Contents

Appendixes

Preface

This text is in a series of texts that includes the following:

Bittinger: *Basic Mathematics,* Eighth Edition

Bittinger: *Fundamental Mathematics,* Second Edition

Bittinger: *Introductory Algebra,* Eighth Edition

Bittinger: *Intermediate Algebra,* Eighth Edition

Bittinger/Beecher: *Introductory and Intermediate Algebra: A Combined Approach*

Introductory and Intermediate Algebra: A Combined Approach, First Edition, is a new text in the Bittinger paperback series. Appropriate for a course combining the study of introductory and intermediate algebra, this text covers both introductory and intermediate algebra topics without the repetition of instruction necessary in two separate texts. Its unique approach blends the following elements in order to bring students success:

- ***Writing style.*** The authors write in a clear easy-to-read style that helps students progress from concepts through examples and margin exercises to section exercises.
- ***Problem-solving approach.*** The basis for solving problems and real-data applications is a five-step process (*Familiarize, Translate, Solve, Check,* and *State*) introduced early in the text and used consistently throughout. This problem-solving approach provides students with a consistent framework for solving applications. (See pages 95, 331, 392, and 515.)
- ***Real data.*** Real-data applications aid in motivating students by connecting the mathematics to their everyday lives. Extensive research was conducted to find new applications that relate mathematics to the real world.

Tornado Touchdowns in Indiana by Time of Day (1950–1994)

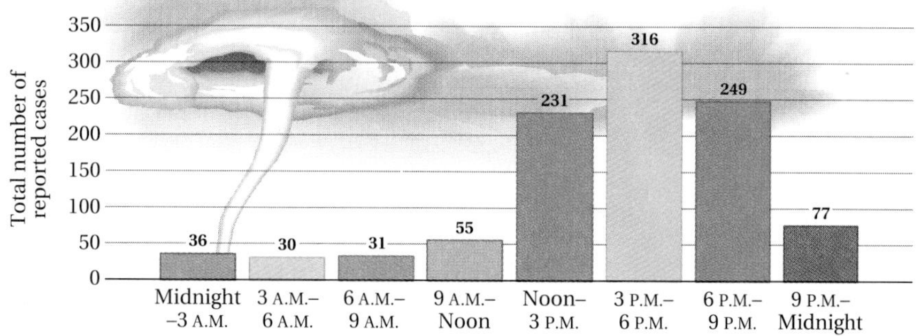

Source: National Weather Service

- ***Art program.*** The art program is designed to improve the visualization of mathematical concepts and to enhance the real-data applications.

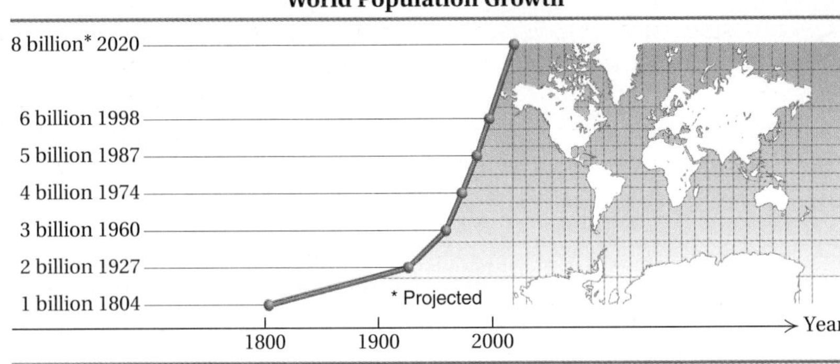

- ***Reviewer feedback.*** The authors solicit feedback from reviewers and students to help fulfill student and instructor needs.
- ***Accuracy.*** The manuscript is subjected to an extensive accuracy-checking process to eliminate errors.
- ***Supplements package.*** All ancillary materials are closely tied with the text and created by members of the author team to provide a complete and consistent package for both students and instructors.

Features

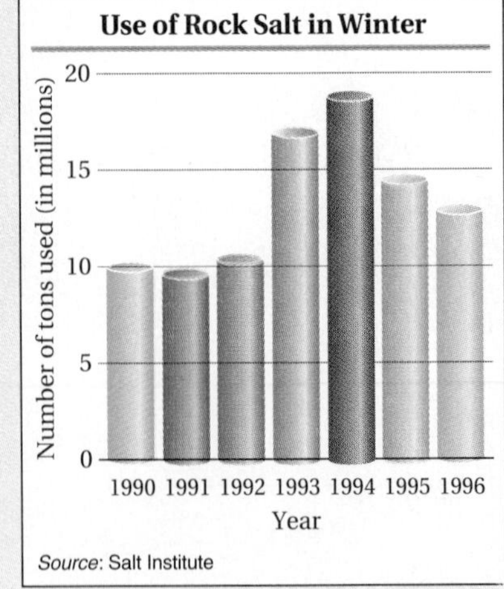

Updated Applications Extensive research has been done to make the applications up to date and realistic. A large number of the applications are drawn from the fields of business and economics, life and physical sciences, social sciences, and areas of general interest such as sports and daily life. To encourage students to understand the relevance of mathematics, many applications are enhanced by graphs and drawings similar to those found in today's newspapers and magazines. Many applications are also titled for quick and easy reference, and most real-data applications are credited with a source line. (See pages 101, 140, 392, 478, and 730.)

Improving Your Math Study Skills Occurring at least once in every chapter, and referenced in the table of contents, these mini-lessons provide students with concrete techniques to improve studying and test-taking. These features can be covered in their entirety at the beginning of the course, encouraging good study habits early on, or they can be used as they occur in the text, allowing students to learn them gradually. These features can also be used in conjunction with Marvin L. Bittinger's "Math Study Skills" Videotape, which is free to adopters. Please contact your Addison Wesley Longman sales consultant for details on how to obtain this videotape. (See pages 16, 62, 382, and 818.)

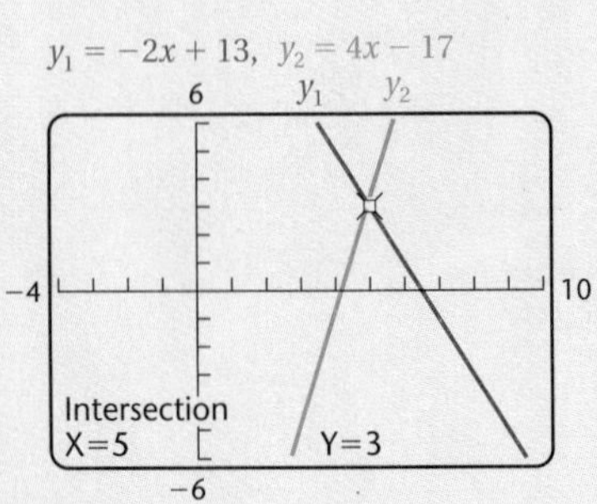

Calculator Spotlights Designed specifically for the beginning algebra student, these optional features include graphing-calculator instruction and practice exercises (see pages 160, 444, 494, 610, and 733). Answers to all Calculator Spotlight exercises appear at the back of the text.

Art and Design To enhance the greater emphasis on real data and applications, we have included a large number of pieces of technical and situational art (see pages 10, 97, 387, 392, 480, and 753). The use of color has been carried out in a methodical and precise manner so that its use carries a consistent meaning, which enhances the readability of the text. For example, the use of both red and blue in mathematical art increases understanding of the concepts. When two lines are graphed using the same set of axes, one is usually red and the other blue. Note that equation labels are the same color as the corresponding line to aid in understanding.

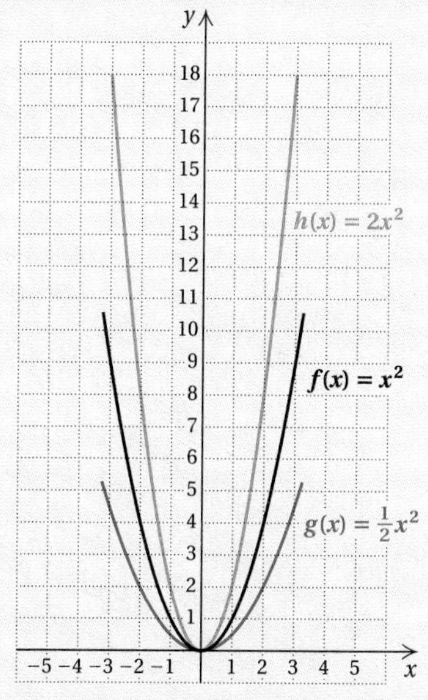

World Wide Web Integration The World Wide Web is a powerful resource available to more and more people every day. In an effort to get students more involved in using this source of information, we have included a World Wide Web address (www.mathmax.com) on every chapter opener (see pages 139, 425, and 751). Students can go to this page on the World Wide Web to further explore the subject matter of the chapter-opening application. Selected exercise sets, marked on the first page of the exercise set with an icon (see pages 161, 295, and 461), have additional practice-problem worksheets that can be downloaded from this site. Additional, more extensive, Summary and Review pages for each chapter, as well as other supplementary material, can also be downloaded for instructor and student use.

Algebraic–Graphical Connections To give students a better visual understanding of algebra, we have included algebraic–graphical connections (see pages 200, 379, 643, and 673). This feature gives the algebra more meaning by connecting the algebra to a graphical interpretation.

Collaborative Learning Features An icon located at the end of an exercise set signals the existence of a Collaborative Learning Activity correlating to that section in Irene Doo's *Collaborative Learning Activities Manual* (see pages 216, 402, 464, and 614). Please contact your Addison Wesley Longman sales consultant for details on ordering this supplement.

Exercises Exercises are paired, meaning that each even-numbered exercise is very much like the odd-numbered one that precedes it. This gives the instructor several options: If an instructor wants the student to have answers available, the odd-numbered exercises are assigned; if an instructor wants the student to practice (perhaps for a test), with no answers available, then the even-numbered exercises are assigned. In this way, each exercise set actually serves as two exercise sets. Answers to all odd-numbered exercises, with the exception of the Thinking and Writing exercises, and *all* Skill Maintenance exercises are provided at the back of the text. If an instructor wants the student to have access to all the answers, a complete answer book is available.

Skill Maintenance Exercises The Skill Maintenance exercises focus on the four Objectives for Retesting listed at the beginning of each chapter, but they also review concepts from other sections of the text in order to prepare students for the final examination. Section and objective codes appear next to each Skill Maintenance exercise for easy reference. Answers to all Skill Maintenance exercises appear at the back of the book (see pages 164, 237, and 360).

Synthesis Exercises These exercises appear in every exercise set, Summary and Review, Chapter Test, and Cumulative Review. Synthesis exercises help build critical thinking skills by requiring students to synthesize or combine learning objectives from the section being studied as well as preceding sections in the book (see pages 322, 594, and 656).

Thinking and Writing Exercises ◈ Two Thinking and Writing exercises (denoted by the maze icon) are included in the Synthesis section of every exercise set and Summary and Review. Designed to develop comprehension of critical concepts, these exercises encourage students to both think and write about key mathematical ideas in the chapter (see pages 114, 206, 628, and 726).

Content

- Combining the content of introductory algebra and intermediate algebra into one text eliminates the overlap of topics that occurs in two separate courses. This approach provides more time to spend on difficult topics and more time to include intermediate topics that often are eliminated from a syllabus because of time constraints.
- One text for both introductory algebra and intermediate algebra gives the student the advantage of learning all algebra from a consistent pedagogy and style.
- A guide to a review of Chapters 1–6 is in Appendix E. This guide outlines a brief algebra review (in syllabus form) for students who begin using this text in Chapter 7.
- Graphing is divided into two chapters. Basic graphing of equations is covered in Chapter 3 (*Graphs of Equations; Data Analysis*) and the concept of slope appears in Chapter 7 (*Graphs, Functions, and Applications*). This early introduction to graphing allows expanded coverage and integration of graphing in the rest of the text.
- To provide a solid background for college algebra, functions are introduced early (Section 7.1, "Functions and Graphs," and Section 7.2, "Finding Domain and Range") and are then integrated throughout. This early function approach allows for many uses of functions such as expansion of graphing applications in the radical, quadratic, and logarithmic topics.
- Modeling data with functions is presented in three sections: Section 7.6 ("Mathematical Modeling with Linear Functions"), Section 11.7 ("Mathematical Modeling with Quadratic Functions"), and Section 12.7 ("Mathematical Modeling with Exponential and Logarithmic Functions"). The study of applications becomes more interactive when students learn to model real data with functions and make predictions from graphs.
- Section 8.6 ("Business and Economics Applications") broadens the base of function applications. This section provides the background for business students who plan to take Business Calculus and Finite Mathematics.
- Section 9.1 ("Sets, Interval Notation, and Inequalities") introduces interval notation. Throughout the rest of the text, solution sets are given in both set-builder notation and interval notation. Interval notation enhances the preparation for later courses.
- Chapter 12 (*Exponential and Logarithmic Functions*) immediately follows the chapter on quadratics. Early presentation of function topics has allowed a smoother transition between types of functions throughout the text. Section 12.2 ("Inverse and Composite Functions") is

placed immediately before the section in which it is needed (Section 12.3, "Logarithmic Functions").

- Five appendixes (*Handling Dimension Symbols, Determinants and Cramer's Rule, Elimination of Matrices, The Algebra of Functions,* and *Introductory Algebra Review*) are included. As reviewers have requested, these topics are placed in a more optional format.

Learning Aids

Interactive Worktext Approach The pedagogy of this text is designed to provide an interactive learning experience between the student and the exposition, annotated examples, art, margin exercises, and exercise sets. This approach provides students with a clear set of learning objectives, involves them with the development of the material, and provides immediate and continual reinforcement and assessment.

Section objectives are keyed by letter not only to section subheadings, but also to exercises in the exercise sets and Summary and Review, as well as answers to the Pretest, Chapter Test, and Cumulative Review questions. This enables students to easily find appropriate review material if they are unable to work a particular exercise.

Throughout the text, students are directed to numerous *margin exercises,* which provide immediate reinforcement of the concepts covered in each section.

Review Material The First Edition of *Introductory and Intermediate Algebra: A Combined Approach* provides many opportunities for students to prepare for final assessment.

The *Summary and Review* (in a two-column format) appears at the end of each chapter and provides an extensive set of review exercises. Reference codes beside each exercise or direction line preceding it allow the student to easily return to the objective being reviewed (see pages 337, 595, and 667).

Also included at the end of Chapters 3, 6, 9, and 12 is a *Cumulative Review,* which reviews material from all preceding chapters. At the back of the text are answers to all Cumulative Review exercises, together with section and objective references, so that students know exactly what material to study if they miss a review exercise (see pages 423, 599, and 827).

Objectives for Retesting are covered in each Summary and Review and Chapter Test, and are also included in the Skill Maintenance exercises and in the Printed Test Bank (see pages 194, 342, and 488).

For Extra Help Many valuable study aids accompany this text. Below the list of objectives found at the beginning of each section are references to appropriate videotape, tutorial software, and CD-ROM programs to make it easy for the student to find the correct support materials.

Objectives

a Determine whether a correspondence is a function.

b Given a function described by an equation, find function values (outputs) for specified values (inputs).

c Draw the graph of a function.

d Determine whether a graph is that of a function using the vertical-line test.

e Solve applied problems involving functions and their graphs.

For Extra Help

TAPE 12 | MAC WIN | CD-ROM

Testing The following assessment opportunities exist in the text.

Chapter Pretests can then be used to place students in a specific section of the chapter, allowing them to concentrate on topics with which they have particular difficulty (see pages 140, 426, and 604).

Chapter Tests allow students to review and test comprehension of chapter skills, as well as the four Objectives for Retesting from earlier chapters (see pages 187, 483, and 749).

Answers to all Chapter Pretest and Chapter Test questions are found at the back of the book, along with appropriate section and objective references.

Supplements for the Instructor

Instructor's Solutions Manual

0-201-51925-9

The *Instructor's Solutions Manual* by Judith A. Penna contains brief worked-out solutions to all even-numbered exercises in the exercise sets and answers to all Thinking and Writing exercises.

Printed Test Bank/Instructor's Resource Guide

by Donna DeSpain

0-201-51924-0

The test-bank section of this supplement contains the following:

- Two alternate test forms for each chapter, modeled after the Chapter Tests in the text
- Two alternate test forms for each chapter, designed for a 50-minute class period
- Two multiple-choice versions of each Chapter Test
- Two cumulative review tests for each chapter, with the exception of Chapters 1 and 12
- Six final examinations: two with questions organized by chapter, two with questions scrambled as in the Cumulative Reviews, and two with multiple-choice questions
- Answers for the Chapter Tests and the final examinations

The resource-guide section contains the following:

- Extra practice exercises (with answers) for the most difficult topics in the text
- Critical Thinking exercises and answers
- Black-line masters of grids and number lines for transparency masters or test preparation
- Indexes to the videotapes that accompany the text

Collaborative Learning Activities Manual

0-201-51927-5

The *Collaborative Learning Activities Manual,* written by Irene Doo of Austin Community College, features group activities that are tied to sections of the text via an icon Collaborative Learning Manual. Instructions for classroom setup are also included in the manual.

Answer Book

0-201-51931-3

The *Answer Book* contains answers to all exercises in the exercise sets in the text. Instructors can make quick reference to all answers or have quantities of these booklets made available for sale if they want students to have access to all the answers.

TestGen-EQ

0-201-38175-3 (Windows), 0-201-38182-6 (Macintosh)

This test generation software is available in Windows and Macintosh versions. TestGen-EQ's friendly graphical interface enables instructors to easily view, edit, and add questions, transfer questions to tests, and print tests in a variety of fonts and forms. Search and sort features help the instructor quickly locate questions and arrange them in a preferred order. Six question formats are available, including short-answer, true–false, multiple-choice, essay, matching, and bimodal formats. A built-in question editor gives the instructor the ability to create graphs, import graphics, insert mathematical symbols and templates, and insert variable numbers or text. Computerized testbanks include algorithmically defined problems organized according to each textbook. An "Export to HTML" feature lets instructors create practice tests for the World Wide Web. TestGen-EQ is free to qualifying adopters.

QuizMaster-EQ

0-201-38147-8 (Windows), 0-201-38154-0 (Macintosh)

QuizMaster-EQ enables instructors to create and save tests and quizzes using TestGen-EQ so students can take them on a computer network. Instructors can set preferences for how and when tests are administered. QuizMaster-EQ automatically grades the exams and allows the instructor to view or print a variety of reports for individual students, classes, or courses. This software is available for both Windows and Macintosh and is fully networkable. QuizMaster-EQ is free to qualifying adopters.

Supplements for the Student

Student's Solutions Manual

0-201-34021-6

The *Student's Solutions Manual* by Judith A. Penna contains fully worked-out solutions with step-by-step annotations for all the odd-numbered exercises in the exercise sets in the text, with the exception of the Thinking and Writing exercises. It may be purchased by your students from Addison Wesley Longman.

"Steps to Success" Videotapes

0-201-57308-3

Steps to Success is a complete revision of the existing series of videotapes, based on extensive input from both students and instructors. These videotapes feature an engaging team of mathematics teachers who present comprehensive coverage of each section of the text in a student-interactive format. The lecturers' presentations include examples and problems from the text and support an approach that emphasizes visualization and problem solving. A video icon at the beginning of each section references the appropriate videotape number. The videotapes are free to qualifying adopters.

"Math Study Skills for Students" Videotape

0-201-84521-0

Designed to help students make better use of their math study time, this videotape help students improve retention of concepts and procedures taught in classes from basic mathematics through intermediate algebra.

Through carefully-crafted graphics and comprehensive on-camera explanation, Marvin L. Bittinger helps viewers focus on study skills that are commonly overlooked.

InterAct Math Tutorial Software

0-201-38119-2 (Windows), 0-201-38126-5 (Macintosh)

InterAct Math Tutorial Software has been developed and designed by professional software engineers working closely with a team of experienced developmental-math teachers. This software includes exercises that are linked one-to-one with the odd-numbered exercises in the text and require the same computational and problem-solving skills as their companion exercises in the text. Each exercise has an example and an interactive guided solution that are designed to involve students in the solution process and to help them identify precisely where they are having trouble. In addition, the software recognizes common student errors and provides students with appropriate customized feedback. With its sophisticated answer recognition capabilities, *InterAct Math Tutorial Software* recognizes equivalent forms of the same answer for any kind of input. It also tracks for each section student activity and scores that can then be printed out. A disk icon 💾 at the beginning of each section identifies section coverage. Available for Windows and Macintosh computers, this software is free to qualifying adopters or can be bundled with books for sale to students.

World Wide Web Supplement (www.mathmax.com)

This on-line supplement provides additional practice and learning resources for the student of introductory algebra. For each book chapter, students can find additional practice exercises, Web links for further exploration, and expanded Summary and Review pages that review and reinforce the concepts and skills learned throughout the chapter. In addition, students can download a plug-in for Addison Wesley Longman's *InterAct Math Tutorial Software* that allows them to access additional tutorial problems directly through their Web browser. Students and instructors can also learn about the other supplements available for the MathMax series via sample audio clips and complete descriptions of other services provided by Addison Wesley Longman.

MathMax Multimedia CD-ROM for Introductory and Intermediate Algebra: A Combined Approach

0-201-51930-5

The Introductory and Intermediate Algebra Combined CD provides an active environment using graphics, animations, and audio narration to build on some of the unique and proven features of the MathMax series. Highlighting key concepts from the book, the content of the CD is tightly and consistently integrated with the *Introductory and Intermediate Algebra: A Combined Approach* text and retains references to the text's numbering scheme so that students can move smoothly between the CD and other *Introductory and Intermediate Algebra: A Combined Approach* supplements. The CD includes Addison Wesley Longman's *InterAct Math Tutorial Software* so that students can practice additional tutorial problems. An interactive Summary and Review section allows students to review and practice what they have learned in each chapter; and multimedia presentations reiterate important study skills described throughout the book. A CD-ROM icon at the beginning of each section indicates section coverage. The Introductory and Intermediate Algebra Combined CD is available for both Windows and Macintosh computers. Contact your Addison Wesley Longman sales consultant for a demonstration.

Your authors and their team have committed themselves to publishing an accessible, clear, accomplishable, error-free book and supplements package that will provide the student with a successful learning experience and will foster appreciation and enjoyment of mathematics. As part of our continual effort to accomplish this goal, we welcome your comments and suggestions at the following email addresses:

Marv Bittinger
exponent@aol.com

Judy Beecher
jabeecher@worldnet.att.net

Acknowledgments

Many of you have helped to shape this text by reviewing, participating in telephone surveys and focus groups, filling out questionnaires, and spending time with us on your campuses. Our deepest appreciation to all of you and in particular to the following:

Dolores Anenson, *Merced College*
Jamie Ashby, *Texarkana College*
James Baglio, *Normandale Community College*
Martin Baker, *Parks College*
Sharon Balk, *Northeast Iowa Community College*
Shirley Beil, *Normandale Community College*
Boyd Benson, *Rio Hondo College*
William Bordeaux, Jr., *Huntington College*
Richard Burns, *Springfield Technical Community College*
Mark Campbell, *Slippery Rock University*
Ben Cornelius, *Oregon Institute of Technology*
Cynthia Coulter, *Catawba Valley Community College*
Sherry Crabtree, *Northwest Shoals Community College*
Debra Caplinger Cross, *Alabama Southern Community College*
Patrick Cross, *University of Oklahoma—Norman*
William Culbertson, *Northwest State Community College*
Laura Davis, *Garland City Community College*
Andres Delgado, *Orange County Community College*
Irene Duranczyk, *Eastern Michigan University*
M. R. Eisfelder, *McHenry County College*
Mimi Elwell, *Lake Michigan College*
Mike Flodin, *Tacoma Community College*
Timothy Foster, *Garrett Community College*
John Fourlis, *Harrisburg Area Community College*
Linda Galloway, *Macon College*
Ormsin Gardner, *Winona State University*
Stephanie Gardner, *College of Eastern Utah*
Catherine Green, *Lawson State Community College*
Phil Green, *Skagit Valley College*
Don Griffin, *Greenville Technical College*
Dennis Houk, *West Shore Community College*
Mary Indelicado, *Normandale Community College*
Cheryl Ingrams, *Crowder College*
Yvonne Jessee, *Mountain Empire Community College*
Juan Carlos Jiminez, *Springfield Technical Community College*
Charles Jordan, *Lawson State University*
Joseph Jordan, *John Tyler Community College*
Judy Kasabian, *El Camino College*

Joann Kelly, *Palm Beach Community College*
Karen Knight, *Jefferson College*
Evelyn Kral, *Morgan Community College*
Lee M. Lacey, *Glendale Community College*
Ira Lansing, *College of Marin*
Christine Ledwith, *Florida Keys Community College*
Sang Lee, *DeVry Institute of Technology*
Linda Long, *Ricks College*
Fred Lemmerhirt, *Waubansee Community College*
Rudy Maglio, *Oakton Community College*
C. Vernon Marlin, *Southeastern Community College*
Ray Maruca, *Delaware County Community College*
Hubert McClure, *Tri-County Technical College*
Jann McInnis, *Florida Community College— Jacksonville*
Sharon McKindrick, *New Mexico State University—Grants Beach*
John Menzie, *Barstow College*
Greg Millican, *Northeast Alabama State Jr. College*
Sandra Miller, *Harrisburg Area Community College*
Steven Mondy, *Normandale Community College*
Michael Montano, *Riverside Community College City Campus*
Charlie Montgomery, *Alabama Southern Community College*
Mika Moteki, *Cardinal Stritch University*
Frank Mulvaney, *Delaware County Community College*
Ellen O'Connell, *Triton College*
Mark Omadt, *Anoka Ramsey Community College*
Linda Padilla, *Joliet Junior College*
Helen Paszek, *Nash Community College*
Barbara Phillips, *Roane State Community College*
Marilyn Platt, *Gaston College*
John Pleasants, *Orange County Community College*
Carol Rardin, *Central Wyoming College*
Margaret Rauch, *Texarkana College*
Eugena Rohrberg, *Los Angeles Valley College*
Suzanne Rosenberger, *Harrisburg Area Community College*
Robert Sampolski, *Oakton Community College*
Bud Sather, *Flathead Valley Community College*
Ned Schillow, *Lehigh Carbon Community College*
Carole Shapero, *Oakton Community College*
Minnie Shuler, *Gulfcoast Community College*
Lynn Siedenstrang, *Gray's Harbor College*
Sam Sochis, *Barstow College*
Lee Ann Spahr, *Durham Technical Community College*
Helen Stewert, *Brunswick Community College*
Roman Sznajder, *Bowie State University*
Sharon Testone, Ph.D., *Onondaga Community College*
Rick Vanamerongen, *Portland Community College*
Sharon Walker, *Catawba Valley Community College*
Tammy Wellick, *McHenry County College*
Joyce Wellington, *Southeastern Community College*
Elaine Werner, *University of the Pacific*
Jaqueline Wing, *Angelina College*
Steve Wittel, *Augusta College*
Gerad Zimmerman, *Tacoma Community College*

We also wish to recognize the following people who wrote scripts, presented lessons on camera, and checked the accuracy of the videotapes:

Beth Burkenstock
Margaret Donlan, *University of Delaware*
David J. Ellenbogen, *Community College of Vermont*
Barbara Johnson, *Indiana University—Purdue University at Indianapolis*
Judith A. Penna, *Indiana University—Purdue University at Indianapolis*
Patricia Schwarzkopf, *University of Delaware*
Clen Vance, *Houston Community College*

We wish to thank Jason Jordan, our publisher and friend at Addison Wesley Longman, for his encouragement, for his marketing insight, and for providing us with the environment of creative freedom. The unwavering support of the Developmental Math group and the endless hours of hard work by Martha Morong and Janet Theurer have led to products of which we are immensely proud.

We also want to thank Judy Penna for writing the Student's and the Instructor's Solutions Manuals and for her strong leadership in the preparation of the printed supplements, videotapes, and interactive CD-ROM. Other strong support has come from Donna DeSpain for the Printed Test Bank; Bill Saler for the audiotapes; Irene Doo for the Collaborative Learning Activities Manual; and Irene Doo, Shane Goodwin, Barbara Johnson, Judy Penna, Vera Preston, and Peggy Reijto for their accuracy checking.

M.L.B.
J.A.B.

1

Introduction to Real Numbers and Algebraic Expressions

An Application

The casino game of blackjack makes use of many card-counting systems to give players a winning edge if the count becomes negative. One such system is called *High–Low*, first developed by Harvey Dubner in 1963. Each card counts as −1, 0, or 1 as follows:

2, 3, 4, 5, 6	count as +1;
7, 8, 9	count as 0;
10, J, Q, K, A	count as −1.

Find the final count on the sequence of cards

K, A, 2, 4, 5, 10, J, 8, Q, K, 5.

***Source*:** Patterson, Jerry L., *Casino Gambling.* New York: Perigee, 1982

The Mathematics

We add the following numbers:

$$(-1) + (-1) + 1 + 1 + 1 + (-1) + (-1) + 0 + (-1) + (-1) + 1 = -2.$$

The numbers in red are negative numbers.

This problem appears as Exercise 120 in Exercise Set 1.4.

World Wide Web For more information, visit us at www.mathmax.com

Introduction

In this chapter, we consider the number system used most often in algebra. It is called the real-number system. We will learn to add, subtract, multiply, and divide real numbers and to manipulate certain expressions. Such manipulation will be important when we solve equations and applied problems in Chapter 2.

Pretest: Chapter 1

1. Evaluate $x/2y$ for $x = 5$ and $y = 8$.

2. Write an algebraic expression: Seventy-eight percent of some number.

3. Find the area of a rectangle when the length is 22.5 ft and the width is 16 ft.

4. Find $-x$ when $x = -12$.

Use either $<$ or $>$ for ▩ to write a true sentence.

5. 0 ▩ -5

6. 10 ▩ -5

7. -35 ▩ -45

8. $-\frac{2}{3}$ ▩ $\frac{4}{5}$

Find the absolute value.

9. $|-12|$

10. $|2.3|$

11. $|0|$

Find the opposite, or additive inverse.

12. 5.4

13. $-\frac{2}{3}$

Find the reciprocal.

14. 10

15. $-\frac{2}{3}$

Compute and simplify.

16. $-9 + (-8)$

17. $20.2 - (-18.4)$

18. $-\frac{5}{6} - \frac{3}{10}$

19. $-11.5 + 6.5$

20. $-9(-7)$

21. $\frac{5}{8}\left(-\frac{2}{3}\right)$

22. $-19.6 \div 0.2$

23. $-56 \div (-7)$

24. $12 - (-6) + 14 - 8$

25. $20 - 10 \div 5 + 2^3$

Multiply.

26. $9(z - 2)$

27. $-2(2a + b - 5c)$

Factor.

28. $4x - 12$

29. $6y - 9z - 18$

Simplify.

30. $3y - 7 - 2(2y + 3)$

31. $\{2[3(y + 1) - 4] - [5(y - 3) - 5]\}$

32. Write an inequality with the same meaning as $x > 12$.

1.1 Introduction to Algebra

Many types of problems require the use of equations in order to be solved effectively. The study of algebra involves the use of equations to solve problems. Equations are constructed from algebraic expressions. The purpose of this section is to introduce you to the types of expressions encountered in algebra.

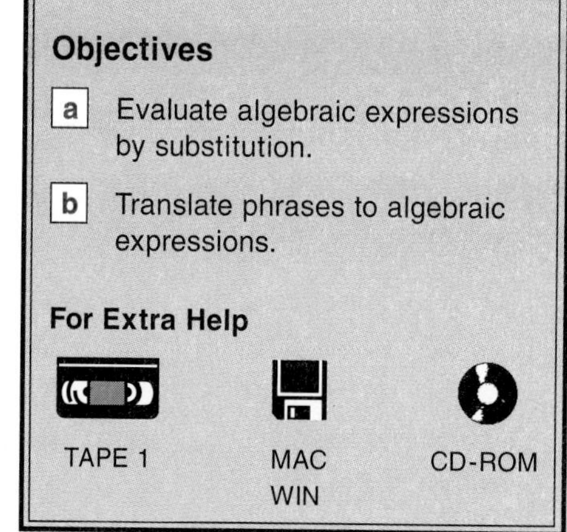

a Evaluating Algebraic Expressions

In arithmetic, you have worked with expressions such as

$$49 + 75, \quad 8 \times 6.07, \quad 29 - 14, \quad \text{and} \quad \frac{5}{6}.$$

In algebra, we use certain letters for numbers and work with *algebraic expressions* such as

$$x + 75, \quad 8 \times y, \quad 29 - t, \quad \text{and} \quad \frac{a}{b}.$$

Sometimes a letter can represent various numbers. In that case, we call the letter a **variable**. Let a = your age. Then a is a variable since a changes from year to year. Sometimes a letter can stand for just one number. In that case, we call the letter a **constant**. Let b = your date of birth. Then b is a constant.

Where do algebraic expressions occur? Most often we encounter them when we are solving applied problems. For example, consider the bar graph shown at right, one that we might find in a book or magazine. Suppose we want to know how much longer the diameter of Earth is than the diameter of Mars.

In algebra, we translate the problem into an equation. It might be done as follows.

Diameter of Mars	plus	How much	is	Diameter of Earth
↓	↓	↓	↓	↓
4217	+	x	=	7927

Note that we have an algebraic expression on the left of the equals sign. To find the number x, we can subtract 4217 on both sides of the equation:

$$4217 + x = 7927$$
$$4217 + x - 4217 = 7927 - 4217$$
$$x = 3710.$$

The value of x gives us the answer, 3710 miles.

In arithmetic, you probably would do this subtraction right away without considering an equation. In algebra, more complex problems are difficult to solve without first solving an equation.

Do Exercise 1.

1. Translate this problem to an equation. Use the graph below.

How much longer is the diameter of Venus than the diameter of Pluto?

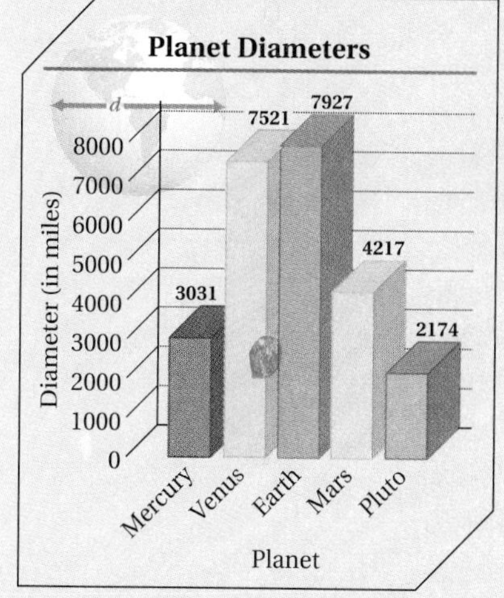

2. Evaluate $a + b$ for $a = 38$ and $b = 26$.

3. Evaluate $x - y$ for $x = 57$ and $y = 29$.

4. Evaluate $4t$ for $t = 15$.

Answers on page A-1

5. Find the area of a rectangle when l is 24 ft and w is 8 ft.

Calculator Spotlight

Evaluating Algebraic Expressions. Eventually in this text, we will be using only graphing calculators in working the Calculator Spotlights. Because of this and because there is extensive variance in keystroke procedures among calculators, we will henceforth focus on the keystrokes for a graphing calculator such as a TI-82 or a TI-83. For the use of any other calculator, consult its particular manual.

To evaluate the expression $12m/n$ for $m = 8$ and $n = 16$, we first write (or think)

$$\frac{12 \cdot 8}{16}.$$

Then we press

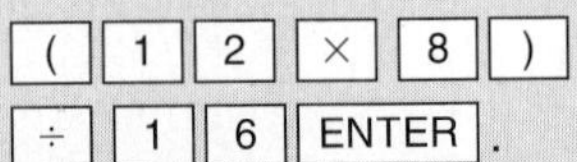

The answer is 6.

Exercises

Evaluate.

1. $\frac{12m}{n}$, for $m = 82.3$ and $n = 16.56$
2. $a + b$, for $a = 8.2$ and $b = 3.7$
3. $b + a$, for $a = 8.2$ and $b = 3.7$
4. $27xy$, for $x = 12.7$ and $y = 100.4$
5. $3a + 2b$, for $a = 8.2$ and $b = 3.7$
6. $2a + 3b$, for $a = 8.2$ and $b = 3.7$

Answer on page A-1

An **algebraic expression** consists of variables, constants, numerals, and operation signs. When we replace a variable with a number, we say that we are **substituting** for the variable. This process is called **evaluating the expression.**

Example 1 Evaluate $x + y$ for $x = 37$ and $y = 29$.

We substitute 37 for x and 29 for y and carry out the addition:

$$x + y = 37 + 29 = 66.$$

The number 66 is called the **value** of the expression.

Algebraic expressions involving multiplication can be written in several ways. For example, "8 times a" can be written as $8 \times a$, $8 \cdot a$, $8(a)$, or simply $8a$. Two letters written together without an operation symbol, such as ab, also indicates a multiplication.

Example 2 Evaluate $3y$ for $y = 14$.

$$3y = 3(14) = 42$$

Do Exercises 2–4 on the preceding page.

Example 3 The area A of a rectangle of length l and width w is given by the formula $A = lw$. Find the area when l is 24.5 in. and w is 16 in.

We substitute 24.5 in. for l and 16 in. for w and carry out the multiplication:

$$A = lw = (24.5 \text{ in.})(16 \text{ in.})$$
$$= (24.5)(16)(\text{in.})(\text{in.})$$
$$= 392 \text{ in}^2, \text{ or } 392 \text{ square inches.}$$

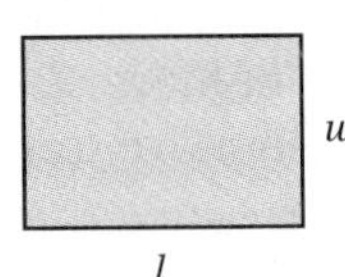

Do Exercise 5.

Algebraic expressions involving division can also be written in several ways. For example, "8 divided by t" can be written as $8 \div t$, $\frac{8}{t}$, or $8/t$, where the fraction bar is a division symbol.

Example 4 Evaluate $\frac{a}{b}$ for $a = 63$ and $b = 9$.

We substitute 63 for a and 9 for b and carry out the division:

$$\frac{a}{b} = \frac{63}{9} = 7.$$

Example 5 Evaluate $\frac{12m}{n}$ for $m = 8$ and $n = 16$.

$$\frac{12m}{n} = \frac{12 \cdot 8}{16} = \frac{96}{16} = 6$$

Do Exercises 6 and 7 on the following page.

Example 6 *Motorcycle Travel.* Ed takes a trip on his motorcycle. He wants to travel 660 mi on a particular day. The time t, in hours, that it takes to travel 660 mi is given by

$$t = \frac{660}{r},$$

where r is the speed of Ed's motorcycle. Find the time of travel if the speed r is 60 mph.

We substitute 60 for r and carry out the division:

$$t = \frac{660}{r} = \frac{660}{60} = 11 \text{ hr}.$$

Do Exercise 8.

b Translating to Algebraic Expressions

In algebra, we translate problems to equations. The different parts of an equation are translations of word phrases to algebraic expressions. It is easier to translate if we know that certain words often translate to certain operation symbols.

KEY WORDS			
Addition (+)	**Subtraction (−)**	**Multiplication (·)**	**Division (÷)**
add sum plus more than increased by	subtract difference minus less than decreased by take from	multiply product times twice of	divide quotient divided by

Example 7 Translate to an algebraic expression:

Twice (or two times) some number.

Think of some number, say, 8. What number is twice 8? It is 16. How did you get 16? You multiplied by 2. Do the same thing using a variable. We can use any variable we wish, such as x, y, m, or n. Let's use y to stand for some number. If we multiply by 2, we get an expression

$y \times 2$, $2 \times y$, $2 \cdot y$, or $2y$.

Example 8 Translate to an algebraic expression:

Thirty-eight percent of some number.

The word "of" translates to a multiplication symbol, so we get the following expressions as a translation:

$38\% \cdot n$, $0.38 \times n$, or $0.38n$.

6. Evaluate a/b for $a = 200$ and $b = 8$.

7. Evaluate $10p/q$ for $p = 40$ and $q = 25$.

8. *Motorcycle Travel.* Find the time it takes to travel 660 mi if the speed is 55 mph.

Answers on page A-1

Translate to an algebraic expression.

9. Eight less than some number

10. Eight more than some number

11. Four less than some number

12. Half of a number

13. Six more than eight times some number

14. The difference of two numbers

15. Fifty-nine percent of some number

16. Two hundred less than the product of two numbers

17. The sum of two numbers

Answers on page A-1

Example 9 Translate to an algebraic expression:

Seven less than some number.

We let

x represent the number.

Now if the number were 23, then the translation would be $23 - 7$. If we knew the number to be 345, then the translation would be $345 - 7$. If the number is x, then the translation is

$x - 7$.

CAUTION! Note that $7 - x$ is *not* a correct translation of the expression in Example 9. The expression $7 - x$ is a translation of "seven minus some number" or "some number less than seven."

Example 10 Translate to an algebraic expression:

Eighteen more than a number.

We let

$t =$ the number.

Now if the number were 26, then the translation would be $26 + 18$. If we knew the number to be 174, then the translation would be $174 + 18$. If the number is t, then the translation is

$t + 18$, or $18 + t$.

Example 11 Translate to an algebraic expression:

A number divided by 5.

We let

$m =$ the number.

Now if the number were 76, then the translation would be $76 \div 5$, or $76/5$, or $\frac{76}{5}$. If the number were 213, then the translation would be $213 \div 5$, or $213/5$, or $\frac{213}{5}$. If the number is m, then the translation is

$$m \div 5, \quad m/5, \quad \text{or} \quad \frac{m}{5}.$$

Example 12 Translate each of the following phrases to an algebraic expression.

Phrase	Algebraic Expression
Five more than some number	$n + 5$, or $5 + n$
Half of a number	$\frac{1}{2}t$, $\frac{t}{2}$, or $t/2$
Five more than three times some number	$3p + 5$, or $5 + 3p$
The difference of two numbers	$x - y$
Six less than the product of two numbers	$mn - 6$
Seventy-six percent of some number	$76\%z$, or $0.76z$

Do Exercises 9–17.

Exercise Set 1.1

a Substitute to find values of the expressions in each of the following applied problems.

1. *Enrollment Costs.* At Emmett Community College, it costs \$600 to enroll in the 8 A.M. section of Elementary Algebra. Suppose that the variable n stands for the number of students who enroll. Then $600n$ stands for the total amount of money collected for this course. How much is collected if 34 students enroll? 78 students? 250 students?

2. *Commuting Time.* It takes Erin 24 min less time to commute to work than it does George. Suppose that the variable x stands for the time it takes George to get to work. Then $x - 24$ stands for the time it takes Erin to get to work. How long does it take Erin to get to work if it takes George 56 min? 93 min? 105 min?

3. The area A of a triangle with base b and height h is given by $A = \frac{1}{2}bh$. Find the area when $b = 45$ m (meters) and $h = 86$ m.

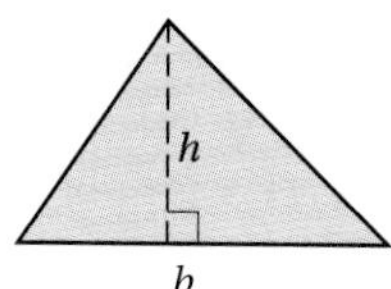

4. The area A of a parallelogram with base b and height h is given by $A = bh$. Find the area of the parallelogram when the height is 15.4 cm (centimeters) and the base is 6.5 cm.

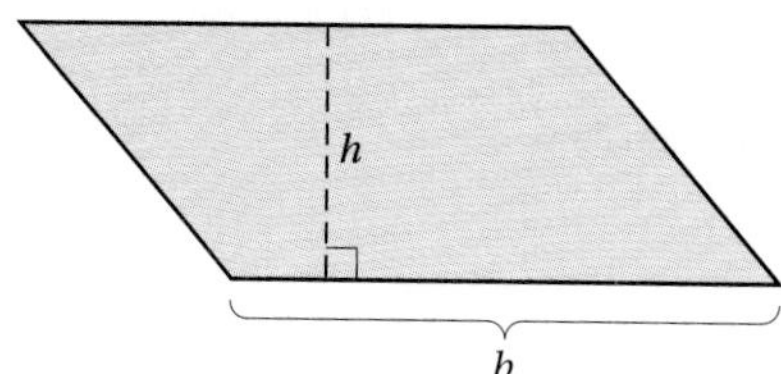

5. *Distance Traveled.* A driver who drives at a speed of r mph for t hr will travel a distance d mi given by $d = rt$ mi. How far will a driver travel at a speed of 65 mph for 4 hr?

6. *Simple Interest.* The simple interest I on a principal of P dollars at interest rate r for time t, in years, is given by $I = Prt$. Find the simple interest on a principal of \$4800 at 9% for 2 yr. (*Hint*: 9% = 0.09.)

Evaluate.

7. $8x$, for $x = 7$

8. $6y$, for $y = 7$

9. $\frac{a}{b}$, for $a = 24$ and $b = 3$

10. $\frac{p}{q}$, for $p = 16$ and $q = 2$

11. $\frac{3p}{q}$, for $p = 2$ and $q = 6$

12. $\frac{5y}{z}$, for $y = 15$ and $z = 25$

13. $\frac{x + y}{5}$, for $x = 10$ and $y = 20$

14. $\frac{p + q}{2}$, for $p = 2$ and $q = 16$

15. $\frac{x - y}{8}$, for $x = 20$ and $y = 4$

16. $\frac{m - n}{5}$, for $m = 16$ and $n = 6$

b Translate to an algebraic expression.

17. 7 more than b

18. 9 more than t

19. 12 less than c

20. 14 less than d

21. 4 increased by q

22. 13 increased by z

23. b more than a

24. c more than d

25. x less than y

26. c less than h

27. x added to w

28. s added to t

29. m subtracted from n

30. p subtracted from q

31. The sum of r and s

32. The sum of a and b

33. Twice z

34. Three times q

35. 3 multiplied by m

36. The product of 8 and t

37. The product of 89% and some number

38. 67% of some number

39. A driver drove at a speed of 55 mph for t hours. How far did the driver travel?

40. An executive assistant has d dollars before going to an office supply store. He bought some fax paper for $18.95. How much did he have after the purchase?

Synthesis

Exercises designated as *Synthesis Exercises* differ from those found in the main body of the exercise set. The icon ◆ denotes synthesis exercises that are writing exercises. Writing exercises are meant to be answered in one or more complete sentences. Because answers to writing exercises often vary, they are not listed at the back of the book. These and the other synthesis exercises will often challenge you to put together two or more objectives at once.

41. ◆ If the length of a rectangle is doubled, does the area double? Why or why not?

42. ◆ If the height and the base of a triangle are doubled, what happens to the area? Explain.

Translate to an algebraic expression.

43. Some number x plus three times y

44. Some number a plus 2 plus b

45. A number that is 3 less than twice x

46. Your age in 5 years, if you are a years old now

1.2 The Real Numbers

A **set** is a collection of objects. For our purposes, we will most often be considering sets of numbers. One way to name a set uses what is called **roster notation.** For example, roster notation for the set containing the numbers 0, 2, and 5 is {0, 2, 5}.

Sets that are parts of other sets are called **subsets**. In this section, we become acquainted with the set of *real numbers* and its various subsets.

Two important subsets of the real numbers are listed below using roster notation.

Natural numbers = {1, 2, 3, ...}. These are the numbers used for counting.

Whole numbers = {0, 1, 2, 3, ...}. This is the set of natural numbers with 0 included.

We can represent these sets on a number line. The natural numbers are those to the right of zero. The whole numbers are the natural numbers and zero.

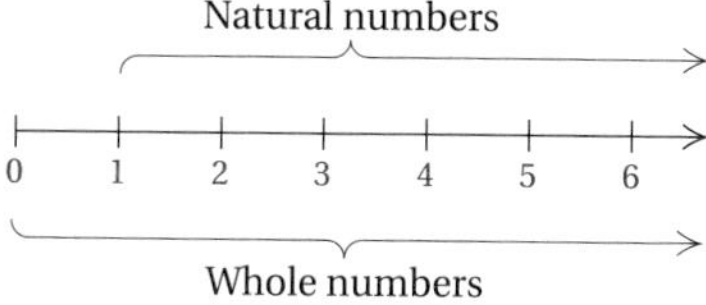

We create a new set, called the *integers,* by starting with the whole numbers, 0, 1, 2, 3, and so on. For each natural number 1, 2, 3, and so on, we obtain a new number to the left of zero on the number line:

For the number 1, there will be an *opposite* number -1 (negative 1).

For the number 2, there will be an *opposite* number -2 (negative 2).

For the number 3, there will be an *opposite* number -3 (negative 3), and so on.

The **integers** consist of the whole numbers and these new numbers. We picture them on a number line as follows.

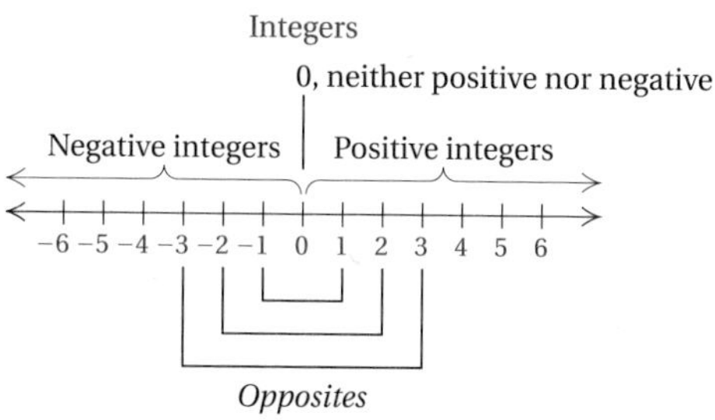

We call these new numbers to the left of 0 **negative integers.** The natural numbers are also called **positive integers.** Zero is neither positive nor negative. We call -1 and 1 **opposites** of each other. Similarly, -2 and 2 are opposites, -3 and 3 are opposites, -100 and 100 are opposites, and 0 is its own opposite. Pairs of opposite numbers like -3 and 3 are equidistant from 0. The integers extend infinitely on the number line to the left and right of zero.

Objectives

a. Name the integer that corresponds to a real-world situation.

b. Graph rational numbers on a number line.

c. Convert from fractional notation to decimal notation for a rational number.

d. Determine which of two real numbers is greater and indicate which, using < or >; given an inequality like $a < b$, write another inequality with the same meaning. Determine whether an inequality like $-3 \le 5$ is true or false.

e. Find the absolute value of a real number.

For Extra Help

TAPE 1 | MAC WIN | CD-ROM

> The set of **integers** = {. . . , −5, −4, −3, −2, −1, 0, 1, 2, 3, 4, 5, . . .}.

a Integers and the Real World

Integers correspond to many real-world problems and situations. The following examples will help you get ready to translate problem situations that involve integers to mathematical language.

Example 1 Tell which integer corresponds to this situation: The temperature is 3 degrees below zero.

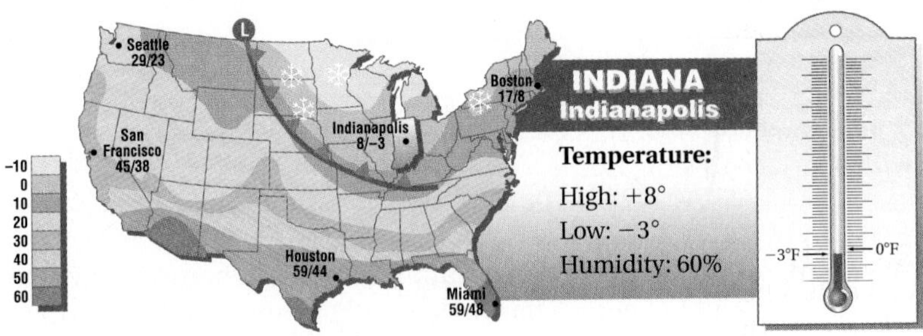

The integer −3 corresponds to the situation. The temperature is −3°.

Example 2 *Jeopardy.* Tell which integer corresponds to this situation: A contestant missed a $600 question on the television game show "Jeopardy."

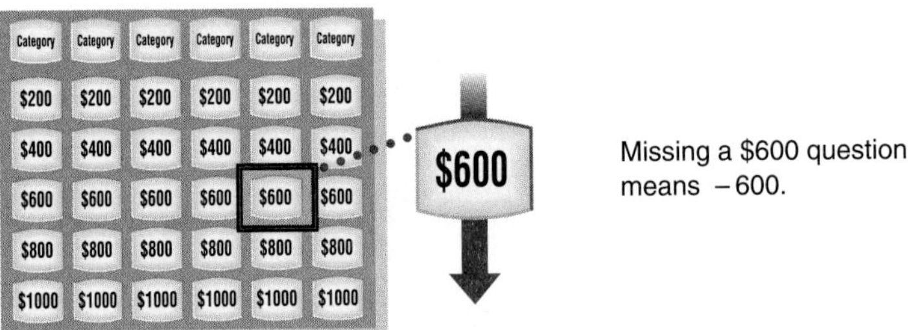

Missing a $600 question causes a $600 loss on the score—that is, the contestant earns −600 dollars.

Example 3 *Elevation.* Tell which integer corresponds to this situation: The lowest point in New Orleans is 8 ft below sea level.

The integer −8 corresponds to the situation. The elevation is −8 ft.

Example 4 *Dow Jones Industrial Average.* Tell which integers correspond to this situation: The largest daily decrease in the Dow Jones Industrial Average was 508 points; the largest increase was 187 (**Source:** *The Guinness Book of Records*).

The integer -508 corresponds to a decrease in the average. The integer 187 corresponds to an increase in the average.

Do Exercises 1–4.

b The Rational Numbers

We created the set of integers by obtaining a negative number for each natural number. To create a larger number system, called the set of **rational numbers,** we consider quotients of integers with nonzero divisors. The following are rational numbers:

$$\frac{2}{3},\quad -\frac{2}{3},\quad \frac{7}{1},\quad 4,\quad -3,\quad 0,\quad \frac{23}{-8},\quad 2.4,\quad -0.17,\quad 10\frac{1}{2}.$$

The number $-\frac{2}{3}$ (read "negative two-thirds") can also be named $\frac{2}{-3}$ or $\frac{-2}{3}$. The number 2.4 can be named $\frac{24}{10}$ or $\frac{12}{5}$, and -0.17 can be named $-\frac{17}{100}$.

Note that this new set of numbers, the rational numbers, contains the whole numbers, the integers, and the arithmetic numbers (also called the nonnegative rational numbers). We can describe the set of rational numbers using **set-builder notation,** as follows.

The set of **rational numbers** $= \left\{\frac{a}{b} \,\middle|\, a \text{ and } b \text{ are integers and } b \neq 0\right\}$.

(This is read "the set of numbers $\frac{a}{b}$, where a and b are integers and $b \neq 0$.")

We picture the rational numbers on a number line as follows. There is a point on the line for every rational number.

Negative numbers | 0 | Positive numbers

$-2.7 \quad -\frac{3}{2} \quad 1.5 \quad 4$

$-4 \quad -3 \quad -2 \quad -1 \quad 0 \quad 1 \quad 2 \quad 3 \quad 4$

To **graph** a number means to find and mark its point on the number line. Some rational numbers are graphed in the preceding figure.

Example 5 Graph: $\frac{5}{2}$.

The number $\frac{5}{2}$ can be named $2\frac{1}{2}$, or 2.5. Its graph is halfway between 2 and 3.

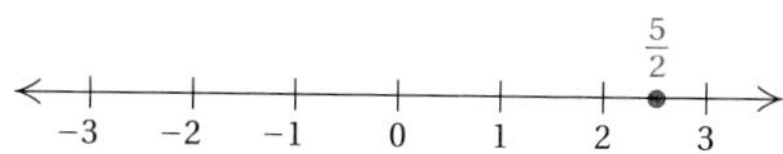

Tell which integers correspond to the given situation.

1. The halfback gained 8 yd on the first down. The quarterback was sacked for a 5-yd loss on the second down.

2. The highest temperature ever recorded in the United States was 134° in Death Valley on July 10, 1913. The coldest temperature ever recorded in the United States was 80° below zero in Prospect Creek, Alaska, in January 1971.

3. At 10 sec before liftoff, ignition occurs. At 156 sec after liftoff, the first stage is detached from the rocket.

4. A submarine dove 120 ft, rose 50 ft, and then dove 80 ft.

Answers on page A-1

Graph on a number line.

5. $-\dfrac{7}{2}$

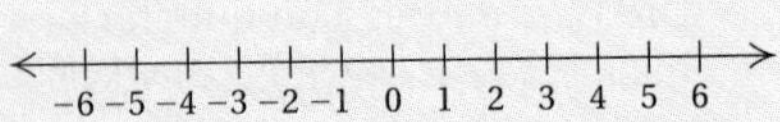

6. -1.4

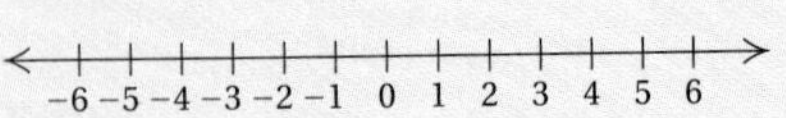

7. $\dfrac{11}{4}$

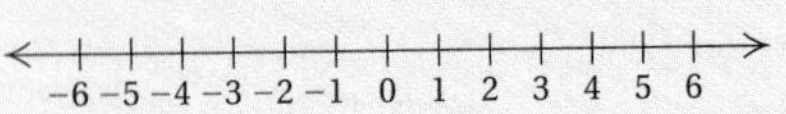

Calculator Spotlight

Negative Numbers on a Calculator. To enter a negative number on a graphing calculator, we use an opposite key (−). Note that this is different from the subtraction key. To enter −5, we press

The display then reads

−5 .

To enter $-\frac{7}{8}$, we press

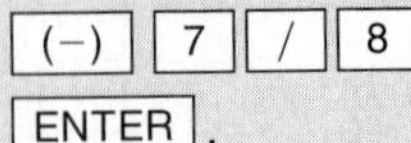

ENTER .

The number is displayed in decimal notation:

−.875 .

Exercises

Press the appropriate keys so that your calculator displays the given number.

1. −3 **2.** −508

3. −0.17 **4.** $-\dfrac{5}{8}$

Answers on page A-1

Example 6 Graph: −3.2.

The graph of −3.2 is $\frac{2}{10}$ of the way from −3 to −4.

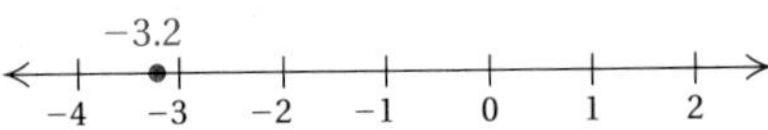

Example 7 Graph: $\frac{13}{8}$.

The number $\frac{13}{8}$ can be named $1\frac{5}{8}$, or 1.625. The graph is about $\frac{6}{10}$ of the way from 1 to 2.

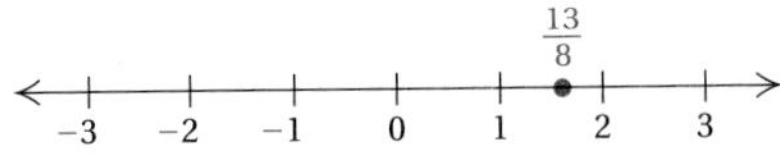

Do Exercises 5–7.

c Notation for Rational Numbers

Each rational number can be named using fractional or decimal notation.

Example 8 Convert to decimal notation: $-\frac{5}{8}$.

We first find decimal notation for $\frac{5}{8}$. Since $\frac{5}{8}$ means 5 ÷ 8, we divide.

```
    0.6 2 5
8 ) 5.0 0 0
    4 8
      2 0
      1 6
        4 0
        4 0
          0
```

Thus, $\frac{5}{8} = 0.625$, so $-\frac{5}{8} = -0.625$.

Decimal notation for $-\frac{5}{8}$ is −0.625. We consider −0.625 to be a **terminating decimal.** Decimal notation for some numbers repeats.

Example 9 Convert to decimal notation: $\frac{7}{11}$.

We divide.

```
     0.6 3 6 3...
1 1 ) 7.0 0 0 0
      6 6
        4 0
        3 3
          7 0
          6 6
            4 0
            3 3
              7
```

$$\frac{7}{11} = 0.\overline{63}$$

We can abbreviate repeating decimal notation by writing a bar over the repeating part—in this case, $0.\overline{63}$.

The following are other examples to show how each rational number can be named using fractional or decimal notation:

$$0 = \frac{0}{8}, \quad \frac{27}{100} = 0.27, \quad -8\frac{3}{4} = -8.75, \quad \frac{-13}{6} = -2.1\overline{6}.$$

Do Exercises 8–10.

d The Real Numbers and Order

Every rational number has a point on the number line. However, there are some points on the line for which there is no rational number. These points correspond to what are called **irrational numbers.**

What kinds of numbers are irrational? One example is the number π, which is used in finding the area and the circumference of a circle: $A = \pi r^2$ and $C = 2\pi r$.

Another example of an irrational number is the square root of 2, named $\sqrt{2}$.

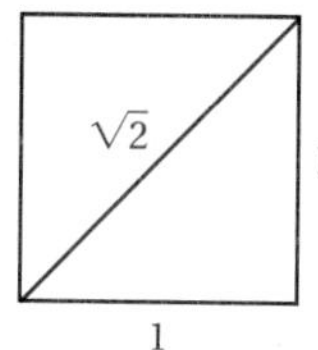

It is the length of the diagonal of a square with sides of length 1. It is also the number that when multiplied by itself gives 2. There is no rational number that can be multiplied by itself to get 2. But the following are rational *approximations*:

1.4 is an approximation of $\sqrt{2}$ because $(1.4)^2 = 1.96$;

1.41 is a better approximation because $(1.41)^2 = 1.9881$;

1.4142 is an even better approximation because $(1.4142)^2 = 1.99996164$.

We can find rational approximations for square roots using a calculator.

Decimal notation for rational numbers *either* terminates *or* repeats. Decimal notation for irrational numbers *neither* terminates *nor* repeats. Some other examples of irrational numbers are $\sqrt{3}$, $-\sqrt{8}$, $\sqrt{11}$, and 0.121221222122221.... Whenever we take the square root of a number that is not a perfect square, we will get an irrational number.

The rational numbers and the irrational numbers together correspond to all the points on a number line and make up what is called the **real-number system.**

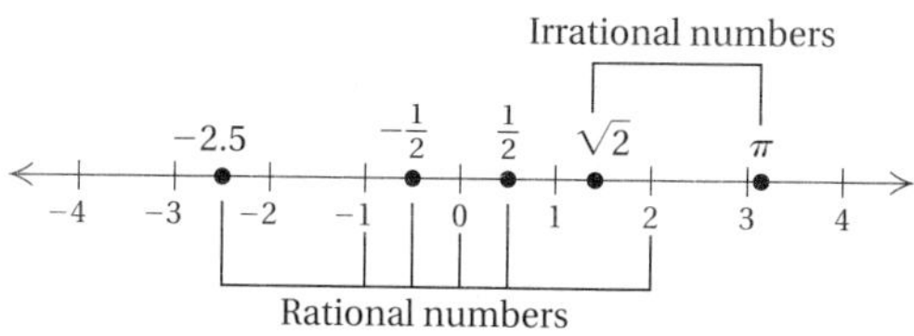

> The set of **real numbers** = The set of all numbers corresponding to points on the number line.

Convert to decimal notation.

8. $-\frac{3}{8}$

9. $-\frac{6}{11}$

10. $\frac{4}{3}$

Calculator Spotlight

Approximating Square Roots and π. Square roots are found by pressing [2nd] [√]. (√ is the second operation associated with the [x^2] key.)

To find an approximation for $\sqrt{48}$, we press

The approximation 6.92820323 is displayed.

To find $8 \cdot \sqrt{13}$, we press

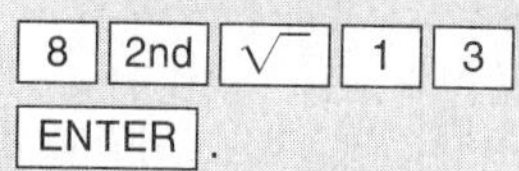

The approximation 28.8444102 is displayed.

The number π is used widely enough to have its own key. (π is the second operation associated with the [^] key.)

To approximate π, we press

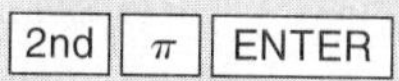

The approximation 3.141592654 is displayed.

Exercises

Approximate.

1. $\sqrt{76}$

2. $\sqrt{317}$

3. $15 \cdot \sqrt{20}$

4. $29 + \sqrt{42}$

5. π

6. $29 \cdot \pi$

7. $\pi \cdot 13^2$

8. $5 \cdot \pi + 8 \cdot \sqrt{237}$

Answers on page A-1

Use either $<$ or $>$ for ■ to write a true sentence.

11. -3 ■ 7

12. -8 ■ -5

13. 7 ■ -10

14. 3.1 ■ -9.5

15. $-\frac{2}{3}$ ■ -1

16. $-\frac{11}{8}$ ■ $\frac{23}{15}$

17. $-\frac{2}{3}$ ■ $-\frac{5}{9}$

18. -4.78 ■ -5.01

Answers on page A-1

The real numbers consist of the rational numbers and the irrational numbers. The following figure shows the relationships among various kinds of numbers.

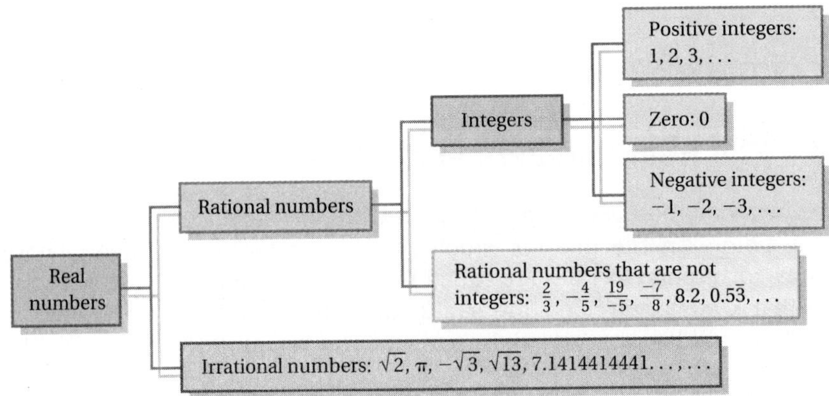

Order

Real numbers are named in order on the number line, with larger numbers named farther to the right. For any two numbers on the line, the one to the left is less than the one to the right.

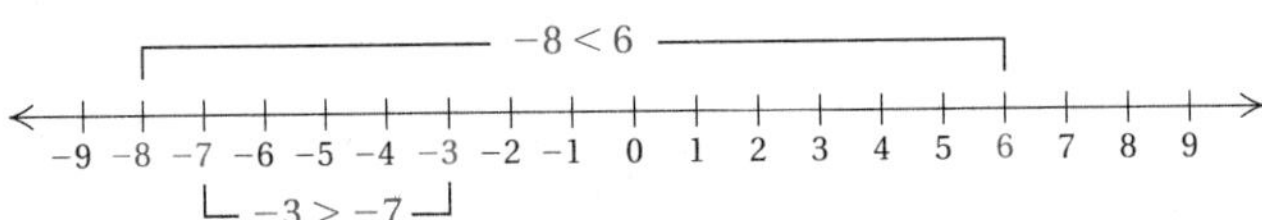

We use the symbol **<** to mean "**is less than.**" The sentence $-8 < 6$ means "-8 is less than 6." The symbol **>** means "**is greater than.**" The sentence $-3 > -7$ means "-3 is greater than -7." The sentences $-8 < 6$ and $-3 > -7$ are **inequalities**.

Examples Use either $<$ or $>$ for ■ to write a true sentence.

10. 2 ■ 9 — Since 2 is to the left of 9, 2 is less than 9, so $2 < 9$.

11. -7 ■ 3 — Since -7 is to the left of 3, we have $-7 < 3$.

12. 6 ■ -12 — Since 6 is to the right of -12, then $6 > -12$.

13. -18 ■ -5 — Since -18 is to the left of -5, we have $-18 < -5$.

14. -2.7 ■ $-\frac{3}{2}$ — The answer is $-2.7 < -\frac{3}{2}$.

15. 1.5 ■ -2.7 — The answer is $1.5 > -2.7$.

16. 1.38 ■ 1.83 — The answer is $1.38 < 1.83$.

17. -3.45 ■ 1.32 — The answer is $-3.45 < 1.32$.

18. -4 ■ 0 — The answer is $-4 < 0$.

19. 5.8 ■ 0 — The answer is $5.8 > 0$.

20. $\frac{5}{8}$ ■ $\frac{7}{11}$ — We convert to decimal notation: $\frac{5}{8} = 0.625$ and $\frac{7}{11} = 0.6363\ldots$. Thus, $\frac{5}{8} < \frac{7}{11}$.

Do Exercises 11–18.

Note that both $-8 < 6$ and $6 > -8$ are true. Every true inequality yields another true inequality when we interchange the numbers or variables and reverse the direction of the inequality sign.

> $a < b$ also has the meaning $b > a$.

Examples Write another inequality with the same meaning.

21. $a < -5$ The inequality $-5 > a$ has the same meaning.

22. $-3 > -8$ The inequality $-8 < -3$ has the same meaning.

A helpful mental device is to think of an inequality sign as an "arrow" with the arrow pointing to the smaller number.

Do Exercises 19 and 20.

Note that all positive real numbers are greater than zero and all negative real numbers are less than zero.

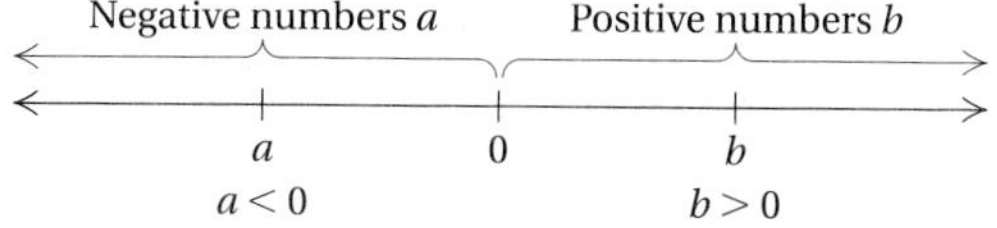

> If b is a positive real number, then $b > 0$.
> If a is a negative real number, then $a < 0$.

Expressions like $a \le b$ and $b \ge a$ are also inequalities. We read $a \le b$ as "***a* is less than or equal to *b*.**" We read $a \ge b$ as "***a* is greater than or equal to *b*.**"

Examples Write true or false for each statement.

23. $-3 \le 5.4$ True since $-3 < 5.4$ is true

24. $-3 \le -3$ True since $-3 = -3$ is true

25. $-5 \ge 1\frac{2}{3}$ False since neither $-5 > 1\frac{2}{3}$ nor $-5 = 1\frac{2}{3}$ is true

Do Exercises 21–23.

e Absolute Value

From the number line, we see that numbers like 4 and -4 are the same distance from zero. Distance is always a nonnegative number. We call the distance from zero on a number line the **absolute value** of the number.

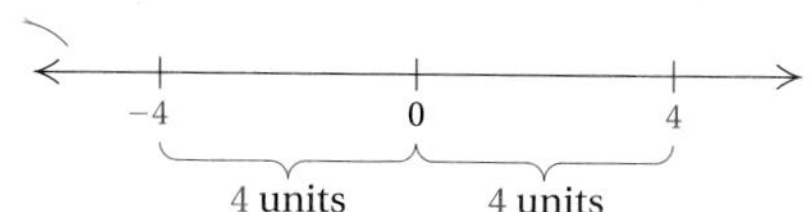

> The **absolute value** of a number is its distance from zero on a number line. We use the symbol $|x|$ to represent the absolute value of a number x.

Write another inequality with the same meaning.

19. $-5 < 7$

20. $x > 4$

Write true or false.

21. $-4 \le -6$

22. $7.8 \ge 7.8$

23. $-2 \le \dfrac{3}{8}$

Answers on page A-1

Find the absolute value.

24. $|8|$

25. $|0|$

26. $|-9|$

27. $\left|-\frac{2}{3}\right|$

28. $|5.6|$

To find absolute value:

a) If a number is negative, make it positive.

b) If a number is positive or zero, leave it alone.

Examples Find the absolute value.

26. $|-7|$ The distance of -7 from 0 is 7, so $|-7| = 7$.

27. $|12|$ The distance of 12 from 0 is 12, so $|12| = 12$.

28. $|0|$ The distance of 0 from 0 is 0, so $|0| = 0$.

29. $\left|\frac{3}{2}\right| = \frac{3}{2}$

30. $|-2.73| = 2.73$

Do Exercises 24–28.

Improving Your Math Study Skills

Tips for Using This Textbook

Throughout this textbook, you will find a feature called "Improving Your Math Study Skills." At least one such topic is included in each chapter. Each topic title is listed in the table of contents beginning on p. iii.

One of the most important ways to improve your math study skills is to learn the proper use of the textbook. Here we highlight a few points that we consider most helpful.

- **Be sure to note the special symbols [a], [b], [c], and so on, that correspond to the objectives you are to be able to perform.** They appear in many places throughout the text. The first time you see them is in the margin at the beginning of each section. The second time is in the subheadings of each section, and the third time is in the exercise set. You will also find them next to the skill maintenance exercises in each exercise set and in the review exercises at the end of the chapter, as well as in the answers to the chapter tests and the cumulative reviews. These objective symbols allow you to refer back whenever you need to review a topic.
- **Note the symbols in the margin under the list of objectives at the beginning of each section.** These refer to the many distinctive study aids that accompany the book.
- **Read and study each step of each example.** The examples include important side comments that explain each step. These carefully chosen examples and notes prepare you for success in the exercise set.
- **Stop and do the margin exercises as you study a section.** When our students come to us troubled about how they are doing in the course, the first question we ask is "Are you doing the margin exercises when directed to do so?" This is one of the most effective ways to enhance your ability to learn mathematics from this text. Don't deprive yourself of its benefits!
- **When you study the book, don't mark the points that you think are important, but mark the points you do not understand!** This book includes many design features that highlight important points. Use your efforts to mark where you are having trouble. Then when you go to class, a math lab, or a tutoring session, you will be prepared to ask questions that home in on your difficulties rather than spending time going over what you already understand.
- **If you are having trouble, consider using the *Student's Solutions Manual,* which contains worked-out solutions to the odd-numbered exercises in the exercise sets.**
- **Try to keep one section ahead of your syllabus.** If you study ahead of your lectures, you can concentrate on what is being explained in them, rather than trying to write everything down. You can then take notes only of special points or of questions related to what is happening in class.

Answers on page A-1

Exercise Set 1.2

a Tell which integers correspond to the situation.

1. *Elevation.* The Dead Sea, between Jordan and Israel, is 1286 ft below sea level; Mt. Rainier in Washington State is 13,804 ft above sea level.

2. Amy's golf score was 3 under par; Juan's was 7 over par.

3. On Wednesday, the temperature was 24° above zero. On Thursday, it was 2° below zero.

4. A student deposited her tax refund of \$750 in a savings account. Two weeks later, she withdrew \$125 to pay sorority fees.

5. *U.S. Public Debt.* Recently, the total public debt of the United States was about \$5.2 trillion (***Source*:** U.S. Department of the Treasury).

6. *Birth and Death Rates.* Recently, the world birth rate was 27 per thousand. The death rate was 9.7 per thousand. (***Source*:** United Nations Population Fund)

b Graph the number on the number line.

7. $\frac{10}{3}$

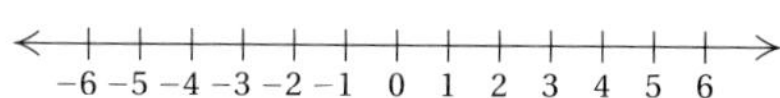

8. $-\frac{17}{4}$

−6 −5 −4 −3 −2 −1 0 1 2 3 4 5 6

9. -5.2

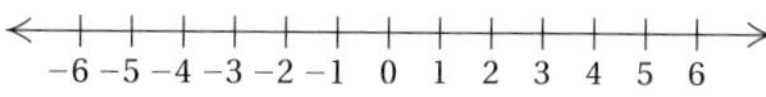

10. 4.78

−6 −5 −4 −3 −2 −1 0 1 2 3 4 5 6

c Convert to decimal notation.

11. $-\frac{7}{8}$ **12.** $-\frac{1}{8}$ **13.** $\frac{5}{6}$ **14.** $\frac{5}{3}$ **15.** $\frac{7}{6}$ **16.** $\frac{5}{12}$

17. $\frac{2}{3}$ **18.** $\frac{1}{4}$ **19.** $-\frac{1}{2}$ **20.** $\frac{5}{8}$ **21.** $\frac{1}{10}$ **22.** $-\frac{7}{20}$

d Use either $<$ or $>$ for $\square$ to write a true sentence.

23. $8 \;\square\; 0$ **24.** $3 \;\square\; 0$ **25.** $-8 \;\square\; 3$ **26.** $6 \;\square\; -6$

27. $-8 \;\square\; 8$ **28.** $0 \;\square\; -9$ **29.** $-8 \;\square\; -5$ **30.** $-4 \;\square\; -3$

31. $-5 \;\square\; -11$ **32.** $-3 \;\square\; -4$ **33.** $-6 \;\square\; -5$ **34.** $-10 \;\square\; -14$

35. $2.14 \;\square\; 1.24$ **36.** $-3.3 \;\square\; -2.2$ **37.** $-14.5 \;\square\; 0.011$ **38.** $17.2 \;\square\; -1.67$

39. $-12.88 \; \square \; -6.45$

40. $-14.34 \; \square \; -17.88$

41. $\frac{5}{12} \; \square \; \frac{11}{25}$

42. $-\frac{13}{16} \; \square \; -\frac{5}{9}$

Write true or false.

43. $-3 \geq -11$

44. $5 \leq -5$

45. $0 \geq 8$

46. $-5 \leq 7$

Write an inequality with the same meaning.

47. $-6 > x$

48. $x < 8$

49. $-10 \leq y$

50. $12 \geq t$

e Find the absolute value.

51. $|-3|$

52. $|-7|$

53. $|10|$

54. $|11|$

55. $|0|$

56. $|-4|$

57. $|-24|$

58. $|325|$

59. $\left|-\frac{2}{3}\right|$

60. $\left|-\frac{10}{7}\right|$

61. $\left|\frac{0}{4}\right|$

62. $|14.8|$

Skill Maintenance

This heading indicates that the exercises that follow are *skill maintenance exercises,* which review any skill previously studied in the text. You can expect such exercises in every exercise set. Answers to *all* skill maintenance exercises are found at the back of the book. If you miss an exercise, restudy the objective shown in blue.

Evaluate. [1.1a]

63. $\frac{5c}{d}$, for $c = 15$ and $d = 25$

64. $\frac{2x + y}{3}$, for $x = 12$ and $y = 9$

65. $\frac{q - r}{8}$, for $q = 30$ and $r = 6$

66. $\frac{w}{4y}$, for $w = 52$ and $y = 13$

Synthesis

Exercises marked with a ▦ are meant to be solved with a calculator.

67. ◆ ▦ When Jennifer's calculator gives a decimal approximation for $\sqrt{2}$ and that approximation is promptly squared, the result is 2. Yet, when that same approximation is entered by hand and then squared, the result is not exactly 2. Why do you suppose this happens?

68. ◆ How many rational numbers are there between 0 and 1? Why?

List in order from the least to the greatest.

69. $-\frac{2}{3}, \frac{1}{2}, -\frac{3}{4}, -\frac{5}{6}, \frac{3}{8}, \frac{1}{6}$

70. $-8\frac{7}{8}, 7^1, -5, |-6|, 4, |3|, -8\frac{5}{8}, -100, 0, 1^7, \frac{14}{4}, \frac{-67}{8}$

Given that $0.3\overline{3} = \frac{1}{3}$ and $0.6\overline{6} = \frac{2}{3}$, express each of the following as a quotient or ratio of two integers.

71. $0.1\overline{1}$

72. $0.9\overline{9}$

73. $5.5\overline{5}$

1.3 Addition of Real Numbers

In this section, we consider addition of real numbers. First, to gain an understanding, we add using a number line. Then we consider rules for addition.

Addition of numbers can be illustrated on a number line. To do the addition $a + b$, we start at a, and then move according to b.

a) If b is positive, we move to the right.

b) If b is negative, we move to the left.

c) If b is 0, we stay at a.

Example 1 Add: $3 + (-5)$.

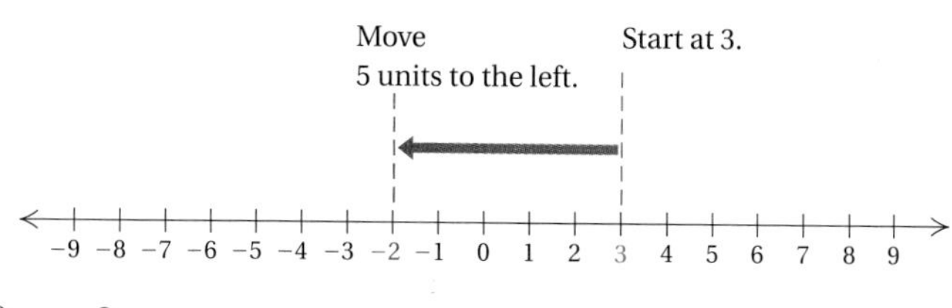

$3 + (-5) = -2$

Example 2 Add: $-4 + (-3)$.

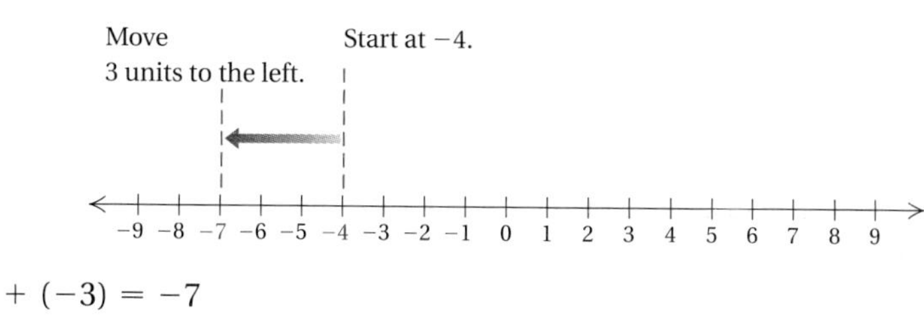

$-4 + (-3) = -7$

Example 3 Add: $-4 + 9$.

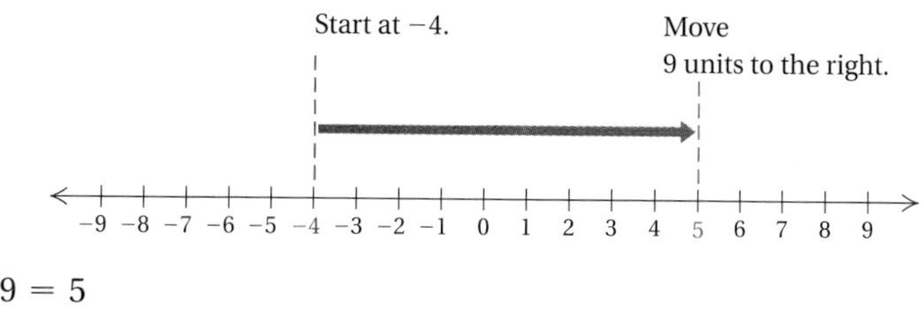

$-4 + 9 = 5$

Example 4 Add: $-5.2 + 0$.

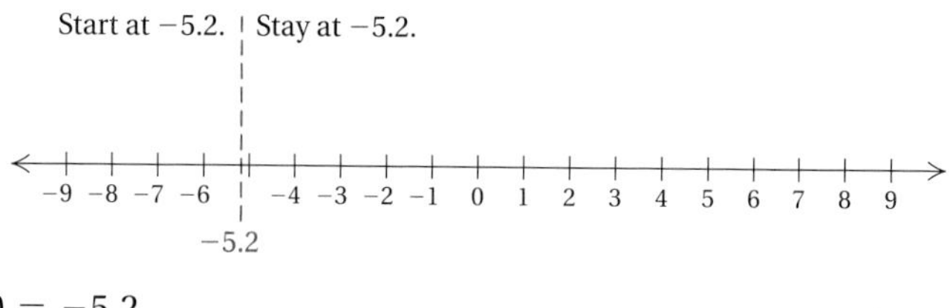

$-5.2 + 0 = -5.2$

Do Exercises 1–6.

Objectives

a Add real numbers without using a number line.

b Find the opposite, or additive inverse, of a real number.

For Extra Help

Add using a number line.

1. $0 + (-6)$

2. $1 + (-4)$

3. $-3 + (-5)$

4. $-3 + 7$

5. $-5.4 + 5.4$

6. $-\frac{5}{2} + \frac{1}{2}$

Answers on page A-2

Add without using a number line.

7. $-5 + (-6)$

8. $-9 + (-3)$

9. $-4 + 6$

10. $-7 + 3$

11. $5 + (-7)$

12. $-20 + 20$

13. $-11 + (-11)$

14. $10 + (-7)$

15. $-0.17 + 0.7$

16. $-6.4 + 8.7$

17. $-4.5 + (-3.2)$

18. $-8.6 + 2.4$

19. $\frac{5}{9} + \left(-\frac{7}{9}\right)$

20. $-\frac{1}{5} + \left(-\frac{3}{4}\right)$

Answers on page A-2

a Adding Without a Number Line

You may have noticed some patterns in the preceding examples. These lead us to rules for adding without using a number line that are more efficient for adding larger numbers.

RULES FOR ADDITION OF REAL NUMBERS

1. *Positive numbers*: Add the same as arithmetic numbers. The answer is positive.
2. *Negative numbers*: Add absolute values. The answer is negative.
3. *A positive and a negative number*: Subtract the smaller absolute value from the larger. Then:
   a) If the positive number has the greater absolute value, the answer is positive.
   b) If the negative number has the greater absolute value, the answer is negative.
   c) If the numbers have the same absolute value, the answer is 0.
4. *One number is zero*: The sum is the other number.

Rule 4 is known as the **identity property of 0.** It says that for any real number a, $a + 0 = a$.

Examples Add without using a number line.

5. $-12 + (-7) = -19$ — Two negatives. *Think*: Add the absolute values, 12 and 7, getting 19. Make the answer *negative*, -19.

6. $-1.4 + 8.5 = 7.1$ — The absolute values are 1.4 and 8.5. The difference is 7.1. The positive number has the larger absolute value, so the answer is *positive*, 7.1.

7. $-36 + 21 = -15$ — The absolute values are 36 and 21. The difference is 15. The negative number has the larger absolute value, so the answer is *negative*, -15.

8. $1.5 + (-1.5) = 0$ — The numbers have the same absolute value. The sum is 0.

9. $-\frac{7}{8} + 0 = -\frac{7}{8}$ — One number is zero. The sum is $-\frac{7}{8}$.

10. $-9.2 + 3.1 = -6.1$

11. $-\frac{3}{2} + \frac{9}{2} = \frac{6}{2} = 3$

12. $-\frac{2}{3} + \frac{5}{8} = -\frac{16}{24} + \frac{15}{24} = -\frac{1}{24}$

Do Exercises 7–20.

Suppose we want to add several numbers, some positive and some negative, as follows. How can we proceed?

$$15 + (-2) + 7 + 14 + (-5) + (-12)$$

We can change grouping and order as we please when adding. For instance, we can group the positive numbers together and the negative numbers together and add them separately. Then we add the two results.

Example 13 Add: $15 + (-2) + 7 + 14 + (-5) + (-12)$.

a) $15 + 7 + 14 = 36$ — **Adding the positive numbers**

b) $-2 + (-5) + (-12) = -19$ — **Adding the negative numbers**

c) $36 + (-19) = 17$ — **Adding the results**

We can also add the numbers in any other order we wish, say, from left to right as follows:

$$\begin{aligned} 15 + (-2) + 7 + 14 + (-5) + (-12) &= 13 + 7 + 14 + (-5) + (-12) \\ &= 20 + 14 + (-5) + (-12) \\ &= 34 + (-5) + (-12) \\ &= 29 + (-12) \\ &= 17 \end{aligned}$$

Do Exercises 21–24.

b Opposites, or Additive Inverses

Suppose we add two numbers that are **opposites**, such as 6 and -6. The result is 0. When opposites are added, the result is always 0. Such numbers are also called **additive inverses.** Every real number has an opposite, or additive inverse.

> Two numbers whose sum is 0 are called **opposites**, or **additive inverses,** of each other.

Examples Find the opposite of each number.

14. 34 — The opposite of 34 is -34 because $34 + (-34) = 0$.

15. -8 — The opposite of -8 is 8 because $-8 + 8 = 0$.

16. 0 — The opposite of 0 is 0 because $0 + 0 = 0$.

17. $-\frac{7}{8}$ — The opposite of $-\frac{7}{8}$ is $\frac{7}{8}$ because $-\frac{7}{8} + \frac{7}{8} = 0$.

Do Exercises 25–30.

To name the opposite, we use the symbol $-$, as follows.

> The opposite, or additive inverse, of a number a can be named $-a$ (read "the opposite of a," or "the additive inverse of a").

Note that if we take a number, say, 8, and find its opposite, -8, and then find the opposite of the result, we will have the original number, 8, again.

Add.

21. $(-15) + (-37) + 25 + 42 + (-59) + (-14)$

22. $42 + (-81) + (-28) + 24 + 18 + (-31)$

23. $-2.5 + (-10) + 6 + (-7.5)$

24. $-35 + 17 + 14 + (-27) + 31 + (-12)$

Find the opposite.

25. -4

26. 8.7

27. -7.74

28. $-\frac{8}{9}$

29. 0

30. 12

Answers on page A-2

Find $-x$ and $-(-x)$ when x is each of the following.

31. 14

32. 1

33. -19

34. -1.6

35. $\frac{2}{3}$

36. $-\frac{9}{8}$

Find the opposite. (Change the sign.)

37. -4

38. -13.4

39. 0

40. $\frac{1}{4}$

Answers on page A-2

> The opposite of the opposite of a number is the number itself. (The additive inverse of the additive inverse of a number is the number itself.) That is, for any number a,
>
> $$-(-a) = a.$$

Example 18 Find $-x$ and $-(-x)$ when $x = 16$.

If $x = 16$, then $-x = -16$. **The opposite of 16 is −16.**

If $x = 16$, then $-(-x) = -(-16) = 16$. **The opposite of the opposite of 16 is 16.**

Example 19 Find $-x$ and $-(-x)$ when $x = -3$.

If $x = -3$, then $-x = -(-3) = 3$.

If $x = -3$, then $-(-x) = -(-(-3)) = -3$.

Note that in Example 19 we used a second set of parentheses to show that we are substituting the negative number -3 for x. Symbolism like $--x$ is not considered meaningful.

Do Exercises 31–36.

A symbol such as -8 is usually read "negative 8." It could be read "the additive inverse of 8," because the additive inverse of 8 is negative 8. It could also be read "the opposite of 8," because the opposite of 8 is -8. Thus a symbol like -8 can be read in more than one way. A symbol like $-x$, which has a variable, should be read "the opposite of x" or "the additive inverse of x" and *not* "negative x," because we do not know whether x represents a positive number, a negative number, or 0. You can check this in Examples 18 and 19.

We can use the symbolism $-a$ to restate the definition of opposite, or additive inverse.

> For any real number a, the **opposite**, or **additive inverse**, of a, which is $-a$, is such that
>
> $$a + (-a) = (-a) + a = 0.$$

Signs of Numbers

A negative number is sometimes said to have a "negative sign." A positive number is said to have a "positive sign." When we replace a number with its opposite, we can say that we have "changed its sign."

Examples Find the opposite. (Change the sign.)

20. -3 $-(-3) = 3$ **The opposite of −3 is 3.**

21. -10 $-(-10) = 10$

22. 0 $-(0) = 0$

23. 14 $-(14) = -14$

Do Exercises 37–40.

Exercise Set 1.3

a Add. Do not use a number line except as a check.

1. $2 + (-9)$
2. $-5 + 2$
3. $-11 + 5$
4. $4 + (-3)$
5. $-6 + 6$
6. $8 + (-8)$
7. $-3 + (-5)$
8. $-4 + (-6)$
9. $-7 + 0$
10. $-13 + 0$
11. $0 + (-27)$
12. $0 + (-35)$
13. $17 + (-17)$
14. $-15 + 15$
15. $-17 + (-25)$
16. $-24 + (-17)$
17. $18 + (-18)$
18. $-13 + 13$
19. $-28 + 28$
20. $11 + (-11)$
21. $8 + (-5)$
22. $-7 + 8$
23. $-4 + (-5)$
24. $10 + (-12)$
25. $13 + (-6)$
26. $-3 + 14$
27. $-25 + 25$
28. $50 + (-50)$
29. $53 + (-18)$
30. $75 + (-45)$
31. $-8.5 + 4.7$
32. $-4.6 + 1.9$
33. $-2.8 + (-5.3)$
34. $-7.9 + (-6.5)$
35. $-\frac{3}{5} + \frac{2}{5}$
36. $-\frac{4}{3} + \frac{2}{3}$
37. $-\frac{2}{9} + \left(-\frac{5}{9}\right)$
38. $-\frac{4}{7} + \left(-\frac{6}{7}\right)$
39. $-\frac{5}{8} + \frac{1}{4}$
40. $-\frac{5}{6} + \frac{2}{3}$
41. $-\frac{5}{8} + \left(-\frac{1}{6}\right)$
42. $-\frac{5}{6} + \left(-\frac{2}{9}\right)$
43. $-\frac{3}{8} + \frac{5}{12}$
44. $-\frac{7}{16} + \frac{7}{8}$

45. $76 + (-15) + (-18) + (-6)$

46. $29 + (-45) + 18 + 32 + (-96)$

47. $-44 + \left(-\frac{3}{8}\right) + 95 + \left(-\frac{5}{8}\right)$

48. $24 + 3.1 + (-44) + (-8.2) + 63$

49. $98 + (-54) + 113 + (-998) + 44 + (-612)$

50. $-458 + (-124) + 1025 + (-917) + 218$

b Find the opposite, or additive inverse.

51. 24

52. -64

53. -26.9

54. 48.2

Find $-x$ when x is each of the following.

55. 8

56. -27

57. $-\frac{13}{8}$

58. $\frac{1}{236}$

Find $-(-x)$ when x is each of the following.

59. -43

60. 39

61. $\frac{4}{3}$

62. -7.1

Find the opposite. (Change the sign.)

63. -24

64. -12.3

65. $-\frac{3}{8}$

66. 10

Skill Maintenance

Convert to decimal notation. [1.2c]

67. $-\frac{5}{8}$

68. $\frac{1}{3}$

69. $-\frac{1}{12}$

70. $\frac{13}{20}$

Find the absolute value. [1.2e]

71. $|2.3|$

72. $|0|$

73. $\left|-\frac{4}{5}\right|$

74. $|-21.4|$

Synthesis

75. ◆ Without actually performing the addition, explain why the sum of all integers from -50 to 50 is 0.

76. ◆ Explain in your own words why the sum of two negative numbers is always negative.

77. For what numbers x is $-x$ negative?

78. For what numbers x is $-x$ positive?

Add.

79. ▦ $-3496 + (-2987)$

80. ▦ $2708 + (-3749)$

Tell whether the sum is positive, negative, or zero.

81. If a is positive and b is negative, $-a + b$ is ____________.

82. If $a = b$ and a and b are negative, $-a + (-b)$ is ____________.

Add integers using a variety of methods.

1.4 Subtraction of Real Numbers

a Subtraction

We now consider subtraction of real numbers. Subtraction is defined as follows.

> The difference $a - b$ is the number that when added to b gives a.

For example, $45 - 17 = 28$ because $28 + 17 = 45$. Let's consider an example whose answer is a negative number.

Example 1 Subtract: $5 - 8$.

Think: $5 - 8$ is the number that when added to 8 gives 5. What number can we add to 8 to get 5? The number must be negative. The number is -3:

$$5 - 8 = -3.$$

That is, $5 - 8 = -3$ because $5 = -3 + 8$.

Do Exercises 1–3.

The definition above does *not* provide the most efficient way to do subtraction. From that definition, however, we can develop a faster way to subtract. Look for a pattern in the following examples.

Subtractions	Adding an Opposite
$5 - 8 = -3$	$5 + (-8) = -3$
$-6 - 4 = -10$	$-6 + (-4) = -10$
$-7 - (-10) = 3$	$-7 + 10 = 3$
$-7 - (-2) = -5$	$-7 + 2 = -5$

Do Exercises 4–7.

Perhaps you have noticed that we can subtract by adding the opposite of the number being subtracted. This can always be done.

> For any real numbers a and b,
>
> $$a - b = a + (-b).$$
>
> (To subtract, add the opposite, or additive inverse, of the number being subtracted.)

This is the method generally used for quick subtraction of real numbers.

Objectives

a Subtract real numbers and simplify combinations of additions and subtractions.

b Solve applied problems involving addition and subtraction of real numbers.

For Extra Help

TAPE 1

MAC
WIN

CD-ROM

Subtract.

1. $-6 - 4$

Think: What number can be added to 4 to get -6?

2. $-7 - (-10)$

Think: What number can be added to -10 to get -7?

3. $-7 - (-2)$

Think: What number can be added to -2 to get -7?

Complete the addition and compare with the subtraction.

4. $4 - 6 = -2$;
$4 + (-6) =$ ______

5. $-3 - 8 = -11$;
$-3 + (-8) =$ ______

6. $-5 - (-9) = 4$;
$-5 + 9 =$ ______

7. $-5 - (-3) = -2$;
$-5 + 3 =$ ______

Answers on page A-2

Subtract.

8. $2 - 8$

9. $-6 - 10$

10. $12.4 - 5.3$

11. $-8 - (-11)$

12. $-8 - (-8)$

13. $\frac{2}{3} - \left(-\frac{5}{6}\right)$

Read each of the following. Then subtract by adding the opposite of the number being subtracted.

14. $3 - 11$

15. $12 - 5$

16. $-12 - (-9)$

17. $-12.4 - 10.9$

18. $-\frac{4}{5} - \left(-\frac{4}{5}\right)$

Simplify.

19. $-6 - (-2) - (-4) - 12 + 3$

20. $9 - (-6) + 7 - 11 - 14 - (-20)$

21. $-9.6 + 7.4 - (-3.9) - (-11)$

Answers on page A-2

Examples Subtract.

2. $2 - 6 = 2 + (-6) = -4$ — **The opposite of 6 is −6. We change the subtraction to addition and add the opposite.**

3. $4 - (-9) = 4 + 9 = 13$ — **The opposite of −9 is 9. We change the subtraction to addition and add the opposite.**

4. $-4.2 - (-3.6) = -4.2 + 3.6 = -0.6$ — **Adding the opposite.** ***Check:*** **$-0.6 + (-3.6) = -4.2$.**

5. $-\frac{1}{2} - \left(-\frac{3}{4}\right) = -\frac{1}{2} + \frac{3}{4} = \frac{1}{4}$ — **Adding the opposite.** ***Check:*** **$\frac{1}{4} + \left(-\frac{3}{4}\right) = -\frac{1}{2}$.**

Do Exercises 8–13.

Examples Read each of the following. Then subtract by adding the opposite of the number being subtracted.

6. $3 - 5$;
$3 - 5 = 3 + (-5) = -2$ — Read "three minus five is three plus the opposite of five"

7. $\frac{1}{8} - \frac{7}{8}$;
$\frac{1}{8} - \frac{7}{8} = \frac{1}{8} + \left(-\frac{7}{8}\right) = -\frac{6}{8}$, or $-\frac{3}{4}$ — Read "one-eighth minus seven-eighths is one-eighth plus the opposite of seven-eighths"

8. $-4.6 - (-9.8)$;
$-4.6 - (-9.8) = -4.6 + 9.8 = 5.2$ — Read "negative four point six minus negative nine point eight is negative four point six plus the opposite of negative nine point eight"

9. $-\frac{3}{4} - \frac{7}{5}$;
$-\frac{3}{4} - \frac{7}{5} = -\frac{3}{4} + \left(-\frac{7}{5}\right) = -\frac{15}{20} + \left(-\frac{28}{20}\right) = -\frac{43}{20}$ — Read "negative three-fourths minus seven-fifths is negative three-fourths plus the opposite of seven-fifths"

Do Exercises 14–18.

When several additions and subtractions occur together, we can make them all additions.

Examples Simplify.

10. $8 - (-4) - 2 - (-4) + 2 = 8 + 4 + (-2) + 4 + 2$
$= 16$ — **Adding the opposites where subtraction is indicated**

11. $8.2 - (-6.1) + 2.3 - (-4) = 8.2 + 6.1 + 2.3 + 4$
$= 20.6$

Do Exercises 19–21.

b Applications and Problem Solving

Let's now see how we can use addition and subtraction of real numbers to solve applied problems.

Example 12 *Home-Run Differential.* In baseball the difference between the number of home runs hit by a team's players and the number given up by its pitchers is called the *home-run differential,* that is,

$$\text{Home run differential} = \begin{matrix}\text{Number of} \\ \text{home runs}\end{matrix} - \begin{matrix}\text{Number of home} \\ \text{runs allowed}\end{matrix}.$$

Teams strive for a positive home-run differential.

a) In a recent year, Atlanta hit 197 home runs and gave up 120. Find its home-run differential.

b) In a recent year, San Francisco hit 153 home runs and gave up 194. Find its home-run differential.

We solve as follows.

a) We subtract 120 from 197 to find the home-run differential for Atlanta:

$$\text{Home-run differential} = 197 - 120 = 77.$$

b) We subtract 194 from 153 to find the home-run differential for San Francisco:

$$\text{Home-run differential} = 153 - 194 = -41.$$

Do Exercises 22 and 23.

22. *Home-Run Differential.* Complete the following table to find the home-run differentials for all the major-league baseball teams.

National League	HRs	HRs allowed	Diff.
Atlanta	197	120	+77
Florida	150	113	
Los Angeles	150	125	
Cincinnati	191	167	
Colorado	221	198	
San Diego	147	138	
Montreal	148	152	
Cubs	175	184	
Mets	147	159	
Houston	129	154	
Philadelphia	132	160	
St. Louis	142	173	
San Francisco	153	194	−41
Pittsburgh	138	183	

American League	HRs	HRs allowed	Diff.
Texas	221	168	
Baltimore	257	209	
Cleveland	218	173	
Oakland	243	205	
Seattle	245	216	
Boston	209	185	
White Sox	195	174	
Yankees	162	143	
Toronto	177	187	
California	192	219	
Milwaukee	178	213	
Detroit	204	241	
Kansas City	123	176	
Minnesota	118	233	

23. *Temperature Extremes.* In Churchill, Manitoba, Canada, the average daily low temperature in January is −31°C. The average daily low temperature in Key West, Florida, is 19°C. How much higher is the average daily low temperature in Key West, Florida?

Answers on page A-2

Improving Your Math Study Skills

Getting Started in a Math Class: The First-Day Handout or Syllabus

There are many ways in which to improve your math study skills. We have already considered some tips on using this book. We now consider some more general tips.

- **Textbook.** On the first day of class, most instructors distribute a handout that lists the textbook and other materials needed in the course. If possible, call the instructor or the department office before the term begins to find out which textbook you will be using and visit the bookstore to pick it up. This way, you can purchase the book before class starts and be ready to begin studying.
- **Attendance.** The handout may also describe the attendance policy for your class. Some instructors take attendance at every class, while others use different methods to track students' attendance. Regardless of the policy, you should plan to attend every class. Missing even one class can cause you to fall behind. If attendance counts toward your course grade, find out if there is a way to make up for missed days.

  If you do miss a class, call the instructor as soon as possible to find out what material was covered and what was assigned for the next class. If you have a study partner, call this person; ask if you can make a copy of his or her notes and find out what the homework assignment was.
- **Homework.** The first-day handout may also detail how homework is handled. Find out when, and how often, homework will be assigned, whether homework is collected or graded, and whether there will be quizzes over the homework material.

If the homework will be graded, find out what part of the final grade it will determine. Ask what the policy is for late homework. If you do miss a homework deadline, be sure to do the assigned homework anyway, as this is the best way to learn the material.

- **Grading.** The handout may also provide information on how your grade will be calculated at the end of the term. Typically, there will be tests during the term and a final exam at the end of the term. Frequently, homework is counted as part of the grade calculation, as are the quizzes. Find out how many tests will be given, if there is an option for make-up tests, or if any test grades will be dropped at the end of the term.

Some instructors keep the class grades on a computer. If this is the case, find out if you can receive current grade reports throughout the term.

- **Get to know your classmates.** It can be a big help in a math class to get to know your fellow students. You might consider forming a study group. If you do so, find out their phone numbers and schedules so that you can coordinate study time for homework or tests.
- **Get to know your instructor.** It can, of course, help immensely to get to know your instructor. Trivial though it may seem, get basic information like his or her name, how he or she can be contacted outside of class, and where the office is.

  Learn about your instructor's teaching style and try to adapt your learning to it.

Exercise Set 1.4

a Subtract.

1. $2 - 9$

2. $3 - 8$

3. $0 - 4$

4. $0 - 9$

5. $-8 - (-2)$

6. $-6 - (-8)$

7. $-11 - (-11)$

8. $-6 - (-6)$

9. $12 - 16$

10. $14 - 19$

11. $20 - 27$

12. $30 - 4$

13. $-9 - (-3)$

14. $-7 - (-9)$

15. $-40 - (-40)$

16. $-9 - (-9)$

17. $7 - 7$

18. $9 - 9$

19. $7 - (-7)$

20. $4 - (-4)$

21. $8 - (-3)$

22. $-7 - 4$

23. $-6 - 8$

24. $6 - (-10)$

25. $-4 - (-9)$

26. $-14 - 2$

27. $1 - 8$

28. $2 - 8$

29. $-6 - (-5)$

30. $-4 - (-3)$

31. $8 - (-10)$

32. $5 - (-6)$

33. $0 - 10$

34. $0 - 18$

35. $-5 - (-2)$

36. $-3 - (-1)$

37. $-7 - 14$

38. $-9 - 16$

39. $0 - (-5)$

40. $0 - (-1)$

41. $-8 - 0$

42. $-9 - 0$

43. $7 - (-5)$

44. $7 - (-4)$

45. $2 - 25$

46. $18 - 63$

47. $-42 - 26$

48. $-18 - 63$

49. $-71 - 2$

50. $-49 - 3$

51. $24 - (-92)$

52. $48 - (-73)$

53. $-50 - (-50)$

54. $-70 - (-70)$

55. $-\frac{3}{8} - \frac{5}{8}$

56. $\frac{3}{9} - \frac{9}{9}$

57. $\frac{3}{4} - \frac{2}{3}$

58. $\frac{5}{8} - \frac{3}{4}$

59. $-\frac{3}{4} - \frac{2}{3}$

60. $-\frac{5}{8} - \frac{3}{4}$

61. $-\frac{5}{8} - \left(-\frac{3}{4}\right)$

62. $-\frac{3}{4} - \left(-\frac{2}{3}\right)$

63. $6.1 - (-13.8)$

64. $1.5 - (-3.5)$

65. $-2.7 - 5.9$

66. $-3.2 - 5.8$

67. $0.99 - 1$

68. $0.87 - 1$

69. $-79 - 114$

70. $-197 - 216$

71. $0 - (-500)$

72. $500 - (-1000)$

73. $-2.8 - 0$

74. $6.04 - 1.1$

75. $7 - 10.53$

76. $8 - (-9.3)$

77. $\frac{1}{6} - \frac{2}{3}$

78. $-\frac{3}{8} - \left(-\frac{1}{2}\right)$

79. $-\frac{4}{7} - \left(-\frac{10}{7}\right)$

80. $\frac{12}{5} - \frac{12}{5}$

81. $-\frac{7}{10} - \frac{10}{15}$

82. $-\frac{4}{18} - \left(-\frac{2}{9}\right)$

83. $\frac{1}{5} - \frac{1}{3}$

84. $-\frac{1}{7} - \left(-\frac{1}{6}\right)$

Simplify.

85. $18 - (-15) - 3 - (-5) + 2$

86. $22 - (-18) + 7 + (-42) - 27$

87. $-31 + (-28) - (-14) - 17$

88. $-43 - (-19) - (-21) + 25$

89. $-34 - 28 + (-33) - 44$

90. $39 + (-88) - 29 - (-83)$

91. $-93 - (-84) - 41 - (-56)$

92. $84 + (-99) + 44 - (-18) - 43$

93. $-5 - (-30) + 30 + 40 - (-12)$

94. $14 - (-50) + 20 - (-32)$

95. $132 - (-21) + 45 - (-21)$

96. $81 - (-20) - 14 - (-50) + 53$

b Solve.

97. *Ocean Depth.* The deepest point in the Pacific Ocean is the Marianas Trench, with a depth of 11,033 m. The deepest point in the Atlantic Ocean is the Puerto Rico Trench, with a depth of 8648 m. What is the difference in the elevation of the two trenches?

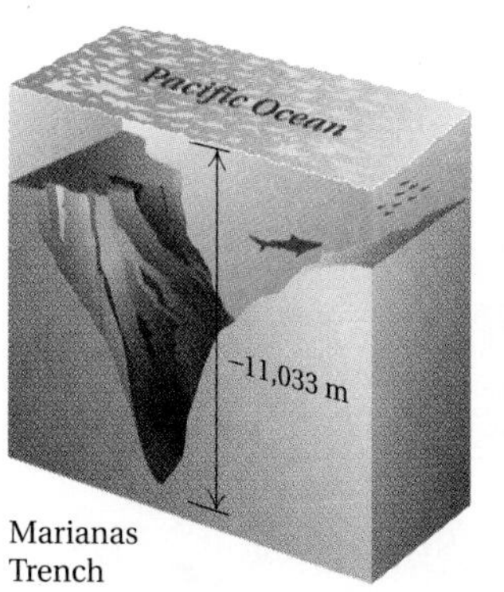

Marianas Trench

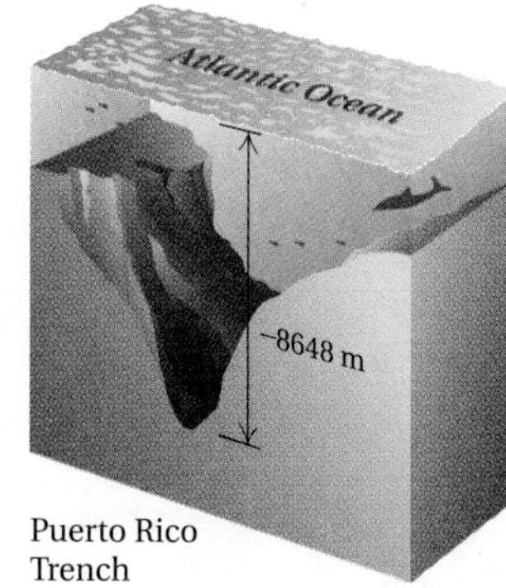

Puerto Rico Trench

98. *Depth of Offshore Oil Wells.* In 1993, the elevation of the world's deepest offshore oil well was −2860 ft. By 1998, the deepest well is expected to be 360 ft deeper. (***Source:*** *New York Times,* 12/7/94, p. D1.) What will be the elevation of the deepest well in 1998?

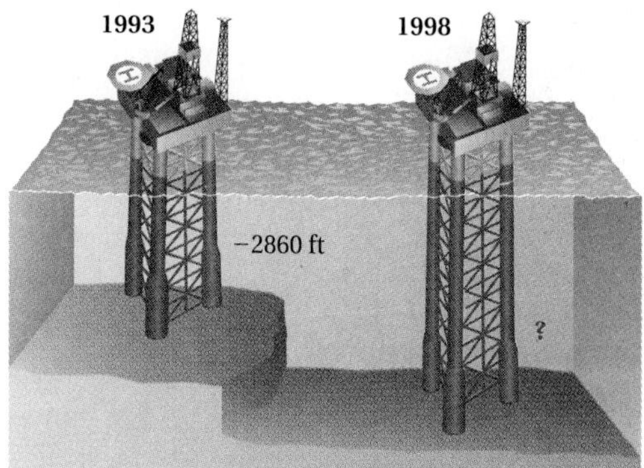

99. Laura has a charge of $476.89 on her credit card, but she then returns a sweater that cost $128.95. How much does she now owe on her credit card?

100. Chris has $720 in a checking account. He writes a check for $970 to pay for a sound system. What is the balance in his checking account?

101. *Temperature Records.* The greatest recorded temperature change in one day occurred in Browning, Montana, where the temperature fell from 44°F to −56°F (***Source:*** *The Guinness Book of Records*). How much did the temperature drop?

102. *Low Points on Continents.* The lowest point in Africa is Lake Assal, which is 515 ft below sea level. The lowest point in South America is the Valdes Peninsula, which is 132 ft below sea level. How much lower is Lake Assal than the Valdes Peninsula?

Skill Maintenance

Translate to an algebraic expression. [1.1b]

103. 7 more than y

104. 41 less than t

105. h subtracted from a

106. The product of 6 and c

107. r more than s

108. x less than y

SYNTHESIS

109. ◆ If a negative number is subtracted from a positive number, will the result always be positive? Why or why not?

110. ◆ Write a problem for a classmate to solve. Design the problem so that the solution is "The temperature dropped to −9°."

Subtract.

111. ▦ 123,907 − 433,789

112. ▦ 23,011 − (−60,432)

Tell whether the statement is true or false for all integers a and b. If false, show why.

113. $a - 0 = 0 - a$

114. $0 - a = a$

115. If $a \neq b$, then $a - b \neq 0$.

116. If $a = -b$, then $a + b = 0$.

117. If $a + b = 0$, then a and b are opposites.

118. If $a - b = 0$, then $a = -b$.

119. Maureen is a stockbroker. She kept track of the changes in the stock market over a period of 5 weeks. By how many points had the market risen or fallen over this time?

Week 1	Week 2	Week 3	Week 4	Week 5
Down 13 pts	Down 16 pts	Up 36 pts	Down 11 pts	Up 19 pts

120. *Blackjack Counting System.* The casino game of blackjack makes use of many card-counting systems to give players a winning edge if the count becomes negative. One such system is called *High–Low*, first developed by Harvey Dubner in 1963. Each card counts as −1, 0, or 1 as follows:

2, 3, 4, 5, 6 count as +1;
7, 8, 9 count as 0;
10, J, Q, K, A count as −1.

(***Source:*** Patterson, Jerry L., *Casino Gambling.* New York: Perigee, 1982)

a) Find the final count on the sequence of cards

K, A, 2, 4, 5, 10, J, 8, Q, K, 5.

b) Does the player have a winning edge?

Collaborative Learning Manual — Subtract integers using tiles.

1.5 Multiplication of Real Numbers

a Multiplication

Multiplication of real numbers is very much like multiplication of arithmetic numbers. The only difference is that we must determine whether the answer is positive or negative.

Multiplication of a Positive Number and a Negative Number

To see how to multiply a positive number and a negative number, consider the pattern of the following.

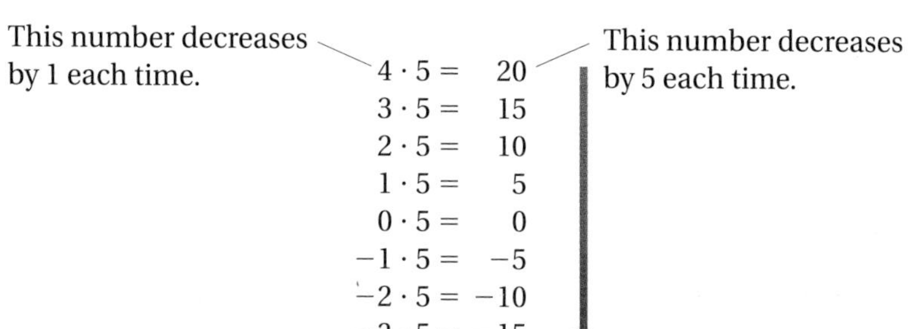

This number decreases by 1 each time.

$$\begin{aligned} 4 \cdot 5 &= 20 \\ 3 \cdot 5 &= 15 \\ 2 \cdot 5 &= 10 \\ 1 \cdot 5 &= 5 \\ 0 \cdot 5 &= 0 \\ -1 \cdot 5 &= -5 \\ -2 \cdot 5 &= -10 \\ -3 \cdot 5 &= -15 \end{aligned}$$

This number decreases by 5 each time.

Do Exercise 1.

According to this pattern, it looks as though the product of a negative number and a positive number is negative. That is the case, and we have the first part of the rule for multiplying numbers.

> To multiply a positive number and a negative number, multiply their absolute values. The answer is negative.

Examples Multiply.

1. $8(-5) = -40$ **2.** $-\frac{1}{3} \cdot \frac{5}{7} = -\frac{5}{21}$ **3.** $(-7.2)5 = -36$

Do Exercises 2–7.

Multiplication of Two Negative Numbers

How do we multiply two negative numbers? Again, we look for a pattern.

This number decreases by 1 each time.

$$\begin{aligned} 4 \cdot (-5) &= -20 \\ 3 \cdot (-5) &= -15 \\ 2 \cdot (-5) &= -10 \\ 1 \cdot (-5) &= -5 \\ 0 \cdot (-5) &= 0 \\ -1 \cdot (-5) &= 5 \\ -2 \cdot (-5) &= 10 \\ -3 \cdot (-5) &= 15 \end{aligned}$$

This number increases by 5 each time.

Do Exercise 8.

Objective

a Multiply real numbers.

For Extra Help

TAPE 2 MAC WIN CD-ROM

1. Complete, as in the example.

$$\begin{aligned} 4 \cdot 10 &= 40 \\ 3 \cdot 10 &= 30 \\ 2 \cdot 10 &= \\ 1 \cdot 10 &= \\ 0 \cdot 10 &= \\ -1 \cdot 10 &= \\ -2 \cdot 10 &= \\ -3 \cdot 10 &= \end{aligned}$$

Multiply.

2. $-3 \cdot 6$

3. $20 \cdot (-5)$

4. $4 \cdot (-20)$

5. $-\frac{2}{3} \cdot \frac{5}{6}$

6. $-4.23(7.1)$

7. $\frac{7}{8}\left(-\frac{4}{5}\right)$

8. Complete, as in the example.

$$\begin{aligned} 3 \cdot (-10) &= -30 \\ 2 \cdot (-10) &= -20 \\ 1 \cdot (-10) &= \\ 0 \cdot (-10) &= \\ -1 \cdot (-10) &= \\ -2 \cdot (-10) &= \\ -3 \cdot (-10) &= \end{aligned}$$

Answers on page A-2

Multiply.

9. $-9 \cdot (-3)$

10. $-16 \cdot (-2)$

11. $-7 \cdot (-5)$

12. $-\frac{4}{7}\left(-\frac{5}{9}\right)$

13. $-\frac{3}{2}\left(-\frac{4}{9}\right)$

14. $-3.25(-4.14)$

Multiply.

15. $5(-6)$

16. $(-5)(-6)$

17. $(-3.2) \cdot 0$

18. $\left(-\frac{4}{5}\right)\left(\frac{10}{3}\right)$

Answers on page A-2

According to the pattern, it appears that the product of two negative numbers is positive. That is actually so, and we have the second part of the rule for multiplying real numbers.

> To multiply two negative numbers, multiply their absolute values. The answer is positive.

Do Exercises 9–14.

The following is another way to consider the rules we have for multiplication.

> To multiply two real numbers:
>
> **a)** Multiply the absolute values.
>
> **b)** If the signs are the same, the answer is positive.
>
> **c)** If the signs are different, the answer is negative.

Multiplication by Zero

The only case that we have not considered is multiplying by zero. As with other numbers, the product of any real number and 0 is 0.

> **THE MULTIPLICATION PROPERTY OF ZERO**
>
> For any real number a,
>
> $$a \cdot 0 = 0.$$
>
> (The product of 0 and any real number is 0.)

Examples Multiply.

4. $(-3)(-4) = 12$
5. $-1.6(2) = -3.2$
6. $-19 \cdot 0 = 0$
7. $\left(-\frac{5}{6}\right)\left(-\frac{1}{9}\right) = \frac{5}{54}$

Do Exercises 15–18.

Multiplying More Than Two Numbers

When multiplying more than two real numbers, we can choose order and grouping as we please.

Examples Multiply.

8. $-8 \cdot 2(-3) = -16(-3)$ Multiplying the first two numbers
$= 48$

9. $-8 \cdot 2(-3) = 24 \cdot 2$ Multiplying the negatives. Every pair of negative numbers gives a positive product.
$= 48$

10. $-3(-2)(-5)(4) = 6(-5)(4)$ Multiplying the first two numbers
$= (-30)4$
$= -120$

11. $\left(-\frac{1}{2}\right)(8)\left(-\frac{2}{3}\right)(-6) = (-4)4$ Multiplying the first two numbers and the last two numbers
$= -16$

12. $-5 \cdot (-2) \cdot (-3) \cdot (-6) = 10 \cdot 18$
$= 180$

13. $(-3)(-5)(-2)(-3)(-6) = (-30)(18)$
$= -540$

We can see the following pattern in the results of Examples 12 and 13.

> The product of an even number of negative numbers is positive.
> The product of an odd number of negative numbers is negative.

Do Exercises 19–24.

Let's compare the expressions $(-x)^2$ and $-x^2$.

Example 14 Evaluate $(-x)^2$ and $-x^2$ for $x = 5$.

$(-x)^2 = (-5)^2 = (-5)(-5) = 25;$ Substitute 5 for x. Then evaluate the power.

$-x^2 = -(5)^2 = -25$ Substitute 5 for x. Evaluate the power. Then find the opposite.

The expressions $(-x)^2$ and $-x^2$ are *not* equivalent. That is, they do not have the same value for every allowable replacement of the variable by a real number. To find $(-x)^2$, we take the opposite and then square. To find $-x^2$, we find the square and then take the opposite.

Example 15 Evaluate $2x^2$ for $x = 3$ and $x = -3$.

$2x^2 = 2(3)^2 = 2(9) = 18;$

$2x^2 = 2(-3)^2 = 2(9) = 18$

Do Exercises 25–27.

Multiply.

19. $5 \cdot (-3) \cdot 2$

20. $-3 \times (-4.1) \times (-2.5)$

21. $-\frac{1}{2} \cdot \left(-\frac{4}{3}\right) \cdot \left(-\frac{5}{2}\right)$

22. $-2 \cdot (-5) \cdot (-4) \cdot (-3)$

23. $(-4)(-5)(-2)(-3)(-1)$

24. $(-1)(-1)(-2)(-3)(-1)(-1)$

25. Evaluate $(-x)^2$ and $-x^2$ for $x = 2$.

26. Evaluate $(-x)^2$ and $-x^2$ for $x = 3$.

27. Evaluate $3x^2$ for $x = 4$ and for $x = -4$.

Answers on page A-2

Improving Your Math Study Skills

Classwork: Before and During Class

Before Class

Textbook

- Check your syllabus (or ask your instructor) to find out which sections will be covered during the next class. Then be sure to read these sections *before* class. Although you may not understand all the concepts, you will at least be familiar with the material, which will help you follow the discussion during class.
- This book makes use of color, shading, and design elements to highlight important concepts, so you do not need to highlight these. Instead, it is more productive for you to note trouble spots with either a highlighter or Post-It notes. Then use these marked points as possible questions for clarification by your instructor at the appropriate time.

Homework

- Review the previous day's homework just before class. This will refresh your memory on the concepts covered in the last class, and again provide you with possible questions to ask your instructor.

During Class

Class Seating

- If possible, choose a seat at the front of the class. In most classes, the more serious students tend to sit there so you will probably be able to concentrate better if you do the same. You should also avoid sitting next to noisy or distracting students.
- If your instructor uses an overhead projector, choose a seat that will give you an unobstructed view of the screen.

Taking Notes

- This textbook has been written and laid out so that it represents a quality set of notes at the same time that it teaches. Thus you might not need to take notes in class. Just watch, listen, and ask yourself questions as the class moves along, rather than racing to keep up your note-taking.

  However, if you still feel more comfortable taking your own notes, consider using the following two-column method. Divide your page in half vertically so that you have two columns side by side. Write down what is on the board in the left column; then, in the right column, write clarifying comments or questions.
- If you have any difficulty keeping up with the instructor, use abbreviations to speed up your note-taking. Consider standard abbreviations like "Ex" for "Example," "$\approx$" for "approximately equal to," or "$\therefore$" for "therefore." Create your own abbreviations as well.
- Another shortcut for note-taking is to write only the beginning of a word, leaving space for the rest. Be sure you write enough of the word to know what it means later on!

Exercise Set 1.5

a Multiply.

1. $-4 \cdot 2$
2. $-3 \cdot 5$
3. $-8 \cdot 6$
4. $-5 \cdot 2$
5. $8 \cdot (-3)$
6. $9 \cdot (-5)$
7. $-9 \cdot 8$
8. $-10 \cdot 3$
9. $-8 \cdot (-2)$
10. $-2 \cdot (-5)$
11. $-7 \cdot (-6)$
12. $-9 \cdot (-2)$
13. $15 \cdot (-8)$
14. $-12 \cdot (-10)$
15. $-14 \cdot 17$
16. $-13 \cdot (-15)$
17. $-25 \cdot (-48)$
18. $39 \cdot (-43)$
19. $-3.5 \cdot (-28)$
20. $97 \cdot (-2.1)$
21. $9 \cdot (-8)$
22. $7 \cdot (-9)$
23. $4 \cdot (-3.1)$
24. $3 \cdot (-2.2)$
25. $-5 \cdot (-6)$
26. $-6 \cdot (-4)$
27. $-7 \cdot (-3.1)$
28. $-4 \cdot (-3.2)$
29. $\frac{2}{3} \cdot \left(-\frac{3}{5}\right)$
30. $\frac{5}{7} \cdot \left(-\frac{2}{3}\right)$
31. $-\frac{3}{8} \cdot \left(-\frac{2}{9}\right)$
32. $-\frac{5}{8} \cdot \left(-\frac{2}{5}\right)$
33. -6.3×2.7
34. -4.1×9.5
35. $-\frac{5}{9} \cdot \frac{3}{4}$
36. $-\frac{8}{3} \cdot \frac{9}{4}$
37. $7 \cdot (-4) \cdot (-3) \cdot 5$
38. $9 \cdot (-2) \cdot (-6) \cdot 7$
39. $-\frac{2}{3} \cdot \frac{1}{2} \cdot \left(-\frac{6}{7}\right)$
40. $-\frac{1}{8} \cdot \left(-\frac{1}{4}\right) \cdot \left(-\frac{3}{5}\right)$
41. $-3 \cdot (-4) \cdot (-5)$
42. $-2 \cdot (-5) \cdot (-7)$
43. $-2 \cdot (-5) \cdot (-3) \cdot (-5)$
44. $-3 \cdot (-5) \cdot (-2) \cdot (-1)$
45. $\frac{1}{5}\left(-\frac{2}{9}\right)$
46. $-\frac{3}{5}\left(-\frac{2}{7}\right)$
47. $-7 \cdot (-21) \cdot 13$
48. $-14 \cdot (34) \cdot 12$
49. $-4 \cdot (-1.8) \cdot 7$
50. $-8 \cdot (-1.3) \cdot (-5)$

51. $-\frac{1}{9}\left(-\frac{2}{3}\right)\left(\frac{5}{7}\right)$

52. $-\frac{7}{2}\left(-\frac{5}{7}\right)\left(-\frac{2}{5}\right)$

53. $4 \cdot (-4) \cdot (-5) \cdot (-12)$

54. $-2 \cdot (-3) \cdot (-4) \cdot (-5)$

55. $0.07 \cdot (-7) \cdot 6 \cdot (-6)$

56. $80 \cdot (-0.8) \cdot (-90) \cdot (-0.09)$

57. $\left(-\frac{5}{6}\right)\left(\frac{1}{8}\right)\left(-\frac{3}{7}\right)\left(-\frac{1}{7}\right)$

58. $\left(\frac{4}{5}\right)\left(-\frac{2}{3}\right)\left(-\frac{15}{7}\right)\left(\frac{1}{2}\right)$

59. $(-14) \cdot (-27) \cdot 0$

60. $7 \cdot (-6) \cdot 5 \cdot (-4) \cdot 3 \cdot (-2) \cdot 1 \cdot 0$

61. $(-8)(-9)(-10)$

62. $(-7)(-8)(-9)(-10)$

63. $(-6)(-7)(-8)(-9)(-10)$

64. $(-5)(-6)(-7)(-8)(-9)(-10)$

65. Evaluate $(-3x)^2$ and $-3x^2$ for $x = 7$.

66. Evaluate $(-2x)^2$ and $-2x^2$ for $x = 3$.

67. Evaluate $5x^2$ for $x = 2$ and for $x = -2$.

68. Evaluate $2x^2$ for $x = 5$ and for $x = -5$.

Skill Maintenance

69. Evaluate $\frac{x - 2y}{3}$ for $x = 20$ and $y = 7$. [1.1a]

70. Subtract: $-\frac{1}{2} - \left(-\frac{1}{6}\right)$. [1.4a]

Write true or false. [1.2d]

71. $-10 > -12$

72. $0 \leq -1$

73. $4 < -8$

74. $-7 \geq -6$

Synthesis

75. ◆ Multiplication can be thought of as repeated addition. Using this concept and a number line, explain why $3 \cdot (-5) = -15$.

76. ◆ What rule have we developed that would tell you the sign of $(-7)^8$ and $(-7)^{11}$ without doing the computations? Explain.

77. After diving 95 m below the surface, a diver rises at a rate of 7 meters per minute for 9 min. What is the diver's new elevation?

78. Jo wrote seven checks for $13 each. If she began with a balance of $68 in her account, what was her balance after having written the checks?

79. What must be true of a and b if $-ab$ is to be (a) positive? (b) zero? (c) negative?

80. Evaluate $-6(3x - 5y) + z$ for $x = -2$, $y = -4$, and $z = 5$.

1.6 Division of Real Numbers

We now consider division of real numbers. The definition of division results in rules for division that are the same as those for multiplication.

a Division of Integers

> The quotient $\frac{a}{b}$ (or $a \div b$) is the number, if there is one, that when multiplied by b gives a.

Let's use the definition to divide integers.

Examples Divide, if possible. Check your answer.

1. $14 \div (-7) = -2$ — *Think*: What number multiplied by -7 gives 14? That number is -2. *Check*: $(-2)(-7) = 14$.

2. $\frac{-32}{-4} = 8$ — *Think*: What number multiplied by -4 gives -32? That number is 8. *Check*: $8(-4) = -32$.

3. $\frac{-10}{7} = -\frac{10}{7}$ — *Think*: What number multiplied by 7 gives -10? That number is $-\frac{10}{7}$. *Check*: $-\frac{10}{7} \cdot 7 = -10$.

4. $\frac{-17}{0}$ **is undefined.** — *Think*: What number multiplied by 0 gives -17? There is no such number because the product of 0 and *any* number is 0.

The rules for division are the same as those for multiplication.

> To multiply or divide two real numbers:
> a) Multiply or divide the absolute values.
> b) If the signs are the same, the answer is positive.
> c) If the signs are different, the answer is negative.

Do Exercises 1–6.

Division by Zero

Example 4 shows why we cannot divide -17 by 0. We can use the same argument to show why we cannot divide any nonzero number b by 0. Consider $b \div 0$. We look for a number that when multiplied by 0 gives b. There is no such number because the product of 0 and any number is 0. Thus we cannot divide a nonzero number b by 0.

On the other hand, if we divide 0 by 0, we look for a number r such that $0 \cdot r = 0$. But $0 \cdot r = 0$ for any number r. Thus it appears that $0 \div 0$ could be any number we choose. Getting any answer we want when we divide 0 by 0 would be very confusing. Thus we agree that division by zero is undefined.

> Division by 0 is undefined.
>
> $a \div 0$ is undefined for all real numbers a.
>
> 0 divided by a nonzero number a is 0.
>
> $0 \div a = 0, \quad a \neq 0.$

Objectives

a Divide integers.

b Find the reciprocal of a real number.

c Divide real numbers.

For Extra Help

TAPE 2

MAC
WIN

CD-ROM

Divide.

1. $6 \div (-3)$
   Think: What number multiplied by -3 gives 6?

2. $\frac{-15}{-3}$
   Think: What number multiplied by -3 gives -15?

3. $-24 \div 8$
   Think: What number multiplied by 8 gives -24?

4. $\frac{-48}{-6}$

5. $\frac{30}{-5}$

6. $\frac{30}{-7}$

Answers on page A-2

Divide, if possible.

7. $\frac{-5}{0}$

8. $\frac{0}{-3}$

Find the reciprocal.

9. $\frac{2}{3}$

10. $-\frac{5}{4}$

11. -3

12. $-\frac{1}{5}$

13. 1.6

14. $\frac{1}{2/3}$

Answers on page A-2

For example, $\frac{0}{4} = 0$, $\frac{4}{0}$ is undefined, and $\frac{0}{0}$ is undefined.

Do Exercises 7 and 8.

b Reciprocals

When two numbers like $\frac{1}{2}$ and 2 are multiplied, the result is 1. Such numbers are called **reciprocals** of each other. Every nonzero real number has a reciprocal, also called a **multiplicative inverse.**

> Two numbers whose product is 1 are called **reciprocals**, or **multiplicative inverses,** of each other.

Examples Find the reciprocal.

5. $\frac{7}{8}$ The reciprocal of $\frac{7}{8}$ is $\frac{8}{7}$ because $\frac{7}{8} \cdot \frac{8}{7} = 1$.

6. -5 The reciprocal of -5 is $-\frac{1}{5}$ because $-5\left(-\frac{1}{5}\right) = 1$.

7. 3.9 The reciprocal of 3.9 is $\frac{1}{3.9}$ because $3.9\left(\frac{1}{3.9}\right) = 1$.

8. $-\frac{1}{2}$ The reciprocal of $-\frac{1}{2}$ is -2 because $\left(-\frac{1}{2}\right)(-2) = 1$.

9. $-\frac{2}{3}$ The reciprocal of $-\frac{2}{3}$ is $-\frac{3}{2}$ because $\left(-\frac{2}{3}\right)\left(-\frac{3}{2}\right) = 1$.

10. $\frac{1}{3/4}$ The reciprocal of $\frac{1}{3/4}$ is $\frac{3}{4}$ because $\left(\frac{1}{3/4}\right)\left(\frac{3}{4}\right) = 1$.

> For $a \neq 0$, the reciprocal of a can be named $\frac{1}{a}$ and the reciprocal of $\frac{1}{a}$ is a.
>
> The reciprocal of a nonzero number $\frac{a}{b}$ can be named $\frac{b}{a}$.
>
> The number 0 has no reciprocal.

Do Exercises 9–14.

The reciprocal of a positive number is also a positive number, because their product must be the positive number 1. The reciprocal of a negative number is also a negative number, because their product must be the positive number 1.

> The reciprocal of a number has the same sign as the number itself.

CAUTION! It is important *not* to confuse *opposite* with *reciprocal.* Keep in mind that the opposite, or additive inverse, of a number is what we add to the number to get 0. The reciprocal, or multiplicative inverse, is what we multiply the number by to get 1.

Compare the following.

Number	Opposite (Change the Sign.)	Reciprocal (Invert But Do Not Change the Sign.)
$-\frac{3}{8}$	$\frac{3}{8}$	$-\frac{8}{3}$
19	-19	$\frac{1}{19}$
$\frac{18}{7}$	$-\frac{18}{7}$	$\frac{7}{18}$
-7.9	7.9	$-\frac{1}{7.9}$, or $-\frac{10}{79}$
0	0	Undefined

$\left(-\frac{3}{8}\right)\left(-\frac{8}{3}\right) = 1$

$-\frac{3}{8} + \frac{3}{8} = 0$

Do Exercise 15.

c Division of Real Numbers

We know that we can subtract by adding an opposite. Similarly, we can divide by multiplying by a reciprocal.

> For any real numbers a and b, $b \neq 0$,
>
> $$a \div b = \frac{a}{b} = a \cdot \frac{1}{b}.$$
>
> (To divide, we can multiply by the reciprocal of the divisor.)

Examples Rewrite the division as a multiplication.

11. $-4 \div 3$ $\qquad$ $-4 \div 3$ is the same as $-4 \cdot \frac{1}{3}$

12. $\frac{6}{-7}$ $\qquad$ $\frac{6}{-7} = 6\left(-\frac{1}{7}\right)$

13. $\frac{x+2}{5}$ $\qquad$ $\frac{x+2}{5} = (x+2)\frac{1}{5}$ $\qquad$ **Parentheses are necessary here.**

14. $\frac{-17}{1/b}$ $\qquad$ $\frac{-17}{1/b} = -17 \cdot b$

15. $\frac{3}{5} \div \left(-\frac{9}{7}\right)$ $\qquad$ $\frac{3}{5} \div \left(-\frac{9}{7}\right) = \frac{3}{5}\left(-\frac{7}{9}\right)$

Do Exercises 16–20.

When actually doing division calculations, we sometimes multiply by a reciprocal and we sometimes divide directly. With fractional notation, it is usually better to multiply by a reciprocal. With decimal notation, it is usually better to divide directly.

15. Complete the following table.

Number	Opposite	Reciprocal
$\frac{2}{3}$		
$-\frac{5}{4}$		
0		
1		
-8		
-4.5		

Rewrite the division as a multiplication.

16. $\frac{4}{7} \div \left(-\frac{3}{5}\right)$

17. $\frac{5}{-8}$

18. $\frac{a-b}{7}$

19. $\frac{-23}{1/a}$

20. $-5 \div 7$

Answers on page A-2

Divide by multiplying by the reciprocal of the divisor.

21. $\frac{4}{7} \div \left(-\frac{3}{5}\right)$

22. $-\frac{8}{5} \div \frac{2}{3}$

23. $-\frac{12}{7} \div \left(-\frac{3}{4}\right)$

24. Divide: $21.7 \div (-3.1)$.

Find two equal expressions for the number with negative signs in different places.

25. $\frac{-5}{6}$

26. $-\frac{8}{7}$

27. $\frac{10}{-3}$

Answers on page A-2

Examples Divide by multiplying by the reciprocal of the divisor.

16. $\frac{2}{3} \div \left(-\frac{5}{4}\right) = \frac{2}{3} \cdot \left(-\frac{4}{5}\right) = -\frac{8}{15}$

17. $-\frac{5}{6} \div \left(-\frac{3}{4}\right) = -\frac{5}{6} \cdot \left(-\frac{4}{3}\right) = \frac{20}{18} = \frac{10 \cdot 2}{9 \cdot 2} = \frac{10}{9} \cdot \frac{2}{2} = \frac{10}{9}$

Caution! Be careful not to change the sign when taking a reciprocal!

18. $-\frac{3}{4} \div \frac{3}{10} = -\frac{3}{4} \cdot \left(\frac{10}{3}\right) = -\frac{30}{12} = -\frac{5}{2} \cdot \frac{6}{6} = -\frac{5}{2}$

With decimal notation, it is easier to carry out long division than to multiply by the reciprocal.

Examples Divide.

19. $-27.9 \div (-3) = \frac{-27.9}{-3} = 9.3$ — Do the long division $3\overline{)27.9}$ (quotient 9.3). The answer is positive.

20. $-6.3 \div 2.1 = -3$ — Do the long division $2.1\overline{)6.3}$ (quotient 3.). The answer is negative.

Do Exercises 21–24.

Consider the following:

1. $\frac{2}{3} = \frac{2}{3} \cdot 1 = \frac{2}{3} \cdot \frac{-1}{-1} = \frac{2(-1)}{3(-1)} = \frac{-2}{-3}$. Thus, $\frac{2}{3} = \frac{-2}{-3}$.

2. $-\frac{2}{3} = -1 \cdot \frac{2}{3} = \frac{-1}{1} \cdot \frac{2}{3} = \frac{-1 \cdot 2}{1 \cdot 3} = \frac{-2}{3}$. Thus, $-\frac{2}{3} = \frac{-2}{3}$.

$\frac{-2}{3} = \frac{-2}{3} \cdot 1 = \frac{-2}{3} \cdot \frac{-1}{-1} = \frac{-2(-1)}{3(-1)} = \frac{2}{-3}$. Thus, $\frac{-2}{3} = \frac{2}{-3}$.

We can use the following properties to make sign changes in fractional notation.

For any numbers a and b, $b \neq 0$:

1. $\frac{-a}{-b} = \frac{a}{b}$

(The opposite of a number a divided by the opposite of another number b is the same as the quotient of the two numbers a and b.)

2. $\frac{-a}{b} = \frac{a}{-b} = -\frac{a}{b}$

(The opposite of a number a divided by another number b is the same as the number a divided by the opposite of the number b, and both are the same as the opposite of *a divided by b*.)

Do Exercises 25–27.

Exercise Set 1.6

a Divide, if possible. Check each answer.

1. $48 \div (-6)$

2. $\frac{42}{-7}$

3. $\frac{28}{-2}$

4. $24 \div (-12)$

5. $\frac{-24}{8}$

6. $-18 \div (-2)$

7. $\frac{-36}{-12}$

8. $-72 \div (-9)$

9. $\frac{-72}{9}$

10. $\frac{-50}{25}$

11. $-100 \div (-50)$

12. $\frac{-200}{8}$

13. $-108 \div 9$

14. $\frac{-63}{-7}$

15. $\frac{200}{-25}$

16. $-300 \div (-16)$

17. $\frac{75}{0}$

18. $\frac{0}{-5}$

19. $\frac{-23}{-2}$

20. $\frac{-23}{0}$

b Find the reciprocal.

21. $\frac{15}{7}$

22. $\frac{3}{8}$

23. $-\frac{47}{13}$

24. $-\frac{31}{12}$

25. 13

26. -10

27. 4.3

28. -8.5

29. $\frac{1}{-7.1}$

30. $\frac{1}{-4.9}$

31. $\frac{p}{q}$

32. $\frac{s}{t}$

33. $\frac{1}{4y}$

34. $\frac{-1}{8a}$

35. $\frac{2a}{3b}$

36. $\frac{-4y}{3x}$

c Rewrite the division as a multiplication.

37. $4 \div 17$

38. $5 \div (-8)$

39. $\frac{8}{-13}$

40. $-\frac{13}{47}$

41. $\frac{13.9}{-1.5}$

42. $-\frac{47.3}{21.4}$

43. $\frac{x}{\frac{1}{y}}$

44. $\frac{13}{x}$

45. $\frac{3x + 4}{5}$

46. $\frac{4y - 8}{-7}$

47. $\frac{5a - b}{5a + b}$

48. $\frac{2x + x^2}{x - 5}$

Divide.

49. $\frac{3}{4} \div \left(-\frac{2}{3}\right)$

50. $\frac{7}{8} \div \left(-\frac{1}{2}\right)$

51. $-\frac{5}{4} \div \left(-\frac{3}{4}\right)$

52. $-\frac{5}{9} \div \left(-\frac{5}{6}\right)$

53. $-\frac{2}{7} \div \left(-\frac{4}{9}\right)$

54. $-\frac{3}{5} \div \left(-\frac{5}{8}\right)$

55. $-\frac{3}{8} \div \left(-\frac{8}{3}\right)$

56. $-\frac{5}{8} \div \left(-\frac{6}{5}\right)$

57. $-6.6 \div 3.3$

58. $-44.1 \div (-6.3)$

59. $\frac{-11}{-13}$

60. $\frac{-1.9}{20}$

61. $\frac{48.6}{-3}$

62. $\frac{-17.8}{3.2}$

63. $\frac{-9}{17 - 17}$

64. $\frac{-8}{-5 + 5}$

Skill Maintenance

Simplify.

65. $\frac{1}{4} - \frac{1}{2}$ [1.4a]

66. $-9 - 3 + 17$ [1.4a]

67. $35 \cdot (-1.2)$ [1.5a]

68. $4 \cdot (-6) \cdot (-2) \cdot (-1)$ [1.5a]

69. $13.4 + (-4.9)$ [1.3a]

70. $-\frac{3}{8} - \left(-\frac{1}{4}\right)$ [1.4a]

Convert to decimal notation. [1.2c]

71. $-\frac{1}{11}$

72. $\frac{11}{12}$

Synthesis

73. ◆ Explain how multiplication can be used to justify why a negative number divided by a positive number is negative.

74. ◆ Explain how multiplication can be used to justify why a negative number divided by a negative number is positive.

75. ◆ 🖩 Find the reciprocal of -10.5. What happens if you take the reciprocal of the result?

76. Determine those real numbers a for which the opposite of a is the same as the reciprocal of a.

Tell whether the expression represents a positive number or a negative number when a and b are negative.

77. $\frac{-a}{b}$

78. $\frac{-a}{-b}$

79. $-\left(\frac{a}{-b}\right)$

80. $-\left(\frac{-a}{b}\right)$

81. $-\left(\frac{-a}{-b}\right)$

1.7 Properties of Real Numbers

a Equivalent Expressions

In solving equations and doing other kinds of work in algebra, we manipulate expressions in various ways. For example, instead of

$$x + x,$$

we might write

$$2x,$$

knowing that the two expressions represent the same number for any allowable replacement of x. In that sense, the expressions $x + x$ and $2x$ are **equivalent**, as are $3/x$ and $3x/x^2$, even though 0 is not an allowable replacement because division by 0 is undefined.

> Two expressions that have the same value for all allowable replacements are called **equivalent**.

The expressions $x + 3x$ and $5x$ are *not* equivalent.

Do Exercises 1 and 2.

In this section, we will consider several laws of real numbers that will allow us to find equivalent expressions. The first two laws are the *identity properties of 0 and 1.*

> **THE IDENTITY PROPERTY OF 0**
>
> For any real number a,
>
> $$a + 0 = 0 + a = a.$$
>
> (The number 0 is the *additive identity.*)

> **THE IDENTITY PROPERTY OF 1**
>
> For any real number a,
>
> $$a \cdot 1 = 1 \cdot a = a.$$
>
> (The number 1 is the *multiplicative identity.*)

We often refer to the use of the identity property of 1 as "multiplying by 1." We can use this method to find equivalent fractional expressions. Recall from arithmetic that to multiply with fractional notation, we multiply numerators and denominators.

Objectives

a Find equivalent fractional expressions and simplify fractional expressions.

b Use the commutative and associative laws to find equivalent expressions.

c Use the distributive laws to multiply expressions like 8 and $x - y$.

d Use the distributive laws to factor expressions like $4x - 12 + 24y$.

e Collect like terms.

For Extra Help

TAPE 2

MAC WIN

CD-ROM

Complete the table by evaluating each expression for the given values.

1.

	$x + x$	$2x$
$x = 3$		
$x = -6$		
$x = 4.8$		

2.

	$x + 3x$	$5x$
$x = 2$		
$x = -6$		
$x = 4.8$		

Answers on page A-2

3. Write a fractional expression equivalent to $\frac{3}{4}$ with a denominator of 8.

4. Write a fractional expression equivalent to $\frac{3}{4}$ with a denominator of $4t$.

Simplify.

5. $\frac{3y}{4y}$

6. $-\frac{16m}{12m}$

7. Evaluate $x + y$ and $y + x$ for $x = -2$ and $y = 3$.

8. Evaluate xy and yx for $x = -2$ and $y = 5$.

Answers on page A-3

Example 1 Write a fractional expression equivalent to $\frac{2}{3}$ with a denominator of $3x$.

Note that $3x = 3 \cdot x$. We want fractional notation for $\frac{2}{3}$ that has a denominator of $3x$, but the denominator 3 is missing a factor of x. Thus we multiply by 1, using x/x as an equivalent expression for 1:

$$\frac{2}{3} = \frac{2}{3} \cdot 1 = \frac{2}{3} \cdot \frac{x}{x} = \frac{2x}{3x}.$$

The expressions $2/3$ and $2x/3x$ are equivalent. They have the same value for any allowable replacement. Note that $2x/3x$ is undefined for a replacement of 0, but for all nonzero real numbers, the expressions $2/3$ and $2x/3x$ have the same value.

Do Exercises 3 and 4.

In algebra, we consider an expression like $2/3$ to be "simplified" from $2x/3x$. To find such simplified expressions, we use the identity property of 1 to remove a factor of 1.

Example 2 Simplify: $-\frac{20x}{12x}$.

$$-\frac{20x}{12x} = -\frac{5 \cdot 4x}{3 \cdot 4x}$$ We look for the largest factor common to both the numerator and the denominator and factor each.

$$= -\frac{5}{3} \cdot \frac{4x}{4x}$$ Factoring the fractional expression

$$= -\frac{5}{3} \cdot 1$$ $\frac{4x}{4x} = 1$

$$= -\frac{5}{3}$$ Removing a factor of 1 using the identity property of 1

Do Exercises 5 and 6.

b The Commutative and Associative Laws

The Commutative Laws

Let's examine the expressions $x + y$ and $y + x$, as well as xy and yx.

Example 3 Evaluate $x + y$ and $y + x$ for $x = 4$ and $y = 3$.

We substitute 4 for x and 3 for y in both expressions:

$$x + y = 4 + 3 = 7; \qquad y + x = 3 + 4 = 7.$$

Example 4 Evaluate xy and yx for $x = 23$ and $y = 12$.

We substitute 23 for x and 12 for y in both expressions:

$$xy = 23 \cdot 12 = 276; \qquad yx = 12 \cdot 23 = 276.$$

Do Exercises 7 and 8.

Note that the expressions

$x + y$ and $y + x$

have the same values no matter what the variables stand for. Thus they are equivalent. Therefore, when we add two numbers, the order in which we add does not matter. Similarly, the expressions xy and yx are equivalent. They also have the same values, no matter what the variables stand for. Therefore, when we multiply two numbers, the order in which we multiply does not matter.

The following are examples of general patterns or laws.

THE COMMUTATIVE LAWS

Addition. For any numbers a and b,

$$a + b = b + a.$$

(We can change the order when adding without affecting the answer.)

Multiplication. For any numbers a and b,

$$ab = ba.$$

(We can change the order when multiplying without affecting the answer.)

Using a commutative law, we know that $x + 2$ and $2 + x$ are equivalent. Similarly, $3x$ and $x(3)$ are equivalent. Thus, in an algebraic expression, we can replace one with the other and the result will be equivalent to the original expression.

Example 5 Use the commutative laws to write an expression equivalent to $y + 5$, ab, and $7 + xy$.

An expression equivalent to $y + 5$ is $5 + y$ by the commutative law of addition.

An expression equivalent to ab is ba by the commutative law of multiplication.

An expression equivalent to $7 + xy$ is $xy + 7$ by the commutative law of addition. Another expression equivalent to $7 + xy$ is $7 + yx$ by the commutative law of multiplication.

Do Exercises 9–11.

The Associative Laws

Now let's examine the expressions $a + (b + c)$ and $(a + b) + c$. Note that these expressions involve the use of parentheses as *grouping* symbols, and they also involve three numbers. Calculations within parentheses are to be done first.

Example 6 Calculate and compare: $3 + (8 + 5)$ and $(3 + 8) + 5$.

$3 + (8 + 5) = 3 + 13$ Calculating within parentheses first; adding the 8 and 5

$= 16;$

$(3 + 8) + 5 = 11 + 5$ Calculating within parentheses first; adding the 3 and 8

$= 16$

Use a commutative law to write an equivalent expression.

9. $x + 9$

10. pq

11. $xy + t$

Answers on page A-3

12. Calculate and compare:

$8 + (9 + 2)$ and $(8 + 9) + 2$.

13. Calculate and compare:

$10 \cdot (5 \cdot 3)$ and $(10 \cdot 5) \cdot 3$.

Use an associative law to write an equivalent expression.

14. $r + (s + 7)$

15. $9(ab)$

Answers on page A-3

The two expressions in Example 6 name the same number. Moving the parentheses to group the additions differently does not affect the value of the expression.

Example 7 Calculate and compare: $3 \cdot (4 \cdot 2)$ and $(3 \cdot 4) \cdot 2$.

$$3 \cdot (4 \cdot 2) = 3 \cdot 8 = 24; \qquad (3 \cdot 4) \cdot 2 = 12 \cdot 2 = 24$$

Do Exercises 12 and 13.

You may have noted that when only addition is involved, parentheses can be placed any way we please without affecting the answer. When only multiplication is involved, parentheses also can be placed any way we please without affecting the answer.

THE ASSOCIATIVE LAWS

Addition. For any numbers a, b, and c,

$$a + (b + c) = (a + b) + c.$$

(Numbers can be grouped in any manner for addition.)

Multiplication. For any numbers a, b, and c,

$$a \cdot (b \cdot c) = (a \cdot b) \cdot c.$$

(Numbers can be grouped in any manner for multiplication.)

Example 8 Use an associative law to write an expression equivalent to $(y + z) + 3$ and $8(xy)$.

An equivalent expression is $y + (z + 3)$ by the associative law of addition.

An equivalent expression is $(8x)y$ by the associative law of multiplication.

Do Exercises 14 and 15.

The associative laws say parentheses can be placed any way we please when only additions or only multiplications are involved. Thus we often omit them. For example,

$$x + (y + 2) \text{ means } x + y + 2, \quad \text{and} \quad (lw)h \text{ means } lwh.$$

Using the Commutative and Associative Laws Together

Example 9 Use the commutative and associative laws to write at least three expressions equivalent to $(x + 5) + y$.

a) $(x + 5) + y = x + (5 + y)$
$= x + (y + 5)$ — Using the associative law first and then using the commutative law

b) $(x + 5) + y = y + (x + 5)$
$= y + (5 + x)$ — Using the commutative law first and then the commutative law again

c) $(x + 5) + y = (5 + x) + y$
$= 5 + (x + y)$ — Using the commutative law first and then the associative law

Example 10 Use the commutative and associative laws to write at least three expressions equivalent to $(3x)y$.

a) $(3x)y = 3(xy)$ **Using the associative law first and then using the commutative law**
$= 3(yx)$

b) $(3x)y = y(3x)$ **Using the commutative law twice**
$= y(x3)$

c) $(3x)y = (x3)y$ **Using the commutative law, and then the associative law, and then the commutative law again**
$= x(3y)$
$= x(y3)$

Do Exercises 16 and 17.

c The Distributive Laws

The *distributive laws* are the basis of many procedures in both arithmetic and algebra. They are probably the most important laws that we use to manipulate algebraic expressions. The distributive law of multiplication over addition involves two operations: addition and multiplication.

Let's begin by considering a multiplication problem from arithmetic:

```
    4 5
  ×   7
  -----
    3 5 <— This is 7 · 5.
  2 8 0 <— This is 7 · 40.
  -----
  3 1 5 <— This is the sum 7 · 40 + 7 · 5.
```

To carry out the multiplication, we actually added two products. That is,

$$7 \cdot 45 = 7(40 + 5) = 7 \cdot 40 + 7 \cdot 5.$$

Let's examine this further. If we wish to multiply a sum of several numbers by a factor, we can either add and then multiply, or multiply and then add.

Example 11 Compute in two ways: $5 \cdot (4 + 8)$.

a) $5 \cdot (4 + 8)$ **Adding within parentheses first, and then multiplying**
$= 5 \cdot 12$
$= 60$

b) $(5 \cdot 4) + (5 \cdot 8)$ **Distributing the multiplication to terms within parentheses first and then adding**
$= 20 + 40$
$= 60$

Do Exercises 18–20.

> **THE DISTRIBUTIVE LAW OF MULTIPLICATION OVER ADDITION**
>
> For any numbers a, b, and c,
>
> $$a(b + c) = ab + ac.$$

Use the commutative and associative laws to write at least three equivalent expressions.

16. $4(tu)$

17. $r + (2 + s)$

Compute.

18. a) $7 \cdot (3 + 6)$

b) $(7 \cdot 3) + (7 \cdot 6)$

19. a) $2 \cdot (10 + 30)$

b) $(2 \cdot 10) + (2 \cdot 30)$

20. a) $(2 + 5) \cdot 4$

b) $(2 \cdot 4) + (5 \cdot 4)$

Answers on page A-3

Calculate.

21. a) $4(5 - 3)$

b) $4 \cdot 5 - 4 \cdot 3$

22. a) $-2 \cdot (5 - 3)$

b) $-2 \cdot 5 - (-2) \cdot 3$

23. a) $5 \cdot (2 - 7)$

b) $5 \cdot 2 - 5 \cdot 7$

What are the terms of the expression?

24. $5x - 8y + 3$

25. $-4y - 2x + 3z$

Multiply.

26. $3(x - 5)$

27. $5(x + 1)$

28. $\frac{3}{5}(p + q - t)$

Answers on page A-3

In the statement of the distributive law, we know that in an expression such as $ab + ac$, the multiplications are to be done first according to the rules for order of operations. So, instead of writing $(4 \cdot 5) + (4 \cdot 7)$, we can write $4 \cdot 5 + 4 \cdot 7$. However, in $a(b + c)$, we cannot omit the parentheses. If we did, we would have $ab + c$, which means $(ab) + c$. For example, $3(4 + 2) = 18$, but $3 \cdot 4 + 2 = 14$.

There is another distributive law that relates multiplication and subtraction. This law says that to multiply by a difference, we can either subtract and then multiply, or multiply and then subtract.

THE DISTRIBUTIVE LAW OF MULTIPLICATION OVER SUBTRACTION

For any numbers a, b, and c,

$$a(b - c) = ab - ac.$$

We often refer to "*the* distributive law" when we mean *either* or *both* of these laws.

Do Exercises 21–23.

What do we mean by the *terms* of an expression? **Terms** are separated by addition signs. If there are subtraction signs, we can find an equivalent expression that uses addition signs.

Example 12 What are the terms of $3x - 4y + 2z$?

We have

$$3x - 4y + 2z = 3x + (-4y) + 2z. \quad \text{Separating parts with + signs}$$

The terms are $3x$, $-4y$, and $2z$.

Do Exercises 24 and 25.

The distributive laws are a basis for a procedure in algebra called **multiplying**. In an expression like $8(a + 2b - 7)$, we multiply each term inside the parentheses by 8:

$$8(a + 2b - 7) = 8 \cdot a + 8 \cdot 2b - 8 \cdot 7 = 8a + 16b - 56.$$

Examples Multiply.

13. $9(x - 5) = 9x - 9(5)$ Using the distributive law of multiplication over subtraction

$= 9x - 45$

14. $\frac{2}{3}(w + 1) = \frac{2}{3} \cdot w + \frac{2}{3} \cdot 1$ Using the distributive law of multiplication over addition

$= \frac{2}{3}w + \frac{2}{3}$

15. $\frac{4}{3}(s - t + w) = \frac{4}{3}s - \frac{4}{3}t + \frac{4}{3}w$ Using both distributive laws

Do Exercises 26–28.

Example 16 Multiply: $-4(x - 2y + 3z)$.

$$-4(x - 2y + 3z) = -4 \cdot x - (-4)(2y) + (-4)(3z) \quad \text{Using both distributive laws}$$
$$= -4x - (-8y) + (-12z) \quad \text{Multiplying}$$
$$= -4x + 8y - 12z$$

We can also do this problem by first finding an equivalent expression with all plus signs and then multiplying:

$$-4(x - 2y + 3z) = -4[x + (-2y) + 3z]$$
$$= -4 \cdot x + (-4)(-2y) + (-4)(3z)$$
$$= -4x + 8y - 12z.$$

Do Exercises 29–31.

Multiply.

29. $-2(x - 3)$

30. $5(x - 2y + 4z)$

31. $-5(x - 2y + 4z)$

d Factoring

Factoring is the reverse of multiplying. To factor, we can use the distributive laws in reverse:

$$ab + ac = a(b + c) \quad \text{and} \quad ab - ac = a(b - c).$$

> To **factor** an expression is to find an equivalent expression that is a product.

Look at Example 13. To *factor* $9x - 45$, we find an equivalent expression that is a product, $9(x - 5)$. When all the terms of an expression have a factor in common, we can "factor it out" using the distributive laws. Note the following.

$9x$ has the factors $9, -9, 3, -3, 1, -1, x, -x, 3x, -3x, 9x, -9x$;

-45 has the factors $1, -1, 3, -3, 5, -5, 9, -9, 15, -15, 45, -45$

We generally remove the largest common factor. In this case, that factor is 9. Thus,

$$9x - 45 = 9 \cdot x - 9 \cdot 5$$
$$= 9(x - 5).$$

Remember that an expression has been factored when we have found an equivalent expression that is a product.

Examples Factor.

17. $5x - 10 = 5 \cdot x - 5 \cdot 2$ Try to do this step mentally.

$= 5(x - 2)$ You can check by multiplying.

18. $ax - ay + az = a(x - y + z)$

19. $9x + 27y - 9 = 9 \cdot x + 9 \cdot 3y - 9 \cdot 1 = 9(x + 3y - 1)$

Answers on page A-3

Factor.

32. $6x - 12$

33. $3x - 6y + 9$

34. $bx + by - bz$

35. $16a - 36b + 42$

36. $\frac{3}{8}x - \frac{5}{8}y + \frac{7}{8}$

37. $-12x + 32y - 16z$

Collect like terms.

38. $6x - 3x$

39. $7x - x$

40. $x - 9x$

41. $x - 0.41x$

42. $5x + 4y - 2x - y$

43. $3x - 7x - 11 + 8y + 4 - 13y$

44. $-\frac{2}{3} - \frac{3}{5}x + y + \frac{7}{10}x - \frac{2}{9}y$

Answers on page A-3

Examples Factor. Try to write just the answer, if you can.

20. $5x - 5y = 5(x - y)$

21. $-3x + 6y - 9z = -3(x - 2y + 3z)$

We usually factor out a negative when the first term is negative. The way we factor can depend on the situation in which we are working. We might also factor the expression in Example 21 as follows:

$$-3x + 6y - 9z = 3(-x + 2y - 3z).$$

22. $18z - 12x - 24 = 6(3z - 2x - 4)$

23. $\frac{1}{2}x + \frac{3}{2}y - \frac{1}{2} = \frac{1}{2}(x + 3y - 1)$

Remember that you can always check factoring by multiplying. Keep in mind that an expression is factored when it is written as a product.

Do Exercises 32–37.

e Collecting Like Terms

Terms such as $5x$ and $-4x$, whose variable factors are exactly the same, are called **like terms.** Similarly, numbers, such as -7 and 13, are like terms. Also, $3y^2$ and $9y^2$ are like terms because the variables are raised to the same power. Terms such as $4y$ and $5y^2$ are not like terms, and $7x$ and $2y$ are not like terms.

The process of **collecting like terms** is also based on the distributive laws. We can apply the distributive law when a factor is on the right because of the commutative law of multiplication.

Examples Collect like terms. Try to write just the answer, if you can.

24. $4x + 2x = (4 + 2)x = 6x$ Factoring out the x using a distributive law

25. $2x + 3y - 5x - 2y = 2x - 5x + 3y - 2y$
$= (2 - 5)x + (3 - 2)y = -3x + y$

26. $3x - x = (3 - 1)x = 2x$

27. $x - 0.24x = 1 \cdot x - 0.24x = (1 - 0.24)x = 0.76x$

28. $x - 6x = 1 \cdot x - 6 \cdot x = (1 - 6)x = -5x$

29. $4x - 7y + 9x - 5 + 3y - 8 = 13x - 4y - 13$

30. $\frac{2}{3}a - b + \frac{4}{5}a + \frac{1}{4}b - 10 = \frac{2}{3}a - 1 \cdot b + \frac{4}{5}a + \frac{1}{4}b - 10$
$= \left(\frac{2}{3} + \frac{4}{5}\right)a + \left(-1 + \frac{1}{4}\right)b - 10$
$= \left(\frac{10}{15} + \frac{12}{15}\right)a + \left(-\frac{4}{4} + \frac{1}{4}\right)b - 10$
$= \frac{22}{15}a - \frac{3}{4}b - 10$

Do Exercises 38–44.

Exercise Set 1.7

a Find an equivalent expression with the given denominator.

1. $\frac{3}{5}$; $5y$ **2.** $\frac{5}{8}$; $8t$ **3.** $\frac{2}{3}$; $15x$ **4.** $\frac{6}{7}$; $14y$

Simplify.

5. $-\frac{24a}{16a}$ **6.** $-\frac{42t}{18t}$ **7.** $-\frac{42ab}{36ab}$ **8.** $-\frac{64pq}{48pq}$

b Write an equivalent expression. Use a commutative law.

9. $y + 8$ **10.** $x + 3$ **11.** mn **12.** ab

13. $9 + xy$ **14.** $11 + ab$ **15.** $ab + c$ **16.** $rs + t$

Write an equivalent expression. Use an associative law.

17. $a + (b + 2)$ **18.** $3(vw)$ **19.** $(8x)y$ **20.** $(y + z) + 7$

21. $(a + b) + 3$ **22.** $(5 + x) + y$ **23.** $3(ab)$ **24.** $(6x)y$

Use the commutative and associative laws to write three equivalent expressions.

25. $(a + b) + 2$ **26.** $(3 + x) + y$ **27.** $5 + (v + w)$ **28.** $6 + (x + y)$

29. $(xy)3$ **30.** $(ab)5$ **31.** $7(ab)$ **32.** $5(xy)$

c Multiply.

33. $2(b + 5)$ **34.** $4(x + 3)$ **35.** $7(1 + t)$ **36.** $4(1 + y)$

37. $6(5x + 2)$ **38.** $9(6m + 7)$ **39.** $7(x + 4 + 6y)$ **40.** $4(5x + 8 + 3p)$

41. $7(x - 3)$

42. $15(y - 6)$

43. $-3(x - 7)$

44. $1.2(x - 2.1)$

45. $\frac{2}{3}(b - 6)$

46. $\frac{5}{8}(y + 16)$

47. $7.3(x - 2)$

48. $5.6(x - 8)$

49. $-\frac{3}{5}(x - y + 10)$

50. $-\frac{2}{3}(a + b - 12)$

51. $-9(-5x - 6y + 8)$

52. $-7(-2x - 5y + 9)$

53. $-4(x - 3y - 2z)$

54. $8(2x - 5y - 8z)$

55. $3.1(-1.2x + 3.2y - 1.1)$

56. $-2.1(-4.2x - 4.3y - 2.2)$

List the terms of the expression.

57. $4x + 3z$

58. $8x - 1.4y$

59. $7x + 8y - 9z$

60. $8a + 10b - 18c$

d Factor. Check by multiplying.

61. $2x + 4$

62. $5y + 20$

63. $30 + 5y$

64. $7x + 28$

65. $14x + 21y$

66. $18a + 24b$

67. $5x + 10 + 15y$

68. $9a + 27b + 81$

69. $8x - 24$

70. $10x - 50$

71. $32 - 4y$

72. $24 - 6m$

73. $8x + 10y - 22$

74. $9a + 6b - 15$

75. $ax - a$

76. $by - 9b$

77. $ax - ay - az$

78. $cx + cy - cz$

79. $18x - 12y + 6$

80. $-14x + 21y + 7$

81. $\frac{2}{3}x - \frac{5}{3}y + \frac{1}{3}$

82. $\frac{3}{5}a + \frac{4}{5}b - \frac{1}{5}$

e Collect like terms.

83. $9a + 10a$

84. $12x + 2x$

85. $10a - a$

86. $-16x + x$

87. $2x + 9z + 6x$

88. $3a - 5b + 7a$

89. $7x + 6y^2 + 9y^2$

90. $12m^2 + 6q + 9m^2$

91. $41a + 90 - 60a - 2$

92. $42x - 6 - 4x + 2$

93. $23 + 5t + 7y - t - y - 27$

94. $45 - 90d - 87 - 9d + 3 + 7d$

95. $\frac{1}{2}b + \frac{1}{2}b$

96. $\frac{2}{3}x + \frac{1}{3}x$

97. $2y + \frac{1}{4}y + y$

98. $\frac{1}{2}a + a + 5a$

99. $11x - 3x$

100. $9t - 17t$

101. $6n - n$

102. $10t - t$

103. $y - 17y$

104. $3m - 9m + 4$

105. $-8 + 11a - 5b + 6a - 7b + 7$

106. $8x - 5x + 6 + 3y - 2y - 4$

107. $9x + 2y - 5x$

108. $8y - 3z + 4y$

109. $11x + 2y - 4x - y$

110. $13a + 9b - 2a - 4b$

111. $2.7x + 2.3y - 1.9x - 1.8y$

112. $6.7a + 4.3b - 4.1a - 2.9b$

113. $\frac{13}{2}a + \frac{9}{5}b - \frac{2}{3}a - \frac{3}{10}b - 42$

114. $\frac{11}{4}x + \frac{2}{3}y - \frac{4}{5}x - \frac{1}{6}y + 12$

Skill Maintenance

115. Evaluate $9w$ for $w = 20$. [1.1a]

116. Find the absolute value: $\left|-\frac{4}{13}\right|$. [1.2e]

Write true or false. [1.2d]

117. $-43 < -40$

118. $-3 \geq 0$

119. $-6 \leq -6$

120. $0 > -4$

Synthesis

121. ◆ The distributive law was introduced before the discussion on collecting like terms. Why do you think this was done?

122. ◆ Find two different expressions for the total area of the two rectangles shown below. Explain the equivalence of the expressions in terms of the distributive law.

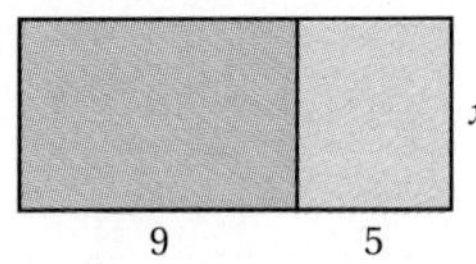

Tell whether the expressions are equivalent. Explain.

123. $3t + 5$ and $3 \cdot 5 + t$

124. $4x$ and $x + 4$

125. $5m + 6$ and $6 + 5m$

126. $(x + y) + z$ and $z + (x + y)$

Collect like terms, if possible, and factor the result.

127. $q + qr + qrs + qrst$

128. $21x + 44xy + 15y - 16x - 8y - 38xy + 2y + xy$

Collaborative Learning Manual

Use the commutative and associative laws to add a series of numbers.

1.8 Simplifying Expressions; Order of Operations

We now expand our ability to manipulate expressions by first considering opposites of sums and differences. Then we simplify expressions involving parentheses.

Objectives

a. Find an equivalent expression for an opposite without parentheses, where an expression has several terms.

b. Simplify expressions by removing parentheses and collecting like terms.

c. Simplify expressions with parentheses inside parentheses.

d. Simplify expressions using rules for order of operations.

For Extra Help

TAPE 2 MAC WIN CD-ROM

a Opposites of Sums

What happens when we multiply a real number by -1? Consider the following products:

$$-1(7) = -7, \qquad -1(-5) = 5, \qquad -1(0) = 0.$$

From these examples, it appears that when we multiply a number by -1, we get the opposite, or additive inverse, of that number.

> **THE PROPERTY OF −1**
>
> For any real number a,
>
> $$-1 \cdot a = -a.$$
>
> (Negative one times a is the opposite, or additive inverse, of a.)

The property of -1 enables us to find certain expressions equivalent to opposites of sums.

Examples Find an equivalent expression without parentheses.

1.
$$\begin{aligned} -(3 + x) &= -1(3 + x) && \text{Using the property of } -1 \\ &= -1 \cdot 3 + (-1)x && \text{Using a distributive law, multiplying each term by } -1 \\ &= -3 + (-x) && \text{Using the property of } -1 \\ &= -3 - x \end{aligned}$$

2.
$$\begin{aligned} -(3x + 2y + 4) &= -1(3x + 2y + 4) && \text{Using the property of } -1 \\ &= -1(3x) + (-1)(2y) + (-1)4 && \text{Using a distributive law} \\ &= -3x - 2y - 4 && \text{Using the property of } -1 \end{aligned}$$

Do Exercises 1 and 2.

Find an equivalent expression without parentheses.

1. $-(x + 2)$

2. $-(5x + 2y + 8)$

Suppose we want to remove parentheses in an expression like

$$-(x - 2y + 5).$$

We can first rewrite any subtractions inside the parentheses as additions. Then we take the opposite of each term:

$$\begin{aligned} -(x - 2y + 5) &= -[x + (-2y) + 5] \\ &= -x + 2y - 5. \end{aligned}$$

The most efficient method for removing parentheses is to replace each term in the parentheses with its opposite ("change the sign of every term"). Doing so for $-(x - 2y + 5)$, we obtain $-x + 2y - 5$ as an equivalent expression.

Answers on page A-3

Find an equivalent expression without parentheses. Try to do this in one step.

3. $-(6 - t)$

4. $-(x - y)$

5. $-(-4a + 3t - 10)$

6. $-(18 - m - 2n + 4z)$

Remove parentheses and simplify.

7. $5x - (3x + 9)$

8. $5y - 2 - (2y - 4)$

Remove parentheses and simplify.

9. $6x - (4x + 7)$

10. $8y - 3 - (5y - 6)$

11. $(2a + 3b - c) - (4a - 5b + 2c)$

Answers on page A-3

Examples Find an equivalent expression without parentheses.

3. $-(5 - y) = -5 + y$ Changing the sign of each term

4. $-(2a - 7b - 6) = -2a + 7b + 6$

5. $-(-3x + 4y + z - 7w - 23) = 3x - 4y - z + 7w + 23$

Do Exercises 3–6.

b Removing Parentheses and Simplifying

When a sum is added, as in $5x + (2x + 3)$, we can simply remove, or drop, the parentheses and collect like terms because of the associative law of addition:

$$5x + (2x + 3) = 5x + 2x + 3 = 7x + 3.$$

On the other hand, when a sum is subtracted, as in $3x - (4x + 2)$, no "associative" law applies. However, we can subtract by adding an opposite. We then remove parentheses by changing the sign of each term inside the parentheses and collecting like terms.

Example 6 Remove parentheses and simplify.

$3x - (4x + 2) = 3x + [-(4x + 2)]$ Adding the opposite of $(4x + 2)$

$= 3x + (-4x - 2)$ Changing the sign of each term inside the parentheses

$= 3x - 4x - 2$

$= -x - 2$ Collecting like terms

Do Exercises 7 and 8.

In practice, the first three steps of Example 6 are usually combined by changing the sign of each term in parentheses and then collecting like terms.

Examples Remove parentheses and simplify.

7. $5y - (3y + 4) = 5y - 3y - 4$ Removing parentheses by changing the sign of every term inside the parentheses

$= 2y - 4$ Collecting like terms

8. $3y - 2 - (2y - 4) = 3y - 2 - 2y + 4 = y + 2$

9. $(3a + 4b - 5) - (2a - 7b + 4c - 8)$
$= 3a + 4b - 5 - 2a + 7b - 4c + 8$
$= a + 11b - 4c + 3$

Do Exercises 9–11.

Next, consider subtracting an expression consisting of several terms multiplied by a number other than 1 or -1.

Example 10 Remove parentheses and simplify.

$x - 3(x + y) = x + [-3(x + y)]$ Adding the opposite of $3(x + y)$

$= x + [-3x - 3y]$ Multiplying $x + y$ by -3

$= x - 3x - 3y$

$= -2x - 3y$ Collecting like terms

Examples Remove parentheses and simplify.

11. $3y - 2(4y - 5) = 3y - 8y + 10$ Multiplying each term in parentheses by -2
$= -5y + 10$

12. $(2a + 3b - 7) - 4(-5a - 6b + 12)$
$= 2a + 3b - 7 + 20a + 24b - 48$
$= 22a + 27b - 55$

13. $2y - \frac{1}{3}(9y - 12) = 2y - 3y + 4$
$= -y + 4$

Do Exercises 12–15.

c Parentheses Within Parentheses

In addition to parentheses, some expressions contain other grouping symbols such as brackets [ ] and braces { }.

> When more than one kind of grouping symbol occurs, do the computations in the innermost ones first. Then work from the inside out.

Examples Simplify.

14. $[3 - (7 + 3)] = [3 - 10]$ Computing 7 + 3
$= -7$

15. $\{8 - [9 - (12 + 5)]\} = \{8 - [9 - 17]\}$ Computing 12 + 5
$= \{8 - [-8]\}$ Computing 9 − 17
$= 8 + 8$
$= 16$

16. $\left[(-4) \div \left(-\frac{1}{4}\right)\right] \div \frac{1}{4} = [(-4) \cdot (-4)] \div \frac{1}{4}$ Working within the brackets computing $(-4) \div \left(-\frac{1}{4}\right)$
$= 16 \div \frac{1}{4}$
$= 16 \cdot 4$
$= 64$

17. $4(2 + 3) - \{7 - [4 - (8 + 5)]\}$
$= 4 \cdot 5 - \{7 - [4 - 13]\}$ Working with the innermost parentheses first
$= 20 - \{7 - [-9]\}$ Computing $4 \cdot 5$ and $4 - 13$
$= 20 - 16$ Computing $7 - [-9]$
$= 4$

Do Exercises 16–19.

Example 18 Simplify.

$[5(x + 2) - 3x] - [3(y + 2) - 7(y - 3)]$
$= [5x + 10 - 3x] - [3y + 6 - 7y + 21]$ Working with the innermost parentheses first
$= [2x + 10] - [-4y + 27]$ Collecting like terms within brackets
$= 2x + 10 + 4y - 27$ Removing brackets
$= 2x + 4y - 17$ Collecting like terms

Do Exercise 20.

Remove parentheses and simplify.

12. $y - 9(x + y)$

13. $5a - 3(7a - 6)$

14. $4a - b - 6(5a - 7b + 8c)$

15. $5x - \dfrac{1}{4}(8x + 28)$

Simplify.

16. $12 - (8 + 2)$

17. $\{9 - [10 - (13 + 6)]\}$

18. $[24 \div (-2)] \div (-2)$

19. $5(3 + 4) - \{8 - [5 - (9 + 6)]\}$

20. Simplify:

$[3(x + 2) + 2x] -$
$[4(y + 2) - 3(y - 2)]$.

Answers on page A-3

Simplify.

21. $23 - 42 \cdot 30$

22. $32 \div 8 \cdot 2$

23. $52 \cdot 5 + 5^3 - (4^2 - 48 \div 4)$

24. $\dfrac{5 - 10 - 5 \cdot 23}{2^3 + 3^2 - 7}$

Answers on page A-3

d Order of Operations

When several operations are to be done in a calculation or a problem, we apply the following rules.

RULES FOR ORDER OF OPERATIONS

1. Do all calculations within parentheses before operations outside.
2. Evaluate all exponential expressions.
3. Do all multiplications and divisions in order from left to right.
4. Do all additions and subtractions in order from left to right.

These rules are consistent with the way in which most computers and scientific calculators perform calculations.

Example 19 Simplify: $-34 \cdot 56 - 17$.

There are no parentheses or powers, so we start with the third step.

$-34 \cdot 56 - 17 = -1904 - 17$ Carrying out all multiplications and divisions in order from left to right

$= -1921$ Carrying out all additions and subtractions in order from left to right

Example 20 Simplify: $2^4 + 51 \cdot 4 - (37 + 23 \cdot 2)$.

$2^4 + 51 \cdot 4 - (37 + 23 \cdot 2)$

$= 2^4 + 51 \cdot 4 - (37 + 46)$ Following the rules for order of operations within the parentheses first

$= 2^4 + 51 \cdot 4 - 83$ Completing the addition inside parentheses

$= 16 + 51 \cdot 4 - 83$ Evaluating exponential expressions

$= 16 + 204 - 83$ Doing all multiplications

$= 220 - 83$ Doing all additions and subtractions in order from left to right

$= 137$

A fraction bar can play the role of a grouping symbol, although such a symbol is not as evident as the others.

Example 21 Simplify: $\dfrac{-64 \div (-16) \div (-2)}{2^3 - 3^2}$.

An equivalent expression with brackets as grouping symbols is

$$[-64 \div (-16) \div (-2)] \div [2^3 - 3^2].$$

This shows, in effect, that we can do the calculations in the numerator and then in the denominator, and divide the results:

$$\frac{-64 \div (-16) \div (-2)}{2^3 - 3^2} = \frac{4 \div (-2)}{8 - 9} = \frac{-2}{-1} = 2.$$

Do Exercises 21–24.

Calculator Spotlight

Calculations Using Order of Operations. To do a calculation like $-8 - (-2.3)$, we press the following keys:

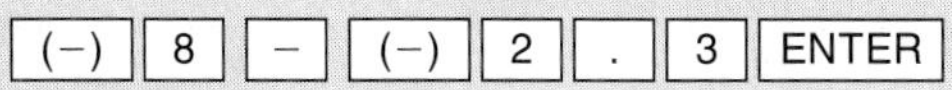

The answer is -5.7.

Note that we did not need to key in grouping symbols. Sometimes we do need the parenthesis keys [( ] and [ )]. For example, to do a calculation like $-7(2 - 9) - 20$, we press the following keys:

[(−)] [7] [(] [2] [−] [9] [)] [−]
[2] [0] [ENTER].

The answer is 29.

To enter a power like $(-39)^4$, we press

[(] [(−)] [3] [9] [)] [^] [4] [ENTER].

The answer is 2,313,441.

To find -39^4, think of the expression as -1×39^4. Then we press

[(−)] [3] [9] [^] [4] [ENTER].

The answer is $-2{,}313{,}441$.

To simplify an expression like

$$\frac{38 + 142}{2 - 47},$$

we first think of it using grouping symbols as

$$(38 + 142) \div (2 - 47).$$

We then press

[(] [3] [8] [+] [1] [4] [2] [)] [÷]
[(] [2] [−] [4] [7] [)] [ENTER].

The answer is -4.

Exercises

Evaluate.

1. $-8 + 4(7 - 9) + 5$
2. $-3[2 + (-5)]$
3. $7[4 - (-3)] + 5[3^2 - (-4)]$
4. $(-7)^6$
5. $(-17)^5$
6. $(-104)^3$
7. -7^6
8. -17^5
9. -104^3

Calculate.

10. $\dfrac{38 - 178}{5 + 30}$
11. $\dfrac{311 - 17^2}{2 - 13}$
12. $785 - \dfrac{285 - 5^4}{17 + 3 \cdot 51}$

In Exercises 13 and 14, place one of $+$, $-$, $\times$, and $\div$ in each blank to make a true sentence.

13. -32 ■ $(88$ ■ $29) = -1888$
14. 3^5 ■ 10^2 ■ $5^2 = -2257$
15. Consider the numbers 2, 4, 6, and 8. Assume that each can be placed in a blank in the following.

    ■ + ■ · ■ − ■ = ?

    What placement of the numbers in the blanks yields the largest number? Explain why there are two answers.
16. Consider the numbers 3, 5, 7, and 9. Assume that each can be placed in a blank in the following.

    ■ + ■^{2} · ■ − ■ = ?

    What placement of the numbers in the blanks yields the largest number? Explain why, unlike Exercise 15, there is just one answer.

Improving Your Math Study Skills

Learning Resources and Time Management

Two other topics to consider in enhancing your math study skills are learning resources and time management.

Learning Resources

- **Textbook supplements.** Are you aware of all the supplements that exist for this textbook? Many details are given in the preface. Now that you are more familiar with the book, let's discuss them.

  1. The *Student's Solutions Manual* contains worked-out solutions to the odd-numbered exercises in the exercise sets. Consider obtaining a copy if you are having trouble. It should be your first choice if you can make an additional purchase.
  2. An extensive set of *videotapes* supplement this text. These may be available to you on your campus at a learning center or math lab. Check with your instructor.
  3. *Tutorial software* also accompanies the text. If not available in the campus learning center, you might order it by calling the number 1-800-322-1377.

- **The Internet.** Our on-line World Wide Web supplement provides additional practice resources. If you have internet access, you can reach this site through the address:

  http://www.mathmax.com

  It contains many helpful ideas as well as many links to other resources for learning mathematics.

- **Your college or university.** Your own college or university probably has resources to enhance your math learning.

  1. For example, is there a learning lab or tutoring center for drop-in tutoring?
  2. Are there special lab classes or formal tutoring sessions tailored for the specific course you are taking?
  3. Perhaps there is a bulletin board or network where you can locate the names of experienced private tutors.

- **Your instructor.** Although it might seem obvious to ask your instructor for help, many students fail to use this valuable resource. Learn what your instructor's office hours are and meet with your instructor at those times when you need extra help.

Time Management

- **Juggling time.** Have reasonable expectations about the time you need to study math. Unreasonable expectations may lead to lower grades and frustrations. Working 40 hours per week and taking 12 hours of credit is equivalent to working two full-time jobs. Can you handle such a load? As a rule of thumb, your ratio of work hours to credit load should be about 40/3, 30/6, 20/9, 10/12, and 5/14. Budget about 2–3 hours of study time outside of class for each hour that you spend in class.
- **Daily schedule.** Make an hour-by-hour schedule of your typical week. Include work, college, home, personal, sleep, study, and leisure times. Be realistic about the amount of time needed for sleep and home duties. If possible, try to schedule time for study when you are most alert.

Exercise Set 1.8

a Find an equivalent expression without parentheses.

1. $-(2x + 7)$
2. $-(8x + 4)$
3. $-(5x - 8)$
4. $-(4x - 3)$
5. $-(4a - 3b + 7c)$
6. $-(x - 4y - 3z)$
7. $-(6x - 8y + 5)$
8. $-(4x + 9y + 7)$
9. $-(3x - 5y - 6)$
10. $-(6a - 4b - 7)$
11. $-(-8x - 6y - 43)$
12. $-(-2a + 9b - 5c)$

b Remove parentheses and simplify.

13. $9x - (4x + 3)$
14. $4y - (2y + 5)$
15. $2a - (5a - 9)$
16. $12m - (4m - 6)$
17. $2x + 7x - (4x + 6)$
18. $3a + 2a - (4a + 7)$
19. $2x - 4y - 3(7x - 2y)$
20. $3a - 9b - 1(4a - 8b)$
21. $15x - y - 5(3x - 2y + 5z)$
22. $4a - b - 4(5a - 7b + 8c)$
23. $(3x + 2y) - 2(5x - 4y)$
24. $(-6a - b) - 5(2b + a)$
25. $(12a - 3b + 5c) - 5(-5a + 4b - 6c)$
26. $(-8x + 5y - 12) - 6(2x - 4y - 10)$

c Simplify.

27. $[9 - 2(5 - 4)]$

28. $[6 - 5(8 - 4)]$

29. $8[7 - 6(4 - 2)]$

30. $10[7 - 4(7 - 5)]$

31. $[4(9 - 6) + 11] - [14 - (6 + 4)]$

32. $[7(8 - 4) + 16] - [15 - (7 + 8)]$

33. $[10(x + 3) - 4] + [2(x - 1) + 6]$

34. $[9(x + 5) - 7] + [4(x - 12) + 9]$

35. $[7(x + 5) - 19] - [4(x - 6) + 10]$

36. $[6(x + 4) - 12] - [5(x - 8) + 14]$

37. $3\{[7(x - 2) + 4] - [2(2x - 5) + 6]\}$

38. $4\{[8(x - 3) + 9] - [4(3x - 2) + 6]\}$

39. $4\{[5(x - 3) + 2] - 3[2(x + 5) - 9]\}$

40. $3\{[6(x - 4) + 5] - 2[5(x + 8) - 3]\}$

d Simplify.

41. $8 - 2 \cdot 3 - 9$

42. $8 - (2 \cdot 3 - 9)$

43. $(8 - 2 \cdot 3) - 9$

44. $(8 - 2)(3 - 9)$

45. $[(-24) \div (-3)] \div \left(-\frac{1}{2}\right)$

46. $[32 \div (-2)] \div (-2)$

47. $16 \cdot (-24) + 50$

48. $10 \cdot 20 - 15 \cdot 24$

49. $2^4 + 2^3 - 10$

50. $40 - 3^2 - 2^3$

51. $5^3 + 26 \cdot 71 - (16 + 25 \cdot 3)$

52. $4^3 + 10 \cdot 20 + 8^2 - 23$

53. $4 \cdot 5 - 2 \cdot 6 + 4$

54. $4 \cdot (6 + 8)/(4 + 3)$

55. $4^3/8$

56. $5^3 - 7^2$

57. $8(-7) + 6(-5)$

58. $10(-5) + 1(-1)$

59. $19 - 5(-3) + 3$

60. $14 - 2(-6) + 7$

61. $9 \div (-3) + 16 \div 8$

62. $-32 - 8 \div 4 - (-2)$

63. $6 - 4^2$

64. $(2 - 5)^2$

65. $(3 - 8)^2$

66. $3 - 3^2$

67. $12 - 20^3$

68. $20 + 4^3 \div (-8)$

69. $2 \cdot 10^3 - 5000$

70. $-7(3^4) + 18$

71. $6[9 - (3 - 4)]$

72. $8[(6 - 13) - 11]$

73. $-1000 \div (-100) \div 10$

74. $256 \div (-32) \div (-4)$

75. $8 - (7 - 9)$

76. $(8 - 7) - 9$

77. $\dfrac{10 - 6^2}{9^2 + 3^2}$

78. $\dfrac{5^2 - 4^3 - 3}{9^2 - 2^2 - 1^5}$

79. $\dfrac{3(6-7)-5\cdot 4}{6\cdot 7-8(4-1)}$

80. $\dfrac{20(8-3)-4(10-3)}{10(2-6)-2(5+2)}$

81. $\dfrac{2^3-3^2+12\cdot 5}{-32\div(-16)\div(-4)}$

82. $\dfrac{|3-5|^2-|7-13|}{|12-9|+|11-14|}$

Skill Maintenance

Multiply. [1.7c]

83. $8(2x-3)$

84. $5(1+a)$

85. $\dfrac{2}{3}(y-9)$

86. $-7(3x+5)$

Factor. [1.7d]

87. $35-5b$

88. $48x-36$

89. $15x-60y$

90. $12x-15y+30z$

Synthesis

91. ◆ Some students use the memory device PEMDAS ("Please Excuse My Dear Aunt Sally") to remember the rules for the order of operations. Explain how this can be done.

92. ◆ Determine whether $|-x|$ and $|x|$ are equivalent. Explain.

Find an equivalent expression by enclosing the last three terms in parentheses preceded by a minus sign.

93. $6y+2x-3a+c$

94. $x-y-a-b$

95. $6m+3n-5m+4b$

Simplify.

96. $z-\{2z-[3z-(4z-5z)-6z]-7z\}-8z$

97. $\{x-[f-(f-x)]+[x-f]\}-3x$

98. $x-\{x-1-[x-2-(x-3-\{x-4-[x-5-(x-6)]\})]\}$

99. Use your calculator to do the following.

a) Evaluate x^2+3 for $x=7$, for $x=-7$, and for $x=-5.013$.

b) Evaluate $1-x^2$ for $x=5$, for $x=-5$, and for $x=-10.455$.

100. Express $3^3+3^3+3^3$ as a power of 3.

Collaborative Learning Manual

Use the order of operations as a group to simplify expressions.

Summary and Review Exercises: Chapter 1

Important Properties and Formulas

Properties of the Real-Number System

The Commutative Laws:	$a + b = b + a, \quad ab = ba$
The Associative Laws:	$a + (b + c) = (a + b) + c, \quad a(bc) = (ab)c$
The Identity Properties:	For every real number a, $a + 0 = a$ and $a \cdot 1 = a$.
The Inverse Properties:	For each real number a, there is an opposite $-a$, such that $a + (-a) = 0$.
	For each nonzero real number a, there is a reciprocal $\frac{1}{a}$, such that $a\left(\frac{1}{a}\right) = 1$.
The Distributive Laws:	$a(b + c) = ab + ac, \quad a(b - c) = ab - ac$

The review exercises that follow are for practice. Answers are at the back of the book. If you miss an exercise, restudy the objective indicated in blue after the exercise or the direction line that precedes it.

1. Evaluate $\frac{x - y}{3}$ for $x = 17$ and $y = 5$. [1.1a]

2. Translate to an algebraic expression: [1.1b]
Nineteen percent of some number.

3. Tell which integers correspond to this situation: [1.2a]
David has a debt of $45 and Joe has $72 in his savings account.

4. Find: $|-38|$. [1.2e]

Graph the number on a number line. [1.2b]

5. -2.5

6. $\frac{8}{9}$

Use either $<$ or $>$ for ▒ to write a true sentence. [1.2d]

7. -3 ▒ 10

8. -1 ▒ -6

9. 0.126 ▒ -12.6

10. $-\frac{2}{3}$ ▒ $-\frac{1}{10}$

Find the opposite. [1.3b]

11. 3.8

12. $-\frac{3}{4}$

Find the reciprocal. [1.6b]

13. $\frac{3}{8}$

14. -7

15. Find $-x$ when $x = -34$. [1.3b]

16. Find $-(-x)$ when $x = 5$. [1.3b]

Compute and simplify.

17. $4 + (-7)$ [1.3a]

18. $6 + (-9) + (-8) + 7$ [1.3a]

19. $-3.8 + 5.1 + (-12) + (-4.3) + 10$ [1.3a]

20. $-3 - (-7)$ [1.4a]

21. $-\frac{9}{10} - \frac{1}{2}$ [1.4a]

22. $-3.8 - 4.1$ [1.4a]

23. $-9 \cdot (-6)$ [1.5a]

24. $-2.7(3.4)$ [1.5a]

25. $\frac{2}{3} \cdot \left(-\frac{3}{7}\right)$ [1.5a]

26. $3 \cdot (-7) \cdot (-2) \cdot (-5)$ [1.5a]

27. $35 \div (-5)$ [1.6a]

28. $-5.1 \div 1.7$ [1.6c]

29. $-\frac{3}{11} \div \left(-\frac{4}{11}\right)$ [1.6c]

30. $(-3.4 - 12.2) - 8(-7)$ [1.8d]

31. $\frac{-12(-3) - 2^3 - (-9)(-10)}{3 \cdot 10 + 1}$ [1.8d]

Solve. [1.4b]

32. On the first, second, and third downs, a football team had these gains and losses: 5-yd gain, 12-yd loss, and 15-yd gain, respectively. Find the total gain (or loss).

33. Kaleb's total assets are $170. He borrows $300. What are his total assets now?

Multiply. [1.7c]

34. $5(3x - 7)$

35. $-2(4x - 5)$

36. $10(0.4x + 1.5)$

37. $-8(3 - 6x)$

Factor. [1.7d]

38. $2x - 14$

39. $6x - 6$

40. $5x + 10$

41. $12 - 3x$

Collect like terms. [1.7e]

42. $11a + 2b - 4a - 5b$

43. $7x - 3y - 9x + 8y$

44. $6x + 3y - x - 4y$

45. $-3a + 9b + 2a - b$

Remove parentheses and simplify.

46. $2a - (5a - 9)$ [1.8b]

47. $3(b + 7) - 5b$ [1.8b]

48. $3[11 - 3(4 - 1)]$ [1.8c]

49. $2[6(y - 4) + 7]$ [1.8c]

50. $[8(x + 4) - 10] - [3(x - 2) + 4]$ [1.8c]

51. $5\{[6(x - 1) + 7] - [3(3x - 4) + 8]\}$ [1.8c]

Write true or false. [1.2d]

52. $-9 \leq 11$

53. $-11 \geq -3$

54. Write another inequality with the same meaning as $-3 < x$. [1.2d]

Synthesis

Simplify. [1.2e], [1.4a], [1.6a], [1.8d]

55. $-\left|\frac{7}{8} - \left(-\frac{1}{2}\right) - \frac{3}{4}\right|$

56. $(|2.7 - 3| + 3^2 - |-3|) \div (-3)$

57. $2000 - 1990 + 1980 - 1970 + \cdots + 20 - 10$

58. Find a formula for the perimeter of the following figure. [1.7e]

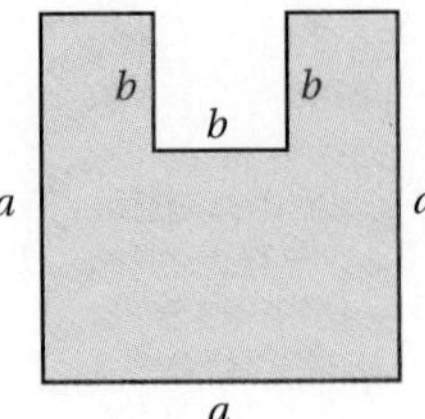

Test: Chapter 1

1. Evaluate $\frac{3x}{y}$ for $x = 10$ and $y = 5$.

2. Write an algebraic expression: Nine less than some number.

3. Find the area of a triangle when the height h is 30 ft and the base b is 16 ft.

Use either $<$ or $>$ for ▒ to write a true sentence.

4. -4 ▒ 0

5. -3 ▒ -8

6. -0.78 ▒ -0.87

7. $-\frac{1}{8}$ ▒ $\frac{1}{2}$

Find the absolute value.

8. $|-7|$

9. $\left|\frac{9}{4}\right|$

10. $|-2.7|$

Find the opposite.

11. $\frac{2}{3}$

12. -1.4

13. Find $-x$ when $x = -8$.

Find the reciprocal.

14. -2

15. $\frac{4}{7}$

Compute and simplify.

16. $3.1 - (-4.7)$

17. $-8 + 4 + (-7) + 3$

18. $-\frac{1}{5} + \frac{3}{8}$

19. $2 - (-8)$

20. $3.2 - 5.7$

21. $\frac{1}{8} - \left(-\frac{3}{4}\right)$

Answers

1. __________

2. __________

3. __________

4. __________

5. __________

6. __________

7. __________

8. __________

9. __________

10. __________

11. __________

12. __________

13. __________

14. __________

15. __________

16. __________

17. __________

18. __________

19. __________

20. __________

21. __________

Answers

22. ______

23. ______

24. ______

25. ______

26. ______

27. ______

28. ______

29. ______

30. ______

31. ______

32. ______

33. ______

34. ______

35. ______

36. ______

37. ______

38. ______

39. ______

40. ______

41. ______

42. ______

22. $4 \cdot (-12)$

23. $-\frac{1}{2} \cdot \left(-\frac{3}{8}\right)$

24. $-45 \div 5$

25. $-\frac{3}{5} \div \left(-\frac{4}{5}\right)$

26. $4.864 \div (-0.5)$

27. $-2(16) - |2(-8) - 5^3|$

28. *Antarctica Highs and Lows.* The continent of Antarctica, which lies in the southern hemisphere, experiences winter in July. The average high temperature is −67°F and the average low temperature is −81°F. How much higher is the average high than the average low?

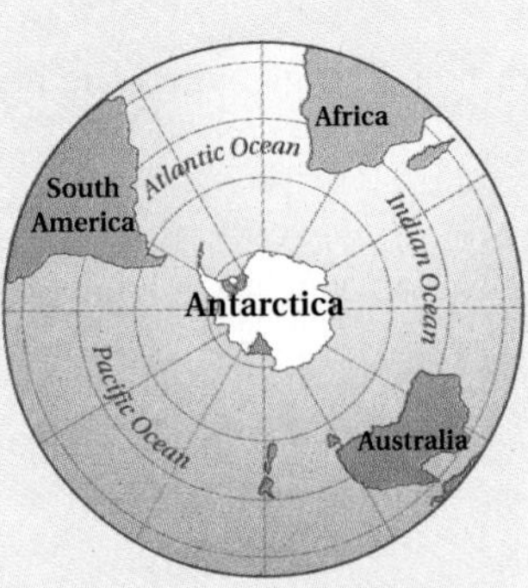

Multiply.

29. $3(6 - x)$

30. $-5(y - 1)$

Factor.

31. $12 - 22x$

32. $7x + 21 + 14y$

Simplify.

33. $6 + 7 - 4 - (-3)$

34. $5x - (3x - 7)$

35. $4(2a - 3b) + a - 7$

36. $4\{3[5(y - 3) + 9] + 2(y + 8)\}$

37. $256 \div (-16) \div 4$

38. $2^3 - 10[4 - (-2 + 18)3]$

39. Write an inequality with the same meaning as $x \leq -2$.

Synthesis

Simplify.

40. $|-27 - 3(4)| - |-36| + |-12|$

41. $a - \{3a - [4a - (2a - 4a)]\}$

42. Find a formula for the perimeter of the following figure.

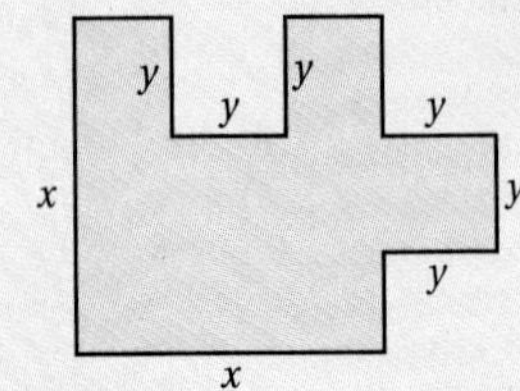

2

Solving Equations and Inequalities

An Application

The perimeter of an NBA basketball court is 288 ft. The length is 44 ft longer than the width. (***Source***: National Basketball Association) Find the dimensions of the court.

This problem appears as Example 5 in Section 2.4.

The Mathematics

The perimeter P of a rectangle is given by the formula $2l + 2w = P$, where $l =$ the length and $w =$ the width. To translate the problem, we substitute $w + 44$ for l and 288 for P:

$$2l + 2w = P$$
$$\underbrace{2(w + 44) + 2w = 288.}$$

To find the dimensions, we first solve this equation.

World Wide Web For more information, visit us at www.mathmax.com

Introduction

In this chapter, we use the manipulations considered in Chapter 1 to solve equations and inequalities. We then use equations and inequalities to solve applied problems.

Pretest: Chapter 2

Solve.

1. $-7x = 49$

2. $4y + 9 = 2y + 7$

3. $6a - 2 = 10$

4. $4 + x = 12$

5. $7 - 3(2x - 1) = 40$

6. $\frac{4}{9}x - 1 = \frac{7}{8}$

7. $1 + 2(a + 3) = 3(2a - 1) + 6$

8. $-3x \leq 18$

9. $y + 5 > 1$

10. $5 - 2a < 7$

11. $3x + 4 \geq 2x + 7$

12. $8y < -18$

13. Solve for G: $P = 3KG$.

14. Solve for a: $A = \dfrac{3a - b}{b}$.

Solve.

15. The perimeter of the ornate frame of an oil painting is 146 in. The width is 5 in. less than the length. Find the dimensions.

16. Money is invested in a savings account at 4.25% simple interest. After 1 year, there is $479.55 in the account. How much was originally invested?

17. The sum of three consecutive integers is 246. Find the integers.

18. When 18 is added to six times a number, the result is less than 120. For what numbers is this possible?

Graph on a number line.

19. $x > -3$

20. $x \leq 4$

Objectives for Retesting

The objectives to be tested in addition to the material in this chapter are as follows.

[1.1a] Evaluate algebraic expressions by substitution.
[1.1b] Translate phrases to algebraic expressions.
[1.3a] Add real numbers.
[1.8b] Simplify expressions by removing parentheses and collecting like terms.

2.1 Solving Equations: The Addition Principle

a Equations and Solutions

In order to solve problems, we must learn to solve equations.

> An **equation** is a number sentence that says that the expressions on either side of the equals sign, =, represent the same number.

Here are some examples:

$$3 + 2 = 5, \quad 14 - 10 = 1 + 3, \quad x + 6 = 13, \quad 3x - 2 = 7 - x.$$

Equations have expressions on each side of the equals sign. The sentence "$14 - 10 = 1 + 3$" asserts that the expressions $14 - 10$ and $1 + 3$ name the same number.

Some equations are true. Some are false. Some are neither true nor false.

Examples Determine whether the equation is true, false, or neither.

1. $3 + 2 = 5$ The equation is *true*.

2. $7 - 2 = 4$ The equation is *false*.

3. $x + 6 = 13$ The equation is *neither* true nor false, because we do not know what number x represents.

Do Exercises 1–3.

> Any replacement for the variable that makes an equation true is called a **solution** of the equation. To solve an equation means to find *all* of its solutions.

One way to determine whether a number is a solution of an equation is to evaluate the expression on each side of the equals sign by substitution. If the values are the same, then the number is a solution.

Example 4 Determine whether 7 is a solution of $x + 6 = 13$.

We have

$x + 6 = 13$		Writing the equation
$7 + 6$	? 13	Substituting 7 for x
13		TRUE

Since the left-hand and the right-hand sides are the same, we have a solution. No other number makes the equation true, so the only solution is the number 7.

Objectives

a Determine whether a given number is a solution of a given equation.

b Solve equations using the addition principle.

For Extra Help

TAPE 3

MAC WIN

CD-ROM

Determine whether the equation is true, false, or neither.

1. $5 - 8 = -4$

2. $12 + 6 = 18$

3. $x + 6 = 7 - x$

Answers on page A-4

Determine whether the given number is a solution of the given equation.

4. 8; $x + 4 = 12$

5. 0; $x + 4 = 12$

6. -3; $7 + x = -4$

7. Solve using the addition principle:

$$x + 2 = 11.$$

Answers on page A-4

Example 5 Determine whether 19 is a solution of $7x = 141$.

We have

$7x = 141$	Writing the equation
$7(19) \; ? \; 141$	Substituting 19 for x
$133 \mid$	FALSE

Since the left-hand and the right-hand sides are not the same, we do not have a solution.

Do Exercises 4–6.

b Using the Addition Principle

Consider the equation

$$x = 7.$$

We can easily see that the solution of this equation is 7. If we replace x with 7, we get

$7 = 7$, which is true.

Now consider the equation of Example 4:

$$x + 6 = 13.$$

In Example 4, we discovered that the solution of this equation is also 7, but the fact that 7 is the solution is not as obvious. We now begin to consider principles that allow us to start with an equation and end up with an *equivalent equation,* like $x = 7$, in which the variable is alone on one side and for which the solution is easy to find.

> Equations with the same solutions are called **equivalent equations.**

One of the principles that we use in solving equations involves adding. An equation $a = b$ says that a and b stand for the same number. Suppose this is true, and we add a number c to the number a. We get the same answer if we add c to b, because a and b are the same number.

> **THE ADDITION PRINCIPLE**
>
> For any real numbers a, b, and c,
>
> $a = b$ is equivalent to $a + c = b + c$.

Let's again solve the equation $x + 6 = 13$ using the addition principle. We want to get x alone on one side. To do so, we use the addition principle, choosing to add -6 because $6 + (-6) = 0$:

$x + 6 = 13$	
$x + 6 + (-6) = 13 + (-6)$	Using the addition principle: adding -6 on both sides
$x + 0 = 7$	Simplifying
$x = 7.$	Identity property of 0: $x + 0 = x$

Do Exercise 7.

When we use the addition principle, we sometimes say that we "add the same number on both sides of the equation." This is also true for subtraction, since we can express every subtraction as an addition. That is, since

$$a - c = b - c \quad \text{is equivalent to} \quad a + (-c) = b + (-c),$$

the addition principle tells us that we can "subtract the same number on both sides of the equation."

Example 6 Solve: $x + 5 = -7$.

We have

$$
\begin{aligned}
x + 5 &= -7 \\
x + 5 - 5 &= -7 - 5 && \text{Using the addition principle: adding } -5 \text{ on both sides or subtracting 5 on both sides} \\
x + 0 &= -12 && \text{Simplifying} \\
x &= -12. && \text{Identity property of 0}
\end{aligned}
$$

We can see that the solution of $x = -12$ is the number -12. To check the answer, we substitute -12 in the original equation.

CHECK:

$$
\begin{array}{r|l}
x + 5 & -7 \\
\hline
-12 + 5 \;?\; & -7 \\
-7 & \quad \text{TRUE}
\end{array}
$$

The solution of the original equation is -12.

In Example 6, to get x alone, we used the addition principle and subtracted 5 on both sides. This eliminated the 5 on the left. We started with $x + 5 = -7$, and, using the addition principle, we found a simpler equation $x = -12$ for which it was easy to "*see*" the solution. The equations $x + 5 = -7$ and $x = -12$ are *equivalent.*

Do Exercise 8.

Now we solve an equation with a subtraction using the addition principle.

Example 7 Solve: $a - 4 = 10$.

We have

$$
\begin{aligned}
a - 4 &= 10 \\
a - 4 + 4 &= 10 + 4 && \text{Using the addition principle: adding 4 on both sides} \\
a + 0 &= 14 && \text{Simplifying} \\
a &= 14. && \text{Identity property of 0}
\end{aligned}
$$

CHECK:

$$
\begin{array}{r|l}
a - 4 & 10 \\
\hline
14 - 4 \;?\; & 10 \\
10 & \quad \text{TRUE}
\end{array}
$$

The solution is 14.

Do Exercise 9.

8. Solve using the addition principle, subtracting 7 on both sides:

$x + 7 = 2.$

9. Solve: $t - 3 = 19$.

Answers on page A-4

Solve.

10. $8.7 = n - 4.5$

11. $y + 17.4 = 10.9$

Solve.

12. $x + \frac{1}{2} = -\frac{3}{2}$

13. $t - \frac{13}{4} = \frac{5}{8}$

Answers on page A-4

Example 8 Solve: $-6.5 = y - 8.4$.

We have

$$-6.5 = y - 8.4$$

$$-6.5 + 8.4 = y - 8.4 + 8.4 \quad \text{Using the addition principle: adding 8.4 to eliminate } -8.4 \text{ on the right}$$

$$1.9 = y.$$

CHECK:

$$\begin{array}{c|c} -6.5 & y - 8.4 \\ \hline -6.5 \;?& 1.9 - 8.4 \\ & -6.5 \end{array} \quad \text{TRUE}$$

The solution is 1.9.

Note that equations are reversible. That is, if $a = b$ is true, then $b = a$ is true. Thus when we solve $-6.5 = y - 8.4$, we can reverse it and solve $y - 8.4 = -6.5$ if we wish.

Do Exercises 10 and 11.

Example 9 Solve: $-\frac{2}{3} + x = \frac{5}{2}$.

We have

$$-\frac{2}{3} + x = \frac{5}{2}$$

$$\frac{2}{3} - \frac{2}{3} + x = \frac{2}{3} + \frac{5}{2} \quad \text{Adding } \frac{2}{3}$$

$$x = \frac{2}{3} \cdot \frac{2}{2} + \frac{5}{2} \cdot \frac{3}{3} \quad \text{Multiplying by 1 to obtain equivalent fractional expressions with the least common denominator 6}$$

$$= \frac{4}{6} + \frac{15}{6}$$

$$= \frac{19}{6}.$$

CHECK:

$$\begin{array}{c|c} -\frac{2}{3} + x & \frac{5}{2} \\ \hline -\frac{2}{3} + \frac{19}{6} \;?& \frac{5}{2} \\ -\frac{4}{6} + \frac{19}{6} & \\ \frac{15}{6} & \\ \frac{5}{2} & \end{array} \quad \text{TRUE}$$

The solution is $\frac{19}{6}$.

Do Exercises 12 and 13.

Exercise Set 2.1

a Determine whether the given number is a solution of the given equation.

1. 15; $x + 17 = 32$

2. 35; $t + 17 = 53$

3. 21; $x - 7 = 12$

4. 36; $a - 19 = 17$

5. -7; $6x = 54$

6. -9; $8y = -72$

7. 30; $\frac{x}{6} = 5$

8. 49; $\frac{y}{8} = 6$

9. 19; $5x + 7 = 107$

10. 9; $9x + 5 = 86$

11. -11; $7(y - 1) = 63$

12. -18; $x + 3 = 3 + x$

b Solve using the addition principle. Don't forget to check!

13. $x + 2 = 6$

CHECK: $x + 2 = 6$?

14. $y + 4 = 11$

CHECK: $y + 4 = 11$?

15. $x + 15 = -5$

CHECK: $x + 15 = -5$?

16. $t + 10 = 44$

CHECK: $t + 10 = 44$?

17. $x + 6 = -8$

CHECK: $x + 6 = -8$?

18. $z + 9 = -14$

19. $x + 16 = -2$

20. $m + 18 = -13$

21. $x - 9 = 6$

22. $x - 11 = 12$

23. $x - 7 = -21$

24. $x - 3 = -14$

25. $5 + t = 7$

26. $8 + y = 12$

27. $-7 + y = 13$

28. $-8 + y = 17$

29. $-3 + t = -9$

30. $-8 + t = -24$

31. $x + \frac{1}{2} = 7$

32. $24 = -\frac{7}{10} + r$

33. $12 = a - 7.9$

34. $2.8 + y = 11$

35. $r + \frac{1}{3} = \frac{8}{3}$

36. $t + \frac{3}{8} = \frac{5}{8}$

37. $m + \frac{5}{6} = -\frac{11}{12}$

38. $x + \frac{2}{3} = -\frac{5}{6}$

39. $x - \frac{5}{6} = \frac{7}{8}$

40. $y - \frac{3}{4} = \frac{5}{6}$

41. $-\frac{1}{5} + z = -\frac{1}{4}$

42. $-\frac{1}{8} + y = -\frac{3}{4}$

43. $7.4 = x + 2.3$

44. $8.4 = 5.7 + y$

45. $7.6 = x - 4.8$

46. $8.6 = x - 7.4$

47. $-9.7 = -4.7 + y$

48. $-7.8 = 2.8 + x$

49. $5\frac{1}{6} + x = 7$

50. $5\frac{1}{4} = 4\frac{2}{3} + x$

51. $q + \frac{1}{3} = -\frac{1}{7}$

52. $52\frac{3}{8} = -84 + x$

Skill Maintenance

53. Add: $-3 + (-8)$. [1.3a]

54. Subtract: $-3 - (-8)$. [1.4a]

55. Multiply: $-\frac{2}{3} \cdot \frac{5}{8}$. [1.5a]

56. Divide: $-\frac{3}{7} \div \left(-\frac{9}{7}\right)$. [1.6c]

57. Divide: $\frac{2}{3} \div \left(-\frac{4}{9}\right)$. [1.6c]

58. Add: $-8.6 + 3.4$. [1.3a]

Translate to an algebraic expression. [1.1b]

59. Liza had \$50 before paying x dollars for a pizza. How much does she have left?

60. Donnie drove his S-10 pickup truck 65 mph for t hours. How far did he drive?

Synthesis

61. ◆ Explain the difference between equivalent expressions and equivalent equations.

62. ◆ When solving an equation using the addition principle, how do you determine which number to add or subtract on both sides of the equation?

Solve.

63. 🖩 $-356.788 = -699.034 + t$

64. $-\frac{4}{5} + \frac{7}{10} = x - \frac{3}{4}$

65. $x + \frac{4}{5} = -\frac{2}{3} - \frac{4}{15}$

66. $8 - 25 = 8 + x - 21$

67. $16 + x - 22 = -16$

68. $x + x = x$

69. $x + 3 = 3 + x$

70. $x + 4 = 5 + x$

71. $-\frac{3}{2} + x = -\frac{5}{17} - \frac{3}{2}$

72. $|x| = 5$

73. $|x| + 6 = 19$

2.2 Solving Equations: The Multiplication Principle

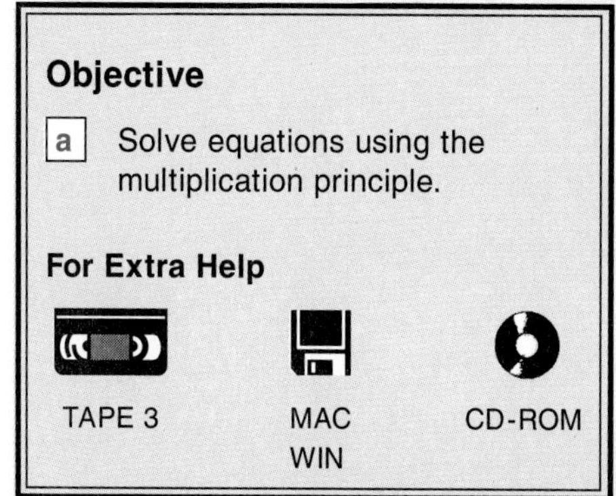

a Using the Multiplication Principle

Suppose that $a = b$ is true, and we multiply a by some number c. We get the same answer if we multiply b by c, because a and b are the same number.

> **THE MULTIPLICATION PRINCIPLE**
>
> For any real numbers a, b, and c, $c \neq 0$,
>
> $$a = b \quad \text{is equivalent to} \quad a \cdot c = b \cdot c.$$

When using the multiplication principle, we sometimes say that we "multiply on both sides of the equation by the same number."

Example 1 Solve: $5x = 70$.

To get x alone, we multiply by the *multiplicative inverse,* or *reciprocal,* of 5. Then we get the *multiplicative identity* 1 times x, or $1 \cdot x$, which simplifies to x. This allows us to eliminate 5 on the left.

$5x = 70$ — The reciprocal of 5 is $\frac{1}{5}$.

$\frac{1}{5} \cdot 5x = \frac{1}{5} \cdot 70$ — Multiplying by $\frac{1}{5}$ to get $1 \cdot x$ and eliminate 5 on the left

$1 \cdot x = 14$ — Simplifying

$x = 14$ — Identity property of 1: $1 \cdot x = x$

CHECK:

$5x = 70$	
$5 \cdot 14 \ ? \ 70$	
70	TRUE

The solution is 14.

The multiplication principle also tells us that we can "divide on both sides of the equation by a nonzero number." This is because division is the same as multiplying by a reciprocal. That is,

$$\frac{a}{c} = \frac{b}{c} \quad \text{is equivalent to} \quad a \cdot \frac{1}{c} = b \cdot \frac{1}{c}, \quad \text{when } c \neq 0.$$

In an expression like $5x$ in Example 1, the number 5 is called the **coefficient**. Example 1 could be done as follows, dividing by 5, the coefficient of x, on both sides.

1. Solve. Multiply on both sides.

$6x = 90$

2. Solve. Divide on both sides.

$4x = -7$

3. Solve: $-6x = 108$.

4. Solve: $-x = -10$.

Answers on page A-4

Example 2 Solve: $5x = 70$.

We have

$$5x = 70$$

$$\frac{5x}{5} = \frac{70}{5} \quad \text{Dividing by 5 on both sides}$$

$$1 \cdot x = 14 \quad \text{Simplifying}$$

$$x = 14. \quad \text{Identity property of 1}$$

Do Exercises 1 and 2.

Example 3 Solve: $-4x = 92$.

We have

$$-4x = 92$$

$$\frac{-4x}{-4} = \frac{92}{-4} \quad \text{Using the multiplication principle. Dividing by } -4 \text{ on both sides is the same as multiplying by } -\tfrac{1}{4}.$$

$$1 \cdot x = -23 \quad \text{Simplifying}$$

$$x = -23. \quad \text{Identity property of 1}$$

CHECK:

$-4x = 92$	
$-4(-23) \;?\; 92$	
$92 \;\vert$	TRUE

The solution is -23.

Do Exercise 3.

Example 4 Solve: $-x = 9$.

We have

$$-x = 9$$

$$-1 \cdot x = 9 \quad \text{Using the property of } -1\text{: } -x = -1 \cdot x$$

$$\frac{-1 \cdot x}{-1} = \frac{9}{-1} \quad \text{Dividing by } -1$$

$$1 \cdot x = -9$$

$$x = -9.$$

CHECK:

$-x = 9$	
$-(-9) \;?\; 9$	
$9 \;\vert$	TRUE

The solution is -9.

Do Exercise 4.

In practice, it is generally more convenient to "divide" on both sides of the equation if the coefficient of the variable is in decimal notation or is an integer. If the coefficient is in fractional notation, it is more convenient to "multiply" by a reciprocal.

Example 5 Solve: $\frac{3}{8} = -\frac{5}{4}x$.

$$\frac{3}{8} = -\frac{5}{4}x$$

The reciprocal of $-\frac{5}{4}$ is $-\frac{4}{5}$. There is no sign change.

$$-\frac{4}{5} \cdot \frac{3}{8} = -\frac{4}{5} \cdot \left(-\frac{5}{4}x\right) \quad \text{Multiplying by } -\tfrac{4}{5} \text{ to get } 1 \cdot x \text{ and eliminate } -\tfrac{5}{4} \text{ on the right}$$

$$-\frac{12}{40} = 1 \cdot x$$

$$-\frac{3}{10} = 1 \cdot x \quad \text{Simplifying}$$

$$-\frac{3}{10} = x \quad \text{Identity property of 1}$$

CHECK:

$$\frac{3}{8} = -\frac{5}{4}x$$

$$\frac{3}{8} \; ? \; -\frac{5}{4}\left(-\frac{3}{10}\right)$$

$$\frac{3}{8} \qquad \text{TRUE}$$

The solution is $-\frac{3}{10}$.

Note that equations are reversible. That is, if $a = b$ is true, then $b = a$ is true. Thus when we solve $\frac{3}{8} = -\frac{5}{4}x$, we can reverse it and solve $-\frac{5}{4}x = \frac{3}{8}$ if we wish.

Do Exercise 5.

Example 6 Solve: $1.16y = 9744$.

$$1.16y = 9744$$

$$\frac{1.16y}{1.16} = \frac{9744}{1.16} \quad \text{Dividing by 1.16}$$

$$y = \frac{9744}{1.16}$$

$$= 8400$$

CHECK:

$$1.16y = 9744$$

$$1.16(8400) \; ? \; 9744$$

$$9744 \qquad \text{TRUE}$$

The solution is 8400.

Do Exercises 6 and 7.

5. Solve: $\frac{2}{3} = -\frac{5}{6}y$.

Solve.

6. $1.12x = 8736$

7. $6.3 = -2.1y$

Answers on page A-4

8. Solve: $-14 = \frac{-y}{2}$.

Now we solve an equation with a division using the multiplication principle. Consider an equation like $-y/9 = 14$. In Chapter 1, we learned that a division can be expressed as multiplication by the reciprocal of the divisor. Thus,

$$\frac{-y}{9} \text{ is equivalent to } \frac{1}{9}(-y).$$

The reciprocal of $\frac{1}{9}$ is 9. Then, using the multiplication principle, we multiply by 9 on both sides. This is shown in the following example.

Example 7 Solve: $\frac{-y}{9} = 14$.

$$\frac{-y}{9} = 14$$

$$\frac{1}{9}(-y) = 14$$

$$9 \cdot \frac{1}{9}(-y) = 9 \cdot 14 \quad \text{Multiplying by 9 on both sides}$$

$$-y = 126$$

$$-1 \cdot (-y) = -1 \cdot 126 \quad \text{Multiplying by } -1 \text{, or dividing by } -1 \text{, on both sides}$$

$$y = -126$$

CHECK:

$$\begin{array}{c|c} \frac{-y}{9} & 14 \\ \hline \frac{-(-126)}{9} & ?\ 14 \\ \frac{126}{9} & \\ 14 & \end{array} \quad \text{TRUE}$$

The solution is -126.

Do Exercise 8.

Answer on page A-4

Exercise Set 2.2

a Solve using the multiplication principle. Don't forget to check!

1. $6x = 36$

CHECK: $6x = 36$ | ?

2. $3x = 51$

CHECK: $3x = 51$ | ?

3. $5x = 45$

CHECK: $5x = 45$ | ?

4. $8x = 72$

CHECK: $8x = 72$ | ?

5. $84 = 7x$

6. $63 = 9x$

7. $-x = 40$

8. $53 = -x$

9. $-x = -1$

10. $-47 = -t$

11. $7x = -49$

12. $8x = -56$

13. $-12x = 72$

14. $-15x = 105$

15. $-21x = -126$

16. $-13x = -104$

17. $\frac{t}{7} = -9$

18. $\frac{y}{-8} = 11$

19. $\frac{3}{4}x = 27$

20. $\frac{4}{5}x = 16$

21. $\frac{-t}{3} = 7$

22. $\frac{-x}{6} = 9$

23. $-\frac{m}{3} = \frac{1}{5}$

24. $\frac{1}{8} = -\frac{y}{5}$

25. $-\frac{3}{5}r = \frac{9}{10}$

26. $\frac{2}{5}y = -\frac{4}{15}$

27. $-\frac{3}{2}r = -\frac{27}{4}$

28. $-\frac{3}{8}x = -\frac{15}{16}$

29. $6.3x = 44.1$

30. $2.7y = 54$

31. $-3.1y = 21.7$

32. $-3.3y = 6.6$

33. $38.7m = 309.6$

34. $29.4m = 235.2$

35. $-\frac{2}{3}y = -10.6$

36. $-\frac{9}{7}y = 12.06$

Skill Maintenance

Collect like terms. [1.7e]

37. $3x + 4x$

38. $6x + 5 - 7x$

39. $-4x + 11 - 6x + 18x$

40. $8y - 16y - 24y$

Remove parentheses and simplify. [1.8b]

41. $3x - (4 + 2x)$

42. $2 - 5(x + 5)$

43. $8y - 6(3y + 7)$

44. $-2a - 4(5a - 1)$

Translate to an algebraic expression. [1.1b]

45. Patty drives her van for 8 hr at a speed of r mph. How far does she drive?

46. A triangle has a height of 10 meters and a base of b meters. What is the area of the triangle?

Synthesis

47. ◆ When solving an equation using the multiplication principle, how do you determine by what number to multiply or divide on both sides of the equation?

48. ◆ Are the equations $x = 5$ and $x^2 = 25$ equivalent? Why or why not?

Solve.

49. ▦ $-0.2344m = 2028.732$

50. $0 \cdot x = 0$

51. $0 \cdot x = 9$

52. $4|x| = 48$

53. $2|x| = -12$

Solve for x.

54. $ax = 5a$

55. $3x = \frac{b}{a}$

56. $cx = a^2 + 1$

57. $\frac{a}{b}x = 4$

58. A student makes a calculation and gets an answer of 22.5. On the last step, the student multiplies by 0.3 when a division by 0.3 should have been done. What is the correct answer?

2.3 Using the Principles Together

a Applying Both Principles

Consider the equation $3x + 4 = 13$. It is more complicated than those we discussed in the preceding two sections. In order to solve such an equation, we first isolate the x-term, $3x$, using the addition principle. Then we apply the multiplication principle to get x by itself.

Example 1 Solve: $3x + 4 = 13$.

$$3x + 4 = 13$$

$$3x + 4 - 4 = 13 - 4 \quad \text{Using the addition principle: subtracting 4 on both sides}$$

First, isolate the x-term. → $3x = 9$ Simplifying

$$\frac{3x}{3} = \frac{9}{3} \quad \text{Using the multiplication principle: dividing by 3 on both sides}$$

Then isolate x. → $x = 3$ Simplifying

CHECK:

$$\begin{array}{c|c} 3x + 4 & 13 \\ \hline 3 \cdot 3 + 4 \ ? & 13 \\ 9 + 4 & \\ 13 & \end{array} \quad \text{TRUE}$$

We use the rules for order of operations to carry out the check. We find the product $3 \cdot 3$. Then we add 4.

The solution is 3.

Do Exercise 1.

Example 2 Solve: $-5x - 6 = 16$.

$$-5x - 6 = 16$$

$$-5x - 6 + 6 = 16 + 6 \quad \text{Adding 6 on both sides}$$

$$-5x = 22$$

$$\frac{-5x}{-5} = \frac{22}{-5} \quad \text{Dividing by } -5 \text{ on both sides}$$

$$x = -\frac{22}{5}, \text{ or } -4\frac{2}{5} \quad \text{Simplifying}$$

CHECK:

$$\begin{array}{c|c} -5x - 6 & 16 \\ \hline -5\left(-\frac{22}{5}\right) - 6 \ ? & 16 \\ 22 - 6 & \\ 16 & \end{array} \quad \text{TRUE}$$

The solution is $-\frac{22}{5}$.

Do Exercises 2 and 3.

Objectives

a Solve equations using both the addition and the multiplication principles.

b Solve equations in which like terms may need to be collected.

c Solve equations by first removing parentheses and collecting like terms.

For Extra Help

TAPE 3

MAC
WIN

CD-ROM

1. Solve: $9x + 6 = 51$.

Solve.

2. $8x - 4 = 28$

3. $-\frac{1}{2}x + 3 = 1$

Answers on page A-4

4. Solve: $-18 - m = -57$.

Example 3 Solve: $45 - t = 13$.

$$45 - t = 13$$
$$-45 + 45 - t = -45 + 13 \quad \text{Adding } -45 \text{ on both sides}$$
$$-t = -32$$
$$-1 \cdot t = -32 \quad \text{Using the property of } -1\text{: } -t = -1 \cdot t$$
$$\frac{-1 \cdot t}{-1} = \frac{-32}{-1}$$

Dividing by -1 on both sides (You could have multiplied by -1 on both sides instead. That would also change the sign on both sides.)

$$t = 32$$

The number 32 checks and is the solution.

Do Exercise 4.

Solve.

5. $-4 - 8x = 8$

6. $41.68 = 4.7 - 8.6y$

Example 4 Solve: $16.3 - 7.2y = -8.18$.

$$16.3 - 7.2y = -8.18$$
$$-16.3 + 16.3 - 7.2y = -16.3 + (-8.18) \quad \text{Adding } -16.3 \text{ on both sides}$$
$$-7.2y = -24.48$$
$$\frac{-7.2y}{-7.2} = \frac{-24.48}{-7.2} \quad \text{Dividing by } -7.2 \text{ on both sides}$$
$$y = 3.4$$

CHECK:

$16.3 - 7.2y = -8.18$		
$16.3 - 7.2(3.4)$	? -8.18	
$16.3 - 24.48$		
-8.18		TRUE

The solution is 3.4.

Do Exercises 5 and 6.

Solve.

7. $4x + 3x = -21$

8. $x - 0.09x = 728$

b Collecting Like Terms

If there are like terms on one side of the equation, we collect them before using the addition or the multiplication principle.

Example 5 Solve: $3x + 4x = -14$.

$$3x + 4x = -14$$
$$7x = -14 \quad \text{Collecting like terms}$$
$$\frac{7x}{7} = \frac{-14}{7} \quad \text{Dividing by 7 on both sides}$$
$$x = -2$$

The number -2 checks, so the solution is -2.

Do Exercises 7 and 8.

Answers on page A-4

If there are like terms on opposite sides of the equation, we get them on the same side by using the addition principle. Then we collect them. In other words, we get all terms with a variable on one side and all numbers on the other.

Example 6 Solve: $2x - 2 = -3x + 3$.

$$2x - 2 = -3x + 3$$

$$2x - 2 + 2 = -3x + 3 + 2 \quad \text{Adding 2}$$

$$2x = -3x + 5 \quad \text{Collecting like terms}$$

$$2x + 3x = -3x + 3x + 5 \quad \text{Adding } 3x$$

$$5x = 5 \quad \text{Simplifying}$$

$$\frac{5x}{5} = \frac{5}{5} \quad \text{Dividing by 5}$$

$$x = 1 \quad \text{Simplifying}$$

CHECK:

$2x - 2$	$= -3x + 3$	
$2 \cdot 1 - 2$?	$-3 \cdot 1 + 3$	
$2 - 2$	$-3 + 3$	
0	0	TRUE

The solution is 1.

Do Exercise 9.

In Example 6, we used the addition principle to get all terms with a variable on one side and all numbers on the other side. Then we collected like terms and proceeded as before. If there are like terms on one side at the outset, they should be collected before proceeding.

Example 7 Solve: $6x + 5 - 7x = 10 - 4x + 3$.

$$6x + 5 - 7x = 10 - 4x + 3$$

$$-x + 5 = 13 - 4x \quad \text{Collecting like terms}$$

$$4x - x + 5 = 13 - 4x + 4x \quad \text{Adding } 4x \text{ to get all terms with a variable on one side}$$

$$3x + 5 = 13 \quad \text{Simplifying; that is, collecting like terms}$$

$$3x + 5 - 5 = 13 - 5 \quad \text{Subtracting 5}$$

$$3x = 8 \quad \text{Simplifying}$$

$$\frac{3x}{3} = \frac{8}{3} \quad \text{Dividing by 3}$$

$$x = \frac{8}{3} \quad \text{Simplifying}$$

The number $\frac{8}{3}$ checks, so it is the solution.

Do Exercises 10–12.

9. Solve: $7y + 5 = 2y + 10$.

Solve.

10. $5 - 2y = 3y - 5$

11. $7x - 17 + 2x = 2 - 8x + 15$

12. $3x - 15 = 5x + 2 - 4x$

Answers on page A-4

13. Solve: $\frac{7}{8}x - \frac{1}{4} + \frac{1}{2}x = \frac{3}{4} + x.$

Answer on page A-4

Clearing Fractions and Decimals

In general, equations are easier to solve if they do not contain fractions or decimals. Consider, for example,

$$\frac{1}{2}x + 5 = \frac{3}{4} \quad \text{and} \quad 2.3x + 7 = 5.4.$$

If we multiply by 4 on both sides of the first equation and by 10 on both sides of the second equation, we have

$$4\left(\frac{1}{2}x + 5\right) = 4 \cdot \frac{3}{4} \quad \text{and} \quad 10(2.3x + 7) = 10 \cdot 5.4$$

or

$$4 \cdot \frac{1}{2}x + 4 \cdot 5 = 4 \cdot \frac{3}{4} \quad \text{and} \quad 10 \cdot 2.3x + 10 \cdot 7 = 10 \cdot 5.4$$

or

$$2x + 20 = 3 \quad \text{and} \quad 23x + 70 = 54.$$

The first equation has been "cleared of fractions" and the second equation has been "cleared of decimals." Both resulting equations are equivalent to the original equations and are easier to solve. *It is your choice* whether to clear fractions or decimals, but doing so often eases computations.

The easiest way to clear an equation of fractions is to multiply *every term on both sides* by the **least common multiple of all the denominators.**

Example 8 Solve: $\frac{2}{3}x - \frac{1}{6} + \frac{1}{2}x = \frac{7}{6} + 2x.$

The number 6 is the least common multiple of all the denominators. We multiply by 6 on both sides.

$$6\left(\frac{2}{3}x - \frac{1}{6} + \frac{1}{2}x\right) = 6\left(\frac{7}{6} + 2x\right)$$ Multiplying by 6 on both sides

$$6 \cdot \frac{2}{3}x - 6 \cdot \frac{1}{6} + 6 \cdot \frac{1}{2}x = 6 \cdot \frac{7}{6} + 6 \cdot 2x$$ Using the distributive law (*Caution!* Be sure to multiply *all* the terms by 6.)

$$4x - 1 + 3x = 7 + 12x$$ Simplifying. Note that the fractions are cleared.

$$7x - 1 = 7 + 12x$$ Collecting like terms

$$7x - 1 - 12x = 7 + 12x - 12x$$ Subtracting $12x$

$$-5x - 1 = 7$$ Collecting like terms

$$-5x - 1 + 1 = 7 + 1$$ Adding 1

$$-5x = 8$$ Collecting like terms

$$\frac{-5x}{-5} = \frac{8}{-5}$$ Dividing by -5

$$x = -\frac{8}{5}$$

The number $-\frac{8}{5}$ checks, so it is the solution.

Do Exercise 13.

To illustrate clearing decimals, we repeat Example 4, but this time we clear the equation of decimals first. Compare both methods.

To clear an equation of decimals, we count the greatest number of decimal places in any one number. If the greatest number of decimal places is 1, we multiply by 10; if it is 2, we multiply by 100; and so on.

Example 9 Solve: $16.3 - 7.2y = -8.18$.

The greatest number of decimal places in any one number is *two*. Multiplying by 100, which has *two* 0's, will clear all decimals.

$$100(16.3 - 7.2y) = 100(-8.18)$$ Multiplying by 100 on both sides

$$100(16.3) - 100(7.2y) = 100(-8.18)$$ Using the distributive law

$$1630 - 720y = -818$$ Simplifying

$$1630 - 720y - 1630 = -818 - 1630$$ Subtracting 1630 on both sides

$$-720y = -2448$$ Collecting like terms

$$\frac{-720y}{-720} = \frac{-2448}{-720}$$ Dividing by −720 on both sides

$$y = 3.4$$

The number 3.4 checks, so it is the solution.

Do Exercise 14.

14. Solve: $41.68 = 4.7 - 8.6y$.

c Equations Containing Parentheses

To solve certain kinds of equations that contain parentheses, we first use the distributive laws to remove the parentheses. Then we proceed as before.

Example 10 Solve: $4x = 2(12 - 2x)$.

$$4x = 2(12 - 2x)$$

$$4x = 24 - 4x$$ Using the distributive law to multiply and remove parentheses

$$4x + 4x = 24 - 4x + 4x$$ Adding $4x$ to get all the x-terms on one side

$$8x = 24$$ Collecting like terms

$$\frac{8x}{8} = \frac{24}{8}$$ Dividing by 8

$$x = 3$$

CHECK:

$4x = 2(12 - 2x)$		
$4 \cdot 3$?	$2(12 - 2 \cdot 3)$	We use the rules for order of operations to carry out the calculations on each side of the equation.
12	$2(12 - 6)$	
	$2 \cdot 6$	
	12	TRUE

The solution is 3.

Do Exercises 15 and 16.

Solve.

15. $2(2y + 3) = 14$

16. $5(3x - 2) = 35$

Answers on page A-4

Solve.

17. $3(7 + 2x) = 30 + 7(x - 1)$

18. $4(3 + 5x) - 4 = 3 + 2(x - 2)$

Here is a procedure for solving the types of equation discussed in this section.

AN EQUATION-SOLVING PROCEDURE

1. Multiply on both sides to clear the equation of fractions or decimals. (This is optional, but it can ease computations.)
2. If parentheses occur, multiply to remove them using the *distributive laws.*
3. Collect like terms on each side, if necessary.
4. Get all terms with variables on one side and all numbers (constant terms) on the other side, using the *addition principle.*
5. Collect like terms again, if necessary.
6. Multiply or divide to solve for the variable, using the *multiplication principle.*
7. Check all possible solutions in the original equation.

Example 11 Solve: $2 - 5(x + 5) = 3(x - 2) - 1$.

$2 - 5(x + 5) = 3(x - 2) - 1$	
$2 - 5x - 25 = 3x - 6 - 1$	Using the distributive laws to multiply and remove parentheses
$-5x - 23 = 3x - 7$	Collecting like terms
$-5x - 23 + 5x = 3x - 7 + 5x$	Adding $5x$
$-23 = 8x - 7$	Collecting like terms
$-23 + 7 = 8x - 7 + 7$	Adding 7
$-16 = 8x$	Collecting like terms
$\frac{-16}{8} = \frac{8x}{8}$	Dividing by 8
$-2 = x$	

CHECK:

$2 - 5(x + 5)$	$= 3(x - 2) - 1$	
$2 - 5(-2 + 5)$?	$3(-2 - 2) - 1$	
$2 - 5(3)$	$3(-4) - 1$	
$2 - 15$	$-12 - 1$	
-13	-13	TRUE

The solution is -2.

Do Exercises 17 and 18.

Calculator Spotlight

Checking Possible Solutions. To check possible solutions on a calculator, we substitute and carry out the calculations on each side. Let's check the solution of Example 11. We first substitute -2 on the left:

$$2 - 5(-2 + 5).$$

We carry out this computation as shown in the Calculator Spotlight in Section 1.8. We get -13. Then we substitute -2 on the right:

$$3(-2 - 2) - 1.$$

Carrying out the calculations, we again get -13.

Exercises

1. Check the solution of Margin Exercise 17.
2. Check the solution of Margin Exercise 18.
3. Check the solution of Example 9.

Answers on page A-4

Exercise Set 2.3

a Solve. Don't forget to check!

1. $5x + 6 = 31$

CHECK: $5x + 6 = 31$?

2. $7x + 6 = 13$

CHECK: $7x + 6 = 13$?

3. $8x + 4 = 68$

CHECK: $8x + 4 = 68$?

4. $4y + 10 = 46$

CHECK: $4y + 10 = 46$?

5. $4x - 6 = 34$

6. $5y - 2 = 53$

7. $3x - 9 = 33$

8. $4x - 19 = 5$

9. $7x + 2 = -54$

10. $5x + 4 = -41$

11. $-45 = 3 + 6y$

12. $-91 = 9t + 8$

13. $-4x + 7 = 35$

14. $-5x - 7 = 108$

15. $-7x - 24 = -129$

16. $-6z - 18 = -132$

b Solve.

17. $5x + 7x = 72$

CHECK: $5x + 7x = 72$?

18. $8x + 3x = 55$

CHECK: $8x + 3x = 55$?

19. $8x + 7x = 60$

CHECK: $8x + 7x = 60$?

20. $8x + 5x = 104$

CHECK: $8x + 5x = 104$?

21. $4x + 3x = 42$

22. $7x + 18x = 125$

23. $-6y - 3y = 27$

24. $-5y - 7y = 144$

25. $-7y - 8y = -15$

26. $-10y - 3y = -39$

27. $x + \frac{1}{3}x = 8$

28. $x + \frac{1}{4}x = 10$

29. $10.2y - 7.3y = -58$

30. $6.8y - 2.4y = -88$

31. $8y - 35 = 3y$

32. $4x - 6 = 6x$

33. $8x - 1 = 23 - 4x$

34. $5y - 2 = 28 - y$

35. $2x - 1 = 4 + x$

36. $4 - 3x = 6 - 7x$

37. $6x + 3 = 2x + 11$

38. $14 - 6a = -2a + 3$

39. $5 - 2x = 3x - 7x + 25$

40. $-7z + 2z - 3z - 7 = 17$

41. $4 + 3x - 6 = 3x + 2 - x$

42. $5 + 4x - 7 = 4x - 2 - x$

43. $4y - 4 + y + 24 = 6y + 20 - 4y$

44. $5y - 7 + y = 7y + 21 - 5y$

Solve. Clear fractions or decimals first.

45. $\frac{7}{2}x + \frac{1}{2}x = 3x + \frac{3}{2} + \frac{5}{2}x$

46. $\frac{7}{8}x - \frac{1}{4} + \frac{3}{4}x = \frac{1}{16} + x$

47. $\frac{2}{3} + \frac{1}{4}t = \frac{1}{3}$

48. $-\frac{3}{2} + x = -\frac{5}{6} - \frac{4}{3}$

49. $\frac{2}{3} + 3y = 5y - \frac{2}{15}$

50. $\frac{1}{2} + 4m = 3m - \frac{5}{2}$

51. $\frac{5}{3} + \frac{2}{3}x = \frac{25}{12} + \frac{5}{4}x + \frac{3}{4}$

52. $1 - \frac{2}{3}y = \frac{9}{5} - \frac{y}{5} + \frac{3}{5}$

53. $2.1x + 45.2 = 3.2 - 8.4x$

54. $0.96y - 0.79 = 0.21y + 0.46$

55. $1.03 - 0.62x = 0.71 - 0.22x$

56. $1.7t + 8 - 1.62t = 0.4t - 0.32 + 8$

57. $\frac{2}{7}x - \frac{1}{2}x = \frac{3}{4}x + 1$

58. $\frac{5}{16}y + \frac{3}{8}y = 2 + \frac{1}{4}y$

c Solve.

59. $3(2y - 3) = 27$

60. $8(3x + 2) = 30$

61. $40 = 5(3x + 2)$

62. $9 = 3(5x - 2)$

63. $2(3 + 4m) - 9 = 45$

64. $5x + 5(4x - 1) = 20$

65. $5r - (2r + 8) = 16$

66. $6b - (3b + 8) = 16$

67. $6 - 2(3x - 1) = 2$

68. $10 - 3(2x - 1) = 1$

69. $5(d + 4) = 7(d - 2)$

70. $3(t - 2) = 9(t + 2)$

71. $8(2t + 1) = 4(7t + 7)$

72. $7(5x - 2) = 6(6x - 1)$

73. $3(r - 6) + 2 = 4(r + 2) - 21$

74. $5(t + 3) + 9 = 3(t - 2) + 6$

75. $19 - (2x + 3) = 2(x + 3) + x$

76. $13 - (2c + 2) = 2(c + 2) + 3c$

77. $2[4 - 2(3 - x)] - 1 = 4[2(4x - 3) + 7] - 25$

78. $5[3(7 - t) - 4(8 + 2t)] - 20 = -6[2(6 + 3t) - 4]$

79. $0.7(3x + 6) = 1.1 - (x + 2)$

80. $0.9(2x + 8) = 20 - (x + 5)$

81. $a + (a - 3) = (a + 2) - (a + 1)$

82. $0.8 - 4(b - 1) = 0.2 + 3(4 - b)$

Skill Maintenance

83. Divide: $-22.1 \div 3.4$. [1.6c]

84. Factor: $7x - 21 - 14y$. [1.7d]

85. Use < or > for ▒ to write a true sentence: [1.2d]
-15 ▒ -13.

86. Find $-(-x)$ when $x = -14$. [1.3b]

87. Add: $-22.1 + 3.4$. [1.3a]

88. Subtract: $-22.1 - 3.4$. [1.4a]

Translate to an algebraic expression. [1.1b]

89. A number c is divided by 8.

90. A parallelogram with height h has a base length of 13.4. What is the area of the parallelogram?

Synthesis

91. ◆ What procedure would you follow to solve an equation like $0.23x + \frac{17}{3} = -0.8 + \frac{3}{4}x$? Could your procedure be streamlined? If so, how?

92. ◆ Consider any equation of the form $ax + b = c$. Describe a procedure that can be used to solve for x.

Solve.

93. 🖩 $0.008 + 9.62x - 42.8 = 0.944x + 0.0083 - x$

94. $\dfrac{y - 2}{3} = \dfrac{2 - y}{5}$

95. $0 = y - (-14) - (-3y)$

96. $3x = 4x$

97. $\dfrac{5 + 2y}{3} = \dfrac{25}{12} + \dfrac{5y + 3}{4}$

98. 🖩 $0.05y - 1.82 = 0.708y - 0.504$

99. $-2y + 5y = 6y$

100. $\dfrac{1}{4}(8y + 4) - 17 = -\dfrac{1}{2}(4y - 8)$

101. $\dfrac{1}{3}(6x + 24) - 20 = -\dfrac{1}{4}(12x - 72)$

102. $\dfrac{2}{3}\left(\dfrac{7}{8} - 4x\right) - \dfrac{5}{8} = \dfrac{3}{8}$

103. $\dfrac{3}{4}\left(3x - \dfrac{1}{2}\right) - \dfrac{2}{3} = \dfrac{1}{3}$

104. $\dfrac{4 - 3x}{7} = \dfrac{2 + 5x}{49} - \dfrac{x}{14}$

105. Solve the equation $4x - 8 = 32$ by first using the addition principle. Then solve it by first using the multiplication principle.

Solve linear equations as a group.

Collaborative Learning Manual

2.4 Applications and Problem Solving

a Five Steps for Solving Problems

We have studied many new equation-solving tools in this chapter. We now use them for applications and problem solving. The following five-step strategy can be very helpful in solving problems.

Five Steps for Problem Solving in Algebra

1. *Familiarize* yourself with the problem situation.
2. *Translate* the problem to an equation.
3. *Solve* the equation.
4. *Check* the answer in the original problem.
5. *State* the answer to the problem clearly.

Of the five steps, the most important is probably the first one: becoming familiar with the problem situation. The table in the margin lists some hints for familiarization.

Example 1 *Subway Sandwich.* Subway is a national restaurant firm that serves sandwiches prepared in buns of length 18 in. (***Source*:** Subway Restaurants). Suppose Jenny, Demi, and Sarah buy one of these sandwiches and take it back to their room. Since they have different appetites, Jenny cuts the sandwich in such a way that Demi gets half of what Jenny gets and Sarah gets three-fourths of what Jenny gets. Find the length of each person's sandwich.

1. **Familiarize.** We first make a drawing. Because the sandwich lengths are expressed in terms of Jenny's sandwich, we let

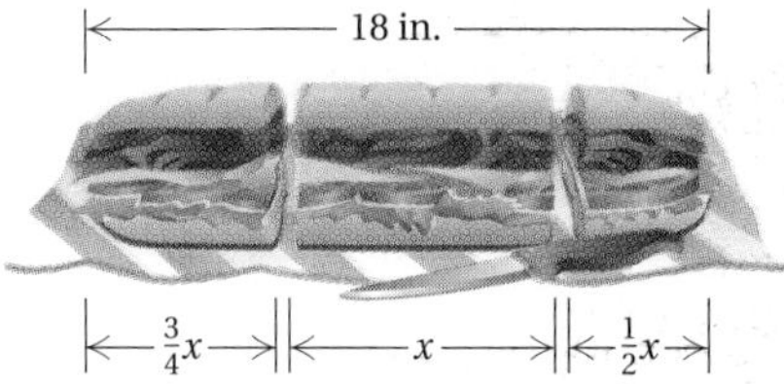

x = the length of Jenny's sandwich.

Then $\frac{1}{2}x$ = the length of Demi's sandwich

and $\frac{3}{4}x$ = the length of Sarah's sandwich.

2. **Translate.** From the statement of the problem and the drawing, we see that the lengths add up to 18 in. That gives us our translation:

Length of Jenny's sandwich	plus	Length of Demi's sandwich	plus	Length of Sarah's sandwich	is	Total length
↓	↓	↓	↓	↓	↓	↓
x	$+$	$\frac{1}{2}x$	$+$	$\frac{3}{4}x$	$=$	18

Objective

a Solve applied problems by translating to equations.

For Extra Help

TAPE 3 | MAC WIN | CD-ROM

To familiarize yourself with a problem situation:

- If a problem is given in words, read it carefully. Reread the problem, perhaps aloud. Try to verbalize the problem as if you were explaining it to someone else.
- Make a drawing and label it with known information, using specific units if given. Also, indicate unknown information.
- Choose a variable (or variables) to represent the unknown and clearly state what the variable represents. Be descriptive! For example, let L = length, d = distance, and so on.
- Find further information. Look up formulas or definitions with which you are not familiar. (Geometric formulas appear on the inside front cover of this text.) Consult a reference librarian or an expert in the field.
- Create a table that lists all the information you have available. Look for patterns that may help in the translation to an equation.
- Guess or estimate the answer.

1. *Rocket Sections.* A rocket is divided into three sections: the payload and navigation section in the top, the fuel section in the middle, and the rocket engine section in the bottom. The top section is one-sixth the length of the bottom section. The middle section is one-half the length of the bottom section. The total length is 240 ft. Find the length of each section.

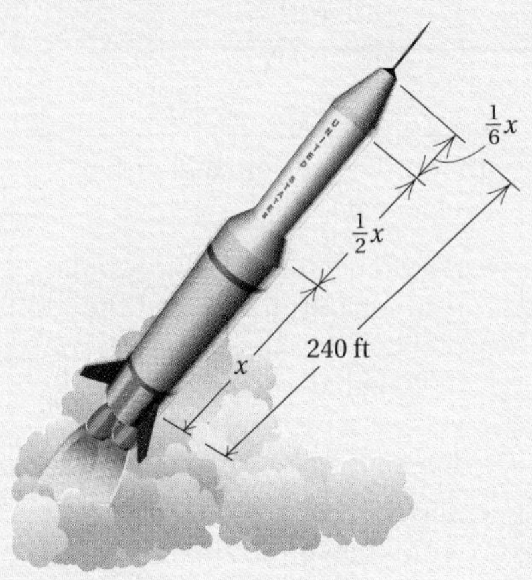

2. If 5 is subtracted from three times a certain number, the result is 10. What is the number?

Answers on page A-4

3. **Solve.** We solve the equation by clearing fractions as follows:

$$x + \frac{1}{2}x + \frac{3}{4}x = 18 \qquad \textbf{The LCM of all the denominators is 4.}$$

$$4\left(x + \frac{1}{2}x + \frac{3}{4}x\right) = 4 \cdot 18 \qquad \textbf{Multiplying by the LCM, 4}$$

$$4 \cdot x + 4 \cdot \frac{1}{2}x + 4 \cdot \frac{3}{4}x = 4 \cdot 18 \qquad \textbf{Using the distributive law}$$

$$4x + 2x + 3x = 72 \qquad \textbf{Simplifying}$$

$$9x = 72 \qquad \textbf{Collecting like terms}$$

$$\frac{9x}{9} = \frac{72}{9} \qquad \textbf{Dividing by 9}$$

$$x = 8.$$

4. **Check.** Do we have an answer to the *problem*? If the length of Jenny's sandwich is 8 in., then the length of Demi's sandwich is $\frac{1}{2} \cdot 8$ in., or 4 in., and the length of Sarah's sandwich is $\frac{3}{4} \cdot 8$ in., or 6 in. These lengths add up to 18 in. Our answer checks.
5. **State.** The length of Jenny's sandwich is 8 in., the length of Demi's sandwich is 4 in., and the length of Sarah's sandwich is 6 in.

Do Exercise 1.

Example 2 Five plus three more than a number is nineteen. What is the number?

1. **Familiarize.** Let x = the number. Then "three more than a number" translates to $x + 3$, and "five plus three more than a number" translates to $5 + (x + 3)$.
2. **Translate.** The familiarization leads us to the following translation:

Five	plus	Three more than a number	is	Nineteen.
↓	↓	↓	↓	↓
5	+	$(x + 3)$	=	19.

3. **Solve.** We solve the equation:

$$5 + (x + 3) = 19$$

$$x + 8 = 19 \qquad \textbf{Collecting like terms}$$

$$x + 8 - 8 = 19 - 8 \qquad \textbf{Subtracting 8}$$

$$x = 11.$$

4. **Check.** Be sure to check your answer in the original wording of the problem, not in the equation that you solved. This will enable you to check for errors in the translation as well. Three more than 11 is 14. Adding 5 to 14, we get 19. This checks.
5. **State.** The number is 11.

Do Exercise 2.

Recall that the

$$\text{Set of integers} = \{\ldots, -5, -4, -3, -2, -1, 0, 1, 2, 3, 4, 5, \ldots\}.$$

Before we solve the next problem, we need to learn some additional terminology regarding integers.

The following are examples of **consecutive integers:** 16, 17, 18, 19, 20; and -31, -30, -29, -28. Note that consecutive integers can be represented in the form x, $x + 1$, $x + 2$, and so on.

The following are examples of **consecutive even integers:** 16, 18, 20, 22, 24; and -52, -50, -48, -46. Note that consecutive even integers can be represented in the form x, $x + 2$, $x + 4$, and so on.

The following are examples of **consecutive odd integers:** 21, 23, 25, 27, 29; and -71, -69, -67, -65. Note that consecutive odd integers can be represented in the form x, $x + 2$, $x + 4$, and so on.

3. *Interstate Mile Markers.* The sum of two consecutive mile markers on I-90 in upstate New York is 627 (***Source:*** New York State Department of Transportation). (On I-90 in New York, the marker numbers *increase* from east to west.) Find the numbers on the markers.

Example 3 *Interstate Mile Markers.* If you are traveling on a U.S. interstate highway, you will notice numbered markers every mile to tell your location in case of an accident or other emergency. In many states, the numbers on the markers increase from west to east. (***Source:*** Federal Highway Administration, Ed Rotalewski) The sum of two consecutive mile markers on I-70 in Kansas is 559. Find the numbers on the markers.

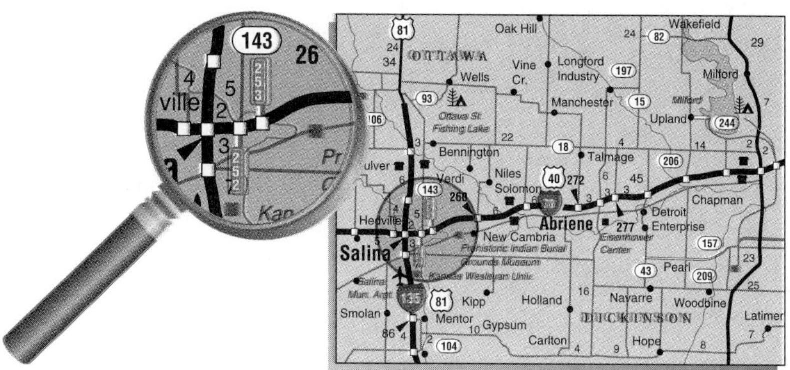

1. **Familiarize.** The numbers on the mile markers are consecutive positive integers. Thus if we let $x =$ the smaller number, then $x + 1 =$ the larger number.

   To become familiar with the problem, we can make a table. First, we guess a value for x; then we find $x + 1$. Finally, we add the two numbers and check the sum. From the table, we see that the first marker should be between 252 and 302. You might actually solve the problem this way, but let's work on developing our algebra skills.

x	$x + 1$	Sum of x and $x + 1$
114	115	229
252	253	505
302	303	605

2. **Translate.** We reword the problem and translate as follows.

First integer	plus	Second integer	is	559	**Rewording**
↓	↓	↓	↓	↓	
x	$+$	$(x + 1)$	$=$	559	**Translating**

3. **Solve.** We solve the equation:

$$x + (x + 1) = 559$$

$$2x + 1 = 559 \quad \text{Collecting like terms}$$

$$2x + 1 - 1 = 559 - 1 \quad \text{Subtracting 1}$$

$$2x = 558$$

$$\frac{2x}{2} = \frac{558}{2} \quad \text{Dividing by 2}$$

$$x = 279.$$

If x is 279, then $x + 1$ is 280.

Answer on page A-4

4. *IKON Copiers.* The law firm in Example 4 decides to raise its budget to \$2400 for the 3-month period. How many copies can they make for \$2400?

Answer on page A-4

4. Check. Our possible answers are 279 and 280. These are consecutive positive integers and $279 + 280 = 559$, so the answers check.

5. State. The mile markers are 279 and 280.

Do Exercise 3 on the preceding page.

Example 4 *IKON Copiers.* IKON Office Solutions rents a Canon GP30F copier for \$240 per month plus 1.8¢ per copy. A law firm needs to lease a copy machine for use during a special case that they anticipate will take 3 months. If they allot a budget of \$1500, how many copies can they make?

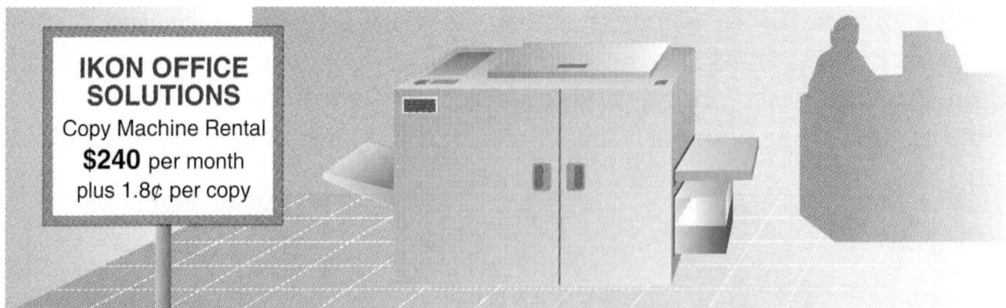

Source: IKON Office Solutions, Keith Palmer

1. Familiarize. Suppose that the law firm makes 20,000 copies. Then the cost is

Monthly charges plus Copy charges

or

3(\$240)	plus	Cost per copy	times	Number of copies
↓	↓	↓	↓	↓
\$720	+	\$0.018	·	20,000,

which is \$1080. This process familiarizes us with the way in which a calculation is made. Note that we convert 1.8¢ to \$0.018 so that all information is in the same unit, dollars. Otherwise, we will not get the correct answer.

We let c = the number of copies that can be made for \$1500.

2. Translate. We reword the problem and translate as follows.

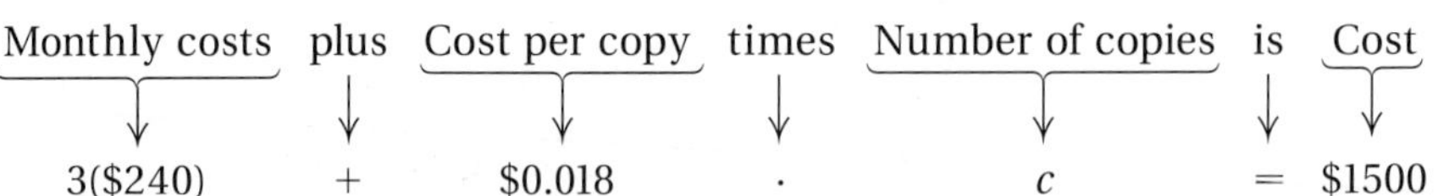

3. Solve. We solve the equation:

$$3(240) + 0.018c = 1500$$
$$720 + 0.018c = 1500$$
$$1000(720 + 0.018c) = 1000 \cdot 1500$$ **Multiplying by 1000 on both sides to clear decimals**
$$1000(720) + 1000(0.018c) = 1{,}500{,}000$$ **Using the distributive law**
$$720{,}000 + 18c = 1{,}500{,}000$$ **Simplifying**
$$720{,}000 + 18c - 720{,}000 = 1{,}500{,}000 - 720{,}000$$ **Subtracting 720,000**
$$18c = 780{,}000$$
$$\frac{18c}{18} = \frac{780{,}000}{18}$$ **Dividing by 18**
$$c \approx 43{,}333.$$ **Rounding to the nearest one. "≈" means "is approximately equal to."**

4. **Check.** We check in the original problem. The cost for 43,333 pages is 43,333($0.018) = $779.994. The rental for 3 months is 3($240) = $720. The total cost is then $779.994 + $720 ≈ $1499.99, which is just about the $1500 allotted.

5. **State.** The law firm can make 43,333 copies on the copy rental allotment of $1500.

Do Exercise 4 on the preceding page.

Example 5 *Perimeter of NBA Court.* The perimeter of an NBA basketball court is 288 ft. The length is 44 ft longer than the width. (***Source:*** National Basketball Association) Find the dimensions of the court.

1. **Familiarize.** We first make a drawing.

We let w = the width of the rectangle. Then $w + 44$ = the length. The perimeter P of a rectangle is the distance around the rectangle and is given by the formula $2l + 2w = P$, where

l = the length and w = the width.

2. **Translate.** To translate the problem, we substitute $w + 44$ for l and 288 for P:

$$2l + 2w = P$$
$$2(w + 44) + 2w = 288.$$

3. **Solve.** We solve the equation:

$$2(w + 44) + 2w = 288$$
$$2 \cdot w + 2 \cdot 44 + 2w = 288 \quad \text{Using the distributive law}$$
$$4w + 88 = 288 \quad \text{Collecting like terms}$$
$$4w + 88 - 88 = 288 - 88 \quad \text{Subtracting 88}$$
$$4w = 200$$
$$\frac{4w}{4} = \frac{200}{4} \quad \text{Dividing by 4}$$
$$w = 50.$$

Thus possible dimensions are

w = 50 ft and $l = w + 44 = 50 + 44$, or 94 ft.

4. **Check.** If the width is 50 ft and the length is 94 ft, then the perimeter is 2(50 ft) + 2(94 ft), or 288 ft. This checks.

5. **State.** The width is 50 ft and the length is 94 ft.

Do Exercise 5.

5. *Perimeter of High School Basketball Court.* The perimeter of a standard high school basketball court is 268 ft. The length is 34 ft longer than the width. Find the dimensions of the court.

Answer on page A-4

6. The second angle of a triangle is three times as large as the first. The third angle measures 30° more than the first angle. Find the measures of the angles.

Answer on page A-4

Example 6 *Cross Section of a Roof.* In a triangular cross section of a roof, the second angle is twice as large as the first angle. The measure of the third angle is 20° greater than that of the first angle. How large are the angles?

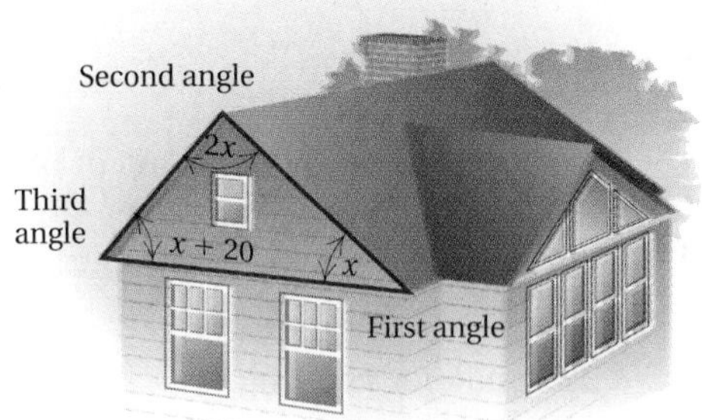

1. **Familiarize.** We first make a drawing as shown above. We let

measure of first angle $= x$.

Then measure of second angle $= 2x$

and measure of third angle $= x + 20$.

2. **Translate.** To translate, we need to recall a geometric fact. (You might, as part of step 1, look it up in a geometry book or in the list of formulas on the inside front cover.) Remember, the measures of the angles of a triangle total 180°.

Measure of first angle	plus	Measure of second angle	plus	Measure of third angle	is	180°
↓	↓	↓	↓	↓	↓	↓
x	$+$	$2x$	$+$	$(x + 20)$	$=$	180°

3. **Solve.** We solve the equation:

$$\begin{aligned} x + 2x + (x + 20) &= 180 \\ 4x + 20 &= 180 \\ 4x + 20 - 20 &= 180 - 20 \\ 4x &= 160 \\ \frac{4x}{4} &= \frac{160}{4} \\ x &= 40. \end{aligned}$$

Possible measures for the angles are as follows:

First angle: $x = 40°$;

Second angle: $2x = 2(40) = 80°$;

Third angle: $x + 20 = 40 + 20 = 60°$.

4. **Check.** Consider our answers: 40°, 80°, and 60°. The second is twice the first and the third is 20° greater than the first. The sum is 180°. The angles check.

5. **State.** The measures of the angles are 40°, 80°, and 60°.

CAUTION! Units are important in answers. Remember to include them, where appropriate.

Do Exercise 6.

Example 7 *Nike, Inc.* The equation

$$y = 0.69606x + 1.68722$$

can be used to approximate the total revenue y, in billions of dollars, of Nike, Inc., in year x, where

$x = 0$ corresponds to 1990,

$x = 1$ corresponds to 1991,

$x = 2$ corresponds to 1992,

$x = 10$ corresponds to 2000,

and so on.

(***Source:*** Nike, Inc.) (This equation was developed from a procedure called *regression*. Its discussion belongs to a later course.)

a) Find the total revenue in 1999 and 2008.

b) In what year will the total revenue be about $12.82418 billion?

Since a formula has been given, we will not use the five-step problem-solving strategy.

a) To find the total revenue for 1999, note first that $1999 - 1990 = 9$. We substitute 9 for x:

$$\begin{aligned} y &= 0.69606x + 1.68722 \\ &= 0.69606(9) + 1.68722 \\ &= \$7.95176. \end{aligned}$$

To find the total revenue for 2008, note that $2008 - 1990 = 18$. We substitute 18 for x:

$$\begin{aligned} y &= 0.69606x + 1.68722 \\ &= 0.69606(18) + 1.68722 \\ &= \$14.2163. \end{aligned}$$

Thus the total revenue will be $7.95176 billion in 1999 and $14.2163 billion in 2008.

b) To determine the year in which the total revenue will be about $12.82418 billion, we first substitute 12.82418 for y. Then we solve for x. Note that we have not cleared decimals because the numbers have the same number of decimal places. (You may choose to do so.)

$$\begin{aligned} y &= 0.69606x + 1.68722 \\ 12.82418 &= 0.69606x + 1.68722 \\ 12.82418 - 1.68722 &= 0.69606x + 1.68722 - 1.68722 \\ 11.13696 &= 0.69606x \\ \frac{11.13696}{0.69606} &= \frac{0.69606x}{0.69606} \\ 16 &= x \end{aligned}$$

The number 16 is the number of years *after* 1990. To find that year, we add 16 to 1990: $1990 + 16 = 2006$. Thus, assuming the equation continues to be valid, the total revenue of Nike, Inc., will be $12.82418 billion in 2006.

Do Exercise 7.

7. *Nike, Inc.* Referring to Example 7:

a) Find the total revenue in 2002 and 2010.

b) In what year will the total revenue be about $9.34388 billion?

Answer on page A-4

Improving Your Math Study Skills

Extra Tips on Problem Solving

The following tips, which are focused on problem solving, summarize some points already considered and propose some new ones.

- Get in the habit of using all five steps for problem solving.

1. ***Familiarize* yourself with the problem situation.** Some suggestions for this are given on p. 95.
2. ***Translate* the problem to an equation.** As you study more mathematics, you will find that the translation may be to some other kind of mathematical language, such as an inequality.
3. ***Solve* the equation.** If the translation is to some other kind of mathematical language, you would carry out some kind of mathematical manipulation—in the case of an inequality, you would solve it.
4. ***Check* the answer in the original problem.** This does not mean to check in the translated equation. It means to go back to the original worded problem.
5. ***State* the answer to the problem clearly.**

For Step 4, some further comment on checking is appropriate. *You may be able to translate to an equation and to solve the equation, but you may find that none of the solutions of the equation is the solution of the original problem.* To see how this can happen, consider this example.

Example

The sum of two consecutive even integers is 537. Find the integers.

1. ***Familiarize.*** Suppose we let x = the first number. Then $x + 2$ = the second number.
2. ***Translate.*** The problem can be translated to the following equation: $x + (x + 2) = 537$.
3. ***Solve.*** We solve the equation as follows:

$$2x + 2 = 537$$
$$2x = 535$$
$$x = \frac{535}{2}, \text{ or } 267.5.$$

4. ***Check.*** Then $x + 2 = 269.5$. However, the numbers are not only not even, but they are not integers.
5. ***State.*** The problem has no solution.

The following are some other tips.

- **To be good at problem solving, do lots of problems.** Learning to solve problems is similar to learning other skills such as golf. At first you may not be successful, but the more you practice and work at improving your skills, the more successful you will become. For problem solving, do more than just two or three odd-numbered assigned problems. Do them all, and if you have time, do the even-numbered problems as well. Then find another book on the same subject and do problems in that book.
- **Look for patterns when solving problems.** You will eventually see patterns in similar kinds of problems. For example, there is a pattern in the way that you solve problems involving consecutive integers.
- **When translating to an equation, or some other mathematical language, consider the dimensions of the variables and the constants in the equation.** The variables that represent length should all be in the same unit, those that represent money should all be in dollars or in cents, and so on.

Exercise Set 2.4

a Solve.

1. Two times a number added to 85 is 117. Find the number.

2. Eight times a number plus 7 is 2559. Find the number.

3. Three less than twice a number is -4. Find the number.

4. Seven less than four times a number is -27. Find the number.

5. When 17 is subtracted from four times a certain number, the result is 211. What is the number?

6. When 36 is subtracted from five times a certain number, the result is 374. What is the number?

7. A 240-in. pipe is cut into two pieces. One piece is three times the length of the other. Find the lengths of the pieces.

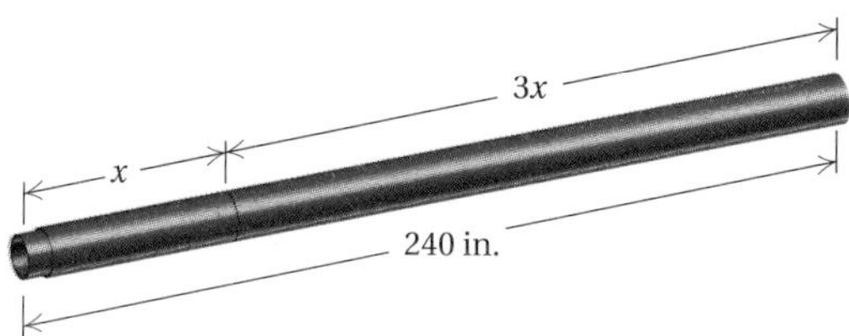

8. A 72-in. board is cut into two pieces. One piece is 2 in. longer than the other. Find the lengths of the pieces.

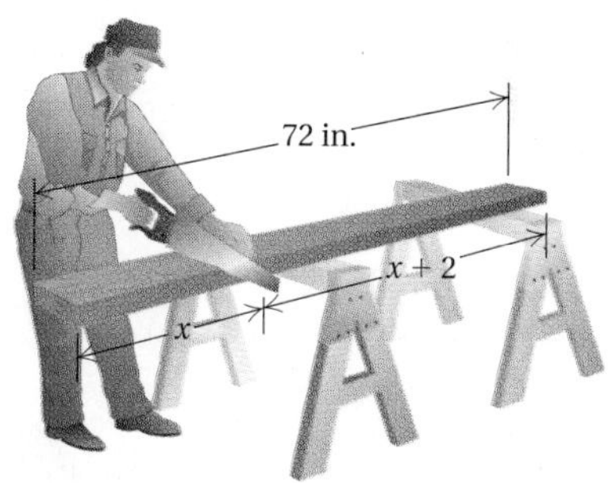

9. *Statue of Liberty.* The height of the Eiffel Tower is 974 ft, which is about 669 ft higher than the Statue of Liberty. What is the height of the Statue of Liberty?

10. *Area of Lake Ontario.* The area of Lake Superior is about four times the area of Lake Ontario. The area of Lake Superior is 30,172 mi². What is the area of Lake Ontario?

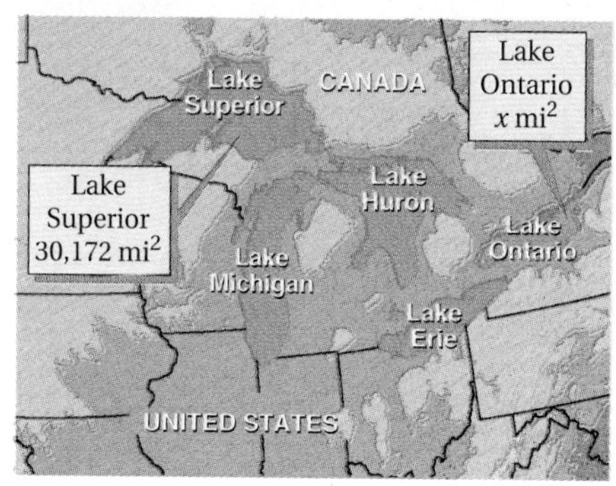

11. *Wheaties.* Recently, the cost of four 18-oz boxes of Wheaties cereal was $11.56. What was the cost of one box?

12. *Women's Dresses.* In a recent year, the total amount spent on women's blouses was $6.5 billion. This was $0.2 billion more than what was spent on women's dresses. How much was spent on women's dresses?

13. If you double a certain number and then add 16, you get $\frac{2}{3}$ of the original number. What is the original number?

14. If you double a certain number and then add 85, you get $\frac{3}{4}$ of the original number. What is the original number?

15. *Iditarod Race.* The Iditarod sled dog race extends for 1049 mi from Anchorage to Nome (***Source:*** Iditarod Trail Commission). If a musher is twice as far from Anchorage as from Nome, how much of the race has the musher completed?

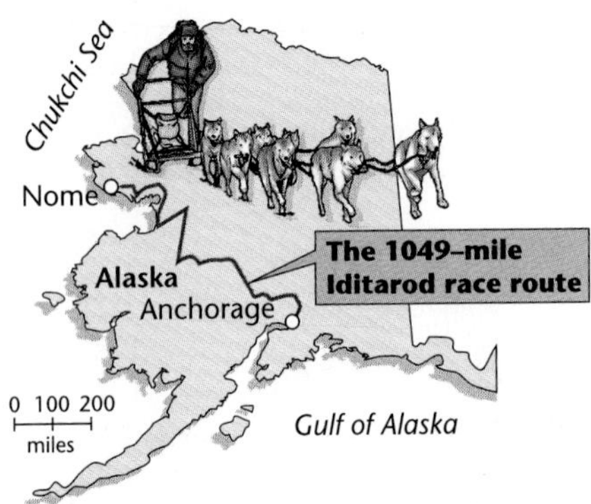

16. *Home Remodeling.* In a recent year, Americans spent a total of $35 billion to remodel bathrooms and kitchens. Twice as much was spent on kitchens as bathrooms. How much was spent on each?

17. *Consecutive Page Numbers.* The sum of the page numbers on the facing pages of a book is 573. What are the page numbers?

18. *Consecutive Post Office Box Numbers.* The sum of the numbers on two consecutive post office boxes is 547. What are the numbers?

19. The numbers on Sam's three raffle tickets are consecutive integers. The sum of the numbers is 126. What are the numbers?

20. The ages of Whitney, Wesley, and Wanda are consecutive integers. The sum of their ages is 108. What are their ages?

21. The sum of three consecutive odd integers is 189. What are the integers?

22. Three consecutive integers are such that the first plus one-half the second plus seven less than twice the third is 2101. What are the integers?

23. *Standard Billboard Sign.* A standard rectangular highway billboard sign has a perimeter of 124 ft. The length is 6 ft more than three times the width. Find the dimensions.

24. *Two-by-Four.* The perimeter of a cross section of a "two-by-four" piece of lumber is $10\frac{1}{2}$ in. The length is twice the width. Find the actual dimensions of the cross section of a two-by-four.

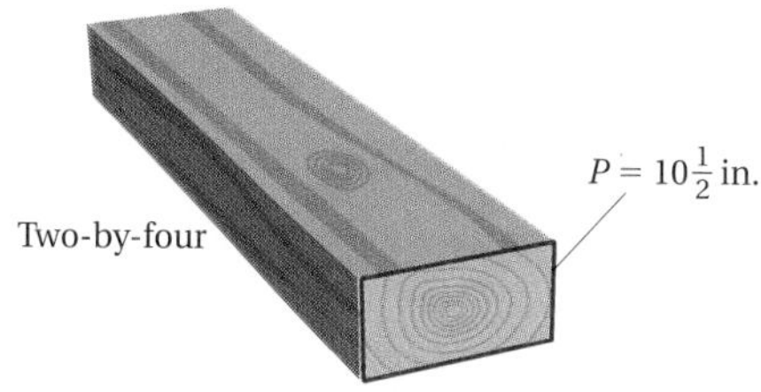

25. *Parking Costs.* A hospital parking lot charges $1.50 for the first hour or part thereof, and $1.00 for each additional hour or part thereof. A weekly pass costs $27.00 and allows unlimited parking for 7 days. Suppose that each visit Ed makes to the hospital lasts $1\frac{1}{2}$ hr. What is the minimum number of times that Ed would have to visit per week to make it worthwhile for him to buy the pass?

26. *Van Rental.* Value Rent-A-Car rents vans at a daily rate of $84.95 plus 60 cents per mile. Molly rents a van to deliver electrical parts to her customers. She is allotted a daily budget of $320. How many miles can she drive for $320?

27. The second angle of a triangular field is three times as large as the first angle. The third angle is 40° greater than the first angle. How large are the angles?

28. *Triangular Parking Lot.* The second angle of a triangular parking lot is four times as large as the first angle. The third angle is 45° less than the sum of the other two angles. How large are the angles?

29. *Triangular Backyard.* A home has a triangular backyard. The second angle of the triangle is 5° more than the first angle. The third angle is 10° more than three times the first angle. Find the angles of the triangular yard.

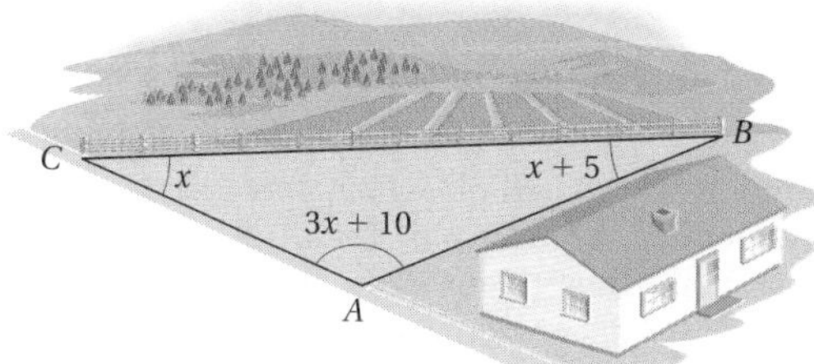

30. *Boarding Stable.* A rancher needs to form a triangular horse pen using ropes next to a stable. The second angle is three times the first angle. The third angle is 15° less than the first angle. Find the angles of the triangular pen.

31. *Coca-Cola Co.* The equation

$$y = 66.2x + 460.2$$

can be used to approximate the total revenue y, in millions of dollars, of the Coca-Cola Co., in year x, where

$x = 0$ corresponds to 1990,

$x = 2$ corresponds to 1992,

$x = 10$ corresponds to 2000,

and so on.

(***Source:*** Coca-Cola Bottling Consolidated)

a) Find the total revenue in 1999, 2000, and 2010.
b) In what year will the total revenue be about $1254.6 million?

32. *Running Records in the 200-m Dash.* The equation

$$R = -0.028t + 20.8$$

can be used to predict the world record in the 200-m dash, where R = the record in seconds, and t = the number of years since 1920 (***Source:*** International Amateur Athletic Federation).

a) Predict the record in 2000 and 2010.
b) In what year will the record be 18.0 sec?

Skill Maintenance

Calculate.

33. $-\frac{4}{5} - \frac{3}{8}$ [1.4a]

34. $-\frac{4}{5} + \frac{3}{8}$ [1.3a]

35. $-\frac{4}{5} \cdot \frac{3}{8}$ [1.5a]

36. $-\frac{4}{5} \div \frac{3}{8}$ [1.6c]

37. $-25.6 \div (-16)$ [1.6c]

38. $-25.6(-16)$ [1.5a]

39. $-25.6 - (-16)$ [1.4a]

40. $-25.6 + (-16)$ [1.3a]

Synthesis

41. ◆ A fellow student claims to be able to solve most of the problems in this section by guessing. Is there anything wrong with this approach? Why or why not?

42. ◆ Write a problem for a classmate to solve so that it can be translated to the equation

$$\tfrac{2}{3}x + (x + 5) + x = 375.$$

43. Apples are collected in a basket for six people. One-third, one-fourth, one-eighth, and one-fifth are given to four people, respectively. The fifth person gets ten apples with one apple remaining for the sixth person. Find the original number of apples in the basket.

44. A student scored 78 on a test that had 4 seven-point fill-ins and 24 three-point multiple-choice questions. The student had one fill-in wrong. How many multiple-choice questions did the student answer correctly?

45. 🖩 The area of this triangle is 2.9047 in^2. Find x.

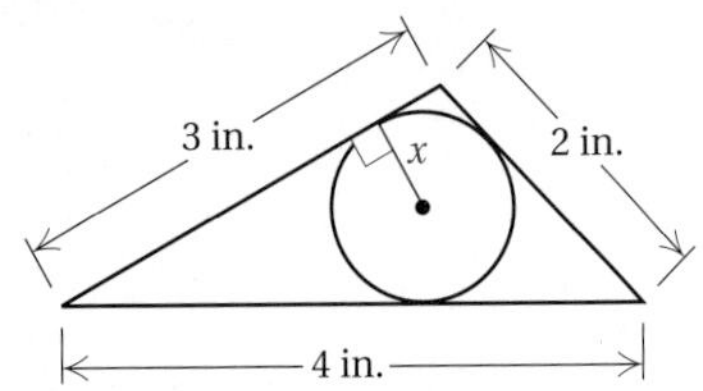

46. A storekeeper goes to the bank to get $10 worth of change. She requests twice as many quarters as half dollars, twice as many dimes as quarters, three times as many nickels as dimes, and no pennies or dollars. How many of each coin did the storekeeper get?

2.5 Applications with Percents

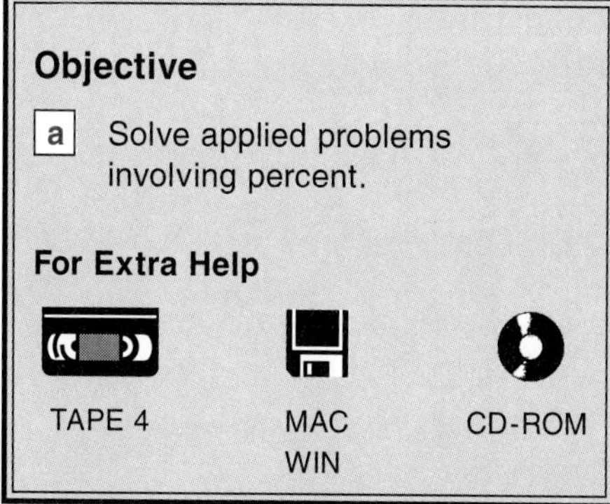

a Many applied problems involve percents. We can use our knowledge of equations and the problem-solving process to solve such problems.

Example 1 What percent of 45 is 15?

1. **Familiarize.** This type of problem is stated so explicitly that we can proceed directly to the translation. We first let $x =$ the percent.
2. **Translate.** We translate as follows:

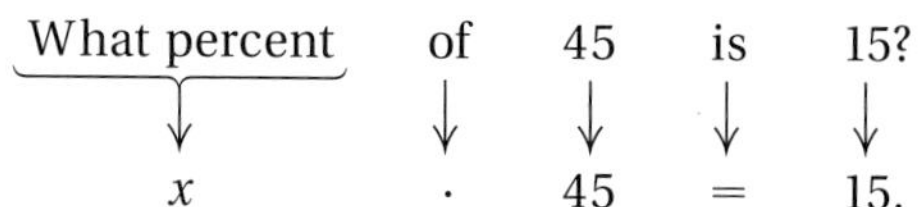

$$x \cdot 45 = 15.$$

3. **Solve.** We solve the equation:

$$x \cdot 45 = 15$$

$$\frac{x \cdot 45}{45} = \frac{15}{45} \quad \text{Dividing by 45}$$

$$x = \frac{1}{3} \quad \text{Simplifying}$$

$$= 33\frac{1}{3}\%. \quad \text{Changing fractional notation to percent notation}$$

4. **Check.** We check by finding $33\frac{1}{3}\%$ of 45:

$$33\frac{1}{3}\% \cdot 45 = \frac{1}{3} \cdot 45 = 15.$$

5. **State.** The answer is $33\frac{1}{3}\%$.

Do Exercises 1 and 2.

Solve.

1. What percent of 50 is 16?

2. 15 is what percent of 60?

Example 2 3 is 16% of what number?

1. **Familiarize.** This problem is stated so explicitly that we can proceed directly to the translation. We let $y =$ the number that we are taking 16% of.
2. **Translate.** The translation is as follows:

3	is	16%	of	what?
↓	↓	↓	↓	↓
3	=	16%	·	y

3. **Solve.** We solve the equation:

$$3 = 16\% \cdot y$$

$$3 = 0.16y \quad \text{Converting to decimal notation}$$

$$0.16y = 3 \quad \text{Reversing the equation}$$

$$\frac{0.16y}{0.16} = \frac{3}{0.16} \quad \text{Dividing by 0.16}$$

$$y = 18.75.$$

Answers on page A-5

Solve.

3. 45 is 20 percent of what number?

4. 120 percent of what number is 60?

Solve.

5. What is 23% of 48?

6. Referring to Example 3, determine how many deaths by lightning occurred near telephone poles.

Answers on page A-5

4. Check. We check by finding 16% of 18.75:

$$16\% \times 18.75 = 0.16 \times 18.75 = 3.$$

5. State. The answer is 18.75.

Do Exercises 3 and 4.

Perhaps you have noticed that to handle percents in problems such as those in Examples 1 and 2, you can convert to decimal notation before continuing.

Example 3 *Locations of Deaths by Lightning.* The circle graph below shows the various locations of people who are struck and killed by lightning. It is known that in the United States, 3327 people were killed by lightning from 1959 to 1996. How many were killed in fields or ballparks?

Number of Deaths Due to Lightning

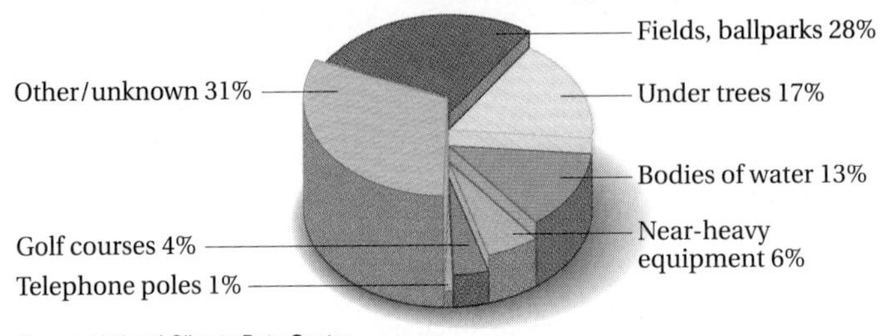

Source: National Climate Data Center

1. Familiarize. We first write down the information.

Total number of lightning strikes: 3327
Percent killed in fields or ballparks: 28%

We let x = the number of people killed in fields or ballparks. It seems reasonable that we would take 28% of 3327. This leads us to rewording and translating the problem.

2. Translate. The translation is as follows.

28%	of	3327	is	what?	**Rewording**
↓	↓	↓	↓	↓	
28%	·	3327	=	x	**Translating**

3. Solve. We solve the equation:

$28\% \cdot 3327 = x$

$0.28 \times 3327 = x$ Converting 28% to decimal notation

$932 \approx x.$ Multiplying and rounding to the nearest one

4. Check. The check is actually the computation we use to solve the equation:

$$28\% \cdot 3327 = 0.28 \times 3327 = 931.56 \approx 932.$$

5. State. About 932 lightning deaths occurred in fields or ballparks.

Do Exercises 5–7. (Exercise 7 is on the following page.)

Example 4 *Simple Interest.* An investment is made at 6% simple interest for 1 year. It grows to \$768.50. How much was originally invested (the principal)?

1. **Familiarize.** Suppose that \$100 was invested. Recalling the formula for simple interest, $I = Prt$, we know that the interest for 1 year on \$100 at 6% simple interest is given by $I = \$100 \cdot 6\% \cdot 1 = \$6$. Then, at the end of the year, the amount in the account is found by adding the principal and the interest:

$$\begin{array}{ccccc} \text{Principal} & + & \text{Interest} & = & \text{Amount} \\ \downarrow & & \downarrow & & \downarrow \\ \$100 & + & \$6 & = & \$106. \end{array}$$

In this problem, we are working backward. We are trying to find the principal, which is the original investment. We let $x =$ the principal.

2. **Translate.** We reword the problem and then translate.

$$\begin{array}{ccccc} \text{Principal} & + & \text{Interest} & = & \text{Amount} \\ \downarrow & & \downarrow & & \downarrow \\ x & + & 6\%x & = & 768.50 \end{array}$$

Interest is 6% of the principal.

3. **Solve.** We solve the equation:

$x + 6\%x = 768.50$

$x + 0.06x = 768.50$ Converting to decimal notation

$1x + 0.06x = 768.50$ Identity property of 1

$1.06x = 768.50$ Collecting like terms

$\frac{1.06x}{1.06} = \frac{768.50}{1.06}$ Dividing by 1.06

$x = 725.$

4. **Check.** We check by taking 6% of \$725 and adding it to \$725:

$6\% \times \$725 = 0.06 \times 725 = \$43.50.$

Then \$725 + \$43.50 = \$768.50, so \$725 checks.

5. **State.** The original investment was \$725.

Do Exercise 8.

7. The area of Arizona is 19% of the area of Alaska. The area of Alaska is 586,400 mi^2. What is the area of Arizona?

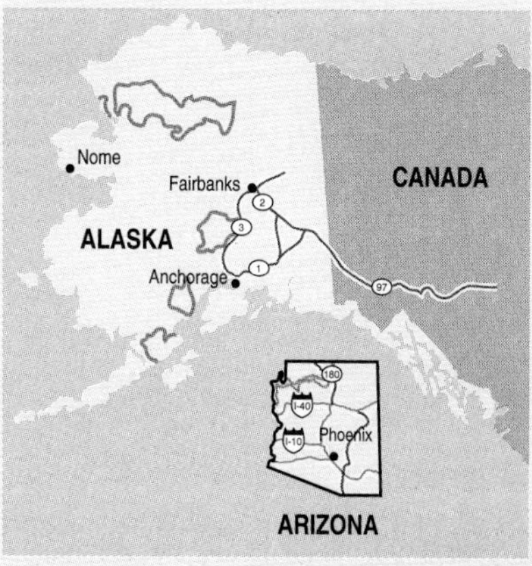

8. An investment is made at 7% simple interest for 1 year. It grows to \$8988. How much was originally invested (the principal)?

Answers on page A-5

Example 5 *Selling a Home.* The Fowlers are selling their home. They want to clear \$115,625 after paying a $7\frac{1}{2}\%$ commission to a realtor. For how much must they sell the house?

1. **Familiarize.** Suppose the Fowlers sold the house for \$120,000. We can determine the $7\frac{1}{2}\%$ commission by taking $7\frac{1}{2}\%$ of \$120,000:

   $$7\tfrac{1}{2}\% \text{ of } \$120{,}000 = 0.075(\$120{,}000) = \$9000.$$

   Subtracting this commission from \$120,000 would leave the Fowlers with

   $$\$120{,}000 - \$9000 = \$111{,}000.$$

   This shows us that in order for the Fowlers to clear \$115,625, the house must be sold for more than \$120,000. To determine exactly what the sale price would need to be, we could check more guesses. Instead, let's take advantage of our algebra skills. We let x = the selling price of the house. Because the commission is $7\frac{1}{2}\%$, the realtor receives $7\frac{1}{2}\%x$.

2. **Translate.** We reword and translate.

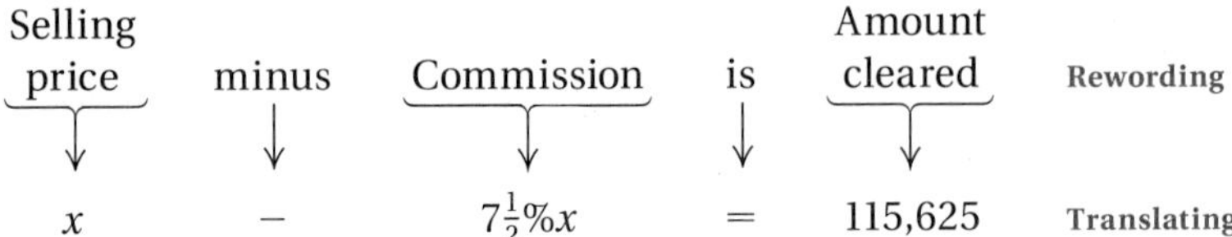

3. **Solve.** We solve the equation:

$$
\begin{aligned}
x - 7\tfrac{1}{2}\%x &= 115{,}625 \\
1x - 0.075x &= 115{,}625 && \text{Converting to decimal notation} \\
(1 - 0.075)x &= 115{,}625 && \text{Collecting like terms} \\
0.925x &= 115{,}625 \\
\frac{0.925x}{0.925} &= \frac{115{,}625}{0.925} && \text{Dividing by 0.925} \\
x &= 125{,}000.
\end{aligned}
$$

4. **Check.** To check, we first find $7\frac{1}{2}\%$ of \$125,000, calculating as we did in the *Familiarize* step:

   $$7\tfrac{1}{2}\% \text{ of } \$125{,}000 = 0.075(\$125{,}000) = \$9375. \quad \text{This is the commission.}$$

   Then we subtract the commission to find the amount cleared:

   $$\$125{,}000 - \$9375 = \$115{,}625.$$

   Since, after the commission, the Fowlers are left with \$115,625, our answer checks. Note that the sale price of \$125,000 is greater than \$120,000, as predicted in the *Familiarize* step.

5. **State.** The Fowlers need to sell their house for \$125,000.

CAUTION! The problem in Example 5 is easy to solve with algebra. Without algebra, it is not. A common error in such a problem is to take $7\frac{1}{2}\%$ of the sale price and then subtract or add. Note that $7\frac{1}{2}\%$ of the selling price $\left(7\frac{1}{2}\% \cdot \$125{,}000 = \$9375\right)$ is not equal to $7\frac{1}{2}\%$ of the price that the Fowlers wanted to clear $\left(7\frac{1}{2}\% \cdot \$115{,}625 \approx \$8671.88\right)$.

Do Exercise 9.

9. The price of a suit was decreased to a sale price of \$526.40. This was a 20% reduction. What was the former price?

Answer on page A-5

Exercise Set 2.5

a Solve.

1. What percent of 180 is 36?

2. What percent of 76 is 19?

3. 45 is 30% of what number?

4. 20.4 is 24% of what number?

5. What number is 65% of 840?

6. What number is 1% of 1,000,000?

7. 30 is what percent of 125?

8. 57 is what percent of 300?

9. 12% of what number is 0.3?

10. 7 is 175% of what number?

11. 2 is what percent of 40?

12. 40 is 2% of what number?

National Hamburger Sales. The circle graph below shows hamburger sales by various restaurants in 1996. The total sales were $39 billion.

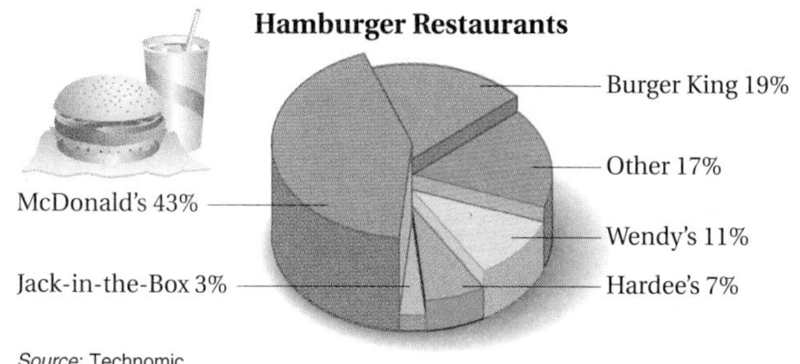

Source: Technomic

13. What was the total amount of hamburger sales, in dollars, by McDonald's?

14. What was the total amount of hamburger sales, in dollars, by Wendy's?

15. *Junk Mail.* The U.S. Postal Service reports that we open and read 78% of the junk mail that we receive (***Source*:** U.S. Postal Service). A sports instructional videotape company sends out 10,500 advertising brochures.
    a) How many of the brochures can it expect to be opened and read?
    b) The company sells videos to 189 of the people who receive the brochure. What percent of the 10,500 people who receive the brochure buy the video?

16. *FBI Applications.* The FBI annually receives 16,000 applications for agents. It accepts 600 of these applicants. (***Source*:** Federal Bureau of Investigation) What percent does it accept?

17. Leon left a $4 tip for a meal that cost $25.

a) What percent of the cost of the meal was the tip?
b) What was the total cost of the meal including the tip?

18. Selena left a $12.76 tip for a meal that cost $58.

a) What percent of the cost of the meal was the tip?
b) What was the total cost of the meal including the tip?

19. Leon left a 15% tip for a meal that cost $25.

a) How much was the tip?
b) What was the total cost of the meal including the tip?

20. Selena left a 15% tip for a meal that cost $58.

a) How much was the tip?
b) What was the total cost of the meal including the tip?

21. Leon left a 15% tip of $4.32 for a meal.

a) What was the cost of the meal before the tip?
b) What was the total cost of the meal including the tip?

22. Selena left a 15% tip of $8.40 for a meal.

a) What was the cost of the meal before the tip?
b) What was the total cost of the meal including the tip?

23. Leon left a 15% tip for a meal. The total cost of the meal, including the tip, was $41.40. What was the cost of the meal before the tip was added?

24. Selena left an 18% tip for a meal. The total cost of the meal, including the tip, was $40.71. What was the cost of the meal before the tip was added?

25. In a medical study of a group of pregnant women with "poor" diets, 16 of the women, or 8%, had babies who were in good or excellent health. How many women were in the original study?

26. In a medical study of a group of pregnant women with "good-to-excellent" diets, 285 of the women, or 95%, had babies who were in good or excellent health. How many women were in the original study?

27. *Life Insurance for Smokers vs. Nonsmokers.* The premium for a $100,000 life insurance policy for a female nonsmoker, age 22, is about $166 per year. The premium for a smoker is 170% of the premium for a nonsmoker. What is the premium for a smoker?

28. *Catching Colds from Kissing.* In a medical study, it was determined that if 800 people kiss someone else who has a cold, only 56 will actually catch the cold. What percent is this?

29. *Body Fat.* The author of this text exercises regularly at a local YMCA that recently offered a body-fat percentage test to its members. The device used measures the passage of a very low voltage of electricity through the body. The author's body-fat percentage was found to be 19.8% and he weighs 214 lb. What part, in pounds, of his body weight is fat?

30. *Calories Burned.* The author of this text exercises regularly at a local YMCA. The readout on a stairmaster machine tells him that if he exercises for 24 min, he will burn 356 calories. He decides to increase his time on the machine in order to lose more weight.

a) By what percent has he increased his time on the stairmaster if he exercises for 30 min?

b) How many calories does he burn if he exercises for 30 min? 40 min?

31. *Lightning Deaths Under Trees.* Referring to Example 3, determine how many deaths by lightning occurred under trees from 1959 to 1996.

32. *Lightning Deaths on Golf Courses.* Referring to Example 3, determine how many deaths by lightning occurred on golf courses from 1959 to 1996.

33. *Major League Baseball.* In 1997, the Boston Red Sox made the highest increase in the average price of a ticket to a baseball game. The price was $17.69 and represented an increase of 15% from the preceding season. (***Source:*** Major League Baseball) What was the average price of a ticket the preceding season?

34. *Major League Baseball.* The Atlanta Braves opened a new ballpark, Turner Field, in 1997. Attending a game represented the costliest outing in major league baseball that year. To take a family of four to a game, buy four small soft drinks, two small beers, and four hot dogs, and add in the cost of parking, two game programs, and two twill caps, the average cost would be $129.16, an increase of 6% over the year before. (***Source:*** Major League Baseball) What did this outing cost the year before?

35. An investment was made at 6% simple interest for 1 year. It grows to \$8268. How much was originally invested?

36. Money is borrowed at 6.2% simple interest. After 1 year, \$6945.48 pays off the loan. How much was originally borrowed?

37. After a 40% reduction, a shirt is on sale for \$34.80. What was the original price (that is, the price before reduction)?

38. After a 34% reduction, a blouse is on sale for \$42.24. What was the original price?

Skill Maintenance

Simplify.

39. $-3 - 8$ [1.4a]

40. $5 \cdot (-20)$ [1.5a]

41. $-\frac{3}{5} + \frac{1}{5}$ [1.3a]

42. $-12 \div (-6)$ [1.6a]

Evaluate. [1.1a]

43. $x - y$, for $x = 58$ and $y = 42$

44. $8t$, for $t = 23.7$

45. $\frac{6a}{b}$, for $a = 25$ and $b = 15$

46. $\frac{a + b}{8}$, for $a = 45.6$ and $b = 102.3$

Synthesis

47. ◆ Comment on the following quote by Yogi Berra, a famous Major League Hall of Fame baseball player: "Ninety percent of hitting is mental. The other half is physical."

48. ◆ Erin returns a tent that she bought during a storewide 35% off sale that has ended. She is offered store credit for 125% of what she paid (not to be used on sale items). Is this fair to Erin? Why or why not?

49. It has been determined that at the age of 15, a boy has reached 96.1% of his final adult height. Jaraan is 6 ft, 4 in. at the age of 15. What will his final adult height be?

50. It has been determined that at the age of 10, a girl has reached 84.4% of her final adult height. Dana is 4 ft, 8 in. at the age of 10. What will her final adult height be?

51. In one city, a sales tax of 9% was added to the price of gasoline as registered on the pump. Suppose a driver asked for \$10 worth of gas. The attendant filled the tank until the pump read \$9.10 and charged the driver \$10. Something was wrong. Use algebra to correct the error.

Calculate the sale price and the original price of discounted items.

2.6 Formulas

a Evaluating and Solving Formulas

A **formula** is a "recipe" for doing a certain type of calculation. Formulas are often given as equations. Here is an example of a formula that has to do with weather: $M = \frac{1}{5}n$. You see a flash of lightning. After a few seconds you hear the thunder associated with that flash. How far away was the lightning?

Your distance from the storm is M miles. You can find that distance by counting the number of seconds n that it takes the sound of the thunder to reach you and then multiplying by $\frac{1}{5}$.

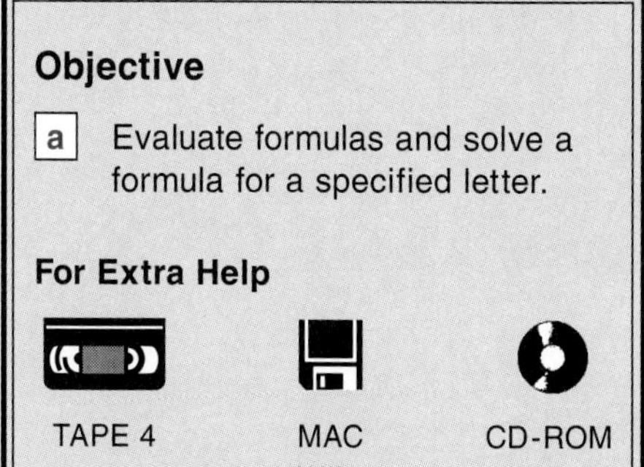

Example 1 *Storm Distance.* Consider the formula $M = \frac{1}{5}n$. It takes 10 sec for the sound of thunder to reach you after you have seen a flash of lightning. How far away is the storm?

We substitute 10 for n and calculate M: $M = \frac{1}{5}n = \frac{1}{5}(10) = 2$. The storm is 2 mi away.

Do Exercise 1.

1. Suppose that it takes the sound of thunder 14 sec to reach you. How far away is the storm?

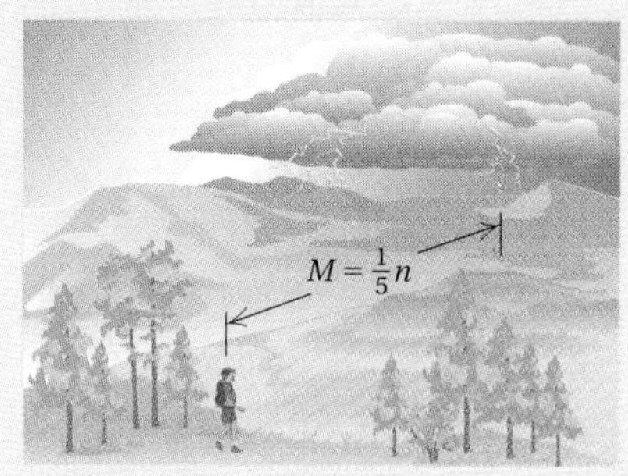

Suppose that we think we know how far we are from the storm and want to check by calculating the number of seconds it should take the sound of the thunder to reach us. We could substitute a number for M—say, 2—and solve for n:

$$2 = \tfrac{1}{5}n$$

$$10 = n. \qquad \text{Multiplying by 5}$$

However, if we wanted to do this repeatedly, it might be easier to solve for n by getting it alone on one side. We "solve" the formula for n.

Example 2 Solve for n: $M = \frac{1}{5}n$.

We have

$$M = \tfrac{1}{5}n \qquad \text{We want this letter alone.}$$

$$5 \cdot M = 5 \cdot \tfrac{1}{5}n \qquad \text{Multiplying by 5 on both sides}$$

$$5M = n.$$

In the above situation for $M = 2$, $n = 5(2)$, or 10.

Do Exercise 2.

2. Solve for I: $E = IR$. (This is a formula from electricity relating voltage E, current I, and resistance R.)

To see how the addition and multiplication principles apply to formulas, compare the following.

A. Solve.

$$5x + 2 = 12$$
$$5x = 12 - 2$$
$$5x = 10$$
$$x = \frac{10}{5} = 2$$

B. Solve.

$$5x + 2 = 12$$
$$5x = 12 - 2$$
$$x = \frac{12 - 2}{5}$$

C. Solve for x.

$$ax + b = c$$
$$ax = c - b$$
$$x = \frac{c - b}{a}$$

In (A), we solved as we did before. In (B), we did not carry out the calculations. In (C), we could not carry out the calculations because we had unknown numbers.

Answers on page A-5

3. Solve for D: $C = \pi D$.

(This is a formula for the circumference C of a circle of diameter D.)

4. *Averages.* Solve for c:

$$A = \frac{a + b + c + d}{4}.$$

5. Use the formula of Example 5.

a) Estimate the weight of a yellow tuna that is 7 ft long and has a girth of about 54 in.

b) Solve the formula for L.

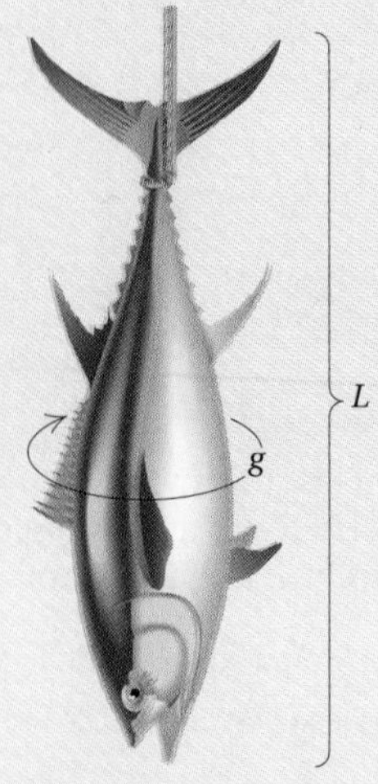

Answers on page A-5

Example 3 *Circumference.* Solve for r: $C = 2\pi r$. This is a formula for the circumference C of a circle of radius r.

$C = 2\pi r$ **We want this letter alone.**

$\frac{C}{2\pi} = \frac{2\pi r}{2\pi}$ **Dividing by 2π**

$\frac{C}{2\pi} = r$

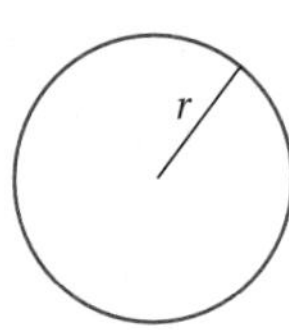

To solve a formula for a given letter, identify the letter and:

1. Multiply on both sides to clear fractions or decimals, if that is needed.
2. Collect like terms on each side, if necessary.
3. Get all terms with the letter to be solved for on one side of the equation and all other terms on the other side.
4. Collect like terms again, if necessary.
5. Solve for the letter in question.

Example 4 Solve for a: $A = \frac{a + b + c}{3}$. This is a formula for the average A of three numbers a, b, and c.

$A = \frac{a + b + c}{3}$ **We want the letter a alone.**

$3A = a + b + c$ **Multiplying by 3 to clear the fraction**

$3A - b - c = a$ **Subtracting b and c**

Do Exercises 3 and 4.

Example 5 *Estimating the Weight of a Fish.* An ancient fisherman's formula for estimating the weight of a fish is

$$W = \frac{Lg^2}{800},$$

where W is the weight in pounds, L is the length in inches, and g is the girth (distance around the midsection) in inches.

a) Estimate the weight of a great bluefin tuna that is 8 ft long and has a girth of about 76 in.

b) Solve the formula for g^2.

We solve as follows:

a) We substitute 96 for L (8 ft = 96 in.) and 76 for g. Then we calculate W.

$$W = \frac{Lg^2}{800} = \frac{96 \cdot 76^2}{800} \approx 693 \text{ lb}$$

The tuna weighs about 693 lb.

b) $W = \frac{Lg^2}{800}$ **We want to get g^2 alone.**

$800W = Lg^2$ **Multiplying by 800**

$\frac{800W}{L} = g^2$ **Dividing by L**

Do Exercise 5.

Exercise Set 2.6

a Solve for the given letter.

1. *Area of a Parallelogram*:

$A = bh$, for h

(Area A, base b, height h)

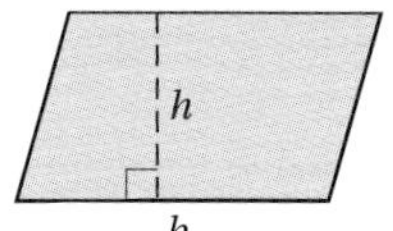

2. *Distance Formula*:

$d = rt$, for r

(Distance d, speed r, time t)

3. *Perimeter of a Rectangle*:

$P = 2l + 2w$, for w

(Perimeter P, length l, width w)

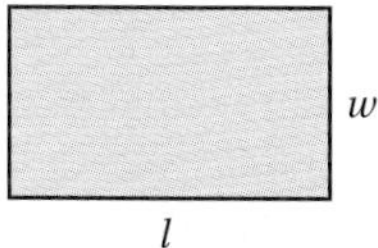

4. *Area of a Circle*:

$A = \pi r^2$, for r^2

(Area A, radius r)

5. *Average of Two Numbers*:

$$A = \frac{a + b}{2}, \quad \text{for } a$$

$a \qquad A = \frac{a+b}{2} \qquad b$

6. *Area of a Triangle*:

$$A = \frac{1}{2}bh, \quad \text{for } b$$

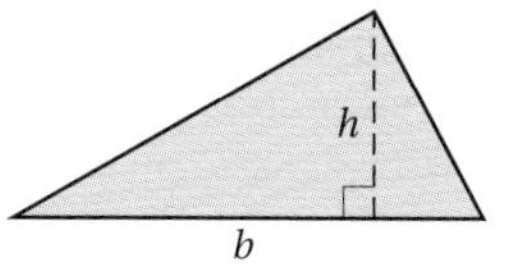

7. *Force*:

$F = ma$, for a

(Force F, mass m, acceleration, a)

8. *Simple Interest*:

$I = Prt$, for P

(Interest I, principal P, interest rate r, time t)

9. *Relativity*:

$E = mc^2$, for c^2

(Energy E, mass m, speed of light c)

10. $Q = \dfrac{p - q}{2}$, for p

11. $Ax + By = c$, for x

12. $Ax + By = c$, for y

13. $v = \dfrac{3k}{t}$, for t

14. $P = \dfrac{ab}{c}$, for c

15. *Furnace Output.* The formula

$b = 30a$

is used in New England to estimate the minimum furnace output, b, in Btu's, for a modern house with a square feet of flooring.

a) Determine the minimum furnace output for a 1900-ft^2 modern house.

b) Solve the formula for a.

16. *Surface Area of a Cube.* The surface area of a cube with side s is given by

$A = 6s^2$.

a) Determine the surface area of a cube with 3-in. sides.

b) Solve the formula for s^2.

17. *Full-Time Equivalent Students.* Colleges accommodate students who need to take different total-credit-hour loads. They determine the number of "full-time-equivalent" students, F, using the formula

$$F = \frac{n}{15},$$

where n is the total number of credits students enroll in for a given semester.

a) Determine the number of full-time equivalent students on a campus in which students register for 21,345 credits.

b) Solve the formula for n.

18. *Young's Rule in Medicine.* Young's rule for determining the amount of a medicine dosage for a child is given by the formula

$$c = \frac{ad}{a + 12},$$

where a is the child's age and d is the usual adult dosage. (*Warning*! Do not apply this formula without checking with a physician!)

a) The usual adult dosage of medication for an adult is 250 mg. Find the dosage for a child of age 2.

b) Solve the formula for d. (***Source*:** Olsen, June L., et al., *Medical Dosage Calculations,* 6th ed. Reading, MA: Addison Wesley Longman, p. A-31.)

19. *Female Caloric Needs.* The number of calories K needed each day by a moderately active woman who weighs w pounds, is h inches tall, and is a years old can be estimated by the formula

$$K = 917 + 6(w + h - a).$$

(***Source*:** Parker, M., *She Does Math.* Mathematical Association of America, p. 96.)

a) Elaine is moderately active, weighs 120 lb, is 67 in. tall, and is 23 yr old. What are her caloric needs?

b) Solve the formula for a, for h, and for w.

20. *Male Caloric Needs.* The number of calories K needed each day by a moderately active man who weighs w kilograms, is h centimeters tall, and is a years old can be estimated by the formula

$$K = 19.18w + 7h - 9.52a + 92.4.$$

(***Source*:** Parker, M., *She Does Math.* Mathematical Association of America, p. 96.)

a) Marv is moderately active, weighs 97 kg, is 185 cm tall, and is 55 yr old. What are his caloric needs?

b) Solve the formula for a, for h, and for w.

Skill Maintenance

21. Evaluate $2a - b$ for $a = 2$ and $b = 3$. [1.1a]

22. Add: $-23 + (-67)$. [1.3a]

23. Subtract: $-45.8 - (-32.6)$. [1.4a]

24. Remove parentheses and simplify: [1.8b]

$4a - 8b - 5(5a - 4b)$.

25. Add: $-\frac{2}{3} + \frac{5}{6}$. [1.3a]

26. Subtract: $-\frac{2}{3} - \frac{5}{6}$. [1.4a]

Synthesis

27. ◆ Devise an application in which it would be useful to solve the equation $d = rt$ for r. (See Exercise 2.)

28. ◆ The equations

$$P = 2l + 2w \quad \text{and} \quad w = \frac{P}{2} - l$$

are equivalent formulas involving the perimeter P, the length l, and the width w of a rectangle. Devise a problem for which the second of the two formulas would be more useful.

Solve.

29. $A = \frac{1}{2}ah + \frac{1}{2}bh$, for b; for h

30. $P = 4m + 7mn$, for m

31. In $A = lw$, l and w both double. What is the effect on A?

32. In $P = 2a + 2b$, P doubles. Do a and b necessarily both double?

33. In $A = \frac{1}{2}bh$, b increases by 4 units and h does not change. What happens to A?

34. Solve for F:

$$D = \frac{1}{E + F}.$$

2.7 Solving Inequalities

We now extend our equation-solving principles to the solving of inequalities.

Objectives

a Determine whether a given number is a solution of an inequality.

b Graph an inequality on a number line.

c Solve inequalities using the addition principle.

d Solve inequalities using the multiplication principle.

e Solve inequalities using the addition and multiplication principles together.

For Extra Help

TAPE 4 MAC WIN CD-ROM

a Solutions of Inequalities

In Section 1.2, we defined the symbols $>$ (greater than), $<$ (less than), $\geq$ (greater than or equal to), and $\leq$ (less than or equal to). For example, $3 \leq 4$ and $3 \leq 3$ are both true, but $-3 \leq -4$ and $0 \geq 2$ are both false.

An **inequality** is a number sentence with $>$, $<$, $\geq$, or $\leq$ as its verb—for example,

$$-4 > t, \quad x < 3, \quad 2x + 5 \geq 0, \quad \text{and} \quad -3y + 7 \leq -8.$$

Some replacements for a variable in an inequality make it true and some make it false.

A replacement that makes an inequality true is called a **solution**. The set of all solutions is called the **solution set.** When we have found the set of all solutions of an inequality, we say that we have **solved** the inequality.

Examples Determine whether the number is a solution of $x < 2$.

1. -2.7 Since $-2.7 < 2$ is true, -2.7 is a solution.

2. 2 Since $2 < 2$ is false, 2 is not a solution.

Examples Determine whether the number is a solution of $y \geq 6$.

3. 6 Since $6 \geq 6$ is true, 6 is a solution.

4. $-\frac{4}{3}$ Since $-\frac{4}{3} \geq 6$ is false, $-\frac{4}{3}$ is not a solution.

Do Exercises 1 and 2.

Determine whether each number is a solution of the inequality.

1. $x > 3$

a) 2 b) 0

c) -5 d) 15.4

e) 3 f) $-\dfrac{2}{5}$

2. $x \leq 6$

a) 6 b) 0

c) -4.3 d) 25

e) -6 f) $\dfrac{5}{8}$

Answers on page A-5

b Graphs of Inequalities

Some solutions of $x < 2$ are 0.45, -8.9, $-\pi$, $\frac{5}{8}$, and so on. In fact, there are infinitely many real numbers that are solutions. Because we cannot list them all individually, it is helpful to make a drawing that represents all the solutions.

A **graph** of an inequality is a drawing that represents its solutions. An inequality in one variable can be graphed on a number line. An inequality in two variables can be graphed on a coordinate plane; we will study such graphs in Chapter 9.

We first graph inequalities in one variable on a number line.

Example 5 Graph: $x < 2$.

The solutions of $x < 2$ are all those numbers less than 2. They are shown on the graph by shading all points to the left of 2. The open circle at 2 indicates that 2 is not part of the graph.

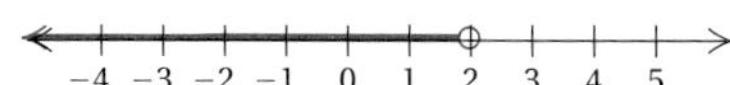

Graph.

3. $x \leq 4$

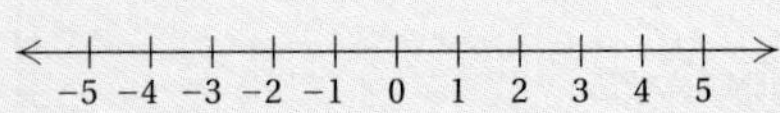

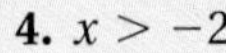

4. $x > -2$

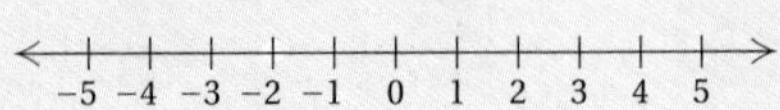

5. $-2 < x \leq 4$

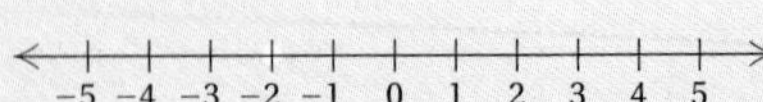

Answers on page A-5

Example 6 Graph: $x \geq -3$.

The solutions of $x \geq -3$ are shown on the number line by shading the point for -3 and all points to the right of -3. The closed circle at -3 indicates that -3 *is* part of the graph.

Example 7 Graph: $-3 \leq x < 2$.

The inequality $-3 \leq x < 2$ is read "-3 is less than or equal to x *and* x is less than 2," or "x is greater than or equal to -3 *and* x is less than 2." In order to be a solution of this inequality, a number must be a solution of both $-3 \leq x$ and $x < 2$. The number 1 is a solution, as are -1.7, 0, 1.5, and $\frac{3}{8}$. We can see from the graphs below that the solution set consists of the numbers that overlap in the two solution sets in Examples 5 and 6:

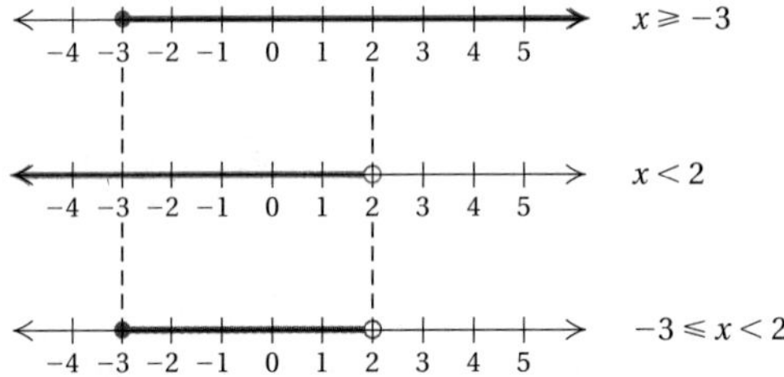

The open circle at 2 means that 2 is *not* part of the graph. The closed circle at -3 means that -3 *is* part of the graph. The other solutions are shaded.

Do Exercises 3–5.

c Solving Inequalities Using the Addition Principle

Consider the true inequality $3 < 7$. If we add 2 on both sides, we get another true inequality:

$$3 + 2 < 7 + 2, \quad \text{or} \quad 5 < 9.$$

Similarly, if we add -4 on both sides of $x + 4 < 10$, we get an *equivalent* inequality:

$$x + 4 + (-4) < 10 + (-4),$$

or

$$x < 6.$$

To say that $x + 4 < 10$ and $x < 6$ are **equivalent** is to say that they have the same solution set. For example, the number 3 is a solution of $x + 4 < 10$. It is also a solution of $x < 6$. The number -2 is a solution of $x < 6$. It is also a solution of $x + 4 < 10$. Any solution of one is a solution of the other—they are equivalent.

> **THE ADDITION PRINCIPLE FOR INEQUALITIES**
>
> For any real numbers a, b, and c:
>
> $a < b$ is equivalent to $a + c < b + c$;
>
> $a > b$ is equivalent to $a + c > b + c$;
>
> $a \le b$ is equivalent to $a + c \le b + c$;
>
> $a \ge b$ is equivalent to $a + c \ge b + c$.
>
> In other words, when we add or subtract the same number on both sides of an inequality, the direction of the inequality symbol is not changed.

As with equation solving, when solving inequalities, our goal is to isolate the variable on one side. Then it is easier to determine the solution set.

Example 8 Solve: $x + 2 > 8$. Then graph.

We use the addition principle, subtracting 2 on both sides:

$$x + 2 - 2 > 8 - 2$$
$$x > 6.$$

From the inequality $x > 6$, we can determine the solutions directly. Any number greater than 6 makes the last sentence true and is a solution of that sentence. Any such number is also a solution of the original sentence. Thus the inequality is solved. The graph is as follows:

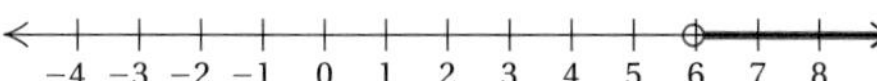

We cannot check all the solutions of an inequality by substitution, as we can check solutions of equations, because there are too many of them. A partial check can be done by substituting a number greater than 6—say, 7—into the original inequality:

$x + 2 > 8$	
$7 + 2$	8
9	TRUE

Since $9 > 8$ is true, 7 is a solution. Any number greater than 6 is a solution.

Example 9 Solve: $3x + 1 \le 2x - 3$. Then graph.

We have

$$3x + 1 \le 2x - 3$$
$$3x + 1 - 1 \le 2x - 3 - 1 \quad \text{Subtracting 1}$$
$$3x \le 2x - 4 \quad \text{Simplifying}$$
$$3x - 2x \le 2x - 4 - 2x \quad \text{Subtracting } 2x$$
$$x \le -4. \quad \text{Simplifying}$$

The graph is as follows:

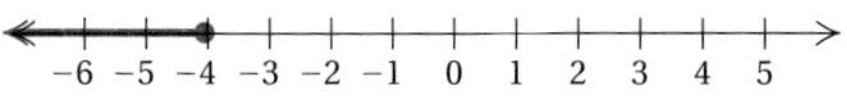

Solve. Then graph.

6. $x + 3 > 5$

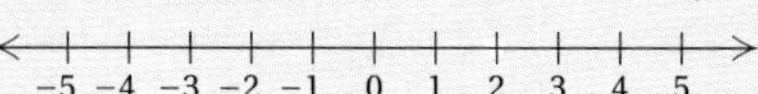

7. $x - 1 \le 2$

−5 −4 −3 −2 −1 0 1 2 3 4 5

8. $5x + 1 < 4x - 2$

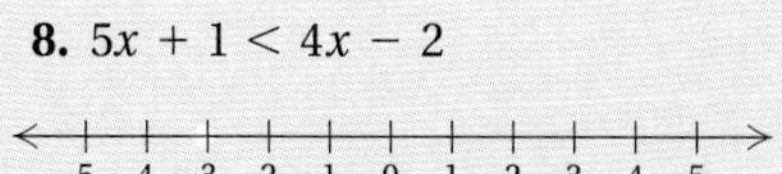

Answers on page A-5

Solve.

9. $x + \frac{2}{3} \geq \frac{4}{5}$

10. $5y + 2 \leq -1 + 4y$

Answers on page A-5

In Example 9, any number less than or equal to -4 is a solution. The following are some solutions:

$$-4, \quad -5, \quad -6, \quad -\frac{13}{3}, \quad -204.5, \quad \text{and} \quad -18\pi.$$

Besides drawing a graph, we can also describe all the solutions of an inequality using **set notation.** We could just begin to list them in a set using roster notation (see p. 9), as follows:

$$\{-4, -5, -6, -4.1, -204.5, -18\pi, \ldots\}.$$

We can never list them all this way, however. Seeing this set without knowing the inequality makes it difficult for us to know what real numbers we are considering. There is, however, another kind of notation that we can use. It is

$$\{x \mid x \leq -4\},$$

which is read

"The set of all x such that x is less than or equal to -4."

This shorter notation for sets is called **set-builder notation** (see Section 1.2). From now on, we will use this notation when solving inequalities.

Do Exercises 6–8 on the preceding page.

Example 10 Solve: $x + \frac{1}{3} > \frac{5}{4}$.

We have

$$x + \tfrac{1}{3} > \tfrac{5}{4}$$

$$x + \tfrac{1}{3} - \tfrac{1}{3} > \tfrac{5}{4} - \tfrac{1}{3} \qquad \textbf{Subtracting } \tfrac{1}{3}$$

$$x > \tfrac{5}{4} \cdot \tfrac{3}{3} - \tfrac{1}{3} \cdot \tfrac{4}{4} \qquad \textbf{Multiplying by 1 to obtain a common denominator}$$

$$x > \tfrac{15}{12} - \tfrac{4}{12}$$

$$x > \tfrac{11}{12}.$$

Any number greater than $\frac{11}{12}$ is a solution. The solution set is

$$\{x \mid x > \tfrac{11}{12}\},$$

which is read

"The set of all x such that x is greater than $\frac{11}{12}$."

When solving inequalities, you may obtain an answer like $7 < x$. Recall from Chapter 1 that this has the same meaning as $x > 7$. Thus the solution set can be described as $\{x \mid 7 < x\}$ or as $\{x \mid x > 7\}$. The latter is used most often.

Do Exercises 9 and 10.

d Solving Inequalities Using the Multiplication Principle

There is a multiplication principle for inequalities that is similar to that for equations, but it must be modified. When we are multiplying on both sides by a negative number, the direction of the inequality symbol must be changed. Let's see what happens. Consider the true inequality $3 < 7$. If we multiply on both sides by a *positive* number, like 2, we get another true inequality:

$3 \cdot 2 < 7 \cdot 2$, or $6 < 14$. **True**

If we multiply on both sides by a *negative* number, like -2, and we do not change the direction of the inequality symbol, we get a *false* inequality:

$3 \cdot (-2) < 7 \cdot (-2)$, or $-6 < -14$. **False**

The fact that $6 < 14$ is true but $-6 < -14$ is false stems from the fact that the negative numbers, in a sense, mirror the positive numbers. That is, whereas 14 is to the *right* of 6 on a number line, the number -14 is to the *left* of -6. Thus, if we reverse (change the direction of) the inequality symbol, we get a *true* inequality: $-6 > -14$.

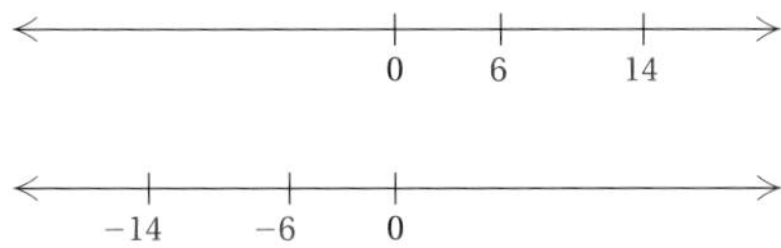

> **THE MULTIPLICATION PRINCIPLE FOR INEQUALITIES**
>
> For any real numbers a and b, and any *positive* number c:
>
> $a < b$ is equivalent to $ac < bc$;
>
> $a > b$ is equivalent to $ac > bc$.
>
> For any real numbers a and b, and any *negative* number c:
>
> $a < b$ is equivalent to $ac > bc$;
>
> $a > b$ is equivalent to $ac < bc$.
>
> Similar statements hold for $\leq$ and $\geq$.
>
> In other words, when we multiply or divide by a positive number on both sides of an inequality, the direction of the inequality symbol stays the same. When we multiply or divide by a negative number on both sides of an inequality, the direction of the inequality symbol is reversed.

Example 11 Solve: $4x < 28$. Then graph.

We have

$4x < 28$

$\frac{4x}{4} < \frac{28}{4}$ **Dividing by 4**

The symbol stays the same.

$x < 7$. **Simplifying**

The solution set is $\{x \mid x < 7\}$. The graph is as follows:

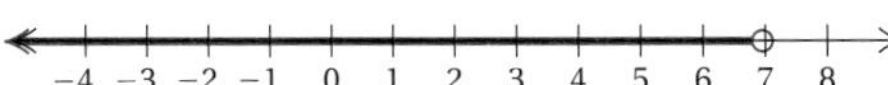

Do Exercises 11 and 12.

Solve. Then graph.

11. $8x < 64$

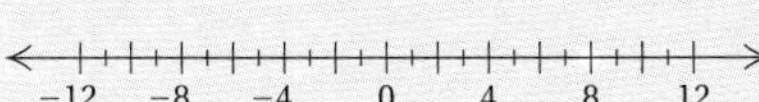

12. $5y \geq 160$

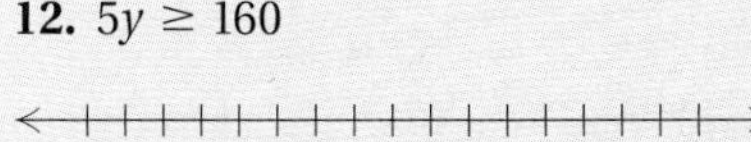

Answers on page A-5

Solve.

13. $-4x \leq 24$

14. $-5y > 13$

15. Solve: $7 - 4x < 8$.

Answers on page A-5

Example 12 Solve: $-2y < 18$. Then graph.

We have

$$-2y < 18$$

$$\frac{-2y}{-2} > \frac{18}{-2} \quad \text{Dividing by } -2$$

The symbol must be reversed!

$$y > -9. \quad \text{Simplifying}$$

The solution set is $\{y \mid y > -9\}$. The graph is as follows:

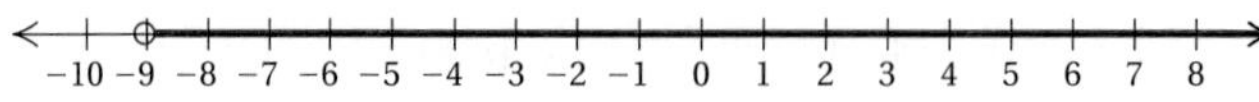

Do Exercises 13 and 14.

e Using the Principles Together

All of the equation-solving techniques used in Sections 2.1–2.3 can be used with inequalities provided we remember to reverse the inequality symbol when multiplying or dividing on both sides by a negative number.

Example 13 Solve: $6 - 5y > 7$.

We have

$$6 - 5y > 7$$

$$-6 + 6 - 5y > -6 + 7 \quad \text{Adding } -6\text{. The symbol stays the same.}$$

$$-5y > 1 \quad \text{Simplifying}$$

$$\frac{-5y}{-5} < \frac{1}{-5} \quad \text{Dividing by } -5$$

The symbol must be reversed.

$$y < -\frac{1}{5}. \quad \text{Simplifying}$$

The solution set is $\left\{y \mid y < -\frac{1}{5}\right\}$.

Do Exercise 15.

Example 14 Solve: $8y - 5 > 17 - 5y$.

$$-17 + 8y - 5 > -17 + 17 - 5y \quad \text{Adding } -17\text{. The symbol stays the same.}$$

$$8y - 22 > -5y \quad \text{Simplifying}$$

$$-8y + 8y - 22 > -8y - 5y \quad \text{Adding } -8y$$

$$-22 > -13y \quad \text{Simplifying}$$

$$\frac{-22}{-13} < \frac{-13y}{-13} \quad \text{Dividing by } -13$$

The symbol must be reversed.

$$\frac{22}{13} < y.$$

The solution set is $\left\{y \mid \frac{22}{13} < y\right\}$, or $\left\{y \mid y > \frac{22}{13}\right\}$.

We can often solve inequalities in such a way as to avoid having to reverse the inequality symbol. We add so that after like terms have been collected, the coefficient of the variable term is positive. We show this by solving the inequality in Example 14 a different way.

Example 15 Solve: $8y - 5 > 17 - 5y$.

Note that if we add $5y$ on both sides, the coefficient of the y-term will be positive after like terms have been collected.

$$8y - 5 + 5y > 17 - 5y + 5y \quad \text{Adding } 5y$$
$$13y - 5 > 17 \quad \text{Simplifying}$$
$$13y - 5 + 5 > 17 + 5 \quad \text{Adding 5}$$
$$13y > 22 \quad \text{Simplifying}$$
$$\frac{13y}{13} > \frac{22}{13} \quad \text{Dividing by 13}$$
$$y > \frac{22}{13}$$

The solution set is $\{y \mid y > \frac{22}{13}\}$.

Do Exercises 16 and 17.

Example 16 Solve: $3(x - 2) - 1 < 2 - 5(x + 6)$.

$$3(x - 2) - 1 < 2 - 5(x + 6)$$
$$3x - 6 - 1 < 2 - 5x - 30 \quad \text{Using the distributive law to multiply and remove parentheses}$$
$$3x - 7 < -5x - 28 \quad \text{Simplifying}$$
$$3x + 5x < -28 + 7 \quad \text{Adding } 5x \text{ and 7 to get all } x\text{-terms on one side and all other terms on the other side}$$
$$8x < -21 \quad \text{Simplifying}$$
$$x < \frac{-21}{8}, \text{ or } -\frac{21}{8} \quad \text{Dividing by 8}$$

The solution set is $\{x \mid x < -\frac{21}{8}\}$.

Do Exercise 18.

Example 17 Solve: $16.3 - 7.2p \leq -8.18$.

The greatest number of decimal places in any one number is *two*. Multiplying by 100, which has two 0's, will clear decimals. Then we proceed as before.

$$16.3 - 7.2p \leq -8.18$$
$$100(16.3 - 7.2p) \leq 100(-8.18) \quad \text{Multiplying by 100}$$
$$100(16.3) - 100(7.2p) \leq 100(-8.18) \quad \text{Using the distributive law}$$
$$1630 - 720p \leq -818 \quad \text{Simplifying}$$
$$1630 - 720p - 1630 \leq -818 - 1630 \quad \text{Subtracting 1630}$$
$$-720p \leq -2448 \quad \text{Simplifying}$$
$$\frac{-720p}{-720} \geq \frac{-2448}{-720} \quad \text{Dividing by } -720$$

The symbol must be reversed.

$$p \geq 3.4$$

The solution set is $\{p \mid p \geq 3.4\}$.

16. Solve: $24 - 7y \leq 11y - 14$.

17. Solve. Use a method like the one used in Example 15.

$24 - 7y \leq 11y - 14$

18. Solve:

$3(7 + 2x) \leq 30 + 7(x - 1)$.

Answers on page A-5

19. Solve:

$2.1x + 43.2 \geq 1.2 - 8.4x.$

Do Exercise 19.

Example 18 Solve: $\frac{2}{3}x - \frac{1}{6} + \frac{1}{2}x > \frac{7}{6} + 2x.$

The number 6 is the least common multiple of all the denominators. Thus we multiply by 6 on both sides.

$$\frac{2}{3}x - \frac{1}{6} + \frac{1}{2}x > \frac{7}{6} + 2x$$

$$6\left(\frac{2}{3}x - \frac{1}{6} + \frac{1}{2}x\right) > 6\left(\frac{7}{6} + 2x\right) \quad \text{Multiplying by 6 on both sides}$$

$$6 \cdot \frac{2}{3}x - 6 \cdot \frac{1}{6} + 6 \cdot \frac{1}{2}x > 6 \cdot \frac{7}{6} + 6 \cdot 2x \quad \text{Using the distributve law}$$

$$4x - 1 + 3x > 7 + 12x \quad \text{Simplifying}$$

$$7x - 1 > 7 + 12x \quad \text{Collecting like terms}$$

$$7x - 1 - 12x > 7 + 12x - 12x \quad \text{Subtracting } 12x$$

$$-5x - 1 > 7 \quad \text{Collecting like terms}$$

$$-5x - 1 + 1 > 7 + 1 \quad \text{Adding 1}$$

$$-5x > 8 \quad \text{Simplifying}$$

$$\frac{-5x}{-5} < \frac{8}{-5} \quad \text{Dividing by } -5$$

The symbol must be reversed.

$$x < -\frac{8}{5}$$

The solution set is $\{x \mid x < -\frac{8}{5}\}$.

20. Solve:

$\frac{3}{4} + x < \frac{7}{8}x - \frac{1}{4} + \frac{1}{2}x.$

Do Exercise 20.

Answers on page A-5

Exercise Set 2.7

a Determine whether each number is a solution of the given inequality.

1. $x > -4$
a) 4
b) 0
c) -4
d) 6
e) 5.6

2. $x \le 5$
a) 0
b) 5
c) -1
d) -5
e) $7\frac{1}{4}$

3. $x \ge 6.8$
a) -6
b) 0
c) 6
d) 8
e) $-3\frac{1}{2}$

4. $x < 8$
a) 8
b) -10
c) 0
d) 11
e) -4.7

b Graph on a number line.

5. $x > 4$

−5 −4 −3 −2 −1 0 1 2 3 4 5

6. $x < 0$

−5 −4 −3 −2 −1 0 1 2 3 4 5

7. $t < -3$

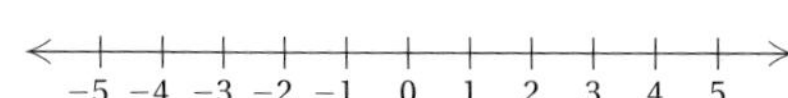

8. $y > 5$

−5 −4 −3 −2 −1 0 1 2 3 4 5

9. $m \ge -1$

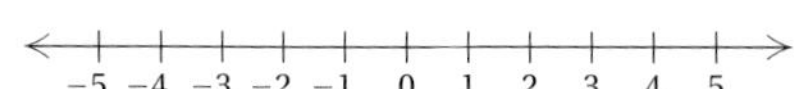

10. $x \le -2$

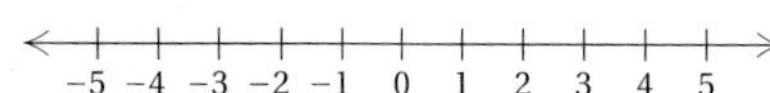

11. $-3 < x \le 4$

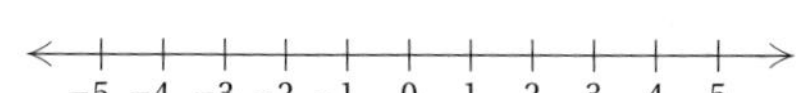

12. $-5 \le x < 2$

−5 −4 −3 −2 −1 0 1 2 3 4 5

13. $0 < x < 3$

−5 −4 −3 −2 −1 0 1 2 3 4 5

14. $-5 \le x \le 0$

−5 −4 −3 −2 −1 0 1 2 3 4 5

c Solve using the addition principle. Then graph.

15. $x + 7 > 2$

−5 −4 −3 −2 −1 0 1 2 3 4 5

16. $x + 5 > 2$

−5 −4 −3 −2 −1 0 1 2 3 4 5

17. $x + 8 \le -10$

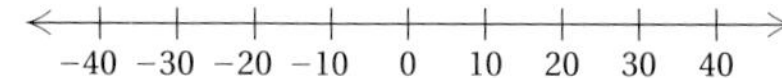

18. $x + 8 \le -11$

−40 −30 −20 −10 0 10 20 30 40

Solve using the addition principle.

19. $y - 7 > -12$

20. $y - 9 > -15$

21. $2x + 3 > x + 5$

22. $2x + 4 > x + 7$

23. $3x + 9 \leq 2x + 6$

24. $3x + 18 \leq 2x + 16$

25. $5x - 6 < 4x - 2$

26. $9x - 8 < 8x - 9$

27. $-9 + t > 5$

28. $-8 + p > 10$

29. $y + \frac{1}{4} \leq \frac{1}{2}$

30. $x - \frac{1}{3} \leq \frac{5}{6}$

31. $x - \frac{1}{3} > \frac{1}{4}$

32. $x + \frac{1}{8} > \frac{1}{2}$

d Solve using the multiplication principle. Then graph.

33. $5x < 35$

−8 −6 −4 −2 0 2 4 6 8

34. $8x \geq 32$

−5 −4 −3 −2 −1 0 1 2 3 4 5

35. $-12x > -36$

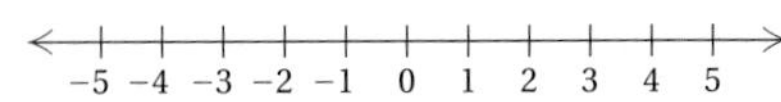

36. $-16x > -64$

−5 −4 −3 −2 −1 0 1 2 3 4 5

Solve using the multiplication principle.

37. $5y \geq -2$

38. $3x < -4$

39. $-2x \leq 12$

40. $-3x \leq 15$

41. $-4y \geq -16$

42. $-7x < -21$

43. $-3x < -17$

44. $-5y > -23$

45. $-2y > \frac{1}{7}$

46. $-4x \leq \frac{1}{9}$

47. $-\frac{6}{5} \leq -4x$

48. $-\frac{7}{9} > 63x$

e Solve using the addition and multiplication principles.

49. $4 + 3x < 28$

50. $3 + 4y < 35$

51. $3x - 5 \leq 13$

52. $5y - 9 \leq 21$

53. $13x - 7 < -46$

54. $8y - 6 < -54$

55. $30 > 3 - 9x$

56. $48 > 13 - 7y$

57. $4x + 2 - 3x \leq 9$

58. $15x + 5 - 14x \leq 9$

59. $-3 < 8x + 7 - 7x$

60. $-8 < 9x + 8 - 8x - 3$

61. $6 - 4y > 4 - 3y$

62. $9 - 8y > 5 - 7y + 2$

63. $5 - 9y \leq 2 - 8y$

64. $6 - 18x \leq 4 - 12x - 5x$

65. $19 - 7y - 3y < 39$

66. $18 - 6y - 4y < 63 + 5y$

67. $2.1x + 45.2 > 3.2 - 8.4x$

68. $0.96y - 0.79 \leq 0.21y + 0.46$

69. $\frac{x}{3} - 2 \leq 1$

70. $\frac{2}{3} + \frac{x}{5} < \frac{4}{15}$

71. $\frac{y}{5} + 1 \leq \frac{2}{5}$

72. $\frac{3x}{4} - \frac{7}{8} \geq -15$

73. $3(2y - 3) < 27$

74. $4(2y - 3) > 28$

75. $2(3 + 4m) - 9 \geq 45$

76. $3(5 + 3m) - 8 \leq 88$

77. $8(2t + 1) > 4(7t + 7)$

78. $7(5y - 2) > 6(6y - 1)$

79. $3(r - 6) + 2 < 4(r + 2) - 21$

80. $5(x + 3) + 9 \leq 3(x - 2) + 6$

81. $0.8(3x + 6) \geq 1.1 - (x + 2)$

82. $0.4(2x + 8) \geq 20 - (x + 5)$

83. $\frac{5}{3} + \frac{2}{3}x < \frac{25}{12} + \frac{5}{4}x + \frac{3}{4}$

84. $1 - \frac{2}{3}y \geq \frac{9}{5} - \frac{y}{5} + \frac{3}{5}$

Skill Maintenance

Add or subtract. [1.3a], [1.4a]

85. $-56 + (-18)$

86. $-2.3 + 7.1$

87. $-\frac{3}{4} + \frac{1}{8}$

88. $8.12 - 9.23$

89. $-56 - (-18)$

90. $-\frac{3}{4} - \frac{1}{8}$

91. $-2.3 - 7.1$

92. $-8.12 + 9.23$

Simplify.

93. $5 - 3^2 + (8 - 2)^2 \cdot 4$ [1.8d]

94. $10 \div 2 \cdot 5 - 3^2 + (-5)^2$ [1.8d]

95. $5(2x - 4) - 3(4x + 1)$ [1.8b]

96. $9(3 + 5x) - 4(7 + 2x)$ [1.8b]

Synthesis

97. ◆ Are the inequalities $3x - 4 < 10 - 4x$ and $2(x - 5) > 3(2x - 6)$ equivalent? Why or why not?

98. ◆ Explain in your own words why it is necessary to reverse the inequality symbol when multiplying on both sides of an inequality by a negative number.

99. Determine whether each number is a solution of the inequality $|x| < 3$.

a) 0
b) −2
c) −3
d) 4
e) 3
f) 1.7
g) −2.8

100. Graph $|x| < 3$ on a number line.

−5 −4 −3 −2 −1 0 1 2 3 4 5

Solve.

101. $x + 3 \leq 3 + x$

102. $x + 4 < 3 + x$

Collaborative Learning Manual

Solve linear inequalities as a group.

2.8 Applications and Problem Solving with Inequalities

We can use inequalities to solve certain types of problems.

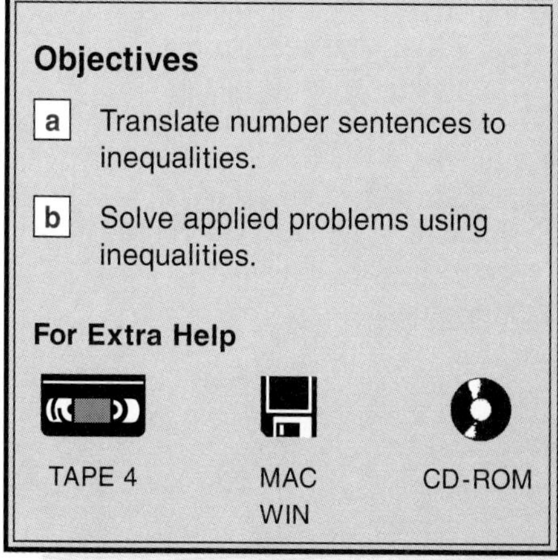

a Translating to Inequalities

First let's practice translating sentences to inequalities.

Examples Translate to an inequality.

1. A number is less than 5.

$x < 5$

2. A number is greater than or equal to $3\frac{1}{2}$.

$y \geq 3\frac{1}{2}$

3. He can earn, at most, \$34,000.

$E \leq \$34{,}000$

4. The number of compact disc players sold in this city in a year is at least 2700.

$C \geq 2700$

5. 12 more than twice a number is less than 37.

$2x + 12 < 37$

Do Exercises 1–5.

b Solving Problems

Example 6 *Test Scores.* A pre-med student is taking a chemistry course in which four tests are to be given. To get an A, she must average at least 90 on the four tests. The student got scores of 91, 86, and 89 on the first three tests. Determine (in terms of an inequality) what scores on the last test will allow her to get an A.

1. **Familiarize.** Let's try some guessing. Suppose the student gets a 92 on the last test. The average of the four scores is their sum divided by the number of tests, 4, and is given by

$$\frac{91 + 86 + 89 + 92}{4} = 89.5.$$

In order for this average to be *at least* 90, it must be greater than or equal to 90. Since $89.5 \geq 90$ is false, a score of 92 will not give the student an A. But there are scores that will give an A. To find them, we translate to an inequality and solve. Let $x =$ the student's score on the last test.

2. **Translate.** The average of the four scores must be *at least* 90. This means that it must be greater than or equal to 90. Thus we can translate the problem to the inequality

$$\frac{91 + 86 + 89 + x}{4} \geq 90.$$

Translate.

1. A number is less than or equal to 8.

2. A number is greater than -2.

3. That car can be driven at most 180 mph.

4. The price of that car is at least \$5800.

5. Twice a number minus 32 is greater than 5.

Answers on page A-5

6. *Test Scores.* A student is taking a literature course in which four tests are to be given. To get a B, he must average at least 80 on the four tests. The student got scores of 82, 76, and 78 on the first three tests. Determine (in terms of an inequality) what scores on the last test will allow him to get at least a B.

7. *Gold Temperatures.* Gold stays solid at Fahrenheit temperatures below 1945.4°. Determine (in terms of an inequality) those Celsius temperatures for which gold stays solid. Use the formula given in Example 7.

Answers on page A-5

3. Solve. We solve the inequality. We first multiply by 4 to clear the fraction.

$$4\left(\frac{91 + 86 + 89 + x}{4}\right) \geq 4 \cdot 90 \qquad \text{Multiplying by 4}$$
$$91 + 86 + 89 + x \geq 360$$
$$266 + x \geq 360 \qquad \text{Collecting like terms}$$
$$x \geq 94 \qquad \text{Subtracting 266}$$

The solution set is $\{x \mid x \geq 94\}$.

4. Check. We can obtain a partial check by substituting a number greater than or equal to 94. We leave it to the student to try 95 in a manner similar to what was done in the *Familiarize* step.

5. State. Any score that is at least 94 will give the student an A.

Do Exercise 6.

Example 7 *Butter Temperatures.* Butter stays solid at Fahrenheit temperatures below 88°. The formula

$$F = \tfrac{9}{5}C + 32$$

can be used to convert Celsius temperatures C to Fahrenheit temperatures F. Determine (in terms of an inequality) those Celsius temperatures for which butter stays solid.

1. Familiarize. Let's make a guess. We try a Celsius temperature of 40°. We substitute and find F:

$$F = \tfrac{9}{5}C + 32 = \tfrac{9}{5}(40) + 32 = 72 + 32 = 104°.$$

This is higher than 88°, so 40° is *not* a solution. To find the solutions, we need to solve an inequality.

2. Translate. The Fahrenheit temperature F is to be less than 88. We have the inequality

$$F < 88.$$

To find the Celsius temperatures C that satisfy this condition, we substitute $\tfrac{9}{5}C + 32$ for F, which gives us the following inequality:

$$\tfrac{9}{5}C + 32 < 88.$$

3. Solve. We solve the inequality:

$$\tfrac{9}{5}C + 32 < 88$$
$$5\left(\tfrac{9}{5}C + 32\right) < 5(88) \qquad \text{Multiplying by 5 to clear the fraction}$$
$$5\left(\tfrac{9}{5}C\right) + 5(32) < 440 \qquad \text{Using a distributive law}$$
$$9C + 160 < 440 \qquad \text{Simplifying}$$
$$9C < 280 \qquad \text{Subtracting 160}$$
$$C < \frac{280}{9} \qquad \text{Dividing by 9}$$
$$C < 31.1. \qquad \text{Dividing and rounding to the nearest tenth}$$

The solution set of the inequality is $\{C \mid C < 31.1°\}$.

4. Check. The check is left to the student.

5. State. Butter stays solid at Celsius temperatures below 31.1°.

Do Exercise 7.

Exercise Set 2.8

a Translate to an inequality.

1. A number is greater than 8.

2. A number is less than 5.

3. A number is less than or equal to -4.

4. A number is greater than or equal to 18.

5. The number of people is at least 1300.

6. The cost is at most \$4857.95.

7. The amount of acid is not to exceed 500 liters.

8. The cost of gasoline is no less than 94 cents per gallon.

9. Two more than three times a number is less than 13.

10. Five less than one-half a number is greater than 17.

b Solve.

11. *Test Scores.* Your quiz grades are 73, 75, 89, and 91. Determine (in terms of an inequality) what scores on the last quiz will allow you to get an average quiz grade of at least 85.

12. *Body Temperatures.* The human body is considered to be fevered when its temperature is higher than 98.6°F. Using the formula given in Example 7, determine (in terms of an inequality) those Celsius temperatures for which the body is fevered.

13. *World Records in the 1500-m Run.* The formula

$$R = -0.075t + 3.85$$

can be used to predict the world record in the 1500-m run t years after 1930. Determine (in terms of an inequality) those years for which the world record will be less than 3.5 min.

14. *World Records in the 200-m Dash.* The formula

$$R = -0.028t + 20.8$$

can be used to predict the world record in the 200-m dash t years after 1920. Determine (in terms of an inequality) those years for which the world record will be less than 19.0 sec.

15. *Sizes of Envelopes.* Rhetoric Advertising is a direct-mail company. It determines that for a particular campaign, it can use any envelope with a fixed width of $3\frac{1}{2}$ in. and an area of at least $17\frac{1}{2}$ in^2. Determine (in terms of an inequality) those lengths that will satisfy the company constraints.

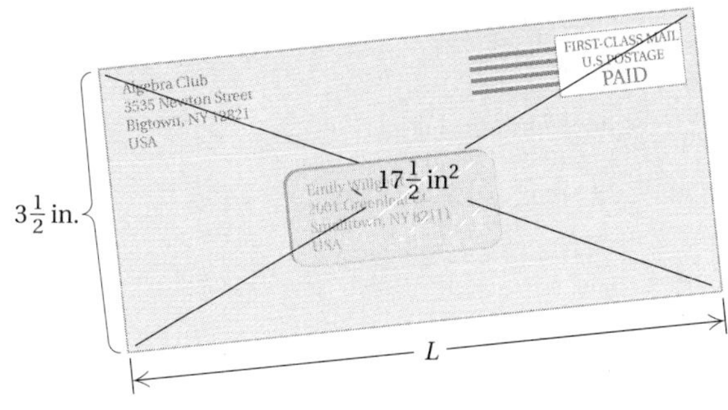

16. *Sizes of Packages.* An overnight delivery service accepts packages of up to 165 in. in length and girth combined. (Girth is the distance around the package.) A package has a fixed girth of 53 in. Determine (in terms of an inequality) those lengths for which a package is acceptable.

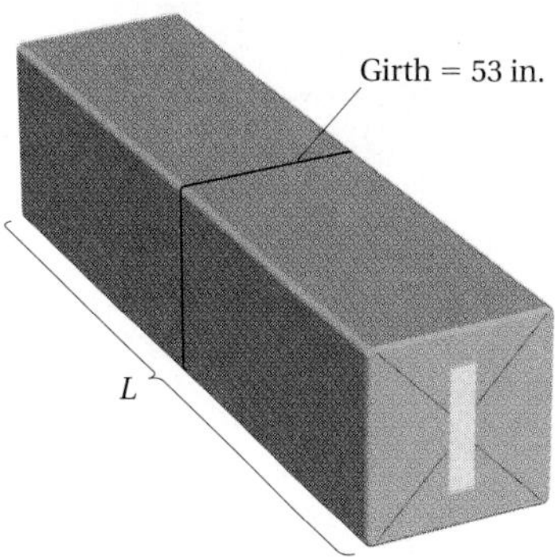

17. Find all numbers such that the sum of the number and 15 is less than four times the number.

18. Find all numbers such that three times the number minus ten times the number is greater than or equal to eight times the number.

19. *Black Angus Calves.* Black Angus calves weigh about 75 lb at birth and gain about 2 lb per day for the first few weeks. Determine (in terms of an inequality) those days for which the calf's weight is more than 125 lb.

20. *IKON Copiers.* IKON Office Solutions rents a Canon GP30F copier for $240 per month plus 1.8¢ per copy (***Source*:** Ikon Office Solutions, Keith Palmer). A catalog publisher needs to lease a copy machine for use during a special project that they anticipate will take 3 months. They decide to rent the copier, but must stay within a budget of $5400 for copies. Determine (in terms of an inequality) the number of copies they can make per month and still remain within budget.

21. One side of a triangle is 2 cm shorter than the base. The other side is 3 cm longer than the base. What lengths of the base will allow the perimeter to be greater than 19 cm?

22. The perimeter of a rectangular swimming pool is not to exceed 70 ft. The length is to be twice the width. What widths will meet these conditions?

23. Dirk's Electric made 17 customer calls last week and 22 calls this week. How many calls must be made next week in order to maintain an average of at least 20 calls for the three-week period?

24. Ginny and Jill do volunteer work at a hospital. Jill worked 3 hr more than Ginny, and together they worked more than 27 hr. What possible number of hours did each work?

25. A family's air conditioner needs freon. The charge for a service call is a flat fee of $70 plus $60 an hour. The freon costs $35. The family has at most $150 to pay for the service call. Determine (in terms of an inequality) those lengths of time of the call that will allow the family to stay within its $150 budget.

26. A student is shopping for a new pair of jeans and two sweaters of the same kind. He is determined to spend no more than $120.00 for the outfit. He buys jeans for $21.95. What is the most that the student can spend for each sweater?

27. *Skippy Reduced-Fat Peanut Butter.* In order for a food to be advertised as "reduced fat," it must have at least 25% less fat than the regular food of that type. Reduced-fat Skippy Peanut Butter contains 12 g of fat per serving. What can you conclude about how much fat is in regular Skippy peanut butter?

28. A landscaping company is laying out a triangular flower bed. The height of the triangle is 16 ft. What lengths of the base will make the area at least 200 ft^2?

Skill Maintenance

Simplify.

29. $-3 + 2(-5)^2(-3) - 7$ [1.8d]

30. $3x + 2[4 - 5(2x - 1)]$ [1.8c]

31. $23(2x - 4) - 15(10 - 3x)$ [1.8b]

32. $256 \div 64 \div 4^2$ [1.8d]

Synthesis

33. ◆ Chassman and Bem booksellers offers a preferred customer card for $25. The card entitles a customer to a 10% discount on all purchases for a period of 1 year. Under what circumstances would an individual save money by buying a card?

34. ◆ After 9 quizzes, Brenda's average is 84. Is it possible for her to improve her average by two points with the next quiz? Why or why not?

Summary and Review Exercises: Chapter 2

Important Properties and Formulas

The Addition Principle for Equations:	For any real numbers a, b, and c: $a = b$ is equivalent to $a + c = b + c$.
The Multiplication Principle for Equations:	For any real numbers a, b, and c, $c \neq 0$: $a = b$ is equivalent to $a \cdot c = b \cdot c$.
The Addition Principle for Inequalities:	For any real numbers a, b, and c: $a < b$ is equivalent to $a + c < b + c$; $a > b$ is equivalent to $a + c > b + c$; $a \le b$ is equivalent to $a + c \le b + c$; $a \ge b$ is equivalent to $a + c \ge b + c$.
The Multiplication Principle for Inequalities:	For any real numbers a and b, and any *positive* number c: $a < b$ is equivalent to $ac < bc$; $a > b$ is equivalent to $ac > bc$. For any real numbers a and b, and any *negative* number c: $a < b$ is equivalent to $ac > bc$; $a > b$ is equivalent to $ac < bc$.

The review exercises that follow are for practice. Answers are at the back of the book. If you miss an exercise, restudy the objective indicated in blue after the exercise or the direction line that precedes it. Beginning with this chapter, certain objectives, from four particular sections of preceding chapters, will be retested on the chapter test. The objectives to be tested in addition to the material in this chapter are [1.1a], [1.1b], [1.3a], and [1.8b].

Solve. [2.1b]

1. $x + 5 = -17$

2. $n - 7 = -6$

3. $x - 11 = 14$

4. $y - 0.9 = 9.09$

Solve. [2.2a]

5. $-\frac{2}{3}x = -\frac{1}{6}$

6. $-8x = -56$

7. $-\frac{x}{4} = 48$

8. $15x = -35$

9. $\frac{4}{5}y = -\frac{3}{16}$

Solve. [2.3a]

10. $5 - x = 13$

11. $\frac{1}{4}x - \frac{5}{8} = \frac{3}{8}$

Solve. [2.3b]

12. $5t + 9 = 3t - 1$

13. $7x - 6 = 25x$

14. $14y = 23y - 17 - 10$

15. $0.22y - 0.6 = 0.12y + 3 - 0.8y$

16. $\frac{1}{4}x - \frac{1}{8}x = 3 - \frac{1}{16}x$

Solve. [2.3c]

17. $4(x + 3) = 36$

18. $3(5x - 7) = -66$

19. $8(x - 2) = 5(x + 4)$

20. $-5x + 3(x + 8) = 16$

Determine whether the given number is a solution of the inequality $x \le 4$. [2.7a]

21. -3

22. 7

23. 4

Solve. Write set notation for the answers. [2.7c, d, e]

24. $y + \frac{2}{3} \ge \frac{1}{6}$

25. $9x \ge 63$

26. $2 + 6y > 14$

27. $7 - 3y \ge 27 + 2y$

28. $3x + 5 < 2x - 6$

29. $-4y < 28$

30. $3 - 4x < 27$

31. $4 - 8x < 13 + 3x$

32. $-3y \ge -21$

33. $-4x \le \frac{1}{3}$

Graph on a number line. [2.7b, e]

34. $4x - 6 < x + 3$

35. $-2 < x \le 5$

36. $y > 0$

Solve. [2.6a]

37. $C = \pi d$, for d

38. $V = \frac{1}{3}Bh$, for B

39. $A = \frac{a + b}{2}$, for a

Solve. [2.4a]

40. *Dimensions of Wyoming.* The state of Wyoming is roughly in the shape of a rectangle whose perimeter is 1280 mi. The length is 90 mi more than the width. Find the dimensions.

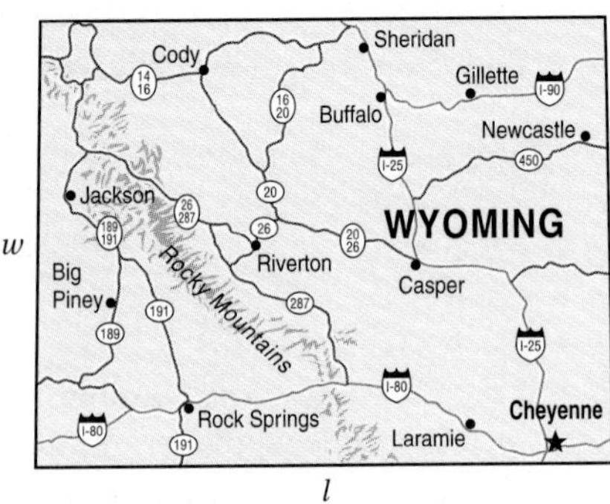

41. If 14 is added to a certain number, the result is 41. Find the number.

42. *Interstate Mile Markers.* The sum of two mile markers on I-5 in California is 691. Find the numbers on the markers.

43. An entertainment center sold for \$2449 in June. This was \$332 more than the cost in February. Find the cost in February.

44. Ty is paid a commission of \$4 for each appliance he sells. One week, he received \$108 in commissions. How many appliances did he sell?

45. The measure of the second angle of a triangle is 50° more than that of the first angle. The measure of the third angle is 10° less than twice the first angle. Find the measures of the angles.

Solve. [2.5a]

46. After a 30% reduction, a bread maker is on sale for \$154. What was the marked price (the price before the reduction)?

47. A hotel manager's salary is \$30,000, which is a 15% increase over the previous year's salary. What was the previous salary (to the nearest dollar)?

48. A tax-exempt charity received a bill of \$145.90 for a sump pump. The bill incorrectly included sales tax of 5%. How much does the charity actually owe?

Solve.

49. *Test Scores.* Your test grades are 71, 75, 82, and 86. What is the lowest grade that you can get on the next test and still have an average test score of at least 80? [2.8b]

50. The length of a rectangle is 43 cm. What widths will make the perimeter greater than 120 cm? [2.8a]

51. *Estimating the Weight of a Fish.* An ancient fisherman's formula for estimating the weight of a fish is

$$W = \frac{Lg^2}{800},$$

where W is the weight in pounds, L is the length in inches, and g is the girth (distance around the midsection) in inches. [2.6a]

a) Estimate the weight of a salmon that is 3 ft long and has a girth of about 13.5 in.

b) Solve the formula for L.

Skill Maintenance

52. Evaluate $\frac{a + b}{4}$ for $a = 16$ and $b = 25$. [1.1a]

53. Translate to an algebraic expression: [1.1b]

Tricia drives her car at 58 mph for t hours. How far has she driven?

54. Add: $-12 + 10 + (-19) + (-24)$. [1.3a]

55. Remove parentheses and simplify: $5x - 8(6x - y)$. [1.8b]

Synthesis

56. ◆ Would it be better to receive a 5% raise and then an 8% raise or the other way around? Why?

57. ◆ Are the inequalities $x > -5$ and $-x < 5$ equivalent? Why or why not?

Solve.

58. $2|x| + 4 = 50$ [1.2e], [2.3a]

59. $|3x| = 60$ [1.2e], [2.2a]

60. $y = 2a - ab + 3$, for a [2.6a]

Test: Chapter 2

Solve.

1. $x + 7 = 15$
2. $t - 9 = 17$
3. $3x = -18$
4. $-\frac{4}{7}x = -28$
5. $3t + 7 = 2t - 5$
6. $\frac{1}{2}x - \frac{3}{5} = \frac{2}{5}$
7. $8 - y = 16$
8. $-\frac{2}{5} + x = -\frac{3}{4}$
9. $3(x + 2) = 27$
10. $-3x - 6(x - 4) = 9$
11. $0.4p + 0.2 = 4.2p - 7.8 - 0.6p$

Solve. Write set notation for the answers.

12. $x + 6 \le 2$
13. $14x + 9 > 13x - 4$
14. $12x \le 60$
15. $-2y \ge 26$
16. $-4y \le -32$
17. $-5x \ge \frac{1}{4}$
18. $4 - 6x > 40$
19. $5 - 9x \ge 19 + 5x$

Graph on a number line.

20. $y \le 9$

−12 −8 −4 0 4 8 12

21. $6x - 3 < x + 2$

−5 −4 −3 −2 −1 0 1 2 3 4 5

22. $-2 \le x \le 2$

−5 −4 −3 −2 −1 0 1 2 3 4 5

Answers

1. ______
2. ______
3. ______
4. ______
5. ______
6. ______
7. ______
8. ______
9. ______
10. ______
11. ______
12. ______
13. ______
14. ______
15. ______
16. ______
17. ______
18. ______
19. ______
20. ______
21. ______
22. ______

Answers

23. ______

24. ______

25. ______

26. ______

27. ______

28. ______

29. a) ______

b) ______

30. ______

31. ______

32. ______

33. ______

34. ______

35. ______

36. ______

37. ______

38. ______

Solve.

23. The perimeter of a rectangular photograph is 36 cm. The length is 4 cm greater than the width. Find the width and the length.

24. If you triple a number and then subtract 14, you get two-thirds of the original number. What is the original number?

25. The numbers on three raffle tickets are consecutive integers whose sum is 7530. Find the integers.

26. Money is invested in a savings account at 5% simple interest. After 1 year, there is $924 in the account. How much was originally invested?

27. An 8-m board is cut into two pieces. One piece is 2 m longer than the other. How long are the pieces?

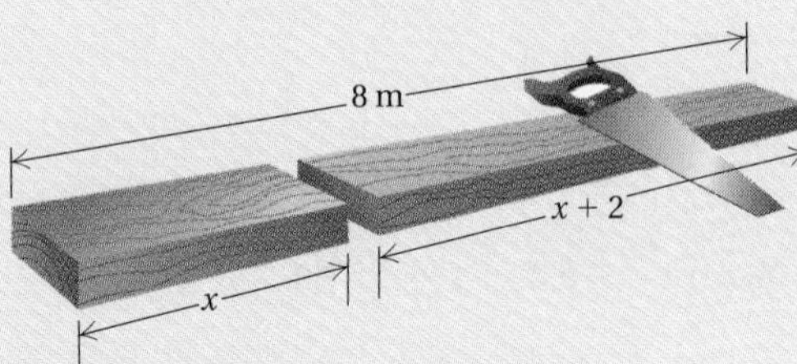

28. Solve $A = 2\pi rh$ for r.

29. *Male Caloric Needs.* The number of calories K needed each day by a moderately active man who weighs w kilograms, is h centimeters tall, and is a years old can be estimated by the formula

$$K = 19.18w + 7h - 9.52a + 92.4.$$

a) David is moderately active, weighs 89 kg, is 180 cm tall, and is 43 yr old. What are his caloric needs?

b) Solve the formula for w.

Source: Parker, M., *She Does Math.* Mathematical Association of America, p. 96.

30. Find all numbers such that six times the number is greater than the number plus 30.

31. The width of a rectangle is 96 yd. Find all possible lengths such that the perimeter of the rectangle will be at least 540 yd.

Skill Maintenance

32. Add: $\frac{2}{3} + \left(-\frac{8}{9}\right)$.

33. Evaluate $\frac{4x}{y}$ for $x = 2$ and $y = 3$.

34. Translate to an algebraic expression: Seventy-three percent of p.

35. Simplify: $2x - 3y - 5(4x - 8y)$.

Synthesis

36. Solve $c = \frac{1}{a - d}$ for d.

37. Solve: $3|w| - 8 = 37$.

38. A movie theater had a certain number of tickets to give away. Five people got the tickets. The first got one-third of the tickets, the second got one-fourth of the tickets, and the third got one-fifth of the tickets. The fourth person got eight tickets, and there were five tickets left for the fifth person. Find the total number of tickets given away.

3

Graphs of Equations; Data Analysis

Introduction

We now begin a study of graphs. First, we examine graphs as they commonly appear in newspapers or magazines and develop some terminology. Following that, we study graphs of certain equations. Finally, we use graphs with applications and data analysis.

An Application

The cost y, in dollars, of mailing a FedEx Priority Overnight package weighing 1 lb or more is given by the equation

$$y = 2.085x + 15.08,$$

where x = the number of pounds (***Source***: Federal Express Corporation). Graph the equation and then use the graph to estimate the cost of mailing a $6\frac{1}{2}$-lb package.

This problem appears as Example 8 in Section 3.2.

The Mathematics

The graph is shown below. It appears that the cost of mailing a $6\frac{1}{2}$-lb package is about \$29.

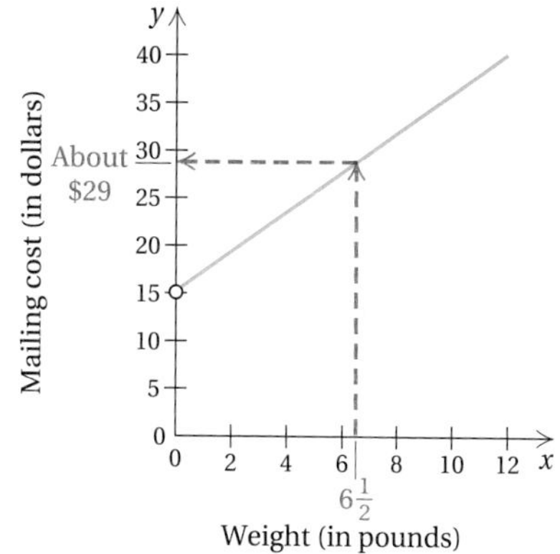

For more information, visit us at www.mathmax.com

Pretest: Chapter 3

Graph on a plane.

1. $y = -x$

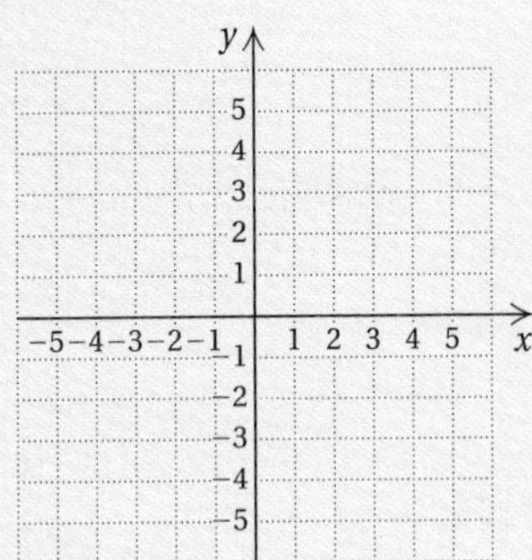

2. $x = -4$

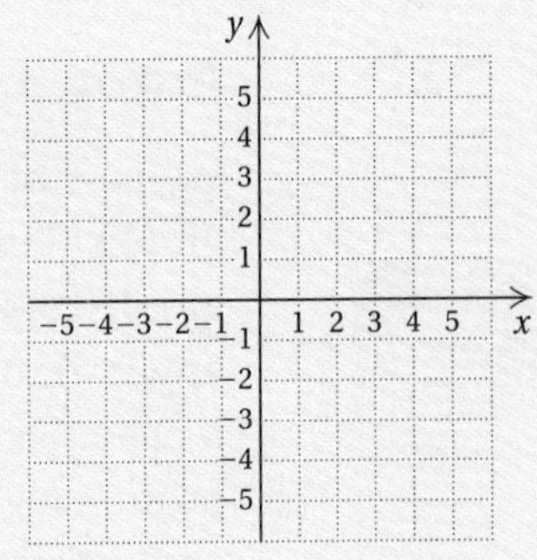

3. $4x - 5y = 20$

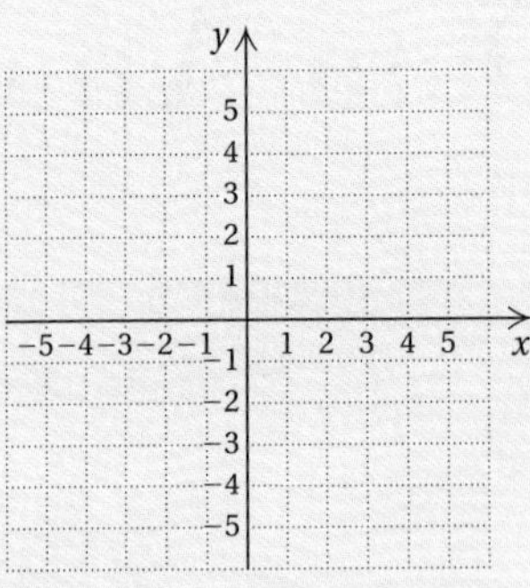

4. $y = \frac{2}{3}x - 1$

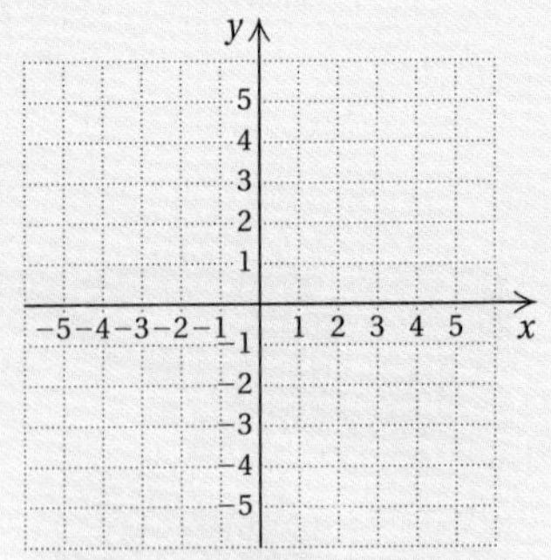

5. In which quadrant is the point $(-4, -1)$ located?

6. Determine whether the ordered pair $(-4, -1)$ is a solution of $4x - 5y = 20$.

7. Find the intercepts of the graph of $4x - 5y = 20$.

8. Find the y-intercept of $y = 3x - 8$.

9. *Price of Printing.* The price P, in cents, of a photocopied and bound lab manual is given by

$$P = \frac{7}{2}n + 20,$$

where $n =$ the number of pages in the manual. Graph the equation and then use the graph to estimate the price of an 85-page manual.

10. *Blood Alcohol Levels of Drivers in Fatal Accidents.* Find the mean, the median, and the mode of this set of data:

0.18, 0.17, 0.21, 0.16, 0.18.

11. *Height of Girls.* Use extrapolation to estimate the missing data.

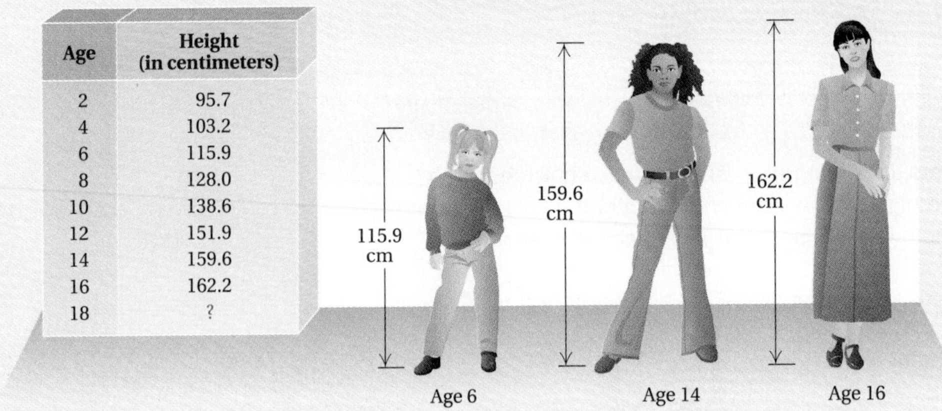

Age	Height (in centimeters)
2	95.7
4	103.2
6	115.9
8	128.0
10	138.6
12	151.9
14	159.6
16	162.2
18	?

Source: Kempe, C. Henry, et al (eds.), *Current Pediatric Diagnosis & Treatment 1987.* Norwalk, CT: Appleton & Lange, 1987

Objectives for Retesting

The objectives to be retested in addition to the material in this chapter are as follows.

[1.2c] Convert from fractional notation to decimal notation for a rational number.
[1.2e] Find the absolute value of a real number.
[2.2a] Solve equations using the multiplication principle.
[2.5a] Solve applied problems involving percent.

3.1 Graphs and Applications

Often data are available regarding an application in mathematics that we are reviewing. We can use graphs to show the data and extract information about the data that can lead to making analyses and predictions.

Today's print and electronic media make extensive use of graphs. This is due in part to the ease with which some graphs can be prepared by computer and in part to the large quantity of information that a graph can display. We first consider applications with circle, bar, and line graphs.

Objectives

a Solve applied problems involving circle, bar, and line graphs.

b Plot points associated with ordered pairs of numbers.

c Determine the quadrant in which a point lies.

d Find the coordinates of a point on a graph.

For Extra Help

TAPE 5

MAC
WIN

CD-ROM

a Applications with Graphs

Circle Graphs

Circle graphs and *pie graphs,* or *charts,* are often used to show what percent of a whole each particular item in a group represents.

Example 1 *U.S. Soft-Drink Retail Sales.* The following circle graph shows the percentages of sales of various soft drinks in the United States in a recent year.

U.S. Soft Drink Retail Sales

Other 11%
Coca-Cola 43%
Dr. Pepper/7Up 15%
Pepsi 31%

Source: Pepsico, Inc.

Total soft-drink sales in the United States that year reached \$54 billion. What were the sales of Pepsi?

1. **Familiarize.** The graph shows that 31% of the soft-drink sales were of Pepsi. We let

   $y =$ the amount spent on Pepsi.

2. **Translate.** We reword and translate the problem as follows.

What	is	31%	of	\$54?	Rewording
↓	↓	↓	↓	↓	
y	$=$	31%	$\cdot$	54	Translating

3. **Solve.** We solve the equation by carrying out the computation on the right:

   $y = 31\% \cdot 54 = 0.31 \cdot 54 = \$16.74.$

4. **Check.** We leave the check to the student.

5. **State.** Pepsi accounted for \$16.74 billion of the soft-drink sales that particular year.

Do Exercise 1.

1. Referring to Example 1, determine the soft-drink sales of Coca-Cola.

Answer on page A-6

2. *Tornado Touchdowns.* Referring to Example 2, determine the following.

a) During which interval did the smallest number of touchdowns occur?

b) During which intervals was the number of touchdowns less than 60?

Answers on page A-6

Bar Graphs

Bar graphs are convenient for showing comparisons. In every bar graph, certain categories are paired with certain numbers. Example 2 pairs intervals of time with the total number of reported cases of tornado touchdowns.

Example 2 *Tornado Touchdowns.* The following bar graph shows the total number of tornado touchdowns by time of day in Indiana from 1950–1994.

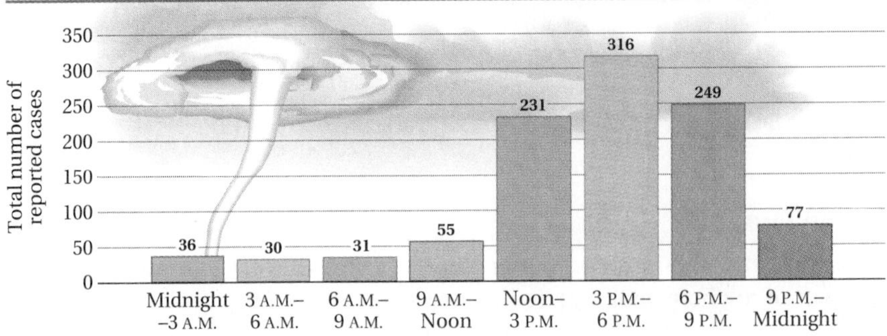

Source: National Weather Service

a) During which interval of time did the greatest number of tornado touchdowns occur?

b) During which intervals was the number of tornado touchdowns greater than 200?

We solve as follows.

a) In this bar graph, the values are written at the top of the bars. We see that 316 is the greatest number. We look at the bottom of that bar on the horizontal scale and see that the time interval of greatest occurrence is 3 P.M.–6 P.M.

b) We locate 200 on the vertical scale and move across the graph or draw a horizontal line. We note that the value on three bars exceeds 200. Then we look down at the horizontal scale and see that the corresponding time intervals are noon–3 P.M., 3 P.M.–6 P.M., and 6 P.M.–9 P.M.

Do Exercise 2.

Line Graphs

Line graphs are often used to show change over time. Certain points are plotted to represent given information. When segments are drawn to connect the points, a line graph is formed.

Sometimes it is impractical to begin the listing of horizontal or vertical values with zero. When this happens, as in Example 3, the symbol ⦚ is used to indicate a break in the list of values.

Example 3 *Exercise and Pulse Rate.* The following line graph shows the relationship between a person's resting pulse rate and months of regular exercise.

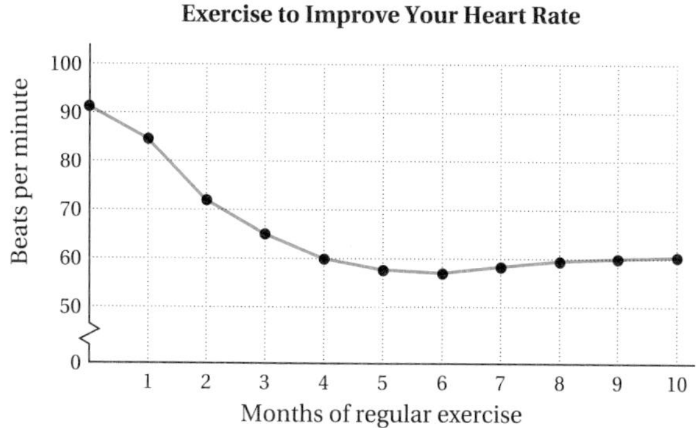

Source: Hughes, Martin, *Body Clock*. New York: Facts on File, Inc., p. 60

a) How many months of regular exercise are required to lower the pulse rate to its lowest point?

b) How many months of regular exercise are needed to achieve a pulse rate of 65 beats per minute?

We solve as follows.

a) The lowest point on the graph occurs above the number 6. Thus after 6 months of regular exercise, the pulse rate has been lowered as much as possible.

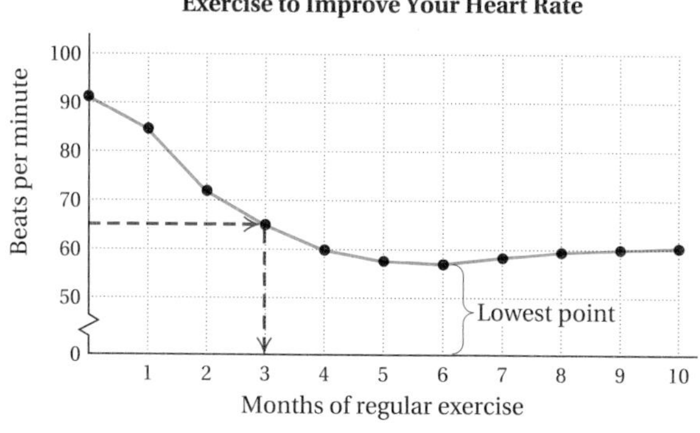

b) We locate 65 on the vertical scale and then move right until we reach the line. At that point, we move down to the horizontal scale and read the information we are seeking. The pulse rate is 65 beats per minute after 3 months of regular exercise.

Do Exercise 3.

3. *Exercise and Pulse Rate.* Referring to Example 3, determine the following.

a) About how many months of regular exercise are needed to achieve a pulse rate of about 72 beats per minute?

b) What pulse rate has been achieved after 10 months of exercise?

Answers on page A-6

Plot these points on the graph below.

4. (4, 5) **5.** (5, 4)

6. (−2, 5) **7.** (−3, −4)

8. (5, −3) **9.** (−2, −1)

10. (0, −3) **11.** (2, 0)

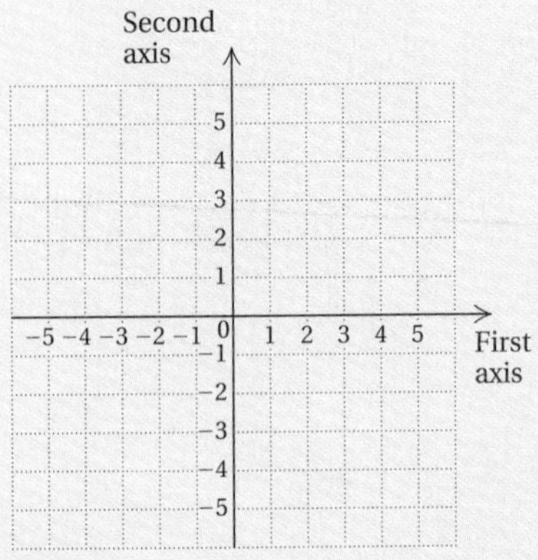

Answers on page A-7

b Plotting Ordered Pairs

The line graph in Example 3 is formed from a collection of points. Each point pairs a number of months of exercise with a pulse rate.

In Chapter 2, we graphed numbers and inequalities in one variable on a line. To enable us to graph an equation that contains two variables, we now learn to graph number pairs on a plane.

On a number line, each point is the graph of a number. On a plane, each point is the graph of a number pair. We use two perpendicular number lines called **axes.** They cross at a point called the **origin**. The arrows show the positive directions.

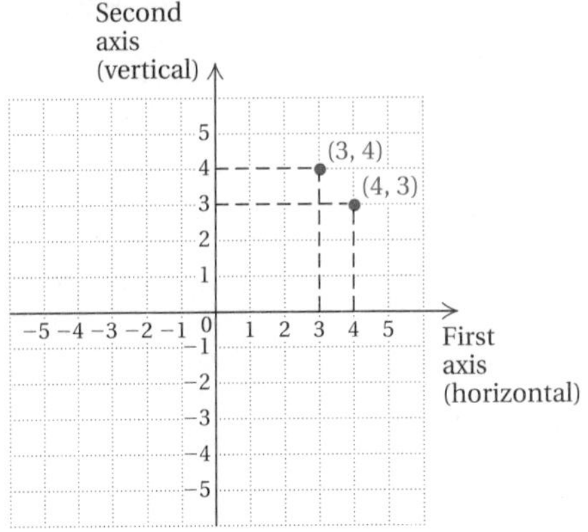

Consider the ordered pair (3, 4). The numbers in an ordered pair are called **coordinates.** In (3, 4), the **first coordinate** is 3 and the **second coordinate** is 4. To plot (3, 4), we start at the origin and move horizontally to the 3. Then we move up vertically 4 units and make a "dot."

The point (4, 3) is also plotted. Note that (3, 4) and (4, 3) give different points. The order of the numbers in the pair is indeed important. They are called **ordered pairs** because it makes a difference which number comes first. The coordinates of the origin are (0, 0) even though it is usually labeled either with the number 0 or not at all.

Example 4 Plot the point (−5, 2).

The first number, −5, is negative. Starting at the origin, we move −5 units in the horizontal direction (5 units to the left). The second number, 2, is positive. We move 2 units in the vertical direction (up).

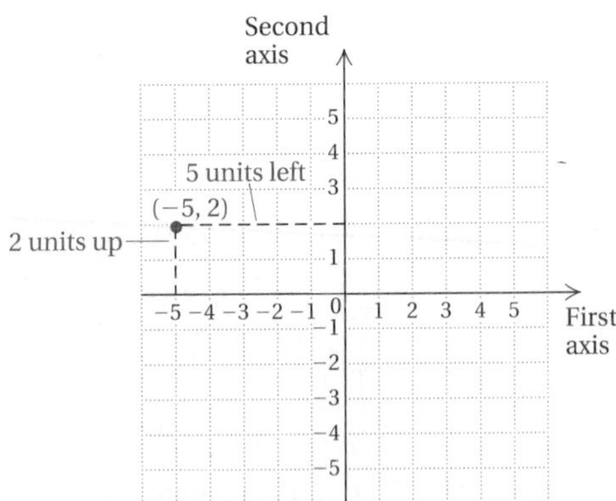

Do Exercises 4–11.

c Quadrants

This figure shows some points and their coordinates. In region I (the *first quadrant*), both coordinates of any point are positive. In region II (the *second quadrant*), the first coordinate is negative and the second positive. In region III (the *third quadrant*), both coordinates are negative. In region IV (the *fourth quadrant*), the first coordinate is positive and the second is negative.

Example 5 In which quadrant, if any, are the points (−4, 5), (5, −5), (2, 4), (−2, −5), and (−5, 0) located?

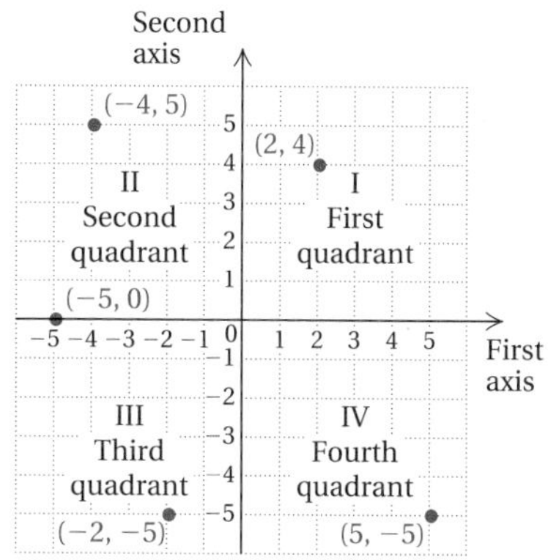

The point (−4, 5) is in the second quadrant. The point (5, −5) is in the fourth quadrant. The point (2, 4) is in the first quadrant. The point (−2, −5) is in the third quadrant. The point (−5, 0) is on an axis and is *not* in any quadrant.

Do Exercises 12–18.

d Finding Coordinates

To find the coordinates of a point, we see how far to the right or left of zero it is located and how far up or down.

Example 6 Find the coordinates of points *A*, *B*, *C*, *D*, *E*, *F*, and *G*.

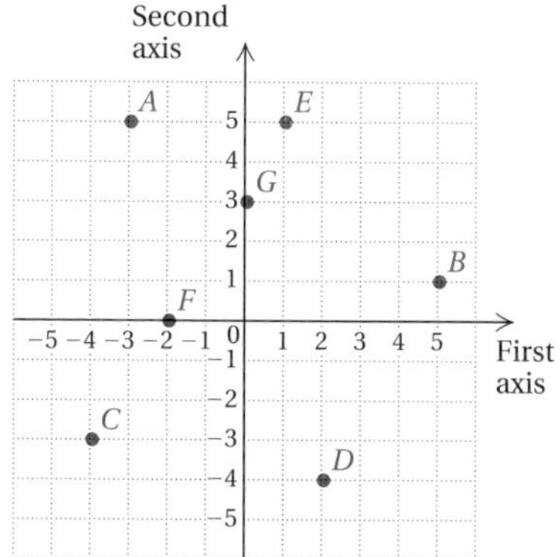

Point *A* is 3 units to the left (horizontal direction) and 5 units up (vertical direction). Its coordinates are (−3, 5). The coordinates of the other points are as follows:

B: (5, 1); *C*: (−4, −3); *D*: (2, −4);

E: (1, 5); *F*: (−2, 0); *G*: (0, 3).

Do Exercise 19.

12. What can you say about the coordinates of a point in the third quadrant?

13. What can you say about the coordinates of a point in the fourth quadrant?

In which quadrant, if any, is the point located?

14. (5, 3)

15. (−6, −4)

16. (10, −14)

17. (−13, 9)

18. (0, −3)

19. Find the coordinates of points *A*, *B*, *C*, *D*, *E*, *F*, and *G* on the graph below.

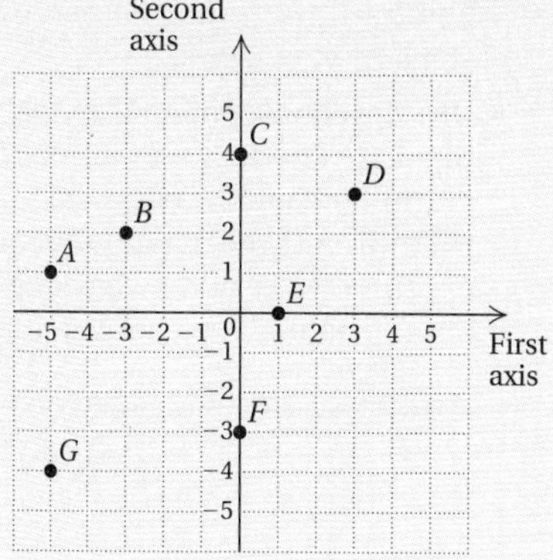

Answers on page A-7

Improving Your Math Study Skills

Homework

Before Doing Your Homework

- **Setting.** Consider doing your homework as soon as possible after class, before you forget what you learned in the lecture. Research has shown that after 24 hours, most people forget about half of what is in their short-term memory. To avoid this "automatic" forgetting, you need to transfer the knowledge into long-term memory.

  Try to set a specific time for your homework. Then choose a location that is quiet and uninterrupted. Some students find it helpful to listen to music when doing homework. Research has shown that classical music creates the best atmosphere for studying: Give it a try!
- **Reading.** Before you begin doing the homework exercises, you should reread the assigned material in the textbook. You may also want to look over your class notes again and rework some of the examples given in class.

  You should not read a math textbook as you would a novel or history textbook. Math texts are not meant to be read passively. Work the examples on your own paper as you read them. Mark what you do not understand. Reread any paragraph as you see the need, and look up any sections referenced.
- **Doing the margin exercises.** Be sure to stop and do the margin exercises when directed. We cannot overemphasize the importance of this.

While Doing Your Homework

- **Write legibly.** Write legibly so that you can easily check over your work. Clearly label each section and each exercise. Your legible writing will also be appreciated should your homework be collected and graded. Tutors and instructors are more helpful if they can see and understand all the steps in your work.
- **Use notebook paper if extra workspace is needed.** The text you are using is a workbook. You might consider tearing out the pages and placing them in a three-ring notebook. Then you can organize your homework and class notes in appropriate places between the text material. You want to be able to go over your homework when studying for a test. Therefore, you need to be able to easily access any problem in your homework notebook.
- **Show all the steps.** Be sure to show all the steps in your work. This avoids the common difficulty of trying to do too much "in your head" and also provides you and your instructor with an effective means of checking your work.
- **Check answers.** When you are finished with your homework, check the answers to the odd-numbered exercises at the back of the book or in the *Student's Solutions Manual* and make corrections. If you do not understand why an answer is wrong, mark it so you can ask questions in class or during the instructor's office hours.
- **Form a study group.** For some students, forming a study group can be helpful. Many times, two heads are better than one. Also, it is true that "to teach is to learn." Thus, when you explain a concept to your classmate, you often gain a better understanding of the concept yourself. If you do study in a group, resist the temptation to waste time by socializing.

  If you work regularly with someone, be careful not to become dependent on that person. Always allow some time to work on your own so that you are able to learn even when your study partner is not available.

After Doing Your Homework

- **Daily review.** If you complete your homework several days before the next class, review your work every day. This will keep the material fresh in your mind. You should also review the work immediately before the next class so that you can ask questions as needed.
- **Extra practice.** The best way to learn math concepts is to perform practice exercises repeatedly. This is the "drill-and-practice" part of learning math that comes when you do your homework. It cannot be overlooked if you want to succeed in your study of math.

Exercise Set 3.1

a Solve.

Driving While Intoxicated (DWI). State laws have determined that a blood alcohol level of at least 0.10% or higher indicates that an individual has consumed too much alcohol to drive safely. The following bar graph shows the number of drinks that a person of a certain weight would need to consume in order to reach a blood alcohol level of 0.10%. A 12-oz beer, a 5-oz glass of wine, or a cocktail containing $1\frac{1}{2}$ oz of distilled liquor all count as one drink. Use the bar graph for Exercises 1–6.

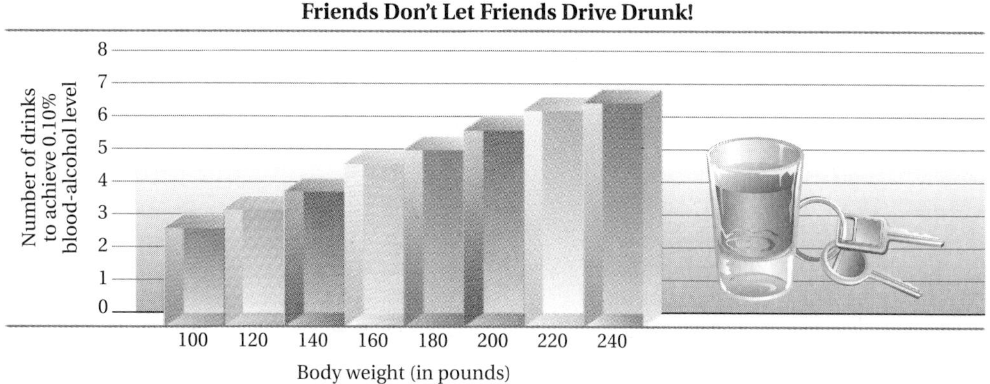

Source: *Neighborhood Digest*, 7, no. 12

1. Approximately how many drinks would a 200-lb person have consumed if he or she had a blood alcohol level of 0.10%?

2. What can be concluded about the weight of someone who can consume 4 drinks without reaching a blood alcohol level of 0.10%?

3. What can be concluded about the weight of someone who can consume 6 drinks without reaching a blood alcohol level of 0.10%?

4. Approximately how many drinks would a 160-lb person have consumed if he or she had a blood alcohol level of 0.10%?

5. What can be concluded about the weight of someone who has consumed $3\frac{1}{2}$ drinks without reaching a blood alcohol level of 0.10%?

6. What can be concluded about the weight of someone who has consumed $4\frac{1}{2}$ drinks without reaching a blood alcohol level of 0.10%?

Cost of Raising a Child. A family is in a \$32,800–\$55,500 income bracket. The following pie chart shows the various costs involved for a family in a \$32,800–\$55,500 income bracket in raising a child to the age of 18. Use the pie chart for Exercises 7–10.

Cost of Raising a Child to the Age of 18

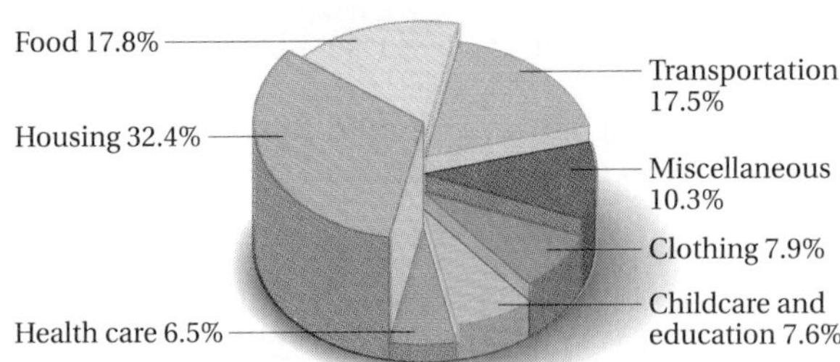

Source: U.S. Department of Agriculture, Food, Nutrition, and Consumer Service

7. What percent of the total expense is for housing?

8. What percent of the total expense is for health care?

9. It costs a total of about \$136,320 to raise a child to the age of 18. How much of this cost is for child care and education?

10. It costs a total of about \$136,320 to raise a child to the age of 18. How much of this cost is for transportation?

MADD (Mothers Against Drunk Driving). Despite efforts by groups such as MADD, the number of alcohol-related deaths is rising after many years of decline. The data in the following graph show the number of deaths from 1989 to 1995. Use this graph for Exercises 11–16.

Number of Alcohol-Related Traffic Deaths

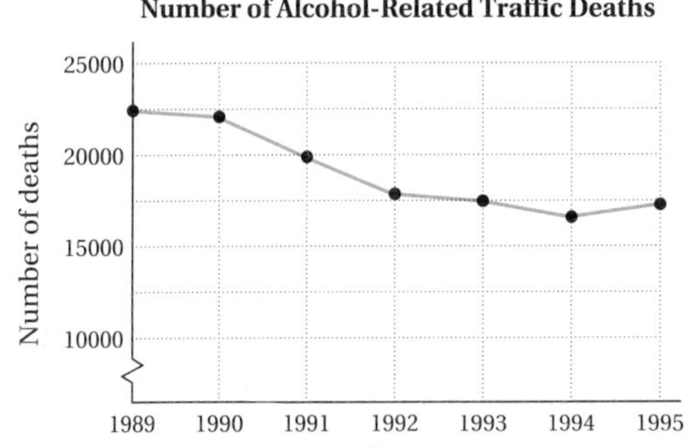

Source: National Highway Traffic Safety Administration

11. About how many alcohol-related deaths occurred in 1991?

12. About how many alcohol-related deaths occurred in 1995?

13. In what year did the lowest number of deaths occur?

14. In what years did fewer than 18,000 deaths occur?

15. By how much did the number of alcohol-related deaths increase from 1994 to 1995?

16. By how much did the number of alcohol-related deaths decrease from 1989 to 1994?

b

17. Plot these points.

(2, 5) (−1, 3) (3, −2) (−2, −4)
(0, 4) (0, −5) (5, 0) (−5, 0)

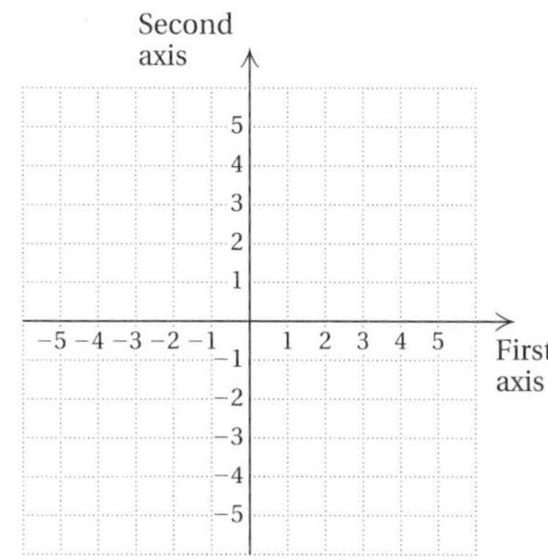

18. Plot these points.

(4, 4) (−2, 4) (5, −3) (−5, −5)
(0, 4) (0, −4) (3, 0) (−4, 0)

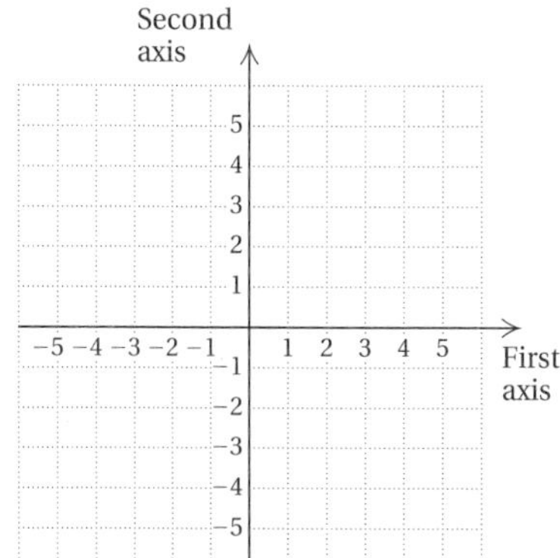

c In which quadrant is the point located?

19. (−5, 3)

20. (1, −12)

21. (100, −1)

22. (−2.5, 35.6)

23. (−6, −29)

24. (3.6, 105.9)

25. (3.8, 9.2)

26. (−895, −492)

27. $\left(-\frac{1}{3}, \frac{15}{7}\right)$

28. $\left(-\frac{2}{3}, -\frac{9}{8}\right)$

29. $\left(12\frac{7}{8}, -1\frac{1}{2}\right)$

30. $\left(23\frac{5}{8}, 81.74\right)$

31. In quadrant III, first coordinates are always __________ and second coordinates are always __________.

32. In quadrant II, __________ coordinates are always positive and __________ coordinates are always negative.

33. In quadrant IV, __________ coordinates are always negative and __________ coordinates are always positive.

34. In quadrant I, first coordinates are always __________ and second coordinates are always __________.

In Exercises 35–38, tell in which quadrant(s) the point can be located.

35. The first coordinate is positive.

36. The second coordinate is negative.

37. The first and second coordinates are equal.

38. The first coordinate is the additive inverse of the second coordinate.

d

39. Find the coordinates of points A, B, C, D, and E.

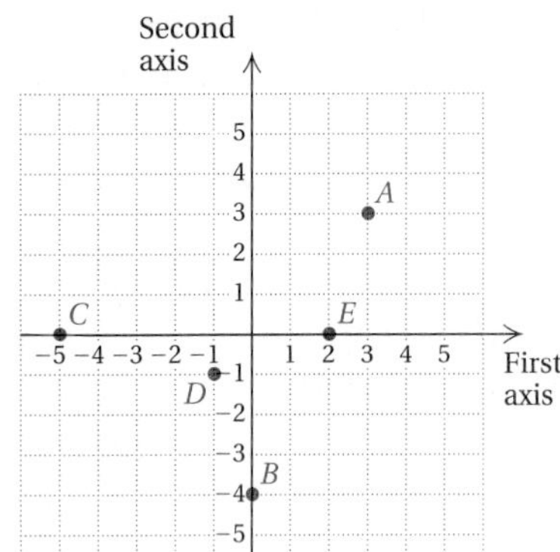

40. Find the coordinates of points A, B, C, D, and E.

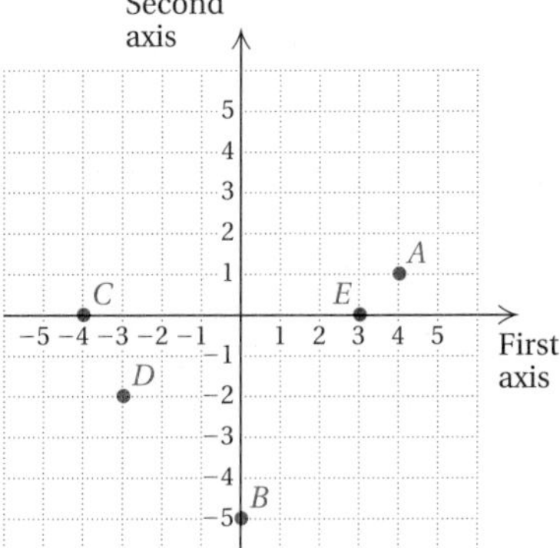

Skill Maintenance

Find the absolute value. [1.2e]

41. $|-12|$

42. $|4.89|$

43. $|0|$

44. $\left|-\frac{4}{5}\right|$

Solve. [2.5a]

45. *Baseball Salaries.* In 1997, the total amount spent on the salaries of major-league baseball players soared to \$1.06 billion. This was a 17.7% increase over the total amount spent in 1996. (***Source:*** Major League Baseball) How much was spent in 1996?

46. Erin left a 15% tip for a meal. The total cost of the meal, including the tip, was \$21.16. What was the cost of the meal before the tip was added?

Synthesis

47. ◆ The sales of snow skis are highest in the winter months and lowest in the summer months. Sketch a line graph that might show the sales of a ski store and explain how an owner might use such a graph in decision making.

48. ◆ The graph in Example 3 tends to flatten out. Explain why the graph does not continue to decrease downward.

49. The points $(-1, 1)$, $(4, 1)$, and $(4, -5)$ are three vertices of a rectangle. Find the coordinates of the fourth vertex.

50. Three parallelograms share the vertices $(-2, -3)$, $(-1, 2)$, and $(4, -3)$. Find the fourth vertex of each parallelogram.

51. Graph eight points such that the sum of the coordinates in each pair is 6.

52. Graph eight points such that the first coordinate minus the second coordinate is 1.

53. Find the perimeter of a rectangle whose vertices have coordinates $(5, 3)$, $(5, -2)$, $(-3, -2)$, and $(-3, 3)$.

54. Find the area of a triangle whose vertices have coordinates $(0, 9)$, $(0, -4)$, and $(5, -4)$.

Collaborative Learning Manual — **Practice finding and plotting ordered pairs by playing a variation of the game Battleship.**

3.2 Graphing Linear Equations

We have seen how circle, bar, and line graphs can be used to represent the data in an application. Now we begin to learn how graphs can be used to represent solutions of equations.

a Solutions of Equations

When an equation contains two variables, the solutions of the equation are *ordered pairs* in which each number in the pair corresponds to a letter in the equation. Unless stated otherwise, to determine whether a pair is a solution, we use the first number in each pair to replace the variable that occurs first alphabetically.

Example 1 Determine whether each of the following pairs is a solution of $4q - 3p = 22$: (2, 7) and (−1, 6).

For (2, 7), we substitute 2 for p and 7 for q (using alphabetical order of variables):

$$\begin{array}{r|l} 4q - 3p & 22 \\ \hline 4 \cdot 7 - 3 \cdot 2 \; ? & 22 \\ 28 - 6 & \\ 22 & \end{array} \qquad \text{TRUE}$$

Thus, (2, 7) is a solution of the equation.

For (−1, 6), we substitute −1 for p and 6 for q:

$$\begin{array}{r|l} 4q - 3p & 22 \\ \hline 4 \cdot 6 - 3 \cdot -1 \; ? & 22 \\ 24 + 3 & \\ 27 & \end{array} \qquad \text{FALSE}$$

Thus, (−1, 6) is *not* a solution of the equation.

Do Exercises 1 and 2.

Example 2 Show that the pairs (3, 7), (0, 1), and (−3, −5) are solutions of $y = 2x + 1$. Then graph the three points and use the graph to determine another pair that is a solution.

To show that a pair is a solution, we substitute, replacing x with the first coordinate and y with the second coordinate of each pair:

$$\begin{array}{l|l} y & 2x + 1 \\ \hline 7 \; ? & 2 \cdot 3 + 1 \\ & 6 + 1 \\ & 7 \end{array} \quad \text{TRUE} \qquad \begin{array}{l|l} y & 2x + 1 \\ \hline 1 \; ? & 2 \cdot 0 + 1 \\ & 0 + 1 \\ & 1 \end{array} \quad \text{TRUE}$$

$$\begin{array}{l|l} y & 2x + 1 \\ \hline -5 \; ? & 2(-3) + 1 \\ & -6 + 1 \\ & -5 \end{array} \quad \text{TRUE}$$

In each of the three cases, the substitution results in a true equation. Thus the pairs are all solutions.

Objectives

a Determine whether an ordered pair is a solution of an equation with two variables.

b Graph linear equations of the type $y = mx + b$ and $Ax + By = C$, identifying the y-intercept.

c Solve applied problems involving graphs of linear equations.

For Extra Help

TAPE 5 MAC WIN CD-ROM

1. Determine whether (2, −4) is a solution of $4q - 3p = 22$.

2. Determine whether (2, −4) is a solution of $7a + 5b = -6$.

Answers on page A-7

3. Use the graph in Example 2 to find at least two more points that are solutions of $y = 2x + 1$.

Answer on page A-7

We plot the points as shown at right. The order of the points follows the alphabetical order of the variables. That is, x comes before y, so x-values are first coordinates and y-values are second coordinates. Similarly, we also label the horizontal axis as the x-axis and the vertical axis as the y-axis.

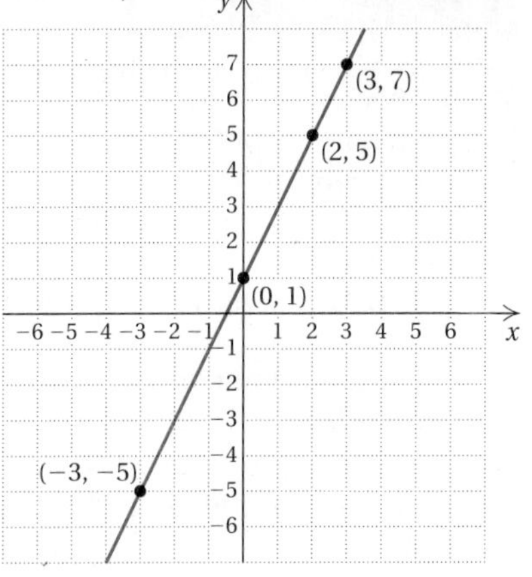

Note that the three points appear to "line up." That is, they appear to be on a straight line. Will other points that line up with these points also represent solutions of $y = 2x + 1$? To find out, we use a straightedge and lightly sketch a line passing through (3, 7), (0, 1), and (−3, −5).

The line appears to also pass through (2, 5). Let's see if this pair is a solution of $y = 2x + 1$:

$$\begin{array}{c|l} y & = 2x + 1 \\ \hline 5 & ?\ 2 \cdot 2 + 1 \\ & 4 + 1 \\ & 5 \qquad\qquad \text{TRUE} \end{array}$$

Thus, (2, 5) is a solution.

Do Exercise 3.

Example 2 leads us to suspect that any point on the line that passes through (3, 7), (0, 1), and (−3, −5) represents a solution of $y = 2x + 1$. In fact, every solution of $y = 2x + 1$ is represented by a point on that line and every point on that line represents a solution. The line is said to be the *graph* of the equation.

The **graph** of an equation is a drawing that represents all its solutions.

b Graphs of Linear Equations

Equations like $y = 2x + 1$ and $4p - 3p = 22$ are said to be **linear** because the graph of each equation is a straight line. In general, any equation equivalent to one of the form $y = mx + b$ or $Ax + By = C$, where m, b, A, B, and C are constants (not variables) and A and B are not both 0, is linear.

To graph a linear equation:

1. Select a value for one variable and calculate the corresponding value of the other variable. Form an ordered pair using alphabetical order as indicated by the variables.
2. Repeat step (1) to obtain at least two other ordered pairs. Two points are essential to determine a straight line. A third point serves as a check.
3. Plot the ordered pairs and draw a straight line passing through the points.

In general, calculating three (or more) ordered pairs is not difficult for equations of the form $y = mx + b$. We simply substitute values for x and calculate the corresponding values for y.

Example 3 Graph: $y = 2x$.

First, we find some ordered pairs that are solutions. We choose *any* number for x and then determine y by substitution. Since $y = 2x$, we find y by doubling x. Suppose that we choose 3 for x. Then

$$y = 2x = 2 \cdot 3 = 6.$$

We get a solution: the ordered pair (3, 6).

Suppose that we choose 0 for x. Then

$$y = 2x = 2 \cdot 0 = 0.$$

We get another solution: the ordered pair (0, 0).

For a third point, we make a negative choice for x. We now have enough points to plot the line, but if we wish, we can compute more. If a number takes us off the graph paper, we either do not use it or we use larger paper or rescale the axes. Continuing in this manner, we create a table like the one shown below.

Now we plot these points. We draw the line, or graph, with a straight-edge and label it $y = 2x$.

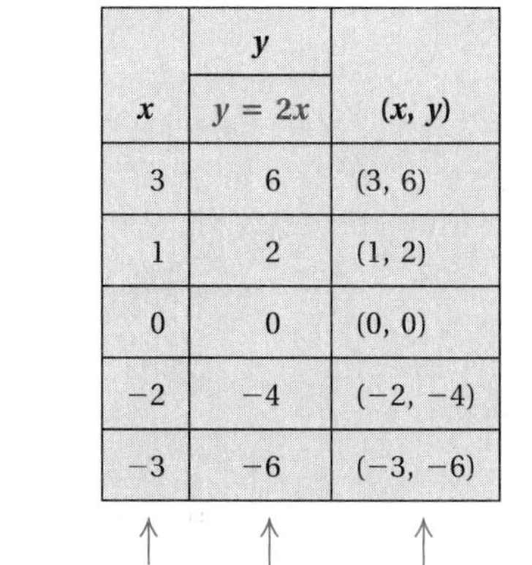

x	y $y = 2x$	(x, y)
3	6	(3, 6)
1	2	(1, 2)
0	0	(0, 0)
−2	−4	(−2, −4)
−3	−6	(−3, −6)

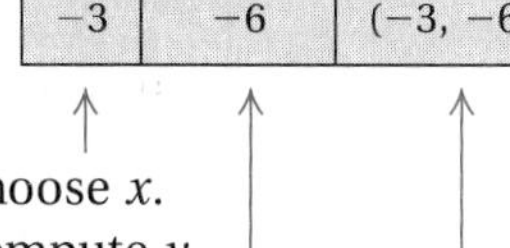

(1) Choose x.
(2) Compute y.
(3) Form the pair (x, y).
(4) Plot the points.

Do Exercises 4 and 5.

Graph.

4. $y = -2x$

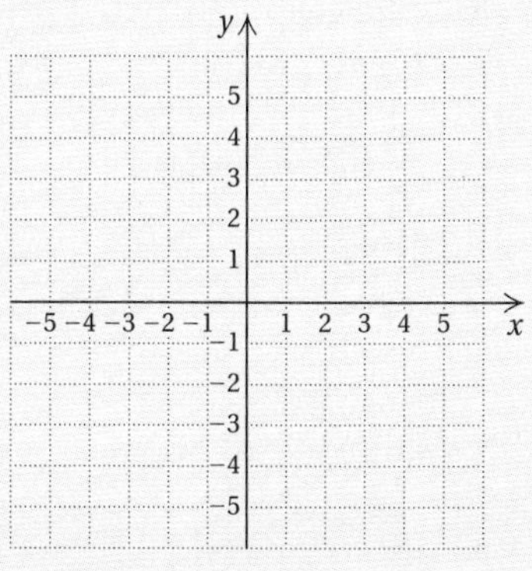

5. $y = \frac{1}{2}x$

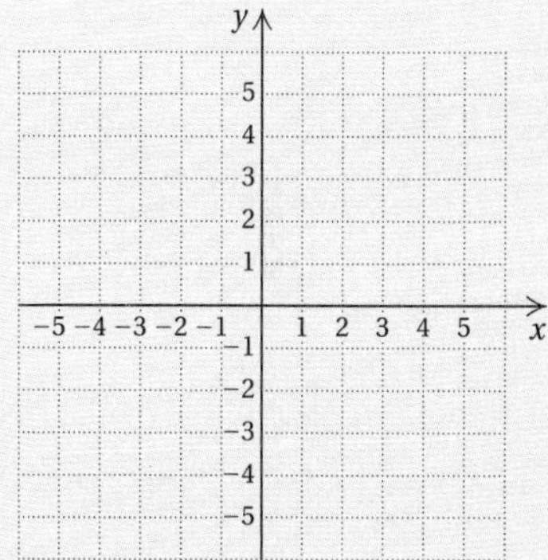

Answers on page A-7

Graph.

6. $y = 2x + 3$

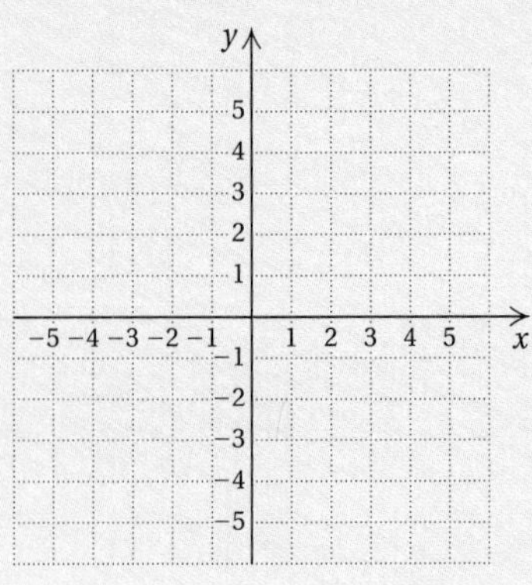

7. $y = -\frac{1}{2}x - 3$

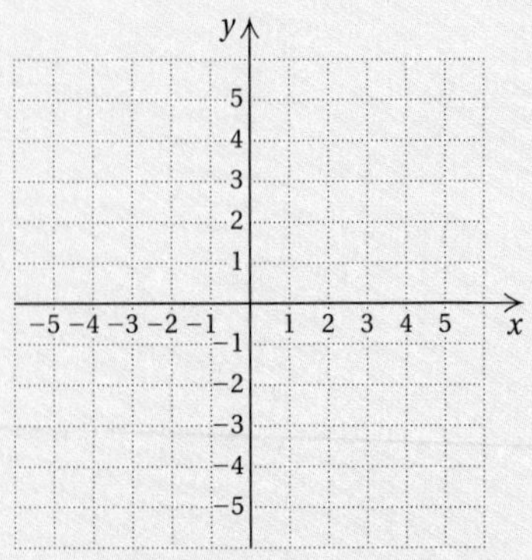

Answers on page A-7

Example 4 Graph: $y = -3x + 1$.

We select a value for x, compute y, and form an ordered pair. Then we repeat the process for other choices of x.

If $x = 2$, then $y = -3 \cdot 2 + 1 = -5$, and $(2, -5)$ is a solution.
If $x = 0$, then $y = -3 \cdot 0 + 1 = 1$, and $(0, 1)$ is a solution.
If $x = -1$, then $y = -3 \cdot (-1) + 1 = 4$, and $(-1, 4)$ is a solution.

Results are often listed in a table, as shown below. The points corresponding to each pair are then plotted.

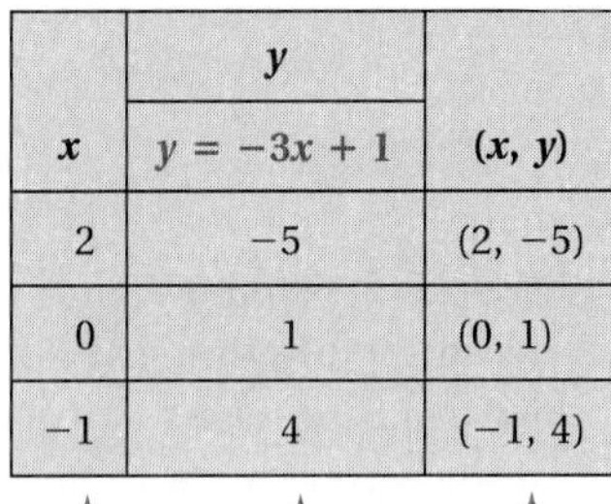

x	y $y = -3x + 1$	(x, y)
2	−5	(2, −5)
0	1	(0, 1)
−1	4	(−1, 4)

(1) Choose x.
(2) Compute y.
(3) Form the pair (x, y).
(4) Plot the points.

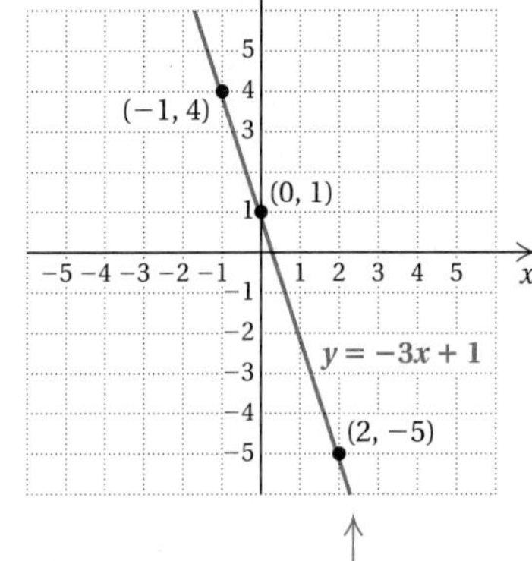

Note that all three points line up. If they did not, we would know that we had made a mistake. When only two points are plotted, a mistake is harder to detect. We use a ruler or other straightedge to draw a line through the points. Every point on the line represents a solution of $y = -3x + 1$.

Do Exercises 6 and 7.

In Example 3, we saw that $(0, 0)$ is a solution of $y = 2x$. It is also the point at which the graph crosses the y-axis. Similarly, in Example 4, we saw that $(0, 1)$ is a solution of $y = -3x + 1$. It is also the point at which the graph crosses the y-axis. A generalization can be made: If x is replaced with 0 in the equation $y = mx + b$, then the corresponding y-value is $m \cdot 0 + b$, or b. Thus any equation of the form $y = mx + b$ has a graph that passes through the point $(0, b)$. Since $(0, b)$ is the point at which the graph crosses the y-axis, it is called the ***y*-intercept**. Sometimes, for convenience, we simply refer to b as the y-intercept.

The graph of the equation $y = mx + b$ passes through the ***y*-intercept** $(0, b)$.

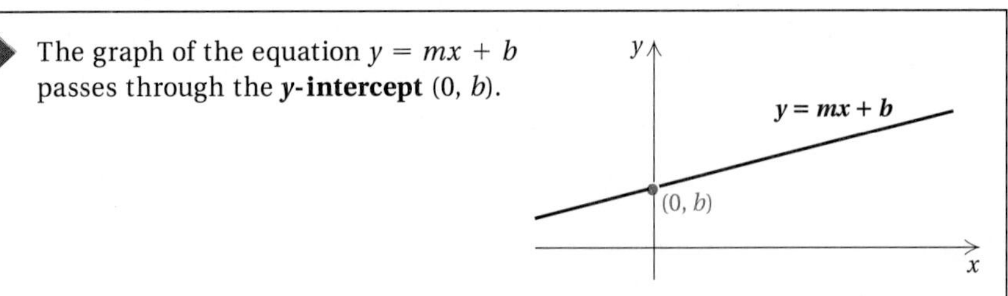

Example 5 Graph $y = \frac{2}{5}x + 4$ and identify the y-intercept.

We select a value for x, compute y, and form an ordered pair. Then we repeat the process for other choices of x. In this case, using multiples of 5 avoids fractions.

If $x = 0$, then $y = \frac{2}{5} \cdot 0 + 4 = 4$, and $(0, 4)$ is a solution.

If $x = 5$, then $y = \frac{2}{5} \cdot 5 + 4 = 6$, and $(5, 6)$ is a solution.

If $x = -5$, then $y = \frac{2}{5} \cdot (-5) + 4 = 2$, and $(-5, 2)$ is a solution.

The following table lists these solutions. Next, we plot the points and see that they form a line. Finally, we draw and label the line.

x	y $y = \frac{2}{5}x + 4$	(x, y)
0	4	(0, 4)
5	6	(5, 6)
−5	2	(−5, 2)

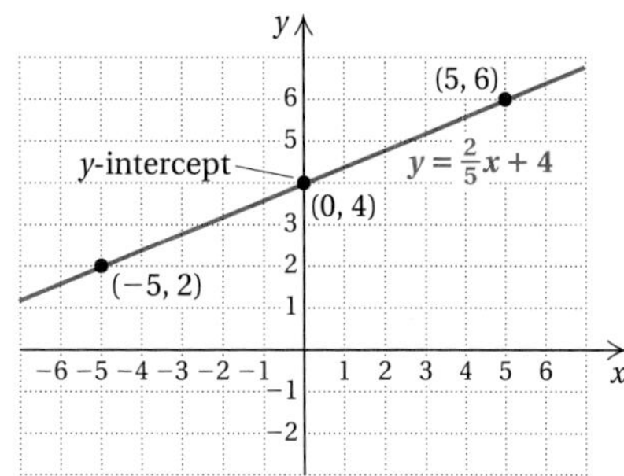

We see that (0, 4) is a solution of $y = \frac{2}{5}x + 4$. It is the y-intercept. Because the equation is in the form $y = mx + b$, we can read the y-intercept directly from the equation as follows:

$$y = \frac{2}{5}x + 4 \qquad (0, 4) \text{ is the } y\text{-intercept.}$$

Do Exercises 8 and 9.

Calculating ordered pairs is generally easiest when y is isolated on one side of the equation, as in $y = mx + b$. To graph an equation in which y is not isolated, we can use the addition and multiplication principles to solve for y (see Sections 2.3 and 2.6).

Example 6 Graph $3y + 5x = 0$ and identify the y-intercept.

To find an equivalent equation in the form $y = mx + b$, we solve for y:

$$3y + 5x = 0$$

$$3y + 5x - 5x = 0 - 5x \qquad \text{Subtracting } 5x$$

$$3y = -5x \qquad \text{Collecting like terms}$$

$$\frac{3y}{3} = \frac{-5x}{3} \qquad \text{Dividing by 3}$$

$$y = -\frac{5}{3}x.$$

Graph the equation and identify the y-intercept.

8. $y = \frac{3}{5}x + 2$

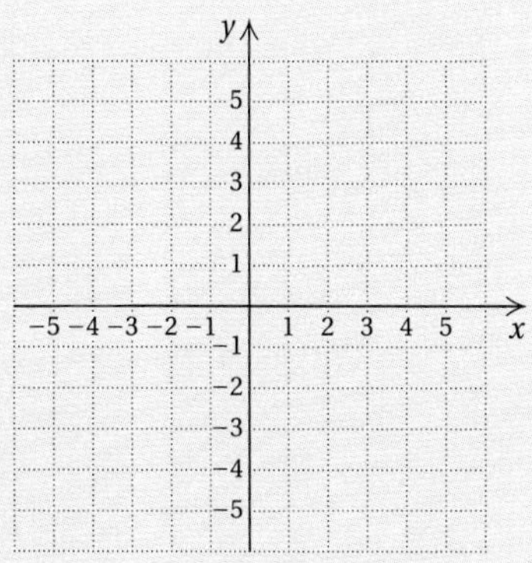

9. $y = -\frac{3}{5}x - 1$

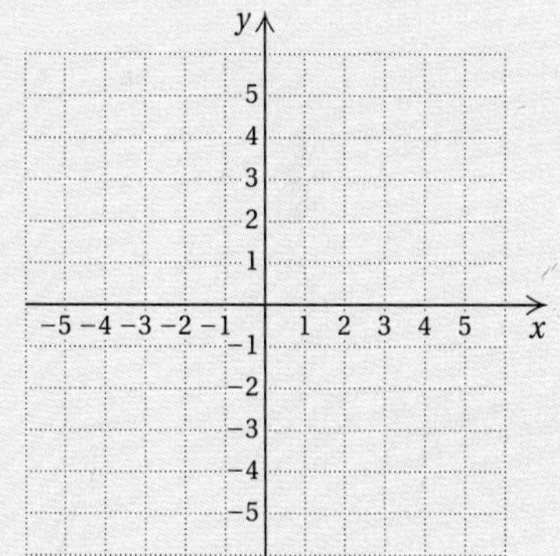

Answers on page A-7

Graph the equation and identify the y-intercept.

10. $5y + 4x = 0$

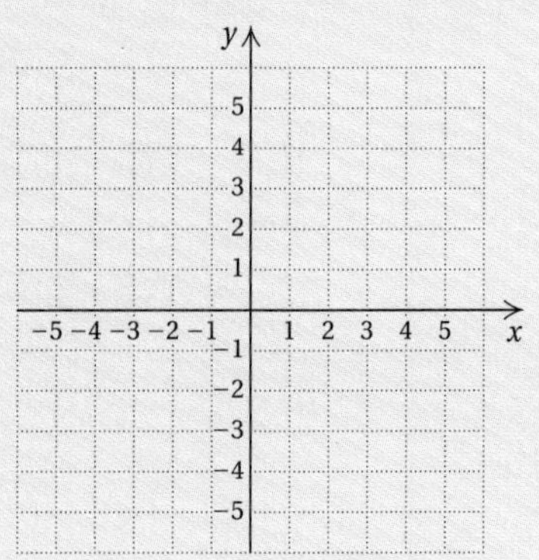

Because all the equations above are equivalent, we can use $y = -\frac{5}{3}x$ to draw the graph of $3y + 5x = 0$. To graph $y = -\frac{5}{3}x$, we select x-values and compute y-values. In this case, if we select multiples of 3, we can avoid fractions.

$$\text{If } x = 0, \quad \text{then } y = -\frac{5}{3} \cdot 0 = 0.$$

$$\text{If } x = 3, \quad \text{then } y = -\frac{5}{3} \cdot 3 = -5.$$

$$\text{If } x = -3, \quad \text{then } y = -\frac{5}{3} \cdot (-3) = 5.$$

We list these solutions in a table. Next, we plot the points and see that they form a line. Finally, we draw and label the line. The y-intercept is (0, 0).

x	y
0	0
3	−5
−3	5

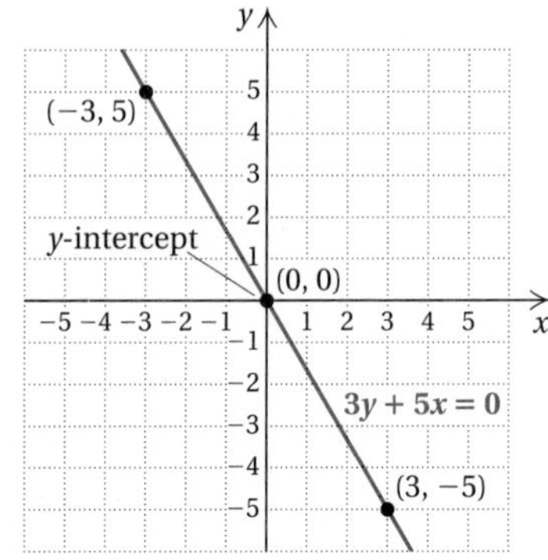

Do Exercises 10 and 11.

11. $4y = 3x$

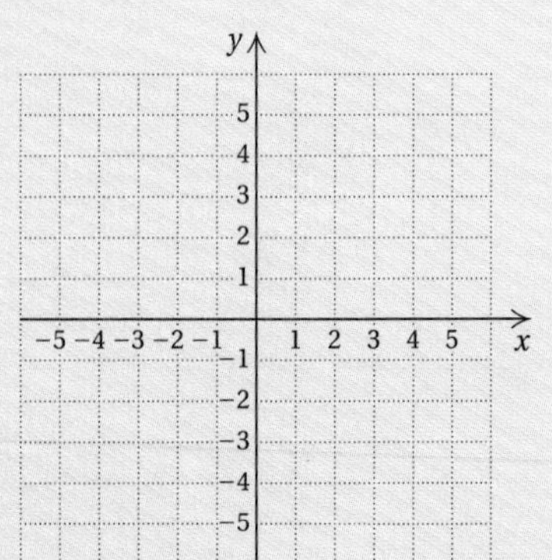

Example 7 Graph $4y + 3x = -8$ and identify the y-intercept.

To find an equivalent equation in the form $y = mx + b$, we solve for y:

$$4y + 3x = -8$$

$$4y + 3x - 3x = -8 - 3x \qquad \text{Subtracting } 3x$$

$$4y = -3x - 8 \qquad \text{Simplifying}$$

$$\frac{1}{4} \cdot 4y = \frac{1}{4} \cdot (-3x - 8) \qquad \text{Multiplying by } \tfrac{1}{4} \text{ or dividing by 4}$$

$$y = \frac{1}{4} \cdot (-3x) - \frac{1}{4} \cdot 8 \qquad \text{Using the distributive law}$$

$$= -\frac{3}{4}x - 2. \qquad \text{Simplifying}$$

Answers on page A-7

Thus, $4y + 3x = -8$ is equivalent to $y = -\frac{3}{4}x - 2$. The y-intercept is $(0, -2)$. We find two other pairs using multiples of 4 for x to avoid fractions. We then complete and label the graph as shown.

x	y
0	−2
4	−5
−4	1

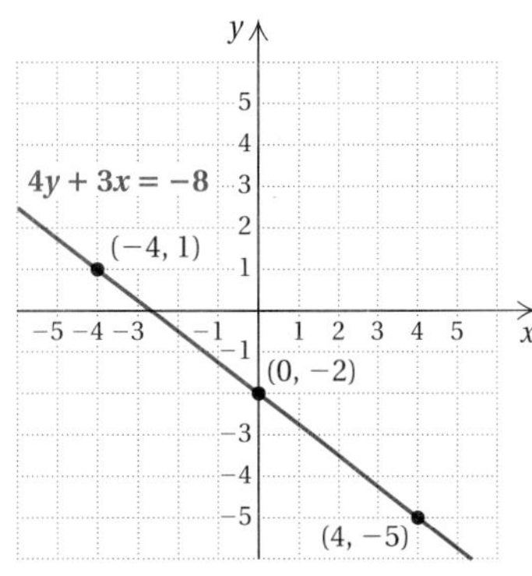

Do Exercises 12 and 13.

Graph the equation and identify the y-intercept.

12. $5y - 3x = -10$

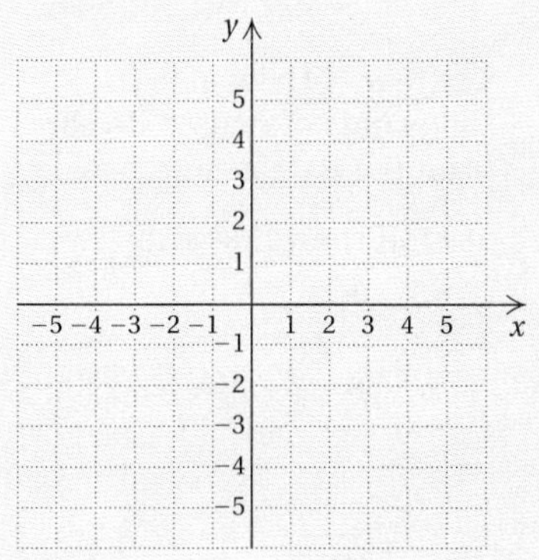

13. $5y + 3x = 20$

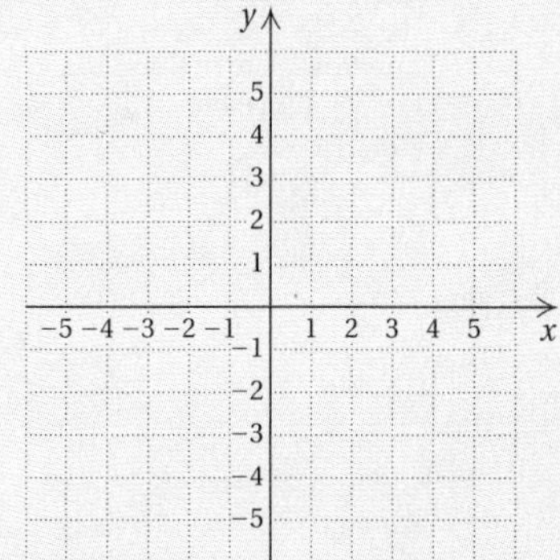

c Applications of Linear Equations

Mathematical concepts become more understandable through visualization. Throughout this text, you will occasionally see the heading **AG** Algebraic–Graphical Connection, as in Example 8, which follows. In this feature, the algebraic approach is enhanced and expanded with a graphical connection. Relating a solution of an equation to a graph can often give added meaning to the algebraic solution.

Example 8 *FedEx Mailing Costs.* The cost y, in dollars, of mailing a FedEx Priority Overnight package weighing 1 lb or more is given by the equation

$$y = 2.085x + 15.08,$$

where x = the number of pounds (***Source*:** Federal Express Corporation).

a) Find the cost of mailing packages weighing 2 lb, 5 lb, and 7 lb.

b) Graph the equation and then use the graph to estimate the cost of mailing a $6\frac{1}{2}$-lb package.

c) If a package costs $177.71 to mail, how much does it weigh?

Answers on page A-7

14. *Value of a Color Copier.* The value of Dupliographic's color copier is given by

$$v = -0.68t + 3.4,$$

where v = the value, in thousands of dollars, t years from the date of purchase.

a) Find the value after 1 yr, 2 yr, 4 yr, and 5 yr.

b) Graph the equation and use the graph to estimate the value of the copier after $2\frac{1}{2}$ yr.

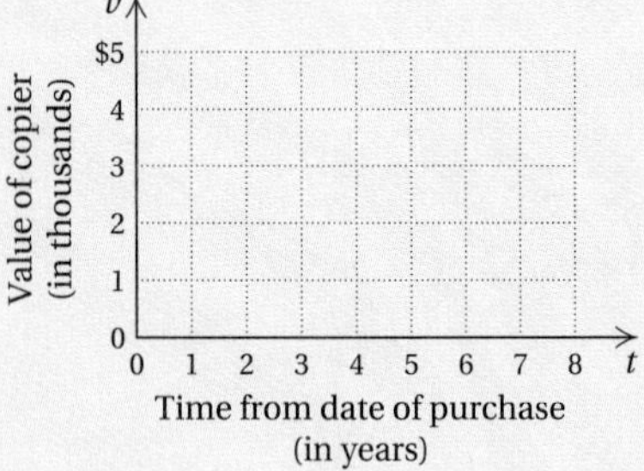

c) After what amount of time is the value of the copier $1500?

Answers on page A-7

We solve as follows.

a) We substitute 2, 5, and 7 for x and then calculate y:

If $x = 2$, then $y = 2.085(2) + 15.08 = \$19.25$.
If $x = 5$, then $y = 2.085(5) + 15.08 \approx \$25.51$.
If $x = 7$, then $y = 2.085(7) + 15.08 \approx \$29.68$.

AG Algebraic–Graphical Connection

b) We have three ordered pairs from (a). We plot these points and see that they line up. Thus our calculations are probably correct. Since zero and negative x-values have no meaning in this problem, we use an open circle at (0, 15.08) when drawing the graph.

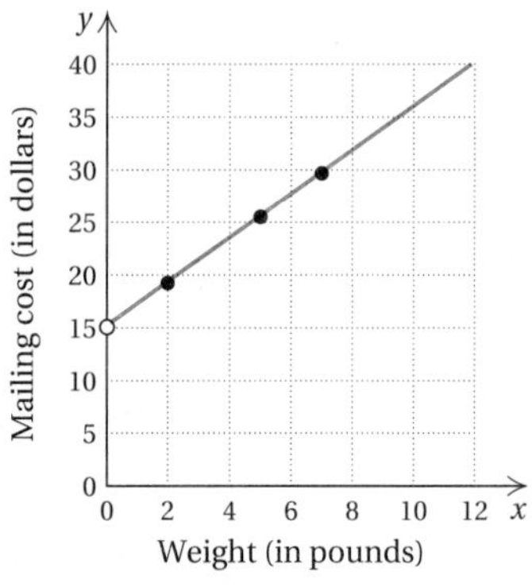

To estimate the cost of mailing a $6\frac{1}{2}$-lb package, we need to determine what y-value is paired with $x = 6\frac{1}{2}$. We locate the point on the line that is above $6\frac{1}{2}$ and then find the value on the y-axis that corresponds to that point. It appears that the cost of mailing a $6\frac{1}{2}$-lb package is about $29.

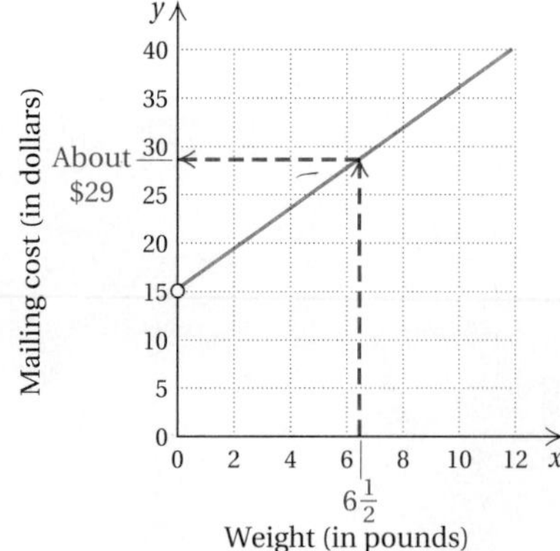

To obtain a more accurate cost, we can simply substitute into the equation:

$$y = 2.085(6.5) + 15.08 \approx \$28.63.$$

c) We substitute \$177.71 for y and then solve for x:

$$
\begin{aligned}
y &= 2.085x + 15.08 \\
177.71 &= 2.085x + 15.08 && \text{Substituting} \\
162.63 &= 2.085x && \text{Subtracting 15.08} \\
78 &= x. && \text{Dividing by 2.085}
\end{aligned}
$$

Do Exercise 14 on the preceding page.

Many equations in two variables have graphs that are not straight lines. Three such graphs are shown below. As before, each graph represents the solutions of the given equation. We are not going to develop methods of doing such graphing at this time, although such *nonlinear graphs* can be created very easily using a graphing calculator. We will cover such graphs in the optional Calculator Spotlights.

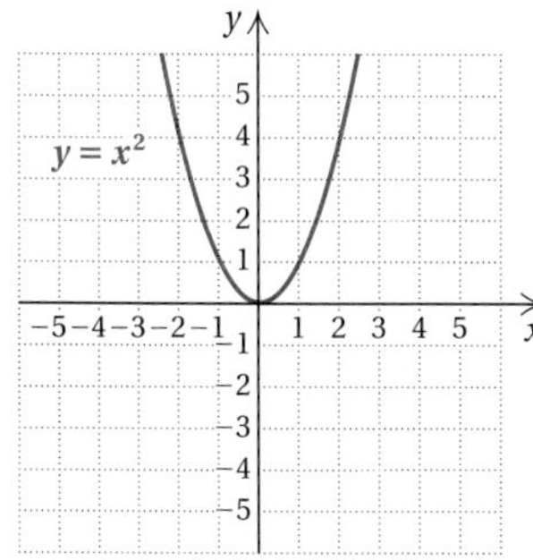

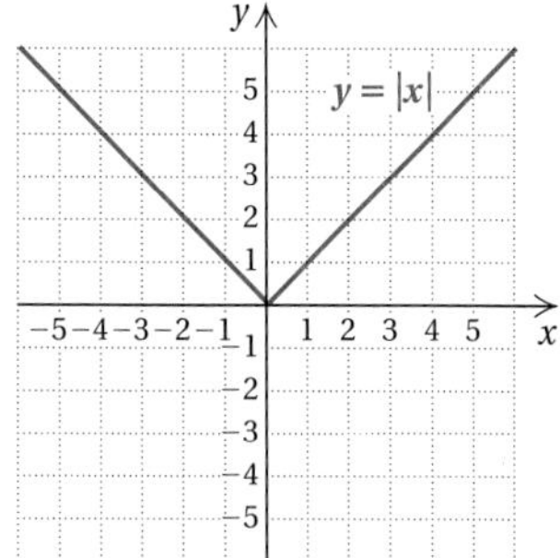

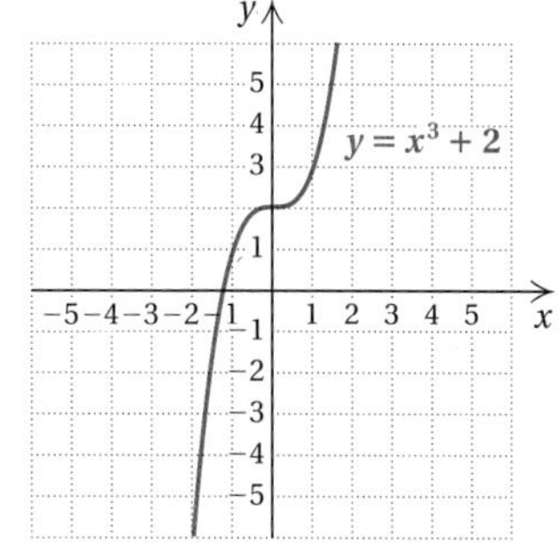

Calculator Spotlight

Introduction to the Use of a Graphing Calculator: Windows and Graphs

Viewing Windows. One feature common to all graphers is the **viewing window.** Windows are described by four numbers, [**L**, **R**, **B**, **T**], which represent the **L**eft and **R**ight endpoints of the x-axis and the **B**ottom and **T**op endpoints of the y-axis. A WINDOW feature is used to set these dimensions. Below is a window setting of $[-20, 20, -5, 5]$ with axis scaling denoted as Xscl = 5 and Yscl = 1. The notation Xres = 1 indicates the number of pixels (black rectangular dots). We will usually leave it at 1 and not refer to it unless needed. On some graphers, a setting of $[-10, 10, -10, 10]$, Xscl = 1, Yscl = 1 is considered **standard**.

WINDOW
Xmin = −20
Xmax = 20
Xscl = 5
Ymin = −5
Ymax = 5
Yscl = 1
Xres = 1

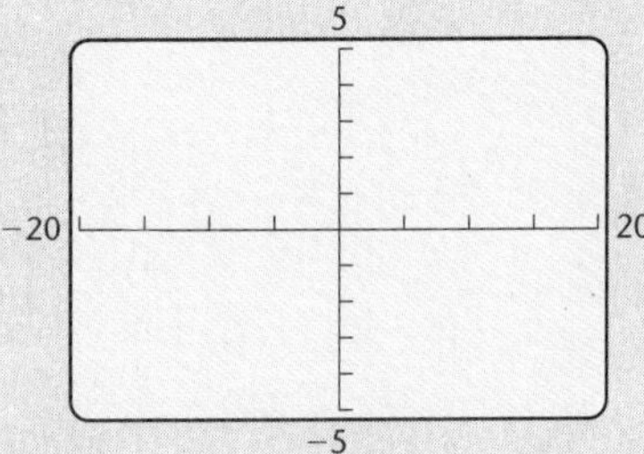

The primary use for a grapher is to graph equations. For example, let's graph the equation $y = x^2 - 3x - 5$. The equation can be entered using the [y=] key.

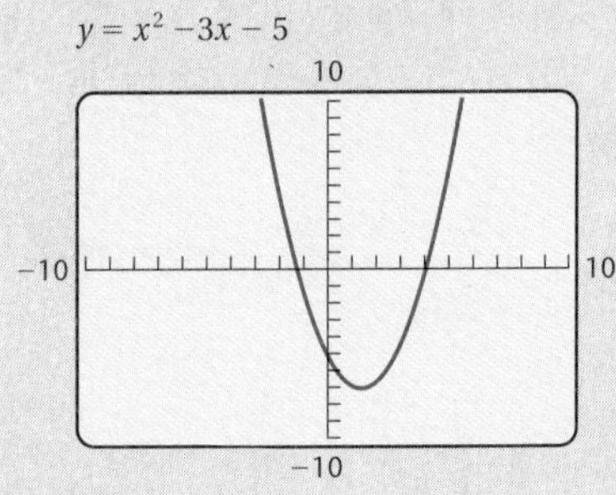

To graph an equation like $4y + 3x = -8$, most graphers require that the equation be solved for y, that is, "$y = \ldots$." Thus we must rewrite the equation $4y + 3x = -8$ as

$$y = \frac{-3x - 8}{4}, \quad \text{or} \quad y = -\frac{3}{4}x - 2,$$

as we did in Example 7. We then enter this equation as $y = -(3/4)x - 2$.

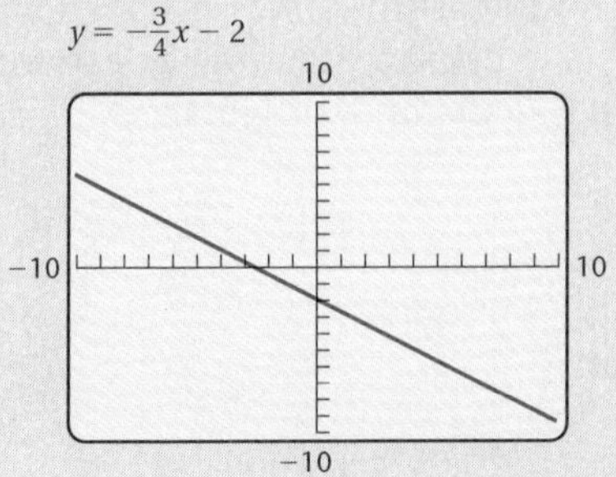

Exercises

Use a grapher to graph each of the following equations. Select the standard window $[-10, 10, -10, 10]$ and axis scaling Xscl = 1, Yscl = 1.

1. $y = 2x + 1$
2. $y = -3x + 1$
3. $y = \frac{2}{5}x + 4$
4. $y = -\frac{3}{5}x - 1$
5. $y = 2.085x + 15.08$
6. $y = -\frac{4}{5}x + \frac{13}{7}$
7. $2x + 3y = 18$
8. $5y + 3x = 4$
9. $y = x^2$
10. $y = 0.5x^2$
11. $y = 8 - x^2$
12. $y = 4 - 3x - x^2$
13. $y = 5x^2 - 3x - 10$
14. $y = x^3 + 2$
15. $y = |x|$ (On most graphers, this is entered as $y = \text{abs}(x)$.)
16. $y = |x - 5|$
17. $y = |x| - 5$
18. $y = 8 - |x|$

Exercise Set 3.2

a Determine whether the given point is a solution of the equation.

1. (2, 9); $y = 3x - 1$

2. (1, 7); $y = 2x + 5$

3. (4, 2); $2x + 3y = 12$

4. (0, 5); $5x - 3y = 15$

5. (3, −1); $3a - 4b = 13$

6. (−5, 1); $2p - 3q = -13$

In Exercises 7–12, an equation and two ordered pairs are given. Show that each pair is a solution. Then use the graph of the two points to determine another solution. Answers may vary.

7. $y = x - 5$; (4, −1) and (1, −4)

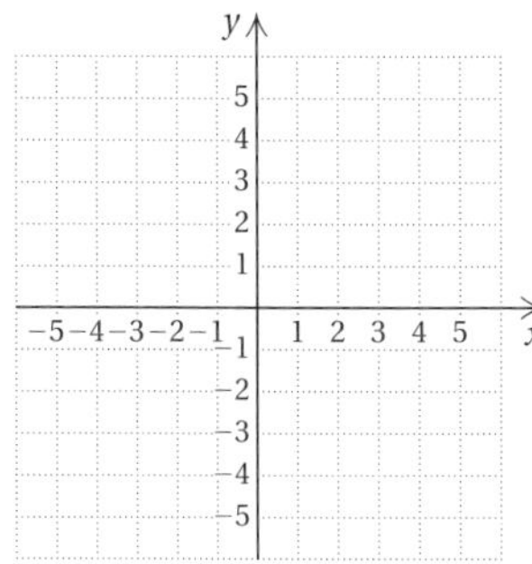

8. $y = x + 3$; (−1, 2) and (3, 6)

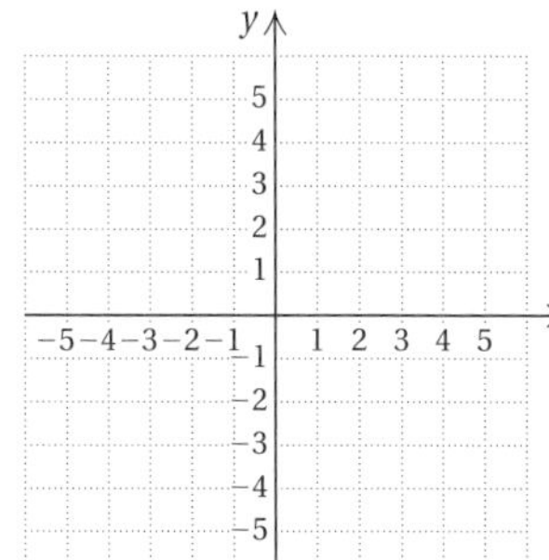

9. $y = \frac{1}{2}x + 3$; (4, 5) and (−2, 2)

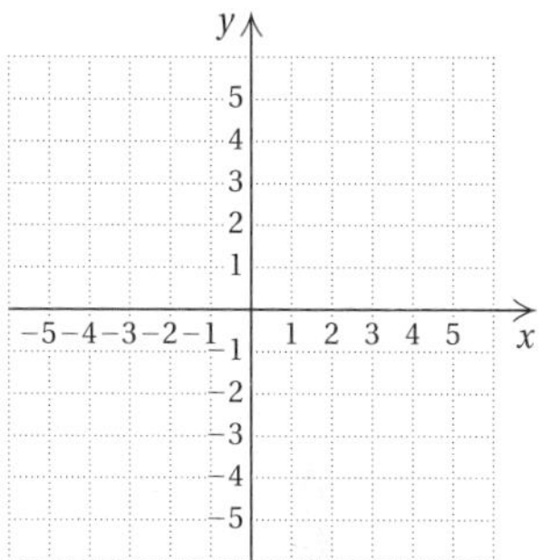

10. $3x + y = 7$; (2, 1) and (4, −5)

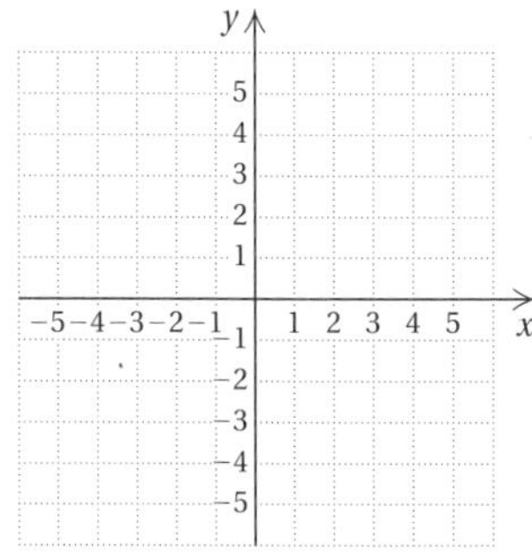

11. $4x - 2y = 10$; (0, −5) and (4, 3)

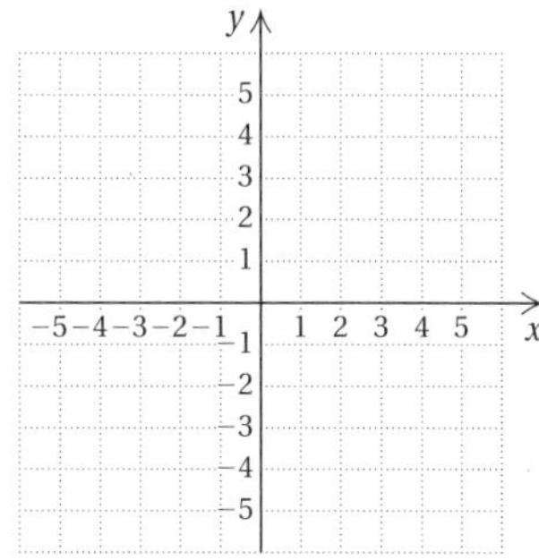

12. $6x - 3y = 3$; (1, 1) and (−1, −3)

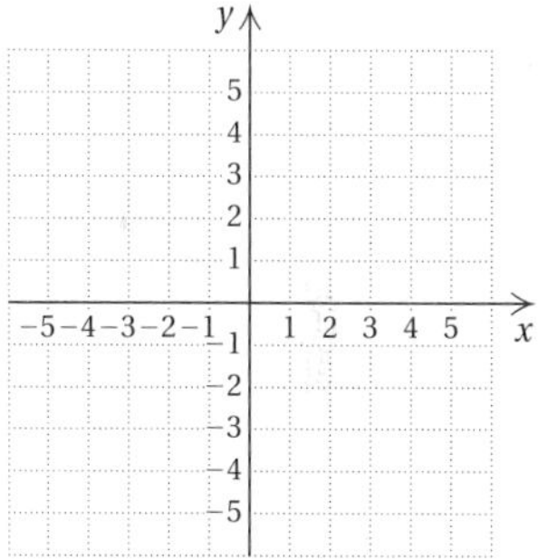

b Graph the equation and identify the y-intercept.

13. $y = x + 1$

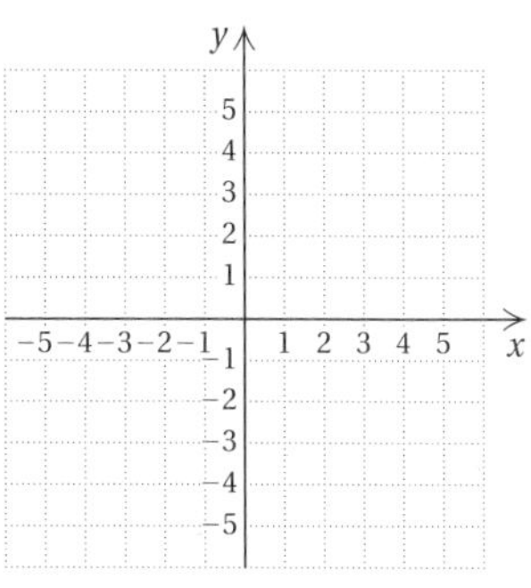

14. $y = x - 1$

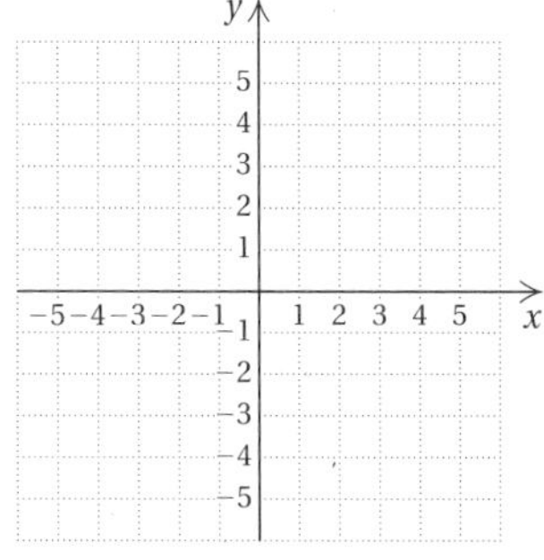

15. $y = x$

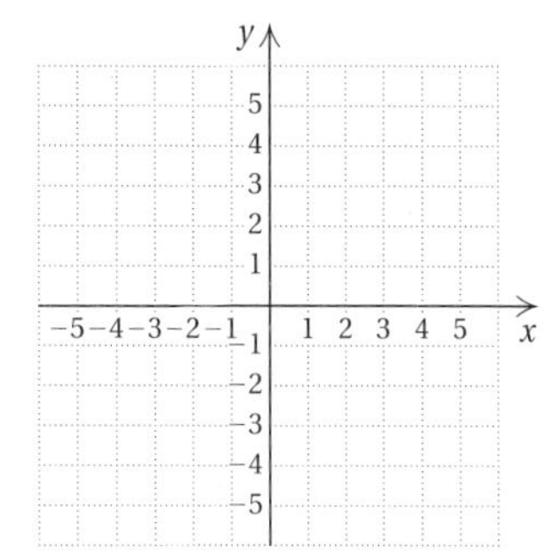

16. $y = -x$

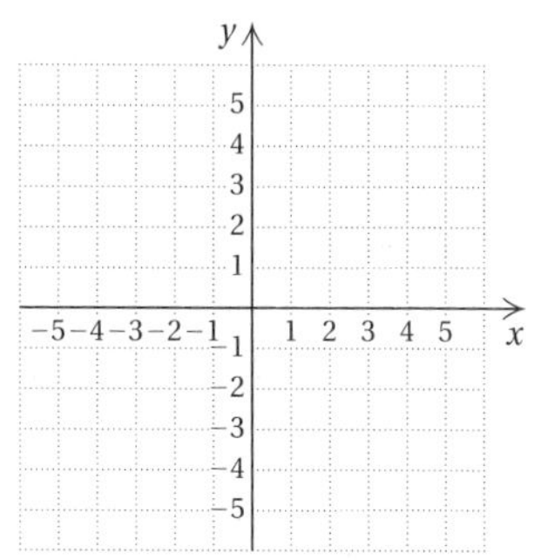

17. $y = \frac{1}{2}x$

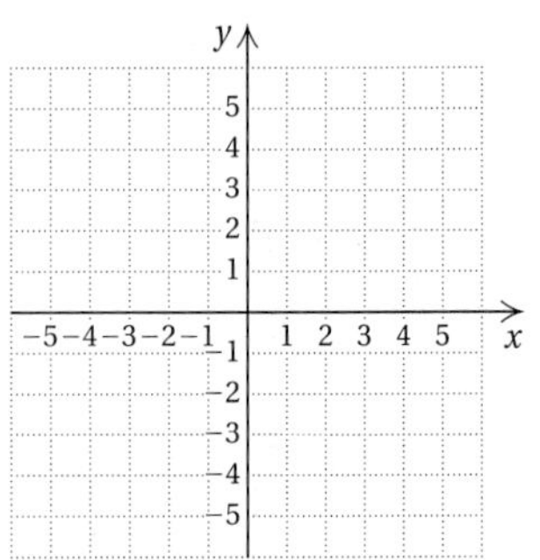

18. $y = \frac{1}{3}x$

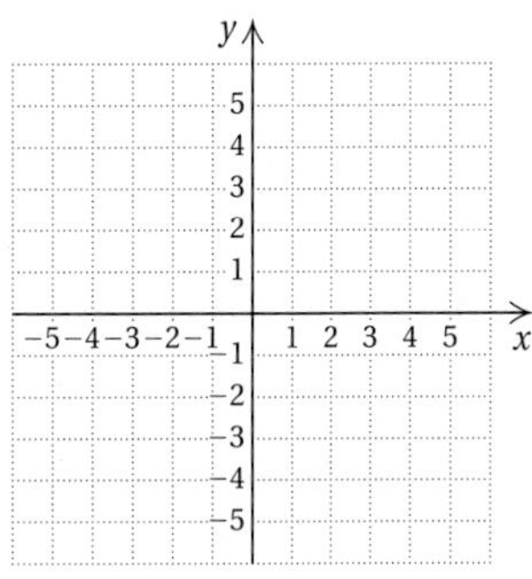

19. $y = x - 3$

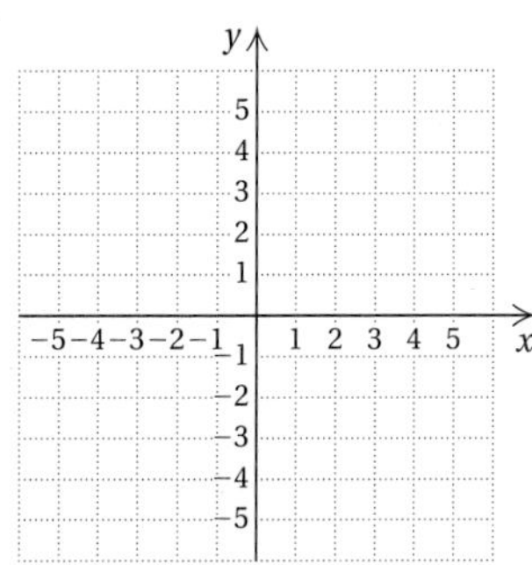

20. $y = x + 3$

21. $y = 3x - 2$

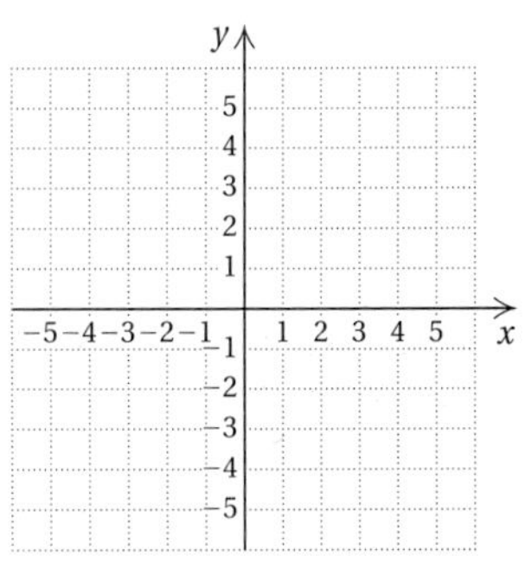

22. $y = 2x + 2$

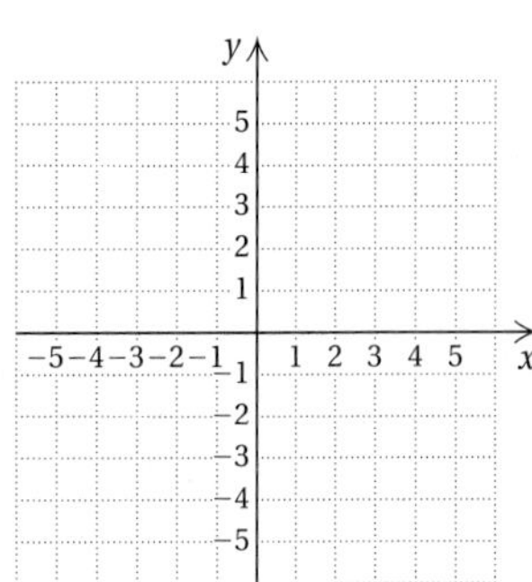

23. $y = \frac{1}{2}x + 1$

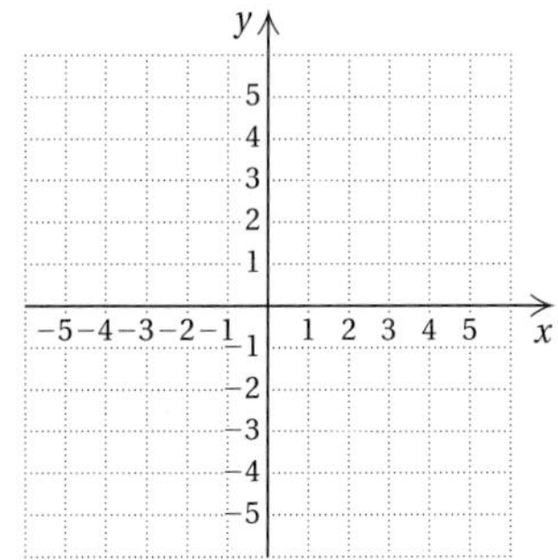

24. $y = \frac{1}{3}x - 4$

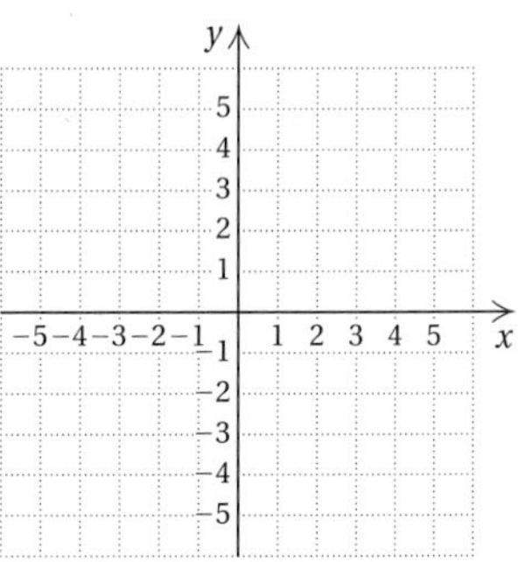

25. $x + y = -5$

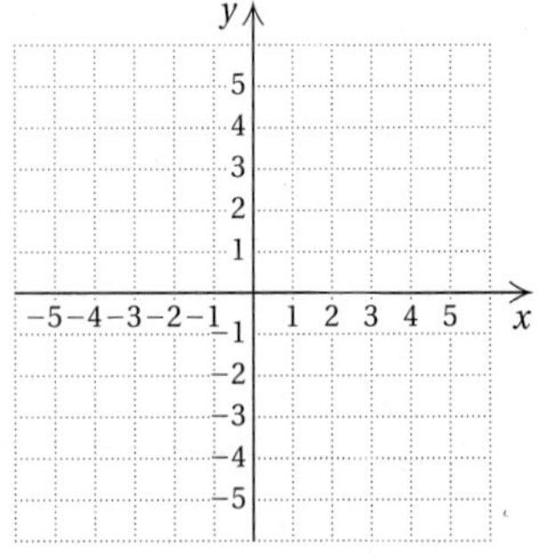

26. $x + y = 4$

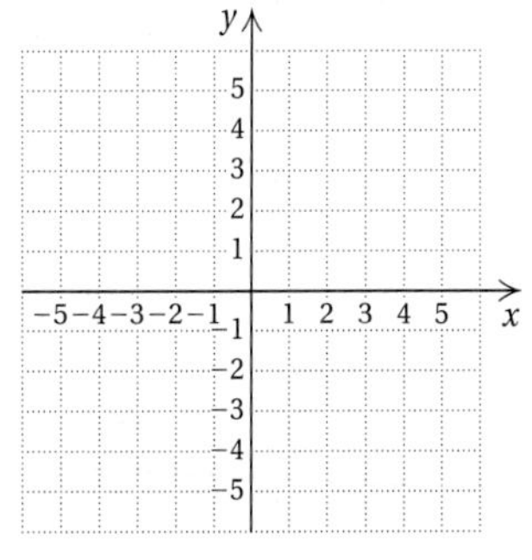

27. $y = \frac{5}{3}x - 2$

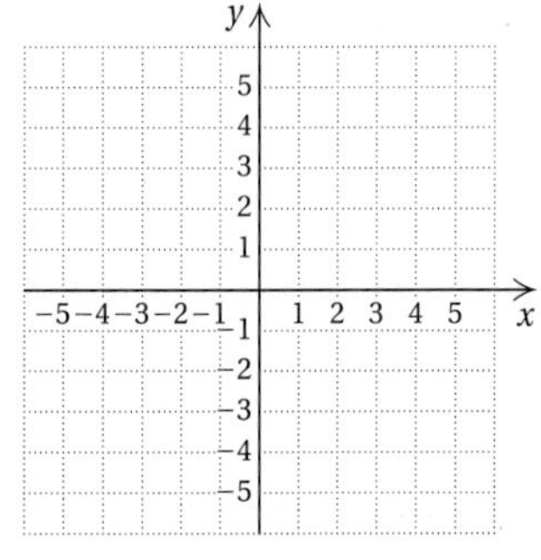

28. $y = \frac{5}{2}x + 3$

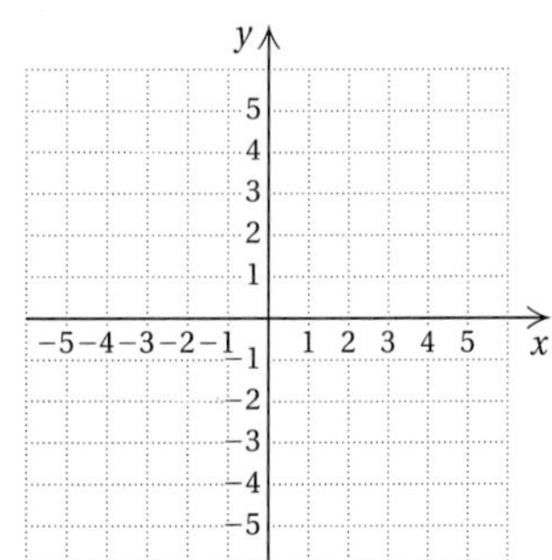

29. $x + 2y = 8$

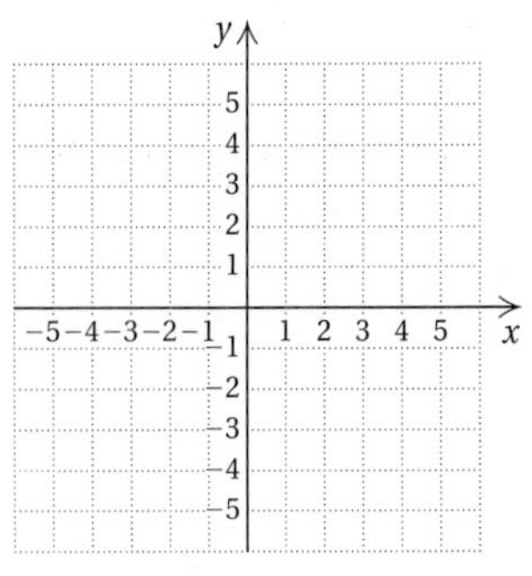

30. $x + 2y = -6$

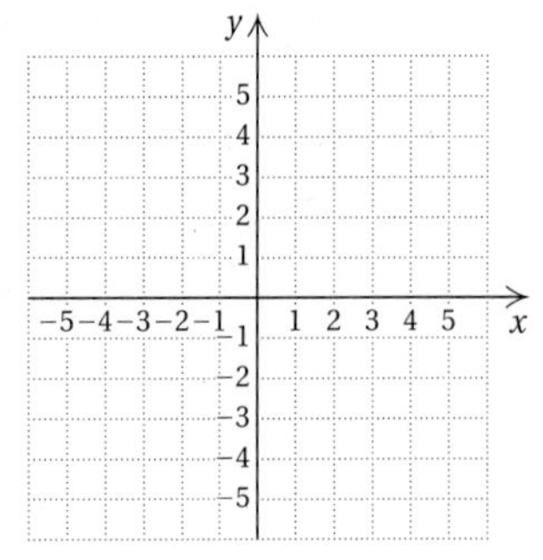

31. $y = \frac{3}{2}x + 1$

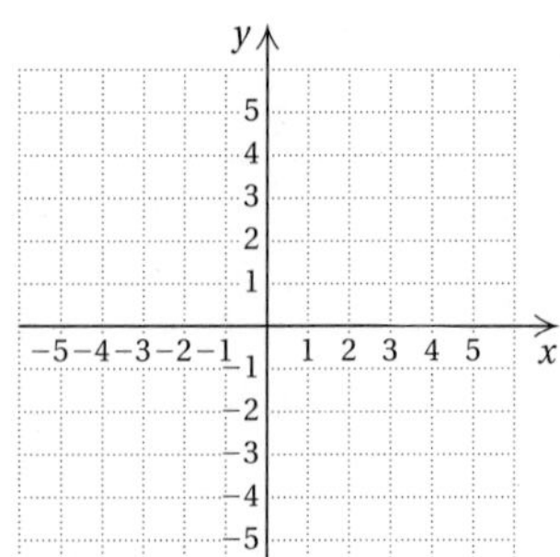

32. $y = -\frac{1}{2}x - 3$

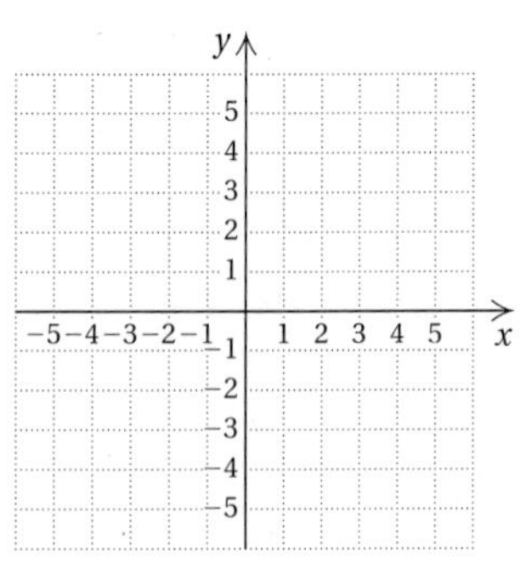

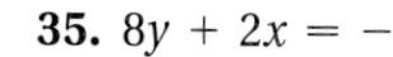

33. $8x - 2y = -10$

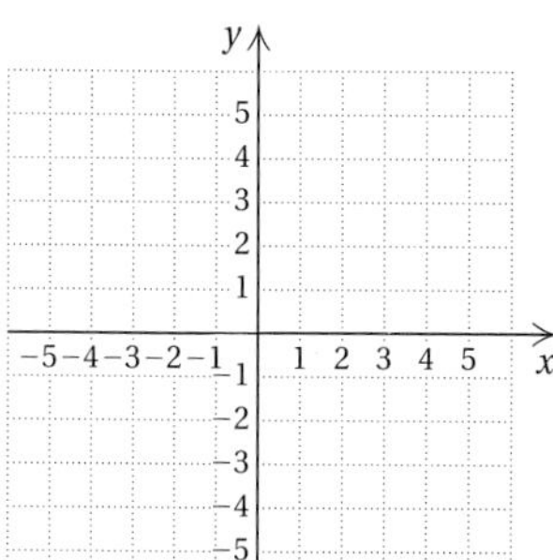

34. $6x - 3y = 9$

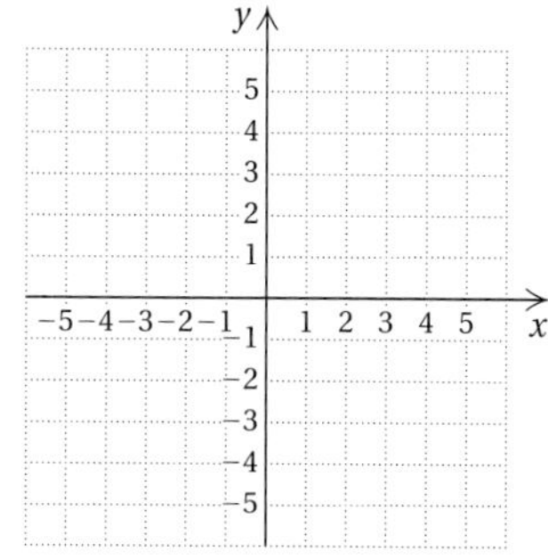

35. $8y + 2x = -4$

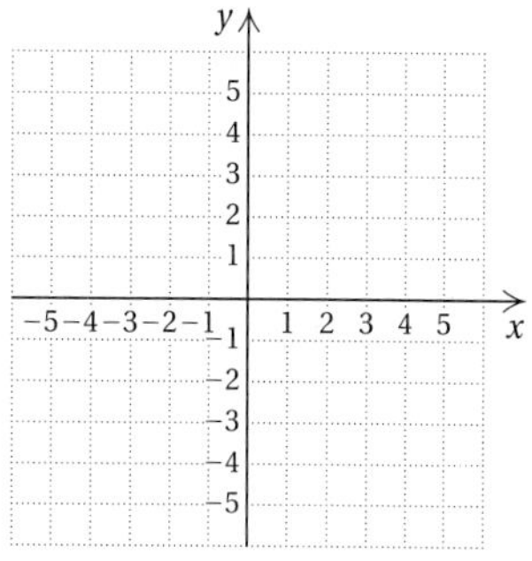

36. $6y + 2x = 8$

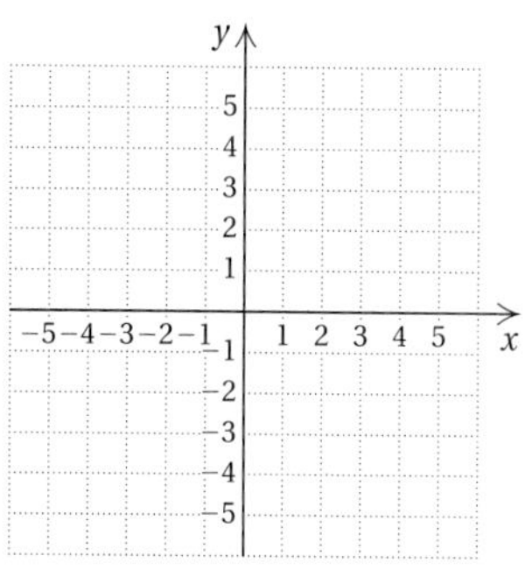

c Solve.

37. *Value of Computer Software.* The value V, in dollars, of a shopkeeper's inventory software program is given by

$$V = -50t + 300,$$

where $t =$ the number of years since the shopkeeper first bought the program.

a) Find the value of the software after 0 yr, 4 yr, and 6 yr.

b) Graph the equation and then use the graph to estimate the value of the software after 5 yr.

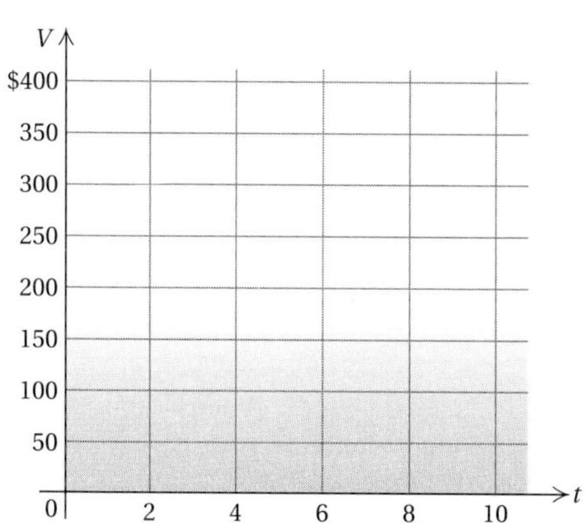

c) After how many years is the value of the software $150?

38. *College Costs.* The cost T, in dollars, of tuition and fees at many community colleges can be approximated by

$$T = 120c + 100,$$

where $c =$ the number of credits for which a student registers (***Source:*** Community College of Vermont).

a) Find the cost of tuition for a student who takes 8 hr, 12 hr, and 15 hr.

b) Graph the equation and then use the graph to estimate the cost of tuition for a student who takes 9 hr.

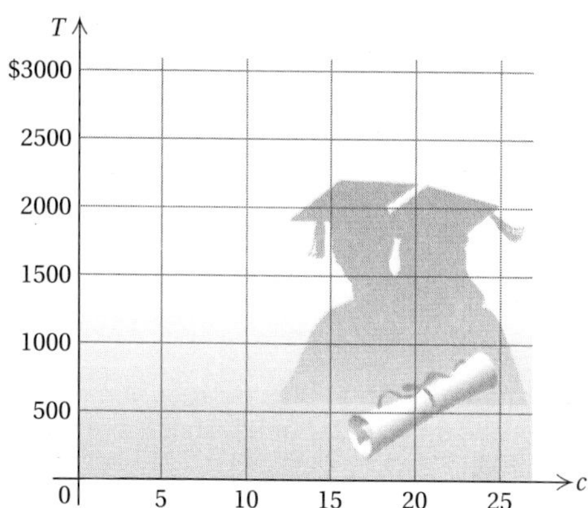

c) Estimate how many hours a student can take for $1420.

39. *Tea Consumption.* The number of gallons N of tea consumed each year by the average U.S. consumer can be approximated by

$$N = 0.1d + 7,$$

where d = the number of years since 1991 (***Source:*** Statistical Abstract of the United States).

a) Find the number of gallons of tea consumed in 1992 ($d = 1$), 1995 ($d = 4$), 1999 ($d = 8$), and 2001 ($d = 10$). 7.1 gal; 7.4 gal; 7.8 gal; 8 gal

b) Graph the equation and use the graph to estimate what the tea consumption was in 1997. 7.6 gal

c) In what year will tea consumption be about 9 gal? 2011

40. *Record Temperature Drop.* On 22 January 1943, the temperature T, in degrees Fahrenheit, in Spearfish, South Dakota, could be approximated by

$$T = -2.15m + 54,$$

where m = the number of minutes since 9:00 that morning (***Source:*** *Information Please Almanac,* 1996).

a) Find the temperature at 9:01 A.M., 9:08 A.M., and 9:20 A.M. 51.85°; 36.8°; 11°

b) Graph the equation and use the graph to estimate the temperature at 9:15 A.M. 22°

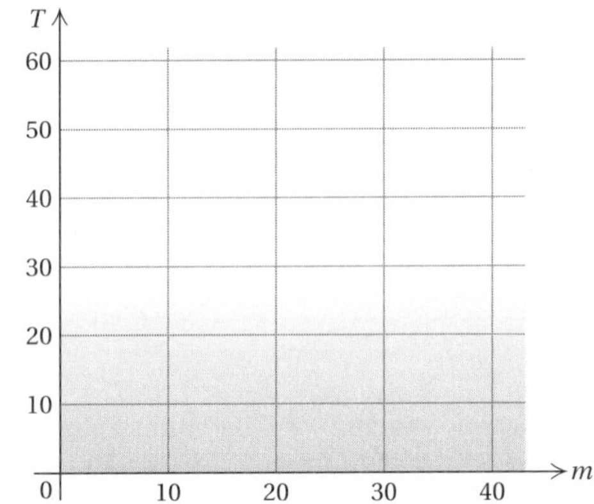

c) The temperature stopped dropping when it reached −4°. At what time did this occur? (*Note:* The linear equation could not be used after that time.) About 9:27 A.M.

Skill Maintenance

Solve. [2.2a]

41. $63 = 9x$

42. $\frac{2}{3}y = -\frac{1}{3}$

43. $13x = -52$

44. $\frac{a}{8} = -2$

Convert to decimal notation. [1.2c]

45. $-\frac{7}{8}$ −0.875

46. $\frac{23}{32}$ 0.71875

47. $\frac{117}{64}$ 1.828125

48. $-\frac{27}{12}$ −2.25

Synthesis

49. ◆ The equations $3x + 4y = 8$ and $y = -\frac{3}{4}x + 2$ are equivalent. Which equation is easier to graph and why?

50. ◆ Referring to Exercise 40, discuss why the linear equation no longer described the temperature after the temperature reached −4°.

In Exercises 51–54, find an equation for the graph shown.

51.

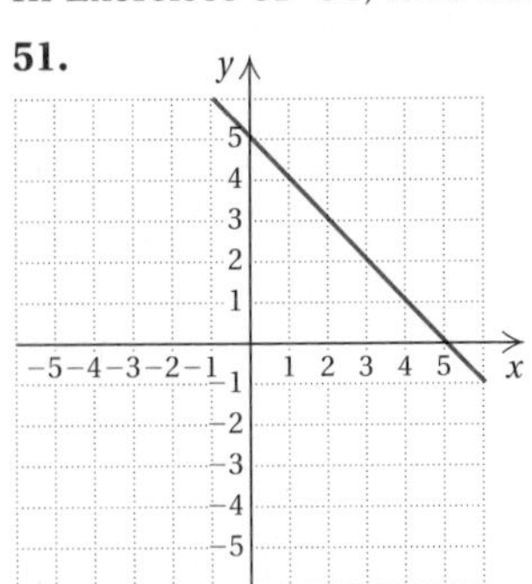

$y = -x + 5$

52.

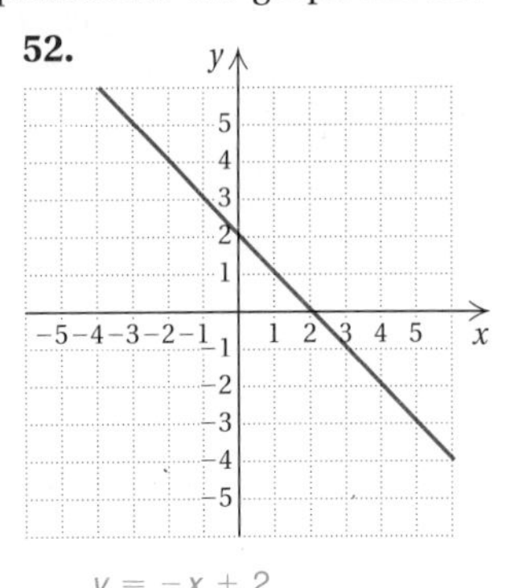

$y = -x + 2$

53.

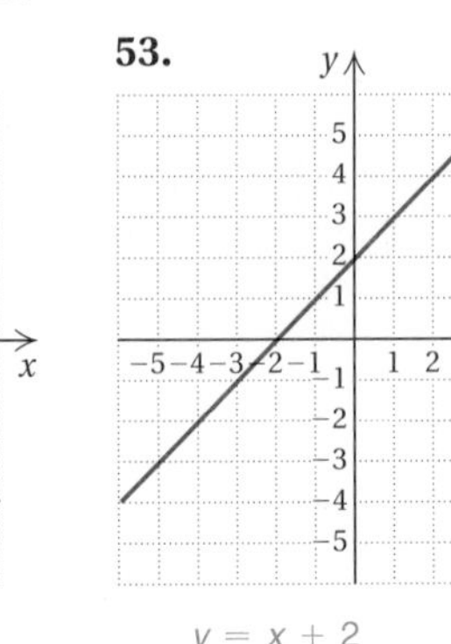

$y = x + 2$

54.

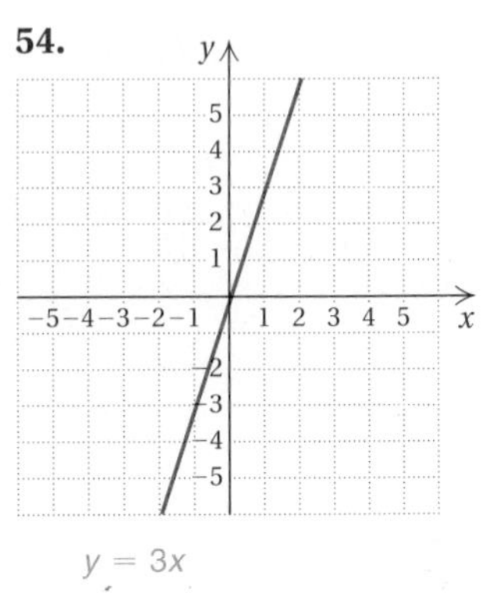

$y = 3x$

3.3 More with Graphing and Intercepts

a Graphing Using Intercepts

In Section 3.2, we graphed linear equations of the form $Ax + By = C$ by first solving for y to find an equivalent equation in the form $y = mx + b$. We did so because it is then easier to calculate the y-value that corresponds to a given x-value. Another convenient way to graph $Ax + By = C$ is to use **intercepts**. Look at the graph of $-2x + y = 4$ shown below.

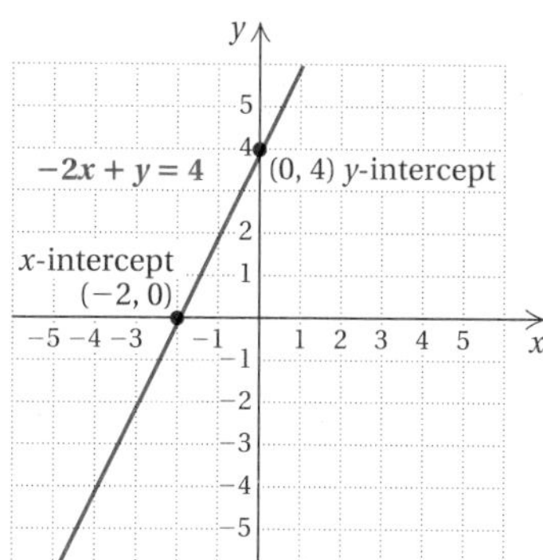

The y-intercept is (0, 4). It occurs where the line crosses the y-axis and thus will always have 0 as the first coordinate. The x-intercept is (−2, 0). It occurs where the line crosses the x-axis and thus will always have 0 as the second coordinate.

Do Exercise 1.

We find intercepts as follows.

> The **y-intercept** is $(0, b)$. To find b, let $x = 0$ and solve the original equation for y.
>
> The **x-intercept** is $(a, 0)$. To find a, let $y = 0$ and solve the original equation for x.

Now let's draw a graph using intercepts.

Example 1 Consider $4x + 3y = 12$. Find the intercepts. Then graph the equation using the intercepts.

To find the y-intercept, we let $x = 0$. Then we solve for y:

$$\begin{aligned} 4 \cdot 0 + 3y &= 12 \\ 3y &= 12 \\ y &= 4. \end{aligned}$$

Thus, (0, 4) is the y-intercept. Note that finding this intercept amounts to covering up the x-term and solving the rest of the equation.

To find the x-intercept, we let $y = 0$. Then we solve for x:

$$\begin{aligned} 4x + 3 \cdot 0 &= 12 \\ 4x &= 12 \\ x &= 3. \end{aligned}$$

Objectives

a Find the intercepts of a linear equation, and graph using intercepts.

b Graph equations equivalent to those of the type $x = a$ and $y = b$.

For Extra Help

TAPE 5 MAC WIN CD-ROM

1. Look at the graph shown below.

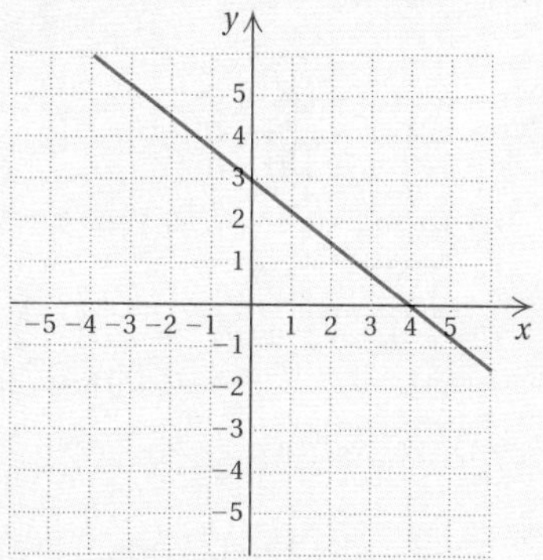

a) Find the coordinates of the y-intercept.

b) Find the coordinates of the x-intercept.

Answers on page A-9

For each equation, find the intercepts. Then graph the equation using the intercepts.

2. $2x + 3y = 6$

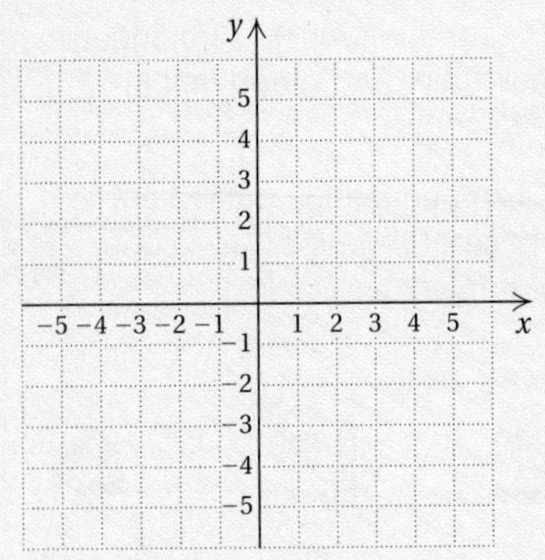

3. $3y - 4x = 12$

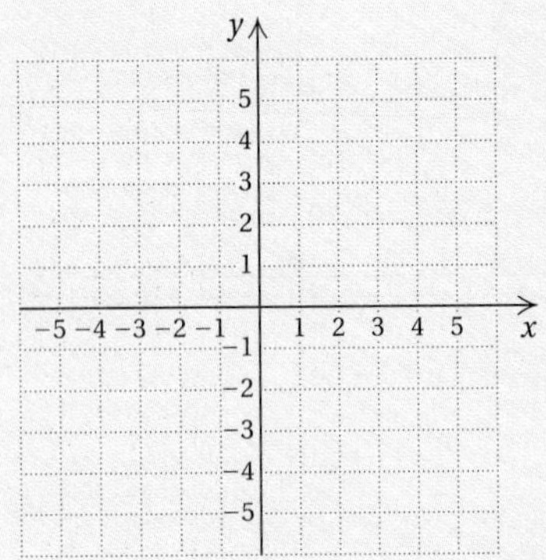

Graph.

4. $x = 5$

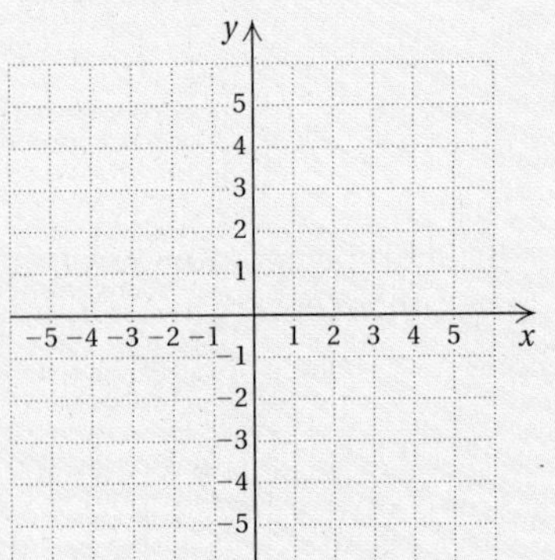

Answers on page A-9

Thus, (3, 0) is the x-intercept. Note that finding this intercept amounts to covering up the y-term and solving the rest of the equation.

We plot these points and draw the line, or graph.

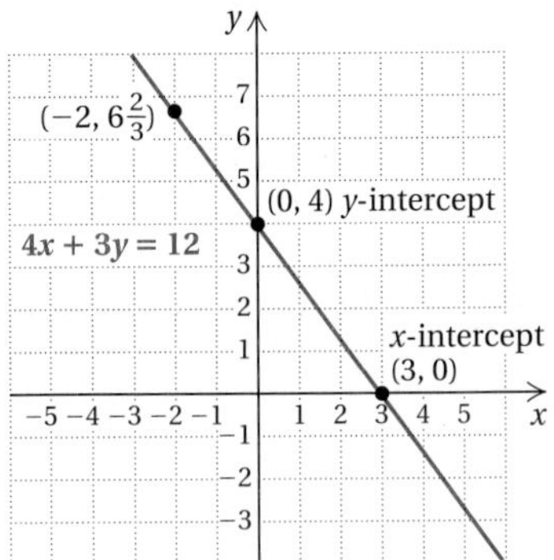

A third point should be used as a check. We substitute any convenient value for x and solve for y. In this case, we choose $x = -2$. Then

$$4(-2) + 3y = 12 \quad \text{Substituting } -2 \text{ for } x$$
$$-8 + 3y = 12$$
$$3y = 12 + 8 = 20$$
$$y = \tfrac{20}{3}, \text{ or } 6\tfrac{2}{3}. \quad \text{Solving for } y$$

It appears that the point $\left(-2, 6\frac{2}{3}\right)$ is on the graph, though graphing fractional values can be inexact. The graph is probably correct.

Graphs of equations of the type $y = mx$ pass through the origin. Thus the x-intercept and the y-intercept are the same, (0, 0). In such cases, we must calculate another point in order to complete the graph. Another point would also have to be calculated if a check is desired.

Do Exercises 2 and 3.

b Equations Whose Graphs Are Horizontal or Vertical Lines

Example 2 Graph: $y = 3$.

Consider $y = 3$. We can also think of this equation as $0 \cdot x + y = 3$. No matter what number we choose for x, we find that y is 3. We make up a table with all 3's in the y-column.

x	y
	3
	3
	3

y must be 3.

Choose any number for x. →

x	y
−2	3
0	3 ← y-intercept
4	3

When we plot the ordered pairs $(-2, 3)$, $(0, 3)$, and $(4, 3)$ and connect the points, we will obtain a horizontal line. Any ordered pair $(x, 3)$ is a solution. So the line is parallel to the x-axis with y-intercept $(0, 3)$.

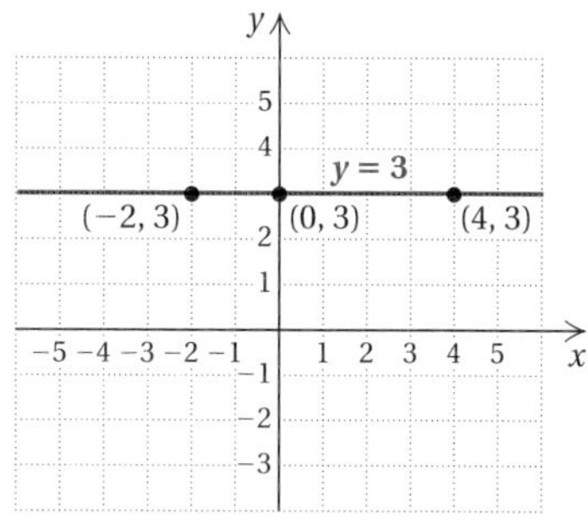

Example 3 Graph: $x = -4$.

Consider $x = -4$. We can also think of this equation as $x + 0 \cdot y = -4$. We make up a table with all -4's in the x-column.

x	y
−4	
−4	
−4	
−4	

x must be −4.

Choose any number for y. →

x	y
−4	−5
−4	1
−4	3
−4	0

x-intercept → (last row)

When we plot the ordered pairs $(-4, -5)$, $(-4, 1)$, $(-4, 3)$, and $(-4, 0)$ and connect the points, we will obtain a vertical line. Any ordered pair $(-4, y)$ is a solution. So the line is parallel to the y-axis with x-intercept $(-4, 0)$.

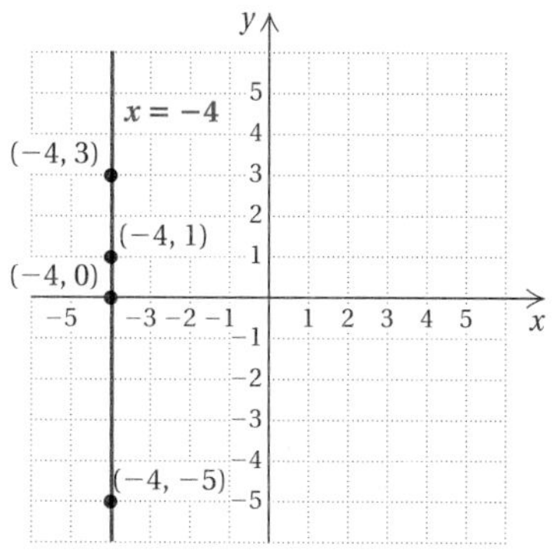

The graph of $y = b$ is a **horizontal line.** The y-intercept is $(0, b)$.

The graph of $x = a$ is a **vertical line.** The x-intercept is $(a, 0)$.

Do Exercises 4–7. (Exercise 4 is on the preceding page.)

Graph.

5. $y = -2$

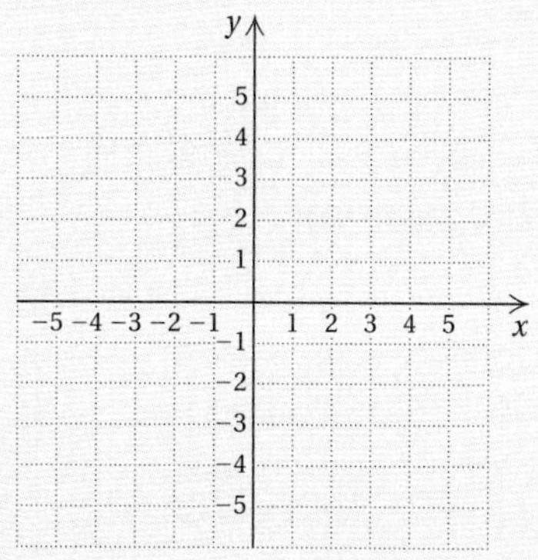

6. $x = 0$

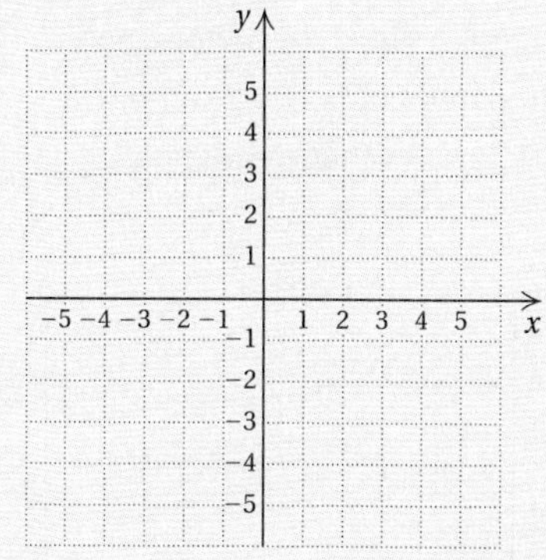

7. $x = -3$

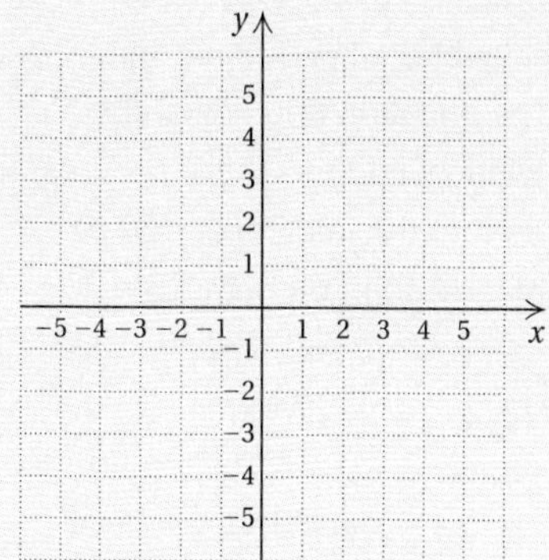

Answers on page A-9

Calculator Spotlight

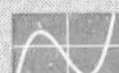

Viewing the Intercepts. Graph the equation $y = -x + 15$ using the standard viewing window.

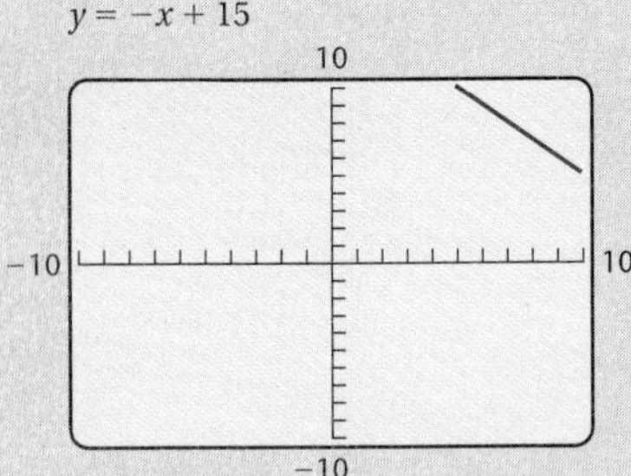

Note that the graph is barely visible in the upper right-hand corner, and neither intercept can be seen. To better view the intercepts, we can try different window settings.

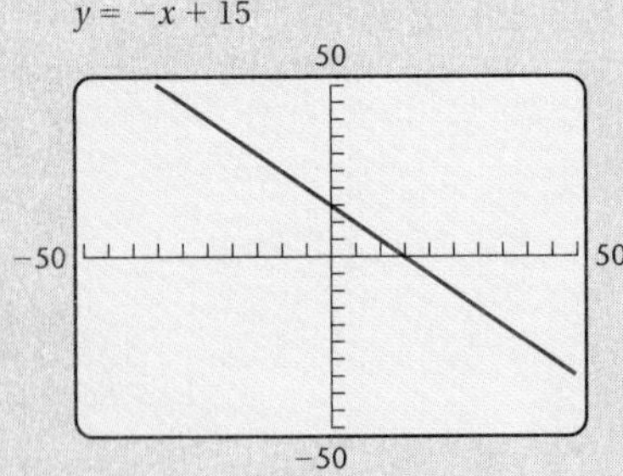

Exercises

Find the intercepts of the equation using algebra (algebraically). Then graph the equation and adjust the window and tick mark settings so that the intercepts can be clearly seen on both axes.

1. $y = -7.2x - 15$
2. $y - 2.13x = 27$
3. $5x + 6y = 84$
4. $2x - 7y = 150$
5. $3x + 2y = 50$
6. $y = 0.2x - 9$
7. $y = 1.3x - 15$
8. $25x - 20y = 1$

The following is a general procedure for graphing linear equations.

GRAPHING LINEAR EQUATIONS

1. If the equation is of the type $x = a$ or $y = b$, the graph will be a line parallel to an axis; $x = a$ is vertical and $y = b$ is horizontal.

Examples.

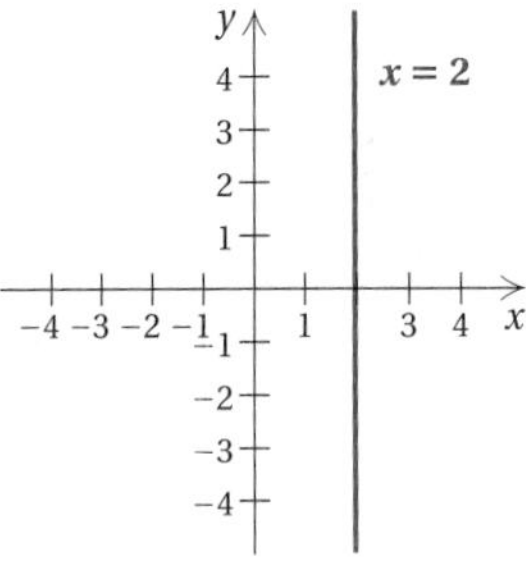

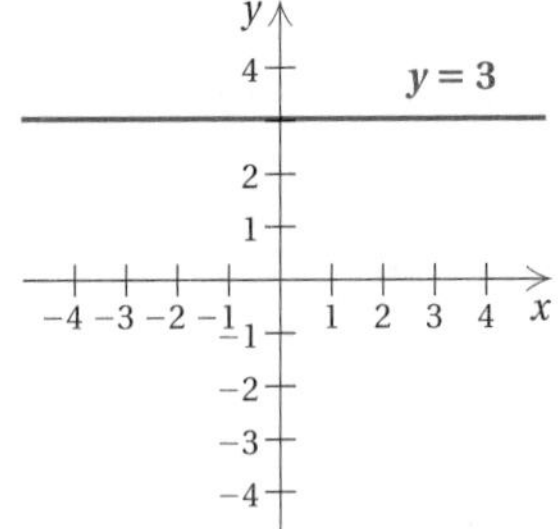

2. If the equation is of the type $y = mx$, both intercepts are the origin, (0, 0). Plot (0, 0) and two other points.

Example.

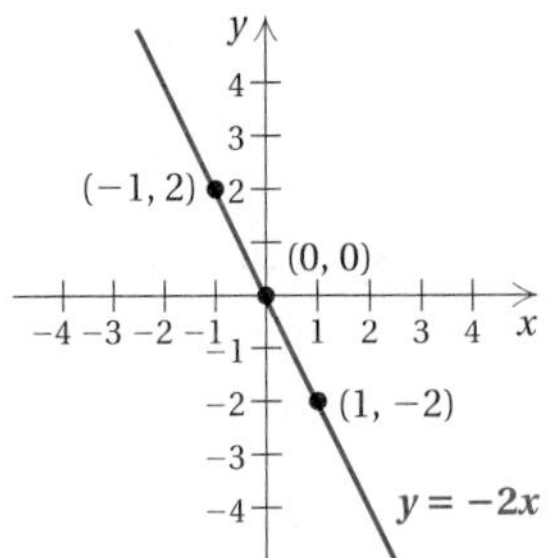

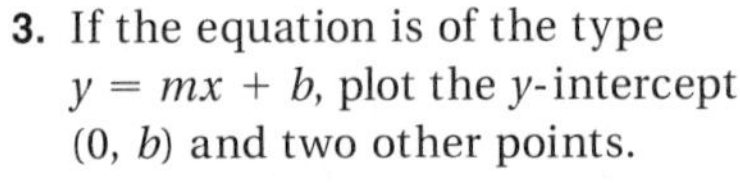

3. If the equation is of the type $y = mx + b$, plot the y-intercept $(0, b)$ and two other points.

Example.

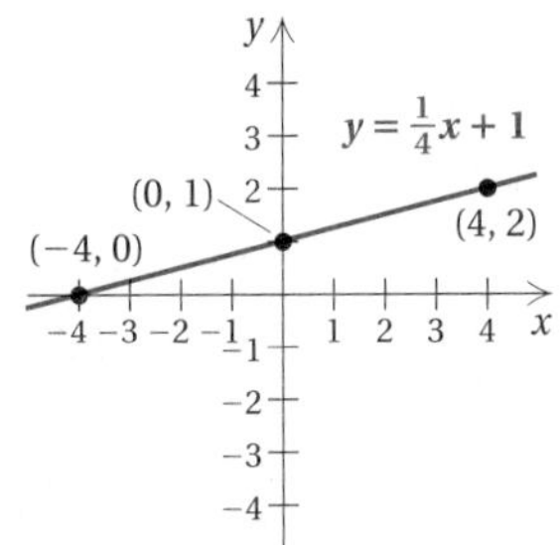

4. If the equation is of the type $Ax + By = C$, but not of the type $x = a$, $y = b$, $y = mx$, or $y = mx + b$, then either solve for y and proceed as with the equation $y = mx + b$, or graph using intercepts. If the intercepts are too close together, choose another point or points farther from the origin.

Examples.

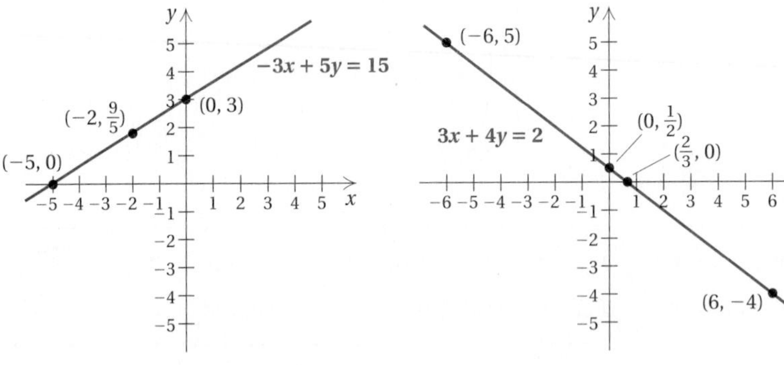

Exercise Set 3.3

a For Exercises 1–4, find (a) the coordinates of the y-intercept and (b) the coordinates of the x-intercept.

1.

2.

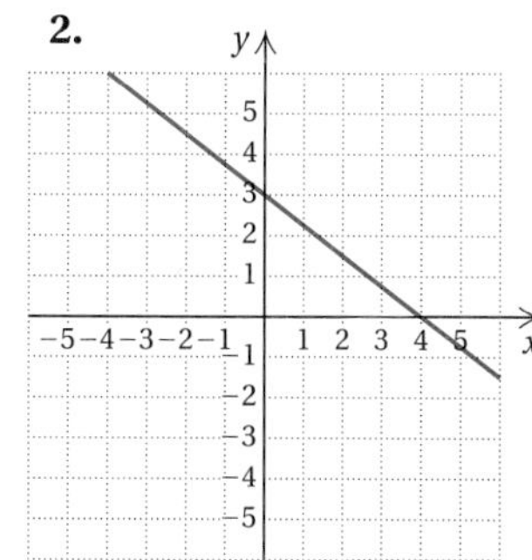

3.

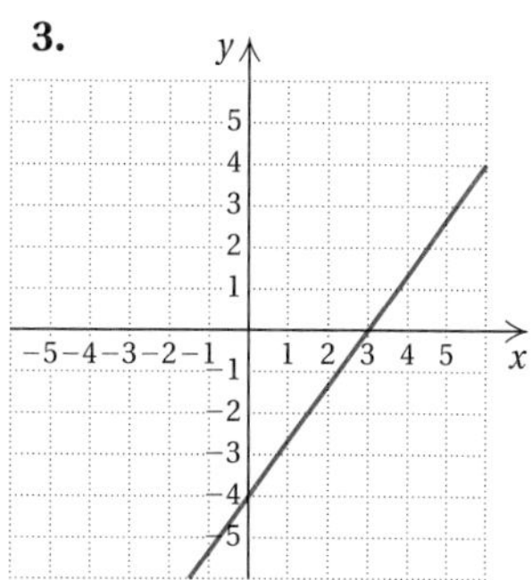

4.

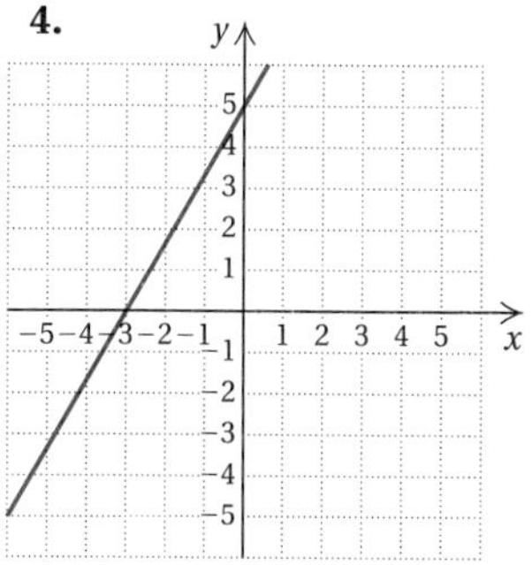

For Exercises 5–12, find (a) the coordinates of the y-intercept and (b) the coordinates of the x-intercept. Do not graph.

5. $3x + 5y = 15$

6. $5x + 2y = 20$

7. $7x - 2y = 28$

8. $3x - 4y = 24$

9. $-4x + 3y = 10$

10. $-2x + 3y = 7$

11. $6x - 3 = 9y$

12. $4y - 2 = 6x$

For each equation, find the intercepts. Then use the intercepts to graph the equation.

13. $x + 3y = 6$

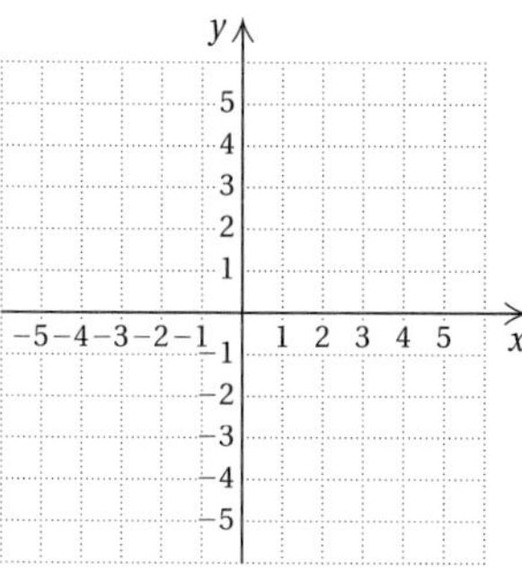

14. $x + 2y = 2$

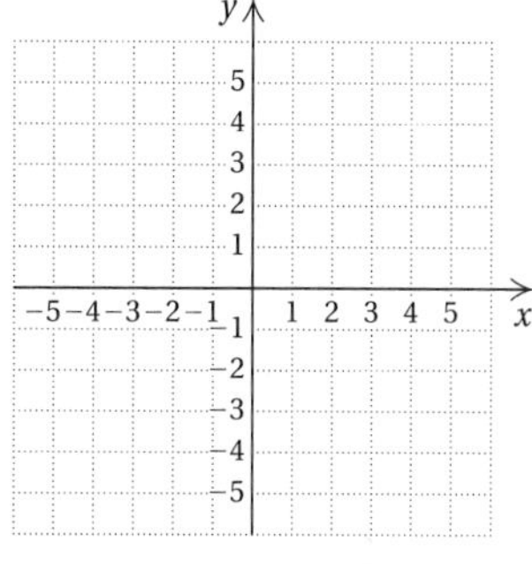

15. $-x + 2y = 4$

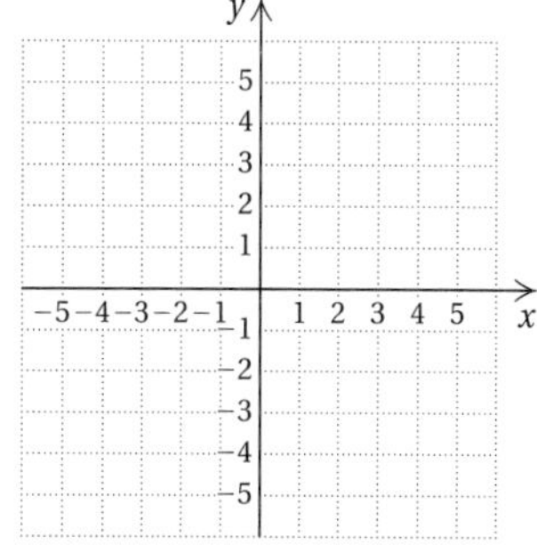

16. $-x + y = 5$

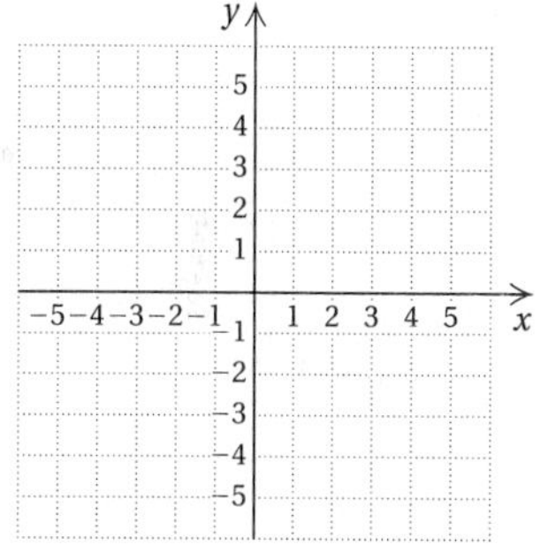

17. $3x + y = 6$

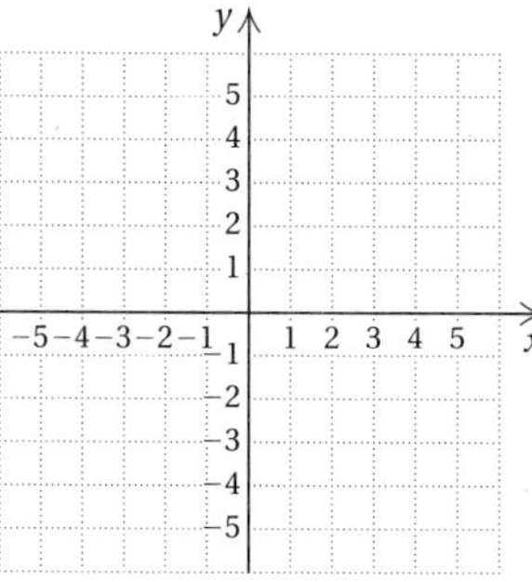

18. $2x + y = 6$

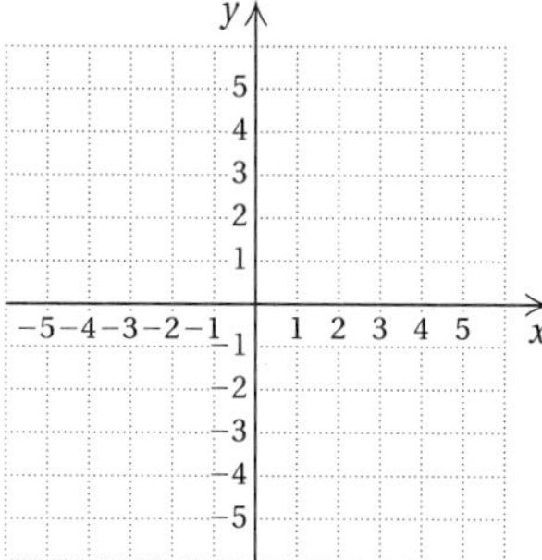

19. $2y - 2 = 6x$

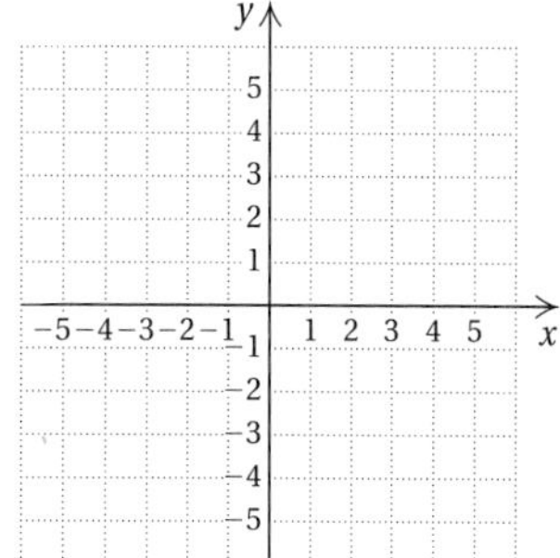

20. $3y - 6 = 9x$

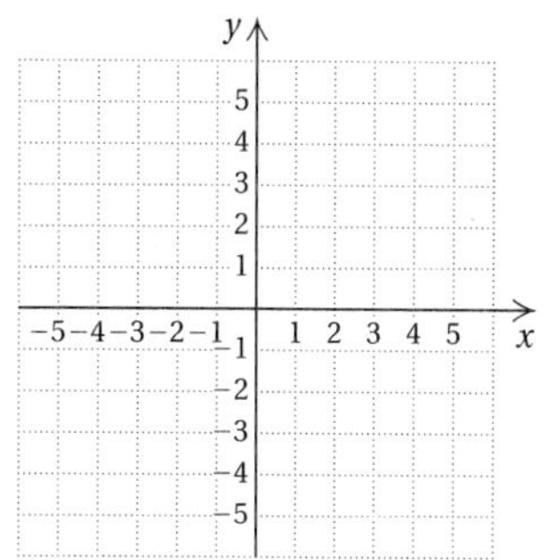

21. $3x - 9 = 3y$

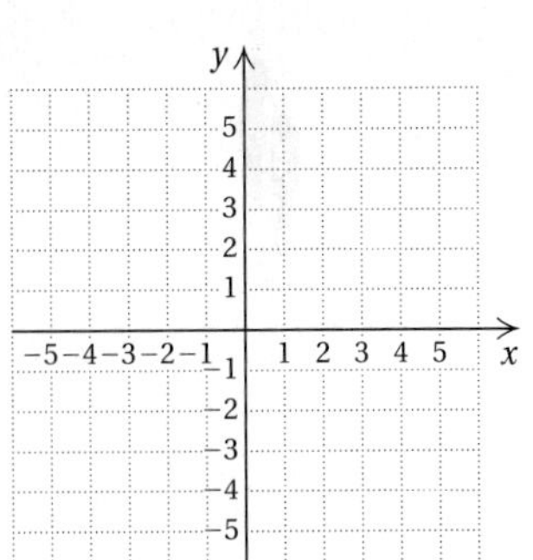

22. $5x - 10 = 5y$

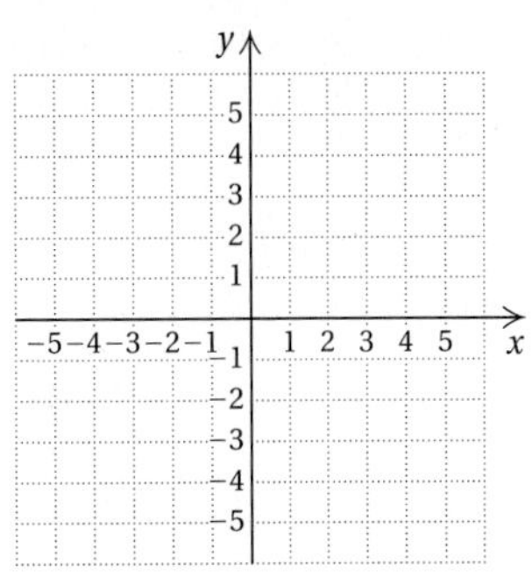

23. $2x - 3y = 6$

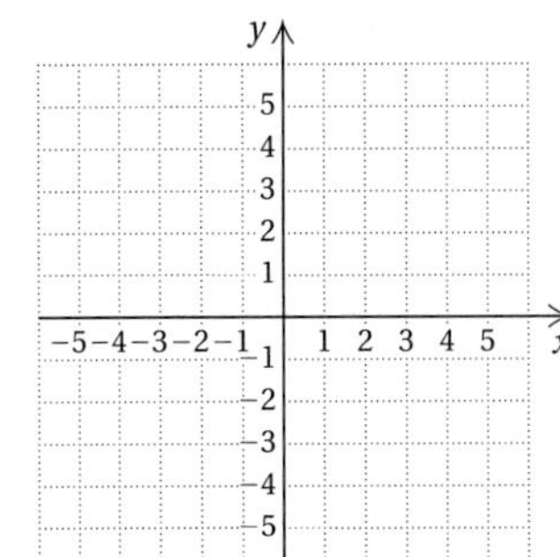

24. $2x - 5y = 10$

25. $4x + 5y = 20$

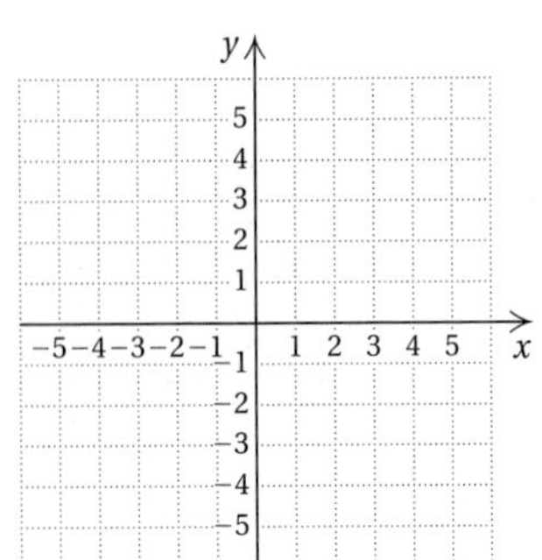

26. $2x + 6y = 12$

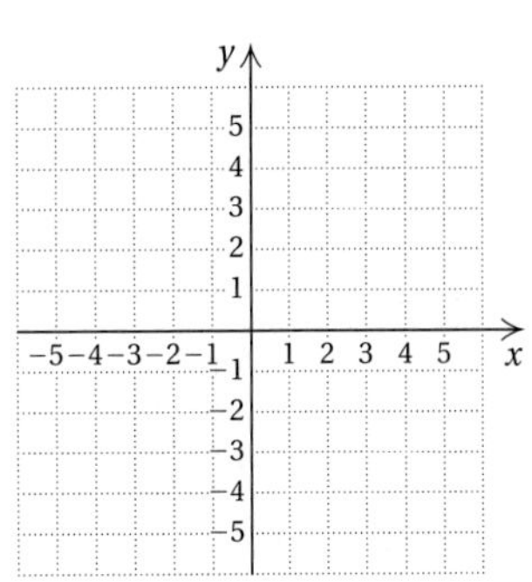

27. $2x + 3y = 8$

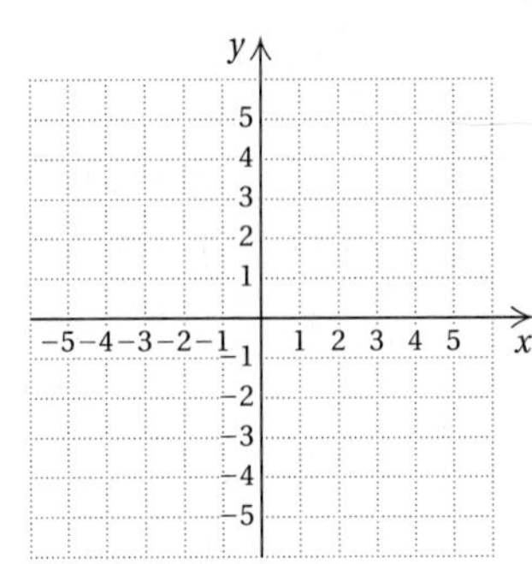

28. $x - 1 = y$

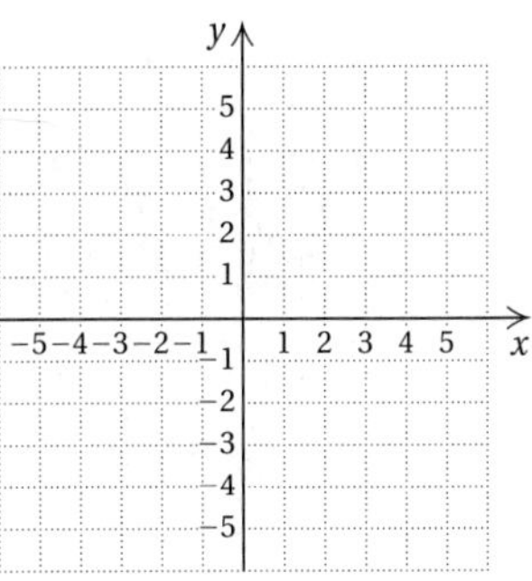

29. $x - 3 = y$

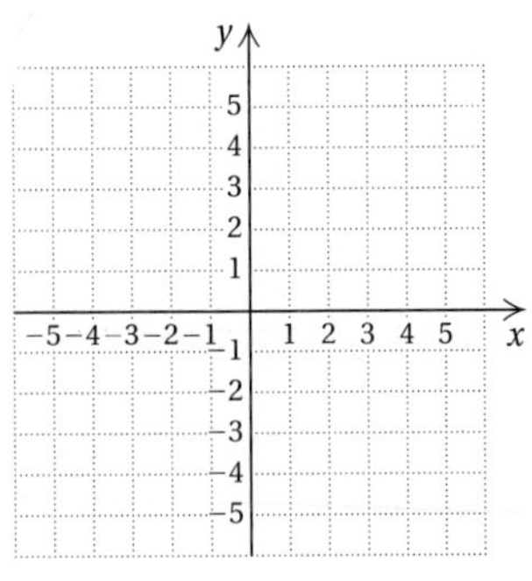

30. $2x - 1 = y$

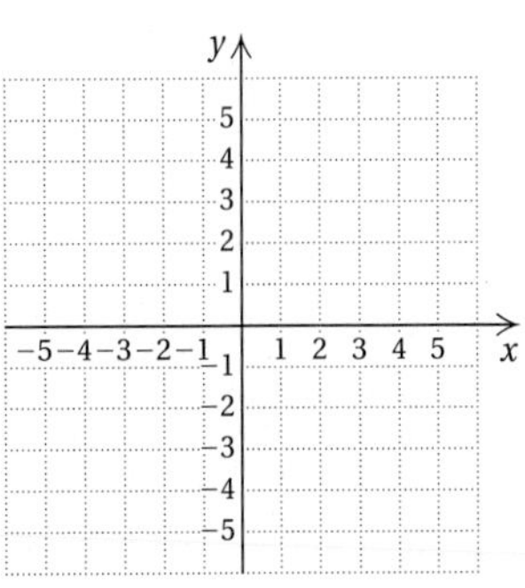

31. $3x - 2 = y$

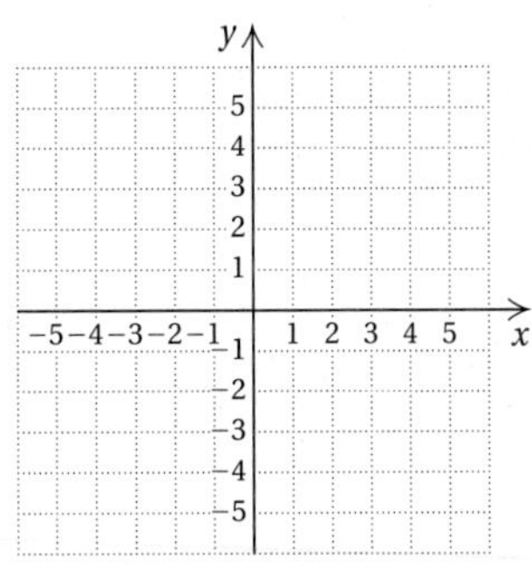

32. $4x - 3y = 12$

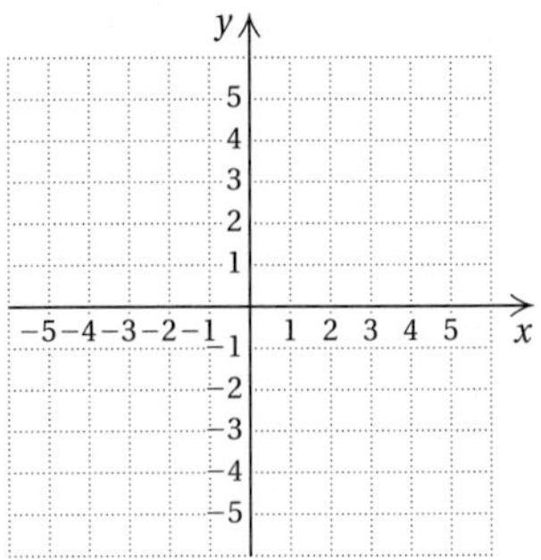

33. $6x - 2y = 12$

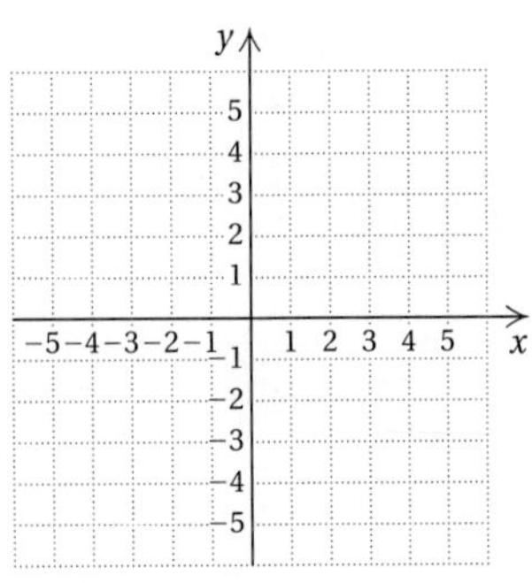

34. $7x + 2y = 6$

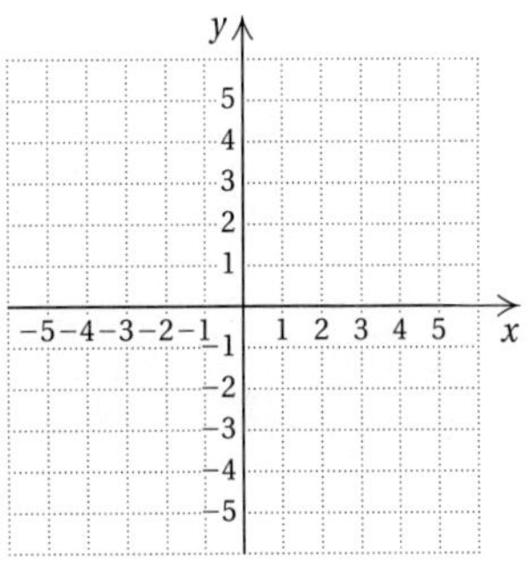

35. $3x + 4y = 5$

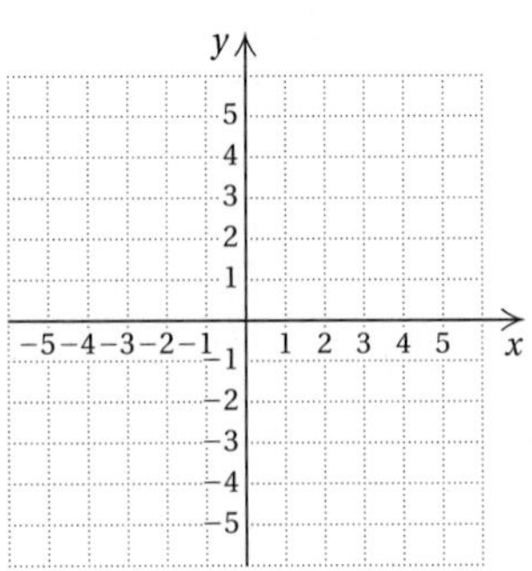

36. $y = -4 - 4x$

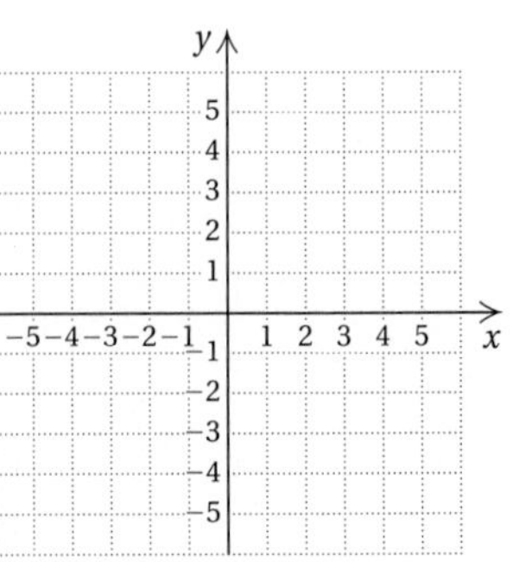

37. $y = -3 - 3x$

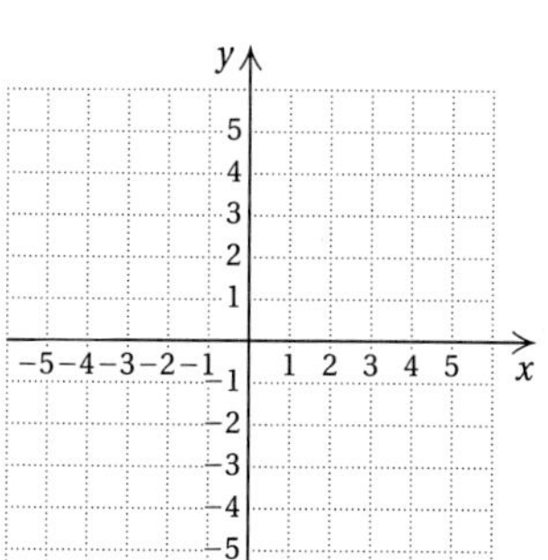

38. $-3x = 6y - 2$

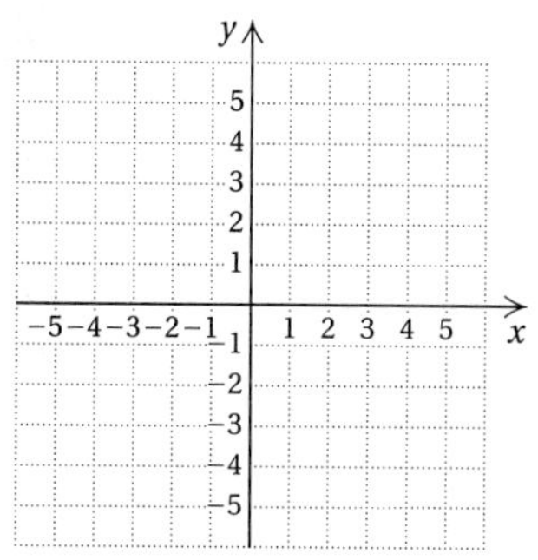

39. $y - 3x = 0$

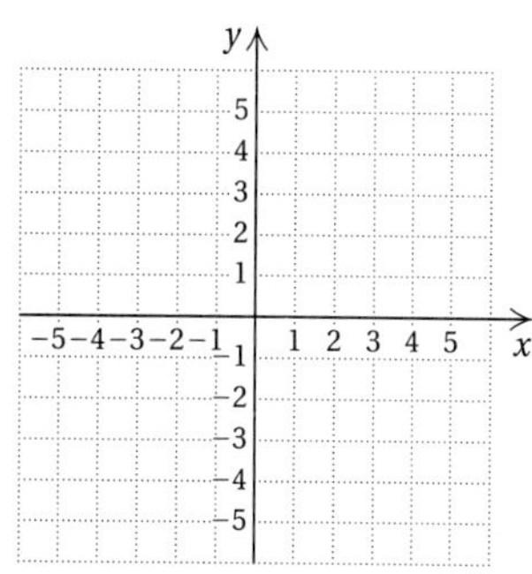

40. $x + 2y = 0$

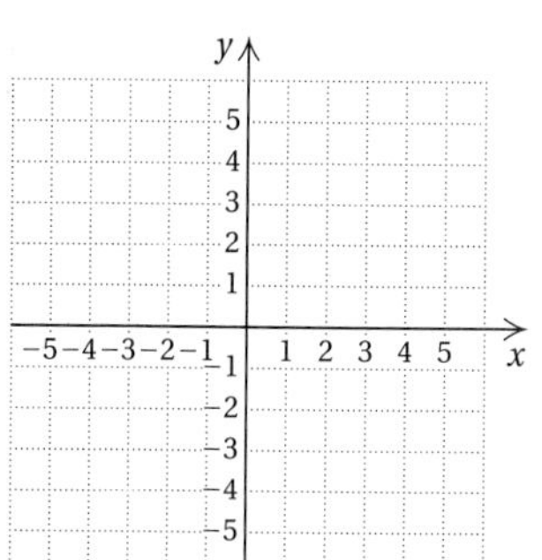

b Graph.

41. $x = -2$

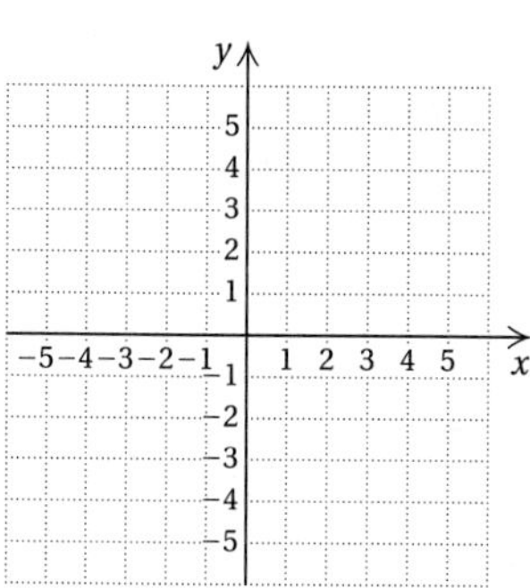

42. $x = 1$

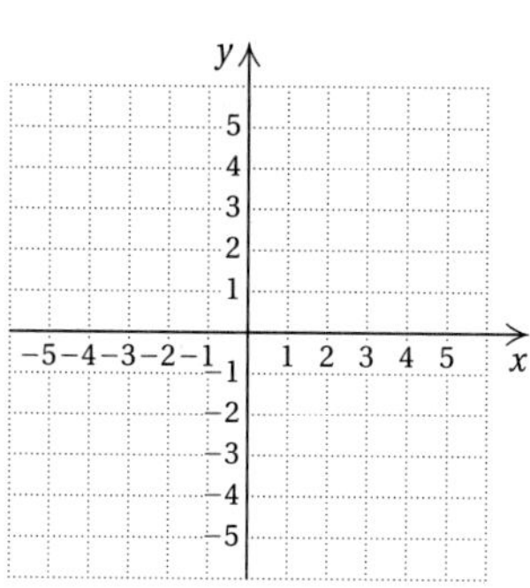

43. $y = 2$

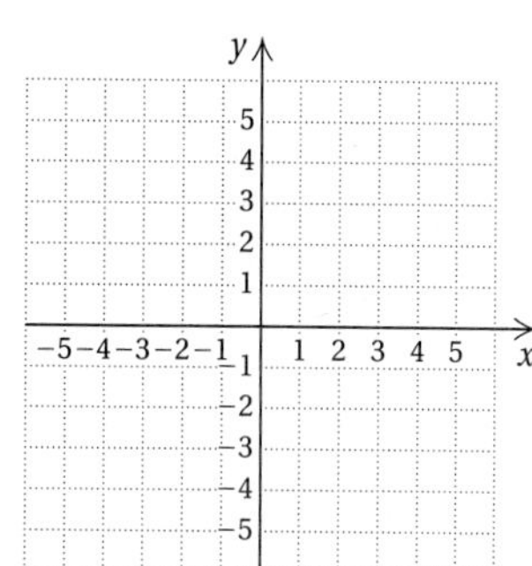

44. $y = -4$

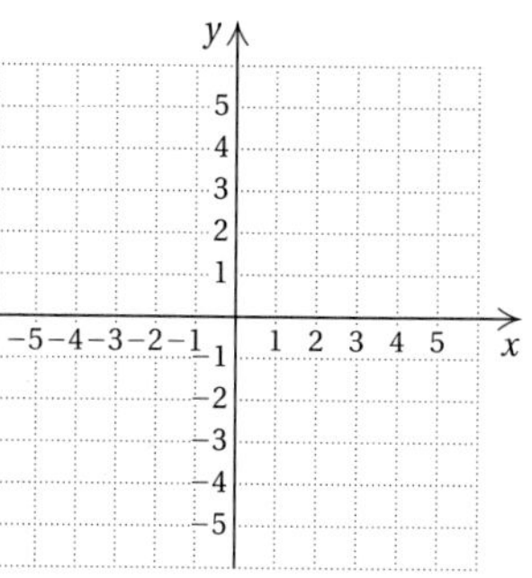

45. $x = 2$

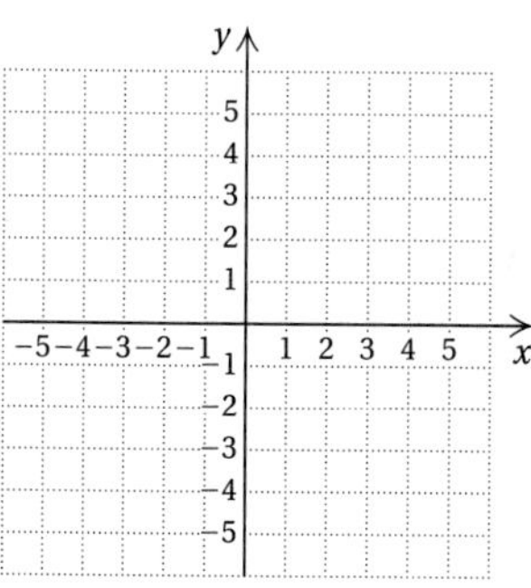

46. $x = 3$

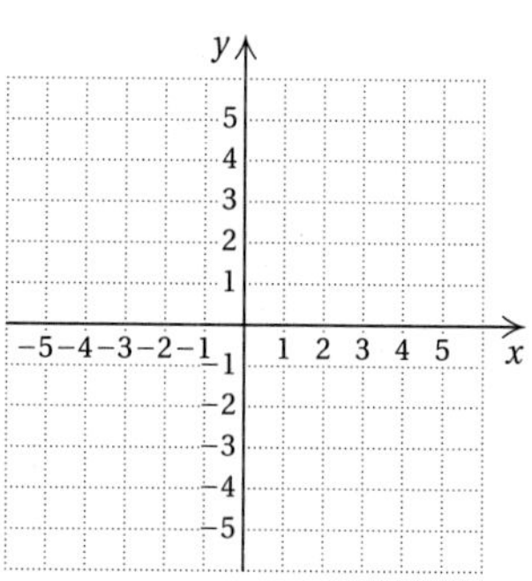

47. $y = 0$

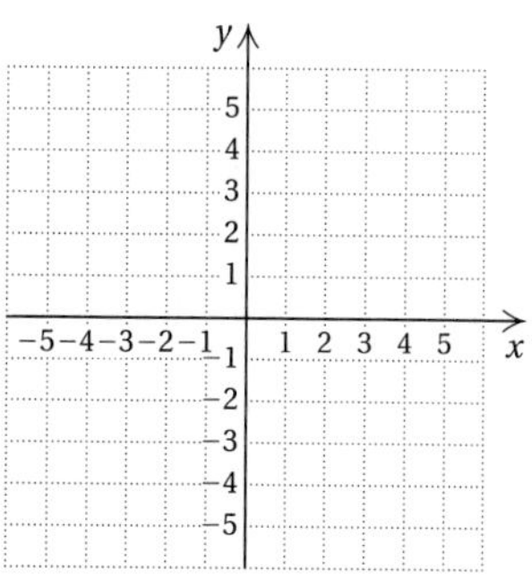

48. $y = -1$

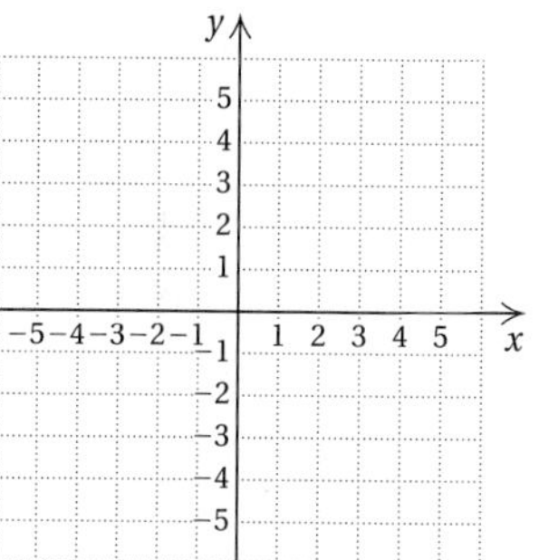

49. $x = \frac{3}{2}$

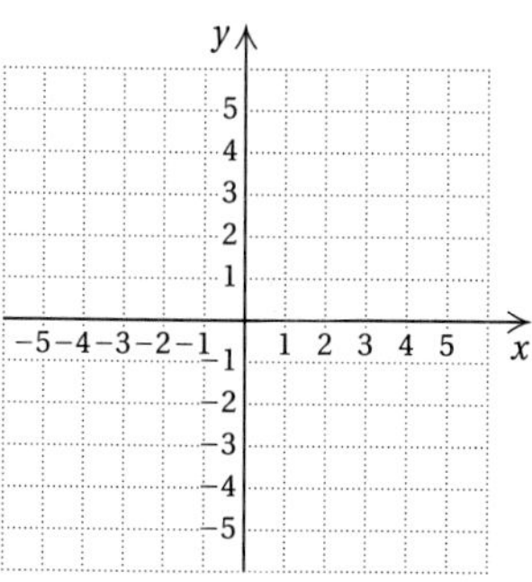

50. $x = -\frac{5}{2}$

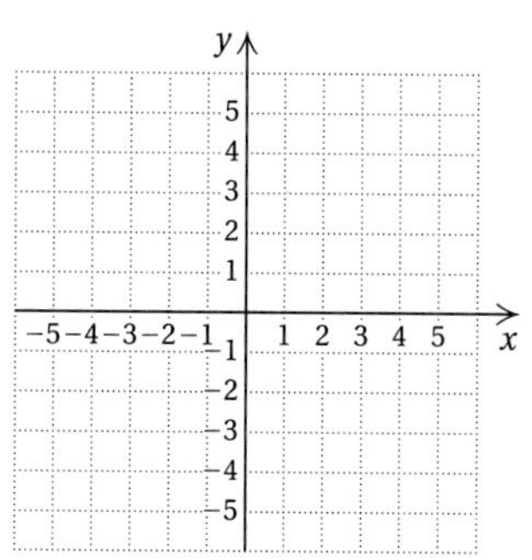

51. $3y = -5$

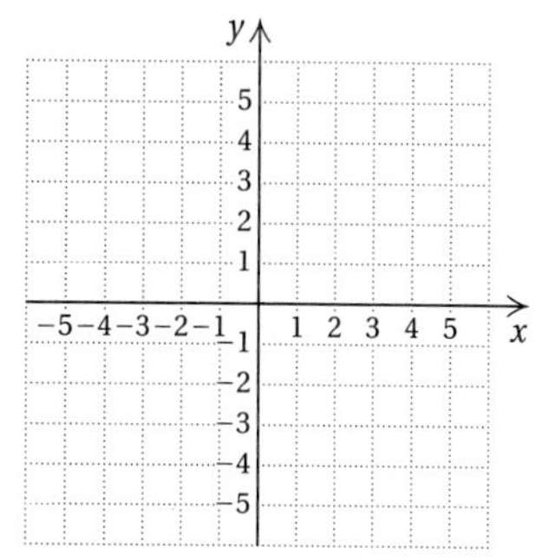

52. $12y = 45$

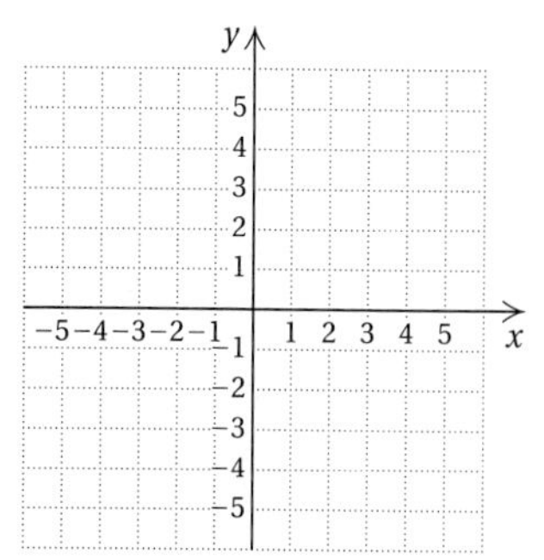

53. $4x + 3 = 0$

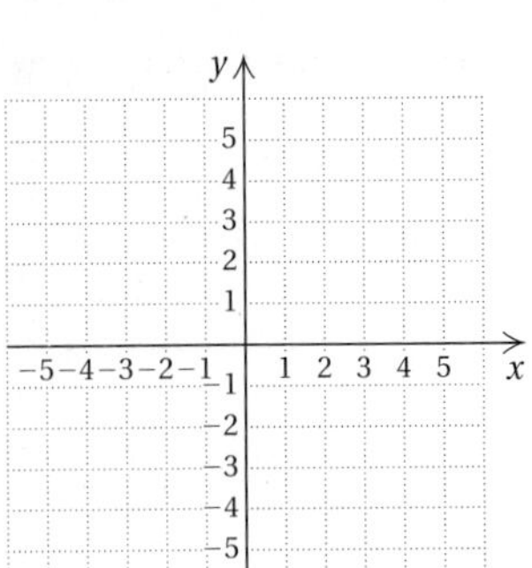

54. $-3x + 12 = 0$

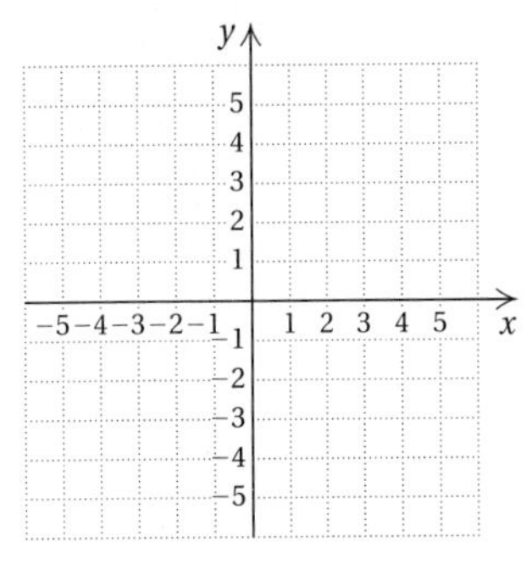

55. $48 - 3y = 0$

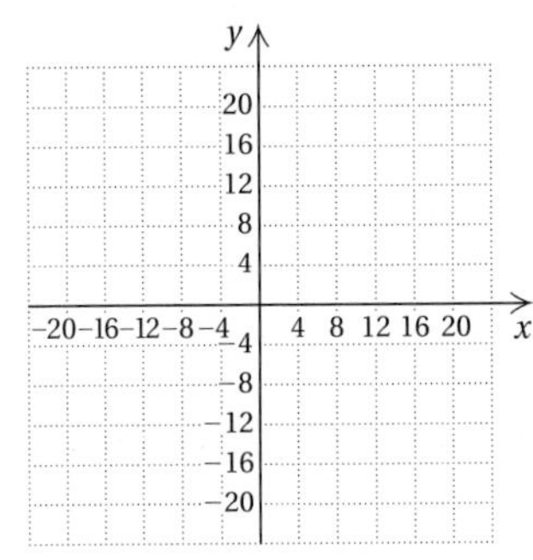

56. $63 + 7y = 0$

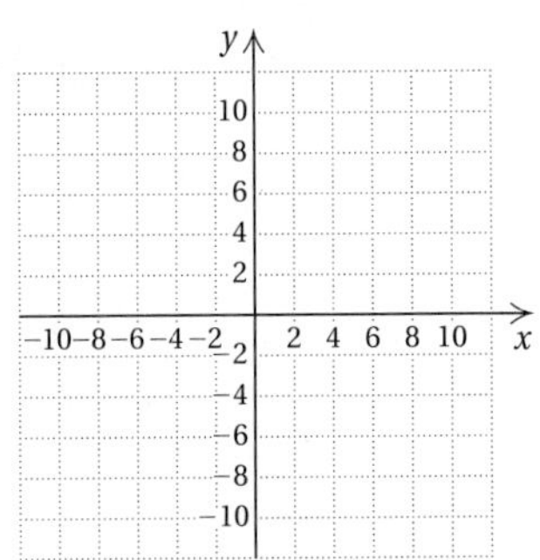

Write an equation for the graph shown.

57.

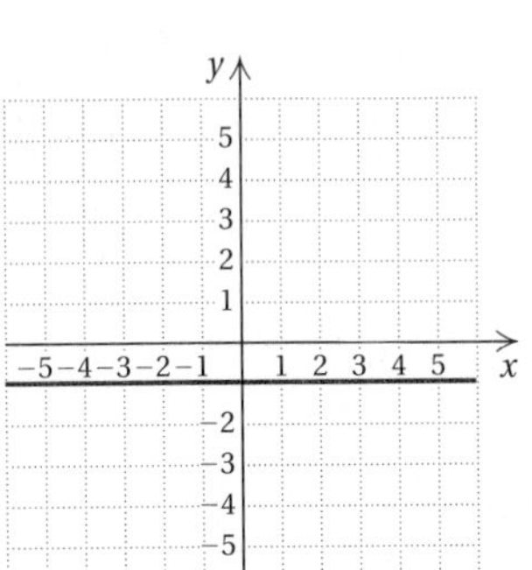

58.

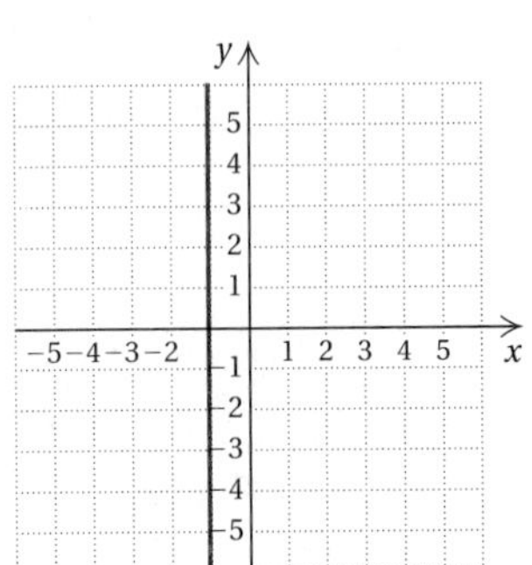

59.

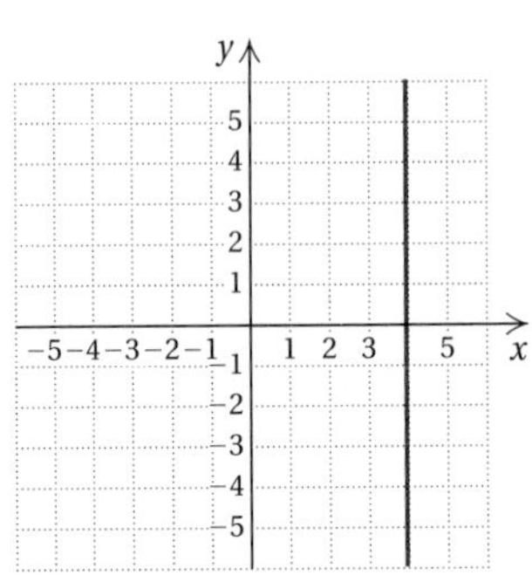

60.

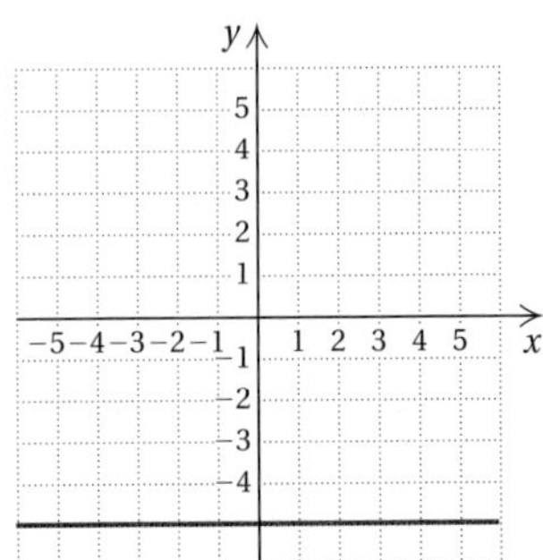

Skill Maintenance

Solve. [2.5a]

61. *Desserts.* If a restaurant sells 250 desserts in an evening, it is typical that 40 of them will be pie. What percent of the desserts sold will be pie?

62. *Desserts.* Of all desserts sold in restaurants, 20% of them are chocolate cake. One evening a restaurant sells 350 desserts. How many were chocolate cake?

63. Harry left a 20% tip of \$6.50 for a meal. What was the cost of the meal before the tip?

64. Rambeau paid \$27.60 for a taxi ride. This included a 20% tip. How much was the fare before the tip?

Solve. [2.7e]

65. $-1.6x < 64$

66. $-12x - 71 \geq 13$

67. $x + (x - 1) < (x + 2) - (x + 1)$

68. $6 - 18x \leq 4 - 12x - 5x$

Synthesis

69. ◆ If the graph of the equation $Ax + By = C$ is a horizontal line, what can you conclude about A? Why?

70. ◆ Explain in your own words why the graph of $x = 7$ is a vertical line.

71. Write an equation for the y-axis.

72. Write an equation for the x-axis.

73. Write an equation of a line parallel to the x-axis and passing through $(-3, -4)$.

74. Find the value of m such that the graph of $y = mx + 6$ has an x-intercept of $(2, 0)$.

75. Find the value of k such that the graph of $3x + k = 5y$ has an x-intercept of $(-4, 0)$.

76. Find the value of k such that the graph of $4x = k - 3y$ has a y-intercept of $(0, -8)$.

3.4 Applications and Data Analysis with Graphs

a Mean, Median, and Mode

One way to analyze data is to look for a single representative number, called a *center point* or *measure of central tendency.* Those most often used are the *mean* (or *average*), the *median,* and the *mode.*

Mean

Let's first consider the *mean,* or average.

> The **mean**, or **average**, of a set of numbers is the sum of the numbers divided by the number of addends.

Example 1 Consider the following data on revenue, in billions of dollars, at McDonald's restaurants in five recent years:

\$12.5, \$13.2, \$14.2, \$14.9, \$15.9.

(***Source***: McDonalds Corporation). What is the mean of the numbers?

First, we add the numbers:

$$12.5 + 13.2 + 14.2 + 14.9 + 15.9 = 70.7.$$

Then we divide by the number of addends, 5:

$$\frac{(12.5 + 13.2 + 14.2 + 14.9 + 15.9)}{5} = \frac{70.7}{5} = 14.14.$$

The mean, or average, revenue of McDonald's for those five years is \$14.14 billion.

Note that

$$14.14 + 14.14 + 14.14 + 14.14 + 14.14 = 70.7.$$

If we use this center point, 14.14, repeatedly as the addend, we get the same sum that we do when adding the individual data numbers.

Do Exercises 1–3.

Median

The *median* is useful when we wish to de-emphasize extreme scores. For example, suppose five workers in a technology company manufactured the following numbers of computers during one day's work:

Sarah:	88	Jen:	94
Matt:	92	Mark:	91
Pat:	66		

Let's first list the scores in order from smallest to largest:

66 88 91 92 94.
↑
Middle number

The middle number—in this case, 91—is the **median.**

Objectives

a Find the mean (average), the median, and the mode of a set of data and solve related applied problems.

b Compare two sets of data using their means.

c Make predictions from a set of data using interpolation or extrapolation.

For Extra Help

TAPE 5

MAC WIN

CD-ROM

Find the mean. Round to the nearest tenth.

1. 28, 103, 39

2. 85, 46, 105.7, 22.1

3. A student scored the following on five tests:

78, 95, 84, 100, 82.

What was the average score?

Answers on page A-11

Find the median.

4. 17, 13, 18, 14, 19

5. 17, 18, 16, 19, 13, 14

6. 122, 102, 103, 91, 83, 81, 78, 119, 88

Find any modes that exist.

7. 33, 55, 55, 88, 55

8. 90, 54, 88, 87, 87, 54

9. 23.7, 27.5, 54.9, 17.2, 20.1

10. In conducting laboratory tests, Carole discovers bacteria in different lab dishes grew to the following areas, in square millimeters:

25, 19, 29, 24, 28.

a) What is the mean?

b) What is the median?

c) What is the mode?

Answers on page A-11

Once a set of data has been arranged from smallest to largest, the **median** of the set of data is the middle number if there is an odd number of data numbers. If there is an even number of data numbers, then there are two middle numbers and the median is the *average* of the two middle numbers.

Example 2 What is the median of the following set of yearly salaries?

\$76,000, \$58,000, \$87,000, \$32,500, \$64,800, \$62,500

We first rearrange the numbers in order from smallest to largest.

\$32,500 \$58,000 \$62,500 ↑ \$64,800 \$76,000 \$87,000

Median

There is an even number of numbers. We look for the middle two, which are \$62,500 and \$64,800. In this case, the median is the average of \$62,500 and \$64,800:

$$\frac{\$62{,}500 + \$64{,}800}{2} = \$63{,}650.$$

Do Exercises 4–6.

Mode

The last center point we consider is called the **mode**. A number that occurs most often in a set of data can be considered a representative number or center point.

The **mode** of a set of data is the number or numbers that occur most often. If each number occurs the same number of times, there is *no* mode.

Example 3 Find the mode of the following data:

23, 24, 27, 18, 19, 27

The number that occurs most often is 27. Thus the mode is 27.

Example 4 Find the mode of the following data:

83, 84, 84, 84, 85, 86, 87, 87, 87, 88, 89, 90.

There are two numbers that occur most often, 84 and 87. Thus the modes are 84 and 87.

Example 5 Find the mode of the following data:

115, 117, 211, 213, 219.

Each number occurs the same number of times. The set of data has *no* mode.

Do Exercises 7–10.

b Comparing Two Sets of Data

We have seen how data are displayed and interpreted using graphs, and we have calculated the mean, the median, and the mode from data. Now we look into using data analysis to solve applied problems.

One goal of analyzing two sets of data is to make a determination about which of two groups is "better." One way to do so is by comparing the means.

Example 6 *Light-Bulb Testing.* An experiment is performed to compare the lives of two types of light bulb. Several bulbs of each type were tested and the results are listed in the following table. On the basis of this test, which bulb is better?

Bulb A: HotLight Life Times (in hours)		
983	964	1214
1417	1211	1521
1084	1075	892
1423	949	

Bulb B: BrightBulb Life Times (in hours)		
979	1083	1344
984	1445	975
1492	1325	1283
1325	1352	1432

Note that it is difficult to analyze the data at a glance because the numbers are close together and there is a different number of data points in each set. We need a way to compare the two groups. Let's compute the average of each set of data.

Bulb A: Average

$$= \frac{(983 + 964 + 1214 + 1417 + 1211 + 1521 + 1084 + 1075 + 892 + 1423 + 949)}{11}$$

$$= \frac{12{,}733}{11} \approx 1157.55;$$

Bulb B: Average

$$= \frac{(979 + 1083 + 1344 + 984 + 1445 + 975 + 1492 + 1325 + 1283 + 1325 + 1352 + 1432)}{12}$$

$$= \frac{15{,}019}{12} \approx 1251.58.$$

We see that the average life of bulb B is higher than that of bulb A and thus conclude that bulb B is "better." (It should be noted that statisticians might question whether these differences are what they call "significant." The answer to that question belongs to a later math course.)

Do Exercise 11.

11. *Quality of Baseballs.* Lauri experiments to see which of two kinds of baseball is better by determining which bounces highest. She drops balls of each kind from a height of 6 ft onto concrete and measures how high they bounce, in inches. Which kind of ball is better?

Ball A Bouncing Heights (in inches)			
16.2	22.3	19.5	15.7
19.6	18.0	15.6	21.7
19.8	16.4	18.4	16.6
21.5	18.7	22.0	18.3

Ball B Bouncing Heights (in inches)			
19.7	18.4	19.7	17.2
19.7	14.6	22.0	23.7
16.5	21.6	22.5	19.8
22.6	17.9	18.7	

Answer on page A-11

12. *Monthly Loan Payment.* The following table lists monthly payments on a loan of $110,000 at 9% interest. Note that there is no data point for a 35-yr loan. Use interpolation to estimate the missing value.

Number of Years	Monthly Payment (in dollars)
5	$2283.42
10	1393.43
15	1115.69
20	989.70
25	923.12
30	885.08
35	?
40	848.50

Answer on page A-11

C Making Predictions

Sometimes we use data to make predictions or estimates of missing data points. One process for doing so is called *interpolation*. It uses graphs and/or averages to guess a missing data point between two data points.

Example 7 *World Bicycle Production.* The following table shows how world bicycle production has grown in recent years. Note that there is no data for 1994. Use interpolation to estimate the missing value.

Year	World Bicycle Production (in millions)
1989	95
1990	90
1991	96
1992	103
1993	108
1994	?
1995	114

Source: United Nations Interbike Directory

First, we analyze the data and look for trends. Note that production decreases for the early years, but in more recent years it increases more like a straight line. It seems reasonable that we can draw a line between the points for 1993 and 1995. We zoom in on that portion of the graph, as shown below.

World Bicycle Production

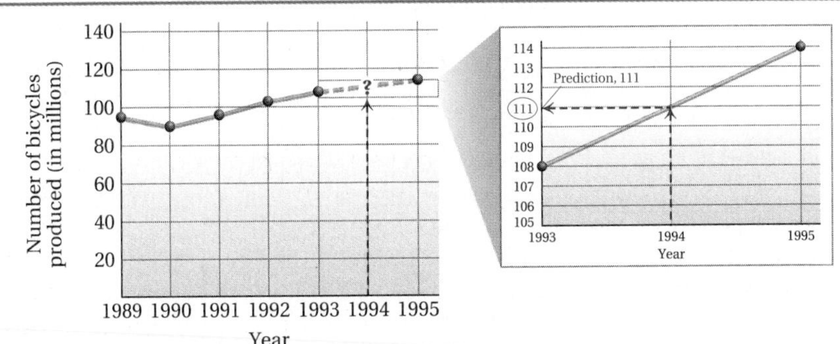

Then we visualize a vertical line up from the point for 1994 and see where the vertical line crosses the line between the data points (1993, 108) and (1995, 114). We move to the left and read off a value—about 111. We can also estimate this value by taking the average of the data values 108 and 114:

$$\frac{(108 + 114)}{2} = 111.$$

When we estimate in this way to find an "in-between value," we are using a process called **interpolation.** Real-world information about the data might tell us that an estimate found in this way is unreliable. For example, data from the stock market might be very erratic.

Do Exercise 12.

We often analyze data with the view of going "beyond" the data. One process for doing so is called *extrapolation*.

Example 8 *World Bicycle Production Extended.* Let's now consider that we know the world production of bicycles in 1994 to be 111 million and add it to the table of data. Use extrapolation to estimate world bicycle production in 1996.

Year	World Bicycle Production (in millions)
1989	95
1990	90
1991	96
1992	103
1993	108
1994	111
1995	114
1996	?

Source: United Nations Interbike Directory

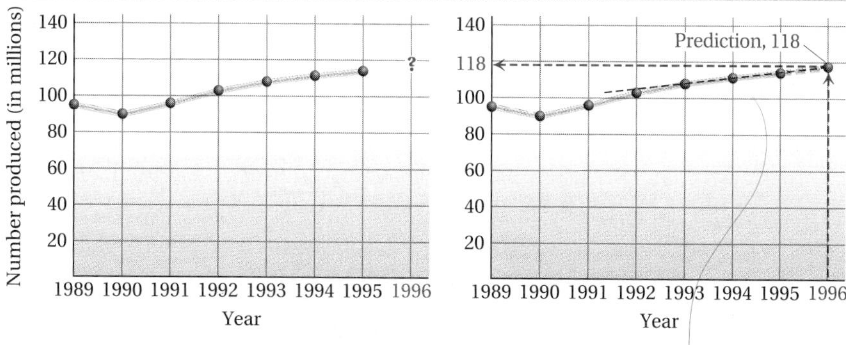

First, we analyze the data and note that they tend to follow a straight line past 1993. Keeping this trend in mind, we draw a "representative" line through the data and beyond. To estimate a value for 1996, we draw a vertical line up from 1996 until it hits the representative line. We go to the left and read off a value—about 118. When we estimate in this way to find a "go-beyond value," we are using a process called **extrapolation.** Answers found with this method can vary greatly depending on the points chosen to determine the "representative" line.

Do Exercise 13.

13. *Study Time and Test Scores.* A professor gathered the following data comparing study time and test scores. Use extrapolation to estimate the test score received when studying for 22 hr.

Study Time (in hours)	Test Grade (in percent)
18	84
19	86
20	89
21	92
22	?

Answer on page A-11

Calculator Spotlight

TRACE and TABLE Features

TRACE Feature. There are two ways in which we can determine the coordinates of points on a graph drawn by a grapher. One approach is to use a TRACE key.

Let's consider the equation for the FedEx mailing costs considered in Example 8 of Section 3.2.

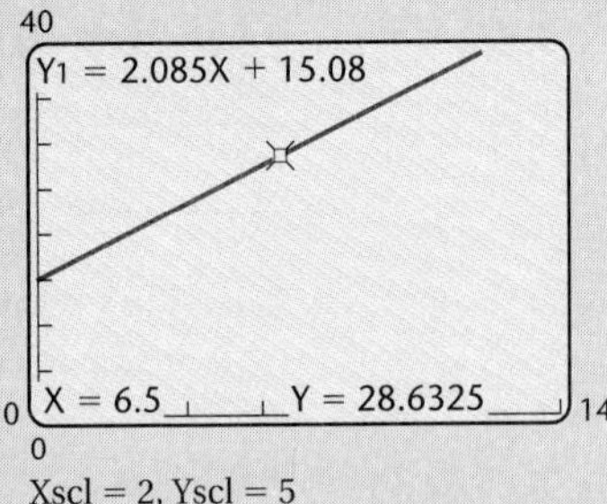

The cursor on the line means that the TRACE feature has been activated. The coordinates at the bottom indicate that the cursor is at the point with coordinates (6.5, 28.6325). By using the arrow keys, we can obtain coordinates of other points. For example, if we press the left arrow key ◁ seven times, we move the cursor to the location shown below, obtaining a point on the graph with coordinates (5.5319149, 26.614043).

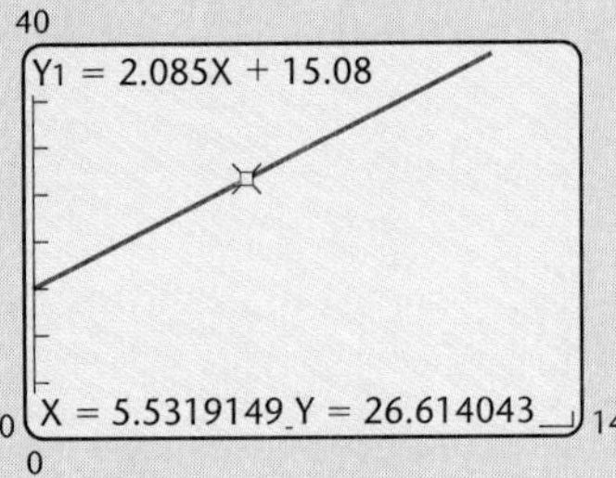

TABLE Feature. Another way to find the coordinates of solutions of equations makes use of the TABLE feature. We first press 2nd TBLSET and set TblStart = 0 and ΔTbl = 10.

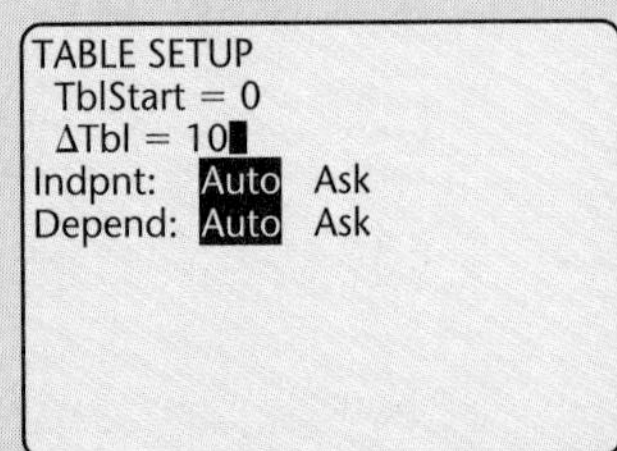

This means that the table's x-values will start at 0 and increase by 10. By setting Indpnt and Depend to Auto, we obtain the following when we press 2nd TABLE: The arrow keys allow us to scroll up and down the table and extend it to other values not initially shown.

X	Y1
0	15.08
10	35.93
20	56.78
30	77.63
40	98.48
50	119.33
60	140.18

X = 0

X	Y1
20	56.78
30	77.63
40	98.48
50	119.33
60	140.18
70	161.03
80	181.88

X = 80

Exercises

1. Use the TRACE feature to find five different ordered-pair solutions of the equation $y = 2.085x + 15.08$.
2. Use the TABLE feature to construct a table of solutions of the equation $y = 2.085x + 15.08$. Set TblStart = 100 with ΔTbl = 50. Find the value of y when x is 100. Then find the value of y when x is 150.
3. Again, use the equation $y = 2.085x + 15.08$ and adjust the table settings to Indpnt: Ask. How does the table change? Enter a number of your choice and see what happens. Use this setting to find the value of y when x is 187.
4. *Value of an Office Machine.* (Refer to Margin Exercise 14 in Section 3.2.) The value of Dupliographic's color copier is given by $v = -0.68t + 3.4$, where v = the value, in thousands of dollars, t years from the date of purchase.
   a) Graph the equation, choosing an appropriate viewing window that shows the intercepts.
   b) Does the equation have meaning for negative x-values?
   c) Find the x-intercept.
   d) For what x-values does the equation have meaning?
   e) Use the TRACE feature to find five ordered-pair solutions for values of x between 0 and 5.
   f) After what amount of time is the value of the copier about \$1500?
   g) Using the TABLE feature, find the value of the copier after 1.7 yr, 2.3 yr, 4.1 yr, and 5 yr.

Exercise Set 3.4

a For each set of numbers, find the mean, the median, and any modes that exist. Round to the nearest tenth.

1. 15, 40, 30, 30, 45, 15, 25

2. 26, 28, 39, 24, 39, 29, 25

3. 81, 93, 96, 98, 102, 94

4. 23.4, 23.4, 22.6, 52.9

5. 23, 42, 35, 37, 23

6. 101.2, 104.3, 107.4, 105.7, 107.4

7. *Coffee Consumption.* The following lists the annual coffee consumption, in cups per person, for various countries.

Germany	1113
United States	610
Switzerland	1215
France	798
Italy	750

(***Source***: Beverage Marketing Corporation)

8. *Calories in Cereal.* The following lists the caloric content of a 2-cup bowl of certain cereals.

Ralston Rice Chex	240
Kellogg's Complete Bran Flakes	240
Kellogg's Special K	220
Honey Nut Cheerios	240
Wheaties	220

9. *NBA Tall Men.* The following is a list of the heights, in inches, of the tallest men in the NBA in a recent year.

Shaquille O'Neal	85
Gheorghe Muresan	91
Shawn Bradley	90
Priest Lauderdale	88
Rik Smits	88
David Robinson	85
Arvydas Sabonis	87

(***Source***: National Basketball Association)

10. *Movie Tickets Sold.* The following lists the number of movie tickets sold, in billions, in six recent years.

1990	1.19
1991	1.14
1992	1.17
1993	1.24
1994	1.29
1995	1.26

(***Source***: Motion Picture Association of America)

b Compare the set of data using their means.

11. *Battery Testing.* An experiment is performed to compare battery quality. Two kinds of battery were tested to see how long, in hours, they kept a portable CD player running. On the basis of this test, which battery is better?

Battery A: EternReady Times (in hours)			Battery B: SturdyCell Times (in hours)		
27.9	28.3	27.4	28.3	27.6	27.8
27.6	27.9	28.0	27.4		27.9
26.8	27.7	28.1	26.9	27.8	28.1
28.2	26.9	27.4	27.9	28.7	27.6

Battery B

12. *Growth of Wheat.* A farmer experiments to see which of two kinds of wheat is better. (In this situation, the shorter wheat is considered "better.") He grows both kinds under similar conditions and measures stalk heights, in inches, as follows. Which kind is better? Wheat B

Wheat A Stalk Heights (in inches)				Wheat B Stalk Heights (in inches)			
16.2	42.3	19.5	25.7	19.7	18.4	32.0	25.7
25.6	18.0	15.6	41.7	19.7	21.6	42.5	32.6
22.6	26.4	18.4	12.6	14.0	10.9	26.7	22.8
41.5	13.7	42.0	21.6	22.6	19.7	17.2	

c Use interpolation or extrapolation to estimate the missing data values. Answers found using extrapolation can vary greatly.

13. *Height of Girls.* 162.4

Age	Height (in centimeters)
2	95.7
4	103.2
6	115.9
8	128.0
10	138.6
12	151.9
14	159.6
16	162.2
17	?
18	162.5

Source: Kempe, C. Henry, M.D., et al., eds., *Current Pediatric Diagnosis & Treatment 1987.* Norwalk, CT: Appleton & Lange, 1987.

14. *Height of Boys.* 167.2

Age	Height (in centimeters)
2	96.2
4	103.4
6	117.5
8	130.0
10	140.3
12	149.6
14	162.7
15	?
16	171.6
18	174.5

Source: Kempe, C. Henry, M.D., et al., eds., *Current Pediatric Diagnosis & Treatment 1987.* Norwalk, CT: Appleton & Lange, 1987.

15. *World Population.*

Population (in billions)	Year in Which Population Is Reached
1	1804
2	1927
3	1960
4	1974
5	1987
6	1998
7	?

Source: U.S. Bureau of the Census

16. *SAT Scores.*

Year	Average SAT Score, Math and Verbal
1991	999
1992	1001
1993	1003
1994	1003
1995	1010
1996	1013
1997	?

Source: The College Board

17. *Movie Tickets Sold.*

Year	Tickets Sold (in billions)
1991	1.19
1992	1.14
1993	1.17
1994	1.24
1995	1.29
1996	1.26
1997	?
2000	?

Source: Motion Picture Association of America

18. *McDonald's Restaurant Revenue in the United States.*

Year	Revenue (in billions)
1990	$12.3
1991	12.5
1992	13.2
1993	14.2
1994	14.9
1995	15.9
1996	?
2000	?

Source: McDonald's Corporation

19. *Average Price of a 30-Second Super Bowl Commercial.*

Year	Cost
1991	$ 800,000
1992	850,000
1993	850,000
1994	900,000
1995	1,000,000
1996	1,300,000
1997	?
2000	?

Source: National Football League

20. *Retail Revenue from Lettuce.*

Year	Revenue (in millions)
1991	$ 106
1992	168
1993	312
1994	577
1995	889
1996	1100
1997	?
2000	?

Source: International Fresh-Cut Produce Association, Information Resources

21. *Study Time vs. Grades.* A mathematics instructor asked her students to keep track of how much time each spent studying the chapter on percent notation in her basic mathematics course. She collected the information together with test scores from that chapter's test. The data are given in the following table. Estimate the missing data value.

Study Time (in hours)	Test Grade (in percent)
9	76
11	94
13	81
15	86
16	87
17	81
18	?
19	87
20	92

22. *Maximum Heart Rate.* A person's maximum heart rate depends on his or her gender, age, and resting heart rate. The following table relates resting heart rate and maximum heart rate for a 20-yr-old man. Estimate the missing data value.

Resting Heart Rate (in beats per minute)	Maximum Heart Rate (in beats per minute)
50	166
60	168
70	170
75	?
80	172

Source: American Heart Association

Skill Maintenance

Solve. [2.2a]

23. $15x = -60$

24. $-\frac{1}{2}x = 10$

25. $-x = 37$

26. $\frac{1}{4} = -\frac{x}{3}$

Solve. [2.5a]

27. Jennifer left an $8.50 tip for a meal that cost $42.50. What percent of the cost of the meal was the tip?

28. Kristen left an 18% tip of $3.24 for a meal. What was the cost of the meal before the tip?

29. Juan left a 15% tip for a meal. The total cost of the meal, including the tip, was $51.92. What was the cost of the meal before the tip was added?

30. After a 25% reduction, a sweater is on sale for $41.25. What was the original price?

Synthesis

31. ◆ In a recent year, the average salary of all players in baseball's American League was $1.3 million and the median was $400,000 (***Source:*** Major League Baseball). Discuss the merits of each value as a measure of central tendency.

32. ◆ Discuss how you might test the estimates that you found in Exercises 13–20.

Graph the equation using the standard viewing window. Then construct a table of y-values for x-values starting at $x = -10$ with $\Delta\text{Tbl} = 0.1$.

33. $y = 0.35x - 7$

34. $y = 5.6 - x^2$

35. $y = x^3 - 5$

36. $y = 4 + 3x - x^2$

Collaborative Learning Manual

Perform a statistical analysis of pulse rates. Make predictions from a set of data.

Summary and Review Exercises: Chapter 3

The objectives to be tested in addition to the material in this chapter are [1.2c], [1.2e], [2.2a], and [2.5a].

1. *Federal Spending.* The following pie chart shows how our federal income tax dollars are used. As a freelance graphic artist, Jennifer pays $3525 in taxes. How much of Jennifer's tax payment goes toward defense? toward social programs? [3.1a]

Where Your Tax Dollars Are Spent

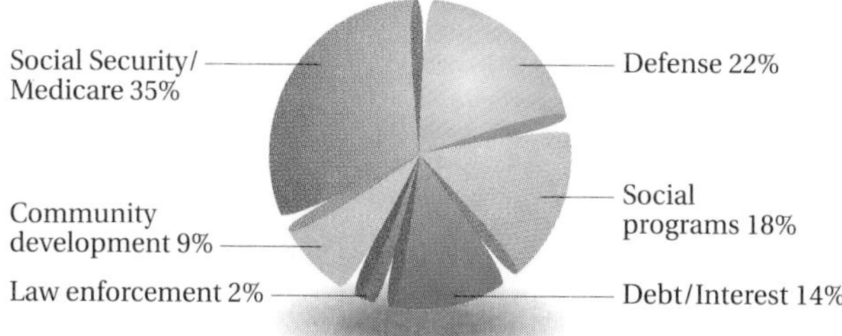

Source: U.S. Department of the Treasury

Chicken Consumption. The following line graph shows average chicken consumption from 1980 to 2000. (The value for the year 2000 is projected.) [3.1a]

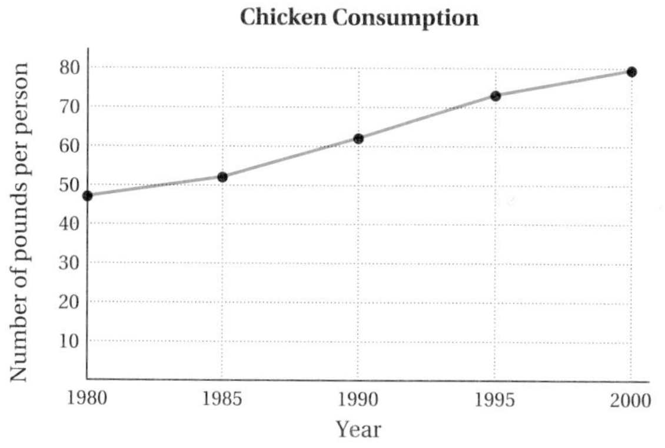

2. About how many pounds of chicken were consumed per person in 1980?

3. About how many pounds of chicken will be consumed per person in 2000?

4. By what amount did chicken consumption increase from 1980 to 2000?

5. In what year did the consumption of chicken exceed 70 lb per person?

6. In what 5-yr period was the difference in consumption the greatest?

Water Usage. The following bar graph shows water usage, in gallons, for various tasks. [3.1a]

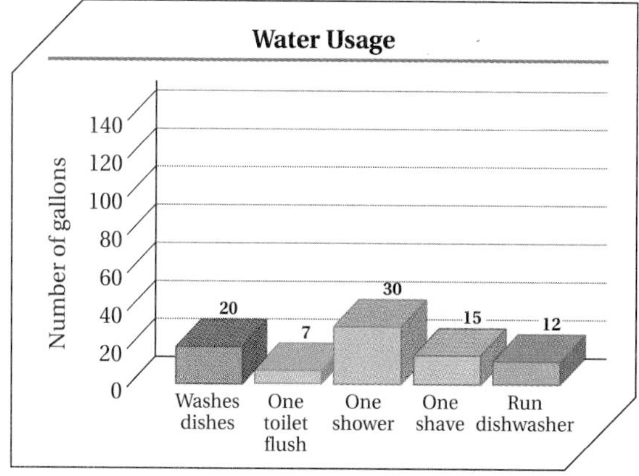

Source: American Water Works Association

7. Which task requires the most water?

8. Which task requires the least water?

9. Which tasks require 15 or more gallons?

10. Which task requires 7 gallons?

Find the coordinates of the point. [3.1d]

11. A **12.** B **13.** C

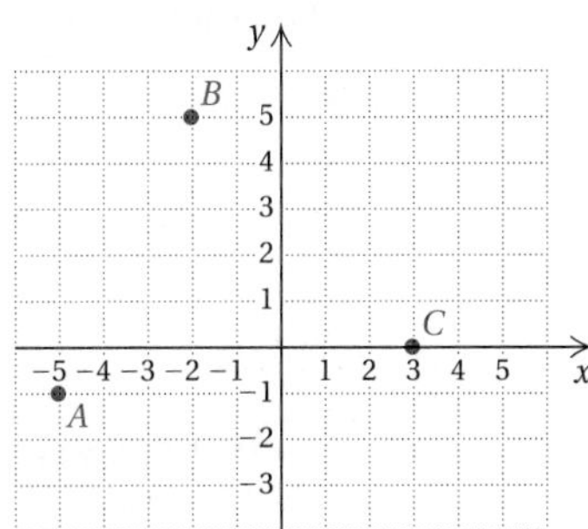

Plot the point. [3.1b]

14. (2, 5) **15.** (0, −3) **16.** (−4, −2)

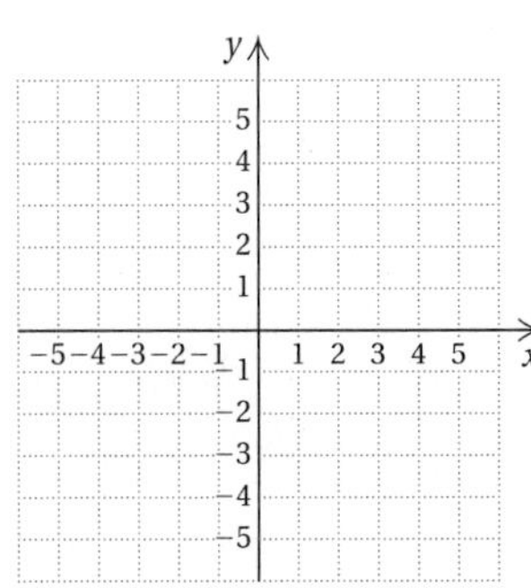

In which quadrant is the point located? [3.1c]

17. (3, −8) **18.** (−20, −14) **19.** (4.9, 1.3)

Determine whether the point is a solution of $2y - x = 10$. [3.2a]

20. (2, −6) **21.** (0, 5)

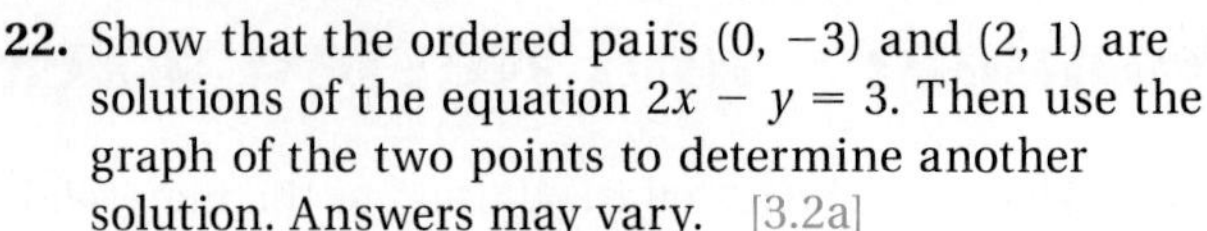

22. Show that the ordered pairs (0, −3) and (2, 1) are solutions of the equation $2x - y = 3$. Then use the graph of the two points to determine another solution. Answers may vary. [3.2a]

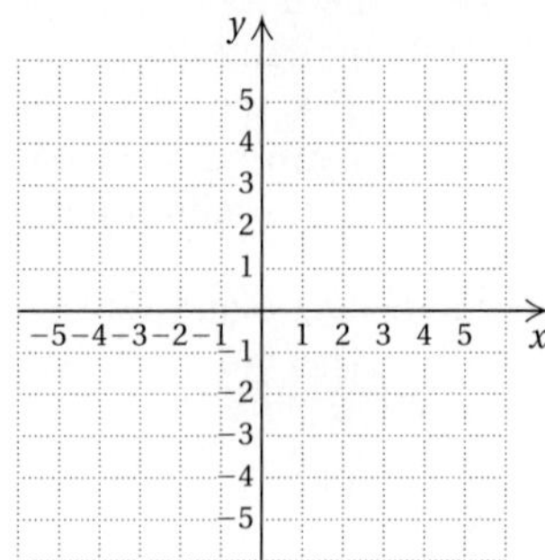

Graph the equation, identifying the y-intercept. [3.2b]

23. $y = 2x - 5$

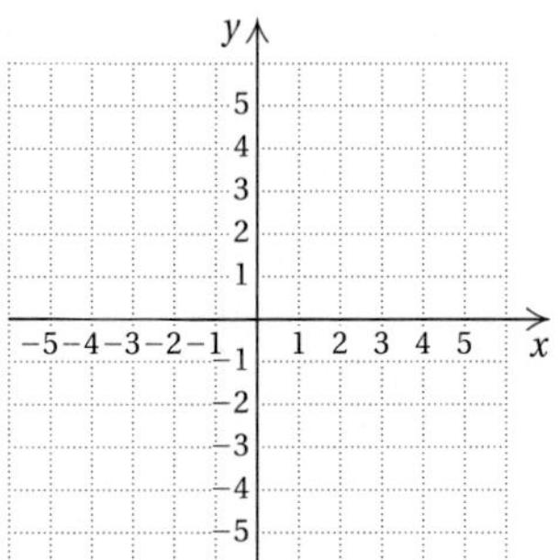

24. $y = -\frac{3}{4}x$

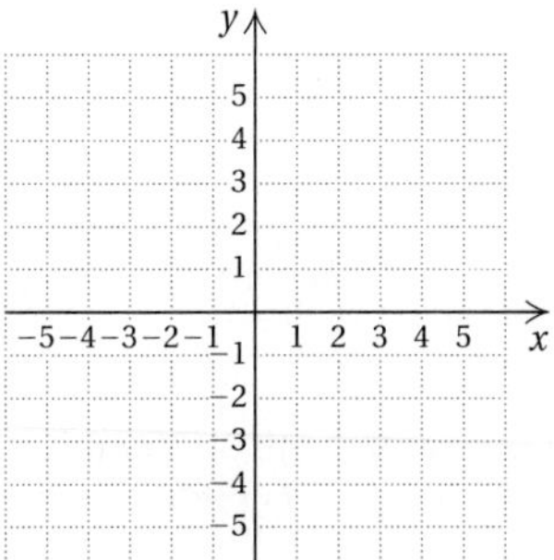

25. $y = -x + 4$

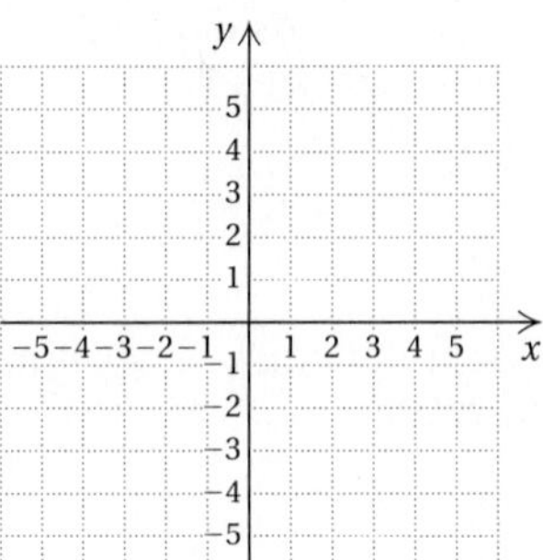

26. $y = 3 - 4x$

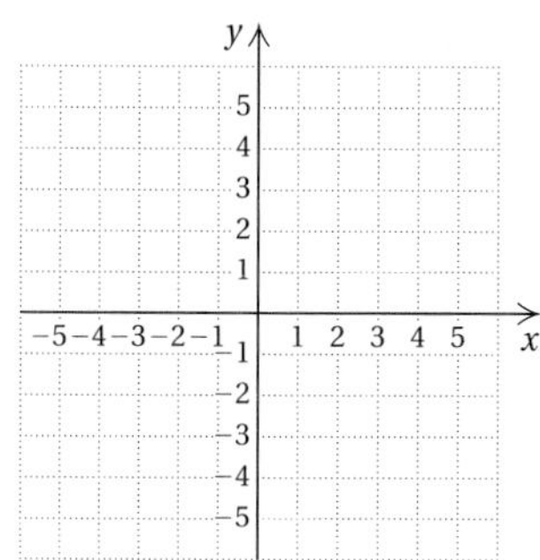

Graph the equation. [3.3b]

27. $y = 3$

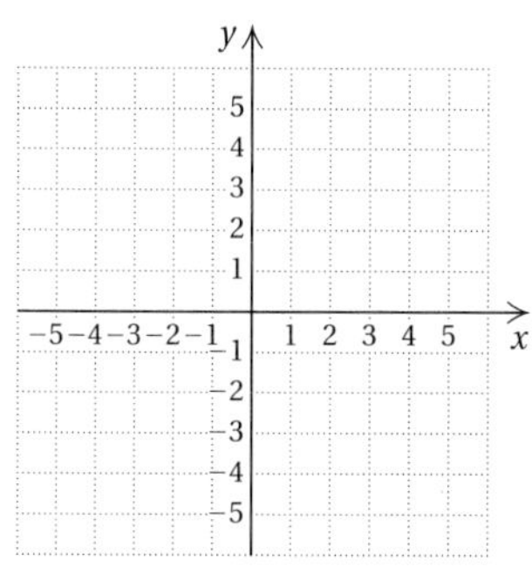

28. $5x - 4 = 0$

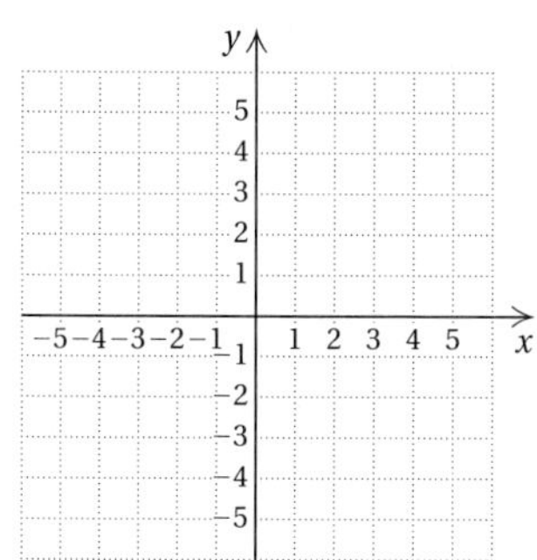

Find the intercepts of the equation. Then graph the equation. [3.3a]

29. $x - 2y = 6$

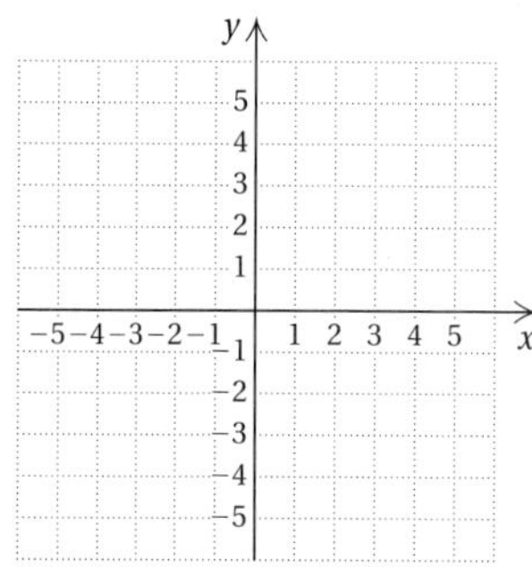

30. $5x - 2y = 10$

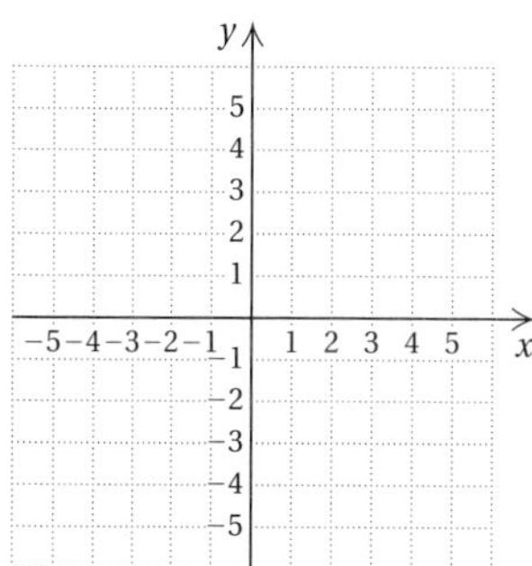

Solve. [3.2c]

31. *Kitchen Design.* Kitchen designers recommend that a refrigerator be selected on the basis of the number of people n in the household. The appropriate size S, in cubic feet, is given by

$$S = \frac{3}{2}n + 13.$$

a) Determine the recommended size of a refrigerator if the number of people is 1, 2, 5, and 10.

b) Graph the equation and use the graph to estimate the recommended size of a refrigerator for 3 people sharing an apartment.

c) A refrigerator is 22 ft^3. For how many residents is it the recommended size?

Find the mean, the median, and the mode. [3.4a]

32. 27, 35, 44, 52

33. 8, 12, 15, 17, 19, 12, 17, 19

34. *Blood Alcohol Levels of Drivers in Fatal Accidents.* 0.18, 0.17, 0.21, 0.16

35. *Bowling Scores of the Author in a Recent Tournament.* 215, 259, 215, 223, 237

36. *Movie Production Costs (in millions of dollars) per Film in Five Recent Years.* $42.3, $44.0, $50.4, $54.1, $59.7

37. *Popcorn Testing.* An experiment is performed to compare the quality of two types of popcorn. The tester puts 200 kernels in a pan, pops them, and counts the number of unpopped kernels. The results are shown in the following table. Determine which type of popcorn is better by comparing their means. [3.4b]

Popcorn A Unpopped Kernels		
20	23	35
10	12	18
18	24	11
19	21	
Popcorn B Unpopped Kernels		
19	25	32
8	22	14
19	13	24
15	22	10

Estimate the missing data values using interpolation or extrapolation. [3.4c]

38. *Height of Girls.*

Age	Height (in centimeters)
2	95.7
4	103.2
5	?
6	115.9
8	128.0
10	138.6
12	151.9
14	159.6
16	162.2
18	162.5

Source: Kempe, C. Henry, M.D., et al., eds., *Current Pediatric Diagnosis & Treatment 1987.* Norwalk, CT: Appleton & Lange, 1987.

39. *Movie Production Costs.*

Year	Cost per Film (in millions)
1992	$42.3
1993	44.0
1994	50.4
1995	54.1
1996	59.7
1997	?
2000	?

Source: Motion Picture Association of America

Skill Maintenance

Convert to decimal notation. [1.2c]

40. $-\dfrac{11}{32}$

41. $\dfrac{8}{9}$

Find the absolute value. [1.2e]

42. $|-3.2|$

43. $\left|\dfrac{17}{19}\right|$

Solve. [2.2a]

44. $-5x = -55$

45. $\dfrac{4}{7}a = \dfrac{3}{7}$

Solve. [2.5a]

46. An investment was made at 6% simple interest for 1 year. It grows to $10,340.40. How much was originally invested?

47. After a 20% reduction, a pair of slacks is on sale for $63.96. What was the original price (that is, the price before reduction)?

Synthesis

48. ◆ Describe two ways in which a small business might make use of graphs. [3.1a], [3.2c]

49. ◆ Explain why the first coordinate of the y-intercept is always 0. [3.2b]

50. Find the value of m in $y = mx + 3$ such that $(-2, 5)$ is on the graph. [3.2a]

51. Find the area and the perimeter of a rectangle for which $(-2, 2)$, $(7, 2)$, and $(7, -3)$ are three of the vertices. [3.1b]

Test: Chapter 3

Toothpaste Sales. The following pie chart shows the percentages of sales of various toothpaste brands in the United States. In a recent year, total sales of toothpaste were $1,500,000,000.

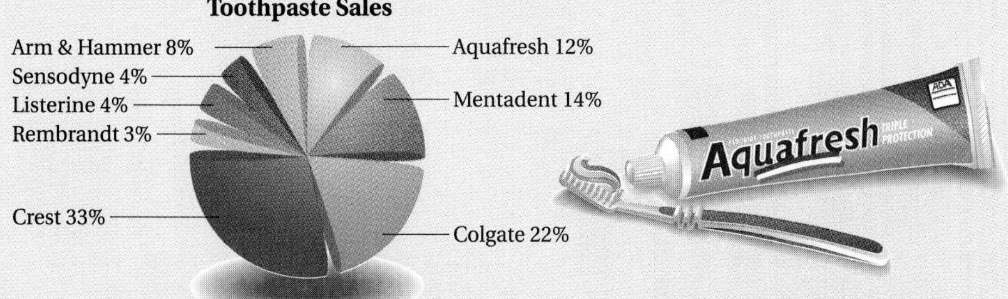

1. What were the total sales of Crest?

2. Which two brands together accounted for over half the sales?

3. Which brand had the greatest sales?

4. Which brand had sales of $120,000,000?

Tornado Touchdowns. The following bar graph shows the total number of tornado touchdowns by month in Indiana from 1950–1994.

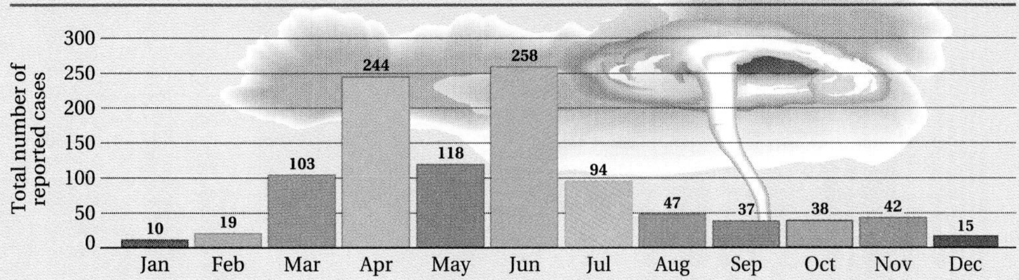

Source: National Weather Service

5. In which month did the greatest number of touchdowns occur?

6. In which month did the least number of touchdowns occur?

7. In which months was the number of touchdowns greater than 90?

8. In which month were there 47 touchdowns?

Answers

1. ______

2. ______

3. ______

4. ______

5. ______

6. ______

7. ______

8. ______

Answers

9. ______

10. ______

11. ______

12. ______

13. ______

14. ______

15. ______

16. ______

17. ______

18. ______

19. ______

20. ______

21. ______

Average Salary of Major-League Baseball Players. The line graph at right shows the average salary of major-league baseball players over a recent seven-year period. Use the graph for Exercises 9–14.

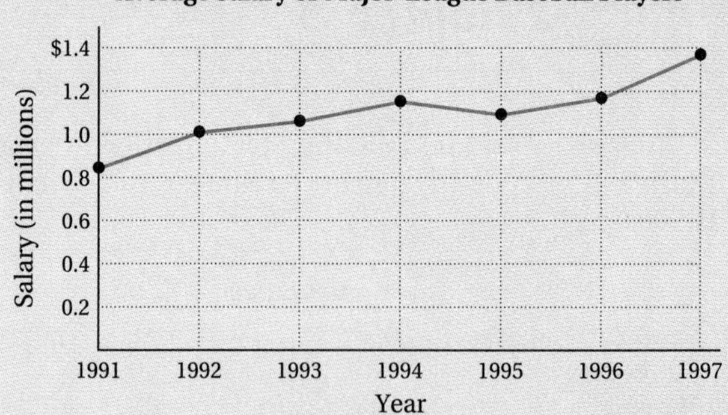

9. In which year was the average salary the highest?

10. In which year was the average salary the lowest?

11. What was the difference in salary between the highest and lowest salaries?

12. Between which two years was the increase in salary the greatest?

13. Between which two years did the salary decrease?

14. By how much did salaries increase between 1991 and 1997?

In which quadrant is the point located?

15. $\left(-\frac{1}{2}, 7\right)$

16. $(-5, -6)$

Find the coordinates of the point.

17. *A*

18. *B*

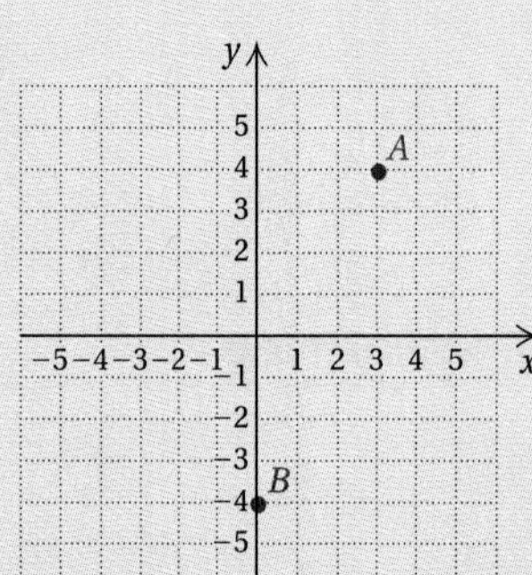

19. Show that the ordered pairs $(-4, -3)$ and $(-1, 3)$ are solutions of the equation $y - 2x = 5$. Then use the graph of the two points to determine another solution. Answers may vary.

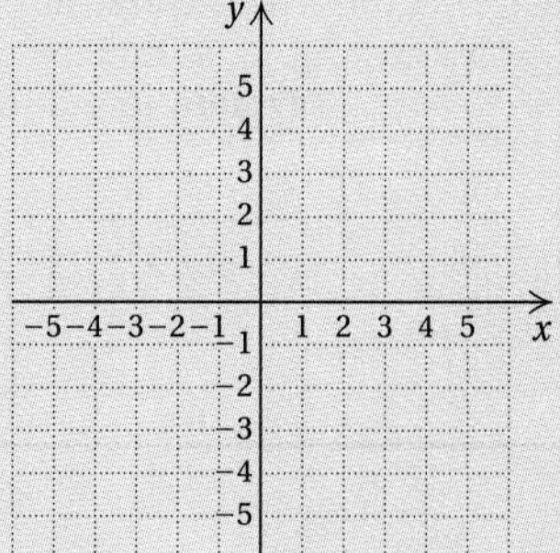

Graph the equation. Identify the y-intercept.

20. $y = 2x - 1$

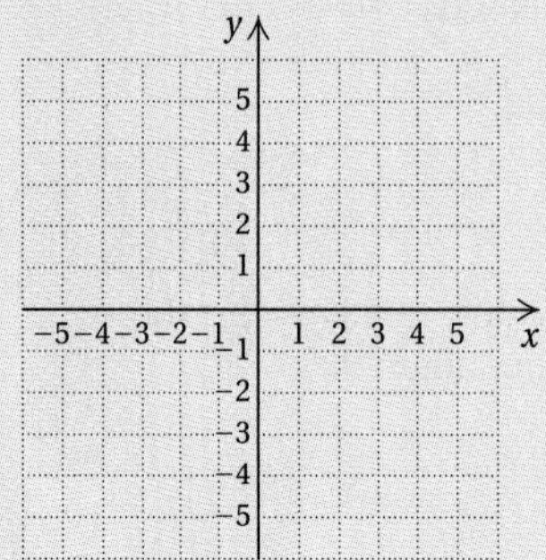

21. $y = -\frac{3}{2}x$

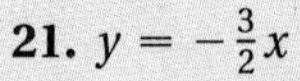

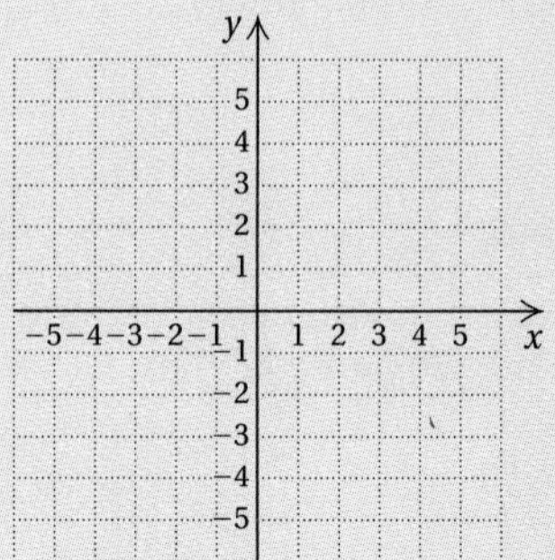

Graph the equation.

22. $2x + 8 = 0$

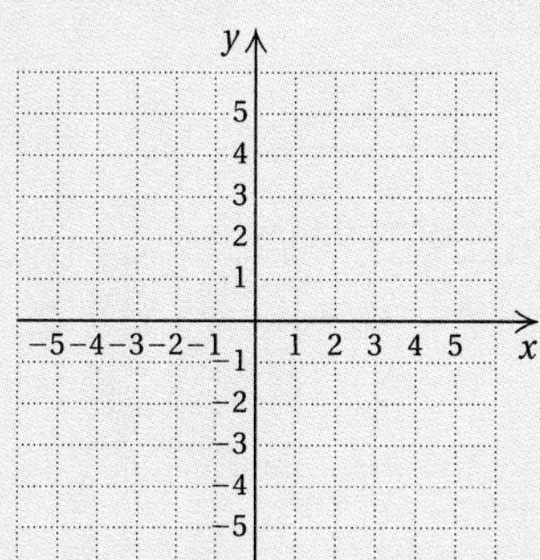

23. $y = 5$

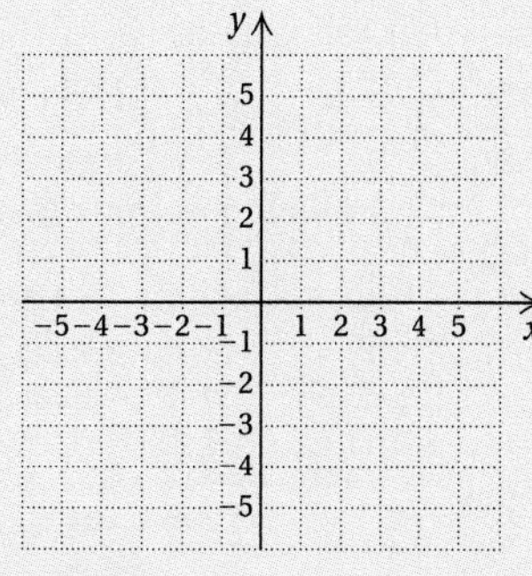

Find the intercepts of the equation. Then graph the equation.

24. $2x - 4y = -8$

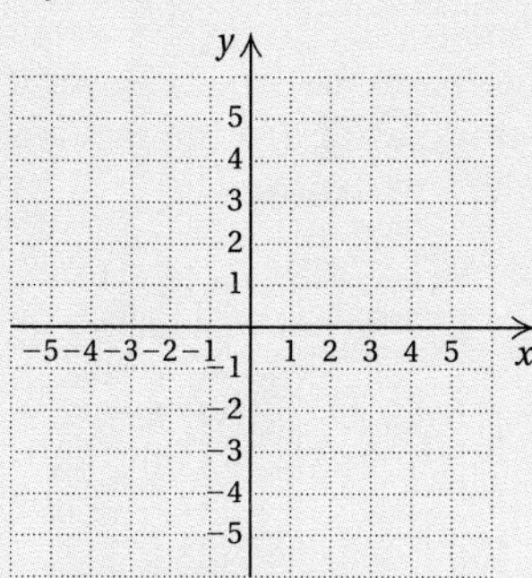

25. $2x - y = 3$

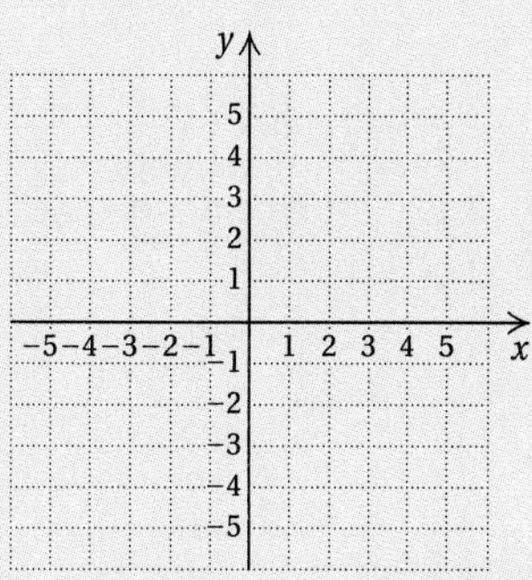

26. *Private-College Costs.* The cost T, in thousands of dollars, of tuition and fees at a private college (all expenses) can be approximated by

$$T = \frac{4}{5}n + 17,$$

where n = the number of years since 1992 (***Source:*** Statistical Abstract of the United States). That is, $n = 0$ corresponds to 1992, $n = 7$ corresponds to 1999, and so on.

a) Find the cost of tuition in 1992, 1995, 1999, and 2001.

b) Graph the equation and then use the graph to estimate the cost of tuition in 2005.

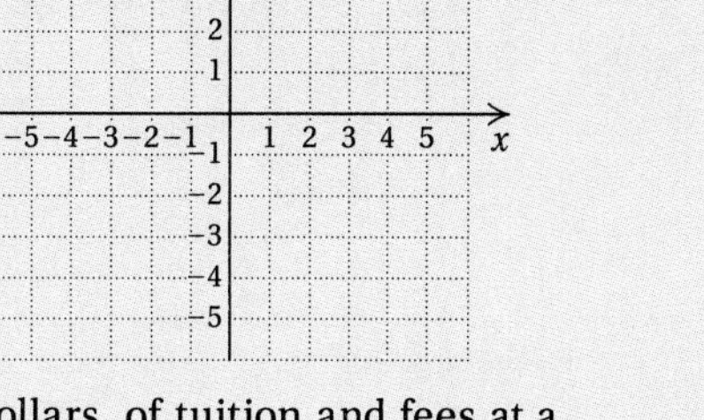

c) Estimate the year in which the cost of tuition will be $25,000.

Find the mean, the median, and the mode.

27. 46, 50, 53, 55

28. 2, 3, 4, 5, 6, 5

29. 4, 19, 20, 18, 19, 18

30. *Animal Speeds.*

Animal	Speed (in miles per hour)
Antelope	61
Bear	30
Cheetah	70
Fastest human	28
Greyhound	39
Lion	50
Zebra	40

Answers

22. ________

23. ________

24. ________

25. ________

26. a) ________

b) ________

c) ________

27. ________

28. ________

29. ________

30. ________

Answers

31. ______

32. ______

33. ______

34. ______

35. ______

36. ______

37. ______

38. ______

39. ______

40. ______

41. ______

42. ______

43. ______

31. *Quality of Golf Balls.* A golf pro experiments to see which of two kinds of golf ball is better. He drops each type of ball from a height of 8 ft onto concrete and measures how high they bounce, in inches. Determine which type of ball is better by comparing their means.

Ball A Bouncing Heights (in inches)				Ball B Bouncing Heights (in inches)			
59.7	58.4	59.7	57.2	56.2	62.3	59.5	65.7
59.7	64.6	62.0	63.7	59.6	58.0	61.6	62.7
66.5	61.6	62.5	59.8	59.8	56.4	58.4	66.6
61.6	57.9	58.7		61.5	58.7	62.0	68.3

Estimate the missing data values using interpolation or extrapolation.

32. *Height of Boys.*

Age	Height (in centimeters)
2	96.2
4	103.4
6	117.5
8	130.0
9	?
10	140.3
12	149.6
14	162.7
16	171.6
18	174.5

Source: Kempe, C. Henry, M.D., et al., eds., *Current Pediatric Diagnosis & Treatment 1987.* Norwalk, CT: Appleton & Lange, 1987.

33. *Deaths from Driving Incidents.* The following table lists for several years the number of driving incidents that resulted in death.

Year	Incidents
1990	1129
1991	1297
1992	1478
1993	1555
1994	1669
1995	1708
1996	?
2000	?

Source: AAA Foundation

Skill Maintenance

Convert to decimal notation.

34. $\frac{39}{40}$

35. $-\frac{13}{12}$

Find the absolute value.

36. $|71.2|$

37. $\left|-\frac{13}{47}\right|$

Solve.

38. $-\frac{5}{8}x = \frac{3}{8}$

39. $39 = 3t$

Solve.

40. After a 24% reduction, a software game is on sale for \$64.22. What was the original price (that is, the price before reduction)?

41. An investment was made at 7% simple interest for 1 year. It grows to \$38,948. How much was originally invested?

Synthesis

42. A diagonal of a square connects the points $(-3, -1)$ and $(2, 4)$. Find the area and the perimeter of the square.

43. Write an equation of a line parallel to the x-axis and 3 units above it.

Cumulative Review: Chapters 1–3

1. Evaluate $\frac{x}{2y}$ for $x = 10$ and $y = 2$.

2. Multiply: $3(4x - 5y + 7)$.

3. Factor: $15x - 9y + 3$.

4. List the terms of the expression $3y - 2x + 4$.

5. Find decimal notation: $\frac{9}{20}$.

6. Find the absolute value: $|-4|$.

7. Find the opposite of -3.08.

8. Find the reciprocal of $-\frac{8}{7}$.

9. Collect like terms: $2x - 5y + (-3x) + 4y$.

10. Find $-x$ when x is 24.6.

Simplify.

11. $\frac{3}{4} - \frac{5}{12}$

12. $3.4 + (-0.8)$

13. $(-2)(-1.4)(2.6)$

14. $\frac{3}{8} \div \left(-\frac{9}{10}\right)$

15. $2 - [32 \div (4 + 2^2)]$

16. $-5 + 16 \div 2 \cdot 4$

17. $y - (3y + 7)$

18. $3(x - 1) - 2[x - (2x + 7)]$

Solve.

19. $1.5 = 2.7 + x$

20. $\frac{2}{7}x = -6$

21. $5x - 9 = 36$

22. $\frac{5}{2}y = \frac{2}{5}$

23. $5.4 - 1.9x = 0.8x$

24. $x - \frac{7}{8} = \frac{3}{4}$

25. $2(2 - 3x) = 3(5x + 7)$

26. $\frac{1}{4}x - \frac{2}{3} = \frac{3}{4} + \frac{1}{3}x$

27. $y + 5 - 3y = 5y - 9$

28. $x - 28 < 20 - 2x$

29. $2(x + 2) \geq 5(2x + 3)$

30. Solve $A = \frac{1}{2}h(b + c)$ for h.

31. In which quadrant is the point $(3, -1)$ located?

32. Graph on a number line: $-1 < x \leq 2$.

Graph.

33. $2x + 5y = 10$

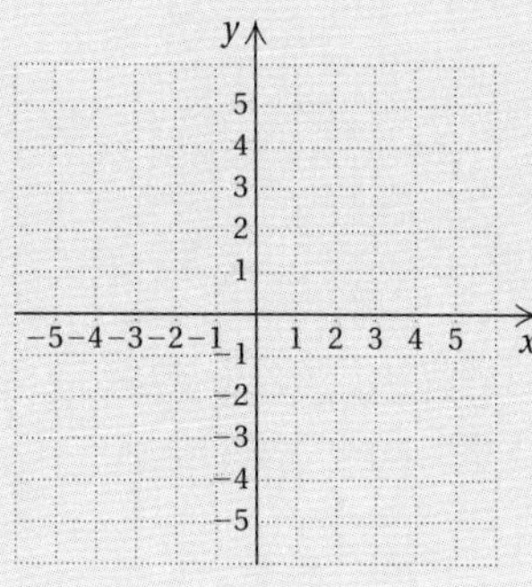

34. $y = -2$

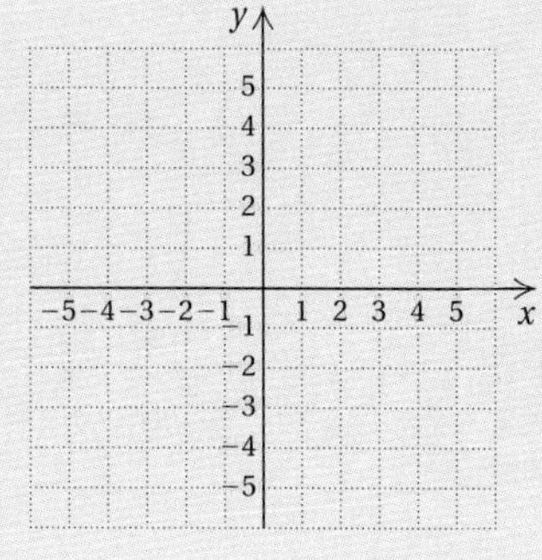

35. $y = -2x + 1$

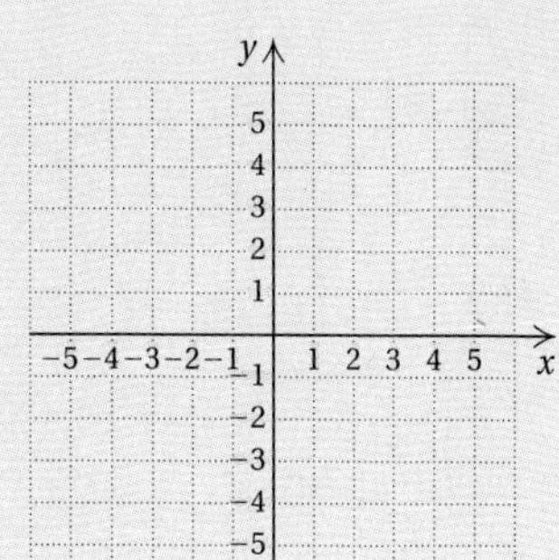

36. $y = \frac{2}{3}x - 2$

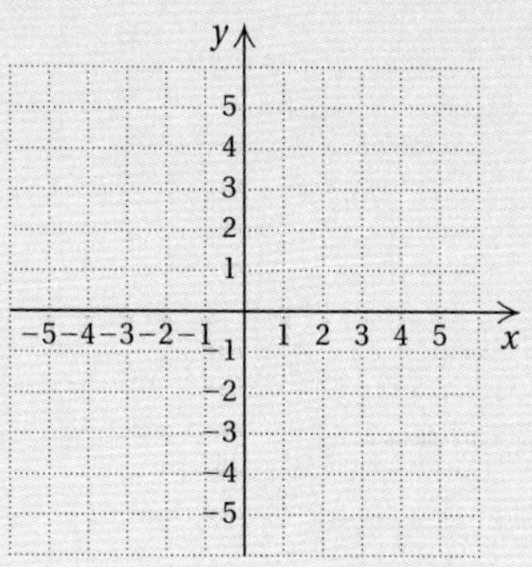

Find the intercepts. Do not graph.

37. $2x - 7y = 21$

38. $y = 4x + 5$

Solve.

39. *Blood Donors.* Each year, 8 million Americans donate blood. This is 5% of those healthy enough to do so. How many Americans are eligible to donate blood?

40. *Blood Types.* There are 117 million Americans with either O-positive or O-negative blood. Those with O-positive blood outnumber those with O-negative blood by 85.8 million. How many Americans have O-negative blood?

41. Tina paid \$126 for a cordless drill. This included 5% for sales tax. How much did the drill cost before tax?

42. A 143-m wire is cut into three pieces. The second piece is 3 m longer than the first. The third is four-fifths as long as the first. How long is each piece?

43. Cory's contract stipulates that he cannot work more than 40 hr per week. For the first four days of one week, he worked 7, 10, 9, and 6 hr. Determine as an inequality the number of hours he can work on the fifth day without violating his contract.

44. *Telephone Line.* The cost P, in hundreds of dollars, of a telephone line for a business is given by $P = \frac{3}{4}n + 3$, where $n =$ the number of months that the line has been in service.

a) Find the cost of the phone line for 1 month, 2 months, 3 months, and 7 months.

b) Graph the equation and use the graph to estimate the cost of the phone line for 10 months.

c) Estimate the number of months it will take for the cost to be \$15,000.

45. Find the mean, the median, and the mode for these medical dosages:

25.4 cc, 31.2 cc, 25.4 cc, 28.7 cc, 32.8 cc, 25.4 cc.

46. *Class Comparisons.* A math instructor conducts an experiment to compare two classes that took the same test. Determine which class performed better by comparing the means.

Class A: 15 students			Class B: 14 students		
78	82	93	79	80	80
100	76	82	74	100	77
66	91	76	64	74	86
78	76	84	88	82	62
92	83	64	93	94	

47. *Height of Girls.* Estimate the missing data value.

Age	Height (in centimeters)
2	95.7
4	103.2
6	115.9
8	128.0
10	138.6
11	?
12	151.9
14	159.6
16	162.2
18	162.5

Synthesis

Solve.

48. $4|x| - 13 = 3$

49. $4(x + 2) = 4(x - 2) + 16$

50. $0(x + 3) + 4 = 0$

51. $\frac{2 + 5x}{4} = \frac{11}{28} + \frac{8x + 3}{7}$

52. $5(7 + x) = (x + 7)5$

53. $p = \frac{2}{m + Q}$, for Q

4

Polynomials: Operations

An Application

The Olympic flame at the 1992 Summer Olympics was lit by a flaming arrow. As the arrow moved d feet horizontally from the archer, its height h, in feet, could be approximated by the polynomial

$$-0.002d^2 + 0.8d + 6.6.$$

Use the polynomial to approximate the height of the arrow after it has traveled 100 ft horizontally.

This problem appears as Exercise 19 in Exercise Set 4.3.

The Mathematics

We substitute 100 for d and evaluate:

$$\underbrace{-0.002d^2 + 0.8d + 6.6}_{\uparrow} = -0.002(100)^2 + 0.8(100) + 6.6 = 66.6 \text{ ft.}$$

This is a polynomial.

World Wide Web For more information, visit us at www.mathmax.com

Introduction

Algebraic expressions like

$$-0.002d^2 + 0.8d + 6.6$$

and

$$x^3 - 5x^2 + 4x - 11$$

are called *polynomials.* One of the most important parts of introductory algebra is the study of polynomials. In this chapter, we learn to add, subtract, multiply, and divide polynomials.

Of particular importance here is the study of the quick ways to multiply certain polynomials. These *special products* will be helpful not only in this text but also in more advanced mathematics.

Pretest: Chapter 4

1. Multiply: $x^{-3} \cdot x^5$.

2. Divide: $\dfrac{x^{-2}}{x^5}$.

3. Simplify: $(-4x^2y^{-3})^2$.

4. Express using a positive exponent: p^{-3}.

5. Convert to scientific notation: 0.000347.

6. Convert to decimal notation: 3.4×10^6.

7. Identify the degree of each term and the degree of the polynomial:

   $2x^3 - 4x^2 + 3x - 5$.

8. Collect like terms:

   $2a^3b - a^2b^2 + ab^3 + 9 - 5a^3b - a^2b^2 + 12b^3$.

9. Add:

   $(5x^2 - 7x + 8) + (6x^2 + 11x - 19)$.

10. Subtract:

    $(5x^2 - 7x + 8) - (6x^2 + 11x - 19)$.

Multiply.

11. $5x^2(3x^2 - 4x + 1)$

12. $(x + 5)^2$

13. $(x - 5)(x + 5)$

14. $(x^3 + 6)(4x^3 - 5)$

15. $(2x - 3y)(2x - 3y)$

16. Divide: $(x^3 - x^2 + x + 2) \div (x - 2)$.

Objectives for Retesting

The objectives to be retested in addition to the material in this chapter are as follows.

[1.4a] Subtract real numbers and simplify combinations of additions and subtractions.

[1.7d] Use the distributive laws to factor expressions like $4x - 12 + 24y$.

[2.3b, c] Solve equations in which like terms may need to be collected, and solve equations by first removing parentheses and collecting like terms.

[2.4a] Solve applied problems by translating to equations.

4.1 Integers as Exponents

a Exponential Notation

An exponent of 2 or greater tells how many times the base is used as a factor. For example,

$$a \cdot a \cdot a \cdot a = a^4.$$

In this case, the **exponent** is 4 and the **base** is a. An expression for a power is called **exponential notation.**

a^n ← This is the exponent.
↑
This is the base.

Example 1 What is the meaning of 3^5? of n^4? of $(2n)^3$? of $50x^2$?

3^5 means $3 \cdot 3 \cdot 3 \cdot 3 \cdot 3$; n^4 means $n \cdot n \cdot n \cdot n$;
$(2n)^3$ means $2n \cdot 2n \cdot 2n$; $50x^2$ means $50 \cdot x \cdot x$

Do Exercises 1–4.

We read exponential notation as follows:

a^n is read the ***n*th power of *a*,** or simply ***a* to the *n*th,** or ***a* to the *n*.**

We often read x^2 as "***x*-squared.**" The reason for this is that the area of a square of side x is $x \cdot x$, or x^2. We often read x^3 as "***x*-cubed.**" The reason for this is that the volume of a cube with length, width, and height x is $x \cdot x \cdot x$, or x^3.

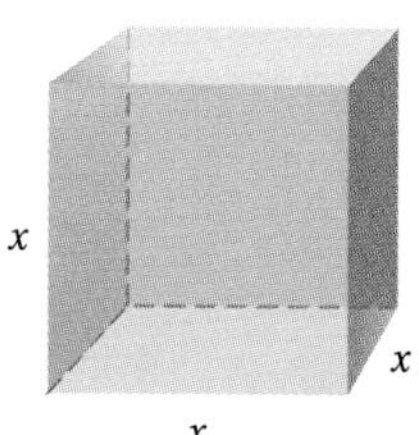

b One and Zero as Exponents

Look for a pattern in the following:

On each side, we divide by 8 at each step.

$$8 \cdot 8 \cdot 8 \cdot 8 = 8^4$$
$$8 \cdot 8 \cdot 8 = 8^3$$
$$8 \cdot 8 = 8^2$$
$$8 = 8^?$$
$$1 = 8^?.$$

On this side, the exponents decrease by 1.

To continue the pattern, we would say that

$$8 = 8^1$$

and $$1 = 8^0.$$

Objectives

a Tell the meaning of exponential notation.

b Evaluate exponential expressions with exponents of 0 and 1.

c Evaluate algebraic expressions containing exponents.

d Use the product rule to multiply exponential expressions with like bases.

e Use the quotient rule to divide exponential expressions with like bases.

f Express an exponential expression involving negative exponents with positive exponents.

For Extra Help

TAPE 6

MAC WIN

CD-ROM

What is the meaning of each of the following?

1. 5^4

2. x^5

3. $(3t)^2$

4. $3t^2$

Answers on page A-14

Evaluate.

5. 6^1

6. 7^0

7. $(8.4)^1$

8. 8654^0

Answers on page A-14

We make the following definition.

> $a^1 = a$, for any number a;
> $a^0 = 1$, for any nonzero number a.

We consider 0^0 to be undefined. We will explain why later in this section.

Example 2 Evaluate 5^1, 8^1, 3^0, $(-7.3)^0$, and $(186{,}892{,}046)^0$.

$5^1 = 5;\quad 8^1 = 8;\quad 3^0 = 1;$

$(-7.3)^0 = 1;\quad (186{,}892{,}046)^0 = 1$

Do Exercises 5–8.

c Evaluating Algebraic Expressions

Algebraic expressions can involve exponential notation. For example, the following are algebraic expressions:

$$x^4,\quad (3x)^3 - 2,\quad a^2 + 2ab + b^2.$$

We evaluate algebraic expressions by replacing variables with numbers and following the rules for order of operations.

Example 3 Evaluate x^4 for $x = 2$.

$x^4 = 2^4$ Substituting

$= 2 \cdot 2 \cdot 2 \cdot 2 = 16$

Example 4 *Area of a Compact Disc.* The standard compact disc used for software and music has a radius of 6 cm. Find the area of such a CD (ignoring the hole in the middle).

$$A = \pi r^2$$
$$= \pi \cdot (6 \text{ cm})^2$$
$$\approx 3.14 \times 36 \text{ cm}^2$$
$$\approx 113.04 \text{ cm}^2$$

$r = 6$ cm

In Example 4, "cm^2" means "square centimeters" and "≈" means "is approximately equal to."

Example 5 Evaluate $(5x)^3$ for $x = -2$.

When we evaluate with a negative number, we often use extra parentheses to show the substitution.

$(5x)^3 = [5 \cdot (-2)]^3$ Substituting

$= [-10]^3$ Multiplying within brackets first

$= -1000$ Evaluating the power

Example 6 Evaluate $5x^3$ for $x = -2$.

$$5x^3 = 5 \cdot (-2)^3 \quad \text{Substituting}$$
$$= 5(-8) \quad \text{Evaluating the power first}$$
$$= -40$$

Recall that two expressions are equivalent if they have the same value for all meaningful replacements. Note that Examples 5 and 6 show that $(5x)^3$ and $5x^3$ are *not* equivalent—that is, $(5x)^3 \neq 5x^3$.

Do Exercises 9–13.

d Multiplying Powers with Like Bases

There are several rules for manipulating exponential notation to obtain equivalent expressions. We first consider multiplying powers with like bases:

$$a^3 \cdot a^2 = \underbrace{(a \cdot a \cdot a)}_{3 \text{ factors}}\underbrace{(a \cdot a)}_{2 \text{ factors}} = \underbrace{a \cdot a \cdot a \cdot a \cdot a}_{5 \text{ factors}} = a^5.$$

Since an integer exponent greater than 1 tells how many times we use a base as a factor, then $(a \cdot a \cdot a)(a \cdot a) = a \cdot a \cdot a \cdot a \cdot a = a^5$ by the associative law. Note that the exponent in a^5 is the sum of those in $a^3 \cdot a^2$. That is, $3 + 2 = 5$. Likewise,

$$b^4 \cdot b^3 = (b \cdot b \cdot b \cdot b)(b \cdot b \cdot b) = b^7, \quad \text{where} \quad 4 + 3 = 7.$$

Adding the exponents gives the correct result.

THE PRODUCT RULE

For any number a and any positive integers m and n,

$$a^m \cdot a^n = a^{m+n}.$$

(When multiplying with exponential notation, if the bases are the same, keep the base and add the exponents.)

Examples Multiply and simplify. By simplify, we mean write the expression as one number to a nonnegative power.

7. $8^4 \cdot 8^3 = 8^{4+3} \quad$ Adding exponents: $a^m \cdot a^n = a^{m+n}$
$= 8^7$

8. $x^2 \cdot x^9 = x^{2+9}$
$= x^{11}$

9. $m^5 m^{10} m^3 = m^{5+10+3}$
$= m^{18}$

10. $x \cdot x^8 = x^1 \cdot x^8 = x^{1+8}$
$= x^9$

11. $(a^3b^2)(a^3b^5) = (a^3a^3)(b^2b^5)$
$= a^6b^7$

Do Exercises 14–18.

9. Evaluate t^3 for $t = 5$.

10. Find the area of a circle when $r = 32$ cm. Use 3.14 for π.

11. Evaluate $200 - a^4$ for $a = 3$.

12. Evaluate $t^1 - 4$ and $t^0 - 4$ for $t = 7$.

13. a) Evaluate $(4t)^2$ for $t = -3$.

b) Evaluate $4t^2$ for $t = -3$.

c) Determine whether $(4t)^2$ and $4t^2$ are equivalent.

Multiply and simplify.

14. $3^5 \cdot 3^5$

15. $x^4 \cdot x^6$

16. $p^4p^{12}p^8$

17. $x \cdot x^4$

18. $(a^2b^3)(a^7b^5)$

Answers on page A-14

Divide and simplify.

19. $\frac{4^5}{4^2}$

20. $\frac{y^6}{y^2}$

21. $\frac{p^{10}}{p}$

22. $\frac{a^7b^6}{a^3b^4}$

Answers on page A-14

e Dividing Powers with Like Bases

The following suggests a rule for dividing powers with like bases, such as a^5/a^2:

$$\frac{a^5}{a^2} = \frac{a \cdot a \cdot a \cdot a \cdot a}{a \cdot a} = \frac{a \cdot a \cdot a \cdot a \cdot a}{1 \cdot a \cdot a} = \frac{a \cdot a \cdot a}{1} \cdot \frac{a \cdot a}{a \cdot a} = \frac{a \cdot a \cdot a}{1} \cdot 1$$
$$= a \cdot a \cdot a = a^3.$$

Note that the exponent in a^3 is the difference of those in $a^5 \div a^2$. If we subtract exponents, we get $5 - 2$, which is 3.

> **THE QUOTIENT RULE**
>
> For any nonzero number a and any positive integers m and n,
>
> $$\frac{a^m}{a^n} = a^{m-n}.$$
>
> (When dividing with exponential notation, if the bases are the same, keep the base and subtract the exponent of the denominator from the exponent of the numerator.)

Examples Divide and simplify. By simplify, we mean write the expression as one number to a nonnegative power.

12. $\frac{6^5}{6^3} = 6^{5-3}$ Subtracting exponents
$= 6^2$

13. $\frac{x^8}{x^2} = x^{8-2}$
$= x^6$

14. $\frac{t^{12}}{t} = \frac{t^{12}}{t^1} = t^{12-1}$
$= t^{11}$

15. $\frac{p^5q^7}{p^2q^5} = \frac{p^5}{p^2} \cdot \frac{q^7}{q^5} = p^{5-2}q^{7-5}$
$= p^3q^2$

The quotient rule can also be used to explain the definition of 0 as an exponent. Consider the expression a^4/a^4, where a is nonzero:

$$\frac{a^4}{a^4} = \frac{a \cdot a \cdot a \cdot a}{a \cdot a \cdot a \cdot a} = 1.$$

This is true because the numerator and the denominator are the same. Now suppose we apply the rule for dividing powers with the same base:

$$\frac{a^4}{a^4} = a^{4-4} = a^0 = 1.$$

Since both expressions a^4/a^4 and a^{4-4} are equivalent to 1, it follows that $a^0 = 1$, when $a \neq 0$.

We can explain why we do not define 0^0 using the quotient rule. We know that 0^0 is 0^{1-1}. But 0^{1-1} is also equal to $0/0$. We have already seen that division by 0 is undefined, so 0^0 is also undefined.

Do Exercises 19–22.

f Negative Integers as Exponents

We can use the rule for dividing powers with like bases to lead us to a definition of exponential notation when the exponent is a negative integer. Consider $5^3/5^7$ and first simplify it using procedures we have learned for working with fractions:

$$\frac{5^3}{5^7} = \frac{5 \cdot 5 \cdot 5}{5 \cdot 5 \cdot 5 \cdot 5 \cdot 5 \cdot 5 \cdot 5} = \frac{5 \cdot 5 \cdot 5 \cdot 1}{5 \cdot 5 \cdot 5 \cdot 5 \cdot 5 \cdot 5 \cdot 5}$$
$$= \frac{5 \cdot 5 \cdot 5}{5 \cdot 5 \cdot 5} \cdot \frac{1}{5 \cdot 5 \cdot 5 \cdot 5} = \frac{1}{5^4}.$$

Now we apply the rule for dividing powers with the same bases. Then

$$\frac{5^3}{5^7} = 5^{3-7} = 5^{-4}.$$

From these two expressions for $5^3/5^7$, it follows that

$$5^{-4} = \frac{1}{5^4}.$$

This leads to our definition of negative exponents:

> For any real number a that is nonzero and any integer n,
>
> $$a^{-n} = \frac{1}{a^n}.$$

In fact, the numbers a^n and a^{-n} are reciprocals of each other because

$$a^n \cdot a^{-n} = a^n \cdot \frac{1}{a^n} = \frac{a^n}{a^n} = 1.$$

Examples Express using positive exponents. Then simplify.

16. $4^{-2} = \frac{1}{4^2} = \frac{1}{16}$

17. $(-3)^{-2} = \frac{1}{(-3)^2} = \frac{1}{(-3)(-3)} = \frac{1}{9}$

18. $m^{-3} = \frac{1}{m^3}$

19. $ab^{-1} = a\left(\frac{1}{b^1}\right) = a\left(\frac{1}{b}\right) = \frac{a}{b}$

20. $\frac{1}{x^{-3}} = x^{-(-3)} = x^3$

21. $3c^{-5} = 3\left(\frac{1}{c^5}\right) = \frac{3}{c^5}$

CAUTION! Note in Example 16 that

$$4^{-2} \neq -16 \quad \text{and} \quad 4^{-2} \neq -\frac{1}{16}.$$

Do Exercises 23–28.

The rules for multiplying and dividing powers with like bases still hold when exponents are 0 or negative. We will state them in a summary at the end of this section.

Express with positive exponents. Then simplify.

23. 4^{-3}

24. 5^{-2}

25. 2^{-4}

26. $(-2)^{-3}$

27. $4p^{-3}$

28. $\frac{1}{x^{-2}}$

Answers on page A-14

Simplify.

29. $5^{-2} \cdot 5^4$

30. $x^{-3} \cdot x^{-4}$

31. $\frac{7^{-2}}{7^3}$

32. $\frac{b^{-2}}{b^{-3}}$

33. $\frac{t}{t^{-5}}$

Answers on page A-14

Examples Simplify. By simplify, we generally mean write the expression as one number to a nonnegative power.

22. $7^{-3} \cdot 7^6 = 7^{-3+6} = 7^3$ Adding exponents

23. $x^4 \cdot x^{-3} = x^{4+(-3)} = x^1 = x$

24. $\frac{5^4}{5^{-2}} = 5^{4-(-2)} = 5^{4+2} = 5^6$ Subtracting exponents

25. $\frac{x}{x^7} = x^{1-7} = x^{-6} = \frac{1}{x^6}$

26. $\frac{b^{-4}}{b^{-5}} = b^{-4-(-5)} = b^{-4+5} = b^1 = b$

27. $y^{-4} \cdot y^{-8} = y^{-4+(-8)} = y^{-12} = \frac{1}{y^{12}}$

In Examples 24–26 (division with exponents), it may help to think as follows: After writing the base, write the top exponent. Then write a subtraction sign. Next write the bottom exponent. Then do the subtraction by adding the opposite. For example,

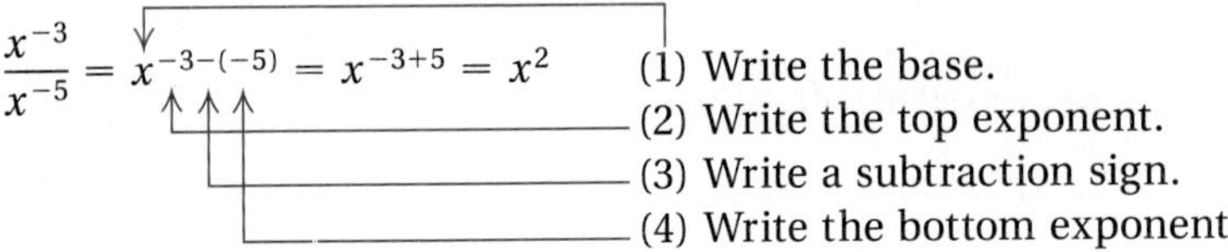

Do Exercises 29–33.

The following is another way to arrive at the definition of negative exponents.

On each side, we divide by 5 at each step.

$5 \cdot 5 \cdot 5 \cdot 5 = 5^4$
$5 \cdot 5 \cdot 5 = 5^3$
$5 \cdot 5 = 5^2$
$5 = 5^1$
$1 = 5^0$
$\frac{1}{5} = 5^?$
$\frac{1}{25} = 5^?$

On this side, the exponents decrease by 1.

To continue the pattern, it should follow that

$$\frac{1}{5} = \frac{1}{5^1} = 5^{-1} \text{ and } \frac{1}{25} = \frac{1}{5^2} = 5^{-2}.$$

The following is a summary of the definitions and rules for exponents that we have considered in this section.

DEFINITIONS AND RULES FOR EXPONENTS

1 as an exponent:	$a^1 = a$;
0 as an exponent:	$a^0 = 1,\ a \neq 0$;
Negative integers as exponents:	$a^{-n} = \frac{1}{a^n},\ \frac{1}{a^{-n}} = a^n;\ a \neq 0$
Product Rule:	$a^m \cdot a^n = a^{m+n}$;
Quotient Rule:	$\frac{a^m}{a^n} = a^{m-n},\ a \neq 0$

Calculator Spotlight

Checking Equivalent Expressions. Let's look at the expressions $x^2 \cdot x^3$ and x^5. We know from the product rule, $x^m \cdot x^n = x^{m+n}$, that these expressions are equivalent. In this case, $x^2 \cdot x^3 = x^{2+3} = x^5$ is true for any real-number substitution. This use of the product rule is an algebraic check of the correctness of the statement $x^2 \cdot x^3 = x^5$. How can we check the result using a grapher? We can do it *graphically* by looking at graphs and *numerically* by looking at a table of values.

Graphical Check. Let's first do a graphical check of $x^2 \cdot x^3 = x^5$. We consider each expression separately and form two equations to be graphed: $y_1 = x^2 \cdot x^3$ and $y_2 = x^5$. The "y_1" is read "y sub 1" and refers simply to a "first" equation. Similarly, "y_2" is read "y sub 2" and refers to a "second" equation. We enter these equations into the grapher using the [y=] key and then graph them, as shown on the left below. Note that the graphs appear to coincide. This is a partial check that the expressions are equivalent. We say "partial check" because most graphs cannot be drawn completely so there is always an element of uncertainty.

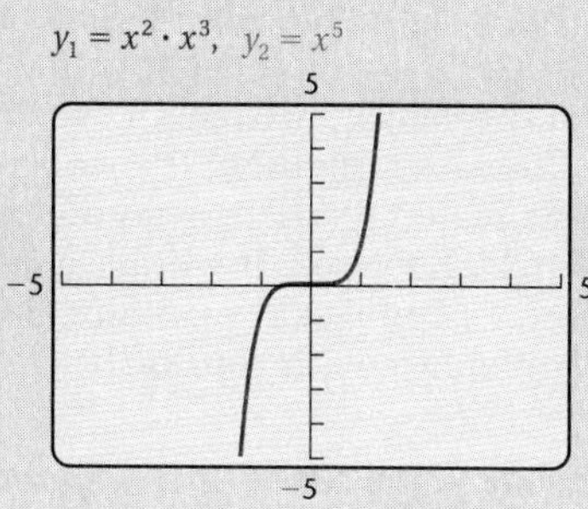

X	Y1	Y2
6	7776	7776
7	16807	16807
8	32768	32768
9	59049	59049
10	100000	100000
11	161051	161051
12	248832	248832

Y1 = 32768

Numerical Check. Now let's use the TABLE feature to check $x^2 \cdot x^3 = x^5$. We already have the equations $y_1 = x^2 \cdot x^3$ and $y_2 = x^5$ entered. The TABLE feature allows us to compare y-values for various x-values. Note in the table on the right above that the y_1- and y_2-values agree. Thus we have a partial check that we have an identity. We say "partial check" because it is impossible to compute all possible y-values and there may be some that disagree.

Let's now consider the equation $x^2 \cdot x^3 = x^6$. Is this a correct result? It seems to violate the product rule, $x^m \cdot x^n = x^{m+n}$. Let's check the equation both graphically and numerically.

Graphical Check. We graph $y_1 = x^2 \cdot x^3$ and $y_2 = x^6$, as shown on the left below. On the TI-83, there is a way to choose a graphing style so that the graphs look different when graphed in the same window. See the window in the middle below. It is obvious that the graphs are different. Thus the equation is not correct.

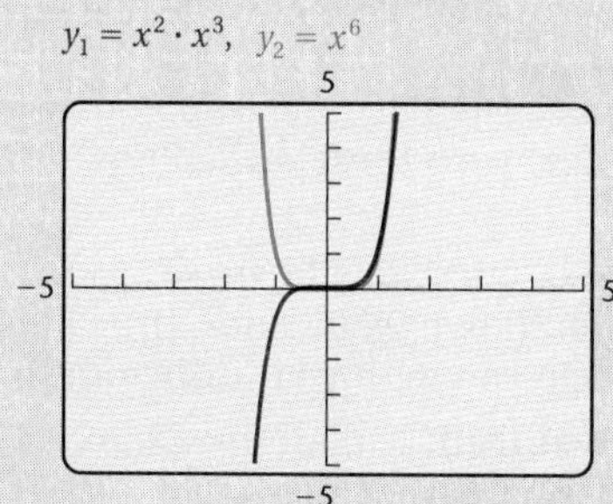

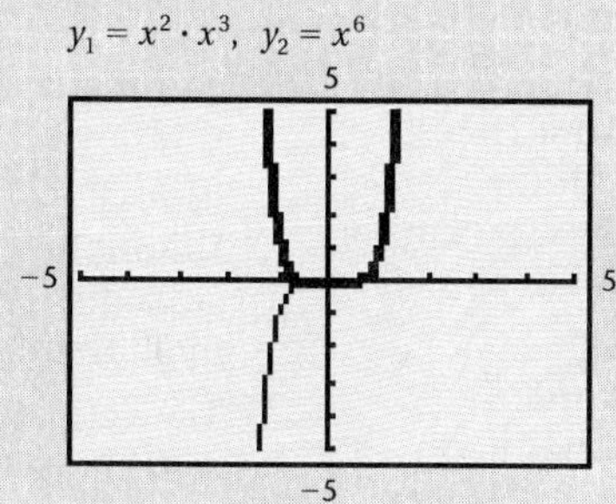

X	Y1	Y2
5	3125	15625
6	7776	46656
7	16807	117649
8	32768	262144
9	59049	531441
10	100000	1E6
11	161051	1.77E6

X = 5

Numerical Check. Let's check a table of y-values. See the table on the right above. Here we note that the y_1- and y_2-values are not the same. Thus, $x^2 \cdot x^3 = x^6$ is not correct.

Exercises Determine whether each of the following equations is correct.

1. $x \cdot x^2 = x^3$
2. $x \cdot x^2 = x^2$
3. $\frac{x^3}{x^2} = x^5$
4. $\frac{x^5}{x^2} = x^3$
5. $\left(\frac{x}{3}\right)^2 = \frac{x^2}{9}$
6. $(5x)^2 = 25x^2$
7. $(x + 2)^2 = x^2 + 4$
8. $(x + 2)^2 = x^2 + 4x + 4$
9. $x + 3 = 3 + x$
10. $3(x - 1) = 3x - 3$
11. $5x - 5 = 5(x - 5)$
12. $10x + 20 = 5(2x + 4)$
13. $2 + (3 + x) = (2 + 3) + x$
14. $5(2x) = 5x$

Improving Your Math Study Skills

Studying for Tests and Making the Most of Tutoring Sessions

This math study skill feature focuses on the very important task of test preparation.

Test-Taking Tips

- **Make up your own test questions as you study.** You have probably become accustomed by now to the section and objective codes that appear throughout the book. After you have done your homework over a particular objective, write one or two questions on your own that you think might be on a test. You will be amazed at the insight this will provide. You are actually carrying out a task similar to what a teacher does in preparing an exam.
- **Do an overall review of the chapter focusing on the objectives and the examples.** This should be accompanied by a study of any class notes you may have taken.
- **Do the review exercises at the end of the chapter.** Check your answers at the back of the book. If you have trouble with an exercise, use the objective symbol as a guide to go back and do further study of that objective. These review exercises are very much like a sample test.
- **Do the chapter test at the end of the chapter.** This is like taking a second sample test. Check the answers and objective symbols at the back of the book.
- **Ask former students for old exams.** Working such exams can be very helpful and allows you to see what various professors think is important.
- **When taking a test, read each question carefully and try to do all the questions the first time through, but pace yourself.** Answer all the questions, and mark those to recheck if you have time at the end. Very often, your first hunch will be correct.
- **Try to write your test in a neat and orderly manner.** Very often, your instructor tries to give you partial credit when grading an exam. If your test paper is sloppy and disorderly, it is difficult to verify the partial credit. Doing your work neatly can ease such a task for the instructor. Try using an erasable pen to make your writing darker and therefore more readable.
- **What about the student who says, "I could do the work at home, but on the test I made silly mistakes"?** Yes, all of us, including instructors, make silly computational mistakes in class, on homework, and on tests. But your instructor, if he or she has taught for some time, is probably aware that 90% of students who make such comments in truth do not have the required depth of knowledge of the subject matter, and such silly mistakes often are a sign that the student has not mastered the material. There is no way we can make that analysis for you. It will have to be unraveled by some careful soul searching on your part or by a conference with your instructor.

Making the Most of Tutoring and Help Sessions

Often you will determine that a tutoring session would be helpful. The following comments may help you to make the most of such sessions.

- **Work on the topics before you go to the help or tutoring session. Do not go to such sessions viewing yourself as an empty cup and the tutor as a magician who will pour in the learning.** The primary source of your ability to learn is within you. We have seen so many students over the years go to help or tutoring sessions with no advanced preparation. You are often wasting your time and perhaps your money if you are paying for such sessions. Go to class, study the textbook, and mark trouble spots. Then use the help and tutoring sessions to deal with these difficulties most efficiently.
- **Do not be afraid to ask questions in these sessions!** The more you talk to your tutor, the more the tutor can help you with your difficulties.
- **Try being a "tutor" yourself.** Explaining a topic to someone else—a classmate, your instructor—is often the best way to learn it.

Exercise Set 4.1

a What is the meaning of each of the following?

1. 3^4

2. 4^3

3. $(1.1)^5$

4. $(87.2)^6$

5. $\left(\frac{2}{3}\right)^4$

6. $\left(-\frac{5}{8}\right)^3$

7. $(7p)^2$

8. $(11c)^3$

9. $8k^3$

10. $17x^2$

b Evaluate.

11. a^0, $a \neq 0$

12. t^0, $t \neq 0$

13. b^1

14. c^1

15. $\left(\frac{2}{3}\right)^0$

16. $\left(-\frac{5}{8}\right)^0$

17. 8.38^0

18. 8.38^1

19. $(ab)^1$

20. $(ab)^0$, $a, b \neq 0$

21. ab^1

22. ab^0

c Evaluate.

23. m^3, for $m = 3$

24. x^6, for $x = 2$

25. p^1, for $p = 19$

26. x^{19}, for $x = 0$

27. x^4, for $x = 4$

28. y^{15}, for $y = 1$

29. $y^2 - 7$, for $y = -10$

30. $z^5 + 5$, for $z = -2$

31. $x^1 + 3$ and $x^0 + 3$, for $x = 7$

32. $y^0 - 8$ and $y^1 - 8$, for $y = -3$

33. Find the area of a circle when $r = 34$ ft. Use 3.14 for π.

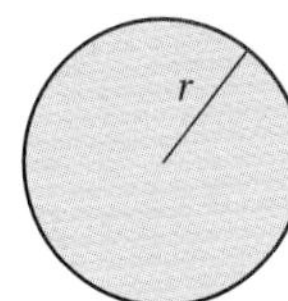

34. The area A of a square with sides of length s is given by $A = s^2$. Find the area of a square with sides of length 24 m.

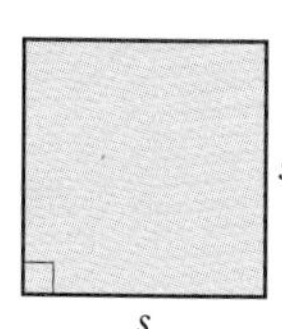

f Express using positive exponents. Then simplify.

35. 3^{-2} **36.** 2^{-3} **37.** 10^{-3} **38.** 5^{-4} **39.** 7^{-3}

40. 5^{-2} **41.** a^{-3} **42.** x^{-2} **43.** $\frac{1}{8^{-2}}$ **44.** $\frac{1}{2^{-5}}$

45. $\frac{1}{y^{-4}}$ **46.** $\frac{1}{t^{-7}}$ **47.** $\frac{1}{z^{-n}}$ **48.** $\frac{1}{h^{-n}}$

Express using negative exponents.

49. $\frac{1}{4^3}$ **50.** $\frac{1}{5^2}$ **51.** $\frac{1}{x^3}$ **52.** $\frac{1}{y^2}$ **53.** $\frac{1}{a^5}$ **54.** $\frac{1}{b^7}$

d, **f** Multiply and simplify.

55. $2^4 \cdot 2^3$ **56.** $3^5 \cdot 3^2$ **57.** $8^5 \cdot 8^9$ **58.** $n^3 \cdot n^{20}$

59. $x^4 \cdot x^3$ **60.** $y^7 \cdot y^9$ **61.** $9^{17} \cdot 9^{21}$ **62.** $t^0 \cdot t^{16}$

63. $(3y)^4(3y)^8$ **64.** $(2t)^8(2t)^{17}$ **65.** $(7y)^1(7y)^{16}$ **66.** $(8x)^0(8x)^1$

67. $3^{-5} \cdot 3^8$ **68.** $5^{-8} \cdot 5^9$ **69.** $x^{-2} \cdot x$ **70.** $x \cdot x^{-1}$

71. $x^{14} \cdot x^3$

72. $x^9 \cdot x^4$

73. $x^{-7} \cdot x^{-6}$

74. $y^{-5} \cdot y^{-8}$

75. $a^{11} \cdot a^{-3} \cdot a^{-18}$

76. $a^{-11} \cdot a^{-3} \cdot a^{-7}$

77. $t^8 \cdot t^{-8}$

78. $m^{10} \cdot m^{-10}$

e, f Divide and simplify.

79. $\frac{7^5}{7^2}$

80. $\frac{5^8}{5^6}$

81. $\frac{8^{12}}{8^6}$

82. $\frac{8^{13}}{8^2}$

83. $\frac{y^9}{y^5}$

84. $\frac{x^{11}}{x^9}$

85. $\frac{16^2}{16^8}$

86. $\frac{7^2}{7^9}$

87. $\frac{m^6}{m^{12}}$

88. $\frac{a^3}{a^4}$

89. $\frac{(8x)^6}{(8x)^{10}}$

90. $\frac{(8t)^4}{(8t)^{11}}$

91. $\frac{(2y)^9}{(2y)^9}$

92. $\frac{(6y)^7}{(6y)^7}$

93. $\frac{x}{x^{-1}}$

94. $\frac{y^8}{y}$

95. $\frac{x^7}{x^{-2}}$

96. $\frac{t^8}{t^{-3}}$

97. $\frac{z^{-6}}{z^{-2}}$

98. $\frac{x^{-9}}{x^{-3}}$

99. $\frac{x^{-5}}{x^{-8}}$

100. $\frac{y^{-2}}{y^{-9}}$

101. $\frac{m^{-9}}{m^{-9}}$

102. $\frac{x^{-7}}{x^{-7}}$

Simplify.

103. 5^2, 5^{-2}, $\left(\frac{1}{5}\right)^2$, $\left(\frac{1}{5}\right)^{-2}$, -5^2, and $(-5)^2$

104. 8^2, 8^{-2}, $\left(\frac{1}{8}\right)^2$, $\left(\frac{1}{8}\right)^{-2}$, -8^2, and $(-8)^2$

Skill Maintenance

105. Translate to an algebraic expression: Sixty-four percent of t. [1.1b]

106. Evaluate $3x/y$ for $x = 4$ and $y = 12$. [1.1a]

107. Divide: $1555.2 \div 24.3$. [1.6c]

108. Add: $1555.2 + 24.3$. [1.3a]

109. Solve: $3x - 4 + 5x - 10x = x - 8$. [2.3b]

110. Factor: $8x - 56$. [1.7d]

Solve. [2.4a]

111. A 12-in. submarine sandwich is cut into two pieces. One piece is twice as long as the other. How long are the pieces?

112. A book is opened. The sum of the page numbers on the facing pages is 457. Find the page numbers.

Synthesis

113. ◆ Under what conditions does a^n represent a negative number? Why?

114. ◆ Explain the errors in each of the following.

a) $2^{-3} = \frac{1}{-8}$

b) $m^{-2}m^5 = m^{10}$

Determine whether each of the following is correct.

115. $(x + 1)^2 = x^2 + 1$

116. $(x - 1)^2 = x^2 - 2x + 1$

117. $(5x)^0 = 5x^0$

118. $\frac{x^3}{x^5} = x^2$

Simplify.

119. $(y^{2x})(y^{3x})$

120. $a^{5k} \div a^{3k}$

121. $\frac{a^{6t}(a^{7t})}{a^{9t}}$

122. $\frac{\left(\frac{1}{2}\right)^4}{\left(\frac{1}{2}\right)^5}$

123. $\frac{(0.8)^5}{(0.8)^3(0.8)^2}$

124. Determine whether $(a + b)^2$ and $a^2 + b^2$ are equivalent. (*Hint*: Choose values for a and b and evaluate.)

Use $>$, $<$, or $=$ for ▒ to write a true sentence.

125. 3^5 ▒ 3^4

126. 4^2 ▒ 4^3

127. 4^3 ▒ 5^3

128. 4^3 ▒ 3^4

Find a value of the variable that shows that the two expressions are *not* equivalent.

129. $3x^2$; $(3x)^2$

130. $\frac{x + 2}{2}$; x

4.2 Exponents and Scientific Notation

We now enhance our ability to manipulate exponential expressions by considering three more rules. The rules are also applied to a new way to name numbers called *scientific notation.*

Objectives

a Use the power rule to raise powers to powers.

b Raise a product to a power and a quotient to a power.

c Convert between scientific notation and decimal notation.

d Multiply and divide using scientific notation.

e Solve applied problems using scientific notation.

For Extra Help

TAPE 6

MAC
WIN

CD-ROM

a Raising Powers to Powers

Consider an expression like $(3^2)^4$. We are raising 3^2 to the fourth power:

$$\begin{aligned}(3^2)^4 &= (3^2)(3^2)(3^2)(3^2)\\ &= (3 \cdot 3)(3 \cdot 3)(3 \cdot 3)(3 \cdot 3)\\ &= 3 \cdot 3 \cdot 3 \cdot 3 \cdot 3 \cdot 3 \cdot 3 \cdot 3\\ &= 3^8.\end{aligned}$$

Note that in this case we could have multiplied the exponents:

$$(3^2)^4 = 3^{2 \cdot 4} = 3^8.$$

Likewise, $(y^8)^3 = (y^8)(y^8)(y^8) = y^{24}$. Once again, we get the same result if we multiply the exponents:

$$(y^8)^3 = y^{8 \cdot 3} = y^{24}.$$

THE POWER RULE

For any real number a and any integers m and n,

$$(a^m)^n = a^{mn}.$$

(To raise a power to a power, multiply the exponents.)

Examples Simplify. Express the answers using positive exponents.

1. $(3^5)^4 = 3^{5 \cdot 4}$ **Multiplying exponents**
   $= 3^{20}$

2. $(2^2)^5 = 2^{2 \cdot 5} = 2^{10}$

3. $(y^{-5})^7 = y^{-5 \cdot 7} = y^{-35} = \dfrac{1}{y^{35}}$

4. $(x^4)^{-2} = x^{4(-2)} = x^{-8} = \dfrac{1}{x^8}$

5. $(a^{-4})^{-6} = a^{(-4)(-6)} = a^{24}$

Do Exercises 1–4.

Calculator Spotlight

Exercises

Determine both graphically and numerically whether each of the following is correct. That is, use the GRAPH and TABLE features on your grapher.

1. $(x^2)^3 = x^6$
2. $(x^2)^3 = x^5$

Simplify. Express the answers using positive exponents.

1. $(3^4)^5$

2. $(x^{-3})^4$

3. $(y^{-5})^{-3}$

4. $(x^4)^{-8}$

Answers on page A-14

b Raising a Product or a Quotient to a Power

When an expression inside parentheses is raised to a power, the inside expression is the base. Let's compare $2a^3$ and $(2a)^3$:

$2a^3 = 2 \cdot a \cdot a \cdot a;$ **The base is a.**

$(2a)^3 = (2a)(2a)(2a)$ **The base is $2a$.**

$= (2 \cdot 2 \cdot 2)(a \cdot a \cdot a)$ **Using the associative and commutative laws of multiplication to regroup the factors**

$= 2^3a^3$

$= 8a^3.$

We see that $2a^3$ and $(2a)^3$ are *not* equivalent. We also see that we can evaluate the power $(2a)^3$ by raising each factor to the power 3. This leads us to the following rule for raising a product to a power.

Simplify.

5. $(2x^5y^{-3})^4$ $\frac{16x^{20}}{y^{12}}$

6. $(5x^5y^{-6}z^{-3})^2$ $\frac{25x^{10}}{y^{12}z^6}$

7. $[(-x)^{37}]^2$ x^{74}

8. $(3y^{-2}x^{-5}z^8)^3$ $\frac{27z^{24}}{y^6x^{15}}$

Simplify.

9. $\left(\frac{x^6}{5}\right)^2$ $\frac{x^{12}}{25}$

10. $\left(\frac{2t^5}{w^4}\right)^3$ $\frac{8t^{15}}{w^{12}}$

11. $\left(\frac{x^4}{3}\right)^{-2}$ $\frac{9}{x^8}$

Calculator Spotlight

Exercises

Determine both graphically and numerically whether each of the following is correct. That is, use the GRAPH and TABLE features on your grapher.

1. $(3x)^2 = 9x^2$ Yes
2. $\left(\frac{x}{4}\right)^2 = \frac{x^2}{16}$ Yes
3. $(2x)^3 = 2x^3$ No
4. $\left(\frac{x}{5}\right)^2 = \frac{x^2}{5}$ No

Answers on page A-14

RAISING A PRODUCT TO A POWER

For any real numbers a and b and any integer n,

$$(ab)^n = a^nb^n.$$

(To raise a product to the nth power, raise each factor to the nth power.)

Examples Simplify.

6. $(4x^2)^3 = 4^3 \cdot (x^2)^3$ Raising each factor to the third power
$= 64x^6$

7. $(5x^3y^5z^2)^4 = 5^4(x^3)^4(y^5)^4(z^2)^4$ Raising each factor to the fourth power
$= 625x^{12}y^{20}z^8$

8. $(-5x^4y^3)^3 = (-5)^3(x^4)^3(y^3)^3$
$= -125x^{12}y^9$

9. $[(-x)^{25}]^2 = (-x)^{50}$ Using the power rule
$= (-1 \cdot x)^{50}$ Using the property of -1 (Section 1.8)
$= (-1)^{50}x^{50}$
$= 1 \cdot x^{50}$ The product of an even number of negative factors is positive.
$= x^{50}$

10. $(5x^2y^{-2})^3 = 5^3(x^2)^3(y^{-2})^3 = 125x^6y^{-6}$ Be sure to raise *each* factor to the third power.
$= \frac{125x^6}{y^6}$

11. $(3x^3y^{-5}z^2)^4 = 3^4(x^3)^4(y^{-5})^4(z^2)^4$
$= 81x^{12}y^{-20}z^8 = \frac{81x^{12}z^8}{y^{20}}$

Do Exercises 5–8.

There is a similar rule for raising a quotient to a power.

RAISING A QUOTIENT TO A POWER

For any real numbers a and b, $b \neq 0$, and any integer n,

$$\left(\frac{a}{b}\right)^n = \frac{a^n}{b^n}.$$

(To raise a quotient to the nth power, raise both the numerator and the denominator to the nth power.)

Examples Simplify.

12. $\left(\frac{x^2}{4}\right)^3 = \frac{(x^2)^3}{4^3} = \frac{x^6}{64}$

13. $\left(\frac{3a^4}{b^3}\right)^2 = \frac{(3a^4)^2}{(b^3)^2} = \frac{3^2(a^4)^2}{b^{3\cdot 2}} = \frac{9a^8}{b^6}$

14. $\left(\frac{y^3}{5}\right)^{-2} = \frac{(y^3)^{-2}}{5^{-2}} = \frac{y^{-6}}{5^{-2}} = \frac{\frac{1}{y^6}}{\frac{1}{5^2}} = \frac{1}{y^6} \div \frac{1}{5^2} = \frac{1}{y^6} \cdot \frac{5^2}{1} = \frac{25}{y^6}$

Do Exercises 9–11.

C Scientific Notation

There are many kinds of symbols, or notation, for numbers. You are already familiar with fractional notation, decimal notation, and percent notation. Now we study another, **scientific notation,** which is especially useful when calculations involve very large or very small numbers. The following are examples of scientific notation:

Niagara Falls: On the Canadian side, during the summer the amount of water that spills over the falls in 1 min is about

1.3088×10^8 L = 130,880,000 L.

The mass of a hydrogen atom:

1.7×10^{-24} g = 0.0000000000000000000000017 g.

> **Scientific notation** for a number is an expression of the type
>
> $M \times 10^n$,
>
> where n is an integer, M is greater than or equal to 1 and less than 10 ($1 \le M < 10$), and M is expressed in decimal notation. 10^n is also considered to be scientific notation when $M = 1$.

You should try to make conversions to scientific notation mentally as much as possible. Here is a handy mental device.

> A positive exponent in scientific notation indicates a large number (greater than 1) and a negative exponent indicates a small number (less than 1).

Examples Convert to scientific notation.

15. $78{,}000 = 7.8 \times 10^4$ 7.8,000. (4 places)

Large number, so the exponent is positive.

16. $0.0000057 = 5.7 \times 10^{-6}$ 0.000005.7 (6 places)

Small number, so the exponent is negative.

Each of the following is *not* scientific notation.

12.46×10^7 — This number is greater than 10.

0.347×10^{-5} — This number is less than 1.

Do Exercises 12 and 13.

Examples Convert mentally to decimal notation.

17. $7.893 \times 10^5 = 789{,}300$ 7.89300. (5 places)

Positive exponent, so the answer is a large number.

18. $4.7 \times 10^{-8} = 0.000000047$ 0.00000004.7 (8 places)

Negative exponent, so the answer is a small number.

Convert to scientific notation.

12. 0.000517

13. 523,000,000

Convert to decimal notation.

14. 6.893×10^{11}

15. 5.67×10^{-5}

Answers on page A-14

Calculator Spotlight

 Scientific Notation.

To enter a number in scientific notation into a graphing calculator, we first enter the decimal portion of the number and then press 2nd EE followed by the exponent.

For example, to enter 1.789×10^{-11}, we press

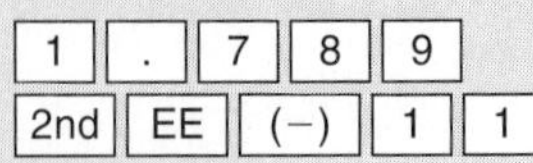

The display reads,

1.789 E ⁻11 .

Exercises

Enter in scientific notation.

1. 260,000,000
2. 0.0000000006709

Multiply and write scientific notation for the result.

16. $(1.12 \times 10^{-8})(5 \times 10^{-7})$

17. $(9.1 \times 10^{-17})(8.2 \times 10^{3})$

Answers on page A-14

Do Exercises 14 and 15 on the preceding page.

d Multiplying and Dividing Using Scientific Notation

Multiplying

Consider the product

$$400 \cdot 2000 = 800{,}000.$$

In scientific notation, this is

$$(4 \times 10^2) \cdot (2 \times 10^3) = (4 \cdot 2)(10^2 \cdot 10^3) = 8 \times 10^5.$$

By applying the commutative and associative laws, we can find this product by multiplying $4 \cdot 2$, to get 8, and $10^2 \cdot 10^3$, to get 10^5 (we do this by adding the exponents).

Example 19 Multiply: $(1.8 \times 10^6) \cdot (2.3 \times 10^{-4})$.

We apply the commutative and associative laws to get

$$\begin{aligned}(1.8 \times 10^6) \cdot (2.3 \times 10^{-4}) &= (1.8 \cdot 2.3) \times (10^6 \cdot 10^{-4}) \\ &= 4.14 \times 10^{6+(-4)} \qquad \text{Adding exponents} \\ &= 4.14 \times 10^2.\end{aligned}$$

Example 20 Multiply: $(3.1 \times 10^5) \cdot (4.5 \times 10^{-3})$.

We have

$$\begin{aligned}(3.1 \times 10^5) \cdot (4.5 \times 10^{-3}) &= (3.1 \times 4.5)(10^5 \cdot 10^{-3}) \\ &= 13.95 \times 10^2.\end{aligned}$$

The answer at this stage is 13.95×10^2, but this is *not* scientific notation, because 13.95 is not a number between 1 and 10. To find scientific notation for the product, we convert 13.95 to scientific notation and simplify:

$$\begin{aligned}13.95 \times 10^2 &= (1.395 \times 10^1) \times 10^2 \qquad \text{Substituting } 1.395 \times 10^1 \text{ for } 13.95 \\ &= 1.395 \times (10^1 \times 10^2) \qquad \text{Associative law} \\ &= 1.395 \times 10^3. \qquad \text{Adding exponents}\end{aligned}$$

The answer is

$$1.395 \times 10^3.$$

Do Exercises 16 and 17.

Dividing

Consider the quotient

$$800{,}000 \div 400 = 2000.$$

In scientific notation, this is

$$(8 \times 10^5) \div (4 \times 10^2) = \frac{8 \times 10^5}{4 \times 10^2} = \frac{8}{4} \times \frac{10^5}{10^2} = 2 \times 10^3.$$

We can find this product by dividing 8 by 4, to get 2, and 10^5 by 10^2, to get 10^3 (we do this by subtracting the exponents).

Example 21 Divide: $(3.41 \times 10^5) \div (1.1 \times 10^{-3})$.

$$\begin{aligned}(3.41 \times 10^5) \div (1.1 \times 10^{-3}) &= \frac{3.41 \times 10^5}{1.1 \times 10^{-3}} \\ &= \frac{3.41}{1.1} \times \frac{10^5}{10^{-3}} \\ &= 3.1 \times 10^{5-(-3)} \\ &= 3.1 \times 10^8\end{aligned}$$

Example 22 Divide: $(6.4 \times 10^{-7}) \div (8.0 \times 10^6)$.

We have

$$\begin{aligned}(6.4 \times 10^{-7}) \div (8.0 \times 10^6) &= \frac{6.4 \times 10^{-7}}{8.0 \times 10^6} \\ &= \frac{6.4}{8.0} \times \frac{10^{-7}}{10^6} \\ &= 0.8 \times 10^{-7-6} \\ &= 0.8 \times 10^{-13}.\end{aligned}$$

The answer at this stage is

$$0.8 \times 10^{-13},$$

but this is *not* scientific notation, because 0.8 is not a number between 1 and 10. To find scientific notation for the quotient, we convert 0.8 to scientific notation and simplify:

$$\begin{aligned}0.8 \times 10^{-13} &= (8.0 \times 10^{-1}) \times 10^{-13} && \text{Substituting } 8.0 \times 10^{-1} \text{ for } 0.8 \\ &= 8.0 \times (10^{-1} \times 10^{-13}) && \text{Associative law} \\ &= 8.0 \times 10^{-14}. && \text{Adding exponents}\end{aligned}$$

The answer is

$$8.0 \times 10^{-14}.$$

Do Exercises 18 and 19.

e Applications with Scientific Notation

Example 23 *Distance from the Sun to Earth.* Light from the sun traveling at a rate of 300,000 kilometers per second (km/s) reaches Earth in 499 sec. Find the distance, expressed in scientific notation, from the sun to Earth.

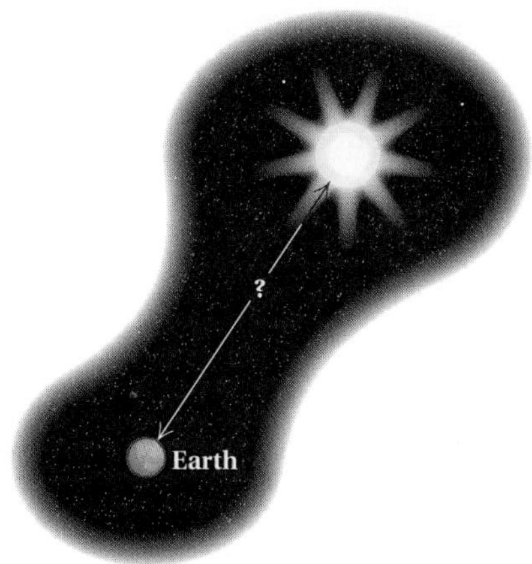

Divide and write scientific notation for the result.

18. $\dfrac{4.2 \times 10^5}{2.1 \times 10^2}$

19. $\dfrac{1.1 \times 10^{-4}}{2.0 \times 10^{-7}}$

Answers on page A-14

20. *Niagara Falls Water Flow.* On the Canadian side, during the summer the amount of water that spills over the falls in 1 min is about

1.3088×10^8 L.

How much water spills over the falls in one day? Express the answer in scientific notation.

21. *Earth vs. Saturn.* The mass of Earth is about 6×10^{21} metric tons. The mass of Saturn is about 5.7×10^{23} metric tons. About how many times the mass of Earth is the mass of Saturn? Express the answer in scientific notation.

Answers on page A-14

The time t that it takes for light to reach Earth from the sun is 4.99×10^2 sec (s). The speed is 3.0×10^5 km/s. Recall that distance can be expressed in terms of speed and time as

$$\text{Distance} = \text{Speed} \cdot \text{Time}$$
$$d = rt.$$

We substitute 3.0×10^5 for r and 4.99×10^2 for t:

$$\begin{aligned} d &= rt \\ &= (3.0 \times 10^5)(4.99 \times 10^2) && \text{Substituting} \\ &= 14.97 \times 10^7 \\ &= 1.497 \times 10^8 \text{ km.} && \text{Converting to scientific notation} \end{aligned}$$

Thus the distance from the sun to Earth is 1.497×10^8 km.

Do Exercise 20.

Example 24 *Earth vs. Jupiter.* The mass of Earth is about 6×10^{21} metric tons. The mass of Jupiter is about 1.908×10^{24} metric tons. About how many times the mass of Earth is the mass of Jupiter? Express the answer in scientific notation.

To determine how many times the mass of Jupiter is of the mass of Earth, we divide the mass of Jupiter by the mass of Earth:

$$\begin{aligned} \frac{1.908 \times 10^{24}}{6 \times 10^{21}} &= \frac{1.908}{6} \times \frac{10^{24}}{10^{21}} \\ &= 0.318 \times 10^3 \\ &= (3.18 \times 10^{-1}) \times 10^3 \\ &= 3.18 \times 10^2. \end{aligned}$$

Thus the mass of Jupiter is 3.18×10^2, or 318, times the mass of Earth.

Do Exercise 21.

The following is a summary of the definitions and rules for exponents that we have considered in this section and the preceding one.

DEFINITIONS AND RULES FOR EXPONENTS

Exponent of 1:	$a^1 = a$
Exponent of 0:	$a^0 = 1,\ a \neq 0$
Negative exponents:	$a^{-n} = \dfrac{1}{a^n},\ a \neq 0$
Product Rule:	$a^m \cdot a^n = a^{m+n}$
Quotient Rule:	$\dfrac{a^m}{a^n} = a^{m-n},\ a \neq 0$
Power Rule:	$(a^m)^n = a^{mn}$
Raising a product to a power:	$(ab)^n = a^n b^n$
Raising a quotient to a power:	$\left(\dfrac{a}{b}\right)^n = \dfrac{a^n}{b^n},\ b \neq 0$
Scientific notation:	$M \times 10^n$, or 10^n, where $1 \leq M < 10$

Exercise Set 4.2

a, b Simplify.

1. $(2^3)^2$

2. $(5^2)^4$

3. $(5^2)^{-3}$

4. $(7^{-3})^5$

5. $(x^{-3})^{-4}$

6. $(a^{-5})^{-6}$

7. $(4x^3)^2$

8. $4(x^3)^2$

9. $(x^4y^5)^{-3}$

10. $(t^5x^3)^{-4}$

11. $(x^{-6}y^{-2})^{-4}$

12. $(x^{-2}y^{-7})^{-5}$

13. $(3x^3y^{-8}z^{-3})^2$

14. $(2a^2y^{-4}z^{-5})^3$

15. $\left(\frac{a^2}{b^3}\right)^4$

16. $\left(\frac{x^3}{y^4}\right)^5$

17. $\left(\frac{y^3}{2}\right)^2$

18. $\left(\frac{a^5}{3}\right)^3$

19. $\left(\frac{y^2}{2}\right)^{-3}$

20. $\left(\frac{a^4}{3}\right)^{-2}$

21. $\left(\frac{x^2y}{z}\right)^3$

22. $\left(\frac{m}{n^4p}\right)^3$

23. $\left(\frac{a^2b}{cd^3}\right)^{-2}$

24. $\left(\frac{2a^2}{3b^4}\right)^{-3}$

c Convert to scientific notation.

25. 28,000,000,000

26. 4,900,000,000,000

27. 907,000,000,000,000,000

28. 168,000,000,000,000

29. 0.00000304

30. 0.000000000865

31. 0.000000018

32. 0.00000000002

33. 100,000,000,000

34. 0.0000001

Convert the number in the sentence to scientific notation.

35. *Niagara Falls Water Flow.* On the American side, during the summer the amount of water that spills over the falls in 1 min is about 11.35 million L (1 million = 10^6).

36. *Proctor & Gamble.* In a recent year, Proctor & Gamble led the nation's advertisers by spending \$2.777 billion on advertising (***Source:*** *Advertising Age*) (1 billion = 10^9).

Convert to decimal notation.

37. 8.74×10^7

38. 1.85×10^8

39. 5.704×10^{-8}

40. 8.043×10^{-4}

41. 10^7

42. 10^6

43. 10^{-5}

44. 10^{-8}

d Multiply or divide and write scientific notation for the result.

45. $(3 \times 10^4)(2 \times 10^5)$

46. $(3.9 \times 10^8)(8.4 \times 10^{-3})$

47. $(5.2 \times 10^5)(6.5 \times 10^{-2})$

48. $(7.1 \times 10^{-7})(8.6 \times 10^{-5})$

49. $(9.9 \times 10^{-6})(8.23 \times 10^{-8})$

50. $(1.123 \times 10^4) \times 10^{-9}$

51. $\dfrac{8.5 \times 10^8}{3.4 \times 10^{-5}}$

52. $\dfrac{5.6 \times 10^{-2}}{2.5 \times 10^5}$

53. $(3.0 \times 10^6) \div (6.0 \times 10^9)$

54. $(1.5 \times 10^{-3}) \div (1.6 \times 10^{-6})$

55. $\dfrac{7.5 \times 10^{-9}}{2.5 \times 10^{12}}$

56. $\dfrac{4.0 \times 10^{-3}}{8.0 \times 10^{20}}$

e Solve.

57. *Total Income of Two-Person Households.* In 1993, there were about 31.2 million two-person households in the United States. The average income of these households was about $42,400. (***Source*:** Statistical Abstract of the United States) Find the total income generated by two-person households in 1993. Express the answer in scientific notation.

58. *Niagara Falls Water Flow.* On the American side, during the summer the amount of water that spills over the falls in 1 min is about 11.35 million L (1 million $= 10^6$). How much water spills over the falls in 1 yr? (Use 365 days for 1 yr.) Express the answer in scientific notation.

59. *Stars.* It is estimated that there are 10 billion trillion stars in the known universe. Express the number of stars in scientific notation.

60. *Closest Star.* Excluding the sun, the closest star to Earth is Proxima Centauri, which is 4.3 light-years away (one light-year $= 5.88 \times 10^{12}$ mi). How far, in miles, is Proxima Centauri from Earth? Express the answer in scientific notation.

61. *Earth vs. Sun.* The mass of Earth is about 6×10^{21} metric tons. The mass of the sun is about 1.998×10^{27} metric tons. About how times the mass of Earth is the mass of the sun? Express the answer in scientific notation.

62. *Red Light.* The wavelength of light is given by the velocity divided by the frequency. The velocity of red light is 300,000,000 m/sec, and its frequency is 400,000,000,000,000 cycles per second. What is the wavelength of red light? Express the answer in scientific notation.

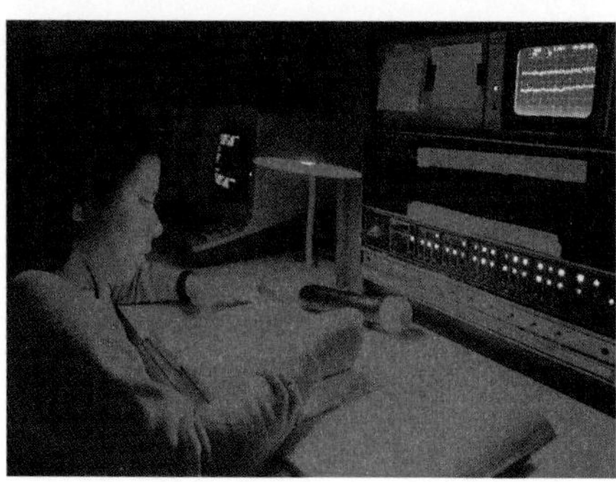

Space Travel. Use the following information for Exercises 63 and 64.

Earth

Moon

$d = 240{,}000$ mi

Approximate Distance from Earth to:	
Moon	240,000 mi
Mars	35,000,000 mi
Pluto	2,670,000,000 mi

63. *Time to Reach Mars.* Suppose that it takes about 3 days for a space vehicle to travel from Earth to the moon. About how long would it take the same space vehicle traveling at the same speed to reach Mars? Express the answer in scientific notation.

64. *Time to Reach Pluto.* Suppose that it takes about 3 days for a space vehicle to travel from Earth to the moon. About how long would it take the same space vehicle traveling at the same speed to reach Pluto? Express the answer in scientific notation.

Skill Maintenance

Factor. [1.7d]

65. $9x - 36$

66. $4x - 2y + 16$

67. $3s + 3t + 24$

68. $-7x - 14$

Solve. [2.3b]

69. $2x - 4 - 5x + 8 = x - 3$

70. $8x + 7 - 9x = 12 - 6x + 5$

Solve. [2.3c]

71. $8(2x + 3) - 2(x - 5) = 10$

72. $4(x - 3) + 5 = 6(x + 2) - 8$

Graph. [3.2b], [3.3a]

73. $y = x - 5$

74. $2x + y = 8$

Synthesis

75. ◆ Using the quotient rule, explain why 9^0 is defined to be 1.

76. ◆ Explain in your own words when exponents should be added and when they should be multiplied.

77. 🖩 Carry out the indicated operations. Express the result in scientific notation.

$$\frac{(5.2 \times 10^6)(6.1 \times 10^{-11})}{1.28 \times 10^{-3}}$$

78. Find the reciprocal and express it in scientific notation.

$$6.25 \times 10^{-3}$$

Simplify.

79. $\dfrac{(5^{12})^2}{5^{25}}$

80. $\dfrac{a^{22}}{(a^2)^{11}}$

81. $\dfrac{(3^5)^4}{3^5 \cdot 3^4}$

82. $\dfrac{49^{18}}{7^{35}}$

(*Hint*: Study Exercise 80.)

83. $\left(\dfrac{1}{a}\right)^{-n}$

84. $\dfrac{(0.4)^5}{[(0.4)^3]^2}$

Determine whether each of the following is true for any pairs of integers m and n and any positive numbers x and y.

85. $x^m \cdot y^n = (xy)^{mn}$

86. $x^m \cdot y^m = (xy)^{2m}$

87. $(x - y)^m = x^m - y^m$

Use exponential and scientific notation to represent the salary for a job.

4.3 Introduction to Polynomials

We have already learned to evaluate and to manipulate certain kinds of algebraic expressions. We will now consider algebraic expressions called *polynomials.*

The following are examples of *monomials in one variable*:

$$3x^2, \quad 2x, \quad -5, \quad 37p^4, \quad 0.$$

Each expression is a constant or a constant times some variable to a nonnegative integer power.

> A **monomial** is an expression of the type ax^n, where a is a real-number constant and n is a nonnegative integer.

Algebraic expressions like the following are **polynomials:**

$$\tfrac{3}{4}y^5, \quad -2, \quad 5y + 3, \quad 3x^2 + 2x - 5, \quad -7a^3 + \tfrac{1}{2}a, \quad 6x, \quad 37p^4, \quad x, \quad 0.$$

> A **polynomial** is a monomial or a combination of sums and/or differences of monomials.

The following algebraic expressions are *not* polynomials:

$$\textbf{(1)}\ \frac{x+3}{x-4}, \qquad \textbf{(2)}\ 5x^3 - 2x^2 + \frac{1}{x}, \qquad \textbf{(3)}\ \frac{1}{x^3 - 2}.$$

Expressions (1) and (3) are not polynomials because they represent quotients, not sums. Expression (2) is not a polynomial because

$$\frac{1}{x} = x^{-1},$$

and this is not a monomial because the exponent is negative.

Do Exercise 1.

Objectives

- a Evaluate a polynomial for a given value of the variable.
- b Identify the terms of a polynomial.
- c Identify the like terms of a polynomial.
- d Identify the coefficients of a polynomial.
- e Collect the like terms of a polynomial.
- f Arrange a polynomial in descending order, or collect the like terms and then arrange in descending order.
- g Identify the degree of each term of a polynomial and the degree of the polynomial.
- h Identify the missing terms of a polynomial.
- i Classify a polynomial as a monomial, binomial, trinomial, or none of these.

For Extra Help

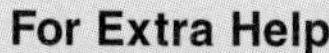

TAPE 6   MAC WIN   CD-ROM

1. Write three polynomials.

a Evaluating Polynomials and Applications

When we replace the variable in a polynomial with a number, the polynomial then represents a number called a **value** of the polynomial. Finding that number, or value, is called **evaluating the polynomial.** We evaluate a polynomial using the rules for order of operations (Section 1.8).

Example 1 Evaluate the polynomial for $x = 2$.

a) $$\begin{aligned} 3x + 5 &= 3 \cdot 2 + 5 \\ &= 6 + 5 \\ &= 11 \end{aligned}$$

b) $$\begin{aligned} 2x^2 - 7x + 3 &= 2 \cdot 2^2 - 7 \cdot 2 + 3 \\ &= 2 \cdot 4 - 7 \cdot 2 + 3 \\ &= 8 - 14 + 3 \\ &= -3 \end{aligned}$$

Answer on page A-14

Evaluate the polynomial for $x = 3$.

2. $-4x - 7$

3. $-5x^3 + 7x + 10$

Evaluate the polynomial for $x = -4$.

4. $5x + 7$

5. $2x^2 + 5x - 4$

6. Referring to Example 3, what is the total number of games to be played in a league of 12 teams?

7. *Perimeter of Baseball Diamond.* The perimeter of a square of side x is given by the polynomial $4x$.

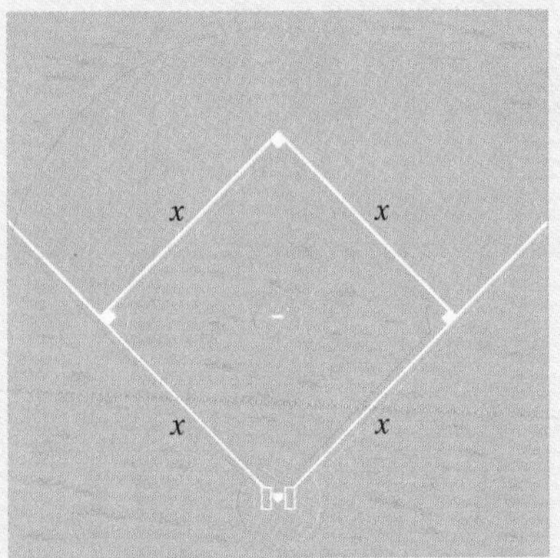

A baseball diamond is a square 90 ft on a side. Find the perimeter of a baseball diamond.

8. Use *only* the graph shown in Example 4 to evaluate the polynomial $2x - 2$ for $x = 4$ and for $x = -1$.

Answers on page A-14

Example 2 Evaluate the polynomial for $x = -4$.

a) $2 - x^3 = 2 - (-4)^3 = 2 - (-64)$
$= 2 + 64$
$= 66$

b) $-x^2 - 3x + 1 = -(-4)^2 - 3(-4) + 1$
$= -16 + 12 + 1$
$= -3$

Do Exercises 2–5.

Polynomials occur in many real-world situations.

Example 3 *Games in a Sports League.* In a sports league of n teams in which each team plays every other team twice, the total number of games to be played is given by the polynomial

$$n^2 - n.$$

A women's slow-pitch softball league has 10 teams. What is the total number of games to be played?

We evaluate the polynomial for $n = 10$:

$$n^2 - n = 10^2 - 10 = 100 - 10 = 90.$$

The league plays 90 games.

Do Exercises 6 and 7.

AG Algebraic–Graphical Connection

An equation like $y = 2x - 2$, which has a polynomial on one side and y on the other, is called a **polynomial equation.** We will here and in many places throughout the book connect graphs to related concepts.

Recall from Chapter 3 that in order to plot points before graphing an equation, we choose values for x and compute the corresponding y-values. If the equation has y on one side and a polynomial involving x on the other, then determining y is the same as evaluating the polynomial. Once the graph of such an equation has been drawn, we can evaluate the polynomial for a given x-value by finding the y-value that is paired with it on the graph.

Example 4 Use *only* the given graph of $y = 2x - 2$ to evaluate the polynomial $2x - 2$ for $x = 3$.

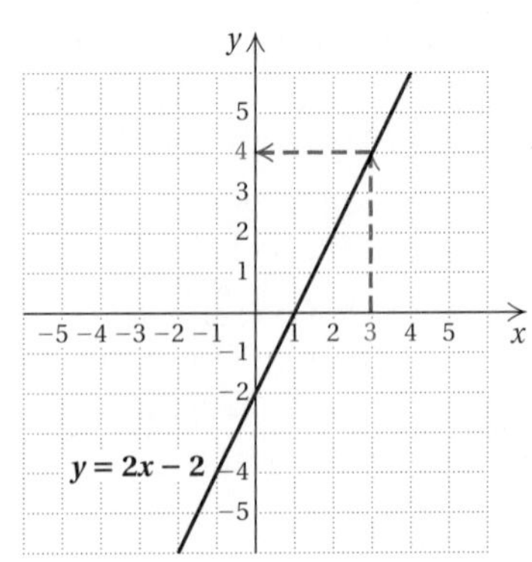

First, we locate 3 on the x-axis. From there we move vertically to the graph of the equation and then horizontally to the y-axis. There we locate the y-value that is paired with 3. Although our drawing may not be precise, it appears that the y-value 4 is paired with 3. Thus the value of $2x - 2$ is 4 when $x = 3$.

Do Exercise 8.

AG

Example 5 *Medical Dosage.* The concentration C, in parts per million, of a certain antibiotic in the bloodstream after t hours is given by the polynomial equation

$$C = -0.05t^2 + 2t + 2.$$

Find the concentration after 2 hr.

To find the concentration after 2 hr, we evaluate the polynomial for $t = 2$:

$$\begin{aligned} -0.05t^2 + 2t + 2 &= -0.05(2)^2 + 2(2) + 2 \\ &= -0.05(4) + 2(2) + 2 \\ &= -0.2 + 4 + 2 \\ &= -0.2 + 6 \\ &= 5.8. \end{aligned}$$

Carrying out the calculation using the rules for order of operations

The concentration after 2 hr is 5.8 parts per million.

Do Exercise 9.

9. Referring to Example 5, find the concentration after 3 hr.

AG Algebraic–Graphical Connection

The polynomial equation in Example 5 can be graphed if we evaluate the polynomial for several values of t. We list the values in a table and show the graph below. Note that the concentration peaks at the 20-hr mark and after a bit more than 40 hr, the concentration is 0. Since neither time nor concentration can be negative, our graph uses only the first quadrant.

t	$-0.05t^2 + 2t + 2$
0	2
2	5.8
10	17
20	22
30	17

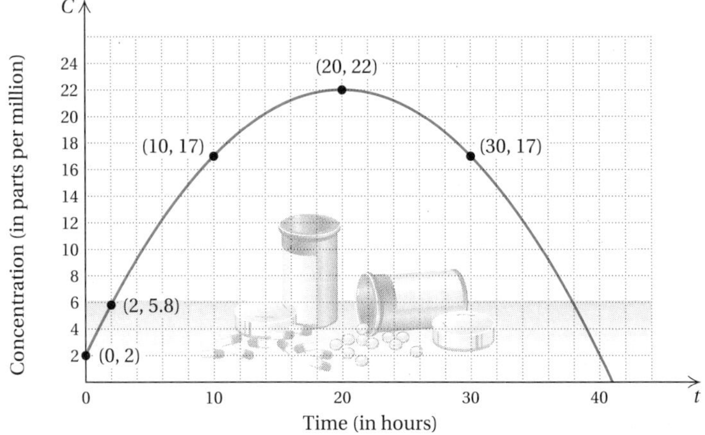

Do Exercise 10.

10. Use *only* the graph showing medical dosage to estimate the value of the polynomial for $t = 26$.

Answers on page A-14

Calculator Spotlight

Evaluating Polynomials. One way to evaluate a polynomial like $-x^2 - 3x + 1$ for $x = -4$ (see Example 2b) is to graph $y_1 = -x^2 - 3x + 1$.

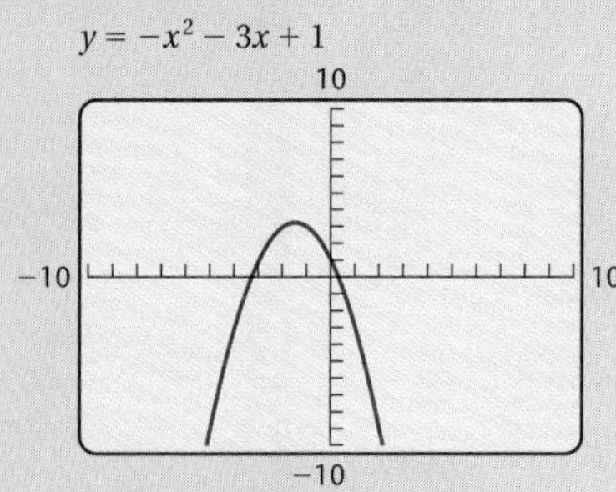

We can also adjust the window to obtain a better view of the graph.

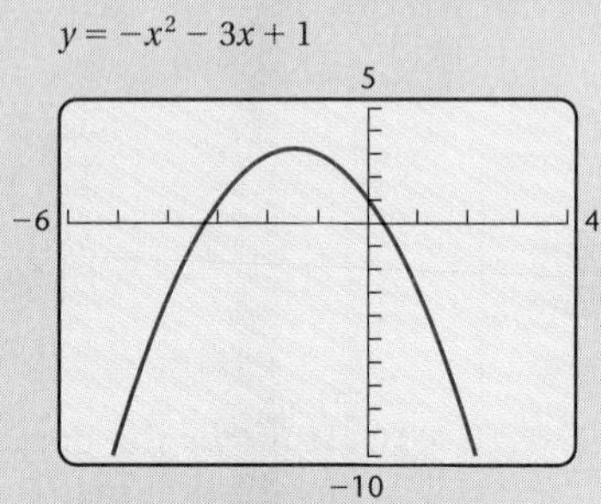

To evaluate the polynomial, we then use the CALC feature and choose VALUE.

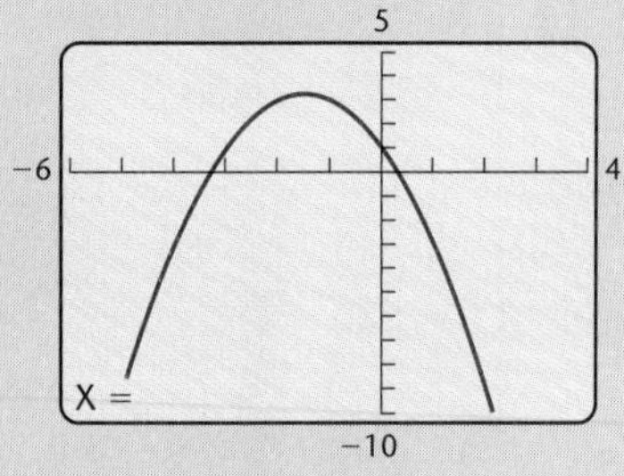

We enter $x = -4$. The y-value, -3, is shown together with a TRACE indicator showing the point $(-4, -3)$.

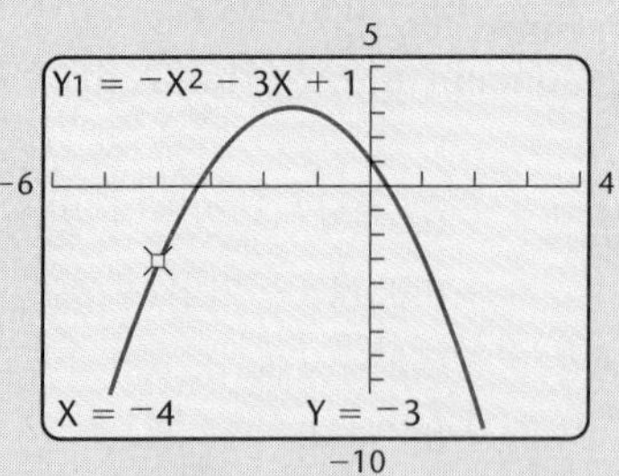

Polynomials can also be evaluated using the TABLE feature as described in Section 3.4.

Exercises

1. Evaluate the polynomial $-x^2 - 3x + 1$ for $x = -1$, for $x = -0.3$, and for $x = 1.7$.
2. Evaluate the polynomial $-0.05x^2 + 2x + 2$ for $x = 0$, for $x = 10$, for $x = 23$, and for $x = 36.4$. Use the viewing window [0, 41, 0, 25].
3. Evaluate the polynomial $2x^2 - x - 8$ for $x = -3$, for $x = -2$, for $x = 0$, for $x = 1.8$, and for $x = 3$.

b Identifying Terms

As we saw in Section 1.4, subtractions can be rewritten as additions. For any polynomial that has some subtractions, we can find an equivalent polynomial using only additions.

Examples Find an equivalent polynomial using only additions.

6. $-5x^2 - x = -5x^2 + (-x)$

7. $4x^5 - 2x^6 - 4x + 7 = 4x^5 + (-2x^6) + (-4x) + 7$

Do Exercises 11 and 12.

When a polynomial has only additions, the monomials being added are called **terms**. In Example 6, the terms are $-5x^2$ and $-x$. In Example 7, the terms are $4x^5$, $-2x^6$, $-4x$, and 7.

Example 8 Identify the terms of the polynomial

$$4x^7 + 3x + 12 + 8x^3 + 5x.$$

Terms: $4x^7$, $3x$, 12, $8x^3$, and $5x$.

If there are subtractions, you can *think* of them as additions without rewriting.

Example 9 Identify the terms of the polynomial

$$3t^4 - 5t^6 - 4t + 2.$$

Terms: $3t^4$, $-5t^6$, $-4t$, and 2.

Do Exercises 13 and 14.

c Like Terms

When terms have the same variable and the variable is raised to the same power, we say that they are **like terms,** or **similar terms.**

Examples Identify the like terms in the polynomials.

10. $4x^3 + 5x - 4x^2 + 2x^3 + x^2$

Like terms: $4x^3$ and $2x^3$ — **Same variable and exponent**

Like terms: $-4x^2$ and x^2 — **Same variable and exponent**

11. $6 - 3a^2 + 8 - a - 5a$

Like terms: 6 and 8 — **Constant terms are like terms because $6 = 6x^0$ and $8 = 8x^0$.**

Like terms: $-a$ and $-5a$

Do Exercises 15–17.

d Coefficients

The coefficient of the term $5x^3$ is 5. In the following polynomial, the color numbers are the **coefficients**:

$$3x^5 - 2x^3 + 5x + 4.$$

Find an equivalent polynomial using only additions.

11. $-9x^3 - 4x^5$

12. $-2y^3 + 3y^7 - 7y$

Identify the terms of the polynomial.

13. $3x^2 + 6x + \frac{1}{2}$

14. $-4y^5 + 7y^2 - 3y - 2$

Identify the like terms in the polynomial.

15. $4x^3 - x^3 + 2$

16. $4t^4 - 9t^3 - 7t^4 + 10t^3$

17. $5x^2 + 3x - 10 + 7x^2 - 8x + 11$

Answers on page A-14

18. Identify the coefficient of each term in the polynomial

$2x^4 - 7x^3 - 8.5x^2 + 10x - 4.$

Collect like terms.

19. $3x^2 + 5x^2$

20. $4x^3 - 2x^3 + 2 + 5$

21. $\frac{1}{2}x^5 - \frac{3}{4}x^5 + 4x^2 - 2x^2$

22. $24 - 4x^3 - 24$

23. $5x^3 - 8x^5 + 8x^5$

24. $-2x^4 + 16 + 2x^4 + 9 - 3x^5$

Collect like terms.

25. $7x - x$

26. $5x^3 - x^3 + 4$

27. $\frac{3}{4}x^3 + 4x^2 - x^3 + 7$

28. $8x^2 - x^2 + x^3 - 1 - 4x^2 + 10$

Answers on page A-14

Example 12 Identify the coefficient of each term in the polynomial

$3x^4 - 4x^3 + 7x^2 + x - 8.$

The coefficient of the first term is 3.

The coefficient of the second term is -4.

The coefficient of the third term is 7.

The coefficient of the fourth term is 1.

The coefficient of the fifth term is -8.

Do Exercise 18.

e Collecting Like Terms

We can often simplify polynomials by **collecting like terms,** or **combining similar terms.** To do this, we use the distributive laws. We factor out the variable expression and add or subtract the coefficients. We try to do this mentally as much as possible.

Examples Collect like terms.

13. $2x^3 - 6x^3 = (2 - 6)x^3 = -4x^3$ Using a distributive law

14. $5x^2 + 7 + 4x^4 + 2x^2 - 11 - 2x^4 = (5 + 2)x^2 + (4 - 2)x^4 + (7 - 11)$
$= 7x^2 + 2x^4 - 4$

Note that using the distributive laws in this manner allows us to collect like terms by adding or subtracting the coefficients. Often the middle step is omitted and we add or subtract mentally, writing just the answer. In collecting like terms, we may get 0.

Examples Collect like terms.

15. $5x^3 - 5x^3 = (5 - 5)x^3 = 0x^3 = 0$

16. $3x^4 + 2x^2 - 3x^4 + 8 = (3 - 3)x^4 + 2x^2 + 8$
$= 0x^4 + 2x^2 + 8 = 2x^2 + 8$

Do Exercises 19–24.

Multiplying a term of a polynomial by 1 does not change the term, but it may make it easier to factor or collect like terms.

Examples Collect like terms.

17. $5x^2 + x^2 = 5x^2 + 1x^2$ Replacing x^2 with $1x^2$
$= (5 + 1)x^2$ Using a distributive law
$= 6x^2$

18. $5x^4 - 6x^3 - x^4 = 5x^4 - 6x^3 - 1x^4$ $x^4 = 1x^4$
$= (5 - 1)x^4 - 6x^3$
$= 4x^4 - 6x^3$

19. $\frac{2}{3}x^4 - x^3 - \frac{1}{6}x^4 + \frac{2}{5}x^3 - \frac{3}{10}x^3 = \left(\frac{2}{3} - \frac{1}{6}\right)x^4 + \left(-1 + \frac{2}{5} - \frac{3}{10}\right)x^3$
$= \left(\frac{4}{6} - \frac{1}{6}\right)x^4 + \left(-\frac{10}{10} + \frac{4}{10} - \frac{3}{10}\right)x^3$
$= \frac{3}{6}x^4 - \frac{9}{10}x^3 = \frac{1}{2}x^4 - \frac{9}{10}x^3$

Do Exercises 25–28.

f Descending and Ascending Order

Note in the following polynomial that the exponents decrease from left to right. We say that the polynomial is arranged in **descending order:**

$$2x^4 - 8x^3 + 5x^2 - x + 3.$$

The term with the largest exponent is first. The term with the next largest exponent is second, and so on. The associative and commutative laws allow us to arrange the terms of a polynomial in descending order.

Examples Arrange the polynomial in descending order.

20. $6x^5 + 4x^7 + x^2 + 2x^3 = 4x^7 + 6x^5 + 2x^3 + x^2$

21. $\frac{2}{3} + 4x^5 - 8x^2 + 5x - 3x^3 = 4x^5 - 3x^3 - 8x^2 + 5x + \frac{2}{3}$

We usually arrange polynomials in descending order, but not always. The opposite order is called **ascending order.** Generally, if an exercise is written in a certain order, we give the answer in that same order.

Do Exercises 29–31.

Example 22 Collect like terms and then arrange in descending order:

$$2x^2 - 4x^3 + 3 - x^2 - 2x^3.$$

We have

$$2x^2 - 4x^3 + 3 - x^2 - 2x^3 = x^2 - 6x^3 + 3 \quad \text{Collecting like terms}$$
$$= -6x^3 + x^2 + 3 \quad \text{Arranging in descending order}$$

Do Exercises 32 and 33.

g Degrees

The **degree** of a term is the exponent of the variable. The degree of the term $5x^3$ is 3.

Example 23 Identify the degree of each term of $8x^4 + 3x + 7$.

The degree of $8x^4$ is 4.

The degree of $3x$ is 1. Recall that $x = x^1$.

The degree of 7 is 0. Think of 7 as $7x^0$. Recall that $x^0 = 1$.

The **degree of a polynomial** is the largest of the degrees of the terms, unless it is the polynomial 0. The polynomial 0 is a special case. We agree that it has *no* degree either as a term or as a polynomial. This is because we can express 0 as $0 = 0x^5 = 0x^7$, and so on, using any exponent we wish.

Example 24 Identify the degree of the polynomial $5x^3 - 6x^4 + 7$.

We have

$$5x^3 - 6x^4 + 7. \quad \text{The largest exponent is 4.}$$

The degree of the polynomial is 4.

Do Exercise 34.

Arrange the polynomial in descending order.

29. $x + 3x^5 + 4x^3 + 5x^2 + 6x^7 - 2x^4$

30. $4x^2 - 3 + 7x^5 + 2x^3 - 5x^4$

31. $-14 + 7t^2 - 10t^5 + 14t^7$

Collect like terms and then arrange in descending order.

32. $3x^2 - 2x + 3 - 5x^2 - 1 - x$

33. $-x + \frac{1}{2} + 14x^4 - 7x - 1 - 4x^4$

34. Identify the degree of each term and the degree of the polynomial

$$-6x^4 + 8x^2 - 2x + 9.$$

Answers on page A-14

Identify the missing terms in the polynomial.

35. $2x^3 + 4x^2 - 2$

36. $-3x^4$

37. $x^3 + 1$

38. $x^4 - x^2 + 3x + 0.25$

Classify the polynomial as a monomial, binomial, trinomial, or none of these.

39. $5x^4$

40. $4x^3 - 3x^2 + 4x + 2$

41. $3x^2 + x$

42. $3x^2 + 2x - 4$

Answers on page A-14

Let's summarize the terminology that we have learned, using the polynomial

$$3x^4 - 8x^3 + 5x^2 + 7x - 6.$$

Term	Coefficient	Degree of the Term	Degree of the Polynomial
$3x^4$	3	4	
$-8x^3$	−8	3	
$5x^2$	5	2	4
$7x$	7	1	
−6	−6	0	

h Missing Terms

If a coefficient is 0, we generally do not write the term. We say that we have a **missing term.**

Example 25 Identify the missing terms in the polynomial

$$8x^5 - 2x^3 + 5x^2 + 7x + 8.$$

There is no term with x^4. We say that the x^4-term (or the *fourth-degree term*) is missing.

For certain skills or manipulations, we can write missing terms with zero coefficients or leave space. For example, we can write the polynomial $3x^2 + 9$ as

$$3x^2 + 0x + 9 \quad \text{or} \quad 3x^2 + \qquad 9.$$

Do Exercises 35–38.

i Classifying Polynomials

Polynomials with just one term are called **monomials**. Polynomials with just two terms are called **binomials**. Those with just three terms are called **trinomials**. Those with more than three terms are generally not specified with a name.

Example 26

Monomials	Binomials	Trinomials	None of These
$4x^2$	$2x + 4$	$3x^3 + 4x + 7$	$4x^3 - 5x^2 + x - 8$
9	$3x^5 + 6x$	$6x^7 - 7x^2 + 4$	
$-23x^{19}$	$-9x^7 - 6$	$4x^2 - 6x - \frac{1}{2}$	

Do Exercises 39–42.

Exercise Set 4.3

a Evaluate the polynomial for $x = 4$ and for $x = -1$.

1. $-5x + 2$

2. $-8x + 1$

3. $2x^2 - 5x + 7$

4. $3x^2 + x - 7$

5. $x^3 - 5x^2 + x$

6. $7 - x + 3x^2$

Evaluate the polynomial for $x = -2$ and for $x = 0$.

7. $3x + 5$

8. $8 - 4x$

9. $x^2 - 2x + 1$

10. $5x + 6 - x^2$

11. $-3x^3 + 7x^2 - 3x - 2$

12. $-2x^3 + 5x^2 - 4x + 3$

13. *Skydiving.* During the first 13 sec of a jump, the number of feet that a skydiver falls in t seconds can be approximated by the polynomial

$11.12t^2$.

Approximately how far has a skydiver fallen 10 sec after having jumped from a plane?

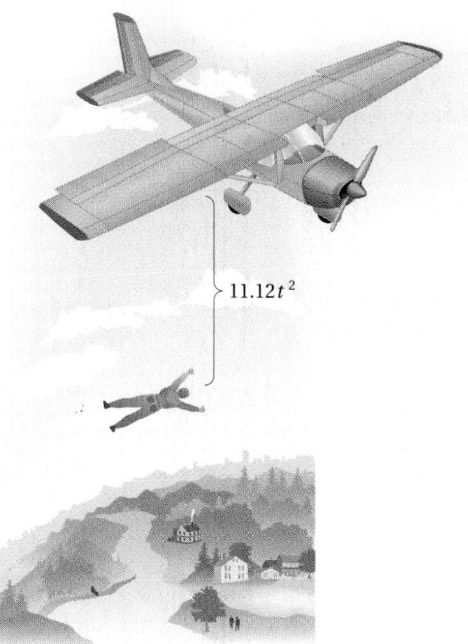

14. *Skydiving.* For jumps that exceed 13 sec, the polynomial

$173t - 369$

can be used to approximate the distance, in feet, that a skydiver has fallen in t seconds. Approximately how far has a skydiver fallen 20 sec after having jumped from a plane?

15. *Total Revenue.* Hadley Electronics is marketing a new kind of high-density TV. The firm determines that when it sells x TVs, its total revenue (the total amount of money taken in) will be

$280x - 0.4x^2$ dollars.

What is the total revenue from the sale of 75 TVs? 100 TVs?

16. *Total Cost.* Hadley Electronics determines that the total cost of producing x high-density TVs is given by

$5000 + 0.6x^2$ dollars.

What is the total cost of producing 500 TVs? 650 TVs?

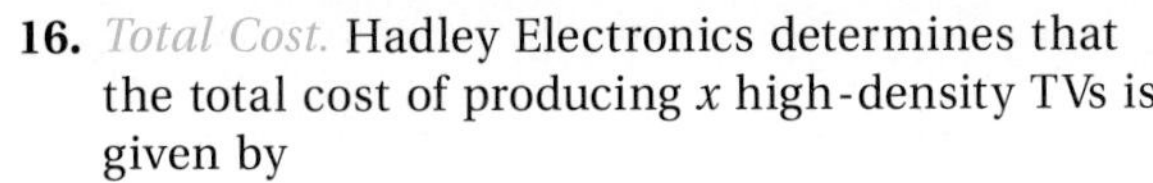

17. The graph of the polynomial equation $y = 5 - x^2$ is shown below. Use *only* the graph to estimate the value of the polynomial for $x = -3$, for $x = -1$, for $x = 0$, for $x = 1.5$, and for $x = 2$.

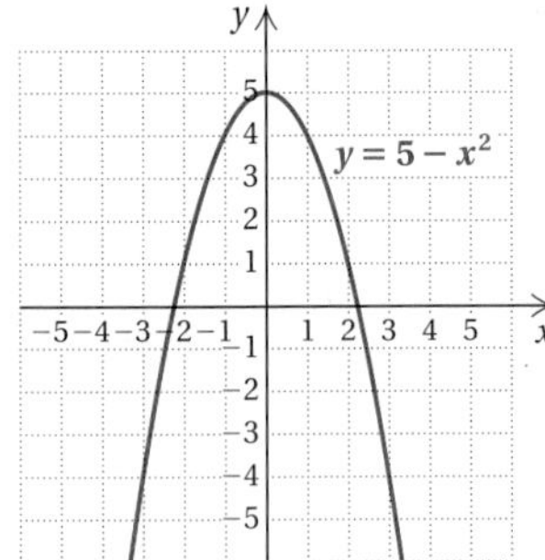

18. The graph of the polynomial equation $y = 6x^3 - 6x$ is shown below. Use *only* the graph to estimate the value of the polynomial for $x = -1$, for $x = -0.5$, for $x = 0.5$, for $x = 1$, and for $x = 1.1$.

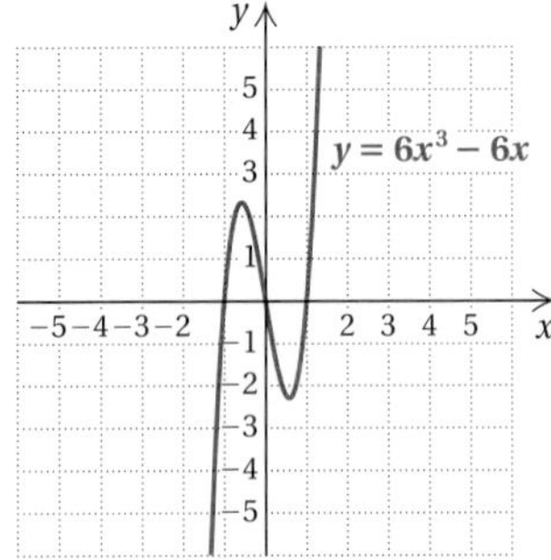

19. *Path of the Olympic Arrow.* The Olympic flame at the 1992 Summer Olympics was lit by a flaming arrow. As the arrow moved d feet horizontally from the archer, its height h, in feet, could be approximated by the polynomial equation

$h = -0.002d^2 + 0.8d + 6.6.$

The graph of this equation is shown at right. Use either the graph or the polynomial to approximate the height of the arrow after it has traveled horizontally for 100 ft, 200 ft, 300 ft, and 350 ft.

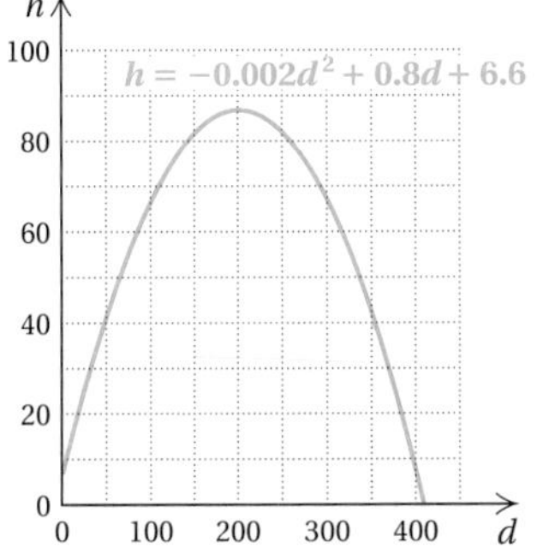

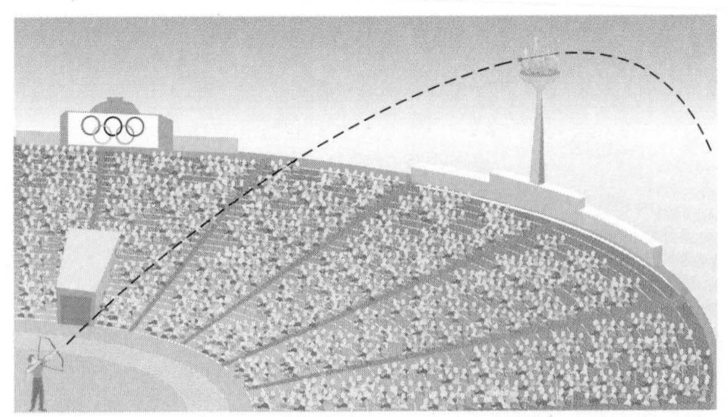

20. *Hearing-Impaired Americans.* The number N, in millions, of hearing-impaired Americans of age x can be approximated by the polynomial equation

$$N = -0.00006x^3 + 0.006x^2 - 0.1x + 1.9$$

(***Source***: American Speech-Language Hearing Association). The graph of this equation is shown at right. Use either the graph or the polynomial to approximate the number of hearing-impaired Americans of ages 20, 40, 50, and 60.

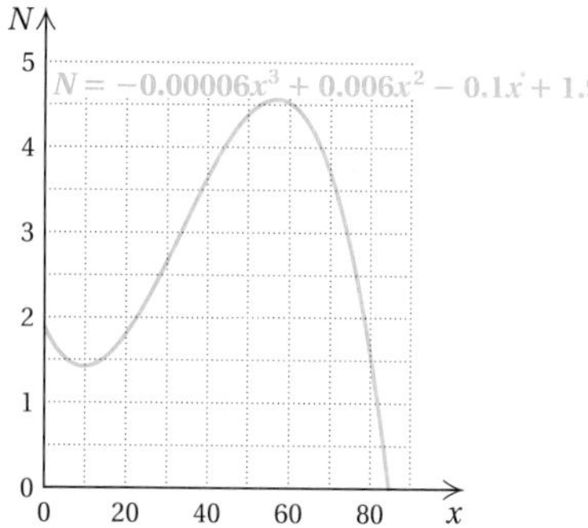

b Identify the terms of the polynomial.

21. $2 - 3x + x^2$

22. $2x^2 + 3x - 4$

c Identify the like terms in the polynomial.

23. $5x^3 + 6x^2 - 3x^2$

24. $3x^2 + 4x^3 - 2x^2$

25. $2x^4 + 5x - 7x - 3x^4$

26. $-3t + t^3 - 2t - 5t^3$

27. $3x^5 - 7x + 8 + 14x^5 - 2x - 9$

28. $8x^3 + 7x^2 - 11 - 4x^3 - 8x^2 - 29$

d Identify the coefficient of each term of the polynomial.

29. $-3x + 6$

30. $2x - 4$

31. $5x^2 + 3x + 3$

32. $3x^2 - 5x + 2$

33. $-5x^4 + 6x^3 - 3x^2 + 8x - 2$

34. $7x^3 - 4x^2 - 4x + 5$

e Collect like terms.

35. $2x - 5x$

36. $2x^2 + 8x^2$

37. $x - 9x$

38. $x - 5x$

39. $5x^3 + 6x^3 + 4$

40. $6x^4 - 2x^4 + 5$

41. $5x^3 + 6x - 4x^3 - 7x$

42. $3a^4 - 2a + 2a + a^4$

43. $6b^5 + 3b^2 - 2b^5 - 3b^2$

44. $2x^2 - 6x + 3x + 4x^2$

45. $\frac{1}{4}x^5 - 5 + \frac{1}{2}x^5 - 2x - 37$

46. $\frac{1}{3}x^3 + 2x - \frac{1}{6}x^3 + 4 - 16$

47. $6x^2 + 2x^4 - 2x^2 - x^4 - 4x^2$

48. $8x^2 + 2x^3 - 3x^3 - 4x^2 - 4x^2$

49. $\frac{1}{4}x^3 - x^2 - \frac{1}{6}x^2 + \frac{3}{8}x^3 + \frac{5}{16}x^3$

50. $\frac{1}{5}x^4 + \frac{1}{5} - 2x^2 + \frac{1}{10} - \frac{3}{15}x^4 + 2x^2 - \frac{3}{10}$

f Arrange the polynomial in descending order.

51. $x^5 + x + 6x^3 + 1 + 2x^2$

52. $3 + 2x^2 - 5x^6 - 2x^3 + 3x$

53. $5y^3 + 15y^9 + y - y^2 + 7y^8$

54. $9p - 5 + 6p^3 - 5p^4 + p^5$

Collect like terms and then arrange in descending order.

55. $3x^4 - 5x^6 - 2x^4 + 6x^6$

56. $-1 + 5x^3 - 3 - 7x^3 + x^4 + 5$

57. $-2x + 4x^3 - 7x + 9x^3 + 8$

58. $-6x^2 + x - 5x + 7x^2 + 1$

59. $3x + 3x + 3x - x^2 - 4x^2$

60. $-2x - 2x - 2x + x^3 - 5x^3$

61. $-x + \frac{3}{4} + 15x^4 - x - \frac{1}{2} - 3x^4$

62. $2x - \frac{5}{6} + 4x^3 + x + \frac{1}{3} - 2x$

g Identify the degree of each term of the polynomial and the degree of the polynomial.

63. $2x - 4$

64. $6 - 3x$

65. $3x^2 - 5x + 2$

66. $5x^3 - 2x^2 + 3$

67. $-7x^3 + 6x^2 + 3x + 7$

68. $5x^4 + x^2 - x + 2$

69. $x^2 - 3x + x^6 - 9x^4$

70. $8x - 3x^2 + 9 - 8x^3$

71. Complete the following table for the polynomial $-7x^4 + 6x^3 - 3x^2 + 8x - 2$.

Term	Coefficient	Degree of the Term	Degree of the Polynomial
$6x^3$	6		
		2	
$8x$		1	
	−2		

72. Complete the following table for the polynomial $3x^2 + 8x^5 - 46x^3 + 6x - 2.4 - \frac{1}{2}x^4$.

Term	Coefficient	Degree of the Term	Degree of the Polynomial
		5	
$-\frac{1}{2}x^4$		4	
	−46		
$3x^2$		2	
	6		
−2.4			

h Identify the missing terms in the polynomial.

73. $x^3 - 27$

74. $x^5 + x$

75. $x^4 - x$

76. $5x^4 - 7x + 2$

77. $2x^3 - 5x^2 + x - 3$

78. $-6x^3$

i Classify the polynomial as a monomial, binomial, trinomial, or none of these.

79. $x^2 - 10x + 25$

80. $-6x^4$

81. $x^3 - 7x^2 + 2x - 4$

82. $x^2 - 9$

83. $4x^2 - 25$

84. $2x^4 - 7x^3 + x^2 + x - 6$

85. $40x$

86. $4x^2 + 12x + 9$

Skill Maintenance

87. Three tired campers stopped for the night. All they had to eat was a bag of apples. During the night, one awoke and ate one-third of the apples. Later, a second camper awoke and ate one-third of the apples that remained. Much later, the third camper awoke and ate one-third of those apples yet remaining after the other two had eaten. When they got up the next morning, 8 apples were left. How many apples did they begin with? [2.4a]

Subtract. [1.4a]

88. $1 - 20$

89. $\frac{1}{8} - \frac{5}{6}$

90. $\frac{3}{8} - \left(-\frac{1}{4}\right)$

91. $5.6 - 8.2$

92. Solve: $3(x + 2) = 5x - 9$. [2.3c]

93. Solve $cx = ab - r$ for b. [2.6a]

94. A nut dealer has 1800 lb of peanuts, 1500 lb of cashews, and 700 lb of almonds. What percent of the total is peanuts? cashews? almonds? [2.5a]

95. Factor: $3x - 15y + 63$. [1.7d]

Synthesis

96. ◆ Is it better to evaluate a polynomial before or after like terms have been collected? Why?

97. ◆ Explain why an understanding of the rules of order of operations is essential when evaluating polynomials.

Collect like terms.

98. $\frac{9}{2}x^8 + \frac{1}{9}x^2 + \frac{1}{2}x^9 + \frac{9}{2}x^1 + \frac{9}{2}x^9 + \frac{8}{9}x^2 + \frac{1}{2}x - \frac{1}{2}x^8$

99. $(3x^2)^3 + 4x^2 \cdot 4x^4 - x^4(2x)^2 + ((2x)^2)^3 - 100x^2(x^2)^2$

100. Construct a polynomial in x (meaning that x is the variable) of degree 5 with four terms and coefficients that are integers.

101. What is the degree of $(5m^5)^2$?

102. A polynomial in x has degree 3. The coefficient of x^2 is 3 less than the coefficient of x^3. The coefficient of x is three times the coefficient of x^2. The remaining coefficient is 2 more than the coefficient of x^3. The sum of the coefficients is -4. Find the polynomial.

Use the CALC feature and choose VALUE on your grapher to find the values in each of the following.

103. Exercise 17

104. Exercise 18

105. Exercise 19

106. Exercise 20

4.4 Addition and Subtraction of Polynomials

a Addition of Polynomials

To add two polynomials, we can write a plus sign between them and then collect like terms. Depending on the situation, you may see polynomials written in descending order, ascending order, or neither. Generally, if an exercise is written in a particular order, we write the answer in that same order.

Example 1 Add: $(-3x^3 + 2x - 4) + (4x^3 + 3x^2 + 2)$.

$$(-3x^3 + 2x - 4) + (4x^3 + 3x^2 + 2)$$
$$= (-3 + 4)x^3 + 3x^2 + 2x + (-4 + 2) \quad \text{Collecting like terms (}\textit{No}\text{ signs are changed.)}$$
$$= x^3 + 3x^2 + 2x - 2$$

Example 2 Add:

$$\left(\tfrac{2}{3}x^4 + 3x^2 - 2x + \tfrac{1}{2}\right) + \left(-\tfrac{1}{3}x^4 + 5x^3 - 3x^2 + 3x - \tfrac{1}{2}\right).$$

We have

$$\left(\tfrac{2}{3}x^4 + 3x^2 - 2x + \tfrac{1}{2}\right) + \left(-\tfrac{1}{3}x^4 + 5x^3 - 3x^2 + 3x - \tfrac{1}{2}\right)$$
$$= \left(\tfrac{2}{3} - \tfrac{1}{3}\right)x^4 + 5x^3 + (3 - 3)x^2 + (-2 + 3)x + \left(\tfrac{1}{2} - \tfrac{1}{2}\right) \quad \text{Collecting like terms}$$
$$= \tfrac{1}{3}x^4 + 5x^3 + x.$$

We can add polynomials as we do because they represent numbers. After some practice, you will be able to add mentally.

Do Exercises 1–4.

Example 3 Add: $(3x^2 - 2x + 2) + (5x^3 - 2x^2 + 3x - 4)$.

$$(3x^2 - 2x + 2) + (5x^3 - 2x^2 + 3x - 4)$$
$$= 5x^3 + (3 - 2)x^2 + (-2 + 3)x + (2 - 4) \quad \text{You might do this step mentally.}$$
$$= 5x^3 + x^2 + x - 2 \quad \text{Then you would write only this.}$$

Do Exercises 5 and 6.

We can also add polynomials by writing like terms in columns.

Example 4 Add: $9x^5 - 2x^3 + 6x^2 + 3$ and $5x^4 - 7x^2 + 6$ and $3x^6 - 5x^5 + x^2 + 5$.

We arrange the polynomials with the like terms in columns.

$$\begin{array}{rrrrrrl}
 & 9x^5 & & -2x^3 & +6x^2 & +3 & \\
 & & 5x^4 & & -7x^2 & +6 & \text{We leave spaces for missing terms.} \\
3x^6 & -5x^5 & & & +x^2 & +5 & \\
\hline
3x^6 & +4x^5 & +5x^4 & -2x^3 & & +14 & \text{Adding}
\end{array}$$

We write the answer as $3x^6 + 4x^5 + 5x^4 - 2x^3 + 14$ without the space.

Objectives

a Add polynomials.

b Find the opposite of a polynomial.

c Subtract polynomials.

d Use polynomials to represent perimeter and area.

For Extra Help

TAPE 6

MAC WIN

CD-ROM

Add.

1. $(3x^2 + 2x - 2) + (-2x^2 + 5x + 5)$

2. $(-4x^5 + x^3 + 4) + (7x^4 + 2x^2)$

3. $(31x^4 + x^2 + 2x - 1) + (-7x^4 + 5x^3 - 2x + 2)$

4. $(17x^3 - x^2 + 3x + 4) + \left(-15x^3 + x^2 - 3x - \dfrac{2}{3}\right)$

Add mentally. Try to write just the answer.

5. $(4x^2 - 5x + 3) + (-2x^2 + 2x - 4)$

6. $(3x^3 - 4x^2 - 5x + 3) + \left(5x^3 + 2x^2 - 3x - \dfrac{1}{2}\right)$

Answers on page A-15

Add.

7.
$$\begin{array}{r} -2x^3 + 5x^2 - 2x + 4 \\ x^4 \qquad\quad + 6x^2 + 7x - 10 \\ -9x^4 + 6x^3 + x^2 \qquad\quad - 2 \\ \hline \end{array}$$

8. $-3x^3 + 5x + 2$ and
$x^3 + x^2 + 5$ and
$x^3 - 2x - 4$

Find two equivalent expressions for the opposite of the polynomial.

9. $12x^4 - 3x^2 + 4x$

10. $-4x^4 + 3x^2 - 4x$

11. $-13x^6 + 2x^4 - 3x^2 + x - \frac{5}{13}$

12. $-7y^3 + 2y^2 - y + 3$

Simplify.

13. $-(4x^3 - 6x + 3)$

14. $-(5x^4 + 3x^2 + 7x - 5)$

15. $-\left(14x^{10} - \frac{1}{2}x^5 + 5x^3 - x^2 + 3x\right)$

Answers on page A-15

Do Exercises 7 and 8.

b Opposites of Polynomials

We now look at subtraction of polynomials. To do so, we first consider the opposite, or additive inverse, of a polynomial.

We know that two numbers are opposites of each other if their sum is zero. For example, 5 and -5 are opposites, since $5 + (-5) = 0$. The same definition holds for polynomials. Two polynomials are **opposites**, or **additive inverses,** of each other if their sum is zero.

To find a way to determine an opposite, look for a pattern in the following examples:

a) $2x + (-2x) = 0$;

b) $-6x^2 + 6x^2 = 0$;

c) $(5t^3 - 2) + (-5t^3 + 2) = 0$;

d) $(7x^3 - 6x^2 - x + 4) + (-7x^3 + 6x^2 + x - 4) = 0$.

Since $(5t^3 - 2) + (-5t^3 + 2) = 0$, we know that the opposite of $(5t^3 - 2)$ is $(-5t^3 + 2)$. To say the same thing with purely algebraic symbolism, consider

$$\underbrace{\text{The opposite of}}_{-}\quad \underbrace{(5t^3 - 2)}_{(5t^3 - 2)}\quad \underbrace{\text{is}}_{=}\quad \underbrace{-5t^3 + 2.}_{-5t^3 + 2.}$$

> We can find an equivalent polynomial for the opposite, or additive inverse, of a polynomial by replacing each term with its opposite—that is, *changing the sign of every term.*

Example 5 Find two equivalent expressions for the opposite of

$$4x^5 - 7x^3 - 8x + \tfrac{5}{6}.$$

The opposite of $4x^5 - 7x^3 - 8x + \frac{5}{6}$ is

$-\left(4x^5 - 7x^3 - 8x + \frac{5}{6}\right)$, or

$-4x^5 + 7x^3 + 8x - \frac{5}{6}$. **Changing the sign of every term**

Thus, $-\left(4x^5 - 7x^3 - 8x + \frac{5}{6}\right)$ is equivalent to $-4x^5 + 7x^3 + 8x - \frac{5}{6}$, and each is the opposite of the original polynomial $4x^5 - 7x^3 - 8x + \frac{5}{6}$.

Do Exercises 9–12.

Example 6 Simplify: $-\left(-7x^4 - \frac{5}{9}x^3 + 8x^2 - x + 67\right)$.

$$-\left(-7x^4 - \tfrac{5}{9}x^3 + 8x^2 - x + 67\right) = 7x^4 + \tfrac{5}{9}x^3 - 8x^2 + x - 67$$

Do Exercises 13–15.

c Subtraction of Polynomials

Recall that we can subtract a real number by adding its opposite, or additive inverse: $a - b = a + (-b)$. This allows us to find an equivalent expression for the difference of two polynomials.

Example 7 Subtract:

$$(9x^5 + x^3 - 2x^2 + 4) - (2x^5 + x^4 - 4x^3 - 3x^2).$$

We have

$(9x^5 + x^3 - 2x^2 + 4) - (2x^5 + x^4 - 4x^3 - 3x^2)$

$= 9x^5 + x^3 - 2x^2 + 4 + [-(2x^5 + x^4 - 4x^3 - 3x^2)]$ **Adding the opposite**

$= 9x^5 + x^3 - 2x^2 + 4 - 2x^5 - x^4 + 4x^3 + 3x^2$ **Finding the opposite by changing the sign of *each* term**

$= 7x^5 - x^4 + 5x^3 + x^2 + 4.$ **Collecting like terms**

Do Exercises 16 and 17.

As with similar work in Section 1.8, we combine steps by changing the sign of each term of the polynomial being subtracted and collecting like terms. Try to do this mentally as much as possible.

Example 8 Subtract: $(9x^5 + x^3 - 2x) - (-2x^5 + 5x^3 + 6)$.

$(9x^5 + x^3 - 2x) - (-2x^5 + 5x^3 + 6)$

$= 9x^5 + x^3 - 2x + 2x^5 - 5x^3 - 6$ **Finding the opposite by changing the sign of each term**

$= 11x^5 - 4x^3 - 2x - 6$ **Collecting like terms**

Do Exercises 18 and 19.

We can use columns to subtract. We replace coefficients with their opposites, as shown in Example 7.

Example 9 Write in columns and subtract:

$$(5x^2 - 3x + 6) - (9x^2 - 5x - 3).$$

a) $\begin{array}{r} 5x^2 - 3x + 6 \\ \underline{-(9x^2 - 5x - 3)} \end{array}$ **Writing similar terms in columns**

b) $\begin{array}{r} 5x^2 - 3x + 6 \\ \underline{-9x^2 + 5x + 3} \end{array}$ **Changing signs**

c) $\begin{array}{r} 5x^2 - 3x + 6 \\ \underline{-9x^2 + 5x + 3} \\ -4x^2 + 2x + 9 \end{array}$ **Adding**

If you can do so without error, you can arrange the polynomials in columns and write just the answer.

Subtract.

16. $(7x^3 + 2x + 4) - (5x^3 - 4)$

17. $(-3x^2 + 5x - 4) - (-4x^2 + 11x - 2)$

Subtract.

18. $(-6x^4 + 3x^2 + 6) - (2x^4 + 5x^3 - 5x^2 + 7)$

19. $\left(\frac{3}{2}x^3 - \frac{1}{2}x^2 + 0.3\right) - \left(\frac{1}{2}x^3 + \frac{1}{2}x^2 + \frac{4}{3}x + 1.2\right)$

Answers on page A-15

Write in columns and subtract.

20. $(4x^3 + 2x^2 - 2x - 3) - (2x^3 - 3x^2 + 2)$

21. $(2x^3 + x^2 - 6x + 2) - (x^5 + 4x^3 - 2x^2 - 4x)$

22. Find a polynomial for the sum of the perimeters and the areas of the rectangles.

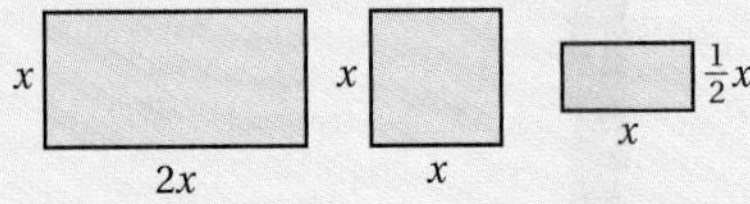

23. Find a polynomial for the shaded area.

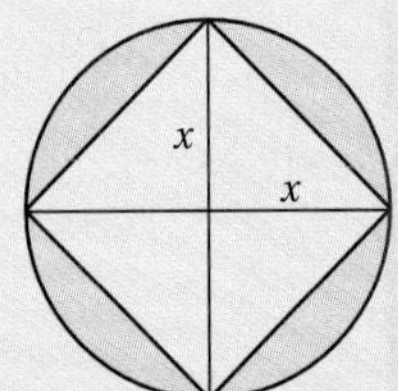

Answers on page A-15

Example 10 Write in columns and subtract:

$$(x^3 + x^2 + 2x - 12) - (-2x^3 + x^2 - 3x).$$

We have

$$\begin{array}{rrrrl} x^3 & + x^2 & + 2x & - 12 & \\ -2x^3 & + x^2 & - 3x & & \text{Changing signs} \\ \hline 3x^3 & & + 5x & - 12. & \text{Adding} \end{array}$$

Do Exercises 20 and 21.

d Polynomials and Geometry

Example 11 Find a polynomial for the sum of the areas of these rectangles.

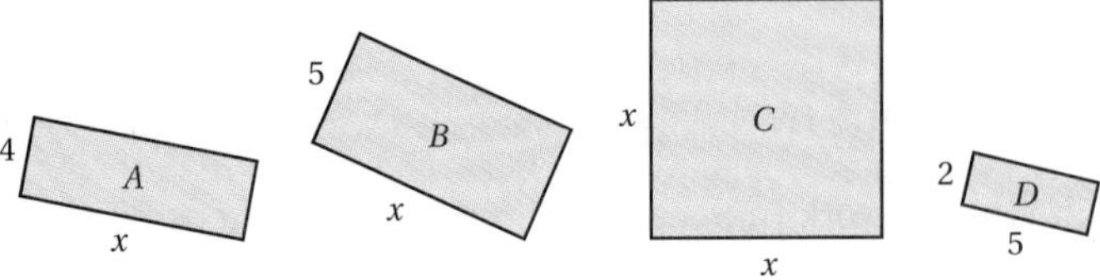

Recall that the area of a rectangle is the product of the length and the width. The sum of the areas is a sum of products. We find these products and then collect like terms.

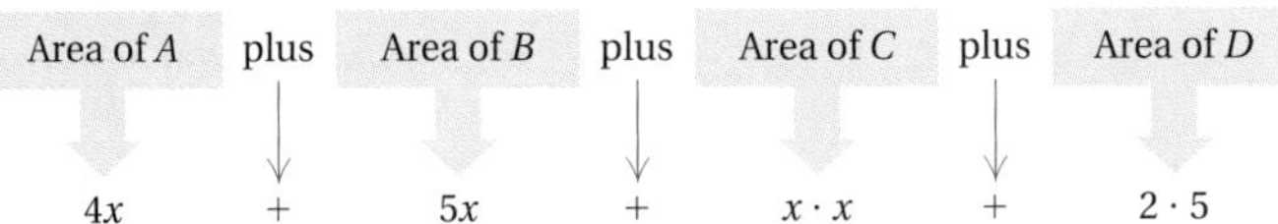

We collect like terms:

$$4x + 5x + x^2 + 10 = x^2 + 9x + 10.$$

Do Exercise 22.

Example 12 A water fountain with a 4-ft by 4-ft square base is placed on a square grassy park area that is x ft on a side. To determine the amount of grass seed needed for the lawn, find a polynomial for the grassy area.

We draw a picture of the situation as shown here. We then reword the problem and write the polynomial as follows.

$$\underbrace{\text{Area of park}} - \underbrace{\text{Area of base of fountain}} = \text{Area left over}$$

$$x \cdot x \quad - \quad 4 \cdot 4 \quad = \quad \text{Area left over}$$

Then $x^2 - 16 =$ Area left over.

Do Exercise 23.

Exercise Set 4.4

a Add.

1. $(3x + 2) + (-4x + 3)$

2. $(6x + 1) + (-7x + 2)$

3. $(-6x + 2) + (x^2 + x - 3)$

4. $(x^2 - 5x + 4) + (8x - 9)$

5. $(x^2 - 9) + (x^2 + 9)$

6. $(x^3 + x^2) + (2x^3 - 5x^2)$

7. $(3x^2 - 5x + 10) + (2x^2 + 8x - 40)$

8. $(6x^4 + 3x^3 - 1) + (4x^2 - 3x + 3)$

9. $(1.2x^3 + 4.5x^2 - 3.8x) + (-3.4x^3 - 4.7x^2 + 23)$

10. $(0.5x^4 - 0.6x^2 + 0.7) + (2.3x^4 + 1.8x - 3.9)$

11. $(1 + 4x + 6x^2 + 7x^3) + (5 - 4x + 6x^2 - 7x^3)$

12. $(3x^4 - 6x - 5x^2 + 5) + (6x^2 - 4x^3 - 1 + 7x)$

13. $\left(\frac{1}{4}x^4 + \frac{2}{3}x^3 + \frac{5}{8}x^2 + 7\right) + \left(-\frac{3}{4}x^4 + \frac{3}{8}x^2 - 7\right)$

14. $\left(\frac{1}{3}x^9 + \frac{1}{5}x^5 - \frac{1}{2}x^2 + 7\right) +$
$\left(-\frac{1}{5}x^9 + \frac{1}{4}x^4 - \frac{3}{5}x^5 + \frac{3}{4}x^2 + \frac{1}{2}\right)$

15. $(0.02x^5 - 0.2x^3 + x + 0.08) +$
$(-0.01x^5 + x^4 - 0.8x - 0.02)$

16. $(0.03x^6 + 0.05x^3 + 0.22x + 0.05) +$
$\left(\frac{7}{100}x^6 - \frac{3}{100}x^3 + 0.5\right)$

17. $(9x^8 - 7x^4 + 2x^2 + 5) + (8x^7 + 4x^4 - 2x) +$
$(-3x^4 + 6x^2 + 2x - 1)$

18. $(4x^5 - 6x^3 - 9x + 1) + (6x^3 + 9x^2 + 9x) +$
$(-4x^3 + 8x^2 + 3x - 2)$

19.
$$\begin{array}{rrrrr}
0.15x^4 & + 0.10x^3 & - \;\; 0.9x^2 & & \\
 & - 0.01x^3 & + 0.01x^2 & + x & \\
1.25x^4 & & + 0.11x^2 & & + 0.01 \\
 & 0.27x^3 & & & + 0.99 \\
\underline{-0.35x^4} & & \underline{+ \;\; 15x^2} & & \underline{- 0.03}
\end{array}$$

20.
$$\begin{array}{rrrrr}
0.05x^4 & + 0.12x^3 & - \;\; 0.5x^2 & & \\
 & - 0.02x^3 & + 0.02x^2 & + 2x & \\
1.5x^4 & & + 0.01x^2 & & + 0.15 \\
 & 0.25x^3 & & & + 0.85 \\
\underline{-0.25x^4} & & \underline{+ \;\; 10x^2} & & \underline{- 0.04}
\end{array}$$

b Find two equivalent expressions for the opposite of the polynomial.

21. $-5x$

22. $x^2 - 3x$

23. $-x^2 + 10x - 2$

24. $-4x^3 - x^2 - x$

25. $12x^4 - 3x^3 + 3$

26. $4x^3 - 6x^2 - 8x + 1$

Simplify.

27. $-(3x - 7)$

28. $-(-2x + 4)$

29. $-(4x^2 - 3x + 2)$

30. $-(-6a^3 + 2a^2 - 9a + 1)$

31. $-(-4x^4 + 6x^2 + \frac{3}{4}x - 8)$

32. $-(-5x^4 + 4x^3 - x^2 + 0.9)$

c Subtract.

33. $(3x + 2) - (-4x + 3)$

34. $(6x + 1) - (-7x + 2)$

35. $(-6x + 2) - (x^2 + x - 3)$

36. $(x^2 - 5x + 4) - (8x - 9)$

37. $(x^2 - 9) - (x^2 + 9)$

38. $(x^3 + x^2) - (2x^3 - 5x^2)$

39. $(6x^4 + 3x^3 - 1) - (4x^2 - 3x + 3)$

40. $(-4x^2 + 2x) - (3x^3 - 5x^2 + 3)$

41. $(1.2x^3 + 4.5x^2 - 3.8x) - (-3.4x^3 - 4.7x^2 + 23)$

42. $(0.5x^4 - 0.6x^2 + 0.7) - (2.3x^4 + 1.8x - 3.9)$

43. $\left(\frac{5}{8}x^3 - \frac{1}{4}x - \frac{1}{3}\right) - \left(-\frac{1}{8}x^3 + \frac{1}{4}x - \frac{1}{3}\right)$

44. $\left(\frac{1}{5}x^3 + 2x^2 - 0.1\right) - \left(-\frac{2}{5}x^3 + 2x^2 + 0.01\right)$

45. $(0.08x^3 - 0.02x^2 + 0.01x) - (0.02x^3 + 0.03x^2 - 1)$

46. $(0.8x^4 + 0.2x - 1) - \left(\frac{7}{10}x^4 + \frac{1}{5}x - 0.1\right)$

Subtract.

47. $\begin{array}{r} x^2 + 5x + 6 \\ \underline{x^2 + 2x } \end{array}$

48. $\begin{array}{r} x^3 + 1 \\ \underline{x^3 + x^2 } \end{array}$

49. $\begin{array}{r} 5x^4 + 6x^3 - 9x^2 \\ \underline{-6x^4 - 6x^3 + 8x + 9} \end{array}$

50. $\begin{array}{r} 5x^4 + 6x^2 - 3x + 6 \\ \underline{6x^3 + 7x^2 - 8x - 9} \end{array}$

51. $\begin{array}{r} x^5 - 1 \\ \underline{x^5 - x^4 + x^3 - x^2 + x - 1} \end{array}$

52. $\begin{array}{r} x^5 + x^4 - x^3 + x^2 - x + 2 \\ \underline{x^5 - x^4 + x^3 - x^2 - x + 2} \end{array}$

d Solve.

53. Find a polynomial for the sum of the areas of these rectangles.

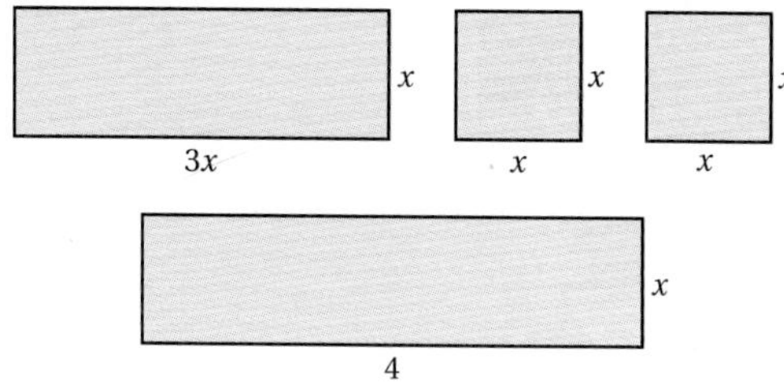

54. Find a polynomial for the sum of the areas of these circles.

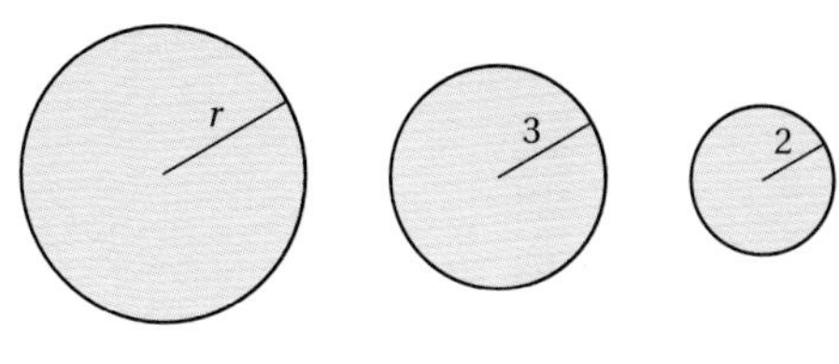

Find a polynomial for the perimeter of the figure.

55.

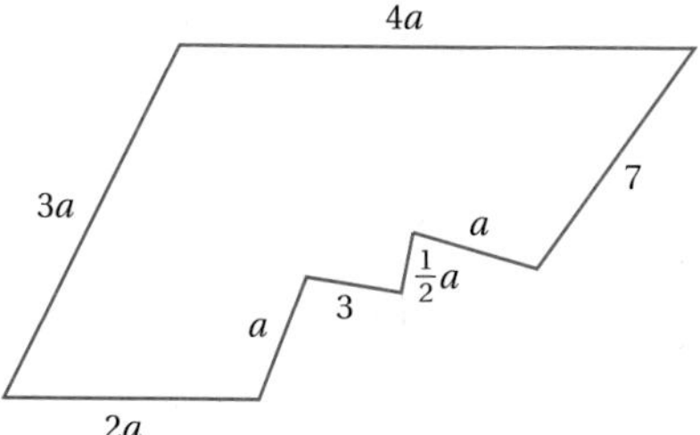

56.

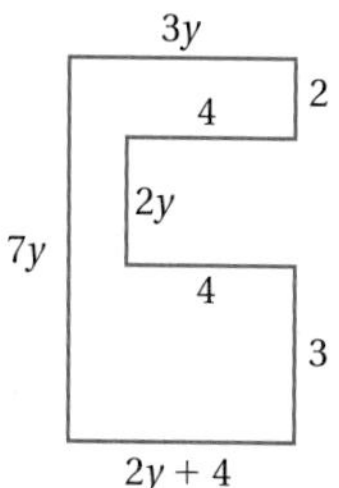

Skill Maintenance

Solve. [2.3b]

57. $8x + 3x = 66$

58. $5x - 7x = 38$

59. $\frac{3}{8}x + \frac{1}{4} - \frac{3}{4}x = \frac{11}{16} + x$

60. $5x - 4 = 26 - x$

61. $1.5x - 2.7x = 22 - 5.6x$

62. $3x - 3 = -4x + 4$

Solve. [2.3c]

63. $6(y - 3) - 8 = 4(y + 2) + 5$

64. $8(5x + 2) = 7(6x - 3)$

Solve. [2.7e]

65. $3x - 7 \leq 5x + 13$

66. $2(x - 4) > 5(x - 3) + 7$

Synthesis

67. ◈ Is the sum of two binomials ever a trinomial? Why or why not?

68. ◈ Which, if any, of the commutative, associative, and distributive laws are needed for adding polynomials? Why?

Find two algebraic expressions for the area of the figure.

69.

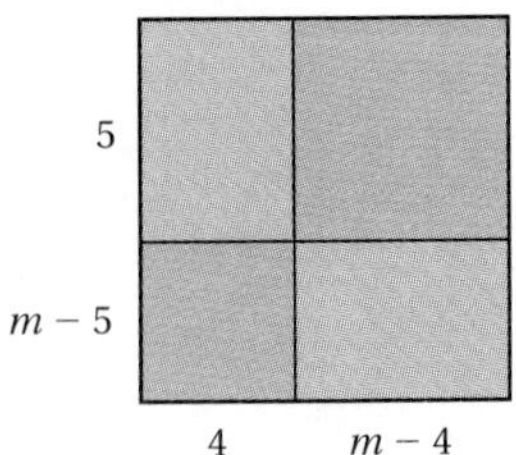

70.

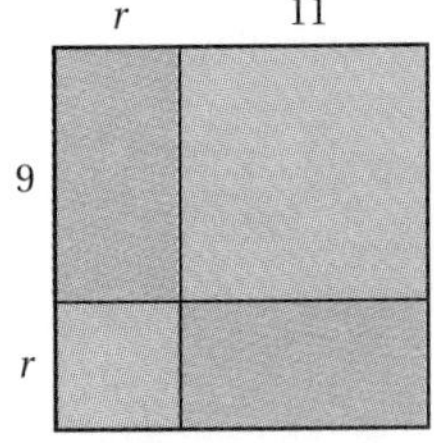

Find a polynomial for the shaded area of the figure.

71.

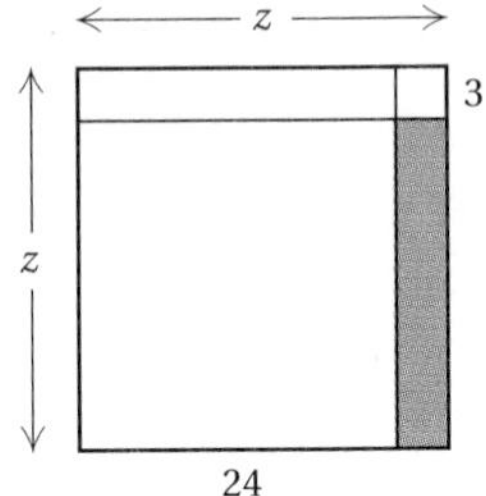

72.

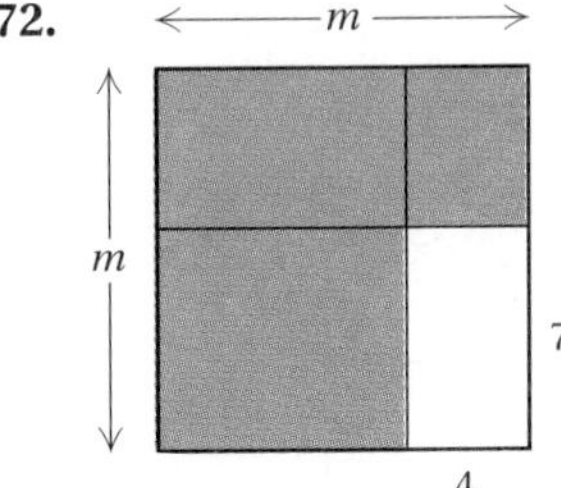

73. Find $(y - 2)^2$ using the four parts of this square.

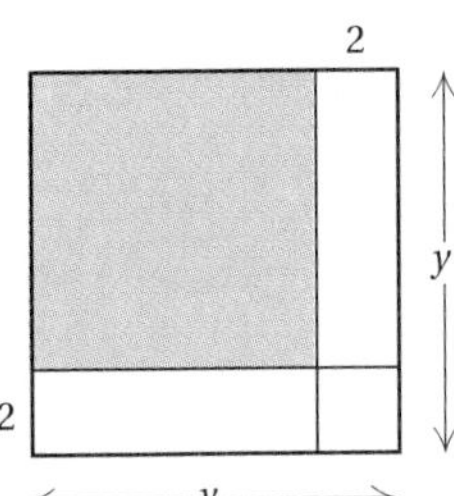

74. Find a polynomial for the surface area of this right rectangular solid.

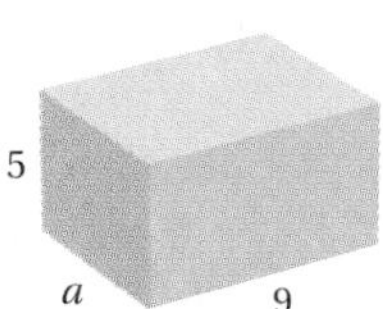

Simplify.

75. $(3x^2 - 4x + 6) - (-2x^2 + 4) + (-5x - 3)$

76. $(7y^2 - 5y + 6) - (3y^2 + 8y - 12) + (8y^2 - 10y + 3)$

77. $(-4 + x^2 + 2x^3) - (-6 - x + 3x^3) - (-x^2 - 5x^3)$

78. $(-y^4 - 7y^3 + y^2) + (-2y^4 + 5y - 2) - (-6y^3 + y^2)$

The TABLE feature can be used as a partial check that polynomials have been added or subtracted correctly. To check Example 3, we enter

$$y_1 = (3x^2 - 2x + 2) + (5x^3 - 2x^2 + 3x - 4)$$
$$\text{and } y_2 = 5x^3 + x^2 + x - 2.$$

If our addition was correct, the y_1- and y_2-values should be the same, regardless of the table settings used.

X	Y1	Y2
4	338	338
5	653	653
6	1120	1120
7	1769	1769
8	2630	2630
9	3733	3733
10	5108	5108

X = 4

79. Use the TABLE feature to check Exercise 7.

80. Use the TABLE feature to check Exercise 9.

4.5 Multiplication of Polynomials

We now multiply polynomials using techniques based, for the most part, on the distributive laws, but also on the associative and commutative laws. As we proceed in this chapter, we will develop special ways to find certain products.

Objectives

a Multiply monomials.

b Multiply a monomial and any polynomial.

c Multiply two binomials.

d Multiply any two polynomials.

For Extra Help

TAPE 7 MAC WIN CD-ROM

a Multiplying Monomials

Consider $(3x)(4x)$. We multiply as follows:

$(3x)(4x) = 3 \cdot x \cdot 4 \cdot x$ By the associative law of multiplication

$= 3 \cdot 4 \cdot x \cdot x$ By the commutative law of multiplication

$= (3 \cdot 4)(x \cdot x)$ By the associative law

$= 12x^2.$ Using the product rule for exponents

To find an equivalent expression for the product of two monomials, multiply the coefficients and then multiply the variables using the product rule for exponents.

Examples Multiply.

1. $5x \cdot 6x = (5 \cdot 6)(x \cdot x)$ By the associative and commutative laws
   $= 30x^2$ Multiplying the coefficients and multiplying the variables
2. $(3x)(-x) = (3x)(-1x)$
   $= (3)(-1)(x \cdot x)$
   $= -3x^2$
3. $(-7x^5)(4x^3) = (-7 \cdot 4)(x^5 \cdot x^3)$
   $= -28x^{5+3}$ Adding the exponents
   $= -28x^8$ Simplifying

After some practice, you can do this mentally. Multiply the coefficients and then the variables by keeping the base and adding the exponents. Write only the answer.

Do Exercises 1–8.

Multiply.

1. $(3x)(-5)$
2. $(-x) \cdot x$
3. $(-x)(-x)$
4. $(-x^2)(x^3)$
5. $3x^5 \cdot 4x^2$
6. $(4y^5)(-2y^6)$
7. $(-7y^4)(-y)$
8. $7x^5 \cdot 0$

Answers on page A-15

b Multiplying a Monomial and Any Polynomial

To find an equivalent expression for the product of a monomial, such as $2x$, and a binomial, such as $5x + 3$, we use a distributive law and multiply each term of $5x + 3$ by $2x$.

Example 4 Multiply: $2x(5x + 3)$.

$2x(5x + 3) = (2x)(5x) + (2x)(3)$ Using a distributive law

$= 10x^2 + 6x$ Multiplying the monomials

Multiply.

9. $4x(2x + 4)$

10. $3t^2(-5t + 2)$

11. $5x^3(x^3 + 5x^2 - 6x + 8)$

Multiply.

12. $(x + 8)(x + 5)$

13. $(x + 5)(x - 4)$

Answers on page A-15

Example 5 Multiply: $5x(2x^2 - 3x + 4)$.

$$5x(2x^2 - 3x + 4) = (5x)(2x^2) - (5x)(3x) + (5x)(4)$$
$$= 10x^3 - 15x^2 + 20x$$

> To multiply a monomial and a polynomial, multiply each term of the polynomial by the monomial.

Example 6 Multiply: $2x^2(x^3 - 7x^2 + 10x - 4)$.

$$2x^2(x^3 - 7x^2 + 10x - 4) = 2x^5 - 14x^4 + 20x^3 - 8x^2$$

Do Exercises 9–11.

c Multiplying Two Binomials

To find an equivalent expression for the product of two binomials, we use the distributive laws more than once. In Example 7, we use a distributive law three times.

Example 7 Multiply: $(x + 5)(x + 4)$.

$(x + 5)(x + 4) = x(x + 4) + 5(x + 4)$ — **Using a distributive law**

$= x \cdot x + x \cdot 4 + 5 \cdot x + 5 \cdot 4$ — **Using a distributive law on each part**

$= x^2 + 4x + 5x + 20$ — **Multiplying the monomials**

$= x^2 + 9x + 20$ — **Collecting like terms**

To visualize the product in Example 7, consider a rectangle of length $x + 5$ and width $x + 4$.

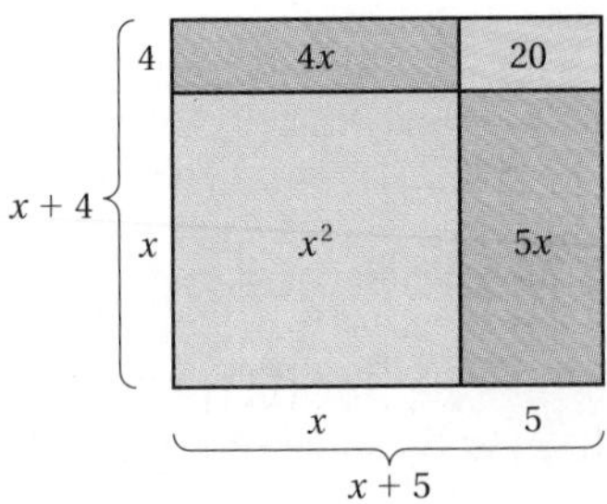

The total area can be expressed as $(x + 5)(x + 4)$ or, by adding the four smaller areas, $x^2 + 5x + 4x + 20$.

Do Exercises 12 and 13.

Example 8 Multiply: $(4x + 3)(x - 2)$.

$$(4x + 3)(x - 2) = 4x(x - 2) + 3(x - 2) \quad \text{Using a distributive law}$$
$$= 4x \cdot x - 4x \cdot 2 + 3 \cdot x - 3 \cdot 2 \quad \text{Using a distributive law on each part}$$
$$= 4x^2 - 8x + 3x - 6 \quad \text{Multiplying the monomials}$$
$$= 4x^2 - 5x - 6 \quad \text{Collecting like terms}$$

Do Exercises 14 and 15.

Multiply.

14. $(5x + 3)(x - 4)$

15. $(2x - 3)(3x - 5)$

d Multiplying Any Two Polynomials

Let's consider the product of a binomial and a trinomial. We use a distributive law four times. You may see ways to skip some steps and do the work mentally.

Example 9 Multiply: $(x^2 + 2x - 3)(x^2 + 4)$.

$$(x^2 + 2x - 3)(x^2 + 4) = x^2(x^2 + 4) + 2x(x^2 + 4) - 3(x^2 + 4)$$
$$= x^2 \cdot x^2 + x^2 \cdot 4 + 2x \cdot x^2 + 2x \cdot 4 - 3 \cdot x^2 - 3 \cdot 4$$
$$= x^4 + 4x^2 + 2x^3 + 8x - 3x^2 - 12$$
$$= x^4 + 2x^3 + x^2 + 8x - 12$$

Do Exercises 16 and 17.

Multiply.

16. $(x^2 + 3x - 4)(x^2 + 5)$

17. $(3y^2 - 7)(2y^3 - 2y + 5)$

Perhaps you have discovered the following in the preceding examples.

> To multiply two polynomials P and Q, select one of the polynomials—say, P. Then multiply each term of P by every term of Q and collect like terms.

We can use columns for long multiplications. We multiply each term at the top by every term at the bottom. We write like terms in columns, and then we add the results. Such multiplication is like multiplying with whole numbers:

4 5 7	4 5 7	= 400 + 50 + 7
× 6 3	× 6 3	= 60 + 3
1 3 7 1	1200 + 150 + 21	= 3(457) = 3(400 + 50 + 7)
2 7 4 2 0	24000 + 3000 + 420	= 60(457) = 60(400 + 50 + 7)
2 8 7 9 1	24000 + 4200 + 570 + 21	= 28,791

Example 10 Multiply: $(4x^2 - 2x + 3)(x + 2)$.

$$\begin{array}{rl}
4x^2 - 2x + 3 & \\
x + 2 & \\
\hline
8x^2 - 4x + 6 & \text{Multiplying the top row by 2} \\
4x^3 - 2x^2 + 3x \quad\;\; & \text{Multiplying the top row by } x \\
\hline
4x^3 + 6x^2 - x + 6 & \text{Collecting like terms}
\end{array}$$

Line up like terms in columns.

Answers on page A-15

Multiply.

18. $\begin{array}{r} 3x^2 - 2x + 4 \\ x + 5 \\ \hline \end{array}$

19. $\begin{array}{r} -5x^2 + 4x + 2 \\ -4x^2 - 8 \\ \hline \end{array}$

20. Multiply.

$\begin{array}{r} 3x^2 - 2x - 5 \\ 2x^2 + x - 2 \\ \hline \end{array}$

Example 11 Multiply: $(5x^3 - 3x + 4)(-2x^2 - 3)$.

When missing terms occur, it helps to leave spaces for them and align like terms as we multiply.

$$\begin{array}{rrrrrl} & 5x^3 & & -\ 3x & +\ 4 & \\ & & -2x^2 & & -\ 3 & \\ \hline & -15x^3 & & +\ 9x & -\ 12 & \text{Multiplying by } -3 \\ -10x^5 & +\ 6x^3 & -\ 8x^2 & & & \text{Multiplying by } -2x^2 \\ \hline -10x^5 & -\ 9x^3 & -\ 8x^2 & +\ 9x & -\ 12 & \text{Collecting like terms} \end{array}$$

Do Exercises 18 and 19.

Example 12 Multiply: $(2x^2 + 3x - 4)(2x^2 - x + 3)$.

$$\begin{array}{rrrrrl} & & 2x^2 & +\ 3x & -\ 4 & \\ & & 2x^2 & -\ x & +\ 3 & \\ \hline & & 6x^2 & +\ 9x & -\ 12 & \text{Multiplying by } 3 \\ & -2x^3 & -\ 3x^2 & +\ 4x & & \text{Multiplying by } -x \\ 4x^4 & +\ 6x^3 & -\ 8x^2 & & & \text{Multiplying by } 2x^2 \\ \hline 4x^4 & +\ 4x^3 & -\ 5x^2 & +\ 13x & -\ 12 & \text{Collecting like terms} \end{array}$$

Do Exercise 20.

Calculator Spotlight

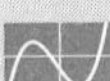 Checking Multiplications with a Table or Graph

Table. The TABLE feature can be used as a partial check that polynomials have been multiplied correctly. To check whether

$(x + 5)(x - 4) = x^2 + 9x - 20$

is correct, we enter

$y_1 = (x + 5)(x - 4)$ and $y_2 = x^2 + 9x - 20$.

X	Y1	Y2
−23	486	302
−13	136	32
−3	−14	−38
7	36	92
17	286	422
27	736	952
37	1386	1682
X = −23		

If our multiplication is correct, the y_1- and y_2-values should be the same, regardless of the table settings used. We see that y_1 and y_2 are not the same, so the multiplication is not correct.

Graph. Multiplication of polynomials can also be checked with the GRAPH feature. In this case, we see that the graphs differ, so the multiplication is not correct.

$y_1 = (x + 5)(x - 4)$, $y_2 = x^2 + 9x - 20$

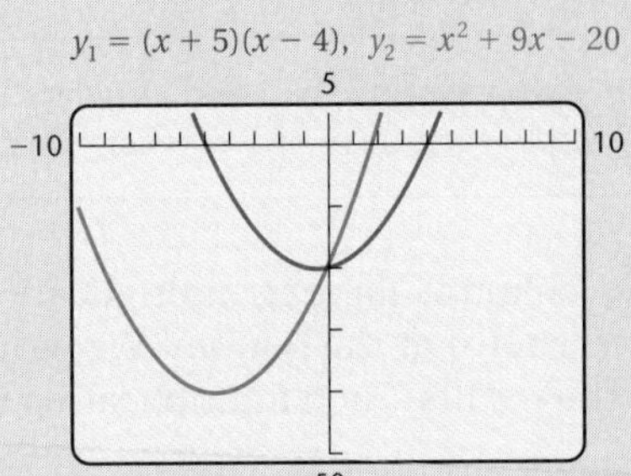

Exercises

Use the TABLE or GRAPH feature to check whether each of the following is correct.

1. $(x + 5)(x + 4) = x^2 + 9x + 20$ (Example 7)
2. $(4x + 3)(x - 2) = 4x^2 - 5x - 6$ (Example 8)
3. $(5x + 3)(x - 4) = 5x^2 + 17x - 12$
4. $(2x - 3)(3x - 5) = 6x^2 - 19x - 15$
5. $(x - 3)(x - 3) = x^2 - 9$
6. $(x - 3)(x + 3) = x^2 - 9$

Answers on page A-15

Exercise Set 4.5

a Multiply.

1. $(8x^2)(5)$

2. $(4x^2)(-2)$

3. $(-x^2)(-x)$

4. $(-x^3)(x^2)$

5. $(8x^5)(4x^3)$

6. $(10a^2)(2a^2)$

7. $(0.1x^6)(0.3x^5)$

8. $(0.3x^4)(-0.8x^6)$

9. $\left(-\frac{1}{5}x^3\right)\left(-\frac{1}{3}x\right)$

10. $\left(-\frac{1}{4}x^4\right)\left(\frac{1}{5}x^8\right)$

11. $(-4x^2)(0)$

12. $(-4m^5)(-1)$

13. $(3x^2)(-4x^3)(2x^6)$

14. $(-2y^5)(10y^4)(-3y^3)$

b Multiply.

15. $2x(-x + 5)$

16. $3x(4x - 6)$

17. $-5x(x - 1)$

18. $-3x(-x - 1)$

19. $x^2(x^3 + 1)$

20. $-2x^3(x^2 - 1)$

21. $3x(2x^2 - 6x + 1)$

22. $-4x(2x^3 - 6x^2 - 5x + 1)$

23. $(-6x^2)(x^2 + x)$

24. $(-4x^2)(x^2 - x)$

25. $(3y^2)(6y^4 + 8y^3)$

26. $(4y^4)(y^3 - 6y^2)$

c Multiply.

27. $(x + 6)(x + 3)$

28. $(x + 5)(x + 2)$

29. $(x + 5)(x - 2)$

30. $(x + 6)(x - 2)$

31. $(x - 4)(x - 3)$

32. $(x - 7)(x - 3)$

33. $(x + 3)(x - 3)$

34. $(x + 6)(x - 6)$

35. $(5 - x)(5 - 2x)$

36. $(3 + x)(6 + 2x)$

37. $(2x + 5)(2x + 5)$

38. $(3x - 4)(3x - 4)$

39. $\left(x - \frac{5}{2}\right)\left(x + \frac{2}{5}\right)$

40. $\left(x + \frac{4}{3}\right)\left(x + \frac{3}{2}\right)$

41. $(x - 2.3)(x + 4.7)$

42. $(2x + 0.13)(2x - 0.13)$

d Multiply.

43. $(x^2 + x + 1)(x - 1)$

44. $(x^2 + x - 2)(x + 2)$

45. $(2x + 1)(2x^2 + 6x + 1)$

46. $(3x - 1)(4x^2 - 2x - 1)$

47. $(y^2 - 3)(3y^2 - 6y + 2)$

48. $(3y^2 - 3)(y^2 + 6y + 1)$

49. $(x^3 + x^2)(x^3 + x^2 - x)$

50. $(x^3 - x^2)(x^3 - x^2 + x)$

51. $(-5x^3 - 7x^2 + 1)(2x^2 - x)$

52. $(-4x^3 + 5x^2 - 2)(5x^2 + 1)$

53. $(1 + x + x^2)(-1 - x + x^2)$

54. $(1 - x + x^2)(1 - x + x^2)$

55. $(2t^2 - t - 4)(3t^2 + 2t - 1)$

56. $(3a^2 - 5a + 2)(2a^2 - 3a + 4)$

57. $(x - x^3 + x^5)(x^2 - 1 + x^4)$

58. $(x - x^3 + x^5)(3x^2 + 3x^6 + 3x^4)$

59. $(x^3 + x^2 + x + 1)(x - 1)$

60. $(x + 2)(x^3 - x^2 + x - 2)$

Skill Maintenance

Simplify.

61. $-\frac{1}{4} - \frac{1}{2}$ [1.4a]

62. $-3.8 - (-10.2)$ [1.4a]

63. $(10 - 2)(10 + 2)$ [1.8d]

64. $10 - 2 + (-6)^2 \div 3 \cdot 2$ [1.8d]

Factor. [1.7d]

65. $15x - 18y + 12$

66. $16x - 24y + 36$

67. $-9x - 45y + 15$

68. $100x - 100y + 1000a$

69. Graph: $y = \frac{1}{2}x - 3$. [3.2b]

70. Solve: $4(x - 3) = 5(2 - 3x) + 1$. [2.3c]

Synthesis

71. ◆ Under what conditions will the product of two binomials be a trinomial?

72. ◆ Is it possible to understand polynomial multiplication without first understanding the distributive law? Why or why not?

73. Find a polynomial for the shaded area.

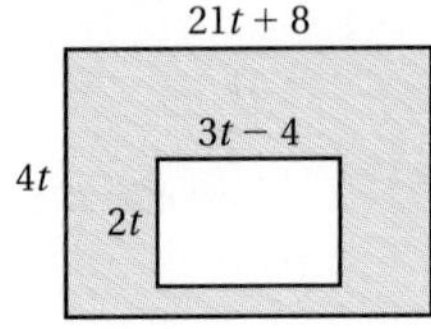

74. A box with a square bottom is to be made from a 12-in.-square piece of cardboard. Squares with side x are cut out of the corners and the sides are folded up. Find polynomials for the volume and the outside surface area of the box.

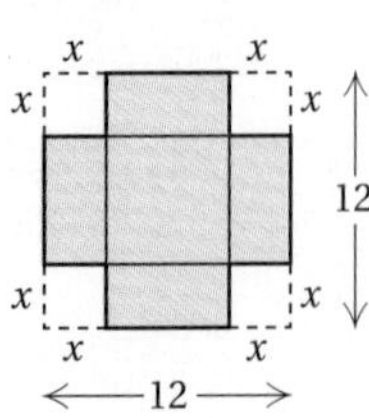

75. The height of a triangle is 4 ft longer than its base. Find a polynomial for the area.

Compute and simplify.

76. $(x + 3)(x + 6) + (x + 3)(x + 6)$

77. $(x - 2)(x - 7) - (x - 2)(x - 7)$

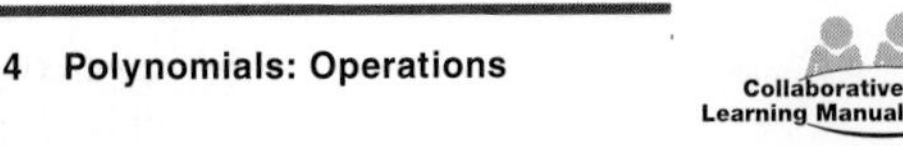

Collaborative Learning Manual

Visualize polynomial multiplication using rectangles.

4.6 Special Products

We encounter certain products so often that it is helpful to have faster methods of computing. We now consider special ways of multiplying any two binomials. Such techniques are called *special products.*

Objectives

a. Multiply two binomials mentally using the FOIL method.
b. Multiply the sum and the difference of two terms mentally.
c. Square a binomial mentally.
d. Find special products when polynomial products are mixed together.

For Extra Help

TAPE 7 | MAC WIN | CD-ROM

a Products of Two Binomials Using FOIL

To multiply two binomials, we can select one binomial and multiply each term of that binomial by every term of the other. Then we collect like terms. Consider the product $(x + 5)(x + 4)$:

$$\begin{aligned}(x + 5)(x + 4) &= x \cdot x + 5 \cdot x + x \cdot 4 + 5 \cdot 4 \\ &= x^2 + 5x + 4x + 20 \\ &= x^2 + 9x + 20.\end{aligned}$$

We can rewrite the first line of this product to show a special technique for finding the product of two binomials:

$$(x + 5)(x + 4) = \underbrace{x \cdot x}_{\text{First terms}} + \underbrace{4 \cdot x}_{\text{Outside terms}} + \underbrace{5 \cdot x}_{\text{Inside terms}} + \underbrace{5 \cdot 4}_{\text{Last terms}}.$$

To remember this method of multiplying, we use the initials **FOIL.**

THE FOIL METHOD

To multiply two binomials, $A + B$ and $C + D$, multiply the First terms AC, the Outside terms AD, the Inside terms BC, and then the Last terms BD. Then collect like terms, if possible.

$$(A + B)(C + D) = AC + AD + BC + BD$$

1. Multiply First terms: AC.
2. Multiply Outside terms: AD.
3. Multiply Inside terms: BC.
4. Multiply Last terms: BD.

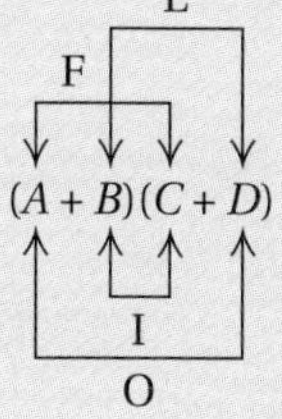

Example 1 Multiply: $(x + 8)(x^2 - 5)$.

We have

$$\begin{aligned}(x + 8)(x^2 - 5) &= \overset{F}{x^3} - \overset{O}{5x} + \overset{I}{8x^2} - \overset{L}{40} \\ &= x^3 + 8x^2 - 5x - 40.\end{aligned}$$

Since each of the original binomials is in descending order, we write the product in descending order, as is customary, but this is not a "must."

Multiply mentally, if possible. If you need extra steps, be sure to use them.

1. $(x + 3)(x + 4)$

2. $(x + 3)(x - 5)$

3. $(2x - 1)(x - 4)$

4. $(2x^2 - 3)(x - 2)$

5. $(6x^2 + 5)(2x^3 + 1)$

6. $(y^3 + 7)(y^3 - 7)$

7. $(t + 5)(t + 3)$

8. $(2x^4 + x^2)(-x^3 + x)$

Multiply.

9. $\left(x + \frac{4}{5}\right)\left(x - \frac{4}{5}\right)$

10. $(x^3 - 0.5)(x^2 + 0.5)$

11. $(2 + 3x^2)(4 - 5x^2)$

12. $(6x^3 - 3x^2)(5x^2 - 2x)$

Answers on page A-16

Often we can collect like terms after we have multiplied.

Examples Multiply.

2. $(x + 6)(x - 6) = x^2 - 6x + 6x - 36$ **Using FOIL**
$= x^2 - 36$ **Collecting like terms**

3. $(x + 7)(x + 4) = x^2 + 4x + 7x + 28$
$= x^2 + 11x + 28$

4. $(y - 3)(y - 2) = y^2 - 2y - 3y + 6$
$= y^2 - 5y + 6$

5. $(x^3 - 5)(x^3 + 5) = x^6 + 5x^3 - 5x^3 - 25$
$= x^6 - 25$

Do Exercises 1–8.

Examples Multiply.

6. $(4t^3 + 5)(3t^2 - 2) = 12t^5 - 8t^3 + 15t^2 - 10$

7. $\left(x - \frac{2}{3}\right)\left(x + \frac{2}{3}\right) = x^2 + \frac{2}{3}x - \frac{2}{3}x - \frac{4}{9}$
$= x^2 - \frac{4}{9}$

8. $(x^2 - 0.3)(x^2 - 0.3) = x^4 - 0.3x^2 - 0.3x^2 + 0.09$
$= x^4 - 0.6x^2 + 0.09$

9. $(3 - 4x)(7 - 5x^3) = 21 - 15x^3 - 28x + 20x^4$
$= 21 - 28x - 15x^3 + 20x^4$

(*Note*: If the original polynomials are in ascending order, it is natural to write the product in ascending order, but this is not a "must.")

10. $(5x^4 + 2x^3)(3x^2 - 7x) = 15x^6 - 35x^5 + 6x^5 - 14x^4$
$= 15x^6 - 29x^5 - 14x^4$

Do Exercises 9–12.

We can show the FOIL method geometrically as follows.

The area of the large rectangle is $(A + B)(C + D)$.

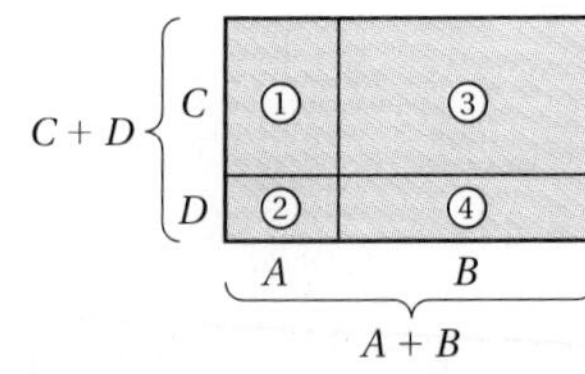

The area of rectangle (1) is AC.
The area of rectangle (2) is AD.
The area of rectangle (3) is BC.
The area of rectangle (4) is BD.

The area of the large rectangle is the sum of the areas of the smaller rectangles. Thus,

$$(A + B)(C + D) = AC + AD + BC + BD.$$

b Multiplying Sums and Differences of Two Terms

Consider the product of the sum and the difference of the same two terms, such as

$$(x + 2)(x - 2).$$

Since this is the product of two binomials, we can use FOIL. This type of product occurs so often, however, that it would be valuable if we could use an even faster method. To find a faster way to compute such a product, look for a pattern in the following:

a) $(x + 2)(x - 2) = x^2 - 2x + 2x - 4$
$= x^2 - 4;$

b) $(3x - 5)(3x + 5) = 9x^2 + 15x - 15x - 25$
$= 9x^2 - 25.$

Do Exercises 13 and 14.

Perhaps you discovered in each case that when you multiply the two binomials, two terms are opposites, or additive inverses, which add to 0 and "drop out."

> The product of the sum and the difference of the same two terms is the square of the first term minus the square of the second term:
> $(A + B)(A - B) = A^2 - B^2.$

It is helpful to memorize this rule in both words and symbols. (If you do forget it, you can, of course, use FOIL.)

Examples Multiply. (Carry out the rule and say the words as you go.)

$(A + B)(A - B) = A^2 - B^2$

11. $(x + 4)(x - 4) = x^2 - 4^2$ "The square of the first term, x^2, minus the square of the second, 4^2"
$= x^2 - 16$ Simplifying

12. $(5 + 2w)(5 - 2w) = 5^2 - (2w)^2$
$= 25 - 4w^2$

13. $(3x^2 - 7)(3x^2 + 7) = (3x^2)^2 - 7^2$
$= 9x^4 - 49$

14. $(-4x - 10)(-4x + 10) = (-4x)^2 - 10^2$
$= 16x^2 - 100$

15. $\left(x + \frac{3}{8}\right)\left(x - \frac{3}{8}\right) = x^2 - \left(\frac{3}{8}\right)^2 = x^2 - \frac{9}{64}$

Do Exercises 15–19.

C Squaring Binomials

Consider the square of a binomial, such as $(x + 3)^2$. This can be expressed as $(x + 3)(x + 3)$. Since this is the product of two binomials, we can again use FOIL. But again, this type of product occurs so often that we would like to use an even faster method. Look for a pattern in the following:

a) $(x + 3)^2 = (x + 3)(x + 3)$
$= x^2 + 3x + 3x + 9$
$= x^2 + 6x + 9;$

b) $(5 + 3p)^2 = (5 + 3p)(5 + 3p)$
$= 25 + 15p + 15p + 9p^2$
$= 25 + 30p + 9p^2;$

Multiply.

13. $(x + 5)(x - 5)$

14. $(2x - 3)(2x + 3)$

Multiply.

15. $(x + 2)(x - 2)$

16. $(x - 7)(x + 7)$

17. $(6 - 4y)(6 + 4y)$

18. $(2x^3 - 1)(2x^3 + 1)$

19. $\left(x - \frac{2}{5}\right)\left(x + \frac{2}{5}\right)$

Answers on page A-16

Multiply.

20. $(x + 8)(x + 8)$

21. $(x - 5)(x - 5)$

Multiply.

22. $(x + 2)^2$

23. $(a - 4)^2$

24. $(2x + 5)^2$

25. $(4x^2 - 3x)^2$

26. $(7.8 + 1.2y)(7.8 + 1.2y)$

27. $(3x^2 - 5)(3x^2 - 5)$

Answers on page A-16

c) $(x - 3)^2 = (x - 3)(x - 3)$
$= x^2 - 3x - 3x + 9$
$= x^2 - 6x + 9;$

d) $(3x - 5)^2 = (3x - 5)(3x - 5)$
$= 9x^2 - 15x - 15x + 25$
$= 9x^2 - 30x + 25.$

Do Exercises 20 and 21.

When squaring a binomial, we multiply a binomial by itself. Perhaps you noticed that two terms are the same and when added give twice their product. The other two terms are squares.

> The square of a sum or a difference of two terms is the square of the first term, plus or minus twice the product of the two terms, plus the square of the last term:
>
> $$(A + B)^2 = A^2 + 2AB + B^2;$$
> $$(A - B)^2 = A^2 - 2AB + B^2.$$

It is helpful to memorize this rule in both words and symbols.

Examples Multiply. (Carry out the rule and say the words as you go.)

$$(A + B)^2 = A^2 + 2 \cdot A \cdot B + B^2$$

16. $(x + 3)^2 = x^2 + 2 \cdot x \cdot 3 + 3^2$ "x^2 plus 2 times x times 3 plus 3^2"
$= x^2 + 6x + 9$

$$(A - B)^2 = A^2 - 2 \cdot A \cdot B + B^2$$

17. $(t - 5)^2 = t^2 - 2 \cdot t \cdot 5 + 5^2$ "t^2 minus 2 times t times 5 plus 5^2"
$= t^2 - 10t + 25$

18. $(2x + 7)^2 = (2x)^2 + 2 \cdot 2x \cdot 7 + 7^2$
$= 4x^2 + 28x + 49$

19. $(5x - 3x^2)^2 = (5x)^2 - 2 \cdot 5x \cdot 3x^2 + (3x^2)^2$
$= 25x^2 - 30x^3 + 9x^4$

20. $(2.3 - 5.4m)^2 = 2.3^2 - 2(2.3)(5.4m) + (5.4m)^2$
$= 5.29 - 24.84m + 29.16m^2$

Do Exercises 22–27.

CAUTION! Note carefully in these examples that the square of a sum is *not* the sum of the squares:

The middle term $2AB$ is missing.

$$(A + B)^2 \neq A^2 + B^2.$$

To see this, note that

$$(20 + 5)^2 = 25^2 = 625,$$

but

$$20^2 + 5^2 = 400 + 25 = 425 \quad \text{and} \quad 425 \neq 625.$$

However, $20^2 + 2(20)(5) + 5^2 = 625$, which illustrates that

$$(A + B)^2 = A^2 + 2AB + B^2.$$

We can look at the rule for finding $(A + B)^2$ geometrically as follows. The area of the large square is

$$(A + B)(A + B) = (A + B)^2.$$

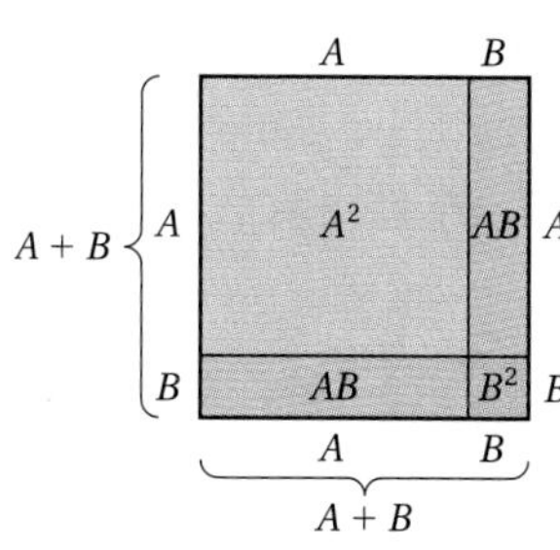

This is equal to the sum of the areas of the smaller rectangles:

$$A^2 + AB + AB + B^2 = A^2 + 2AB + B^2.$$

Thus,

$$(A + B)^2 = A^2 + 2AB + B^2.$$

d Multiplication of Various Types

We have considered how to quickly multiply certain kinds of polynomials. Let's now try several types of multiplications mixed together so that we can learn to sort them out. When you multiply, first see what kind of multiplication you have. Then use the best method. The formulas you should know and the questions you should ask yourself are as follows.

MULTIPLYING TWO POLYNOMIALS

1. Is the product the square of a binomial? If so, use the following:

$$(A + B)(A + B) = (A + B)^2 = A^2 + 2AB + B^2,$$

or $(A - B)(A - B) = (A - B)^2 = A^2 - 2AB + B^2.$

The square of a binomial is the square of the first term, plus or minus *twice* the product of the two terms, plus the square of the last term.

[The answer has 3 terms.]

Example: $(x + 7)(x + 7) = (x + 7)^2$
$= x^2 + 2 \cdot x \cdot 7 + 7^2 = x^2 + 14x + 49$

2. Is it the product of the sum and the difference of the *same* two terms? If so, use the following:

$$(A + B)(A - B) = A^2 - B^2.$$

The product of the sum and the difference of the same two terms is the difference of the squares.

[The answer has 2 terms.]

Example: $(x + 7)(x - 7) = x^2 - 7^2 = x^2 - 49$

3. Is it the product of two binomials other than those above? If so, use FOIL.

[The answer will have 3 or 4 terms.]

Example: $(x + 7)(x - 4) = x^2 - 4x + 7x - 28 = x^2 + 3x - 28$

4. Is it the product of a monomial and a polynomial? If so, multiply each term of the polynomial by the monomial.

Example: $5x(x + 7) = 5x \cdot x + 5x \cdot 7 = 5x^2 + 35x$

5. Is it the product of two polynomials other than those above? If so, multiply each term of one by every term of the other. Use columns if you wish.

[The answer will have 2 or more terms, usually more than 2 terms.]

Example: $(x^2 - 3x + 2)(x + 7) = x^2(x + 7) - 3x(x + 7) + 2(x + 7)$
$= x^2 \cdot x + x^2 \cdot 7 - 3x \cdot x - 3x \cdot 7$
$+ 2 \cdot x + 2 \cdot 7$
$= x^3 + 7x^2 - 3x^2 - 21x + 2x + 14$
$= x^3 + 4x^2 - 19x + 14$

Remember that FOIL will *always* work for two binomials. You can use it instead of either of the first two rules, but those rules will make your work go faster.

Multiply.

28. $(x + 5)(x + 6)$

29. $(t - 4)(t + 4)$

30. $4x^2(-2x^3 + 5x^2 + 10)$

31. $(9x^2 + 1)^2$

32. $(2a - 5)(2a + 8)$

33. $\left(5x + \frac{1}{2}\right)^2$

34. $\left(2x - \frac{1}{2}\right)^2$

35. $(x^2 - x + 4)(x - 2)$

Answers on page A-16

Example 21 Multiply: $(x + 3)(x - 3)$.

$(x + 3)(x - 3) = x^2 - 9$ — Using method 2 (the product of the sum and the difference of two terms)

Example 22 Multiply: $(t + 7)(t - 5)$.

$(t + 7)(t - 5) = t^2 + 2t - 35$ — Using method 3 (the product of two binomials, but neither the square of a binomial nor the product of the sum and the difference of two terms)

Example 23 Multiply: $(x + 6)(x + 6)$.

$(x + 6)(x + 6) = x^2 + 2(6)x + 36$ — Using method 1 (the square of a binomial sum)

$= x^2 + 12x + 36$

Example 24 Multiply: $2x^3(9x^2 + x - 7)$.

$2x^3(9x^2 + x - 7) = 18x^5 + 2x^4 - 14x^3$ — Using method 4 (the product of a monomial and a trinomial; multiplying each term of the trinomial by the monomial)

Example 25 Multiply: $(5x^3 - 7x)^2$.

$(5x^3 - 7x)^2 = 25x^6 - 2(5x^3)(7x) + 49x^2$ — Using method 1 (the square of a binomial difference)

$= 25x^6 - 70x^4 + 49x^2$

Example 26 Multiply: $\left(3x + \frac{1}{4}\right)^2$.

$\left(3x + \frac{1}{4}\right)^2 = 9x^2 + 2(3x)\left(\frac{1}{4}\right) + \frac{1}{16}$ — Using method 1 (the square of a binomial sum. To get the middle term, we multiply $3x$ by $\frac{1}{4}$ and double.)

$= 9x^2 + \frac{3}{2}x + \frac{1}{16}$

Example 27 Multiply: $\left(4x - \frac{3}{4}\right)^2$.

$\left(4x - \frac{3}{4}\right)^2 = 16x^2 - 2(4x)\left(\frac{3}{4}\right) + \frac{9}{16}$ — Using method 1 (the square of a binomial difference)

$= 16x^2 - 6x + \frac{9}{16}$

Example 28 Multiply: $(p + 3)(p^2 + 2p - 1)$.

$$
\begin{array}{r l}
p^2 + 2p - 1 & \text{Using method 5 (the product of two polynomials)} \\
p + 3 & \\
\hline
3p^2 + 6p - 3 & \text{Multiplying by 3} \\
p^3 + 2p^2 - p & \text{Multiplying by } p \\
\hline
p^3 + 5p^2 + 5p - 3 &
\end{array}
$$

Do Exercises 28–35.

Exercise Set 4.6

a Multiply. Try to write only the answer. If you need more steps, be sure to use them.

1. $(x + 1)(x^2 + 3)$

2. $(x^2 - 3)(x - 1)$

3. $(x^3 + 2)(x + 1)$

4. $(x^4 + 2)(x + 10)$

5. $(y + 2)(y - 3)$

6. $(a + 2)(a + 3)$

7. $(3x + 2)(3x + 2)$

8. $(4x + 1)(4x + 1)$

9. $(5x - 6)(x + 2)$

10. $(x - 8)(x + 8)$

11. $(3t - 1)(3t + 1)$

12. $(2m + 3)(2m + 3)$

13. $(4x - 2)(x - 1)$

14. $(2x - 1)(3x + 1)$

15. $\left(p - \frac{1}{4}\right)\left(p + \frac{1}{4}\right)$

16. $\left(q + \frac{3}{4}\right)\left(q + \frac{3}{4}\right)$

17. $(x - 0.1)(x + 0.1)$

18. $(x + 0.3)(x - 0.4)$

19. $(2x^2 + 6)(x + 1)$

20. $(2x^2 + 3)(2x - 1)$

21. $(-2x + 1)(x + 6)$

22. $(3x + 4)(2x - 4)$

23. $(a + 7)(a + 7)$

24. $(2y + 5)(2y + 5)$

25. $(1 + 2x)(1 - 3x)$

26. $(-3x - 2)(x + 1)$

27. $(x^2 + 3)(x^3 - 1)$

28. $(x^4 - 3)(2x + 1)$

29. $(3x^2 - 2)(x^4 - 2)$

30. $(x^{10} + 3)(x^{10} - 3)$

31. $(2.8x - 1.5)(4.7x + 9.3)$

32. $\left(x - \frac{3}{8}\right)\left(x + \frac{4}{7}\right)$

33. $(3x^5 + 2)(2x^2 + 6)$

34. $(1 - 2x)(1 + 3x^2)$

35. $(8x^3 + 1)(x^3 + 8)$

36. $(4 - 2x)(5 - 2x^2)$

37. $(4x^2 + 3)(x - 3)$

38. $(7x - 2)(2x - 7)$

39. $(4y^4 + y^2)(y^2 + y)$

40. $(5y^6 + 3y^3)(2y^6 + 2y^3)$

b Multiply mentally, if possible. If you need extra steps, be sure to use them.

41. $(x + 4)(x - 4)$

42. $(x + 1)(x - 1)$

43. $(2x + 1)(2x - 1)$

44. $(x^2 + 1)(x^2 - 1)$

45. $(5m - 2)(5m + 2)$

46. $(3x^4 + 2)(3x^4 - 2)$

47. $(2x^2 + 3)(2x^2 - 3)$

48. $(6x^5 - 5)(6x^5 + 5)$

49. $(3x^4 - 4)(3x^4 + 4)$

50. $(t^2 - 0.2)(t^2 + 0.2)$

51. $(x^6 - x^2)(x^6 + x^2)$

52. $(2x^3 - 0.3)(2x^3 + 0.3)$

53. $(x^4 + 3x)(x^4 - 3x)$

54. $\left(\frac{3}{4} + 2x^3\right)\left(\frac{3}{4} - 2x^3\right)$

55. $(x^{12} - 3)(x^{12} + 3)$

56. $(12 - 3x^2)(12 + 3x^2)$

57. $(2y^8 + 3)(2y^8 - 3)$

58. $\left(m - \frac{2}{3}\right)\left(m + \frac{2}{3}\right)$

59. $\left(\frac{5}{8}x - 4.3\right)\left(\frac{5}{8}x + 4.3\right)$

60. $(10.7 - x^3)(10.7 + x^3)$

c Multiply mentally, if possible. If you need extra steps, be sure to use them.

61. $(x + 2)^2$

62. $(2x - 1)^2$

63. $(3x^2 + 1)^2$

64. $\left(3x + \frac{3}{4}\right)^2$

65. $\left(a - \frac{1}{2}\right)^2$

66. $\left(2a - \frac{1}{5}\right)^2$

67. $(3 + x)^2$

68. $(x^3 - 1)^2$

69. $(x^2 + 1)^2$

70. $(8x - x^2)^2$

71. $(2 - 3x^4)^2$

72. $(6x^3 - 2)^2$

73. $(5 + 6t^2)^2$

74. $(3p^2 - p)^2$

75. $\left(x - \frac{5}{8}\right)^2$

76. $(0.3y + 2.4)^2$

d Multiply mentally, if possible.

77. $(3 - 2x^3)^2$

78. $(x - 4x^3)^2$

79. $4x(x^2 + 6x - 3)$

80. $8x(-x^5 + 6x^2 + 9)$

81. $\left(2x^2 - \frac{1}{2}\right)\left(2x^2 - \frac{1}{2}\right)$

82. $(-x^2 + 1)^2$

83. $(-1 + 3p)(1 + 3p)$

84. $(-3q + 2)(3q + 2)$

85. $3t^2(5t^3 - t^2 + t)$

86. $-6x^2(x^3 + 8x - 9)$

87. $(6x^4 + 4)^2$

88. $(8a + 5)^2$

89. $(3x + 2)(4x^2 + 5)$

90. $(2x^2 - 7)(3x^2 + 9)$

91. $(8 - 6x^4)^2$

92. $\left(\frac{1}{5}x^2 + 9\right)\left(\frac{3}{5}x^2 - 7\right)$

93. $(t - 1)(t^2 + t + 1)$

94. $(y + 5)(y^2 - 5y + 25)$

Compute each of the following and compare.

95. $3^2 + 4^2$; $(3 + 4)^2$

96. $6^2 + 7^2$; $(6 + 7)^2$

97. $9^2 - 5^2$; $(9 - 5)^2$

98. $11^2 - 4^2$; $(11 - 4)^2$

Skill Maintenance

99. In apartment 3B, lamps, an air conditioner, and a television set are all operating at the same time. The lamps use 10 times as many watts of electricity as the television set, and the air conditioner uses 40 times as many watts as the television set. The total wattage used in the apartment is 2550. How many watts are used by each appliance? [2.4a]

Solve. [2.3c]

100. $3x - 8x = 4(7 - 8x)$

101. $3(x - 2) = 5(2x + 7)$

102. $5(2x - 3) - 2(3x - 4) = 20$

Solve. [2.6a]

103. $3x - 2y = 12$, for y

104. $ab - cd = 4$, for a

Synthesis

105. ◆ Under what conditions is the product of two binomials a binomial?

106. ◆ Todd feels that since the FOIL method can be used to find the product of any two binomials, he needn't study the other special products. What advice would you give him?

Multiply.

107. $5x(3x - 1)(2x + 3)$

108. $[(2x - 3)(2x + 3)](4x^2 + 9)$

109. $[(a - 5)(a + 5)]^2$

110. $(a - 3)^2(a + 3)^2$
(*Hint*: Examine Exercise 109.)

111. $(3t^4 - 2)^2(3t^4 + 2)^2$
(*Hint*: Examine Exercise 109.)

112. $[3a - (2a - 3)][3a + (2a - 3)]$

Solve.

113. $(x + 2)(x - 5) = (x + 1)(x - 3)$

114. $(2x + 5)(x - 4) = (x + 5)(2x - 4)$

115. *Factors and Sums.* To *factor* a number is to express it as a product. Since $12 = 4 \cdot 3$, we say that 12 is *factored* and that 4 and 3 are *factors* of 12. In the following table, the top number has been factored in such a way that the sum of the factors is the bottom number. For example, in the first column, 40 has been factored as $5 \cdot 8$, and $5 + 8 = 13$, the bottom number. Such thinking is important in algebra when we factor trinomials of the type $x^2 + bx + c$. Find the missing numbers in the table.

Product	40	63	36	72	−140	−96	48	168	110			
Factor	5									−9	−24	−3
Factor	8									−10	18	
Sum	13	16	−20	−38	−4	4	−14	−29	−21			18

Find the total shaded area.

116.

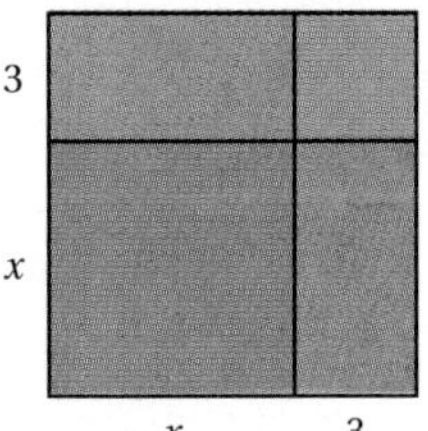

117.

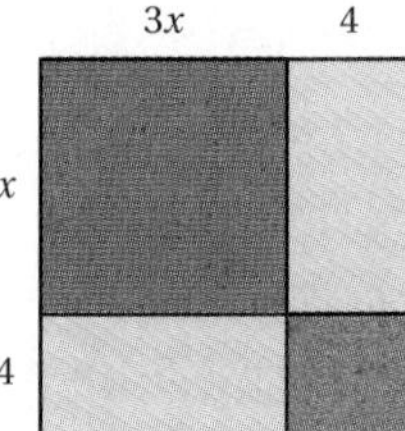

118. A factored polynomial for the shaded area in this rectangle is $(A + B)(A - B)$.

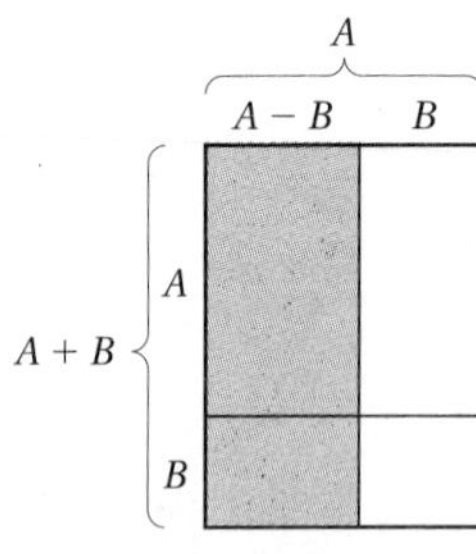

a) Find a polynomial for the area of the entire rectangle.
b) Find a polynomial for the sum of the areas of the two small unshaded rectangles.
c) Find a polynomial for the area in part (a) minus the area in part (b).
d) Find a polynomial for the area of the shaded region and compare this with the polynomial found in part (c).

Use the TABLE or GRAPH feature to check whether each of the following is correct.

119. $(x - 1)^2 = x^2 - 2x + 1$

120. $(x - 2)^2 = x^2 - 4x - 4$

121. $(x - 3)(x + 3) = x^2 - 6$

122. $(x - 3)(x + 2) = x^2 - x - 6$

Collaborative Learning Manual

Derive the special-product formulas.

4.7 Operations with Polynomials in Several Variables

The polynomials that we have been studying have only one variable. A **polynomial in several variables** is an expression like those you have already seen, but with more than one variable. Here are two examples:

$$3x + xy^2 + 5y + 4, \qquad 8xy^2z - 2x^3z - 13x^4y^2 + 15.$$

Objectives

a Evaluate a polynomial in several variables for given values of the variables.

b Identify the coefficients and the degrees of the terms of a polynomial and the degree of a polynomial.

c Collect like terms of a polynomial.

d Add polynomials.

e Subtract polynomials.

f Multiply polynomials.

For Extra Help

TAPE 7

MAC WIN

CD-ROM

a Evaluating Polynomials

Example 1 Evaluate the polynomial $4 + 3x + xy^2 + 8x^3y^3$ for $x = -2$ and $y = 5$.

We replace x with -2 and y with 5:

$$\begin{aligned} 4 + 3x + xy^2 + 8x^3y^3 &= 4 + 3(-2) + (-2) \cdot 5^2 + 8(-2)^3 \cdot 5^3 \\ &= 4 - 6 - 50 - 8000 \\ &= -8052. \end{aligned}$$

Example 2 *Male Caloric Needs.* The number of calories needed each day by a moderately active man who weighs w kilograms, is h centimeters tall, and is a years old can be estimated by the polynomial

$$19.18w + 7h - 9.52a + 92.4$$

(***Source:*** Parker, M., *She Does Math.* Mathematical Association of America, p. 96). The author of this text is moderately active, weighs 97 kg, is 185 cm tall, and is 55 yr old. What are his daily caloric needs?

We evaluate the polynomial for $w = 97$, $h = 185$, and $a = 55$:

$$\begin{aligned} &19.18w + 7h - 9.52a + 92.4 \\ &= 19.18(97) + 7(185) - 9.52(55) + 92.4 \quad \text{Substituting} \\ &= 2724.26. \end{aligned}$$

His daily caloric need is about 2724 calories.

Do Exercises 1–3.

1. Evaluate the polynomial

$$4 + 3x + xy^2 + 8x^3y^3$$

for $x = 2$ and $y = -5$.

2. Evaluate the polynomial

$$8xy^2 - 2x^3z - 13x^4y^2 + 5$$

for $x = -1$, $y = 3$, and $z = 4$.

3. *Female Caloric Needs.* The number of calories needed each day by a moderately active woman who weighs w pounds, is h inches tall, and is a years old can be estimated by the polynomial

$$917 + 6w + 6h - 6a$$

(***Source:*** Parker, M., *She Does Math.* Mathematical Association of America, p. 96). Christine is moderately active, weighs 125 lb, is 64 in. tall, and is 27 yr old. What are her daily caloric needs?

Answers on page A-17

4. Identify the coefficient of each term:

$-3xy^2 + 3x^2y - 2y^3 + xy + 2.$

5. Identify the degree of each term and the degree of the polynomial

$4xy^2 + 7x^2y^3z^2 - 5x + 2y + 4.$

Collect like terms.

6. $4x^2y + 3xy - 2x^2y$

7. $-3pq - 5pqr^3 - 12 + 8pq + 5pqr^3 + 4$

Answers on page A-17

b Coefficients and Degrees

The **degree** of a term is the sum of the exponents of the variables. The **degree of a polynomial** is the degree of the term of highest degree.

Example 3 Identify the coefficient and the degree of each term and the degree of the polynomial

$$9x^2y^3 - 14xy^2z^3 + xy + 4y + 5x^2 + 7.$$

Term	Coefficient	Degree	Degree of the Polynomial
$9x^2y^3$	9	5	
$-14xy^2z^3$	-14	6	6
xy	1	2	
$4y$	4	1	Think: $4y = 4y^1$.
$5x^2$	5	2	
7	7	0	Think: $7 = 7x^0$, or $7x^0y^0z^0$.

Do Exercises 4 and 5.

c Collecting Like Terms

Like terms (or **similar terms**) have exactly the same variables with exactly the same exponents. For example,

$3x^2y^3$ and $-7x^2y^3$ are like terms;

$9x^4z^7$ and $12x^4z^7$ are like terms.

But

$13xy^5$ and $-2x^2y^5$ are *not* like terms, because the x-factors have different exponents;

and

$3xyz^2$ and $4xy$ are *not* like terms, because there is no factor of z^2 in the second expression.

Collecting like terms is based on the distributive laws.

Examples Collect like terms.

4. $5x^2y + 3xy^2 - 5x^2y - xy^2 = (5 - 5)x^2y + (3 - 1)xy^2 = 2xy^2$

5. $8a^2 - 2ab + 7b^2 + 4a^2 - 9ab - 17b^2 = 12a^2 - 11ab - 10b^2$

6. $7xy - 5xy^2 + 3xy^2 - 7 + 6x^3 + 9xy - 11x^3 + y - 1$
$= -2xy^2 + 16xy - 5x^3 + y - 8$

Do Exercises 6 and 7.

d Addition

We can find the sum of two polynomials in several variables by writing a plus sign between them and then collecting like terms.

Example 7 Add: $(-5x^3 + 3y - 5y^2) + (8x^3 + 4x^2 + 7y^2)$.

$$(-5x^3 + 3y - 5y^2) + (8x^3 + 4x^2 + 7y^2)$$
$$= (-5 + 8)x^3 + 4x^2 + 3y + (-5 + 7)y^2$$
$$= 3x^3 + 4x^2 + 3y + 2y^2$$

Example 8 Add:

$$(5xy^2 - 4x^2y + 5x^3 + 2) + (3xy^2 - 2x^2y + 3x^3y - 5).$$

We first look for like terms. They are $5xy^2$ and $3xy^2$, $-4x^2y$ and $-2x^2y$, and 2 and -5. We collect these. Since there are no more like terms, the answer is

$$8xy^2 - 6x^2y + 5x^3 + 3x^3y - 3.$$

Do Exercises 8–10.

e Subtraction

We subtract a polynomial by adding its opposite, or additive inverse. The opposite of the polynomial

$$4x^2y - 6x^3y^2 + x^2y^2 - 5y$$

can be represented by

$$-(4x^2y - 6x^3y^2 + x^2y^2 - 5y).$$

We find an equivalent expression for the opposite of a polynomial by replacing each coefficient with its opposite, or by changing the sign of each term. Thus,

$$-(4x^2y - 6x^3y^2 + x^2y^2 - 5y) = -4x^2y + 6x^3y^2 - x^2y^2 + 5y.$$

Example 9 Subtract:

$$(4x^2y + x^3y^2 + 3x^2y^3 + 6y + 10) - (4x^2y - 6x^3y^2 + x^2y^2 - 5y - 8).$$

We have

$$(4x^2y + x^3y^2 + 3x^2y^3 + 6y + 10) - (4x^2y - 6x^3y^2 + x^2y^2 - 5y - 8)$$
$$= 4x^2y + x^3y^2 + 3x^2y^3 + 6y + 10 - 4x^2y + 6x^3y^2 - x^2y^2 + 5y + 8$$

Finding the opposite by changing the sign of each term

$$= 7x^3y^2 + 3x^2y^3 - x^2y^2 + 11y + 18.$$

Collecting like terms. (Try to write just the answer!)

Do Exercises 11 and 12.

Add.

8. $(4x^3 + 4x^2 - 8y - 3) + (-8x^3 - 2x^2 + 4y + 5)$

9. $(13x^3y + 3x^2y - 5y) + (x^3y + 4x^2y - 3xy + 3y)$

10. $(-5p^2q^4 + 2p^2q^2 + 3q) + (6pq^2 + 3p^2q + 5)$

Subtract.

11. $(-4s^4t + s^3t^2 + 2s^2t^3) - (4s^4t - 5s^3t^2 + s^2t^2)$

12. $(-5p^4q + 5p^3q^2 - 3p^2q^3 - 7q^4 - 2) - (4p^4q - 4p^3q^2 + p^2q^3 + 2q^4 - 7)$

Answers on page A-17

Multiply.

13. $(x^2y^3 + 2x)(x^3y^2 + 3x)$

14. $(p^4q - 2p^3q^2 + 3q^3)(p + 2q)$

Multiply.

15. $(3xy + 2x)(x^2 + 2xy^2)$

16. $(x - 3y)(2x - 5y)$

17. $(4x + 5y)^2$

18. $(3x^2 - 2xy^2)^2$

19. $(2xy^2 + 3x)(2xy^2 - 3x)$

20. $(3xy^2 + 4y)(-3xy^2 + 4y)$

21. $(3y + 4 - 3x)(3y + 4 + 3x)$

22. $(2a + 5b + c)(2a - 5b - c)$

Answers on page A-17

f Multiplication

To multiply polynomials in several variables, we can multiply each term of one by every term of the other. We can use columns for long multiplications as with polynomials in one variable. We multiply each term at the top by every term at the bottom. We write like terms in columns, and then we add the results.

Example 10 Multiply: $(3x^2y - 2xy + 3y)(xy + 2y)$.

$$\begin{array}{rrrrl} & 3x^2y & -\ 2xy & +\ 3y & \\ & & xy & +\ 2y & \\ \hline & 6x^2y^2 & -\ 4xy^2 & +\ 6y^2 & \text{Multiplying by } 2y \\ 3x^3y^2 & -\ 2x^2y^2 & +\ 3xy^2 & & \text{Multiplying by } xy \\ \hline 3x^3y^2 & +\ 4x^2y^2 & -\ xy^2 & +\ 6y^2 & \text{Adding} \end{array}$$

Do Exercises 13 and 14.

Where appropriate, we use the special products that we have learned.

Examples Multiply.

11. $(x^2y + 2x)(xy^2 + y^2) = \overset{F}{x^3y^3} + \overset{O}{x^2y^3} + \overset{I}{2x^2y^2} + \overset{L}{2xy^2}$

12. $(p + 5q)(2p - 3q) = 2p^2 - 3pq + 10pq - 15q^2$
$= 2p^2 + 7pq - 15q^2$

$(A + B)^2 = A^2 + 2 \cdot A \cdot B + B^2$

13. $(3x + 2y)^2 = (3x)^2 + 2(3x)(2y) + (2y)^2$
$= 9x^2 + 12xy + 4y^2$

$(A - B)^2 = A^2 - 2 \cdot A \cdot B + B^2$

14. $(2y^2 - 5x^2y)^2 = (2y^2)^2 - 2(2y^2)(5x^2y) + (5x^2y)^2$
$= 4y^4 - 20x^2y^3 + 25x^4y^2$

$(A + B)(A - B) = A^2 - B^2$

15. $(3x^2y + 2y)(3x^2y - 2y) = (3x^2y)^2 - (2y)^2$
$= 9x^4y^2 - 4y^2$

16. $(-2x^3y^2 + 5t)(2x^3y^2 + 5t) = (5t - 2x^3y^2)(5t + 2x^3y^2)$
$= (5t)^2 - (2x^3y^2)^2$
$= 25t^2 - 4x^6y^4$

$(A - B)(A + B) = A^2 - B^2$

17. $(\boxed{2x + 3} - 2y)(\boxed{2x + 3} + 2y) = (\boxed{2x + 3})^2 - (2y)^2$
$= 4x^2 + 12x + 9 - 4y^2$

Do Exercises 15–22.

Exercise Set 4.7

a Evaluate the polynomial for $x = 3$, $y = -2$, and $z = -5$.

1. $x^2 - y^2 + xy$

2. $x^2 + y^2 - xy$

3. $x^2 - 3y^2 + 2xy$

4. $x^2 - 4xy + 5y^2$

5. $8xyz$

6. $-3xyz^2$

7. $xyz^2 - z$

8. $xy - xz + yz$

Lung Capacity. The polynomial

$$0.041h - 0.018A - 2.69$$

can be used to estimate the lung capacity, in liters, of a female of height h, in centimeters, and age A, in years.

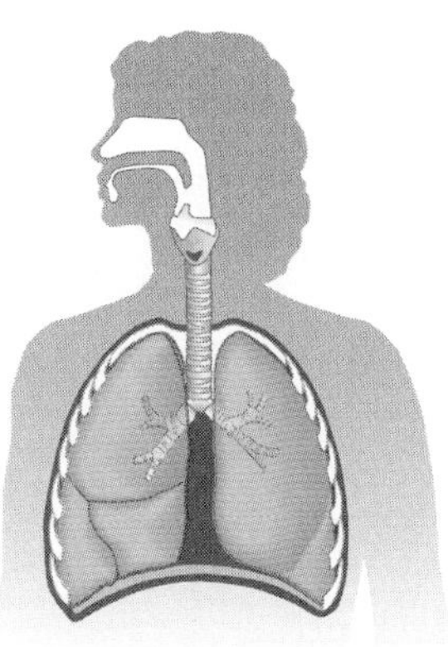

9. Find the lung capacity of a 20-yr-old woman who is 165 cm tall.

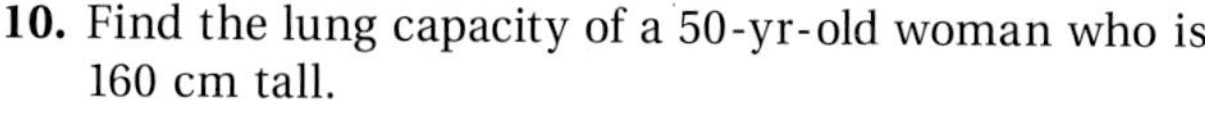

10. Find the lung capacity of a 50-yr-old woman who is 160 cm tall.

Altitude of a Launched Object. The altitude, in meters, of a launched object is given by the polynomial

$$h + vt - 4.9t^2,$$

where h is the height, in meters, from which the launch occurs, v is the initial upward speed (or velocity), in meters per second (m/s), and t is the number of seconds for which the object is airborne.

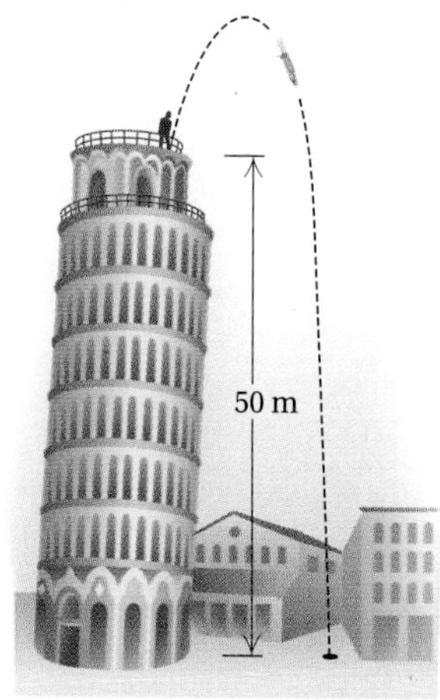

11. A model rocket is launched from the top of the Leaning Tower of Pisa, 50 m above the ground. The upward speed is 40 m/s. How high will the rocket be 2 sec after the blastoff?

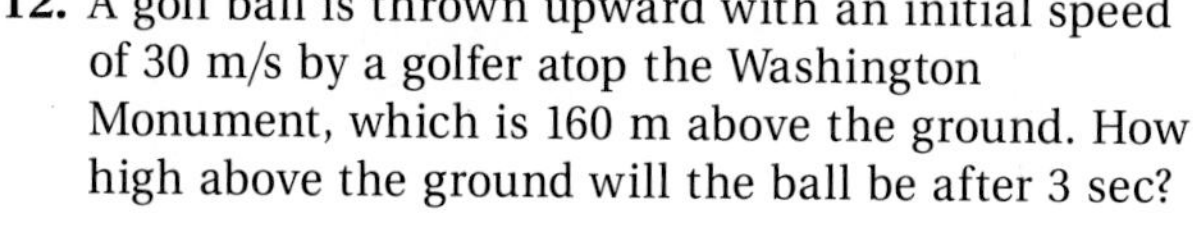

12. A golf ball is thrown upward with an initial speed of 30 m/s by a golfer atop the Washington Monument, which is 160 m above the ground. How high above the ground will the ball be after 3 sec?

Surface Area of a Right Circular Cylinder. The area of a right circular cylinder is given by the polynomial

$$2\pi rh + 2\pi r^2,$$

where h is the height and r is the radius of the base.

13. A 16-oz beverage can has a height of 6.3 in. and a radius of 1.2 in. Evaluate the polynomial for $h = 6.3$ and $r = 1.2$ to find the area of the can. Use 3.14 for π.

14. A 26-oz coffee can has a height of 6.5 in. and a radius of 2.5 in. Evaluate the polynomial for $h = 6.5$ and $r = 2.5$ to find the area of the can. Use 3.14 for π.

b Identify the coefficient and the degree of each term of the polynomial. Then find the degree of the polynomial.

15. $x^3y - 2xy + 3x^2 - 5$

16. $5y^3 - y^2 + 15y + 1$

17. $17x^2y^3 - 3x^3yz - 7$

18. $6 - xy + 8x^2y^2 - y^5$

c Collect like terms.

19. $a + b - 2a - 3b$

20. $y^2 - 1 + y - 6 - y^2$

21. $3x^2y - 2xy^2 + x^2$

22. $m^3 + 2m^2n - 3m^2 + 3mn^2$

23. $6au + 3av + 14au + 7av$

24. $3x^2y - 2z^2y + 3xy^2 + 5z^2y$

25. $2u^2v - 3uv^2 + 6u^2v - 2uv^2$

26. $3x^2 + 6xy + 3y^2 - 5x^2 - 10xy - 5y^2$

d Add.

27. $(2x^2 - xy + y^2) + (-x^2 - 3xy + 2y^2)$

28. $(2z - z^2 + 5) + (z^2 - 3z + 1)$

29. $(r - 2s + 3) + (2r + s) + (s + 4)$

30. $(ab - 2a + 3b) + (5a - 4b) + (3a + 7ab - 8b)$

31. $(b^3a^2 - 2b^2a^3 + 3ba + 4) + (b^2a^3 - 4b^3a^2 + 2ba - 1)$

32. $(2x^2 - 3xy + y^2) + (-4x^2 - 6xy - y^2) + (x^2 + xy - y^2)$

e Subtract.

33. $(a^3 + b^3) - (a^2b - ab^2 + b^3 + a^3)$

34. $(x^3 - y^3) - (-2x^3 + x^2y - xy^2 + 2y^3)$

35. $(xy - ab - 8) - (xy - 3ab - 6)$

36. $(3y^4x^2 + 2y^3x - 3y - 7) - (2y^4x^2 + 2y^3x - 4y - 2x + 5)$

37. $(-2a + 7b - c) - (-3b + 4c - 8d)$

38. Find the sum of $2a + b$ and $3a - b$. Then subtract $5a + 2b$.

f Multiply.

39. $(3z - u)(2z + 3u)$

40. $(a - b)(a^2 + b^2 + 2ab)$

41. $(a^2b - 2)(a^2b - 5)$

42. $(xy + 7)(xy - 4)$

43. $(a^3 + bc)(a^3 - bc)$

44. $(m^2 + n^2 - mn)(m^2 + mn + n^2)$

45. $(y^4x + y^2 + 1)(y^2 + 1)$

46. $(a - b)(a^2 + ab + b^2)$

47. $(3xy - 1)(4xy + 2)$

48. $(m^3n + 8)(m^3n - 6)$

49. $(3 - c^2d^2)(4 + c^2d^2)$

50. $(6x - 2y)(5x - 3y)$

51. $(m^2 - n^2)(m + n)$

52. $(pq + 0.2)(0.4pq - 0.1)$

53. $(xy + x^5y^5)(x^4y^4 - xy)$

54. $(x - y^3)(2y^3 + x)$

55. $(x + h)^2$

56. $(3a + 2b)^2$

57. $(r^3t^2 - 4)^2$

58. $(3a^2b - b^2)^2$

59. $(p^4 + m^2n^2)^2$

60. $(2ab - cd)^2$

61. $\left(2a^3 - \frac{1}{2}b^3\right)^2$

62. $-3x(x + 8y)^2$

63. $3a(a - 2b)^2$

64. $(a^2 + b + 2)^2$

65. $(2a - b)(2a + b)$

66. $(x - y)(x + y)$

67. $(c^2 - d)(c^2 + d)$

68. $(p^3 - 5q)(p^3 + 5q)$

69. $(ab + cd^2)(ab - cd^2)$

70. $(xy + pq)(xy - pq)$

71. $(x + y - 3)(x + y + 3)$

72. $(p + q + 4)(p + q - 4)$

73. $[x + y + z][x - (y + z)]$

74. $[a + b + c][a - (b + c)]$

75. $(a + b + c)(a - b - c)$

76. $(3x + 2 - 5y)(3x + 2 + 5y)$

Skill Maintenance

In which quadrant is the point located? [3.1c]

77. $(2, -5)$

78. $(-8, -9)$

79. $(16, 23)$

80. $(-3, 2)$

Graph. [3.3b]

81. $2x = -10$

82. $y = -4$

83. $8y - 16 = 0$

84. $x = 4$

Find the mean, the median, and the mode, if it exists. [3.4a]

85. 23, 31, 24, 31, 25, 28, 31

86. 5.2, 5.6, 5.8, 6.1, 5.6, 5.2, 6.3

Synthesis

87. ◆ Is it possible for a polynomial in four variables to have a degree less than 4? Why or why not?

88. ◆ Can the sum of two trinomials in several variables be a trinomial in one variable? Why or why not?

Find a polynomial for the shaded area. (Leave results in terms of π where appropriate.)

89.

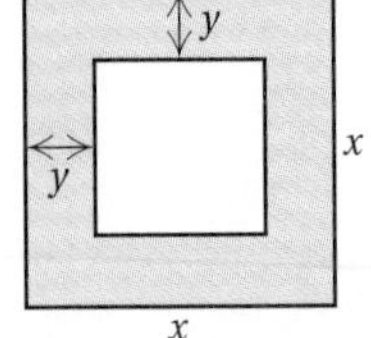

90.

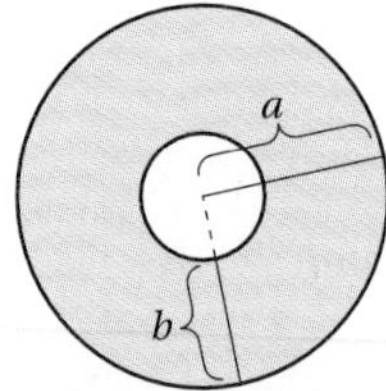

91.

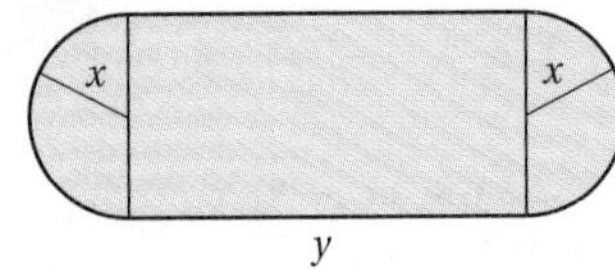

92.

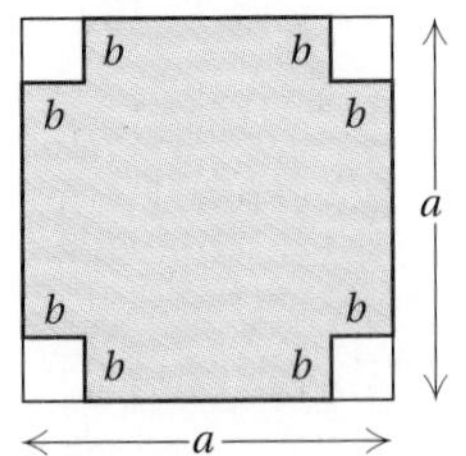

Find a formula for the surface area of the solid object. Leave results in terms of π.

93.

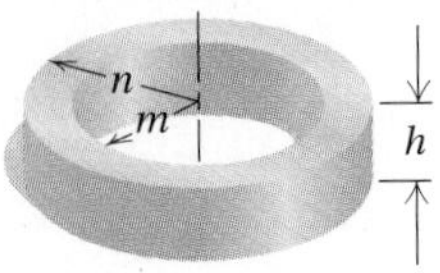

94.

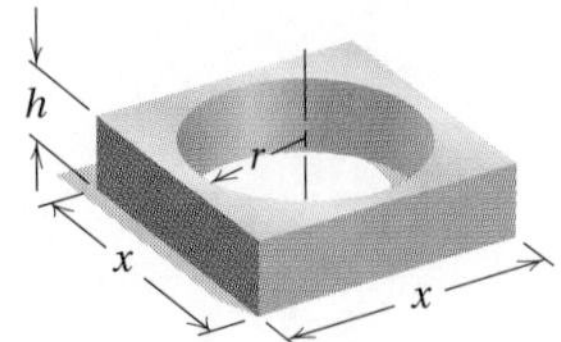

4.8 Division of Polynomials

A rational expression represents division. "Long" division of polynomials, like division of real numbers, relies on our multiplication and subtraction skills.

Objectives

a Divide a polynomial by a monomial.

b Divide a polynomial by a divisor that is not a monomial, and if there is a remainder, express the result in two ways.

c Use synthetic division to divide a polynomial by a binomial of the type $x - a$.

For Extra Help

TAPE 7

MAC WIN

CD-ROM

a Divisor a Monomial

We first consider division by a monomial. When we are dividing a monomial by a monomial, we can use the rules of exponents and subtract exponents when the bases are the same. For example,

$$\frac{45x^{10}}{3x^4} = \frac{45}{3}x^{10-4} = 15x^6, \qquad \frac{48a^2b^5}{-3ab^2} = \frac{48}{-3}a^{2-1}b^{5-2} = -16ab^3.$$

When we divide a polynomial by a monomial, we break up the division into a sum of quotients of monomials. To do so, we use the rule for addition using fractional notation in reverse. That is, since

$$\frac{A}{C} + \frac{B}{C} = \frac{A+B}{C}, \quad \text{we know that} \quad \frac{A+B}{C} = \frac{A}{C} + \frac{B}{C}.$$

Example 1 Divide $12x^3 + 8x^2 + x + 4$ by $4x$.

$\dfrac{12x^3 + 8x^2 + x + 4}{4x}$ — Writing a fractional expression

$= \dfrac{12x^3}{4x} + \dfrac{8x^2}{4x} + \dfrac{x}{4x} + \dfrac{4}{4x}$ — Dividing each term of the numerator by the monomial: This is the reverse of addition.

$= 3x^2 + 2x + \dfrac{1}{4} + \dfrac{1}{x}$ — Doing the four indicated divisions

Do Exercise 1.

1. Divide: $\dfrac{x^3 + 16x^2 + 6x}{2x}$.

Example 2 Divide: $(8x^4y^5 - 3x^3y^4 + 5x^2y^3) \div x^2y^3$.

$$\frac{8x^4y^5 - 3x^3y^4 + 5x^2y^3}{x^2y^3} = \frac{8x^4y^5}{x^2y^3} - \frac{3x^3y^4}{x^2y^3} + \frac{5x^2y^3}{x^2y^3}$$
$$= 8x^2y^2 - 3xy + 5$$

You should try to write only the answer.

> To divide a polynomial by a monomial, divide each term by the monomial.

Do Exercises 2 and 3.

Divide.

2. $(15y^5 - 6y^4 + 18y^3) \div 3y^2$

3. $(x^4y^3 + 10x^3y^2 + 16x^2y) \div 2x^2y$

b Divisor Not a Monomial

When the divisor is not a monomial, we use a procedure very much like long division in arithmetic.

Answers on page A-17

4. Divide and check:

$$x - 2\overline{)x^2 + 3x - 10}.$$

5. Divide and check:

$(2x^4 + 3x^3 - x^2 - 7x + 9) \div (x + 4)$.

Answers on page A-17

Example 3 Divide $x^2 + 5x + 8$ by $x + 3$.

We have

$$\begin{array}{r l}
x \phantom{{}+3x} & \leftarrow \text{Divide the first term by the first term: } x^2/x = x. \\
x + 3\overline{)x^2 + 5x + 8} & \\
x^2 + 3x \phantom{{}+8} & \leftarrow \text{Multiply } x \text{ above by the divisor, } x + 3. \\
\hline
2x \phantom{{}+8} & \leftarrow \text{Subtract: } (x^2 + 5x) - (x^2 + 3x) = x^2 + 5x - x^2 - 3x = 2x.
\end{array}$$

We now "bring down" the other terms of the dividend—in this case, 8.

$$\begin{array}{r l}
x + 2 & \leftarrow \text{Divide the first term by the first term: } 2x/x = 2. \\
x + 3\overline{)x^2 + 5x + 8} & \\
x^2 + 3x \phantom{{}+8} & \\
\hline
2x + 8 & \leftarrow \text{The 8 has been "brought down."} \\
2x + 6 & \leftarrow \text{Multiply 2 above by the divisor, } x + 3. \\
\hline
2 & \leftarrow \text{Subtract: } (2x + 8) - (2x + 6) = 2x + 8 - 2x - 6 = 2.
\end{array}$$

The answer is $x + 2$, R 2, or

$$x + 2 + \frac{2}{x + 3}.$$

This expression is the remainder over the divisor.

Note that the answer is not a polynomial unless the remainder is 0.

To check, we multiply the quotient by the divisor and add the remainder to see if we get the dividend:

$$\underbrace{(x + 3)}_{\text{Divisor}} \cdot \underbrace{(x + 2)}_{\text{Quotient}} + \underbrace{2}_{\text{Remainder}} = \underbrace{(x^2 + 5x + 6) + 2}_{\text{Dividend}}$$
$$= x^2 + 5x + 8$$

The answer checks.

Example 4 Divide: $(5x^4 + x^3 - 3x^2 - 6x - 8) \div (x - 1)$.

$$\begin{array}{r l}
5x^3 + 6x^2 + 3x - 3 & \\
x - 1\overline{)5x^4 + x^3 - 3x^2 - 6x - 8} & \\
5x^4 - 5x^3 \phantom{{} - 3x^2 - 6x - 8} & \\
\hline
6x^3 - 3x^2 \phantom{{} - 6x - 8} & \leftarrow \text{Subtract: } (5x^4 + x^3) - (5x^4 - 5x^3) = 6x^3. \\
6x^3 - 6x^2 \phantom{{} - 6x - 8} & \\
\hline
3x^2 - 6x \phantom{{} - 8} & \leftarrow \text{Subtract: } (6x^3 - 3x^2) - (6x^3 - 6x^2) = 3x^2. \\
3x^2 - 3x \phantom{{} - 8} & \\
\hline
-3x - 8 & \leftarrow \text{Subtract: } (3x^2 - 6x) - (3x^2 - 3x) = -3x. \\
-3x + 3 & \\
\hline
-11 & \leftarrow \text{Subtract: } (-3x - 8) - (-3x + 3) = -11.
\end{array}$$

The answer is $5x^3 + 6x^2 + 3x - 3$, R -11; or

$$5x^3 + 6x^2 + 3x - 3 + \frac{-11}{x - 1}.$$

Do Exercises 4 and 5.

Always remember when dividing polynomials to arrange the polynomials in descending order. In a polynomial division, if there are *missing* terms in the dividend, either write them with 0 coefficients or leave space for them. For example, in $125y^3 - 8$, we say that "the y^2- and y-terms are **missing**." We could write them in as follows: $125y^3 + 0y^2 + 0y - 8$.

Example 5 Divide: $(125y^3 - 8) \div (5y - 2)$.

$$\begin{array}{r l}
 & 25y^2 + 10y + 4 \\
5y - 2 \,\big)\! & 125y^3 + 0y^2 + 0y - 8 \\
 & 125y^3 - 50y^2 \\
 & \phantom{125y^3 - {}} 50y^2 + 0y \\
 & \phantom{125y^3 - {}} 50y^2 - 20y \\
 & \phantom{125y^3 - 50y^2 + {}} 20y - 8 \\
 & \phantom{125y^3 - 50y^2 + {}} 20y - 8 \\
 & \phantom{125y^3 - 50y^2 + 20y - {}} 0
\end{array}$$

← When there are missing terms, we can write them in.

← Subtract: $125y^3 - (125y^3 - 50y^2) = 50y^2$.

The answer is $25y^2 + 10y + 4$.

Do Exercise 6.

Another way to deal with missing terms is to leave space for them, as we see in Example 6.

Example 6 Divide: $(x^4 - 9x^2 - 5) \div (x - 2)$.

Note that the x^3- and x-terms are missing in the dividend.

$$\begin{array}{r l}
 & x^3 + 2x^2 - 5x - 10 \\
x - 2 \,\big)\! & x^4 \qquad\quad - 9x^2 \qquad\quad - 5 \\
 & x^4 - 2x^3 \\
 & \phantom{x^4 - {}} 2x^3 - 9x^2 \\
 & \phantom{x^4 - {}} 2x^3 - 4x^2 \\
 & \phantom{x^4 - 2x^3 {}} - 5x^2 \\
 & \phantom{x^4 - 2x^3 {}} - 5x^2 + 10x \\
 & \phantom{x^4 - 2x^3 - 5x^2 {}} - 10x - 5 \\
 & \phantom{x^4 - 2x^3 - 5x^2 {}} - 10x + 20 \\
 & \phantom{x^4 - 2x^3 - 5x^2 - 10x {}} - 25
\end{array}$$

← We leave spaces for missing terms.

← Subtract: $x^4 - (x^4 - 2x^3) = 2x^3$.

← Subtract: $(2x^3 - 9x^2) - (2x^3 - 4x^2) = -5x^2$.

The answer is $x^3 + 2x^2 - 5x - 10$, R -25, or

$$x^3 + 2x^2 - 5x - 10 + \frac{-25}{x - 2}.$$

Do Exercises 7 and 8.

6. Divide and check:

$(9y^4 + 14y^2 - 8) \div (3y + 2)$.

Divide and check.

7. $(y^3 - 11y^2 + 6) \div (y - 3)$

8. $(x^3 + 9x^2 - 5) \div (x - 1)$

Answers on page A-17

9. Divide and check:

$(y^3 - 11y^2 + 6) \div (y^2 - 3)$.

Answer on page A-17

When dividing, we may "come out even" (have a remainder of 0) or we may not. If not, how long should we keep working? We continue until the degree of the remainder is less than the degree of the divisor, as in the next example.

Example 7 Divide: $(6x^3 + 9x^2 - 5) \div (x^2 - 2x)$.

$$\begin{array}{r|l} & 6x + 21 \\ x^2 - 2x & 6x^3 + 9x^2 + 0x - 5 \\ & \underline{6x^3 - 12x^2} \\ & \quad 21x^2 + 0x \\ & \quad \underline{21x^2 - 42x} \\ & \qquad 42x - 5 \end{array}$$

Again we have a missing term, so we can write it in.

The degree of the remainder is less than the degree of the divisor, so we are finished.

The answer is $6x + 21$, R $42x - 5$, or

$$6x + 21 + \frac{42x - 5}{x^2 - 2x}.$$

Do Exercise 9.

c Synthetic Division

To divide a polynomial by a binomial of the type $x - a$, we can streamline the general procedure by a process called **synthetic division.**

Compare the following. In **A** we perform a division. In **B** we also divide but we do not write the variables.

A.

$$\begin{array}{r|l} & 4x^2 + 5x + 11 \\ x - 2 & 4x^3 - 3x^2 + x + 7 \\ & \underline{4x^3 - 8x^2} \\ & \quad 5x^2 + x \\ & \quad \underline{5x^2 - 10x} \\ & \qquad 11x + 7 \\ & \qquad \underline{11x - 22} \\ & \qquad\qquad 29 \end{array}$$

B.

$$\begin{array}{r|l} & 4 + 5 + 11 \\ 1 - 2 & 4 - 3 + 1 + 7 \\ & \underline{4 - 8} \\ & \quad 5 + 1 \\ & \quad \underline{5 - 10} \\ & \qquad 11 + 7 \\ & \qquad \underline{11 - 22} \\ & \qquad\qquad 29 \end{array}$$

In **B** there is still some duplication of writing. Also, since we can subtract by adding the opposite, we can use 2 instead of -2 and then add instead of subtracting.

C. *Synthetic Division*

a) $\underline{2}|\,4 \;-\; 3 \;+\; 1 \;+\; 7$ — **Write the 2, the opposite of −2 in the divisor $x - 2$, and the coefficients of the dividend.**

$$\begin{array}{r|rrrr} 2 & 4 & -3 & +1 & +7 \\ \hline & 4 & & & \end{array}$$

Bring down the first coefficient.

b) **Multiply 4 by 2 to get 8. Add 8 and −3.**

$$\begin{array}{r|rrrr} 2 & 4 & -3 & +1 & +7 \\ & & 8 & & \\ \hline & 4 & 5 & & \end{array}$$

c) **Multiply 5 by 2 to get 10. Add 10 and 1.**

$$\begin{array}{r|rrrr} 2 & 4 & -3 & +1 & +7 \\ & & 8 & 10 & \\ \hline & 4 & 5 & 11 & \end{array}$$

d) **Multiply 11 by 2 to get 22. Add 22 and 7.**

$$\begin{array}{r|rrr|r} 2 & 4 & -3 & +1 & +7 \\ & & 8 & 10 & 22 \\ \hline & 4 & 5 & 11 & 29 \end{array}$$

Quotient Remainder

The last number, 29, is the remainder. The other numbers are the coefficients of the quotient, with that of the term of highest degree first, as follows.

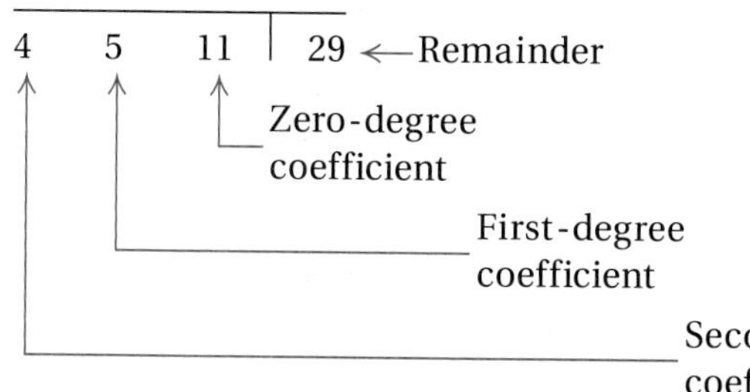

The answer is $4x^2 + 5x + 11$, R 29, or $4x^2 + 5x + 11 + \dfrac{29}{x - 2}$.

> It is important to remember that in order for this method to work, the divisor must be of the form $x - a$, that is, a variable minus a constant. The coefficient of the variable must be 1.

Example 8 Use synthetic division to divide:

$$(x^3 + 6x^2 - x - 30) \div (x - 2).$$

We have

$$\begin{array}{r|rrr|r} 2 & 1 & 6 & -1 & -30 \\ & & 2 & 16 & 30 \\ \hline & 1 & 8 & 15 & 0 \end{array}$$

The answer is $x^2 + 8x + 15$, R 0, or just $x^2 + 8x + 15$.

Do Exercise 10.

10. Use synthetic division to divide:

$(2x^3 - 4x^2 + 8x - 8) \div (x - 3).$

Answer on page A-17

Use synthetic division to divide.

11. $(x^3 - 2x^2 + 5x - 4) \div (x + 2)$

12. $(y^3 + 1) \div (y + 1)$

Answers on page A-17

When there are missing terms, be sure to write 0's for their coefficients.

Examples Use synthetic division to divide.

9. $(2x^3 + 7x^2 - 5) \div (x + 3)$

There is no x-term, so we must write a 0 for its coefficient. Note that $x + 3 = x - (-3)$, so we write -3 in the left corner.

$$\begin{array}{r|rrrr} -3 & 2 & 7 & 0 & -5 \\ & & -6 & -3 & 9 \\ \hline & 2 & 1 & -3 & 4 \end{array}$$

The answer is $2x^2 + x - 3$, R 4, or $2x^2 + x - 3 + \dfrac{4}{x + 3}$.

10. $(x^3 + 4x^2 - x - 4) \div (x + 4)$

$$\begin{array}{r|rrrr} -4 & 1 & 4 & -1 & -4 \\ & & -4 & 0 & 4 \\ \hline & 1 & 0 & -1 & 0 \end{array}$$

The answer is $x^2 - 1$.

11. $(x^4 - 1) \div (x - 1)$

$$\begin{array}{r|rrrrr} 1 & 1 & 0 & 0 & 0 & -1 \\ & & 1 & 1 & 1 & 1 \\ \hline & 1 & 1 & 1 & 1 & 0 \end{array}$$

The answer is $x^3 + x^2 + x + 1$.

12. $(8x^5 - 6x^3 + x - 8) \div (x + 2)$

$$\begin{array}{r|rrrrrr} -2 & 8 & 0 & -6 & 0 & 1 & -8 \\ & & -16 & 32 & -52 & 104 & -210 \\ \hline & 8 & -16 & 26 & -52 & 105 & -218 \end{array}$$

The answer is $8x^4 - 16x^3 + 26x^2 - 52x + 105$, R -218, or

$$8x^4 - 16x^3 + 26x^2 - 52x + 105 + \frac{-218}{x + 2}.$$

Do Exercises 11 and 12.

Exercise Set 4.8

a Divide and check.

1. $\dfrac{24x^6 + 18x^5 - 36x^2}{6x^2}$

2. $\dfrac{30y^8 - 15y^6 + 40y^4}{5y^4}$

3. $\dfrac{45y^7 - 20y^4 + 15y^2}{5y^2}$

4. $\dfrac{60x^8 + 44x^5 - 28x^3}{4x^3}$

5. $(32a^4b^3 + 14a^3b^2 - 22a^2b) \div 2a^2b$

6. $(7x^3y^4 - 21x^2y^3 + 28xy^2) \div 7xy$

b Divide.

7. $(x^2 + 10x + 21) \div (x + 3)$

8. $(y^2 - 8y + 16) \div (y - 4)$

9. $(a^2 - 8a - 16) \div (a + 4)$

10. $(y^2 - 10y - 25) \div (y - 5)$

11. $(x^2 + 7x + 14) \div (x + 5)$

12. $(t^2 - 7t - 9) \div (t - 3)$

13. $(4y^3 + 6y^2 + 14) \div (2y + 4)$

14. $(6x^3 - x^2 - 10) \div (3x + 4)$

15. $(10y^3 + 6y^2 - 9y + 10) \div (5y - 2)$

16. $(6x^3 - 11x^2 + 11x - 2) \div (2x - 3)$

17. $(2x^4 - x^3 - 5x^2 + x - 6) \div (x^2 + 2)$

18. $(3x^4 + 2x^3 - 11x^2 - 2x + 5) \div (x^2 - 2)$

19. $(2x^5 - x^4 + 2x^3 - x) \div (x^2 - 3x)$

20. $(2x^5 + 3x^3 + x^2 - 4) \div (x^2 + x)$

c Use synthetic division to divide.

21. $(x^3 - 2x^2 + 2x - 5) \div (x - 1)$

22. $(x^3 - 2x^2 + 2x - 5) \div (x + 1)$

23. $(a^2 + 11a - 19) \div (a + 4)$

24. $(a^2 + 11a - 19) \div (a - 4)$

25. $(x^3 - 7x^2 - 13x + 3) \div (x - 2)$

26. $(x^3 - 7x^2 - 13x + 3) \div (x + 2)$

27. $(3x^3 + 7x^2 - 4x + 3) \div (x + 3)$

28. $(3x^3 + 7x^2 - 4x + 3) \div (x - 3)$

29. $(y^3 - 3y + 10) \div (y - 2)$

30. $(x^3 - 2x^2 + 8) \div (x + 2)$

31. $(3x^4 - 25x^2 - 18) \div (x - 3)$

32. $(6y^4 + 15y^3 + 28y + 6) \div (y + 3)$

33. $(x^3 - 8) \div (x - 2)$

34. $(y^3 + 125) \div (y + 5)$

35. $(y^4 - 16) \div (y - 2)$

36. $(x^5 - 32) \div (x - 2)$

Skill Maintenance

Multiply or divide and simplify. [4.1d, e, f]

37. $w^4 \cdot w^5$

38. $\frac{w^4}{w^5}$

39. $x^{-4} \cdot x^{17}$

40. $\frac{x^{-12}}{x^{-4}}$

Solve. [2.3b]

41. $15x + 9x = 72$

42. $-2y - 9y = -55$

43. $a + \frac{1}{2}a = 21$

44. $7x - 3 = 24 + x$

Convert to scientific notation. [4.2c]

45. 0.000213

46. 84,000,000

Convert to decimal notation. [4.2c]

47. 3.8×10^6

48. 2.527×10^{-5}

Synthesis

49. ◆ Do addition, subtraction, multiplication, and division of polynomials always result in a polynomial? Why or why not?

50. ◆ Explain how synthetic division can be useful when factoring a polynomial.

51. Let $f(x) = 4x^3 + 16x^2 - 3x - 45$. Find $f(-3)$ and then solve $f(x) = 0$.

52. Use the TRACE feature on a grapher to check your answer to Exercise 51.

53. Let $f(x) = 6x^3 - 13x^2 - 79x + 140$. Find $f(4)$ and then solve $f(x) = 0$.

54. Use the TRACE feature on a grapher to check your answer to Exercise 53.

Summary and Review Exercises: Chapter 4

Important Properties and Formulas

FOIL:	$(A + B)(C + D) = AC + AD + BC + BD$
Square of a Sum:	$(A + B)(A + B) = (A + B)^2 = A^2 + 2AB + B^2$
Square of a Difference:	$(A - B)(A - B) = (A - B)^2 = A^2 - 2AB + B^2$
Product of a Sum and a Difference:	$(A + B)(A - B) = A^2 - B^2$

Definitions and Rules for Exponents

See p. 200.

The objectives to be tested in addition to the material in this chapter are [1.4a], [1.7d], [2.3b, c], and [2.4a].

Multiply and simplify. [4.1d, f]

1. $7^2 \cdot 7^{-4}$

2. $y^7 \cdot y^3 \cdot y$

3. $(3x)^5 \cdot (3x)^9$

4. $t^8 \cdot t^0$

Divide and simplify. [4.1e, f]

5. $\dfrac{4^5}{4^2}$

6. $\dfrac{a^5}{a^8}$

7. $\dfrac{(7x)^4}{(7x)^4}$

Simplify.

8. $(3t^4)^2$ [4.2a, b]

9. $(2x^3)^2(-3x)^2$ [4.1d], [4.2a, b]

10. $\left(\dfrac{2x}{y}\right)^{-3}$ [4.2b]

11. Express using a negative exponent: $\dfrac{1}{t^5}$. [4.1f]

12. Express using a positive exponent: y^{-4}. [4.1f]

13. Convert to scientific notation: 0.0000328. [4.2c]

14. Convert to decimal notation: 8.3×10^6. [4.2c]

Multiply or divide and write scientific notation for the result. [4.2d]

15. $(3.8 \times 10^4)(5.5 \times 10^{-1})$

16. $\dfrac{1.28 \times 10^{-8}}{2.5 \times 10^{-4}}$

17. *Diet-Drink Consumption.* It has been estimated that there will be 275 million people in the United States by the year 2000 and that on average, each of them will drink 15.3 gal of diet drinks that year (***Source:*** U.S. Department of Agriculture). How many gallons of diet drinks will be consumed by the entire population in 2000? Express the answer in scientific notation. [4.2e]

18. Evaluate the polynomial $x^2 - 3x + 6$ for $x = -1$. [4.3a]

19. Identify the terms of the polynomial $-4y^5 + 7y^2 - 3y - 2$. [4.3b]

20. Identify the missing terms in $x^3 + x$. [4.3h]

21. Identify the degree of each term and the degree of the polynomial $4x^3 + 6x^2 - 5x + \frac{5}{3}$. [4.3g]

Classify the polynomial as a monomial, binomial, trinomial, or none of these. [4.3i]

22. $4x^3 - 1$

23. $4 - 9t^3 - 7t^4 + 10t^2$

24. $7y^2$

Collect like terms and then arrange in descending order. [4.3f]

25. $3x^2 - 2x + 3 - 5x^2 - 1 - x$

26. $-x + \frac{1}{2} + 14x^4 - 7x^2 - 1 - 4x^4$

Add. [4.4a]

27. $(3x^4 - x^3 + x - 4) + (x^5 + 7x^3 - 3x^2 - 5) + (-5x^4 + 6x^2 - x)$

28. $(3x^5 - 4x^4 + x^3 - 3) + (3x^4 - 5x^3 + 3x^2) + (-5x^5 - 5x^2) + (-5x^4 + 2x^3 + 5)$

Subtract. [4.4c]

29. $(5x^2 - 4x + 1) - (3x^2 + 1)$

30. $(3x^5 - 4x^4 + 3x^2 + 3) - (2x^5 - 4x^4 + 3x^3 + 4x^2 - 5)$

31. Find a polynomial for the perimeter and for the area. [4.4d], [4.5b]

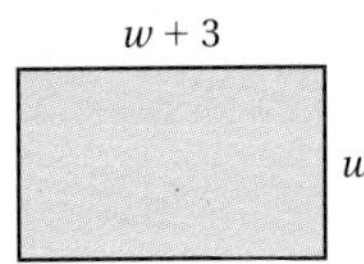

Multiply.

32. $\left(x + \frac{2}{3}\right)\left(x + \frac{1}{2}\right)$ [4.6a]

33. $(7x + 1)^2$ [4.6c]

34. $(4x^2 - 5x + 1)(3x - 2)$ [4.5d]

35. $(3x^2 + 4)(3x^2 - 4)$ [4.6b]

36. $5x^4(3x^3 - 8x^2 + 10x + 2)$ [4.5b]

37. $(x + 4)(x - 7)$ [4.6a]

38. $(3y^2 - 2y)^2$ [4.6c]

39. $(2t^2 + 3)(t^2 - 7)$ [4.6a]

40. Evaluate the polynomial

$$2 - 5xy + y^2 - 4xy^3 + x^6$$

for $x = -1$ and $y = 2$. [4.7a]

41. Identify the coefficient and the degree of each term of the polynomial

$$x^5y - 7xy + 9x^2 - 8.$$

Then find the degree of the polynomial. [4.7b]

Collect like terms. [4.7c]

42. $y + w - 2y + 8w - 5$

43. $m^6 - 2m^2n + m^2n^2 + n^2m - 6m^3 + m^2n^2 + 7n^2m$

44. Add: [4.7d]
$(5x^2 - 7xy + y^2) + (-6x^2 - 3xy - y^2) +$
$(x^2 + xy - 2y^2)$.

45. Subtract: [4.7e]
$(6x^3y^2 - 4x^2y - 6x) - (-5x^3y^2 + 4x^2y + 6x^2 - 6)$.

Multiply. [4.7f]

46. $(p - q)(p^2 + pq + q^2)$ **47.** $\left(3a^4 - \frac{1}{3}b^3\right)^2$

Divide.

48. $(10x^3 - x^2 + 6x) \div (2x)$ [4.8a]

49. $(6x^3 - 5x^2 - 13x + 13) \div (2x + 3)$ [4.8b]

50. The graph of the polynomial equation $y = 10x^3 - 10x$ is shown below. Use *only* the graph to estimate the value of the polynomial for $x = -1$, for $x = -0.5$, for $x = 0.5$, for $x = 1$, and for $x = 1.1$. [4.3a]

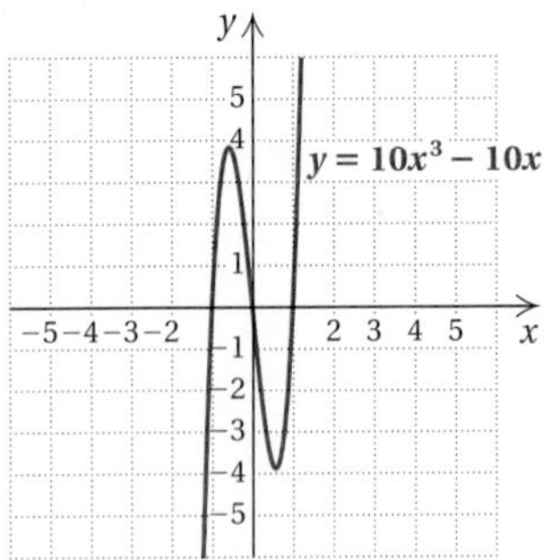

Skill Maintenance

51. Factor: $25t - 50 + 100m$. [1.7d]

52. Solve: $7x + 6 - 8x = 11 - 5x + 4$. [2.3b]

53. Solve: $3(x - 2) + 6 = 5(x + 3) + 9$. [2.3c]

54. Subtract: $-3.4 - 7.8$. [1.4a]

55. The perimeter of a rectangle is 540 m. The width is 19 m less than the length. Find the width and the length. [2.4a]

Synthesis

56. ◈ Explain why the expression 578.6×10^{-7} is not in scientific notation. [4.2c]

57. ◈ Write a short explanation of the difference between a monomial, a binomial, a trinomial, and a general polynomial. [4.3i]

Find a polynomial for the shaded area. [4.4d], [4.6b]

58.

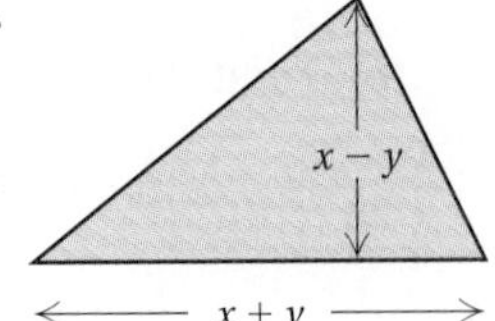

59.

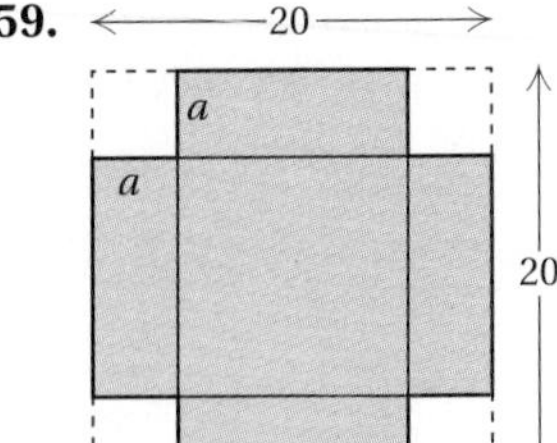

60. Collect like terms: [4.1d], [4.2a], [4.3e]
$-3x^5 \cdot 3x^3 - x^6(2x)^2 + (3x^4)^2 + (2x^2)^4 - 40x^2(x^3)^2$.

61. Solve: [4.6a]

$$(x - 7)(x + 10) = (x - 4)(x - 6).$$

62. The product of two polynomials is $x^5 - 1$. One of the polynomials is $x - 1$. Find the other. [4.8b]

Test: Chapter 4

Multiply and simplify.

1. $6^{-2} \cdot 6^{-3}$
2. $x^6 \cdot x^2 \cdot x$
3. $(4a)^3 \cdot (4a)^8$

Divide and simplify.

4. $\frac{3^5}{3^2}$
5. $\frac{x^3}{x^8}$
6. $\frac{(2x)^5}{(2x)^5}$

Simplify.

7. $(x^3)^2$
8. $(-3y^2)^3$
9. $(2a^3b)^4$
10. $\left(\frac{ab}{c}\right)^3$
11. $(3x^2)^3(-2x^5)^3$
12. $3(x^2)^3(-2x^5)^3$
13. $2x^2(-3x^2)^4$
14. $(2x)^2(-3x^2)^4$

15. Express using a positive exponent: 5^{-3}.

16. Express using a negative exponent: $\frac{1}{y^8}$.

17. Convert to scientific notation: 3,900,000,000.

18. Convert to decimal notation: 5×10^{-8}.

Multiply or divide and write scientific notation for the answer.

19. $\frac{5.6 \times 10^6}{3.2 \times 10^{-11}}$

20. $(2.4 \times 10^5)(5.4 \times 10^{16})$

21. A CD-ROM can contain about 600 million pieces of information (bytes). How many sound files, each containing 40,000 bytes, can a CD-ROM hold? Express the answer in scientific notation.

22. Evaluate the polynomial $x^5 + 5x - 1$ for $x = -2$.

23. Identify the coefficient of each term of the polynomial $\frac{1}{3}x^5 - x + 7$.

24. Identify the degree of each term and the degree of the polynomial $2x^3 - 4 + 5x + 3x^6$.

25. Classify the polynomial $7 - x$ as a monomial, binomial, trinomial, or none of these.

Collect like terms.

26. $4a^2 - 6 + a^2$

27. $y^2 - 3y - y + \frac{3}{4}y^2$

Answers

1. ____
2. ____
3. ____
4. ____
5. ____
6. ____
7. ____
8. ____
9. ____
10. ____
11. ____
12. ____
13. ____
14. ____
15. ____
16. ____
17. ____
18. ____
19. ____
20. ____
21. ____
22. ____
23. ____
24. ____
25. ____
26. ____
27. ____

Answers

28. ______

29. ______

30. ______

31. ______

32. ______

33. ______

34. ______

35. ______

36. ______

37. ______

38. ______

39. ______

40. ______

41. ______

42. ______

43. ______

44. ______

45. ______

46. ______

47. ______

48. ______

49. ______

50. ______

51. ______

52. ______

53. ______

28. Collect like terms and then arrange in descending order:
$3 - x^2 + 2x^3 + 5x^2 - 6x - 2x + x^5$.

Add.

29. $(3x^5 + 5x^3 - 5x^2 - 3) + (x^5 + x^4 - 3x^3 - 3x^2 + 2x - 4)$

30. $\left(x^4 + \frac{2}{3}x + 5\right) + \left(4x^4 + 5x^2 + \frac{1}{3}x\right)$

Subtract.

31. $(2x^4 + x^3 - 8x^2 - 6x - 3) - (6x^4 - 8x^2 + 2x)$

32. $(x^3 - 0.4x^2 - 12) - (x^5 + 0.3x^3 + 0.4x^2 + 9)$

Multiply.

33. $-3x^2(4x^2 - 3x - 5)$

34. $\left(x - \frac{1}{3}\right)^2$

35. $(3x + 10)(3x - 10)$

36. $(3b + 5)(b - 3)$

37. $(x^6 - 4)(x^8 + 4)$

38. $(8 - y)(6 + 5y)$

39. $(2x + 1)(3x^2 - 5x - 3)$

40. $(5t + 2)^2$

41. Collect like terms: $x^3y - y^3 + xy^3 + 8 - 6x^3y - x^2y^2 + 11$.

42. Subtract: $(8a^2b^2 - ab + b^3) - (-6ab^2 - 7ab - ab^3 + 5b^3)$.

43. Multiply: $(3x^5 - 4y^5)(3x^5 + 4y^5)$.

Divide.

44. $(12x^4 + 9x^3 - 15x^2) \div (3x^2)$

45. $(6x^3 - 8x^2 - 14x + 13) \div (3x + 2)$

46. The graph of the polynomial equation $y = x^3 - 5x - 1$ is shown at right. Use *only* the graph to estimate the value of the polynomial for $x = -1$, for $x = -0.5$, for $x = 0.5$, for $x = 1$, and for $x = 1.1$.

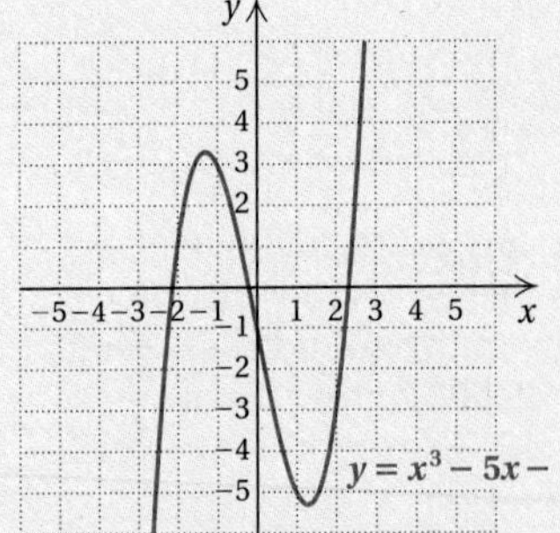

Skill Maintenance

47. Solve: $7x - 4x - 2 = 37$.

48. Solve: $4(x + 2) - 21 = 3(x - 6) + 2$.

49. Factor: $64t - 32m + 16$.

50. Subtract: $\frac{2}{5} - \left(-\frac{3}{4}\right)$.

51. The first angle of a triangle is four times as large as the second. The measure of the third angle is 30° greater than that of the second. How large are the angles?

Synthesis

52. The height of a box is 1 less than its length, and the length is 2 more than its width. Find the volume in terms of the length.

53. Solve: $(x - 5)(x + 5) = (x + 6)^2$.

5

Polynomials: Factoring

An Application

A ladder of length 13 ft is placed against a building in such a way that the distance from the top of the ladder to the ground is 7 ft more than the distance from the bottom of the ladder to the building. Find both distances.

This problem appears as Example 6 in Section 5.9.

The Mathematics

If we visualize this as a triangle, we can let x = the length of the side (leg) across the bottom. Then $x + 7$ = the length of the other side (leg). The hypotenuse has length 13 ft. Using the Pythagorean theorem, we translate the problem to

$$x^2 + (x + 7)^2 = 13^2$$
$$x^2 + x^2 + 14x + 49 = 169$$
$$\underbrace{2x^2 + 14x - 120 = 0.}$$

This is a second-degree, or quadratic, equation.

World Wide Web For more information, visit us at www.mathmax.com

Introduction

Factoring is the reverse of multiplying. To *factor* a polynomial, or other algebraic expression, is to find an equivalent expression that is a product. In this chapter, we study factoring polynomials. To learn to factor quickly, we use the quick methods for multiplication that we learned in Chapter 4.

At the end of this chapter, we find the payoff for learning to factor. We can solve certain new equations containing second-degree polynomials. This in turn allows us to solve applied problems, like the one below, that we could not have solved before.

Pretest: Chapter 5

1. Find three factorizations of $-20x^6$.

Factor.

2. $2x^2 + 4x + 2$

3. $x^2 + 6x + 8$

4. $8a^5 + 4a^3 - 20a$

5. $-6 + 5x^2 - 13x$

6. $81 - z^4$

7. $y^6 - 4y^3 + 4$

8. $3x^3 + 2x^2 + 12x + 8$

9. $p^2 - p - 30$

10. $x^4y^2 - 64$

11. $2p^2 + 7pq - 4q^2$

Solve.

12. $x^2 - 5x = 0$

13. $(x - 4)(5x - 3) = 0$

14. $3x^2 + 10x - 8 = 0$

Solve.

15. Six less than the square of a number is five times the number. Find all such numbers.

16. The height of a triangle is 3 cm longer than the base. The area of the triangle is 44 cm^2. Find the base and the height.

Objectives for Retesting

The objectives to be tested in addition to the material in this chapter are as follows.

[1.6c] Divide real numbers.
[2.7e] Solve inequalities using the addition and multiplication principles together.
[3.3a] Find the intercepts of a linear equation, and graph using intercepts.
[4.6d] Find special products when polynomial products are mixed together.

5.1 Introduction to Factoring

To solve certain types of algebraic equations involving polynomials of second degree, we must learn to factor polynomials.

Consider the product $15 = 3 \cdot 5$. We say that 3 and 5 are **factors** of 15 and that $3 \cdot 5$ is a **factorization** of 15. Since $15 = 15 \cdot 1$, we also know that 15 and 1 are factors of 15 and that $15 \cdot 1$ is a factorization of 15.

> To **factor** a polynomial is to express it as a product.
>
> A **factor** of a polynomial P is a polynomial that can be used to express P as a product.
>
> A **factorization** of a polynomial is an expression that names that polynomial as a product.

Objectives

a. Factor monomials.

b. Factor polynomials when the terms have a common factor, factoring out the largest common factor.

c. Factor certain expressions with four terms using factoring by grouping.

For Extra Help

TAPE 8

MAC WIN

CD-ROM

a Factoring Monomials

To factor a monomial, we find two monomials whose product is equivalent to the original monomial. Compare.

Multiplying	*Factoring*
a) $(4x)(5x) = 20x^2$	$20x^2 = (4x)(5x)$
b) $(2x)(10x) = 20x^2$	$20x^2 = (2x)(10x)$
c) $(-4x)(-5x) = 20x^2$	$20x^2 = (-4x)(-5x)$
d) $(x)(20x) = 20x^2$	$20x^2 = (x)(20x)$

You can see that the monomial $20x^2$ has many factorizations. There are still other ways to factor $20x^2$.

Do Exercises 1 and 2.

Example 1 Find three factorizations of $15x^3$.

a) $15x^3 = (3 \cdot 5)(x \cdot x^2)$
$= (3x)(5x^2)$

b) $15x^3 = (3 \cdot 5)(x^2 \cdot x)$
$= (3x^2)(5x)$

c) $15x^3 = (-15)(-1)x^3$
$= (-15)(-x^3)$

Do Exercises 3–5.

b Factoring When Terms Have a Common Factor

To factor polynomials quickly, we consider the special-product rules studied in Chapter 4, but we first factor out the largest common factor.

To multiply a monomial and a polynomial with more than one term, we multiply each term of the polynomial by the monomial using the distributive laws,

$$a(b + c) = ab + ac \quad \text{and} \quad a(b - c) = ab - ac.$$

1. a) Multiply: $(3x)(4x)$.

b) Factor: $12x^2$.

2. a) Multiply: $(2x)(8x^2)$.

b) Factor: $16x^3$.

Find three factorizations of the monomial.

3. $8x^4$

4. $21x^2$

5. $6x^5$

Answers on page A-18

6. a) Multiply: $3(x + 2)$.

b) Factor: $3x + 6$.

7. a) Multiply: $2x(x^2 + 5x + 4)$.

b) Factor: $2x^3 + 10x^2 + 8x$.

Answers on page A-18

To factor, we do the reverse. We express a polynomial as a product using the distributive laws in reverse:

$$ab + ac = a(b + c) \quad \text{and} \quad ab - ac = a(b - c).$$

Compare.

Multiply	*Factor*
$3x(x^2 + 2x - 4)$	$3x^3 + 6x^2 - 12x$
$= 3x \cdot x^2 + 3x \cdot 2x - 3x \cdot 4$	$= 3x \cdot x^2 + 3x \cdot 2x - 3x \cdot 4$
$= 3x^3 + 6x^2 - 12x$	$= 3x(x^2 + 2x - 4)$

Do Exercises 6 and 7.

CAUTION! Consider the following:

$$3x^3 + 6x^2 - 12x = 3 \cdot x \cdot x \cdot x + 2 \cdot 3 \cdot x \cdot x - 2 \cdot 2 \cdot 3 \cdot x.$$

The terms of the polynomial, $3x^3$, $6x^2$, and $-12x$, have been factored but the polynomial itself has not been factored. This is not what we mean by a factorization of the polynomial. The *factorization* is

$$3x(x^2 + 2x - 4).$$

The expressions $3x$ and $x^2 + 2x - 4$ are *factors* of $3x^3 + 6x^2 - 12x$.

To factor, we first try to find a factor common to all terms. There may not always be one other than 1. When there is, we generally use the factor with the largest possible coefficient and the largest possible exponent.

Example 2 Factor: $7x^2 + 14$.

We have

$$7x^2 + 14 = 7 \cdot x^2 + 7 \cdot 2 \qquad \text{Factoring each term}$$
$$= 7(x^2 + 2). \qquad \text{Factoring out the common factor 7}$$

CHECK: We multiply to check:

$$7(x^2 + 2) = 7 \cdot x^2 + 7 \cdot 2 = 7x^2 + 14.$$

Example 3 Factor: $16x^3 + 20x^2$.

$$16x^3 + 20x^2 = (4x^2)(4x) + (4x^2)(5) \qquad \text{Factoring each term}$$
$$= 4x^2(4x + 5) \qquad \text{Factoring out the common factor } 4x^2$$

Suppose in Example 3 that you had not recognized the largest common factor and removed only part of it, as follows:

$$16x^3 + 20x^2 = (2x^2)(8x) + (2x^2)(10)$$
$$= 2x^2(8x + 10).$$

Note that $8x + 10$ still has a common factor of 2. You need not begin again. Just continue factoring out common factors, as follows, until finished:

$$= 2x^2[2(4x + 5)]$$
$$= 4x^2(4x + 5).$$

Example 4 Factor: $15x^5 - 12x^4 + 27x^3 - 3x^2$.

$$15x^5 - 12x^4 + 27x^3 - 3x^2 = (3x^2)(5x^3) - (3x^2)(4x^2) + (3x^2)(9x) - (3x^2)(1)$$
$$= 3x^2(5x^3 - 4x^2 + 9x - 1) \quad \text{Factoring out } 3x^2$$

CAUTION! Don't forget the term -1.

CHECK: We multiply to check:

$$3x^2(5x^3 - 4x^2 + 9x - 1)$$
$$= (3x^2)(5x^3) - (3x^2)(4x^2) + (3x^2)(9x) - (3x^2)(1)$$
$$= 15x^5 - 12x^4 + 27x^3 - 3x^2.$$

As you become more familiar with factoring, you will be able to spot the largest common factor without factoring each term. Then you can write just the answer.

Examples Factor.

5. $8m^3 - 16m = 8m(m^2 - 2)$
6. $14p^2y^3 - 8py^2 + 2py = 2py(7py^2 - 4y + 1)$
7. $\frac{4}{5}x^2 + \frac{1}{5}x + \frac{2}{5} = \frac{1}{5}(4x^2 + x + 2)$
8. $2.4x^2 + 1.2x - 3.6 = 1.2(2x^2 + x - 3)$

Do Exercises 8–13.

There are two important points to keep in mind as we study this chapter.

- Before doing any other kind of factoring, first try to factor out the largest common factor.
- Always check the result of factoring by multiplying.

C Factoring by Grouping: Four Terms

Certain polynomials with four terms can be factored using a method called *factoring by grouping.*

Example 9 Factor: $x^2(x + 1) + 2(x + 1)$.

The binomial $x + 1$ is common to both terms:

$$x^2(x + 1) + 2(x + 1) = (x^2 + 2)(x + 1).$$

The factorization is $(x^2 + 2)(x + 1)$.

Do Exercises 14 and 15.

Factor. Check by multiplying.

8. $x^2 + 3x$

9. $3y^6 - 5y^3 + 2y^2$

10. $9x^4 - 15x^3 + 3x^2$

11. $\frac{3}{4}t^3 + \frac{5}{4}t^2 + \frac{7}{4}t + \frac{1}{4}$

12. $35x^7 - 49x^6 + 14x^5 - 63x^3$

13. $8.4x^2 - 5.6x + 2.8$

Factor.

14. $x^2(x + 7) + 3(x + 7)$

15. $x^2(a + b) + 2(a + b)$

Answers on page A-18

Factor by grouping.

16. $x^3 + 7x^2 + 3x + 21$

17. $8t^3 + 2t^2 + 12t + 3$

18. $3m^5 - 15m^3 + 2m^2 - 10$

19. $3x^3 - 6x^2 - x + 2$

20. $4x^3 - 6x^2 - 6x + 9$

21. $y^4 - 2y^3 - 2y - 10$

Answers on page A-18

Consider the four-term polynomial

$$x^3 + x^2 + 2x + 2.$$

There is no factor other than 1 that is common to all the terms. We can, however, factor $x^3 + x^2$ and $2x + 2$ separately:

$x^3 + x^2 = x^2(x + 1);$ Factoring $x^3 + x^2$

$2x + 2 = 2(x + 1).$ Factoring $2x + 2$

We have grouped certain terms and factored each polynomial separately:

$$\begin{aligned} x^3 + x^2 + 2x + 2 &= (x^3 + x^2) + (2x + 2) \\ &= x^2(x + 1) + 2(x + 1) \\ &= (x^2 + 2)(x + 1), \end{aligned}$$

as in Example 9. This method is called **factoring by grouping.** We began with a polynomial with four terms. After grouping and removing common factors, we obtained a polynomial with two parts, each having a common factor $x + 1$. Not all polynomials with four terms can be factored by this procedure, but it does give us a method to try.

Examples Factor by grouping.

10. $6x^3 - 9x^2 + 4x - 6$

$= (6x^3 - 9x^2) + (4x - 6)$

$= 3x^2(2x - 3) + 2(2x - 3)$ Factoring each binomial

$= (3x^2 + 2)(2x - 3)$ Factoring out the common factor $2x - 3$

We think through this process as follows:

$$6x^3 - 9x^2 + 4x - 6 = 3x^2(2x - 3) \quad \blacksquare \quad (2x - 3)$$

(1) Factor the first two terms.

(2) This factor, $2x - 3$, gives us a hint to the factorization on the right.

(3) Now we ask ourselves, "What needs to be here to enable us to get $4x - 6$ when we multiply?"

11. $x^3 + x^2 + x + 1 = (x^3 + x^2) + (x + 1)$

$= x^2(x + 1) + 1(x + 1)$ Factoring each binomial

$= (x^2 + 1)(x + 1)$ Factoring out the common factor $x + 1$

12. $2x^3 - 6x^2 - x + 3$

$= (2x^3 - 6x^2) + (-x + 3)$

$= 2x^2(x - 3) - 1(x - 3)$ *Check:* $-1(x - 3) = -x + 3.$

$= (2x^2 - 1)(x - 3)$ Factoring out the common factor $x - 3$

13. $12x^5 + 20x^2 - 21x^3 - 35 = 4x^2(3x^3 + 5) - 7(3x^3 + 5)$

$= (4x^2 - 7)(3x^3 + 5)$

14. $x^3 + x^2 + 2x - 2 = x^2(x + 1) + 2(x - 1)$

This polynomial is not factorable using factoring by grouping. It may be factorable, but not by methods that we will consider in this text.

Do Exercises 16–21.

Exercise Set 5.1

a Find three factorizations for the monomial.

1. $8x^3$ **2.** $6x^4$ **3.** $-10a^6$ **4.** $-8y^5$ **5.** $24x^4$ **6.** $15x^5$

b Factor. Check by multiplying.

7. $x^2 - 6x$ **8.** $x^2 + 5x$ **9.** $2x^2 + 6x$

10. $8y^2 - 8y$ **11.** $x^3 + 6x^2$ **12.** $3x^4 - x^2$

13. $8x^4 - 24x^2$ **14.** $5x^5 + 10x^3$ **15.** $2x^2 + 2x - 8$

16. $8x^2 - 4x - 20$ **17.** $17x^5y^3 + 34x^3y^2 + 51xy$ **18.** $16p^6q^4 + 32p^5q^3 - 48pq^2$

19. $6x^4 - 10x^3 + 3x^2$ **20.** $5x^5 + 10x^2 - 8x$ **21.** $x^5y^5 + x^4y^3 + x^3y^3 - x^2y^2$

22. $x^9y^6 - x^7y^5 + x^4y^4 + x^3y^3$ **23.** $2x^7 - 2x^6 - 64x^5 + 4x^3$ **24.** $8y^3 - 20y^2 + 12y - 16$

25. $1.6x^4 - 2.4x^3 + 3.2x^2 + 6.4x$ **26.** $2.5x^6 - 0.5x^4 + 5x^3 + 10x^2$

27. $\frac{5}{3}x^6 + \frac{4}{3}x^5 + \frac{1}{3}x^4 + \frac{1}{3}x^3$ **28.** $\frac{5}{9}x^7 + \frac{2}{9}x^5 - \frac{4}{9}x^3 - \frac{1}{9}x$

c Factor.

29. $x^2(x + 3) + 2(x + 3)$ **30.** $3z^2(2z + 1) + (2z + 1)$

31. $5a^3(2a - 7) - (2a - 7)$ **32.** $m^4(8 - 3m) - 7(8 - 3m)$

Factor by grouping.

33. $x^3 + 3x^2 + 2x + 6$

34. $6z^3 + 3z^2 + 2z + 1$

35. $2x^3 + 6x^2 + x + 3$

36. $3x^3 + 2x^2 + 3x + 2$

37. $8x^3 - 12x^2 + 6x - 9$

38. $10x^3 - 25x^2 + 4x - 10$

39. $12x^3 - 16x^2 + 3x - 4$

40. $18x^3 - 21x^2 + 30x - 35$

41. $5x^3 - 5x^2 - x + 1$

42. $7x^3 - 14x^2 - x + 2$

43. $x^3 + 8x^2 - 3x - 24$

44. $2x^3 + 12x^2 - 5x - 30$

45. $2x^3 - 8x^2 - 9x + 36$

46. $20g^3 - 4g^2 - 25g + 5$

Skill Maintenance

Solve.

47. $-2x < 48$ [2.7d]

48. $4x - 8x + 16 \geq 6(x - 2)$ [2.7e]

49. Divide: $\frac{-108}{-4}$. [1.6a]

50. Solve $A = \frac{p + q}{2}$ for p. [2.6a]

Multiply. [4.6d]

51. $(y + 5)(y + 7)$

52. $(y + 7)^2$

53. $(y + 7)(y - 7)$

54. $(y - 7)^2$

Find the intercepts of the equation. Then graph the equation. [3.3a]

55. $x + y = 4$

56. $x - y = 3$

57. $5x - 3y = 15$

58. $y - 3x = 6$

Synthesis

59. ◈ Josh says that there is no need to print answers for Exercises 1–46 at the back of the book. Is he correct in saying this? Why or why not?

60. ◈ Explain how one could construct a polynomial with four terms that can be factored by grouping.

Factor.

61. $4x^5 + 6x^3 + 6x^2 + 9$

62. $x^6 + x^4 + x^2 + 1$

63. $x^{12} + x^7 + x^5 + 1$

64. $x^3 - x^2 - 2x + 5$

65. $p^3 + p^2 - 3p + 10$

5.2 Factoring Trinomials of the Type $x^2 + bx + c$

Objective

a Factor trinomials of the type $x^2 + bx + c$ by examining the constant term c.

For Extra Help

TAPE 8 | MAC WIN | CD-ROM

a We now begin a study of the factoring of trinomials. We first factor trinomials like

$x^2 + 5x + 6$ and $x^2 + 3x - 10$

by a refined *trial-and-error* process. In this section, we restrict our attention to trinomials of the type $ax^2 + bx + c$, where $a = 1$. The coefficient a is often called the **leading coefficient.**

Constant Term Positive

Recall the FOIL method of multiplying two binomials:

$$\begin{aligned}&\quad\ \ \text{F}\quad\ \text{O}\quad\ \text{I}\quad\ \text{L}\\(x + 2)(x + 5) &= x^2 + 5x + 2x + 10\\&= x^2 \quad + 7x \quad + 10.\end{aligned}$$

The product above is a trinomial. The term of highest degree, x^2, called the leading term, has a coefficient of 1. The constant term, 10, is positive. To factor $x^2 + 7x + 10$, we think of FOIL in reverse. We multiplied x times x to get the first term of the trinomial, so we know that the first term of each binomial factor is x. Next, we look for numbers p and q such that

$$x^2 + 7x + 10 = (x + p)(x + q).$$

To get the middle term and the last term of the trinomial, we look for two numbers p and q whose product is 10 and whose sum is 7. Those numbers are 2 and 5. Thus the factorization is

$$(x + 2)(x + 5).$$

Example 1 Factor: $x^2 + 5x + 6$.

Think of FOIL in reverse. The first term of each factor is x: $(x + \quad)(x + \quad)$. Next, we look for two numbers whose product is 6 and whose sum is 5. All the pairs of factors of 6 are shown in the table on the left below. Since both the product, 6, and the sum, 5, of the pair of numbers must be positive, we need consider only the positive factors, listed in the table on the right.

Pairs of Factors	Sums of Factors
1, 6	7
−1, −6	−7
2, 3	5
−2, −3	−5

Pairs of Factors	Sums of Factors
1, 6	7
2, 3	5

The numbers we need are 2 and 3.

The factorization is $(x + 2)(x + 3)$. We can check by multiplying to see whether we get the original trinomial.

CHECK: $(x + 2)(x + 3) = x^2 + 3x + 2x + 6 = x^2 + 5x + 6$.

Do Exercises 1 and 2.

1. Consider the trinomial $x^2 + 7x + 12$.

a) Complete the following table.

Pairs of Factors	Sums of Factors
1, 12	13
−1, −12	
2, 6	
−2, −6	
3, 4	
−3, −4	

b) Explain why you need to consider only positive factors, as in the following table.

Pairs of Factors	Sums of Factors
1, 12	
2, 6	
3, 4	

c) Factor: $x^2 + 7x + 12$.

2. Factor: $x^2 + 13x + 36$.

Answers on page A-19

3. Explain why you would not consider the pairs of factors listed below in factoring $y^2 - 8y + 12$.

Pairs of Factors	Sums of Factors
1, 12	
2, 6	
3, 4	

Factor.

4. $x^2 - 8x + 15$

5. $t^2 - 9t + 20$

Answers on page A-19

Consider this multiplication:

$$\begin{aligned}(x-2)(x-5) &= \overset{F}{x^2} \overset{O}{- 5x} \overset{I}{- 2x} \overset{L}{+ 10}\\ &= x^2 - 7x + 10.\end{aligned}$$

When the constant term of a trinomial is positive, we look for two numbers with the same sign (both negative or both positive). The sign is that of the middle term:

$$x^2 - 7x + 10 = (x-2)(x-5), \quad \text{or} \quad x^2 + 7x + 10 = (x+2)(x+5).$$

Example 2 Factor: $y^2 - 8y + 12$.

Since the constant term, 12, is positive and the coefficient of the middle term, -8, is negative, we look for a factorization of 12 in which both factors are negative. Their sum must be -8.

Pairs of Factors	Sums of Factors
−1, −12	−13
−2, −6	−8 ← The numbers we need are −2 and −6.
−3, −4	−7

The factorization is $(y - 2)(y - 6)$.

Do Exercises 3–5.

Constant Term Negative

Sometimes when we use FOIL, the product has a negative constant term. Consider these multiplications:

a) $$\begin{aligned}(x-5)(x+2) &= \overset{F}{x^2} \overset{O}{+ 2x} \overset{I}{- 5x} \overset{L}{- 10}\\ &= x^2 - 3x - 10;\end{aligned}$$

b) $$\begin{aligned}(x+5)(x-2) &= \overset{F}{x^2} \overset{O}{- 2x} \overset{I}{+ 5x} \overset{L}{- 10}\\ &= x^2 + 3x - 10.\end{aligned}$$

Reversing the signs of the factors changes the sign of the middle term.

When the constant term of a trinomial is negative, we look for two factors whose product is negative. One of them must be positive and the other negative. Their sum must be the coefficient of the middle term:

$$x^2 - 3x - 10 = (x-5)(x+2), \quad \text{or} \quad x^2 + 3x - 10 = (x+5)(x-2).$$

Example 3 Factor: $x^3 - 8x^2 - 20x$.

Always look first for a common factor. This time there is one, x. We first factor it out: $x^3 - 8x^2 - 20x = x(x^2 - 8x - 20)$. Now consider the expression $x^2 - 8x - 20$. Since the constant term, -20, is negative, we look for a factorization of -20 in which one factor is positive and one factor is negative. The sum must be -8, so the negative factor must have the larger absolute value. Thus we consider only pairs of factors in which the negative factor has the larger absolute value.

Pairs of Factors	Sums of Factors
1, −20	−19
2, −10	−8 ← The numbers we need are 2 and −10.
4, −5	−1

The factorization of $x^2 - 8x - 20$ is $(x + 2)(x - 10)$. But we must also remember to include the common factor. The factorization of the original polynomial is

$$x(x + 2)(x - 10).$$

Do Exercise 6.

6. Explain why you would not consider the pairs of factors listed below in factoring $x^2 - 8x - 20$.

Pairs of Factors	Sums of Factors
−1, 20	
−2, 10	
−4, 5	

Example 4 Factor: $t^2 - 24 + 5t$.

It helps to first write the trinomial in descending order: $t^2 + 5t - 24$. Since the constant term, -24, is negative, we look for a factorization of -24 in which one factor is positive and one factor is negative. Their sum must be 5, so the positive factor must have the larger absolute value. Thus we consider only pairs of factors in which the positive term has the larger absolute value.

Pairs of Factors	Sums of Factors
−1, 24	23
−2, 12	10
−3, 8	5 ← The numbers we need are −3 and 8.
−4, 6	2

The factorization is $(t - 3)(t + 8)$.

Do Exercise 7.

7. Explain why you would not consider the pairs of factors listed below in factoring $t^2 + 5t - 24$.

Pairs of Factors	Sums of Factors
1, −24	
2, −12	
3, −8	
4, −6	

Example 5 Factor: $x^4 - x^2 - 110$.

Consider this trinomial as $(x^2)^2 - x^2 - 110$. We look for numbers p and q such that

$$x^4 - x^2 - 110 = (x^2 + p)(x^2 + q).$$

Since the constant term, -110, is negative, we look for a factorization of -110 in which one factor is positive and one factor is negative. Their sum must be -1. The middle-term coefficient, -1, is small compared to -110. This tells us that the desired factors are close to each other in absolute value. The numbers we want are 10 and -11. The factorization is

$$(x^2 + 10)(x^2 - 11).$$

Answers on page A-19

Factor.

8. $x^3 + 4x^2 - 12x$

9. $y^2 - 12 - 4y$

10. $t^4 + 5t^2 - 14$

11. $p^2 - pq - 3pq^2$

12. $x^2 + 2x + 7$

13. Factor: $x^2 + 8x + 16$.

Answers on page A-19

Example 6 Factor: $a^2 + 4ab - 21b^2$.

We consider the trinomial in the equivalent form

$$a^2 + 4ba - 21b^2.$$

We think of $4b$ as a "coefficient" of a. Then we look for factors of $-21b^2$ whose sum is $4b$. Those factors are $-3b$ and $7b$. The factorization is

$$(a - 3b)(a + 7b).$$

There are polynomials that are not factorable.

Example 7 Factor: $x^2 - x + 5$.

Since 5 has very few factors, we can easily check all possibilities.

Pairs of Factors	Sums of Factors
5, 1	6
−5, −1	−6

There are no factors whose sum is -1. Thus the polynomial is *not* factorable into binomials.

Do Exercises 8–12.

We can factor a trinomial that is a perfect square using this method.

Example 8 Factor: $x^2 - 10x + 25$.

Since the constant term, 25, is positive and the coefficient of the middle term, -10, is negative, we look for a factorization of 25 in which both factors are negative. Their sum must be -10.

Pairs of Factors	Sums of Factors
−25, −1	−26
−5, −5	−10 ← The numbers we need are −5 and −5.

The factorization is $(x - 5)(x - 5)$, or $(x - 5)^2$.

Do Exercise 13.

The following is a summary of our procedure for factoring $x^2 + bx + c$.

To factor $x^2 + bx + c$:

1. First arrange in descending order.
2. Use a trial-and-error process that looks for factors of c whose sum is b.
3. If c is positive, the signs of the factors are the same as the sign of b.
4. If c is negative, one factor is positive and the other is negative. If the sum of two factors is the opposite of b, changing the sign of each factor will give the desired factors whose sum is b.
5. Check by multiplying.

Exercise Set 5.2

a Factor. Remember that you can check by multiplying.

1. $x^2 + 8x + 15$

2. $x^2 + 5x + 6$

3. $x^2 + 7x + 12$

4. $x^2 + 9x + 8$

5. $x^2 - 6x + 9$

6. $y^2 - 11y + 28$

7. $x^2 + 9x + 14$

8. $a^2 + 11a + 30$

9. $b^2 + 5b + 4$

10. $z^2 - 8z + 7$

11. $x^2 + \frac{2}{3}x + \frac{1}{9}$

12. $x^2 - \frac{2}{5}x + \frac{1}{25}$

13. $d^2 - 7d + 10$

14. $t^2 - 12t + 35$

15. $y^2 - 11y + 10$

16. $x^2 - 4x - 21$

17. $x^2 + x - 42$

18. $x^2 + 2x - 15$

19. $x^2 - 7x - 18$

20. $y^2 - 3y - 28$

21. $x^3 - 6x^2 - 16x$

22. $x^3 - x^2 - 42x$

23. $y^3 - 4y^2 - 45y$

24. $x^3 - 7x^2 - 60x$

25. $-2x - 99 + x^2$

26. $x^2 - 72 + 6x$

27. $c^4 + c^2 - 56$

28. $b^4 + 5b^2 - 24$

29. $a^4 + 2a^2 - 35$

30. $x^4 - x^2 - 6$

31. $x^2 + x + 1$

32. $x^2 + 5x + 3$

33. $7 - 2p + p^2$

34. $11 - 3w + w^2$

35. $x^2 + 20x + 100$

36. $a^2 + 19a + 88$

37. $x^4 - 21x^3 - 100x^2$

38. $x^4 - 20x^3 + 96x^2$

39. $x^2 - 21x - 72$

40. $4x^2 + 40x + 100$

41. $x^2 - 25x + 144$

42. $y^2 - 21y + 108$

43. $a^2 + a - 132$

44. $a^2 + 9a - 90$

45. $120 - 23x + x^2$

46. $96 + 22d + d^2$

47. $108 - 3x - x^2$

48. $112 + 9y - y^2$

49. $y^2 - 0.2y - 0.08$

50. $t^2 - 0.3t - 0.10$

51. $p^2 + 3pq - 10q^2$

52. $a^2 + 2ab - 3b^2$

53. $m^2 + 5mn + 4n^2$

54. $x^2 + 11xy + 24y^2$

55. $s^2 - 2st - 15t^2$

56. $p^2 + 5pq - 24q^2$

Skill Maintenance

Multiply. [4.6d]

57. $8x(2x^2 - 6x + 1)$

58. $(7w + 6)(4w - 11)$

59. $(7w + 6)^2$

60. $(4w - 11)^2$

61. $(4w - 11)(4w + 11)$

62. Simplify: $(3x^4)^3$. [4.2b]

Solve. [2.3a]

63. $3x - 8 = 0$

64. $2x + 7 = 0$

Solve.

65. *Arrests for Counterfeiting.* In a recent year, 29,200 people were arrested for counterfeiting. This number was down 1.2% from the preceding year. How many people were arrested the preceding year? [2.5a]

66. The first angle of a triangle is four times as large as the second. The measure of the third angle is 30° greater than that of the second. Find the angle measures. [2.4a]

Synthesis

67. ◆ Without doing the multiplication $(x - 17)(x - 18)$, explain why it cannot possibly be a factorization of $x^2 + 35x + 306$.

68. ◆ When searching for a factorization of $x^2 + bx + c$, why do we list pairs of numbers with the specified product c instead of pairs of numbers with the specified sum b?

69. Find all integers m for which $y^2 + my + 50$ can be factored.

70. Find all integers b for which $a^2 + ba - 50$ can be factored.

Factor completely.

71. $x^2 - \frac{1}{2}x - \frac{3}{16}$

72. $x^2 - \frac{1}{4}x - \frac{1}{8}$

73. $x^2 + \frac{30}{7}x - \frac{25}{7}$

74. $\frac{1}{3}x^3 + \frac{1}{3}x^2 - 2x$

75. $b^{2n} + 7b^n + 10$

76. $a^{2m} - 11a^m + 28$

Find a polynomial in factored form for the shaded area. (Leave answers in terms of π.)

77.

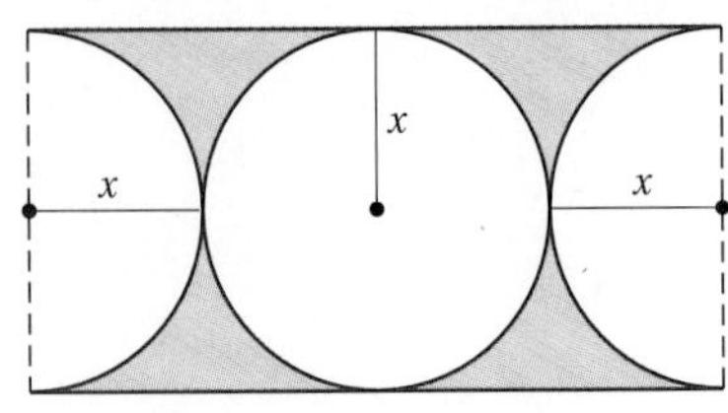

78.

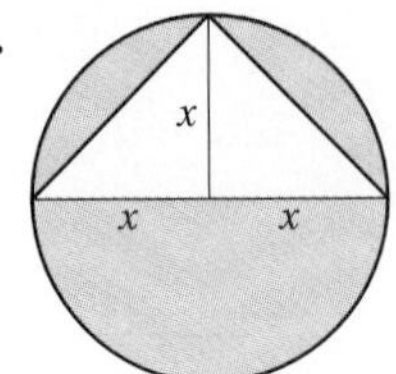

5.3 Factoring $ax^2 + bx + c,\ a \neq 1$, Using FOIL

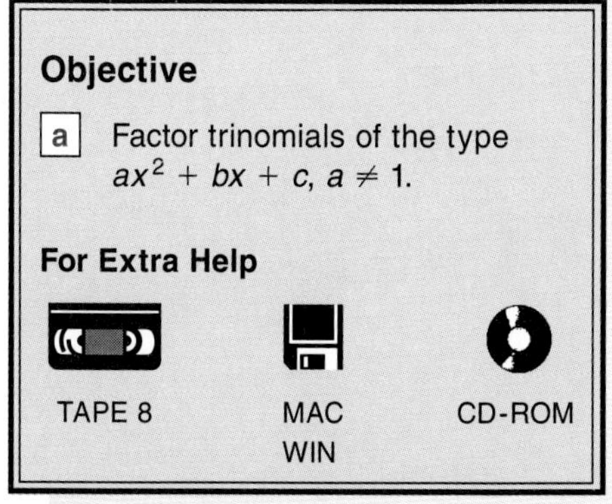

In Section 5.2, we learned a trial-and-error method to factor trinomials of the type $x^2 + bx + c$. In this section, we factor trinomials in which the coefficient of the leading term x^2 is not 1. The procedure we learn is a refined trial-and-error method. (In Section 5.4, we will consider an alternative method for the same kind of factoring. It involves *factoring by grouping.*)

a We want to factor trinomials of the type $ax^2 + bx + c$. Consider the following multiplication:

$$\begin{aligned}&\qquad\quad\ \ \text{F}\qquad\ \ \text{O}\qquad\ \ \text{I}\qquad\ \ \text{L}\\(2x + 5)(3x + 4) &= 6x^2 + 8x + 15x + 20\\&= 6x^2 + \quad 23x \quad + 20\end{aligned}$$

F	O + I		L
$2 \cdot 3$	$2 \cdot 4$	$5 \cdot 3$	$5 \cdot 4$

To factor $6x^2 + 23x + 20$, we reverse the above multiplication, using what we might call an "unFOIL" process. We look for two binomials $rx + p$ and $sx + q$ whose product is this trinomial. The product of the First terms must be $6x^2$. The product of the Outside terms plus the product of the Inside terms must be $23x$. The product of the Last terms must be 20. We know from the preceding discussion that the answer is $(2x + 5)(3x + 4)$. Generally, however, finding such an answer is a refined trial-and-error process. It turns out that $(-2x - 5)(-3x - 4)$ is also a correct answer, but we usually choose an answer in which the first coefficients are positive.

We will use the following trial-and-error method.

To factor $ax^2 + bx + c$, $a \neq 1$, using FOIL:

1. Factor out the largest common factor, if any.
2. Find the First terms whose product is ax^2.

   $(\blacksquare x + \quad)(\blacksquare x + \quad) = ax^2 + bx + c.$

   FOIL
3. Find two Last terms whose product is c:

   $(\ \ x + \blacksquare)(\ \ x + \blacksquare) = ax^2 + bx + c.$

   FOIL
4. Repeat steps (2) and (3) until a combination is found for which the sum of the Outer and Inner products is bx:

   $(\blacksquare x + \blacksquare)(\blacksquare x + \blacksquare) = ax^2 + bx + c.$

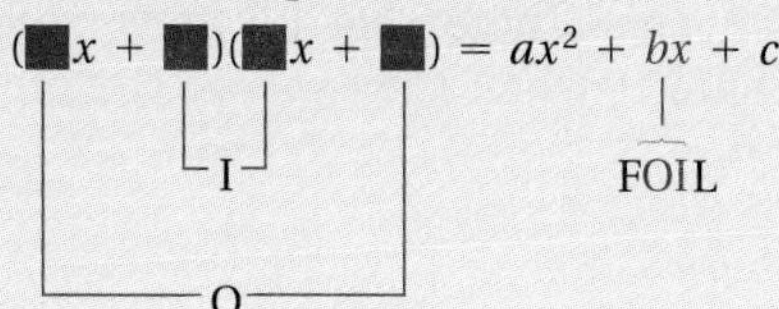

FOIL

Factor.

1. $2x^2 - x - 15$

2. $12x^2 - 17x - 5$

Answers on page A-19

Example 1 Factor: $3x^2 - 10x - 8$.

1) First, we check for a common factor. Here there is none (other than 1 or -1).

2) Find two **F**irst terms whose product is $3x^2$.

The only possibilities for the **F**irst terms are $3x$ and x, so any factorization must be of the form

$(3x + \quad)(x + \quad)$.

3) Find two **L**ast terms whose product is -8.

Possible factorizations of -8 are

$(-8) \cdot 1$, $8 \cdot (-1)$, $(-2) \cdot 4$, and $2 \cdot (-4)$.

Since the First terms are not identical, we must also consider

$1 \cdot (-8)$, $(-1) \cdot 8$, $4 \cdot (-2)$, and $(-4) \cdot 2$.

4) Inspect the **O**uter and **I**nner products resulting from steps (2) and (3). Look for a combination in which the sum of the products is the middle term, $-10x$:

Trial	*Product*	
$(3x - 8)(x + 1)$	$3x^2 + 3x - 8x - 8$	
	$= 3x^2 - 5x - 8$	← Wrong middle term
$(3x + 8)(x - 1)$	$3x^2 - 3x + 8x - 8$	
	$= 3x^2 + 5x - 8$	← Wrong middle term
$(3x - 2)(x + 4)$	$3x^2 + 12x - 2x - 8$	
	$= 3x^2 + 10x - 8$	← Wrong middle term
$(3x + 2)(x - 4)$	$3x^2 - 12x + 2x - 8$	
	$= 3x^2 - 10x - 8$	← Correct middle term!
$(3x + 1)(x - 8)$	$3x^2 - 24x + x - 8$	
	$= 3x^2 - 23x - 8$	← Wrong middle term
$(3x - 1)(x + 8)$	$3x^2 + 24x - x - 8$	
	$= 3x^2 + 23x - 8$	← Wrong middle term
$(3x + 4)(x - 2)$	$3x^2 - 6x + 4x - 8$	
	$= 3x^2 - 2x - 8$	← Wrong middle term
$(3x - 4)(x + 2)$	$3x^2 + 6x - 4x - 8$	
	$= 3x^2 + 2x - 8$	← Wrong middle term

The correct factorization is $(3x + 2)(x - 4)$.

CHECK: $(3x + 2)(x - 4) = 3x^2 - 10x - 8$.

Two observations can be made from Example 1. First, we listed all possible trials even though we could have stopped after having found the correct factorization. We did this to show that each trial differs only in the middle term of the product. Second, note that as in Section 5.2, only the sign of the middle term changes when the signs in the binomials are reversed.

Do Exercises 1 and 2.

Example 2 Factor: $24x^2 - 76x + 40$.

1) First, we factor out the largest common factor, 4:

$$4(6x^2 - 19x + 10).$$

Now we factor the trinomial $6x^2 - 19x + 10$.

2) Because $6x^2$ can be factored as $3x \cdot 2x$ or $6x \cdot x$, we have these possibilities for factorizations:

$$(3x + \square)(2x + \square) \quad \text{or} \quad (6x + \square)(x + \square).$$

3) There are four pairs of factors of 10 and they each can be listed in two ways:

$$10, 1 \qquad -10, -1 \qquad 5, 2 \qquad -5, -2$$

and

$$1, 10 \qquad -1, -10 \qquad 2, 5 \qquad -2, -5.$$

4) The two possibilities from step (2) and the eight possibilities from step (3) give $2 \cdot 8$, or 16 possibilities for factorizations. We look for **O**uter and **I**nner products resulting from steps (2) and (3) for which the sum is the middle term, $-19x$. Since the sign of the middle term is negative, but the sign of the last term, 10, is positive, the two factors of 10 must both be negative. This means only four pairings from step (3) need be considered. We first try these factors with $(3x + \square)(2x + \square)$. If none gives the correct factorization, we will consider $(6x + \square)(x + \square)$.

Trial	*Product*	
$(3x - 10)(2x - 1)$	$6x^2 - 3x - 20x + 10$	
	$= 6x^2 - 23x + 10$	← **Wrong middle term**
$(3x - 1)(2x - 10)$	$6x^2 - 30x - 2x + 10$	
	$= 6x^2 - 32x + 10$	← **Wrong middle term**
$(3x - 5)(2x - 2)$	$6x^2 - 6x - 10x + 10$	
	$= 6x^2 - 16x + 10$	← **Wrong middle term**
$(3x - 2)(2x - 5)$	$6x^2 - 15x - 4x + 10$	
	$= 6x^2 - 19x + 10$	← **Correct middle term!**

Since we have a correct factorization, we need not consider

$$(6x + \square)(x + \square).$$

The factorization of $6x^2 - 19x + 10$ is $(3x - 2)(2x - 5)$, but *do not forget the common factor*! We must include it in order to factor the original trinomial:

$$\begin{aligned} 24x^2 - 76x + 40 &= 4(6x^2 - 19x + 10) \\ &= 4(3x - 2)(2x - 5). \end{aligned}$$

Caution! When factoring any polynomial, always look for a common factor. Failure to do so is such a common error that this caution bears repeating.

Factor.

3. $3x^2 - 19x + 20$

4. $20x^2 - 46x + 24$

5. Factor: $6x^2 + 7x + 2$.

Answers on page A-19

In Example 2, look again at the possibility $(3x - 5)(2x - 2)$. Without multiplying, we can reject such a possibility. To see why, consider the following:

$$(3x - 5)(2x - 2) = 2(3x - 5)(x - 1).$$

The expression $2x - 2$ has a common factor, 2. But we removed the *largest* common factor in the first step. If $2x - 2$ were one of the factors, then 2 would have to be a common factor in addition to the original 4. Thus, $(2x - 2)$ cannot be part of the factorization of the original trinomial.

> Given that the largest common factor is factored out at the outset, we need not consider factorizations that have a common factor.

Do Exercises 3 and 4.

Example 3 Factor: $10x^2 + 37x + 7$.

1) There is no common factor (other than 1 or −1).
2) Because $10x^2$ factors as $10x \cdot x$ or $5x \cdot 2x$, we have these possibilities for factorizations:

$$(10x + \quad)(x + \quad) \quad \text{or} \quad (5x + \quad)(2x + \quad).$$

3) There are two pairs of factors of 7 and they each can be listed in two ways:

1, 7   −1, −7   and   7, 1   −7, −1.

4) From steps (2) and (3), we see that there are 8 possibilities for factorizations. Look for **O**uter and **I**nner products for which the sum is the middle term. Because all coefficients in $10x^2 + 37x + 7$ are positive, we need consider only positive factors of 7. The possibilities are

$$(10x + 1)(x + 7) = 10x^2 + 71x + 7,$$
$$(10x + 7)(x + 1) = 10x^2 + 17x + 7,$$
$$(5x + 7)(2x + 1) = 10x^2 + 19x + 7,$$
$$(5x + 1)(2x + 7) = 10x^2 + 37x + 7.$$

The factorization is $(5x + 1)(2x + 7)$.

Keep in mind that this method of factoring trinomials of the type $ax^2 + bx + c$ involves *trial and error.* As you practice, you will find that you can make better and better guesses.

Do Exercise 5.

TIPS FOR FACTORING $ax^2 + bx + c$, $a \neq 1$

1. Always factor out the largest common factor, if one exists. Once the common factor has been factored out of the original trinomial, no binomial factor can contain a common factor (other than 1 or −1).
2. If c is positive, then the signs in both binomial factors must match the sign of b. (This assumes that $a > 0$.)
3. Reversing the signs in the binomials reverses the sign of the middle term of their product.
4. Be systematic about your trials. Keep track of those pairs you have tried and those you have not.
5. Always check by multiplying.

Example 4 Factor: $6p^2 - 13pq - 28q^2$.

1) Factor out a common factor, if any. There is none (other than 1 or -1).
2) Factor the first term, $6p^2$. Possibilities are $2p$, $3p$ and $6p$, p. We have these as possibilities for factorizations:

$$(2p + \square)(3p + \square) \quad \text{or} \quad (6p + \square)(p + \square).$$

3) Factor the last term, $-28q^2$, which has a negative coefficient. The possibilities are $-14q$, $2q$ and $14q$, $-2q$; $-28q$, q and $28q$, $-q$; and $-7q$, $4q$ and $7q$, $-4q$.
4) The coefficient of the middle term is negative, so we look for combinations of factors from steps (2) and (3) such that the sum of their products has a negative coefficient. We try some possibilities:

$$(2p + q)(3p - 28q) = 6p^2 - 53pq - 28q^2,$$
$$(2p - 7q)(3p + 4q) = 6p^2 - 13pq - 28q^2.$$

The factorization of $6p^2 - 13pq - 28q^2$ is $(2p - 7q)(3p + 4q)$.

Do Exercises 6 and 7.

Factor.

6. $6a^2 - 5ab + b^2$

7. $6x^2 + 15xy + 9y^2$

Calculator Spotlight

 Checking Factorizations with a Table or a Graph

Table. The TABLE feature can be used as a partial check that polynomials have been factored correctly. To check whether the factoring of Example 1,

$$3x^2 - 10x - 8 = (3x + 2)(x - 4),$$

is correct, we enter

$$y_1 = 3x^2 - 10x - 8 \quad \text{and} \quad y_2 = (3x + 2)(x - 4).$$

If our factoring is correct, the y_1- and y_2-values should be the same, regardless of the table settings used.

X	Y1	Y2
0	−8	−8
1	−15	−15
2	−16	−16
3	−11	−11
4	0	0
5	17	17
6	40	40

X = 0

We see that y_1 and y_2 are the same, so the factoring seems to be correct. Remember, though, that this is only a partial check.

Graph. This factorization of Example 1 can also be checked with the GRAPH feature. We see that the graphs of y_1 and y_2 are the same, so the factoring seems to be correct.

$y_1 = 3x^2 - 10x - 8,\ y_2 = (3x + 2)(x - 4)$

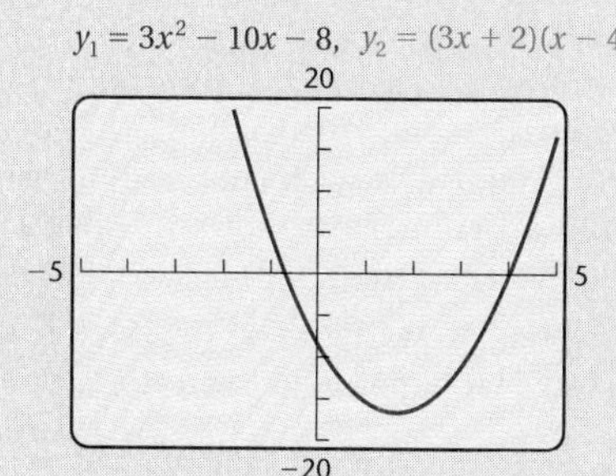

Exercises

Use the TABLE or the GRAPH feature to check whether the factorization is correct.

1. $24x^2 - 76x + 40 = 4(3x - 2)(2x - 5)$ (Example 2)
2. $4x^2 - 5x - 6 = (4x + 3)(x - 2)$
3. $5x^2 + 17x - 12 = (5x + 3)(x - 4)$
4. $10x^2 + 37x + 7 = (5x - 1)(2x + 7)$
5. $12x^2 - 17x - 5 = (6x + 1)(2x - 5)$
6. $12x^2 - 17x - 5 = (4x + 1)(3x - 5)$
7. $x^2 - 4 = (x - 2)(x - 2)$
8. $x^2 - 4 = (x + 2)(x - 2)$

Answers on page A-19

Improving Your Math Study Skills

Forming Math Study Groups, by James R. Norton

Dr. James Norton has taught at the University of Phoenix and Scottsdale Community College. He has extensive experience with the use of study groups to learn mathematics.

The use of math study groups for learning has become increasingly common in recent years. Some instructors regard them as a primary source of learning, while others let students form groups on their own.

A study group generally consists of study partners who help each other learn the material and do the homework. You will probably meet outside of class at least once or twice a week. Here are some do's and don'ts to make your study group more valuable.

- DO make the group up of no more than four or five people. Research has shown clearly that this size works best.
- DO trade phone numbers so that you can get in touch with each other for help between team meetings.
- DO make sure that everyone in the group has a chance to contribute.
- DON'T let a group member copy from others without contributing. If this should happen, one member should speak with that student privately; if the situation continues, that student should be asked to leave the group.
- DON'T let the "A" students be passive. The group needs them! The benefits to even the best students are twofold: (1) Other students will benefit from their expertise and (2) the bright students will learn the material better by teaching it to someone else.
- DON'T let the slower students be passive either. *Everyone* can contribute something, and being in a group will actually improve their self-esteem as well as their performance.

How do you form study groups if the instructor has not already done so? A good place to begin is to get together with three or four friends and arrange a study time. If you don't know anyone, start getting acquainted with other people in the class during the first week of the semester.

What should you look for in a study partner?

- Do you live near each other to make it easy to get together?
- What are your class schedules like? Are you both on campus? Do you have free time?
- What about work schedules, athletic practice, and other out-of-school commitments that you might have to work around?

Making use of a study group is not a form of "cheating." You are merely helping each other learn. So long as everyone in the group is both contributing and doing the work, this method will bring you great success!

Exercise Set 5.3

a Factor.

1. $2x^2 - 7x - 4$
2. $3x^2 - x - 4$
3. $5x^2 - x - 18$
4. $4x^2 - 17x + 15$
5. $6x^2 + 23x + 7$
6. $6x^2 - 23x + 7$
7. $3x^2 + 4x + 1$
8. $7x^2 + 15x + 2$
9. $4x^2 + 4x - 15$
10. $9x^2 + 6x - 8$
11. $2x^2 - x - 1$
12. $15x^2 - 19x - 10$
13. $9x^2 + 18x - 16$
14. $2x^2 + 5x + 2$
15. $3x^2 - 5x - 2$
16. $18x^2 - 3x - 10$
17. $12x^2 + 31x + 20$
18. $15x^2 + 19x - 10$
19. $14x^2 + 19x - 3$
20. $35x^2 + 34x + 8$
21. $9x^2 + 18x + 8$
22. $6 - 13x + 6x^2$
23. $49 - 42x + 9x^2$
24. $16 + 36x^2 + 48x$
25. $24x^2 + 47x - 2$
26. $16p^2 - 78p + 27$
27. $35x^2 - 57x - 44$
28. $9a^2 + 12a - 5$
29. $20 + 6x - 2x^2$
30. $15 + x - 2x^2$
31. $12x^2 + 28x - 24$
32. $6x^2 + 33x + 15$
33. $30x^2 - 24x - 54$
34. $18t^2 - 24t + 6$
35. $4y + 6y^2 - 10$
36. $-9 + 18x^2 - 21x$
37. $3x^2 - 4x + 1$
38. $6t^2 + 13t + 6$
39. $12x^2 - 28x - 24$
40. $6x^2 - 33x + 15$
41. $-1 + 2x^2 - x$
42. $-19x + 15x^2 + 6$
43. $9x^2 - 18x - 16$
44. $14y^2 + 35y + 14$
45. $15x^2 - 25x - 10$
46. $18x^2 + 3x - 10$
47. $12p^3 + 31p^2 + 20p$
48. $15x^3 + 19x^2 - 10x$

49. $14x^4 + 19x^3 - 3x^2$

50. $70x^4 + 68x^3 + 16x^2$

51. $168x^3 - 45x^2 + 3x$

52. $144x^5 + 168x^4 + 48x^3$

53. $15x^4 - 19x^2 + 6$

54. $9x^4 + 18x^2 + 8$

55. $25t^2 + 80t + 64$

56. $9x^2 - 42x + 49$

57. $6x^3 + 4x^2 - 10x$

58. $18x^3 - 21x^2 - 9x$

59. $25x^2 + 79x + 64$

60. $9y^2 + 42y + 47$

61. $6x^2 - 19x - 5$

62. $2x^2 + 11x - 9$

63. $12m^2 - mn - 20n^2$

64. $12a^2 - 17ab + 6b^2$

65. $6a^2 - ab - 15b^2$

66. $3p^2 - 16pq - 12q^2$

67. $9a^2 + 18ab + 8b^2$

68. $10s^2 + 4st - 6t^2$

69. $35p^2 + 34pq + 8q^2$

70. $30a^2 + 87ab + 30b^2$

71. $18x^2 - 6xy - 24y^2$

72. $15a^2 - 5ab - 20b^2$

Skill Maintenance

Solve. [2.6a]

73. $A = pq - 7$, for q

74. $y = mx + b$, for x

75. $3x + 2y = 6$, for y

76. $p - q + r = 2$, for q

Solve. [2.7e]

77. $5 - 4x < -11$

78. $2x - 4(x + 3x) \geq 6x - 8 - 9x$

79. Graph: $y = \frac{2}{5}x - 1$. [3.2b]

80. Divide: $\frac{y^{12}}{y^4}$. [4.1e]

Multiply. [4.6d]

81. $(3x - 5)(3x + 5)$

82. $(4a - 3)^2$

Synthesis

83. ◆ Explain how the factoring in Exercise 21 can be used to aid the factoring in Exercise 67.

84. ◆ A student presents the following work:

$$4x^2 + 28x + 48 = (2x + 6)(2x + 8)$$
$$= 2(x + 3)(x + 4).$$

Is it correct? Explain.

Factor.

85. $20x^{2n} + 16x^n + 3$

86. $-15x^{2m} + 26x^m - 8$

87. $3x^{6a} - 2x^{3a} - 1$

88. $x^{2n+1} - 2x^{n+1} + x$

5.4 Factoring $ax^2 + bx + c$, $a \neq 1$, Using Grouping

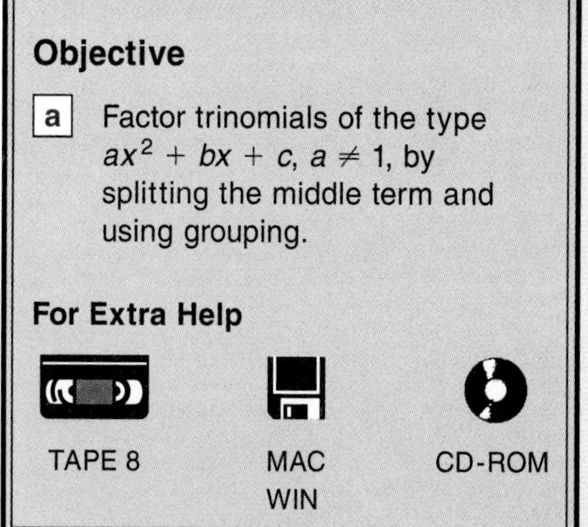

a Another method of factoring trinomials of the type $ax^2 + bx + c$, $a \neq 1$, is known as the **grouping method.** It involves factoring by grouping. We know how to factor the trinomial $x^2 + 5x + 6$. We look for factors of the constant term, 6, whose sum is the coefficient of the middle term, 5:

$$x^2 + 5x + 6.$$

(1) Factor: $6 = 2 \cdot 3$

(2) Sum: $2 + 3 = 5$

What happens when the leading coefficient is not 1? To factor a trinomial like $3x^2 - 10x - 8$, we can use a method similar to what we used for the preceding trinomial, but we need two more steps. That method is outlined as follows.

To factor $ax^2 + bx + c$, $a \neq 1$, using the grouping method:

1. Factor out a common factor, if any.
2. Multiply the leading coefficient a and the constant c.
3. Try to factor the product ac so that the sum of the factors is b. That is, find integers p and q such that $pq = ac$ and $p + q = b$.
4. Split the middle term. That is, write it as a sum using the factors found in step (3).
5. Then factor by grouping.

Example 1 Factor: $3x^2 - 10x - 8$.

1) First, we factor out a common factor, if any. There is none (other than 1 or -1).

2) We multiply the leading coefficient, 3, and the constant, -8:

$$3(-8) = -24.$$

3) Then we look for a factorization of -24 in which the sum of the factors is the coefficient of the middle term, -10.

Pairs of Factors	Sums of Factors
−1, 24	23
1, −24	−23
−2, 12	10
2, −12	−10 ← $2 + (-12) = -10$
−3, 8	5
3, −8	−5
−4, 6	2
4, −6	−2

4) Next, we split the middle term as a sum or a difference using the factors found in step (3):

$$-10x = 2x - 12x.$$

Factor.

1. $6x^2 + 7x + 2$

2. $12x^2 - 17x - 5$

Factor.

3. $6x^2 + 15x + 9$

4. $20x^2 - 46x + 24$

Answers on page A-19

5) Finally, we factor by grouping, as follows:

$$3x^2 - 10x - 8 = 3x^2 + 2x - 12x - 8 \quad \text{Substituting } 2x - 12x \text{ for } -10x$$
$$= x(3x + 2) - 4(3x + 2) \quad \text{Factoring by grouping; see Section 5.1}$$
$$= (x - 4)(3x + 2).$$

We can also split the middle term as $-12x + 2x$. We still get the same factorization, although the factors may be in a different order. Note the following:

$$3x^2 - 10x - 8 = 3x^2 - 12x + 2x - 8 \quad \text{Substituting } -12x + 2x \text{ for } -10x$$
$$= 3x(x - 4) + 2(x - 4) \quad \text{Factoring by grouping; see Section 5.1}$$
$$= (3x + 2)(x - 4).$$

Check by multiplying: $(3x + 2)(x - 4) = 3x^2 - 10x - 8.$

Do Exercises 1 and 2.

Example 2 Factor: $8x^2 + 8x - 6$.

1) First, we factor out a common factor, if any. The number 2 is common to all three terms, so we factor it out:

$$2(4x^2 + 4x - 3).$$

2) Next, we factor the trinomial $4x^2 + 4x - 3$. We multiply the leading coefficient and the constant, 4 and -3:

$$4(-3) = -12.$$

3) We try to factor -12 so that the sum of the factors is 4.

Pairs of Factors	Sums of Factors
−1, 12	11
1, −12	−11
−2, 6	4 ← −2 + 6 = 4
2, −6	−4
−3, 4	1
3, −4	−1

4) Then we split the middle term, $4x$, as follows:

$$4x = -2x + 6x.$$

5) Finally, we factor by grouping:

$$4x^2 + 4x - 3 = 4x^2 - 2x + 6x - 3 \quad \text{Substituting } -2x + 6x \text{ for } 4x$$
$$= 2x(2x - 1) + 3(2x - 1) \quad \text{Factoring by grouping}$$
$$= (2x + 3)(2x - 1).$$

The factorization of $4x^2 + 4x - 3$ is $(2x + 3)(2x - 1)$. But don't forget the common factor! We must include it to get a factorization of the original trinomial:

$$8x^2 + 8x - 6 = 2(2x + 3)(2x - 1).$$

Do Exercises 3 and 4.

Exercise Set 5.4

a Factor. Note that the middle term has already been split.

1. $x^2 + 2x + 7x + 14$

2. $x^2 + 3x + x + 3$

3. $x^2 - 4x - x + 4$

4. $a^2 + 5a - 2a - 10$

5. $6x^2 + 4x + 9x + 6$

6. $3x^2 - 2x + 3x - 2$

7. $3x^2 - 4x - 12x + 16$

8. $24 - 18y - 20y + 15y^2$

9. $35x^2 - 40x + 21x - 24$

10. $8x^2 - 6x - 28x + 21$

11. $4x^2 + 6x - 6x - 9$

12. $2x^4 - 6x^2 - 5x^2 + 15$

13. $2x^4 + 6x^2 + 5x^2 + 15$

14. $9x^4 - 6x^2 - 6x^2 + 4$

Factor by grouping.

15. $2x^2 - 7x - 4$

16. $5x^2 - x - 18$

17. $3x^2 + 4x - 15$

18. $3x^2 + x - 4$

19. $6x^2 + 23x + 7$

20. $6x^2 + 13x + 6$

21. $3x^2 + 4x + 1$

22. $7x^2 + 15x + 2$

23. $4x^2 + 4x - 15$

24. $9x^2 + 6x - 8$

25. $2x^2 + x - 1$

26. $15x^2 + 19x - 10$

27. $9x^2 - 18x - 16$

28. $2x^2 - 5x + 2$

29. $3x^2 + 5x - 2$

30. $18x^2 + 3x - 10$

31. $12x^2 - 31x + 20$

32. $15x^2 - 19x - 10$

33. $14x^2 + 19x - 3$

34. $35x^2 + 34x + 8$

35. $9x^2 + 18x + 8$

36. $6 - 13x + 6x^2$

37. $49 - 42x + 9x^2$

38. $25x^2 + 40x + 16$

39. $24x^2 + 47x - 2$

40. $16a^2 + 78a + 27$

41. $35x^5 - 57x^4 - 44x^3$

42. $18a^3 + 24a^2 - 10a$

43. $60x + 18x^2 - 6x^3$

44. $60x + 4x^2 - 8x^3$

45. $15x^3 + 33x^4 + 6x^5$

46. $8x^2 + 2x + 6x^3$

Skill Maintenance

Solve. [2.7d, e]

47. $-10x > 1000$

48. $-3.8x \leq -824.6$

49. $6 - 3x \geq -18$

50. $3 - 2x - 4x > -9$

51. $\frac{1}{2}x - 6x + 10 \leq x - 5x$

52. $-2(x + 7) > -4(x - 5)$

53. $3x - 6x + 2(x - 4) > 2(9 - 4x)$

54. $-6(x - 4) + 8(4 - x) \leq 3(x - 7)$

Synthesis

55. ◆ If you have studied both the FOIL and the grouping methods of factoring $ax^2 + bx + c$, $a \neq 1$, decide which method you think is better and explain why.

56. ◆ Explain factoring $ax^2 + bx + c$, $a \neq 1$, by grouping as though you were teaching a fellow student.

Factor.

57. $9x^{10} - 12x^5 + 4$

58. $24x^{2n} + 22x^n + 3$

59. $16x^{10} + 8x^5 + 1$

60. $(a + 4)^2 - 2(a + 4) + 1$

61.–70. Use the TABLE feature to check the factoring in Exercises 15–24.

5.5 Factoring Trinomial Squares and Differences of Squares

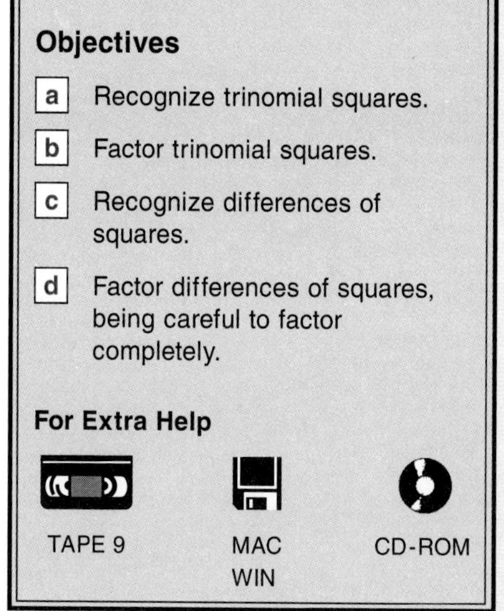

In this section, we first learn to factor trinomials that are squares of binomials. Then we factor binomials that are differences of squares.

a Recognizing Trinomial Squares

Some trinomials are squares of binomials. For example, the trinomial $x^2 + 10x + 25$ is the square of the binomial $x + 5$. To see this, we can calculate $(x + 5)^2$. It is $x^2 + 2 \cdot x \cdot 5 + 5^2$, or $x^2 + 10x + 25$. A trinomial that is the square of a binomial is called a **trinomial square.**

In Chapter 4, we considered squaring binomials as special-product rules:

$$(A + B)^2 = A^2 + 2AB + B^2;$$
$$(A - B)^2 = A^2 - 2AB + B^2.$$

We can use these equations in reverse to factor trinomial squares.

$$A^2 + 2AB + B^2 = (A + B)^2;$$
$$A^2 - 2AB + B^2 = (A - B)^2$$

How can we recognize when an expression to be factored is a trinomial square? Look at $A^2 + 2AB + B^2$ and $A^2 - 2AB + B^2$. In order for an expression to be a trinomial square:

a) Two terms, A^2 and B^2, must be squares, such as

$$4, \quad x^2, \quad 25x^4, \quad 16t^2.$$

When the coefficient is a perfect square and the power(s) of the variable(s) is (are) even, then the expression is a perfect square.

b) There must be no minus sign before A^2 or B^2.

c) If we multiply A and B (expressions whose squares are A^2 and B^2) and double the result, we get either the remaining term $2 \cdot A \cdot B$, or its opposite, $-2 \cdot A \cdot B$.

Example 1 Determine whether $x^2 + 6x + 9$ is a trinomial square.

a) We know that x^2 and 9 are squares.

b) There is no minus sign before x^2 or 9.

c) If we multiply the square roots, x and 3, and double the product, we get the remaining term: $2 \cdot x \cdot 3 = 6x$.

Thus, $x^2 + 6x + 9$ is the square of a binomial. In fact, $x^2 + 6x + 9 = (x + 3)^2$.

Example 2 Determine whether $x^2 + 6x + 11$ is a trinomial square.

The answer is no, because only one term is a square.

Determine whether each is a trinomial square. Write "yes" or "no."

1. $x^2 + 8x + 16$

2. $25 - x^2 + 10x$

3. $t^2 - 12t + 4$

4. $25 + 20y + 4y^2$

5. $5x^2 + 16 - 14x$

6. $16x^2 + 40x + 25$

7. $p^2 + 6p - 9$

8. $25a^2 + 9 - 30a$

Factor.

9. $x^2 + 2x + 1$

10. $1 - 2x + x^2$

11. $4 + t^2 + 4t$

12. $25x^2 - 70x + 49$

13. $49 - 56y + 16y^2$

Answers on page A-19

Example 3 Determine whether $16x^2 + 49 - 56x$ is a trinomial square.

It helps to first write the trinomial in descending order:

$$16x^2 - 56x + 49.$$

a) We know that $16x^2$ and 49 are squares.

b) There is no minus sign before $16x^2$ or 49.

c) If we multiply the square roots, $4x$ and 7, and double the product, we get the opposite of the remaining term: $2 \cdot 4x \cdot 7 = 56x$; $56x$ is the opposite of $-56x$.

Thus, $16x^2 + 49 - 56x$ is a trinomial square. In fact, $16x^2 - 56x + 49 = (4x - 7)^2$.

Do Exercises 1–8.

b Factoring Trinomial Squares

We can use the trial-and-error or grouping methods from Sections 5.2–5.4 to factor such trinomial squares, but there is a faster method using the following equations:

$$A^2 + 2AB + B^2 = (A + B)^2;$$
$$A^2 - 2AB + B^2 = (A - B)^2.$$

We consider 3 to be a square root of 9 because $3^2 = 9$. Similarly, A is a square root of A^2. We use square roots of the squared terms and the sign of the remaining term to factor a trinomial square.

Example 4 Factor: $x^2 + 6x + 9$.

$$x^2 + 6x + 9 = x^2 + 2 \cdot x \cdot 3 + 3^2 = (x + 3)^2$$
$$A^2 + 2\ A\ B + B^2 = (A + B)^2$$

The sign of the middle term is positive.

Example 5 Factor: $x^2 + 49 - 14x$.

$x^2 + 49 - 14x = x^2 - 14x + 49$ **Changing order**

$= x^2 - 2 \cdot x \cdot 7 + 7^2$ **The sign of the middle term is negative.**

$= (x - 7)^2$

Example 6 Factor: $16x^2 - 40x + 25$.

$$16x^2 - 40x + 25 = (4x)^2 - 2 \cdot 4x \cdot 5 + 5^2 = (4x - 5)^2$$
$$A^2 - 2\ A\ B + B^2 = (A - B)^2$$

Do Exercises 9–13.

Example 7 Factor: $t^4 + 20t^2 + 100$.

$$t^4 + 20t^2 + 100 = (t^2)^2 + 2(t^2)(10) + 10^2$$
$$= (t^2 + 10)^2$$

Example 8 Factor: $75m^3 + 210m^2 + 147m$.

Always look first for a common factor. This time there is one, $3m$:

$$75m^3 + 210m^2 + 147m = 3m[25m^2 + 70m + 49]$$
$$= 3m[(5m)^2 + 2(5m)(7) + 7^2]$$
$$= 3m(5m + 7)^2.$$

Example 9 Factor: $4p^2 - 12pq + 9q^2$.

$$4p^2 - 12pq + 9q^2 = (2p)^2 - 2(2p)(3q) + (3q)^2$$
$$= (2p - 3q)^2$$

Do Exercises 14–17.

c Recognizing Differences of Squares

The following polynomials are *differences of squares*:

$$x^2 - 9, \quad 4t^2 - 49, \quad a^2 - 25b^2.$$

To factor a difference of squares such as $x^2 - 9$, think about the formula we used in Chapter 4:

$$(A + B)(A - B) = A^2 - B^2.$$

Equations are reversible, so we also know that

$$A^2 - B^2 = (A + B)(A - B).$$

Thus,

$$x^2 - 9 = (x + 3)(x - 3).$$

To use this formula, we must be able to recognize when it applies. A **difference of squares** is an expression like the following:

$$A^2 - B^2.$$

How can we recognize such expressions? Look at $A^2 - B^2$. In order for a binomial to be a difference of squares:

a) There must be two expressions, both squares, such as

$$4x^2, \quad 9, \quad 25t^4, \quad 1, \quad x^6, \quad 49y^8.$$

b) The terms must have different signs.

Factor.

14. $48m^2 + 75 + 120m$

15. $p^4 + 18p^2 + 81$

16. $4z^5 - 20z^4 + 25z^3$

17. $9a^2 + 30ab + 25b^2$

Answers on page A-19

Determine whether each is a difference of squares. Write "yes" or "no."

18. $x^2 - 25$

19. $t^2 - 24$

20. $y^2 + 36$

21. $4x^2 - 15$

22. $16x^4 - 49$

23. $9w^6 - 1$

24. $-49 + 25t^2$

Answers on page A-19

Example 10 Is $9x^2 - 64$ a difference of squares?

a) The first expression is a square: $9x^2 = (3x)^2$.
The second expression is a square: $64 = 8^2$.

b) The terms have different signs.

Thus we have a difference of squares, $(3x)^2 - 8^2$.

Example 11 Is $25 - t^3$ a difference of squares?

a) The expression t^3 is not a square.

The expression is not a difference of squares.

Example 12 Is $-4x^2 + 16$ a difference of squares?

a) The expressions $4x^2$ and 16 are squares: $4x^2 = (2x)^2$ and $16 = 4^2$.

b) The terms have different signs.

Thus we have a difference of squares. We can also see this by rewriting in the equivalent form: $16 - 4x^2$.

Do Exercises 18–24.

d Factoring Differences of Squares

To factor a difference of squares, we use the following equation:

$$A^2 - B^2 = (A + B)(A - B).$$

To factor a difference of squares $A^2 - B^2$, we find A and B, which are square roots of the expressions A^2 and B^2. We then use A and B to form two factors. One is the sum $A + B$, and the other is the difference $A - B$.

Example 13 Factor: $x^2 - 4$.

$$x^2 - 4 = \underset{A^2}{x^2} - \underset{B^2}{2^2} = (\underset{A}{x} + \underset{B}{2})(\underset{A}{x} - \underset{B}{2})$$

Example 14 Factor: $9 - 16t^4$.

$$9 - 16t^4 = \underset{A^2}{3^2} - \underset{B^2}{(4t^2)^2} = (\underset{A}{3} + \underset{B}{4t^2})(\underset{A}{3} - \underset{B}{4t^2})$$

Example 15 Factor: $m^2 - 4p^2$.

$$m^2 - 4p^2 = m^2 - (2p)^2 = (m + 2p)(m - 2p)$$

Example 16 Factor: $x^2 - \frac{1}{9}$.

$$x^2 - \frac{1}{9} = x^2 - \left(\frac{1}{3}\right)^2 = \left(x + \frac{1}{3}\right)\left(x - \frac{1}{3}\right)$$

Example 17 Factor: $18x^2 - 50x^6$.

Always look first for a factor common to all terms. This time there is one, $2x^2$.

$$\begin{aligned} 18x^2 - 50x^6 &= 2x^2(9 - 25x^4) \\ &= 2x^2[3^2 - (5x^2)^2] \\ &= 2x^2(3 + 5x^2)(3 - 5x^2) \end{aligned}$$

Example 18 Factor: $49x^4 - 9x^6$.

$$49x^4 - 9x^6 = x^4(49 - 9x^2) = x^4(7 + 3x)(7 - 3x)$$

Do Exercises 25–29.

CAUTION! Note carefully in these examples that a difference of squares is *not* the square of the difference; that is,

$$A^2 - B^2 \neq (A - B)^2 = A^2 - 2AB + B^2.$$

For example,

$$(45 - 5)^2 = 40^2 = 1600,$$

but

$$45^2 - 5^2 = 2025 - 25 = 2000.$$

Factoring Completely

If a factor with more than one term can still be factored, you should do so. When no factor can be factored further, you have **factored completely.** Always factor completely whenever told to factor.

Example 19 Factor: $p^4 - 16$.

$$\begin{aligned} p^4 - 16 &= (p^2)^2 - 4^2 \\ &= (p^2 + 4)(p^2 - 4) && \text{Factoring a difference of squares} \\ &= (p^2 + 4)(p + 2)(p - 2) && \text{Factoring further. The factor } p^2 - 4 \text{ is a difference of squares.} \end{aligned}$$

The polynomial $p^2 + 4$ cannot be factored further into polynomials with real coefficients.

CAUTION! If the greatest common factor has been removed, then you cannot factor a sum of squares further. In particular,

$$(A + B)^2 \neq A^2 + B^2.$$

Consider $25x^2 + 100$. This is a case in which we have a sum of squares, but there is a common factor, 25. Factoring, we get $25(x^2 + 4)$. Now $x^2 + 4$ cannot be factored further.

Example 20 Factor: $y^4 - 16x^{12}$.

$$\begin{aligned} y^4 - 16x^{12} &= (y^2 + 4x^6)(y^2 - 4x^6) && \text{Factoring a difference of squares} \\ &= (y^2 + 4x^6)(y + 2x^3)(y - 2x^3) && \text{Factoring further. The factor } y^2 - 4x^6 \text{ is a difference of squares.} \end{aligned}$$

Factor.

25. $x^2 - 9$

26. $64 - 4t^2$

27. $a^2 - 25b^2$

28. $64x^4 - 25x^6$

29. $5 - 20t^6$
[*Hint*: $1 = 1^2$, $t^6 = (t^3)^2$.]

Answers on page A-19

Factor completely.

30. $81x^4 - 1$

31. $49p^4 - 25q^6$

FACTORING HINTS

1. Always look first for a common factor. If there is one, factor out the largest common factor.
2. Always factor completely.
3. Check by multiplying.

Do Exercises 30 and 31.

Improving Your Math Study Skills

A Checklist of Your Study Skills

You are now about halfway through this textbook as well as the course. How are you doing? If you are struggling, we might ask if you are making full use of the study skills that we have suggested in these inserts. To determine this, review the following list of all study skill suggestions made so far and answer the questions "yes" or "no."

Study Skill Questions	Yes	No
1. Are you doing a thorough job of reading the book?		
2. Are you stopping and working the margin exercises when directed to do so?		
3. Are you doing your homework as soon as possible after class?		
4. Are you doing your homework at a specified time and in a quiet setting?		
5. Have you found a study group in which to work?		
6. Are you consistently trying to apply the five-step problem-solving strategy when working applied problems?		
7. Are you asking questions in class and in tutoring sessions?		
8. Are you doing lots of even-numbered exercises for which answers are not available?		

Study Skill Questions	Yes	No
9. Are you keeping one section ahead of your syllabus?		
10. Are you using the book supplements, such as the *Student's Solutions Manual* and the *InterAct Math Tutorial Software*?		
11. When you study the book, are you marking the points that you do not understand as a source for in-class questions?		
12. Are you reading and studying each step of each example?		
13. Are you using the objective code symbols (a, b, c, etc.) that appear at the beginning of each section, throughout the section, and in the exercise sets, the summary–reviews, and the answers for the chapter tests?		

If you have answered "no" seven or more times and are struggling in the course, you need to improve your study skills by following more of these suggestions.

A consultation with your instructor regarding your situation is strongly advised.

Answers on page A-20

Exercise Set 5.5

a Determine whether each of the following is a trinomial square.

1. $x^2 - 14x + 49$

2. $x^2 - 16x + 64$

3. $x^2 + 16x - 64$

4. $x^2 - 14x - 49$

5. $x^2 - 2x + 4$

6. $x^2 + 3x + 9$

7. $9x^2 - 36x + 24$

8. $36x^2 - 24x + 16$

b Factor completely. Remember to look first for a common factor and to check by multiplying.

9. $x^2 - 14x + 49$

10. $x^2 - 20x + 100$

11. $x^2 + 16x + 64$

12. $x^2 + 20x + 100$

13. $x^2 - 2x + 1$

14. $x^2 + 2x + 1$

15. $4 + 4x + x^2$

16. $4 + x^2 - 4x$

17. $q^4 - 6q^2 + 9$

18. $64 + 16a^2 + a^4$

19. $49 + 56y + 16y^2$

20. $75 + 48a^2 - 120a$

21. $2x^2 - 4x + 2$

22. $2x^2 - 40x + 200$

23. $x^3 - 18x^2 + 81x$

24. $x^3 + 24x^2 + 144x$

25. $12q^2 - 36q + 27$

26. $20p^2 + 100p + 125$

27. $49 - 42x + 9x^2$

28. $64 - 112x + 49x^2$

29. $5y^4 + 10y^2 + 5$

30. $a^4 + 14a^2 + 49$

31. $1 + 4x^4 + 4x^2$

32. $1 - 2a^5 + a^{10}$

33. $4p^2 + 12pq + 9q^2$

34. $25m^2 + 20mn + 4n^2$

35. $a^2 - 6ab + 9b^2$

36. $x^2 - 14xy + 49y^2$

37. $81a^2 - 18ab + b^2$

38. $64p^2 + 16pq + q^2$

39. $36a^2 + 96ab + 64b^2$

40. $16m^2 - 40mn + 25n^2$

c Determine whether each of the following is a difference of squares.

41. $x^2 - 4$

42. $x^2 - 36$

43. $x^2 + 25$

44. $x^2 + 9$

45. $x^2 - 45$

46. $x^2 - 80y^2$

47. $16x^2 - 25y^2$

48. $-1 + 36x^2$

d Factor completely. Remember to look first for a common factor.

49. $y^2 - 4$

50. $q^2 - 1$

51. $p^2 - 9$

52. $x^2 - 36$

53. $-49 + t^2$

54. $-64 + m^2$

55. $a^2 - b^2$

56. $p^2 - q^2$

57. $25t^2 - m^2$

58. $w^2 - 49z^2$

59. $100 - k^2$

60. $81 - w^2$

61. $16a^2 - 9$

62. $25x^2 - 4$

63. $4x^2 - 25y^2$

64. $9a^2 - 16b^2$

65. $8x^2 - 98$

66. $24x^2 - 54$

67. $36x - 49x^3$

68. $16x - 81x^3$

69. $49a^4 - 81$

70. $25a^4 - 9$

71. $a^4 - 16$

72. $y^4 - 1$

73. $5x^4 - 405$

74. $4x^4 - 64$

75. $1 - y^8$

76. $x^8 - 1$

77. $x^{12} - 16$

78. $x^8 - 81$

79. $y^2 - \frac{1}{16}$

80. $x^2 - \frac{1}{25}$

81. $25 - \frac{1}{49}x^2$

82. $\frac{1}{4} - 9q^2$

83. $16m^4 - t^4$

84. $p^4q^4 - 1$

Skill Maintenance

Divide. [1.6c]

85. $(-110) \div 10$

86. $-1000 \div (-2.5)$

87. $\left(-\frac{2}{3}\right) \div \frac{4}{5}$

88. $8.1 \div (-9)$

89. $-64 \div (-32)$

90. $-256 \div 1.6$

Find a polynomial for the shaded area. (Leave results in terms of π where appropriate.) [4.4d]

91.

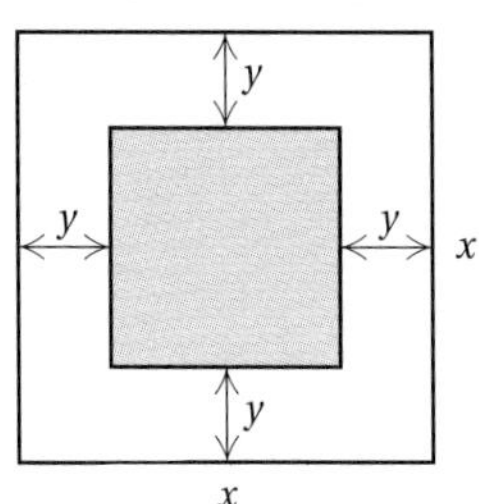

92.

Simplify.

93. $y^5 \cdot y^7$ [4.1d]

94. $(5a^2b^3)^2$ [4.2a]

Find the intercepts. Then graph the equation. [3.3a]

95. $y - 6x = 6$

96. $3x - 5y = 15$

Synthesis

97. ◆ Explain in your own words how to determine whether a polynomial is a trinomial square.

98. ◆ A student concludes that since $x^2 - 9 = (x - 3)(x + 3)$, it must follow that $x^2 + 9 = (x + 3)(x + 3)$. What mistake is the student making? How would you go about correcting the misunderstanding?

Factor completely, if possible.

99. $49x^2 - 216$

100. $27x^3 - 13x$

101. $x^2 + 22x + 121$

102. $x^2 - 5x + 25$

103. $18x^3 + 12x^2 + 2x$

104. $162x^2 - 82$

105. $x^8 - 2^8$

106. $4x^4 - 4x^2$

107. $3x^5 - 12x^3$

108. $3x^2 - \frac{1}{3}$

109. $18x^3 - \frac{8}{25}x$

110. $x^2 - 2.25$

111. $0.49p - p^3$

112. $3.24x^2 - 0.81$

113. $0.64x^2 - 1.21$

114. $1.28x^2 - 2$

115. $(x + 3)^2 - 9$

116. $(y - 5)^2 - 36q^2$

117. $x^2 - \left(\frac{1}{x}\right)^2$

118. $a^{2n} - 49b^{2n}$

119. $81 - b^{4k}$

120. $9x^{18} + 48x^9 + 64$

121. $9b^{2n} + 12b^n + 4$

122. $(x + 7)^2 - 4x - 24$

123. $(y + 3)^2 + 2(y + 3) + 1$

124. $49(x + 1)^2 - 42(x + 1) + 9$

Find c such that the polynomial is the square of a binomial.

125. $cy^2 + 6y + 1$

126. $cy^2 - 24y + 9$

Use the TABLE feature to determine whether the factorization is correct.

127. $x^2 + 9 = (x + 3)(x + 3)$

128. $x^2 - 49 = (x - 7)(x + 7)$

129. $x^2 + 9 = (x + 3)^2$

130. $x^2 - 49 = (x - 7)^2$

5.6 Factoring Sums or Differences of Cubes

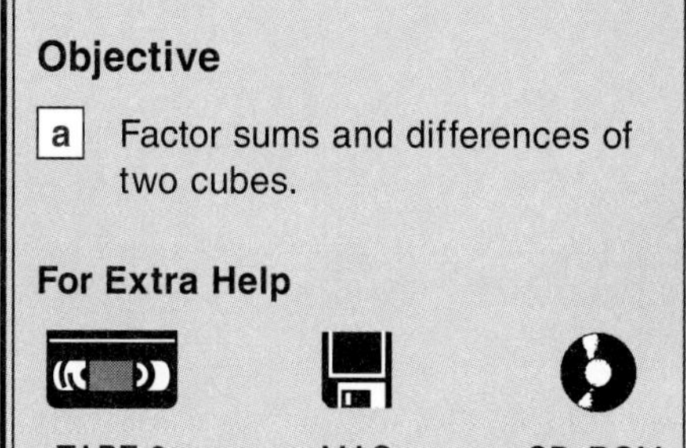

a We can factor the sum or the difference of two expressions that are cubes.

Consider the following products:

$$(A + B)(A^2 - AB + B^2) = A(A^2 - AB + B^2) + B(A^2 - AB + B^2)$$
$$= A^3 - A^2B + AB^2 + A^2B - AB^2 + B^3$$
$$= A^3 + B^3$$

and

$$(A - B)(A^2 + AB + B^2) = A(A^2 + AB + B^2) - B(A^2 + AB + B^2)$$
$$= A^3 + A^2B + AB^2 - A^2B - AB^2 - B^3$$
$$= A^3 - B^3.$$

The above equations (reversed) show how we can factor a sum or a difference of two cubes.

$$A^3 + B^3 = (A + B)(A^2 - AB + B^2),$$
$$A^3 - B^3 = (A - B)(A^2 + AB + B^2)$$

Note that what we are considering here is a sum or a difference of cubes. We are not cubing a binomial. For example, $(A + B)^3$ is *not* the same as $A^3 + B^3$. The table of cubes in the margin is helpful.

N	N^3
0.2	0.008
0.1	0.001
0	0
1	1
2	8
3	27
4	64
5	125
6	216
7	343
8	512
9	729
10	1000

Example 1 Factor: $x^3 - 27$.

We have

$$x^3 - 27 = x^3 - 3^3.$$

In one set of parentheses, we write the cube root of the first term, x. Then we write the cube root of the second term, -3. This gives us the expression $x - 3$:

$$(x - 3)(\qquad\qquad).$$

To get the next factor, we think of $x - 3$ and do the following:

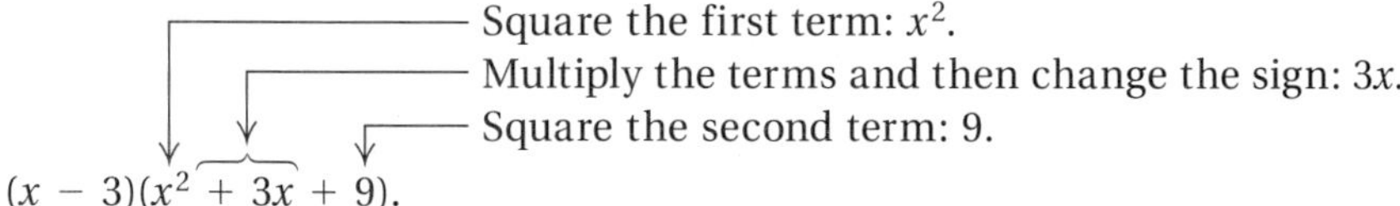

Note that we cannot factor $x^2 + 3x + 9$. It is not a trinomial square nor can it be factored by trial and error.

Do Exercises 1 and 2.

Factor.

1. $x^3 - 8$

2. $64 - y^3$

Example 2 Factor: $125x^3 + y^3$.

We have

$$125x^3 + y^3 = (5x)^3 + y^3.$$

In one set of parentheses, we write the cube root of the first term, $5x$. Then we write a plus sign, and then the cube root of the second term, y:

$$(5x + y)(\qquad\qquad).$$

Answers on page A-20

Factor.

3. $27x^3 + y^3$

4. $8y^3 + z^3$

Factor.

5. $m^6 - n^6$

6. $16x^7y + 54xy^7$

7. $729x^6 - 64y^6$

8. $x^3 - 0.027$

Answers on page A-20

To get the next factor, we think of $5x + y$ and do the following:

Square the first term: $25x^2$.
Multiply the terms and then change the sign: $-5xy$.
Square the second term: y^2.

$(5x + y)(25x^2 - 5xy + y^2)$.

Do Exercises 3 and 4.

Example 3 Factor: $128y^7 - 250x^6y$.

We first look for a common factor:

$$128y^7 - 250x^6y = 2y(64y^6 - 125x^6) = 2y[(4y^2)^3 - (5x^2)^3]$$
$$= 2y(4y^2 - 5x^2)(16y^4 + 20x^2y^2 + 25x^4).$$

Example 4 Factor: $a^6 - b^6$.

We can express this polynomial as a difference of squares:

$$(a^3)^2 - (b^3)^2.$$

We factor as follows:

$$a^6 - b^6 = (a^3 + b^3)(a^3 - b^3).$$

One factor is a sum of two cubes, and the other factor is a difference of two cubes. We factor them:

$$(a + b)(a^2 - ab + b^2)(a - b)(a^2 + ab + b^2).$$

We have now factored completely.

In Example 4, had we thought of factoring first as a difference of two cubes, we would have had

$$(a^2)^3 - (b^2)^3 = (a^2 - b^2)(a^4 + a^2b^2 + b^4)$$
$$= (a + b)(a - b)(a^4 + a^2b^2 + b^4).$$

In this case, we might have missed some factors; $a^4 + a^2b^2 + b^4$ can be factored as $(a^2 - ab + b^2)(a^2 + ab + b^2)$, but we probably would not have known to do such factoring.

Example 5 Factor: $64a^6 - 729b^6$.

We have

$$64a^6 - 729b^6 = (8a^3 - 27b^3)(8a^3 + 27b^3) \quad \text{Factoring a difference of squares}$$
$$= [(2a)^3 - (3b)^3][(2a)^3 + (3b)^3].$$

Each factor is a sum or a difference of cubes. We factor each:

$$= (2a - 3b)(4a^2 + 6ab + 9b^2)(2a + 3b)(4a^2 - 6ab + 9b^2).$$

Sum of cubes:	$A^3 + B^3 = (A + B)(A^2 - AB + B^2)$;
Difference of cubes:	$A^3 - B^3 = (A - B)(A^2 + AB + B^2)$;
Difference of squares:	$A^2 - B^2 = (A + B)(A - B)$;
Sum of squares:	$A^2 + B^2$ cannot be factored using real numbers if the largest common factor has been removed.

Do Exercises 5–8.

Exercise Set 5.6

a Factor.

1. $z^3 + 27$

2. $a^3 + 8$

3. $x^3 - 1$

4. $c^3 - 64$

5. $y^3 + 125$

6. $x^3 + 1$

7. $8a^3 + 1$

8. $27x^3 + 1$

9. $y^3 - 8$

10. $p^3 - 27$

11. $8 - 27b^3$

12. $64 - 125x^3$

13. $64y^3 + 1$

14. $125x^3 + 1$

15. $8x^3 + 27$

16. $27y^3 + 64$

17. $a^3 - b^3$

18. $x^3 - y^3$

19. $a^3 + \frac{1}{8}$

20. $b^3 + \frac{1}{27}$

21. $2y^3 - 128$

22. $3z^3 - 3$

23. $24a^3 + 3$

24. $54x^3 + 2$

25. $rs^3 + 64r$

26. $ab^3 + 125a$

27. $5x^3 - 40z^3$

28. $2y^3 - 54z^3$

29. $x^3 + 0.001$

30. $y^3 + 0.125$

31. $64x^6 - 8t^6$

32. $125c^6 - 8d^6$

33. $2y^4 - 128y$

34. $3z^5 - 3z^2$

35. $z^6 - 1$

36. $t^6 + 1$

37. $t^6 + 64y^6$

38. $p^6 - q^6$

Synthesis

Consider these polynomials:

$(a + b)^3$; $a^3 + b^3$; $(a + b)(a^2 - ab + b^2)$;

$(a + b)(a^2 + ab + b^2)$; $(a + b)(a + b)(a + b)$.

39. Evaluate each polynomial for $a = -2$ and $b = 3$.

40. Evaluate each polynomial for $a = 4$ and $b = -1$.

Factor. Assume that variables in exponents represent natural numbers.

41. $x^{6a} + y^{3b}$

42. $a^3x^3 - b^3y^3$

43. $3x^{3a} + 24y^{3b}$

44. $\frac{8}{27}x^3 + \frac{1}{64}y^3$

45. $\frac{1}{24}x^3y^3 + \frac{1}{3}z^3$

46. $7x^3 - \frac{7}{8}$

47. $(x + y)^3 - x^3$

48. $(1 - x)^3 + (x - 1)^6$

49. $(a + 2)^3 - (a - 2)^3$

50. $y^4 - 8y^3 - y + 8$

5.7 Factoring: A General Strategy

Objective

[a] Factor polynomials completely using any of the methods considered in this chapter.

For Extra Help

TAPE 9 | MAC WIN | CD-ROM

[a] We now combine all of our factoring techniques and consider a general strategy for factoring polynomials. Here we will encounter polynomials of all the types we have considered, in random order, so you will have the opportunity to determine which method to use.

> To factor a polynomial:
>
> a) Always look first for a common factor. If there is one, factor out the largest common factor.
>
> b) Then look at the number of terms.
>
> *Two terms*: Try factoring as a difference of squares first. Next, try factoring as a sum or a difference of cubes. Do *not* try to factor a *sum* of squares: $A^2 + B^2$.
>
> *Three terms*: Determine whether the trinomial is a square. If it is, you know how to factor. If not, try trial and error, using FOIL or grouping.
>
> *Four terms*: Try factoring by grouping.
>
> c) *Always factor completely.* If a factor with more than one term can still be factored, you should factor it. When no factor can be factored further, you have finished.

Example 1 Factor: $5t^4 - 80$.

a) We look for a common factor:

$$5t^4 - 80 = 5(t^4 - 16).$$

b) The factor $t^4 - 16$ has only two terms. It is a difference of squares: $(t^2)^2 - 4^2$. We factor it, being careful to include the common factor:

$$5(t^2 + 4)(t^2 - 4).$$

c) We see that one of the factors is again a difference of squares. We factor it:

$$5(t^2 + 4)(t + 2)(t - 2).$$

↑ This is a sum of squares. It cannot be factored!

We have factored completely because no factor with more than one term can be factored further.

Example 2 Factor: $2x^3 + 10x^2 + x + 5$.

a) We look for a common factor. There isn't one.

b) There are four terms. We try factoring by grouping:

$2x^3 + 10x^2 + x + 5$

$= (2x^3 + 10x^2) + (x + 5)$ Separating into two binomials

$= 2x^2(x + 5) + 1(x + 5)$ Factoring each binomial

$= (2x^2 + 1)(x + 5).$ Factoring out the common factor $x + 5$

c) None of these factors can be factored further, so we have factored completely.

Factor.

1. $3m^4 - 3$

2. $x^6 + 8x^3 + 16$

3. $2x^4 + 8x^3 + 6x^2$

4. $3x^3 + 12x^2 - 2x - 8$

5. $8x^3 - 200x$

Answers on page A-20

Example 3 Factor: $x^5 - 2x^4 - 35x^3$.

a) We look first for a common factor. This time there is one, x^3:

$$x^5 - 2x^4 - 35x^3 = x^3(x^2 - 2x - 35).$$

b) The factor $x^2 - 2x - 35$ has three terms, but it is not a trinomial square. We factor it using trial and error (FOIL or grouping):

$$x^5 - 2x^4 - 35x^3 = x^3(x^2 - 2x - 35) = x^3(x - 7)(x + 5).$$

Don't forget to include the common factor in the final answer!

c) No factor with more than one term can be factored further, so we have factored completely.

Example 4 Factor: $x^4 - 10x^2 + 25$.

a) We look first for a common factor. There isn't one.

b) There are three terms. We see that this polynomial is a trinomial square. We factor it:

$$x^4 - 10x^2 + 25 = (x^2)^2 - 2 \cdot x^2 \cdot 5 + 5^2 = (x^2 - 5)^2.$$

c) Since $x^2 - 5$ cannot be factored further, we have factored completely.

Do Exercises 1–5.

Example 5 Factor: $6x^2y^4 - 21x^3y^5 + 3x^2y^6$.

a) We look first for a common factor:

$$6x^2y^4 - 21x^3y^5 + 3x^2y^6 = 3x^2y^4(2 - 7xy + y^2).$$

b) There are three terms in $2 - 7xy + y^2$. We determine whether the trinomial is a square. Since only y^2 is a square, we do not have a trinomial square. Can the trinomial be factored by trial and error? A key to the answer is that x is only in the term $-7xy$. The polynomial might be in a form like $(1 - y)(2 + y)$, but there would be no x in the middle term. Thus, $2 - 7xy + y^2$ cannot be factored.

c) Have we factored completely? Yes, because no factor with more than one term can be factored further.

Example 6 Factor: $(p + q)(x + 2) + (p + q)(x + y)$.

a) We look for a common factor:

$$\begin{aligned}(p + q)(x + 2) + (p + q)(x + y) &= (p + q)[(x + 2) + (x + y)]\\ &= (p + q)(2x + y + 2).\end{aligned}$$

b) There are three terms in $2x + y + 2$, but this trinomial cannot be factored further.

c) Neither factor can be factored further, so we have factored completely.

Example 7 Factor: $px + py + qx + qy$.

a) We look first for a common factor. There isn't one.

b) There are four terms. We try factoring by grouping:

$$\begin{aligned} px + py + qx + qy &= p(x + y) + q(x + y) \\ &= (p + q)(x + y). \end{aligned}$$

c) Have we factored completely? Since neither factor can be factored further, we have factored completely.

Example 8 Factor: $25x^2 + 20xy + 4y^2$.

a) We look first for a common factor. There isn't one.

b) There are three terms. We determine whether the trinomial is a square. The first term and the last term are squares:

$$25x^2 = (5x)^2 \quad \text{and} \quad 4y^2 = (2y)^2.$$

Since twice the product of $5x$ and $2y$ is the other term,

$$2 \cdot 5x \cdot 2y = 20xy,$$

the trinomial is a perfect square.

We factor by writing the square roots of the square terms and the sign of the middle term:

$$25x^2 + 20xy + 4y^2 = (5x + 2y)^2.$$

We can check by squaring $5x + 2y$.

c) Since $5x + 2y$ cannot be factored further, we have factored completely.

Example 9 Factor: $p^2q^2 + 7pq + 12$.

a) We look first for a common factor. There isn't one.

b) There are three terms. We determine whether the trinomial is a square. The first term is a square, but neither of the other terms is a square, so we do not have a trinomial square. We use the trial-and-error or grouping method, thinking of the product pq as a single variable. We consider this possibility for factorization:

$$(pq + \square)(pq + \square).$$

We factor the last term, 12. All the signs are positive, so we consider only positive factors. Possibilities are 1, 12 and 2, 6 and 3, 4. The pair 3, 4 gives a sum of 7 for the coefficient of the middle term. Thus,

$$p^2q^2 + 7pq + 12 = (pq + 3)(pq + 4).$$

c) No factor with more than one term can be factored further, so we have factored completely.

Factor.

6. $x^4y^2 + 2x^3y + 3x^2y$

7. $10p^6q^2 + 4p^5q^3 + 2p^4q^4$

8. $(a - b)(x + 5) + (a - b)(x + y^2)$

9. $ax^2 + ay + bx^2 + by$

10. $x^4 + 2x^2y^2 + y^4$

11. $x^2y^2 + 5xy + 4$

12. $p^4 - 81q^4$

13. $15a^3 - 120b^3$

Answers on page A-20

Example 10 Factor: $8x^4 - 20x^2y - 12y^2$.

a) We look first for a common factor:

$$8x^4 - 20x^2y - 12y^2 = 4(2x^4 - 5x^2y - 3y^2).$$

b) There are three terms in $2x^4 - 5x^2y - 3y^2$. We determine whether the trinomial is a square. Since none of the terms is a square, we do not have a trinomial square. We factor $2x^4$. Possibilities are $2x^2$, x^2 and $2x$, x^3 and others. We also factor the last term, $-3y^2$. Possibilities are $3y$, $-y$ and $-3y$, y and others. We look for factors such that the sum of their products is the middle term. We try some possibilities:

$$(2x - y)(x^3 + 3y) = 2x^4 + 6xy - x^3y - 3y^2,$$
$$(2x^2 - y)(x^2 + 3y) = 2x^4 + 5x^2y - 3y^2,$$
$$(2x^2 + y)(x^2 - 3y) = 2x^4 - 5x^2y - 3y^2.$$

c) No factor with more than one term can be factored further, so we have factored completely. The factorization, including the common factor, is

$$4(2x^2 + y)(x^2 - 3y).$$

Example 11 Factor: $a^4 - 16b^4$.

a) We look first for a common factor. There isn't one.

b) There are two terms. Since $a^4 = (a^2)^2$ and $16b^4 = (4b^2)^2$, we see that we do have a difference of squares. Thus,

$$a^4 - 16b^4 = (a^2 + 4b^2)(a^2 - 4b^2).$$

c) The last factor can be factored further. It is also a difference of squares. Thus,

$$a^4 - 16b^4 = (a^2 + 4b^2)(a + 2b)(a - 2b).$$

Example 12 Factor: $40t^3 - 5s^3$.

a) We look first for a common factor:

$$40t^3 - 5s^3 = 5(8t^3 - s^3).$$

b) The factor $8t^3 - s^3$ has only two terms. It is a difference of two cubes. We factor as follows:

$$(2t - s)(4t^2 + 2ts + s^2).$$

c) No factor with more than one term can be factored further, so we have factored completely. The factorization, including the common factor, is

$$5(2t - s)(4t^2 + 2ts + s^2).$$

Do Exercises 6–13.

Exercise Set 5.7

a Factor completely.

1. $3x^2 - 192$

2. $2t^2 - 18$

3. $a^2 + 25 - 10a$

4. $y^2 + 49 + 14y$

5. $2x^2 - 11x + 12$

6. $8y^2 - 18y - 5$

7. $x^3 + 24x^2 + 144x$

8. $x^3 - 18x^2 + 81x$

9. $x^3 + 3x^2 - 4x - 12$

10. $x^3 - 5x^2 - 25x + 125$

11. $48x^2 - 3$

12. $50x^2 - 32$

13. $9x^3 + 12x^2 - 45x$

14. $20x^3 - 4x^2 - 72x$

15. $x^2 + 4$

16. $t^2 + 25$

17. $x^4 + 7x^2 - 3x^3 - 21x$

18. $m^4 + 8m^3 + 8m^2 + 64m$

19. $x^5 - 14x^4 + 49x^3$

20. $2x^6 + 8x^5 + 8x^4$

21. $20 - 6x - 2x^2$

22. $45 - 3x - 6x^2$

23. $x^2 - 6x + 1$

24. $x^2 + 8x + 5$

25. $4x^4 - 64$

26. $5x^5 - 80x$

27. $1 - y^8$

28. $t^8 - 1$

29. $x^5 - 4x^4 + 3x^3$

30. $x^6 - 2x^5 + 7x^4$

31. $\frac{1}{81}x^6 - \frac{8}{27}x^3 + \frac{16}{9}$

32. $36a^2 - 15a + \frac{25}{16}$

33. $27w^3 + 1000z^3$

34. $125a^3 - 8b^3$

35. $9x^2y^2 - 36xy$

36. $x^2y - xy^2$

37. $2\pi rh + 2\pi r^2$

38. $10p^4q^4 + 35p^3q^3 + 10p^2q^2$

39. $(a + b)(x - 3) + (a + b)(x + 4)$

40. $5c(a^3 + b) - (a^3 + b)$

41. $(x - 1)(x + 1) - y(x + 1)$

42. $3(p - q) - q^2(p - q)$

43. $n^2 + 2n + np + 2p$

44. $a^2 - 3a + ay - 3y$

45. $6q^2 - 3q + 2pq - p$

46. $2x^2 - 4x + xy - 2y$

47. $4b^2 + a^2 - 4ab$

48. $x^2 + y^2 - 2xy$

49. $16x^2 + 24xy + 9y^2$

50. $9c^2 + 6cd + d^2$

51. $49m^4 - 112m^2n + 64n^2$

52. $4x^2y^2 + 12xyz + 9z^2$

53. $y^4 + 10y^2z^2 + 25z^4$

54. $0.01x^4 - 0.1x^2y^2 + 0.25y^4$

55. $\frac{1}{4}a^2 + \frac{1}{3}ab + \frac{1}{9}b^2$

56. $4p^2q + pq^2 + 4p^3$

57. $a^2 - ab - 2b^2$

58. $3b^2 - 17ab - 6a^2$

59. $2mn - 360n^2 + m^2$

60. $15 + x^2y^2 + 8xy$

61. $m^2n^2 - 4mn - 32$

62. $p^2q^2 + 7pq + 6$

63. $a^2b^6 + 4ab^5 - 32b^4$

64. $p^5q^2 + 3p^4q - 10p^3$

65. $a^5 + 4a^4b - 5a^3b^2$

66. $2s^6t^2 + 10s^3t^3 + 12t^4$

67. $a^2 - \frac{1}{25}b^2$

68. $p^2 - \frac{1}{49}b^2$

69. $7x^6 - 7y^6$

70. $16p^3 + 54q^3$

71. $16 - p^4q^4$

72. $15a^4 - 15b^4$

73. $1 - 16x^{12}y^{12}$

74. $81a^4 - b^4$

75. $q^3 + 8q^2 - q - 8$

76. $m^3 - 7m^2 - 4m + 28$

77. $112xy + 49x^2 + 64y^2$

78. $4ab^5 - 32b^4 + a^2b^6$

Skill Maintenance

Sports-Car Sales. The sales of sports cars rise and fall over the years due often to new or redesigned models, such as the 1997 Corvette. Sales for recent years are shown in the following line graph. Use the graph for Exercises 79–84. [3.1a]

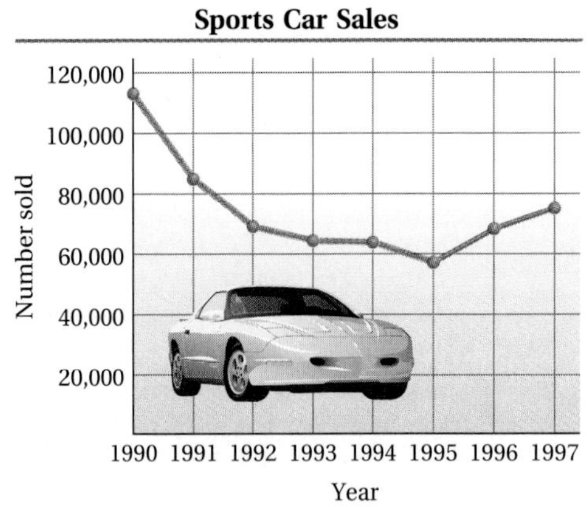

Source: Autodata

79. In which year were sports-car sales the highest?

80. In which year were sports-car sales the lowest?

81. In which year were sports-car sales about 68,000?

82. What were sports-car sales in 1997?

83. By how much did sales increase from 1995 to 1997?

84. By how much did sales decrease from 1990 to 1995?

85. Divide: $\frac{7}{5} \div \left(-\frac{11}{10}\right)$. [1.6c]

86. Multiply: $(5x - t)^2$. [4.6d]

87. Solve $A = aX + bX - 7$ for X. [2.6a]

88. Solve: $4(x - 9) - 2(x + 7) < 14$. [2.7e]

Synthesis

89. ◆ Kelly factored $16 - 8x + x^2$ as $(x - 4)^2$, while Tony factored it as $(4 - x)^2$. Evaluate each expression for several values of x. Then explain why both answers are correct.

90. ◆ Describe in your own words a strategy that can be used to factor polynomials.

Factor completely.

91. $a^4 - 2a^2 + 1$

92. $x^4 + 9$

93. $12.25x^2 - 7x + 1$

94. $\frac{1}{5}x^2 - x + \frac{4}{5}$

95. $5x^2 + 13x + 7.2$

96. $x^3 - (x - 3x^2) - 3$

97. $18 + y^3 - 9y - 2y^2$

98. $-(x^4 - 7x^2 - 18)$

99. $a^3 + 4a^2 + a + 4$

100. $x^3 + x^2 - (4x + 4)$

101. $x^4 - 7x^2 - 18$

102. $3x^4 - 15x^2 + 12$

103. $x^3 - x^2 - 4x + 4$

104. $y^2(y + 1) - 4y(y + 1) - 21(y + 1)$

105. $y^2(y - 1) - 2y(y - 1) + (y - 1)$

106. $6(x - 1)^2 + 7y(x - 1) - 3y^2$

107. $(y + 4)^2 + 2x(y + 4) + x^2$

108. $a^4 - 81$

Create polynomials for factoring.

Collaborative Learning Manual

5.8 Solving Quadratic Equations by Factoring

In this section, we introduce a new equation-solving method and use it along with factoring to solve certain equations like $x^2 + x - 156 = 0$.

Objectives

a Solve equations (already factored) using the principle of zero products.

b Solve quadratic equations by factoring and then using the principle of zero products.

For Extra Help

TAPE 9

MAC WIN

CD-ROM

> A **quadratic equation** is an equation equivalent to an equation of the type
>
> $$ax^2 + bx + c = 0, \quad \text{where } a > 0.$$
>
> The trinomial on the left is of second degree.

a The Principle of Zero Products

The product of two numbers is 0 if one or both of the numbers is 0. Furthermore, *if any product is* 0, *then a factor must be* 0. For example:

If $7x = 0$, then we know that $x = 0$.

If $x(2x - 9) = 0$, then we know that $x = 0$ or $2x - 9 = 0$.

If $(x + 3)(x - 2) = 0$, then we know that $x + 3 = 0$ or $x - 2 = 0$.

In a product such as $ab = 24$, we cannot conclude with certainty that a is 24 or that b is 24, but if $ab = 0$, we can conclude that $a = 0$ or $b = 0$.

Example 1 Solve: $(x + 3)(x - 2) = 0$.

We have a product of 0. This equation will be true when either factor is 0. Thus it is true when

$$x + 3 = 0 \quad \text{or} \quad x - 2 = 0.$$

Here we have two simple equations that we know how to solve:

$$x = -3 \quad \text{or} \quad x = 2.$$

Each of the numbers -3 and 2 is a solution of the original equation, as we can see in the following checks.

CHECK: For -3:

$$\begin{array}{r|l} (x + 3)(x - 2) & 0 \\ \hline (-3 + 3)(-3 - 2) & 0 \\ 0(-5) & \\ 0 & \end{array} \quad \text{TRUE}$$

For 2:

$$\begin{array}{r|l} (x + 3)(x - 2) & 0 \\ \hline (2 + 3)(2 - 2) & 0 \\ 5(0) & \\ 0 & \end{array} \quad \text{TRUE}$$

We now have a principle to help in solving quadratic equations.

> **THE PRINCIPLE OF ZERO PRODUCTS**
>
> An equation $ab = 0$ is true if and only if $a = 0$ is true or $b = 0$ is true, or both are true. (A product is 0 if and only if one or both of the factors is 0.)

Solve using the principle of zero products.

1. $(x - 3)(x + 4) = 0$

2. $(x - 7)(x - 3) = 0$

3. $(4t + 1)(3t - 2) = 0$

4. Solve: $y(3y - 17) = 0$.

Answers on page A-21

Calculator Spotlight

Checking Solutions on a Grapher. We can check solutions using the TABLE or the CALC feature. Consider the equation $(x + 3)(x - 2) = 0$ of Example 1. We enter $y_1 = (x + 3)(x - 2)$ into the grapher.

Table. First, we set up the table with TblStart $= -4$ and ΔTbl $= 1$. By scrolling through the table, we see that when $x = -3$, $y = 0$ and when $x = 2$, $y = 0$.

X	Y1	
−4	6	
−3	0	
−2	−4	
−1	−6	
0	−6	
1	−4	
2	0	

X = −4

Calc. We graph the equation $y = (x + 3)(x - 2)$ and use the CALC and VALUE features to check the solutions to the equation $(x + 3)(x - 2) = 0$. By entering -3 for x, we see that when $x = -3$, $y = 0$ (that is, when $x = -3$, the graph crosses the x-axis). It can also be shown that when $x = 2$, $y = 0$.

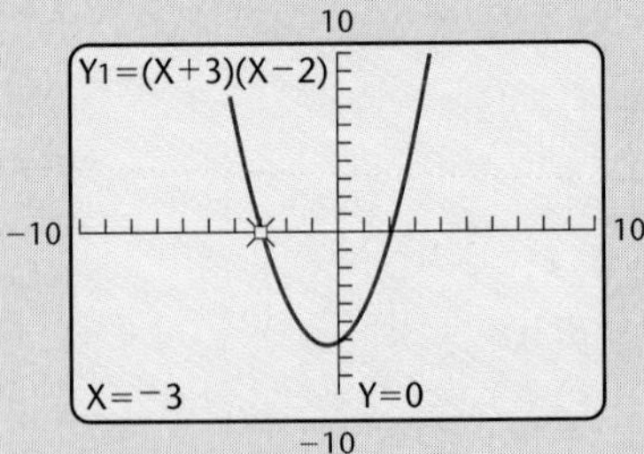

Exercises

1. Check the solutions of the equations in Examples 1–3.

Example 2 Solve: $(5x + 1)(x - 7) = 0$.

$$(5x + 1)(x - 7) = 0$$

$$5x + 1 = 0 \quad or \quad x - 7 = 0 \qquad \text{Using the principle of zero products}$$

$$5x = -1 \quad or \quad x = 7 \qquad \text{Solving the two equations separately}$$

$$x = -\tfrac{1}{5} \quad or \quad x = 7$$

CHECK: For $-\frac{1}{5}$:

$$\frac{(5x + 1)(x - 7) = 0}{\left(5\left(-\tfrac{1}{5}\right) + 1\right)\left(-\tfrac{1}{5} - 7\right) \; ? \; 0}$$

$$(-1 + 1)\left(-7\tfrac{1}{5}\right)$$

$$0\left(-7\tfrac{1}{5}\right)$$

$$0 \qquad \text{TRUE}$$

For 7:

$$\frac{(5x + 1)(x - 7) = 0}{(5(7) + 1)(7 - 7) \; ? \; 0}$$

$$(35 + 1) \cdot 0$$

$$36 \cdot 0$$

$$0 \qquad \text{TRUE}$$

The solutions are $-\frac{1}{5}$ and 7.

When you solve an equation using the principle of zero products, you may wish to check by substitution, as in Examples 1 and 2. Such a check will detect errors in solving.

Do Exercises 1–3 on the preceding page.

When some factors have only one term, you can still use the principle of zero products.

Example 3 Solve: $x(2x - 9) = 0$.

$$x(2x - 9) = 0$$

$$x = 0 \quad or \quad 2x - 9 = 0 \qquad \text{Using the principle of zero products}$$

$$x = 0 \quad or \quad 2x = 9$$

$$x = 0 \quad or \quad x = \frac{9}{2}$$

The solutions are 0 and $\frac{9}{2}$. The check is left to the student.

Do Exercise 4 on the preceding page.

b Using Factoring to Solve Equations

Using factoring and the principle of zero products, we can solve some new kinds of equations. Thus we have extended our equation-solving abilities.

Example 4 Solve: $x^2 + 5x + 6 = 0$.

Compare this equation to those that we know how to solve from Chapter 2. There are no like terms to collect, and we have a squared term. We first factor the polynomial. Then we use the principle of zero products.

$$x^2 + 5x + 6 = 0$$

$$(x + 2)(x + 3) = 0 \qquad \text{Factoring}$$

$$x + 2 = 0 \quad or \quad x + 3 = 0 \qquad \text{Using the principle of zero products}$$

$$x = -2 \quad or \quad x = -3$$

CHECK: For −2:

$$\begin{array}{c|c} x^2 + 5x + 6 & 0 \\ \hline (-2)^2 + 5(-2) + 6 & ?\ 0 \\ 4 - 10 + 6 & \\ -6 + 6 & \\ 0 & \end{array} \quad \text{TRUE}$$

For −3:

$$\begin{array}{c|c} x^2 + 5x + 6 & 0 \\ \hline (-3)^2 + 5(-3) + 6 & ?\ 0 \\ 9 - 15 + 6 & \\ -6 + 6 & \\ 0 & \end{array} \quad \text{TRUE}$$

The solutions are −2 and −3.

CAUTION! Keep in mind that you *must* have 0 on one side of the equation before you can use the principle of zero products. Get all nonzero terms on one side and 0 on the other.

Do Exercise 5.

Example 5 Solve: $x^2 - 8x = -16$.

We first add 16 to get a 0 on one side:

$$x^2 - 8x = -16$$

$$x^2 - 8x + 16 = 0 \quad \text{Adding 16}$$

$$(x - 4)(x - 4) = 0 \quad \text{Factoring}$$

$$x - 4 = 0 \quad or \quad x - 4 = 0 \quad \text{Using the principle of zero products}$$

$$x = 4 \quad or \quad x = 4$$

There is only one solution, 4. The check is left to the student.

Do Exercises 6 and 7.

Example 6 Solve: $x^2 + 5x = 0$.

$$x^2 + 5x = 0$$

$$x(x + 5) = 0 \quad \text{Factoring out a common factor}$$

$$x = 0 \quad or \quad x + 5 = 0 \quad \text{Using the principle of zero products}$$

$$x = 0 \quad or \quad x = -5$$

The solutions are 0 and −5. The check is left to the student.

Example 7 Solve: $4x^2 = 25$.

$$4x^2 = 25$$

$$4x^2 - 25 = 0 \quad \text{Subtracting 25 on both sides to get 0 on one side}$$

$$(2x - 5)(2x + 5) = 0 \quad \text{Factoring a difference of squares}$$

$$2x - 5 = 0 \quad or \quad 2x + 5 = 0$$

$$2x = 5 \quad or \quad 2x = -5$$

$$x = \frac{5}{2} \quad or \quad x = -\frac{5}{2}$$

The solutions are $\frac{5}{2}$ and $-\frac{5}{2}$.

Do Exercises 8 and 9.

5. Solve: $x^2 - x - 6 = 0$.

Solve.

6. $x^2 - 3x = 28$

7. $x^2 = 6x - 9$

Solve.

8. $x^2 - 4x = 0$

9. $9x^2 = 16$

Answers on page A-21

10. Solve: $(x + 1)(x - 1) = 8$.

11. Find the x-intercepts of the graph shown below.

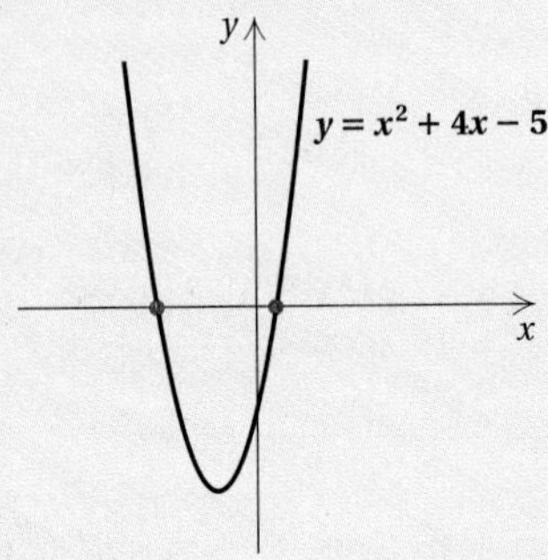

12. Use *only* the graph shown below to solve $3x - x^2 = 0$.

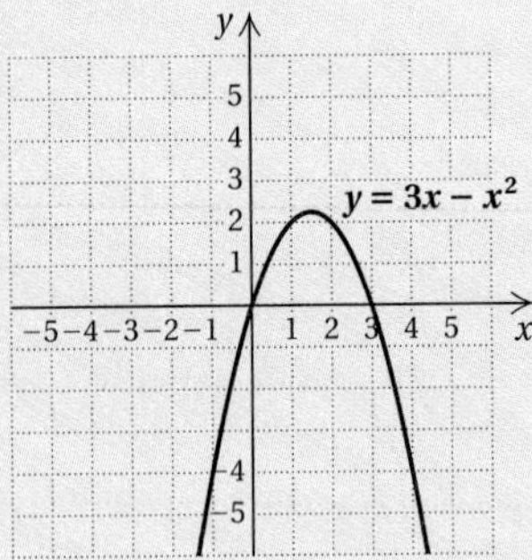

Answer on page A-21

Example 8 Solve: $(x + 2)(x - 2) = 5$.

Be careful with an equation like this one! It might be tempting to set each factor equal to 5. Remember: We must have a 0 on one side. We first carry out the product on the left. Then we subtract 5 on both sides to get 0 on one side. Then we proceed with the principle of zero products.

$$(x + 2)(x - 2) = 5$$

$$x^2 - 4 = 5 \qquad \text{Multiplying on the left}$$

$$x^2 - 4 - 5 = 5 - 5 \qquad \text{Subtracting 5}$$

$$x^2 - 9 = 0 \qquad \text{Simplifying}$$

$$(x + 3)(x - 3) = 0 \qquad \text{Factoring}$$

$$x + 3 = 0 \quad or \quad x - 3 = 0 \qquad \text{Using the principle of zero products}$$

$$x = -3 \quad or \quad x = 3$$

The solutions are -3 and 3. The check is left to the student.

Do Exercise 10.

AG Algebraic–Graphical Connection

In Chapter 3, we graphed linear equations of the type $y = mx + b$ and $Ax + By = C$. Recall that to find the x-intercept, we replaced y with 0 and solved for x. This procedure can also be used to find the x-intercepts when an equation of the form $y = ax^2 + bx + c$, $a \neq 0$, is to be graphed. Although the details of creating such graphs will be left to Chapter 11, we consider them briefly here from the standpoint of finding the x-intercepts. The graphs are shaped like the following curves. Note that each x-intercept represents a solution of $ax^2 + bx + c = 0$.

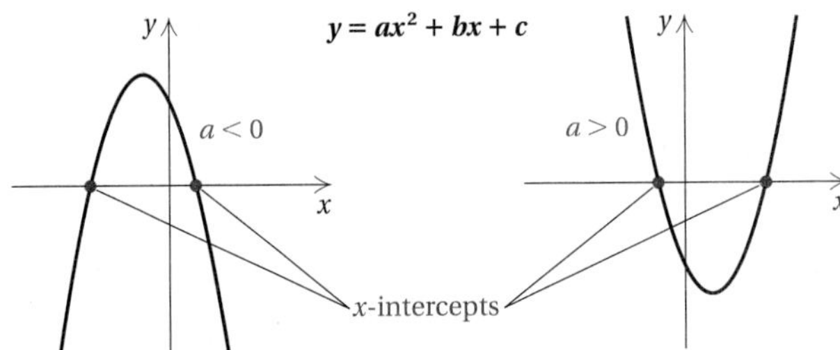

Example 9 Find the x-intercepts of the graph of $y = x^2 - 4x - 5$ shown at right.

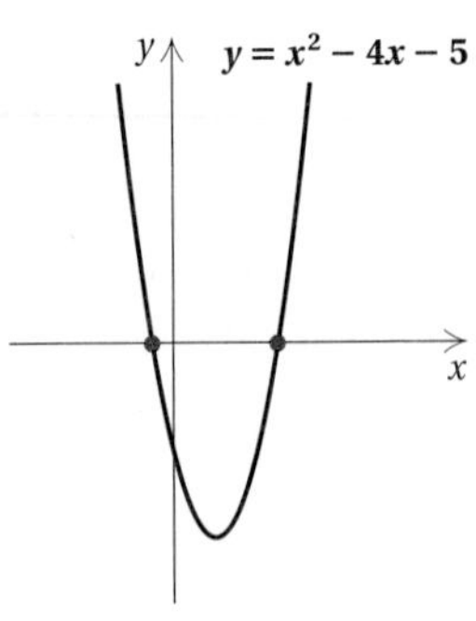

To find the x-intercepts, we let $y = 0$ and solve for x:

$$0 = x^2 - 4x - 5 \qquad \text{Substituting 0 for } y$$

$$0 = (x - 5)(x + 1) \qquad \text{Factoring}$$

$$x - 5 = 0 \quad or \quad x + 1 = 0 \qquad \text{Using the principle of zero products}$$

$$x = 5 \quad or \quad x = -1.$$

The x-intercepts are $(5, 0)$ and $(-1, 0)$.

Do Exercises 11 and 12.

Exercise Set 5.8

a Solve using the principle of zero products.

1. $(x + 4)(x + 9) = 0$

2. $(x + 2)(x - 7) = 0$

3. $(x + 3)(x - 8) = 0$

4. $(x + 6)(x - 8) = 0$

5. $(x + 12)(x - 11) = 0$

6. $(x - 13)(x + 53) = 0$

7. $x(x + 3) = 0$

8. $y(y + 5) = 0$

9. $0 = y(y + 18)$

10. $0 = x(x - 19)$

11. $(2x + 5)(x + 4) = 0$

12. $(2x + 9)(x + 8) = 0$

13. $(5x + 1)(4x - 12) = 0$

14. $(4x + 9)(14x - 7) = 0$

15. $(7x - 28)(28x - 7) = 0$

16. $(13x + 14)(6x - 5) = 0$

17. $2x(3x - 2) = 0$

18. $55x(8x - 9) = 0$

19. $\left(\frac{1}{5} + 2x\right)\left(\frac{1}{9} - 3x\right) = 0$

20. $\left(\frac{7}{4}x - \frac{1}{16}\right)\left(\frac{2}{3}x - \frac{16}{15}\right) = 0$

21. $(0.3x - 0.1)(0.05x + 1) = 0$

22. $(0.1x + 0.3)(0.4x - 20) = 0$

23. $9x(3x - 2)(2x - 1) = 0$

24. $(x + 5)(x - 75)(5x - 1) = 0$

b Solve by factoring and using the principle of zero products. Remember to check.

25. $x^2 + 6x + 5 = 0$

26. $x^2 + 7x + 6 = 0$

27. $x^2 + 7x - 18 = 0$

28. $x^2 + 4x - 21 = 0$

29. $x^2 - 8x + 15 = 0$

30. $x^2 - 9x + 14 = 0$

31. $x^2 - 8x = 0$

32. $x^2 - 3x = 0$

33. $x^2 + 18x = 0$

34. $x^2 + 16x = 0$

35. $x^2 = 16$

36. $100 = x^2$

37. $9x^2 - 4 = 0$

38. $4x^2 - 9 = 0$

39. $0 = 6x + x^2 + 9$

40. $0 = 25 + x^2 + 10x$

41. $x^2 + 16 = 8x$

42. $1 + x^2 = 2x$

43. $5x^2 = 6x$

44. $7x^2 = 8x$

45. $6x^2 - 4x = 10$

46. $3x^2 - 7x = 20$

47. $12y^2 - 5y = 2$

48. $2y^2 + 12y = -10$

49. $t(3t + 1) = 2$

50. $x(x - 5) = 14$

51. $100y^2 = 49$

52. $64a^2 = 81$

53. $x^2 - 5x = 18 + 2x$

54. $3x^2 + 8x = 9 + 2x$

55. $10x^2 - 23x + 12 = 0$

56. $12x^2 + 17x - 5 = 0$

Find the x-intercepts for the graph of the equation.

57.

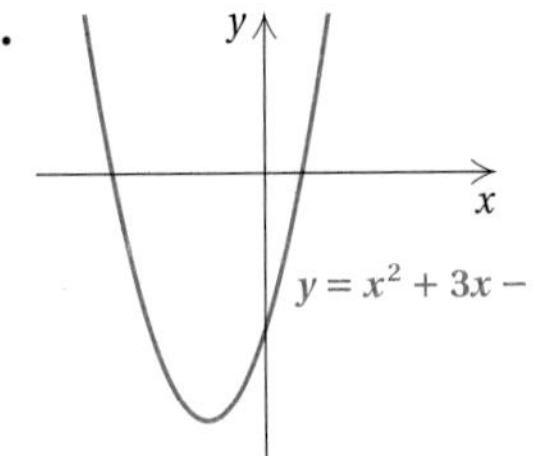

58.

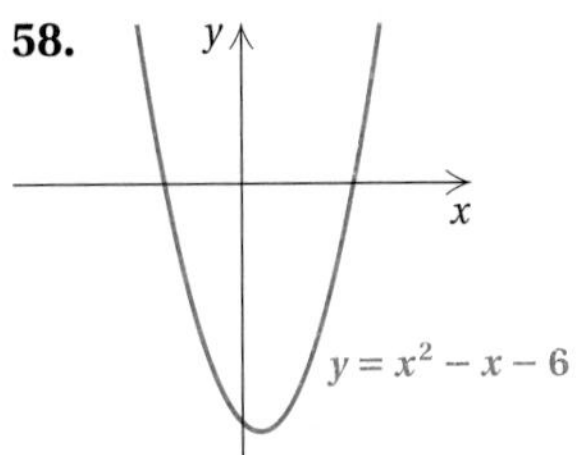

59.

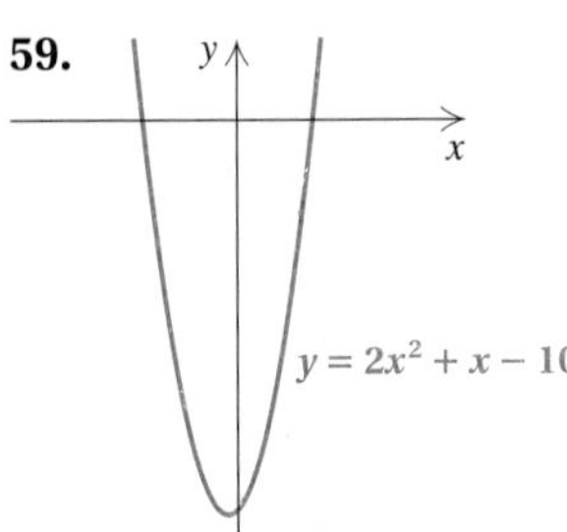

60.

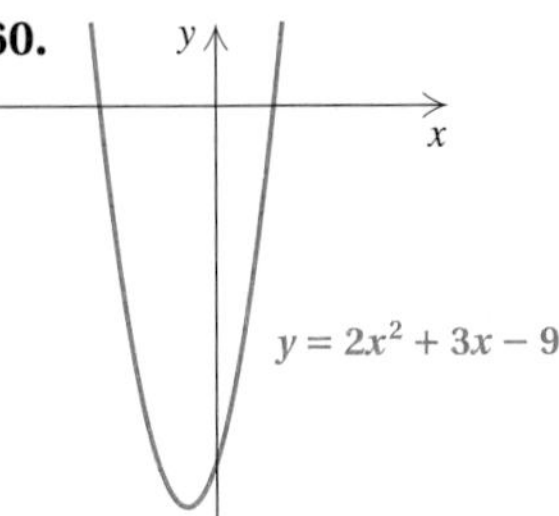

Skill Maintenance

Translate to an algebraic expression. [1.1b]

61. The square of the sum of a and b

62. The sum of the squares of a and b

Divide. [1.6c]

63. $144 \div (-9)$

64. $-24.3 \div 5.4$

65. $-\frac{5}{8} \div \frac{3}{16}$

66. $-\frac{3}{16} \div \left(-\frac{5}{8}\right)$

Synthesis

67. ◆ What is wrong with the following? Explain the correct method of solution.

$$(x - 3)(x + 4) = 8$$
$$x - 3 = 8 \quad or \quad x + 4 = 8$$
$$x = 11 \quad or \quad x = 4$$

68. ◆ What is incorrect about solving $x^2 = 3x$ by dividing by x on both sides?

Solve.

69. $b(b + 9) = 4(5 + 2b)$

70. $y(y + 8) = 16(y - 1)$

71. $(t - 3)^2 = 36$

72. $(t - 5)^2 = 2(5 - t)$

73. $x^2 - \frac{1}{64} = 0$

74. $x^2 - \frac{25}{36} = 0$

75. $\frac{5}{16}x^2 = 5$

76. $\frac{27}{25}x^2 = \frac{1}{3}$

77. Find an equation that has the given numbers as solutions. For example, 3 and -2 are solutions to $x^2 - x - 6 = 0$.

a) $-3, 4$ b) $-3, -4$ c) $\frac{1}{2}, \frac{1}{2}$ d) $5, -5$ e) $0, 0.1, \frac{1}{4}$

Create and solve quadratic equations.

5.9 Applications and Problem Solving

a We can now use our new method for solving quadratic equations and the five steps for solving problems.

Objective

a Solve applied problems involving quadratic equations that can be solved by factoring.

For Extra Help

TAPE 9 MAC WIN CD-ROM

Example 1 One more than a number times one less than the number is 8. Find all such numbers.

1. **Familiarize.** Let's make a guess. Try 5. One more than 5 is 6. One less than the number is 4. The product of one more than the number and one less than the number is 6(4), or 24, which is too large. We could continue to guess, but let's use our algebraic skills to find the numbers. Let $x =$ the number (there could be more than one).
2. **Translate.** From the familiarization, we can translate as follows:

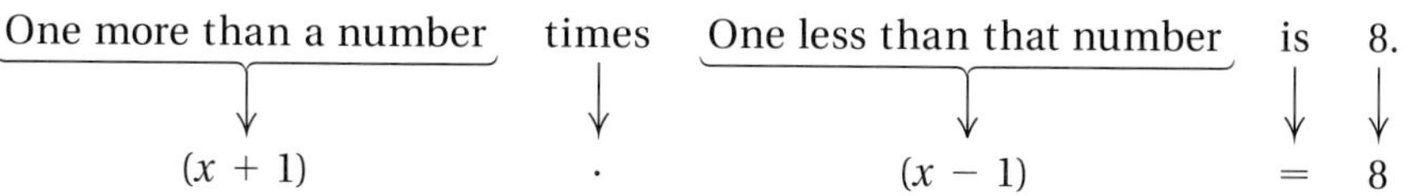

3. **Solve.** We solve the equation as follows:

$$(x + 1)(x - 1) = 8$$
$$x^2 - 1 = 8 \quad \text{Multiplying}$$
$$x^2 - 1 - 8 = 8 - 8 \quad \text{Subtracting 8 on both sides to get 0 on one side}$$
$$x^2 - 9 = 0 \quad \text{Simplifying}$$
$$(x - 3)(x + 3) = 0 \quad \text{Factoring}$$
$$x - 3 = 0 \quad or \quad x + 3 = 0 \quad \text{Using the principle of zero products}$$
$$x = 3 \quad or \quad x = -3.$$

4. **Check.** One more than 3 (this is 4) times one less than 3 (this is 2) is 8. Thus, 3 checks. One more than -3 (this is -2) times one less than -3 (this is -4) is 8. Thus, -3 also checks.
5. **State.** There are two such numbers, 3 and -3.

Do Exercises 1 and 2.

1. One more than a number times one less than the number is 24. Find all such numbers.

2. Seven less than a number times eight less than the number is 0. Find all such numbers.

Example 2 The square of a number minus twice the number is 48. Find all such numbers.

1. **Familiarize.** Let's make a guess to help understand the problem and ease the translation. Try 6. The square of 6 is 36, and twice the number is 12. Then $36 - 12 = 24$, so 6 is not a number we want. We find the numbers using our algebraic skills. Let $x =$ the number, or numbers.
2. **Translate.** We translate as follows:

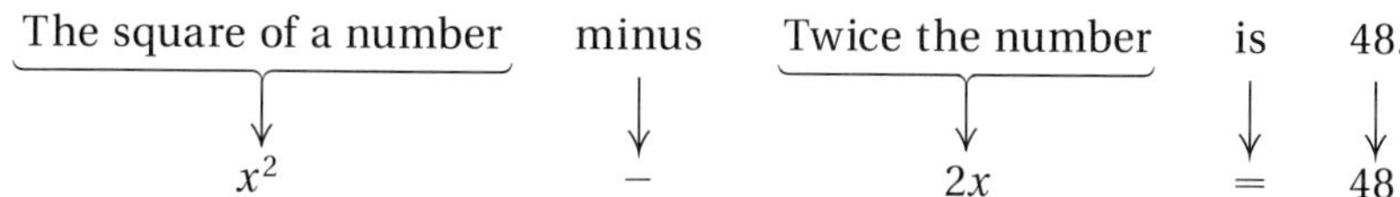

3. **Solve.** We solve the equation as follows:

$$x^2 - 2x = 48$$
$$x^2 - 2x - 48 = 48 - 48 \quad \text{Subtracting 48 to get 0 on one side}$$
$$x^2 - 2x - 48 = 0 \quad \text{Simplifying}$$
$$(x - 8)(x + 6) = 0 \quad \text{Factoring}$$
$$x - 8 = 0 \quad or \quad x + 6 = 0 \quad \text{Using the principle of zero products}$$
$$x = 8 \quad or \quad x = -6.$$

Answers on page A-21

3. The square of a number minus the number is 20. Find all such numbers.

4. The width of a rectangle is 2 cm less than the length. The area is 15 cm^2. Find the length and the width.

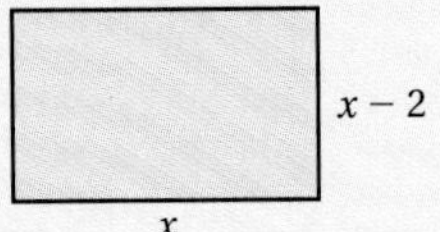

Answers on page A-21

4. **Check.** The square of 8 is 64, and twice the number 8 is 16. Then $64 - 16$ is 48, so 8 checks. The square of -6 is $(-6)^2$, or 36, and twice -6 is -12. Then $36 - (-12)$ is 48, so -6 checks.

5. **State.** There are two such numbers, 8 and -6.

Do Exercise 3.

Example 3 *Sailing.* The height of a triangular foresail on a racing yacht is 7 ft more than the base. The area of the triangle is 30 ft^2. Find the height and the base.

1. **Familiarize.** We first make a drawing. If you don't remember the formula for the area of a triangle, look it up at the back of the book or in a geometry book. The area is $\frac{1}{2}$(base)(height).

   We let b = the base of the triangle. Then $b + 7$ = the height.

2. **Translate.** It helps to reword this problem before translating:

$\frac{1}{2}$	times	Base	times	Height	is	30.	Rewording
↓	↓	↓	↓	↓	↓	↓	
$\frac{1}{2}$	$\cdot$	b	$\cdot$	$(b + 7)$	$=$	30	Translating

3. **Solve.** We solve the equation as follows:

$$\frac{1}{2} \cdot b \cdot (b + 7) = 30$$

$$\frac{1}{2}(b^2 + 7b) = 30 \quad \text{Multiplying}$$

$$2 \cdot \frac{1}{2}(b^2 + 7b) = 2 \cdot 30 \quad \text{Multiplying by 2}$$

$$b^2 + 7b = 60 \quad \text{Simplifying}$$

$$b^2 + 7b - 60 = 60 - 60 \quad \text{Subtracting 60 to get 0 on one side}$$

$$b^2 + 7b - 60 = 0$$

$$(b + 12)(b - 5) = 0 \quad \text{Factoring}$$

$$b + 12 = 0 \quad or \quad b - 5 = 0 \quad \text{Using the principle of zero products}$$

$$b = -12 \quad or \quad b = 5.$$

4. **Check.** The base of a triangle cannot have a negative length, so -12 cannot be a solution. Suppose the base is 5 ft. Then the height is 7 ft more than the base, so the height is 12 ft and the area is $\frac{1}{2}(5)(12)$, or 30 ft^2. These numbers check in the original problem.

5. **State.** The height is 12 ft and the base is 5 ft.

Do Exercise 4.

Example 4 *Games in a Sports League.* In a sports league of n teams in which each team plays every other team twice, the total number N of games to be played is given by

$$n^2 - n = N.$$

If a basketball league plays a total of 240 games, how many teams are in the league?

1., 2. **Familiarize** and **Translate.** We are given that $n =$ the number of teams in a league and $N =$ the number of games. To familiarize yourself with this problem, reread Example 3 in Section 4.3 where we first considered it. To find the number of teams n in a league in which 240 games are played, we substitute 240 for N in the equation:

$$n^2 - n = 240. \quad \text{Substituting 240 for } N$$

3. **Solve.** We solve the equation as follows:

$$
\begin{aligned}
n^2 - n &= 240 \\
n^2 - n - 240 &= 240 - 240 && \text{Subtracting 240 to get 0 on one side} \\
n^2 - n - 240 &= 0 \\
(n - 16)(n + 15) &= 0 && \text{Factoring} \\
n - 16 = 0 \quad &or \quad n + 15 = 0 && \text{Using the principle of zero products} \\
n = 16 \quad &or \quad n = -15.
\end{aligned}
$$

4. **Check.** The solutions of the equation are 16 and -15. Since the number of teams cannot be negative, -15 cannot be a solution. But 16 checks, since $16^2 - 16 = 256 - 16 = 240$.

5. **State.** There are 16 teams in the league.

Do Exercise 5.

Example 5 The product of the numbers of two consecutive entrants in a marathon race is 156. Find the numbers.

1. **Familiarize.** The numbers are consecutive integers. Recall that consecutive integers are next to each other, such as 49 and 50, or -6 and -5. Let $x =$ the smaller integer; then $x + 1 =$ the larger integer.

2. **Translate.** It helps to reword the problem before translating:

First integer	times	Second integer	is	156.	Rewording
↓	↓	↓	↓	↓	
x	$\cdot$	$(x + 1)$	$=$	156	Translating

3. **Solve.** We solve the equation as follows:

$$
\begin{aligned}
x(x + 1) &= 156 \\
x^2 + x &= 156 && \text{Multiplying} \\
x^2 + x - 156 &= 156 - 156 && \text{Subtracting 156 to get 0 on one side} \\
x^2 + x - 156 &= 0 && \text{Simplifying} \\
(x - 12)(x + 13) &= 0 && \text{Factoring} \\
x - 12 = 0 \quad &or \quad x + 13 = 0 && \text{Using the principle of zero products} \\
x = 12 \quad &or \quad x = -13.
\end{aligned}
$$

4. **Check.** The solutions of the equation are 12 and -13. When x is 12, then $x + 1$ is 13, and $12 \cdot 13 = 156$. The numbers 12 and 13 are consecutive integers that are solutions to the problem. When x is -13, then $x + 1$ is -12, and $(-13)(-12) = 156$. The numbers -13 and -12 are also consecutive integers, but they are not solutions of the problem because negative numbers are not used as entry numbers.

5. **State.** The entry numbers are 12 and 13.

5. Use $N = n^2 - n$ for the following.

a) *Volleyball League.* A women's volleyball league has 19 teams. What is the total number of games to be played?

b) *Softball League.* A slow-pitch softball league plays a total of 72 games. How many teams are in the league?

Answers on page A-21

6. The product of the page numbers on two facing pages of a book is 506. Find the page numbers.

7. The length of one leg of a right triangle is 1 m longer than the other. The length of the hypotenuse is 5 m. Find the lengths of the legs.

Answers on page A-21

Do Exercise 6.

The following example involves the Pythagorean theorem, which relates the lengths of the sides of a right triangle. A **right triangle** has a 90° angle. The side opposite the 90° angle is called the **hypotenuse**. The other sides are called **legs**.

THE PYTHAGOREAN THEOREM

The sum of the squares of the legs of a right triangle is equal to the square of the hypotenuse:

$$a^2 + b^2 = c^2.$$

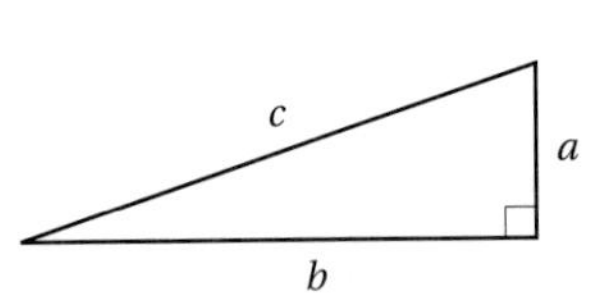

Example 6 *Ladder Settings.* A ladder of length 13 ft is placed against a building in such a way that the distance from the top of the ladder to the ground is 7 ft more than the distance from the bottom of the ladder to the building. Find both distances.

1. **Familiarize.** We first make a drawing. The ladder and the missing distances form the hypotenuse and legs of a right triangle. We let x = the length of the side (leg) across the bottom. Then $x + 7$ = the length of the other side (leg). The hypotenuse has length 13 ft.

2. **Translate.** Since a right triangle is formed, we can use the Pythagorean theorem:

$$a^2 + b^2 = c^2$$
$$x^2 + (x + 7)^2 = 13^2. \quad \text{Substituting}$$

3. **Solve.** We solve the equation as follows:

$$x^2 + (x^2 + 14x + 49) = 169 \quad \text{Squaring the binomial and 13}$$
$$2x^2 + 14x + 49 = 169 \quad \text{Collecting like terms}$$
$$2x^2 + 14x + 49 - 169 = 169 - 169 \quad \text{Subtracting 169 to get 0 on one side}$$
$$2x^2 + 14x - 120 = 0 \quad \text{Simplifying}$$
$$2(x^2 + 7x - 60) = 0 \quad \text{Factoring out a common factor}$$
$$x^2 + 7x - 60 = 0 \quad \text{Dividing by 2}$$
$$(x + 12)(x - 5) = 0 \quad \text{Factoring}$$
$$x + 12 = 0 \quad or \quad x - 5 = 0$$
$$x = -12 \quad or \quad x = 5.$$

4. **Check.** The negative integer −12 cannot be the length of a side. When $x = 5$, $x + 7 = 12$, and $5^2 + 12^2 = 13^2$. So 5 and 12 check.

5. **State.** The distance from the top of the ladder to the ground is 12 ft. The distance from the bottom of the ladder to the building is 5 ft.

Do Exercise 7.

Exercise Set 5.9

a Solve.

1. If 7 is added to the square of a number, the result is 32. Find all such numbers.

2. If you subtract a number from four times its square, the result is 3. Find all such numbers.

3. Fifteen more than the square of a number is eight times the number. Find all such numbers.

4. Eight more than the square of a number is six times the number. Find all such numbers.

5. *Calculator Dimensions.* The length of a rectangular calculator is 5 cm greater than the width. The area of the calculator is 84 cm^2. Find the length and the width.

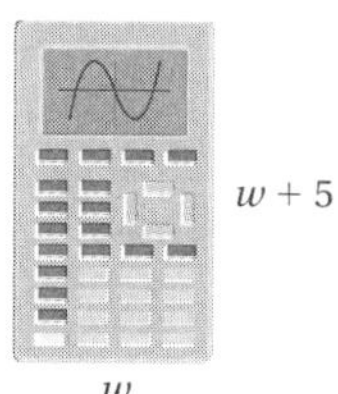

6. *Garden Dimensions.* The length of a rectangular garden is 4 m greater than the width. The area of the garden is 96 m^2. Find the length and the width.

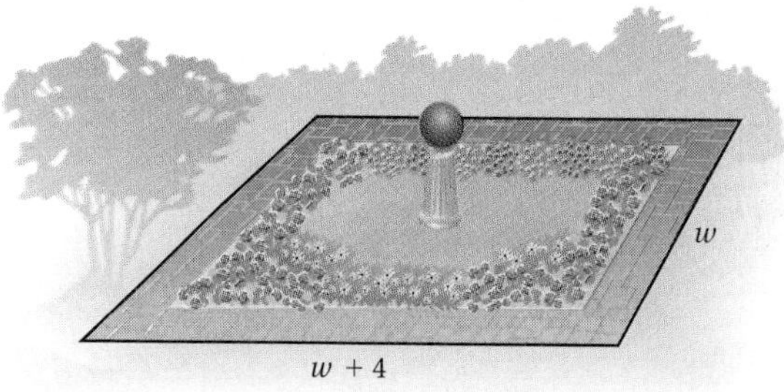

7. *Consecutive Page Numbers.* The product of the page numbers on two facing pages of a book is 210. Find the page numbers.

8. *Consecutive Page Numbers.* The product of the page numbers on two facing pages of a book is 420. Find the page numbers.

9. The product of two consecutive even integers is 168. Find the integers.

10. The product of two consecutive even integers is 224. Find the integers.

11. The product of two consecutive odd integers is 255. Find the integers.

12. The product of two consecutive odd integers is 143. Find the integers.

13. The area of a square bookcase is 5 more than the perimeter. Find the length of a side.

14. The perimeter of a square porch is 3 more than the area. Find the length of a side.

15. *Sharks' Teeth.* Sharks' teeth are shaped like triangles. The height of a tooth of a great white shark is 1 cm longer than the base. The area is 15 cm^2. Find the height and the base.

16. The base of a triangle is 6 cm greater than twice the height. The area is 28 cm^2. Find the height and the base.

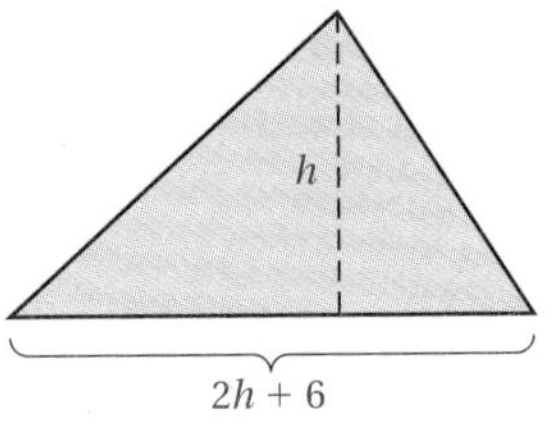

17. If the sides of a square are lengthened by 3 km, the area becomes 81 km^2. Find the length of a side of the original square.

18. The base and the height of a triangle are the same length. If the length of the base is increased by 4 in., the area becomes 96 in^2. Find the length of the base of the original triangle.

Rocket Launch. A model water rocket is launched with an initial velocity of 180 ft/sec. Its height h, in feet, after t seconds is given by the formula

$$h = 180t - 16t^2.$$

Use this formula for Exercises 19 and 20.

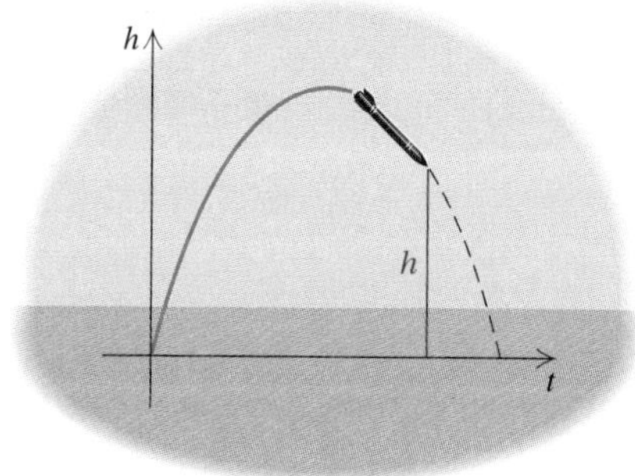

19. After how many seconds will the rocket first reach a height of 464 ft?

20. After how many seconds will the rocket again be at that same height of 464 ft? (See Exercise 19.)

21. The sum of the squares of two consecutive odd positive integers is 74. Find the integers.

22. The sum of the squares of two consecutive odd positive integers is 130. Find the integers.

Games in a League. Use $n^2 - n = N$ for Exercises 23–26.

23. A chess league has 14 teams. What is the total number of games to be played?

24. A women's volleyball league has 23 teams. What is the total number of games to be played?

25. A slow-pitch softball league plays a total of 132 games. How many teams are in the league?

26. A basketball league plays a total of 90 games. How many teams are in the league?

Handshakes. A researcher wants to investigate the potential spread of germs by contact. She knows that the number of possible handshakes within a group of n people is given by

$$N = \frac{1}{2}(n^2 - n).$$

27. There are 100 people at a party. How many handshakes are possible?

28. There are 40 people at a meeting. How many handshakes are possible?

29. Everyone at a meeting shook hands. There were 300 handshakes in all. How many people were at the meeting?

30. Everyone at a party shook hands. There were 190 handshakes in all. How many people were at the party?

31. The length of one leg of a right triangle is 8 ft. The length of the hypotenuse is 2 ft longer than the other leg. Find the length of the hypotenuse and the other leg.

32. The length of one leg of a right triangle is 24 ft. The length of the other leg is 16 ft shorter than the hypotenuse. Find the length of the hypotenuse and the other leg.

Skill Maintenance

Multiply. [4.6d], [4.7f]

33. $(3x - 5y)(3x + 5y)$

34. $(3x - 5y)^2$

35. $(3x + 5y)^2$

36. $(3x - 5y)(2x + 7y)$

Find the intercepts of the equation. [3.3a]

37. $4x - 16y = 64$

38. $4x + 16y = 64$

39. $x - 1.3y = 6.5$

40. $\frac{2}{3}x + \frac{5}{8}y = \frac{5}{12}$

Synthesis

41. ◆ Write a problem in which a quadratic equation must be solved.

42. ◆ Write a problem for a classmate to solve such that only one of the two solutions of a quadratic equation can be used as an answer.

43. A cement walk of constant width is built around a 20-ft by 40-ft rectangular pool. The total area of the pool and the walk is 1500 ft^2. Find the width of the walk.

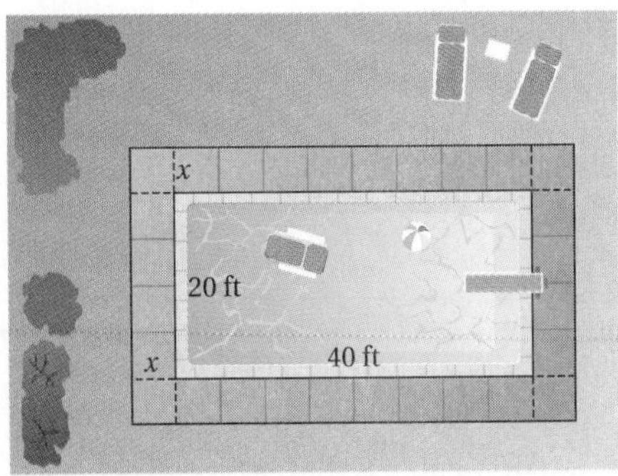

44. An open rectangular gutter is made by turning up the sides of a piece of metal 20 in. wide. The area of the cross-section of the gutter is 50 in^2. Find the depth of the gutter.

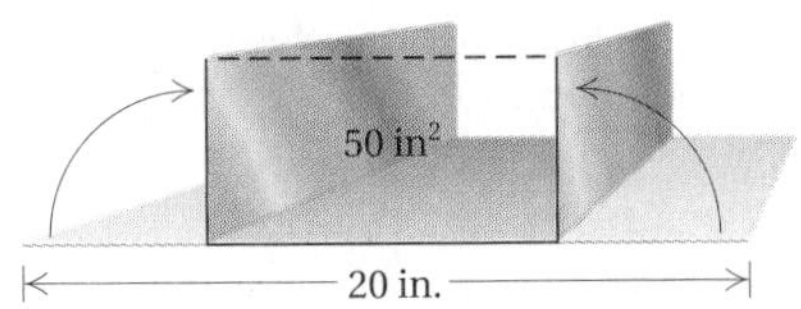

45. The ones digit of a number less than 100 is 4 greater than the tens digit. The sum of the number and the product of the digits is 58. Find the number.

46. The total surface area of a closed box is 350 m^2. The box is 9 m high and has a square base and lid. Find the length of the side of the base.

47. A rectangular piece of cardboard is twice as long as it is wide. A 4-cm square is cut out of each corner, and the sides are turned up to make a box with an open top. The volume of the box is 616 cm^3. Find the original dimensions of the cardboard.

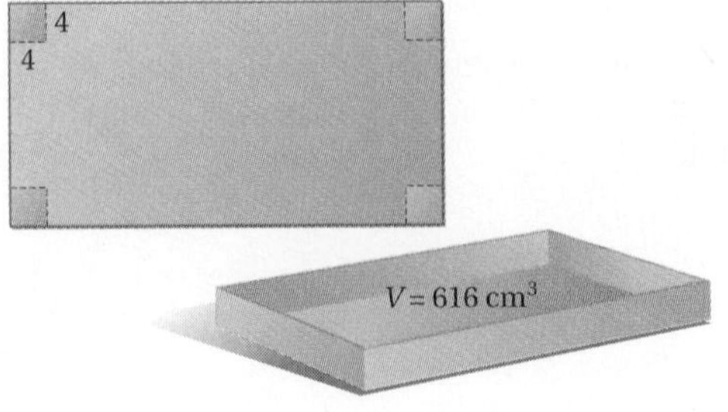

Summary and Review Exercises: Chapter 5

Important Properties and Formulas

Factoring Formulas:

$A^2 - B^2 = (A + B)(A - B)$, $A^3 + B^3 = (A + B)(A^2 - AB + B^2)$,
$A^2 + 2AB + B^2 = (A + B)^2$, $A^3 - B^3 = (A - B)(A^2 + AB + B^2)$
$A^2 - 2AB + B^2 = (A - B)^2$,

The Principle of Zero Products: **An equation $ab = 0$ is true if and only if $a = 0$ is true or $b = 0$ is true, or both are true.**

The objectives to be tested in addition to the material in this chapter are [1.6c], [2.7e], [3.3a], and [4.6d].

Find three factorizations of the monomial. [5.1a]

1. $-10x^2$

2. $36x^5$

Factor completely. [5.7a]

3. $5 - 20x^6$

4. $x^2 - 3x$

5. $9x^2 - 4$

6. $x^2 + 4x - 12$

7. $x^2 + 14x + 49$

8. $6x^3 + 12x^2 + 3x$

9. $x^3 + x^2 + 3x + 3$

10. $6x^2 - 5x + 1$

11. $x^4 - 81$

12. $9x^3 + 12x^2 - 45x$

13. $2x^2 - 50$

14. $x^4 + 4x^3 - 2x - 8$

15. $16x^4 - 1$

16. $8x^6 - 32x^5 + 4x^4$

17. $75 + 12x^2 + 60x$

18. $x^2 + 9$

19. $x^3 - x^2 - 30x$

20. $8x^3 - 125$

21. $9x^2 + 25 - 30x$

22. $6x^2 - 28x - 48$

23. $x^2 - 6x + 9$

24. $2x^2 - 7x - 4$

25. $18x^2 - 12x + 2$

26. $3x^2 - 27$

27. $15 - 8x + x^2$

28. $25x^2 - 20x + 4$

29. $49b^{10} + 4a^8 - 28a^4b^5$

30. $x^2y^2 + xy - 12$

31. $12a^2 + 84ab + 147b^2$

32. $m^2 + 5m + mt + 5t$

33. $32x^4 - 128y^4z^4$

Solve. [5.8a], [5.8b]

34. $(x - 1)(x + 3) = 0$

35. $x^2 + 2x - 35 = 0$

36. $x^2 + x - 12 = 0$

37. $3x^2 + 2 = 5x$

38. $2x^2 + 5x = 12$

39. $16 = x(x - 6)$

Solve. [5.9a]

40. The square of a number is 6 more than the number. Find all such numbers.

41. The product of two consecutive even integers is 288. Find the integers.

42. Twice the square of a number is 10 more than the number. Find all such numbers.

43. The product of two consecutive odd integers is 323. Find the integers.

44. *House Plan.* An architect has allocated a rectangular space of 264 ft^2 for a square dining room and a 10-ft wide kitchen. Find the dimensions of each room.

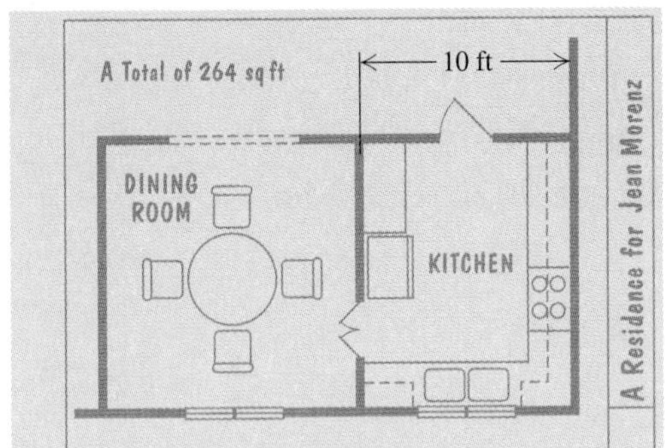

45. *Antenna Guy Wire.* The guy wires for a television antenna are 1 m longer than the height of the antenna. The guy wires are anchored 3 m from the foot of the antenna. How tall is the antenna?

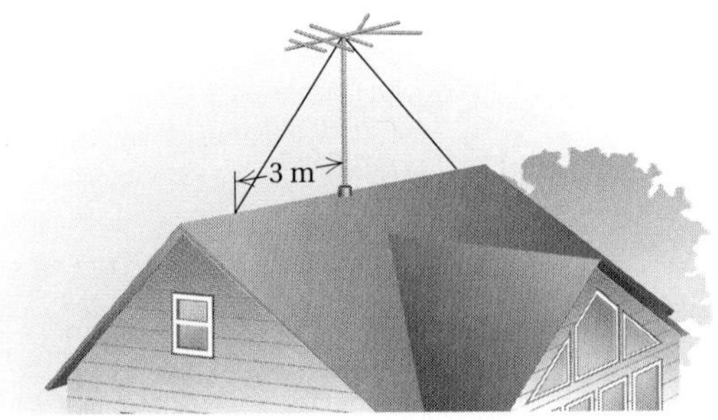

Find the x-intercepts for the graph of the equation. [5.8b]

46. $y = x^2 + 9x + 20$

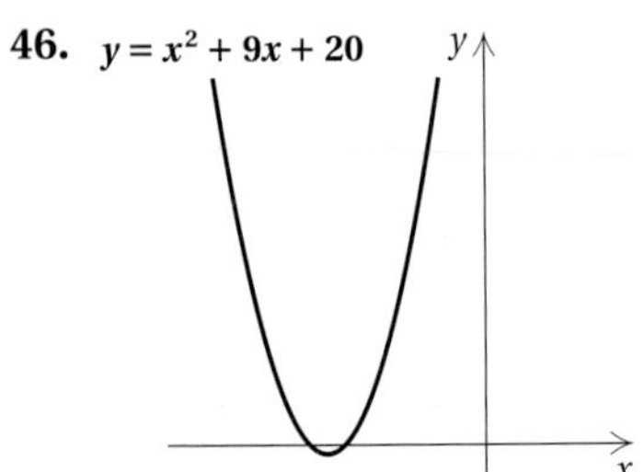

47. $y = 2x^2 - 7x - 15$

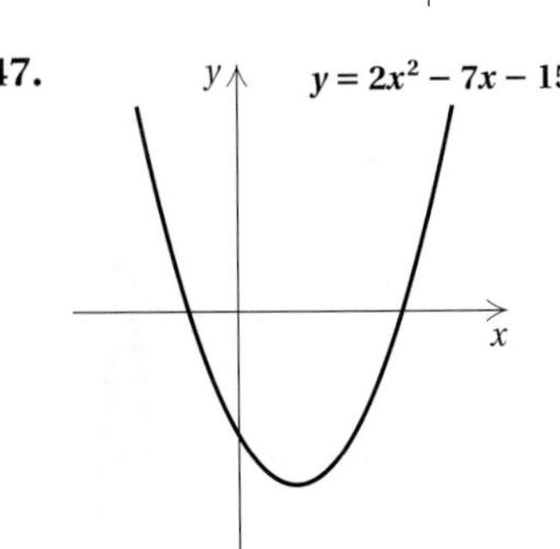

Skill Maintenance

48. Divide: $-\frac{12}{25} \div \left(-\frac{21}{10}\right)$. [1.6c]

49. Solve: $20 - (3x + 2) \geq 2(x + 5) + x$. [2.7e]

50. Multiply: $(2a - 3)(2a + 3)$. [4.6d]

51. Find the intercepts. Then graph the equation. [3.3a]

$3y - 4x = -12$

Synthesis

52. ◆ Compare the types of equations that we are able to solve after having studied this chapter with those we have studied earlier. [2.3a, b, c], [5.8a, b]

53. ◆ Describe as many procedures as you can for checking the result of factoring a polynomial. [5.3a], [5.8a]

Solve. [5.9a]

54. The pages of a book measure 15 cm by 20 cm. Margins of equal width surround the printing on each page and constitute one-half of the area of the page. Find the width of the margins.

55. The cube of a number is the same as twice the square of the number. Find all such numbers.

56. The length of a rectangle is two times its width. When the length is increased by 20 and the width decreased by 1, the area is 160. Find the original length and width.

Solve. [5.8b]

57. $x^2 + 25 = 0$

58. $(x - 2)(x + 3)(2x - 5) = 0$

59. For each equation in group A, find an equivalent equation in group B. [5.7a]

A. a) $3x^2 - 4x + 8 = 0$
b) $(x - 6)(x + 3) = 0$
c) $x^2 + 2x + 9 = 0$
d) $(2x - 5)(x + 4) = 0$
e) $5x^2 - 5 = 0$
f) $x^2 + 10x - 2 = 0$

B. g) $4x^2 + 8x + 36 = 0$
h) $(2x + 8)(2x - 5) = 0$
i) $9x^2 - 12x + 24 = 0$
j) $(x + 1)(5x - 5) = 0$
k) $x^2 - 3x - 18 = 0$
l) $2x^2 + 20x - 4 = 0$

60. Which is greater, $2^{90} + 2^{90}$, or 2^{100}? Why? [5.1b]

Test: Chapter 5

1. Find three factorizations of $4x^3$.

Factor completely.

2. $x^2 - 7x + 10$

3. $x^2 + 25 - 10x$

4. $6y^2 - 8y^3 + 4y^4$

5. $x^3 + x^2 + 2x + 2$

6. $x^2 - 5x$

7. $x^3 + 2x^2 - 3x$

8. $28x - 48 + 10x^2$

9. $4x^2 - 9$

10. $x^2 - x - 12$

11. $6m^3 + 9m^2 + 3m$

12. $3w^2 - 75$

13. $60x + 45x^2 + 20$

14. $3x^4 - 48$

15. $49x^2 - 84x + 36$

16. $5x^2 - 26x + 5$

17. $x^4 + 2x^3 - 3x - 6$

18. $80 - 5x^4$

19. $4x^2 - 4x - 15$

20. $6t^3 + 9t^2 - 15t$

21. $3m^2 - 9mn - 30n^2$

22. $1000a^3 - 27b^3$

Answers

1. ________
2. ________
3. ________
4. ________
5. ________
6. ________
7. ________
8. ________
9. ________
10. ________
11. ________
12. ________
13. ________
14. ________
15. ________
16. ________
17. ________
18. ________
19. ________
20. ________
21. ________
22. ________

Answers

23. ______

24. ______

25. ______

26. ______

27. ______

28. ______

29. ______

30. ______

31. ______

32. ______

33. ______

34. ______

35. ______

36. ______

37. ______

Solve.

23. $x^2 - x - 20 = 0$

24. $2x^2 + 7x = 15$

25. $x(x - 3) = 28$

Solve.

26. The square of a number is 24 more than five times the number. Find all such numbers.

27. *Dimensions of a Sail.* The height of the jib sail on a Lightning sailboat is 5 ft greater than the length of its "foot." If the area of the sail is 42 ft^2, find the length of the foot and the height of the sail.

Find the x-intercepts for the graph of the equation.

28.

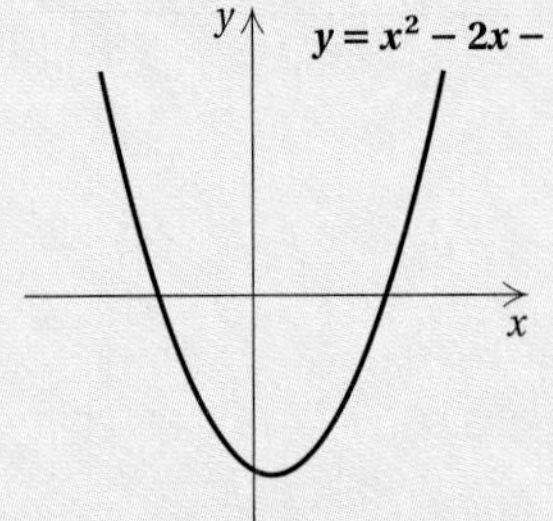

29.

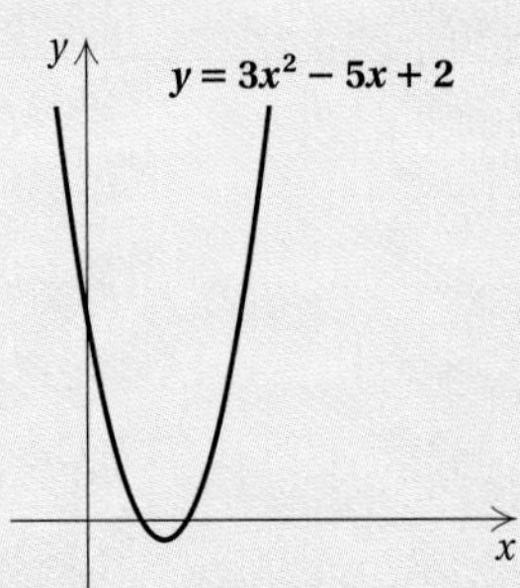

Skill Maintenance

30. Divide: $\frac{5}{8} \div \left(-\frac{11}{16}\right)$.

31. Solve: $10(x - 3) < 4(x + 2)$.

32. Find the intercepts. Then graph the equation.
$2y - 5x = 10$

33. Multiply: $(5x^2 - 7)^2$.

Synthesis

34. The length of a rectangle is five times its width. When the length is decreased by 3 and the width is increased by 2, the area of the new rectangle is 60. Find the original length and width.

35. Factor: $(a + 3)^2 - 2(a + 3) - 35$.

36. If $x^2 - 4 = (14)(18)$, then one possibility for x is which of the following?

a) 12 b) 14
c) 16 d) 18

37. If $x + y = 4$ and $x - y = 6$, then $x^2 - y^2 = ?$

a) 2 b) 10
c) 34 d) 24

6

Rational Expressions and Equations

An Application

Erin and Nick work as volunteers at a community recycling depot. Erin can sort a morning's accumulation of recyclables in 4 hr, while Nick requires 6 hr to do the same job. How long would it take them, working together, to sort the recyclables?

This problem appears as Example 3 in Section 6.6.

The Mathematics

We let t = the time it takes them, working together, to sort the recyclables. The problem then translates to the equation

These are *rational expressions.*

$$\frac{t}{4} + \frac{t}{6} = 1.$$

This is a *rational equation.*

For more information, visit us at www.mathmax.com

Introduction

In this chapter, we learn to manipulate rational expressions. We learn to simplify rational expressions as well as to add, subtract, multiply, and divide them. Then we use these skills to solve equations, formulas, and applied problems.

Pretest: Chapter 6

1. Find the LCM of $x^2 + 5x + 6$ and $x^2 + 6x + 9$.

Perform the indicated operations and simplify.

2. $\dfrac{b-1}{2-b} + \dfrac{b^2-3}{b^2-4}$

3. $\dfrac{4y-4}{y^2-y-2} - \dfrac{3y-5}{y^2-y-2}$

4. $\dfrac{4}{a+2} + \dfrac{3}{a}$

5. $\dfrac{x}{x+1} - \dfrac{x}{x-1} + \dfrac{2x^2}{x^2-1}$

6. $\dfrac{4x+8}{x+1} \cdot \dfrac{x^2-2x-3}{2x^2-8}$

7. $\dfrac{x+3}{x^2-9} \div \dfrac{x+3}{x^2-6x+9}$

8. Simplify: $\dfrac{\frac{1}{x}+\frac{1}{y}}{\frac{1}{x}-\frac{1}{y}}$.

Solve.

9. $\dfrac{1}{x+4} = \dfrac{5}{x}$

10. $\dfrac{3}{x-2} + \dfrac{x}{2} = \dfrac{6}{2x-4}$

11. Solve $R = \dfrac{1}{3}M(a-b)$ for M.

12. It takes 6 hr for a paper carrier to deliver 200 papers. At this rate, how long would it take to deliver 350 papers?

13. One data-entry clerk can key in a report in 6 hr. Another can key in the same report in 5 hr. How long would it take them, working together, to key in the same report?

14. One car travels 20 mph faster than another. While one car travels 300 mi, the other travels 400 mi. Find their speeds.

15. Find an equation of variation in which y varies jointly as x and the square of z and inversely as w, and $y = 1$ when $x = 5$, $z = 4$, and $w = 20$.

16. *Patio Concrete.* The amount of concrete necessary to construct a patio varies directly as the area of the patio. You note on the package that it takes 4.5 yd^3 of concrete to construct a 200-ft^2 patio. How many cubic yards of concrete will you need to construct a 250-ft^2 patio?

Objectives for Retesting

The objectives to be tested in addition to the material in this chapter are as follows.

[4.2b] Raise a product to a power and a quotient to a power.
[4.4c] Subtract polynomials.
[5.7a] Factor polynomials completely.
[5.9a] Solve applied problems involving quadratic equations that can be solved by factoring.

6.1 Rational Expressions: Multiplying, Dividing, and Simplifying

a Rational Expressions and Replacements

Rational numbers are quotients of integers. Some examples are

$$\frac{2}{3}, \quad \frac{4}{-5}, \quad \frac{-8}{17}, \quad \frac{563}{1}.$$

The following are called **rational expressions** or **fractional expressions.** They are quotients, or ratios, of polynomials:

$$\frac{3}{4}, \quad \frac{z}{6}, \quad \frac{5}{x+2}, \quad \frac{t^2+3t-10}{7t^2-4}.$$

A rational expression is also a division. For example,

$$\frac{3}{4} \text{ means } 3 \div 4 \quad \text{and} \quad \frac{x-8}{x+2} \text{ means } (x-8) \div (x+2).$$

Because rational expressions indicate division, we must be careful to avoid denominators of zero. When a variable is replaced with a number that produces a denominator equal to zero, the rational expression is undefined. For example, in the expression

$$\frac{x-8}{x+2},$$

when x is replaced with -2, the denominator is 0, and the expression is undefined:

$$\frac{x-8}{x+2} = \frac{-2-8}{-2+2} = \frac{-10}{0} \longleftarrow \text{Undefined}$$

When x is replaced with a number other than -2, such as 3, the expression *is* defined because the denominator is nonzero:

$$\frac{x-8}{x+2} = \frac{3-8}{3+2} = \frac{-5}{5} = -1.$$

Example 1 Find all numbers for which the rational expression

$$\frac{x+4}{x^2-3x-10}$$

is undefined.

To determine which numbers make the rational expression undefined, we set the denominator equal to 0 and solve:

$$x^2 - 3x - 10 = 0$$
$$(x-5)(x+2) = 0 \qquad \text{Factoring}$$
$$x - 5 = 0 \quad or \quad x + 2 = 0 \qquad \text{Using the principle of zero products}$$
$$x = 5 \quad or \quad x = -2.$$

The expression is undefined for the replacement numbers 5 and -2.

Do Exercises 1–3.

Objectives

a Find all numbers for which a rational expression is undefined.

b Multiply a rational expression by 1, using an expression such as A/A.

c Simplify rational expressions by factoring the numerator and the denominator and removing factors of 1.

d Multiply rational expressions and simplify.

e Find the reciprocal of a rational expression.

f Divide rational expressions and simplify.

For Extra Help

TAPE 10 | MAC WIN | CD-ROM

Find all numbers for which the rational expression is undefined.

1. $\dfrac{16}{x-3}$

2. $\dfrac{2x-7}{x^2+5x-24}$

3. $\dfrac{x+5}{8}$

Answers on page A-22

Multiply.

4. $\frac{2x+1}{3x-2} \cdot \frac{x}{x}$

5. $\frac{x+1}{x-2} \cdot \frac{x+2}{x+2}$

6. $\frac{x-8}{x-y} \cdot \frac{-1}{-1}$

Answers on page A-22

b Multiplying by 1

We multiply rational expressions in the same way that we multiply fractional notation in arithmetic. Note that

$$\frac{3}{7} \cdot \frac{2}{5} = \frac{3 \cdot 2}{7 \cdot 5} = \frac{6}{35}.$$

> To multiply rational expressions, multiply the numerators and multiply the denominators.

For example,

$$\frac{x-2}{3} \cdot \frac{x+2}{x+7} = \frac{(x-2)(x+2)}{3(x+7)}.$$ **Multiplying the numerators and the denominators**

Note that we leave the numerator, $(x - 2)(x + 2)$, and the denominator, $3(x + 7)$, in factored form because it is easier to simplify if we do not multiply. In order to learn to simplify, we first need to consider multiplying the rational expression by 1.

Any rational expression with the same numerator and denominator is a symbol for 1:

$$\frac{19}{19} = 1, \quad \frac{x+8}{x+8} = 1, \quad \frac{3x^2-4}{3x^2-4} = 1, \quad \frac{-1}{-1} = 1.$$

> Expressions that have the same value for all allowable (or meaningful) replacements are called **equivalent expressions.**

We can multiply by 1 to obtain an equivalent expression. At this point, we select expressions for 1 arbitrarily. Later, we will have a system for our choices when we add and subtract.

Examples Multiply.

2. $\frac{3x+2}{x+1} \cdot 1 = \frac{3x+2}{x+1} \cdot \frac{2x}{2x} = \frac{(3x+2)2x}{(x+1)2x}$ **Identity property of 1**

3. $\frac{x+2}{x-7} \cdot \frac{x+3}{x+3} = \frac{(x+2)(x+3)}{(x-7)(x+3)}$

4. $\frac{2+x}{2-x} \cdot \frac{-1}{-1} = \frac{(2+x)(-1)}{(2-x)(-1)}$

Do Exercises 4–6.

C Simplifying Rational Expressions

Simplifying rational expressions is similar to simplifying fractional expressions in arithmetic. For example, that an expression like $\frac{15}{40}$ can be simplified as follows:

$$\frac{15}{40} = \frac{3 \cdot 5}{8 \cdot 5}$$ Factoring the numerator and the denominator. Note the common factor of 5.

$$= \frac{3}{8} \cdot \frac{5}{5}$$ Factoring the fractional expression

$$= \frac{3}{8} \cdot 1$$ $\frac{5}{5} = 1$

$$= \frac{3}{8}.$$ Using the identity property of 1, or "removing a factor of 1"

In algebra, instead of simplifying

$$\frac{15}{40},$$

we may need to simplify an expression like

$$\frac{x^2 - 16}{x + 4}.$$

Just as factoring is important in simplifying in arithmetic, so too is it important in simplifying rational expressions. The factoring we use most is the factoring of polynomials, which we studied in Chapter 5.

To simplify, we can do the reverse of multiplying. We factor the numerator and the denominator and "remove" a factor of 1.

Example 5 Simplify by removing a factor of 1: $\frac{8x^2}{24x}$.

$$\frac{8x^2}{24x} = \frac{8 \cdot x \cdot x}{3 \cdot 8 \cdot x}$$ Factoring the numerator and the denominator

$$= \frac{8x}{8x} \cdot \frac{x}{3}$$ Factoring the rational expression

$$= 1 \cdot \frac{x}{3}$$ $\frac{8x}{8x} = 1$

$$= \frac{x}{3}$$ We removed a factor of 1.

Do Exercises 7 and 8.

Examples Simplify by removing a factor of 1.

6. $$\frac{5a + 15}{10} = \frac{5(a + 3)}{5 \cdot 2}$$ Factoring the numerator and the denominator

$$= \frac{5}{5} \cdot \frac{a + 3}{2}$$ Factoring the rational expression

$$= 1 \cdot \frac{a + 3}{2}$$ $\frac{5}{5} = 1$

$$= \frac{a + 3}{2}$$ Removing a factor of 1

Simplify by removing a factor of 1.

7. $\frac{5y}{y}$

8. $\frac{9x^2}{36x}$

Answers on page A-22

Simplify by removing a factor of 1.

9. $\dfrac{2x^2 + x}{3x^2 + 2x}$

10. $\dfrac{x^2 - 1}{2x^2 - x - 1}$

11. $\dfrac{7x + 14}{7}$

12. $\dfrac{12y + 24}{48}$

Answers on page A-22

7. $$\frac{6a + 12}{7a + 14} = \frac{6(a + 2)}{7(a + 2)}$$ Factoring the numerator and the denominator

$$= \frac{6}{7} \cdot \frac{a + 2}{a + 2}$$ Factoring the rational expression

$$= \frac{6}{7} \cdot 1$$ $\frac{a + 2}{a + 2} = 1$

$$= \frac{6}{7}$$ Removing a factor of 1

8. $$\frac{6x^2 + 4x}{2x^2 + 2x} = \frac{2x(3x + 2)}{2x(x + 1)}$$ Factoring the numerator and the denominator

$$= \frac{2x}{2x} \cdot \frac{3x + 2}{x + 1}$$ Factoring the rational expression

$$= 1 \cdot \frac{3x + 2}{x + 1}$$ $\frac{2x}{2x} = 1$

$$= \frac{3x + 2}{x + 1}$$ Removing a factor of 1. Note in this step that you *cannot* remove the x's because x is not a factor of the entire numerator and the entire denominator.

9. $$\frac{x^2 + 3x + 2}{x^2 - 1} = \frac{(x + 2)(x + 1)}{(x + 1)(x - 1)}$$

$$= \frac{x + 1}{x + 1} \cdot \frac{x + 2}{x - 1}$$

$$= 1 \cdot \frac{x + 2}{x - 1}$$

$$= \frac{x + 2}{x - 1}$$

Canceling

You may have encountered canceling when working with rational expressions. With great concern, we mention it as a possible way to speed up your work. Our concern is that canceling be done with care and understanding. Example 9 might have been done faster as follows:

$$\frac{x^2 + 3x + 2}{x^2 - 1} = \frac{(x + 2)(x + 1)}{(x + 1)(x - 1)}$$ Factoring the numerator and the denominator

$$= \frac{(x + 2)\cancel{(x + 1)}}{\cancel{(x + 1)}(x - 1)}$$ When a factor of 1 is noted, it is canceled, as shown: $\frac{x + 1}{x + 1} = 1$.

$$= \frac{x + 2}{x - 1}.$$ Simplifying

CAUTION! The difficulty with canceling is that it is often applied incorrectly, as in the following situations:

$$\frac{\cancel{x} + 3}{\cancel{x}} = 3; \qquad \frac{\cancel{4} + 1}{\cancel{4} + 2} = \frac{1}{2}; \qquad \frac{1\cancel{5}}{\cancel{5}4} = \frac{1}{4}.$$

Wrong! Wrong! Wrong!

In each of these situations, the expressions canceled were *not* factors of 1. Factors are parts of products. For example, in $2 \cdot 3$, 2 and 3 are factors, but in $2 + 3$, 2 and 3 are *not* factors. **If you can't factor, you can't cancel. If in doubt, don't cancel!**

Do Exercises 9–12.

Factors That Are Opposites

Consider

$$\frac{x-4}{4-x}.$$

At first glance, the numerator and the denominator do not appear to have any common factors other than 1. But $x - 4$ and $4 - x$ are opposites, or additive inverses, of each other. Thus we can rewrite one as the opposite of the other by factoring out a -1.

Example 10 Simplify: $\frac{x-4}{4-x}$.

$$\frac{x-4}{4-x} = \frac{x-4}{-(-4+x)} = \frac{x-4}{-1(x-4)}$$
$$= -1 \cdot \frac{x-4}{x-4}$$
$$= -1 \cdot 1$$
$$= -1$$

Do Exercises 13–15.

Simplify.

13. $\frac{x-8}{8-x}$

14. $\frac{c-d}{d-c}$

15. $\frac{-x-7}{x+7}$

d Multiplying and Simplifying

We try to simplify after we multiply. That is why we leave the numerator and the denominator in factored form.

Example 11 Multiply and simplify: $\frac{5a^3}{4} \cdot \frac{2}{5a}$.

$$\frac{5a^3}{4} \cdot \frac{2}{5a} = \frac{5a^3(2)}{4(5a)}$$ Multiplying the numerators and the denominators

$$= \frac{2 \cdot 5 \cdot a \cdot a \cdot a}{2 \cdot 2 \cdot 5 \cdot a}$$ Factoring the numerator and the denominator

$$= \frac{\cancel{2} \cdot \cancel{5} \cdot \cancel{a} \cdot a \cdot a}{\cancel{2} \cdot 2 \cdot \cancel{5} \cdot \cancel{a}}$$ Removing a factor of 1: $\frac{2 \cdot 5 \cdot a}{2 \cdot 5 \cdot a} = 1$

$$= \frac{a^2}{2}$$ Simplifying

Example 12 Multiply and simplify: $\frac{x^2+6x+9}{x^2-4} \cdot \frac{x-2}{x+3}$.

$$\frac{x^2+6x+9}{x^2-4} \cdot \frac{x-2}{x+3} = \frac{(x^2+6x+9)(x-2)}{(x^2-4)(x+3)}$$ Multiplying the numerators and the denominators

$$= \frac{(x+3)(x+3)(x-2)}{(x+2)(x-2)(x+3)}$$ Factoring the numerator and the denominator

$$= \frac{\cancel{(x+3)}(x+3)\cancel{(x-2)}}{(x+2)\cancel{(x-2)}\cancel{(x+3)}}$$ Removing a factor of 1: $\frac{(x+3)(x-2)}{(x+3)(x-2)} = 1$

$$= \frac{x+3}{x+2}$$ Simplifying

Do Exercise 16.

16. Multiply and simplify:
$$\frac{a^2-4a+4}{a^2-9} \cdot \frac{a+3}{a-2}.$$

Answers on page A-22

17. Multiply and simplify:

$$\frac{x^2 - 25}{6} \cdot \frac{3}{x + 5}.$$

Example 13 Multiply and simplify: $\frac{x^2 + x - 2}{15} \cdot \frac{5}{2x^2 - 3x + 1}$.

$$\frac{x^2 + x - 2}{15} \cdot \frac{5}{2x^2 - 3x + 1} = \frac{(x^2 + x - 2)5}{15(2x^2 - 3x + 1)}$$ Multiplying the numerators and the denominators

$$= \frac{(x + 2)(x - 1)5}{5(3)(x - 1)(2x - 1)}$$ Factoring the numerator and the denominator

$$= \frac{(x + 2)\cancel{(x - 1)}\cancel{5}}{\cancel{5}(3)\cancel{(x - 1)}(2x - 1)}$$ Removing a factor of 1: $\frac{(x - 1)5}{(x - 1)5} = 1$

$$= \frac{x + 2}{3(2x - 1)}$$ Simplifying

You need not carry out this multiplication.

Do Exercise 17.

Calculator Spotlight

Checking Simplifications Using a Table. We can use the TABLE feature as a partial check that rational expressions have been multiplied and/or simplified correctly. To check whether the simplifying of Example 9,

$$\frac{x^2 + 3x + 2}{x^2 - 1} = \frac{x + 2}{x - 1},$$

is correct, we enter

$$y_1 = \frac{x^2 + 3x + 2}{x^2 - 1} \quad \text{as} \quad y_1 = (x^2 + 3x + 2)/(x^2 - 1)$$

and

$$y_2 = \frac{x + 2}{x - 1} \quad \text{as} \quad y_2 = (x + 2)/(x - 1).$$

If our simplifying is correct, the y_1- and y_2-values should be the same for all allowable replacements, regardless of the table settings used. Let TblStart = −4 and ΔTbl = 1.

X	Y1	Y2
−4	.4	.4
−3	.25	.25
−2	0	0
−1	ERROR	−.5
0	−2	−2
1	ERROR	ERROR
2	4	4

X = −4

Note that −1 and 1 are not allowable replacements in the first expression, and 1 is not an allowable replacement in the second. These facts are indicated by the ERROR messages. For all other numbers, we see that y_1 and y_2 are the same, so the simplifying seems to be correct. Remember, this is only a partial check.

Exercises

Use the TABLE feature to check whether each of the following is correct.

1. $\frac{8x^2}{24x} = \frac{x}{3}$
2. $\frac{5a + 15}{10} = \frac{a + 3}{2}$
3. $\frac{x + 3}{x} = 3$
4. $\frac{x^2 - 3x + 2}{x^2 - 1} = \frac{x + 2}{x - 1}$
5. $\frac{x^2 - 9}{x^2 - 3} = 3$
6. $\frac{x^2 + 6x + 9}{x^2 - 4} \cdot \frac{x - 2}{x + 3} = \frac{x + 3}{x + 2}$
7. $\frac{x^2 - 25}{6} \cdot \frac{3}{x + 5} = \frac{x - 5}{3}$

Answer on page A-22

After variables have been replaced with rational numbers, a rational expression represents a rational number. There is a similarity between what we do with rational expressions and what we do with rational numbers.

e Finding Reciprocals

Two expressions are reciprocals of each other if their product is 1. The reciprocal of a rational expression is found by interchanging the numerator and the denominator.

Examples

14. The reciprocal of $\frac{2}{5}$ is $\frac{5}{2}$. $\left(\text{This is because } \frac{2}{5} \cdot \frac{5}{2} = \frac{10}{10} = 1.\right)$

15. The reciprocal of $\frac{2x^2 - 3}{x + 4}$ is $\frac{x + 4}{2x^2 - 3}$.

16. The reciprocal of $x + 2$ is $\frac{1}{x + 2}$. $\left(\text{Think of } x + 2 \text{ as } \frac{x + 2}{1}.\right)$

Do Exercises 18–21.

Find the reciprocal.

18. $\frac{7}{2}$

19. $\frac{x^2 + 5}{2x^3 - 1}$

20. $x - 5$

21. $\frac{1}{x^2 - 3}$

f Division

We divide rational expressions in the same way that we divide fractional notation in arithmetic.

> To divide rational expressions, multiply by the reciprocal of the divisor. Then factor and simplify the result.

Examples Divide.

17. $\frac{3}{4} \div \frac{2}{5} = \frac{3}{4} \cdot \frac{5}{2}$ **Multiplying by the reciprocal of the divisor**

$= \frac{3 \cdot 5}{4 \cdot 2}$

$= \frac{15}{8}$

18. $\frac{2}{x} \div \frac{x}{3} = \frac{2}{x} \cdot \frac{3}{x}$ **Multiplying by the reciprocal of the divisor**

$= \frac{2 \cdot 3}{x \cdot x}$

$= \frac{6}{x^2}$

Do Exercises 22 and 23.

Divide.

22. $\frac{3}{5} \div \frac{7}{2}$

23. $\frac{x}{8} \div \frac{5}{x}$

Answers on page A-22

24. Divide:

$$\frac{x-3}{x+5} \div \frac{x+5}{x-2}.$$

Example 19 Divide: $\dfrac{x+1}{x+2} \div \dfrac{x-1}{x+3}$.

$$\frac{x+1}{x+2} \div \frac{x-1}{x+3} = \frac{x+1}{x+2} \cdot \frac{x+3}{x-1}$$ Multiplying by the reciprocal of the divisor

$$= \frac{(x+1)(x+3)}{(x+2)(x-1)}$$ ← We usually do not carry out the multiplication in the numerator or the denominator. It is not wrong to do so, but the factored form is often more useful.

Do Exercise 24.

Divide and simplify.

25. $\dfrac{x-3}{x+5} \div \dfrac{x+2}{x+5}$

26. $\dfrac{x^2-5x+6}{x+5} \div \dfrac{x+2}{x+5}$

27. $\dfrac{y^2-1}{y+1} \div \dfrac{y^2-2y+1}{y+1}$

Example 20 Divide and simplify: $\dfrac{x+1}{x^2-1} \div \dfrac{x+1}{x^2-2x+1}$.

$$\frac{x+1}{x^2-1} \div \frac{x+1}{x^2-2x+1}$$

$$= \frac{x+1}{x^2-1} \cdot \frac{x^2-2x+1}{x+1}$$ Multiplying by the reciprocal

$$= \frac{(x+1)(x^2-2x+1)}{(x^2-1)(x+1)}$$

$$= \frac{(x+1)(x-1)(x-1)}{(x-1)(x+1)(x+1)}$$ Factoring the numerator and the denominator

$$= \frac{\cancel{(x+1)}\cancel{(x-1)}(x-1)}{\cancel{(x-1)}\cancel{(x+1)}(x+1)}$$ Removing a factor of 1: $\dfrac{(x+1)(x-1)}{(x+1)(x-1)} = 1$

$$= \frac{x-1}{x+1}$$

Example 21 Divide and simplify: $\dfrac{x^2-2x-3}{x^2-4} \div \dfrac{x+1}{x+5}$.

$$\frac{x^2-2x-3}{x^2-4} \div \frac{x+1}{x+5}$$

$$= \frac{x^2-2x-3}{x^2-4} \cdot \frac{x+5}{x+1}$$ Multiplying by the reciprocal

$$= \frac{(x^2-2x-3)(x+5)}{(x^2-4)(x+1)}$$

$$= \frac{(x-3)(x+1)(x+5)}{(x-2)(x+2)(x+1)}$$ Factoring the numerator and the denominator

$$= \frac{(x-3)\cancel{(x+1)}(x+5)}{(x-2)(x+2)\cancel{(x+1)}}$$ Removing a factor of 1: $\dfrac{x+1}{x+1} = 1$

$$= \frac{(x-3)(x+5)}{(x-2)(x+2)}$$ ← You need not carry out the multiplications in the numerator and the denominator.

Do Exercises 25–27.

Calculator Spotlight

Use the TABLE feature to check the divisions in Examples 18–21.

Answers on page A-22

Exercise Set 6.1

a Find all numbers for which the rational expression is undefined.

1. $\frac{-3}{2x}$

2. $\frac{24}{-8y}$

3. $\frac{5}{x - 8}$

4. $\frac{y - 4}{y + 6}$

5. $\frac{3}{2y + 5}$

6. $\frac{x^2 - 9}{4x - 12}$

7. $\frac{x^2 + 11}{x^2 - 3x - 28}$

8. $\frac{p^2 - 9}{p^2 - 7p + 10}$

9. $\frac{m^3 - 2m}{m^2 - 25}$

10. $\frac{7 - 3x + x^2}{49 - x^2}$

11. $\frac{x - 4}{3}$

12. $\frac{x^2 - 25}{14}$

b Multiply. Do not simplify. Note that in each case you are multiplying by 1.

13. $\frac{4x}{4x} \cdot \frac{3x^2}{5y}$

14. $\frac{5x^2}{5x^2} \cdot \frac{6y^3}{3z^4}$

15. $\frac{2x}{2x} \cdot \frac{x - 1}{x + 4}$

16. $\frac{2a - 3}{5a + 2} \cdot \frac{a}{a}$

17. $\frac{3 - x}{4 - x} \cdot \frac{-1}{-1}$

18. $\frac{x - 5}{5 - x} \cdot \frac{-1}{-1}$

19. $\frac{y + 6}{y + 6} \cdot \frac{y - 7}{y + 2}$

20. $\frac{x^2 + 1}{x^3 - 2} \cdot \frac{x - 4}{x - 4}$

c Simplify.

21. $\frac{8x^3}{32x}$

22. $\frac{4x^2}{20x}$

23. $\frac{48p^7q^5}{18p^5q^4}$

24. $\frac{-76x^8y^3}{-24x^4y^3}$

25. $\frac{4x - 12}{4x}$

26. $\frac{5a - 40}{5}$

27. $\dfrac{3m^2 + 3m}{6m^2 + 9m}$

28. $\dfrac{4y^2 - 2y}{5y^2 - 5y}$

29. $\dfrac{a^2 - 9}{a^2 + 5a + 6}$

30. $\dfrac{t^2 - 25}{t^2 + t - 20}$

31. $\dfrac{a^2 - 10a + 21}{a^2 - 11a + 28}$

32. $\dfrac{x^2 - 2x - 8}{x^2 - x - 6}$

33. $\dfrac{x^2 - 25}{x^2 - 10x + 25}$

34. $\dfrac{x^2 + 8x + 16}{x^2 - 16}$

35. $\dfrac{a^2 - 1}{a - 1}$

36. $\dfrac{t^2 - 1}{t + 1}$

37. $\dfrac{x^2 + 1}{x + 1}$

38. $\dfrac{m^2 + 9}{m + 3}$

39. $\dfrac{6x^2 - 54}{4x^2 - 36}$

40. $\dfrac{8x^2 - 32}{4x^2 - 16}$

41. $\dfrac{6t + 12}{t^2 - t - 6}$

42. $\dfrac{4x + 32}{x^2 + 9x + 8}$

43. $\dfrac{2t^2 + 6t + 4}{4t^2 - 12t - 16}$

44. $\dfrac{3a^2 - 9a - 12}{6a^2 + 30a + 24}$

45. $\dfrac{t^2 - 4}{(t + 2)^2}$

46. $\dfrac{m^2 - 10m + 25}{m^2 - 25}$

47. $\dfrac{6 - x}{x - 6}$

48. $\dfrac{t - 3}{3 - t}$

49. $\dfrac{a - b}{b - a}$

50. $\dfrac{y - x}{-x + y}$

51. $\dfrac{6t - 12}{2 - t}$

52. $\dfrac{5a - 15}{3 - a}$

53. $\dfrac{x^2 - 1}{1 - x}$

54. $\dfrac{a^2 - b^2}{b^2 - a^2}$

d Multiply and simplify.

55. $\dfrac{4x^3}{3x} \cdot \dfrac{14}{x}$

56. $\dfrac{18}{x^3} \cdot \dfrac{5x^2}{6}$

57. $\dfrac{3c}{d^2} \cdot \dfrac{4d}{6c^3}$

58. $\dfrac{3x^2y}{2} \cdot \dfrac{4}{xy^3}$

59. $\dfrac{x^2 - 3x - 10}{x^2 - 4x + 4} \cdot \dfrac{x - 2}{x - 5}$

60. $\dfrac{t^2}{t^2 - 4} \cdot \dfrac{t^2 - 5t + 6}{t^2 - 3t}$

61. $\dfrac{a^2 - 9}{a^2} \cdot \dfrac{a^2 - 3a}{a^2 + a - 12}$

62. $\dfrac{x^2 + 10x - 11}{x^2 - 1} \cdot \dfrac{x + 1}{x + 11}$

63. $\dfrac{4a^2}{3a^2 - 12a + 12} \cdot \dfrac{3a - 6}{2a}$

64. $\dfrac{5v + 5}{v - 2} \cdot \dfrac{v^2 - 4v + 4}{v^2 - 1}$

65. $\dfrac{t^4 - 16}{t^4 - 1} \cdot \dfrac{t^2 + 1}{t^2 + 4}$

66. $\dfrac{x^4 - 1}{x^4 - 81} \cdot \dfrac{x^2 + 9}{x^2 + 1}$

67. $\dfrac{(x+4)^3}{(x+2)^3} \cdot \dfrac{x^2+4x+4}{x^2+8x+16}$

68. $\dfrac{(t-2)^3}{(t-1)^3} \cdot \dfrac{t^2-2t+1}{t^2-4t+4}$

69. $\dfrac{5a^2-180}{10a^2-10} \cdot \dfrac{20a+20}{2a-12}$

70. $\dfrac{2t^2-98}{4t^2-4} \cdot \dfrac{8t+8}{16t-112}$

e Find the reciprocal.

71. $\dfrac{4}{x}$

72. $\dfrac{a+3}{a-1}$

73. $x^2 - y^2$

74. $x^2 - 5x + 7$

75. $\dfrac{1}{a+b}$

76. $\dfrac{x^2}{x^2-3}$

77. $\dfrac{x^2+2x-5}{x^2-4x+7}$

78. $\dfrac{(a-b)(a+b)}{(a+4)(a-5)}$

f Divide and simplify.

79. $\dfrac{2}{5} \div \dfrac{4}{3}$

80. $\dfrac{3}{10} \div \dfrac{3}{2}$

81. $\dfrac{2}{x} \div \dfrac{8}{x}$

82. $\dfrac{t}{3} \div \dfrac{t}{15}$

83. $\dfrac{a}{b^2} \div \dfrac{a^2}{b^3}$

84. $\dfrac{x^2}{y} \div \dfrac{x^3}{y^3}$

85. $\dfrac{a+2}{a-3} \div \dfrac{a-1}{a+3}$

86. $\dfrac{x-8}{x+9} \div \dfrac{x+2}{x-1}$

87. $\dfrac{x^2-1}{x} \div \dfrac{x+1}{x-1}$

88. $\frac{4y - 8}{y + 2} \div \frac{y - 2}{y^2 - 4}$

89. $\frac{x + 1}{6} \div \frac{x + 1}{3}$

90. $\frac{a}{a - b} \div \frac{b}{a - b}$

91. $\frac{5x - 5}{16} \div \frac{x - 1}{6}$

92. $\frac{4y - 12}{12} \div \frac{y - 3}{3}$

93. $\frac{-6 + 3x}{5} \div \frac{4x - 8}{25}$

94. $\frac{-12 + 4x}{4} \div \frac{-6 + 2x}{6}$

95. $\frac{a + 2}{a - 1} \div \frac{3a + 6}{a - 5}$

96. $\frac{t - 3}{t + 2} \div \frac{4t - 12}{t + 1}$

97. $\frac{x^2 - 4}{x} \div \frac{x - 2}{x + 2}$

98. $\frac{x + y}{x - y} \div \frac{x^2 + y}{x^2 - y^2}$

99. $\frac{x^2 - 9}{4x + 12} \div \frac{x - 3}{6}$

100. $\frac{a - b}{2a} \div \frac{a^2 - b^2}{8a^3}$

101. $\frac{c^2 + 3c}{c^2 + 2c - 3} \div \frac{c}{c + 1}$

102. $\frac{y + 5}{2y} \div \frac{y^2 - 25}{4y^2}$

103. $\frac{2y^2 - 7y + 3}{2y^2 + 3y - 2} \div \frac{6y^2 - 5y + 1}{3y^2 + 5y - 2}$

104. $\frac{x^2 + x - 20}{x^2 - 7x + 12} \div \frac{x^2 + 10x + 25}{x^2 - 6x + 9}$

105. $\frac{x^2 - 1}{4x + 4} \div \frac{2x^2 - 4x + 2}{8x + 8}$

106. $\frac{5t^2 + 5t - 30}{10t + 30} \div \frac{2t^2 - 8}{6t^2 + 36t + 54}$

Skill Maintenance

Solve.

107. The product of two consecutive even integers is 360. Find the integers. [5.9a]

108. Sixteen more than the square of a number is eight times the number. Find the number. [5.9a]

Subtract. [4.4c]

109. $(8x^3 - 3x^2 + 7) - (8x^2 + 3x - 5)$

110. $(3p^2 - 6pq + 7q^2) - (5p^2 - 10pq + 11q^2)$

Simplify. [4.2b]

111. $(2x^{-3}y^4)^2$

112. $(5x^6y^{-4})^3$

113. $\left(\frac{2x^3}{y^5}\right)^2$

114. $\left(\frac{a^{-3}}{b^4}\right)^5$

Synthesis

115. ◆ Explain why 5, −1, and 7 are *not* allowable replacements in the division

$$\frac{x+3}{x-5} \div \frac{x-7}{x+1}.$$

116. ◆ Is the reciprocal of a product the product of the reciprocals? Why or why not?

Simplify.

117. $\dfrac{x^4 - 16y^4}{(x^2 + 4y^2)(x - 2y)}$

118. $\dfrac{(a - b)^2}{b^2 - a^2}$

119. $\dfrac{t^4 - 1}{t^4 - 81} \cdot \dfrac{t^2 - 9}{t^2 + 1} \cdot \dfrac{(t - 9)^2}{(t + 1)^2}$

120. $\dfrac{(t + 2)^3}{(t + 1)^3} \cdot \dfrac{t^2 + 2t + 1}{t^2 + 4t + 4} \cdot \dfrac{t + 1}{t + 2}$

121. $\dfrac{x^2 - y^2}{(x - y)^2} \cdot \dfrac{x^2 - 2xy + y^2}{x^2 - 4xy - 5y^2}$

122. $\dfrac{x - 1}{x^2 + 1} \cdot \dfrac{x^4 - 1}{(x - 1)^2} \cdot \dfrac{x^2 - 1}{x^4 - 2x^2 + 1}$

123. $\dfrac{3a^2 - 5ab - 12b^2}{3ab + 4b^2} \div (3b^2 - ab)$

124. $\dfrac{3x + 3y + 3}{9x} \div \left(\dfrac{x^2 + 2xy + y^2 - 1}{x^4 + x^2}\right)$

125. The volume of this rectangular solid is $x - 3$. What is its height?

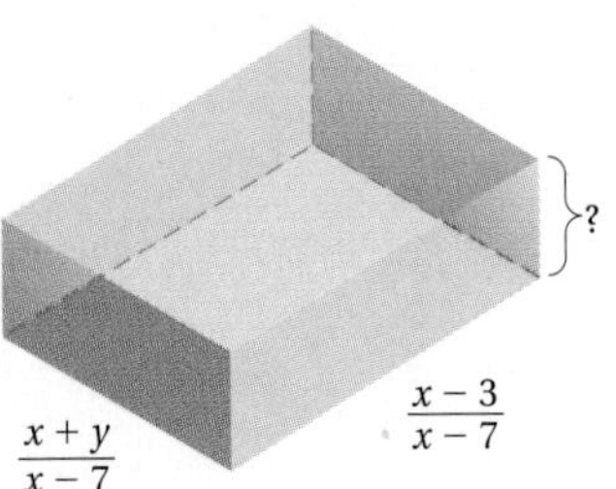

6.2 Least Common Multiples and Denominators

Objectives

a. Find the LCM of several numbers by factoring.

b. Add fractions, first finding the LCD.

c. Find the LCM of algebraic expressions by factoring.

For Extra Help

TAPE 10 | MAC WIN | CD-ROM

a Least Common Multiples

To add when denominators are different, we first find a common denominator. We saw there, for example, that to add $\frac{5}{12}$ and $\frac{7}{30}$, we first look for the **least common multiple, LCM,** of both 12 and 30. That number becomes the **least common denominator, LCD.** To find the LCM of 12 and 30, we factor:

$$12 = 2 \cdot 2 \cdot 3;$$

$$30 = 2 \cdot 3 \cdot 5.$$

The LCM is the number that has 2 as a factor twice, 3 as a factor once, and 5 as a factor once:

$$\text{LCM} = 2 \cdot 2 \cdot 3 \cdot 5, \text{ or } 60.$$

> To find the LCM, use each factor the greatest number of times that it appears in any one factorization.

Example 1 Find the LCM of 24 and 36.

$$\left.\begin{array}{l} 24 = 2 \cdot 2 \cdot 2 \cdot 3 \\ 36 = 2 \cdot 2 \cdot 3 \cdot 3 \end{array}\right\} \quad \text{LCM} = 2 \cdot 2 \cdot 2 \cdot 3 \cdot 3, \text{ or } 72$$

Do Exercises 1–4.

Find the LCM by factoring.

1. 16, 18

2. 6, 12

3. 2, 5

4. 24, 30, 20

b Adding Using the LCD

Let's finish adding $\frac{5}{12}$ and $\frac{7}{30}$:

$$\frac{5}{12} + \frac{7}{30} = \frac{5}{2 \cdot 2 \cdot 3} + \frac{7}{2 \cdot 3 \cdot 5}.$$

The least common denominator, LCD, is $2 \cdot 2 \cdot 3 \cdot 5$. To get the LCD in the first denominator, we need a 5. To get the LCD in the second denominator, we need another 2. We get these numbers by multiplying by 1:

$$\frac{5}{12} + \frac{7}{30} = \frac{5}{2 \cdot 2 \cdot 3} \cdot \frac{5}{5} + \frac{7}{2 \cdot 3 \cdot 5} \cdot \frac{2}{2} \qquad \text{Multiplying by 1}$$

$$= \frac{25}{2 \cdot 2 \cdot 3 \cdot 5} + \frac{14}{2 \cdot 3 \cdot 5 \cdot 2} \qquad \text{The denominators are now the LCD.}$$

$$= \frac{39}{2 \cdot 2 \cdot 3 \cdot 5} \qquad \text{Adding the numerators and keeping the LCD}$$

$$= \frac{\not{3} \cdot 13}{2 \cdot 2 \cdot \not{3} \cdot 5} \qquad \text{Factoring the numerator and removing a factor of 1: } \frac{3}{3} = 1$$

$$= \frac{13}{20}. \qquad \text{Simplifying}$$

Answers on page A-22

Add, first finding the LCD. Simplify, if possible.

5. $\frac{3}{16}+\frac{1}{18}$

6. $\frac{1}{6}+\frac{1}{12}$

7. $\frac{1}{2}+\frac{3}{5}$

8. $\frac{1}{24}+\frac{1}{30}+\frac{3}{20}$

Find the LCM.

9. $12xy^2,\quad 15x^3y$

10. $y^2+5y+4,\quad y^2+2y+1$

11. $t^2+16,\quad t-2,\quad 7$

12. $x^2+2x+1,\quad 3x^2-3x,\quad x^2-1$

Answers on page A-22

Example 2 Add: $\frac{5}{12}+\frac{11}{18}$.

$$\left.\begin{aligned}12 &= 2\cdot 2\cdot 3\\ 18 &= 2\cdot 3\cdot 3\end{aligned}\right\}\quad \text{LCD} = 2\cdot 2\cdot 3\cdot 3, \text{ or } 36$$

$$\frac{5}{12}+\frac{11}{18}=\frac{5}{2\cdot 2\cdot 3}\cdot\frac{3}{3}+\frac{11}{2\cdot 3\cdot 3}\cdot\frac{2}{2}=\frac{15+22}{2\cdot 2\cdot 3\cdot 3}=\frac{37}{36}$$

Do Exercises 5–8.

c LCMs of Algebraic Expressions

To find the LCM of two or more algebraic expressions, we factor them. Then we use each factor the greatest number of times that it occurs in any one expression.

Example 3 Find the LCM of $12x$, $16y$, and $8xyz$.

$$\left.\begin{aligned}12x &= 2\cdot 2\cdot 3\cdot x\\ 16y &= 2\cdot 2\cdot 2\cdot 2\cdot y\\ 8xyz &= 2\cdot 2\cdot 2\cdot x\cdot y\cdot z\end{aligned}\right\}\quad \begin{aligned}\text{LCM} &= 2\cdot 2\cdot 2\cdot 2\cdot 3\cdot x\cdot y\cdot z\\ &= 48xyz\end{aligned}$$

Example 4 Find the LCM of x^2+5x-6 and x^2-1.

$$\left.\begin{aligned}x^2+5x-6 &= (x+6)(x-1)\\ x^2-1 &= (x+1)(x-1)\end{aligned}\right\}\quad \text{LCM} = (x+6)(x-1)(x+1)$$

Example 5 Find the LCM of x^2+4, $x+1$, and 5.

These expressions do not share a common factor other than 1, so the LCM is their product:

$$5(x^2+4)(x+1).$$

Example 6 Find the LCM of x^2-25 and $2x-10$.

$$\left.\begin{aligned}x^2-25 &= (x+5)(x-5)\\ 2x-10 &= 2(x-5)\end{aligned}\right\}\quad \text{LCM} = 2(x+5)(x-5)$$

Example 7 Find the LCM of x^2-4y^2, $x^2-4xy+4y^2$, and $x-2y$.

$$\left.\begin{aligned}x^2-4y^2 &= (x-2y)(x+2y)\\ x^2-4xy+4y^2 &= (x-2y)(x-2y)\\ x-2y &= x-2y\end{aligned}\right\}\quad \begin{aligned}\text{LCM} &= (x+2y)(x-2y)(x-2y)\\ &= (x+2y)(x-2y)^2\end{aligned}$$

Do Exercises 9–12.

Exercise Set 6.2

a Find the LCM.

1. 12, 27

2. 10, 15

3. 8, 9

4. 12, 18

5. 6, 9, 21

6. 8, 36, 40

7. 24, 36, 40

8. 4, 5, 20

9. 10, 100, 500

10. 28, 42, 60

b Add, first finding the LCD. Simplify, if possible.

11. $\frac{7}{24} + \frac{11}{18}$

12. $\frac{7}{60} + \frac{2}{25}$

13. $\frac{1}{6} + \frac{3}{40}$

14. $\frac{5}{24} + \frac{3}{20}$

15. $\frac{1}{20} + \frac{1}{30} + \frac{2}{45}$

16. $\frac{2}{15} + \frac{5}{9} + \frac{3}{20}$

c Find the LCM.

17. $6x^2$, $12x^3$

18. $2a^2b$, $8ab^3$

19. $2x^2$, $6xy$, $18y^2$

20. p^3q, p^2q, pq^2

21. $2(y - 3)$, $6(y - 3)$

22. $5(m + 2)$, $15(m + 2)$

23. t, $t + 2$, $t - 2$

24. y, $y - 5$, $y + 5$

25. $x^2 - 4$, $x^2 + 5x + 6$

26. $x^2 - 4$, $x^2 - x - 2$

27. $t^3 + 4t^2 + 4t$, $t^2 - 4t$

28. $m^4 - m^2$, $m^3 - m^2$

29. $a + 1$, $(a - 1)^2$, $a^2 - 1$

30. $a^2 - 2ab + b^2$, $a^2 - b^2$, $3a + 3b$

31. $m^2 - 5m + 6$, $m^2 - 4m + 4$

32. $2x^2 + 5x + 2$, $2x^2 - x - 1$

33. $2 + 3x$, $4 - 9x^2$, $2 - 3x$

34. $9 - 4x^2$, $3 + 2x$, $3 - 2x$

35. $10v^2 + 30v, \quad 5v^2 + 35v + 60$

36. $12a^2 + 24a, \quad 4a^2 + 20a + 24$

37. $9x^3 - 9x^2 - 18x, \quad 6x^5 - 24x^4 + 24x^3$

38. $x^5 - 4x^3, \quad x^3 + 4x^2 + 4x$

39. $x^5 + 4x^4 + 4x^3, \quad 3x^2 - 12, \quad 2x + 4$

40. $x^5 + 2x^4 + x^3, \quad 2x^3 - 2x, \quad 5x - 5$

Skill Maintenance

Factor. [5.7a]

41. $x^2 - 6x + 9$

42. $6x^2 + 4x$

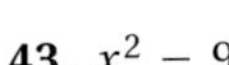

43. $x^2 - 9$

44. $x^2 + 4x - 21$

45. $x^2 + 6x + 9$

46. $x^2 - 4x - 21$

Divorce Rate. The graph at right is that of the equation

$$D = 0.00509x^2 - 19.17x + 18{,}065.305$$

for values of x ranging from 1900 to 2010. It shows the percentage of couples who are married in a given year, x, whose marriages, it is predicted, will end in divorce. Use *only* the graph to answer the questions in Exercises 47–52. [3.1a], [4.3a]

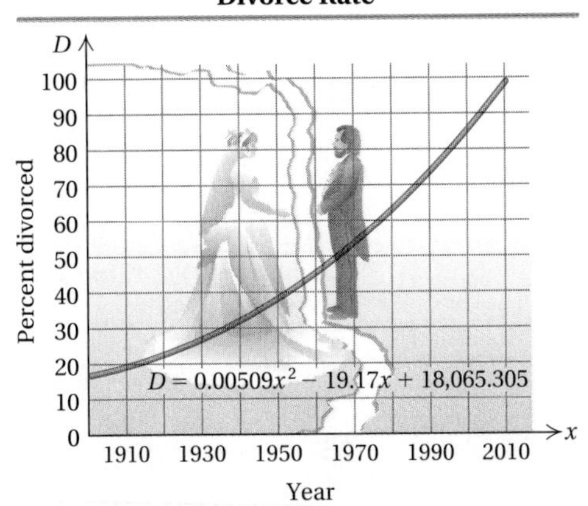

Source: Gottman, John, *What Predicts Divorce: The Relationship Between Marital Processes and Marital Outcomes.* New Jersey: Lawrence Erlbaum Associates, 1993.

47. Estimate the divorce percentage of those married in 1970.

48. Estimate the divorce percentage of those married in 1980.

49. Estimate the divorce percentage of those married in 1990.

50. Estimate the divorce percentage of those married in 2010.

51. In what year was the divorce percentage about 50%?

52. In what year was the divorce percentage about 84%?

Synthesis

53. ◆ If the LCM of a binomial and a trinomial is the trinomial, what relationship exists between the two expressions?

54. ◆ Explain how you might find the LCD of these two expressions:

$$\frac{x + 1}{x^2 - 4}, \quad \frac{x - 2}{x^2 + 5x + 6}.$$

6.3 Adding Rational Expressions

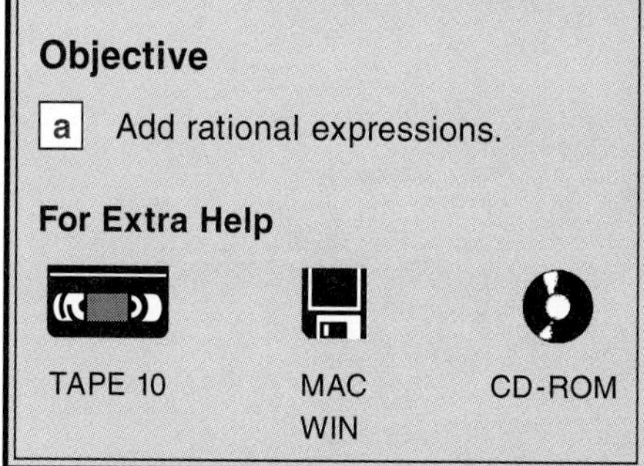

a We add rational expressions as we do rational numbers.

> To add when the denominators are the same, add the numerators and keep the same denominator.

Examples Add.

1. $\dfrac{x}{x+1} + \dfrac{2}{x+1} = \dfrac{x+2}{x+1}$

2. $\dfrac{2x^2+3x-7}{2x+1} + \dfrac{x^2+x-8}{2x+1} = \dfrac{(2x^2+3x-7)+(x^2+x-8)}{2x+1}$

$= \dfrac{3x^2+4x-15}{2x+1}$

3. $\dfrac{x-5}{x^2-9} + \dfrac{2}{x^2-9} = \dfrac{(x-5)+2}{x^2-9} = \dfrac{x-3}{x^2-9}$

$= \dfrac{x-3}{(x-3)(x+3)}$ Factoring

$= \dfrac{\cancel{x-3}}{\cancel{(x-3)}(x+3)}$ Removing a factor of 1: $\dfrac{x-3}{x-3} = 1$

$= \dfrac{1}{x+3}$ Simplifying

As in Example 3, simplifying should be done if possible after adding.

Do Exercises 1–3.

Add.

1. $\dfrac{5}{9} + \dfrac{2}{9}$

2. $\dfrac{3}{x-2} + \dfrac{x}{x-2}$

3. $\dfrac{4x+5}{x-1} + \dfrac{2x-1}{x-1}$

When denominators are not the same, we multiply by 1 to obtain equivalent expressions with the same denominator. When one denominator is the opposite of the other, we can first multiply either expression by 1 using $-1/-1$.

Examples

4. $\dfrac{x}{2} + \dfrac{3}{-2} = \dfrac{x}{2} + \dfrac{3}{-2} \cdot \dfrac{-1}{-1}$ Multiplying by 1 using $\dfrac{-1}{-1}$

$= \dfrac{x}{2} + \dfrac{-3}{2}$ The denominators are now the same.

$= \dfrac{x+(-3)}{2} = \dfrac{x-3}{2}$

5. $\dfrac{3x+4}{x-2} + \dfrac{x-7}{2-x} = \dfrac{3x+4}{x-2} + \dfrac{x-7}{2-x} \cdot \dfrac{-1}{-1}$

We could have chosen to multiply this expression by $-1/-1$. We multiply only one expression, *not* both.

$= \dfrac{3x+4}{x-2} + \dfrac{-x+7}{x-2}$ *Note*: $(2-x)(-1) = -2+x = x-2.$

$= \dfrac{(3x+4)+(-x+7)}{x-2} = \dfrac{2x+11}{x-2}$

Do Exercises 4 and 5.

Add.

4. $\dfrac{x}{4} + \dfrac{5}{-4}$

5. $\dfrac{2x+1}{x-3} + \dfrac{x+2}{3-x}$

Answers on page A-23

Add.

6. $\frac{3x}{16} + \frac{5x^2}{24}$

7. $\frac{3}{16x} + \frac{5}{24x^2}$

Answers on page A-23

When denominators are different, we find the least common denominator, LCD. The procedure we will use is as follows.

> To add rational expressions with different denominators:
>
> 1. Find the LCM of the denominators. This is the least common denominator (LCD).
> 2. For each rational expression, find an equivalent expression with the LCD. To do so, multiply by 1 using an expression for 1 made up of factors of the LCD that are missing from the original denominator.
> 3. Add the numerators. Write the sum over the LCD.
> 4. Simplify, if possible.

Example 6 Add: $\frac{5x^2}{8} + \frac{7x}{12}$.

First, we find the LCD:

$$\left.\begin{aligned} 8 &= 2 \cdot 2 \cdot 2 \\ 12 &= 2 \cdot 2 \cdot 3 \end{aligned}\right\} \quad \text{LCD} = 2 \cdot 2 \cdot 2 \cdot 3, \text{ or } 24.$$

Compare the factorization $8 = 2 \cdot 2 \cdot 2$ with the factorization of the LCD, $24 = 2 \cdot 2 \cdot 2 \cdot 3$. The factor of the LCD missing from 8 is 3. Compare $12 = 2 \cdot 2 \cdot 3$ and $24 = 2 \cdot 2 \cdot 2 \cdot 3$. The factor of the LCD missing from 12 is 2. We multiply by 1 to get the LCD in each expression, and then add and simplify, if possible:

$$\begin{aligned} \frac{5x^2}{8} + \frac{7x}{12} &= \frac{5x^2}{2 \cdot 2 \cdot 2} + \frac{7x}{2 \cdot 2 \cdot 3} \\ &= \frac{5x^2}{2 \cdot 2 \cdot 2} \cdot \frac{3}{3} + \frac{7x}{2 \cdot 2 \cdot 3} \cdot \frac{2}{2} \qquad \text{Multiplying by 1 to get the same denominators} \\ &= \frac{15x^2}{24} + \frac{14x}{24} \\ &= \frac{15x^2 + 14x}{24}. \end{aligned}$$

Example 7 Add: $\frac{3}{8x} + \frac{5}{12x^2}$.

First, we find the LCD:

$$\left.\begin{aligned} 8x &= 2 \cdot 2 \cdot 2 \cdot x \\ 12x^2 &= 2 \cdot 2 \cdot 3 \cdot x \cdot x \end{aligned}\right\} \quad \text{LCD} = 2 \cdot 2 \cdot 2 \cdot 3 \cdot x \cdot x, \text{ or } 24x^2.$$

The factors of the LCD missing from $8x$ are 3 and x. The factor of the LCD missing from $12x^2$ is 2. We multiply by 1 to get the LCD in each expression, and then add and simplify, if possible:

$$\begin{aligned} \frac{3}{8x} + \frac{5}{12x^2} &= \frac{3}{8x} \cdot \frac{3 \cdot x}{3 \cdot x} + \frac{5}{12x^2} \cdot \frac{2}{2} \\ &= \frac{9x}{24x^2} + \frac{10}{24x^2} \\ &= \frac{9x + 10}{24x^2}. \end{aligned}$$

Do Exercises 6 and 7.

Example 8 Add: $\dfrac{2a}{a^2 - 1} + \dfrac{1}{a^2 + a}$.

First, we find the LCD:

$$\left.\begin{aligned} a^2 - 1 &= (a - 1)(a + 1) \\ a^2 + a &= a(a + 1) \end{aligned}\right\} \quad \text{LCD} = a(a - 1)(a + 1).$$

We multiply by 1 to get the LCD in each expression, and then add and simplify:

$$\frac{2a}{(a - 1)(a + 1)} \cdot \frac{a}{a} + \frac{1}{a(a + 1)} \cdot \frac{a - 1}{a - 1}$$

$$= \frac{2a^2}{a(a - 1)(a + 1)} + \frac{a - 1}{a(a - 1)(a + 1)}$$

$$= \frac{2a^2 + a - 1}{a(a - 1)(a + 1)}$$

$$= \frac{(a + 1)(2a - 1)}{a(a - 1)(a + 1)} \qquad \text{Factoring the numerator in order to simplify}$$

$$= \frac{\cancel{(a + 1)}(2a - 1)}{a(a - 1)\cancel{(a + 1)}} \qquad \text{Removing a factor of 1: } \frac{a + 1}{a + 1} = 1$$

$$= \frac{2a - 1}{a(a - 1)}.$$

Do Exercise 8.

Example 9 Add: $\dfrac{x + 4}{x - 2} + \dfrac{x - 7}{x + 5}$.

First, we find the LCD. It is just the product of the denominators:

$$\text{LCD} = (x - 2)(x + 5).$$

We multiply by 1 to get the LCD in each expression, and then add and simplify:

$$\frac{x + 4}{x - 2} \cdot \frac{x + 5}{x + 5} + \frac{x - 7}{x + 5} \cdot \frac{x - 2}{x - 2} = \frac{(x + 4)(x + 5)}{(x - 2)(x + 5)} + \frac{(x - 7)(x - 2)}{(x - 2)(x + 5)}$$

$$= \frac{x^2 + 9x + 20}{(x - 2)(x + 5)} + \frac{x^2 - 9x + 14}{(x - 2)(x + 5)}$$

$$= \frac{x^2 + 9x + 20 + x^2 - 9x + 14}{(x - 2)(x + 5)}$$

$$= \frac{2x^2 + 34}{(x - 2)(x + 5)}.$$

Do Exercise 9.

8. Add:

$$\frac{3}{x^3 - x} + \frac{4}{x^2 + 2x + 1}.$$

9. Add:

$$\frac{x - 2}{x + 3} + \frac{x + 7}{x + 8}.$$

Calculator Spotlight

Use the TABLE feature to check the additions in Examples 7–9. Then check your answers to Margin Exercises 7–9.

Answers on page A-23

10. Add:

$$\frac{5}{x^2 + 17x + 16} + \frac{3}{x^2 + 9x + 8}.$$

Example 10 Add: $\dfrac{x}{x^2 + 11x + 30} + \dfrac{-5}{x^2 + 9x + 20}$.

$$\frac{x}{x^2 + 11x + 30} + \frac{-5}{x^2 + 9x + 20}$$

$$= \frac{x}{(x + 5)(x + 6)} + \frac{-5}{(x + 5)(x + 4)}$$ Factoring the denominators in order to find the LCD. The LCD is $(x + 4)(x + 5)(x + 6)$.

$$= \frac{x}{(x + 5)(x + 6)} \cdot \frac{x + 4}{x + 4} + \frac{-5}{(x + 5)(x + 4)} \cdot \frac{x + 6}{x + 6}$$ Multiplying by 1

$$= \frac{x(x + 4) + (-5)(x + 6)}{(x + 4)(x + 5)(x + 6)} = \frac{x^2 + 4x - 5x - 30}{(x + 4)(x + 5)(x + 6)}$$

$$= \frac{x^2 - x - 30}{(x + 4)(x + 5)(x + 6)}$$

$$= \frac{(x - 6)\cancel{(x + 5)}}{(x + 4)\cancel{(x + 5)}(x + 6)}$$

$$= \frac{(x - 6)}{(x + 4)(x + 6)}$$

→ Always simplify at the end if possible: $\dfrac{x + 5}{x + 5} = 1$.

Do Exercise 10.

Suppose that after we factor to find the LCD, we find factors that are opposites. There are several ways to handle this, but the easiest is to first go back and multiply by $-1/-1$ appropriately to change factors so that they are not opposites.

11. Add:

$$\frac{x + 3}{x^2 - 16} + \frac{5}{12 - 3x}.$$

Example 11 Add: $\dfrac{x}{x^2 - 25} + \dfrac{3}{10 - 2x}$.

First, we factor as though we are going to find the LCD:

$$x^2 - 25 = (x - 5)(x + 5);$$

$$10 - 2x = 2(5 - x).$$

We note that there is an $x - 5$ as one factor and a $5 - x$ as another factor. If the denominator of the second expression were $2x - 10$, this situation would not occur. To rewrite the second expression with a denominator of $2x - 10$, we multiply by 1 using $-1/-1$, and then continue as before:

$$\frac{x}{x^2 - 25} + \frac{3}{10 - 2x} = \frac{x}{(x - 5)(x + 5)} + \frac{3}{10 - 2x} \cdot \frac{-1}{-1}$$

$$= \frac{x}{(x - 5)(x + 5)} + \frac{-3}{2x - 10}$$

$$= \frac{x}{(x - 5)(x + 5)} + \frac{-3}{2(x - 5)}$$ LCD $= 2(x - 5)(x + 5)$

$$= \frac{x}{(x - 5)(x + 5)} \cdot \frac{2}{2} + \frac{-3}{2(x - 5)} \cdot \frac{x + 5}{x + 5}$$

$$= \frac{2x - 3(x + 5)}{2(x - 5)(x + 5)} = \frac{2x - 3x - 15}{2(x - 5)(x + 5)}$$

$$= \frac{-x - 15}{2(x - 5)(x + 5)}.$$ Collecting like terms

Do Exercise 11.

Answers on page A-23

Exercise Set 6.3

a Add. Simplify, if possible.

1. $\frac{5}{8} + \frac{3}{8}$

2. $\frac{3}{16} + \frac{5}{16}$

3. $\frac{1}{3+x} + \frac{5}{3+x}$

4. $\frac{4x+6}{2x-1} + \frac{5-8x}{-1+2x}$

5. $\frac{x^2+7x}{x^2-5x} + \frac{x^2-4x}{x^2-5x}$

6. $\frac{4}{x+y} + \frac{9}{y+x}$

7. $\frac{7}{8} + \frac{5}{-8}$

8. $\frac{5}{-3} + \frac{11}{3}$

9. $\frac{3}{t} + \frac{4}{-t}$

10. $\frac{5}{-a} + \frac{8}{a}$

11. $\frac{2x+7}{x-6} + \frac{3x}{6-x}$

12. $\frac{2x-7}{5x-8} + \frac{6+10x}{8-5x}$

13. $\frac{y^2}{y-3} + \frac{9}{3-y}$

14. $\frac{t^2}{t-2} + \frac{4}{2-t}$

15. $\frac{b-7}{b^2-16} + \frac{7-b}{16-b^2}$

16. $\frac{a-3}{a^2-25} + \frac{a-3}{25-a^2}$

17. $\frac{a^2}{a-b} + \frac{b^2}{b-a}$

18. $\frac{x^2}{x-7} + \frac{49}{7-x}$

19. $\frac{x+3}{x-5} + \frac{2x-1}{5-x} + \frac{2(3x-1)}{x-5}$

20. $\frac{3(x-2)}{2x-3} + \frac{5(2x+1)}{2x-3} + \frac{3(x+1)}{3-2x}$

21. $\frac{2(4x+1)}{5x-7} + \frac{3(x-2)}{7-5x} + \frac{-10x-1}{5x-7}$

22. $\frac{5(x-2)}{3x-4} + \frac{2(x-3)}{4-3x} + \frac{3(5x+1)}{4-3x}$

23. $\dfrac{x+1}{(x+3)(x-3)} + \dfrac{4(x-3)}{(x-3)(x+3)} + \dfrac{(x-1)(x-3)}{(3-x)(x+3)}$

24. $\dfrac{2(x+5)}{(2x-3)(x-1)} + \dfrac{3x+4}{(2x-3)(1-x)} + \dfrac{x-5}{(3-2x)(x-1)}$

25. $\dfrac{2}{x} + \dfrac{5}{x^2}$

26. $\dfrac{3}{y^2} + \dfrac{6}{y}$

27. $\dfrac{5}{6r} + \dfrac{7}{8r}$

28. $\dfrac{13}{18x} + \dfrac{7}{24x}$

29. $\dfrac{4}{xy^2} + \dfrac{6}{x^2y}$

30. $\dfrac{8}{ab^3} + \dfrac{3}{a^2b}$

31. $\dfrac{2}{9t^3} + \dfrac{1}{6t^2}$

32. $\dfrac{5}{c^2d^3} + \dfrac{-4}{7cd^2}$

33. $\dfrac{x+y}{xy^2} + \dfrac{3x+y}{x^2y}$

34. $\dfrac{2c-d}{c^2d} + \dfrac{c+d}{cd^2}$

35. $\dfrac{3}{x-2} + \dfrac{3}{x+2}$

36. $\dfrac{2}{y+1} + \dfrac{2}{y-1}$

37. $\dfrac{3}{x+1} + \dfrac{2}{3x}$

38. $\dfrac{4}{5y} + \dfrac{7}{y-2}$

39. $\dfrac{2x}{x^2-16} + \dfrac{x}{x-4}$

40. $\dfrac{4x}{x^2 - 25} + \dfrac{x}{x + 5}$

41. $\dfrac{5}{z + 4} + \dfrac{3}{3z + 12}$

42. $\dfrac{t}{t - 3} + \dfrac{5}{4t - 12}$

43. $\dfrac{3}{x - 1} + \dfrac{2}{(x - 1)^2}$

44. $\dfrac{8}{(y + 3)^2} + \dfrac{5}{y + 3}$

45. $\dfrac{4a}{5a - 10} + \dfrac{3a}{10a - 20}$

46. $\dfrac{9x}{6x - 30} + \dfrac{3x}{4x - 20}$

47. $\dfrac{x + 4}{x} + \dfrac{x}{x + 4}$

48. $\dfrac{a}{a - 3} + \dfrac{a - 3}{a}$

49. $\dfrac{4}{a^2 - a - 2} + \dfrac{3}{a^2 + 4a + 3}$

50. $\dfrac{a}{a^2 - 2a + 1} + \dfrac{1}{a^2 - 5a + 4}$

51. $\dfrac{x + 3}{x - 5} + \dfrac{x - 5}{x + 3}$

52. $\dfrac{3x}{2y - 3} + \dfrac{2x}{3y - 2}$

53. $\dfrac{a}{a^2 - 1} + \dfrac{2a}{a^2 - a}$

54. $\dfrac{3x + 2}{3x + 6} + \dfrac{x - 2}{x^2 - 4}$

55. $\dfrac{6}{x - y} + \dfrac{4x}{y^2 - x^2}$

56. $\dfrac{a - 2}{3 - a} + \dfrac{4 - a^2}{a^2 - 9}$

57. $\dfrac{4 - a}{25 - a^2} + \dfrac{a + 1}{a - 5}$

58. $\dfrac{x + 2}{x - 7} + \dfrac{3 - x}{49 - x^2}$

59. $\dfrac{2}{t^2 + t - 6} + \dfrac{3}{t^2 - 9}$

60. $\dfrac{10}{a^2 - a - 6} + \dfrac{3a}{a^2 + 4a + 4}$

Skill Maintenance

Subtract. [4.4c]

61. $(x^2 + x) - (x + 1)$

62. $(4y^3 - 5y^2 + 7y - 24) - (-9y^3 + 9y^2 - 5y + 49)$

Simplify. [4.2b]

63. $(2x^4y^3)^{-3}$

64. $\left(\dfrac{x^3}{5y}\right)^2$

65. $\left(\dfrac{x^{-4}}{y^7}\right)^3$

66. $(5x^{-2}y^{-3})^2$

Graph.

67. $y = \dfrac{1}{2}x - 5$ [3.2b], [3.3a]

68. $2y + x + 10 = 0$ [3.2b], [3.3a]

69. $y = 3$ [3.3b]

70. $x = -5$ [3.3b]

Solve.

71. $3x - 7 = 5x + 9$ [2.3b]

72. $2a + 8 = 13 - 4a$ [2.3b]

73. $x^2 - 8x + 15 = 0$ [5.8b]

74. $x^2 - 7x = 18$ [5.8b]

Synthesis

75. ◆ Explain why the expressions

$$\frac{1}{3 - x} \quad \text{and} \quad \frac{1}{x - 3}$$

are opposites.

76. ◆ Why is it better to use the *least* common denominator, rather than *any* common denominator, when adding rational expressions?

Find the perimeter and the area of the figure.

77.

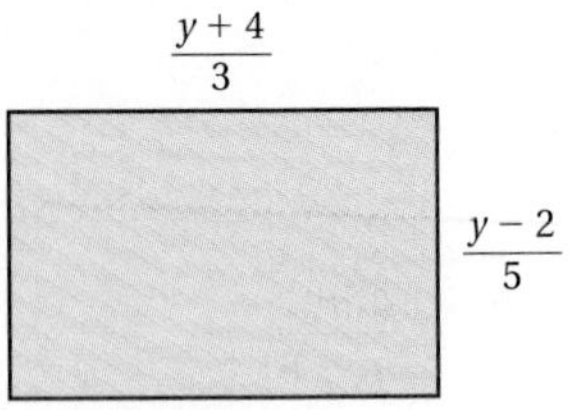

$\dfrac{y + 4}{3}$ (top), $\dfrac{y - 2}{5}$ (side)

78.

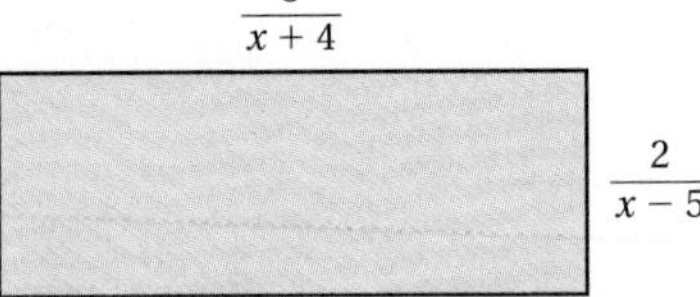

$\dfrac{3}{x + 4}$ (top), $\dfrac{2}{x - 5}$ (side)

Add. Simplify, if possible.

79. $\dfrac{5}{z + 2} + \dfrac{4z}{z^2 - 4} + 2$

80. $\dfrac{-2}{y^2 - 9} + \dfrac{4y}{(y - 3)^2} + \dfrac{6}{3 - y}$

81. $\dfrac{3z^2}{z^4 - 4} + \dfrac{5z^2 - 3}{2z^4 + z^2 - 6}$

82. Find an expression equivalent to

$$\frac{a - 3b}{a - b}$$

that is a sum of two fractional expressions. Answers may vary.

83.–88. Use the TABLE feature to check the additions in Exercises 47–52.

6.4 Subtracting Rational Expressions

a We subtract rational expressions as we do rational numbers.

> To subtract when the denominators are the same, subtract the numerators and keep the same denominator.

Example 1 Subtract: $\frac{8}{x} - \frac{3}{x}$.

$$\frac{8}{x} - \frac{3}{x} = \frac{8-3}{x} = \frac{5}{x}$$

Example 2 Subtract: $\frac{3x}{x+2} - \frac{x-2}{x+2}$.

$$\frac{3x}{x+2} - \frac{x-2}{x+2} = \frac{3x-(x-2)}{x+2}$$

The parentheses are important to make sure that you subtract the entire numerator.

$$= \frac{3x-x+2}{x+2} = \frac{2x+2}{x+2}$$

Do Exercises 1–3.

When one denominator is the opposite of the other, we can first multiply one expression by $-1/-1$ to obtain a common denominator.

Example 3 Subtract: $\frac{x}{5} - \frac{3x-4}{-5}$.

$$\frac{x}{5} - \frac{3x-4}{-5} = \frac{x}{5} - \frac{3x-4}{-5}\cdot\frac{-1}{-1}$$

Multiplying by 1 using $\frac{-1}{-1}$ ← This is equal to 1 (not −1).

$$= \frac{x}{5} - \frac{(3x-4)(-1)}{(-5)(-1)}$$

$$= \frac{x}{5} - \frac{4-3x}{5}$$

Remember the parentheses!

$$= \frac{x-(4-3x)}{5}$$

$$= \frac{x-4+3x}{5} = \frac{4x-4}{5}$$

Example 4 Subtract: $\frac{5y}{y-5} - \frac{2y-3}{5-y}$.

$$\frac{5y}{y-5} - \frac{2y-3}{5-y} = \frac{5y}{y-5} - \frac{2y-3}{5-y}\cdot\frac{-1}{-1}$$

$$= \frac{5y}{y-5} - \frac{(2y-3)(-1)}{(5-y)(-1)} = \frac{5y}{y-5} - \frac{3-2y}{y-5}$$

Remember the parentheses!

$$= \frac{5y-(3-2y)}{y-5}$$

Objectives

a Subtract rational expressions.

b Simplify combined additions and subtractions of rational expressions.

For Extra Help

TAPE 10

MAC WIN

CD-ROM

Subtract.

1. $\frac{7}{11} - \frac{3}{11}$

2. $\frac{7}{y} - \frac{2}{y}$

3. $\frac{2x^2+3x-7}{2x+1} - \frac{x^2+x-8}{2x+1}$

Answers on page A-23

Subtract.

4. $\dfrac{x}{3} - \dfrac{2x - 1}{-3}$

5. $\dfrac{3x}{x - 2} - \dfrac{x - 3}{2 - x}$

6. Subtract:

$$\frac{x - 2}{3x} - \frac{2x - 1}{5x}.$$

Answers on page A-23

Then $$= \frac{5y - 3 + 2y}{y - 5}$$
$$= \frac{7y - 3}{y - 5}.$$

Do Exercises 4 and 5.

To subtract rational expressions with different denominators, we use a procedure similar to what we used for addition, except that we subtract numerators and write the difference over the LCD.

To subtract rational expressions with different denominators:

1. Find the LCM of the denominators. This is the least common denominator (LCD).
2. For each rational expression, find an equivalent expression with the LCD. To do so, multiply by 1 using a symbol for 1 made up of factors of the LCD that are missing from the original denominator.
3. Subtract the numerators. Write the difference over the LCD.
4. Simplify, if possible.

Example 5 Subtract: $\dfrac{x + 2}{x - 4} - \dfrac{x + 1}{x + 4}$.

The LCD = $(x - 4)(x + 4)$.

$$\frac{x + 2}{x - 4} \cdot \frac{x + 4}{x + 4} - \frac{x + 1}{x + 4} \cdot \frac{x - 4}{x - 4}$$ Multiplying by 1

$$= \frac{(x + 2)(x + 4)}{(x - 4)(x + 4)} - \frac{(x + 1)(x - 4)}{(x - 4)(x + 4)}$$

$$= \frac{x^2 + 6x + 8}{(x - 4)(x + 4)} - \frac{x^2 - 3x - 4}{(x - 4)(x + 4)}$$

Subtracting this numerator. Don't forget the parentheses.

$$= \frac{x^2 + 6x + 8 - (x^2 - 3x - 4)}{(x - 4)(x + 4)}$$

$$= \frac{x^2 + 6x + 8 - x^2 + 3x + 4}{(x - 4)(x + 4)}$$

$$= \frac{9x + 12}{(x - 4)(x + 4)}$$

Do Exercise 6.

Example 6 Subtract: $\dfrac{x}{x^2 + 5x + 6} - \dfrac{2}{x^2 + 3x + 2}$.

$$\frac{x}{x^2 + 5x + 6} - \frac{2}{x^2 + 3x + 2}$$

$$= \frac{x}{(x + 2)(x + 3)} - \frac{2}{(x + 2)(x + 1)}$$ LCD = $(x + 1)(x + 2)(x + 3)$

$$= \frac{x}{(x + 2)(x + 3)} \cdot \frac{x + 1}{x + 1} - \frac{2}{(x + 2)(x + 1)} \cdot \frac{x + 3}{x + 3}$$

$$= \frac{x^2 + x}{(x + 1)(x + 2)(x + 3)} - \frac{2x + 6}{(x + 1)(x + 2)(x + 3)}$$

Then

$$= \frac{x^2 + x - (2x + 6)}{(x+1)(x+2)(x+3)}$$ Subtracting this numerator. Don't forget the parentheses.

$$= \frac{x^2 + x - 2x - 6}{(x+1)(x+2)(x+3)}$$

$$= \frac{x^2 - x - 6}{(x+1)(x+2)(x+3)}$$

$$= \frac{(x+2)(x-3)}{(x+1)(x+2)(x+3)}$$

$$= \frac{\cancel{(x+2)}(x-3)}{(x+1)\cancel{(x+2)}(x+3)}$$ Simplifying by removing a factor of 1: $\frac{x+2}{x+2} = 1$

$$= \frac{x-3}{(x+1)(x+3)}.$$

Do Exercise 7.

Suppose that after we factor to find the LCD, we find factors that are opposites. Then we multiply by $-1/-1$ appropriately to change factors so that they are not opposites.

Example 7 Subtract: $\frac{p}{64 - p^2} - \frac{5}{p - 8}$.

Factoring $64 - p^2$, we get $(8 - p)(8 + p)$. Note that the factors $8 - p$ in the first denominator and $p - 8$ in the second denominator are opposites. We multiply the first expression by $-1/-1$ to avoid this situation. Then we proceed as before.

$$\frac{p}{64 - p^2} - \frac{5}{p-8} = \frac{p}{64 - p^2} \cdot \frac{-1}{-1} - \frac{5}{p-8}$$

$$= \frac{-p}{p^2 - 64} - \frac{5}{p-8}$$

$$= \frac{-p}{(p-8)(p+8)} - \frac{5}{p-8}$$ LCD = $(p-8)(p+8)$

$$= \frac{-p}{(p-8)(p+8)} - \frac{5}{p-8} \cdot \frac{p+8}{p+8}$$

$$= \frac{-p}{(p-8)(p+8)} - \frac{5p+40}{(p-8)(p+8)}$$

$$= \frac{-p - (5p + 40)}{(p-8)(p+8)}$$ Subtracting this numerator. Don't forget the parentheses.

$$= \frac{-p - 5p - 40}{(p-8)(p+8)}$$

$$= \frac{-6p - 40}{(p-8)(p+8)}$$

Do Exercise 8.

7. Subtract:

$$\frac{x}{x^2 + 15x + 56} - \frac{6}{x^2 + 13x + 42}.$$

8. Subtract:

$$\frac{y}{16 - y^2} - \frac{7}{y - 4}.$$

Calculator Spotlight

Use the TABLE feature to check the subtractions in Examples 5–7. Then check your answers to Margin Exercises 6–8.

Answers on page A-23

9. Perform the indicated operations and simplify:

$$\frac{x+2}{x^2-9} - \frac{x-7}{9-x^2} + \frac{-8-x}{x^2-9}.$$

10. Perform the indicated operations and simplify:

$$\frac{1}{x} - \frac{5}{3x} + \frac{2x}{x+1}.$$

Answers on page A-23

b Combined Additions and Subtractions

Now let's look at some combined additions and subtractions.

Example 8 Perform the indicated operations and simplify:

$$\frac{x+9}{x^2-4} + \frac{5-x}{4-x^2} - \frac{2+x}{x^2-4}.$$

We have

$$\begin{aligned}
\frac{x+9}{x^2-4} + \frac{5-x}{4-x^2} - \frac{2+x}{x^2-4} &= \frac{x+9}{x^2-4} + \frac{5-x}{4-x^2} \cdot \frac{-1}{-1} - \frac{2+x}{x^2-4} \\
&= \frac{x+9}{x^2-4} + \frac{x-5}{x^2-4} - \frac{2+x}{x^2-4} \\
&= \frac{(x+9)+(x-5)-(2+x)}{x^2-4} \\
&= \frac{x+9+x-5-2-x}{x^2-4} \\
&= \frac{x+2}{x^2-4} \\
&= \frac{\cancel{(x+2)} \cdot 1}{\cancel{(x+2)}(x-2)} \qquad \frac{x+2}{x+2} = 1 \\
&= \frac{1}{x-2}.
\end{aligned}$$

Do Exercise 9.

Example 9 Perform the indicated operations and simplify:

$$\frac{1}{x} - \frac{1}{x^2} + \frac{2}{x+1}.$$

The LCD $= x \cdot x(x+1)$, or $x^2(x+1)$.

$$\begin{aligned}
&\frac{1}{x} \cdot \frac{x(x+1)}{x(x+1)} - \frac{1}{x^2} \cdot \frac{(x+1)}{(x+1)} + \frac{2}{x+1} \cdot \frac{x^2}{x^2} \\
&= \frac{x(x+1)}{x^2(x+1)} - \frac{x+1}{x^2(x+1)} + \frac{2x^2}{x^2(x+1)} \\
&= \frac{x(x+1)-(x+1)+2x^2}{x^2(x+1)} \qquad \text{Subtracting this numerator. Don't forget the parentheses.} \\
&= \frac{x^2+x-x-1+2x^2}{x^2(x+1)} \\
&= \frac{3x^2-1}{x^2(x+1)}
\end{aligned}$$

Do Exercise 10.

Exercise Set 6.4

a Subtract. Simplify, if possible.

1. $\frac{7}{x} - \frac{3}{x}$

2. $\frac{5}{a} - \frac{8}{a}$

3. $\frac{y}{y - 4} - \frac{4}{y - 4}$

4. $\frac{t^2}{t + 5} - \frac{25}{t + 5}$

5. $\frac{2x - 3}{x^2 + 3x - 4} - \frac{x - 7}{x^2 + 3x - 4}$

6. $\frac{x + 1}{x^2 - 2x + 1} - \frac{5 - 3x}{x^2 - 2x + 1}$

7. $\frac{11}{6} - \frac{5}{-6}$

8. $\frac{5}{9} - \frac{7}{-9}$

9. $\frac{5}{a} - \frac{8}{-a}$

10. $\frac{8}{x} - \frac{3}{-x}$

11. $\frac{4}{y - 1} - \frac{4}{1 - y}$

12. $\frac{5}{a - 2} - \frac{3}{2 - a}$

13. $\frac{3 - x}{x - 7} - \frac{2x - 5}{7 - x}$

14. $\frac{t^2}{t - 2} - \frac{4}{2 - t}$

15. $\frac{a - 2}{a^2 - 25} - \frac{6 - a}{25 - a^2}$

16. $\frac{x - 8}{x^2 - 16} - \frac{x - 8}{16 - x^2}$

17. $\frac{4 - x}{x - 9} - \frac{3x - 8}{9 - x}$

18. $\frac{4x - 6}{x - 5} - \frac{7 - 2x}{5 - x}$

19. $\frac{2(x-1)}{2x-3} - \frac{3(x+2)}{2x-3} - \frac{x-1}{3-2x}$

20. $\frac{5(2y+1)}{2y-3} - \frac{3(y-1)}{3-2y} - \frac{3(y-2)}{2y-3}$

21. $\frac{a-2}{10} - \frac{a+1}{5}$

22. $\frac{y+3}{2} - \frac{y-4}{4}$

23. $\frac{4z-9}{3z} - \frac{3z-8}{4z}$

24. $\frac{a-1}{4a} - \frac{2a+3}{a}$

25. $\frac{4x+2t}{3xt^2} - \frac{5x-3t}{x^2t}$

26. $\frac{5x+3y}{2x^2y} - \frac{3x+4y}{xy^2}$

27. $\frac{5}{x+5} - \frac{3}{x-5}$

28. $\frac{3t}{t-1} - \frac{8t}{t+1}$

29. $\frac{3}{2t^2-2t} - \frac{5}{2t-2}$

30. $\frac{11}{x^2-4} - \frac{8}{x+2}$

31. $\frac{2s}{t^2-s^2} - \frac{s}{t-s}$

32. $\frac{3}{12+x-x^2} - \frac{2}{x^2-9}$

33. $\frac{y-5}{y} - \frac{3y-1}{4y}$

34. $\frac{3x-2}{4x} - \frac{3x+1}{6x}$

35. $\frac{a}{x+a} - \frac{a}{x-a}$

36. $\frac{a}{a-b} - \frac{a}{a+b}$

37. $\dfrac{5x}{x^2 - 9} - \dfrac{4}{3 - x}$

38. $\dfrac{8x}{16 - x^2} - \dfrac{5}{x - 4}$

39. $\dfrac{t^2}{2t^2 - 2t} - \dfrac{1}{2t - 2}$

40. $\dfrac{4}{5a^2 - 5a} - \dfrac{2}{5a - 5}$

41. $\dfrac{x}{x^2 + 5x + 6} - \dfrac{2}{x^2 + 3x + 2}$

42. $\dfrac{a}{a^2 + 11a + 30} - \dfrac{5}{a^2 + 9a + 20}$

b Perform the indicated operations and simplify.

43. $\dfrac{3(2x + 5)}{x - 1} - \dfrac{3(2x - 3)}{1 - x} + \dfrac{6x - 1}{x - 1}$

44. $\dfrac{a - 2b}{b - a} - \dfrac{3a - 3b}{a - b} + \dfrac{2a - b}{a - b}$

45. $\dfrac{x - y}{x^2 - y^2} + \dfrac{x + y}{x^2 - y^2} - \dfrac{2x}{x^2 - y^2}$

46. $\dfrac{x - 3y}{2(y - x)} + \dfrac{x + y}{2(x - y)} - \dfrac{2x - 2y}{2(x - y)}$

47. $\dfrac{10}{2y - 1} - \dfrac{6}{1 - 2y} + \dfrac{y}{2y - 1} + \dfrac{y - 4}{1 - 2y}$

48. $\dfrac{(x + 1)(2x - 1)}{(2x - 3)(x - 3)} - \dfrac{(x - 3)(x + 1)}{(3 - x)(3 - 2x)} + \dfrac{(2x + 1)(x + 3)}{(3 - 2x)(x - 3)}$

49. $\dfrac{a + 6}{4 - a^2} - \dfrac{a + 3}{a + 2} + \dfrac{a - 3}{2 - a}$

50. $\dfrac{4t}{t^2 - 1} - \dfrac{2}{t} - \dfrac{2}{t + 1}$

51. $\frac{2z}{1-2z} + \frac{3z}{2z+1} - \frac{3}{4z^2-1}$

52. $\frac{1}{x-y} - \frac{2x}{x^2-y^2} + \frac{1}{x+y}$

53. $\frac{1}{x+y} - \frac{1}{x-y} + \frac{2x}{x^2-y^2}$

54. $\frac{2b}{a^2-b^2} - \frac{1}{a+b} + \frac{1}{a-b}$

Skill Maintenance

Simplify.

55. $\frac{x^8}{x^3}$ [4.1e]

56. $3x^4 \cdot 10x^8$ [4.1d]

57. $(a^2b^{-5})^{-4}$ [4.2b]

58. $\frac{54x^{10}}{3x^7}$ [4.1e]

59. $\frac{66x^2}{11x^5}$ [4.1e]

60. $5x^{-7} \cdot 2x^4$ [4.1d]

Find a polynomial for the shaded area of the figure. [4.4d]

61.

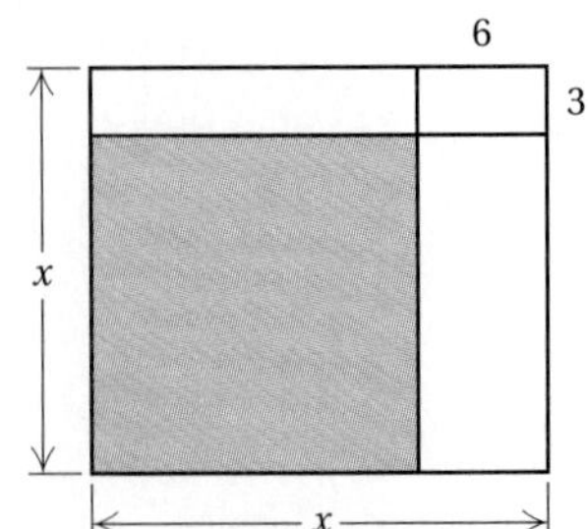

62.

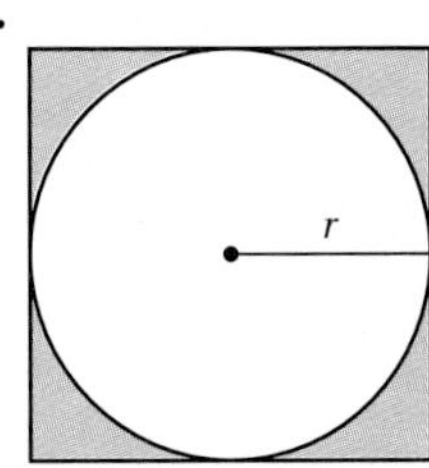

Synthesis

63. ◆ Are parentheses as important when adding rational expressions as they are when subtracting? Why or why not?

64. ◆ Is it possible to add or subtract rational expressions without knowing how to factor? Why or why not?

Perform the indicated operations and simplify.

65. $\frac{2x+11}{x-3} \cdot \frac{3}{x+4} + \frac{2x+1}{4+x} \cdot \frac{3}{3-x}$

66. $\frac{x^2}{3x^2-5x-2} - \frac{2x}{3x+1} \cdot \frac{1}{x-2}$

67. $\frac{x}{x^4-y^4} - \left(\frac{1}{x+y}\right)^2$

68. $\left(\frac{a}{a-b} + \frac{b}{a+b}\right)\left(\frac{1}{3a+b} + \frac{2a+6b}{9a^2-b^2}\right)$

69. The perimeter of the following right triangle is $2a + 5$. Find the length of the missing side and the area.

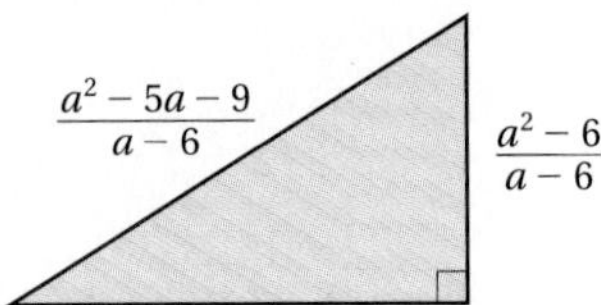

70.–75. Use the TABLE feature to check the subtractions in Exercises 29–34.

6.5 Solving Rational Equations

a Rational Equations

In Sections 6.1–6.4, we studied operations with *rational expressions.* These expressions have no equals signs. We can perform the operations and simplify, but we cannot solve if there are no equals signs—as, for example, in

$$\frac{x^2+6x+9}{x^2-4} \cdot \frac{x-2}{x+3}, \quad \frac{x+y}{x-y} \div \frac{x^2+y}{x^2-y^2}, \quad \text{and} \quad \frac{a+3}{a^2-16} + \frac{5}{12-3a}.$$

Operation signs occur. There are no equals signs!

Most often, the result of our calculation is another rational expression that has not been cleared of fractions.

Equations *do have* equals signs, and we can clear them of fractions as we did in Section 2.3. A **rational**, or **fractional, equation** is an equation containing one or more rational expressions. Here are some examples:

$$\frac{2}{3} + \frac{5}{6} = \frac{x}{9}, \quad x + \frac{6}{x} = -5, \quad \text{and} \quad \frac{x^2}{x-1} = \frac{1}{x-1}.$$

There are equals signs as well as operation signs.

> To solve a rational equation, the first step is to clear the equation of fractions. To do this, multiply both sides of the equation by the LCM of all the denominators. Then carry out the equation-solving process as we learned it in Chapter 2.

When clearing an equation of fractions, we use the terminology LCM instead of LCD because we are *not* adding or subtracting rational expressions.

Example 1 Solve: $\frac{2}{3} + \frac{5}{6} = \frac{x}{9}$.

The LCM of all denominators is $2 \cdot 3 \cdot 3$, or 18. We multiply by 18 on both sides:

$$18\left(\frac{2}{3} + \frac{5}{6}\right) = 18 \cdot \frac{x}{9}$$ Multiplying by the LCM on both sides

$$18 \cdot \frac{2}{3} + 18 \cdot \frac{5}{6} = 18 \cdot \frac{x}{9}$$ Multiplying to remove parentheses

When clearing an equation of fractions, be sure to multiply *each* term by the LCM.

$$12 + 15 = 2x$$ Simplifying. Note that we have now cleared fractions.

$$27 = 2x$$

$$\frac{27}{2} = x.$$

The solution is $\frac{27}{2}$.

Do Exercise 1.

Objective

a Solve rational equations.

For Extra Help

TAPE 10 | MAC WIN | CD-ROM

1. Solve: $\frac{3}{4} + \frac{5}{8} = \frac{x}{12}$.

Answer on page A-23

2. Solve: $\frac{1}{x} = \frac{1}{6 - x}$.

3. Solve: $\frac{x}{4} - \frac{x}{6} = \frac{1}{8}$.

Answers on page A-23

Example 2 Solve: $\frac{1}{x} = \frac{1}{4 - x}$.

The LCM is $x(4 - x)$. We multiply by $x(4 - x)$ on both sides:

$$\frac{1}{x} = \frac{1}{4 - x}$$

$$x(4 - x) \cdot \frac{1}{x} = x(4 - x) \cdot \frac{1}{4 - x} \quad \text{Multiplying by the LCM on both sides}$$

$$4 - x = x \quad \text{Simplifying}$$

$$4 = 2x$$

$$x = 2.$$

CHECK:

$$\frac{1}{x} = \frac{1}{4 - x}$$

$\frac{1}{2}$	$\frac{1}{4 - 2}$
	$\frac{1}{2}$ TRUE

This checks, so the solution is 2.

Do Exercise 2.

Example 3 Solve: $\frac{x}{6} - \frac{x}{8} = \frac{1}{12}$.

The LCM is 24. We multiply by 24 on both sides:

$$\frac{x}{6} - \frac{x}{8} = \frac{1}{12}$$

$$24\left(\frac{x}{6} - \frac{x}{8}\right) = 24 \cdot \frac{1}{12} \quad \text{Multiplying by the LCM on both sides}$$

$$24 \cdot \frac{x}{6} - 24 \cdot \frac{x}{8} = 24 \cdot \frac{1}{12} \quad \text{Multiplying to remove parentheses}$$

Be sure to multiply *each* term by the LCM.

$$4x - 3x = 2 \quad \text{Simplifying}$$

$$x = 2.$$

CHECK:

$$\frac{x}{6} - \frac{x}{8} = \frac{1}{12}$$

$\frac{2}{6} - \frac{2}{8}$	$\frac{1}{12}$
$\frac{1}{3} - \frac{1}{4}$	
$\frac{4}{12} - \frac{3}{12}$	
$\frac{1}{12}$	TRUE

This checks, so the solution is 2.

Do Exercise 3.

AG Algebraic–Graphical Connection

We can obtain a visual check of the solutions of a rational equation by graphing. For example, consider the equation

$$\frac{x}{4} + \frac{x}{2} = 6.$$

We can examine the solution by graphing the equations

$$y = \frac{x}{4} + \frac{x}{2} \quad \text{and} \quad y = 6$$

using the same set of axes.

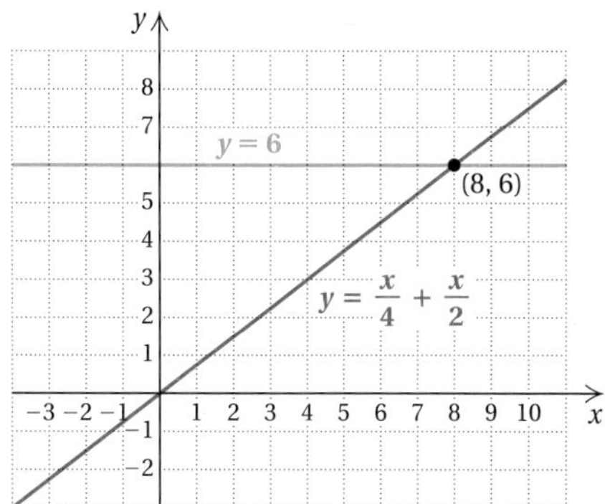

The y-values for each equation will be the same where the graphs intersect. The x-value of that point will yield that value, so it will be the solution of the equation. It appears from the graph that when $x = 8$, the value of $x/4 + x/2$ is 6. We can check by substitution:

$$\frac{x}{4} + \frac{x}{2} = \frac{8}{4} + \frac{8}{2} = 2 + 4 = 6.$$

Thus the solution is 8.

Example 4 Solve: $\frac{2}{3x} + \frac{1}{x} = 10.$

The LCM is $3x$. We multiply by $3x$ on both sides:

$$\frac{2}{3x} + \frac{1}{x} = 10$$

$$3x\left(\frac{2}{3x} + \frac{1}{x}\right) = 3x \cdot 10 \qquad \text{Multiplying by the LCM on both sides}$$

$$3x \cdot \frac{2}{3x} + 3x \cdot \frac{1}{x} = 3x \cdot 10 \qquad \text{Multiplying to remove parentheses}$$

$$2 + 3 = 30x \qquad \text{Simplifying}$$

$$5 = 30x$$

$$\frac{5}{30} = x$$

$$\frac{1}{6} = x.$$

We leave the check to the student. The solution is $\frac{1}{6}$.

Do Exercise 4.

4. Solve: $\frac{1}{2x} + \frac{1}{x} = -12.$

Answer on page A-23

5. Solve: $x + \frac{1}{x} = 2$.

Example 5 Solve: $x + \frac{6}{x} = -5$.

The LCM is x. We multiply by x on both sides:

$$x + \frac{6}{x} = -5$$

$$x\left(x + \frac{6}{x}\right) = -5x \quad \text{Multiplying by } x \text{ on both sides}$$

$$x \cdot x + x \cdot \frac{6}{x} = -5x \quad \text{Note that each rational expression on the left is now multiplied by } x.$$

$$x^2 + 6 = -5x \quad \text{Simplifying}$$

$$x^2 + 5x + 6 = 0 \quad \text{Adding } 5x \text{ to get a 0 on one side}$$

$$(x + 3)(x + 2) = 0 \quad \text{Factoring}$$

$$x + 3 = 0 \quad or \quad x + 2 = 0 \quad \text{Using the principle of zero products}$$

$$x = -3 \quad or \quad x = -2.$$

CHECK:

For -3:

$$x + \frac{6}{x} = -5$$

$-3 + \frac{6}{-3}$	-5
$-3 - 2$	
-5	TRUE

For -2:

$$x + \frac{6}{x} = -5$$

$-2 + \frac{6}{-2}$	-5
$-2 - 3$	
-5	TRUE

Both of these check, so there are two solutions, -3 and -2.

Answer on page A-23

Do Exercise 5.

Calculator Spotlight

Checking Solutions Graphically. A grapher can be used to check the solutions of the equation in Example 5:

$$x + \frac{6}{x} = -5.$$

To do so, we graph

$$y_1 = x + \frac{6}{x} \quad \text{and} \quad y_2 = -5.$$

$y_1 = x + \frac{6}{x}$, $y_2 = -5$

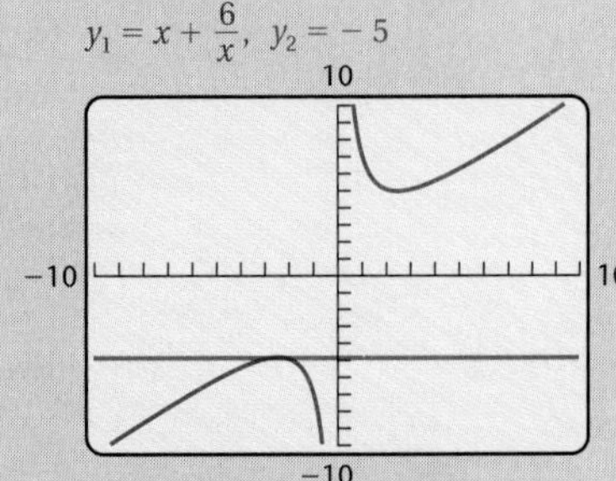

We see that the graphs appear to cross each other in the third quadrant. To get a better look, we change window settings and obtain the following graph.

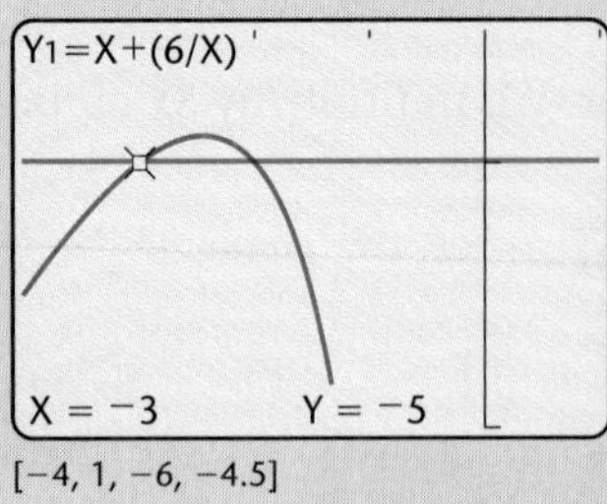

[−4, 1, −6, −4.5]

Next, we can use the CALC and VALUE features to confirm that the points of intersection occur at $x = -3$ and $x = -2$.

Exercises

Use a grapher to check the solutions in each of the following.

1. Example 1

2. Margin Exercise 1

3. Example 3

4. Margin Exercise 5

When we multiply on both sides of an equation by the LCM, the resulting equation might have solutions that are *not* solutions of the original equation. Thus we must *always* check possible solutions in the original equation.

1. If you have carried out all algebraic procedures correctly, you need only check to see if a number makes a denominator 0 in the original equation. If it does make a denominator 0, it is *not* a solution.
2. To be sure that no computational errors have been made and that you indeed have a solution, a complete check is necessary, as we did in Chapter 2.

The next example illustrates the importance of checking all possible solutions.

Example 6 Solve: $\dfrac{x^2}{x-1} = \dfrac{1}{x-1}$.

The LCM is $x - 1$. We multiply by $x - 1$ on both sides:

$$\frac{x^2}{x-1} = \frac{1}{x-1}$$

$$(x-1)\cdot\frac{x^2}{x-1} = (x-1)\cdot\frac{1}{x-1} \quad \text{Multiplying by } x-1 \text{ on both sides}$$

$$x^2 = 1 \quad \text{Simplifying}$$

$$x^2 - 1 = 0 \quad \text{Subtracting 1 to get a 0 on one side}$$

$$(x-1)(x+1) = 0 \quad \text{Factoring}$$

$$x - 1 = 0 \quad or \quad x + 1 = 0 \quad \text{Using the principle of zero products}$$

$$x = 1 \quad or \quad x = -1.$$

The numbers 1 and -1 are possible solutions. We look at the original equation and see that 1 makes a denominator 0 and is therefore not a solution. The number -1 checks and is a solution.

Do Exercise 6.

Example 7 Solve: $\dfrac{3}{x-5} + \dfrac{1}{x+5} = \dfrac{2}{x^2-25}$.

The LCM is $(x-5)(x+5)$. We multiply by $(x-5)(x+5)$ on both sides:

$$(x-5)(x+5)\left(\frac{3}{x-5} + \frac{1}{x+5}\right) = (x-5)(x+5)\left(\frac{2}{x^2-25}\right)$$

Multiplying on both sides by the LCM

$$(x-5)(x+5)\cdot\frac{3}{x-5} + (x-5)(x+5)\cdot\frac{1}{x+5} = (x-5)(x+5)\cdot\frac{2}{x^2-25}$$

$$3(x+5) + (x-5) = 2 \quad \text{Simplifying}$$

$$3x + 15 + x - 5 = 2 \quad \text{Removing parentheses}$$

$$4x + 10 = 2$$

$$4x = -8$$

$$x = -2.$$

The check is left to the student. The number -2 checks and is the solution.

Do Exercise 7.

6. Solve: $\dfrac{x^2}{x+2} = \dfrac{4}{x+2}$.

7. Solve: $\dfrac{4}{x-2} + \dfrac{1}{x+2} = \dfrac{26}{x^2-4}$.

Calculator Spotlight

Use a grapher to check the solution to Example 6.

CAUTION! We have introduced a new use of the LCM in this section. We previously used the LCM in adding or subtracting rational expressions. *Now* we have equations with equals signs. We clear fractions by multiplying on both sides of the equation by the LCM. This eliminates the denominators. Do *not* make the mistake of trying to clear fractions when you do not have an equation.

Answers on page A-23

Improving Your Math Study Skills

Are You Calculating or Solving?

At the beginning of this section, we noted that one of the common difficulties with this chapter is knowing for sure the task at hand. Are you combining expressions using operations to get another *rational expression,* or are you solving equations for which the results are numbers that are *solutions* of an equation? To learn to make these decisions, complete the following list by writing in the blank the type of answer you should get: "Rational expression" or "Solutions." You do not need to complete the mathematical operations.

Task	Answer (Just write "Rational expression" or "Solutions.")
1. Add: $\frac{4}{x-2} + \frac{1}{x+2}$.	
2. Solve: $\frac{4}{x-2} = \frac{1}{x+2}$.	
3. Subtract: $\frac{4}{x-2} - \frac{1}{x+2}$.	
4. Multiply: $\frac{4}{x-2} \cdot \frac{1}{x+2}$.	
5. Divide: $\frac{4}{x-2} \div \frac{1}{x+2}$.	
6. Solve: $\frac{4}{x-2} + \frac{1}{x+2} = \frac{26}{x^2-4}$.	
7. Perform the indicated operations and simplify: $\frac{4}{x-2} + \frac{1}{x+2} - \frac{26}{x^2-4}$.	
8. Solve: $\frac{x^2}{x-1} = \frac{1}{x-1}$.	
9. Solve: $\frac{2}{y^2-25} = \frac{3}{y-5} + \frac{1}{y-5}$.	
10. Solve: $\frac{x}{x+4} - \frac{4}{x-4} = \frac{x^2+16}{x^2-16}$.	
11. Perform the indicated operations and simplify: $\frac{x}{x+4} - \frac{4}{x-4} - \frac{x^2+16}{x^2-16}$.	
12. Solve: $\frac{5}{y-3} - \frac{30}{y^2-9} = 1$.	
13. Add: $\frac{5}{y-3} + \frac{30}{y^2-9} + 1$.	

Exercise Set 6.5

a Solve. Don't forget to check!

1. $\frac{4}{5} - \frac{2}{3} = \frac{x}{9}$

2. $\frac{x}{20} = \frac{3}{8} - \frac{4}{5}$

3. $\frac{3}{5} + \frac{1}{8} = \frac{1}{x}$

4. $\frac{2}{3} + \frac{5}{6} = \frac{1}{x}$

5. $\frac{3}{8} + \frac{4}{5} = \frac{x}{20}$

6. $\frac{3}{5} + \frac{2}{3} = \frac{x}{9}$

7. $\frac{1}{x} = \frac{2}{3} - \frac{5}{6}$

8. $\frac{1}{x} = \frac{1}{8} - \frac{3}{5}$

9. $\frac{1}{6} + \frac{1}{8} = \frac{1}{t}$

10. $\frac{1}{8} + \frac{1}{12} = \frac{1}{t}$

11. $x + \frac{4}{x} = -5$

12. $\frac{10}{x} - x = 3$

13. $\frac{x}{4} - \frac{4}{x} = 0$

14. $\frac{x}{5} - \frac{5}{x} = 0$

15. $\frac{5}{x} = \frac{6}{x} - \frac{1}{3}$

16. $\frac{4}{x} = \frac{5}{x} - \frac{1}{2}$

17. $\frac{5}{3x} + \frac{3}{x} = 1$

18. $\frac{5}{2y} + \frac{8}{y} = 1$

19. $\frac{t - 2}{t + 3} = \frac{3}{8}$

20. $\frac{x - 7}{x + 2} = \frac{1}{4}$

21. $\frac{2}{x + 1} = \frac{1}{x - 2}$

22. $\frac{8}{y - 3} = \frac{6}{y + 4}$

23. $\frac{x}{6} - \frac{x}{10} = \frac{1}{6}$

24. $\frac{x}{8} - \frac{x}{12} = \frac{1}{8}$

25. $\frac{t + 2}{5} - \frac{t - 2}{4} = 1$

26. $\frac{x + 1}{3} - \frac{x - 1}{2} = 1$

27. $\frac{5}{x - 1} = \frac{3}{x + 2}$

28. $\frac{x - 7}{x - 9} = \frac{2}{x - 9}$

29. $\frac{a - 3}{3a + 2} = \frac{1}{5}$

30. $\frac{x + 7}{8x - 5} = \frac{2}{3}$

31. $\dfrac{x-1}{x-5} = \dfrac{4}{x-5}$

32. $\dfrac{y+11}{y+8} = \dfrac{3}{y+8}$

33. $\dfrac{2}{x+3} = \dfrac{5}{x}$

34. $\dfrac{6}{y} = \dfrac{5}{y-8}$

35. $\dfrac{x-2}{x-3} = \dfrac{x-1}{x+1}$

36. $\dfrac{t+5}{t-2} = \dfrac{t-2}{t+4}$

37. $\dfrac{1}{x+3} + \dfrac{1}{x-3} = \dfrac{1}{x^2-9}$

38. $\dfrac{4}{x-3} + \dfrac{2x}{x^2-9} = \dfrac{1}{x+3}$

39. $\dfrac{x}{x+4} - \dfrac{4}{x-4} = \dfrac{x^2+16}{x^2-16}$

40. $\dfrac{5}{y-3} - \dfrac{30}{y^2-9} = 1$

41. $\frac{4-a}{8-a} = \frac{4}{a-8}$

42. $\frac{3}{x-7} = \frac{x+10}{x-7}$

43. $2 - \frac{a-2}{a+3} = \frac{a^2-4}{a+3}$

44. $\frac{5}{x-1} + x + 1 = \frac{5x+4}{x-1}$

Skill Maintenance

Simplify.

45. $(a^2b^5)^{-3}$ [4.2b]

46. $(x^{-2}y^{-3})^{-4}$ [4.2b]

47. $\left(\frac{2x}{t^2}\right)^4$ [4.2b]

48. $\left(\frac{y^3}{w^2}\right)^{-2}$ [4.2b]

49. $4x^{-5} \cdot 8x^{11}$ [4.1d]

50. $(8x^5y^{-4})^2$ [4.2b]

Find the intercepts. Then graph the equation. [3.3a]

51. $5x + 10y = 20$

52. $2x - 4y = 8$

53. $10y - 4x = -20$

54. $y - 5x = 5$

Synthesis

55. ◆ Why is it especially important to check the possible solutions to a rational equation?

56. ◆ How can a graph be used to determine how many solutions an equation has?

Solve.

57. $\frac{4}{y-2} - \frac{2y-3}{y^2-4} = \frac{5}{y+2}$

58. $\frac{x}{x^2+3x-4} + \frac{x+1}{x^2+6x+8} = \frac{2x}{x^2+x-2}$

59. $\frac{x+1}{x+2} = \frac{x+3}{x+4}$

60. $\frac{x^2}{x^2-4} = \frac{x}{x+2} - \frac{2x}{2-x}$

61. $4a - 3 = \frac{a+13}{a+1}$

62. $\frac{3x-9}{x-3} = \frac{5x-4}{2}$

63. $\frac{y^2-4}{y+3} = 2 - \frac{y-2}{y+3}$

64. $\frac{3a-5}{a^2+4a+3} + \frac{2a+2}{a+3} = \frac{a-3}{a+1}$

65. Use a grapher to check the solutions to Exercises 1–4.

66. Use a grapher to check the solutions to Exercises 13, 15, and 25.

6.6 Applications, Proportions, and Problem Solving

a Solving Applied Problems

Example 1 If 2 is subtracted from a number and then the reciprocal is found, the result is twice the reciprocal of the number itself. What is the number?

1. **Familiarize.** Let's try to guess such a number. Try 10: $10 - 2$ is 8, and the reciprocal of 8 is $\frac{1}{8}$. Two times the reciprocal of 10 is $2\left(\frac{1}{10}\right)$, or $\frac{1}{5}$. Since $\frac{1}{8} \neq \frac{1}{5}$, the number 10 does not check, but the process helps us understand the translation. Let $x =$ the number.
2. **Translate.** From the *Familiarize* step, we get the following translation. Subtracting 2 from the number gives us $x - 2$. Twice the reciprocal of the original number is $2(1/x)$.

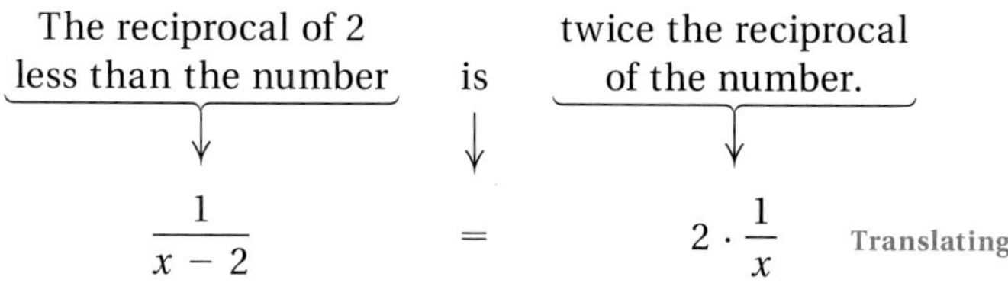

$$\frac{1}{x-2} = 2 \cdot \frac{1}{x} \qquad \text{Translating}$$

3. **Solve.** We solve the equation. The LCM is $x(x - 2)$.

$$x(x-2) \cdot \frac{1}{x-2} = x(x-2) \cdot \frac{2}{x} \qquad \text{Multiplying by the LCM}$$
$$x = 2(x-2) \qquad \text{Simplifying}$$
$$x = 2x - 4$$
$$-x = -4$$
$$x = 4$$

4. **Check.** We go back to the original problem. The number to be checked is 4. Two from 4 is 2. The reciprocal of 2 is $\frac{1}{2}$. The reciprocal of the number itself is $\frac{1}{4}$. Since $\frac{1}{2}$ is twice $\frac{1}{4}$, the conditions are satisfied.
5. **State.** The number is 4.

Do Exercise 1.

Example 2 *Animal Speeds.* A cheetah can run 20 mph faster than a lion. A cheetah can run 7 mi in the same time that a lion can run 5 mi. Find the speed of each animal.

1. **Familiarize.** We first make a drawing. Let $r =$ the speed of the lion. Then $r + 20 =$ the speed of the cheetah.

Objectives

a Solve applied problems using rational equations.

b Solve proportion problems.

For Extra Help

TAPE 11 | MAC WIN | CD-ROM

1. The reciprocal of 2 more than a number is three times the reciprocal of the number. Find the number.

Answer on page A-24

2. *Car Speeds.* One car travels 20 km/h faster than another. While one car travels 240 km, the other travels 160 km. Find the speed of each car.

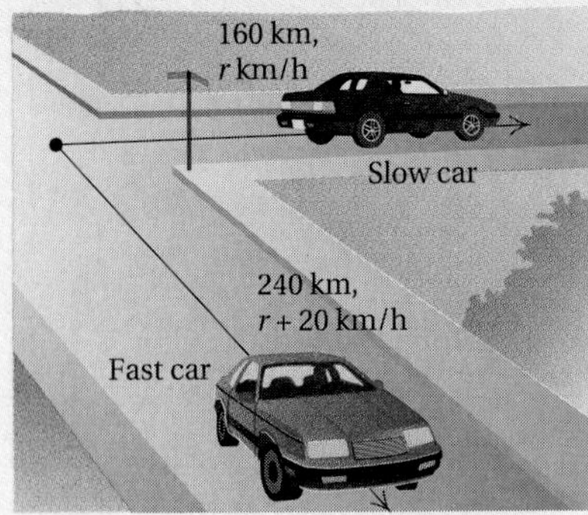

Answer on page A-24

Recall that sometimes we need to find a formula in order to solve an application. A formula that relates the notions of distance, speed, and time is $d = rt$, or

Distance = Speed · Time.

(Indeed, you may need to look up such a formula.)

Since each animal travels the same length of time, we can use just t for time. We organize the information in a chart, as follows.

$d = r \cdot t$

	Distance	Speed	Time	
Lion	5	r	t	$\to 5 = rt$
Cheetah	7	$r + 20$	t	$\to 7 = (r + 20)t$

2. **Translate.** We can apply the formula $d = rt$ along the rows of the table to obtain two equations:

$$5 = rt, \qquad (1)$$

$$7 = (r + 20)t. \qquad (2)$$

We know that the animals travel for the same length of time. Thus if we solve each equation for t and set the results equal to each other, we get an equation in terms of r.

Solving $5 = rt$ for t: $t = \frac{5}{r}$

Solving $7 = (r + 20)t$ for t: $t = \frac{7}{r + 20}$

Since the times are the same, we have the following equation:

$$\frac{5}{r} = \frac{7}{r + 20}.$$

3. **Solve.** To solve the equation, we first multiply on both sides by the LCM, which is $r(r + 20)$:

$$r(r + 20) \cdot \frac{5}{r} = r(r + 20) \cdot \frac{7}{r + 20} \quad \text{Multiplying on both sides by the LCM, which is } r(r + 20)$$

$$5(r + 20) = 7r \quad \text{Simplifying}$$

$$5r + 100 = 7r \quad \text{Removing parentheses}$$

$$100 = 2r$$

$$50 = r.$$

We now have a possible solution. The speed of the lion is 50 mph, and the speed of the cheetah is $r = 50 + 20$, or 70 mph.

4. **Check.** We first reread the problem to see what we were to find. We check the speeds of 50 for the lion and 70 for the cheetah. The cheetah does travel 20 mph faster than the lion and will travel farther than the lion, which runs at a slower speed. If the cheetah runs 7 mi at 70 mph, the time it has traveled is $\frac{7}{70}$, or $\frac{1}{10}$ hr. If the lion runs 5 mi at 50 mph, the time it has traveled is $\frac{5}{50}$, or $\frac{1}{10}$ hr. Since the times are the same, the speeds check.

5. **State.** The speed of the lion is 50 mph and the speed of the cheetah is 70 mph.

Do Exercise 2.

Example 3 *Recyclable Work.* Erin and Nick work as volunteers at a community recycling depot. Erin can sort a morning's accumulation of recyclables in 4 hr, while Nick requires 6 hr to do the same job. How long would it take them, working together, to sort the recyclables?

1. **Familiarize.** We familiarize ourselves with the problem by considering two *incorrect* ways of translating the problem to mathematical language.

   a) A common *incorrect* way to translate the problem is to add the two times: 4 hr + 6 hr = 10 hr. Let's think about this. Erin can do the job alone in 4 hr. If Erin and Nick work together, whatever time it takes them should be *less* than 4 hr. Thus we reject 10 hr as a solution, but we do have a partial check on any answer we get. The answer should be less than 4 hr.

   b) Another *incorrect* way to translate the problem is as follows. Suppose the two people split up the sorting job in such a way that Erin does half the sorting and Nick does the other half. Then

   Erin sorts $\frac{1}{2}$ the recyclables in $\frac{1}{2}$(4 hr), or 2 hr,

   and Nick sorts $\frac{1}{2}$ the recyclables in $\frac{1}{2}$(6 hr), or 3 hr.

   But time is wasted since Erin would finish 1 hr earlier than Nick. In effect, they have not worked together to get the job done as fast as possible. If Erin helps Nick after completing her half, the entire job could be done in a time somewhere between 2 hr and 3 hr.

   We proceed to a translation by considering how much of the job is finished in 1 hr, 2 hr, 3 hr, and so on. It takes Erin 4 hr to do the sorting job alone. Then, in 1 hr, she can do $\frac{1}{4}$ of the job. It takes Nick 6 hr to do the job alone. Then, in 1 hr, he can do $\frac{1}{6}$ of the job. Working together, they can do

   $$\frac{1}{4} + \frac{1}{6}, \text{ or } \frac{5}{12} \text{ of the job in 1 hr.}$$

   In 2 hr, Erin can do $2\left(\frac{1}{4}\right)$ of the job and Nick can do $2\left(\frac{1}{6}\right)$ of the job. Working together, they can do

   $$2\left(\frac{1}{4}\right) + 2\left(\frac{1}{6}\right), \text{ or } \frac{5}{6} \text{ of the job in 2 hr.}$$

   Continuing this reasoning, we can create a table like the following one.

	Fraction of the Job Completed		
Time	**Erin**	**Nick**	**Together**
1 hr	$\frac{1}{4}$	$\frac{1}{6}$	$\frac{1}{4} + \frac{1}{6}$, or $\frac{5}{12}$
2 hr	$2\left(\frac{1}{4}\right)$	$2\left(\frac{1}{6}\right)$	$2\left(\frac{1}{4}\right) + 2\left(\frac{1}{6}\right)$, or $\frac{5}{6}$
3 hr	$3\left(\frac{1}{4}\right)$	$3\left(\frac{1}{6}\right)$	$3\left(\frac{1}{4}\right) + 3\left(\frac{1}{6}\right)$, or $1\frac{1}{4}$
t hr	$t\left(\frac{1}{4}\right)$	$t\left(\frac{1}{6}\right)$	$t\left(\frac{1}{4}\right) + t\left(\frac{1}{6}\right)$

3. By checking work records, a contractor finds that it takes Eduardo 6 hr to construct a wall of a certain size. It takes Yolanda 8 hr to construct the same wall. How long would it take if they worked together?

Answer on page A-24

Calculator Spotlight

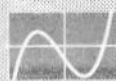 Work Problems on a Calculator.
Using a reciprocal key $\boxed{x^{-1}}$ on a calculator can make it easier to solve problems when using the work formula.

Example. It takes Bianca 6.75 hr, working alone, to keyboard a financial statement. It takes Arturo 7 hr, working alone, and Alani 8 hr, working alone, to produce the same statement. How long does it take if they all work together?

Let t = the time it takes them to do the job working together. Then

$$\frac{t}{6.75} + \frac{t}{7} + \frac{t}{8} = 1, \quad \text{or}$$

$$\frac{1}{6.75} + \frac{1}{7} + \frac{1}{8} = \frac{1}{t}.$$

We calculate $1/t$ using the reciprocal key:

$$\frac{1}{t} = 6.75^{-1} + 7^{-1} + 8^{-1}$$

$$\frac{1}{t} = 0.416005291.$$

This is $1/t$, so we press the reciprocal key $\boxed{x^{-1}}$ again to obtain

$$t = 2.40381558.$$

The financial statement will take about 2.4 hr to complete if they all work together.

Exercises

Solve.

1. Irwin can edge a lawn in 45 min if he works alone. It takes Hannah 38 min to edge the same lawn if she works alone. How long would it take if they work together?
2. Five sorority sisters work on invitations to a dance. Working alone, it would take them, respectively, 2.8, 3.1, 2.2, 3.4, and 2.4 hr to do the job. How long would it take if they work together?

From the table, we see that if they work 3 hr, the fraction of the job completed is $1\frac{1}{4}$, which is more of the job than needs to be done. We see again that the answer is somewhere between 2 hr and 3 hr. What we want is a number t such that the fraction of the job that gets completed is 1; that is, the job is just completed.

2. **Translate.** From the table, we see that the time we want is some number t for which

$$t\left(\frac{1}{4}\right) + t\left(\frac{1}{6}\right) = 1, \quad \text{or} \quad \frac{t}{4} + \frac{t}{6} = 1,$$

where 1 represents the idea that the entire job is completed in time t.

3. **Solve.** We solve the equation:

$$12\left(\frac{t}{4} + \frac{t}{6}\right) = 12 \cdot 1 \qquad \text{Multiplying by the LCM, which is } 2 \cdot 2 \cdot 3\text{, or } 12$$

$$12 \cdot \frac{t}{4} + 12 \cdot \frac{t}{6} = 12$$

$$3t + 2t = 12$$

$$5t = 12$$

$$t = \frac{12}{5}, \text{ or } 2\frac{2}{5} \text{ hr.}$$

4. **Check.** The check can be done by recalculating:

$$\frac{12}{5}\left(\frac{1}{4}\right) + \frac{12}{5}\left(\frac{1}{6}\right) = \frac{3}{5} + \frac{2}{5} = \frac{5}{5} = 1.$$

We also have another check in what we learned from the *Familiarize* step. The answer, $2\frac{2}{5}$ hr, is between 2 hr and 3 hr (see the table), and it is less than 4 hr, the time it takes Erin working alone.

5. **State.** It takes $2\frac{2}{5}$ hr for them to do the sorting, working together.

THE WORK PRINCIPLE

Suppose a = the time it takes A to do a job, b = the time it takes B to do the same job, and t = the time it takes them to do the same job working together. Then

$$\frac{t}{a} + \frac{t}{b} = 1, \quad \text{or} \quad \frac{1}{a} + \frac{1}{b} = \frac{1}{t}.$$

Do Exercise 3 on the preceding page.

b Applications Involving Proportions

We now consider applications with proportions. A **proportion** involves ratios. A **ratio** of two quantities is their quotient. For example, 73% is the ratio of 73 to 100, $\frac{73}{100}$. The ratio of two different kinds of measure is called a **rate**. Suppose an animal travels 720 ft in 2.5 hr. Its **rate**, or **speed**, is then

$$\frac{720 \text{ ft}}{2.5 \text{ hr}} = 288 \frac{\text{ft}}{\text{hr}}.$$

Do Exercises 4–7 on the following page.

An equality of ratios, $A/B = C/D$, is called a **proportion**. The numbers named in a proportion are said to be **proportional**.

Proportions can be used to solve applications by expressing a single ratio in two ways.

Example 4 *Gas Mileage.* A Ford Taurus can travel 135 mi of city driving on 6 gal of gas (***Source:*** Ford Motor Company). How much gas would be required for 360 mi of city driving?

1. **Familiarize.** We know that the Taurus can travel 135 mi on 6 gal of gas. Thus we can set up ratios, letting $x =$ the amount of gas required to drive 360 mi.
2. **Translate.** We assume that the car uses gas at the same rate throughout the 360 miles. Thus the ratios are the same and we can write a proportion. Note that the units of *mileage* are in the numerators and the units of *gasoline* are in the denominators.

$$\begin{array}{r} \text{Miles} \longrightarrow \\ \text{Gas} \longrightarrow \end{array} \frac{135}{6} = \frac{360}{x} \begin{array}{l} \longleftarrow \text{Miles} \\ \longleftarrow \text{Gas} \end{array}$$

3. **Solve.** To solve for x, we multiply on both sides by the LCM, which is $6x$:

$$6x \cdot \frac{135}{6} = 6x \cdot \frac{360}{x}$$

$$135x = 2160 \quad \text{Simplifying}$$

$$\frac{135x}{135} = \frac{2160}{135} \quad \text{Dividing by 135}$$

$$x = 16. \quad \text{Simplifying}$$

We can also use **cross products** to solve the proportion:

$$\frac{135}{6} = \frac{360}{x} \qquad 135x \text{ and } 6 \cdot 360 \text{ are called cross products.}$$

$$135x = 6 \cdot 360 \quad \text{Equating the cross products}$$

$$\frac{135x}{135} = \frac{6 \cdot 360}{135} \quad \text{Dividing by 135}$$

$$x = 16.$$

4. **Check.** We leave the check to the student.
5. **State.** The Taurus will require 16 gal of gas for 360 mi of city driving.

Do Exercise 8.

4. Find the ratio of 145 km to 2.5 liters (L).

5. *Batting Average.* Recently, a baseball player got 7 hits in 25 times at bat. What was the rate, or batting average, in hits per times at bat?

6. Impulses in nerve fibers travel 310 km in 2.5 hr. What is the rate, or speed, in kilometers per hour?

7. A lake of area 550 yd^2 contains 1320 fish. What is the population density of the lake in fish per square yard?

8. *Gas Mileage.* An Oldsmobile Achieva can travel 576 mi of interstate driving on 18 gal of gas (***Source:*** General Motors Corporation). How much gas would be required for 2592 mi of interstate driving?

Answers on page A-24

9. In 1997, Mark McGwire of the Oakland Athletics (and later with the St. Louis Cardinals) had 27 home runs after 77 games.

a) At this rate, how many home runs could McGwire hit in 162 games?

b) Could it be predicted that he would break Maris's record? (McGwire actually completed the season hitting a major-league high of 58 home runs.) (***Source***: Major League Baseball)

10. A sample of 184 light bulbs contained 6 defective bulbs. How many would you expect to find in a sample of 1288 bulbs?

Answers on page A-24

Proportions can be used in many types of applications. In the following example, we predict whether an important home-run record can be broken.

Example 5 *Home-Run Record.* Baseball fans enjoy speculating about records being broken. Roger Maris hit 61 home runs in 1961 to claim the major-league season home-run record. In 1997, Ken Griffey, Jr., had 20 home runs after 44 games. The season consists of 162 games. At this rate, could it be predicted that Griffey would break Maris's record? (***Source***: Major League Baseball)

1. **Familiarize.** Let's assume that Griffey's rate of hitting 20 home runs in 44 games will continue for the 162-game season. We let H = the number of home runs that Griffey can hit in 162 games.

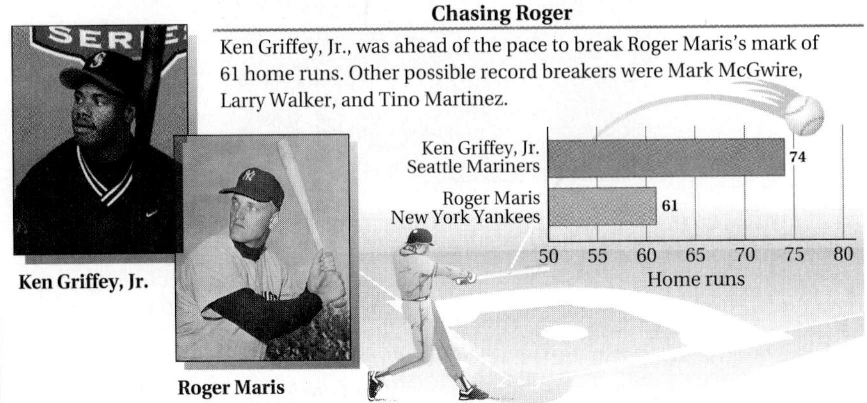

2. **Translate.** Assuming the rate of hitting home runs continues, the ratios are the same, and we have the proportion

$$\text{Number of home runs} \rightarrow \frac{H}{162} = \frac{20}{44}. \begin{array}{l}\leftarrow \text{Number of home runs}\\ \leftarrow \text{Number of games}\end{array}$$

(Number of games $\rightarrow$ 162)

3. **Solve.** We solve the equation:

$$\frac{H}{162} = \frac{20}{44}$$

$$44H = 162 \cdot 20 \quad \text{Equating cross products}$$

$$\frac{44H}{44} = \frac{162 \cdot 20}{44} \quad \text{Dividing by 44}$$

$$H \approx 73.64.$$

4. **Check.** We leave the check to the student.

5. **State.** We can indeed predict that Griffey, Jr., will hit about 74 home runs and break Maris's record. (Griffey actually completed the season with 56 home runs, having hit only 8 home runs in June and July.)

Do Exercises 9 and 10.

Example 6 *Estimating Wildlife Populations.* To determine the number of fish in a lake, a park ranger catches 225 fish, tags them, and throws them back into the lake. Later, 108 fish are caught, and 15 of them are found to be tagged. Estimate how many fish are in the lake.

1. **Familiarize.** The ratio of fish tagged to the total number of fish in the lake, F, is $\frac{225}{F}$. Of the 108 fish caught later, 15 fish were tagged. The ratio of fish tagged to fish caught is $\frac{15}{108}$.
2. **Translate.** Assuming that the two ratios are the same, we can translate to a proportion.

$$\text{Fish tagged originally} \rightarrow \frac{225}{F} \leftarrow \text{Fish in lake} = \frac{15}{108} \begin{matrix} \leftarrow \text{Tagged fish caught later} \\ \leftarrow \text{Fish caught later} \end{matrix}$$

3. **Solve.** We solve the proportion. We multiply by the LCM, which is $108F$:

$$108F \cdot \frac{225}{F} = 108F \cdot \frac{15}{108} \qquad \text{Multiplying by } 108F$$

$$108 \cdot 225 = F \cdot 15$$

$$\frac{108 \cdot 225}{15} = F \qquad \text{Dividing by 15}$$

$$1620 = F.$$

4. **Check.** We leave the check to the student.
5. **State.** We estimate that there are about 1620 fish in the lake.

Do Exercise 11.

11. To determine the number of deer in a forest, a conservationist catches 612 deer, tags them, and lets them loose. Later, 244 deer are caught, 72 of which are tagged. Estimate how many deer are in the forest.

Similar Triangles

Proportions also occur geometrically with *similar triangles.* Although similar triangles have the same shape, their sizes may be different.

$$\frac{a}{r} = \frac{b}{s} = \frac{c}{t}$$

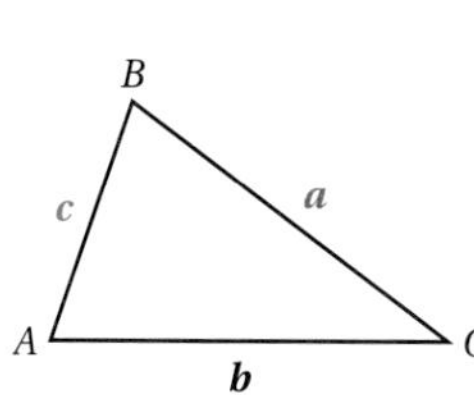

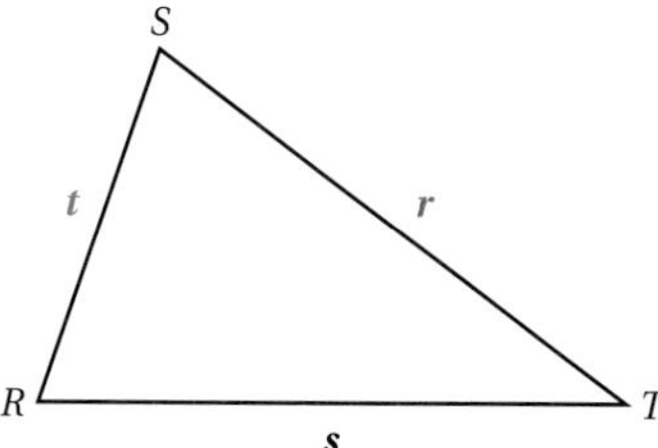

> In **similar triangles,** corresponding angles have the same measure and the lengths of corresponding sides are proportional.

Answer on page A-24

12. *Height of a Flagpole.* How high is a flagpole that casts a 45-ft shadow at the same time that a 5.5-ft woman casts a 10-ft shadow?

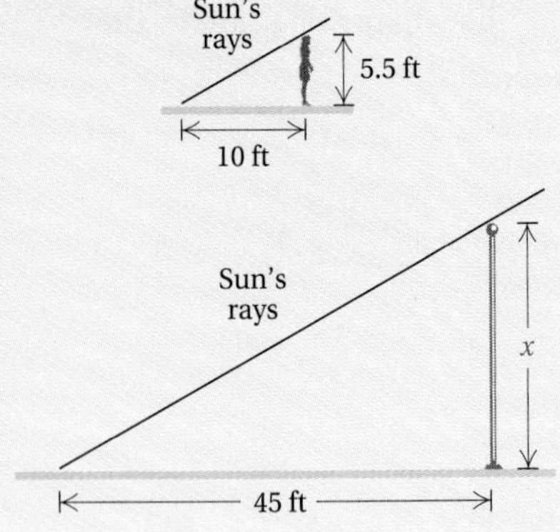

13. *F-106 Blueprint.* Referring to Example 8, find the length x on the plane.

Answers on page A-24

Example 7 Triangles ABC and XYZ below are similar triangles. Solve for z if $x = 10$, $a = 8$, and $c = 5$.

We make a sketch, write a proportion, and then solve. Note that side a is always opposite angle A, side x is always opposite angle X, and so on.

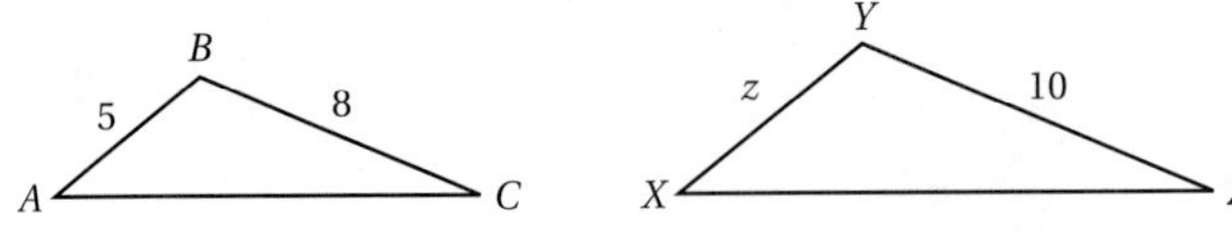

We have

$$\frac{z}{5} = \frac{10}{8}$$ The proportion $\frac{5}{z} = \frac{8}{10}$ could also be used.

$$40 \cdot \frac{z}{5} = 40 \cdot \frac{10}{8}$$ Multiplying by 40

$$8z = 50$$

$$z = \frac{50}{8} \text{ or } 6.25.$$ Dividing by 8

Example 8 *F-106 Blueprint.* A blueprint for an F-106 Delta Dart fighter plane is a scale drawing, as shown below. Each wing has a triangular shape. The blueprint shows similar triangles. Find the length of side a of the wing.

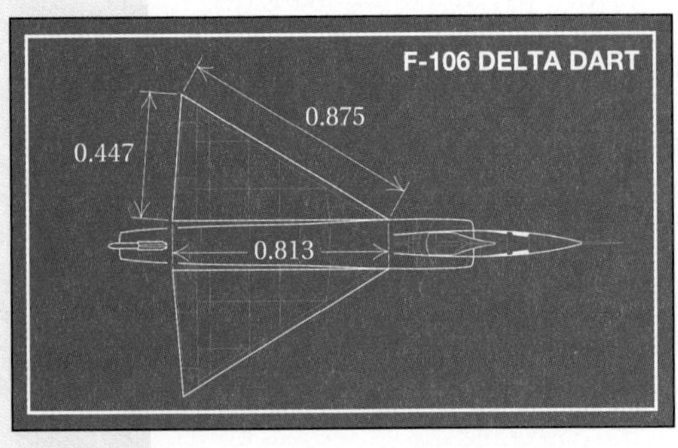

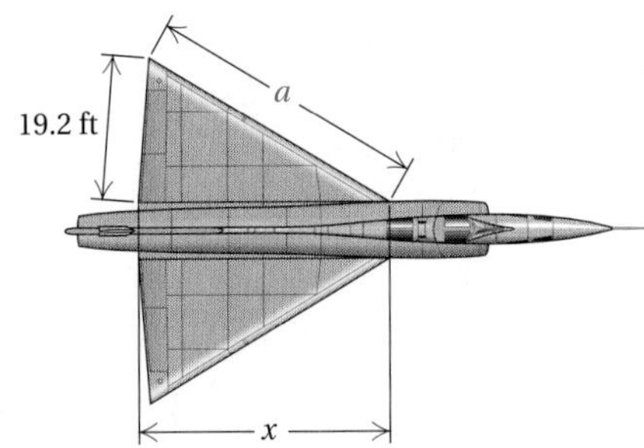

We let a = the length of the wing. Thus we have the proportion

$$\begin{array}{r} \text{Length on the blueprint} \rightarrow \\ \text{Length of the wing} \rightarrow \end{array} \frac{0.447}{19.2} = \frac{0.875}{a} \begin{array}{l} \leftarrow \text{Length on the blueprint} \\ \leftarrow \text{Length of the wing} \end{array}$$

Solve: $0.447 \cdot a = 19.2 \cdot 0.875$ Equating cross products

$$a = \frac{19.2 \cdot 0.875}{0.447}$$ Dividing by 0.447

$$a \approx 37.6 \text{ ft}.$$

The length of side a of the wing is about 37.6 ft.

Do Exercises 12 and 13.

Exercise Set 6.6

a Solve.

1. The reciprocal of 6 plus the reciprocal of 8 is the reciprocal of what number?

2. The reciprocal of 5 plus the reciprocal of 4 is the reciprocal of what number?

3. One number is 5 more than another. The quotient of the larger divided by the smaller is $\frac{4}{3}$. Find the numbers.

4. One number is 4 more than another. The quotient of the larger divided by the smaller is $\frac{5}{2}$. Find the numbers.

5. *Car Speeds.* Rick drives his four-wheel-drive truck 40 km/h faster than Sarah drives her Saturn. While Sarah travels 150 km, Rick travels 350 km. Find their speeds.

   Complete this table and the equations as part of the *Familiarize* step.

$d = r \cdot t$

	Distance	Speed	Time	
Car	150	r		→ $150 = r(\quad)$
Truck	350		t	→ $350 = (\quad)t$

6. *Car Speeds.* A passenger car travels 30 km/h faster than a delivery truck. While the car goes 400 km, the truck goes 250 km. Find their speeds.

7. *Train Speeds.* The speed of a freight train is 14 mph slower than the speed of a passenger train. The freight train travels 330 mi in the same time that it takes the passenger train to travel 400 mi. Find the speed of each train.

   Complete this table and the equations as part of the *Familiarize* step.

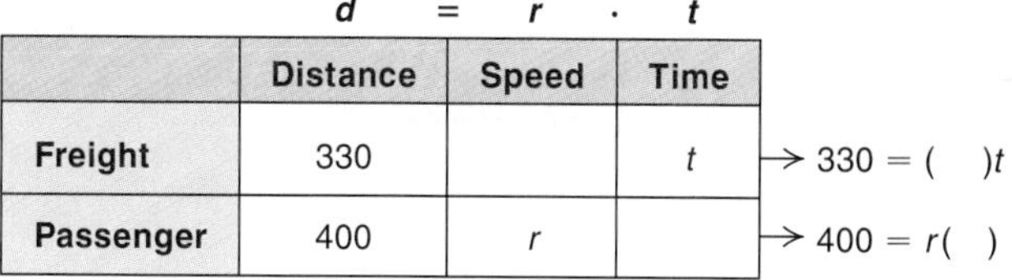

$d = r \cdot t$

	Distance	Speed	Time	
Freight	330		t	→ $330 = (\quad)t$
Passenger	400	r		→ $400 = r(\quad)$

8. *Train Speeds.* The speed of a freight train is 15 mph slower than the speed of a passenger train. The freight train travels 390 mi in the same time that it takes the passenger train to travel 480 mi. Find the speed of each train.

9. A long-distance trucker traveled 120 mi in one direction during a snowstorm. The return trip in rainy weather was accomplished at double the speed and took 3 hr less time. Find the speed going.

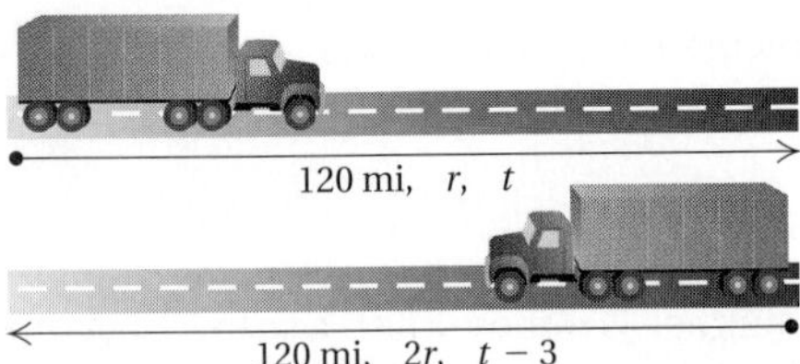

10. After making a trip of 126 mi, a person found that the trip would have taken 1 hr less time by increasing the speed by 8 mph. What was the actual speed?

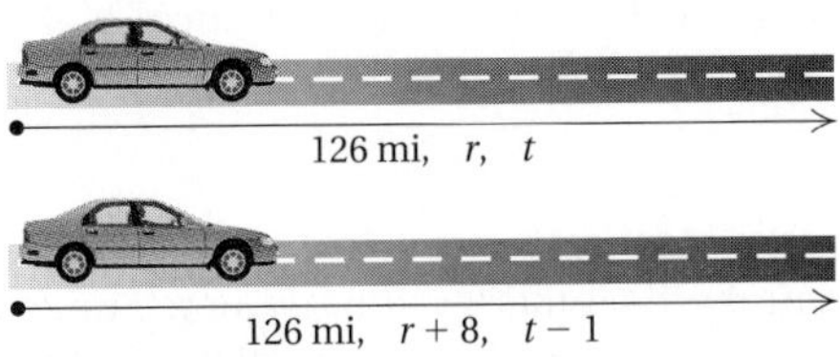

11. The Brother MFC4500 can fax a year-end report in 10 min while the Xerox 850 can fax the same report in 8 min. How long would it take the two machines, working together, to fax the report? (Assume that the recipient has two machines for incoming faxes.)

12. Zack mows the backyard in 40 min, while Angela can mow the same yard in 50 min. How long would it take them, working together with two mowers, to mow the yard?

13. By checking work records, a plumber finds that Rory can fit a kitchen in 12 hr. Mira can do the same job in 9 hr. How long would it take if they worked together?

14. Morgan can proofread 25 pages in 40 min. Shelby can proofread the same 25 pages in 30 min. How long would it take them, working together, to proofread 25 pages?

b Find the ratio of the following. Simplify, if possible.

15. 54 days, 6 days

16. 800 mi, 50 gal

17. A black racer snake travels 4.6 km in 2 hr. What is the speed in kilometers per hour?

18. *Speed of Light.* Light travels 558,000 mi in 3 sec. What is the speed in miles per second?

Solve.

19. A 120-lb person should eat a minimum of 44 g of protein each day. How much protein should a 180-lb person eat each day?

20. *Coffee Beans.* The coffee beans from 14 trees are required to produce 7.7 kg of coffee (this is the average amount that each person in the United States drinks each year). How many trees are required to produce 320 kg of coffee?

21. A student traveled 234 km in 14 days. At this same rate, how far would the student travel in 42 days?

22. In a potato bread recipe, the ratio of milk to flour is $\frac{3}{13}$. If 5 cups of milk are used, how many cups of flour are used?

23. A sample of 144 firecrackers contained 9 "duds." How many duds would you expect in a sample of 3200 firecrackers?

24. *Grass Seed.* It takes 60 oz of grass seed to seed 3000 ft^2 of lawn. At this rate, how much would be needed to seed 5000 ft^2 of lawn?

25. *Home Runs.* In 1997, Tino Martinez of the New York Yankees had 17 home runs after 44 games (***Source:*** Major League Baseball).

a) At this rate, how many home runs could Martinez hit in 162 games?
b) Could it be predicted that Martinez would break Maris's record of 61 home runs in a season?

26. *Home Runs.* In 1997, Larry Walker of the Colorado Rockies had 14 home runs after 40 games (***Source:*** Major League Baseball).

a) At this rate, how many home runs could Walker hit in 162 games?
b) Could it be predicted that Walker would break Maris's record of 61 home runs in a season?

27. *Estimating Whale Population.* To determine the number of blue whales in the world's oceans, marine biologists tag 500 blue whales in various parts of the world. Later, 400 blue whales are checked, and it is found that 20 of them are tagged. Estimate the blue whale population.

28. *Estimating Trout Population.* To determine the number of trout in a lake, a conservationist catches 112 trout, tags them, and throws them back into the lake. Later, 82 trout are caught; 32 of them are tagged. Estimate the number of trout in the lake.

29. *Weight on Mars.* The ratio of the weight of an object on Mars to the weight of an object on Earth is 0.4 to 1.

a) How much would a 12-ton rocket weigh on Mars?
b) How much would a 120-lb astronaut weigh on Mars?

30. *Weight on Moon.* The ratio of the weight of an object on the moon to the weight of an object on Earth is 0.16 to 1.

a) How much would a 12-ton rocket weigh on the moon?
b) How much would a 180-lb astronaut weigh on the moon?

31. A basketball team has 12 more games to play. They have won 25 of the 36 games they have played. How many more games must they win in order to finish with a 0.750 record?

32. Simplest fractional notation for a rational number is $\frac{9}{17}$. Find an equal ratio in which the sum of the numerator and the denominator is 104.

For each pair of similar triangles, find the length of the indicated letter.

33. b:

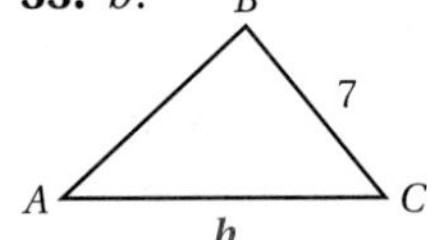

34. a:

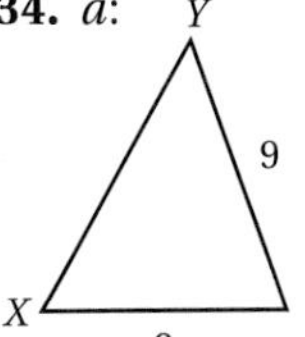

35. f:

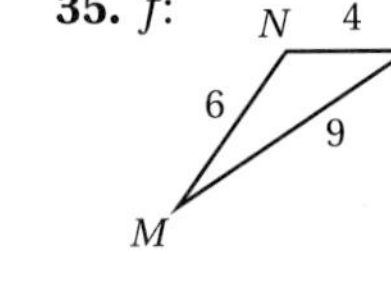

36. r:

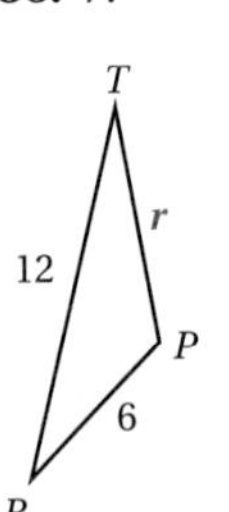

37. h:

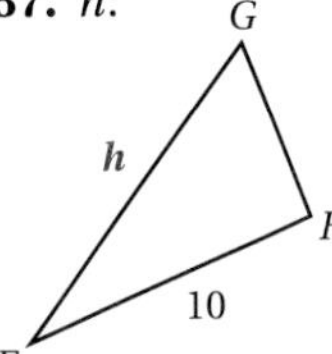

38. n:

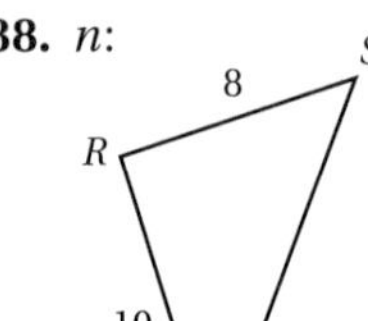

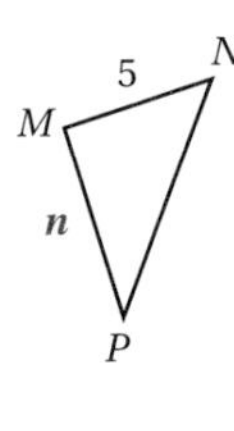

Skill Maintenance

Simplify. [4.1d]

39. $x^5 \cdot x^6$

40. $x^{-5} \cdot x^6$

41. $x^{-5} \cdot x^{-6}$

42. $x^5 \cdot x^{-6}$

Synthesis

43. ◆ Explain why it is incorrect to assume that two workers can complete a task twice as quickly as one person working alone.

44. ◆ Write a problem similar to Example 3 or Margin Exercise 3 for a classmate to solve. Design the problem so that the translation step is

$$\frac{t}{7} + \frac{t}{5} = 1.$$

45. Larry, Moe, and Curly are accountants who can complete a financial report together in 3 days. Larry can do the job in 8 days and Moe can do it in 10 days. How many days will it take Curly to complete the job?

46. Ann and Betty work together and complete a sales report in 4 hr. It would take Betty 6 hr longer, working alone, to do the job than it would Ann. How long would it take each of them to do the job working alone?

47. The denominator of a fraction is 1 more than the numerator. If 2 is subtracted from both the numerator and the denominator, the resulting fraction is $\frac{1}{2}$. Find the original fraction.

48. Express 100 as the sum of two numbers for which the ratio of one number, increased by 5, to the other number, decreased by 5, is 4.

49. How soon after 5 o'clock will the hands on a clock first be together?

50. Rachel allows herself 1 hr to reach a sales appointment 50 mi away. After she has driven 30 mi, she realizes that she must increase her speed by 15 mph in order to get there on time. What was her speed for the first 30 mi?

6.7 Formulas and Applications

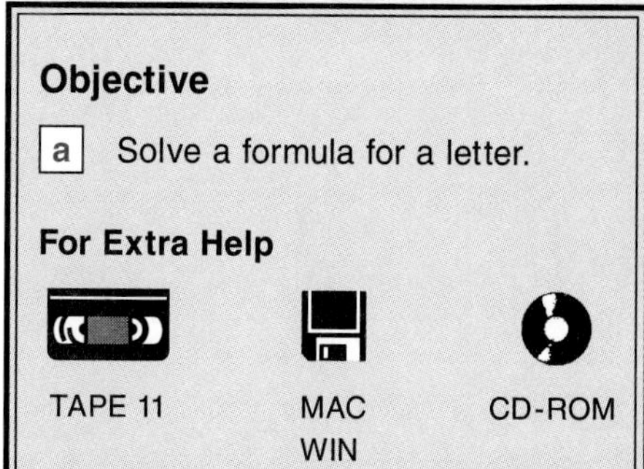

Objective

a Solve a formula for a letter.

For Extra Help

TAPE 11 MAC WIN CD-ROM

a The use of formulas is important in many applications of mathematics. We use the following procedure to solve a rational formula for a letter.

> To solve a rational formula for a given letter, identify the letter, and:
>
> 1. Multiply on both sides to clear fractions or decimals, if that is needed.
> 2. Multiply to remove parentheses, if necessary.
> 3. Get all terms with the letter to be solved for on one side of the equation and all other terms on the other side, using the addition principle.
> 4. Factor out the unknown.
> 5. Solve for the letter in question, using the multiplication principle.

1. Solve for M: $f = \dfrac{kMm}{d^2}$.

Example 1 *Gravitational Force.* The gravitational force f between planets of mass M and m, at a distance d from each other, is given by

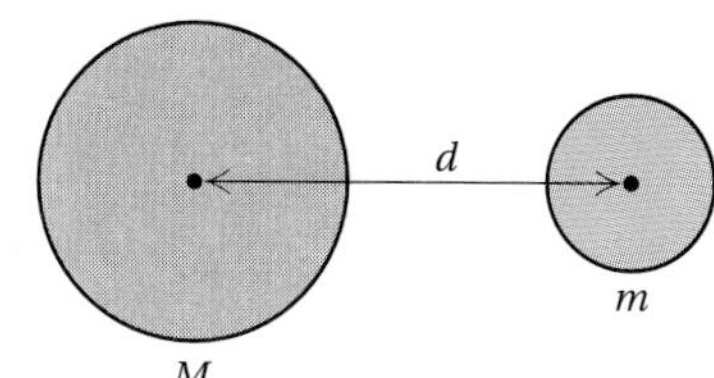

$$f = \frac{kMm}{d^2},$$

where k represents a fixed number constant. Solve for m.

We have

$$f \cdot d^2 = \frac{kMm}{d^2} \cdot d^2 \qquad \text{Multiplying by the LCM, } d^2$$

$$fd^2 = kMm \qquad \text{Simplifying}$$

$$\frac{fd^2}{kM} = m. \qquad \text{Dividing by } kM$$

Do Exercise 1.

Example 2 *Area of a Trapezoid.* The area A of a trapezoid is half the product of the height h and the sum of the lengths b_1 and b_2 of the parallel sides. Solve for b_2.

$$A = \frac{1}{2}h(b_1 + b_2)$$

We consider b_1 and b_2 to be different variables (or constants). The letter b_1 represents the length of the first parallel side and b_2 represents the length of the second parallel side. The small numbers 1 and 2 are called **subscripts**. Subscripts are used to identify different variables with related meanings.

$$2 \cdot A = 2 \cdot \frac{1}{2}h(b_1 + b_2) \qquad \text{Multiplying by 2 to clear fractions}$$

$$2A = h(b_1 + b_2) \qquad \text{Simplifying}$$

Answer on page A-24

2. Solve for b_1: $A = \frac{1}{2}h(b_1 + b_2)$.

3. Solve for f: $\frac{1}{p} + \frac{1}{q} = \frac{1}{f}$.
(This is an optics formula.)

4. Solve for b: $Q = \frac{a - b}{2b}$.

Answers on page A-24

Then

$$2A = hb_1 + hb_2 \quad \text{Using a distributive law to remove parentheses}$$

$$2A - hb_1 = hb_2 \quad \text{Subtracting } hb_1$$

$$\frac{2A - hb_1}{h} = b_2. \quad \text{Dividing by } h$$

Do Exercise 2.

Example 3 *A Work Formula.* The following work formula was considered in Section 6.6. Solve it for t.

$$\frac{t}{a} + \frac{t}{b} = 1$$

We multiply by the LCM, which is ab:

$$ab \cdot \left(\frac{t}{a} + \frac{t}{b}\right) = ab \cdot 1 \quad \text{Multiplying by } ab$$

$$ab \cdot \frac{t}{a} + ab \cdot \frac{t}{b} = ab \quad \text{Using a distributive law to remove parentheses}$$

$$bt + at = ab \quad \text{Simplifying}$$

$$(b + a)t = ab \quad \text{Factoring out } t$$

$$t = \frac{ab}{b + a}. \quad \text{Dividing by } b + a$$

Do Exercise 3.

In Examples 1 and 2, the letter for which we solved was on the right side of the equation. In Example 3, the letter was on the left. Since all equations are reversible, the location of the letter is a matter of choice.

TIP FOR FORMULA SOLVING

The variable to be solved for should be alone on one side of the equation, with *no* occurrence of that variable on the other side.

Example 4 Solve for b: $S = \frac{a + b}{3b}$.

We multiply by the LCM, which is $3b$:

$$3b \cdot S = 3b \cdot \frac{a + b}{3b} \quad \text{Multiplying by } 3b$$

$$3bS = a + b \quad \text{Simplifying}$$

If we divide by $3S$, we will have b alone on the left, but we will still have a term with b on the right.

$$3bS - b = a \quad \text{Subtracting } b \text{ to get all terms involving } b \text{ on one side}$$

$$b(3S - 1) = a$$

$$b = \frac{a}{3S - 1}. \quad \text{Dividing by } 3S - 1$$

Do Exercise 4.

Exercise Set 6.7

a Solve.

1. $S = 2\pi rh$, for r

2. $A = P(1 + rt)$, for t
(An interest formula)

3. $A = \frac{1}{2}bh$, for b
(The area of a triangle)

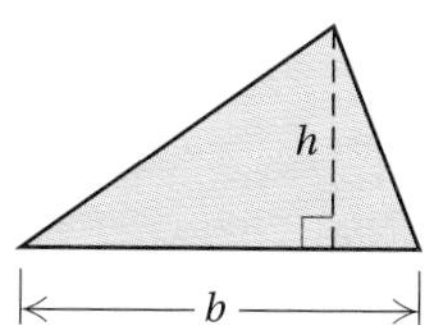

4. $s = \frac{1}{2}gt^2$, for g

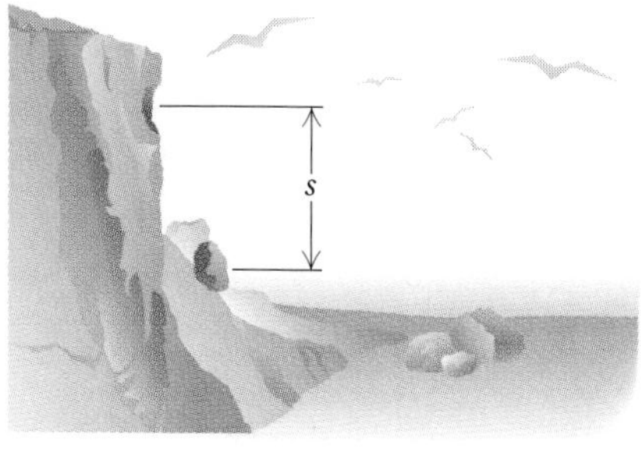

5. $S = 180(n - 2)$, for n

6. $S = \frac{n}{2}(a + l)$, for a

7. $V = \frac{1}{3}k(B + b + 4M)$, for b

8. $A = P + Prt$, for P
(*Hint*: Factor the right-hand side.)

9. $S(r - 1) = rl - a$, for r

10. $T = mg - mf$, for m
(*Hint*: Factor the right-hand side.)

11. $A = \frac{1}{2}h(b_1 + b_2)$, for h

12. $S = 2\pi r(r + h)$, for h
(The surface area of a right circular cylinder)

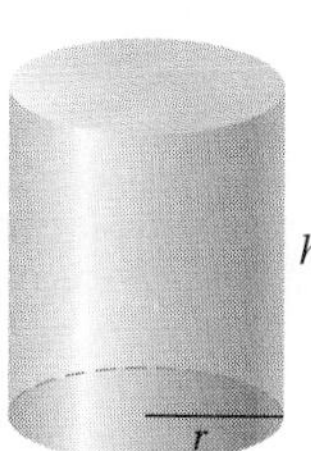

13. $\frac{A - B}{AB} = Q$, for B

14. $L = \frac{Mt + g}{t}$, for t

15. $\frac{1}{p} + \frac{1}{q} = \frac{1}{f}$, for p

16. $\frac{1}{a} + \frac{1}{b} = \frac{1}{t}$, for b

17. $\frac{A}{P} = 1 + r$, for A

18. $\frac{2A}{h} = a + b$, for h

19. $\frac{1}{R} = \frac{1}{r_1} + \frac{1}{r_2}$, for R
(An electricity formula)

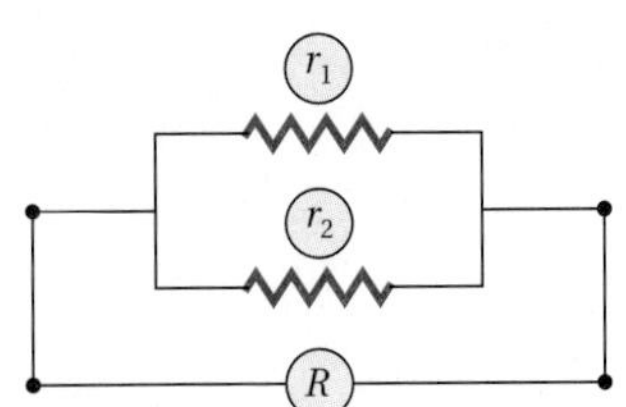

20. $\frac{1}{R} = \frac{1}{r_1} + \frac{1}{r_2}$, for r_1

21. $\frac{A}{B} = \frac{C}{D}$, for D

22. $q = \frac{VQ}{I}$, for I
(An engineering formula)

23. $h_1 = q\left(1 + \frac{h_2}{p}\right)$, for h_2

24. $S = \frac{a - ar^n}{1 - r}$, for a

25. $C = \frac{Ka - b}{a}$, for a

26. $Q = \frac{Pt - h}{t}$, for t

Skill Maintenance

Subtract. [4.4c]

27. $(5x^3 - 7x^2 + 9) - (8x^3 - 2x^2 + 4)$

28. $(5x^4 - 6x^3 + 23x^2 - 79x + 24) - (-18x^4 - 56x^3 + 84x - 17)$

Factor. [5.7a]

29. $x^2 - 4$

30. $30y^4 + 9y^2 - 12$

31. $49m^2 - 112mn + 64n^2$

32. $y^2 + 2y - 35$

33. $y^4 - 1$

34. $a^2 - 100b^2$

Divide and check. [4.8b]

35. $(x^3 + 4x - 4) \div (x - 2)$

36. $(x^4 - 6x^2 + 9) \div (x^2 - 3)$

Synthesis

37. ◆ Describe a situation in which the result of Example 3,

$$t = \frac{ab}{a + b},$$

would be especially useful.

38. ◆ Which of the following is easier to solve for x?

$$\frac{1}{23} + \frac{1}{25} = \frac{1}{x} \quad \text{or} \quad \frac{1}{a} + \frac{1}{b} = \frac{1}{x}$$

Explain the reasons for your choice.

Solve.

39. $u = -F\left(E - \frac{P}{T}\right)$, for T

40. $l = a + (n - 1)d$, for d

41. The formula

$$N = \frac{(b + d)f_1 - v}{(b - v)f_2}$$

is used when monitoring the water in fisheries. Solve for v.

42. In

$$N = \frac{a}{c},$$

what is the effect on N when c increases? when c decreases? Assume that a, c, and N are positive.

Collaborative Learning Manual

Develop a formula for calculating the time required to complete a task when two or more people are working together.

6.8 Complex Rational Expressions

a A **complex rational expression,** or **complex fractional expression,** is a rational expression that has one or more rational expressions within its numerator or denominator. Here are some examples:

$$\frac{1+\frac{2}{x}}{3}, \quad \frac{\frac{x+y}{2}}{\frac{2x}{x+1}}, \quad \frac{\frac{1}{3}+\frac{1}{5}}{\frac{2}{x}-\frac{x}{y}}.$$

These are rational expressions within the complex rational expression.

Objective

a Simplify complex rational expressions.

For Extra Help

TAPE 11 MAC WIN CD-ROM

There are two methods to simplify complex rational expressions. We will consider them both. Use the one that works best for you or the one that your instructor directs you to use.

Multiplying by the LCM of All the Denominators: Method 1

METHOD 1

To simplify a complex rational expression:

1. First, find the LCM of all the denominators of all the rational expressions occurring *within* both the numerator and the denominator of the complex rational expression.
2. Then multiply by 1 using LCM/LCM.
3. If possible, simplify by removing a factor of 1.

Example 1 Simplify: $\dfrac{\frac{1}{2}+\frac{3}{4}}{\frac{5}{6}-\frac{3}{8}}$.

We have

$$\frac{\frac{1}{2}+\frac{3}{4}}{\frac{5}{6}-\frac{3}{8}}$$

The denominators *within* the complex rational expression are 2, 4, 6, and 8. The LCM of these denominators is 24. We multiply by 1 using $\frac{24}{24}$.

$$= \frac{\frac{1}{2}+\frac{3}{4}}{\frac{5}{6}-\frac{3}{8}} \cdot \frac{24}{24}$$

Multiplying by 1

$$= \frac{\left(\frac{1}{2}+\frac{3}{4}\right)24}{\left(\frac{5}{6}-\frac{3}{8}\right)24}$$

← Multiplying the numerator by 24

← Multiplying the denominator by 24

1. Simplify. Use method 1.

$$\frac{\frac{1}{3}+\frac{4}{5}}{\frac{7}{8}-\frac{5}{6}}$$

2. Simplify. Use method 1.

$$\frac{\frac{x}{2}+\frac{2x}{3}}{\frac{1}{x}-\frac{x}{2}}$$

3. Simplify. Use method 1.

$$\frac{1+\frac{1}{x}}{1-\frac{1}{x^2}}$$

Answers on page A-24

Using the distributive laws, we carry out the multiplications:

$$= \frac{\frac{1}{2}(24)+\frac{3}{4}(24)}{\frac{5}{6}(24)-\frac{3}{8}(24)}$$

$$= \frac{12+18}{20-9} \quad \text{Simplifying}$$

$$= \frac{30}{11}.$$

Multiplying in this manner has the effect of clearing fractions in both the top and the bottom of the complex rational expression.

Do Exercise 1.

Example 2 Simplify: $\dfrac{\frac{3}{x}+\frac{1}{2x}}{\frac{1}{3x}-\frac{3}{4x}}$.

The denominators within the complex expression are x, $2x$, $3x$, and $4x$. The LCM of these denominators is $12x$. We multiply by 1 using $12x/12x$.

$$\frac{\frac{3}{x}+\frac{1}{2x}}{\frac{1}{3x}-\frac{3}{4x}}\cdot\frac{12x}{12x} = \frac{\left(\frac{3}{x}+\frac{1}{2x}\right)12x}{\left(\frac{1}{3x}-\frac{3}{4x}\right)12x} = \frac{\frac{3}{x}(12x)+\frac{1}{2x}(12x)}{\frac{1}{3x}(12x)-\frac{3}{4x}(12x)}$$

$$= \frac{36+6}{4-9} = -\frac{42}{5}$$

Do Exercise 2.

Example 3 Simplify: $\dfrac{1-\frac{1}{x}}{1-\frac{1}{x^2}}$.

The denominators within the complex expression are x and x^2. The LCM of these denominators is x^2. We multiply by 1 using x^2/x^2. Then, after obtaining a single rational expression, we simplify:

$$\frac{1-\frac{1}{x}}{1-\frac{1}{x^2}}\cdot\frac{x^2}{x^2} = \frac{\left(1-\frac{1}{x}\right)x^2}{\left(1-\frac{1}{x^2}\right)x^2} = \frac{1(x^2)-\frac{1}{x}(x^2)}{1(x^2)-\frac{1}{x^2}(x^2)} = \frac{x^2-x}{x^2-1}$$

$$= \frac{x\cancel{(x-1)}}{(x+1)\cancel{(x-1)}} = \frac{x}{x+1}.$$

Do Exercise 3.

Adding in the Numerator and the Denominator: Method 2

METHOD 2

To simplify a complex rational expression:

1. Add or subtract, as necessary, to get a single rational expression in the numerator.
2. Add or subtract, as necessary, to get a single rational expression in the denominator.
3. Divide the numerator by the denominator.
4. If possible, simplify by removing a factor of 1.

We will redo Examples 1–3 using this method.

Example 4 Simplify: $\dfrac{\frac{1}{2}+\frac{3}{4}}{\frac{5}{6}-\frac{3}{8}}$.

We have

$$\frac{\frac{1}{2}+\frac{3}{4}}{\frac{5}{6}-\frac{3}{8}} = \frac{\frac{1}{2}\cdot\frac{2}{2}+\frac{3}{4}}{\frac{5}{6}\cdot\frac{4}{4}-\frac{3}{8}\cdot\frac{3}{3}}$$

Multiplying the $\frac{1}{2}$ by 1 to get a common denominator

Multiplying the $\frac{5}{6}$ and the $\frac{3}{8}$ by 1 to get a common denominator

$$= \frac{\frac{2}{4}+\frac{3}{4}}{\frac{20}{24}-\frac{9}{24}}$$

$$= \frac{\frac{5}{4}}{\frac{11}{24}}$$ Adding in the numerator; subtracting in the denominator

$$= \frac{5}{4}\cdot\frac{24}{11}$$ Multiplying by the reciprocal of the divisor

$$= \frac{5\cdot 3\cdot 2\cdot 2\cdot 2}{2\cdot 2\cdot 11}$$ Factoring

$$= \frac{5\cdot 3\cdot 2\cdot \cancel{2}\cdot \cancel{2}}{\cancel{2}\cdot \cancel{2}\cdot 11}$$ Removing a factor of 1: $\frac{2\cdot 2}{2\cdot 2} = 1$

$$= \frac{30}{11}.$$

Do Exercise 4.

4. Simplify. Use method 2.

$$\frac{\frac{1}{3}+\frac{4}{5}}{\frac{7}{8}-\frac{5}{6}}$$

Answer on page A-24

5. Simplify. Use method 2.

$$\frac{\frac{x}{2}+\frac{2x}{3}}{\frac{1}{x}-\frac{x}{2}}$$

6. Simplify. Use method 2.

$$\frac{1+\frac{1}{x}}{1-\frac{1}{x^2}}$$

Answers on page A-24

Example 5 Simplify: $\dfrac{\frac{3}{x}+\frac{1}{2x}}{\frac{1}{3x}-\frac{3}{4x}}$.

We have

$$\frac{\frac{3}{x}+\frac{1}{2x}}{\frac{1}{3x}-\frac{3}{4x}} = \frac{\frac{3}{x}\cdot\frac{2}{2}+\frac{1}{2x}}{\frac{1}{3x}\cdot\frac{4}{4}-\frac{3}{4x}\cdot\frac{3}{3}}$$

Finding the LCD, $2x$, and multiplying by 1 in the numerator

Finding the LCD, $12x$, and multiplying by 1 in the denominator

$$= \frac{\frac{6}{2x}+\frac{1}{2x}}{\frac{4}{12x}-\frac{9}{12x}} = \frac{\frac{7}{2x}}{\frac{-5}{12x}}$$

Adding in the numerator and subtracting in the denominator

$$= \frac{7}{2x}\cdot\frac{12x}{-5}$$

Multiplying by the reciprocal of the divisor

$$= \frac{7}{2x}\cdot\frac{6(2x)}{-5}$$

Factoring

$$= \frac{7}{\cancel{2x}}\cdot\frac{6(\cancel{2x})}{-5}$$

Removing a factor of 1: $\frac{2x}{2x}=1$

$$= \frac{42}{-5} = -\frac{42}{5}.$$

Do Exercise 5.

Example 6 Simplify: $\dfrac{1-\frac{1}{x}}{1-\frac{1}{x^2}}$.

We have

$$\frac{1-\frac{1}{x}}{1-\frac{1}{x^2}} = \frac{\frac{x}{x}-\frac{1}{x}}{\frac{x^2}{x^2}-\frac{1}{x^2}}$$

Finding the LCD, x, and multiplying by 1 in the numerator

Finding the LCD, x^2, and multiplying by 1 in the denominator

$$= \frac{\frac{x-1}{x}}{\frac{x^2-1}{x^2}}$$

Subtracting in the numerator and subtracting in the denominator

$$= \frac{x-1}{x}\cdot\frac{x^2}{x^2-1}$$

Multiplying by the reciprocal of the divisor

$$= \frac{(x-1)x\cdot x}{x(x-1)(x+1)}$$

Factoring

$$= \frac{\cancel{(x-1)}\cancel{x}\cdot x}{\cancel{x}\cancel{(x-1)}(x+1)}$$

Removing a factor of 1: $\frac{x(x-1)}{x(x-1)}=1$

$$= \frac{x}{x+1}.$$

Do Exercise 6.

Exercise Set 6.8

a Simplify.

1. $\dfrac{1+\dfrac{9}{16}}{1-\dfrac{3}{4}}$

2. $\dfrac{6-\dfrac{3}{8}}{4+\dfrac{5}{6}}$

3. $\dfrac{1-\dfrac{3}{5}}{1+\dfrac{1}{5}}$

4. $\dfrac{2+\dfrac{2}{3}}{2-\dfrac{2}{3}}$

5. $\dfrac{\dfrac{1}{2}+\dfrac{3}{4}}{\dfrac{5}{8}-\dfrac{5}{6}}$

6. $\dfrac{\dfrac{3}{4}+\dfrac{7}{8}}{\dfrac{2}{3}-\dfrac{5}{6}}$

7. $\dfrac{\dfrac{1}{x}+3}{\dfrac{1}{x}-5}$

8. $\dfrac{2-\dfrac{1}{a}}{4+\dfrac{1}{a}}$

9. $\dfrac{4-\dfrac{1}{x^2}}{2-\dfrac{1}{x}}$

10. $\dfrac{\dfrac{2}{y}+\dfrac{1}{2y}}{y+\dfrac{y}{2}}$

11. $\dfrac{8+\dfrac{8}{d}}{1+\dfrac{1}{d}}$

12. $\dfrac{3+\dfrac{2}{t}}{3-\dfrac{2}{t}}$

13. $\dfrac{\dfrac{x}{8}-\dfrac{8}{x}}{\dfrac{1}{8}+\dfrac{1}{x}}$

14. $\dfrac{\dfrac{2}{m}+\dfrac{m}{2}}{\dfrac{m}{3}-\dfrac{3}{m}}$

15. $\dfrac{1+\dfrac{1}{y}}{1-\dfrac{1}{y^2}}$

16. $\dfrac{\dfrac{1}{q^2}-1}{\dfrac{1}{q}+1}$

17. $\dfrac{\dfrac{1}{5}-\dfrac{1}{a}}{\dfrac{5-a}{5}}$

18. $\dfrac{\dfrac{4}{t}}{4+\dfrac{1}{t}}$

19. $\dfrac{\dfrac{1}{a}+\dfrac{1}{b}}{\dfrac{1}{a^2}-\dfrac{1}{b^2}}$

20. $\dfrac{\dfrac{1}{x^2}-\dfrac{1}{y^2}}{\dfrac{2}{x}-\dfrac{2}{y}}$

21. $\dfrac{\dfrac{p}{q}+\dfrac{q}{p}}{\dfrac{1}{p}+\dfrac{1}{q}}$

22. $\dfrac{x-3+\dfrac{2}{x}}{x-4+\dfrac{3}{x}}$

Skill Maintenance

Add. [4.4a]

23. $(2x^3 - 4x^2 + x - 7) + (4x^4 + x^3 + 4x^2 + x)$

24. $(2x^3 - 4x^2 + x - 7) + (-2x^3 + 4x^2 - x + 7)$

Factor. [5.7a]

25. $p^2 - 10p + 25$

26. $p^2 + 10p + 25$

27. $50p^2 - 100$

28. $5p^2 - 40p - 100$

Solve. [5.9a]

29. The length of a rectangle is 3 yd greater than the width. The area of the rectangle is 10 yd^2. Find the perimeter.

30. A ladder of length 13 ft is placed against a building in such a way that the distance from the top of the ladder to the ground is 7 ft more than the distance from the bottom of the ladder to the building. Find these distances.

Synthesis

31. ◆ Why is factoring an important skill when simplifying complex rational expressions?

32. ◆ Why is the distributive law especially important when using method 1 of this section?

33. Find the reciprocal of $\dfrac{2}{x-1} - \dfrac{1}{3x-2}$.

Simplify.

34. $\dfrac{\dfrac{a}{b}+\dfrac{c}{d}}{\dfrac{b}{a}+\dfrac{d}{c}}$

35. $\dfrac{\dfrac{a}{b}-\dfrac{c}{d}}{\dfrac{b}{a}-\dfrac{d}{c}}$

36. $\left[\dfrac{\dfrac{x+1}{x-1}+1}{\dfrac{x+1}{x-1}-1}\right]^5$

37. $1+\dfrac{1}{1+\dfrac{1}{1+\dfrac{1}{1+\dfrac{1}{x}}}}$

38. $\dfrac{\dfrac{z}{1-\dfrac{z}{2+2z}}-2z}{\dfrac{2z}{5z-2}-3}$

6.9 Variation and Applications

We now extend our study of formulas and functions by considering applications involving variation.

Objectives

a. Find an equation of direct variation given a pair of values of the variables.

b. Solve applied problems involving direct variation.

c. Find an equation of inverse variation given a pair of values of the variables.

d. Solve applied problems involving inverse variation.

e. Find equations of other kinds of variation given values of the variables.

f. Solve applied problems involving other kinds of variation.

For Extra Help

TAPE 11

MAC
WIN

CD-ROM

a Equations of Direct Variation

An electrician earns \$21 per hour. In 1 hr, \$21 is earned; in 2 hr, \$42 is earned; in 3 hr, \$63 is earned; and so on. We plot this information on a graph, using the number of hours as the first coordinate and the amount earned as the second coordinate to form a set of ordered pairs:

(1, 21), (2, 42),
(3, 63), (4, 84),
and so on.

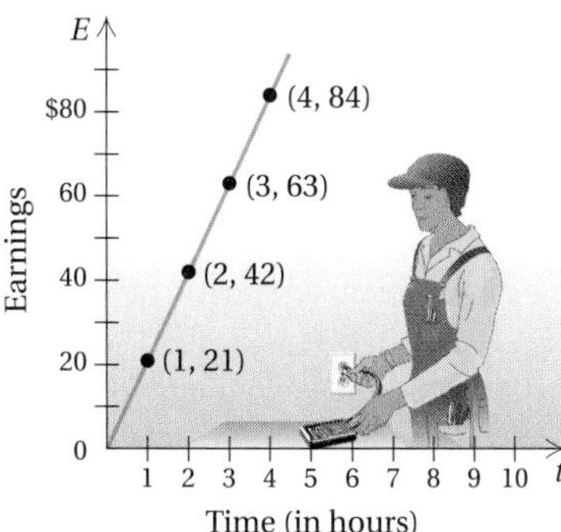

Note that the ratio of the second coordinate to the first is the same number for each point:

$$\frac{21}{1} = 21, \quad \frac{42}{2} = 21,$$

$$\frac{63}{3} = 21, \quad \frac{84}{4} = 21,$$

and so on.

Whenever a situation produces pairs of numbers in which the *ratio is constant,* we say that there is **direct variation.** Here the amount earned varies directly as the time:

$$\frac{E}{t} = 21 \text{ (a constant)}, \quad \text{or} \quad E = 21t,$$

or, using function notation, $E(t) = 21t$. The equation is an equation of **direct variation.** The coefficient, 21 in the situation above, is called the **variation constant.** In this case, it is the rate of change of earnings with respect to time.

DIRECT VARIATION

If a situation gives rise to a linear function $f(x) = kx$, or $y = kx$, where k is a positive constant, we say that we have **direct variation,** or that **y varies directly as x,** or that **y is directly proportional to x.** The number k is called the **variation constant,** or **constant of proportionality.**

Example 1 Find the variation constant and an equation of variation in which y varies directly as x, and $y = 32$ when $x = 2$.

We know that (2, 32) is a solution of $y = kx$. Thus,

$$y = kx$$
$$32 = k \cdot 2 \qquad \text{Substituting}$$
$$\frac{32}{2} = k, \text{ or } k = 16. \qquad \text{Solving for } k$$

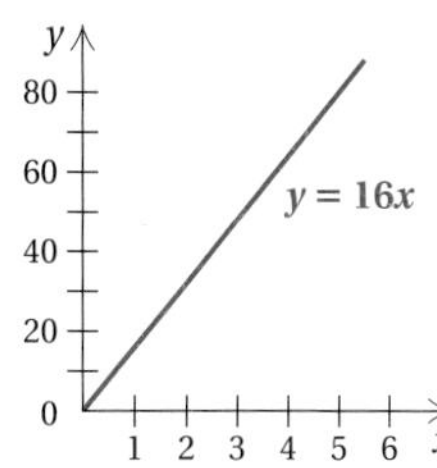

The variation constant, 16, is the rate of change of y with respect to x. The equation of variation is $y = 16x$.

Do Exercises 1 and 2.

1. Find the variation constant and an equation of variation in which y varies directly as x, and $y = 8$ when $x = 20$.

2. Find the variation constant and an equation of variation in which y varies directly as x, and $y = 5.6$ when $x = 8$.

The graph of $y = kx$, $k > 0$, always goes through the origin and rises from left to right. Note that as x increases, y increases. The constant k is also the slope of the line.

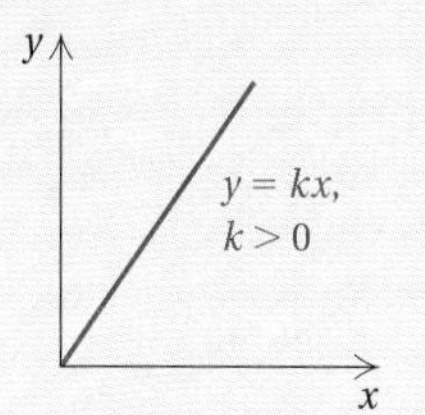

Answers on page A-25

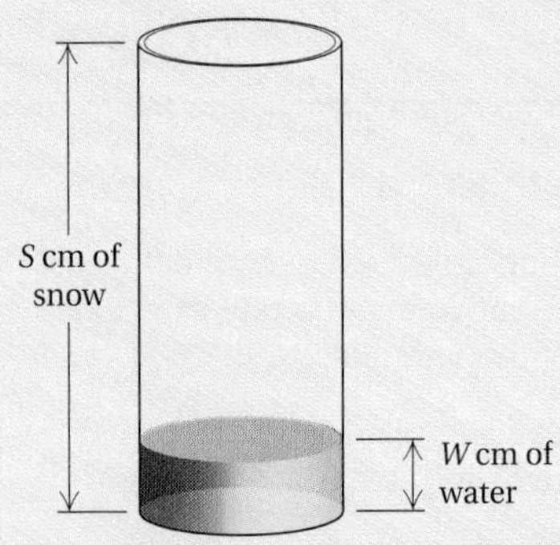

b Applications of Direct Variation

Example 2 *Water from Melting Snow.* The number of centimeters W of water produced from melting snow varies directly as S, the number of centimeters of snow. Meteorologists have found that 150 cm of snow will melt to 16.8 cm of water. To how many centimeters of water will 200 cm of snow melt?

We first find the variation constant using the data and then find an equation of variation:

$W = kS$ — **W varies directly as S.**

$16.8 = k \cdot 150$ — **Substituting**

$\dfrac{16.8}{150} = k$ — **Solving for k**

$0.112 = k.$ — **This is the variation constant.**

The equation of variation is $W = 0.112S$.

Next, we use the equation to find how many centimeters of water will result from melting 200 cm of snow:

$$\begin{aligned} W &= 0.112S \\ &= 0.112(200) \quad \textbf{Substituting} \\ &= 22.4. \end{aligned}$$

Thus, 200 cm of snow will melt to 22.4 cm of water.

3. *Ohm's Law.* Ohm's Law states that the voltage V in an electric circuit varies directly as the number of amperes I of electric current in the circuit. If the voltage is 10 volts when the current is 3 amperes, what is the voltage when the current is 15 amperes?

4. *An Ecology Problem.* The amount of garbage G produced in the United States varies directly as the number of people N who produce the garbage. It is known that 50 tons of garbage is produced by 200 people in 1 year. The population of the San Francisco–Oakland–San Jose area is 6,300,000. How much garbage is produced by this area in 1 year?

Do Exercises 3 and 4.

c Equations of Inverse Variation

A bus is traveling a distance of 20 mi. At a speed of 5 mph, the trip will take 4 hr; at 20 mph, it will take 1 hr; at 40 mph, it will take $\frac{1}{2}$ hr; and so on. We plot this information on a graph, using speed as the first coordinate and time as the second coordinate to determine a set of ordered pairs:

$(5, 4)$, $(20, 1)$, $\left(40, \frac{1}{2}\right)$, and so on.

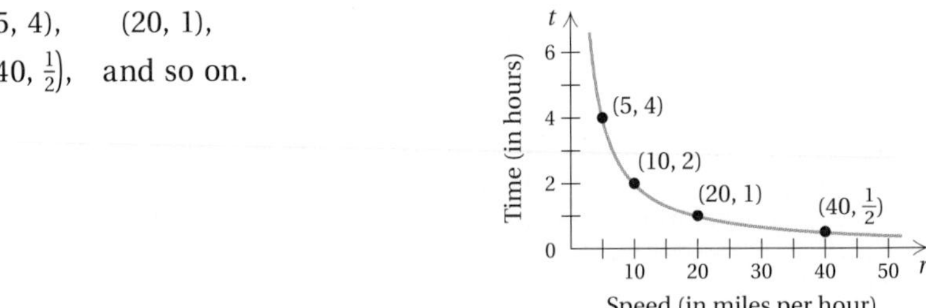

Note that the products of the coordinates are all the same number:

$5 \cdot 4 = 20, \quad 20 \cdot 1 = 20, \quad 40 \cdot \frac{1}{2} = 20,$ and so on.

Whenever a situation produces pairs of numbers in which the *product is constant,* we say that there is **inverse variation.** Here the time varies inversely as the speed:

$$rt = 20 \text{ (a constant)}, \quad \text{or} \quad t = \frac{20}{r}.$$

The equation is an equation of **inverse variation.** The coefficient, 20 in the situation above, is called the **variation constant.** Note that as the first number increases, the second number decreases.

Answers on page A-25

INVERSE VARIATION

If a situation gives rise to a function $f(x) = k/x$, or $y = k/x$, where k is a positive constant, we say that we have **inverse variation,** or that ***y* varies inversely as *x*,** or that ***y* is inversely proportional to *x*.** The number k is called the **variation constant,** or **constant of proportionality.**

It is helpful to look at the graph of $y = k/x$, $k > 0$. The graph is like the one shown below for positive values of x. Note that as x increases, y decreases.

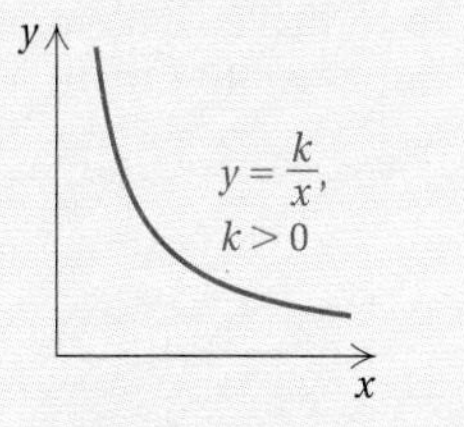

Example 3 Find the variation constant and an equation of variation in which y varies inversely as x, and $y = 32$ when $x = 0.2$.

We know that (0.2, 32) is a solution of $y = k/x$. We substitute:

$$y = \frac{k}{x}$$

$$32 = \frac{k}{0.2} \quad \text{Substituting}$$

$$(0.2)32 = k \quad \text{Solving for } k$$

$$6.4 = k.$$

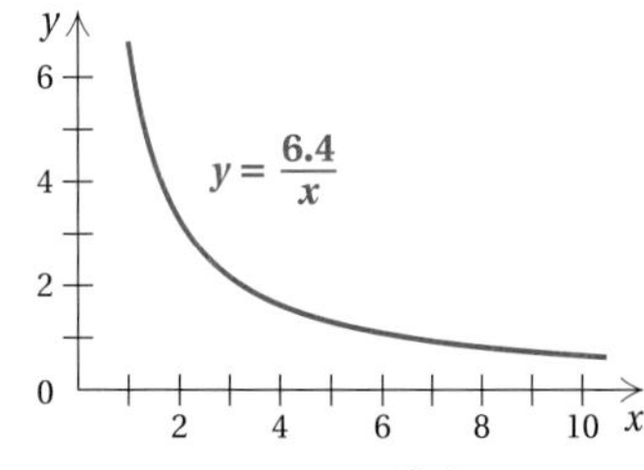

The variation constant is 6.4. The equation of variation is $y = \dfrac{6.4}{x}$.

Do Exercise 5.

5. Find the variation constant and an equation of variation in which y varies inversely as x, and $y = 0.012$ when $x = 50$.

d Applications of Inverse Variation

Example 4 *Building a Shed.* The time t required to do a job varies inversely as the number of people P who work on the job (assuming that all work at the same rate). It takes 4 hr for 12 people to build a woodshed. How long would it take 3 people to complete the same job?

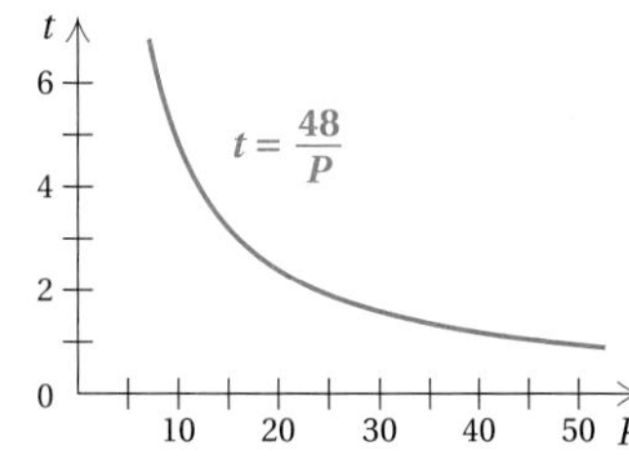

We first find the variation constant using the data and then find an equation of variation:

$$t = \frac{k}{P} \quad t \text{ varies inversely as } P.$$

$$4 = \frac{k}{12} \quad \text{Substituting}$$

$$48 = k. \quad \text{Solving for } k\text{, the variation constant}$$

The equation of variation is $t = 48/P$. In this case, 48 is the total number of hours required to build the shed.

Answer on page A-25

6. *Time and Speed.* The time t required to drive a fixed distance varies inversely as the speed r. It takes 5 hr at 60 km/h to drive a fixed distance. How long would it take to drive that same distance at 40 km/h?

7. Find an equation of variation in which y varies directly as the square of x, and $y = 175$ when $x = 5$.

Answers on page A-25

Next, we use the equation to find the time that it would take 3 people to do the job:

$$t = \frac{48}{P}$$

$$= \frac{48}{3} \quad \text{Substituting}$$

$$= 16.$$

It would take 16 hr for 3 people to do the job.

Do Exercise 6.

e Other Kinds of Variation

We now look at other kinds of variation. Consider the equation for the area of a circle, in which A and r are variables and π is a constant:

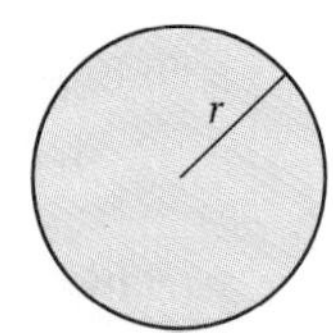

$$A = \pi r^2, \quad \text{or, as a function,} \quad A(r) = \pi r^2.$$

We say that the area *varies directly* as the square of the radius.

> y varies directly as the nth power of x if there is some positive constant k such that $y = kx^n$.

Example 5 Find an equation of variation in which y varies directly as the square of x, and $y = 12$ when $x = 2$.

We write an equation of variation and find k:

$$y = kx^2$$

$$12 = k \cdot 2^2 \quad \text{Substituting}$$

$$12 = k \cdot 4$$

$$3 = k.$$

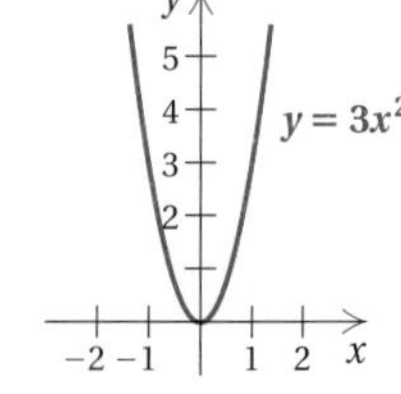

Thus, $y = 3x^2$.

Do Exercise 7.

From the law of gravity, we know that the weight W of an object *varies inversely* as the square of its distance d from the center of the earth:

$$W = \frac{k}{d^2}.$$

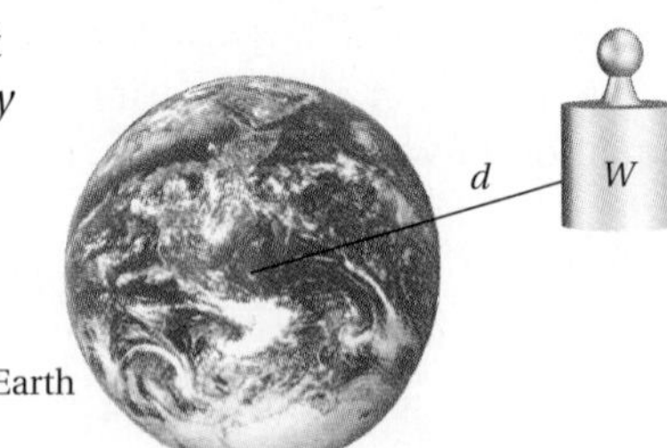

> y varies inversely as the nth power of x if there is some positive constant k such that
>
> $$y = \frac{k}{x^n}.$$

Example 6 Find an equation of variation in which W varies inversely as the square of d, and $W = 3$ when $d = 5$.

$$W = \frac{k}{d^2}$$

$$3 = \frac{k}{5^2} \quad \text{Substituting}$$

$$3 = \frac{k}{25}$$

$$75 = k$$

Thus, $W = \frac{75}{d^2}$.

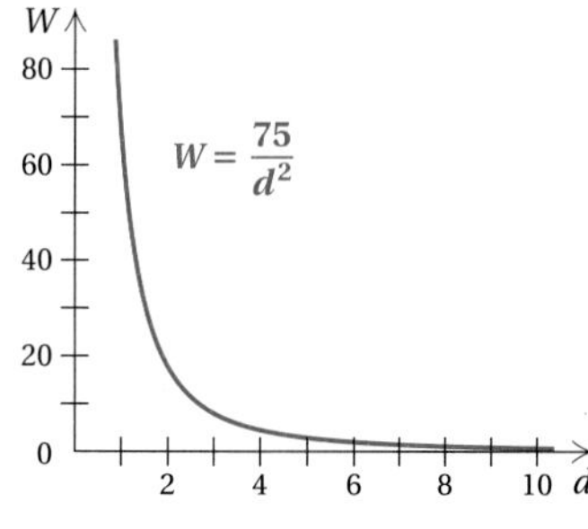

Do Exercise 8.

Consider the equation for the area A of a triangle with height h and base b: $A = \frac{1}{2}bh$. We say that the area *varies jointly* as the height and the base.

> y varies jointly as x and z if there is some positive constant k such that
> $y = kxz$.

Example 7 Find an equation of variation in which y varies jointly as x and z, and $y = 42$ when $x = 2$ and $z = 3$.

$$y = kxz$$

$$42 = k \cdot 2 \cdot 3 \quad \text{Substituting}$$

$$42 = k \cdot 6$$

$$7 = k$$

Thus, $y = 7xz$.

Do Exercise 9.

The equation

$$y = k \cdot \frac{xz^2}{w}$$

asserts that y varies jointly as x and the square of z, and inversely as w.

Example 8 Find an equation of variation in which y varies jointly as x and z and inversely as the square of w, and $y = 105$ when $x = 3$, $z = 20$, and $w = 2$.

$$y = k \cdot \frac{xz}{w^2}$$

$$105 = k \cdot \frac{3 \cdot 20}{2^2} \quad \text{Substituting}$$

$$105 = k \cdot 15$$

$$7 = k$$

Thus, $y = 7 \cdot \frac{xz}{w^2}$.

Do Exercise 10.

8. Find an equation of variation in which y varies inversely as the square of x, and $y = \frac{1}{4}$ when $x = 6$.

9. Find an equation of variation in which y varies jointly as x and z, and $y = 65$ when $x = 10$ and $z = 13$.

10. Find an equation of variation in which y varies jointly as x and the square of z and inversely as w, and $y = 80$ when $x = 4$, $z = 10$, and $w = 25$.

Answers on page A-25

11. *Distance of a Dropped Object.* The distance s that an object falls when dropped from some point above the ground varies directly as the square of the time t that it falls. If the object falls 19.6 m in 2 sec, how far will the object fall in 10 sec?

12. *Electrical Resistance.* At a fixed temperature, the resistance R of a wire varies directly as the length l and inversely as the square of its diameter d. If the resistance is 0.1 ohm when the diameter is 1 mm and the length is 50 cm, what is the resistance when the length is 2000 cm and the diameter is 2 mm?

Answers on page A-25

f Other Applications of Variation

Many problem situations can be described with equations of variation.

Example 9 *Volume of a Tree.* The volume of wood V in a tree varies jointly as the height h and the square of the girth g (girth is distance around). If the volume of a redwood tree is 216 m^3 when the height is 30 m and the girth is 1.5 m, what is the height of a tree whose volume is 960 m^3 and girth is 2 m?

We first find k using the first set of data. Then we solve for h using the second set of data.

$$V = khg^2$$
$$216 = k \cdot 30 \cdot 1.5^2$$
$$3.2 = k$$

Then the equation of variation is $V = 3.2hg^2$. We substitute the second set of data into the equation:

$$960 = 3.2 \cdot h \cdot 2^2$$
$$75 = h.$$

Therefore, the height of the tree is 75 m.

Example 10 *TV Signal.* The intensity I of a TV signal varies inversely as the square of the distance d from the transmitter. If the intensity is 23 watts per square meter (W/m^2) at a distance of 2 km, what is the intensity at a distance of 6 km?

We first find k using the first set of data. Then we solve for I using the second set of data.

$$I = \frac{k}{d^2}$$
$$23 = \frac{k}{2^2}$$
$$92 = k$$

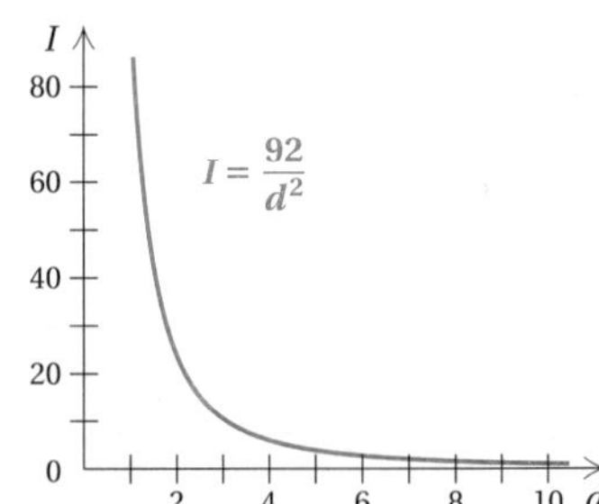

Then the equation of variation is $I = 92/d^2$. We substitute the second distance into the equation:

$$I = \frac{92}{d^2} = \frac{92}{6^2} \approx 2.56. \quad \text{Rounded to the nearest hundredth}$$

Therefore, at 6 km, the intensity is about 2.56 W/m^2.

Do Exercises 11 and 12.

Exercise Set 6.9

a Find the variation constant and an equation of variation in which y varies directly as x and the following are true.

1. $y = 40$ when $x = 8$

2. $y = 54$ when $x = 12$

3. $y = 4$ when $x = 30$

4. $y = 3$ when $x = 33$

5. $y = 0.9$ when $x = 0.4$

6. $y = 0.8$ when $x = 0.2$

b Solve.

7. *Aluminum Usage.* The number N of aluminum cans used each year varies directly as the number of people using the cans. If 250 people use 60,000 cans in one year, how many cans are used each year in Dallas, which has a population of 1,008,000?

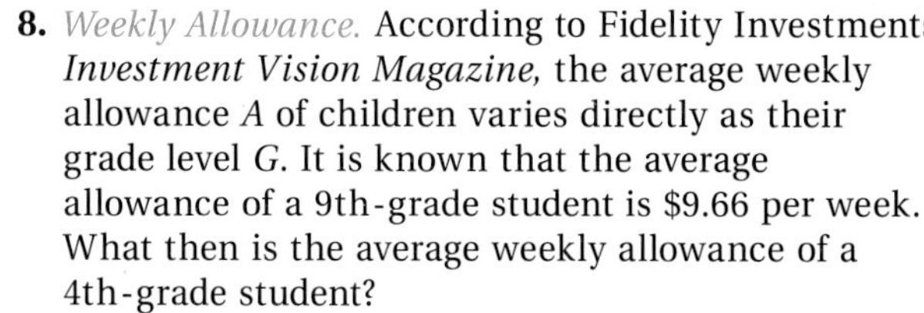

8. *Weekly Allowance.* According to Fidelity Investments *Investment Vision Magazine,* the average weekly allowance A of children varies directly as their grade level G. It is known that the average allowance of a 9th-grade student is $9.66 per week. What then is the average weekly allowance of a 4th-grade student?

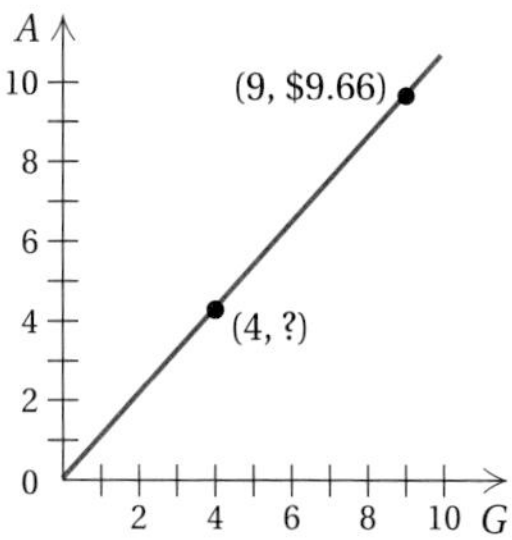

9. *Hooke's Law.* Hooke's law states that the distance d that a spring is stretched by a hanging object varies directly as the weight w of the object. If a spring is stretched 40 cm by a 3-kg barbell, what is the distance stretched by a 5-kg barbell?

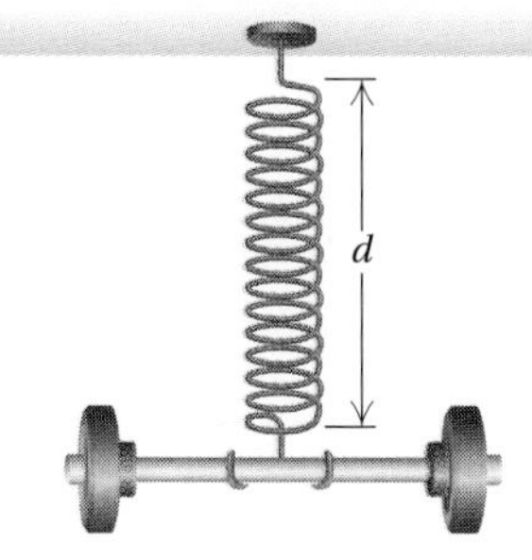

10. *Lead Pollution.* The average U.S. community of population 12,500 released about 385 tons of lead into the environment in a recent year (***Source:*** *Conservation Matters,* Autumn 1995 issue. Boston: Conservation Law Foundation, p. 30). How many tons were released nationally? Use 250,000,000 as the U.S. population.

11. *Fat Intake.* The maximum number of grams of fat that should be in a diet varies directly as a person's weight. A person weighing 120 lb should have no more than 60 g of fat per day. What is the maximum daily fat intake for a person weighing 180 lb?

12. *Relative Aperture.* The relative aperture, or f-stop, of a 23.5-mm diameter lens is directly proportional to the focal length F of the lens. If a 150-mm focal length has an f-stop of 6.3, find the f-stop of a 23.5-mm diameter lens with a focal length of 80 mm.

13. *Mass of Water in Body.* The number of kilograms W of water in a human body varies directly as the mass of the body. A 96-kg person contains 64 kg of water. How many kilograms of water are in a 60-kg person?

14. *Weight on Mars.* The weight M of an object on Mars varies directly as its weight E on Earth. A person who weighs 95 lb on Earth weighs 38 lb on Mars. How much would a 100-lb person weigh on Mars?

c Find the variation constant and an equation of variation in which y varies inversely as x and the following are true.

15. $y = 14$ when $x = 7$

16. $y = 1$ when $x = 8$

17. $y = 3$ when $x = 12$

18. $y = 12$ when $x = 5$

19. $y = 0.1$ when $x = 0.5$

20. $y = 1.8$ when $x = 0.3$

d Solve.

21. *Work Rate.* The time T required to do a job varies inversely as the number of people P working. It takes 5 hr for 7 bricklayers to build a park wall. How long will it take 10 bricklayers to complete the job?

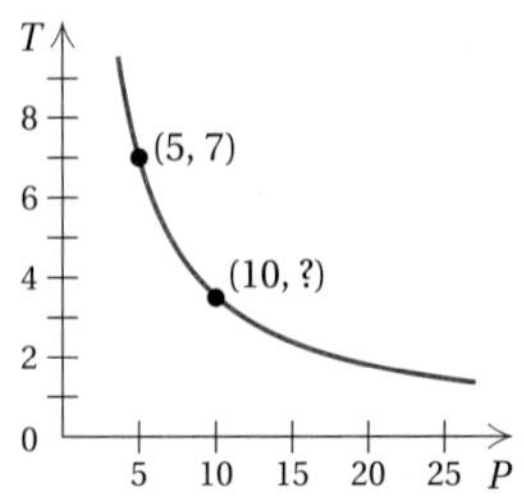

22. *Pumping Rate.* The time t required to empty a tank varies inversely as the rate r of pumping. If a pump can empty a tank in 45 min at the rate of 600 kL/min, how long will it take the pump to empty the same tank at the rate of 1000 kL/min?

23. *Current and Resistance.* The current I in an electrical conductor varies inversely as the resistance R of the conductor. If the current is $\frac{1}{2}$ ampere when the resistance is 240 ohms, what is the current when the resistance is 540 ohms?

24. *Wavelength and Frequency.* The wavelength W of a radio wave varies inversely as its frequency F. A wave with a frequency of 1200 kilohertz has a length of 300 meters. What is the length of a wave with a frequency of 800 kilohertz?

25. *Musical Pitch.* The pitch P of a musical tone varies inversely as its wavelength W. One tone has a pitch of 330 vibrations per second and a wavelength of 3.2 ft. Find the wavelength of another tone that has a pitch of 550 vibrations per second.

26. *Beam Weight.* The weight W that a horizontal beam can support varies inversely as the length L of the beam. Suppose that an 8-m beam can support 1200 kg. How many kilograms can a 14-m beam support?

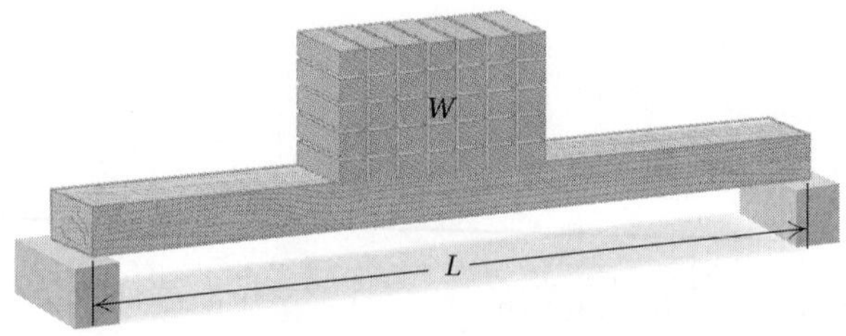

27. *Volume and Pressure.* The volume V of a gas varies inversely as the pressure P upon it. The volume of a gas is 200 cm^3 under a pressure of 32 kg/cm^2. What will be its volume under a pressure of 40 kg/cm^2?

28. *Rate of Travel.* The time t required to drive a fixed distance varies inversely as the speed r. It takes 5 hr at a speed of 80 km/h to drive a fixed distance. How long will it take to drive the same distance at a speed of 70 km/h?

e Find an equation of variation in which the following are true.

29. y varies directly as the square of x, and $y = 0.15$ when $x = 0.1$

30. y varies directly as the square of x, and $y = 6$ when $x = 3$

31. y varies inversely as the square of x, and $y = 0.15$ when $x = 0.1$

32. y varies inversely as the square of x, and $y = 6$ when $x = 3$

33. y varies jointly as x and z, and $y = 56$ when $x = 7$ and $z = 8$

34. y varies directly as x and inversely as z, and $y = 4$ when $x = 12$ and $z = 15$

35. y varies jointly as x and the square of z, and $y = 105$ when $x = 14$ and $z = 5$

36. y varies jointly as x and z and inversely as w, and $y = \frac{3}{2}$ when $x = 2$, $z = 3$, and $w = 4$

37. y varies jointly as x and z and inversely as the product of w and p, and $y = \frac{3}{28}$ when $x = 3$, $z = 10$, $w = 7$, and $p = 8$

38. y varies jointly as x and z and inversely as the square of w, and $y = \frac{12}{5}$ when $x = 16$, $z = 3$, and $w = 5$

f Solve.

39. *Stopping Distance of a Car.* The stopping distance d of a car after the brakes have been applied varies directly as the square of the speed r. If a car traveling 60 mph can stop in 200 ft, how fast can a car travel and still stop in 72 ft?

40. *Boyle's Law.* The volume V of a given mass of a gas varies directly as the temperature T and inversely as the pressure P. If $V = 231$ cm^3 when $T = 42°$ and $P = 20$ kg/cm^2, what is the volume when $T = 30°$ and $P = 15$ kg/cm^2?

41. *Intensity of Light.* The intensity I of light from a light bulb varies inversely as the square of the distance d from the bulb. Suppose that I is 90 W/m^2 (watts per square meter) when the distance is 5 m. How much *further* would it be to a point where the intensity is 40 W/m^2?

42. *Weight of an Astronaut.* The weight W of an object varies inversely as the square of the distance d from the center of the earth. At sea level (3978 mi from the center of the earth), an astronaut weighs 220 lb. Find his weight when he is 200 mi above the surface of the earth and the spacecraft is not in motion.

43. *Earned-Run Average.* A pitcher's earned-run average E varies directly as the number R of earned runs allowed and inversely as the number I of innings pitched. In a recent year, Shawn Estes of the San Francisco Giants had an earned-run average of 3.18. He gave up 71 earned runs in 201 innings. How many earned runs would he have given up had he pitched 300 innings with the same average? Round to the nearest whole number.

44. *Atmospheric Drag.* Wind resistance, or atmospheric drag, tends to slow down moving objects. Atmospheric drag varies jointly as an object's surface area A and velocity v. If a car traveling at a speed of 40 mph with a surface area of 37.8 ft^2 experiences a drag of 222 N (Newtons), how fast must a car with 51 ft^2 of surface area travel in order to experience a drag force of 430 N?

45. *Water Flow.* The amount Q of water emptied by a pipe varies directly as the square of the diameter d. A pipe 5 in. in diameter will empty 225 gal of water over a fixed time period. If we assume the same kind of flow, how many gallons of water are emptied in the same amount of time by a pipe that is 9 in. in diameter?

46. *Weight of a Sphere.* The weight W of a sphere of a given material varies directly as its volume V, and its volume V varies directly as the cube of its diameter.

a) Find an equation of variation relating the weight W to the diameter d.

b) An iron ball that is 5 in. in diameter is known to weigh 25 lb. Find the weight of an iron ball that is 8 in. in diameter.

Skill Maintenance

Factor completely. [5.7a]

47. $x^2 - x - 56$

48. $a^2 - 16a + 64$

49. $x^5 - 2x^4 - 35x^3$

50. $2y^3 - 10y^2 + y - 5$

51. $16 - t^4$

52. $10x^2 + 80x + 70$

53. $x^2 - 9x + 14$

54. $x^2 + x + 7$

55. $16x^2 - 40xy + 25y^2$

56. $a^2 - 9ab + 14b^2$

57. $3x^3 - 3y^3$

58. $w^6 - t^6$

Synthesis

59. ◆ If y varies directly as x and x varies inversely as z, how does y vary with regard to z? Why?

60. ◆ Write a variation problem for a classmate to solve. Design the problem so that the answer is "When Simone studies for 8 hr a week, her quiz score is 92."

61. *Area of a Circle.* The area of a circle varies directly as the square of the length of a diameter. What is the variation constant?

62. In each of the following equations, state whether y varies directly as x, inversely as x, or neither directly nor inversely as x.

a) $7xy = 14$

b) $x - 2y = 12$

c) $-2x + 3y = 0$

d) $x = \frac{3}{4}y$

e) $\frac{x}{y} = 2$

Describe, in words, the variation given by the equation.

63. $Q = \frac{kp^2}{q^3}$

64. $W = \frac{km_1M_1}{d^2}$

65. *Volume and Cost.* A peanut butter jar in the shape of a right circular cylinder is 4 in. high and 3 in. in diameter and sells for $1.20. If we assume that cost is proportional to volume, how much should a jar 6 in. high and 6 in. in diameter cost?

Collaborative Learning Manual

Model the height of a bouncing ball with an equation of direct variation.

Summary and Review Exercises: Chapter 6

Important Properties and Formulas

Direct Variation: $y = kx$ *Inverse Variation:* $y = \frac{k}{x}$ *Joint Variation:* $y = kxz$

The objectives to be tested in addition to the material in this chapter are [4.2b], [4.4c], [5.7a], and [5.9a].

Find all numbers for which the rational expression is undefined. [6.1a]

1. $\frac{3}{x}$

2. $\frac{4}{x-6}$

3. $\frac{x+5}{x^2-36}$

4. $\frac{x^2-3x+2}{x^2+x-30}$

5. $\frac{-4}{(x+2)^2}$

6. $\frac{x-5}{x^3-8x^2+15x}$

Simplify. [6.1c]

7. $\frac{4x^2-8x}{4x^2+4x}$

8. $\frac{14x^2-x-3}{2x^2-7x+3}$

9. $\frac{(y-5)^2}{y^2-25}$

Multiply and simplify. [6.1d]

10. $\frac{a^2-36}{10a} \cdot \frac{2a}{a+6}$

11. $\frac{6t-6}{2t^2+t-1} \cdot \frac{t^2-1}{t^2-2t+1}$

Divide and simplify. [6.1f]

12. $\frac{10-5t}{3} \div \frac{t-2}{12t}$

13. $\frac{4x^4}{x^2-1} \div \frac{2x^3}{x^2-2x+1}$

Find the LCM. [6.2c]

14. $3x^2$, $10xy$, $15y^2$

15. $a-2$, $4a-8$

16. y^2-y-2, y^2-4

Add and simplify. [6.3a]

17. $\frac{x+8}{x+7} + \frac{10-4x}{x+7}$

18. $\frac{3}{3x-9} + \frac{x-2}{3-x}$

19. $\frac{2a}{a+1} + \frac{4a}{a^2-1}$

20. $\frac{d^2}{d-c} + \frac{c^2}{c-d}$

Subtract and simplify. [6.4a]

21. $\frac{6x-3}{x^2-x-12} - \frac{2x-15}{x^2-x-12}$

22. $\frac{3x-1}{2x} - \frac{x-3}{x}$

23. $\frac{x+3}{x-2} - \frac{x}{2-x}$

24. $\frac{1}{x^2-25} - \frac{x-5}{x^2-4x-5}$

25. Perform the indicated operations and simplify: [6.4b]

$$\frac{3x}{x+2} - \frac{x}{x-2} + \frac{8}{x^2-4}.$$

Simplify. [6.8a]

26. $\dfrac{\frac{1}{z}+1}{\frac{1}{z^2}-1}$

27. $\dfrac{\frac{c}{d}-\frac{d}{c}}{\frac{1}{c}+\frac{1}{d}}$

Solve. [6.5a]

28. $\frac{3}{y} - \frac{1}{4} = \frac{1}{y}$

29. $\frac{15}{x} - \frac{15}{x+2} = 2$

Solve. [6.6a]

30. In checking records, a contractor finds that crew A can pave a certain length of highway in 9 hr, while crew B can do the same job in 12 hr. How long would it take if they worked together?

31. *Train Speeds.* A manufacturer is testing two high-speed trains. One train travels 40 km/h faster than the other. While one train travels 70 km, the other travels 60 km. Find the speed of each train.

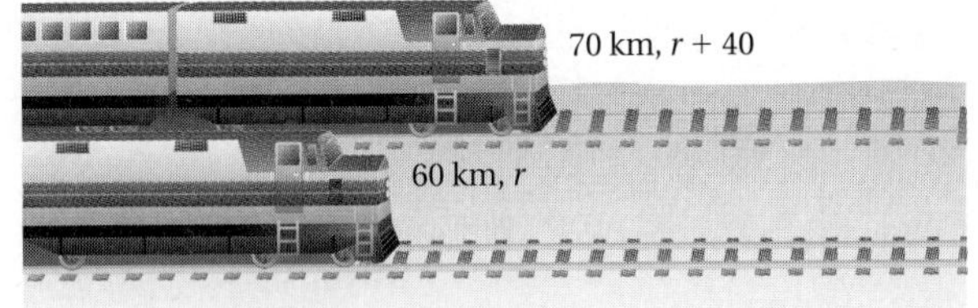

32. The reciprocal of 1 more than a number is twice the reciprocal of the number itself. What is the number?

33. *Airplane Speeds.* One plane travels 80 mph faster than another. While one travels 1750 mi, the other travels 950 mi. Find the speed of each plane.

Solve. [6.6b]

34. A sample of 250 calculators contained 8 defective calculators. How many defective calculators would you expect to find in a sample of 5000?

35. It is known that 10 cm^3 of a normal specimen of human blood contains 1.2 g of hemoglobin. How many grams of hemoglobin would 16 cm^3 of the same blood contain?

36. Triangles ABC and XYZ below are similar. Find the value of x.

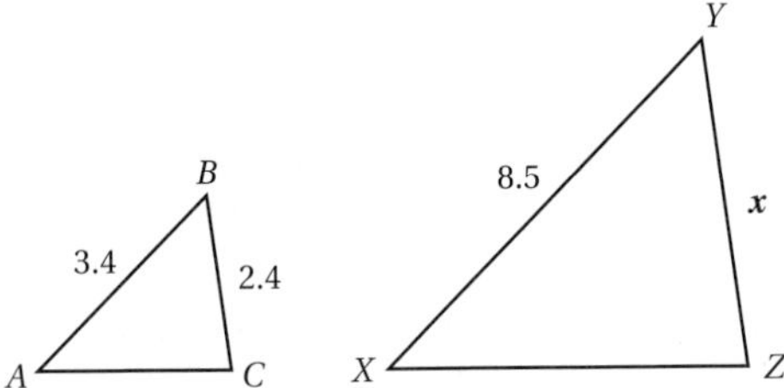

Solve for the letter indicated. [6.7a]

37. $\frac{1}{r} + \frac{1}{s} = \frac{1}{t}$, for s

38. $F = \frac{9C + 160}{5}$, for C

39. $V = \frac{4}{3}\pi r^3$, for r^3
(The volume of a sphere)

40. Find an equation of variation in which y varies directly as x, and $y = 100$ when $x = 25$. [6.9a]

41. Find an equation of variation in which y varies inversely as x, and $y = 100$ when $x = 25$. [6.9c]

42. *Pumping Time.* The time t required to empty a tank varies inversely as the rate r of pumping. If a pump can empty a tank in 35 min at the rate of 800 kL per minute, how long will it take the pump to empty the same tank at the rate of 1400 kL per minute? [6.9d]

Skill Maintenance

43. Factor: $5x^3 + 20x^2 - 3x - 12$. [5.7a]

44. Simplify: $(5x^3y^2)^{-3}$. [4.2b]

45. Subtract: [4.4c]
$(5x^3 - 4x^2 + 3x - 4) - (7x^3 - 7x^2 - 9x + 14)$.

46. The width of a rectangle is 2 cm less than the length. The area is 15 cm^2. Find the dimensions and the perimeter of the rectangle. [5.9a]

Synthesis

◆ Carry out the direction for each of the following. Explain the use of the LCM in each case.

47. Add: $\frac{4}{x-2} + \frac{1}{x+2}$. [6.3a]

48. Subtract: $\frac{4}{x-2} - \frac{1}{x+2}$. [6.4a]

49. Solve: $\frac{4}{x-2} + \frac{1}{x+2} = \frac{26}{x^2-4}$. [6.5a]

50. Simplify: $\dfrac{1 - \frac{2}{x}}{1 + \frac{x}{4}}$. [6.8a]

Simplify.

51. $\frac{2a^2 + 5a - 3}{a^2} \cdot \frac{5a^3 + 30a^2}{2a^2 + 7a - 4} \div \frac{a^2 + 6a}{a^2 + 7a + 12}$
[6.1d, f]

52. $\frac{12a}{(a-b)(b-c)} - \frac{2a}{(b-a)(c-b)}$ [6.4a]

53. Compare
$$\frac{A+B}{B} = \frac{C+D}{D}$$
with the proportion
$$\frac{A}{B} = \frac{C}{D}.$$
[6.6b]

Test: Chapter 6

Find all numbers for which the rational expression is undefined.

1. $\frac{8}{2x}$

2. $\frac{5}{x + 8}$

3. $\frac{x - 7}{x^2 - 49}$

4. $\frac{x^2 + x - 30}{x^2 - 3x + 2}$

5. $\frac{11}{(x - 1)^2}$

6. $\frac{x + 2}{x^3 + 8x^2 + 15x}$

7. Simplify:

$$\frac{6x^2 + 17x + 7}{2x^2 + 7x + 3}.$$

8. Multiply and simplify:

$$\frac{a^2 - 25}{6a} \cdot \frac{3a}{a - 5}.$$

9. Divide and simplify:

$$\frac{25x^2 - 1}{9x^2 - 6x} \div \frac{5x^2 + 9x - 2}{3x^2 + x - 2}.$$

10. Find the LCM:

$y^2 - 9,\ y^2 + 10y + 21,\ y^2 + 4y - 21.$

Add or subtract. Simplify, if possible.

11. $\frac{16 + x}{x^3} + \frac{7 - 4x}{x^3}$

12. $\frac{5 - t}{t^2 + 1} - \frac{t - 3}{t^2 + 1}$

13. $\frac{x - 4}{x - 3} + \frac{x - 1}{3 - x}$

14. $\frac{x - 4}{x - 3} - \frac{x - 1}{3 - x}$

15. $\frac{5}{t - 1} + \frac{3}{t}$

16. $\frac{1}{x^2 - 16} - \frac{x + 4}{x^2 - 3x - 4}$

17. $\frac{1}{x - 1} + \frac{4}{x^2 - 1} - \frac{2}{x^2 - 2x + 1}$

18. Simplify: $\dfrac{9 - \frac{1}{y^2}}{3 - \frac{1}{y}}.$

Answers

1. ________

2. ________

3. ________

4. ________

5. ________

6. ________

7. ________

8. ________

9. ________

10. ________

11. ________

12. ________

13. ________

14. ________

15. ________

16. ________

17. ________

18. ________

Answers

19. ____________

20. ____________

21. ____________

22. ____________

23. ____________

24. ____________

25. ____________

26. ____________

27. ____________

28. ____________

29. ____________

30. ____________

31. ____________

32. ____________

33. ____________

34. ____________

Solve.

19. $\frac{7}{y} - \frac{1}{3} = \frac{1}{4}$

20. $\frac{15}{x} - \frac{15}{x-2} = -2$

Solve.

21. The reciprocal of 3 less than a number is four times the reciprocal of the number itself. What is the number?

22. A sample of 125 spark plugs contained 4 defective spark plugs. How many defective spark plugs would you expect to find in a sample of 500?

23. One car travels 20 mph faster than another on a freeway. While one goes 225 mi, the other goes 325 mi. Find the speed of each car.

24. *Income vs. Time.* Dean's income I varies directly as the time t worked. He gets a job that pays \$275 for 40 hr of work. What is he paid for working 72 hr, assuming that there is no change in payscale for overtime?

25. Find an equation of variation in which Q varies jointly as x and y, and $Q = 25$ when $x = 2$ and $y = 5$.

26. Find an equation of variation in which y varies inversely as x, and $y = 10$ when $x = 25$.

27. Solve $L = \frac{Mt - g}{t}$ for t.

28. This pair of triangles is similar. Find the missing length x.

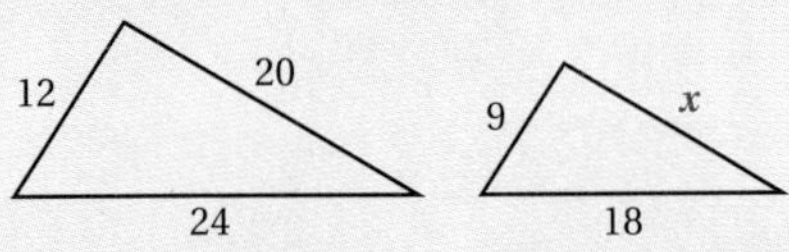

Skill Maintenance

29. Factor: $16a^2 - 49$.

30. Simplify: $\left(\frac{3x^2}{y^3}\right)^{-4}$.

31. Subtract:
$(5x^2 - 19x + 34) - (-8x^2 + 10x - 42)$.

32. The product of two consecutive integers is 462. Find the integers.

Synthesis

33. Team A and team B work together and complete a job in $2\frac{6}{7}$ hr. It would take team B 6 hr longer, working alone, to do the job than it would team A. How long would it take each of them to do the job working alone?

34. Simplify: $1 + \cfrac{1}{1 + \cfrac{1}{1 + \cfrac{1}{a}}}$.

Cumulative Review: Chapters 1–6

Evaluate.

1. $\dfrac{2x+5}{y-10}$, for $x = 2$ and $y = 5$

2. $4 - x^3$, for $x = -2$

Simplify.

3. $x - [x - 2(x+3)]$

4. $(2x^{-2})^{-2}(3x)^3$

5. $\dfrac{24x^8}{18x^{-2}}$

6. $\dfrac{2t^2 + 8t - 42}{2t^2 + 13t - 7}$

7. $\dfrac{\frac{2}{x} + 1}{\frac{x}{x+2}}$

8. $\dfrac{a^2 - 16}{a^2 - 8a + 16}$

Add. Simplify, if possible.

9. $\dfrac{9}{14} + \left(-\dfrac{5}{21}\right)$

10. $\dfrac{2x+y}{x^2y} + \dfrac{x+2y}{xy^2}$

11. $\dfrac{z}{z^2-1} + \dfrac{2}{z+1}$

12. $(2x^4 + 5x^3 + 4) + (3x^3 - 2x + 5)$

Subtract. Simplify, if possible.

13. $1.53 - (-0.8)$

14. $(x^2 - xy - y^2) - (x^2 - y^2)$

15. $\dfrac{3}{x^2-9} - \dfrac{x}{9-x^2}$

16. $\dfrac{2x}{x^2 - x - 20} - \dfrac{4}{x^2 - 10x + 25}$

Multiply. Simplify, if possible.

17. $(1.3)(-0.5)(2)$

18. $3x^2(2x^2 + 4x - 5)$

19. $\left(3t + \frac{1}{2}\right)\left(3t - \frac{1}{2}\right)$

20. $(2p - q)^2$

21. $(3x+5)(x-4)$

22. $(2x^2+1)(2x^2-1)$

23. $\dfrac{6t+6}{t^3 - 2t^2} \cdot \dfrac{t^3 - 3t^2 + 2t}{3t+3}$

24. $\dfrac{a^2-1}{a^2} \cdot \dfrac{2a}{1-a}$

Divide. Simplify, if possible.

25. $(3x^3 - 7x^2 + 9x - 5) \div (x - 1)$

26. $-\dfrac{21}{25} \div \dfrac{28}{15}$

27. $\dfrac{x^2 - x - 2}{4x^3 + 8x^2} \div \dfrac{x^2 - 2x - 3}{2x^2 + 4x}$

28. $\dfrac{3 - 3x}{x^2} \div \dfrac{x-1}{4x}$

Factor completely.

29. $4x^3 + 12x^2 - 9x - 27$

30. $x^2 + 7x - 8$

31. $3x^2 - 14x - 5$

32. $16y^2 + 40xy + 25x^2$

33. $3x^3 + 24x^2 + 45x$

34. $2x^2 - 2$

35. $x^2 - 28x + 196$

36. $4y^3 + 10y^2 + 12y + 30$

Solve.

37. $2(x - 3) = 5(x + 3)$

38. $2x(3x + 4) = 0$

39. $x^2 = 8x$

40. $x^2 + 16 = 8x$

41. $x - 5 \leq 2x + 4$

42. $3x^2 = 27$

43. $\frac{1}{3}x - \frac{2}{5} = \frac{4}{5}x + \frac{1}{3}$

44. $\frac{x}{3} = \frac{3}{x}$

45. $\frac{x + 5}{2x + 1} = \frac{x - 7}{2x - 1}$

46. $\frac{1}{3}x\left(2x - \frac{1}{5}\right) = 0$

47. $\frac{3 - x}{x - 1} = \frac{2}{x - 1}$

48. $\frac{3}{2x + 5} = \frac{2}{5 - x}$

49. $\frac{1}{x} + \frac{1}{y} = \frac{1}{z}$, for z

50. $\frac{3N}{T} = D$, for N

51. Find the intercepts. Then graph the equation.

$5y - 2x = -10$

Solve.

52. The sum of three consecutive integers is 99. What are the integers?

53. The speed of one bicyclist is 2 km/h faster than the speed of another bicyclist. The first bicyclist travels 60 km in the same time that it takes the second to travel 50 km. Find the speed of each bicyclist.

54. A swimming pool can be filled in 5 hr by hose A alone and in 6 hr by hose B alone. How long would it take to fill the tank if both hoses were working?

55. The sum of the page numbers on the facing pages of a book is 69. What are the page numbers?

56. The product of the page numbers on two facing pages of a book is 272. Find the page numbers.

57. *Insurance Costs.* The cost c of an insurance policy varies directly as the age a of the insured. A 32-year-old person pays an annual premium of \$152. What is the age of a person who pays \$285?

58. The area of a circle is 35π more than the circumference. Find the length of the radius.

59. The sum of the squares of two consecutive odd positive integers is 202. Find the integers.

Synthesis

Solve.

60. $(2x - 1)^2 = (x + 3)^2$

61. $\frac{x + 2}{3x + 2} = \frac{1}{x}$

62. $\frac{2 + \frac{2}{x}}{x + 2 + \frac{1}{x}} = \frac{x + 2}{3}$

63. $\frac{x^6x^4}{x^9x^{-1}} = \frac{5^{14}}{25^6}$

64. Find the reciprocal of $\frac{1 - x}{x + 3} + \frac{x + 1}{2 - x}$.

65. Find the reciprocal of 2.0×10^{-8} and express in scientific notation.

7

Graphs, Functions, and Applications

An Application

The graph at right approximates the weekly U.S. revenue, in millions of dollars, from the recent movie *Air Force One*. The revenue is a function f of the number of weeks since the movie was released. No equation is given for the function. What was the movie revenue for week 5? In other words, we need to find $f(5)$.

This problem appears as Example 9 in Section 7.1.

The Mathematics

We locate 5 on the horizontal axis and move directly up until we reach the graph. Then we move across to the vertical axis. We estimate that value to be about \$13 million—that is, $f(5) = 13$.

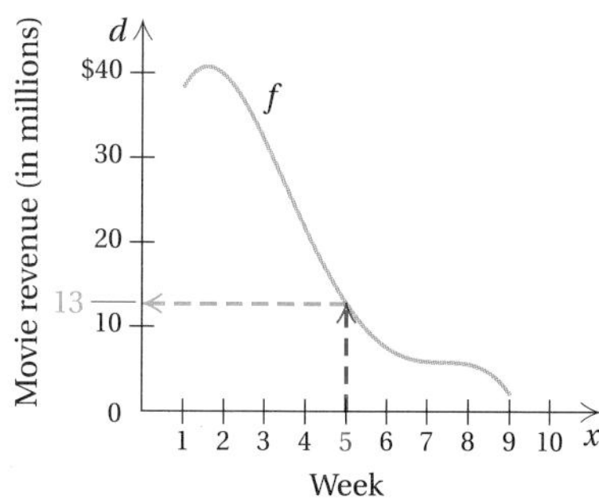

Source: Exhibitor Relations Co., Inc.

World Wide Web For more information, visit us at www.mathmax.com

Introduction

Graphs help us to *see* relationships among quantities. In this chapter, we graph many kinds of equations, emphasizing those whose graphs are straight lines. These are called *linear equations.*

We also consider the concept of a *function*. Functions are the backbone of any future study in mathematics. Here we introduce graphs, as well as applications, of functions.

For linear equations and functions, we develop the concepts of slope and intercepts. Finally, we bring the ideas of this chapter together in a study of data analysis and applications.

Pretest: Chapter 7

Graph on a plane.

1. $2x - 5y = 20$
2. $x = 4$
3. $y = x - 2$
4. $f(x) = -2$
5. $f(x) = 3 - x^2$

6. Find the slope and the y-intercept: $y = 5x - 3$.

7. Find the slope, if it exists, of the line containing the points $(7, 4)$ and $(-5, 4)$.

8. Find an equation of the line containing the points $(-3, 7)$ and $(-8, 2)$.

9. Find an equation of the line having the given slope and containing the given point.

   $m = 2; \quad (-2, 3)$

10. Find an equation of the line containing the given point and parallel to the given line.

    $(2, 5); \quad 2x - 7y = 10$

11. Find an equation of the line containing the given point and perpendicular to the given line.

    $(2, 5); \quad 2x - 7y = 10$

Determine whether the graphs of the pair of lines are parallel or perpendicular.

12. $3y - 2x = 21,$
    $3x + 2y = 8$

13. $y = 3x + 7,$
    $y = 3x - 4$

14. Find the intercepts of $2x - 3y = 12$.

15. For the function f given by $f(x) = |x| - 3$, find $f(0)$, $f(-2)$, and $f(4)$.

16. Find the domain:

    $$f(x) = \frac{3}{2x - 5}.$$

17. Determine whether each of the following is the graph of a function.

a)

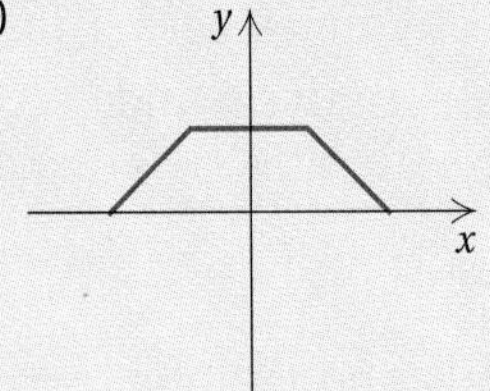

b)

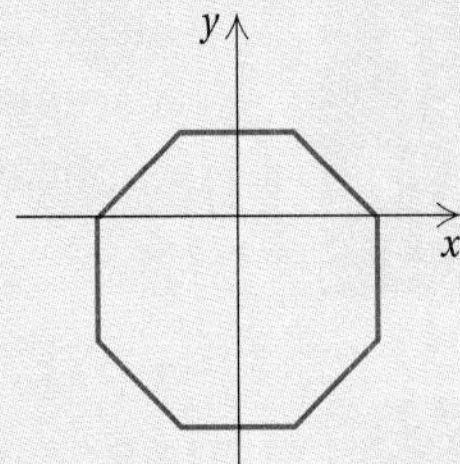

18. *Spending on Recorded Music.* Consider the following graph showing the average amount of spending per person per year on recorded music.

Average Amount Spent per Person
(Data for 1997–2001 are projections.)

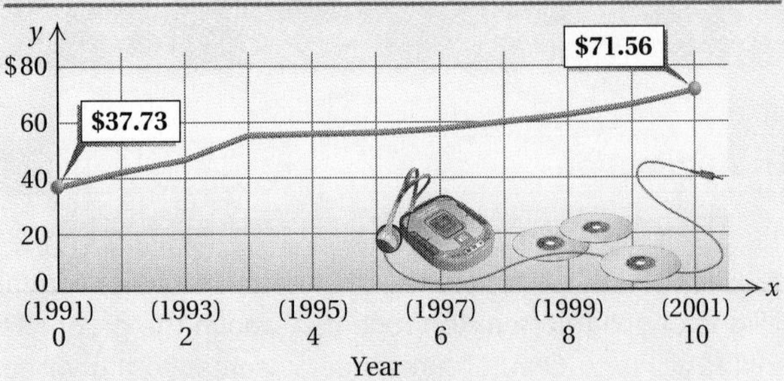

Source: Veronis, Suhler & Associates, Industry Sources

a) Use the two points (0, \$37.73) and (10, \$71.56) to find a linear function that fits the data.

b) Use the function to predict the average amount of spending on recorded music in 2011.

Objectives for Retesting

The objectives to be tested in addition to the material in this chapter are as follows.

[4.2c] Convert between scientific notation and decimal notation.

[4.8b] Divide a polynomial by a divisor that is not a monomial, and if there is a remainder, express the result in two ways.

[5.6a] Factor sums and differences of two cubes.

[6.1c] Simplify rational expressions by factoring the numerator and the denominator and removing factors of 1.

7.1 Functions and Graphs

a Identifying Functions

We now develop one of the most important concepts in mathematics, **functions**. We have actually been studying functions all through this text; we just haven't identified them as such. In much the same way that ordered pairs form correspondences between first and second coordinates, a *function* is a correspondence from one set to another. For example:

> To each student in a college, there corresponds his or her student ID.
>
> To each item in a store, there corresponds its price.
>
> To each real number, there corresponds the cube of that number.

In each case, the first set is called the **domain** and the second set is called the **range**. This kind of correspondence is called a **function**. Given a member of the domain, there is *just one* member of the range to which it corresponds.

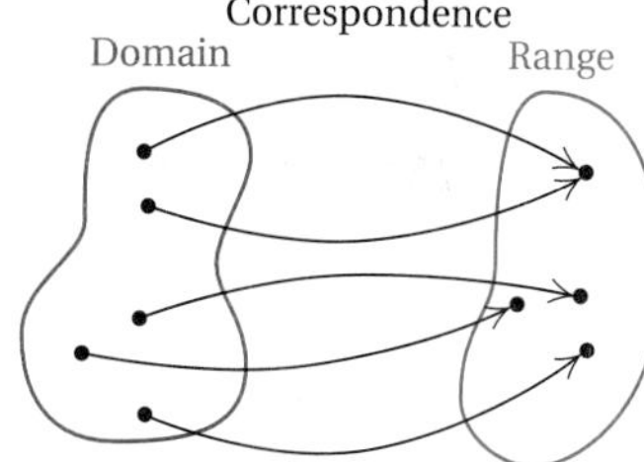

Objectives

a Determine whether a correspondence is a function.

b Given a function described by an equation, find function values (outputs) for specified values (inputs).

c Draw the graph of a function.

d Determine whether a graph is that of a function using the vertical-line test.

e Solve applied problems involving functions and their graphs.

For Extra Help

TAPE 12 | MAC WIN | CD-ROM

Example 1 Determine whether the correspondence is a function.

f:

Domain		*Range*
1	→	\$107.4
2	→	\$ 34.1
3	→	\$ 29.6
4	→	\$ 19.6

g:

Domain		*Range*
3	→	5
4	→	9
5	→	−7
6	→	−7

h:

Domain		*Range*
Chicago	→	Cubs
Chicago	→	White Sox
Baltimore	→	Orioles
San Diego	→	Padres

p:

Domain		*Range*
Cubs	→	Chicago
White Sox	→	Chicago
Orioles	→	Baltimore
Padres	→	San Diego

The correspondence f is a function because each member of the domain is matched to *only one* member of the range.

The correspondence g *is* also a function because each member of the domain is matched to *only one* member of the range.

The correspondence h *is not* a function because one member of the domain, Chicago, is matched to *more than one* member of the range.

The correspondence p *is* a function because each member of the domain is matched to *only one* member of the range.

> A **function** is a correspondence between a first set, called the **domain**, and a second set, called the **range**, such that each member of the domain corresponds to *exactly one* member of the range.

Do Exercises 1–4.

Determine whether the correspondence is a function.

1.

Domain		*Range*
Cheetah	→	70 mph
Human	→	28 mph
Lion	→	50 mph
Chicken	→	9 mph

2.

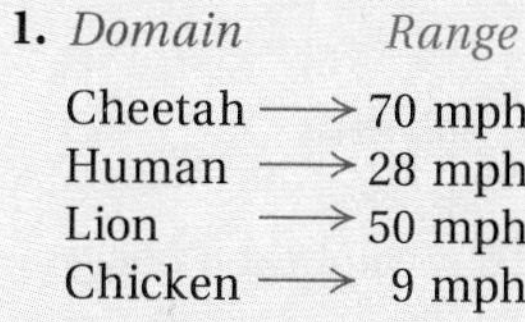

3.

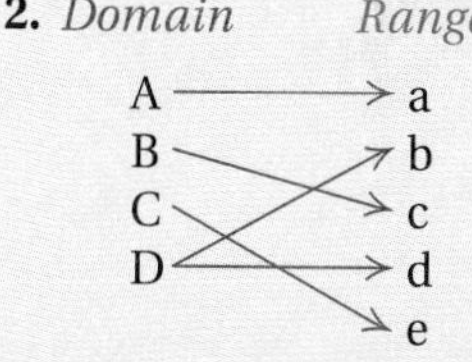

4.

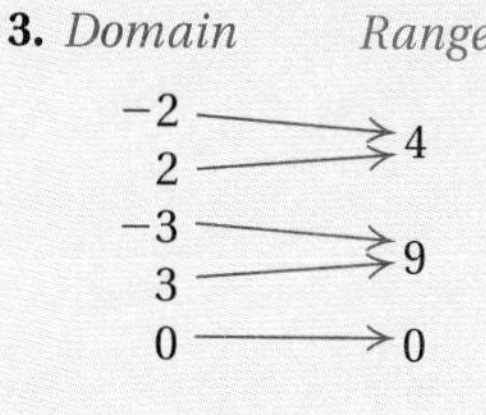

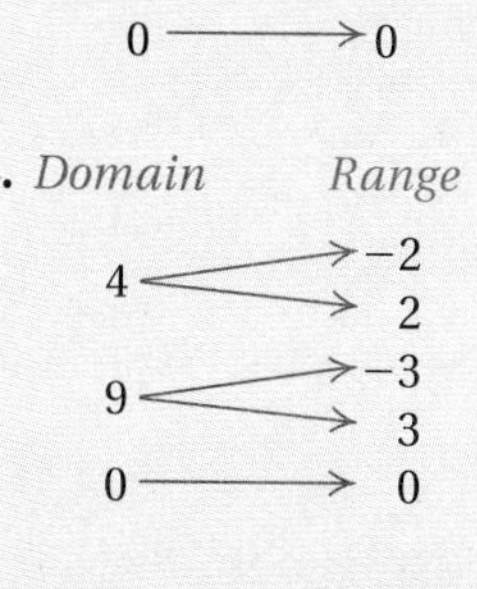

Answers on page A-26

Determine whether each of the following is a function.

5. *Domain*
A set of numbers
Correspondence
Square each number and subtract 10.
Range
A set of numbers

6. *Domain*
A set of polygons
Correspondence
Find the area of each polygon.
Range
A set of numbers

Answers on page A-26

Example 2 Determine whether the correspondence is a function.

Domain	*Correspondence*	*Range*
a) A family	Each person's weight	A set of positive numbers
b) The integers	Each number's square	A set of nonnegative integers
c) The set of all states	Each state's members of the U.S. Senate	A set of U.S. Senators

a) The correspondence *is* a function because each person has *only one* weight.

b) The correspondence *is* a function because each integer has *only one* square.

c) The correspondence *is not* a function because each state has two U.S. Senators.

Do Exercises 5 and 6.

When a correspondence between two sets is not a function, it is still an example of a **relation**.

> A **relation** is a correspondence between a first set, called the **domain**, and a second set, called the **range**, such that each member of the domain corresponds to *at least one* member of the range.

Thus, although the correspondences of Examples 1 and 2 are not all functions, they *are* all relations. A function is a special type of relation—one in which each member of the domain is paired with *exactly one* member of the range.

b Finding Function Values

Most functions considered in mathematics are described by equations like $y = 2x + 3$ or $y = 4 - x^2$. We graph the function $y = 2x + 3$ by first performing calculations like the following:

for $x = 4$, $y = 2x + 3 = 2 \cdot 4 + 3 = 11$;

for $x = -5$, $y = 2x + 3 = 2 \cdot (-5) + 3 = -7$;

for $x = 0$, $y = 2x + 3 = 2 \cdot 0 + 3 = 3$; and so on.

For $y = 2x + 3$, the **inputs** (members of the domain) are values of x substituted into the equation. The **outputs** (members of the range) are the resulting values of y. If we call the function f, we can use x to represent an arbitrary *input* and $f(x)$—read "f of x," or "f at x," or "the value of f at x"—to represent the corresponding *output*. In this notation, the function given by $y = 2x + 3$ is written as $f(x) = 2x + 3$ and the calculations above can be written more concisely as follows:

$f(4) = 2 \cdot 4 + 3 = 11$;

$f(-5) = 2 \cdot (-5) + 3 = -7$;

$f(0) = 2 \cdot 0 + 3 = 3$; and so on.

Thus instead of writing "when $x = 4$, the value of y is 11," we can simply write "$f(4) = 11$," which can also be read as "f of 4 is 11" or "for the input 4, the output of f is 11."

It helps to think of a function as a machine. Think of $f(4) = 11$ as putting a member of the domain (an input), 4, into the machine. The machine knows the correspondence $f(x) = 2x + 3$, multiplies 4 by 2 and adds 3, and gives out a member of the range (the output), 11.

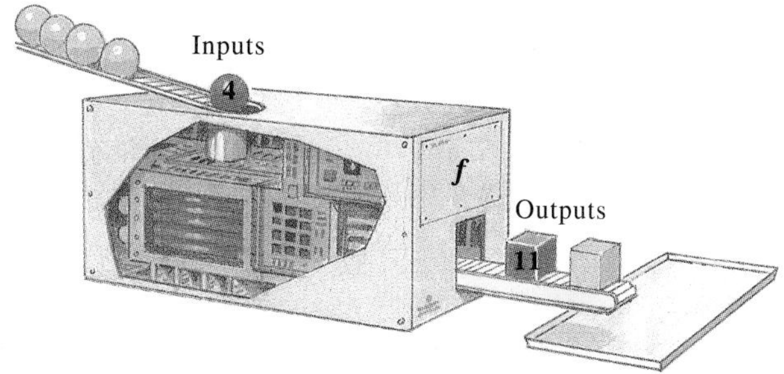

CAUTION! The notation $f(x)$ *does not mean "f times x"* and should not be read that way.

Example 3 A function f is given by $f(x) = 3x^2 - 2x + 8$. Find each of the indicated function values.

a) $f(0)$

b) $f(1)$

c) $f(-5)$

d) $f(7a)$

One way to find function values when a formula is given is to think of the formula with blanks, or placeholders, as follows:

$$f(\quad) = 3\quad^2 - 2\quad + 8.$$

To find an output for a given input, we think: "Whatever goes in the blank on the left goes in the blank(s) on the right." With this in mind, let's complete the example.

a) $f(0) = 3 \cdot 0^2 - 2 \cdot 0 + 8 = 8$

b) $f(1) = 3 \cdot 1^2 - 2 \cdot 1 + 8 = 3 \cdot 1 - 2 + 8 = 3 - 2 + 8 = 9$

c) $f(-5) = 3(-5)^2 - 2 \cdot (-5) + 8 = 3 \cdot 25 + 10 + 8 = 75 + 10 + 8 = 93$

d) $f(7a) = 3(7a)^2 - 2(7a) + 8 = 3 \cdot 49a^2 - 14a + 8 = 147a^2 - 14a + 8$

Do Exercise 7.

Example 4 Find the indicated function value.

a) $f(5)$, for $f(x) = 3x + 2$

b) $g(-2)$, for $g(r) = 5r^2 + 3r$

c) $h(4)$, for $h(x) = 7$

d) $F(a + 1)$, for $F(x) = 5x - 8$

a) $f(5) = 3 \cdot 5 + 2 = 15 + 2 = 17$

b) $g(-2) = 5(-2)^2 + 3(-2) = 5(4) - 6 = 20 - 6 = 14$

c) For the function given by $h(x) = 7$, all inputs share the same output, 7. Thus, $h(4) = 7$. The function h is an example of a **constant function.**

d) $F(a + 1) = 5(a + 1) - 8 = 5a + 5 - 8 = 5a - 3$

Do Exercise 8.

7. Find the indicated function values for the following function:

$$f(x) = 2x^2 + 3x - 4.$$

a) $f(0)$
b) $f(8)$
c) $f(-5)$
d) $f(7a)$

8. Find the indicated function value.

a) $f(-6)$, for $f(x) = 5x - 3$
b) $g(-1)$, for $g(r) = 5r^2 + 3r$
c) $h(55)$, for $h(x) = 7$
d) $F(a + 2)$, for $F(x) = 5x - 8$

Answers on page A-26

Graph.

9. $f(x) = x - 4$

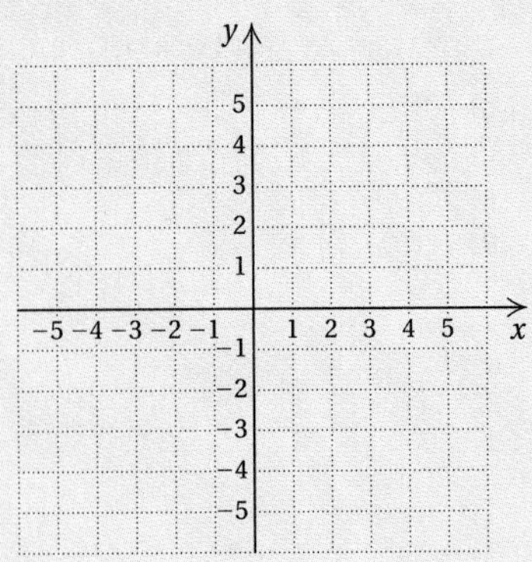

10. $g(x) = 5 - x^2$

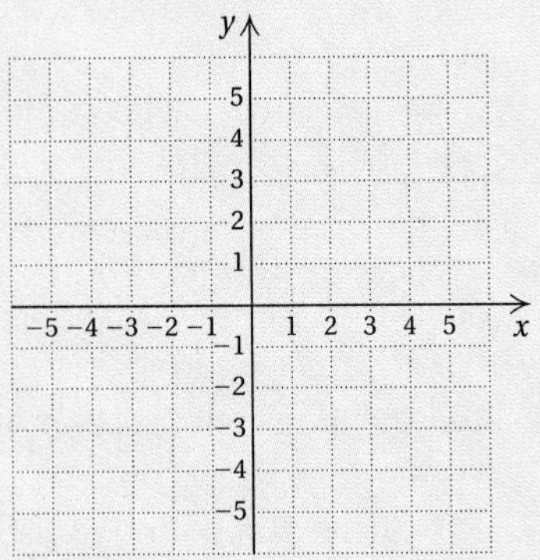

11. $t(x) = 3 - |x|$

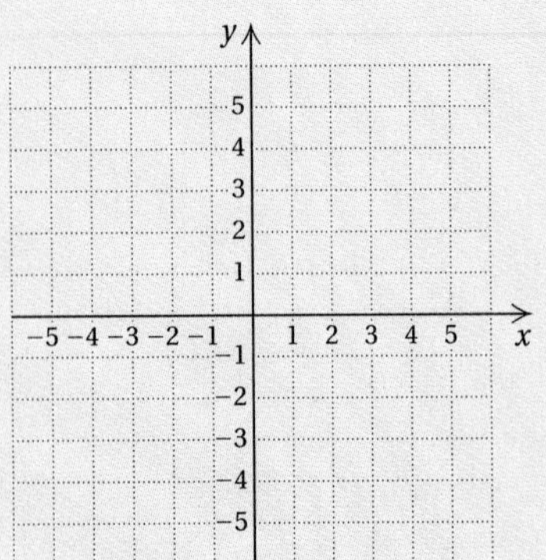

Answers on page A-27

c Graphs of Functions

To graph a function, we find ordered pairs (x, y) or $(x, f(x))$, plot them, and connect the points. Note that y and $f(x)$ are used interchangeably—that is, $y = f(x)$—when working with functions and their graphs.

Example 5 Graph: $f(x) = x + 2$.

A list of some function values is shown in this table. We plot the points and connect them. The graph is a straight line.

x	$f(x)$
−4	−2
−3	−1
−2	0
−1	1
0	2
1	3
2	4
3	5
4	6

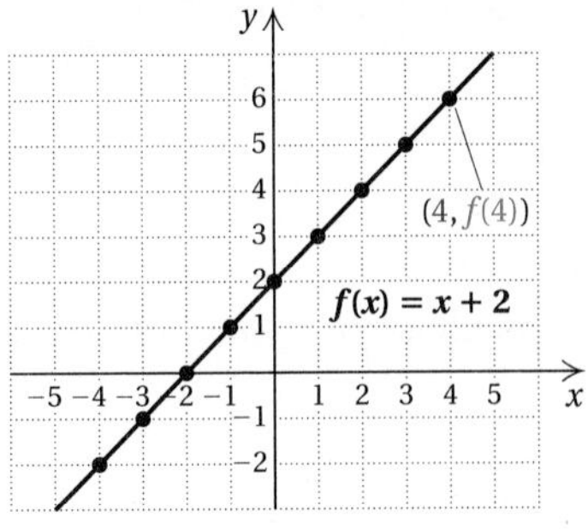

Example 6 Graph: $g(x) = 4 - x^2$.

We calculate some function values and draw the curve.

$$g(0) = 4 - 0^2 = 4 - 0 = 4,$$
$$g(-1) = 4 - (-1)^2 = 4 - 1 = 3,$$
$$g(2) = 4 - 2^2 = 4 - 4 = 0,$$
$$g(-3) = 4 - (-3)^2 = 4 - 9 = -5$$

x	$g(x)$
−3	−5
−2	0
−1	3
0	4
1	3
2	0
3	−5

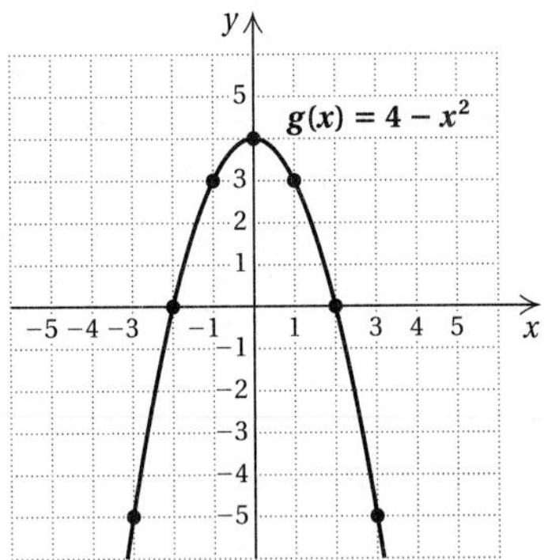

Example 7 Graph: $h(x) = |x|$.

A list of some function values is shown in the following table. We plot the points and connect them. The graph is a V-shaped "curve" that rises on either side of the vertical axis.

x	$h(x)$
−3	3
−2	2
−1	1
0	0
1	1
2	2
3	3

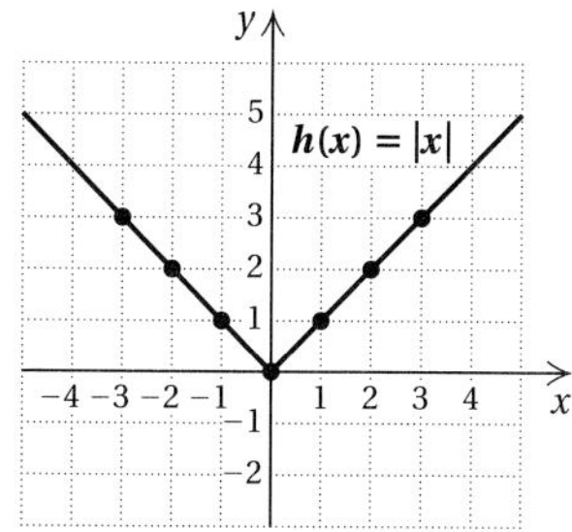

Do Exercises 9–11 on the preceding page.

d The Vertical-Line Test

Consider the graph of the function f described by $f(x) = x^2 - 5$. Its graph is shown at right. It is also the graph of the equation $y = x^2 - 5$.

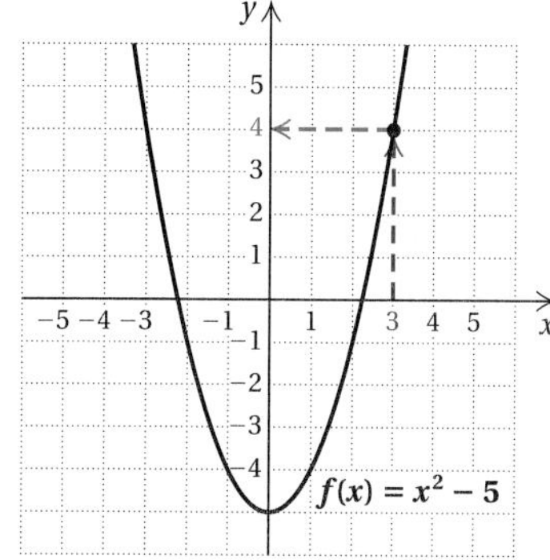

To find a function value, like $f(3)$, from a graph, we locate the input on the horizontal axis, move directly up or down to the graph of the function, and then move left or right to find the output on the vertical axis. Thus, $f(3) = 4$. Keep in mind that members of the domain are found on the horizontal axis and members of the range are found on the vertical axis.

When one member of the domain is paired with two or more different members of the range, the correspondence is not a function. Thus, when a graph contains two or more different points with the same first coordinate, the graph cannot represent a function. Points sharing a common first coordinate are vertically above or below each other (see the following graph).

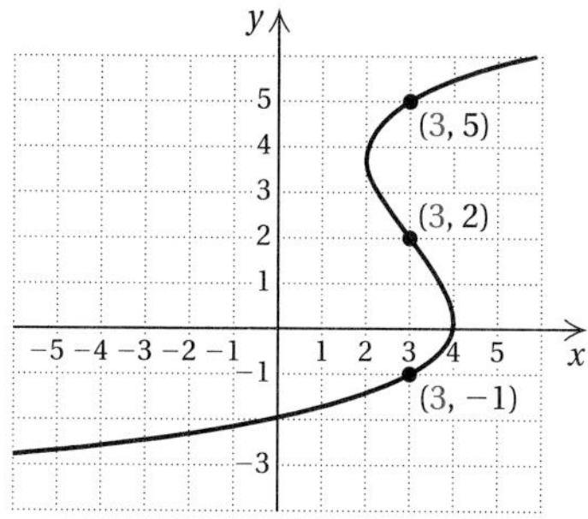

Since 3 is paired with more than one member of the range, the graph does not represent a function.

This observation leads to the *vertical-line test.*

> **THE VERTICAL-LINE TEST**
>
> A graph represents a function if it is impossible to draw a vertical line that intersects the graph more than once.

Calculator Spotlight

Graphs and Function Values. To graph the function $f(x) = 2x^2 + x$, we enter it into the grapher as $y_1 = 2x^2 + x$ and use the GRAPH feature.

To find function values, we can use the CALC menu and choose Value. To evaluate the function $f(x) = 2x^2 + x$ at $x = -2$, that is, to find $f(-2)$, we press [2nd] [CALC] [1]. We enter $x = -2$. The function value, or y-value, $y = -6$ appears together with a trace indicator showing the point $(-2, 6)$.

Functions can also be evaluated using the TABLE feature or the Y-VARS feature. Consult the manual for your particular grapher for specific keystrokes.

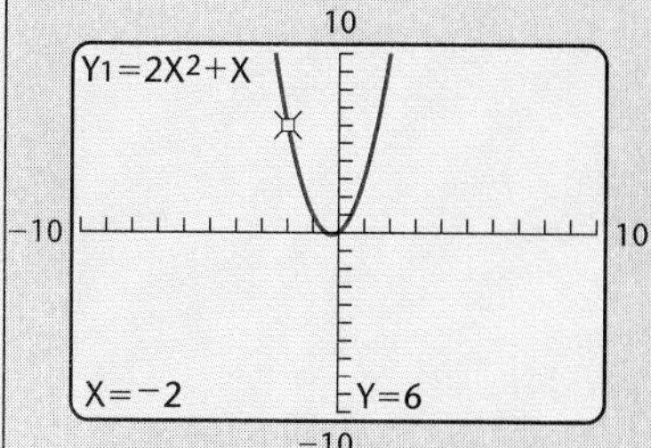

Exercises

1. Graph $f(x) = x^2 + 3x - 4$. Then find $f(-5)$, $f(-4.7)$, $f(11)$, and $f(2/3)$.
2. Graph $f(x) = 3.7 - x^2$. Then find $f(-5)$, $f(-4.7)$, $f(11)$, and $f(2/3)$.
3. Graph $f(x) = 4 - 1.2x - 3.4x^2$. Then find $f(-5)$, $f(-4.7)$, $f(11)$, and $f(2/3)$.

Determine whether each of the following is a graph of a function.

12.

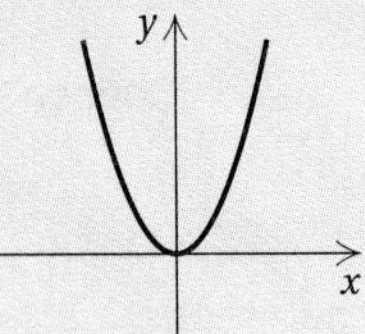

13.

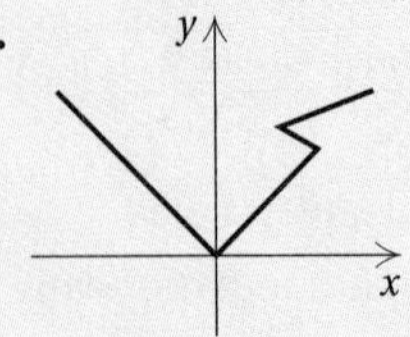

14.

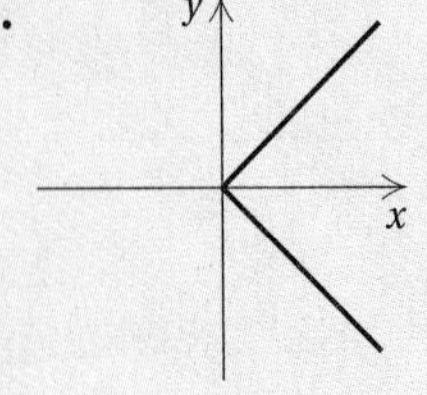

15.

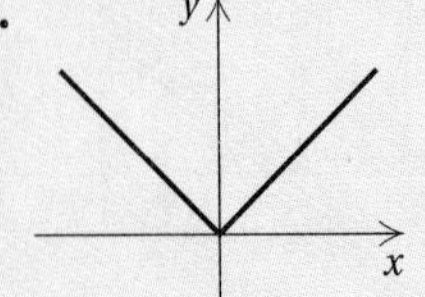

Answers on page A-27

Example 8 Determine whether each of the following is the graph of a function.

a)

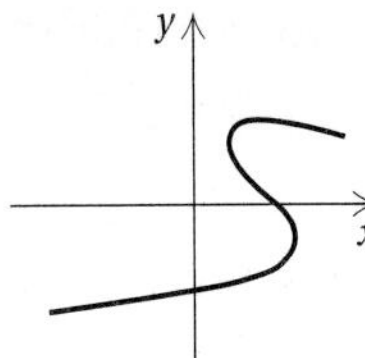

b)

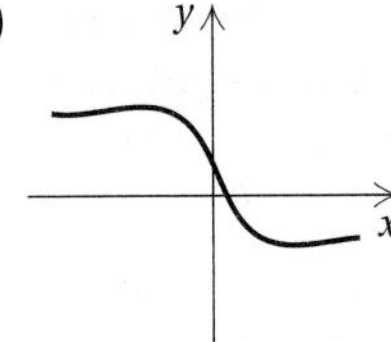

c)

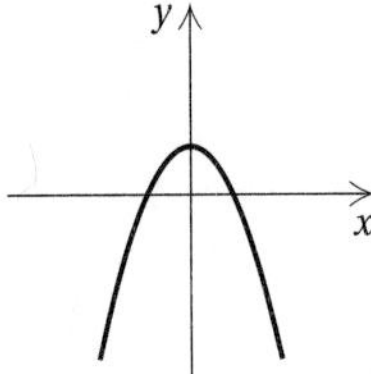

d)

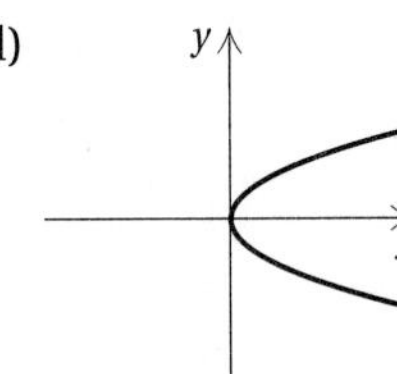

a) The graph *is not* that of a function because a vertical line can cross the graph at more than one point.

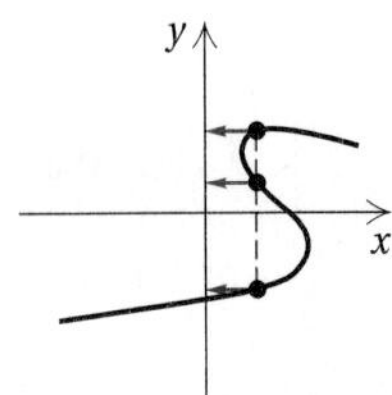

b) The graph *is* that of a function because no vertical line can cross the graph at more than one point. This can be confirmed with a ruler or straightedge.

c) The graph *is* that of a function.

d) The graph *is not* that of a function. There is a vertical line that can cross the graph more than once.

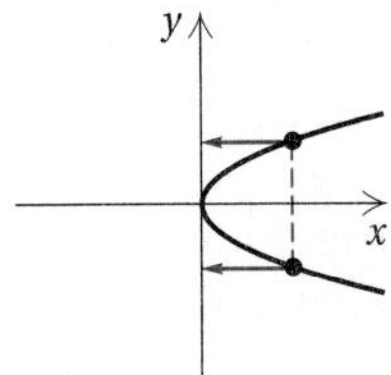

Do Exercises 12–15.

e Applications of Functions and Their Graphs

Functions are often described by graphs, whether or not an equation is given. To use a graph in an application, we note that each point on the graph represents a pair of values.

Example 9 *Movie Revenue.* The following graph approximates the weekly U.S. revenue, in millions of dollars, from the recent movie *Air Force One*. The revenue is a function f of the number of weeks x since the movie was released. No equation is given for the function.

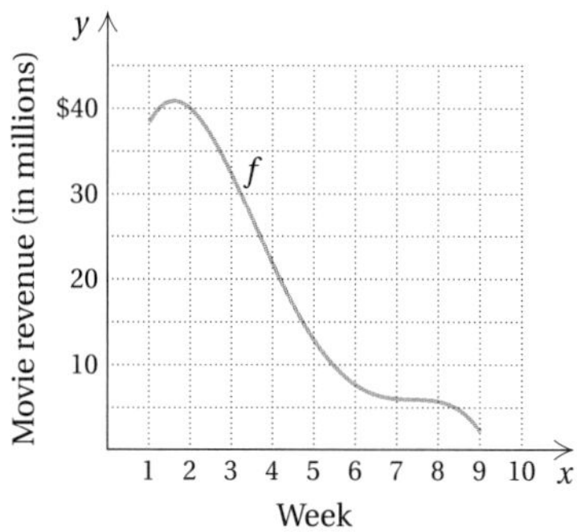

Use the graph to answer the following.

a) What was the movie revenue for week 1? That is, find $f(1)$.

b) What was the movie revenue for week 5? That is, find $f(5)$.

a) To estimate the revenue for week 1, we locate 1 on the horizontal axis and move directly up until we reach the graph. Then we move across to the vertical axis. We estimate that value to be about $38 million—that is, $f(1) = 38$.

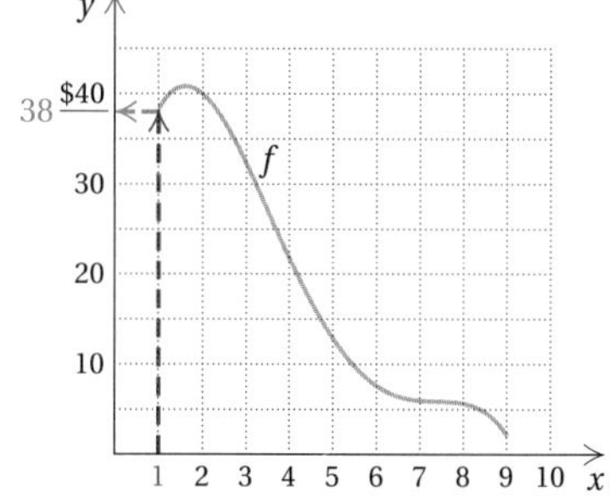

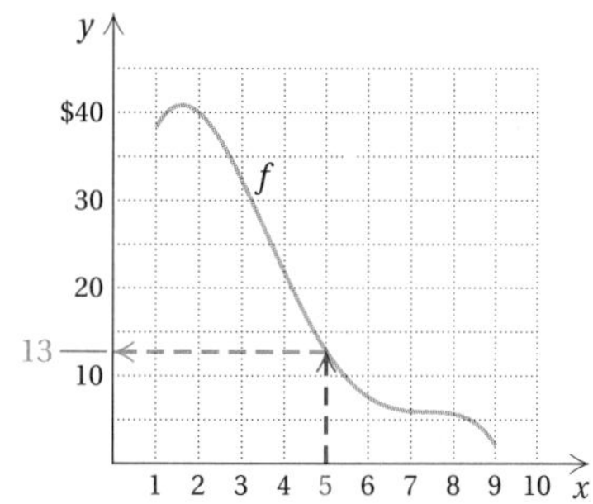

b) To estimate the revenue for week 5, we locate 5 on the horizontal axis and move directly up until we reach the graph. Then we move across to the vertical axis. We estimate that value to be about $13 million—that is, $f(5) = 13$.

Do Exercises 16 and 17.

Referring to the graph in Example 9:

16. What was the movie revenue for week 2?

17. What was the movie revenue for week 6?

Answers on page A-27

Calculator Spotlight

TRACE and TABLE Features

TRACE Feature. There are two ways in which we can determine the coordinates of points on a graph drawn by a grapher. One approach is to use the TRACE key. When the TRACE feature is activated, the cursor appears on the line (or curve) that has been graphed and the coordinates at that point are displayed.

Let's consider the equation or function $y = x^3 - 5x + 1$. We graph the equation in the window $[-5, 5, -10, 10]$ and press TRACE:

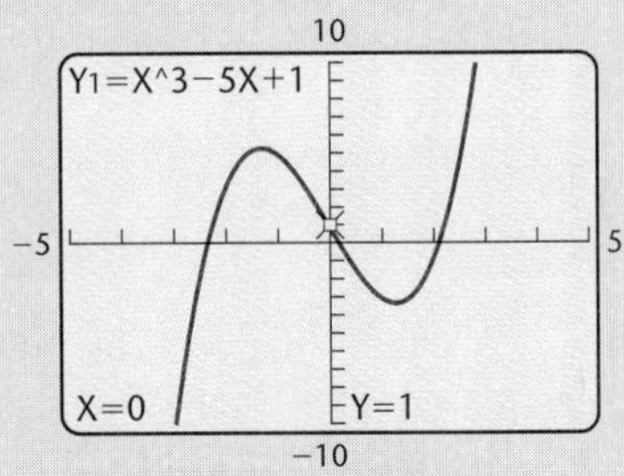

The coordinates at the bottom indicate that the cursor is at the point with coordinates (0, 1). By using the arrow keys, we can obtain coordinates of other points. For example, if we press the left arrow key ◁ seven times, we move the cursor to the location shown below, obtaining a point on the graph with coordinates (−0.7446809, 4.3104418).

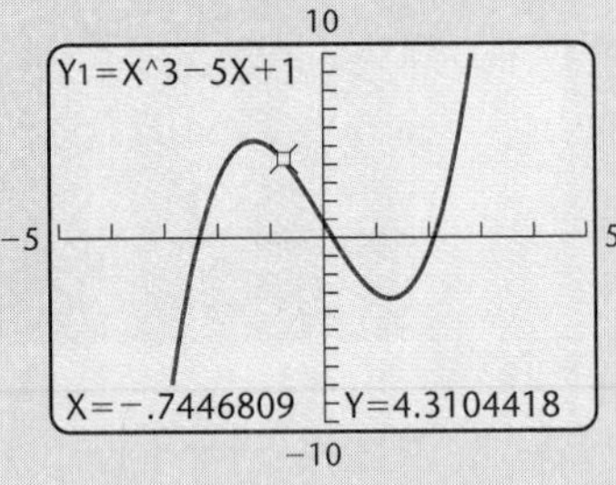

TABLE Feature. Another way to find the coordinates of solutions of equations makes use of the TABLE feature. We first press 2nd TBLSET. For the equation above, let's set TblStart = 0.3 and ΔTbl = 1.

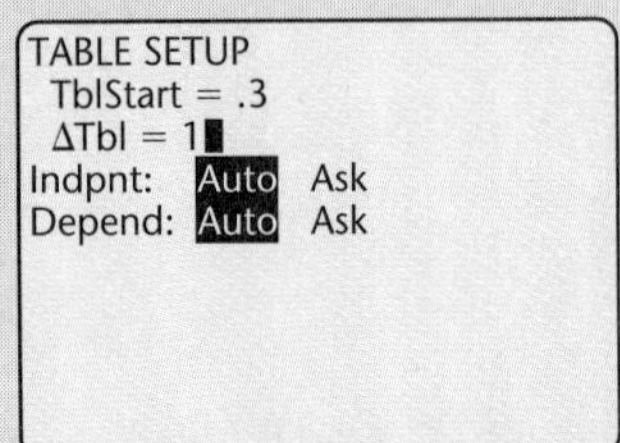

This means that the table's x-values will start at 0.3 and increase by 1. (We could choose other values for TblStart and ΔTbl.) By setting Indpnt and Depend to Auto, we obtain the following when we press 2nd TABLE:

X	Y1
.3	−.473
1.3	−3.303
2.3	1.667
3.3	20.437
4.3	59.007
5.3	123.38
6.3	219.55

X = .3

The arrow keys allow us to scroll up and down the table and extend it to other values not initially shown.

X	Y1
12.3	1800.4
13.3	2287.1
14.3	2853.7
15.3	3506.1
16.3	4250.2
17.3	5092.2
18.3	6038

X = 18.3

Exercises

1. Use the TRACE feature to find five different ordered-pair solutions of the equation $y = x^3 - 5x + 1$.
2. For $y = x^3 - 5x + 1$, use the TABLE feature to construct a table starting with $x = 10$ and $\Delta\text{Tbl} = 5$. Find the value of y when x is 10. Then find the value of y when x is 35.
3. Adjust the table settings to Indpnt: Ask. How does the table change? Enter a number of your choice and see what happens. Use this setting to find the value of y when $x = 28$.

Exercise Set 7.1

a Determine whether the correspondence is a function.

1.

Domain	*Range*
2	9
5	8
19	

2.

Domain	*Range*
5	3
−3	7
7	
−7	

3.

Domain	*Range*
−5	1
5	
8	

4.

Domain	*Range*
6	−6
7	−7
3	−3

5.

Domain (Girl's Age, in months)	*Range* (Average Daily Weight Gain, in grams)
2	21.8
9	11.7
16	8.5
23	7.0

Source: *American Family Physician*, December 1993, p. 1435.

6.

Domain (Year)	*Range* (Consumption of Diet Cola, in gallons per person)
1991	11.7
1992	11.6
1993	
1994	11.9

Source: U.S. Department of Agriculture Economic Research Service

7.

Domain	*Range*
	Austin
Texas	Houston
	Dallas
	Cleveland
Ohio	Toledo
	Cincinnati

8.

Domain	*Range*
Austin	
Houston	Texas
Dallas	
Cleveland	
Toledo	Ohio
Cincinnati	

	Domain	*Correspondence*	*Range*
9.	A family	Each person's eye color	A set of colors
10.	A textbook	An even-numbered page in the book	A set of pages
11.	A set of avenues	An intersecting road	A set of cross streets
12.	A math class	Each person's seat number	A set of numbers
13.	A set of numbers	Square each number and then add 4.	A set of positive numbers
14.	A set of shapes	The perimeter of each shape	A set of positive numbers

b Find the function values.

15. $f(x) = x + 5$

a) $f(4)$ b) $f(7)$
c) $f(-3)$ d) $f(0)$
e) $f(2.4)$ f) $f\left(\frac{2}{3}\right)$

16. $g(t) = t - 6$

a) $g(0)$ b) $g(6)$
c) $g(13)$ d) $g(-1)$
e) $g(-1.08)$ f) $g\left(\frac{7}{8}\right)$

17. $h(p) = 3p$

a) $h(-7)$ b) $h(5)$
c) $h(14)$ d) $h(0)$
e) $h\left(\frac{2}{3}\right)$ f) $h(a + 1)$

18. $f(x) = -4x$

a) $f(6)$ b) $f\left(-\frac{1}{2}\right)$
c) $f(a - 1)$ d) $f(11.8)$
e) $f(0)$ f) $f(-1)$

19. $g(s) = 3s + 4$

a) $g(1)$ b) $g(-7)$
c) $g(6.7)$ d) $g(0)$
e) $g(-10)$ f) $g\left(\frac{2}{3}\right)$

20. $h(x) = 19$, a constant function

a) $h(4)$ b) $h(-6)$
c) $h(12.5)$ d) $h(0)$
e) $h\left(\frac{2}{3}\right)$ f) $h(1234)$

21. $f(x) = 2x^2 - 3x$

a) $f(0)$ b) $f(-1)$
c) $f(2)$ d) $f(10)$
e) $f(-5)$ f) $f(4a)$

22. $f(x) = 3x^2 - 2x + 1$

a) $f(0)$ b) $f(1)$
c) $f(-1)$ d) $f(10)$
e) $f(2a)$ f) $f(-3)$

23. $f(x) = |x| + 1$

a) $f(0)$ b) $f(-2)$
c) $f(2)$ d) $f(-3)$
e) $f(-10)$ f) $f(a - 1)$

24. $g(t) = |t - 1|$

a) $g(4)$ b) $g(-2)$
c) $g(-1)$ d) $g(100)$
e) $g(-50)$ f) $g(a + 1)$

25. $f(x) = x^3$

a) $f(0)$ b) $f(-1)$
c) $f(2)$ d) $f(10)$
e) $f(-5)$ f) $f(-10)$

26. $f(x) = x^4 - 3$

a) $f(1)$ b) $f(-1)$
c) $f(0)$ d) $f(2)$
e) $f(-2)$ f) $f(10)$

27. *Archaeology.* The function H described by

$$H(x) = 2.75x + 71.48$$

can be used to predict the height, in centimeters, of a woman whose *humerus* (the bone from the elbow to the shoulder) is x cm long. If a humerus is known to be from a female, how tall was she if the bone is **(a)** 32 cm long? **(b)** 35 cm long?

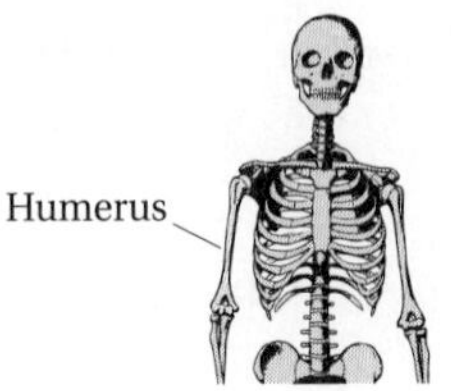

28. Refer to Exercise 27. When a humerus is from a male, the function

$$M(x) = 2.89x + 70.64$$

can be used to find the male's height, in centimeters. If a humerus is known to be from a male, how tall was the male if the bone is **(a)** 30 cm long? **(b)** 35 cm long?

29. *Pressure at Sea Depth.* The function $P(d) = 1 + (d/33)$ gives the pressure, in *atmospheres* (atm), at a depth of d feet in the sea. Note that $P(0) = 1$ atm, $P(33) = 2$ atm, and so on. Find the pressure at 20 ft, 30 ft, and 100 ft.

30. *Temperature as a Function of Depth.* The function $T(d) = 10d + 20$ gives the temperature, in degrees Celsius, inside the earth as a function of the depth d, in kilometers. Find the temperature at 5 km, 20 km, and 1000 km.

31. *Melting Snow.* The function $W(d) = 0.112d$ approximates the amount, in centimeters, of water that results from d centimeters of snow melting. Find the amount of water that results from snow melting from depths of 16 cm, 25 cm, and 100 cm.

32. *Temperature Conversions.* The function $C(F) = \frac{5}{9}(F - 32)$ determines the Celsius temperature that corresponds to F degrees Fahrenheit. Find the Celsius temperature that corresponds to 62°F, 77°F, and 23°F.

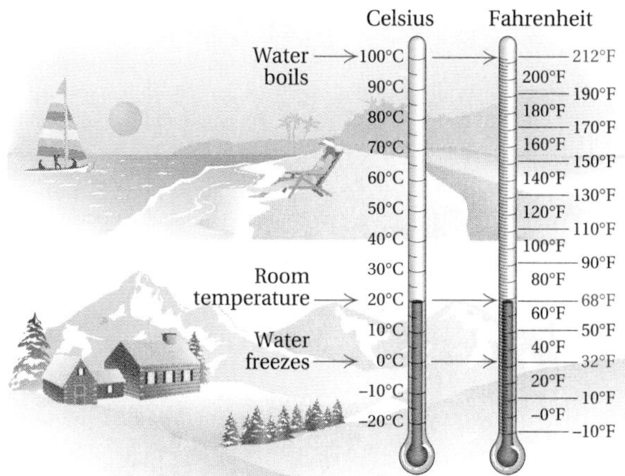

c Graph the function.

33. $f(x) = 3x - 1$

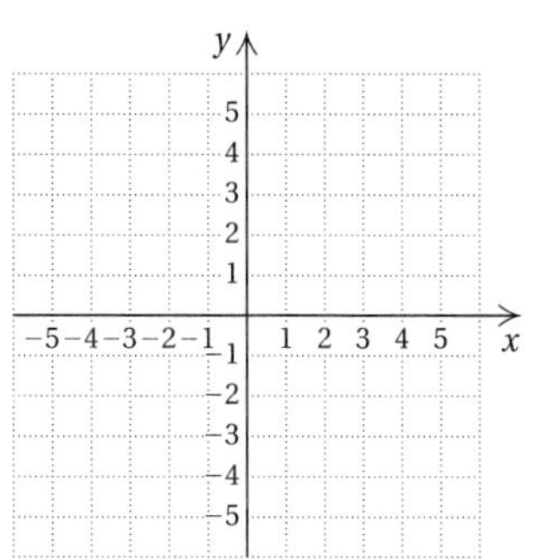

34. $g(x) = 2x + 5$

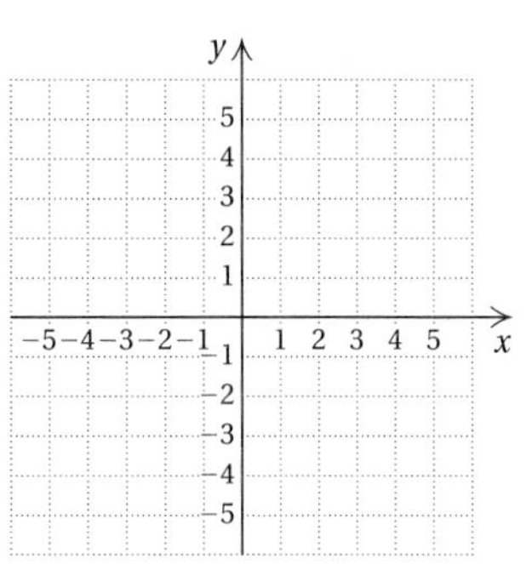

35. $g(x) = -2x + 3$

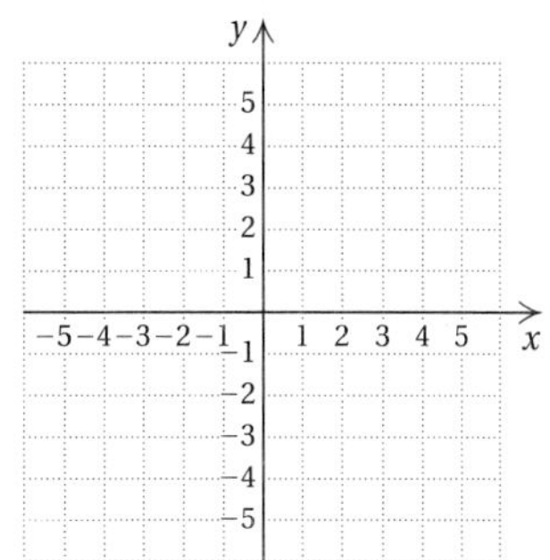

36. $f(x) = -\frac{1}{2}x + 2$

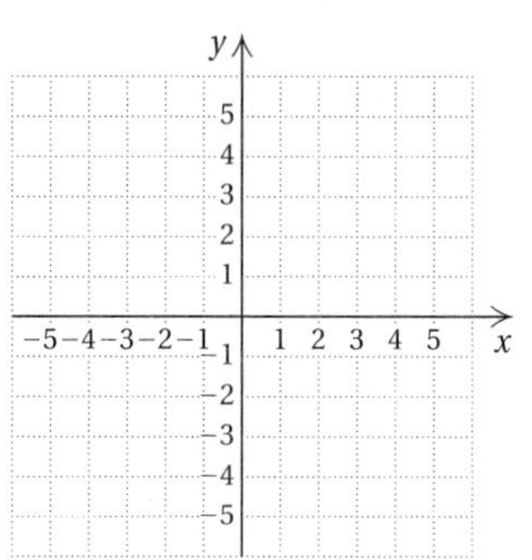

37. $f(x) = \frac{1}{2}x + 1$

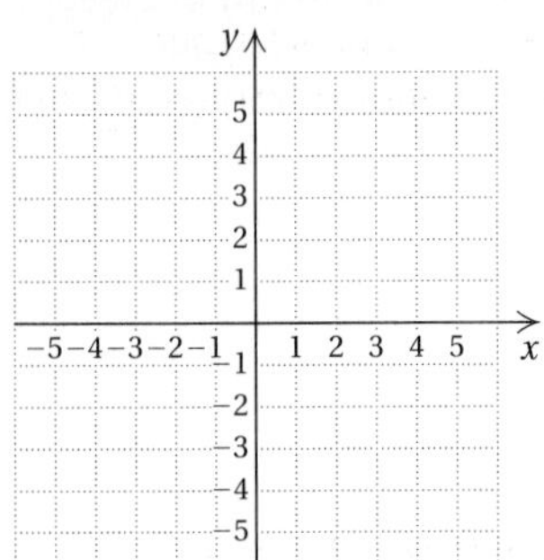

38. $f(x) = -\frac{3}{4}x - 2$

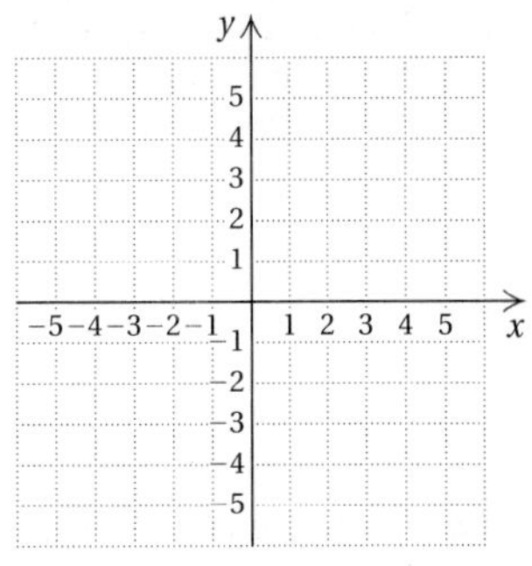

39. $f(x) = 2 - |x|$

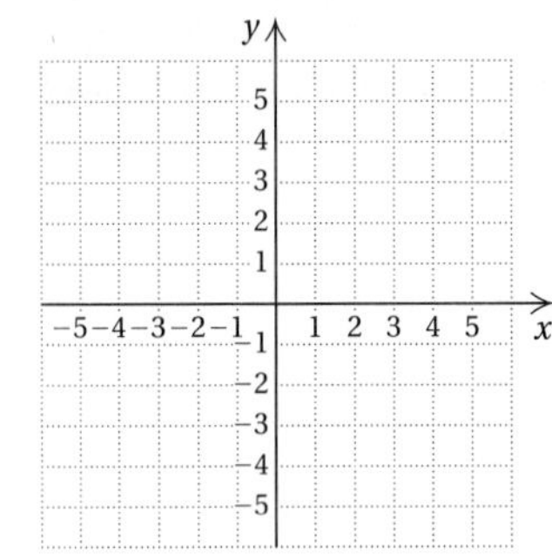

40. $f(x) = |x| - 4$

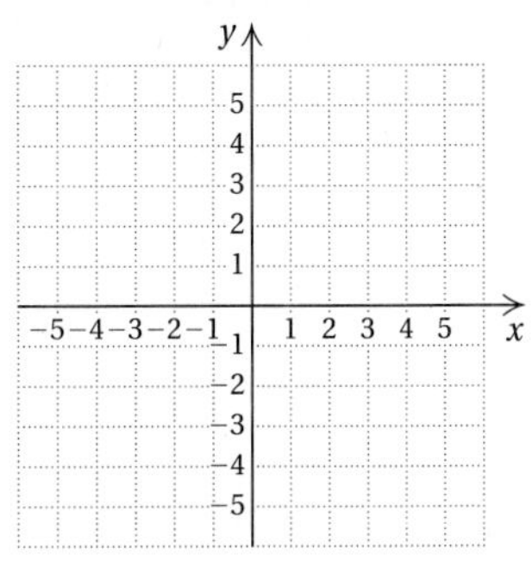

41. $f(x) = x^2$

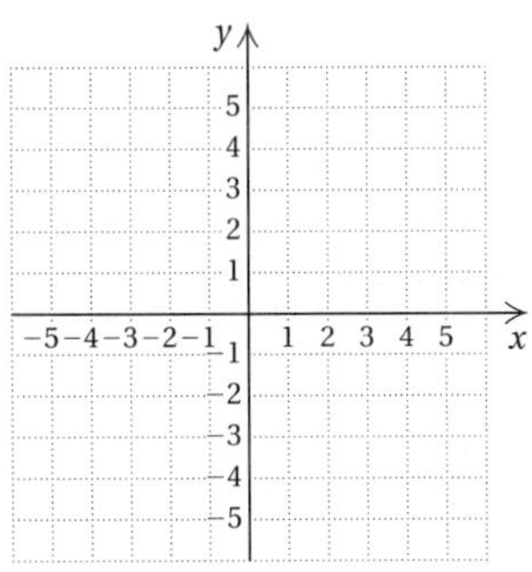

42. $f(x) = x^2 - 1$

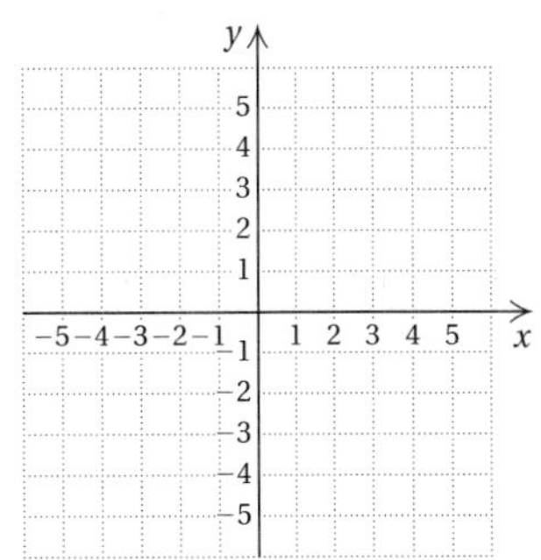

43. $f(x) = x^2 - x - 2$

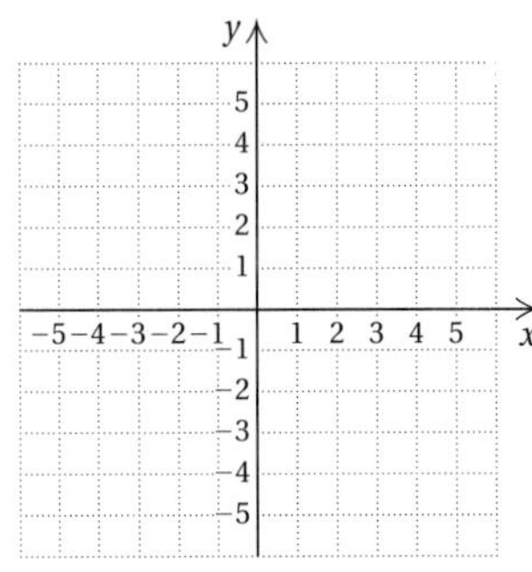

44. $f(x) = x^2 + 6x + 5$

d Determine whether each of the following is the graph of a function.

45.

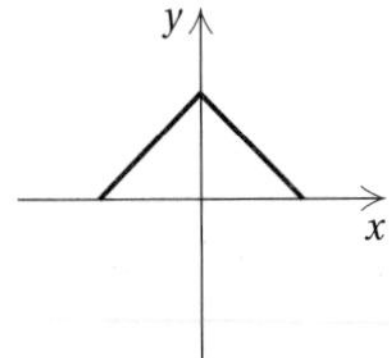

46.

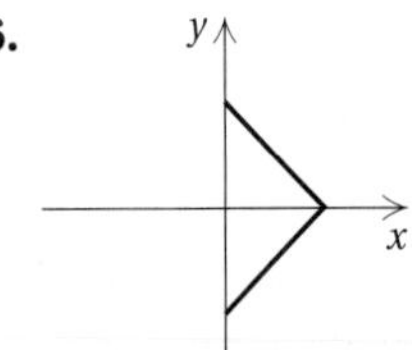

47.

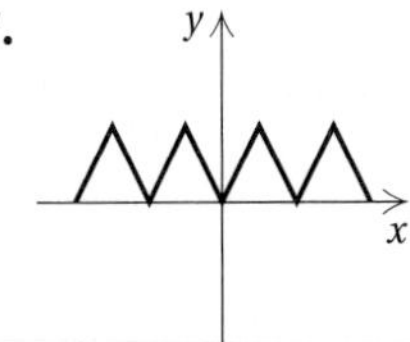

48.

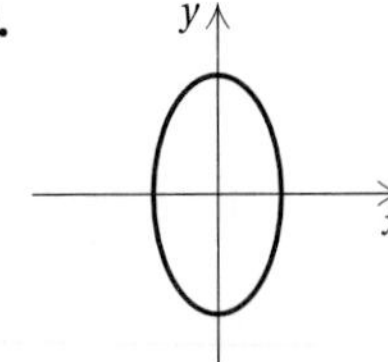

49.

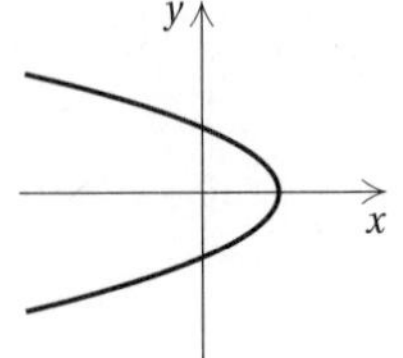

50.

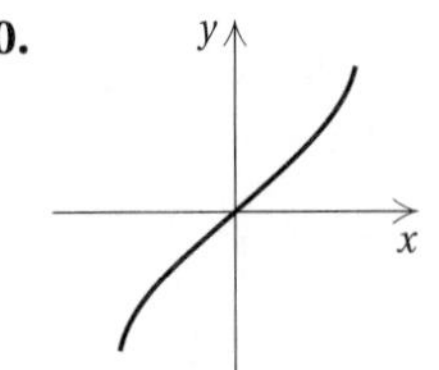

51.

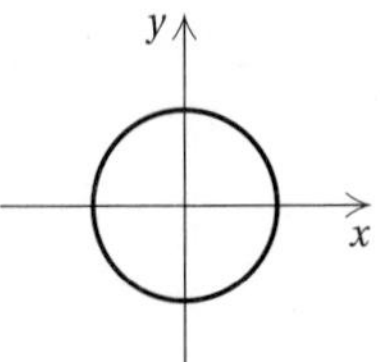

52.

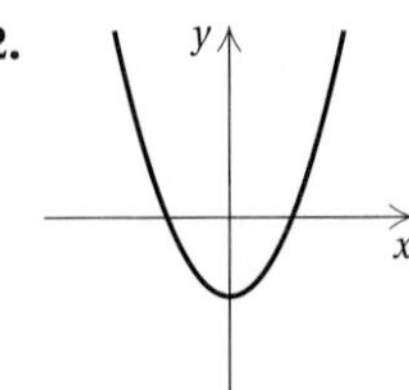

e

Cholesterol Level and Risk of a Heart Attack. The following graph shows the annual heart attack rate per 10,000 men as a function of blood cholesterol level.

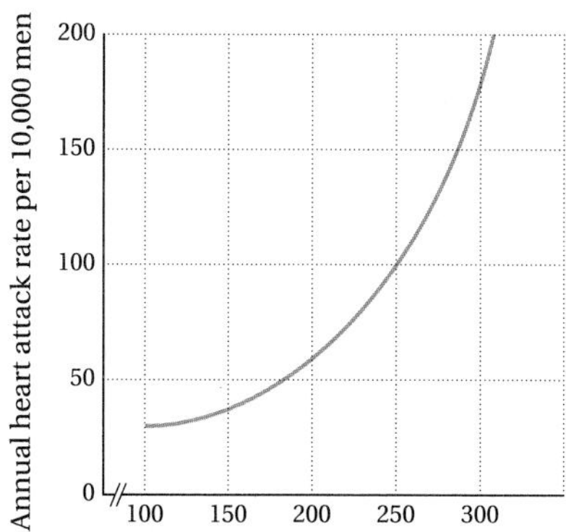

Source: Copyright 1989, CSPI. Adapted from *Nutrition Action Healthletter* (1875 Connecticut Avenue, N.W., Suite 300, Washington, DC 20009-5728)

53. Approximate the annual heart attack rate per 10,000 men for those whose blood cholesterol level is 225 mg/dl.

54. Approximate the annual heart attack rate per 10,000 men for those whose blood cholesterol level is 275 mg/dl.

Videotape Rentals. For Exercises 55–58, use the following graph, which shows the number of videotape rentals as a function of time.

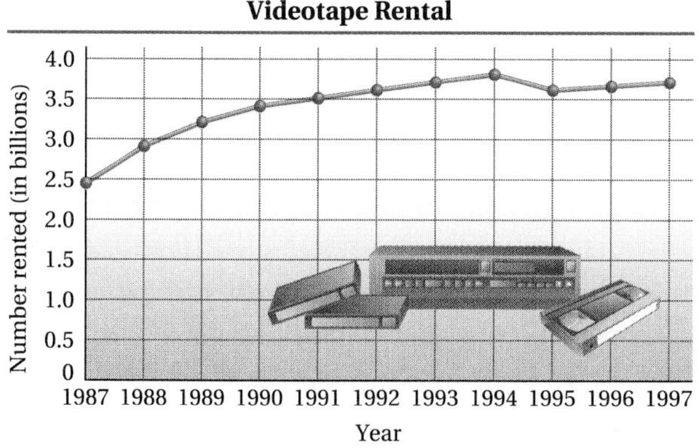

Source: Adams Media Research, Carmel Valley, California

55. Approximate the number of videotape rentals in 1997.

56. Approximate the number of videotape rentals in 1994.

57. In what year did the greatest number of rentals occur?

58. In what year did the least number of rentals occur?

Skill Maintenance

Solve.

59. Find three consecutive even integers such that the sum of the first, twice the second, and three times the third is 124. [2.4a]

60. A piece of wire 32.8 ft long is to be cut into two pieces and each of those pieces is to be bent into a square. The length of a side of one square is to be 2.2 ft longer than the length of a side of the other square. How should the wire be cut? [2.4a]

61. The surface area of a rectangular solid of length l, width w, and height h is given by

$S = 2lh + 2lw + 2wh.$

Solve for l. [2.6a]

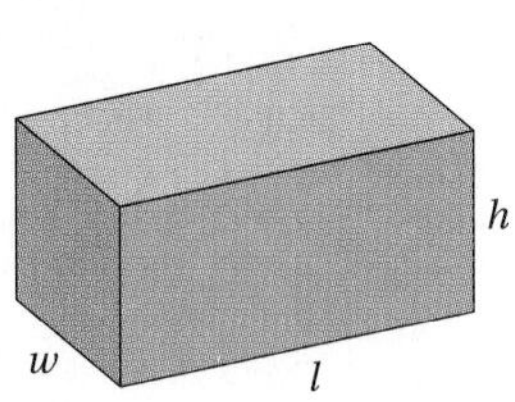

62. Solve the formula in Exercise 61 for w. [1.2a]

Convert to decimal notation. [4.2c]

63. 9.3×10^{-9}

64. 3.04×10^{6}

Convert to scientific notation. [4.2c]

65. 1,075,000,000

66. 0.00703

Factor. [5.6a]

67. $w^3 + \frac{1}{27}$

68. $a^3 - b^3$

69. $xy^3 + 64x$

70. $c^3 - 0.008$

Synthesis

Researchers at Yale University have suggested that the following graphs may represent three different aspects of love.

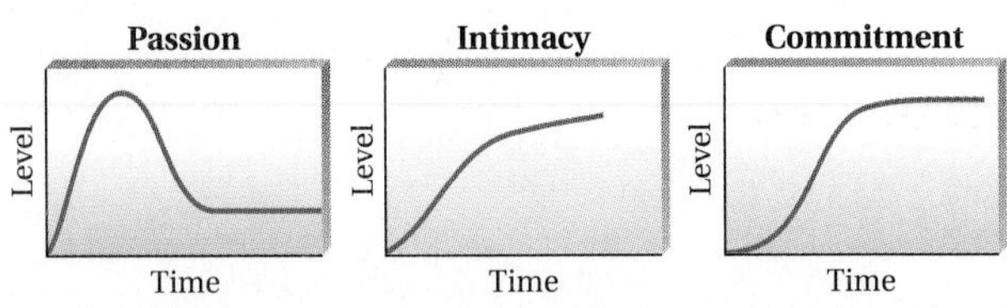

Source: "A Triangular Theory of Love," by R. J. Sternberg, 1986, *Psychological Review*, **93**(2), 119–135. Copyright 1986 by the American Psychological Association, Inc. Reprinted by permission.

71. ◆ In what unit would you measure time if the horizontal length of each graph were 10 units? Why?

72. ◆ Do you agree with the researchers that these graphs should be shaped as they are? Why or why not?

73. ◆ Is it possible for a function to have more numbers as outputs than as inputs? Why or why not?

74. ◆ Look up the word "function" in a dictionary. Explain how that definition might be related to the mathematical one given in this section.

For Exercises 75 and 76, let $f(x) = 3x^2 - 1$ and $g(x) = 2x + 5$.

75. Find $f(g(-4))$ and $g(f(-4))$.

76. Find $f(g(-1))$ and $g(f(-1))$.

77. Suppose that a function g is such that $g(-1) = -7$ and $g(3) = 8$. Find a formula for g if $g(x)$ is of the form $g(x) = mx + b$, where m and b are constants.

7.2 Finding Domain and Range

a Finding Domain and Range

The solutions of an equation in two variables consist of a set of ordered pairs. An arbitrary set of ordered pairs is called a **relation**. When a set of ordered pairs is such that no two different pairs share a common first coordinate, we have a function. The **domain** is the set of all first coordinates and the **range** is the set of all second coordinates.

Example 1 Find the domain and the range of the function f whose graph is shown below.

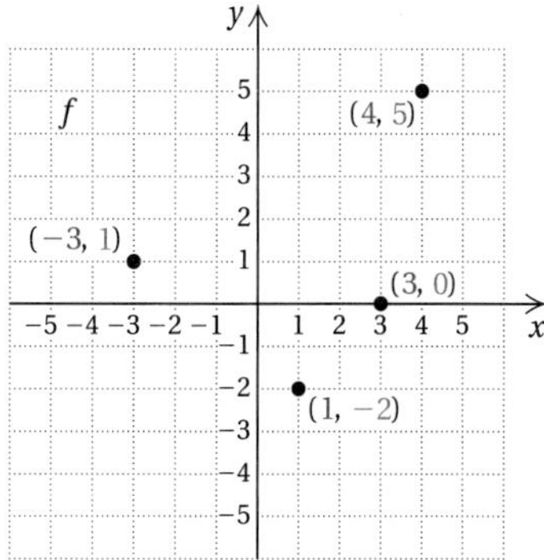

This is a rather simple function. Its graph contains just four ordered pairs and it can be written as

$$\{(-3, 1), (1, -2), (3, 0), (4, 5)\}.$$

We can determine the domain and the range by reading the x- and y-values directly from the graph.

The domain is the set of all first coordinates, $\{-3, 1, 3, 4\}$. The range is the set of all second coordinates, $\{1, -2, 0, 5\}$.

Do Exercise 1.

Example 2 For the function f whose graph is shown below, determine each of the following.

a) The number in the range that is paired with 1 (from the domain). That is, find $f(1)$.

b) The domain of f

c) The numbers in the domain that are paired with 1 (from the range). That is, find all x such that $f(x) = 1$.

d) The range of f

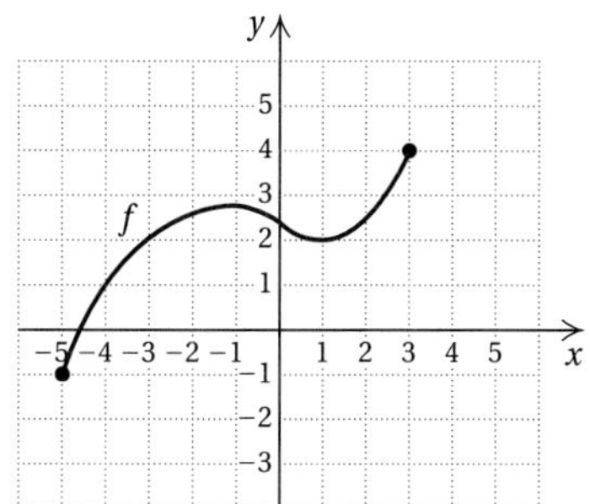

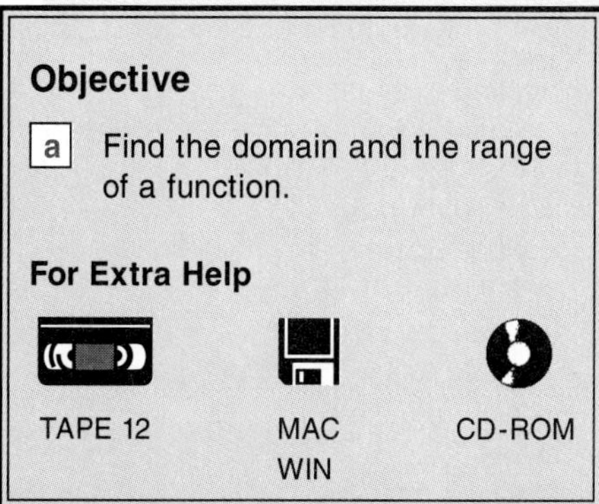

1. Find the domain and the range of the function f whose graph is shown below.

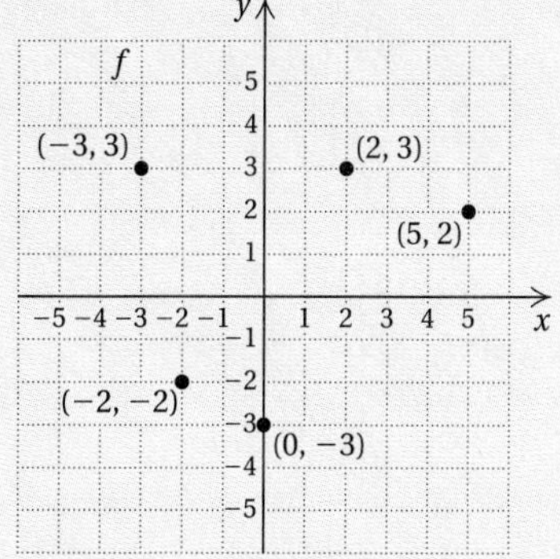

Answer on page A-27

2. For the function f whose graph is shown below, determine each of the following.

a) The number in the range that is paired with the input 1. That is, find $f(1)$.
b) The domain of f
c) The number in the domain that is paired with 4
d) The range of f

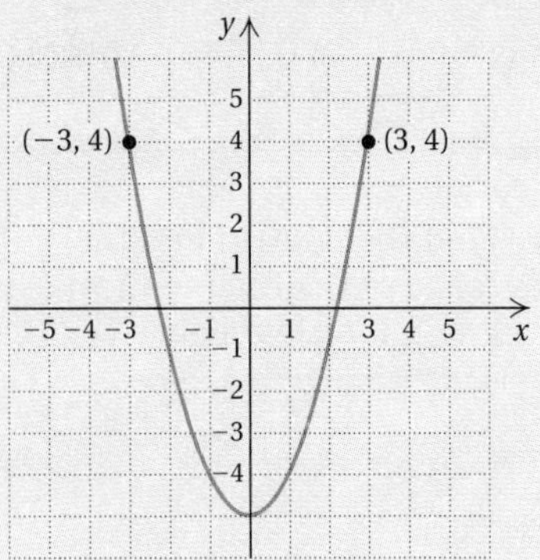

Answers on page A-27

a) To determine which number in the range is paired with 1 in the domain, we locate 1 on the horizontal axis. Next, we find the point on the graph of f for which 1 is the first coordinate. From that point, we can look to the vertical axis to find the corresponding y-coordinate, 2. The input 1 has the output 2—that is, $f(1) = 2$.

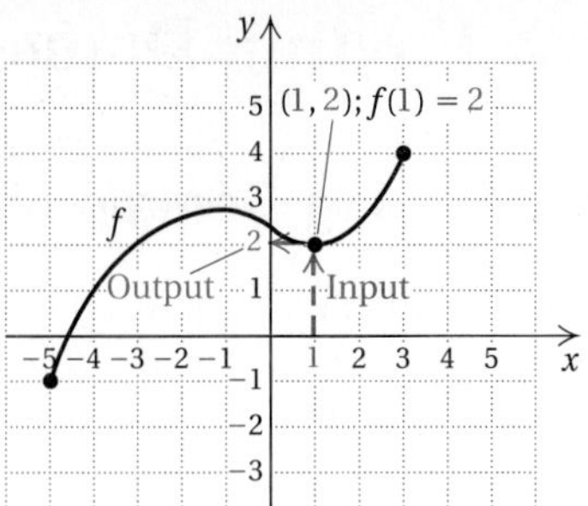

b) The domain of the function is the set of all x-values, or inputs, of the points on the graph. These extend from -5 to 3 and can be viewed as the curve's shadow, or projection, onto the x-axis. Thus the domain is the set $\{x \mid -5 \leq x \leq 3\}$.

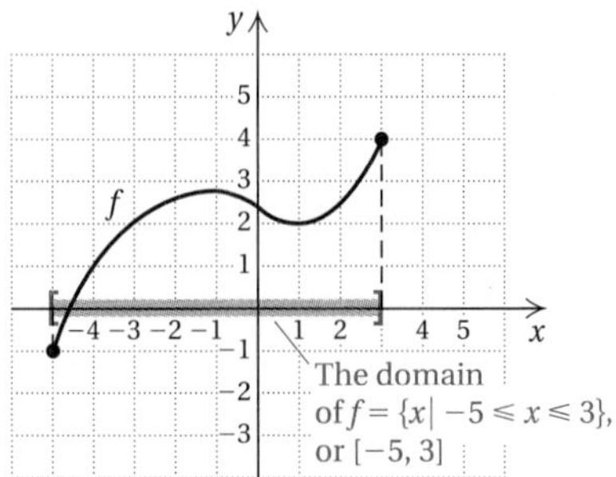

c) To determine which numbers in the domain are paired with 1 in the range, we locate 1 on the vertical axis. From there, we look left and right to the graph of f to find any points (inputs) for which 1 is the second coordinate (output). One such point exists, $(-4, 1)$. For this function, we note that $x = -4$ is the only member of the domain paired with 1. For other functions, there might be more than one member of the domain paired with a member of the range.

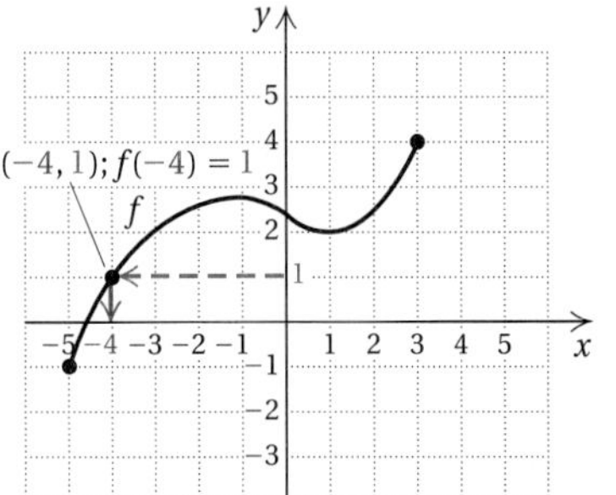

d) The range of the function is the set of all y-values, or outputs, of the points on the graph. These extend from -1 to 4 and can be viewed as the curve's shadow, or projection, onto the y-axis. Thus the range is the set $\{y \mid -1 \leq y \leq 4\}$.

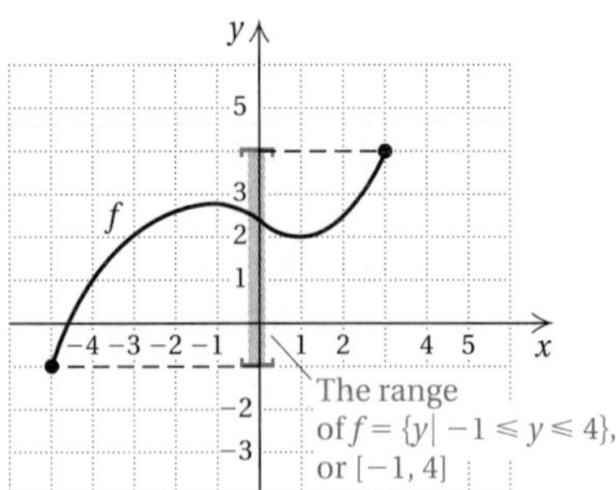

Do Exercise 2.

When a function is given by an equation or formula, the domain is understood to be the largest set of real numbers (inputs) for which function values (outputs) can be calculated. That is, the domain is the set of all possible allowable inputs into the formula. To find the domain, think, "What can we substitute?"

Example 3 Find the domain: $f(x) = |x|$.

We ask, "What can we substitute?" Is there any number x for which we cannot calculate $|x|$? The answer is no. Thus the domain of f is the set of all real numbers.

Example 4 Find the domain: $f(x) = \dfrac{3}{2x - 5}$.

We ask, "What can we substitute?" Is there any number x for which we cannot calculate $3/(2x - 5)$? Since $3/(2x - 5)$ cannot be calculated when the denominator $2x - 5$ is 0, we solve the following equation to find those real numbers that must be excluded from the domain of f:

$2x - 5 = 0$ **Setting the denominator equal to 0**

$2x = 5$ **Adding 5**

$x = \frac{5}{2}$. **Dividing by 2**

Thus $\frac{5}{2}$ is not in the domain, whereas all other real numbers are.
The domain of f is $\{x \mid x \text{ is a real number and } x \neq \frac{5}{2}\}$.

Do Exercises 3 and 4.

The task of determining the domain and the range of a function is one that we will return to several times as we consider other types of functions in this book.

The following is a review of the function concepts considered in Sections 7.1 and 7.2.

Function Concepts

- Formula for f: $f(x) = x^2 - 7$
- For every input of f, there is exactly one output.
- 1 is an input; -6 is an output.
- $f(1) = -6$
- $(1, -6)$ is on the graph.
- Domain = The set of all inputs = The set of all real numbers
- Range = The set of all outputs = $\{y \mid y \geq -7\}$

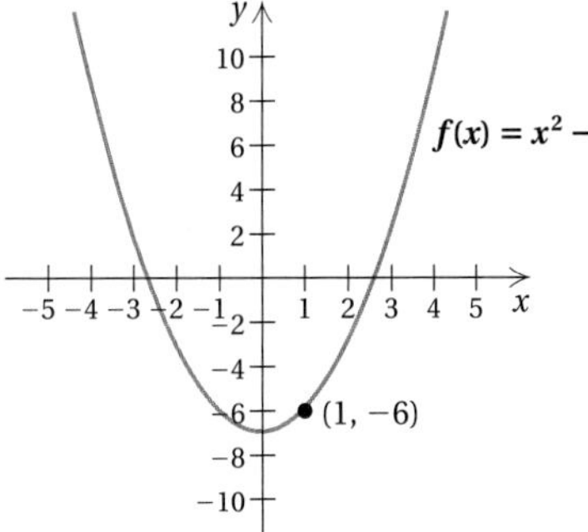

3. Find the domain:
$f(x) = x^3 - |x|$.

4. Find the domain:
$f(x) = \dfrac{4}{3x + 2}$.

Answers on page A-27

Calculator Spotlight

Determining Domain and Range. Use a grapher to graph the function in the given viewing window. Then determine the domain and the range.

a) $f(x) = 3 - |x|$, $[-10, 10, -10, 10]$
b) $f(x) = x^3 - x$, $[-3, 3, -4, 4]$
c) $f(x) = \dfrac{12}{x}$, or $12x^{-1}$, $[-14, 14, -14, 14]$
d) $f(x) = x^4 - 2x^2 - 3$, $[-4, 4, -6, 6]$

We have the following.

a)

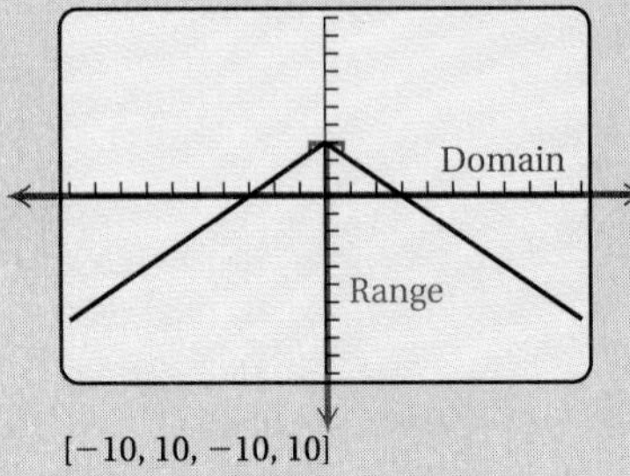

Domain = all real numbers;
range = $\{y \mid y \leq 3\}$

b)

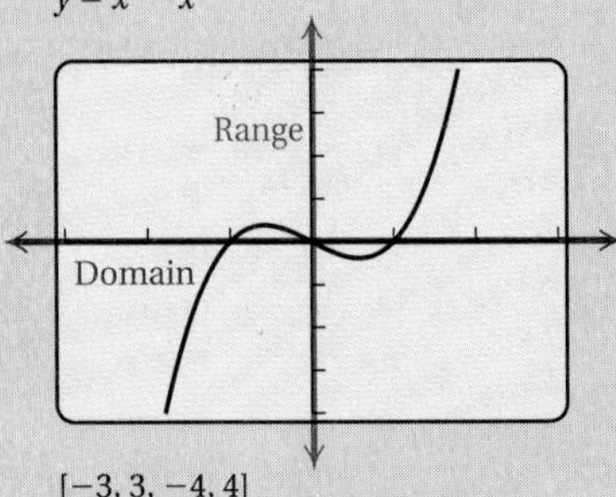

Domain = all real numbers;
range = all real numbers

c)

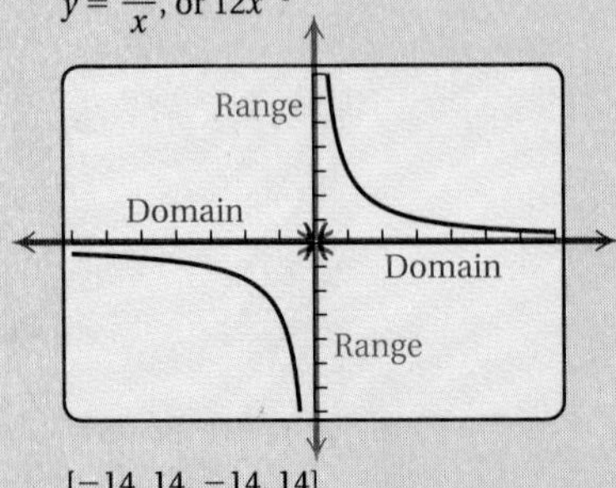

The number 0 is excluded as an input.
Domain = $\{x \mid x \text{ is a real number } and\ x \neq 0\}$
range = $\{y \mid y \text{ is a real number } and\ y \neq 0\}$

d)

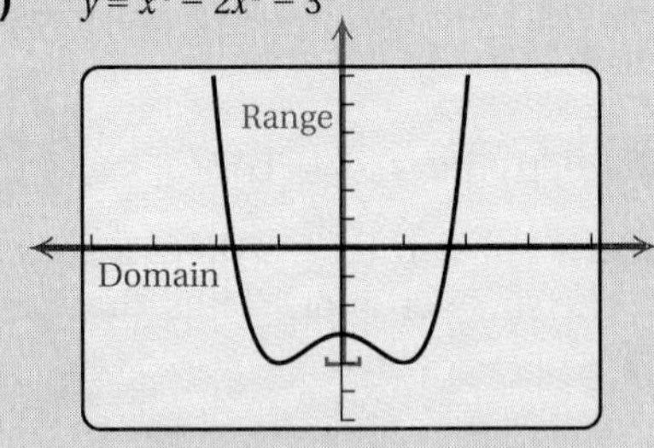

Domain = all real numbers;
range = $\{y \mid y \geq -4\}$

We can confirm our results using the TRACE feature, moving the cursor from left to right along the curve. We can also use the TABLE feature. In Example (d), it might not appear as though the domain is all real numbers because the graph seems "thin," but reexamining the formula shows that we can indeed substitute any real number.

Exercises

Use a grapher to graph the function in the given viewing window. Then determine the domain and the range.

1. $f(x) = |x| - 7$, $[-10, 10, -10, 10]$
2. $f(x) = 2 + 3x - x^3$, $[-5, 5, -5, 5]$
3. $f(x) = \dfrac{-16}{x}$, or $-16x^{-1}$, $[-20, 20, -20, 20]$
4. $f(x) = x^4 - 2x^2 - 7$, $[-4, 4, -9, 9]$

Exercise Set 7.2

a In Exercises 1–12, the graph is that of a function. Determine for each one **(a)** $f(1)$; **(b)** the domain; **(c)** all x-values such that $f(x) = 2$; and **(d)** the range. An open dot indicates that the point does not belong to the graph.

1.

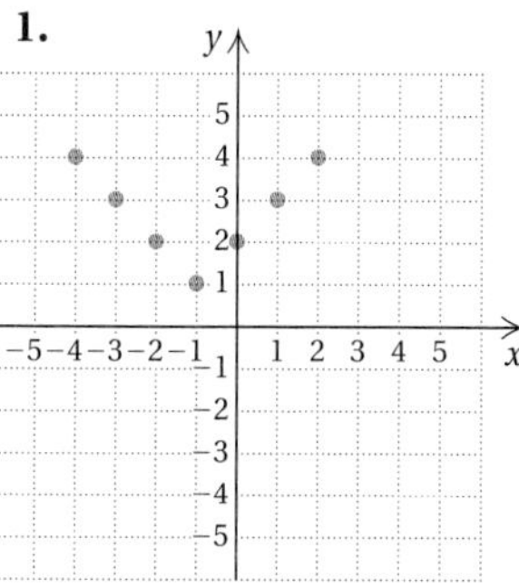

2.

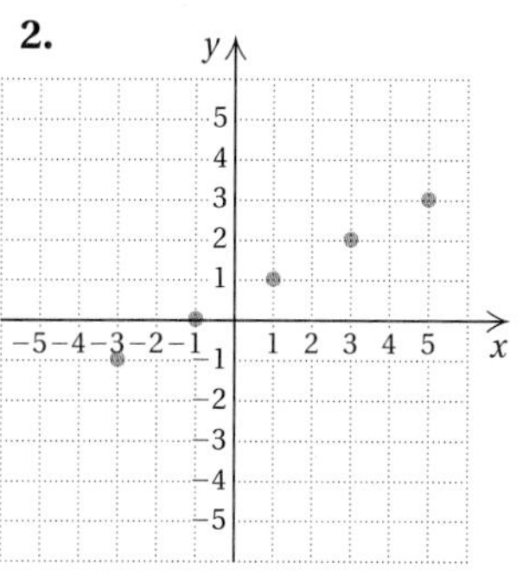

3.

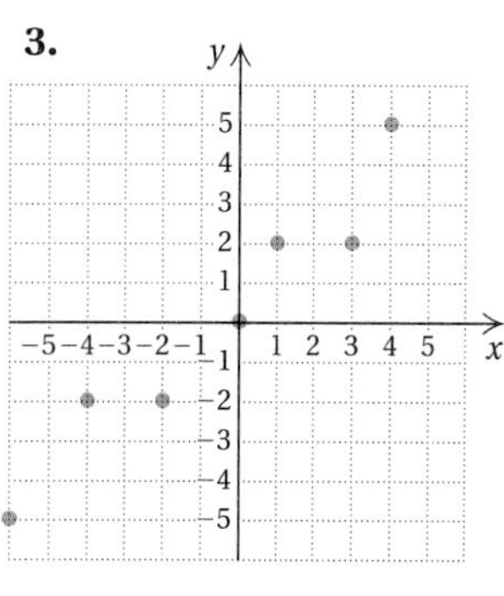

4.

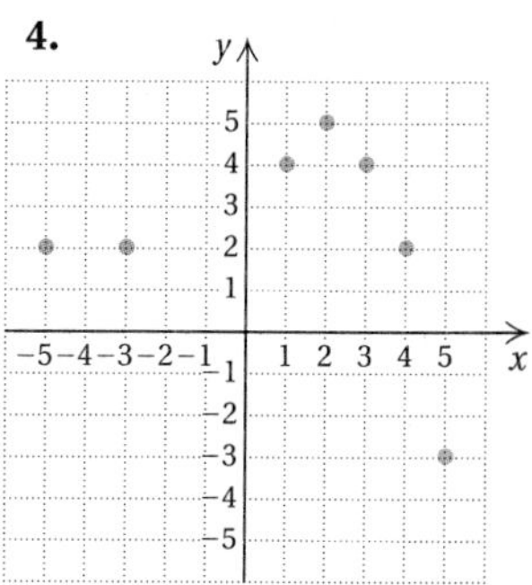

5.

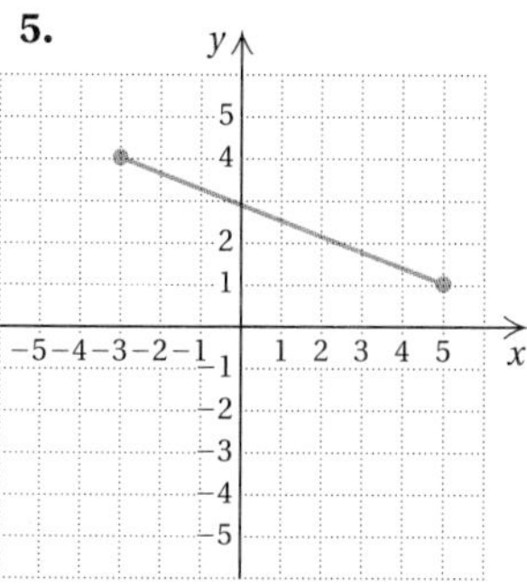

6.

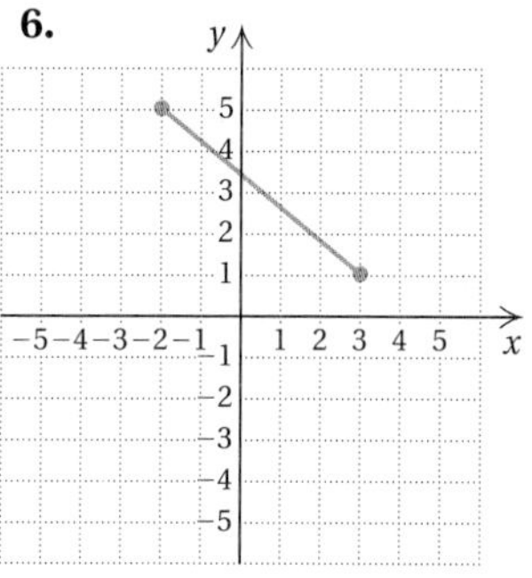

7.

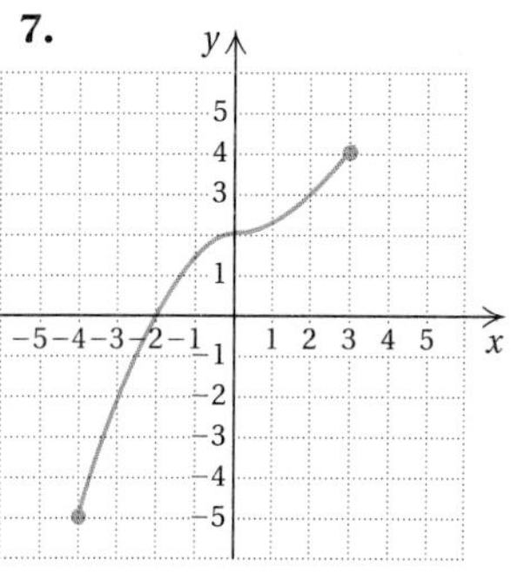

8.

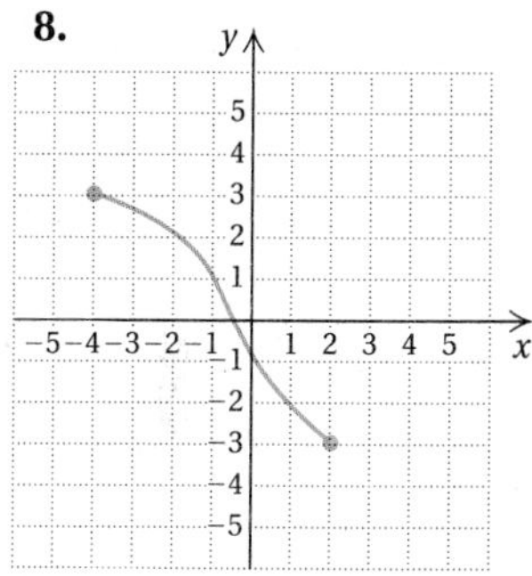

9.

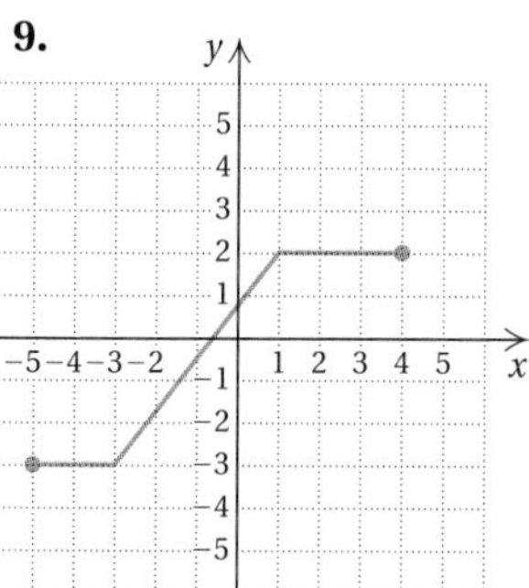

10.

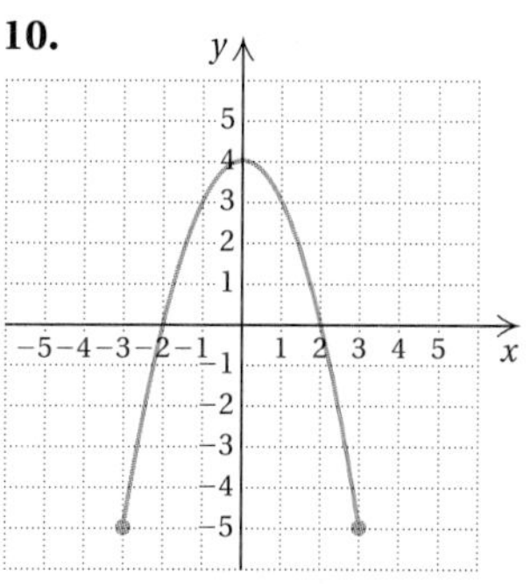

11.

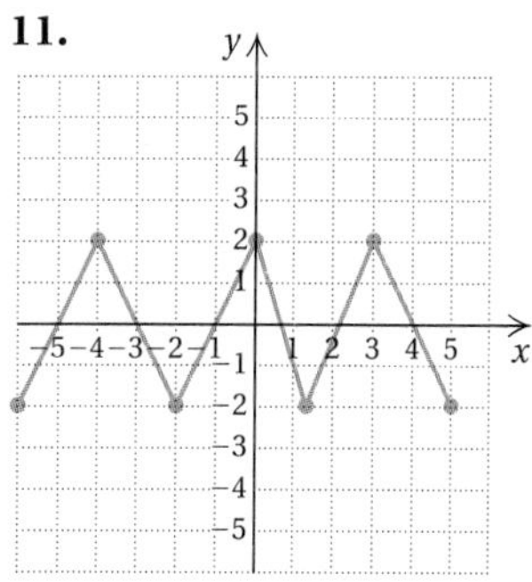

12.

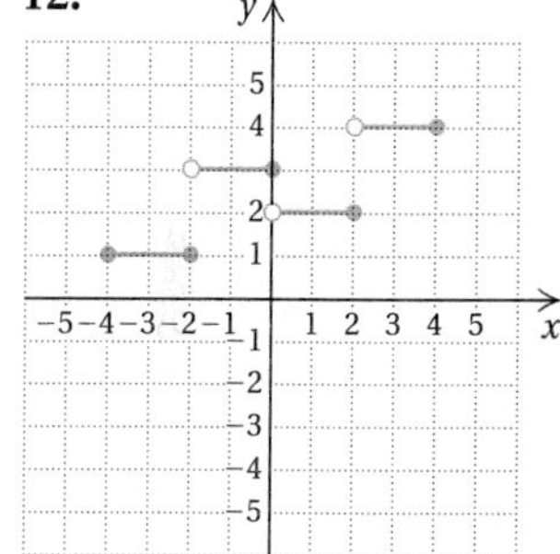

Find the domain.

13. $f(x) = \dfrac{2}{x + 3}$

14. $f(x) = \dfrac{7}{5 - x}$

15. $f(x) = 2x + 1$

16. $f(x) = 4 - 5x$

17. $f(x) = x^2 + 3$

18. $f(x) = x^2 - 2x + 3$

19. $f(x) = \dfrac{8}{5x - 14}$

20. $f(x) = \dfrac{x - 2}{3x + 4}$

21. $f(x) = |x| - 4$

22. $f(x) = |x - 4|$

23. $f(x) = \dfrac{4}{|2x - 3|}$

24. $f(x) = \dfrac{x^2 - 3x}{|4x - 7|}$

25. $g(x) = \dfrac{1}{x - 1}$

26. $g(x) = \dfrac{-11}{4 + x}$

27. $g(x) = x^2 - 2x + 1$

28. $g(x) = 8 - x^2$

29. $g(x) = x^3 - 1$

30. $g(x) = 4x^3 + 5x^2 - 2x$

31. $g(x) = \dfrac{7}{20 - 8x}$

32. $g(x) = \dfrac{2x - 3}{6x - 12}$

33. $g(x) = |x + 7|$

34. $g(x) = |x| + 1$

35. $g(x) = \dfrac{-2}{|4x + 5|}$

36. $g(x) = \dfrac{x^2 + 2x}{|10x - 20|}$

37. For the function f whose graph is shown below, find $f(-1)$, $f(0)$, and $f(1)$.

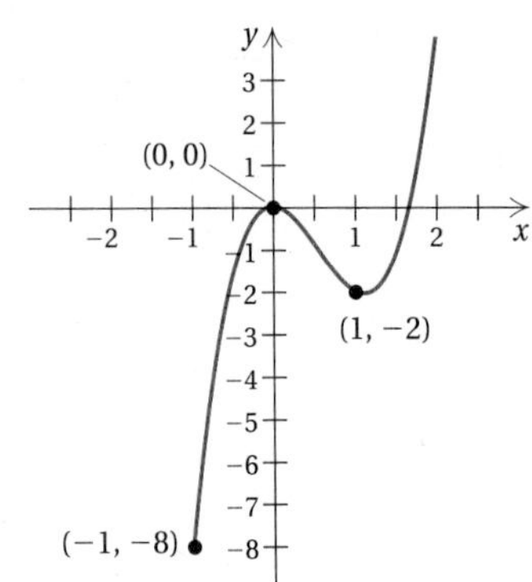

38. For the function g whose graph is shown below, find all the x-values for which $g(x) = 1$.

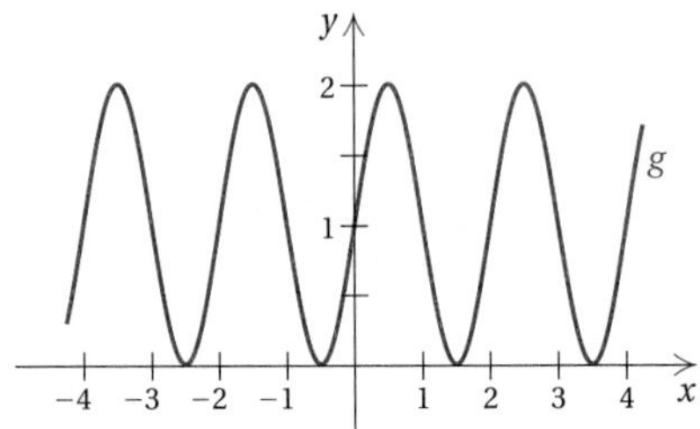

Skill Maintenance

Simplify. [6.1c]

39. $\dfrac{a^2 - 1}{a + 1}$

40. $\dfrac{10y^2 + 10y - 20}{35y^2 + 210y - 245}$

41. $\dfrac{5x - 15}{x^2 - x - 6}$

Divide. [4.8b]

42. $(t^2 + 3t - 28) \div (t + 7)$

43. $(w^2 + 4w + 5) \div (w + 3)$

44. $(x^6 + x^5 - 3x^4 + x + 5) \div (x^2 - 1)$

Multiply. [4.6d]

45. $(7x - 3)(2x + 9)$

46. $(a - 1)(a + 1)$

47. $(9y + 10)^2$

48. $\left(2w - \frac{1}{2}\right)\left(4w + \frac{1}{2}\right)$

Synthesis

49. ◈ Explain the difference between the domain and the range of a function.

50. ◈ For a given function f, it is known that $f(2) = -3$. Give as many interpretations of this fact as you can.

7.3 Linear Functions: Graphs and Slope

We now turn our attention to functions whose graphs are straight lines. Such functions are called **linear** and can be written in the form $f(x) = mx + b$.

> **LINEAR FUNCTION**
>
> A **linear function** f is any function that can be described by $f(x) = mx + b$.

In this section, we consider the effects of the constants m and b on the graphs of linear functions.

a The Constant b: The y-Intercept

Let's first explore the effect of the constant b.

Example 1 Graph $y = 2x$ and $y = 2x + 3$ using the same set of axes. Compare the graphs.

We first make a table of solutions of both equations.

	y	y
x	$y = 2x$	$y = 2x + 3$
0	0	3
1	2	5
−1	−2	1
2	4	7
−2	−4	−1

Next, we plot these points. Drawing a red line for $y = 2x$ and a blue line for $y = 2x + 3$, we note that the graph of $y = 2x + 3$ is simply the graph of $y = 2x$ shifted, or *translated,* up 3 units. The lines are parallel.

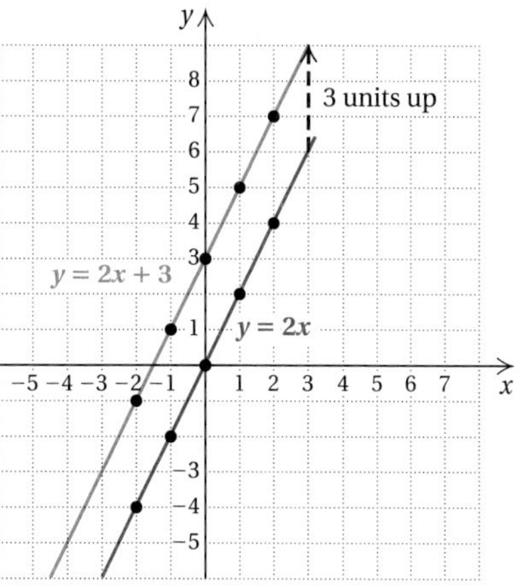

Do Exercises 1 and 2.

Example 2 Graph $f(x) = \frac{1}{3}x$ and $g(x) = \frac{1}{3}x - 2$ using the same set of axes. Compare the graphs.

We first make a table of solutions of both equations. By choosing multiples of 3, we can avoid fractions.

	$f(x)$	$g(x)$
x	$f(x) = \frac{1}{3}x$	$g(x) = \frac{1}{3}x - 2$
0	0	−2
3	1	−1
−3	−1	−3
6	2	0

Objectives

a Find the y-intercept of a line from the equation $y = mx + b$ or $f(x) = mx + b$.

b Given two points on a line, find the slope; given a linear equation, derive the equivalent slope–intercept equation and determine the slope and the y-intercept.

c Solve applied problems involving slope.

For Extra Help

 TAPE 12

 MAC WIN

 CD-ROM

1. Graph $y = 3x$ and $y = 3x - 6$ using the same set of axes. Compare the graphs.

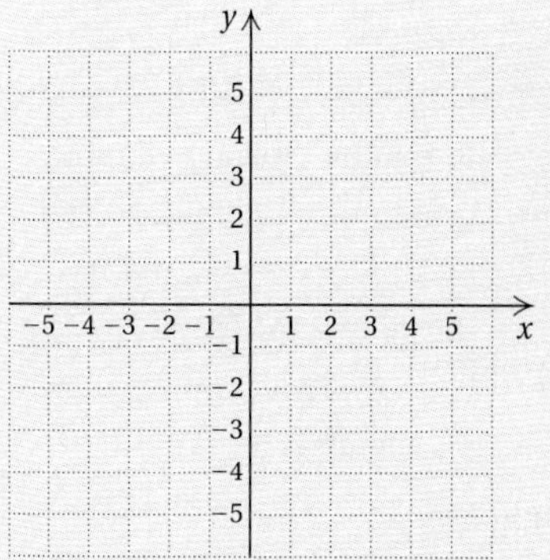

2. Graph $y = -2x$ and $y = -2x + 3$ using the same set of axes. Compare the graphs.

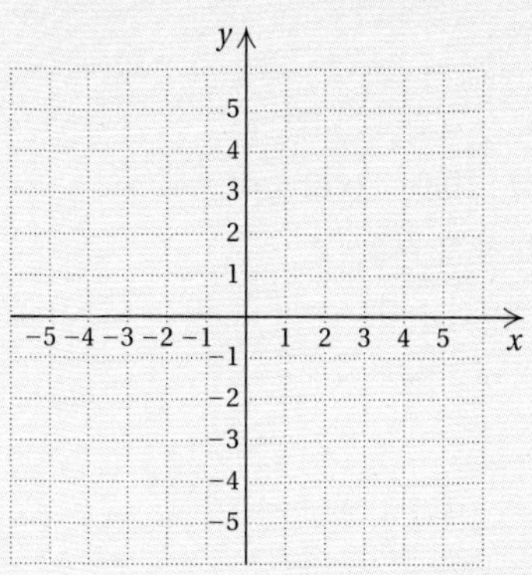

Answers on page A-28

3. Graph $f(x) = \frac{1}{3}x$ and $g(x) = \frac{1}{3}x + 2$ using the same set of axes. Compare the graphs.

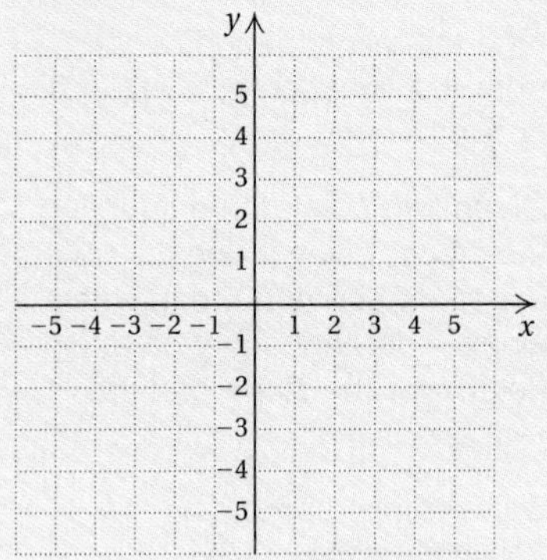

We then plot these points. Drawing a line for $f(x) = \frac{1}{3}x$ and a line for $g(x) = \frac{1}{3}x - 2$, we see that the graph of $g(x) = \frac{1}{3}x - 2$ is simply the graph of $f(x) = \frac{1}{3}x$ shifted, or translated, down 2 units.

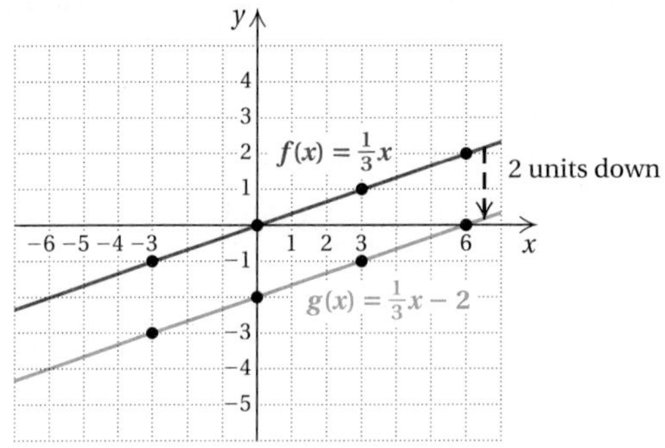

Note that in Example 1, the graph of $y = 2x + 3$ passed through the point (0, 3) and in Example 2, the graph of $g(x) = \frac{1}{3}x - 2$ passed through the point (0, −2). In general, the graph of $y = mx + b$ is a line parallel to $y = mx$, passing through the point $(0, b)$. The point $(0, b)$ is called the **y-intercept** because it is the point at which the graph crosses the y-axis. Often it is convenient to refer to the number b as the y-intercept. The constant b has the effect of moving the graph of $y = mx$ up or down $|b|$ units to obtain the graph of $y = mx + b$.

Do Exercise 3.

Calculator Spotlight

Exploring b. We can use a grapher to explore the effect of b when we are graphing $f(x) = mx + b$.

Exercises

1. Graph $y_1 = x$. Then, using the same viewing window, graph $y_2 = x + 3$ followed by $y_3 = x - 4$. Compare the graphs of y_2 and y_3 with y_1. Then without drawing the graphs, compare the graph of $y = x - 5$ with the graph of y_1.
2. Use the TABLE feature to compare the values of y_1, y_2, and y_3.

The **y-intercept** of the graph of $f(x) = mx + b$ is the point $(0, b)$ or, simply, b.

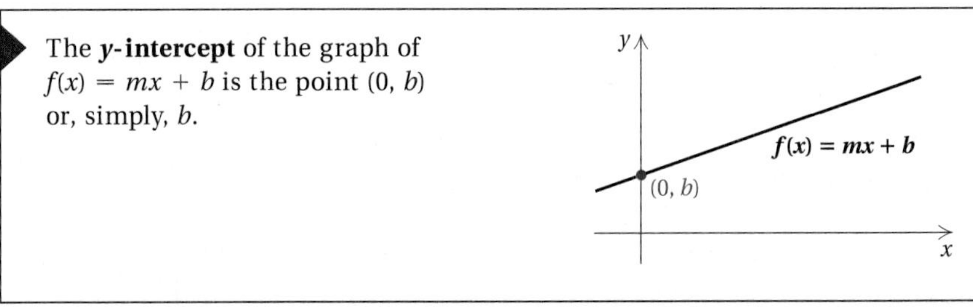

Example 3 Find the y-intercept: $y = -5x + 4$.

$y = -5x + 4$ (0, 4), or simply 4, is the y-intercept.

Example 4 Find the y-intercept: $f(x) = 6.3x - 7.8$.

$f(x) = 6.3x - 7.8$ (0, −7.8), or simply −7.8, is the y-intercept.

Do Exercises 4 and 5.

Find the y-intercept.

4. $y = 7x + 8$

5. $f(x) = -6x - \frac{2}{3}$

Answers on page A-28

b The Constant m: Slope

Look again at the graphs in Examples 1 and 2. Note that the slant of each red line seems to match the slant of each blue line. This leads us to believe that the number m in the equation $y = mx + b$ is related to the slant of the line. The following definition enables us to visualize this slant and attach a number, a geometric ratio, or *slope*, to the line.

> The **slope** of a line containing points (x_1, y_1) and (x_2, y_2) is given by
>
> 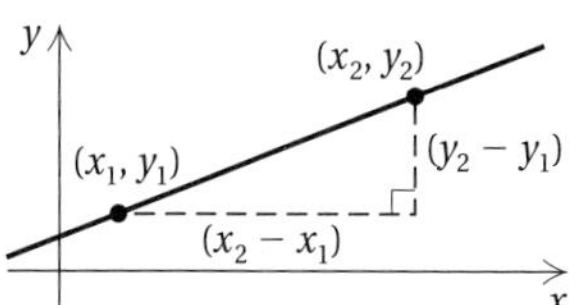
>
>
> $$m = \frac{\text{rise}}{\text{run}} = \frac{\text{change in } y}{\text{change in } x} = \frac{y_2 - y_1}{x_2 - x_1} = \frac{y_1 - y_2}{x_1 - x_2}.$$

Consider a line with two points marked P_1 and P_2. As we move from P_1 to P_2, the y-coordinate changes from 1 to 3 and the x-coordinate changes from 2 to 7. The change in y is $3 - 1$, or 2. The change in x is $7 - 2$, or 5.

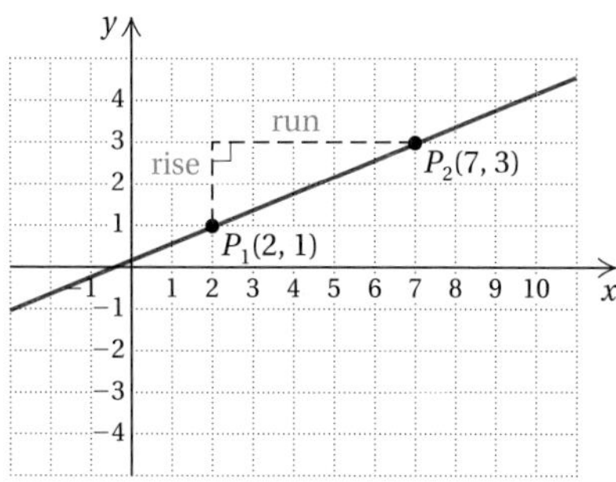

We call the change in y the **rise** and the change in x the **run**. The ratio rise/run is the same for any two points on a line. We call this ratio the **slope**. Slope describes the slant of a line. The slope of the line in the graph above is given by

$$\frac{\text{rise}}{\text{run}}, \quad \text{or} \quad \frac{\text{change in } y}{\text{change in } x}, \quad \text{or} \quad \frac{2}{5}.$$

Whenever x increases by 5 units, y increases by 2 units. Equivalently, whenever x increases by 1 unit, y increases by $\frac{2}{5}$ unit.

Example 5 Graph the line containing the points $(-4, 3)$ and $(2, -5)$ and find the slope.

The graph is shown at right. Going from $(-4, 3)$ to $(2, -5)$, we see that the change in y, or the rise, is $-5 - 3$, or -8. The change in x, or the run, is $2 - (-4)$, or 6.

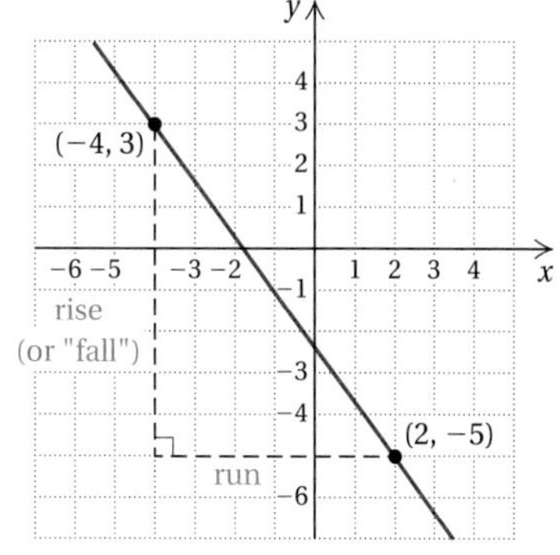

$$\text{Slope} = \frac{\text{rise}}{\text{run}} = \frac{\text{change in } y}{\text{change in } x} = \frac{-5 - 3}{2 - (-4)} = \frac{-8}{6} = -\frac{8}{6}, \text{ or } -\frac{4}{3}$$

The formula

$$m = \frac{y_2 - y_1}{x_2 - x_1} = \frac{y_1 - y_2}{x_1 - x_2}$$

tells us that we can subtract in two ways. We must remember, however, to subtract the x-coordinates in the same order that we subtract the y-coordinates.

Calculator Spotlight

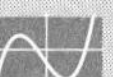
Visualizing Slope

Exercises

Use the window settings $[-6, 6, -4, 4]$, Xscl = 1, Yscl = 1 for Exercises 1–4.

1. Graph
   $y = x + 1$, $y = 2x + 1$,
   $y = 3x + 1$, $y = 10x + 1$.
   What do you think the graph of $y = 247x + 1$ will look like?
2. Graph
   $y = x$, $y = \frac{7}{8}x$,
   $y = 0.47x$, $y = \frac{2}{31}x$.
   What do you think the graph of $y = 0.000018x$ will look like?
3. Graph
   $y = -x$, $y = -2x$,
   $y = -5x$, $y = -10x$.
   What do you think the graph of $y = -247x$ will look like?
4. Graph
   $y = -x - 1$,
   $y = -\frac{3}{4}x - 1$,
   $y = -0.38x - 1$,
   $y = -\frac{5}{32}x - 1$.
   What do you think the graph of $y = -0.000043x - 1$ will look like?

Graph the line through the given points and find its slope.

6. $(-1, -1)$ and $(2, -4)$

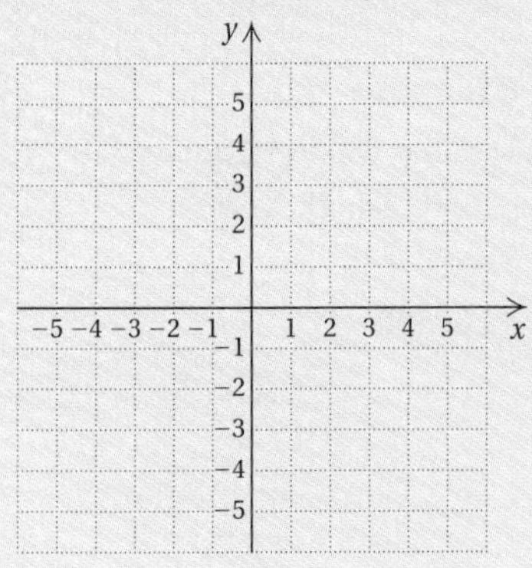

7. $(0, 2)$ and $(3, 1)$

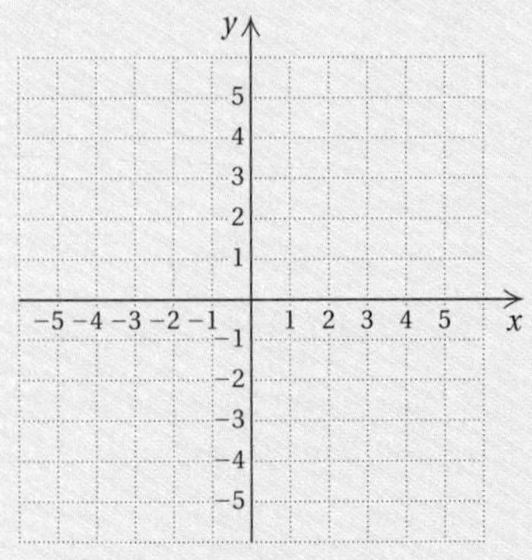

8. Find the slope of the line $f(x) = -\frac{2}{3}x + 1$. Use the points $(9, -5)$ and $(3, -1)$.

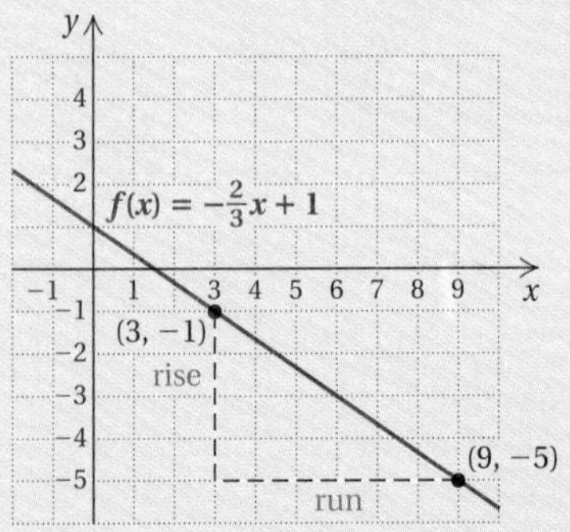

Answers on page A-28

Let's do Example 5 again:

$$\text{Slope} = \frac{\text{change in } y}{\text{change in } x} = \frac{3 - (-5)}{-4 - 2} = \frac{8}{-6} = -\frac{8}{6} = -\frac{4}{3}.$$

We see that both ways give the same slope value.

The slope of a line tells how it slants. A line with positive slope slants up from left to right. The larger the positive number, the steeper the slant. A line with negative slope slants downward from left to right. The smaller the negative number, the steeper the line.

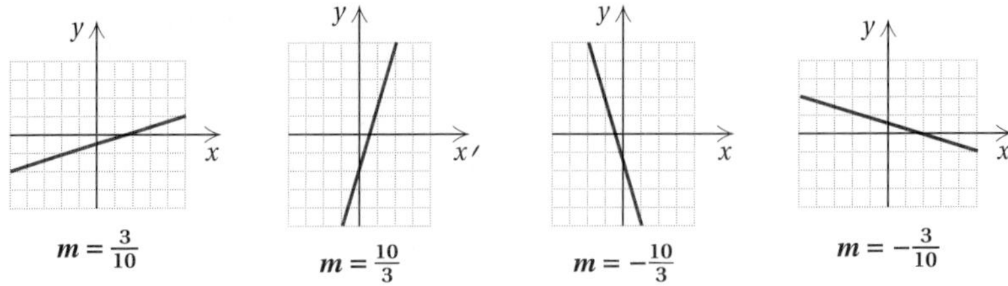

Do Exercises 6 and 7.

How can we find the slope from a given equation? Let's consider the equation $y = 2x + 3$, which is in the form $y = mx + b$. We can find two points by choosing convenient values for x, say 0 and 1, and substituting to find the corresponding y-values. We find two points on the line to be $(0, 3)$ and $(1, 5)$. The slope of the line is found as follows, using the definition of slope:

$$m = \frac{\text{change in } y}{\text{change in } x} = \frac{5 - 3}{1 - 0} = \frac{2}{1} = 2.$$

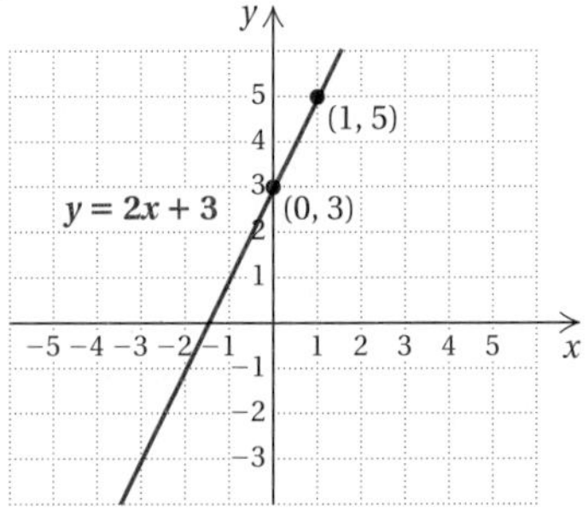

The slope is 2. Note that this is also the coefficient of the x-term in the equation $y = 2x + 3$.

Do Exercise 8.

We see that the slope of the line $y = mx + b$ is indeed the constant m, the coefficient of x.

> The **slope** of the line $y = mx + b$ is m.

From a linear equation in the form $y = mx + b$, we can read directly the slope and the y-intercept of the graph.

> The equation $y = mx + b$ is called the **slope–intercept equation.** The slope is m and the y-intercept is $(0, b)$.

Note that any graph of an equation $y = mx + b$ passes the vertical-line test and thus represents a function.

Example 6 Find the slope and the y-intercept of $y = 5x - 4$.

Since the equation is already in the form $y = mx + b$, we simply read the slope and the y-intercept from the equation:

$$y = 5x - 4.$$

The slope is 5. The y-intercept is $(0, -4)$.

Example 7 Find the slope and the y-intercept of $2x + 3y = 8$.

We first solve for y so we can easily read the slope and the y-intercept:

$$2x + 3y = 8$$

$$3y = -2x + 8 \quad \text{Subtracting } 2x$$

$$\frac{3y}{3} = \frac{-2x + 8}{3} \quad \text{Dividing by 3}$$

$$y = -\frac{2}{3}x + \frac{8}{3}. \quad \text{Finding the form } y = mx + b$$

The slope is $-\frac{2}{3}$. The y-intercept is $\left(0, \frac{8}{3}\right)$.

Do Exercises 9 and 10.

c Applications

Slope has many real-world applications. For example, numbers like 2%, 3%, and 6% are often used to represent the *grade* of a road, a measure of how steep a road on a hill or mountain is. A 3% grade $\left(3\% = \frac{3}{100}\right)$ means that for every horizontal distance of 100 ft, the road rises 3 ft, and a −3% grade means that for every horizontal distance of 100 ft, the road drops 3 ft. An athlete might change the grade of a treadmill during a workout. An escape ramp on an airliner might have a slope of about −0.6.

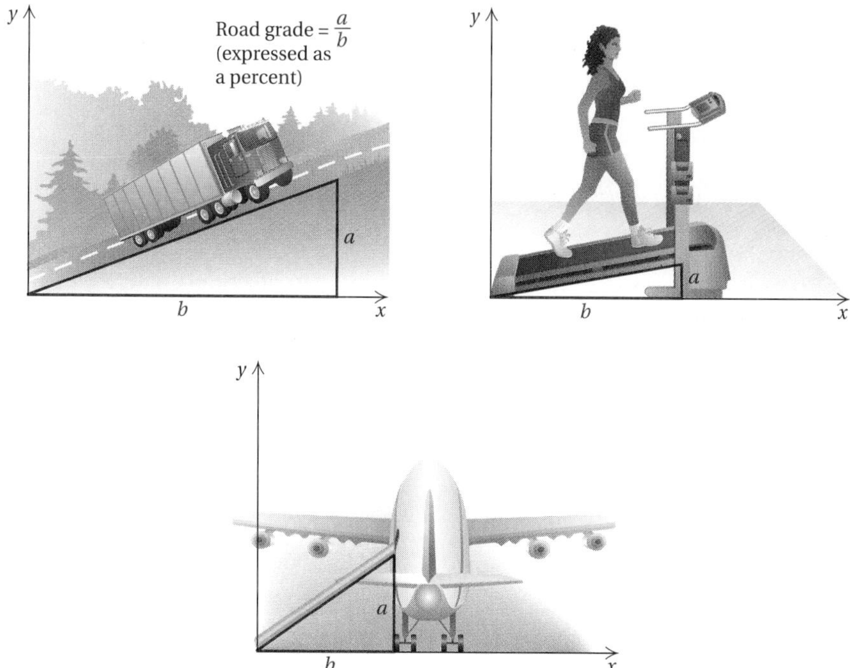

Find the slope and the y-intercept.

9. $f(x) = -8x + 23$

10. $5x - 10y = 25$

Answers on page A-28

11. *Capital Outlay for U.S. Defense.* Find the rate of change of capital outlay for U.S. defense.

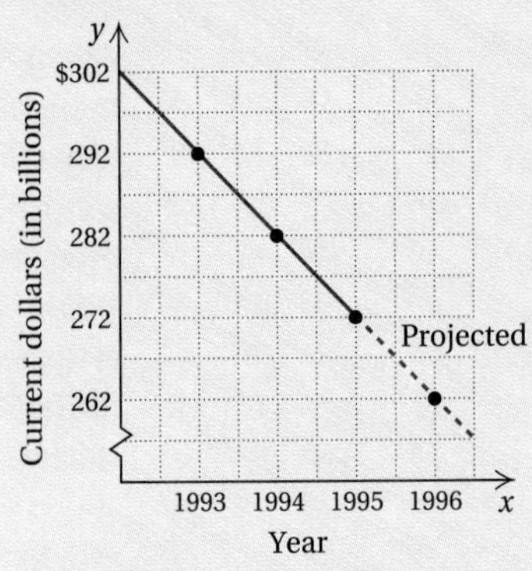

Source: Statistical Abstract of the United States

Answer on page A-28

Architects and carpenters use slope when designing and building stairs, ramps, or roof pitches. Another application occurs in hydrology. When a river flows, the strength or force of the river depends on how far the river falls vertically compared to how far it flows horizontally. Slope can also be considered as a **rate of change.**

Example 8 *Amount Spent on Cancer Research.* The amount spent on cancer research has increased steadily over the years, as shown in the following graph. Find the rate of change of that amount.

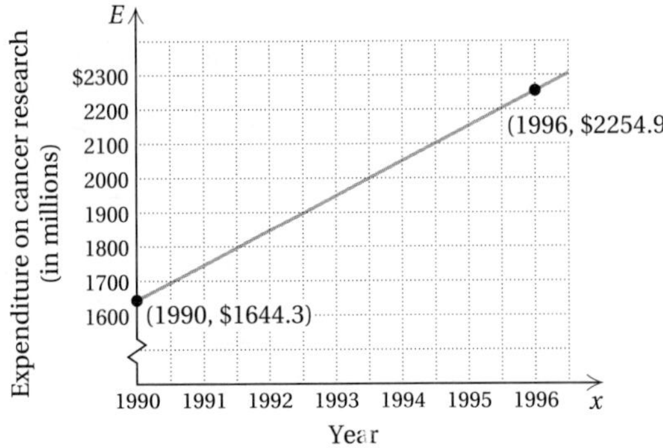

Source: The New England Journal of Medicine

First, we determine the coordinates of two points on the graph. In this case, they are given as (1990, $1644.3) and (1996, $2254.9). Then we compute the slope, or rate of change, as follows:

$$\text{Slope} = \text{Rate of change} = \frac{\text{change in } y}{\text{change in } x}$$

$$= \frac{\$2254.9 - \$1644.3}{1996 - 1990} = \frac{\$610.6}{6} \approx 101.8\,\frac{\$}{\text{yr}}.$$

This result tells us that each year the amount spent on cancer research has increased by about $101.8 million.

Do Exercise 11.

Exercise Set 7.3

a, b Find the slope and the y-intercept.

1. $y = 4x + 5$

2. $y = -5x + 10$

3. $f(x) = -2x - 6$

4. $g(x) = -5x + 7$

5. $y = -\frac{3}{8}x - \frac{1}{5}$

6. $y = \frac{15}{7}x + \frac{16}{5}$

7. $g(x) = 0.5x - 9$

8. $f(x) = -3.1x + 5$

9. $2x - 3y = 8$

10. $-8x - 7y = 24$

11. $9x = 3y + 6$

12. $9y + 36 - 4x = 0$

13. $3 - \frac{1}{4}y = 2x$

14. $5x = \frac{2}{3}y - 10$

15. $17y + 4x + 3 = 7 + 4x$

16. $3y - 2x = 5 + 9y - 2x$

b Find the slope of the line.

17.

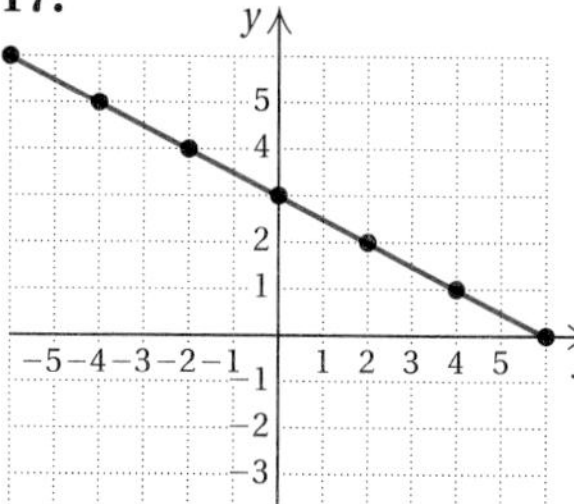

18.

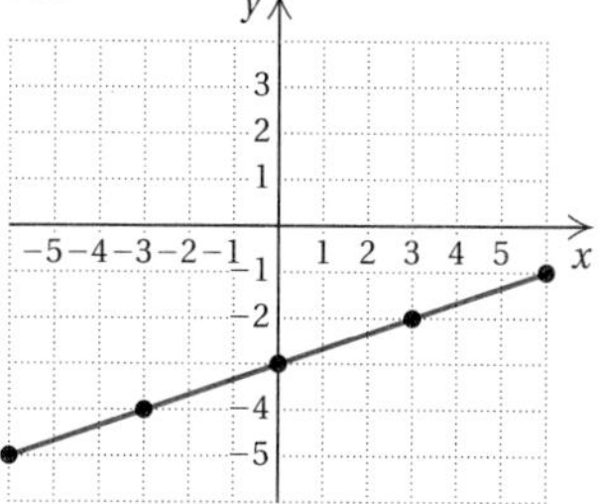

19.

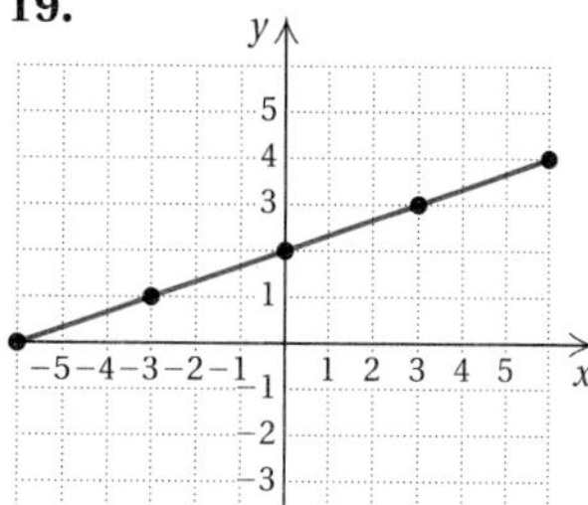

20.

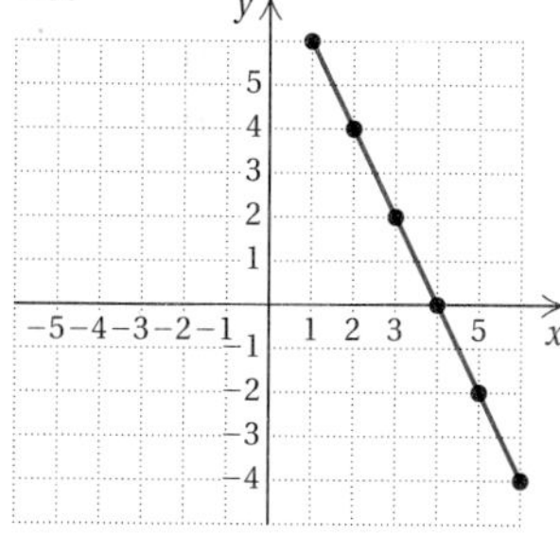

Find the slope of the line containing the given pair of points.

21. (6, 9) and (4, 5)

22. (8, 7) and (2, −1)

23. (9, −4) and (3, −8)

24. (17, −12) and (−9, −15)

25. (−16.3, 12.4) and (−5.2, 8.7)

26. (14.4, −7.8) and (−12.5, −17.6)

c Find the slope (or rate of change).

27. Find the slope (or grade) of the treadmill.

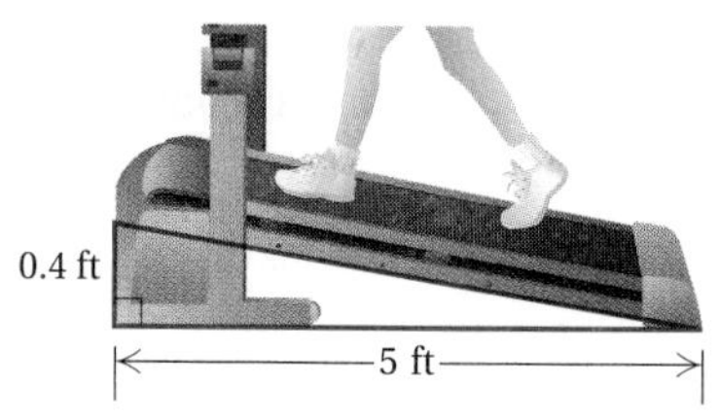

28. Find the slope (or pitch) of the roof.

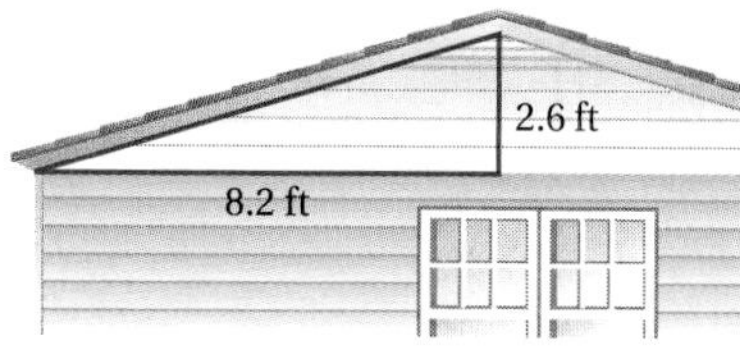

29. Find the slope (or head) of the river.

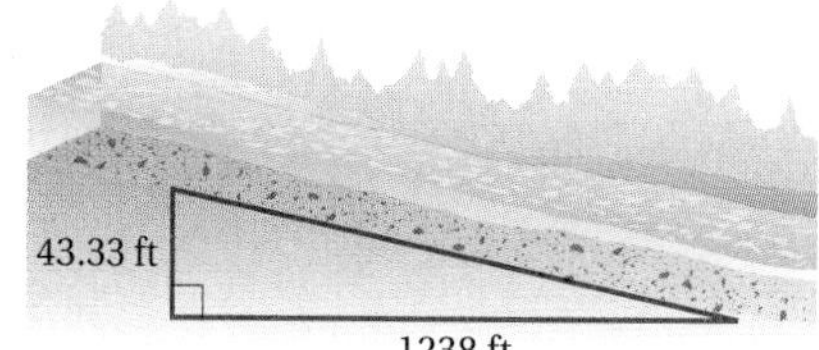

30. Public buildings regularly include steps with 7-in. risers and 11-in. treads. Find the grade of such a stairway.

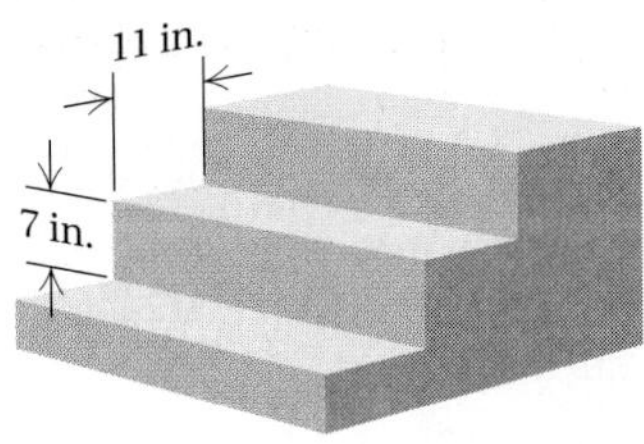

31. Find the rate of change of the tuition and fees at public two-year colleges.

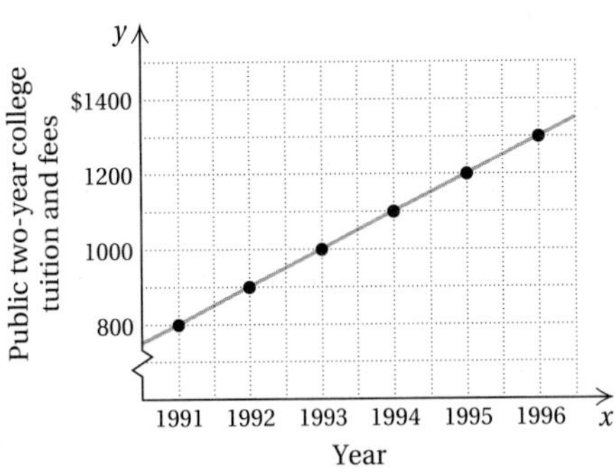

Source: *Statistical Abstract of the United States*

32. Find the rate of change of the cost of a formal wedding.

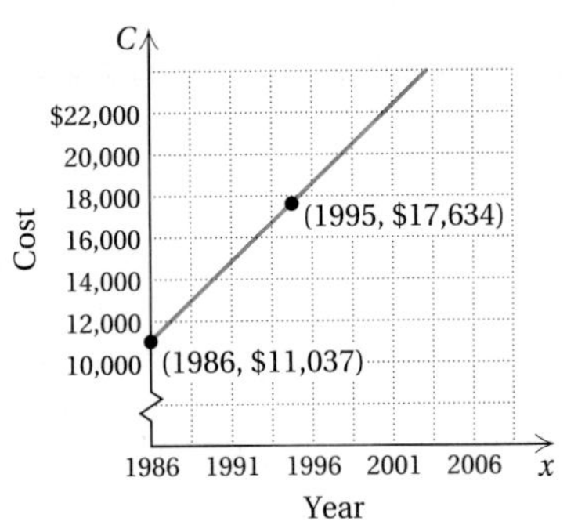

Source: *Modern Bride Magazine*

Skill Maintenance

Simplify. [1.8d], [4.1b]

33. $3^2 - 24 \cdot 56 + 144 \div 12$

34. $9\{2x - 3[5x + 2(-3x + y^0 - 2)]\}$

35. $10\{2x + 3[5x - 2(-3x + y^1 - 2)]\}$

36. $5^4 \div 625 \div 5^2 \cdot 5^7 \div 5^3$

Solve. [2.4a]

37. One side of a square is 5 yd less than a side of an equilateral triangle. If the perimeter of the square is the same as the perimeter of the triangle, what is the length of a side of the square? of the triangle?

Factor. [5.6a]

38. $8 - 125x^3$

39. $c^6 - d^6$

40. $56x^3 - 7$

41. Divide: $(a^2 - 11a + 6) \div (a - 1)$. [4.8b]

Synthesis

42. ◆ A student makes a mistake when using a grapher to draw $4x + 5y = 12$ and the following screen appears. Use algebra to show that a mistake has been made. What do you think the mistake was?

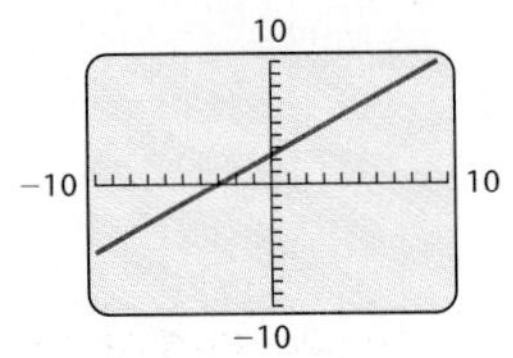

43. ◆ A student makes a mistake when using a grapher to draw $5x - 2y = 3$ and the following screen appears. Use algebra to show that a mistake has been made. What do you think the mistake was?

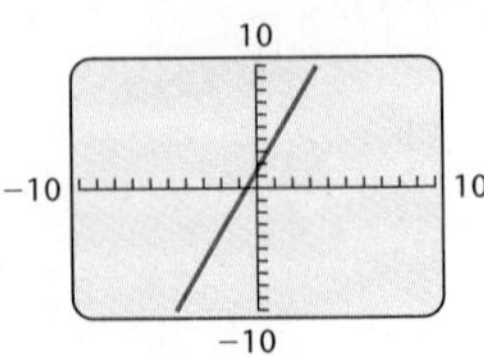

7.4 More on Graphing Linear Equations

a Graphing Using Intercepts

The **x-intercept** of the graph of a linear equation or function is the point at which the graph crosses the x-axis. The **y-intercept** is the point at which the graph crosses the y-axis. We know from geometry that only one line can be drawn through two given points. Thus, if we know the intercepts, we can graph the line. To ensure that a computation error has not been made, it is a good idea to calculate a third point as a check.

Many equations of the type $Ax + By = C$ can be graphed conveniently using intercepts.

> A **y-intercept** is a point $(0, b)$. To find b, let $x = 0$ and solve for y.
> An **x-intercept** is a point $(a, 0)$. To find a, let $y = 0$ and solve for x.

Example 1 Find the intercepts of $3x + 2y = 12$ and then graph the line.

To find the y-intercept, we let $x = 0$ and see what y must be. Covering up the $3x$ or ignoring it amounts to letting x be 0. So we cover up $3x$ and see that $2y = 12$. Then y must be 6. The y-intercept is $(0, 6)$.

To find the x-intercept, we can cover up the y-term and look at the rest of the equation. We have $3x = 12$, or $x = 4$. The x-intercept is $(4, 0)$.

We plot these points and draw the line, using a third point as a check. We choose $x = 6$ and solve for y:

$$
\begin{aligned}
3(6) + 2y &= 12 \\
18 + 2y &= 12 \\
2y &= -6 \\
y &= -3.
\end{aligned}
$$

We plot $(6, -3)$ and note that it is on the line.

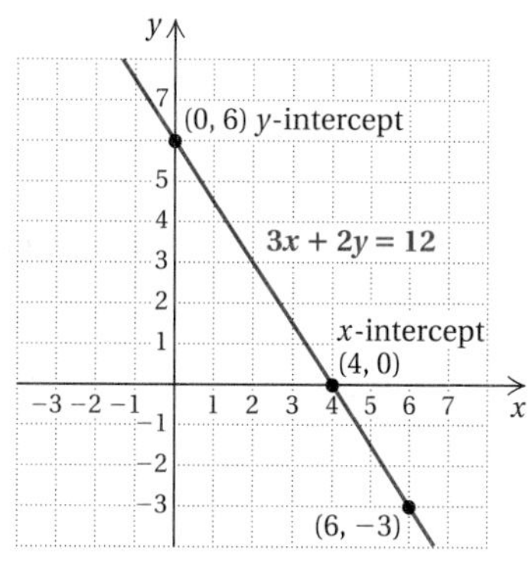

Do Exercise 1.

b Graphing Using the Slope and the y-Intercept

We can also graph a line using its slope and y-intercept.

Example 2 Graph: $y = -\frac{2}{3}x + 1$.

This equation is in slope–intercept form, $y = mx + b$. The y-intercept is $(0, 1)$. We plot $(0, 1)$. We can think of the slope as $\frac{-2}{3}$. Starting at the y-intercept and using the slope, we find another point by moving 2 units down (since the numerator is *negative* and corresponds to the change in y) and 3 units to the right (since the denominator is *positive* and corresponds to the change in x). We get to a new point, $(3, -1)$. In a similar manner, we can move from the point $(3, -1)$ to find another point, $(6, -3)$.

Objectives

a. Graph linear equations using intercepts.
b. Given a linear equation in slope–intercept form, use the slope and the y-intercept to graph the line.
c. Graph linear equations of the form $x = a$ or $y = b$.
d. Given the equations of two lines, determine whether their graphs are parallel or whether they are perpendicular.

For Extra Help

TAPE 12 MAC WIN CD-ROM

1. Find the intercepts of $4y - 12 = -6x$ and then graph the line.

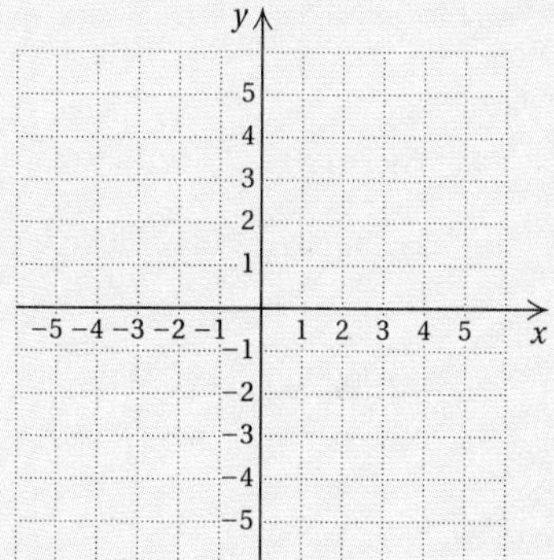

Answer on page A-29

Graph using the slope and the y-intercept.

2. $y = \frac{3}{2}x + 1$

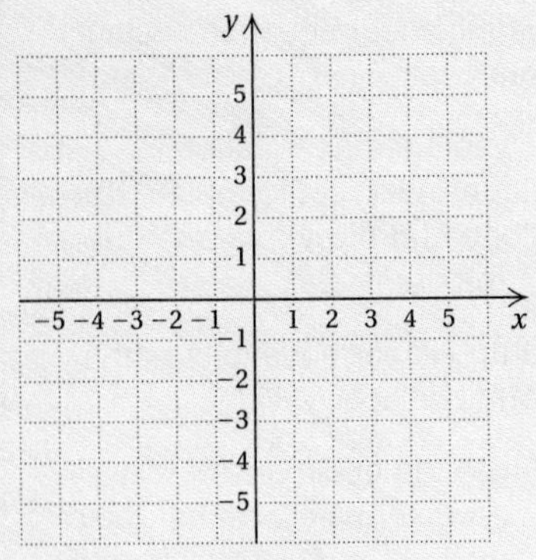

3. $f(x) = \frac{3}{4}x - 2$

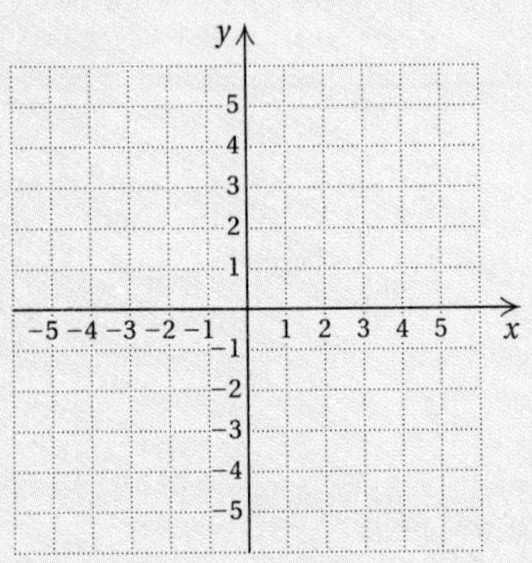

4. $g(x) = -\frac{3}{5}x + 5$

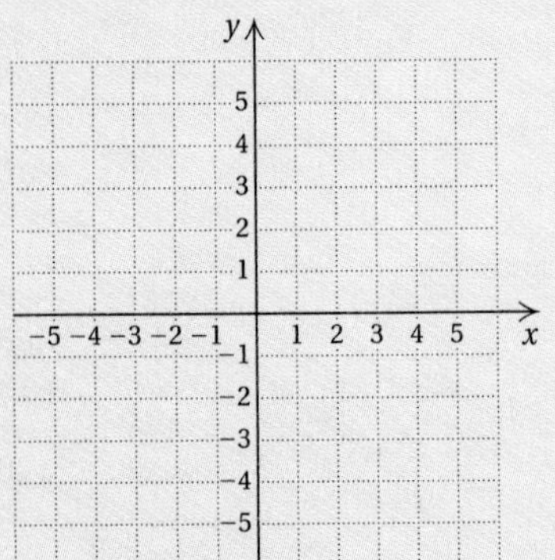

Answers on page A-29

Suppose we think of the slope $-\frac{2}{3}$ as $\frac{2}{-3}$. Then we can start again at (0, 1), but this time we move 2 units up (since the numerator is *positive* and corresponds to the change in y) and 3 units to the left (since the denominator is *negative* and corresponds to the change in x). We get another point on the graph, $(-3, 3)$, and from it we can obtain $(-6, 5)$ and others in a similar manner. We plot the points and draw the line.

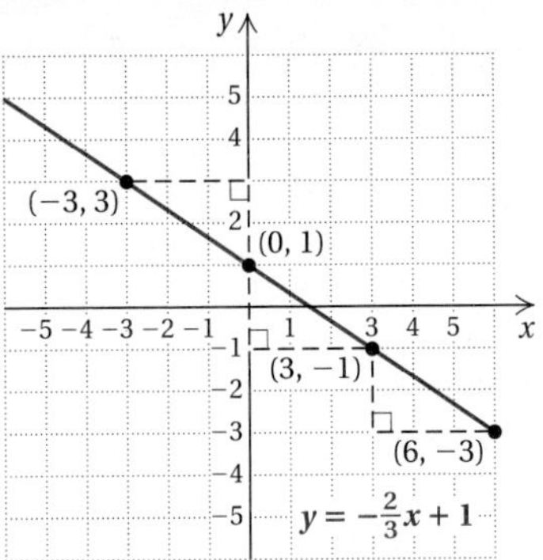

Example 3 Graph: $f(x) = \frac{2}{5}x + 4$.

First, we plot the y-intercept, (0, 4). We then consider the slope $\frac{2}{5}$. Starting at the y-intercept and using the slope, we find another point by moving 2 units up (since the numerator is *positive* and corresponds to the change in y) and 5 units to the right (since the denominator is *positive* and corresponds to the change in x). We get to a new point, (5, 6).

We can also think of the slope $\frac{2}{5}$ as $\frac{-2}{-5}$. We again start at the y-intercept, (0, 4). We move 2 units down (since the numerator is *negative* and corresponds to the change in y) and 5 units to the left (since the denominator is *negative* and corresponds to the change in x). We get to another new point, $(-5, 2)$. We plot the points and draw the line.

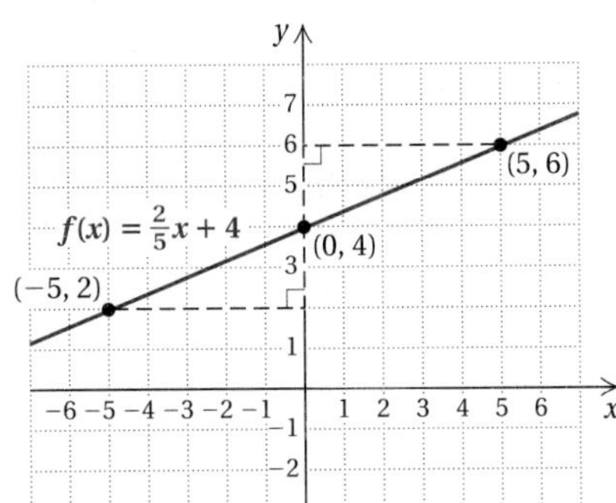

Do Exercises 2–5. (Exercise 5 is on the following page.)

c Horizontal and Vertical Lines

Some equations have graphs that are parallel to one of the axes. This happens when either A or B is 0 in $Ax + By = C$. These equations have a missing variable. In the following example, x is missing.

Example 4 Graph: $y = 3$.

Since x is missing, any number for x will do. Thus all ordered pairs $(x, 3)$ are solutions. The graph is a **horizontal line** parallel to the x-axis.

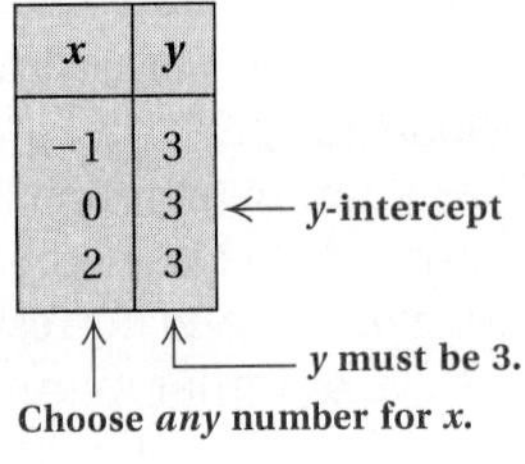

x	y
−1	3
0	3
2	3

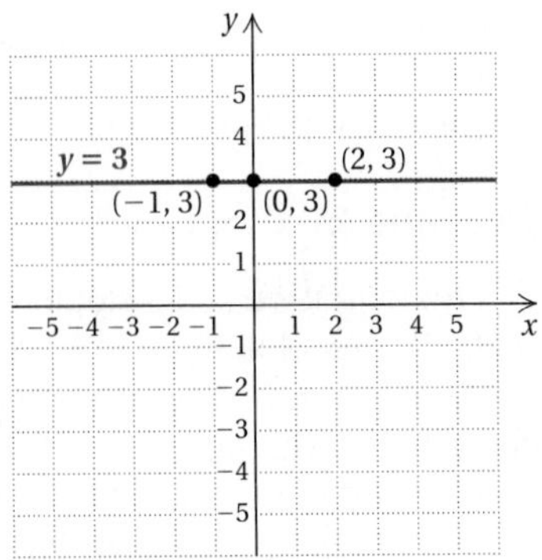

What about the slope of a horizontal line? In Example 4, consider the points $(-1, 3)$ and $(2, 3)$, which are on the line $y = 3$. The change in y is $3 - 3$, or 0. The change in x is $-1 - 2$, or -3. Thus,

$$m = \frac{3 - 3}{-1 - 2} = \frac{0}{-3} = 0.$$

Any two points on a horizontal line have the same y-coordinate. Thus the change in y is always 0, so the slope is 0.

We can also determine the slope by noting that $y = 3$ can be written in slope–intercept form as $y = 0x + 3$, or $f(x) = 0x + 3$. From this equation, we read that the slope is 0. A function of this type is called a **constant function.** We can express it in the form $y = b$, or $f(x) = b$. Its graph is a horizontal line that crosses the y-axis at $(0, b)$.

Do Exercises 6 and 7.

In the following example, y is missing and the graph is parallel to the y-axis.

Example 5 Graph: $x = -2$.

Since y is missing, any number for y will do. Thus all ordered pairs $(-2, y)$ are solutions. The graph is a **vertical line** parallel to the y-axis.

x	y	
-2	0	← x-intercept
-2	3	
-2	-4	

↑ x must be -2. ↑ Choose *any* number for y.

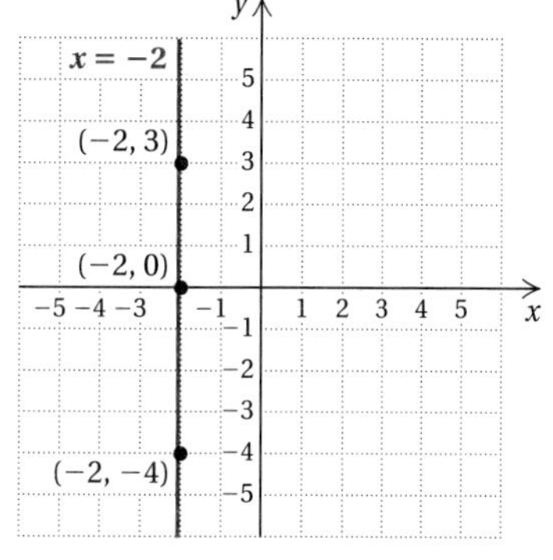

This graph is not the graph of a function because it fails the vertical-line test. There is a vertical line (itself) that crosses the graph more than once.

What about the slope of a vertical line? In Example 5, consider the points $(-2, 3)$ and $(-2, -4)$, which are on the line $x = -2$. The change in y is $3 - (-4)$, or 7. The change in x is $-2 - (-2)$, or 0. Thus,

$$m = \frac{3 - (-4)}{-2 - (-2)} = \frac{7}{0}. \quad \text{Undefined}$$

Since division by 0 is not defined, the slope of this line is not defined. Any two points on a vertical line have the same x-coordinate. Thus the change in x is always 0, so the slope of any vertical line is undefined.

The following summarizes horizontal and vertical lines and their equations.

> **HORIZONTAL LINE**
>
> The graph of $y = b$, or $f(x) = b$, is a horizontal line with y-intercept $(0, b)$. It is the graph of a constant function with slope 0.
>
> **VERTICAL LINE**
>
> The graph of $x = a$ is a vertical line through the point $(a, 0)$. The slope is undefined. It is not the graph of a function.

Do Exercises 8–10 on the following page.

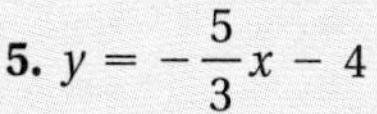

5. $y = -\frac{5}{3}x - 4$

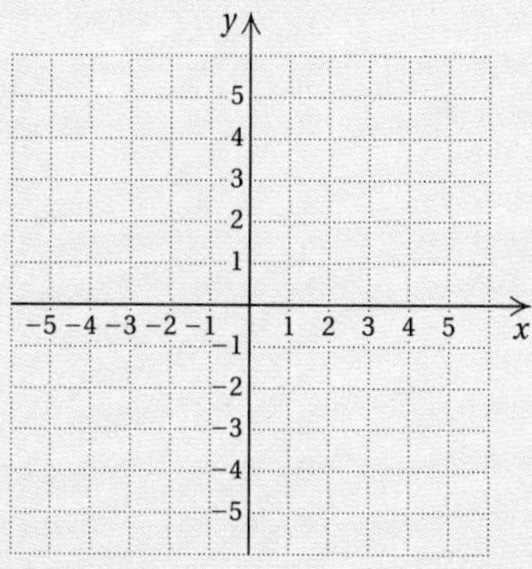

Graph and determine the slope.

6. $f(x) = -4$

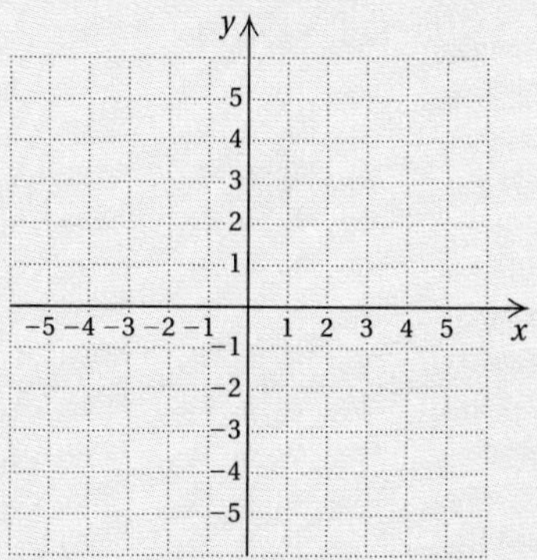

7. $y = 3.6$

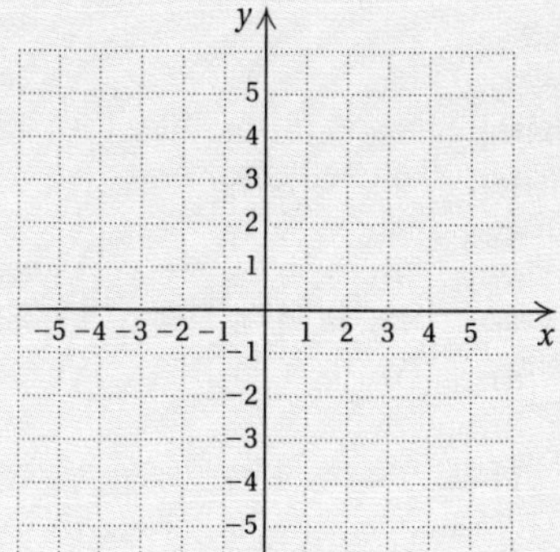

Answers on page A-29

Graph.

8. $x = -5$

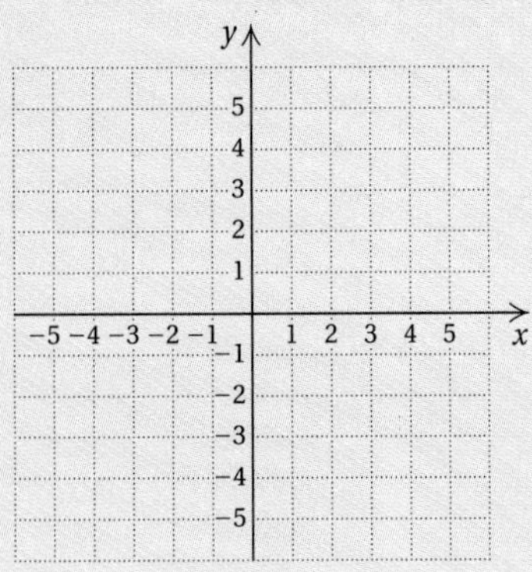

9. $8x - 5 = 19$
(*Hint*: Solve for x.)

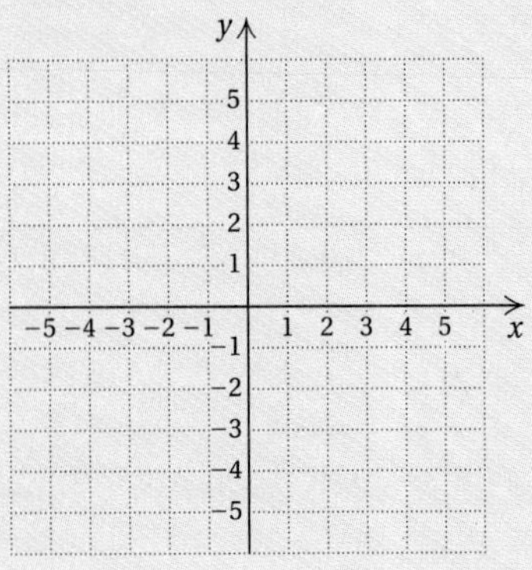

10. Determine, if possible, the slope of each line.

a) $x = -12$
b) $y = 6$
c) $2y + 7 = 11$
d) $x = 0$
e) $y = -\frac{3}{4}$
f) $10 - 5x = 15$

Answers on page A-29

We have graphed linear equations in several ways in this chapter. Although, in general, you can use any method that works best for you, the following are some guidelines.

> To graph a linear equation:
>
> 1. Is the equation of the type $x = a$ or $y = b$? If so, the graph will be a line parallel to an axis; $x = a$ is vertical and $y = b$ is horizontal.
> 2. If the line is of the type $y = mx$, both intercepts are the origin (0, 0). Plot (0, 0) and one other point.
> 3. If the line is of the type $y = mx + b$, plot the y-intercept and one other point.
> 4. If the equation is of the form $Ax + By = C$, but not of the form $x = a$, $y = b$, $y = mx$, or $y = mx + b$, graph using intercepts. If the intercepts are too close together, choose another point farther from the origin.
> 5. In all cases, use a third point as a check.

d Parallel and Perpendicular Lines

Parallel Lines

If two lines are vertical, they are parallel. How can we tell whether nonvertical lines are parallel? We examine their slopes and y-intercepts.

> Two nonvertical lines are **parallel** if they have the same slope and different y-intercepts.

Example 6 Determine whether the graphs of

$$y - 3x = 1 \quad \text{and} \quad 3x + 2y = -2$$

are parallel.

To determine whether lines are parallel, we first find their slopes. Thus we first find the slope–intercept form of each equation by solving for y:

$$\begin{aligned} y - 3x &= 1 \\ y &= 3x + 1; \end{aligned}$$

$$\begin{aligned} 3x + 2y &= -2 \\ 2y &= -3x - 2 \\ y &= \tfrac{1}{2}(-3x - 2) \\ &= \tfrac{1}{2}(-3x) - \tfrac{1}{2}(2) \\ &= -\tfrac{3}{2}x - 1. \end{aligned}$$

The slopes, 3 and $-\frac{3}{2}$, are different. Thus the lines are not parallel, as the graphs shown at right confirm.

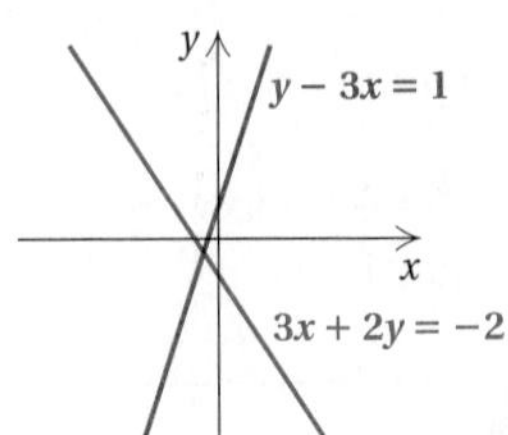

Example 7 Determine whether the graphs of

$$3x - y = -5 \quad \text{and} \quad y - 3x = -2$$

are parallel.

We first find the slope–intercept form of each equation by solving for y:

$$3x - y = -5 \qquad y - 3x = -2$$
$$-y = -3x - 5 \qquad y = 3x - 2.$$
$$y = 3x + 5;$$

The slopes, 3, are the same. The y-intercepts are different. Thus the lines are parallel, as the graphs seem to confirm.

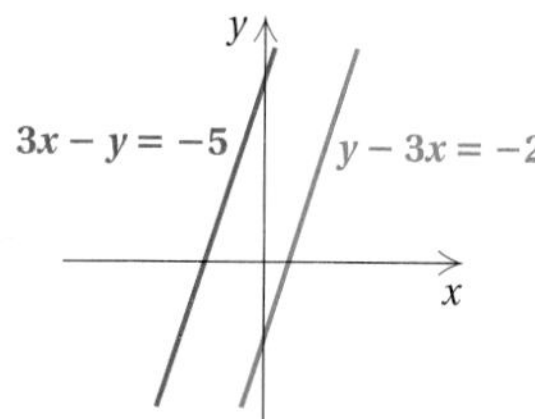

Determine whether the graphs of the given pair of lines are parallel.

11. $x + 4 = y$,
$y - x = -3$

12. $y + 4 = 3x$,
$4x - y = -7$

13. $y = 4x + 5$,
$2y = 8x + 10$

Answers on page A-29

Do Exercises 11–13.

If one line is vertical and another is horizontal, they are perpendicular. Otherwise, how can we tell whether two lines are perpendicular?

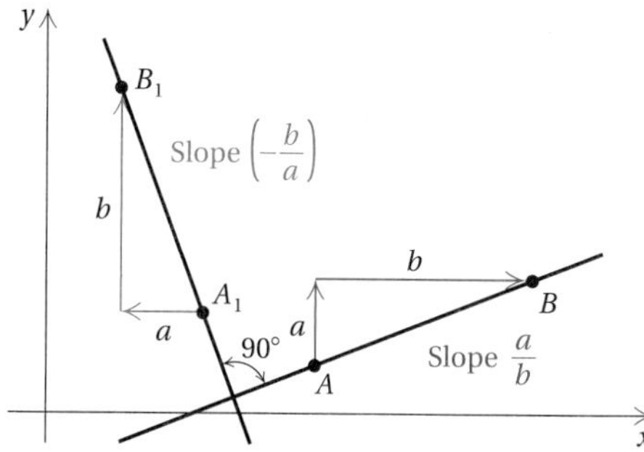

Consider a line $\overleftrightarrow{AB}$, as shown in the figure above, with slope a/b. Then think of rotating the line 90° to get a line $\overleftrightarrow{A_1B_1}$ perpendicular to $\overleftrightarrow{AB}$. For the new line, the rise and the run are interchanged, but the run is now negative. Thus the slope of the new line is $-b/a$, which is the opposite of the reciprocal of the slope of the first line. Also note that when we multiply the slopes, we get

$$\frac{a}{b}\left(-\frac{b}{a}\right) = -1.$$

This is the condition under which lines will be perpendicular.

> Two nonvertical lines are **perpendicular** if the product of their slopes is -1. (If one line has slope m, the slope of a line perpendicular to it is $-1/m$. That is, to find the slope of a line perpendicular to a given line, we take the reciprocal of the given slope and change the sign.)
>
> Lines are also perpendicular if one of them is vertical ($x = a$) and one of them is horizontal ($y = b$).

Example 8 Determine whether the graphs of $5y = 4x + 10$ and $4y = -5x + 4$ are perpendicular.

To determine whether the lines are perpendicular, we determine whether the product of their slopes is -1. We first find the slope–intercept form of each equation by solving for y.

Determine whether the graphs of the given pair of lines are perpendicular.

14. $2y - x = 2,$
$y + 2x = 4$

15. $3y = 2x + 15,$
$2y = 3x + 10$

Answers on page A-29

We have

$$\begin{aligned} 5y &= 4x + 10 \\ y &= \tfrac{1}{5}(4x + 10) \\ &= \tfrac{1}{5}(4x) + \tfrac{1}{5}(10) \\ &= \tfrac{4}{5}x + 2; \end{aligned} \qquad \begin{aligned} 4y &= -5x + 4 \\ y &= \tfrac{1}{4}(-5x + 4) \\ &= \tfrac{1}{4}(-5x) + \tfrac{1}{4}(4) \\ &= -\tfrac{5}{4}x + 1. \end{aligned}$$

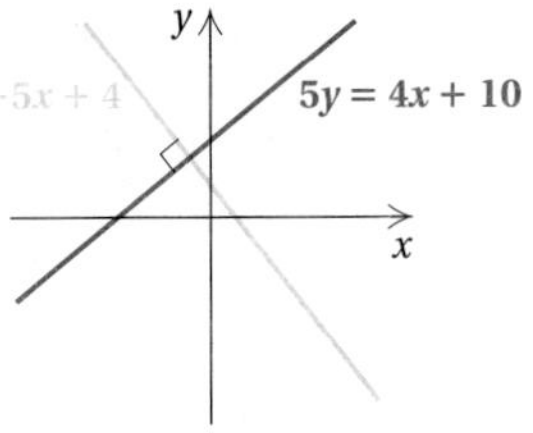

The slope of the first line is $\frac{4}{5}$, and the slope of the second line is $-\frac{5}{4}$. The product of the slopes is -1; that is, $\frac{4}{5} \cdot \left(-\frac{5}{4}\right) = -1$. Thus the lines are perpendicular.

Do Exercises 14 and 15.

Calculator Spotlight

Squaring Viewing Windows; Visualizing Parallel and Perpendicular Lines

Squaring a Viewing Window. Consider the [−10, 10, −10, 10] viewing window on the left below. Note that the distance between units is not visually the same on both axes. In this case, the length of the interval shown on the y-axis is about two-thirds of the length of the interval on the x-axis. If we change the dimensions of the window to [−6, 6, −4, 4], we get a graph for which the units are visually about the same on both axes. Creating such a window is called **squaring**. On a TI-83 grapher, there is a ZSQUARE feature for automatic window squaring. This feature alters the standard window dimensions to [−15.1613, 15.1613, −10, 10].

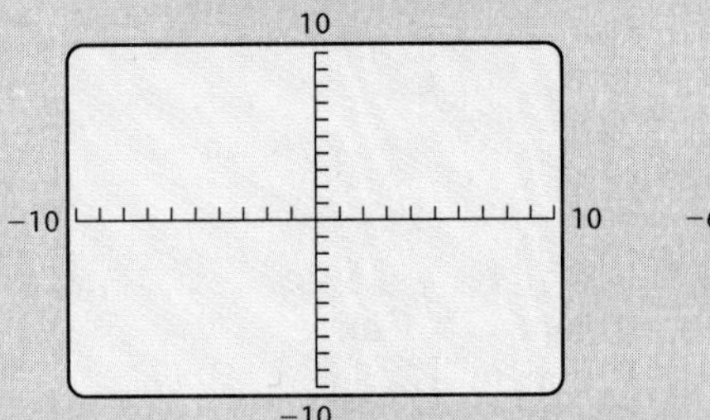

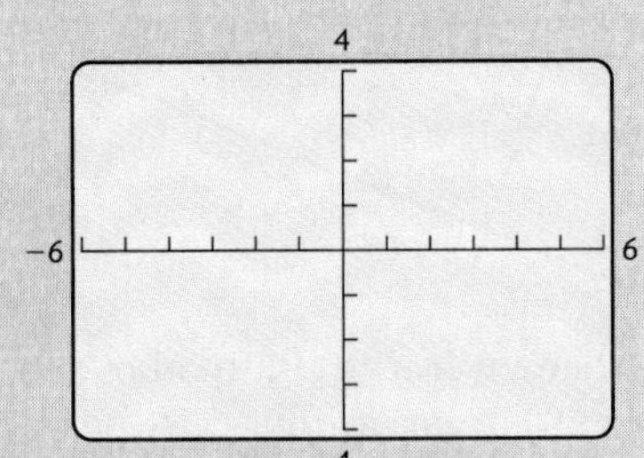

Each of the following is a graph of the line $y = 2x - 3$, but the viewing windows are different. When the window is square, as shown on the right, we get an accurate representation of the *slope* of the line. This is important when we need to visualize perpendicular lines.

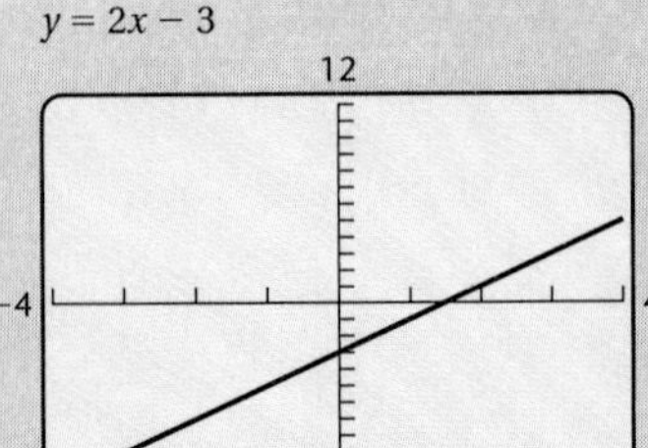

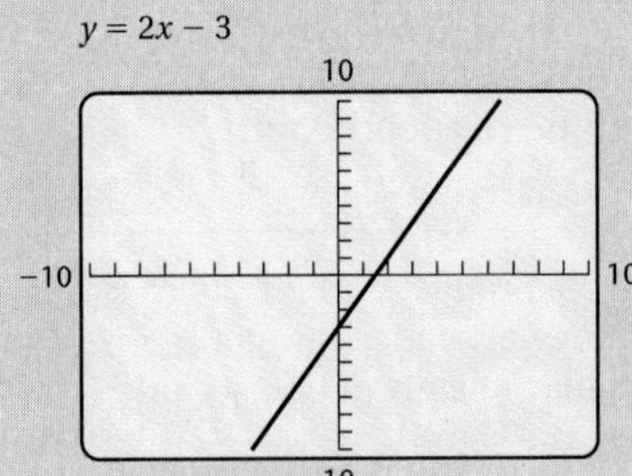

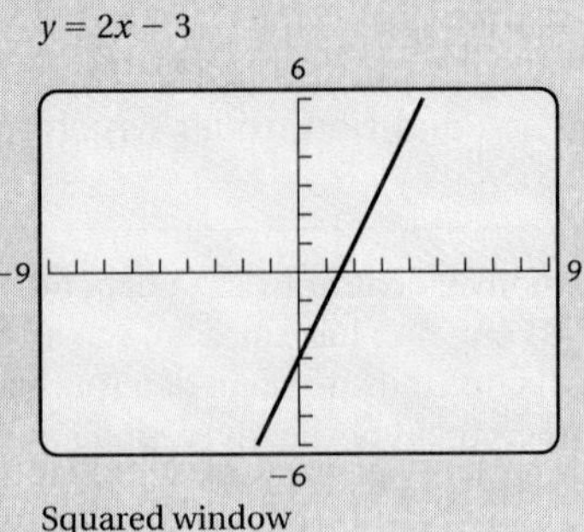

Squared window

Exercises

1. Graph each pair of equations in Margin Exercises 11–13. Check visually whether the lines appear to be parallel.
2. Graph each pair of equations in Margin Exercises 14 and 15 using a squared viewing window. Check visually whether the lines appear to be perpendicular.

Exercise Set 7.4

a Find the intercepts and then graph the line.

1. $x - 2 = y$

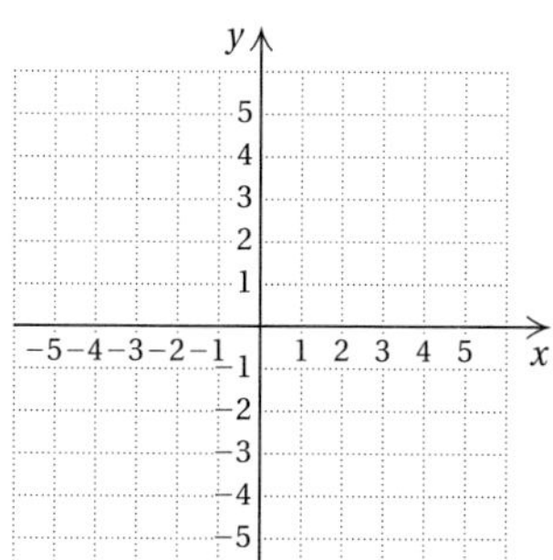

2. $x + 3 = y$

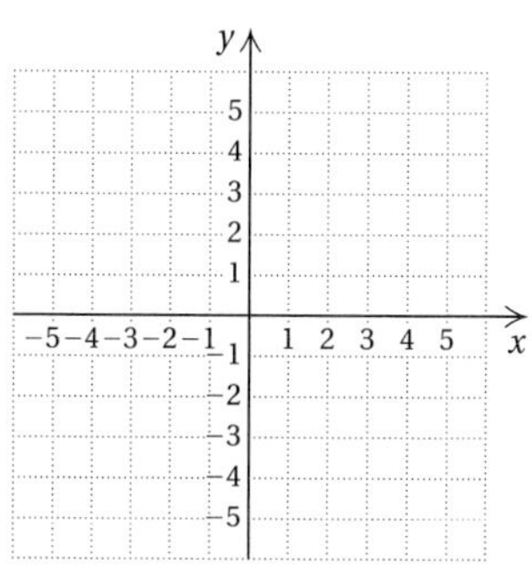

3. $x + 3y = 6$

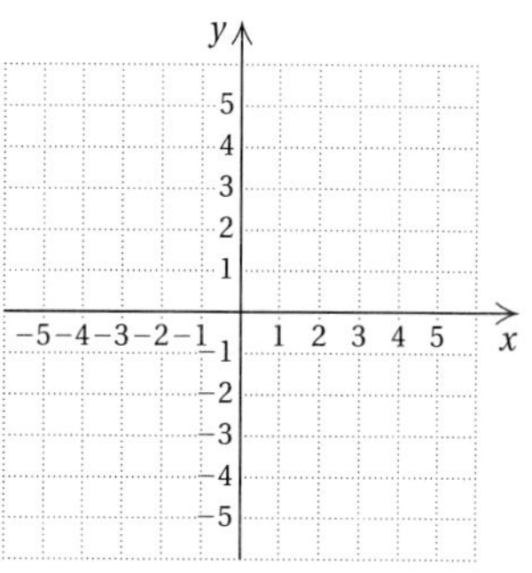

4. $x - 2y = 4$

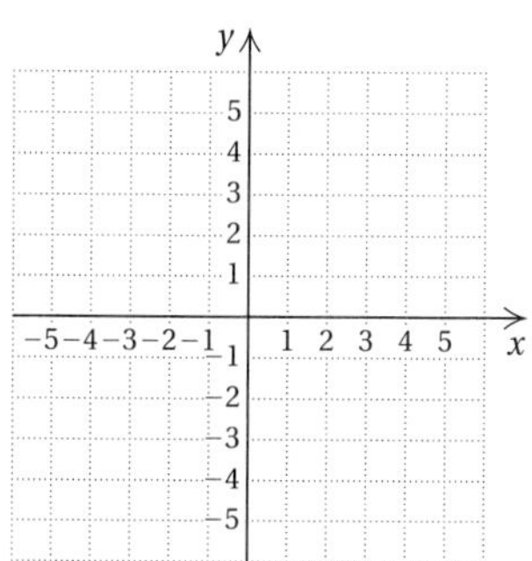

5. $2x + 3y = 6$

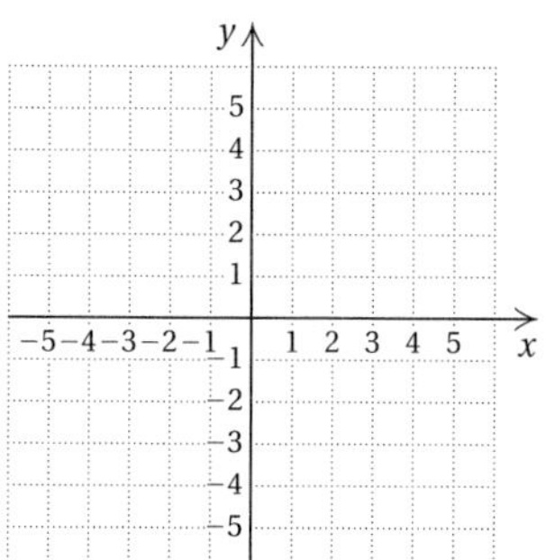

6. $5x - 2y = 10$

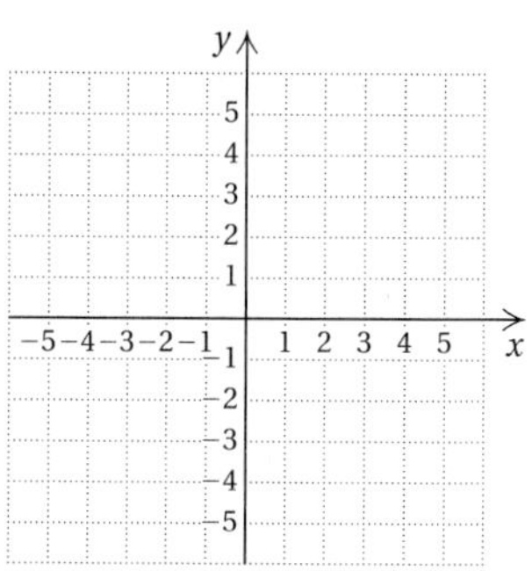

7. $f(x) = -2 - 2x$

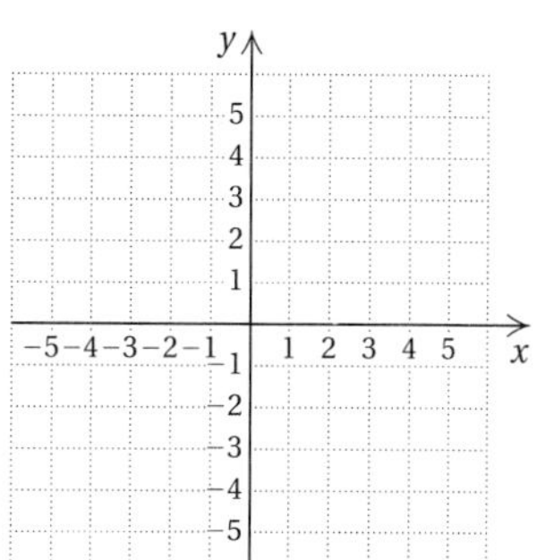

8. $g(x) = 5x - 5$

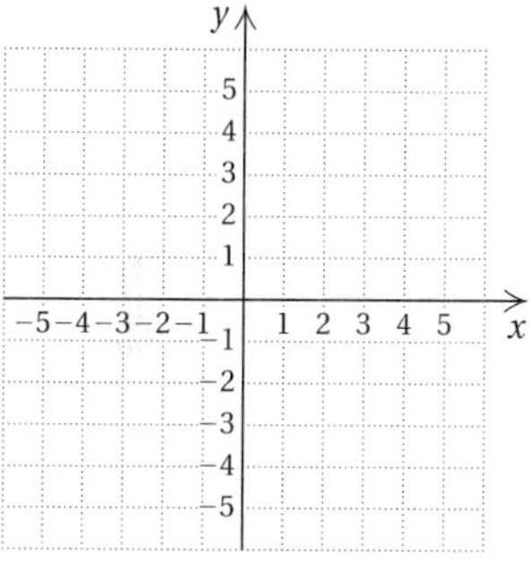

9. $5y = -15 + 3x$

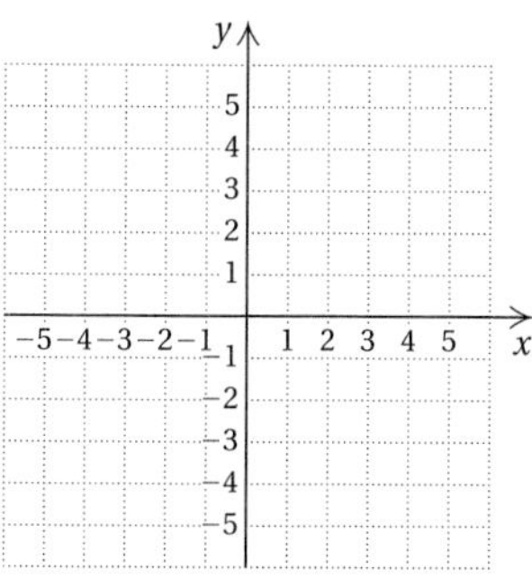

10. $5x - 10 = 5y$

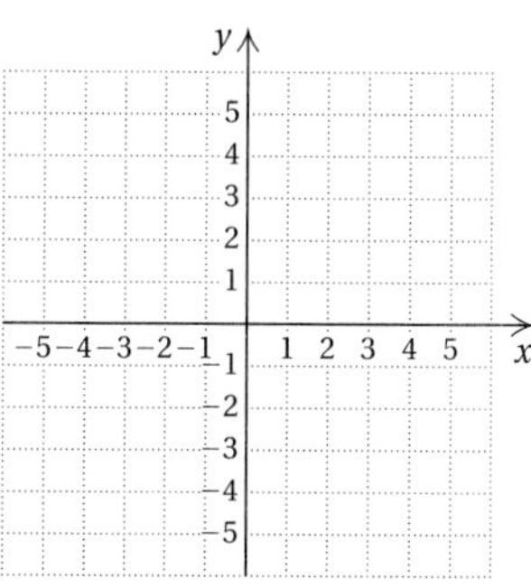

11. $2x - 3y = 6$

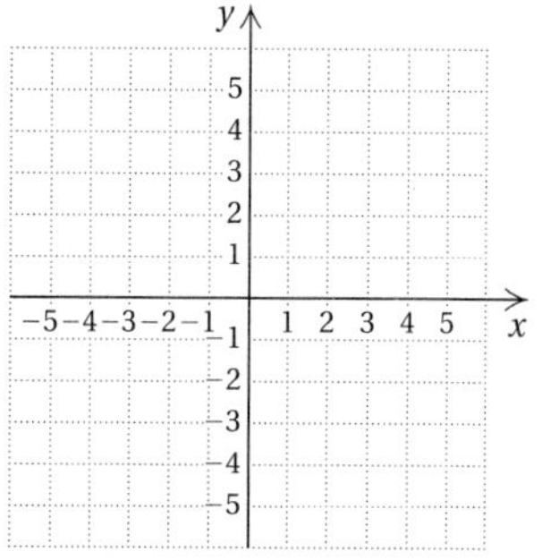

12. $4x + 5y = 20$

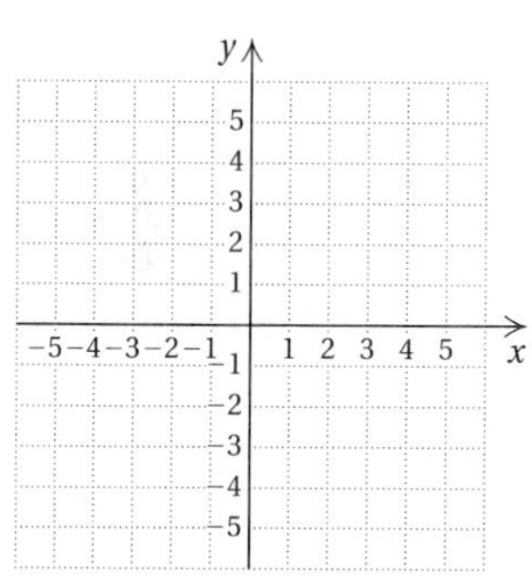

13. $2.8y - 3.5x = -9.8$

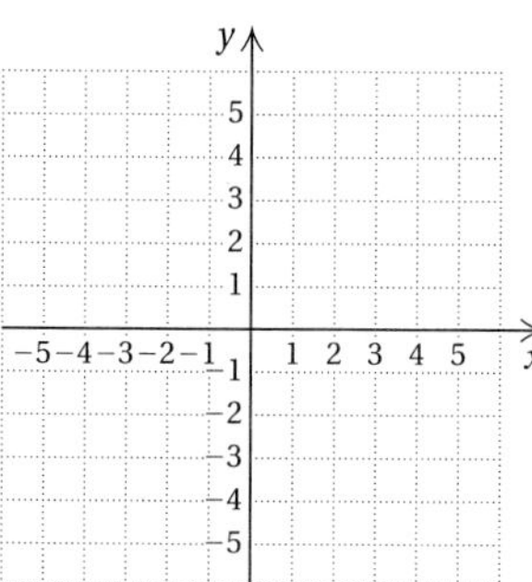

14. $10.8x - 22.68 = 4.2y$

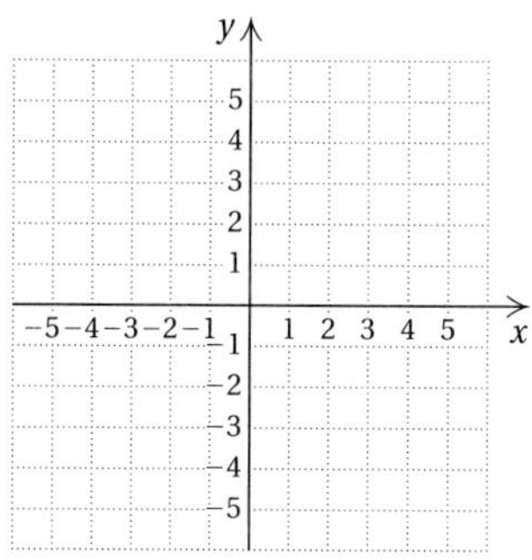

15. $5x + 2y = 7$

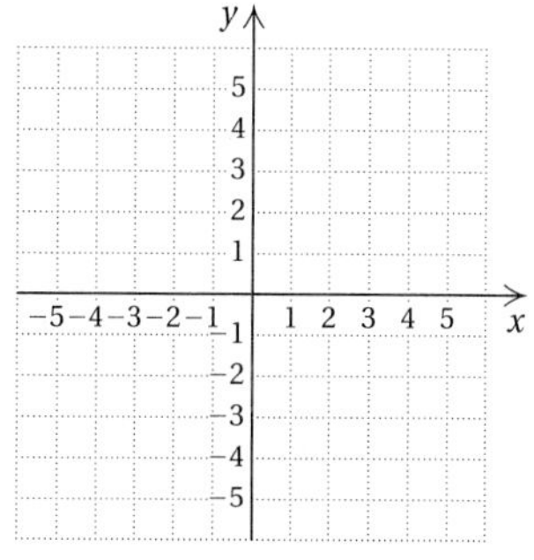

16. $3x - 4y = 10$

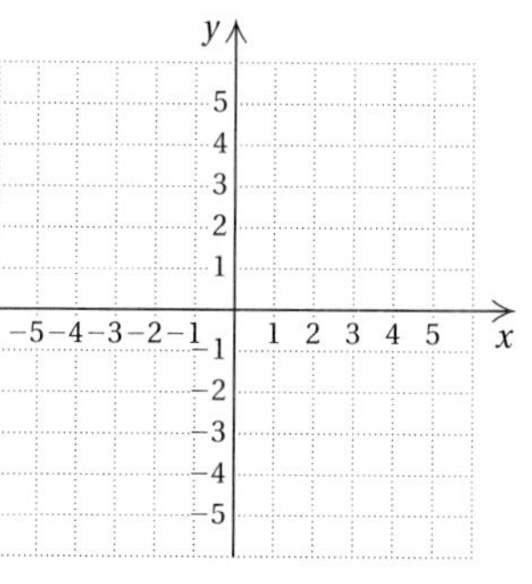

b Graph using the slope and the y-intercept.

17. $y = \frac{5}{2}x + 1$

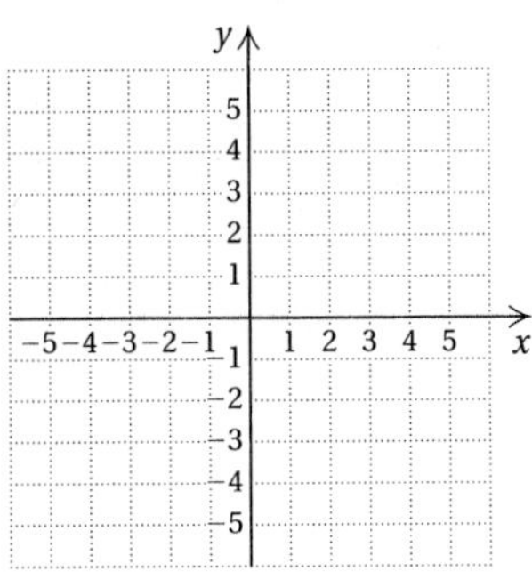

18. $y = \frac{2}{5}x - 4$

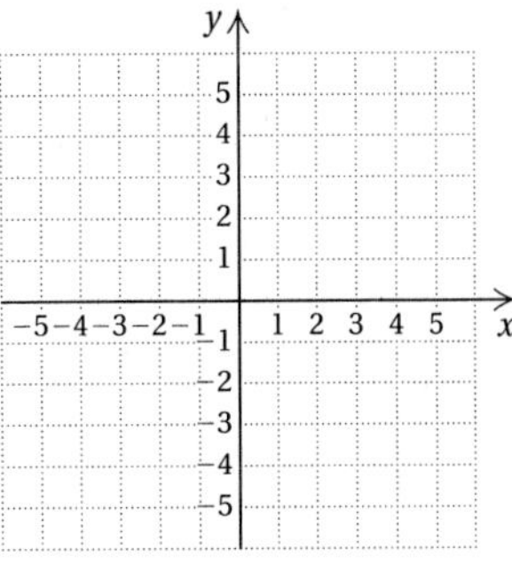

19. $f(x) = -\frac{5}{2}x - 4$

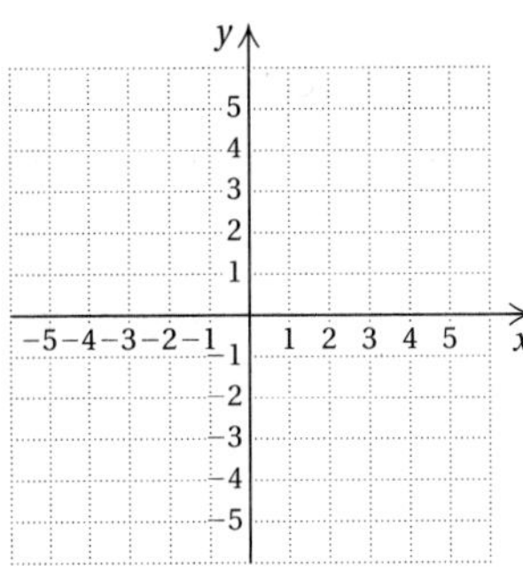

20. $f(x) = \frac{2}{5}x + 3$

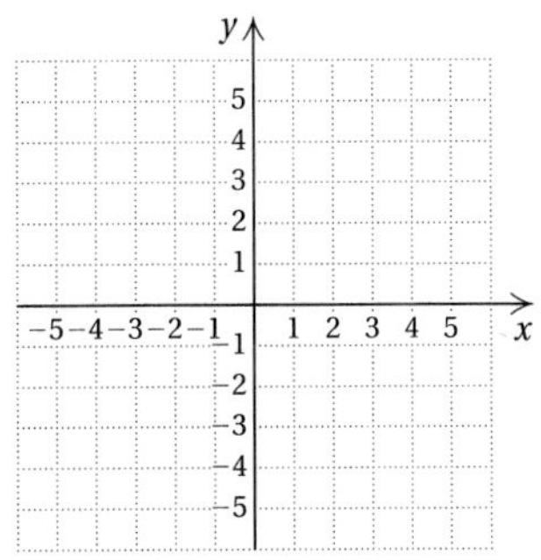

21. $x + 2y = 4$

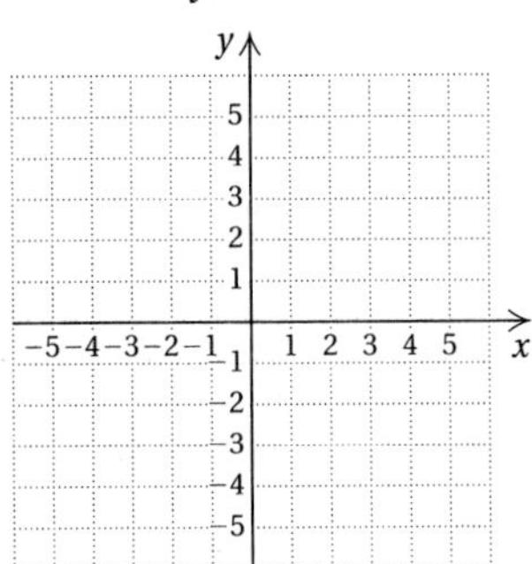

22. $x - 3y = 6$

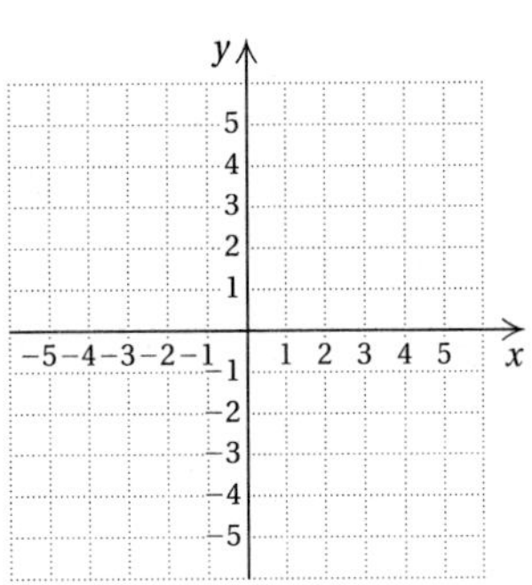

23. $4x - 3y = 12$

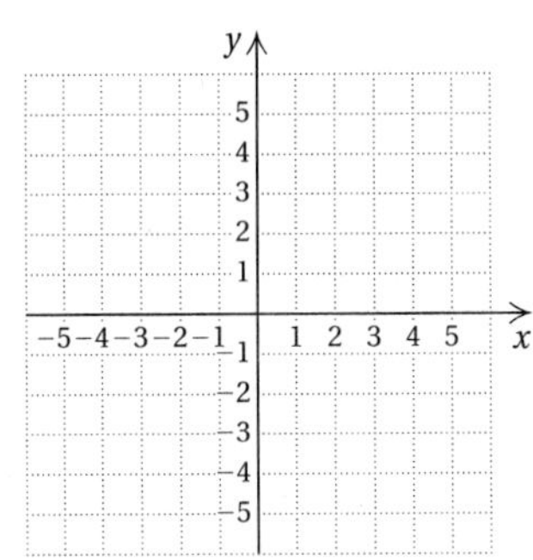

24. $2x + 6y = 12$

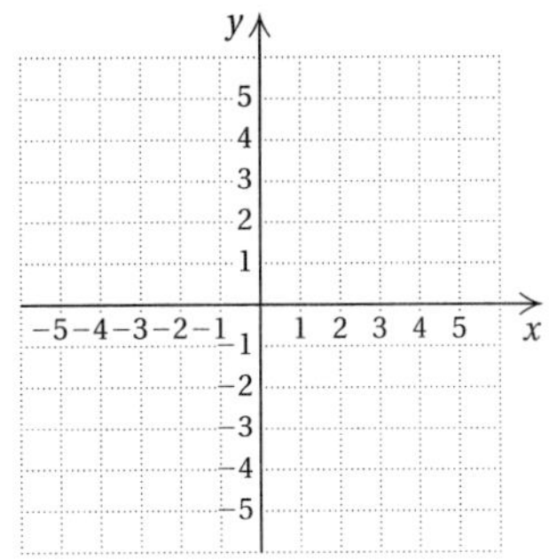

25. $f(x) = \frac{1}{3}x - 4$

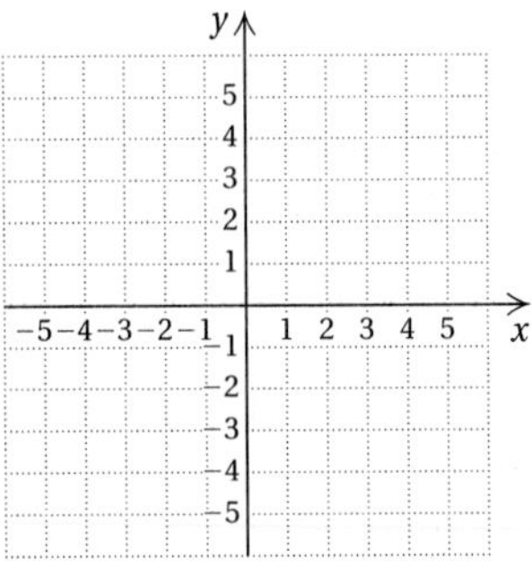

26. $g(x) = -0.25x + 2$

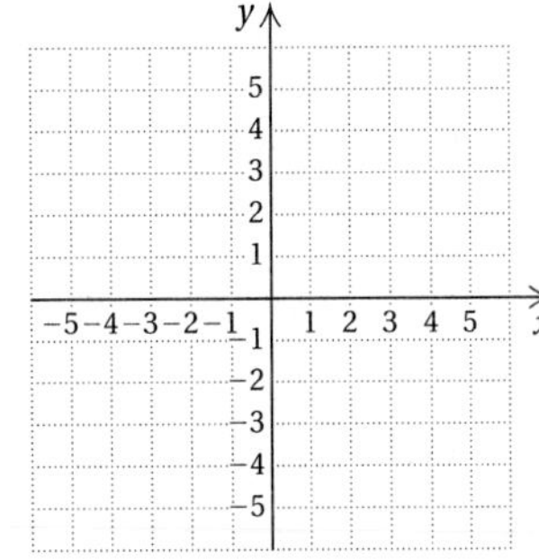

27. $5x + 4 \cdot f(x) = 4$
(*Hint*: Solve for $f(x)$.)

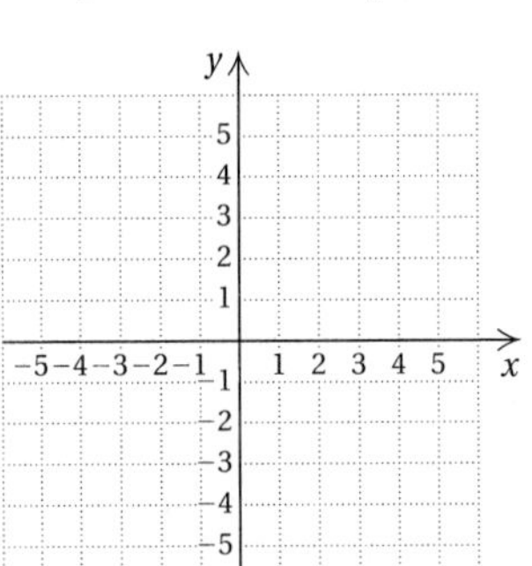

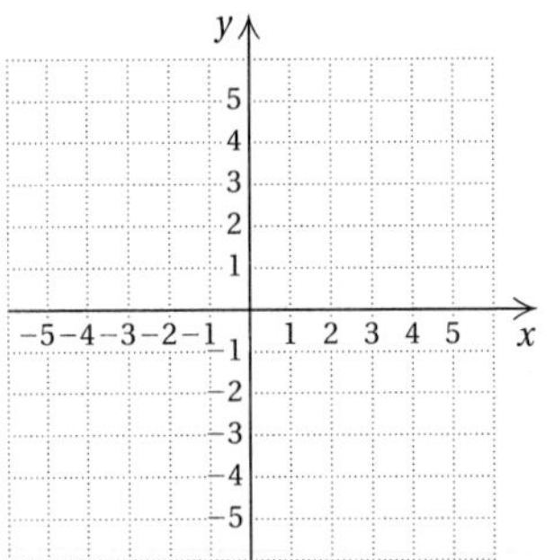

28. $3 \cdot f(x) = 4x + 6$

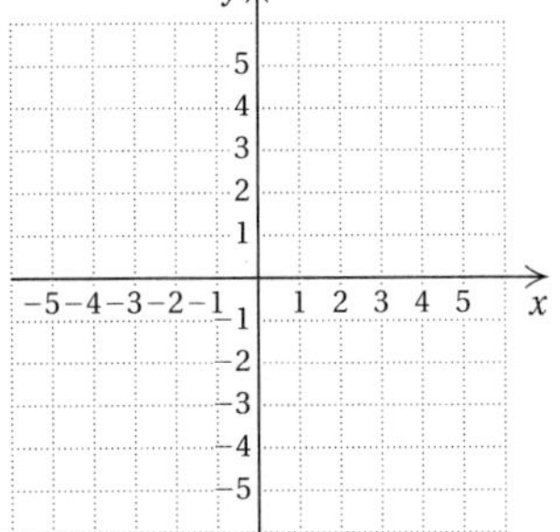

c Graph and, if possible, determine the slope.

29. $x = 1$

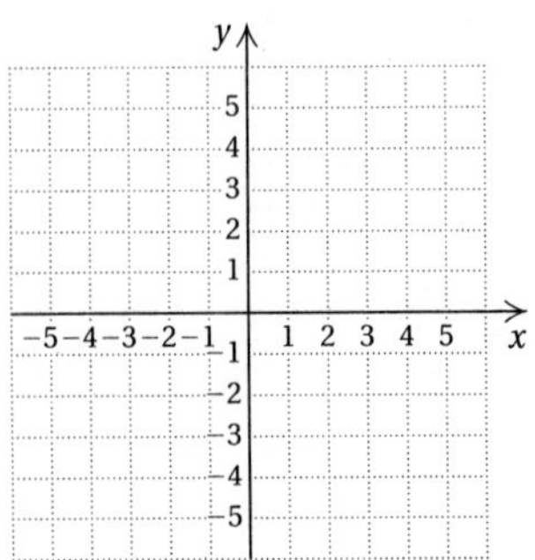

30. $x = -4$

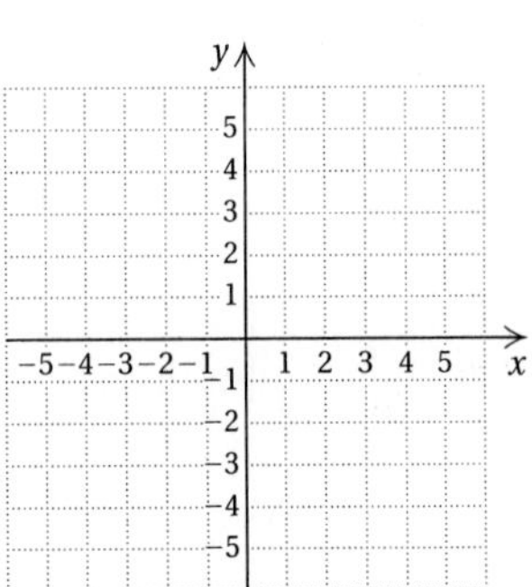

31. $y = -1$

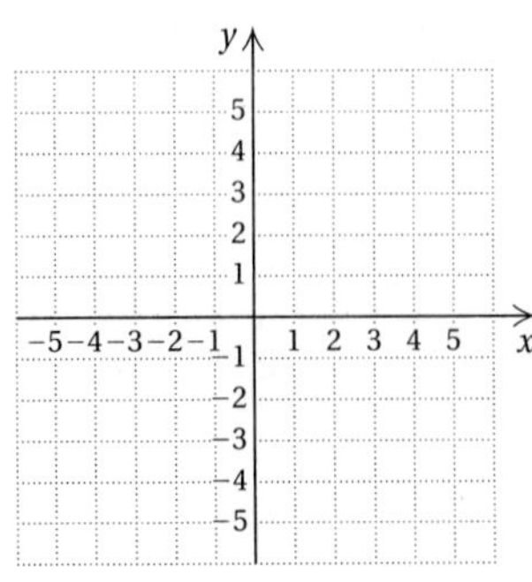

32. $y = \frac{3}{2}$

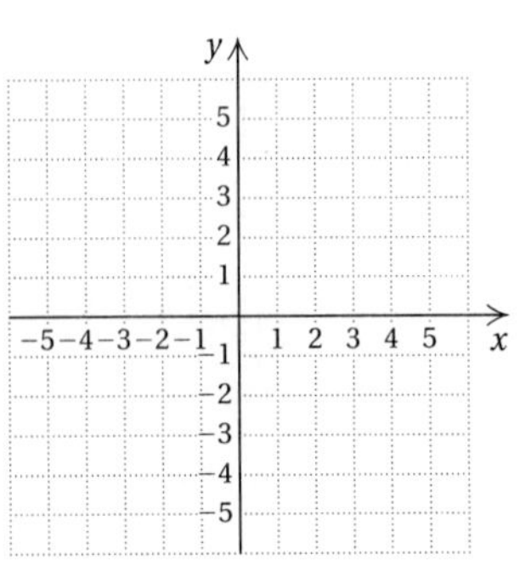

33. $f(x) = -6$

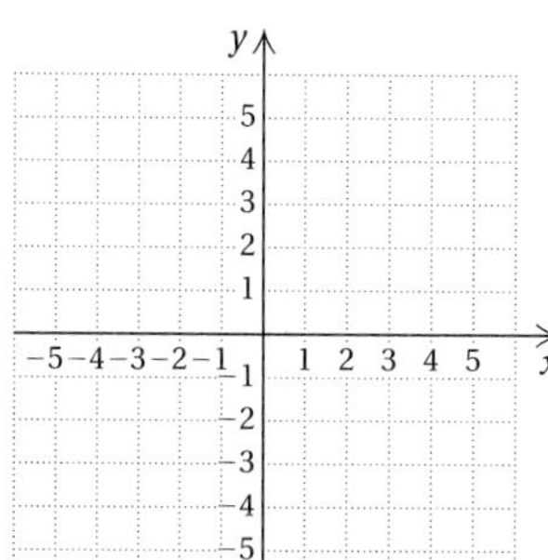

34. $f(x) = 2$

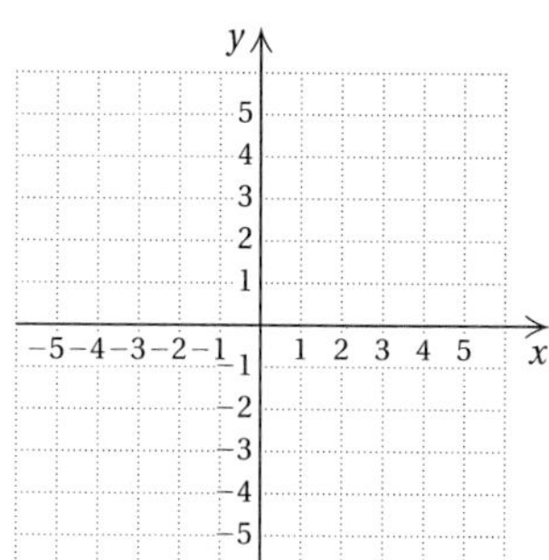

35. $y = 0$

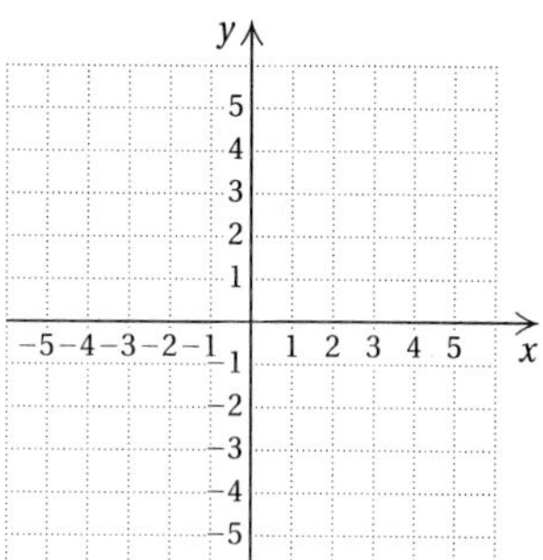

36. $x = 0$

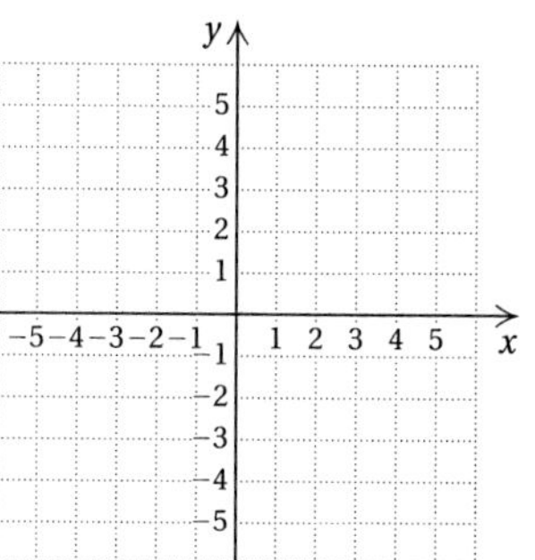

37. $2 \cdot f(x) + 5 = 0$

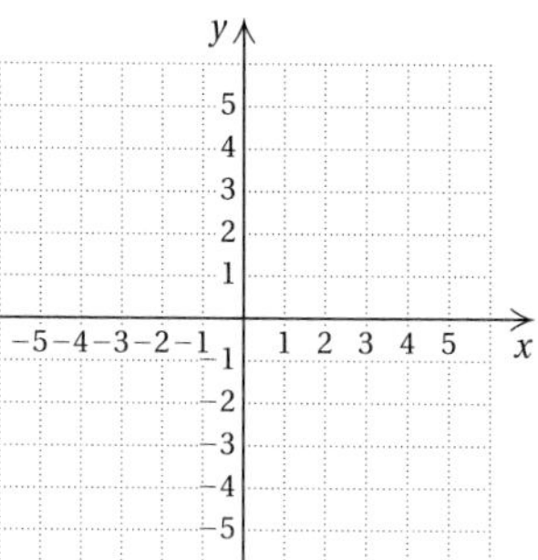

38. $4 \cdot g(x) + 3x = 12 + 3x$

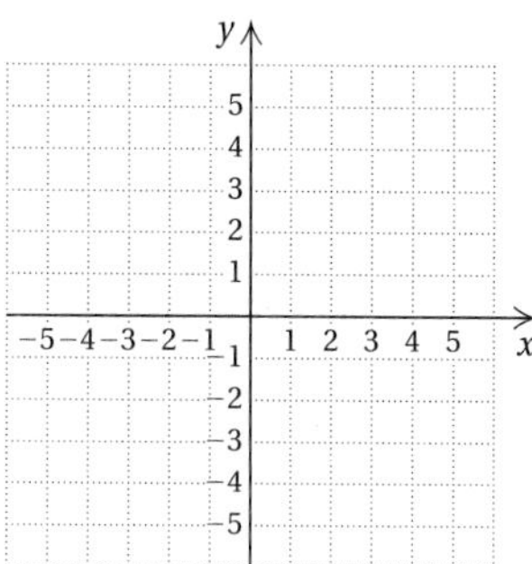

39. $7 - 3x = 4 + 2x$

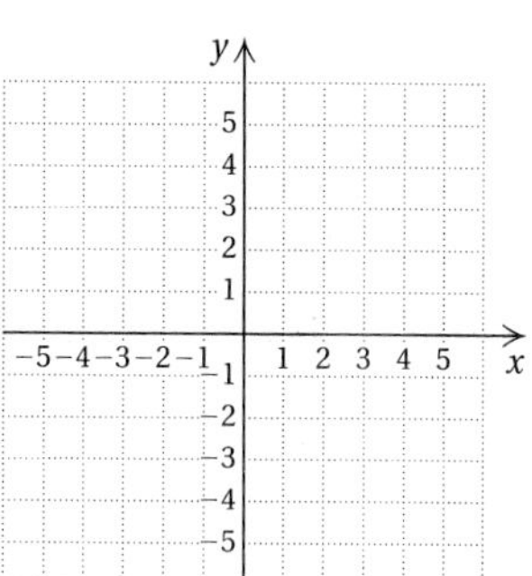

40. $3 - f(x) = 2$

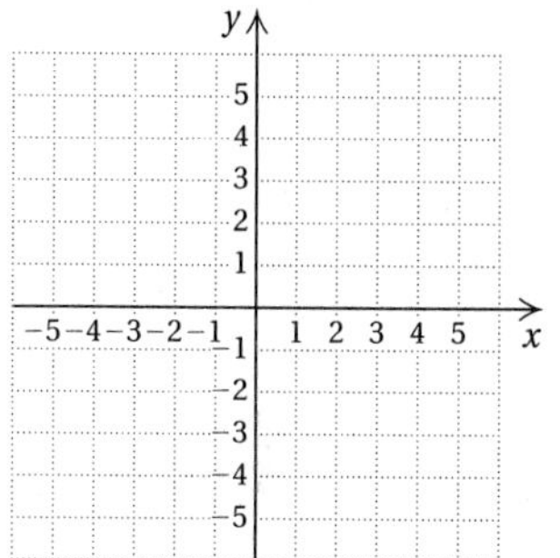

d Determine whether the graphs of the given pair of lines are parallel.

41. $x + 6 = y,$
$y - x = -2$

42. $2x - 7 = y,$
$y - 2x = 8$

43. $y + 3 = 5x,$
$3x - y = -2$

44. $y + 8 = -6x,$
$-2x + y = 5$

45. $y = 3x + 9,$
$2y = 6x - 2$

46. $y + 7x = -9,$
$-3y = 21x + 7$

47. $12x = 3,$
$-7x = 10$

48. $5y = -2,$
$\frac{3}{4}x = 16$

Determine whether the graphs of the given pair of lines are perpendicular.

49. $y = 4x - 5,$
$4y = 8 - x$

50. $2x - 5y = -3,$
$2x + 5y = 4$

51. $x + 2y = 5,$
$2x + 4y = 8$

52. $y = -x + 7,$
$y = x + 3$

53. $2x - 3y = 7,$
$2y - 3x = 10$

54. $x = y,$
$y = -x$

55. $2x = 3,$
$-3y = 6$

56. $-5y = 10,$
$y = -\frac{4}{9}$

Skill Maintenance

Write in scientific notation. [4.2c]

57. 53,000,000,000

58. 0.000047

59. 0.018

60. 99,902,000

Write in decimal notation. [4.2c]

61. 2.13×10^{-5}

62. 9.01×10^{8}

63. 2×10^{4}

64. 8.5677×10^{-2}

Factor. [5.7a]

65. $25x^2 - 10x + 1$

66. $9b^2 - c^2$

67. $77x^2 + 41x - 10$

68. $6y^2 + 33y - 18$

Synthesis

69. ◆ Under what conditions will the x- and the y-intercepts of a line be the same? What would the equation for such a line look like?

70. ◆ Explain why the slope of a vertical line is undefined but the slope of a horizontal line is 0.

71. Find an equation of a horizontal line that passes through the point $(-2, 3)$.

72. Find an equation of a vertical line that passes through the point $(-2, 3)$.

73. Find the value of a such that the graphs of $5y = ax + 5$ and $\frac{1}{4}y = \frac{1}{10}x - 1$ are parallel.

74. Find the value of k such that the graphs of $x + 7y = 70$ and $y + 3 = kx$ are perpendicular.

75. Write an equation of the line that has x-intercept $(-3, 0)$ and y-intercept $\left(0, \frac{2}{5}\right)$.

76. Find the coordinates of the point of intersection of the graphs of the equations $x = -4$ and $y = 5$.

77. Write an equation for the x-axis. Is this equation a function?

78. Write an equation for the y-axis. Is this equation a function?

79. Find the value of m in $y = mx + 3$ so that the x-intercept of its graph will be $(4, 0)$.

80. Find the value of b in $2y = -7x + 3b$ so that the y-intercept of its graph will be $(0, -13)$.

81. Match each sentence with the most appropriate graph.

a) The rate at which fluids were given intravenously was doubled after 3 hr.

b) The rate at which fluids were given intravenously was gradually reduced to 0.

c) The rate at which fluids were given intravenously remained constant for 5 hr.

d) The rate at which fluids were given intravenously was gradually increased.

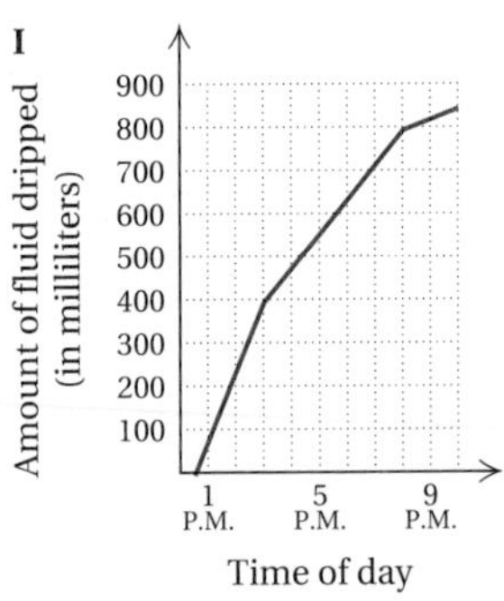

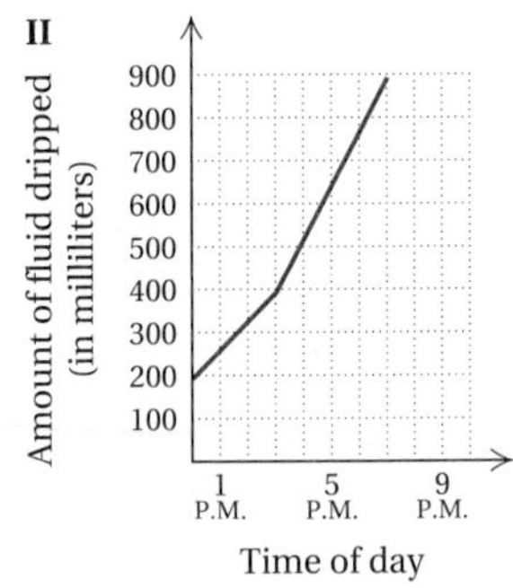

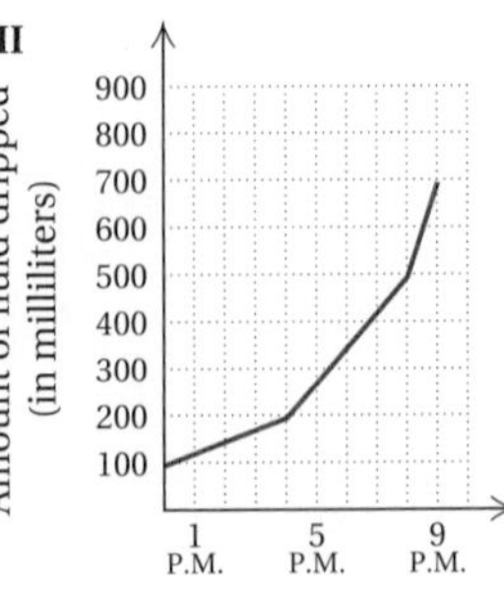

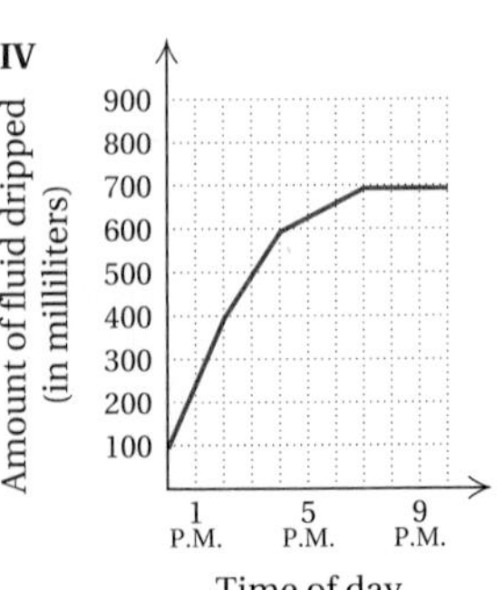

Collaborative Learning Manual

Practice graphing and identifying the graphs of linear equations.

7.5 Finding Equations of Lines

In this section, we will learn to find a linear equation for a line for which we have been given two pieces of information.

Objectives

a Find an equation of a line when the slope and the y-intercept are given.

b Find an equation of a line when the slope and a point are given.

c Find an equation of a line when two points are given.

d Given a line and a point not on the given line, find an equation of the line parallel to the line and containing the point, and find an equation of the line perpendicular to the line and containing the point.

e Solve applied problems involving linear functions.

For Extra Help

TAPE 12

MAC WIN

CD-ROM

a Finding an Equation of a Line When the Slope and the y-Intercept Are Given

If we know the slope and the y-intercept of a line, we can find an equation of the line using the slope–intercept equation $y = mx + b$.

Example 1 A line has slope -0.7 and y-intercept $(0, 13)$. Find an equation of the line.

We use the slope–intercept equation and substitute -0.7 for m and 13 for b:

$$y = mx + b$$
$$y = -0.7x + 13.$$

Do Exercise 1.

1. A line has slope 3.4 and y-intercept $(0, -8)$. Find an equation of the line.

b Finding an Equation of a Line When the Slope and a Point Are Given

If we know the slope of a line and a certain point on that line, we can find an equation of the line using the slope–intercept equation

$$y = mx + b.$$

Example 2 Find an equation of the line with slope 5 and containing the point $\left(\frac{1}{2}, -1\right)$.

The point $\left(\frac{1}{2}, -1\right)$ is on the line, so it is a solution. Thus we can substitute $\frac{1}{2}$ for x and -1 for y in $y = mx + b$. We also substitute 5 for m, the slope. Then we solve for b:

$$\begin{aligned} y &= mx + b \\ -1 &= 5 \cdot \left(\tfrac{1}{2}\right) + b \quad \text{Substituting} \\ -1 &= \tfrac{5}{2} + b \\ -1 - \tfrac{5}{2} &= b \\ -\tfrac{2}{2} - \tfrac{5}{2} &= b \\ -\tfrac{7}{2} &= b. \quad \text{Solving for } b \end{aligned}$$

We then use the equation $y = mx + b$ and substitute 5 for m and $-\frac{7}{2}$ for b:

$$y = 5x - \tfrac{7}{2}.$$

Do Exercises 2–5.

Find an equation of the line with the given slope and containing the given point.

2. $m = -5$, $(-4, 2)$

3. $m = 3$, $(1, -2)$

4. $m = 8$, $(3, 5)$

5. $m = -\dfrac{2}{3}$, $(1, 4)$

Answers on page A-30

6. Find an equation of the line containing the points (4, −3) and (1, 2).

7. Find an equation of the line containing the points (−3, −5) and (−4, 12).

Answers on page A-30

c Finding an Equation of a Line When Two Points Are Given

We can also use the slope–intercept equation to find an equation of a line when two points are given.

Example 3 Find an equation of the line containing the points (2, 3) and (−6, 1).

First, we find the slope:

$$m = \frac{3-1}{2-(-6)} = \frac{2}{8}, \text{ or } \frac{1}{4}.$$

Now we have the slope and two points. We then proceed as we did in Example 2, using either point. We choose (2, 3) and substitute 2 for x, 3 for y, and $\frac{1}{4}$ for m:

$$\begin{aligned} y &= mx + b \\ 3 &= \tfrac{1}{4} \cdot 2 + b && \text{Substituting} \\ 3 &= \tfrac{1}{2} + b \\ 3 - \tfrac{1}{2} &= \tfrac{1}{2} + b - \tfrac{1}{2} \\ \tfrac{6}{2} - \tfrac{1}{2} &= b \\ \tfrac{5}{2} &= b. && \text{Solving for } b \end{aligned}$$

Finally, we use the equation $y = mx + b$ and substitute $\frac{1}{4}$ for m and $\frac{5}{2}$ for b:

$$y = \tfrac{1}{4}x + \tfrac{5}{2}.$$

Do Exercises 6 and 7.

d Finding an Equation of a Line Parallel or Perpendicular to a Given Line Through a Point Off the Line

We can also use the method of Example 2 to find equations of lines through a point off the line parallel and perpendicular to a given line.

Example 4 Find an equation of the line containing the point (−1, 3) and parallel to the line $2x + y = 10$.

An equation parallel to the given line $2x + y = 10$ must have the same slope. To find that slope, we first find the slope–intercept equation by solving for y:

$$\begin{aligned} 2x + y &= 10 \\ y &= -2x + 10. \end{aligned}$$

Thus the new line through (−1, 3) must also have slope −2.

We then proceed as in Example 2, substituting −1 for x and 3 for y in $y = mx + b$. We also substitute −2 for m, the slope. Then we solve for b:

$$\begin{aligned} y &= mx + b \\ 3 &= -2(-1) + b && \text{Substituting} \\ 3 &= 2 + b \\ 1 &= b. && \text{Solving for } b \end{aligned}$$

We then use the equation $y = mx + b$ and substitute -2 for m and 1 for b:

$$y = -2x + 1.$$

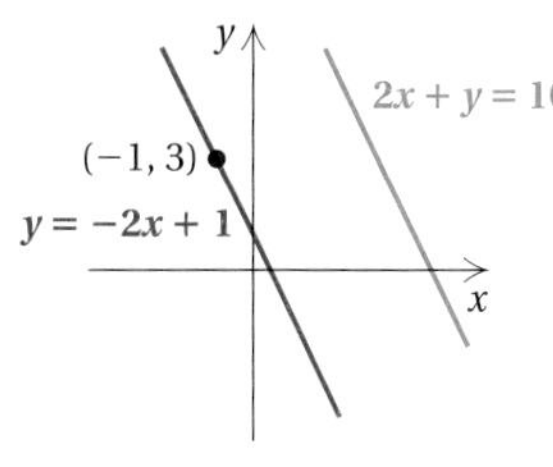

The given line $y = -2x + 10$ and the new line $y = -2x + 1$ have the same slope but different y-intercepts. Thus their graphs are parallel.

Do Exercise 8.

Example 5 Find an equation of the line containing the point $(2, -3)$ and perpendicular to the line $4y - x = 20$.

To find the slope of the given line, we first find its slope–intercept form by solving for y:

$$4y - x = 20$$
$$4y = x + 20$$
$$\frac{4y}{4} = \frac{x + 20}{4}$$
$$y = \tfrac{1}{4}x + 5.$$

We know that the slope of the perpendicular line must be the opposite of the reciprocal of $\frac{1}{4}$. Thus the new line through $(2, -3)$ must have slope -4.

We now substitute 2 for x and -3 for y in $y = mx + b$. We also substitute -4 for m, the slope. Then we solve for b:

$$y = mx + b$$
$$-3 = -4(2) + b \qquad \text{Substituting}$$
$$-3 = -8 + b$$
$$5 = b. \qquad \text{Solving for } b$$

Finally, we use the equation $y = mx + b$ and substitute -4 for m and 5 for b:

$$y = -4x + 5.$$

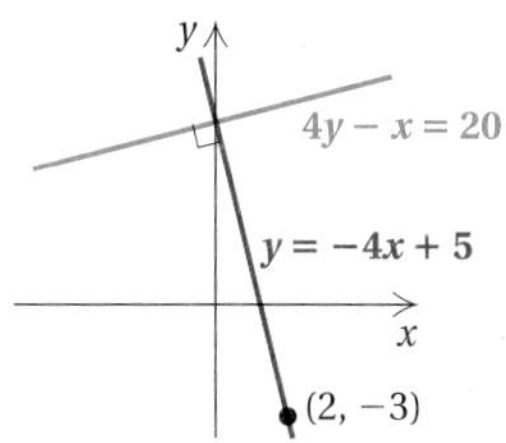

Do Exercise 9.

e Applications of Linear Functions

When the essential parts of a problem are described in mathematical language, we say that we have a **mathematical model.** We have already studied many kinds of mathematical models in this text—for example, the formulas in Section 2.6 and the functions in Section 7.1. Here we study linear functions as models.

8. Find an equation of the line containing the point $(2, -1)$ and parallel to the line $8x = 7y - 24$.

9. Find an equation of the line containing the point $(5, 4)$ and perpendicular to the line $2x - 4y = 12$.

Answers on page A-30

10. *Cable TV Service.* Clear County Cable TV Service charges a \$25 installation fee and \$20 per month for basic service.

a) Formulate a linear function for cost $C(t)$ for t months of cable TV service.

b) Graph the model.

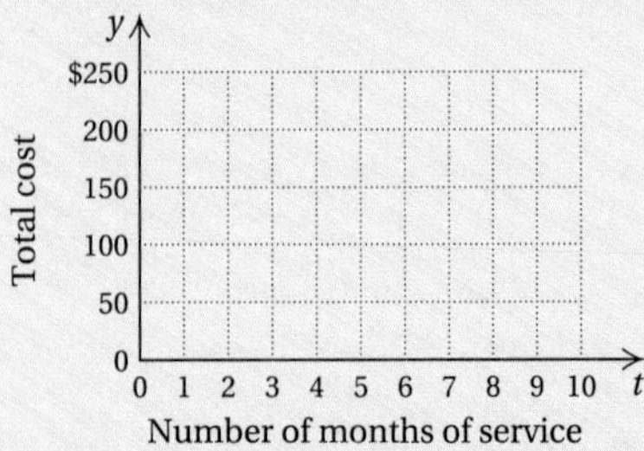

c) Use the model to determine the cost of $8\frac{1}{2}$ months of service.

Answers on page A-30

Example 6 *Cost Projections.* Cleartone Communications charges \$50 for a cellular telephone and \$40 per month for phone calls under its economy plan.

a) Formulate a linear function for total cost $C(t)$, where t is the number of months of telephone usage.

b) Graph the model.

c) Use the model to determine the cost of $3\frac{1}{2}$ months of service.

a) The problem describes a situation in which a monthly fee is charged after an initial purchase has been made. After 1 month of service, the total cost is

$$\$50 + \$40 \cdot 1 = \$90.$$

After 2 months of service, the total cost is

$$\$50 + \$40 \cdot 2 = \$130.$$

We can generalize that after t months of service, the total cost $C(t)$ is $C(t) = 50 + 40t$, where $t \geq 0$ (since there cannot be a negative number of months). Note that $C(t)$ is a way of saying that the cost C of the phone is a function of time t.

AG Algebraic–Graphical Connection

b) Before graphing, we rewrite the model in slope–intercept form:

$$C(t) = 40t + 50.$$

We note that the y-intercept is $(0, 50)$ and the slope, or rate of change, is \$40 per month. We plot $(0, 50)$, and from there we count \$40 *up* and 1 month *to the right*. This takes us to $(1, 90)$. We then draw a line through the points, calculating a third value as a check:

$$C(4) = 40 \cdot 4 + 50 = 210.$$

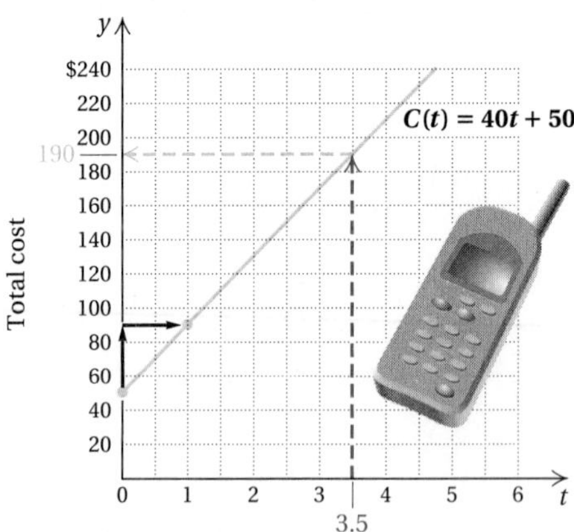

c) To find the cost for $3\frac{1}{2}$ months, we determine $C(3.5)$:

$$C(3.5) = 40(3.5) + 50 = 190.$$

Thus it would cost \$190 for $3\frac{1}{2}$ months of service.

Do Exercise 10.

Exercise Set 7.5

a Find an equation of the line having the given slope and y-intercept.

1. Slope: -8; y-intercept: $(0, 4)$

2. Slope: 5; y-intercept: $(0, -3)$

3. Slope: 2.3; y-intercept: $(0, -1)$

4. Slope: -9.1; y-intercept: $(0, 2)$

Find a linear function $f(x) = mx + b$ whose graph has the given slope and y-intercept.

5. Slope: $-\frac{7}{3}$; y-intercept: $(0, -5)$

6. Slope: $\frac{4}{5}$; y-intercept: $(0, 28)$

7. Slope: $\frac{2}{3}$; y-intercept: $\left(0, \frac{5}{8}\right)$

8. Slope: $-\frac{7}{8}$; y-intercept: $\left(0, -\frac{7}{11}\right)$

b Find an equation of the line having the given slope and containing the given point.

9. $m = 5$, $(4, 3)$

10. $m = 4$, $(5, 2)$

11. $m = -3$, $(9, 6)$

12. $m = -2$, $(2, 8)$

13. $m = 1$, $(-1, -7)$

14. $m = 3$, $(-2, -2)$

15. $m = -2$, $(8, 0)$

16. $m = -3$, $(-2, 0)$

17. $m = 0$, $(0, -7)$

18. $m = 0$, $(0, 4)$

19. $m = \frac{2}{3}$, $(1, -2)$

20. $m = -\frac{4}{5}$, $(2, 3)$

c Find an equation of the line containing the given pair of points.

21. $(1, 4)$ and $(5, 6)$

22. $(2, 5)$ and $(4, 7)$

23. $(-3, -3)$ and $(2, 2)$

24. $(-1, -1)$ and $(9, 9)$

25. $(-4, 0)$ and $(0, 7)$

26. $(0, -5)$ and $(3, 0)$

27. $(-2, -3)$ and $(-4, -6)$

28. $(-4, -7)$ and $(-2, -1)$

29. $(0, 0)$ and $(6, 1)$

30. $(0, 0)$ and $(-4, 7)$

31. $\left(\frac{1}{4}, -\frac{1}{2}\right)$ and $\left(\frac{3}{4}, 6\right)$

32. $\left(\frac{2}{3}, \frac{3}{2}\right)$ and $\left(-3, \frac{5}{6}\right)$

d Write an equation of the line containing the given point and parallel to the given line.

33. $(3, 7)$; $x + 2y = 6$

34. $(0, 3)$; $2x - y = 7$

35. $(2, -1)$; $5x - 7y = 8$

36. $(-4, -5)$; $2x + y = -3$

37. $(-6, 2)$; $3x = 9y + 2$

38. $(-7, 0)$; $2y + 5x = 6$

Write an equation of the line containing the given point and perpendicular to the given line.

39. (2, 5); $2x + y = 3$

40. (4, 1); $x - 3y = 9$

41. (3, −2); $3x + 4y = 5$

42. (−3, −5); $5x - 2y = 4$

43. (0, 9); $2x + 5y = 7$

44. (−3, −4); $6y - 3x = 2$

e Solve.

45. *Moving Costs.* Musclebound Movers charges $85 plus $40 an hour to move households across town.

a) Formulate a linear function for total cost $C(t)$ for t hours of moving.
b) Graph the model.
c) Use the model to determine the cost of $6\frac{1}{2}$ hr of moving service.

46. *Deluxe Cable TV Service.* Twin Cities Cable TV Service charges a $35 installation fee and $20 per month for basic service.

a) Formulate a linear function for total cost $C(t)$ for t months of cable TV service.
b) Graph the model.
c) Use the model to determine the cost of 9 months of service.

47. *Value of a Fax Machine.* FaxMax bought a multifunction fax machine for $750. The value $V(t)$ of the machine depreciates (declines) at a rate of $25 per month.

a) Formulate a linear function for the value $V(t)$ of the machine after t months.
b) Graph the model.
c) Use the model to determine the value of the machine after 13 months.

48. *Value of a Computer.* SendUp Graphics bought a computer for $3800. The value $V(t)$ of the computer depreciates at a rate of $50 per month.

a) Formulate a linear function for the value $V(t)$ of the computer after t months.
b) Graph the model.
c) Use the model to determine the value of the computer after $10\frac{1}{2}$ months.

Skill Maintenance

Simplify. [6.1c]

49. $\dfrac{w - t}{t - w}$

50. $\dfrac{b^2 - 1}{b - 1}$

51. $\dfrac{3x^2 + 15x - 72}{6x^2 + 18x - 240}$

52. $\dfrac{4y + 32}{y^2 - y - 72}$

Synthesis

Determine m and b in each application and explain their meaning.

53. ◆ *Cost of a Movie Ticket.* The average price $P(t)$, in dollars, of a movie ticket can be estimated by the function

$$P(t) = 0.1522t + 4.29,$$

where t is the number of years since 1990 (***Source:*** Motion Picture Association of America).

54. ◆ *Cost of a Taxi Ride.* The cost $C(d)$, in dollars, of a taxi ride in Pelham is given by

$$C(d) = 0.75d + 2,$$

where d is the number of miles traveled.

7.6 Mathematical Modeling with Linear Functions

We have considered many linear functions or models in this chapter. Many of them were derived from data gathered in real-world situations. How can we find a linear model from the data? In this section, we consider two methods in the examples and another, called **regression**, in a Calculator Spotlight.

Objectives

a Using a set of data, draw a representative graph of a linear function and make predictions from the graph.

b Using a set of data, choose two representative points, find a linear function using the two points, and make predictions from the function.

For Extra Help

TAPE 12 | MAC WIN | CD-ROM

a Data Analysis: Drawing a Linear Model

Often we will encounter data in an application. To determine whether a linear function fits the data, we graph ordered pairs of data, forming a **scatterplot**. Then we decide whether we can draw a straight line that fits the data well.

Let's look at two sets of data and their scatterplots. Does it appear that a linear function could fit either set of data?

MEDIA USAGE BY EIGHTEEN-YEAR-OLDS

Year, x	Media Usage, y (in hours per day)	Scatterplot
0. 1990	8.944	It appears that the data points can be represented or fitted by a straight line. The graph is linear.
1. 1991	8.917	
2. 1992	9.114	
3. 1993	9.040	
4. 1994	9.314	
5. 1995	9.402	
6. 1996	9.465	
7. 1997	9.552	
8. 1998	9.659	
9. 1999	9.747	

Media Usage
y
9.8
9.6
9.4
9.2
9.0
8.8
8.6
Number of hours per day
0
5
10 x
Number of years since 1990

Source: Veronis, Suhler & Associates, Inc., New York

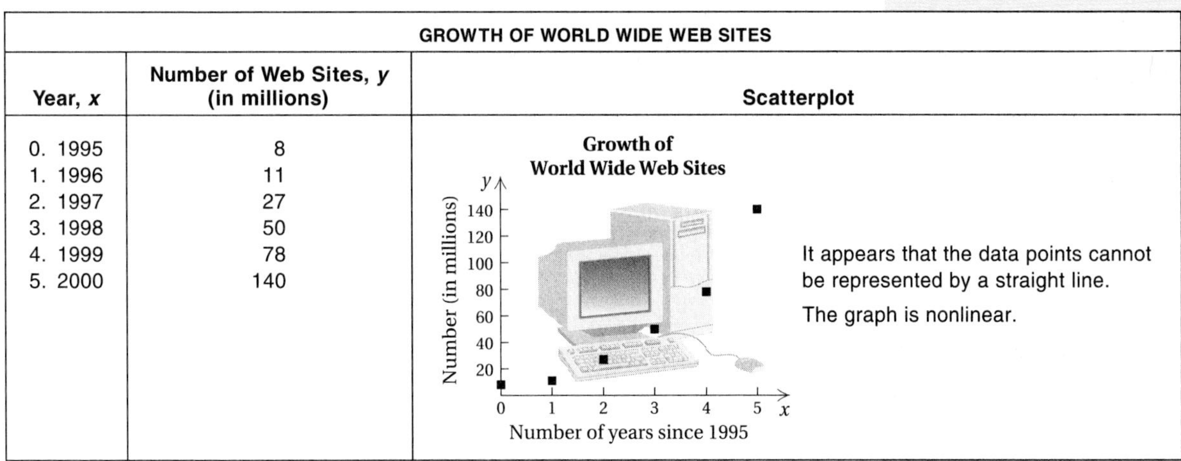

GROWTH OF WORLD WIDE WEB SITES

Year, x	Number of Web Sites, y (in millions)	Scatterplot
0. 1995	8	It appears that the data points cannot be represented by a straight line. The graph is nonlinear.
1. 1996	11	
2. 1997	27	
3. 1998	50	
4. 1999	78	
5. 2000	140	

Source: International Data Corporation, 1996

1. *Study Time and Test Scores.* A professor gathered the following data comparing study time and test scores.

a) Make a scatterplot of the data (graph the ordered pairs).

b) Use extrapolation to estimate the test score received if one has studied for 23 hr.

Study Time (in hours)	Test Score (in percent)
19	83
20	85
21	88
22	91
23	?

Answers on page A-31

Looking at the scatterplots, we see that the media usage data seem to be rising in a manner to suggest that a linear function might fit, although a "perfect" straight line cannot be drawn through the data points. However, a linear function does not seem to fit the web site data.

Let's try to use the data in the first table to make a prediction.

Example 1 *Media Usage by Eighteen-Year-Olds.* Consider the preceding data and scatterplot. Use extrapolation to estimate media usage in the year 2000 ($x = 10$).

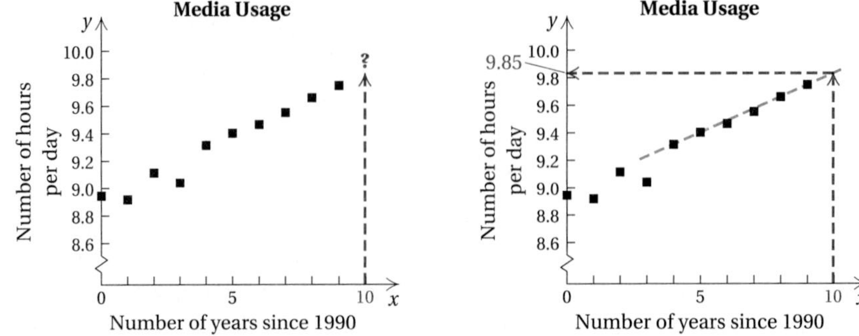

We analyze the data and note that they tend to follow a straight line past 1994 ($x = 4$). Keeping this in mind, we draw a "representative" line through the data and beyond. To estimate a value for the year 2000 ($x = 10$), we draw a vertical line up from 10 to the representative line. We then move to the left and read a value from the vertical axis. That value is about 9.85.

When we estimate to find a "go-beyond value," as we did in Example 1, we are using a process called **extrapolation**. Note that the process of drawing a representative line is arbitrary. Thus the answer we provide might differ from the one you find. This is not important in our work here. The idea is to use a line to go beyond the data to make an estimated prediction.

Do Exercise 1.

b Data Analysis: Using Two Points to Find a Linear Model

In the following example, we refine the process of finding a linear model. First, we choose two points and find an equation for the linear function through these points. Then we use the equation to make a prediction.

Example 2 *Media Usage by Eighteen-Year-Olds.* Choose two points from the data (this can vary) regarding media usage.

a) Use the two points to find a linear function that fits the data.

b) Use the function to predict media usage in the year 2000 ($x = 10$).

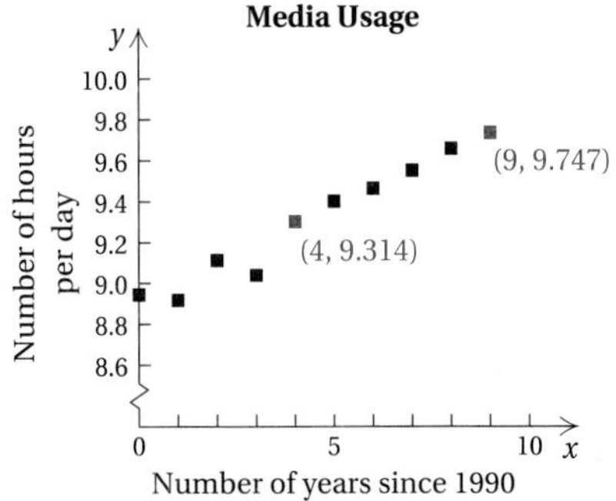

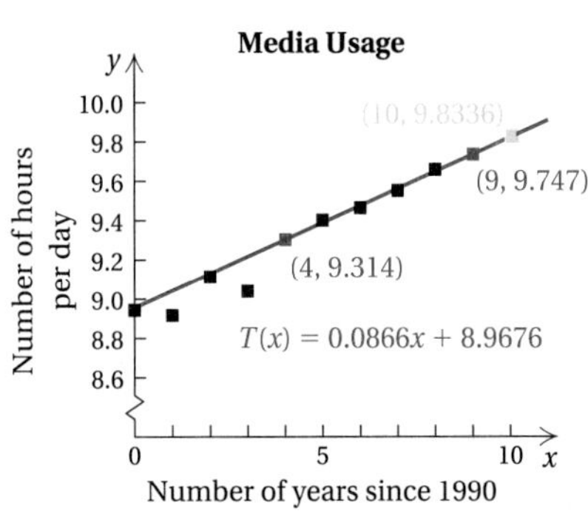

a) We note in the data chart on p. 209 that media usage in 1994 ($x = 4$) was about 9.314 hr per day, while in 1999 it is projected to be about 9.747. We let $T(x) =$ the number of hours per day. We determine a slope–intercept equation $T(x) = mx + b$ using the procedure of Section 7.5(c) with the data points (4, 9.314) and (9, 9.747).

We first find the slope, or rate of change:

$$m = \frac{9.747 \text{ hr/day} - 9.314 \text{ hr/day}}{9 \text{ yr} - 4 \text{ yr}} = \frac{0.433 \text{ hr/day}}{5 \text{ yr}} = 0.0866 \text{ hr/day per year.}$$

Now we know the slope and two points. We proceed to find b using the slope and either one of the points. We choose to use (4, 9.314) and substitute 4 for x, 9.314 for y, and 0.0866 for m:

$$y = mx + b$$
$$9.314 = 0.0866(4) + b \quad \text{Substituting}$$
$$9.314 = 0.3464 + b$$
$$8.9676 = b. \quad \text{Solving for } b$$

Next, we use the equation $y = mx + b$, substituting $T(x)$ for y, 0.0866 for m, and 8.9676 for b. This gives us a linear function:

$$T(x) = 0.0866x + 8.9676.$$

b) To predict media usage in the year 2000 ($x = 10$), we find $T(10)$:

$$T(10) = 0.0866(10) + 8.9676 = 9.8336.$$

Thus media usage in 2000 will be about 9.83 hr per day.

Note that this prediction differs slightly from the number 9.85 found in Example 1. Such is the nature of the process of making predictions. How much you rely on such information depends on the type of application. Nevertheless, modeling applications have extensive use in areas such as life science, physical science, business, and social science. To see this, look through an economics book or a medical journal.

Do Exercise 2.

2. *Media Usage.* Repeat Example 2 using the data points (5, 9.402) and (9, 9.747). Compare the results with those of Example 2.

Answers on page A-31

Linear Regression

Another procedure, called **linear regression,** can be used to fit a linear function to data. The strength of this procedure is that it makes use of ***all*** the data, not just two points. Most graphing calculators have a REGRESSION feature, which we discuss in the following Calculator Spotlight. The mathematical basis for regression belongs to a course in statistics and/or calculus.

Calculator Spotlight

Linear Regression: Fitting a Linear Function to Data. We now consider **linear regression,** a procedure that can be used to fit a linear function to a set of data. Although the complete basis for this method belongs to a statistics and/or calculus course, we consider it here because we can carry out the procedure easily using technology. The grapher gives us the powerful capability to find linear models and make predictions.

Example Consider the data (on p. 471) on media usage by eighteen-year-olds. If we look at the scatterplot, it appears that the data points can be modeled by a linear function.

a) Fit a regression line to the data using the REGRESSION feature on a grapher.

b) Make a scatterplot of the data. Then graph the regression line with the scatterplot.

c) Use the linear model to predict media usage in the year 2000 ($x = 10$).

We proceed as follows.

a) We can fit a linear function to the data using linear regression. We show this using a TI-83 grapher.

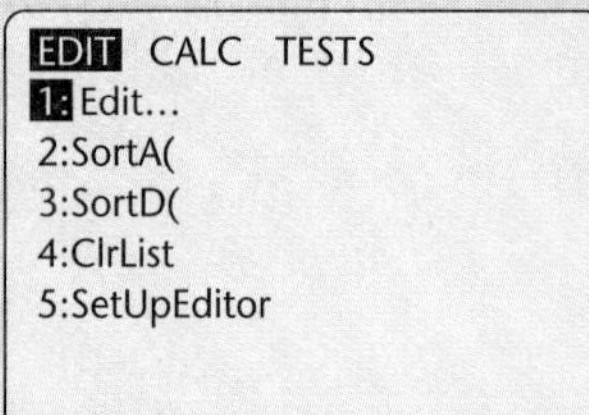

1. Press [STAT] and then [1] or [ENTER] to enter the data.
2. If there are any data in column one, clear the numbers by pressing the arrow keys until L_1 is highlighted. Then press [CLEAR] and [▽]. Repeat for each column in which data appear, using [▷] or [◁] to move between columns.
3. Go back to the first column and enter the years by typing the number and then the down arrow key. Then move to the second column and enter the hours of media usage in a similar manner. You will not be able to see all the data points at once on the display.

L1	L2	L3 1
0		
1		
2		
3		
4		
5		

L1 (7) = 6

L1	L2	L3 2
3	9.04	
4	9.314	
5	9.402	
6	9.465	
7	9.552	
8	9.659	
9		

L2 (10) = 9.747

4. Equations are calculated using the STAT-CALC menu. Press [STAT] [▷] [4] to choose LinReg($ax + b$). Then press [VARS] [▷] [1] [1] to copy Y_1 on the screen, as shown below. The regression equation will now automatically be copied as Y_1 on the Y= screen. If an equation is currently entered as Y_1, it must be cleared *before* the regression equation can be found.

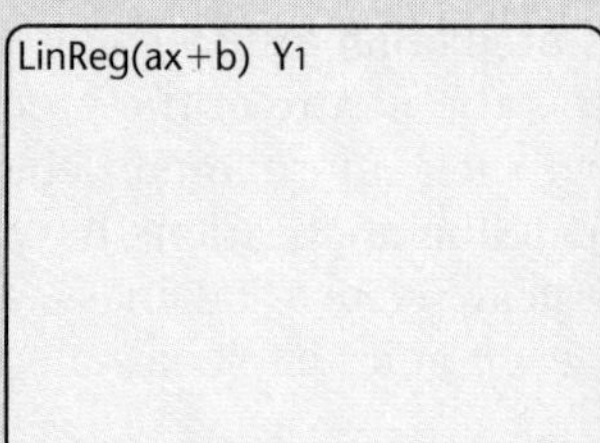

(*continued*)

5. Next, press ENTER. We now have a linear regression equation that fits the data.

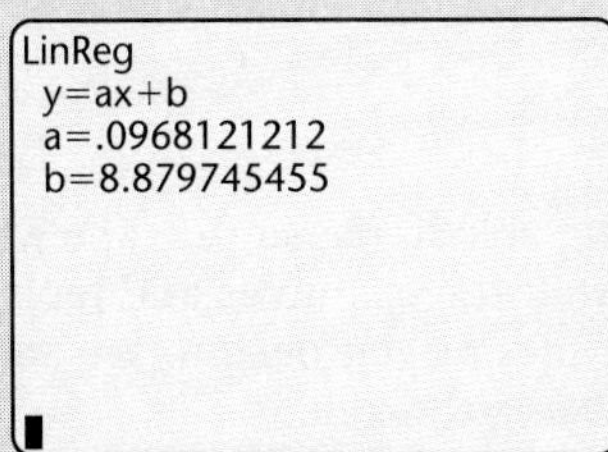

Rounding the coefficients to four decimal places, we find that the regression line is $y = 0.0968x + 8.8797$.

b) To make a scatterplot, or *xy*-graph, of the data, we can use the STAT PLOT feature. (If the scatterplot had not been shown on p. 471, we would begin this example by creating a scatterplot to determine whether it appears as though a linear function could fit the data points.)

1. Press 2nd STAT PLOT. Turn on Plot 1 by pressing ENTER twice. Note that the highlighting indicates that the data in the first list L1 are related to the independent variable, *x*, and that the data in the second list L2 are related to the dependent variable, y_1.

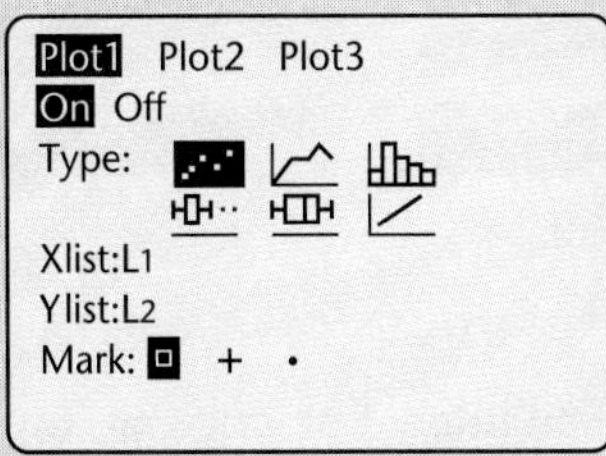

2. Press the Y= screen. You will see that the regression function has already been entered. Using the ZOOMSTAT feature—that is, pressing ZOOM 9, we get the graph of the data points and the regression function.

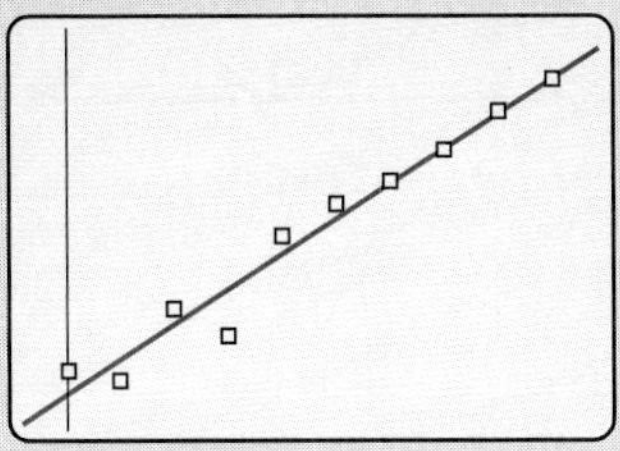

c) To predict media usage in the year 2000, we evaluate Y_1 for $x = 10$. Press VARS ▷ 1 1. We get Y_1 on the screen. Then press (1 0) ENTER. This gives $Y_1(10) = 9.847866667$, as shown below.

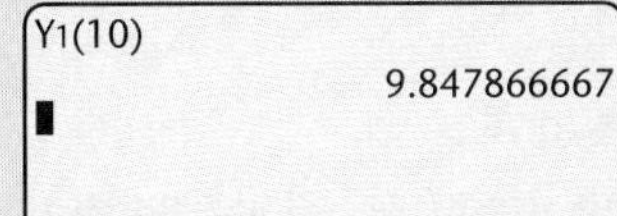

Thus we predict media usage in the year 2000 to be about 9.85 hr/day. Note this agrees exactly with the estimate found by "eyeballing" the data in Example 1 and is close to that found with a linear function formed by using only two data points in Example 2. This may not always be the case.

Exercises

1. Use the data on study time and test scores in Margin Exercise 1.
   a) Fit a regression line to the data using the REGRESSION feature on a grapher.
   b) Make a scatterplot of the data. Then graph the regression line with the scatterplot.
   c) Use the linear model to predict the test score received if one has studied for 23 hr.
   d) Compare your answers with those found in Margin Exercise 1.

Improving Your Math Study Skills

Classwork: During and After Class

During Class

Asking Questions

Many students are afraid to ask questions in class. You will find that most instructors are not only willing to answer questions during class, but often encourage students to ask questions. In fact, some instructors would like more questions than are offered. Probably your question is one that other students in the class might have been afraid to ask!

Wait for an appropriate time to ask questions. Some instructors will pause to ask the class if they have questions. Use this opportunity to get clarification on any concept you do not understand.

After Class

Restudy Examples and Class Notes

As soon as possible after class, find some time to go over your notes. Read the appropriate sections from the textbook and try to correlate the text with your class notes. You may also want to restudy the examples in the textbook for added comprehension.

Often students make the mistake of doing the homework exercises without reading their notes or textbook. This is not a good idea, since you may lose the opportunity for a complete understanding of the concepts. Simply being able to work the exercises does not ensure that you know the material well enough to work problems on a test.

Videotapes

If you can find the time, visit the library, math lab, or media center to view the videotapes on the textbook. Look on the first page of each section in the textbook for the appropriate tape reference.

The videotapes provide detailed explanations of each objective and they may give you a different presentation than the one offered by your instructor. Being able to pause the tape while you take notes or work the examples and replaying the tape as many times as you need to are additional advantages to using the videos.

Also, consider studying the special tapes *Math Problem Solving in the Real World* and *Math Study Skills* prepared by the author. If these are not available in the media center, contact your instructor.

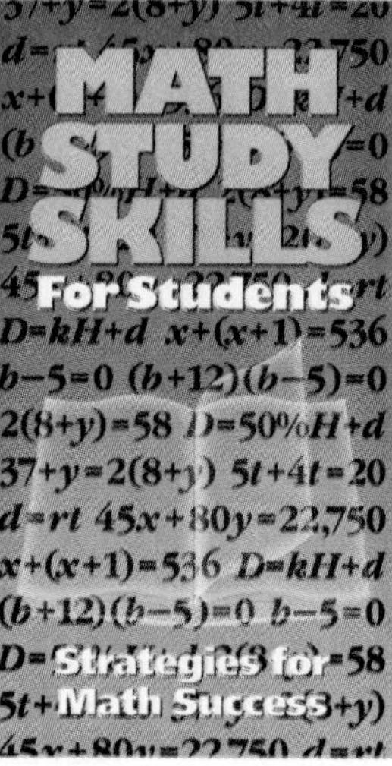

Software

If you would like additional practice on any section of the textbook, you can use the accompanying Interact Math Tutorial Software. This software can generate many different versions of the basic odd-numbered exercises for added practice. You can also ask the software to work out each problem step by step.

Ask your instructor about the availability of this software.

Exercise Set 7.6

1. *Net Sales of the Gap.*

Years, x (since 1990)	Net Sales, S (in billions)
0. 1990	$1.9
1. 1991	2.5
2. 1992	3.0
3. 1993	3.3
4. 1994	3.7
5. 1995	3.4
6. 1996	5.3

Source: The Gap, Inc.

a) Make a scatterplot of the data.
b) Draw a representative graph of a linear function.
c) Predict net sales for The Gap in 1999 and 2001.

2. *Earnings Per Share of Toys "R" Us.*

Years, x (since 1992)	Earnings per Share, E
0. 1992	$1.15
1. 1993	1.47
2. 1994	1.63
3. 1995	1.85
4. 1996	1.73

Source: Toys Я Us

a) Make a scatterplot of the data.
b) Draw a representative graph of a linear function.
c) Predict earnings per share for Toys "R" Us in 1999 and 2001.

For each set of data, **(a)** make a scatterplot of the data, **(b)** draw a representative graph of a linear function, and **(c)** predict the number of home runs in 1998 and 2000.

3. *Home Runs Per Game in the National League.*

Years, x (since 1992)	Average Number of Home Runs per Game, H
0. 1992	1.30
1. 1993	1.72
2. 1994	1.91
3. 1995	1.90
4. 1996	2.17
5. 1997	1.83

Source: Major League Baseball

4. *Home Runs Per Game in the American League.*

Years, x (since 1992)	Average Number of Home Runs per Game, H
0. 1992	1.57
1. 1993	1.83
2. 1994	2.23
3. 1995	2.14
4. 1996	2.48
5. 1997	2.09

Source: Major League Baseball

b Solve.

5. *Bird Watching.* Bird watching has been on the increase in recent years, as shown by the following graph.

The American Birding Association

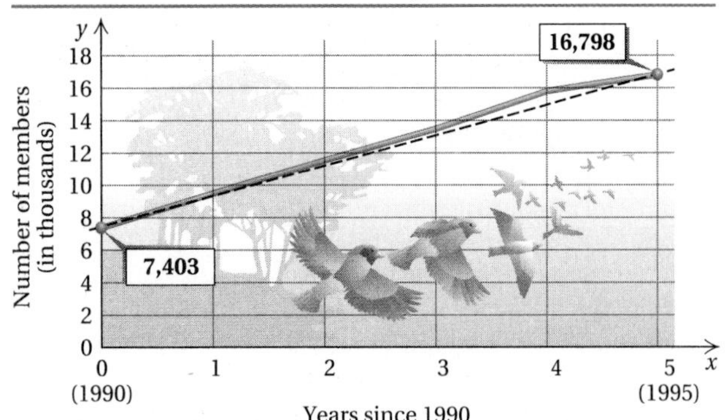

Source: American Birding Association

a) Use the two points (0, 7403) and (5, 16,798) to find a linear function that fits the data.
b) Graph the function.
c) Use the function to predict the number of members of the American Birding Association in 2000 and 2010.

6. *The National Debt.* The national debt has been increasing for several years, as shown by the following graph.

Growth of the National Debt

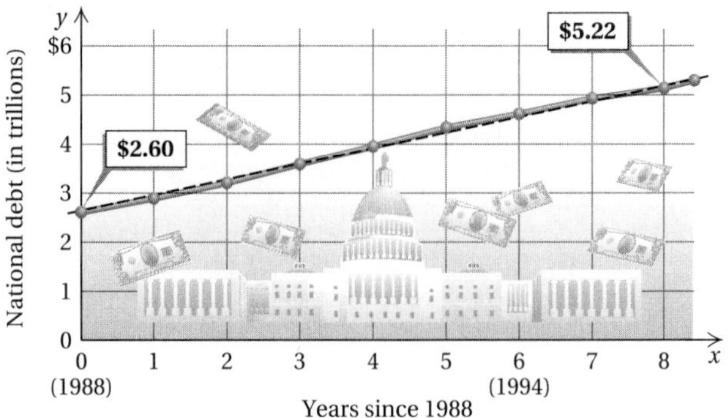

Source: U.S. Department of the Treasury

a) Use the two points (0, \$2.6) and (8, \$5.22) to find a linear function that fits the data.
b) Graph the function.
c) Use the function to predict the national debt in 2000 and 2010.

7. *Home Runs Per Game in the National League.* Use the data from Exercise 3.

a) Choose two points and find a linear function that fits the data. Answers may vary depending on the two points chosen.
b) Graph the function on the scatterplot.
c) Use the function to predict the average number of home runs in 1999 and 2002.

8. *Home Runs Per Game in the American League.* Use the data from Exercise 4.

a) Choose two points and find a linear function that fits the data. Answers may vary.
b) Graph the function on the scatterplot.
c) Use the function to predict the average number of home runs in 1999 and 2002.

9. *Net Sales of The Gap.* Use the data from Exercise 1.

a) Choose two points and find a linear function that fits the data. Answers may vary.
b) Graph the function on the scatterplot.
c) Use the function to predict net sales of The Gap in 2000 and 2005.

10. *Earnings Per Share of Toys "R" Us.* Use the data from Exercise 2.

a) Choose two points and find a linear function that fits the data. Answers may vary.
b) Graph the function on the scatterplot.
c) Use the function to predict earnings per share of Toys "R" Us in 2000 and 2010.

11. *Study Time vs. Grades.* A mathematics instructor asked her students to keep track of how much time each spent studying the chapter on percent notation in her basic mathematics course. She collected the information, together with test scores from that chapter's test, in the table below.

Study Time, x (in hours)	Test Grade, y (in percent)
9	74
11	94
13	81
15	86
16	87
17	81
21	87
23	92

a) Choose two points from the data (this can vary) and find a linear function that fits the data.
b) Use the function to predict the test scores of someone who has studied for 18 hr; for 25 hr.

12. *Maximum Heart Rate.* A person exercising should not exceed a maximum heart rate, which depends on his or her gender, age, and resting heart rate. The following table shows data relating resting heart rate and maximum heart rate for a 20-yr-old woman.

Resting Heart Rate, r (in beats per minute)	Maximum Heart Rate, M (in beats per minute)
50	170
60	172
70	174
80	176

Source: American Heart Association

a) Choose two points from the data and find a linear function that fits the data.
b) Use the function to predict the maximum heart rate of a woman whose resting heart rate is 62; whose resting heart rate is 75.

Skill Maintenance

Solve. [2.4a]

13. The price of a radio, including 5% sales tax, is \$36.75. Find the price of the radio before the tax was added.

14. A basketball team increases its final score by 7 points in each of three games. The total of the three scores was 228. What was the score in the last game?

Solve for the indicated letter. [2.6a]

15. $Ax + By = C$, for y

16. $3x - 7y = 12$, for y

17. $P = \frac{2}{3}q - y$, for q

18. $A = \frac{1}{2}bh$, for b

Divide. [4.8b]

19. $(a^2 + 2a - 63) \div (a - 7)$

20. $(x^2 + 4x + 1) \div (x - 7)$

21. $(w^2 - 3w + 8) \div (w - 2)$

22. $(x^5 - 2x^2 + x - 2) \div (x^2 + 2)$

Synthesis

Determine whether the graph might be modeled by a linear function. Give reasons why or why not.

23. ◈

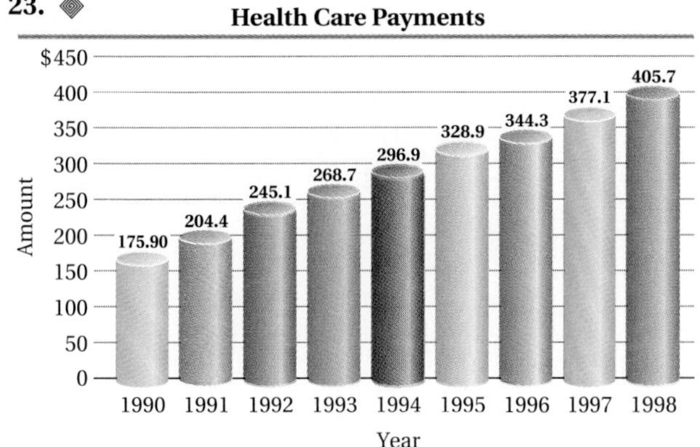

Source: Healthcare Financing Administration, U.S. Department of Health and Human Resources

24. ◈

Government's exit from milk price support business has helped create more volatile farm-paid prices for milk.

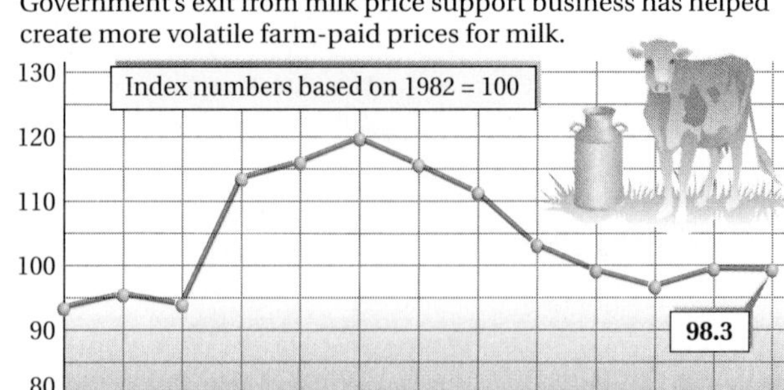

Source: U.S. Department of Agriculture

Use the REGRESSION feature on a grapher for Exercises 25–28.

25. *Home Runs Per Game in the National League.*

a) Fit a regression line to the data in Exercise 3.
b) Make a scatterplot of the data. Then graph the regression line with the scatterplot.
c) Use the linear model to predict the average number of home runs per game in 2002 ($x = 10$).

26. *Net Sales of The Gap.*

a) Fit a regression line to the data in Exercise 1.
b) Make a scatterplot of the data. Then graph the regression line with the scatterplot.
c) Use the linear model to predict the net sales of The Gap in 2001 ($x = 11$).

The following table contains data relating infant mortality rate, life expectancy, and average daily caloric intake. Use it for Exercises 27 and 28.

Country	Average Daily Caloric Intake, *c*, in 1992	Life Expectancy, *L*, Projected in 2000	Infant Mortality Rate, *M* (per 1000 births)
Argentina	2880	72.3	26.1
Bolivia	2100	62.0	60.2
Canada	3482	80.0	5.5
Dominican Republic	2359	70.4	40.8
Germany	3443	76.7	22.2
Haiti	1707	50.2	98.4
Mexico	3181	75.0	20.7
United States	3671	76.3	6.2

Source: *Universal Almanac*; *Statistical Abstract of the United States*

27. *Infant Mortality Rate as a Function of Daily Caloric Intake.*

a) Fit a regression line, $M(c) = mc + b$, to the data using the REGRESSION feature on a grapher.
b) Make a scatterplot of the data. Then graph the regression line with the scatterplot.
c) Use the linear model to estimate the infant mortality rate for Australia, which has an average daily caloric intake of 3216.
d) Use the linear model to estimate the infant mortality rate for Venezuela, which has an average daily caloric intake of 2622.

28. *Life Expectancy as a Function of Daily Caloric Intake.*

a) Fit a regression line, $L(c) = mc + b$, to the data using the REGRESSION feature on a grapher.
b) Make a scatterplot of the data. Then graph the regression line with the scatterplot.
c) Use the linear model to estimate life expectancy for Australia, which has an average daily caloric intake of 3216.
d) Use the linear model to estimate life expectancy for Venezuela, which has an average daily caloric intake of 2622.

Make predictions from a set of data.

Summary and Review Exercises: Chapter 7

Important Properties and Formulas

Slope $= m = \dfrac{y_2 - y_1}{x_2 - x_1}$, or $\dfrac{y_1 - y_2}{x_1 - x_2}$

Equations of Lines and Linear Functions

Horizontal Line: $f(x) = b$, or $y = b$; slope $= 0$

Vertical Line: $x = a$, slope is undefined.

Slope–Intercept Equation: $f(x) = mx + b$, or $y = mx + b$

Parallel Lines: $m_1 = m_2$, $b_1 \neq b_2$

Perpendicular Lines: $m_1 = -\dfrac{1}{m_2}$

The objectives to be tested in addition to the material in this chapter are [4.2c], [4.8b], [5.6a], and [6.1c].

Determine whether the correspondence is a function. [7.1a]

1.

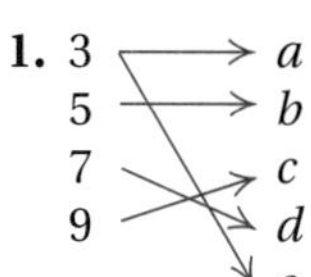

2. 1, 2, 3, 4, 5 → a, b, c, d

Find the function values. [7.1b]

3. $g(x) = -2x + 5$; $g(0)$ and $g(-1)$

4. $f(x) = 3x^2 - 2x + 7$; $f(0)$ and $f(-1)$

5. The function described by $C(t) = 645t + 9800$ can be used to estimate the average cost of tuition at a state university t years after 1997. Estimate the average cost of tuition at a state university in 2010. [7.1b]

Determine whether each of the following is the graph of a function. [7.1d]

6.

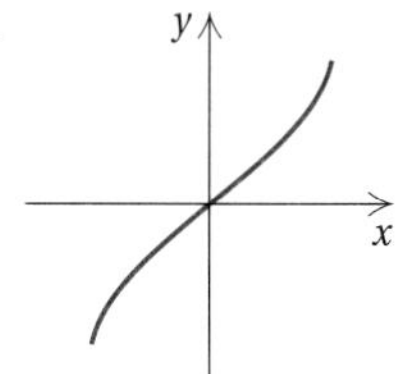

7.

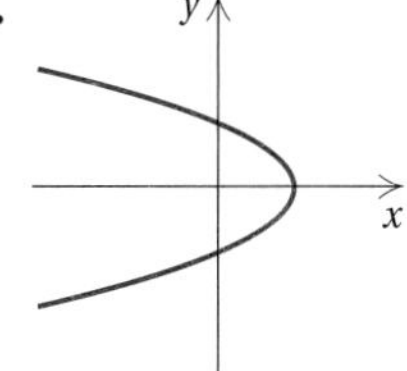

8. For the following graph of a function f, determine **(a)** $f(2)$; **(b)** the domain; **(c)** all x-values such that $f(x) = 2$; and **(d)** the range. [7.2a]

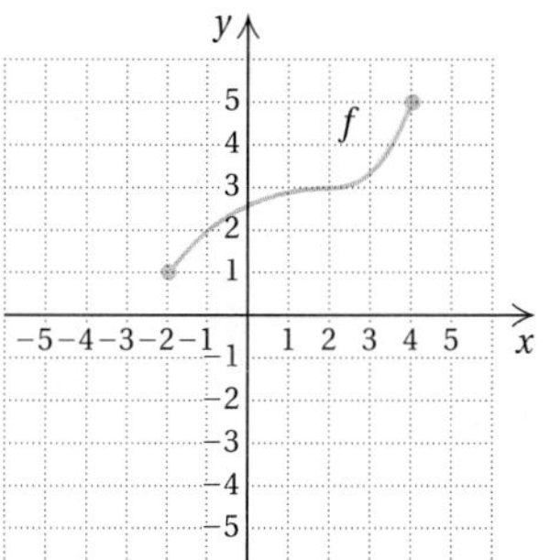

Find the domain. [7.2a]

9. $f(x) = \dfrac{5}{x - 4}$

10. $g(x) = x - x^2$

Find the slope and the y-intercept. [7.3a, b]

11. $y = -3x + 2$

12. $4y + 2x = 8$

13. Find the slope, if it exists, of the line containing the points (13, 7) and (10, −4). [7.3b]

Find the intercepts. Then graph the equation. [7.4a]

14. $2y + x = 4$

15. $2y = 6 - 3x$

Graph using the slope and the y-intercept. [7.4b]

16. $g(x) = -\frac{2}{3}x - 4$

17. $f(x) = \frac{5}{2}x + 3$

Graph. [7.4c]

18. $x = -3$

19. $f(x) = 4$

Determine whether the graphs of the given pair of lines are parallel or perpendicular. [7.4d]

20. $y + 5 = -x,$
$x - y = 2$

21. $3x - 5 = 7y,$
$7y - 3x = 7$

22. $4y + x = 3,$
$2x + 8y = 5$

23. $x = 4,$
$y = -3$

24. Find a linear function $f(x) = mx + b$ whose graph has the given slope and y-intercept: [7.5a]

slope: 4.7; y-intercept: $(0, -23)$.

25. Find an equation of the line having the given slope and containing the given point: [7.5b]

$m = -3$; $(3, -5)$.

26. Find an equation of the line containing the given pair of points: [7.5c]

$(-2, 3)$, and $(-4, 6)$.

27. Find an equation of the line containing the given point and parallel to the given line: [7.5d]

$(14, -1)$; $5x + 7y = 8$.

28. Find an equation of the line containing the given point and perpendicular to the given line: [7.5d]

$(5, 2)$; $3x + y = 5$.

Use the following table of data for Exercise 29.

Year, x, since 1990	Health Care Payments, H (in billions)
0. 1990	\$175.9
1. 1991	204.4
2. 1992	245.1
3. 1993	268.7
4. 1994	328.9
5. 1995	344.3
6. 1996	377.1
7. 1997	405.7

Source: Adams Media Research, Carmel, California

29. a) Use the two points (0, 175.9) and (7, 405.7) to find a linear function that fits the data.
b) Use the function to predict health care payments in 2000 and 2005. [7.6b]

Skill Maintenance

30. Convert 3.3×10^{-8} to decimal notation. [4.2c]

31. Convert 13,700,000,000 to scientific notation. [4.2c]

Divide. [4.8b]

32. $(6x^2 + 13x - 5) \div (2x + 5)$

33. $(r^2 - 3r + 6) \div (r - 4)$

Factor. [5.6a]

34. $81y^3 - 3000$

35. $9 + 72z^3$

36. Simplify: $\dfrac{2a^2 - 5a - 12}{6a^2 + 7a - 3}$. [6.1c]

Synthesis

37. ◆ Explain the usefulness of the slope concept when describing a line. [7.3b, c], [7.4b], [7.5a, b, c, d]

38. ◆ Explain the meaning of the notation $f(x)$ when considering a function in as many ways as you can. [7.1b]

Use the REGRESSION feature on a grapher for Exercise 39. [7.6b]

39. a) Fit a regression line, $H(x) = mx + b$, to the data in Exercise 29.
b) Make a scatterplot of the data. Then graph the regression line with the scatterplot.
c) Use the linear model to predict health care payments in 2005.

Test: Chapter 7

Answers

Graph.

1. $f(x) = -\frac{3}{5}x$

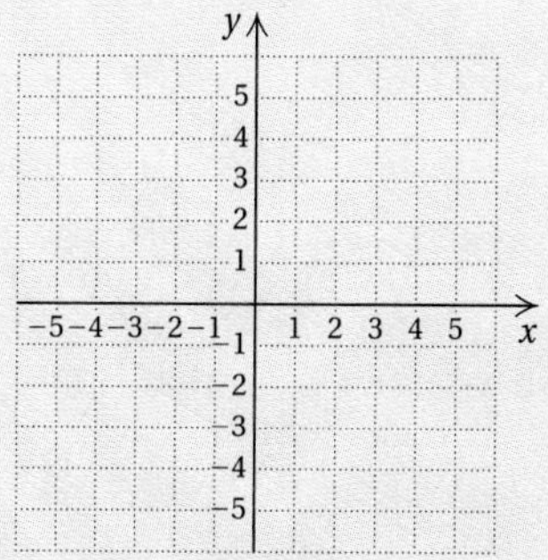

2. $g(x) = 2 - |x|$

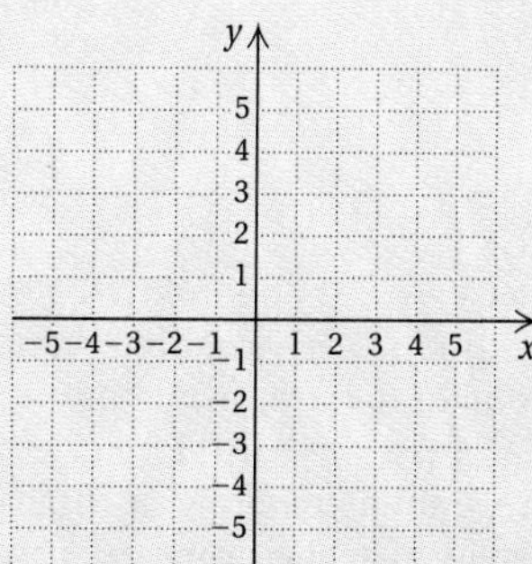

3. **Median Age of Cars.** The function

$$A(t) = 0.233t + 5.87$$

can be used to estimate the median age of cars in the United States t years after 1990 (***Source:*** The Polk Co.). (In this context, we mean that if the median age of cars is 3 yr, then half the cars are older than 3 yr and half are younger.)

a) Find the median age of cars in 2002.

b) In what year will the median age of cars be 7.734 yr?

Determine whether the correspondence is a function.

4. cat → fish; fish → worm; dog → cat; tiger → dog; teacher → student

5. Lake Placid → 1980, 1932; Oslo → 1952; Squaw Valley → 1960; Innsbruck → 1976

Find the function values.

6. $f(x) = -3x - 4$; $f(0)$ and $f(-2)$

7. $g(x) = x^2 + 7$; $g(0)$ and $g(-1)$

1. ______

2. ______

3. a) ______

b) ______

4. ______

5. ______

6. ______

7. ______

Answers

8. ______

9. ______

10. a) ______

b) ______

11. a) ______

b) ______

c) ______

d) ______

12. ______

13. ______

14. ______

15. ______

16. ______

17. ______

Determine whether each of the following is the graph of a function.

8.

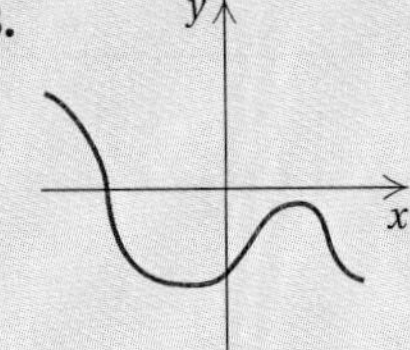

9.

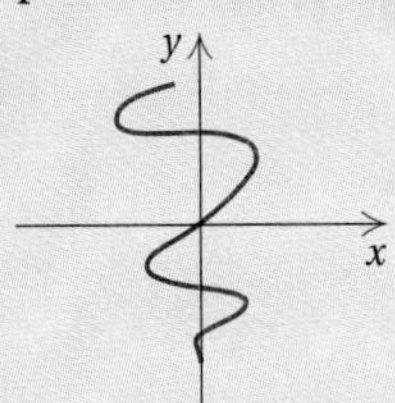

10. *Movie Revenue.* The following graph approximates the weekly revenue, in millions of dollars, from the recent movie *Jurassic Park—The Lost World.* The revenue is given as a function of the week. Use the graph to answer the following.

a) What was the movie revenue for week 1?

b) What was the movie revenue for week 5?

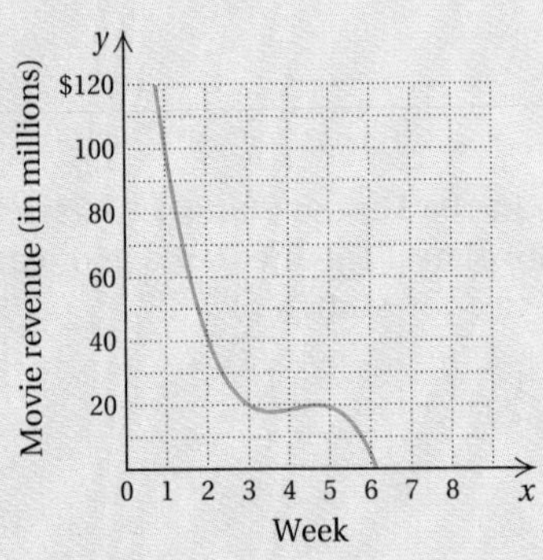

Source: Exhibitor Relations Co., Inc.

11. For the following graph of function f, determine **(a)** $f(2)$; **(b)** the domain; **(c)** all x-values such that $f(x) = 2$; and **(d)** the range.

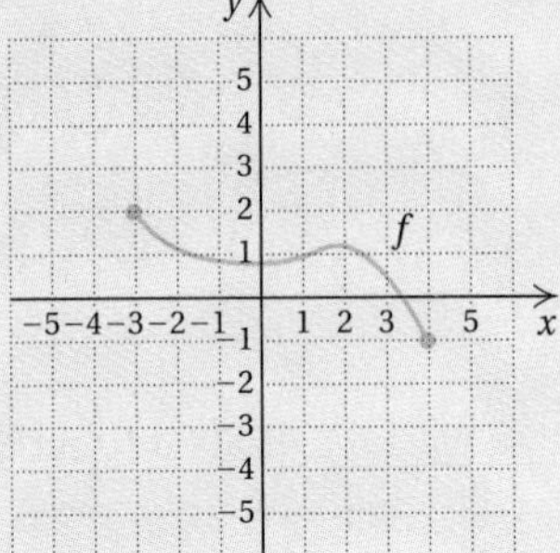

Find the domain.

12. $g(x) = 5 - x^2$

13. $f(x) = \dfrac{8}{2x + 3}$

Find the slope and the y-intercept.

14. $f(x) = -\frac{3}{5}x + 12$

15. $-5y - 2x = 7$

Find the slope, if it exists, of the line containing the following points.

16. $(-2, -2)$ and $(6, 3)$

17. $(-3.1, 5.2)$ and $(-4.4, 5.2)$

18. Find the slope, or rate of change, of the graph at right.

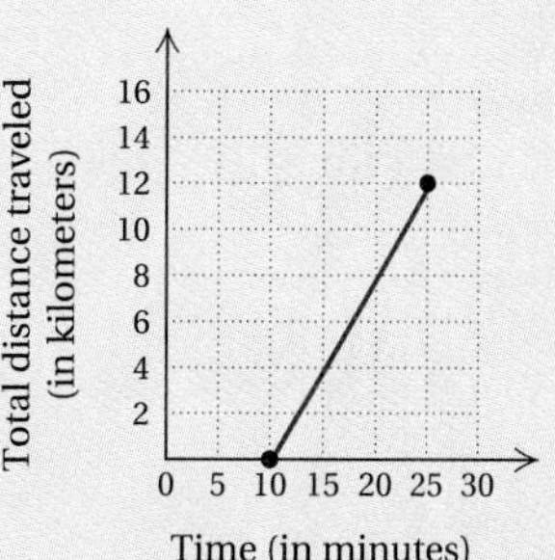

19. Find the intercepts. Then graph the equation.

$2x + 3y = 6$

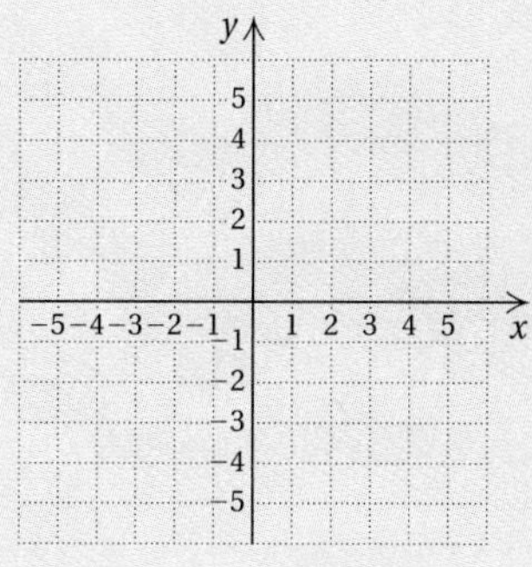

20. Graph using the slope and the y-intercept:

$$f(x) = -\frac{2}{3}x - 1.$$

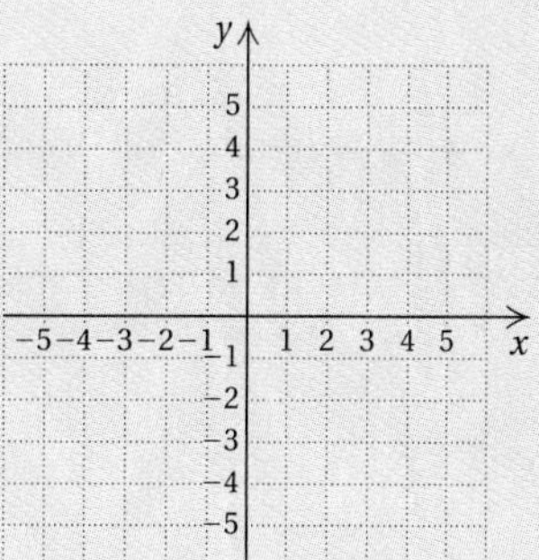

Graph.

21. $y = f(x) = -3$

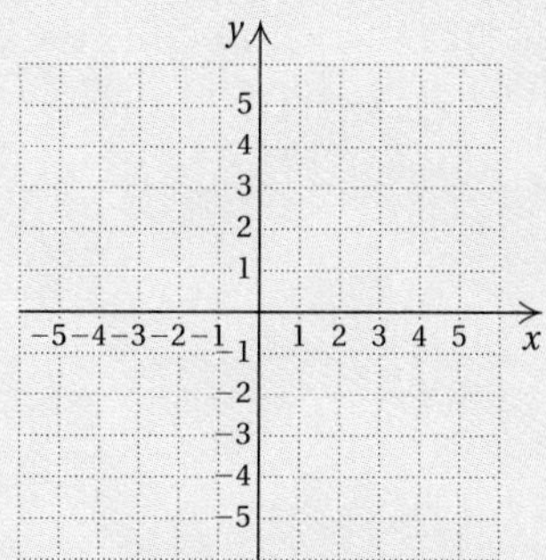

22. $2x = -4$

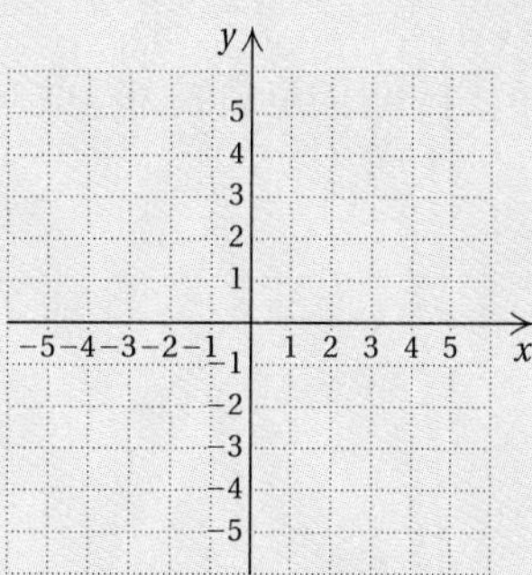

Determine whether the graphs of the given pair of lines are parallel or perpendicular.

23. $4y + 2 = 3x,$
$-3x + 4y = -12$

24. $y = -2x + 5,$
$2y - x = 6$

25. Find an equation of the line that has the given characteristics:

slope: -3; y-intercept: $(0, 4.8)$.

26. Find a linear function $f(x) = mx + b$ whose graph has the given slope and y-intercept:

slope: 5.2; y-intercept: $\left(0, -\frac{5}{8}\right)$.

27. Find an equation of the line having the given slope and containing the given point:

$m = -4$; $(1, -2)$.

28. Find an equation of the line containing the given pair of points:

$(4, -6)$ and $(-10, 15)$.

Answers

18. ______

19. ______

20. ______

21. ______

22. ______

23. ______

24. ______

25. ______

26. ______

27. ______

28. ______

Answers

29. ____________

30. ____________

31. a) ____________

b) ____________

32. ____________

33. ____________

34. ____________

35. ____________

36. ____________

37. ____________

29. Find an equation of the line containing the given point and parallel to the given line:

$(4, -1)$; $x - 2y = 5$.

30. Find an equation of the line containing the given point and perpendicular to the given line:

$(2, 5)$; $x + 3y = 2$.

Sales of Books on Tape. Sales of books on audiotape have increased in recent years. Use the following table of data for Exercise 31.

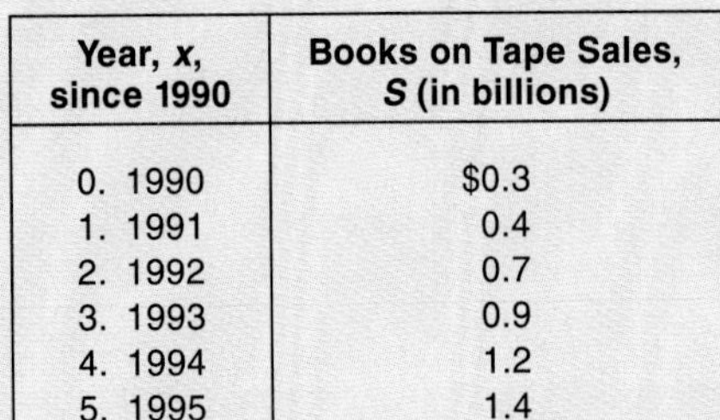

Year, x, since 1990	Books on Tape Sales, S (in billions)
0. 1990	\$0.3
1. 1991	0.4
2. 1992	0.7
3. 1993	0.9
4. 1994	1.2
5. 1995	1.4
6. 1996	1.6

Source: Audio Book Club

31. a) Use the two points $(0, 0.3)$ and $(6, 1.6)$ to find a linear function that fits the data.
b) Use the function to predict sales of books on audiotape in 1998 and 2000.

Skill Maintenance

32. Convert 0.0000008204 to scientific notation.

33. Divide: $(y^2 + 6y - 3) \div (y + 4)$.

34. Factor: $125a^3 + 27$.

35. Simplify: $\dfrac{3m^2 + m - 10}{7m^2 + 18m + 8}$.

Synthesis

36. Find k such that the line $3x + ky = 17$ is perpendicular to the line $8x - 5y = 26$.

37. Find a formula for a function f for which $f(-2) = 3$.

8

Systems of Equations

An Application

A nontoxic floor wax can be made by combining lemon juice and food-grade linseed oil. The amount of oil should be twice the amount of lemon juice. How much of each ingredient is needed in order to make 32 oz of floor wax?

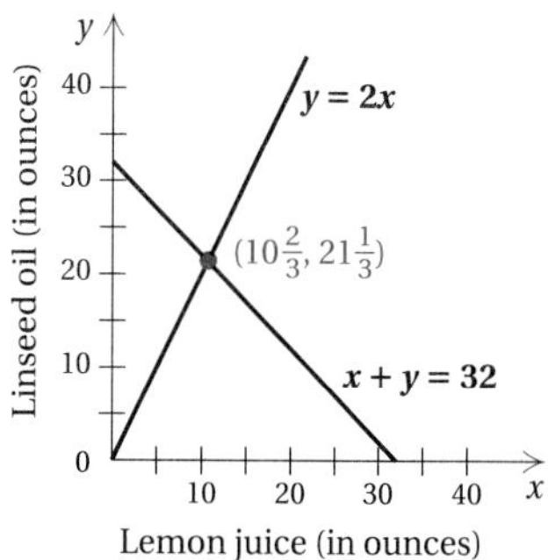

The Mathematics

We let $x =$ the amount of lemon juice, in ounces, and $y =$ the amount of linseed oil, in ounces. We can then translate the problem as:

$$\left.\begin{aligned} x + y &= 32, \\ y &= 2x. \end{aligned}\right\}$$

This is a *system of equations.*

This problem appears as Exercise 6 in Exercise Set 8.3.

For more information, visit us at www.mathmax.com

Introduction

As you have probably noted, the most difficult and time-consuming step in solving an applied problem is generally the translation of the situation to mathematical language. When you can translate using more than one variable and more than one equation, thus obtaining a *system of equations,* the translation step is often easier than if you translate to a single equation. In this chapter, you will learn to solve systems of equations and then apply that skill to the solving of applied problems.

Pretest: Chapter 8

1. Solve this system by graphing:

$$y = x + 1,$$
$$y + x = 3.$$

2. Solve the system of equations in Exercise 1 by the substitution method.

3. Solve this system by the elimination method:

$$3x + 5y = 1,$$
$$4x + 3y = -6.$$

4. Classify the system of equations in Exercise 1 as consistent or inconsistent.

5. Classify the system of equations in Exercise 1 as dependent or independent.

6. Solve:

$$3x + 5y - 2z = 7,$$
$$2x + y - 3z = -5,$$
$$4x - 2y + z = 3.$$

Solve.

7. *Mixed Nuts.* The Nutty Professor sells cashews for \$6.75 per pound and Brazil nuts for \$5.00 per pound. How much of each type should be used to make a 50-lb mixture that sells for \$5.70 per pound?

8. *Investments.* Two investments are made totaling \$7500. For a certain year, the investments yielded \$516 in simple interest. Part of the \$7500 is invested at 8% and part at 6%. Find the amount invested at each rate.

9. *Salt-Water Mixtures.* Mixture A is 32% salt and the rest water. Mixture B is 58% salt and the rest water. How many pounds of each mixture should be combined in order to obtain 120 lb of a mixture that is 44% salt?

10. *Marine Travel.* A motorboat took 4 hr to make a trip downstream with a 6-mph current. The return trip against the same current took 5 hr. Find the speed of the boat in still water.

11. For the following total-cost and total-revenue functions, find **(a)** the total-profit function and **(b)** the break-even point.

$$C(x) = 90{,}000 + 15x,$$
$$R(x) = 26x.$$

Objectives for Retesting

The objectives to be tested in addition to the material in this chapter are as follows.

[2.6a] Evaluate formulas and solve a formula for a specified letter.

[7.1b] Given a function described by an equation, find function values (outputs) for specified values (inputs).

[7.3b] Given a linear equation, derive the equivalent slope–intercept equation and determine the slope and the y-intercept.

[7.5d] Given a line and a point not on the given line, find an equation of the line parallel to the line and containing the point, and find an equation of the line perpendicular to the line and containing the point.

8.1 Systems of Equations in Two Variables

We can solve many applied problems more easily by translating to two or more equations in two or more variables than by translating to a single equation. Let's look at such a problem.

Real Estate Merger

In 1996, the Simon Property Group and the DeBartolo Realty Corporation merged to form the largest real estate company in the United States, owning 183 shopping centers in 32 states (***Source*:** Simon Property Group; DeBartolo Realty Corporation). Before the merger, Simon owned twice as many properties as DeBartolo. How many properties did each own originally?

The Indianapolis Star
REAL ESTATE GIANTS MERGE

DeBartolo shareholders would receive 0.68 share of Simon common stock for each share of DeBartolo common stock. Simon also would agree to repay $1.5 billion in DeBartolo debt. At Tuesday's closing price of $23.625 a share for common stock, the transaction is valued at roughly $3 billion.

Executives say the proposed company, Simon DeBartolo Group, would be the largest real estate company in the United States, worth $7.5 billion.

Not included in the deal: DeBartolo's ownership stake in the San Francisco 49ers, or the Indiana Pacers, owned separately by the Simon

To solve, we let $x =$ the number of properties originally owned by Simon and $y =$ the number of properties originally owned by DeBartolo. There are two statements to translate.

First we look at the total number of properties involved:

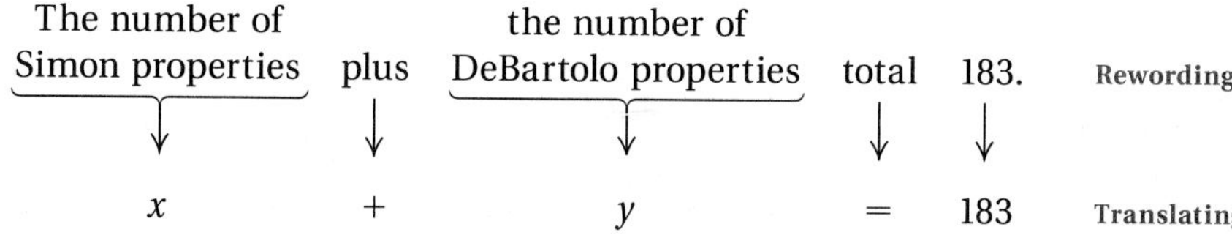

The second statement compares the number of properties that each company held before merging:

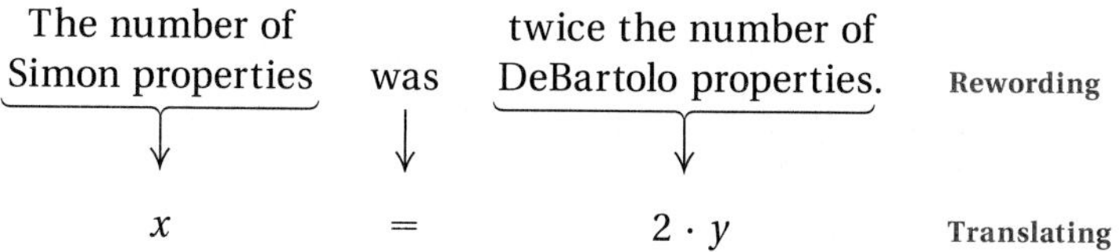

We have now translated the problem to a **pair**, or **system, of equations**:

$$x + y = 183,$$
$$x = 2y.$$

We can also write this system in function notation by solving each equation for y:

$$x + y = 183 \longrightarrow y = 183 - x \longrightarrow f(x) = 183 - x,$$
$$x = 2y \longrightarrow y = \tfrac{1}{2}x \longrightarrow g(x) = \tfrac{1}{2}x.$$

Objective

a Solve a system of two linear equations or two functions by graphing and determine whether a system is consistent or inconsistent and whether it is dependent or independent.

For Extra Help

TAPE 13

MAC WIN

CD-ROM

Solve the system graphically.

1. $-2x + y = 1,$
$3x + y = 1$

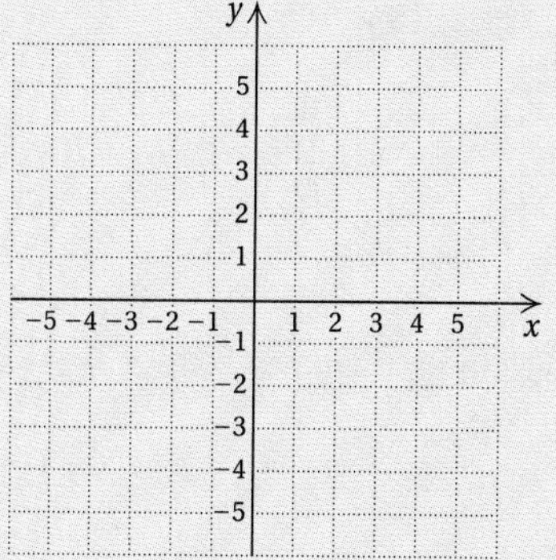

2. $f(x) = \frac{1}{2}x,$
$g(x) = -\frac{1}{4}x + \frac{3}{2}$

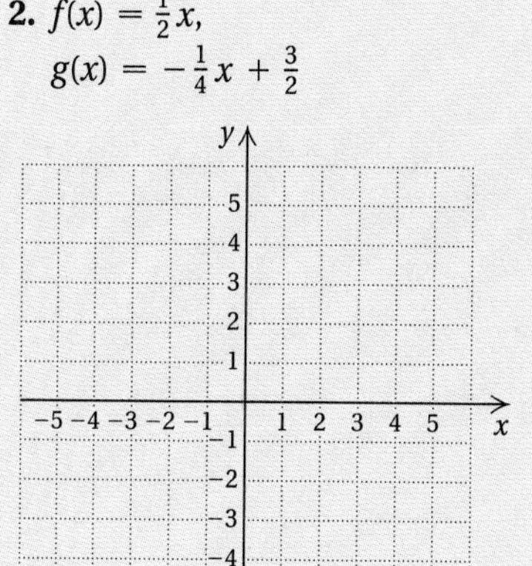

Answers on page A-32

Calculator Spotlight

Checking with the TABLE feature. We can use the TABLE feature to check a possible solution of a system. Consider the system in Example 1. We enter each equation in the "y =" form:

$$y_1 = x + 1,$$
$$y_2 = 3 - x.$$

After pressing 2nd TBLSET and selecting Indpnt:Ask, we press 2nd TABLE 1 ENTER and obtain the following table.

X	Y1	Y2
1	2	2

X = 1

Note that when $x = 1$, $y_1 = 2$ and $y_2 = 2$.

To solve this system, whether it is written in function notation or not, we graph each equation and look for the point of intersection of the graphs. (We may need to use detailed graphing paper to determine the point of intersection exactly.)

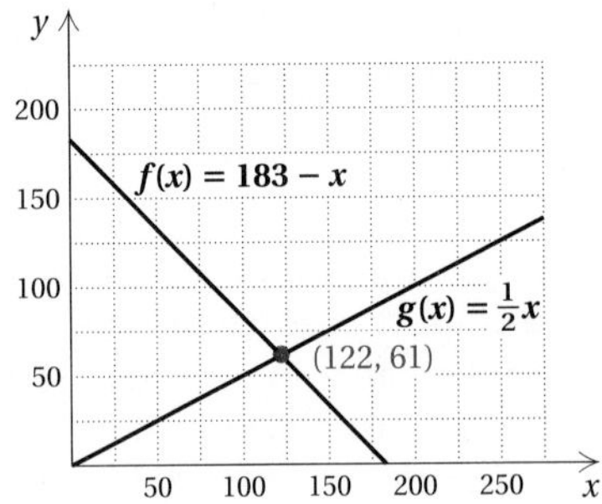

As we see in the graph above, the ordered pair (122, 61) is the intersection and thus the solution—that is, $x = 122$ and $y = 61$. This tells us that Simon originally owned 122 properties and DeBartolo owned 61.

a Solving Systems of Equations Graphically

One Solution

A **solution** of a system of two equations in two variables is an ordered pair that makes *both* equations true. If we graph a system of equations or functions, the point at which the graphs intersect will be a solution of *both* equations or functions.

Example 1 Solve this system graphically:

$$y - x = 1,$$
$$y + x = 3.$$

We draw the graph of each equation using any method studied in Chapter 2.

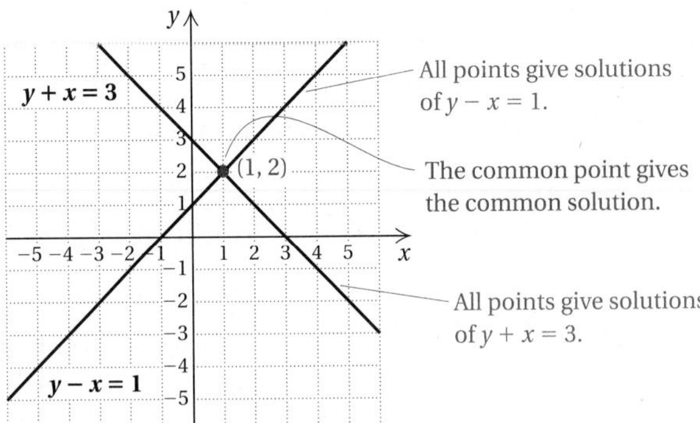

The point of intersection has coordinates that make *both* equations true. The solution seems to be the point (1, 2). However, since graphing alone is not perfectly accurate, solving by graphing may give only approximate answers. Thus we check the pair (1, 2) as follows.

CHECK:

$y - x = 1$		$y + x = 3$	
$2 - 1 \ ? \ 1$		$2 + 1 \ ? \ 3$	
1	TRUE	3	TRUE

The solution is (1, 2).

Do Exercises 1 and 2 on the preceding page.

No Solution

Sometimes the equations in a system have graphs that are parallel lines.

Example 2 Solve graphically:

$f(x) = -3x + 5,$

$g(x) = -3x - 2.$

We graph the functions. The graphs have the same slope, -3, and different y-intercepts, so they are parallel. There is no point at which they cross, so the system has no solution. No matter what point we try, it will *not* check in *both* equations. The solution set is thus the empty set, denoted $\varnothing$ or $\{\ \}$.

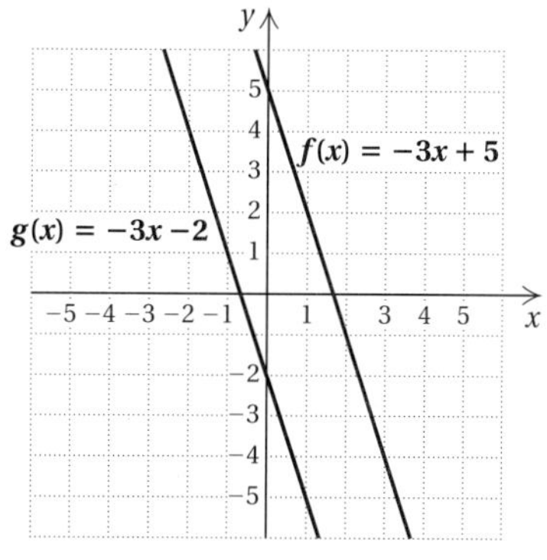

CONSISTENT AND INCONSISTENT SYSTEMS

A system of equations has a solution. ⟹ It is **consistent**.

A system of equations has no solution. ⟹ It is **inconsistent**.

The system in Example 1 is consistent. The system in Example 2 is inconsistent.

Do Exercises 3 and 4.

Infinitely Many Solutions

Sometimes the equations in a system have the same graph.

Example 3 Solve graphically:

$3y - 2x = 6,$

$-12y + 8x = -24.$

We graph the equations and see that the graphs are the same. Thus any solution of one of the equations is a solution of the other. Each equation has an infinite number of solutions, two of which are shown on the graph.

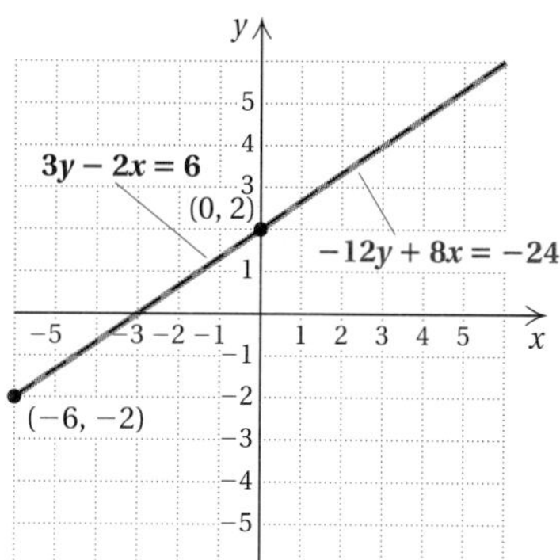

We check one such solution, (0, 2), which is the y-intercept of each equation.

3. Solve graphically:

$y + 2x = 3,$

$y + 2x = -4.$

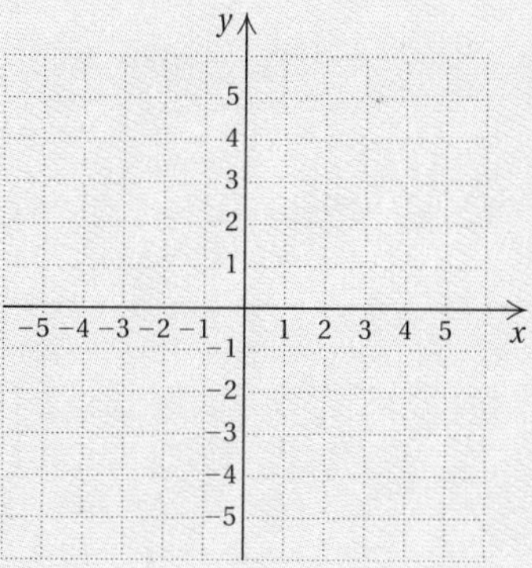

4. Classify each of the systems in Margin Exercises 1–3 as consistent or inconsistent.

Calculator Spotlight

Use the TABLE feature to check the results of Examples 2 and 3.

Answers on page A-32

5. Solve graphically:

$2x - 5y = 10,$

$-6x + 15y = -30.$

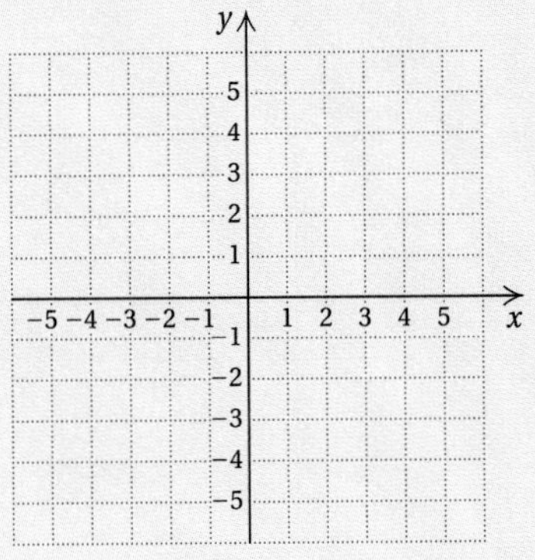

6. Classify each system in Margin Exercises 1, 2, 3, and 5 as dependent or independent.

7. a) Solve $x + 1 = \frac{2}{3}x$ algebraically.

b) Solve $x + 1 = \frac{2}{3}x$ graphically using Method 1.

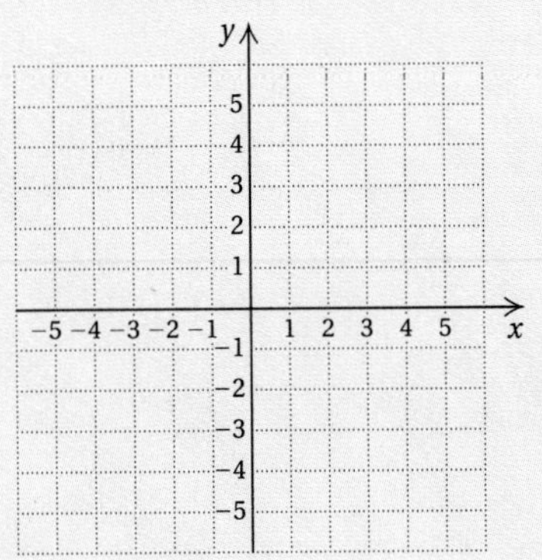

c) Compare your answers to parts (a) and (b).

Answers on page A-32

CHECK:

$3y - 2x = 6$	
$3(2) - 2(0)$	? 6
$6 - 0$	
6	TRUE

$-12y + 8x = -24$	
$-12(2) + 8(0)$	? -24
$-24 + 0$	
-24	TRUE

On your own, check that $(-6, -2)$ is a solution of both equations. If $(0, 2)$ and $(-6, -2)$ are solutions, then all points on the line containing them will be solutions. The system has an infinite number of solutions.

DEPENDENT AND INDEPENDENT SYSTEMS

A system of two equations in two variables:

has infinitely many solutions. ⟶ It is **dependent**.

has one solution or no solutions. ⟶ It is **independent**.

The system in Example 3 is dependent. The systems in Examples 1 and 2 are independent.

When we graph a system of two equations, one of the following three things can happen.

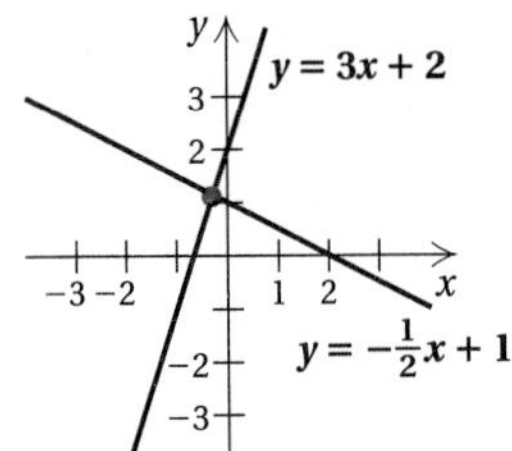

One solution.
Graphs intersect.
Equations are *consistent* and *independent.*

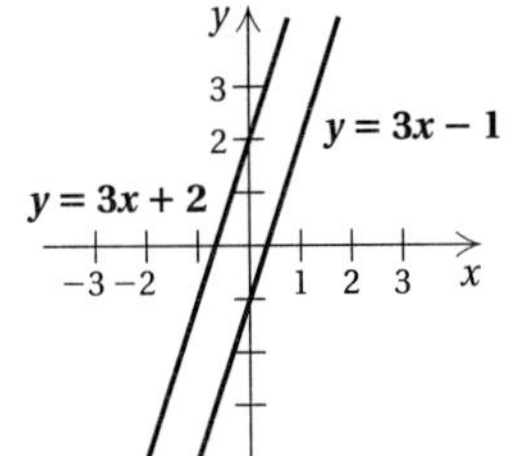

No solution.
Graphs are parallel.
Equations are *inconsistent* and *independent.*

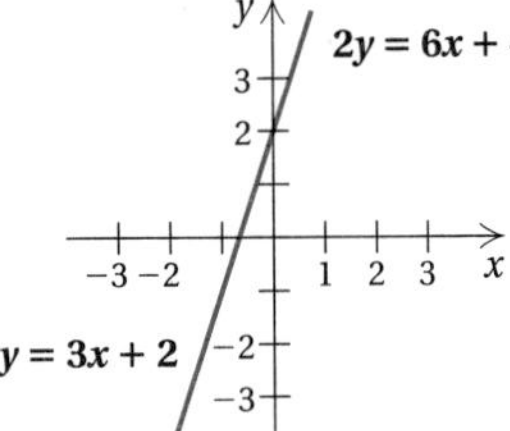

Infinitely many solutions.
Equations have the same graph. Equations are *consistent* and *dependent.*

Do Exercises 5 and 6.

AG Algebraic–Graphical Connection

Consider the equation $-2x + 13 = 4x - 17$. Let's solve it algebraically as we did in Chapter 2:

$$-2x + 13 = 4x - 17$$

$$13 = 6x - 17 \quad \text{Adding } 2x$$

$$30 = 6x \quad \text{Adding 17}$$

$$5 = x. \quad \text{Dividing by 6}$$

Could we also solve the equation graphically? The answer is yes, as we see in the following two methods.

METHOD 1 Solve $-2x + 13 = 4x - 17$ graphically.

We let

$f(x) = -2x + 13$ and $g(x) = 4x - 17$.

Graphing the system of equations, we get the graph shown at right.

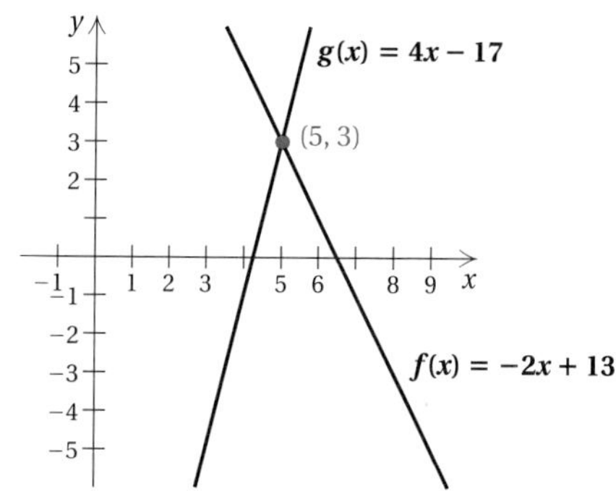

The point of intersection of the two graphs is (5, 3). Note that the x-coordinate of the intersection is 5. This value for x is the solution of the equation $-2x + 13 = 4x - 17$.

Do Exercises 7 and 8. (Exercise 7 is on the preceding page.)

METHOD 2 Solve $-2x + 13 = 4x - 17$ graphically.

Adding $-4x$ and 17 on both sides, we obtain the form

$-6x + 30 = 0$.

This time we let

$f(x) = -6x + 30$ and $g(x) = 0$.

Since $g(x) = 0$, or $y = 0$, is the x-axis, we need only graph $f(x) = -6x + 30$ and see where it crosses the x-axis.

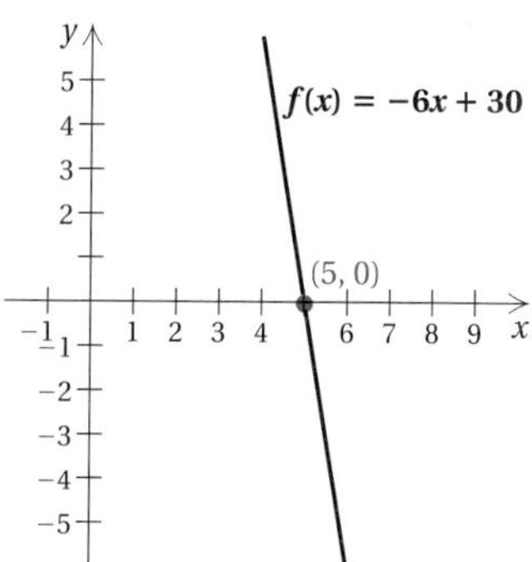

Note that the x-intercept of $f(x) = -6x + 30$ is (5, 0), or just 5. This x-value is the solution of the equation $-2x + 13 = 4x - 17$.

Do Exercise 9.

Let's compare the two methods. Using Method 1, we graph two functions. The solution of the original equation is the x-coordinate of the point of intersection. Using Method 2, we find that the solution of the original equation is the x-intercept of the graph.

Do Exercise 10.

Methods 1 and 2 are helpful when we are using a graphing calculator or computer graphing software. See the Calculator Spotlight that follows.

8. Solve $\frac{1}{2}x + 3 = 2$ graphically using Method 1.

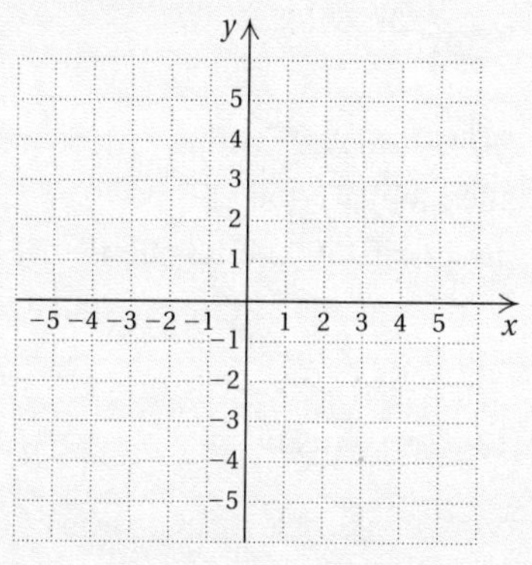

9. a) Solve $x + 1 = \frac{2}{3}x$ graphically using Method 2.

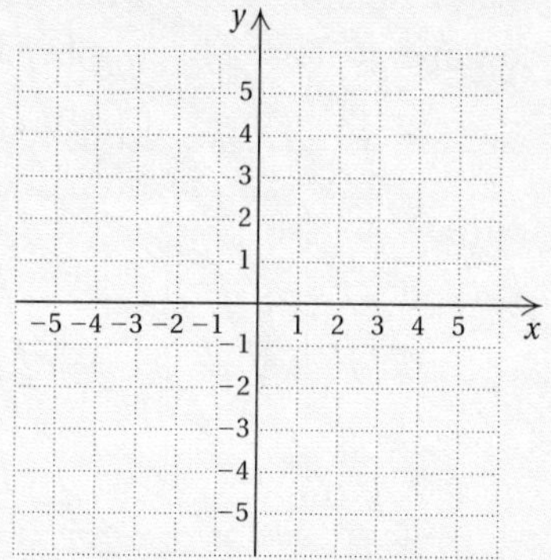

b) Compare your answers to Margin Exercises 7(a), 7(b), and 9(a).

10. Solve $\frac{1}{2}x + 3 = 2$ graphically using Method 2.

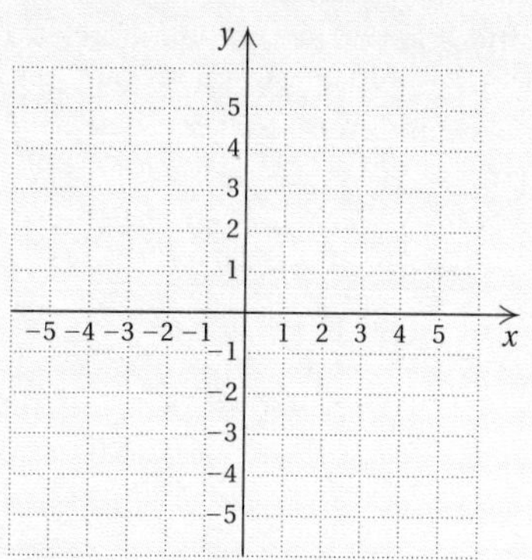

Answers on page A-33

Calculator Spotlight

 The INTERSECT and ZERO Features

INTERSECT Feature. Most graphers have an INTERSECT feature that can be used to find the intersection of the graphs of two functions. Let's use it to find the point of intersection of the two functions

$$f(x) = -2x + 13,$$
$$g(x) = 4x - 17.$$

We enter the functions as

$$y_1 = -2x + 13,$$
$$y_2 = 4x - 17.$$

Then we graph the functions, adjusting the viewing window in order to see the point of intersection.

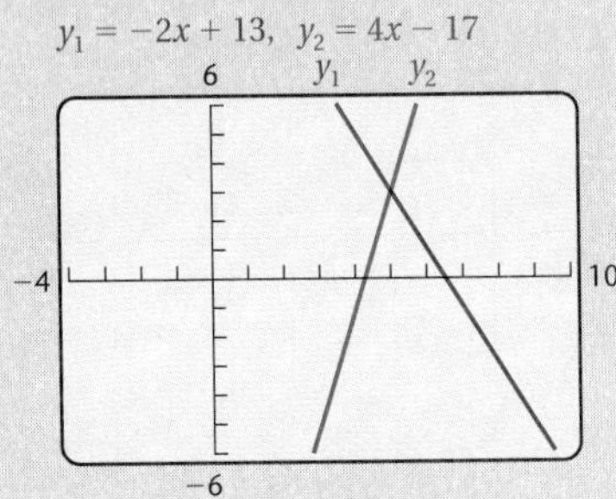

Next, we use the INTERSECT feature (see the CALC menu) to obtain the point of intersection (5, 3).

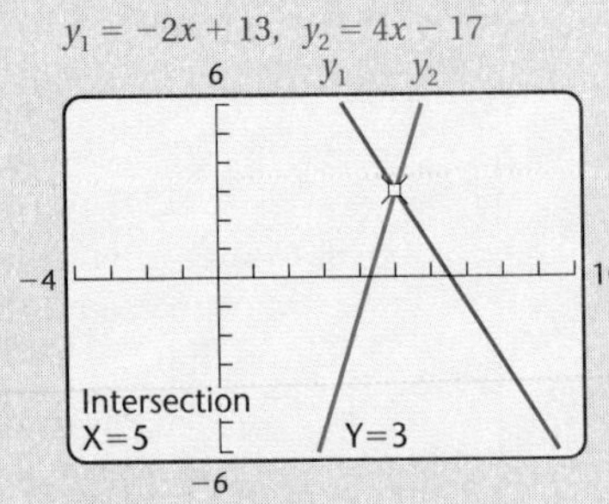

The first coordinate of the point of intersection, 5, is the solution of the equation $-2x + 13 = 4x - 17$.

ZERO Feature. Most graphers have a ZERO, or ROOT, feature that allows us to solve an equation quickly. In this context, the word "zero" refers to an input for which the output of a function is 0.

To use this feature, we must first have a 0 on one side of the equation. So to solve

$$-2x + 13 = 4x - 17,$$

we subtract $4x$ and add 17 on both sides to get $-6x + 30 = 0$. Graphing $y = -6x + 30$ and using the ZERO feature, we obtain a screen like the following.

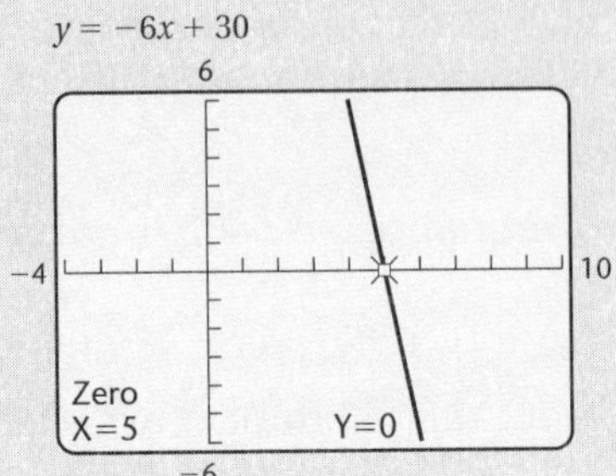

We see that $-6x + 30 = 0$ when $x = 5$, so 5 is the solution of the equation $-2x + 13 = 4x - 17$.

Exercises

Use the INTERSECT feature on your grapher to solve the equation.

1. $x + 1 = \frac{1}{2}x$

2. $\frac{1}{2}x + 3 = 2$

3. $-\frac{3}{4}x + 6 = 2x - 1$

4. $-3x + 4 = 3x - 4$

5. $2.4 - 1.8x = 6.8 - x^2$

6. $x^3 - 3x - 5 = 0$

7.–12. Use the ZERO, or ROOT, feature on your grapher to solve each of the equations in Exercises 1–6.

Exercise Set 8.1

[a] Solve the system of equations graphically. Then classify the system as consistent or inconsistent and as dependent or independent.

1. $x + y = 4,$
$x - y = 2$

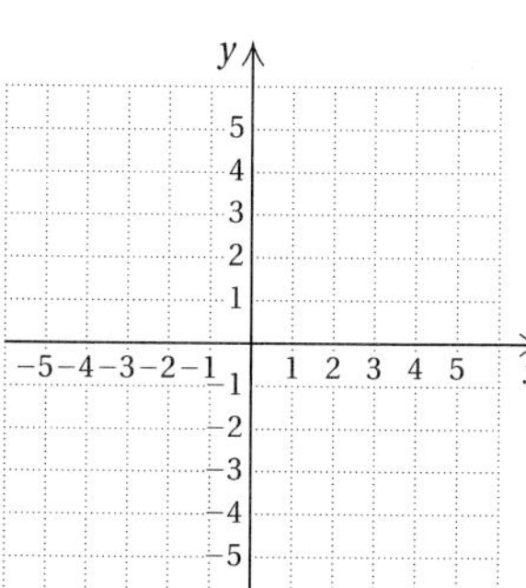

CHECK:

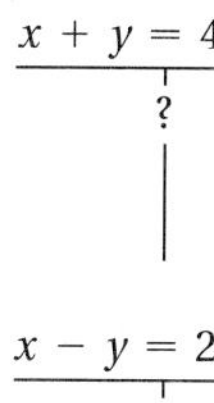

$x + y = 4$ | ?

$x - y = 2$ | ?

2. $x - y = 3,$
$x + y = 5$

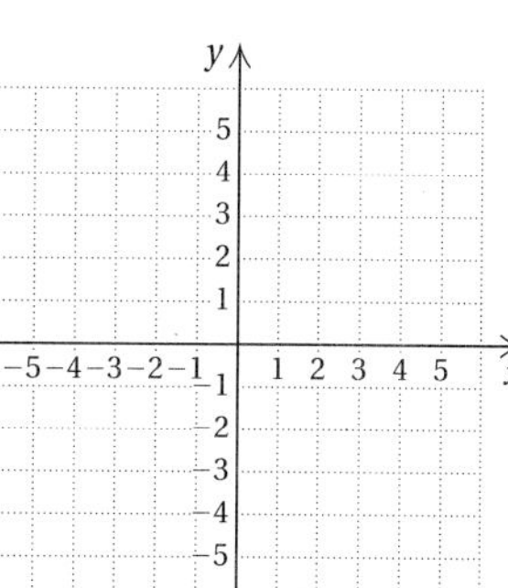

CHECK:

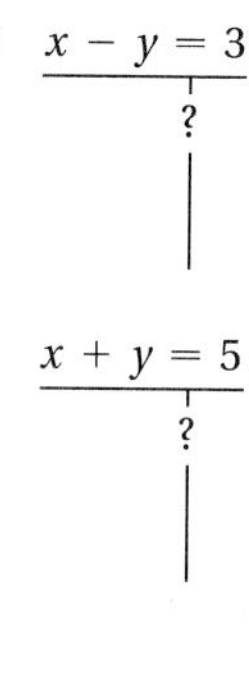

$x - y = 3$ | ?

$x + y = 5$ | ?

3. $2x - y = 4,$
$2x + 3y = -4$

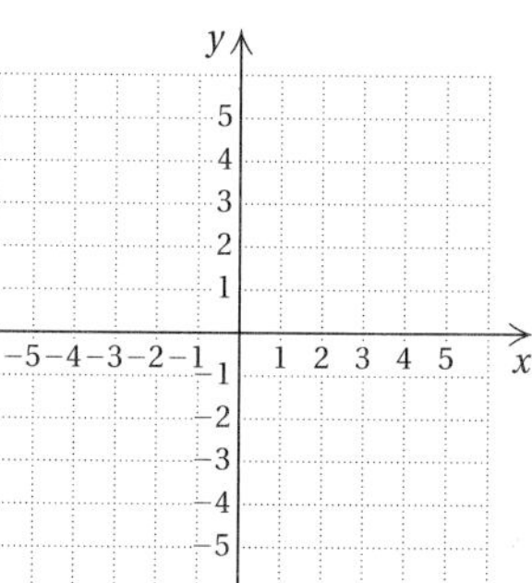

CHECK: $2x - y = 4$ | ?

$2x + 3y = -4$ | ?

4. $3x + y = 5,$
$x - 2y = 4$

CHECK: $3x + y = 5$ | ?

$x - 2y = 4$ | ?

5. $2x + y = 6,$
$3x + 4y = 4$

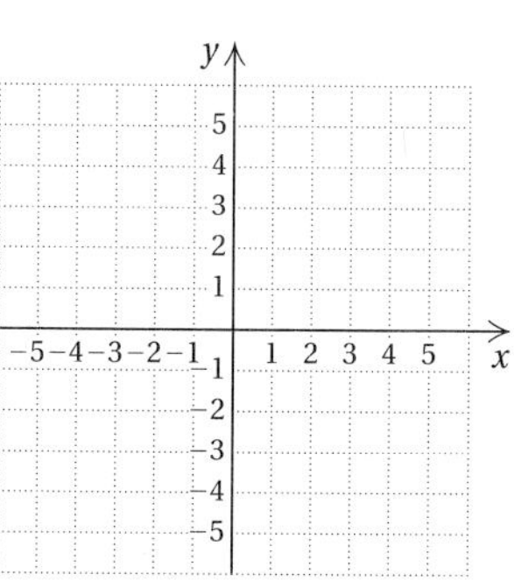

6. $2y = 6 - x,$
$3x - 2y = 6$

7. $f(x) = x - 1,$
$g(x) = -2x + 5$

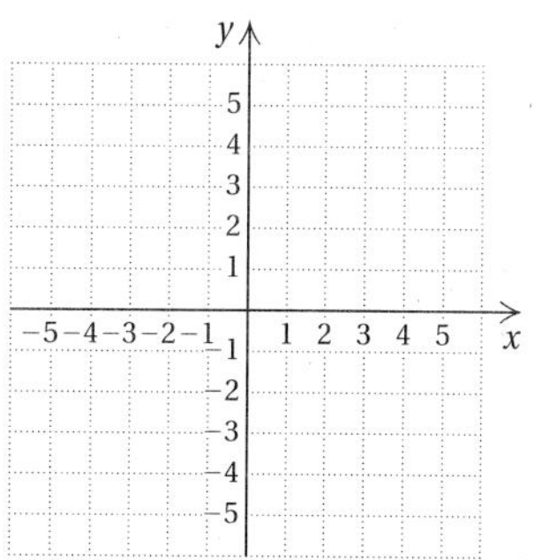

8. $f(x) = x + 1,$
$g(x) = \frac{2}{3}x$

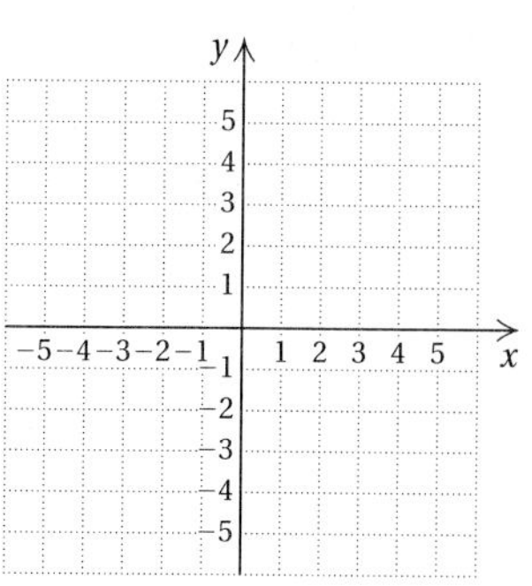

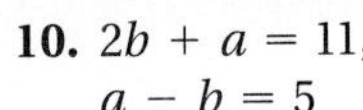

9. $2u + v = 3,$
$2u = v + 7$

10. $2b + a = 11,$
$a - b = 5$

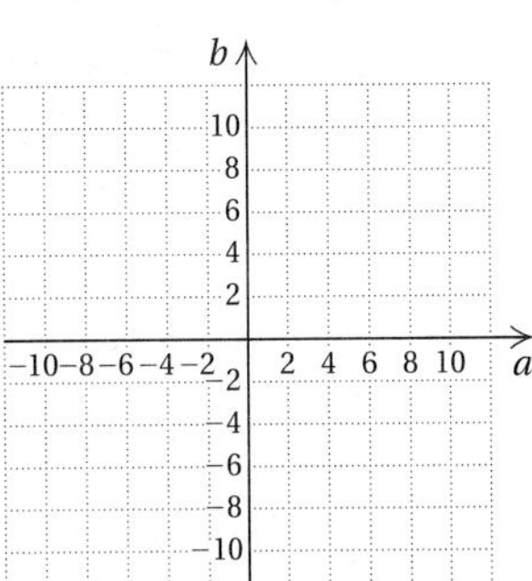

11. $f(x) = -\frac{1}{3}x - 1,$
$g(x) = \frac{4}{3}x - 6$

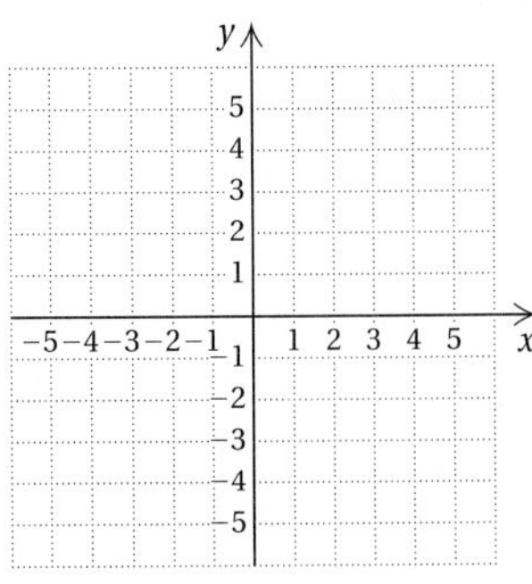

12. $f(x) = -\frac{1}{4}x + 1,$
$g(x) = \frac{1}{2}x - 2$

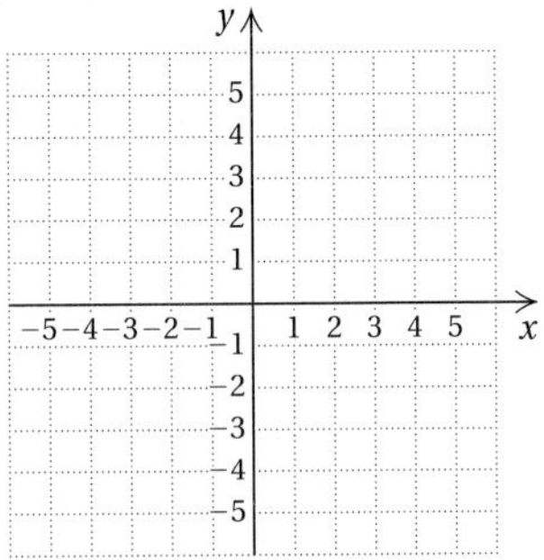

13. $6x - 2y = 2,$
$9x - 3y = 1$

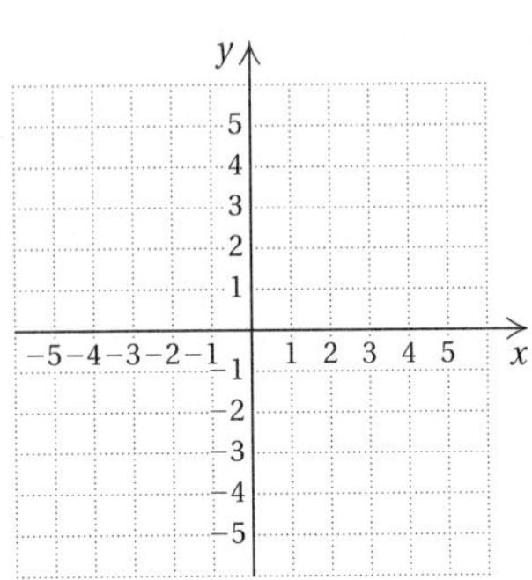

14. $y - x = 5,$
$2x - 2y = 10$

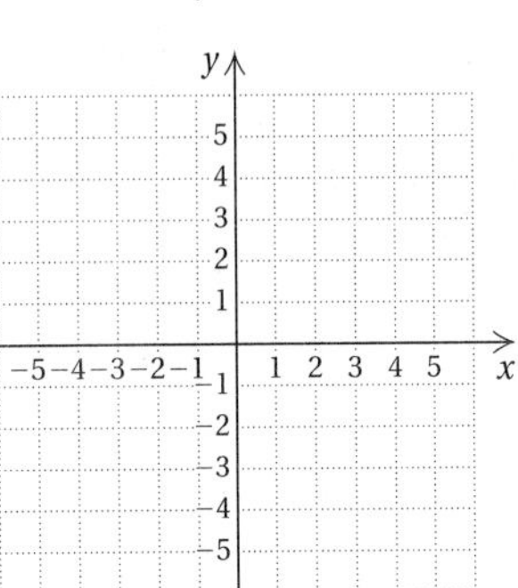

15. $2x - 3y = 6,$
$3y - 2x = -6$

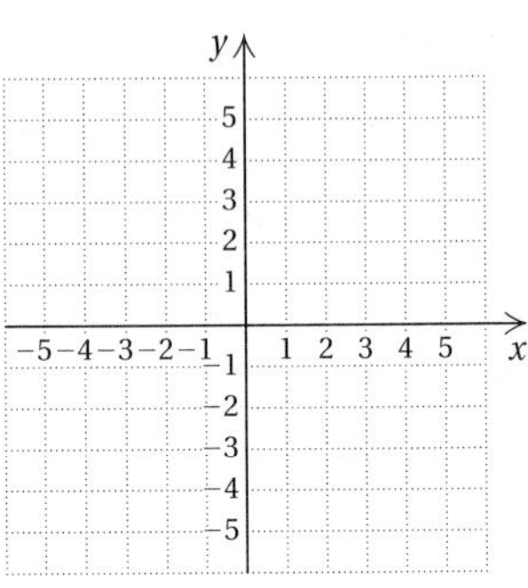

16. $y = 3 - x,$
$2x + 2y = 6$

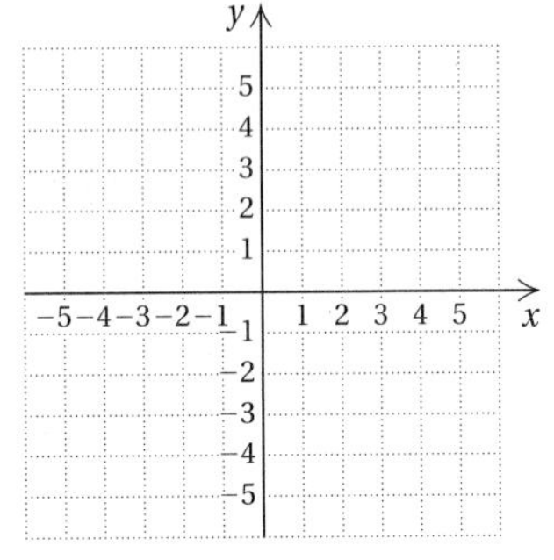

17. $x = 4,$
$y = -5$

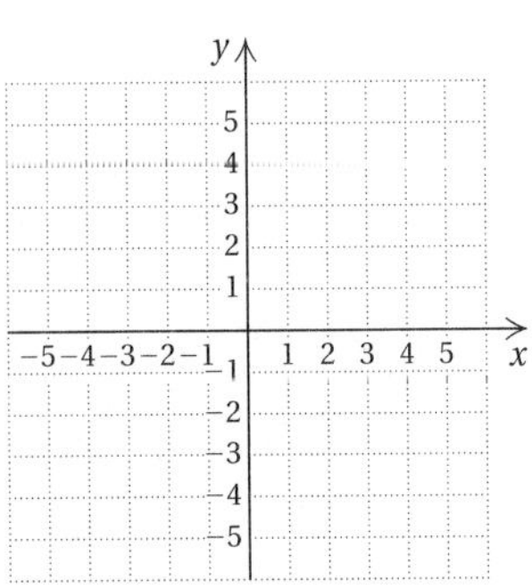

18. $x = -3,$
$y = 2$

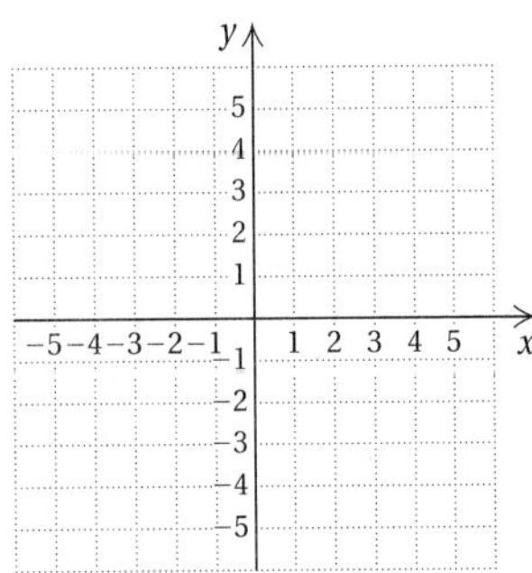

19. $y = -x - 1,$
$4x - 3y = 17$

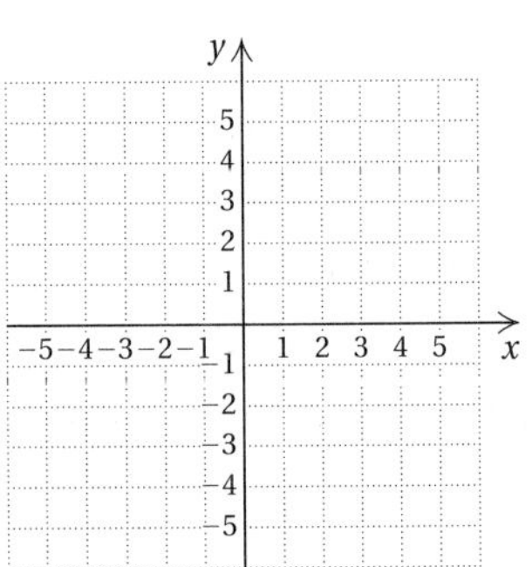

20. $a + 2b = -3,$
$b - a = 6$

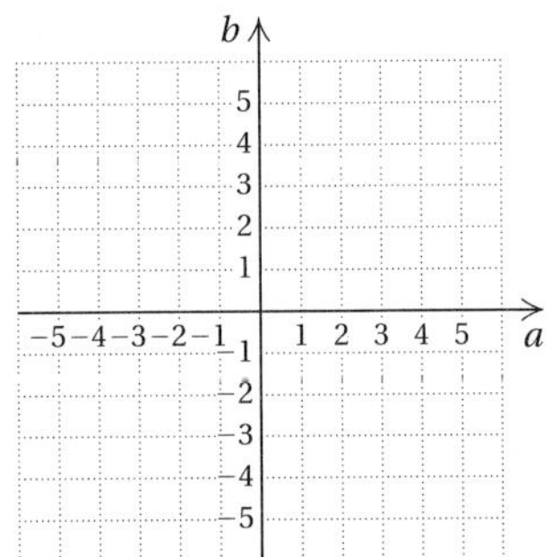

Skill Maintenance

Write an equation of the line containing the given point and parallel to the given line. [7.5d]

21. $(-4, 2);\ 3x = 5y - 4$

22. $(-6, 0);\ 8y - 3x = 2$

Synthesis

23. ◆ Explain how to find the solution of $\frac{3}{4}x + 2 = \frac{2}{5}x - 5$ in two ways graphically and in two ways algebraically.

24. ◆ Write a system of equations with the given solution. Answers may vary.

a) $(4, -3)$ **b)** No solution

c) Infinitely many solutions

Use a grapher to find the point of intersection of the pair of equations. Round all answers to the nearest hundredth. You may need to solve for y first.

25. $2.18x + 7.81y = 13.78,$
$5.79x - 3.45y = 8.94$

26. $f(x) = 123.52x + 89.32,$
$g(x) = -89.22x + 33.76$

8.2 Solving by Substitution or Elimination

Consider this system of equations:

$$5x + 9y = 2,$$
$$4x - 9y = 10.$$

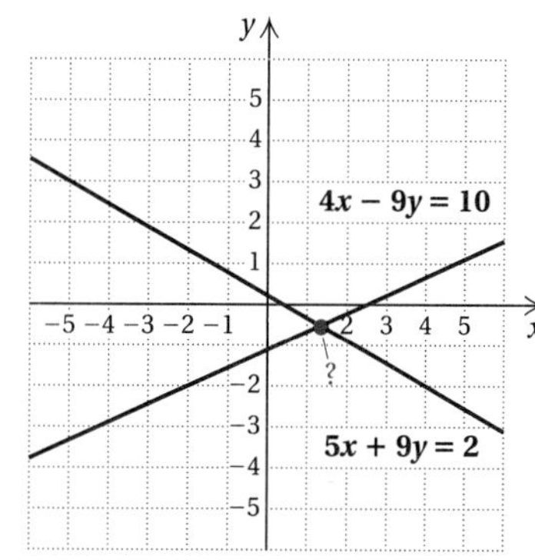

What is the solution? It is rather difficult to tell exactly by graphing. It would appear that fractions are involved. It turns out that the solution is

$$\left(\frac{4}{3}, -\frac{14}{27}\right).$$

Solving by graphing, though useful in many applied situations, is not always fast or accurate in cases where solutions are not integers. We need techniques involving algebra to determine the solution exactly. Because they use algebra, they are called **algebraic methods.**

Objectives

a Solve systems of equations in two variables by the substitution method.

b Solve systems of equations in two variables by the elimination method.

c Solve applied problems by solving systems of two equations using substitution or elimination.

For Extra Help

TAPE 13 MAC WIN CD-ROM

a The Substitution Method

One nongraphical method for solving systems is known as the **substitution method.**

Example 1 Solve this system:

$$x + y = 4, \quad (1)$$
$$x = y + 1. \quad (2)$$

Equation (2) says that x and $y + 1$ name the same number. Thus we can substitute $y + 1$ for x in equation (1):

$$x + y = 4 \quad \text{Equation (1)}$$
$$(y + 1) + y = 4. \quad \text{Substituting } y + 1 \text{ for } x$$

Since this equation has only one variable, we can solve for y using methods learned earlier:

$$(y + 1) + y = 4$$
$$2y + 1 = 4 \quad \text{Removing parentheses and collecting like terms}$$
$$2y = 3 \quad \text{Subtracting 1}$$
$$y = \tfrac{3}{2}. \quad \text{Dividing by 2}$$

We return to the original pair of equations and substitute $\frac{3}{2}$ for y in *either* equation so that we can solve for x. Calculation will be easier if we choose equation (2) since it is already solved for x:

$$x = y + 1 \quad \text{Equation (2)}$$
$$= \tfrac{3}{2} + 1 \quad \text{Substituting } \tfrac{3}{2} \text{ for } y$$
$$= \tfrac{3}{2} + \tfrac{2}{2} = \tfrac{5}{2}.$$

We obtain the ordered pair $\left(\frac{5}{2}, \frac{3}{2}\right)$. Even though we solved for y *first,* it is still the *second* coordinate. We check to be sure that the ordered pair is a solution.

Solve by the substitution method.

1. $x + y = 6,$
   $y = x + 2$

2. $y = 7 - x,$
   $2x - y = 8$

   (*Caution*: Use parentheses when you substitute, being careful about removing them. Remember to solve for both variables.)

Solve by the substitution method.

3. $2y + x = 1,$
   $y - 2x = 8$

4. $8x + 5y = 184,$
   $x - y = -3$

Calculator Spotlight

Use the TABLE feature to check the solutions of the systems in Examples 1 and 2 and Margin Exercises 1–4.

Answers on page A-33

CHECK:

$x + y = 4$		
$\frac{5}{2} + \frac{3}{2}$	?	4
$\frac{8}{2}$		
4		TRUE

$x = y + 1$		
$\frac{5}{2}$	?	$\frac{3}{2} + 1$
		$\frac{3}{2} + \frac{2}{2}$
		$\frac{5}{2}$ TRUE

Since $\left(\frac{5}{2}, \frac{3}{2}\right)$ checks, it is the solution. Even though exact fractional solutions are difficult to determine graphically, a graph can help us to visualize whether the solution is reasonable.

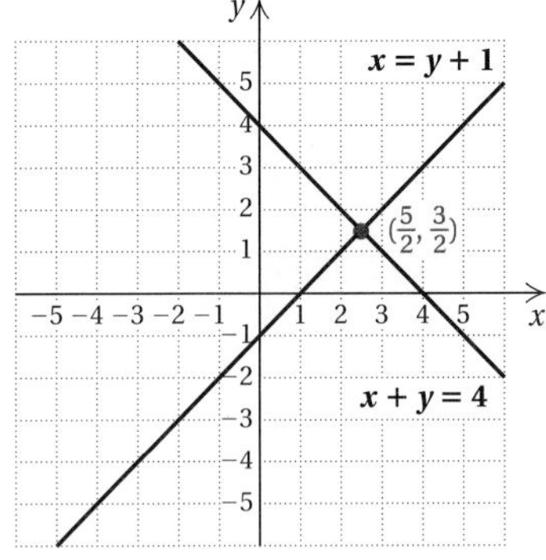

Do Exercises 1 and 2.

Suppose neither equation of a pair has a variable alone on one side. We then solve one equation for one of the variables.

Example 2 Solve this system:

$$2x + y = 6, \quad (1)$$
$$3x + 4y = 4. \quad (2)$$

First, we solve one equation for one variable. Since the coefficient of y is 1 in equation (1), it is the easier one to solve for y:

$$y = 6 - 2x. \quad (3)$$

Next, we substitute $6 - 2x$ for y in equation (2) and solve for x:

$$3x + 4(6 - 2x) = 4 \quad \text{Substituting } 6 - 2x \text{ for } y$$
$$3x + 24 - 8x = 4 \quad \text{Multiplying to remove parentheses}$$
$$24 - 5x = 4 \quad \text{Collecting like terms}$$
$$-5x = -20 \quad \text{Subtracting 24}$$
$$x = 4. \quad \text{Dividing by } -5$$

Remember to use parentheses when you substitute. Then remove them carefully.

We return to either of the original equations, (1) or (2), or equation (3), which we solved for y. It is generally easier to use an equation like (3), where we have solved for the specific variable. We substitute 4 for x in equation (3) and solve for y:

$$y = 6 - 2x$$
$$= 6 - 2(4) = 6 - 8 = -2.$$

We obtain the ordered pair $(4, -2)$.

CHECK:

$2x + y = 6$		
$2(4) + (-2)$	?	6
$8 - 2$		
6		TRUE

$3x + 4y = 4$		
$3(4) + 4(-2)$	?	4
$12 - 8$		
4		TRUE

Since $(4, -2)$ checks, it is the solution.

Do Exercises 3 and 4.

b The Elimination Method

The **elimination method** for solving systems of equations makes use of the *addition principle* for equations. Some systems are much easier to solve using the elimination method rather than the substitution method.

Example 3 Solve this system:

$$2x - 3y = 0, \quad (1)$$
$$-4x + 3y = -1. \quad (2)$$

The key to the advantage of the elimination method for solving this system involves the $-3y$ in one equation and the $3y$ in the other. These terms are opposites. If we add them, these terms will add to 0, and in effect, the variable y will have been "eliminated."

We will use the addition principle for equations, adding the same number on both sides of the equation. According to equation (2), $-4x + 3y$ and -1 are the same number. Thus we can use a vertical form and add $-4x + 3y$ to the left side of equation (1) and -1 to the right side:

$$\begin{aligned} 2x - 3y &= 0 && (1) \\ -4x + 3y &= -1 && (2) \\ \hline -2x + 0y &= -1 && \text{Adding} \\ -2x + 0 &= -1 \\ -2x &= -1. \end{aligned}$$

We have eliminated the variable y, which is why we call this the *elimination method.* We now have an equation with just one variable, which we solve for x:

$$\begin{aligned} -2x &= -1 \\ x &= \tfrac{1}{2}. \end{aligned}$$

Next, we substitute $\frac{1}{2}$ for x in either equation and solve for y:

$$\begin{aligned} 2 \cdot \tfrac{1}{2} - 3y &= 0 && \text{Substituting in equation (1)} \\ 1 - 3y &= 0 \\ -3y &= -1 && \text{Subtracting 1} \\ y &= \tfrac{1}{3}. && \text{Dividing by } -3 \end{aligned}$$

CHECK:

$2x - 3y = 0$			$-4x + 3y = -1$		
$2\left(\frac{1}{2}\right) - 3\left(\frac{1}{3}\right)$	? 0		$-4\left(\frac{1}{2}\right) + 3\left(\frac{1}{3}\right)$	? -1	
$1 - 1$			$-2 + 1$		
0		TRUE	-1		TRUE

Since $\left(\frac{1}{2}, \frac{1}{3}\right)$ checks, it is the solution. We can also see this in the graph shown at right.

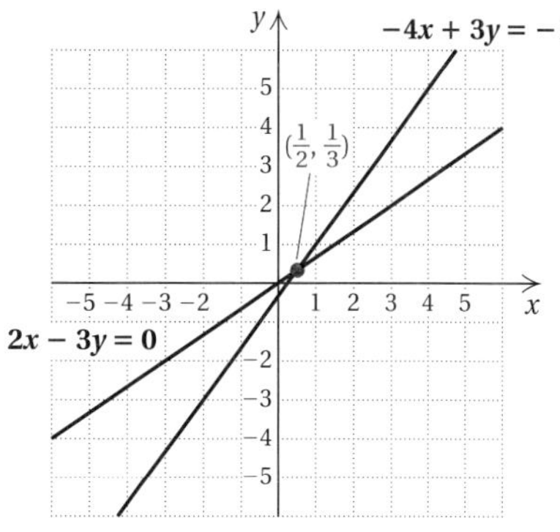

Do Exercises 5 and 6.

Solve by the elimination method.

5. $5x + 3y = 17,$
$-5x + 2y = 3$

6. $-3a + 2b = 0,$
$3a - 4b = -1$

Answers on page A-33

7. Solve by the elimination method:

$$2y + 3x = 12,$$
$$-4y + 5x = -2.$$

Calculator Spotlight

Use the INTERSECT feature to check the solutions of the systems in Examples 3–5 and Margin Exercises 5–9.

Answer on page A-33

In order to eliminate a variable, we sometimes use the multiplication principle to multiply one or both of the equations by a particular number before adding.

Example 4 Solve this system:

$$3x + 3y = 15, \quad (1)$$
$$2x + 6y = 22. \quad (2)$$

If we add directly, we get $5x + 9y = 37$, but we have not eliminated a variable. However, note that if the $3y$ in equation (1) were $-6y$, we could eliminate y. Thus we multiply by -2 on both sides of equation (1) and add:

$$-6x - 6y = -30 \quad \text{Multiplying equation (1) by } -2 \text{ on both sides}$$
$$2x + 6y = 22 \quad \text{Equation (2)}$$
$$-4x + 0 = -8 \quad \text{Adding}$$
$$-4x = -8$$
$$x = 2. \quad \text{Solving for } x$$

Then

$$2 \cdot 2 + 6y = 22 \quad \text{Substituting 2 for } x \text{ in Equation (2)}$$
$$\left.\begin{aligned} 4 + 6y &= 22 \\ 6y &= 18 \\ y &= 3. \end{aligned}\right\} \quad \text{Solving for } y$$

We obtain (2, 3), or $x = 2$, $y = 3$. This checks, so it is the solution.

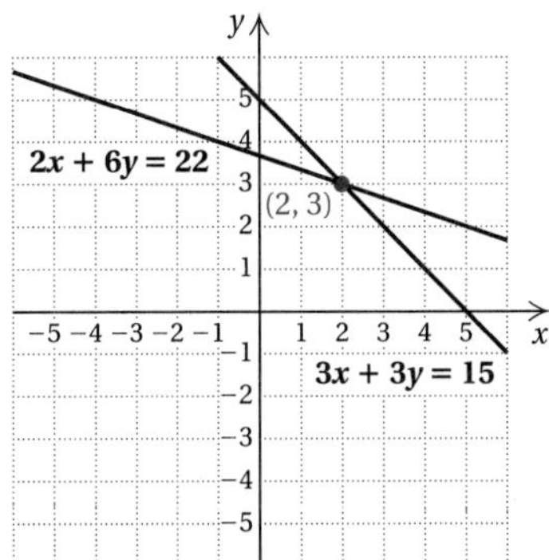

Do Exercise 7.

Sometimes we must multiply twice in order to make two terms opposites.

Example 5 Solve this system:

$$2x + 3y = 17, \quad (1)$$
$$5x + 7y = 29. \quad (2)$$

We must first multiply in order to make one pair of terms with the same variable opposites. We decide to do this with the x-terms in each equation. We multiply equation (1) by 5 and equation (2) by -2. Then we get $10x$ and $-10x$, which are opposites.

From Equation (1): $\quad 10x + 15y = 85 \quad$ Multiplying by 5
From Equation (2): $\quad -10x - 14y = -58 \quad$ Multiplying by -2

$$0 + y = 27 \quad \text{Adding}$$
$$y = 27. \quad \text{Solving for } y$$

Then

$$2x + 3 \cdot 27 = 17 \quad \text{Substituting 27 for } y \text{ in equation (1)}$$

$$\left.\begin{aligned} 2x + 81 &= 17 \\ 2x &= -64 \\ x &= -32. \end{aligned}\right\} \quad \text{Solving for } x$$

CHECK:

$$\begin{array}{c|c} 2x + 3y & 17 \\ \hline 2(-32) + 3(27) & ?\ 17 \\ -64 + 81 & \\ 17 & \text{TRUE} \end{array} \qquad \begin{array}{c|c} 5x + 7y & 29 \\ \hline 5(-32) + 7(27) & ?\ 29 \\ -160 + 189 & \\ 29 & \text{TRUE} \end{array}$$

We obtain $(-32, 27)$, or $x = -32$, $y = 27$, as the solution.

Do Exercises 8 and 9.

Some systems have no solution, as we saw graphically in Section 8.1. How do we recognize such systems if we are solving using an algebraic method?

Example 6 Solve this system:

$$y + 3x = 5, \qquad (1)$$
$$y + 3x = -2. \qquad (2)$$

If we find the slope–intercept equations for this system, we get

$$y = -3x + 5,$$
$$y = -3x - 2.$$

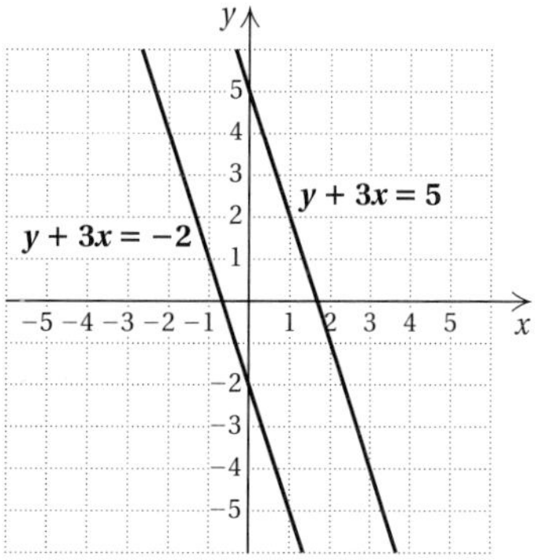

The graphs are parallel lines. The system has no solution.

Let's see what happens if we attempt to solve the system by the elimination method. We multiply by -1 on both sides of equation (2) and add:

$$\begin{aligned} y + 3x &= 5 && \text{Equation (1)} \\ -y - 3x &= 2 && \text{Multiplying equation (2) by } -1 \\ \hline 0 &= 7. && \text{Adding, we obtain a false equation.} \end{aligned}$$

The x-terms and the y-terms are eliminated and we end up with a *false* equation. Thus, if we obtain a false equation when solving algebraically, we know that the system has no solution. The system is inconsistent and independent.

Do Exercise 10.

Solve by the elimination method.

8. $4x + 5y = -8,$
$7x + 9y = 11$

9. $4x - 5y = 38,$
$7x - 8y = -22$

10. Solve by the elimination method:

$$y + 2x = 3,$$
$$y + 2x = -1.$$

Answers on page A-33

11. Solve by the elimination method:

$$2x - 5y = 10,$$
$$-6x + 15y = -30.$$

12. Clear the decimals. Then solve.

$$0.02x + 0.03y = 0.01,$$
$$0.3x - 0.1y = 0.7$$

(*Hint*: Multiply the first equation by 100 and the second one by 10.)

13. Clear the fractions. Then solve.

$$\frac{3}{5}x + \frac{2}{3}y = \frac{1}{3},$$
$$\frac{3}{4}x - \frac{1}{3}y = \frac{1}{4}$$

Answers on page A-33

Some systems have infinitely many solutions. How can we recognize such a situation when we are solving systems using an algebraic method?

Example 7 Solve this system:

$$3y - 2x = 6, \quad (1)$$
$$-12y + 8x = -24. \quad (2)$$

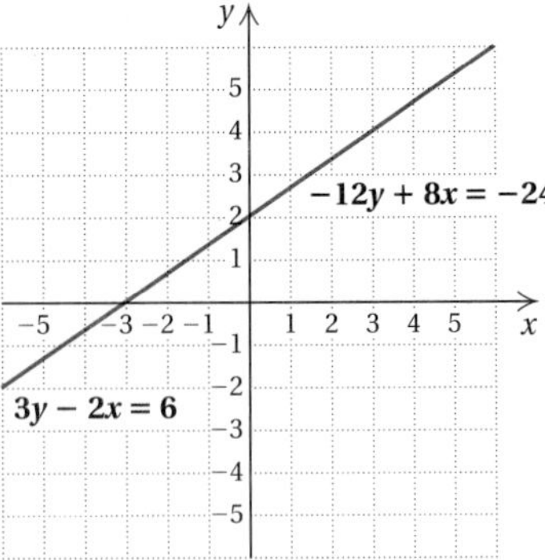

The graphs are the same line. The system has an infinite number of solutions.

Suppose we try to solve this system by the elimination method:

$12y - 8x = 24$	**Multiplying equation (1) by 4**
$-12y + 8x = -24$	**Equation (2)**
$0 = 0.$	**Adding, we obtain a true equation.**

We have eliminated both variables, and what remains is a true equation. It can be expressed as $0 \cdot x + 0 \cdot y = 0$, and is true for all numbers x and y. If a pair is a solution of one of the original equations, then it will be a solution of the other. The system has an infinite number of solutions. The system is consistent and dependent.

When solving a system of two linear equations in two variables:

1. If a false equation is obtained, such as $0 = 7$, then the system has no solution. The system is inconsistent and independent.
2. If a true equation is obtained, such as $0 = 0$, then the system has an infinite number of solutions. The system is consistent and dependent.

Do Exercise 11.

In carrying out the elimination method, it helps to first write the equations in the form $Ax + By = C$. When decimals or fractions occur, it also helps to *clear* before solving.

Example 8 Solve this system:

$$0.2x + 0.3y = 1.7,$$
$$\tfrac{1}{7}x + \tfrac{1}{5}y = \tfrac{29}{35}.$$

We have

$$0.2x + 0.3y = 1.7, \longrightarrow \textbf{Multiplying by 10} \longrightarrow 2x + 3y = 17,$$
$$\tfrac{1}{7}x + \tfrac{1}{5}y = \tfrac{29}{35} \longrightarrow \textbf{Multiplying by 35} \longrightarrow 5x + 7y = 29.$$

We multiplied by 10 to clear the decimals. Multiplication by 35, the least common denominator, clears the fractions. The problem is now identical to Example 5. The solution is $(-32, 27)$, or $x = -32$, $y = 27$.

Do Exercises 12 and 13.

To use the elimination method to solve systems of two equations:

1. Write both equations in the form $Ax + By = C$.
2. Clear any decimals or fractions.
3. Choose a variable to eliminate.
4. Make the chosen variable's terms opposites by multiplying one or both equations by appropriate numbers if necessary.
5. Eliminate a variable by adding the sides of the equations and then solve for the remaining variable.
6. Substitute in either of the original equations to find the value of the other variable.

Comparing Methods

The following table is a summary that compares the graphical, substitution, and elimination methods for solving systems of equations.

When deciding which method to use, consider this table and directions from your instructor. The situation is analogous to having a piece of wood to cut and three different types of saws available. Although all three saws can cut the wood, the "best" choice depends on the particular piece of wood, the type of cut being made, and your level of skill with each saw.

Method	Strengths	Weaknesses
Graphical	Can "see" solutions.	Inexact when solutions involve numbers that are not integers. Solution may not appear on the part of the graph drawn.
Substitution	Yields exact solutions. Convenient to use when a variable has a coefficient of 1.	Can introduce extensive computations with fractions. Cannot "see" solutions quickly.
Elimination	Yields exact solutions. Convenient to use when no variable has a coefficient of 1. The preferred method for systems of 3 or more equations in 3 or more variables (see Section 8.4).	Cannot "see" solutions quickly.

C Solving Applied Problems Involving Two Equations

Many applied problems are easier to solve if we first translate to a system of two equations rather than to a single equation. Using substitution or elimination, we will solve some fairly easy problems here; then in Section 8.3 we will consider more complicated problems. Let's begin with the real-world application involving the Simon–DeBartolo merger that we solved graphically in Section 8.1.

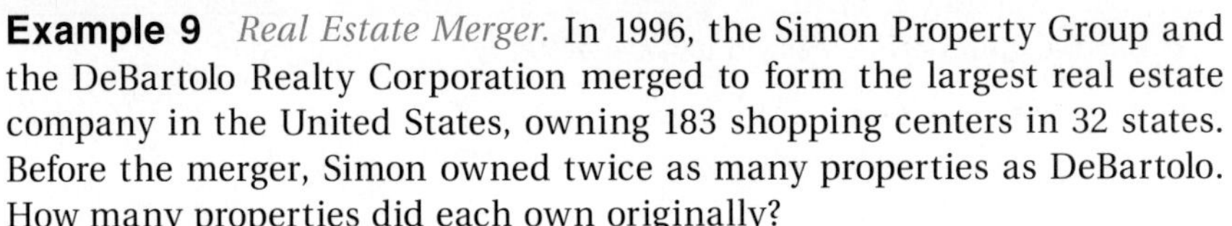

14. *Basketball Scoring.* The Central College Cougars made 40 field goals in a recent basketball game, some 2-pointers and the rest 3-pointers. Altogether the 40 baskets counted for 89 points. How many of each type of field goal was made?

Example 9 *Real Estate Merger.* In 1996, the Simon Property Group and the DeBartolo Realty Corporation merged to form the largest real estate company in the United States, owning 183 shopping centers in 32 states. Before the merger, Simon owned twice as many properties as DeBartolo. How many properties did each own originally?

1., 2. Familiarize and **Translate.** These steps were actually completed at the beginning of Section 8.1. The resulting system of equations is

$$x + y = 183, \quad (1)$$
$$x = 2y, \quad (2)$$

where x is the number of properties originally owned by Simon and y is the number of properties originally owned by DeBartolo.

3. Solve. We solve the system of equations. We note that one equation already has a variable by itself on one side, so substitution can be used. (Keep in mind that unless you are directed otherwise, you can use the method you prefer.)

$$x + y = 183$$
$$2y + y = 183 \quad \text{Substituting } 2y \text{ for } x \text{ in equation (1)}$$
$$3y = 183 \quad \text{Collecting like terms}$$
$$y = 61 \quad \text{Solving for } y$$

Then we substitute 61 for y in equation (2) and solve for x:

$$x = 2y = 2(61) = 122.$$

4. Check. The sum of 122 and 61 is 183, so the total number of properties is correct. Since 122 is twice 61, the numbers check.

5. State. Before the merger, Simon owned 122 properties and DeBartolo owned 61.

Do Exercise 14.

Example 10 *Architecture.* The architects who designed the John Hancock Building in Chicago created a visually appealing building that slants on the sides. Thus the ground floor is a rectangle that is larger than the rectangle formed by the top floor. The ground floor has a perimeter of 860 ft. The length is 100 ft more than the width. Find the length and the width.

1. Familiarize. We first make a drawing and label it. We recall, or look up, the formula for perimeter: $P = 2l + 2w$. This formula can be found at the back of the book.

2. Translate. We translate as follows:

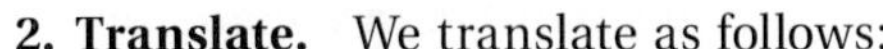

The perimeter	is	860 ft.
↓	↓	↓
$2l + 2w$	$=$	860

We then translate the second statement:

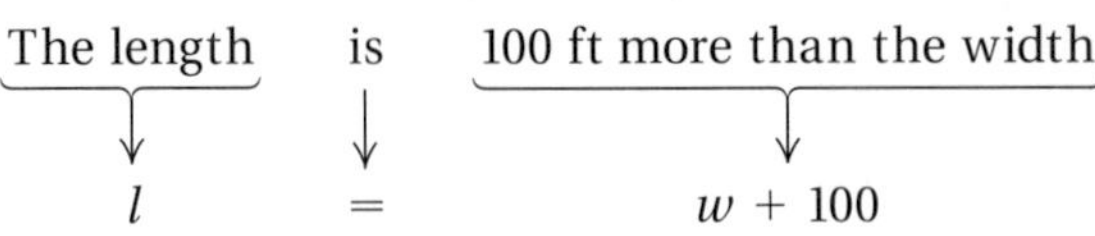

The length	is	100 ft more than the width.
↓	↓	↓
l	$=$	$w + 100$

Answer on page A-33

We now have a system of equations:

$$2l + 2w = 860, \quad (1)$$
$$l = w + 100. \quad (2)$$

3. **Solve.** For comparison, we use both the substitution and elimination methods. Let's first use *substitution,* substituting $w + 100$ for l in equation (1):

$2(w + 100) + 2w = 860$	Substituting in equation (1)
$2w + 200 + 2w = 860$	Multiplying to remove parentheses on the left
$4w + 200 = 860$	Collecting like terms
$4w = 660$	Solving for w
$w = 165.$	

Next, we substitute 165 for w in equation (2) and solve for l:

$$l = 165 + 100 = 265.$$

To use *elimination,* we first rewrite both equations in the form $Ax + By = C$. Since equation (1) is already in this form, we need only rewrite equation (2):

$l = w + 100$	Equation (2)
$l - w = 100.$	Subtracting w

Now we solve the system

$$2l + 2w = 860, \quad (1)$$
$$l - w = 100. \quad (2)$$

We multiply by 2 on both sides of equation (2) and add:

$2l + 2w = 860$	Equation (1)
$2l - 2w = 200$	Multiplying by 2 on both sides of equation (2)
$4l = 1060$	Adding
$l = 265.$	Solving for l

Next, we substitute 265 for l in the equation $l = w + 100$ and solve for w:

$$265 = w + 100$$
$$165 = w.$$

Which method do you prefer to use?

4. **Check.** Consider the dimensions 265 ft and 165 ft. The length is 100 ft more than the width. The perimeter is 2(265 ft) + 2(165 ft), or 860 ft. The dimensions 265 ft and 165 ft check in the original problem.

5. **State.** The length is 265 ft and the width is 165 ft.

Do Exercise 15.

15. *Architecture.* The top floor of the John Hancock Building is also a rectangle, but its perimeter is 520 ft. The width is 60 ft less than the length. Find the length and the width.

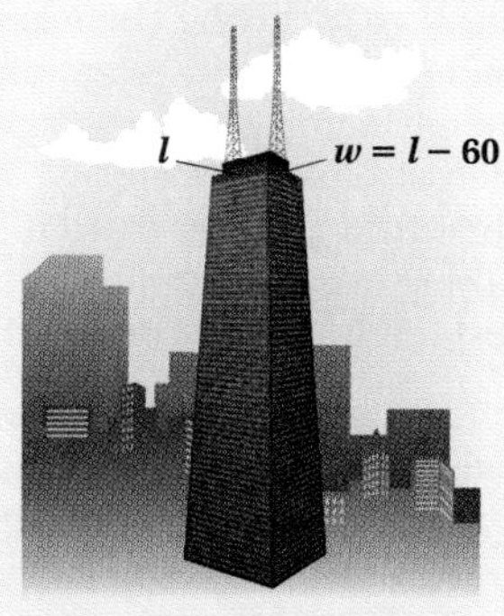

a) First, use substitution to solve the resulting system.

b) Second, use elimination to solve the resulting system.

Answers on page A-33

Improving Your Math Study Skills

Better Test Taking

How often do you make the following statement after taking a test: "I was able to do the homework, but I froze during the test"? Instructors have heard this comment for years, and in most cases, it is merely a coverup for a lack of proper study habits. Here are two related tips, however, to help you with this difficulty. Both are intended to make test taking less stressful by getting you to practice good test-taking habits on a daily basis.

- **Treat *every* homework exercise as if it were a test question.** If you had to work a problem at your job with no backup answer provided, what would you do? You would probably work it very deliberately, checking and rechecking every step. You might work it more than one time, or you might try to work it another way to check the result. Try to use this approach when doing your homework. Treat every exercise as though it were a test question and no answers were provided at the back of the book.
- **Be sure that you do questions without answers as part of every homework assignment whether or not the instructor has assigned them!** One reason a test may seem such a different task from homework is that questions on a test lack answers. That is the reason for taking a test: to see if you can answer the questions without assistance. As part of your test preparation, be sure you do some exercises for which you do not have the answers. Thus when you take a test, you are doing a more familiar task.

The purpose of doing your homework using these approaches is to give you more test-taking practice beforehand. Let's make a sports analogy here. At a basketball game, the players take lots of practice shots before the game. They play the first half, go to the locker room, and come out for the second half. What do they do before the second half, even though they have just played 20 minutes of basketball? They shoot baskets again! We suggest the same approach here. Create more and more situations in which you practice taking test questions by treating each homework exercise like a test question and by doing exercises for which you have no answers. Good luck! Please send me an e-mail (exponent@aol.com) and let me know how it works for you.

Exercise Set 8.2

a Solve the system by the substitution method.

1. $2x + 5y = 4,$
$x = 3 - 3y$

2. $5x - 2y = 23,$
$y = 8 - 4x$

3. $9x - 2y = -6,$
$7x + 8 = y$

4. $x = 3y - 3,$
$x + 2y = 9$

5. $5m + n = 8,$
$3m - 4n = 14$

6. $4x + y = -1,$
$x - 2y = 11$

7. $4x + 13y = 5,$
$-6x + y = 13$

8. $-5a + b = -23,$
$6a + 7b = 3$

b Solve the system by the elimination method.

9. $x + 3y = 7,$
$-x + 4y = 7$

10. $x + y = 9,$
$2x - y = -3$

11. $9x + 5y = 6,$
$2x - 5y = -17$

12. $8x - 3y = 16,$
$8x + 3y = -8$

13. $5x + 3y = 19,$
$2x - 5y = 11$

14. $3x + 2y = 3,$
$9x - 8y = -2$

15. $5r - 3s = 24,$
$3r + 5s = 28$

16. $5x - 7y = -16,$
$2x + 8y = 26$

17. $0.3x - 0.2y = 4,$
$0.2x + 0.3y = 1$

18. $0.7x - 0.3y = 0.5,$
$-0.4x + 0.7y = 1.3$

19. $\frac{1}{2}x + \frac{1}{3}y = 4,$
$\frac{1}{4}x + \frac{1}{3}y = 3$

20. $\frac{2}{3}x + \frac{1}{7}y = -11,$
$\frac{1}{7}x - \frac{1}{3}y = -10$

21. $\frac{2}{5}x + \frac{1}{2}y = 2,$
$\frac{1}{2}x - \frac{1}{6}y = 3$

22. $\frac{1}{3}x + \frac{1}{5}y = 7,$
$\frac{1}{6}x - \frac{2}{5}y = -4$

23. $2x + 3y = 1,$
$4x + 6y = 2$

24. $3x - 2y = 1,$
$-6x + 4y = -2$

25. $2x - 4y = 5,$
$2x - 4y = 6$

26. $3x - 5y = -2,$
$5y - 3x = 7$

27. $5x - 9y = 7,$
$7y - 3x = -5$

28. $a - 2b = 16,$
$b + 3 = 3a$

29. $3(a - b) = 15,$
$4a = b + 1$

30. $10x + y = 306,$
$10y + x = 90$

31. $x - \frac{1}{10}y = 100,$
$y - \frac{1}{10}x = -100$

32. $\frac{1}{8}x + \frac{3}{5}y = \frac{19}{2},$
$-\frac{3}{10}x - \frac{7}{20}y = -1$

33. $0.05x + 0.25y = 22,$
$0.15x + 0.05y = 24$

34. $1.3x - 0.2y = 12,$
$0.4x + 17y = 89$

Solve.

35. *Racquetball Court.* A regulation racquetball court has a perimeter of 120 ft, with a length that is twice the width. Find the length and the width of such a court.

36. *Soccer Field.* The perimeter of a soccer field is 340 m. The length exceeds the width by 50 m. Find the length and the width.

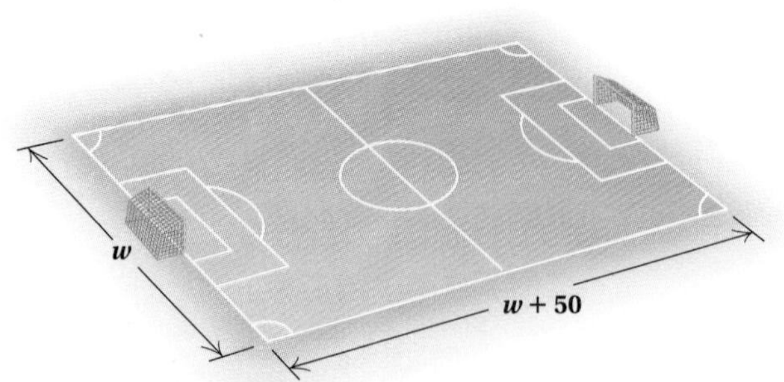

37. *Supplementary Angles.* **Supplementary angles** are angles whose sum is 180°. Two supplementary angles are such that one angle is 12° less than three times the other. Find the measures of the angles.

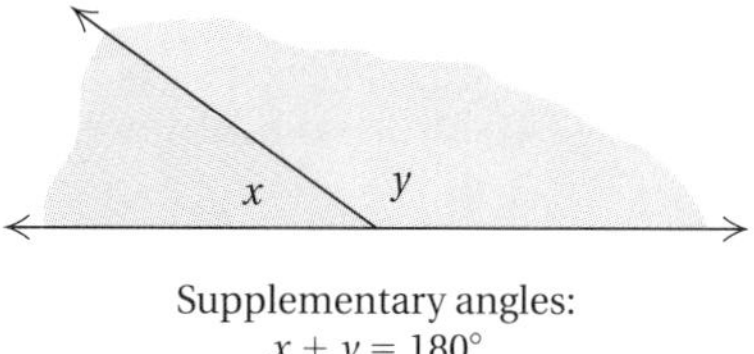

Supplementary angles:
$x + y = 180°$

38. *Complementary Angles.* **Complementary angles** are angles whose sum is 90°. Two complementary angles are such that one angle is 6° more than five times the other. Find the measures of the angles.

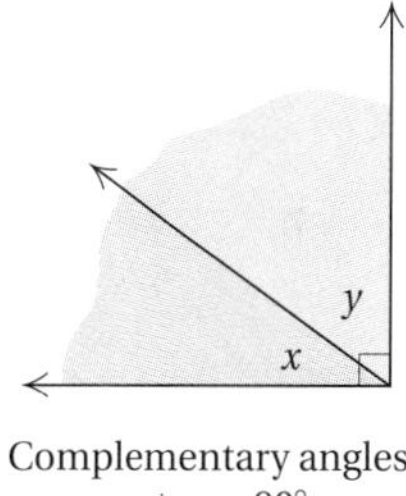

Complementary angles:
$x + y = 90°$

39. The sum of two numbers is -42. The first number minus the second is 52. What are the numbers?

40. The difference between two numbers is 11. Twice the smaller plus three times the larger is 123. What are the numbers?

41. *Hockey Points.* Hockey teams receive two points when they win a game and one point when they tie. One season, a team won a championship with 60 points. They won 9 more games than they tied. How many wins and how many ties did the team have?

42. *Airplane Seating.* An airplane has a total of 152 seats. The number of coach-class seats is 5 more than six times the number of first-class seats. How many of each type of seat are there on the plane?

Skill Maintenance

43. Find the slope of the line $y = 1.3x - 7$. [7.3b]

44. Simplify: $-9(y + 7) - 6(y - 4)$. [1.8b]

45. Solve $A = \frac{pq}{7}$ for p. [2.6a]

46. Find the slope of the line containing the points $(-2, 3)$ and $(-5, -4)$. [7.3b]

Write an equation of the line containing the given point and perpendicular to the given line. [7.5d]

47. $(10, 1)$; $2x - 7y = 3$

48. $(-2, -3)$; $7y - 5x = 1$

Given the function $f(x) = 3x^2 - x + 1$, find each of the following function values. [7.1b]

49. $f(0)$

50. $f(-1)$

51. $f(1)$

52. $f(10)$

53. $f(-2)$

54. $f(2a)$

55. $f(-4)$

56. $f(1.8)$

Synthesis

57. ◆ Describe a method that could be used to create inconsistent systems of equations.

58. ◆ Describe a method that could be used to create dependent systems of equations.

59. Use the INTERSECT feature to solve the following system of equations. You may need to first solve for y. Round answers to the nearest hundredth.

$3.5x - 2.1y = 106.2,$

$4.1x + 16.7y = -106.28$

60. Solve:

$$\frac{x + y}{2} - \frac{x - y}{5} = 1,$$

$$\frac{x - y}{2} + \frac{x + y}{6} = -2.$$

61. Solve for x and y in terms of a and b:

$5x + 2y = a,$

$x - y = b.$

62. Determine a and b for which $(-4, -3)$ will be a solution of the system

$ax + by = -26,$

$bx - ay = 7.$

63. The points $(0, -3)$ and $\left(-\frac{3}{2}, 6\right)$ are two of the solutions of the equation $px - qy = -1$. Find p and q.

64. For $y = mx + b$, two solutions are $(1, 2)$ and $(-3, 4)$. Find m and b.

65. The solution of this system is $(-5, -1)$. Find A and B.

$Ax - 7y = -3,$

$x - By = -1$

66. Find an equation to pair with $6x + 7y = -4$ such that $(-3, 2)$ is a solution of the system.

Collaborative Learning Manual

Compare the three methods for solving systems of equations in two variables.

8.3 Solving Applied Problems: Systems of Two Equations

Objectives

[a] Solve applied problems involving total value and mixture using systems of two equations.

[b] Solve applied problems involving motion using systems of two equations.

For Extra Help

TAPE 13 MAC WIN CD-ROM

a Total-Value and Mixture Problems

Systems of equations can be a useful tool in solving applied problems. Using systems often makes the *Translate* step easier than using a single equation. The first kind of problem we consider involves quantities of items sold and the total value of the items. We refer to this type of problem as a **total-value problem.**

Example 1 *Retail Sales of Gloves.* In one day, Glovers, Inc., sold 20 pairs of gloves. Fleece gloves sold for \$24.95 a pair and Gore-Tex gloves for \$37.50. Receipts totaled \$687.25. How many of each kind of glove were sold?

1. **Familiarize.** To familiarize ourselves with the problem situation, let's guess that Glovers sold 12 pairs of fleece gloves and 8 pairs of Gore-Tex gloves. The total is 20. How much money was taken in? Since fleece gloves sold for \$24.95 a pair and Gore-Tex for \$37.50, the total received would then be

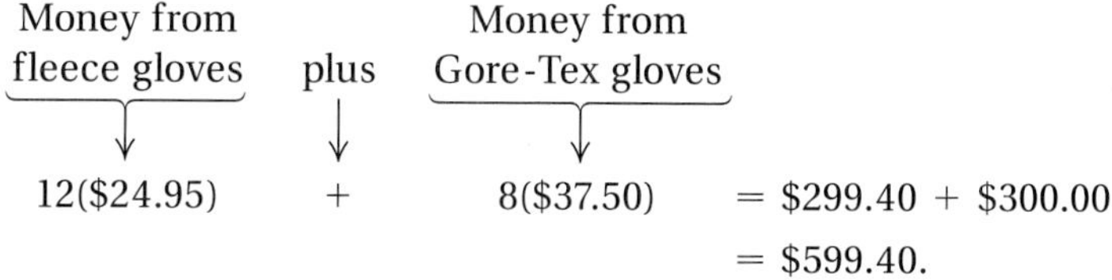

Although the total number of pairs is correct, our guess is incorrect because the problem states that the total amount received was \$687.25. Since \$599.40 is less than \$687.25, more of the expensive gloves were sold than we had guessed. We could now adjust our guess accordingly. Instead, let's use an algebraic approach that avoids guessing.

We let p = the number of pairs of fleece gloves sold and q = the number of pairs of Gore-Tex gloves sold. It helps to organize the information in a table as follows:

Kind of Glove	Fleece	Gore-Tex	Total	
Number Sold	p	q	20	→ $p + q = 20$
Price	\$24.95	\$37.50		
Amount Taken In	$24.95p$	$37.50q$	687.25	→ $24.95p + 37.50q = 687.25$

2. **Translate.** The first row of the table and the first sentence of the problem tell us that a total of 20 pairs of gloves was sold. Thus we have one equation:

$$p + q = 20.$$

Since each pair of fleece gloves costs \$24.95 and p pairs were sold, $24.95p$ is the amount taken in from the sale of fleece gloves. Similarly, $37.50q$ is the amount taken in from the sale of q pairs of the Gore-Tex gloves. From the third row of the table and the third sentence of the problem, we get the second equation:

$$24.95p + 37.50q = 687.25.$$

1. *Retail Sales of Sweatshirts.* Sandy's Sweatshirt Shop sells college sweatshirts. White sweatshirts sell for \$18.95 each and red ones sell for \$19.50 each. If receipts for 30 shirts total \$572.90, how many of each color did the shop sell? Complete the following table, letting w = the number of white sweatshirts and r = the number of red sweatshirts.

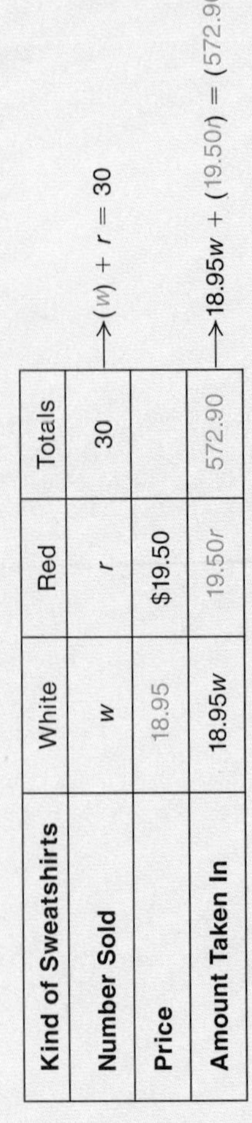

Kind of Sweatshirts	White	Red	Totals	
Number Sold	w	r	30	$\rightarrow (w) + r = 30$
Price	18.95	\$19.50		
Amount Taken In	$18.95w$	$19.50r$	572.90	$\rightarrow 18.95w + (19.50r) = (572.90)$

White: 22; red: 8

Answer on page A-33

We can multiply by 100 on both sides of this equation in order to clear the decimals. This gives us the following system of equations as a translation:

$$p + q = 20, \quad (1)$$
$$2495p + 3750q = 68{,}725. \quad (2)$$

3. **Solve.** We choose to use the elimination method to solve the system. We eliminate p by multiplying equation (1) by -2495 and adding it to equation (2):

$$\begin{aligned} -2495p - 2495q &= -49{,}900 && \text{Multiplying equation (1) by } -2495 \\ 2495p + 3750q &= 68{,}725 \\ \hline 1255q &= 18{,}825 && \text{Adding} \\ q &= 15. && \text{Solving for } q \end{aligned}$$

To find p, we substitute 15 for q in equation (1) and solve for p:

$$\begin{aligned} p + q &= 20 && \text{Equation (1)} \\ p + 15 &= 20 && \text{Substituting 15 for } q \\ p &= 5. && \text{Solving for } p \end{aligned}$$

We obtain (5, 15), or $p = 5$, $q = 15$.

4. **Check.** We check in the original problem. Remember that p is the number of pairs of fleece gloves and q the number of pairs of Gore-Tex gloves:

Number of gloves: $p + q = 5 + 15 = 20$
Money from fleece gloves: $\$24.95p = 24.95 \times 5 = \$124.75$
Money from Gore-Tex gloves: $\$37.50q = 37.50 \times 15 = \$562.50$
Total = \$687.25

The numbers check.

5. **State.** The store sold 5 pairs of fleece gloves and 15 pairs of Gore-Tex gloves.

Do Exercise 1.

The following problem, similar to Example 1, is called a **mixture problem.**

Example 2 *Blending Teas.* Tara's Tea Terrace sells loose Black tea for 95¢ per ounce and Lapsang Souchong tea for \$1.43 per ounce. Tara wants to mix the two teas to get a 1-lb mixture, called Imperial Blend, that sells for \$1.10 per ounce. How many ounces of each type of tea should Tara use?

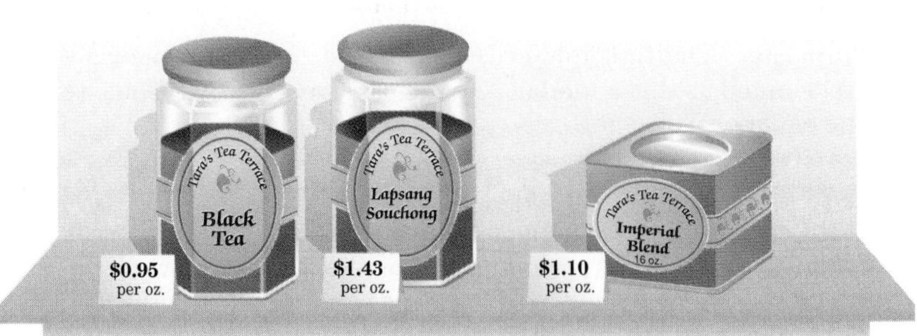

1. **Familiarize.** To familiarize ourselves with the problem situation, we make a guess and do some calculations. The total amount of tea is to be 1 lb, or 16 oz. Let's try 12 oz of Black tea and 4 oz of Lapsang Souchong.

The sum of the amounts of tea is $12 + 4$, or 16.

The values of these amounts of tea are found by multiplying the cost per ounce, in dollars, by the number of ounces and adding:

$0.95(12) + $1.43(4), or $17.12.

The desired cost is $1.10 per ounce. If we multiply $1.10 by 16, we get 16($1.10), or $17.60. This does not agree with $17.12, but these calculations help us to translate.

We let a = the number of ounces of Black tea and b = the number of ounces of Lapsang Souchong tea. Next, we organize the information in a table, as follows.

	Black Tea	Lapsang Souchong	Imperial Blend	
Number of Ounces	a	b	16	→ $a + b = 16$
Price per Ounce	$0.95	$1.43	$1.10	
Value of Tea	$0.95a$	$1.43b$	$16 \cdot 1.10$, or 17.60	→ $0.95a + 1.43b = 17.60$

2. **Translate.** The total amount of tea is 16 oz, so we have

$$a + b = 16.$$

The value of the Black tea is $0.95a$ and the value of the Lapsang Souchong tea is $1.43b$. These amounts are in dollars. Since the total is to be 16($1.10), or $17.60, we have

$$0.95a + 1.43b = 17.60.$$

We can multiply by 100 on both sides of this equation to clear the decimals. Thus we have the translation, a system of equations:

$$a + b = 16, \qquad (1)$$
$$95a + 143b = 1760. \qquad (2)$$

3. **Solve.** We decide to use substitution, although elimination could be used as we did in Example 1. When equation (1) is solved for b, we get $b = 16 - a$. Substituting $16 - a$ for b in equation (2) and solving gives us

$$95a + 143(16 - a) = 1760 \quad \text{Substituting}$$
$$95a + 2288 - 143a = 1760 \quad \text{Using the distributive law}$$
$$-48a = -528 \quad \text{Subtracting 2288 and collecting like terms}$$
$$a = 11.$$

We have $a = 11$. Substituting this value in the equation $b = 16 - a$, we obtain $b = 16 - 11$, or 5.

2. *Blending Coffees.* The Coffee Counter charges $9.00 per pound for Kenyan French Roast coffee and $8.00 per pound for Sumatran coffee. How much of each type should be used to make a 20-lb blend that sells for $8.40 per pound?

Answer on page A-33

3. *Client Investments.* Kaufman Financial Corporation makes investments for corporate clients. It makes an investment of \$3700 for one year at simple interest, yielding \$297. Part of the money is invested at 7% and the rest at 9%. How much was invested at each rate?

Do the *Familiarize* and *Translate* steps by completing the following table. Let x = the number of dollars invested at 7% and y = the number of dollars invested at 9%.

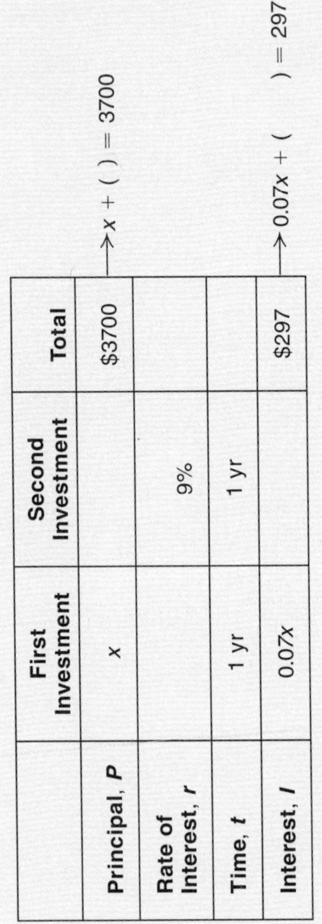

	First Investment	Second Investment	Total	
Principal, P	x		\$3700	→ $x + (\ ) = 3700$
Rate of Interest, r		9%		
Time, t	1 yr	1 yr		
Interest, I	$0.07x$		\$297	→ $0.07x + (\ \ ) = 297$

Answer on page A-33

4. **Check.** We check in a manner similar to our guess in the *Familiarize* step. The total amount of tea is 11 + 5, or 16 oz. The value of the tea is

$$\$0.95(11) + \$1.43(5), \text{ or } \$17.60.$$

Thus the amounts of tea check.

5. **State.** The Imperial Blend can be made by mixing 11 oz of Black tea with 5 oz of Lapsang Souchong tea.

Do Exercise 2 on the preceding page.

Example 3 *Student Loans.* Enid's student loans totaled \$9600. Part was a Perkins loan made at 5% interest and the rest was a Federal Education Loan made at 8% interest. After one year, Enid's loans accumulated \$633 in interest. What was the amount of each loan?

1. **Familiarize.** Listing the given information in a table will help. The columns in the table come from the formula for simple interest: $I = Prt$. We let x = the number of dollars in the Perkins loan and y = the number of dollars in the Federal Education loan.

	Perkins Loan	Federal Loan	Total	
Principal	x	y	\$9600	→ $x + y = 9600$
Rate of Interest	5%	8%		
Time	1 yr	1 yr		
Interest	$0.05x$	$0.08y$	\$633	→ $0.05x + 0.08y = 633$

2. **Translate.** The total of the amounts of the loans is found in the first row of the table. This gives us one equation:

$$x + y = 9600.$$

Look at the last row of the table. The interest, or **yield**, totals \$633. This gives us a second equation:

$$5\%x + 8\%y = 633, \quad \text{or} \quad 0.05x + 0.08y = 633.$$

After we multiply on both sides to clear the decimals, we have

$$5x + 8y = 63{,}300.$$

3. **Solve.** Using either elimination or substitution, we solve the resulting system:

$$x + y = 9600,$$
$$5x + 8y = 63{,}300.$$

We find that $x = 4500$ and $y = 5100$.

4. **Check.** The sum is \$4500 + \$5100, or \$9600. The interest from \$4500 at 5% for one year is 5%(\$4500), or \$225. The interest from \$5100 at 8% for one year is 8%(\$5100), or \$408. The total interest is \$225 + \$408, or \$633. The numbers check in the problem.

5. **State.** The Perkins loan was for \$4500 and the Federal Education loan was for \$5100.

Do Exercise 3.

Example 4 *Mixing Fertilizers.* Yardbird Gardening carries two kinds of fertilizer containing nitrogen and water. "Gently Green" is 5% nitrogen and "Sun Saver" is 15% nitrogen. Yardbird Gardening needs to combine the two types of solution to make 90 L of a solution that is 12% nitrogen. How much of each brand should be used?

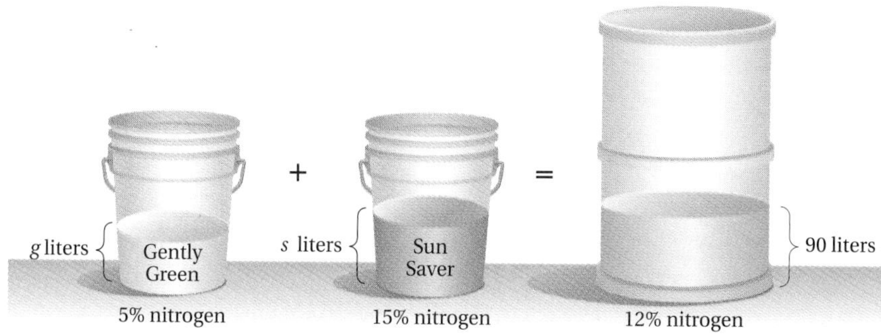

1. **Familiarize.** We first make a drawing and a guess to become familiar with the problem.

   We choose two numbers that total 90 L—say, 40 L of Gently Green and 50 L of Sun Saver—for the amounts of each fertilizer. Will the resulting mixture have the correct percentage of fertilizer? To find out, we multiply as follows:

   $5\%(40 \text{ L}) = 2 \text{ L of nitrogen}$ and $15\%(50 \text{ L}) = 7.5 \text{ L of nitrogen}.$

   Thus the total amount of nitrogen in the mixture is 2 L + 7.5 L, or 9.5 L. The final mixture of 90 L is supposed to have 12% nitrogen. Now

   $12\%(90 \text{ L}) = 10.8 \text{ L}.$

   Since 9.5 L and 10.8 L are not the same, our guess is incorrect. But these calculations help us to become familiar with the problem and to make the translation.

   We let g = the number of liters of Gently Green and s = the number of liters of Sun Saver.

   The information can be organized in a table, as follows.

	Gently Green	Sun Saver	Mixture	
Number of Liters	g	s	90	$\rightarrow g + s = 90$
Percent of Nitrogen	5%	15%	12%	
Amount of Nitrogen	$0.05g$	$0.15s$	0.12×90, or 10.8 liters	$\rightarrow 0.05g + 0.15s = 10.8$

2. **Translate.** If we add g and s in the first row, we get 90, and this gives us one equation:

   $g + s = 90.$

   If we add the amounts of nitrogen listed in the third row, we get 10.8, and this gives us another equation:

   $5\%g + 15\%s = 10.8,$ or $0.05g + 0.15s = 10.8.$

4. *Mixing Cleaning Solutions.* King's Service Station uses two kinds of cleaning solution containing acid and water. "Attack" is 2% acid and "Blast" is 6% acid. They want to mix the two to get 60 qt of a solution that is 5% acid. How many quarts of each should they use?

Do the *Familiarize* and *Translate* steps by completing the following table. Let a = the number of quarts of Attack and b = the number of quarts of Blast.

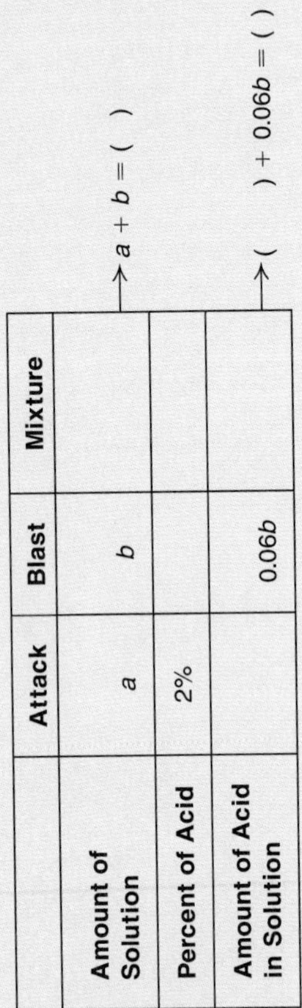

	Attack	Blast	Mixture	
Amount of Solution	a	b		→ $a + b = (\ \ )$
Percent of Acid	2%			
Amount of Acid in Solution		$0.06b$		→ $(\ \ \ ) + 0.06b = (\ \ )$

Answer on page A-33

After clearing the decimals, we have the following system:

$$g + s = 90, \quad (1)$$
$$5g + 15s = 1080. \quad (2)$$

3. Solve. We solve the system using elimination. We multiply equation (1) by -5 and add the result to equation (2):

$$\begin{aligned} -5g - \ \ 5s &= -450 && \text{Multiplying equation (1) by } -5 \\ 5g + 15s &= 1080 \\ \hline 10s &= 630 && \text{Adding} \\ s &= 63; && \text{Dividing by 10} \end{aligned}$$

$$\begin{aligned} g + 63 &= 90 && \text{Substituting in equation (1) of the system} \\ g &= 27. && \text{Solving for g} \end{aligned}$$

4. Check. Remember that g is the number of liters of Gently Green, with 5% nitrogen, and s is the number of liters of Sun Saver, with 15% nitrogen.

Total number of liters of mixture: $g + s = 27 + 63 = 90$

Amount of nitrogen: $5\%(27) + 15\%(63) = 1.35 + 9.45 = 10.8$ L

Percentage of nitrogen in mixture: $\dfrac{10.8}{90} = 0.12 = 12\%$

The numbers check in the original problem.

5. State. Yardbird Gardening should mix 27 L of Gently Green with 63 L of Sun Saver.

Do Exercise 4.

b Motion Problems

When a problem deals with speed, distance, and time, we can expect to use the following *motion formula*.

THE MOTION FORMULA

Distance = Rate (or speed) · Time

$$d = rt$$

We have five steps for problem solving. The following tips are also helpful when solving motion problems.

TIPS FOR SOLVING MOTION PROBLEMS

1. Draw a diagram using an arrow or arrows to represent distance and the direction of each object in motion.
2. Organize the information in a table or chart.
3. Look for as many things as you can that are the same, so you can write equations.

Example 5 *Auto Travel.* Your brother leaves on a trip, forgetting his suitcase. You know that he normally drives at a speed of 55 mph. You do not discover the suitcase until 1 hr after he has left. If you follow him at a speed of 65 mph, how long will it take you to catch up with him?

1. Familiarize. We first make a drawing.

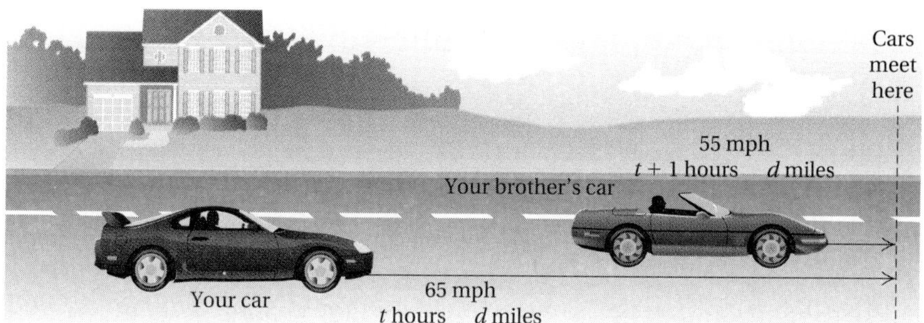

From the drawing, we see that when you catch up with your brother, the distances from home are the same. Let's call the distance d. If we let $t =$ the time for you to catch your brother, then $t + 1 =$ the time traveled by your brother at a slower speed. We organize the information in a table.

	d Distance	$=$ r Rate	$\cdot$ t Time	
Brother	d	55	$t + 1$	$\rightarrow d = 55(t + 1)$
You	d	65	t	$\rightarrow d = 65t$

2. Translate. Using $d = rt$ in each row of the table, we get an equation. Thus we have a system of equations:

$$d = 55(t + 1), \quad (1)$$
$$d = 65t. \quad (2)$$

3. Solve. We solve the system using the substitution method:

$$65t = 55(t + 1) \quad \text{Substituting } 65t \text{ for } d \text{ in equation (1)}$$
$$65t = 55t + 55 \quad \text{Multiplying to remove parentheses on the right}$$
$$\left.\begin{aligned} 10t &= 55 \\ t &= 5.5. \end{aligned}\right\} \quad \text{Solving for } t$$

Your time is 5.5 hr, which means that your brother's time is $5.5 + 1$, or 6.5 hr.

4. Check. At 65 mph, you will travel $65 \cdot 5.5$, or 357.5 mi, in 5.5 hr. At 55 mph, your brother will travel $55 \cdot 6.5$, or the same 357.5 mi, in 6.5 hr. The numbers check.

5. State. You will overtake your brother in 5.5 hr.

Do Exercise 5.

5. *Train Travel.* A train leaves Barstow traveling east at 35 km/h. One hour later, a faster train leaves Barstow, also traveling east on a parallel track at 40 km/h. How far from Barstow will the faster train catch up with the slower one?

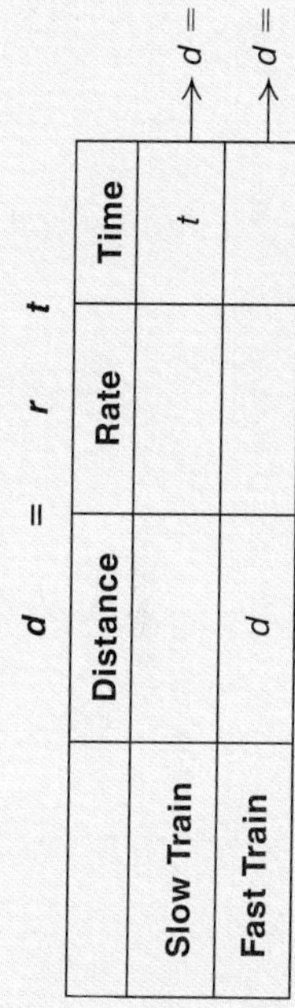

	d Distance	$=$ r Rate	$\cdot$ t Time	
Slow Train			t	$\rightarrow d =$
Fast Train	d			$\rightarrow d =$

Answer on page A-34

6. *Air Travel.* An airplane flew for 4 hr with a 20-mph tailwind. The return flight against the same wind took 5 hr. Find the speed of the plane in still air.

$d = r \cdot t$

	Distance	Rate	Time	
With Wind		$r + 20$		$\to d =$
Against Wind	d			$\to d =$

Answer on page A-34

Example 6 *Marine Travel.* A Coast-Guard patrol boat travels 4 hr on a trip downstream with a 6-mph current. The return trip against the same current takes 5 hr. Find the speed of the boat in still water.

1. **Familiarize.** We first make a drawing. From the drawing, we see that the distances are the same. We let d = the distance, in miles, and r = the speed of the boat in still water, in miles per hour. Then, when the boat is traveling downstream, its speed is $r + 6$ (the current helps the boat along). When it is traveling upstream, its speed is $r - 6$ (the current holds the boat back). We can organize the information in a table. We use the formula $d = rt$.

$d = r \cdot t$

	Distance	Rate	Time	
Downstream	d	$r + 6$	4	$\to d = (r + 6)4$
Upstream	d	$r - 6$	5	$\to d = (r - 6)5$

2. **Translate.** From each row of the table, we get an equation, $d = rt$:

$$d = 4r + 24, \quad (1)$$
$$d = 5r - 30. \quad (2)$$

3. **Solve.** We solve the system using the substitution method:

$$4r + 24 = 5r - 30 \qquad \text{Substituting } 4r + 24 \text{ for } d \text{ in equation (2)}$$
$$\left.\begin{aligned} 24 &= r - 30 \\ 54 &= r. \end{aligned}\right\} \qquad \text{Solving for } r$$

4. **Check.** If $r = 54$, then $r + 6 = 60$; and $60 \cdot 4 = 240$, the distance traveled downstream. If $r = 54$, then $r - 6 = 48$; and $48 \cdot 5 = 240$, the distance traveled upstream. The distances are the same. In this type of problem, a problem-solving tip to keep in mind is "Have I found what the problem asked for?" We could solve for a certain variable but still have not answered the question of the original problem. For example, we might have found speed when the problem wanted distance. In this problem, we want the speed of the boat in still water, and that is r.

5. **State.** The speed in still water is 54 mph.

Do Exercise 6.

Exercise Set 8.3

a Solve.

1. *Retail Sales.* Paint Town sold 45 paintbrushes, one kind at \$8.50 each and another at \$9.75 each. In all, \$398.75 was taken in for the brushes. How many of each kind were sold?

2. *Retail Sales.* Mountainside Fleece sold 40 neckwarmers. Solid-color neckwarmers sold for \$9.90 each and print ones sold for \$12.75 each. In all, \$421.65 was taken in for the neckwarmers. How many of each type were sold?

3. *Sales of Pharmaceuticals.* The Diabetic Express recently charged \$15.75 for a vial of Humulin insulin and \$12.95 for a vial of Novolin insulin. If a total of \$959.35 was collected for 65 vials of insulin, how many vials of each type were sold?

4. *Fundraising.* The St. Mark's Community Barbecue served 250 dinners. A child's plate cost \$3.50 and an adult's plate cost \$7.00. A total of \$1347.50 was collected. How many of each type of plate was served?

5. *Radio Airplay.* Omar must play 12 commercials during his 1-hr radio show. Each commercial is either 30 sec or 60 sec long. If the total commercial time during that hour is 10 min, how many commercials of each type does Omar play?

6. *Nontoxic Floor Wax.* A nontoxic floor wax can be made by combining lemon juice and food-grade linseed oil. The amount of oil should be twice the amount of lemon juice. How much of each ingredient is needed in order to make 32 oz of floor wax? (The mix should be spread with a rag and buffed when dry.)

7. *Catering.* Casella's Catering is planning a wedding reception. The bride and groom would like to serve a nut mixture containing 25% peanuts. Casella has available mixtures that are either 40% or 10% peanuts. How much of each type should be mixed to get a 10-lb mixture that is 25% peanuts?

8. *Blending Granola.* Deep Thought Granola is 25% nuts and dried fruit. Oat Dream Granola is 10% nuts and dried fruit. How much of Deep Thought and how much of Oat Dream should be mixed to form a 20-lb batch of granola that is 19% nuts and dried fruit?

9. *Ink Remover.* Etch Clean Graphics uses one cleanser that is 25% acid and a second that is 50% acid. How many liters of each should be mixed to get 10 L of a solution that is 40% acid?

10. *Livestock Feed.* Soybean meal is 16% protein and corn meal is 9% protein. How many pounds of each should be mixed to get a 350-lb mixture that is 12% protein?

11. *Student Loans.* Lomasi's two student loans totaled \$12,000. One of her loans was at 6% simple interest and the other at 9%. After one year, Lomasi owed \$855 in interest. What was the amount of each loan?

12. *Investments.* An executive nearing retirement made two investments totaling \$15,000. In one year, these investments yielded \$1432 in simple interest. Part of the money was invested at 9% and the rest at 10%. How much was invested at each rate?

13. *Food Science.* The following bar graph shows the milk fat percentages in three dairy products. How many pounds each of whole milk and cream should be mixed to form 200 lb of milk for cream cheese?

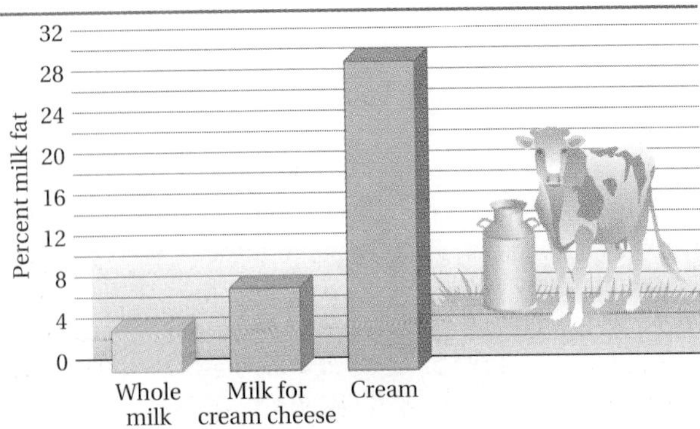

14. *Automotive Maintenance.* "Arctic Antifreeze" is 18% alcohol and "Frost No-More" is 10% alcohol. How many liters of each should be mixed to get 20 L of a mixture that is 15% alcohol?

15. *Teller Work.* Ashford goes to a bank and gets change for a \$50 bill consisting of all \$5 bills and \$1 bills. There are 22 bills in all. How many of each kind are there?

16. *Making Change.* Cecilia makes a \$9.25 purchase at the bookstore with a \$20 bill. The store has no bills and gives her the change in quarters and fifty-cent pieces. There are 30 coins in all. How many of each kind are there?

b Solve.

17. *Train Travel.* A train leaves Danville Junction and travels north at a speed of 75 mph. Two hours later, a second train leaves on a parallel track and travels north at 125 mph. How far from the station will they meet?

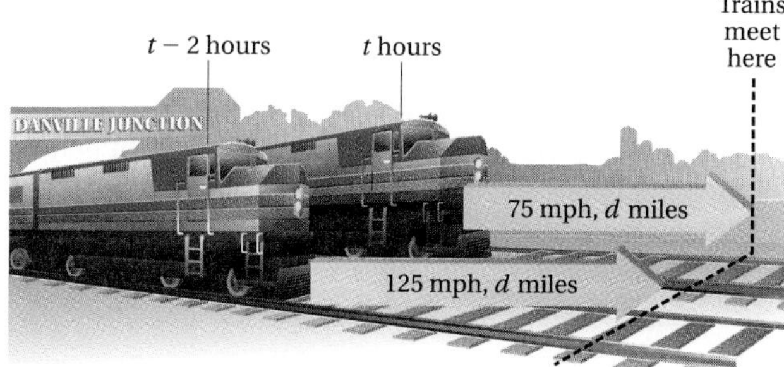

18. *Car Travel.* Two cars leave Denver traveling in opposite directions. One car travels at a speed of 80 km/h and the other at 96 km/h. In how many hours will they be 528 km part?

19. *Canoeing.* Alvin paddled for 4 hr with a 6-km/h current to reach a campsite. The return trip against the same current took 10 hr. Find the speed of Alvin's canoe in still water.

20. *Boating.* Mia's motorboat took 3 hr to make a trip downstream with a 6-mph current. The return trip against the same current took 5 hr. Find the speed of the boat in still water.

21. *Car Travel.* Donna is late for a sales meeting after traveling from one town to another at a speed of 32 mph. If she had traveled 4 mph faster, she could have made the trip in $\frac{1}{2}$ hr less time. How far apart are the towns?

22. *Air Travel.* Rod is a pilot for Crossland Airways. He computes his flight time against a headwind for a trip of 2900 mi at 5 hr. The flight would take 4 hr and 50 min if the headwind were half as great. Find the headwind and the plane's air speed.

23. *Air Travel.* Two planes travel toward each other from cities that are 780 km apart at rates of 190 km/h and 200 km/h. They started at the same time. In how many hours will they meet?

24. *Motorcycle Travel.* Sally and Rocky travel on motorcycles toward each other from Chicago and Indianapolis, which are about 350 km apart, and they are biking at rates of 110 km/h and 90 km/h. They started at the same time. In how many hours will they meet?

25. *Air Travel.* Two airplanes start at the same time and fly toward each other from points 1000 km apart at rates of 420 km/h and 330 km/h. After how many hours will they meet?

26. *Truck and Car Travel.* A truck and a car leave a service station at the same time and travel in the same direction. The truck travels at 55 mph and the car at 40 mph. They can maintain CB radio contact within a range of 10 mi. When will they lose contact?

27. *Point of No Return.* A plane flying the 3458-mi trip from New York City to London has a 50-mph tailwind. The flight's *point of no return* is the point at which the flight time required to return to New York is the same as the time required to continue to London. If the speed of the plane in still air is 360 mph, how far is New York from the point of no return?

28. *Point of No Return.* A plane is flying the 2553-mi trip from Los Angeles to Honolulu into a 60-mph headwind. If the speed of the plane in still air is 310 mph, how far from Los Angeles is the plane's point of no return? (See Exercise 27.)

Skill Maintenance

Given the function $f(x) = 4x - 7$, find each of the following function values. [7.1b]

29. $f(0)$	30. $f(-1)$	31. $f(1)$	32. $f(10)$
33. $f(-2)$	34. $f(2a)$	35. $f(-4)$	36. $f(1.8)$
37. $f\left(\frac{3}{4}\right)$	38. $f(-2.5)$	39. $f(-3h)$	40. $f(1000)$

Synthesis

41. ◆ List three or four study tips for someone beginning this exercise set.

42. ◆ Write a problem similar to Example 1 for a classmate to solve. Design the problem so the answer is "The florist sold 14 hanging plants and 9 flats of petunias."

43. *Automotive Maintenance.* The radiator in Michelle's car contains 16 L of antifreeze and water. This mixture is 30% antifreeze. How much of this mixture should she drain and replace with pure antifreeze so that there will be a mixture of 50% antifreeze?

44. *Physical Exercise.* Natalie jogs and walks to school each day. She averages 4 km/h walking and 8 km/h jogging. The distance from home to school is 6 km and Natalie makes the trip in 1 hr. How far does she jog in a trip?

45. *Fuel Economy.* Ellen Jordan's station wagon gets 18 miles per gallon (mpg) in city driving and 24 mpg in highway driving. The car is driven 465 mi on 23 gal of gasoline. How many miles were driven in the city and how many were driven on the highway?

46. *Gender.* Phil and Phyllis are siblings. Phyllis has twice as many brothers as she has sisters. Phil has the same number of brothers as sisters. How many girls and how many boys are in the family?

47. *Wood Stains.* Bennett Custom Flooring has 0.5 gal of stain that is 20% brown and 80% neutral. A customer orders 1.5 gal of a stain that is 60% brown and 40% neutral. How much pure brown stain and how much neutral stain should be added to the original 0.5 gal in order to make up the order?

48. See Exercise 47. Let $x =$ the amount of pure brown stain added to the original 0.5 gal. Find a function $P(x)$ that can be used to determine the percentage of brown stain in the 1.5-gal mixture. On a grapher, draw the graph of P and use ZOOM and TRACE or the TABLE feature to confirm the answer to Exercise 47.

8.4 Systems of Equations in Three Variables

Objective

a Solve systems of three equations in three variables.

For Extra Help

TAPE 14 MAC WIN CD-ROM

a Solving Systems in Three Variables

A **linear equation in three variables** is an equation equivalent to one of the type $Ax + By + Cz = D$. A solution of a system of three equations in three variables is an ordered triple (x, y, z) that makes *all three* equations true.

The substitution method can be used to solve systems of three equations, but it is not efficient unless a variable has already been eliminated from one or more of the equations. Therefore, we will use only the elimination method—essentially the same procedure for systems of three equations as for systems of two equations. The first step is to eliminate a variable and obtain a system of two equations in two variables.

Example 1 Solve the following system of equations:

$$\begin{aligned} x + y + z &= 4, && (1)\\ x - 2y - z &= 1, && (2)\\ 2x - y - 2z &= -1. && (3) \end{aligned}$$

a) We first use *any* two of the three equations to get an equation in two variables. In this case, let's use equations (1) and (2) and add to eliminate z:

$$\begin{aligned} x + y + z &= 4 && (1)\\ x - 2y - z &= 1 && (2)\\ \hline 2x - y \quad &= 5. && (4) \quad \text{Adding} \end{aligned}$$

b) We use a different pair of equations and eliminate the *same variable* that we did in part (a). Let's use equations (1) and (3) and again eliminate z. Be careful! A common error is to eliminate a different variable the second time.

$$\begin{aligned} x + y + z &= 4 && (1)\\ 2x - y - 2z &= -1 && (3) \end{aligned}$$

$$\begin{aligned} 2x + 2y + 2z &= 8 && \text{Multiplying equation (1) by 2}\\ 2x - y - 2z &= -1 && (3)\\ \hline 4x + y \quad &= 7 && (5) \quad \text{Adding} \end{aligned}$$

c) Now we solve the resulting system of equations, (4) and (5). That solution will give us two of the numbers. Note that we now have two equations in two variables. Had we eliminated two different variables in parts (a) and (b), this would not be the case.

$$\begin{aligned} 2x - y &= 5 && (4)\\ 4x + y &= 7 && (5)\\ \hline 6x \quad &= 12 && \text{Adding}\\ x &= 2 \end{aligned}$$

We can use either equation (4) or (5) to find y. We choose equation (5):

$$\begin{aligned} 4x + y &= 7 && (5)\\ 4(2) + y &= 7 && \text{Substituting 2 for } x\\ 8 + y &= 7\\ y &= -1. \end{aligned}$$

1. Solve. Don't forget to check.

$$4x - y + z = 6,$$
$$-3x + 2y - z = -3,$$
$$2x + y + 2z = 3$$

Answer on page A-34

d) We now have $x = 2$ and $y = -1$. To find the value for z, we use any of the original three equations and substitute to find the third number, z. Let's use equation (1) and substitute our two numbers in it:

$$x + y + z = 4 \quad (1)$$
$$2 + (-1) + z = 4 \quad \text{Substituting 2 for } x \text{ and } -1 \text{ for } y$$
$$\left.\begin{aligned} 1 + z &= 4 \\ z &= 3. \end{aligned}\right\} \quad \text{Solving for } z$$

We have obtained the ordered triple $(2, -1, 3)$. We check as follows, substituting $(2, -1, 3)$ into each of the three equations using alphabetical order.

CHECK:

$$\begin{array}{c|c} x + y + z & 4 \\ \hline 2 + (-1) + 3 \ ? & 4 \\ 4 & \end{array} \quad \text{TRUE}$$

$$\begin{array}{c|c} x - 2y - z & 1 \\ \hline 2 - 2(-1) - 3 \ ? & 1 \\ 2 + 2 - 3 & \\ 1 & \end{array} \quad \text{TRUE}$$

$$\begin{array}{c|c} 2x - y - 2z & -1 \\ \hline 2(2) - (-1) - 2 \cdot 3 \ ? & -1 \\ 4 + 1 - 6 & \\ -1 & \end{array} \quad \text{TRUE}$$

The triple $(2, -1, 3)$ checks and is the solution.

To use the elimination method to solve systems of three equations:

1. Write all equations in the standard form $Ax + By + Cz = D$.
2. Clear any decimals or fractions.
3. Choose a variable to eliminate. Then use *any* two of the three equations to eliminate that variable, getting an equation in two variables.
4. Next, use a different pair of equations and get another equation in *the same two variables.* That is, eliminate the same variable that you did in step (3).
5. Solve the resulting system (pair) of equations. That will give two of the numbers.
6. Then use any of the original three equations to find the third number.

Do Exercise 1.

Example 2 Solve this system:

$$4x - 2y - 3z = 5, \quad (1)$$
$$-8x - y + z = -5, \quad (2)$$
$$2x + y + 2z = 5. \quad (3)$$

a) The equations are in standard form and do not contain decimals or fractions.

b) We decide to eliminate the variable y since the y-terms are opposites in equations (2) and (3). We add:

$$\begin{aligned} -8x - y + z &= -5 \quad (2) \\ 2x + y + 2z &= \ \ 5 \quad (3) \\ \hline -6x \qquad + 3z &= \ \ 0. \quad (4) \quad \text{Adding} \end{aligned}$$

c) We use another pair of equations to get an equation in the same two variables, x and z. That is, we eliminate the same variable y that we did in step (b). We use equations (1) and (3) and eliminate y:

$$4x - 2y - 3z = 5 \qquad (1)$$
$$2x + y + 2z = 5 \qquad (3)$$

$$\begin{array}{rll} 4x - 2y - 3z = 5 & (1) & \\ 4x + 2y + 4z = 10 & & \text{Multiplying equation (3) by 2} \\ \hline 8x \qquad + z = 15. & (5) & \text{Adding} \end{array}$$

d) Now we solve the resulting system of equations (4) and (5). That will give us two of the numbers:

$$-6x + 3z = 0, \qquad (4)$$
$$8x + z = 15. \qquad (5)$$

We multiply equation (5) by -3. (We could also have multiplied equation (4) by $-\frac{1}{3}$.)

$$\begin{array}{rl} -6x + 3z = 0 & (4) \\ -24x - 3z = -45 & \text{Multiplying equation (5) by } -3 \\ \hline -30x \qquad = -45 & \text{Adding} \\ x = \frac{-45}{-30} = \frac{3}{2} & \end{array}$$

We now use equation (5) to find z:

$$\begin{array}{rl} 8x + z = 15 & (5) \\ 8\left(\frac{3}{2}\right) + z = 15 & \text{Substituting } \frac{3}{2} \text{ for } x \\ 12 + z = 15 & \\ z = 3. & \text{Solving for } z \end{array}$$

e) Next, we use any of the original equations and substitute to find the third number, y. We choose equation (3) since the coefficient of y there is 1:

$$\begin{array}{rl} 2x + y + 2z = 5 & (3) \\ 2\left(\frac{3}{2}\right) + y + 2(3) = 5 & \text{Substituting } \frac{3}{2} \text{ for } x \text{ and 3 for } z \\ 3 + y + 6 = 5 & \\ y + 9 = 5 & \text{Solving for } y \\ y = -4. & \end{array}$$

The solution is $\left(\frac{3}{2}, -4, 3\right)$. The check is as follows.

CHECK:

$$\begin{array}{c|l} 4x - 2y - 3z = 5 & \\ \hline 4 \cdot \frac{3}{2} - 2(-4) - 3(3) \;?\; 5 & \\ 6 + 8 - 9 & \\ 5 & \text{TRUE} \end{array}$$

$$\begin{array}{c|l} -8x - y + z = -5 & \\ \hline -8 \cdot \frac{3}{2} - (-4) + 3 \;?\; -5 & \\ -12 + 4 + 3 & \\ -5 & \text{TRUE} \end{array}$$

$$\begin{array}{c|l} 2x + y + 2z = 5 & \\ \hline 2 \cdot \frac{3}{2} + (-4) + 2(3) \;?\; 5 & \\ 3 - 4 + 6 & \\ 5 & \text{TRUE} \end{array}$$

Do Exercise 2.

2. Solve. Don't forget to check.

$$2x + y - 4z = 0,$$
$$x - y + 2z = 5,$$
$$3x + 2y + 2z = 3$$

Answer on page A-34

3. Solve. Don't forget to check.

$$\begin{aligned} x + y + z &= 100, \\ x - y \quad &= -10, \\ x \quad - z &= -30 \end{aligned}$$

Answer on page A-34

In Example 3, two of the equations have a missing variable.

Example 3 Solve this system:

$$\begin{aligned} x + y + z &= 180, && (1) \\ x \quad - z &= -70, && (2) \\ 2y - z &= 0. && (3) \end{aligned}$$

We note that there is no y in equation (2). In order to have a system of two equations in the variables x and z, we need to find another equation without a y. We use equations (1) and (3) to eliminate y:

$$\begin{aligned} x + y + z &= 180 && (1) \\ 2y - z &= 0 && (3) \end{aligned}$$

$$\begin{aligned} -2x - 2y - 2z &= -360 && \text{Multiplying equation (1) by } -2 \\ 2y - z &= 0 && (3) \\ \hline -2x \qquad - 3z &= -360. && (4) \quad \text{Adding} \end{aligned}$$

Now we solve the resulting system of equations (2) and (4):

$$\begin{aligned} x - z &= -70 && (2) \\ -2x - 3z &= -360 && (4) \end{aligned}$$

$$\begin{aligned} 2x - 2z &= -140 && \text{Multiplying equation (2) by 2} \\ -2x - 3z &= -360 && (4) \\ \hline -5z &= -500 && \text{Adding} \\ z &= 100. \end{aligned}$$

To find x, we substitute 100 for z in equation (2) and solve for x:

$$\begin{aligned} x - z &= -70 \\ x - 100 &= -70 \\ x &= 30. \end{aligned}$$

To find y, we substitute 100 for z in equation (3) and solve for y:

$$\begin{aligned} 2y - z &= 0 \\ 2y - 100 &= 0 \\ 2y &= 100 \\ y &= 50. \end{aligned}$$

The triple (30, 50, 100) is the solution. The check is left to the student.

Do Exercise 3.

It is possible for a system of three equations to have no solution, that is, to be inconsistent. An example is the system

$$\begin{aligned} x + y + z &= 14, \\ x + y + z &= 11, \\ 2x - 3y + 4z &= -3. \end{aligned}$$

Note the first two equations. It is not possible for a sum of three numbers to be both 14 and 11. Thus the system has no solution. We will not consider such systems here, nor will we consider systems with infinitely many solutions, which also exist.

Exercise Set 8.4

a Solve.

1. $x + y + z = 2,$
$2x - y + 5z = -5,$
$-x + 2y + 2z = 1$

2. $2x - y - 4z = -12,$
$2x + y + z = 1,$
$x + 2y + 4z = 10$

3. $2x - y + z = 5,$
$6x + 3y - 2z = 10,$
$x - 2y + 3z = 5$

4. $x - y + z = 4,$
$3x + 2y + 3z = 7,$
$2x + 9y + 6z = 5$

5. $2x - 3y + z = 5,$
$x + 3y + 8z = 22,$
$3x - y + 2z = 12$

6. $6x - 4y + 5z = 31,$
$5x + 2y + 2z = 13,$
$x + y + z = 2$

7. $3a - 2b + 7c = 13,$
$a + 8b - 6c = -47,$
$7a - 9b - 9c = -3$

8. $x + y + z = 0,$
$2x + 3y + 2z = -3,$
$-x + 2y - 3z = -1$

9. $2x + 3y + z = 17,$
$x - 3y + 2z = -8,$
$5x - 2y + 3z = 5$

10. $2x + y - 3z = -4,$
$4x - 2y + z = 9,$
$3x + 5y - 2z = 5$

11. $2x + y + z = -2,$
$2x - y + 3z = 6,$
$3x - 5y + 4z = 7$

12. $2x + y + 2z = 11,$
$3x + 2y + 2z = 8,$
$x + 4y + 3z = 0$

13. $x - y + z = 4,$
$5x + 2y - 3z = 2,$
$3x - 7y + 4z = 8$

14. $2x + y + 2z = 3,$
$x + 6y + 3z = 4,$
$3x - 2y + z = 0$

15. $4x - y - z = 4,$
$2x + y + z = -1,$
$6x - 3y - 2z = 3$

16. $a + 2b + c = 1,$
$7a + 3b - c = -2,$
$a + 5b + 3c = 2$

17. $2r + 3s + 12t = 4,$
$4r - 6s + 6t = 1,$
$r + s + t = 1$

18. $10x + 6y + z = 7,$
$5x - 9y - 2z = 3,$
$15x - 12y + 2z = -5$

19. $4a + 9b = 8,$
$8a + 6c = -1,$
$6b + 6c = -1$

20. $3p + 2r = 11,$
$q - 7r = 4,$
$p - 6q = 1$

21. $x + y + z = 57,$
$-2x + y = 3,$
$x - z = 6$

22. $x + y + z = 105,$
$10y - z = 11,$
$2x - 3y = 7$

23. $r + s = 5,$
$3s + 2t = -1,$
$4r + t = 14$

24. $a - 5c = 17,$
$b + 2c = -1,$
$4a - b - 3c = 12$

Skill Maintenance

Solve for the indicated letter. [2.6a]

25. $F = 3ab$, for a

26. $Q = 4(a + b)$, for a

27. $F = \frac{1}{2}t(c - d)$, for c

28. $F = \frac{1}{2}t(c - d)$, for d

29. $Ax - By = c$, for y

30. $Ax + By = c$, for y

Find the slope and the y-intercept. [7.3b]

31. $y = -\frac{2}{3}x - \frac{5}{4}$

32. $y = 5 - 4x$

33. $2x - 5y = 10$

34. $7x - 6.4y = 20$

Synthesis

35. ◆ Explain a procedure that could be used to solve a system of four equations in four variables.

36. ◆ Is it possible for a system of three equations to have exactly two ordered triples in its solution set? Why or why not?

Solve.

37. $w + x + y + z = 2,$
$w + 2x + 2y + 4z = 1,$
$w - x + y + z = 6,$
$w - 3x - y + z = 2$

38. $w + x - y + z = 0,$
$w - 2x - 2y - z = -5,$
$w - 3x - y + z = 4,$
$2w - x - y + 3z = 7$

8.5 Solving Applied Problems: Systems of Three Equations

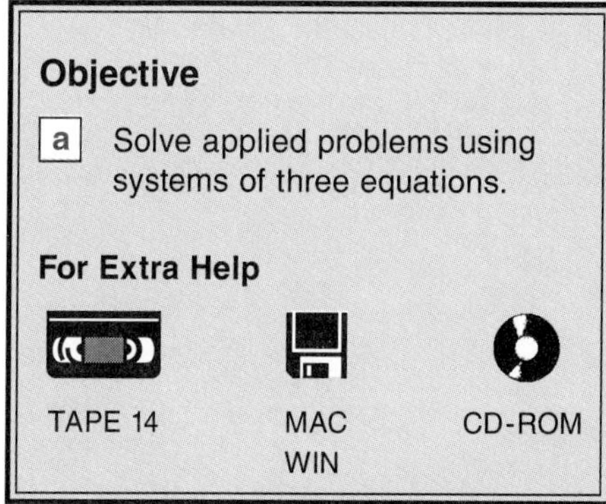

a Using Systems of Three Equations

Solving systems of three or more equations is important in many applications occurring in the natural and social sciences, business, and engineering.

Example 1 *Architecture.* In a triangular cross-section of a roof, the largest angle is 70° greater than the smallest angle. The largest angle is twice as large as the remaining angle. Find the measure of each angle.

1. **Familiarize.** We first make a drawing. Since we do not know the size of any angle, we use x, y, and z for the measures of the angles. We let $x =$ the smallest angle, $z =$ the largest angle, and $y =$ the remaining angle.

2. **Translate.** In order to translate the problem, we need to make use of a geometric fact—that is, the sum of the measures of the angles of a triangle is 180°. This fact about triangles gives us one equation:

$$x + y + z = 180.$$

There are two statements in the problem that we can translate directly.

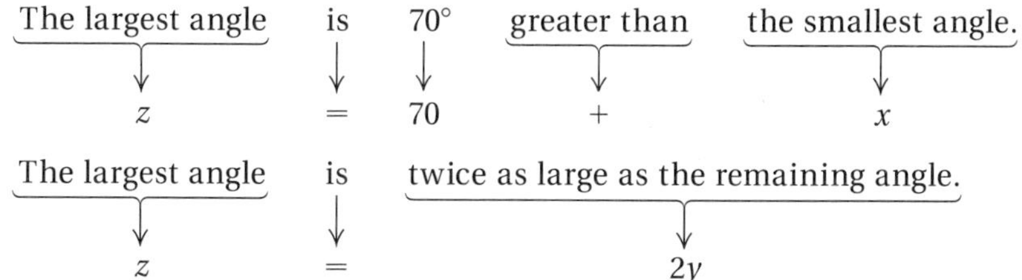

We now have a system of three equations:

$$\begin{aligned} x + y + z &= 180, \\ x + 70 &= z, \\ 2y &= z; \end{aligned} \qquad \text{or} \qquad \begin{aligned} x + y + z &= 180, \\ x \qquad - z &= -70, \\ 2y - z &= 0. \end{aligned}$$

3. **Solve.** The system was solved in Example 3 of Section 8.4. The solution is (30, 50, 100).

4. **Check.** The sum of the numbers is 180. The largest angle measures 100° and the smallest measures 30°. The largest angle is 70° greater than the smallest. The remaining angle measures 50°. The largest angle is twice as large as the remaining angle. We do have an answer to the problem.

5. **State.** The measures of the angles of the triangle are 30°, 50°, and 100°.

Do Exercise 1.

1. *Triangle Measures.* One angle of a triangle is twice as large as a second angle. The remaining angle is 20° greater than the first angle. Find the measure of each angle.

Answer on page A-34

Example 2 *Cholesterol Levels.* Americans have become very conscious of their cholesterol levels. Recent studies indicate that a child's intake of cholesterol should be no more than 300 mg per day. By eating 1 egg, 1 cupcake, and 1 slice of pizza, a child consumes 302 mg of cholesterol. If the child eats 2 cupcakes and 3 slices of pizza, he or she takes in 65 mg of cholesterol. By eating 2 eggs and 1 cupcake, a child consumes 567 mg of cholesterol. How much cholesterol is in each item?

1. **Familiarize.** After we have read the problem a few times, it becomes clear that an egg contains considerably more cholesterol than the other foods. Let's guess that one egg contains 200 mg of cholesterol and one cupcake contains 50 mg. Because of the third sentence in the problem, it would follow that a slice of pizza contains 52 mg of cholesterol since $200 + 50 + 52 = 302$.

   To see if our guess satisfies the other statements in the problem, we find the amount of cholesterol that 2 cupcakes and 3 slices of pizza would contain: $2 \cdot 50 + 3 \cdot 52 = 256$. Since this does not match the 65 mg listed in the fourth sentence of the problem, our guess was incorrect. Rather than guess again, we examine how we checked our guess and let e, c, and s = the number of milligrams of cholesterol in an egg, a cupcake, and a slice of pizza, respectively.

2. **Translate.** By rewording some of the sentences in the problem, we can translate it into three equations.

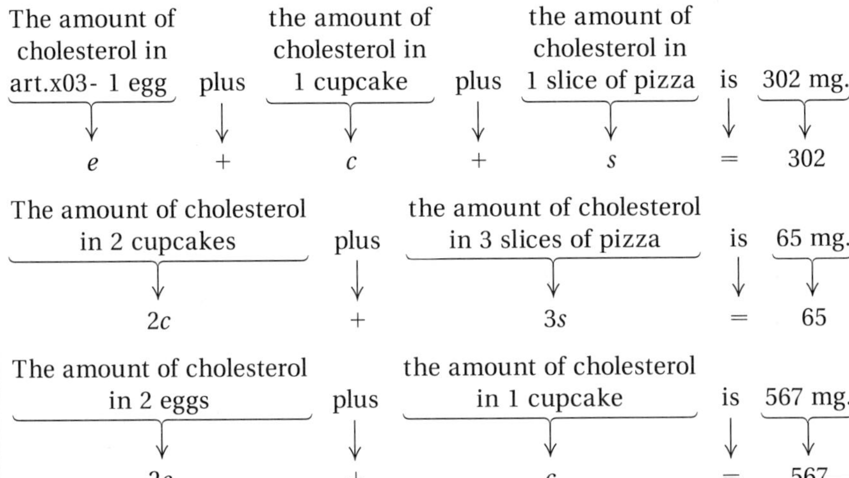

   We now have a system of three equations:

$$
\begin{aligned}
e + c + s &= 302, \\
2c + 3s &= 65, \\
2e + c &= 567.
\end{aligned}
$$

3. **Solve.** We solve and get $e = 274$, $c = 19$, $s = 9$, or $(274, 19, 9)$.

4. **Check.** The sum of 274, 19, and 9 is 302 so the total cholesterol in 1 egg, 1 cupcake, and 1 slice of pizza checks. Two cupcakes and three slices of pizza would contain $2 \cdot 19 + 3 \cdot 9$, or 65 mg, while two eggs and one cupcake would contain $2 \cdot 274 + 19$, or 567 mg of cholesterol. The answer checks.

5. **State.** An egg contains 274 mg of cholesterol, a cupcake contains 19 mg of cholesterol, and a slice of pizza contains 9 mg of cholesterol.

Do Exercise 2.

2. *Client Investments.* Kaufman Financial Corporation makes investments for corporate clients. One year, a client receives $1620 in simple interest from three investments that total $25,000. Part is invested at 5%, part at 6%, and part at 7%. There is $11,000 more invested at 7% than at 6%. How much was invested at each rate?

Answer on page A-34

Exercise Set 8.5

a Solve.

1. *Restaurant Management.* Kyle works at Dunkin Donuts®, where a 10-oz cup of coffee costs 95¢, a 14-oz cup costs $1.15, and a 20-oz cup costs $1.50. During one busy period, Kyle served 34 cups of coffee, emptying five 96-oz pots while collecting a total of $39.60. How many cups of each size did Kyle fill?

10 oz $0.95 | 14 oz $1.15 | 20 oz $1.50

2. *Restaurant Management.* McDonald's® recently sold small soft drinks for 89¢, medium soft drinks for 99¢, and large soft drinks for $1.19. During a lunch-time rush, Chris sold 55 soft drinks for a total of $54.95. The number of small and large drinks, combined, was 5 fewer than the number of medium drinks. How many drinks of each size were sold?

small $0.89 | medium $0.99 | large $1.19

3. *Triangle Measures.* In triangle *ABC*, the measure of angle *B* is three times that of angle *A*. The measure of angle *C* is 20° more than that of angle *A*. Find the measure of each angle.

4. *Triangle Measures.* In triangle *ABC*, the measure of angle *B* is twice the measure of angle *A*. The measure of angle *C* is 80° more than that of angle *A*. Find the measure of each angle.

5. *Automobile Pricing.* A recent basic model of a particular automobile had a price of $12,685. The basic model with the added features of automatic transmission and power door locks was $14,070. The basic model with air conditioning (AC) and power door locks was $13,580. The basic model with AC and automatic transmission was $13,925. What was the individual cost of each of the three options?

6. *Telemarketing.* Sven, Tillie, and Isaiah can process 740 telephone orders per day. Sven and Tillie together can process 470 orders, while Tillie and Isaiah together can process 520 orders per day. How many orders can each person process alone?

7. *Lens Production.* When Sight-Rite's three polishing machines, A, B, and C, are all working, 5700 lenses can be polished in one week. When only A and B are working, 3400 lenses can be polished in one week. When only B and C are working, 4200 lenses can be polished in one week. How many lenses can be polished in a week by each machine alone?

8. *Welding Rates.* Elrod, Dot, and Wendy can weld 74 linear feet per hour when working together. Elrod and Dot together can weld 44 linear feet per hour, while Elrod and Wendy can weld 50 linear feet per hour. How many linear feet per hour can each weld alone?

9. *Investments.* A business class divided an imaginary investment of $80,000 among three mutual funds. The first fund grew by 10%, the second by 6%, and the third by 15%. Total earnings were $8850. The earnings from the first fund were $750 more than the earnings from the third. How much was invested in each fund?

10. *Advertising.* In a recent year, companies spent a total of $84.8 billion on newspaper, television, and radio ads. The total amount spent on television and radio ads was only $2.6 billion more than the amount spent on newspaper ads alone. The amount spent on newspaper ads was $5.1 billion more than what was spent on television ads. How much was spent on each form of advertising? (*Hint*: Let the variables represent numbers of billions of dollars.)

11. *Twin Births.* In the United States, the highest incidence of fraternal twin births occurs among Asian-Americans, then African-Americans, and then Caucasians. Of every 15,400 births, the total number of fraternal twin births for all three is 739, where there are 185 more for Asian-Americans than African-Americans and 231 more for Asian-Americans than Caucasians. How many births of fraternal twins are there for each group out of every 15,400 births?

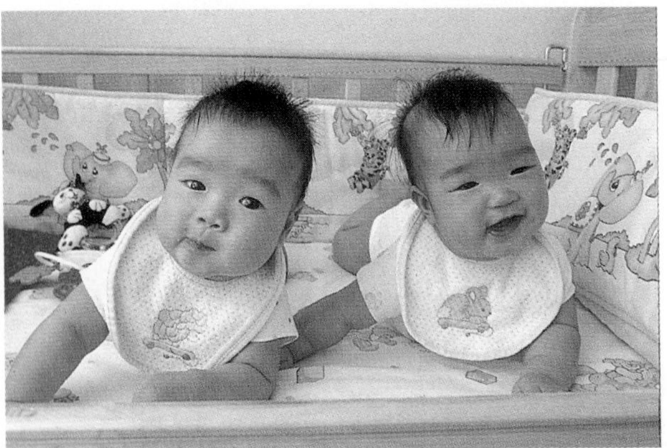

12. *Crying Rate.* The sum of the average number of times a man, a woman, and a one-year-old child cry each month is 71.7. A one-year-old cries 46.4 more times than a man. The average number of times a one-year-old cries per month is 28.3 more than the average number of times combined that a man and a woman cry. What is the average number of times per month that each cries?

13. *Nutrition.* A dietician in a hospital prepares meals under the guidance of a physician. Suppose that for a particular patient a physician prescribes a meal to have 800 calories, 55 g of protein, and 220 mg of vitamin C. The dietician prepares a meal of roast beef, baked potatoes, and broccoli according to the data in the following table.

	Calories	Protein (in grams)	Vitamin C (in milligrams)
Roast Beef, 3 oz	300	20	0
Baked Potato	100	5	20
Broccoli, 156 g	50	5	100

How many servings of each food are needed in order to satisfy the doctor's orders?

14. *Nutrition.* Repeat Exercise 13 but replace the broccoli with asparagus, for which one 180-g serving contains 50 calories, 5 g of protein, and 44 mg of vitamin C. Which meal would you prefer eating?

15. *Golf.* On an 18-hole golf course, there are par-3 holes, par-4 holes, and par-5 holes. A golfer who shoots par on every hole has a total of 70. There are twice as many par-4 holes as there are par-5 holes. How many of each type of hole are there on the golf course?

16. *Golf.* On an 18-hole golf course, there are par-3 holes, par-4 holes, and par-5 holes. A golfer who shoots par on every hole has a total of 72. The sum of the number of par-3 holes and the number of par-5 holes is 8. How many of each type of hole are there on the golf course?

17. The sum of three numbers is 5. The first number minus the second plus the third is 1. The first minus the third is 3 more than the second. Find the numbers.

18. The sum of three numbers is 26. Twice the first minus the second is 2 less than the third. The third is the second minus three times the first. Find the numbers.

19. *Basketball Scoring.* The New York Knicks recently scored a total of 92 points on a combination of 2-point field goals, 3-point field goals, and 1-point foul shots. Altogether, the Knicks made 50 baskets and 19 more 2-pointers than foul shots. How many shots of each kind were made?

20. *History.* Find the year in which the first U.S. transcontinental railroad was completed. The following are some facts about the number. The sum of the digits in the year is 24. The ones digit is 1 more than the hundreds digit. Both the tens and the ones digits are multiples of 3.

Skill Maintenance

Determine whether the correspondence is a function. [7.1a]

21. *Domain* (State) → *Range* (City)

California → Los Angeles, San Francisco, San Diego
Kansas → Kansas City, Topeka, Wichita

22.

Domain (Boy's Age, in months)	*Range* (Average Daily Weight Gain, in grams)
2	24.3
9	11.7
16	8.2
23	7.0

Source: *American Family Physician,* December 1993, p. 1435

23. Find the domain of the function: [7.2a]

$$f(x) = \frac{x-5}{x+7}.$$

24. Find the domain and the range of the function: [7.2a]

$$g(x) = 5 - x^2.$$

25. Find an equation of the line with slope $-\frac{3}{5}$ and y-intercept $(0, -7)$. [7.5a]

26. Simplify: $\dfrac{(a^2b^3)^5}{a^7b^{16}}$. [4.1e], [4.2a]

Synthesis

27. ◆ Exercise 8 can be solved mentally after a careful reading of the problem. How is this possible?

28. ◆ A theater audience of 100 people consists of adults, senior citizens, and children. The ticket prices are \$10 each for adults, \$3 each for senior citizens, and \$0.50 each for children. The total amount of money taken in is \$100. How many adults, senior citizens, and children are in attendance? Does there seem to be some information missing? Do some careful reasoning and explain.

29. Find the sum of the angle measures at the tips of the star in this figure.

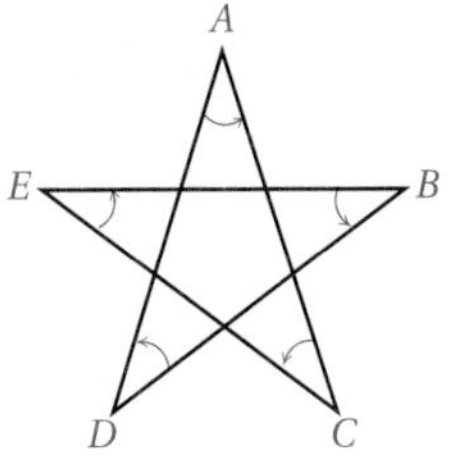

30. *Sharing Raffle Tickets.* Hal gives Tom as many raffle tickets as Tom has and Gary as many as Gary has. In like manner, Tom then gives Hal and Gary as many tickets as each then has. Similarly, Gary gives Hal and Tom as many tickets as each then has. If each finally has 40 tickets, with how many tickets does Tom begin?

Collaborative Learning Manual

Solve a system of equations in three variables by choosing a different variable to eliminate.

8.6 Business and Economics Applications

a Break-Even Analysis

When a company manufactures x units of a product, it invests money. This is **total cost** and can be thought of as a function C, where $C(x)$ is the total cost of producing x units. When the company sells x units of the product, it takes in money. This is **total revenue** and can be thought of as a function R, where $R(x)$ is the total revenue from the sale of x units. **Total profit** is the money taken in less the money spent, or total revenue minus total cost. Total profit from the production and sale of x units is a function P given by

Profit = Revenue − Cost, or $P(x) = R(x) - C(x)$.

If $R(x)$ is greater than $C(x)$, the company has a profit. If $C(x)$ is greater than $R(x)$, the company has a loss. When $R(x) = C(x)$, the company breaks even.

There are two kinds of costs. First, there are costs like rent, insurance, machinery, and so on. These costs, which must be paid whether a product is produced or not, are called **fixed costs.** When a product is being produced, there are costs for labor, materials, marketing, and so on. These are called **variable costs,** because they vary according to the amount of the product being produced. The sum of the fixed costs and the variable costs gives the **total cost** of producing a product.

Objectives

a Given total-cost and total-revenue functions, find the total-profit function and the break-even point.

b Given supply and demand functions, find the equilibrium point.

For Extra Help

TAPE 14 MAC WIN CD-ROM

Example 1 *Manufacturing Radios.* Ergs, Inc., is planning to make a new kind of radio. Fixed costs will be \$90,000, and it will cost \$15 to produce each radio (variable costs). Each radio sells for \$26.

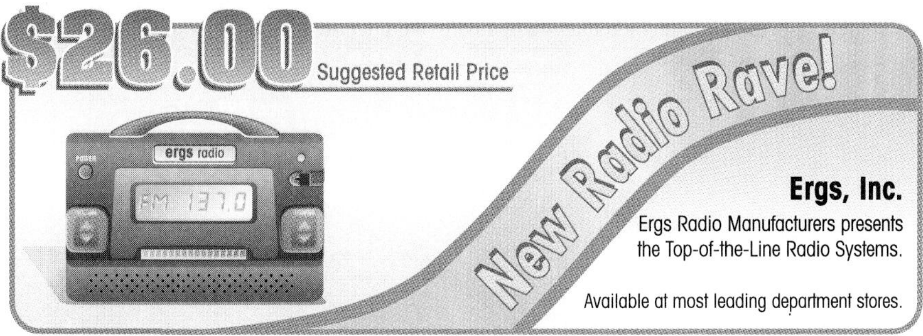

a) Find the total cost $C(x)$ of producing x radios.

b) Find the total revenue $R(x)$ from the sale of x radios.

c) Find the total profit $P(x)$ from the production and sale of x radios.

d) What profit or loss will the company realize from the production and sale of 3000 radios? of 14,000 radios?

e) Graph the total-cost, total-revenue, and total-profit functions using the same set of axes. Determine the break-even point.

a) Total cost is given by

$C(x) =$ (Fixed costs) plus (Variable costs),

or $C(x) = 90{,}000 + 15x$,

where x is the number of radios produced.

1. *Manufacturing Radios.* Refer to Example 1. Suppose that fixed costs are \$80,000, and it costs \$20 to produce each radio. Each radio sells for \$36.

a) Find the total cost $C(x)$ of producing x radios.

b) Find the total revenue $R(x)$ from the sale of x radios.

c) Find the total profit $P(x)$ from the production and sale of x radios.

d) What profit or loss will the company realize from the production and sale of 4000 radios? of 16,000 radios?

e) Graph the total-cost, total-revenue, and total-profit functions using the same set of axes. Determine the break-even point.

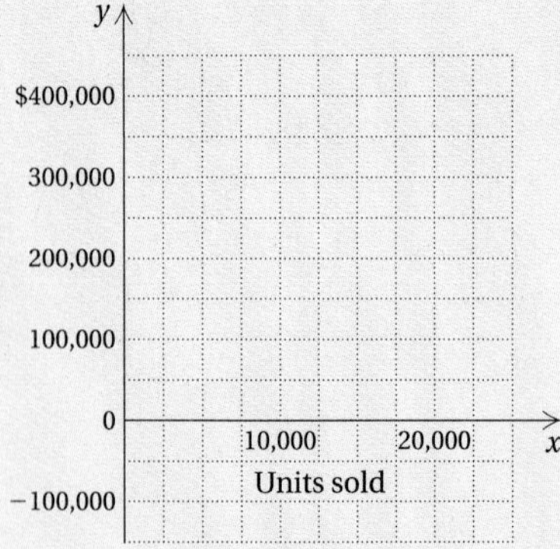

Answers on page A-34

b) Total revenue is given by

$$R(x) = 26x. \quad \text{\$26 times the number of radios sold. We assume that every radio produced is sold.}$$

c) Total profit is given by

$$P(x) = R(x) - C(x)$$
$$= 26x - (90{,}000 + 15x)$$
$$= 11x - 90{,}000.$$

d) Profits will be

$$P(3000) = 11 \cdot 3000 - 90{,}000 = -\$57{,}000$$

when 3000 radios are produced and sold, and

$$P(14{,}000) = 11 \cdot 14{,}000 - 90{,}000 = \$64{,}000$$

when 14,000 radios are produced and sold. Thus the company loses \$57,000 if only 3000 radios are sold, but makes \$64,000 if 14,000 are sold.

e) The graphs of each of the three functions are shown below:

$$C(x) = 90{,}000 + 15x, \quad (1)$$
$$R(x) = 26x, \quad (2)$$
$$P(x) = 11x - 90{,}000. \quad (3)$$

$R(x)$, $C(x)$, and $P(x)$ are all in dollars.

Equation (2) has a graph that goes through the origin and has a slope of 26. Equation (1) has an intercept on the y-axis of 90,000 and has a slope of 15. Equation (3) has an intercept on the \$-axis of −90,000 and has a slope of 11. It is shown by the dashed line. The red dashed line shows a "negative" profit, which is a loss. (That is what is known as "being in the red.") The black dashed line shows a "positive" profit, or gain. (That is what is known as "being in the black.")

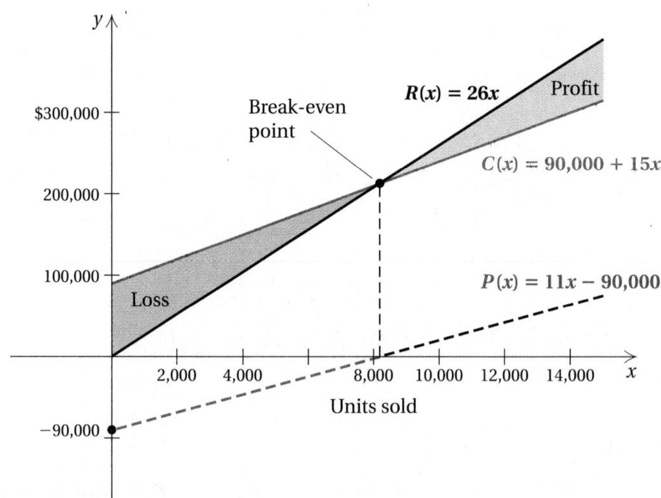

Profits occur when the revenue is greater than the cost. Losses occur when the revenue is less than the cost. The **break-even point** occurs where the graphs of R and C cross. Thus to find the break-even point, we solve a system:

$$C(x) = 90{,}000 + 15x,$$
$$R(x) = 26x.$$

Since both revenue and cost are in *dollars* and they are equal at the break-even point, the system can be rewritten as

$$d = 90{,}000 + 15x, \qquad (1)$$
$$d = 26x \qquad (2)$$

and solved using substitution:

$$26x = 90{,}000 + 15x \qquad \text{Substituting } 26x \text{ for } d \text{ in equation (1)}$$
$$11x = 90{,}000$$
$$x \approx 8181.8.$$

The firm will break even if it produces and sells about 8182 radios (8181 will yield a tiny loss and 8182 a tiny gain), and takes in a total of $R(8182) = 26 \cdot 8182 = \$212{,}732$ in revenue. Note that the x-coordinate of the break-even point is also the x-intercept of the profit function. It can also be found by solving $P(x) = 0$.

Do Exercise 1 on the preceding page.

b Supply and Demand

As the price of coffee varies, the amount sold varies. The table and graph below both show that consumer demand goes down as the price goes up and the demand goes up as the price goes down.

DEMAND FUNCTION, *D*

Price, *p*, per Kilogram	Quantity, *D*(*p*) (in millions of kilograms)
$ 8.00	25
9.00	20
10.00	15
11.00	10
12.00	5

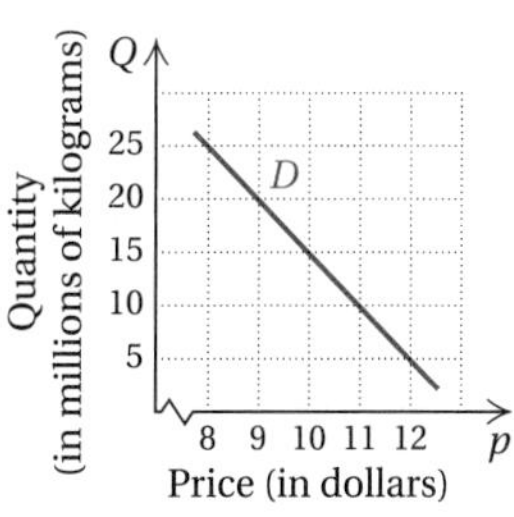

As the price of coffee varies, the amount available varies. The table and graph below both show that sellers will supply less as the price goes down, but will supply more as the price goes up.

SUPPLY FUNCTION, *S*

Price, *p*, per Kilogram	Quantity, *S*(*p*) (in millions of kilograms)
$ 9.00	5
9.50	10
10.00	15
10.50	20
11.00	25

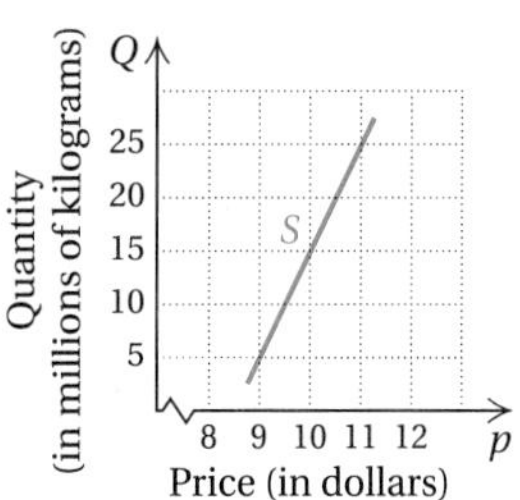

Calculator Spotlight

Linear regression can be used to find equations for supply and demand. See the Calculator Spotlight in Section 7.6.

2. Find the equilibrium point for the following supply and demand functions:

$$D(p) = 1000 - 46p,$$
$$S(p) = 300 + 4p.$$

Let's look at the above graphs together. We see that as price increases, demand decreases. As price increases, supply increases. The point of intersection of the demand and supply functions is called the **equilibrium point.** At the equilibrium point, the amount that the seller will supply is the same amount that the consumer will buy. The situation is analogous to a buyer and a seller negotiating the price of an item. The equilibrium point is the price and quantity on which they finally agree.

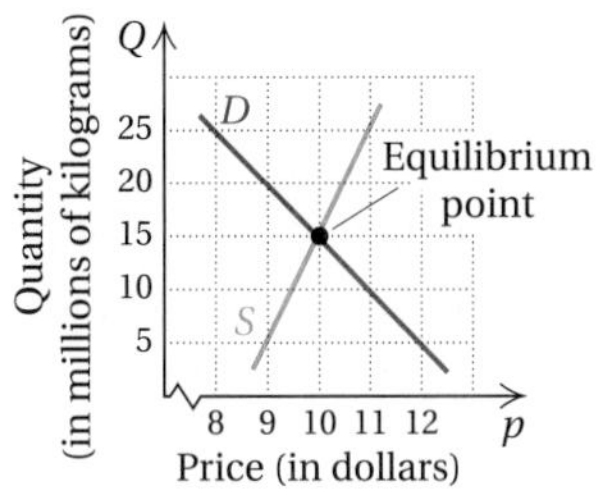

Any ordered pair of coordinates from the graph is (price, quantity), because the horizontal axis is the price axis and the vertical axis is the quantity axis. If D is a demand function and S is a supply function, then the equilibrium point is where demand equals supply:

$$D(p) = S(p).$$

Example 2 Find the equilibrium point for the following demand and supply functions:

$$D(p) = 1000 - 60p, \quad (1)$$
$$S(p) = 200 + 4p. \quad (2)$$

Since both demand and supply are *quantities* and they are equal at the equilibrium point, we rewrite the system as

$$q = 1000 - 60p, \quad (1)$$
$$q = 200 + 4p. \quad (2)$$

We substitute $200 + 4p$ for q in equation (1) and solve:

$$200 + 4p = 1000 - 60p$$
$$200 + 64p = 1000 \quad \text{Adding } 60p \text{ on both sides}$$
$$64p = 800 \quad \text{Subtracting 200 on both sides}$$
$$p = \frac{800}{64} = 12.5.$$

Thus the equilibrium price is \$12.50 per unit.

To find the equilibrium quantity, we substitute \$12.50 into either $D(p)$ or $S(p)$. We use $S(p)$:

$$S(12.5) = 200 + 4(12.5)$$
$$= 200 + 50 = 250.$$

Thus the equilibrium quantity is 250 units, and the equilibrium point is (\$12.50, 250).

Do Exercise 2.

Calculator Spotlight

Use the INTERSECT feature on your grapher to solve Example 2 and Margin Exercise 2.

Answer on page A-34

Exercise Set 8.6

a For each of the following pairs of total-cost and total-revenue functions, find **(a)** the total-profit function and **(b)** the break-even point.

1. $C(x) = 25x + 270{,}000$;
$R(x) = 70x$

2. $C(x) = 45x + 300{,}000$;
$R(x) = 65x$

3. $C(x) = 10x + 120{,}000$;
$R(x) = 60x$

4. $C(x) = 30x + 49{,}500$;
$R(x) = 85x$

5. $C(x) = 20x + 10{,}000$;
$R(x) = 100x$

6. $C(x) = 40x + 22{,}500$;
$R(x) = 85x$

7. $C(x) = 22x + 16{,}000$;
$R(x) = 40x$

8. $C(x) = 15x + 75{,}000$;
$R(x) = 55x$

9. $C(x) = 50x + 195{,}000$;
$R(x) = 125x$

10. $C(x) = 34x + 928{,}000$;
$R(x) = 128x$

Solve.

11. *Manufacturing Lamps.* City Lights is planning to manufacture a new type of lamp. For the first year, the fixed costs for setting up production are $22,500. The variable costs for producing each lamp are $40. The revenue from each lamp is $85. Find the following.

a) The total cost $C(x)$ of producing x lamps
b) The total revenue $R(x)$ from the sale of x lamps
c) The total profit $P(x)$ from the production and sale of x lamps
d) The profit or loss from the production and sale of 3000 lamps; of 400 lamps
e) The break-even point

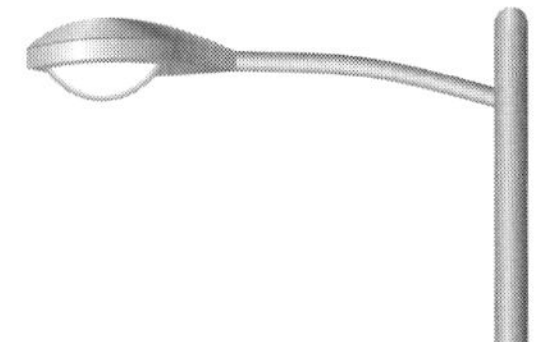

12. *Computer Manufacturing.* Sky View Electronics is planning to introduce a new line of computers. For the first year, the fixed costs for setting up production are $125,100. The variable costs for producing each computer are $750. The revenue from each computer is $1050. Find the following.

a) The total cost $C(x)$ of producing x computers
b) The total revenue $R(x)$ from the sale of x computers
c) The total profit $P(x)$ from the production and sale of x computers
d) The profit or loss from the production and sale of 400 computers; of 700 computers
e) The break-even point

13. *Manufacturing Caps.* Martina's Custom Printing is planning on adding painter's caps to its product line. For the first year, the fixed costs for setting up production are \$16,404. The variable costs for producing a dozen caps are \$6.00. The revenue on each dozen caps is \$18.00. Find the following.

a) The total cost $C(x)$ of producing x dozen caps
b) The total revenue $R(x)$ from the sale of x dozen caps
c) The total profit $P(x)$ from the production and sale of x dozen caps
d) The profit or loss from the production and sale of 3000 dozen caps; of 1000 dozen caps
e) The break-even point

14. *Sport Coat Production.* Sarducci's is planning a new line of sport coats. For the first year, the fixed costs for setting up production are \$10,000. The variable costs for producing each coat are \$20. The revenue from each coat is \$100. Find the following.

a) The total cost $C(x)$ of producing x coats
b) The total revenue $R(x)$ from the sale of x coats
c) The total profit $P(x)$ from the production and sale of x coats
d) The profit or loss from the production and sale of 2000 coats; of 50 coats
e) The break-even point

b Find the equilibrium point for each of the following pairs of demand and supply functions.

15. $D(p) = 1000 - 10p;$ $S(p) = 230 + p$

16. $D(p) = 2000 - 60p;$ $S(p) = 460 + 94p$

17. $D(p) = 760 - 13p;$ $S(p) = 430 + 2p$

18. $D(p) = 800 - 43p;$ $S(p) = 210 + 16p$

19. $D(p) = 7500 - 25p;$ $S(p) = 6000 + 5p$

20. $D(p) = 8800 - 30p;$ $S(p) = 7000 + 15p$

21. $D(p) = 1600 - 53p;$ $S(p) = 320 + 75p$

22. $D(p) = 5500 - 40p;$ $S(p) = 1000 + 85p$

Skill Maintenance

Find the slope and the y-intercept. [7.3b]

23. $5y - 3x = 8$

24. $6x + 7y - 9 = 4$

25. $2y = 3.4x + 98$

26. $\frac{x}{3} + \frac{y}{4} = 1$

Synthesis

27. ◆ Variable costs and fixed costs are often compared to the slope and the y-intercept, respectively, of an equation of a line. Explain why this analogy is valid.

28. ◆ In this section, we examined supply and demand functions for coffee. Does it seem realistic to you for the graph of D to have a constant slope? Why or why not?

Summary and Review Exercises: Chapter 8

The objectives to be tested in addition to the material in this chapter are [2.6a], [7.1b], [7.3b], and [7.5d].

Solve graphically. Then classify the system as consistent or inconsistent and as dependent or independent. [8.1a]

1. $4x - y = -9,$
$x - y = -3$

2. $15x + 10y = -20,$
$3x + 2y = -4$

3. $y - 2x = 4,$
$y - 2x = 5$

Solve by the substitution method. [8.2a]

4. $7x - 4y = 6,$
$y - 3x = -2$

5. $y = x + 2,$
$y - x = 8$

6. $9x - 6y = 2,$
$x = 4y + 5$

Solve by the elimination method. [8.2b]

7. $8x - 2y = 10,$
$-4y - 3x = -17$

8. $4x - 7y = 18,$
$9x + 14y = 40$

9. $3x - 5y = -4,$
$5x - 3y = 4$

10. $1.5x - 3 = -2y,$
$3x + 4y = 6$

11. *Music Spending.* Sean has \$37 to spend. He can spend all of it on two compact discs and a cassette, or he can buy one CD and two cassettes and have \$5.00 left over. What is the price of a CD? of a cassette? [8.3a]

12. *Orange Drink Mixtures.* "Orange Thirst" is 15% orange juice and "Quencho" is 5% orange juice. How many liters of each should be combined in order to get 10 L of a mixture that is 10% orange juice? [8.3a]

13. *Train Travel.* A train leaves Watsonville at noon traveling north at 44 mph. One hour later, another train, going 52 mph, travels north on a parallel track. How many hours will the second train travel before it overtakes the first train? [8.3b]

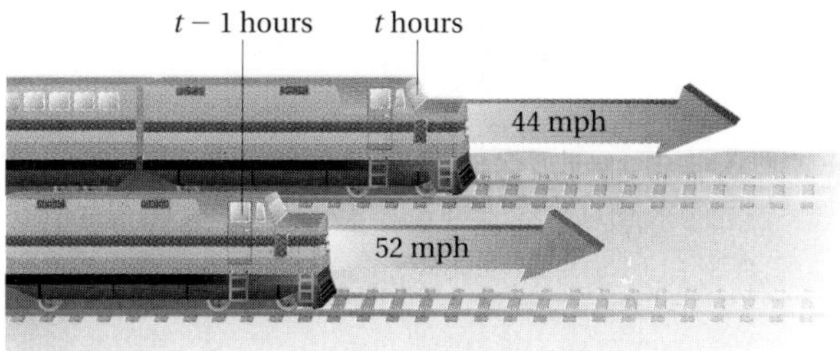

Solve. [8.4a]

14. $x + 2y + z = 10,$
$2x - y + z = 8,$
$3x + y + 4z = 2$

15. $3x + 2y + z = 3,$
$6x - 4y - 2z = -34,$
$-x + 3y - 3z = 14$

16. $2x - 5y - 2z = -4,$
$7x + 2y - 5z = -6,$
$-2x + 3y + 2z = 4$

17. $x + y + 2z = 1,$
$x - y + z = 1,$
$x + 2y + z = 2$

18. *Triangle Measures.* In triangle ABC, the measure of angle A is four times the measure of angle C, and the measure of angle B is 45° more than the measure of angle C. What are the measures of the angles of the triangle? [8.5a]

19. *Money Mixtures.* Elaine has \$194, consisting of \$20, \$5, and \$1 bills. The number of \$1 bills is 1 less than the total number of \$20 and \$5 bills. If she has 39 bills in her purse, how many of each denomination does she have? [8.5a]

20. *Bed Manufacturing.* Kregel Furniture is planning to produce a new type of bed. For the first year, the fixed costs for setting up production are \$35,000. The variable costs for producing each bed are \$175. The revenue from each bed is \$300. Find the following. [8.6a]

a) The total cost $C(x)$ of producing x beds
b) The total revenue $R(x)$ from the sale of x beds
c) The total profit from the production and sale of x beds
d) The profit or loss from the production and sale of 1200 beds; of 200 beds
e) The break-even point

21. Find the equilibrium point for the following demand and supply functions: [8.6b]

$D(p) = 120 - 13p,$
$S(p) = 60 + 7p.$

Skill Maintenance

Write an equation of the line containing the given point and parallel to the given line. [7.5d]

22. $(0, 5)$; $x - 2y = -3$

23. $(1, -1)$; $10x - 5y = 7$

Write an equation of the line containing the given point and perpendicular to the given line. [7.5d]

24. $(8, 0)$; $4x + y = 10$

25. $(2, -2)$; $y = 4 - x$

26. Solve $Q = at - 4t$ for t. [2.6a]

27. Given the function $f(x) = 8 - 3x$, find $f(0)$ and $f(-2)$. [7.1b]

28. For $5x - 8y = 40$, find the slope and the y-intercept. [7.3b]

Synthesis

29. ◆ Briefly compare the strengths and the weaknesses of the graphical, substitution, and elimination methods as applied to the solution of two equations in two variables. [8.1a], [8.2a, b]

30. ◆ Explain the advantages of using a system of equations to solve an applied problem. [8.3a, b]

31. Solve graphically:

$y = x + 2,$
$y = x^2 + 2.$ [8.1a]

32. *Height Estimation in Anthropology.* An anthropologist can use linear functions to estimate the height of a male or female, given the length of certain bones. The *femur* is the large bone from the hip to the knee, as shown below. Let $x =$ the length of the femur, in centimeters. Then the height, in centimeters, of a male with a femur of length x is given by the function

$M(x) = 1.88x + 81.31.$

The height, in centimeters, of a female with a femur of length x is given by the function

$F(x) = 1.95x + 72.85.$

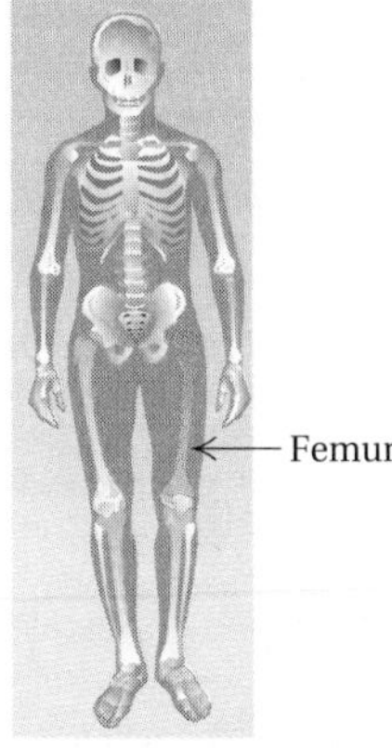

A 45-cm femur was uncovered at an anthropological dig. [8.1a]

a) If we assume that it was from a male, how tall was he?
b) If we assume that it was from a female, how tall was she?
c) Graph each equation and find the point of intersection of the graphs of the equations.
d) For what length of a male femur and a female femur, if any, would the height be the same?

Test: Chapter 8

Answers

Solve graphically. Then classify the system as consistent or inconsistent and as dependent or independent.

1. $y = 3x + 7,$
   $3x + 2y = -4$

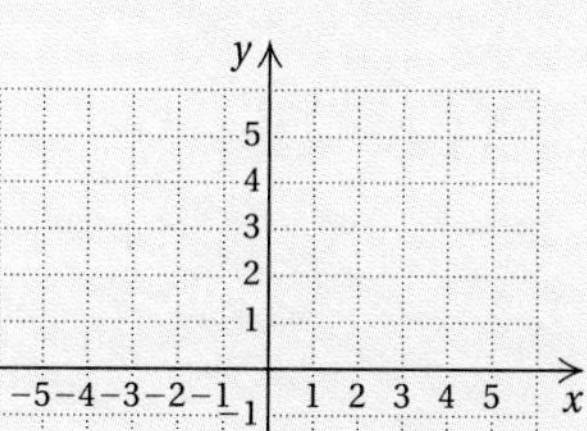

2. $y = 3x + 4,$
   $y = 3x - 2$

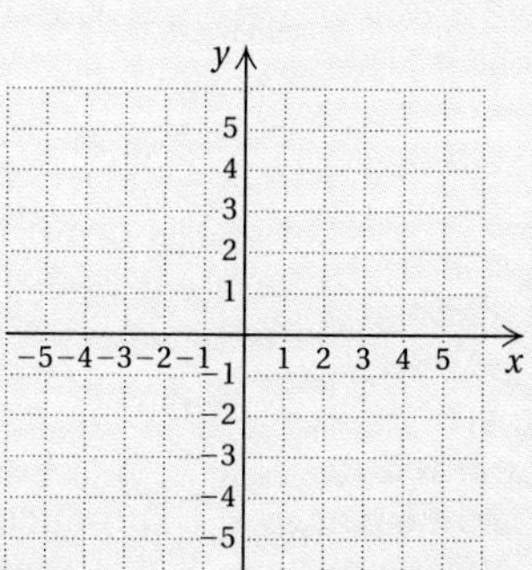

3. $y - 3x = 6,$
   $6x - 2y = -12$

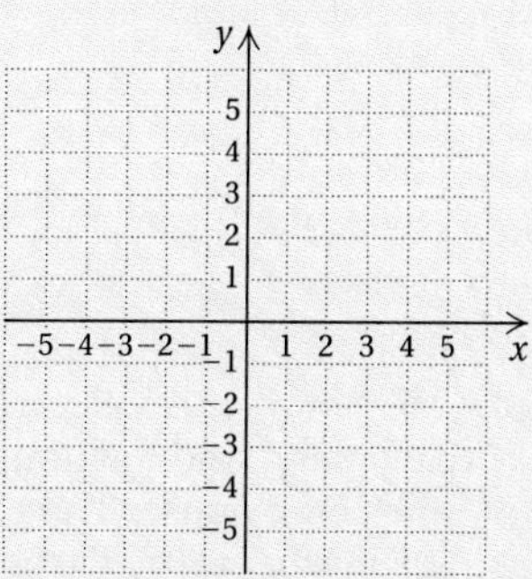

Solve by the substitution method.

4. $x + 3y = -8,$
   $4x - 3y = 23$

5. $2x + 4y = -6,$
   $y = 3x - 9$

Solve by the elimination method.

6. $4x - 6y = 3,$
   $6x - 4y = -3$

7. $4y + 2x = 18,$
   $3x + 6y = 26$

Solve.

8. *Saline Solutions.* Saline (saltwater) solutions are often used for sore throats. A nurse has a saline solution that is 34% salt and the rest water. She also has a 61% saline solution. She wants to use a 50% solution with a particular patient without wasting the existing solutions. How many milliliters (mL) of each solution would be needed in order to obtain 120 mL of a mixture that is 50% salt?

9. *Chicken Dinners.* High Flyin' Wings charges $12 for a bucket of chicken wings and $7 for a chicken dinner. After filling 28 orders for buckets and dinners during a football game, the waiters had collected $281. How many buckets and how many dinners did they sell?

10. *Air Travel.* An airplane flew for 5 hr with a 20-km/h tailwind and returned in 7 hr against the same wind. Find the speed of the plane in still air.

11. *Tennis Court.* The perimeter of a standard tennis court used for playing doubles is 288 ft. The width of the court is 42 ft less than the length. Find the length and the width.

1. ______
2. ______
3. ______
4. ______
5. ______
6. ______
7. ______
8. ______
9. ______
10. ______
11. ______

Answers

12. ______

13. ______

14. ______

15. a) ______

b) ______

c) ______

d) ______

e) ______

16. ______

17. ______

18. ______

19. ______

20. ______

21. ______

12. Solve:

$$6x + 2y - 4z = 15,$$
$$-3x - 4y + 2z = -6,$$
$$4x - 6y + 3z = 8.$$

13. Find the equilibrium point for the following demand and supply functions:

$$D(p) = 79 - 8p,$$
$$S(p) = 37 + 6p.$$

Solve.

14. *Repair Rates.* An electrician, a carpenter, and a plumber are hired to work on a house. The electrician earns \$21 per hour, the carpenter \$19.50 per hour, and the plumber \$24 per hour. The first day on the job, they worked a total of 21.5 hr and earned a total of \$469.50. If the plumber worked 2 hr more than the carpenter did, how many hours did the electrician work?

15. *Manufacturing Tennis Rackets.* Sweet Spot Manufacturing is planning to produce a new type of tennis racket. For the first year, the fixed costs for setting up production are \$40,000. The variable costs for producing each racket are \$30. The sales department predicts that 1500 rackets can be sold during the first year. The revenue from each racket is \$80. Find the following.

a) The total cost $C(x)$ of producing x rackets
b) The total revenue $R(x)$ from the sale of x rackets
c) The total profit from the production and sale of x rackets
d) The profit or loss from the production and sale of 1200 rackets; of 200 rackets
e) The break-even point

Skill Maintenance

16. Write an equation of the line containing the given point and parallel to the given line.

$(-1, -2)$; $3x + y = -4$

17. Write an equation of the line containing the given point and perpendicular to the given line.

$(0, 4)$; $5x + 4y = -2$

18. Solve $P = 4a - 3b$ for a.

19. For $7x = 14 - 2y$, find the slope and the y-intercept.

20. Given the function $f(x) = x^2 - 8$, find $f(0)$ and $f(-3)$.

Synthesis

21. The graph of the function $f(x) = mx + b$ contains the points $(-1, 3)$ and $(-2, -4)$. Find m and b.

9

More on Inequalities

Introduction

In this chapter, we expand our coverage of inequalities by expressing solution sets in interval notation. We solve and graph conjunctions and disjunctions of inequalities and solve systems of inequalities. We also solve equations and inequalities with absolute-value expressions.

An Application

Michael Johnson set a world record of 19.32 sec in the men's 200-m dash in the 1996 Olympics. The equation

$$R = -0.0513t + 19.32$$

can be used to predict the world record in the men's 200-m dash t years after 1996. Determine those years for which the world record will be less than 19.0 sec.

The Mathematics

We can translate the problem to the following sentence:

$$\underbrace{-0.0513t + 19.32 < 19.0.}_{\uparrow}$$

This is an inequality.

This problem appears as Example 15 in Section 9.1.

World Wide Web For more information, visit us at www.mathmax.com

Pretest: Chapter 9

Solve.

1. $-7x < 21$

2. $2x - 1 \leq 5x + 8$

3. $-1 \leq 3x + 2 \leq 4$

4. $|2y - 7| = 9$

5. $4a + 5 < -3$ *or* $4a + 5 > 3$

6. $|3a + 5| > 1$

7. $|3x + 4| = |x - 7|$

8. $-8(x - 7) \geq 10(2x + 3) - 12$

9. Find the distance between 8.2 and 10.6.

10. *Records in the Women's 100-m Dash.* Florence Griffith Joyner set a world record of 10.49 sec in the women's 100-m dash in 1988. The equation

$$R = -0.0433t + 10.49$$

can be used to predict the world record in the women's 100-m dash t years after 1988. Determine those years for which the world record will be less than 10.35 sec.

11. Find the intersection:
$\{-4, -3, 1, 4, 9\} \cap \{-3, -2, 2, 4, 5\}$.

12. Find the union: $\{-1, 1\} \cup \{0\}$.

13. Simplify: $\left|\frac{-5y}{2x}\right|$.

14. Graph on a plane: $6x - 2y < 12$.

15. Graph the following system of inequalities. Find the coordinates of any vertices formed.

$$x + y \leq 16,$$
$$3x + 6y \leq 60,$$
$$x \geq 0,$$
$$y \geq 0$$

Objectives for Retesting

The objectives to be tested in addition to the material in this chapter are as follows.

[4.6a] Multiply two binomials mentally using the FOIL method.
[7.2a] Find the domain and the range of a function.
[7.5c] Find an equation of a line when two points are given.
[8.2b] Solve systems of equations in two variables by the elimination method.

9.1 Sets, Interval Notation, and Inequalities

We begin this chapter with a review of solving inequalities. In Chapter 2, we wrote solution sets of inequalities using *set-builder notation.* Here we will also write solution sets using *interval notation.*

Objectives

a. Determine whether a given number is a solution of an inequality.

b. Write interval notation for the solution set or graph of an inequality.

c. Solve an inequality using the addition and multiplication principles and then graph the inequality.

d. Solve applied problems by translating to inequalities.

For Extra Help

TAPE 15

MAC
WIN

CD-ROM

a Inequalities

An **inequality** is any sentence containing $<$, $>$, $\leq$, $\geq$, or $\neq$.

Examples of inequalities are the following:

$-2 < a$, $\quad x > 4$, $\quad x + 3 \leq 6$, $\quad 6 - 7y \geq 10y - 4$, and $5x \neq 10$.

Any replacement for the variables that makes an inequality true is called a **solution**. The set of all solutions is called the **solution set.** When all the solutions of an inequality have been found, we say that we have **solved** the inequality.

Examples Determine whether the given number is a solution of the inequality.

1. $x + 3 < 6$; 5

We substitute and get $5 + 3 < 6$, or $8 < 6$, a false sentence. Therefore, 5 is not a solution.

2. $2x - 3 > -3$; 1

We substitute and get $2(1) - 3 > -3$, or $-1 > -3$, a true sentence. Therefore, 1 is a solution.

3. $4x - 1 \leq 3x + 2$; 3

We substitute and get $4(3) - 1 \leq 3(3) + 2$, or $11 \leq 11$, a true sentence. Therefore, 3 is a solution.

Do Exercises 1–3.

Determine whether the given number is a solution of the inequality.

1. $3 - x < 2$; 8

2. $3x + 2 > -1$; -2

3. $3x + 2 \leq 4x - 3$; 5

Answers on page A-35

b Inequalities and Interval Notation

The **graph** of an inequality is a drawing that represents its solutions. An inequality in one variable can be graphed on the number line.

Example 4 Graph $x < 4$ on the number line.

The solutions are all real numbers less than 4, so we shade all numbers less than 4 on the number line. To indicate that 4 is not a solution, we use a right parenthesis ")" at 4.

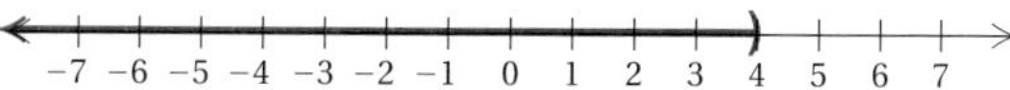

We can write the solution set for $x < 4$ using **set-builder notation:**

$\{x \mid x < 4\}$.

This is read

"The set of all x such that x is less than 4."

Another way to write solutions of an inequality in one variable is to use **interval notation.** Interval notation uses parentheses ( ) and brackets [ ].

If a and b are real numbers such that $a < b$, we define the interval (a, b) as the set of all numbers between but not including a and b—that is, the set of all x for which $a < x < b$. Thus,

$$(a, b) = \{x \mid a < x < b\}.$$

(a, b)

a b

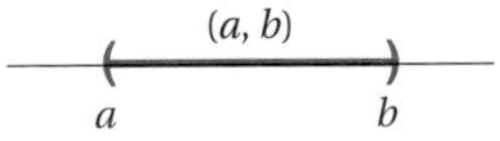

The points a and b are the **endpoints**. The parentheses indicate that the endpoints are *not* included in the graph.

The interval $[a, b]$ is defined as the set of all numbers x for which $a \leq x \leq b$. Thus,

$$[a, b] = \{x \mid a \leq x \leq b\}.$$

[a, b]

a b

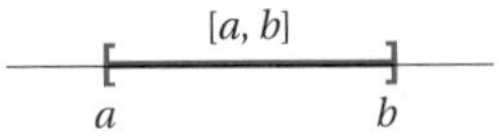

The brackets indicate that the endpoints *are* included in the graph.*

CAUTION! Do not confuse the *interval* (a, b) with the *ordered pair* (a, b) used in connection with an equation in two variables in the plane, as in Chapter 3. The context in which the notation appears usually makes the meaning clear.

The following intervals include one endpoint and exclude the other:

$(a, b] = \{x \mid a < x \leq b\}$. The graph excludes a and includes b.

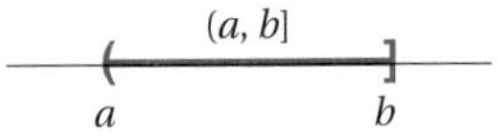

(a, b]

a b

$[a, b) = \{x \mid a \leq x < b\}$. The graph includes a and excludes b.

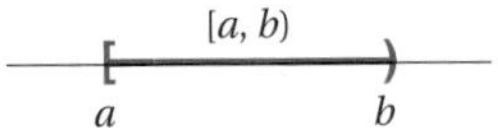

[a, b)

a b

Some intervals extend without bound in one or both directions. We use the symbols ∞, read "infinity," and $-\infty$, read "negative infinity," to name these intervals. The notation (a, ∞) represents the set of all numbers greater than a—that is,

$$(a, \infty) = \{x \mid x > a\}.$$

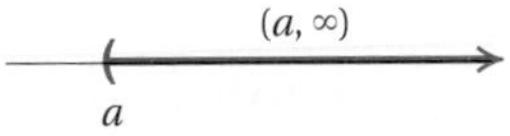

(a, ∞)

a

Similarly, the notation $(-\infty, a)$ represents the set of all numbers less than a—that is,

$$(-\infty, a) = \{x \mid x < a\}.$$

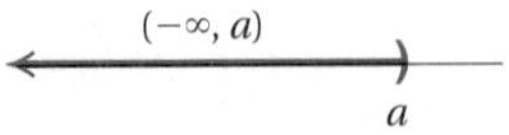

(−∞, a)

a

The notations $[a, \infty)$ and $(-\infty, a]$ are used when we want to include the endpoints. The interval $(-\infty, \infty)$ names the set of all real numbers.

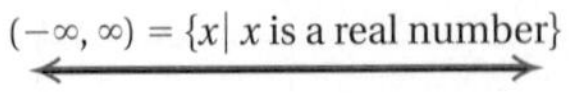

$(-\infty, \infty) = \{x \mid x \text{ is a real number}\}$

Interval notation is summarized in the following table.

---

*Some books use the representations (open circles at a and b) and (solid dots at a and b) instead of, respectively, (parentheses at a and b) and (brackets at a and b).

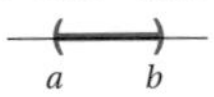

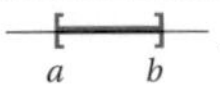

INTERVALS: NOTATION AND GRAPHS

Interval Notation	*Set Notation*	*Graph*
(a, b)	$\{x \mid a < x < b\}$	a b
$[a, b]$	$\{x \mid a \le x \le b\}$	a b
$[a, b)$	$\{x \mid a \le x < b\}$	a b
$(a, b]$	$\{x \mid a < x \le b\}$	a b
(a, ∞)	$\{x \mid x > a\}$	a
$[a, \infty)$	$\{x \mid x \ge a\}$	a
$(-\infty, b)$	$\{x \mid x < b\}$	b
$(-\infty, b]$	$\{x \mid x \le b\}$	b
$(-\infty, \infty)$	$\{x \mid x$ is a real number$\}$	

Examples Write interval notation for the given set or graph.

5. $\{x \mid -4 < x < 5\} = (-4, 5)$

6. $\{x \mid x \ge -2\} = [-2, \infty)$

7.

$(-2, 4]$

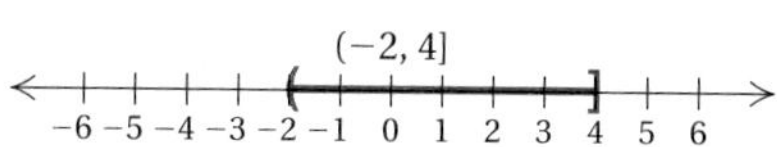

8.

$(-\infty, -1)$

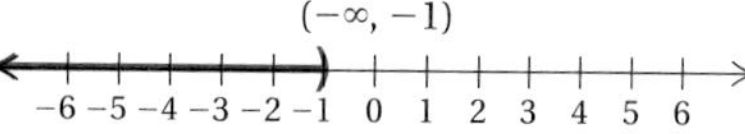

Do Exercises 4–7.

c Solving Inequalities

Two inequalities are **equivalent** if they have the same solution set. For example, the inequalities $x > 4$ and $4 < x$ are equivalent. Just as the addition principle for equations gives us equivalent equations, the addition principle for inequalities gives us equivalent inequalities.

THE ADDITION PRINCIPLE FOR INEQUALITIES

For any real numbers a, b, and c:

$a < b$ is equivalent to $a + c < b + c$;

$a > b$ is equivalent to $a + c > b + c$.

Similar statements hold for $\le$ and $\ge$.

Since subtracting c is the same as adding $-c$, there is no need for a separate subtraction principle.

Write interval notation for the given set or graph.

4. $\{x \mid -4 \le x < 5\}$

5. $\{x \mid x \le -2\}$

6.

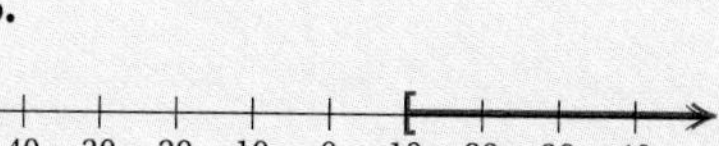

7.

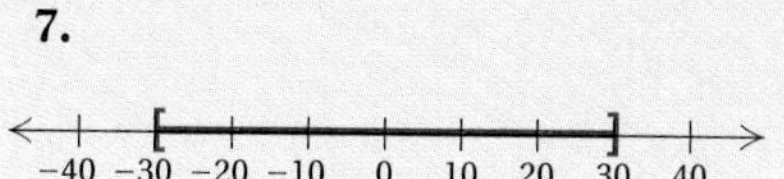

Answers on page A-35

Solve and graph.

8. $x + 6 > 9$

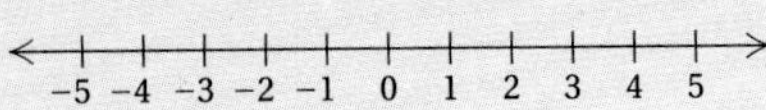

9. $x + 4 \le 7$

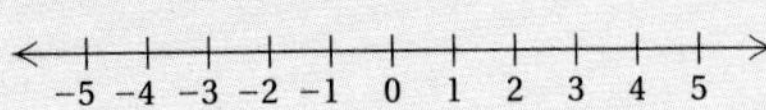

10. Solve and graph:

$2x - 3 \ge 3x - 1.$

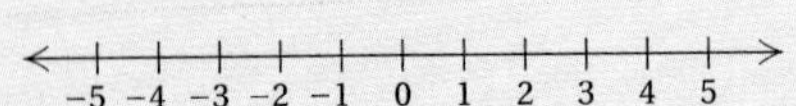

Answers on page A-35

Example 9 Solve and graph: $x + 5 > 1$.

We have

$$x + 5 > 1$$

$$x + 5 - 5 > 1 - 5 \quad \text{Using the addition principle: adding } -5 \text{ or subtracting } 5$$

$$x > -4.$$

We used the addition principle to show that the inequalities $x + 5 > 1$ and $x > -4$ are equivalent. The solution set is $\{x \mid x > -4\}$ and consists of an infinite number of solutions. We cannot possibly check them all. Instead, we can perform a partial check by substituting one member of the solution set (here we use -1) into the original inequality:

$$\begin{array}{c|c} x + 5 & 1 \\ \hline -1 + 5 \; ? & 1 \\ 4 & \end{array} \quad \text{TRUE}$$

Since $4 > 1$ is true, we have our check. The solution set is $\{x \mid x > -4\}$, or $(-4, \infty)$. The graph is as follows:

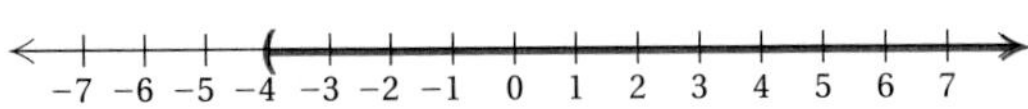

Do Exercises 8 and 9.

Example 10 Solve and graph: $4x - 1 \ge 5x - 2$.

We have

$$4x - 1 \ge 5x - 2$$

$$4x - 1 + 2 \ge 5x - 2 + 2 \quad \text{Adding 2}$$

$$4x + 1 \ge 5x \quad \text{Simplifying}$$

$$4x + 1 - 4x \ge 5x - 4x \quad \text{Subtracting } 4x$$

$$1 \ge x. \quad \text{Simplifying}$$

We know that $1 \ge x$ has the same meaning as $x \le 1$. You can check that any number less than or equal to 1 is a solution. The solution set is $\{x \mid 1 \ge x\}$ or, more commonly, $\{x \mid x \le 1\}$. Using interval notation, we write that the solution set is $(-\infty, 1]$. The graph is as follows:

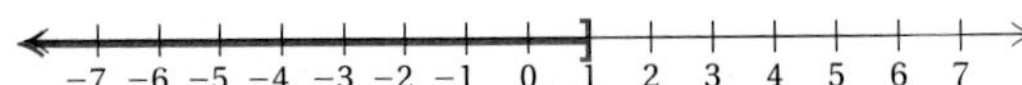

Do Exercise 10.

The multiplication principle for inequalities is somewhat different from the multiplication principle for equations. Consider this true inequality:

$$-4 < 9. \quad \text{True}$$

If we multiply both numbers by 2, we get another true inequality:

$$-4(2) < 9(2), \quad \text{or} \quad -8 < 18. \quad \text{True}$$

If we multiply both numbers by -3, we get a false inequality:

$$-4(-3) < 9(-3), \quad \text{or} \quad 12 < -27. \quad \text{False}$$

However, if we now *reverse* the inequality symbol above, we get a true inequality:

$$12 > -27. \quad \text{True}$$

THE MULTIPLICATION PRINCIPLE FOR INEQUALITIES

For any real numbers a and b, and any *positive* number c:

$a < b$ is equivalent to $ac < bc$;

$a > b$ is equivalent to $ac > bc$.

For any real numbers a and b, and any *negative* number c:

$a < b$ is equivalent to $ac > bc$;

$a > b$ is equivalent to $ac < bc$.

Similar statements hold for $\leq$ and $\geq$.

Since division by c is the same as multiplication by $1/c$, there is no need for a separate division principle.

Example 11 Solve and graph: $3y < \frac{3}{4}$.

We have

$$3y < \frac{3}{4}$$

$$\frac{1}{3} \cdot 3y < \frac{1}{3} \cdot \frac{3}{4}$$ Multiplying by $\frac{1}{3}$. The symbol stays the same.

$$y < \frac{1}{4}.$$

Any number less than $\frac{1}{4}$ is a solution. The solution set is $\{y \mid y < \frac{1}{4}\}$, or $\left(-\infty, \frac{1}{4}\right)$. The graph is as follows:

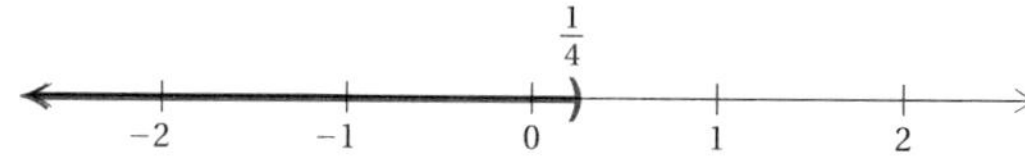

Example 12 Solve and graph: $-5x \geq -80$.

We have

$$-5x \geq -80$$

$$\frac{-5x}{-5} \leq \frac{-80}{-5}$$ Dividing by -5. The symbol must be reversed.

$$x \leq 16.$$

The solution set is $\{x \mid x \leq 16\}$, or $(-\infty, 16]$. The graph is as follows:

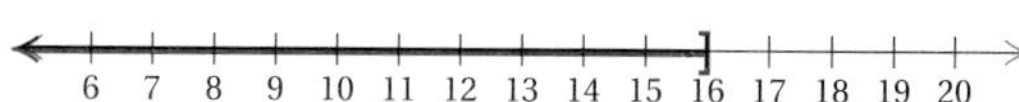

Do Exercises 11–13.

Solve and graph.

11. $5y \leq \frac{3}{2}$

−5 −4 −3 −2 −1 0 1 2 3 4 5

12. $-2y > 10$

−5 −4 −3 −2 −1 0 1 2 3 4 5

13. $-\frac{1}{3}x \leq -4$

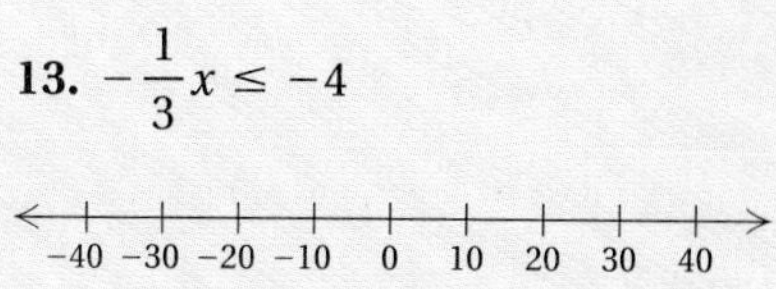

Answers on page A-35

Solve.

14. $6 - 5y \geq 7$

15. $3x + 5x < 4$

16. $17 - 5(y - 2) \leq 45y + 8(2y - 3) - 39y$

Answers on page A-35

We use the addition and multiplication principles together in solving inequalities in much the same way as in solving equations.

Example 13 Solve: $16 - 7y \geq 10y - 4$.

We have

$$\begin{aligned} 16 - 7y &\geq 10y - 4 \\ -16 + 16 - 7y &\geq -16 + 10y - 4 && \text{Adding } -16 \\ -7y &\geq 10y - 20 && \text{Collecting like terms} \\ -10y + (-7y) &\geq -10y + 10y - 20 && \text{Adding } -10y \\ -17y &\geq -20 && \text{Collecting like terms} \\ \frac{-17y}{-17} &\leq \frac{-20}{-17} && \text{Dividing by } -17. \text{ The symbol must be reversed.} \\ y &\leq \frac{20}{17}. && \text{Simplifying} \end{aligned}$$

The solution set is $\{y \mid y \leq \frac{20}{17}\}$, or $\left(-\infty, \frac{20}{17}\right]$.

In some cases, you can avoid the concern about multiplying or dividing by a negative number by using the addition principle in a different way. Let's rework Example 13 by adding $7y$ instead of $-10y$:

$$\begin{aligned} 16 - 7y &\geq 10y - 4 \\ 16 - 7y + 7y &\geq 10y - 4 + 7y && \text{Adding } 7y \\ 16 &\geq 17y - 4 && \text{Collecting like terms} \\ 16 + 4 &\geq 17y - 4 + 4 && \text{Adding } 4 \\ 20 &\geq 17y && \text{Collecting like terms} \\ \frac{20}{17} &\geq \frac{17y}{17} && \text{Dividing by } 17 \\ \frac{20}{17} &\geq y. \end{aligned}$$

Note that $\frac{20}{17} \geq y$ is equivalent to $y \leq \frac{20}{17}$.

Example 14 Solve: $-3(x + 8) - 5x > 4x - 9$.

We have

$$\begin{aligned} -3(x + 8) - 5x &> 4x - 9 \\ -3x - 24 - 5x &> 4x - 9 && \text{Using the distributive law} \\ -24 - 8x &> 4x - 9 && \text{Collecting like terms} \\ -24 - 8x + 8x &> 4x - 9 + 8x && \text{Adding } 8x \\ -24 &> 12x - 9 && \text{Collecting like terms} \\ -24 + 9 &> 12x - 9 + 9 && \text{Adding } 9 \\ -15 &> 12x && \text{Simplifying} \\ \frac{-15}{12} &> \frac{12x}{12} && \text{Dividing by 12. The symbol stays the same.} \\ -\frac{5}{4} &> x. \end{aligned}$$

The solution set is $\{x \mid -\frac{5}{4} > x\}$, or $\{x \mid x < -\frac{5}{4}\}$, or $\left(-\infty, -\frac{5}{4}\right)$.

Do Exercises 14–16.

d Applications and Problem Solving

Many problem-solving and applied situations translate to inequalities. In addition to "is less than" and "is more than," other phrases are commonly used.

Phrase	Translation
a "is at most" 17	$a \leq 17$
a "is at least" 5	$a \geq 5$
a "can't exceed" 12	$a \leq 12$
a "is a better price than" b	$a < b$

Example 15 *Records in the Men's 200-m Dash.* Michael Johnson set a world record of 19.32 sec in the men's 200-m dash in the 1996 Olympics (***Source***: International Amateur Athletic Foundation). The equation

$$R = -0.0513t + 19.32$$

can be used to predict the world record in the men's 200-m dash t years after 1996. Determine (in terms of an inequality) those years for which the world record will be less than 19.0 sec.

1. **Familiarize.** We already have a formula. To become more familiar with it, we might make a substitution for t. Suppose we want to know the record after 20 yr, in the year 2016. We substitute 20 for t:

   $$R = -0.0513(20) + 19.32 = 18.294 \text{ sec.}$$

   We see that by 2016, the record will be less than 19.0 sec. To predict the exact year in which the 19.0-sec mark will be broken, we could make other guesses that are less than 20. Instead, we proceed to the next step.
2. **Translate.** The record R is to be *less than* 19.0 sec. Thus we have

   $$R < 19.0.$$

   We replace R with $-0.0513t + 19.32$ to find the times t that solve the inequality:

   $$-0.0513t + 19.32 < 19.0.$$
3. **Solve.** We solve the inequality:

   $$-0.0513t + 19.32 < 19.0$$
   $$-0.0513t < -0.32 \quad \text{Subtracting 19.32}$$
   $$t > 6.24. \quad \text{Dividing by } -0.0513 \text{ and rounding}$$
4. **Check.** A partial check is to substitute a value for t greater than 6.24. We did that in the *Familiarize* step.
5. **State.** The record will be less than 19.0 sec for races occurring more than 6.24 yr after 1996, or approximately after 2002.

Do Exercise 17.

17. *Records in the Men's 200-m Dash.* Refer to Example 15. Determine (in terms of an inequality) those years for which the world record will be less than 19.1 sec.

Answer on page A-35

18. *Salary Plans.* A painter can be paid in one of two ways:

Plan A: $500 plus $4 per hour;
Plan B: Straight $9 per hour.

Suppose that the job takes n hours. For what values of n is plan A better for the painter?

Answer on page A-35

Example 16 *Salary Plans.* On a new job, Rose can be paid in one of two ways: *Plan A* is a salary of $600 per month, plus a commission of 4% of sales; and *Plan B* is a salary of $800 per month, plus a commission of 6% of sales in excess of $10,000. For what amount of monthly sales is plan A better than plan B, if we assume that sales are always more than $10,000?

1. Familiarize. Listing the given information in a table will be helpful.

Plan A: Monthly Income	Plan B: Monthly Income
$600 salary 4% of sales *Total*: $600 + 4% of sales	$800 salary 6% of sales over $10,000 *Total*: $800 + 6% of sales over $10,000

Next, suppose that Rose sold $12,000 in one month. Which plan would be better? Under plan A, she would earn $600 plus 4% of $12,000, or

$$600 + 0.04(12{,}000) = \$1080.$$

Since with plan B commissions are paid only on sales in excess of $10,000, Rose would earn $800 plus 6% of ($12,000 − $10,000), or

$$800 + 0.06(12{,}000 - 10{,}000) = \$920.$$

This shows that for monthly sales of $12,000, plan A is better. Similar calculations will show that for sales of $30,000 a month, plan B is better. To determine *all* values for which plan A pays more money, we must solve an inequality that is based on the calculations above.

2. Translate. We let S represent the amount of monthly sales. If we examine the calculations in the *Familiarize* step, we see that the monthly income from plan A is $600 + 0.04S$ and from plan B is $800 + 0.06(S - 10{,}000)$. Thus we want to find all values of S for which

Income from plan A	is greater than	Income from plan B
$600 + 0.04S$	$>$	$800 + 0.06(S - 10{,}000)$.

3. Solve. We solve the inequality:

$$600 + 0.04S > 800 + 0.06(S - 10{,}000)$$
$$600 + 0.04S > 800 + 0.06S - 600 \quad \text{Using the distributive law}$$
$$600 + 0.04S > 200 + 0.06S \quad \text{Collecting like terms}$$
$$400 > 0.02S \quad \text{Subtracting both 200 and } 0.04S$$
$$20{,}000 > S, \text{ or } S < 20{,}000. \quad \text{Dividing by 0.02}$$

4. Check. For $S = 20{,}000$, the income from plan A is

$600 + 4\% \cdot 20{,}000$, or $1400.

The income from plan B is

$800 + 6\% \cdot (20{,}000 - 10{,}000)$, or $1400.

This confirms that for sales of $20,000, Rose's pay is the same under either plan.

In the *Familiarize* step, we saw that for sales of $12,000, plan A pays more. Since $12{,}000 < 20{,}000$, this is a partial check. Since we cannot check all possible values of S, we will stop here.

5. State. For monthly sales of less than $20,000, plan A is better.

Do Exercise 18.

Exercise Set 9.1

a Determine whether the given numbers are solutions of the inequality.

1. $x - 2 \geq 6$; $-4, 0, 4, 8$

2. $3x + 5 \leq -10$; $-5, -10, 0, 27$

3. $t - 8 > 2t - 3$; $0, -8, -9, -3, -\frac{7}{8}$

4. $5y - 7 < 8 - y$; $2, -3, 0, 3, \frac{2}{3}$

b Write interval notation for the given set or graph.

5. $\{x \mid x < 5\}$

6. $\{t \mid t \geq -5\}$

7. $\{x \mid -3 \leq x \leq 3\}$

8. $\{t \mid -10 < t \leq 10\}$

9.

−6 −5 −4 −3 −2 −1 0 1 2 3 4 5 6

10.

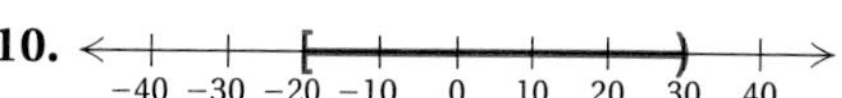

11.

$-\sqrt{2}$

−2 −1 0 1 2

12.

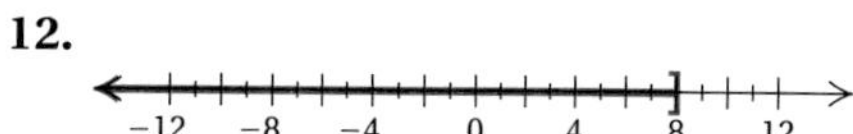

c Solve and graph.

13. $x + 2 > 1$

−6 −5 −4 −3 −2 −1 0 1 2 3 4 5 6

14. $x + 8 > 4$

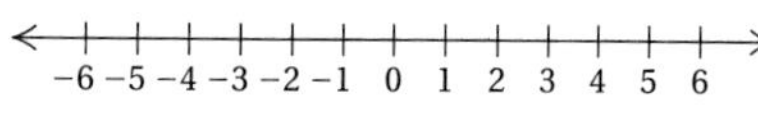

15. $y + 3 < 9$

−6 −5 −4 −3 −2 −1 0 1 2 3 4 5 6

16. $y + 4 < 10$

−6 −5 −4 −3 −2 −1 0 1 2 3 4 5 6

17. $a - 9 \leq -31$

−40 −30 −20 −10 0 10 20 30 40

18. $a + 6 \leq -14$

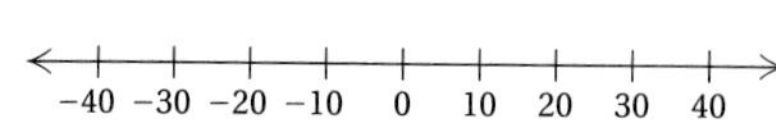

19. $t + 13 \geq 9$

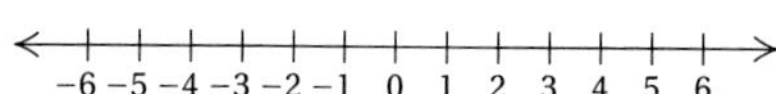

20. $x - 8 \leq 17$

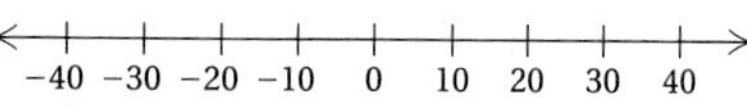

21. $y - 8 > -14$

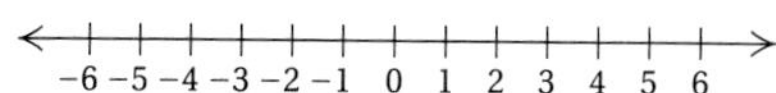

22. $y - 9 > -18$

−12 −8 −4 0 4 8 12

23. $x - 11 \le -2$

−12 −8 −4 0 4 8 12

24. $y - 18 \le -4$

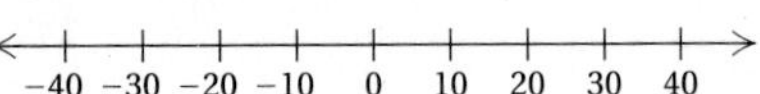

25. $8x \ge 24$

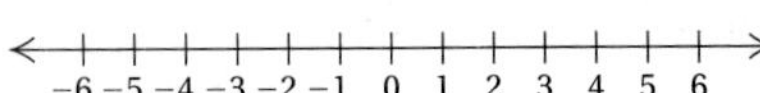

26. $8t < -56$

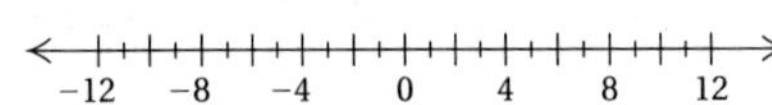

27. $0.3x < -18$

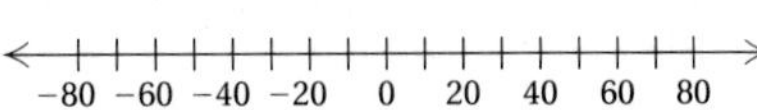

28. $0.6x < 30$

−80 −60 −40 −20 0 20 40 60 80

Solve.

29. $-9x \ge -8.1$

30. $-5y \le 3.5$

31. $-\frac{3}{4}x \ge -\frac{5}{8}$

32. $-\frac{1}{8}y \le -\frac{9}{8}$

33. $2x + 7 < 19$

34. $5y + 13 > 28$

35. $5y + 2y \le -21$

36. $-9x + 3x \ge -24$

37. $2y - 7 < 5y - 9$

38. $8x - 9 < 3x - 11$

39. $0.4x + 5 \le 1.2x - 4$

40. $0.2y + 1 > 2.4y - 10$

41. $5x - \frac{1}{12} \le \frac{5}{12} + 4x$

42. $2x - 3 < \frac{13}{4}x + 10 - 1.25x$

43. $4(4y - 3) \geq 9(2y + 7)$

44. $2m + 5 \geq 16(m - 4)$

45. $3(2 - 5x) + 2x < 2(4 + 2x)$

46. $2(0.5 - 3y) + y > (4y - 0.2)8$

47. $5[3m - (m + 4)] > -2(m - 4)$

48. $[8x - 3(3x + 2)] - 5 \geq 3(x + 4) - 2x$

49. $3(r - 6) + 2 > 4(r + 2) - 21$

50. $5(t + 3) + 9 < 3(t - 2) + 6$

51. $19 - (2x + 3) \leq 2(x + 3) + x$

52. $13 - (2c + 2) \geq 2(c + 2) + 3c$

53. $\frac{1}{4}(8y + 4) - 17 < -\frac{1}{2}(4y - 8)$

54. $\frac{1}{3}(6x + 24) - 20 > -\frac{1}{4}(12x - 72)$

55. $2[4 - 2(3 - x)] - 1 \geq 4[2(4x - 3) + 7] - 25$

56. $5[3(7 - t) - 4(8 + 2t)] - 20 \leq -6[2(6 + 3t) - 4]$

57. $\frac{4}{5}(7x - 6) < 40$

58. $\frac{2}{3}(4x - 3) > 30$

59. $\frac{3}{4}(3 + 2x) + 1 \geq 13$

60. $\frac{7}{8}(5 - 4x) - 17 \geq 38$

61. $\frac{3}{4}\left(3x - \frac{1}{2}\right) - \frac{2}{3} < \frac{1}{3}$

62. $\frac{2}{3}\left(\frac{7}{8} - 4x\right) - \frac{5}{8} < \frac{3}{8}$

63. $0.7(3x + 6) \geq 1.1 - (x + 2)$

64. $0.9(2x + 8) < 20 - (x + 5)$

65. $a + (a - 3) \leq (a + 2) - (a + 1)$

66. $0.8 - 4(b - 1) > 0.2 + 3(4 - b)$

d Solve.

Body Mass Index. The *body mass index I* can be used to determine an individual's risk of heart disease. An index less than 25 indicates a low risk. The body mass index is given by the formula, or model,

$$I = \frac{700W}{H^2},$$

where W is the weight, in pounds, and H is the height, in inches.

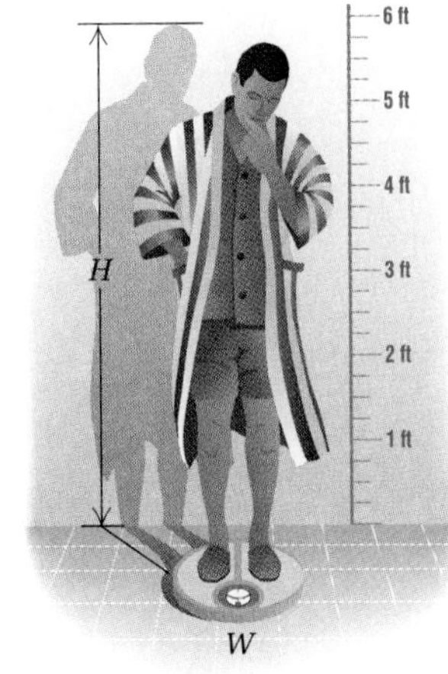

67. a) Marv weighs 214 lb and his height is 73 in. What is his body mass index?
b) Determine (in terms of an inequality) those weights W for Marv that will keep him in the low-risk category.

68. a) Elaine weighs 110 lb and her height is 67 in. What is her body mass index?
b) Determine (in terms of an inequality) those weights W for Elaine that will keep her in the low-risk category.

69. *Grades.* You are taking a European history course in which there will be 4 tests, each worth 100 points. You have scores of 89, 92, and 95 on the first three tests. You must make a total of at least 360 in order to get an A. What scores on the last test will give you an A?

70. *Grades.* You are taking a literature course in which there will be 5 tests, each worth 100 points. You have scores of 94, 90, and 89 on the first three tests. You must make a total of at least 450 in order to get an A. What scores on the fourth test will keep you eligible for an A?

71. *Insurance Claims.* After a serious automobile accident, most insurance companies will replace the damaged car with a new one if repair costs exceed 80% of the N.A.D.A., or "blue-book," value of the car. Miguel's car recently sustained \$9200 worth of damage but was not replaced. What was the blue-book value of his car?

72. *Phone Rates.* A long-distance telephone call using Down East Calling costs 20 cents for the first minute and 16 cents for each additional minute. The same call, placed on Long Call Systems, costs 19 cents for the first minute and 18 cents for each additional minute. For what length phone calls is Down East Calling less expensive?

73. *Salary Plans.* Toni can be paid in one of two ways:

Plan A: A salary of \$400 per month plus a commission of 8% of gross sales;

Plan B: A salary of \$610 per month, plus a commission of 5% of gross sales.

For what amount of gross sales should Toni select plan A?

74. *Salary Plans.* Branford can be paid for his masonry work in one of two ways:

Plan A: \$300 plus \$9.00 per hour;

Plan B: Straight \$12.50 per hour.

Suppose that the job takes n hours. For what values of n is plan B better for Branford?

75. *Checking-Account Rates.* The Hudson Bank offers two checking-account plans. Their Anywhere plan charges 20¢ per check whereas their Acu-checking plan costs \$2 per month plus 12¢ per check. For what numbers of checks per month will the Acu-checking plan cost less?

76. *Insurance Benefits.* Bayside Insurance offers two plans. Under plan A, Giselle would pay the first \$50 of her medical bills and 20% of all bills after that. Under plan B, Giselle would pay the first \$250 of bills, but only 10% of the rest. For what amount of medical bills will plan B save Giselle money? (Assume that her bills will exceed \$250.)

77. *Wedding Costs.* The Arnold Inn offers two plans for wedding parties. Under plan A, the inn charges \$30 for each person in attendance. Under plan B, the inn charges \$1300 plus \$20 for each person in excess of the first 25 who attend. For what size parties will plan B cost less? (Assume that more than 25 guests will attend.)

78. *Investing.* Lillian is about to invest \$20,000, part at 6% and the rest at 8%. What is the most that she can invest at 6% and still be guaranteed at least \$1500 in interest per year?

79. *Converting Dress Sizes.* The formula

$I = 2(s + 10)$

can be used to convert dress sizes s in the United States to dress sizes I in Italy. For what dress sizes in the United States will dress sizes in Italy be larger than 36?

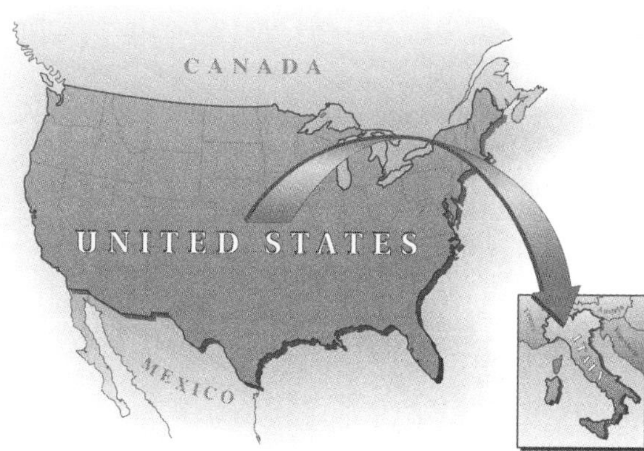

80. *Temperatures of Solids.* The formula

$C = \frac{5}{9}(F - 32)$

can be used to convert Fahrenheit temperatures F to Celsius temperatures C.

a) Gold is a solid at Celsius temperatures less than 1063°C. Find the Fahrenheit temperatures for which gold is a solid.

b) Silver is a solid at Celsius temperatures less than 960.8°C. Find the Fahrenheit temperatures for which silver is a solid.

81. *Consumption of Bottled Water.* People are consuming more bottled water each year. The number N of gallons per year that each person drinks t years after 1990 is approximated by

$N = 0.733t + 8.398.$

a) How many gallons of bottled water did each person drink in 1990 ($t = 0$)? in 1995 ($t = 5$)? in 2000?
b) For what years will the amount of bottled water drunk be at least 15 gal?

82. *Dewpoint Spread.* Pilots use the **dewpoint spread,** or the difference between the current temperature and the dewpoint (the temperature at which dew occurs), to estimate the height of the cloud cover. Each 3° of dewpoint spread corresponds to an increased height of cloud cover of 1000 ft. A plane, flying with limited instruments, must have a cloud cover greater than 3500 ft. What dewpoint spreads will allow the plane to fly?

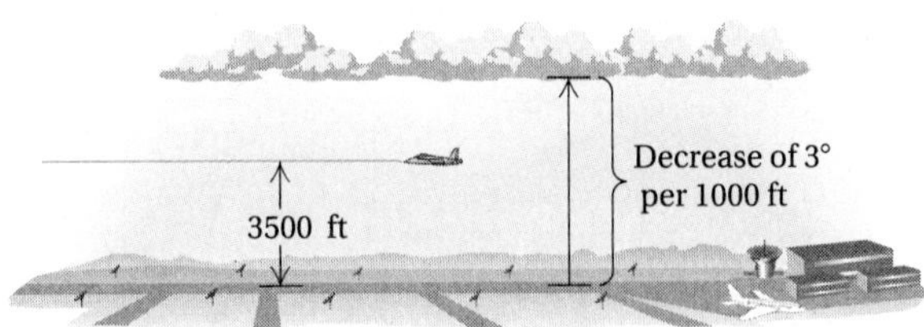

Skill Maintenance

Multiply. [4.6a]

83. $(3x - 4)(x + 8)$

84. $(r - 4s)(6r + s)$

85. $(2a - 5)(3a + 11)$

86. $(t + 2s)(t - 9s)$

Find the domain. [7.2a]

87. $f(x) = \dfrac{-3}{x + 8}$

88. $f(x) = 3x - 5$

89. $f(x) = |x| - 4$

90. $f(x) = \dfrac{x + 7}{3x - 2}$

Synthesis

91. ◆ Explain in your own words why the inequality symbol must be reversed when both sides of an inequality are multiplied or divided by a negative number.

92. ◆ Graph the solution set of each of the following on a number line and compare:

$x < 2$, $x > 2$, $x = 2$, $x \leq 2$, and $x \geq 2$.

93. *Supply and Demand.* The supply S and demand D for a certain product are given by

$S = 460 + 94p$ and $D = 2000 - 60p$.

a) Find those values of p for which supply exceeds demand.
b) Find those values of p for which supply is less than demand.

Determine whether the statement is true or false. If false, give a counterexample.

94. For any real numbers x and y, if $x < y$, then $x^2 < y^2$.

95. For any real numbers a, b, c, and d, if $a < b$ and $c < d$, then $a + c < b + d$.

96. Determine whether the inequalities

$x < 3$ and $0 \cdot x < 0 \cdot 3$

are equivalent. Give reasons to support your answer.

Solve.

97. $x + 5 \leq 5 + x$

98. $x + 8 < 3 + x$

99. $x^2 + 1 > 0$

9.2 Intersections, Unions, and Compound Inequalities

The steroid cholesterol is a fat-related compound in tissues of cells. One commonly used measure of cholesterol is *total cholesterol.* The following table shows the health-risk levels of this measure of cholesterol.

Type of Cholesterol	Normal Risk	Borderline Risk	High Risk
Total	Total $\leq$ 200	200 $<$ Total $<$ 240	Total $\geq$ 240

A total-cholesterol level T between 200 and 240 is considered borderline risk. We can express this by the sentence

$$200 < T \quad and \quad T < 240$$

or more simply by

$$200 < T < 240.$$

This is an example of a *compound inequality.* **Compound inequalities** consist of two or more inequalities joined by the word *and* or the word *or.* We now "solve" such sentences—that is, we find the set of all solutions.

Objectives

a Find the intersection of two sets. Solve and graph conjunctions of inequalities.

b Find the union of two sets. Solve and graph disjunctions of inequalities.

c Solve applied problems involving conjunctions and disjunctions of inequalities.

For Extra Help

TAPE 15

MAC WIN

CD-ROM

a Intersections of Sets and Conjunctions of Inequalities

The **intersection** of two sets A and B is the set of all members that are common to A and B. We denote the intersection of sets A and B as

$$A \cap B.$$

The intersection of two sets is often illustrated as shown at right.

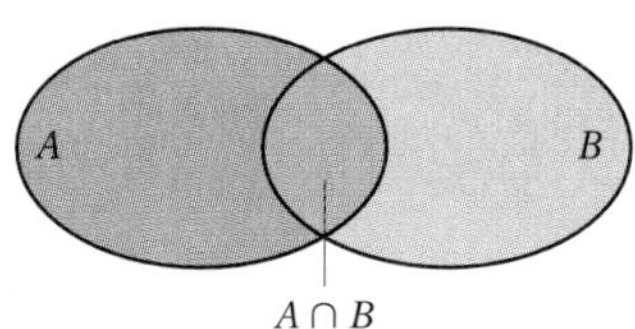

Example 1 Find the intersection:

$$\{1, 2, 3, 4, 5\} \cap \{-2, -1, 0, 1, 2, 3\}.$$

The numbers 1, 2, and 3 are common to the two sets, so the intersection is $\{1, 2, 3\}$.

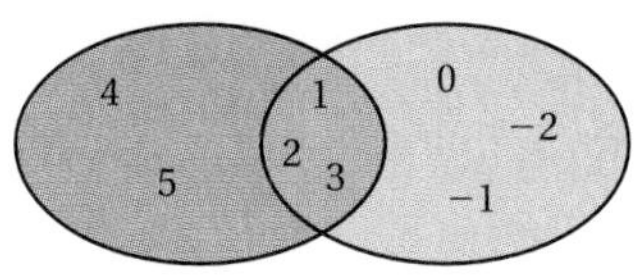

Do Exercises 1 and 2.

When two or more sentences are joined by the word *and* to make a compound sentence, the new sentence is called a **conjunction** of the sentences. The following is a conjunction of inequalities:

$$-2 < x \quad and \quad x < 1.$$

1. Find the intersection:

$$\{0, 3, 5, 7\} \cap \{0, 1, 3, 11\}.$$

2. Shade the intersection of sets A and B.

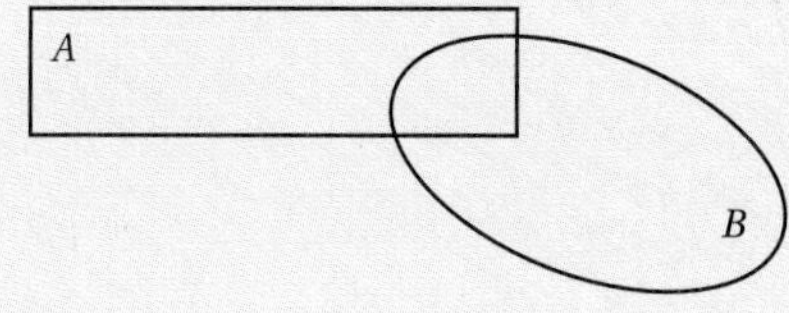

Answers on page A-36

3. Graph and write interval notation:

$-1 < x$ *and* $x < 4$.

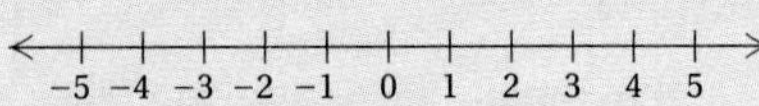

Answer on page A-36

In order for a conjunction to be true, each individual sentence must be true. *The solution set of a conjunction is the intersection of the solution sets of the individual sentences.* Consider the conjunction

$$-2 < x \quad and \quad x < 1.$$

The graphs of each separate sentence are shown below, and the intersection is the last graph. We use both set-builder and interval notations.

$\{x \mid -2 < x\}$ $(-2, \infty)$

$\{x \mid x < 1\}$ $(-\infty, 1)$

$\{x \mid -2 < x\} \cap \{x \mid x < 1\}$
$= \{x \mid -2 < x \text{ and } x < 1\}$ $(-2, 1)$

Because there are numbers that are both greater than -2 and less than 1, the conjunction $-2 < x$ *and* $x < 1$ can be abbreviated by $-2 < x < 1$. Thus the interval $(-2, 1)$ can be represented as $\{x \mid -2 < x < 1\}$, the set of all numbers that are *simultaneously* greater than -2 *and* less than 1. Note that, in general, for $a < b$,

$a < x$ *and* $x < b$ can be abbreviated $a < x < b$;

and $b > x$ *and* $x > a$ can be abbreviated $b > x > a$.

> The word "and" corresponds to "intersection" and to the symbol "$\cap$". In order for a number to be a solution of a conjunction, it must be in *both* solution sets.

Do Exercise 3.

Example 2 Solve and graph: $-1 \le 2x + 5 < 13$.

This inequality is an abbreviation for the conjunction

$$-1 \le 2x + 5 \quad and \quad 2x + 5 < 13.$$

The word *and* corresponds to set *intersection*, $\cap$. To solve the conjunction, we solve each of the two inequalities separately and then find the intersection of the solution sets:

$-1 \le 2x + 5$ *and* $2x + 5 < 13$

$-6 \le 2x$ *and* $2x < 8$ **Subtracting 5**

$-3 \le x$ *and* $x < 4$. **Dividing by 2**

We now abbreviate the answer:

$$-3 \le x < 4.$$

The solution set is $\{x \mid -3 \le x < 4\}$, or, in interval notation, $[-3, 4)$. The graph is the intersection of the two separate solution sets.

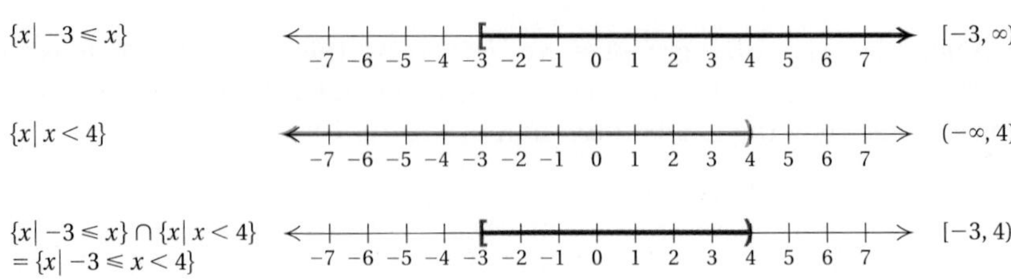

The steps above are generally combined as follows:

$$-1 \le 2x + 5 < 13 \qquad \textbf{2x + 5 appears in both inequalities.}$$
$$-6 \le 2x < 8 \qquad \textbf{Subtracting 5}$$
$$-3 \le x < 4. \qquad \textbf{Dividing by 2}$$

Such an approach saves some writing and will prove useful in Section 9.3.

Do Exercise 4.

Sometimes there is no way to solve both parts of a conjunction at the same time.

Example 3 Solve and graph: $2x - 5 \ge -3$ *and* $5x + 2 \ge 17$.

We first solve each inequality separately:

$$2x - 5 \ge -3 \quad \textit{and} \quad 5x + 2 \ge 17$$
$$2x \ge 2 \quad \textit{and} \quad 5x \ge 15$$
$$x \ge 1 \quad \textit{and} \quad x \ge 3.$$

Next, we find the intersection of the two separate solution sets:

$\{x | x \ge 1\}$ — $[1, \infty)$

$\{x | x \ge 3\}$ — $[3, \infty)$

$\{x | x \ge 1\} \cap \{x | x \ge 3\}$
$= \{x | x \ge 3\}$ — $[3, \infty)$

The numbers common to both sets are those that are greater than or equal to 3. Thus the solution set is $\{x | x \ge 3\}$, or, in interval notation, $[3, \infty)$. You should check that any number in $[3, \infty)$ satisfies the conjunction whereas numbers outside $[3, \infty)$ do not.

Do Exercise 5.

Sometimes two sets have no elements in common. In such a case, we say that the intersection of the two sets is the empty set, denoted { } or $\varnothing$. Two sets with an empty intersection are said to be **disjoint**.

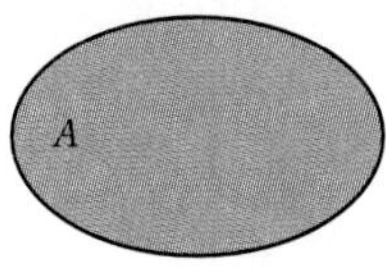

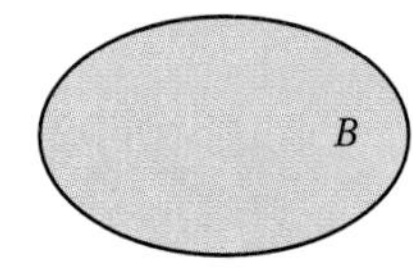

$A \cap B = \varnothing$.

Example 4 Solve and graph: $2x - 3 > 1$ *and* $3x - 1 < 2$.

We solve each inequality separately:

$$2x - 3 > 1 \quad \textit{and} \quad 3x - 1 < 2$$
$$2x > 4 \quad \textit{and} \quad 3x < 3$$
$$x > 2 \quad \textit{and} \quad x < 1.$$

The solution set is the intersection of the solution sets of the individual inequalities.

4. Solve and graph:

$$-22 < 3x - 7 \le 23.$$

5. Solve and graph:

$3x + 4 < 10$ *and* $2x - 7 < -13$.

Answers on page A-36

6. Solve and graph.

$3x - 7 \leq -13$ *and* $4x + 3 > 8$.

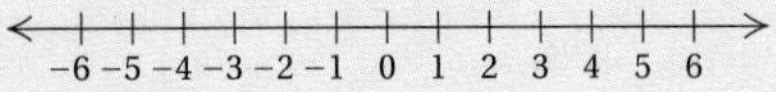

7. Solve: $-4 \leq 8 - 2x \leq 4$.

8. Find the union:

$\{0, 1, 3, 4\} \cup \{0, 1, 7, 9\}$.

9. Shade the union of sets A and B.

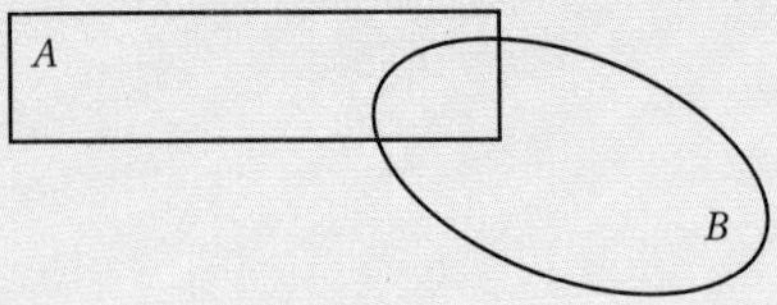

Answers on page A-36

$\{x \mid x > 2\}$

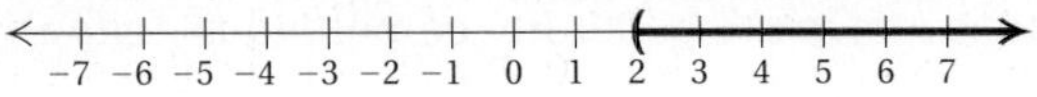

$(2, \infty)$

$\{x \mid x < 1\}$

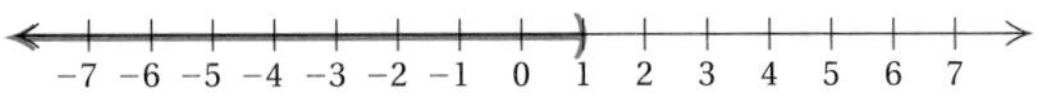

$(-\infty, 1)$

$\{x \mid x > 2\} \cap \{x \mid x < 1\}$
$= \{x \mid x > 2 \text{ and } x < 1\}$
$= \varnothing$

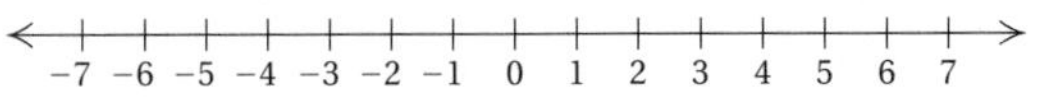

$\varnothing$

Since no number is both greater than 2 and less than 1, the solution set is the empty set, $\varnothing$.

Do Exercise 6.

Example 5 Solve: $3 \leq 5 - 2x < 7$.

We have

$$3 \leq 5 - 2x < 7$$

$$3 - 5 \leq 5 - 2x - 5 < 7 - 5 \qquad \textbf{Subtracting 5}$$

$$-2 \leq \quad -2x \quad < 2 \qquad \textbf{Simplifying}$$

$$\frac{-2}{-2} \geq \frac{-2x}{-2} > \frac{2}{-2} \qquad \textbf{Dividing by −2. The symbols must be reversed.}$$

$$1 \geq x > -1. \qquad \textbf{Simplifying}$$

The solution set is $\{x \mid 1 \geq x > -1\}$, or $\{x \mid -1 < x \leq 1\}$, since the inequalities $1 \geq x > -1$ and $-1 < x \leq 1$ are equivalent. The solution, in interval notation, is $(-1, 1]$.

Do Exercise 7.

b Unions of Sets and Disjunctions of Inequalities

The **union** of two sets A and B is the collection of elements belonging to A and/or B. We denote the union of A and B by

$A \cup B$.

The union of two sets is often pictured as shown below.

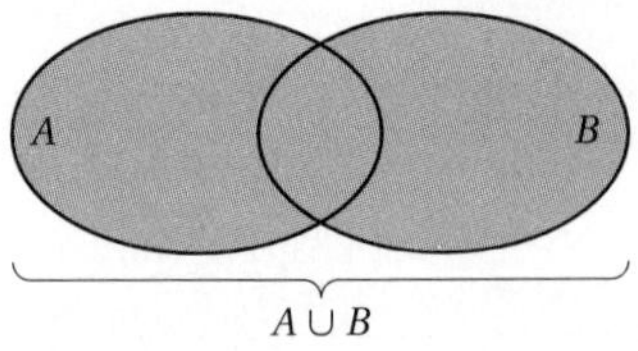

Example 6 Find the union: $\{2, 3, 4\} \cup \{3, 5, 7\}$.

The numbers in either or both sets are 2, 3, 4, 5, and 7, so the union is $\{2, 3, 4, 5, 7\}$.

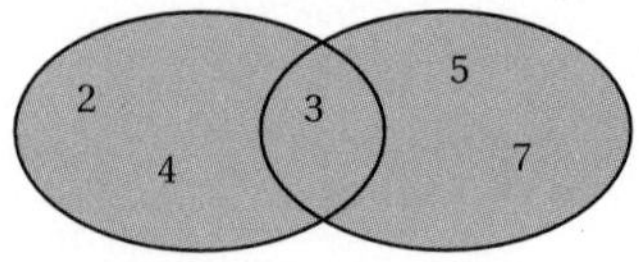

Do Exercises 8 and 9.

When two or more sentences are joined by the word *or* to make a compound sentence, the new sentence is called a **disjunction** of the sentences. Here are three examples:

$x < -3$ *or* $x > 3$;

y is an odd number *or* y is a prime number;

$x < 0$ *or* $x = 0$ *or* $x > 0$.

In order for a disjunction to be true, at least one of the individual sentences must be true. *The solution set of a disjunction is the union of the individual solution sets.* Consider the disjunction

$x < -3$ *or* $x > 3$.

The graphs of each separate sentence are shown below, and the union is the last graph. Again, we use both set-builder and interval notations.

$\{x \mid x < -3\}$ — $(-\infty, -3)$

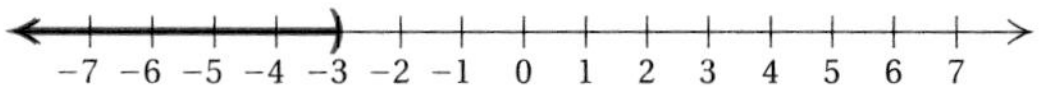

$\{x \mid x > 3\}$ — $(3, \infty)$

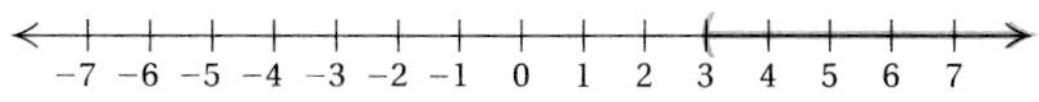

$\{x \mid x < -3\} \cup \{x \mid x > 3\}$
$= \{x \mid x < -3 \text{ or } x > 3\}$ — $(-\infty, -3) \cup (3, \infty)$

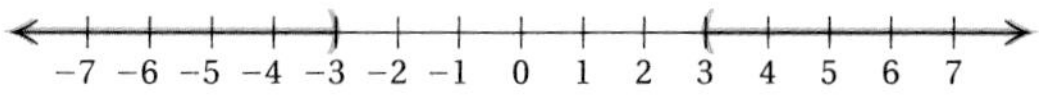

Answers to disjunctions can rarely be written in a shorter manner. The solution set of $x < -3$ *or* $x > 3$ is simply written $\{x \mid x < -3 \text{ or } x > 3\}$, or $(-\infty, -3) \cup (3, \infty)$.

> The word "or" corresponds to "union" and the symbol "∪". In order for a number to be in the solution set of a disjunction, it must be in *at least one* of the solution sets.

Do Exercise 10.

Example 7 Solve and graph: $7 + 2x < -1$ *or* $13 - 5x \leq 3$.

We solve each inequality separately, retaining the word *or*:

$7 + 2x < -1$ *or* $13 - 5x \leq 3$

$2x < -8$ *or* $-5x \leq -10$

Dividing by −5. The symbol must be reversed.

$x < -4$ *or* $x \geq 2$.

To find the solution set of the disjunction, we consider the individual graphs. We graph $x < -4$ and then $x \geq 2$. Then we take the union of the graphs.

$\{x \mid x < -4\}$ — $(-\infty, -4)$

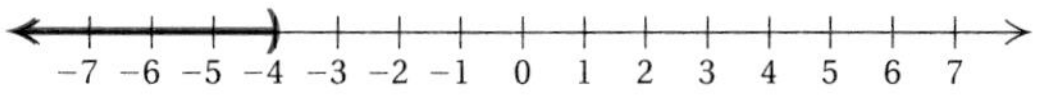

$\{x \mid x \geq 2\}$ — $[2, \infty)$

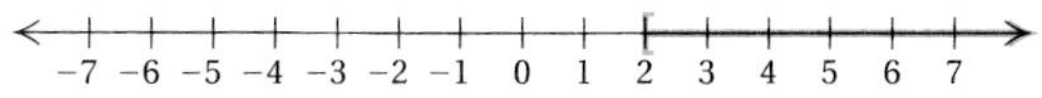

$\{x \mid x < -4 \text{ or } x \geq 2\}$ — $(-\infty, -4) \cup [2, \infty)$

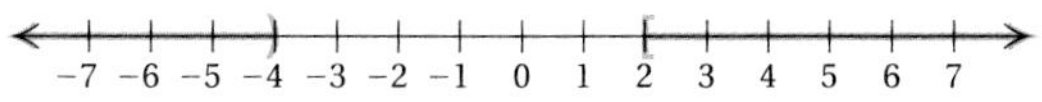

The solution set is $\{x \mid x < -4 \text{ or } x \geq 2\}$, or, in interval notation, $(-\infty, -4) \cup [2, \infty)$.

10. Graph and write interval notation:

$x \leq -2$ *or* $x > 4$.

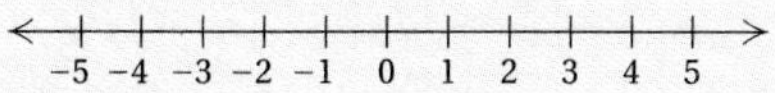

-5 -4 -3 -2 -1 0 1 2 3 4 5

Answer on page A-36

Solve and graph.

11. $x - 4 < -3$ *or* $x - 3 \geq 3$

12. $-2x + 4 \leq -3$ *or* $x + 5 < 3$

13. Solve:

$-3x - 7 < -1$ *or* $x + 4 < -1$.

14. Solve and graph:

$5x - 7 \leq 13$ *or* $2x - 1 \geq -7$.

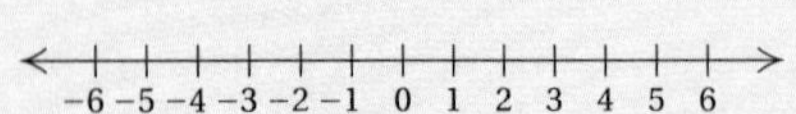

Answers on page A-36

CAUTION! A compound inequality like

$$x < -4 \quad or \quad x \geq 2,$$

as in Example 7, *cannot* be expressed as $2 \leq x < -4$ because to do so would be to say that x is *simultaneously* less than -4 and greater than or equal to 2. No number is both less than -4 *and* greater than or equal to 2, but many are less than -4 *or* greater than or equal to 2.

Do Exercises 11 and 12.

Example 8 Solve: $-2x - 5 < -2$ *or* $x - 3 < -10$.

We solve the individual inequalities separately, retaining the word *or*:

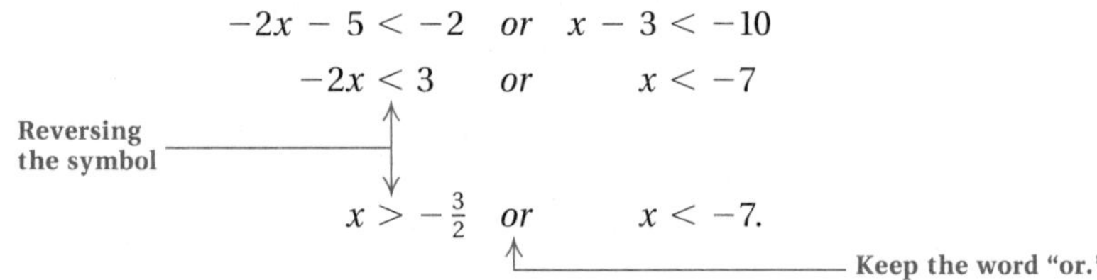

The solution set is $\{x \mid x < -7 \text{ or } x > -\frac{3}{2}\}$, or, in interval notation, $(-\infty, -7) \cup \left(-\frac{3}{2}, \infty\right)$.

Do Exercise 13.

Example 9 Solve: $3x - 11 < 4$ *or* $4x + 9 \geq 1$.

We solve the individual inequalities separately, retaining the word *or*:

$$\begin{aligned} 3x - 11 < 4 \quad &or \quad 4x + 9 \geq 1 \\ 3x < 15 \quad &or \quad 4x \geq -8 \\ x < 5 \quad &or \quad x \geq -2. \end{aligned}$$

To find the solution set, we first look at the individual graphs.

$\{x \mid x < 5\}$ — $(-\infty, 5)$

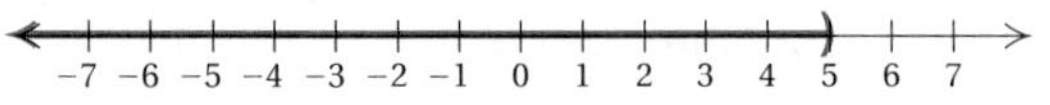

$\{x \mid x \geq -2\}$ — $[-2, \infty)$

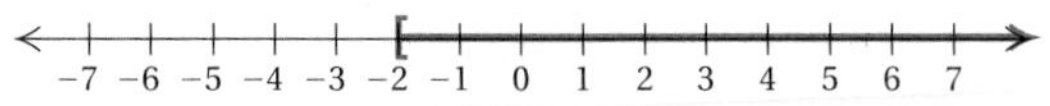

$\{x \mid x < 5\} \cup \{x \mid x \geq -2\} = \{x \mid x < 5 \text{ or } x \geq -2\}$ — $(-\infty, \infty)$ = The set of all real numbers

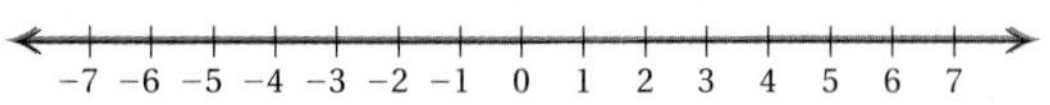

(Number lines labeled −7 −6 −5 −4 −3 −2 −1 0 1 2 3 4 5 6 7)

Since any number is either less than 5 or greater than or equal to -2, the two sets fill the entire number line. Thus the solution set is the set of all real numbers.

Do Exercise 14.

C Applications and Problem Solving

Example 10 *Converting Dress Sizes.* The equation

$$I = 2(s + 10)$$

can be used to convert dress sizes s in the United States to dress sizes I in Italy. Which dress sizes in the United States correspond to dress sizes between 32 and 46 in Italy?

1. **Familiarize.** We have a formula for converting the dress sizes. Thus we can substitute a value into the formula. For a dress of size 6 in the United States, we get the corresponding dress size in Italy as follows:

   $$I = 2(6 + 10) = 2 \cdot 16 = 32.$$

   This familiarizes us with the formula and also tells us that the United States sizes that we are looking for must be larger than size 6.

2. **Translate.** We want the Italian sizes *between* 32 and 46, so we want to find those values of s for which

   $$32 < I < 46 \qquad I \text{ is between 32 and 46}$$

   or

   $$32 < 2(s + 10) < 46. \qquad \text{Substituting } 2(s + 10) \text{ for } I$$

   Thus we have translated the problem to an inequality.

3. **Solve.** We solve the inequality:

   $$32 < 2(s + 10) < 46$$
   $$\frac{32}{2} < \frac{2(s + 10)}{2} < \frac{46}{2} \qquad \text{Dividing by 2}$$
   $$16 < s + 10 < 23 \qquad \text{Simplifying}$$
   $$6 < s < 13. \qquad \text{Subtracting 10}$$

4. **Check.** We substitute some values as we did in the *Familiarize* step.

5. **State.** Dress sizes between 6 and 13 in the United States correspond to dress sizes between 32 and 46 in Italy.

Do Exercise 15.

15. *Converting Dress Sizes.* Refer to Example 10. Which dress sizes in the United States correspond to dress sizes between 36 and 58 in Italy?

Answer on page A-36

Improving Your Math Study Skills

Study Tips for Trouble Spots

By now you have probably encountered certain topics that gave you more difficulty than others. It is important to know that this happens to every person who studies mathematics. Unfortunately, frustration is often part of the learning process and it is important not to give up when difficulty arises.

One source of frustration for many students is not being able to set aside sufficient time for studying. Family commitments, work schedules, and athletics are just a few of the time demands that many students face. Couple these demands with a math lesson that seems to require a greater than usual amount of study time, and it is no wonder that many students often feel frustrated. Below are some study tips that might be useful if troubles arise.

- **Realize that everyone—even your instructor—has been stymied at times when studying math.** You are not the first person, nor will you be the last, to encounter a "roadblock."
- **Whether working alone or with a classmate, try to allow enough study time so that you won't need to constantly glance at a clock.** Difficult material is best mastered when your mind is completely focused on the subject matter. Thus, if you are tired, it is usually best to study early the next morning or to take a ten-minute "power-nap" in order to make the most productive use of your time.
- **Talk about your trouble spot with a classmate.** It is possible that she or he is also having difficulty with the same material. If that is the case, perhaps the majority of your class is confused and the class can ask the instructor to go over the material again. If your classmate *does* understand the topic that is troubling you, patiently allow him or her to explain it to you. By verbalizing the math in question, your classmate may help clarify the material for both of you. Perhaps you will be able to return the favor for your classmate when he or she is struggling with a topic that you understand.
- **Try to study in a "controlled" environment.** What we mean by this is to put yourself in a setting that will enable you to maximize your powers of concentration. For example, whereas some students may succeed in studying at home or in a dorm room, for many these settings are filled with distractions. Consider a trip to a library, classroom building, or perhaps the attic or basement if such a setting is more conducive to studying. If you plan on working with a classmate, try to find a location in which conversation will not be bothersome to others.
- **When working on difficult material, it is often helpful to first "back up" and review the most recent material that *did* make sense.** This can build your confidence and create a momentum that can often carry you through the roadblock. Sometimes a small piece of information that appeared in a previous section is all that is needed for your problem spot to disappear. When the difficult material is finally mastered, try to make use of what is fresh in your mind by taking a "sneak preview" of what your next topic for study will be.

Exercise Set 9.2

a, b Find the intersection or union.

1. $\{9, 10, 11\} \cap \{9, 11, 13\}$

2. $\{1, 5, 10, 15\} \cap \{5, 15, 20\}$

3. $\{a, b, c, d\} \cap \{b, f, g\}$

4. $\{m, n, o, p\} \cap \{m, o, p\}$

5. $\{9, 10, 11\} \cup \{9, 11, 13\}$

6. $\{1, 5, 10, 15\} \cup \{5, 15, 20\}$

7. $\{a, b, c, d\} \cup \{b, f, g\}$

8. $\{m, n, o, p\} \cup \{m, o, p\}$

9. $\{2, 5, 7, 9\} \cap \{1, 3, 4\}$

10. $\{a, e, i, o, u\} \cap \{m, q, w, s, t\}$

11. $\{3, 5, 7\} \cup \varnothing$

12. $\{3, 5, 7\} \cap \varnothing$

a Graph and write interval notation.

13. $-4 < a$ *and* $a \leq 1$

14. $-\frac{5}{2} \leq m$ *and* $m < \frac{3}{2}$

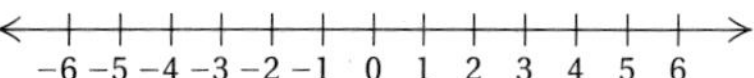

15. $1 < x < 6$

−6 −5 −4 −3 −2 −1 0 1 2 3 4 5 6

16. $-3 \leq y \leq 4$

−6 −5 −4 −3 −2 −1 0 1 2 3 4 5 6

Solve and graph.

17. $-10 \leq 3x + 2$ *and* $3x + 2 < 17$

−6 −5 −4 −3 −2 −1 0 1 2 3 4 5 6

18. $-11 < 4x - 3$ *and* $4x - 3 \leq 13$

−6 −5 −4 −3 −2 −1 0 1 2 3 4 5 6

19. $3x + 7 \geq 4$ *and* $2x - 5 \geq -1$

−6 −5 −4 −3 −2 −1 0 1 2 3 4 5 6

20. $4x - 7 < 1$ *and* $7 - 3x > -8$

−6 −5 −4 −3 −2 −1 0 1 2 3 4 5 6

21. $4 - 3x \geq 10$ *and* $5x - 2 > 13$

−6 −5 −4 −3 −2 −1 0 1 2 3 4 5 6

22. $5 - 7x > 19$ *and* $2 - 3x < -4$

−6 −5 −4 −3 −2 −1 0 1 2 3 4 5 6

Solve.

23. $-4 < x + 4 < 10$

24. $-6 < x + 6 \leq 8$

25. $6 > -x \geq -2$

26. $3 > -x \geq -5$

27. $1 < 3y + 4 \le 19$

28. $5 \le 8x + 5 \le 21$

29. $-10 \le 3x - 5 \le -1$

30. $-18 \le -2x - 7 < 0$

31. $2 < x + 3 \le 9$

32. $-6 \le x + 1 < 9$

33. $-6 \le 2x - 3 < 6$

34. $4 > -3m - 7 \ge 2$

35. $-\frac{1}{2} < \frac{1}{4}x - 3 \le \frac{1}{2}$

36. $-\frac{2}{3} \le 4 - \frac{1}{4}x < \frac{2}{3}$

37. $-3 < \frac{2x - 5}{4} < 8$

38. $-4 \le \frac{7 - 3x}{5} \le 4$

b Graph and write interval notation.

39. $x < -2$ *or* $x > 1$

40. $x < -4$ *or* $x > 0$

41. $x \le -3$ *or* $x > 1$

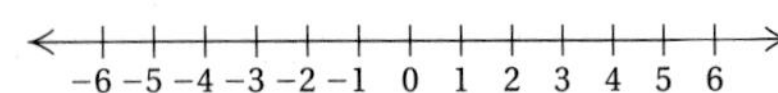

42. $x \le -1$ *or* $x > 3$

Solve and graph.

43. $x + 3 < -2$ *or* $x + 3 > 2$

44. $x - 2 < -1$ *or* $x - 2 > 3$

45. $2x - 8 \le -3$ *or* $x - 1 \ge 3$

46. $x - 5 \le -4$ *or* $2x - 7 \ge 3$

47. $7x + 4 \ge -17$ *or* $6x + 5 \ge -7$

48. $4x - 4 < -8$ *or* $4x - 4 < 12$

Solve.

49. $7 > -4x + 5$ *or* $10 \le -4x + 5$

50. $6 > 2x - 1$ *or* $-4 \le 2x - 1$

51. $3x - 7 > -10$ *or* $5x + 2 \leq 22$

52. $3x + 2 < 2$ *or* $4 - 2x < 14$

53. $-2x - 2 < -6$ *or* $-2x - 2 > 6$

54. $-3m - 7 < -5$ *or* $-3m - 7 > 5$

55. $\frac{2}{3}x - 14 < -\frac{5}{6}$ *or* $\frac{2}{3}x - 14 > \frac{5}{6}$

56. $\frac{1}{4} - 3x \leq -3.7$ *or* $\frac{1}{4} - 5x \geq 4.8$

57. $\frac{2x - 5}{6} \leq -3$ *or* $\frac{2x - 5}{6} \geq 4$

58. $\frac{7 - 3x}{5} < -4$ *or* $\frac{7 - 3x}{5} > 4$

c Solve.

59. *Pressure at Sea Depth.* The equation

$$P = 1 + \frac{d}{33}$$

gives the pressure P, in atmospheres (atm), at a depth of d feet in the sea. For what depths d is the pressure at least 1 atm and at most 7 atm?

60. *Temperatures of Liquids.* The formula

$$C = \frac{5}{9}(F - 32)$$

can be used to convert Fahrenheit temperatures F to Celsius temperatures C.

a) Gold is a liquid for Celsius temperatures C such that $1063° \leq C < 2660°$. Find such an inequality for the corresponding Fahrenheit temperatures.

b) Silver is a liquid for Celsius temperatures C such that $960.8° \leq C < 2180°$. Find such an inequality for the corresponding Fahrenheit temperatures.

61. *Solid-Waste Generation.* The equation

$$W = 0.05t + 4.3$$

can be used to estimate the average number of pounds W of solid waste produced daily by each person in the United States, t years after 1991. For what years will waste production range from 5.0 to 5.25 lb per person per day?

62. *Aerobic Exercise.* In order to achieve maximum results from aerobic exercise, one should maintain one's heart rate at a certain level. A 30-yr-old woman with a resting heart rate of 60 beats per minute should keep her heart rate between 138 and 162 beats per minute while exercising. She checks her pulse for 10 sec while exercising. What should the number of beats be?

63. *Body Mass Index.* See Exercises 67 and 68 in Exercise Set 9.1. Marv's height is 73 in. What weights W will allow Marv to keep his body mass index I between 20 and 30?

64. *Young's Rule in Medicine.* Young's rule for determining the amount of a medicine dosage for a child is given by

$$c = \frac{ad}{a + 12},$$

where a is the child's age and d is the usual adult dosage, in milligrams. (*Warning*! Do not apply this formula without checking with a physician!) (***Source*:** Olsen, June L., et al., *Medical Dosage Calculations,* 6th ed. Reading, MA: Addison Wesley Longman, p. A-31.) An 8-yr-old child needs medication. What adult dosage can be used if a child's dosage must stay between 100 mg and 200 mg?

Skill Maintenance

Solve. [8.2b]

65. $3x - 2y = -7,$
$2x + 5y = 8$

66. $4x - 7y = 23,$
$x + 6y = -33$

67. $x + y = 0,$
$x - y = 8$

Find an equation of the line containing the given pair of points. [7.5c]

68. $(2, 7), (3, -4)$

69. $(0, 7), (2, -1)$

70. $(4, -2), (-2, 4)$

Multiply. [4.6a]

71. $(2a - b)(3a + 5b)$

72. $(5y + 6)(5y + 1)$

73. $(7x - 8)(3x - 5)$

74. $(13x - 2y)(x + 3y)$

Synthesis

75. ◆ Explain why the conjunction $3 < x$ *and* $x < 5$ is equivalent to $3 < x < 5$, but the disjunction $3 < x$ *or* $x < 5$ is not.

76. ◆ Describe the circumstances under which $[a, b] \cup [c, d] = [a, d]$.

77. What is the union of the set of all rational numbers with the set of all irrational numbers? the intersection?

78. *Minimizing Tolls.* A \$3.00 toll is charged to cross the bridge from Sanibel Island to mainland Florida. A six-month pass, costing \$15.00, reduces the toll to \$0.50 per crossing. A one-year pass, costing \$60, allows for free crossings. How many crossings per month does it take, on average, for the six-month pass to be the more economical choice?

Solve.

79. $x - 10 < 5x + 6 \le x + 10$

80. $4m - 8 > 6m + 5$ *or* $5m - 8 < -2$

81. $-\frac{2}{15} \le \frac{2}{3}x - \frac{2}{5} \le \frac{2}{15}$

82. $2[5(3 - y) - 2(y - 2)] > y + 4$

83. $3x < 4 - 5x < 5 + 3x$

84. $2x - \frac{3}{4} < -\frac{1}{10}$ *or* $2x - \frac{3}{4} > \frac{1}{10}$

85. $x + 4 < 2x - 6 \le x + 12$

86. $2x + 3 \le x - 6$ *or* $3x - 2 \le 4x + 5$

9.3 Absolute-Value Equations and Inequalities

a Properties of Absolute Value

We can think of the **absolute value** of a number as its distance from zero on the number line. Recall the formal definition from Section 1.2.

> The absolute value of x, denoted $|x|$, is defined as follows:
>
> $x \geq 0 \longrightarrow |x| = x;$ $x < 0 \longrightarrow |x| = -x.$

Some simple properties of absolute value allow us to manipulate or simplify algebraic expressions.

> **PROPERTIES OF ABSOLUTE VALUE**
>
> a) For any real numbers a and b, $|ab| = |a| \cdot |b|$.
>
> (The absolute value of a product is the product of the absolute values.)
>
> b) $\left|\frac{a}{b}\right| = \frac{|a|}{|b|}$, provided that $b \neq 0$.
>
> (The absolute value of a quotient is the quotient of the absolute values.)
>
> c) $|-a| = |a|$
>
> (The absolute value of the opposite of a number is the same as the absolute value of the number.)

Examples Simplify, leaving as little as possible inside the absolute-value signs.

1. $|5x| = |5| \cdot |x| = 5|x|$
2. $|-3y| = |-3| \cdot |y| = 3|y|$
3. $|7x^2| = |7| \cdot |x^2| = 7|x^2| = 7x^2$ Since x^2 is never negative for any number x
4. $\left|\frac{6x}{-3x^2}\right| = \left|\frac{2}{-x}\right| = \frac{|2|}{|-x|} = \frac{2}{|x|}$

Do Exercises 1–5.

b Distance on a Number Line

The number line below shows that the distance between -3 and 2 is 5.

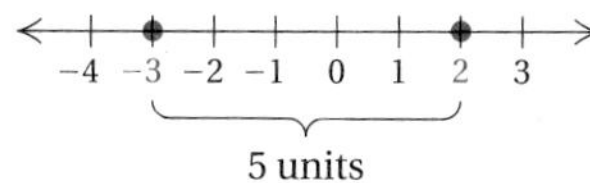

Another way to find the distance between two numbers on a number line is to take the absolute value of the difference, as follows:

$$|-3 - 2| = |-5| = 5, \quad \text{or} \quad |2 - (-3)| = |5| = 5.$$

Note that the order in which we subtract does not matter because we are taking the absolute value after we have subtracted.

Objectives

a Simplify expressions containing absolute-value symbols.

b Find the distance between two points on a number line.

c Solve equations with absolute-value expressions.

d Solve equations with two absolute-value expressions.

e Solve inequalities with absolute-value expressions.

For Extra Help

TAPE 15  MAC WIN  CD-ROM

Simplify, leaving as little as possible inside the absolute-value signs.

1. $|7x|$

2. $|x^8|$

3. $|5a^2b|$

4. $\left|\frac{7a}{b^2}\right|$

5. $|-9x|$

Answers on page A-37

Find the distance between the points.

6. $-6, -35$

7. $19, 14$

8. $0, p$

9. Solve: $|x| = 6$. Then graph using a number line.

10. Solve: $|x| = -6$.

11. Solve: $|p| = 0$.

Answers on page A-37

> For any real numbers a and b, the **distance** between them is $|a - b|$.

We should note that the distance is also $|b - a|$, because $a - b$ and $b - a$ are opposites and hence have the same absolute value.

Example 5 Find the distance between -8 and -92 on a number line.

$$|-8 - (-92)| = |84| = 84, \quad \text{or} \quad |-92 - (-8)| = |-84| = 84$$

Example 6 Find the distance between x and 0 on the number line.

$$|x - 0| = |x|$$

Do Exercises 6–8.

c Equations with Absolute Value

Example 7 Solve: $|x| = 4$. Then graph using the number line.

Note that $|x| = |x - 0|$, so that $|x - 0|$ is the distance from x to 0. Thus solutions of the equation $|x| = 4$, or $|x - 0| = 4$, are those numbers x whose distance from 0 is 4. Those numbers are -4 and 4. The solution set is $\{-4, 4\}$. The graph consists of just two points, as shown.

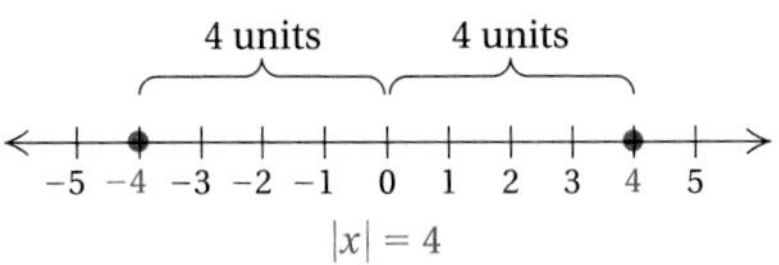

Example 8 Solve: $|x| = 0$.

The only number whose absolute value is 0 is 0 itself. Thus the solution is 0. The solution set is $\{0\}$.

Example 9 Solve: $|x| = -7$.

The absolute value of a number is always nonnegative. There is no number whose absolute value is -7. Thus there is no solution. The solution set is $\varnothing$.

Examples 7–9 lead us to the following principle for solving linear equations with absolute value.

> **THE ABSOLUTE-VALUE PRINCIPLE**
>
> For any positive number p and any algebraic expression X:
>
> a) The solutions of $|X| = p$ are those numbers that satisfy $X = -p$ *or* $X = p$.
>
> b) The equation $|X| = 0$ is equivalent to the equation $X = 0$.
>
> c) The equation $|X| = -p$ has no solution.

Do Exercises 9–11.

We can use the absolute-value principle with the addition and multiplication principles to solve equations with absolute value.

Example 10 Solve: $2|x| + 5 = 9$.

We first use the addition and multiplication principles to get $|x|$ by itself. Then we use the absolute-value principle.

$$2|x| + 5 = 9$$
$$2|x| = 4 \quad \text{Subtracting 5}$$
$$|x| = 2 \quad \text{Dividing by 2}$$
$$x = -2 \quad or \quad x = 2 \quad \text{Using the absolute-value principle}$$

The solutions are -2 and 2. The solution set is $\{-2, 2\}$.

Do Exercises 12–14.

Example 11 Solve: $|x - 2| = 3$.

We can consider solving this equation in two different ways.

METHOD 1 This method allows us to see the meaning of the solutions graphically. The solution set consists of those numbers that are 3 units from 2 on the number line.

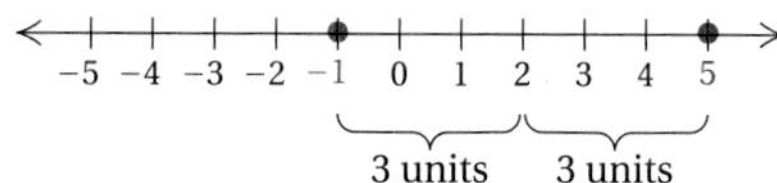

The solutions of $|x - 2| = 3$ are -1 and 5. The solution set is $\{-1, 5\}$.

METHOD 2 This method is more efficient. We use the absolute-value principle, replacing X with $x - 2$ and p with 3. Then we solve each equation separately.

$$|X| = p$$
$$|x - 2| = 3$$
$$x - 2 = -3 \quad or \quad x - 2 = 3 \quad \text{Absolute-value principle}$$
$$x = -1 \quad or \quad x = 5$$

The solutions are -1 and 5. The solution set is $\{-1, 5\}$.

Do Exercise 15.

Example 12 Solve: $|2x + 5| = 13$.

We use the absolute-value principle, replacing X with $2x + 5$ and p with 13:

$$|X| = p$$
$$|2x + 5| = 13$$
$$2x + 5 = -13 \quad or \quad 2x + 5 = 13 \quad \text{Absolute-value principle}$$
$$2x = -18 \quad or \quad 2x = 8$$
$$x = -9 \quad or \quad x = 4.$$

The solutions are -9 and 4. The solution set is $\{-9, 4\}$.

Do Exercise 16.

Solve.

12. $|3x| = 6$

13. $4|x| + 10 = 27$

14. $3|x| - 2 = 10$

15. Solve: $|x - 4| = 1$. Use two methods as in Example 11.

16. Solve: $|3x - 4| = 17$.

Answers on page A-37

17. Solve: $|6 + 2x| = -3$.

Solve.

18. $|5x - 3| = |x + 4|$

19. $|x - 3| = |x + 10|$

20. Solve: $|x| = 5$. Then graph using a number line.

Answers on page A-37

Example 13 Solve: $|4 - 7x| = -8$.

Since absolute value is always nonnegative, this equation has no solution. The solution set is $\varnothing$.

Do Exercise 17.

d Equations with Two Absolute-Value Expressions

Sometimes equations have two absolute-value expressions. Consider $|a| = |b|$. This means that a and b are the same distance from 0. If a and b are the same distance from 0, then either they are the same number or they are opposites of each other.

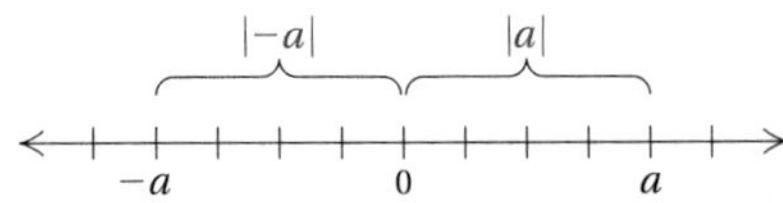

Example 14 Solve: $|2x - 3| = |x + 5|$.

Either $2x - 3 = x + 5$ or $2x - 3 = -(x + 5)$. We solve each equation:

$$
\begin{aligned}
2x - 3 &= x + 5 &\quad \text{or} \quad 2x - 3 &= -(x + 5) \\
x - 3 &= 5 &\quad \text{or} \quad 2x - 3 &= -x - 5 \\
x &= 8 &\quad \text{or} \quad 3x - 3 &= -5 \\
x &= 8 &\quad \text{or} \quad 3x &= -2 \\
x &= 8 &\quad \text{or} \quad x &= -\tfrac{2}{3}.
\end{aligned}
$$

The solutions are 8 and $-\frac{2}{3}$. The solution set is $\{8, -\frac{2}{3}\}$.

Example 15 Solve: $|x + 8| = |x - 5|$.

$$
\begin{aligned}
x + 8 &= x - 5 &\quad \text{or} \quad x + 8 &= -(x - 5) \\
8 &= -5 &\quad \text{or} \quad x + 8 &= -x + 5 \\
8 &= -5 &\quad \text{or} \quad 2x &= -3 \\
8 &= -5 &\quad \text{or} \quad x &= -\tfrac{3}{2}
\end{aligned}
$$

The first equation has no solution. The second equation has $-\frac{3}{2}$ as a solution. The solution set is $\{-\frac{3}{2}\}$.

Do Exercises 18 and 19.

e Inequalities with Absolute Value

We can extend our methods for solving equations with absolute value to those for solving inequalities with absolute value.

Example 16 Solve: $|x| = 4$. Then graph using a number line.

From Example 7, we know that the solutions are −4 and 4. The solution set is $\{-4, 4\}$. The graph consists of just two points, as shown here.

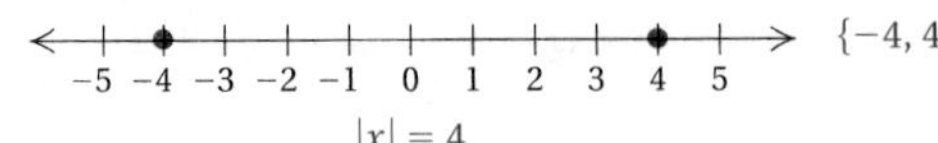

Do Exercise 20.

Example 17 Solve: $|x| < 4$. Then graph.

The solutions of $|x| < 4$ are the solutions of $|x - 0| < 4$ and are those numbers x whose distance from 0 is less than 4. We can check by substituting or by looking at the number line that numbers like $-3, -2, -1, -\frac{1}{2}, -\frac{1}{4}, 0, \frac{1}{4}, \frac{1}{2}, 1, 2$, and 3 are all solutions. In fact, the solutions are all the real numbers x between -4 and 4, such that $-4 < x < 4$. The solution set is $\{x \mid -4 < x < 4\}$ or, in interval notation, $(-4, 4)$. The graph is as follows.

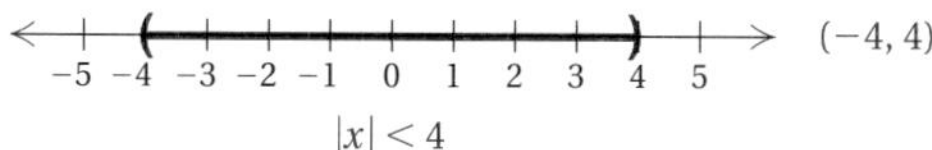

Do Exercise 21.

Example 18 Solve: $|x| \geq 4$. Then graph.

The solutions of $|x| \geq 4$ are solutions of $|x - 0| \geq 4$ and are those numbers whose distance from 0 is greater than or equal to 4—in other words, those numbers x such that $x \leq -4$ *or* $x \geq 4$. The solution set is $\{x \mid x \leq -4$ *or* $x \geq 4\}$, or $(-\infty, -4] \cup [4, \infty)$. The graph is as follows.

(−∞, −4] ∪ [4, ∞)

−5 −4 −3 −2 −1 0 1 2 3 4 5

|x| ≥ 4

Do Exercise 22.

Examples 16–18 illustrate three cases of solving equations and inequalities with absolute value. The expression inside the absolute-value signs can be something besides a single variable. The following is a general principle for solving.

> For any positive number p and any algebraic expression X:
>
> a) The solutions of $|X| = p$ are those numbers that satisfy $X = -p$ or $X = p$.
>
> As an example, replacing X with $5x - 1$ and p with 8, we see that the solutions of $|5x - 1| = 8$ are those numbers x for which
>
> $$5x - 1 = -8 \quad or \quad 5x - 1 = 8$$
> $$5x = -7 \quad or \quad 5x = 9$$
> $$x = -\tfrac{7}{5} \quad or \quad x = \tfrac{9}{5}.$$
>
> The solution set is $\left\{-\frac{7}{5}, \frac{9}{5}\right\}$.
>
> 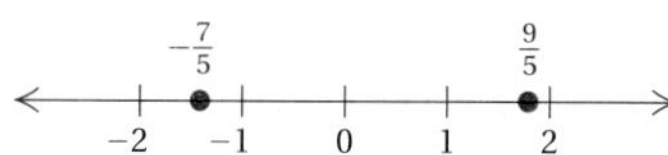
>
>
> b) The solutions of $|X| < p$ are those numbers that satisfy $-p < X < p$.
>
> As an example, replacing X with $6x + 7$ and p with 5, we see that the solutions of $|6x + 7| < 5$ are those numbers x for which
>
> $$-5 < 6x + 7 < 5$$
> $$-12 < 6x < -2$$
> $$-2 < x < -\tfrac{1}{3}.$$
>
> The solution set is $\left\{x \mid -2 < x < -\frac{1}{3}\right\}$, or $\left(-2, -\frac{1}{3}\right)$.
>
> 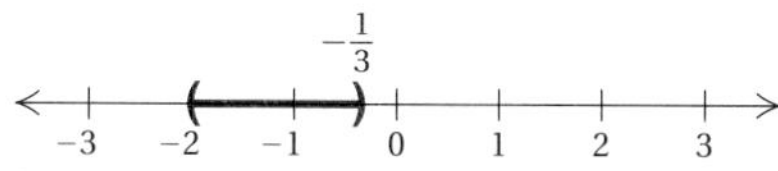
>
>
> (continued)

21. Solve: $|x| < 5$. Then graph.

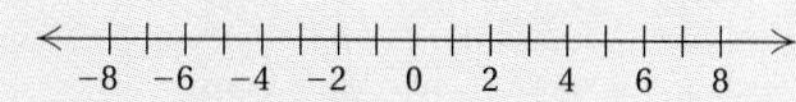

22. Solve: $|x| \geq 5$. Then graph.

Answers on page A-37

23. Solve: $|2x - 3| < 7$. Then graph.

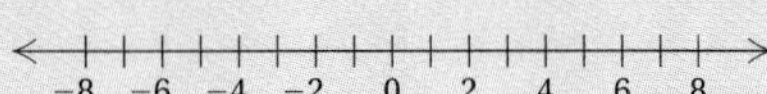

24. Solve: $|7 - 3x| \le 4$.

25. Solve: $|3x + 2| \ge 5$. Then graph.

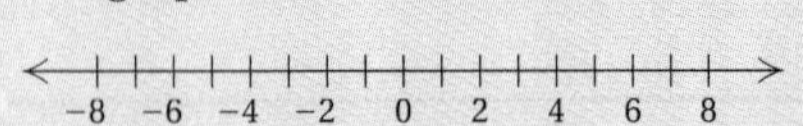

Answers on page A-37

c) The solutions of $|X| > p$ are those numbers that satisfy $X < -p$ or $X > p$.

As an example, replacing X with $2x - 9$ and p with 4, we see that the solutions of $|2x - 9| > 4$ are those numbers x for which

$$2x - 9 < -4 \quad or \quad 2x - 9 > 4$$
$$2x < 5 \quad or \quad 2x > 13$$
$$x < \frac{5}{2} \quad or \quad x > \frac{13}{2}.$$

The solution set is $\left\{x \mid x < \frac{5}{2} \text{ or } x > \frac{13}{2}\right\}$, or $\left(-\infty, \frac{5}{2}\right) \cup \left(\frac{13}{2}, \infty\right)$.

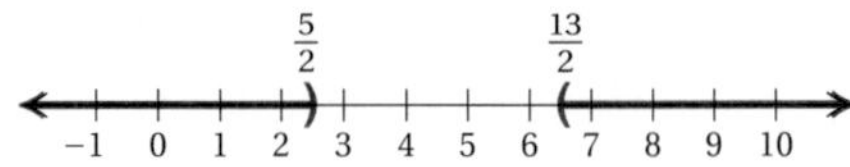

Example 19 Solve: $|3x - 2| < 4$. Then graph.

We use part (b). In this case, X is $3x - 2$ and p is 4:

$$|X| < p$$
$$|3x - 2| < 4 \qquad \textbf{Replacing } X \textbf{ with } 3x - 2 \textbf{ and } p \textbf{ with } 4$$
$$-4 < 3x - 2 < 4$$
$$-2 < 3x < 6$$
$$-\frac{2}{3} < x < 2.$$

The solution set is $\left\{x \middle| -\frac{2}{3} < x < 2\right\}$, or $\left(-\frac{2}{3}, 2\right)$. The graph is as follows.

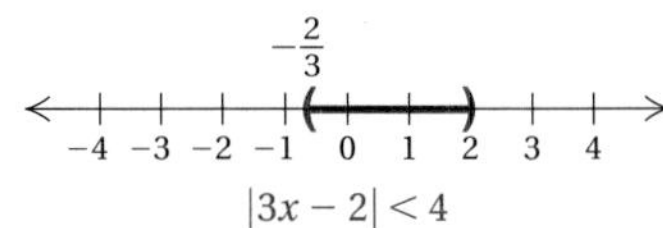

Example 20 Solve: $|8 - 4x| \le 5$.

We use part (b). In this case, X is $8 - 4x$ and p is 5:

$$|X| \le p$$
$$|8 - 4x| \le 5 \qquad \textbf{Replacing } X \textbf{ with } 8 - 4x \textbf{ and } p \textbf{ with } 5$$
$$-5 \le 8 - 4x \le 5$$
$$-13 \le -4x \le -3$$
$$\frac{13}{4} \ge x \ge \frac{3}{4}. \qquad \textbf{Dividing by } -4 \textbf{ and reversing the inequality symbols}$$

The solution set is $\left\{x \middle| \frac{13}{4} \ge x \ge \frac{3}{4}\right\}$, or $\left\{x \middle| \frac{3}{4} \le x \le \frac{13}{4}\right\}$, or $\left[\frac{3}{4}, \frac{13}{4}\right]$.

Example 21 Solve: $|4x + 2| \ge 6$.

We use part (c). In this case, X is $4x + 2$ and p is 6:

$$|X| \ge p$$
$$|4x + 2| \ge 6 \qquad \textbf{Replacing } X \textbf{ with } 4x + 2 \textbf{ and } p \textbf{ with } 6$$
$$4x + 2 \le -6 \quad or \quad 4x + 2 \ge 6$$
$$4x \le -8 \quad or \quad 4x \ge 4$$
$$x \le -2 \quad or \quad x \ge 1.$$

The solution set is $\{x \mid x \le -2 \text{ or } x \ge 1\}$, or $(-\infty, -2] \cup [1, \infty)$.

Do Exercises 23–25.

Exercise Set 9.3

a Simplify, leaving as little as possible inside absolute-value signs.

1. $|9x|$

2. $|26x|$

3. $|2x^2|$

4. $|8x^2|$

5. $|-2x^2|$

6. $|-20x^2|$

7. $|-6y|$

8. $|-17y|$

9. $\left|\frac{-2}{x}\right|$

10. $\left|\frac{y}{3}\right|$

11. $\left|\frac{x^2}{-y}\right|$

12. $\left|\frac{x^4}{-y}\right|$

13. $\left|\frac{-8x^2}{2x}\right|$

14. $\left|\frac{9y}{3y^2}\right|$

b Find the distance between the points on a number line.

15. $-8, -46$

16. $-7, -32$

17. $36, 17$

18. $52, 18$

19. $-3.9, 2.4$

20. $-1.8, -3.7$

21. $-5, 0$

22. $\frac{2}{3}, -\frac{5}{6}$

c Solve.

23. $|x| = 3$

24. $|x| = 5$

25. $|x| = -3$

26. $|x| = -9$

27. $|q| = 0$

28. $|y| = 7.4$

29. $|x - 3| = 12$

30. $|3x - 2| = 6$

31. $|2x - 3| = 4$

32. $|5x + 2| = 3$

33. $|4x - 9| = 14$

34. $|9y - 2| = 17$

35. $|x| + 7 = 18$

36. $|x| - 2 = 6.3$

37. $574 = 283 + |t|$

38. $-562 = -2000 + |x|$

39. $|5x| = 40$

40. $|2y| = 18$

41. $|3x| - 4 = 17$

42. $|6x| + 8 = 32$

43. $7|w| - 3 = 11$

44. $5|x| + 10 = 26$

45. $\left|\frac{2x - 1}{3}\right| = 5$

46. $\left|\frac{4 - 5x}{6}\right| = 7$

47. $|m + 5| + 9 = 16$

48. $|t - 7| - 5 = 4$

49. $10 - |2x - 1| = 4$

50. $2|2x - 7| + 11 = 25$

51. $|3x - 4| = -2$

52. $|x - 6| = -8$

53. $\left|\frac{5}{9} + 3x\right| = \frac{1}{6}$

54. $\left|\frac{2}{3} - 4x\right| = \frac{4}{5}$

d Solve.

55. $|3x + 4| = |x - 7|$

56. $|2x - 8| = |x + 3|$

57. $|x + 3| = |x - 6|$

58. $|x - 15| = |x + 8|$

59. $|2a + 4| = |3a - 1|$

60. $|5p + 7| = |4p + 3|$

61. $|y - 3| = |3 - y|$

62. $|m - 7| = |7 - m|$

63. $|5 - p| = |p + 8|$

64. $|8 - q| = |q + 19|$

65. $\left|\frac{2x - 3}{6}\right| = \left|\frac{4 - 5x}{8}\right|$

66. $\left|\frac{6 - 8x}{5}\right| = \left|\frac{7 + 3x}{2}\right|$

67. $\left|\frac{1}{2}x - 5\right| = \left|\frac{1}{4}x + 3\right|$

68. $\left|2 - \frac{2}{3}x\right| = \left|4 + \frac{7}{8}x\right|$

e Solve.

69. $|x| < 3$

70. $|x| \leq 5$

71. $|x| \geq 2$

72. $|y| > 12$

73. $|x - 1| < 1$

74. $|x + 4| \leq 9$

75. $|x + 4| \leq 1$

76. $|x - 2| > 6$

77. $|2x - 3| \leq 4$

78. $|5x + 2| \leq 3$

79. $|2y - 7| > 10$

80. $|3y - 4| > 8$

81. $|4x - 9| \geq 14$

82. $|9y - 2| \geq 17$

83. $|y - 3| < 12$

84. $|p - 2| < 6$

85. $|2x + 3| \leq 4$

86. $|5x + 2| \leq 13$

87. $|4 - 3y| > 8$

88. $|7 - 2y| > 5$

89. $|9 - 4x| \geq 14$

90. $|2 - 9p| \geq 17$

91. $|3 - 4x| < 21$

92. $|-5 - 7x| \leq 30$

93. $\left|\frac{1}{2} + 3x\right| \geq 12$

94. $\left|\frac{1}{4}y - 6\right| > 24$

95. $\left|\frac{x - 7}{3}\right| < 4$

96. $\left|\frac{x + 5}{4}\right| \leq 2$

97. $\left|\frac{2 - 5x}{4}\right| \geq \frac{2}{3}$

98. $\left|\frac{1 + 3x}{5}\right| > \frac{7}{8}$

99. $|m + 5| + 9 \leq 16$

100. $|t - 7| + 3 \geq 4$

101. $7 - |3 - 2x| \geq 5$

102. $16 \leq |2x - 3| + 9$

103. $\left|\frac{2x - 1}{0.0059}\right| \leq 1$

104. $\left|\frac{3x - 2}{5}\right| \geq 1$

Skill Maintenance

Find the domain. [7.2a]

105. $f(x) = |x - 2|$

106. $f(x) = \dfrac{2x - 4}{2x + 5}$

107. $f(x) = \dfrac{1}{x - 5}$

108. $f(x) = 10 - 2x$

Solve. [8.2b]

109. $2x + y = 7,$
$x - 3y = 21$

110. $10x + y = 16,$
$x - 10y = 42$

Synthesis

111. ◆ Explain in your own words why the solutions of the inequality $|x + 5| \leq 2$ can be interpreted as "all those numbers x whose distance from -5 is at most 2 units."

112. ◆ Explain in your own words why the interval $[6, \infty)$ is only part of the solution set of $|x| \geq 6$.

113. *Motion of a Spring.* A weighted spring is bouncing up and down so that its distance d above the ground satisfies the inequality $|d - 6 \text{ ft}| \leq \frac{1}{2}$ ft. Find all possible distances d.

114. *Container Sizes.* A container company is manufacturing rectangular boxes of various sizes. The length of any box must exceed the width by at least 3 in., but the perimeter cannot exceed 24 in. What widths are possible?

$l \geq w + 3,$
$2l + 2w \leq 24$

Solve.

115. $|x + 5| = x + 5$

116. $1 - \left|\frac{1}{4}x + 8\right| = \frac{3}{4}$

117. $|7x - 2| = x + 4$

118. $|x - 1| = x - 1$

119. $|x - 6| \leq -8$

120. $|3x - 4| > -2$

121. $|x + 5| > x$

122. $\left|\frac{5}{9} + 3x\right| < -\frac{1}{6}$

123. $|x| \geq 0$

124. $2 \leq |x - 1| \leq 5$

Find an equivalent inequality with absolute value.

125. $-3 < x < 3$

126. $-5 \leq y \leq 5$

127. $x \leq -6$ *or* $x \geq 6$

128. $-5 < x < 1$

129. $x < -8$ *or* $x > 2$

Collaborative Learning Manual — Create and solve equations with absolute value as a group.

9.4 Systems of Linear Inequalities in Two Variables

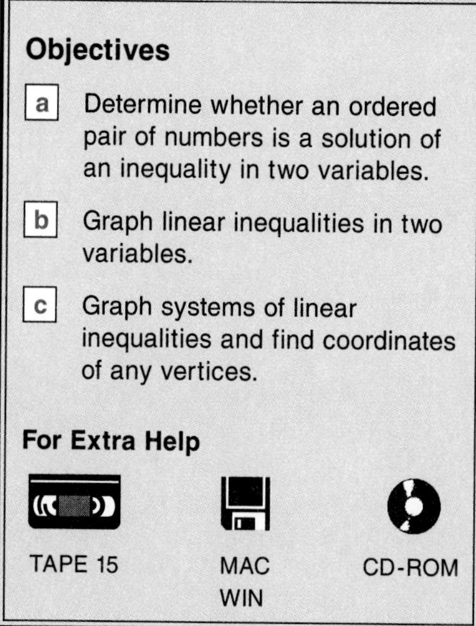

A **graph** of an inequality is a drawing that represents its solutions. In Sections 2.7 and 9.1, we graphed inequalities in one variable on the number line. Inequalities in two variables can be graphed on a coordinate plane.

A **linear inequality** is one that we can get from a related linear equation by changing the equals symbol to an inequality symbol. The graph of a linear inequality is a region on one side of a line. This region is called a **half-plane**. The graph sometimes includes the graph of the related line at the boundary of the half-plane.

a Solutions of Inequalities in Two Variables

The solutions of an inequality in two variables are ordered pairs.

Examples Determine whether the ordered pair is a solution of the inequality $5x - 4y > 13$.

1. $(-3, 2)$

We have

$$\begin{array}{r|l} 5x - 4y > 13 & \\ \hline 5(-3) - 4 \cdot 2 \ ? & 13 \\ -15 - 8 & \\ -23 & \text{FALSE} \end{array}$$

We use alphabetical order to replace x with -3 and y with 2.

Since $-23 > 13$ is false, $(-3, 2)$ is not a solution.

2. $(6, -7)$

We have

$$\begin{array}{r|l} 5x - 4y > 13 & \\ \hline 5(6) - 4(-7) \ ? & 13 \\ 30 + 28 & \\ 58 & \text{TRUE} \end{array}$$

Replacing x with 6 and y with -7

Since $58 > 13$ is true, $(6, -7)$ is a solution.

Do Exercises 1 and 2.

1. Determine whether $(1, -4)$ is a solution of $4x - 5y < 12$.

2. Determine whether $(4, -3)$ is a solution of $3y - 2x \leq 6$.

Answers on page A-37

b Graphing Inequalities in Two Variables

Example 3 Graph: $y < x$.

We first graph the line $y = x$ for comparison. Every solution of $y = x$ is an ordered pair like (3, 3), where the first and second coordinates are the same. The graph of $y = x$ is shown on the left below. We draw it dashed because these points are *not* solutions of $y < x$.

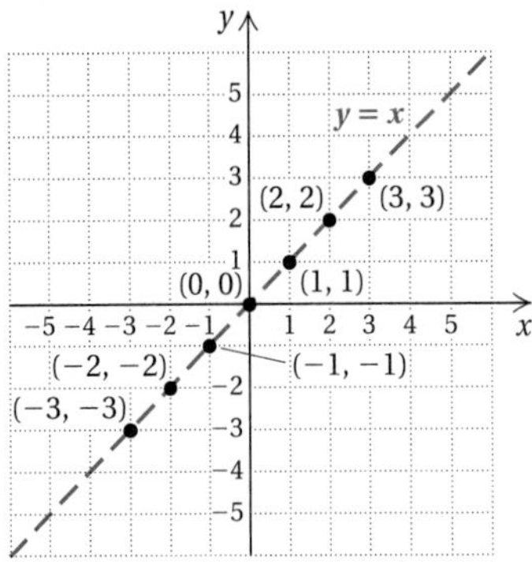

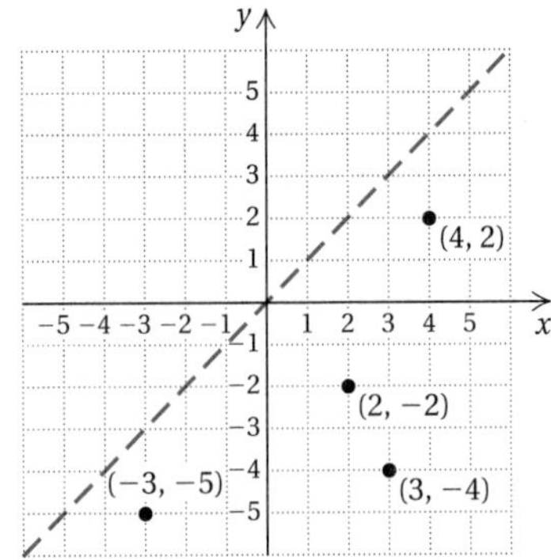

Now look at the graph on the right above. Several ordered pairs are plotted on the half-plane below $y = x$. Each is a solution of $y < x$. We can check a pair (4, 2) as follows:

$$\frac{y < x}{2 \ ? \ 4} \quad \text{TRUE}$$

It turns out that any point on the same side of $y = x$ as (4, 2) is also a solution. Thus, if you know that one point in a half-plane is a solution, then all points in that half-plane are solutions. In this text, we will usually indicate this by color shading. We shade the half-plane below $y = x$.

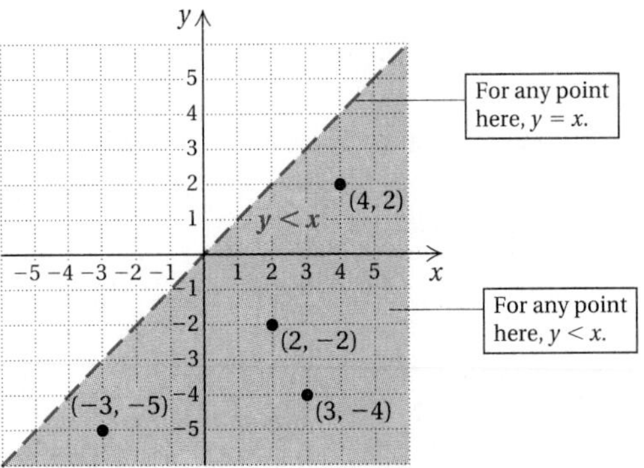

Example 4 Graph: $8x + 3y \geq 24$.

First, we sketch the line $8x + 3y = 24$. Points on the line $8x + 3y = 24$ are also in the graph of $8x + 3y \geq 24$, so we draw the line solid. This indicates that all points on the line are solutions. The rest of the solutions are in the half-plane either to the left or to the right of the line. To determine which, we select a point that is not on the line and determine whether it is a solution of $8x + 3y \geq 24$. We try $(-3, 4)$ as a test point:

$$\begin{array}{r|l} 8x + 3y \geq 24 & \\ \hline 8(-3) + 3(4) \ ? \ 24 & \\ -24 + 12 & \\ -12 & \quad \text{FALSE} \end{array}$$

We see that $-12 \geq 24$ is *false*. Since $(-3, 4)$ is not a solution, none of the points in the half-plane containing $(-3, 4)$ is a solution. Thus the points in the opposite half-plane are solutions. We shade that half-plane and obtain the graph shown below.

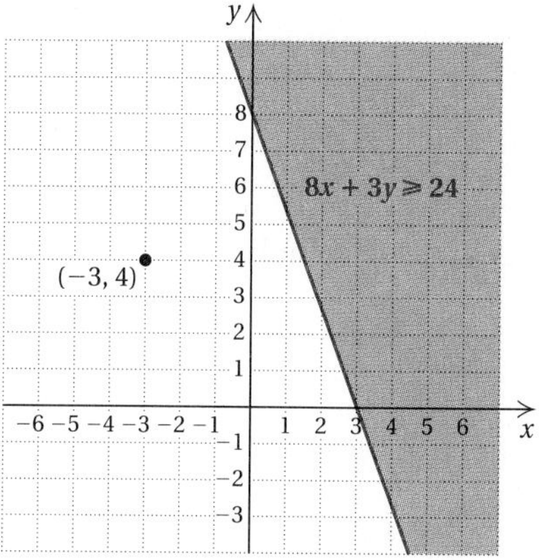

To graph an inequality in two variables:

1. Replace the inequality symbol with an equals sign and graph this related equation.
2. If the inequality symbol is $<$ or $>$, draw the line dashed. If the inequality symbol is $\leq$ or $\geq$, draw the line solid.
3. The graph consists of a half-plane that is either above or below or to the left or right of the line and, if the line is solid, the line as well. To determine which half-plane to shade, choose a point not on the line as a test point. Substitute to determine whether that point is a solution. If so, shade the half-plane containing that point. If not, shade the opposite half-plane.

Example 5 Graph: $6x - 2y < 12$.

1. We first graph the related equation $6x - 2y = 12$.
2. Since the inequality uses the symbol $<$, points on the line are not solutions of the inequality, so we draw a dashed line.
3. To determine which half-plane to shade, we consider a test point *not* on the line. We try $(0, 0)$ and substitute:

$$\begin{array}{r|l} 6x - 2y < 12 & \\ \hline 6(0) - 2(0) \ ? \ 12 & \\ 0 - 0 & \\ 0 & \quad \text{TRUE} \end{array}$$

Graph.

3. $6x - 3y < 18$

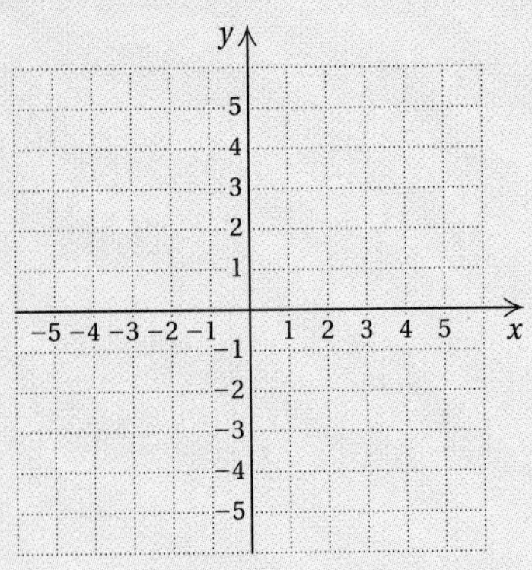

4. $4x + 3y \geq 12$

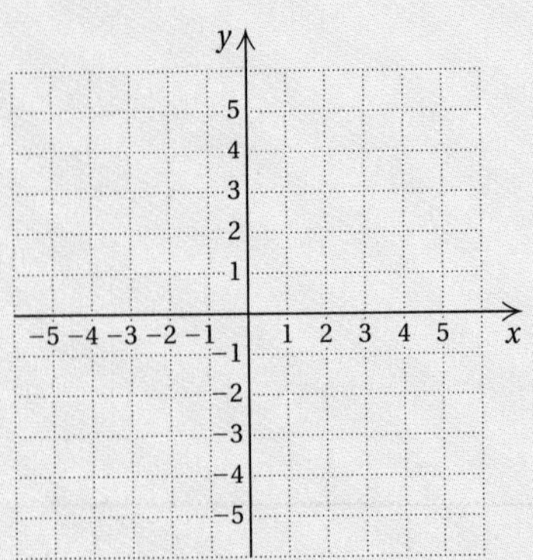

Answers on page A-37

Since the inequality $0 < 12$ is *true,* the point (0, 0) is a solution; each point in the half-plane containing (0, 0) is a solution. Thus each point in the opposite half-plane is *not* a solution. The graph is shown below.

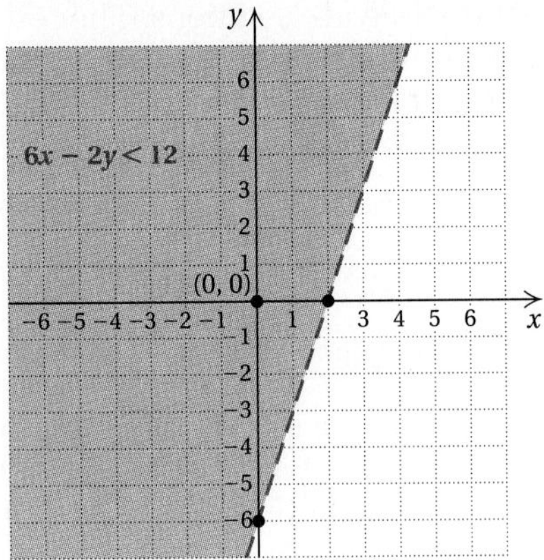

Do Exercises 3 and 4.

Example 6 Graph $x > -3$ on a plane.

There is a missing variable in this inequality. If we graph the inequality on the number line, its graph is as follows:

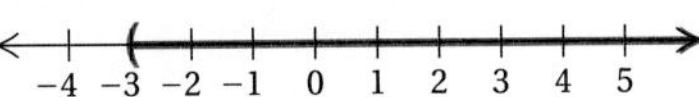

However, we can also write this inequality as $x + 0y > -3$ and consider graphing it in the plane. We use the same technique that we have used with the other examples. We first graph the related equation $x = -3$ in the plane. We draw the boundary with a dashed line. The rest of the graph is a half-plane to the right or left of the line $x = -3$. To determine which, we consider a test point, (2, 5):

$$\begin{array}{c|c} x + 0y & > -3 \\ \hline 2 + 0(5) & ?\ -3 \\ 2 & \end{array} \quad \text{TRUE}$$

Since (2, 5) is a solution, all the pairs in the half-plane containing (2, 5) are solutions. We shade that half-plane.

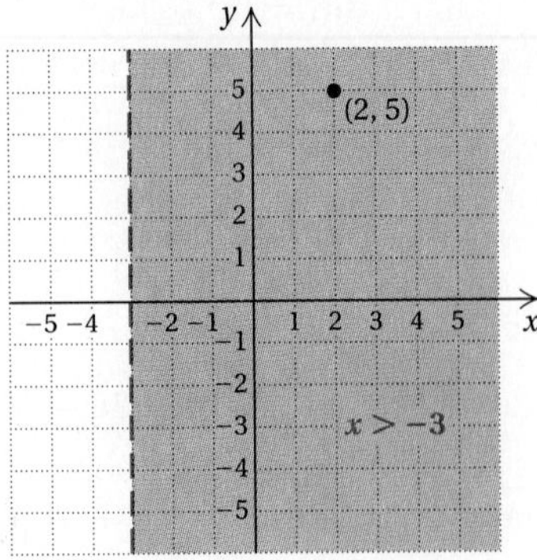

We see that the solutions of $x > -3$ are all those ordered pairs whose first coordinates are greater than -3.

Example 7 Graph $y \leq 4$ on a plane.

We first graph $y = 4$ using a solid line. We then use $(2, -3)$ as a test point and substitute:

$$\begin{array}{c|c} 0x + y & \leq 4 \\ \hline 0(2) + (-3) & ?\ 4 \\ -3 & \end{array} \quad \text{TRUE}$$

We see that $(2, -3)$ is a solution, so all points in the half-plane containing $(2, -3)$ are solutions. Note that this half-plane consists of all ordered pairs whose second coordinate is less than or equal to 4.

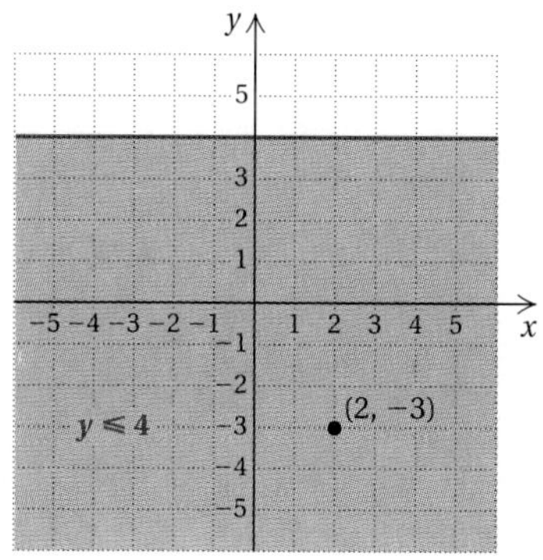

Do Exercises 5 and 6.

Graph on a plane.

5. $x < 3$

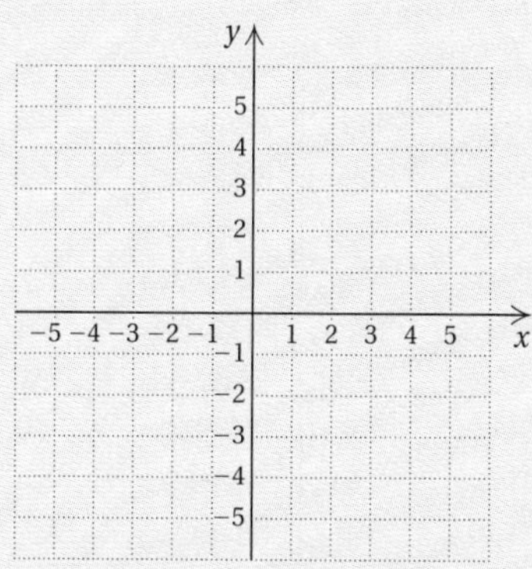

6. $y \geq -4$

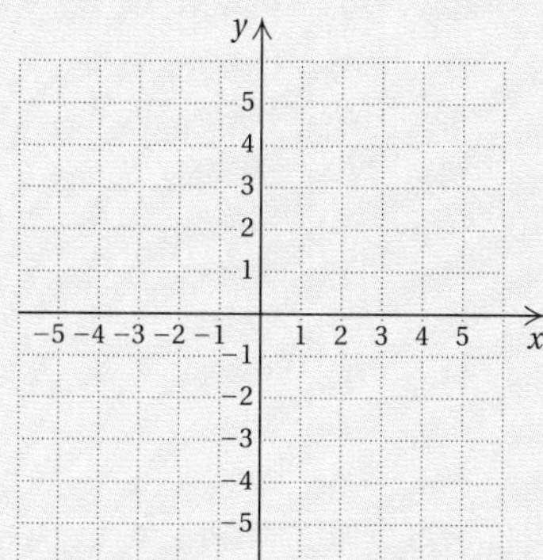

c Systems of Linear Inequalities

The following is an example of a system of two linear inequalities in two variables:

$$x + y \leq 4,$$
$$x - y < 4.$$

A **solution** of a system of linear inequalities is an ordered pair that is a solution of *both* inequalities. We now graph solutions of systems of linear inequalities. To do so, we graph each inequality and determine where the graphs overlap, or intersect. That will be a region in which the ordered pairs are solutions of both inequalities.

Example 8 Graph the solutions of the system

$$x + y \leq 4,$$
$$x - y < 4.$$

We graph the inequality $x + y \leq 4$ by first graphing the equation $x + y = 4$ using a solid red line. We consider $(0, 0)$ as a test point and find that it is a solution, so we shade all points on that side of the line using red shading. The arrows at the ends of the line also indicate the half-plane, or region, that contains the solutions.

Answers on page A-37

7. Graph:

$x + y \geq 1,$

$y - x \geq 2.$

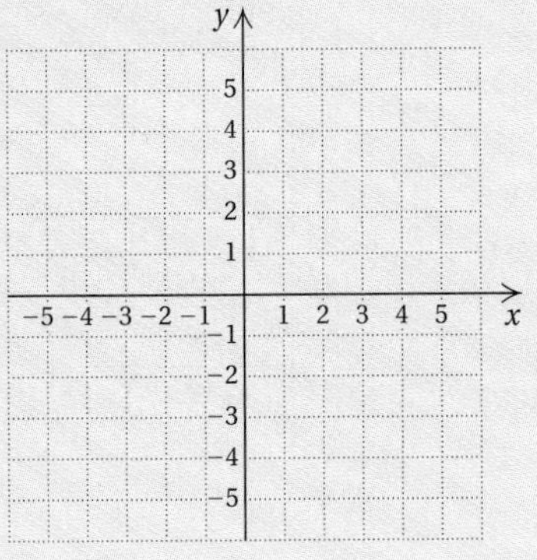

8. Graph: $-3 \leq y < 4$.

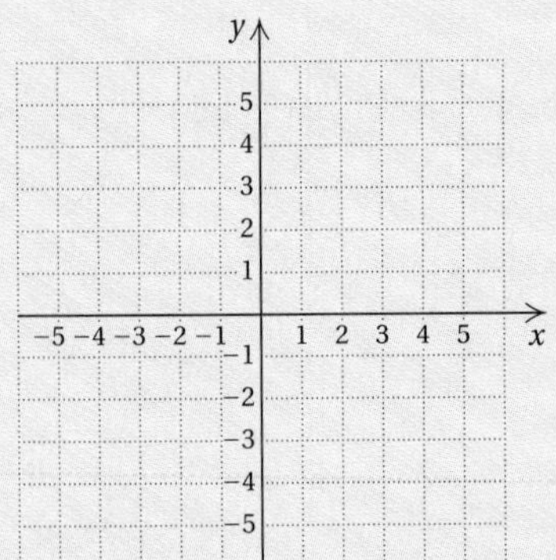

Answers on page A-37

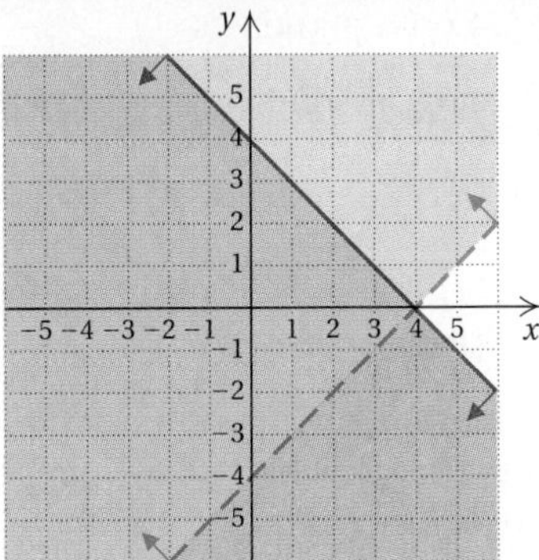

Next, we graph $x - y < 4$. We begin by graphing the equation $x - y = 4$ using a dashed blue line and consider (0, 0) as a test point. Again, (0, 0) is a solution so we shade that side of the line using blue shading. The solution set of the system is the region that is shaded both red and blue and part of the line $x + y = 4$.

Do Exercise 7.

Example 9 Graph: $-2 < x \leq 5$.

This is actually a system of inequalities:

$-2 < x,$

$x \leq 5.$

We graph the equation $-2 = x$ and see that the graph of the first inequality is the half-plane to the right of the line $-2 = x$ (see the graph on the left below).

Next, we graph the second inequality, starting with the line $x = 5$, and find that its graph is the line and also the half-plane to the left of it (see the graph on the right below).

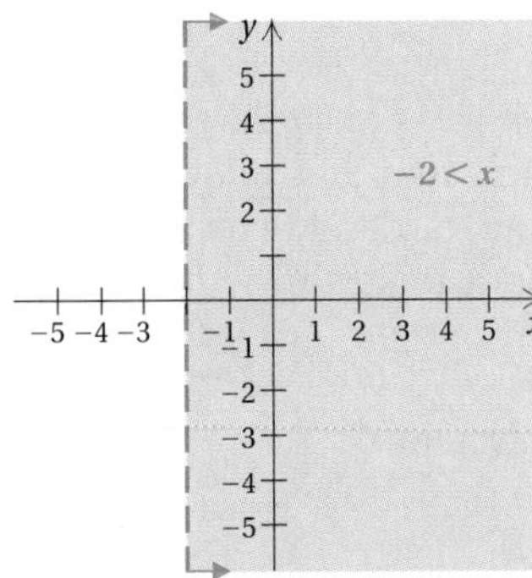

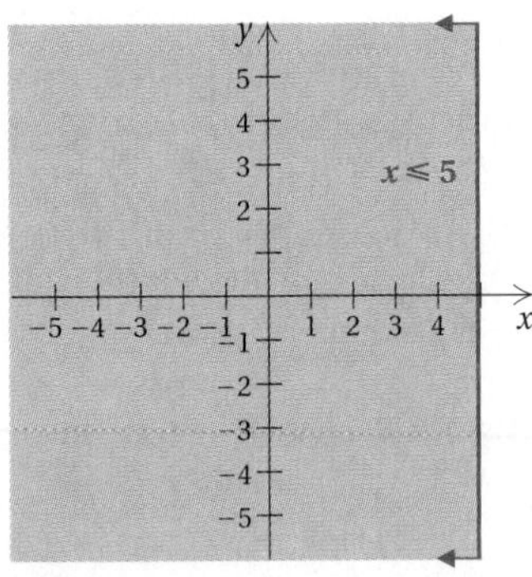

We shade the intersection of these graphs.

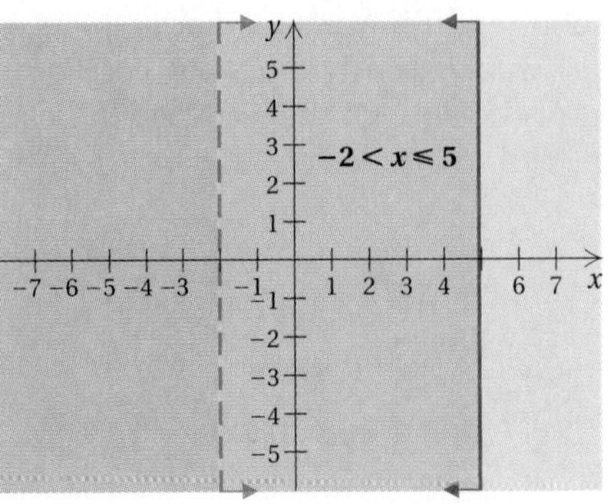

Do Exercise 8.

A system of inequalities may have a graph that consists of a polygon and its interior. In *linear programming,* which is a topic rich in application that you may study in a later course, it is important to be able to find the vertices of such a polygon.

Example 10 Graph the following system of inequalities. Find the coordinates of any vertices formed.

$$6x - 2y \leq 12, \quad (1)$$
$$y - 3 \leq 0, \quad (2)$$
$$x + y \geq 0 \quad (3)$$

We graph the lines $6x - 2y = 12$, $y - 3 = 0$, and $x + y = 0$ using solid lines. The regions for each inequality are indicated by the arrows at the ends of the lines. We then note where the regions overlap and shade the region of solutions using one color.

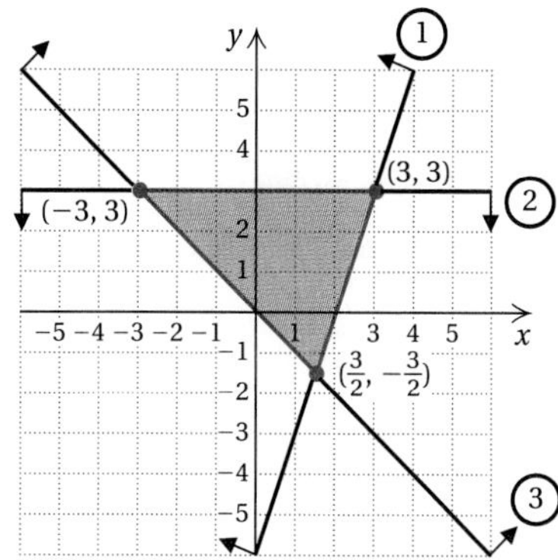

To find the vertices, we solve three different systems of equations. The system of equations from inequalities (1) and (2) is

$$6x - 2y = 12,$$
$$y - 3 = 0.$$

Solving, we obtain the vertex (3, 3).

The system of equations from inequalities (1) and (3) is

$$6x - 2y = 12,$$
$$x + y = 0.$$

Solving, we obtain the vertex $\left(\frac{3}{2}, -\frac{3}{2}\right)$.

The system of equations from inequalities (2) and (3) is

$$y - 3 = 0,$$
$$x + y = 0.$$

Solving, we obtain the vertex $(-3, 3)$.

Do Exercise 9.

9. Graph the system of inequalities. Find the coordinates of any vertices formed.

$$5x + 6y \leq 30,$$
$$0 \leq y \leq 3,$$
$$0 \leq x \leq 4$$

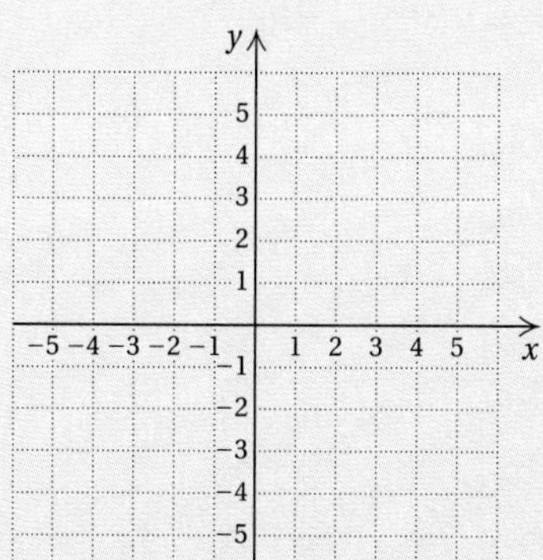

Answer on page A-38

10. Graph the system of inequalities. Find the coordinates of any vertices formed.

$$2x + 4y \le 8,$$
$$x + y \le 3,$$
$$x \ge 0,$$
$$y \ge 0$$

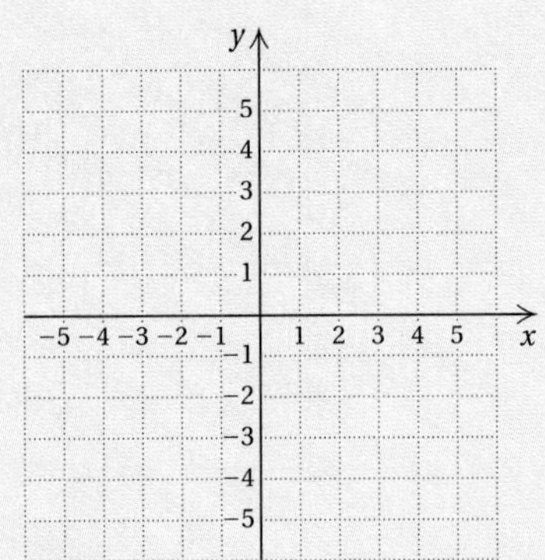

Answer on page A-38

Example 11 Graph the following system of inequalities. Find the coordinates of any vertices formed.

$$x + y \le 16, \quad (1)$$
$$3x + 6y \le 60, \quad (2)$$
$$x \ge 0, \quad (3)$$
$$y \ge 0 \quad (4)$$

We graph each inequality using solid lines. The regions for each inequality are indicated by the arrows at the ends of the lines. We then note where the regions overlap and shade the region of solutions using one color.

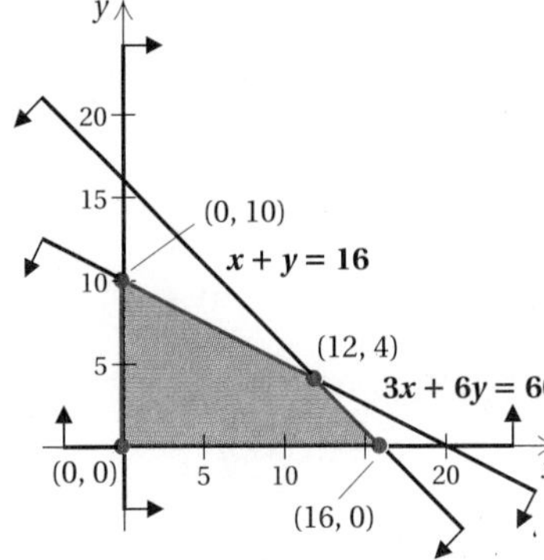

To find the vertices, we solve four different systems of equations. The system of equations from inequalities (1) and (2) is

$$x + y = 16,$$
$$3x + 6y = 60.$$

Solving, we obtain the vertex (12, 4).

The system of equations from inequalities (1) and (4) is

$$x + y = 16,$$
$$y = 0.$$

Solving, we obtain the vertex (16, 0).

The system of equations from inequalities (3) and (4) is

$$x = 0,$$
$$y = 0.$$

The vertex is (0, 0).

The system of equations from inequalities (2) and (3) is

$$3x + 6y = 60,$$
$$x = 0.$$

Solving, we obtain the vertex (0, 10).

Do Exercise 10.

Exercise Set 9.4

a Determine whether the given ordered pair is a solution of the given inequality.

1. $(-3, 3)$; $3x + y < -5$

2. $(6, -8)$; $4x + 3y \geq 0$

3. $(5, 9)$; $2x - y > -1$

4. $(5, -2)$; $6y - x > 2$

b Graph the inequality on a plane.

5. $y > 2x$

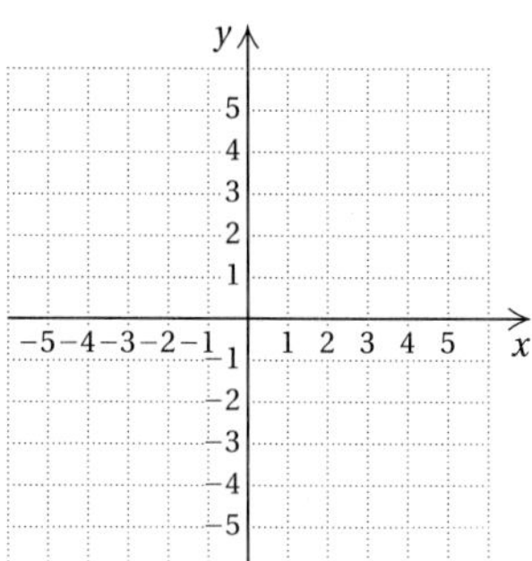

6. $y < 3x$

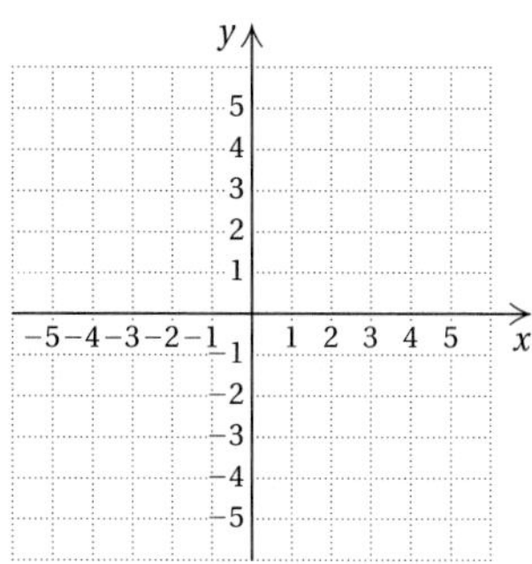

7. $y < x + 1$

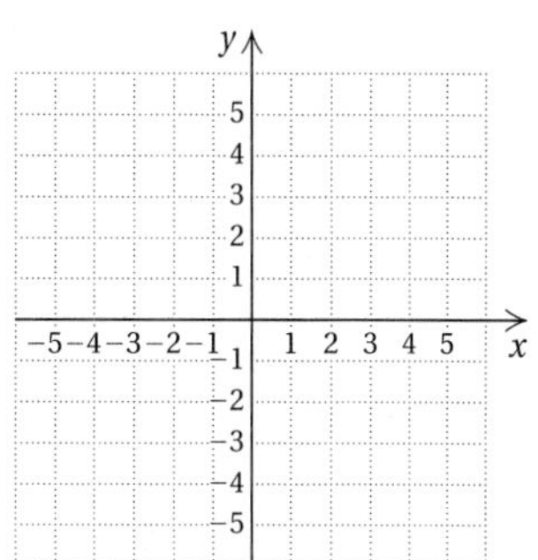

8. $y \leq x - 3$

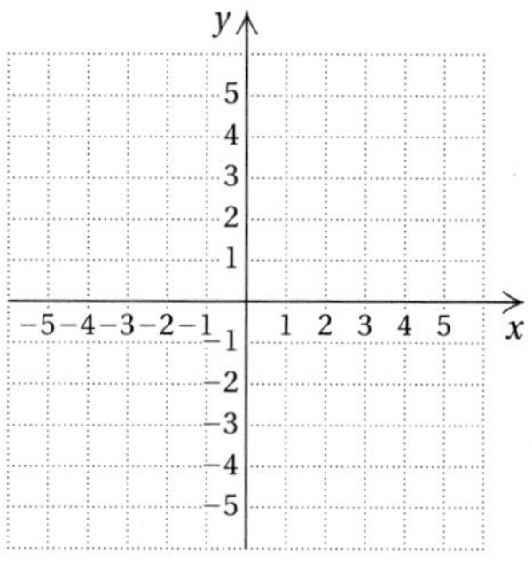

9. $y > x - 2$

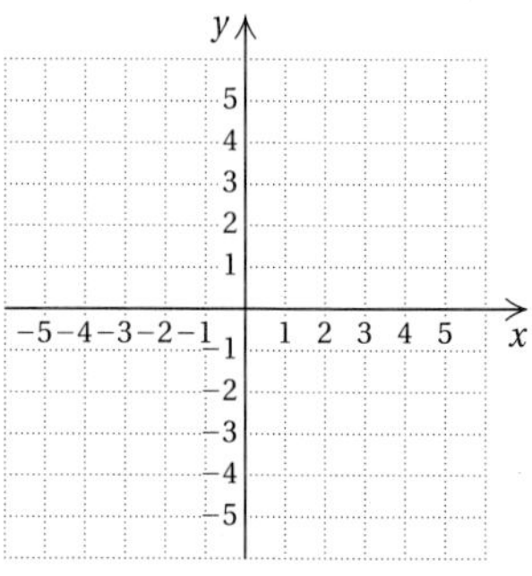

10. $y \geq x + 4$

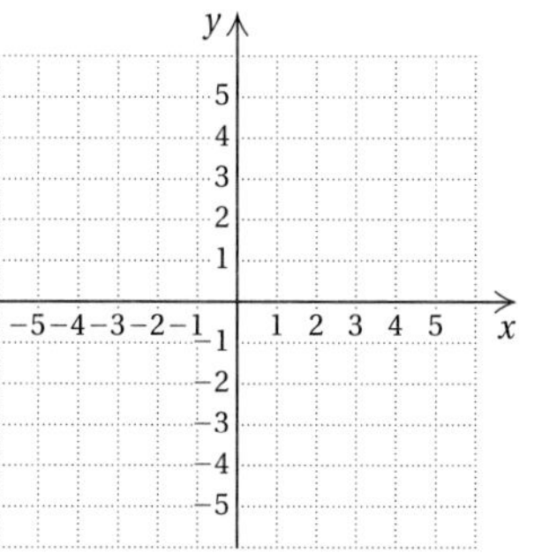

11. $x + y < 4$

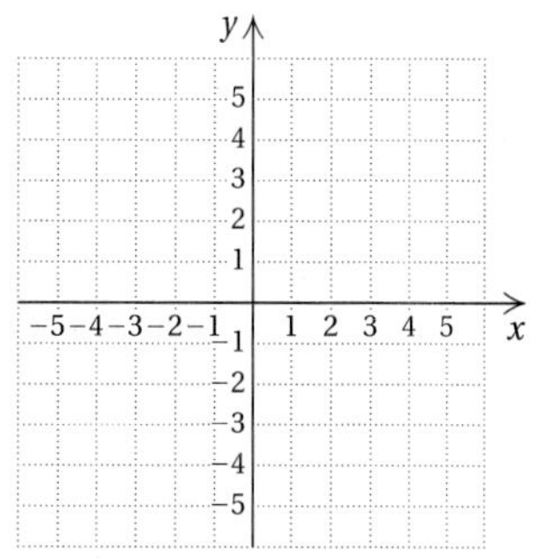

12. $x - y \geq 3$

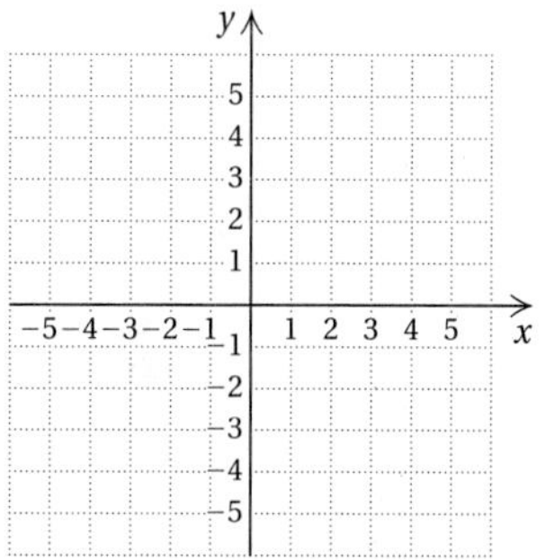

13. $3x + 4y \leq 12$

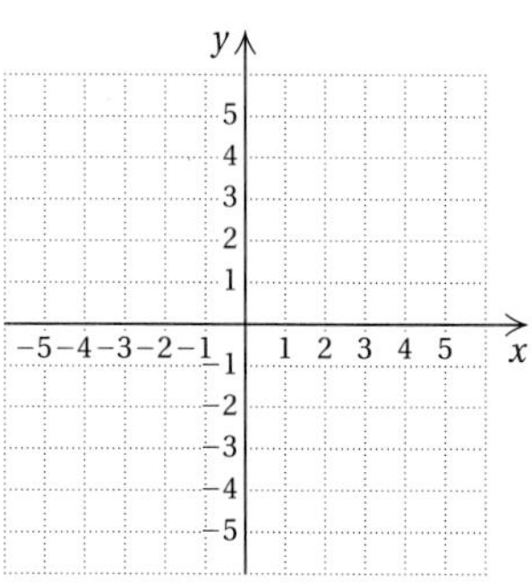

14. $2x + 3y < 6$

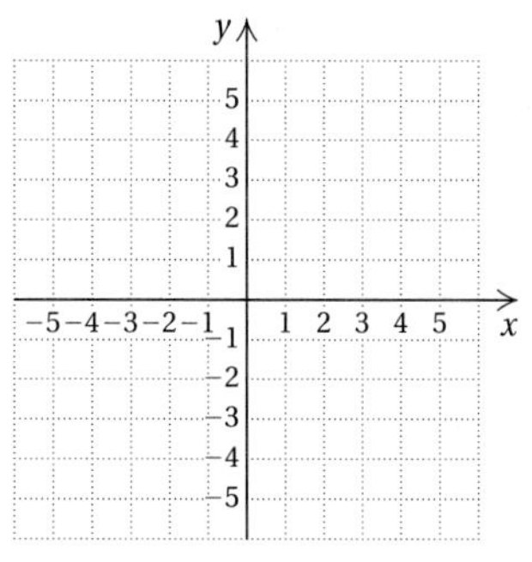

15. $2y - 3x > 6$

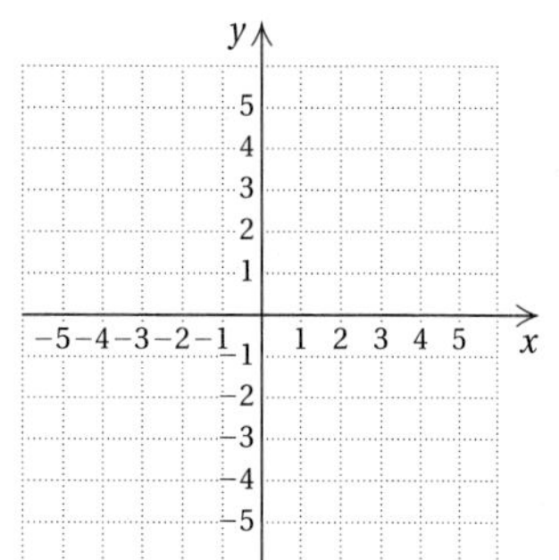

16. $2y - x \leq 4$

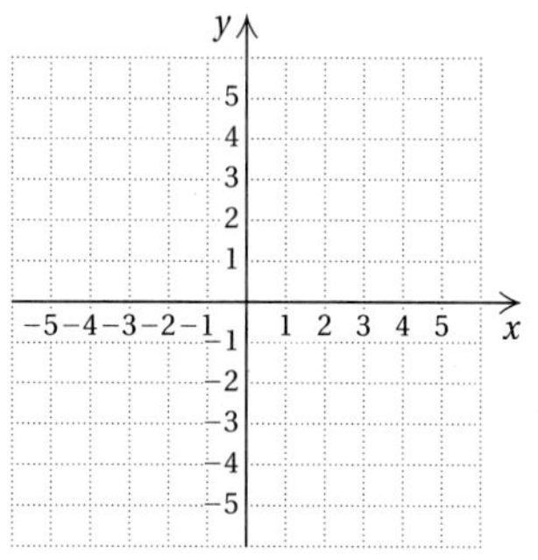

17. $3x - 2 \leq 5x + y$

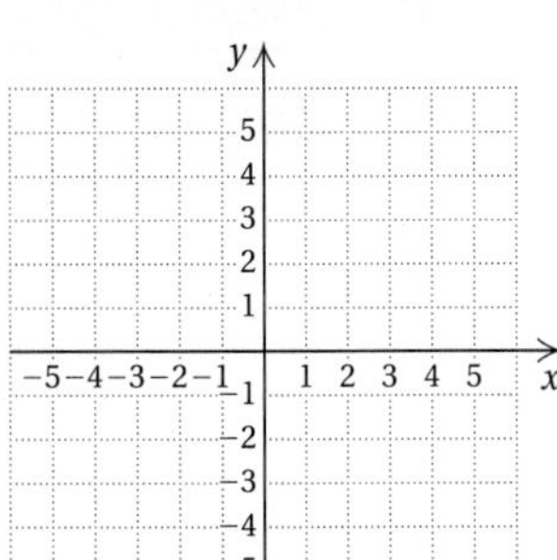

18. $2x - 2y \geq 8 + 2y$

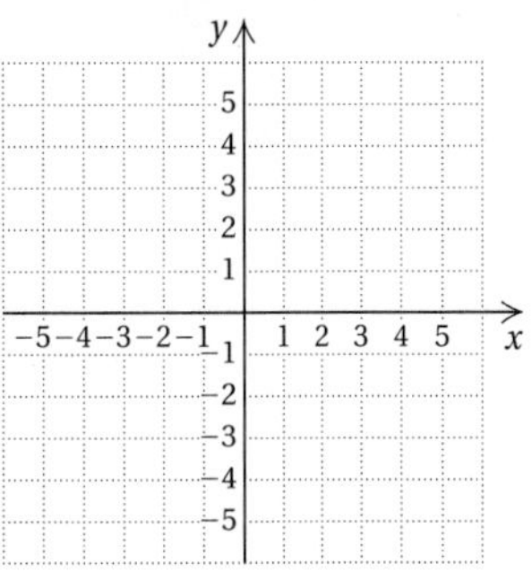

19. $x < 5$

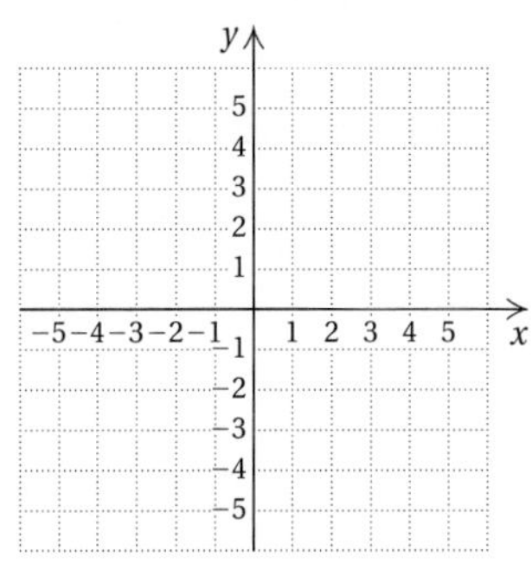

20. $y \geq -2$

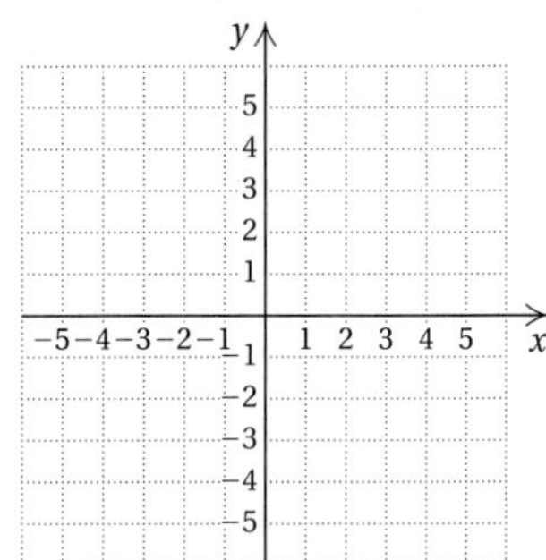

21. $y > 2$

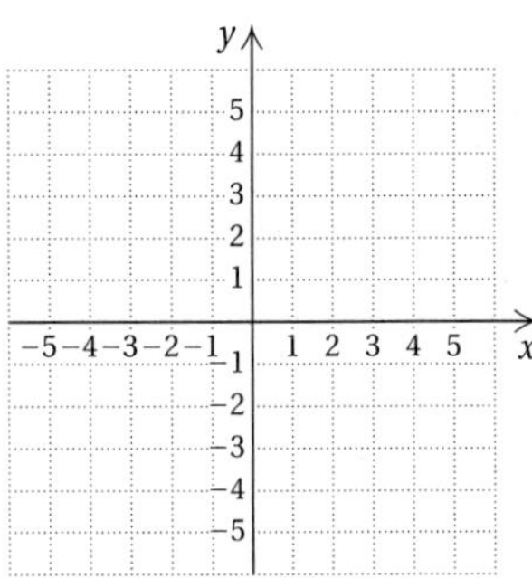

22. $x \leq -4$

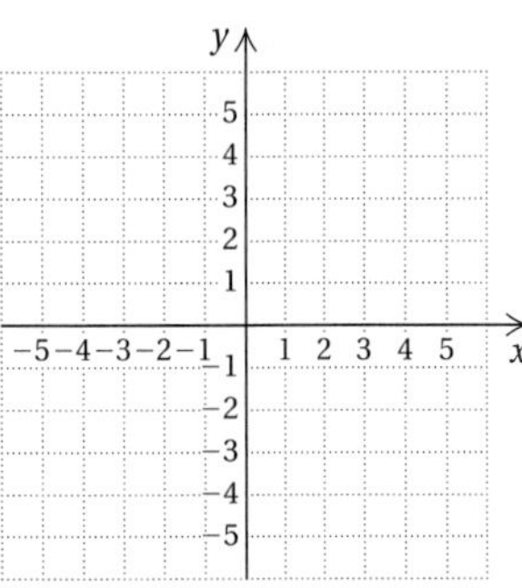

23. $2x + 3y \leq 6$

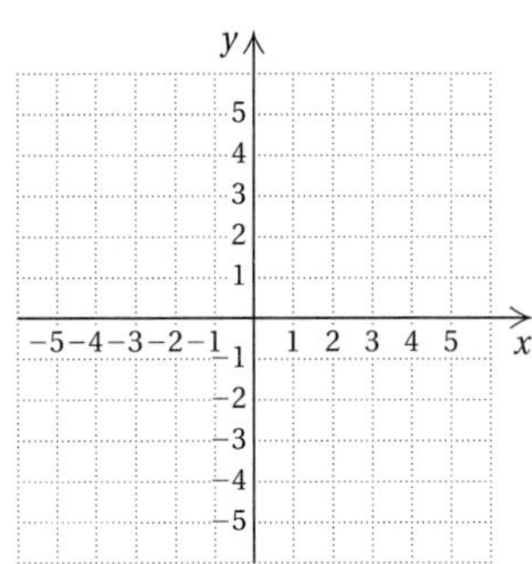

24. $7x + 2y \geq 21$

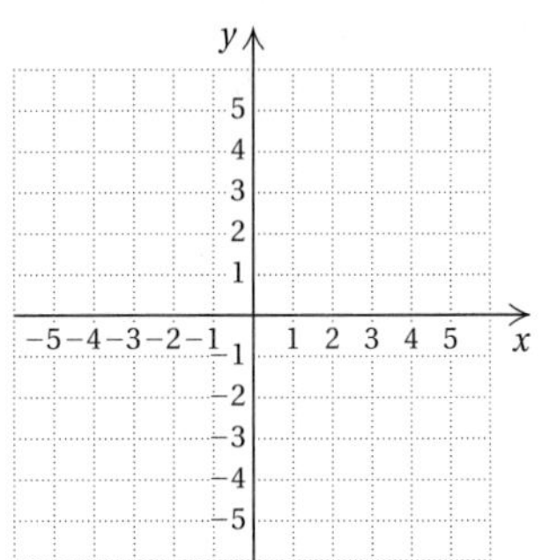

c Graph the system of inequalities. Find the coordinates of any vertices formed.

25. $y \geq x,$
$y \leq -x + 2$

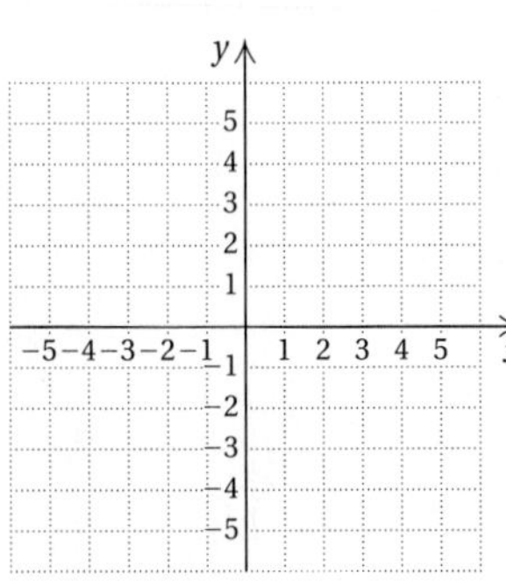

26. $y \geq x,$
$y \leq -x + 4$

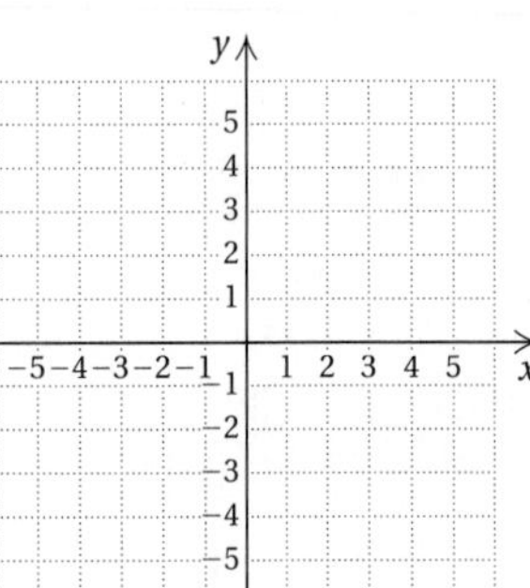

27. $y > x,$
$y < -x + 1$

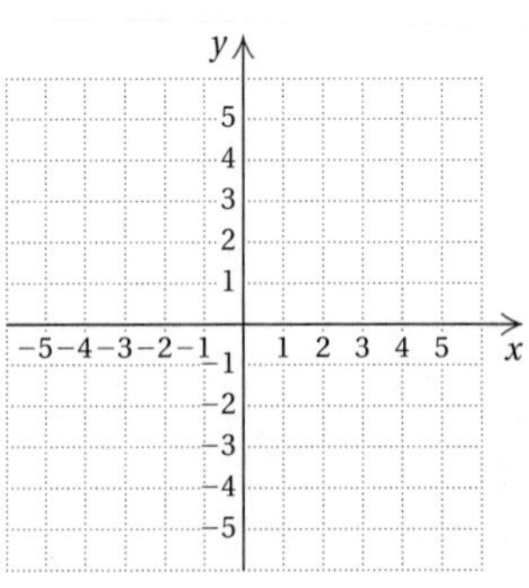

28. $y < x,$
$y > -x + 3$

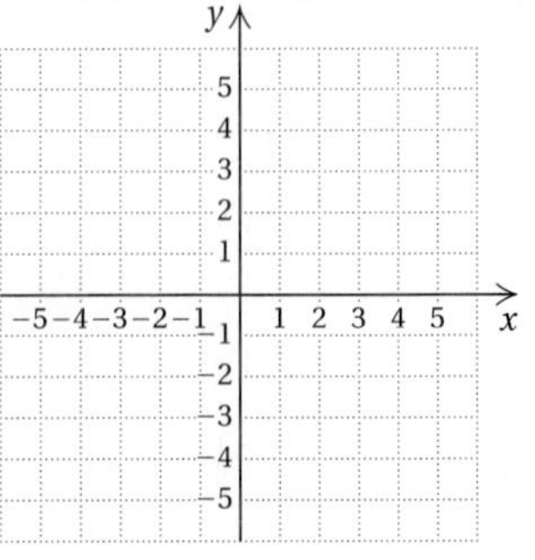

29. $y \geq -2,$
$x \geq 1$

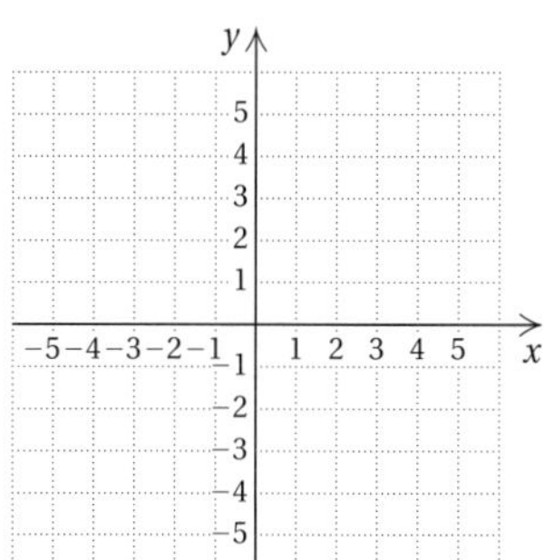

30. $y \leq -2,$
$x \geq 2$

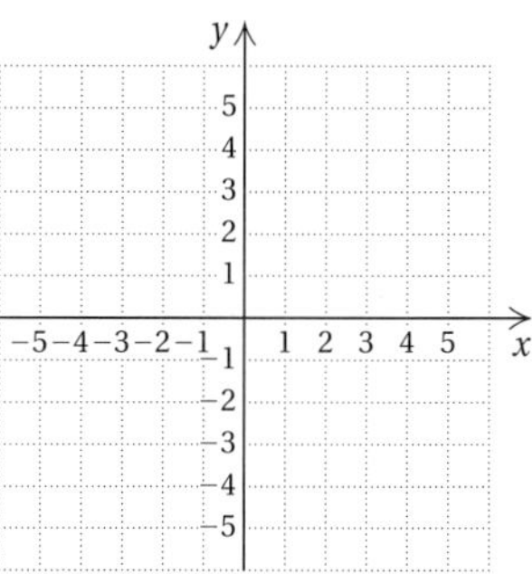

31. $x \leq 3,$
$y \geq -3x + 2$

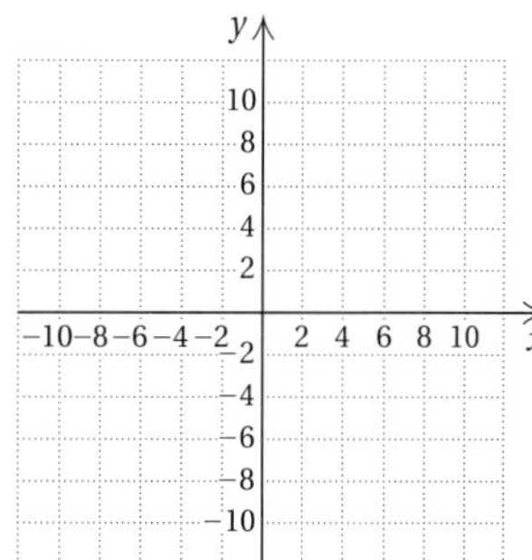

32. $x \geq -2,$
$y \leq -2x + 3$

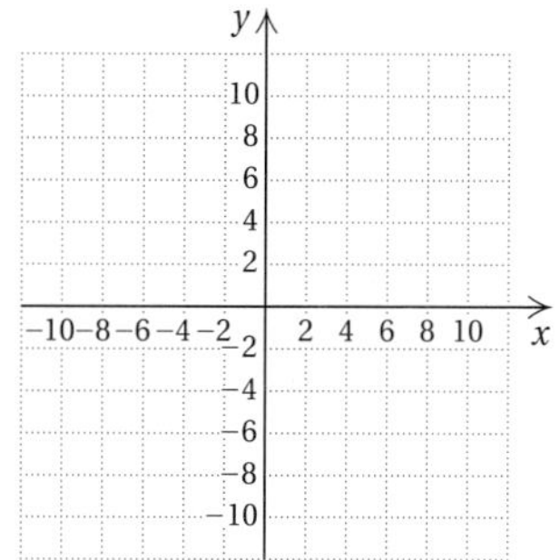

33. $y \geq -2,$
$y \geq x + 3$

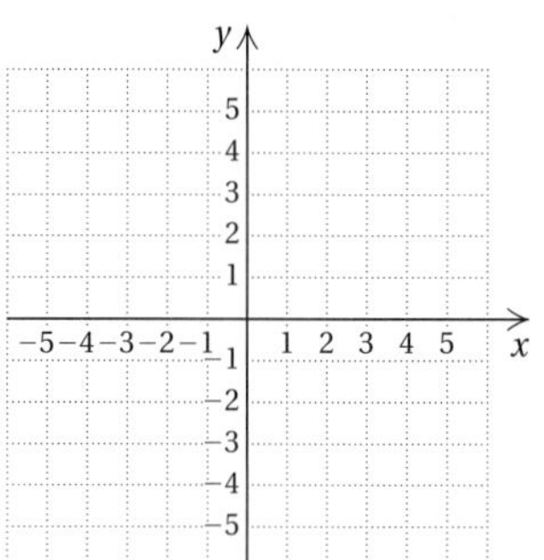

34. $y \leq 4,$
$y \geq -x + 2$

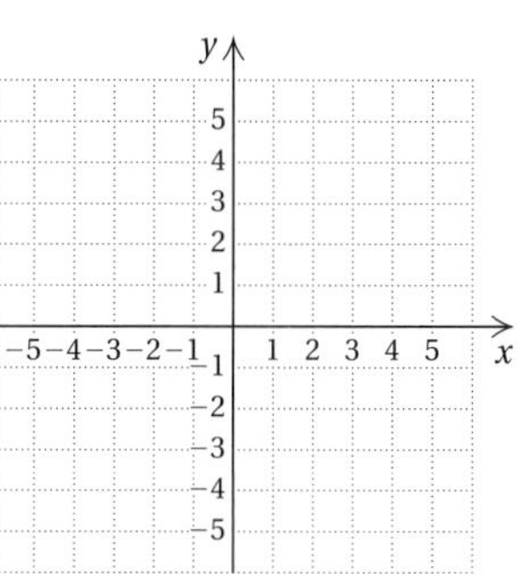

35. $x + y \leq 1,$
$x - y \leq 2$

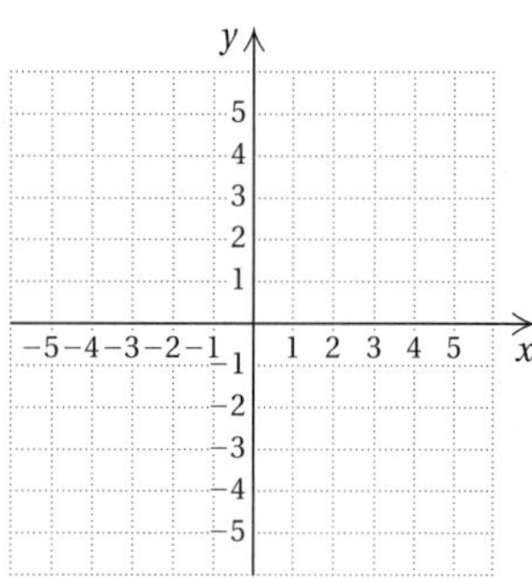

36. $x + y \leq 3,$
$x - y \leq 4$

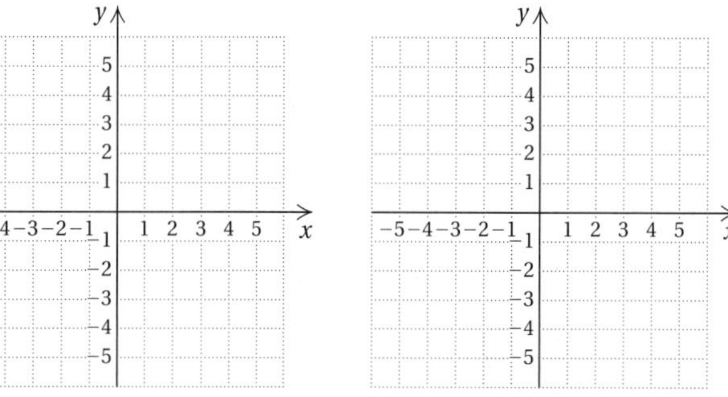

37. $y - 2x \geq 1,$
$y - 2x \leq 3$

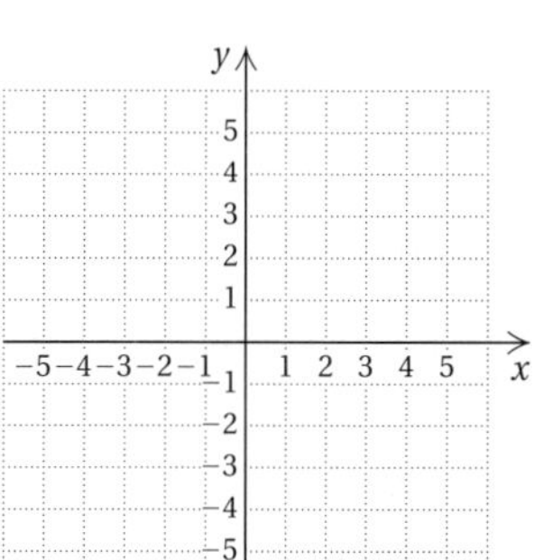

38. $y + 3x \geq 0,$
$y + 3x \leq 2$

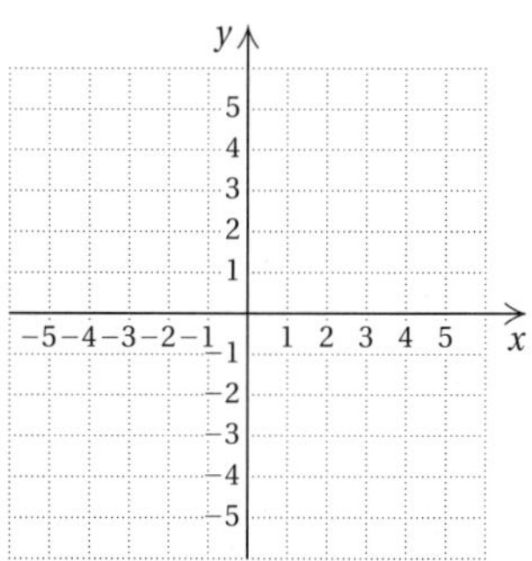

39. $y \leq 2x + 1,$
$y \geq -2x + 1,$
$x \leq 2$

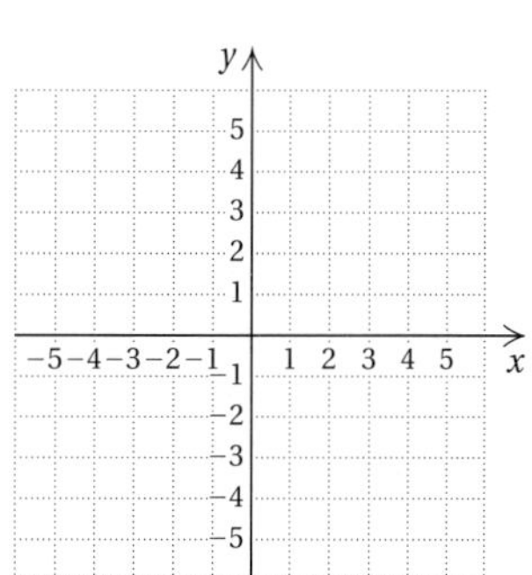

40. $x - y \leq 2,$
$x + 2y \geq 8,$
$y \leq 4$

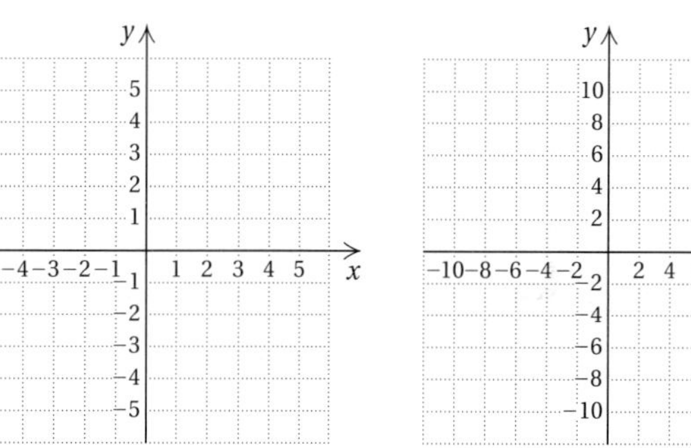

41. $x + 2y \leq 12,$
$2x + y \leq 12,$
$x \geq 0,$
$y \geq 0$

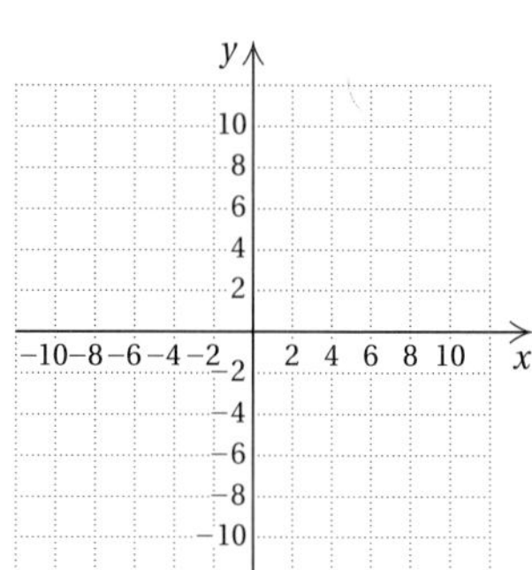

42. $4y - 3x \geq -12,$
$4y + 3x \geq -36,$
$y \leq 0,$
$x \leq 0$

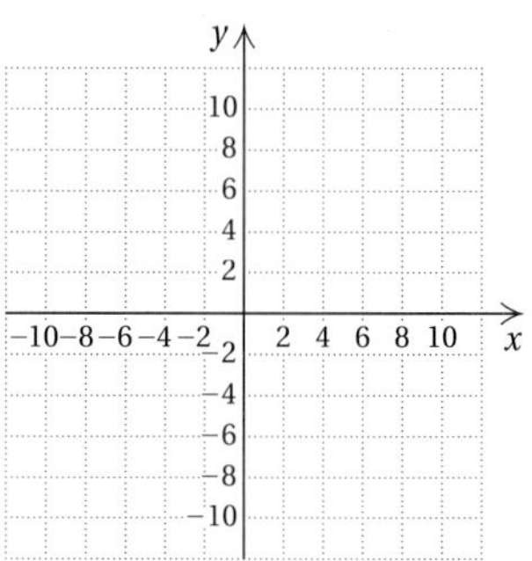

43. $8x + 5y \leq 40,$
$x + 2y \leq 8,$
$x \geq 0,$
$y \geq 0$

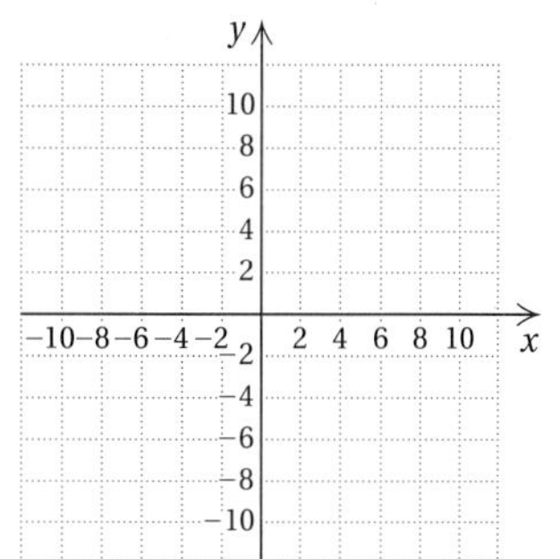

44. $y - x \geq 1,$
$y - x \leq 3,$
$2 \leq x \leq 5$

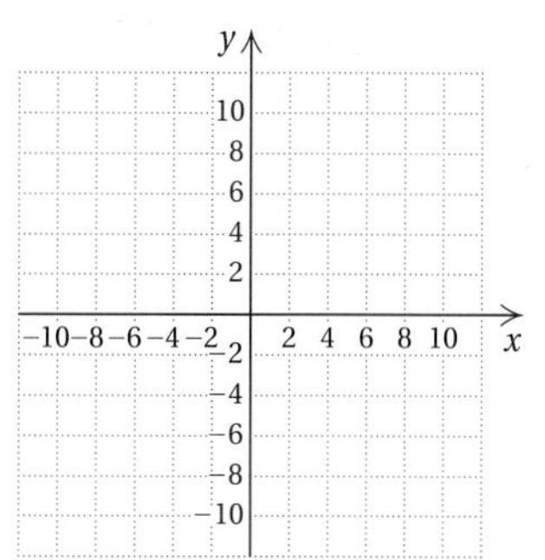

Skill Maintenance

Find an equation of the line containing the given pair of points. [7.5c]

45. $(-5, -6), (3, -2)$

46. $(0, 0), (8, 3)$

47. $(3, 6), (1, 12)$

48. $(-2, 0), (0, 9)$

49. $(-5, 5), (-1, -1)$

50. $(2, 6), (-3, 6)$

Given the function $f(x) = |2 - x|$, find each of the following function values. [7.1b]

51. $f(0)$

52. $f(-1)$

53. $f(1)$

54. $f(10)$

55. $f(-2)$

56. $f(2a)$

57. $f(-4)$

58. $f(1.8)$

Synthesis

59. ◆ Do all systems of linear inequalities have solutions? Why or why not?

60. ◆ When graphing linear inequalities, Ron always shades above the line when he sees a $\geq$ symbol. Is this wise? Why or why not?

61. *Luggage Size.* Unless an additional fee is paid, most major airlines will not check any luggage that is more than 62 in. long. The U.S. Postal Service will ship a package only if the sum of the package's length and girth (distance around its midsection) does not exceed 108 in. Concert Productions is ordering several 62-in. long trunks that will be both mailed and checked as luggage. Using w and h for width and height (in inches), respectively, write and graph an inequality that represents all acceptable combinations of width and height.

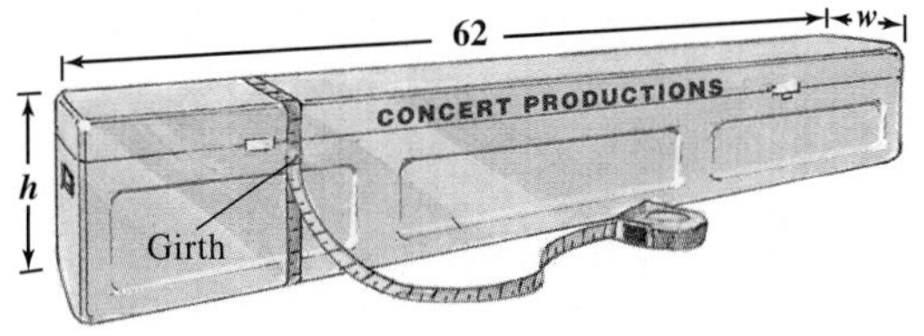

62. *Exercise Danger Zone.* It is dangerous to exercise when the weather is hot and humid. The solutions of the following system of inequalities give a "danger zone" for which it is dangerous to exercise intensely:

$$4H - 3F < 70,$$
$$F + H > 160,$$
$$2F + 3H > 390,$$

where F is the temperature, in degrees Fahrenheit, and H is the humidity.

a) Draw the danger zone by graphing the system of inequalities.

b) Is it dangerous to exercise when $F = 80°$ and $H = 80\%$?

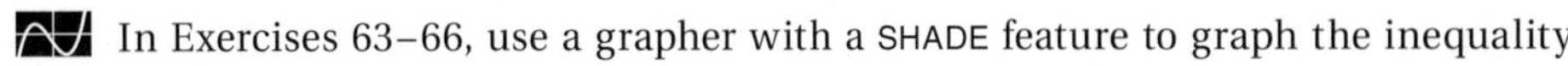

In Exercises 63–66, use a grapher with a SHADE feature to graph the inequality.

63. $3x + 6y > 2$

64. $x - 5y \leq 10$

65. $13x - 25y + 10 \leq 0$

66. $2x + 5y > 0$

67. Use a grapher with a SHADE feature to check your answers to Exercises 25–38. Then use the INTERSECT feature to determine any point(s) of intersection.

Summary and Review Exercises: Chapter 9

Important Properties and Formulas

The Addition Principle for Inequalities: For any real numbers a, b, and c: $a < b$ is equivalent to $a + c < b + c$; $a > b$ is equivalent to $a + c > b + c$.

The Multiplication Principle for Inequalities: For any real numbers a and b, and any *positive* number c: $a < b$ is equivalent to $ac < bc$; $a > b$ is equivalent to $ac > bc$.
For any real numbers a and b, and any *negative* number c: $a < b$ is equivalent to $ac > bc$; $a > b$ is equivalent to $ac < bc$.
Similar statements hold for $\leq$ and $\geq$.

Set Intersection: $A \cap B = \{x \mid x \text{ is in } A \text{ and } x \text{ is in } B\}$

Set Union: $A \cup B = \{x \mid x \text{ is in } A \text{ or in } B, \text{ or both}\}$

"$a < x$ *and* $x < b$" is equivalent to "$a < x < b$"

Properties of Absolute Value

$|ab| = |a| \cdot |b|$, $\quad \left|\dfrac{a}{b}\right| = \dfrac{|a|}{|b|}$, $\quad |-a| = |a|$, $\quad$ The distance between a and b is $|a - b|$.

Principles for Solving Equations and Inequalities Involving Absolute Value:

For any positive number p and any algebraic expression X:

a) The solutions of $|X| = p$ are those numbers that satisfy $X = -p$ *or* $X = p$.

b) The solutions of $|X| < p$ are those numbers that satisfy $-p < X < p$.

c) The solutions of $|X| > p$ are those numbers that satisfy $X < -p$ or $X > p$.

The objectives to be tested in addition to the material in this chapter are [4.6a], [7.2a], [7.5c], and [8.2b].

Write interval notation for the given set or graph. [9.1b]

1. $\{x \mid -8 \leq x < 9\}$

2. 

Solve and graph. Write interval notation for the solution set. [9.1c]

3. $x - 2 \leq -4$

4. $x + 5 > 6$

Solve. [9.1c]

5. $a + 7 \leq -14$

6. $y - 5 \geq -12$

7. $4y > -16$

8. $-0.3y < 9$

9. $-6x - 5 < 13$

10. $4y + 3 \leq -6y - 9$

11. $-\frac{1}{2}x - \frac{1}{4} > \frac{1}{2} - \frac{1}{4}x$

12. $0.3y - 8 < 2.6y + 15$

13. $-2(x - 5) \geq 6(x + 7) - 12$

14. *Moving Costs.* Musclebound Movers charges \$85 plus \$40 an hour to move households across town. Champion Moving charges \$60 an hour for cross-town moves. For what lengths of time is Champion more expensive? [9.1d]

15. *Investments.* You are going to invest \$30,000, part at 13% and part at 15%, for one year. What is the most that can be invested at 13% in order to make at least \$4300 interest in one year? [9.1d]

Graph and write interval notation. [9.2a, b]

16. $-2 \leq x < 5$

17. $x \leq -2$ *or* $x > 5$

Solve. [9.2a, b]

18. $2x - 5 < -7$ *and* $3x + 8 \geq 14$

19. $-4 < x + 3 \leq 5$

20. $-15 < -4x - 5 < 0$

21. $3x < -9$ *or* $-5x < -5$

22. $2x + 5 < -17$ *or* $-4x + 10 \leq 34$

23. $2x + 7 \leq -5$ *or* $x + 7 \geq 15$

24. *Records in the Women's 100-m Dash.* Florence Griffith Joyner set a world record of 10.49 sec in the women's 100-m dash in 1988. The equation

$$R = -0.0433t + 10.49$$

can be used to predict the world record in the women's 100-m dash t years after 1988. For what years was the record between 10.15 and 10.35 sec? [9.2c]

Simplify. [9.3a]

25. $\left|-\frac{3}{x}\right|$ **26.** $\left|\frac{2x}{y^2}\right|$ **27.** $\left|\frac{12y}{-3y^2}\right|$

28. Find the distance between -23 and 39. [9.3b]

Solve. [9.3c, d]

29. $|x| = 6$ **30.** $|x - 2| = 7$

31. $|2x + 5| = |x - 9|$ **32.** $|5x + 6| = -8$

Solve. [9.3e]

33. $|2x + 5| < 12$ **34.** $|x| \geq 3.5$

35. $|3x - 4| \geq 15$ **36.** $|x| < 0$

37. Find the intersection: [9.2a]
$\{1, 2, 5, 6, 9\} \cap \{1, 3, 5, 9\}$.

38. Find the union: [9.2b]
$\{1, 2, 5, 6, 9\} \cup \{1, 3, 5, 9\}$.

Graph. [9.4b]

39. $2x + 3y < 12$ **40.** $y \leq 0$

41. $x + y \geq 1$

Graph. Find the coordinates of any vertices formed. [9.4c]

42. $y \geq -3,$
$x \geq 2$

43. $x + 3y \geq -1,$
$x + 3y \leq 4$

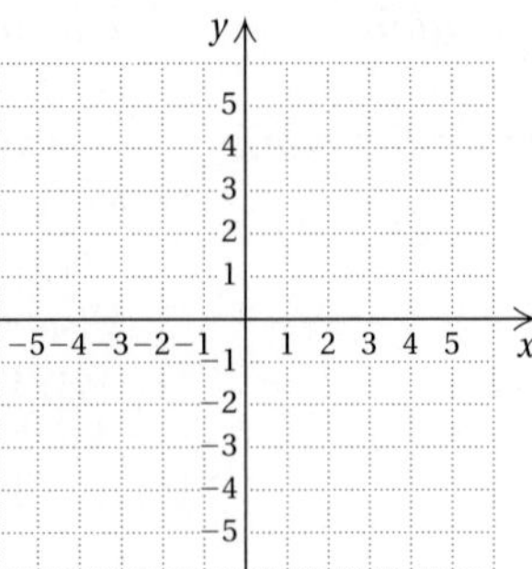

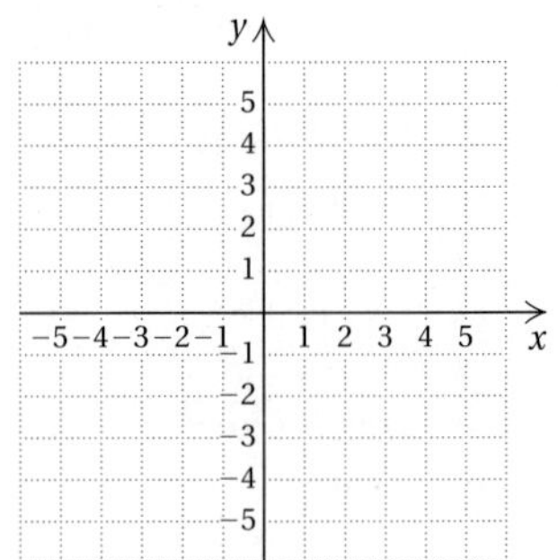

44. $x - 3y \leq 3,$
$x + 3y \geq 9,$
$y \leq 6$

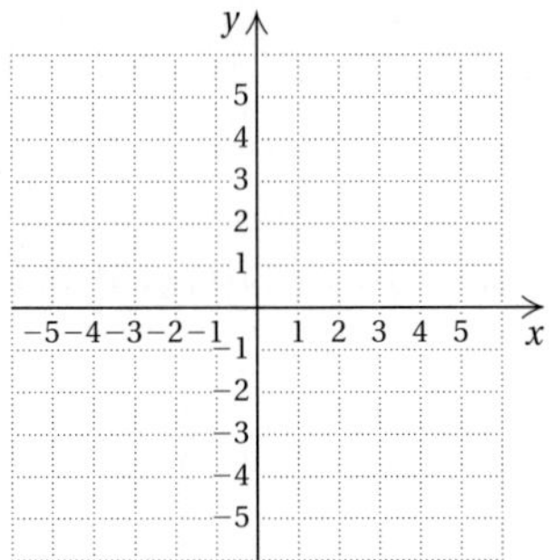

Skill Maintenance

45. Multiply: $(8y + 5)(3y - 7)$. [4.6a]

46. Find the domain: $f(x) = \frac{4}{x + 9}$. [7.2a]

47. Solve: $6x - y = 13,$
$y - 5x = -11.$

48. Find an equation of the line containing the points $(3, -3)$ and $(-2, -1)$.

Synthesis

49. ◈ Find the error or errors in each of the following steps: [9.1c]

$$\begin{aligned} 7 - 9x + 6x &< -9(x + 2) + 10x & \\ 7 - 9x + 6x &< -9x + 2 + 10x & (1) \\ 7 + 6x &> 2 + 10x & (2) \\ -4x &> 8 & (3) \\ x &> -2. & (4) \end{aligned}$$

50. ◈ How does the word "solve" vary in meaning in this chapter?

51. Solve: $|2x + 5| \leq |x + 3|$. [9.3d, e]

Test: Chapter 9

Write interval notation for the given set or graph.

1. $\{x \mid -3 < x \leq 2\}$

2. 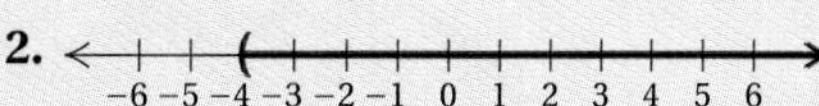

Solve and graph. Write interval notation for the solution set.

3. $x - 2 \leq 4$

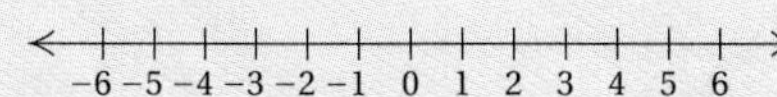

4. $-4y - 3 \geq 5$

−6 −5 −4 −3 −2 −1 0 1 2 3 4 5 6

Solve.

5. $-0.6y < 30$

6. $3a - 5 \leq -2a + 6$

7. $-5y - 1 > -9y + 3$

8. $4(5 - x) < 2x + 5$

9. $-8(2x + 3) + 6(4 - 5x) \geq 2(1 - 7x) - 4(4 + 6x)$

Solve.

10. *Moving Costs.* Motivated Movers charges $105 plus $30 an hour to move households across town. Quick-Pak Moving charges $80 an hour for cross-town moves. For what lengths of time is Quick-Pak more expensive?

11. *Pressure at Sea Depth.* The equation

$$P = 1 + \frac{d}{33}$$

gives the pressure P, in atmospheres (atm), at a depth of d feet in the sea. For what depths d is the pressure at least 2 atm and at most 8 atm?

Graph and write interval notation.

12. $-3 \leq x \leq 4$

−6 −5 −4 −3 −2 −1 0 1 2 3 4 5 6

13. $x < -3$ *or* $x > 4$

−6 −5 −4 −3 −2 −1 0 1 2 3 4 5 6

Solve.

14. $5 - 2x \leq 1$ *and* $3x + 2 \geq 14$

15. $-3 < x - 2 < 4$

16. $-11 \leq -5x - 2 < 0$

17. $-3x > 12$ *or* $4x > -10$

18. $x - 7 \leq -5$ *or* $x - 7 \geq -10$

19. $3x - 2 < 7$ *or* $x - 2 > 4$

Simplify.

20. $\left|\frac{7}{x}\right|$

21. $\left|\frac{-6x^2}{3x}\right|$

Answers

1. ______
2. ______
3. ______
4. ______
5. ______
6. ______
7. ______
8. ______
9. ______
10. ______
11. ______
12. ______
13. ______
14. ______
15. ______
16. ______
17. ______
18. ______
19. ______
20. ______
21. ______

Answers

22. ______

23. ______

24. ______

25. ______

26. ______

27. ______

28. ______

29. ______

30. ______

31. ______

32. ______

33. ______

34. ______

35. ______

36. ______

37. ______

38. ______

39. ______

40. ______

22. Find the distance between 4.8 and −3.6.

Solve.

23. $|x| = 9$

24. $|x| > 3$

25. $|4x - 1| < 4.5$

26. $|-5x - 3| \geq 10$

27. $|x + 10| = |x - 12|$

28. $|2 - 5x| = -10$

29. $\left|\dfrac{6 - x}{7}\right| \leq 15$

30. Find the intersection:
$\{1, 3, 5, 7, 9\} \cap \{3, 5, 11, 13\}$.

31. Find the union:
$\{1, 3, 5, 7, 9\} \cup \{3, 5, 11, 13\}$.

Graph. Find the coordinates of any vertices formed.

32. $x - 6y < -6$

33. $x + y \geq 3,$
$x - y \geq 5$

34. $2y - x \geq -7,$
$2y + 3x \leq 15,$
$y \leq 0,$
$x \leq 0$

Skill Maintenance

35. Multiply: $(9x - 7)(8x + 5)$.

36. Find an equation of the line containing the points $(6, -8)$ and $(1, 2)$.

37. Solve: $4x - 5y = -12,$
$2x - 9y = 20.$

38. Find the domain: 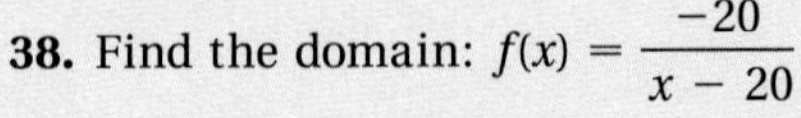 $f(x) = \dfrac{-20}{x - 20}$.

Synthesis

Solve.

39. $|3x - 4| \leq -3$

40. $7x < 8 - 3x < 6 + 7x$

Cumulative Review: Chapters 1–9

1. Evaluate $\frac{4a}{7b}$ for $a = 14$ and $b = 6$.

Simplify.

2. $|-12|$

3. $-\frac{1}{3} \div \frac{3}{8}$

4. $12b - [9 - 7(5b - 6)]$

5. $5^3 \div \frac{1}{2} - 2(3^2 + 2^2)$

6. $\left(\frac{2x^3y^{-6}}{-4y^{-2}}\right)^2$

7. $(6p^2 - 2p + 5) - (-10p^2 + 6p + 5)$

8. $(6m - n)^2$

9. $(3a - 4b)(5a + 2b)$

10. $\frac{y^2 - 4}{3y + 33} \cdot \frac{y + 11}{y + 2}$

11. $\frac{9x^2 - 25}{x^2 - 16} \div \frac{3x + 5}{x - 4}$

12. $\frac{2x + 1}{4x - 12} - \frac{x - 2}{5x - 15}$

13. $\frac{2}{x + 2} + \frac{3}{x - 2} - \frac{x + 1}{x^2 - 4}$

14. $\dfrac{\frac{1}{x} + \frac{2}{y}}{\frac{3}{x} - \frac{1}{y}}$

15. $\dfrac{1 - \frac{2}{y^2}}{1 - \frac{1}{y^3}}$

16. $(2x^3 - 7x^2 + x - 3) \div (x + 2)$

Solve.

17. $9y - (5y - 3) = 33$

18. $F = \frac{9}{5}C + 32$, for C

19. $-3 < -2x - 6 < 0$

20. $|x| \geq 2.1$

21. $4x - 2y = 6,$
$6x - 3y = 9$

22. $x + 2y - 2z = 9,$
$2x - 3y + 4z = -4,$
$5x - 4y + 2z = 5$

23. $8x = 1 + 16x^2$

24. $14 + 3x = 2x^2$

25. $\frac{3x}{x - 2} - \frac{6}{x + 2} = \frac{24}{x^2 - 4}$

26. $\frac{6}{x - 5} = \frac{2}{2x}$

27. $P = \frac{3a}{a + b}$, for a

28. $\frac{1}{2}x - 3 = \frac{7}{2}$

29. $6x - 1 \leq 3(5x + 2)$

30. $|2x - 3| < 7$

31. $|3x - 6| = 2$

Solve.

32. $4x + 5y = -3,$
$x = 1 - 3y$

33. $x + 6y + 4z = -2,$
$4x + 4y + z = 2,$
$3x + 2y - 4z = 5$

Factor.

34. $4x^3 + 18x^2$

35. $8a^3 - 4a^2 - 6a + 3$

36. $x^2 + 8x - 84$

37. $6x^2 + 11x - 10$

38. $16y^2 - 81$

39. $t^2 - 16t + 64$

40. $64x^3 + 8$

41. $0.027b^3 - 0.008c^3$

42. $x^6 - x^2$

43. $20x^2 + 7x - 3$

Graph.

44. $y = -5x + 4$

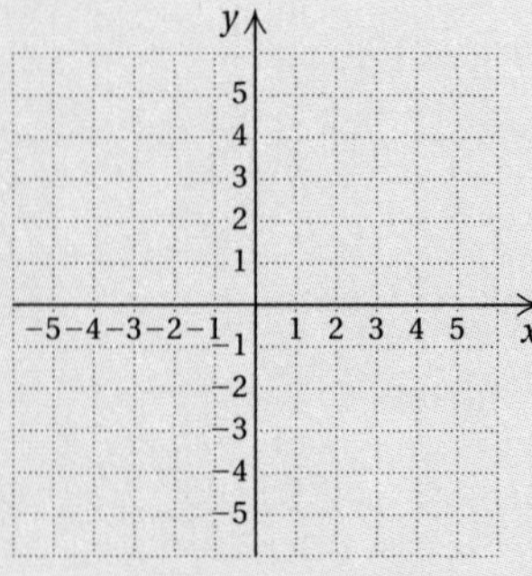

45. $3x - 18 = 0$

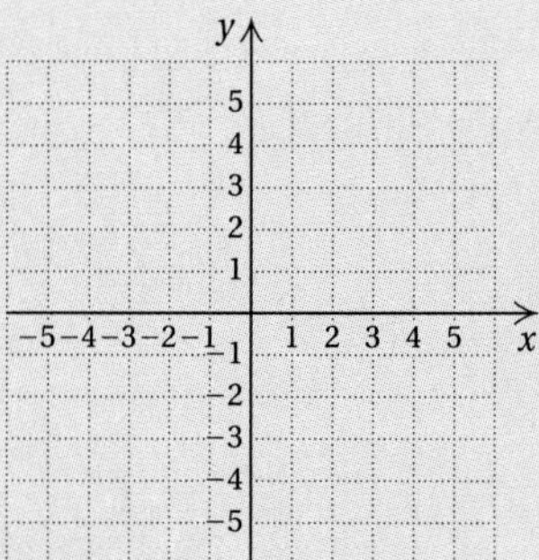

46. $x + 3y < 4$

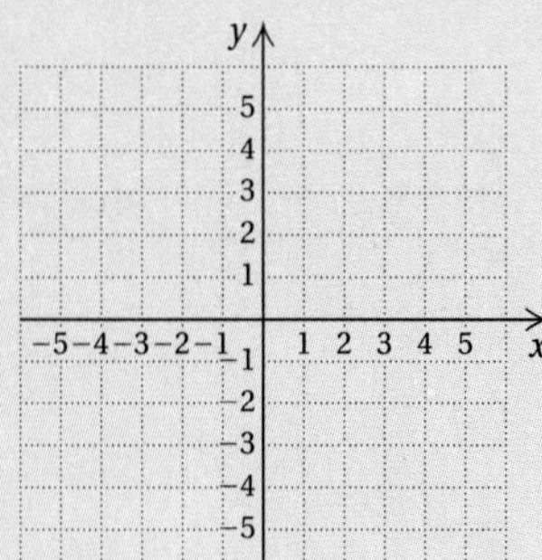

47. $x + y \geq 4,$
$x - y > 1$

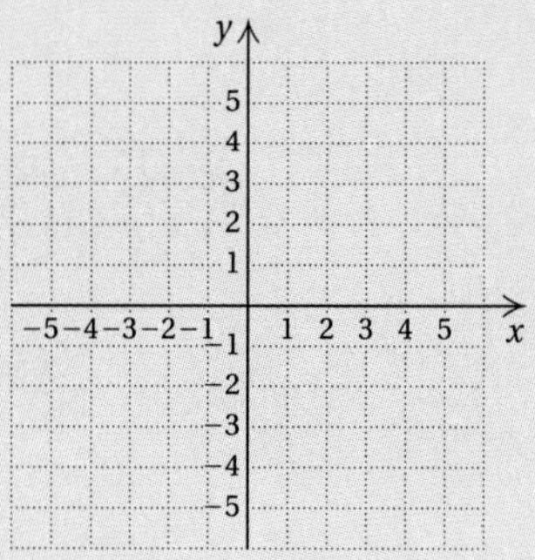

48. Find an equation of the line containing the point (3, 7) and parallel to the line $x + 2y = 6$.

49. Find an equation of the line containing the point (3, −2) and perpendicular to the line $3x + 4y = 5$.

50. Find an equation of the line containing the points (−1, 4) and (−1, 0).

51. Find an equation of the line with slope −3 and through the point (2, 1).

52. Find the slope and the y-intercept of the line $3x - 2y = 8$.

53. *Movie Costs.* The average cost of a movie ticket in Finland is $1.54 more than the average cost of a movie ticket in Sweden. The sum of the average costs is $14.44. How much does a movie ticket cost in each country?

54. *Diaper Changes.* A family with twins did an experiment on the cost of diapering their two children. Disposables were used on one child and cloth diapers from a diaper service were used on the other child. After the twins were out of diapers, calculations showed that for both children, $2886 was spent on diapering. Using disposables cost $546 more than using the diaper service. How much did each cost?

55. *Casualty Loss.* When filing your income tax, you may claim a casualty loss only if it exceeds 10% of your adjusted gross income plus $100. If the loss was $1500, can you claim it if your adjusted gross income is $13,500?

56. *Medical Expenses.* Medical expenses can be deducted on your tax return only if they exceed $7\frac{1}{2}$% of your adjusted gross income. If your expenses were $3400, what is the largest adjusted gross income for which no medical expenses can be deducted?

57. *Housing.* According to Census Bureau statistics, 20.4% of housing is in the West, 24.0% in the North Central, and 35.3% in the South. What percent is in the Northeast?

58. *Residential Taxes.* The residential taxes per $100 of property value are $2.43 more in Detroit than in Chicago. The taxes for $10,000 of property in Detroit and $20,000 of property in Chicago total $741. What are the residential property taxes per $100 of property value in each city?

59. *Hockey Results.* A hockey team played 81 games in a season. They won 1 fewer game than three times the number of ties and lost 8 fewer games than they won. How many games did they win? lose? tie?

60. *Painting Time.* Dave can paint the outside of his house in 15 hr. Bill can paint the same house in 12 hr. How long would it take them to paint the house together?

61. *Insurance Costs.* The cost c of an insurance policy varies directly as the age a of the insured. A 32-year-old person pays an annual premium of \$152. What is the age of a person who pays \$285?

62. *Solid-Waste Generation.* The function

$$W(t) = 0.05t + 4.3$$

can be used to estimate the number of pounds W of solid waste produced daily, on average, by each person in the United States for t years since 1991.

a) Estimate the number of pounds of solid waste that will be produced in 2001 by finding $W(10)$.
b) In what year will solid-waste production be 6.25 lb per day?
c) For what years will waste production range from 6.0 to 6.25 lb per person per day?

63. *Toys "R" Us.* Toys "R" Us, Inc., an international company specializing in toys, has experienced tremendous growth in recent years. Its net sales S, in billions of dollars, are shown in the graph at right.

a) Use the data points (1, 5.5) and (7, 9.9) to fit a linear function to the data.
b) Use the linear function to predict net sales in 2000 and in 2004.

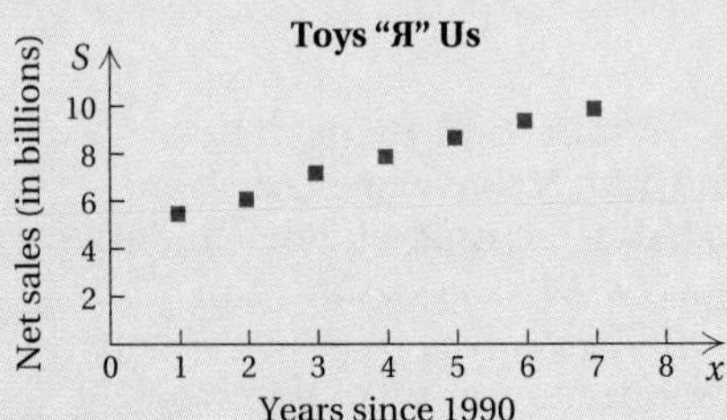

Source: Toys "Я" Us Annual Report, 1997

Find the domain and the range of the function.

64.

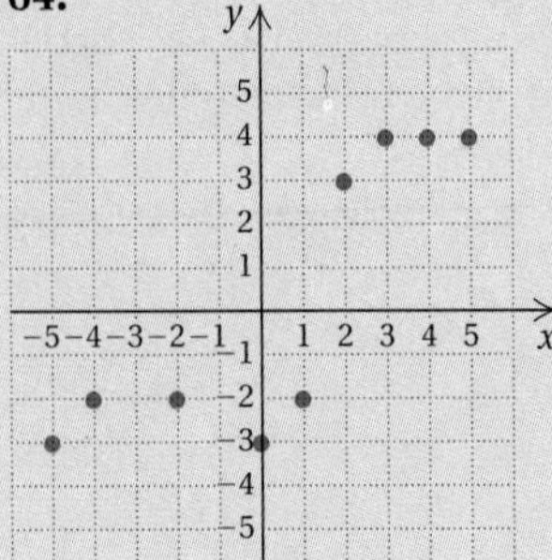

65.

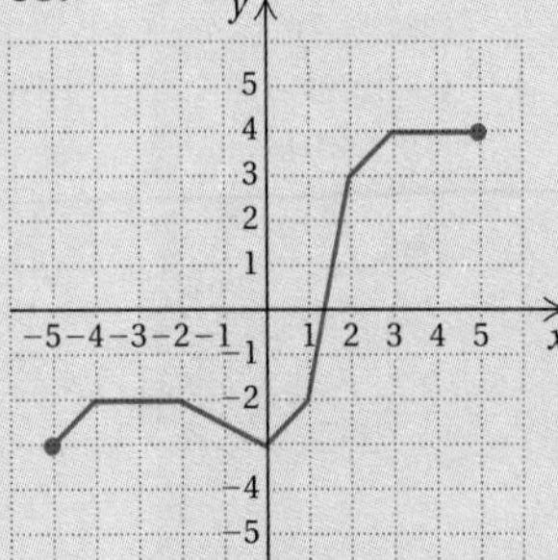

66.

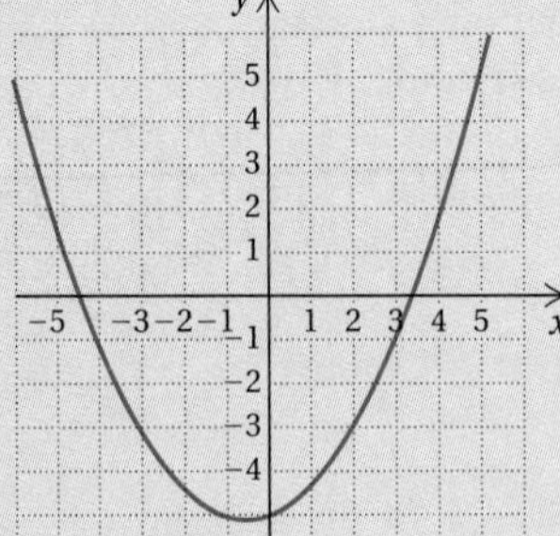

67.

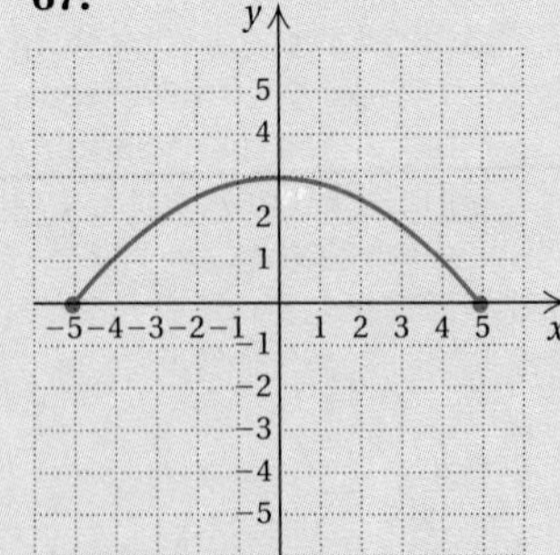

Synthesis

68. The graph of $y = ax^2 + bx + c$ contains the three points (4, 2), (2, 0), and (1, 2). Find a, b, and c.

69. Solve:

$$\frac{18}{x - 9} + \frac{10}{x + 5} = \frac{28x}{x^2 - 4x - 45}.$$

70. Solve: $16x^3 = x$.

10

Radical Expressions, Equations, and Functions

An Application

The *wind chill temperature* T_w is what the temperature would have to be with no wind in order to give the same chilling effect as with wind. A formula for wind chill temperature is given by

$$T_w = 91.4 - \frac{(10.45 + 6.68\sqrt{v} - 0.447v)(457 - 5T)}{110},$$

where T is the actual temperature, in degrees Fahrenheit, and v is the wind speed, in miles per hour. Find the wind chill when $T = 10°F$ and $v = 20$ mph.

The Mathematics

We substitute 10 for T and 20 for v in the formula:

$$T_w = 91.4 - \frac{(10.45 + 6.68\sqrt{20} - 0.447 \cdot 20)(457 - 5 \cdot 10)}{110} \approx -24.7°.$$

This is a *radical equation.*

This problem appears as Exercise 44 in Exercise Set 10.7.

World Wide Web For more information, visit us at www.mathmax.com

Introduction

In this chapter, we introduce square roots, cube roots, fourth roots, fifth roots, and so on. We study them in relation to radical expressions and functions and use them to solve applied problems like the one below. We also consider expressions with rational exponents and their related functions.

Pretest: Chapter 10

Simplify. Assume that letters can represent *any* real number.

1. $\sqrt{t^2}$

2. $\sqrt[3]{27x^3}$

3. $\sqrt[12]{y^{12}}$

In Questions 4–20, assume that all expressions under the radical represent positive numbers.

Simplify.

4. $(\sqrt[3]{9a^2b})^2$

5. $\sqrt{45} - 3\sqrt{125} + 4\sqrt{80}$

Multiply and simplify.

6. $\sqrt{18x^2}\sqrt{8x^3}$

7. $(2\sqrt{6} - 1)^2$

8. Divide and simplify:

$$\frac{\sqrt{52a^4}}{\sqrt{13a^3}}.$$

9. Rationalize the denominator:

$$\frac{3}{2 + \sqrt{5}}.$$

Solve.

10. $\sqrt{2x - 1} = 5$

11. $\sqrt[3]{1 - 6x} = 2$

12. $\sqrt{3x + 1} - \sqrt{2x} = 1$

In Questions 13 and 14, give an exact answer and an approximation to three decimal places.

13. In a right triangle with leg $b = 5$ and hypotenuse $c = 8$, find the length of leg a.

14. The diagonal of a square has length 8 ft. Find the length of a side.

15. Use rational exponents to write a single radical expression:

$$\sqrt{x}\sqrt[3]{x}.$$

16. Determine whether $-1 + i$ is a solution of $x^2 + 2x + 4 = 0$.

17. Subtract: $(-2 + 7i) - (3 - 6i)$.

18. Multiply: $(4 + 3i)(3 - 4i)$.

19. Divide: $\dfrac{-4 + i}{3 - 2i}$.

20. Simplify: i^{87}.

Objectives for Retesting

The objectives to be tested in addition to the material in this chapter are as follows.

[5.8b] Solve quadratic and other polynomial equations by first factoring and then using the principle of zero products.

[6.1d, f] Multiply and divide rational expressions and simplify.

[6.5a] Solve rational equations.

[6.6a, b] Solve applied problems involving work, proportions, and motion.

10.1 Radical Expressions and Functions

In this section, we consider roots, such as square roots and cube roots. We define the symbolism and consider methods of manipulating symbols to get equivalent expressions.

Objectives

a Find principal square roots and their opposites, approximate square roots, find outputs of square-root functions, graph square-root functions, and find the domains of square-root functions.

b Simplify radical expressions with perfect-square radicands.

c Find cube roots, simplifying certain expressions, and find outputs of cube-root functions.

d Simplify expressions involving odd and even roots.

For Extra Help

TAPE 16 | MAC WIN | CD-ROM

a Square Roots and Square-Root Functions

When we raise a number to the second power, we say that we have **squared** the number. Sometimes we may need to find the number that was squared. We call this process **finding a square root** of a number.

> The number c is a **square root** of a if $c^2 = a$.

For example:

5 is a square root of 25 because $5^2 = 5 \cdot 5 = 25$;

-5 is a square root of 25 because $(-5)^2 = (-5)(-5) = 25$.

The number -4 does not have a real-number square root because there is no real number b such that $b^2 = -4$.

> **SQUARE ROOTS**
>
> Every positive real number has two real-number square roots.
>
> The number 0 has just one square root, 0 itself.
>
> Negative numbers do not have real-number square roots.*

Example 1 Find the two square roots of 64.

The square roots of 64 are 8 and -8 because $8^2 = 64$ and $(-8)^2 = 64$.

Do Exercises 1–3.

> The **principal square root** of a nonnegative number is its nonnegative square root. The symbol $\sqrt{a}$ represents the principal square root of a. To name the negative square root of a, we can write $-\sqrt{a}$.

Examples Simplify.

2. $\sqrt{25} = 5$ — *Remember*: $\sqrt{}$ indicates the principal (nonnegative) square root.

3. $\sqrt{\dfrac{81}{64}} = \dfrac{9}{8}$

4. $\sqrt{0.0049} = 0.07$

5. $\sqrt{0} = 0$

6. $-\sqrt{25} = -5$

7. $\sqrt{-25}$ Does not exist as a real number. Negative numbers do not have real-number square roots.

Do Exercises 4–10.

Find the square roots.

1. 9
2. 36
3. 121

Simplify.

4. $\sqrt{1}$
5. $\sqrt{36}$
6. $\sqrt{\dfrac{81}{100}}$
7. $\sqrt{0.0064}$

Find the following.

8. a) $\sqrt{16}$
   b) $-\sqrt{16}$
   c) $\sqrt{-16}$
9. a) $\sqrt{49}$
   b) $-\sqrt{49}$
   c) $\sqrt{-49}$
10. a) $\sqrt{144}$
    b) $-\sqrt{144}$
    c) $\sqrt{-144}$

Answers on page A-41

*In Section 10.8, we will consider an expansion of the real-number system, in which negative numbers do have square roots.

It would be helpful to memorize the following table of exact square roots.

Table of Common Square Roots

$\sqrt{1} = 1$	$\sqrt{196} = 14$
$\sqrt{4} = 2$	$\sqrt{225} = 15$
$\sqrt{9} = 3$	$\sqrt{256} = 16$
$\sqrt{16} = 4$	$\sqrt{289} = 17$
$\sqrt{25} = 5$	$\sqrt{324} = 18$
$\sqrt{36} = 6$	$\sqrt{361} = 19$
$\sqrt{49} = 7$	$\sqrt{400} = 20$
$\sqrt{64} = 8$	$\sqrt{441} = 21$
$\sqrt{81} = 9$	$\sqrt{484} = 22$
$\sqrt{100} = 10$	$\sqrt{529} = 23$
$\sqrt{121} = 11$	$\sqrt{576} = 24$
$\sqrt{144} = 12$	$\sqrt{625} = 25$
$\sqrt{169} = 13$	

Use a calculator to approximate each of the following square roots to three decimal places.

11. $\sqrt{17}$ **12.** $\sqrt{40}$

13. $\sqrt{1138}$ **14.** $-\sqrt{867.6}$

15. $\sqrt{\dfrac{22}{35}}$

16. $-\sqrt{\dfrac{2103.4}{67.82}}$

Identify the radicand.

17. $\sqrt{28 + x}$

18. $\sqrt{\dfrac{y}{y + 3}}$

Answers on page A-41

We found exact square roots in Examples 1–6. We often need to use rational numbers to *approximate* square roots that are irrational. Such expressions can be found using a calculator with a square-root key.

Examples Use a calculator to approximate each of the following.

	Using a calculator with a 10-digit readout	*Rounded to three decimal places*
8. $\sqrt{11}$	3.316624790	3.317
9. $\sqrt{487}$	22.06807649	22.068
10. $-\sqrt{7297.8}$	-85.42716196	-85.427
11. $\sqrt{\dfrac{463}{557}}$	.9117229728	0.912

Do Exercises 11–16.

The symbol $\sqrt{}$ is called a **radical**.

An expression written with a radical is called a **radical expression.**

The expression written under the radical is called the **radicand**.

These are radical expressions:

$$\sqrt{5}, \quad \sqrt{a}, \quad -\sqrt{5x}, \quad \sqrt{y^2 + 7}.$$

The radicands in these expressions are 5, a, $5x$, and $y^2 + 7$, respectively.

Example 12 Identify the radicand in $\sqrt{x^2 - 9}$.

The radicand in $\sqrt{x^2 - 9}$ is $x^2 - 9$.

Do Exercises 17 and 18.

Since each nonnegative real number x has exactly one principal square root, the symbol $\sqrt{x}$ represents exactly one real number and thus can be used to define a square-root function:

$$f(x) = \sqrt{x}.$$

The domain of this function is the set of nonnegative real numbers. In interval notation, the domain is $[0, \infty)$.

Example 13 For the given function, find the indicated function values:

$$f(x) = \sqrt{3x - 2}; \quad f(1), f(5), \text{ and } f(0).$$

We have

$$f(1) = \sqrt{3 \cdot 1 - 2} \qquad \textbf{Substituting}$$
$$= \sqrt{3 - 2} = \sqrt{1} = 1; \qquad \textbf{Simplifying and taking the square root}$$

$$f(5) = \sqrt{3 \cdot 5 - 2} \qquad \textbf{Substituting}$$
$$= \sqrt{13} \approx 3.606; \qquad \textbf{Simplifying and approximating}$$

$$f(0) = \sqrt{3 \cdot 0 - 2} \qquad \textbf{Substituting}$$
$$= \sqrt{-2}. \qquad \textbf{Negative radicand. No real-number function value exists.}$$

Do Exercises 19 and 20 on the following page.

Example 14 Find the domain of $f(x) = \sqrt{x + 2}$.

The expression $\sqrt{x + 2}$ is a real number only when $x + 2$ is nonnegative. Thus the domain of $f(x) = \sqrt{x + 2}$ is the set of all x-values for which $x + 2 \geq 0$. We solve as follows:

$$x + 2 \geq 0$$
$$x \geq -2. \quad \text{Adding } -2$$

The domain of $f = \{x \mid x \geq -2\} = [-2, \infty)$.

Example 15 Graph: $f(x) = \sqrt{x}$.

We first find outputs as we did in Example 13. We can either select inputs that have exact outputs or use a calculator to make approximations. Once ordered pairs have been calculated, a smooth curve can be drawn.

x	$f(x) = \sqrt{x}$	$(x, f(x))$
0	0	(0, 0)
1	1	(1, 1)
3	1.7	(3, 1.7)
4	2	(4, 2)
7	2.6	(7, 2.6)
9	3	(9, 3)

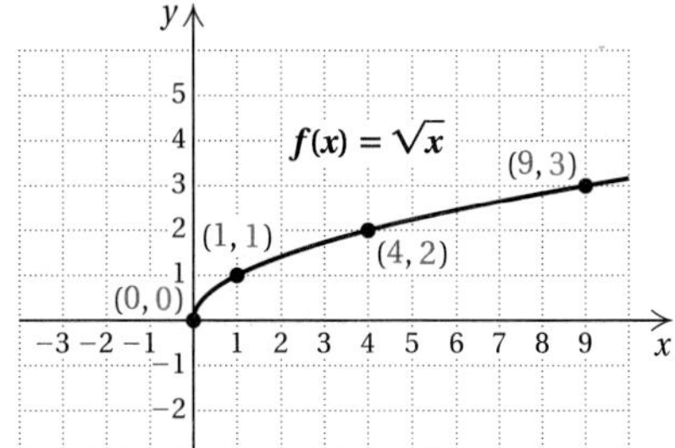

We can see from the table and the graph that the domain is $[0, \infty)$. The range is also the set of nonnegative real numbers $[0, \infty)$.

Do Exercises 21–24.

b Finding $\sqrt{a^2}$

In the expression $\sqrt{a^2}$, the radicand is a perfect square. It is tempting to think that $\sqrt{a^2} = a$, but we see below that this is not the case.

Suppose $a = 5$. Then we have $\sqrt{5^2}$, which is $\sqrt{25}$, or 5.
Suppose $a = -5$. Then we have $\sqrt{(-5)^2}$, which is $\sqrt{25}$, or 5.
Suppose $a = 0$. Then we have $\sqrt{0^2}$, which is $\sqrt{0}$, or 0.

The symbol $\sqrt{a^2}$ never represents a negative number. It represents the principal square root of a^2. Note that if a represents a positive number or 0, then $\sqrt{a^2}$ represents a. If a is negative, then $\sqrt{a^2}$ represents the opposite of a. In all cases, the radical expression represents the absolute value of a.

> For any real number a, $\sqrt{a^2} = |a|$. The principal (nonnegative) square root of a^2 is the absolute value of a.

The absolute value is used to ensure that the principal square root is nonnegative, which is as it is defined.

Examples Find the following. Assume that letters can represent any real number.

16. $\sqrt{(-16)^2} = |-16|$, or 16

For the given function, find the indicated function values.

19. $g(x) = \sqrt{6x + 4}$; $g(0)$, $g(3)$, and $g(-5)$

20. $f(x) = -\sqrt{x}$; $f(4)$, $f(7)$, and $f(-3)$

Find the domain of the function.

21. $f(x) = \sqrt{x - 5}$

22. $g(x) = \sqrt{2x + 3}$

Graph.

23. $g(x) = -\sqrt{x}$

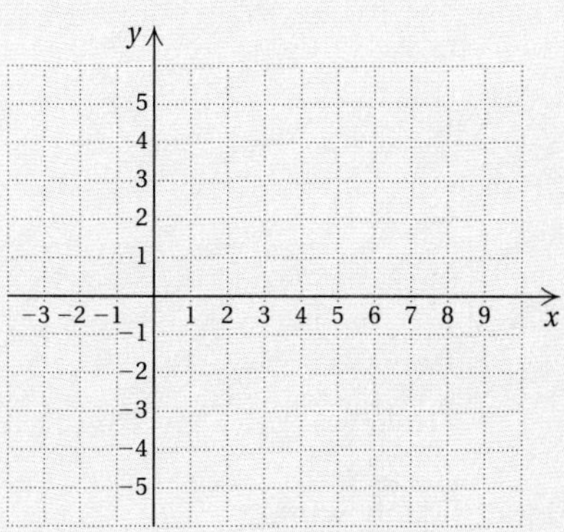

24. $f(x) = 2\sqrt{x + 3}$

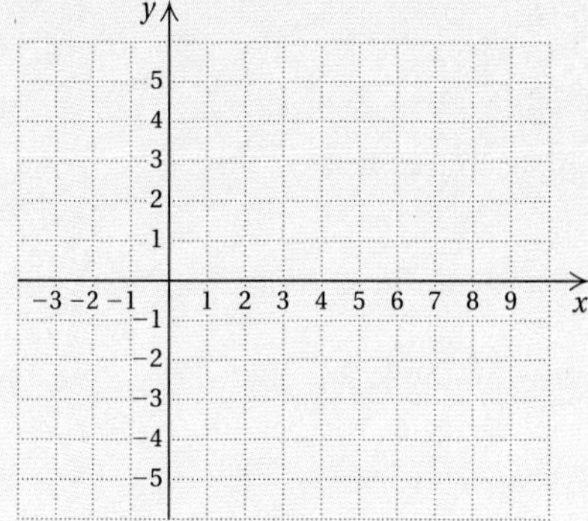

Answers on page A-41

Find the following. Assume that letters can represent *any* real number.

25. $\sqrt{y^2}$

26. $\sqrt{(-24)^2}$

27. $\sqrt{(5y)^2}$

28. $\sqrt{16y^2}$

29. $\sqrt{(x+7)^2}$

30. $\sqrt{4(x-2)^2}$

31. $\sqrt{49(y+5)^2}$

32. $\sqrt{x^2-6x+9}$

Find the following.

33. $\sqrt[3]{-64}$

34. $\sqrt[3]{27y^3}$

35. $\sqrt[3]{8(x+2)^3}$

36. $\sqrt[3]{-\dfrac{343}{64}}$

Answers on page A-41

17. $\sqrt{(3b)^2} = |3b| = |3| \cdot |b| = 3|b|$

> $|3b|$ can be simplified to $3|b|$ because the absolute value of any product is the product of the absolute values. That is, $|a \cdot b| = |a| \cdot |b|$.

18. $\sqrt{(x-1)^2} = |x-1|$

19. $\sqrt{x^2+8x+16} = \sqrt{(x+4)^2}$
$= |x+4|$ ←

> CAUTION! $|x+4|$ is *not* the same as $|x|+4$.

Do Exercises 25–32.

c Cube Roots

> The number c is the **cube root** of a if its third power is a—that is, if $c^3 = a$.

For example:

2 is the cube root of 8 because $2^3 = 2 \cdot 2 \cdot 2 = 8$;

-4 is the cube root of -64 because $(-4)^3 = (-4)(-4)(-4) = -64$.

We talk about *the* cube root of a number because of the following.

> Every real number has exactly one cube root in the system of real numbers. The symbol $\sqrt[3]{a}$ represents the cube root of a.

Examples Find the following.

20. $\sqrt[3]{8} = 2$

21. $\sqrt[3]{-27} = -3$

22. $\sqrt[3]{-\dfrac{216}{125}} = -\dfrac{6}{5}$

23. $\sqrt[3]{0.001} = 0.1$

24. $\sqrt[3]{x^3} = x$

25. $\sqrt[3]{-8} = -2$

26. $\sqrt[3]{0} = 0$

27. $\sqrt[3]{-8y^3} = -2y$

When we are determining a cube root, no absolute-value signs are needed because a real number has just one cube root. The real-number cube root of a positive number is positive. The real-number cube root of a negative number is negative. The cube root of 0 is 0. That is, $\sqrt[3]{a^3} = a$ whether $a > 0$, $a < 0$, or $a = 0$.

Do Exercises 33–36.

Since the symbol $\sqrt[3]{x}$ represents exactly one real number, it can be used to define a function.

Example 28 For the given function, find the indicated function values.

$f(x) = \sqrt[3]{x}$; $f(125)$, $f(-8)$, $f(0)$, and $f(-10)$.

We have

$f(125) = \sqrt[3]{125} = 5$; $\quad f(-8) = \sqrt[3]{-8} = -2$;

$f(0) = \sqrt[3]{0} = 0$; $\quad f(-10) = \sqrt[3]{-10} \approx -2.1544$.

Do Exercise 37 on the following page.

The graph of $f(x) = \sqrt[3]{x}$ is shown below for reference. Note that the domain and the range *each* consists of the entire set of real numbers, $(-\infty, \infty)$.

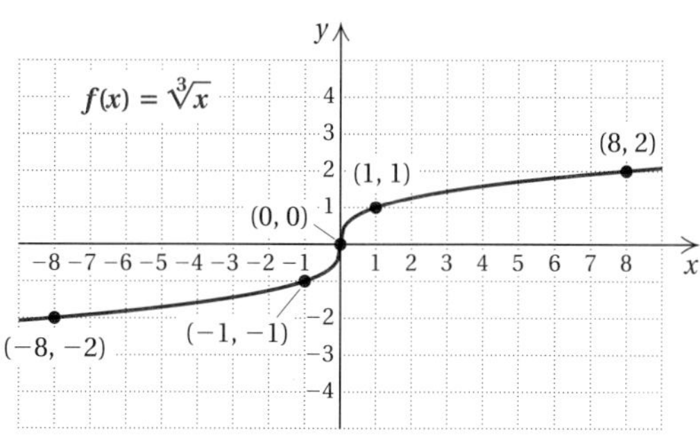

d Odd and Even *k*th Roots

In the expression $\sqrt[k]{a}$, we call k the **index** and assume $k \geq 2$.

Odd Roots

The 5th root of a number a is the number c for which $c^5 = a$. There are also 7th roots, 9th roots, and so on. Whenever the number k in $\sqrt[k]{}$ is an odd number, we say that we are taking an **odd root.**

Every number has just one real-number odd root. If the number is positive, then the root is positive. If the number is negative, then the root is negative. If the number is 0, then the root is 0.

> If k is an *odd* natural number, then for any real number a,
> $$\sqrt[k]{a^k} = a.$$

Absolute-value signs are not needed when we are finding odd roots.

Examples Find the following.

29. $\sqrt[5]{32} = 2$
30. $\sqrt[5]{-32} = -2$
31. $-\sqrt[5]{32} = -2$
32. $-\sqrt[5]{-32} = -(-2) = 2$
33. $\sqrt[7]{x^7} = x$
34. $\sqrt[7]{128} = 2$
35. $\sqrt[7]{-128} = -2$
36. $\sqrt[7]{0} = 0$
37. $\sqrt[5]{a^5} = a$
38. $\sqrt[9]{(x-1)^9} = x - 1$

Do Exercises 38–44.

Even Roots

When the index k in $\sqrt[k]{}$ is an even number, we say that we are taking an **even root.** Every positive real number has two real-number kth roots when k is even. One of those roots is positive and one is negative. Negative real numbers do not have real-number kth roots when k is even. When we are finding even kth roots, absolute-value signs are sometimes necessary, as they are with square roots. When the index is 2, we do not write it. For example,

$$\sqrt{64} = 8, \quad \sqrt[6]{64} = 2, \quad -\sqrt[6]{64} = -2, \quad \sqrt[6]{64x^6} = |2x| = 2|x|.$$

Note that in $\sqrt[6]{64x^6}$, we need absolute-value signs because a variable is involved.

37. For the given function, find the indicated function values.

$g(x) = \sqrt[3]{x-4}$; $g(-23)$, $g(4)$, $g(-1)$, and $g(11)$

Calculator Spotlight

We can approximate cube roots by using the MATH feature to access a $\sqrt[3]{}$ key. For example,

$$\sqrt[3]{19} \approx 2.668401649.$$

Exercises

Approximate each of the following.

1. $\sqrt[3]{22}$
2. $\sqrt[3]{-43}$
3. $\sqrt[3]{46.7}$
4. $\sqrt[3]{-2040}$

Find the following.

38. $\sqrt[5]{243}$

39. $\sqrt[5]{-243}$

40. $\sqrt[5]{x^5}$

41. $\sqrt[7]{y^7}$

42. $\sqrt[5]{0}$

43. $\sqrt[5]{-32x^5}$

44. $\sqrt[7]{(3x+2)^7}$

Answers on page A-41

Find the following. Assume that letters can represent any real number.

45. $\sqrt[4]{81}$ **46.** $-\sqrt[4]{81}$

47. $\sqrt[4]{-81}$ **48.** $\sqrt[4]{0}$

49. $\sqrt[4]{16(x-2)^4}$

50. $\sqrt[6]{x^6}$

51. $\sqrt[8]{(x+3)^8}$

Examples Find the following. Assume that letters can represent any real number.

39. $\sqrt[4]{16} = 2$

40. $-\sqrt[4]{16} = -2$

41. $\sqrt[4]{-16}$ Does not exist as a real number.

42. $\sqrt[4]{81x^4} = 3|x|$

43. $\sqrt[6]{(y+7)^6} = |y+7|$

44. $\sqrt{81y^2} = 9|y|$

The following is a summary of how absolute value is used when we are taking even or odd roots.

> For any real number a:
> a) $\sqrt[k]{a^k} = |a|$ when k is an *even* natural number. We use absolute value when k is even unless a is nonnegative.
> b) $\sqrt[k]{a^k} = a$ when k is an *odd* natural number greater than 1. We do not use absolute value when k is odd.

Do Exercises 45–51.

Calculator Spotlight

Graphing Radical Functions. Graphing functions defined by radical expressions involves approximating roots. Since the square root of a negative number is not a real number, y-values may not exist for some x-values. For example, y-values for the graph of $f(x) = \sqrt{x-1}$ do not exist for x-values that are less than 1 because square roots of negative numbers would result.

We must enter $y = \sqrt{x-1}$ using parentheses around the radicand as $y_1 = \sqrt{(x-1)}$. Some graphers supply the left parenthesis automatically.

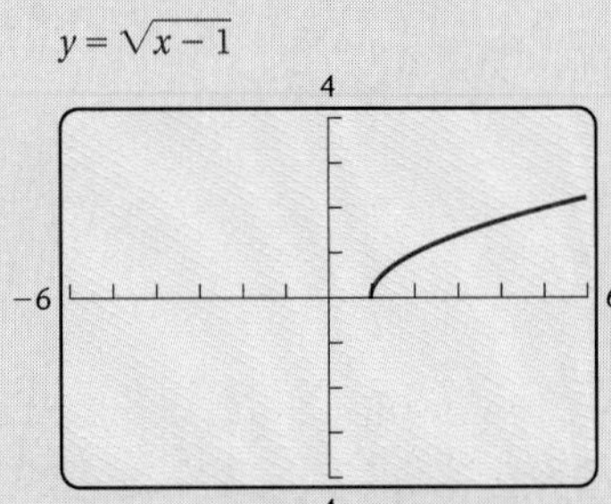

Similarly, y-values for the graph of $f(x) = \sqrt{2-x}$ do not exist for x-values that are greater than 2.

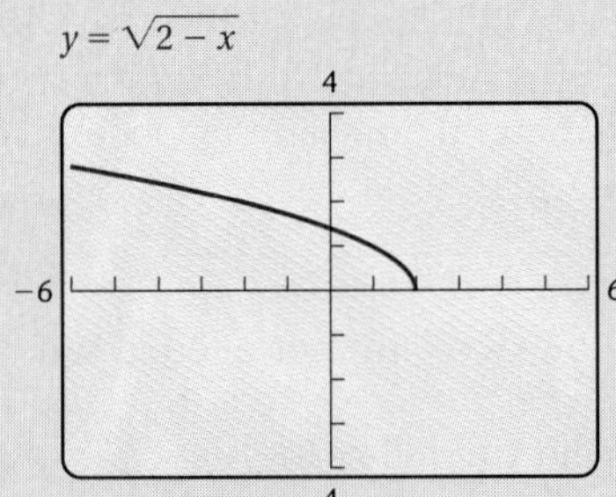

Exercises

Graph each of the following functions. Then use the TABLE and TRACE features to determine the domain and the range of each function. (See also the Calculator Spotlight in Section 7.2.) The MATH feature contains $\sqrt[3]{\ }$ and $\sqrt[x]{\ }$ keys to enter kth roots.

1. $f(x) = \sqrt{x}$ **2.** $g(x) = \sqrt{x+2}$

3. $f(x) = \sqrt[3]{x}$ **4.** $f(x) = \sqrt[3]{x-2}$

5. $f(x) = \sqrt[4]{x-1}$ **6.** $F(x) = \sqrt[5]{6-x}$

7. $g(x) = 5 - \sqrt{x+3}$ **8.** $f(x) = 4 - \sqrt[3]{x}$

Use the GRAPH and TABLE features to determine whether each of the following is correct.

9. $\sqrt{x+4} = \sqrt{x} + 2$

10. $\sqrt{25x} = 5\sqrt{x}$

Answers on page A-41

Exercise Set 10.1

a Find the square roots.

1. 16

2. 225

3. 144

4. 9

5. 400

6. 81

Simplify.

7. $-\sqrt{\frac{49}{36}}$

8. $-\sqrt{\frac{361}{9}}$

9. $\sqrt{196}$

10. $\sqrt{441}$

11. $\sqrt{0.0036}$

12. $\sqrt{0.04}$

Use a calculator to approximate to three decimal places.

13. $\sqrt{347}$

14. $-\sqrt{1839.2}$

15. $\sqrt{\frac{285}{74}}$

16. $\sqrt{\frac{839.4}{19.7}}$

Identify the radicand.

17. $9\sqrt{y^2 + 16}$

18. $-3\sqrt{p^2 - 10}$

19. $x^4y^5\sqrt{\frac{x}{y-1}}$

20. $a^2b^2\sqrt{\frac{a^2 - b}{b}}$

For the given function, find the indicated function values.

21. $f(x) = \sqrt{5x - 10}$; $f(6)$, $f(2)$, $f(1)$, and $f(-1)$

22. $t(x) = -\sqrt{2x + 1}$; $t(4)$, $t(0)$, $t(-1)$, and $t\left(-\frac{1}{2}\right)$

23. $g(x) = \sqrt{x^2 - 25}$; $g(-6)$, $g(3)$, $g(6)$, and $g(13)$

24. $F(x) = \sqrt{x^2 + 1}$; $F(0)$, $F(-1)$, and $F(-10)$

25. Find the domain of the function f in Exercise 21.

26. Find the domain of the function t in Exercise 22.

27. *Speed of a Skidding Car.* How do police determine how fast a car had been traveling after an accident has occurred? The function

$$S(x) = 2\sqrt{5x}$$

can be used to approximate the speed S, in miles per hour, of a car that has left a skid mark of length x, in feet. What was the speed of a car that left skid marks of length 30 ft? 150 ft?

28. *Parking-Lot Arrival Spaces.* The attendants at a parking lot park cars in temporary spaces before the cars are taken to permanent parking stalls. The number N of such spaces needed is approximated by the function

$$N(a) = 2.5\sqrt{a},$$

where a is the average number of arrivals in peak hours. What is the number of spaces needed when the average number of arrivals is 66? 100?

Graph.

29. $f(x) = 2\sqrt{x}$

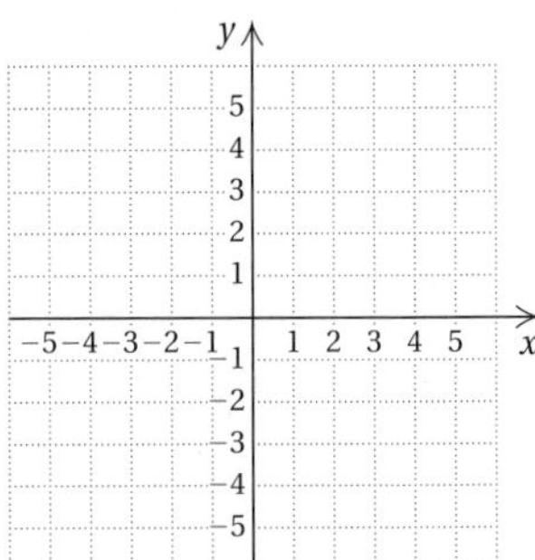

30. $g(x) = 3 - \sqrt{x}$

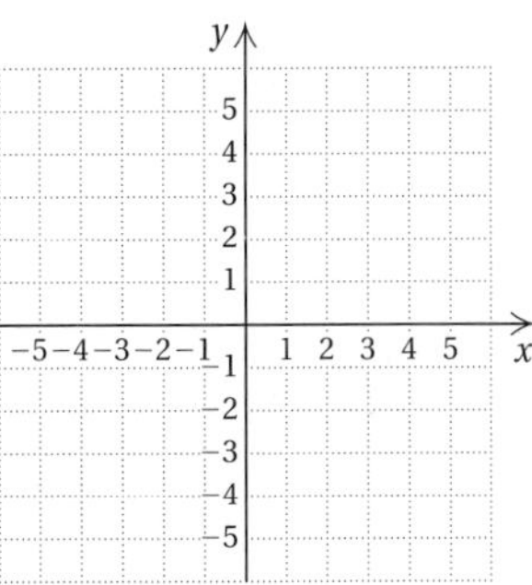

31. $F(x) = -3\sqrt{x}$

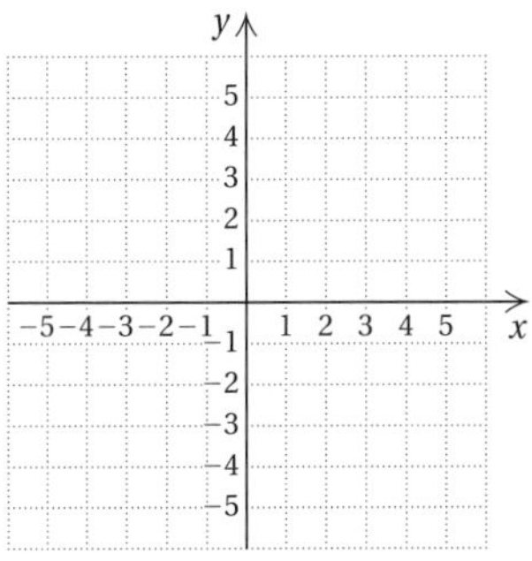

32. $f(x) = 2 + \sqrt{x - 1}$

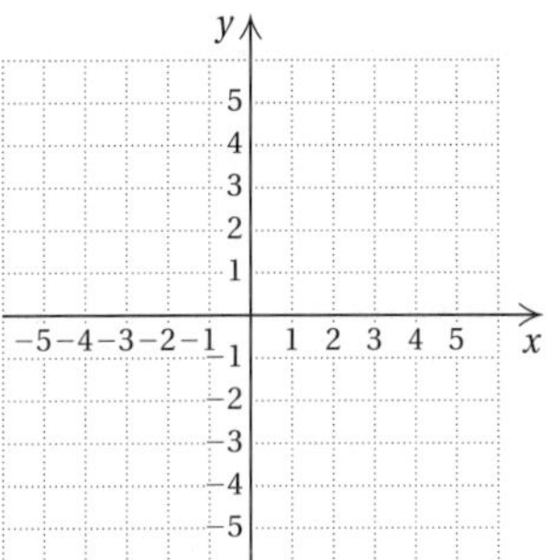

33. $f(x) = \sqrt{x}$

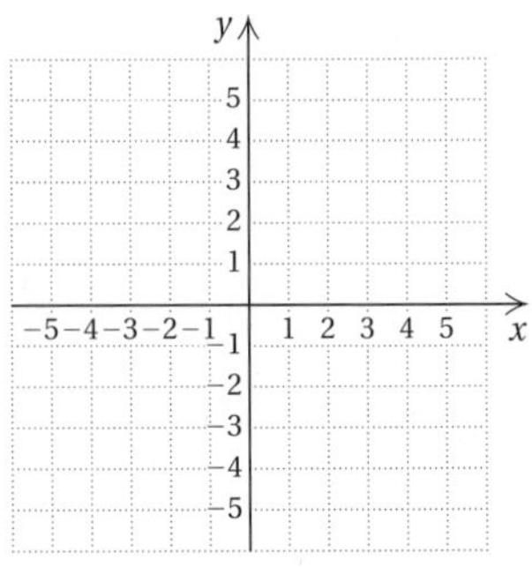

34. $g(x) = -\sqrt{x}$

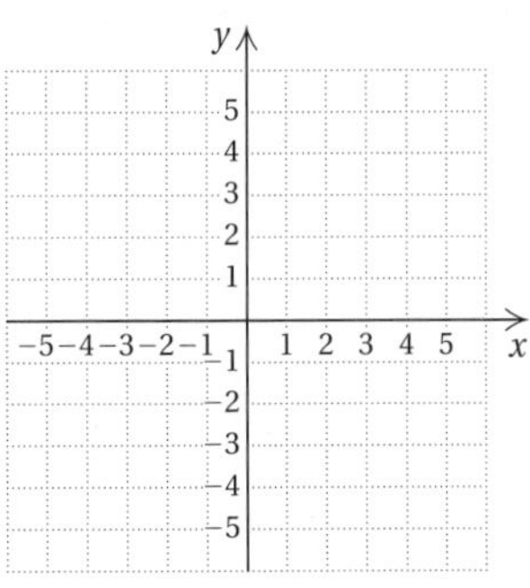

35. $f(x) = \sqrt{x - 2}$

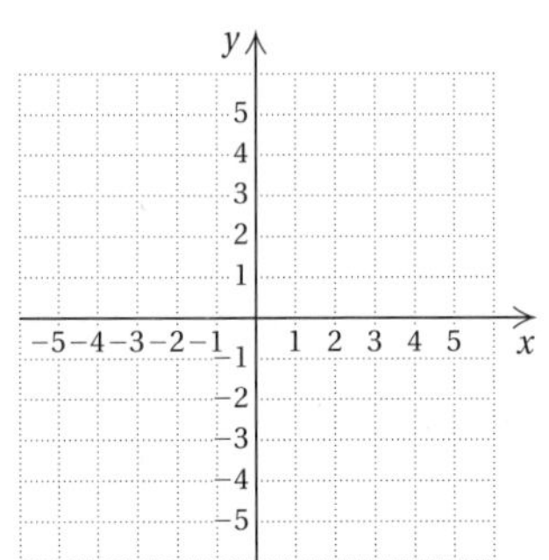

36. $g(x) = \sqrt{x + 3}$

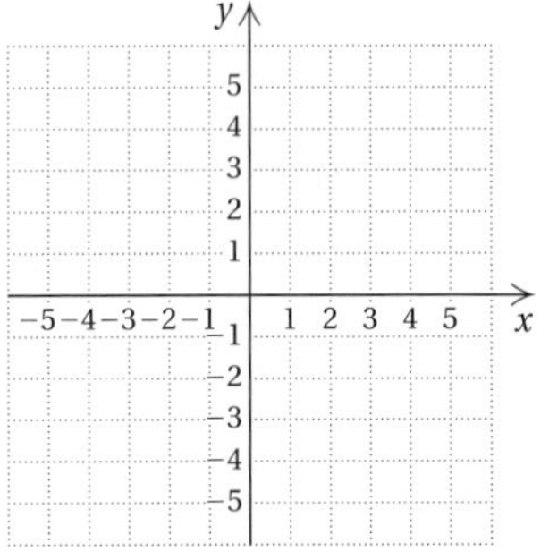

37. $f(x) = \sqrt{12 - 3x}$

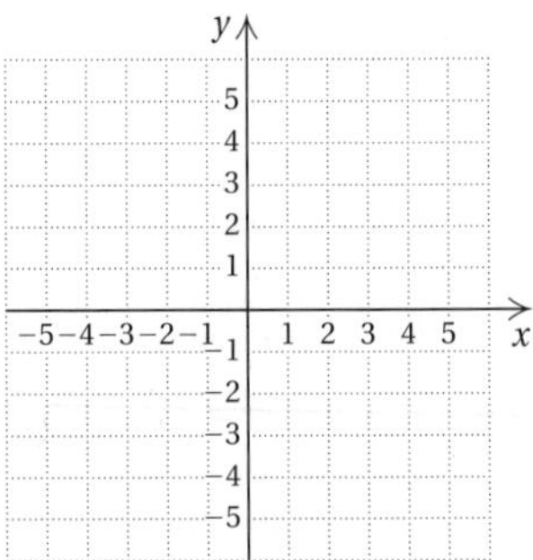

38. $g(x) = \sqrt{8 - 4x}$

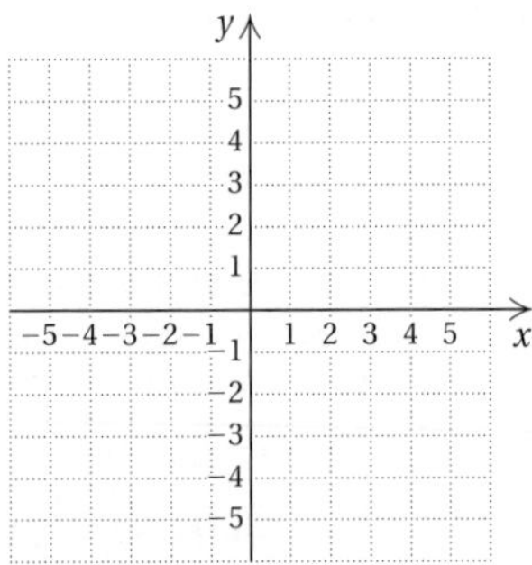

39. $g(x) = \sqrt{3x + 9}$

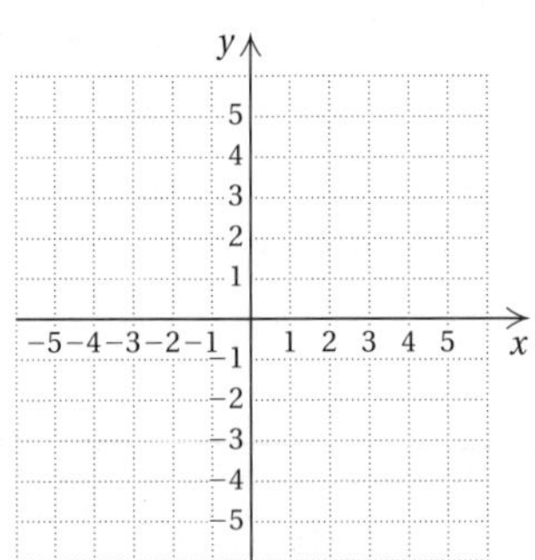

40. $f(x) = \sqrt{3x - 6}$

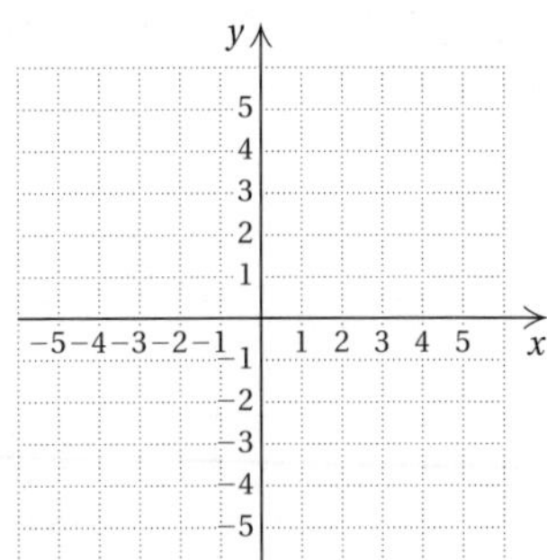

b Find the following. Assume that letters can represent *any* real number.

41. $\sqrt{16x^2}$

42. $\sqrt{25t^2}$

43. $\sqrt{(-12c)^2}$

44. $\sqrt{(-9d)^2}$

45. $\sqrt{(p + 3)^2}$

46. $\sqrt{(2 - x)^2}$

47. $\sqrt{x^2 - 4x + 4}$

48. $\sqrt{9t^2 - 30t + 25}$

c Simplify.

49. $\sqrt[3]{27}$

50. $-\sqrt[3]{64}$

51. $\sqrt[3]{-64x^3}$

52. $\sqrt[3]{-125y^3}$

53. $\sqrt[3]{-216}$

54. $-\sqrt[3]{-1000}$

55. $\sqrt[3]{0.343(x + 1)^3}$

56. $\sqrt[3]{0.000008(y - 2)^3}$

For the given function, find the indicated function values.

57. $f(x) = \sqrt[3]{x + 1}$; $f(7)$, $f(26)$, $f(-9)$, and $f(-65)$

58. $g(x) = -\sqrt[3]{2x - 1}$; $g(-62)$, $g(0)$, $g(-13)$, and $g(63)$

59. $f(x) = -\sqrt[3]{3x + 1}$; $f(0)$, $f(-7)$, $f(21)$, and $f(333)$

60. $g(t) = \sqrt[3]{t - 3}$; $g(30)$, $g(-5)$, $g(1)$, and $g(67)$

d Find the following. Assume that letters can represent *any* real number.

61. $\sqrt[4]{625}$

62. $-\sqrt[4]{256}$

63. $\sqrt[5]{-1}$

64. $\sqrt[5]{-32}$

65. $\sqrt[5]{-\frac{32}{243}}$

66. $\sqrt[5]{-\frac{1}{32}}$

67. $\sqrt[6]{x^6}$

68. $\sqrt[8]{y^8}$

69. $\sqrt[4]{(5a)^4}$

70. $\sqrt[4]{(7b)^4}$

71. $\sqrt[10]{(-6)^{10}}$

72. $\sqrt[12]{(-10)^{12}}$

73. $\sqrt[414]{(a + b)^{414}}$

74. $\sqrt[1999]{(2a + b)^{1999}}$

75. $\sqrt[7]{y^7}$

76. $\sqrt[3]{(-6)^3}$

77. $\sqrt[5]{(x - 2)^5}$

78. $\sqrt[9]{(2xy)^9}$

Skill Maintenance

Solve. [5.8b]

79. $x^2 + x - 2 = 0$

80. $x^2 + x = 0$

81. $4x^2 - 49 = 0$

82. $2x^2 - 26x + 72 = 0$

83. $3x^2 + x = 10$

84. $4x^2 - 20x + 25 = 0$

85. $4x^3 - 20x^2 + 25x = 0$

86. $x^3 - x^2 = 0$

Simplify.

87. $(a^3b^2c^5)^3$ [4.2a]

88. $(5a^7b^8)(2a^3b)$ [4.1d]

Synthesis

89. ◆ Does the nth root of x^2 always exist? Why or why not?

90. ◆ Explain how to formulate a radical expression that can be used to define a function f with a domain of $\{x \mid x \leq 5\}$.

91. Find the domain of

$$f(x) = \frac{\sqrt{x + 3}}{\sqrt{2 - x}}.$$

92. Use a grapher to check your answers to Exercises 33, 37, and 39.

93. Use only the graph of $f(x) = \sqrt{x}$, shown below, to approximate $\sqrt{3}$, $\sqrt{5}$, and $\sqrt{10}$. Answers may vary.

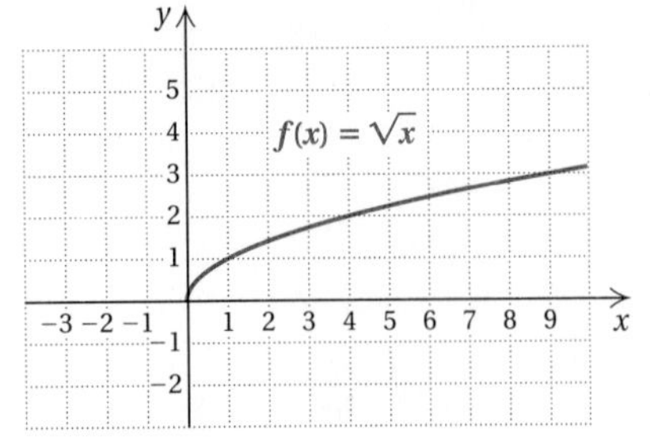

94. Use only the graph of $f(x) = \sqrt[3]{x}$, shown below, to approximate $\sqrt[3]{4}$, $\sqrt[3]{6}$, and $\sqrt[3]{-5}$. Answers may vary.

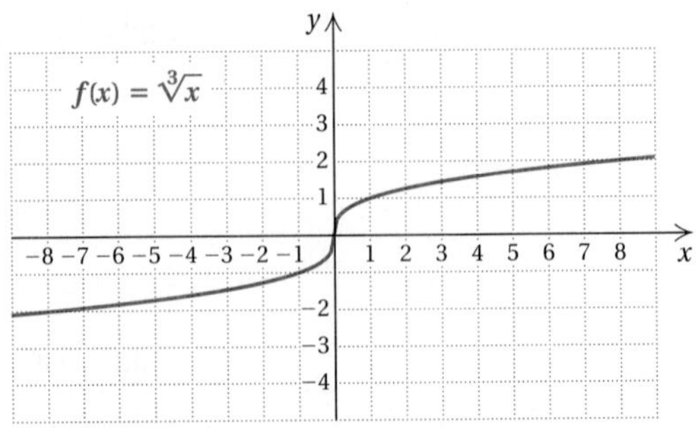

95. Use the TABLE, TRACE, and GRAPH features of a grapher to find the domain and the range of each of the following functions.

a) $f(x) = \sqrt[3]{x}$

b) $g(x) = \sqrt[3]{4x - 5}$

c) $q(x) = 2 - \sqrt{x + 3}$

d) $h(x) = \sqrt[4]{x}$

e) $t(x) = \sqrt[4]{x - 3}$

Use the function concerning skid length to determine a safe distance from which to follow another vehicle.

10.2 Rational Numbers as Exponents

In this section, we give meaning to expressions such as $a^{1/3}$, $7^{-1/2}$, and $(3x)^{0.84}$, which have rational numbers as exponents. We will see that using such notation can help simplify certain radical expressions.

Objectives

a Write expressions with or without rational exponents, and simplify, if possible.

b Write expressions without negative exponents, and simplify, if possible.

c Use the laws of exponents with rational exponents.

d Use rational exponents to simplify radical expressions.

For Extra Help

TAPE 16 MAC WIN CD-ROM

a Rational Exponents

Expressions like $a^{1/2}$, $5^{-1/4}$, and $(2y)^{4/5}$ have not yet been defined. We will define such expressions so that the general properties of exponents hold.

Consider $a^{1/2} \cdot a^{1/2}$. If we want to multiply by adding exponents, it must follow that $a^{1/2} \cdot a^{1/2} = a^{1/2+1/2}$, or a^1. Thus we should define $a^{1/2}$ to be a square root of a. Similarly, $a^{1/3} \cdot a^{1/3} \cdot a^{1/3} = a^{1/3+1/3+1/3}$, or a^1, so $a^{1/3}$ should be defined to mean $\sqrt[3]{a}$.

> For any nonnegative real number a and any natural number index n ($n \neq 1$), $a^{1/n}$ means $\sqrt[n]{a}$ (the nonnegative nth root of a).

Whenever we use rational exponents, we assume that the bases are nonnegative.

Examples Rewrite without rational exponents, and simplify, if possible.

1. $x^{1/2} = \sqrt{x}$
2. $27^{1/3} = \sqrt[3]{27} = 3$
3. $(abc)^{1/5} = \sqrt[5]{abc}$

Do Exercises 1–5.

Examples Rewrite with rational exponents.

4. $\sqrt[5]{7xy} = (7xy)^{1/5}$ — We need parentheses around the radicand here.

5. $8\sqrt[3]{xy} = 8(xy)^{1/3}$

6. $\sqrt[7]{\dfrac{x^3y}{9}} = \left(\dfrac{x^3y}{9}\right)^{1/7}$

Do Exercises 6–9.

How should we define $a^{2/3}$? If the general properties of exponents are to hold, we have $a^{2/3} = (a^{1/3})^2$, or $(\sqrt[3]{a})^2$, or $\sqrt[3]{a^2}$. We define this accordingly.

> For any natural numbers m and n ($n \neq 1$) and any nonnegative real number a,
>
> $a^{m/n}$ means $\sqrt[n]{a^m}$, or $(\sqrt[n]{a})^m$.

Examples Rewrite without rational exponents, and simplify, if possible.

7. $(27)^{2/3} = \sqrt[3]{27^2} = (\sqrt[3]{27})^2 = 3^2 = 9$

8. $4^{3/2} = \sqrt[2]{4^3} = (\sqrt[2]{4})^3 = 2^3 = 8$

Do Exercises 10–12.

Rewrite without rational exponents, and simplify, if possible.

1. $y^{1/4}$
2. $(3a)^{1/2}$
3. $16^{1/4}$
4. $(125)^{1/3}$
5. $(a^3b^2c)^{1/5}$

Rewrite with rational exponents.

6. $\sqrt[3]{19ab}$
7. $19\sqrt[3]{ab}$
8. $\sqrt[5]{\dfrac{x^2y}{16}}$
9. $7\sqrt[4]{2ab}$

Rewrite without rational exponents, and simplify, if possible.

10. $x^{3/2}$
11. $8^{2/3}$
12. $4^{5/2}$

Rewrite with rational exponents.

13. $(\sqrt[3]{7abc})^4$
14. $\sqrt[5]{6^7}$

Answers on page A-42

Calculator Spotlight

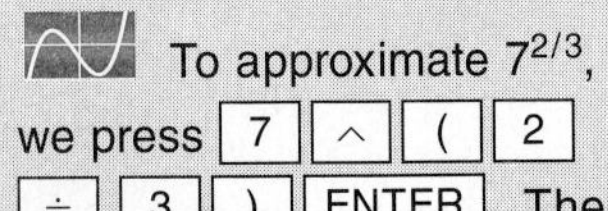

To approximate $7^{2/3}$, we press 7 ^ (2 ÷ 3) ENTER. The display will give an approximation like 3.65930571.

On some graphers, such as the TI-82, to graph radical functions of the form $f(x) = x^{m/n}$, where $m > 1$, we must enter $y = (x^m)^{1/n}$ or $y = (x^{1/n})^m$ in order to see the graph for negative values of x.

Exercises

Approximate each of the following.

1. $5^{2/3}$
2. $8^{4/5}$
3. $29^{-3/8}$
4. $73^{0.56}$
5. $34^{-2.78}$

Graph each of the following and compare.

6. $f(x) = x^{1/2}$
7. $g(x) = x$
8. $t(x) = x^{3/2}$
9. $f(x) = x^{-1/2}$
10. $t(x) = x^{-3/2}$
11. $f(x) = x^{2/3}$

Rewrite with positive exponents, and simplify, if possible.

15. $16^{-1/4}$

16. $(3xy)^{-7/8}$

17. $81^{-3/4}$

18. $7p^{3/4}q^{-6/5}$

19. $\left(\dfrac{11m}{7n}\right)^{-2/3}$

Answers on page A-42

Examples Rewrite with rational exponents.

The index becomes the denominator of the rational exponent.

9. $\sqrt[3]{9^4} = 9^{4/3}$

10. $(\sqrt[4]{7xy})^5 = (7xy)^{5/4}$

Do Exercises 13 and 14 on the preceding page.

b Negative Rational Exponents

Negative rational exponents have a meaning similar to that of negative integer exponents.

> For any rational number m/n and any positive real number a,
>
> $$a^{-m/n} \text{ means } \frac{1}{a^{m/n}},$$
>
> that is, $a^{m/n}$ and $a^{-m/n}$ are reciprocals.

Examples Rewrite with positive exponents, and simplify, if possible.

11. $9^{-1/2} = \dfrac{1}{9^{1/2}} = \dfrac{1}{\sqrt{9}} = \dfrac{1}{3}$

12. $(5xy)^{-4/5} = \dfrac{1}{(5xy)^{4/5}}$ $(5xy)^{-4/5}$ **is the reciprocal of** $(5xy)^{4/5}$.

13. $64^{-2/3} = \dfrac{1}{64^{2/3}} = \dfrac{1}{(\sqrt[3]{64})^2} = \dfrac{1}{4^2} = \dfrac{1}{16}$

14. $4x^{-2/3}y^{1/5} = 4 \cdot \dfrac{1}{x^{2/3}} \cdot y^{1/5} = \dfrac{4y^{1/5}}{x^{2/3}}$

15. $\left(\dfrac{3r}{7s}\right)^{-5/2} = \left(\dfrac{7s}{3r}\right)^{5/2}$ **Since** $\left(\dfrac{a}{b}\right)^{-n} = \left(\dfrac{b}{a}\right)^n$

Do Exercises 15–19.

c Laws of Exponents

The same laws hold for rational-number exponents as for integer exponents. We list them for review.

> For any real number a and any rational exponents m and n:
>
> 1. $a^m \cdot a^n = a^{m+n}$ — In multiplying, we can add exponents if the bases are the same.
> 2. $\dfrac{a^m}{a^n} = a^{m-n}$ — In dividing, we can subtract exponents if the bases are the same.
> 3. $(a^m)^n = a^{m \cdot n}$ — To raise a power to a power, we can multiply the exponents.
> 4. $(ab)^m = a^m b^m$ — To raise a product to a power, we can raise each factor to the power.
> 5. $\left(\dfrac{a}{b}\right)^n = \dfrac{a^n}{b^n}$ — To raise a quotient to a power, we can raise both the numerator and the denominator to the power.

Examples Use the laws of exponents to simplify.

16. $3^{1/5} \cdot 3^{3/5} = 3^{1/5+3/5} = 3^{4/5}$ — Adding exponents

17. $\dfrac{7^{1/4}}{7^{1/2}} = 7^{1/4-1/2} = 7^{1/4-2/4} = 7^{-1/4} = \dfrac{1}{7^{1/4}}$ — Subtracting exponents

18. $(7.2^{2/3})^{3/4} = 7.2^{2/3 \cdot 3/4} = 7.2^{6/12}$ — Multiplying exponents

$= 7.2^{1/2}$

19. $(a^{-1/3}b^{2/5})^{1/2} = a^{-1/3 \cdot 1/2} \cdot b^{2/5 \cdot 1/2}$ — Raising a product to a power and multiplying exponents

$= a^{-1/6}b^{1/5} = \dfrac{b^{1/5}}{a^{1/6}}$

Do Exercises 20–23.

d Simplifying Radical Expressions

Rational exponents can be used to simplify some radical expressions. The procedure is as follows.

1. Convert radical expressions to exponential expressions.
2. Use arithmetic and the laws of exponents to simplify.
3. Convert back to radical notation when appropriate.

Important: This procedure works only when all expressions under radicals are nonnegative since rational exponents are not defined otherwise. No absolute-value signs will be needed.

Examples Use rational exponents to simplify.

20. $\sqrt[6]{x^3} = x^{3/6}$ — Converting to an exponential expression

$= x^{1/2}$ — Simplifying the exponent

$= \sqrt{x}$ — Converting back to radical notation

21. $\sqrt[6]{4} = 4^{1/6}$ — Converting to exponential notation

$= (2^2)^{1/6}$ — Renaming 4 as 2^2

$= 2^{2/6}$ — Using $(a^m)^n = a^{mn}$; multiplying exponents

$= 2^{1/3}$ — Simplifying the exponent

$= \sqrt[3]{2}$ — Converting back to radical notation

Do Exercises 24–26.

Example 22 Use rational exponents to simplify: $\sqrt[8]{a^2b^4}$.

$\sqrt[8]{a^2b^4} = (a^2b^4)^{1/8}$ — Converting to exponential notation

$= a^{2/8} \cdot b^{4/8}$ — Using $(ab)^n = a^nb^n$

$= a^{1/4} \cdot b^{1/2}$ — Simplifying the exponents

$= a^{1/4} \cdot b^{2/4}$ — Rewriting $\frac{1}{2}$ with a denominator of 4

$= (ab^2)^{1/4}$ — Using $a^nb^n = (ab)^n$

$= \sqrt[4]{ab^2}$ — Converting back to radical notation

Do Exercises 27–29.

Use the laws of exponents to simplify.

20. $7^{1/3} \cdot 7^{3/5}$

21. $\dfrac{5^{7/6}}{5^{5/6}}$

22. $(9^{3/5})^{2/3}$

23. $(p^{-2/3}q^{1/4})^{1/2}$

Use rational exponents to simplify.

24. $\sqrt[4]{a^2}$

25. $\sqrt[4]{x^4}$

26. $\sqrt[6]{8}$

Use rational exponents to simplify.

27. $\sqrt[12]{x^3y^6}$

28. $\sqrt[6]{a^{12}b^3}$

29. $\sqrt[5]{a^5b^{10}}$

Answers on page A-42

30. Use rational exponents to write a single radical expression:

$\sqrt[4]{7} \cdot \sqrt{3}$.

Write a single radical expression.

31. $x^{2/3}y^{1/2}z^{5/6}$

32. $\dfrac{a^{1/2}b^{3/8}}{a^{1/4}b^{1/8}}$

Use rational exponents to simplify.

33. $\sqrt[14]{(5m)^2}$

34. $\sqrt[18]{m^3}$

35. $(\sqrt[6]{a^5b^3c})^{24}$

36. $\sqrt[5]{\sqrt{x}}$

Answers on page A-42

We can use properties of rational exponents to write a single radical expression for a product or a quotient.

Example 23 Use rational exponents to write a single radical expression for $\sqrt[3]{5} \cdot \sqrt{2}$.

$\sqrt[3]{5} \cdot \sqrt{2} = 5^{1/3} \cdot 2^{1/2}$ Converting to exponential notation

$= 5^{2/6} \cdot 2^{3/6}$ Rewriting so that exponents have a common denominator

$= (5^2 \cdot 2^3)^{1/6}$ Using $a^nb^n = (ab)^n$

$= \sqrt[6]{5^2 \cdot 2^3}$ Converting back to radical notation

$= \sqrt[6]{200}$ Multiplying under the radical

Do Exercise 30.

Example 24 Write a single radical expression for $a^{1/2}b^{-1/2}c^{5/6}$.

$a^{1/2}b^{-1/2}c^{5/6} = a^{3/6}b^{-3/6}c^{5/6}$ Rewriting so that exponents have a common denominator

$= (a^3b^{-3}c^5)^{1/6}$ Using $a^nb^n = (ab)^n$

$= \sqrt[6]{a^3b^{-3}c^5}$ Converting to radical notation

Do Exercises 31 and 32.

Examples Use rational exponents to simplify.

25. $\sqrt[6]{(5x)^3} = (5x)^{3/6}$ Converting to exponential notation

$= (5x)^{1/2}$ Simplifying the exponent

$= \sqrt{5x}$ Converting back to radical notation

26. $\sqrt[5]{t^{20}} = t^{20/5}$ Converting to exponential notation

$= t^4$ Simplifying the exponent

27. $(\sqrt[3]{pq^2c})^{12} = (pq^2c)^{12/3}$ Converting to exponential notation

$= (pq^2c)^4$ Simplifying the exponent

$= p^4q^8c^4$ Using $(ab)^n = a^nb^n$

28. $\sqrt{\sqrt[3]{x}} = \sqrt{x^{1/3}}$ Converting the radicand to exponential notation

$= (x^{1/3})^{1/2}$ Try to go directly to this step.

$= x^{1/6}$ Multiplying exponents

$= \sqrt[6]{x}$ Converting back to radical notation

Do Exercises 33–36.

Exercise Set 10.2

a Rewrite without rational exponents, and simplify, if possible.

1. $y^{1/7}$ **2.** $x^{1/6}$ **3.** $8^{1/3}$ **4.** $16^{1/2}$ **5.** $(a^3b^3)^{1/5}$

6. $(x^2y^2)^{1/3}$ **7.** $16^{3/4}$ **8.** $4^{7/2}$ **9.** $49^{3/2}$ **10.** $27^{4/3}$

Rewrite with rational exponents.

11. $\sqrt{17}$ **12.** $\sqrt{x^3}$ **13.** $\sqrt[3]{18}$ **14.** $\sqrt[3]{23}$ **15.** $\sqrt[5]{xy^2z}$

16. $\sqrt[7]{x^3y^2z^2}$ **17.** $(\sqrt{3mn})^3$ **18.** $(\sqrt[3]{7xy})^4$ **19.** $(\sqrt[7]{8x^2y})^5$ **20.** $(\sqrt[6]{2a^5b})^7$

b Rewrite with positive exponents, and simplify, if possible.

21. $27^{-1/3}$ **22.** $100^{-1/2}$ **23.** $100^{-3/2}$ **24.** $16^{-3/4}$ **25.** $x^{-1/4}$

26. $y^{-1/7}$ **27.** $(2rs)^{-3/4}$ **28.** $(5xy)^{-5/6}$ **29.** $2a^{3/4}b^{-1/2}c^{2/3}$ **30.** $5x^{-2/3}y^{4/5}$

31. $\left(\frac{7x}{8yz}\right)^{-3/5}$ **32.** $\left(\frac{2ab}{3c}\right)^{-5/6}$ **33.** $\frac{1}{x^{-2/3}}$ **34.** $\frac{1}{a^{-7/8}}$ **35.** $2^{-1/3}x^4y^{-2/7}$

36. $3^{-5/2}a^3b^{-7/3}$ **37.** $\frac{7x}{\sqrt[3]{z}}$ **38.** $\frac{6a}{\sqrt[4]{b}}$ **39.** $\frac{5a}{3c^{-1/2}}$ **40.** $\frac{2z}{5x^{-1/3}}$

c Use the laws of exponents to simplify. Write the answers with positive exponents.

41. $5^{3/4} \cdot 5^{1/8}$ **42.** $11^{2/3} \cdot 11^{1/2}$ **43.** $\frac{7^{5/8}}{7^{3/8}}$ **44.** $\frac{3^{5/8}}{3^{-1/8}}$ **45.** $\frac{4.9^{-1/6}}{4.9^{-2/3}}$

46. $\frac{2.3^{-3/10}}{2.3^{-1/5}}$ **47.** $(6^{3/8})^{2/7}$ **48.** $(3^{2/9})^{3/5}$ **49.** $a^{2/3} \cdot a^{5/4}$ **50.** $x^{3/4} \cdot x^{2/3}$

51. $(a^{2/3} \cdot b^{5/8})^4$

52. $(x^{-1/3} \cdot y^{-2/5})^{-15}$

53. $(x^{2/3})^{-3/7}$

54. $(a^{-3/2})^{2/9}$

d Use rational exponents to simplify. Write the answer in radical notation if appropriate.

55. $\sqrt[6]{a^2}$

56. $\sqrt[6]{t^4}$

57. $\sqrt[3]{x^{15}}$

58. $\sqrt[4]{a^{12}}$

59. $\sqrt[6]{x^{-18}}$

60. $\sqrt[5]{a^{-10}}$

61. $(\sqrt[3]{ab})^{15}$

62. $(\sqrt[7]{cd})^{14}$

63. $\sqrt[4]{32}$

64. $\sqrt[6]{81}$

65. $\sqrt[6]{4x^2}$

66. $\sqrt[3]{8y^6}$

67. $\sqrt{x^4y^6}$

68. $\sqrt[4]{16x^4y^2}$

69. $\sqrt[5]{32c^{10}d^{15}}$

Use rational exponents to write a single radical expression.

70. $\sqrt[3]{3}\sqrt{3}$

71. $\sqrt[3]{7} \cdot \sqrt[4]{5}$

72. $\sqrt[7]{11} \cdot \sqrt[6]{13}$

73. $\sqrt[4]{5} \cdot \sqrt[5]{7}$

74. $\sqrt[3]{y}\sqrt[5]{3y}$

75. $\sqrt{x}\sqrt[3]{2x}$

76. $(\sqrt[3]{x^2y^5})^{12}$

77. $(\sqrt[5]{a^2b^4})^{15}$

78. $\sqrt[4]{\sqrt{x}}$

79. $\sqrt[3]{\sqrt[6]{m}}$

80. $a^{2/3} \cdot b^{3/4}$

81. $x^{1/3} \cdot y^{1/4} \cdot z^{1/6}$

82. $\dfrac{x^{8/15} \cdot y^{7/5}}{x^{1/3} \cdot y^{-1/5}}$

83. $\left(\dfrac{c^{-4/5}d^{5/9}}{c^{3/10}d^{1/6}}\right)^3$

84. $\sqrt[3]{\sqrt[4]{xy}}$

Skill Maintenance

Solve. [6.7a]

85. $A = \dfrac{ab}{a+b}$, for a

86. $Q = \dfrac{st}{s-t}$, for s

87. $Q = \dfrac{st}{s-t}$, for t

88. $\dfrac{1}{t} = \dfrac{1}{a} - \dfrac{1}{b}$, for b

Synthesis

89. ◆ Find the domain of
$f(x) = (x+5)^{1/2}(x+7)^{-1/2}$
and explain how you found your answer.

90. ◆ Explain why $\sqrt[3]{x^6} = x^2$ for any value of x, but $\sqrt[2]{x^6} = x^3$ only when $x \geq 0$.

91. Use the SIMULTANEOUS mode to graph
$y_1 = x^{1/2}, \quad y_2 = 3x^{2/5}, \quad y_3 = x^{4/7}, \quad y_4 = \frac{1}{5}x^{3/4}.$
Then, looking only at coordinates, match each graph with its equation.

92. Simplify:
$(\sqrt[10]{\sqrt[5]{x^{15}}})^5(\sqrt[5]{\sqrt[10]{x^{15}}})^5.$

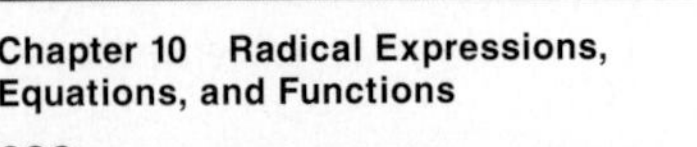

Investigate the effect of the order of rational exponents on exponential functions.

10.3 Simplifying Radical Expressions

a Multiplying and Simplifying Radical Expressions

Note that $\sqrt{4}\sqrt{25} = 2 \cdot 5 = 10$. Also $\sqrt{4 \cdot 25} = \sqrt{100} = 10$. Likewise,

$$\sqrt[3]{27}\sqrt[3]{8} = 3 \cdot 2 = 6 \quad \text{and} \quad \sqrt[3]{27 \cdot 8} = \sqrt[3]{216} = 6.$$

These examples suggest the following.

THE PRODUCT RULE FOR RADICALS

For any nonnegative real numbers a and b and any index k,

$$\sqrt[k]{a} \cdot \sqrt[k]{b} = \sqrt[k]{a \cdot b}.$$

(To multiply, multiply the radicands.)

Note that the index k must be the same throughout.

Examples Multiply.

1. $\sqrt{3} \cdot \sqrt{5} = \sqrt{3 \cdot 5} = \sqrt{15}$
2. $\sqrt{5a}\sqrt{2b} = \sqrt{5a \cdot 2b} = \sqrt{10ab}$
3. $\sqrt[3]{4}\sqrt[3]{5} = \sqrt[3]{4 \cdot 5} = \sqrt[3]{20}$
4. $\sqrt[4]{\frac{y}{5}}\sqrt[4]{\frac{7}{x}} = \sqrt[4]{\frac{y}{5} \cdot \frac{7}{x}} = \sqrt[4]{\frac{7y}{5x}}$

A common error is to omit the index in the answer.

Do Exercises 1–4.

Keep in mind that the product rule can be used only when the indexes are the same. When indexes differ, we can use rational exponents as we did in Section 10.2.

Example 5 Multiply: $\sqrt{5x} \cdot \sqrt[4]{3y}$.

$\sqrt{5x} \cdot \sqrt[4]{3y} = (5x)^{1/2}(3y)^{1/4}$ — Converting to exponential notation

$= (5x)^{2/4}(3y)^{1/4}$ — Rewriting so that exponents have a common denominator

$= [(5x)^2(3y)]^{1/4}$ — Using $a^nb^n = (ab)^n$

$= \sqrt[4]{(25x^2)(3y)}$ — Squaring $5x$ and converting back to radical notation

$= \sqrt[4]{75x^2y}$ — Multiplying under the radical

Do Exercises 5 and 6.

We can reverse the product rule to simplify a product. We simplify the root of a product by taking the root of each factor separately.

For any nonnegative real numbers a and b and any index k,

$$\sqrt[k]{ab} = \sqrt[k]{a} \cdot \sqrt[k]{b}.$$

(Take the kth root of each factor separately.)

Objectives

a Multiply and simplify radical expressions.

b Divide and simplify radical expressions.

For Extra Help

TAPE 16

MAC
WIN

CD-ROM

Multiply.

1. $\sqrt{19}\sqrt{7}$

2. $\sqrt{3p}\sqrt{7q}$

3. $\sqrt[4]{403}\sqrt[4]{7}$

4. $\sqrt[3]{\frac{5}{p}} \cdot \sqrt[3]{\frac{2}{q}}$

Multiply.

5. $\sqrt{5}\sqrt[3]{2}$

6. $\sqrt{x}\sqrt[3]{5y}$

Answers on page A-42

Simplify by factoring.

7. $\sqrt{32}$

8. $\sqrt[3]{80}$

Answers on page A-42

This process shows a way to factor and thus simplify radical expressions. Consider $\sqrt{20}$. The number 20 has the factor 4, which is a perfect square. Therefore,

$$\sqrt{20} = \sqrt{4 \cdot 5} \quad \text{Factoring the radicand (4 is a perfect square)}$$
$$= \sqrt{4} \cdot \sqrt{5} \quad \text{Factoring into two radicals}$$
$$= 2\sqrt{5}. \quad \text{Taking the square root of 4}$$

> To simplify a radical expression by factoring:
>
> 1. Look for the largest factors of the radicand that are perfect kth powers (where k is the index).
> 2. Then take the kth root of the resulting factors.
> 3. A radical expression, with index k, is *simplified* when its radicand has no factors that are perfect kth powers.

Examples Simplify by factoring.

6. $\sqrt{50} = \sqrt{25 \cdot 2} = \sqrt{25} \cdot \sqrt{2} = 5\sqrt{2}$

This factor is a perfect square.

7. $\sqrt[3]{32} = \sqrt[3]{8 \cdot 4} = \sqrt[3]{8} \cdot \sqrt[3]{4} = 2\sqrt[3]{4}$

This factor is a perfect cube (third power).

8. $\sqrt[4]{48} = \sqrt[4]{16 \cdot 3} = \sqrt[4]{16} \cdot \sqrt[4]{3} = 2\sqrt[4]{3}$

Do Exercises 7 and 8.

> In many situations, expressions under radicals never represent negative numbers. In such cases, absolute-value notation is not necessary. For this reason, we will henceforth assume that *all expressions under radicals are nonnegative.*

Examples Simplify by factoring. Assume that all expressions under radicals represent nonnegative numbers.

9. $\sqrt{5x^2} = \sqrt{x^2 \cdot 5}$ Factoring the radicand

$= \sqrt{x^2} \cdot \sqrt{5}$ Factoring into two radicals

$= x \cdot \sqrt{5}$ Taking the square root of x^2

10. $\sqrt{18x^2y} = \sqrt{9x^2 \cdot 2y}$

$= \sqrt{9x^2} \cdot \sqrt{2y}$

$= 3x\sqrt{2y}$

Absolute-value notation is not needed because expressions under radicals are not negative.

11. $\sqrt{216x^5y^3} = \sqrt{36 \cdot 6 \cdot x^4 \cdot x \cdot y^2 \cdot y}$ Factoring the radicand

$= \sqrt{36 \cdot x^4 \cdot y^2 \cdot 6 \cdot x \cdot y}$

$= \sqrt{36}\sqrt{x^4}\sqrt{y^2}\sqrt{6xy}$ Factoring into several radicals

$= 6x^2y\sqrt{6xy}$ Taking square roots

Note: Had we not seen in Example 11 that $216 = 36 \cdot 6$, where 36 is the largest square factor of 216, we could have found the prime factorization

$$2 \cdot 2 \cdot 2 \cdot 3 \cdot 3 \cdot 3.$$

Each pair of factors makes a square, so

$$\sqrt{2 \cdot 2 \cdot 2 \cdot 3 \cdot 3 \cdot 3} = \sqrt{2^2 \cdot 3^2 \cdot 2 \cdot 3} = 2 \cdot 3\sqrt{2 \cdot 3} = 6\sqrt{6}.$$

12. $\sqrt[3]{16a^7b^{11}} = \sqrt[3]{8 \cdot 2 \cdot a^6 \cdot a \cdot b^9 \cdot b^2}$ **Factoring the radicand. The index is 3, so we look for the largest powers that are multiples of 3 because these are perfect cubes.**

$= \sqrt[3]{8} \cdot \sqrt[3]{a^6} \cdot \sqrt[3]{b^9} \cdot \sqrt[3]{2ab^2}$ **Factoring into radicals**

$= 2a^2b^3\sqrt[3]{2ab^2}$ **Taking cube roots**

Do Exercises 9–14.

Sometimes after we have multiplied, we can then simplify by factoring.

Examples Multiply and simplify. Assume that all expressions under radicals represent nonnegative numbers.

13. $\sqrt{20}\sqrt{8} = \sqrt{20 \cdot 8} = \sqrt{4 \cdot 5 \cdot 4 \cdot 2} = 4\sqrt{10}$

14. $3\sqrt[3]{25} \cdot 2\sqrt[3]{5} = 6 \cdot \sqrt[3]{25 \cdot 5} = 6 \cdot \sqrt[3]{125}$

$= 6 \cdot 5 = 30$

15. $\sqrt[3]{18y^3}\sqrt[3]{4x^2} = \sqrt[3]{18y^3 \cdot 4x^2}$ **Multiplying radicands**

$= \sqrt[3]{72y^3x^2}$

$= \sqrt[3]{8y^3 \cdot 9x^2}$ **Factoring the radicand**

$= \sqrt[3]{8y^3}\sqrt[3]{9x^2}$ **Factoring into two radicals**

$= 2y\sqrt[3]{9x^2}$ **Taking the cube root**

Do Exercises 15–18.

b Dividing and Simplifying Radical Expressions

Note that $\dfrac{\sqrt[3]{27}}{\sqrt[3]{8}} = \dfrac{3}{2}$ and that $\sqrt[3]{\dfrac{27}{8}} = \dfrac{3}{2}$. This example suggests the following.

> **THE QUOTIENT RULE FOR RADICALS**
>
> For any nonnegative number a, any positive number b, and any index k,
>
> $$\frac{\sqrt[k]{a}}{\sqrt[k]{b}} = \sqrt[k]{\frac{a}{b}}.$$
>
> (To divide, divide the radicands. After doing this, you can sometimes simplify by taking roots.)

Examples Divide and simplify. Assume that all expressions under radicals represent positive numbers.

16. $\dfrac{\sqrt{80}}{\sqrt{5}} = \sqrt{\dfrac{80}{5}} = \sqrt{16} = 4$ We divide the radicands.

Simplify by factoring. Assume that all expressions under radicals represent nonnegative numbers.

9. $\sqrt{300}$

10. $\sqrt{36y^2}$

11. $\sqrt{12a^2b}$

12. $\sqrt{12ab^3c^2}$

13. $\sqrt[3]{16}$

14. $\sqrt[3]{81x^4y^8}$

Multiply and simplify. Assume that all expressions under radicals represent nonnegative numbers.

15. $\sqrt{3}\sqrt{6}$

16. $\sqrt{18y}\sqrt{14y}$

17. $\sqrt[3]{3x^2y}\sqrt[3]{36x}$

18. $\sqrt{7a}\sqrt{21b}$

Answers on page A-42

Divide and simplify. Assume that all expressions under radicals represent positive numbers.

19. $\dfrac{\sqrt{75}}{\sqrt{3}}$

20. $\dfrac{14\sqrt{128xy}}{2\sqrt{2}}$

21. $\dfrac{\sqrt{50a^3}}{\sqrt{2a}}$

22. $\dfrac{4\sqrt[3]{250}}{7\sqrt[3]{2}}$

Simplify by taking the roots of the numerator and the denominator. Assume that all expressions under radicals represent positive numbers.

23. $\sqrt{\dfrac{25}{36}}$

24. $\sqrt{\dfrac{x^2}{100}}$

25. $\sqrt[3]{\dfrac{54x^5}{125}}$

26. Divide and simplify:

$$\frac{\sqrt[4]{x^3y^2}}{\sqrt[3]{x^2y}}.$$

Answers on page A-42

17. $\dfrac{5\sqrt[3]{32}}{\sqrt[3]{2}} = 5\sqrt[3]{\dfrac{32}{2}} = 5\sqrt[3]{16} = 5\sqrt[3]{8 \cdot 2} = 5\sqrt[3]{8}\sqrt[3]{2} = 5 \cdot 2\sqrt[3]{2} = 10\sqrt[3]{2}$

18. $\dfrac{\sqrt{72xy}}{2\sqrt{2}} = \dfrac{1}{2}\dfrac{\sqrt{72xy}}{\sqrt{2}} = \dfrac{1}{2}\sqrt{\dfrac{72xy}{2}} = \dfrac{1}{2}\sqrt{36xy} = \dfrac{1}{2}\sqrt{36}\sqrt{xy}$

$= \dfrac{1}{2} \cdot 6\sqrt{xy} = 3\sqrt{xy}$

Do Exercises 19–22.

We can reverse the quotient rule to simplify a quotient. We simplify the root of a quotient by taking the roots of the numerator and of the denominator separately.

For any nonnegative number a, any positive number b, and any index k,

$$\sqrt[k]{\frac{a}{b}} = \frac{\sqrt[k]{a}}{\sqrt[k]{b}}.$$

(Take the kth roots of the numerator and of the denominator separately.)

Examples Simplify by taking the roots of the numerator and the denominator. Assume that all expressions under radicals represent positive numbers.

19. $\sqrt[3]{\dfrac{27}{125}} = \dfrac{\sqrt[3]{27}}{\sqrt[3]{125}} = \dfrac{3}{5}$ — We take the cube root of the numerator and of the denominator.

20. $\sqrt{\dfrac{25}{y^2}} = \dfrac{\sqrt{25}}{\sqrt{y^2}} = \dfrac{5}{y}$ — We take the square root of the numerator and of the denominator.

21. $\sqrt{\dfrac{16x^3}{y^4}} = \dfrac{\sqrt{16x^3}}{\sqrt{y^4}} = \dfrac{\sqrt{16x^2 \cdot x}}{\sqrt{y^4}} = \dfrac{\sqrt{16x^2} \cdot \sqrt{x}}{\sqrt{y^4}} = \dfrac{4x\sqrt{x}}{y^2}$

22. $\sqrt[3]{\dfrac{27y^5}{343x^3}} = \dfrac{\sqrt[3]{27y^5}}{\sqrt[3]{343x^3}} = \dfrac{\sqrt[3]{27y^3 \cdot y^2}}{\sqrt[3]{343x^3}} = \dfrac{\sqrt[3]{27y^3} \cdot \sqrt[3]{y^2}}{\sqrt[3]{343x^3}} = \dfrac{3y\sqrt[3]{y^2}}{7x}$

We are assuming that no expression represents 0 or a negative number. Thus we need not be concerned about zero denominators.

Do Exercises 23–25.

When indexes differ, we can use rational exponents.

Example 23 Divide and simplify: $\dfrac{\sqrt[3]{a^2b^4}}{\sqrt{ab}}$.

$\dfrac{\sqrt[3]{a^2b^4}}{\sqrt{ab}} = \dfrac{(a^2b^4)^{1/3}}{(ab)^{1/2}}$ — Converting to exponential notation

$= \dfrac{a^{2/3}b^{4/3}}{a^{1/2}b^{1/2}}$ — Using the product and power rules

$= a^{2/3-1/2}b^{4/3-1/2}$ — Subtracting exponents

$= a^{4/6-3/6}b^{8/6-3/6} = a^{1/6}b^{5/6}$

$= (ab^5)^{1/6} = \sqrt[6]{ab^5}$

Do Exercise 26.

Exercise Set 10.3

a Simplify by factoring. Assume that all expressions under radicals represent nonnegative numbers.

1. $\sqrt{24}$

2. $\sqrt{20}$

3. $\sqrt{90}$

4. $\sqrt{18}$

5. $\sqrt[3]{250}$

6. $\sqrt[3]{108}$

7. $\sqrt{180x^4}$

8. $\sqrt{175y^6}$

9. $\sqrt[3]{54x^8}$

10. $\sqrt[3]{40y^3}$

11. $\sqrt[3]{80t^8}$

12. $\sqrt[3]{108x^5}$

13. $\sqrt[4]{80}$

14. $\sqrt[4]{32}$

15. $\sqrt{32a^2b}$

16. $\sqrt{75p^3q^4}$

17. $\sqrt[4]{243x^8y^{10}}$

18. $\sqrt[4]{162c^4d^6}$

19. $\sqrt[5]{96x^7y^{15}}$

20. $\sqrt[5]{p^{14}q^9r^{23}}$

Multiply and simplify. Assume that all expressions under radicals represent nonnegative numbers.

21. $\sqrt{10}\sqrt{5}$

22. $\sqrt{6}\sqrt{3}$

23. $\sqrt{15}\sqrt{6}$

24. $\sqrt{2}\sqrt{32}$

25. $\sqrt[3]{2}\sqrt[3]{4}$

26. $\sqrt[3]{9}\sqrt[3]{3}$

27. $\sqrt{45}\sqrt{60}$

28. $\sqrt{24}\sqrt{75}$

29. $\sqrt{3x^3}\sqrt{6x^5}$

30. $\sqrt{5a^7}\sqrt{15a^3}$

31. $\sqrt{5b^3}\sqrt{10c^4}$

32. $\sqrt{2x^3y}\sqrt{12xy}$

33. $\sqrt[3]{5a^2}\sqrt[3]{2a}$

34. $\sqrt[3]{7x}\sqrt[3]{3x^2}$

35. $\sqrt[3]{y^4}\sqrt[3]{16y^5}$

36. $\sqrt[3]{s^2t^4}\sqrt[3]{s^4t^6}$

37. $\sqrt[4]{16}\sqrt[4]{64}$

38. $\sqrt[5]{64}\sqrt[5]{16}$

39. $\sqrt{12a^3b}\sqrt{8a^4b^2}$

40. $\sqrt{30x^3y^4}\sqrt{18x^2y^5}$

41. $\sqrt{2}\sqrt[3]{5}$

42. $\sqrt{6}\sqrt[3]{5}$

43. $\sqrt[4]{3}\sqrt{2}$

44. $\sqrt[3]{5}\sqrt[4]{2}$

45. $\sqrt{a}\sqrt[4]{a^3}$

46. $\sqrt[3]{x^2}\sqrt[6]{x^5}$

47. $\sqrt[5]{b^2}\sqrt{b^3}$

48. $\sqrt[4]{a^3}\sqrt[3]{a^2}$

49. $\sqrt{xy^3}\sqrt[3]{x^2y}$

50. $\sqrt[4]{9ab^3}\sqrt{3a^4b}$

b Divide and simplify. Assume that all expressions under radicals represent positive numbers.

51. $\dfrac{\sqrt{90}}{\sqrt{5}}$

52. $\dfrac{\sqrt{98}}{\sqrt{2}}$

53. $\dfrac{\sqrt{35q}}{\sqrt{7q}}$

54. $\dfrac{\sqrt{30x}}{\sqrt{10x}}$

55. $\dfrac{\sqrt[3]{54}}{\sqrt[3]{2}}$

56. $\dfrac{\sqrt[3]{40}}{\sqrt[3]{5}}$

57. $\dfrac{\sqrt{56xy^3}}{\sqrt{8x}}$

58. $\dfrac{\sqrt{52ab^3}}{\sqrt{13a}}$

59. $\dfrac{\sqrt[3]{96a^4b^2}}{\sqrt[3]{12a^2b}}$

60. $\dfrac{\sqrt[3]{189x^5y^7}}{\sqrt[3]{7x^2y^2}}$

61. $\dfrac{\sqrt{128xy}}{2\sqrt{2}}$

62. $\dfrac{\sqrt{48ab}}{2\sqrt{3}}$

63. $\dfrac{\sqrt[4]{48x^9y^{13}}}{\sqrt[4]{3xy^5}}$

64. $\dfrac{\sqrt[5]{64a^{11}b^{28}}}{\sqrt[5]{2ab^2}}$

65. $\dfrac{\sqrt[3]{a}}{\sqrt{a}}$

66. $\dfrac{\sqrt{x}}{\sqrt[4]{x}}$

67. $\dfrac{\sqrt[3]{a^2}}{\sqrt[4]{a}}$

68. $\dfrac{\sqrt[3]{x^2}}{\sqrt[5]{x}}$

69. $\dfrac{\sqrt[4]{x^2y^3}}{\sqrt[3]{xy}}$

70. $\dfrac{\sqrt[5]{a^4b^2}}{\sqrt[3]{ab^2}}$

Simplify.

71. $\sqrt{\dfrac{25}{36}}$

72. $\sqrt{\dfrac{49}{64}}$

73. $\sqrt{\dfrac{16}{49}}$

74. $\sqrt{\dfrac{100}{81}}$

75. $\sqrt[3]{\dfrac{125}{27}}$

76. $\sqrt[3]{\dfrac{343}{1000}}$

77. $\sqrt{\dfrac{49}{y^2}}$

78. $\sqrt{\dfrac{121}{x^2}}$

79. $\sqrt{\dfrac{25y^3}{x^4}}$

80. $\sqrt{\dfrac{36a^5}{b^6}}$

81. $\sqrt[3]{\dfrac{27a^4}{8b^3}}$

82. $\sqrt[3]{\dfrac{64x^7}{216y^6}}$

83. $\sqrt[4]{\frac{81x^4}{16}}$

84. $\sqrt[4]{\frac{81x^4}{y^8z^4}}$

85. $\sqrt[5]{\frac{32x^8}{y^{10}}}$

86. $\sqrt[5]{\frac{32b^{10}}{243a^{20}}}$

87. $\sqrt[6]{\frac{x^{13}}{y^6z^{12}}}$

88. $\sqrt[6]{\frac{p^9q^{24}}{r^{18}}}$

Skill Maintenance

Solve. [5.8b]

89. The sum of a number and its square is 90. Find the number.

90. The base of a triangle is 2 in. longer than the height. The area is 12 in^2. Find the height and the base.

Solve. [6.5a]

91. $\frac{12x}{x-4} - \frac{3x^2}{x+4} = \frac{384}{x^2-16}$

92. $\frac{2}{3} + \frac{1}{t} = \frac{4}{5}$

93. $\frac{18}{x^2-3x} = \frac{2x}{x-3} - \frac{6}{x}$

94. $\frac{4x}{x+5} + \frac{20}{x} = \frac{100}{x^2+5x}$

Synthesis

95. ◆ Is the quotient of two irrational numbers always an irrational number? Why or why not?

96. ◆ Ron is puzzled. When he uses a grapher to graph $y = \sqrt{x} \cdot \sqrt{x}$, he gets the following screen. Explain why Ron did not get the complete line $y_1 = x$.

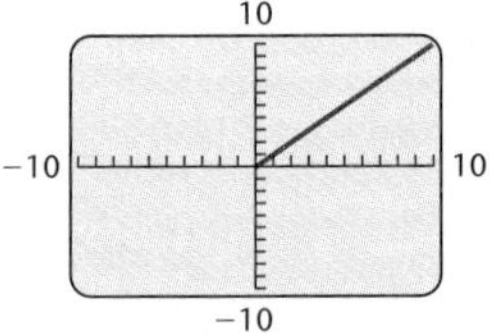

97. *Pendulums.* The *period* of a pendulum is the time it takes to complete one cycle, swinging to and fro. For a pendulum that is L centimeters long, the period T is given by the function

$$T(L) = 2\pi\sqrt{\frac{L}{980}},$$

where T is in seconds. Find, to the nearest hundredth of a second, the period of a pendulum of length **(a)** 65 cm; **(b)** 98 cm; **(c)** 120 cm. Use a calculator's $\boxed{\pi}$ key if possible.

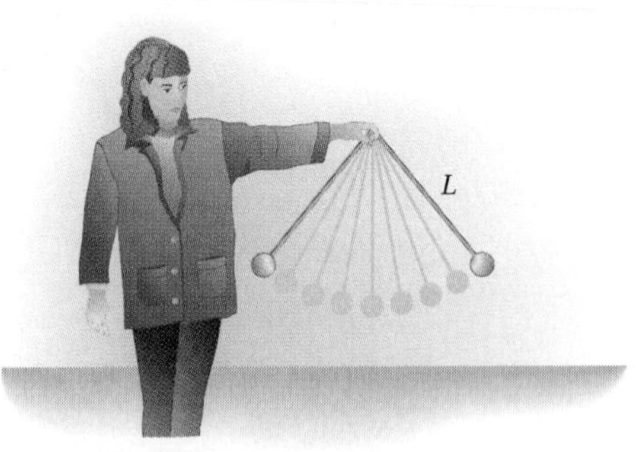

Simplify.

98. $\frac{\sqrt[3]{x^3 - y^3}}{\sqrt[3]{x - y}}$

99. $\frac{\sqrt{44x^2y^9z}\sqrt{22y^9z^6}}{(\sqrt{11xy^8z^2})^2}$

100. Use a grapher to check your answers to Exercises 7, 12, 30, and 54.

10.4 Addition, Subtraction, and More Multiplication

a Addition and Subtraction

Any two real numbers can be added. For example, the sum of 7 and $\sqrt{3}$ can be expressed as $7 + \sqrt{3}$. We cannot simplify this sum. However, when we have **like radicals** (radicals having the same index and radicand), we can use the distributive laws to simplify by collecting like radical terms. For example,

$$7\sqrt{3} + \sqrt{3} = 7\sqrt{3} + 1 \cdot \sqrt{3} = (7 + 1)\sqrt{3} = 8\sqrt{3}.$$

Examples Add or subtract. Simplify by collecting like radical terms, if possible.

1. $6\sqrt{7} + 4\sqrt{7} = (6 + 4)\sqrt{7}$ **Using a distributive law (factoring out $\sqrt{7}$)**
$= 10\sqrt{7}$

2. $8\sqrt[3]{2} - 7x\sqrt[3]{2} + 5\sqrt[3]{2} = (8 - 7x + 5)\sqrt[3]{2}$ **Factoring out $\sqrt[3]{2}$**
$= (13 - 7x)\sqrt[3]{2}$

These parentheses *are* necessary!

3. $6\sqrt[5]{4x} + 4\sqrt[5]{4x} - \sqrt[3]{4x} = (6 + 4)\sqrt[5]{4x} - \sqrt[3]{4x}$
$= 10\sqrt[5]{4x} - \sqrt[3]{4x}$

Note that these expressions have the *same* radicand, but they are *not* like radicals because they do not have the same index.

Do Exercises 1 and 2.

Sometimes we need to simplify radicals by factoring in order to obtain terms with like radicals.

Examples Add or subtract. Simplify by collecting like radical terms, if possible.

4. $3\sqrt{8} - 5\sqrt{2} = 3(\sqrt{4 \cdot 2}) - 5\sqrt{2}$ **Factoring 8**
$= 3\sqrt{4} \cdot \sqrt{2} - 5\sqrt{2}$ **Factoring $\sqrt{4 \cdot 2}$ into two radicals**
$= 3 \cdot 2\sqrt{2} - 5\sqrt{2}$ **Taking the square root of 4**
$= 6\sqrt{2} - 5\sqrt{2}$
$= (6 - 5)\sqrt{2}$ **Collecting like radical terms**
$= \sqrt{2}$

5. $5\sqrt{2} - 4\sqrt{3}$ No simplification possible

6. $5\sqrt[3]{16y^4} + 7\sqrt[3]{2y} = 5\sqrt[3]{8y^3 \cdot 2y} + 7\sqrt[3]{2y}$
$= 5\sqrt[3]{8y^3} \cdot \sqrt[3]{2y} + 7\sqrt[3]{2y}$ **Factoring the first radical**
$= 5 \cdot 2y \cdot \sqrt[3]{2y} + 7\sqrt[3]{2y}$ **Taking the cube root of $8y^3$**
$= 10y\sqrt[3]{2y} + 7\sqrt[3]{2y}$
$= (10y + 7)\sqrt[3]{2y}$ **Collecting like radical terms**

Do Exercises 3–5.

Objectives

a Add or subtract with radical notation and simplify.

b Multiply expressions involving radicals in which some factors contain more than one term.

For Extra Help

TAPE 16

MAC
WIN

CD-ROM

Add or subtract. Simplify by collecting like radical terms, if possible.

1. $5\sqrt{2} + 8\sqrt{2}$

2. $7\sqrt[4]{5x} + 3\sqrt[4]{5x} - \sqrt{7}$

Add or subtract. Simplify by collecting like radical terms, if possible.

3. $7\sqrt{45} - 2\sqrt{5}$

4. $3\sqrt[3]{y^5} + 4\sqrt[3]{y^2} + \sqrt[3]{8y^6}$

5. $\sqrt{25x - 25} - \sqrt{9x - 9}$

Answers on page A-43

Multiply. Assume that all expressions under radicals represent nonnegative numbers.

6. $\sqrt{2}(5\sqrt{3} + 3\sqrt{7})$

7. $\sqrt[3]{a^2}(\sqrt[3]{3a} - \sqrt[3]{2})$

Multiply. Assume that all expressions under radicals represent nonnegative numbers.

8. $(\sqrt{3} - 5\sqrt{2})(2\sqrt{3} + \sqrt{2})$

9. $(\sqrt{a} + 2\sqrt{3})(3\sqrt{b} - 4\sqrt{3})$

Multiply. Assume that all expressions under radicals represent nonnegative numbers.

10. $(\sqrt{2} + \sqrt{5})(\sqrt{2} - \sqrt{5})$

11. $(\sqrt{p} - \sqrt{q})(\sqrt{p} + \sqrt{q})$

Multiply.

12. $(2\sqrt{5} - y)^2$

13. $(3\sqrt{6} + 2)^2$

Answers on page A-43

b More Multiplication

To multiply expressions in which some factors contain more than one term, we use the procedures for multiplying polynomials.

Examples Multiply.

7. $\sqrt{3}(x - \sqrt{5}) = \sqrt{3} \cdot x - \sqrt{3} \cdot \sqrt{5}$ **Using a distributive law**

$= x\sqrt{3} - \sqrt{15}$ **Multiplying radicals**

8. $\sqrt[3]{y}(\sqrt[3]{y^2} + \sqrt[3]{2}) = \sqrt[3]{y} \cdot \sqrt[3]{y^2} + \sqrt[3]{y} \cdot \sqrt[3]{2}$ **Using a distributive law**

$= \sqrt[3]{y^3} + \sqrt[3]{2y}$ **Multiplying radicals**

$= y + \sqrt[3]{2y}$ **Simplifying $\sqrt[3]{y^3}$**

Do Exercises 6 and 7.

Example 9 Multiply: $(4\sqrt{3} + \sqrt{2})(\sqrt{3} - 5\sqrt{2})$.

F O I L

$$(4\sqrt{3} + \sqrt{2})(\sqrt{3} - 5\sqrt{2}) = 4(\sqrt{3})^2 - 20\sqrt{3} \cdot \sqrt{2} + \sqrt{2} \cdot \sqrt{3} - 5(\sqrt{2})^2$$

$$= 4 \cdot 3 - 20\sqrt{6} + \sqrt{6} - 5 \cdot 2$$

$$= 12 - 20\sqrt{6} + \sqrt{6} - 10$$

$$= 2 - 19\sqrt{6}$$ **Collecting like terms**

Example 10 Multiply: $(\sqrt{a} + \sqrt{3})(\sqrt{b} + \sqrt{3})$. Assume that all expressions under radicals represent nonnegative numbers.

$$(\sqrt{a} + \sqrt{3})(\sqrt{b} + \sqrt{3}) = \sqrt{a}\sqrt{b} + \sqrt{a}\sqrt{3} + \sqrt{3}\sqrt{b} + \sqrt{3}\sqrt{3}$$

$$= \sqrt{ab} + \sqrt{3a} + \sqrt{3b} + 3$$

Do Exercises 8 and 9.

Example 11 Multiply: $(\sqrt{5} + \sqrt{7})(\sqrt{5} - \sqrt{7})$.

$$(\sqrt{5} + \sqrt{7})(\sqrt{5} - \sqrt{7}) = (\sqrt{5})^2 - (\sqrt{7})^2$$ **This is now a difference of two squares.**

$$= 5 - 7 = -2$$

Example 12 Multiply: $(\sqrt{a} + \sqrt{b})(\sqrt{a} - \sqrt{b})$. Assume that all expressions under radicals represent nonnegative numbers.

$$(\sqrt{a} + \sqrt{b})(\sqrt{a} - \sqrt{b}) = (\sqrt{a})^2 - (\sqrt{b})^2$$

$$= a - b \leftarrow$$ No radicals

Expressions of the form $\sqrt{a} + \sqrt{b}$ and $\sqrt{a} - \sqrt{b}$ are called **conjugates**. Their product is always an expression that has no radicals.

Do Exercises 10 and 11.

Example 13 Multiply: $(\sqrt{3} + x)^2$.

$$(\sqrt{3} + x)^2 = (\sqrt{3})^2 + 2x\sqrt{3} + x^2$$ **Squaring a binomial**

$$= 3 + 2x\sqrt{3} + x^2$$

Do Exercises 12 and 13.

Exercise Set 10.4

a Add or subtract. Then simplify by collecting like radical terms, if possible. Assume that all expressions under radicals represent nonnegative numbers.

1. $7\sqrt{5} + 4\sqrt{5}$

2. $2\sqrt{3} + 9\sqrt{3}$

3. $6\sqrt[3]{7} - 5\sqrt[3]{7}$

4. $13\sqrt[5]{3} - 8\sqrt[5]{3}$

5. $4\sqrt[3]{y} + 9\sqrt[3]{y}$

6. $6\sqrt[4]{t} - 3\sqrt[4]{t}$

7. $5\sqrt{6} - 9\sqrt{6} - 4\sqrt{6}$

8. $3\sqrt{10} - 8\sqrt{10} + 7\sqrt{10}$

9. $4\sqrt[3]{3} - \sqrt{5} + 2\sqrt[3]{3} + \sqrt{5}$

10. $5\sqrt{7} - 8\sqrt[4]{11} + \sqrt{7} + 9\sqrt[4]{11}$

11. $8\sqrt{27} - 3\sqrt{3}$

12. $9\sqrt{50} - 4\sqrt{2}$

13. $8\sqrt{45} + 7\sqrt{20}$

14. $9\sqrt{12} + 16\sqrt{27}$

15. $18\sqrt{72} + 2\sqrt{98}$

16. $12\sqrt{45} - 8\sqrt{80}$

17. $3\sqrt[3]{16} + \sqrt[3]{54}$

18. $\sqrt[3]{27} - 5\sqrt[3]{8}$

19. $2\sqrt{128} - \sqrt{18} + 4\sqrt{32}$

20. $5\sqrt{50} - 2\sqrt{18} + 9\sqrt{32}$

21. $\sqrt{5a} + 2\sqrt{45a^3}$

22. $4\sqrt{3x^3} - \sqrt{12x}$

23. $\sqrt[3]{24x} - \sqrt[3]{3x^4}$

24. $\sqrt[3]{54x} - \sqrt[3]{2x^4}$

25. $5\sqrt[3]{32} - \sqrt[3]{108} + 2\sqrt[3]{256}$

26. $3\sqrt[3]{8x} - 4\sqrt[3]{27x} + 2\sqrt[3]{64x}$

b Multiply.

27. $\sqrt{5}(4 - 2\sqrt{5})$

28. $\sqrt{6}(2 + \sqrt{6})$

29. $\sqrt{3}(\sqrt{2} - \sqrt{7})$

30. $\sqrt{2}(\sqrt{5} - \sqrt{2})$

31. $\sqrt{3}(2\sqrt{5} - 3\sqrt{4})$

32. $\sqrt{2}(3\sqrt{10} - 2\sqrt{2})$

33. $\sqrt[3]{2}(\sqrt[3]{4} - 2\sqrt[3]{32})$

34. $\sqrt[3]{3}(\sqrt[3]{9} - 4\sqrt[3]{21})$

35. $\sqrt[3]{a}(\sqrt[3]{2a^2} + \sqrt[3]{16a^2})$

36. $\sqrt[3]{x}(\sqrt[3]{3x^2} - \sqrt[3]{81x^2})$

37. $(\sqrt{3} - \sqrt{2})(\sqrt{3} + \sqrt{2})$

38. $(\sqrt{5} + \sqrt{6})(\sqrt{5} - \sqrt{6})$

39. $(\sqrt{8} + 2\sqrt{5})(\sqrt{8} - 2\sqrt{5})$

40. $(\sqrt{18} + 3\sqrt{7})(\sqrt{18} - 3\sqrt{7})$

41. $(7 + \sqrt{5})(7 - \sqrt{5})$

42. $(4 - \sqrt{3})(4 + \sqrt{3})$

43. $(2 - \sqrt{3})(2 + \sqrt{3})$

44. $(11 - \sqrt{2})(11 + \sqrt{2})$

45. $(\sqrt{8} + \sqrt{5})(\sqrt{8} - \sqrt{5})$

46. $(\sqrt{6} - \sqrt{7})(\sqrt{6} + \sqrt{7})$

47. $(3 + 2\sqrt{7})(3 - 2\sqrt{7})$

48. $(6 - 3\sqrt{2})(6 + 3\sqrt{2})$

For the following exercises, assume that all expressions under radicals represent nonnegative numbers.

49. $(\sqrt{a} + \sqrt{b})(\sqrt{a} - \sqrt{b})$

50. $(\sqrt{x} - \sqrt{y})(\sqrt{x} + \sqrt{y})$

51. $(3 - \sqrt{5})(2 + \sqrt{5})$

52. $(2 + \sqrt{6})(4 - \sqrt{6})$

53. $(\sqrt{3} + 1)(2\sqrt{3} + 1)$

54. $(4\sqrt{3} + 5)(\sqrt{3} - 2)$

55. $(2\sqrt{7} - 4\sqrt{2})(3\sqrt{7} + 6\sqrt{2})$

56. $(4\sqrt{5} + 3\sqrt{3})(3\sqrt{5} - 4\sqrt{3})$

57. $(\sqrt{a} + \sqrt{2})(\sqrt{a} + \sqrt{3})$

58. $(2 - \sqrt{x})(1 - \sqrt{x})$

59. $(2\sqrt[3]{3} + \sqrt[3]{2})(\sqrt[3]{3} - 2\sqrt[3]{2})$

60. $(3\sqrt[4]{7} + \sqrt[4]{6})(2\sqrt[4]{9} - 3\sqrt[4]{6})$

61. $(2 + \sqrt{3})^2$

62. $(\sqrt{5} + 1)^2$

63. $(\sqrt[5]{9} - \sqrt[5]{3})(\sqrt[5]{8} + \sqrt[5]{27})$

64. $(\sqrt[3]{8x} - \sqrt[3]{5y})^2$

Skill Maintenance

Multiply or divide and simplify. [6.1d, f]

65. $\dfrac{x^3 + 4x}{x^2 - 16} \div \dfrac{x^2 + 8x + 15}{x^2 + x - 20}$

66. $\dfrac{a^2 - 4}{a} \div \dfrac{a - 2}{a + 4}$

67. $\dfrac{a^3 + 8}{a^2 - 4} \cdot \dfrac{a^2 - 4a + 4}{a^2 - 2a + 4}$

68. $\dfrac{y^3 - 27}{y^2 - 9} \cdot \dfrac{y^2 - 6y + 9}{y^2 + 3y + 9}$

Simplify. [6.8a]

69. $\dfrac{x - \frac{1}{3}}{x + \frac{1}{4}}$

70. $\dfrac{1 - \frac{1}{x}}{1 - \frac{1}{x^2}}$

71. $\dfrac{\frac{1}{p} - \frac{1}{q}}{\frac{1}{p^2} - \frac{1}{q^2}}$

Synthesis

72. ◆ Why do we need to know how to simplify radical expressions before we learn to add them?

73. ◆ In what way(s) is collecting like radical terms the same as collecting like monomial terms?

74. Graph the function $f(x) = \sqrt{(x - 2)^2}$. What is the domain?

75. Use a grapher to check your answers to Exercises 5, 22, and 58.

Multiply and simplify.

76. $\sqrt{9 + 3\sqrt{5}}\sqrt{9 - 3\sqrt{5}}$

77. $(\sqrt{x + 2} - \sqrt{x - 2})^2$

78. $(\sqrt{3} + \sqrt{5} - \sqrt{6})^2$

79. $\sqrt[3]{y}(1 - \sqrt[3]{y})(1 + \sqrt[3]{y})$

80. $(\sqrt[3]{9} - 2)(\sqrt[3]{9} + 4)$

81. $[\sqrt{3 + \sqrt{2 + \sqrt{1}}}]^4$

10.5 More on Division of Radical Expressions

a Rationalizing Denominators

Sometimes in mathematics it is useful to find an equivalent expression without a radical in the denominator. This provides a standard notation for expressing results. The procedure for finding such an expression is called **rationalizing the denominator.** We carry this out by multiplying by 1.

Example 1 Rationalize the denominator: $\sqrt{\frac{7}{3}}$.

We multiply by 1, using $\sqrt{3}/\sqrt{3}$. We do this so that the denominator of the radicand will be a perfect square.

$$\sqrt{\frac{7}{3}} = \frac{\sqrt{7}}{\sqrt{3}} \cdot \frac{\sqrt{3}}{\sqrt{3}} = \frac{\sqrt{7} \cdot \sqrt{3}}{\sqrt{3} \cdot \sqrt{3}} = \frac{\sqrt{21}}{\sqrt{3^2}} = \frac{\sqrt{21}}{3}$$

The radicand is a perfect square.

Do Exercise 1.

Example 2 Rationalize the denominator: $\sqrt[3]{\frac{7}{25}}$.

We first factor the denominator:

$$\sqrt[3]{\frac{7}{25}} = \sqrt[3]{\frac{7}{5 \cdot 5}}.$$

To get a perfect cube in the denominator, we consider the index 3 and the factors. We have 2 factors of 5, and we need 3 factors of 5. We achieve this by multiplying by 1, using $\sqrt[3]{5}/\sqrt[3]{5}$.

$$\sqrt[3]{\frac{7}{25}} = \sqrt[3]{\frac{7}{5 \cdot 5}} \cdot \frac{\sqrt[3]{5}}{\sqrt[3]{5}}$$

Multiplying by $\frac{\sqrt[3]{5}}{\sqrt[3]{5}}$ to make the denominator of the radicand a perfect cube

$$= \frac{\sqrt[3]{7} \cdot \sqrt[3]{5}}{\sqrt[3]{5 \cdot 5} \cdot \sqrt[3]{5}}$$

$$= \frac{\sqrt[3]{35}}{\sqrt[3]{5^3}}$$

The radicand is a perfect cube.

$$= \frac{\sqrt[3]{35}}{5}.$$

Do Exercise 2.

Objectives

a. Rationalize the denominator of a radical expression having one term in the denominator.

b. Rationalize the denominator of a radical expression having two terms in the denominator.

For Extra Help

TAPE 17

MAC WIN

CD-ROM

1. Rationalize the denominator:

$\sqrt{\frac{2}{5}}$.

2. Rationalize the denominator:

$\sqrt[3]{\frac{5}{4}}$.

Answers on page A-43

3. Rationalize the denominator:

$$\sqrt{\frac{4a}{3b}}.$$

Rationalize the denominator.

4. $\dfrac{\sqrt[4]{7}}{\sqrt[4]{2}}$

5. $\sqrt[3]{\dfrac{3x^5}{2y}}$

6. Rationalize the denominator:

$$\frac{7x}{\sqrt[3]{4xy^5}}.$$

Answers on page A-43

Example 3 Rationalize the denominator: $\sqrt{\dfrac{2a}{5b}}$. Assume that all expressions under radicals represent positive numbers.

$$\sqrt{\frac{2a}{5b}} = \frac{\sqrt{2a}}{\sqrt{5b}} \qquad \text{Converting to a quotient of radicals}$$

$$= \frac{\sqrt{2a}}{\sqrt{5b}} \cdot \frac{\sqrt{5b}}{\sqrt{5b}} \qquad \text{Multiplying by 1}$$

$$= \frac{\sqrt{10ab}}{\sqrt{(5b)^2}} \qquad \text{The denominator is a perfect square.}$$

$$= \frac{\sqrt{10ab}}{5b}$$

Do Exercise 3.

Example 4 Rationalize the denominator: $\dfrac{\sqrt[3]{a}}{\sqrt[3]{9x}}$.

We factor the denominator:

$$\frac{\sqrt[3]{a}}{\sqrt[3]{9x}} = \frac{\sqrt[3]{a}}{\sqrt[3]{3 \cdot 3 \cdot x}}.$$

To choose the symbol for 1, we look at $3 \cdot 3 \cdot x$. To make it a cube, we need another 3 and two more x's. Thus we multiply by 1, using $\sqrt[3]{3x^2}/\sqrt[3]{3x^2}$:

$$\frac{\sqrt[3]{a}}{\sqrt[3]{9x}} = \frac{\sqrt[3]{a}}{\sqrt[3]{9x}} \cdot \frac{\sqrt[3]{3x^2}}{\sqrt[3]{3x^2}} \qquad \text{Multiplying by 1}$$

$$= \frac{\sqrt[3]{3ax^2}}{\sqrt[3]{(3x)^3}} \qquad \text{The denominator is a perfect cube.}$$

$$= \frac{\sqrt[3]{3ax^2}}{3x}.$$

Do Exercises 4 and 5.

Example 5 Rationalize the denominator: $\dfrac{3x}{\sqrt[5]{2x^2y^3}}$.

$$\frac{3x}{\sqrt[5]{2x^2y^3}} = \frac{3x}{\sqrt[5]{2 \cdot x \cdot x \cdot y \cdot y \cdot y}}$$

$$= \frac{3x}{\sqrt[5]{2x^2y^3}} \cdot \frac{\sqrt[5]{2^4x^3y^2}}{\sqrt[5]{2^4x^3y^2}}$$

$$= \frac{3x\sqrt[5]{16x^3y^2}}{\sqrt[5]{32x^5y^5}} \qquad \text{The denominator is a perfect fifth power.}$$

$$= \frac{3x\sqrt[5]{16x^3y^2}}{2xy}$$

$$= \frac{3\sqrt[5]{16x^3y^2}}{2y}$$

Do Exercise 6.

b Rationalizing When There Are Two Terms

Do Exercises 7 and 8.

Certain pairs of expressions containing square roots, such as $c - \sqrt{b}$, $c + \sqrt{b}$ and $\sqrt{a} - \sqrt{b}$, $\sqrt{a} + \sqrt{b}$, are called **conjugates**. The product of such a pair of conjugates has no radicals in it. (See Example 12 of Section 10.4). Thus when we wish to rationalize a denominator that has two terms and one or more of them involves a square-root radical, we multiply by 1 using the conjugate of the denominator to write a symbol for 1.

Examples What symbol for 1 would you use to rationalize the denominator?

	Expression	*Symbol for 1*	
6.	$\dfrac{3}{x + \sqrt{7}}$	$\dfrac{x - \sqrt{7}}{x - \sqrt{7}}$	Change the operation sign to obtain the conjugate. Use the conjugate for the numerator and denominator of the symbol for 1.
7.	$\dfrac{\sqrt{7} + 4}{3 - 2\sqrt{5}}$	$\dfrac{3 + 2\sqrt{5}}{3 + 2\sqrt{5}}$	

Do Exercises 9 and 10.

Example 8 Rationalize the denominator: $\dfrac{4 + \sqrt{2}}{\sqrt{5} - \sqrt{2}}$.

$$\frac{4 + \sqrt{2}}{\sqrt{5} - \sqrt{2}} = \frac{4 + \sqrt{2}}{\sqrt{5} - \sqrt{2}} \cdot \frac{\sqrt{5} + \sqrt{2}}{\sqrt{5} + \sqrt{2}}$$

Multiplying by 1, using the conjugate of $\sqrt{5} - \sqrt{2}$, which is $\sqrt{5} + \sqrt{2}$

$$= \frac{(4 + \sqrt{2})(\sqrt{5} + \sqrt{2})}{(\sqrt{5} - \sqrt{2})(\sqrt{5} + \sqrt{2})}$$

Multiplying numerators and denominators

$$= \frac{4\sqrt{5} + 4\sqrt{2} + \sqrt{2}\sqrt{5} + (\sqrt{2})^2}{(\sqrt{5})^2 - (\sqrt{2})^2}$$

$$= \frac{4\sqrt{5} + 4\sqrt{2} + \sqrt{10} + 2}{5 - 2}$$

$$= \frac{4\sqrt{5} + 4\sqrt{2} + \sqrt{10} + 2}{3}$$

Example 9 Rationalize the denominator: $\dfrac{4}{\sqrt{3} + x}$.

$$\frac{4}{\sqrt{3} + x} = \frac{4}{\sqrt{3} + x} \cdot \frac{\sqrt{3} - x}{\sqrt{3} - x}$$

$$= \frac{4(\sqrt{3} - x)}{(\sqrt{3} + x)(\sqrt{3} - x)}$$

$$= \frac{4\sqrt{3} - 4x}{3 - x^2}$$

Do Exercises 11 and 12.

Multiply.

7. $(c - \sqrt{b})(c + \sqrt{b})$

8. $(\sqrt{a} + \sqrt{b})(\sqrt{a} - \sqrt{b})$

What symbol for 1 would you use to rationalize the denominator?

9. $\dfrac{\sqrt{5} + 1}{\sqrt{3} - y}$

10. $\dfrac{1}{\sqrt{2} + \sqrt{3}}$

Rationalize the denominator.

11. $\dfrac{5 + \sqrt{2}}{1 - \sqrt{2}}$

12. $\dfrac{14}{3 + \sqrt{2}}$

Answers on page A-43

Improving Your Math Study Skills

On Reading and Writing Mathematics

Mike Rosenborg is a former student of Marv Bittinger. He went on to receive a Master's Degree in mathematics and is now a math teacher. Here are some of his study tips regarding the reading and writing of mathematics.

Why read your math text? This is a legitimate question when you consider that the instructor usually covers most of the material in the text. I have a reason: I don't want to be spoon-fed the material; I want to learn on my own. It's a lot more fun, and it builds my self-confidence to know that I can learn the material without the need for a teacher or a classroom.

It's a good idea to read your math text regularly for several reasons. When you read mathematics, you rely exclusively on the written word, and this is where mathematics derives much of its power. Mathematics is very precise, and it depends on writing to maintain and communicate this precision. Definitions and theorems in mathematics are stated in precise terms, and mathematical manipulations (such as solving equations) are performed by writing in a precise way.

In general, math texts develop the concepts in mathematics in a clear, tightly reasoned format, showing many examples along the way. Remember: The authors are mathematicians, and the way they write reflects their extensive mathematical training and thought processes. If you carefully read through the text, you will experience what it is like to think in a mathematical, rigorous, and precise way. This will not happen if you rely exclusively on oral lectures, because oral presentations are intrinsically "loose."

Reading your math text has other benefits. You will often find how to solve a difficult problem in the exercise set by looking at the text; in fact, there may be an example developed for you in the text that is much like your problem. Often your instructor will not have the time to cover everything in the text, or may want to cover something a little different. In these cases, reading your text will fill in the gaps.

But how do you read a math text? There is, of course, a difference between reading mathematics and a novel. Mathematics is like a chain with each link being developed in sequence and in order, and each link demands careful thought and attention before one can proceed to the next link—each link depends on the link before it. This is why you will experience troubles throughout an entire math course if you miss a single concept. Here, then, is a list of math reading tips and techniques:

- **Always read with a pencil and a piece of paper nearby.** When you find a section in your text that is difficult to understand, stop and work it out on your scratch paper.
- **Make notes in the margins of your text.** For instance, if you come across a word you don't understand, look it up and write its definition in the margin. Also, if the book refers to something covered previously that you have forgotten, find where it was originally covered, write the page number down in the margin, and go back and review the word.
- **Proceed slowly and carefully, making sure you understand what you read before continuing.** If there are worked-out examples in the book, read one and then try the next ones on your own. If you make a mistake, the details on the worked examples in the text will enable you to find your mistake quickly.
- **Do the Thinking and Writing Exercises.** The benefits of writing mathematics are that writing forces you to think through what you are writing about in a step-by-step manner, the act of writing itself helps to reinforce the concepts in your mind, and you have a ready, easy-to-read reference for further study or review. For instance, when studying for a test, if you have written up all your homework problems in a complete way, it will be easy for you to study for the test directly from your homework.

Exercise Set 10.5

a Rationalize the denominator. Assume that all expressions under radicals represent positive numbers.

1. $\sqrt{\frac{5}{3}}$

2. $\sqrt{\frac{8}{7}}$

3. $\sqrt{\frac{11}{2}}$

4. $\sqrt{\frac{17}{6}}$

5. $\frac{2\sqrt{3}}{7\sqrt{5}}$

6. $\frac{3\sqrt{5}}{8\sqrt{2}}$

7. $\sqrt[3]{\frac{16}{9}}$

8. $\sqrt[3]{\frac{3}{9}}$

9. $\frac{\sqrt[3]{3a}}{\sqrt[3]{5c}}$

10. $\frac{\sqrt[3]{7x}}{\sqrt[3]{3y}}$

11. $\frac{\sqrt[3]{2y^4}}{\sqrt[3]{6x^4}}$

12. $\frac{\sqrt[3]{3a^4}}{\sqrt[3]{7b^2}}$

13. $\frac{1}{\sqrt[4]{st}}$

14. $\frac{1}{\sqrt[3]{yz}}$

15. $\sqrt{\frac{3x}{20}}$

16. $\sqrt{\frac{7a}{32}}$

17. $\sqrt[3]{\frac{4}{5x^5y^2}}$

18. $\sqrt[3]{\frac{7c}{100ab^5}}$

19. $\sqrt[4]{\frac{1}{8x^7y^3}}$

20. $\frac{2x}{\sqrt[5]{18x^8y^6}}$

b Rationalize the denominator. Assume that all expressions under radicals represent positive numbers.

21. $\frac{9}{6 - \sqrt{10}}$

22. $\frac{3}{8 + \sqrt{5}}$

23. $\frac{-4\sqrt{7}}{\sqrt{5} - \sqrt{3}}$

24. $\frac{34\sqrt{5}}{2\sqrt{5} - \sqrt{3}}$

25. $\frac{\sqrt{5} - 2\sqrt{6}}{\sqrt{3} - 4\sqrt{5}}$

26. $\frac{\sqrt{6} - 3\sqrt{5}}{\sqrt{3} - 2\sqrt{7}}$

27. $\frac{2 - \sqrt{a}}{3 + \sqrt{a}}$

28. $\frac{5 + \sqrt{x}}{8 - \sqrt{x}}$

29. $\frac{5\sqrt{3} - 3\sqrt{2}}{3\sqrt{2} - 2\sqrt{3}}$

30. $\frac{7\sqrt{2} + 4\sqrt{3}}{4\sqrt{3} - 3\sqrt{2}}$

31. $\frac{\sqrt{x} - \sqrt{y}}{\sqrt{x} + \sqrt{y}}$

32. $\frac{\sqrt{a} + \sqrt{b}}{\sqrt{a} - \sqrt{b}}$

Skill Maintenance

Solve. [6.5a]

33. $\frac{1}{2} - \frac{1}{3} = \frac{5}{t}$

34. $\frac{5}{x - 1} + \frac{9}{x^2 + x + 1} = \frac{15}{x^3 - 1}$

Divide and simplify. [6.1f]

35. $\frac{1}{x^3 - y^3} \div \frac{1}{(x - y)(x^2 + xy + y^2)}$

36. $\frac{2x^2 - x - 6}{x^2 + 4x + 3} \div \frac{2x^2 + x - 3}{x^2 - 1}$

Synthesis

37. ◆ A student *incorrectly* claims that

$$\frac{5 + \sqrt{2}}{\sqrt{18}} = \frac{5 + \sqrt{1}}{\sqrt{9}} = \frac{5 + 1}{3} = 2.$$

How could you convince the student that a mistake has been made? How would you explain the correct way of rationalizing the denominator?

38. ◆ A student considers the radical expression

$$\frac{11}{\sqrt[3]{4} - \sqrt[3]{5}}$$

and tries to rationalize the denominator by multiplying by

$$\frac{\sqrt[3]{4} + \sqrt[3]{5}}{\sqrt[3]{4} + \sqrt[3]{5}}.$$

Discuss the difficulties of such a plan.

39. Use a grapher to check your answers to Exercises 15, 16, and 28.

40. Express each of the following as the product of two radical expressions.

a) $x - 5$ **b)** $x - a$

Simplify. (*Hint*: Rationalize the denominator.)

41. $\sqrt{a^2 - 3} - \frac{a^2}{\sqrt{a^2 - 3}}$

42. $\frac{1}{4 + \sqrt{3}} + \frac{1}{\sqrt{3}} + \frac{1}{\sqrt{3} - 4}$

10.6 Solving Radical Equations

a The Principle of Powers

Objectives

a Solve radical equations with one radical term.

b Solve radical equations with two radical terms.

c Solve applied problems involving radical equations.

For Extra Help

TAPE 17 MAC WIN CD-ROM

A **radical equation** has variables in one or more radicands—for example,

$$\sqrt[3]{2x} + 1 = 5, \qquad \sqrt{x} + \sqrt{4x - 2} = 7.$$

To solve such an equation, we need a new principle. Suppose that an equation $a = b$ is true. If we square both sides, we get another true equation: $a^2 = b^2$. This can be generalized.

THE PRINCIPLE OF POWERS

For any natural number n, if an equation $a = b$ is true, then $a^n = b^n$ is true.

However, if an equation $a^n = b^n$ is true, it *may not* be true that $a = b$. For example, $3^2 = (-3)^2$ is true, but $3 = -3$ is not true. Thus we must make a check when we solve an equation using the principle of powers.

Example 1 Solve: $\sqrt{x} - 3 = 4$.

$$\sqrt{x} - 3 = 4$$

$$\sqrt{x} = 7 \qquad \text{Adding to isolate the radical}$$

$$(\sqrt{x})^2 = 7^2 \qquad \text{Using the principle of powers (squaring)}$$

$$x = 49. \qquad \sqrt{x} \cdot \sqrt{x} = x$$

The number 49 is a possible solution. But we *must* make a check in order to be sure!

CHECK:

$\sqrt{x} - 3$	$= 4$
$\sqrt{49} - 3$	$?\ 4$
$7 - 3$	
4	

TRUE

The solution is 49.

The principle of powers does not always give equivalent equations. For this reason, a check is a must!

Example 2 Solve: $\sqrt{x} = -3$.

We might observe at the outset that this equation has no solution because the principal square root of a number is never negative. Let's continue as above for comparison.

$$\sqrt{x} = -3$$

$$(\sqrt{x})^2 = (-3)^2$$

$$x = 9$$

CHECK:

$\sqrt{x}$	$= -3$
$\sqrt{9}$	$?\ -3$
3	

FALSE

The number 9 does *not* check. Thus the equation $\sqrt{x} = -3$ has no real-number solution. Note that the equation $x = 9$ has solution 9, but that $\sqrt{x} = -3$ has *no* solution. Thus the equations $x = 9$ and $\sqrt{x} = -3$ are *not* equivalent. That is, $\sqrt{9} \neq -3$.

Solve.

1. $\sqrt{x} - 7 = 3$

2. $\sqrt{x} = -2$

Solve.

3. $x + 2 = \sqrt{2x + 7}$

4. $x + 1 = 3\sqrt{x - 1}$

Answers on page A-43

Do Exercises 1 and 2.

To solve an equation with a radical term, we first isolate the radical term on one side of the equation. Then we use the principle of powers.

Example 3 Solve: $x - 7 = 2\sqrt{x + 1}$.

The radical term is already isolated. We proceed with the principle of powers:

$$x - 7 = 2\sqrt{x + 1}$$

$$(x - 7)^2 = (2\sqrt{x + 1})^2 \quad \text{Using the principle of powers (squaring)}$$

$$x^2 - 14x + 49 = 2^2(\sqrt{x + 1})^2 \quad \text{Squaring the binomial on the left; raising a product to a power on the right}$$

$$x^2 - 14x + 49 = 4(x + 1)$$

$$x^2 - 14x + 49 = 4x + 4$$

$$x^2 - 18x + 45 = 0$$

$$(x - 3)(x - 15) = 0 \quad \text{Factoring}$$

$$x - 3 = 0 \quad or \quad x - 15 = 0 \quad \text{Using the principle of zero products}$$

$$x = 3 \quad or \quad x = 15.$$

The possible solutions are 3 and 15. We check.

For 3:

$x - 7 = 2\sqrt{x + 1}$		
$3 - 7$?	$2\sqrt{3 + 1}$	
-4	$2\sqrt{4}$	
	$2(2)$	
	4	FALSE

For 15:

$x - 7 = 2\sqrt{x + 1}$		
$15 - 7$?	$2\sqrt{15 + 1}$	
8	$2\sqrt{16}$	
	$2(4)$	
	8	TRUE

The number 3 does *not* check, but the number 15 does check. The solution is 15.

The number 3 in Example 3 is sometimes called an *extraneous solution,* but such terminology is risky at best because the number 3 is in *no way* a solution of the original equation.

Do Exercises 3 and 4.

Example 4 Solve: $x = \sqrt{x + 7} + 5$.

$$x = \sqrt{x + 7} + 5$$

$$x - 5 = \sqrt{x + 7} \quad \text{Subtracting 5 to isolate the radical term}$$

$$(x - 5)^2 = (\sqrt{x + 7})^2 \quad \text{Using the principle of powers (squaring both sides)}$$

$$x^2 - 10x + 25 = x + 7$$

$$x^2 - 11x + 18 = 0$$

$$(x - 9)(x - 2) = 0 \quad \text{Factoring}$$

$$x = 9 \quad or \quad x = 2 \quad \text{Using the principle of zero products}$$

The possible solutions are 9 and 2. Let's check.

For 9:

$x = \sqrt{x + 7} + 5$		
9 ?	$\sqrt{9 + 7} + 5$	
	9	TRUE

For 2:

$x = \sqrt{x + 7} + 5$		
2 ?	$\sqrt{2 + 7} + 5$	
	8	FALSE

Since 9 checks but 2 does not, the solution is 9.

Algebraic–Graphical Connection

We can visualize or check the solutions of a radical equation graphically. Consider the equation of Example 3:

$$x - 7 = 2\sqrt{x + 1}.$$

We can examine the solutions by graphing the equations

$$y = x - 7 \quad \text{and} \quad y = 2\sqrt{x + 1}$$

using the same set of axes. A hand-drawn graph of $y = 2\sqrt{x + 1}$ would involve approximating square roots on a calculator.

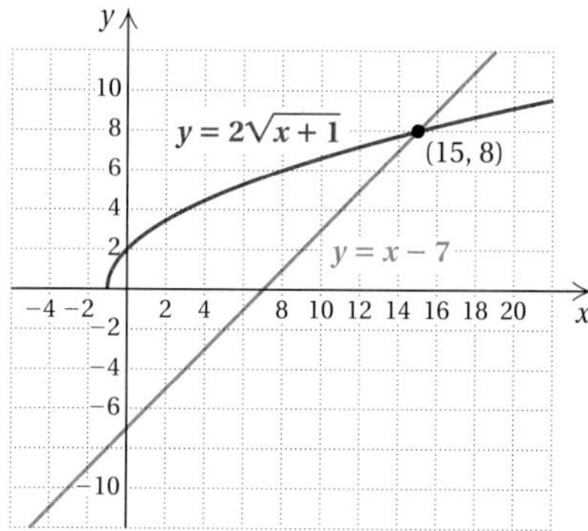

It appears from the graph that when $x = 15$, the values of $y = x - 7$ and $y = 2\sqrt{x + 1}$ are the same, 8. We can check this as we did in Example 3. Note too that the graphs *do not* intersect at $x = 3$, the "extraneous" solution.

Calculator Spotlight

Solving Radical Equations. Consider the equation in Example 3. From each side of the equation, we form two new equations in the "y=" form:

$$y_1 = x - 7, \quad (1)$$
$$y_2 = 2\sqrt{x + 1}. \quad (2)$$

The graphs of y_1 and y_2 are as shown below.

$y_1 = x - 7, \ y_2 = 2\sqrt{x + 1}$

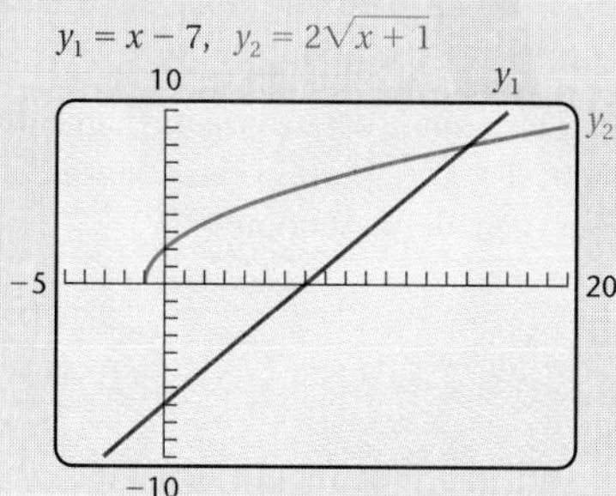

In some cases, the viewing window may need to be adjusted to view the intersection.

To find the solution, we use the CALC-INTERSECT features. The solution of the equation $x - 7 = 2\sqrt{x + 1}$ is the first coordinate at the bottom of the screen. Note that there is no point of intersection at $x = 3$, the "extraneous" solution. The number 3 is *not* a solution of the original equation.

$y_1 = x - 7, \ y_2 = 2\sqrt{x + 1}$

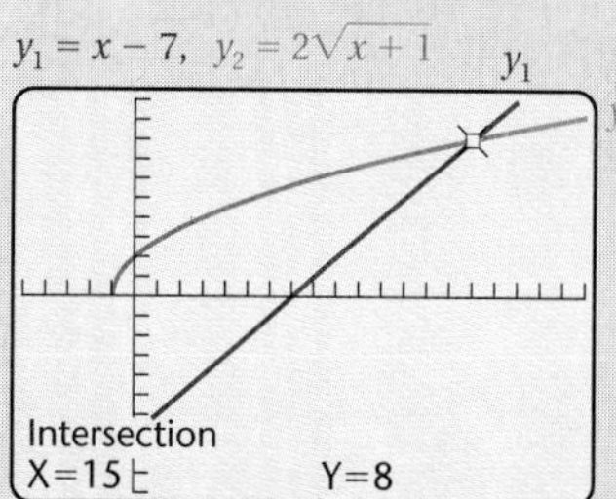

Exercises

Solve.

1. $x = \sqrt{x + 7} + 5$ (Example 4)

2. $x + 2 = \sqrt{2x + 7}$ (Margin Exercise 3)
What happens to the equations at $x = -3$? Do the graphs intersect?

3. $x + 1 = 3\sqrt{x - 1}$ (Margin Exercise 4)

Approximate the solutions.

4. $x - 7.4 = 2\sqrt{x + 1.3}$

5. $x - 5.3 = \sqrt{x + 8.1}$

6. $x + 1.92 = 3.6\sqrt{x - 0.8}$

Solve.

5. $x = \sqrt{x+5} + 1$

6. $\sqrt[4]{x-1} - 2 = 0$

Answers on page A-43

Example 5 Solve: $\sqrt[3]{2x+1} + 5 = 0$.

$$\sqrt[3]{2x+1} + 5 = 0$$

$$\sqrt[3]{2x+1} = -5 \quad \text{Subtracting 5; this isolates the radical term}$$

$$(\sqrt[3]{2x+1})^3 = (-5)^3 \quad \text{Using the principle of powers (raising to the third power)}$$

$$2x + 1 = -125$$

$$2x = -126 \quad \text{Subtracting 1}$$

$$x = -63$$

CHECK:

$\sqrt[3]{2x+1} + 5 = 0$		
$\sqrt[3]{2\cdot(-63)+1} + 5$	? 0	
$\sqrt[3]{-125} + 5$		
$-5 + 5$		
0		TRUE

The solution is -63.

Do Exercises 5 and 6.

b Equations with Two Radical Terms

A general strategy for solving equations with two radical terms is as follows.

> To solve an equation with two radical terms:
> 1. Isolate one of the radical terms.
> 2. Use the principle of powers.
> 3. If a radical remains, perform steps (1) and (2) again.
> 4. Check possible solutions.

Example 6 Solve: $\sqrt{x-3} + \sqrt{x+5} = 4$.

$$\sqrt{x-3} + \sqrt{x+5} = 4$$

$$\sqrt{x-3} = 4 - \sqrt{x+5} \quad \text{Subtracting } \sqrt{x+5}\text{; this isolates one of the radical terms}$$

$$(\sqrt{x-3})^2 = (4 - \sqrt{x+5})^2 \quad \text{Using the principle of powers (squaring both sides)}$$

Here we are squaring a binomial. We square 4, then subtract twice the product of 4 and $\sqrt{x+5}$, and then add the square of $\sqrt{x+5}$. (See the rule in Section 4.6.)

$$x - 3 = 16 - 8\sqrt{x+5} + (x+5)$$

$$-3 = 21 - 8\sqrt{x+5} \quad \text{Subtracting } x \text{ and collecting like terms}$$

$$-24 = -8\sqrt{x+5} \quad \text{Isolating the remaining radical term}$$

$$3 = \sqrt{x+5} \quad \text{Dividing by } -8$$

$$3^2 = (\sqrt{x+5})^2 \quad \text{Squaring}$$

$$9 = x + 5$$

$$4 = x$$

The number 4 checks and is the solution.

CAUTION! A common error in solving equations like $\sqrt{x-3}+\sqrt{x+5}=4$ is to square the left side, obtaining $(x-3)+(x+5)$. That is wrong because the square of a sum is *not* the sum of the squares.

Example: $\sqrt{9}+\sqrt{16}=7$, but $9+16=25$, not 7^2.

Example 7 Solve: $\sqrt{2x-5}=1+\sqrt{x-3}$.

$$\sqrt{2x-5}=1+\sqrt{x-3}$$

$$(\sqrt{2x-5})^2=(1+\sqrt{x-3})^2 \qquad \text{One radical is already isolated; we square both sides.}$$

$$2x-5=1+2\sqrt{x-3}+(\sqrt{x-3})^2$$

$$2x-5=1+2\sqrt{x-3}+(x-3)$$

$$x-3=2\sqrt{x-3} \qquad \text{Isolating the remaining radical term}$$

$$(x-3)^2=(2\sqrt{x-3})^2 \qquad \text{Squaring both sides}$$

$$x^2-6x+9=4(x-3)$$

$$x^2-6x+9=4x-12$$

$$x^2-10x+21=0$$

$$(x-7)(x-3)=0 \qquad \text{Factoring}$$

$$x=7 \quad or \quad x=3 \qquad \text{Using the principle of zero products}$$

The numbers 7 and 3 check and are the solutions.

Do Exercises 7 and 8.

Solve.

7. $\sqrt{x}-\sqrt{x-5}=1$

8. $\sqrt{2x-5}-2=\sqrt{x-2}$

Example 8 Solve: $\sqrt{x+2}-\sqrt{2x+2}+1=0$.

We first isolate one radical.

$$\sqrt{x+2}-\sqrt{2x+2}+1=0$$

$$\sqrt{x+2}=\sqrt{2x+2}-1 \qquad \text{Adding } \sqrt{2x+2} \text{ and subtracting 1 to isolate a radical}$$

$$(\sqrt{x+2})^2=(\sqrt{2x+2}-1)^2 \qquad \text{Squaring both sides}$$

$$x+2=(\sqrt{2x+2})^2-2\sqrt{2x+2}+1$$

$$x+2=2x+2-2\sqrt{2x+2}+1$$

$$-x-1=-2\sqrt{2x+2} \qquad \text{Isolating the remaining radical}$$

$$x+1=2\sqrt{2x+2} \qquad \text{Multiplying by } -1$$

$$(x+1)^2=(2\sqrt{2x+2})^2 \qquad \text{Squaring both sides}$$

$$x^2+2x+1=4(2x+2)$$

$$x^2+2x+1=8x+8$$

$$x^2-6x-7=0$$

$$(x-7)(x+1)=0 \qquad \text{Factoring}$$

$$x-7=0 \quad or \quad x+1=0 \qquad \text{Using the principle of zero products}$$

$$x=7 \quad or \quad x=-1$$

The check is left to the student. The number 7 checks, but -1 does not. The solution is 7.

Do Exercise 9.

9. Solve:

$\sqrt{3x+1}-1-\sqrt{x+4}=0$.

Answers on page A-43

10. How far to the horizon can you see through an airplane window at a height, or altitude, of 38,000 ft?

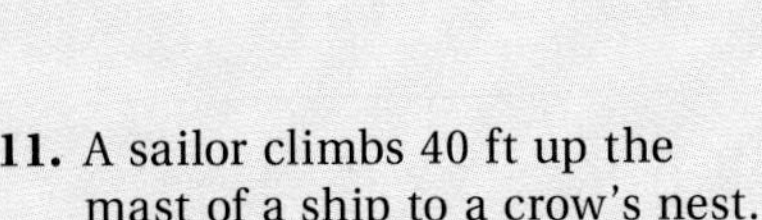

11. A sailor climbs 40 ft up the mast of a ship to a crow's nest. How far can he see to the horizon?

12. How far above sea level must a sailor climb on the mast of a ship in order to see 18.4 mi out to sea?

Answers on page A-43

c Applications

Sighting to the Horizon. How far can you see from a given height? The function

$$D(h) = \sqrt{2h} \qquad \textbf{(1)}$$

can be used to approximate the distance D, in miles, that a person can see to the horizon from a height h, in feet.

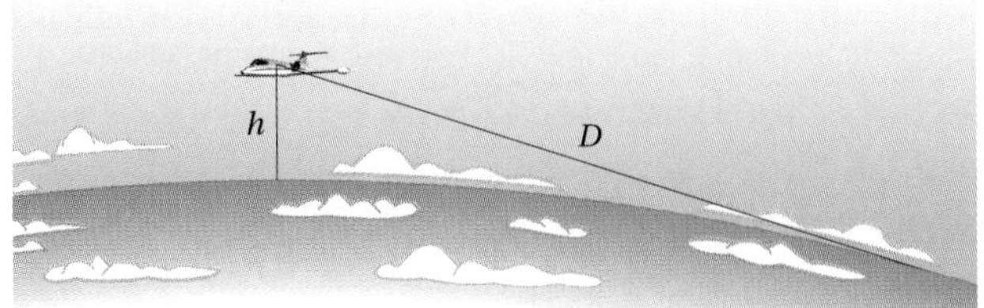

Example 9 How far to the horizon can you see through an airplane window at a height, or altitude, of 30,000 ft?

We substitute 30,000 for h in equation (1) and find an approximation using a calculator:

$$\begin{aligned} D(30{,}000) &= \sqrt{2 \cdot 30{,}000} \\ &\approx 245 \text{ mi}. \end{aligned}$$

You can see for about 245 mi to the horizon.

Do Exercises 10 and 11.

Example 10 How far above sea level must a sailor climb on the mast of a ship in order to see 10.2 mi out to an iceberg?

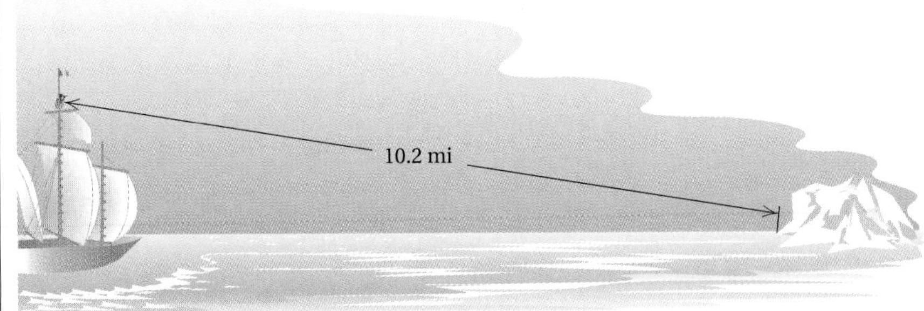

We substitute 10.2 for $D(h)$ in equation (1) and solve:

$$\begin{aligned} 10.2 &= \sqrt{2h} \\ (10.2)^2 &= (\sqrt{2h})^2 \\ 104.04 &= 2h \\ \frac{104.04}{2} &= h \\ 52.02 &= h. \end{aligned}$$

The sailor must climb to a height of about 52 ft in order to see 10.2 mi out to the iceberg.

Do Exercise 12.

Exercise Set 10.6

a Solve.

1. $\sqrt{2x - 3} = 4$

2. $\sqrt{5x + 2} = 7$

3. $\sqrt{6x} + 1 = 8$

4. $\sqrt{3x} - 4 = 6$

5. $\sqrt{y + 7} - 4 = 4$

6. $\sqrt{x - 1} - 3 = 9$

7. $\sqrt{5y + 8} = 10$

8. $\sqrt{2y + 9} = 5$

9. $\sqrt[3]{x} = -1$

10. $\sqrt[3]{y} = -2$

11. $\sqrt{x + 2} = -4$

12. $\sqrt{y - 3} = -2$

13. $\sqrt[3]{x + 5} = 2$

14. $\sqrt[3]{x - 2} = 3$

15. $\sqrt[4]{y - 3} = 2$

16. $\sqrt[4]{x + 3} = 3$

17. $\sqrt[3]{6x + 9} + 8 = 5$

18. $\sqrt[3]{3y + 6} + 2 = 3$

19. $8 = \frac{1}{\sqrt{x}}$

20. $\frac{1}{\sqrt{y}} = 3$

b Solve.

21. $\sqrt{3y + 1} = \sqrt{2y + 6}$

22. $\sqrt{5x - 3} = \sqrt{2x + 3}$

23. $\sqrt{y - 5} + \sqrt{y} = 5$

24. $\sqrt{x - 9} + \sqrt{x} = 1$

25. $3 + \sqrt{z - 6} = \sqrt{z + 9}$

26. $\sqrt{4x - 3} = 2 + \sqrt{2x - 5}$

27. $\sqrt{20 - x} + 8 = \sqrt{9 - x} + 11$

28. $4 + \sqrt{10 - x} = 6 + \sqrt{4 - x}$

29. $\sqrt{4y + 1} - \sqrt{y - 2} = 3$

30. $\sqrt{y + 15} - \sqrt{2y + 7} = 1$

31. $\sqrt{x + 2} + \sqrt{3x + 4} = 2$

32. $\sqrt{6x + 7} - \sqrt{3x + 3} = 1$

33. $\sqrt{3x - 5} + \sqrt{2x + 3} + 1 = 0$

34. $\sqrt{2m - 3} + 2 - \sqrt{m + 7} = 0$

35. $2\sqrt{t - 1} - \sqrt{3t - 1} = 0$

36. $3\sqrt{2y + 3} - \sqrt{y + 10} = 0$

c Solve.

Use the formula $D(h) = \sqrt{2h}$ for Exercises 37–40.

37. How far to the horizon can you see through an airplane window at a height, or altitude, of 27,000 ft?

38. How far to the horizon can you see through an airplane window at a height, or altitude, of 32,000 ft?

39. How far above sea level must a pilot fly in order to see to a horizon that is 180 mi away?

40. A person can see 220 mi to the horizon through an airplane window. How high above sea level is the airplane?

Speed of a Skidding Car. How do police determine how fast a car had been traveling after an accident has occurred? The formula

$$S(x) = 2\sqrt{5x}$$

can be used to approximate the speed S, in miles per hour, of a car that has left a skid mark of length x, in feet. (See Exercise 27 in Section 10.1.) Use this formula for Exercises 41 and 42.

41. How far will a car skid at 65 mph? at 75 mph?

42. How far will a car skid at 55 mph? at 90 mph?

43. Find the number such that twice its square root is 14.

44. Find the number such that the square root of 4 more than five times the number is 8.

45. Find the number such that the square root of twice the number minus 1, all added to 1, is the square root of the number plus 11.

46. Find the number such that the square root of 4 less than the number plus the square root of 1 more than the number is 5.

Skill Maintenance

Solve. [6.6a]

47. *Painting a Room.* Julia can paint a room in 8 hr. George can paint the same room in 10 hr. How long will it take them, working together, to paint the same room?

48. *Delivering Leaflets.* Jeff can drop leaflets in mailboxes three times as fast as Grace can. If they work together, it takes them 1 hr to complete the job. How long would it take each to deliver the leaflets alone?

Solve. [6.6a]

49. *Bicycle Travel.* A cyclist traveled 702 mi in 14 days. At this same ratio, how far would the cyclists have traveled in 56 days?

50. *Earnings.* Dharma earned \$696.64 working for 56 hr at a fruit stand. How many hours must she work in order to earn \$1044.96?

Solve. [5.8b]

51. $x^2 + 2.8x = 0$

52. $3x^2 - 5x = 0$

53. $x^2 - 64 = 0$

54. $2x^2 = x + 21$

Synthesis

55. ◆ The principle of powers contains an "if–then" statement that becomes false when the parts are interchanged. Find another mathematical example of such an "if–then" statement.

56. ◆ Is checking necessary when the principle of powers is used with an odd power n? Why or why not?

57. Use a grapher to check your answers to Exercises 4, 9, 25, and 30.

58. ◆ Consider the equation
$\sqrt{2x + 1} + \sqrt{5x - 4} = \sqrt{10x + 9}.$
a) Use a grapher to solve the equation.
b) Solve the equation algebraically.
c) Explain the advantages and disadvantages of using each method. Which do you prefer?

Solve.

59. $\sqrt[3]{\frac{z}{4}} - 10 = 2$

60. $\sqrt[4]{z^2 + 17} = 3$

61. $\sqrt{\sqrt{y + 49} - \sqrt{y}} = \sqrt{7}$

62. $\sqrt[3]{x^2 + x + 15} - 3 = 0$

63. $\sqrt{\sqrt{x^2 + 9x + 34}} = 2$

64. $\sqrt{8 - b} = b\sqrt{8 - b}$

65. $\sqrt{x - 2} - \sqrt{x + 2} + 2 = 0$

66. $6\sqrt{y} + 6y^{-1/2} = 37$

67. $\sqrt{a^2 + 30a} = a + \sqrt{5a}$

68. $\sqrt{\sqrt{x} + 4} = \sqrt{x} - 2$

69. $\frac{x - 1}{\sqrt{x^2 + 3x + 6}} = \frac{1}{4}$

70. $\sqrt{x + 1} - \frac{2}{\sqrt{x + 1}} = 1$

71. $\sqrt{y^2 + 6} + y - 3 = 0$

72. $2\sqrt{x - 1} - \sqrt{3x - 5} = \sqrt{x - 9}$

73. $\sqrt{y + 1} - \sqrt{2y - 5} = \sqrt{y - 2}$

74. Evaluate: $\sqrt{7 + 4\sqrt{3}} - \sqrt{7 - 4\sqrt{3}}.$

10.7 Applications Involving Powers and Roots

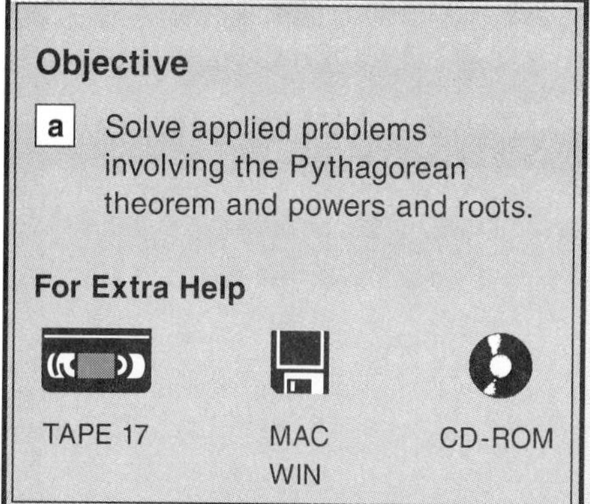

a Applications

There are many kinds of applied problems that involve powers and roots. Many also make use of right triangles and the Pythagorean theorem: $a^2 + b^2 = c^2$.

Example 1 *Vegetable Garden.* Benito and Dominique are planting a vegetable garden in the backyard. They decide that it will be a 30-ft by 40-ft rectangle and begin to lay it out using string. They soon realize that it is difficult to form the right angles and that it would be helpful to know the length of a diagonal. Find the length of a diagonal.

Using the Pythagorean theorem, $a^2 + b^2 = c^2$, we substitute 30 for a and 40 for b and then solve for c:

$$a^2 + b^2 = c^2$$
$$30^2 + 40^2 = c^2 \quad \textbf{Substituting}$$
$$900 + 1600 = c^2$$
$$2500 = c^2$$
$$\sqrt{2500} = c$$
$$50 = c.$$

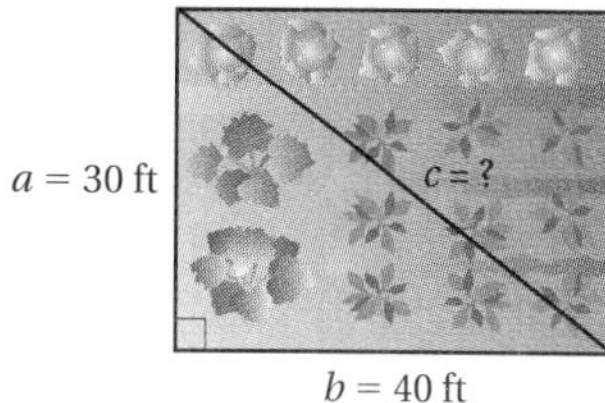

The length of the hypotenuse, or the diagonal, is 50 ft. Knowing this measurement would help in laying out the garden. Construction workers often use a procedure like this to lay out a right angle.

Example 2 Find the length of the hypotenuse of this right triangle. Give an exact answer and an approximation to three decimal places.

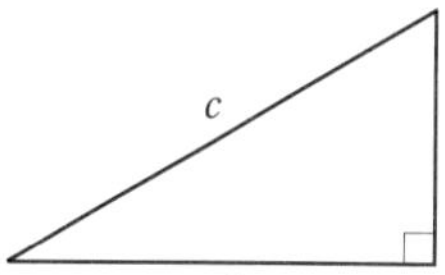

$$7^2 + 4^2 = c^2 \quad \textbf{Substituting in the Pythagorean theorem}$$
$$49 + 16 = c^2$$
$$65 = c^2$$

Exact answer: $c = \sqrt{65}$

Approximation: $c \approx 8.062$ **Using a calculator**

Example 3 Find the missing length b in this right triangle. Give an exact answer and an approximation to three decimal places.

$$1^2 + b^2 = (\sqrt{11})^2 \quad \textbf{Substituting in the Pythagorean theorem}$$
$$1 + b^2 = 11$$
$$b^2 = 10$$

Exact answer: $b = \sqrt{10}$

Approximation: $b \approx 3.162$

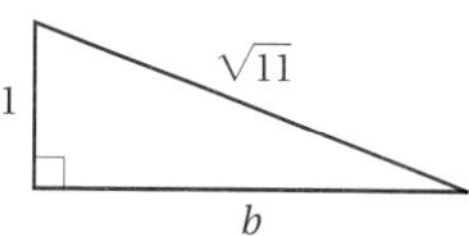

Do Exercises 1–3.

1. Find the length of the hypotenuse of this right triangle. Give an exact answer and an approximation to three decimal places.

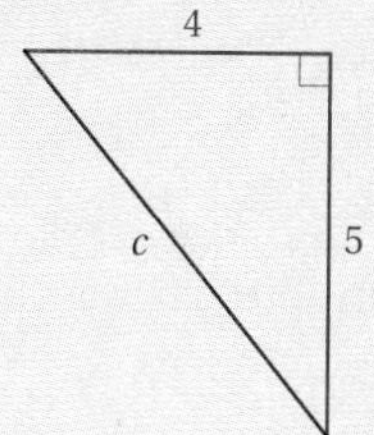

2. Find the length of the leg of this right triangle. Give an exact answer and an approximation to three decimal places.

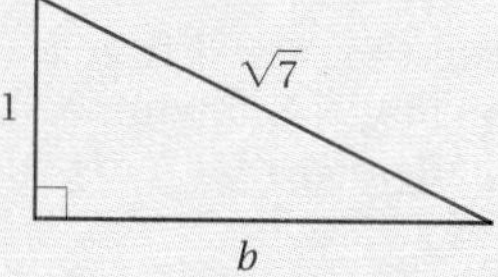

3. Find the length of the hypotenuse of this right triangle. Give an exact answer and an approximation to three decimal places.

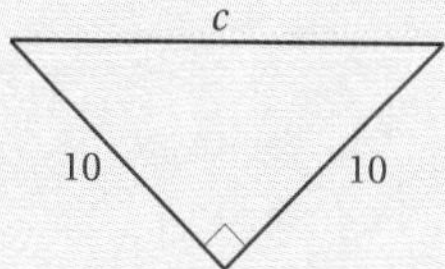

Answers on page A-43

4. *Baseball Diamond.* A baseball diamond is actually a square 90 ft on a side. Suppose a catcher fields a bunt along the third-base line 10 ft from home plate. How far would the catcher have to throw the ball to first base? Give an exact answer and an approximation to three decimal places.

5. Referring to Example 5, find L given that $h = 3$ ft and $d = 180$ ft. You will need a calculator with an exponential key $\boxed{y^x}$, or $\boxed{\wedge}$.

Answers on page A-43

Example 4 *Ramps for the Handicapped.* Laws regarding access ramps for the handicapped state that a ramp must be in the form of a right triangle, where every vertical length (leg) of 1 ft has a horizontal length (leg) of 12 ft. What is the length of a ramp with a 12-ft horizontal leg and a 1-ft vertical leg? Give an exact answer and an approximation to three decimal places.

We make a drawing and let $h =$ the length of the ramp. It is the length of the hypotenuse of a right triangle whose legs are 12 ft and 1 ft. We substitute these values into the Pythagorean theorem to find h.

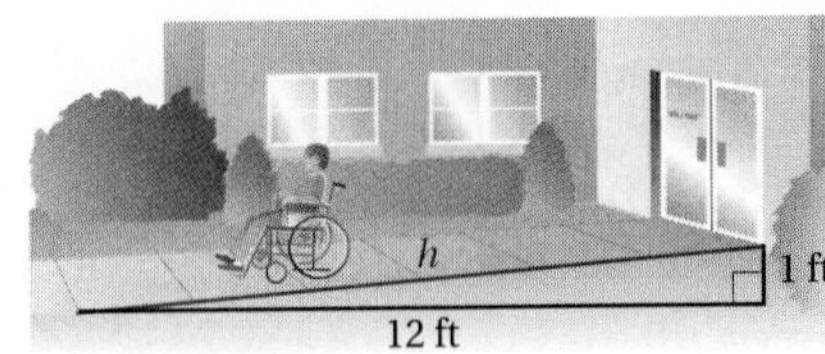

$$h^2 = 12^2 + 1^2$$
$$h^2 = 144 + 1$$
$$h^2 = 145$$
$$h = \sqrt{145}$$

Exact answer: $h = \sqrt{145}$ ft

Approximation: $h \approx 12.042$ ft

Do Exercise 4.

Example 5 *Road-Pavement Messages.* In a psychological study, it was determined that the proper length L of the letters of a word painted on pavement is given by

$$L = \frac{0.000169d^{2.27}}{h},$$

where d is the distance of a car from the lettering and h is the height of the eye above the road. All units are in feet. In other words, for a person h feet above the road, a message d feet away will be the most readable if the length of the letters is L.

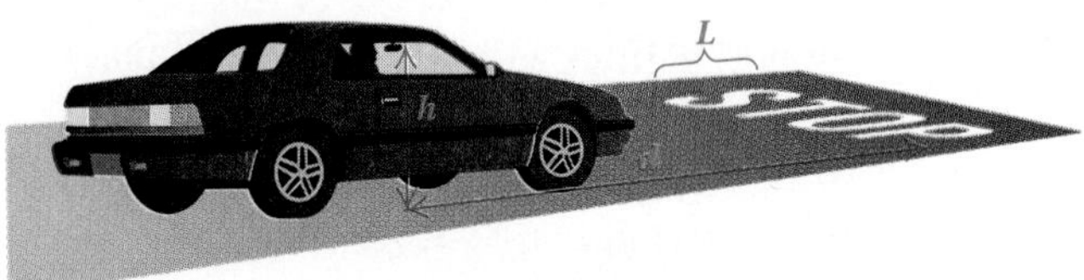

Find L, given that $h = 4$ ft and $d = 180$ ft.

We substitute 4 for h and 180 for d and calculate L using a calculator with an exponential key $\boxed{y^x}$, or $\boxed{\wedge}$:

$$L = \frac{0.000169(180)^{2.27}}{4}$$
$$\approx 5.6 \text{ ft.}$$

Do Exercise 5.

Exercise Set 10.7

a In a right triangle, find the length of the side not given. Give an exact answer and an approximation to three decimal places.

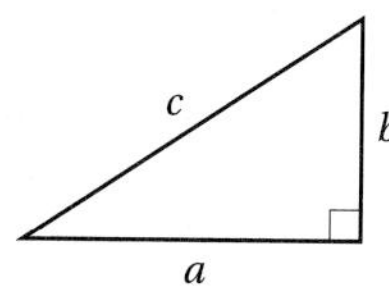

1. $a = 3, \quad b = 5$

2. $a = 8, \quad b = 10$

3. $a = 15, \quad b = 15$

4. $a = 8, \quad b = 8$

5. $b = 12, \quad c = 13$

6. $a = 5, \quad c = 12$

7. $c = 7, \quad a = \sqrt{6}$

8. $c = 10, \quad a = 4\sqrt{5}$

9. $b = 1, \quad c = \sqrt{13}$

10. $a = 1, \quad c = \sqrt{12}$

11. $a = 1, \quad c = \sqrt{n}$

12. $c = 2, \quad a = \sqrt{n}$

In the following problems, give an exact answer and, where appropriate, an approximation to three decimal places.

13. *Bridge Expansion.* During the summer heat, a 2-mi bridge expands 2 ft in length. If we assume that the bulge occurs straight up the middle, how high is the bulge? (The answer may surprise you. In reality, bridges are built with expansion spaces to avoid such buckling.)

14. Triangle ABC has sides of lengths 25 ft, 25 ft, and 30 ft. Triangle PQR has sides of lengths 25 ft, 25 ft, and 40 ft. Which triangle has the greater area and by how much?

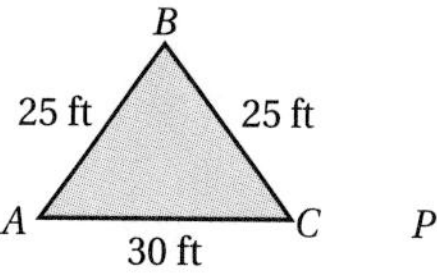

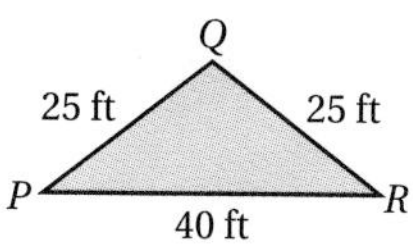

15. *Guy Wire.* How long is a guy wire reaching from the top of a 10-ft pole to a point on the ground 4 ft from the pole?

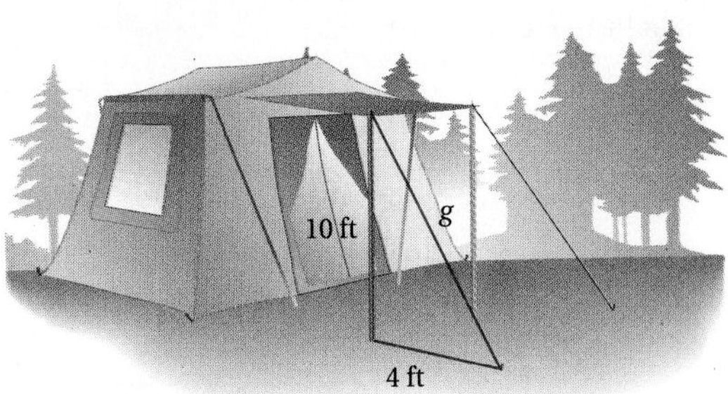

16. *Softball Diamond.* A slow-pitch softball diamond is actually a square 65 ft on a side. How far is it from home to second base?

17. Each side of a regular octagon has length s. Find a formula for the distance d between the parallel sides of the octagon.

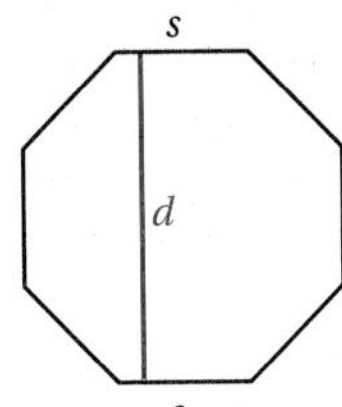

18. The two equal sides of an isosceles right triangle are of length s. Find a formula for the length of the hypotenuse.

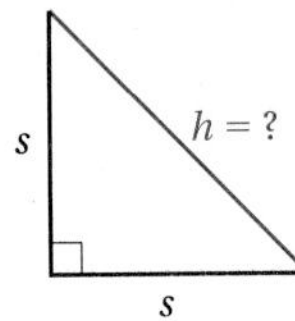

19. *Road-Pavement Messages.* Using the formula of Example 5, find the length L of a road-pavement message when $h = 4$ ft and $d = 200$ ft.

20. *Road-Pavement Messages.* Using the formula of Example 5, find the length L of a road-pavement message when $h = 8$ ft and $d = 300$ ft.

21. The length and the width of a rectangle are given by consecutive integers. The area of the rectangle is 90 cm^2. Find the length of a diagonal of the rectangle.

22. The diagonal of a square has length $8\sqrt{2}$ ft. Find the length of a side of the square.

23. *Television Sets.* What does it mean to refer to a 20-in. TV set or a 25-in. TV set? Such units refer to the diagonal of the screen. A 20-in. TV set also has a width of 16 in. What is its height?

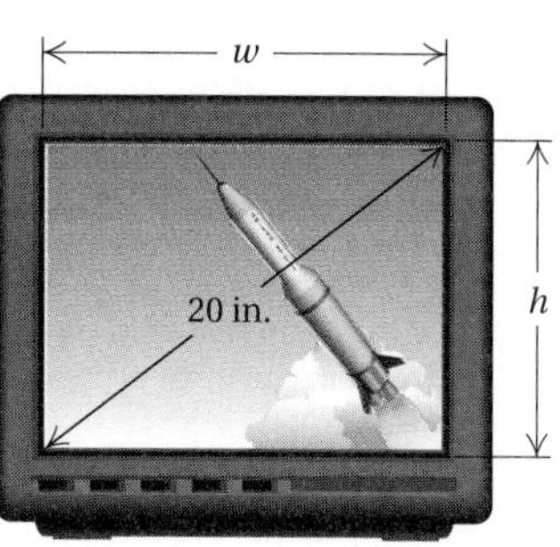

24. *Television Sets.* A 25-in. TV set has a screen with a height of 15 in. What is its width?

25. Find all points on the x-axis of a Cartesian coordinate system that are 5 units from the point (0, 4).

26. Find all points on the y-axis of a Cartesian coordinate system that are 5 units from the point (3, 0).

27. *Speaker Placement.* A stereo receiver is in a corner of a 12-ft by 14-ft room. Speaker wire will run under a rug, diagonally, to a speaker in the far corner. If 4 ft of slack is required on each end, how long a piece of wire should be purchased?

28. *Distance Over Water.* To determine the width of a pond, a surveyor locates two stakes at either end of the pond and uses instrumentation to place a third stake so that the distance across the pond is the length of a hypotenuse. If the third stake is 90 m from one stake and 70 m from the other, how wide is the pond?

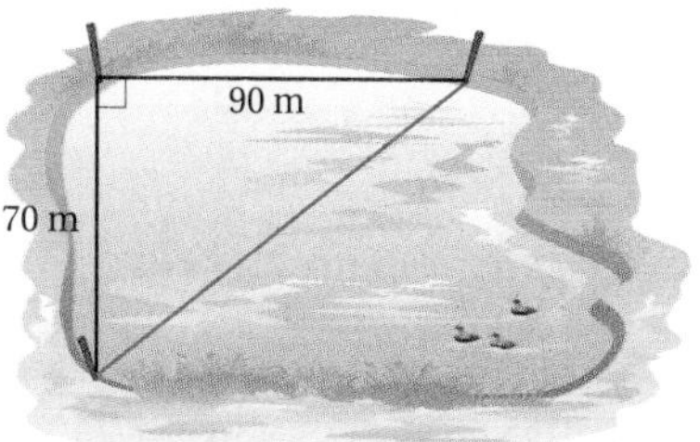

29. *Plumbing.* Plumbers use the Pythagorean theorem to calculate pipe length. If a pipe is to be offset, as shown in the figure, the *travel,* or length, of the pipe, is calculated using the lengths of the *advance* and *offset.*

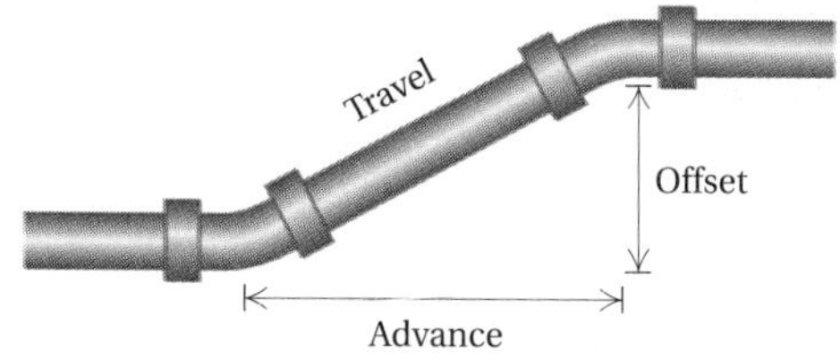

Find the travel if the offset is 17.75 in. and the advance is 10.25 in.

30. *Carpentry.* Dale is laying out the footer of a house. To see if the corner is square, she measures 16 ft from the corner along one wall and 20 ft from the corner along the other wall. How long should the diagonal be between those two points if the corner is a right angle?

Skill Maintenance

Solve. [6.6a]

31. *Commuter Travel.* The speed of the Zionsville Flash commuter train is 14 mph faster than that of the Carmel Crawler. The Flash travels 290 mi in the same time that it takes the Crawler to travel 230 mi. Find the speed of each train.

32. *Marine Travel.* A motor boat travels three times as fast as the current in the Saskatee River. A trip up the river and back takes 10 hr, and the total distance of the trip is 100 mi. Find the speed of the current.

Solve. [5.8b], [6.5a]

33. $2x^2 + 11x - 21 = 0$

34. $x^2 + 24 = 11x$

35. $\dfrac{x+2}{x+3} = \dfrac{x-4}{x-5}$

36. $3x^2 - 12 = 0$

37. $\dfrac{x-5}{x-7} = \dfrac{4}{3}$

38. $\dfrac{x-1}{x-3} = \dfrac{6}{x-3}$

Synthesis

39. ◆ Write a problem for a classmate to solve in which the solution is "The height of the tepee is $5\sqrt{3}$ yd."

40. ◆ Write a problem for a classmate to solve in which the solution is "The height of the window is $15\sqrt{3}$ yd."

41. *Roofing.* Kit's cottage, which is 24 ft wide and 32 ft long, needs a new roof. By counting clapboards that are 4 in. apart, Kit determines that the peak of the roof is 6 ft higher than the sides. If one packet of shingles covers 100 square feet, how many packets will the job require?

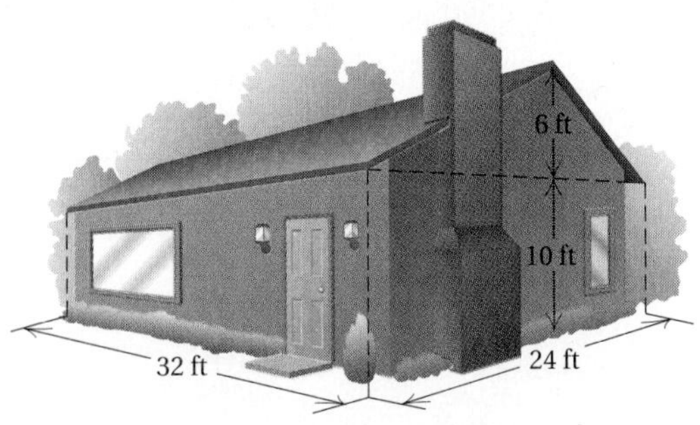

42. *Painting.* (Refer to Exercise 41.) A gallon of paint covers about 275 square feet. If Kit's first floor is 10 ft high, how many gallons of paint should be bought to paint the house? What assumption(s) is made in your answer?

43. A cube measures 5 cm on each side. How long is the diagonal that connects two opposite corners of the cube? Give an exact answer.

44. *Wind Chill Temperature.* We can use square roots to consider an application involving the effect of wind on the feeling of cold in the winter. Because wind enhances the loss of heat from the skin, we feel colder when there is wind than when there is not. The *wind chill temperature* is what the temperature would have to be with no wind in order to give the same chilling effect. A formula for finding the wind chill temperature, T_W, is

$$T_W = 91.4 - \frac{(10.45 + 6.68\sqrt{v} - 0.447v)(457 - 5T)}{110},$$

where T is the actual temperature given by a thermometer, in degrees Fahrenheit, and v is the wind speed, in miles per hour. Use a calculator to find the wind chill temperature in each case. Round to the nearest degree.

a) $T = 40°\text{F}$, $v = 25$ mph

b) $T = 20°\text{F}$, $v = 25$ mph

c) $T = 10°\text{F}$, $v = 20$ mph

d) $T = 10°\text{F}$, $v = 40$ mph

e) $T = -5°\text{F}$, $v = 35$ mph

f) $T = -16°\text{F}$, $v = 40$ mph

Collaborative Learning Manual — Develop a formula for the swing time of a pendulum.

10.8 The Complex Numbers

a Imaginary and Complex Numbers

Negative numbers do not have square roots in the real-number system. However, mathematicians have invented a larger number system that contains the real-number system, but is such that negative numbers have square roots. That system is called the **complex-number system.** We begin by creating a number that is a square root of -1. We call this new number i.

> We define the number i to be $\sqrt{-1}$. That is, $i = \sqrt{-1}$ and $i^2 = -1$.

To express roots of negative numbers in terms of i, we can use the fact that in the complex numbers, $\sqrt{-p} = \sqrt{-1 \cdot p} = \sqrt{-1}\sqrt{p}$ when p is a positive real number.

Examples Express in terms of i.

1. $\sqrt{-7} = \sqrt{-1 \cdot 7} = \sqrt{-1} \cdot \sqrt{7} = i\sqrt{7}$, or $\sqrt{7}\,i$ (i is *not* under the radical.)
2. $\sqrt{-16} = \sqrt{-1 \cdot 16} = \sqrt{-1} \cdot \sqrt{16} = i \cdot 4 = 4i$
3. $-\sqrt{-13} = -\sqrt{-1 \cdot 13} = -\sqrt{-1} \cdot \sqrt{13} = -i\sqrt{13}$, or $-\sqrt{13}\,i$
4. $-\sqrt{-64} = -\sqrt{-1 \cdot 64} = -\sqrt{-1} \cdot \sqrt{64} = -i \cdot 8 = -8i$
5. $\sqrt{-48} = \sqrt{-1 \cdot 48} = \sqrt{-1} \cdot \sqrt{48} = i\sqrt{48} = i \cdot 4\sqrt{3} = 4\sqrt{3}\,i$, or $4i\sqrt{3}$

Do Exercises 1–5.

> An **imaginary*** **number** is a number that can be named
>
> bi,
>
> where b is some real number and $b \neq 0$.

To form the system of **complex numbers,** we take the imaginary numbers and the real numbers and all possible sums of real and imaginary numbers. These are complex numbers:

$$7 - 4i, \qquad -\pi + 19i, \qquad 37, \qquad i\sqrt{8}.$$

> A **complex number** is any number that can be named
>
> $a + bi$,
>
> where a and b are any real numbers. (Note that either a or b or both can be 0.)

*Don't let the name "imaginary" fool you. The imaginary numbers are very important in such fields as engineering and the physical sciences.

Objectives

- a Express imaginary numbers as bi, where b is a nonzero real number, and complex numbers as $a + bi$, where a and b are real numbers.
- b Add and subtract complex numbers.
- c Multiply complex numbers.
- d Write expressions involving powers of i in the form $a + bi$.
- e Find conjugates of complex numbers and divide complex numbers.
- f Determine whether a given complex number is a solution of an equation.

For Extra Help

TAPE 17

MAC WIN

CD-ROM

Express in terms of i.

1. $\sqrt{-5}$
2. $\sqrt{-25}$
3. $-\sqrt{-11}$
4. $-\sqrt{-36}$
5. $\sqrt{-54}$

Answers on page A-43

Add or subtract.

6. $(7 + 4i) + (8 - 7i)$

7. $(-5 - 6i) + (-7 + 12i)$

8. $(8 + 3i) - (5 + 8i)$

9. $(5 - 4i) - (-7 + 3i)$

Answers on page A-43

Since $0 + bi = bi$, every imaginary number is a complex number. Similarly, $a + 0i = a$, so every real number is a complex number. The relationships among various real and complex numbers are shown in the following diagram.

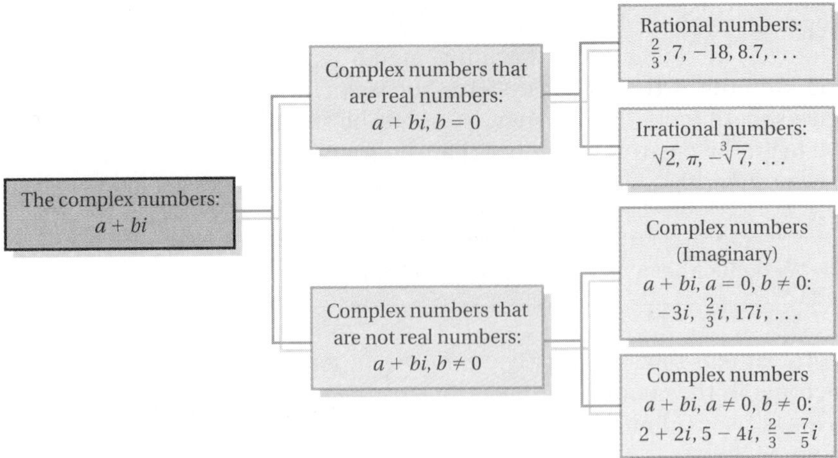

It is important to keep in mind some comparisons between numbers that have real-number roots and those that have complex-number roots that are not real. For example, $\sqrt{-48}$ is a complex number that is not a real number because we are taking the square root of a negative number. *But,* $\sqrt[3]{-125}$ is a real number because we are taking the cube root of a negative number and *any* real number has a cube root that is a real number.

b Addition and Subtraction

The complex numbers follow the commutative and associative laws of addition. Thus we can add and subtract them as we do binomials with real-number coefficients, that is, we collect like terms.

Examples Add or subtract.

6. $(8 + 6i) + (3 + 2i) = (8 + 3) + (6 + 2)i = 11 + 8i$

7. $(3 + 2i) - (5 - 2i) = (3 - 5) + [2 - (-2)]i = -2 + 4i$

Do Exercises 6–9.

c Multiplication

The complex numbers obey the commutative, associative, and distributive laws. But although the property $\sqrt{a}\sqrt{b} = \sqrt{ab}$ does *not* hold for complex numbers in general, it does hold when $a = -1$ and b is a positive real number. To multiply square roots of negative real numbers, we first express them in terms of i. For example,

$$\sqrt{-2} \cdot \sqrt{-5} = \sqrt{-1} \cdot \sqrt{2} \cdot \sqrt{-1} \cdot \sqrt{5} = i\sqrt{2} \cdot i\sqrt{5}$$
$$= i^2\sqrt{10} = -\sqrt{10} \quad \text{is correct!}$$

But $\sqrt{-2} \cdot \sqrt{-5} = \sqrt{(-2)(-5)} = \sqrt{10}$ is wrong!

Keeping this and the fact that $i^2 = -1$ in mind, we multiply in much the same way that we do with real numbers.

Examples Multiply.

8. $\sqrt{-49} \cdot \sqrt{-16} = \sqrt{-1} \cdot \sqrt{49} \cdot \sqrt{-1} \cdot \sqrt{16}$
$= i \cdot 7 \cdot i \cdot 4$
$= i^2(28)$
$= (-1)(28)$ **$i^2 = -1$**
$= -28$

9. $\sqrt{-3} \cdot \sqrt{-7} = \sqrt{-1} \cdot \sqrt{3} \cdot \sqrt{-1} \cdot \sqrt{7}$
$= i \cdot \sqrt{3} \cdot i \cdot \sqrt{7}$
$= i^2(\sqrt{21})$
$= (-1)\sqrt{21}$ **$i^2 = -1$**
$= -\sqrt{21}$

10. $-2i \cdot 5i = -10 \cdot i^2$
$= (-10)(-1)$ **$i^2 = -1$**
$= 10$

11. $(-4i)(3 - 5i) = (-4i) \cdot 3 - (-4i)(5i)$ **Using a distributive law**
$= -12i + 20i^2$
$= -12i + 20(-1)$ **$i^2 = -1$**
$= -12i - 20$
$= -20 - 12i$

12. $(1 + 2i)(1 + 3i) = 1 + 3i + 2i + 6i^2$ **Multiplying each term of one number by every term of the other (FOIL)**
$= 1 + 3i + 2i - 6$ **$i^2 = -1$**
$= -5 + 5i$ **Collecting like terms**

13. $(3 - 2i)^2 = 3^2 - 2(3)(2i) + (2i)^2$ **Squaring the binomial**
$= 9 - 12i + 4i^2$
$= 9 - 12i - 4$
$= 5 - 12i$

Do Exercises 10–17.

d Powers of *i*

We now want to simplify certain expressions involving powers of i. To do so, we first see how to simplify powers of i. Simplifying powers of i can be done by using the fact that $i^2 = -1$ and expressing the given power of i in terms of even powers, and then in terms of powers of i^2. Consider the following:

$i,$
$i^2 = -1,$
$i^3 = i^2 \cdot i = (-1)i = -i,$
$i^4 = (i^2)^2 = (-1)^2 = 1,$
$i^5 = i^4 \cdot i = (i^2)^2 \cdot i = (-1)^2 \cdot i = i,$
$i^6 = (i^2)^3 = (-1)^3 = -1.$

Note that the powers of i cycle themselves through the values i, -1, $-i$, and 1.

Multiply.

10. $\sqrt{-25} \cdot \sqrt{-4}$

11. $\sqrt{-2} \cdot \sqrt{-17}$

12. $-6i \cdot 7i$

13. $-3i(4 - 3i)$

14. $5i(-5 + 7i)$

15. $(1 + 3i)(1 + 5i)$

16. $(3 - 2i)(1 + 4i)$

17. $(3 + 2i)^2$

Answers on page A-43

Simplify.

18. i^{47}

19. i^{68}

20. i^{85}

21. i^{90}

Simplify.

22. $8 - i^5$

23. $7 + 4i^2$

24. $6i^{11} + 7i^{14}$

25. $i^{34} - i^{55}$

Find the conjugate.

26. $6 + 3i$

27. $-9 - 5i$

28. $\pi - \frac{1}{4}i$

Answers on page A-43

Examples Simplify.

14. $i^{37} = i^{36} \cdot i = (i^2)^{18} \cdot i = (-1)^{18} \cdot i = 1 \cdot i = i$

15. $i^{58} = (i^2)^{29} = (-1)^{29} = -1$

16. $i^{75} = i^{74} \cdot i = (i^2)^{37} \cdot i = (-1)^{37} \cdot i = -1 \cdot i = -i$

17. $i^{80} = (i^2)^{40} = (-1)^{40} = 1$

Do Exercises 18–21.

Now let's simplify other expressions.

Examples Simplify to the form $a + bi$.

18. $8 - i^2 = 8 - (-1) = 8 + 1 = 9$

19. $17 + 6i^3 = 17 + 6 \cdot i^2 \cdot i = 17 + 6(-1)i = 17 - 6i$

20. $i^{22} - 67i^2 = (i^2)^{11} - 67(-1) = (-1)^{11} + 67 = -1 + 67 = 66$

21. $i^{23} + i^{48} = (i^{22}) \cdot i + (i^2)^{24} = (i^2)^{11} \cdot i + (-1)^{24} = (-1)^{11} \cdot i + (-1)^{24}$
$= -i + 1 = 1 - i$

Do Exercises 22–25.

e Conjugates and Division

Conjugates of complex numbers are defined as follows.

The **conjugate** of a complex number $a + bi$ is $a - bi$, and the **conjugate** of $a - bi$ is $a + bi$.

Examples Find the conjugate.

22. $5 + 7i$ The conjugate is $5 - 7i$.

23. $14 - 3i$ The conjugate is $14 + 3i$.

24. $-3 - 9i$ The conjugate is $-3 + 9i$.

25. $4i$ The conjugate is $-4i$.

Do Exercises 26–28.

When we multiply a complex number by its conjugate, we get a real number.

Examples Multiply.

26.
$$\begin{aligned}(5 + 7i)(5 - 7i) &= 5^2 - (7i)^2 && \text{Using } (A + B)(A - B) = A^2 - B^2\\ &= 25 - 49i^2\\ &= 25 - 49(-1) && i^2 = -1\\ &= 25 + 49\\ &= 74\end{aligned}$$

27.
$$\begin{aligned}(2 - 3i)(2 + 3i) &= 2^2 - (3i)^2\\ &= 4 - 9i^2\\ &= 4 - 9(-1) && i^2 = -1\\ &= 4 + 9\\ &= 13\end{aligned}$$

Do Exercises 29 and 30.

We use conjugates in dividing complex numbers.

Example 28 Divide and simplify to the form $a + bi$: $\dfrac{-5 + 9i}{1 - 2i}$.

$$\frac{-5 + 9i}{1 - 2i} \cdot \frac{1 + 2i}{1 + 2i} = \frac{(-5 + 9i)(1 + 2i)}{(1 - 2i)(1 + 2i)}$$ Multiplying by 1 using the conjugate of the denominator in the symbol for 1

$$= \frac{-5 - 10i + 9i + 18i^2}{1^2 - 4i^2}$$

$$= \frac{-5 - i + 18(-1)}{1 - 4(-1)}$$ $i^2 = -1$

$$= \frac{-5 - i - 18}{1 + 4}$$

$$= \frac{-23 - i}{5} = -\frac{23}{5} - \frac{1}{5}i$$

Note the similarity between the preceding example and rationalizing denominators. In both cases, we used the conjugate of the denominator to write another name for 1. In Example 28, the symbol for the number 1 was chosen using the conjugate of the divisor, $1 - 2i$.

Example 29 What symbol for 1 would you use to divide?

Division to be done	*Symbol for 1*
$\dfrac{3 + 5i}{4 + 3i}$	$\dfrac{4 - 3i}{4 - 3i}$

Example 30 Divide and simplify to the form $a + bi$: $\dfrac{3 + 5i}{4 + 3i}$.

$$\frac{3 + 5i}{4 + 3i} \cdot \frac{4 - 3i}{4 - 3i} = \frac{(3 + 5i)(4 - 3i)}{(4 + 3i)(4 - 3i)}$$ Multiplying by 1

$$= \frac{12 - 9i + 20i - 15i^2}{4^2 - 9i^2}$$

$$= \frac{12 + 11i - 15(-1)}{16 - 9(-1)}$$ $i^2 = -1$

$$= \frac{27 + 11i}{25} = \frac{27}{25} + \frac{11}{25}i$$

Do Exercises 31 and 32.

Multiply.

29. $(7 - 2i)(7 + 2i)$

30. $(-3 - i)(-3 + i)$

Divide and simplify to the form $a + bi$.

31. $\dfrac{6 + 2i}{1 - 3i}$

32. $\dfrac{2 + 3i}{-1 + 4i}$

33. Determine whether $-i$ is a solution of $x^2 + 1 = 0$.

34. Determine whether $1 - i$ is a solution of $x^2 - 2x + 2 = 0$.

Answers on page A-43

Calculator Spotlight

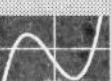 Complex Numbers. Many graphers can be used to handle complex numbers. The TI-83 (but not the TI-82) can do complex operations. We first set the grapher in the complex $a + bi$ mode by pressing MODE, positioning the cursor over $a + bi$, and pressing ENTER. Then we can carry out operations by using the i key (press 2nd i). When finding $\sqrt{-1}$, we get i on the display if we are in complex $a + bi$ mode. We get an error message if we are in real mode.

Some graphers (the TI-85, TI-86, and TI-92, for example) will give complex solutions to equations.

Exercises

Carry out each of the following in complex $a + bi$ mode.

1. $(9 + 4i) + (-11 - 13i)$
2. $(9 + 4i) - (-11 - 13i)$
3. $(9 + 4i) \cdot (-11 - 13i)$
4. $(9 + 4i) \div (-11 - 13i)$
5. $\sqrt{-16}$
6. $\sqrt{-16} \cdot \sqrt{-25}$
7. $\sqrt{-23} \cdot \sqrt{-35}$
8. $\dfrac{4 - 5i}{-6 + 8i}$
9. $(-3i)^4$
10. i^{11}
11. $i^{86} - i^{100}$
12. $\left(\frac{2}{3} - \frac{4}{5}i\right)^3 - \left(\frac{1}{2} + \frac{3}{5}i\right)^4$
13. $\dfrac{i^{47} - i^{38}}{1 + i}$

f Solutions of Equations

The equation $x^2 + 1 = 0$ has no real-number solution, but it has two nonreal complex solutions.

Example 31 Determine whether i is a solution of the equation $x^2 + 1 = 0$.

We substitute i for x in the equation.

$$\begin{array}{r|l} x^2 + 1 & 0 \\ \hline i^2 + 1 & ?\ 0 \\ -1 + 1 & \\ 0 & \qquad \text{TRUE} \end{array}$$

The number i is a solution.

Do Exercise 33 on the preceding page.

Any equation consisting of a polynomial in one variable on one side and 0 on the other has complex-number solutions (some may be real). It is not always easy to find the solutions, but they always exist.

Example 32 Determine whether $1 + i$ is a solution of the equation $x^2 - 2x + 2 = 0$.

We substitute $1 + i$ for x in the equation.

$$\begin{array}{r|l} x^2 - 2x + 2 & 0 \\ \hline (1 + i)^2 - 2(1 + i) + 2 & ?\ 0 \\ 1 + 2i + i^2 - 2 - 2i + 2 & \\ 1 + 2i - 1 - 2 - 2i + 2 & \\ (1 - 1 - 2 + 2) + (2 - 2)i & \\ 0 + 0i & \\ 0 & \qquad \text{TRUE} \end{array}$$

The number $1 + i$ is a solution.

Example 33 Determine whether $2i$ is a solution of $x^2 + 3x - 4 = 0$.

$$\begin{array}{r|l} x^2 + 3x - 4 & 0 \\ \hline (2i)^2 + 3(2i) - 4 & ?\ 0 \\ 4i^2 + 6i - 4 & \\ -4 + 6i - 4 & \\ -8 + 6i & \qquad \text{FALSE} \end{array}$$

The number $2i$ is not a solution.

Do Exercise 34 on the preceding page.

Exercise Set 10.8

a Express in terms of i.

1. $\sqrt{-35}$
2. $\sqrt{-21}$
3. $\sqrt{-16}$
4. $\sqrt{-36}$
5. $-\sqrt{-12}$
6. $-\sqrt{-20}$
7. $\sqrt{-3}$
8. $\sqrt{-4}$
9. $\sqrt{-81}$
10. $\sqrt{-27}$
11. $\sqrt{-98}$
12. $-\sqrt{-18}$
13. $-\sqrt{-49}$
14. $-\sqrt{-125}$
15. $4 - \sqrt{-60}$
16. $6 - \sqrt{-84}$
17. $\sqrt{-4} + \sqrt{-12}$
18. $-\sqrt{-76} + \sqrt{-125}$

b Add or subtract and simplify.

19. $(7 + 2i) + (5 - 6i)$
20. $(-4 + 5i) + (7 + 3i)$
21. $(4 - 3i) + (5 - 2i)$
22. $(-2 - 5i) + (1 - 3i)$
23. $(9 - i) + (-2 + 5i)$
24. $(6 + 4i) + (2 - 3i)$
25. $(6 - i) - (10 + 3i)$
26. $(-4 + 3i) - (7 + 4i)$
27. $(4 - 2i) - (5 - 3i)$
28. $(-2 - 3i) - (1 - 5i)$
29. $(9 + 5i) - (-2 - i)$
30. $(6 - 3i) - (2 + 4i)$

c Multiply.

31. $\sqrt{-36} \cdot \sqrt{-9}$

32. $\sqrt{-16} \cdot \sqrt{-64}$

33. $\sqrt{-7} \cdot \sqrt{-2}$

34. $\sqrt{-11} \cdot \sqrt{-3}$

35. $-3i \cdot 7i$

36. $8i \cdot 5i$

37. $-3i(-8 - 2i)$

38. $4i(5 - 7i)$

39. $(3 + 2i)(1 + i)$

40. $(4 + 3i)(2 + 5i)$

41. $(2 + 3i)(6 - 2i)$

42. $(5 + 6i)(2 - i)$

43. $(6 - 5i)(3 + 4i)$

44. $(5 - 6i)(2 + 5i)$

45. $(7 - 2i)(2 - 6i)$

46. $(-4 + 5i)(3 - 4i)$

47. $(3 - 2i)^2$

48. $(5 - 2i)^2$

49. $(1 + 5i)^2$

50. $(6 + 2i)^2$

51. $(-2 + 3i)^2$

52. $(-5 - 2i)^2$

d Simplify.

53. i^7

54. i^{11}

55. i^{24}

56. i^{35}

57. i^{42}

58. i^{64}

59. i^9

60. $(-i)^{71}$

61. i^6

62. $(-i)^4$

63. $(5i)^3$

64. $(-3i)^5$

Simplify to the form $a + bi$.

65. $7 + i^4$

66. $-18 + i^3$

67. $i^{28} - 23i$

68. $i^{29} + 33i$

69. $i^2 + i^4$

70. $5i^5 + 4i^3$

71. $i^5 + i^7$

72. $i^{84} - i^{100}$

73. $1 + i + i^2 + i^3 + i^4$

74. $i - i^2 + i^3 - i^4 + i^5$

75. $5 - \sqrt{-64}$

76. $\sqrt{-12} + 36i$

77. $\dfrac{8 - \sqrt{-24}}{4}$

78. $\dfrac{9 + \sqrt{-9}}{3}$

e Divide and simplify to the form $a + bi$.

79. $\dfrac{4 + 3i}{3 - i}$

80. $\dfrac{5 + 2i}{2 + i}$

81. $\dfrac{3 - 2i}{2 + 3i}$

82. $\dfrac{6 - 2i}{7 + 3i}$

83. $\dfrac{8 - 3i}{7i}$

84. $\dfrac{3 + 8i}{5i}$

85. $\dfrac{4}{3 + i}$

86. $\dfrac{6}{2 - i}$

87. $\dfrac{2i}{5 - 4i}$

88. $\dfrac{8i}{6 + 3i}$

89. $\dfrac{4}{3i}$

90. $\dfrac{5}{6i}$

91. $\dfrac{2 - 4i}{8i}$

92. $\dfrac{5 + 3i}{i}$

93. $\dfrac{6 + 3i}{6 - 3i}$

94. $\dfrac{4 - 5i}{4 + 5i}$

f Determine whether the complex number is a solution of the equation.

95. $1 - 2i$;
$x^2 - 2x + 5 = 0$

96. $1 + 2i$;
$x^2 - 2x + 5 = 0$

97. $2 + i$;
$x^2 - 4x - 5 = 0$

98. $1 - i$;
$x^2 + 2x + 2 = 0$

Skill Maintenance

Solve. [6.5a]

99. $\dfrac{196}{x^2 - 7x + 49} - \dfrac{2x}{x + 7} = \dfrac{2058}{x^3 + 343}$

100. $\dfrac{5}{t} - \dfrac{3}{2} = \dfrac{4}{7}$

Solve. [9.3c, d, e]

101. $|3x + 7| = 22$

102. $|3x + 7| < 22$

103. $|3x + 7| \geq 22$

104. $|3x + 7| = |2x - 5|$

Synthesis

105. ◆ How are conjugates of complex numbers similar to the conjugates used in Section 10.5?

106. ◆ Is every real number a complex number? Why or why not?

107. A complex function g is given by

$$g(z) = \frac{z^4 - z^2}{z - 1}.$$

Find $g(2i)$, $g(1 + i)$, and $g(-1 + 2i)$.

108. Evaluate $\dfrac{1}{w - w^2}$ for $w = \dfrac{1 - i}{10}$.

Express in terms of i.

109. $\dfrac{1}{8}(-24 - \sqrt{-1024})$

110. $12\sqrt{-\dfrac{1}{32}}$

111. $7\sqrt{-64} - 9\sqrt{-256}$

Simplify.

112. $\dfrac{i^5 + i^6 + i^7 + i^8}{(1 - i)^4}$

113. $(1 - i)^3(1 + i)^3$

114. $\dfrac{5 - \sqrt{5}i}{\sqrt{5}i}$

115. $\dfrac{6}{1 + \dfrac{3}{i}}$

116. $\left(\dfrac{1}{2} - \dfrac{1}{3}i\right)^2 - \left(\dfrac{1}{2} + \dfrac{1}{3}i\right)^2$

117. $\dfrac{i - i^{38}}{1 + i}$

118. Find all numbers a for which the opposite of a is the same as the reciprocal of a.

Summary and Review Exercises: Chapter 10

Important Properties and Formulas

$\sqrt{a^2} = |a|$; $\sqrt[k]{a^k} = |a|$, when k is even; $\sqrt[k]{a^k} = a$, when k is odd;

$\sqrt[k]{ab} = \sqrt[k]{a} \cdot \sqrt[k]{b}$; $\sqrt[k]{\dfrac{a}{b}} = \dfrac{\sqrt[k]{a}}{\sqrt[k]{b}}$; $\sqrt[k]{a^m} = (\sqrt[k]{a})^m$;

$a^{1/n} = \sqrt[n]{a}$; $a^{m/n} = \sqrt[n]{a^m} = (\sqrt[n]{a})^m$; $a^{-m/n} = \dfrac{1}{a^{m/n}}$

Principle of Powers: If $a = b$ is true, then $a^n = b^n$ is true.
Pythagorean Theorem: $a^2 + b^2 = c^2$, in a right triangle.

$i = \sqrt{-1}$, $i^2 = -1$, $i^3 = -i$, $i^4 = 1$

Imaginary Numbers: bi, $i^2 = -1$, $b \neq 0$

Complex Numbers: $a + bi$, $i^2 = -1$

Conjugates: $a + bi$, $a - bi$

The objectives to be tested in addition to the material in this chapter are [5.8b], [6.1d, f], [6.5a], and [6.6a, b].

Use a calculator to approximate to three decimal places. [10.1a]

1. $\sqrt{778}$ **2.** $\sqrt{\dfrac{963.2}{23.68}}$

3. For the given function, find the indicated function values. [10.1a]
$f(x) = \sqrt{3x - 16}$; $f(0), f(-1), f(1)$, and $f\left(\frac{41}{3}\right)$

4. Find the domain of the function f in Exercise 3. [10.1a]

Simplify. Assume that letters represent *any* real number. [10.1b]

5. $\sqrt{81a^2}$ **6.** $\sqrt{(-7z)^2}$

7. $\sqrt{(c-3)^2}$ **8.** $\sqrt{x^2 - 6x + 9}$

Simplify. [10.1c]

9. $\sqrt[3]{-1000}$ **10.** $\sqrt[3]{-\dfrac{1}{27}}$

11. For the given function, find the indicated function values. [10.1c]
$f(x) = \sqrt[3]{x + 2}$; $f(6), f(-10)$, and $f(25)$

Simplify. Assume that letters represent *any* real number. [10.1d]

12. $\sqrt[10]{x^{10}}$ **13.** $-\sqrt[13]{(-3)^{13}}$

Rewrite without rational exponents, and simplify, if possible. [10.2a]

14. $a^{1/5}$ **15.** $64^{3/2}$

Rewrite with rational exponents. [10.2a]

16. $\sqrt{31}$ **17.** $\sqrt[5]{a^2b^3}$

Rewrite with positive exponents, and simplify, if possible. [10.2b]

18. $49^{-1/2}$ **19.** $(8xy)^{-2/3}$

20. $5a^{-3/4}b^{1/2}c^{-2/3}$ **21.** $\dfrac{3a}{\sqrt[4]{t}}$

Use the laws of exponents to simplify. Write answers with positive exponents. [10.2c]

22. $(x^{-2/3})^{3/5}$ **23.** $\dfrac{7^{-1/3}}{7^{-1/2}}$

Use rational exponents to simplify. Write the answer in radical notation if appropriate. [10.2d]

24. $\sqrt[3]{x^{21}}$ **25.** $\sqrt[3]{27x^6}$

Use rational exponents to write a single radical expression. [10.2d]

26. $x^{1/3}y^{1/4}$ **27.** $\sqrt[4]{x}\sqrt[3]{x}$

Simplify by factoring. Assume that all expressions under radicals represent nonnegative numbers. [10.3a]

28. $\sqrt{245}$ **29.** $\sqrt[3]{-108}$

30. $\sqrt[3]{250a^2b^6}$

Simplify. Assume that all expressions under radicals represent positive numbers. [10.3b]

31. $\sqrt{\dfrac{49}{36}}$ **32.** $\sqrt[3]{\dfrac{64x^6}{27}}$

33. $\sqrt[4]{\dfrac{16x^8}{81y^{12}}}$

Perform the indicated operations and simplify. Assume that all expressions under radicals represent positive numbers. [10.3a, b]

34. $\sqrt{5x}\sqrt{3y}$

35. $\sqrt[3]{a^5b}\sqrt[3]{27b}$

36. $\sqrt[3]{a}\sqrt[5]{b^3}$

37. $\dfrac{\sqrt[3]{60xy^3}}{\sqrt[3]{10x}}$

38. $\dfrac{\sqrt{75x}}{2\sqrt{3}}$

39. $\dfrac{\sqrt[3]{x^2}}{\sqrt[4]{x}}$

Add or subtract. Assume that all expressions under radicals represent nonnegative numbers. [10.4a]

40. $5\sqrt[3]{x} + 2\sqrt[3]{x}$

41. $2\sqrt{75} - 7\sqrt{3}$

42. $\sqrt[3]{8x^4} + \sqrt[3]{xy^6}$

43. $\sqrt{50} + 2\sqrt{18} + \sqrt{32}$

Multiply. [10.4b]

44. $(\sqrt{5} - 3\sqrt{8})(\sqrt{5} + 2\sqrt{8})$

45. $(1 - \sqrt{7})^2$

46. $(\sqrt[3]{27} - \sqrt[3]{2})(\sqrt[3]{27} + \sqrt[3]{2})$

Rationalize the denominator. [10.5a, b]

47. $\sqrt{\dfrac{8}{3}}$

48. $\dfrac{2}{\sqrt{a} + \sqrt{b}}$

Solve. [10.6a, b]

49. $\sqrt[4]{x + 3} = 2$

50. $1 + \sqrt{x} = \sqrt{3x - 3}$

51. The diagonal of a square has length $9\sqrt{2}$ cm. Find the length of a side of the square. [10.7a]

52. A bookcase is 5 ft tall and has a 7-ft diagonal brace, as shown. How wide is the bookcase? [10.7a]

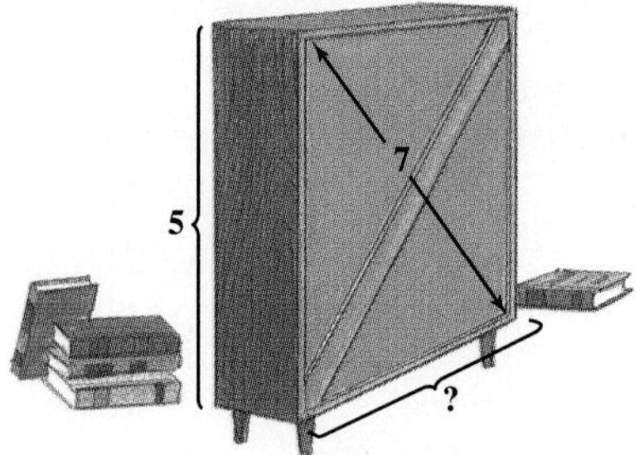

In a right triangle, find the length of the side not given. Give an exact answer and an answer to three decimal places. [10.7a]

53. $a = 7, \quad b = 24$

54. $a = 2, \quad c = 5\sqrt{2}$

55. Express in terms of i: $\sqrt{-25} + \sqrt{-8}$. [10.8a]

Add or subtract. [10.8b]

56. $(-4 + 3i) + (2 - 12i)$

57. $(4 - 7i) - (3 - 8i)$

Multiply. [10.8c, d]

58. $(2 + 5i)(2 - 5i)$

59. i^{13}

60. $(6 - 3i)(2 - i)$

Divide. [10.8e]

61. $\dfrac{-3 + 2i}{5i}$

62. $\dfrac{6 - 3i}{2 - i}$

63. Determine whether $1 + i$ is a solution of $x^2 + x + 2 = 0$. [10.8f]

64. Graph: $f(x) = \sqrt{x}$. [10.1a]

Skill Maintenance

Solve.

65. $\dfrac{7}{x + 2} + \dfrac{5}{x^2 - 2x + 4} = \dfrac{84}{x^3 + 8}$ [6.5a]

66. $3x^2 - 5x - 12 = 0$ [5.8b]

67. Multiply and simplify: [6.1d]

$$\frac{x^2 + 3x}{x^2 - y^2} \cdot \frac{x^2 - xy + 2x - 2y}{x^2 - 9}.$$

68. A motorcycle and a sport utility vehicle (SUV) are driven out of a car dealership at the same time. The SUV travels 8 mph faster than the motorcycle. The SUV travels 105 mi in the same time that the motorcycle travels 93 mi. Find the speed of each vehicle. [6.6a]

Synthesis

69. ◆ We learned a new method of equation solving in this chapter. Explain how this procedure differs from others we have used. [10.6a, b]

70. Simplify: $i \cdot i^2 \cdot i^3 \cdots i^{99} \cdot i^{100}$. [10.8c, d]

71. Solve: $\sqrt{11x + \sqrt{6 + x}} = 6$. [10.6a]

Test: Chapter 10

1. Use a calculator to approximate $\sqrt{148}$ to three decimal places.

2. For the given function, find the indicated function values.
   $f(x) = \sqrt{8 - 4x}$; $f(1)$ and $f(3)$

3. Find the domain of the function f in Exercise 2.

Simplify. Assume that letters represent *any* real number.

4. $\sqrt{(-3q)^2}$

5. $\sqrt{x^2 + 10x + 25}$

6. $\sqrt[3]{-\frac{1}{1000}}$

7. $\sqrt[5]{x^5}$

8. $\sqrt[10]{(-4)^{10}}$

Rewrite without rational exponents, and simplify, if possible.

9. $a^{2/3}$

10. $32^{3/5}$

Rewrite with rational exponents.

11. $\sqrt{37}$

12. $(\sqrt{5xy^2})^5$

Rewrite with positive exponents, and simplify, if possible.

13. $1000^{-1/3}$

14. $8a^{3/4}b^{-3/2}c^{-2/5}$

Use the laws of exponents to simplify. Write answers with positive exponents.

15. $(x^{2/3}y^{-3/4})^{12/5}$

16. $\frac{2.9^{-5/8}}{2.9^{2/3}}$

Use rational exponents to simplify. Write the answer in radical notation if appropriate.

17. $\sqrt[8]{x^2}$

18. $\sqrt[4]{16x^6}$

Use rational exponents to write a single radical expression.

19. $a^{2/5}b^{1/3}$

20. $\sqrt[4]{2y}\,\sqrt[3]{y}$

Simplify by factoring. Assume that all expressions under radicals represent nonnegative numbers.

21. $\sqrt{148}$

22. $\sqrt[4]{80}$

23. $(\sqrt[3]{16a^2b})^2$

Simplify. Assume that all expressions under radicals represent positive numbers.

24. $\sqrt[3]{-\frac{8}{x^6}}$

25. $\sqrt{\frac{25x^2}{36y^4}}$

Answers

1. ______
2. ______
3. ______
4. ______
5. ______
6. ______
7. ______
8. ______
9. ______
10. ______
11. ______
12. ______
13. ______
14. ______
15. ______
16. ______
17. ______
18. ______
19. ______
20. ______
21. ______
22. ______
23. ______
24. ______
25. ______

Answers

26. ______

27. ______

28. ______

29. ______

30. ______

31. ______

32. ______

33. ______

34. ______

35. ______

36. ______

37. ______

38. ______

39. ______

40. ______

41. ______

42. ______

43. ______

44. ______

45. ______

46. ______

47. ______

48. ______

49. ______

50. ______

51. ______

Perform the indicated operations and simplify. Assume that all expressions under radicals represent positive numbers.

26. $\sqrt[3]{2x}\sqrt[3]{5y^2}$

27. $\sqrt[4]{x^3y^2}\sqrt{xy}$

28. $\dfrac{\sqrt[5]{x^3y^4}}{\sqrt[5]{xy^2}}$

29. $\dfrac{\sqrt{300a}}{5\sqrt{3}}$

30. Add: $3\sqrt{128} + 2\sqrt{18} + 2\sqrt{32}$.

Multiply.

31. $(\sqrt{20} + 2\sqrt{5})(\sqrt{20} - 3\sqrt{5})$

32. $(3 + \sqrt{x})^2$

33. Rationalize the denominator: $\dfrac{1 + \sqrt{2}}{3 - 5\sqrt{2}}$.

Solve.

34. $\sqrt[5]{x - 3} = 2$

35. $\sqrt{x - 6} = \sqrt{x + 9} - 3$

36. The diagonal of a square has length $7\sqrt{2}$ ft. Find the length of a side of the square.

In a right triangle, find the length of the side not given. Give an exact answer and an answer to three decimal places.

37. $a = 7, \quad b = 7$

38. $a = 1, \quad c = \sqrt{5}$

39. Express in terms of i: $\sqrt{-9} + \sqrt{-64}$.

40. Subtract: $(5 + 8i) - (-2 + 3i)$.

Multiply.

41. $(3 - 4i)(3 + 7i)$

42. i^{95}

43. Divide: $\dfrac{-7 + 14i}{6 - 8i}$.

44. Determine whether $1 + 2i$ is a solution of $x^2 + 2x + 5 = 0$.

45. Graph: $f(x) = 4 - \sqrt{x}$.

Skill Maintenance

Solve.

46. $\dfrac{11x}{x + 3} + \dfrac{33}{x} + 12 = \dfrac{99}{x^2 + 3x}$

47. $6x^2 = 13x + 5$

48. Divide and simplify:

$$\frac{x^3 - 27}{x^2 - 16} \div \frac{x^2 + 3x + 9}{x + 4}.$$

49. *Mowing Time.* Fran and Juan do all the mowing for the local community college. It takes Juan 9 hr more than Fran to do the mowing. Working together, they can complete the job in 20 hr. How long would it take each, working alone, to do the mowing?

Synthesis

50. Simplify: $\dfrac{1 - 4i}{4i(1 + 4i)^{-1}}$.

51. Solve:
$\sqrt{2x - 2} + \sqrt{7x + 4} = \sqrt{13x + 10}$.

11

Quadratic Equations and Functions

An Application

Typically rivers are deepest in the middle, with the depth decreasing to 0 at the edges. A hydrologist measures the depths D, in feet, of a river at distances x, in feet, from one bank. Some of the data points found by the hydrologist are (0, 0), (50, 20), and (100, 0). Find a quadratic function that fits the data points and graph the function.

This problem appears as Example 7 in Section 11.7.

The Mathematics

The quadratic function is

$$D(x) = -0.008x^2 + 0.8x.$$

The graph is as shown below.

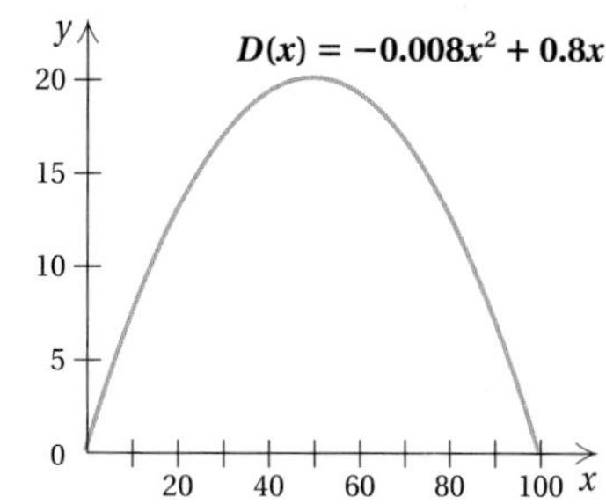

For more information, visit us at www.mathmax.com

Introduction

We began our study of quadratic equations in Chapter 4 in connection with polynomials. Here we extend our equation-solving skills to those quadratic equations that do not lend themselves to being solved easily by factoring. We consider both real-number and complex-number solutions found using the quadratic formula.

We also learn to graph quadratic functions in various forms and to solve related applied problems.

Pretest: Chapter 11

Solve.

1. $5x^2 + 15x = 0$

2. $y^2 + 4y + 8 = 0$

3. $x^2 - 10x + 25 = 0$

4. $3x^4 - 7x^2 + 2 = 0$

5. $\dfrac{2x}{2x+1} + \dfrac{x+1}{2x-1} = \dfrac{6}{4x^2-1}$

6. $x + 3\sqrt{x} - 4 = 0$

7. Solve. Give the exact solution and approximate the solutions to three decimal places.

   $2x^2 + 4x - 1 = 0$

8. Solve for T:

   $W = \sqrt{\dfrac{1}{RT}}$.

9. Write a quadratic equation having solutions $\frac{2}{3}$ and -2.

10. For $f(x) = 2x^2 - 12x + 16$:
    a) Find the vertex.
    b) Find the line of symmetry.
    c) Graph the function.

11. Find the x-intercepts: $f(x) = x^2 - 6x + 4$.

12. Find the quadratic function that fits the data points (0, 1), (1, 0), and (2, 7).

13. Find three consecutive even integers such that twice the product of the first two is equal to the square of the third plus five times the third.

14. *Car Travel.* During the first part of a trip, a car travels 100 mi at a certain speed. It travels 120 mi on the second part of the trip at a speed that is 8 mph faster. The total time for the trip is 5 hr. Find the speed of the car during each part of the trip.

15. *Corral Design.* What is the area of the largest rectangular horse corral that a rancher can enclose with 300 ft of fencing?

Solve.

16. $x(x-3)(x+5) < 0$

17. $\dfrac{x-2}{x+3} \geq 0$

Objectives for Retesting

The objectives to be tested in addition to the material in this chapter are as follows.

[6.3a], [6.4a] Add and subtract rational expressions.

[7.6b] Using a set of data, choose two representative points, find a linear function using the two points, and make predictions from the function.

[10.3a] Multiply and simplify radical expressions.

[10.6a, b] Solve radical equations.

11.1 The Basics of Solving Quadratic Equations

 Algebraic–Graphical Connection

Let's reexamine the graphical connections to the algebraic equation-solving concepts we have studied before.

In Chapter 7, we introduced the graph of a quadratic function:

$$f(x) = ax^2 + bx + c, \quad a \neq 0.$$

For example, the graph of the function $f(x) = x^2 + 6x + 8$ and its x-intercepts are shown below.

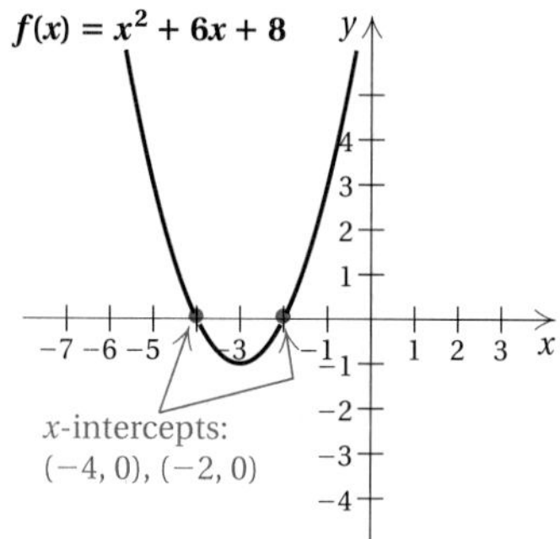

The x-intercepts are $(-4, 0)$ and $(-2, 0)$. These pairs are also the points of intersection of the graphs of $f(x) = x^2 + 6x + 8$ and $g(x) = 0$ (the x-axis). We will analyze the graphs of quadratic functions in greater detail in Sections 11.5–11.7.

In Chapter 5, we solved quadratic equations like $x^2 + 6x + 8 = 0$ using factoring, as here:

$$x^2 + 6x + 8 = 0$$

$$(x + 4)(x + 2) = 0 \qquad \text{Factoring}$$

$$x + 4 = 0 \quad or \quad x + 2 = 0 \qquad \text{Using the principle of zero products}$$

$$x = -4 \quad or \quad x = -2.$$

We see that the solutions of $0 = x^2 + 6x + 8$, -4 and -2, are the first coordinates of the x-intercepts, $(-4, 0)$ and $(-2, 0)$, of the graph of $f(x) = x^2 + 6x + 8$.

Do Exercise 1.

We now enhance our ability to solve quadratic equations.

a The Principle of Square Roots

The quadratic equation

$$5x^2 + 8x - 2 = 0$$

is said to be in **standard form.** The quadratic equation

$$5x^2 = 2 - 8x$$

is equivalent to the preceding, but it is *not* in standard form.

Objectives

a Solve quadratic equations using the principle of square roots and find the x-intercepts of the graph of a related function.

b Solve quadratic equations by completing the square.

c Solve applied problems using quadratic equations.

For Extra Help

 TAPE 18

 MAC WIN

 CD-ROM

1. Consider solving the equation

$$x^2 - 6x + 8 = 0.$$

Below is the graph of

$$f(x) = x^2 - 6x + 8.$$

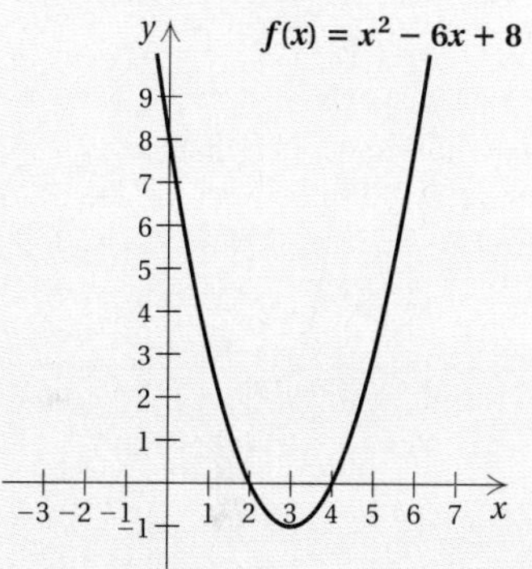

a) What are the x-intercepts of the graph?

b) What are the solutions of $x^2 - 6x + 8 = 0$?

c) What relationship exists between the answers to parts (a) and (b)?

Answers on page A-45

We require $a > 0$ in the standard form to allow a smoother proof of the quadratic formula, which we consider later. This restriction also eases related factoring.

> An equation of the type $ax^2 + bx + c = 0$, where a, b, and c are real-number constants and $a > 0$, is called the **standard form of a quadratic equation.**

For example, the equation $-5x^2 + 4x - 7 = 0$ is not in standard form. However, we can find an equivalent equation that is in standard form by multiplying by -1 on both sides:

$$-1(-5x^2 + 4x - 7) = -1(0)$$
$$5x^2 - 4x + 7 = 0.$$

In Section 5.8, we studied the use of factoring and the principle of zero products to solve certain quadratic equations. Let's review that procedure and introduce a new one.

2. a) Solve: $5x^2 = 8x - 3$.

b) Find the x-intercepts of $f(x) = 5x^2 - 8x + 3$.

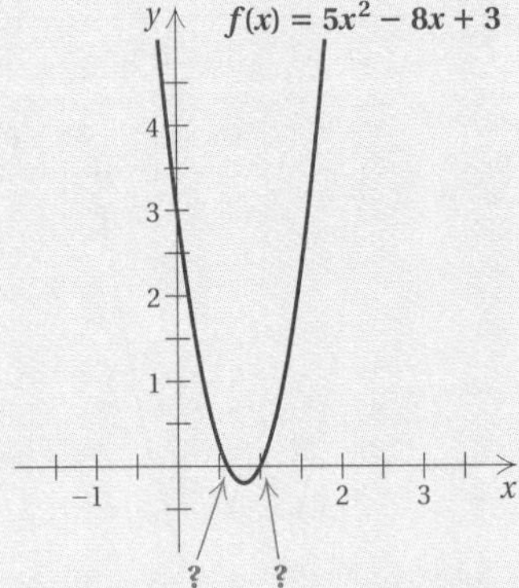

Example 1

a) Solve: $3x^2 = 2 - x$.

b) Find the x-intercepts of $f(x) = 3x^2 + x - 2$.

a) We first find standard form. Then we factor and use the principle of zero products.

$$3x^2 = 2 - x$$
$$3x^2 + x - 2 = 0 \quad \text{Adding } x \text{ and subtracting 2 to get the standard form}$$
$$(3x - 2)(x + 1) = 0 \quad \text{Factoring}$$
$$3x - 2 = 0 \quad or \quad x + 1 = 0 \quad \text{Using the principle of zero products}$$
$$3x = 2 \quad or \quad x = -1$$
$$x = \tfrac{2}{3} \quad or \quad x = -1$$

CHECK:

For $\frac{2}{3}$:

$3x^2$	$= 2 - x$
$3\left(\frac{2}{3}\right)^2$	? $2 - \left(\frac{2}{3}\right)$
$3 \cdot \frac{4}{9}$	$\frac{6}{3} - \frac{2}{3}$
$\frac{4}{3}$	$\frac{4}{3}$ TRUE

For -1:

$3x^2$	$= 2 - x$
$3(-1)^2$	? $2 - (-1)$
$3 \cdot 1$	$2 + 1$
3	3 TRUE

The solutions are -1 and $\frac{2}{3}$.

b) The x-intercepts of $f(x) = 3x^2 + x - 2$ are $(-1, 0)$ and $\left(\frac{2}{3}, 0\right)$. The solutions of the equation $3x^2 = 2 - x$ are the first coordinates of the x-intercepts of the graph of $f(x) = 3x^2 + x - 2$.

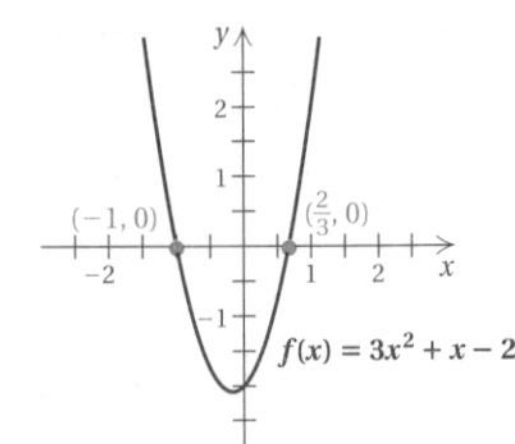

Do Exercise 2.

Calculator Spotlight

Use the SOLVE or INTERSECT feature to check the results of Example 1 and Margin Exercise 2.

Answers on page A-45

Example 2

a) Solve: $x^2 = 25$.

b) Find the x-intercepts of $f(x) = x^2 - 25$.

a) We first find standard form and then factor:

$$x^2 - 25 = 0 \qquad \textbf{Subtracting 25}$$

$$(x - 5)(x + 5) = 0 \qquad \textbf{Factoring}$$

$$x - 5 = 0 \quad or \quad x + 5 = 0 \qquad \textbf{Using the principle of zero products}$$

$$x = 5 \quad or \quad x = -5.$$

The solutions are 5 and -5.

b) The x-intercepts of $f(x) = x^2 - 25$ are $(-5, 0)$ and $(5, 0)$.

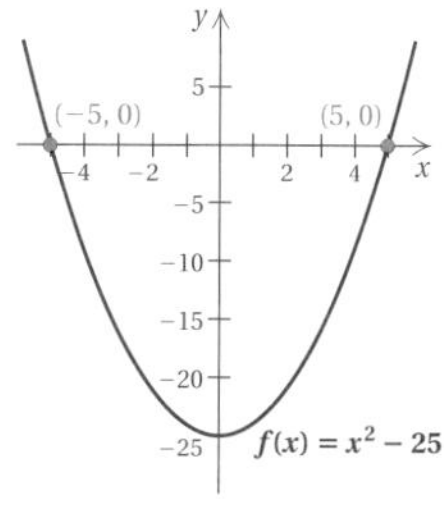

Solve.

3. $x^2 = 16$

Example 3

Solve: $6x^2 - 15x = 0$.

We factor and use the principle of zero products:

$$6x^2 - 15x = 0$$

$$3x(2x - 5) = 0$$

$$3x = 0 \quad or \quad 2x - 5 = 0$$

$$x = 0 \quad or \quad x = \tfrac{5}{2}.$$

The solutions are 0 and $\frac{5}{2}$. We leave the check to the student.

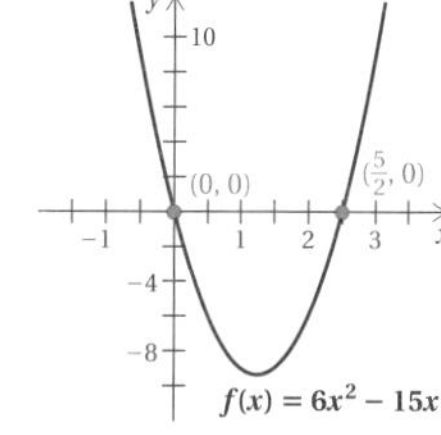

4. $4x^2 + 14x = 0$

Do Exercises 3 and 4.

Solving Equations of the Type $x^2 = d$

Consider the equation $x^2 = 25$ again. We know from Chapter 10 that the number 25 has two real-number square roots, namely, 5 and -5. Note that these are the solutions of the equation in Example 2. This exemplifies the principle of square roots, which provides a quick method for solving equations of the type $x^2 = d$.

> **THE PRINCIPLE OF SQUARE ROOTS**
>
> The equation $x^2 = d$ has two real-number solutions when $d > 0$. The solutions are $\sqrt{d}$ and $-\sqrt{d}$.
>
> The equation $x^2 = 0$ has 0 as its only solution.
>
> The equation $x^2 = d$ has two imaginary-number solutions when $d < 0$.

Calculator Spotlight

Use the SOLVE or INTERSECT feature to check the results of Examples 2 and 3 and Margin Exercises 3 and 4.

Answers on page A-45

5. Solve: $5x^2 = 15$. Give the exact solution and approximate the solutions to three decimal places.

6. Solve: $-3x^2 + 8 = 0$. Give the exact solution and approximate the solutions to three decimal places.

Calculator Spotlight

Referring to Example 5, we note that once the algebraic work yields $\sqrt{10}/5$, we can find an approximation using the keystrokes

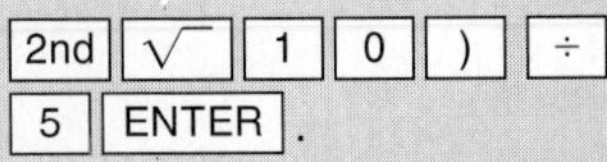

We obtain the answer 0.632.

The SOLVE or INTERSECT feature can also be used to approximate the solutions of $-5x^2 + 2 = 0$ graphically.

Exercises

Approximate the solutions.

1. $-3x^2 + 8 = 0$
2. $-5.13x^2 + 2.17 = 0$

Answers on page A-45

Example 4 Solve: $3x^2 = 6$. Give the exact solution and approximate the solutions to three decimal places.

We have

$$3x^2 = 6$$
$$x^2 = 2$$
$$x = \sqrt{2} \quad or \quad x = -\sqrt{2}.$$

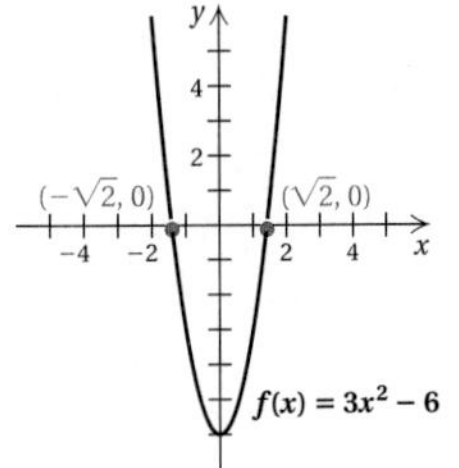

We often use the symbol $\pm\sqrt{2}$ to represent both of the solutions.

CHECK: For $\sqrt{2}$:

$$\begin{array}{r|l} 3x^2 & = 6 \\ \hline 3(\sqrt{2})^2 & ?\ 6 \\ 3 \cdot 2 & \\ 6 & \quad \text{TRUE} \end{array}$$

For $-\sqrt{2}$:

$$\begin{array}{r|l} 3x^2 & = 6 \\ \hline 3(-\sqrt{2})^2 & ?\ 6 \\ 3 \cdot 2 & \\ 6 & \quad \text{TRUE} \end{array}$$

The solutions are $\sqrt{2}$ and $-\sqrt{2}$, or $\pm\sqrt{2}$, which are about 1.414 and -1.414 when written to three decimal places.

Do Exercise 5.

Sometimes we rationalize denominators to simplify answers.

Example 5 Solve: $-5x^2 + 2 = 0$. Give the exact solution and approximate the solutions to three decimal places.

$$-5x^2 + 2 = 0$$
$$x^2 = \frac{2}{5} \qquad \text{Subtracting 2 and dividing by } -5$$
$$x = \sqrt{\frac{2}{5}} \quad or \quad x = -\sqrt{\frac{2}{5}} \qquad \text{Using the principle of square roots}$$
$$x = \sqrt{\frac{2}{5} \cdot \frac{5}{5}} \quad or \quad x = -\sqrt{\frac{2}{5} \cdot \frac{5}{5}} \qquad \text{Rationalizing the denominators}$$
$$x = \frac{\sqrt{10}}{5} \quad or \quad x = -\frac{\sqrt{10}}{5}$$

CHECK: We check both numbers at once, since there is no x-term in the equation.

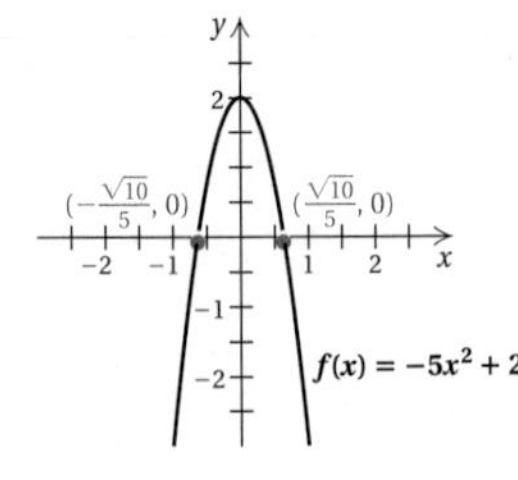

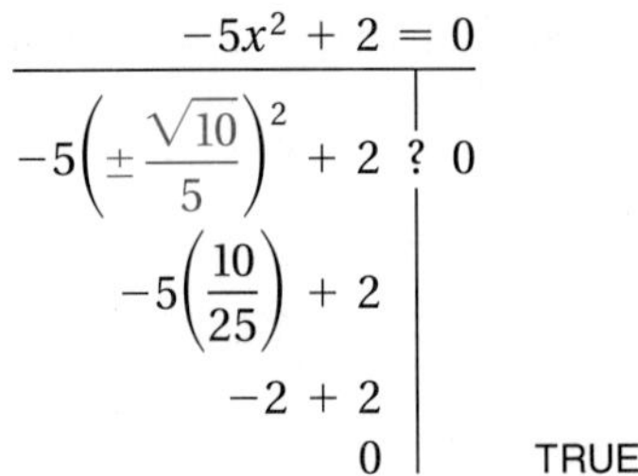

$$\begin{array}{r|l} -5x^2 + 2 & = 0 \\ \hline -5\left(\pm\frac{\sqrt{10}}{5}\right)^2 + 2 & ?\ 0 \\ -5\left(\frac{10}{25}\right) + 2 & \\ -2 + 2 & \\ 0 & \quad \text{TRUE} \end{array}$$

The solutions are $\frac{\sqrt{10}}{5}$ and $-\frac{\sqrt{10}}{5}$, or $\pm\frac{\sqrt{10}}{5}$. We can use a calculator for an approximation:

$$\pm\frac{\sqrt{10}}{5} \approx \pm 0.632.$$

Do Exercise 6 on the preceding page.

Sometimes we get solutions that are imaginary numbers.

Example 6 Solve: $4x^2 + 9 = 0$.

$$4x^2 + 9 = 0$$

$$x^2 = -\frac{9}{4} \qquad \text{Subtracting 9 and dividing by 4}$$

$$x = \sqrt{-\frac{9}{4}} \quad or \quad x = -\sqrt{-\frac{9}{4}} \qquad \text{Using the principle of square roots}$$

$$x = \frac{3}{2}i \quad or \quad x = -\frac{3}{2}i \qquad \text{Simplifying}$$

CHECK:

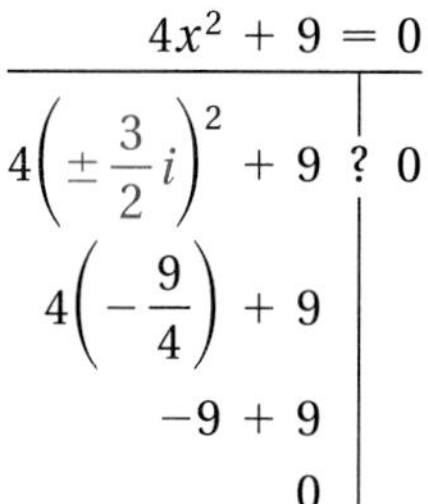

$$\begin{array}{c|c} 4x^2 + 9 = 0 & \\ \hline 4\left(\pm\frac{3}{2}i\right)^2 + 9 \; ? \; 0 & \\ 4\left(-\frac{9}{4}\right) + 9 & \\ -9 + 9 & \\ 0 & \text{TRUE} \end{array}$$

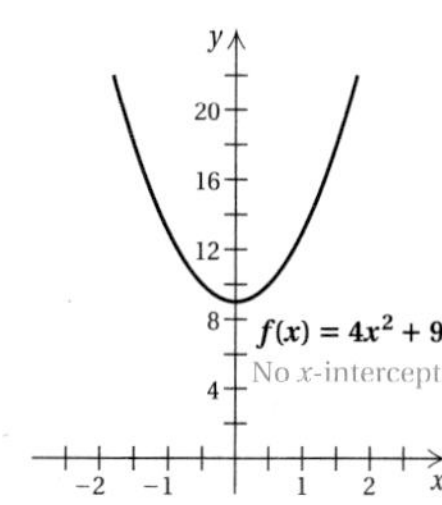

The solutions are $\frac{3}{2}i$ and $-\frac{3}{2}i$, or $\pm\frac{3}{2}i$.

We see that the graph of $f(x) = 4x^2 + 9$ does not cross the x-axis. This is true because the equation $4x^2 + 9 = 0$ has imaginary solutions.

Do Exercise 7.

Solving Equations of the Type $(x + c)^2 = d$

The equation $(x - 2)^2 = 7$ can also be solved using the principle of square roots.

Example 7

a) Solve: $(x - 2)^2 = 7$.

b) Find the x-intercepts of $f(x) = (x - 2)^2 - 7$.

a) We have

$$(x - 2)^2 = 7$$

$$x - 2 = \sqrt{7} \quad or \quad x - 2 = -\sqrt{7} \qquad \text{Using the principle of square roots}$$

$$x = 2 + \sqrt{7} \quad or \quad x = 2 - \sqrt{7}.$$

The solutions are $2 + \sqrt{7}$ and $2 - \sqrt{7}$, or $2 \pm \sqrt{7}$.

b) The x-intercepts of $f(x) = (x - 2)^2 - 7$ are $(2 - \sqrt{7}, 0)$ and $(2 + \sqrt{7}, 0)$.

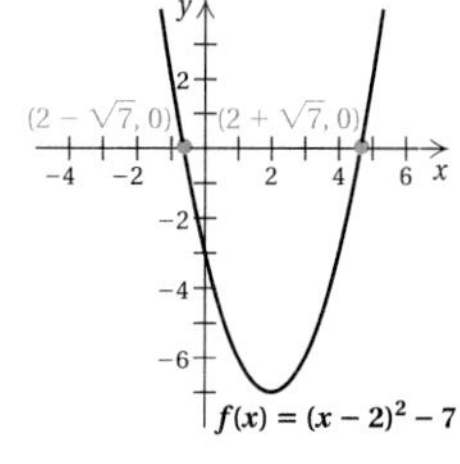

Do Exercise 8.

7. Solve: $2x^2 + 1 = 0$.

Calculator Spotlight

What happens when we use the SOLVE or INTERSECT feature to solve the equations in Example 6 and Margin Exercise 7?

8. a) Solve: $(x - 1)^2 = 5$.

b) Find the x-intercepts of $f(x) = (x - 1)^2 - 5$.

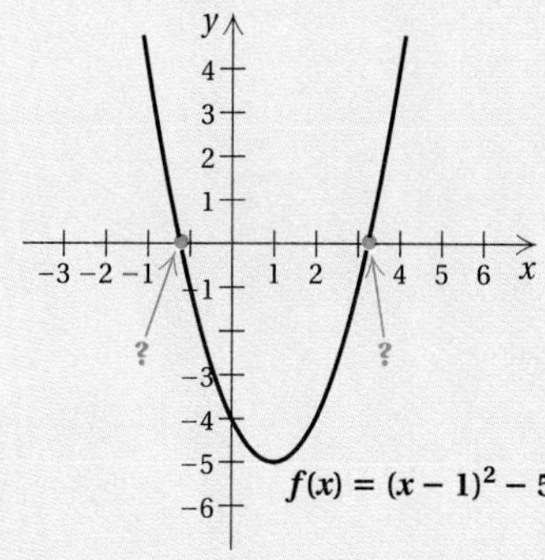

Answers on page A-45

9. Solve: $x^2 + 16x + 64 = 11$.

Answer on page A-45

If we can express the left side of an equation as the square of a binomial, we can proceed as we did in Example 7.

Example 8 Solve: $x^2 + 6x + 9 = 2$.

We have

$$x^2 + 6x + 9 = 2 \quad \text{The left side is the square of a binomial.}$$

$$(x + 3)^2 = 2$$

$$x + 3 = \sqrt{2} \quad or \quad x + 3 = -\sqrt{2} \quad \text{Using the principle of square roots}$$

$$x = -3 + \sqrt{2} \quad or \quad x = -3 - \sqrt{2}.$$

The solutions are $-3 + \sqrt{2}$ and $-3 - \sqrt{2}$, or $-3 \pm \sqrt{2}$.

Do Exercise 9.

b Completing the Square

We can solve quadratic equations like $3x^2 = 6$ and $(x - 2)^2 = 7$ by using the principle of square roots. We can also solve an equation such as $x^2 + 6x + 9 = 2$ in like manner because the expression on the left side is the square of a binomial, $(x + 3)^2$. This second procedure is the basis for a method called **completing the square.** *It can be used to solve any quadratic equation.*

Suppose we have the following quadratic equation:

$$x^2 + 14x = 4.$$

If we could add on both sides of the equation a constant that would make the expression on the left the square of a binomial, we could then solve the equation using the principle of square roots.

How can we determine what to add to $x^2 + 14x$ to construct the square of a binomial? We want to find a number a such that the following equation is satisfied:

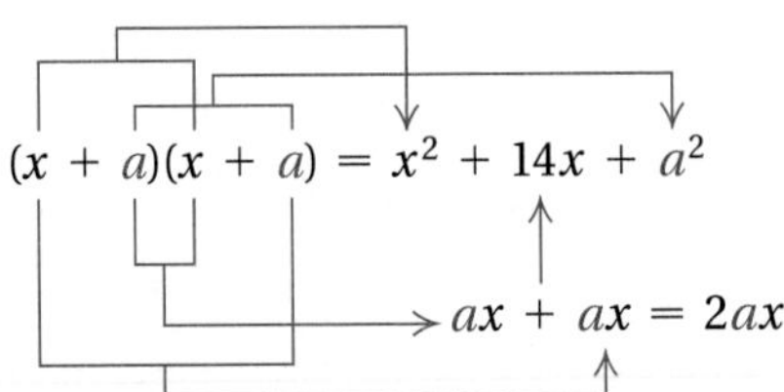

Thus a is such that $2ax = 14x$. Solving for a, we get

$$a = \frac{14x}{2x} = \frac{14}{2} = 7.$$

That is, a is half of the coefficient of x in $x^2 + 14x$. Since $a^2 = \left(\frac{14}{2}\right)^2 = 7^2 = 49$, we add 49 to our original expression:

$$x^2 + 14x + 49 \text{ is the square of } x + 7;$$

that is,

$$x^2 + 14x + 49 = (x + 7)^2.$$

When solving an equation, to **complete the square** of an expression like $x^2 + bx$, we take half the x-coefficient, which is $b/2$, and square. Then we add that number, $(b/2)^2$, on both sides.

Returning to solving our original equation, we first add 49 on *both* sides to *complete the square.* Then we solve:

$$x^2 + 14x = 4 \qquad \text{Original equation}$$

$$x^2 + 14x + 49 = 4 + 49 \qquad \text{Adding 49: } \left(\frac{14}{2}\right)^2 = 7^2 = 49$$

$$(x + 7)^2 = 53$$

$$x + 7 = \sqrt{53} \quad or \quad x + 7 = -\sqrt{53} \qquad \text{Using the principle of square roots}$$

$$x = -7 + \sqrt{53} \quad or \quad x = -7 - \sqrt{53}.$$

The solutions are $-7 \pm \sqrt{53}$.

We have seen that a quadratic equation $(x + c)^2 = d$ can be solved using the principle of square roots. Any equation, such as $x^2 - 6x + 8 = 0$, can be put in this form by completing the square. Then we can solve as before.

Example 9 Solve: $x^2 - 6x + 8 = 0$.

We have

$$x^2 - 6x + 8 = 0$$

$$x^2 - 6x = -8. \qquad \text{Subtracting 8}$$

We take half of -6 and square it, to get 9. Then we add 9 on *both* sides of the equation. This makes the left side the square of a binomial, $x - 3$. We have now *completed the square.*

$$x^2 - 6x + 9 = -8 + 9 \qquad \text{Adding 9}$$

$$(x - 3)^2 = 1$$

$$x - 3 = 1 \quad or \quad x - 3 = -1 \qquad \text{Using the principle of square roots}$$

$$x = 4 \quad or \quad x = 2$$

The solutions are 2 and 4.

Do Exercises 10 and 11.

Example 10 Solve $x^2 + 4x - 7 = 0$ by completing the square.

$$x^2 + 4x - 7 = 0$$

$$x^2 + 4x = 7 \qquad \text{Adding 7}$$

$$x^2 + 4x + 4 = 7 + 4 \qquad \text{Adding 4: } \left(\frac{4}{2}\right)^2 = (2)^2 = 4$$

$$(x + 2)^2 = 11$$

$$x + 2 = \sqrt{11} \quad or \quad x + 2 = -\sqrt{11} \qquad \text{Using the principle of square roots}$$

$$x = -2 + \sqrt{11} \quad or \quad x = -2 - \sqrt{11}$$

The solutions are $-2 \pm \sqrt{11}$.

Do Exercise 12.

Solve.

10. $x^2 + 6x + 8 = 0$

11. $x^2 - 8x - 20 = 0$

12. Solve by completing the square:

$x^2 + 6x - 1 = 0.$

Answers on page A-45

13. Solve by completing the square:

$3x^2 - 2x = 7.$

Answer on page A-45

When the coefficient of x^2 is not 1, we can make it 1, as shown in the following example.

Example 11 Solve $2x^2 = 3x - 7$ by completing the square.

$$2x^2 = 3x - 7$$

$$2x^2 - 3x = -7$$

$$\frac{1}{2}(2x^2 - 3x) = \frac{1}{2} \cdot (-7)$$ Multiplying by $\frac{1}{2}$ to make the x^2-coefficient 1

$$x^2 - \frac{3}{2}x = -\frac{7}{2}$$ Multiplying and simplifying

$$x^2 - \frac{3}{2}x + \frac{9}{16} = -\frac{7}{2} + \frac{9}{16}$$ Adding $\frac{9}{16}$: $\left[\frac{1}{2}\left(-\frac{3}{2}\right)\right]^2 = \left[-\frac{3}{4}\right]^2 = \frac{9}{16}$

$$\left(x - \frac{3}{4}\right)^2 = -\frac{56}{16} + \frac{9}{16}$$ Finding a common denominator

$$\left(x - \frac{3}{4}\right)^2 = -\frac{47}{16}$$

$$x - \frac{3}{4} = \sqrt{-\frac{47}{16}} \quad or \quad x - \frac{3}{4} = -\sqrt{-\frac{47}{16}}$$ Using the principle of square roots

$$x - \frac{3}{4} = i\sqrt{\frac{47}{16}} \quad or \quad x - \frac{3}{4} = -i\sqrt{\frac{47}{16}}$$ $i\sqrt{-1} = i$

$$x = \frac{3}{4} + \frac{i\sqrt{47}}{4} \quad or \quad x = \frac{3}{4} - \frac{i\sqrt{47}}{4}$$

The solutions are $\frac{3}{4} \pm i\frac{\sqrt{47}}{4}$.

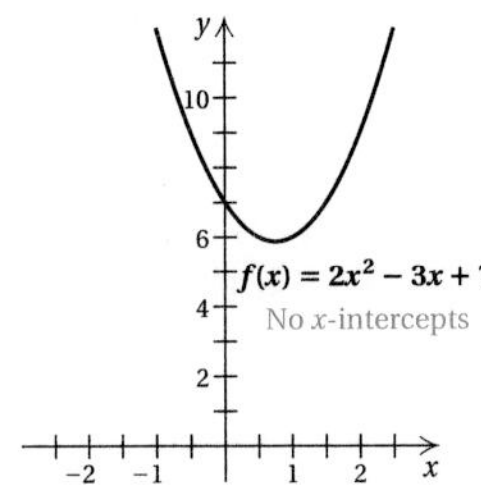

SOLVING BY COMPLETING THE SQUARE

To solve an equation $ax^2 + bx + c = 0$ by completing the square:

1. If $a \neq 1$, multiply by $1/a$ so that the x^2-coefficient is 1.
2. If the x^2-coefficient is 1, add or subtract so that the equation is in the form

$$x^2 + bx = -c, \quad \text{or} \quad x^2 + \frac{b}{a}x = -\frac{c}{a} \quad \text{if step (1) has been applied.}$$

3. Take half of the x-coefficient and square it. Add the result on both sides of the equation.
4. Express the side with the variables as the square of a binomial.
5. Use the principle of square roots and complete the solution.

Do Exercise 13.

c Applications and Problem Solving

If you put money in a savings account, the bank will pay you interest. If interest is **compounded annually,** the bank, at the end of a year, will start paying interest on both the original amount and the interest that has been earned.

> **THE COMPOUND-INTEREST FORMULA**
>
> If an amount of money P is invested at interest rate r, compounded annually, then in t years, it will grow to the amount A given by
>
> $$A = P(1 + r)^t.$$

We can use quadratic equations to solve certain interest problems.

Example 12 *Compound Interest.* $1000 invested at 8.4%, compounded annually, for 2 yr will grow to what amount?

$$\begin{aligned} A &= P(1 + r)^t \\ &= 1000(1 + 0.084)^2 && \textbf{Substituting into the formula; 8.4\% = 0.0084} \\ &= 1000(1.084)^2 \\ &= 1000(1.175056) \\ &\approx 1175.06. && \textbf{Computing and rounding} \end{aligned}$$

The amount is $1175.06.

Example 13 *Compound Interest.* $4000 is invested at interest rate r, compounded annually. In 2 yr, it grows to $4410. What is the interest rate?

We know that $4000 is originally invested. Thus P is $4000. That amount grows to $4410 in 2 yr. Thus when t is 2, A is $4410. We substitute 4000 for P, 4410 for A, and 2 for t in the formula, and solve the resulting equation for r:

$$\begin{aligned} A &= P(1 + r)^t \\ 4410 &= 4000(1 + r)^2 \\ \frac{4410}{4000} &= (1 + r)^2 \\ \frac{441}{400} &= (1 + r)^2. \end{aligned}$$

We then have

$$\begin{aligned} \sqrt{\frac{441}{400}} &= 1 + r \quad \text{or} \quad -\sqrt{\frac{441}{400}} = 1 + r && \textbf{Using the principle of square roots} \\ \frac{21}{20} &= 1 + r \quad \text{or} \quad -\frac{21}{20} = 1 + r && \textbf{Simplifying} \\ -\frac{20}{20} + \frac{21}{20} &= r \quad \text{or} \quad -\frac{20}{20} - \frac{21}{20} = r \\ \frac{1}{20} &= r \quad \text{or} \quad -\frac{41}{20} = r. \end{aligned}$$

Since the interest rate cannot be negative, we have

$$\begin{aligned} \frac{1}{20} &= r \\ r &= 0.05, \text{ or } 5\%. \end{aligned}$$

The interest rate must be 5%.

Do Exercises 14 and 15.

14. *Compound Interest.* Suppose that $3000 is invested at 9.6%, compounded annually, for 2 yr. To what amount will the investment grow?

15. *Compound Interest.* Suppose that $2500 is invested at interest rate r, compounded annually. In 2 yr, the investment grows to $3600. What is the interest rate?

Answers on page A-45

16. *Hang Time.* Anfernee Hardaway, of the Orlando Magic, has a hang time of about 0.866 sec. What is his vertical leap?

17. *Hang Time.* The record for the greatest vertical leap in the NBA is held by Darryl Griffith of the Utah Jazz. It was 48 in. What was his hang time?

Answers on page A-45

Example 14 *Hang Time.* One of the most exciting plays in basketball is the dunk shot. The amount of time T that passes from the moment a player leaves the ground, goes up, makes the shot, and arrives back on the ground is called the *hang time.* A function relating an athlete's vertical leap V, in inches, to hang time T, in seconds, is given by

$$V(t) = 48T^2.$$

a) Michael Jordan, of the Chicago Bulls, has a hang time of about 0.889 sec. What is his vertical leap?

b) Although his height is only 5 ft 7 in., Spud Webb, formerly of the Sacramento Kings, had a vertical leap of about 44 in. What is his hang time? (***Source:*** Peter Brancazio, "The Mechanics of a Slam Dunk," *Popular Mechanics,* November 1991. Courtesy of Professor Peter Brancazio, Brooklyn College.)

a) To find Jordan's vertical leap, we substitute 0.889 for T in the function and compute V:

$$V(0.889) = 48(0.889)^2 \approx 37.9 \text{ in.}$$

Jordan's vertical leap is about 37.9 in. Surprisingly, Jordan does not have the vertical leap most fans would expect.

b) To find Webb's hang time, we substitute 44 for V and solve for T:

$$44 = 48T^2 \qquad \text{Substituting 44 for } V$$

$$\frac{44}{48} = T^2 \qquad \text{Solving for } T^2$$

$$0.91\overline{6} = T^2$$

$$\sqrt{0.91\overline{6}} = T \qquad \text{Hang time is positive.}$$

$$0.957 \approx T. \qquad \text{Using a calculator}$$

Webb's hang time is 0.957 sec. Note that his hang time is greater than Jordan's.

Do Exercises 16 and 17.

Exercise Set 11.1

a

1. a) Solve:
   $6x^2 = 30.$
   b) Find the x-intercepts of $f(x) = 6x^2 - 30.$

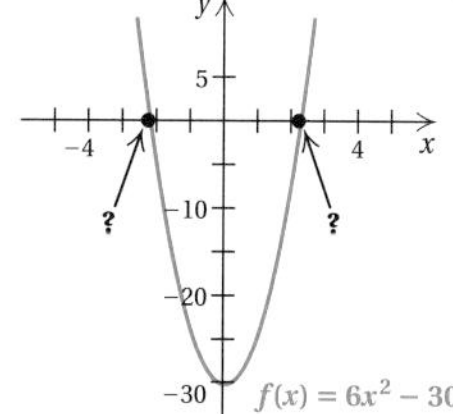

2. a) Solve:
   $5x^2 = 35.$
   b) Find the x-intercepts of $f(x) = 5x^2 - 35.$

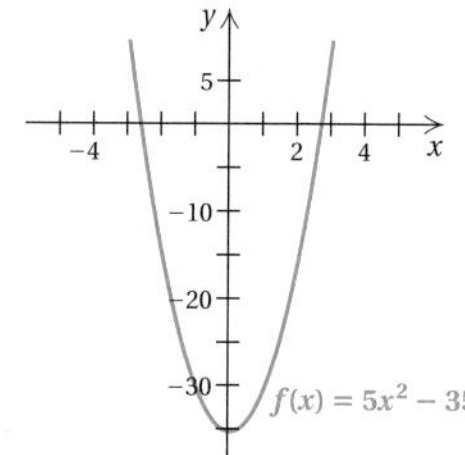

3. a) Solve:
   $9x^2 + 25 = 0.$
   b) Find the x-intercepts of $f(x) = 9x^2 + 25.$

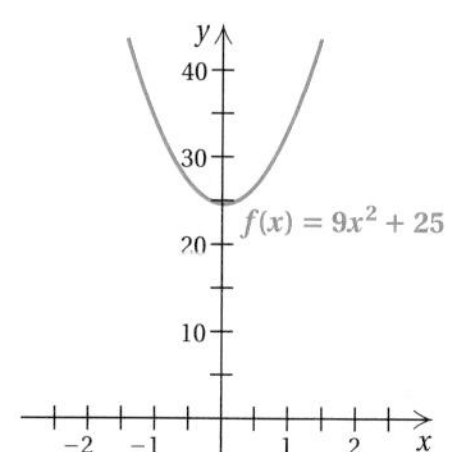

4. a) Solve:
   $36x^2 + 49 = 0.$
   b) Find the x-intercepts of $f(x) = 36x^2 + 49.$

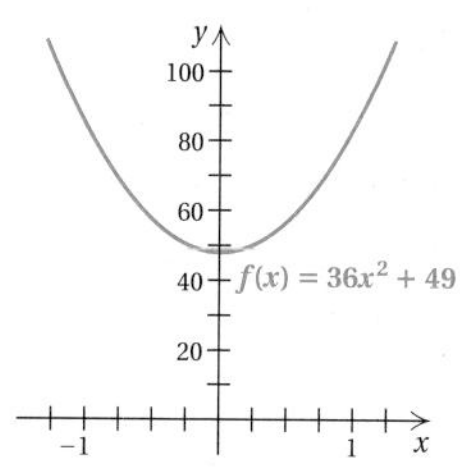

Solve. Give the exact solution and approximate solutions to three decimal places, when appropriate.

5. $2x^2 - 3 = 0$

6. $3x^2 - 7 = 0$

7. $(x + 2)^2 = 49$

8. $(x - 1)^2 = 6$

9. $(x - 4)^2 = 16$

10. $(x + 3)^2 = 9$

11. $(x - 11)^2 = 7$

12. $(x - 9)^2 = 34$

13. $(x - 7)^2 = -4$

14. $(x + 1)^2 = -9$

15. $(x - 9)^2 = 81$

16. $(t - 2)^2 = 25$

17. $\left(x - \frac{3}{2}\right)^2 = \frac{7}{2}$

18. $\left(y + \frac{3}{4}\right)^2 = \frac{17}{16}$

19. $x^2 + 6x + 9 = 64$

20. $x^2 + 10x + 25 = 100$

21. $y^2 - 14y + 49 = 4$

22. $p^2 - 8p + 16 = 1$

b Solve by completing the square. Show your work.

23. $x^2 + 4x = 2$

24. $x^2 + 2x = 5$

25. $x^2 - 22x = 11$

26. $x^2 - 18x = 10$

27. $x^2 + x = 1$

28. $x^2 - x = 3$

29. $t^2 - 5t = 7$

30. $y^2 + 9y = 8$

31. $x^2 + \frac{3}{2}x = 3$

32. $x^2 - \frac{4}{3}x = \frac{2}{3}$

33. $m^2 - \frac{9}{2}m = \frac{3}{2}$

34. $r^2 + \frac{2}{5}r = \frac{4}{5}$

35. $x^2 + 6x - 16 = 0$

36. $x^2 - 8x + 15 = 0$

37. $x^2 + 22x + 102 = 0$

38. $x^2 + 18x + 74 = 0$

39. $x^2 - 10x - 4 = 0$

40. $x^2 + 10x - 4 = 0$

41. a) Solve: $x^2 + 7x - 2 = 0$.
b) Find the x-intercepts of $f(x) = x^2 + 7x - 2$.

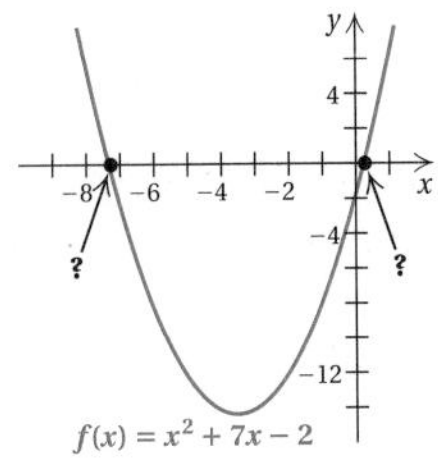

42. a) Solve: $x^2 - 7x - 2 = 0$.
b) Find the x-intercepts of $f(x) = x^2 - 7x - 2$.

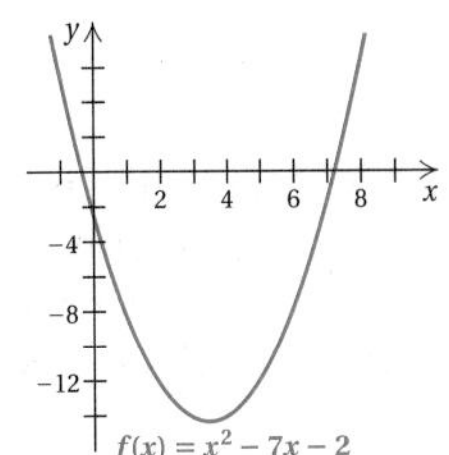

43. a) Solve: $2x^2 - 5x + 8 = 0$.
b) Find the x-intercepts of $f(x) = 2x^2 - 5x + 8$.

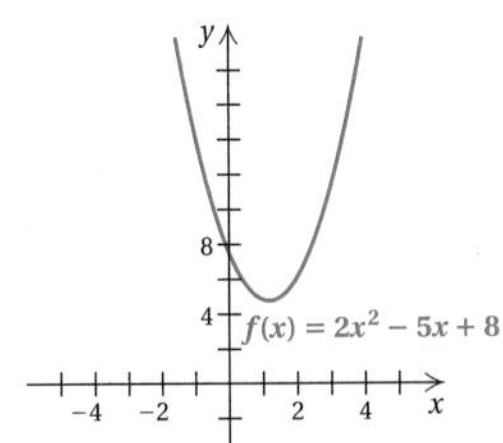

44. a) Solve: $2x^2 - 3x + 9 = 0$.
b) Find the x-intercepts of $f(x) = 2x^2 - 3x + 9$.

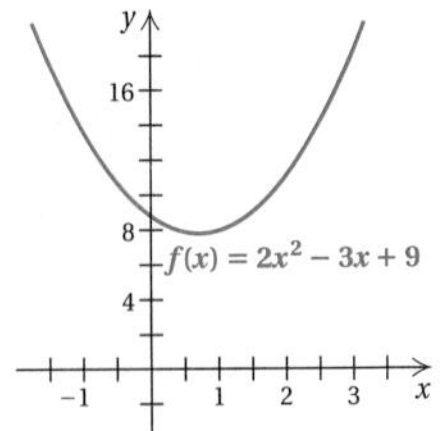

Solve by completing the square. Show your work.

45. $x^2 - \frac{3}{2}x - \frac{1}{2} = 0$

46. $x^2 + \frac{3}{2}x - 2 = 0$

47. $2x^2 - 3x - 17 = 0$

48. $2x^2 + 3x - 1 = 0$

49. $3x^2 - 4x - 1 = 0$

50. $3x^2 + 4x - 3 = 0$

51. $x^2 + x + 2 = 0$

52. $x^2 - x + 1 = 0$

53. $x^2 - 4x + 13 = 0$

54. $x^2 - 6x + 13 = 0$

c *Compound Interest.* For Exercises 55–60, use the compound-interest formula $A = P(1 + r)^t$ to determine the interest rate.

55. \$1000 grows to \$1210 in 2 yr

56. \$2560 grows to \$2890 in 2 yr

57. \$8000 grows to \$10,125 in 2 yr

58. \$10,000 grows to \$10,404 in 2 yr

59. \$6250 grows to \$6760 in 2 yr

60. \$6250 grows to \$7290 in 2 yr

Hang Time. For Exercises 61 and 62, use the hang-time function $V(T) = 48T^2$, relating vertical leap to hang time.

61. A basketball player has a vertical leap of 36 in. What is his hang time?

62. Tracy McGrady, an NBA player, has a vertical leap of 40 in. What is his hang time?

Free-Falling Objects. The function $s(t) = 16t^2$ is used to approximate the distance s, in feet, that an object falls freely from rest in t seconds. Use the formula for Exercises 63–66.

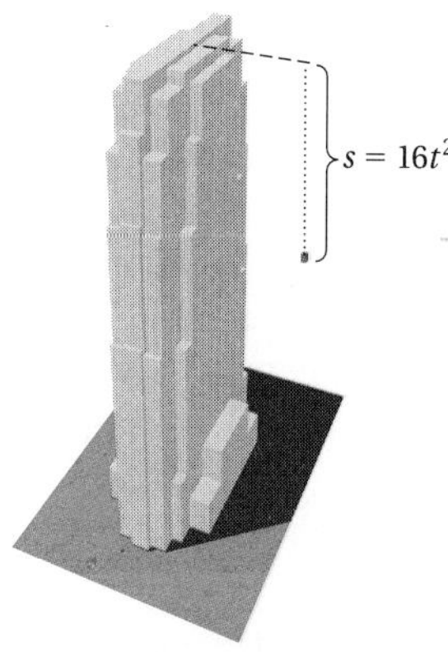

63. The RCA Building in New York City is 850 ft tall. How long will it take an object to fall from the top?

64. The CN Tower in Toronto, at 1815 ft, is the world's tallest self-supporting tower (no guy wires) (***Source:*** *The Guinness Book of Records*). How long would it take an object to fall freely from the top?

65. Reaching 745 ft above the water, the towers of California's Golden Gate Bridge are the world's tallest bridge towers (***Source:*** *The Guinness Book of Records*). How long would it take an object to fall freely from the top?

66. The Gateway Arch in St. Louis is 640 ft high. How long would it take an object to fall freely from the top?

Skill Maintenance

Solve. [7.6b]

67. *Computer Usage in Schools.* Schools are spending more and more on technology. The following graph shows the ratio of number of students to number of computers for recent years.

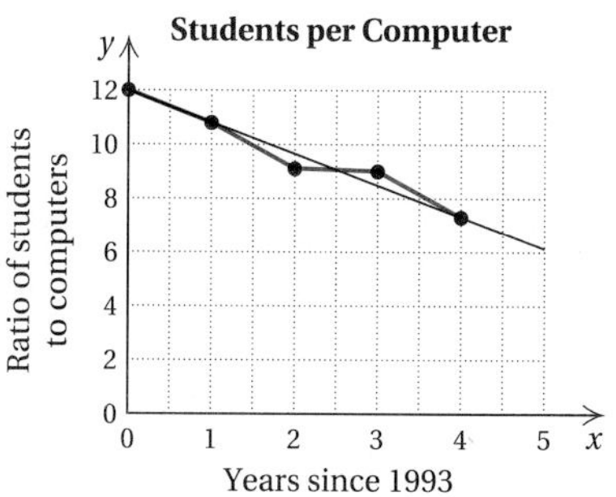

Source: Market Data Retrieval, Inc.

a) Use the two points (0, 12) and (4, 7.3) to find a linear function that fits the data.
b) Graph the function.
c) Use the function to predict the ratio of number of students to number of computers in 1998 and in 2000.

68. *Computers in Schools.* The inventory of computers in schools has soared in recent years. The following graph shows the number of computers, in millions, in schools for recent years.

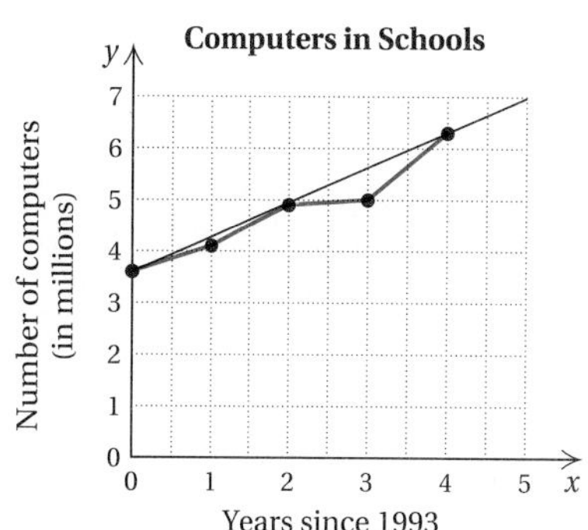

Source: Market Data Retrieval, Inc.

a) Use the two points (0, 3.6) and (4, 6.3) to find a linear function that fits the data.
b) Graph the function.
c) Use the function to predict the number of computers in schools in 1998 and in 2000.

Graph. [7.1c], [7.4a]

69. $f(x) = 5 - 2x$

70. $f(x) = 5 - 2x^2$

71. $f(x) = |5 - 2x|$

72. $2x - 5y = 10$

73. Simplify: $\sqrt{88}$. [10.3a]

74. Rationalize the denominator: $\sqrt{\frac{2}{5}}$. [10.5a]

Synthesis

75. ◆ Explain in your own words a sequence of steps that you might follow to solve any quadratic equation.

76. ◆ Example 13 can be solved with a grapher by graphing each side of

$$4410 = 4000(1 + r)^2$$

and using the INTERSECT feature. How could you determine, from a reading of the problem, a suitable viewing window? What might that window be?

77. Problems such as those in Exercises 17, 21, and 25 can be solved without first finding standard form by using the INTERSECT feature of a grapher. We let $y_1 =$ the left side of the equation and $y_2 =$ the right side. Use a grapher to solve Exercises 17, 21, and 25 in this manner.

78. Use a grapher to solve each of the following equations.

a) $25.55x^2 - 1635.2 = 0$
b) $-0.0644x^2 + 0.0936x + 4.56 = 0$
c) $2.101x + 3.121 = 0.97x^2$

Find b such that the trinomial is a square.

79. $x^2 + bx + 64$

80. $x^2 + bx + 75$

Solve.

81. $x(2x^2 + 9x - 56)(3x + 10) = 0$

82. $\left(x - \frac{1}{3}\right)\left(x - \frac{1}{3}\right) + \left(x - \frac{1}{3}\right)\left(x + \frac{2}{9}\right) = 0$

83. *Boating.* A barge and a fishing boat leave a dock at the same time, traveling at right angles to each other. The barge travels 7 km/h slower than the fishing boat. After 4 hr, the boats are 68 km apart. Find the speed of each vessel.

68 km

84. Find three consecutive integers such that the square of the first plus the product of the other two is 67.

Collaborative Learning Manual

Discover the rule for completing the square.

11.2 The Quadratic Formula

There are at least two reasons for learning to complete the square. One is to enhance your ability to graph certain equations that are needed to solve problems later in this chapter. The other is to prove a general formula for solving quadratic equations.

Objective

a Solve quadratic equations using the quadratic formula, and approximate solutions using a calculator.

For Extra Help

TAPE 18 MAC WIN CD-ROM

a Solving Using the Quadratic Formula

Each time you solve by completing the square, the procedure is the same. When we do the same kind of procedure many times, we look for a formula to speed up our work. Consider

$$ax^2 + bx + c = 0, \quad a > 0.$$

Let's solve by *completing the square.* As we carry out the steps, compare them with Example 11 in the preceding section.

$$x^2 + \frac{b}{a}x + \frac{c}{a} = 0 \qquad \text{Multiplying by } \frac{1}{a}$$

$$x^2 + \frac{b}{a}x = -\frac{c}{a} \qquad \text{Subtracting } \frac{c}{a}$$

Half of $\frac{b}{a}$ is $\frac{b}{2a}$. The square is $\frac{b^2}{4a^2}$. We add $\frac{b^2}{4a^2}$ on both sides:

$$x^2 + \frac{b}{a}x + \frac{b^2}{4a^2} = -\frac{c}{a} + \frac{b^2}{4a^2} \qquad \text{Adding } \frac{b^2}{4a^2}$$

$$\left(x + \frac{b}{2a}\right)^2 = -\frac{4ac}{4a^2} + \frac{b^2}{4a^2} \qquad \text{Factoring the left side and finding a common denominator on the right}$$

$$\left(x + \frac{b}{2a}\right)^2 = \frac{b^2 - 4ac}{4a^2}$$

$$x + \frac{b}{2a} = \sqrt{\frac{b^2 - 4ac}{4a^2}} \quad or \quad x + \frac{b}{2a} = -\sqrt{\frac{b^2 - 4ac}{4a^2}} \qquad \text{Using the principle of square roots}$$

Since $a > 0$, $\sqrt{4a^2} = 2a$, so we can simplify as follows:

$$x + \frac{b}{2a} = \frac{\sqrt{b^2 - 4ac}}{2a} \quad \text{or} \quad x + \frac{b}{2a} = -\frac{\sqrt{b^2 - 4ac}}{2a}.$$

Thus,

$$x = -\frac{b}{2a} \pm \frac{\sqrt{b^2 - 4ac}}{2a}, \quad \text{or} \quad x = \frac{-b \pm \sqrt{b^2 - 4ac}}{2a}.$$

Note that the formula also holds when $a < 0$. A similar proof would show this, but we will not consider it here.

AG Algebraic–Graphical Connection

The Quadratic Formula (Algebraic).
The solutions of
$ax^2 + bx + c = 0, a \neq 0,$
are given by

$$x = \frac{-b \pm \sqrt{b^2 - 4ac}}{2a}.$$

The Quadratic Formula (Graphical).
The x-intercepts of the graph of the function
$f(x) = ax^2 + bx + c, a \neq 0,$
are given by

$$\left(\frac{-b \pm \sqrt{b^2 - 4ac}}{2a}, 0\right).$$

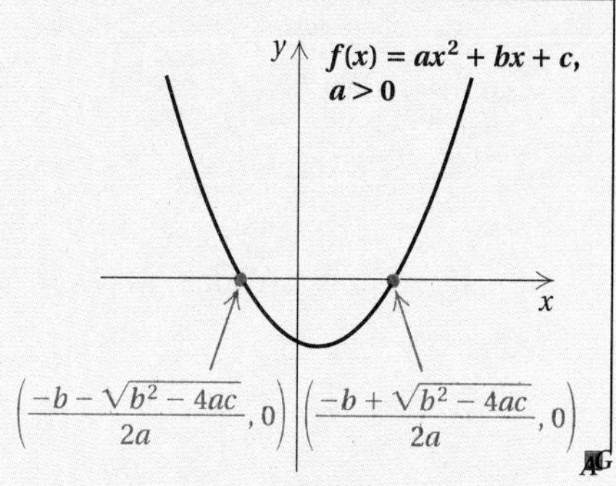

Example 1 Solve $5x^2 + 8x = -3$ using the quadratic formula.

We first find standard form and determine a, b, and c:

$$5x^2 + 8x + 3 = 0;$$
$$a = 5, \quad b = 8, \quad c = 3.$$

We then use the quadratic formula:

$$x = \frac{-b \pm \sqrt{b^2 - 4ac}}{2a}$$

$$x = \frac{-8 \pm \sqrt{8^2 - 4 \cdot 5 \cdot 3}}{2 \cdot 5} \quad \text{Substituting}$$

Be sure to write the fraction bar all the way across.

$$x = \frac{-8 \pm \sqrt{64 - 60}}{10}$$

$$x = \frac{-8 \pm \sqrt{4}}{10}$$

$$x = \frac{-8 \pm 2}{10}$$

$$x = \frac{-8 + 2}{10} \quad \text{or} \quad x = \frac{-8 - 2}{10}$$

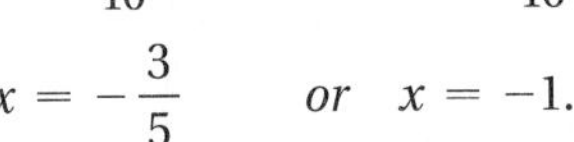

$$x = \frac{-6}{10} \quad \text{or} \quad x = \frac{-10}{10}$$

$$x = -\frac{3}{5} \quad \text{or} \quad x = -1.$$

The solutions are $-\frac{3}{5}$ and -1.

It turns out that we could have solved the equation in Example 1 more easily by factoring as follows:

$$5x^2 + 8x + 3 = 0$$
$$(5x + 3)(x + 1) = 0$$
$$5x + 3 = 0 \quad \text{or} \quad x + 1 = 0$$
$$5x = -3 \quad \text{or} \quad x = -1$$
$$x = -\tfrac{3}{5} \quad \text{or} \quad x = -1.$$

We will see in Example 2 that we cannot always rely on factoring.

To solve a quadratic equation:

1. Check for the form $x^2 = d$ or $(x + c)^2 = d$. If it is in this form, use the principle of square roots as in Section 11.1.
2. If it is not in the form of step (1), write it in standard form $ax^2 + bx + c = 0$ with a and b nonzero.
3. Then try factoring.
4. If it is not possible to factor or if factoring seems difficult, use the quadratic formula.

The solutions of a quadratic equation cannot always be found by factoring. They can always be found using the quadratic formula.

Do Exercise 1.

1. Consider the equation

$$2x^2 = 4 + 7x.$$

a) Solve using the quadratic formula.

b) Solve by factoring.

Calculator Spotlight

Use the SOLVE or INTERSECT feature to check the results of Example 1 and Margin Exercise 1.

Answers on page A-46

Example 2 Solve: $5x^2 - 8x = 3$. Give the exact solution and approximate the solutions to three decimal places.

We first find standard form and determine a, b, and c:

$$5x^2 - 8x - 3 = 0;$$
$$a = 5, \quad b = -8, \quad c = -3.$$

We then use the quadratic formula:

$$x = \frac{-(-8) \pm \sqrt{(-8)^2 - 4 \cdot 5 \cdot (-3)}}{2 \cdot 5} \quad \text{Substituting}$$
$$= \frac{8 \pm \sqrt{64 + 60}}{10} = \frac{8 \pm \sqrt{124}}{10} = \frac{8 \pm \sqrt{4 \cdot 31}}{10}$$
$$= \frac{8 \pm 2\sqrt{31}}{10} = \frac{2(4 \pm \sqrt{31})}{2 \cdot 5} = \frac{4 \pm \sqrt{31}}{5}.$$

CAUTION! To avoid a common error in simplifying, remember to *factor the numerator and the denominator* and then remove a factor of 1.

We can use a calculator to approximate the solutions:

$$\frac{4 + \sqrt{31}}{5} \approx 1.914; \qquad \frac{4 - \sqrt{31}}{5} \approx -0.314.$$

CHECK: Checking the exact solutions $(4 \pm \sqrt{31})/5$ can be quite cumbersome. It could be done on a calculator or by using the approximations. Here we check 1.914; the check for -0.314 is left to the student.

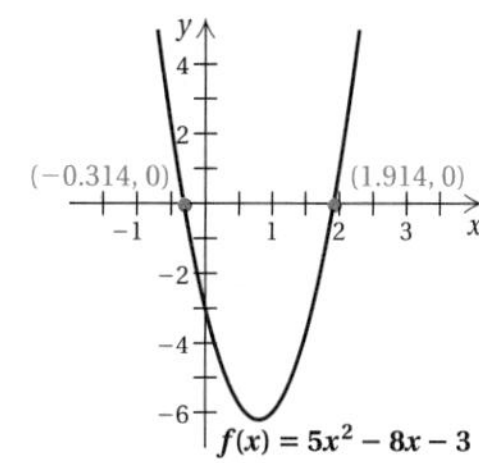

For 1.914:

$$\begin{array}{r|l} 5x^2 - 8x & = 3 \\ \hline 5(1.914)^2 - 8(1.914) & ?\ 3 \\ 5(3.663396) - 15.312 & \\ 18.31698 - 15.312 & \\ 3.00498 & \end{array}$$

Approximately TRUE

We do not have a perfect check due to the rounding error. But our check seems to confirm the solutions.

Do Exercise 2.

Some quadratic equations have solutions that are nonreal complex numbers.

Example 3 Solve: $x^2 + x + 1 = 0$.

We have $a = 1$, $b = 1$, $c = 1$. We use the quadratic formula:

$$x = \frac{-1 \pm \sqrt{1^2 - 4 \cdot 1 \cdot 1}}{2 \cdot 1}$$
$$= \frac{-1 \pm \sqrt{1 - 4}}{2}$$
$$= \frac{-1 \pm \sqrt{-3}}{2}$$
$$= \frac{-1 \pm i\sqrt{3}}{2}.$$

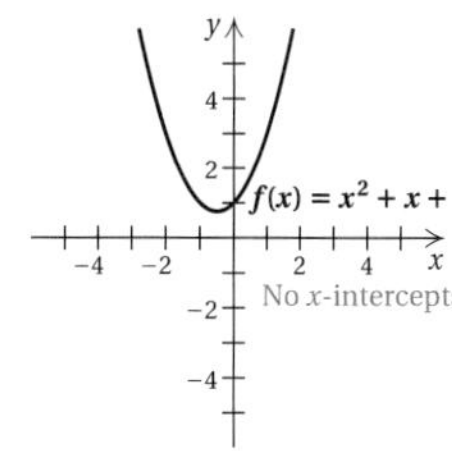

2. Solve using the quadratic formula:

$$3x^2 + 2x = 7.$$

Give the exact solution and approximate solutions to three decimal places.

Calculator Spotlight

Referring to Example 2, we note that once the algebraic work yields $(4 + \sqrt{31})/5$, we can find an approximation using the keystrokes

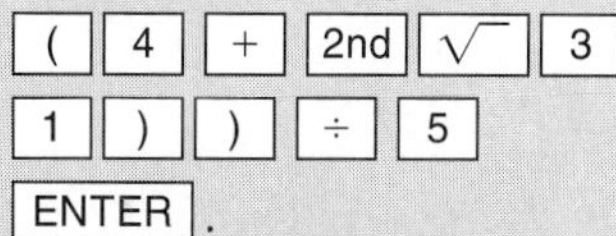

On a scientific calculator (nongrapher), the keystrokes might be

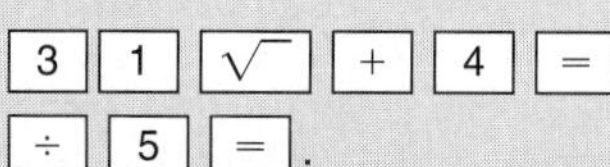

On a grapher, the SOLVE or INTERSECT feature can also be used to find the approximate solutions of $5x^2 - 8x = 3$ graphically.

Exercises

Approximate the solutions.

1. $3x^2 + 2x = 7$
2. $5.13x^2 - 8.4x - 3.08 = 0$

Answers on page A-46

3. Solve: $x^2 - x + 2 = 0$.

The solutions are

$$\frac{-1 + i\sqrt{3}}{2} \quad \text{and} \quad \frac{-1 - i\sqrt{3}}{2}.$$

The solutions can also be expressed in the form

$$-\frac{1}{2} + i\frac{\sqrt{3}}{2} \quad \text{and} \quad -\frac{1}{2} - i\frac{\sqrt{3}}{2}.$$

Do Exercise 3.

Example 4 Solve: $2 + \frac{7}{x} = \frac{5}{x^2}$. Give the exact solution and approximate solutions to three decimal places.

We first find standard form:

$$x^2\left(2 + \frac{7}{x}\right) = x^2 \cdot \frac{5}{x^2} \qquad \text{Multiplying by } x^2 \text{ to clear fractions, noting that } x \neq 0$$

$$2x^2 + 7x = 5$$

$$2x^2 + 7x - 5 = 0 \qquad \text{Subtracting 5}$$

$$a = 2, \quad b = 7, \quad c = -5$$

$$x = \frac{-7 \pm \sqrt{7^2 - 4 \cdot 2 \cdot (-5)}}{2 \cdot 2}$$

$$x = \frac{-7 \pm \sqrt{49 + 40}}{4} = \frac{-7 \pm \sqrt{89}}{4}$$

$$x = \frac{-7 + \sqrt{89}}{4} \quad \text{or} \quad x = \frac{-7 - \sqrt{89}}{4}.$$

4. Solve: $3 = \frac{5}{x} + \frac{4}{x^2}$.
Give the exact solution and approximate solutions to three decimal places.

The quadratic formula always gives correct results when we begin with the standard form. In such cases, we need check only to detect errors in substitution and computation. In this case, since we began with a rational equation, which was *not* in standard form, we *do* need to check. We cleared the fractions before obtaining standard form, and this step could introduce numbers that do not check in the original equation. At the very least, we need to show that neither of the numbers makes a denominator 0. Since neither of them does, the solutions are

$$\frac{-7 + \sqrt{89}}{4} \quad \text{and} \quad \frac{-7 - \sqrt{89}}{4}.$$

We can use a calculator to approximate the solutions:

$$\frac{-7 + \sqrt{89}}{4} \approx 0.608;$$

$$\frac{-7 - \sqrt{89}}{4} \approx -4.108.$$

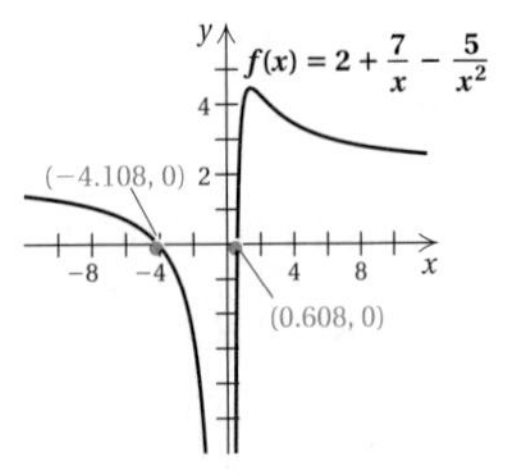

Do Exercise 4.

Calculator Spotlight

Use the SOLVE or INTERSECT feature to check the results of Example 4 and Margin Exercise 4.

Answers on page A-46

Exercise Set 11.2

a Solve.

1. $x^2 + 6x + 4 = 0$

2. $x^2 - 6x - 4 = 0$

3. $3p^2 = -8p - 1$

4. $3u^2 = 18u - 6$

5. $x^2 - x + 1 = 0$

6. $x^2 + x + 2 = 0$

7. $x^2 + 13 = 4x$

8. $x^2 + 13 = 6x$

9. $r^2 + 3r = 8$

10. $h^2 + 4 = 6h$

11. $1 + \frac{2}{x} + \frac{5}{x^2} = 0$

12. $1 + \frac{5}{x^2} = \frac{2}{x}$

13. a) Solve: $3x + x(x - 2) = 0$.
b) Find the x-intercepts of $f(x) = 3x + x(x - 2)$.

14. a) Solve: $4x + x(x - 3) = 0$.
b) Find the x-intercepts of $f(x) = 4x + x(x - 3)$.

15. a) Solve: $11x^2 - 3x - 5 = 0$.
b) Find the x-intercepts of $f(x) = 11x^2 - 3x - 5$.

16. a) Solve: $7x^2 + 8x = -2$.
b) Find the x-intercepts of $f(x) = 7x^2 + 8x + 2$.

17. a) Solve: $25x^2 = 20x - 4$.
b) Find the x-intercepts of $f(x) = 25x^2 - 20x + 4$.

18. a) Solve: $49x^2 - 14x + 1 = 0$.
b) Find the x-intercepts of $f(x) = 49x^2 - 14x + 1$.

Solve.

19. $4x(x - 2) - 5x(x - 1) = 2$

20. $3x(x + 1) - 7x(x + 2) = 6$

21. $14(x - 4) - (x + 2) = (x + 2)(x - 4)$

22. $11(x - 2) + (x - 5) = (x + 2)(x - 6)$

23. $5x^2 = 17x - 2$

24. $15x = 2x^2 + 16$

25. $x^2 + 5 = 4x$

26. $x^2 + 5 = 2x$

27. $x + \frac{1}{x} = \frac{13}{6}$

28. $\frac{3}{x} + \frac{x}{3} = \frac{5}{2}$

29. $\frac{1}{y} + \frac{1}{y + 2} = \frac{1}{3}$

30. $\frac{1}{x} + \frac{1}{x + 4} = \frac{1}{7}$

31. $(2t - 3)^2 + 17t = 15$

32. $2y^2 - (y + 2)(y - 3) = 12$

33. $(x - 2)^2 + (x + 1)^2 = 0$

34. $(x + 3)^2 + (x - 1)^2 = 0$

35. $x^3 - 1 = 0$
(*Hint*: Factor the difference of cubes. Then use the quadratic formula.)

36. $x^3 + 27 = 0$

Solve. Give the exact solution and approximate solutions to three decimal places.

37. $x^2 + 6x + 4 = 0$

38. $x^2 + 4x - 7 = 0$

39. $x^2 - 6x + 4 = 0$

40. $x^2 - 4x + 1 = 0$

41. $2x^2 - 3x - 7 = 0$

42. $3x^2 - 3x - 2 = 0$

43. $5x^2 = 3 + 8x$

44. $2y^2 + 2y - 3 = 0$

Skill Maintenance

Solve. [10.6a, b]

45. $x = \sqrt{x + 2}$

46. $x = \sqrt{15 - 2x}$

47. $\sqrt{x + 2} = \sqrt{2x - 8}$

48. $\sqrt{x + 1} + 2 = \sqrt{3x + 1}$

49. $\sqrt{x + 5} = -7$

50. $\sqrt{2x - 6} + 11 = 2$

51. $\sqrt[3]{4x - 7} = 2$

52. $\sqrt[4]{3x - 1} = 2$

Synthesis

53. ◆ The list of steps on p. 688 does not mention completing the square as a method of solving quadratic equations. Why not?

54. ◆ Given the solutions of a quadratic equation, is it possible to reconstruct the original equation? Why or why not?

55. Use the SOLVE or INTERSECT feature to solve the equations in Exercises 3, 16, 17, and 37. Then solve $2.2x^2 + 0.5x - 1 = 0$.

56. Use a grapher to solve the equations in Exercises 9, 27, and 30 using the INTERSECT feature, letting y_1 = the left side and y_2 = the right side. Then solve $5.33x^2 = 8.23x + 3.24$.

Solve.

57. $2x^2 - x - \sqrt{5} = 0$

58. $\frac{5}{x} + \frac{x}{4} = \frac{11}{7}$

59. $ix^2 - x - 1 = 0$

60. $\sqrt{3}x^2 + 6x + \sqrt{3} = 0$

61. $\frac{x}{x + 1} = 4 + \frac{1}{3x^2 - 3}$

62. $(1 + \sqrt{3})x^2 - (3 + 2\sqrt{3})x + 3 = 0$

63. Let $f(x) = (x - 3)^2$. Find all inputs x such that $f(x) = 13$.

64. Let $f(x) = x^2 + 14x + 49$. Find all inputs x such that $f(x) = 36$.

11.3 Applications Involving Quadratic Equations

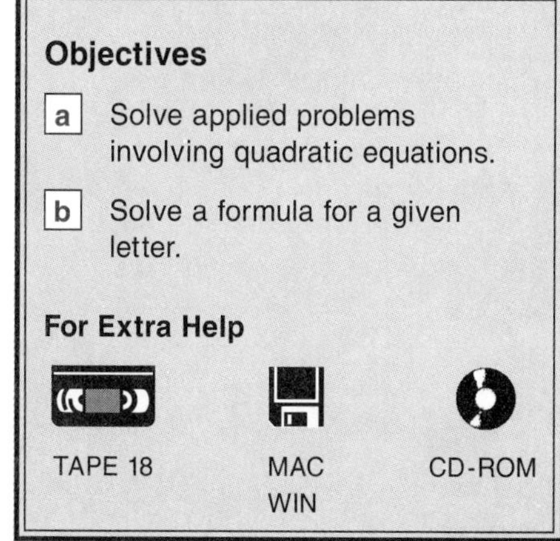

a Applications and Problem Solving

Sometimes when we translate a problem to mathematical language, the result is a quadratic equation.

Example 1 *Landscaping.* A rectangular garden is 60 ft by 80 ft. Part of the garden is torn up to install a sidewalk of uniform width around it. The area of the new garden is $\frac{1}{2}$ of the old area. How wide is the sidewalk?

1. **Familiarize.** We first make a drawing and label it with the known information. We don't know how wide the sidewalk is, so we have called its width x.

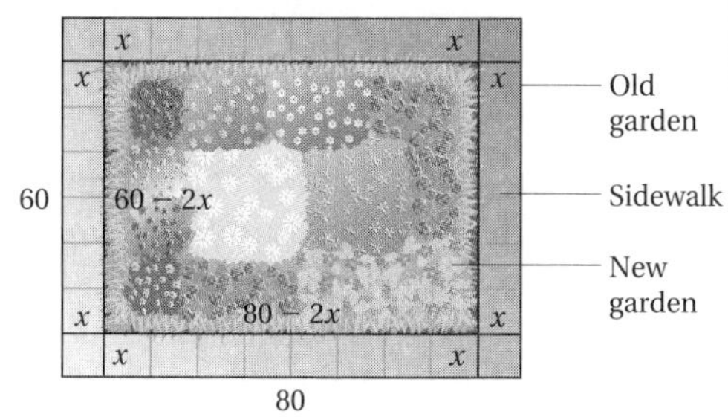

2. **Translate.** Remember, the area of a rectangle is lw (length times width). Then:

   Area of old garden $= 60 \cdot 80$;
   Area of new garden $= (60 - 2x)(80 - 2x)$.

   Since the area of the new garden is $\frac{1}{2}$ of the area of the old, we have

   $$(60 - 2x)(80 - 2x) = \tfrac{1}{2} \cdot 60 \cdot 80.$$

3. **Solve.** We solve the equation:

   $$4800 - 120x - 160x + 4x^2 = 2400 \quad \text{Using FOIL on the left}$$
   $$4x^2 - 280x + 2400 = 0 \quad \text{Collecting like terms}$$
   $$x^2 - 70x + 600 = 0 \quad \text{Dividing by 4}$$
   $$(x - 10)(x - 60) = 0 \quad \text{Factoring}$$
   $$x = 10 \quad or \quad x = 60. \quad \text{Using the principle of zero products}$$

4. **Check.** We check in the original problem. We see that 60 is not a solution because when $x = 60$, $60 - 2x = -60$, and the width of the garden cannot be negative.

   If the sidewalk is 10 ft wide, then the garden itself will have length $80 - 2 \cdot 10$, or 60 ft. The width will be $60 - 2 \cdot 10$, or 40 ft. The new area is thus $60 \cdot 40$, or 2400 ft^2. The old area was $60 \cdot 80$, or 4800 ft^2. The new area of 2400 ft^2 is $\frac{1}{2}$ of 4800 ft^2, so the number 10 checks.

5. **State.** The sidewalk is 10 ft wide.

Do Exercise 1.

1. *Box Construction.* An open box is to be made from a 10-ft by 20-ft rectangular piece of cardboard by cutting a square from each corner. The area of the bottom of the box is to be 96 ft^2. What is the length of the sides of the squares that are cut from the corners?

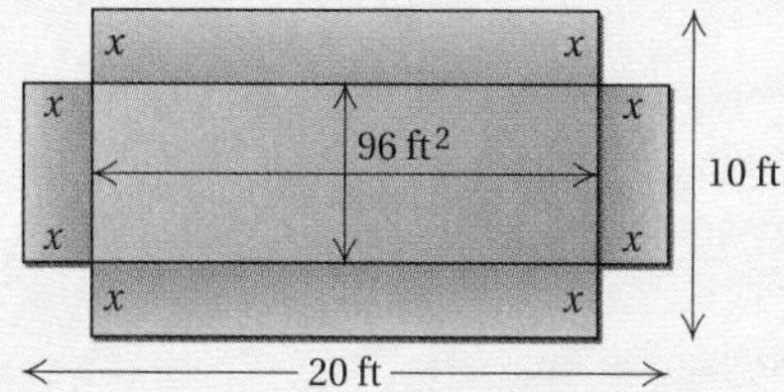

Answer on page A-46

2. *Town Planning.* Three towns A, B, and C are situated as shown. The roads at A form a right angle. The distance from A to B is 2 mi less than the distance from A to C. The distance from B to C is 10 mi. Find the distance from A to B and the distance from A to C.

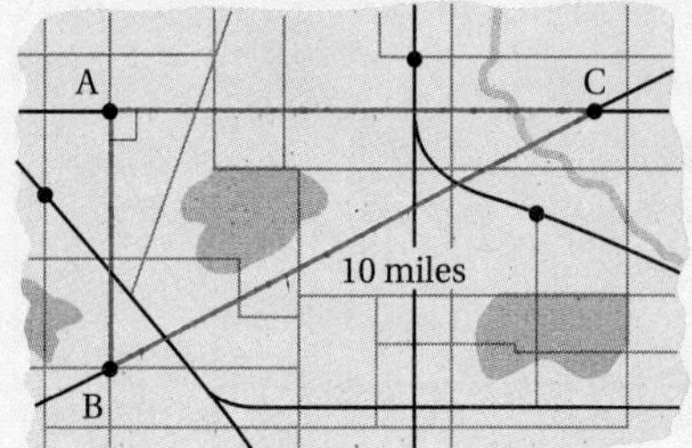

Answers on page A-46

Example 2 *Ladder Location.* A ladder leans against a building, as shown below. The ladder is 20 ft long. The distance to the top of the ladder is 4 ft greater than the distance d from the building. Find the distance d and the distance to the top of the ladder.

1. **Familiarize.** We first make a drawing and label it. We want to find d and $d + 4$.

2. **Translate.** As we look at the figure, we see that a right triangle is formed. We can use the Pythagorean equation, which we studied in Chapter 10:

$$c^2 = a^2 + b^2.$$

In this problem then, we have

$$20^2 = d^2 + (d + 4)^2.$$

3. **Solve.** We solve the equation:

$400 = d^2 + d^2 + 8d + 16$	Squaring
$2d^2 + 8d - 384 = 0$	Finding standard form
$d^2 + 4d - 192 = 0$	Dividing by 2
$(d + 16)(d - 12) = 0$	Factoring
$d + 16 = 0 \quad or \quad d - 12 = 0$	Using the principle of zero products
$d = -16 \quad or \quad d = 12.$	

4. **Check.** We know that -16 is not an answer because distances are not negative. The number 12 checks (we leave the check to the student).

5. **State.** The distance d is 12 ft, and the distance to the top of the ladder is $12 + 4$, or 16 ft.

Do Exercise 2.

Example 3 *Ladder Location.* Suppose that the ladder in Example 2 has length 10 ft. Find the distance d and the distance $d + 4$.

Using the same reasoning that we did in Example 2, we translate the problem to the equation

$$10^2 = d^2 + (d + 4)^2.$$

We solve as follows. Note that the quadratic equation we get is not easily factored, so we use the quadratic formula:

$$100 = d^2 + d^2 + 8d + 16 \quad \text{Squaring}$$

$$2d^2 + 8d - 84 = 0 \quad \text{Finding standard form}$$

$$d^2 + 4d - 42 = 0 \quad \text{Multiplying by } \tfrac{1}{2}\text{, or dividing by 2}$$

$$d = \frac{-b \pm \sqrt{b^2 - 4ac}}{2a}$$

$$= \frac{-4 \pm \sqrt{4^2 - 4(1)(-42)}}{2(1)}$$

$$= \frac{-4 \pm \sqrt{16 + 168}}{2} = \frac{-4 \pm \sqrt{184}}{2}$$

$$= \frac{-4 \pm \sqrt{4(46)}}{2} = \frac{-4 \pm 2\sqrt{46}}{2}$$

$$= -2 \pm \sqrt{46}.$$

Since $-2 - \sqrt{46} < 0$ and $\sqrt{46} > 2$, it follows that d is given by $d = -2 + \sqrt{46}$. Using a calculator, we find that $d \approx -2 + 6.782 \approx 4.782$ ft, and that $d + 4 \approx 4.782 + 4$, or 8.782 ft.

3. *Town Planning.* Three towns A, B, and C are situated as shown. The roads at A form a right angle. The distance from A to B is 2 mi less than the distance from A to C. The distance from B to C is 8 mi. Find the distance from A to B and the distance from A to C. Find exact and approximate answers to the nearest hundredth of a mile.

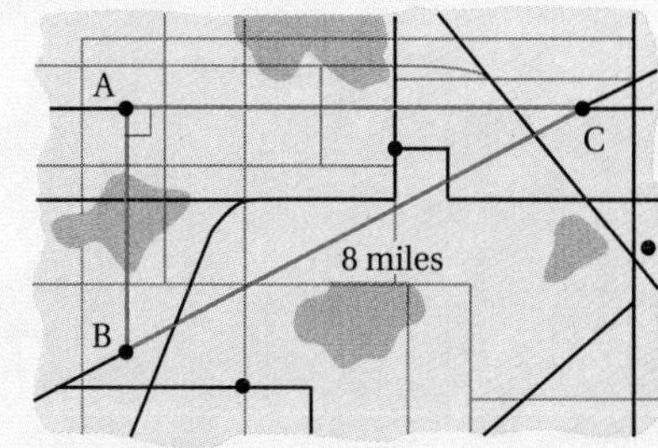

Answers on page A-46

Do Exercise 3.

Some problems translate to rational equations. The solution of such rational equations can involve quadratic equations as well.

Example 4 *Motorcycle Travel.* Karin's motorcycle traveled 300 mi at a certain speed. Had she gone 10 mph faster, she could have made the trip in 1 hr less time. Find her speed.

1. **Familiarize.** We first make a drawing, labeling it with known and unknown information. We can also organize the information, in a table. We let r = the speed, in miles per hour, and t = the time, in hours.

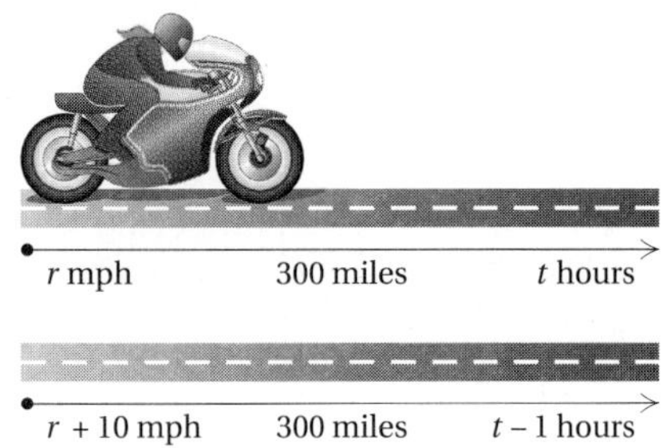

Distance	Speed	Time	
300	r	t	$\to r = \frac{300}{t}$
300	$r + 10$	$t - 1$	$\to r + 10 = \frac{300}{t - 1}$

Recalling the motion formula $d = rt$ and solving for r, we get $r = d/t$. From the rows of the table, we obtain

$$r = \frac{300}{t} \quad \text{and} \quad r + 10 = \frac{300}{t - 1}.$$

2. **Translate.** We substitute for r from the first equation into the second and get a translation:

$$\frac{300}{t} + 10 = \frac{300}{t - 1}.$$

4. *Marine Travel.* Two ships make the same voyage to a destination of 3000 nautical miles. The faster ship travels 10 knots faster than the slower one (a *knot* is 1 nautical mile per hour). The faster ship makes the voyage in 50 hr less time than the slower one. Find the speeds of the two ships.

Complete this table to help with the familiarization.

	Distance	Speed	Time
Faster Ship	3000		$t - 50$
Slower Ship	3000		t

3. Solve. We solve as follows:

$$t(t-1)\left[\frac{300}{t} + 10\right] = t(t-1)\cdot\frac{300}{t-1} \quad \text{Multiplying by the LCM}$$

$$t(t-1)\cdot\frac{300}{t} + t(t-1)\cdot 10 = t(t-1)\cdot\frac{300}{t-1}$$

$$300(t-1) + 10(t^2 - t) = 300t$$

$$10t^2 - 10t - 300 = 0 \quad \text{Standard form}$$

$$t^2 - t - 30 = 0 \quad \text{Dividing by 10}$$

$$(t-6)(t+5) = 0 \quad \text{Factoring}$$

$$t = 6 \quad or \quad t = -5 \quad \text{Using the principle of zero products}$$

4. Check. Since negative time has no meaning in this problem, we try 6 hr. Remembering that $r = d/t$, we get $r = 300/6 = 50$ mph.

To check, we take the speed 10 mph faster, which is 60 mph, and see how long the trip would have taken at that speed:

$$t = \frac{d}{r} = \frac{300}{60} = 5 \text{ hr}.$$

This is 1 hr less than the trip actually took, so we have an answer.

5. State. Karin's speed was 50 mph.

Do Exercise 4.

b Solving Formulas

Recall that to solve a formula for a certain letter, we use the principles for solving equations to get that letter alone on one side.

Example 5 *Period of a Pendulum.* The time T required for a pendulum of length l to swing back and forth (complete one period) is given by the formula $T = 2\pi\sqrt{L/g}$, where g is the gravitational constant. Solve for L.

We have

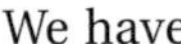

$$T = 2\pi\sqrt{\frac{L}{g}} \quad \text{This is a radical equation (see Section 6.6).}$$

$$T^2 = \left(2\pi\sqrt{\frac{L}{g}}\right)^2 \quad \text{Principle of powers (squaring)}$$

$$T^2 = 2^2\pi^2\frac{L}{g}$$

$$gT^2 = 4\pi^2 L \quad \text{Clearing fractions}$$

$$\frac{gT^2}{4\pi^2} = L. \quad \text{Multiplying by } \frac{1}{4\pi^2}$$

We now have L alone on one side and L does not appear on the other side, so the formula is solved for L.

5. Solve $A = \sqrt{\frac{w_1}{w_2}}$ for w_2.

Do Exercise 5.

Answers on page A-46

In most formulas, variables represent nonnegative numbers, so we do not need to use absolute-value signs when taking square roots.

Example 6 *Hang Time.* An athlete's *hang time* is the amount of time that the athlete can remain airborne when jumping. A formula relating an athlete's vertical leap V, in inches, to hang time T, in seconds, is $V = 48T^2$. (See Example 14 in Section 11.1.) Solve for T.

We have

$$48T^2 = V$$

$$T^2 = \frac{V}{48} \qquad \text{Multiplying by } \tfrac{1}{48} \text{ to get } T^2 \text{ alone}$$

$$T = \sqrt{\frac{V}{48}} \qquad \text{Using the principle of square roots; note that } T \geq 0.$$

$$T = \sqrt{\frac{V}{16 \cdot 3} \cdot \frac{3}{3}} \qquad \text{Factoring and multiplying by 1 to rationalize the denominator. Note that 16 is a perfect square.}$$

$$T = \sqrt{\frac{3V}{144}} = \frac{\sqrt{3V}}{12}.$$

Do Exercise 6.

Example 7 *Falling Distance.* An object tossed downward with an initial speed (velocity) of v_0 will travel a distance of s meters, where $s = 4.9t^2 + v_0t$ and t is measured in seconds. Solve for t.

Since t is squared in one term and raised to the first power in the other term, the equation is quadratic in t.

$$4.9t^2 + v_0t = s$$

$$4.9t^2 + v_0t - s = 0 \qquad \text{Writing standard form}$$

$$a = 4.9, \quad b = v_0, \quad c = -s$$

$$t = \frac{-v_0 \pm \sqrt{v_0^2 - 4(4.9)(-s)}}{2(4.9)} \qquad \text{Using the quadratic formula}$$

Since the negative square root would yield a negative value for t, we use only the positive root:

$$t = \frac{-v_0 + \sqrt{v_0^2 + 19.6s}}{9.8}.$$

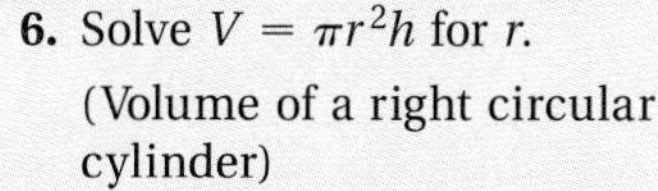

6. Solve $V = \pi r^2 h$ for r.

(Volume of a right circular cylinder)

Answer on page A-46

7. Solve $s = gt + 16t^2$ for t.

8. Solve $\dfrac{b}{\sqrt{a^2 - b^2}} = p$ for b.

Answers on page A-46

The following list of steps should help you when solving formulas for a given letter. Try to remember that when solving a formula, you do the same things you would do to solve any equation.

To solve a formula for a letter, say, b:

1. Clear the fractions and use the principle of powers, as needed, until b does not appear in any radicand or denominator. (In some cases, you may clear the fractions first, and in some cases you may use the principle of powers first.)
2. Collect all terms with b^2 in them. Also collect all terms with b in them.
3. If b^2 does not appear, you can finish by using just the addition and multiplication principles.
4. If b^2 appears but b does not, solve the equation for b^2. Then take square roots on both sides.
5. If there are terms containing both b and b^2, write the equation in standard form and use the quadratic formula.

Do Exercise 7.

Example 8 Solve $q = \dfrac{a}{\sqrt{a^2 + b^2}}$ for a.

In this case, we could either clear the fractions first or use the principle of powers first. Let's clear the fractions. We then have

$$q\sqrt{a^2 + b^2} = a.$$

Now we square both sides and then continue:

$$(q\sqrt{a^2 + b^2})^2 = a^2 \quad \text{Squaring}$$

Don't forget to square both q and $\sqrt{a^2 + b^2}$.

$$q^2(a^2 + b^2) = a^2$$

$$q^2a^2 + q^2b^2 = a^2$$

$$q^2b^2 = a^2 - q^2a^2 \quad \text{Getting all } a^2\text{-terms together}$$

$$q^2b^2 = a^2(1 - q^2) \quad \text{Factoring out } a^2$$

$$\frac{q^2b^2}{1 - q^2} = a^2 \quad \text{Dividing by } 1 - q^2$$

$$\sqrt{\frac{q^2b^2}{1 - q^2}} = a \quad \text{Taking the square root}$$

$$\frac{qb}{\sqrt{1 - q^2}} = a. \quad \text{Simplifying}$$

You need not rationalize denominators in situations such as this.

Do Exercise 8.

Exercise Set 11.3

a Solve.

1. *Flower Bed.* The width of a rectangular flower bed is 7 ft less than the length. The area is 18 ft^2. Find the length and the width.

2. *Feed Lot.* The width of a rectangular feed lot is 8 m less than the length. The area is 20 m^2. Find the length and the width.

3. *Parking Lot.* The length of a rectangular parking lot is twice the width. The area is 162 yd^2. Find the length and the width.

4. *Computer Part.* The length of a rectangular computer part is twice the width. The area is 242 cm^2. Find the length and the width.

5. *Picture Framing.* The outside of a picture frame measures 12 cm by 20 cm; 84 cm^2 of picture shows. Find the width of the frame.

6. *Picture Framing.* The outside of a picture frame measures 14 in. by 20 in.; 160 in^2 of picture shows. Find the width of the frame.

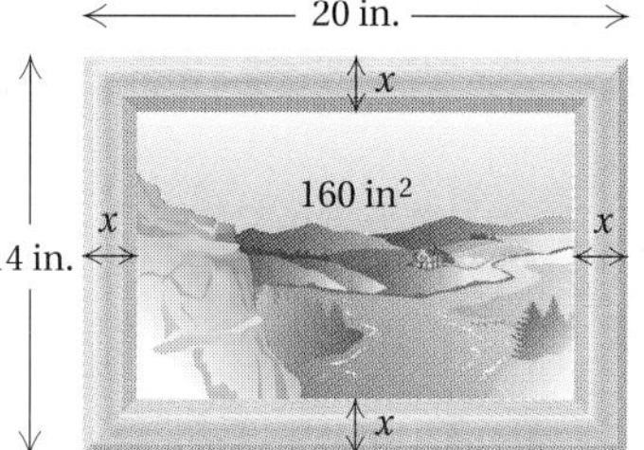

7. *Landscaping.* A landscaper is designing a flower garden in the shape of a right triangle. She wants 10 ft of a perennial border to form the hypotenuse of the triangle, and one leg is to be 2 ft longer than the other. Find the lengths of the legs.

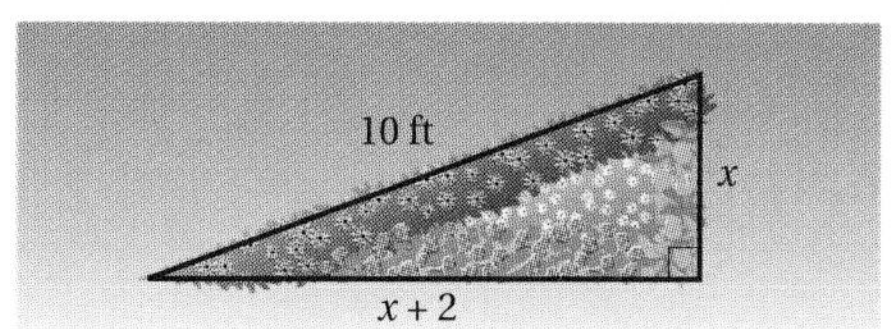

8. The hypotenuse of a right triangle is 25 m long. The length of one leg is 17 m less than the other. Find the lengths of the legs.

9. *Page Numbers.* A student opens a literature book to two facing pages. The product of the page numbers is 812. Find the page numbers.

10. *Page Numbers.* A student opens a mathematics book to two facing pages. The product of the page numbers is 1980. Find the page numbers.

Solve. Find exact and approximate answers rounded to three decimal places.

11. The width of a rectangle is 4 ft less than the length. The area is 10 ft^2. Find the length and the width.

12. The length of a rectangle is twice the width. The area is 328 cm^2. Find the length and the width.

13. *Page Dimensions.* The outside of an oversized book page measures 14 in. by 20 in.; 100 in^2 of printed text shows. Find the width of the margin.

14. *Picture Framing.* The outside of a picture frame measures 12 cm by 20 cm; 80 cm^2 of picture shows. Find the width of the frame.

15. The hypotenuse of a right triangle is 24 ft long. The length of one leg is 14 ft more than the other. Find the lengths of the legs.

16. The hypotenuse of a right triangle is 22 m long. The length of one leg is 10 m less than the other. Find the lengths of the legs.

17. *Car Trips.* During the first part of a trip, Meira's Honda traveled 120 mi at a certain speed. Meira then drove another 100 mi at a speed that was 10 mph slower. If Meira's total trip time was 4 hr, what was her speed on each part of the trip?

18. *Canoeing.* During the first part of a canoe trip, Tim covered 60 km at a certain speed. He then traveled 24 km at a speed that was 4 km/h slower. If the total time for the trip was 8 hr, what was the speed on each part of the trip?

19. *Car Trips.* Petra's Plymouth travels 200 mi at a certain speed. If the car had gone 10 mph faster, the trip would have taken 1 hr less. Find Petra's speed.

20. *Car Trips.* Sandi's Subaru travels 280 mi at a certain speed. If the car had gone 5 mph faster, the trip would have taken 1 hr less. Find Sandi's speed.

21. *Air Travel.* A Cessna flies 600 mi at a certain speed. A Beechcraft flies 1000 mi at a speed that is 50 mph faster, but takes 1 hr longer. Find the speed of each plane.

22. *Air Travel.* A turbo-jet flies 50 mph faster than a super-prop plane. If a turbo-jet goes 2000 mi in 3 hr less time than it takes the super-prop to go 2800 mi, find the speed of each plane.

23. *Bicycling.* Naoki bikes the 40 mi to Hillsboro at a certain speed. The return trip is made at a speed that is 6 mph slower. Total time for the round trip is 14 hr. Find Naoki's speed on each part of the trip.

24. *Car Speed.* On a sales trip, Gail drives the 600 mi to Richmond at a certain speed. The return trip is made at a speed that is 10 mph slower. Total time for the round trip was 22 hr. How fast did Gail travel on each part of the trip?

25. *Navigation.* The current in a typical Mississippi River shipping route flows at a rate of 4 mph. In order for a barge to travel 24 mi upriver and then return in a total of 5 hr, approximately how fast must the barge be able to travel in still water?

26. *Navigation.* The Hudson River flows at a rate of 3 mph. A patrol boat travels 60 mi upriver and returns in a total time of 9 hr. What is the speed of the boat in still water?

b Solve the formula for the given letter. Assume that all variables represent nonnegative numbers.

27. $A = 6s^2$, for s
(Surface area of a cube)

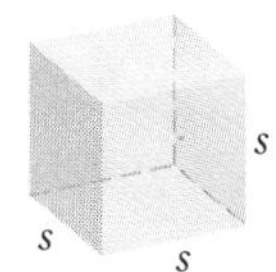

28. $A = 4\pi r^2$, for r
(Surface area of a sphere)

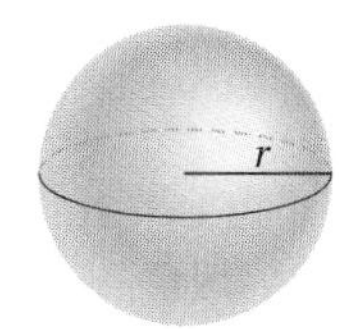

29. $F = \dfrac{Gm_1m_2}{r^2}$, for r
(Newton's law of gravity)

30. $N = \dfrac{kQ_1Q_2}{s^2}$, for s
(Number of phone calls between two cities)

31. $E = mc^2$, for c
(Einstein's energy–mass relationship)

32. $V = \frac{1}{3}s^2h$, for s
(Volume of a pyramid)

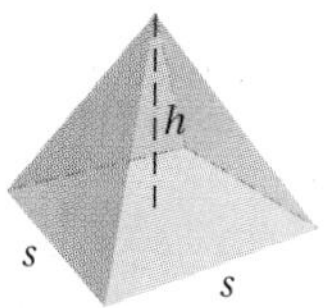

33. $a^2 + b^2 = c^2$, for b
(Pythagorean formula in two dimensions)

34. $a^2 + b^2 + c^2 = d^2$, for c
(Pythagorean formula in three dimensions)

35. $N = \dfrac{k^2 - 3k}{2}$, for k
(Number of diagonals of a polygon of k sides)

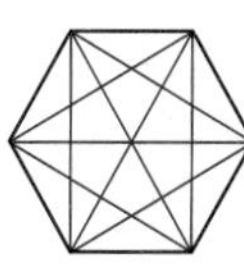

36. $s = v_0t + \dfrac{gt^2}{2}$, for t
(A motion formula)

37. $A = 2\pi r^2 + 2\pi rh$, for r
(Surface area of a cylinder)

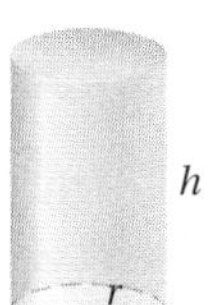

38. $A = \pi r^2 + \pi rs$, for r
(Surface area of a cone)

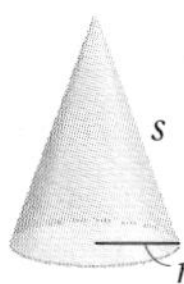

39. $T = 2\pi\sqrt{\dfrac{L}{g}}$, for g
(A pendulum formula)

40. $W = \sqrt{\dfrac{1}{LC}}$, for L
(An electricity formula)

41. $I = \dfrac{700W}{H^2}$, for H
(Body mass index)

42. $N + p = \frac{6.2A^2}{pR^2}$, for R

43. $m = \frac{m_0}{\sqrt{1 - \frac{v^2}{c^2}}}$, for v

(A relativity formula)

44. Solve the formula given in Exercise 43 for c.

Skill Maintenance

Add or subtract.

45. $\frac{1}{x-1} + \frac{1}{x^2 - 3x + 2}$ [6.3a]

46. $\frac{x+1}{x-1} - \frac{x+1}{x^2 + x + 1}$ [6.4a]

47. $\frac{2}{x+3} - \frac{x}{x-1} + \frac{x^2+2}{x^2 + 2x - 3}$ [6.4b]

48. Multiply and simplify: $\sqrt{3x^2}\,\sqrt{3x^3}$. [10.3a]

49. Express in terms of i: $\sqrt{-20}$. [10.8a]

Simplify. [6.8a]

50. $\dfrac{\frac{3}{x-1}}{\frac{1}{x+1} + \frac{2}{x-1}}$

51. $\dfrac{\frac{4}{a^2b}}{\frac{3}{a} - \frac{4}{b^2}}$

Synthesis

52. ◆ Explain how Exercises 1–26 can be solved without using factoring, completing the square, or the quadratic formula.

53. ◆ Explain how the quadratic formula can be used to factor a quadratic polynomial into two binomials. Use it to factor $5x^2 + 8x - 3$.

54. Solve: $\frac{4}{2x+i} - \frac{1}{x-i} = \frac{2}{x+i}$.

55. Find a when the reciprocal of $a - 1$ is $a + 1$.

56. *Bungee Jumping.* Jesse is tied to one end of a 40-m elasticized (bungee) cord. The other end of the cord is tied to the middle of a train trestle. If Jesse jumps off the bridge, for how long will he fall before the cord begins to stretch? (See Example 7 and let $v_0 = 0$.)

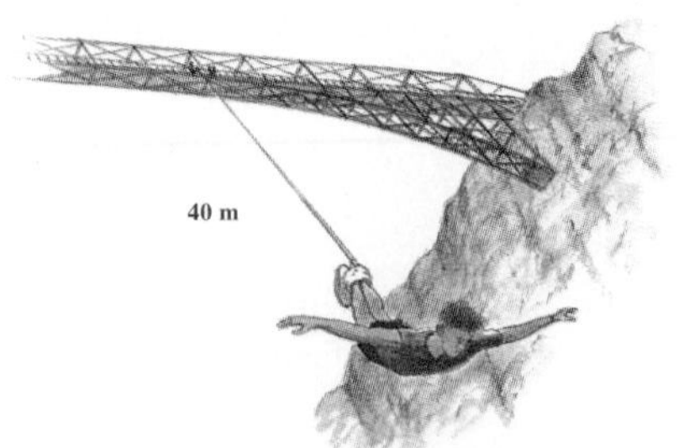

57. *Surface Area.* A sphere is inscribed in a cube as shown in the figure below. Express the surface area of the sphere as a function of the surface area S of the cube.

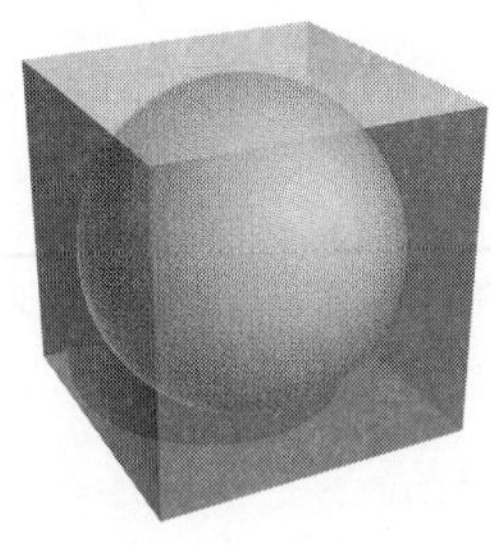

58. *Pizza Crusts.* At Pizza Perfect, Ron can make 100 large pizza crusts in 1.2 hr less than Chad. Together they can do the job in 1.8 hr. How long does it take each to do the job alone?

59. *The Golden Rectangle.* For over 2000 yr, the proportions of a "golden" rectangle have been considered visually appealing. A rectangle of width w and length l is considered "golden" if

$$\frac{w}{l} = \frac{l}{w+l}.$$

Solve for l.

11.4 More on Quadratic Equations

Objectives

a Determine the nature of the solutions of a quadratic equation.

b Write a quadratic equation having two numbers specified as solutions.

c Solve equations that are quadratic in form.

For Extra Help

TAPE 18

MAC WIN

CD-ROM

a The Discriminant

From the quadratic formula, we know that the solutions x_1 and x_2 of a quadratic equation are given by

$$x_1 = \frac{-b + \sqrt{b^2 - 4ac}}{2a} \quad \text{and} \quad x_2 = \frac{-b - \sqrt{b^2 - 4ac}}{2a}.$$

The expression $b^2 - 4ac$ is called the **discriminant**. When using the quadratic formula, it is helpful to compute the discriminant first. If it is 0, there will be just one real solution. If it is positive, there will be two real solutions. If it is negative, we will be taking the square root of a negative number; hence there will be two nonreal complex-number solutions, and they will be complex conjugates.

Discriminant $b^2 - 4ac$	Nature of Solutions	x-intercepts
0	Only one solution; it is a real number	Only one
Positive	Two different real-number solutions	Two different
Negative	Two different nonreal complex-number solutions (complex conjugates)	None

If the discriminant is a perfect square, we can solve the equation by factoring, not needing the quadratic formula.

Example 1 Determine the nature of the solutions of $9x^2 - 12x + 4 = 0$.

We have

$$a = 9, \quad b = -12, \quad c = 4.$$

We compute the discriminant:

$$\begin{aligned} b^2 - 4ac &= (-12)^2 - 4 \cdot 9 \cdot 4 \\ &= 144 - 144 \\ &= 0. \end{aligned}$$

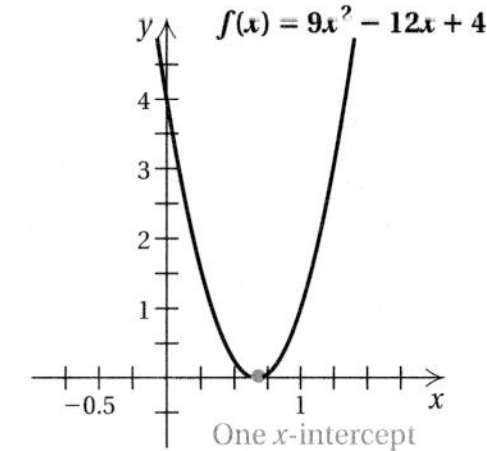

There is just one solution, and it is a real number. Since 0 is a perfect square, the equation can be solved by factoring.

Example 2 Determine the nature of the solutions of $x^2 + 5x + 8 = 0$.

We have

$$a = 1, \quad b = 5, \quad c = 8.$$

We compute the discriminant:

$$\begin{aligned} b^2 - 4ac &= 5^2 - 4 \cdot 1 \cdot 8 \\ &= 25 - 32 \\ &= -7. \end{aligned}$$

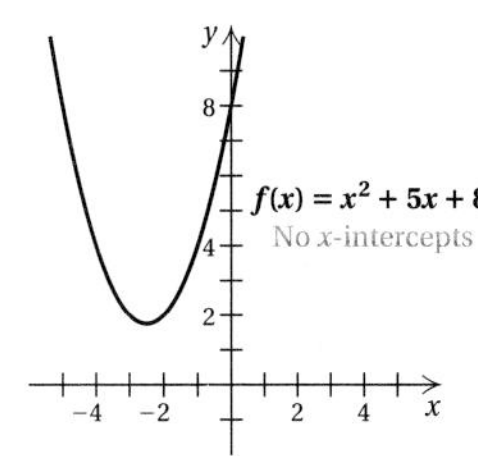

Since the discriminant is negative, there are two nonreal complex-number solutions. The equation cannot be solved by factoring because -7 is not a perfect square.

Determine the nature of the solutions without solving.

1. $x^2 + 5x - 3 = 0$

2. $9x^2 - 6x + 1 = 0$

3. $3x^2 - 2x + 1 = 0$

Answers on page A-46

Example 3 Determine the nature of the solutions of $x^2 + 5x + 6 = 0$.

We have

$$a = 1, \quad b = 5, \quad c = 6;$$
$$b^2 - 4ac = 5^2 - 4 \cdot 1 \cdot 6 = 1.$$

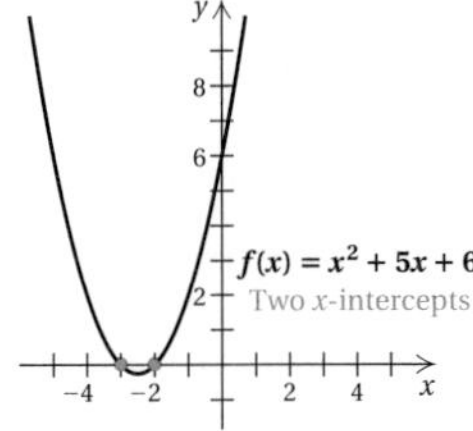

Since the discriminant is positive, there are two solutions, and they are real numbers. The equation can be solved by factoring since the discriminant is a perfect square.

The discriminant, $b^2 - 4ac$, tells us how many real-number solutions the equation $0 = ax^2 + bx + c$ has, so it also indicates how many x-intercepts the graph of $f(x) = ax^2 + bx + c$ has. Compare the following.

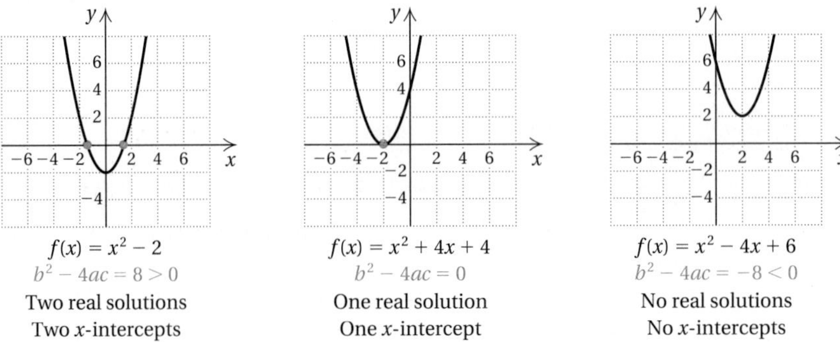

$f(x) = x^2 - 2$
$b^2 - 4ac = 8 > 0$
Two real solutions
Two x-intercepts

$f(x) = x^2 + 4x + 4$
$b^2 - 4ac = 0$
One real solution
One x-intercept

$f(x) = x^2 - 4x + 6$
$b^2 - 4ac = -8 < 0$
No real solutions
No x-intercepts

Do Exercises 1–3.

b Writing Equations from Solutions

We know by the principle of zero products that $(x - 2)(x + 3) = 0$ has solutions 2 and -3. If we know the solutions of an equation, we can write the equation, using this principle in reverse.

Example 4 Find a quadratic equation whose solutions are 3 and $-\frac{2}{5}$.

We have

$x = 3$ *or* $x = -\frac{2}{5}$

$x - 3 = 0$ *or* $x + \frac{2}{5} = 0$ **Getting the 0's on one side**

$x - 3 = 0$ *or* $5x + 2 = 0$ **Clearing the fraction**

$(x - 3)(5x + 2) = 0$ **Using the principle of zero products in reverse**

$5x^2 - 13x - 6 = 0.$ **Using FOIL**

Example 5 Write a quadratic equation whose solutions are $2i$ and $-2i$.

We have

$$x = 2i \quad or \quad x = -2i$$

$$x - 2i = 0 \quad or \quad x + 2i = 0 \qquad \text{Getting the 0's on one side}$$

$$(x - 2i)(x + 2i) = 0 \qquad \text{Using the principle of zero products in reverse}$$

$$x^2 - (2i)^2 = 0 \qquad \text{Using } (A - B)(A + B) = A^2 - B^2$$

$$x^2 - 4i^2 = 0$$

$$x^2 - 4(-1) = 0$$

$$x^2 + 4 = 0.$$

Example 6 Write a quadratic equation whose solutions are $\sqrt{3}$ and $-2\sqrt{3}$.

We have

$$x = \sqrt{3} \quad or \quad x = -2\sqrt{3}$$

$$x - \sqrt{3} = 0 \quad or \quad x + 2\sqrt{3} = 0 \qquad \text{Getting the 0's on one side}$$

$$(x - \sqrt{3})(x + 2\sqrt{3}) = 0 \qquad \text{Using the principle of zero products}$$

$$x^2 + 2\sqrt{3}x - \sqrt{3}x - 2(\sqrt{3})^2 = 0 \qquad \text{Using FOIL}$$

$$x^2 + \sqrt{3}x - 6 = 0. \qquad \text{Collecting like terms}$$

Do Exercises 4–7.

Find a quadratic equation having the following solutions.

4. 7 and -2

5. -4 and $\frac{5}{3}$

6. $5i$ and $-5i$

7. $-2\sqrt{2}$ and $\sqrt{2}$

Answers on page A-46

c Equations Quadratic in Form

Certain equations that are not really quadratic can still be solved as quadratic. Consider this fourth-degree equation.

$$x^4 - 9x^2 + 8 = 0$$

$$\downarrow \quad \downarrow \quad \downarrow \quad \downarrow$$

$$(x^2)^2 - 9(x^2) + 8 = 0 \qquad \text{Thinking of } x^4 \text{ as } (x^2)^2$$

$$\downarrow \quad \downarrow \quad \downarrow \quad \downarrow$$

$$u^2 - 9u + 8 = 0 \qquad \text{To make this clearer, write } u \text{ instead of } x^2.$$

The equation $u^2 - 9u + 8 = 0$ can be solved by factoring or by the quadratic formula. After that, we can find x by remembering that $x^2 = u$. Equations that can be solved like this are said to be **quadratic in form,** or **reducible to quadratic.**

Example 7 Solve: $x^4 - 9x^2 + 8 = 0$.

Let $u = x^2$. Then we solve the equation found by substituting u for x^2:

$$u^2 - 9u + 8 = 0$$

$$(u - 8)(u - 1) = 0 \qquad \text{Factoring}$$

$$u - 8 = 0 \quad or \quad u - 1 = 0 \qquad \text{Using the principle of zero products}$$

$$u = 8 \quad or \quad u = 1.$$

Next, we substitute x^2 for u and solve these equations:

$$x^2 = 8 \quad or \quad x^2 = 1$$

$$x = \pm\sqrt{8} \quad or \quad x = \pm 1$$

$$x = \pm 2\sqrt{2} \quad or \quad x = \pm 1.$$

8. Solve: $x^4 - 10x^2 + 9 = 0$.

To check, first note that when $x = 2\sqrt{2}$, $x^2 = 8$ and $x^4 = 64$. Also, when $x = -2\sqrt{2}$, $x^2 = 8$ and $x^4 = 64$. Similarly, when $x = 1$, $x^2 = 1$ and $x^4 = 1$, and when $x = -1$, $x^2 = 1$ and $x^4 = 1$. Thus, instead of making four checks, we need make only two.

CHECK:

For $\pm 2\sqrt{2}$:

$x^4 - 9x^2 + 8 = 0$		
$(\pm 2\sqrt{2})^4 - 9(\pm 2\sqrt{2})^2 + 8$	0	
$64 - 9 \cdot 8 + 8$		
0		TRUE

For ± 1:

$x^4 - 9x^2 + 8 = 0$		
$(\pm 1)^4 - 9(\pm 1)^2 + 8$	0	
$1 - 9 + 8$		
0		TRUE

The solutions are $1, -1, 2\sqrt{2}$, and $-2\sqrt{2}$.

CAUTION! A common error is to solve for u and then forget to solve for x. Remember that you *must* find values for the *original* variable!

Solving equations quadratic in form can sometimes introduce numbers that are not solutions of the original equation. Thus a check by substituting is necessary.

Do Exercise 8.

Example 8 Solve: $x - 3\sqrt{x} - 4 = 0$.

9. Solve $x + 3\sqrt{x} - 10 = 0$. Be sure to check.

Let $u = \sqrt{x}$. Then we solve the equation found by substituting u for $\sqrt{x}$ (and, of course, u^2 for x):

$$u^2 - 3u - 4 = 0$$

$$(u - 4)(u + 1) = 0$$

$$u = 4 \quad or \quad u = -1.$$

Next, we substitute $\sqrt{x}$ for u and solve these equations:

$$\sqrt{x} = 4 \quad or \quad \sqrt{x} = -1.$$

Squaring the first equation, we get $x = 16$, and 16 checks. Squaring the second equation, we get $x = 1$. But the number 1 does not check. The number 16 is the solution.

Do Exercise 9.

Answers on page A-46

Example 9 Find the x-intercepts of the graph of $f(x) = (x^2 - 1)^2 - (x^2 - 1) - 2$.

The x-intercepts occur where $f(x) = 0$ so we must have

$$(x^2 - 1)^2 - (x^2 - 1) - 2 = 0.$$

Let $u = x^2 - 1$. Then we solve the equation found by substituting u for $x^2 - 1$:

$$u^2 - u - 2 = 0$$
$$(u - 2)(u + 1) = 0$$
$$u = 2 \quad or \quad u = -1.$$

Next, we substitute $x^2 - 1$ for u and solve these equations:

$$x^2 - 1 = 2 \quad or \quad x^2 - 1 = -1$$
$$x^2 = 3 \quad or \quad x^2 = 0$$
$$x = \pm\sqrt{3} \quad or \quad x = 0.$$

The numbers $\sqrt{3}$, $-\sqrt{3}$, and 0 check. They are the solutions of $(x^2 - 1)^2 - (x^2 - 1) - 2 = 0$. Thus the x-intercepts of the graph of $f(x)$ are $(-\sqrt{3}, 0)$, $(0, 0)$, and $(\sqrt{3}, 0)$.

Do Exercise 10.

10. Find the x-intercepts of

$f(x) = (x^2 - x)^2 - 14(x^2 - x) + 24$.

Example 10 Solve: $y^{-2} - y^{-1} - 2 = 0$.

Let $u = y^{-1}$. Then we solve the equation found by substituting u for y^{-1} and u^2 for y^{-2}:

$$u^2 - u - 2 = 0$$
$$(u - 2)(u + 1) = 0$$
$$u = 2 \quad or \quad u = -1.$$

Next, we substitute y^{-1} or $1/y$ for u and solve these equations:

$$\frac{1}{y} = 2 \quad or \quad \frac{1}{y} = -1.$$

Solving, we get

$$y = \frac{1}{2} \quad \text{or} \quad y = \frac{1}{(-1)} = -1.$$

The numbers $\frac{1}{2}$ and -1 both check. They are the solutions.

Do Exercise 11.

11. Solve: $x^{-2} + x^{-1} - 6 = 0$.

Answers on page A-46

Improving Your Math Study Skills

How Many Women Have Won the Ultimate Math Contest?

Although this Study Skill feature does not contain specific tips on studying mathematics, we hope that you will find this article both challenging and encouraging.

Every year on college campuses across the United States and Canada, the most brilliant math students face the ultimate challenge. For six hours, they struggle with problems from the merely intractable to the seemingly impossible.

Every spring, five are chosen winners of the William Lowell Putnam Mathematical Competition, the Olympics of college mathematics. Every year for 56 years, all have been men.

Until this year.

This spring, Ioana Dumitriu (pronounced yo-AHN-na doo-mee-TREE-oo), 20, a New York University sophomore from Romania, became the first woman to win the award.

Ms. Dumitriu, the daughter of two electrical engineering professors in Romania, who as a girl solved math puzzles for fun, was identified as a math talent early in her schooling in Bucharest. At 11, Ms. Dumitriu was steered into years of math training camps as preparation for the Romanian entry in the International Mathematics Olympiad.

It was this training, and a handsome young coach, that led her to New York City. He was several years older. They fell in love. He chose N.Y.U. for its graduate school in mathematics, and at 19 she joined him in New York.

The test Ms. Dumitriu won is dauntingly difficult, even for math majors. About half of the 2,407 test-takers scored 2 or less of a possible 120, and a third scored 0. Some students simply walk out after staring at the questions for a while.

Ms. Dumitriu said that in the six hours allotted, she had time to do 8 of the 12 problems, each worth a maximum of 10 points. The last one she did in 10 minutes. This year, Ms. Dumitriu and her five co-winners (there was a tie for fifth place) scored between 76 and 98. She does not know her exact score or rank because the organizers do not announce them.

"I didn't ever tell myself that I was unlikely to win, that no woman before had ever won and therefore I couldn't," she said. "It is not that I forget that I'm a woman. It's just that I don't see it as an obstacle or a —."

Her English is near-perfect, but she paused because she could not find the right word. "The mathematics community is made up of persons, and that is what I am primarily."

Prof. Joel Spencer, who was a Putnam winner himself, said her work for his class in problem solving last year was remarkable. "What really got me was her fearlessness," he said. "To be good at math, you have to go right at it and start playing around with it, and she had that from the start."

In the graduate lounge in the Courant Institute of Mathematical Sciences at N.Y.U., Ms. Dumitriu, a tall, striking redhead, stands out. Instead of jeans and T-shirts, she wears gray pin-striped slacks and a rust-colored turtleneck and vest.

"There is a social perception of women and math, a stereotype," Ms. Dumitriu said during an interview. "What's happening right now is that the stereotype is defied. It starts breaking."

Still, even as women began to flock to sciences, math has remained largely a male bastion.

"Math remains the bottom line of sex differences for many," said Sheila Tobias, author of "Overcoming Math Anxiety" (W.W. Norton & Company, 1994). "It's one thing for women to write books, negotiate bills through Congress, litigate, fire missiles; quite another for them to do math."

Besides collecting the $1,000 awarded to each Putnam fellow, Ms. Dumitriu also won the $500 Elizabeth Lowell Putman prize for the top woman finisher for the second year in a row, a prize created five years ago to encourage women to take the test. This year 414 did.

In her view, there are never too many problems, never too much practice.

Besides, each new problem holds its own allure: "When you have all the pieces and you put them together and you see the puzzle, that moment always amazes me."

Exercise Set 11.4

a Determine the nature of the solutions of the equation.

1. $x^2 - 8x + 16 = 0$

2. $x^2 + 12x + 36 = 0$

3. $x^2 + 1 = 0$

4. $x^2 + 6 = 0$

5. $x^2 - 6 = 0$

6. $x^2 - 3 = 0$

7. $4x^2 - 12x + 9 = 0$

8. $4x^2 + 8x - 5 = 0$

9. $x^2 - 2x + 4 = 0$

10. $x^2 + 3x + 4 = 0$

11. $9t^2 - 3t = 0$

12. $4m^2 + 7m = 0$

13. $y^2 = \frac{1}{2}y + \frac{3}{5}$

14. $y^2 + \frac{9}{4} = 4y$

15. $4x^2 - 4\sqrt{3}x + 3 = 0$

16. $6y^2 - 2\sqrt{3}y - 1 = 0$

b Write a quadratic equation having the given numbers as solutions.

17. -4 and 4

18. -11 and 9

19. -2 and -7

20. 3 and 10

21. 8, only solution [*Hint*: It must be a double solution, that is, $(x - 8)(x - 8) = 0$.]

22. -3, only solution

23. $-\frac{2}{5}$ and $\frac{6}{5}$

24. $-\frac{1}{4}$ and $-\frac{1}{2}$

25. $\frac{k}{3}$ and $\frac{m}{4}$

26. $\frac{c}{2}$ and $\frac{d}{2}$

27. $-\sqrt{3}$ and $2\sqrt{3}$

28. $\sqrt{2}$ and $3\sqrt{2}$

c Solve.

29. $x^4 - 6x^2 + 9 = 0$

30. $x^4 - 7x^2 + 12 = 0$

31. $x - 10\sqrt{x} + 9 = 0$

32. $2x - 9\sqrt{x} + 4 = 0$

33. $(x^2 - 6x)^2 - 2(x^2 - 6x) - 35 = 0$

34. $(x^2 + 5x)^2 + 2(x^2 + 5x) - 24 = 0$

35. $x^{-2} - 5x^{-1} - 36 = 0$

36. $3x^{-2} - x^{-1} - 14 = 0$

37. $(1 + \sqrt{x})^2 + (1 + \sqrt{x}) - 6 = 0$

38. $(2 + \sqrt{x})^2 - 3(2 + \sqrt{x}) - 10 = 0$

39. $(y^2 - 5y)^2 - 2(y^2 - 5y) - 24 = 0$

40. $(2t^2 + t)^2 - 4(2t^2 + t) + 3 = 0$

41. $t^4 - 6t^2 - 4 = 0$

42. $w^4 - 7w^2 + 7 = 0$

43. $2x^{-2} + x^{-1} - 1 = 0$

44. $m^{-2} + 9m^{-1} - 10 = 0$

45. $6x^4 - 19x^2 + 15 = 0$

46. $6x^4 - 17x^2 + 5 = 0$

47. $x^{2/3} - 4x^{1/3} - 5 = 0$

48. $x^{2/3} + 2x^{1/3} - 8 = 0$

49. $\left(\frac{x-4}{x+1}\right)^2 - 2\left(\frac{x-4}{x+1}\right) - 35 = 0$

50. $\left(\frac{x+3}{x-3}\right)^2 - \left(\frac{x+3}{x-3}\right) - 6 = 0$

51. $9\left(\frac{x+2}{x+3}\right)^2 - 6\left(\frac{x+2}{x+3}\right) + 1 = 0$

52. $16\left(\frac{x-1}{x-8}\right)^2 + 8\left(\frac{x-1}{x-8}\right) + 1 = 0$

53. $\left(\frac{x^2-2}{x}\right)^2 - 7\left(\frac{x^2-2}{x}\right) - 18 = 0$

54. $\left(\frac{y^2-1}{y}\right)^2 - 4\left(\frac{y^2-1}{y}\right) - 12 = 0$

Find the x-intercepts of the function.

55. $f(x) = 5x + 13\sqrt{x} - 6$

56. $f(x) = 3x + 10\sqrt{x} - 8$

57. $f(x) = (x^2 - 3x)^2 - 10(x^2 - 3x) + 24$

58. $f(x) = x^{2/5} + x^{1/5} - 6$

Skill Maintenance

Solve. [8.3a]

59. *Coffee Beans.* Twin Cities Roasters has Kenyan coffee worth \$6.75 per pound and Peruvian coffee worth \$11.25 per pound. How many pounds of each kind should be mixed in order to obtain a 50-lb mixture that is worth \$8.55 per pound?

60. *Solution Mixtures.* Solution A is 18% alcohol and solution B is 45% alcohol. How many liters of each should be mixed in order to get 12 L of a solution that is 36% alcohol?

Multiply and simplify. [10.3a]

61. $\sqrt{8x}\,\sqrt{2x}$

62. $\sqrt[3]{x^2}\,\sqrt[3]{27x^4}$

63. $\sqrt[4]{9a^2}\,\sqrt[4]{18a^3}$

64. $\sqrt[5]{16}\,\sqrt[5]{64}$

Graph. [7.1c], [7.4a, c]

65. $f(x) = -\frac{3}{5}x + 4$

66. $5x - 2y = 8$

67. $y = 4$

68. $f(x) = -x - 3$

Synthesis

69. ◆ Describe a procedure that could be used to write an equation having the first seven natural numbers as solutions.

70. ◆ Describe a procedure that could be used to write an equation that is quadratic in $3x^2 + 1$ and has real-number solutions.

71. Use a grapher to check your answers to Exercises 30, 32, 34, and 37.

72. Use a grapher to solve each of the following equations.

a) $6.75x - 35\sqrt{x} - 5.26 = 0$
b) $\pi x^4 - \pi^2 x^2 = \sqrt{99.3}$
c) $x^4 - x^3 - 13x^2 + x + 12 = 0$

For each equation under the given condition, **(a)** find k and **(b)** find the other solution.

73. $kx^2 - 2x + k = 0$; one solution is -3.

74. $kx^2 - 17x + 33 = 0$; one solution is 3.

75. Find a quadratic equation for which the sum of the solutions is $\sqrt{3}$ and the product is 8.

76. Find k given that $kx^2 - 4x + (2k - 1) = 0$ and the product of the solutions is 3.

77. The graph of a function of the form

$$f(x) = ax^2 + bx + c$$

is a curve similar to the one shown below. Determine a, b, and c from the information given.

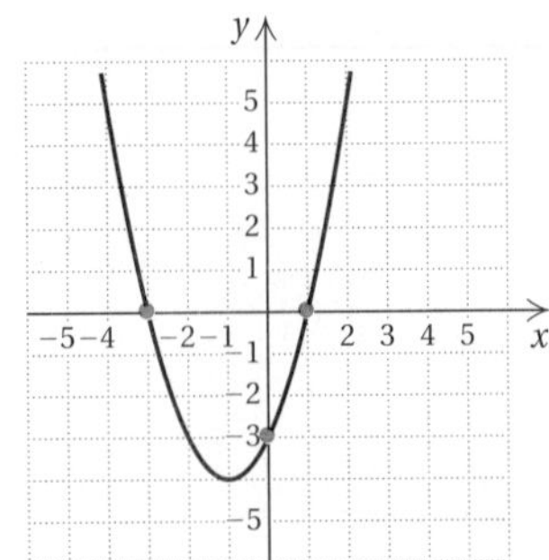

78. ◆ While solving a quadratic equation of the form $ax^2 + bx + c = 0$ with a grapher, Shawn-Marie gets the following screen.

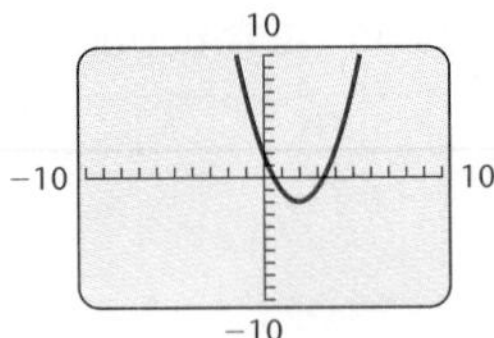

How could the discriminant help her check the graph?

Solve.

79. $\dfrac{x}{x-1} - 6\sqrt{\dfrac{x}{x-1}} - 40 = 0$

80. $\left(\sqrt{\dfrac{x}{x-3}}\right)^2 - 24 = 10\sqrt{\dfrac{x}{x-3}}$

81. $\sqrt{x-3} - \sqrt[4]{x-3} = 12$

82. $a^3 - 26a^{3/2} - 27 = 0$

83. $x^6 - 28x^3 + 27 = 0$

84. $x^6 + 7x^3 - 8 = 0$

11.5 Graphs of Quadratic Functions of the Type $f(x) = a(x - h)^2 + k$

In this section and the next, we develop techniques for graphing quadratic functions.

a Graphs of $f(x) = ax^2$

The most basic quadratic function is $f(x) = x^2$.

Example 1 Graph: $f(x) = x^2$.

We choose some values for x and compute $f(x)$ for each. Then we plot the ordered pairs and connect them with a smooth curve.

x	$f(x) = x^2$	$(x, f(x))$
−3	9	(−3, 9)
−2	4	(−2, 4)
−1	1	(−1, 1)
0	0	(0, 0)
1	1	(1, 1)
2	4	(2, 4)
3	9	(3, 9)

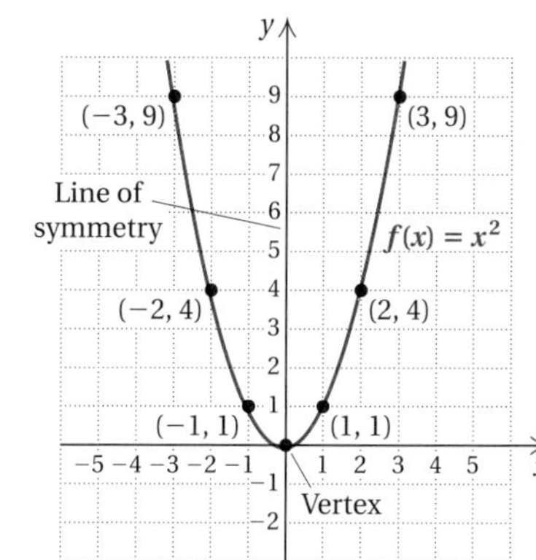

All quadratic functions have graphs similar to the one in Example 1. Such curves are called **parabolas**. They are cup-shaped curves that are symmetric with respect to a vertical line known as the parabola's **line of symmetry,** or **axis of symmetry.** In the graph of $f(x) = x^2$, shown above, the y-axis (or the line $x = 0$) is the line of symmetry. If the paper were to be folded on this line, the two halves of the curve would match. The point (0, 0) is the **vertex** of this parabola.

Let's compare the graphs of $g(x) = \frac{1}{2}x^2$ and $h(x) = 2x^2$ with the graph of $f(x) = x^2$. We choose x-values and plot points for both functions.

x	$g(x) = \frac{1}{2}x^2$
−3	$\frac{9}{2}$
−2	2
−1	$\frac{1}{2}$
0	0
1	$\frac{1}{2}$
2	2
3	$\frac{9}{2}$

x	$h(x) = 2x^2$
−3	18
−2	8
−1	2
0	0
1	2
2	8
3	18

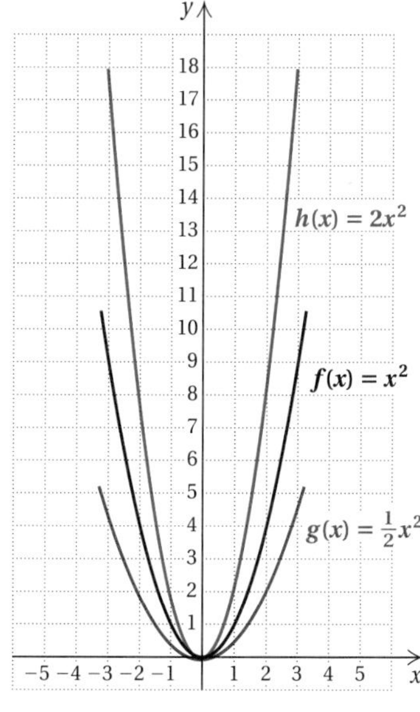

Note that the graph of $g(x) = \frac{1}{2}x^2$ is a wider parabola than the graph of $f(x) = x^2$, and the graph of $h(x) = 2x^2$ is narrower. The vertex and the line of symmetry, however, remain (0, 0) and $x = 0$, respectively.

Objectives

a Graph quadratic functions of the type $f(x) = ax^2$ and then label the vertex and the line of symmetry.

b Graph quadratic functions of the type $f(x) = a(x - h)^2$ and then label the vertex and the line of symmetry.

c Graph quadratic functions of the type $f(x) = a(x - h)^2 + k$, finding the vertex, the line of symmetry, and the maximum or minimum y-value.

For Extra Help

TAPE 19

MAC WIN

CD-ROM

1. Graph: $f(x) = -\frac{1}{3}x^2$.

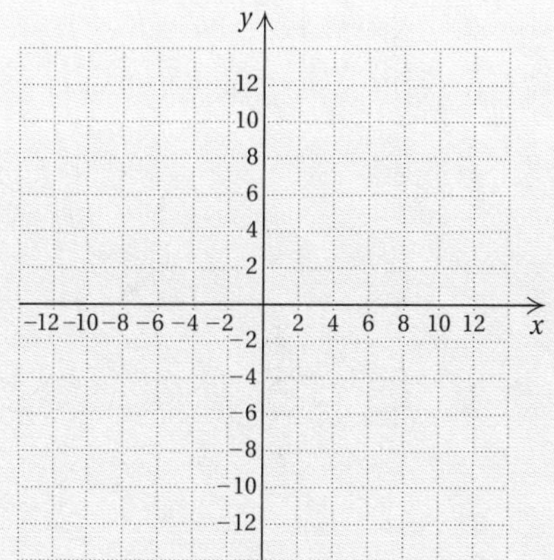

Answer on page A-47

Calculator Spotlight

Exercises

1. a) Graph each of the following equations using the viewing window [−5, 5, −10, 10]:

$y_1 = x^2$; $y_2 = 3x^2$; $y_3 = \frac{1}{3}x^2$.

b) See if you can determine a rule that describes the effect of multiplying x^2 by a, in the graph of $y = ax^2$, when $a > 1$ and $0 < a < 1$. Try some other graphs to test your assertion.

2. a) Graph each of the following equations using the viewing window [−5, 5, −10, 10]:

$y_1 = -x^2$; $y_2 = -4x^2$; $y_3 = -\frac{2}{3}x^2$.

b) See if you can determine a rule that describes the effect of multiplying x^2 by a, in the graph of $y = ax^2$, when $a < -1$ and $-1 < a < 0$. Try some other graphs to test your assertion.

Graph.

2. $f(x) = 3x^2$

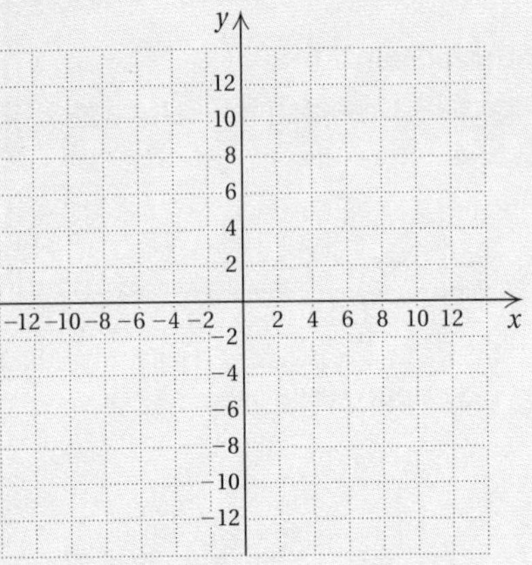

3. $f(x) = -2x^2$

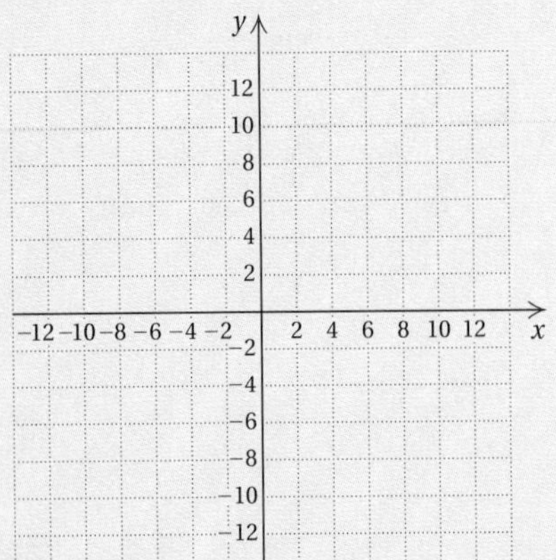

When we consider the graph of $k(x) = -\frac{1}{2}x^2$, we see that the parabola opens down and is the same shape as the graph of $g(x) = \frac{1}{2}x^2$.

x	$k(x) = -\frac{1}{2}x^2$
−3	$-\frac{9}{2}$
−2	−2
−1	$-\frac{1}{2}$
0	0
1	$-\frac{1}{2}$
2	−2
3	$-\frac{9}{2}$

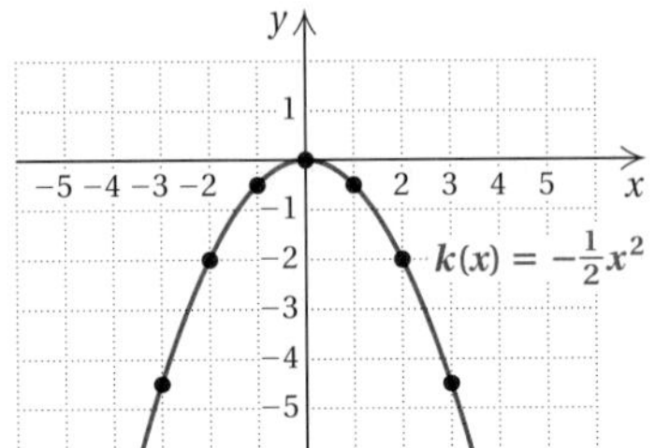

The graph of $f(x) = ax^2$, or $y = ax^2$, is a parabola with $x = 0$ as its line of symmetry; its vertex is the origin.

For $a > 0$, the parabola opens up; for $a < 0$, the parabola opens down.

If $|a|$ is greater than 1, the parabola is narrower than $y = x^2$.

If $|a|$ is between 0 and 1, the parabola is wider than $y = x^2$.

Do Exercises 1–3. (Exercise 1 is on the preceding page.)

b Graphs of $f(x) = a(x - h)^2$

It would seem logical now to consider functions of the type

$$f(x) = ax^2 + bx + c.$$

We are heading in that direction, but it is convenient to first consider graphs of $f(x) = a(x - h)^2$ and then $f(x) = a(x - h)^2 + k$, where a, h, and k are constants.

Answers on page A-47

Example 2 Graph: $g(x) = (x - 3)^2$.

We choose some values for x and compute $g(x)$. Then we plot the points and draw the curve.

x	$g(x) = (x - 3)^2$	
-1	16	
0	9	
1	4	
2	1	
3	0	← Vertex
4	1	
5	4	
6	9	

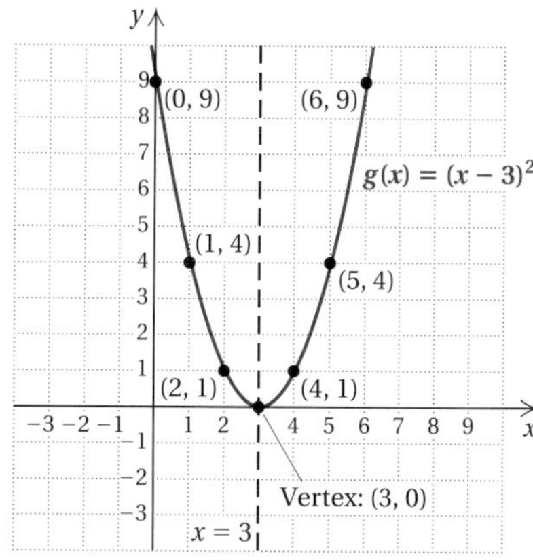

Note that $g(x) = 16$ when $x = -1$ and $g(x)$ gets larger as x gets more negative. Thus we use positive values to fill out the table. Note that the line $x = 3$ is the line of symmetry and the point $(3, 0)$ is the vertex. Had we known earlier that $x = 3$ is the line of symmetry, we could have computed some values on one side, such as $(4, 1)$, $(5, 4)$, and $(6, 9)$, and then used symmetry to get their mirror images $(2, 1)$, $(1, 4)$, and $(0, 9)$ without further computation.

The graph of $g(x) = (x - 3)^2$ in Example 2 looks just like the graph of $f(x) = x^2$ in Example 1, except that it is moved, or translated, 3 units to the right. Comparing the pairs for $g(x)$ with those for $f(x)$, we see that when an input for $g(x)$ is 3 more than an input for $f(x)$, the outputs match.

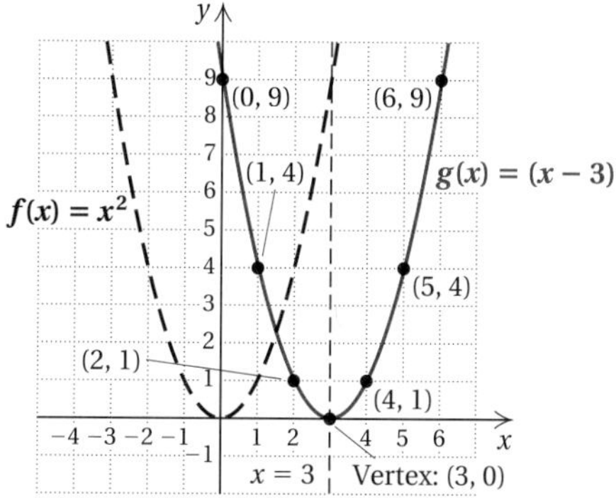

The graph of $f(x) = a(x - h)^2$ has the same shape as the graph of $y = ax^2$.
If h is positive, the graph of $y = ax^2$ is shifted h units to the right.
If h is negative, the graph of $y = ax^2$ is shifted $|h|$ units to the left.
The vertex is $(h, 0)$ and the axis of symmetry is $x = h$.

Calculator Spotlight

Exercises

1. a) Graph each of the following equations using the viewing window $[-5, 5, -5, 5]$:

   $y_1 = 7x^2$;
   $y_2 = 7(x - 1)^2$;
   $y_3 = 7(x - 2)^2$.

   Use the TABLE feature with TblStart $= -2$ and ΔTbl $= 1$ to compare y-values.

   b) See if you can determine a rule that describes the effect of subtracting h from x, in the graph of $y = a(x - h)^2$. Try some other graphs to test your assertion.

2. a) Graph each of the following equations using the viewing window $[-5, 5, -5, 5]$:

   $y_1 = 7x^2$;
   $y_2 = 7(x + 1)^2$;
   $y_3 = 7(x + 2)^2$.

   b) See if you can determine a rule that describes the effect of adding h to x, in the graph of $y = a(x + h)^2$. Try some other graphs to test your assertion.

Example 3 Graph: $f(x) = -2(x + 3)^2$.

We first rewrite the equation as $f(x) = -2[x - (-3)]^2$. In this case, $a = -2$ and $h = -3$, so the graph looks like that of $g(x) = 2x^2$ translated 3 units to the left and, since $-2 < 0$, the graph opens down. The vertex is $(-3, 0)$, and the line of symmetry is $x = -3$. Plotting points as needed, we obtain the graph shown below.

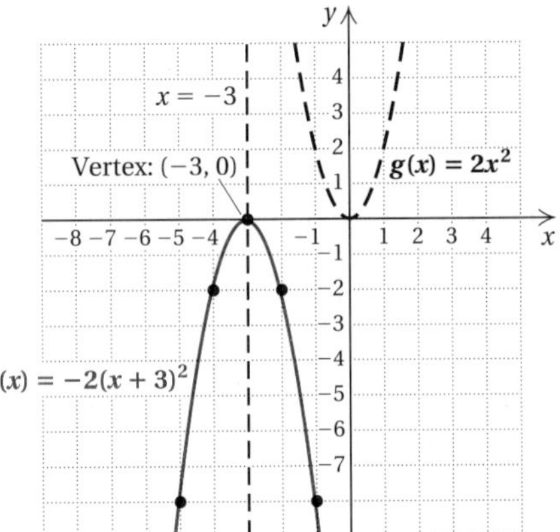

Do Exercises 4 and 5 on the following page.

Calculator Spotlight

Exercises

1. a) Graph each of the following equations using the viewing window [−5, 5, −5, 5]:

   $y_1 = 7(x - 1)^2$;
   $y_2 = 7(x - 1)^2 + 2$.

   Use the TABLE feature with TblStart = −2 and ΔTbl = 1 to compare y-values.

   b) See if you can determine a rule that describes the effect of adding k, in the graph of $y = a(x - h)^2 + k$. Try some other graphs to test your assertion.

2. a) Graph each of the following equations using the viewing window [−5, 5, −5, 5]:

   $y_1 = 7(x - 1)^2$;
   $y_3 = 7(x - 1)^2 - 4$.

   b) See if you can determine a rule that describes the effect of adding k, where k is negative, in the graph of $y = a(x - h)^2 + k$. Try some other graphs to test your assertion.

c Graphs of $f(x) = a(x - h)^2 + k$

Given a graph of $f(x) = a(x - h)^2$, what happens if we add a constant k? Suppose that we add 2. This increases each function value $f(x)$ by 2, so the curve is moved up. If k is negative, the curve is moved down. The line of symmetry for the parabola remains $x = h$, but the vertex will be at (h, k), or, equivalently, $(h, f(h))$.

Note that if a parabola opens up $(a > 0)$, the function value, or y-value, at the vertex is a least, or **minimum**, value. That is, it is less than the y-value at any other point on the graph. If the parabola opens down $(a < 0)$, the function value at the vertex is a greatest, or **maximum**, value.

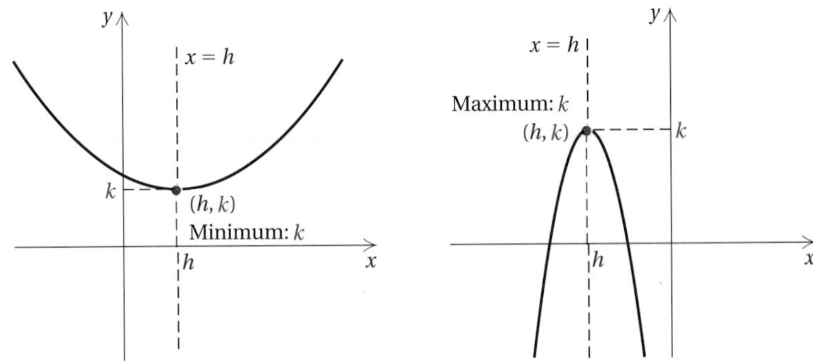

The graph of $f(x) = a(x - h)^2 + k$ has the same shape as the graph of $y = a(x - h)^2$.

If k is positive, the graph of $y = a(x - h)^2$ is shifted k units up.

If k is negative, the graph of $y = a(x - h)^2$ is shifted $|k|$ units down.

The vertex is (h, k), and the line of symmetry is $x = h$.

For $a > 0$, k is the minimum function value. For $a < 0$, k is the maximum function value.

Example 4 Graph $f(x) = (x - 3)^2 - 5$, and find the minimum function value.

The graph will look like that of $g(x) = (x - 3)^2$ (see Example 2) but translated 5 units down. You can confirm this by plotting some points. For instance, $f(4) = (4 - 3)^2 - 5 = -4$, whereas in Example 2, $g(4) = (4 - 3)^2 = 1$. Note here that $h = 3$, so we calculate points on both sides of $x = 3$.

The vertex is now $(3, -5)$, and the minimum function value is -5.

x	$f(x) = (x - 3)^2 - 5$	
0	4	
1	-1	
2	-4	
3	-5	←Vertex
4	-4	
5	-1	
6	4	

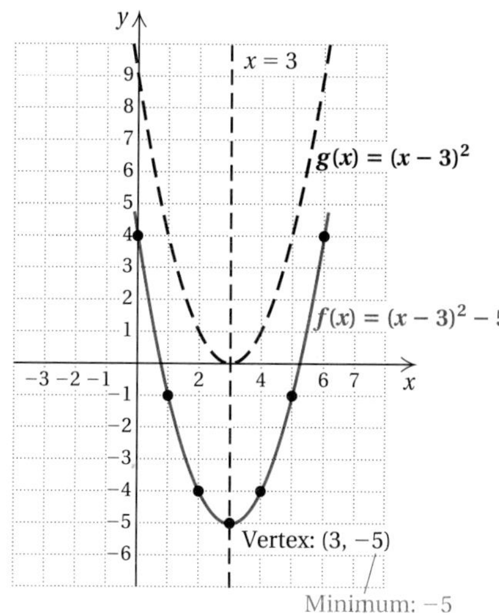

Example 5 Graph $h(x) = \frac{1}{2}(x - 3)^2 + 5$, and find the minimum function value.

The graph looks just like that of $f(x) = \frac{1}{2}x^2$ but moved 3 units to the right and 5 units up. The vertex is $(3, 5)$, and the line of symmetry is $x = 3$. We draw $f(x) = \frac{1}{2}x^2$ and then shift the curve over and up. The minimum function value is 5. By plotting some points, we have a check.

x	$h(x) = \frac{1}{2}(x - 3)^2 + 5$	
0	$9\frac{1}{2}$	
1	7	
3	5	←Vertex
5	7	
6	$9\frac{1}{2}$	

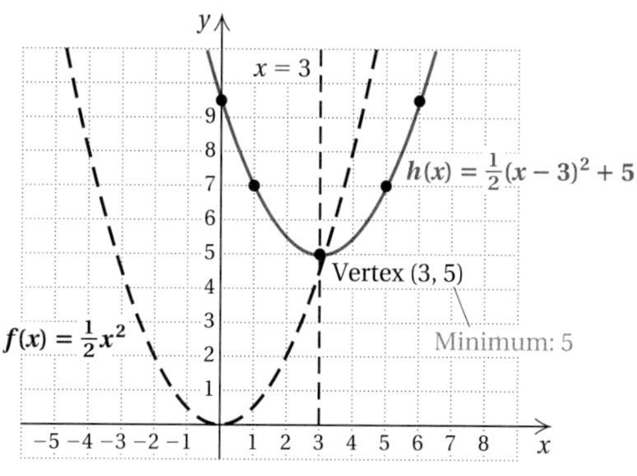

Graph. Find and label the vertex and the line of symmetry.

4. $f(x) = \frac{1}{2}(x - 4)^2$

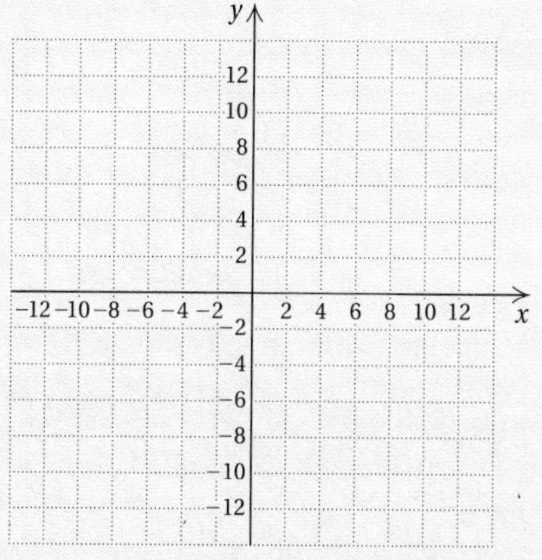

5. $f(x) = -\frac{1}{2}(x - 4)^2$

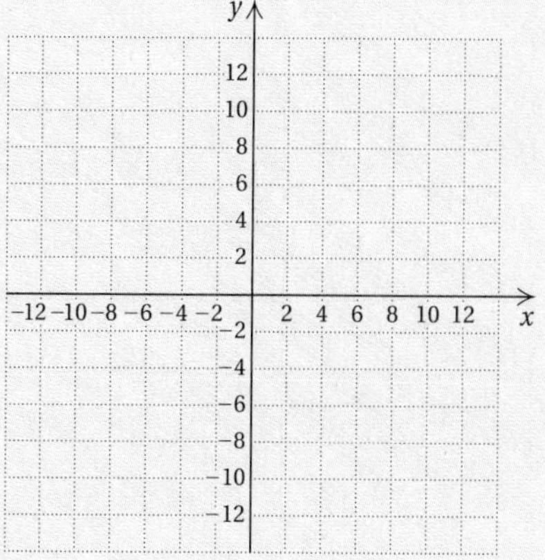

Answers on page A-47

Graph. Find the vertex, the line of symmetry, and the maximum or minimum y-value.

6. $f(x) = \frac{1}{2}(x + 2)^2 - 4$

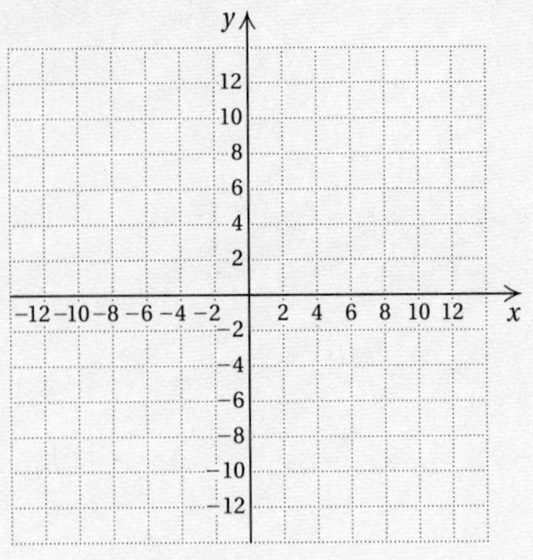

7. $f(x) = -2(x - 5)^2 + 3$

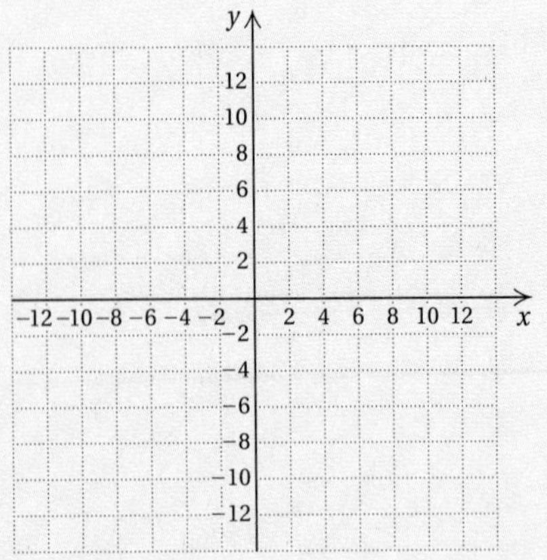

Answers on page A-47

Example 6 Graph $f(x) = -2(x + 3)^2 + 5$. Find the vertex, the line of symmetry, and the maximum or minimum value.

We first express the equation in the equivalent form

$$f(x) = -2[x - (-3)]^2 + 5.$$

The graph looks like that of $g(x) = -2x^2$ translated 3 units to the left and 5 units up. The vertex is $(-3, 5)$, and the line of symmetry is $x = -3$. Since $-2 < 0$, we know that the graph opens down so 5, the second coordinate of the vertex, is the maximum y-value.

We compute a few points as needed. The graph is shown here.

x	$f(x) = -2(x + 3)^2 + 5$	
-4	3	
-3	5	← Vertex
-2	3	

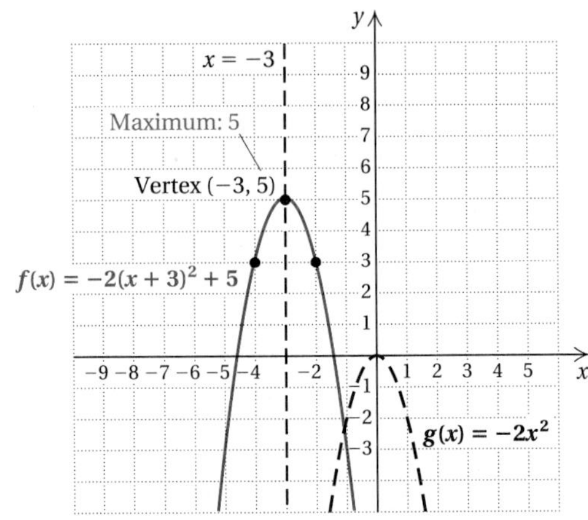

Do Exercises 6 and 7.

Exercise Set 11.5

a, b Graph. Find and label the vertex and the line of symmetry.

1. $f(x) = 4x^2$

x	$f(x)$
0	
1	
−1	
2	
−2	

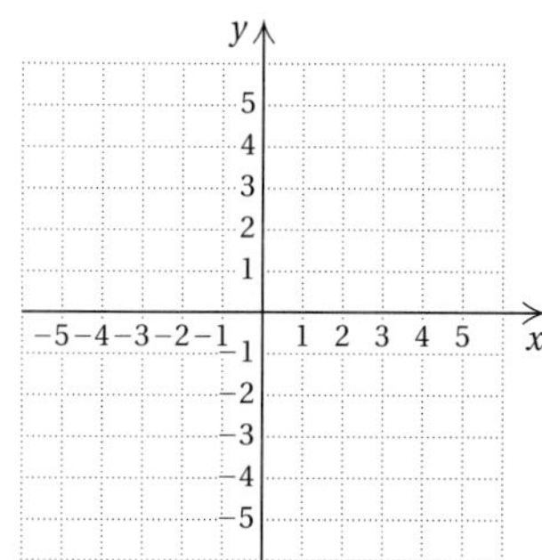

2. $f(x) = 5x^2$

x	$f(x)$
0	
1	
−1	
2	
−2	

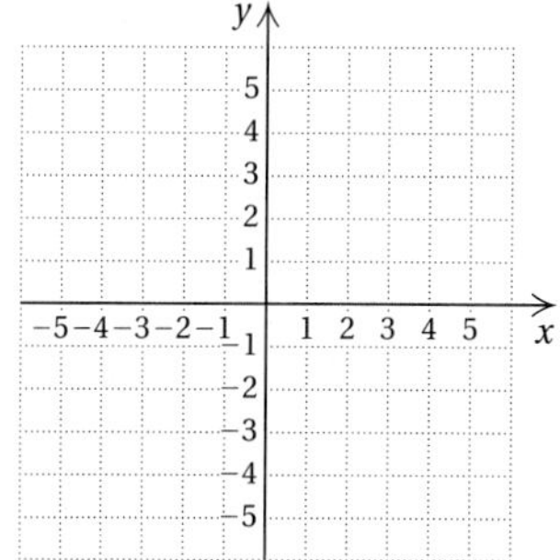

3. $f(x) = \frac{1}{3}x^2$

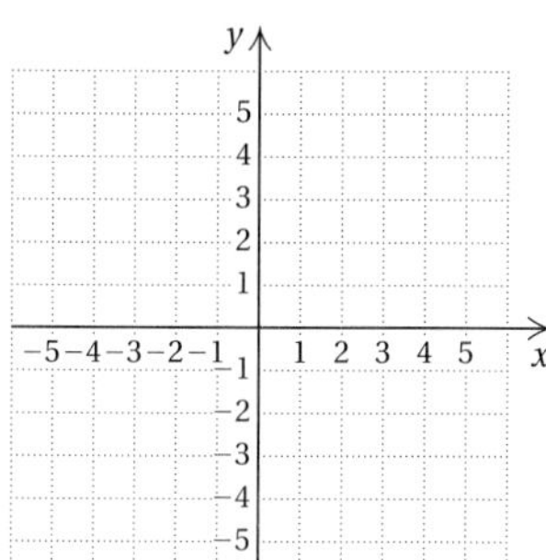

4. $f(x) = \frac{1}{4}x^2$

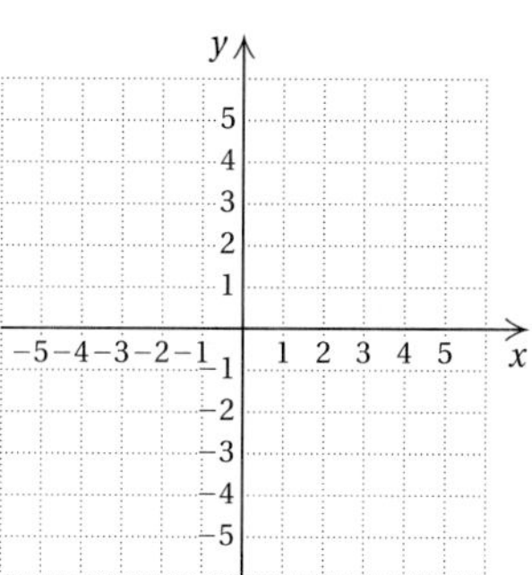

5. $f(x) = -\frac{1}{2}x^2$

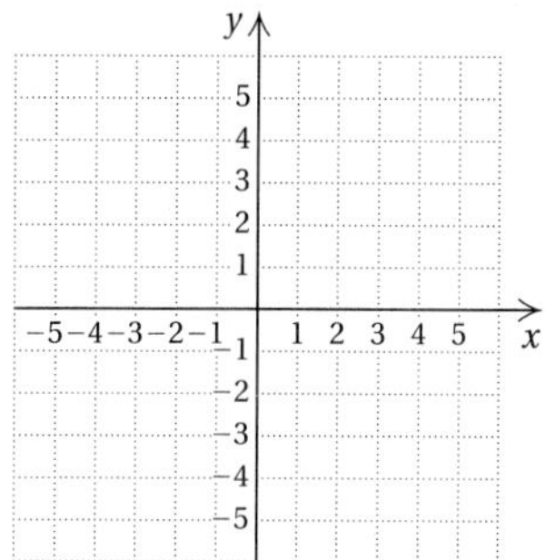

6. $f(x) = -\frac{1}{4}x^2$

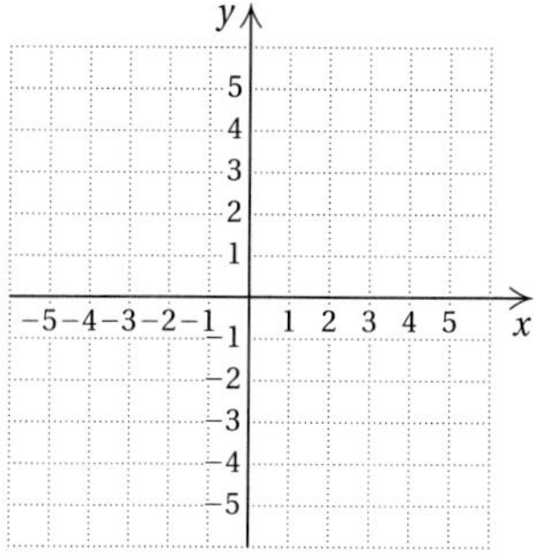

7. $f(x) = -4x^2$

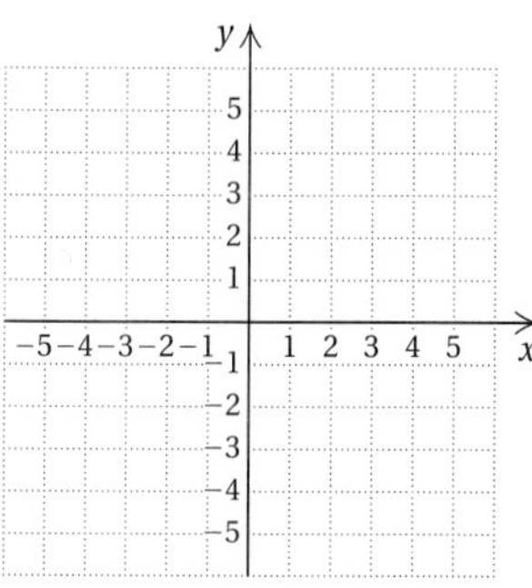

8. $f(x) = -3x^2$

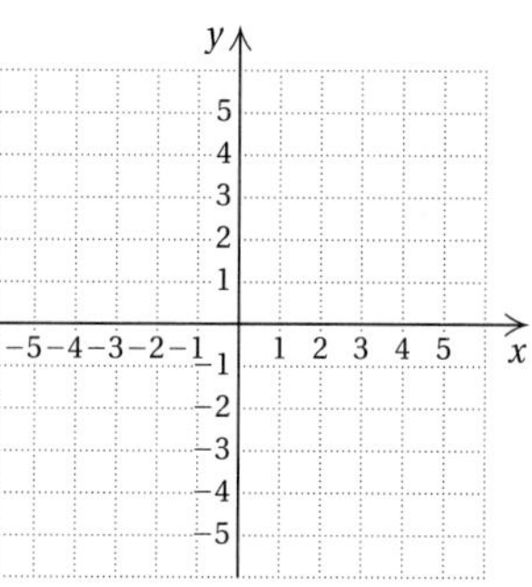

9. $f(x) = (x + 3)^2$

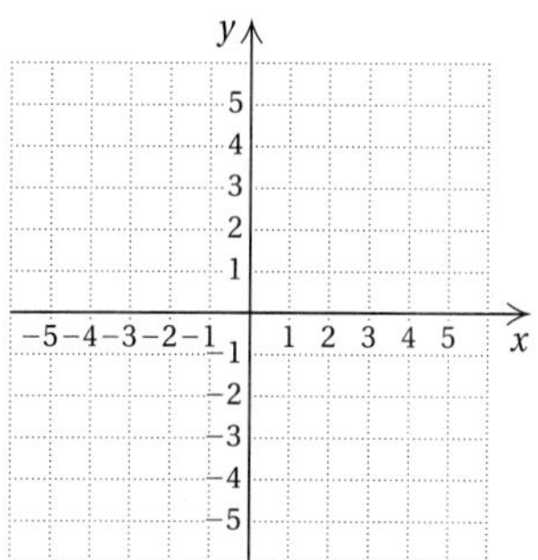

10. $f(x) = (x + 1)^2$

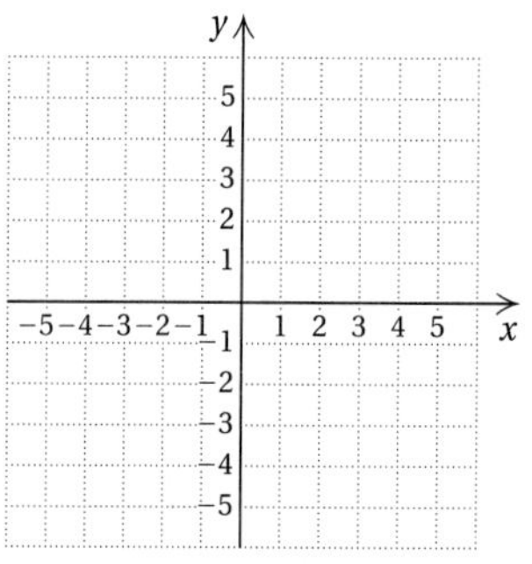

11. $f(x) = 2(x - 4)^2$

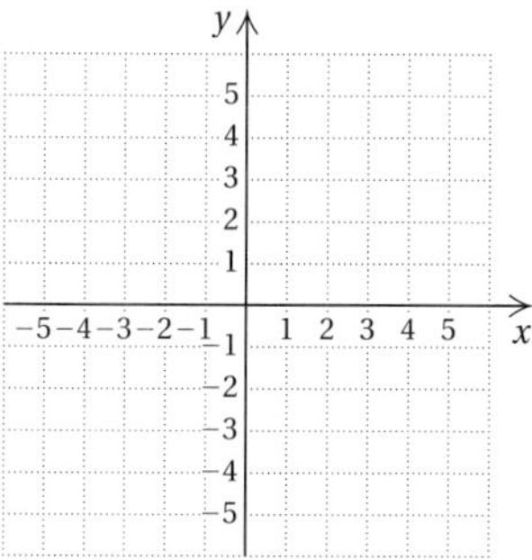

12. $f(x) = 4(x - 1)^2$

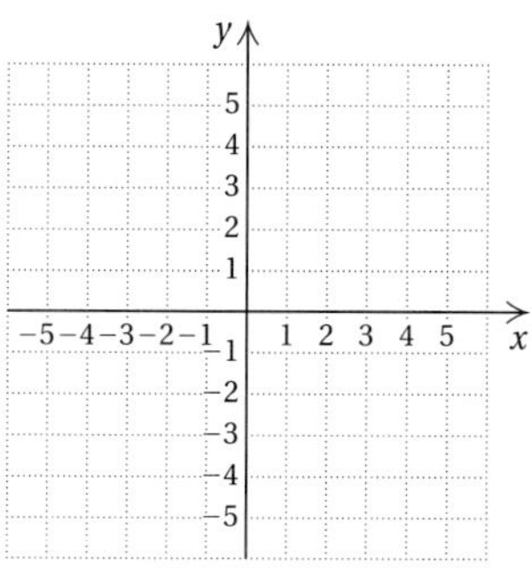

13. $f(x) = -2(x + 2)^2$

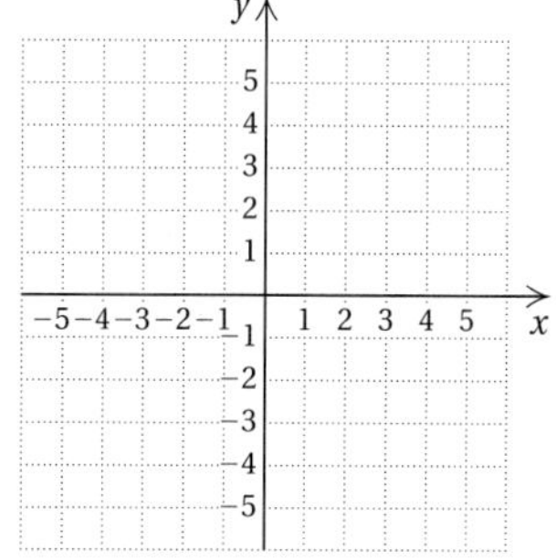

14. $f(x) = -2(x + 4)^2$

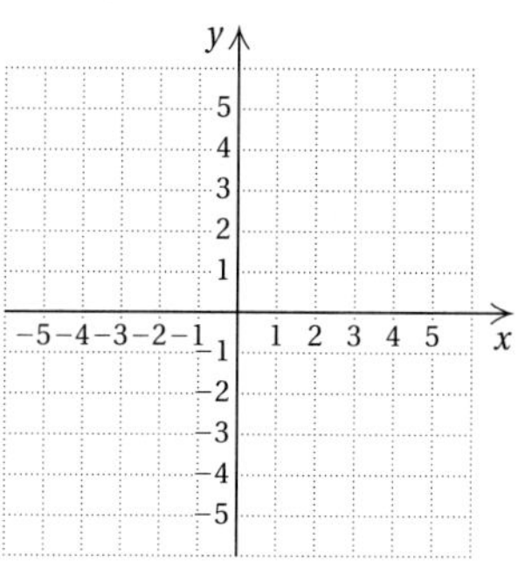

15. $f(x) = 3(x - 1)^2$

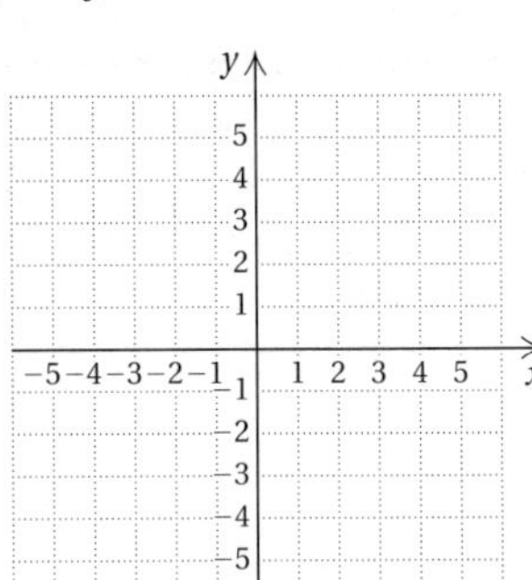

16. $f(x) = 4(x - 2)^2$

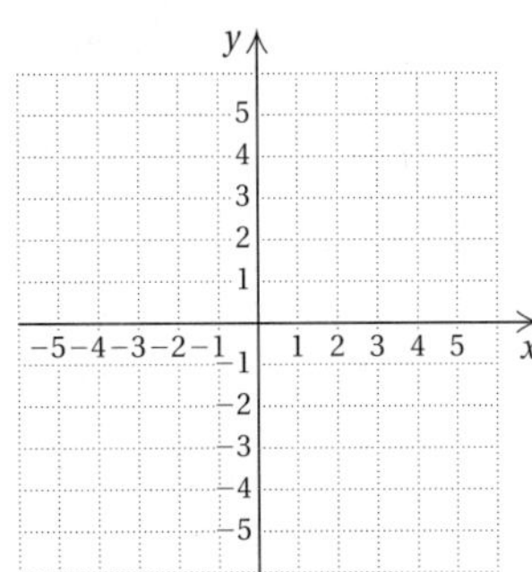

17. $f(x) = -\frac{3}{2}(x + 2)^2$

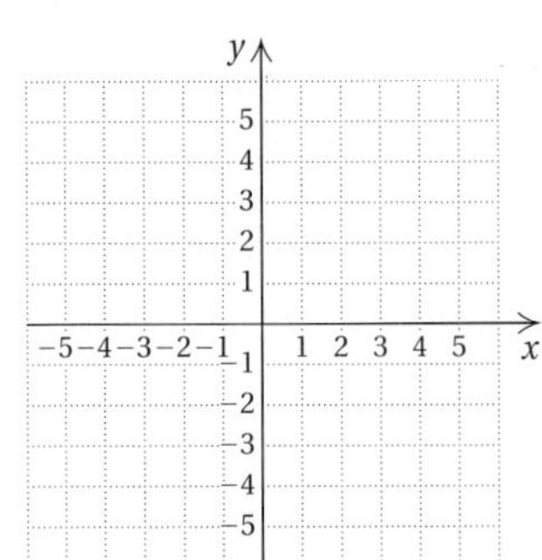

18. $f(x) = -\frac{5}{2}(x + 3)^2$

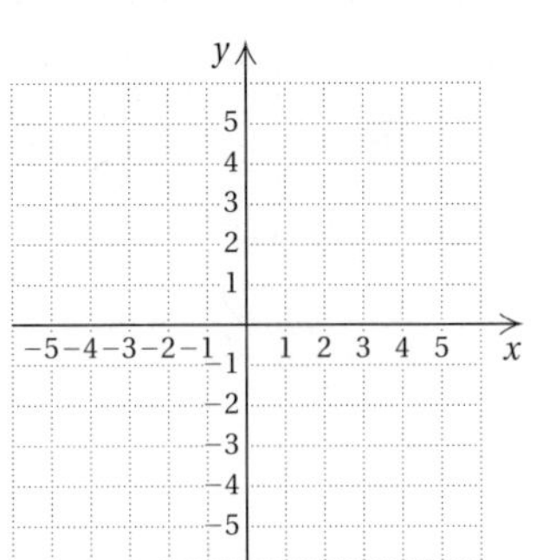

c Graph. Find and label the vertex and the line of symmetry. Find the maximum or minimum value.

19. $f(x) = (x - 3)^2 + 1$

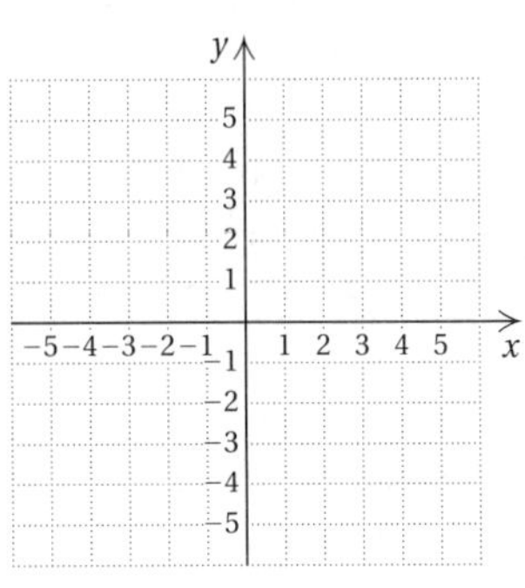

20. $f(x) = (x + 2)^2 - 3$

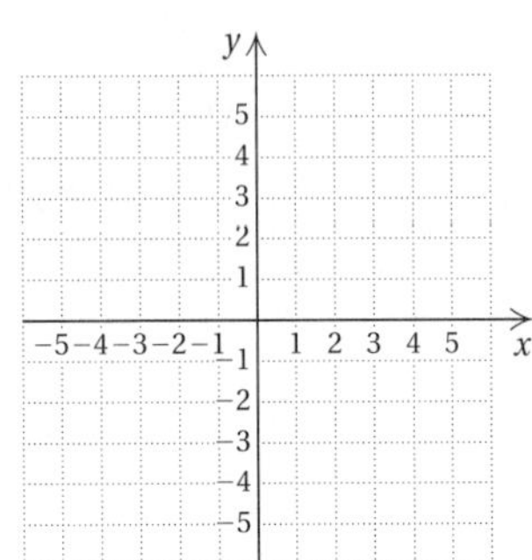

21. $f(x) = -3(x + 4)^2 + 1$

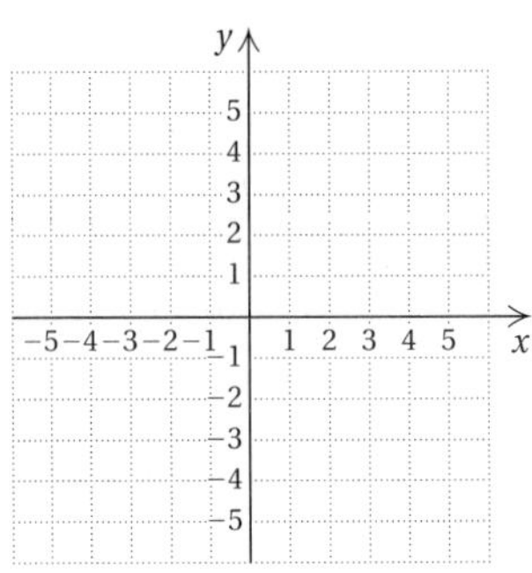

22. $f(x) = \frac{1}{2}(x - 1)^2 - 3$

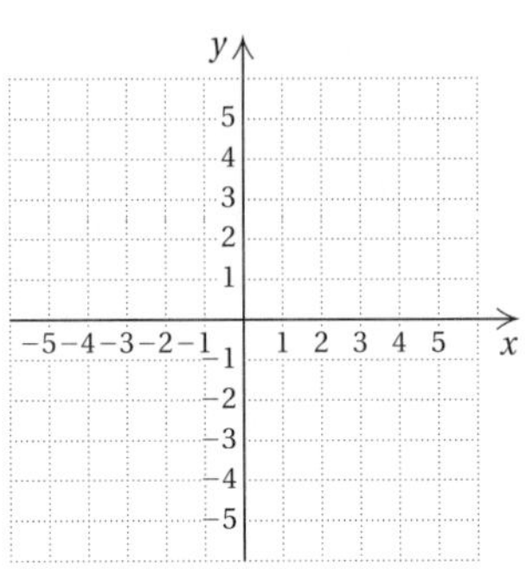

23. $f(x) = \frac{1}{2}(x + 1)^2 + 4$

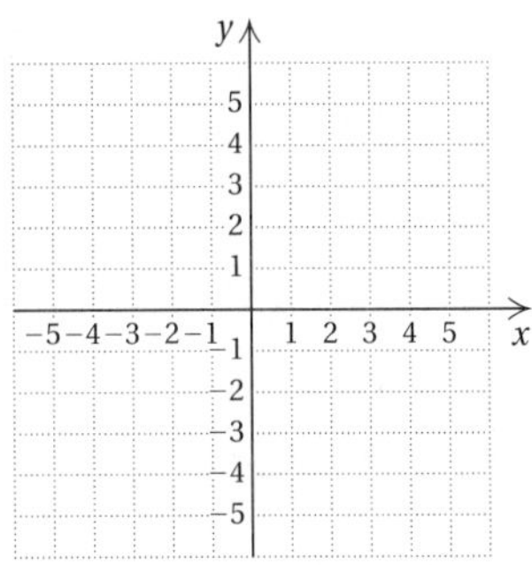

24. $f(x) = -2(x - 5)^2 - 3$

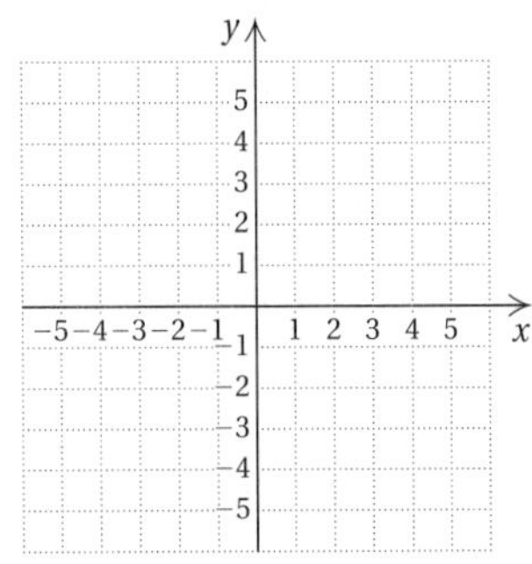

25. $f(x) = -(x + 1)^2 - 2$

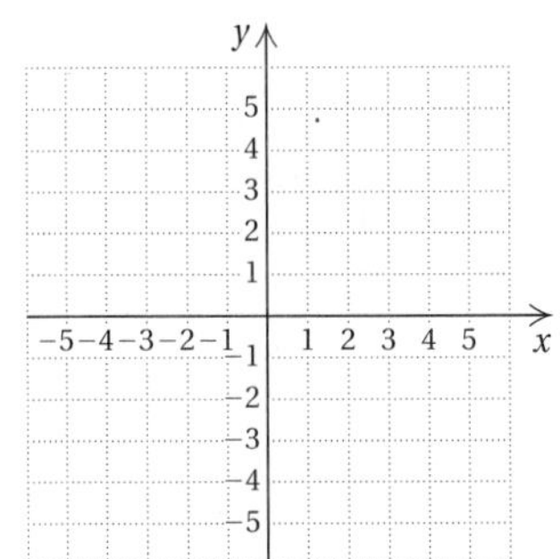

26. $f(x) = 3(x - 4)^2 + 2$

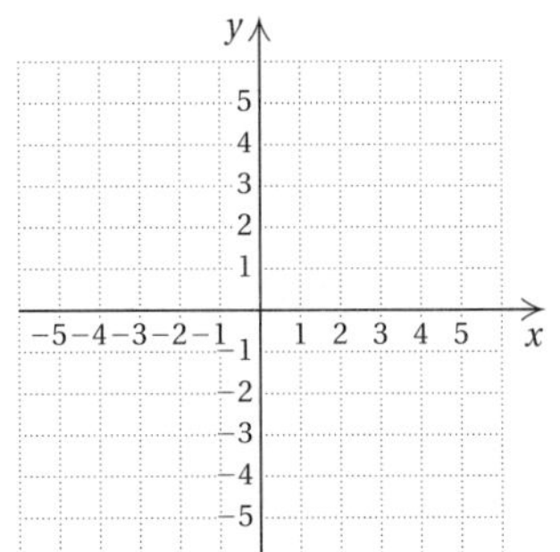

Skill Maintenance

Solve. [10.6a, b]

27. $x - 5 = \sqrt{x + 7}$

28. $\sqrt{2x + 7} = \sqrt{5x - 4}$

29. $\sqrt{x + 4} = -11$

30. $x = 7 + 2\sqrt{x + 1}$

Synthesis

31. ◆ Explain, without plotting points, why the graph of $f(x) = (x + 3)^2$ looks like the graph of $f(x) = x^2$ translated 3 units to the left.

32. ◆ Explain, without plotting points, why the graph of $f(x) = (x + 3)^2 - 4$ looks like the graph of $f(x) = x^2$ translated 3 units to the left and 4 units down.

33. Use the TRACE and/or TABLE features on a grapher to confirm the maximum or minimum values given in Exercises 23, 24, and 26. Also, confirm the results using the MINIMUM or MAXIMUM feature under the CALC key.

Collaborative Learning Manual

Practice graphing and identifying the graphs of quadratic functions.

11.6 Graphs of Quadratic Functions of the Type $f(x) = ax^2 + bx + c$

a Graphing and Analyzing $f(x) = ax^2 + bx + c$

By *completing the square,* we can begin with any quadratic polynomial $ax^2 + bx + c$ and find an equivalent expression $a(x - h)^2 + k$. This allows us to combine the skills of Sections 11.1 and 11.5 to graph and analyze any quadratic function $f(x) = ax^2 + bx + c$.

Example 1 For $f(x) = x^2 - 6x + 4$, find the vertex, the line of symmetry, and the minimum value. Then graph.

We first find the vertex and the line of symmetry. To do so, we find the equivalent form $a(x - h)^2 + k$ by completing the square, beginning as follows:

$$f(x) = x^2 - 6x + 4 = (x^2 - 6x \quad\quad) + 4.$$

We complete the square inside the parentheses, but in a different manner than we did before. We take half the x-coefficient, $-6/2 = -3$, and square it: $(-3)^2 = 9$. Then we add 0, or $9 - 9$, inside the parentheses:

$$\begin{aligned} f(x) &= (x^2 - 6x + 0) + 4 && \text{Adding 0} \\ &= (x^2 - 6x + 9 - 9) + 4 && \text{Substituting } 9 - 9 \text{ for } 0 \\ &= (x^2 - 6x + 9) + (-9 + 4) && \text{Using the associative law of addition to regroup} \\ &= (x - 3)^2 - 5. && \text{Factoring and simplifying} \end{aligned}$$

(This equation was graphed in Example 4 of Section 11.5.) The vertex is $(3, -5)$, and the line of symmetry is $x = 3$. The coefficient of x^2 is 1, which is positive, so the graph opens up. This tells us that -5 is a minimum. We plot the vertex and draw the line of symmetry. We choose some x-values on both sides of the vertex and graph the parabola. Suppose we compute the pair $(5, -1)$:

$$f(5) = 5^2 - 6(5) + 4 = 25 - 30 + 4 = -1.$$

We note that it is 2 units to the right of the line of symmetry. There will also be a pair with the same y-coordinate on the graph 2 units to the *left* of the line of symmetry. Thus we get a second point, $(1, -1)$, without making another calculation.

x	$f(x)$	
3	−5	← Vertex
4	−4	
2	−4	
5	−1	
1	−1	
6	4	
0	4	← y-intercept

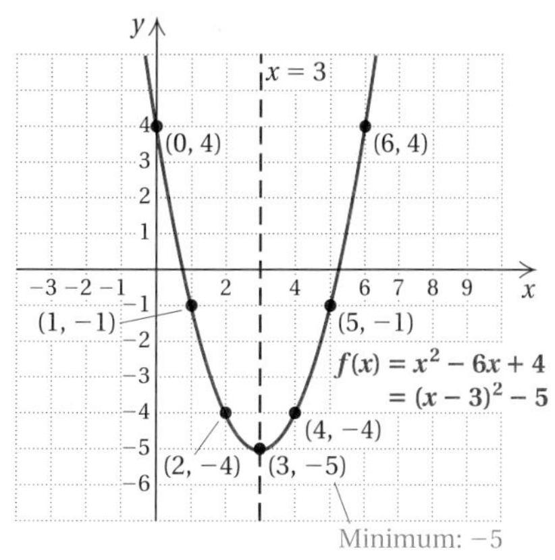

Do Exercise 1.

Objectives

a For a quadratic function, find the vertex, the line of symmetry, and the maximum or minimum value, and graph the function.

b Find the intercepts of a quadratic function.

For Extra Help

TAPE 19

MAC WIN

CD-ROM

1. For $f(x) = x^2 - 4x + 7$, find the vertex, the line of symmetry, and the minimum value. Then graph.

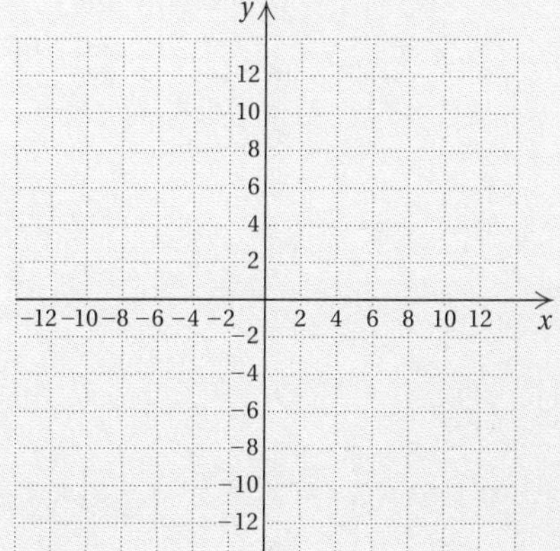

Vertex: ____________

Line of symmetry: ____________

Minimum value: ____________

Answers on page A-48

2. For $f(x) = 3x^2 - 24x + 43$, find the vertex, the line of symmetry, and the minimum value. Then graph.

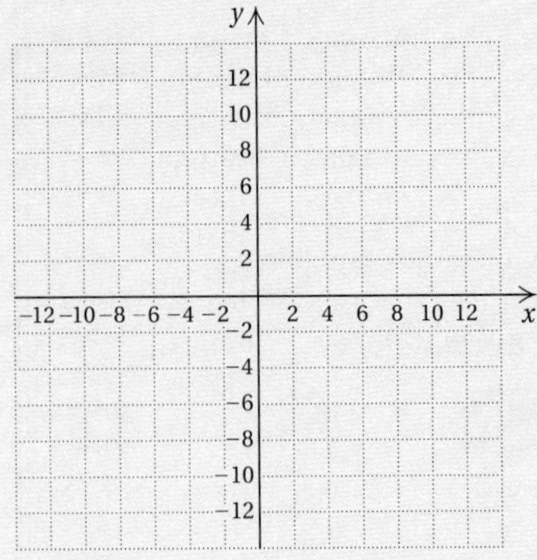

Vertex: ____________

Line of symmetry: ____________

Minimum value: ____________

Answer on page A-48

Example 2 For $f(x) = 3x^2 + 12x + 13$, find the vertex, the line of symmetry, and the minimum value. Then graph.

Since the coefficient of x^2 is not 1, we factor out 3 from only the *first two* terms of the expression. Remember that we want to get to the form $f(x) = a(x - h)^2 + k$:

$$f(x) = 3x^2 + 12x + 13$$
$$= 3(x^2 + 4x) + 13. \quad \text{Factoring 3 out of the first two terms}$$

Next, we complete the square inside the parentheses. We take half the x-coefficient, $\frac{1}{2} \cdot 4 = 2$, and square it: $2^2 = 4$. Then we add 0, or $4 - 4$, inside the parentheses:

$$f(x) = 3(x^2 + 4x + 0) + 13 \quad \text{Adding 0}$$
$$= 3(x^2 + 4x + 4 - 4) + 13 \quad \text{Substituting } 4 - 4 \text{ for } 0$$
$$= 3(x^2 + 4x + 4) + 3(-4) + 13 \quad \text{Using the distributive law to separate } -4 \text{ from the trinomial}$$
$$= 3(x + 2)^2 + 1. \quad \text{Factoring and simplifying}$$

The vertex is $(-2, 1)$, and the line of symmetry is $x = -2$. The coefficient of x^2 is 3, so the graph is narrow and opens up. This tells us that 1 is a minimum. We choose a few x-values on one side of the line of symmetry, compute y-values, and use the resulting coordinates to find more points on the other side of the line of symmetry. We plot points and graph the parabola.

x	$f(x)$	
-2	1	← Vertex
-3	4	
-1	4	
0	13	← y-intercept
-4	13	

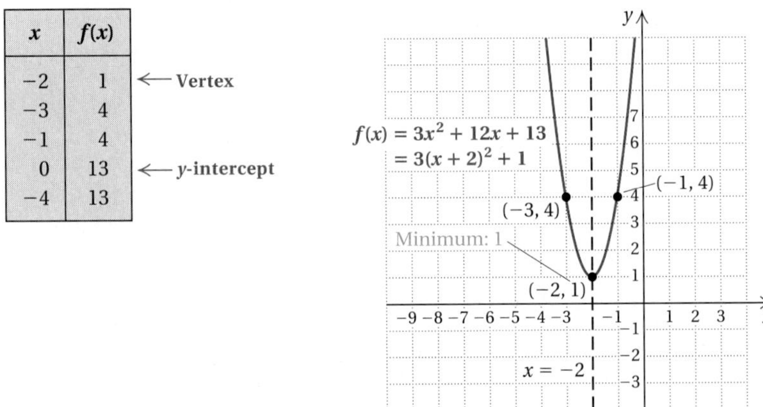

Do Exercise 2.

Example 3 For $f(x) = -2x^2 + 10x - 7$, find the vertex, the line of symmetry, and the maximum value. Then graph.

Again, the coefficient x^2 is not 1. We factor out -2 from only the *first two* terms of the expression. This makes the coefficient of x^2 inside the parentheses 1:

$$f(x) = -2x^2 + 10x - 7$$
$$= -2(x^2 - 5x) - 7.$$

Next, we complete the square as before. We take half the x-coefficient, $\frac{1}{2}(-5) = -\frac{5}{2}$, and square it: $\left(-\frac{5}{2}\right)^2 = \frac{25}{4}$. Then we add 0, or $\frac{25}{4} - \frac{25}{4}$, inside the parentheses:

$$f(x) = -2\left(x^2 - 5x + \tfrac{25}{4} - \tfrac{25}{4}\right) - 7$$

$$= -2\left(x^2 - 5x + \tfrac{25}{4}\right) + (-2)\left(-\tfrac{25}{4}\right) - 7 \qquad \text{Using the distributive law to separate the } -\tfrac{25}{4} \text{ from the trinomial}$$

$$= -2\left(x^2 - 5x + \tfrac{25}{4}\right) + \tfrac{25}{2} - 7$$

$$= -2\left(x - \tfrac{5}{2}\right)^2 + \tfrac{11}{2}.$$

The vertex is $\left(\frac{5}{2}, \frac{11}{2}\right)$, and the line of symmetry is $x = \frac{5}{2}$. The coefficient of x^2 is -2, so the graph is narrow and opens down. This tells us that $\frac{11}{2}$ is a maximum. We choose a few x-values on one side of the line of symmetry, compute y-values, and use the resulting coordinates to find more points on the other side of the line of symmetry. We plot points and graph the parabola.

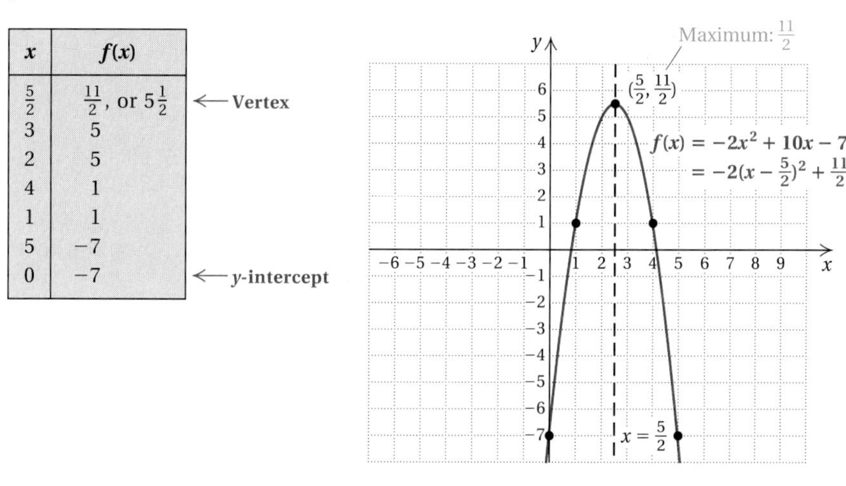

x	$f(x)$	
$\frac{5}{2}$	$\frac{11}{2}$, or $5\frac{1}{2}$	← Vertex
3	5	
2	5	
4	1	
1	1	
5	−7	
0	−7	← y-intercept

3. For $f(x) = -4x^2 + 12x - 5$, find the vertex, the line of symmetry, and the maximum value. Then graph.

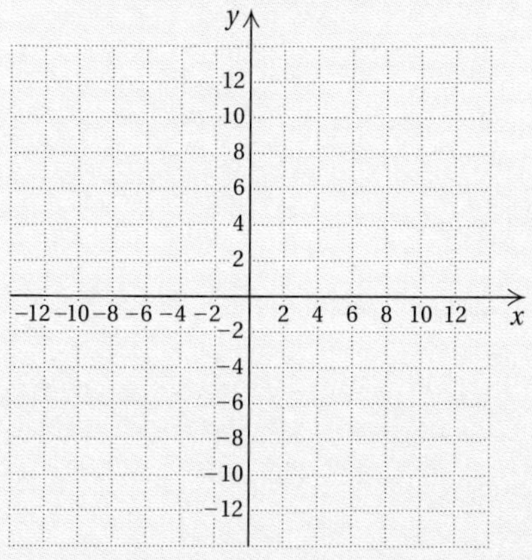

Vertex: ____________

Line of symmetry: ____________

Maximum value: ____________

Answer on page A-48

Do Exercise 3.

The method used in Examples 1–3 can be generalized to find a formula for locating the vertex. We complete the square as follows:

$$f(x) = ax^2 + bx + c$$

$$= a\left(x^2 + \frac{b}{a}x\right) + c. \qquad \text{Factoring } a \text{ out of the first two terms. Check by multiplying.}$$

Half of the x-coefficient, $\dfrac{b}{a}$, is $\dfrac{b}{2a}$. We square it to get $\dfrac{b^2}{4a^2}$ and add $\dfrac{b^2}{4a^2} - \dfrac{b^2}{4a^2}$ inside the parentheses. Then we distribute the a:

$$f(x) = a\left(x^2 + \frac{b}{a}x + \frac{b^2}{4a^2} - \frac{b^2}{4a^2}\right) + c$$

$$= a\left(x^2 + \frac{b}{a}x + \frac{b^2}{4a^2}\right) + a\left(-\frac{b^2}{4a^2}\right) + c \qquad \text{Using the distributive law}$$

$$= a\left(x + \frac{b}{2a}\right)^2 + \frac{-b^2}{4a} + \frac{4ac}{4a} \qquad \text{Factoring and finding a common denominator}$$

$$= a\left[x - \left(-\frac{b}{2a}\right)\right]^2 + \frac{4ac - b^2}{4a}.$$

Find the vertex of the parabola using the formula.

4. $f(x) = x^2 - 6x + 4$

5. $f(x) = 3x^2 - 24x + 43$

6. $f(x) = -4x^2 + 12x - 5$

Find the intercepts.

7. $f(x) = x^2 + 2x - 3$

8. $f(x) = x^2 + 8x + 16$

9. $f(x) = x^2 - 4x + 1$

Answers on page A-48

Thus we have the following.

> The **vertex** of the parabola given by $f(x) = ax^2 + bx + c$ is
>
> $$\left(-\frac{b}{2a}, \frac{4ac - b^2}{4a}\right), \quad \text{or} \quad \left(-\frac{b}{2a}, f\left(-\frac{b}{2a}\right)\right).$$
>
> The x-coordinate of the vertex is $-b/(2a)$. The **line of symmetry** is $x = -b/(2a)$. The second coordinate of the vertex is easiest to find by computing $f\left(-\frac{b}{2a}\right)$.

Let's reexamine Example 3 to see how we could have found the vertex directly. From the formula above,

$$\text{the } x\text{-coordinate of the vertex is } -\frac{b}{2a} = -\frac{10}{2(-2)} = \frac{5}{2}.$$

Substituting $\frac{5}{2}$ into $f(x) = -2x^2 + 10x - 7$, we find the second coordinate of the vertex:

$$\begin{aligned} f\left(\tfrac{5}{2}\right) &= -2\left(\tfrac{5}{2}\right)^2 + 10\left(\tfrac{5}{2}\right) - 7 \\ &= -2\left(\tfrac{25}{4}\right) + 25 - 7 \\ &= -\tfrac{25}{2} + 18 = -\tfrac{25}{2} + \tfrac{36}{2} = \tfrac{11}{2}. \end{aligned}$$

The vertex is $\left(\frac{5}{2}, \frac{11}{2}\right)$. The line of symmetry is $x = \frac{5}{2}$.

We have developed two methods for finding the vertex. One is by completing the square and the other is by using a formula. You should check with your instructor about which method to use.

Do Exercises 4–6.

b Finding the Intercepts of a Quadratic Function

The points at which a graph crosses an axis are called **intercepts**. We determine the y-intercept by finding $f(0)$. For $f(x) = ax^2 + bx + c$, the y-intercept is $(0, c)$.

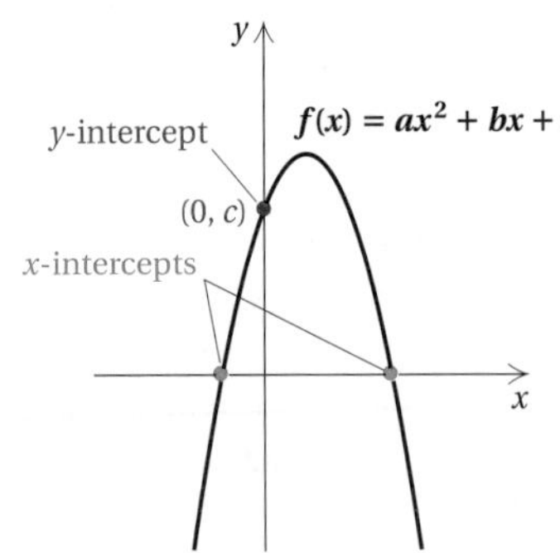

To find the x-intercepts, we look for values of x for which $f(x) = 0$. For $f(x) = ax^2 + bx + c$, we solve

$$0 = ax^2 + bx + c.$$

Example 4 Find the intercepts of $f(x) = x^2 - 2x - 2$.

The y-intercept is $(0, f(0))$. Since $f(0) = 0^2 - 2 \cdot 0 - 2 = -2$, the y-intercept is $(0, -2)$. To find the x-intercepts, we solve

$$0 = x^2 - 2x - 2.$$

Using the quadratic formula gives us $x = 1 \pm \sqrt{3}$. Thus the x-intercepts are $(1 - \sqrt{3}, 0)$ and $(1 + \sqrt{3}, 0)$, or, approximately, $(-0.732, 0)$ and $(2.732, 0)$.

Do Exercises 7–9.

Exercise Set 11.6

a For each quadratic function, find **(a)** the vertex, **(b)** the line of symmetry, and **(c)** the maximum or minimum value. Then **(d)** graph the function.

1. $f(x) = x^2 - 2x - 3$

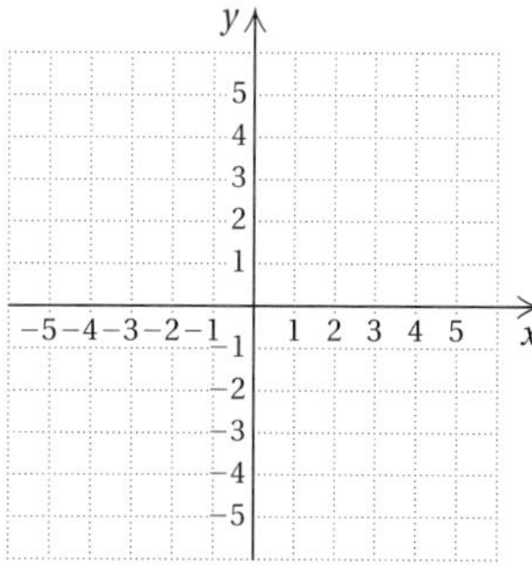

2. $f(x) = x^2 + 2x - 5$

3. $f(x) = -x^2 - 4x - 2$

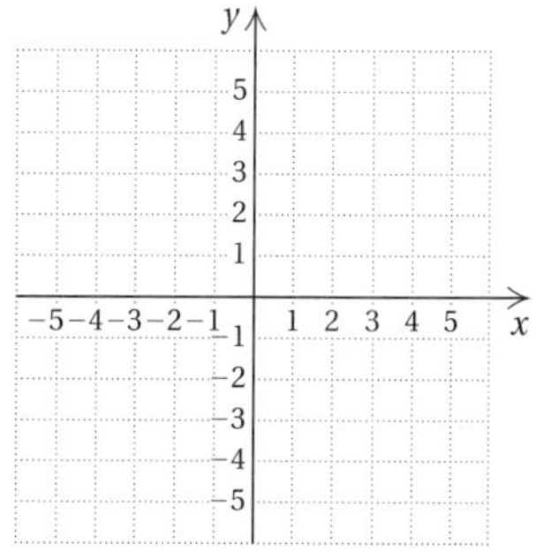

4. $f(x) = -x^2 + 4x + 1$

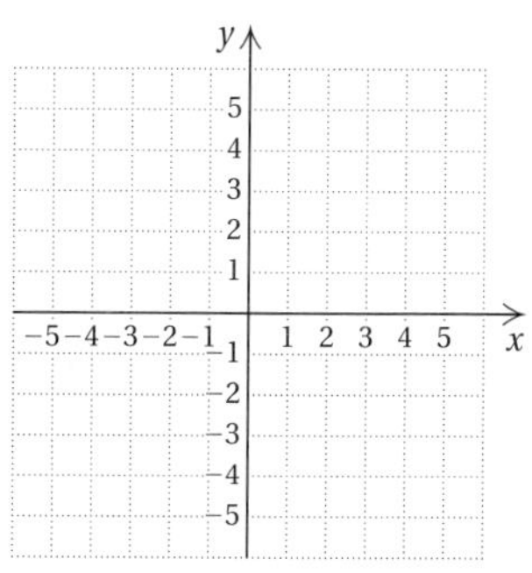

5. $f(x) = 3x^2 - 24x + 50$

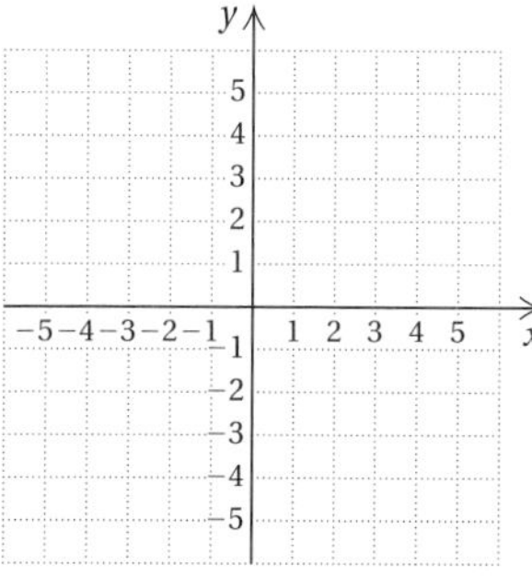

6. $f(x) = 4x^2 + 8x + 1$

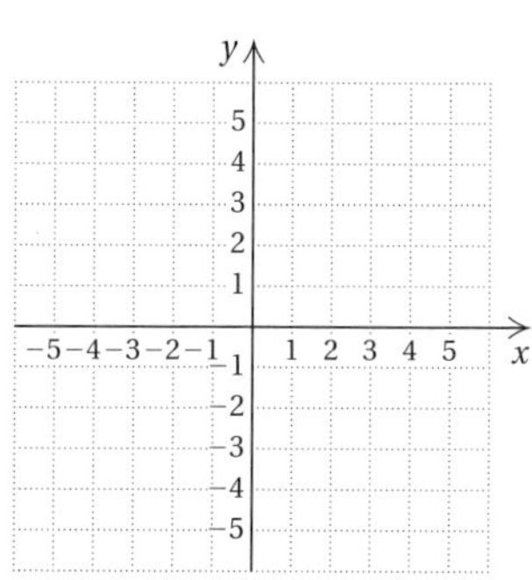

7. $f(x) = -2x^2 - 2x + 3$

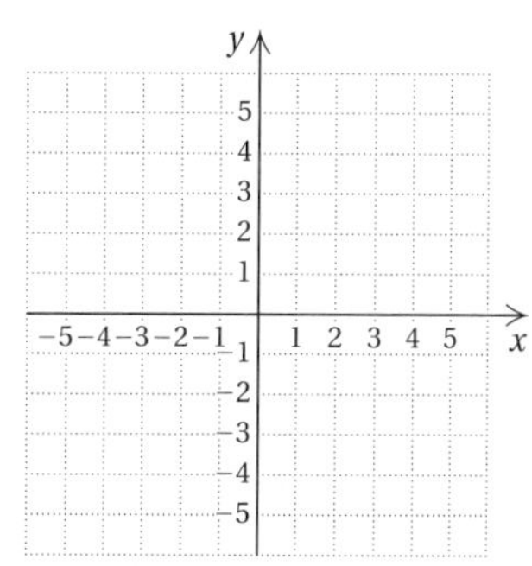

8. $f(x) = -2x^2 + 2x + 1$

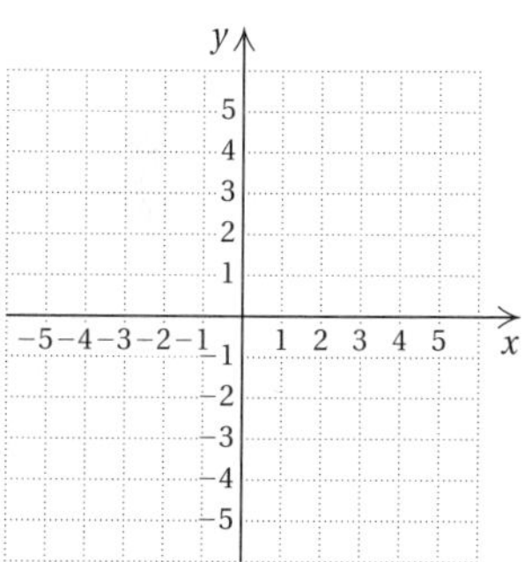

9. $f(x) = 5 - x^2$

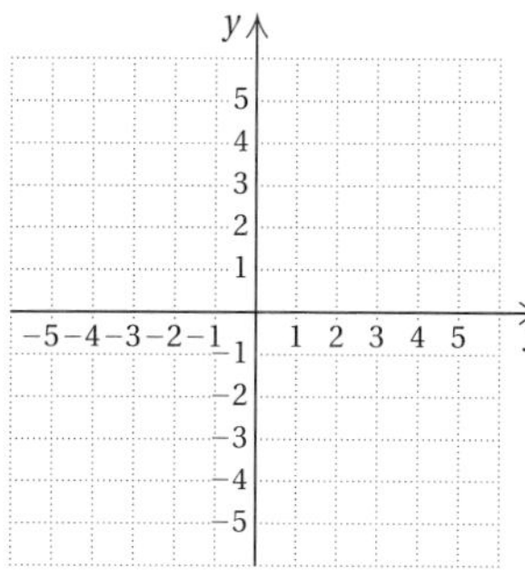

10. $f(x) = x^2 - 3x$

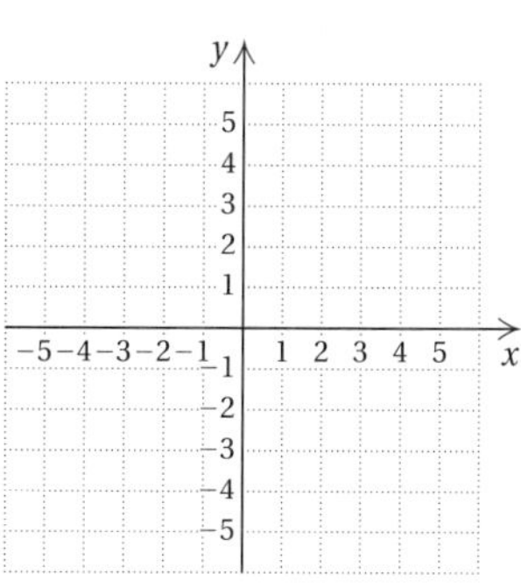

11. $f(x) = 2x^2 + 5x - 2$

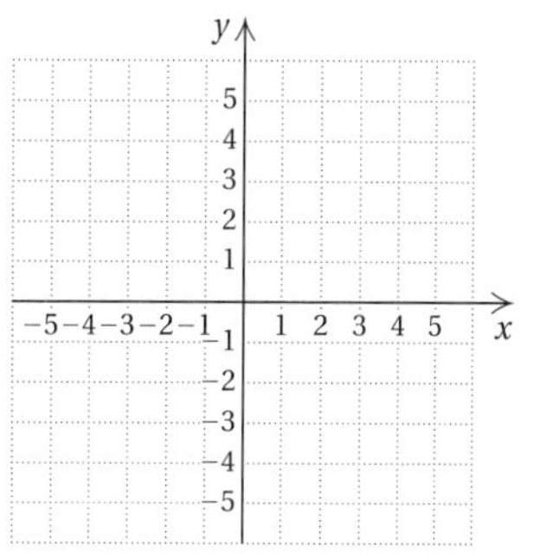

12. $f(x) = -4x^2 - 7x + 2$

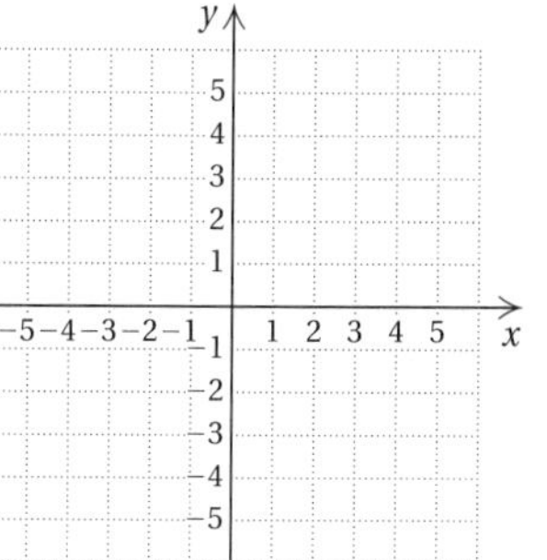

b Find the x- and y-intercepts.

13. $f(x) = x^2 - 6x + 1$

14. $f(x) = x^2 + 2x + 12$

15. $f(x) = -x^2 + x + 20$

16. $f(x) = -x^2 + 5x + 24$

17. $f(x) = 4x^2 + 12x + 9$

18. $f(x) = 3x^2 - 6x + 1$

19. $f(x) = 4x^2 - x + 8$

20. $f(x) = 2x^2 + 4x - 1$

Skill Maintenance

Solve. [6.9a, b]

21. *Determining Medication Dosage.* A child's dosage D, in milligrams, of a medication varies directly as the child's weight w, in kilograms. To control a fever, a doctor suggests that a child who weighs 28 kg be given 420 mg of Tylenol.

a) Find an equation of variation.
b) How much Tylenol would be recommended for a child who weighs 42 kg?

22. *Calories Burned.* The number C of calories burned while exercising varies directly as the time t, in minutes, spent exercising. Harold exercises for 24 min on a stairmaster and burns 356 calories.

a) Find an equation of variation.
b) How many calories would he burn if he were to exercise for 48 min?

Find the variation constant and an equation of variation in which y varies inversely as x and the following are true. [6.9c]

23. $y = 125$ when $x = 2$

24. $y = 2$ when $x = 125$

Find the variation constant and an equation of variation in which y varies directly as x and the following are true. [6.9a]

25. $y = 125$ when $x = 2$

26. $y = 2$ when $x = 125$

Synthesis

27. ◆ Does the graph of every quadratic function have a y-intercept? Why or why not?

28. ◆ Is it possible for the graph of a quadratic function to have only one x-intercept if the vertex is off the x-axis? Why or why not?

29. Use the TRACE and/or TABLE features of a grapher to estimate the maximum or minimum values of the following functions. Confirm the results using the MINIMUM or MAXIMUM feature under the CALC key.

a) $f(x) = 2.31x^2 - 3.135x - 5.89$
b) $f(x) = -18.8x^2 + 7.92x + 6.18$

30. Use the TRACE and/or TABLE features of a grapher to confirm the maximum or minimum values given in Exercises 8, 11, and 12. Confirm the results using the MINIMUM or MAXIMUM feature under the CALC key.

Graph.

31. $f(x) = |x^2 - 1|$

32. $f(x) = |x^2 + 6x + 4|$

33. $f(x) = |x^2 - 3x - 4|$

34. $f(x) = |2(x - 3)^2 - 5|$

35. A quadratic function has $(-1, 0)$ as one of its intercepts and $(3, -5)$ as its vertex. Find an equation for the function.

36. A quadratic function has $(4, 0)$ as one of its intercepts and $(-1, 7)$ as its vertex. Find an equation for the function.

37. Consider

$$f(x) = \frac{x^2}{8} + \frac{x}{4} - \frac{3}{8}.$$

Find **(a)** the vertex, **(b)** the line of symmetry, and **(c)** the maximum or minimum value. Then **(d)** draw the graph.

38. Use only the graph in Exercise 37 to approximate the solutions of each of the following equations.

a) $\frac{x^2}{8} + \frac{x}{4} - \frac{3}{8} = 0$

b) $\frac{x^2}{8} + \frac{x}{4} - \frac{3}{8} = 1$

c) $\frac{x^2}{8} + \frac{x}{4} - \frac{3}{8} = 2$

Use the INTERSECT feature of a grapher to find the points of intersection of the graphs of each pair of functions.

39. $f(x) = x^2 - 4x + 2, \quad g(x) = 2 + x$

40. $f(x) = x^2 + 2x + 1, \quad g(x) = -2x^2 - 4x + 1$

11.7 Mathematical Modeling with Quadratic Functions

We now consider some of the many situations in which quadratic functions can serve as mathematical models.

Objectives

a. Solve maximum–minimum problems involving quadratic functions.

b. Fit a quadratic function to a set of data to form a mathematical model, and solve related applied problems.

For Extra Help

TAPE 19

MAC WIN

CD-ROM

a Maximum–Minimum Problems

We have seen that for any quadratic function $f(x) = ax^2 + bx + c$, the value of $f(x)$ at the vertex is either a maximum or a minimum, meaning that either all outputs are smaller than that value for a maximum or larger than that value for a minimum.

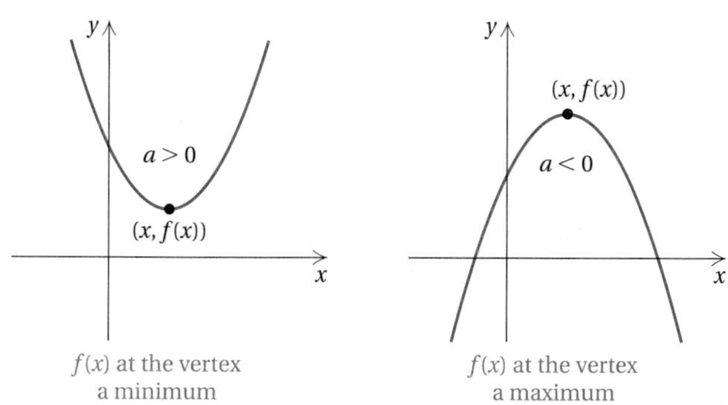

$f(x)$ at the vertex a minimum

$f(x)$ at the vertex a maximum

There are many types of applied problems in which we want to find a maximum or minimum value of a quantity. If a quadratic function can be used as a model, we can find such maximums or minimums by finding coordinates of the vertex.

Example 1 *Fenced-In Land.* A farmer has 64 yd of fencing. What are the dimensions of the largest rectangular pen that the farmer can enclose?

1. **Familiarize.** We first make a drawing and label it. We let l = the length of the pen and w = the width. Recall the following formulas:

Perimeter: $2l + 2w$;

Area: $l \cdot w$.

To become familiar with the problem, let's choose some dimensions for which $2l + 2w = 64$ and then calculate the corresponding areas.

l	*w*	*A*
22	10	220
20	12	240
18	14	252
18.5	13.5	249.75
12.4	19.6	243.04
15	17	255

What choice of l and w will maximize A?

Calculator Spotlight

Refer to Example 1. We can use the STAT feature and then EDIT to create an extended table to check for various values of w. The variable L_1 would represent length, L_2 would represent width and be defined in terms of L_1, and L_3 would represent area and be defined in terms of L_1 and L_2.

L1	L2	L3 3
22	10	220
20	12	240
18	14	252
18.5	13.5	249.75
12.4	19.6	243.04
10.6	21.4	226.84
15	17	255

L3 = {220, 240, 252...

Create such a table, check many values, and try to make an assertion about the dimensions that yield the largest area.

1. *Fenced-In Land.* A farmer has 100 yd of fencing. What are the dimensions of the largest rectangular pen that the farmer can enclose?

To familiarize yourself with the problem, complete the following table.

l	w	A
12	38	456
15	35	
24	26	
25	25	
26.2	23.8	

Calculator Spotlight

Use the TRACE and/or TABLE features to confirm the maximum or minimum values given in Example 1 and Margin Exercise 1. Also, confirm the results using the MAXIMUM feature under the CALC key.

Answer on page A-49

2. Translate. We have two equations, one for perimeter and one for area:

$$2l + 2w = 64,$$
$$A = l \cdot w.$$

Let's use them to express A as a function of l or w, but not both. To express A in terms of w, for example, we solve for l in the first equation:

$$2l + 2w = 64$$
$$2l = 64 - 2w$$
$$l = \frac{64 - 2w}{2}$$
$$= 32 - w.$$

Substituting $32 - w$ for l, we get a quadratic function $A(w)$, or just A:

$$A = lw = (32 - w)w = 32w - w^2 = -w^2 + 32w.$$

3. Carry out. Note here that we are altering the third step of our five-step problem-solving strategy to "carry out" some kind of mathematical manipulation, because we are going to find the vertex rather than solve an equation. To do so, we complete the square as in Section 11.6:

$$A = -w^2 + 32w \quad \text{This is a parabola opening down, so a maximum exists.}$$
$$= -1(w^2 - 32w) \quad \text{Factoring out } -1$$
$$= -1(w^2 - 32w + 256 - 256) \quad \tfrac{1}{2}(-32) = -16;\ (-16)^2 = 256$$
$$= -1(w^2 - 32w + 256) + (-1)(-256)$$
$$= -(w - 16)^2 + 256.$$

The vertex is (16, 256). Thus the maximum value is 256. It occurs when $w = 16$ and $l = 32 - w = 32 - 16 = 16$.

4. Check. We note that 256 is larger than any of the values found in the *Familiarize* step. To be more certain, we could make more calculations. We leave this to the student. We can also use the graph of the function to check the maximum value.

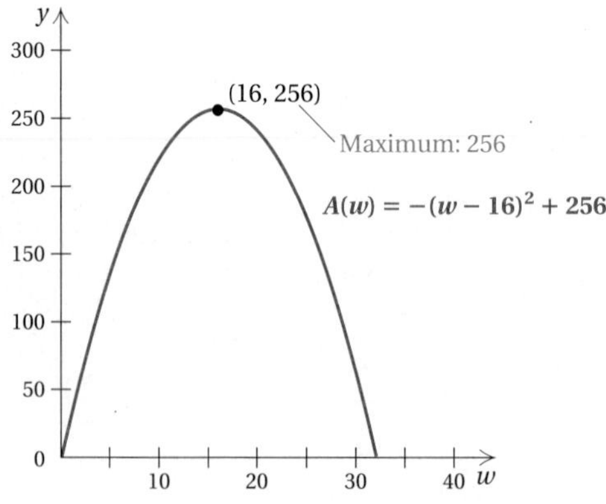

5. State. The largest rectangular pen that can be enclosed is 16 yd by 16 yd; that is, a square.

Do Exercise 1.

b Fitting Quadratic Functions to Data

As we move through our study of mathematics, we develop a library of functions. These functions can serve as models for many applications. Some of them are graphed below. We have not considered the cubic or quartic functions in detail other than in the Calculator Spotlights (we leave that discussion to a later course), but we show them here for reference.

Linear function:
$f(x) = mx + b$

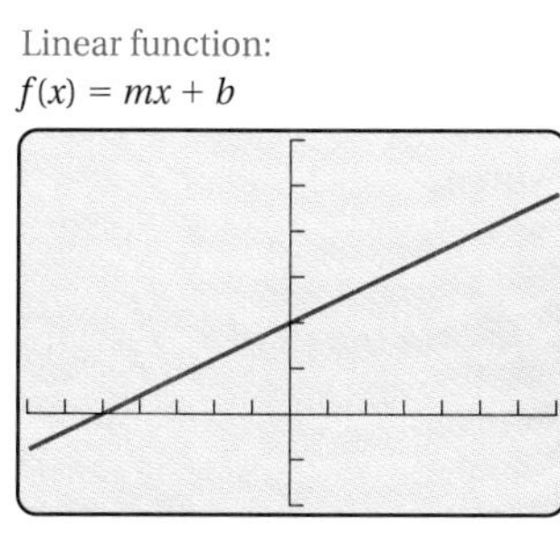

Quadratic function:
$f(x) = ax^2 + bx + c,\ a > 0$

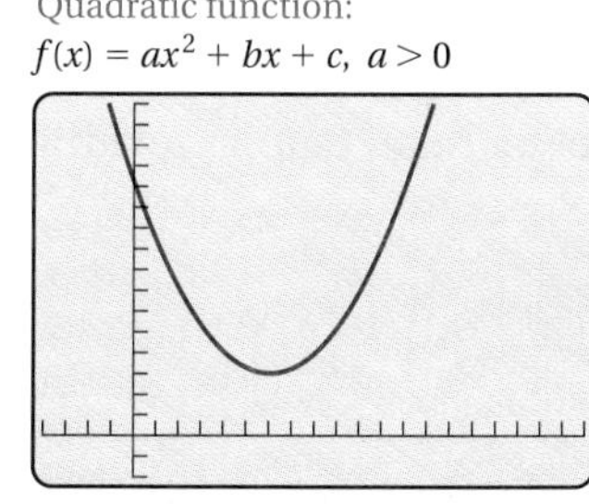

Quadratic function:
$f(x) = ax^2 + bx + c,\ a < 0$

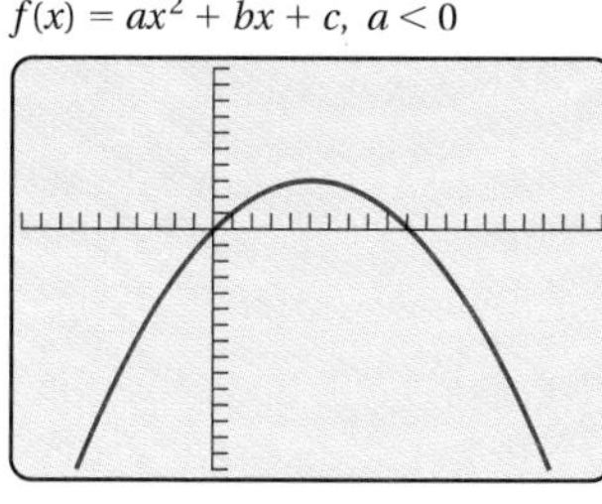

Absolute-value function:
$f(x) = |x|$

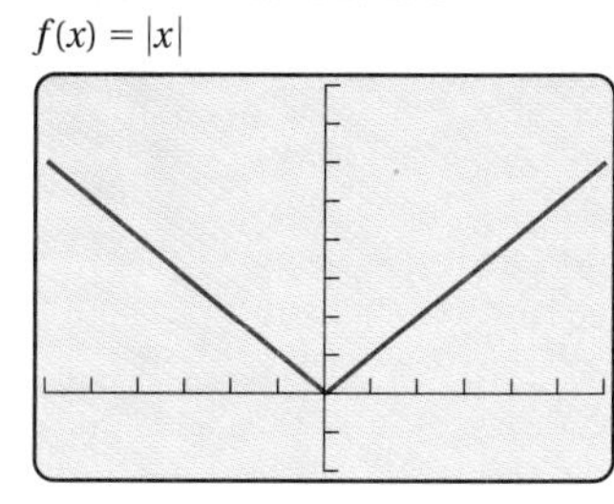

Cubic function:
$f(x) = ax^3 + bx^2 + cx + d,\ a > 0$

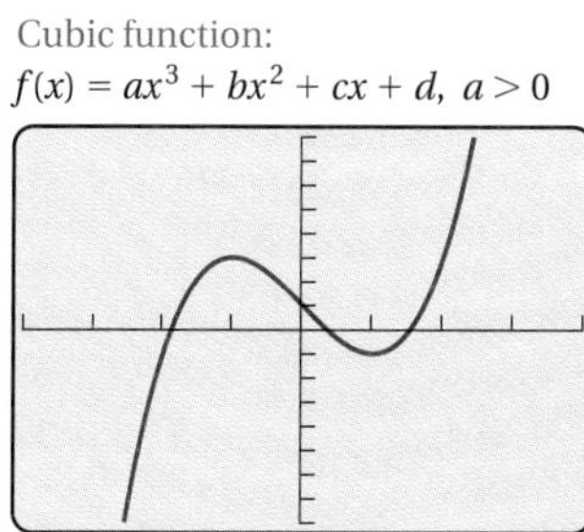

Quartic function:
$f(x) = ax^4 + bx^3 + cx^2 + dx + e,\ a > 0$

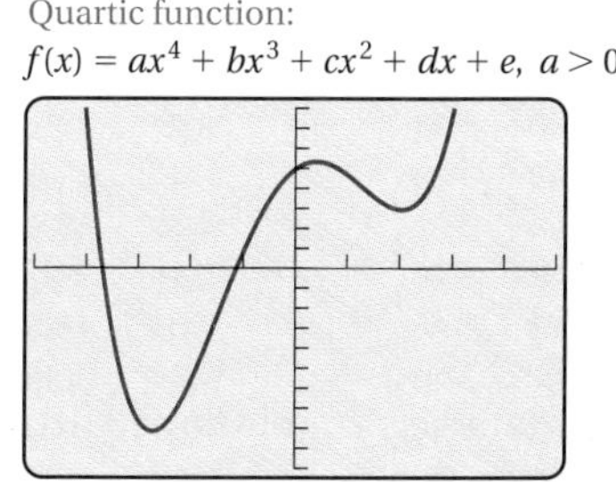

Now let's consider some real-world data. How can we decide which type of function might fit the data of a particular application? One simple way is to graph the data and look for a pattern resembling one of the graphs above. For example, data might be modeled by a linear function if the graph resembles a straight line. The data might be modeled by a quadratic function if the graph rises and then falls, or falls and then rises, in a curved manner resembling a parabola. For a quadratic, it might also just rise or fall in a curved manner as if following only one part of the parabola.

Choosing Models. For the scatterplots and graphs in Margin Exercises 2–5, determine which, if any, of the following functions might be used as a model for the data.

Linear, $f(x) = mx + b$;

Quadratic, $f(x) = ax^2 + bx + c$, $a > 0$;

Quadratic, $f(x) = ax^2 + bx + c$, $a < 0$;

Polynomial, neither quadratic nor linear

2.

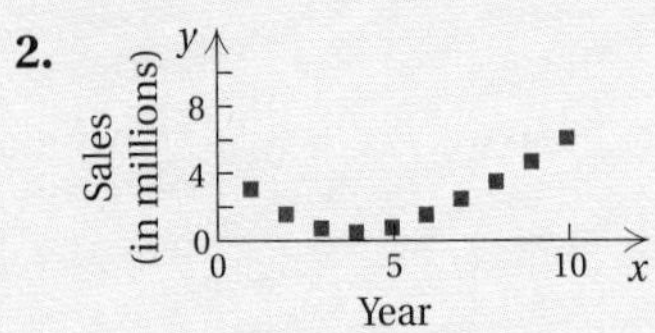

3.

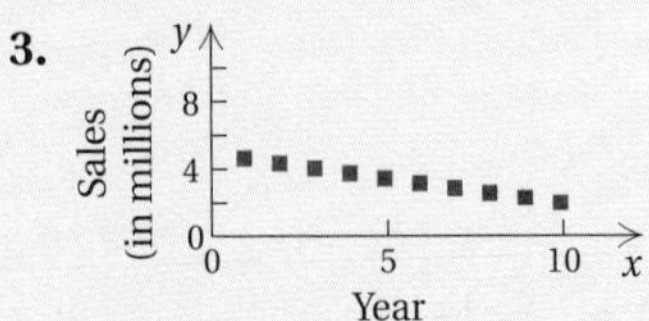

4.

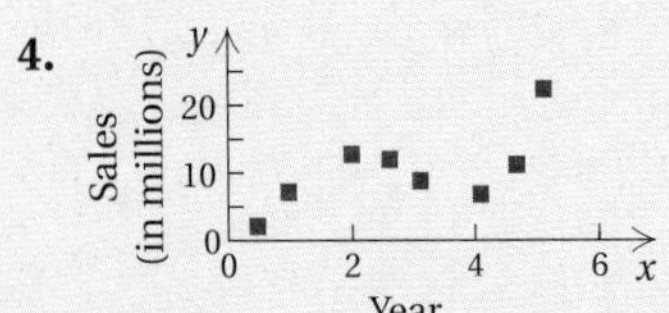

5.

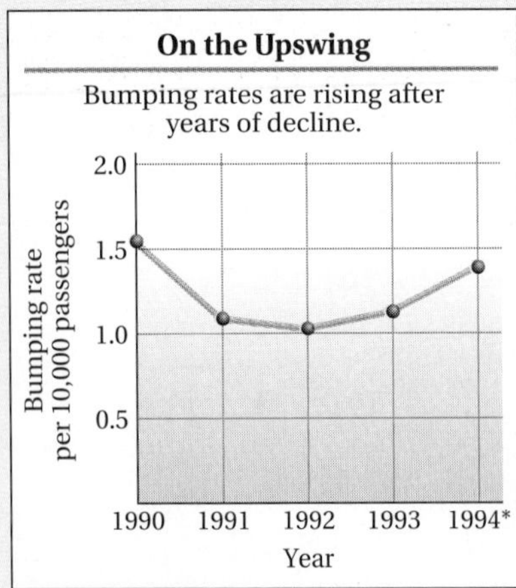

* January–June figure
Source: Department of Transportation

Answers on page A-49

Let's now use our library of functions to see which, if any, might fit certain data situations.

Examples *Choosing Models.* For the scatterplots and graphs below, determine which, if any, of the following functions might be used as a model for the data.

Linear, $f(x) = mx + b$;

Quadratic, $f(x) = ax^2 + bx + c,\ a > 0$;

Quadratic, $f(x) = ax^2 + bx + c,\ a < 0$;

Polynomial, neither quadratic nor linear

2.

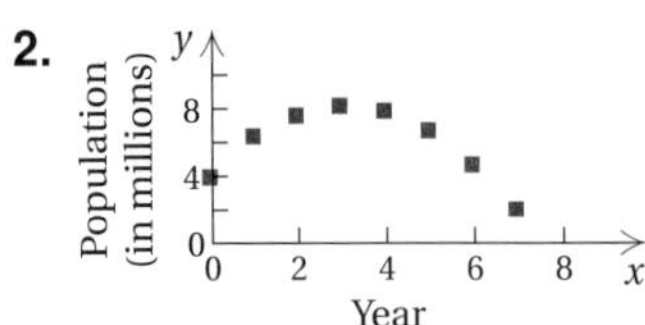

The data rise and then fall in a curved manner fitting a quadratic function $f(x) = ax^2 + bx + c,\ a < 0$.

3.

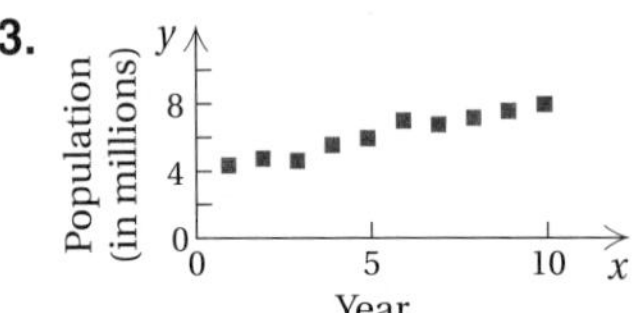

The data seem to fit a linear function $f(x) = mx + b$.

4.

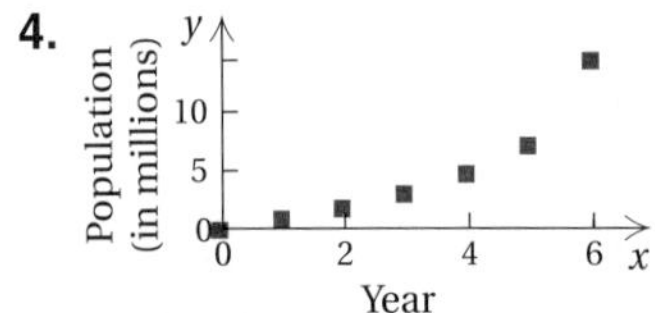

The data rise in a manner fitting the right side of a quadratic function $f(x) = ax^2 + bx + c,\ a > 0$.

5.

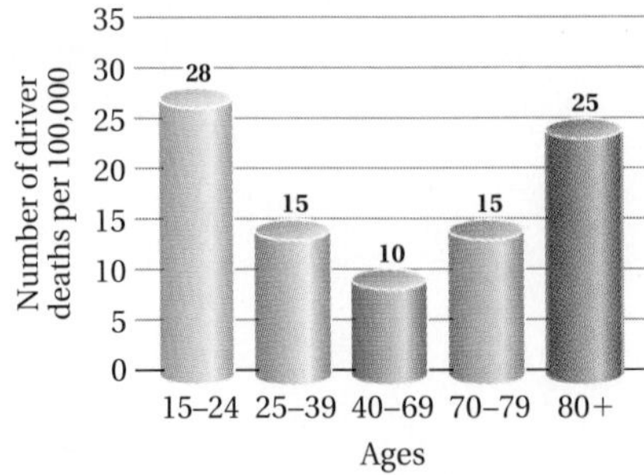

Source: National Highway Traffic Administration

The data fall and then rise in a curved manner fitting a quadratic function $f(x) = ax^2 + bx + c,\ a > 0$.

6.

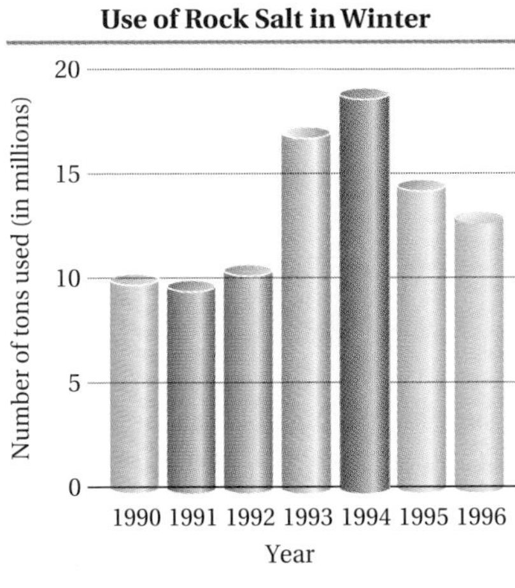

Source: Salt Institute

The data fall, then rise, then fall again, so they do not fit a linear or quadratic function but might fit a polynomial function that is neither quadratic nor linear.

Do Exercises 2–5 on the preceding page.

Whenever a quadratic function seems to fit a data situation, that function can be determined if at least three inputs and their outputs are known.

Example 7 *River Depth.* The drawing below shows the cross section of a river. Typically rivers are deepest in the middle, with the depth decreasing to 0 at the edges. A hydrologist measures the depths D, in feet, of a river at distances x, in feet, from one bank. The results are listed in the table below.

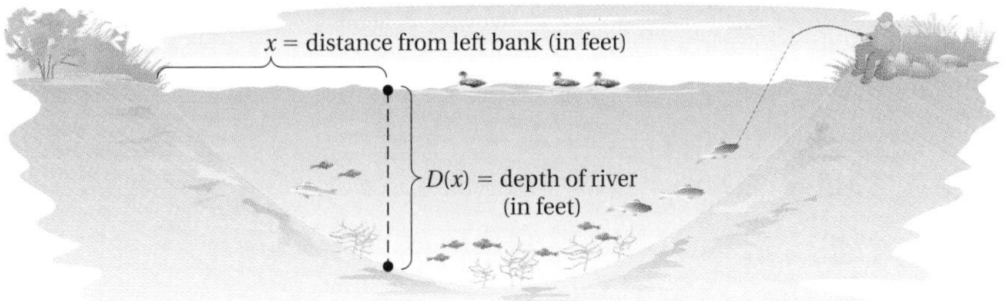

Distance, x, from the Riverbank (in feet)	Depth, D, of the River (in feet)
0	0
15	10.2
25	17
50	20
90	7.2
100	0

a) Make a scatterplot of the data.

b) Decide whether the data seem to fit a quadratic function.

c) Use the data points (0, 0), (50, 20), and (100, 0) to find a quadratic function that fits the data.

d) Use the function to estimate the depth of the river at 75 ft.

6. *Ticket Profits.* Valley Community College is presenting a play. The profit P, in dollars, after x days is given in the following table. (Profit can be negative when costs exceed revenue. See Section 8.6.)

Days, x	Profit, P
0	\$−100
90	560
180	872
270	870
360	548
450	−100

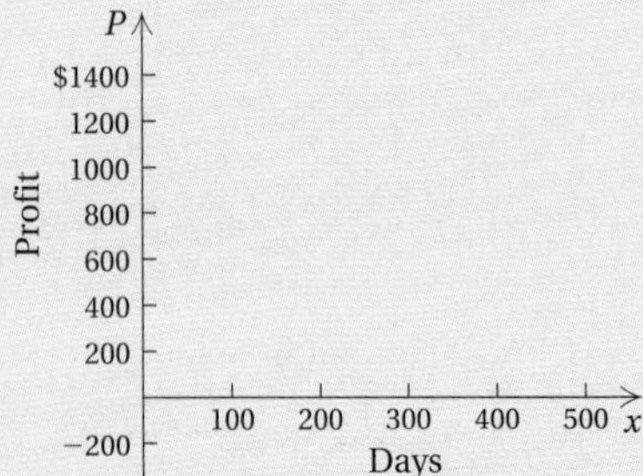

a) Make a scatterplot of the data.
b) Decide whether the data can be modeled by a quadratic function.
c) Use the data points (0, −100), (180, 872), and (360, 548) to find a quadratic function that fits the data.
d) Use the function to estimate the profits after 225 days.

Answers on page A-49

a) The scatterplot is as follows.

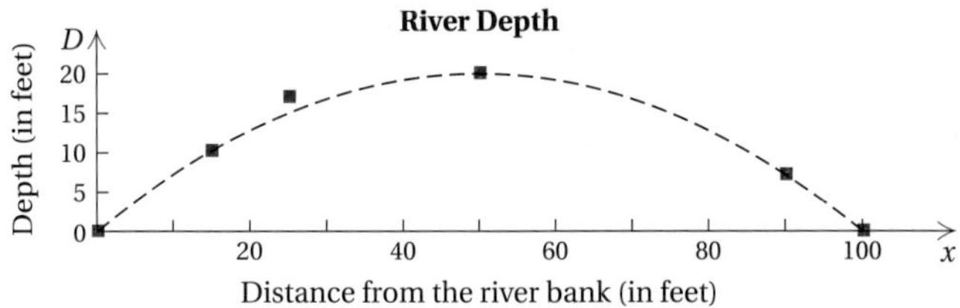

b) The data seem to rise and fall in a manner similar to a quadratic function. The dashed black line in the graph represents a sample quadratic function of fit. Note that it may not necessarily go through each point.

c) We are looking for a quadratic function

$$D(x) = ax^2 + bx + c.$$

We need to determine the constants a, b, and c. We use the three data points (0, 0), (50, 20), and (100, 0) and substitute as follows:

$$0 = a \cdot 0^2 + b \cdot 0 + c,$$
$$20 = a \cdot 50^2 + b \cdot 50 + c,$$
$$0 = a \cdot 100^2 + b \cdot 100 + c.$$

After simplifying, we see that we need to solve the system

$$0 = c,$$
$$20 = 2{,}500a + 50b + c,$$
$$0 = 10{,}000a + 100b + c.$$

Since $c = 0$, the system reduces to a system of two equations in two variables:

$$20 = 2{,}500a + 50b, \qquad (1)$$
$$0 = 10{,}000a + 100b. \qquad (2)$$

We multiply equation (1) by −2, add, and solve for a (see Section 8.2):

$$-40 = -5{,}000a - 100b,$$
$$0 = 10{,}000a + 100b$$
$$-40 = 5000a \qquad \text{Adding}$$
$$\frac{-40}{5000} = a \qquad \text{Solving for } a$$
$$-0.008 = a.$$

Next, we substitute −0.008 for a in equation (2) and solve for b:

$$0 = 10{,}000(-0.008) + 100b$$
$$0 = -80 + 100b$$
$$80 = 100b$$
$$0.8 = b.$$

This gives us the quadratic function:

$$D(x) = -0.008x^2 + 0.8x.$$

d) To find the depth 75 ft from the riverbank, we substitute:

$$D(75) = -0.008(75)^2 + 0.8(75) = 15.$$

At a distance of 75 ft from the riverbank, the depth of the river is 15 ft.

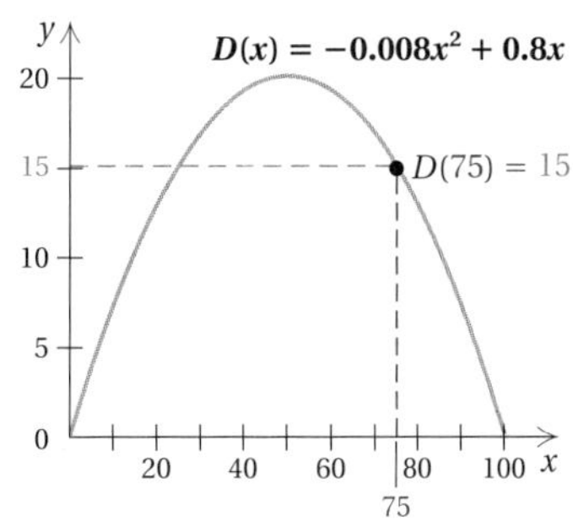

Do Exercise 6 on the preceding page.

Calculator Spotlight

Mathematical Modeling Using Regression: Fitting Quadratic and Other Polynomial Functions to Data. In the Calculator Spotlight of Section 7.6, we considered a method of fitting a linear function to a set of data, linear regression. Regression can be extended to quadratic, cubic, and quartic polynomial functions. The grapher gives us the powerful capability to find these polynomial models, select which seems best, and make predictions.

Example *Live Births to Women of Age x.* The following chart relates the number of live births to women of a particular age.

Age, x	Average Number of Live Births per 1000 Women
16	34
18.5	86.5
22	111.1
27	113.9
32	84.5
37	35.4
42	6.8

Source: Centers for Disease Control and Prevention

a) Fit a quadratic function to the data using the REGRESSION feature on a grapher.

b) Make a scatterplot of the data. Then graph the quadratic function with the scatterplot.

c) Fit a cubic function to the data using the REGRESSION feature on a grapher.

d) Make a scatterplot of the data. Then graph the cubic function with the scatterplot.

e) Decide which function seems to fit the data better.

f) Use the function from part (e) to estimate the average number of live births by women of ages 20 and 30.

We proceed as follows.

a) We fit a quadratic function to the data using the REGRESSION feature. The procedure is similar to what is outlined in the Calculator Spotlight of Section 7.6. We just choose QuadReg instead of LinReg(*ax* + *b*). (Consult your manual for further details.) We obtain the following function:

$$y_1 = f(x) = -0.49x^2 + 25.95x - 238.49.$$

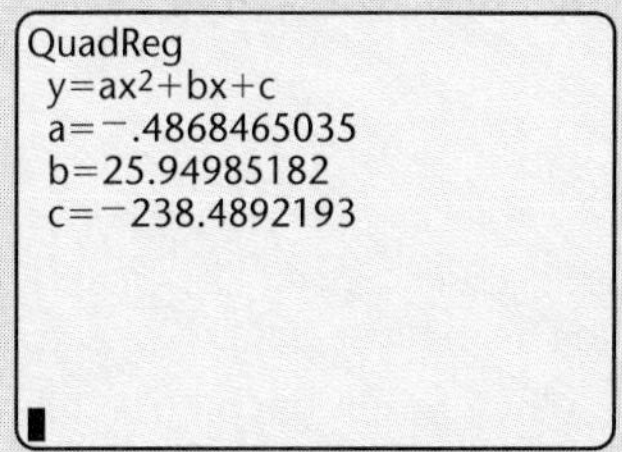

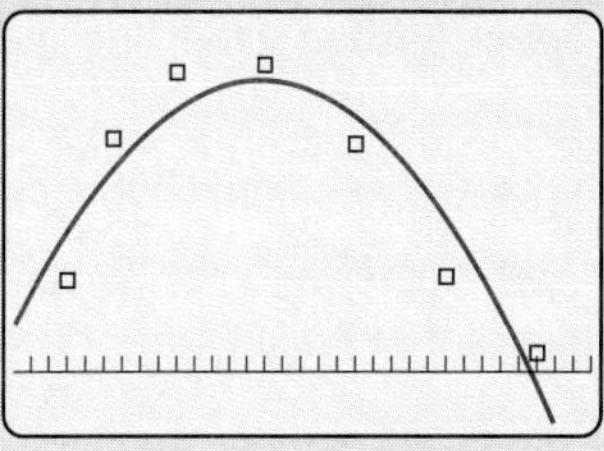

b) The quadratic function is graphed with the scatterplot above.

(continued)

c) We fit a cubic function to the data using the REGRESSION feature. We choose CubicReg and obtain the following function:

$$y_2 = f(x) = 0.03x^3 - 3.22x^2 + 101.18x - 886.93.$$

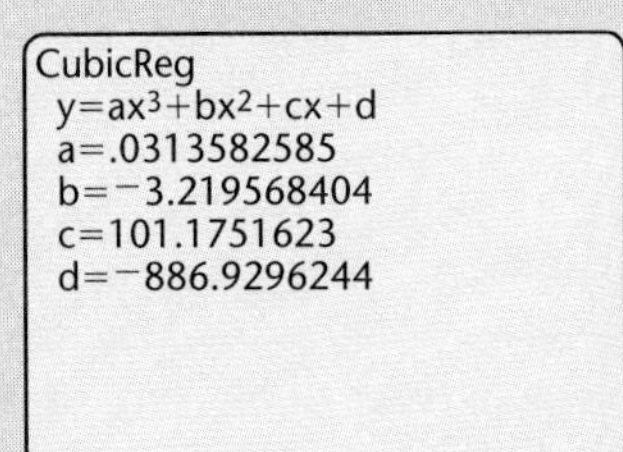
CubicReg
y=ax³+bx²+cx+d
a=.0313582585
b=−3.219568404
c=101.1751623
d=−886.9296244

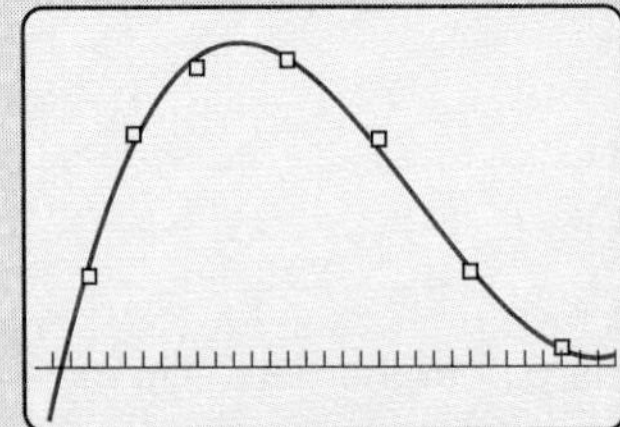

d) The cubic function is graphed with the scatterplot above.

e) The graph of the cubic function seems to fit closer to the data points. Thus we choose it as a model.

f) Using the grapher to do the calculation, we get $Y_2(20) \approx 99.6$ and $Y_2(30) \approx 97.4$ as shown.

```
Y2(20)
                99.61232795
Y2(30)
                97.38665968
```

Thus the average number of live births is 99.6 per 1000 by women age 20 and 97.4 per 1000 by women age 30.

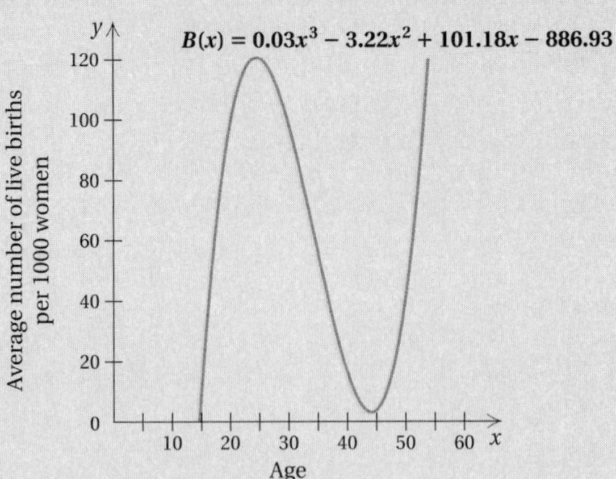

Exercises

1. **a)** Use the REGRESSION feature to fit a quartic equation to the live-birth data. Make a scatterplot of the data. Then graph the quartic function with the scatterplot. Decide whether the quartic function gives a better fit than either the quadratic or the cubic function.

   b) Explain why the domain of the cubic live-birth function should probably be restricted to women whose ages are in the interval [15, 45].

2. **MEDIAN HOUSEHOLD INCOME BY AGE**

Age, x	Median Income in 1996
19.5	$21,438
29.5	35,888
39.5	44,420
49.5	50,472
59.5	39,815
65	19,448

Source: U.S. Bureau of the Census; The Conference Board: Simmons Bureau of Labor Statistics

   a) Fit a quadratic function to the data using a REGRESSION feature on a grapher.

   b) Make a scatterplot of the data. Then graph the quadratic function with the scatterplot.

   c) Fit a cubic function to the data using a REGRESSION feature on a grapher.

   d) Make a scatterplot of the data. Then graph the cubic function with the scatterplot.

   e) Fit a quartic function to the data using a REGRESSION feature on a grapher.

   f) Make a scatterplot of the data. Then graph the quartic function with the scatterplot.

   g) Decide which of the quadratic, cubic, or quartic functions seems to fit the data best.

   h) Use the function from part (e) to estimate the median household income of people ages 25 and 45.

3. *River Depth.* Use your grapher to rework Example 3.

Exercise Set 11.7

a Solve.

1. *Architecture.* An architect is designing the floors for a hotel. Each floor is to be rectangular and is allotted 720 ft of security piping around walls outside the rooms. What dimensions of the floors will allow an atrium at the bottom to have maximum area?

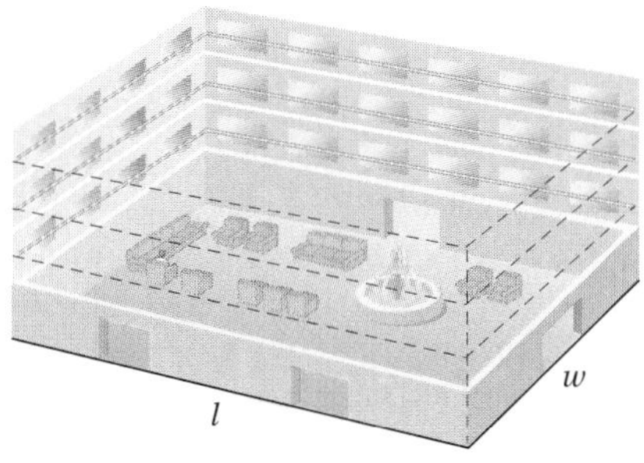

2. *Stained-Glass Window Design.* An artist is designing a rectangular stained-glass window with a perimeter of 84 in. What dimensions will yield the maximum area?

3. What is the maximum product of two numbers whose sum is 22? What numbers yield this product?

4. What is the maximum product of two numbers whose sum is 45? What numbers yield this product?

5. What is the minimum product of two numbers whose difference is 4? What are the numbers?

6. What is the minimum product of two numbers whose difference is 6? What are the numbers?

7. What is the maximum product of two numbers that add to -12? What numbers yield this product?

8. What is the minimum product of two numbers that differ by 9? What are the numbers?

9. *Garden Design.* A farmer decides to enclose a rectangular garden, using the side of a barn as one side of the rectangle. What is the maximum area that the farmer can enclose with 40 ft of fence? What should the dimensions of the garden be in order to yield this area?

10. *Patio Design.* A stone mason has enough stones to enclose a rectangular patio with 60 ft of perimeter, assuming that the attached house forms one side of the rectangle. What is the maximum area that the mason can enclose? What should the dimensions of the patio be in order to yield this area?

11. *Molding Plastics.* Economite Plastics plans to produce a one-compartment vertical file by bending the long side of an 8-in. by 14-in. sheet of plastic along two lines to form a U shape. How tall should the file be in order to maximize the volume that the file can hold?

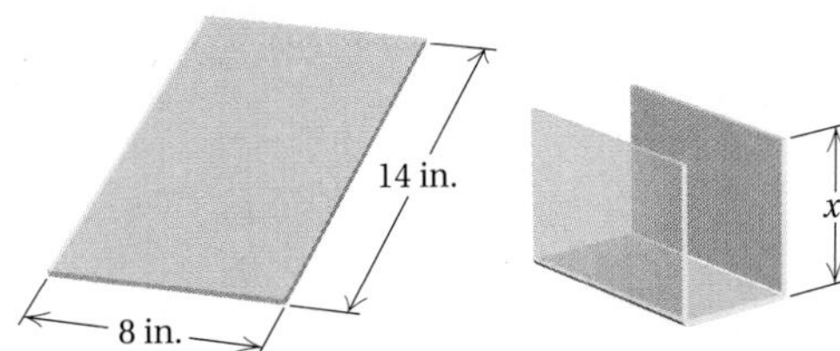

12. *Composting.* A rectangular compost container is to be formed in a corner of a fenced yard, with 8 ft of chicken wire completing the other two sides of the rectangle. If the chicken wire is 3 ft high, what dimensions of the base will maximize the volume of the container?

13. *Minimizing Cost.* Aki's Bicycle Designs has determined that when x hundred bicycles are built, the average cost per bicycle is given by

$$C(x) = 0.1x^2 - 0.7x + 2.425,$$

where $C(x)$ is in hundreds of dollars. How many bicycles should the shop build in order to minimize the average cost per bicycle?

14. *Corral Design.* A rancher needs to enclose two adjacent rectangular corrals, one for sheep and one for cattle. If a river forms one side of the corrals and 180 yd of fencing is available, what is the largest total area that can be enclosed?

Maximizing Profit. Recall (Section 8.6) that total profit P is the difference between total revenue R and total cost C. Given the following total-revenue and total-cost functions, find the total profit, the maximum value of the total profit, and the value of x at which it occurs.

15. $R(x) = 1000x - x^2$,
$C(x) = 3000 + 20x$

16. $R(x) = 200x - x^2$,
$C(x) = 5000 + 8x$

b *Choosing Models.* For the scatterplots and graphs in Exercises 17–24, determine which, if any, of the following functions might be used as a model for the data: Linear, $f(x) = mx + b$; quadratic, $f(x) = ax^2 + bx + c, a > 0$; quadratic, $f(x) = ax^2 + bx + c, a < 0$; polynomial, neither quadratic nor linear.

17.

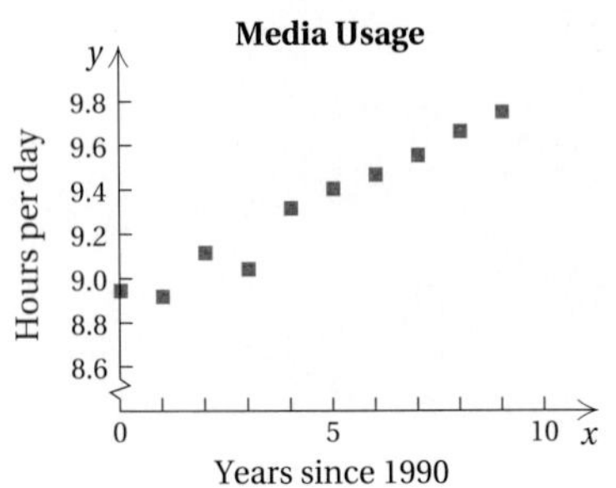

18.

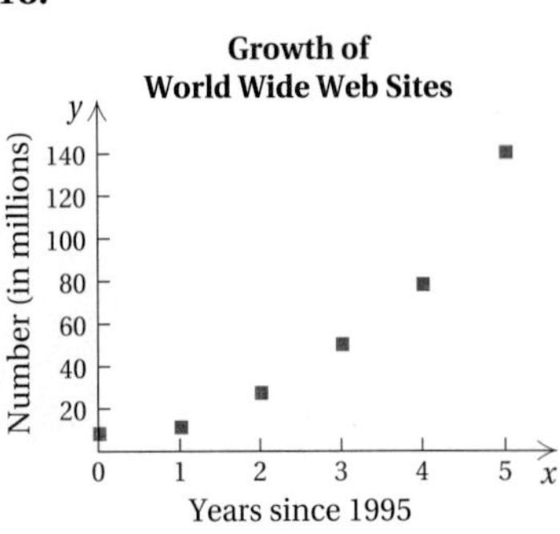

19.

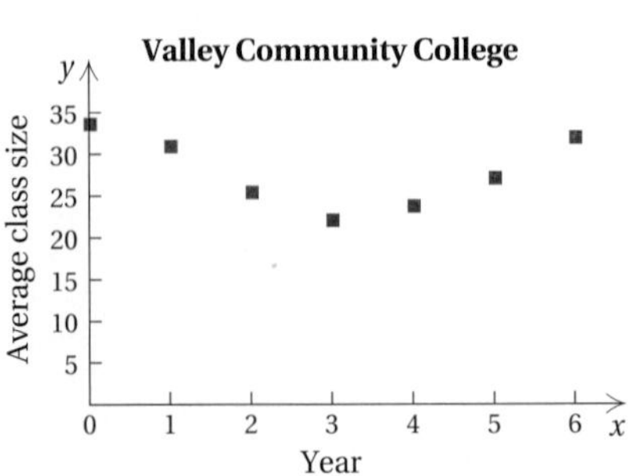

20.

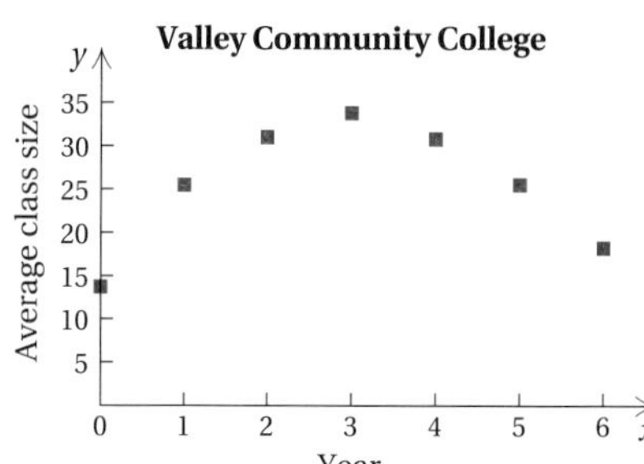

21.

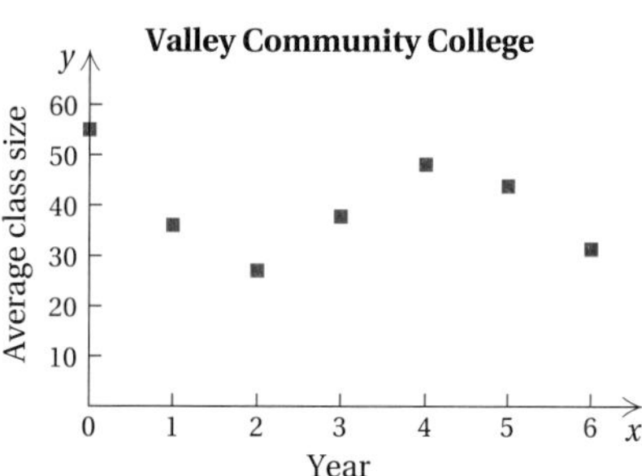

22.

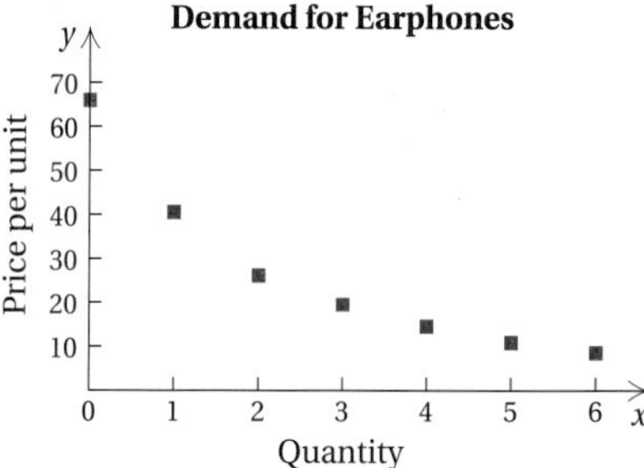

23.

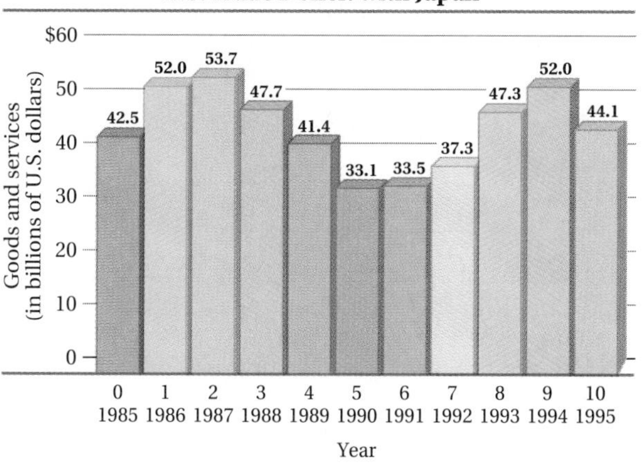

24.

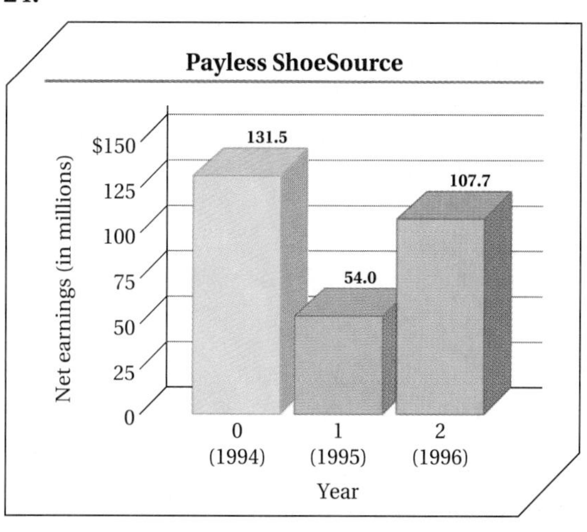

Source: Payless 1996 Annual Report

Find a quadratic function that fits the set of data points.

25. $(1, 4), (-1, -2), (2, 13)$

26. $(1, 4), (-1, 6), (-2, 16)$

27. $(2, 0), (4, 3), (12, -5)$

28. $(-3, -30), (3, 0), (6, 6)$

29. *Nighttime Accidents.*

a) Find a quadratic function that fits the following data.

Travel Speed (in kilometers per hour)	Number of Nighttime Accidents (for every 200 million kilometers driven)
60	400
80	250
100	250

b) Use the function to estimate the number of nighttime accidents that occur at 50 km/h.

30. *Daytime Accidents.*

a) Find a quadratic function that fits the following data.

Travel Speed (in kilometers per hour)	Number of Daytime Accidents (for every 200 million kilometers driven)
60	100
80	130
100	200

b) Use the function to estimate the number of daytime accidents that occur at 50 km/h.

31. *Payless Shoe Source Net Earnings.* Use the data from Exercise 24.

a) Find a quadratic function that fits the data.
b) Use the quadratic function to predict net earnings in 1999 and in 2000.

32. *Rocket Height.* The data in the following table give the height H, in feet, of a rocket t seconds after blastoff.

Time, t (in seconds)	Height, H (in feet)
2	2304
6	2048
10	1280
12	704

a) Use the data points (2, 2304), (6, 2048), and (12, 704) to find a quadratic function that fits the data.
b) Use the function to estimate the height of the rocket after 1 sec, 8 sec, and 11 sec.
c) Determine the maximum height and when it is attained.
d) Determine when the rocket reaches the ground.

Skill Maintenance

Multiply and simplify. [10.3a]

33. $\sqrt[4]{5x^3y^5}\,\sqrt[4]{125x^2y^3}$

34. $\sqrt{9a^3}\,\sqrt{16ab^4}$

Solve. [10.6a, b]

35. $\sqrt{4x-4} = \sqrt{x+4}+1$

36. $\sqrt{5x-4}+\sqrt{13-x} = 7$

37. $-35 = \sqrt{2x+5}$

38. $\sqrt{7x-5} = \sqrt{4x+7}$

Synthesis

39. ◆ Explain the restrictions that should be placed on the domains of the quadratic functions found in Exercises 15 and 29 and why such restrictions are needed.

40. ◆ Explain how the leading coefficient of a quadratic function can be used to determine whether a maximum or minimum function value exists.

41. *Japanese Trade Deficit.* Use the REGRESSION feature of a grapher to fit a quartic function to the data in Exercise 23.

42. The sum of the base and the height of a triangle is 38 cm. Find the dimensions for which the area is a maximum, and find the maximum area.

Collaborative Learning Manual

Fit a quadratic function to a set of data.

11.8 Polynomial and Rational Inequalities

a Quadratic and Other Polynomial Inequalities

Inequalities like the following are called **quadratic inequalities:**

$$x^2 + 3x - 10 < 0, \qquad 5x^2 - 3x + 2 \geq 0.$$

In each case, we have a polynomial of degree 2 on the left. We will solve such inequalities in two ways. The first method provides understanding and the second yields the more efficient method.

The first method for solving a quadratic inequality, such as $ax^2 + bx + c > 0$, is by considering the graph of a related function, $f(x) = ax^2 + bx + c$.

Example 1 Solve: $x^2 + 3x - 10 > 0$.

Consider the function $f(x) = x^2 + 3x - 10$ and its graph. The graph opens up since the leading coefficient ($a = 1$) is positive. We find the intercepts by setting the polynomial equal to 0 and solving:

$$x^2 + 3x - 10 = 0$$
$$(x + 5)(x - 2) = 0$$
$$x + 5 = 0 \quad or \quad x - 2 = 0$$
$$x = -5 \quad or \quad x = 2.$$

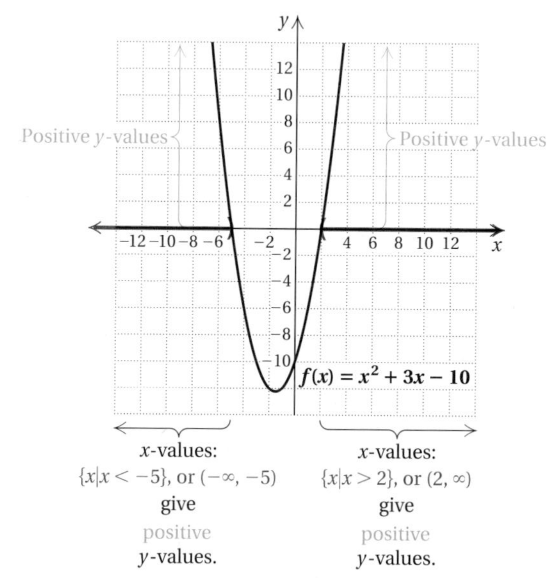

Values of y will be positive to the left and right of the intercepts, as shown. Thus the solution set of the inequality is

$$\{x \mid x < -5 \text{ or } x > 2\}, \quad \text{or} \quad (-\infty, -5) \cup (2, \infty).$$

Do Exercise 1.

We can solve any inequality by considering the graph of a related function and finding intercepts as in Example 1. In some cases, we may need to use the quadratic formula to find the intercepts.

Objectives

a Solve quadratic and other polynomial inequalities.

b Solve rational inequalities.

For Extra Help

 TAPE 19

 MAC WIN

 CD-ROM

1. Solve by graphing:

$$x^2 + 2x - 3 > 0.$$

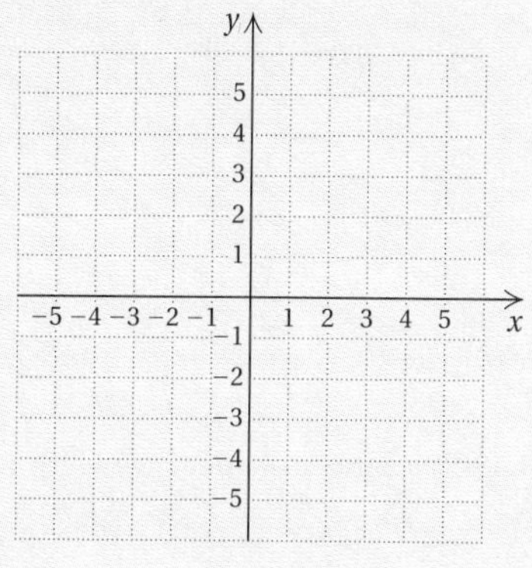

Answer on page A-50

2. Solve by graphing:

$x^2 + 2x - 3 < 0.$

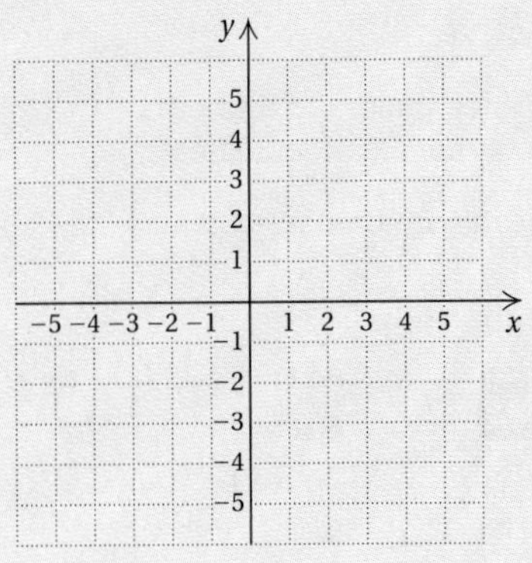

3. Solve by graphing:

$x^2 + 2x - 3 \le 0.$

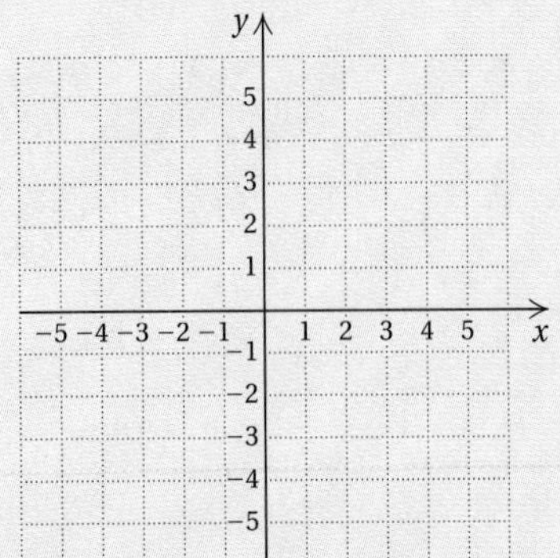

Answers on page A-50

Example 2 Solve: $x^2 + 3x - 10 < 0$.

Looking again at the graph of $f(x) = x^2 + 3x - 10$ or at least visualizing it tells us that y-values are negative for those x-values between -5 and 2.

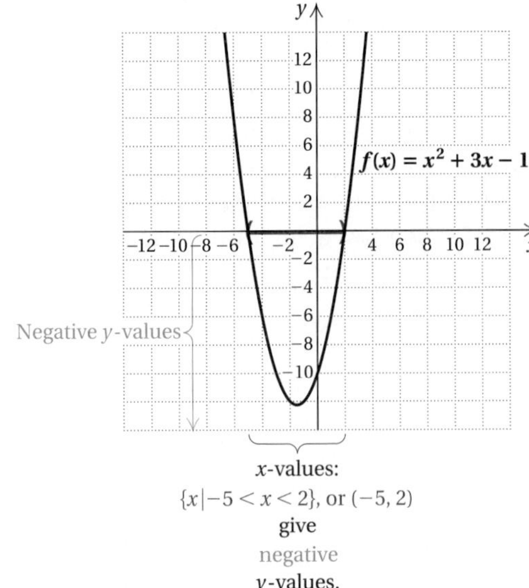

That is, the solution set is $\{x \mid -5 < x < 2\}$, or $(-5, 2)$.

Do Exercise 2.

When an inequality contains $\le$ or $\ge$, the x-values of the intercepts must be included. Thus the solution set of the inequality $x^2 + 3x - 10 \le 0$ is

$$\{x \mid -5 \le x \le 2\}, \quad \text{or} \quad [-5, 2].$$

Do Exercise 3.

We now consider a more efficient method for solving polynomial inequalities. The preceding discussion provides the understanding for this method. In Examples 1 and 2, we see that the intercepts divide the number line into intervals.

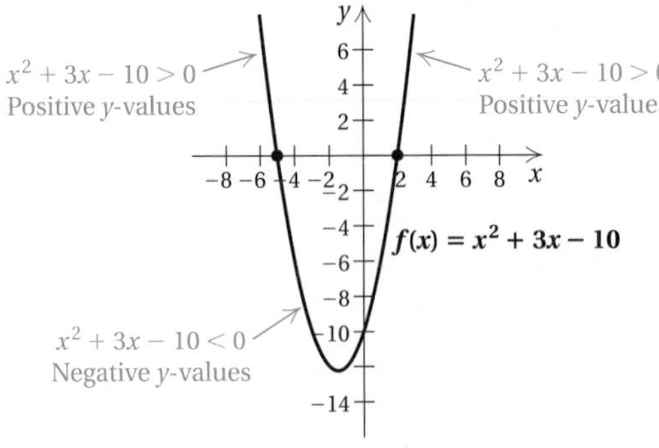

If a function has a positive output for one number in an interval, it will be positive for all the numbers in the interval. The same is true for negative outputs. Thus we can merely make a test substitution in each interval to solve the inequality. This is very similar to our method of using test points to graph a linear inequality in a plane.

Example 3 Solve: $x^2 + 3x - 10 < 0$.

We set the polynomial equal to 0 and solve. The solutions of $x^2 + 3x - 10 = 0$, or $(x + 5)(x - 2) = 0$, are -5 and 2. We then locate them on a number line as follows. Note that the numbers divide the number line into three intervals, which we will call A, B, and C.

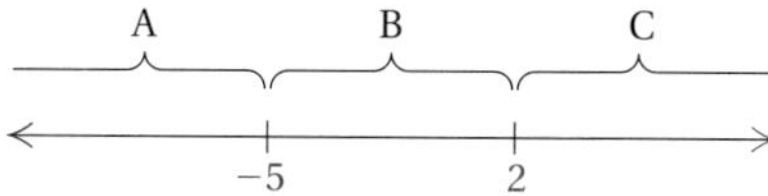

We choose a test number in interval A, say -7, and substitute -7 for x in the function $f(x) = x^2 + 3x - 10$:

$$\begin{aligned} f(-7) &= (-7)^2 + 3(-7) - 10 \\ &= 49 - 21 - 10 = 18. \qquad \text{Thus, } f(-7) > 0. \end{aligned}$$

Note that $18 > 0$, so the function values will be positive for any number in interval A.

Next, we try a test number in interval B, say 1, and find the corresponding function value:

$$\begin{aligned} f(1) &= 1^2 + 3(1) - 10 \\ &= 1 + 3 - 10 = -6. \qquad \text{Thus, } f(1) < 0. \end{aligned}$$

Note that $-6 < 0$, so the function values will be negative for any number in interval B.

Next, we try a test number in interval C, say 4, and find the corresponding function value:

$$\begin{aligned} f(4) &= 4^2 + 3(4) - 10 \\ &= 16 + 12 - 10 = 18. \qquad \text{Thus, } f(4) > 0. \end{aligned}$$

Note that $18 > 0$, so the function values will be positive for any number in interval C.

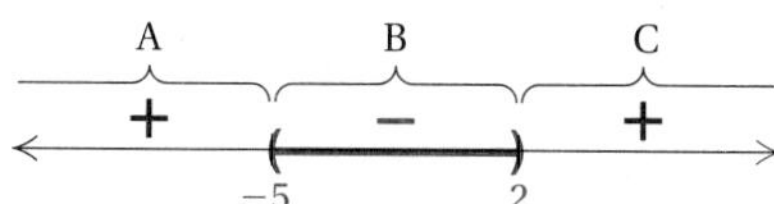

We are looking for numbers x for which $f(x) = x^2 + 3x - 10 < 0$. Thus any number x in interval B is a solution. If the inequality had been $\leq$, it would have been necessary to include the intercepts -5 and 2 in the solution set as well. The solution set is $\{x \mid -5 < x < 2\}$, or the interval $(-5, 2)$.

To solve a polynomial inequality:

1. Get 0 on one side, set the expression on the other side equal to 0, and solve to find the intercepts.
2. Use the numbers found in step (1) to divide the number line into intervals.
3. Substitute a number from each interval into the related function. If the function value is positive, then the expression will be positive for all numbers in the interval. If the function value is negative, then the expression will be negative for all numbers in the interval.
4. Select the intervals for which the inequality is satisfied and write set-builder or interval notation for the solution set.

Do Exercises 4 and 5.

Solve using the method of Example 3.

4. $x^2 + 3x > 4$

5. $x^2 + 3x \leq 4$

Answers on page A-50

Calculator Spotlight

Solving Polynomial Inequalities on a Grapher. Consider solving $2.3x^2 + 2.94x - 9.11 \leq 0$ on a grapher. We first graph the related function

$$f(x) = 2.3x^2 + 2.94x - 9.11.$$

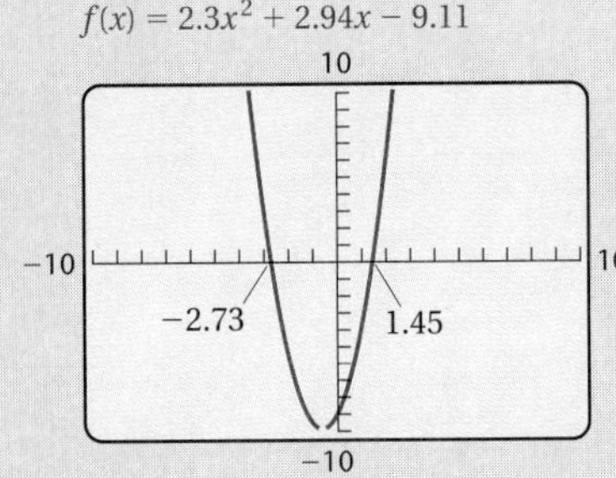

Then we solve the equation $f(x) = 0$ or $2.3x^2 + 2.94x - 9.11 = 0$ using the ZERO feature. The solutions are -2.73 and 1.45 to two decimal places. They divide the number line into three intervals:

$$(-\infty, -2.73), \quad (-2.73, 1.45), \quad \text{and} \quad (1.45, \infty).$$

We note on the graph where the function is negative.

We can check this using the TRACE or TABLE feature. Then, including appropriate endpoints, we find the solution set to be

$$\{x \mid -2.73 \leq x \leq 1.45\}, \quad \text{or} \quad [-2.73, 1.45].$$

To solve $2.3x^2 + 2.94x - 9.11 > 0$, we look for portions of the graph above the x-axis and the solution set would be

$$\{x \mid x < -2.73 \quad or \quad x > 1.45\},$$

or $(-\infty, -2.73) \cup (1.45, \infty)$.

Exercises

Use a grapher to solve the inequality.

1. $x^2 + 3x - 10 > 0$
2. $x^2 + 3x - 10 \geq 0$
3. $x^2 + 3x - 10 \leq 0$
4. $x^2 + 3x - 10 < 0$
5. $4.32x^2 - 3.54x - 5.34 \leq 0$
6. $7.34x^2 \geq 16.55x + 3.89$
7. $10.85x^2 + 4.28x + 4.44 > 7.91x^2 + 7.43x + 13.03$
8. $5.79x^3 - 5.68x^2 + 10.68x > 2.11x^3 + 16.90x - 11.69$

6. Solve: $6x(x + 1)(x - 1) < 0$.

Example 4 Solve: $5x(x + 3)(x - 2) \geq 0$.

The solutions of $f(x) = 0$, or $5x(x + 3)(x - 2) = 0$, are -3, 0, and 2. They divide the real-number line into four intervals, as shown below.

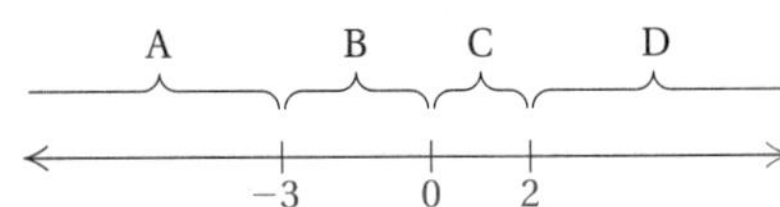

We try test numbers in each interval:

A: Test -5, $\quad f(-5) = 5(-5)(-5 + 3)(-5 - 2) = -350 < 0.$

B: Test -2, $\quad f(-2) = 5(-2)(-2 + 3)(-2 - 2) = 40 > 0.$

C: Test 1, $\quad f(1) = 5(1)(1 + 3)(1 - 2) = -20 < 0.$

D: Test 3, $\quad f(3) = 5(3)(3 + 3)(3 - 2) = 90 > 0.$

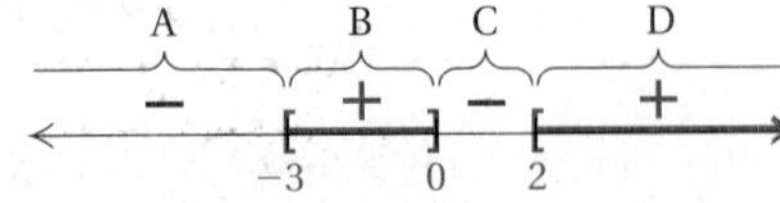

The expression is positive for values of x in intervals B and D. Since the inequality symbol is $\geq$, we need to include the intercepts. The solution set of the inequality is

$$\{x \mid -3 \leq x \leq 0 \ or\ 2 \leq x\}, \quad \text{or} \quad [-3, 0] \cup [2, \infty).$$

We visualize this with the graph at left.

f(x) > 0
f(x) > 0
f(x) < 0
f(x) < 0

$f(x) = 5x(x + 3)(x - 2)$

Answer on page A-50

Do Exercise 6.

b Rational Inequalities

We adapt the preceding method when an inequality involves rational expressions. We call these **rational inequalities.**

Example 5 Solve: $\dfrac{x-3}{x+4} \geq 2$.

We write a related equation by changing the $\geq$ symbol to $=$:

$$\frac{x-3}{x+4} = 2.$$

Then we solve this related equation. First, we multiply on both sides of the equation by the LCM, which is $x + 4$:

$$(x+4) \cdot \frac{x-3}{x+4} = (x+4) \cdot 2$$
$$x - 3 = 2x + 8$$
$$-11 = x.$$

With rational inequalities, we also need to determine those numbers for which the rational expression is not defined—that is, those numbers that make the denominator 0. We set the denominator equal to 0 and solve: $x + 4 = 0$, or $x = -4$. Next, we use the numbers -11 and -4 to divide the number line into intervals, as shown below.

A B C
−11 −4

We try test numbers in each interval to see if each satisfies the original inequality.

A: Test -15,

$$\frac{x-3}{x+4} \geq 2$$
$$\frac{-15-3}{-15+4} \;?\; 2$$
$$\frac{18}{11} \quad \text{FALSE}$$

Since the inequality is false for $x = -15$, the number -15 is not a solution of the inequality. Interval A is *not* part of the solution set.

B: Test -8,

$$\frac{x-3}{x+4} \geq 2$$
$$\frac{-8-3}{-8+4} \;?\; 2$$
$$\frac{11}{4} \quad \text{TRUE}$$

Since the inequality is true for $x = -8$, the number -8 is a solution of the inequality. Interval B *is* part of the solution set.

C: Test 1,

$$\frac{x-3}{x+4} \geq 2$$
$$\frac{1-3}{1+4} \;?\; 2$$
$$-\frac{2}{5} \quad \text{FALSE}$$

Since the inequality is false for $x = 1$, the number 1 is not a solution of the inequality. Interval C is *not* part of the solution set.

Solve.

7. $\dfrac{x+1}{x-2} \geq 3$

8. $\dfrac{x}{x-5} < 2$

The solution set includes the interval B. The number -11 is also included since the inequality symbol is $\geq$ and -11 is a solution of the related equation. The number -4 is not included; it is not an allowable replacement because it results in division by 0. Thus the solution set of the original inequality is

$$\{x \mid -11 \leq x < -4\}, \quad \text{or} \quad [-11, -4).$$

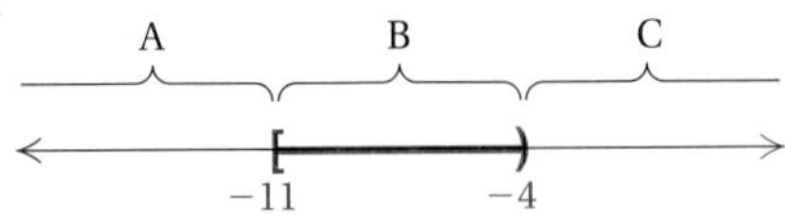

To solve a rational inequality:

1. Change the inequality symbol to an equals sign and solve the related equation.
2. Find the numbers for which any rational expression in the inequality is not defined.
3. Use the numbers found in steps (1) and (2) to divide the number line into intervals.
4. Substitute a number from each interval into the inequality. If the number is a solution, then the interval to which it belongs is part of the solution set.
5. Select the intervals for which the inequality is satisfied and write set-builder or interval notation for the solution set.

Do Exercises 7 and 8.

AG Algebraic–Graphical Connection

Let's compare the algebraic solution of Example 5 to a graphical solution. Although it is not a goal of this text to instruct you on graphing rational functions, let's look at the graph of the function

$$f(x) = \frac{x-3}{x+4}.$$

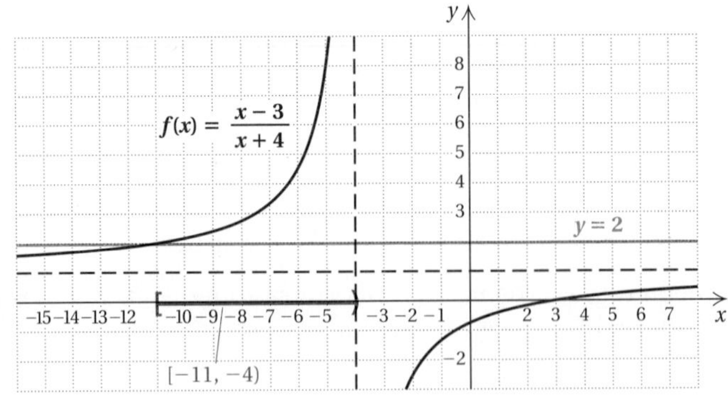

We see that the solutions of the inequality

$$\frac{x-3}{x+4} \geq 2$$

can be found by drawing the line $y = 2$ and determining all x-values for which $f(x) \geq 2$.

Calculator Spotlight

Check the results of Example 5 and Margin Exercises 7 and 8 on your grapher.

Answers on page A-50

Exercise Set 11.8

a Solve algebraically and verify results from the graph.

1. $(x - 6)(x + 2) > 0$

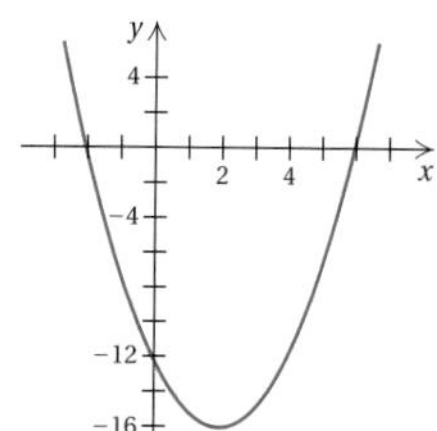

2. $(x - 5)(x + 1) > 0$

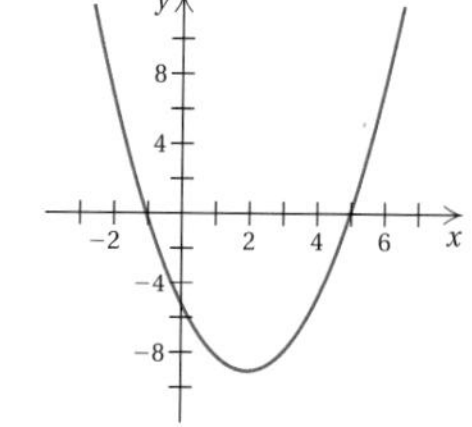

3. $4 - x^2 \geq 0$

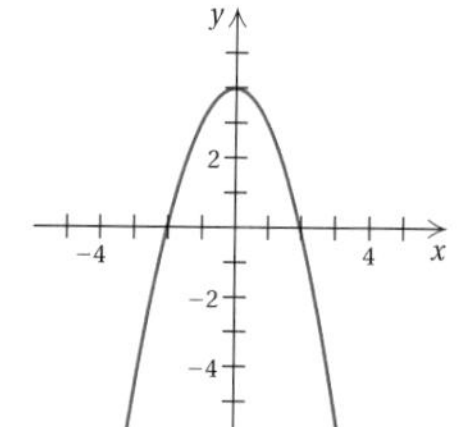

4. $9 - x^2 \leq 0$

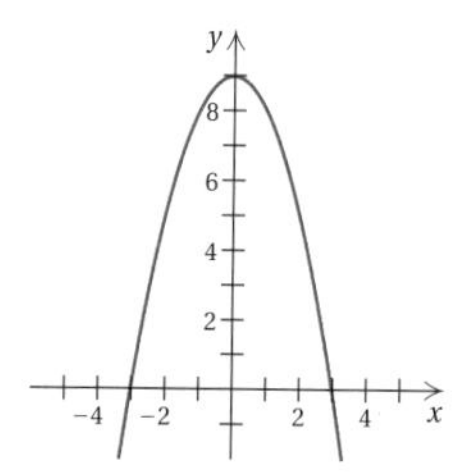

Solve.

5. $3(x + 1)(x - 4) \leq 0$

6. $(x - 7)(x + 3) \leq 0$

7. $x^2 - x - 2 < 0$

8. $x^2 + x - 2 < 0$

9. $x^2 - 2x + 1 \geq 0$

10. $x^2 + 6x + 9 < 0$

11. $x^2 + 8 < 6x$

12. $x^2 - 12 > 4x$

13. $3x(x + 2)(x - 2) < 0$

14. $5x(x + 1)(x - 1) > 0$

15. $(x + 9)(x - 4)(x + 1) > 0$

16. $(x - 1)(x + 8)(x - 2) < 0$

17. $(x + 3)(x + 2)(x - 1) < 0$

18. $(x - 2)(x - 3)(x + 1) < 0$

b Solve.

19. $\frac{1}{x - 6} < 0$

20. $\frac{1}{x + 4} > 0$

21. $\frac{x + 1}{x - 3} > 0$

22. $\frac{x - 2}{x + 5} < 0$

23. $\frac{3x + 2}{x - 3} \leq 0$

24. $\frac{5 - 2x}{4x + 3} \leq 0$

25. $\frac{x - 1}{x - 2} > 3$

26. $\frac{x + 1}{2x - 3} < 1$

27. $\dfrac{(x-2)(x+1)}{x-5} < 0$

28. $\dfrac{(x+4)(x-1)}{x+3} > 0$

29. $\dfrac{x+3}{x} \leq 0$

30. $\dfrac{x}{x-2} \geq 0$

31. $\dfrac{x}{x-1} > 2$

32. $\dfrac{x-5}{x} < 1$

33. $\dfrac{x-1}{(x-3)(x+4)} < 0$

34. $\dfrac{x+2}{(x-2)(x+7)} > 0$

35. $3 < \dfrac{1}{x}$

36. $\dfrac{1}{x} \leq 2$

37. $\dfrac{(x-1)(x+2)}{(x+3)(x-4)} > 0$

38. $\dfrac{x^2-11x+30}{x^2-8x-9} \geq 0$

Skill Maintenance

Simplify. [10.3b]

39. $\sqrt[3]{\dfrac{125}{27}}$

40. $\sqrt{\dfrac{25}{4a^2}}$

41. $\sqrt{\dfrac{16a^3}{b^4}}$

42. $\sqrt[3]{\dfrac{27c^5}{343d^3}}$

Add or subtract. [10.4a]

43. $3\sqrt{8} - 5\sqrt{2}$

44. $7\sqrt{45} - 2\sqrt{20}$

45. $5\sqrt[3]{16a^4} + 7\sqrt[3]{2a}$

46. $3\sqrt{10} + 8\sqrt{20} - 5\sqrt{80}$

Synthesis

47. ◆ Describe a method that could be used to create quadratic inequalities that have no solution.

48. ◆ Describe a method that could be used to create quadratic inequalities that have all real numbers as solutions.

49. Use a grapher to solve Exercises 11, 22, and 25 by graphing two curves, one for each side of an inequality.

50. Use a grapher to solve each of the following.

a) $x + \dfrac{1}{x} < 0$

b) $x - \sqrt{x} \geq 0$

c) $\frac{1}{3}x^3 - x + \frac{2}{3} \leq 0$

Solve.

51. $x^2 - 2x \leq 2$

52. $x^2 + 2x > 4$

53. $x^4 + 2x^2 > 0$

54. $x^4 + 3x^2 \leq 0$

55. $\left|\dfrac{x+2}{x-1}\right| < 3$

56. *Total Profit.* A company determines that its total profit on the production and sale of x units of a product is given by

$$P(x) = -x^2 + 812x - 9600.$$

a) A company makes a profit for those nonnegative values of x for which $P(x) > 0$. Find the values of x for which the company makes a profit.

b) A company loses money for those nonnegative values of x for which $P(x) < 0$. Find the values of x for which the company loses money.

57. *Height of a Thrown Object.* The function

$$H(t) = -16t^2 + 32t + 1920$$

gives the height S of an object thrown from a cliff 1920 ft high, after time t seconds.

a) For what times is the height greater than 1920 ft?

b) For what times is the height less than 640 ft?

Summary and Review Exercises: Chapter 11

Important Properties and Formulas

Principle of Square Roots: $x^2 = d$ has solutions $\sqrt{d}$ and $-\sqrt{d}$.

Quadratic Formula: $x = \dfrac{-b \pm \sqrt{b^2 - 4ac}}{2a}$; *Discriminant:* $b^2 - 4ac$

The vertex of the graph of $f(x) = ax^2 + bx + c$ is $\left(-\dfrac{b}{2a}, \dfrac{4ac - b^2}{4a}\right)$, or $\left(-\dfrac{b}{2a}, f\left(-\dfrac{b}{2a}\right)\right)$.

The line of symmetry of the graph of $f(x) = ax^2 + bx + c$ is $x = -\dfrac{b}{2a}$.

The objectives to be tested in addition to the material in this chapter are [6.3a], [6.4a], [7.6b], [10.3a], and [10.6a, b].

1. a) Solve: $2x^2 - 7 = 0$. [11.1a]
b) Find the x-intercepts of $f(x) = 2x^2 - 7$.

Solve. [11.2a]

2. $14x^2 + 5x = 0$

3. $x^2 - 12x + 27 = 0$

4. $4x^2 + 3x + 1 = 0$

5. $x^2 - 7x + 13 = 0$

6. $4x(x - 1) + 15 = x(3x + 4)$

7. $x^2 + 4x + 1 = 0$. Give exact solutions and approximate solutions to three decimal places.

8. $\dfrac{x}{x - 2} + \dfrac{4}{x - 6} = 0$

9. $\dfrac{x}{4} - \dfrac{4}{x} = 2$

10. $15 = \dfrac{8}{x + 2} - \dfrac{6}{x - 2}$

11. Solve $x^2 + 4x + 1 = 0$ by completing the square. Show your work. [11.1b]

12. *Hang Time.* Use the function $V(T) = 48T^2$. A basketball player has a vertical leap of 39 in. What is his hang time? [11.1c]

13. *VCR Screen.* The width of a rectangular screen on a portable VCR is 5 cm less than the length. The area is 126 cm^2. Find the length and the width. [11.3a]

14. *Picture Matting.* A picture mat measures 12 in. by 16 in.; 140 in^2 of picture shows. Find the width of the mat. [11.3a]

15. *Motorcycle Travel.* During the first part of a trip, a motorcyclist travels 50 mi at a certain speed. The rider travels 80 mi on the second part of the trip at a speed that is 10 mph slower. The total time for the trip is 3 hr. What is the speed on each part of the trip? [11.3a]

Determine the nature of the solutions of the equation. [11.4a]

16. $x^2 + 3x - 6 = 0$

17. $x^2 + 2x + 5 = 0$

Write a quadratic equation having the given solutions. [11.4b]

18. $\frac{1}{5}, -\frac{3}{5}$

19. -4, only solution

Solve for the indicated letter. [11.3b]

20. $N = 3\pi\sqrt{\dfrac{1}{p}}$, for p

21. $2A = \dfrac{3B}{T^2}$, for T

Solve. [11.4c]

22. $x^4 - 13x^2 + 36 = 0$

23. $15x^{-2} - 2x^{-1} - 1 = 0$

24. $(x^2 - 4)^2 - (x^2 - 4) - 6 = 0$

25. $x - 13\sqrt{x} + 36 = 0$

For each quadratic function, find and label **(a)** the vertex, **(b)** the line of symmetry, and **(c)** the maximum or minimum value. Then **(d)** graph the function. [11.5c], [11.6a]

26. $f(x) = -\frac{1}{2}(x - 1)^2 + 3$

27. $f(x) = x^2 - x + 6$

28. $f(x) = -3x^2 - 12x - 8$

29. Find the x- and y-intercepts: [11.6b]
$f(x) = x^2 - 9x + 14.$

30. What is the minimum product of two numbers whose difference is 22? What numbers yield this product? [11.7a]

31. Find the quadratic function that fits the data points $(0, -2)$, $(1, 3)$, and $(3, 7)$. [11.7b]

32. *Hotel Occupancy.* Hotel occupancy has risen and fallen in recent years. Percentages of rooms occupied in various recent years are shown in the graph below. [11.7b]

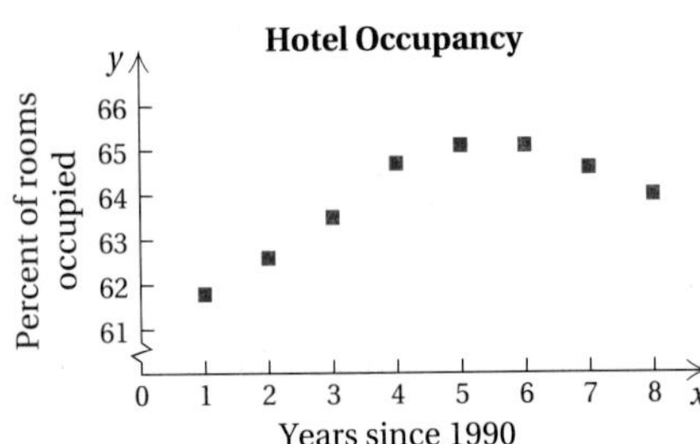

Source: Coopers & Lybrand Lodging Research Network

a) Use the data points $(1, 61.8)$, $(5, 65.1)$, and $(8, 64)$ to fit a quadratic function to the data.
b) Use the quadratic function to predict the percentage of rooms occupied in 2000 and in 2004.

Solve. [11.8a, b]

33. $(x + 2)(x - 1)(x - 2) > 0$

34. $\dfrac{(x + 4)(x - 1)}{(x + 2)} < 0$

Skill Maintenance

35. *SAT Verbal Scores.* The following table lists SAT verbal scores for recent years. [7.6b]

Years Since 1990, x	SAT Verbal Score
1	499
2	500
3	500
4	499
5	504
6	505

Source: The College Board

a) Use the data points $(1, 499)$ and $(6, 505)$ to fit a linear function to the data.
b) Use the linear function to predict SAT verbal scores in 2000 and in 2004.

36. Add and simplify: [6.3a]
$$\frac{x}{x^2 - 3x + 2} + \frac{2}{x^2 - 5x + 6}.$$

37. Multiply and simplify: [10.3a]
$\sqrt[3]{9t^6}\,\sqrt[3]{3s^4t^9}.$

38. Solve: $\sqrt{5x - 1} + \sqrt{2x} = 5.$ [10.6b]

Synthesis

39. Explain as many characteristics as you can of the graph of the quadratic function $f(x) = ax^2 + bx + c$. [11.5a, b, c], [11.6a, b]

40. Explain how the x-intercepts of a quadratic function can be used to help find the vertex of the function. What piece of information would still be missing? [11.6a, b]

41. A quadratic function has x-intercepts $(-3, 0)$ and $(5, 0)$ and y-intercept $(0, -7)$. Find an equation for the function. What is its maximum or minimum value? [11.6a, b]

42. Find h and k such that $3x^2 - hx + 4k = 0$, the sum of the solutions is 20, and the product of the solutions is 80. [11.2a], [11.7b]

43. The average of two numbers is 171. One of the numbers is the square root of the other. Find the numbers. [11.3a]

Test: Chapter 11

1. a) Solve: $3x^2 - 4 = 0$.
   b) Find the x-intercepts of $f(x) = 3x^2 - 4$.

Solve.

2. $x^2 + x + 1 = 0$

3. $x - 8\sqrt{x} + 7 = 0$

4. $4x(x - 2) - 3x(x + 1) = -18$

5. $x^4 - 5x^2 + 5 = 0$

6. $x^2 + 4x = 2$. Give exact solutions and approximate solutions to three decimal places.

7. $\dfrac{1}{4 - x} + \dfrac{1}{2 + x} = \dfrac{3}{4}$

8. Solve $x^2 - 4x + 1 = 0$ by completing the square. Show your work.

9. *Free-Falling Objects.* The Peachtree Plaza in Atlanta, Georgia, is 723 ft tall. Use the function $s(t) = 16t^2$ to approximate how long it would take an object to fall from the top.

10. *Marine Travel.* The Columbia River flows at a rate of 2 mph for the length of a popular boating route. In order for a motorized dinghy to travel 3 mi upriver and then return in a total of 4 hr, how fast must the boat be able to travel in still water?

11. *Memory Board.* A computer-parts company wants to make a rectangular memory board that has a perimeter of 28 cm. What dimensions will allow the board to have a maximum area?

12. *Hang Time.* Use the function $V(T) = 48T^2$. A basketball player has a vertical leap of 35 in. What is his hang time?

13. Determine the nature of the solutions of the equation $x^2 + 5x + 17 = 0$.

14. Write a quadratic equation having the solutions $\sqrt{3}$ and $3\sqrt{3}$.

15. Solve $V = 48T^2$ for T.

For each quadratic function, find and label **(a)** the vertex, **(b)** the line of symmetry, and **(c)** the maximum or minimum value. Then **(d)** graph the function.

16. $f(x) = -x^2 - 2x$

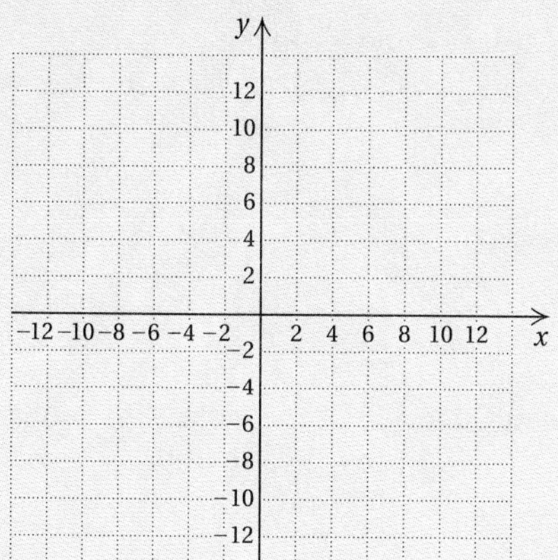

17. $f(x) = 4x^2 - 24x + 41$

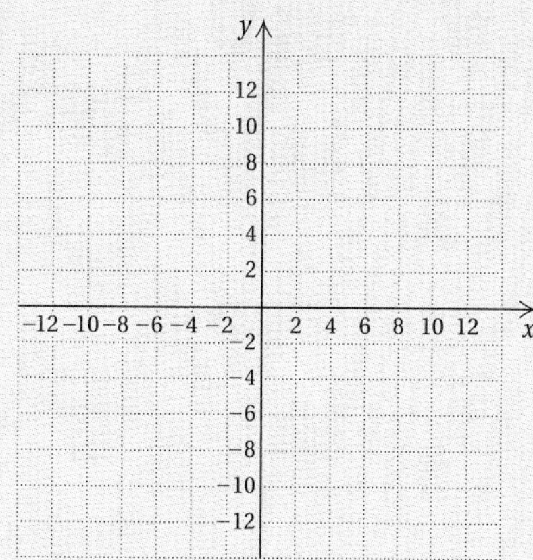

Answers

1. a) ______

b) ______

2. ______

3. ______

4. ______

5. ______

6. ______

7. ______

8. ______

9. ______

10. ______

11. ______

12. ______

13. ______

14. ______

15. ______

16. a) ______

b) ______

c) ______

d) ______

17. a) ______

b) ______

c) ______

d) ______

Answers

18. ______

19. ______

20. ______

21. a) ______

b) ______

22. ______

23. ______

24. a) ______

b) ______

25. ______

26. ______

27. ______

28. ______

29. ______

30. ______

18. Find the x- and y-intercepts:
$f(x) = -x^2 + 4x - 1$.

19. What is the maximum value of two numbers whose difference is 8? What numbers yield this product?

20. Find the quadratic function that fits the data points (0, 0), (3, 0), and (5, 2).

21. *Hotel Construction.* Hotel-room construction has risen and fallen in recent years. The numbers of rooms constructed in various recent years are shown in the graph at right.

a) Use the data points (5, 84.8), (7, 127.5), and (9, 114.2) to fit a quadratic function to the data.

b) Use the quadratic function to predict the number of rooms constructed in 2000 and in 2001.

Construction of Hotel Rooms

Source: Coopers & Lybrand Lodging Research Network

Solve.

22. $x^2 < 6x + 7$

23. $\frac{x - 5}{x + 3} < 0$

Skill Maintenance

24. *SAT Math Scores.* The following table lists SAT math scores for recent years.

Years Since 1990, x	SAT Math Score
1	500
2	501
3	503
4	504
5	506
6	508

Source: The College Board

a) Use the data points (1, 500) and (6, 508) to fit a linear function to the data.

b) Use the linear function to predict SAT math scores in 2000 and in 2004.

25. Subtract and simplify:

$$\frac{x}{x^2 + 15x + 56} - \frac{7}{x^2 + 13x + 42}.$$

26. Multiply and simplify:
$\sqrt[4]{2a^2b^3}\,\sqrt[4]{a^4b}$.

27. Solve: $\sqrt{x + 3} = x - 3$.

Synthesis

28. A quadratic function has x-intercepts $(-2, 0)$ and $(7, 0)$ and y-intercept $(0, 8)$. Find an equation for the function. What is its maximum or minimum value?

29. One solution of $kx^2 + 3x - k = 0$ is -2. Find the other solution.

30. Solve: $x^8 - 20x^4 + 64 = 0$.

12

Exponential and Logarithmic Functions

An Application

The world is experiencing an exponential demand for lumber. The amount of timber N, in billions of cubic feet, consumed t years after 1997, can be approximated by

$$N(t) = 62(1.018)^t,$$

where $t = 0$ corresponds to 1997 (***Source***: U. N. Food and Agricultural Organization; American Forest and Paper Association). Estimate the amount of timber to be consumed in 2000.

This problem appears as Exercise 27 in Exercise Set 12.1.

The Mathematics

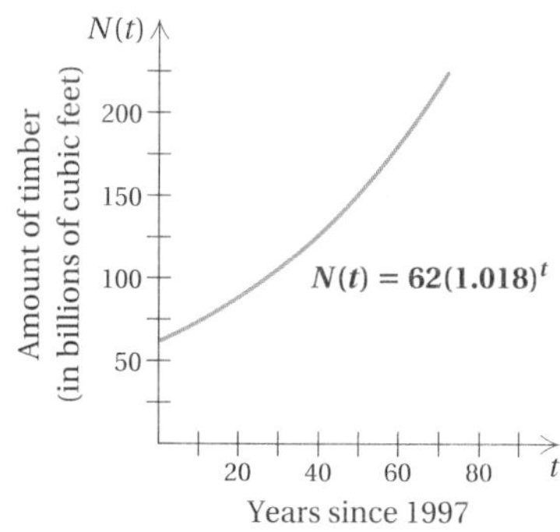

Since the year 2000 is 3 yr after 1997, we substitute 3 for t:

$$N(3) = 62(1.018)^3 \approx 65.4.$$

We estimate that in 2000, 65.4 billion cubic feet of timber will be consumed.

World Wide Web For more information, visit us at www.mathmax.com

Introduction

The functions that we consider in this chapter are important for their rich applications to many fields. We will look at such applications as compound interest, population growth, and radioactive decay, but there are many others.

Exponents are the basis of the theory in this chapter. We define some functions having variable exponents, called exponential functions. The logarithmic functions follow from the exponential functions.

Pretest: Chapter 12

Graph.

1. $f(x) = 2^x$

2. $f(x) = \log_2 x$

3. Convert to a logarithmic equation:

$a^3 = 1000.$

4. Convert to an exponential equation:

$\log_2 32 = t.$

5. Express as a single logarithm:

$2 \log_a M + \log_a N - \frac{1}{3} \log_a Q.$

6. Express in terms of logarithms of x, y, and z:

$\log_a \sqrt[4]{\frac{x^3 y}{z^2}}.$

Solve.

7. $\log_x \frac{1}{4} = -2$

8. $\log_5 x = 2$

9. $3^x = 8.6$

10. $16^{x-3} = 4$

11. $\log (x^2 - 4) - \log (x - 2) = 1$

12. $\ln x = 0$

Find each of the following using a calculator.

13. $\log 714$

14. $\ln 0.0008464$

15. $\ln 1$

16. $e^{4.2}$

17. $e^{-2.13}$

18. $\log (0.0234)$

19. *Consumer Price Index.* The consumer price index compares the cost of goods and services over various years using \$100 worth of goods and services in 1967 as a base. It is approximated by the function

$P(t) = \$100e^{0.06t},$

where t is the number of years since 1967. Goods and services that cost \$100 in 1967 will cost how much in 2000?

20. *Carbon Dating.* How old is an animal bone that has lost 76% of its carbon-14?

Objectives for Retesting

The objectives to be tested in addition to the material in this chapter are as follows.

[10.8b, c, d, e] Add, subtract, multiply, and divide complex numbers. Write expressions involving powers of i in the form $a + bi$.

[11.4c] Solve equations that are quadratic in form.

[11.6a] For a quadratic function, find the vertex, the line of symmetry, and the maximum or minimum value, and graph the function.

[11.7b] Fit a quadratic function to a set of data to form a mathematical model, and solve related applied problems.

12.1 Exponential Functions

The rapidly rising graph shown below approximates the graph of an *exponential function.* We will consider such functions and some of their applications.

World Population Growth

8 billion* 2020
6 billion 1998
5 billion 1987
4 billion 1974
3 billion 1960
2 billion 1927
1 billion 1804

* Projected

1800 1900 2000 Year

Source: U.S. Bureau of the Census

Objectives

a Graph exponential equations and functions.

b Graph exponential equations in which x and y have been interchanged.

c Solve applied problems involving applications of exponential functions and their graphs.

For Extra Help

TAPE 20 | MAC WIN | CD-ROM

a Graphing Exponential Functions

We now develop the meaning of exponential expressions with irrational exponents. In Chapter 10, we gave meaning to exponential expressions with rational-number exponents such as

$$8^{1/4},\quad 3^{-3/4},\quad 7^{2.34},\quad 5^{1.73}.$$

For example, $5^{1.73}$, or $5^{173/100}$, or $\sqrt[100]{5^{173}}$, means to raise 5 to the 173rd power and then take the 100th root. Examples of expressions with irrational exponents are

$$5^{\sqrt{3}},\quad 7^{\pi},\quad 9^{-\sqrt{2}}.$$

Since we can approximate irrational numbers with decimal approximations, we can also approximate expressions with irrational exponents. For example, consider $5^{\sqrt{3}}$. We know that $5^{\sqrt{3}} \approx 5^{1.73} = \sqrt[100]{5^{173}}$. As rational values of r get close to $\sqrt{3}$, 5^r gets close to some real number. Note the following:

r closes in on $\sqrt{3}$.	5^r closes in on some real number p.
r	5^r
$1 < \sqrt{3} < 2$	$5 = 5^1 < p < 5^2 = 25$
$1.7 < \sqrt{3} < 1.8$	$15.426 = 5^{1.7} < p < 5^{1.8} = 18.119$
$1.73 < \sqrt{3} < 1.74$	$16.189 = 5^{1.73} < p < 5^{1.74} = 16.452$
$1.732 < \sqrt{3} < 1.733$	$16.241 = 5^{1.732} < p < 5^{1.733} = 16.267$

As r closes in on $\sqrt{3}$, 5^r closes in on some real number p. We define $5^{\sqrt{3}}$ to be that number p. To seven decimal places, we have

$$5^{\sqrt{3}} \approx 16.2424508.$$

Any positive irrational exponent can be defined in a similar way. Negative irrational exponents are then defined in the same way as negative integer exponents. Then the expression a^x has meaning for any real number x. The general laws of exponents still hold, but we will not prove that here.

Calculator Spotlight

We can approximate $5^{\sqrt{3}}$ using the scientific keys.

Exercises

Approximate.

1. $5^{\sqrt{3}}$ **2.** 7^{π} **3.** $9^{-\sqrt{2}}$

1. Graph: $f(x) = 3^x$.

a) Complete this table of solutions.

x	$f(x)$
0	
1	
2	
3	
−1	
−2	
−3	

b) Plot the points from the table and connect them with a smooth curve.

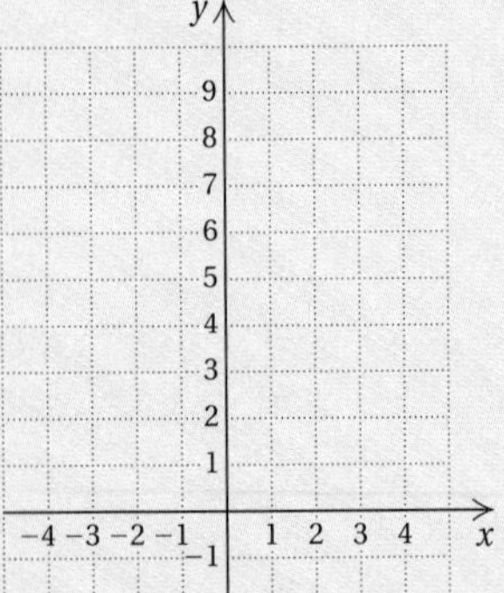

Answers on page A-52

We now define exponential functions.

> **EXPONENTIAL FUNCTION**
>
> The function $f(x) = a^x$, where a is a positive constant different from 1, is called the **exponential function,** base a.

We restrict the base a to be positive to avoid the possibility of taking even roots of negative numbers such as the square root of -1, $(-1)^{1/2}$, which is not a real number. We restrict the base from being 1 because for $a = 1$, $f(x) = 1^x = 1$, which is a constant.

The following are examples of exponential functions:

$$f(x) = 2^x, \qquad f(x) = \left(\tfrac{1}{2}\right)^x, \qquad f(x) = (0.4)^x.$$

Note that in contrast to polynomial functions like $f(x) = x^2$ and $f(x) = x^3$, the variable is *in the exponent.* Let's consider graphs of exponential functions.

Example 1 Graph the exponential function $f(x) = 2^x$.

We compute some function values and list the results in a table. It is a good idea to begin by letting $x = 0$.

$$f(0) = 2^0 = 1;$$
$$f(1) = 2^1 = 2;$$
$$f(2) = 2^2 = 4;$$
$$f(3) = 2^3 = 8;$$
$$f(-1) = 2^{-1} = \frac{1}{2^1} = \frac{1}{2};$$
$$f(-2) = 2^{-2} = \frac{1}{2^2} = \frac{1}{4};$$
$$f(-3) = 2^{-3} = \frac{1}{2^3} = \frac{1}{8}.$$

x	$f(x)$
0	1
1	2
2	4
3	8
−1	$\frac{1}{2}$
−2	$\frac{1}{4}$
−3	$\frac{1}{8}$

Next, we plot these points and connect them with a smooth curve.

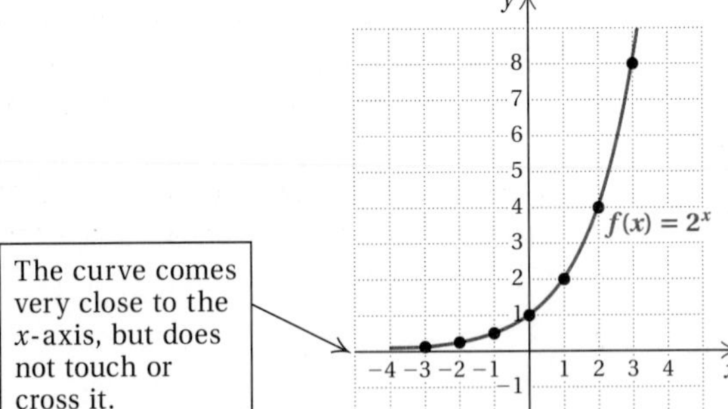

The curve comes very close to the x-axis, but does not touch or cross it.

In graphing, be sure to plot enough points to determine how steeply the curve rises.

Note that as x increases, the function values increase indefinitely. As x decreases, the function values decrease, getting very close to 0. The x-axis, or the line $y = 0$, is an *asymptote,* meaning here that as x gets very small, the curve comes very close to but never touches the axis.

Do Exercise 1.

Example 2 Graph the exponential function $f(x) = \left(\frac{1}{2}\right)^x$.

We compute some function values and list the results in a table. Before we do so, note that

$$f(x) = \left(\tfrac{1}{2}\right)^x = (2^{-1})^x = 2^{-x}.$$

Then we have

$$f(0) = 2^{-0} = 1;$$

$$f(1) = 2^{-1} = \frac{1}{2^1} = \frac{1}{2};$$

$$f(2) = 2^{-2} = \frac{1}{2^2} = \frac{1}{4};$$

$$f(3) = 2^{-3} = \frac{1}{2^3} = \frac{1}{8};$$

$$f(-1) = 2^{-(-1)} = 2^1 = 2;$$

$$f(-2) = 2^{-(-2)} = 2^2 = 4;$$

$$f(-3) = 2^{-(-3)} = 2^3 = 8.$$

x	$f(x)$
0	1
1	$\frac{1}{2}$
2	$\frac{1}{4}$
3	$\frac{1}{8}$
−1	2
−2	4
−3	8

Next, we plot these points and draw the curve. Note that this graph is a reflection across the y-axis of the graph in Example 1. The line $y = 0$ is again an asymptote.

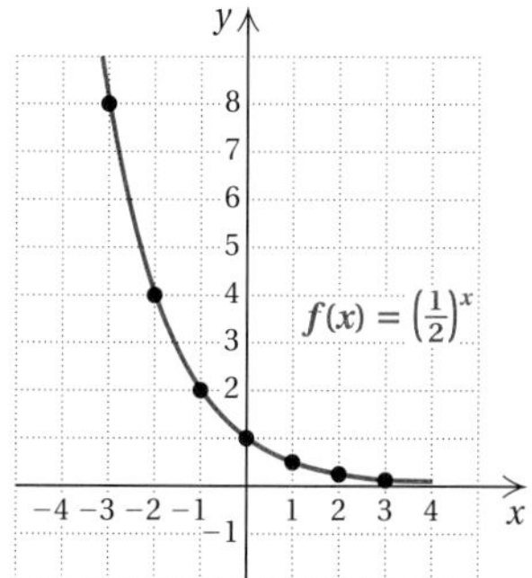

Do Exercise 2.

The preceding examples illustrate exponential functions with various bases. Let's list some of their characteristics. Keep in mind that the definition of an exponential function, $f(x) = a^x$, requires that the base be positive and different from 1.

When $a > 1$, the function $f(x) = a^x$ increases from left to right. The greater the value of a, the steeper the curve. As x gets smaller and smaller, the curve gets closer to the line $y = 0$: It is an asymptote.

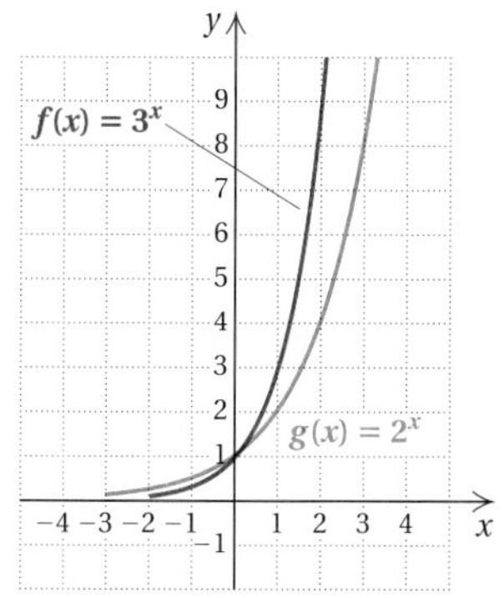

2. Graph: $f(x) = \left(\frac{1}{3}\right)^x$.

a) Complete this table of solutions.

x	$f(x)$
0	
1	
2	
3	
−1	
−2	
−3	

b) Plot the points from the table and connect them with a smooth curve.

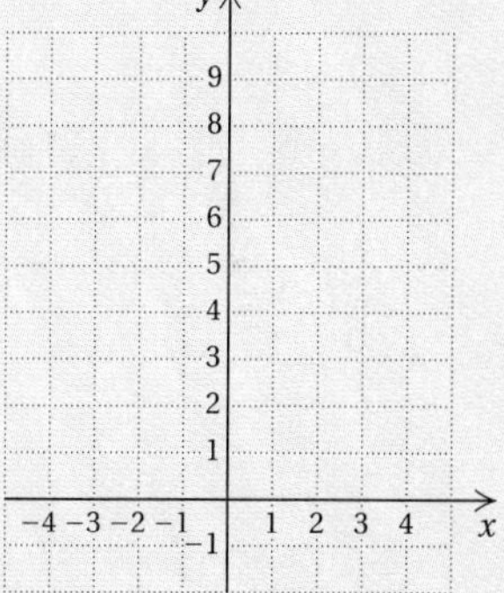

Answers on page A-52

Graph.

3. $f(x) = 4^x$

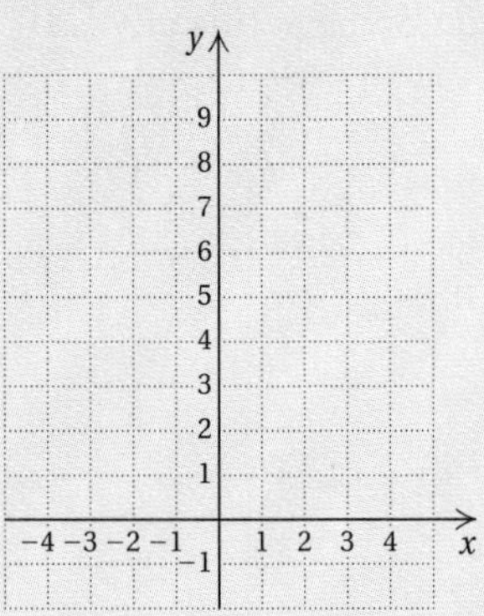

4. $f(x) = \left(\frac{1}{4}\right)^x$

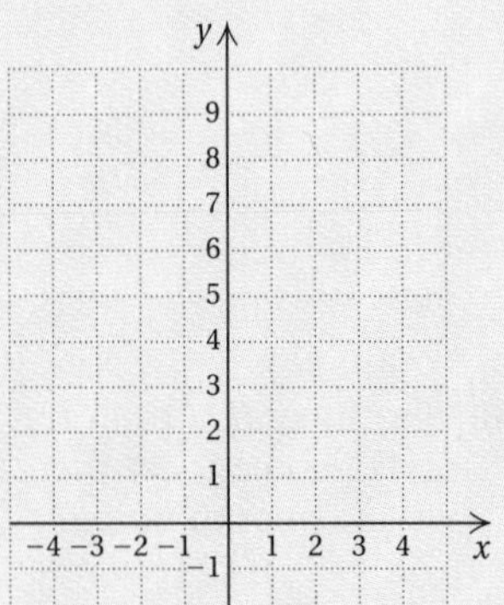

5. Graph: $f(x) = 2^{x+2}$.

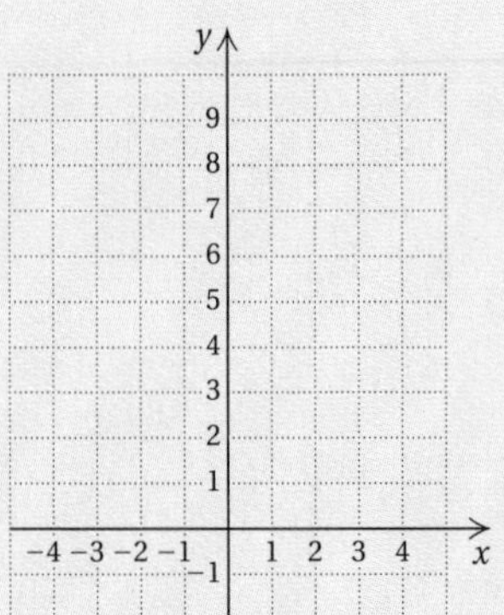

Answers on page A-52

When $0 < a < 1$, the function $f(x) = a^x$ decreases from left to right. As a approaches 1, the curve becomes less steep. As x gets larger and larger, the curve gets closer to the line $y = 0$: It is an asymptote.

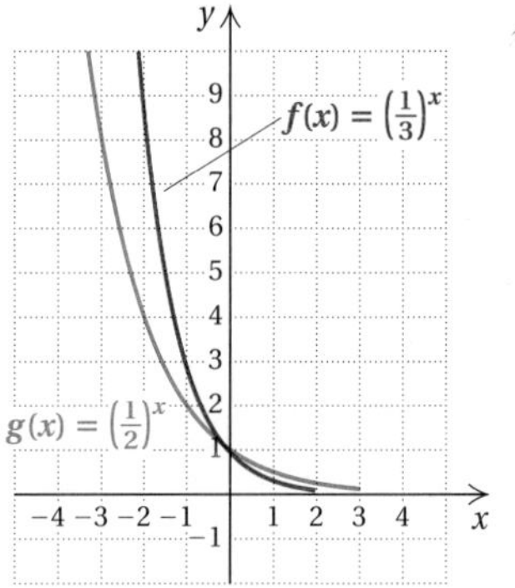

Note that all functions $f(x) = a^x$ go through the point (0, 1). That is, the y-intercept is (0, 1).

Do Exercises 3 and 4.

Example 3 Graph: $f(x) = 2^{x-2}$.

We construct a table of values. Then we plot the points and connect them with a smooth curve. Be sure to note that $x - 2$ is the *exponent.*

$$f(0) = 2^{0-2} = 2^{-2} = \frac{1}{2^2} = \frac{1}{4};$$

$$f(1) = 2^{1-2} = 2^{-1} = \frac{1}{2^1} = \frac{1}{2};$$

$$f(2) = 2^{2-2} = 2^0 = 1;$$

$$f(3) = 2^{3-2} = 2^1 = 2;$$

$$f(4) = 2^{4-2} = 2^2 = 4;$$

$$f(-1) = 2^{-1-2} = 2^{-3} = \frac{1}{2^3} = \frac{1}{8};$$

$$f(-2) = 2^{-2-2} = 2^{-4} = \frac{1}{2^4} = \frac{1}{16}.$$

x	$f(x)$
0	$\frac{1}{4}$
1	$\frac{1}{2}$
2	1
3	2
4	4
−1	$\frac{1}{8}$
−2	$\frac{1}{16}$

The graph looks just like the graph of $g(x) = 2^x$, but it is translated 2 units to the right. The y-intercept of $g(x) = 2^x$ is (0, 1). The y-intercept of $f(x) = 2^{x-2}$ is $\left(0, \frac{1}{4}\right)$. The line $y = 0$ is still an asymptote.

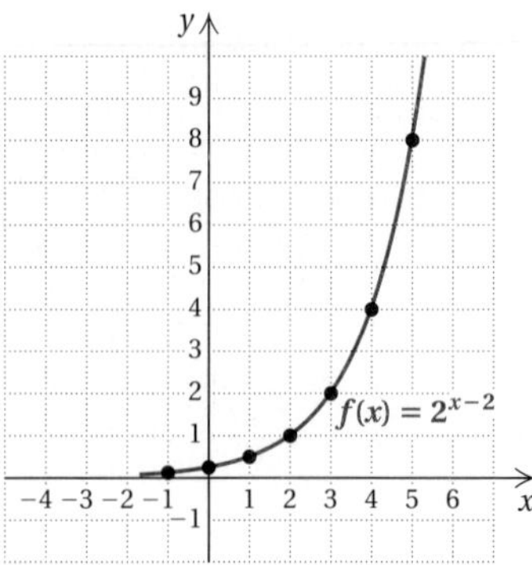

Do Exercise 5.

Example 4 Graph: $f(x) = 2^x - 3$.

We construct a table of values. Then we plot the points and connect them with a smooth curve. Note that the only expression in the exponent is x.

$$f(0) = 2^0 - 3 = 1 - 3 = -2;$$
$$f(1) = 2^1 - 3 = 2 - 3 = -1;$$
$$f(2) = 2^2 - 3 = 4 - 3 = 1;$$
$$f(3) = 2^3 - 3 = 8 - 3 = 5;$$
$$f(4) = 2^4 - 3 = 16 - 3 = 13;$$
$$f(-1) = 2^{-1} - 3 = \frac{1}{2} - 3 = -\frac{5}{2};$$
$$f(-2) = 2^{-2} - 3 = \frac{1}{4} - 3 = -\frac{11}{4}.$$

x	$f(x)$
0	-2
1	-1
2	1
3	5
4	13
-1	$-\frac{5}{2}$
-2	$-\frac{11}{4}$

The graph looks just like the graph of $g(x) = 2^x$, but it is translated down 3 units. The y-intercept is $(0, -2)$. The line $y = -3$ is an asymptote. The curve gets closer to this line as x gets smaller and smaller.

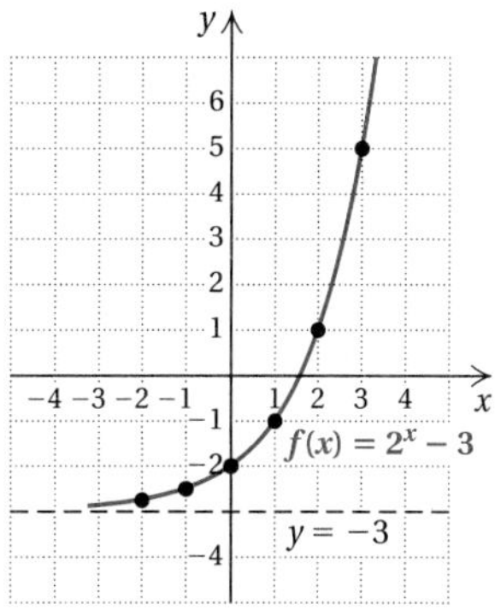

Do Exercise 6.

6. Graph: $f(x) = 2^x - 4$.

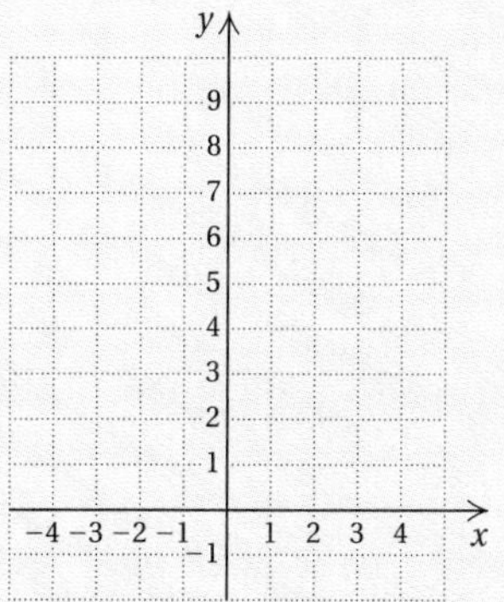

Answer on page A-52

Calculator Spotlight

Graphers are especially helpful when we are graphing exponential functions because bases may not be whole numbers and function values can quickly become very large or very small. To graph

$$f(x) = 5000(1.075)^x,$$

we enter $y_1 = 5000(1.075) \wedge x$. Because y-values are *always* positive and will become quite large, we use the viewing window $[-10, 10, 0, 15000]$, with Yscl = 1000.

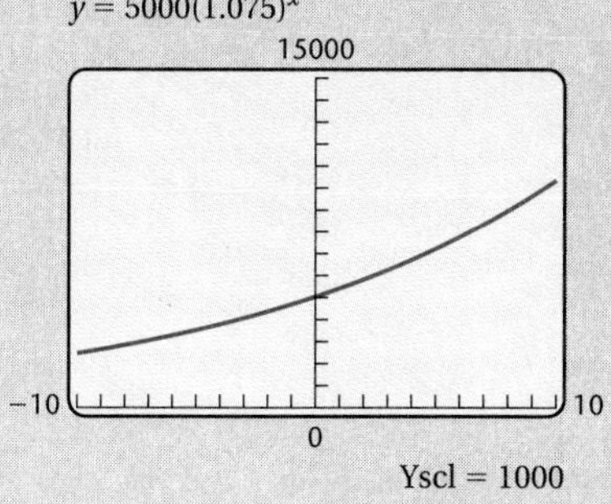

Exercises

Graph the given pair of exponential functions. Adjust the viewing window and scale as needed. Then compare each pair of graphs.

1. $y_1 = \left(\frac{5}{2}\right)^x$, $y_2 = \left(\frac{2}{5}\right)^x$
2. $y_1 = 3.2^x$, $y_2 = 3.2^{-x}$
3. $y_1 = 9.34^x$, $y_2 = 9.34^{-x}$
4. $y_1 = \left(\frac{3}{7}\right)^x$, $y_2 = \left(\frac{7}{3}\right)^x$
5. $y_1 = 5000(1.08)^x$, $y_2 = 5000(1.08)^{x-3}$
6. $y_1 = 2000(1.09)^x$, $y_2 = 2000(1.09)^{x+3}$

7. Graph: $x = 3^y$.

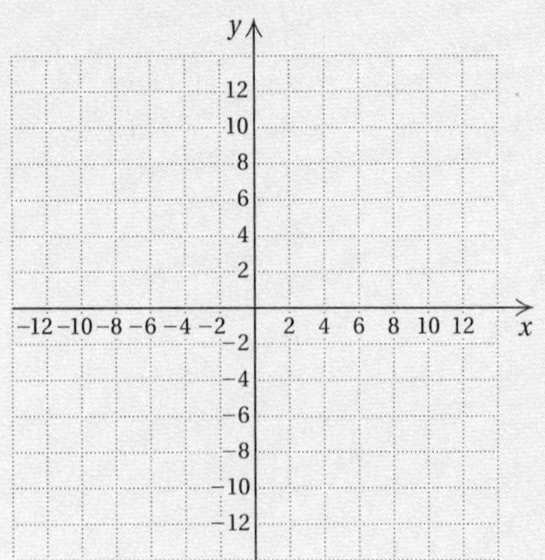

Answer on page A-52

b Equations with *x* and *y* Interchanged

It will be helpful in later work to be able to graph an equation in which the x and the y in $y = a^x$ are interchanged.

Example 5 Graph: $x = 2^y$.

Note that x is alone on one side of the equation. We can find ordered pairs that are solutions more easily by choosing values for y and then computing the x-values.

For $y = 0$, $x = 2^0 = 1$.

For $y = 1$, $x = 2^1 = 2$.

For $y = 2$, $x = 2^2 = 4$.

For $y = 3$, $x = 2^3 = 8$.

For $y = -1$, $x = 2^{-1} = \dfrac{1}{2^1} = \dfrac{1}{2}$.

For $y = -2$, $x = 2^{-2} = \dfrac{1}{2^2} = \dfrac{1}{4}$.

For $y = -3$, $x = 2^{-3} = \dfrac{1}{2^3} = \dfrac{1}{8}$.

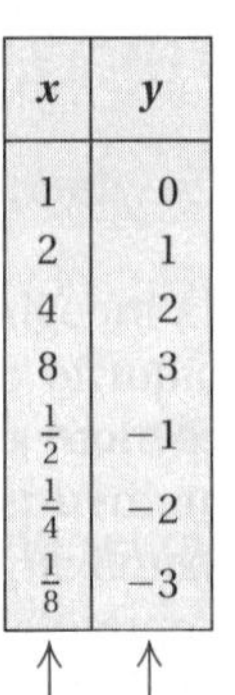

x	y
1	0
2	1
4	2
8	3
$\frac{1}{2}$	−1
$\frac{1}{4}$	−2
$\frac{1}{8}$	−3

(1) Choose values for y.
(2) Compute values for x.

We plot the points and connect them with a smooth curve.

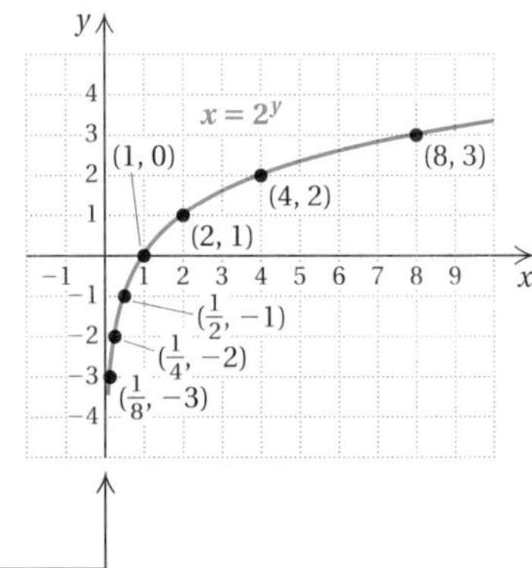

This curve does not touch or cross the y-axis.

Note that this curve looks just like the graph of $y = 2^x$, except that it is reflected, or flipped, across the line $y = x$, as shown below.

$y = 2^x$	
x	y
0	1
1	2
2	4
3	8
−1	$\frac{1}{2}$
−2	$\frac{1}{4}$
−3	$\frac{1}{8}$

$x = 2^y$	
x	y
1	0
2	1
4	2
8	3
$\frac{1}{2}$	−1
$\frac{1}{4}$	−2
$\frac{1}{8}$	−3

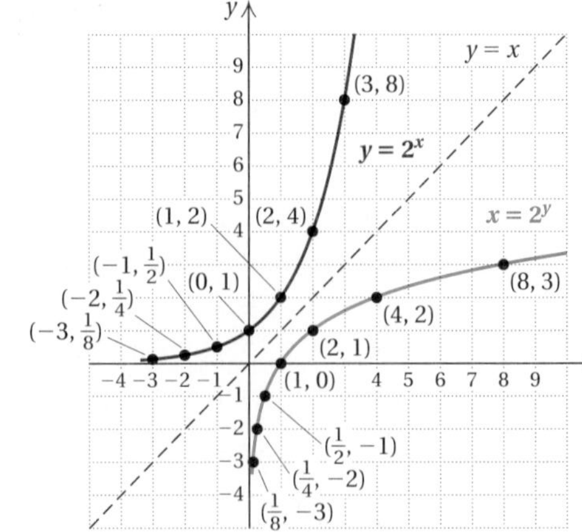

Do Exercise 7.

C Applications of Exponential Functions

Example 6 *Interest Compounded Annually.* The amount of money A that a principal P will grow to after t years at interest rate r, compounded annually, is given by the formula

$$A = P(1 + r)^t.$$

Suppose that \$100,000 is invested at 8% interest, compounded annually.

a) Find a function for the amount in the account after t years.

b) Find the amount of money in the account at $t = 0$, $t = 4$, $t = 8$, and $t = 10$.

c) Graph the function.

We solve as follows:

a) If $P = \$100{,}000$ and $r = 8\% = 0.08$, we can substitute these values and form the following function:

$$A(t) = \$100{,}000(1 + 0.08)^t = \$100{,}000(1.08)^t.$$

b) To find the function values, you might find a calculator with a power key helpful.

$$A(0) = \$100{,}000(1.08)^0 = \$100{,}000(1) = \$100{,}000;$$
$$A(4) = \$100{,}000(1.08)^4 = \$100{,}000(1.36048896) \approx \$136{,}048.90;$$
$$A(8) = \$100{,}000(1.08)^8 = \$100{,}000(1.85093021) \approx \$185{,}093.02;$$
$$A(10) = \$100{,}000(1.08)^{10} = \$100{,}000(2.158924997) \approx \$215{,}892.50$$

c) We use the function values computed in (b) with others, if we wish, to draw the graph as follows. Note that the axes are scaled differently because of the large numbers.

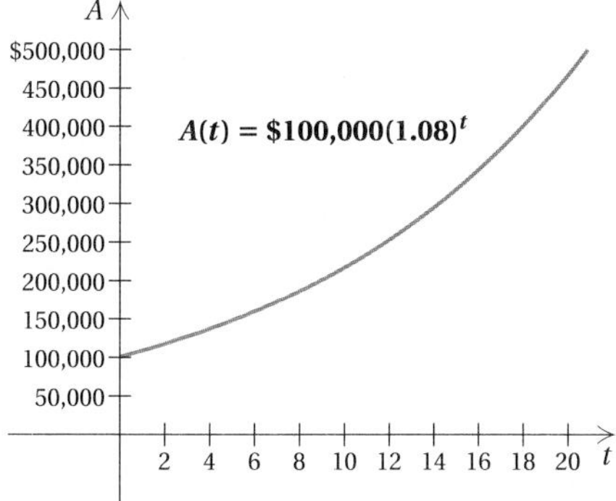

Do Exercise 8.

8. *Interest Compounded Annually.* Suppose that \$40,000 is invested at 5% interest, compounded annually.

a) Find a function for the amount in the account after t years.

b) Find the amount of money in the account at $t = 0$, $t = 4$, $t = 8$, and $t = 10$.

c) Graph the function.

Calculator Spotlight

 Graph

$y_1 = 100{,}000(1.08)^x$.

Then use the TABLE feature or the $Y_1(\ )$ notation to check Example 6(b).

Answers on page A-52

Calculator Spotlight

The Compound-Interest Formula. When interest is compounded quarterly, we can find a formula like the one considered in this section as follows:

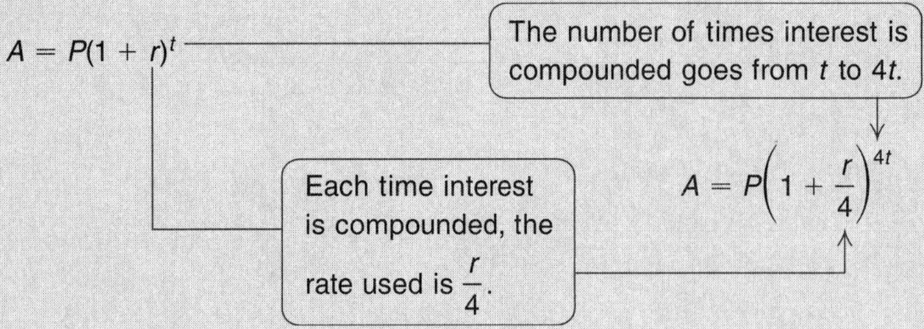

In general, the following formula for compound interest can be used. The amount of money A in an account is given by

$$A = P\left(1 + \frac{r}{n}\right)^{nt},$$

where P is the principal, r is the annual interest rate, n is the number of times per year that interest is compounded, and t is the time, in years.

Example Suppose that \$1000 is invested at 8%, compounded quarterly. How much is in the account at the end of 2 yr?

We use the equation above, substituting 1000 for P, 0.08 for r, 4 for n (compounding quarterly), and 2 for t. Then we get

$$A = P\left(1 + \frac{r}{n}\right)^{nt} = 1000\left(1 + \frac{0.08}{4}\right)^{4\cdot 2} = 1000(1.02)^8 \approx \$1171.66.$$

Exercises

1. Suppose that \$1000 is invested at 6%, compounded semiannually. How much is in the account at the end of 2 yr?
2. Suppose that \$1000 is invested at 6%, compounded monthly. How much is in the account at the end of 2 yr?
3. Suppose that \$20,000 is invested at 4.2%, compounded quarterly. How much is in the account at the end of 10 yr?
4. Suppose that \$420,000 is invested at 4.2%, compounded daily. How much is in the account at the end of 30 yr?
5. Suppose that \$10,000 is invested at 5.4%. How much is in the account at the end of 1 yr, if interest is compounded **(a)** annually? **(b)** semiannually? **(c)** quarterly? **(d)** daily? **(e)** hourly?
6. Suppose that \$10,000 is invested at 3%. How much is in the account at the end of 1 yr, if interest is compounded **(a)** annually? **(b)** semiannually? **(c)** quarterly? **(d)** daily? **(e)** hourly?

Exercise Set 12.1

a Graph.

1. $f(x) = 2^x$

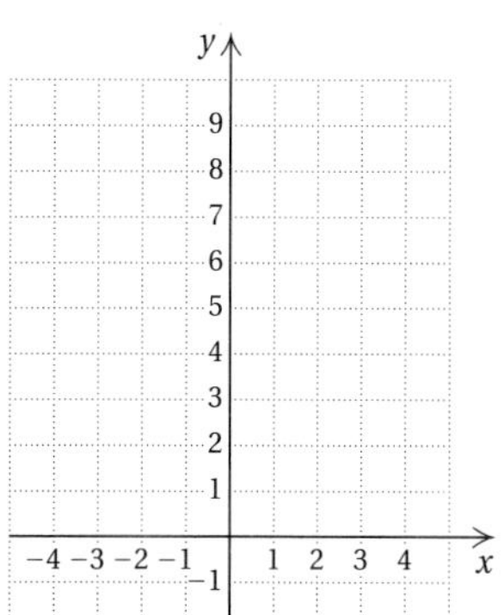

2. $f(x) = 3^x$

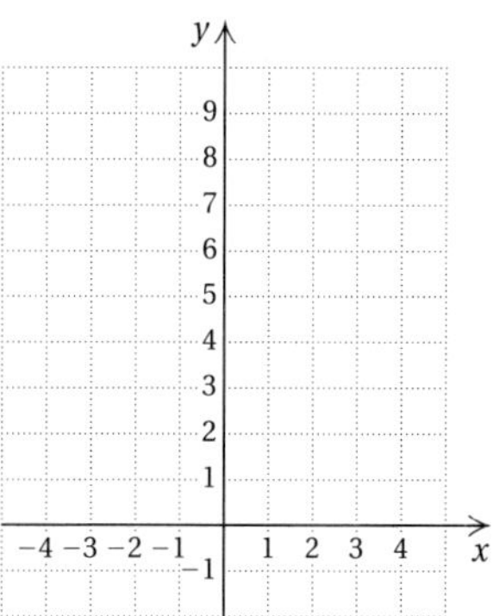

3. $f(x) = 5^x$

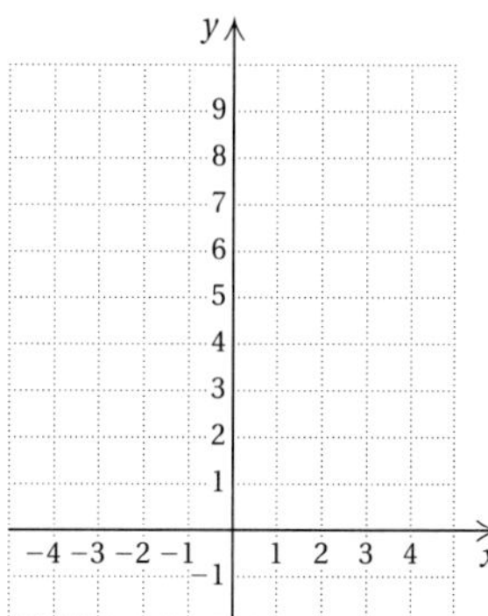

4. $f(x) = 6^x$

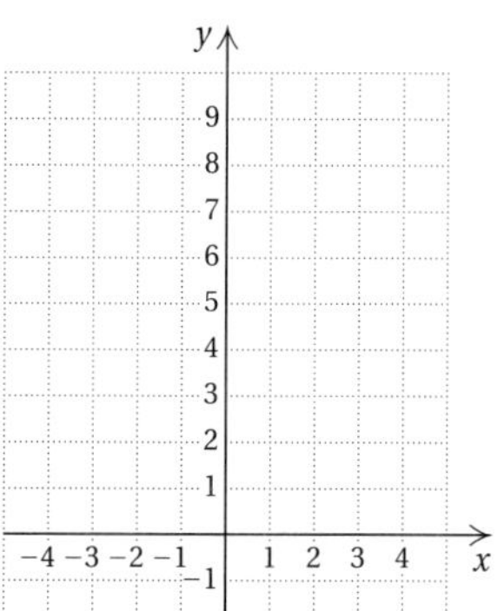

5. $f(x) = 2^{x+1}$

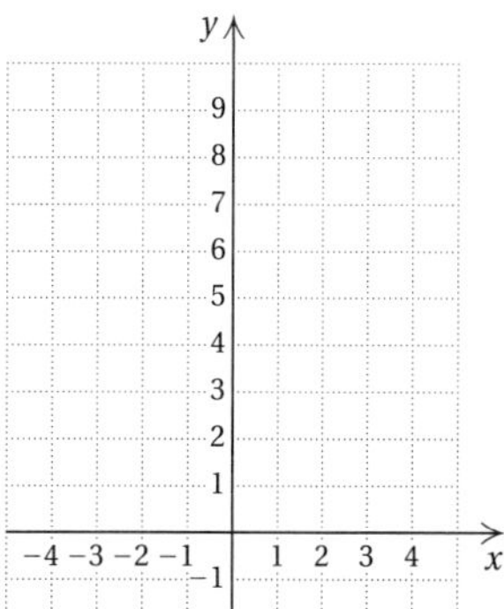

6. $f(x) = 2^{x-1}$

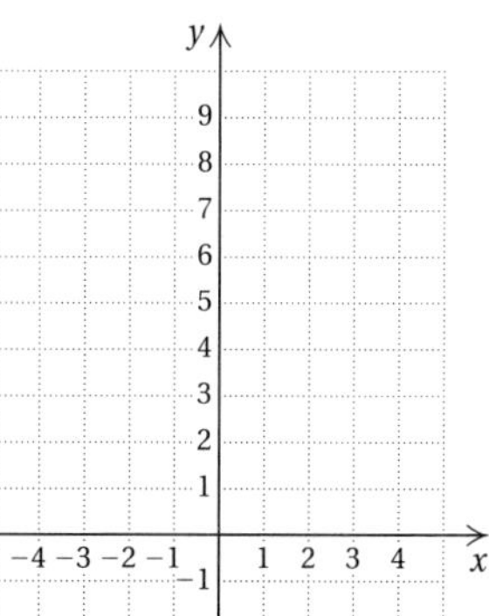

7. $f(x) = 3^{x-2}$

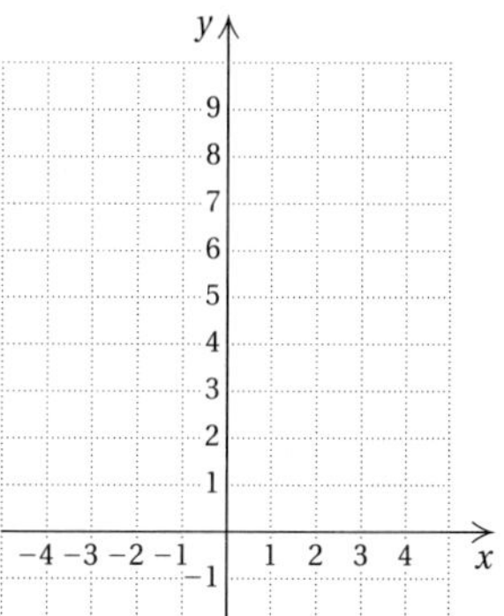

8. $f(x) = 3^{x+2}$

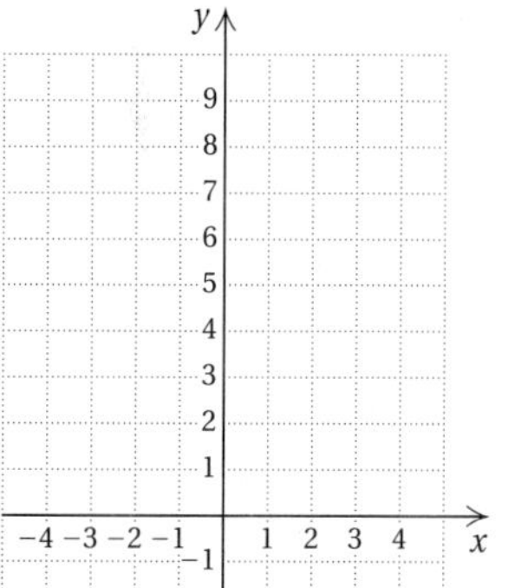

9. $f(x) = 2^x - 3$

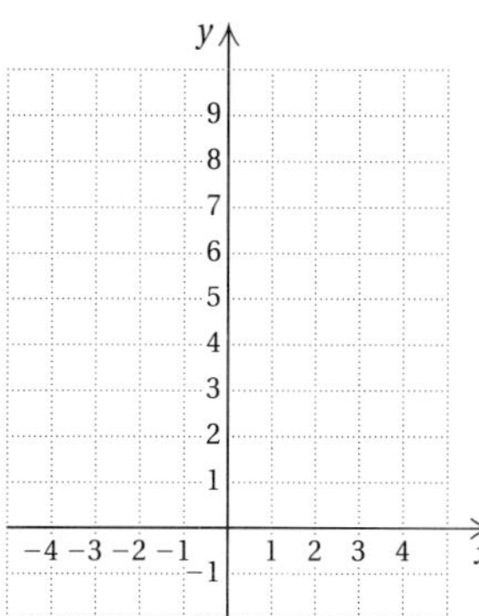

10. $f(x) = 2^x + 1$

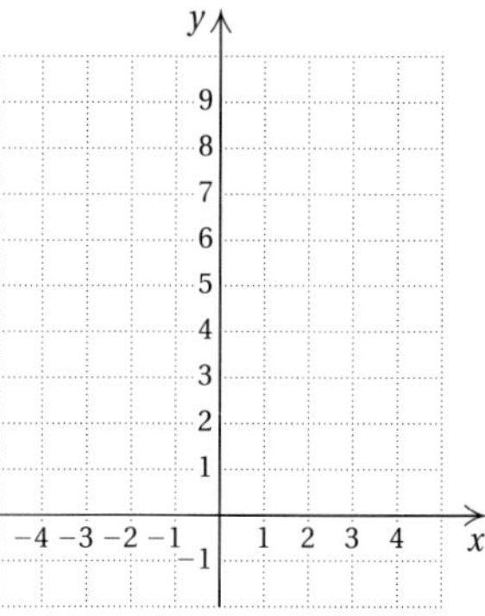

11. $f(x) = 5^{x+3}$

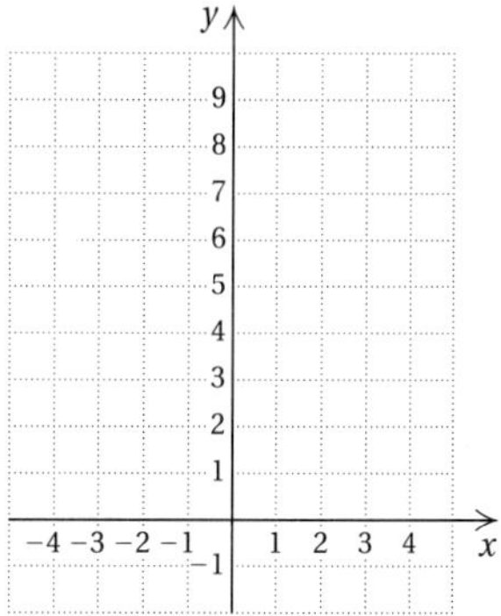

12. $f(x) = 6^{x-4}$

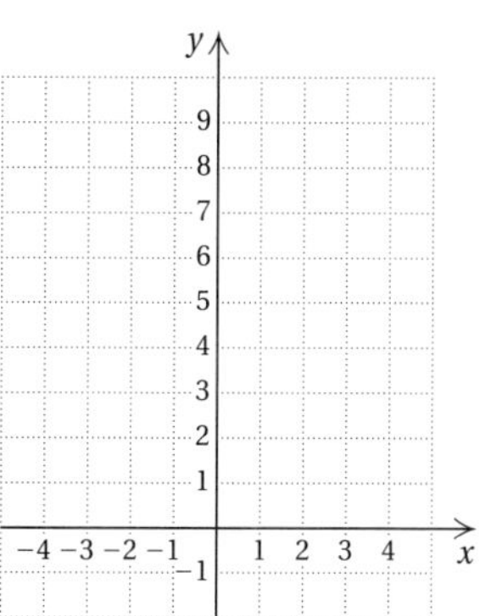

13. $f(x) = \left(\frac{1}{2}\right)^x$

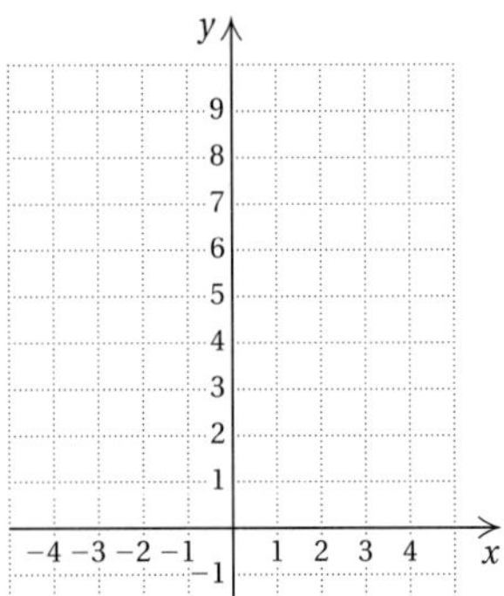

14. $f(x) = \left(\frac{1}{3}\right)^x$

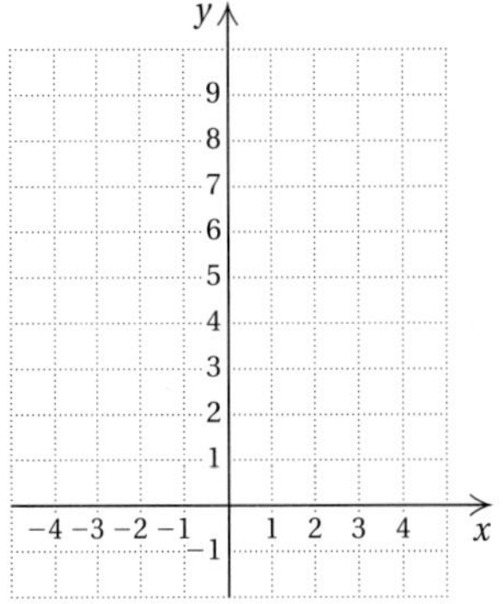

15. $f(x) = \left(\frac{1}{5}\right)^x$

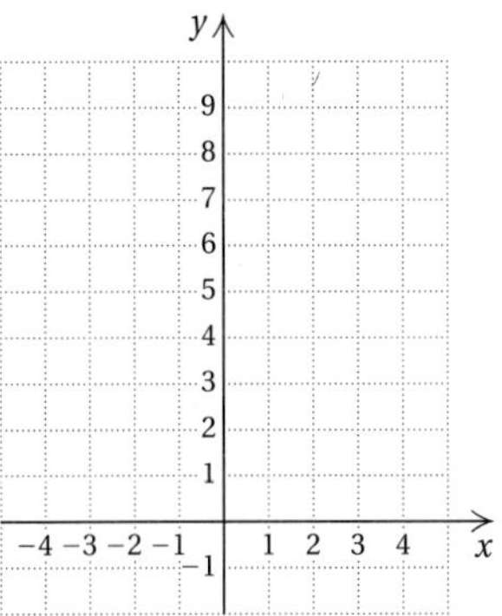

16. $f(x) = \left(\frac{1}{4}\right)^x$

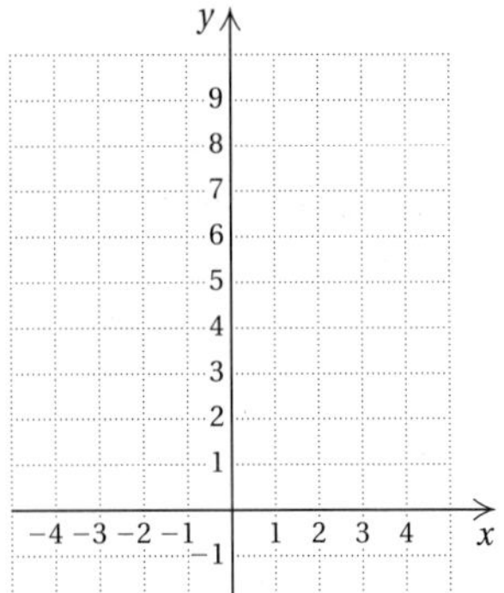

17. $f(x) = 2^{2x-1}$

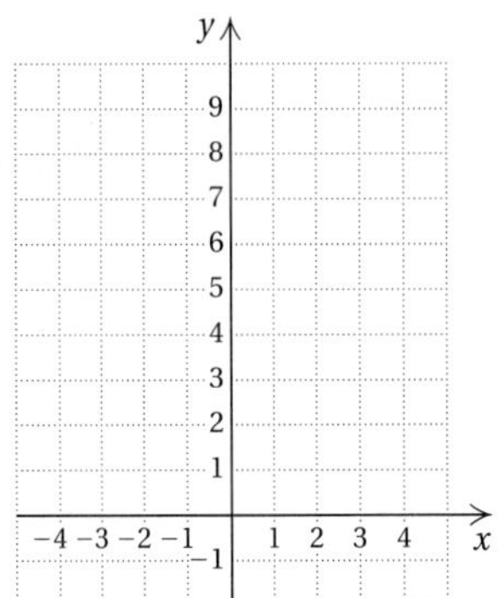

18. $f(x) = 3^{3-x}$

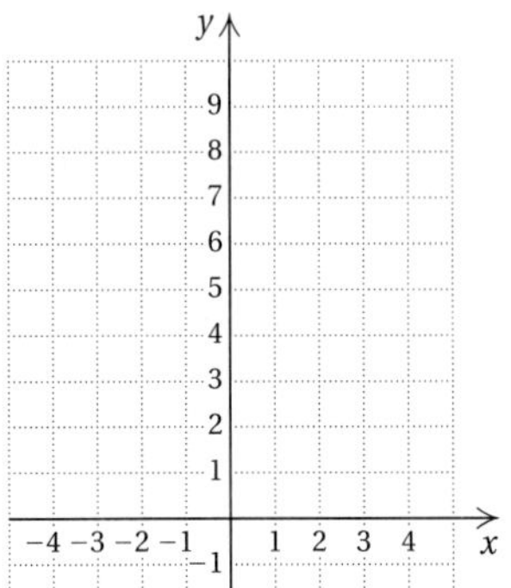

b Graph.

19. $x = 2^y$

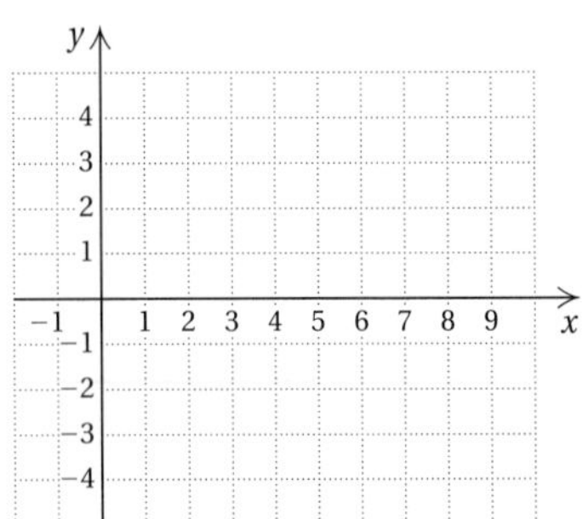

20. $x = 6^y$

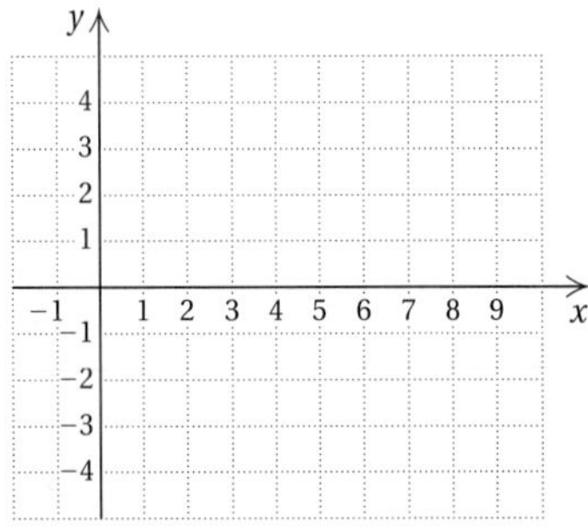

21. $x = \left(\frac{1}{2}\right)^y$

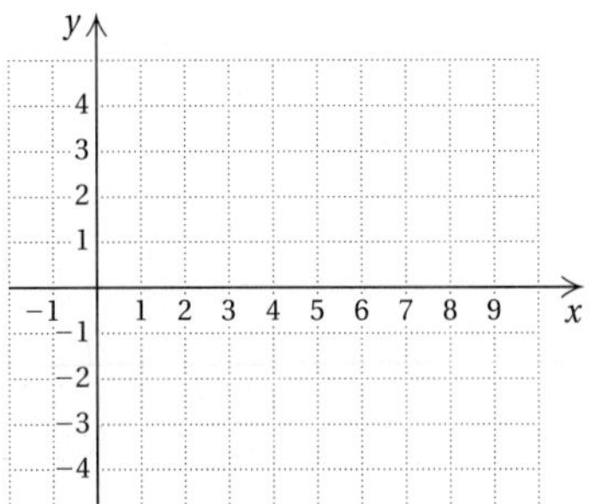

22. $x = \left(\frac{1}{3}\right)^y$

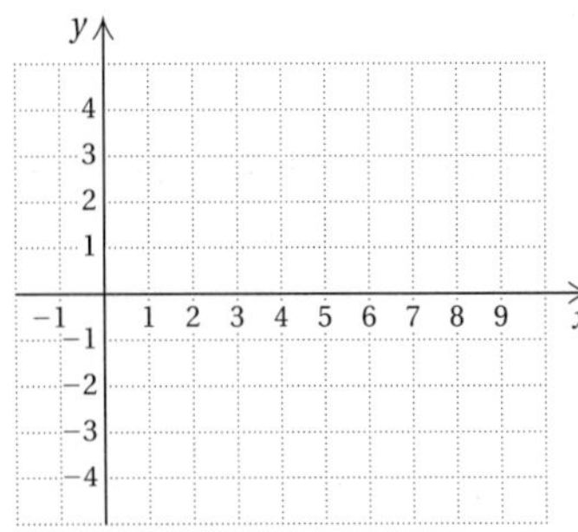

23. $x = 5^y$

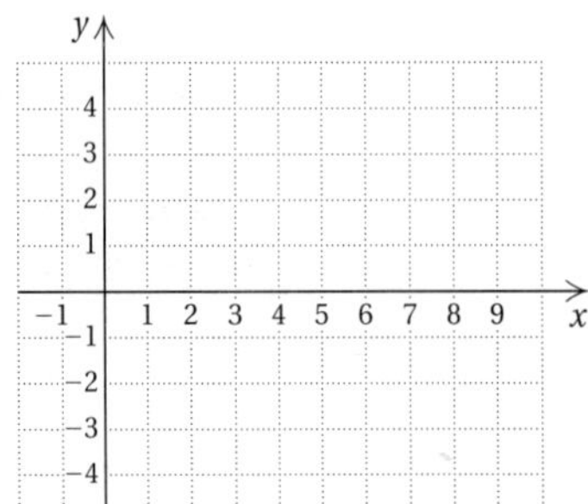

24. $x = \left(\frac{2}{3}\right)^y$

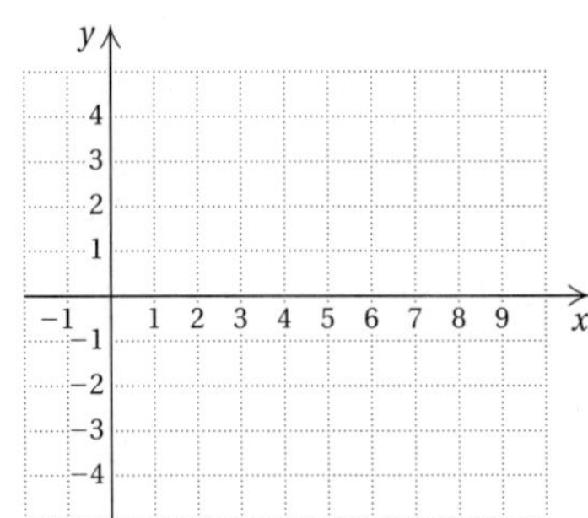

Graph both equations using the same set of axes.

25. $y = 2^x, \quad x = 2^y$

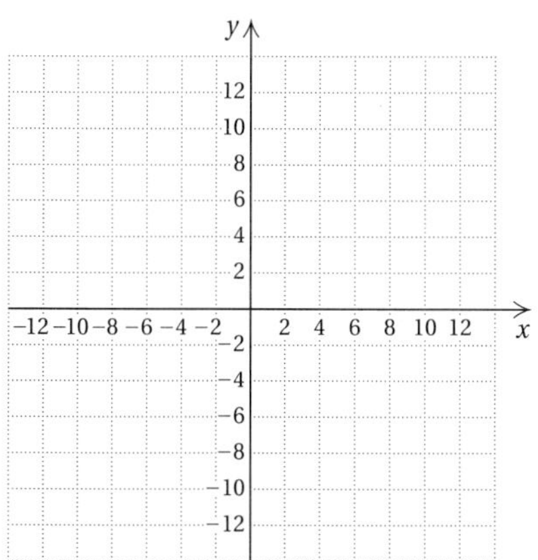

26. $y = \left(\frac{1}{2}\right)^x, \quad x = \left(\frac{1}{2}\right)^y$

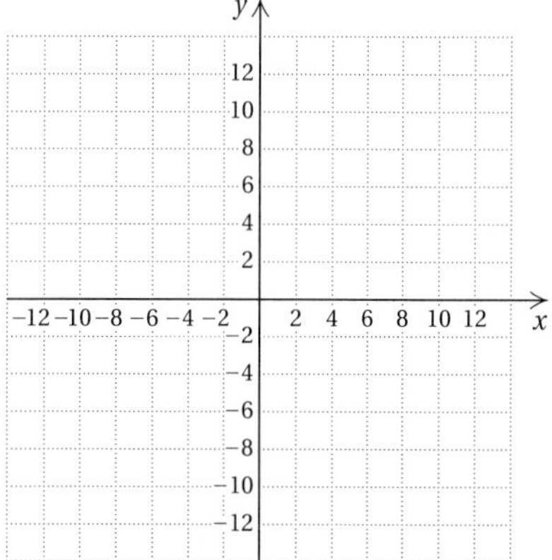

c Solve.

27. *World Demand for Lumber.* The world is experiencing an exponential demand for lumber. The amount of timber N, in billions of cubic feet, consumed t years after 1997, can be approximated by

$$N(t) = 62(1.018)^t,$$

where $t = 0$ corresponds to 1997 (***Source:*** U. N. Food and Agricultural Organization; American Forest and Paper Association).

a) How much timber was consumed in 1997? in 1998?

b) Estimate the amount of timber to be consumed in 2000 and in 2010.

c) Graph the function.

28. *Interest Compounded Annually.* Suppose that \$50,000 is invested at 8% interest, compounded annually.

a) Find a function for the amount in the account after t years.

b) Find the amount of money in the account at $t = 0$, $t = 4$, $t = 8$, and $t = 10$.

c) Graph the function.

29. *Growth of Bacteria.* Bladder infections are often caused when the bacteria *Escherichia coli* reach the human bladder. Suppose that 3000 of the bacteria are present at time $t = 0$. Then t min later, the number of bacteria present will be

$$N(t) = 3000(2)^{t/20}$$

(***Source:*** Chris Hayes, "Detecting a Human Health Risk: *E. coli*," *Laboratory Medicine* 29, no. 6, June 1998: 347–355).

a) How many bacteria will be present after 10 min? 20 min? 30 min? 40 min? 60 min?

b) Graph the function.

30. *Salvage Value.* An office machine is purchased for \$5200. Its value each year is about 80% of the value the preceding year. Its value after t years is given by the exponential function

$$V(t) = \$5200(0.8)^t.$$

a) Find the value of the machine after 0 yr, 1 yr, 2 yr, 5 yr, and 10 yr.

b) Graph the function.

31. *Spread of Zebra Mussels.* Beginning in 1988, infestations of zebra mussels started spreading throughout North American waters. These mussels spread with such speed that water treatment facilities, power plants, and entire ecosystems can become threatened. The function

$$A(t) = 10 \cdot 34^t$$

can be used to estimate the number of square centimeters of lake bottom that will be covered with mussels t years after an infestation covering 10 cm^2 first occurs. (***Source*:** Many thanks to Dr. Gerald Mackie of the Department of Zoology at the University of Guelph in Ontario for the background information for this exercise.)

a) How many square centimeters of lake bottom will be covered with mussels 5 yr after an infestation covering 10 cm^2 first appears? 7 yr after the infestation first appears?

b) Graph the function.

32. *Cases of AIDS.* The total number of Americans who have contracted AIDS, in thousands, can be approximated by the exponential function

$$N(t) = 1476(1.4)^t,$$

where $t = 0$ corresponds to 1997.

a) According to the function, how many Americans had been infected as of 1998?

b) Estimate the total number of Americans who will have been infected as of 2010.

c) Graph the function.

Skill Maintenance

33. Multiply and simplify: $x^{-5} \cdot x^3$. [4.1f]

34. Simplify: $(x^{-3})^4$. [4.2a]

Simplify. [4.1b]

35. 9^0

36. $\left(\frac{2}{3}\right)^0$

37. $\left(\frac{2}{3}\right)^1$

38. 2.7^1

Divide and simplify. [4.1e, f]

39. $\frac{x^{-3}}{x^4}$

40. $\frac{x}{x^{11}}$

41. $\frac{x}{x^0}$

42. $\frac{x^{-3}}{x^{-4}}$

Synthesis

43. ◆ Why was it necessary to discuss irrational exponents before graphing exponential functions?

44. ◆ Suppose that \$1000 is invested for 5 yr at 6% interest, compounded annually. In what year will the greatest amount of interest be earned? Why?

45. Simplify: $(5^{\sqrt{2}})^{2\sqrt{2}}$.

46. Which is larger: $\pi^{\sqrt{2}}$ or $(\sqrt{2})^{\pi}$?

Graph.

47. $y = 2^x + 2^{-x}$

48. $y = |2^x - 2|$

49. $y = \left|\left(\frac{1}{2}\right)^x - 1\right|$

50. $y = 2^{-x^2}$

Graph both equations using the same set of axes.

51. $y = 3^{-(x-1)}$, $x = 3^{-(y-1)}$

52. $y = 1^x$, $x = 1^y$

53. Use a grapher to graph each of the equations in Exercises 47–50.

12.2 Inverse and Composite Functions

When we go from an output of a function back to its input or inputs, we get what is called an *inverse relation.* When that relation is a function, we have what is called an *inverse function.* We now study such inverse functions and how to find formulas when the original function has a formula. We do so to understand the relationships among the special functions that we study in this chapter.

Objectives

a. Find the inverse of a relation if it is described as a set of ordered pairs or as an equation.

b. Given a function, determine whether it is one-to-one and has an inverse that is a function.

c. Find a formula for the inverse of a function, if it exists.

d. Graph inverse relations and functions.

e. Find the composition of functions and express certain functions as a composition of functions.

f. Determine whether a function is an inverse by checking its composition with the original function.

For Extra Help

TAPE 20 MAC WIN CD-ROM

a Inverses

A set of ordered pairs is called a **relation**. When we consider the graph of a function, we are thinking of a set of ordered pairs. Thus a function can be thought of as a special kind of relation, in which to each first coordinate there corresponds one and only one second coordinate.

Consider the relation h given as follows:

$$h = \{(-7, 4), (3, -1), (-6, 5), (0, 2)\}.$$

Suppose we *interchange* the first and second coordinates. The relation we obtain is called the **inverse** of the relation h and is given as follows:

$$\text{Inverse of } h = \{(4, -7), (-1, 3), (5, -6), (2, 0)\}.$$

> Interchanging the ordered pairs in a relation produces the **inverse relation.**

Example 1 Consider the relation g given by

$$g = \{(2, 4), (-1, 3), (-2, 0)\}.$$

In the figure below, the relation g is shown in red. The inverse of the relation is

$$\{(4, 2), (3, -1), (0, -2)\}$$

and is shown in blue.

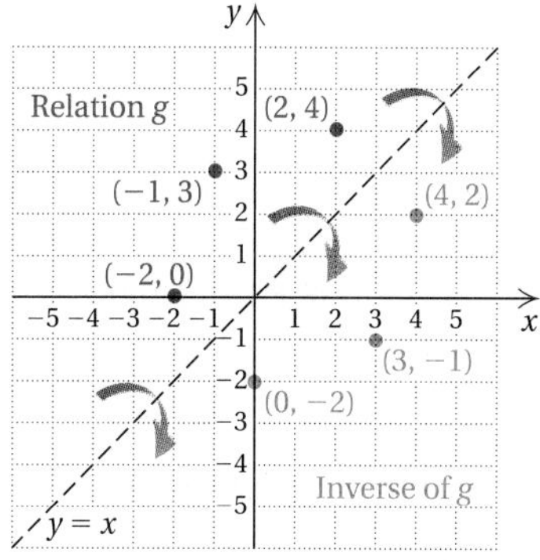

Do Exercise 1.

1. Consider the relation g given by

$$g = \{(2, 5), (-1, 4), (-2, 1)\}.$$

The graph of the relation is shown below in red. Find the inverse and draw its graph in blue.

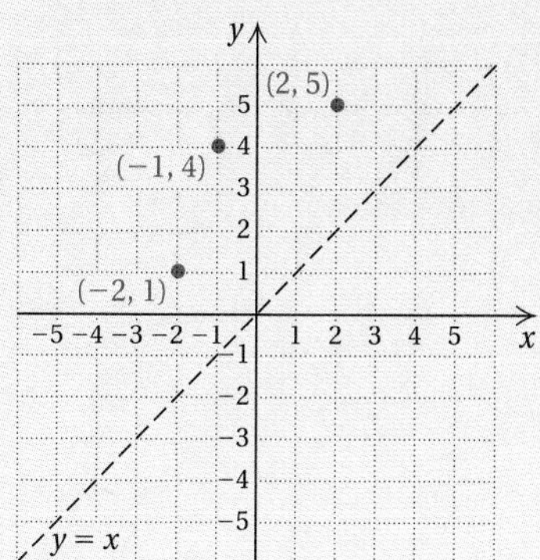

Answer on page A-54

2. Find an equation of the inverse relation. Then graph both the original relation and its inverse.

Relation:
$y = 6 - 2x$

x	y
0	6
2	2
3	0
5	−4

Inverse:

____

x	y
6	
2	
0	
−4	

Complete the table and draw the graphs using the same set of axes.

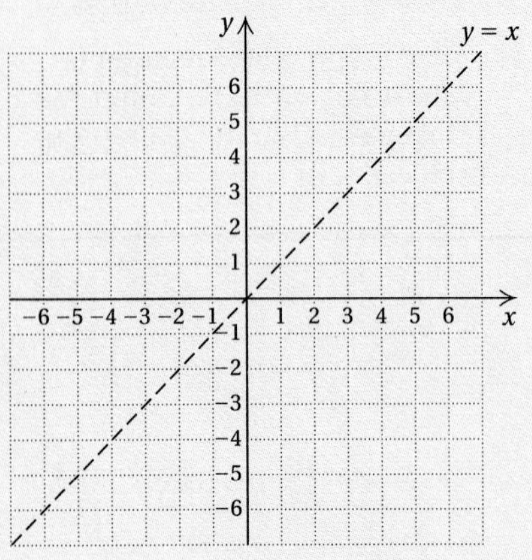

Answers on page A-54

> If a relation is defined by an equation, interchanging the variables produces an equation of the **inverse relation.**

Example 2 Find an equation of the inverse of the relation

$$y = 3x - 4.$$

Then graph both the relation and its inverse.

We interchange x and y and obtain an equation of the inverse:

$$x = 3y - 4.$$

Relation: $y = 3x - 4$ *Inverse*: $x = 3y - 4$

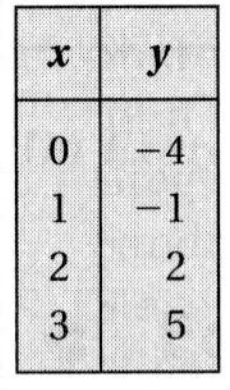

x	y
0	−4
1	−1
2	2
3	5

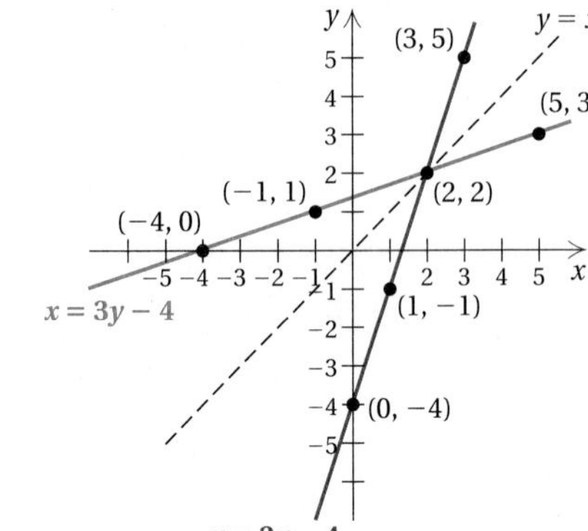

x	y
−4	0
−1	1
2	2
5	3

Example 3 Find an equation of the inverse of the relation

$$y = 6x - x^2.$$

Then graph both the original relation and its inverse.

We interchange x and y and obtain an equation of the inverse:

$$x = 6y - y^2.$$

Relation: $y = 6x - x^2$ *Inverse*: $x = 6y - y^2$

x	y
−1	−7
0	0
1	5
3	9
5	5

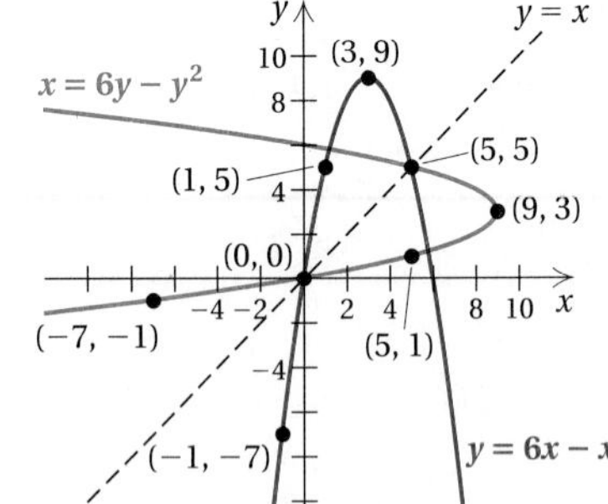

x	y
−7	−1
0	0
5	1
9	3
5	5

Do Exercises 2 and 3. (Exercise 3 is on the following page.)

b Inverses and One-To-One Functions

Let's consider the following two functions.

Number (Domain)		Cube (Range)
−3	→	−27
−2	→	−8
−1	→	−1
0	→	0
1	→	1
2	→	8
3	→	27

Year (Domain)		First-class Postage Cost, in Cents (Range)
1978	→	15
1983	→	20
1984	→	20
1989	→	25
1991	→	29
1995	→	32

Source: U.S. Postal Service

Suppose we reverse the arrows. Are these inverse relations functions?

Cube Root (Range)		Number (Domain)
−3	←	−27
−2	←	−8
−1	←	−1
0	←	0
1	←	1
2	←	8
3	←	27

Year (Range)		First-class Postage Cost, in Cents (Domain)
1978	←	15
1983	←	20
1984	←	20
1989	←	25
1991	←	29
1995	←	32

We see that the inverse of the cubing function is a function. The inverse of the postage function is not a function, however, because the input 20 has *two* outputs, 1983 and 1984. Recall that for a function, each input has exactly one output. However, it can happen that the same output comes from two or more different inputs. If this is the case, the inverse cannot be a function. When this possibility is excluded, the inverse is also a function.

In the cubing function, different inputs have different outputs. Thus its inverse is also a function. The cubing function is what is called a **one-to-one function.** If the inverse of a function f is also a function, it is named f^{-1} (read "f-inverse").

CAUTION! The -1 in f^{-1} is *not* an exponent and f^{-1} does not represent a reciprocal!

ONE-TO-ONE FUNCTION AND INVERSES

A function f is **one-to-one** if different inputs have different outputs—that is,

if $a \neq b$, then $f(a) \neq f(b)$. Or,

A function f is **one-to-one** if when the outputs are the same, the inputs are the same—that is,

if $f(a) = f(b)$, then $a = b$.

If a function is one-to-one, then its inverse is a function.

The domain of a one-to-one function f is the range of the inverse f^{-1}.

The range of a one-to-one function f is the domain of the inverse f^{-1}.

3. Find an equation of the inverse relation. Then graph both the original relation and its inverse.

Relation:
$y = x^2 - 4x + 7$

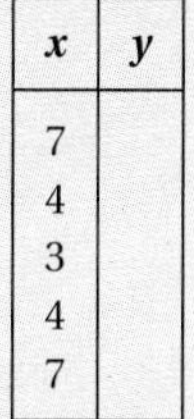

x	y
0	7
1	4
2	3
3	4
4	7

Inverse:

____

x	y
7	
4	
3	
4	
7	

Complete the table and draw the graphs using the same set of axes.

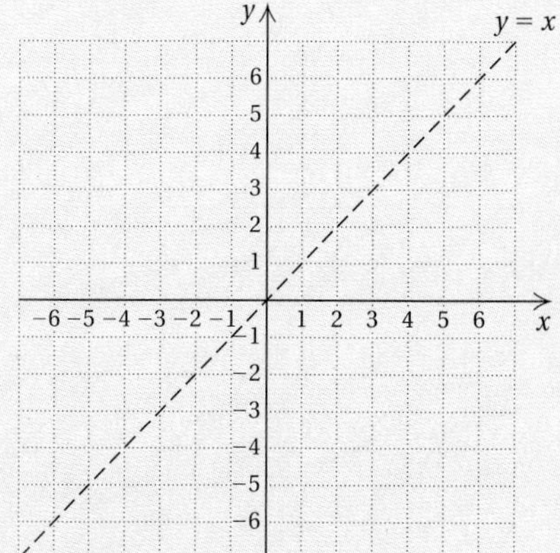

4. Determine whether the function is one-to-one and thus has an inverse that is also a function.

$f(x) = 4 - x$

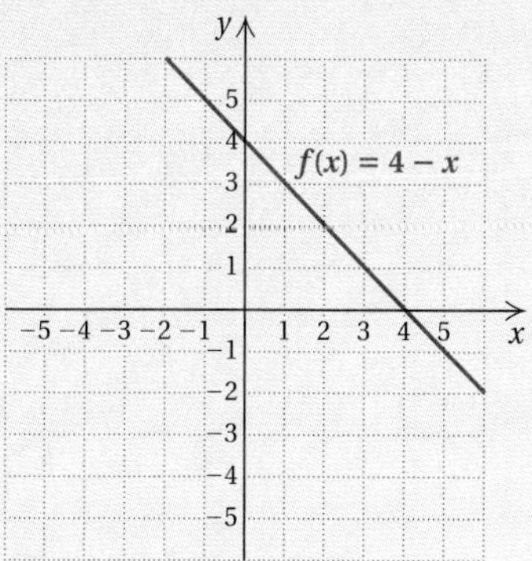

Answers on page A-54

Determine whether the function is one-to-one and thus has an inverse that is also a function.

5. $f(x) = x^2 - 1$

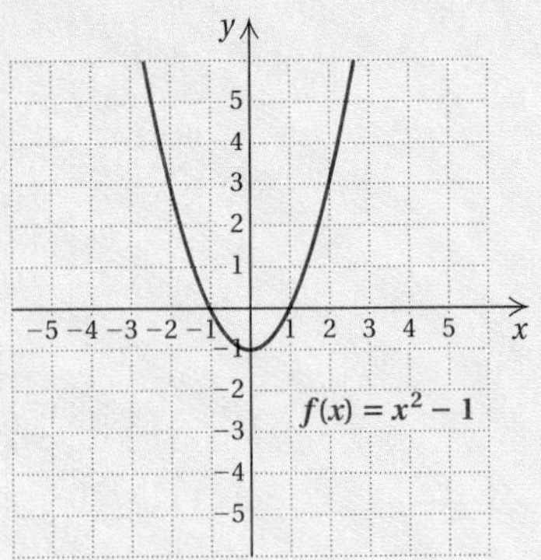

6. $f(x) = 4^x$
(Sketch this graph yourself.)

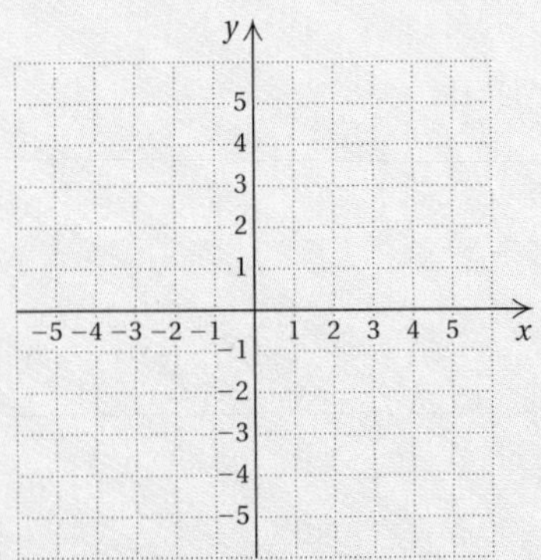

7. $f(x) = |x| - 3$
(Sketch this graph yourself.)

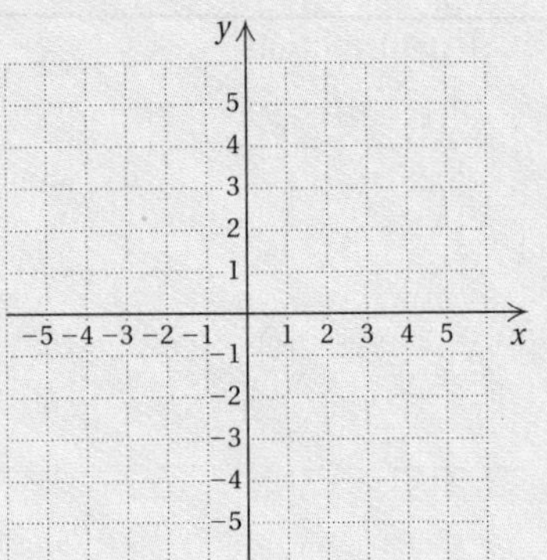

Answers on page A-54

How can we tell graphically whether a function is one-to-one and thus has an inverse that is a function?

Example 4 Here is the graph of the exponential function $f(x) = 2^x$. Determine whether the function is one-to-one and thus has an inverse that is a function.

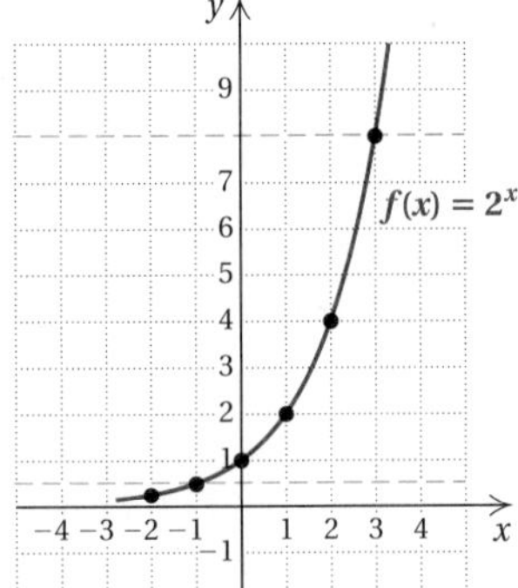

A function is one-to-one if different inputs have different outputs. In other words, no two x-values will have the same y-value. For this function, we cannot find two x-values that have the same y-value. Note also that no horizontal line can be drawn that will cross the graph more than once. The function is thus one-to-one and its inverse is a function.

THE HORIZONTAL-LINE TEST

If it is possible for a horizontal line to intersect the graph of a function more than once, then the function is not one-to-one and therefore its inverse is not a function.

Recall that a graph is that of a function if no vertical line crosses the graph more than once. A function has an inverse that is also a function if no horizontal line crosses the graph more than once.

Example 5 Determine whether the function $f(x) = x^2$ is one-to-one and has an inverse that is also a function.

The graph is shown below. There are many horizontal lines that cross the graph more than once, as shown, so this function is not one-to-one and does not have an inverse that is a function.

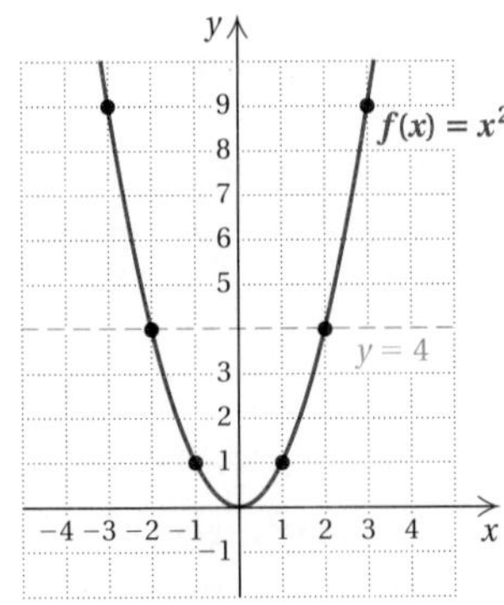

Do Exercises 4–7. (Exercise 4 is on the preceding page.)

C Finding a Formula for an Inverse

Suppose that a function is described by a formula. If it has an inverse that is a function, how do we find a formula for its inverse? If for any equation with two variables such as x and y we interchange the variables, we obtain an equation of the inverse relation. We proceed as follows to find a formula for f^{-1}.

If a function f is one-to-one, a formula for its inverse f^{-1} can be found as follows:

1. Replace $f(x)$ with y.
2. Interchange x and y. (This gives the inverse relation.)
3. Solve for y.
4. Replace y with $f^{-1}(x)$.

Example 6 Given $f(x) = x + 1$:

a) Determine whether the function is one-to-one.

b) If it is one-to-one, find a formula for $f^{-1}(x)$.

We solve as follows.

a) The graph of $f(x) = x + 1$ is shown below. It passes the horizontal-line test, so it is one-to-one. Thus its inverse is a function.

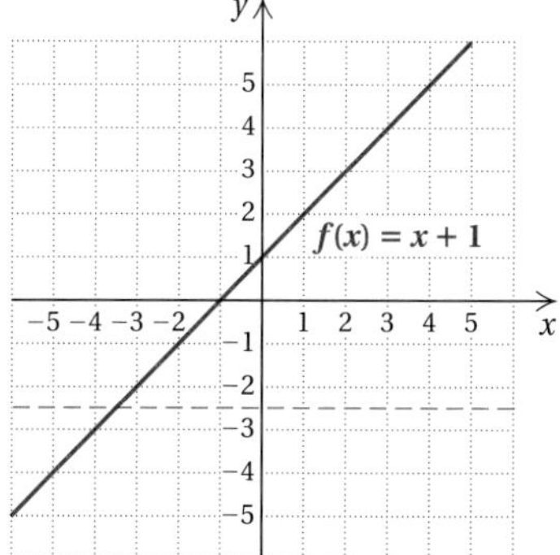

b) **1.** Replace $f(x)$ with y: $y = x + 1.$

2. Interchange x and y: $x = y + 1.$ This gives the inverse relation.

3. Solve for y: $x - 1 = y.$

4. Replace y with $f^{-1}(x)$: $f^{-1}(x) = x - 1.$

Given each function:

a) Determine whether it is one-to-one.

b) If it is one-to-one, find a formula for the inverse.

8. $f(x) = 3 - x$

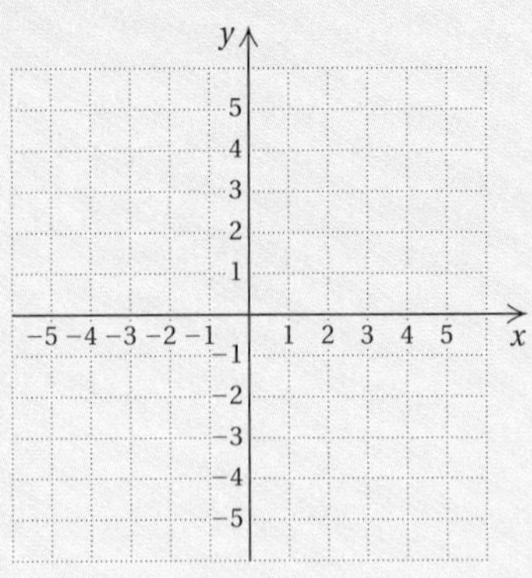

9. $g(x) = 3x - 2$

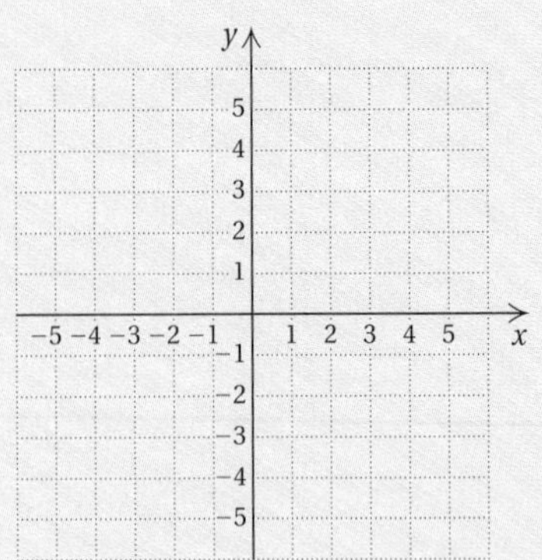

Answers on page A-54

Example 7 Given $f(x) = 2x - 3$:

a) Determine whether the function is one-to-one.

b) If it is one-to-one, find a formula for $f^{-1}(x)$.

We solve as follows.

a) The graph of $f(x) = 2x - 3$ is shown below. It passes the horizontal-line test and is one-to-one.

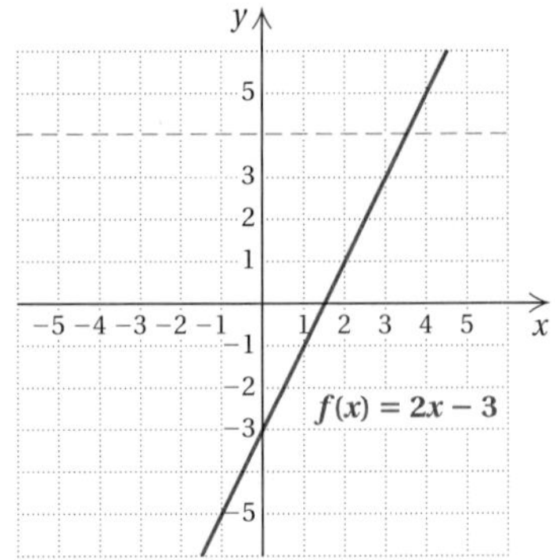

b) **1.** Replace $f(x)$ with y: $y = 2x - 3.$

2. Interchange x and y: $x = 2y - 3.$

3. Solve for y: $x + 3 = 2y$

$$\frac{x+3}{2} = y.$$

4. Replace y with $f^{-1}(x)$: $f^{-1}(x) = \dfrac{x+3}{2}.$

Do Exercises 8 and 9.

Let's now consider inverses of functions in terms of a function machine. Suppose that a one-to-one function f is programmed into a machine. If the machine has a reverse switch, when the switch is thrown, the machine performs the inverse function f^{-1}. Inputs then enter at the opposite end, and the entire process is reversed.

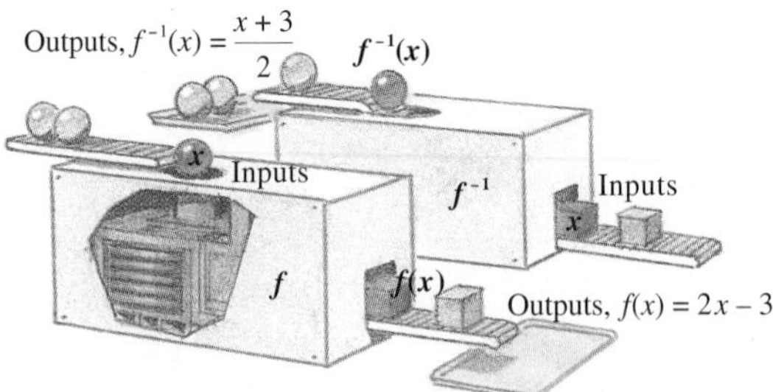

Consider $f(x) = 2x - 3$ and $f^{-1}(x) = (x + 3)/2$ from Example 7. For the input 5,

$$f(5) = 2 \cdot 5 - 3 = 10 - 3 = 7.$$

The output is 7. Now we use 7 for the input in the inverse:

$$f^{-1}(7) = \frac{7 + 3}{2} = \frac{10}{2} = 5.$$

The function f takes 5 to 7. The inverse function f^{-1} takes the number 7 back to 5.

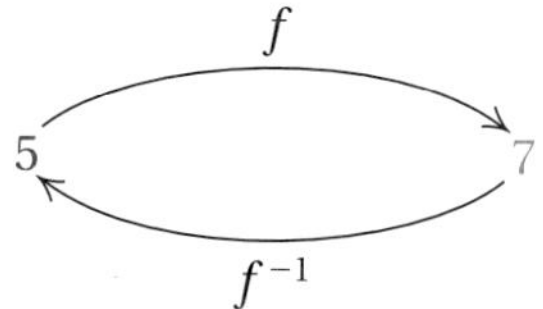

10. Graph $g(x) = 3x - 2$ and $g^{-1}(x) = (x + 2)/3$ using the same set of axes.

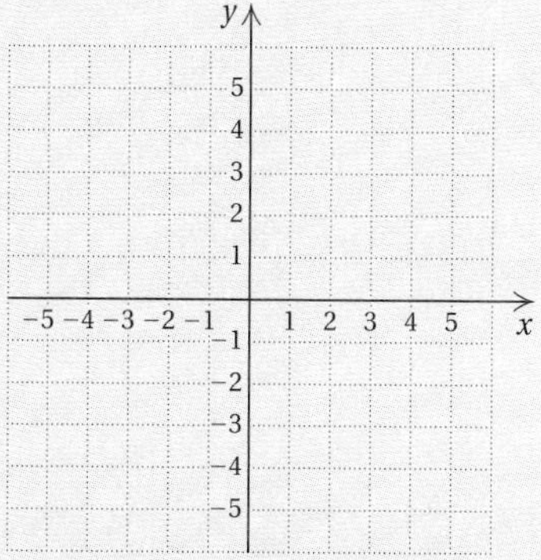

Answer on page A-54

d Graphing Functions and Their Inverses

How do the graphs of a function and its inverse compare?

Example 8 Graph $f(x) = 2x - 3$ and $f^{-1}(x) = (x + 3)/2$ using the same set of axes. Then compare.

The graph of each function follows. Note that the graph of f^{-1} can be drawn by reflecting the graph of f across the line $y = x$. That is, if we graph $f(x) = 2x - 3$ in wet ink and fold the paper along the line $y = x$, the graph of $f^{-1}(x) = (x + 3)/2$ will appear as the impression made by f.

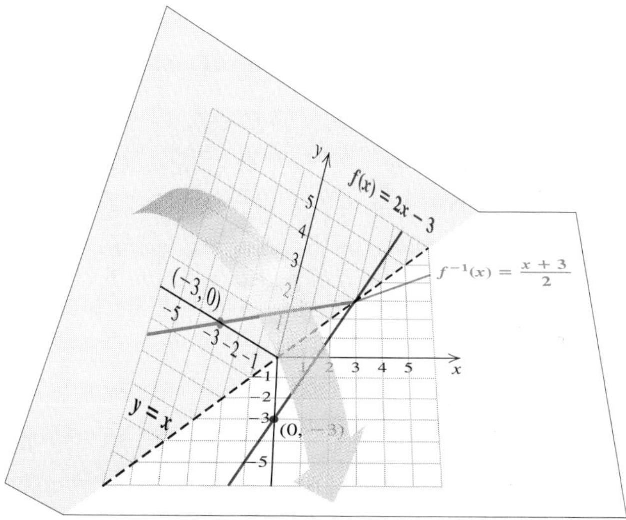

When x and y are interchanged to find a formula for the inverse, we are, in effect, flipping the graph of $f(x) = 2x - 3$ over the line $y = x$. For example, when the coordinates of the y-intercept of the graph of f, $(0, -3)$, are reversed, we get the x-intercept of the graph of f^{-1}, $(-3, 0)$.

11. Given $f(x) = x^3 + 1$:

a) Determine whether the function is one-to-one.

b) If it is one-to-one, find a formula for its inverse.

c) Graph the function and its inverse using the same set of axes.

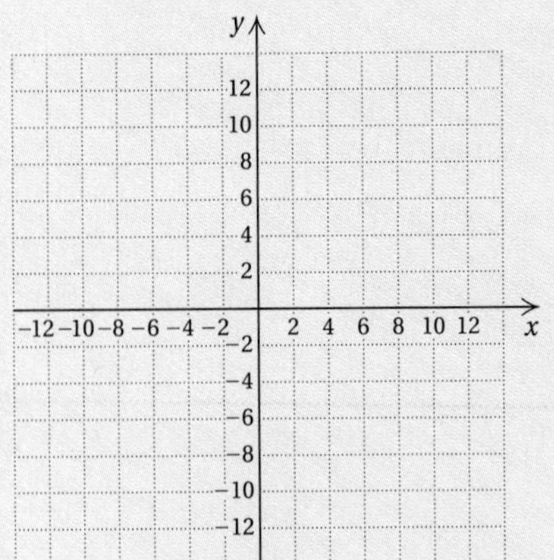

Answers on page A-54

> The graph of f^{-1} is a reflection of the graph of f across the line $y = x$.

Do Exercise 10 on the preceding page.

Example 9 Consider $g(x) = x^3 + 2$.

a) Determine whether the function is one-to-one.

b) If it is one-to-one, find a formula for its inverse.

c) Graph the inverse, if it exists.

We solve as follows.

a) The graph of $g(x) = x^3 + 2$ is shown below. It passes the horizontal-line test and thus is one-to-one.

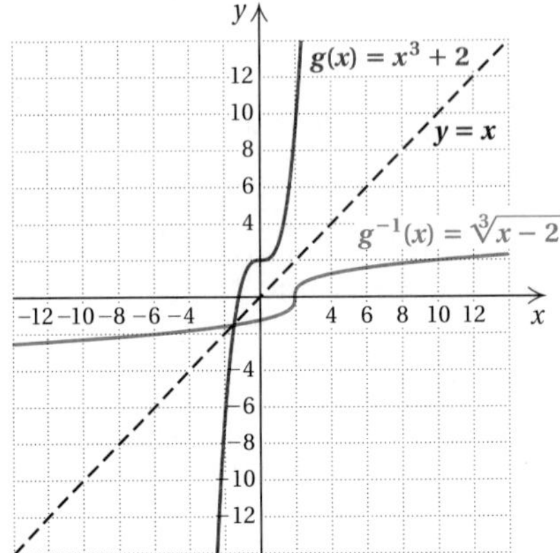

b) **1.** Replace $g(x)$ with y: $\quad y = x^3 + 2.$

2. Interchange x and y: $\quad x = y^3 + 2.$

3. Solve for y: $\quad x - 2 = y^3$

$$\sqrt[3]{x-2} = y.$$ **Since a number has only one cube root, we can solve for y.**

4. Replace y with $g^{-1}(x)$: $\quad g^{-1}(x) = \sqrt[3]{x-2}.$

c) To find the graph, we reflect the graph of $g(x) = x^3 + 2$ across the line $y = x$, as we did in Example 8. It can also be found by substituting into $g^{-1}(x) = \sqrt[3]{x-2}$ and plotting points. The graphs of g and g^{-1} are shown together above.

Do Exercise 11.

We can now see why we exclude 1 as a base for an exponential function. Consider

$$f(x) = a^x = 1^x = 1.$$

The graph of f is the horizontal line $y = 1$. The graph is not one-to-one. The function does not have an inverse that is a function. All other positive bases yield exponential functions that are one-to-one.

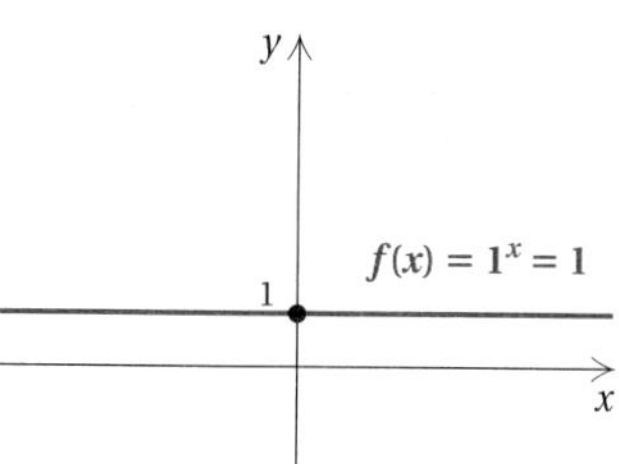

Calculator Spotlight

Graphing Inverses. The TI-82 and TI-83 graphers have the capability to automatically draw the graph of the inverse of a function. Consider the function $g(x) = x^3 + 2$ of Example 9. We enter $y_1 = x^3 + 2$. Then we press 2nd DRAW 8 to select the DRAWINV feature. On the TI-83, y_1 is selected by pressing VARS ▷ 1 1. Next, we press ENTER. The graph of the function and its inverse follows.

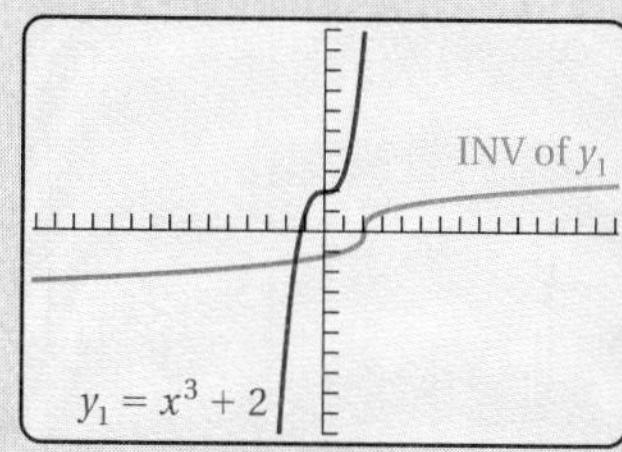

If you press the y= key, you will note that the DRAWINV feature does not give you a formula for the inverse. Suppose we solved for the inverse and incorrectly got the function $h(x) = \sqrt[3]{x + 2}$. Then we could graph $h(x)$ and use the DRAWINV feature again to graph the inverse of $g(x) = x^3 + 2$. We would see all three graphs and could tell that a mistake had been made.

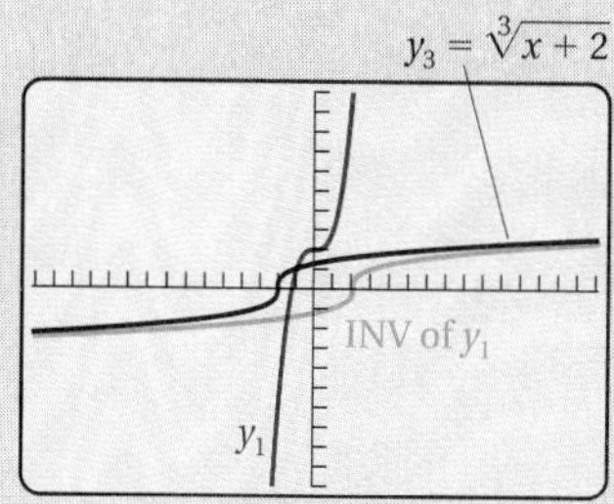

Exercises

1. Use your grapher to check the results of Examples 7–9 and Margin Exercises 8–11.

Use your grapher to determine whether the given function g is the inverse of the given function f.

2. $f(x) = \dfrac{x^2 - 10}{3}$, $g(x) = \sqrt{3x + 12}$
3. $f(x) = \frac{2}{3}x$, $g(x) = \frac{3}{2}x$
4. $f(x) = 3x - 8$, $g(x) = \dfrac{x + 8}{3}$
5. $f(x) = x^3 - 5$, $g(x) = \sqrt[3]{x + 5}$

e Composite Functions

In the real world, functions frequently occur in which some quantity depends on a variable that, in turn, depends on another variable. For instance, the number of employees hired by a firm may depend on the firm's profits, which may in turn depend on the number of items the firm produces. Functions like this are called **composite functions.**

For example, the function g that gives a correspondence between women's shoe sizes in the United States and those in Italy is given by $g(x) = 2x + 24$, where x is the U.S. size and $g(x)$ is the Italian size. Thus a U.S. size 4 corresponds to a shoe size of $g(4) = 2 \cdot 4 + 24$, or 32, in Italy.

There is also a function that gives a correspondence between women's shoe sizes in Italy and those in Britain. The function is given by $f(x) = \frac{1}{2}x - 14$, where x is the Italian size and $f(x)$ is the corresponding British size. Thus an Italian size 32 corresponds to a British size $f(32) = \frac{1}{2}(32) - 14$, or 2.

It seems reasonable to conclude that a shoe size of 4 in the United States corresponds to a size of 2 in Britain and that some function h describes this correspondence. Can we find a formula for h? If we look at the following tables, we might guess that such a formula is $h(x) = x - 2$, and that is indeed correct. But, for more complicated formulas, we would need to use algebra.

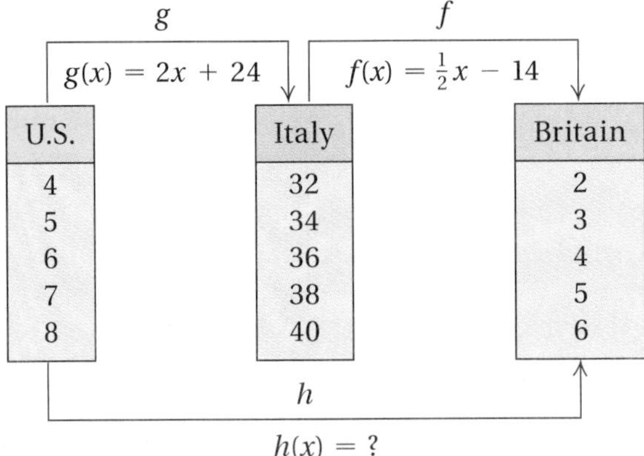

A shoe size x in the United States corresponds to a shoe size $g(x)$ in Italy, where

$$g(x) = 2x + 24.$$

Now $2x + 24$ is a shoe size in Italy. If we replace x in $f(x)$ with $2x + 24$, we can find the corresponding shoe size in Britain:

$$\begin{aligned} f(g(x)) &= \tfrac{1}{2}[2x + 24] - 14 \\ &= x + 12 - 14 = x - 2. \end{aligned}$$

This gives a formula for h: $h(x) = x - 2$. Thus a shoe size of 4 in the United States corresponds to a shoe size of $h(4) = 4 - 2$, or 2 in Britain. The function h is the **composition** of f and g, symbolized by $f \circ g$. To find $f \circ g(x)$, we substitute $g(x)$ for x in $f(x)$.

> The **composite function** $f \circ g$, the **composition** of f and g, is defined as
>
> $f \circ g(x) = f(g(x))$, or $(f \circ g)(x) = f[g(x)]$.

Example 10 Given $f(x) = 3x$ and $g(x) = 1 + x^2$:

a) Find $f \circ g(5)$ and $g \circ f(5)$.

b) Find $f \circ g(x)$ and $g \circ f(x)$.

We consider each function separately:

$f(x) = 3x$ — This function multiplies each input by 3.

and $g(x) = 1 + x^2$. — This function adds 1 to the square of each input.

a) $f \circ g(5) = f(g(5)) = f(1 + 5^2) = f(26) = 3(26) = 78;$
$g \circ f(5) = g(f(5)) = g(3 \cdot 5) = g(15) = 1 + 15^2 = 1 + 225 = 226$

b) $f \circ g(x) = f(g(x))$
$= f(1 + x^2)$ — Substituting $1 + x^2$ for x
$= 3(1 + x^2)$
$= 3 + 3x^2;$

$g \circ f(x) = g(f(x))$
$= g(3x)$ — Substituting $3x$ for x
$= 1 + (3x)^2$
$= 1 + 9x^2$

As a check, note that $g \circ f(5) = 1 + 9 \cdot 5^2 = 1 + 9 \cdot 25 = 226$, as expected from part (a) above.

Example 10 shows that $f \circ g(5) \neq g \circ f(5)$ and, in general, $f \circ g(x) \neq g \circ f(x)$.

Do Exercise 12.

Example 11 Given $f(x) = \sqrt{x}$ and $g(x) = x - 1$, find $f \circ g(x)$ and $g \circ f(x)$.

$f \circ g(x) = f(g(x)) = f(x - 1) = \sqrt{x - 1};$

$g \circ f(x) = g(f(x)) = g(\sqrt{x}) = \sqrt{x} - 1$

Do Exercise 13.

It is important to be able to recognize how a function can be expressed as a composition. Such a situation can occur in a study of calculus.

Example 12 Find $f(x)$ and $g(x)$ such that $h(x) = f \circ g(x)$:

$$h(x) = (7x + 3)^2.$$

This is $7x + 3$ to the 2nd power. Two functions that can be used for the composition are $f(x) = x^2$ and $g(x) = 7x + 3$. We can check by forming the composition:

$$h(x) = f \circ g(x) = f(g(x)) = f(7x + 3) = (7x + 3)^2.$$

This is the most "obvious" answer to the question. There can be other less obvious answers. For example, if

$$f(x) = (x - 1)^2 \quad \text{and} \quad g(x) = 7x + 4,$$

then

$$h(x) = f \circ g(x) = f(g(x)) = f(7x + 4) = (7x + 4 - 1)^2 = (7x + 3)^2.$$

Do Exercise 14.

12. Given $f(x) = x + 5$ and $g(x) = x^2 - 1$, find $f \circ g(x)$ and $g \circ f(x)$.

13. Given $f(x) = 4x + 5$ and $g(x) = \sqrt[3]{x}$, find $f \circ g(x)$ and $g \circ f(x)$.

14. Find $f(x)$ and $g(x)$ such that $h(x) = f \circ g(x)$. Answers may vary.

a) $h(x) = \sqrt[3]{x^2 + 1}$

b) $h(x) = \dfrac{1}{(x + 5)^4}$

Answers on page A-54

15. Let $f(x) = \frac{2}{3}x - 4$.
Use composition to show that

$$f^{-1}(x) = \frac{3x + 12}{2}.$$

Answer on page A-54

f Inverse Functions and Composition

Suppose that we used some input x for the function f and found its output, $f(x)$. The function f^{-1} would then take that output back to x. Similarly, if we began with an input x for the function f^{-1} and found its output, $f^{-1}(x)$, the original function f would then take that output back to x.

> If a function f is one-to-one, then f^{-1} is the unique function for which
> $f^{-1} \circ f(x) = x$ and $f \circ f^{-1}(x) = x$.

Example 13 Let $f(x) = 2x - 3$. Use composition to show that

$$f^{-1}(x) = \frac{x + 3}{2}. \qquad \text{(See Example 7.)}$$

We find $f^{-1} \circ f(x)$ and $f \circ f^{-1}(x)$ and check to see that each is x.

$$\begin{aligned} f^{-1} \circ f(x) &= f^{-1}(f(x)) \\ &= f^{-1}(2x - 3) \\ &= \frac{(2x - 3) + 3}{2} \\ &= \frac{2x}{2} \\ &= x; \end{aligned}$$

$$\begin{aligned} f \circ f^{-1}(x) &= f(f^{-1}(x)) \\ &= f\left(\frac{x + 3}{2}\right) \\ &= 2 \cdot \frac{x + 3}{2} - 3 \\ &= x + 3 - 3 \\ &= x \end{aligned}$$

Do Exercise 15.

Calculator Spotlight

Checking Inverses Using Composites. Consider $f(x) = 2x + 6$ and $g(x) = \frac{1}{2}x - 3$. Suppose that we want to check that g is the inverse of f. We enter

$$y_1 = 2x + 6, \quad y_2 = \frac{1}{2}x - 3,$$

and use the VARS key to enter the composition of y_1 and y_2 as $y_3 = y_1(y_2)$ and then press GRAPH.

$y_1 = 2x + 6,\ y_2 = \frac{1}{2}x - 3,\ y_3 = x$

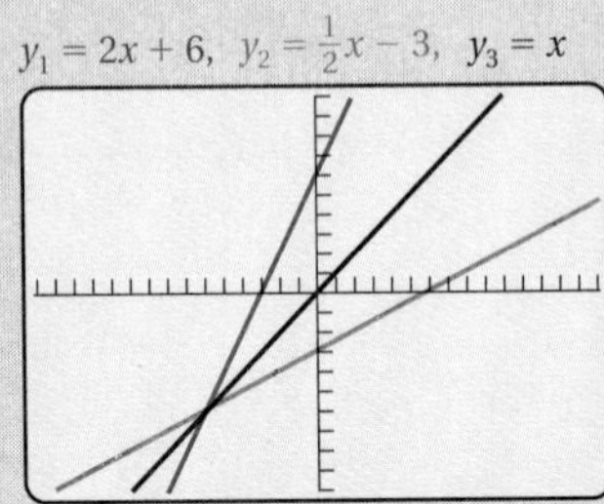

If g is the inverse of f, then the graph of $y_3 = y_1(y_2)$ should be $y_3 = x$. We can check this by using the TRACE feature along y_3 or by comparing the values of x and y_3 with the TABLE feature. The x-values and y_3-values should agree.

X	Y2	Y3
0	−3	0
1	−2.5	1
2	−2	2
3	−1.5	3
4	−1	4
5	−.5	5
6	0	6

Y3 = 0

Exercises

1. Use your grapher to check the results of Example 13 and Margin Exercise 15.

Use composites to determine whether the given function g is the inverse of the given function f.

2. $f(x) = \dfrac{x^2 - 12}{3}$, $g(x) = \sqrt{3x + 12}$
3. $f(x) = \frac{2}{3}x$, $g(x) = -\frac{2}{3}x$
4. $f(x) = 3x - 7$, $g(x) = \dfrac{x + 7}{3}$
5. $f(x) = x^3 - 5$, $g(x) = \sqrt[3]{x - 5}$

Exercise Set 12.2

a Find the inverse of the relation. Graph the original relation and then graph the inverse relation in blue.

1. $\{(1, 2), (6, -3), (-3, -5)\}$

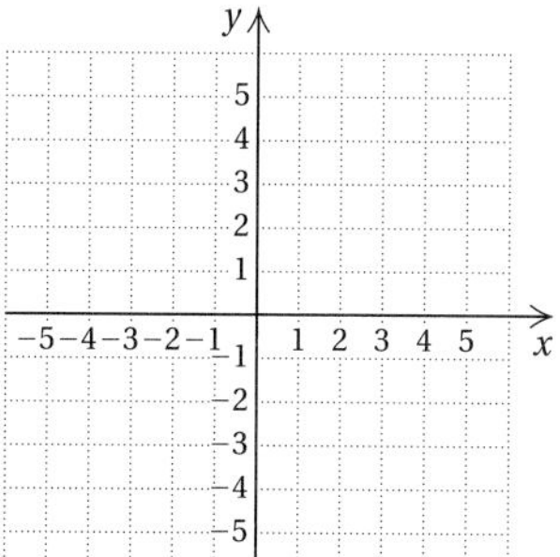

2. $\{(3, -1), (5, 2), (5, -3), (2, 0)\}$

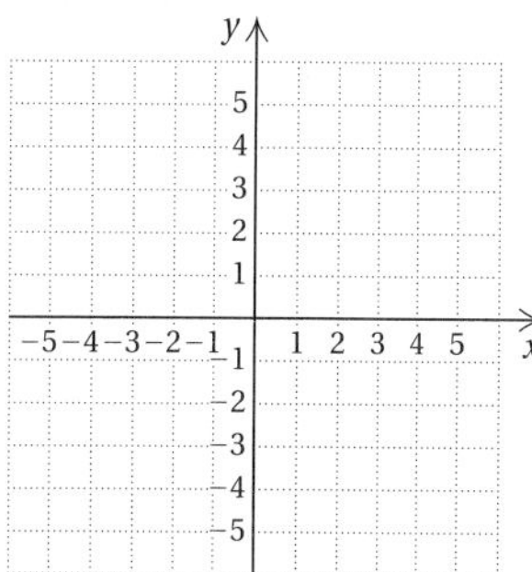

Find an equation of the inverse of the relation. Then complete the second table and graph both the original relation and its inverse.

3. $y = 2x + 6$

x	y
−1	4
0	6
1	8
2	10
3	12

x	y
4	
6	
8	
10	
12	

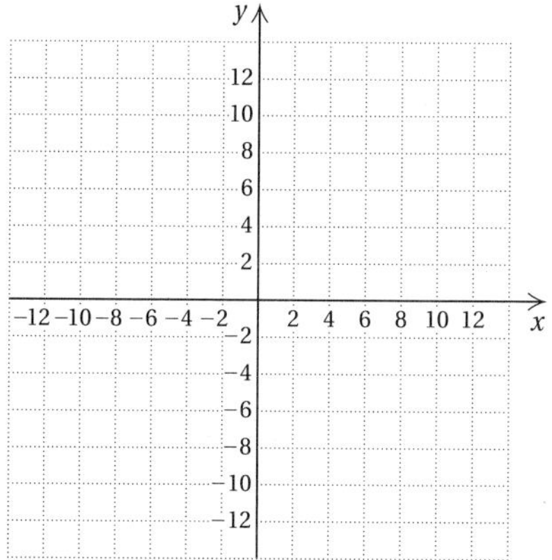

4. $y = \frac{1}{2}x^2 - 8$

x	y
−4	0
−2	−6
0	−8
2	−6
4	0

x	y
0	
−6	
−8	
−6	
0	

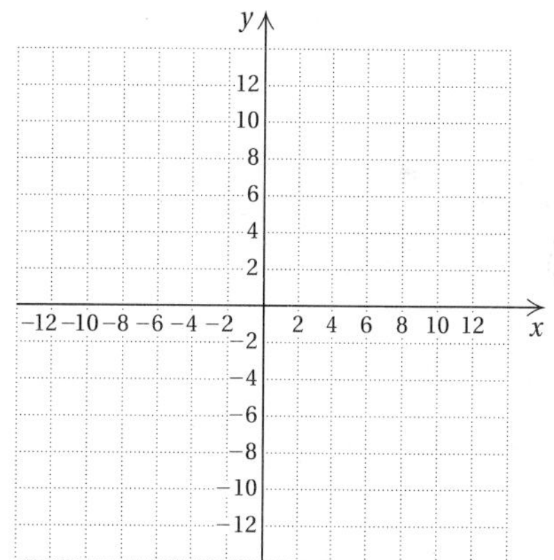

b Determine whether the function is one-to-one.

5. $f(x) = x - 5$

6. $f(x) = 3 - 6x$

7. $f(x) = x^2 - 2$

8. $f(x) = 4 - x^2$

9. $f(x) = |x| - 3$

10. $f(x) = |x - 2|$

11. $f(x) = 3^x$

12. $f(x) = \left(\frac{1}{2}\right)^x$

c Determine whether the function is one-to-one. If it is, find a formula for its inverse.

13. $f(x) = 5x - 2$

14. $f(x) = 4 + 7x$

15. $f(x) = \frac{-2}{x}$

16. $f(x) = \frac{1}{x}$

17. $f(x) = \frac{4}{3}x + 7$

18. $f(x) = -\frac{7}{8}x + 2$

19. $f(x) = \frac{2}{x + 5}$

20. $f(x) = \frac{1}{x - 8}$

21. $f(x) = 5$

22. $f(x) = -2$

23. $f(x) = \frac{2x + 1}{5x + 3}$

24. $f(x) = \frac{2x - 1}{5x + 3}$

25. $f(x) = x^3 - 1$

26. $f(x) = x^3 + 5$

27. $f(x) = \sqrt[3]{x}$

28. $f(x) = \sqrt[3]{x - 4}$

d Graph the function and its inverse using the same set of axes.

29. $f(x) = \frac{1}{2}x - 3$

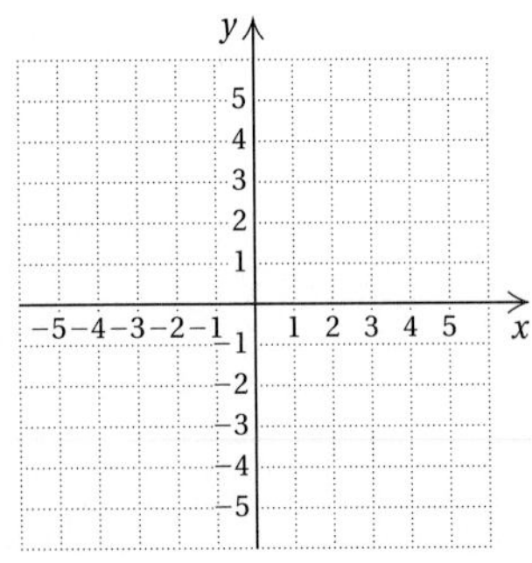

30. $g(x) = x + 4$

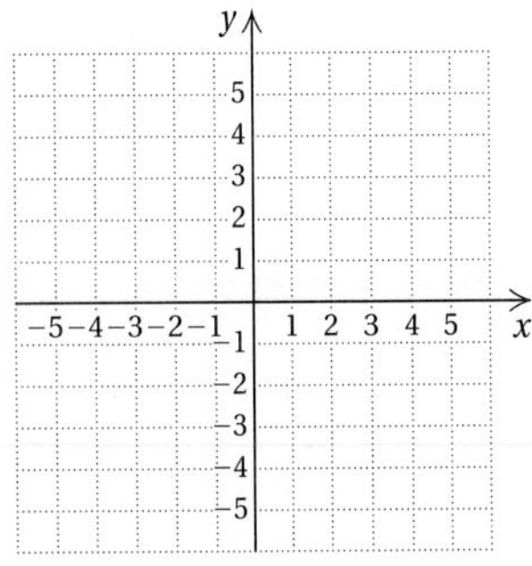

31. $f(x) = x^3$

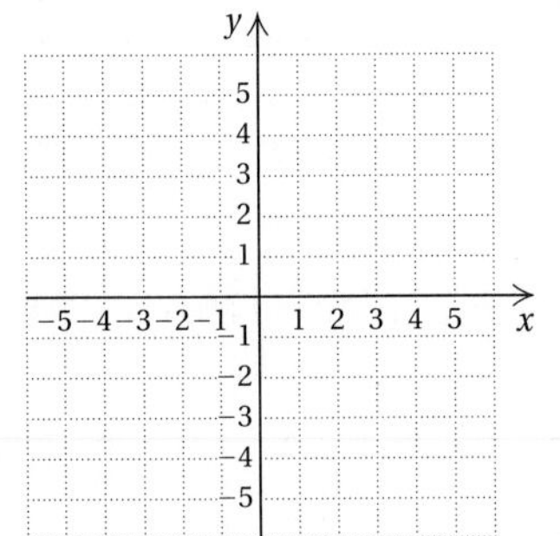

32. $f(x) = x^3 - 1$

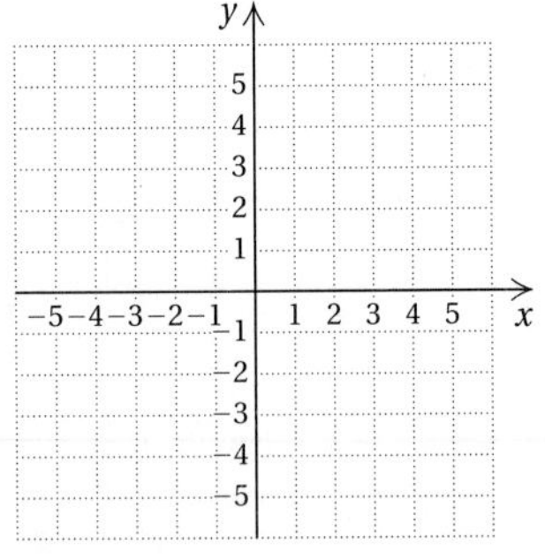

e Find $f \circ g(x)$ and $g \circ f(x)$.

33. $f(x) = 2x - 3,$
$g(x) = 6 - 4x$

34. $f(x) = 9 - 6x,$
$g(x) = 0.37x + 4$

35. $f(x) = 3x^2 + 2,$
$g(x) = 2x - 1$

36. $f(x) = 4x + 3,$
$g(x) = 2x^2 - 5$

37. $f(x) = 4x^2 - 1$, $g(x) = \dfrac{2}{x}$

38. $f(x) = \dfrac{3}{x}$, $g(x) = 2x^2 + 3$

39. $f(x) = x^2 + 5$, $g(x) = x^2 - 5$

40. $f(x) = \dfrac{1}{x^2}$, $g(x) = x - 1$

Find $f(x)$ and $g(x)$ such that $h(x) = f \circ g(x)$. Answers may vary.

41. $h(x) = (5 - 3x)^2$

42. $h(x) = 4(3x - 1)^2 + 9$

43. $h(x) = \sqrt{5x + 2}$

44. $h(x) = (3x^2 - 7)^5$

45. $h(x) = \dfrac{1}{x - 1}$

46. $h(x) = \dfrac{3}{x} + 4$

47. $h(x) = \dfrac{1}{\sqrt{7x + 2}}$

48. $h(x) = \sqrt{x - 7} - 3$

49. $h(x) = (\sqrt{x} + 5)^4$

50. $h(x) = \dfrac{x^3 + 1}{x^3 - 1}$

f For each function, use composition to show that the inverse is correct.

51. $f(x) = \frac{4}{5}x$, $f^{-1}(x) = \frac{5}{4}x$

52. $f(x) = \dfrac{x + 7}{2}$, $f^{-1}(x) = 2x - 7$

53. $f(x) = \dfrac{1 - x}{x}$, $f^{-1}(x) = \dfrac{1}{x + 1}$

54. $f(x) = x^3 - 5$, $f^{-1}(x) = \sqrt[3]{x + 5}$

Find the inverse of the given function by thinking about the operations of the function and then reversing, or undoing, them. Then use composition to show whether the inverse is correct.

Function	*Inverse*
55. $f(x) = 3x$	$f^{-1}(x) =$ ________
56. $f(x) = \frac{1}{4}x + 7$	$f^{-1}(x) =$ ________
57. $f(x) = -x$	$f^{-1}(x) =$ ________
58. $f(x) = \sqrt[3]{x} - 5$	$f^{-1}(x) =$ ________
59. $f(x) = \sqrt[3]{x - 5}$	$f^{-1}(x) =$ ________
60. $f(x) = x^{-1}$	$f^{-1}(x) =$ ________

61. *Dress Sizes in the United States and France.* A size-6 dress in the United States is size 38 in France. A function that converts dress sizes in the United States to those in France is

$$f(x) = x + 32.$$

a) Find the dress sizes in France that correspond to sizes of 8, 10, 14, and 18 in the United States.
b) Determine whether this function has an inverse that is a function. If so, find a formula for the inverse.
c) Use the inverse function to find dress sizes in the United States that correspond to sizes of 40, 42, 46, and 50 in France.

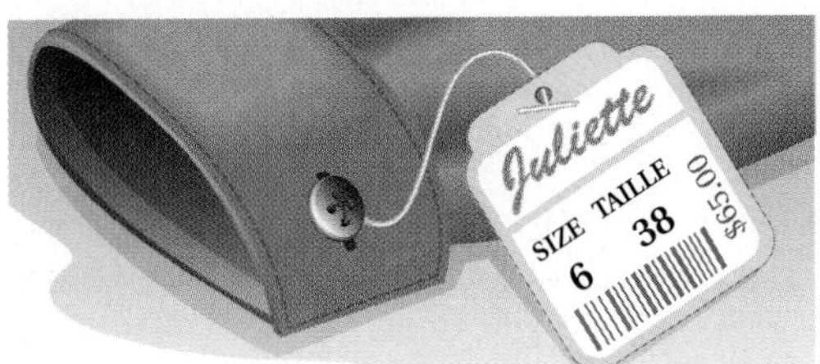

62. *Dress Sizes in the United States and Italy.* A size-6 dress in the United States is size 36 in Italy. A function that converts dress sizes in the United States to those in Italy is

$$f(x) = 2(x + 12).$$

a) Find the dress sizes in Italy that correspond to sizes of 8, 10, 14, and 18 in the United States.
b) Determine whether this function has an inverse that is a function. If so, find a formula for the inverse.
c) Use the inverse function to find dress sizes in the United States that correspond to sizes of 40, 44, 52, and 60 in Italy.

Skill Maintenance

Use rational exponents to simplify. [10.2d]

63. $\sqrt[6]{a^2}$

64. $\sqrt[6]{x^4}$

65. $\sqrt{a^4b^6}$

66. $\sqrt[3]{8t^6}$

67. $\sqrt[8]{81}$

68. $\sqrt[4]{32}$

69. $\sqrt[12]{64x^6y^6}$

70. $\sqrt[8]{p^4t^2}$

71. $\sqrt[5]{32a^{15}b^{40}}$

72. $\sqrt[3]{1000x^9y^{18}}$

73. $\sqrt[4]{81a^8b^8}$

74. $\sqrt[3]{27p^3q^9}$

Synthesis

75. ◆ How can a grapher be used to determine whether a function is one-to-one?

76. ◆ The function $V(t) = 750(1.2)^t$ is used to predict the value V of a certain rare stamp t years from 1999. Do not calculate $V^{-1}(t)$ but explain how V^{-1} could be used.

In Exercises 77–80, use a grapher to help determine whether or not the given functions are inverses of each other.

77. $f(x) = 0.75x^2 + 2;\quad g(x) = \sqrt{\dfrac{4(x-2)}{3}}$

78. $f(x) = 1.4x^3 + 3.2;\quad g(x) = \sqrt[3]{\dfrac{x-3.2}{1.4}}$

79. $f(x) = \sqrt{2.5x + 9.25};\quad g(x) = 0.4x^2 - 3.7,\ x \geq 0$

80. $f(x) = 0.8x^{1/2} + 5.23;\quad g(x) = 1.25(x^2 - 5.23),\ x \geq 0$

81. Use a grapher to help match each function in Column A with its inverse from Column B.

Column A	*Column B*
(1) $y = 5x^3 + 10$	**A.** $y = \dfrac{\sqrt[3]{x} - 10}{5}$
(2) $y = (5x + 10)^3$	**B.** $y = \sqrt[3]{\dfrac{x}{5}} - 10$
(3) $y = 5(x + 10)^3$	**C.** $y = \sqrt[3]{\dfrac{x-10}{5}}$
(4) $y = (5x)^3 + 10$	**D.** $y = \dfrac{\sqrt[3]{x-10}}{5}$

Collaborative Learning Manual: Create composite functions and deduce the original functions.

12.3 Logarithmic Functions

We are now ready to study inverses of exponential functions. These functions have many applications and are referred to as *logarithm,* or *logarithmic, functions.*

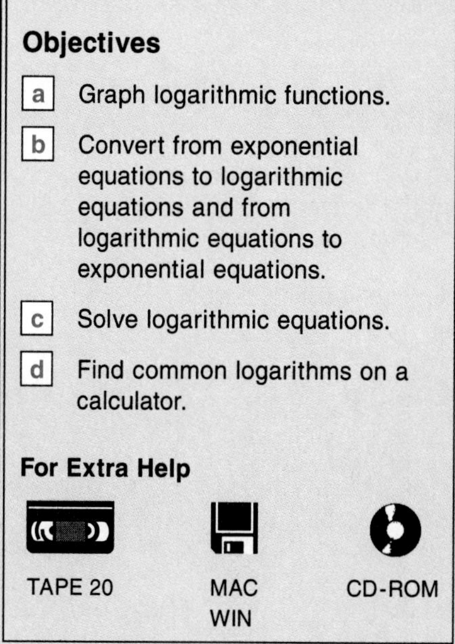

a Graphing Logarithmic Functions

Consider the exponential function $f(x) = 2^x$. Like all exponential functions, f is one-to-one. Can a formula for f^{-1} be found? To answer this, we use the method of Section 12.2:

1. Replace $f(x)$ with y: $\quad y = 2^x$.
2. Interchange x and y: $\quad x = 2^y$.
3. Solve for y: $\quad y =$ the power to which we raise 2 to get x.
4. Replace y with $f^{-1}(x)$: $\quad f^{-1}(x) =$ the power to which we raise 2 to get x.

We now define a new symbol to replace the words "the power to which we raise 2 to get x".

> $\log_2 x$, read "the logarithm, base 2, of x", or "log, base 2, of x," means "the power to which we raise 2 to get x."

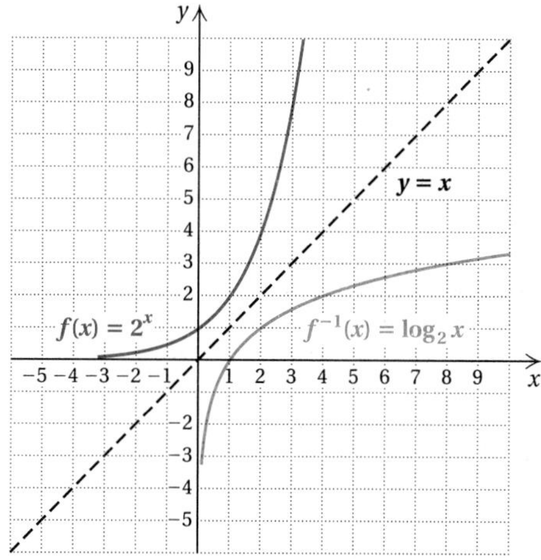

Thus if $f(x) = 2^x$, then $f^{-1}(x) = \log_2 x$. Note that $f^{-1}(8) = \log_2 8 = 3$, because 3 *is the power to which we raise* 2 *to get* 8; that is, $2^3 = 8$.

Although expressions like $\log_2 13$ can only be approximated, remember that $\log_2 13$ represents *the power to which we raise* 2 *to get* 13. That is, $2^{\log_2 13} = 13$.

Do Exercise 1.

1. Write the meaning of $\log_2 64$. Then find $\log_2 64$.

Answer on page A-55

2. Graph: $y = f(x) = \log_3 x$.

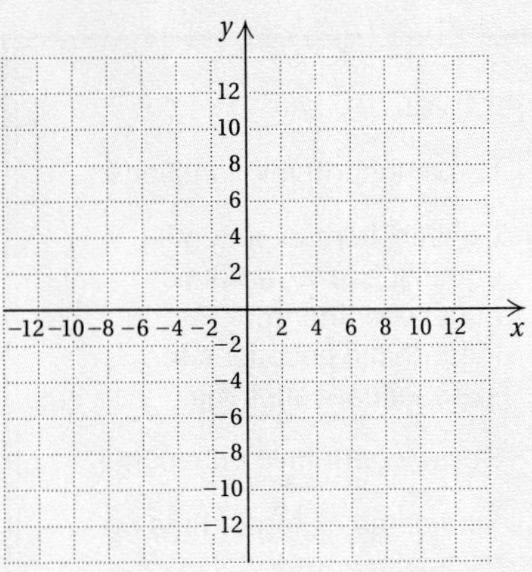

Answer on page A-55

For any exponential function $f(x) = a^x$, the inverse is called a **logarithmic function, base *a*.** The graph of the inverse can, of course, be drawn by reflecting the graph of $f(x) = a^x$ across the line $y = x$. It will be helpful to remember that the inverse of $f(x) = a^x$ is given by $f^{-1}(x) = \log_a x$. Normally, we use a number a that is greater than 1 for the logarithm base.

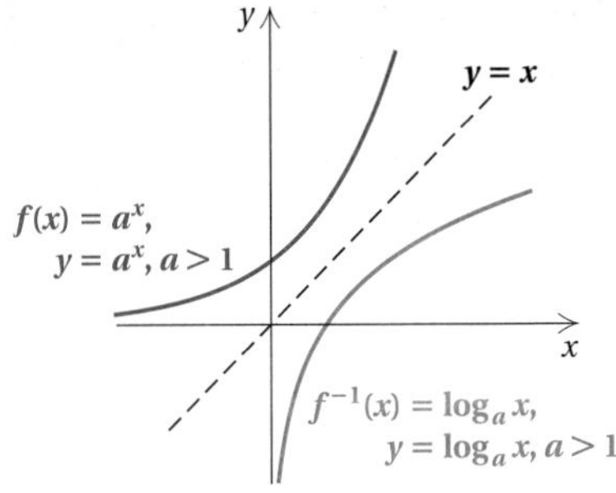

DEFINITION OF LOGARITHMS

The inverse of $f(x) = a^x$ is given by

$$f^{-1}(x) = \log_a x.$$

We read "$\log_a x$" as "the logarithm, base a, of x." We define $y = \log_a x$ as that number y such that $a^y = x$, where $x > 0$ and a is a positive constant other than 1.

It is helpful in dealing with logarithmic functions to remember that the logarithm of a number is an **exponent**. It is the exponent y in $x = a^y$. Keep thinking, "The logarithm, base a, of a number x is the power to which a must be raised in order to get x."

A logarithm is an exponent.

Exponential Function	Logarithmic Function
$y = a^x$ $f(x) = a^x$ $a > 0, a \neq 1$ Domain = The set of real numbers Range = The set of positive numbers	$x = a^y$ $f^{-1}(x) = \log_a x$ $a > 0, a \neq 1$ Range = The set of real numbers Domain = The set of positive numbers

Why do we exclude 1 from being a logarithm base? If we did include it, we would be considering $x = 1^y = 1$. The graph of this equation is a vertical line, which is not a function. It does not pass the vertical-line test.

Example 1 Graph: $y = f(x) = \log_5 x$.

The equation $y = \log_5 x$ is equivalent to $5^y = x$. We can find ordered pairs that are solutions by choosing values for y and computing the corresponding x-values.

For $y = 0$, $x = 5^0 = 1$.

For $y = 1$, $x = 5^1 = 5$.

For $y = 2$, $x = 5^2 = 25$.

For $y = 3$, $x = 5^3 = 125$.

For $y = -1$, $x = 5^{-1} = \dfrac{1}{5}$.

For $y = -2$, $x = 5^{-2} = \dfrac{1}{25}$.

x, or 5^y	y
1	0
5	1
25	2
125	3
$\frac{1}{5}$	−1
$\frac{1}{25}$	−2

(1) Select y.
(2) Compute x.

We plot the ordered pairs and connect them with a smooth curve. The graph of $y = 5^x$ has been shown only for reference.

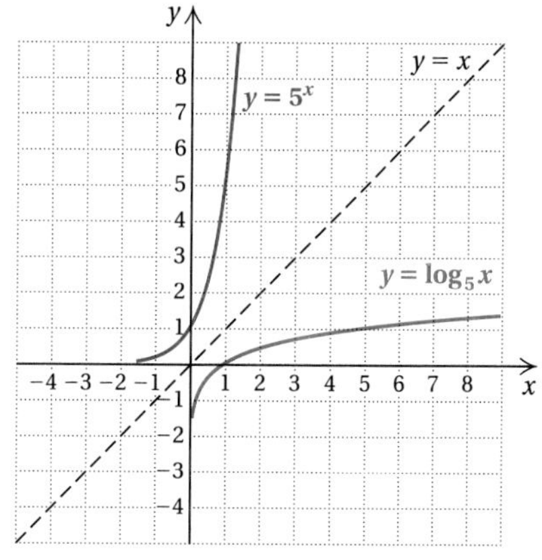

Do Exercise 2 on the preceding page.

b Converting Between Exponential Equations and Logarithmic Equations

We use the definition of logarithms to convert from exponential equations to logarithmic equations.

> $y = \log_a x \Rightarrow a^y = x$; $\quad a^y = x \Rightarrow y = \log_a x$
>
> Be sure to memorize this relationship! It is probably the most important definition in the chapter. Many times this definition will be a justification for a proof or a procedure that we are considering.

Examples Convert to a logarithmic equation.

2. $8 = 2^x \Rightarrow x = \log_2 8$ The exponent is the logarithm. The base remains the base.
3. $y^{-1} = 4 \Rightarrow -1 = \log_y 4$
4. $a^b = c \Rightarrow b = \log_a c$

Do Exercises 3–6.

We also use the definition of logarithms to convert from logarithmic equations to exponential equations.

Examples Convert to an exponential equation.

5. $y = \log_3 5 \Rightarrow 3^y = 5$ The logarithm is the exponent. The base does not change.
6. $-2 = \log_a 7 \Rightarrow a^{-2} = 7$
7. $a = \log_b d \Rightarrow b^a = d$

Do Exercises 7–10.

Convert to a logarithmic equation.

3. $6^0 = 1$

4. $10^{-3} = 0.001$

5. $16^{0.25} = 2$

6. $m^T = P$

Convert to an exponential equation.

7. $\log_2 32 = 5$

8. $\log_{10} 1000 = 3$

9. $\log_a Q = 7$

10. $\log_t M = x$

Answers on page A-55

Solve.

11. $\log_{10} x = 4$

12. $\log_x 81 = 4$

13. $\log_2 x = -2$

Answers on page A-55

C Solving Certain Logarithmic Equations

Certain equations involving logarithms can be solved by first converting to exponential equations. We will solve more complicated equations later.

Example 8 Solve: $\log_2 x = -3$.

$$\log_2 x = -3$$

$$2^{-3} = x \quad \text{Converting to an exponential equation}$$

$$\frac{1}{2^3} = x$$

$$\frac{1}{8} = x$$

CHECK: $\log_2 \frac{1}{8}$ is the exponent to which we raise 2 to get $\frac{1}{8}$. Since $2^{-3} = \frac{1}{8}$, we know that $\frac{1}{8}$ checks and is the solution.

Example 9 Solve: $\log_x 16 = 2$.

$$\log_x 16 = 2$$

$$x^2 = 16 \quad \text{Converting to an exponential equation}$$

$$x = 4 \quad \text{or} \quad x = -4 \quad \text{Using the principle of square roots}$$

CHECK: $\log_4 16 = 2$ because $4^2 = 16$. Thus, 4 is a solution. Since all logarithm bases must be positive, $\log_{-4} 16$ is not defined. Therefore, -4 is not a solution.

Do Exercises 11–13.

To think of finding logarithms as solving equations may help in some cases.

Example 10 Find $\log_{10} 1000$.

METHOD 1. Let $\log_{10} 1000 = x$. Then

$$10^x = 1000 \quad \text{Converting to an exponential equation}$$

$$10^x = 10^3$$

$$x = 3. \quad \text{The exponents are the same.}$$

Therefore, $\log_{10} 1000 = 3$.

METHOD 2. Think of the meaning of $\log_{10} 1000$. It is the exponent to which we raise 10 to get 1000. That exponent is 3. Therefore, $\log_{10} 1000 = 3$.

Example 11 Find $\log_{10} 0.01$.

METHOD 1. Let $\log_{10} 0.01 = x$. Then

$$10^x = 0.01 \quad \text{Converting to an exponential equation}$$

$$10^x = \frac{1}{100}$$

$$10^x = 10^{-2}$$

$$x = -2. \quad \text{The exponents are the same.}$$

Therefore, $\log_{10} 0.01 = -2$.

METHOD 2. $\log_{10} 0.01$ is the exponent to which we raise 10 to get 0.01. Noting that

$$0.01 = \frac{1}{100} = \frac{1}{10^2} = 10^{-2},$$

we see that the exponent is -2. Therefore, $\log_{10} 0.01 = -2$.

Example 12 Find $\log_5 1$.

METHOD 1. Let $\log_5 1 = x$. Then

$5^x = 1$ Converting to an exponential equation

$5^x = 5^0$

$x = 0.$ The exponents are the same.

Therefore, $\log_5 1 = 0$.

METHOD 2. $\log_5 1$ is the exponent to which we raise 5 to get 1. That exponent is 0. Therefore, $\log_5 1 = 0$.

Do Exercises 14–16.

Example 12 illustrates an important property of logarithms.

> For any base a,
>
> $\log_a 1 = 0.$
>
> The logarithm, base a, of 1 is always 0.

The proof follows from the fact that $a^0 = 1$. This is equivalent to the logarithmic equation $\log_a 1 = 0$.

Another property follows similarly. We know that $a^1 = a$ for any real number a. In particular, it holds for any positive number a. This is equivalent to the logarithmic equation $\log_a a = 1$.

> For any base a,
>
> $\log_a a = 1.$

Do Exercises 17–20.

d Finding Common Logarithms on a Calculator

Base-10 logarithms are called **common logarithms.** Before calculators became so widely available, common logarithms were used extensively to do complicated calculations. In fact, that is why logarithms were invented. The abbreviation **log**, with no base written, is used for the common logarithm, base-10. Thus,

$\log 29$ means $\log_{10} 29$.

We can approximate log 29. Note the following:

$\log 100 = \log_{10} 100 = 2;$

$\log 29 = ?;$ It seems reasonable that log 29 is between 1 and 2.

$\log 10 = \log_{10} 10 = 1.$

Find each of the following.

14. $\log_{10} 10{,}000$

15. $\log_{10} 0.0001$

16. $\log_7 1$

Simplify.

17. $\log_3 1$

18. $\log_3 3$

19. $\log_c c$

20. $\log_c 1$

Answers on page A-55

On a scientific calculator or grapher, the scientific key for common logarithms is generally marked [LOG]. We find that

$$\log 29 \approx 1.462397998 \approx 1.4624,$$

rounded to four decimal places. This also tells us that $10^{1.4624} \approx 29$.

On some scientific calculators, the keystrokes for doing such a calculation might be

[2] [9] [LOG] [=]. The display would then read `1.462398`.

Using the scientific keys on a grapher, the keystrokes might be

[LOG] [2] [9] [ENTER]. The display would then read `1.462397998`.

Examples Find each common logarithm, to four decimal places, on a scientific calculator or grapher.

	Function Value	*Readout*	*Rounded*
13.	log 287,523	5.458672591	5.4587
14.	log 0.000486	-3.313363731	-3.3134
15.	log (−5)	NONREAL ANS	NONREAL ANS

In Example 15, log (−5) does not exist as a real number because there is no real-number power to which we can raise 10 to get −5. The number 10 raised to any power is nonnegative. The logarithm of a negative number does not exist.

Find the common logarithm, to four decimal places, on a scientific calculator or a grapher.

21. log 78,235.4

22. log 0.0000309

23. log (−3)

24. Find

log 1000 and log 10,000

without using a calculator. Between what two whole numbers is log 9874? Then approximate log 9874 on a calculator rounded to four decimal places.

Do Exercises 21–24.

The inverse of a logarithmic function is an exponential function. Thus, if $f(x) = \log x$, then $f^{-1}(x) = 10^x$. Because of this, on many calculators, the [LOG] key doubles as the [10^x] key after a [2nd] or [SHIFT] key has been pressed. To find $10^{5.4587}$ on a scientific calculator, we might enter 5.4587 and press [10^x]. On a grapher, we press [2nd] [10^x], followed by 5.4587. In either case, we get the approximation

$$10^{5.4587} \approx 287{,}541.1465.$$

Compare this computation to Example 13. Note that, apart from the rounding error, $10^{5.4587}$ takes us back to about 287,523.

25. Find $10^{4.8934}$ using a calculator. (Compare your computation to that of Margin Exercise 21.)

Do Exercise 25.

Using the scientific keys on a calculator or grapher would allow us to construct a graph of $f(x) = \log_{10} x = \log x$ by finding function values directly, rather than converting to exponential form as we did in Example 1.

x	$f(x)$
0.5	−0.3010
1	0
2	0.3010
3	0.4771
5	0.6990
9	0.9542
10	1

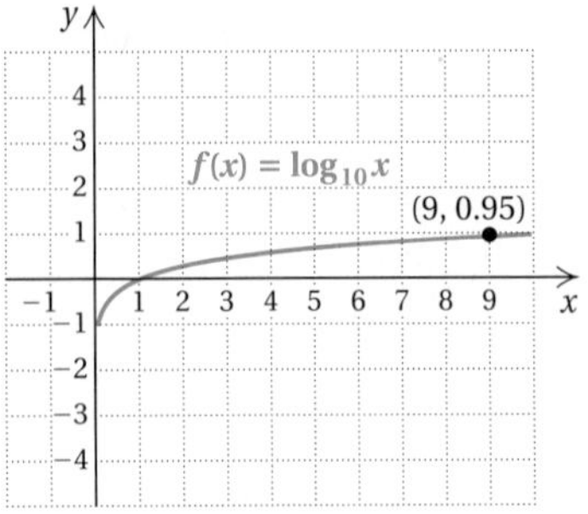

Calculator Spotlight

To graph $f(x) = \log_{10} x = \log x$, we press [y=] [LOG] [X]. You will see $y_1 = \log (x$ on the display. Note that there is no right parenthesis. It does not need to be included unless you are entering a function like

$$g(x) = 2 \cdot \log (x - 3) + 1.$$

Exercises

Graph.

1. $f(x) = \log_{10} x = \log x$

2. $f(x) = 2 \cdot \log x$

3. $g(x) = 2 \cdot \log (x - 3) + 1$

Answers on page A-55

Exercise Set 12.3

a Graph.

1. $f(x) = \log_2 x$

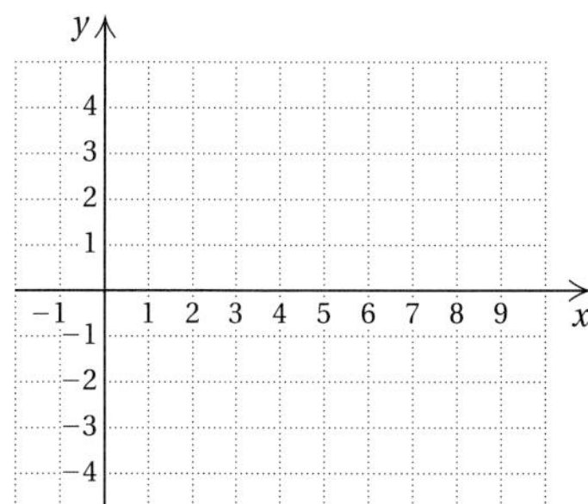

2. $f(x) = \log_{10} x$

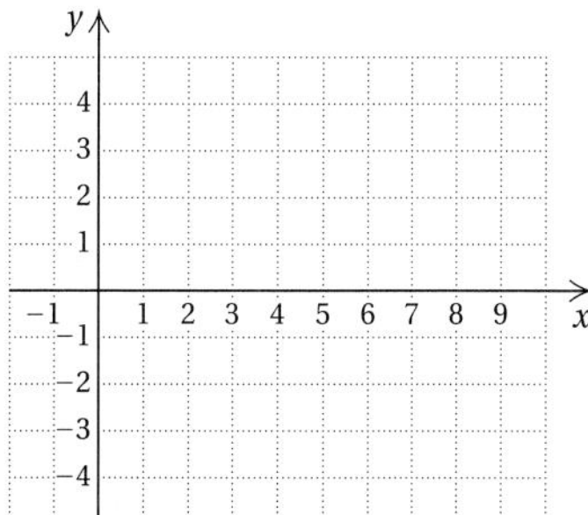

3. $f(x) = \log_{1/3} x$

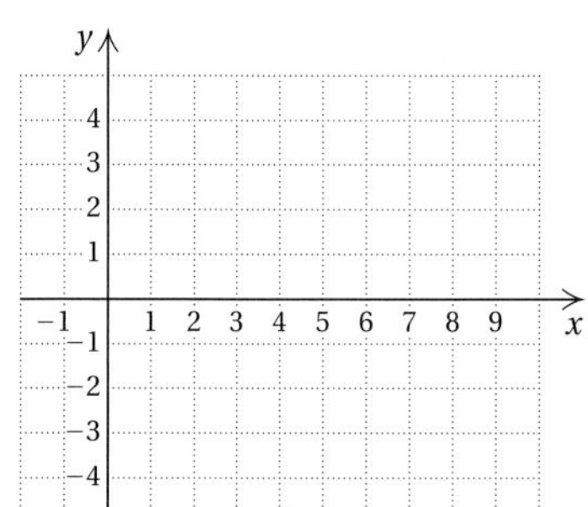

4. $f(x) = \log_{1/2} x$

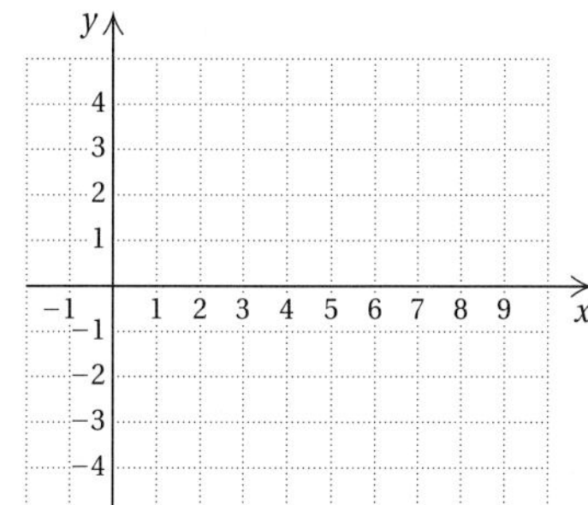

Graph both functions using the same set of axes.

5. $f(x) = 3^x, \quad f^{-1}(x) = \log_3 x$

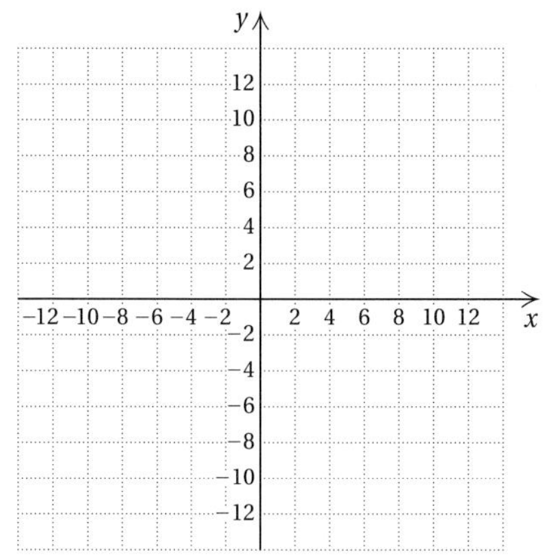

6. $f(x) = 4^x, \quad f^{-1}(x) = \log_4 x$

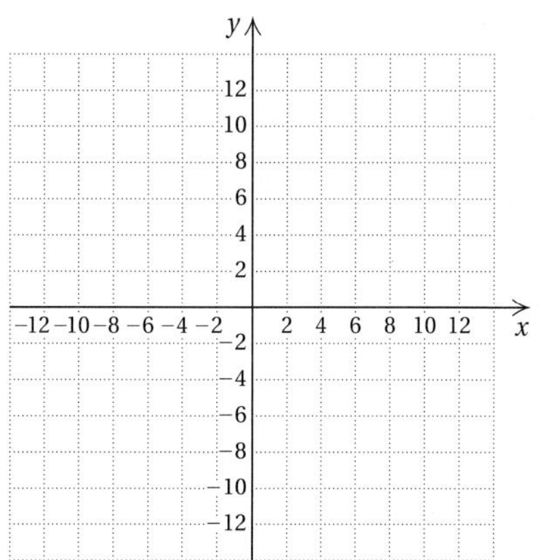

b Convert to a logarithmic equation.

7. $10^3 = 1000$

8. $10^2 = 100$

9. $5^{-3} = \frac{1}{125}$

10. $4^{-5} = \frac{1}{1024}$

11. $8^{1/3} = 2$

12. $16^{1/4} = 2$

13. $10^{0.3010} = 2$

14. $10^{0.4771} = 3$

15. $e^2 = t$

16. $p^k = 3$

17. $Q^t = x$

18. $P^m = V$

19. $e^2 = 7.3891$

20. $e^3 = 20.0855$

21. $e^{-2} = 0.1353$

22. $e^{-4} = 0.0183$

Convert to an exponential equation.

23. $w = \log_4 10$

24. $t = \log_5 9$

25. $\log_6 36 = 2$

26. $\log_7 7 = 1$

27. $\log_{10} 0.01 = -2$

28. $\log_{10} 0.001 = -3$

29. $\log_{10} 8 = 0.9031$

30. $\log_{10} 2 = 0.3010$

31. $\log_e 100 = 4.6052$

32. $\log_e 10 = 2.3026$

33. $\log_t Q = k$

34. $\log_m P = a$

c Solve.

35. $\log_3 x = 2$

36. $\log_4 x = 3$

37. $\log_x 16 = 2$

38. $\log_x 64 = 3$

39. $\log_2 16 = x$

40. $\log_5 25 = x$

41. $\log_3 27 = x$

42. $\log_4 16 = x$

43. $\log_x 25 = 1$

44. $\log_x 9 = 1$

45. $\log_3 x = 0$

46. $\log_2 x = 0$

47. $\log_2 x = -1$

48. $\log_3 x = -2$

49. $\log_8 x = \frac{1}{3}$

50. $\log_{32} x = \frac{1}{5}$

Find each of the following.

51. $\log_{10} 100$

52. $\log_{10} 100{,}000$

53. $\log_{10} 0.1$

54. $\log_{10} 0.001$

55. $\log_{10} 1$

56. $\log_{10} 10$

57. $\log_5 625$

58. $\log_2 64$

59. $\log_7 49$

60. $\log_5 125$

61. $\log_2 8$

62. $\log_8 64$

63. $\log_9 \frac{1}{81}$

64. $\log_5 \frac{1}{125}$

65. $\log_8 1$

66. $\log_6 6$

67. $\log_e e$

68. $\log_e 1$

69. $\log_{27} 9$

70. $\log_8 2$

d Find the common logarithm, to four decimal places, on a calculator.

71. $\log 78{,}889.2$

72. $\log 9{,}043{,}788$

73. $\log 0.67$

74. $\log 0.0067$

75. $\log (-97)$

76. $\log 0$

77. $\log \left(\frac{289}{32.7}\right)$

78. $\log \left(\frac{23}{86.2}\right)$

Skill Maintenance

For each quadratic function, find and label **(a)** the vertex, **(b)** the line of symmetry, and **(c)** the maximum or minimum value. Then **(d)** graph the function. [11.6a]

79. $f(x) = 4 - x^2$

80. $f(x) = (x + 3)^2$

81. $f(x) = -2(x - 1)^2 - 3$

82. $f(x) = x^2 - 2x - 5$

Solve for the given letter. [11.3b]

83. $E = mc^2$, for c

84. $B = 3a^2 - 4a$, for a

85. $A = \sqrt{3ab}$, for b

86. $T = 2\pi\sqrt{\frac{L}{g}}$, for L

87. *Automobile Accidents by Age.* The following table lists the percentages of drivers of age a involved in automobile accidents in a recent year. [11.7b]

Age, a	Percentage Involved in an Accident
20	31
24	34
34	22
44	18

Source: National Safety Council

a) Use the data points (20, 31), (24, 34), and (34, 22) to fit a quadratic function to the data.
b) Use the quadratic function to predict the percentage of drivers involved in accidents at age 30 and at age 37.

Synthesis

88. ◆ Explain in your own words what is meant by $\log_a b = c$.

89. ◆ John Napier (1550–1617) of Scotland is credited by mathematicians as the inventor of logarithms. Make a report on Napier and his work.

Graph.

90. $f(x) = \log_2 (x - 1)$

91. $f(x) = \log_3 |x + 1|$

Solve.

92. $|\log_3 x| = 3$

93. $\log_{125} x = \frac{2}{3}$

94. $\log_4 (3x - 2) = 2$

95. $\log_8 (2x + 1) = -1$

96. $\log_{10} (x^2 + 21x) = 2$

Simplify.

97. $\log_{1/4} \frac{1}{64}$

98. $\log_{81} 3 \cdot \log_3 81$

99. $\log_{10} (\log_4 (\log_3 81))$

100. $\log_2 (\log_2 (\log_4 256))$

101. $\log_{1/5} 25$

Collaborative Learning Manual: Practice graphing logarithmic functions and their inverses.

12.4 Properties of Logarithmic Functions

The ability to manipulate logarithmic expressions is important in many applications and in more advanced mathematics. We now establish some basic properties that are useful in manipulating logarithmic expressions.

Objectives

a Express the logarithm of a product as a sum of logarithms, and conversely.

b Express the logarithm of a power as a product, and conversely.

c Express the logarithm of a quotient as a difference of logarithms, and conversely.

d Convert from logarithms of products, quotients, and powers to expressions in terms of individual logarithms, and conversely.

e Simplify expressions of the type $\log_a a^k$.

For Extra Help

TAPE 21

MAC
WIN

CD-ROM

a Logarithms of Products

PROPERTY 1: THE PRODUCT RULE

For any positive numbers M and N,

$$\log_a (M \cdot N) = \log_a M + \log_a N.$$

(The logarithm of a product is the sum of the logarithms of the factors. The number a can be any logarithm base.)

Example 1 Express as a sum of logarithms: $\log_2 (4 \cdot 16)$.

$$\log_2 (4 \cdot 16) = \log_2 4 + \log_2 16 \qquad \text{By Property 1}$$

Example 2 Express as a single logarithm: $\log_{10} 0.01 + \log_{10} 1000$.

$$\log_{10} 0.01 + \log_{10} 1000 = \log_{10} (0.01 \times 1000) \qquad \text{By Property 1}$$
$$= \log_{10} 10$$

Do Exercises 1–4.

Express as a sum of logarithms.

1. $\log_5 25 \cdot 5$

2. $\log_b PQ$

Express as a single logarithm.

3. $\log_3 7 + \log_3 5$

4. $\log_a J + \log_a A + \log_a M$

A PROOF OF PROPERTY 1 (*OPTIONAL*). Let $\log_a M = x$ and $\log_a N = y$. Converting to exponential equations, we have $a^x = M$ and $a^y = N$. Then we multiply to obtain

$$M \cdot N = a^x \cdot a^y = a^{x+y}.$$

Converting $M \cdot N = a^{x+y}$ back to a logarithmic equation, we get

$$\log_a M \cdot N = x + y.$$

Remembering what x and y represent, we get

$$\log_a M \cdot N = \log_a M + \log_a N.$$

b Logarithms of Powers

PROPERTY 2: THE POWER RULE

For any positive number M and any real number k,

$$\log_a M^k = k \cdot \log_a M.$$

(The logarithm of a power of M is the exponent times the logarithm of M. The number a can be any logarithm base.)

Calculator Spotlight

Exercises

Use the TABLE and GRAPH features to determine whether each of the following is correct.

1. $\log (5x) = \log 5 + \log x$

2. $\log x^2 = 2 \log x$

3. $\log \left(\frac{x}{3}\right) = \log x - \log 3$

4. $\log (x + 2) = \log x + \log 2$

Answers on page A-56

Express as a product.

5. $\log_7 4^5$

6. $\log_a \sqrt{5}$

7. Express as a difference of logarithms:

$\log_b \frac{P}{Q}$.

8. Express as a single logarithm:

$\log_2 125 - \log_2 25$.

Answers on page A-56

Examples Express as a product.

3. $\log_a 9^{-5} = -5 \log_a 9$ — By Property 2

4. $\log_a \sqrt[4]{5} = \log_a 5^{1/4}$ — Writing exponential notation

$= \frac{1}{4} \log_a 5$ — By Property 2

Do Exercises 5 and 6.

A PROOF OF PROPERTY 2 (*OPTIONAL*). Let $x = \log_a M$. Then we convert to an exponential equation to get $a^x = M$. Raising both sides to the kth power, we obtain

$$(a^x)^k = M^k, \quad \text{or} \quad a^{xk} = M^k.$$

Converting back to a logarithmic equation with base a, we get $\log_a M^k = xk$. But $x = \log_a M$, so

$$\log_a M^k = (\log_a M)k = k \cdot \log_a M.$$

C Logarithms of Quotients

PROPERTY 3: THE QUOTIENT RULE

For any positive numbers M and N,

$$\log_a \frac{M}{N} = \log_a M - \log_a N.$$

(The logarithm of a quotient is the logarithm of the numerator minus the logarithm of the denominator. The number a can be any logarithm base.)

Example 5 Express as a difference of logarithms: $\log_t \frac{6}{U}$.

$\log_t \frac{6}{U} = \log_t 6 - \log_t U$ — By Property 3

Example 6 Express as a single logarithm: $\log_b 17 - \log_b 27$.

$\log_b 17 - \log_b 27 = \log_b \frac{17}{27}$ — By Property 3

Example 7 Express as a single logarithm: $\log_{10} 10{,}000 - \log_{10} 100$.

$$\log_{10} 10{,}000 - \log_{10} 100 = \log_{10} \frac{10{,}000}{100} = \log_{10} 100$$

Do Exercises 7 and 8.

A PROOF OF PROPERTY 3 (*OPTIONAL*). The proof makes use of Property 1 and Property 2.

$$\log_a \frac{M}{N} = \log_a MN^{-1}$$

$= \log_a M + \log_a N^{-1}$ — By Property 1

$= \log_a M + (-1) \log_a N$ — By Property 2

$= \log_a M - \log_a N$

d Using the Properties Together

Examples Express in terms of logarithms.

8. $\log_a \dfrac{x^2y^3}{z^4} = \log_a (x^2y^3) - \log_a z^4$ Using Property 3

$= \log_a x^2 + \log_a y^3 - \log_a z^4$ Using Property 1

$= 2\log_a x + 3\log_a y - 4\log_a z$ Using Property 2

9. $\log_a \sqrt[4]{\dfrac{xy}{z^3}} = \log_a \left(\dfrac{xy}{z^3}\right)^{1/4}$ Writing exponential notation

$= \frac{1}{4}\log_a \dfrac{xy}{z^3}$ Using Property 2

$= \frac{1}{4}(\log_a xy - \log_a z^3)$ Using Property 3 (note the parentheses)

$= \frac{1}{4}(\log_a x + \log_a y - 3\log_a z)$ Using Properties 1 and 2

$= \frac{1}{4}\log_a x + \frac{1}{4}\log_a y - \frac{3}{4}\log_a z$ Distributive law

10. $\log_b \dfrac{xy}{m^3n^4} = \log_b xy - \log_b m^3n^4$ Using Property 3

$= (\log_b x + \log_b y) - (\log_b m^3 + \log_b n^4)$ Using Property 1

$= \log_b x + \log_b y - \log_b m^3 - \log_b n^4$ Removing parentheses

$= \log_b x + \log_b y - 3\log_b m - 4\log_b n$ Using Property 2

Do Exercises 9–11.

Examples Express as a single logarithm.

11. $\dfrac{1}{2}\log_a x - 7\log_a y + \log_a z$

$= \log_a x^{1/2} - \log_a y^7 + \log_a z$ Using Property 2

$= \log_a \dfrac{\sqrt{x}}{y^7} + \log_a z$ Using Property 3

$= \log_a \dfrac{z\sqrt{x}}{y^7}$ Using Property 1

12. $\log_a \dfrac{b}{\sqrt{x}} + \log_a \sqrt{bx}$

$= \log_a b - \log_a \sqrt{x} + \log_a \sqrt{bx}$ Using Property 3

$= \log_a b - \frac{1}{2}\log_a x + \frac{1}{2}\log_a (bx)$ Using Property 2

$= \log_a b - \frac{1}{2}\log_a x + \frac{1}{2}(\log_a b + \log_a x)$ Using Property 1

$= \log_a b - \frac{1}{2}\log_a x + \frac{1}{2}\log_a b + \frac{1}{2}\log_a x$

$= \frac{3}{2}\log_a b$ Collecting like terms

$= \log_a b^{3/2}$ Using Property 2

Example 12 could also be done as follows:

$\log_a \dfrac{b}{\sqrt{x}} + \log_a \sqrt{bx} = \log_a \dfrac{b}{\sqrt{x}}\sqrt{bx}$ Using Property 1

$= \log_a \dfrac{b}{\sqrt{x}} \cdot \sqrt{b} \cdot \sqrt{x}$

$= \log_a b\sqrt{b}$, or $\log_a b^{3/2}$.

Do Exercises 12 and 13.

Express in terms of logarithms of x, y, z, and w.

9. $\log_a \sqrt{\dfrac{z^3}{xy}}$

10. $\log_a \dfrac{x^2}{y^3z}$

11. $\log_a \dfrac{x^3y^4}{z^5w^9}$

Express as a single logarithm.

12. $5\log_a x - \log_a y + \dfrac{1}{4}\log_a z$

13. $\log_a \dfrac{\sqrt{x}}{b} - \log_a \sqrt{bx}$

Answers on page A-56

Given

$\log_a 2 = 0.301,$
$\log_a 5 = 0.699,$

find each of the following.

14. $\log_a 4$

15. $\log_a 10$

16. $\log_a \frac{2}{5}$

17. $\log_a \frac{5}{2}$

18. $\log_a \frac{1}{5}$

19. $\log_a \sqrt{a^3}$

20. $\log_a 5a$

21. $\log_a 16$

Simplify.

22. $\log_2 2^6$

23. $\log_{10} 10^{3.2}$

24. $\log_e e^{12}$

CAUTION! Keep in mind that, in general,

$\log_a (M + N) \neq \log_a M + \log_a N,$

$\log_a (M - N) \neq \log_a M - \log_a N,$

$\log_a (MN) \neq (\log_a M)(\log_a N),$

and

$\log_a (M/N) \neq (\log_a M) \div (\log_a N).$

Answers on page A-56

Examples Given $\log_a 2 = 0.301$ and $\log_a 3 = 0.477$, find each of the following.

13. $\log_a 6 = \log_a (2 \cdot 3) = \log_a 2 + \log_a 3$ **Property 1**

$= 0.301 + 0.477 = 0.778$

14. $\log_a \frac{2}{3} = \log_a 2 - \log_a 3$ **Property 3**

$= 0.301 - 0.477 = -0.176$

15. $\log_a 81 = \log_a 3^4 = 4 \log_a 3$ **Property 2**

$= 4(0.477) = 1.908$

16. $\log_a \frac{1}{3} = \log_a 1 - \log_a 3$ **Property 3**

$= 0 - 0.477 = -0.477$

17. $\log_a \sqrt{a} = \log_a a^{1/2} = \frac{1}{2} \log_a a = \frac{1}{2} \cdot 1 = \frac{1}{2}$ **Property 2**

18. $\log_a 2a = \log_a 2 + \log_a a$ **Property 1**

$= 0.301 + 1 = 1.301$

19. $\log_a 5$ **No way to find using these properties. ($\log_a 5 \neq \log_a 2 + \log_a 3$)**

20. $\frac{\log_a 3}{\log_a 2} = \frac{0.477}{0.301} \approx 1.58$ **We simply divide the logarithms, not using any property.**

Do Exercises 14–21.

e The Logarithm of the Base to a Power

> **PROPERTY 4**
>
> For any base a,
>
> $$\log_a a^k = k.$$
>
> (The logarithm, base a, of a to a power is the power.)

A PROOF OF PROPERTY 4 (*OPTIONAL*). The proof involves Property 2 and the fact that $\log_a a = 1$:

$\log_a a^k = k(\log_a a)$ **Using Property 2**

$= k \cdot 1$ **Using $\log_a a = 1$**

$= k.$

Examples Simplify.

21. $\log_3 3^7 = 7$

22. $\log_{10} 10^{5.6} = 5.6$

23. $\log_e e^{-t} = -t$

Do Exercises 22–24.

Exercise Set 12.4

a Express as a sum of logarithms.

1. $\log_2 (32 \cdot 8)$

2. $\log_3 (27 \cdot 81)$

3. $\log_4 (64 \cdot 16)$

4. $\log_5 (25 \cdot 125)$

5. $\log_a Qx$

6. $\log_r 8Z$

Express as a single logarithm.

7. $\log_b 3 + \log_b 84$

8. $\log_a 75 + \log_a 5$

9. $\log_c K + \log_c y$

10. $\log_t H + \log_t M$

b Express as a product.

11. $\log_c y^4$

12. $\log_a x^3$

13. $\log_b t^6$

14. $\log_{10} y^7$

15. $\log_b C^{-3}$

16. $\log_c M^{-5}$

c Express as a difference of logarithms.

17. $\log_a \frac{67}{5}$

18. $\log_t \frac{T}{7}$

19. $\log_b \frac{2}{5}$

20. $\log_a \frac{z}{y}$

Express as a single logarithm.

21. $\log_c 22 - \log_c 3$

22. $\log_d 54 - \log_d 9$

d Express in terms of logarithms.

23. $\log_a x^2y^3z$

24. $\log_a 5xy^4z^3$

25. $\log_b \frac{xy^2}{z^3}$

26. $\log_b \frac{p^2q^5}{m^4n^7}$

27. $\log_c \sqrt[3]{\frac{x^4}{y^3z^2}}$

28. $\log_a \sqrt{\frac{x^6}{p^5q^8}}$

29. $\log_a \sqrt[4]{\frac{m^8n^{12}}{a^3b^5}}$

30. $\log_a \sqrt{\frac{a^6b^8}{a^2b^5}}$

Express as a single logarithm and simplify if possible.

31. $\frac{2}{3} \log_a x - \frac{1}{2} \log_a y$

32. $\frac{1}{2} \log_a x + 3 \log_a y - 2 \log_a x$

33. $\log_a 2x + 3(\log_a x - \log_a y)$

34. $\log_a x^2 - 2 \log_a \sqrt{x}$

35. $\log_a \frac{a}{\sqrt{x}} - \log_a \sqrt{ax}$

36. $\log_a (x^2 - 4) - \log_a (x - 2)$

Given $\log_b 3 = 1.099$ and $\log_b 5 = 1.609$, find each of the following.

37. $\log_b 15$

38. $\log_b \frac{3}{5}$

39. $\log_b \frac{5}{3}$

40. $\log_b \frac{1}{3}$

41. $\log_b \frac{1}{5}$

42. $\log_b \sqrt{b}$

43. $\log_b \sqrt{b^3}$

44. $\log_b 3b$

45. $\log_b 5b$

46. $\log_b 9$

e Simplify.

47. $\log_e e^t$

48. $\log_w w^8$

49. $\log_p p^5$

50. $\log_Y Y^{-4}$

Solve for x.

51. $\log_2 2^7 = x$

52. $\log_9 9^4 = x$

53. $\log_e e^x = -7$

54. $\log_a a^x = 2.7$

Skill Maintenance

Compute and simplify. Express answers in the form $a + bi$, where $i^2 = -1$. [10.8b, c, d, e]

55. i^{29}

56. i^{34}

57. $(2 + i)(2 - i)$

58. $\frac{2 + i}{2 - i}$

59. $(7 - 8i) - (-16 + 10i)$

60. $2i^2 \cdot 5i^3$

61. $(8 + 3i)(-5 - 2i)$

62. $(2 - i)^2$

Synthesis

63. ◆ Find a way to express $\log_a (x/5)$ as a difference of logarithms without using the quotient rule. Explain your work.

64. ◆ A student incorrectly reasons that

$$\log_a \frac{1}{x} = \log_a \frac{x}{x \cdot x}$$
$$= \log_a x - \log_a x + \log_a x$$
$$= \log_a x.$$

What mistake has the student made? Explain what the answer should be.

65. Use the TABLE and GRAPH features to show that $\log x^2 \neq (\log x)(\log x)$.

66. Use the TABLE and GRAPH features to show that $\frac{\log x}{\log 4} \neq \log x - \log 4$.

Express as a single logarithm and, if possible, simplify.

67. $\log_a (x^8 - y^8) - \log_a (x^2 + y^2)$

68. $\log_a (x + y) + \log_a (x^2 - xy + y^2)$

Express as a sum or a difference of logarithms.

69. $\log_a \sqrt{1 - s^2}$

70. $\log_a \frac{c - d}{\sqrt{c^2 - d^2}}$

Determine whether each is true or false.

71. $\frac{\log_a P}{\log_a Q} = \log_a \frac{P}{Q}$

72. $\frac{\log_a P}{\log_a Q} = \log_a P - \log_a Q$

73. $\log_a 3x = \log_a 3 + \log_a x$

74. $\log_a 3x = 3 \log_a x$

75. $\log_a (P + Q) = \log_a P + \log_a Q$

76. $\log_a x^2 = 2 \log_a x$

12.5 Natural Logarithmic Functions

Any positive number other than 1 can serve as the base of a logarithmic function. Common, or base-10, logarithms, which were introduced in Section 12.3, are useful because they have the same base as our "commonly" used decimal system of naming numbers.

Today, another base is widely used. It is an irrational number named e. We now consider e and base-e, or **natural, logarithms.**

Objectives

a. Find logarithms or powers, base e, using a calculator.

b. Use the change-of-base formula to find logarithms to bases other than e or 10.

c. Graph exponential and logarithmic functions, base e.

For Extra Help

TAPE 21 MAC WIN CD-ROM

a The Base e and Natural Logarithms

When interest is computed n times per year, the compound-interest formula is

$$A = P\left(1 + \frac{r}{n}\right)^{nt},$$

where A is the amount that an initial investment P will grow to after t years at interest rate r. Suppose that \$1 could be invested at 100% interest for 1 year. (In reality, no financial institution would pay such an interest rate.) The preceding formula becomes a function A defined in terms of the number of compounding periods n:

$$A(n) = \left(1 + \frac{1}{n}\right)^n.$$

Let's find some function values, using a calculator and rounding to six decimal places. The numbers in this table approach a very important number called e. It is an irrational number, so its decimal representation neither terminates nor repeats.

n	$A(n) = \left(1 + \frac{1}{n}\right)^n$
1 (compounded annually)	\$2.00
2 (compounded semiannually)	\$2.25
3	\$2.370370
4 (compounded quarterly)	\$2.441406
5	\$2.488320
100	\$2.704814
365 (compounded daily)	\$2.714567
8760 (compounded hourly)	\$2.718127

The number e: $e \approx 2.7182818284\ldots$

Logarithms, base e, are called **natural logarithms,** or **Naperian logarithms,** in honor of John Napier (1550–1617), a Scotsman who invented logarithms.

The abbreviation **ln** is commonly used with natural logarithms. Thus,

$$\ln 29 \quad \text{means} \quad \log_e 29.$$

We usually read "ln 29" as "the natural log of 29," or simply "el en of 29."

On a scientific calculator or grapher, the scientific key for natural logarithms is generally marked [LN]. Using that key, we find that

$$\ln 29 \approx 3.36729583 \approx 3.3673,$$

rounded to four decimal places. This also tells us that $e^{3.3673} \approx 29$.

On some scientific calculators, the keystrokes for doing such a calculation might be

[2] [9] [LN] [=].

The display would then read `3.3672958`.

If we were to use the scientific keys on a grapher, the keystrokes might be

[LN] [2] [9] [ENTER]

The display would then read `3.36729583`.

Find the natural logarithm, to four decimal places, on a scientific calculator or a grapher.

1. ln 78,235.4

2. ln 0.0000309

3. ln (−3)

4. ln 0

5. ln 10

6. Find $e^{11.2675}$ using a calculator. (Compare this computation to that of Margin Exercise 1.)

7. Find e^{-2} using a calculator.

Answers on page A-56

Examples Find each natural logarithm, to four decimal places, on a scientific calculator or grapher.

	Function Value	*Readout*	*Rounded*
1.	ln 287,523	12.56905814	12.5691
2.	ln 0.000486	-7.629301934	-7.6293
3.	ln (−5)	NONREAL ANS	NONREAL ANS
4.	ln (*e*)	1	1
5.	ln 1	0	0

Do Exercises 1–5.

The inverse of a logarithmic function is an exponential function. Thus, if $f(x) = \ln x$, then $f^{-1}(x) = e^x$. Because of this, on many calculators, the [LN] key doubles as the [e^x] key after a [2nd] or [SHIFT] key has been pressed.

Example 6 Find $e^{12.5691}$ using a calculator.

On a scientific calculator, we might enter 12.5691 and press [e^x]. On a grapher, we might press [2nd] [e^x], followed by 12.5691. In either case, we get the approximation

$$e^{12.5691} \approx 287{,}535.0371.$$

Compare this computation to Example 1. Note that, apart from the rounding error, $e^{12.5691}$ takes us back to about 287,523.

Example 7 Find $e^{-1.524}$ using a calculator.

On a scientific calculator, we might enter −1.524 and press [e^x]. On a grapher, we might press [2nd] [e^x], followed by −1.524. In either case, we get the approximation

$$e^{-1.524} \approx 0.2178.$$

Do Exercises 6 and 7.

b Changing Logarithm Bases

Most calculators give the values of both common logarithms and natural logarithms. To find a logarithm with some other base, we can use the following conversion formula.

THE CHANGE-OF-BASE FORMULA

For any logarithm bases a and b and any positive number M,

$$\log_b M = \frac{\log_a M}{\log_a b}.$$

A PROOF OF THE CHANGE-OF-BASE FORMULA (*OPTIONAL*). Let $x = \log_b M$. Then, writing an equivalent exponential equation, we have $b^x = M$. Next, we take the logarithm base a on both sides. This gives us

$$\log_a b^x = \log_a M.$$

By Property 2,

$$x \log_a b = \log_a M,$$

and solving for x, we obtain

$$x = \frac{\log_a M}{\log_a b}.$$

But $x = \log_b M$, so we have

$$\log_b M = \frac{\log_a M}{\log_a b},$$

which is the change-of-base formula.

Example 8 Find $\log_4 7$ using common logarithms.

Let $a = 10$, $b = 4$, and $M = 7$. Then we substitute into the change-of-base formula:

$$\log_b M = \frac{\log_a M}{\log_a b}$$

$$\log_4 7 = \frac{\log_{10} 7}{\log_{10} 4} \qquad \text{Substituting 10 for } a\text{, 4 for } b\text{, and 7 for } M$$

$$\approx 1.4036.$$

To check, we use a calculator with a power key $\boxed{y^x}$ to verify that

$$4^{1.4036} \approx 7.$$

We can also use base e for a conversion.

Example 9 Find $\log_5 29$ using natural logarithms.

Substituting e for a, 5 for b, and 29 for M, we have

$$\log_5 29 = \frac{\log_e 29}{\log_e 5} \qquad \text{Using the change-of-base formula}$$

$$= \frac{\ln 29}{\ln 5} \approx 2.0923.$$

Do Exercises 8 and 9.

8. Find $\log_6 7$ using common logarithms.

9. Find $\log_2 46$ using natural logarithms.

c Graphs of Exponential and Logarithmic Functions, Base e

Example 10 Graph $f(x) = e^x$ and $g(x) = e^{-x}$.

We use a calculator with an $\boxed{e^x}$ key to find approximate values of e^x and e^{-x}. Using these values, we can graph the functions.

x	e^x	e^{-x}
0	1	1
1	2.7	0.4
2	7.4	0.1
−1	0.4	2.7
−2	0.1	7.4

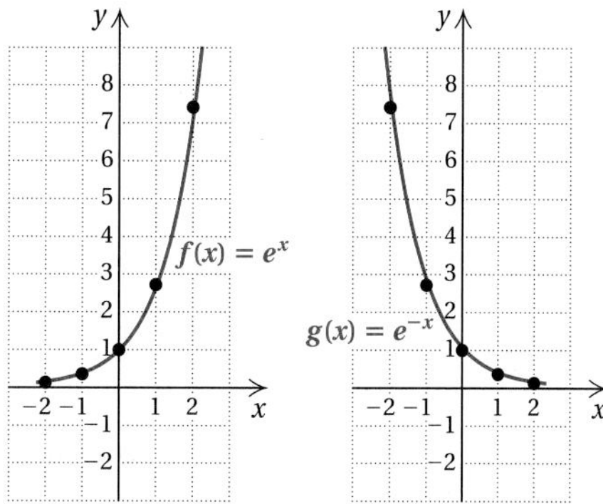

Calculator Spotlight

To graph $f(x) = e^x$, we press $\boxed{y=}$ $\boxed{2nd}$ $\boxed{e^x}$ $\boxed{X}$. To graph $f(x) = \ln_e x = \ln x$, we press $\boxed{y=}$ $\boxed{LN}$ $\boxed{X}$. You will see $y_1 = \ln(x$ on the display. Note that there is no right parenthesis. It does not need to be included unless you are entering a function like

$$g(x) = 2 \cdot \ln(x - 3) + 1.$$

To graph $y = \log_5 x$, we use the change-of-base formula and graph

$$y = \log_5 x = \frac{\log_e x}{\log_e 5} = \frac{\ln x}{\ln 5},$$

$$\text{or } y = \frac{\log_{10} x}{\log_{10} 5} = \frac{\log x}{\log 5}.$$

Exercises

Graph.

1. $f(x) = e^x$
2. $f(x) = e^{-0.06x}$
3. $f(x) = \log_e x = \ln x$
4. $f(x) = 2 \cdot \ln x$
5. $g(x) = 2 \cdot \ln(x + 3) - 1$
6. $y = \log_5 x$
7. $f(x) = \log_7 x$
8. $y = \log_5(x + 2)$
9. $g(x) = \log_7 x + 2$

Answers on page A-56

Graph.

10. $f(x) = e^{2x}$

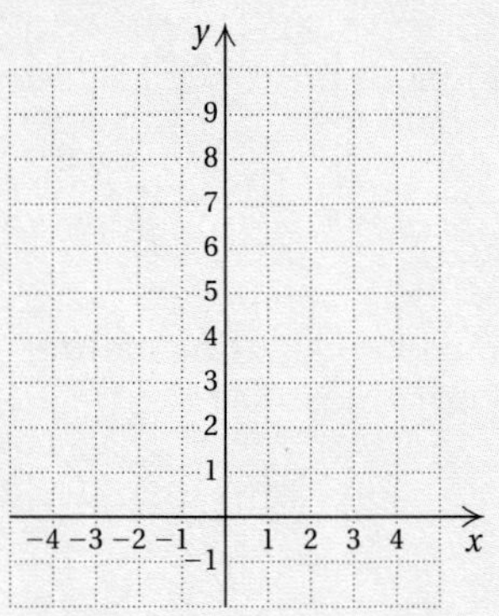

11. $g(x) = \frac{1}{2}e^{-x}$

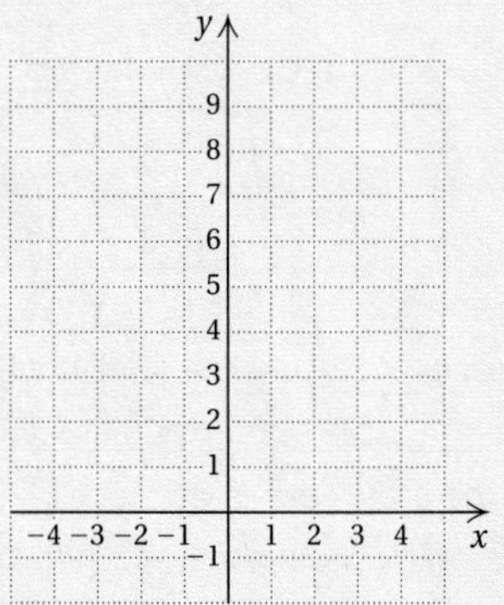

Graph.

12. $f(x) = 2 \ln x$

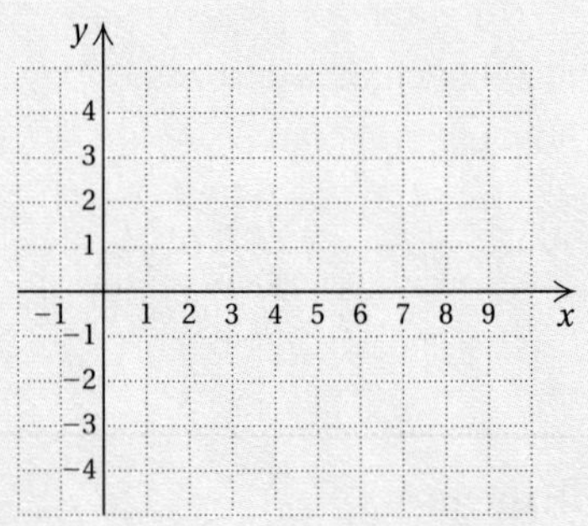

13. $g(x) = \ln (x - 2)$

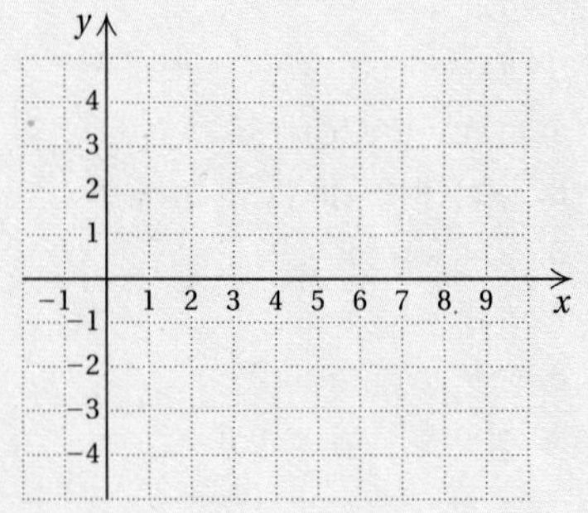

Answers on page A-56

Example 11 Graph: $f(x) = e^{-0.5x}$.

We find some solutions with a calculator, plot them, and then draw the graph. For example,

$$f(2) = e^{-0.5(2)} = e^{-1} \approx 0.4.$$

x	$e^{-0.5x}$
0	1
1	0.6
2	0.4
3	0.2
−1	1.6
−2	2.7
−3	4.5

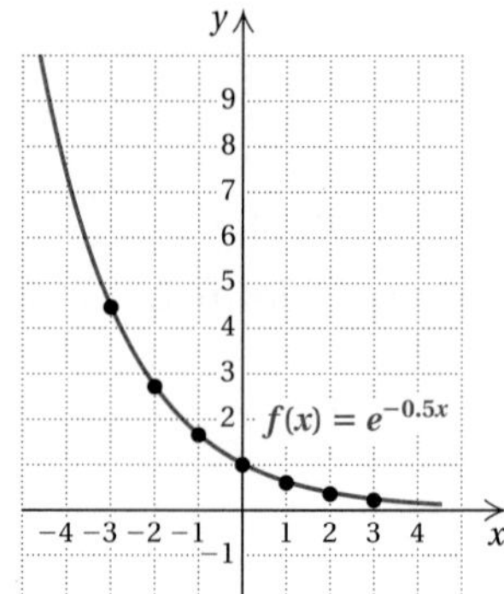

Do Exercises 10 and 11.

Example 12 Graph: $g(x) = \ln x$.

We find some solutions with a calculator and then draw the graph. As expected, the graph is a reflection across the line $y = x$ of the graph of $y = e^x$.

x	$\ln x$
1	0
4	1.4
7	1.9
0.5	−0.7

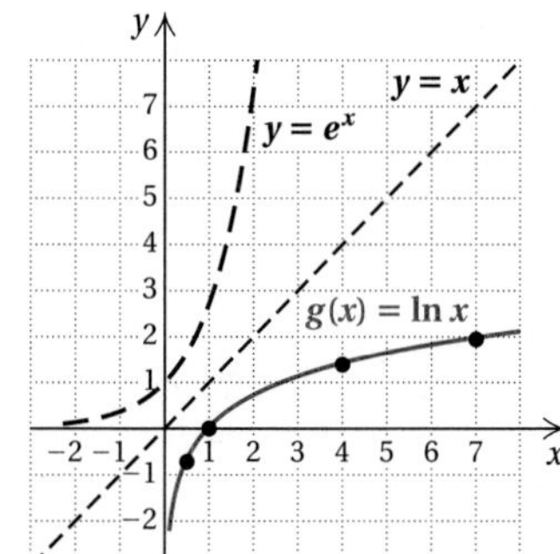

Example 13 Graph: $f(x) = \ln (x + 3)$.

We find some solutions with a calculator, plot them, and then draw the graph.

x	$\ln (x + 3)$
0	1.1
1	1.4
2	1.6
3	1.8
4	1.9
−1	0.7
−2	0
−2.5	−0.7

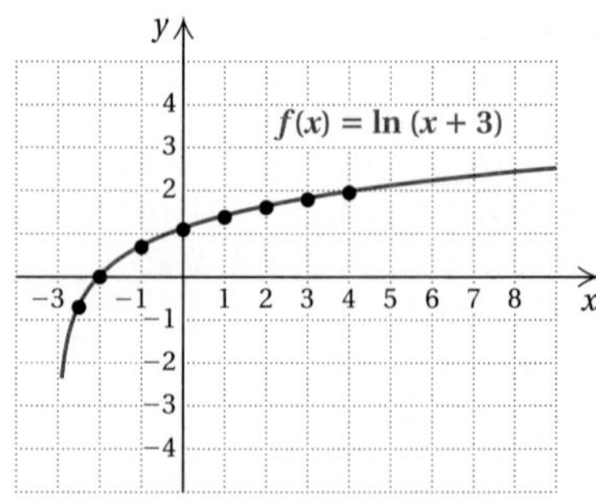

The graph of $y = \ln (x + 3)$ is the graph of $y = \ln x$ translated 3 units to the left.

Do Exercises 12 and 13.

Exercise Set 12.5

a Find each of the following logarithms or powers, base e, using a calculator. Round answers to four decimal places.

1. $\ln 2$ **2.** $\ln 5$ **3.** $\ln 62$ **4.** $\ln 30$ **5.** $\ln 4365$ **6.** $\ln 901.2$

7. $\ln 0.0062$ **8.** $\ln 0.00073$ **9.** $\ln 0.2$ **10.** $\ln 0.04$ **11.** $\ln 0$ **12.** $\ln(-4)$

13. $\ln\left(\frac{97.4}{558}\right)$ **14.** $\ln\left(\frac{786.2}{77.2}\right)$ **15.** $\ln e$ **16.** $\ln e^2$ **17.** $e^{2.71}$ **18.** $e^{3.06}$

19. $e^{-3.49}$ **20.** $e^{-2.64}$ **21.** $e^{4.7}$ **22.** $e^{1.23}$ **23.** $\ln e^5$ **24.** $e^{\ln 7}$

b Find each of the following logarithms using the change-of-base formula.

25. $\log_6 100$ **26.** $\log_3 100$ **27.** $\log_2 100$ **28.** $\log_7 100$ **29.** $\log_7 65$ **30.** $\log_5 42$

31. $\log_{0.5} 5$ **32.** $\log_{0.1} 3$ **33.** $\log_2 0.2$ **34.** $\log_2 0.08$ **35.** $\log_\pi 200$ **36.** $\log_\pi \pi$

c Graph.

37. $f(x) = e^x$

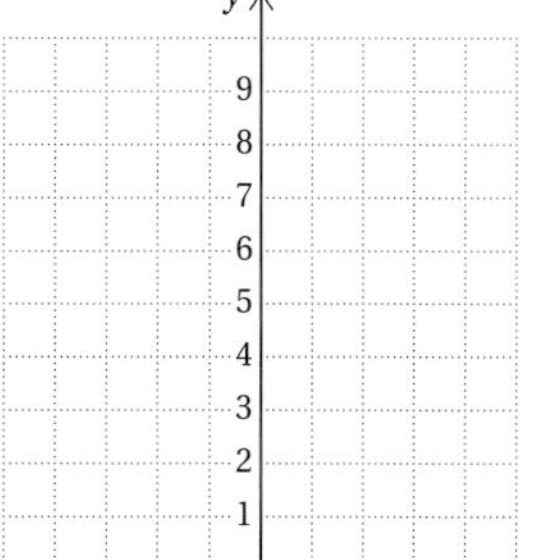

38. $f(x) = e^{0.5x}$

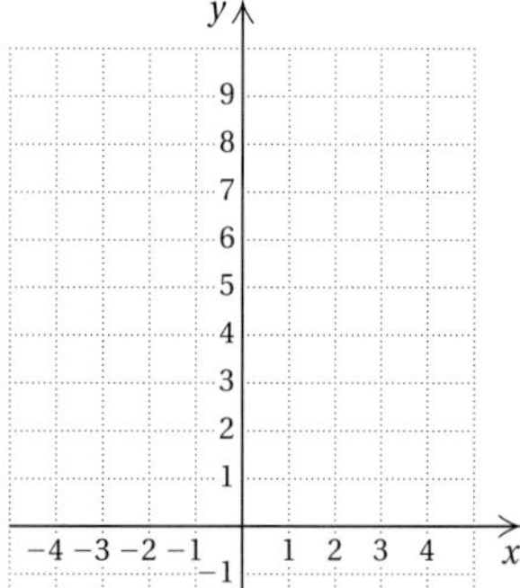

39. $f(x) = e^{-0.5x}$

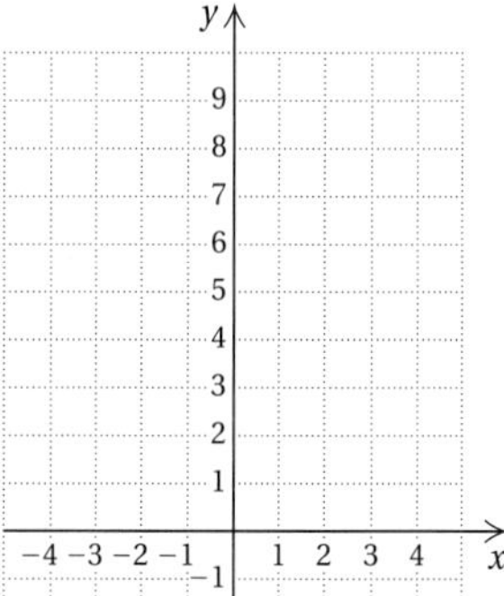

40. $f(x) = e^{-x}$

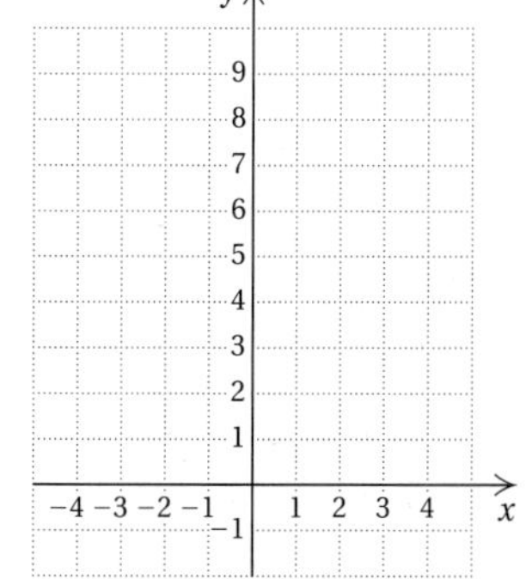

41. $f(x) = e^{x-1}$

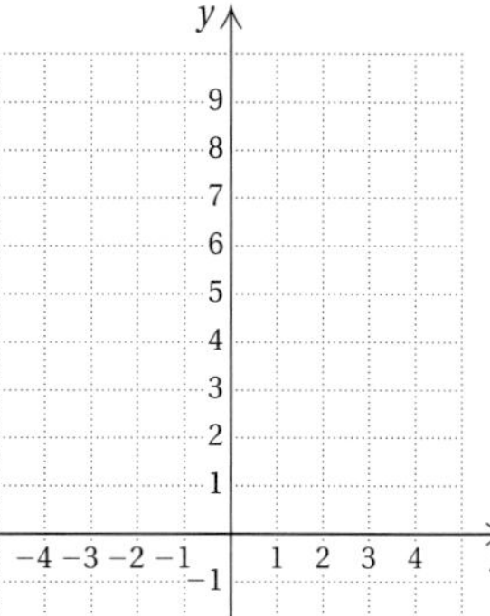

42. $f(x) = e^{-x} + 3$

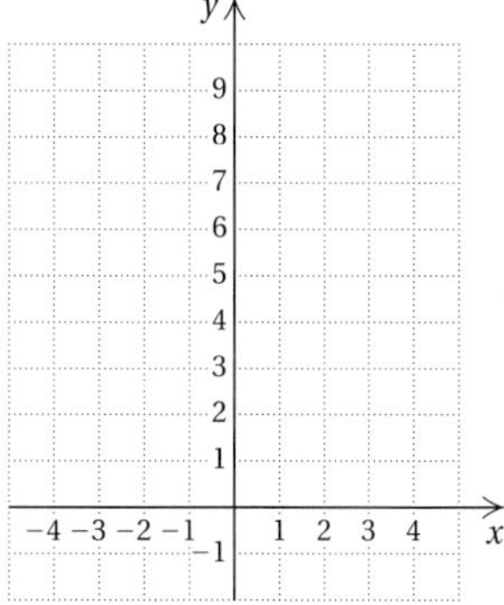

43. $f(x) = e^{x+2}$

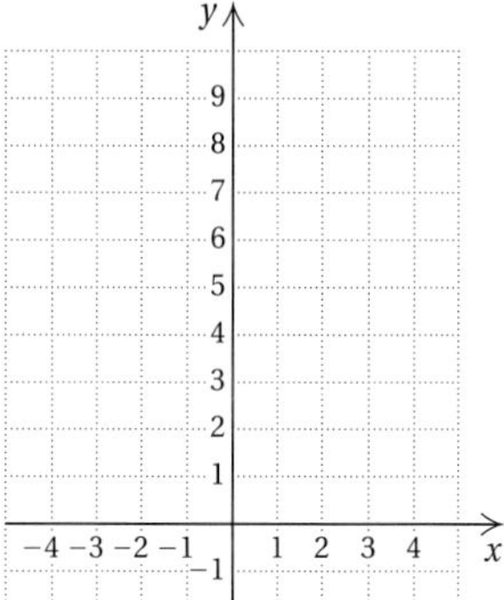

44. $f(x) = e^{x-2}$

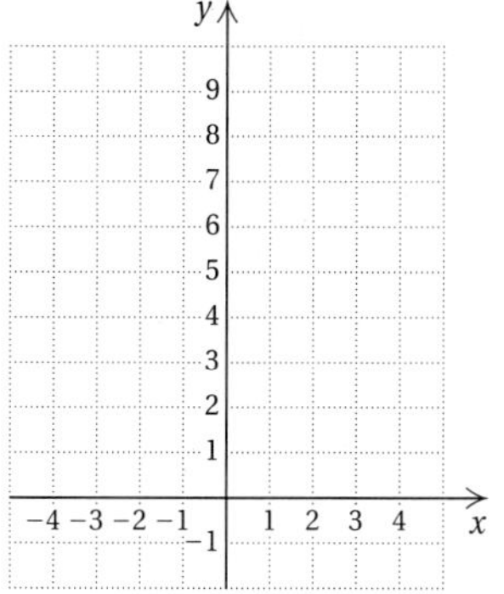

45. $f(x) = e^x - 1$

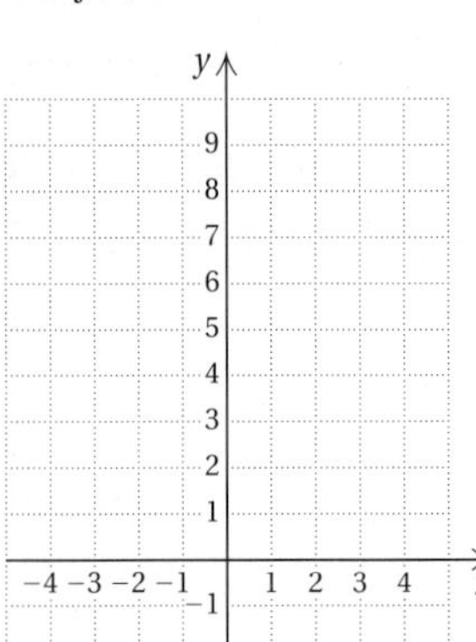

46. $f(x) = 2e^{0.5x}$

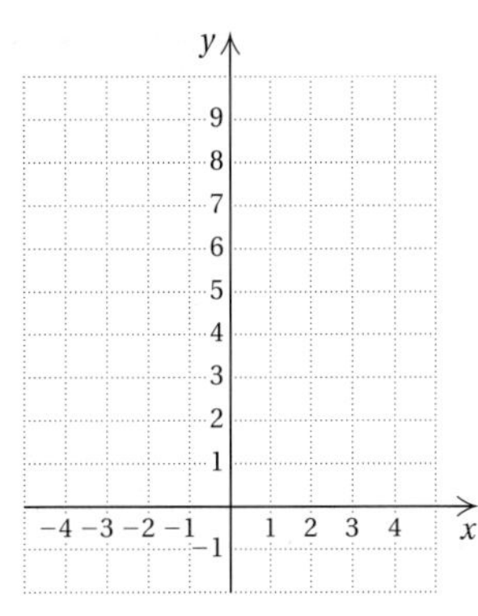

47. $f(x) = \ln (x + 2)$

48. $f(x) = \ln (x + 1)$

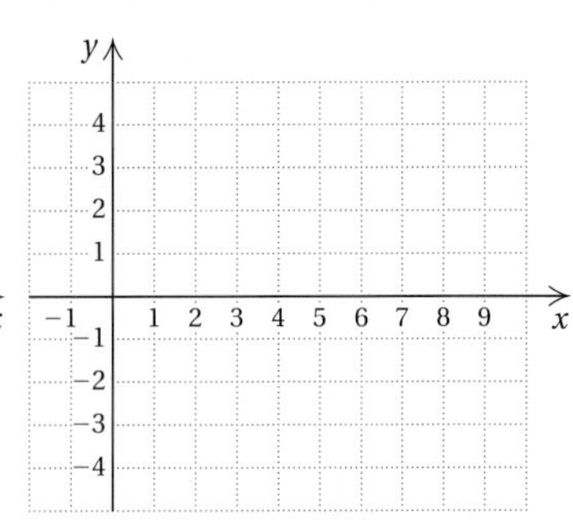

49. $f(x) = \ln (x - 3)$

50. $f(x) = 2 \ln (x - 2)$

51. $f(x) = 2 \ln x$

52. $f(x) = \ln x - 3$

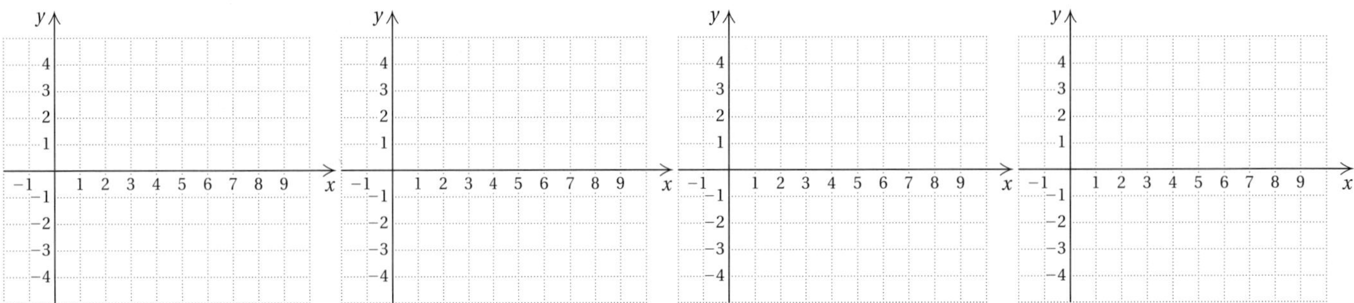

53. $f(x) = \frac{1}{2} \ln x + 1$

54. $f(x) = \ln x^2$

55. $f(x) = |\ln x|$

56. $f(x) = \ln |x|$

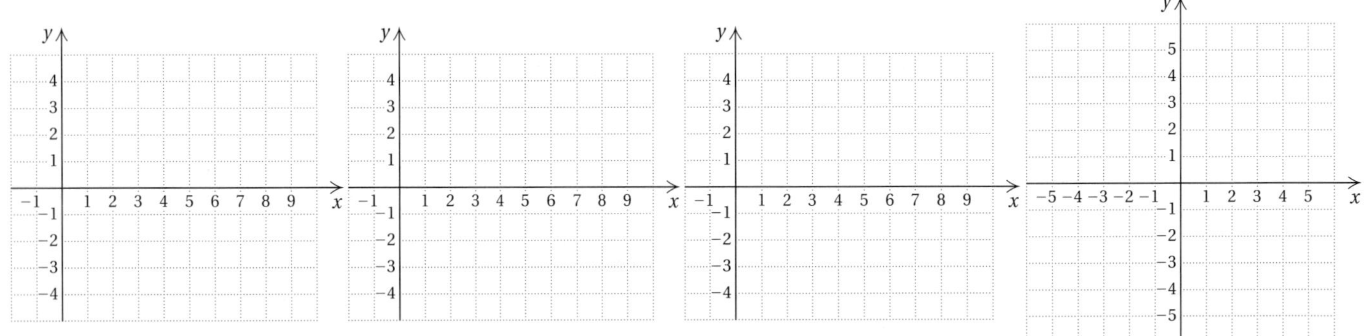

Skill Maintenance

Solve. [11.4c]

57. $x^{1/2} - 6x^{1/4} + 8 = 0$

58. $2y - 7\sqrt{y} + 3 = 0$

59. $x - 18\sqrt{x} + 77 = 0$

60. $x^4 - 25x^2 + 144 = 0$

Synthesis

61. ◆ Explain how the graph of $f(x) = e^x$ could be used to obtain the graph of $g(x) = 1 + \ln x$.

62. Find a formula for converting common logarithms to natural logarithms. Then find a formula for converting natural logarithms to common logarithms.

Graph and find the domain and the range.

63. $f(x) = 10x^2e^{-x}$

64. $f(x) = 7.4e^x \ln x$

65. $f(x) = 100(1 - e^{-0.3x})$

Find the domain.

66. $f(x) = \log_3 x^2$

67. $f(x) = \log (2x - 5)$

12.6 Solving Exponential and Logarithmic Equations

a Solving Exponential Equations

Equations with variables in exponents, such as $5^x = 12$ and $2^{7x} = 64$, are called **exponential equations.** Sometimes, as is the case with $2^{7x} = 64$, we can write each side as a power of the same number:

$$2^{7x} = 2^6.$$

Since the base is the same, 2, the exponents are the same. We can set them equal and solve:

$$7x = 6$$
$$x = \frac{6}{7}.$$

We use the following property, which is true because exponential functions are one-to-one.

> For any $a > 0$, $a \neq 1$,
> $$a^x = a^y \Rightarrow x = y.$$

Example 1 Solve: $2^{3x-5} = 16$.

Note that $16 = 2^4$. Thus we can write each side as a power of the same number:

$$2^{3x-5} = 2^4.$$

Since the base is the same, 2, the exponents must be the same. Thus,

$$3x - 5 = 4$$
$$3x = 9$$
$$x = 3.$$

CHECK:

2^{3x-5}	$= 16$
$2^{3\cdot 3-5}$	? 16
2^{9-5}	
2^4	
16	TRUE

The solution is 3.

Do Exercises 1 and 2.

AG Algebraic–Graphical Connection

The solution, 3, of the equation $2^{3x-5} = 16$ in Example 1 is the x-coordinate of the point of intersection of the graphs of $y = 2^{3x-5}$ and $y = 16$, as we see in the graph on the left below.

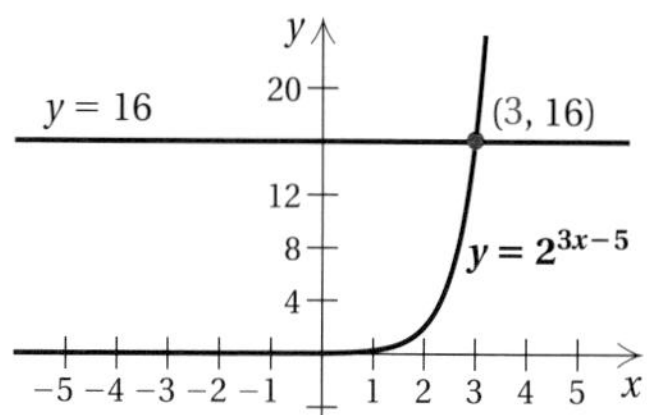

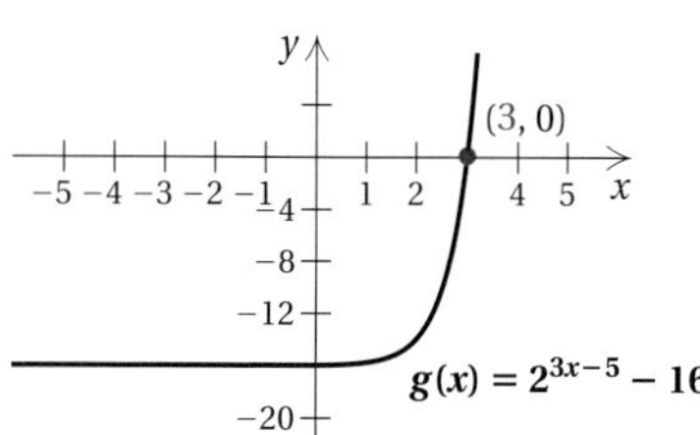

If we subtract 16 on both sides of $2^{3x-5} = 16$, we get $2^{3x-5} - 16 = 0$. The solution, 3, is then the x-coordinate of the x-intercept of the function $g(x) = 2^{3x-5} - 16$, as we see in the graph on the right above.

Objectives

a Solve exponential equations.

b Solve logarithmic equations.

For Extra Help

TAPE 21 MAC WIN CD-ROM

Solve.

1. $3^{2x} = 9$

2. $4^{2x-3} = 64$

Answers on page A-57

3. Solve: $7^x = 20$.

4. Solve: $e^{0.3t} = 80$.

When it does not seem possible to write each side as a power of the same base, we can take the common or natural logarithm on each side and then use Property 2.

Example 2 Solve: $5^x = 12$.

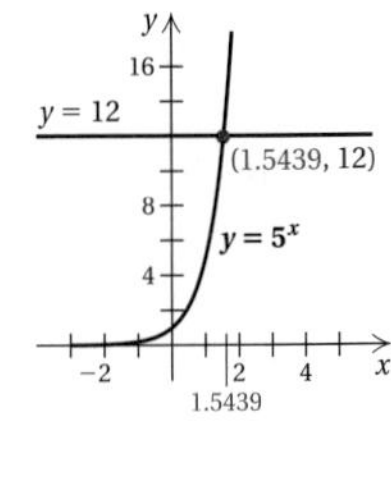

$$5^x = 12$$

$$\log 5^x = \log 12 \quad \text{Taking the common logarithm on both sides}$$

$$x \log 5 = \log 12 \quad \text{Property 2}$$

$$x = \frac{\log 12}{\log 5}$$

Caution! This is not $\log \frac{12}{5}$, or $\log 12 - \log 5$!

This is an exact answer. We cannot simplify further, but we can approximate using a calculator:

$$x = \frac{\log 12}{\log 5} \approx 1.5439.$$

We can also partially check this answer by finding $5^{1.5439}$ using a calculator.

CHECK:

$5^x = 12$	
$5^{1.5439}$?	12
11.99885457	TRUE

We get an answer close to 12, due to the rounding. This checks.

Do Exercise 3.

If the base is e, we can make our work easier by taking the logarithm, base e, on both sides.

Example 3 Solve: $e^{0.06t} = 1500$.

We take the natural logarithm on both sides:

$$e^{0.06t} = 1500$$

$$\ln e^{0.06t} = \ln 1500 \quad \text{Taking ln on both sides}$$

$$\log_e e^{0.06t} = \ln 1500 \quad \text{Definition of natural logarithms}$$

$$0.06t = \ln 1500 \quad \text{Here we use Property 4: } \log_a a^k = k.$$

$$t = \frac{\ln 1500}{0.06}.$$

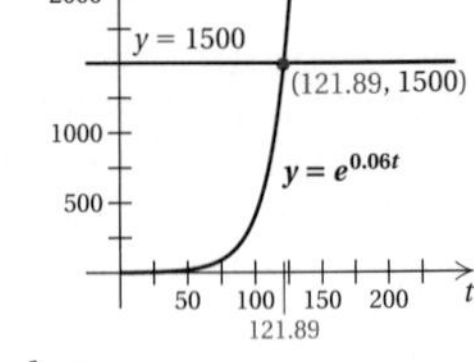

We can approximate using a calculator:

$$t = \frac{\ln 1500}{0.06} \approx \frac{7.3132}{0.06} \approx 121.89.$$

We can also partially check this answer using a calculator.

CHECK:

$e^{0.06t} = 1500$	
$e^{0.06(121.89)}$?	1500
$e^{7.3134}$	
1500.269444	TRUE

The solution is about 121.89.

Do Exercise 4.

Calculator Spotlight

Use the SOLVE or INTERSECT feature to check the results of Examples 1–3 and Margin Exercises 1–4.

Answers on page A-57

b Solving Logarithmic Equations

Equations containing logarithmic expressions are called **logarithmic equations.** We solved some logarithmic equations in Section 12.3 by converting to equivalent exponential equations.

Example 4 Solve: $\log_2 x = 3$.

We obtain an equivalent exponential equation:

$$x = 2^3$$
$$x = 8.$$

The solution is 8.

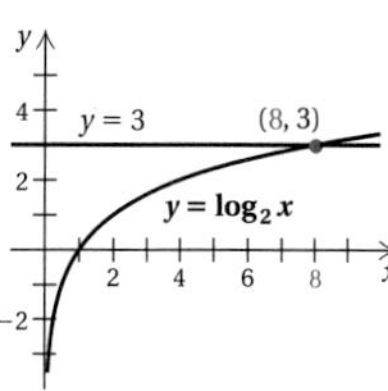

Do Exercise 5.

5. Solve: $\log_5 x = 2$.

To solve a logarithmic equation, first try to obtain a single logarithmic expression on one side and then write an equivalent exponential equation.

Example 5 Solve: $\log_4 (8x - 6) = 3$.

We already have a single logarithmic expression, so we write an equivalent exponential equation:

$8x - 6 = 4^3$ Writing an equivalent exponential equation

$8x - 6 = 64$

$8x = 70$

$x = \frac{70}{8}$, or $\frac{35}{4}$

CHECK:

$\log_4 (8x - 6) = 3$	
$\log_4 \left(8 \cdot \frac{35}{4} - 6\right)$	? 3
$\log_4 (70 - 6)$	
$\log_4 64$	
3	TRUE

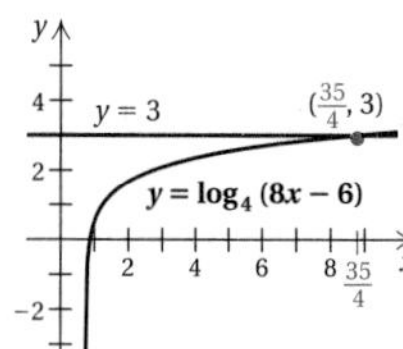

The solution is $\frac{35}{4}$.

Do Exercise 6.

6. Solve: $\log_3 (5x + 7) = 2$.

Example 6 Solve: $\log x + \log (x - 3) = 1$.

Here we have common logarithms. It will help us follow the solution to first write in the 10's and obtain a single logarithmic expression on the left.

$\log_{10} x + \log_{10} (x - 3) = 1$

$\log_{10} [x(x - 3)] = 1$ Using Property 1 to obtain a single logarithm

$x(x - 3) = 10^1$ Writing an equivalent exponential expression

$x^2 - 3x = 10$

$x^2 - 3x - 10 = 0$

$(x + 2)(x - 5) = 0$ Factoring

$x + 2 = 0$ *or* $x - 5 = 0$ Using the principle of zero products

$x = -2$ *or* $x = 5$

Answers on page A-57

7. Solve: $\log x + \log (x + 3) = 1$.

8. Solve:

$\log_3 (2x - 1) - \log_3 (x - 4) = 2$.

Answers on page A-57

CHECK: For -2:

$$\begin{array}{c|l} \log x + \log (x - 3) & 1 \\ \hline \log (-2) + \log (-2 - 3) & ?\ 1 \end{array}$$

The number -2 does *not* check because negative numbers do not have logarithms.

For 5:

$$\begin{array}{r|l} \log x + \log (x - 3) & 1 \\ \hline \log 5 + \log (5 - 3) & ?\ 1 \\ \log 5 + \log 2 & \\ \log (5 \cdot 2) & \\ \log 10 & \\ 1 & \quad \text{TRUE} \end{array}$$

The solution is 5.

Do Exercise 7.

Example 7 Solve: $\log_2 (x + 7) - \log_2 (x - 7) = 3$.

$\log_2 (x + 7) - \log_2 (x - 7) = 3$

$$\log_2 \frac{x + 7}{x - 7} = 3$$ Using Property 3 to obtain a single logarithm

$$\frac{x + 7}{x - 7} = 2^3$$ Writing an equivalent exponential expression

$$\frac{x + 7}{x - 7} = 8$$

$$x + 7 = 8(x - 7)$$ Multiplying by the LCM, $x - 7$

$$x + 7 = 8x - 56$$ Using a distributive law

$$63 = 7x$$

$$\frac{63}{7} = x$$

$$9 = x$$

CHECK:

$$\begin{array}{r|l} \log_2 (x + 7) - \log_2 (x - 7) & 3 \\ \hline \log_2 (9 + 7) - \log_2 (9 - 7) & ?\ 3 \\ \log_2 16 - \log_2 2 & \\ \log_2 \frac{16}{2} & \\ \log_2 8 & \\ 3 & \quad \text{TRUE} \end{array}$$

The solution is 9.

Do Exercise 8.

Exercise Set 12.6

a Solve.

1. $2^x = 8$

2. $3^x = 81$

3. $4^x = 256$

4. $5^x = 125$

5. $2^{2x} = 32$

6. $4^{3x} = 64$

7. $3^{5x} = 27$

8. $5^{7x} = 625$

9. $2^x = 11$

10. $2^x = 20$

11. $2^x = 43$

12. $2^x = 55$

13. $5^{4x-7} = 125$

14. $4^{3x+5} = 16$

15. $3^{x^2} \cdot 3^{4x} = \frac{1}{27}$

16. $3^{5x} \cdot 9^{x^2} = 27$

17. $4^x = 8$

18. $6^x = 10$

19. $e^t = 100$

20. $e^t = 1000$

21. $e^{-t} = 0.1$

22. $e^{-t} = 0.01$

23. $e^{-0.02t} = 0.06$

24. $e^{0.07t} = 2$

25. $2^x = 3^{x-1}$

26. $3^{x+2} = 5^{x-1}$

27. $(3.6)^x = 62$

28. $(5.2)^x = 70$

b Solve.

29. $\log_4 x = 4$

30. $\log_7 x = 3$

31. $\log_2 x = -5$

32. $\log_9 x = \frac{1}{2}$

33. $\log x = 1$

34. $\log x = 3$

35. $\log x = -2$

36. $\log x = -3$

37. $\ln x = 2$

38. $\ln x = 1$

39. $\ln x = -1$

40. $\ln x = -3$

41. $\log_3 (2x + 1) = 5$

42. $\log_2 (8 - 2x) = 6$

43. $\log x + \log (x - 9) = 1$

44. $\log x + \log (x + 9) = 1$

45. $\log x - \log (x + 3) = -1$

46. $\log (x + 9) - \log x = 1$

47. $\log_2 (x + 1) + \log_2 (x - 1) = 3$

48. $\log_2 x + \log_2 (x - 2) = 3$

49. $\log_4 (x + 6) - \log_4 x = 2$

50. $\log_4 (x + 3) - \log_4 (x - 5) = 2$

51. $\log_4 (x + 3) + \log_4 (x - 3) = 2$

52. $\log_5 (x + 4) + \log_5 (x - 4) = 2$

53. $\log_3 (2x - 6) - \log_3 (x + 4) = 2$

54. $\log_4 (2 + x) - \log_4 (3 - 5x) = 3$

Skill Maintenance

Solve. [11.4c]

55. $x^4 + 400 = 104x^2$

56. $x^{2/3} + 2x^{1/3} = 8$

57. $(x^2 + 5x)^2 + 2(x^2 + 5x) = 24$

58. $10 = x^{-2} + 9x^{-1}$

59. Simplify: $(125x^3y^{-2}z^6)^{-2/3}$. [10.2c]

60. Simplify: i^{79}. [10.8d]

Synthesis

61. ◆ Explain how Exercises 37–40 could be solved using only the graph of $f(x) = \ln x$.

62. ◆ Christina first determines that the solution of $\log_3 (x + 4) = 1$ is -1, but then rejects it. What mistake do you think she might be making?

63. Find the value of x for which the natural logarithm is the same as the common logarithm.

64. Use a grapher to check your answers to Exercises 4, 20, 36, and 54.

65. Use a grapher to solve each of the following equations.

a) $e^{7x} = 14$
b) $8e^{0.5x} = 3$
c) $xe^{3x-1} = 5$
d) $4 \ln (x + 3.4) = 2.5$

66. Use the INTERSECT feature of a grapher to find the points of intersection of the graphs of each pair of functions.

a) $f(x) = e^{0.5x-7}$, $g(x) = 2x + 6$
b) $f(x) = \ln 3x$, $g(x) = 3x - 8$
c) $f(x) = \ln x^2$, $g(x) = -x^2$

Solve.

67. $2^{2x} + 128 = 24 \cdot 2^x$

68. $27^x = 81^{2x-3}$

69. $8^x = 16^{3x+9}$

70. $\log_x (\log_3 27) = 3$

71. $\log_6 (\log_2 x) = 0$

72. $x \log \frac{1}{8} = \log 8$

73. $\log_5 \sqrt{x^2 - 9} = 1$

74. $2^{x^2+4x} = \frac{1}{8}$

75. $\log (\log x) = 5$

76. $\log_5 |x| = 4$

77. $\log x^2 = (\log x)^2$

78. $\log_3 |5x - 7| = 2$

79. $\log_a a^{x^2+4x} = 21$

80. $\sqrt{x} \cdot \sqrt[3]{x} \cdot \sqrt[4]{x} \cdot \sqrt[5]{x} = 146$

81. $3^{2x} - 8 \cdot 3^x + 15 = 0$

82. If $x = (\log_{125} 5)^{\log_5 125}$, what is the value of $\log_3 x$?

12.7 Mathematical Modeling with Exponential and Logarithmic Functions

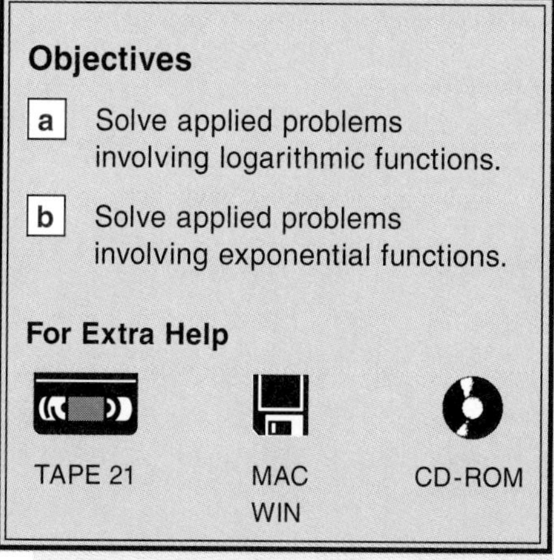

Exponential and logarithmic functions can now be added to our library of functions that can serve as models for many kinds of applications. Let's review some of their graphs.

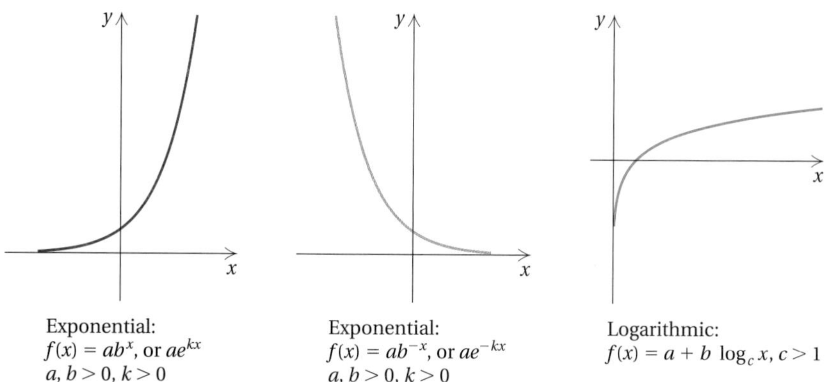

Exponential: $f(x) = ab^x$, or ae^{kx}, $a, b > 0, k > 0$

Exponential: $f(x) = ab^{-x}$, or ae^{-kx}, $a, b > 0, k > 0$

Logarithmic: $f(x) = a + b \log_c x, c > 1$

a Applications of Logarithmic Functions

Example 1 *Sound Levels.* To measure the "loudness" of any particular sound, the decibel scale is used. The loudness L, in decibels (dB), of a sound is given by

$$L = 10 \cdot \log \frac{I}{I_0},$$

where I is the intensity of the sound, in watts per square meter (W/m^2), and $I_0 = 10^{-12}$ W/m^2. (I_0 is approximately the intensity of the softest sound that can be heard.)

a) It is common for the intensity of sound at live performances of rock music to reach 10^{-1} W/m^2 (even higher close to the stage). How loud, in decibels, is this sound level?

b) Audiologists and physicians recommend that earplugs be worn when one is exposed to sounds in excess of 90 db. What is the intensity of such sounds?

1. *Acoustics.* The intensity of sound in normal conversation is about 3.2×10^{-6} W/m^2. How high is this sound level in decibels?

2. *Audiology.* Overexposure to excessive sound levels can diminish one's hearing to the point where the softest sound that is audible is 28 dB. What is the intensity of such a sound?

Answers on page A-57

a) To find the loudness, in decibels, we use the above formula:

$$\begin{aligned} L &= 10 \cdot \log \frac{I}{I_0} \\ &= 10 \cdot \log \frac{10^{-1}}{10^{-12}} && \text{Substituting} \\ &= 10 \cdot \log 10^{11} && \text{Subtracting exponents} \\ &= 10 \cdot 11 && \log 10^a = a \\ &= 110. \end{aligned}$$

The volume of the music is 110 decibels.

b) We substitute and solve for I:

$$\begin{aligned} L &= 10 \cdot \log \frac{I}{I_0} \\ 90 &= 10 \cdot \log \frac{I}{10^{-12}} && \text{Substituting} \\ 9 &= \log \frac{I}{10^{-12}} && \text{Dividing by 10} \\ 9 &= \log I - \log 10^{-12} && \text{Using Property 3} \\ 9 &= \log I - (-12) && \log 10^a = a \\ -3 &= \log I && \text{Adding } -12 \\ 10^{-3} &= I. && \text{Converting to an exponential equation} \end{aligned}$$

Earplugs are recommended for sounds with intensities exceeding 10^{-3} W/m^2.

Do Exercises 1 and 2.

Example 2 *Chemistry: pH of Liquids.* In chemistry the pH of a liquid is defined as:

$$\text{pH} = -\log\,[\text{H}^+],$$

where $[\text{H}^+]$ is the hydrogen ion concentration in moles per liter.

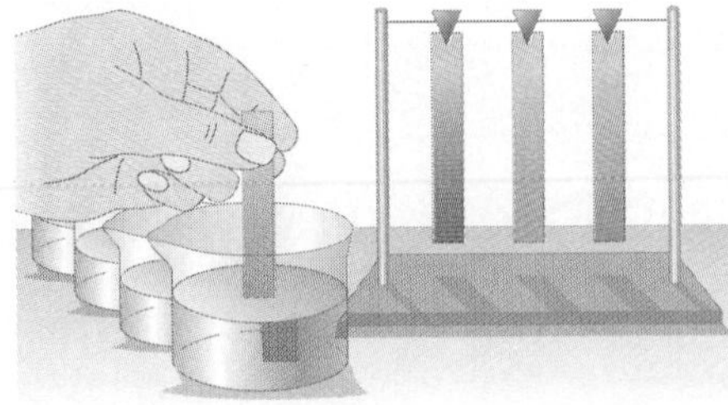

a) The hydrogen ion concentration of human blood is normally about 3.98×10^{-8} moles per liter. Find the pH.

b) The pH of seawater is about 8.3. Find the hydrogen ion concentration.

a) To find the pH of human blood, we use the above formula:

$$\begin{aligned} \text{pH} &= -\log\,[\text{H}^+] \\ &= -\log\,[3.98 \times 10^{-8}] \\ &\approx -(-7.400117) && \text{Using a calculator} \\ &\approx 7.4. \end{aligned}$$

The pH of human blood is normally about 7.4.

b) We substitute and solve for $[H^+]$:

$$8.3 = -\log [H^+] \quad \text{Using pH} = -\log [H^+]$$
$$-8.3 = \log [H^+] \quad \text{Dividing by } -1$$
$$10^{-8.3} = [H^+] \quad \text{Converting to an exponential equation}$$
$$5.01 \times 10^{-9} \approx [H^+]. \quad \text{Using a calculator; writing scientific notation}$$

The hydrogen ion concentration of seawater is about 5.01×10^{-9} moles per liter.

Do Exercises 3 and 4.

b Applications of Exponential Functions

Example 3 *Interest Compounded Annually.* Suppose that \$30,000 is invested at 8% interest, compounded annually. In t years, it will grow to the amount A given by the function

$$A(t) = 30{,}000(1.08)^t.$$

(See Example 6 in Section 12.1.)

a) How long will it take to accumulate \$150,000 in the account?

b) Let T = the amount of time it takes for the \$30,000 to double itself; T is called the **doubling time.** Find the doubling time.

a) We set $A(t) = 150{,}000$ and solve for t:

$$150{,}000 = 30{,}000(1.08)^t$$
$$\frac{150{,}000}{30{,}000} = (1.08)^t \quad \text{Dividing by 30,000}$$
$$5 = (1.08)^t$$
$$\log 5 = \log (1.08)^t \quad \text{Taking the common logarithm on both sides}$$
$$\log 5 = t \log 1.08 \quad \text{Using Property 2}$$
$$\frac{\log 5}{\log 1.08} = t \quad \text{Dividing by log 1.08}$$
$$20.9 \approx t. \quad \text{Using a calculator}$$

It will take about 20.9 yr for the \$30,000 to grow to \$150,000.

b) To find the *doubling time T,* we replace $A(t)$ with 60,000 and t with T and solve for T:

$$60{,}000 = 30{,}000(1.08)^T$$
$$2 = (1.08)^T \quad \text{Dividing by 30,000}$$
$$\log 2 = \log (1.08)^T \quad \text{Taking the common logarithm on both sides}$$
$$\log 2 = T \log 1.08 \quad \text{Using Property 2}$$
$$T = \frac{\log 2}{\log 1.08} \approx 9.0. \quad \text{Using a calculator}$$

The doubling time is about 9 yr.

Do Exercise 5.

3. *Coffee.* The hydrogen ion concentration of freshly brewed coffee is about 1.3×10^{-5} moles per liter. Find the pH.

4. *Acidosis.* When the pH of a patient's blood drops below 7.4, a condition called *acidosis* sets in. Acidosis can be fatal at a pH level of 7.0. What would the hydrogen ion concentration of the patient's blood be at that point?

5. *Interest Compounded Annually.* Suppose that \$40,000 is invested at 7% interest, compounded annually.

a) After what amount of time will there be \$250,000 in the account?

b) Find the doubling time.

Answers on page A-57

6. *Population Growth of the United States.* What will the population of the United States be in 2000? in 2005?

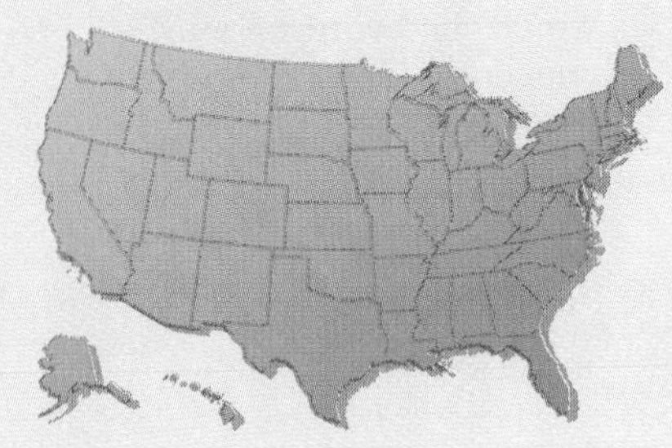

Answer on page A-57

The function in Example 3 illustrates exponential growth. Populations often grow exponentially according to the following model.

An **exponential growth model** is a function of the form

$$P(t) = P_0 e^{kt}, \quad k > 0,$$

where P_0 is the population at time 0, $P(t)$ is the population at time t, and k is the **exponential growth rate** for the situation. The **doubling time** is the amount of time necessary for the population to double in size.

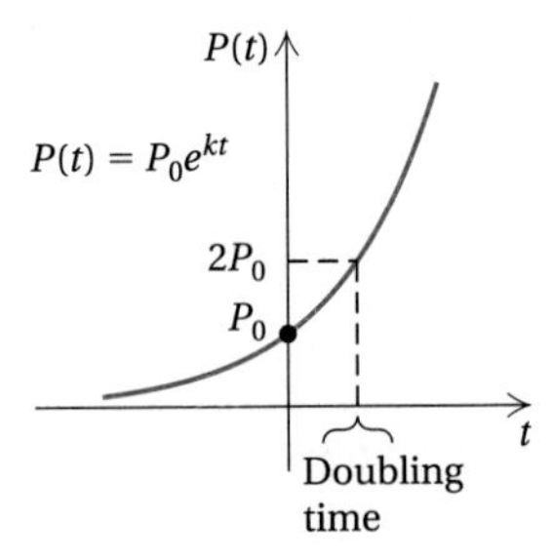

The exponential growth rate is the rate of growth of a population at any *instant* in time. Since the population is continually growing, the percent of total growth after one year will exceed the exponential growth rate.

Example 4 *Population Growth of the United States.* In 1998, the population of the United States was 267 million, and the exponential growth rate was 0.8% per year.

a) Find the exponential growth function.

b) What will the population be in 2010?

a) We are trying to find a model. The given information allows us to create one. At $t = 0$ (1998), the population was 267 million. We substitute 267 for P_0 and 0.8%, or 0.008, for k to obtain the exponential growth function:

$$P(t) = P_0 e^{kt}$$
$$P(t) = 267e^{0.008t}.$$

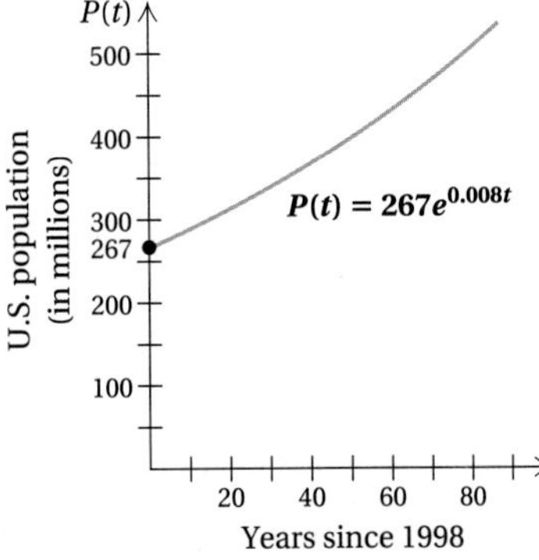

b) In 2010, we have $t = 12$. That is, 12 yr have passed since 1998. To find the population in 2010, we substitute 12 for t:

$$P(12) = 267e^{0.008(12)} \quad \text{Substituting 12 for } t$$
$$\approx 294 \text{ million.} \quad \text{Using a calculator}$$

The population of the United States will be about 294 million in 2010.

Do Exercise 6.

Example 5 *Interest Compounded Continuously.* Suppose that an amount of money P_0 is invested in a savings account at interest rate k, compounded continuously. That is, suppose that interest is computed every "instant" and added to the amount in the account. The balance $P(t)$, after t years, is given by the exponential growth model

$$P(t) = P_0 e^{kt}.$$

a) Suppose that \$30,000 is invested and grows to \$44,754.75 in 5 yr. Find the interest rate and then the exponential growth function.

b) What is the balance after 10 yr?

c) What is the doubling time?

a) We have $P(0) = 30{,}000$. Thus the exponential growth function is

$$P(t) = 30{,}000e^{kt}, \quad \text{where } k \text{ must still be determined.}$$

We know that at $t = 5$, $P(5) = 44{,}754.75$. We substitute and solve for k:

$$44{,}754.75 = 30{,}000e^{k(5)} = 30{,}000e^{5k}$$

$$\frac{44{,}754.75}{30{,}000} = e^{5k} \qquad \text{Dividing by 30,000}$$

$$1.491825 = e^{5k}$$

$$\ln 1.491825 = \ln e^{5k} \qquad \text{Taking the natural logarithm on both sides}$$

$$0.4 \approx 5k \qquad \text{Finding ln 1.491825 on a calculator and simplifying } \ln e^{5k}$$

$$\frac{0.4}{5} = 0.08 \approx k.$$

The interest rate is about 0.08, or 8%, compounded continuously. Note that since interest is being compounded continuously, the interest earned each year is more than 8%. The exponential growth function is

$$P(t) = 30{,}000e^{0.08t}.$$

b) We substitute 10 for t:

$$P(10) = 30{,}000e^{0.08(10)} = 66{,}766.23.$$

The balance in the account after 10 yr will be \$66,766.23.

c) To find the doubling time T, we replace $P(t)$ with 60,000 and solve for T:

$$60{,}000 = 30{,}000e^{0.08T}$$

$$2 = e^{0.08T} \qquad \text{Dividing by 30,000}$$

$$\ln 2 = \ln e^{0.08T} \qquad \text{Taking the natural logarithm on both sides}$$

$$\ln 2 = 0.08T$$

$$\frac{\ln 2}{0.08} = T \qquad \text{Dividing}$$

$$8.7 \approx T.$$

Thus the original investment of \$30,000 will double in about 8.7 yr, as shown in the following graph of the growth function.

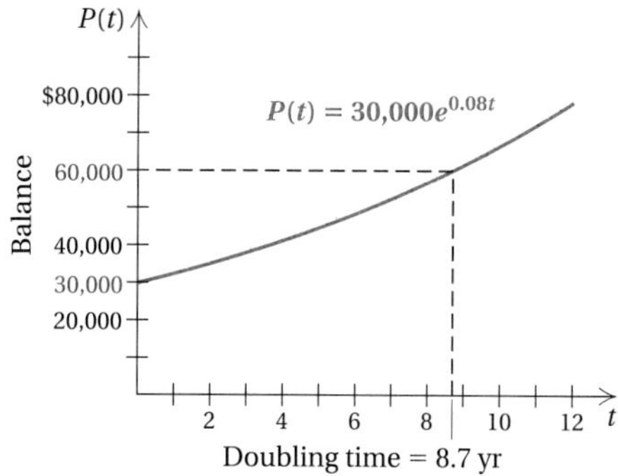

7. *Interest Compounded Continuously.*

a) Suppose that \$5000 is invested and grows to \$6356.25 in 4 yr. Find the interest rate and then the exponential growth function.

b) What is the balance after 1 yr? 2 yr? 10 yr?

c) What is the doubling time?

Do Exercise 7.

Compare the results of Examples 3(b) and 5(c). Note that continuous compounding gives a higher yield and shorter doubling time. Greater understanding about continuous compounding belongs to a course in business calculus.

Answers on page A-57

8. *Bagged Lettuce.* The convenience of bagged lettuce has resulted in huge increases in the sales of lettuce in recent years, as shown in the graph below (***Source:*** International Fresh-Cut Produce Association, Information Resources).

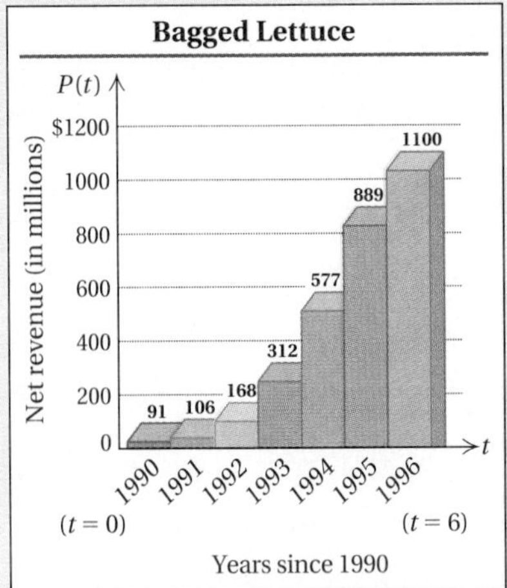

a) Let $t = 0$ correspond to 1990 and $t = 6$ correspond to 1996. Then t is the number of years since 1990. Use the data points (0, 91) and (6, 1100) to find the exponential growth rate and fit an exponential growth function $P(t) = P_0e^{kt}$ to the data, where $P(t)$ is the revenue, in millions, t years after 1990.

b) Use the function found in part (a) to predict the net revenue in 2004.

c) Assuming exponential growth continues, predict the year in which revenue from the sale of lettuce will reach $8000 million.

Answers on page A-57

Take a look at the graph that follows in Example 6. Note that the data tend to be rising steeply in a manner that might fit an exponential function. To fit an exponential function to the data, we choose a pair of data points and use them to determine P_0 and k.

Example 6 *Intel Corporation.* The Intel Corporation manufactures computer chips such as the Pentium II microprocessor. Its net revenues have skyrocketed in recent years, as shown in the graph.

a) If we let $t = 0$ correspond to 1992 and $t = 4$ to 1996, then t is the numbers of years since 1992. Use the data points (0, 5844) and (4, 20,847) to find the exponential growth rate and fit an exponential growth function to the data.

b) Use the function found in part (a) to predict the net revenue in 2000.

c) Assuming that exponential growth continues, predict the year in which the net revenue of Intel will reach $100,000 million.

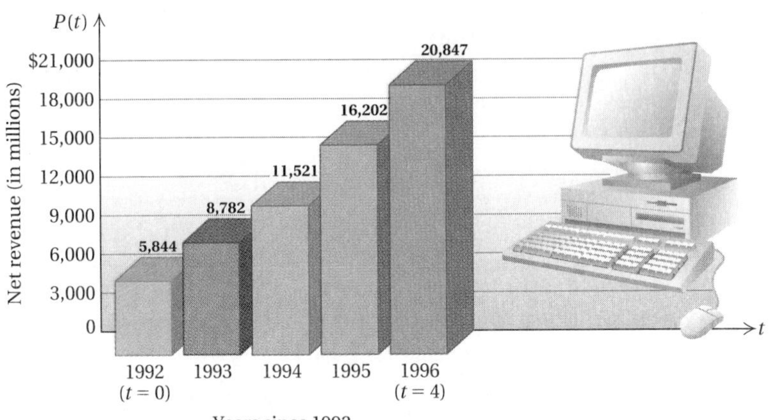

Source: Intel Corporation, 1996 Annual Report

a) We use the equation $P(t) = P_0e^{kt}$, where $P(t)$ is the net revenue of Intel, in millions of dollars, t years after 1992. In 1992, at $t = 0$, the net revenue was $5844 million. Thus we substitute 5844 for P_0:

$$P(t) = 5844e^{kt}.$$

To find the exponential growth rate, k, we note that 4 yr later, at the end of 1996, the net revenue was $20,847 million. We substitute and solve for k:

$$\left.\begin{aligned} P(4) &= 5844e^{k(4)} \\ 20{,}847 &= 5844e^{4k} \end{aligned}\right\} \quad \textbf{Substituting}$$

$$3.5672 \approx e^{4k} \quad \textbf{Dividing by 5844}$$

$$\ln 3.5672 = \ln e^{4k} \quad \textbf{Taking the natural logarithm on both sides}$$

$$1.2718 = 4k \quad \ln e^a = a$$

$$0.3180 \approx k. \quad \textbf{Dividing by 4}$$

The exponential growth rate was 0.318, or 31.8%, and the exponential growth function is $P(t) = 5844e^{0.318t}$.

b) Since 2000 is 8 yr from 1992, we substitute 8 for t:

$$P(8) = 5844e^{0.318(8)} \approx 74{,}397.$$

The net revenue in 2000 will be about $74,397 million, or $74.397 billion.

c) To predict when net revenues will reach $100,000 million, we substitute 100,000 for $P(t)$ and solve for t:

$$100{,}000 = 5844e^{0.318t}$$

$$17.1116 \approx e^{0.318t} \qquad \text{Dividing by 5844}$$

$$\ln 17.1116 \approx \ln e^{0.318t} \qquad \text{Taking the natural logarithm on both sides}$$

$$2.8398 \approx 0.318t \qquad \ln e^a = a$$

$$8.93 \approx t. \qquad \text{Dividing by 0.318}$$

Rounding to 9 yr, we see that, according to this model, by the end of 1992 + 9, or 2001, the net revenue of Intel will reach $100,000 million, or $100 billion.

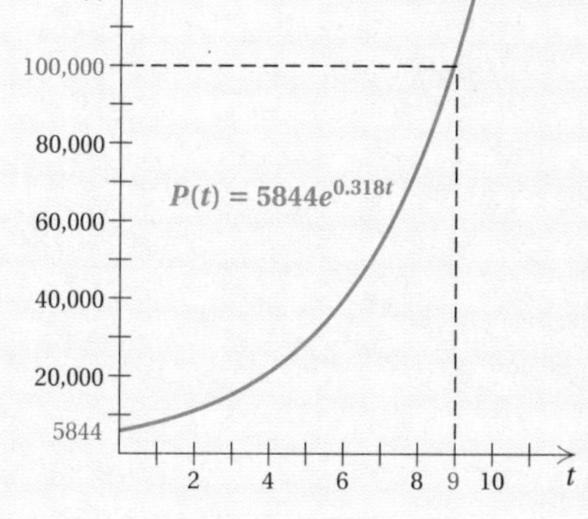

Do Exercise 8 on the preceding page.

In some real-life situations, a quantity or population is *decreasing* or *decaying* exponentially.

An **exponential decay model** is a function of the form

$$P(t) = P_0e^{-kt}, \quad k > 0,$$

where P_0 is the quantity present at time 0, $P(t)$ is the amount present at time t, and k is the **decay rate.** The **half-life** is the amount of time necessary for half of the quantity to decay.

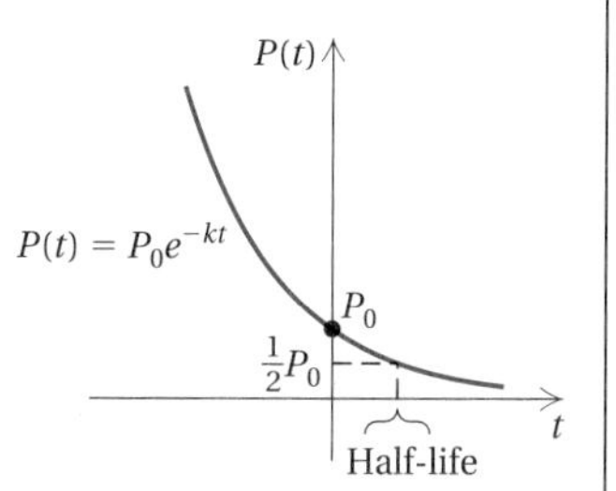

Example 7 *Carbon Dating.* The radioactive element carbon-14 has a half-life of 5750 yr. The percentage of carbon-14 present in the remains of organic matter can be used to determine the age of that organic matter. Recently, while digging in Chaco Canyon, New Mexico, archaeologists found corn pollen that had lost 38.1% of its carbon-14. The age of this corn pollen was evidence that Indians had been cultivating crops in the Southwest centuries earlier than scientists had thought. What was the age of the pollen? (***Source:*** *American Anthropologist*)

We first find k. To do so, we use the concept of half-life. When $t = 5750$ (the half-life), $P(t)$ will be half of P_0. Then

$$0.5P_0 = P_0e^{-k(5750)}$$

$$0.5 = e^{-5750k} \qquad \text{Dividing by } P_0$$

$$\ln 0.5 = \ln e^{-5750k} \qquad \text{Taking the natural logarithm on both sides}$$

$$\ln 0.5 = -5750k$$

$$\frac{\ln 0.5}{-5750} = k$$

$$0.00012 \approx k.$$

Now we have a function for the decay of carbon-14:

$$P(t) = P_0e^{-0.00012t}. \qquad \text{This completes the first part of our solution.}$$

(*Note*: This equation can be used for any subsequent carbon-dating problem.)

9. *Carbon Dating.* How old is an animal bone that has lost 30% of its carbon-14?

If the corn pollen has lost 38.1% of its carbon-14 from an initial amount P_0, then 100% − 38.1%, or 61.9%, of P_0 is still present. To find the age t of the pollen, we solve the following equation for t:

$$61.9\%P_0 = P_0e^{-0.00012t} \qquad \text{We want to find } t \text{ for which } P(t) = 0.619P_0.$$

$$0.619 = e^{-0.00012t} \qquad \text{Dividing by } P_0$$

$$\ln 0.619 = \ln e^{-0.00012t} \qquad \text{Taking the natural logarithm on both sides}$$

$$-0.4797 = -0.00012t \qquad \ln e^a = a$$

$$\frac{-0.4797}{-0.00012} \approx t \qquad \text{Dividing by } -0.00012$$

$$4000 \approx t. \qquad \text{Rounding}$$

The pollen is about 4000 yr old.

Do Exercise 9.

Calculator Spotlight

Mathematical Modeling Using Regression: Fitting Exponential Functions to Data. In the Calculator Spotlight of Section 7.6, we considered a method to fit a linear function to a set of data, *linear regression.* In the Calculator Spotlight of Section 11.7, we extended that regression to quadratic, cubic, and quartic polynomial functions. The grapher also gives us the powerful capability to find exponential models and make predictions. One of the advantages of using regression is that it takes into account all the data, whereas the method we used in Example 6 used only two data points.

Example *Intel Corporation.* The following data relate the net revenue of the Intel Corporation to the number of years t since 1992. They are illustrated in the graph of Example 6.

Years, t, Since 1992	Net Revenue, $P(t)$ (in millions)
0	$ 5,844
1	8,782
2	11,521
3	16,202
4	20,847

Source: Intel Corporation, 1996 Annual Report

a) Fit an exponential function to the data using the REGRESSION feature on a grapher.

b) Make a scatterplot of the data. Then graph the exponential function with the scatterplot.

c) Use the function from part (a) to estimate the net revenue of Intel in 2000. Compare your answer to that found in Example 6(c).

a) We fit an exponential function to the data using the REGRESSION feature on a TI-83 grapher. The procedure is similar to what is outlined in the Calculator Spotlight of Section 7.6. The only difference is that we choose [ExpReg] instead of [LinReg($ax + b$)]. (Consult your manual for further details.) We obtain the following function.

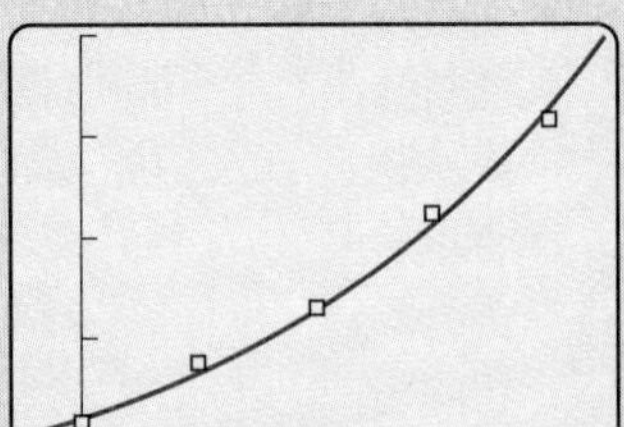

(continued)

Answer on page A-57

On some graphers, there may be a REGRESSION feature that yields an exponential function, base e. If not, and you wish to find such a function, a conversion can be done using $b^x = e^{x(\ln b)}$. For the equation given here, we would have

$$\begin{aligned} y &= 6108.76(1.371)^x \\ &= 6108.76(e^{x(\ln 1.371)}) \\ &= 6108.76e^{0.316x}. \end{aligned}$$

b) The exponential function is graphed with the scatterplot shown on the preceding page.

c) The year 2000 is 8 yr from 1992. We can use the grapher to do the calculation. We get Y1(8) ≈ 76,290, as shown. Thus the net revenue of Intel in 2000 will be about \$76,290 million, or \$76.29 billion. In Example 6(b), we predict a net of \$74,906 million.

```
Y1(8)
                76289.77784
```

Exercises

1. *Bagged Lettuce.* The following data relate the net revenue from bagged lettuce to the number of years t since 1990. They are illustrated in the graph of Margin Exercise 8.

Years, t, Since 1990	Net Revenue, $P(t)$ (in millions)
0	\$ 91
1	106
2	168
3	312
4	577
5	889
6	1100

Source: International Fresh-Cut Produce Association, Information Resources

a) Use the REGRESSION feature to fit an exponential function to the data.

b) Make a scatterplot of the data. Then graph the exponential function with the scatterplot.

c) Use the function from part (a) to estimate the net revenue from bagged lettuce in 2004. Compare your answer to that found in Margin Exercise 8(b).

2. *Average Price of a 30-Second Super Bowl Commercial.* The following data relate the cost of a Super Bowl commercial to the number of years t since 1990.

Years, t, Since 1990	Cost
1	\$ 800,000
3	850,000
4	900,000
5	1,000,000
8	1,300,000

Source: National Football League

a) Use the REGRESSION feature to fit an exponential function to the Super Bowl data.

b) Make a scatterplot of the data. Then graph the exponential function with the scatterplot.

c) Use the function from part (a) to estimate the cost of a 30-second commercial in 2000 and in 2010.

d) Predict the year in which the cost of a Super Bowl commercial will be \$5 million.

Improving Your Math Study Skills

Preparing for a Final Exam

Best Scenario: Two Weeks of Study Time

The best scenario for preparing for a final exam is to do so over a period of at least two weeks. Work in a diligent, disciplined manner, doing some final-exam preparation *each* day. Here is a detailed plan that many find useful.

1. **Begin by browsing through each chapter, reviewing the highlighted or boxed information regarding important formulas in both the text and the Summary and Review.** There may be some formulas that you will need to memorize.
2. **Retake each chapter test that you took in class, assuming your instructor has returned it. Otherwise, use the chapter test in the book.** Restudy the objectives in the text that correspond to each question you missed.
3. **Then work the Cumulative Review that covers all chapters up to that point.** Be careful to avoid any questions corresponding to objectives not covered. Again, restudy the objectives in the text that correspond to each question you missed.
4. **If you are still missing questions, use supplements for extra review.** For example, you might check out the video- or audiotapes, the *Student's Solutions Manual,* or the Interact Math Tutorial Software.
5. **For remaining difficulties, see your instructor, go to a tutoring session, or participate in a study group.**
6. **Check for former final exams that may be on file in the math department or a study center, or with students who have already taken the course.** Use them for practice, being alert to trouble spots.
7. **Take the Final Examination in the text during the last couple of days before the final.** Set aside the same amount of time that you will have for the final. See how much of the final exam you can complete under test-like conditions.

Moderate Scenario: Three Days to Two Weeks of Study Time

1. **Begin by browsing through each chapter, reviewing the highlighted or boxed information regarding important formulas in both the text and the Summary and Review.** There may be some formulas that you will need to memorize.
2. **Retake each chapter test that you took in class, assuming your instructor has returned it. Otherwise, use the chapter test in the book.** Restudy the objectives in the text that correspond to each question you missed.
3. **Then work the last Cumulative Review in the text.** Be careful to avoid any questions corresponding to objectives not covered. Again, restudy the objectives in the text that correspond to each question you missed.
4. **For remaining difficulties, see your instructor, go to a tutoring session, or participate in a study group.**
5. **Take the Final Examination in the text during the last couple of days before the final.** Set aside the same amount of time that you will have for the final. See how much of the final exam you can complete under test-like conditions.

Worst Scenario: One or Two Days of Study Time

1. **Begin by browsing through each chapter, reviewing the highlighted or boxed information regarding important formulas in both the text and the Summary and Review.** There may be some formulas that you will need to memorize.
2. **Then work the last Cumulative Review in the text.** Be careful to avoid any questions corresponding to objectives not covered. Restudy the objectives in the text that correspond to each question you missed.
3. **Attend a final-exam review session if one is available.**
4. **Take the Final Examination in the text during the last couple of days before the final.** Set aside the same amount of time that you will have for the final. See how much of the final exam you can complete under test-like conditions.

   Promise yourself that next semester you will allow a more appropriate amount of time for final exam preparation.

Other "Improving Your Math Study Skills" concerning test preparation appear in Sections 4.1 and 8.2.

Exercise Set 12.7

a Solve.

Sound Levels. Use the decibel formula from Example 1 for Exercises 1–4.

1. *Acoustics.* The intensity of sound of a riveter at work is about 3.2×10^{-3} W/m^2. How high is this sound level in decibels?

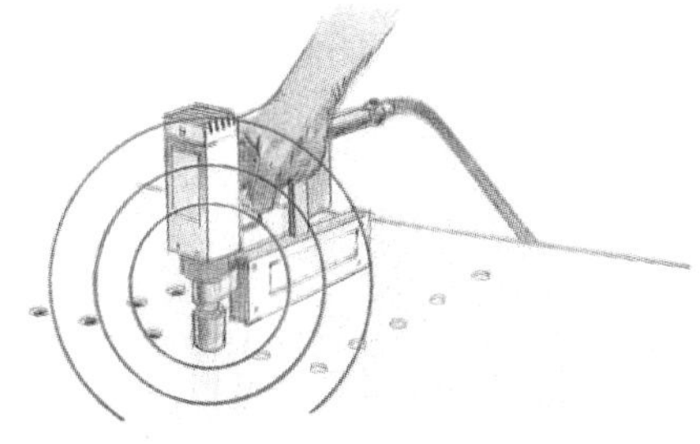

2. *Acoustics.* The intensity of sound of a dishwasher is about 2.5×10^{-6} W/m^2. How high is this sound level in decibels?

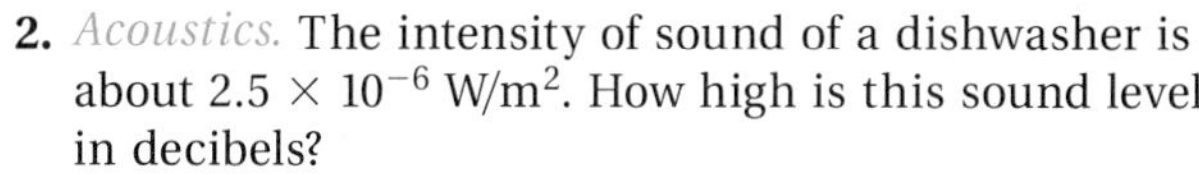

3. *Music.* At a recent performance of the rock group Phish, sound measurements of 105 dB were recorded. What is the intensity of such sounds?

4. *Jet Plane at Takeoff.* A jet plane at takeoff can generate sound measurements of 130 dB. What is the intensity of such sounds?

pH. Use the pH formula from Example 2 for Exercises 5–8.

5. *Milk.* The hydrogen ion concentration of milk is about 1.6×10^{-7} moles per liter. Find the pH.

6. *Mouthwash.* The hydrogen ion concentration of mouthwash is about 6.3×10^{-7} moles per liter. Find the pH.

7. *Alkalosis.* When the pH of a patient's blood rises above 7.4, a condition called *alkalosis* sets in. Alkalosis can be fatal at a pH level above 7.8. What would the hydrogen ion concentration of the patient's blood be at that point?

8. *Orange Juice.* The pH of orange juice is 3.2. What is its hydrogen ion concentration?

Walking Speed. In a study by psychologists Bornstein and Bornstein, it was found that the average walking speed w, in feet per second, of a person living in a city of population P, in thousands, is given by the function

$$w(P) = 0.37 \ln P + 0.05$$

(**Source:** *International Journal of Psychology*). In Exercises 9–12, various cities and their populations are given. Find the walking speed of people in each city.

9. Phoenix, Arizona: 2,892,000

10. Salt Lake City, Utah: 1,178,000

11. Savannah, Georgia: 139,000

12. Plano, Texas: 142,000

b Solve.

13. *Spread of Rumor.* The number of people who hear a rumor increases exponentially. If 20 people start a rumor and if each person who hears the rumor repeats it to two people a day, the number of people N who have heard the rumor after t days is given by the function

$$N(t) = 20(3)^t.$$

a) How many people have heard the rumor after 5 days?
b) After what amount of time will 1000 people have heard the rumor?
c) What is the doubling time for the number of people who have heard the rumor?

14. *Salvage Value.* A color photocopier is purchased for \$5200. Its value each year is about 80% of its value in the preceding year. Its value in dollars after t years is given by the exponential function

$$V(t) = 5200(0.8)^t.$$

a) Find the salvage value of the copier after 3 yr.
b) After what amount of time will the salvage value be \$1200?
c) After what amount of time will the salvage value be half the original value?

15. *Credit-Card Spending.* The amount S, in billions, charged on credit cards between Thanksgiving and Christmas each year has been increasing exponentially according to the function

$$S(t) = 52.4(1.16)^t,$$

where t is the number of years since 1990 (***Source:*** RAM Research Group, National Credit Counseling Services).

a) Estimate the amount spent between Thanksgiving and Christmas in 2000.
b) In what year will the spending be \$1 trillion?
c) What is the doubling time?

16. *Projected College Costs.* In the new millennium, the cost of tuition, books, room, and board at a state university is projected to follow the exponential function

$$C(t) = 11{,}054(1.06)^t,$$

where C is the cost, in dollars, and t is the number of years after 2000 (***Source:*** College Board, Senate Labor Committee).

a) Find the college costs in 2005.
b) In what year will the cost be \$21,000?
c) What is the doubling time of the costs?

Growth. Use the exponential growth model $P(t) = P_0e^{kt}$ for Exercises 17–24.

17. *World Population Growth.* In 1998, the population of the world reached 6 billion, and the exponential growth rate was 1.5% per year.

a) Find the exponential growth function.
b) What will the population be in 2010?
c) In what year will the population be 10 billion?
d) What is the doubling time?

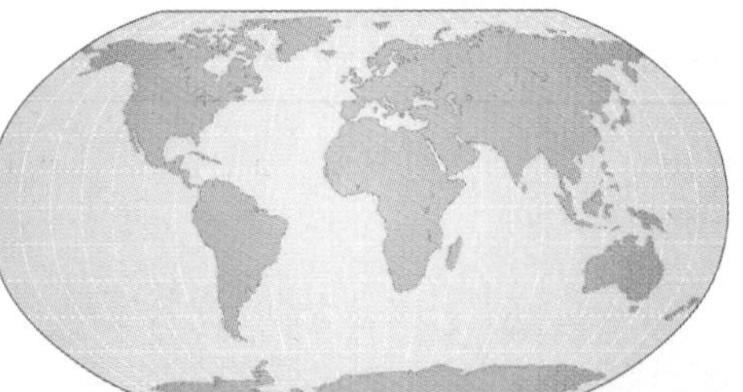

18. *Anchorage Population Growth.* In 1998, the population of Anchorage, Alaska, reached 253,750, and the exponential growth rate was 2.9% per year.

a) Find the exponential growth function.
b) What will the population be in 2010?
c) In what year will the population be 600,000?
d) What is the doubling time?

19. *Interest Compounded Continuously.* Suppose that P_0 is invested in a savings account in which interest is compounded continuously at 6% per year.

a) Express $P(t)$ in terms of P_0 and 0.06.
b) Suppose that \$5000 is invested. What is the balance after 1 yr? 2 yr? 10 yr?
c) When will the investment of \$5000 double itself?

20. *Interest Compounded Continuously.* Suppose that P_0 is invested in a savings account in which interest is compounded continuously at 5.4% per year.

a) Express $P(t)$ in terms of P_0 and 0.054.
b) Suppose that \$10,000 is invested. What is the balance after 1 yr? 2 yr? 10 yr?
c) When will the investment of \$10,000 double itself?

21. *Interest Compounded Continuously.* Suppose that P_0 is invested in a savings account in which interest is compounded continuously.

a) Suppose that \$40,000 is invested and grows to \$56,762.70 in 5 yr. Find the interest rate and then the exponential growth function.
b) What is the balance after 1 yr? 2 yr? 10 yr?
c) What is the doubling time?

22. *Interest Compounded Continuously.* Suppose that P_0 is invested in a savings account in which interest is compounded continuously.

a) Suppose that \$100,000 is invested and grows to \$164,213.96 in 8 yr. Find the interest rate and then the exponential growth function.
b) What is the balance after 1 yr? 3 yr? 12 yr?
c) What is the doubling time?

23. *First-Class Postage.* First-class postage (for the first ounce) was 20¢ in 1981 and 32¢ in 1995 (***Source:*** U.S. Postal Service). Assume that the cost increases according to an exponential growth function.

a) Let $t = 0$ correspond to 1981 and $t = 14$ correspond to 1995. Then t is the number of years since 1981. Use the data points (0, 20) and (14, 32) to find the exponential growth rate and fit an exponential growth function $P(t) = P_0e^{kt}$ to the data, where $P(t)$ is the cost of first-class postage, in cents, t years after 1995.
b) Use the function found in part (a) to predict the cost of first-class postage in 2000.
c) When will the cost of first-class postage be \$1.00 or 100¢?

24. *Books on Tape.* Sales of books on audiotape totaled \$1.6 billion in 1996, increasing rapidly from \$0.25 billion in 1989. Assume that sales increase according to an exponential growth function (***Source:*** Audio Book Club).

a) Let $t = 0$ correspond to 1989 and $t = 7$ correspond to 1996. Then $t =$ the number of years since 1989. Use the data points (0, 0.25) and (7, 1.6) to find the exponential growth rate and fit an exponential growth function $P(t) = P_0e^{kt}$ to the data, where $P(t)$ is the sales, in billions of dollars, t years after 1989.
b) Use the function found in part (a) to predict the sales of books on audiotape in 2000.
c) When will sales reach \$4.0 billion?

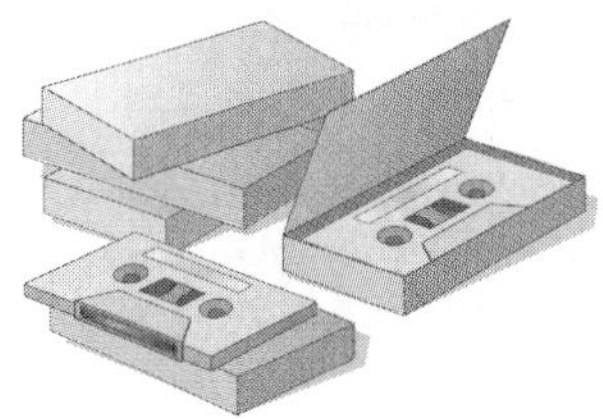

Carbon Dating. Use the carbon-14 decay function $P(t) = P_0e^{-0.00012t}$ for Exercises 25 and 26.

25. *Carbon Dating.* When archaeologists found the Dead Sea scrolls, they determined that the linen wrapping had lost 22.3% of its carbon-14. How old was the linen wrapping?

26. *Carbon Dating.* In 1998, researchers found an ivory tusk that had lost 18% of its carbon-14. How old was the tusk?

Decay. Use the exponential decay function $P(t) = P_0e^{-kt}$ for Exercises 27 and 28.

27. *Chemistry.* The decay rate of iodine-131 is 9.6% per day. What is the half-life?

28. *Chemistry.* The decay rate of krypton-85 is 6.3% per day . What is the half-life?

29. *Home Construction.* The chemical urea formaldehyde was found in some insulation used in houses built during the mid to late 1960s. Unknown at the time was the fact that urea formaldehyde emitted toxic fumes as it decayed. The half-life of urea formaldehyde is 1 yr. What is its decay rate?

30. *Plumbing.* Lead pipes and solder are often found in older buildings. Unfortunately, as lead decays, toxic chemicals can get in the water resting in the pipes. The half-life of lead is 22 yr. What is its decay rate?

31. *Value of a Sports Card.* Because he objected to smoking, and because his first baseball card was issued in cigarette packs, the great shortstop Honus Wagner halted production of his card before many were produced. One of these cards was sold in 1991 for $451,000 and again in 1996 for $640,500. For the following questions, assume that the card's value increases exponentially.

a) Find an exponential function $V(t)$, where t is the number of years after 1991, if $V_0 = 451{,}000$.
b) Predict the card's value in 2005.
c) What is the doubling time for the value of the card?
d) In what year will the value of the card first exceed $1,000,000?

32. *Value of a Van Gogh Painting.* The Van Gogh painting *Irises*, shown below, sold for $84,000 in 1947 and was sold again for $53,900,000 in 1987. Assume that the growth in the value V of the painting is exponential.

Van Gogh's *Irises*, a 28-in. by 32-in. oil on canvas.

a) Find the exponential growth rate k, and determine the exponential growth function, where t is the number of years after 1947, assuming $V_0 = 84{,}000$.
b) Estimate the value of the painting in 2001.
c) What is the doubling time for the value of the painting?
d) How long after 1947 will the value of the painting be $1 billion?

Skill Maintenance

Compute and simplify. Express answers in the form $a + bi$, where $i^2 = -1$. [10.8c, d, e]

33. i^{46}

34. i^{53}

35. $i^{14} + i^{15}$

36. $i^{18} - i^{16}$

37. $\dfrac{8 - i}{8 + i}$

38. $\dfrac{2 + 3i}{5 - 4i}$

39. $(5 - 4i)(5 + 4i)$

40. $(-10 - 3i)^2$

Synthesis

41. ◆ Write a problem for a classmate to solve in which data that seem to fit an exponential growth function are provided. Try to find data in a newspaper to make the problem as realistic as possible.

42. ◆ Do some research or consult with a chemist to determine how carbon dating is carried out. Write a report.

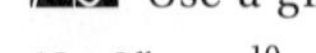 Use a grapher to solve each of the following equations.

43. $2^x = x^{10}$

44. $(\ln 2)x = 10 \ln x$

45. $x^2 = 2^x$

Fit an exponential function to a set of data.

Collaborative Learning Manual

Summary and Review Exercises: Chapter 12

Important Properties and Formulas

Exponential Functions: $f(x) = a^x, \quad f(x) = e^x$

Composition of Functions: $f \circ g(x) = f(g(x))$

Definition of Logarithms: $y = \log_a x$ is that number y such that $x = a^y$, where $x > 0$ and a is a positive constant other than 1.

Properties of Logarithms:

$\log M = \log_{10} M$, $\quad \log_a 1 = 0$, $\quad \ln M = \log_e M$, $\quad \log_a a = 1$,

$\log_a MN = \log_a M + \log_a N$, $\quad \log_a a^k = k$, $\quad \log_a M^k = k \cdot \log_a M$, $\quad \log_b M = \dfrac{\log_a M}{\log_a b}$,

$\log_a \dfrac{M}{N} = \log_a M - \log_a N$, $\quad e \approx 2.7182818284\ldots$

Growth: $P(t) = P_0 e^{kt}$

Decay: $P(t) = P_0 e^{-kt}$

Carbon Dating: $P(t) = P_0 e^{-0.00012t}$

Interest Compounded Annually: $A = P(1 + i)^t$

Interest Compounded n Times per Year: $A = P\left(1 + \dfrac{r}{n}\right)^{nt}$

Interest Compounded Continuously: $P(t) = P_0 e^{kt}$, where P_0 dollars are invested for t years at interest rate k

The objectives to be tested in addition to the material in this chapter are [10.8b, c, d, e], [11.4c], [11.6a], and [11.7b].

Graph.

1. $f(x) = 3^{x-1}$ [12.1a]

2. $f(x) = \log_3 x$ [12.3a]

3. $f(x) = e^{x+1}$ [12.5c]

4. $f(x) = \ln(x - 1)$ [12.5c]

5. Find the inverse of the relation [12.2a]
$\{(-4, 2), (5, -7), (-1, -2), (10, 11)\}$.

Determine whether the function is one-to-one. If it is, find a formula for its inverse. [12.2b, c]

6. $f(x) = 4 - x^2$

7. $g(x) = \dfrac{2x - 3}{7}$

8. $f(x) = 8x^3$

9. $f(x) = \dfrac{4}{3 - 2x}$

10. Graph the function $f(x) = x^3 + 1$ and its inverse using the same set of axes. [12.2d]

11. Find $f \circ g(x)$ and $g \circ f(x)$ if $f(x) = x^2$ and $g(x) = 3x - 5$. [12.2e]

12. If $h(x) = \sqrt{4 - 7x}$, find $f(x)$ and $g(x)$ such that $h(x) = f \circ g(x)$. [12.2e]

Convert to a logarithmic equation. [12.3b]

13. $10^4 = 10{,}000$

14. $25^{1/2} = 5$

Convert to an exponential equation. [12.3b]

15. $\log_4 16 = x$

16. $\log_{1/2} 8 = -3$

Find each of the following. [12.3c]

17. $\log_3 9$

18. $\log_{10} \frac{1}{10}$

19. $\log_m m$

20. $\log_m 1$

Find the common logarithm, to four decimal places, using a calculator. [12.3e]

21. $\log\left(\dfrac{78}{43{,}112}\right)$

22. $\log(-4)$

Express in terms of logarithms of x, y, and z. [12.4d]

23. $\log_a x^4y^2z^3$

24. $\log \sqrt[4]{\dfrac{z^2}{x^3y}}$

Express as a single logarithm. [12.4d]

25. $\log_a 8 + \log_a 15$

26. $\frac{1}{2}\log a - \log b - 2\log c$

Simplify. [12.4e]

27. $\log_m m^{17}$

28. $\log_m m^{-7}$

Given $\log_a 2 = 1.8301$ and $\log_a 7 = 5.0999$, find each of the following. [12.4d]

29. $\log_a 28$

30. $\log_a 3.5$

31. $\log_a \sqrt{7}$

32. $\log_a \frac{1}{4}$

Find each of the following, to four decimal places, using a calculator. [12.5a]

33. $\ln 0.06774$

34. $e^{-0.98}$

35. $e^{2.91}$

36. $\ln 1$

37. $\ln 0$

38. $\ln e$

Find the logarithm using the change-of-base formula. [12.5b]

39. $\log_5 2$

40. $\log_{12} 70$

Solve. Where appropriate, give approximations to four decimal places. [12.6a, b]

41. $\log_3 x = -2$

42. $\log_x 32 = 5$

43. $\log x = -4$

44. $3 \ln x = -6$

45. $4^{2x-5} = 16$

46. $2^{x^2} \cdot 2^{4x} = 32$

47. $4^x = 8.3$

48. $e^{-0.1t} = 0.03$

49. $\log_4 16 = x$

50. $\log_4 x + \log_4 (x - 6) = 2$

51. $\log x + \log (x - 15) = 2$

52. $\log_3 (x - 4) = 3 - \log_3 (x + 4)$

Solve. [12.7a, b]

53. *Acoustics.* The intensity of the sound at the foot of Niagara Falls is about 10^{-3} W/m^2. How high is this sound level in decibels? [Use $L = 10 \cdot \log (I/I_0)$.]

54. *Stock Index Funds.* Standard & Poor's 500 Stock Index is an indicator of the stock-price performance of 500 selected stocks. Many people purchase mutual funds that include these stocks. In recent years, the amount A, in billions of dollars, invested in such funds has grown exponentially according to the function

$A(t) = 4.55(1.54)^t$,

where t is the number of years since 1990 (***Source:*** Lipper Analytical Services).

a) Find the amount invested in these funds in 2000.
b) In what year did the amount invested reach $80 billion?
c) Find the doubling time.

55. *Aeronautics.* There were 6 commercial space launches in 1994 and 12 in 1996. Assume that the number N is growing exponentially according to the function

$N(t) = N_0 e^{kt}$.

(***Source:*** Federal Aviation Administration, U.S. Department of Transportation)

a) Find k and write the exponential growth function.
b) Predict the number of launches in 2003.
c) When will there be 192 annual scheduled launches?

56. The population of Riverton doubled in 16 yr. Find the exponential growth rate.

57. How long will it take $7600 to double itself if it is invested at 8.4%, compounded continuously?

58. How old is a skeleton that has lost 34% of its carbon-14? (Use $P(t) = P_0 e^{-0.00012t}$.)

Skill Maintenance

59. *Pizza Prices.* Pizza Unlimited has the following prices for pizza.

Diameter	Price
8 in.	$ 6.00
12 in.	8.50
16 in.	11.50

a) Use the data points (8, 6), (12, 8.5), and (16, 11.5) to fit a quadratic function to the data.
b) Use the function to estimate the price of a 24-in. pizza. [11.7b]

60. Solve: $x^4 - 11x^2 - 80 = 0$. [11.4c]

61. For $f(x) = x^2 + 2x + 3$, find and label **(a)** the vertex, **(b)** the line of symmetry, and **(c)** the maximum or minimum value. Then **(d)** graph the function. [11.6a]

Compute and simplify. Express answers in the form $a + bi$, where $i^2 = -1$. [10.8b, c, d, e]

62. $(2 - 3i)(4 + i)$

63. i^{20}

64. $\dfrac{4 - 5i}{1 + 3i}$

65. $(4 - 5i) + (1 + 3i)$

Synthesis

66. ◆ Explain why you cannot take the logarithm of a negative number. [12.3a]

67. ◆ Explain why $\log_a 1 = 0$. [12.3a]

Solve. [12.6a, b]

68. $\ln (\ln x) = 3$

69. $5^{x+y} = 25,\ 2^{2x-y} = 64$

Test: Chapter 12

Graph.

1. $f(x) = 2^{x+1}$

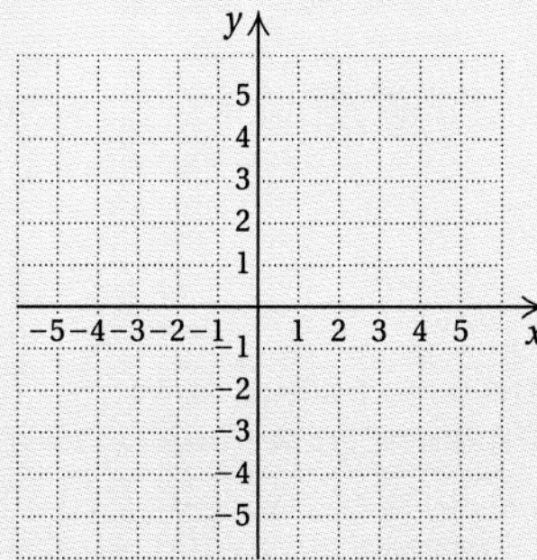

2. $y = \log_2 x$

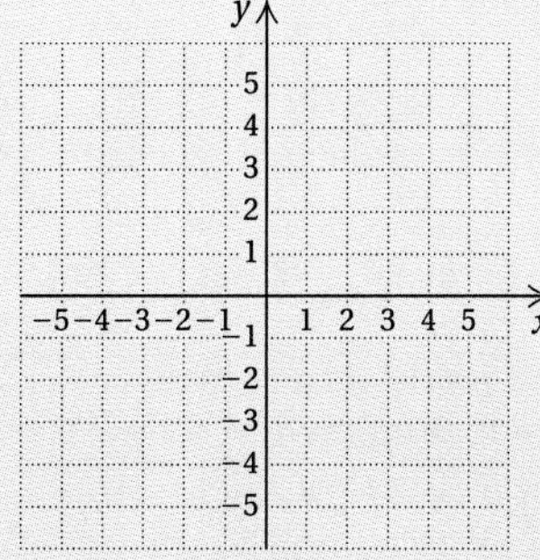

3. $f(x) = e^{x-2}$

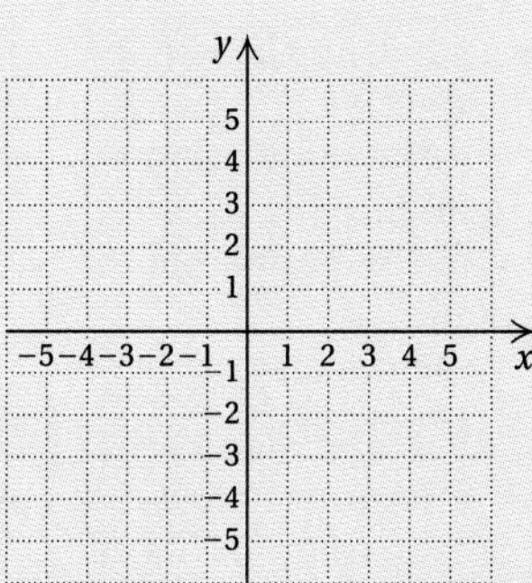

4. $f(x) = \ln (x - 4)$

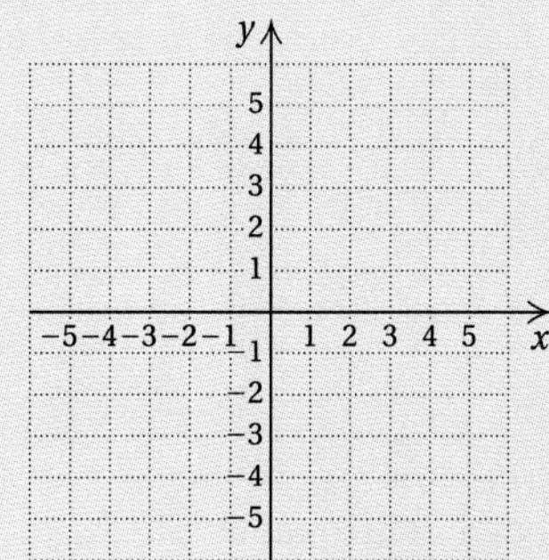

5. Find the inverse of the relation $\{(-4, 3), (5, -8), (-1, -3), (10, 12)\}$.

Determine whether the function is one-to-one. If it is, find a formula for its inverse.

6. $f(x) = 4x - 3$

7. $f(x) = (x + 1)^3$

8. $f(x) = 2 - |x|$

9. Find $f \circ g(x)$ and $g \circ f(x)$ if $f(x) = x + x^2$ and $g(x) = 5x - 2$.

10. Convert to a logarithmic equation: $256^{1/2} = 16$.

11. Convert to an exponential equation: $m = \log_7 49$.

Find each of the following.

12. $\log_5 125$

13. $\log_t t^{23}$

14. $\log_p 1$

Find the common logarithm, to four decimal places, using a calculator.

15. $\log 0.0123$

16. $\log (-5)$

17. Express in terms of logarithms of x, y, and z:
$$\log \frac{a^3 b^{1/2}}{c^2}.$$

18. Express as a single logarithm: $\frac{1}{3} \log_a x - 3 \log_a y + 2 \log_a z$.

Given $\log_a 2 = 0.301$, $\log_a 6 = 0.778$, and $\log_a 7 = 0.845$, find each of the following.

19. $\log_a \frac{2}{7}$

20. $\log_a 12$

Find each of the following, to four decimal places, using a calculator.

21. $\ln 807.39$

22. $e^{4.68}$

23. $\ln 1$

24. Find $\log_{18} 31$ using the change-of-base formula.

Answers

1. ________
2. ________
3. ________
4. ________
5. ________
6. ________
7. ________
8. ________
9. ________
10. ________
11. ________
12. ________
13. ________
14. ________
15. ________
16. ________
17. ________
18. ________
19. ________
20. ________
21. ________
22. ________
23. ________
24. ________

Answers

25. ______

26. ______

27. ______

28. ______

29. ______

30. ______

31. ______

32. ______

33. a) ______

b) ______

c) ______

34. a) ______

b) ______

c) ______

d) ______

35. ______

36. ______

37. a) ______

b) ______

38. ______

39. ______

40. ______

41. ______

42. ______

43. ______

Solve. Where appropriate, give approximations to four decimal places.

25. $\log_x 25 = 2$

26. $\log_4 x = \frac{1}{2}$

27. $\log x = 4$

28. $\ln x = \frac{1}{4}$

29. $7^x = 1.2$

30. $\log (x^2 - 1) - \log (x - 1) = 1$

31. $\log_5 x + \log_5 (x + 4) = 1$

32. *Tomatoes.* What is the pH of tomatoes if the hydrogen ion concentration is 6.3×10^{-5} moles per liter? (Use $\text{pH} = -\log [\text{H}^+]$.)

33. *Projected College Costs.* In the new millennium, the cost of tuition, books, room, and board at a private university is projected to follow the exponential function

$$C(t) = 24{,}534(1.06)^t,$$

where C is the cost, in dollars, and t is the number of years after 2000 (***Source***: College Board, Senate Labor Committee).

a) Find the college costs in 2008.
b) In what year will the cost be \$43,937?
c) What is the doubling time of the costs?

34. *Population Growth of Canada.* The population of Canada was 30 million in 1996 and is expected to be 30.753 million in 2000 (***Source***: Census Bureau of Canada). Assume that the number N is growing exponentially according to the equation

$$N(t) = N_0 e^{kt}.$$

a) Find k and write the exponential growth function.
b) What will the population be in 2003? in 2010?
c) When will the population be 50 million?
d) What is the doubling time?

35. An investment with interest compounded continuously doubled in 15 yr. What is the interest rate?

36. How old is an animal bone that has lost 43% of its carbon-14? (Use $P(t) = P_0 e^{-0.00012t}$.)

Skill Maintenance

37. *Audio Equipment.* The owner's manual for a top-rated cassette deck includes a table relating the time that a tape has to run to the counter reading.

a) Use the data points (0, 0), (400, 25), and (700, 50) to fit a quadratic function to the data.
b) Use the function to estimate how long a tape has run when the counter reading is 200.

Counter Reading	Time Tape Has Run (in minutes)
000	0
400	25
700	50

38. Solve: $y - 9\sqrt{y} + 8 = 0$.

39. For $f(x) = x^2 - 4x - 2$, find and label **(a)** the vertex, **(b)** the line of symmetry, and **(c)** the maximum or minimum value. Then **(d)** graph the function.

Compute and simplify. Express answers in the form $a + bi$, where $i^2 = -1$.

40. i^{29}

41. $\dfrac{4 - i}{1 + i}$

Synthesis

42. Solve: $\log_3 |2x - 7| = 4$.

43. If $\log_a x = 2$, $\log_a y = 3$, and $\log_a z = 4$, find

$$\log_a \frac{\sqrt[3]{x^2 z}}{\sqrt[3]{y^2 z^{-1}}}.$$

Cumulative Review: Chapters 1–12

Simplify.

1. $\left|\frac{2}{3} - \frac{4}{5}\right|$

2. $\dfrac{63x^2y^3}{-7x^{-4}y}$

3. $1000 \div 10^2 \cdot 25 \div 4$

4. $5x - 3[4(x - 2) - 2(x + 1)]$

Solve.

5. $\dfrac{1}{3}x - \dfrac{1}{5} \geq \dfrac{1}{5}x - \dfrac{1}{3}$

6. $|x| > 6.4$

7. $3 \leq 4x + 7 < 31$

8. $3x + y = 4,$
   $-6x - y = -3$

9. $x - y + 2z = 3,$
   $-x + z = 4,$
   $2x + y - z = -3$

10. $2x^2 = x + 3$

11. $3x - \dfrac{6}{x} = 7$

12. $\sqrt{x + 5} = x - 1$

13. $x(x + 10) = -21$

14. $2x^2 + x + 1 = 0$

15. $x^4 - 13x^2 + 36 = 0$

16. $\dfrac{3}{x - 3} - \dfrac{x + 2}{x^2 + 2x - 15} = \dfrac{1}{x + 5}$

17. $\log_2 x + \log_2 (x + 7) = 3$

18. $7^x = 30$

19. $\log_3 x = 2$

20. $x^2 - 1 \geq 0$

21. $\dfrac{x + 1}{x - 2} > 0$

22. $P = \dfrac{3}{4}(M + 2N)$, for N

23. $\dfrac{1}{p} + \dfrac{1}{q} = \dfrac{1}{f}$, for p

Simplify.

24. $(2x + 3)(x^2 - 2x - 1)$

25. $(3x^2 + x^3 - 1) - (2x^3 + x + 5)$

26. $\dfrac{2m^2 + 11m - 6}{m^3 + 1} \cdot \dfrac{m^2 - m + 1}{m + 6}$

27. $\dfrac{x}{x - 1} + \dfrac{2}{x + 1} - \dfrac{2x}{x^2 - 1}$

28. $\dfrac{1 - \dfrac{5}{x}}{x - 4 - \dfrac{5}{x}}$

29. $(x^4 + 3x^3 - x + 4) \div (x + 1)$

30. $\dfrac{\sqrt{75x^5y^2}}{\sqrt{3xy}}$

31. $4\sqrt{50} - 3\sqrt{18}$

32. $(16^{3/2})^{1/2}$

33. $(2 - i\sqrt{2})(5 + 3i\sqrt{2})$

34. $\dfrac{5 + i}{2 - 4i}$

Graph.

35. $4y - 3x = 12$

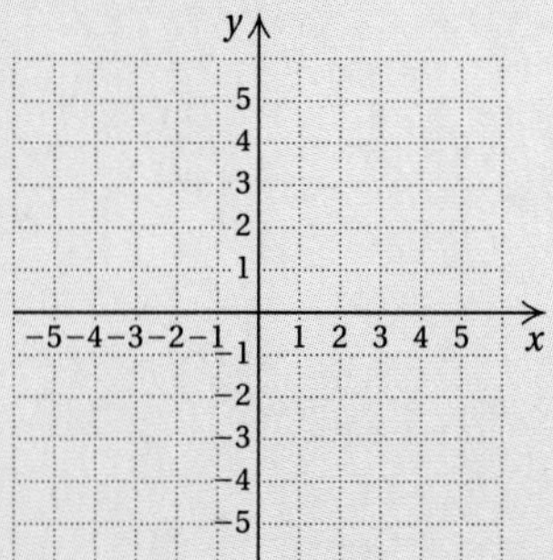

36. $y < -2$

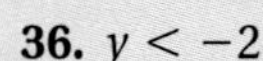

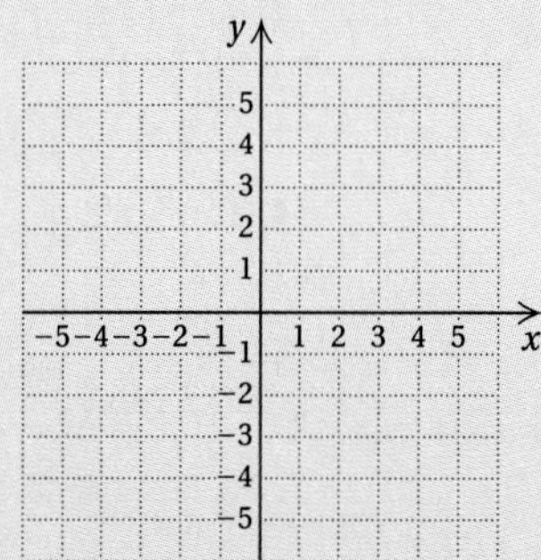

37. $x + y \leq 0,$
$x \geq -4,$
$y \geq -1$

38. $f(x) = 2x^2 - 8x + 9$

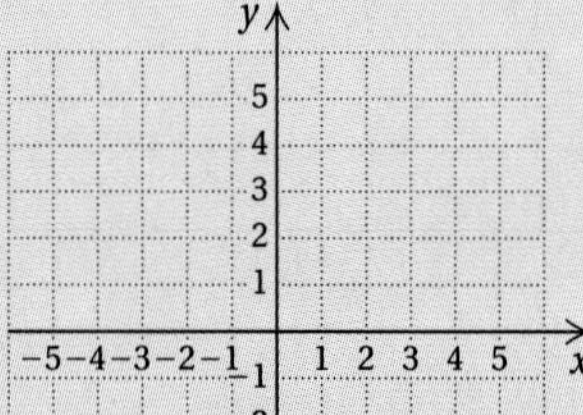

39. $f(x) = e^{-x}$

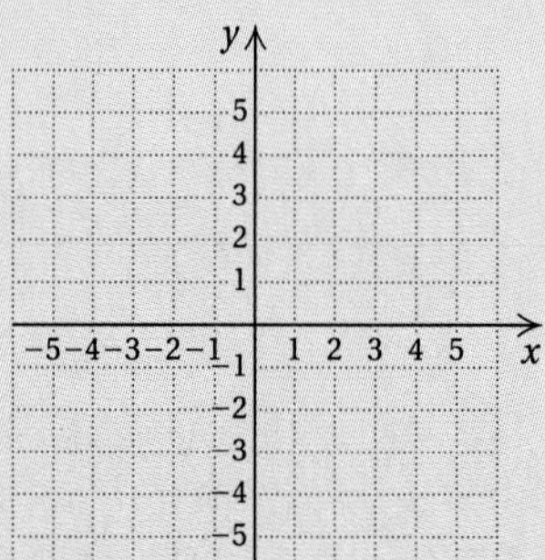

40. $f(x) = \log_2 x$

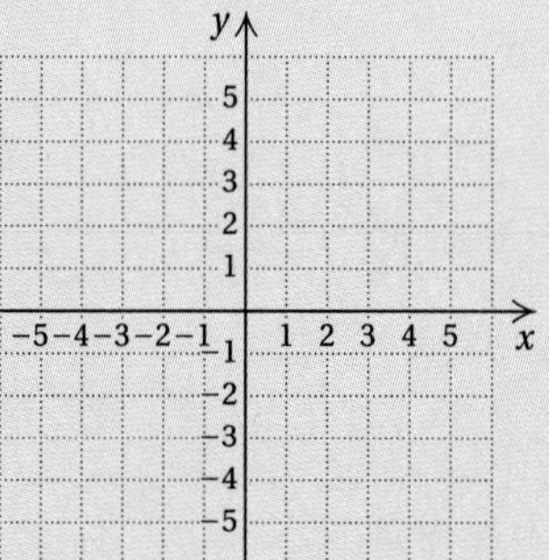

Factor.

41. $2x^4 - 12x^3 + x - 6$

42. $3a^2 - 12ab - 135b^2$

43. $x^2 - 17x + 72$

44. $81m^4 - n^4$

45. $16x^2 - 16x + 4$

46. $81a^3 - 24$

47. $10x^2 + 66x - 28$

48. $6x^3 + 27x^2 - 15x$

49. Find an equation of the line containing the points $(1, 4)$ and $(-1, 0)$.

50. Find an equation of the line containing the point $(1, 2)$ and perpendicular to the line whose equation is $2x - y = 3$.

51. Find $f^{-1}(x)$ when $f(x) = 2x - 3$.

52. z varies directly as x and inversely as the cube of y, and $z = 5$ when $x = 4$ and $y = 2$. What is z when $x = 10$ and $y = 5$?

53. Given the function f described by $f(x) = x^3 - 2$, find $f(-2)$.

54. Rationalize the denominator: $\dfrac{5 + \sqrt{a}}{3 - \sqrt{a}}$.

55. Find the domain: $f(x) = x^2 - 2x + 3$.

56. Find the domain: $f(x) = \dfrac{-43}{2x + 1}$.

Solve.

57. The square of a number plus the number is 156. Find the number.

58. *Work.* Jane can finish a quilted pillow sham in 3 hr. Laura can finish the same pillow sham in $1\frac{1}{2}$ hr. How long would it take to finish the pillow sham if they work together?

59. The sum of the squares of three consecutive even integers is equal to 8 more than three times the square of the second number. Find the integers.

60. *America Online.* America Online is an internet, e-mail, and information service. In addition to monthly service fees, it makes money through on-line sales of various products and through advertising. In 1996, such sales were \$102 million. In 1997, sales were \$256 million. (***Source:*** America Online)

a) Find an exponential growth function $P(t) = P_0e^{kt}$.
b) Predict the sales in the year 2000.
c) What is the doubling time?

61. *Train Travel.* A passenger train travels at twice the speed of a freight train. The freight train leaves a station at 2 A.M. and travels north at 34 mph. The passenger train leaves the station at 11 A.M. traveling north on a parallel track. How far from the station will the passenger train overtake the freight train?

62. *Carbon Dating.* Use the function $P(t) = P_0e^{-0.00012t}$ to find the age of a bone that has lost 25% of its carbon-14.

63. *Beam Load.* The weight W that a horizontal beam can support varies inversely as the length L of the beam. If a 14-m beam can support 1440 kg, what weight can a 6-m beam support?

64. Fit a linear function to the data points (2, −3) and (5, −4).

65. Fit a quadratic function to the data points (−2, 4), (−5, −6), and (1, −3).

66. Convert to a logarithmic equation: $10^6 = r$.

67. Convert to an exponential equation: $\log_3 Q = x$.

68. Express as a single logarithm:
$\frac{1}{5}(7 \log_b x - \log_b y - 8 \log_b z)$.

69. Express in terms of logarithms of x, y, and z:
$$\log_b \left(\frac{xy^5}{z}\right)^{-6}.$$

70. What is the maximum product of two numbers whose sum is 26?

71. Determine whether the function $f(x) = 4 - x^2$ is one-to-one.

72. For the graph of function f shown here, determine **(a)** $f(2)$; **(b)** the domain; **(c)** all x-values such that $f(x) = -5$; and **(d)** the range.

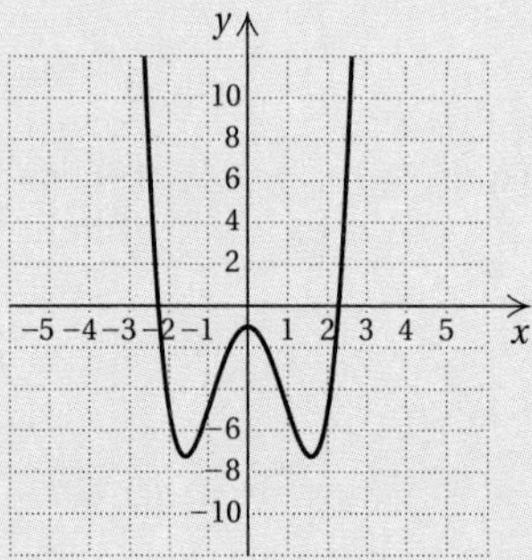

Synthesis

73. Solve $ax^2 - bx = 0$ for x.

74. Simplify: $\left[\frac{1}{(-3)^{-2}} - (-3)^1\right] \cdot [(-3)^2 + (-3)^{-2}]$.

75. Solve: $\frac{2x + 1}{x} = 3 + 7\sqrt{\frac{2x + 1}{x}}$.

76. Factor: $\frac{a^3}{8} + \frac{8b^3}{729}$.

77. Pam can do a certain job in a hours working alone. Elaine can do the same job in b hours working alone. Working together, it takes them t hours to do the job.

a) Find a formula for t.
b) Solve the formula for a.
c) Solve the formula for b.

Appendix A Handling Dimension Symbols

a Calculating with Dimension Symbols

In many applications, we add, subtract, multiply, and divide quantities having units, or dimensions, such as ft, km, sec, and hr. For example, to find average speed, we divide total distance by total time. What results is notation very much like a rational expression.

Example 1 A car travels 150 km in 2 hr. What is its average speed?

$$\text{Speed} = \frac{150\text{ km}}{2\text{ hr}}, \text{ or } 75\,\frac{\text{km}}{\text{hr}}$$

(The standard abbreviation for km/hr is km/h, but it does not suit our present discussion well.)

The symbol km/hr makes it look as if we are dividing kilometers by hours. It can be argued that we can divide only numbers. Nevertheless, we treat dimension symbols, such as km, ft, and hr, as if they were numerals or variables, obtaining correct results mechanically.

Do Exercise 1.

Examples Compare the following.

2. $\dfrac{150x}{2y} = \dfrac{150}{2} \cdot \dfrac{x}{y} = 75\dfrac{x}{y}$ with $\dfrac{150\text{ km}}{2\text{ hr}} = \dfrac{150}{2}\dfrac{\text{km}}{\text{hr}} = 75\dfrac{\text{km}}{\text{hr}}$

3. $3x + 2x = (3 + 2)x = 5x$ with $3\text{ ft} + 2\text{ ft} = (3 + 2)\text{ ft} = 5\text{ ft}$

4. $5x \cdot 3x = 15x^2$ with $5\text{ ft} \cdot 3\text{ ft} = 15\text{ ft}^2$ (square feet)

Do Exercises 2–4.

If 5 men work 8 hours, the total amount of labor is 40 man-hours.

Example 5 Compare

$5x \cdot 8y = 40xy$ with 5 men · 8 hours = 40 man-hours.

Do Exercise 5.

Objectives

- a Perform calculations with dimension symbols.
- b Make unit changes.

1. A truck travels 210 mi in 3 hr. What is its average speed?

Perform these calculations.

2. $\dfrac{100\text{ m}}{4\text{ sec}}$

3. 7 yd + 9 yd

4. 24 in. · 3 in.

5. Calculate: 6 men · 11 hours.

Answers on page A-60

6. Calculate:

$$\frac{200 \text{ kW} \cdot 140 \text{ hr}}{35 \text{ da}}.$$

7. Convert 7 ft to inches.

8. Convert 90 mi/hr to ft/sec.

Answers on page A-60

Example 6 Compare

$$\frac{300x \cdot 240y}{15t} = 4800\frac{xy}{t} \quad \text{with} \quad \frac{300 \text{ kW} \cdot 240 \text{ hr}}{15 \text{ da}} = 4800\frac{\text{kW-hr}}{\text{da}}.$$

If an electrical device uses 300 kW (kilowatts) for 240 hr over a period of 15 days, its rate of usage of energy is 4800 kilowatt-hours per day. The standard abbreviation for kilowatt-hours is kWh.

Do Exercise 6.

b Making Unit Changes

We can treat dimension symbols much like numerals or variables, because we obtain correct results that way. We can change units by substituting or by multiplying by 1, as shown below.

Example 7 Convert 3 ft to inches.

METHOD 1. We have 3 ft. We know that 1 ft = 12 in., so we substitute 12 in. for ft:

$$3 \text{ ft} = 3 \cdot 12 \text{ in.} = 36 \text{ in.}$$

METHOD 2. We want to convert from "ft" to "in." We multiply by 1 using a symbol for 1 with "ft" on the bottom since we are converting from "ft," and with "in." on the top since we are converting to "in."

$$\begin{aligned} 3 \text{ ft} &= 3 \text{ ft} \cdot \frac{12 \text{ in.}}{1 \text{ ft}} \\ &= \frac{3 \cdot 12}{1} \cdot \frac{\text{ft}}{\text{ft}} \cdot \text{in.} = 36 \text{ in.} \end{aligned}$$

Do Exercise 7.

We can multiply by 1 several times to make successive conversions. In the following example, we convert mi/hr to ft/sec by converting successively from mi/hr to ft/hr to ff/min to ft/sec.

Example 8 Convert 60 mi/hr to ft/sec.

$$\begin{aligned} 60\frac{\text{mi}}{\text{hr}} &= 60\frac{\text{mi}}{\text{hr}} \cdot \frac{5280 \text{ ft}}{1 \text{ mi}} \cdot \frac{1 \text{ hr}}{60 \text{ min}} \cdot \frac{1 \text{ min}}{60 \text{ sec}} \\ &= \frac{60 \cdot 5280}{60 \cdot 60} \cdot \frac{\text{mi}}{\text{mi}} \cdot \frac{\text{hr}}{\text{hr}} \cdot \frac{\text{min}}{\text{min}} \cdot \frac{\text{ft}}{\text{sec}} = 88\frac{\text{ft}}{\text{sec}}. \end{aligned}$$

Do Exercise 8.

Exercise Set A

a Add these measures.

1. 45 ft + 23 ft

2. 55 km/hr + 27 km/hr

3. 17 g + 28 g

4. 3.4 lb + 5.2 lb

Find the average speeds, given total distance and total time.

5. 90 mi, 6 hr

6. 640 km, 20 hr

7. 9.9 m, 3 sec

8. 76 ft, 4 min

Perform these calculations.

9. $\frac{3 \text{ in.} \cdot 8 \text{ lb}}{6 \text{ sec}}$

10. $\frac{60 \text{ men} \cdot 8 \text{ hr}}{20 \text{ da}}$

11. $36 \text{ ft} \cdot \frac{1 \text{ yd}}{3 \text{ ft}}$

12. $55 \frac{\text{mi}}{\text{hr}} \cdot 4 \text{ hr}$

13. $5 \text{ ft}^3 + 11 \text{ ft}^3$

14. $\frac{3 \text{ lb}}{14 \text{ ft}} \cdot \frac{7 \text{ lb}}{6 \text{ ft}}$

15. Divide $4850 by 5 days.

16. Divide $25.60 by 8 hr.

b Make these unit changes.

17. Change 3.2 lb to oz (16 oz = 1 lb).

18. Change 6.2 km to m.

19. Change 35 mi/hr to ft/min.

20. Change $375 per day to dollars per minute.

21. Change 8 ft to in.

22. Change 25 yd to ft.

23. How many years ago is 1 million sec ago? Let 365 days = 1 yr.

24. How many years ago is 1 billion sec ago?

25. How many years ago is 1 trillion sec ago?

26. Change 20 lb to oz.

27. Change $60 \frac{\text{lb}}{\text{ft}}$ to $\frac{\text{oz}}{\text{in.}}$.

28. Change $44 \frac{\text{ft}}{\text{sec}}$ to $\frac{\text{mi}}{\text{hr}}$.

29. Change 2 days to seconds.

30. Change 128 hr to days.

31. Change 216 in^2 to ft^2.

32. Change 1440 man-hours to man-days.

33. Change $80 \frac{\text{lb}}{\text{ft}^3}$ to $\frac{\text{ton}}{\text{yd}^3}$.

34. Change the speed of light, 186,000 mi/sec, to mi/yr.

Appendix B Determinants and Cramer's Rule

Objectives

a. Evaluate second-order determinants.

b. Evaluate third-order determinants.

c. Solve systems of equations using Cramer's Rule.

In Chapter 8, you probably noticed that the elimination method concerns itself primarily with the coefficients and constants of the equations. Here we learn a method for solving a system of equations using just the coefficients and constants. This method involves *determinants.*

a Evaluating Determinants

The following symbolism represents a **second-order determinant:**

$$\begin{vmatrix} a_1 & b_1 \\ a_2 & b_2 \end{vmatrix}.$$

To evaluate a determinant, we do two multiplications and subtract.

Example 1 Evaluate:

$$\begin{vmatrix} 2 & -5 \\ 6 & 7 \end{vmatrix}.$$

We multiply and subtract as follows:

$$\begin{vmatrix} 2 & -5 \\ 6 & 7 \end{vmatrix} = 2 \cdot 7 - 6 \cdot (-5) = 14 + 30 = 44.$$

Determinants are defined according to the pattern shown in Example 1.

> The determinant $\begin{vmatrix} a_1 & b_1 \\ a_2 & b_2 \end{vmatrix}$ is defined to mean $a_1b_2 - a_2b_1$.

The value of a determinant is a *number*. In Example 1, the value is 44.

Do Exercises 1 and 2.

Evaluate.

1. $\begin{vmatrix} 3 & 2 \\ 4 & 1 \end{vmatrix}$

2. $\begin{vmatrix} 5 & -2 \\ -1 & -1 \end{vmatrix}$

b Third-Order Determinants

A **third-order determinant** is defined as follows.

$$\begin{vmatrix} a_1 & b_1 & c_1 \\ a_2 & b_2 & c_2 \\ a_3 & b_3 & c_3 \end{vmatrix} = a_1\begin{vmatrix} b_2 & c_2 \\ b_3 & c_3 \end{vmatrix} - a_2\begin{vmatrix} b_1 & c_1 \\ b_3 & c_3 \end{vmatrix} + a_3\begin{vmatrix} b_1 & c_1 \\ b_2 & c_2 \end{vmatrix}$$

Note the minus sign here.

Note that the a's come from the first column.

Answers on page A-60

Note that the second-order determinants on the right can be obtained by crossing out the row and the column in which each a occurs.

For a_1: $\begin{vmatrix} a_1 & b_1 & c_1 \\ a_2 & b_2 & c_2 \\ a_3 & b_3 & c_3 \end{vmatrix}$ For a_2: $\begin{vmatrix} a_1 & b_1 & c_1 \\ a_2 & b_2 & c_2 \\ a_3 & b_3 & c_3 \end{vmatrix}$

For a_3: $\begin{vmatrix} a_1 & b_1 & c_1 \\ a_2 & b_2 & c_2 \\ a_3 & b_3 & c_3 \end{vmatrix}$

Example 2 Evaluate this third-order determinant:

$$\begin{vmatrix} -1 & 0 & 1 \\ -5 & 1 & -1 \\ 4 & 8 & 1 \end{vmatrix} = -1\begin{vmatrix} 1 & -1 \\ 8 & 1 \end{vmatrix} - (-5)\begin{vmatrix} 0 & 1 \\ 8 & 1 \end{vmatrix} + 4\begin{vmatrix} 0 & 1 \\ 1 & -1 \end{vmatrix}.$$

We calculate as follows:

$$-1\begin{vmatrix} 1 & -1 \\ 8 & 1 \end{vmatrix} - (-5)\begin{vmatrix} 0 & 1 \\ 8 & 1 \end{vmatrix} + 4\begin{vmatrix} 0 & 1 \\ 1 & -1 \end{vmatrix}$$
$$= -1[1 \cdot 1 - 8(-1)] + 5(0 \cdot 1 - 8 \cdot 1) + 4[0 \cdot (-1) - 1 \cdot 1]$$
$$= -1(9) + 5(-8) + 4(-1)$$
$$= -9 - 40 - 4$$
$$= -53.$$

Do Exercises 3 and 4.

c Solving Systems by Determinants

Here is a system of two equations in two variables:

$$a_1x + b_1y = c_1,$$
$$a_2x + b_2y = c_2.$$

We form three determinants, which we call D, D_x, and D_y.

$D = \begin{vmatrix} a_1 & b_1 \\ a_2 & b_2 \end{vmatrix}$ In D, we have the coefficients of x and y.

$D_x = \begin{vmatrix} c_1 & b_1 \\ c_2 & b_2 \end{vmatrix}$ To form D_x, we replace the x-coefficients in D with the constants on the right side of the equations.

$D_y = \begin{vmatrix} a_1 & c_1 \\ a_2 & c_2 \end{vmatrix}$ To form D_y, we replace the y-coefficients in D with the constants on the right.

It is important that the replacement be done *without changing the order of the columns.* Then the solution of the system can be found as follows. This is known as **Cramer's rule.**

Evaluate.

3. $\begin{vmatrix} 2 & -1 & 1 \\ 1 & 2 & -1 \\ 3 & 4 & -3 \end{vmatrix}$

4. $\begin{vmatrix} 3 & 2 & 2 \\ -2 & 1 & 4 \\ 4 & -3 & 3 \end{vmatrix}$

Answers on page A-60

5. Solve using Cramer's rule.

$$4x - 3y = 15,$$
$$x + 3y = 0$$

Answer on page A-60

CRAMER'S RULE

$$x = \frac{D_x}{D}, \quad y = \frac{D_y}{D}$$

Example 3 Solve using Cramer's rule:

$$3x - 2y = 7,$$
$$3x + 2y = 9.$$

We compute D, D_x, and D_y:

$$D = \begin{vmatrix} 3 & -2 \\ 3 & 2 \end{vmatrix} = 3 \cdot 2 - 3 \cdot (-2) = 6 + 6 = 12;$$

$$D_x = \begin{vmatrix} 7 & -2 \\ 9 & 2 \end{vmatrix} = 7 \cdot 2 - 9(-2) = 14 + 18 = 32;$$

$$D_y = \begin{vmatrix} 3 & 7 \\ 3 & 9 \end{vmatrix} = 3 \cdot 9 - 3 \cdot 7 = 27 - 21 = 6.$$

Then

$$x = \frac{D_x}{D} = \frac{32}{12}, \text{ or } \frac{8}{3} \quad \text{and} \quad y = \frac{D_y}{D} = \frac{6}{12} = \frac{1}{2}.$$

The solution is $\left(\frac{8}{3}, \frac{1}{2}\right)$.

Do Exercise 5.

Cramer's rule for three equations is very similar to that for two.

$$a_1x + b_1y + c_1z = d_1$$
$$a_2x + b_2y + c_2z = d_2$$
$$a_3x + b_3y + c_3z = d_3$$

$$D = \begin{vmatrix} a_1 & b_1 & c_1 \\ a_2 & b_2 & c_2 \\ a_3 & b_3 & c_3 \end{vmatrix} \qquad D_x = \begin{vmatrix} d_1 & b_1 & c_1 \\ d_2 & b_2 & c_2 \\ d_3 & b_3 & c_3 \end{vmatrix}$$

$$D_y = \begin{vmatrix} a_1 & d_1 & c_1 \\ a_2 & d_2 & c_2 \\ a_3 & d_3 & c_3 \end{vmatrix}$$

D is again the determinant of the coefficients of x, y, and z. This time we have one more determinant, D_z. We get it by replacing the z-coefficients in D with the constants on the right:

$$D_z = \begin{vmatrix} a_1 & b_1 & d_1 \\ a_2 & b_2 & d_2 \\ a_3 & b_3 & d_3 \end{vmatrix}.$$

The solution of the system is given by

CRAMER'S RULE

$$x = \frac{D_x}{D}, \quad y = \frac{D_y}{D}, \quad z = \frac{D_z}{D}$$

Example 4 Solve using Cramer's rule:

$$x - 3y + 7z = 13,$$
$$x + y + z = 1,$$
$$x - 2y + 3z = 4.$$

We compute D, D_x, D_y, and D_z:

$$D = \begin{vmatrix} 1 & -3 & 7 \\ 1 & 1 & 1 \\ 1 & -2 & 3 \end{vmatrix} = -10; \qquad D_x = \begin{vmatrix} 13 & -3 & 7 \\ 1 & 1 & 1 \\ 4 & -2 & 3 \end{vmatrix} = 20;$$

$$D_y = \begin{vmatrix} 1 & 13 & 7 \\ 1 & 1 & 1 \\ 1 & 4 & 3 \end{vmatrix} = -6; \qquad D_z = \begin{vmatrix} 1 & -3 & 13 \\ 1 & 1 & 1 \\ 1 & -2 & 4 \end{vmatrix} = -24.$$

Then

$$x = \frac{D_x}{D} = \frac{20}{-10} = -2;$$

$$y = \frac{D_y}{D} = \frac{-6}{-10} = \frac{3}{5};$$

$$z = \frac{D_z}{D} = \frac{-24}{-10} = \frac{12}{5}.$$

The solution is $\left(-2, \frac{3}{5}, \frac{12}{5}\right)$.

In Example 4, we would not have needed to evaluate D_z. Once we found x and y, we could have substituted them into one of the equations to find z. In practice, it is faster to use determinants to find only two of the numbers; then we find the third by substitution into an equation.

Do Exercise 6.

In using Cramer's rule, we divide by D. If D were 0, we could not do so.

> If $D = 0$ and at least one of the other determinants is not 0, then the system does not have a solution, and we say that it is *inconsistent.*
>
> If $D = 0$ and all the other determinants are also 0, then there is an infinite set of solutions. In that case, we say that the system is *dependent.*

6. Solve using Cramer's rule:

$$x - 3y - 7z = 6,$$
$$2x + 3y + z = 9,$$
$$4x + y = 7.$$

Answer on page A-60

Exercise Set B

a Evaluate.

1. $\begin{vmatrix} 3 & 7 \\ 2 & 8 \end{vmatrix}$

2. $\begin{vmatrix} 5 & 4 \\ 4 & -5 \end{vmatrix}$

3. $\begin{vmatrix} -3 & -6 \\ -5 & -10 \end{vmatrix}$

4. $\begin{vmatrix} 4 & 5 \\ -7 & 9 \end{vmatrix}$

5. $\begin{vmatrix} 8 & 2 \\ 12 & -3 \end{vmatrix}$

6. $\begin{vmatrix} 1 & 1 \\ 9 & 8 \end{vmatrix}$

7. $\begin{vmatrix} 2 & -7 \\ 0 & 0 \end{vmatrix}$

8. $\begin{vmatrix} 0 & -4 \\ 0 & -6 \end{vmatrix}$

b Evaluate.

9. $\begin{vmatrix} 0 & 2 & 0 \\ 3 & -1 & 1 \\ 1 & -2 & 2 \end{vmatrix}$

10. $\begin{vmatrix} 3 & 0 & -2 \\ 5 & 1 & 2 \\ 2 & 0 & -1 \end{vmatrix}$

11. $\begin{vmatrix} -1 & -2 & -3 \\ 3 & 4 & 2 \\ 0 & 1 & 2 \end{vmatrix}$

12. $\begin{vmatrix} 1 & 2 & 2 \\ 2 & 1 & 0 \\ 3 & 3 & 1 \end{vmatrix}$

13. $\begin{vmatrix} 3 & 2 & -2 \\ -2 & 1 & 4 \\ -4 & -3 & 3 \end{vmatrix}$

14. $\begin{vmatrix} 2 & -1 & 1 \\ 1 & 2 & -1 \\ 3 & 4 & -3 \end{vmatrix}$

15. $\begin{vmatrix} 3 & 2 & 4 \\ 1 & 1 & 1 \\ 1 & 1 & 1 \end{vmatrix}$

16. $\begin{vmatrix} -1 & 6 & -5 \\ 2 & 4 & 4 \\ 5 & 3 & 10 \end{vmatrix}$

c Solve using Cramer's rule.

17. $3x - 4y = 6,$
$5x + 9y = 10$

18. $5x + 8y = 1,$
$3x + 7y = 5$

19. $-2x + 4y = 3,$
$3x - 7y = 1$

20. $5x - 4y = -3,$
$7x + 2y = 6$

21. $4x + 2y = 11,$
$3x - y = 2$

22. $3x - 3y = 11,$
$9x - 2y = 5$

23. $x + 4y = 8,$
$3x + 5y = 3$

24. $x + 4y = 5,$
$-3x + 2y = 13$

25. $2x - 3y + 5z = 27,$
$x + 2y - z = -4,$
$5x - y + 4z = 27$

26. $x - y + 2z = -3,$
$x + 2y + 3z = 4,$
$2x + y + z = -3$

27. $r - 2s + 3t = 6,$
$2r - s - t = -3,$
$r + s + t = 6$

28. $a - 3c = 6,$
$b + 2c = 2,$
$7a - 3b - 5c = 14$

29. $4x - y - 3z = 1,$
$8x + y - z = 5,$
$2x + y + 2z = 5$

30. $3x + 2y + 2z = 3,$
$x + 2y - z = 5,$
$2x - 4y + z = 0$

31. $p + q + r = 1,$
$p - 2q - 3r = 3,$
$4p + 5q + 6r = 4$

32. $x + 2y - 3z = 9,$
$2x - y + 2z = -8,$
$3x - y - 4z = 3$

Appendix C Elimination Using Matrices

Objective

a Solve systems of two or three equations using matrices.

The elimination method concerns itself primarily with the coefficients and constants of the equations. In what follows, we learn a method for solving systems using just the coefficients and the constants. This procedure involves what are called *matrices*.

a In solving systems of equations, we perform computations with the constants. The variables play no important role until the end. Thus we can simplify writing a system by omitting the variables. For example, the system

$$\begin{aligned} 3x + 4y &= 5, \\ x - 2y &= 1 \end{aligned} \quad \text{simplifies to} \quad \begin{matrix} 3 & 4 & 5 \\ 1 & -2 & 1 \end{matrix}$$

if we omit the variables, the operation of addition, and the equals signs, the result is a rectangular array of numbers. Such an array is called a **matrix** (plural, **matrices**). We ordinarily write brackets around matrices. The following are matrices:

$$\begin{bmatrix} 4 & 1 & 3 & 5 \\ 1 & 0 & 1 & 2 \\ 6 & 3 & -2 & 0 \end{bmatrix}, \quad \begin{bmatrix} 6 & 2 & 1 & 4 & 7 \\ 1 & 2 & 1 & 3 & 1 \\ 4 & 0 & -2 & 0 & -3 \end{bmatrix}, \quad \begin{bmatrix} 1 & 2 \\ 145 & 0 \\ -7 & 9 \\ 8 & 1 \\ 0 & 0 \end{bmatrix}.$$

The **rows** of a matrix are horizontal, and the **columns** are vertical.

$$\begin{bmatrix} 5 & -2 & 2 \\ 1 & 0 & 1 \\ 0 & 1 & 2 \end{bmatrix} \begin{matrix} \longleftarrow \text{row 1} \\ \longleftarrow \text{row 2} \\ \longleftarrow \text{row 3} \end{matrix}$$

column 1 column 2 column 3

Let's now use matrices to solve systems of linear equations.

Example 1 Solve the system

$$\begin{aligned} 5x - 4y &= -1, \\ -2x + 3y &= 2. \end{aligned}$$

We write a matrix using only the coefficients and the constants, keeping in mind that x corresponds to the first column and y to the second. A dashed line separates the coefficients from the constants at the end of each equation:

$$\left[\begin{array}{cc:c} 5 & -4 & -1 \\ -2 & 3 & 2 \end{array}\right].$$

The individual numbers are called *elements* or *entries*.

Our goal is to transform this matrix into one of the form

$$\left[\begin{array}{cc:c} a & b & c \\ 0 & d & e \end{array}\right].$$

The variables can then be reinserted to form equations from which we can complete the solution.

1. Solve using matrices:

$$5x - 2y = -44,$$
$$2x + 5y = -6.$$

We do calculations that are similar to those that we would do if we wrote the entire equations. The first step, if possible, is to multiply and/or interchange the rows so that each number in the first column below the first number is a multiple of that number. In this case, we do so by multiplying Row 2 by 5. This corresponds to multiplying the second equation by 5.

$$\left[\begin{array}{rr|r} 5 & -4 & -1 \\ -10 & 15 & 10 \end{array}\right] \qquad \text{New Row 2} = 5(\text{Row 2})$$

Next, we multiply the first row by 2 and add the result to the second row. This corresponds to multiplying the first equation by 2 and adding the result to the second equation. Although we write these equations out here, we generally try to do them mentally:

$$2 \cdot 5 + (-10) = 0; \qquad 2(-4) + 15 = 7; \qquad 2(-1) + 10 = 8.$$

$$\left[\begin{array}{rr|r} 5 & -4 & -1 \\ 0 & 7 & 8 \end{array}\right] \qquad \text{New Row 2} = 2(\text{Row 1}) + (\text{Row 2})$$

If we now reinsert the variables, we have

$$5x - 4y = -1, \qquad (1)$$
$$7y = 8. \qquad (2)$$

We can now proceed as before, solving equation (2) for y:

$$7y = 8 \qquad (2)$$
$$y = \tfrac{8}{7}.$$

Next, we substitute $\frac{8}{7}$ for y back in equation (1). This procedure is called *back-substitution.*

$$5x - 4y = -1 \qquad (1)$$
$$5x - 4 \cdot \tfrac{8}{7} = -1 \qquad \text{Substituting } \tfrac{8}{7} \text{ for } y \text{ in equation (1)}$$
$$x = \tfrac{5}{7} \qquad \text{Solving for } x$$

The solution is $\left(\frac{5}{7}, \frac{8}{7}\right)$.

Do Exercise 1.

Example 2 Solve the system

$$2x - y + 4z = -3,$$
$$x - 4z = 5,$$
$$6x - y + 2z = 10.$$

We first write a matrix, using only the coefficients and the constants. Where there are missing terms, we must write 0's:

$$\left[\begin{array}{rrr|r} 2 & -1 & 4 & -3 \\ 1 & 0 & -4 & 5 \\ 6 & -1 & 2 & 10 \end{array}\right] \begin{array}{l} \text{(P1)} \\ \text{(P2)} \\ \text{(P3)} \end{array}$$

(P1), (P2), and (P3) designate the equations that are in the first, second, and third position, respectively.

Our goal is to find an equivalent matrix of the form

$$\left[\begin{array}{ccc|c} a & b & c & d \\ 0 & e & f & g \\ 0 & 0 & h & i \end{array}\right].$$

A matrix of this form can be rewritten as a system of equations from which a solution can be found easily.

Answer on page A-60

The first step, if possible, is to interchange the rows so that each number in the first column below the first number is a multiple of that number. In this case, we do so by interchanging Rows 1 and 2:

$$\left[\begin{array}{rrr|r} 1 & 0 & -4 & 5 \\ 2 & -1 & 4 & -3 \\ 6 & -1 & 2 & 10 \end{array}\right]$$

This corresponds to interchanging the first two equations.

Next, we multiply the first row by -2 and add it to the second row:

$$\left[\begin{array}{rrr|r} 1 & 0 & -4 & 5 \\ 0 & -1 & 12 & -13 \\ 6 & -1 & 2 & 10 \end{array}\right].$$

This corresponds to multiplying new equation (P1) by -2 and adding it to new equation (P2). The result replaces the former (P2). We perform the calculations mentally.

Now we multiply the first row by -6 and add it to the third row:

$$\left[\begin{array}{rrr|r} 1 & 0 & -4 & 5 \\ 0 & -1 & 12 & -13 \\ 0 & -1 & 26 & -20 \end{array}\right].$$

This corresponds to multiplying equation (P1) by -6 and adding it to equation (P3).

Next, we multiply Row 2 by -1 and add it to the third row:

$$\left[\begin{array}{rrr|r} 1 & 0 & -4 & 5 \\ 0 & -1 & 12 & -13 \\ 0 & 0 & 14 & -7 \end{array}\right].$$

This corresponds to multiplying equation (P2) by -1 and adding it to equation (P3).

Reinserting the variables gives us

$$\begin{aligned} x \qquad - 4z &= 5, && \text{(P1)} \\ -y + 12z &= -13, && \text{(P2)} \\ 14z &= -7. && \text{(P3)} \end{aligned}$$

We now solve (P3) for z:

$$\begin{aligned} 14z &= -7 && \text{(P3)} \\ z &= -\tfrac{7}{14} && \text{Solving for } z \\ z &= -\tfrac{1}{2}. \end{aligned}$$

Next, we back-substitute $-\frac{1}{2}$ for z in (P2) and solve for y:

$$\begin{aligned} -y + 12z &= -13 && \text{(P2)} \\ -y + 12\left(-\tfrac{1}{2}\right) &= -13 && \text{Substituting } -\tfrac{1}{2} \text{ for } z \text{ in equation (P2)} \\ -y - 6 &= -13 \\ -y &= -7 \\ y &= 7. && \text{Solving for } y \end{aligned}$$

Since there is no y-term in (P1), we need only substitute $-\frac{1}{2}$ for z in (P1) and solve for x:

$$\begin{aligned} x - 4z &= 5 && \text{(P1)} \\ x - 4\left(-\tfrac{1}{2}\right) &= 5 && \text{Substituting } -\tfrac{1}{2} \text{ for } z \text{ in equation (P1)} \\ x + 2 &= 5 \\ x &= 3. && \text{Solving for } x \end{aligned}$$

The solution is $\left(3, 7, -\frac{1}{2}\right)$.

Do Exercise 2.

2. Solve using matrices:

$$\begin{aligned} x - 2y + 3z &= 4, \\ 2x - y + z &= -1, \\ 4x + y + z &= 1. \end{aligned}$$

Answer on page A-60

All the operations used in the preceding example correspond to operations with the equations and produce equivalent systems of equations. We call the matrices **row-equivalent** and the operations that produce them **row-equivalent operations.**

ROW-EQUIVALENT OPERATIONS

Each of the following row-equivalent operations produces an equivalent matrix:

a) Interchanging any two rows.

b) Multiplying each element of a row by the same nonzero number.

c) Multiplying each element of a row by a nonzero number and adding the result to another row.

The best overall method for solving systems of equations is by row-equivalent matrices; graphers and computers are programmed to use them. Matrices are part of a branch of mathematics known as linear algebra. They are also studied in more detail in many courses in finite mathematics.

Exercise Set C

a Solve using matrices.

1. $4x + 2y = 11,$
$3x - y = 2$

2. $3x - 3y = 11,$
$9x - 2y = 5$

3. $x + 4y = 8,$
$3x + 5y = 3$

4. $x + 4y = 5,$
$-3x + 2y = 13$

5. $5x - 3y = -2,$
$4x + 2y = 5$

6. $3x + 4y = 7,$
$-5x + 2y = 10$

7. $2x - 3y = 50,$
$5x + y = 40$

8. $4x + 5y = -8,$
$7x + 9y = 11$

9. $4x - y - 3z = 1,$
$8x + y - z = 5,$
$2x + y + 2z = 5$

10. $3x + 2y + 2z = 3,$
$x + 2y - z = 5,$
$2x - 4y + z = 0$

11. $p + q + r = 1,$
$p - 2q - 3r = 3,$
$4p + 5q + 6r = 4$

12. $x + 2y - 3z = 9,$
$2x - y + 2z = -8,$
$3x - y - 4z = 3$

13. $x - y + 2z = 0,$
$x - 2y + 3z = -1,$
$2x - 2y + z = -3$

14. $4a + 9b = 8,$
$8a + 6c = -1,$
$6b + 6c = -1$

15. $3p + 2r = 11,$
$q - 7r = 4,$
$p - 6q = 1$

16. $m + n + t = 6,$
$m - n - t = 0,$
$m + 2n + t = 5$

17. $2x + 2y - 2z - 2w = -10,$
$x + y + z + w = -5,$
$x - y + 4z + 3w = -2,$
$3x - 2y + 2z + w = -6$

18. $2x - 3y + z - w = -8,$
$x + y - z - w = -4,$
$x + y + z + w = 22,$
$x - y - z - w = -14$

Appendix D The Algebra of Functions

Objective

a Given two functions f and g, find their sum, difference, product, and quotient.

a The Sum, Difference, Product, and Quotient of Functions

Suppose that a is in the domain of two functions, f and g. The input a is paired with $f(a)$ by f and with $g(a)$ by g. The outputs can then be added to get $f(a) + g(a)$.

Example 1 Let $f(x) = x + 4$ and $g(x) = x^2 + 1$. Find $f(2) + g(2)$.

We visualize two function machines. Because 2 is in the domain of each function, we can compute $f(2)$ and $g(2)$.

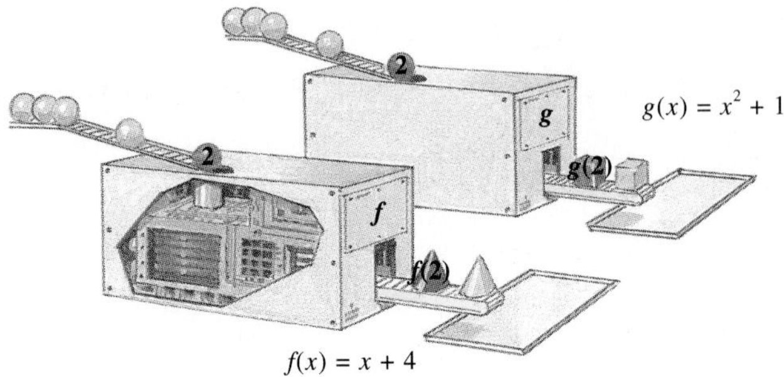

Since

$$f(2) = 2 + 4 = 6 \quad \text{and} \quad g(2) = 2^2 + 1 = 5,$$

we have

$$f(2) + g(2) = 6 + 5 = 11.$$

In Example 1, suppose that we were to write $f(x) + g(x)$ as $(x + 4) + (x^2 + 1)$, or $f(x) + g(x) = x^2 + x + 5$. This could then be regarded as a "new" function: $(f + g)(x) = x^2 + x + 5$. We can alternatively find $f(2) + g(2)$ with $(f + g)(x)$:

$$(f + g)(x) = x^2 + x + 5$$

$$(f + g)(2) = 2^2 + 2 + 5 \quad \text{Substituting 2 for } x$$

$$= 4 + 2 + 5$$

$$= 11.$$

Similar notations exist for subtraction, multiplication, and division of functions.

From any functions f and g, we can form new functions defined as:

1. The **sum** $f + g$: $(f + g)(x) = f(x) + g(x)$;
2. The **difference** $f - g$: $(f - g)(x) = f(x) - g(x)$;
3. The **product** fg: $f \cdot g(x) = f(x) \cdot g(x)$;
4. The **quotient** f/g: $(f/g)(x) = f(x)/g(x)$, where $g(x) \neq 0$.

Example 2 Given f and g described by $f(x) = x^2 - 5$ and $g(x) = x + 7$, find $(f + g)(x)$, $(f - g)(x)$, $(f \cdot g)(x)$, $(f/g)(x)$, and $(g \cdot g)(x)$.

$$(f + g)(x) = f(x) + g(x) = (x^2 - 5) + (x + 7) = x^2 + x + 2;$$
$$(f - g)(x) = f(x) - g(x) = (x^2 - 5) - (x + 7) = x^2 - x - 12;$$
$$(f \cdot g)(x) = f(x) \cdot g(x) = (x^2 - 5)(x + 7) = x^3 + 7x^2 - 5x - 35;$$
$$(f/g)(x) = f(x)/g(x) = \frac{x^2 - 5}{x + 7};$$
$$(g \cdot g)(x) = g(x) \cdot g(x) = (x + 7)(x + 7) = x^2 + 14x + 49$$

Note that the sum, difference, and product of polynomials are also polynomial functions, but the quotient may not be.

Do Exercise 1.

Example 3 For $f(x) = x^2 - x$ and $g(x) = x + 2$, find $(f + g)(3)$, $(f - g)(-1)$, $(f \cdot g)(5)$, and $(f/g)(-4)$.

We first find $(f + g)(x)$, $(f - g)(x)$, $(f \cdot g)(x)$, and $(f/g)(x)$.

$$(f + g)(x) = f(x) + g(x) = x^2 - x + x + 2$$
$$= x^2 + 2;$$

$$(f - g)(x) = f(x) - g(x) = x^2 - x - (x + 2)$$
$$= x^2 - x - x - 2$$
$$= x^2 - 2x - 2;$$

$$(f \cdot g)(x) = f(x) \cdot g(x) = (x^2 - x)(x + 2)$$
$$= x^3 + 2x^2 - x^2 - 2x$$
$$= x^3 + x^2 - 2x;$$

$$(f/g)(x) = \frac{f(x)}{g(x)} = \frac{x^2 - x}{x + 2}.$$

Then we substitute.

$$(f + g)(3) = 3^2 + 2 \qquad \text{Using } (f + g)(x) = x^2 + 2$$
$$= 9 + 2 = 11;$$

$$(f - g)(-1) = (-1)^2 - 2(-1) - 2 \qquad \text{Using } (f - g)(x) = x^2 - 2x - 2$$
$$= 1 + 2 - 2 = 1;$$

$$(f \cdot g)(5) = 5^3 + 5^2 - 2 \cdot 5 \qquad \text{Using } (f \cdot g)(x) = x^3 + x^2 - 2x$$
$$= 125 + 25 - 10 = 140;$$

$$(f/g)(-4) = \frac{(-4)^2 - (-4)}{-4 + 2} \qquad \text{Using } (f/g)(x) = (x^2 - x)/(x + 2)$$
$$= \frac{16 + 4}{-2} = \frac{20}{-2} = -10$$

Do Exercise 2.

1. Given $f(x) = x^2 + 3$ and $g(x) = x^2 - 3$, find each of the following.

a) $(f + g)(x)$

b) $(f - g)(x)$

c) $(f \cdot g)(x)$

d) $(f/g)(x)$

e) $(f \cdot f)(x)$

2. Given $f(x) = x^2 + x$ and $g(x) = 2x - 3$, find each of the following.

a) $(f + g)(-2)$

b) $(f - g)(4)$

c) $(f \cdot g)(-3)$

d) $(f/g)(2)$

Answers on page A-60

Exercise Set D

Let $f(x) = -3x + 1$ and $g(x) = x^2 + 2$. Find the following.

1. $f(2) + g(2)$

2. $f(-1) + g(-1)$

3. $f(5) - g(5)$

4. $f(4) - g(4)$

5. $f(-1) \cdot g(-1)$

6. $f(-2) \cdot g(-2)$

7. $f(-4)/g(-4)$

8. $f(3)/g(3)$

9. $g(1) - f(1)$

10. $g(2)/f(2)$

11. $g(0)/f(0)$

12. $g(6) - f(6)$

Let $f(x) = x^2 - 3$ and $g(x) = 4 - x$. find the following.

13. $(f + g)(x)$

14. $(f - g)(x)$

15. $(f + g)(-4)$

16. $(f + g)(-5)$

17. $(f - g)(3)$

18. $(f - g)(2)$

19. $(f \cdot g)(x)$

20. $(f/g)(x)$

21. $(f \cdot g)(-3)$

22. $(f \cdot g)(-4)$

23. $(f/g)(0)$

24. $(f/g)(1)$

25. $(f/g)(-2)$

26. $(f/g)(-1)$

For each pair of functions f and g, find $(f + g)(x)$, $(f - g)(x)$, $(f \cdot g)(x)$, and $(f/g)(x)$.

27. $f(x) = x^2$, $g(x) = 3x - 4$

28. $f(x) = 5x - 1$, $g(x) = 2x^2$

29. $f(x) = \dfrac{1}{x - 2}$, $g(x) = 4x^3$

30. $f(x) = 3x^2$, $g(x) = \dfrac{1}{x - 4}$

31. $f(x) = \dfrac{3}{x - 2}$, $g(x) = \dfrac{5}{4 - x}$

32. $f(x) = \dfrac{5}{x - 3}$, $g(x) = \dfrac{1}{x - 2}$

Appendix E Introductory Algebra Review

This text is appropriate for a two-semester course that combines the study of introductory and intermediate algebra. Students who take only the second-semester course (which generally begins with Chapter 7) often need a review of the topics covered in the first-semester course. This appendix is a guide for a review of the first six chapters of this text. Below is a syllabus of selected exercises that can be used as a condensed review of the main objectives in the first half of the text. For extra help, consult the *Student's Solutions Manual,* which contains fully worked-out solutions with step-by-step annotations for all the odd-numbered exercises in the exercise sets.

Section/Objective	Examples	Exercises in Exercise Set
1.3a	5–13	3, 11, 19, 25, 33, 39, 43
1.4a	6–11	7, 11, 19, 23, 25, 59, 63, 77, 85
1.5a	4–10, 14	17, 29, 33, 39, 53, 65
1.6c	16–20	49, 53, 55, 57
1.7c	13–16	43, 45, 55
1.7d	20–23	69, 77, 79
1.7e	24–30	85, 93, 97, 103
1.8b	7, 11, 12	15, 21
1.8c	15	29, 35
1.8d	19–21	41, 51, 61, 81
2.1b	6, 7	19, 39, 47
2.2a	2, 3, 5	3, 7, 27, 33
2.3a	2	5, 15
2.3b	7, 9	23, 43, 51
2.3c	11	67, 75
2.4a	1, 2, 5	3, 15, 23
2.5a	1, 4	1, 3, 23
2.7e	13, 16	53, 61, 75
3.2b	4, 5, 7	17, 25, 33
3.3a	1	21, 25
3.3b	2, 3	45, 51
4.1b	2	13, 15
4.1c	5, 6	23, 29
4.1d, e, f	8, 11, 13, 15, 22–27	59, 63, 75, 83, 87, 95
4.2a, b	1–4, 10, 13, 14	3, 13, 17, 19
4.4a	3, 4	7, 17
4.4c	8, 9	33, 41, 49
4.5d	10	43
4.6a	2–6	9, 27
4.6b	11–13	41, 45
4.6c	16–18	61, 65, 69
4.7c	4–6	19, 25
4.7f	11, 12	39, 55
4.8b	3, 4	7, 9, 13

(continued)

Section/Objective	Examples	Exercises in Exercise Set
5.1b	4–6	13, 21
5.1c	10, 13	33, 41
5.2a	1, 2, 3	1, 21, 27
5.3a	1, 2	3, 11, 25, 39, 67
5.5b	4, 6	13, 19, 29
5.5d	13, 17, 19	51, 55, 63, 73
5.6a	1, 2	1, 9, 23, 27
5.8b	4	25, 29, 37
5.9a	6	31
6.1d	11, 12	57, 59
6.1f	19, 20	83, 101
6.2c	4, 6	25, 31
6.3a	7, 8, 10	29, 35, 45, 59
6.4a	5, 6	23, 31, 41
6.5a	1, 2, 5	5, 11, 21, 35
6.6a	2	7
6.6b	4	21

Two other features of the text that can be used for review of the first six chapters are as follows.

- At the end of each chapter is a *Summary and Review* that provides an extensive set of review exercises. Reference codes beside each exercise or direction line preceding it allow the student to easily return to the objective being reviewed. Answers to all of these exercises appear at the back of the text.
- The *Cumulative Review* that follows both Chapter 3 and Chapter 6 can also be used for review. Each reviews material from all preceding chapters. At the back of the text are answers to all Cumulative Review exercises, together with section and objective references, so that students know exactly what material to study if they miss an exercise.

The extensive supplements package that accompanies this text also includes material appropriate for a structured review of the first six chapters. Consult the preface in the text for detailed descriptions of each of the following.

- "Steps to Success" Videotapes: A videotape index can be found on the inside back cover of this text.
- InterAct Math Tutorial Software
- MathMax Multimedia CD-ROM

Answers

Chapter 1

Pretest: Chapter 1, p. 2

1. [1.1a] $\frac{5}{16}$ **2.** [1.1b] 78%x, or $0.78x$ **3.** [1.1a] 360 ft^2 **4.** [1.3b] 12 **5.** [1.2d] > **6.** [1.2d] > **7.** [1.2d] > **8.** [1.2d] < **9.** [1.2e] 12 **10.** [1.2e] 2.3 **11.** [1.2e] 0 **12.** [1.3b] −5.4 **13.** [1.3b] $\frac{2}{3}$ **14.** [1.6b] $\frac{1}{10}$ **15.** [1.6b] $-\frac{3}{2}$ **16.** [1.3a] −17 **17.** [1.4a] 38.6 **18.** [1.4a] $-\frac{17}{15}$ **19.** [1.3a] −5 **20.** [1.5a] 63 **21.** [1.5a] $-\frac{5}{12}$ **22.** [1.6c] −98 **23.** [1.6a] 8 **24.** [1.4a] 24 **25.** [1.8d] 26 **26.** [1.7c] $9z - 18$ **27.** [1.7c] $-4a - 2b + 10c$ **28.** [1.7d] $4(x - 3)$ **29.** [1.7d] $3(2y - 3z - 6)$ **30.** [1.8b] $-y - 13$ **31.** [1.8c] $y + 18$ **32.** [1.2d] $12 < x$

Margin Exercises, Section 1.1, pp. 3–6

1. $2174 + x = 7521$ **2.** 64 **3.** 28 **4.** 60 **5.** 192 ft^2 **6.** 25 **7.** 16 **8.** 12 hr **9.** $x - 8$ **10.** $y + 8$, or $8 + y$ **11.** $m - 4$ **12.** $\frac{1}{2}p$ **13.** $6 + 8x$, or $8x + 6$ **14.** $a - b$ **15.** 59%x, or $0.59x$ **16.** $xy - 200$ **17.** $p + q$

Calculator Spotlight, p. 4

1. 59.63768116 **2.** 11.9 **3.** 11.9 **4.** 34,427.16 **5.** 32 **6.** 27.5

Exercise Set 1.1, p. 7

1. $20,400; $46,800; $150,000 **3.** 1935 m^2 **5.** 260 mi **7.** 56 **9.** 8 **11.** 1 **13.** 6 **15.** 2 **17.** $b + 7$, or $7 + b$ **19.** $c - 12$ **21.** $4 + q$, or $q + 4$ **23.** $a + b$, or $b + a$ **25.** $y - x$ **27.** $w + x$, or $x + w$ **29.** $n - m$ **31.** $r + s$, or $s + r$ **33.** $2z$ **35.** $3m$ **37.** 89%x, or $0.89x$ **39.** $55t$ miles **41.** ◈ **43.** $x + 3y$ **45.** $2x - 3$

Margin Exercises, Section 1.2, pp. 11–16

1. 8; −5 **2.** 134; −80 **3.** −10; 156 **4.** −120; 50; −80

5. $-\frac{7}{2}$ 0

6. −1.4 0

7. $\frac{11}{4}$ 0

8. −0.375 **9.** $-0.\overline{54}$ **10.** $1.\overline{3}$ **11.** < **12.** < **13.** > **14.** > **15.** > **16.** < **17.** < **18.** > **19.** $7 > -5$ **20.** $4 < x$ **21.** False **22.** True **23.** True **24.** 8 **25.** 0 **26.** 9 **27.** $\frac{2}{3}$ **28.** 5.6

Calculator Spotlight, p. 12

1. [(−)] [3] [ENTER]; −3
2. [(−)] [5] [0] [8] [ENTER]; −508
3. [(−)] [.] [1] [7] [ENTER]; −.17
4. [(−)] [5] [÷] [8] [ENTER]; −.625

Calculator Spotlight, p. 13

1. 8.717797887 **2.** 17.80449381 **3.** 67.08203932 **4.** 35.4807407 **5.** 3.141592654 **6.** 91.10618695 **7.** 530.9291585 **8.** 138.8663978

Exercise Set 1.2, p. 17

1. −1286; 13,804 **3.** 24; −2 **5.** −5,200,000,000,000

7. $\frac{10}{3}$ 0

9. $-\frac{17}{4}$ 0

11. −0.875 **13.** $0.8\overline{3}$ **15.** $1.1\overline{6}$ **17.** $0.\overline{6}$ **19.** −0.5 **21.** 0.1 **23.** > **25.** < **27.** < **29.** < **31.** > **33.** < **35.** > **37.** < **39.** < **41.** < **43.** True **45.** False **47.** $x < -6$ **49.** $y \geq -10$ **51.** 3 **53.** 10 **55.** 0 **57.** 24 **59.** $\frac{2}{3}$ **61.** 0 **63.** 3 **64.** 11 **65.** 3 **66.** 1 **67.** ◈ **69.** $-\frac{5}{6}, -\frac{3}{4}, -\frac{2}{3}, \frac{1}{6}, \frac{3}{8}, \frac{1}{2}$ **71.** $\frac{1}{9}$ **73.** $5\frac{5}{9}$, or $\frac{50}{9}$

Margin Exercises, Section 1.3, pp. 19–22

1. −6 **2.** −3 **3.** −8 **4.** 4 **5.** 0 **6.** −2 **7.** −11 **8.** −12 **9.** 2 **10.** −4 **11.** −2 **12.** 0 **13.** −22 **14.** 3 **15.** 0.53 **16.** 2.3 **17.** −7.7 **18.** −6.2 **19.** $-\frac{2}{9}$ **20.** $-\frac{19}{20}$ **21.** −58 **22.** −56 **23.** −14 **24.** −12 **25.** 4 **26.** −8.7 **27.** 7.74 **28.** $\frac{8}{9}$ **29.** 0 **30.** −12 **31.** −14; 14 **32.** −1; 1 **33.** 19; −19 **34.** 1.6; −1.6 **35.** $-\frac{2}{3}$; $\frac{2}{3}$ **36.** $\frac{9}{8}$; $-\frac{9}{8}$ **37.** 4 **38.** 13.4 **39.** 0 **40.** $-\frac{1}{4}$

Exercise Set 1.3, p. 23

1. −7 **3.** −6 **5.** 0 **7.** −8 **9.** −7 **11.** −27 **13.** 0 **15.** −42 **17.** 0 **19.** 0 **21.** 3 **23.** −9 **25.** 7 **27.** 0 **29.** 35 **31.** −3.8 **33.** −8.1 **35.** $-\frac{1}{5}$ **37.** $-\frac{7}{9}$ **39.** $-\frac{3}{8}$ **41.** $-\frac{19}{24}$ **43.** $\frac{1}{24}$ **45.** 37 **47.** 50 **49.** −1409 **51.** −24 **53.** 26.9 **55.** −8 **57.** $\frac{13}{8}$ **59.** −43 **61.** $\frac{4}{3}$ **63.** 24 **65.** $\frac{3}{8}$ **67.** −0.625 **68.** $0.\overline{3}$ **69.** $-0.08\overline{3}$ **70.** 0.65 **71.** 2.3 **72.** 0 **73.** $\frac{4}{5}$ **74.** 21.4 **75.** ◆ **77.** All positive **79.** −6483 **81.** Negative

Margin Exercises, Section 1.4, pp. 25–27

1. −10 **2.** 3 **3.** −5 **4.** −2 **5.** −11 **6.** 4 **7.** −2 **8.** −6 **9.** −16 **10.** 7.1 **11.** 3 **12.** 0 **13.** $\frac{3}{2}$ **14.** −8 **15.** 7 **16.** −3 **17.** −23.3 **18.** 0 **19.** −9 **20.** 17 **21.** 12.7 **22.** 77, 37, 25, 24, 23, 9, −4, −9, −12, −25, −28, −31, −41, −45; 53, 48, 45, 38, 29, 24, 21, 19, −10, −27, −35, −37, −53, −115 **23.** 50°C

Exercise Set 1.4, p. 29

1. −7 **3.** −4 **5.** −6 **7.** 0 **9.** −4 **11.** −7 **13.** −6 **15.** 0 **17.** 0 **19.** 14 **21.** 11 **23.** −14 **25.** 5 **27.** −7 **29.** −1 **31.** 18 **33.** −10 **35.** −3 **37.** −21 **39.** 5 **41.** −8 **43.** 12 **45.** −23 **47.** −68 **49.** −73 **51.** 116 **53.** 0 **55.** −1 **57.** $\frac{1}{12}$ **59.** $-\frac{17}{12}$ **61.** $\frac{1}{8}$ **63.** 19.9 **65.** −8.6 **67.** −0.01 **69.** −193 **71.** 500 **73.** −2.8 **75.** −3.53 **77.** $-\frac{1}{2}$ **79.** $\frac{6}{7}$ **81.** $-\frac{41}{30}$ **83.** $-\frac{2}{15}$ **85.** 37 **87.** −62 **89.** −139 **91.** 6 **93.** 107 **95.** 219 **97.** 2385 m **99.** $347.94 **101.** 100°F **103.** $y + 7$, or $7 + y$ **104.** $t - 41$ **105.** $a - h$ **106.** $6c$, or $c \cdot 6$ **107.** $r + s$, or $s + r$ **108.** $y - x$ **109.** ◆ **111.** −309,882 **113.** False; $3 - 0 \neq 0 - 3$ **115.** True **117.** True **119.** Up 15 points

Margin Exercises, Section 1.5, pp. 33–35

1. 20; 10; 0; −10; −20; −30 **2.** −18 **3.** −100 **4.** −80 **5.** $-\frac{5}{9}$ **6.** −30.033 **7.** $-\frac{7}{10}$ **8.** −10; 0; 10; 20; 30 **9.** 27 **10.** 32 **11.** 35 **12.** $\frac{20}{63}$ **13.** $\frac{2}{3}$ **14.** 13.455 **15.** −30 **16.** 30 **17.** 0 **18.** $-\frac{8}{3}$ **19.** −30 **20.** −30.75 **21.** $-\frac{5}{3}$ **22.** 120 **23.** −120 **24.** 6 **25.** 4; −4 **26.** 9; −9 **27.** 48; 48

Exercise Set 1.5, p. 37

1. −8 **3.** −48 **5.** −24 **7.** −72 **9.** 16 **11.** 42 **13.** −120 **15.** −238 **17.** 1200 **19.** 98 **21.** −72 **23.** −12.4 **25.** 30 **27.** 21.7 **29.** $-\frac{2}{5}$ **31.** $\frac{1}{12}$ **33.** −17.01 **35.** $-\frac{5}{12}$ **37.** 420 **39.** $\frac{2}{7}$ **41.** −60 **43.** 150 **45.** $-\frac{2}{45}$ **47.** 1911 **49.** 50.4 **51.** $\frac{10}{189}$ **53.** −960 **55.** 17.64 **57.** $-\frac{5}{784}$ **59.** 0 **61.** −720 **63.** −30,240 **65.** 441; −147 **67.** 20; 20 **69.** 2 **70.** $-\frac{1}{3}$ **71.** True **72.** False **73.** False **74.** False **75.** ◆ **77.** 32 m below the surface **79.** **(a)** One must be negative, and one must be positive. **(b)** Either or both must be zero. **(c)** Both must be negative or both must be positive.

Margin Exercises, Section 1.6, pp. 39–42

1. −2 **2.** 5 **3.** −3 **4.** 8 **5.** −6 **6.** $-\frac{30}{7}$ **7.** Undefined **8.** 0 **9.** $\frac{3}{2}$ **10.** $-\frac{4}{5}$ **11.** $-\frac{1}{3}$ **12.** −5 **13.** $\frac{1}{1.6}$ **14.** $\frac{2}{3}$ **15.** First row: $-\frac{2}{3}$, $\frac{3}{2}$; second row: $\frac{5}{4}$, $-\frac{4}{5}$; third row: 0, undefined; fourth row: −1, 1; fifth row: 8, $-\frac{1}{8}$; sixth row: 4.5, $-\frac{1}{4.5}$ **16.** $\frac{4}{7} \cdot \left(-\frac{5}{3}\right)$ **17.** $5 \cdot \left(-\frac{1}{8}\right)$ **18.** $(a - b) \cdot \left(\frac{1}{7}\right)$ **19.** $-23 \cdot a$ **20.** $-5 \cdot \left(\frac{1}{7}\right)$ **21.** $-\frac{20}{21}$ **22.** $-\frac{12}{5}$ **23.** $\frac{16}{7}$ **24.** −7 **25.** $\frac{5}{-6}$, $-\frac{5}{6}$ **26.** $\frac{-8}{7}$, $\frac{8}{-7}$ **27.** $\frac{-10}{3}$, $-\frac{10}{3}$

Exercise Set 1.6, p. 43

1. −8 **3.** −14 **5.** −3 **7.** 3 **9.** −8 **11.** 2 **13.** −12 **15.** −8 **17.** Undefined **19.** $\frac{23}{2}$ **21.** $\frac{7}{15}$ **23.** $-\frac{13}{47}$ **25.** $\frac{1}{13}$ **27.** $\frac{1}{4.3}$ **29.** −7.1 **31.** $\frac{q}{p}$ **33.** $4y$ **35.** $\frac{3b}{2a}$ **37.** $4 \cdot \left(\frac{1}{17}\right)$ **39.** $8 \cdot \left(-\frac{1}{13}\right)$ **41.** $13.9 \cdot \left(-\frac{1}{1.5}\right)$ **43.** $x \cdot y$ **45.** $(3x + 4)\left(\frac{1}{5}\right)$ **47.** $(5a - b)\left(\frac{1}{5a + b}\right)$ **49.** $-\frac{9}{8}$ **51.** $\frac{5}{3}$ **53.** $\frac{9}{14}$ **55.** $\frac{9}{64}$ **57.** −2 **59.** $\frac{11}{13}$ **61.** −16.2 **63.** Undefined **65.** $-\frac{1}{4}$ **66.** 5 **67.** −42 **68.** −48 **69.** 8.5 **70.** $-\frac{1}{8}$ **71.** $-0.\overline{09}$ **72.** $0.91\overline{6}$ **73.** ◆ **75.** ◆ **77.** Negative **79.** Positive **81.** Negative

Margin Exercises, Section 1.7, pp. 45–52

1.

	$x + x$	$2x$
$x = 3$	6	6
$x = -6$	−12	−12
$x = 4.8$	9.6	9.6

2.

	$x + 3x$	$5x$
$x = 2$	8	10
$x = -6$	-24	-30
$x = 4.8$	19.2	24

3. $\frac{6}{8}$ **4.** $\frac{3t}{4t}$ **5.** $\frac{3}{4}$ **6.** $-\frac{4}{3}$ **7.** 1; 1 **8.** -10; -10
9. $9 + x$ **10.** qp **11.** $t + xy$, or $yx + t$, or $t + yx$
12. 19; 19 **13.** 150; 150 **14.** $(r + s) + 7$ **15.** $(9a)b$
16. $(4t)u$, $(tu)4$, $t(4u)$; answers may vary
17. $(2 + r) + s$, $(r + s) + 2$, $s + (r + 2)$; answers may vary **18. (a)** 63; **(b)** 63 **19. (a)** 80; **(b)** 80
20. (a) 28; **(b)** 28 **21. (a)** 8; **(b)** 8 **22. (a)** -4; **(b)** -4
23. (a) -25; **(b)** -25 **24.** $5x, -8y, 3$ **25.** $-4y, -2x, 3z$ **26.** $3x - 15$ **27.** $5x + 5$ **28.** $\frac{3}{5}p + \frac{3}{5}q - \frac{3}{5}t$
29. $-2x + 6$ **30.** $5x - 10y + 20z$
31. $-5x + 10y - 20z$ **32.** $6(x - 2)$
33. $3(x - 2y + 3)$ **34.** $b(x + y - z)$
35. $2(8a - 18b + 21)$ **36.** $\frac{1}{8}(3x - 5y + 7)$
37. $-4(3x - 8y + 4z)$ **38.** $3x$ **39.** $6x$ **40.** $-8x$
41. $0.59x$ **42.** $3x + 3y$ **43.** $-4x - 5y - 7$
44. $-\frac{2}{3} + \frac{1}{10}x + \frac{7}{9}y$

Exercise Set 1.7, p. 53

1. $\frac{3y}{5y}$ **3.** $\frac{10x}{15x}$ **5.** $-\frac{3}{2}$ **7.** $-\frac{7}{6}$ **9.** $8 + y$ **11.** nm
13. $xy + 9$, or $9 + yx$ **15.** $c + ab$, or $ba + c$
17. $(a + b) + 2$ **19.** $8(xy)$ **21.** $a + (b + 3)$
23. $(3a)b$ **25.** $2 + (b + a)$, $(2 + a) + b$, $(b + 2) + a$; answers may vary **27.** $(5 + w) + v$, $(v + 5) + w$, $(w + v) + 5$; answers may vary **29.** $(3x)y$, $y(x \cdot 3)$, $3(yx)$; answers may vary **31.** $a(7b)$, $b(7a)$, $(7b)a$; answers may vary **33.** $2b + 10$ **35.** $7 + 7t$
37. $30x + 12$ **39.** $7x + 28 + 42y$ **41.** $7x - 21$
43. $-3x + 21$ **45.** $\frac{2}{3}b - 4$ **47.** $7.3x - 14.6$
49. $-\frac{3}{5}x + \frac{3}{5}y - 6$ **51.** $45x + 54y - 72$
53. $-4x + 12y + 8z$ **55.** $-3.72x + 9.92y - 3.41$
57. $4x, 3z$ **59.** $7x, 8y, -9z$ **61.** $2(x + 2)$
63. $5(6 + y)$ **65.** $7(2x + 3y)$ **67.** $5(x + 2 + 3y)$
69. $8(x - 3)$ **71.** $4(8 - y)$ **73.** $2(4x + 5y - 11)$
75. $a(x - 1)$ **77.** $a(x - y - z)$ **79.** $6(3x - 2y + 1)$
81. $\frac{1}{3}(2x - 5y + 1)$ **83.** $19a$ **85.** $9a$ **87.** $8x + 9z$
89. $7x + 15y^2$ **91.** $-19a + 88$ **93.** $4t + 6y - 4$
95. b **97.** $\frac{13}{4}y$ **99.** $8x$ **101.** $5n$ **103.** $-16y$
105. $17a - 12b - 1$ **107.** $4x + 2y$ **109.** $7x + y$
111. $0.8x + 0.5y$ **113.** $\frac{35}{6}a + \frac{3}{2}b - 42$ **115.** 180
116. $\frac{4}{13}$ **117.** True **118.** False **119.** True
120. True **121.** ◈ **123.** Not equivalent; $3 \cdot 2 + 5 \neq 3 \cdot 5 + 2$ **125.** Equivalent; commutative law of addition **127.** $q(1 + r + rs + rst)$

Margin Exercises, Section 1.8, pp. 57–60

1. $-x - 2$ **2.** $-5x - 2y - 8$ **3.** $-6 + t$ **4.** $-x + y$
5. $4a - 3t + 10$ **6.** $-18 + m + 2n - 4z$ **7.** $2x - 9$
8. $3y + 2$ **9.** $2x - 7$ **10.** $3y + 3$
11. $-2a + 8b - 3c$ **12.** $-9x - 8y$ **13.** $-16a + 18$
14. $-26a + 41b - 48c$ **15.** $3x - 7$ **16.** 2 **17.** 18
18. 6 **19.** 17 **20.** $5x - y - 8$ **21.** -1237 **22.** 8
23. 381 **24.** -12

Calculator Spotlight, p. 61

1. -11 **2.** 9 **3.** 114 **4.** 117,649 **5.** $-1{,}419{,}857$
6. $-1{,}124{,}864$ **7.** $-117{,}649$ **8.** $-1{,}419{,}857$
9. $-1{,}124{,}864$ **10.** -4 **11.** -2 **12.** 787
13. $-32 \times (88 - 29) = -1888$
14. $3^5 - 10^2 \times 5^2 = -2257$
15. $4 + 6 \cdot 8 - 2 = 4 + 8 \cdot 6 - 2 = 50$; the commutative law of multiplication
16. $5 + 9^2 \cdot 7 - 3 = 569$; because $a^2 \cdot b \neq b^2 \cdot a$, although students might phrase this verbally and not symbolically.

Exercise Set 1.8, p. 63

1. $-2x - 7$ **3.** $-5x + 8$ **5.** $-4a + 3b - 7c$
7. $-6x + 8y - 5$ **9.** $-3x + 5y + 6$ **11.** $8x + 6y + 43$
13. $5x - 3$ **15.** $-3a + 9$ **17.** $5x - 6$
19. $-19x + 2y$ **21.** $9y - 25z$ **23.** $-7x + 10y$
25. $37a - 23b + 35c$ **27.** 7 **29.** -40 **31.** 19
33. $12x + 30$ **35.** $3x + 30$ **37.** $9x - 18$
39. $-4x - 64$ **41.** -7 **43.** -7 **45.** -16
47. -334 **49.** 14 **51.** 1880 **53.** 12 **55.** 8
57. -86 **59.** 37 **61.** -1 **63.** -10 **65.** 25
67. -7988 **69.** -3000 **71.** 60 **73.** 1 **75.** 10
77. $-\frac{13}{45}$ **79.** $-\frac{23}{18}$ **81.** -118 **83.** $16x - 24$
84. $5 + 5a$ **85.** $\frac{2}{3}y - 6$ **86.** $-21x - 35$
87. $5(7 - b)$ **88.** $12(4x - 3)$ **89.** $15(x - 4y)$
90. $3(4x - 5y + 10z)$ **91.** ◈
93. $6y - (-2x + 3a - c)$
95. $6m - (-3n + 5m - 4b)$ **97.** $-2x - f$
99. (a) 52, 52, 28.130169; **(b)** -24, -24, -108.307025

Summary and Review: Chapter 1, p. 67

1. 4 **2.** 19%x, or $0.19x$ **3.** -45, 72 **4.** 38
5.
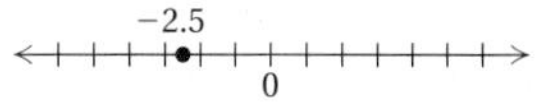

6.
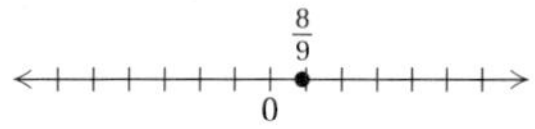

7. $<$ **8.** $>$ **9.** $>$ **10.** $<$ **11.** -3.8 **12.** $\frac{3}{4}$
13. $\frac{8}{3}$ **14.** $-\frac{1}{7}$ **15.** 34 **16.** 5 **17.** -3 **18.** -4
19. -5 **20.** 4 **21.** $-\frac{7}{5}$ **22.** -7.9 **23.** 54
24. -9.18 **25.** $-\frac{2}{7}$ **26.** -210 **27.** -7 **28.** -3
29. $\frac{3}{4}$ **30.** 40.4 **31.** -2 **32.** 8-yd gain
33. $-\$130$ **34.** $15x - 35$ **35.** $-8x + 10$
36. $4x + 15$ **37.** $-24 + 48x$ **38.** $2(x - 7)$
39. $6(x - 1)$ **40.** $5(x + 2)$ **41.** $3(4 - x)$
42. $7a - 3b$ **43.** $-2x + 5y$ **44.** $5x - y$
45. $-a + 8b$ **46.** $-3a + 9$ **47.** $-2b + 21$

48. 6 **49.** $12y - 34$ **50.** $5x + 24$ **51.** $-15x + 25$ **52.** True **53.** False **54.** $x > -3$ **55.** $-\frac{5}{8}$ **56.** -2.1 **57.** 1000 **58.** $4a + 2b$

Test: Chapter 1, p. 69

1. [1.1a] 6 **2.** [1.1b] $x - 9$ **3.** [1.1a] 240 ft^2 **4.** [1.2d] $<$ **5.** [1.2d] $>$ **6.** [1.2d] $>$ **7.** [1.2d] $<$ **8.** [1.2e] 7 **9.** [1.2e] $\frac{9}{4}$ **10.** [1.2e] 2.7 **11.** [1.3b] $-\frac{2}{3}$ **12.** [1.3b] 1.4 **13.** [1.3b] 8 **14.** [1.6b] $-\frac{1}{2}$ **15.** [1.6b] $\frac{7}{4}$ **16.** [1.4a] 7.8 **17.** [1.3a] -8 **18.** [1.3a] $\frac{7}{40}$ **19.** [1.4a] 10 **20.** [1.4a] -2.5 **21.** [1.4a] $\frac{7}{8}$ **22.** [1.5a] -48 **23.** [1.5a] $\frac{3}{16}$ **24.** [1.6a] -9 **25.** [1.6c] $\frac{3}{4}$ **26.** [1.6c] -9.728 **27.** [1.8d] -173 **28.** [1.4b] 14°F **29.** [1.7c] $18 - 3x$ **30.** [1.7c] $-5y + 5$ **31.** [1.7d] $2(6 - 11x)$ **32.** [1.7d] $7(x + 3 + 2y)$ **33.** [1.4a] 12 **34.** [1.8b] $2x + 7$ **35.** [1.8b] $9a - 12b - 7$ **36.** [1.8c] $68y - 8$ **37.** [1.8d] -4 **38.** [1.8d] 448 **39.** [1.2d] $-2 \geq x$ **40.** [1.2e], [1.8d] 15 **41.** [1.8c] $4a$ **42.** [1.7e] $4x + 4y$

Chapter 2

Pretest: Chapter 2, p. 72

1. [2.2a] -7 **2.** [2.3b] -1 **3.** [2.3a] 2 **4.** [2.1b] 8 **5.** [2.3c] -5 **6.** [2.3a] $\frac{135}{32}$ **7.** [2.3c] 1 **8.** [2.7d] $\{x \mid x \geq -6\}$ **9.** [2.7c] $\{y \mid y > -4\}$ **10.** [2.7e] $\{a \mid a > -1\}$ **11.** [2.7c] $\{x \mid x \geq 3\}$ **12.** [2.7d] $\{y \mid y < -\frac{9}{4}\}$ **13.** [2.6a] $G = \dfrac{P}{3K}$ **14.** [2.6a] $a = \dfrac{Ab + b}{3}$ **15.** [2.4a] Width: 34 in.; length: 39 in. **16.** [2.5a] \$460 **17.** [2.4a] 81, 82, 83 **18.** [2.8b] Numbers less than 17 **19.** [2.7b]

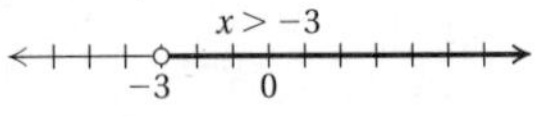

20. [2.7b]

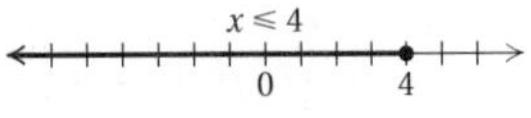

Margin Exercises, Section 2.1, pp. 73–76

1. False **2.** True **3.** Neither **4.** Yes **5.** No **6.** No **7.** 9 **8.** -5 **9.** 22 **10.** 13.2 **11.** -6.5 **12.** -2 **13.** $\frac{31}{8}$

Exercise Set 2.1, p. 77

1. Yes **3.** No **5.** No **7.** Yes **9.** No **11.** No **13.** 4 **15.** -20 **17.** -14 **19.** -18 **21.** 15 **23.** -14 **25.** 2 **27.** 20 **29.** -6 **31.** $6\frac{1}{2}$ **33.** 19.9 **35.** $\frac{7}{3}$ **37.** $-\frac{7}{4}$ **39.** $\frac{41}{24}$ **41.** $-\frac{1}{20}$ **43.** 5.1 **45.** 12.4 **47.** -5 **49.** $1\frac{5}{6}$ **51.** $-\frac{10}{21}$ **53.** -11 **54.** 5 **55.** $-\frac{5}{12}$ **56.** $\frac{1}{3}$ **57.** $-\frac{3}{2}$ **58.** -5.2 **59.** $50 - x$ **60.** $65t$ **61.** ◆ **63.** 342.246 **65.** $-\frac{26}{15}$ **67.** -10 **69.** All real numbers **71.** $-\frac{5}{17}$ **73.** 13, -13

Margin Exercises, Section 2.2, pp. 80–82

1. 15 **2.** $-\frac{7}{4}$ **3.** -18 **4.** 10 **5.** $-\frac{4}{5}$ **6.** 7800 **7.** -3 **8.** 28

Exercise Set 2.2, p. 83

1. 6 **3.** 9 **5.** 12 **7.** -40 **9.** 1 **11.** -7 **13.** -6 **15.** 6 **17.** -63 **19.** 36 **21.** -21 **23.** $-\frac{3}{5}$ **25.** $-\frac{3}{2}$ **27.** $\frac{9}{2}$ **29.** 7 **31.** -7 **33.** 8 **35.** 15.9 **37.** $7x$ **38.** $-x + 5$ **39.** $8x + 11$ **40.** $-32y$ **41.** $x - 4$ **42.** $-23 - 5x$ **43.** $-10y - 42$ **44.** $-22a + 4$ **45.** $8r$ **46.** $\frac{1}{2}b \cdot 10$, or $5b$ **47.** ◆ **49.** -8655 **51.** No solution **53.** No solution **55.** $\dfrac{b}{3a}$ **57.** $\dfrac{4b}{a}$

Margin Exercises, Section 2.3, pp. 85–90

1. 5 **2.** 4 **3.** 4 **4.** 39 **5.** $-\frac{3}{2}$ **6.** -4.3 **7.** -3 **8.** 800 **9.** 1 **10.** 2 **11.** 2 **12.** $\frac{17}{2}$ **13.** $\frac{8}{3}$ **14.** -4.3 **15.** 2 **16.** 3 **17.** -2 **18.** $-\frac{1}{2}$

Calculator Spotlight, p. 90

1. Both sides equal 9. **2.** Both sides equal -2. **3.** Both sides equal -8.18.

Exercise Set 2.3, p. 91

1. 5 **3.** 8 **5.** 10 **7.** 14 **9.** -8 **11.** -8 **13.** -7 **15.** 15 **17.** 6 **19.** 4 **21.** 6 **23.** -3 **25.** 1 **27.** 6 **29.** -20 **31.** 7 **33.** 2 **35.** 5 **37.** 2 **39.** 10 **41.** 4 **43.** 0 **45.** -1 **47.** $-\frac{4}{3}$ **49.** $\frac{2}{5}$ **51.** -2 **53.** -4 **55.** $\frac{4}{5}$ **57.** $-\frac{28}{27}$ **59.** 6 **61.** 2 **63.** 6 **65.** 8 **67.** 1 **69.** 17 **71.** $-\frac{5}{3}$ **73.** -3 **75.** 2 **77.** $\frac{4}{7}$ **79.** $-\frac{51}{31}$ **81.** 2 **83.** -6.5 **84.** $7(x - 3 - 2y)$ **85.** $<$ **86.** -14 **87.** -18.7 **88.** -25.5 **89.** $c \div 8$, or $\dfrac{c}{8}$ **90.** $13.4h$ **91.** ◆ **93.** 4.4233464 **95.** $-\frac{7}{2}$ **97.** -2 **99.** 0 **101.** 6 **103.** $\frac{11}{18}$ **105.** 10

Margin Exercises, Section 2.4, pp. 96–101

1. Top: 24 ft; middle: 72 ft; bottom: 144 ft **2.** 5 **3.** 313 and 314 **4.** 93,333 **5.** Length: 84 ft; width: 50 ft **6.** First: 30°; second: 90°; third: 60° **7.** **(a)** \$10.03994 billion, \$15.60842 billion; **(b)** 2001

Exercise Set 2.4, p. 103

1. 16 **3.** $-\frac{1}{2}$ **5.** 57 **7.** 180 in., 60 in. **9.** 305 ft **11.** \$2.89 **13.** -12 **15.** $699\frac{1}{3}$ mi **17.** 286, 287 **19.** 41, 42, 43 **21.** 61, 63, 65 **23.** Length: 48 ft; width: 14 ft **25.** 11 **27.** 28°, 84°, 68° **29.** 33°, 38°, 109° **31.** **(a)** \$1056 million, \$1122.2 million, \$1784.2 million; **(b)** 2002 **33.** $-\frac{47}{40}$ **34.** $-\frac{17}{40}$ **35.** $-\frac{3}{10}$ **36.** $-\frac{32}{15}$ **37.** 1.6 **38.** 409.6 **39.** -9.6 **40.** -41.6 **41.** ◆ **43.** 120 **45.** About 0.65 in.

Margin Exercises, Section 2.5, pp. 107–110

1. 32% **2.** 25% **3.** 225 **4.** 50 **5.** 11.04
6. About 33 **7.** 111,416 mi^2 **8.** \$8400 **9.** \$658

Exercise Set 2.5, p. 111

1. 20% **3.** 150 **5.** 546 **7.** 24% **9.** 2.5 **11.** 5%
13. \$16.77 billion **15.** **(a)** 8190; **(b)** 1.8% **17.** **(a)** 16%;
(b) \$29 **19.** **(a)** \$3.75; **(b)** \$28.75 **21.** **(a)** \$28.80;
(b) \$33.12 **23.** \$36 **25.** 200 **27.** \$282.20
29. About 42.4 lb **31.** About 566 **33.** \$15.38
35. \$7800 **37.** \$58 **39.** −11 **40.** −100 **41.** $-\frac{2}{5}$
42. 2 **43.** 16 **44.** 189.6 **45.** 10 **46.** 18.4875
47. ◆ **49.** 6 ft, 7 in. **51.** \$9.17, not \$9.10

Margin Exercises, Section 2.6, pp. 115–116

1. 2.8 mi **2.** $I = \dfrac{E}{R}$ **3.** $D = \dfrac{C}{\pi}$
4. $c = 4A - a - b - d$ **5.** **(a)** About 306 lb;
(b) $L = \dfrac{800W}{g^2}$

Exercise Set 2.6, p. 117

1. $h = \dfrac{A}{b}$ **3.** $w = \dfrac{P - 2l}{2}$, or $\dfrac{1}{2}P - l$ **5.** $a = 2A - b$
7. $a = \dfrac{F}{m}$ **9.** $c^2 = \dfrac{E}{m}$ **11.** $x = \dfrac{c - By}{A}$ **13.** $t = \dfrac{3k}{v}$
15. **(a)** 57,000 Btu; **(b)** $a = \dfrac{b}{30}$ **17.** **(a)** 1423;
(b) $n = 15F$ **19.** **(a)** 1901 calories;
(b) $a = \dfrac{917 + 6w + 6h - K}{6}$; $h = \dfrac{K - 917 - 6w + 6a}{6}$;
$w = \dfrac{K - 917 - 6h + 6a}{6}$ **21.** 1 **22.** −90 **23.** −13.2
24. $-21a + 12b$ **25.** $\frac{1}{6}$ **26.** $-\frac{3}{2}$ **27.** ◆
29. $b = \dfrac{2A - ah}{h}$; $h = \dfrac{2A}{a + b}$ **31.** A quadruples.
33. A increases by $2h$ units.

Margin Exercises, Section 2.7, pp. 119–126

1. **(a)** No; **(b)** no; **(c)** no; **(d)** yes; **(e)** no; **(f)** no
2. **(a)** Yes; **(b)** yes; **(c)** yes; **(d)** no; **(e)** yes; **(f)** yes
3.

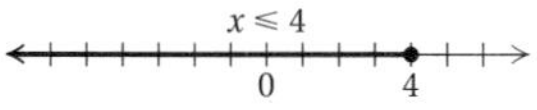

4.

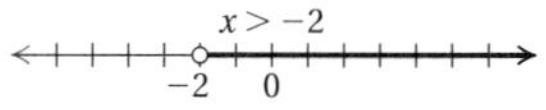

5.

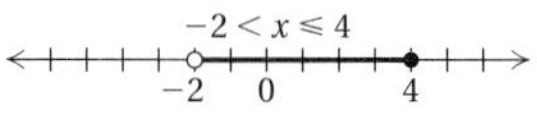

6. $\{x \mid x > 2\}$;

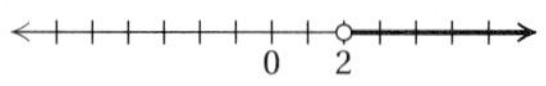

7. $\{x \mid x \le 3\}$;

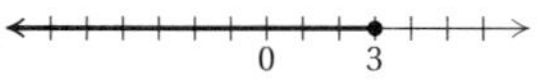

8. $\{x \mid x < -3\}$;

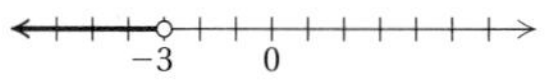

9. $\left\{x \mid x \ge \frac{2}{15}\right\}$ **10.** $\{y \mid y \le -3\}$

11. $\{x \mid x < 8\}$;

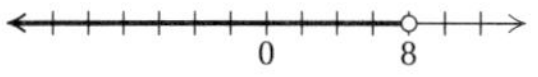

12. $\{y \mid y \ge 32\}$;

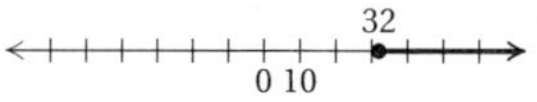

13. $\{x \mid x \ge -6\}$ **14.** $\left\{y \mid y < -\frac{13}{5}\right\}$ **15.** $\left\{x \mid x > -\frac{1}{4}\right\}$
16. $\left\{y \mid y \ge \frac{19}{9}\right\}$ **17.** $\left\{y \mid y \ge \frac{19}{9}\right\}$ **18.** $\{x \mid x \ge -2\}$
19. $\{x \mid x \ge -4\}$ **20.** $\left\{x \mid x > \frac{8}{3}\right\}$

Exercise Set 2.7, p. 127

1. **(a)** Yes; **(b)** yes; **(c)** no; **(d)** yes; **(e)** yes
3. **(a)** No; **(b)** no; **(c)** no; **(d)** yes; **(e)** no
5.

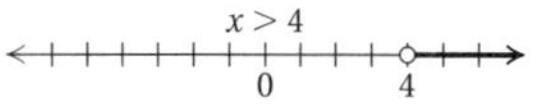

7.

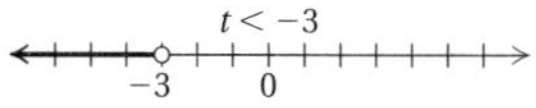

9.

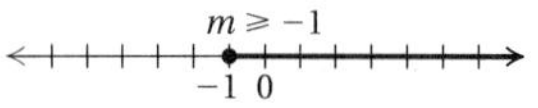

11.

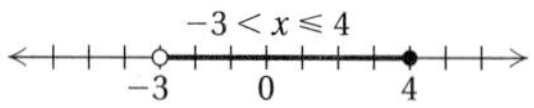

13.

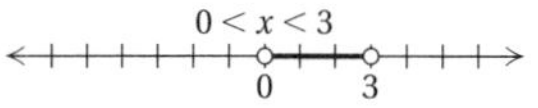

15. $\{x \mid x > -5\}$;

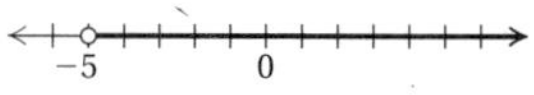

17. $\{x \mid x \le -18\}$;

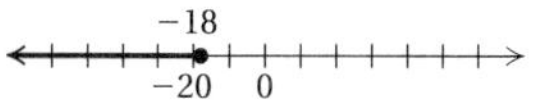

19. $\{y \mid y > -5\}$ **21.** $\{x \mid x > 2\}$ **23.** $\{x \mid x \le -3\}$
25. $\{x \mid x < 4\}$ **27.** $\{t \mid t > 14\}$ **29.** $\left\{y \mid y \le \frac{1}{4}\right\}$
31. $\left\{x \mid x > \frac{7}{12}\right\}$
33. $\{x \mid x < 7\}$;

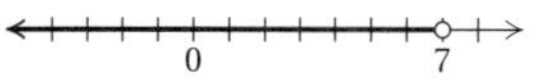

35. $\{x \mid x < 3\}$;

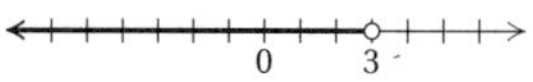

37. $\left\{y \mid y \ge -\frac{2}{5}\right\}$ **39.** $\{x \mid x \ge -6\}$ **41.** $\{y \mid y \le 4\}$
43. $\left\{x \mid x > \frac{17}{3}\right\}$ **45.** $\left\{y \mid y < -\frac{1}{14}\right\}$ **47.** $\left\{x \mid x \le \frac{3}{10}\right\}$
49. $\{x \mid x < 8\}$ **51.** $\{x \mid x \le 6\}$ **53.** $\{x \mid x < -3\}$
55. $\{x \mid x > -3\}$ **57.** $\{x \mid x \le 7\}$ **59.** $\{x \mid x > -10\}$
61. $\{y \mid y < 2\}$ **63.** $\{y \mid y \ge 3\}$ **65.** $\{y \mid y > -2\}$
67. $\{x \mid x > -4\}$ **69.** $\{x \mid x \le 9\}$ **71.** $\{y \mid y \le -3\}$
73. $\{y \mid y < 6\}$ **75.** $\{m \mid m \ge 6\}$ **77.** $\left\{t \mid t < -\frac{5}{3}\right\}$
79. $\{r \mid r > -3\}$ **81.** $\left\{x \mid x \ge -\frac{57}{34}\right\}$ **83.** $\{x \mid x > -2\}$
85. −74 **86.** 4.8 **87.** $-\frac{5}{8}$ **88.** −1.11 **89.** −38
90. $-\frac{7}{8}$ **91.** −9.4 **92.** 1.11 **93.** 140 **94.** 41
95. $-2x - 23$ **96.** $37x - 1$ **97.** ◆ **99.** **(a)** Yes;
(b) yes; **(c)** no; **(d)** no; **(e)** no; **(f)** yes; **(g)** yes
101. All real numbers

Margin Exercises, Section 2.8, pp. 131–132

1. $x \le 8$ **2.** $y > -2$ **3.** $s \le 180$ **4.** $p \ge \$5800$
5. $2x - 32 > 5$ **6.** $\{x \mid x \ge 84\}$ **7.** $\{C \mid C < 1063°\}$

Exercise Set 2.8, p. 133

1. $x > 8$ **3.** $y \le -4$ **5.** $n \ge 1300$ **7.** $a \le 500$
9. $2 + 3x < 13$ **11.** $\{x \mid x \ge 97\}$ **13.** $\{Y \mid Y \ge 1935\}$
15. $\{L \mid L \ge 5 \text{ in.}\}$ **17.** $\{x \mid x > 5\}$ **19.** $\{d \mid d > 25\}$
21. $\{b \mid b > 6 \text{ cm}\}$ **23.** $\{x \mid x \ge 21\}$
25. $\{t \mid t \le 0.75 \text{ hr}\}$ **27.** $\{f \mid f \ge 16 \text{ g}\}$ **29.** −160

30. $-17x + 18$ **31.** $91x - 242$ **32.** 0.25 **33.** ◈

Summary and Review: Chapter 2, p. 135

1. -22 **2.** 1 **3.** 25 **4.** 9.99 **5.** $\frac{1}{4}$ **6.** 7
7. -192 **8.** $-\frac{7}{3}$ **9.** $-\frac{15}{64}$ **10.** -8 **11.** 4 **12.** -5
13. $-\frac{1}{3}$ **14.** 3 **15.** 4 **16.** 16 **17.** 6 **18.** -3
19. 12 **20.** 4 **21.** Yes **22.** No **23.** Yes
24. $\{y \mid y \geq -\frac{1}{2}\}$ **25.** $\{x \mid x \geq 7\}$ **26.** $\{y \mid y > 2\}$
27. $\{y \mid y \leq -4\}$ **28.** $\{x \mid x < -11\}$ **29.** $\{y \mid y > -7\}$
30. $\{x \mid x > -6\}$ **31.** $\{x \mid x > -\frac{9}{11}\}$ **32.** $\{y \mid y \leq 7\}$
33. $\{x \mid x \geq -\frac{1}{12}\}$
34.

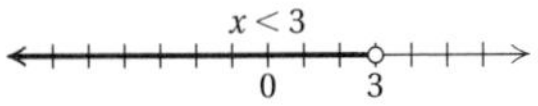

35.

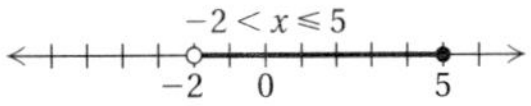

36.

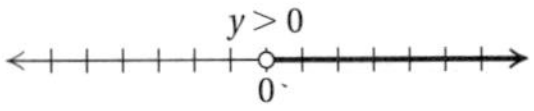

37. $d = \dfrac{C}{\pi}$ **38.** $B = \dfrac{3V}{h}$ **39.** $a = 2A - b$
40. Length: 365 mi; width: 275 mi **41.** 27
42. 345, 346 **43.** \$2117 **44.** 27 **45.** 35°, 85°, 60°
46. \$220 **47.** \$26,087 **48.** \$138.95 **49.** 86
50. $\{w \mid w > 17 \text{ cm}\}$ **51.** **(a)** 8.2 lb; **(b)** $L = \dfrac{800W}{g^2}$
52. $\frac{41}{4}$ **53.** $58t$ **54.** -45 **55.** $-43x + 8y$
56. ◈ The end result is the same either way. If s is the original salary, the new salary after a 5% raise followed by an 8% raise is $1.08(1.05s)$. If the raises occur the other way around, the new salary is $1.05(1.08s)$. By the commutative and associative laws of multiplication, we see that these are equal. However, it would be better to receive the 8% raise first, because this increase yields a higher salary the first year than a 5% raise.
57. ◈ The inequalities are equivalent by the multiplication principle for inequalities. If we multiply on both sides of one inequality by -1, the other inequality results. **58.** 23, -23 **59.** 20, -20
60. $a = \dfrac{y - 3}{2 - b}$

Test: Chapter 2, p. 137

1. [2.1b] 8 **2.** [2.1b] 26 **3.** [2.2a] -6 **4.** [2.2a] 49
5. [2.3b] -12 **6.** [2.3a] 2 **7.** [2.3a] -8
8. [2.1b] $-\frac{7}{20}$ **9.** [2.3c] 7 **10.** [2.3c] $\frac{5}{3}$ **11.** [2.3b] 2.5
12. [2.7c] $\{x \mid x \leq -4\}$ **13.** [2.7c] $\{x \mid x > -13\}$
14. [2.7d] $\{x \mid x \leq 5\}$ **15.** [2.7d] $\{y \mid y \leq -13\}$
16. [2.7d] $\{y \mid y \geq 8\}$ **17.** [2.7d] $\{x \mid x \leq -\frac{1}{20}\}$
18. [2.7e] $\{x \mid x < -6\}$ **19.** [2.7e] $\{x \mid x \leq -1\}$
20. [2.7b]

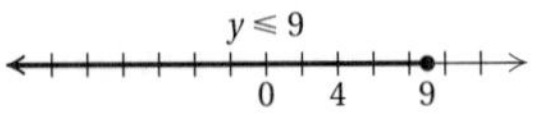

21. [2.7b, e]

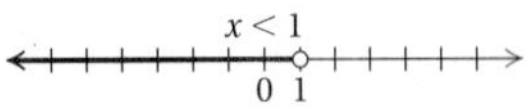

22. [2.7b]

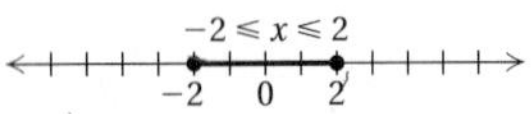

23. [2.4a] Width: 7 cm; length: 11 cm **24.** [2.4a] 6
25. [2.4a] 2509, 2510, 2511 **26.** [2.5a] \$880
27. [2.4a] 3 m, 5 m **28.** [2.6a] $r = \dfrac{A}{2\pi h}$
29. [2.6a] **(a)** 2650; **(b)** $w = \dfrac{K - 7h + 9.52a - 92.4}{19.18}$
30. [2.8b] $\{x \mid x > 6\}$ **31.** [2.8b] $\{l \mid l \geq 174 \text{ yd}\}$
32. [1.3a] $-\frac{2}{9}$ **33.** [1.1a] $\frac{8}{3}$ **34.** [1.1b] 73%p, or $0.73p$
35. [1.8b] $-18x + 37y$ **36.** [2.6a] $d = \dfrac{1 - ca}{-c}$, or $\dfrac{ca - 1}{c}$
37. [1.2e], [2.3a] 15, -15 **38.** [2.4a] 60

Chapter 3

Pretest: Chapter 3, p. 140

1. [3.2b]

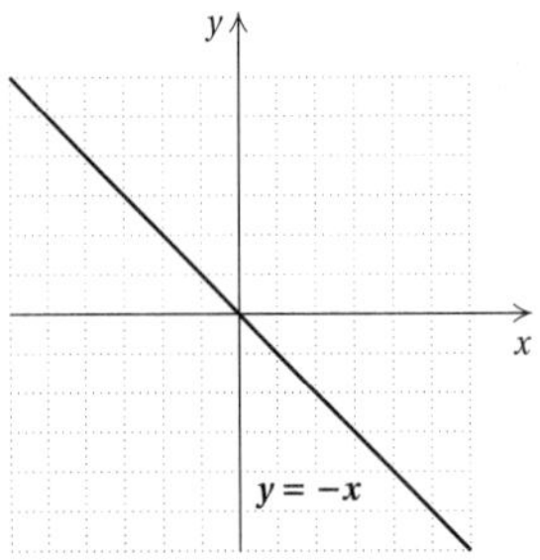

2. [3.3b]

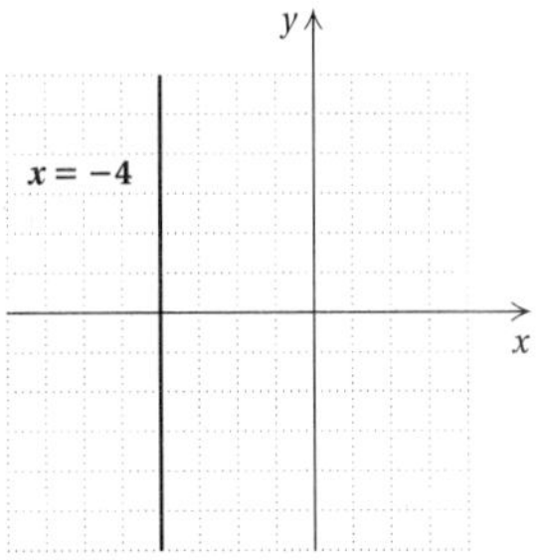

3. [3.3a]

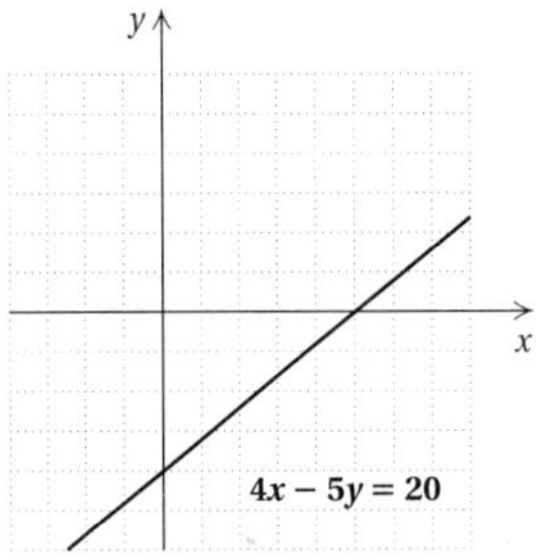

4. [3.2b]

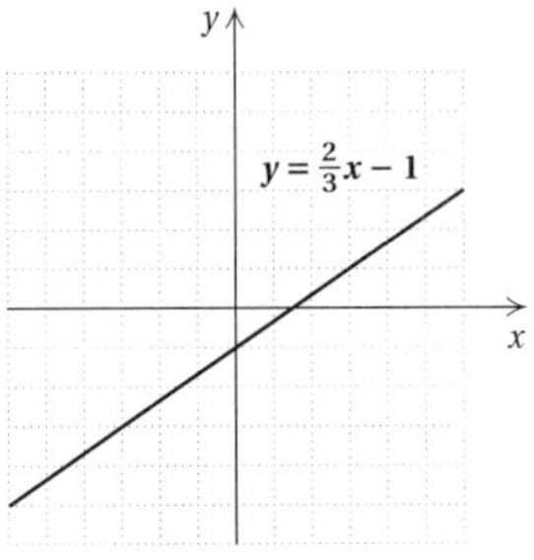

5. [3.1c] III **6.** [3.2a] No **7.** [3.3a] y-intercept: $(0, -4)$; x-intercept: $(5, 0)$ **8.** [3.2b] $(0, -8)$
9. [3.2c]

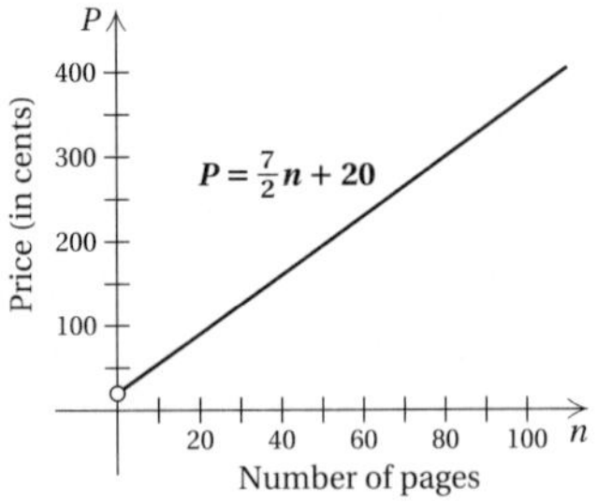

320¢, or \$3.20

10. [3.4a] Mean: 0.18; median: 0.18; mode: 0.18
11. [3.4c] 164 cm

Margin Exercises, Section 3.1, pp. 141–145

1. \$23.22 billion **2.** **(a)** 3 A.M.–6 A.M.; **(b)** midnight–3 A.M., 3 A.M.–6 A.M., 6 A.M.–9 A.M., 9 A.M.–noon **3.** **(a)** 2; **(b)** 60 beats per minute

4.–11.

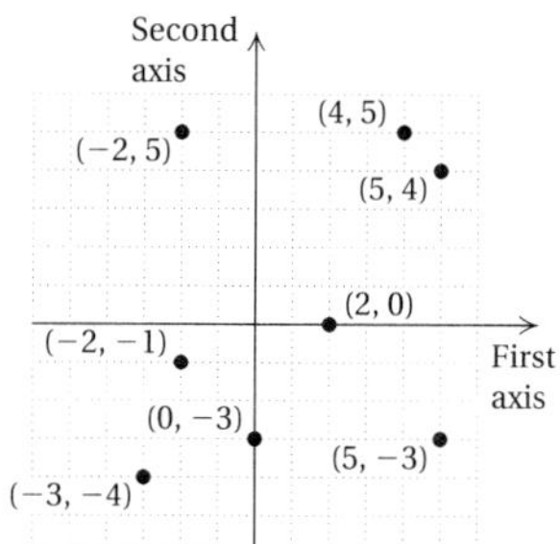

12. Both are negative numbers. **13.** First, positive; second, negative **14.** I **15.** III **16.** IV **17.** II **18.** Not in any quadrant **19.** *A*: (−5, 1); *B*: (−3, 2); *C*: (0, 4); *D*: (3, 3); *E*: (1, 0); *F*: (0, −3); *G*: (−5, −4)

Exercise Set 3.1, p. 147

1. 6 **3.** The weight is greater than 200 lb. **5.** The weight is greater than or equal to 120 lb. **7.** 32.4% **9.** $10,360.32 **11.** 20,000 **13.** 1994 **15.** About 500 **17.**

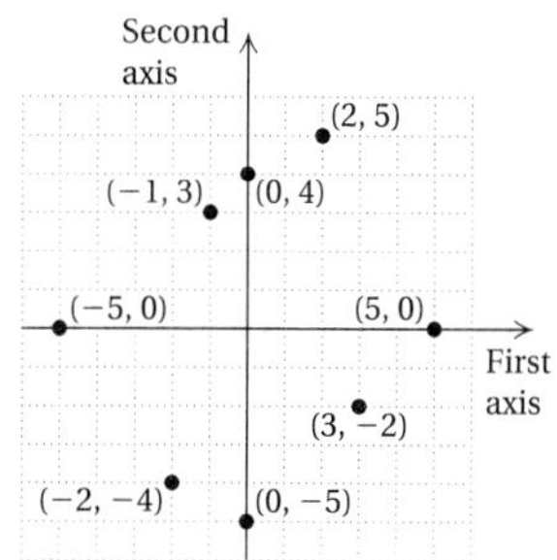

19. II **21.** IV **23.** III **25.** I **27.** II **29.** IV **31.** Negative; negative **33.** Second; first **35.** I, IV **37.** I, III **39.** *A*: (3, 3); *B*: (0, −4); *C*: (−5, 0); *D*: (−1, −1); *E*: (2, 0) **41.** 12 **42.** 4.89 **43.** 0 **44.** $\frac{4}{5}$ **45.** $0.9 billion **46.** $18.40 **47.** ◆ **49.** (−1, −5) **51.** Answers may vary. **53.** 26

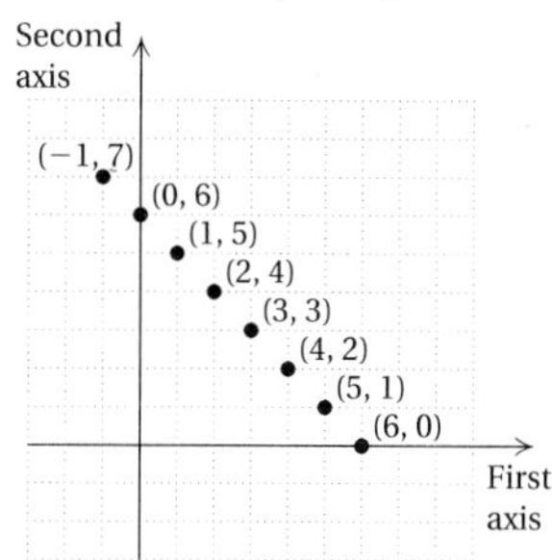

Margin Exercises, Section 3.2, pp. 151–158

1. No **2.** Yes **3.** (−2, −3), (1, 3); answers may vary

4.

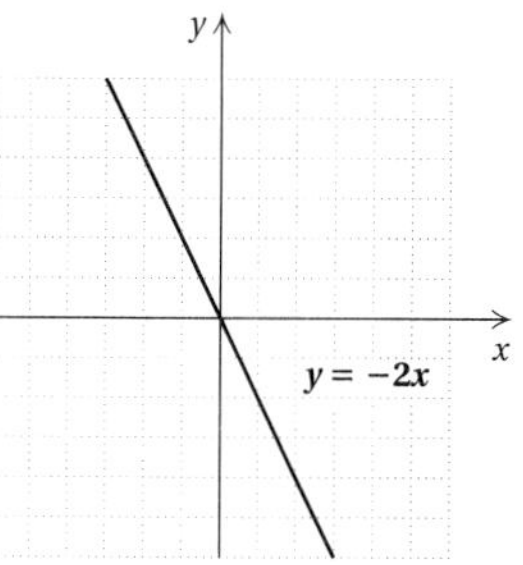

5.

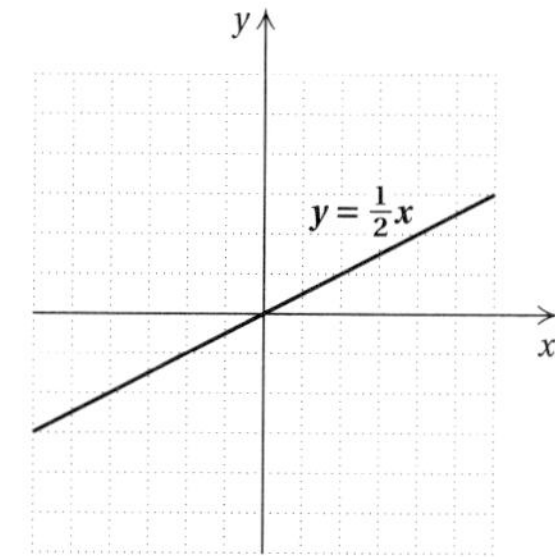

6.

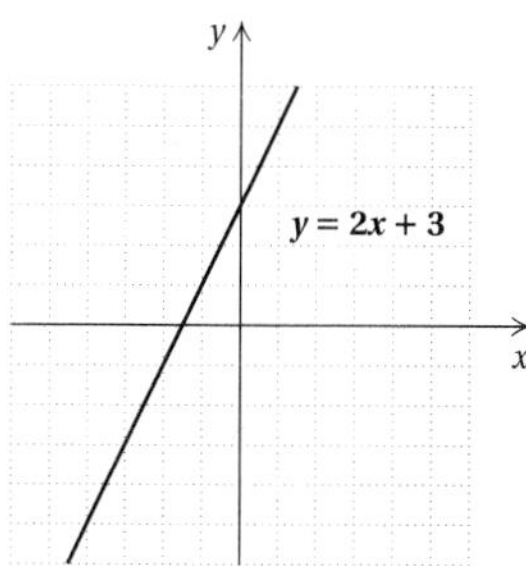

7.

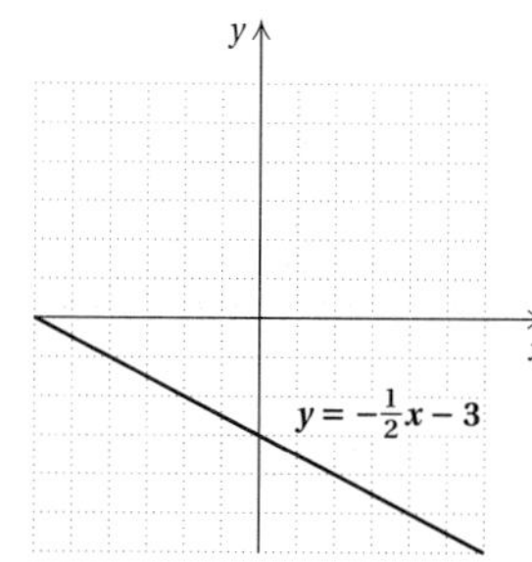

8.

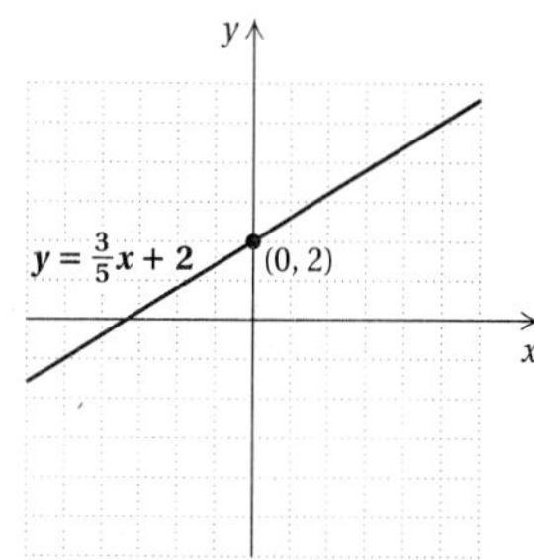

9.

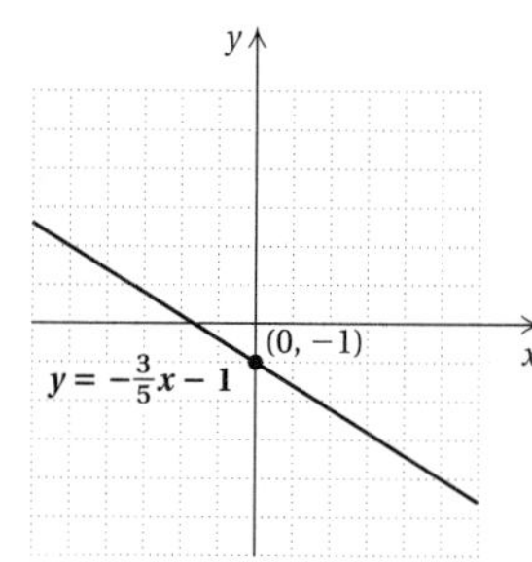

10.

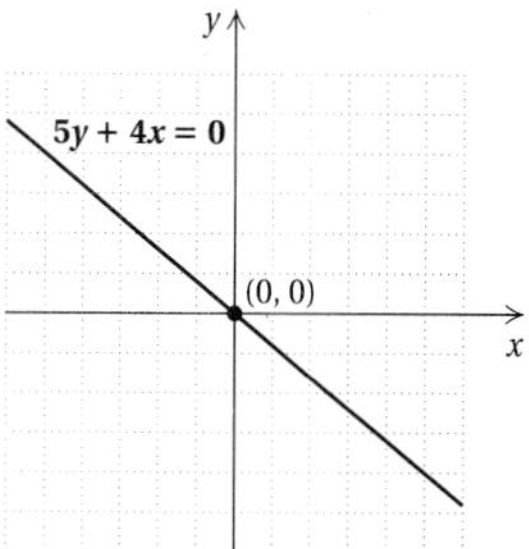

11.

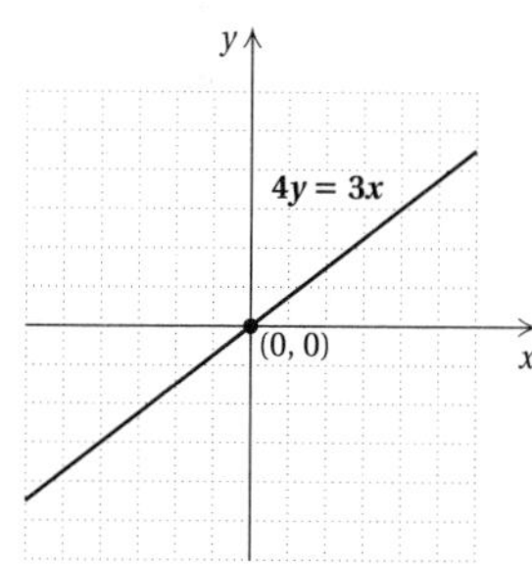

12.

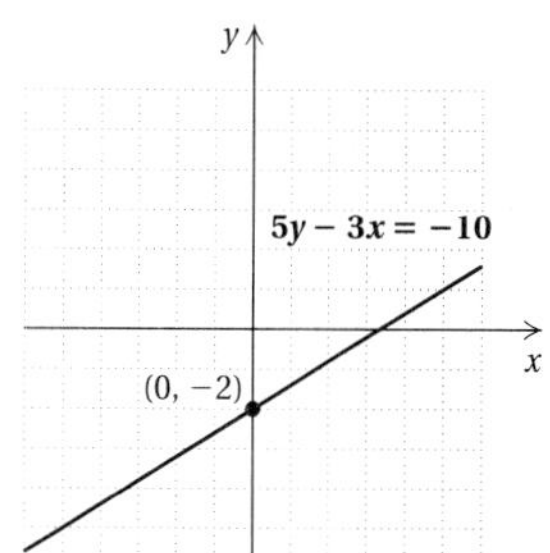

13.

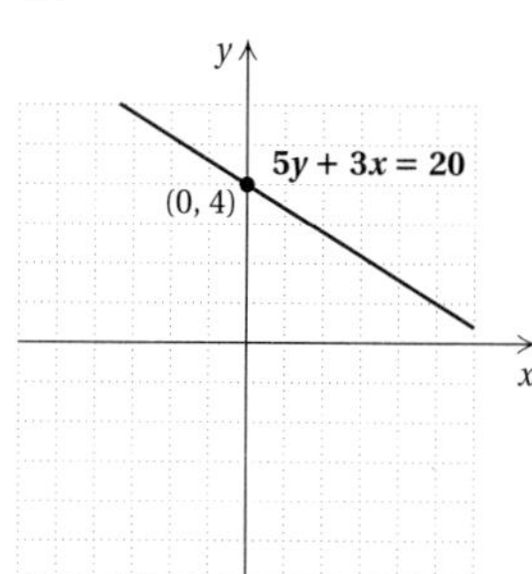

14. **(a)** $2720, $2040, $680, $0;

(b) \$1700 **(c)** About 2.8 yr

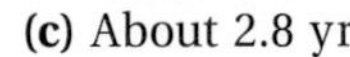

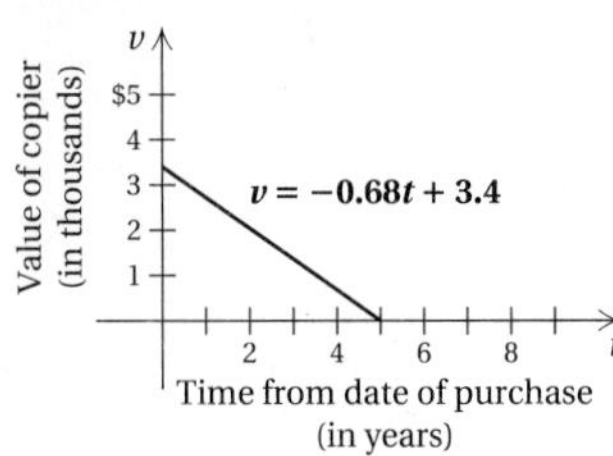

Calculator Spotlight, p. 160

1. 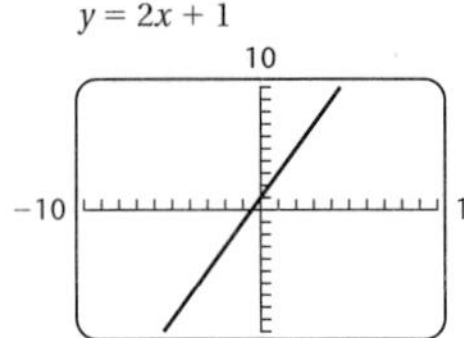

2. 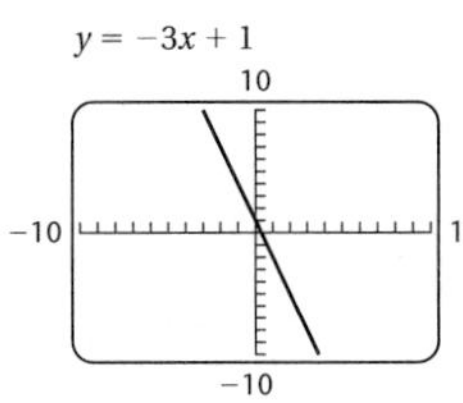

3. 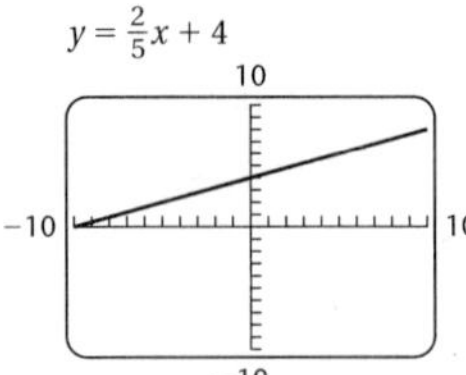

4. 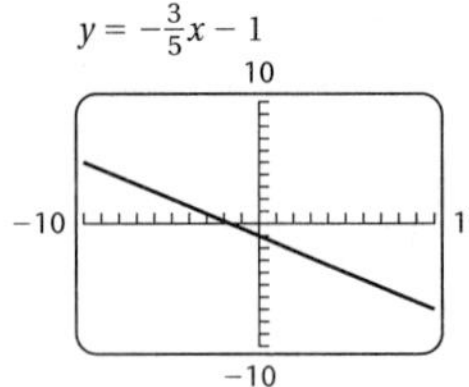

5. 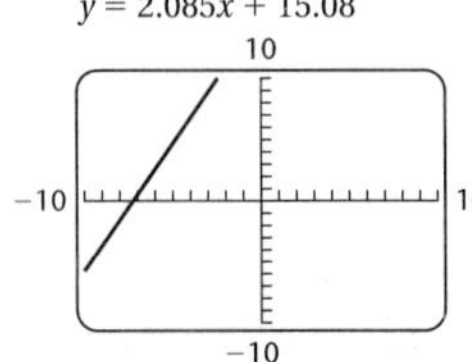

6. 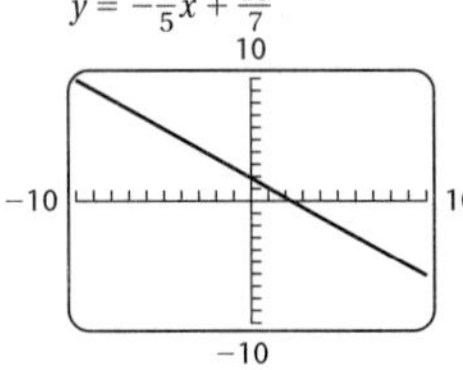

7. 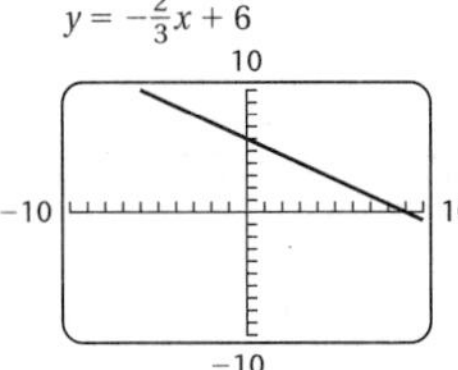

8. 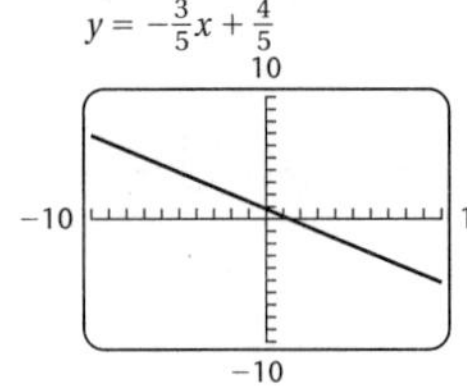

9. 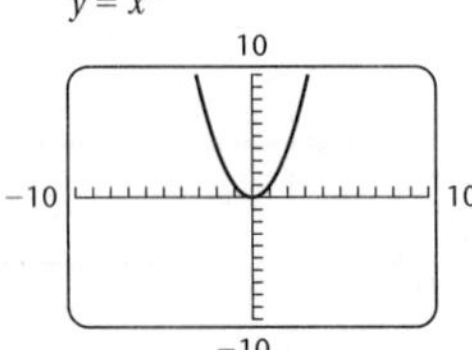

10. 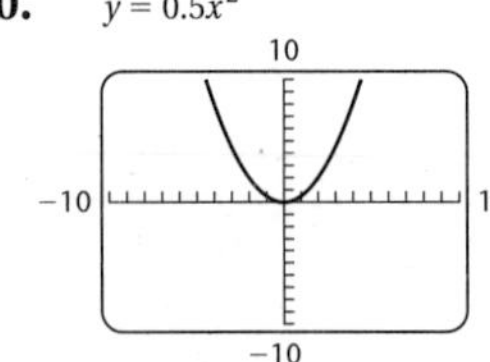

11. 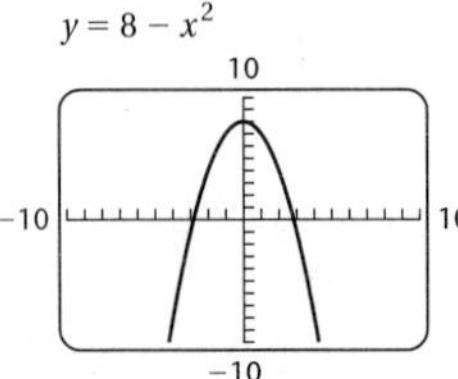

12. 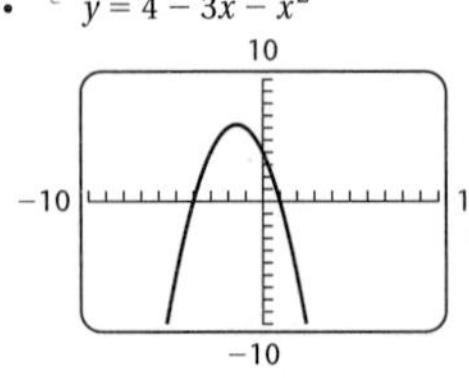

13. 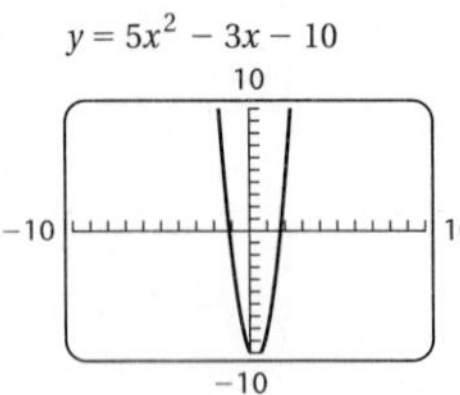

14. 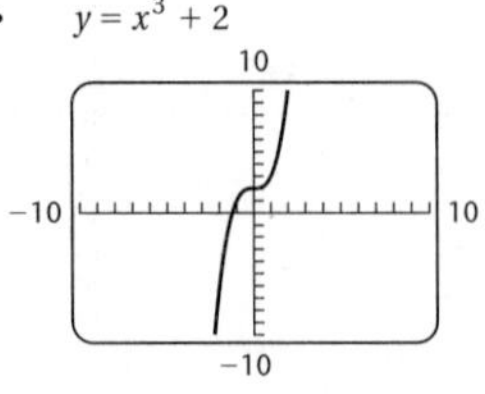

15. 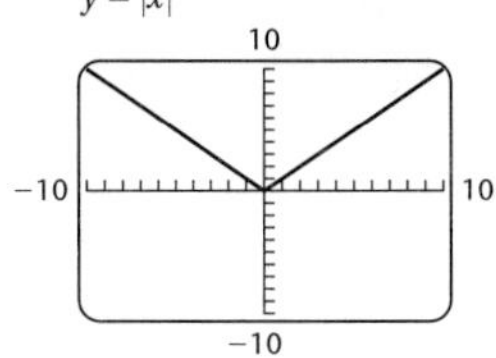

16. 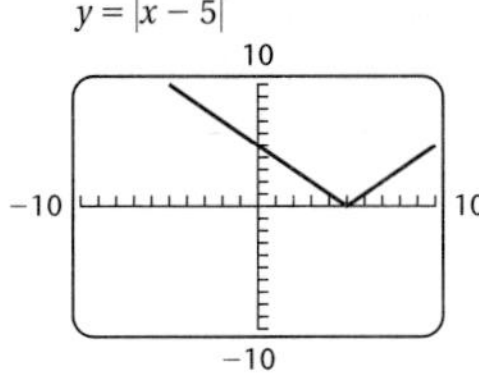

17. 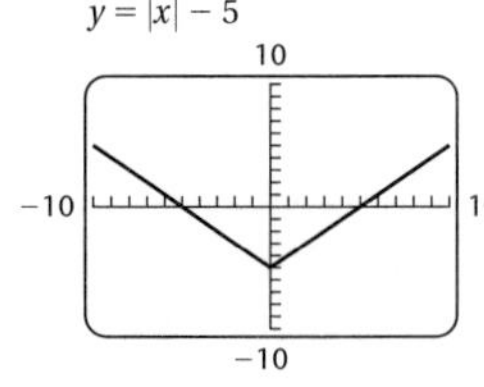

18. 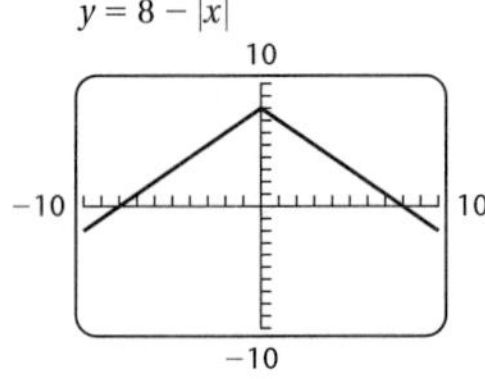

Exercise Set 3.2, p. 161

1. No **3.** No **5.** Yes

7.
$$\begin{array}{c|c} y & x - 5 \\ \hline -1 & ?\ 4 - 5 \\ & -1 \end{array} \quad \text{TRUE}$$

$$\begin{array}{c|c} y & x - 5 \\ \hline -4 & ?\ 1 - 5 \\ & -4 \end{array} \quad \text{TRUE}$$

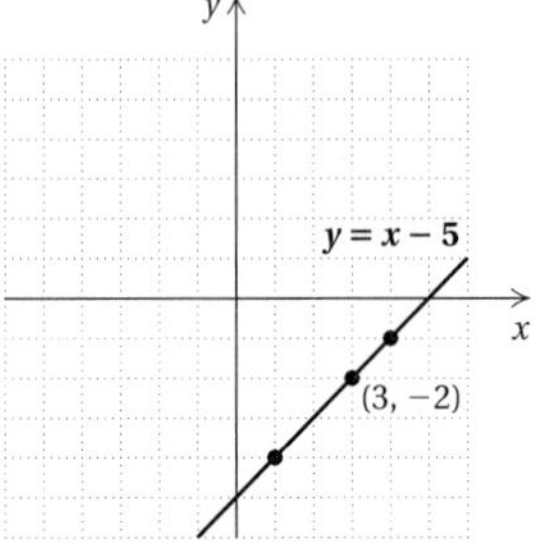

9.
$$\begin{array}{c|c} y & \frac{1}{2}x + 3 \\ \hline 5 & ?\ \frac{1}{2} \cdot 4 + 3 \\ & 2 + 3 \\ & 5 \end{array} \quad \text{TRUE}$$

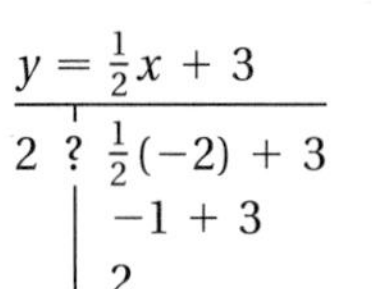

$$\begin{array}{c|c} y & \frac{1}{2}x + 3 \\ \hline 2 & ?\ \frac{1}{2}(-2) + 3 \\ & -1 + 3 \\ & 2 \end{array} \quad \text{TRUE}$$

$y = \frac{1}{2}x + 3$; (−4, 1)

11.
$$\begin{array}{c|c} 4x - 2y & 10 \\ \hline 4 \cdot 0 - 2(-5) & ?\ 10 \\ 0 + 10 & \\ 10 & \end{array} \quad \text{TRUE}$$

$$\begin{array}{c|c} 4x - 2y & 10 \\ \hline 4 \cdot 4 - 2 \cdot 3 & ?\ 10 \\ 16 - 6 & \\ 10 & \end{array} \quad \text{TRUE}$$

$4x - 2y = 10$; (1, −3)

13.

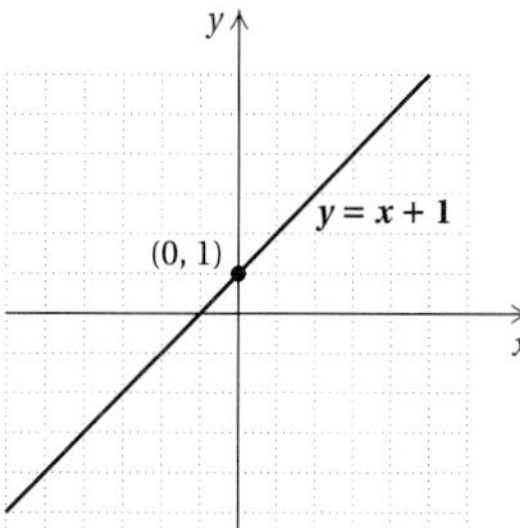

15.

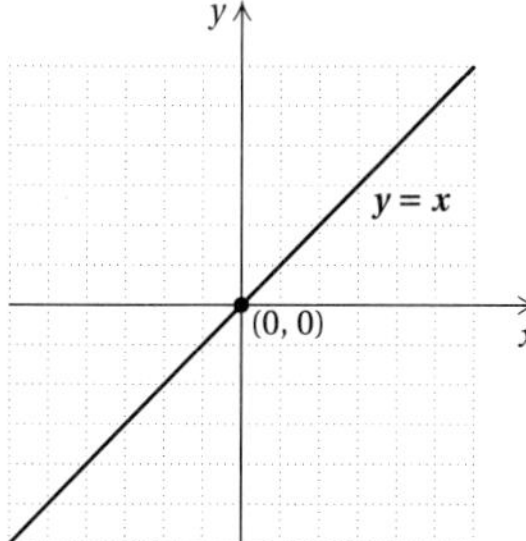

17.

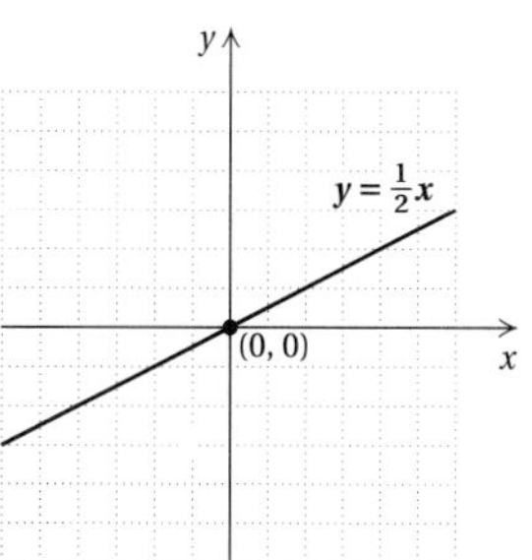

19.

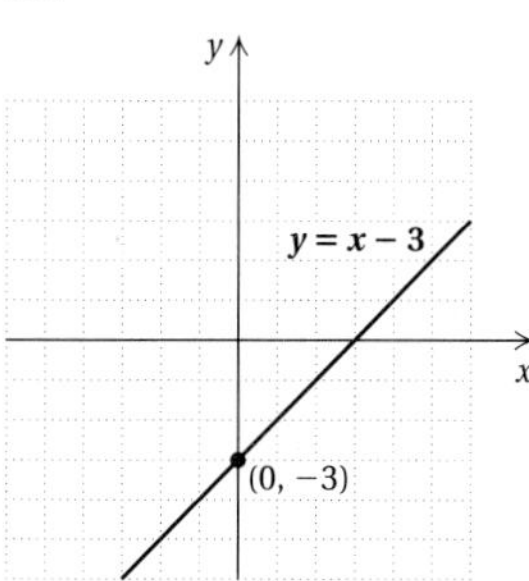

21.

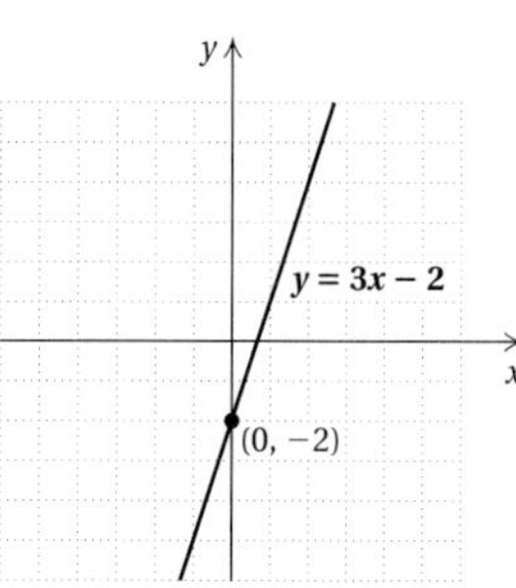

23.

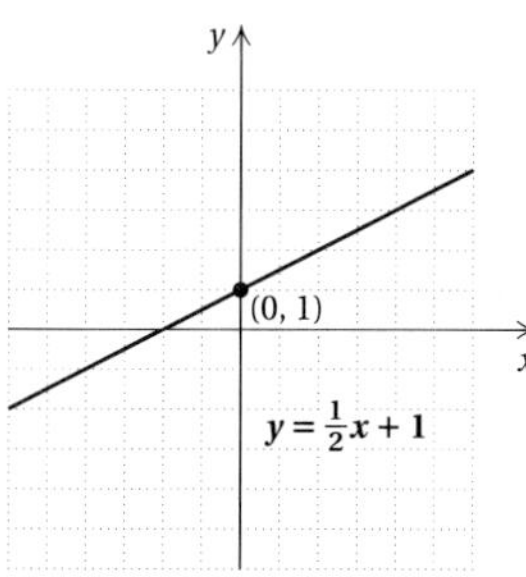

25.

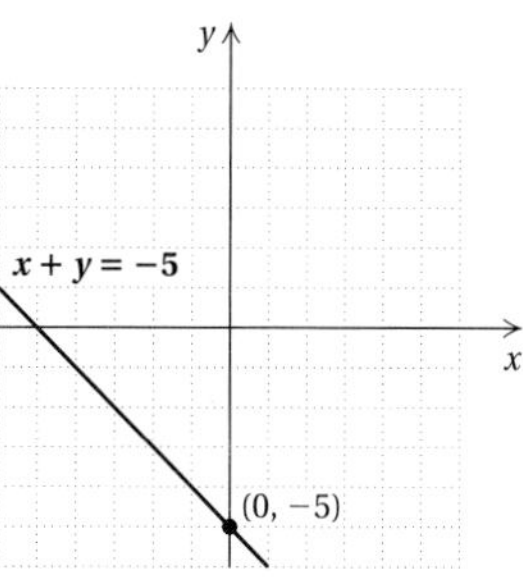

27.

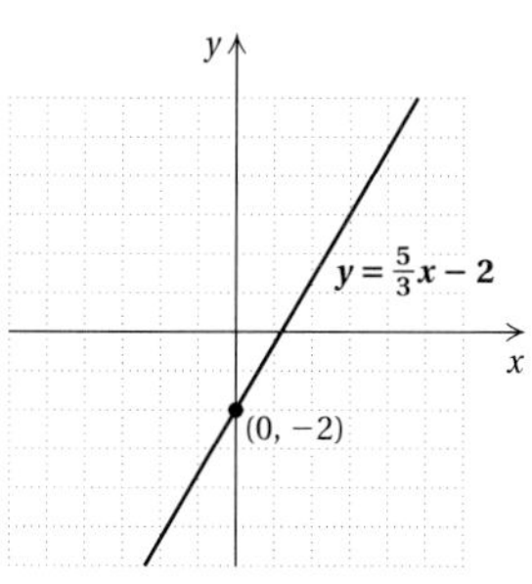

29.

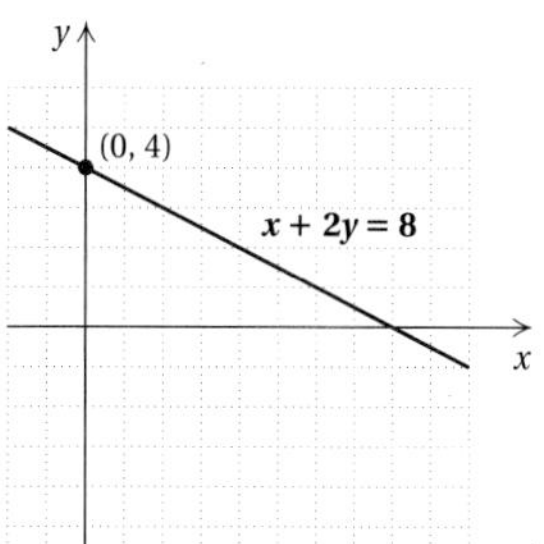

31.

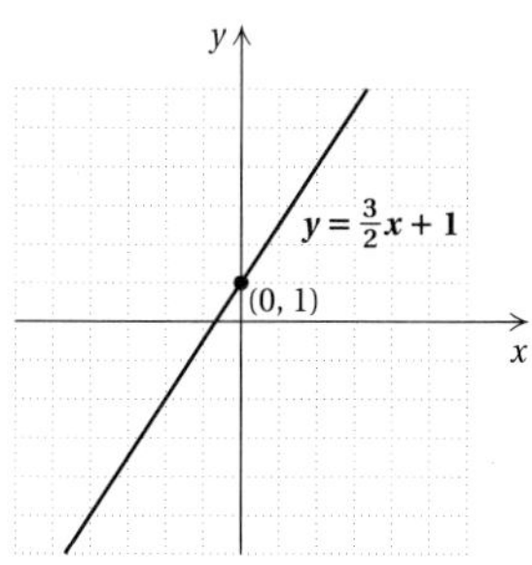

33.

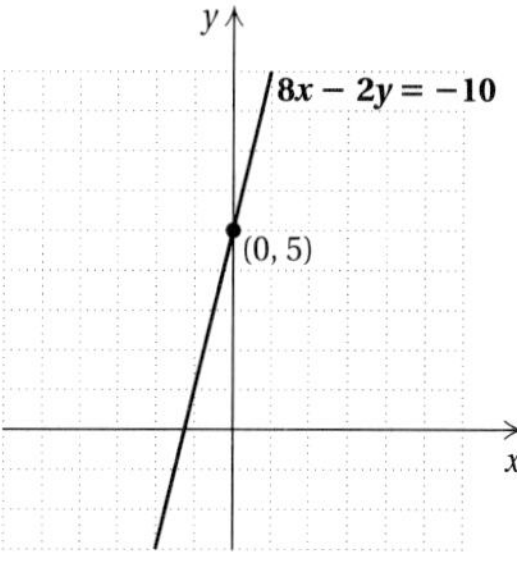

35.

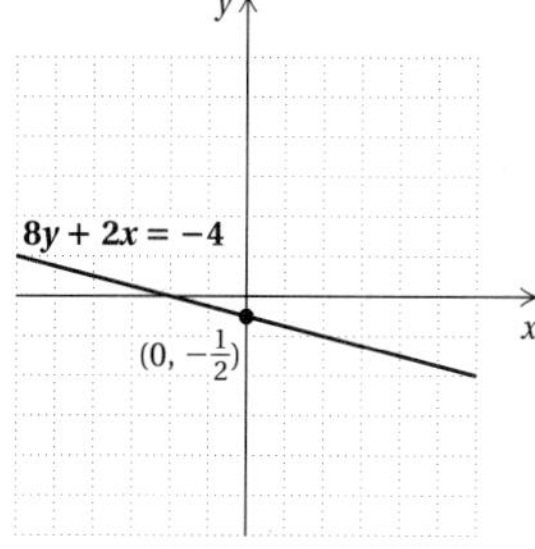

37. **(a)** \$300, \$100, \$0; **(b)**

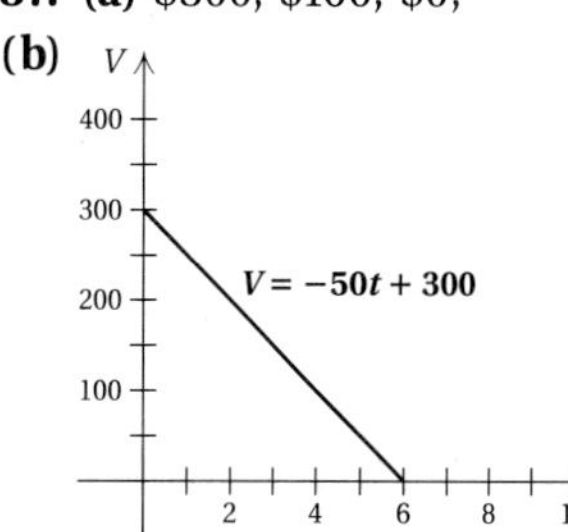
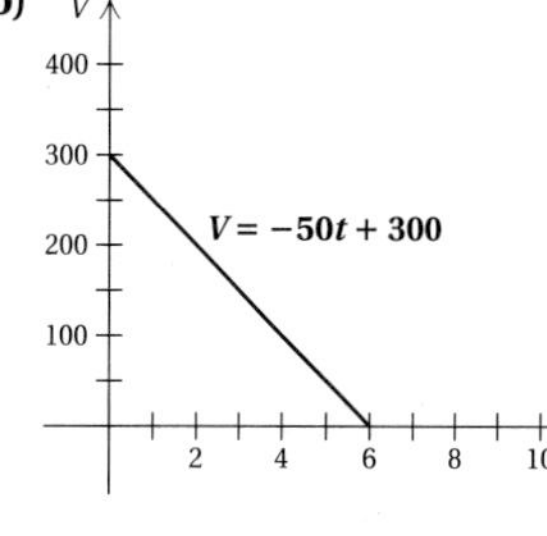

\$50; **(c)** 3

39. **(a)** 7.1 gal, 7.4 gal, 7.8 gal, 8 gal; **(b)**

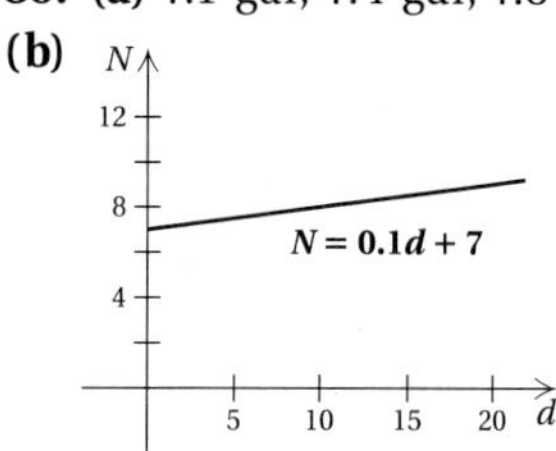

7.6 gal; **(c)** 2011

41. 7 **42.** $-\frac{1}{2}$ **43.** −4 **44.** −16 **45.** −0.875 **46.** 0.71875 **47.** 1.828125 **48.** −2.25 **49.** ◈ **51.** $y = -x + 5$ **53.** $y = x + 2$

Margin Exercises, Section 3.3, pp. 165–167

1. **(a)** (0, 3); **(b)** (4, 0)

2.

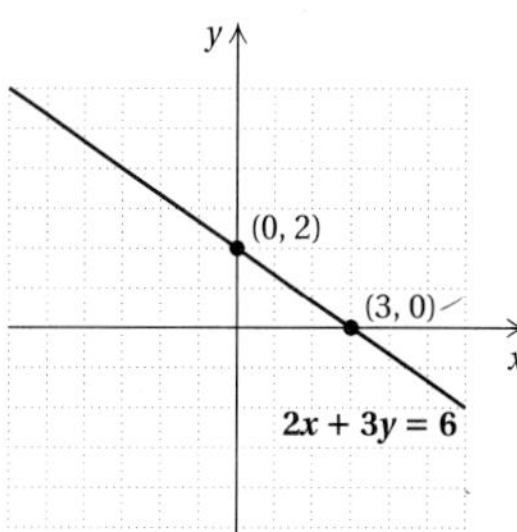

3.

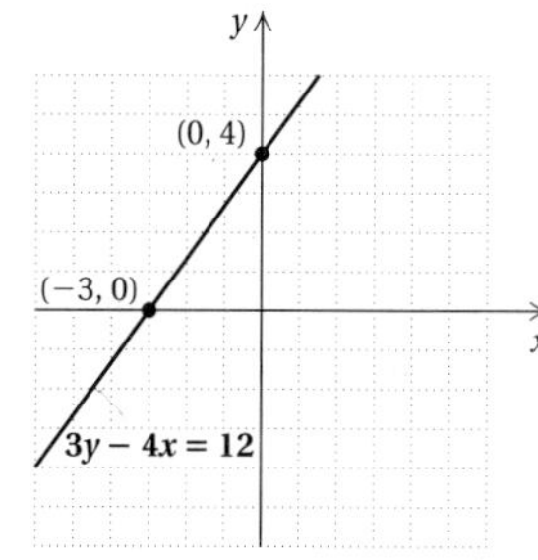

4.

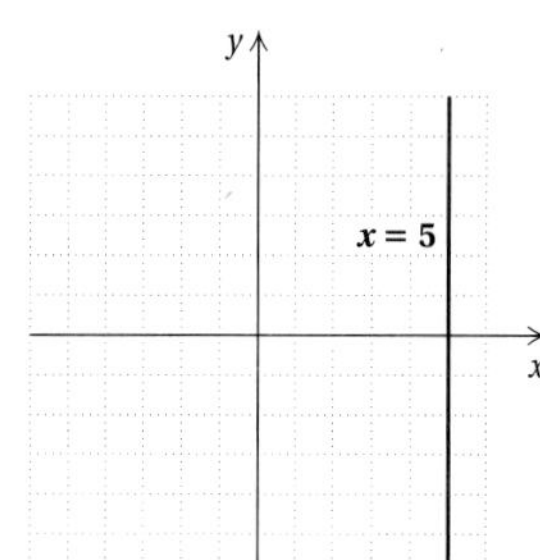

5.

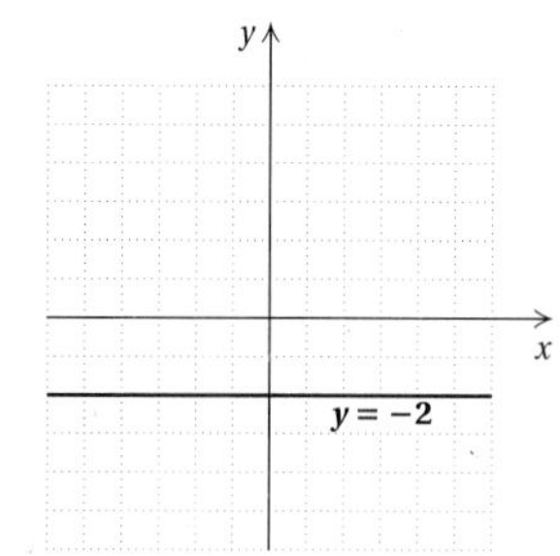

6.

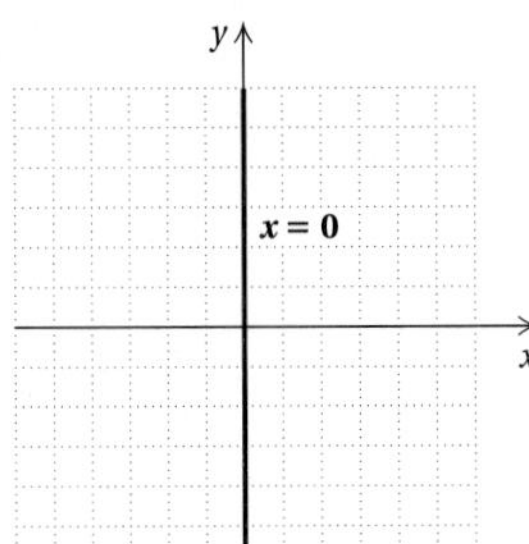

7.

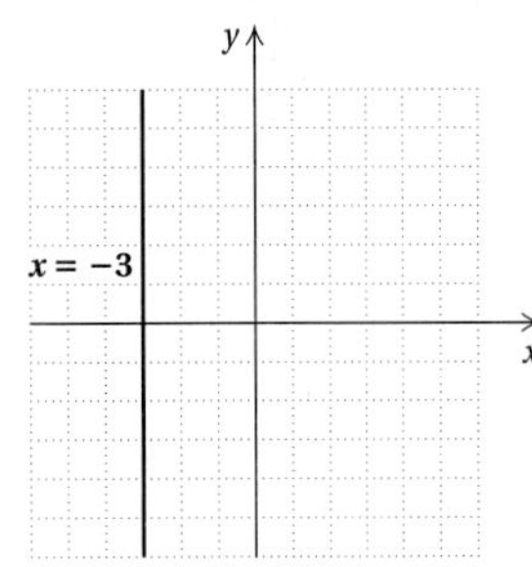

Calculator Spotlight, p. 168

1. y-intercept: (0, −15); x-intercept: (−2.08, 0)
2. y-intercept: (0, 27); x-intercept: (−12.68, 0)
3. y-intercept: (0, 14); x-intercept: (16.8, 0)
4. y-intercept: (0, −21.43); x-intercept: (75, 0)
5. y-intercept: (0, 25); x-intercept: (16.67, 0)
6. y-intercept: (0, −9); x-intercept: (45, 0)
7. y-intercept: (0, −15); x-intercept: (11.54, 0)
8. y-intercept: (0, −0.05); x-intercept: (0.04, 0)

Exercise Set 3.3, p. 169

1. **(a)** (0, 5); **(b)** (2, 0) **3.** **(a)** (0, −4); **(b)** (3, 0)
5. **(a)** (0, 3); **(b)** (5, 0) **7.** **(a)** (0, −14); **(b)** (4, 0)
9. **(a)** $\left(0, \frac{10}{3}\right)$; **(b)** $\left(-\frac{5}{2}, 0\right)$ **11.** **(a)** $\left(0, -\frac{1}{3}\right)$; **(b)** $\left(\frac{1}{2}, 0\right)$

13.

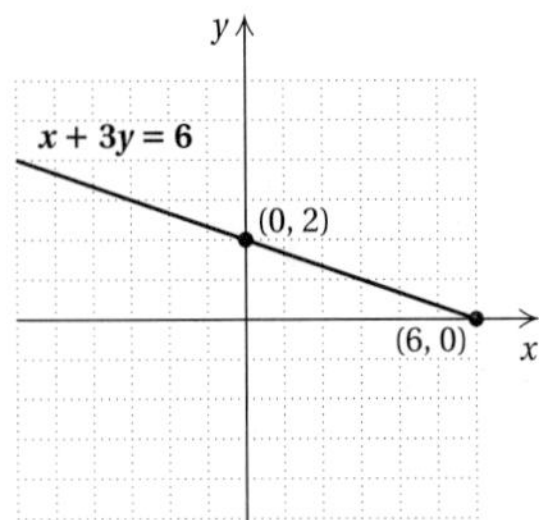

15.

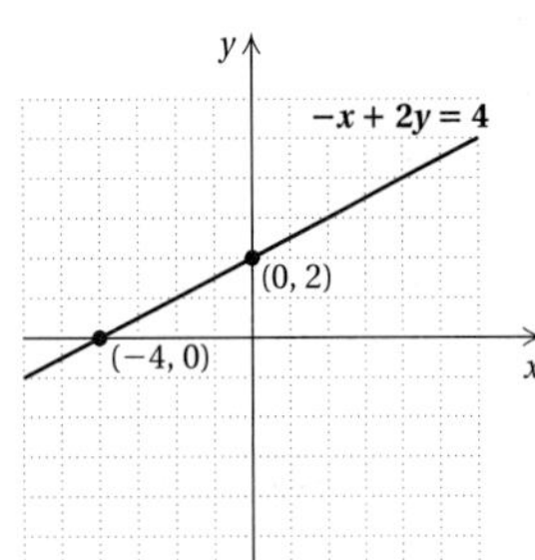

17.

19.

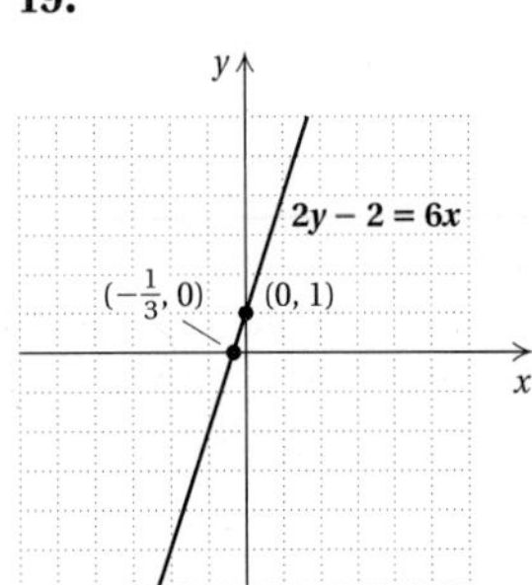

21.

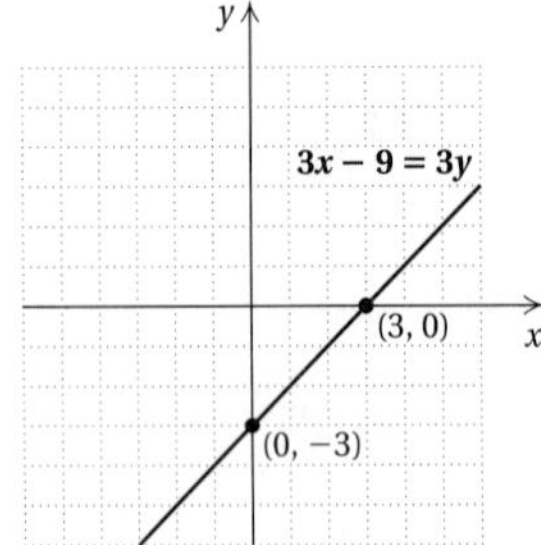

23.

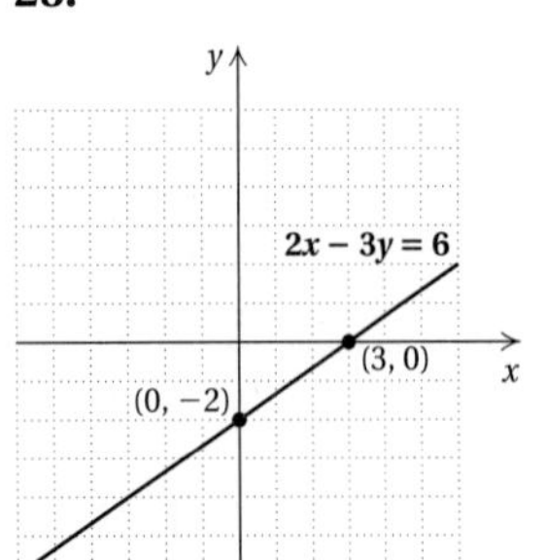

25.

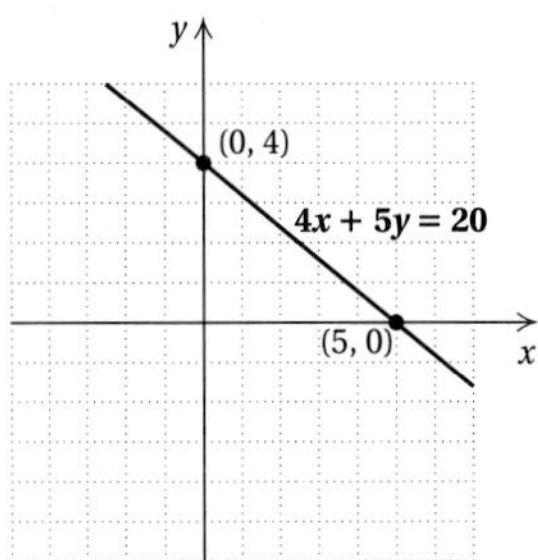

27.

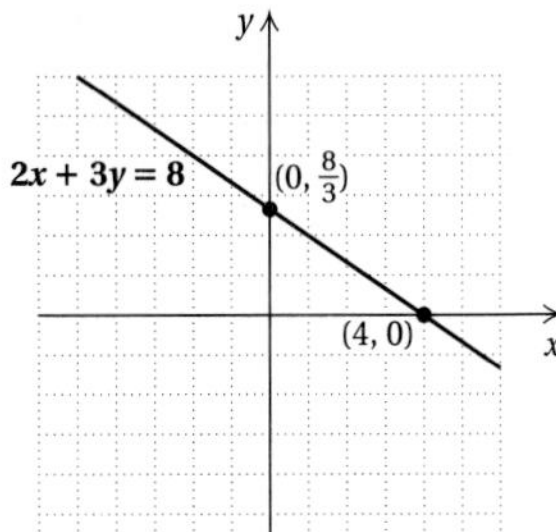

29.

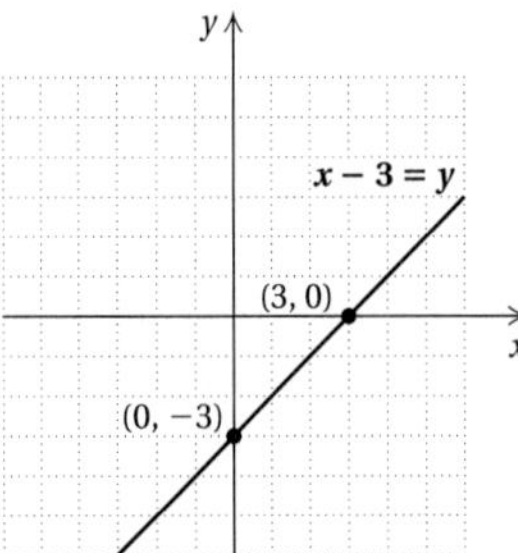

31.

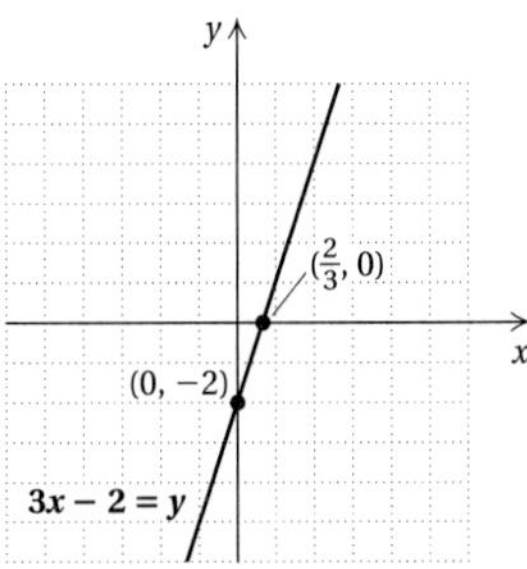

33.

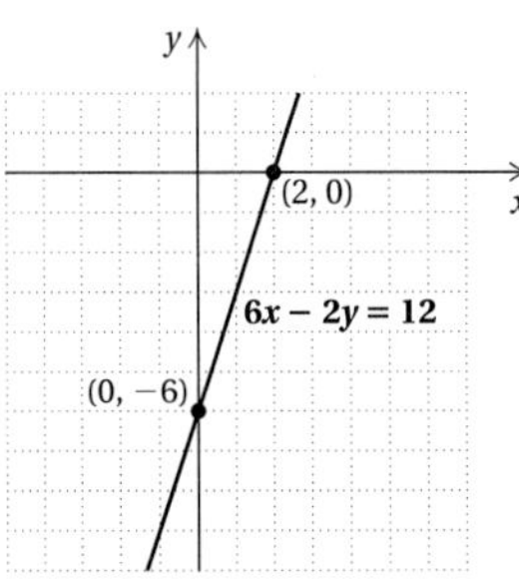

35.

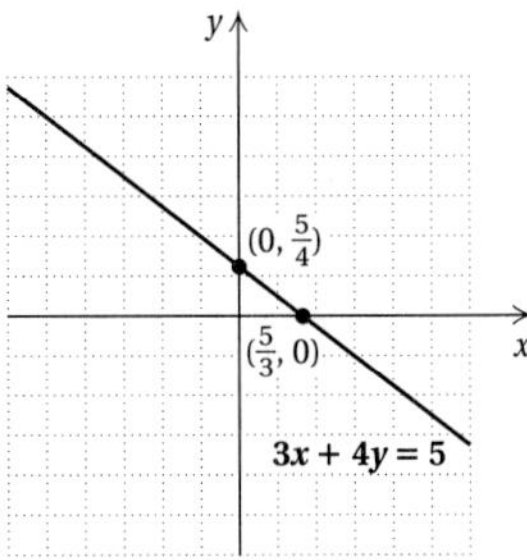

37.

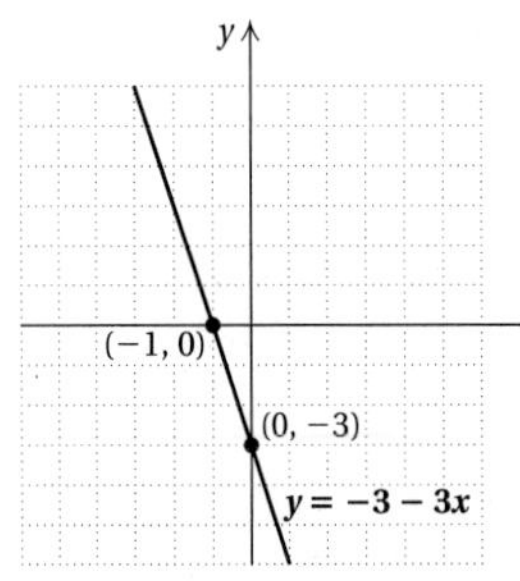

39.

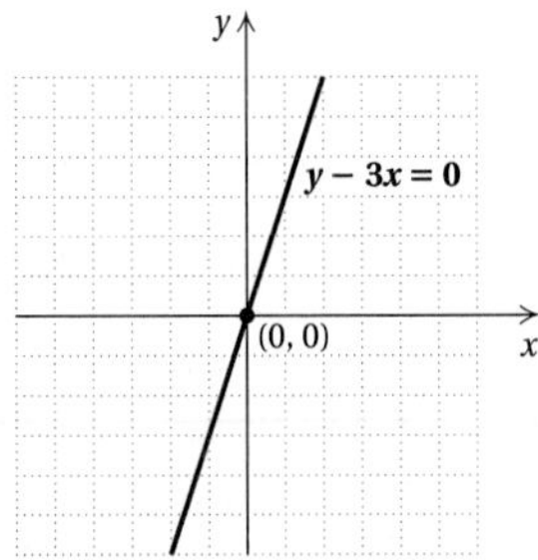

41.

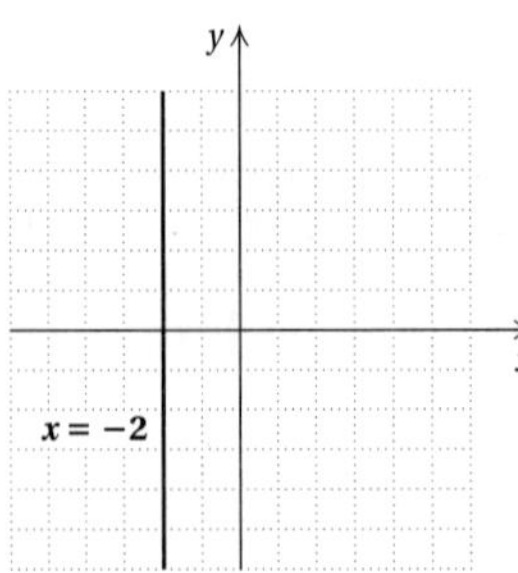

43.

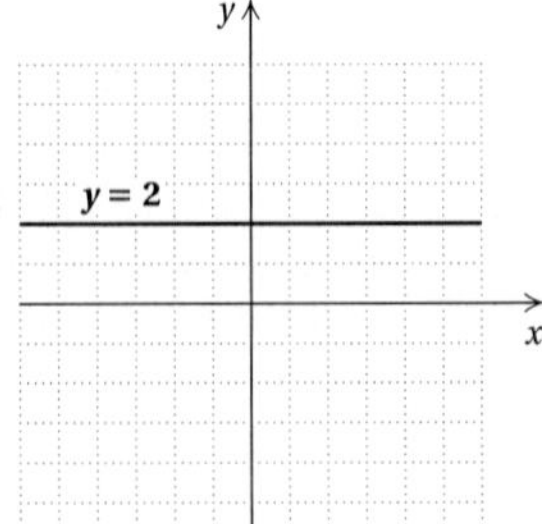

45.

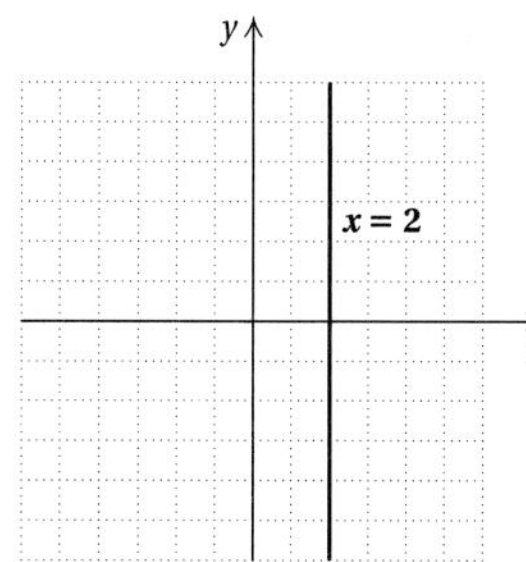

47.

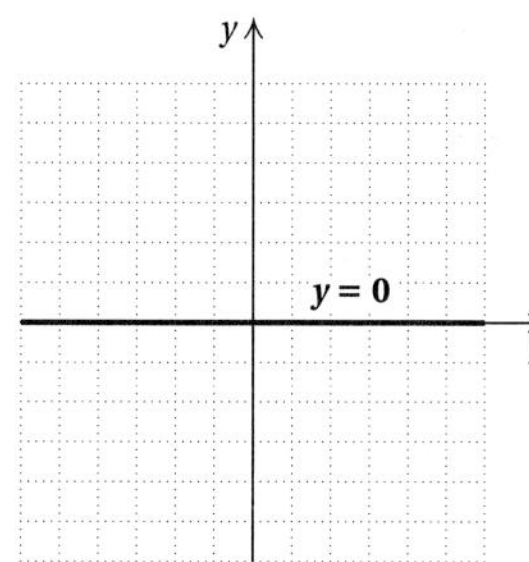

49.

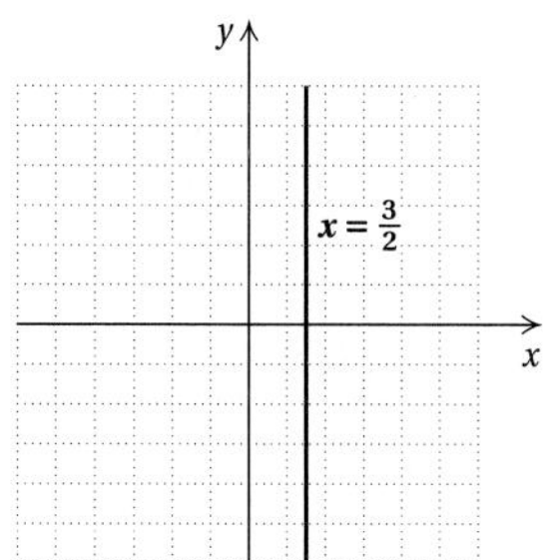

51.

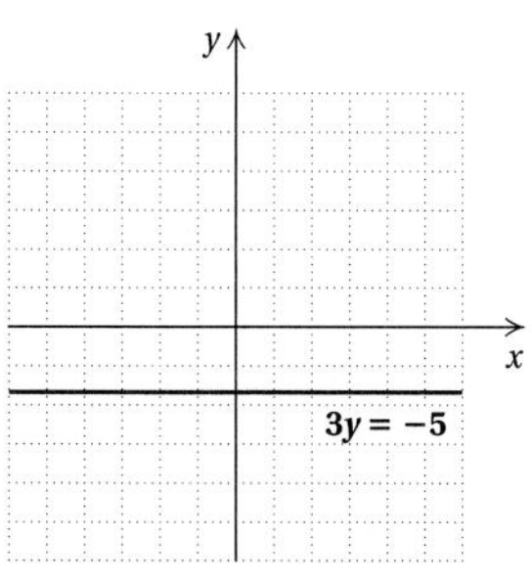

53.

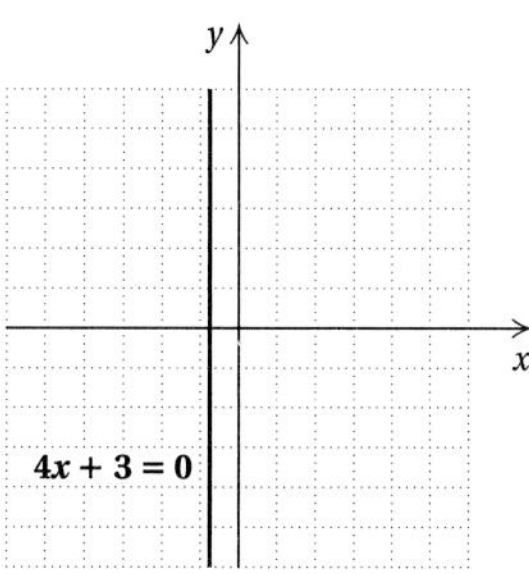

55.

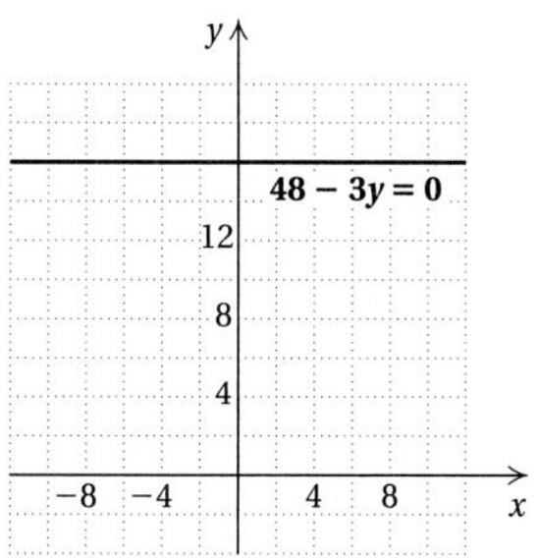

57. $y = -1$ **59.** $x = 4$ **61.** 16% **62.** 70
63. \$32.50 **64.** \$23.00 **65.** $\{x \mid x > -40\}$
66. $\{x \mid x \leq -7\}$ **67.** $\{x \mid x < 1\}$ **68.** $\{x \mid x \geq 2\}$
69. ◆ **71.** $x = 0$ **73.** $y = -4$ **75.** $k = 12$

Margin Exercises, Section 3.4, pp. 173–177

1. 56.7 **2.** 64.7 **3.** 87.8 **4.** 17 **5.** 16.5 **6.** 91
7. 55 **8.** 54, 87 **9.** No mode exists.
10. **(a)** 25 mm^2; **(b)** 25 mm^2; **(c)** no mode exists.
11. Ball A: 18.8 in.; Ball B: 19.6 in.; Ball B is better.
12. \$866.79 **13.** 95

Calculator Spotlight, p. 178

1. (1.106383, 17.386809), (3.0425532, 21.423723), (5.6702128, 26.902394), (8.712766, 33.246117), (10.234043, 36.417979); answers may vary
2. 223.58; 327.83 **3.** 404.98

4. **(a)**

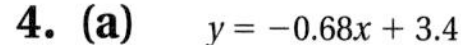

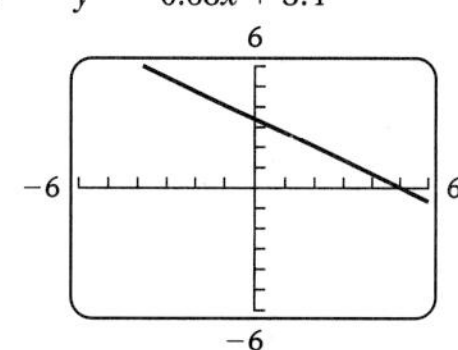

(b) no; **(c)** (5, 0); **(d)** $0 \leq x \leq 5$;
(e) (0.76595745, 2.8791489), (1.787234, 2.1846809), (2.2978723, 1.8374468), (2.9361702, 1.4034043), (4.212766, 0.53531915); answers may vary; **(f)** 2.79 yr;
(g) \$2244, \$1836, \$612, \$0

Exercise Set 3.4, p. 179

1. Mean: 28.6; median: 30; modes: 15, 30
3. Mean: 94; median: 95; no mode exists.
5. Mean: 32; median: 35; mode: 23
7. Mean: 897.2; median: 798; no mode exists.
9. Mean: 87.7; median: 88; modes: 85, 88
11. Battery B **13.** 162.4 **15.** 2008 **17.** 1997: 1.39; 2000: 1.54 **19.** 1997: \$1,500,000; 2000: \$1,700,000
21. 84 **23.** −4 **24.** −20 **25.** −37 **26.** $-\frac{3}{4}$
27. 20% **28.** \$18 **29.** \$45.15 **30.** \$55 **31.** ◆
33. $y = 0.35x - 7$

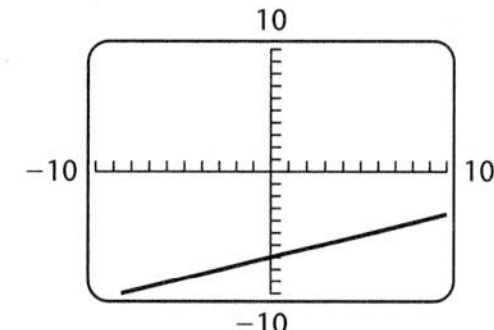

X	Y_1
−10	−10.5
−9.9	−10.47
−9.8	−10.43
−9.7	−10.4
−9.6	−10.36
−9.5	−10.33
−9.4	−10.29

X = −10

35. $y = x^3 - 5$

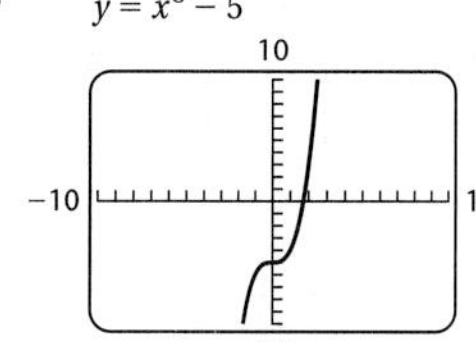

X	Y_1
−10	−1005
−9.9	−975.3
−9.8	−946.2
−9.7	−917.7
−9.6	−889.7
−9.5	−862.4
−9.4	−835.6

X = −10

Summary and Review: Chapter 3, p. 183

1. \$775.50; \$634.50 **2.** 47 lb **3.** 80 lb **4.** 33 lb
5. 1995 **6.** 1990–1995 **7.** One shower
8. One toilet flush **9.** One shave, wash dishes, one shower **10.** One toilet flush **11.** (−5, −1)
12. (−2, 5) **13.** (3, 0)
14.–16.

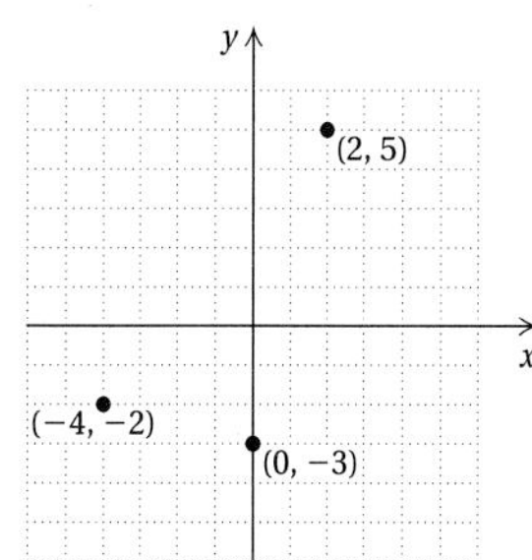

17. IV **18.** III **19.** I **20.** No **21.** Yes

22.

$$\begin{array}{c|c} 2x - y & = 3 \\ \hline 2 \cdot 0 - (-3) & ?\ 3 \\ 0 + 3 & \\ 3 & \end{array} \quad \text{TRUE}$$

$$\begin{array}{c|c} 2x - y & = 3 \\ \hline 2 \cdot 2 - 1 & ?\ 3 \\ 4 - 1 & \\ 3 & \end{array} \quad \text{TRUE}$$

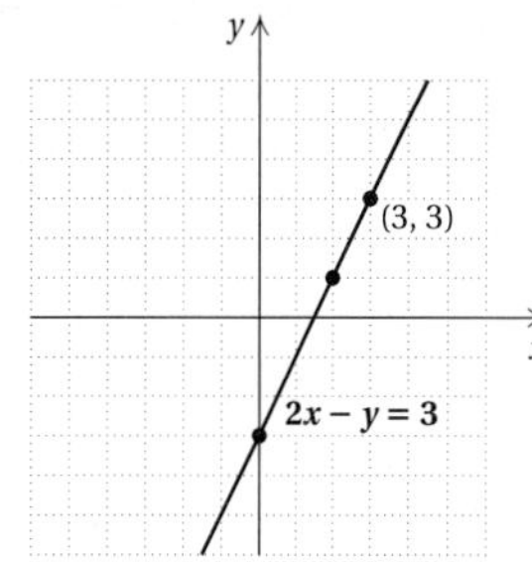

23.

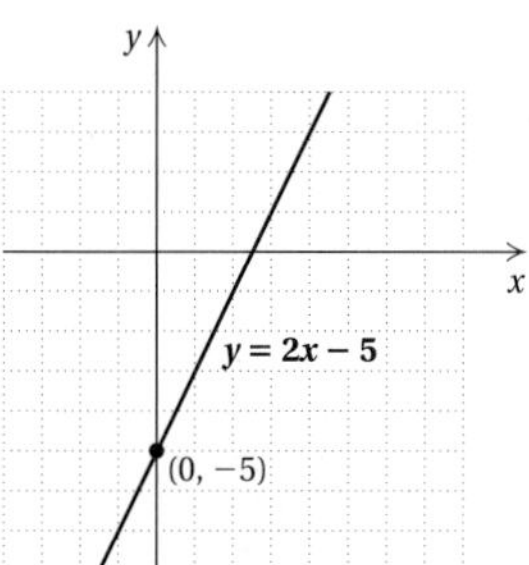

24.

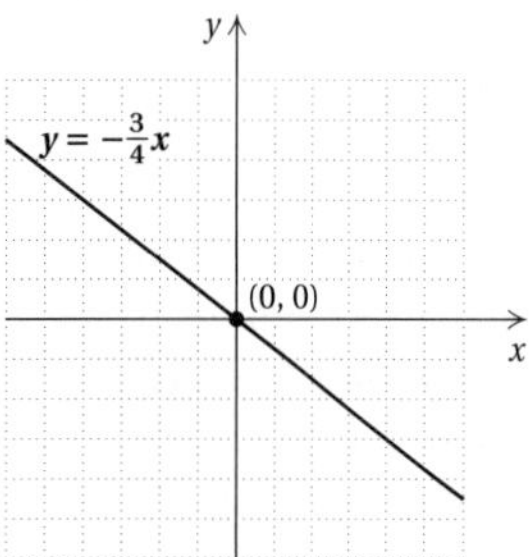

25.

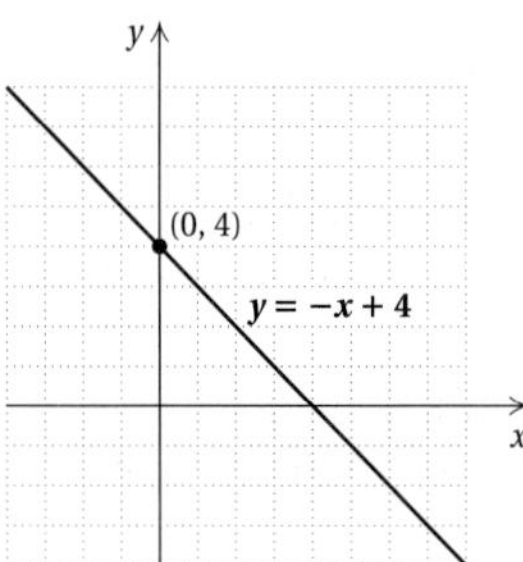

26.

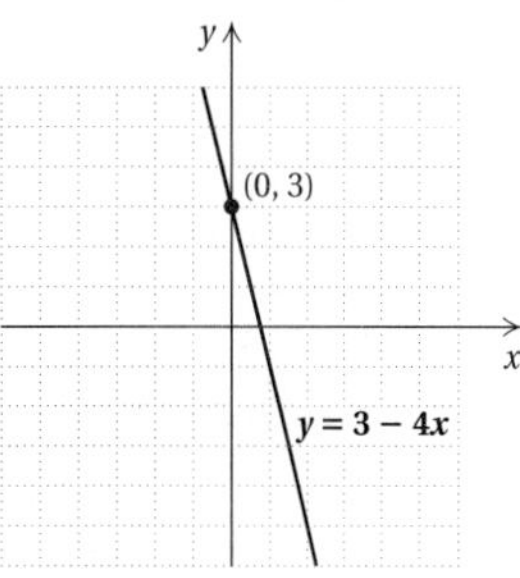

27.

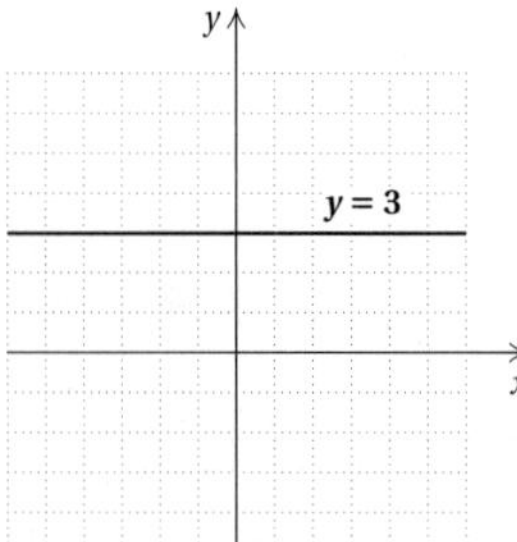

28.

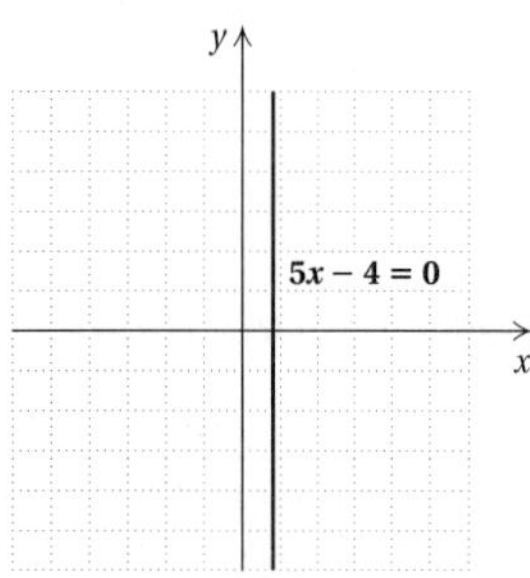

29.

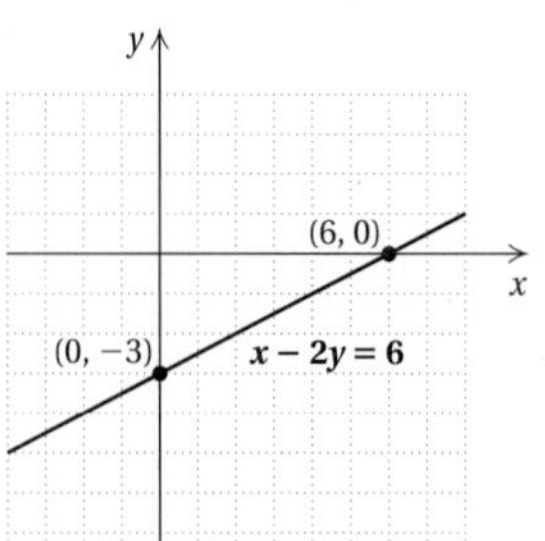

30.

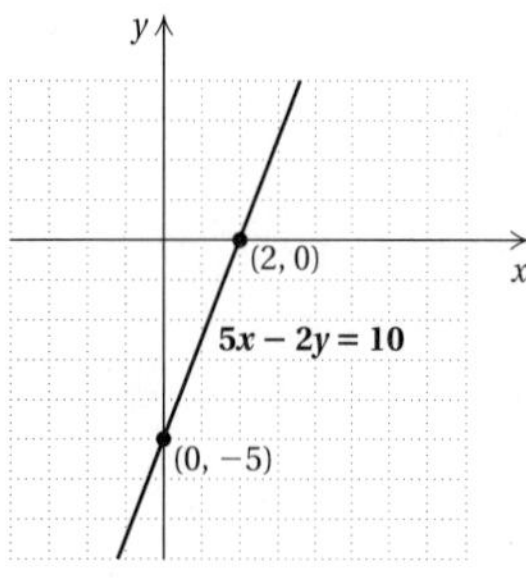

31. **(a)** $14\frac{1}{2}$ ft^3, 16 ft^3, $20\frac{1}{2}$ ft^3, 28 ft^3; $17\frac{1}{2}$ ft^3; **(b)**

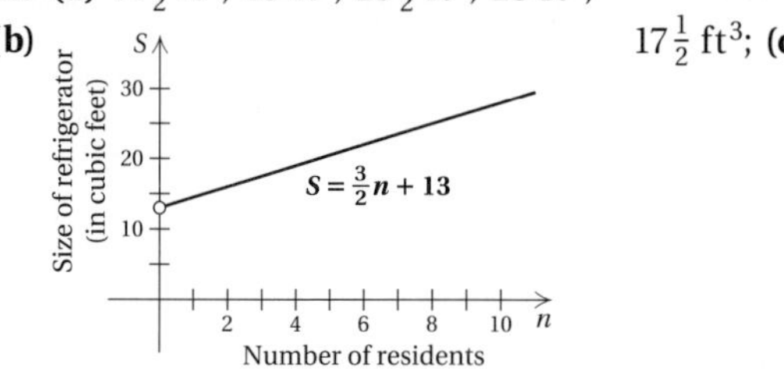

(c) 6

32. Mean: 39.5; median: 39.5; no mode exists.
33. Mean: 14.875; median: 16; modes: 12, 17, 19
34. Mean: 0.18; median: 0.175; no mode exists.
35. Mean: 229.8; median: 223; mode: 215
36. Mean: \$50.1; median: \$50.4; no mode exists.
37. Popcorn A **38.** 109.55 **39.** 1997: 64.4; 2000: 78.3
40. −0.34375 **41.** $0.\overline{8}$ **42.** 3.2 **43.** $\frac{17}{19}$ **44.** 11
45. $\frac{3}{4}$ **46.** \$9755.09 **47.** \$79.95
48. ◈ A small business might use a graph to look up prices quickly (as in the FedEx mailing costs example) or to plot change in sales over a period of time. Many other applications exist. **49.** ◈ The y-intercept is the point at which the graph crosses the y-axis. Since a point on the y-axis is neither left nor right of the origin, the first or x-coordinate of the point is 0.
50. $m = -1$ **51.** 45 square units; 28 linear units

Test: Chapter 3, p. 187

1. [3.1a] \$495,000,000 **2.** [3.1a] Crest and Colgate
3. [3.1a] Crest **4.** [3.1a] Arm & Hammer
5. [3.1a] June **6.** [3.1a] January **7.** [3.1a] March, April, May, June, July **8.** [3.1a] August
9. [3.1a] 1997 **10.** [3.1a] 1991 **11.** [3.1a] About \$500,000 **12.** [3.1a] 1996–1997
13. [3.1a] 1994–1995 **14.** [3.1a] About \$500,000
15. [3.1c] II **16.** [3.1c] III **17.** [3.1d] (3, 4)
18. [3.1d] (0, −4)
19. [3.2a]

$$\begin{array}{c|c} y - 2x & = 5 \\ \hline -3 - 2(-4) & ?\ 5 \\ -3 + 8 & \\ 5 & \end{array} \quad \text{TRUE}$$

$$\begin{array}{c|c} y - 2x & = 5 \\ \hline 3 - 2(-1) & ?\ 5 \\ 3 + 2 & \\ 5 & \end{array} \quad \text{TRUE}$$

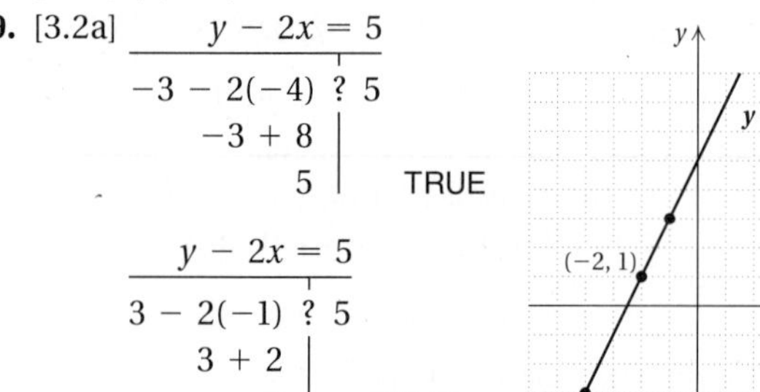

20. [3.2b]

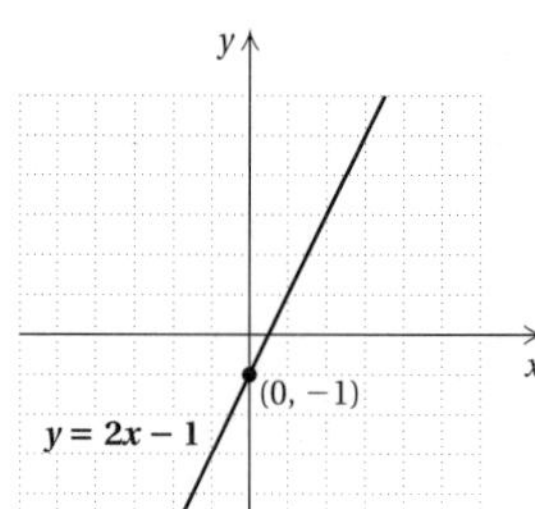

21. [3.2b]

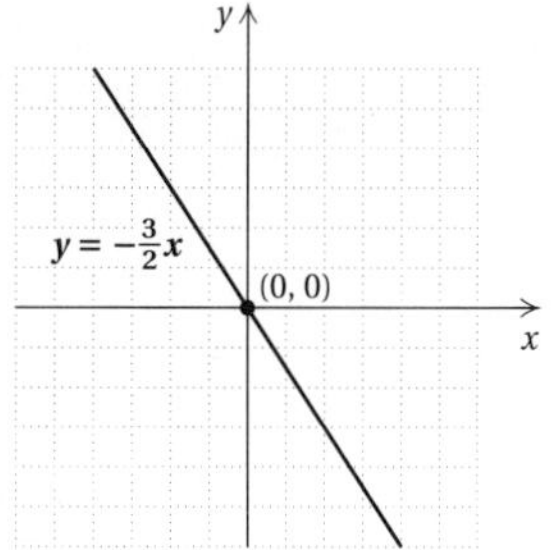

22. [3.3b]

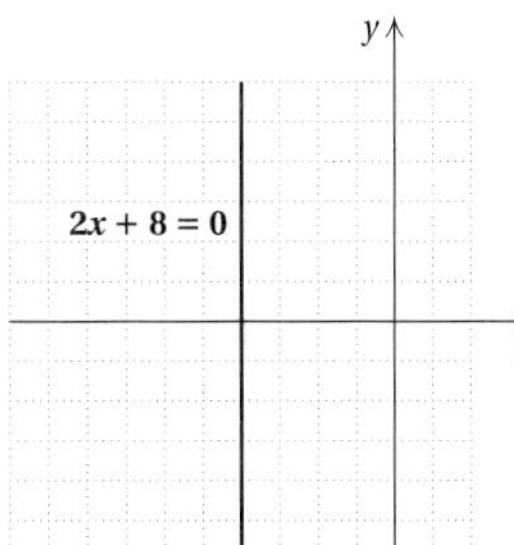

23. [3.3b]

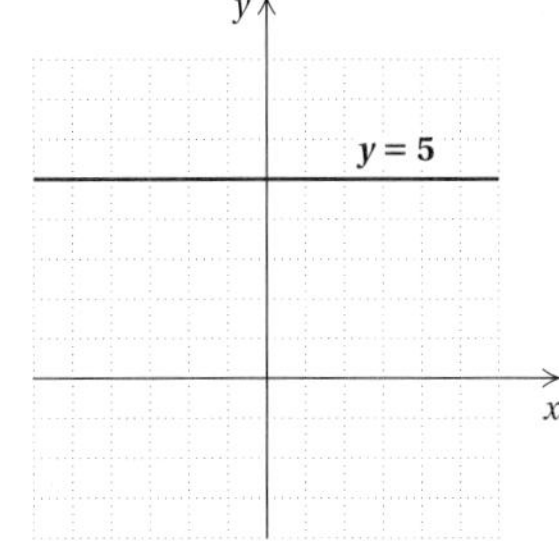

24. [3.3a] **25.** [3.3a]

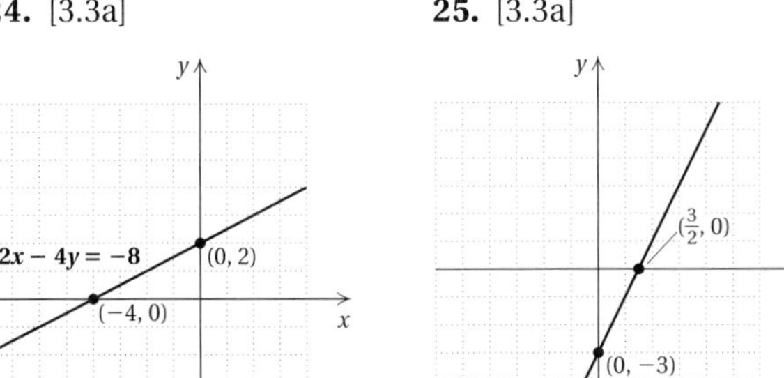

26. [3.2c] **(a)** \$17,000; \$19,400; \$22,600; \$24,200; \$27,500; **(c)** 2002
(b)

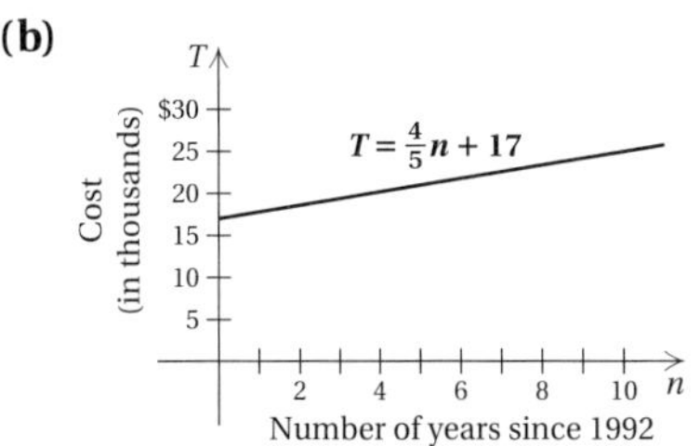

27. [3.4a] Mean: 51; median: 51.5; no mode exists.
28. [3.4a] Mean: 4.2; median: 4.5; mode: 5
29. [3.4a] Mean: 16.3; median: 18.5; modes: 18, 19
30. [3.4a] Mean: 45.4; median: 40; no mode exists.
31. [3.4b] Ball B **32.** [3.4c] 135.15 **33.** [3.4c] 1996: 1785; 2000: 2093 **34.** [1.2c] 0.975 **35.** [1.2c] $-1.08\overline{3}$
36. [1.2e] 71.2 **37.** [1.2e] $\frac{13}{47}$ **38.** [2.2a] $-\frac{3}{5}$
39. [2.2a] 13 **40.** [2.5a] \$84.50 **41.** [2.5a] \$36,400
42. [3.1b] 25 square units, 20 linear units
43. [3.3b] $y = 3$

Cumulative Review: Chapters 1–3, p. 191

1. [1.1a] $\frac{5}{2}$ **2.** [1.7c] $12x - 15y + 21$
3. [1.7d] $3(5x - 3y + 1)$ **4.** [1.7c] $3y, -2x, 4$
5. [1.2c] 0.45 **6.** [1.2e] 4 **7.** [1.3b] 3.08
8. [1.6b] $-\frac{7}{8}$ **9.** [1.7e] $-x - y$ **10.** [1.3b] -24.6
11. [1.4a] $\frac{1}{3}$ **12.** [1.3a] 2.6 **13.** [1.5a] 7.28
14. [1.6c] $-\frac{5}{12}$ **15.** [1.8d] -2 **16.** [1.8d] 27
17. [1.8b] $-2y - 7$ **18.** [1.8c] $5x + 11$
19. [2.1b] -1.2 **20.** [2.2a] -21 **21.** [2.3a] 9
22. [2.2a] $\frac{4}{25}$ **23.** [2.3b] 2 **24.** [2.1b] $\frac{13}{8}$
25. [2.3c] $-\frac{17}{21}$ **26.** [2.3b] -17 **27.** [2.3b] 2
28. [2.7e] $\{x \mid x < 16\}$ **29.** [2.7e] $\left\{x \mid x \leq -\frac{11}{8}\right\}$
30. [2.6a] $h = \dfrac{2A}{b + c}$ **31.** [3.1c] IV

32. [2.7b] $-1 < x \leq 2$

33. [3.3a] **34.** [3.3b]

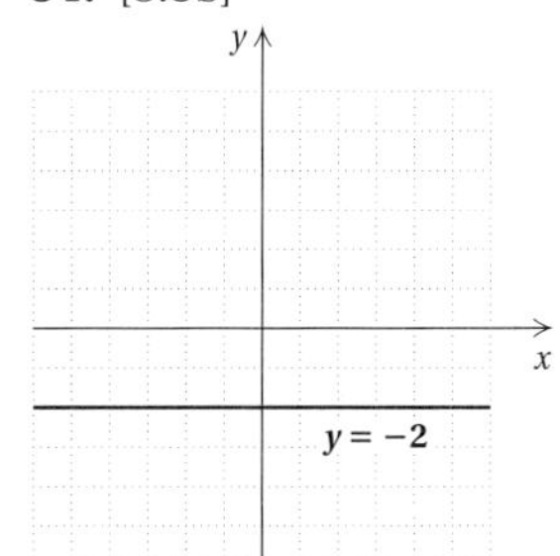

35. [3.2b] **36.** [3.2b]

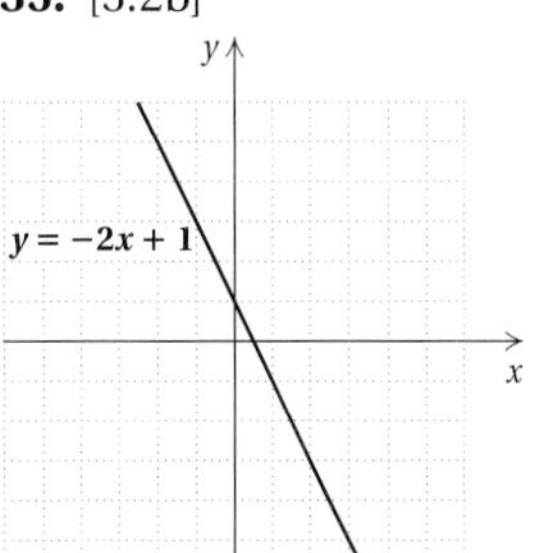

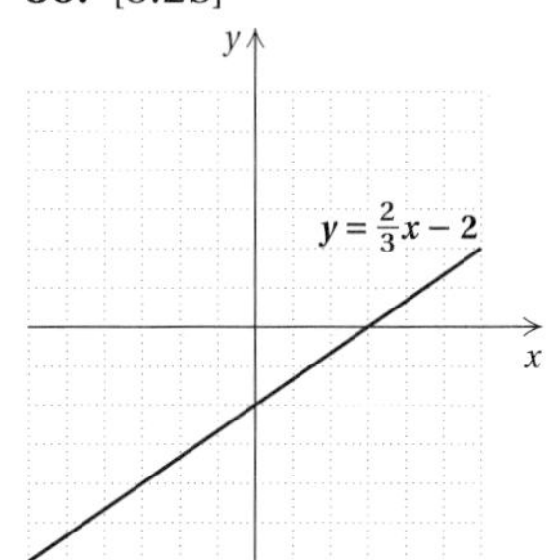

37. [3.3a] y-intercept: $(0, -3)$; x-intercept: $(10.5, 0)$
38. [3.3a] y-intercept: $(0, 5)$; x-intercept: $\left(-\frac{5}{4}, 0\right)$
39. [2.5a] 160 million **40.** [2.4a] 15.6 million
41. [2.5a] \$120 **42.** [2.4a] First: 50 m; second: 53 m; third: 40 m **43.** [2.8b] $\{x \mid x \leq 8\}$
44. [3.2c] **(a)** \$375, \$450, \$525, \$825; \$1050; **(c)** 196 months
(b)

45. [3.4a] Mean: 28.15 cc; median: 27.05 cc; mode: 25.4 cc **46.** [3.4b] Class A **47.** [3.4c] 145.25
48. [2.3a], [1.2e] $-4, 4$ **49.** [2.3c] All real numbers
50. [2.3c] No solution **51.** [2.3b] 3 **52.** [2.3c] All real numbers **53.** [2.6a] $Q = \dfrac{2 - pm}{p}$, or $\dfrac{2}{p} - m$

Chapter 4

Pretest: Chapter 4, p. 194

1. [4.1d, f] x^2 **2.** [4.1e, f] $\dfrac{1}{x^7}$ **3.** [4.2b] $\dfrac{16x^4}{y^6}$
4. [4.1f] $\dfrac{1}{p^3}$ **5.** [4.2c] 3.47×10^{-4} **6.** [4.2c] 3,400,000
7. [4.3g] 3, 2, 1, 0; 3
8. [4.3e] $-3a^3b - 2a^2b^2 + ab^3 + 12b^3 + 9$
9. [4.4a] $11x^2 + 4x - 11$ **10.** [4.4c] $-x^2 - 18x + 27$
11. [4.5b] $15x^4 - 20x^3 + 5x^2$ **12.** [4.6c] $x^2 + 10x + 25$

13. [4.6b] $x^2 - 25$ **14.** [4.6a] $4x^6 + 19x^3 - 30$ **15.** [4.7f] $4x^2 - 12xy + 9y^2$ **16.** [4.8b] $x^2 + x + 3$, R 8; or $x^2 + x + 3 + \frac{8}{x - 2}$

Margin Exercises, Section 4.1, pp. 195–200

1. $5 \cdot 5 \cdot 5 \cdot 5$ **2.** $x \cdot x \cdot x \cdot x \cdot x$ **3.** $3t \cdot 3t$ **4.** $3 \cdot t \cdot t$ **5.** 6 **6.** 1 **7.** 8.4 **8.** 1 **9.** 125 **10.** 3215.36 cm^2 **11.** 119 **12.** 3; −3 **13. (a)** 144; **(b)** 36; **(c)** no **14.** 3^{10} **15.** x^{10} **16.** p^{24} **17.** x^5 **18.** a^9b^8 **19.** 4^3 **20.** y^4 **21.** p^9 **22.** a^4b^2 **23.** $\frac{1}{4^3} = \frac{1}{64}$ **24.** $\frac{1}{5^2} = \frac{1}{25}$ **25.** $\frac{1}{2^4} = \frac{1}{16}$ **26.** $\frac{1}{(-2)^3} = -\frac{1}{8}$ **27.** $\frac{4}{p^3}$ **28.** x^2 **29.** 5^2 **30.** $\frac{1}{x^7}$ **31.** $\frac{1}{7^5}$ **32.** b **33.** t^6

Calculator Spotlight, p. 201

1. Yes **2.** No **3.** No **4.** Yes **5.** Yes **6.** Yes **7.** No **8.** Yes **9.** Yes **10.** Yes **11.** No **12.** Yes **13.** Yes **14.** No

Exercise Set 4.1, p. 203

1. $3 \cdot 3 \cdot 3 \cdot 3$ **3.** (1.1)(1.1)(1.1)(1.1)(1.1) **5.** $\left(\frac{2}{3}\right)\left(\frac{2}{3}\right)\left(\frac{2}{3}\right)\left(\frac{2}{3}\right)$ **7.** $(7p)(7p)$ **9.** $8 \cdot k \cdot k \cdot k$ **11.** 1 **13.** b **15.** 1 **17.** 1 **19.** ab **21.** ab **23.** 27 **25.** 19 **27.** 256 **29.** 93 **31.** 10; 4 **33.** 3629.84 ft^2 **35.** $\frac{1}{3^2} = \frac{1}{9}$ **37.** $\frac{1}{10^3} = \frac{1}{1000}$ **39.** $\frac{1}{7^3} = \frac{1}{343}$ **41.** $\frac{1}{a^3}$ **43.** $8^2 = 64$ **45.** y^4 **47.** z^n **49.** 4^{-3} **51.** x^{-3} **53.** a^{-5} **55.** 2^7 **57.** 8^{14} **59.** x^7 **61.** 9^{38} **63.** $(3y)^{12}$ **65.** $(7y)^{17}$ **67.** 3^3 **69.** $\frac{1}{x}$ **71.** x^{17} **73.** $\frac{1}{x^{13}}$ **75.** $\frac{1}{a^{10}}$ **77.** 1 **79.** 7^3 **81.** 8^6 **83.** y^4 **85.** $\frac{1}{16^6}$ **87.** $\frac{1}{m^6}$ **89.** $\frac{1}{(8x)^4}$ **91.** 1 **93.** x^2 **95.** x^9 **97.** $\frac{1}{z^4}$ **99.** x^3 **101.** 1 **103.** 25, $\frac{1}{25}$, $\frac{1}{25}$, 25, −25, .25 **105.** 64%t, or $0.64t$ **106.** 1 **107.** 64 **108.** 1579.5 **109.** $\frac{4}{3}$ **110.** $8(x - 7)$ **111.** 8 in., 4 in. **112.** 228, 229 **113.** ◆ **115.** No **117.** No **119.** y^{5x} **121.** a^{4t} **123.** 1 **125.** > **127.** < **129.** Let $x = 2$; then $3x^2 = 12$, but $(3x)^2 = 36$.

Margin Exercises, Section 4.2, pp. 207–212

1. 3^{20} **2.** $\frac{1}{x^{12}}$ **3.** y^{15} **4.** $\frac{1}{x^{32}}$ **5.** $\frac{16x^{20}}{y^{12}}$ **6.** $\frac{25x^{10}}{y^{12}z^6}$ **7.** x^{74} **8.** $\frac{27z^{24}}{y^6x^{15}}$ **9.** $\frac{x^{12}}{25}$ **10.** $\frac{8t^{15}}{w^{12}}$ **11.** $\frac{9}{x^8}$ **12.** 5.17×10^{-4} **13.** 5.23×10^8 **14.** 689,300,000,000 **15.** 0.0000567 **16.** 5.6×10^{-15} **17.** 7.462×10^{-13} **18.** 2.0×10^3 **19.** 5.5×10^2 **20.** 1.884672×10^{11} **21.** 9.5×10

Calculator Spotlight, p. 207

1. Yes **2.** No

Calculator Spotlight, p. 208

1. Yes **2.** Yes **3.** No **4.** No

Calculator Spotlight, p. 210

1. 2.6 E 8 **2.** 6.709 E $^{-}$11

Exercise Set 4.2, p. 213

1. 2^6 **3.** $\frac{1}{5^6}$ **5.** x^{12} **7.** $16x^6$ **9.** $\frac{1}{x^{12}y^{15}}$ **11.** $x^{24}y^8$ **13.** $\frac{9x^6}{y^{16}z^6}$ **15.** $\frac{a^8}{b^{12}}$ **17.** $\frac{y^6}{4}$ **19.** $\frac{8}{y^6}$ **21.** $\frac{x^6y^3}{z^3}$ **23.** $\frac{c^2d^6}{a^4b^2}$ **25.** 2.8×10^{10} **27.** 9.07×10^{17} **29.** 3.04×10^{-6} **31.** 1.8×10^{-8} **33.** 10^{11} **35.** 1.135×10^7 **37.** 87,400,000 **39.** 0.00000005704 **41.** 10,000,000 **43.** 0.00001 **45.** 6×10^9 **47.** 3.38×10^4 **49.** 8.1477×10^{-13} **51.** 2.5×10^{13} **53.** 5.0×10^{-4} **55.** 3.0×10^{-21} **57.** $\$1.32288 \times 10^{12}$ **59.** 1×10^{22} **61.** 3.3×10^5 **63.** 4.375×10^2 days **65.** $9(x - 4)$ **66.** $2(2x - y + 8)$ **67.** $3(s + t + 8)$ **68.** $-7(x + 2)$ **69.** $\frac{7}{4}$ **70.** 2 **71.** $-\frac{12}{7}$ **72.** $-\frac{11}{2}$ **73.**

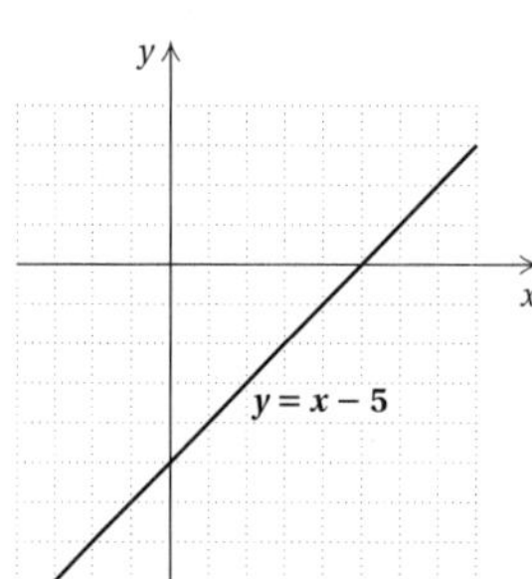

74.

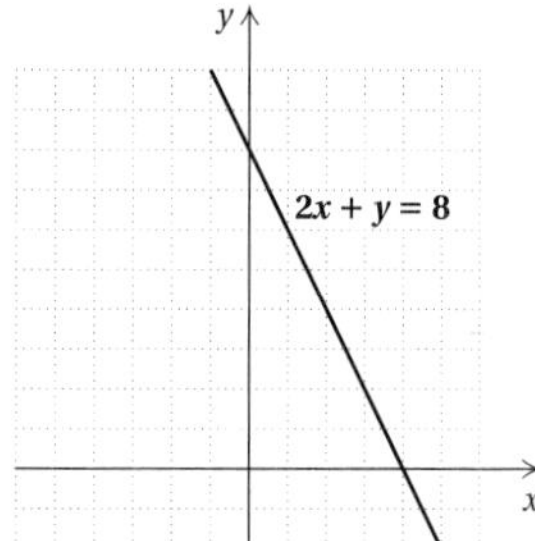

75. ◆ **77.** 2.478125×10^{-1} **79.** $\frac{1}{5}$ **81.** 3^{11} **83.** a^n **85.** False **87.** False

Margin Exercises, Section 4.3, pp. 217–224

1. $4x^2 - 3x + \frac{5}{4}$; $15y^3$; $-7x^3 + 1.1$; answers may vary **2.** −19 **3.** −104 **4.** −13 **5.** 8 **6.** 132 **7.** 360 ft **8.** 6; −4 **9.** 7.55 parts per million **10.** 20 **11.** $-9x^3 + (-4x^5)$ **12.** $-2y^3 + 3y^7 + (-7y)$ **13.** $3x^2$, $6x$, $\frac{1}{2}$ **14.** $-4y^5$, $7y^2$, $-3y$, -2 **15.** $4x^3$ and $-x^3$ **16.** $4t^4$ and $-7t^4$; $-9t^3$ and $10t^3$ **17.** $5x^2$ and $7x^2$; $3x$ and $-8x$; −10 and 11 **18.** 2, −7, −8.5, 10, −4 **19.** $8x^2$ **20.** $2x^3 + 7$ **21.** $-\frac{1}{4}x^5 + 2x^2$ **22.** $-4x^3$ **23.** $5x^3$ **24.** $25 - 3x^5$ **25.** $6x$ **26.** $4x^3 + 4$ **27.** $-\frac{1}{4}x^3 + 4x^2 + 7$ **28.** $3x^2 + x^3 + 9$ **29.** $6x^7 + 3x^5 - 2x^4 + 4x^3 + 5x^2 + x$ **30.** $7x^5 - 5x^4 + 2x^3 + 4x^2 - 3$ **31.** $14t^7 - 10t^5 + 7t^2 - 14$ **32.** $-2x^2 - 3x + 2$ **33.** $10x^4 - 8x - \frac{1}{2}$ **34.** 4, 2, 1, 0; 4 **35.** x **36.** x^3, x^2, x, x^0 **37.** x^2, x **38.** x^3 **39.** Monomial

40. None of these **41.** Binomial **42.** Trinomial

Calculator Spotlight, p. 220

1. 3, 1.81, −6.99 **2.** 2, 17, 21.55, 8.552
3. 13, 2, −8, −3.32, 7

Exercise Set 4.3, p. 225

1. −18, 7 **3.** 19, 14 **5.** −12, −7 **7.** −1, 5 **9.** 9, 1 **11.** 56, −2 **13.** 1112 ft **15.** \$18,750; \$24,000 **17.** −4, 4, 5, 2.75, 1 **19.** 66.6 ft, 86.6 ft, 66.6 ft, 41.6 ft **21.** 2, $-3x$, x^2 **23.** $6x^2$ and $-3x^2$ **25.** $2x^4$ and $-3x^4$; $5x$ and $-7x$ **27.** $3x^5$ and $14x^5$; $-7x$ and $-2x$; 8 and −9 **29.** −3, 6 **31.** 5, 3, 3 **33.** −5, 6, −3, 8, −2 **35.** $-3x$ **37.** $-8x$ **39.** $11x^3 + 4$ **41.** $x^3 - x$ **43.** $4b^5$ **45.** $\frac{3}{4}x^5 - 2x - 42$ **47.** x^4 **49.** $\frac{15}{16}x^3 - \frac{7}{6}x^2$ **51.** $x^5 + 6x^3 + 2x^2 + x + 1$ **53.** $15y^9 + 7y^8 + 5y^3 - y^2 + y$ **55.** $x^6 + x^4$ **57.** $13x^3 - 9x + 8$ **59.** $-5x^2 + 9x$ **61.** $12x^4 - 2x + \frac{1}{4}$ **63.** 1, 0; 1 **65.** 2, 1, 0; 2 **67.** 3, 2, 1, 0; 3 **69.** 2, 1, 6, 4; 6

71.

Term	Coefficient	Degree of Term	Degree of Polynomial
$-7x^4$	−7	4	
$6x^3$	6	3	
$-3x^2$	−3	2	4
$8x$	8	1	
−2	−2	0	

73. x^2, x **75.** x^3, x^2, x^0 **77.** None missing **79.** Trinomial **81.** None of these **83.** Binomial **85.** Monomial **87.** 27 **88.** −19 **89.** $-\frac{17}{24}$ **90.** $\frac{5}{8}$ **91.** −2.6 **92.** $\frac{15}{2}$ **93.** $b = \dfrac{cx + r}{a}$ **94.** 45%, 37.5%, 17.5% **95.** $3(x - 5y + 21)$ **97.** ◆ **99.** $3x^6$ **101.** 10 **103.** −4, 4, 5, 2.75, 1 **105.** 66.6 ft, 86.6 ft, 66.6 ft, 41.6 ft

Margin Exercises, Section 4.4, pp. 231–234

1. $x^2 + 7x + 3$ **2.** $-4x^5 + 7x^4 + x^3 + 2x^2 + 4$ **3.** $24x^4 + 5x^3 + x^2 + 1$ **4.** $2x^3 + \frac{10}{3}$ **5.** $2x^2 - 3x - 1$ **6.** $8x^3 - 2x^2 - 8x + \frac{5}{2}$ **7.** $-8x^4 + 4x^3 + 12x^2 + 5x - 8$ **8.** $-x^3 + x^2 + 3x + 3$ **9.** $-(12x^4 - 3x^2 + 4x)$; $-12x^4 + 3x^2 - 4x$ **10.** $-(-4x^4 + 3x^2 - 4x)$; $4x^4 - 3x^2 + 4x$ **11.** $-\left(-13x^6 + 2x^4 - 3x^2 + x - \frac{5}{13}\right)$; $13x^6 - 2x^4 + 3x^2 - x + \frac{5}{13}$ **12.** $-(-7y^3 + 2y^2 - y + 3)$; $7y^3 - 2y^2 + y - 3$ **13.** $-4x^3 + 6x - 3$ **14.** $-5x^4 - 3x^2 - 7x + 5$ **15.** $-14x^{10} + \frac{1}{2}x^5 - 5x^3 + x^2 - 3x$ **16.** $2x^3 + 2x + 8$ **17.** $x^2 - 6x - 2$ **18.** $-8x^4 - 5x^3 + 8x^2 - 1$ **19.** $x^3 - x^2 - \frac{4}{3}x - 0.9$ **20.** $2x^3 + 5x^2 - 2x - 5$ **21.** $-x^5 - 2x^3 + 3x^2 - 2x + 2$ **22.** Sum of perimeters: $13x$; sum of areas: $\frac{7}{2}x^2$ **23.** $\pi x^2 - 2x^2$, or $(\pi - 2)x^2$

Exercise Set 4.4, p. 235

1. $-x + 5$ **3.** $x^2 - 5x - 1$ **5.** $2x^2$ **7.** $5x^2 + 3x - 30$ **9.** $-2.2x^3 - 0.2x^2 - 3.8x + 23$ **11.** $12x^2 + 6$ **13.** $-\frac{1}{2}x^4 + \frac{2}{3}x^3 + x^2$ **15.** $0.01x^5 + x^4 - 0.2x^3 + 0.2x + 0.06$ **17.** $9x^8 + 8x^7 - 6x^4 + 8x^2 + 4$ **19.** $1.05x^4 + 0.36x^3 + 14.22x^2 + x + 0.97$ **21.** $-(-5x)$; $5x$ **23.** $-(-x^2 + 10x - 2)$; $x^2 - 10x + 2$ **25.** $-(12x^4 - 3x^3 + 3)$; $-12x^4 + 3x^3 - 3$ **27.** $-3x + 7$ **29.** $-4x^2 + 3x - 2$ **31.** $4x^4 - 6x^2 - \frac{3}{4}x + 8$ **33.** $7x - 1$ **35.** $-x^2 - 7x + 5$ **37.** −18 **39.** $6x^4 + 3x^3 - 4x^2 + 3x - 4$ **41.** $4.6x^3 + 9.2x^2 - 3.8x - 23$ **43.** $\frac{3}{4}x^3 - \frac{1}{2}x$ **45.** $0.06x^3 - 0.05x^2 + 0.01x + 1$ **47.** $3x + 6$ **49.** $11x^4 + 12x^3 - 9x^2 - 8x - 9$ **51.** $x^4 - x^3 + x^2 - x$ **53.** $5x^2 + 4x$ **55.** $\frac{23}{2}a + 10$ **57.** 6 **58.** −19 **59.** $-\frac{7}{22}$ **60.** 5 **61.** 5 **62.** 1 **63.** $\frac{39}{2}$ **64.** $\frac{37}{2}$ **65.** $\{x \mid x \geq -10\}$ **66.** $\{x \mid x < 0\}$ **67.** ◆ **69.** $20 + 5(m - 4) + 4(m - 5) + (m - 5)(m - 4)$; m^2 **71.** $z^2 - 27z + 72$ **73.** $y^2 - 4y + 4$ **75.** $5x^2 - 9x - 1$ **77.** $4x^3 + 2x^2 + x + 2$ **79.** Both columns are equal.

Margin Exercises, Section 4.5, pp. 239–242

1. $-15x$ **2.** $-x^2$ **3.** x^2 **4.** $-x^5$ **5.** $12x^7$ **6.** $-8y^{11}$ **7.** $7y^5$ **8.** 0 **9.** $8x^2 + 16x$ **10.** $-15t^3 + 6t^2$ **11.** $5x^6 + 25x^5 - 30x^4 + 40x^3$ **12.** $x^2 + 13x + 40$ **13.** $x^2 + x - 20$ **14.** $5x^2 - 17x - 12$ **15.** $6x^2 - 19x + 15$ **16.** $x^4 + 3x^3 + x^2 + 15x - 20$ **17.** $6y^5 - 20y^3 + 15y^2 + 14y - 35$ **18.** $3x^3 + 13x^2 - 6x + 20$ **19.** $20x^4 - 16x^3 + 32x^2 - 32x - 16$ **20.** $6x^4 - x^3 - 18x^2 - x + 10$

Calculator Spotlight, p. 242

1. Yes **2.** Yes **3.** No **4.** No **5.** No **6.** Yes

Exercise Set 4.5, p. 243

1. $40x^2$ **3.** x^3 **5.** $32x^8$ **7.** $0.03x^{11}$ **9.** $\frac{1}{15}x^4$ **11.** 0 **13.** $-24x^{11}$ **15.** $-2x^2 + 10x$ **17.** $-5x^2 + 5x$ **19.** $x^5 + x^2$ **21.** $6x^3 - 18x^2 + 3x$ **23.** $-6x^4 - 6x^3$ **25.** $18y^6 + 24y^5$ **27.** $x^2 + 9x + 18$ **29.** $x^2 + 3x - 10$ **31.** $x^2 - 7x + 12$ **33.** $x^2 - 9$ **35.** $25 - 15x + 2x^2$ **37.** $4x^2 + 20x + 25$ **39.** $x^2 - \frac{21}{10}x - 1$ **41.** $x^2 + 2.4x - 10.81$ **43.** $x^3 - 1$ **45.** $4x^3 + 14x^2 + 8x + 1$ **47.** $3y^4 - 6y^3 - 7y^2 + 18y - 6$ **49.** $x^6 + 2x^5 - x^3$ **51.** $-10x^5 - 9x^4 + 7x^3 + 2x^2 - x$ **53.** $x^4 - x^2 - 2x - 1$ **55.** $6t^4 + t^3 - 16t^2 - 7t + 4$ **57.** $x^9 - x^5 + 2x^3 - x$ **59.** $x^4 - 1$ **61.** $-\frac{3}{4}$ **62.** 6.4 **63.** 96 **64.** 32 **65.** $3(5x - 6y + 4)$

66. $4(4x - 6y + 9)$ **67.** $-3(3x + 15y - 5)$
68. $100(x - y + 10a)$
69.

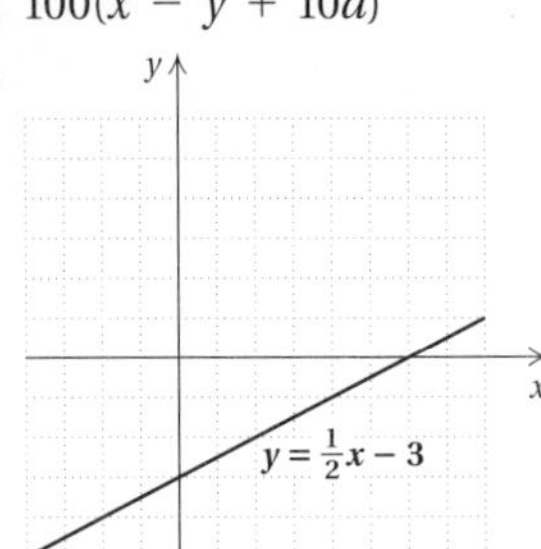

70. $\frac{23}{19}$ **71.** ◈
73. $78t^2 + 40t$
75. $A = \frac{1}{2}b^2 + 2b$
77. 0

Margin Exercises, Section 4.6, pp. 246–250

1. $x^2 + 7x + 12$ **2.** $x^2 - 2x - 15$ **3.** $2x^2 - 9x + 4$
4. $2x^3 - 4x^2 - 3x + 6$ **5.** $12x^5 + 10x^3 + 6x^2 + 5$
6. $y^6 - 49$ **7.** $t^2 + 8t + 15$ **8.** $-2x^7 + x^5 + x^3$
9. $x^2 - \frac{16}{25}$ **10.** $x^5 + 0.5x^3 - 0.5x^2 - 0.25$
11. $8 + 2x^2 - 15x^4$ **12.** $30x^5 - 27x^4 + 6x^3$
13. $x^2 - 25$ **14.** $4x^2 - 9$ **15.** $x^2 - 4$ **16.** $x^2 - 49$
17. $36 - 16y^2$ **18.** $4x^6 - 1$ **19.** $x^2 - \frac{4}{25}$
20. $x^2 + 16x + 64$ **21.** $x^2 - 10x + 25$
22. $x^2 + 4x + 4$ **23.** $a^2 - 8a + 16$
24. $4x^2 + 20x + 25$ **25.** $16x^4 - 24x^3 + 9x^2$
26. $60.84 + 18.72y + 1.44y^2$ **27.** $9x^4 - 30x^2 + 25$
28. $x^2 + 11x + 30$ **29.** $t^2 - 16$
30. $-8x^5 + 20x^4 + 40x^2$ **31.** $81x^4 + 18x^2 + 1$
32. $4a^2 + 6a - 40$ **33.** $25x^2 + 5x + \frac{1}{4}$
34. $4x^2 - 2x + \frac{1}{4}$ **35.** $x^3 - 3x^2 + 6x - 8$

Exercise Set 4.6, p. 251

1. $x^3 + x^2 + 3x + 3$ **3.** $x^4 + x^3 + 2x + 2$
5. $y^2 - y - 6$ **7.** $9x^2 + 12x + 4$ **9.** $5x^2 + 4x - 12$
11. $9t^2 - 1$ **13.** $4x^2 - 6x + 2$ **15.** $p^2 - \frac{1}{16}$
17. $x^2 - 0.01$ **19.** $2x^3 + 2x^2 + 6x + 6$
21. $-2x^2 - 11x + 6$ **23.** $a^2 + 14a + 49$
25. $1 - x - 6x^2$ **27.** $x^5 + 3x^3 - x^2 - 3$
29. $3x^6 - 2x^4 - 6x^2 + 4$ **31.** $13.16x^2 + 18.99x - 13.95$
33. $6x^7 + 18x^5 + 4x^2 + 12$ **35.** $8x^6 + 65x^3 + 8$
37. $4x^3 - 12x^2 + 3x - 9$ **39.** $4y^6 + 4y^5 + y^4 + y^3$
41. $x^2 - 16$ **43.** $4x^2 - 1$ **45.** $25m^2 - 4$
47. $4x^4 - 9$ **49.** $9x^8 - 16$ **51.** $x^{12} - x^4$
53. $x^8 - 9x^2$ **55.** $x^{24} - 9$ **57.** $4y^{16} - 9$
59. $\frac{25}{64}x^2 - 18.49$ **61.** $x^2 + 4x + 4$
63. $9x^4 + 6x^2 + 1$ **65.** $a^2 - a + \frac{1}{4}$ **67.** $9 + 6x + x^2$
69. $x^4 + 2x^2 + 1$ **71.** $4 - 12x^4 + 9x^8$
73. $25 + 60t^2 + 36t^4$ **75.** $x^2 - \frac{5}{4}x + \frac{25}{64}$
77. $9 - 12x^3 + 4x^6$ **79.** $4x^3 + 24x^2 - 12x$
81. $4x^4 - 2x^2 + \frac{1}{4}$ **83.** $9p^2 - 1$ **85.** $15t^5 - 3t^4 + 3t^3$
87. $36x^8 + 48x^4 + 16$ **89.** $12x^3 + 8x^2 + 15x + 10$
91. $64 - 96x^4 + 36x^8$ **93.** $t^3 - 1$ **95.** 25; 49
97. 56; 16 **99.** Lamps: 500 watts; air conditioner: 2000 watts; television: 50 watts **100.** $\frac{28}{27}$ **101.** $-\frac{41}{7}$
102. $\frac{27}{4}$ **103.** $y = \dfrac{3x - 12}{2}$, or $y = \dfrac{3}{2}x - 6$
104. $a = \dfrac{4 + cd}{b}$ **105.** ◈
107. $30x^3 + 35x^2 - 15x$ **109.** $a^4 - 50a^2 + 625$
111. $81t^{16} - 72t^8 + 16$ **113.** -7 **115.** First row: 90, -432, -63; second row: 7, -18, -36, -14, 12, -6, -21, -11; third row: 9, -2, -2, 10, -8, -8, -8, -10, 21; fourth row: -19, -6 **117.** $9x^2 + 24x + 16$ **119.** Yes
121. No

Margin Exercises, Section 4.7, pp. 255–258

1. -7940 **2.** -176 **3.** 1889 **4.** $-3, 3, -2, 1, 2$
5. 3, 7, 1, 1, 0; 7 **6.** $2x^2y + 3xy$ **7.** $5pq - 8$
8. $-4x^3 + 2x^2 - 4y + 2$ **9.** $14x^3y + 7x^2y - 3xy - 2y$
10. $-5p^2q^4 + 2p^2q^2 + 3p^2q + 6pq^2 + 3q + 5$
11. $-8s^4t + 6s^3t^2 + 2s^2t^3 - s^2t^2$
12. $-9p^4q + 9p^3q^2 - 4p^2q^3 - 9q^4 + 5$
13. $x^5y^5 + 2x^4y^2 + 3x^3y^3 + 6x^2$
14. $p^5q - 4p^3q^3 + 3pq^3 + 6q^4$
15. $3x^3y + 6x^2y^3 + 2x^3 + 4x^2y^2$
16. $2x^2 - 11xy + 15y^2$ **17.** $16x^2 + 40xy + 25y^2$
18. $9x^4 - 12x^3y^2 + 4x^2y^4$ **19.** $4x^2y^4 - 9x^2$
20. $16y^2 - 9x^2y^4$ **21.** $9y^2 + 24y + 16 - 9x^2$
22. $4a^2 - 25b^2 - 10bc - c^2$

Exercise Set 4.7, p. 259

1. -1 **3.** -15 **5.** 240 **7.** -145 **9.** 3.715 liters
11. 110.4 m **13.** 56.52 in^2 **15.** Coefficients: 1, -2, 3, -5; degrees: 4, 2, 2, 0; 4 **17.** Coefficients: 17, -3, -7; degrees: 5, 5, 0; 5 **19.** $-a - 2b$
21. $3x^2y - 2xy^2 + x^2$ **23.** $20au + 10av$
25. $8u^2v - 5uv^2$ **27.** $x^2 - 4xy + 3y^2$ **29.** $3r + 7$
31. $-a^3b^2 - 3a^2b^3 + 5ab + 3$ **33.** $ab^2 - a^2b$
35. $2ab - 2$ **37.** $-2a + 10b - 5c + 8d$
39. $6z^2 + 7zu - 3u^2$ **41.** $a^4b^2 - 7a^2b + 10$
43. $a^6 - b^2c^2$ **45.** $y^6x + y^4x + y^4 + 2y^2 + 1$
47. $12x^2y^2 + 2xy - 2$ **49.** $12 - c^2d^2 - c^4d^4$
51. $m^3 + m^2n - mn^2 - n^3$
53. $x^9y^9 - x^6y^6 + x^5y^5 - x^2y^2$ **55.** $x^2 + 2xh + h^2$
57. $r^6t^4 - 8r^3t^2 + 16$ **59.** $p^8 + 2m^2n^2p^4 + m^4n^4$
61. $4a^6 - 2a^3b^3 + \frac{1}{4}b^6$ **63.** $3a^3 - 12a^2b + 12ab^2$
65. $4a^2 - b^2$ **67.** $c^4 - d^2$ **69.** $a^2b^2 - c^2d^4$
71. $x^2 + 2xy + y^2 - 9$ **73.** $x^2 - y^2 - 2yz - z^2$
75. $a^2 - b^2 - 2bc - c^2$ **77.** IV **78.** III **79.** I
80. II
81.

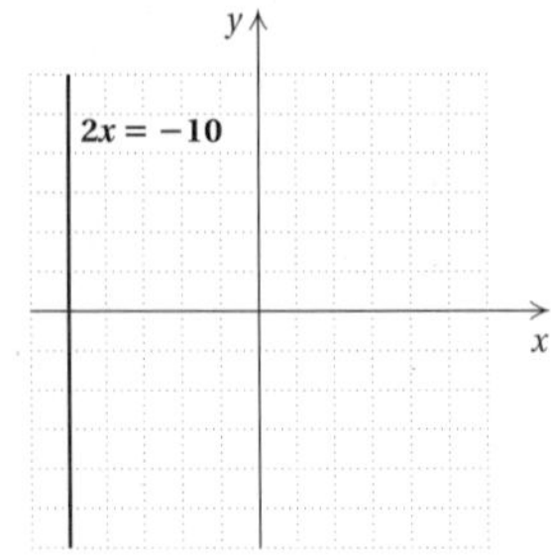

82.

83.

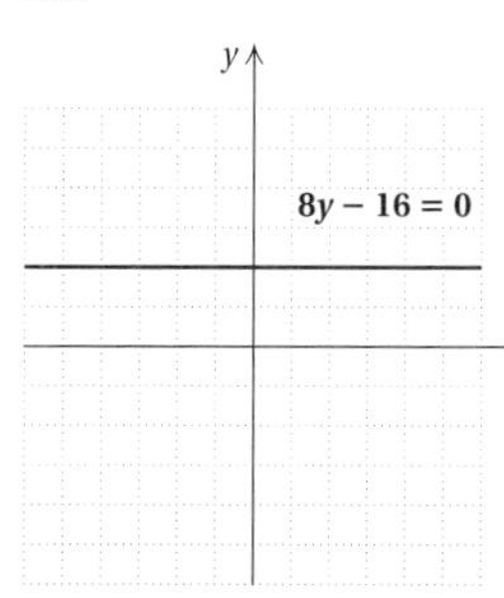

84.

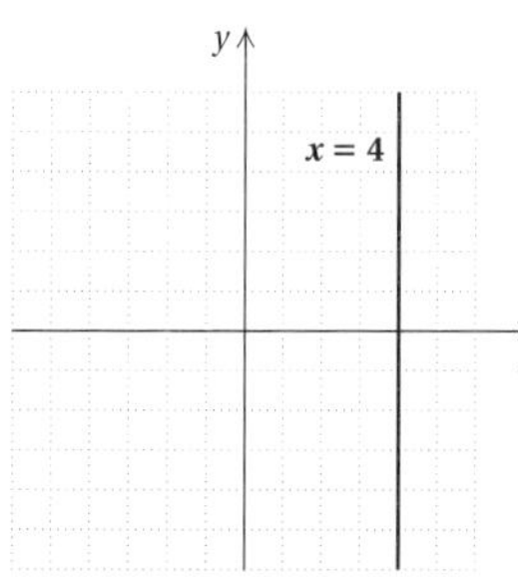

85. Mean: 27.57; median: 28; mode: 31
86. Mean: 5.69; median: 5.6; modes: 5.2, 5.6 **87.** ◆
89. $4xy - 4y^2$ **91.** $2xy + \pi x^2$
93. $2\pi nh + 2\pi mh + 2\pi n^2 - 2\pi m^2$

Margin Exercises, Section 4.8, pp. 263–268

1. $\frac{x^2}{2} + 8x + 3$ **2.** $5y^3 - 2y^2 + 6y$
3. $\frac{x^2y^2}{2} + 5xy + 8$ **4.** $x + 5$
5. $2x^3 - 5x^2 + 19x - 83$, R 341; or
$2x^3 - 5x^2 + 19x - 83 + \frac{341}{x + 4}$
6. $3y^3 - 2y^2 + 6y - 4$ **7.** $y^2 - 8y - 24$, R -66; or
$y^2 - 8y - 24 + \frac{-66}{y - 3}$ **8.** $x^2 + 10x + 10$, R 5; or
$x^2 + 10x + 10 + \frac{5}{x - 1}$ **9.** $y - 11$, R $3y - 27$; or
$y - 11 + \frac{3y - 27}{y^2 - 3}$ **10.** $2x^2 + 2x + 14$, R 34; or
$2x^2 + 2x + 14 + \frac{34}{x - 3}$ **11.** $x^2 - 4x + 13$, R -30; or
$x^2 - 4x + 13 + \frac{-30}{x + 2}$ **12.** $y^2 - y + 1$

Exercise Set 4.8, p. 269

1. $4x^4 + 3x^3 - 6$ **3.** $9y^5 - 4y^2 + 3$
5. $16a^2b^2 + 7ab - 11$ **7.** $x + 7$
9. $a - 12$, R 32; or $a - 12 + \frac{32}{a + 4}$ **11.** $x + 2$, R 4; or
$x + 2 + \frac{4}{x + 5}$ **13.** $2y^2 - y + 2$, R 6; or
$2y^2 - y + 2 + \frac{6}{2y + 4}$ **15.** $2y^2 + 2y - 1$, R 8; or
$2y^2 + 2y - 1 + \frac{8}{5y - 2}$ **17.** $2x^2 - x - 9$, R $(3x + 12)$;
or $2x^2 - x - 9 + \frac{3x + 12}{x^2 + 2}$ **19.** $2x^3 + 5x^2 + 17x + 51$,
R $152x$; or $2x^3 + 5x^2 + 17x + 51 + \frac{152x}{x^2 - 3x}$
21. $x^2 - x + 1$, R -4; or $x^2 - x + 1 + \frac{-4}{x - 1}$
23. $a + 7$, R -47; or $a + 7 + \frac{-47}{a + 4}$ **25.** $x^2 - 5x - 23$,
R -43; or $x^2 - 5x - 23 + \frac{-43}{x - 2}$ **27.** $3x^2 - 2x + 2$,
R -3; or $3x^2 - 2x + 2 + \frac{-3}{x + 3}$ **29.** $y^2 + 2y + 1$, R 12;
or $y^2 + 2y + 1 + \frac{12}{y - 2}$ **31.** $3x^3 + 9x^2 + 2x + 6$
33. $x^2 + 2x + 4$ **35.** $y^3 + 2y^2 + 4y + 8$
37. w^9 **38.** $\frac{1}{w}$ **39.** x^{13} **40.** $\frac{1}{x^8}$ **41.** 3 **42.** 5
43. 14 **44.** $\frac{9}{2}$ **45.** 2.13×10^{-4} **46.** 8.4×10^7
47 3,800,000 **48.** 0.00002527 **49.** ◆
51. $0; -3, -\frac{5}{2}, \frac{3}{2}$ **53.** $0; -\frac{7}{2}, \frac{5}{3}, 4$

Summary and Review: Chapter 4, p. 271

1. $\frac{1}{7^2}$ **2.** y^{11} **3.** $(3x)^{14}$ **4.** t^8 **5.** 4^3 **6.** $\frac{1}{a^3}$ **7.** 1
8. $9t^8$ **9.** $36x^8$ **10.** $\frac{y^3}{8x^3}$ **11.** t^{-5} **12.** $\frac{1}{y^4}$
13. 3.28×10^{-5} **14.** 8,300,000 **15.** 2.09×10^4
16. 5.12×10^{-5} **17.** 4.2075×10^9 **18.** 10
19. $-4y^5, 7y^2, -3y, -2$ **20.** x^2, x^0 **21.** 3, 2, 1, 0; 3
22. Binomial **23.** None of these **24.** Monomial
25. $-2x^2 - 3x + 2$ **26.** $10x^4 - 7x^2 - x - \frac{1}{2}$
27. $x^5 - 2x^4 + 6x^3 + 3x^2 - 9$
28. $-2x^5 - 6x^4 - 2x^3 - 2x^2 + 2$ **29.** $2x^2 - 4x$
30. $x^5 - 3x^3 - x^2 + 8$ **31.** Perimeter: $4w + 6$; area:
$w^2 + 3w$ **32.** $x^2 + \frac{7}{6}x + \frac{1}{3}$ **33.** $49x^2 + 14x + 1$
34. $12x^3 - 23x^2 + 13x - 2$ **35.** $9x^4 - 16$
36. $15x^7 - 40x^6 + 50x^5 + 10x^4$ **37.** $x^2 - 3x - 28$
38. $9y^4 - 12y^3 + 4y^2$ **39.** $2t^4 - 11t^2 - 21$ **40.** 49
41. Coefficients: 1, −7, 9, −8; degrees: 6, 2, 2, 0; 6
42. $-y + 9w - 5$
43. $m^6 - 2m^2n + 2m^2n^2 + 8n^2m - 6m^3$
44. $-9xy - 2y^2$ **45.** $11x^3y^2 - 8x^2y - 6x^2 - 6x + 6$
46. $p^3 - q^3$ **47.** $9a^8 - 2a^4b^3 + \frac{1}{9}b^6$
48. $5x^2 - \frac{1}{2}x + 3$ **49.** $3x^2 - 7x + 4 + \frac{1}{2x + 3}$
50. 0, 3.75, −3.75, 0, 2.25 **51.** $25(t - 2 + 4m)$ **52.** $\frac{9}{4}$
53. −12 **54.** −11.2 **55.** Width: 125.5 m; length: 144.5 m **56.** ◆ 578.6×10^{-7} is not in scientific notation because 578.6 is larger than 10. **57.** ◆ A monomial is an expression of the type ax^n, where n is a whole number and a is a real number. A binomial is a sum of two monomials and has two terms. A trinomial is a sum of three monomials and has three terms. A general polynomial is a monomial or a sum of monomials and has one or more terms.
58. $\frac{1}{2}x^2 - \frac{1}{2}y^2$ **59.** $400 - 4a^2$ **60.** $-28x^8$
61. $\frac{94}{13}$ **62.** $x^4 + x^3 + x^2 + x + 1$

Test: Chapter 4, p. 273

1. [4.1d, f] $\frac{1}{6^5}$ **2.** [4.1d] x^9 **3.** [4.1d] $(4a)^{11}$

4. [4.1e] 3^3 **5.** [4.1e, f] $\frac{1}{x^5}$ **6.** [4.1b, e] 1 **7.** [4.2a] x^6 **8.** [4.2a, b] $-27y^6$ **9.** [4.2a, b] $16a^{12}b^4$ **10.** [4.2b] $\frac{a^3b^3}{c^3}$ **11.** [4.1d], [4.2a, b] $-216x^{21}$ **12.** [4.1d], [4.2a, b] $-24x^{21}$ **13.** [4.1d], [4.2a, b] $162x^{10}$ **14.** [4.1d], [4.2a, b] $324x^{10}$ **15.** [4.1f] $\frac{1}{5^3}$ **16.** [4.1f] y^{-8} **17.** [4.2c] 3.9×10^9 **18.** [4.2c] 0.00000005 **19.** [4.2d] 1.75×10^{17} **20.** [4.2d] 1.296×10^{22} **21.** [4.2e] 1.5×10^4 **22.** [4.3a] -43 **23.** [4.3d] $\frac{1}{3}, -1, 7$ **24.** [4.3g] 3, 0, 1, 6; 6 **25.** [4.3i] Binomial **26.** [4.3e] $5a^2 - 6$ **27.** [4.3e] $\frac{7}{4}y^2 - 4y$ **28.** [4.3f] $x^5 + 2x^3 + 4x^2 - 8x + 3$ **29.** [4.4a] $4x^5 + x^4 + 2x^3 - 8x^2 + 2x - 7$ **30.** [4.4a] $5x^4 + 5x^2 + x + 5$ **31.** [4.4c] $-4x^4 + x^3 - 8x - 3$ **32.** [4.4c] $-x^5 + 0.7x^3 - 0.8x^2 - 21$ **33.** [4.5b] $-12x^4 + 9x^3 + 15x^2$ **34.** [4.6c] $x^2 - \frac{2}{3}x + \frac{1}{9}$ **35.** [4.6b] $9x^2 - 100$ **36.** [4.6a] $3b^2 - 4b - 15$ **37.** [4.6a] $x^{14} - 4x^8 + 4x^6 - 16$ **38.** [4.6a] $48 + 34y - 5y^2$ **39.** [4.5d] $6x^3 - 7x^2 - 11x - 3$ **40.** [4.6c] $25t^2 + 20t + 4$ **41.** [4.7c] $-5x^3y - y^3 + xy^3 - x^2y^2 + 19$ **42.** [4.7e] $8a^2b^2 + 6ab - 4b^3 + 6ab^2 + ab^3$ **43.** [4.7f] $9x^{10} - 16y^{10}$ **44.** [4.8a] $4x^2 + 3x - 5$ **45.** [4.8b] $2x^2 - 4x - 2 + \frac{17}{3x + 2}$ **46.** [4.3a] 3, 1.5, −3.5, −5, −5.25 **47.** [2.3b] 13 **48.** [2.3c] −3 **49.** [1.7d] $16(4t - 2m + 1)$ **50.** [1.4a] $\frac{23}{20}$ **51.** [2.4a] 100°, 25°, 55° **52.** [4.5b], [4.6a] $V = l^3 - 3l^2 + 2l$ **53.** [2.3b], [4.6b, c] $-\frac{61}{12}$

Chapter 5

Pretest: Chapter 5, p. 276

1. [5.1a] $4(-5x^6)$, $(-2x^3)(10x^3)$, $x^2(-20x^4)$; answers may vary **2.** [5.5b] $2(x + 1)^2$ **3.** [5.2a] $(x + 4)(x + 2)$ **4.** [5.1b] $4a(2a^4 + a^2 - 5)$ **5.** [5.3a], [5.4a] $(5x + 2)(x - 3)$ **6.** [5.5d] $(9 + z^2)(3 + z)(3 - z)$ **7.** [5.5b] $(y^3 - 2)^2$ **8.** [5.1c] $(x^2 + 4)(3x + 2)$ **9.** [5.2a] $(p - 6)(p + 5)$ **10.** [5.5d] $(x^2y + 8)(x^2y - 8)$ **11.** [5.3a], [5.4a] $(2p - q)(p + 4q)$ **12.** [5.8b] 0, 5 **13.** [5.8a] 4, $\frac{3}{5}$ **14.** [5.8b] $\frac{2}{3}, -4$ **15.** [5.9a] 6, −1 **16.** [5.9a] Base: 8 cm; height: 11 cm

Margin Exercises, Section 5.1, pp. 277–280

1. **(a)** $12x^2$; **(b)** $(3x)(4x)$, $(2x)(6x)$; answers may vary **2.** **(a)** $16x^3$; **(b)** $(2x)(8x^2)$, $(4x)(4x^2)$; answers may vary **3.** $(8x)(x^3)$, $(4x^2)(2x^2)$, $(2x^3)(4x)$; answers may vary **4.** $(7x)(3x)$, $(-7x)(-3x)$, $(21x)(x)$; answers may vary **5.** $(6x^4)(x)$, $(-2x^3)(-3x^2)$, $(3x^3)(2x^2)$; answers may vary **6.** **(a)** $3x + 6$; **(b)** $3(x + 2)$ **7.** **(a)** $2x^3 + 10x^2 + 8x$; **(b)** $2x(x^2 + 5x + 4)$ **8.** $x(x + 3)$ **9.** $y^2(3y^4 - 5y + 2)$ **10.** $3x^2(3x^2 - 5x + 1)$ **11.** $\frac{1}{4}(3t^3 + 5t^2 + 7t + 1)$ **12.** $7x^3(5x^4 - 7x^3 + 2x^2 - 9)$ **13.** $2.8(3x^2 - 2x + 1)$ **14.** $(x^2 + 3)(x + 7)$ **15.** $(x^2 + 2)(a + b)$ **16.** $(x^2 + 3)(x + 7)$ **17.** $(2t^2 + 3)(4t + 1)$ **18.** $(3m^3 + 2)(m^2 - 5)$ **19.** $(3x^2 - 1)(x - 2)$ **20.** $(2x^2 - 3)(2x - 3)$ **21.** Not factorable using factoring by grouping

Exercise Set 5.1, p. 281

1. $(4x^2)(2x)$, $(-8)(-x^3)$, $(2x^2)(4x)$; answers may vary **3.** $(-5a^5)(2a)$, $(10a^3)(-a^3)$, $(-2a^2)(5a^4)$; answers may vary **5.** $(8x^2)(3x^2)$, $(-8x^2)(-3x^2)$, $(4x^3)(6x)$; answers may vary **7.** $x(x - 6)$ **9.** $2x(x + 3)$ **11.** $x^2(x + 6)$ **13.** $8x^2(x^2 - 3)$ **15.** $2(x^2 + x - 4)$ **17.** $17xy(x^4y^2 + 2x^2y + 3)$ **19.** $x^2(6x^2 - 10x + 3)$ **21.** $x^2y^2(x^3y^3 + x^2y + xy - 1)$ **23.** $2x^3(x^4 - x^3 - 32x^2 + 2)$ **25.** $0.8x(2x^3 - 3x^2 + 4x + 8)$ **27.** $\frac{1}{3}x^3(5x^3 + 4x^2 + x + 1)$ **29.** $(x^2 + 2)(x + 3)$ **31.** $(5a^3 - 1)(2a - 7)$ **33.** $(x^2 + 2)(x + 3)$ **35.** $(2x^2 + 1)(x + 3)$ **37.** $(4x^2 + 3)(2x - 3)$ **39.** $(4x^2 + 1)(3x - 4)$ **41.** $(5x^2 - 1)(x - 1)$ **43.** $(x^2 - 3)(x + 8)$ **45.** $(2x^2 - 9)(x - 4)$ **47.** $\{x \mid x > -24\}$ **48.** $\left\{x \mid x \le \frac{14}{5}\right\}$ **49.** 27 **50.** $p = 2A - q$ **51.** $y^2 + 12y + 35$ **52.** $y^2 + 14y + 49$ **53.** $y^2 - 49$ **54.** $y^2 - 14y + 49$

55.

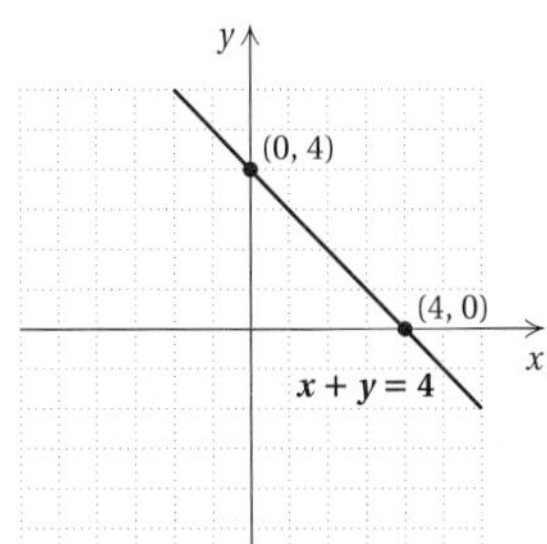

56.

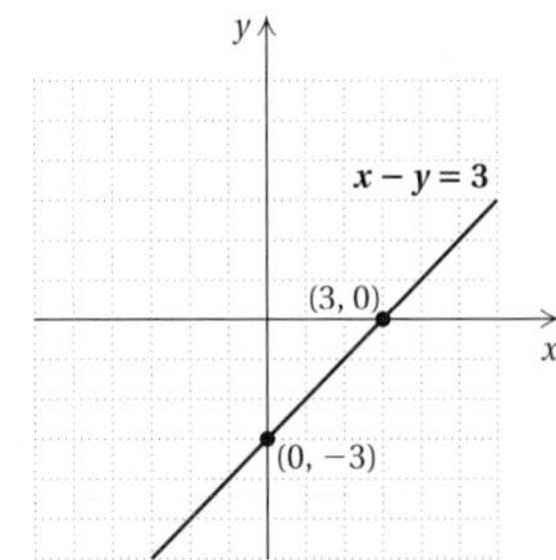

57.

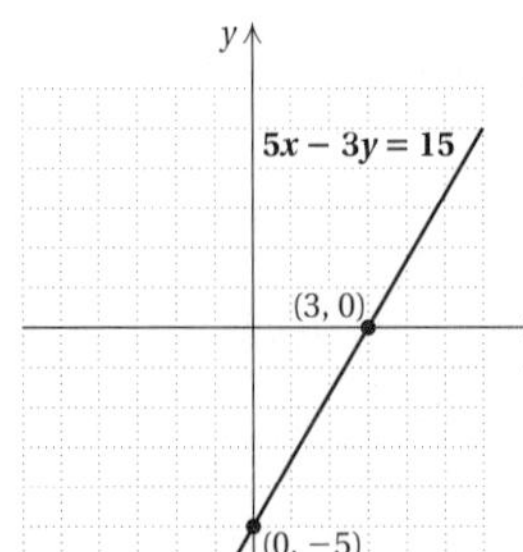

58.

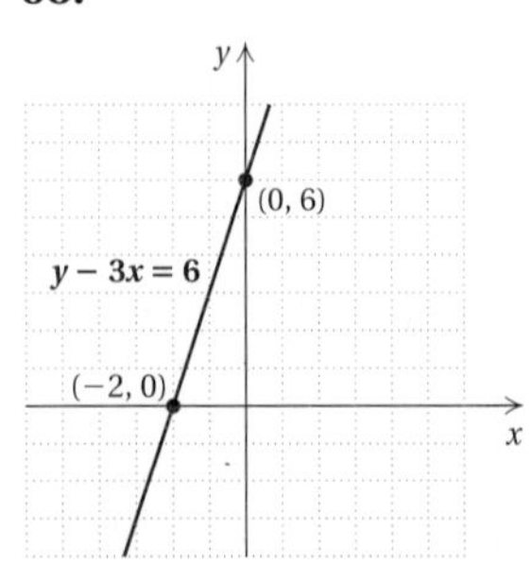

59. ◆ **61.** $(2x^3 + 3)(2x^2 + 3)$ **63.** $(x^7 + 1)(x^5 + 1)$ **65.** Not factorable by grouping

Margin Exercises, Section 5.2, pp. 283–286

1. **(a)** $-13, 8, -8, 7, -7$; **(b)** $13, 8, 7$; both 7 and 12 are positive; **(c)** $(x + 3)(x + 4)$ **2.** $(x + 9)(x + 4)$ **3.** The coefficient of the middle term, -8, is negative. **4.** $(x - 5)(x - 3)$ **5.** $(t - 5)(t - 4)$ **6.** 19, 8, 1; the positive factor has the larger absolute value. **7.** -23, $-10, -5, -2$; the negative factor has the larger absolute value. **8.** $x(x + 6)(x - 2)$ **9.** $(y - 6)(y + 2)$ **10.** $(t^2 + 7)(t^2 - 2)$ **11.** $p(p - q - 3q^2)$ **12.** Not factorable **13.** $(x + 4)^2$

Exercise Set 5.2, p. 287

1. $(x + 3)(x + 5)$ **3.** $(x + 3)(x + 4)$ **5.** $(x - 3)^2$ **7.** $(x + 2)(x + 7)$ **9.** $(b + 1)(b + 4)$ **11.** $\left(x + \frac{1}{3}\right)^2$ **13.** $(d - 2)(d - 5)$ **15.** $(y - 1)(y - 10)$ **17.** $(x - 6)(x + 7)$ **19.** $(x - 9)(x + 2)$ **21.** $x(x - 8)(x + 2)$ **23.** $y(y - 9)(y + 5)$ **25.** $(x - 11)(x + 9)$ **27.** $(c^2 + 8)(c^2 - 7)$ **29.** $(a^2 + 7)(a^2 - 5)$ **31.** Not factorable **33.** Not factorable **35.** $(x + 10)^2$ **37.** $x^2(x - 25)(x + 4)$ **39.** $(x - 24)(x + 3)$ **41.** $(x - 9)(x - 16)$ **43.** $(a + 12)(a - 11)$ **45.** $(x - 15)(x - 8)$ **47.** $(12 + x)(9 - x)$, or $-(x + 12)(x - 9)$ **49.** $(y - 0.4)(y + 0.2)$ **51.** $(p + 5q)(p - 2q)$ **53.** $(m + 4n)(m + n)$ **55.** $(s + 3t)(s - 5t)$ **57.** $16x^3 - 48x^2 + 8x$ **58.** $28w^2 - 53w - 66$ **59.** $49w^2 + 84w + 36$ **60.** $16w^2 - 88w + 121$ **61.** $16w^2 - 121$ **62.** $27x^{12}$ **63.** $\frac{8}{3}$ **64.** $-\frac{7}{2}$ **65.** 29,555 **66.** 100°, 25°, 55° **67.** ◆ **69.** 15, −15, 27, −27, 51, −51 **71.** $\left(x + \frac{1}{4}\right)\left(x - \frac{3}{4}\right)$ **73.** $(x + 5)\left(x - \frac{5}{7}\right)$ **75.** $(b^n + 5)(b^n + 2)$ **77.** $2x^2(4 - \pi)$

Margin Exercises, Section 5.3, pp. 290–293

1. $(2x + 5)(x - 3)$ **2.** $(4x + 1)(3x - 5)$ **3.** $(3x - 4)(x - 5)$ **4.** $2(5x - 4)(2x - 3)$ **5.** $(2x + 1)(3x + 2)$ **6.** $(2a - b)(3a - b)$ **7.** $3(2x + 3y)(x + y)$

Calculator Spotlight, p. 293

1. Correct **2.** Correct **3.** Not correct **4.** Not correct **5.** Not correct **6.** Correct **7.** Not correct **8.** Correct

Exercise Set 5.3, p. 295

1. $(2x + 1)(x - 4)$ **3.** $(5x + 9)(x - 2)$ **5.** $(3x + 1)(2x + 7)$ **7.** $(3x + 1)(x + 1)$ **9.** $(2x - 3)(2x + 5)$ **11.** $(2x + 1)(x - 1)$ **13.** $(3x - 2)(3x + 8)$ **15.** $(3x + 1)(x - 2)$ **17.** $(3x + 4)(4x + 5)$ **19.** $(7x - 1)(2x + 3)$ **21.** $(3x + 2)(3x + 4)$ **23.** $(3x - 7)^2$ **25.** $(24x - 1)(x + 2)$ **27.** $(5x - 11)(7x + 4)$ **29.** $2(5 - x)(2 + x)$ **31.** $4(3x - 2)(x + 3)$ **33.** $6(5x - 9)(x + 1)$ **35.** $2(3y + 5)(y - 1)$ **37.** $(3x - 1)(x - 1)$ **39.** $4(3x + 2)(x - 3)$ **41.** $(2x + 1)(x - 1)$ **43.** $(3x + 2)(3x - 8)$ **45.** $5(3x + 1)(x - 2)$ **47.** $p(3p + 4)(4p + 5)$ **49.** $x^2(7x - 1)(2x + 3)$ **51.** $3x(8x - 1)(7x - 1)$ **53.** $(5x^2 - 3)(3x^2 - 2)$ **55.** $(5t + 8)^2$ **57.** $2x(3x + 5)(x - 1)$ **59.** Not factorable **61.** Not factorable **63.** $(4m + 5n)(3m - 4n)$ **65.** $(2a + 3b)(3a - 5b)$ **67.** $(3a + 2b)(3a + 4b)$ **69.** $(5p + 2q)(7p + 4q)$ **71.** $6(3x - 4y)(x + y)$ **73.** $q = \frac{A + 7}{p}$ **74.** $x = \frac{y - b}{m}$ **75.** $y = \frac{6 - 3x}{2}$ **76.** $q = p + r - 2$ **77.** $\{x \mid x > 4\}$ **78.** $\left\{x \mid x \le \frac{8}{11}\right\}$
79.

$y = \frac{2}{5}x - 1$

80. y^8 **81.** $9x^2 - 25$ **82.** $16a^2 - 24a + 9$ **83.** ◆ **85.** $(2x^n + 1)(10x^n + 3)$ **87.** $(x^{3a} - 1)(3x^{3a} + 1)$

Margin Exercises, Section 5.4, p. 298

1. $(2x + 1)(3x + 2)$ **2.** $(4x + 1)(3x - 5)$ **3.** $3(2x + 3)(x + 1)$ **4.** $2(5x - 4)(2x - 3)$

Exercise Set 5.4, p. 299

1. $(x + 7)(x + 2)$ **3.** $(x - 1)(x - 4)$ **5.** $(2x + 3)(3x + 2)$ **7.** $(x - 4)(3x - 4)$ **9.** $(5x + 3)(7x - 8)$ **11.** $(2x - 3)(2x + 3)$ **13.** $(2x^2 + 5)(x^2 + 3)$ **15.** $(2x + 1)(x - 4)$ **17.** $(3x - 5)(x + 3)$ **19.** $(2x + 7)(3x + 1)$ **21.** $(3x + 1)(x + 1)$ **23.** $(2x - 3)(2x + 5)$ **25.** $(2x - 1)(x + 1)$ **27.** $(3x + 2)(3x - 8)$ **29.** $(3x - 1)(x + 2)$ **31.** $(3x - 4)(4x - 5)$ **33.** $(7x - 1)(2x + 3)$ **35.** $(3x + 2)(3x + 4)$ **37.** $(3x - 7)^2$ **39.** $(24x - 1)(x + 2)$ **41.** $x^3(5x - 11)(7x + 4)$ **43.** $6x(5 - x)(2 + x)$ **45.** $3x^3(5 + x)(1 + 2x)$ **47.** $\{x \mid x < -100\}$ **48.** $\{x \mid x \ge 217\}$ **49.** $\{x \mid x \le 8\}$ **50.** $\{x \mid x < 2\}$ **51.** $\left\{x \mid x \ge \frac{20}{3}\right\}$ **52.** $\{x \mid x > 17\}$ **53.** $\left\{x \mid x > \frac{26}{7}\right\}$ **54.** $\left\{x \mid x \ge \frac{77}{17}\right\}$ **55.** ◆ **57.** $(3x^5 - 2)^2$ **59.** $(4x^5 + 1)^2$ **61.–69.** Left to the student

Margin Exercises, Section 5.5, pp. 302–306

1. Yes **2.** No **3.** No **4.** Yes **5.** No **6.** Yes **7.** No **8.** Yes **9.** $(x + 1)^2$ **10.** $(x - 1)^2$ **11.** $(t + 2)^2$ **12.** $(5x - 7)^2$ **13.** $(7 - 4y)^2$ **14.** $3(4m + 5)^2$ **15.** $(p^2 + 9)^2$ **16.** $z^3(2z - 5)^2$ **17.** $(3a + 5b)^2$ **18.** Yes **19.** No **20.** No **21.** No **22.** Yes **23.** Yes **24.** Yes **25.** $(x + 3)(x - 3)$ **26.** $4(4 + t)(4 - t)$ **27.** $(a + 5b)(a - 5b)$

28. $x^4(8 + 5x)(8 - 5x)$ **29.** $5(1 + 2t^3)(1 - 2t^3)$
30. $(9x^2 + 1)(3x + 1)(3x - 1)$
31. $(7p^2 + 5q^3)(7p^2 - 5q^3)$

Exercise Set 5.5, p. 307

1. Yes **3.** No **5.** No **7.** No **9.** $(x - 7)^2$
11. $(x + 8)^2$ **13.** $(x - 1)^2$ **15.** $(x + 2)^2$
17. $(q^2 - 3)^2$ **19.** $(4y + 7)^2$ **21.** $2(x - 1)^2$
23. $x(x - 9)^2$ **25.** $3(2q - 3)^2$ **27.** $(7 - 3x)^2$
29. $5(y^2 + 1)^2$ **31.** $(1 + 2x^2)^2$ **33.** $(2p + 3q)^2$
35. $(a - 3b)^2$ **37.** $(9a - b)^2$ **39.** $4(3a + 4b)^2$
41. Yes **43.** No **45.** No **47.** Yes
49. $(y + 2)(y - 2)$ **51.** $(p + 3)(p - 3)$
53. $(t + 7)(t - 7)$ **55.** $(a + b)(a - b)$
57. $(5t + m)(5t - m)$ **59.** $(10 + k)(10 - k)$
61. $(4a + 3)(4a - 3)$ **63.** $(2x + 5y)(2x - 5y)$
65. $2(2x + 7)(2x - 7)$ **67.** $x(6 + 7x)(6 - 7x)$
69. $(7a^2 + 9)(7a^2 - 9)$ **71.** $(a^2 + 4)(a + 2)(a - 2)$
73. $5(x^2 + 9)(x + 3)(x - 3)$
75. $(1 + y^4)(1 + y^2)(1 + y)(1 - y)$
77. $(x^6 + 4)(x^3 + 2)(x^3 - 2)$ **79.** $\left(y + \frac{1}{4}\right)\left(y - \frac{1}{4}\right)$
81. $\left(5 + \frac{1}{7}x\right)\left(5 - \frac{1}{7}x\right)$ **83.** $(4m^2 + t^2)(2m + t)(2m - t)$
85. -11 **86.** 400 **87.** $-\frac{5}{6}$ **88.** -0.9 **89.** 2
90. -160 **91.** $x^2 - 4xy + 4y^2$ **92.** $2xy$ **93.** y^{12}
94. $25a^4b^6$
95.

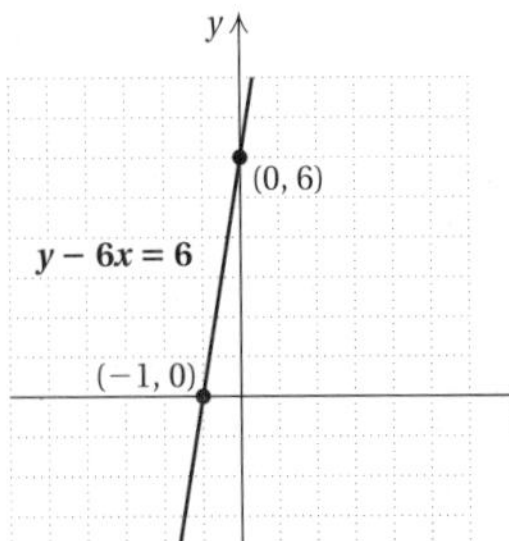

96.

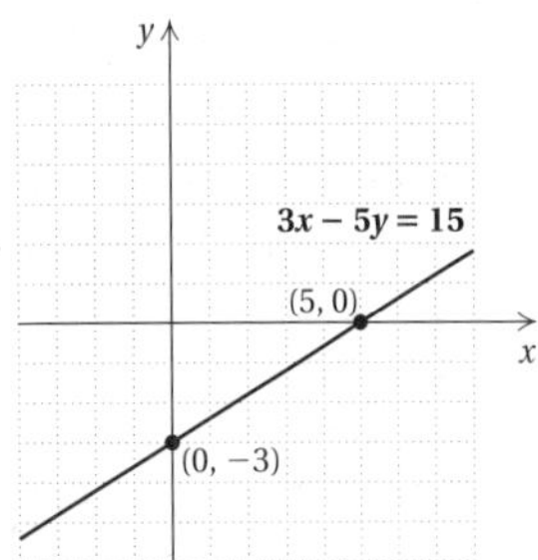

97. ◈ **99.** Not factorable **101.** $(x + 11)^2$
103. $2x(3x + 1)^2$ **105.** $(x^4 + 2^4)(x^2 + 2^2)(x + 2)(x - 2)$
107. $3x^3(x + 2)(x - 2)$ **109.** $2x\left(3x + \frac{2}{5}\right)\left(3x - \frac{2}{5}\right)$
111. $p(0.7 + p)(0.7 - p)$ **113.** $(0.8x + 1.1)(0.8x - 1.1)$
115. $x(x + 6)$ **117.** $\left(x + \dfrac{1}{x}\right)\left(x - \dfrac{1}{x}\right)$
119. $(9 + b^{2k})(3 - b^k)(3 + b^k)$ **121.** $(3b^n + 2)^2$
123. $(y + 4)^2$ **125.** 9 **127.** Not correct
129. Not correct

Margin Exercises, Section 5.6, pp. 311–312

1. $(x - 2)(x^2 + 2x + 4)$ **2.** $(4 - y)(16 + 4y + y^2)$
3. $(3x + y)(9x^2 - 3xy + y^2)$
4. $(2y + z)(4y^2 - 2yz + z^2)$
5. $(m + n)(m^2 - mn + n^2)(m - n)(m^2 + mn + n^2)$
6. $2xy(2x^2 + 3y^2)(4x^4 - 6x^2y^2 + 9y^4)$
7. $(3x + 2y)(9x^2 - 6xy + 4y^2)(3x - 2y) \times (9x^2 + 6xy + 4y^2)$
8. $(x - 0.3)(x^2 + 0.3x + 0.09)$

Exercise Set 5.6, p. 313

1. $(z + 3)(z^2 - 3z + 9)$ **3.** $(x - 1)(x^2 + x + 1)$
5. $(y + 5)(y^2 - 5y + 25)$ **7.** $(2a + 1)(4a^2 - 2a + 1)$
9. $(y - 2)(y^2 + 2y + 4)$ **11.** $(2 - 3b)(4 + 6b + 9b^2)$
13. $(4y + 1)(16y^2 - 4y + 1)$
15. $(2x + 3)(4x^2 - 6x + 9)$ **17.** $(a - b)(a^2 + ab + b^2)$
19. $\left(a + \frac{1}{2}\right)\left(a^2 - \frac{1}{2}a + \frac{1}{4}\right)$ **21.** $2(y - 4)(y^2 + 4y + 16)$
23. $3(2a + 1)(4a^2 - 2a + 1)$
25. $r(s + 4)(s^2 - 4s + 16)$
27. $5(x - 2z)(x^2 + 2xz + 4z^2)$
29. $(x + 0.1)(x^2 - 0.1x + 0.01)$
31. $8(2x^2 - t^2)(4x^4 + 2x^2t^2 + t^4)$
33. $2y(y - 4)(y^2 + 4y + 16)$
35. $(z - 1)(z^2 + z + 1)(z + 1)(z^2 - z + 1)$
37. $(t^2 + 4y^2)(t^4 - 4t^2y^2 + 16y^4)$ **39.** 1; 19; 19; 7; 1
41. $(x^{2a} + y^b)(x^{4a} - x^{2a}y^b + y^{2b})$
43. $3(x^a + 2y^b)(x^{2a} - 2x^ay^b + 4y^{2b})$
45. $\frac{1}{3}\left(\frac{1}{2}xy + z\right)\left(\frac{1}{4}x^2y^2 - \frac{1}{2}xyz + z^2\right)$
47. $y(3x^2 + 3xy + y^2)$ **49.** $4(3a^2 + 4)$

Margin Exercises, Section 5.7, pp. 316–318

1. $3(m^2 + 1)(m + 1)(m - 1)$ **2.** $(x^3 + 4)^2$
3. $2x^2(x + 1)(x + 3)$ **4.** $(3x^2 - 2)(x + 4)$
5. $8x(x - 5)(x + 5)$ **6.** $x^2y(x^2y + 2x + 3)$
7. $2p^4q^2(5p^2 + 2pq + q^2)$ **8.** $(a - b)(2x + 5 + y^2)$
9. $(a + b)(x^2 + y)$ **10.** $(x^2 + y^2)^2$
11. $(xy + 1)(xy + 4)$ **12.** $(p^2 + 9q^2)(p + 3q)(p - 3q)$
13. $15(a - 2b)(a^2 + 2ab + 4b^2)$

Exercise Set 5.7, p. 319

1. $3(x + 8)(x - 8)$ **3.** $(a - 5)^2$ **5.** $(2x - 3)(x - 4)$
7. $x(x + 12)^2$ **9.** $(x + 2)(x - 2)(x + 3)$
11. $3(4x + 1)(4x - 1)$ **13.** $3x(3x - 5)(x + 3)$
15. Not factorable **17.** $x(x^2 + 7)(x - 3)$
19. $x^3(x - 7)^2$ **21.** $2(2 - x)(5 + x)$, or $-2(x - 2)(x + 5)$ **23.** Not factorable
25. $4(x^2 + 4)(x + 2)(x - 2)$
27. $(1 + y^4)(1 + y^2)(1 + y)(1 - y)$ **29.** $x^3(x - 3)(x - 1)$
31. $\frac{1}{81}(x^3 - 12)^2$ **33.** $(3w + 10z)(9w^2 - 30wz + 100z^2)$
35. $9xy(xy - 4)$ **37.** $2\pi r(h + r)$ **39.** $(a + b)(2x + 1)$
41. $(x + 1)(x - 1 - y)$ **43.** $(n + p)(n + 2)$
45. $(3q + p)(2q - 1)$ **47.** $(2b - a)^2$, or $(a - 2b)^2$
49. $(4x + 3y)^2$ **51.** $(7m^2 - 8n)^2$ **53.** $(y^2 + 5z^2)^2$
55. $\left(\frac{1}{2}a + \frac{1}{3}b\right)^2$ **57.** $(a + b)(a - 2b)$
59. $(m + 20n)(m - 18n)$ **61.** $(mn - 8)(mn + 4)$
63. $b^4(ab + 8)(ab - 4)$ **65.** $a^3(a - b)(a + 5b)$
67. $\left(a + \frac{1}{5}b\right)\left(a - \frac{1}{5}b\right)$
69. $7(x + y)(x^2 - xy + y^2)(x - y)(x^2 + xy + y^2)$
71. $(4 + p^2q^2)(2 + pq)(2 - pq)$
73. $(1 + 4x^6y^6)(1 + 2x^3y^3)(1 - 2x^3y^3)$
75. $(q + 1)(q - 1)(q + 8)$ **77.** $(7x + 8y)^2$ **79.** 1990
80. 1995 **81.** 1992 and 1996 **82.** 75,000 **83.** 18,000
84. 59,000 **85.** $-\frac{14}{11}$ **86.** $25x^2 - 10xt + t^2$
87. $X = \dfrac{A + 7}{a + b}$ **88.** $\{x \mid x < 32\}$ **89.** ◈

91. $(a + 1)^2(a - 1)^2$ **93.** $(3.5x - 1)^2$
95. $(5x + 4)(x + 1.8)$ **97.** $(y + 3)(y - 3)(y - 2)$
99. $(a^2 + 1)(a + 4)$ **101.** $(x + 3)(x - 3)(x^2 + 2)$
103. $(x + 2)(x - 2)(x - 1)$ **105.** $(y - 1)^3$
107. $(y + 4 + x)^2$

Margin Exercises, Section 5.8, pp. 323–326

1. 3, −4 **2.** 7, 3 **3.** $-\frac{1}{4}, \frac{2}{3}$ **4.** 0, $\frac{17}{3}$ **5.** −2, 3
6. 7, −4 **7.** 3 **8.** 0, 4 **9.** $\frac{4}{3}, -\frac{4}{3}$ **10.** 3, −3
11. (−5, 0), (1, 0) **12.** 0, 3

Exercise Set 5.8, p. 327

1. −4, −9 **3.** −3, 8 **5.** −12, 11 **7.** 0, −3
9. 0, −18 **11.** $-\frac{5}{2}, -4$ **13.** $-\frac{1}{5}, 3$ **15.** 4, $\frac{1}{4}$ **17.** 0, $\frac{2}{3}$
19. $-\frac{1}{10}, \frac{1}{27}$ **21.** $\frac{1}{3}, -20$ **23.** 0, $\frac{2}{3}, \frac{1}{2}$ **25.** −1, −5
27. −9, 2 **29.** 3, 5 **31.** 0, 8 **33.** 0, −18
35. 4, −4 **37.** $-\frac{2}{3}, \frac{2}{3}$ **39.** −3 **41.** 4 **43.** 0, $\frac{6}{5}$
45. $\frac{5}{3}, -1$ **47.** $\frac{2}{3}, -\frac{1}{4}$ **49.** $\frac{2}{3}, -1$ **51.** $\frac{7}{10}, -\frac{7}{10}$
53. 9, −2 **55.** $\frac{4}{5}, \frac{3}{2}$ **57.** (−4, 0), (1, 0)
59. $\left(-\frac{5}{2}, 0\right)$, (2, 0) **61.** $(a + b)^2$ **62.** $a^2 + b^2$
63. −16 **64.** −4.5 **65.** $-\frac{10}{3}$ **66.** $\frac{3}{10}$ **67.** ◈
69. 4, −5 **71.** 9, −3 **73.** $\frac{1}{8}, -\frac{1}{8}$ **75.** 4, −4
77. **(a)** $x^2 - x - 12 = 0$; **(b)** $x^2 + 7x + 12 = 0$;
(c) $4x^2 - 4x + 1 = 0$; **(d)** $x^2 - 25 = 0$;
(e) $40x^3 - 14x^2 + x = 0$

Margin Exercises, Section 5.9, pp. 329–332

1. 5 and −5 **2.** 7 and 8 **3.** −4 and 5
4. Length: 5 cm; width: 3 cm **5.** **(a)** 342; **(b)** 9
6. 22 and 23 **7.** 3 m, 4 m

Exercise Set 5.9, p. 333

1. 5 and −5 **3.** 3 and 5 **5.** Length: 12 cm; width: 7 cm **7.** 14 and 15 **9.** 12 and 14; −12 and −14
11. 15 and 17; −15 and −17 **13.** 5
15. Height: 6 cm; base: 5 cm **17.** 6 km **19.** 4 sec
21. 5 and 7 **23.** 182 **25.** 12 **27.** 4950 **29.** 25
31. Hypotenuse: 17 ft; leg: 15 ft **33.** $9x^2 - 25y^2$
34. $9x^2 - 30xy + 25y^2$ **35.** $9x^2 + 30xy + 25y^2$
36. $6x^2 + 11xy - 35y^2$ **37.** y-intercept: (0, −4); x-intercept: (16, 0) **38.** y-intercept: (0, 4); x-intercept: (16, 0) **39.** y-intercept: (0, −5); x-intercept: (6.5, 0)
40. y-intercept: $\left(0, \frac{2}{3}\right)$; x-intercept: $\left(\frac{5}{8}, 0\right)$ **41.** ◈
43. 5 ft **45.** 37 **47.** 30 cm by 15 cm

Summary and Review: Chapter 5, p. 337

1. $(-10x)(x)$; $(-5x)(2x)$; $(5x)(-2x)$; answers may vary
2. $(6x)(6x^4)$; $(4x^2)(9x^3)$; $(-2x^4)(-18x)$; answers may vary
3. $5(1 + 2x^3)(1 - 2x^3)$ **4.** $x(x - 3)$
5. $(3x + 2)(3x - 2)$ **6.** $(x + 6)(x - 2)$ **7.** $(x + 7)^2$
8. $3x(2x^2 + 4x + 1)$ **9.** $(x^2 + 3)(x + 1)$
10. $(3x - 1)(2x - 1)$ **11.** $(x^2 + 9)(x + 3)(x - 3)$
12. $3x(3x - 5)(x + 3)$ **13.** $2(x + 5)(x - 5)$
14. $(x^3 - 2)(x + 4)$ **15.** $(4x^2 + 1)(2x + 1)(2x - 1)$
16. $4x^4(2x^2 - 8x + 1)$ **17.** $3(2x + 5)^2$
18. Not factorable **19.** $x(x - 6)(x + 5)$
20. $(2x - 5)(4x^2 + 10x + 25)$ **21.** $(3x - 5)^2$
22. $2(3x + 4)(x - 6)$ **23.** $(x - 3)^2$
24. $(2x + 1)(x - 4)$ **25.** $2(3x - 1)^2$
26. $3(x + 3)(x - 3)$ **27.** $(x - 5)(x - 3)$
28. $(5x - 2)^2$ **29.** $(7b^5 - 2a^4)^2$ **30.** $(xy + 4)(xy - 3)$
31. $3(2a + 7b)^2$ **32.** $(m + t)(m + 5)$
33. $32(x^2 - 2y^2z^2)(x^2 + 2y^2z^2)$ **34.** 1, −3 **35.** −7, 5
36. −4, 3 **37.** $\frac{2}{3}$, 1 **38.** $\frac{3}{2}$, −4 **39.** 8, −2
40. 3 and −2 **41.** −18 and −16; 16 and 18
42. $\frac{5}{2}$ and −2 **43.** −19 and −17; 17 and 19
44. Dining room: 12 ft by 12 ft; kitchen: 12 ft by 10 ft
45. 4 ft **46.** (−5, 0), (−4, 0) **47.** $\left(-\frac{3}{2}, 0\right)$, (5, 0)
48. $\frac{8}{35}$ **49.** $\left\{x \mid x \le \frac{4}{3}\right\}$ **50.** $4a^2 - 9$
51.

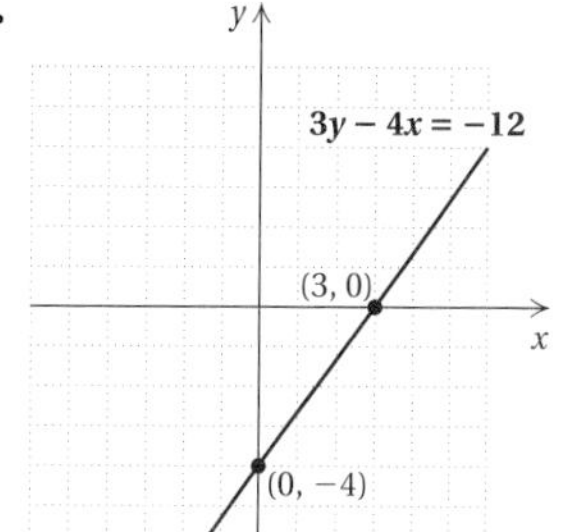

52. ◈ In this chapter, we learned to solve equations of the type $ax^2 + bx + c = 0$ (quadratic equations). Previously, we could solve only first-degree, or linear, equations (equations equivalent to those of the form $ax + b = 0$). The principle of zero products is used to solve quadratic equations, but it is not used to solve linear equations. **53.** ◈ Multiplying can be used to check factoring because factoring is the reverse of multiplying. The TABLE feature of a grapher can provide a partial check of factoring. When a polynomial and its factorization are entered as y_1 and y_2, the factorization is probably correct if corresponding values in the Y1 and Y2 columns are the same. The GRAPH feature of a grapher can also provide a partial check. If the graphs of y_1 and y_2 (entered as described above) coincide, then the factorization is probably correct. **54.** 2.5 cm
55. 0, 2 **56.** Length: 12; width: 6 **57.** No solution
58. 2, −3, $\frac{5}{2}$ **59.** a, i; b, k; c, g; d, h; e, j; f, l
60. 2^{100}; $2^{90} + 2^{90} = 2 \cdot 2^{90} = 2^{91} < 2^{100}$

Test: Chapter 5, p. 339

1. [5.1a] $(4x)(x^2)$; $(2x^2)(2x)$; $(-2x)(-2x^2)$; answers may vary **2.** [5.2a] $(x - 5)(x - 2)$ **3.** [5.5b] $(x - 5)^2$
4. [5.1b] $2y^2(2y^2 - 4y + 3)$ **5.** [5.1c] $(x^2 + 2)(x + 1)$
6. [5.1b] $x(x - 5)$ **7.** [5.2a] $x(x + 3)(x - 1)$
8. [5.3a], [5.4a] $2(5x - 6)(x + 4)$
9. [5.5d] $(2x + 3)(2x - 3)$ **10.** [5.2a] $(x - 4)(x + 3)$
11. [5.3a], [5.4a] $3m(2m + 1)(m + 1)$

12. [5.5d] $3(w + 5)(w - 5)$ **13.** [5.5b] $5(3x + 2)^2$
14. [5.5d] $3(x^2 + 4)(x + 2)(x - 2)$ **15.** [5.5b] $(7x - 6)^2$
16. [5.3a], [5.4a] $(5x - 1)(x - 5)$
17. [5.1c] $(x^3 - 3)(x + 2)$
18. [5.5d] $5(4 + x^2)(2 + x)(2 - x)$
19. [5.3a], [5.4a] $(2x + 3)(2x - 5)$
20. [5.3a], [5.4a] $3t(2t + 5)(t - 1)$
21. [5.2a] $3(m + 2n)(m - 5n)$
22. [5.6a] $(10a - 3b)(100a^2 + 30ab + 9b^2)$
23. [5.8b] 5, −4 **24.** [5.8b] $\frac{3}{2}, -5$ **25.** [5.8b] 7, −4
26. [5.9a] 8, −3 **27.** [5.9a] Length of foot is 7 ft; height of sail is 12 ft **28.** [5.8b] (7, 0), (−5, 0)
29. [5.8b] $\left(\frac{2}{3}, 0\right)$, (1, 0) **30.** [1.6c] $-\frac{10}{11}$
31. [2.7e] $\left\{x \mid x < \frac{19}{3}\right\}$
32. [3.3a]

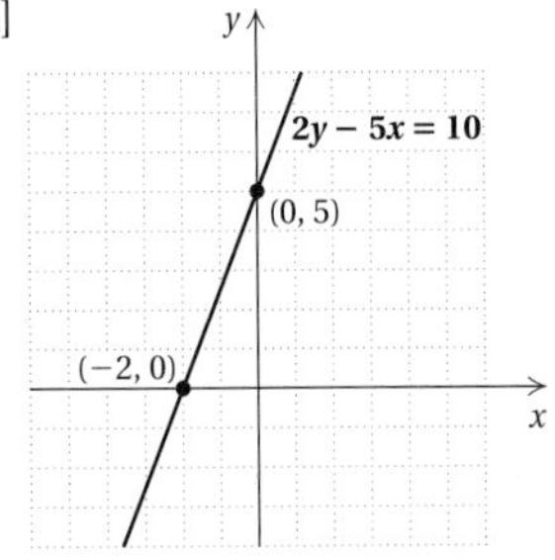

33. [4.6d] $25x^4 - 70x^2 + 49$
34. [5.9a] Length: 15; width: 3
35. [5.2a] $(a - 4)(a + 8)$ **36.** [5.5d], [5.8a] (c)
37. [4.6b], [5.5d] (d)

Chapter 6

Pretest: Chapter 6, p. 342

1. [6.2c] $(x + 2)(x + 3)^2$ **2.** [6.3a] $\frac{-b - 1}{b^2 - 4}$, or $\frac{b + 1}{4 - b^2}$
3. [6.4a] $\frac{1}{y - 2}$ **4.** [6.3a] $\frac{7a + 6}{a(a + 2)}$ **5.** [6.4b] $\frac{2x}{x + 1}$
6. [6.1d] $\frac{2(x - 3)}{x - 2}$ **7.** [6.1f] $\frac{x - 3}{x + 3}$ **8.** [6.8a] $\frac{y + x}{y - x}$
9. [6.5a] −5 **10.** [6.5a] 0 **11.** [6.7a] $M = \frac{3R}{a - b}$
12. [6.6b] 10.5 hr **13.** [6.6a] $\frac{30}{11}$ hr
14. [6.6a] 60 mph, 80 mph **15.** [6.9e] $y = \frac{1}{4} \cdot \frac{xz^2}{w}$
16. [6.9b] 5.625 yd^3

Margin Exercises, Section 6.1, pp. 343–350

1. 3 **2.** −8, 3 **3.** None **4.** $\frac{x(2x + 1)}{x(3x - 2)}$
5. $\frac{(x + 1)(x + 2)}{(x - 2)(x + 2)}$ **6.** $\frac{-1(x - 8)}{-1(x - y)}$ **7.** 5 **8.** $\frac{x}{4}$
9. $\frac{2x + 1}{3x + 2}$ **10.** $\frac{x + 1}{2x + 1}$ **11.** $x + 2$ **12.** $\frac{y + 2}{4}$
13. −1 **14.** −1 **15.** −1 **16.** $\frac{a - 2}{a - 3}$ **17.** $\frac{x - 5}{2}$

18. $\frac{2}{7}$ **19.** $\frac{2x^3 - 1}{x^2 + 5}$ **20.** $\frac{1}{x - 5}$ **21.** $x^2 - 3$
22. $\frac{6}{35}$ **23.** $\frac{x^2}{40}$ **24.** $\frac{(x - 3)(x - 2)}{(x + 5)(x + 5)}$ **25.** $\frac{x - 3}{x + 2}$
26. $\frac{(x - 3)(x - 2)}{x + 2}$ **27.** $\frac{y + 1}{y - 1}$

Calculator Spotlight, p. 348

1. Correct **2.** Correct **3.** Not correct
4. Not correct **5.** Not correct **6.** Correct
7. Not correct

Exercise Set 6.1, p. 351

1. 0 **3.** 8 **5.** $-\frac{5}{2}$ **7.** 7, −4 **9.** 5, −5 **11.** None
13. $\frac{(4x)(3x^2)}{(4x)(5y)}$ **15.** $\frac{2x(x - 1)}{2x(x + 4)}$ **17.** $\frac{-1(3 - x)}{-1(4 - x)}$
19. $\frac{(y + 6)(y - 7)}{(y + 6)(y + 2)}$ **21.** $\frac{x^2}{4}$ **23.** $\frac{8p^2q}{3}$ **25.** $\frac{x - 3}{x}$
27. $\frac{m + 1}{2m + 3}$ **29.** $\frac{a - 3}{a + 2}$ **31.** $\frac{a - 3}{a - 4}$ **33.** $\frac{x + 5}{x - 5}$
35. $a + 1$ **37.** $\frac{x^2 + 1}{x + 1}$ **39.** $\frac{3}{2}$ **41.** $\frac{6}{t - 3}$
43. $\frac{t + 2}{2(t - 4)}$ **45.** $\frac{t - 2}{t + 2}$ **47.** −1 **49.** −1 **51.** −6
53. $-x - 1$ **55.** $\frac{56x}{3}$ **57.** $\frac{2}{dc^2}$ **59.** $\frac{x + 2}{x - 2}$
61. $\frac{(a + 3)(a - 3)}{a(a + 4)}$ **63.** $\frac{2a}{a - 2}$ **65.** $\frac{(t + 2)(t - 2)}{(t + 1)(t - 1)}$
67. $\frac{x + 4}{x + 2}$ **69.** $\frac{5(a + 6)}{a - 1}$ **71.** $\frac{x}{4}$ **73.** $\frac{1}{x^2 - y^2}$
75. $a + b$ **77.** $\frac{x^2 - 4x + 7}{x^2 + 2x - 5}$ **79.** $\frac{3}{10}$ **81.** $\frac{1}{4}$
83. $\frac{b}{a}$ **85.** $\frac{(a + 2)(a + 3)}{(a - 3)(a - 1)}$ **87.** $\frac{(x - 1)^2}{x}$ **89.** $\frac{1}{2}$
91. $\frac{15}{8}$ **93.** $\frac{15}{4}$ **95.** $\frac{a - 5}{3(a - 1)}$ **97.** $\frac{(x + 2)^2}{x}$ **99.** $\frac{3}{2}$
101. $\frac{c + 1}{c - 1}$ **103.** $\frac{y - 3}{2y - 1}$ **105.** $\frac{x + 1}{x - 1}$
107. 18 and 20; −18 and −20 **108.** 4
109. $8x^3 - 11x^2 - 3x + 12$ **110.** $-2p^2 + 4pq - 4q^2$
111. $\frac{4y^8}{x^6}$ **112.** $\frac{125x^{18}}{y^{12}}$ **113.** $\frac{4x^6}{y^{10}}$ **114.** $\frac{1}{a^{15}b^{20}}$
115. ◆ **117.** $x + 2y$ **119.** $\frac{(t - 9)^2(t - 1)}{(t^2 + 9)(t + 1)}$
121. $\frac{x - y}{x - 5y}$ **123.** $-\frac{1}{b^2}$ **125.** $\frac{(x - 7)^2}{x + y}$

Margin Exercises, Section 6.2, pp. 357–358

1. 144 **2.** 12 **3.** 10 **4.** 120 **5.** $\frac{35}{144}$ **6.** $\frac{1}{4}$ **7.** $\frac{11}{10}$
8. $\frac{9}{40}$ **9.** $60x^3y^2$ **10.** $(y + 1)^2(y + 4)$
11. $7(t^2 + 16)(t - 2)$ **12.** $3x(x + 1)^2(x - 1)$

Exercise Set 6.2, p. 359

1. 108 **3.** 72 **5.** 126 **7.** 360 **9.** 500 **11.** $\frac{65}{72}$

13. $\frac{29}{120}$ 15. $\frac{23}{180}$ 17. $12x^3$ 19. $18x^2y^2$ 21. $6(y-3)$
23. $t(t+2)(t-2)$ 25. $(x+2)(x-2)(x+3)$
27. $t(t+2)^2(t-4)$ 29. $(a+1)(a-1)^2$
31. $(m-3)(m-2)^2$ 33. $(2+3x)(2-3x)$
35. $10v(v+4)(v+3)$ 37. $18x^3(x-2)^2(x+1)$
39. $6x^3(x+2)^2(x-2)$ 41. $(x-3)^2$ 42. $2x(3x+2)$
43. $(x+3)(x-3)$ 44. $(x+7)(x-3)$ 45. $(x+3)^2$
46. $(x-7)(x+3)$ 47. 54% 48. 64% 49. 74%
50. 98% 51. 1965 52. 1999 53. ◈

Margin Exercises, Section 6.3, pp. 361–364

1. $\frac{7}{9}$ 2. $\frac{3+x}{x-2}$ 3. $\frac{6x+4}{x-1}$ 4. $\frac{x-5}{4}$ 5. $\frac{x-1}{x-3}$
6. $\frac{10x^2+9x}{48}$ 7. $\frac{9x+10}{48x^2}$ 8. $\frac{4x^2-x+3}{x(x-1)(x+1)^2}$
9. $\frac{2x^2+16x+5}{(x+3)(x+8)}$ 10. $\frac{8x+88}{(x+16)(x+1)(x+8)}$
11. $\frac{-2x-11}{3(x+4)(x-4)}$

Exercise Set 6.3, p. 365

1. 1 3. $\frac{6}{3+x}$ 5. $\frac{2x+3}{x-5}$ 7. $\frac{1}{4}$ 9. $-\frac{1}{t}$
11. $\frac{-x+7}{x-6}$ 13. $y+3$ 15. $\frac{2b-14}{b^2-16}$ 17. $a+b$
19. $\frac{5x+2}{x-5}$ 21. -1 23. $\frac{-x^2+9x-14}{(x-3)(x+3)}$
25. $\frac{2x+5}{x^2}$ 27. $\frac{41}{24r}$ 29. $\frac{4x+6y}{x^2y^2}$ 31. $\frac{4+3t}{18t^3}$
33. $\frac{x^2+4xy+y^2}{x^2y^2}$ 35. $\frac{6x}{(x-2)(x+2)}$ 37. $\frac{11x+2}{3x(x+1)}$
39. $\frac{x^2+6x}{(x+4)(x-4)}$ 41. $\frac{6}{z+4}$ 43. $\frac{3x-1}{(x-1)^2}$
45. $\frac{11a}{10(a-2)}$ 47. $\frac{2x^2+8x+16}{x(x+4)}$
49. $\frac{7a+6}{(a-2)(a+1)(a+3)}$ 51. $\frac{2x^2-4x+34}{(x-5)(x+3)}$
53. $\frac{3a+2}{(a+1)(a-1)}$ 55. $\frac{2x+6y}{(x+y)(x-y)}$
57. $\frac{a^2+7a+1}{(a+5)(a-5)}$ 59. $\frac{5t-12}{(t+3)(t-3)(t-2)}$
61. x^2-1 62. $13y^3-14y^2+12y-73$ 63. $\frac{1}{8x^{12}y^9}$
64. $\frac{x^6}{25y^2}$ 65. $\frac{1}{x^{12}y^{21}}$ 66. $\frac{25}{x^4y^6}$
67. 68.

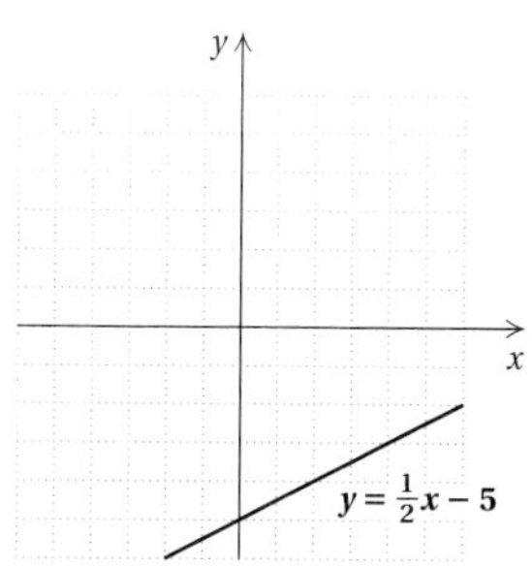

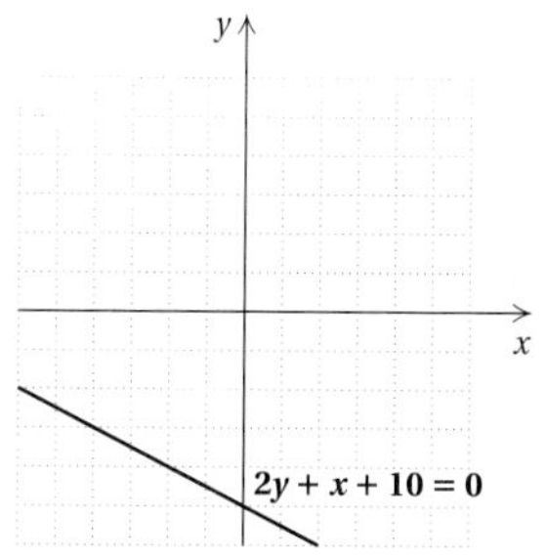

69. 70.

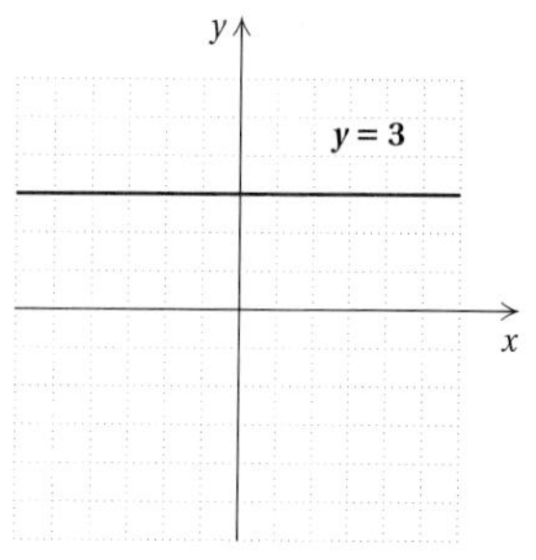

(70.)
x = −5

71. −8 72. $\frac{5}{6}$ 73. 5, 3 74. 9, −2 75. ◈
77. Perimeter: $\frac{16y+28}{15}$; area: $\frac{y^2+2y-8}{15}$
79. $\frac{(z+6)(2z-3)}{(z+2)(z-2)}$ 81. $\frac{11z^4-22z^2+6}{(z^2+2)(z^2-2)(2z^2-3)}$

Margin Exercises, Section 6.4, pp. 369–372

1. $\frac{4}{11}$ 2. $\frac{5}{y}$ 3. $\frac{x^2+2x+1}{2x+1}$ 4. $\frac{3x-1}{3}$ 5. $\frac{4x-3}{x-2}$
6. $\frac{-x-7}{15x}$ 7. $\frac{x^2-48}{(x+7)(x+8)(x+6)}$
8. $\frac{-8y-28}{(y+4)(y-4)}$ 9. $\frac{x-13}{(x+3)(x-3)}$
10. $\frac{6x^2-2x-2}{3x(x+1)}$

Exercise Set 6.4, p. 373

1. $\frac{4}{x}$ 3. 1 5. $\frac{1}{x-1}$ 7. $\frac{8}{3}$ 9. $\frac{13}{a}$ 11. $\frac{8}{y-1}$
13. $\frac{x-2}{x-7}$ 15. $\frac{4}{a^2-25}$ 17. $\frac{2x-4}{x-9}$ 19. $\frac{-9}{2x-3}$
21. $\frac{-a-4}{10}$ 23. $\frac{7z-12}{12z}$ 25. $\frac{4x^2-13xt+9t^2}{3x^2t^2}$
27. $\frac{2x-40}{(x+5)(x-5)}$ 29. $\frac{3-5t}{2t(t-1)}$ 31. $\frac{2s-st-s^2}{(t+s)(t-s)}$
33. $\frac{y-19}{4y}$ 35. $\frac{-2a^2}{(x+a)(x-a)}$ 37. $\frac{9x+12}{(x+3)(x-3)}$
39. $\frac{1}{2}$ 41. $\frac{x-3}{(x+3)(x+1)}$ 43. $\frac{18x+5}{x-1}$ 45. 0
47. $\frac{20}{2y-1}$ 49. $\frac{2a-3}{2-a}$ 51. $\frac{z-3}{2z-1}$ 53. $\frac{2}{x+y}$
55. x^5 56. $30x^{12}$ 57. $\frac{b^{20}}{a^8}$ 58. $18x^3$ 59. $\frac{6}{x^3}$
60. $\frac{10}{x^3}$ 61. $x^2-9x+18$ 62. $(4-\pi)r^2$ 63. ◈
65. $\frac{30}{(x-3)(x+4)}$ 67. $\frac{x^2+xy-x^3+x^2y-xy^2+y^3}{(x^2+y^2)(x+y)^2(x-y)}$
69. $\frac{-2a-15}{a-6}$; area $=\frac{-2a^3-15a^2+12a+90}{2(a-6)^2}$

Margin Exercises, Section 6.5, pp. 377–381

1. $\frac{33}{2}$ 2. 3 3. $\frac{3}{2}$ 4. $-\frac{1}{8}$ 5. 1 6. 2 7. 4

Improving Your Math Study Skills, p. 382

1. Rational expression **2.** Solutions **3.** Rational expression **4.** Rational expression **5.** Rational expression **6.** Solutions **7.** Rational expression **8.** Solutions **9.** Solutions **10.** Solutions **11.** Rational expression **12.** Solutions **13.** Rational expression

Exercise Set 6.5, p. 383

1. $\frac{6}{5}$ **3.** $\frac{40}{29}$ **5.** $\frac{47}{2}$ **7.** -6 **9.** $\frac{24}{7}$ **11.** $-4, -1$ **13.** $4, -4$ **15.** 3 **17.** $\frac{14}{3}$ **19.** 5 **21.** 5 **23.** $\frac{5}{2}$ **25.** -2 **27.** $-\frac{13}{2}$ **29.** $\frac{17}{2}$ **31.** No solution **33.** -5 **35.** $\frac{5}{3}$ **37.** $\frac{1}{2}$ **39.** No solution **41.** No solution **43.** 4 **45.** $\frac{1}{a^6b^{15}}$ **46.** x^8y^{12} **47.** $\frac{16x^4}{t^8}$ **48.** $\frac{w^4}{y^6}$ **49.** $32x^6$ **50.** $\frac{64x^{10}}{y^8}$

51.

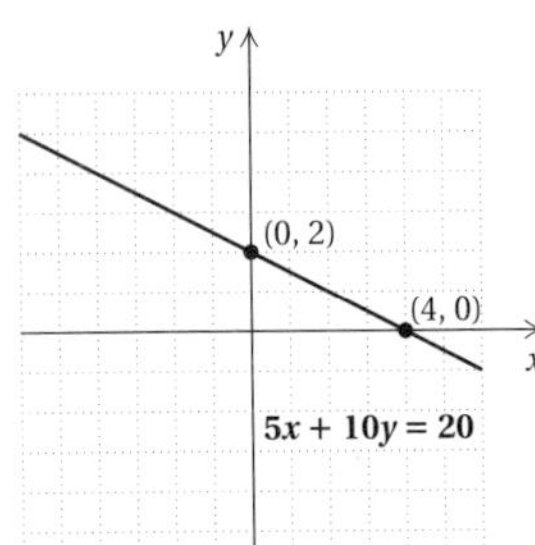

52.

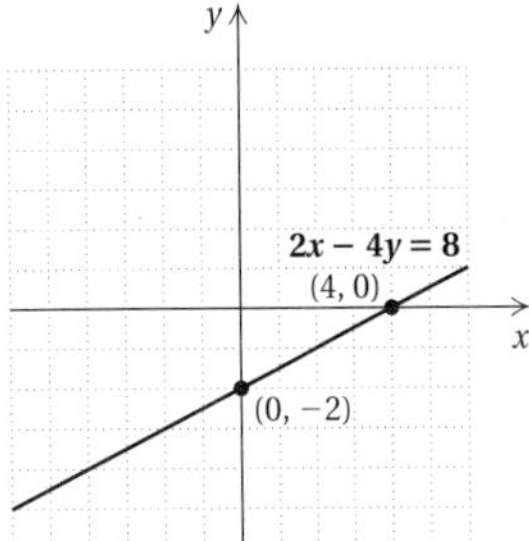

53.

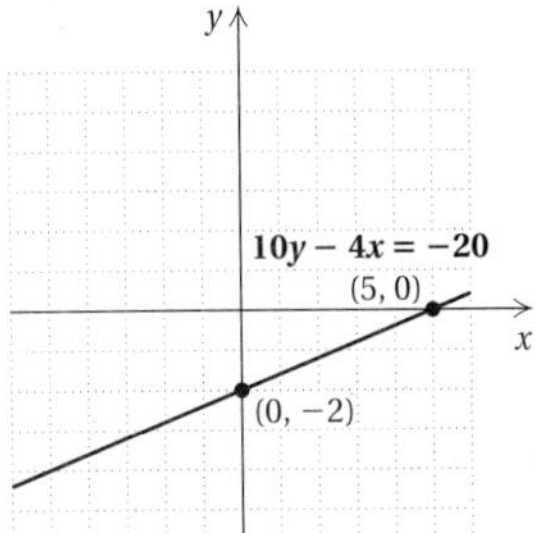

54.

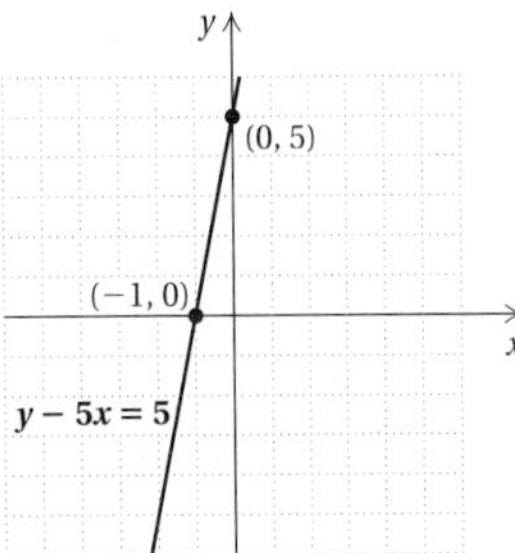

55. ◈ **57.** 7 **59.** No solution **61.** $2, -2$ **63.** 4

Margin Exercises, Section 6.6, pp. 387–394

1. -3 **2.** 40 km/h, 60 km/h **3.** $\frac{24}{7}$, or $3\frac{3}{7}$ hr **4.** 58 km/L **5.** 0.280 **6.** 124 km/h **7.** 2.4 fish/yd^2 **8.** 81 gal **9.** **(a)** About 57; **(b)** Since $57 < 61$, it could not be predicted that McGwire would break Maris's record. **10.** 42 **11.** 2074 **12.** 24.75 ft **13.** 34.9 ft

Calculator Spotlight, p. 390

1. 20.6 min **2.** 0.54 hr

Exercise Set 6.6, p. 395

1. $\frac{24}{7}$ **3.** 20 and 15 **5.** 30 km/h, 70 km/h **7.** Passenger: 80 mph; freight: 66 mph **9.** 20 mph **11.** $4\frac{4}{9}$ min **13.** $5\frac{1}{7}$ hr **15.** 9 **17.** 2.3 km/h **19.** 66 g **21.** 702 km **23.** 200 **25.** **(a)** About 63; **(b)** yes **27.** 10,000 **29.** **(a)** 4.8 tons; **(b)** 48 lb **31.** 11 **33.** $\frac{21}{2}$ **35.** $\frac{8}{3}$ **37.** $\frac{35}{3}$ **39.** x^{11} **40.** x **41.** $\frac{1}{x^{11}}$ **42.** $\frac{1}{x}$ **43.** ◈ **45.** $9\frac{3}{13}$ days **47.** $\frac{3}{4}$ **49.** $27\frac{3}{11}$ min

Margin Exercises, Section 6.7, pp. 399–400

1. $M = \frac{fd^2}{km}$ **2.** $b_1 = \frac{2A - hb_2}{h}$ **3.** $f = \frac{pq}{p + q}$ **4.** $b = \frac{a}{2Q + 1}$

Exercise Set 6.7, p. 401

1. $r = \frac{S}{2\pi h}$ **3.** $b = \frac{2A}{h}$ **5.** $n = \frac{S + 360}{180}$ **7.** $b = \frac{3V - kB - 4kM}{k}$ **9.** $r = \frac{S - a}{S - l}$ **11.** $h = \frac{2A}{b_1 + b_2}$ **13.** $B = \frac{A}{AQ + 1}$ **15.** $p = \frac{qf}{q - f}$ **17.** $A = P(1 + r)$ **19.** $R = \frac{r_1r_2}{r_1 + r_2}$ **21.** $D = \frac{BC}{A}$ **23.** $h_2 = \frac{p(h_1 - q)}{q}$ **25.** $a = \frac{b}{K - C}$ **27.** $-3x^3 - 5x^2 + 5$ **28.** $23x^4 + 50x^3 + 23x^2 - 163x + 41$ **29.** $(x + 2)(x - 2)$ **30.** $3(2y^2 - 1)(5y^2 + 4)$ **31.** $(7m - 8n)^2$ **32.** $(y + 7)(y - 5)$ **33.** $(y^2 + 1)(y + 1)(y - 1)$ **34.** $(a - 10b)(a + 10b)$ **35.** $x^2 + 2x + 8 + \frac{12}{x - 2}$ **36.** $x^2 - 3$ **37.** ◈ **39.** $T = \frac{FP}{u + EF}$ **41.** $v = \frac{Nbf_2 - bf_1 - df_1}{Nf_2 - 1}$

Margin Exercises, Section 6.8, pp. 404–406

1. $\frac{136}{5}$ **2.** $\frac{7x^2}{3(2 - x^2)}$ **3.** $\frac{x}{x - 1}$ **4.** $\frac{136}{5}$ **5.** $\frac{7x^2}{3(2 - x^2)}$ **6.** $\frac{x}{x - 1}$

Exercise Set 6.8, p. 407

1. $\frac{25}{4}$ **3.** $\frac{1}{3}$ **5.** -6 **7.** $\frac{1 + 3x}{1 - 5x}$ **9.** $\frac{2x + 1}{x}$ **11.** 8 **13.** $x - 8$ **15.** $\frac{y}{y - 1}$ **17.** $-\frac{1}{a}$ **19.** $\frac{ab}{b - a}$ **21.** $\frac{p^2 + q^2}{q + p}$ **23.** $4x^4 + 3x^3 + 2x - 7$ **24.** 0 **25.** $(p - 5)^2$ **26.** $(p + 5)^2$ **27.** $50(p^2 - 2)$ **28.** $5(p + 2)(p - 10)$ **29.** 14 yd **30.** 12 ft, 5 ft **31.** ◈ **33.** $\frac{(x - 1)(3x - 2)}{5x - 3}$ **35.** $-\frac{ac}{bd}$ **37.** $\frac{5x + 3}{3x + 2}$

Margin Exercises, Section 6.9, pp. 409–414

1. $\frac{2}{5}$; $y = \frac{2}{5}x$ **2.** 0.7; $y = 0.7x$ **3.** 50 volts **4.** 1,575,000 tons **5.** 0.6; $y = \frac{0.6}{x}$ **6.** $7\frac{1}{2}$hr **7.** $y = 7x^2$ **8.** $y = \frac{9}{x^2}$ **9.** $y = \frac{1}{2}xz$ **10.** $y = \frac{5xz^2}{w}$ **11.** 490 m **12.** 1 ohm

Exercise Set 6.9, p. 415

1. 5; $y = 5x$ **3.** $\frac{2}{15}$; $y = \frac{2}{15}x$ **5.** $\frac{9}{4}$; $y = \frac{9}{4}x$ **7.** 241,920,000 **9.** $66\frac{2}{3}$ cm **11.** 90 g **13.** 40 kg **15.** 98; $y = \frac{98}{x}$ **17.** 36; $y = \frac{36}{x}$ **19.** 0.05; $y = \frac{0.05}{x}$ **21.** 3.5 hr **23.** $\frac{2}{9}$ ampere **25.** 1.92 ft **27.** 160 cm^3 **29.** $y = 15x^2$ **31.** $y = \frac{0.0015}{x^2}$ **33.** $y = xz$ **35.** $y = \frac{3}{10}xz^2$ **37.** $y = \frac{xz}{5wp}$ **39.** 36 mph **41.** 2.5 m **43.** About 106 **45.** 729 gal **47.** $(x - 8)(x + 7)$ **48.** $(a - 8)^2$ **49.** $x^3(x - 7)(x + 5)$ **50.** $(2y^2 + 1)(y - 5)$ **51.** $(2 - t)(2 + t)(4 + t^2)$ **52.** $10(x + 7)(x + 1)$ **53.** $(x - 7)(x - 2)$ **54.** Not factorable **55.** $(4x - 5y)^2$ **56.** $(a - 7b)(a - 2b)$ **57.** $3(x - y)(x^2 + xy + y^3)$ **58.** $(w + t)(w^2 - wt + t^2)(w - t)(w^2 + wt + t^2)$ **59.** ◈ **61.** $\frac{\pi}{4}$ **63.** Q varies directly as the square of p and inversely as the cube of q. **65.** \$7.20

Summary and Review: Chapter 6, p. 419

1. 0 **2.** 6 **3.** 6, −6 **4.** −6, 5 **5.** −2 **6.** 0, 3, 5 **7.** $\frac{x - 2}{x + 1}$ **8.** $\frac{7x + 3}{x - 3}$ **9.** $\frac{y - 5}{y + 5}$ **10.** $\frac{a - 6}{5}$ **11.** $\frac{6}{2t - 1}$ **12.** $-20t$ **13.** $\frac{2x^2 - 2x}{x + 1}$ **14.** $30x^2y^2$ **15.** $4(a - 2)$ **16.** $(y - 2)(y + 2)(y + 1)$ **17.** $\frac{-3x + 18}{x + 7}$ **18.** −1 **19.** $\frac{2a}{a - 1}$ **20.** $d + c$ **21.** $\frac{4}{x - 4}$ **22.** $\frac{x + 5}{2x}$ **23.** $\frac{2x + 3}{x - 2}$ **24.** $\frac{-x^2 + x + 26}{(x - 5)(x + 5)(x + 1)}$ **25.** $\frac{2(x - 2)}{x + 2}$ **26.** $\frac{z}{1 - z}$ **27.** $c - d$ **28.** 8 **29.** 3, −5 **30.** $5\frac{1}{7}$ hr **31.** 240 km/h, 280 km/h **32.** −2 **33.** 95 mph, 175 mph **34.** 160 **35.** 1.92 g **36.** 6 **37.** $s = \frac{rt}{r - t}$ **38.** $C = \frac{5}{9}(F - 32)$, or $C = \frac{5F - 160}{9}$ **39.** $r^3 = \frac{3V}{4\pi}$ **40.** $y = 4x$ **41.** $y = \frac{2500}{x}$ **42.** 20 min **43.** $(5x^2 - 3)(x + 4)$ **44.** $\frac{1}{125x^9y^6}$ **45.** $-2x^3 + 3x^2 + 12x - 18$ **46.** Length: 5 cm; width: 3 cm; perimeter: 16 cm **47.** $\frac{5x + 6}{(x + 2)(x - 2)}$; used to find an equivalent expression for each rational expression with the LCM as the least common denominator **48.** $\frac{3x + 10}{(x - 2)(x + 2)}$; used to find an equivalent expression for each rational expression with the LCM as the least common denominator **49.** 4; used to clear fractions **50.** $\frac{4(x - 2)}{x(x + 4)}$; Method 1: used to multiply by 1 using LCM/LCM; Method 2: used the LCM of the denominators in the numerator to subtract in the numerator and used the LCM of the denominators in the denominator to add in the denominator. **51.** $\frac{5(a + 3)^2}{a}$ **52.** $\frac{10a}{(a - b)(b - c)}$ **53.** They are equivalent equations.

Test: Chapter 6, p. 421

1. [6.1a] 0 **2.** [6.1a] −8 **3.** [6.1a] 7, −7 **4.** [6.1a] 1, 2 **5.** [6.1a] 1 **6.** [6.1a] 0, −3, −5 **7.** [6.1c] $\frac{3x + 7}{x + 3}$ **8.** [6.1d] $\frac{a + 5}{2}$ **9.** [6.1f] $\frac{(5x + 1)(x + 1)}{3x(x + 2)}$ **10.** [6.2c] $(y - 3)(y + 3)(y + 7)$ **11.** [6.3a] $\frac{23 - 3x}{x^3}$ **12.** [6.4a] $\frac{8 - 2t}{t^2 + 1}$ **13.** [6.3a] $\frac{-3}{x - 3}$ **14.** [6.4a] $\frac{2x - 5}{x - 3}$ **15.** [6.3a] $\frac{8t - 3}{t(t - 1)}$ **16.** [6.4a] $\frac{-x^2 - 7x - 15}{(x + 4)(x - 4)(x + 1)}$ **17.** [6.4b] $\frac{x^2 + 2x - 7}{(x - 1)^2(x + 1)}$ **18.** [6.8a] $\frac{3y + 1}{y}$ **19.** [6.5a] 12 **20.** [6.5a] 5, −3 **21.** [6.6a] 4 **22.** [6.6b] 16 **23.** [6.6a] 45 mph, 65 mph **24.** [6.9b] \$495 **25.** [6.9e] $Q = \frac{5}{2}xy$ **26.** [6.9c] $y = \frac{250}{x}$ **27.** [6.7a] $t = \frac{g}{M - L}$ **28.** [6.6b] 15 **29.** [5.7a] $(4a + 7)(4a - 7)$ **30.** [4.2a, b] $\frac{y^{12}}{81x^8}$ **31.** [4.4c] $13x^2 - 29x + 76$ **32.** [5.9a] 21 and 22; −22 and −21 **33.** [6.6a] Team A: 4 hr; team B: 10 hr **34.** [6.8a] $\frac{3a + 2}{2a + 1}$

Cumulative Review: Chapters 1–6, p. 423

1. [1.1a] $-\frac{9}{5}$ **2.** [4.3a] 12 **3.** [1.8c] $2x + 6$ **4.** [4.1d], [4.2a, b] $\frac{27x^7}{4}$ **5.** [4.1e] $\frac{4x^{10}}{3}$ **6.** [6.1c] $\frac{2(t - 3)}{2t - 1}$ **7.** [6.8a] $\frac{(x + 2)^2}{x^2}$ **8.** [6.1c] $\frac{a + 4}{a - 4}$

9. [1.3a] $\frac{17}{42}$ 10. [6.3a] $\frac{x^2 + 4xy + y^2}{x^2y^2}$

11. [6.3a] $\frac{3z - 2}{z^2 - 1}$ 12. [4.4a] $2x^4 + 8x^3 - 2x + 9$

13. [1.4a] 2.33 14. [4.7e] $-xy$ 15. [6.4a] $\frac{1}{x - 3}$

16. [6.4a] $\frac{2x^2 - 14x - 16}{(x + 4)(x - 5)^2}$ 17. [1.5a] -1.3

18. [4.5b] $6x^4 + 12x^3 - 15x^2$ 19. [4.6b] $9t^2 - \frac{1}{4}$

20. [4.6c] $4p^2 - 4pq + q^2$ 21. [4.6a] $3x^2 - 7x - 20$

22. [4.6b] $4x^4 - 1$ 23. [6.1d] $\frac{2(t - 1)}{t}$

24. [6.1d] $-\frac{2(a + 1)}{a}$ 25. [4.8b] $3x^2 - 4x + 5$

26. [1.6c] $-\frac{9}{20}$ 27. [6.1f] $\frac{x - 2}{2x(x - 3)}$ 28. [6.1f] $-\frac{12}{x}$

29. [5.7a] $(2x + 3)(2x - 3)(x + 3)$

30. [5.2a] $(x + 8)(x - 1)$

31. [5.3a], [5.4a] $(3x + 1)(x - 5)$ 32. [5.5b] $(4y + 5x)^2$

33. [5.2a] $3x(x + 5)(x + 3)$ 34. [5.5d] $2(x + 1)(x - 1)$

35. [5.5b] $(x - 14)^2$ 36. [5.1b, c] $2(y^2 + 3)(2y + 5)$

37. [2.3c] -7 38. [5.8a] $0, -\frac{4}{3}$ 39. [5.8b] 0, 8

40. [5.8b] 4 41. [2.7e] $\{x \mid x \geq -9\}$ 42. [5.8b] $3, -3$

43. [2.3b] $-\frac{11}{7}$ 44. [6.5a] $3, -3$ 45. [6.5a] $-\frac{1}{11}$

46. [5.8a] $0, \frac{1}{10}$ 47. [6.5a] No solution 48. [6.5a] $\frac{5}{7}$

49. [6.7a] $z = \frac{xy}{x + y}$ 50. [6.7a] $N = \frac{DT}{3}$

51. [3.3a]

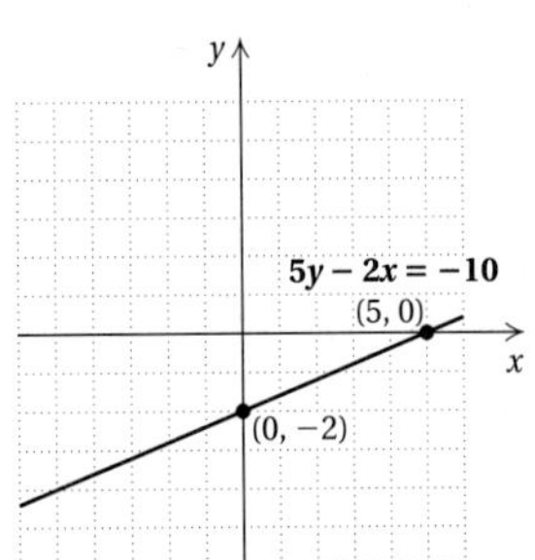

52. [2.4a] 32, 33, 34 53. [6.6a] 12 km/h, 10 km/h

54. [6.6a] $\frac{30}{11}$ hr 55. [2.4a] 34 and 35

56. [5.9a] 16 and 17 57. [6.9b] 60 yr old

58. [5.9a] 7 59. [5.9a] 9 and 11 60. [5.8b] $4, -\frac{2}{3}$

61. [5.8b], [6.5a] $2, -1$ 62. [5.8b], [6.5a], [6.8a] $1, -4$

63. [4.1d, e], [5.8b] $5, -5$

64. [6.1e], [6.3a] $\frac{-x^2 - x + 6}{2x^2 + x + 5}$ 65. [4.2d] 5×10^7

Chapter 7

Pretest: Chapter 7, p. 426

1. [7.4a]

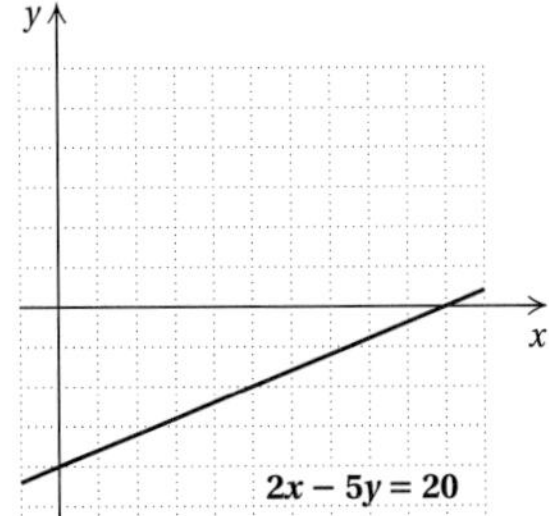

2. [7.4c]

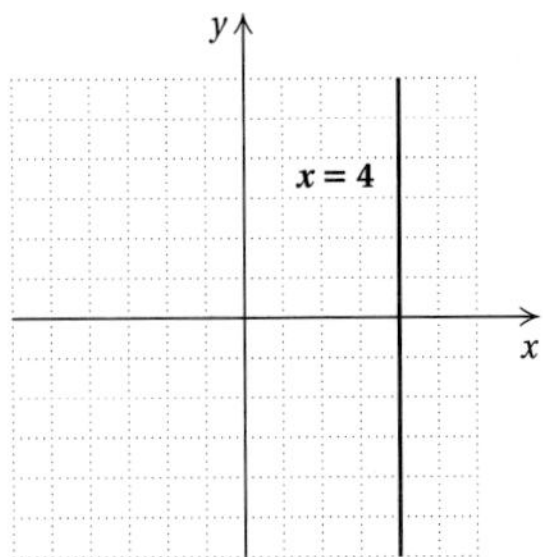

3. [7.4b]

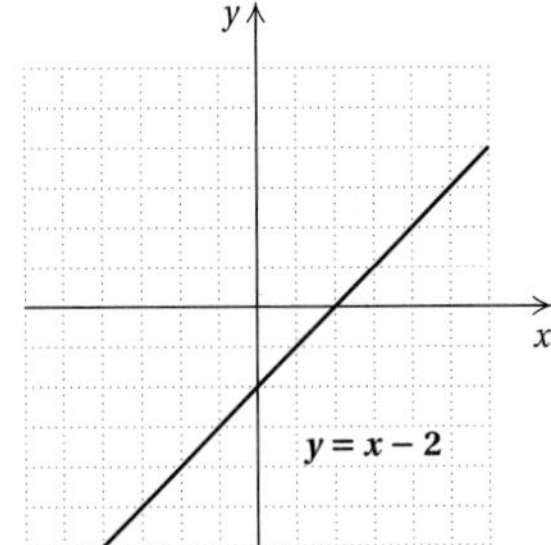

4. [7.4c]

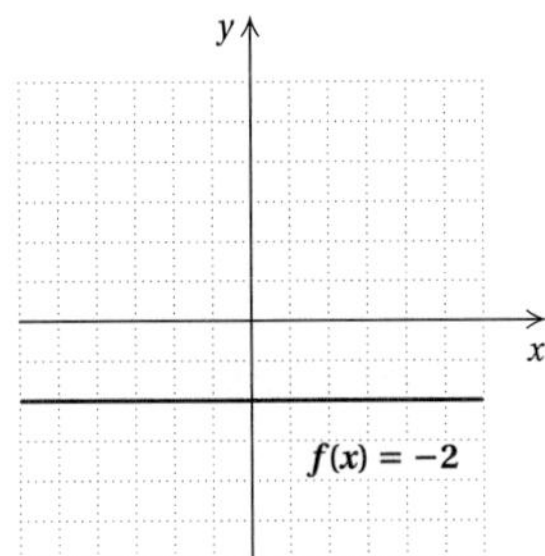

5. [7.1c]

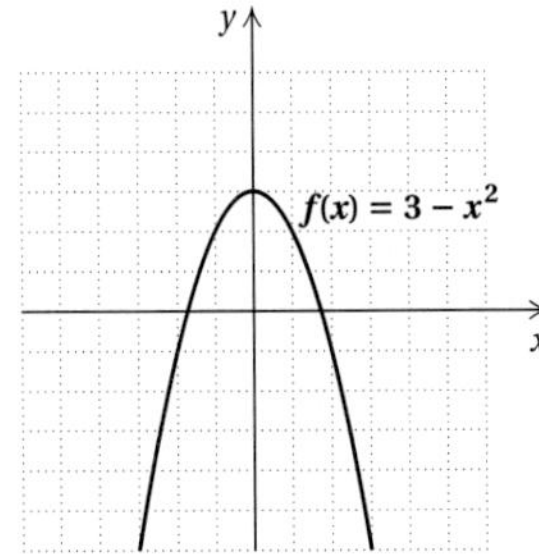

6. [7.3b] 5; $(0, -3)$ 7. [7.3b] 0 8. [7.5c] $y = x + 10$

9. [7.5b] $y = 2x + 7$ 10. [7.5d] $y = \frac{2}{7}x + \frac{31}{7}$

11. [7.5d] $y = -\frac{7}{2}x + 12$ 12. [7.4d] Perpendicular

13. [7.4d] Parallel 14. [7.4a] y-intercept: $(0, -4)$; x-intercept: $(6, 0)$ 15. [7.1b] -3; -1; 1

16. [7.2a] $\{x \mid x \text{ is a real number } and\ x \neq \frac{5}{2}\}$

17. [7.1d] **(a)** Yes; **(b)** no

18. [7.6b] **(a)** $m(t) = 3.383t + 37.73$, where t is the number of years since 1991; **(b)** \$105.39

Margin Exercises, Section 7.1, pp. 427–433

1. Yes 2. No 3. Yes 4. No 5. Yes 6. Yes

7. **(a)** -4; **(b)** 148; **(c)** 31; **(d)** $98a^2 + 21a - 4$

8. **(a)** -33; **(b)** 2; **(c)** 7; **(d)** $5a + 2$

9.

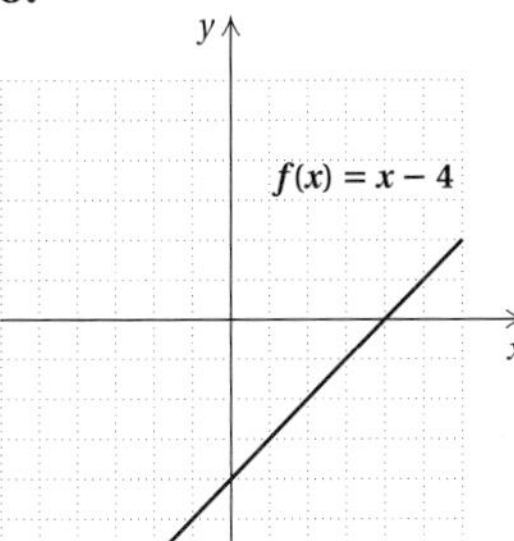

10.

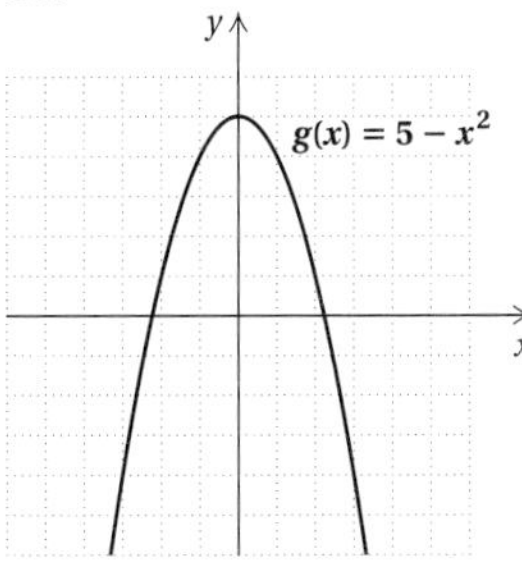

11.

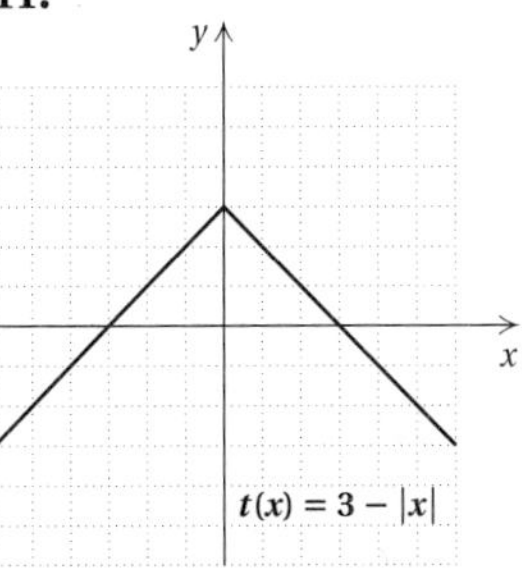

12. Yes **13.** No **14.** No **15.** Yes **16.** \$40 million **17.** \$7 million

Calculator Spotlight, p. 431

1. 6, 3.99, 150, $-1.\overline{5}$ **2.** -21.3, -18.39, -117.3, $3.2\overline{5}$ **3.** -75, -65.466, -420.6, $1.\overline{68}$

Calculator Spotlight, p. 434

1. (0, 1), $(-2.765957, -6.331227)$, $(-1.914894, 3.5529025)$, $(1.2765957, -3.302515)$, $(2.7659574, 8.3312272)$, answers may vary **2.** 951; 42,701 **3.** The table is cleared. Any number can be entered for x; 21,813

Exercise Set 7.1, p. 435

1. Yes **3.** Yes **5.** Yes **7.** No **9.** Function **11.** A relation but not a function **13.** Function **15.** (a) 9; (b) 12; (c) 2; (d) 5; (e) 7.4; (f) $5\frac{2}{3}$ **17.** (a) -21; (b) 15; (c) 42; (d) 0; (e) 2; (f) $3a + 3$ **19.** (a) 7; (b) -17; (c) 24.1; (d) 4; (e) -26; (f) 6 **21.** (a) 0; (b) 5; (c) 2; (d) 170; (e) 65; (f) $32a^2 - 12a$ **23.** (a) 1; (b) 3; (c) 3; (d) 4; (e) 11; (f) $|a - 1| + 1$ **25.** (a) 0; (b) -1; (c) 8; (d) 1000; (e) -125; (f) -1000 **27.** (a) 159.48 cm; (b) 167.73 cm **29.** $1\frac{20}{33}$ atm; $1\frac{10}{11}$ atm; $4\frac{1}{33}$ atm **31.** 1.792 cm; 2.8 cm; 11.2 cm

33.

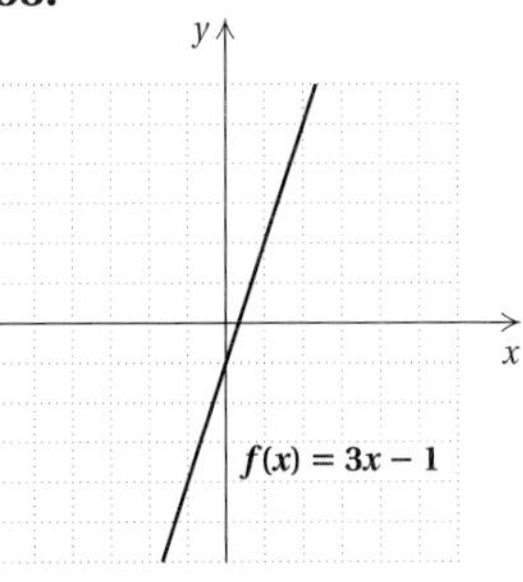

35.

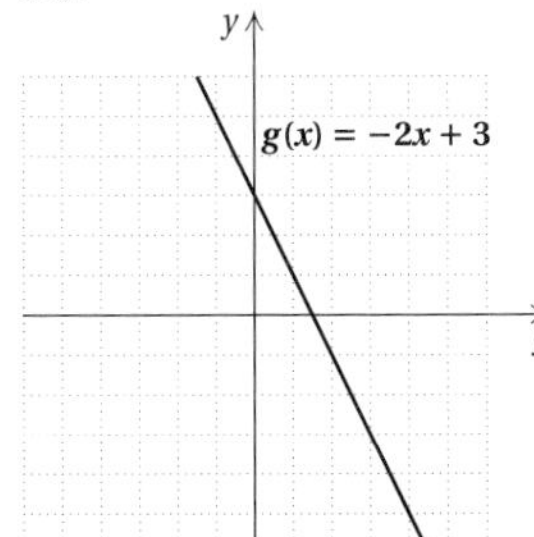

37.

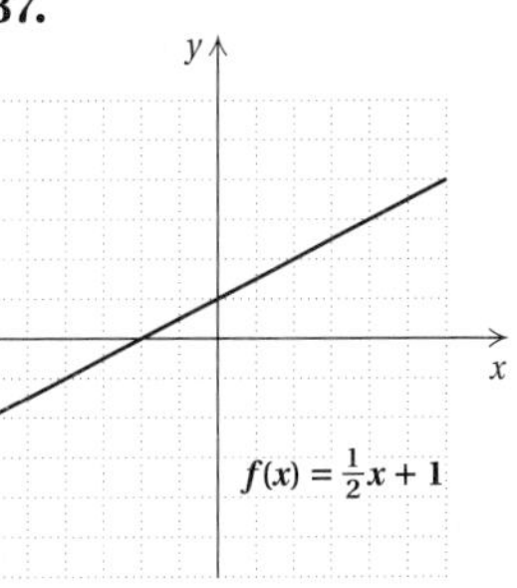

39.

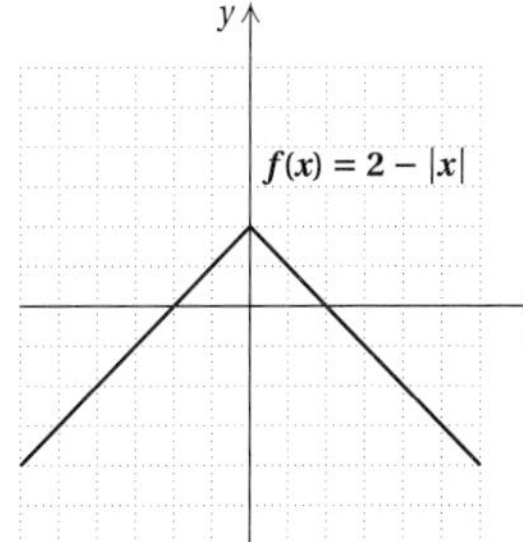

41.

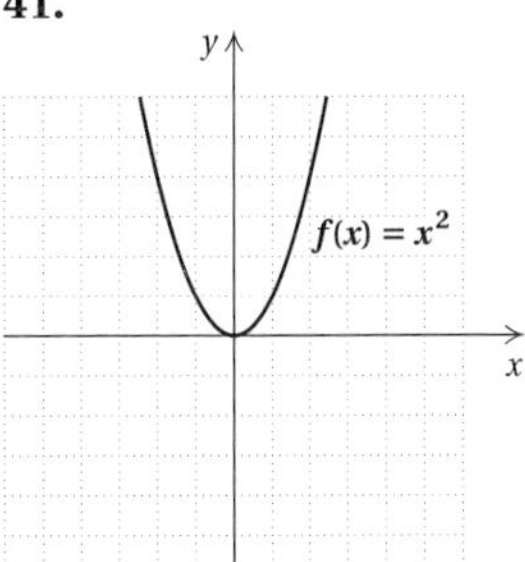

43.

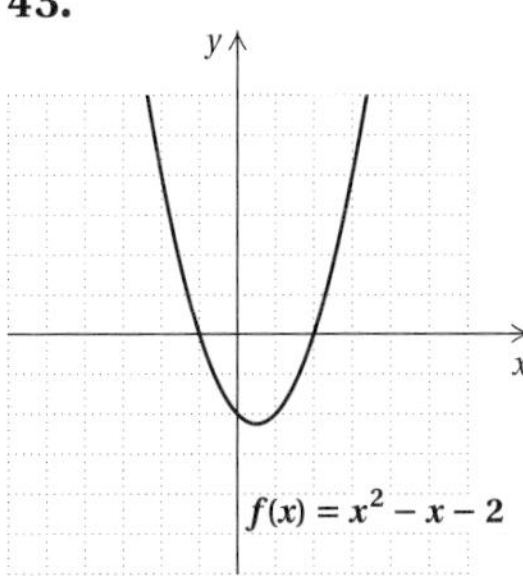

45. Yes **47.** Yes **49.** No **51.** No **53.** 75 **55.** 3.7 billion **57.** 1994 **59.** 18, 20, 22 **60.** 12 ft; 20.8 ft **61.** $l = \dfrac{S - 2wh}{2h + 2w}$ **62.** $w = \dfrac{S - 2lh}{2l + 2h}$ **63.** 0.0000000093 **64.** 3,040,000 **65.** 1.075×10^9 **66.** 7.03×10^{-3} **67.** $\left(w + \frac{1}{3}\right)\left(w^2 - \frac{1}{3}w + \frac{1}{9}\right)$ **68.** $(a - b)(a^2 + ab + b^2)$ **69.** $x(y + 4)(y^2 - 4y + 16)$ **70.** $(c - 0.2)(c^2 + 0.2c + 0.04)$ **71.** ◆ **73.** ◆ **75.** 26; 99 **77.** $g(x) = \frac{15}{4}x - \frac{13}{4}$

Margin Exercises, Section 7.2, pp. 441–443

1. Domain = $\{-3, -2, 0, 2, 5\}$; range = $\{-3, -2, 2, 3\}$ **2.** (a) -4; (b) all real numbers; (c) -3, 3; (d) $\{y \mid y \geq -5\}$ **3.** All real numbers **4.** $\left\{x \mid x \text{ is a real number } and\ x \neq -\frac{2}{3}\right\}$

Calculator Spotlight, p. 444

1. Domain = all real numbers; range = all real numbers ≥ -7 **2.** Domain = all real numbers; range = all real numbers **3.** Domain = all real numbers except 0; range = all real numbers except 0 **4.** Domain = all real numbers; range = all real numbers ≥ -8

Exercise Set 7.2, p. 445

1. (a) 3; (b) $\{-4, -3, -2, -1, 0, 1, 2\}$; (c) -2, 0; (d) $\{1, 2, 3, 4\}$ **3.** (a) 2; (b) $\{-6, -4, -2, 0, 1, 3, 4\}$; (c) 1, 3; (d) $\{-5, -2, 0, 2, 5\}$ **5.** (a) $2\frac{1}{2}$; (b) $\{x \mid -3 \leq x \leq 5\}$; (c) $2\frac{1}{4}$; (d) $\{y \mid 1 \leq y \leq 4\}$ **7.** (a) $2\frac{1}{4}$; (b) $\{x \mid -4 \leq x \leq 3\}$; (c) 0; (d) $\{y \mid -5 \leq y \leq 4\}$ **9.** (a) 2; (b) $\{x \mid -5 \leq x \leq 4\}$; (c) [1, 4]; (d) $\{y \mid -3 \leq y \leq 2\}$ **11.** (a) -1; (b) $\{x \mid -6 \leq x \leq 5\}$; (c) -4, 0, 3; (d) $\{y \mid -2 \leq y \leq 2\}$

13. $\{x \mid x \text{ is a real number } and\ x \neq -3\}$ **15.** All real numbers **17.** All real numbers **19.** $\left\{x \mid x \text{ is a real number } and\ x \neq \frac{14}{5}\right\}$ **21.** All real numbers
23. $\left\{x \mid x \text{ is a real number } and\ x \neq \frac{3}{2}\right\}$
25. $\{x \mid x \text{ is a real number } and\ x \neq 1\}$
27. All real numbers **29.** All real numbers
31. $\left\{x \mid x \text{ is a real number } and\ x \neq \frac{5}{2}\right\}$
33. All real numbers
35. $\left\{x \mid x \text{ is a real number } and\ x \neq -\frac{5}{4}\right\}$ **37.** -8; 0; -2
39. $a - 1$ **40.** $\dfrac{2(y+2)}{7(y+7)}$ **41.** $\dfrac{5}{x+2}$ **42.** $t - 4$
43. $w + 1$, R 2; or $w + 1 + \dfrac{2}{w+3}$
44. $x^4 + x^3 - 2x^2 + x - 2$, R $2x + 3$; or $x^4 + x^3 - 2x^2 + x - 2 + \dfrac{2x+3}{x^2-1}$
45. $14x^2 + 57x - 27$ **46.** $a^2 - 1$
47. $81y^2 + 180y + 100$ **48.** $8w^2 - w - \frac{1}{4}$ **49.** ◆

Margin Exercises, Section 7.3, pp. 447–452

1. The graph of $y = 3x - 6$ looks just like the graph of $y = 3x$, but it is moved down 6 units.

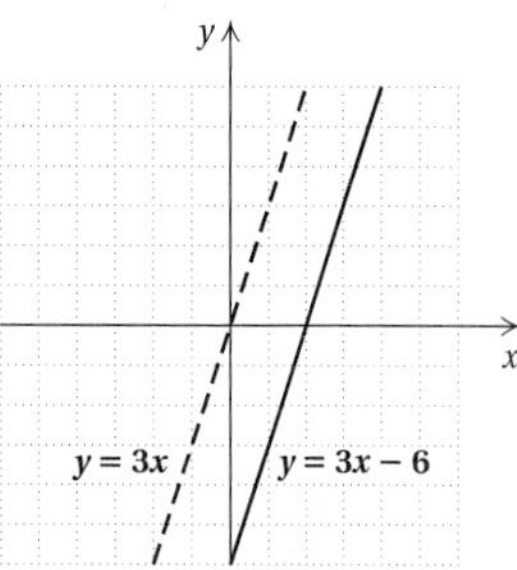

2. The graph of $y = -2x + 3$ looks just like the graph of $y = -2x$, but it is moved up 3 units.

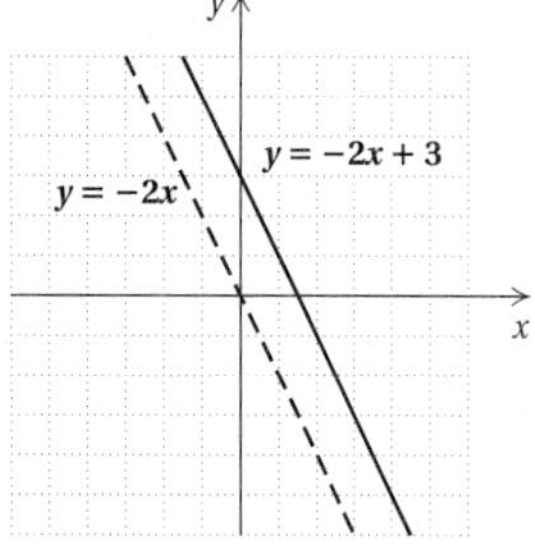

3. The graph of $g(x)$ looks just like the graph of $f(x)$, but it is moved up 2 units.

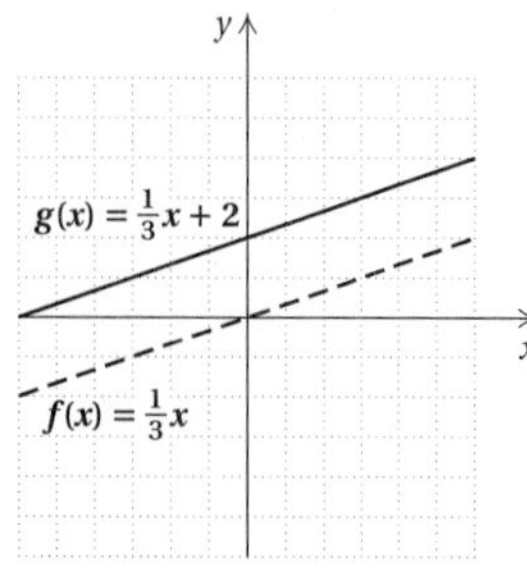

4. (0, 8) **5.** $\left(0, -\frac{2}{3}\right)$
6. $m = -1$

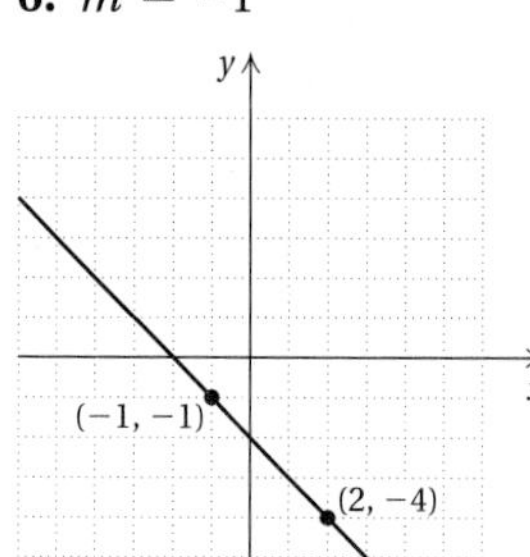

7. $m = -\frac{1}{3}$

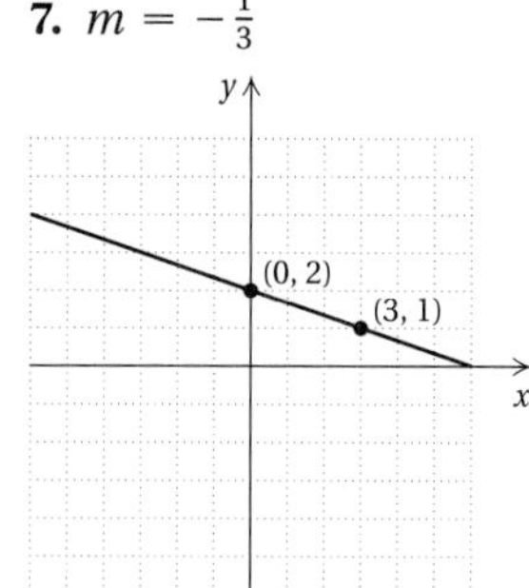

8. $m = -\frac{2}{3}$ **9.** Slope: -8; y-intercept: (0, 23)
10. Slope: $\frac{1}{2}$; y-intercept: $\left(0, -\frac{5}{2}\right)$ **11.** The rate of change is $-10 \dfrac{\$\text{ billion}}{\text{yr}}$.

Calculator Spotlight, p. 448

1. The graph of $y = x - 5$ is the same as the graph of $y = x$, but it is moved down 5 units. **2.** The values of y_2 are 3 more than the values of y_1. The values of y_3 are 4 less than the values of y_1.

Calculator Spotlight, p. 449

1. The graph of $y = 247x + 1$ will be steeper than the others. **2.** The graph of $y = 0.000018x$ will be less steep than the others. **3.** The graph of $y = -247x$ will be steeper than the others. **4.** The graph of $y = -0.000043x - 1$ will be less steep than the others.

Exercise Set 7.3, p. 453

1. $m = 4$; y-intercept: (0, 5) **3.** $m = -2$; y-intercept: (0, -6) **5.** $m = -\frac{3}{8}$; y-intercept: $\left(0, -\frac{1}{5}\right)$ **7.** $m = 0.5$; y-intercept: (0, -9) **9.** $m = \frac{2}{3}$; y-intercept: $\left(0, -\frac{8}{3}\right)$ **11.** $m = 3$; y-intercept: (0, -2) **13.** $m = -8$; y-intercept: (0, 12) **15.** $m = 0$; y-intercept: $\left(0, \frac{4}{17}\right)$ **17.** $m = -\frac{1}{2}$ **19.** $m = \frac{1}{3}$ **21.** $m = 2$ **23.** $m = \frac{2}{3}$ **25.** $m = -\frac{1}{3}$ **27.** $\frac{2}{25}$, or 8% **29.** 3.5% **31.** $100 \dfrac{\$}{\text{yr}}$ **33.** -1323 **34.** $45x + 54$ **35.** $350x - 60y + 120$ **36.** 25 **37.** Square: 15 yd; triangle: 20 yd **38.** $(2 - 5x)(4 + 10x + 25x^2)$ **39.** $(c - d)(c^2 + cd + d^2)(c + d)(c^2 - cd + d^2)$ **40.** $7(2x - 1)(4x^2 + 2x + 1)$ **41.** $a - 10$, R -4; or $a - 10 + \dfrac{-4}{a-1}$ **43.** ◆

Margin Exercises, Section 7.4, pp. 455–460

1.

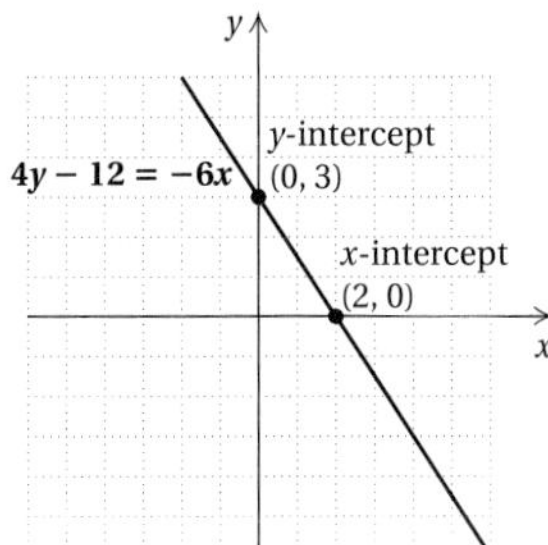

2.

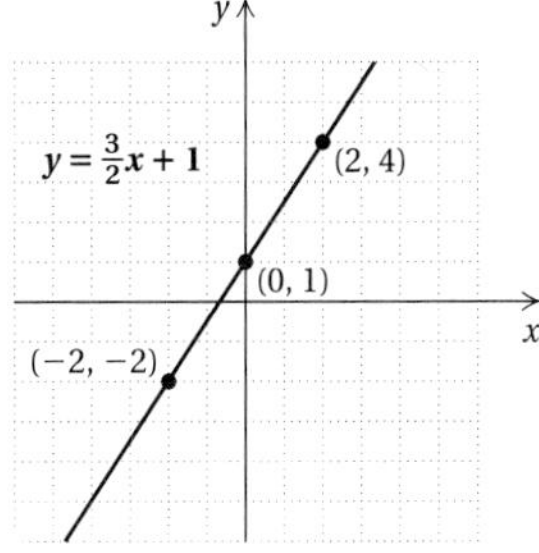

3.

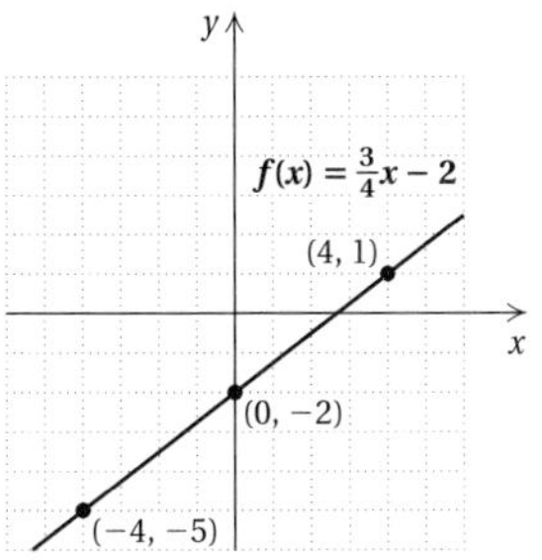

4.

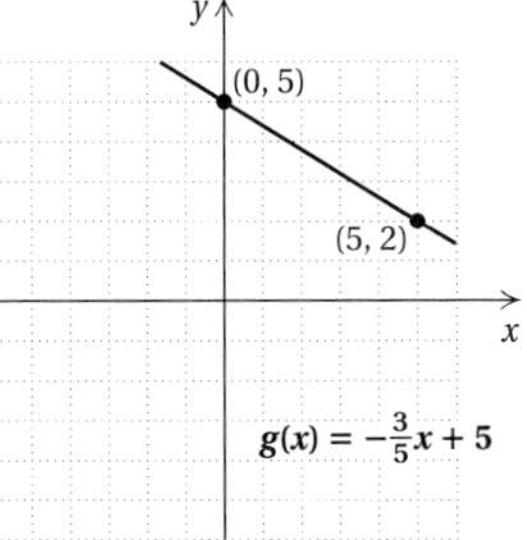

5.

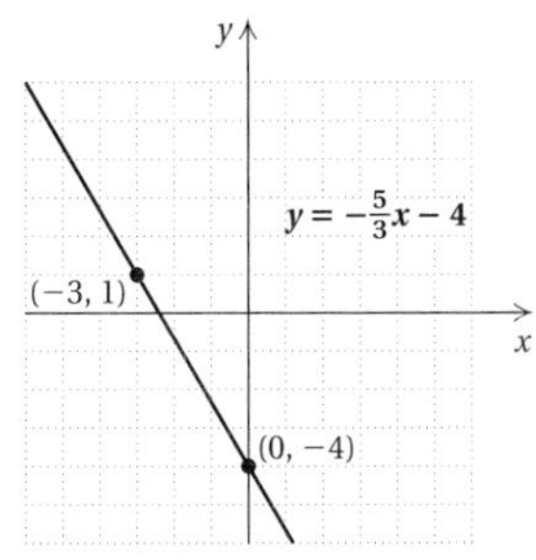

6. $m = 0$

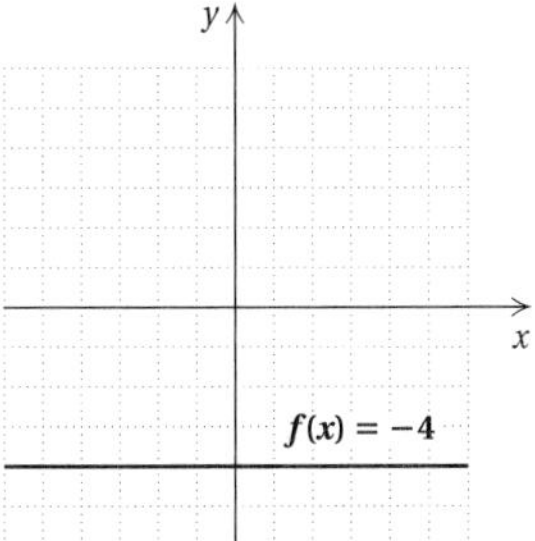

7. $m = 0$

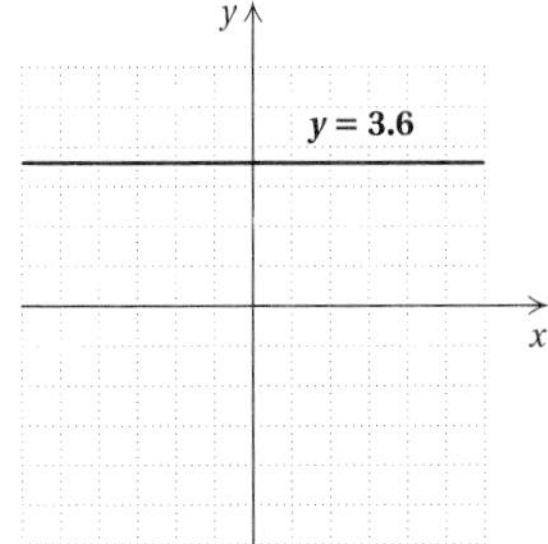

8.

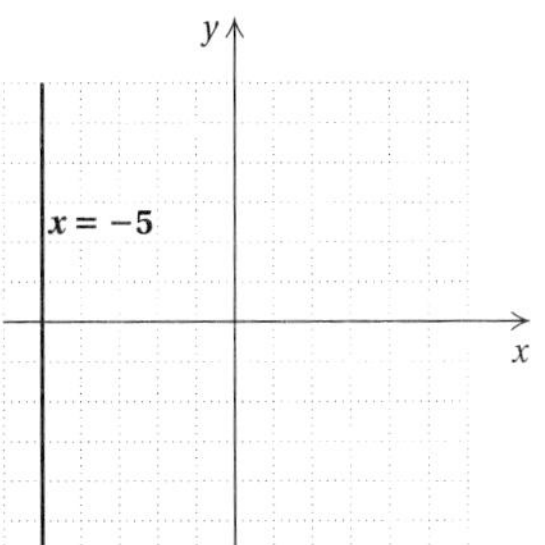

9.

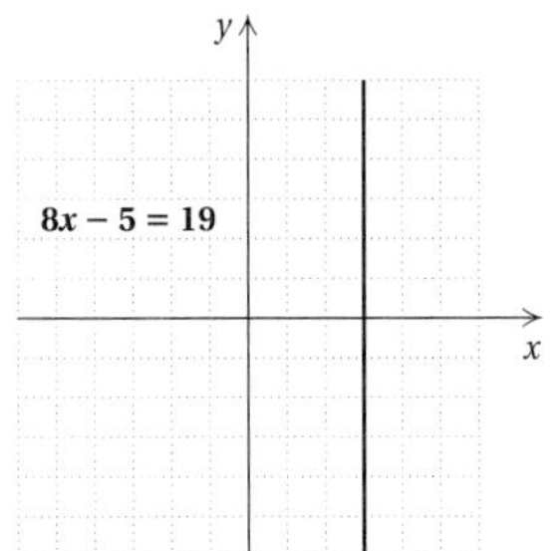

10. **(a)** Undefined; **(b)** $m = 0$; **(c)** $m = 0$; **(d)** undefined; **(e)** $m = 0$; **(f)** undefined **11.** Yes **12.** No **13.** No **14.** Yes **15.** No

Calculator Spotlight, p. 460

1.–2. Left to the student.

Exercise Set 7.4, p. 461

1.

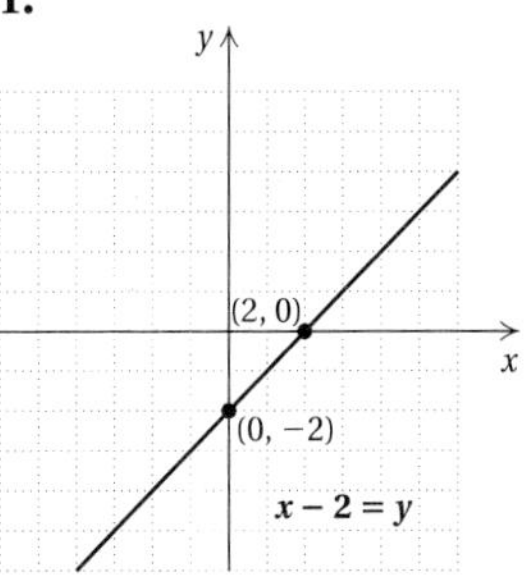

3.

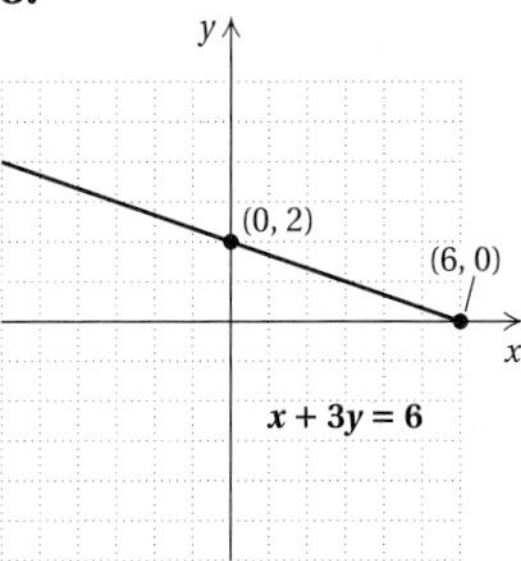

5.

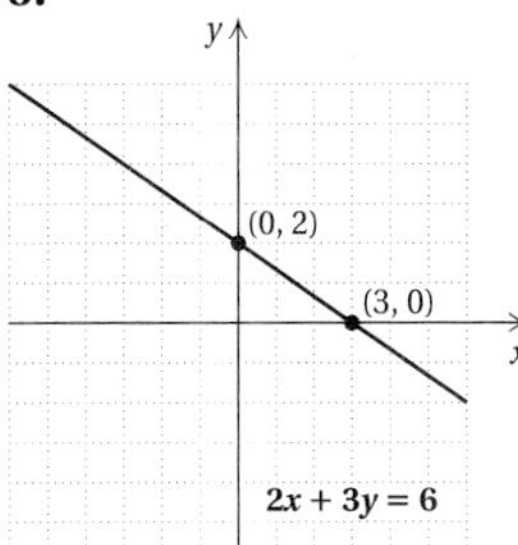

7.

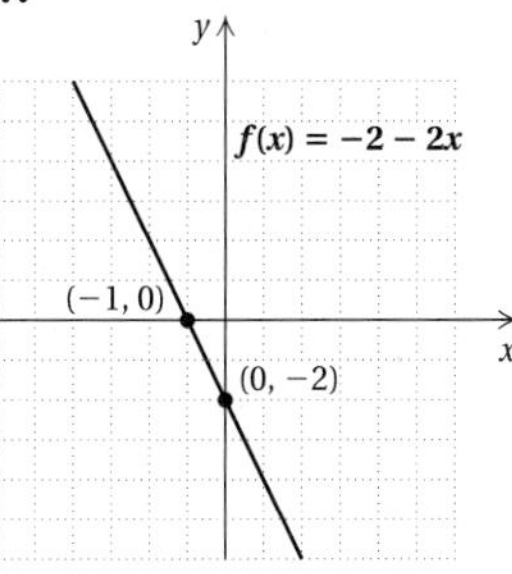

9.

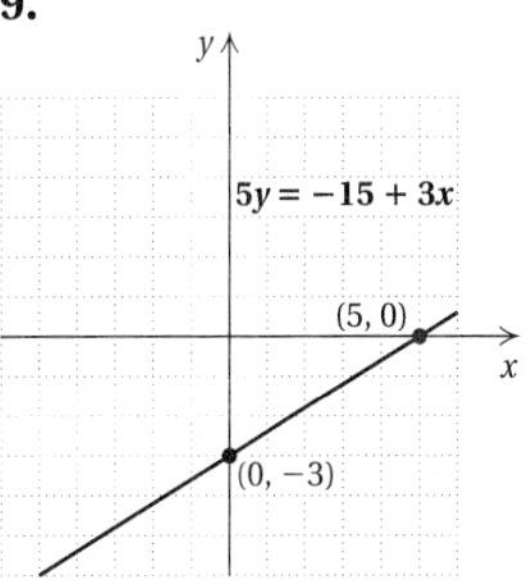

11.

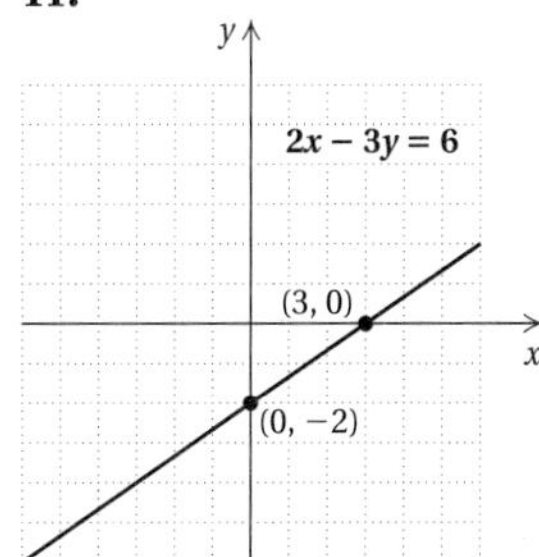

13.

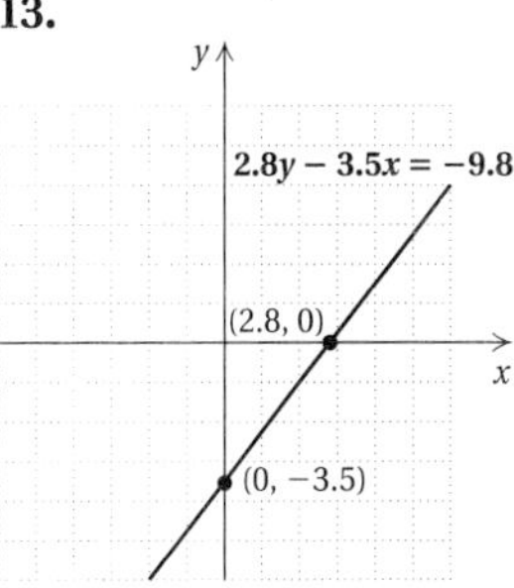

15.

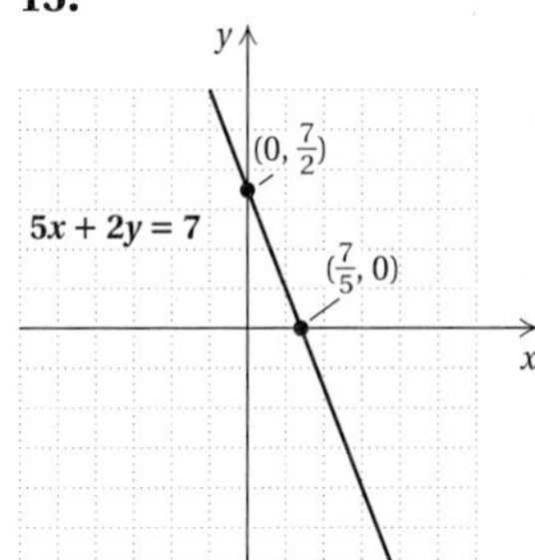

17.

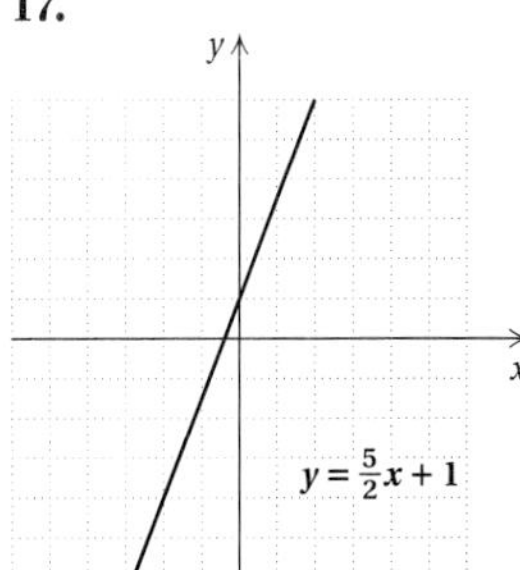

19.

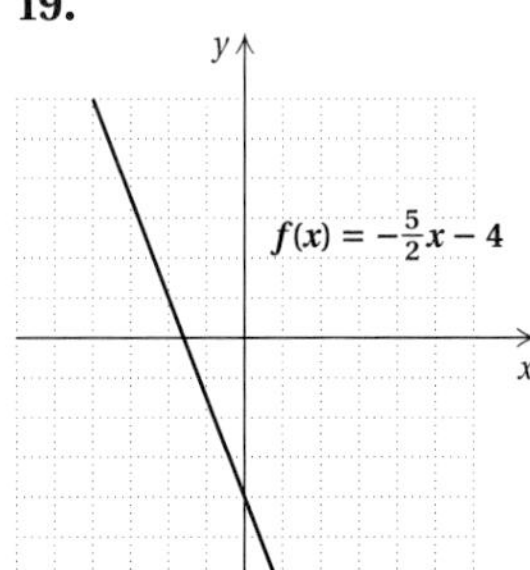

21.

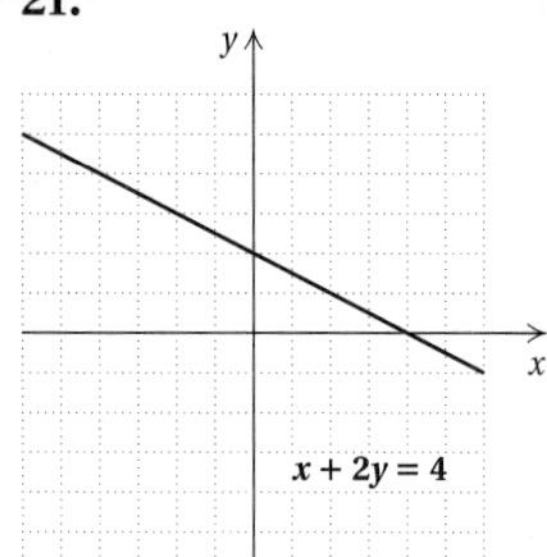

23.

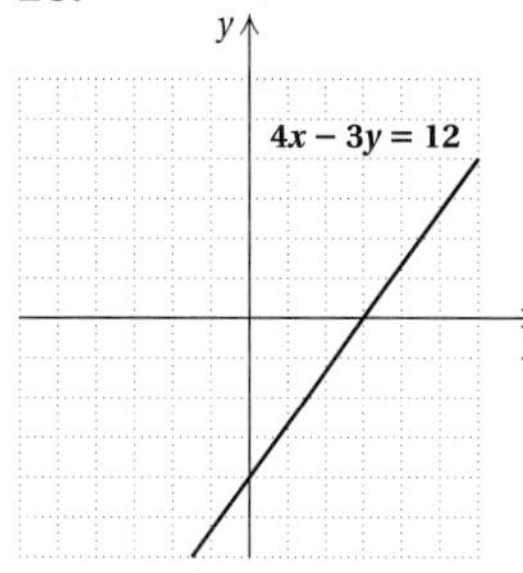

25.

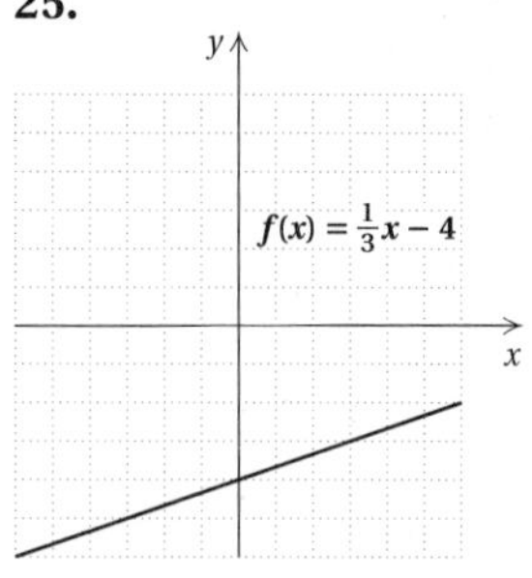

27.

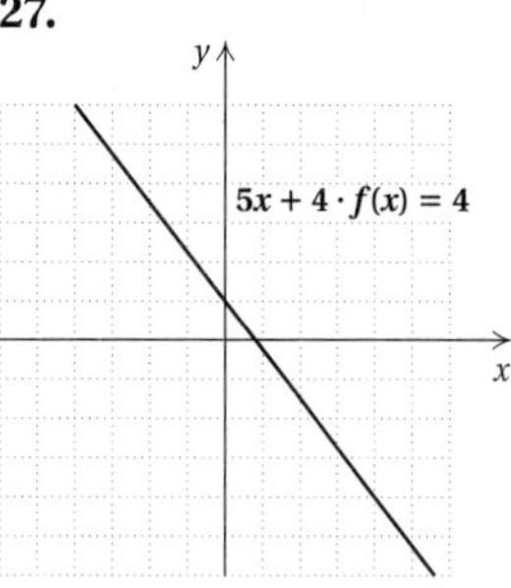

29. Undefined

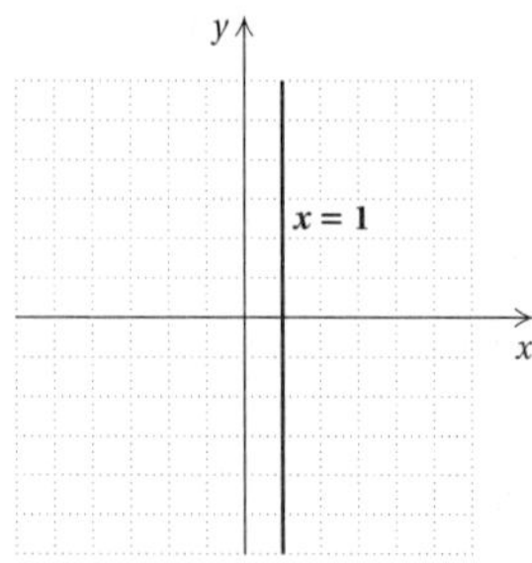

31. $m = 0$

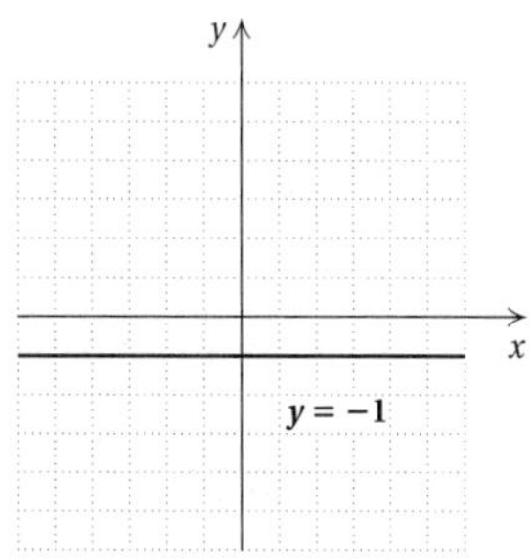

33. $m = 0$

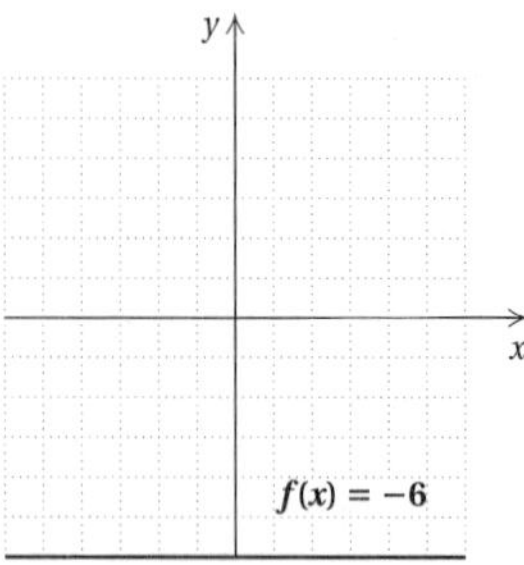

35. $m = 0$

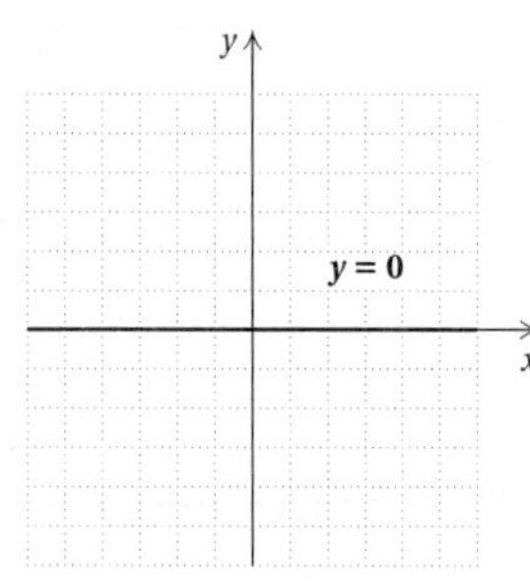

37. $m = 0$

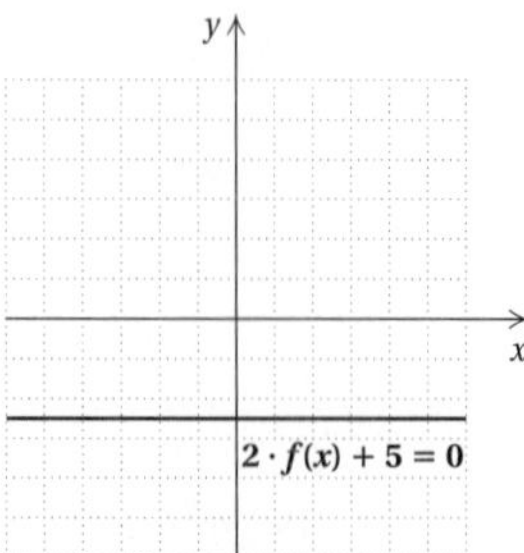

39. Undefined

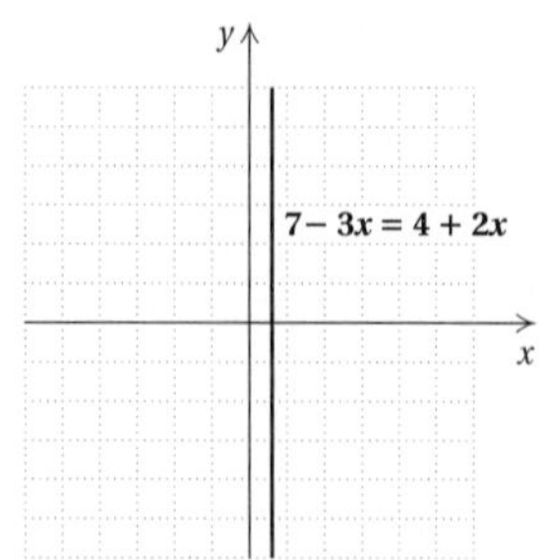

41. Yes **43.** No **45.** Yes **47.** Yes **49.** Yes **51.** No **53.** No **55.** Yes **57.** 5.3×10^{10} **58.** 4.7×10^{-5} **59.** 1.8×10^{-2} **60.** 9.9902×10^{7} **61.** 0.0000213 **62.** 901,000,000 **63.** 20,000 **64.** 0.085677 **65.** $(5x - 1)^2$ **66.** $(3b - c)(3b + c)$ **67.** $(11x - 2)(7x + 5)$ **68.** $3(2y - 1)(y + 6)$ **69.** ◆ **71.** $y = 3$ **73.** $a = 2$ **75.** $y = \frac{2}{15}x + \frac{2}{5}$ **77.** $y = 0$; yes **79.** $m = -\frac{3}{4}$ **81.** **(a)** II; **(b)** IV; **(c)** I; **(d)** III

Margin Exercises, Section 7.5, pp. 465–468

1. $y = 3.4x - 8$ **2.** $y = -5x - 18$ **3.** $y = 3x - 5$ **4.** $y = 8x - 19$ **5.** $y = -\frac{2}{3}x + \frac{14}{3}$ **6.** $y = -\frac{5}{3}x + \frac{11}{3}$ **7.** $y = -17x - 56$ **8.** $y = \frac{8}{7}x - \frac{23}{7}$ **9.** $y = -2x + 14$ **10.** **(a)** $C(t) = 20t + 25$;
(b)

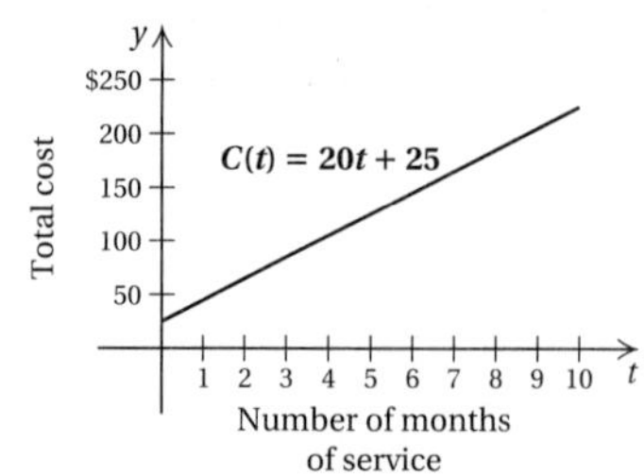

(c) $195

Exercise Set 7.5, p. 469

1. $y = -8x + 4$ **3.** $y = 2.3x - 1$ **5.** $f(x) = -\frac{7}{3}x - 5$ **7.** $f(x) = \frac{2}{3}x + \frac{5}{8}$ **9.** $y = 5x - 17$ **11.** $y = -3x + 33$ **13.** $y = x - 6$ **15.** $y = -2x + 16$ **17.** $y = -7$ **19.** $y = \frac{2}{3}x - \frac{8}{3}$ **21.** $y = \frac{1}{2}x + \frac{7}{2}$ **23.** $y = x$ **25.** $y = \frac{7}{4}x + 7$ **27.** $y = \frac{3}{2}x$ **29.** $y = \frac{1}{6}x$ **31.** $y = 13x - \frac{15}{4}$ **33.** $y = -\frac{1}{2}x + \frac{17}{2}$ **35.** $y = \frac{5}{7}x - \frac{17}{7}$ **37.** $y = \frac{1}{3}x + 4$ **39.** $y = \frac{1}{2}x + 4$ **41.** $y = \frac{4}{3}x - 6$ **43.** $y = \frac{5}{2}x + 9$ **45.** **(a)** $C(t) = 40t + 85$;
(b)

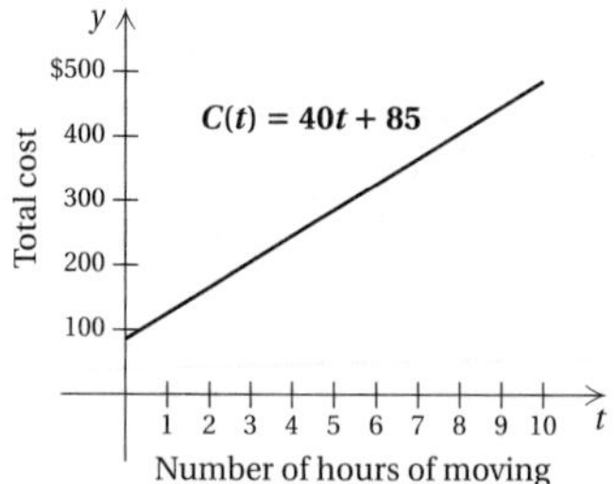

(c) $345

47. **(a)** $V(t) = 750 - 25t$;
(b)

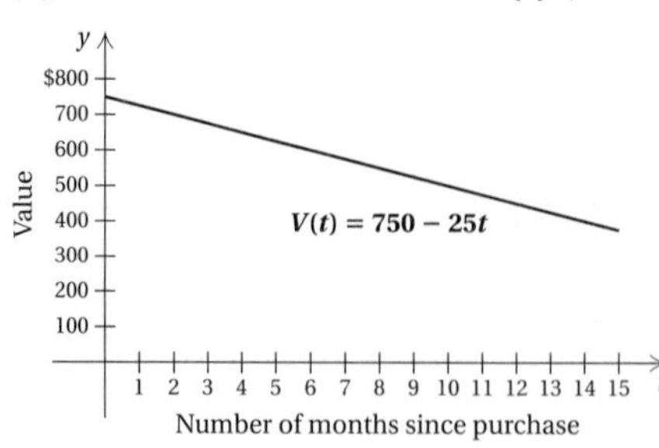

(c) $425

49. -1 **50.** $b + 1$ **51.** $\frac{x - 3}{2(x - 5)}$ **52.** $\frac{4}{y - 9}$ **53.** ◆

Margin Exercises, Section 7.6, pp. 472–473

1. (a)

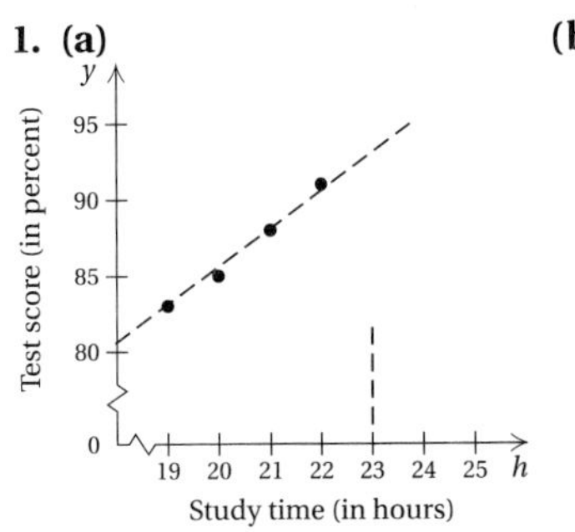

(b) 94%

2. (a) $T(x) = 0.08625x + 8.97075$; **(b)** $T(10) = 9.83325$. This prediction differs slightly from the number found in Example 2 and the number 9.85 found in Example 1.

Calculator Spotlight, pp. 474–475

1. (a) $y = 2.7x + 31.4$; **(b)**

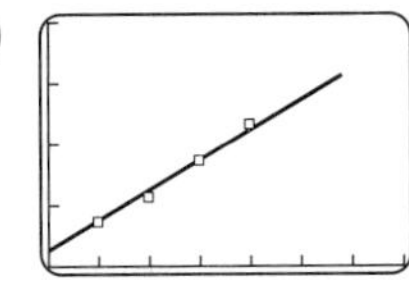

(c) 93.5%; **(d)** This prediction differs by 0.5% from the score of 94% found in Margin Exercise 1.

Exercise Set 7.6, p. 477

Answers may vary in Exercises 1–12.

1. (a) and **(b)**

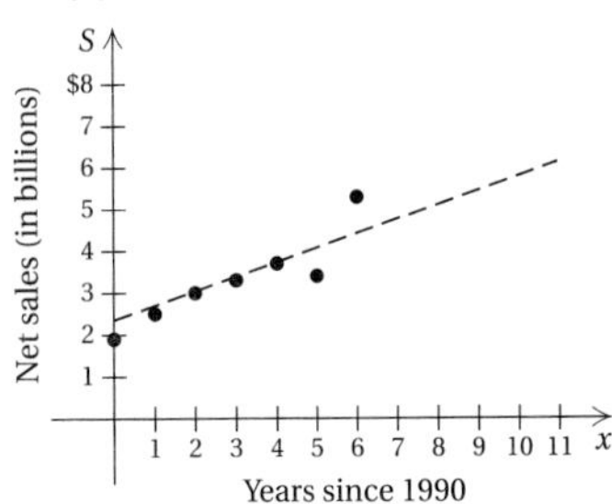

(c) \$5.5 billion; \$6.2 billion

3. (a) and **(b)**

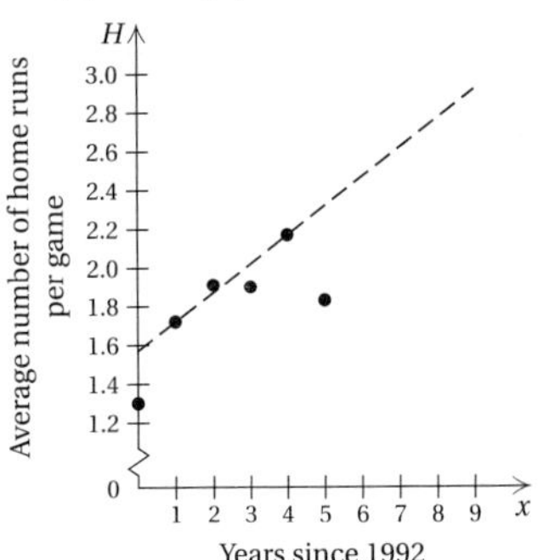

(c) 2.5; 2.8

5. (a) $B(x) = 1879x + 7403$;
(b)

(c) 26,193; 44,983 **7. (a)** $H(x) = 0.15x + 1.57$; **(b)** see graph in Exercise 3; **(c)** 2.62; 3.07 **9. (a)** $S(x) = 0.35x + 2.3$; **(b)** see graph in Exercise 1; **(c)** \$5.8 billion; \$7.55 billion **11. (a)** $y = 0.75x + 71.25$; **(b)** 84.75; 90 **13.** \$35 **14.** 83 **15.** $y = \dfrac{C - Ax}{B}$ **16.** $y = \dfrac{3x - 12}{7}$ **17.** $q = \frac{3}{2}(P + y)$ **18.** $b = \dfrac{2A}{h}$ **19.** $a + 9$ **20.** $x + 11$, R 78; or $x + 11 + \dfrac{78}{x - 7}$ **21.** $w - 1$, R 6; or $w - 1 + \dfrac{6}{w - 2}$ **22.** $x^3 - 2x - 2$, R $5x + 2$; or $x^3 - 2x - 2 + \dfrac{5x + 2}{x^2 + 2}$ **23.** ◆

25. (a) $y = 0.114x + 1.52$; **(b)** left to the student; **(c)** 2.66 **27. (a)** $M(c) = -0.0408975991x + 151.6882381$; **(b)** left to the student; **(c)** 20.2; **(d)** 44.5

Summary and Review: Chapter 7, p. 481

1. No **2.** Yes **3.** $g(0) = 5$; $g(-1) = 7$ **4.** $f(0) = 7$; $f(-1) = 12$ **5.** \$18,185 **6.** Yes **7.** No **8. (a)** $f(2) = 3$; **(b)** $[-2, 4]$; **(c)** -1; **(d)** $[1, 5]$ **9.** $\{x \mid x \text{ is a real number } and\ x \neq 4\}$ **10.** All real numbers **11.** Slope: -3; y-intercept: $(0, 2)$ **12.** Slope: $-\frac{1}{2}$; y-intercept: $(0, 2)$ **13.** $m = \frac{11}{3}$

14.

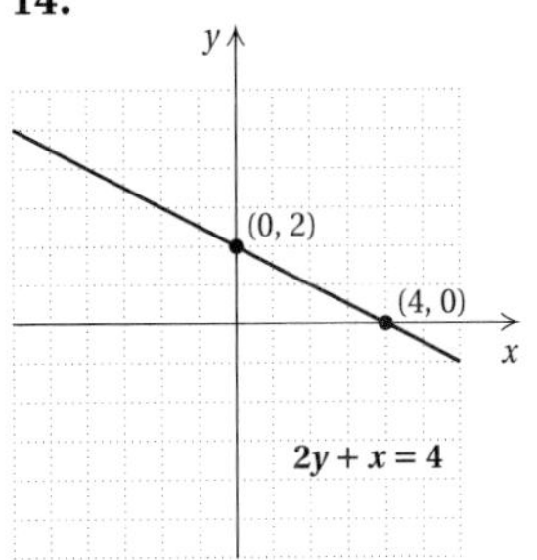

15.

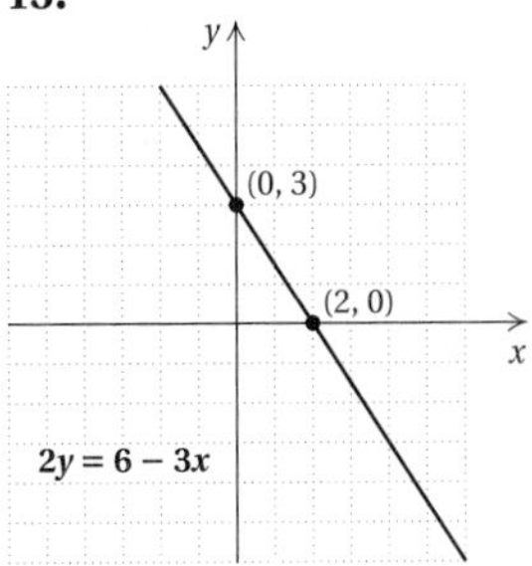

16.

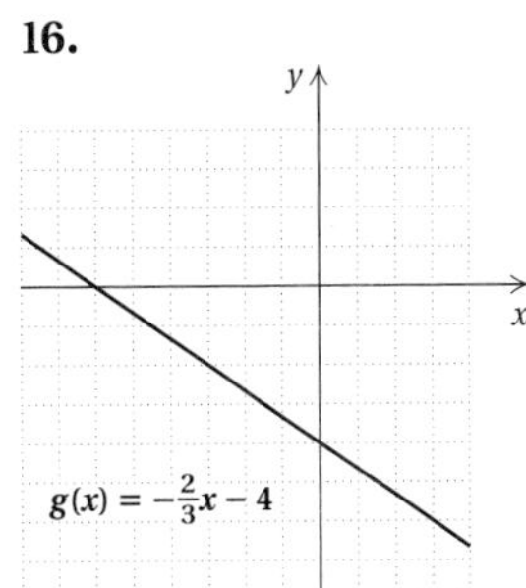

17.

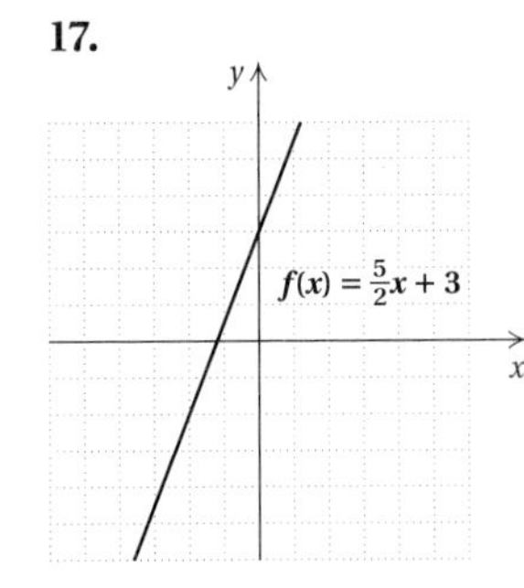

18.

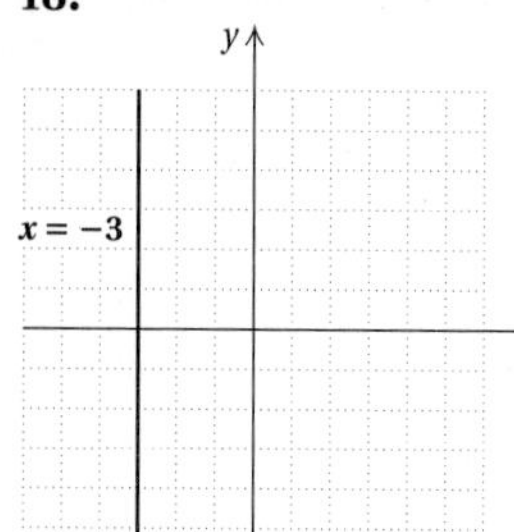

19.

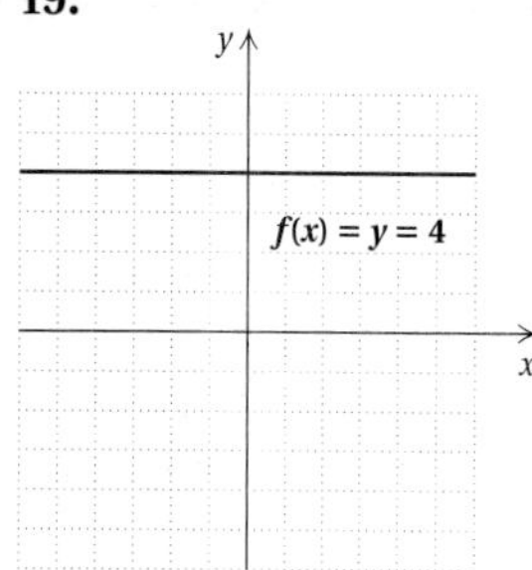

20. Perpendicular **21.** Parallel **22.** Parallel
23. Perpendicular **24.** $f(x) = 4.7x - 23$
25. $y = -3x + 4$ **26.** $y = -\frac{3}{2}x$
27. $y = -\frac{5}{7}x + 9$ **28.** $y = \frac{1}{3}x + \frac{1}{3}$
29. **(a)** $y = 32.8286x + 175.9$;
(b) \$504.2 billion; \$668.3 billion **30.** 0.000000033
31. 1.37×10^{10} **32.** $3x - 1$ **33.** $r + 1$, R 10; or $r + 1 + \dfrac{10}{r - 4}$ **34.** $3(3y - 10)(9y^2 + 30y + 100)$
35. $9(1 + 2z)(1 - 2z + 4z^2)$ **36.** $\dfrac{a - 4}{3a - 1}$
37. ◆ The concept of slope is useful in describing how a line slants. A line with positive slope slants up from left to right. A line with negative slope slants down from left to right. The larger the absolute value of the slope, the steeper the slant.
38. ◆ The notation $f(x)$ can be read "f of x" or "f at x" or "the value of f at x." It represents the output of the function f for the input x. The notation $f(a) = b$ provides a concise way to indicate that for the input a, the output of the function f is b.
39. **(a)** $H(x) = 33.6893x + 175.85$; **(b)** left to the student; **(c)** \$680.5 billion

Test: Chapter 7, p. 483

1. [7.1c]

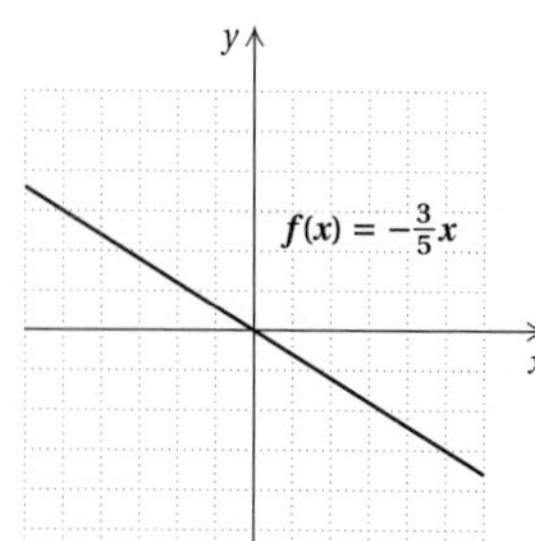

2. [7.1c]

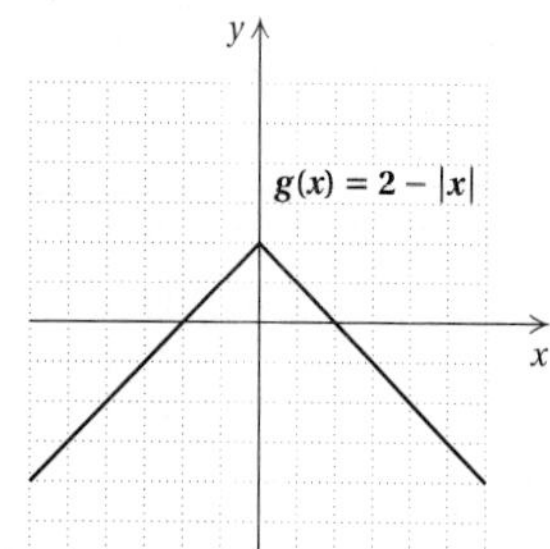

3. [7.1b] **(a)** 8.666; **(b)** 1998 **4.** [7.1a] Yes
5. [7.1a] No **6.** [7.1b] −4; 2 **7.** [7.1b] 7; 8
8. [7.1d] Yes **9.** [7.1d] No
10. [7.1e] **(a)** \$120 million; **(b)** \$20 million
11. [7.2a] **(a)** 1.2; **(b)** $\{x \mid -1 \le x \le 2\}$; **(c)** −3;
(d) $\{x \mid -3 \le x \le 4\}$ **12.** [7.2a] All real numbers
13. [7.2a] $\left\{x \mid x \text{ is a real number } and\ x \ne -\frac{3}{2}\right\}$
14. [7.3b] Slope: $-\frac{3}{5}$; y-intercept: (0, 12)
15. [7.3b] Slope: $-\frac{2}{5}$; y-intercept: $\left(0, -\frac{7}{5}\right)$ **16.** [7.3b] $m = \frac{5}{8}$
17. [7.3b] $m = 0$ **18.** [7.3c] m (or rate of change) $= \frac{4}{5}$
19. [7.4a]

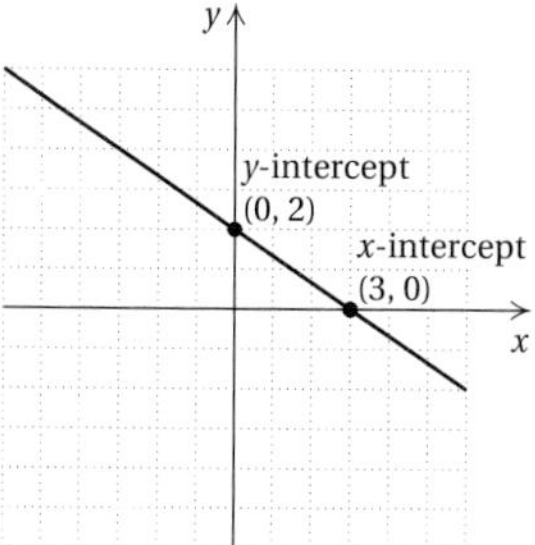

20. [7.4b]

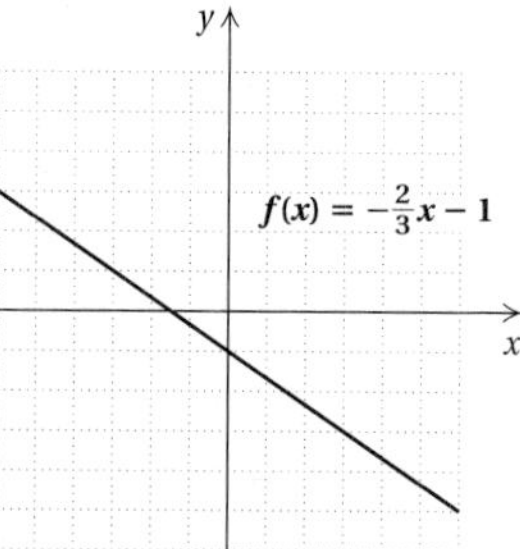

21. [7.4c]

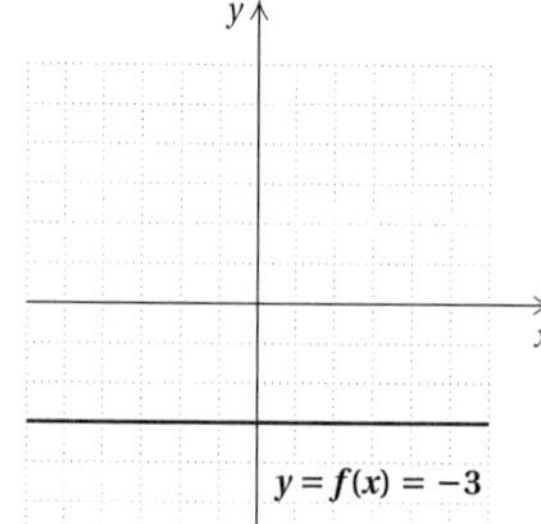

22. [7.4c]

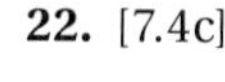

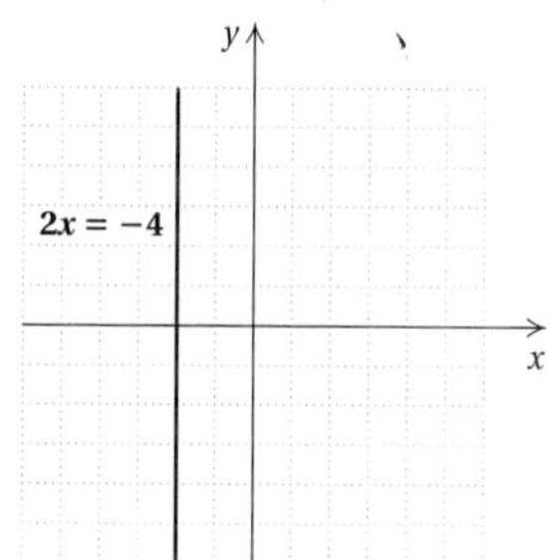

23. [7.4d] Parallel **24.** [7.4d] Perpendicular
25. [7.5a] $y = -3x + 4.8$ **26.** [7.5a] $f(x) = 5.2x - \frac{5}{8}$
27. [7.5b] $y = -4x + 2$ **28.** [7.5c] $y = -\frac{3}{2}x$
29. [7.5d] $y = \frac{1}{2}x - 3$ **30.** [7.5d] $y = 3x - 1$
31. [7.6b] **(a)** $S(x) = 0.2167x + 0.3$;
(b) \$2.0 billion; \$2.5 billion **32.** [4.2c] 8.204×10^{-7}
33. [4.8b] $y + 2$, R −11; or $y + 2 + \dfrac{-11}{y + 4}$
34. [5.6a] $(5a + 3)(25a^2 - 15a + 9)$
35. [6.1c] $\dfrac{3m - 5}{7m + 4}$ **36.** [7.5d] $\frac{24}{5}$ **37.** [7.1b] $f(x) = 3$; answers may vary

CHAPTER 8

Pretest: Chapter 8, p. 488

1. [8.1a] (1, 2) **2.** [8.2a] (1, 2) **3.** [8.2b] (−3, 2)
4. [8.1a] Consistent **5.** [8.1a] Independent
6. [8.4a] (1, 2, 3) **7.** [8.3a] Cashews: 20 lb;
Brazil nuts: 30 lb **8.** [8.3a] \$3300 at 8%; \$4200 at 6%
9. [8.3a] A: $64\frac{8}{13}$ lb; B: $55\frac{5}{13}$ lb **10.** [8.3b] 54 mph
11. [8.6a] **(a)** $P(x) = 11x - 90{,}000$; **(b)** (8182, \$212,732)

Margin Exercises, Section 8.1, pp. 489–493

1. (0, 1) **2.** (2, 1) **3.** No solution
4. Consistent: 1, 2; inconsistent: 3 **5.** Infinitely many solutions **6.** Independent: 1, 2, 3; dependent: 5

7. **(a)** −3; **(b)** ; −3

8. 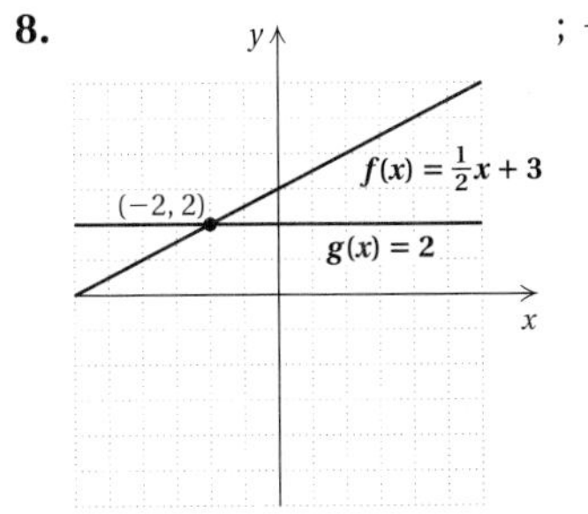

; −2

9. **(a)** 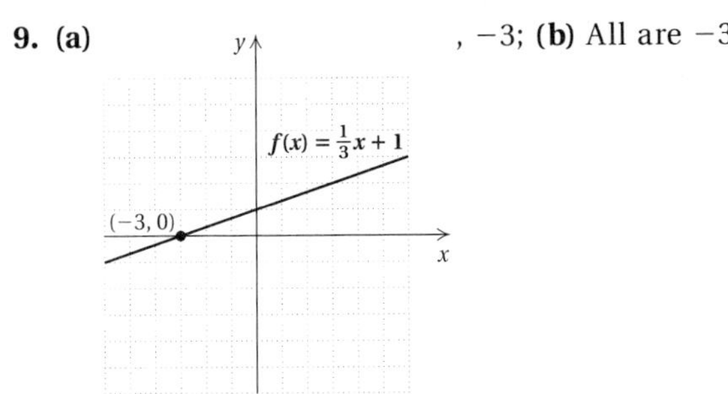

, −3; **(b)** All are −3.

10. 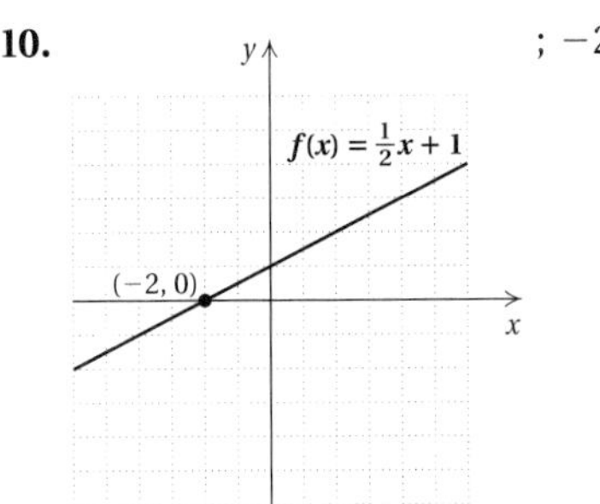

; −2

Calculator Spotlight, p. 494

1. −2 **2.** −2 **3.** 2.5455 **4.** 1.3333 **5.** −1.3825, 3.1825 **6.** 2.2790 **7.–12.** See answers for Exercises 1–6.

Exercise Set 8.1, p. 495

1. (3, 1); consistent; independent **3.** (1, −2); consistent; independent **5.** (4, −2); consistent; independent **7.** (2, 1); consistent; independent **9.** $\left(\frac{5}{2}, -2\right)$; consistent; independent **11.** (3, −2); consistent; independent **13.** No solution; inconsistent; independent **15.** Infinitely many solutions; consistent; dependent **17.** (4, −5); consistent; independent **19.** (2, −3); consistent; independent **21.** $y = \frac{3}{5}x + \frac{22}{5}$ **22.** $y = \frac{3}{8}x + \frac{9}{4}$ **23.** ◆ **25.** (2.23, 1.14)

Margin Exercises, Section 8.2, pp. 498–505

1. (2, 4) **2.** (5, 2) **3.** (−3, 2) **4.** (13, 16) **5.** (1, 4) **6.** $\left(\frac{1}{3}, \frac{1}{2}\right)$ **7.** (2, 3) **8.** (−127, 100) **9.** (−138, −118) **10.** No solution **11.** Infinitely many solutions **12.** $2x + 3y = 1$, $3x - y = 7$; (2, −1) **13.** $9x + 10y = 5$, $9x - 4y = 3$; $\left(\frac{25}{63}, \frac{1}{7}\right)$ **14.** 2-pointers: 31; 3-pointers: 9 **15.** **(a)** and **(b)** Length: 160 ft; width: 100 ft

Exercise Set 8.2, p. 507

1. (−3, 2) **3.** (−2, −6) **5.** (2, −2) **7.** (−2, 1) **9.** (1, 2) **11.** (−1, 3) **13.** $\left(\frac{128}{31}, -\frac{17}{31}\right)$ **15.** (6, 2) **17.** $\left(\frac{140}{13}, -\frac{50}{13}\right)$ **19.** (4, 6) **21.** $\left(\frac{110}{19}, -\frac{12}{19}\right)$ **23.** Infinitely many solutions **25.** No solution **27.** $\left(\frac{1}{2}, -\frac{1}{2}\right)$ **29.** $\left(-\frac{4}{3}, -\frac{19}{3}\right)$ **31.** $\left(\frac{1000}{11}, -\frac{1000}{11}\right)$ **33.** (140, 60) **35.** Length: 40 ft; width: 20 ft **37.** 48° and 132° **39.** 5 and −47 **41.** 23 wins; 14 ties **43.** 1.3 **44.** $-15y - 39$ **45.** $p = \frac{7A}{q}$ **46.** $\frac{7}{3}$ **47.** $y = -\frac{7}{2}x + 36$ **48.** $y = -\frac{7}{5}x - \frac{29}{5}$ **49.** 1 **50.** 5 **51.** 3 **52.** 291 **53.** 15 **54.** $12a^2 - 2a + 1$ **55.** 53 **56.** 8.92 **57.** ◆ **59.** (23.12, −12.04) **61.** $\left(\frac{a + 2b}{7}, \frac{a - 5b}{7}\right)$ **63.** $p = 2$; $q = -\frac{1}{3}$ **65.** $A = 2$; $B = 4$

Margin Exercises, Section 8.3, pp. 512–518

1. White: 22; red: 8

White	Red	Totals	
w	r	30	→ $w + r = 30$
18.95	19.50		
$18.95w$	$19.50r$	572.90	→ $18.95w + 19.50r = 572.90$

2. Kenyan: 8 lb; Sumatran: 12 lb
3. \$1800 at 7%; \$1900 at 9%

First Investment	Second Investment	Total	
x	y	\$3700	→ $x + y = 3700$
7%	9%		
1 yr	1 yr		
$0.07x$	$0.09y$	\$297	→ $0.07x + 0.09y = 297$

4. Attack: 15 qt; Blast: 45 qt

Attack	Blast	Mixture	
a	b	60	→ $a + b = 60$
2%	6%	5%	
$0.02a$	$0.06b$	0.05×60, or 3	→ $0.02a + 0.06b = 3$

5. 280 km

Distance	Rate	Time	
d	35 km/h	t	→ $d = 35t$
d	40 km/h	$t - 1$	→ $d = 40(t - 1)$

6. 180 mph

Distance	Rate	Time	
d	$r + 20$	4 hr	→ $d = 4(r + 20)$
d	$r - 20$	5 hr	→ $d = 5(r - 20)$

Exercise Set 8.3, p. 519

1. 32 at \$8.50; 13 at \$9.75 **3.** Humulin: 42; Novolin: 23 **5.** 30-sec: 4; 60-sec: 8 **7.** 5 lb of each **9.** 4 L of 25%; 6 L of 50% **11.** \$7500 at 6%; \$4500 at 9% **13.** Whole milk: $169\frac{3}{13}$ lb; cream: $30\frac{10}{13}$ lb **15.** \$5 bills: 7; \$1 bills: 15 **17.** 375 mi **19.** 14 km/h **21.** 144 mi **23.** 2 hr **25.** $1\frac{1}{3}$ hr **27.** About 1489 mi **29.** -7 **30.** -11 **31.** -3 **32.** 33 **33.** -15 **34.** $8a - 7$ **35.** -23 **36.** 0.2 **37.** -4 **38.** -17 **39.** $-12h - 7$ **40.** 3993 **41.** ◆ **43.** $4\frac{4}{7}$ L **45.** City: 261 mi; highway: 204 mi **47.** Brown: 0.8 gal; neutral: 0.2 gal

Margin Exercises, Section 8.4, pp. 524–526

1. $(2, 1, -1)$ **2.** $\left(2, -2, \frac{1}{2}\right)$ **3.** $(20, 30, 50)$

Exercise Set 8.4, p. 527

1. $(1, 2, -1)$ **3.** $(2, 0, 1)$ **5.** $(3, 1, 2)$ **7.** $(-3, -4, 2)$ **9.** $(2, 4, 1)$ **11.** $(-3, 0, 4)$ **13.** $(2, 2, 4)$ **15.** $\left(\frac{1}{2}, 4, -6\right)$ **17.** $\left(\frac{1}{2}, \frac{1}{3}, \frac{1}{6}\right)$ **19.** $\left(\frac{1}{2}, \frac{2}{3}, -\frac{5}{6}\right)$ **21.** $(15, 33, 9)$ **23.** $(4, 1, -2)$ **25.** $a = \dfrac{F}{3b}$ **26.** $a = \dfrac{Q - 4b}{4}$, or $\dfrac{Q}{4} - b$ **27.** $c = \dfrac{2F + td}{t}$, or $\dfrac{2F}{t} + d$ **28.** $d = \dfrac{tc - 2F}{t}$, or $c - \dfrac{2F}{t}$ **29.** $y = \dfrac{Ax - c}{B}$ **30.** $y = \dfrac{c - Ax}{B}$ **31.** Slope: $-\frac{2}{3}$; y-intercept: $\left(0, -\frac{5}{4}\right)$ **32.** Slope: -4; y-intercept: $(0, 5)$ **33.** Slope: $\frac{2}{5}$; y-intercept: $(0, -2)$ **34.** Slope: 1.09375; y-intercept: $(0, -3.125)$ **35.** ◆ **37.** $(1, -2, 4, -1)$

Margin Exercises, Section 8.5, pp. 529–530

1. 64°, 32°, 84° **2.** \$4000 at 5%; \$5000 at 6%; \$16,000 at 7%

Exercise Set 8.5, p. 531

1. 10-oz: 8; 14-oz: 20; 20-oz: 6 **3.** 32°, 96°, 52° **5.** Automatic transmission: \$865; power door locks: \$520; air conditioning: \$375 **7.** A: 1500; B: 1900; C: 2300 **9.** First fund: \$45,000; second fund: \$10,000; third fund: \$25,000 **11.** Asian-American: 385; African-American: 200; Caucasian: 154 **13.** Roast beef: 2; baked potato: 1; broccoli: 2 **15.** Par-3: 6; par-4: 8; par-5: 4 **17.** 4, 2, -1 **19.** 2-point: 32; 3-point: 5; foul shots: 13 **21.** No **22.** Yes **23.** $\{x \mid x \text{ is a real number } and\ x \neq -7\}$, or $(-\infty, -7) \cup (-7, \infty)$ **24.** Domain: all real numbers; range: $\{y \mid y \leq 5\}$, or $(-\infty, 5]$ **25.** $y = -\frac{3}{5}x - 7$ **26.** $\dfrac{a^3}{b}$ **27.** ◆ **29.** 180°

Margin Exercises, Section 8.6, pp. 536–538

1. (a) $C(x) = 80{,}000 + 20x$; **(b)** $R(x) = 36x$; **(c)** $P(x) = 16x - 80{,}000$; **(d)** a loss of \$16,000; a profit of \$176,000
(e)

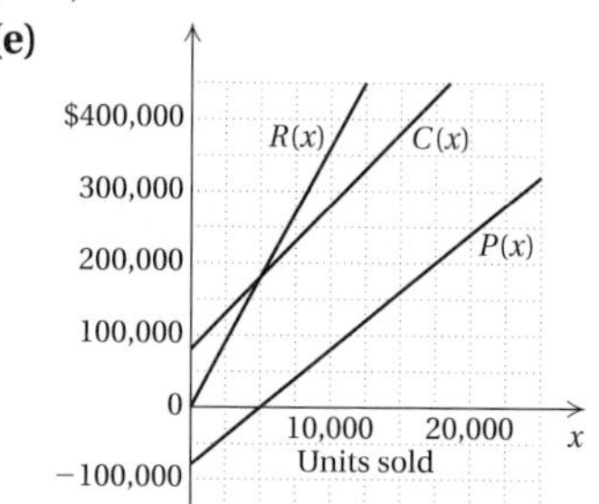

Break-even point: (5000, \$180,000)
2. (\$14, 356)

Exercise Set 8.6, p. 539

1. (a) $P(x) = 45x - 270{,}000$; **(b)** (6000, \$420,000) **3. (a)** $P(x) = 50x - 120{,}000$; **(b)** (2400, \$144,000) **5. (a)** $P(x) = 80x - 10{,}000$; **(b)** (125, \$12,500) **7. (a)** $P(x) = 18x - 16{,}000$; **(b)** (889, \$35,560) **9. (a)** $P(x) = 75x - 195{,}000$; **(b)** (2600, \$325,000) **11. (a)** $C(x) = 22{,}500 + 40x$; **(b)** $R(x) = 85x$; **(c)** $P(x) = 45x - 22{,}500$; **(d)** \$112,500 profit; \$4500 loss; **(e)** (500, \$42,500) **13. (a)** $C(x) = 16{,}404 + 6x$; **(b)** $R(x) = 18x$; **(c)** $P(x) = 12x - 16{,}404$; **(d)** \$19,596 profit; \$4404 loss; **(e)** (1367, \$24,606) **15.** (\$70, 300) **17.** (\$22, 474) **19.** (\$50, 6250) **21.** (\$10, 1070) **23.** Slope: $\frac{3}{5}$; y-intercept: $\left(0, \frac{8}{5}\right)$ **24.** Slope: $-\frac{6}{7}$; y-intercept: $\left(0, \frac{13}{7}\right)$ **25.** Slope: 1.7; y-intercept: $(0, 49)$ **26.** Slope: $-\frac{4}{3}$; y-intercept: $(0, 4)$ **27.** ◆

Summary and Review: Chapter 8, p. 541

1. $(-2, 1)$; consistent; independent
2. Infinitely many solutions; consistent; dependent
3. No solution; inconsistent; independent **4.** $\left(\frac{2}{5}, -\frac{4}{5}\right)$ **5.** No solution **6.** $\left(-\frac{11}{15}, -\frac{43}{30}\right)$ **7.** $\left(\frac{37}{19}, \frac{53}{19}\right)$ **8.** $\left(\frac{76}{17}, -\frac{2}{119}\right)$ **9.** $(2, 2)$ **10.** Infinitely many solutions **11.** CD: \$14; cassette: \$9 **12.** 5 L of each **13.** $5\frac{1}{2}$ hr **14.** $(10, 4, -8)$ **15.** $\left(-\frac{7}{3}, \frac{125}{27}, \frac{20}{27}\right)$ **16.** $(2, 0, 4)$ **17.** $\left(2, \frac{1}{3}, -\frac{2}{3}\right)$ **18.** 90°, $67\frac{1}{2}$°, $22\frac{1}{2}$° **19.** \$20 bills: 5; \$5 bills: 15; \$1 bills: 19

20. **(a)** $C(x) = 35{,}000 + 175x$; **(b)** $R(x) = 300x$; **(c)** $P(x) = 125x - 35{,}000$; **(d)** \$115,000 profit; \$10,000 loss **(e)** (280, \$84,000) **21.** (\$3, 81) **22.** $y = \frac{1}{2}x + 5$ **23.** $y = 2x - 3$ **24.** $y = \frac{1}{4}x - 2$ **25.** $y = x - 4$ **26.** $t = \dfrac{Q}{a - 4}$ **27.** $f(0) = 8$; $f(-2) = 14$ **28.** Slope: $\frac{5}{8}$; y-intercept: $(0, -5)$
29. ◈ The comparison is summarized in the table in Section 8.2.
30. ◈ Many problems that deal with more than one unknown quantity are often easier to translate to a system of equations than to a single equation. Problems involving complementary or supplementary angles, the dimensions of a geometric figure, mixtures, and the measures of the angles of a triangle are examples.
31. (0, 2) and (1, 3) **32.** **(a)** 165.91 cm; **(b)** 160.60 cm; **(c)** (120.857, 308.521); **(d)** 120.857 cm

Test: Chapter 8, p. 543

1. [8.1a] $(-2, 1)$; consistent; independent
2. [8.1a] No solution; inconsistent; independent
3. [8.1a] Infinitely many solutions; consistent; dependent **4.** [8.2a] $\left(3, -\frac{11}{3}\right)$ **5.** [8.2a] $\left(\frac{15}{7}, -\frac{18}{7}\right)$
6. [8.2b] $\left(-\frac{3}{2}, -\frac{3}{2}\right)$ **7.** [8.2b] No solution
8. [8.3a] 34% solution: $48\frac{8}{9}$ mL; 61% solution: $71\frac{1}{9}$ mL **9.** [8.3a] Buckets: 17; dinners: 11 **10.** [8.3b] 120 km/h
11. [8.2c] Length: 93 ft; width: 51 ft
12. [8.4a] $\left(2, -\frac{1}{2}, -1\right)$ **13.** [8.6b] (\$3, 55)
14. [8.5a] 3.5 hr **15.** [8.6a] **(a)** $C(x) = 40{,}000 + 30x$; **(b)** $R(x) = 80x$; **(c)** $P(x) = 50x - 40{,}000$; **(d)** \$20,000 profit; \$30,000 loss; **(e)** (800, \$64,000)
16. [7.5d] $y = -3x - 5$ **17.** [7.5d] $y = \frac{4}{5}x + 4$
18. [2.6a] $a = \dfrac{P + 3b}{4}$
19. [7.3b] Slope: $-\frac{7}{2}$; y-intercept: $(0, 7)$
20. [7.1b] $f(0) = -8$; $f(-3) = 1$ **21.** [8.2b] $m = 7$; $b = 10$

Chapter 9

Pretest: Chapter 9, p. 546

1. [9.1c] $\{x \mid x > -3\}$ **2.** [9.1c] $\{x \mid x \geq -3\}$
3. [9.2a] $\left\{x \mid -1 \leq x \leq \frac{2}{3}\right\}$ **4.** [9.3c] $-1, 8$
5. [9.2b] $\left\{a \mid a < -2 \text{ or } a > -\frac{1}{2}\right\}$
6. [9.3e] $\left\{a \mid a < -2 \text{ or } a > -\frac{4}{3}\right\}$ **7.** [9.3d] $-\frac{11}{2}, \frac{3}{4}$
8. [9.1c] $\left\{x \mid x \leq \frac{19}{14}\right\}$ **9.** [9.3b] 2.4
10. [9.1d] $\{t \mid t > 3.23\}$, 3.23 yr after 1988
11. [9.2a] $\{-3, 4\}$ **12.** [9.2b] $\{-1, 0, 1\}$
13. [9.3a] $\dfrac{5}{2} \,\Big|\, \dfrac{y}{x} \,\Big|$

14. [9.4b]

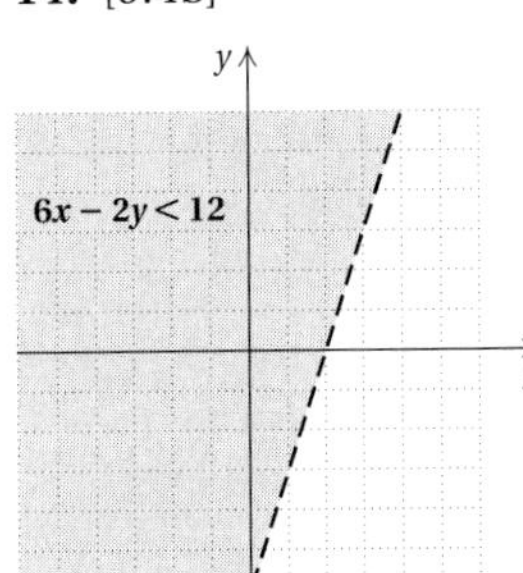

15. [9.4c]

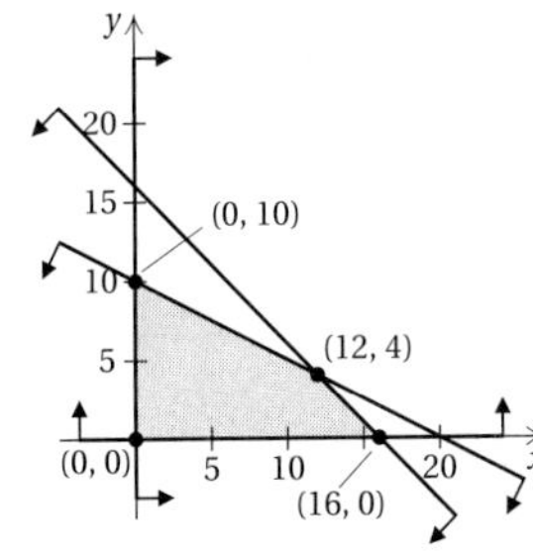

Margin Exercises, Section 9.1, pp. 547–554

1. Yes **2.** No **3.** Yes **4.** $[-4, 5)$ **5.** $(-\infty, -2]$
6. $[10, \infty)$ **7.** $[-30, 30]$
8. $\{x \mid x > 3\}$, or $(3, \infty)$; **9.** $\{x \mid x \leq 3\}$, or $(-\infty, 3]$;
10. $\{x \mid x \leq -2\}$, or $(-\infty, -2]$;
11. $\left\{y \mid y \leq \frac{3}{10}\right\}$, or $\left(-\infty, \frac{3}{10}\right]$;
12. $\{y \mid y < -5\}$, or $(-\infty, -5)$
13. $\{x \mid x \geq 12\}$, or $[12, \infty)$
14. $\left\{y \mid y \leq -\frac{1}{5}\right\}$, or $\left(-\infty, -\frac{1}{5}\right]$
15. $\left\{x \mid x < \frac{1}{2}\right\}$, or $\left(-\infty, \frac{1}{2}\right)$ **16.** $\left\{y \mid y \geq \frac{17}{9}\right\}$, or $\left[\frac{17}{9}, \infty\right)$
17. $\{t \mid t > 4.29\}$, or approximately after 2000
18. For $\{n \mid n < 100 \text{ hr}\}$, plan A is better.

Exercise Set 9.1, p. 555

1. No, no, no, yes **3.** No, yes, yes, no, no
5. $(-\infty, 5)$ **7.** $[-3, 3]$ **9.** $(-2, 5)$ **11.** $(-\sqrt{2}, \infty)$
13. $\{x \mid x > -1\}$, or $(-1, \infty)$ **15.** $\{y \mid y < 6\}$, or $(-\infty, 6)$
17. $\{a \mid a \leq -22\}$, or $(-\infty, -22]$
19. $\{t \mid t \geq -4\}$, or $[-4, \infty)$
21. $\{y \mid y > -6\}$, or $(-6, \infty)$ **23.** $\{x \mid x \leq 9\}$, or $(-\infty, 9]$
25. $\{x \mid x \geq 3\}$, or $[3, \infty)$
27. $\{x \mid x < -60\}$, or $(-\infty, -60)$

29. $\{x \mid x \le 0.9\}$, or $(-\infty, 0.9]$ **31.** $\left\{x \mid x \le \frac{5}{6}\right\}$, or $\left(-\infty, \frac{5}{6}\right]$
33. $\{x \mid x < 6\}$, or $(-\infty, 6)$ **35.** $\{y \mid y \le -3\}$, or $(-\infty, -3]$
37. $\left\{y \mid y > \frac{2}{3}\right\}$, or $\left(\frac{2}{3}, \infty\right)$ **39.** $\left\{x \mid x \ge \frac{45}{4}\right\}$, or $\left[\frac{45}{4}, \infty\right)$
41. $\left\{x \mid x \le \frac{1}{2}\right\}$, or $\left(-\infty, \frac{1}{2}\right]$ **43.** $\left\{y \mid y \le -\frac{75}{2}\right\}$, or $\left(-\infty, -\frac{75}{2}\right]$
45. $\left\{x \mid x > -\frac{2}{17}\right\}$, or $\left(-\frac{2}{17}, \infty\right)$ **47.** $\left\{m \mid m > \frac{7}{3}\right\}$, or $\left(\frac{7}{3}, \infty\right)$
49. $\{r \mid r < -3\}$, or $(-\infty, -3)$ **51.** $\{x \mid x \ge 2\}$, or $[2, \infty)$
53. $\{y \mid y < 5\}$, or $(-\infty, 5)$ **55.** $\left\{x \mid x \le \frac{4}{7}\right\}$, or $\left(-\infty, \frac{4}{7}\right]$
57. $\{x \mid x < 8\}$, or $(-\infty, 8)$ **59.** $\left\{x \mid x \ge \frac{13}{2}\right\}$, or $\left[\frac{13}{2}, \infty\right)$
61. $\left\{x \mid x < \frac{11}{18}\right\}$, or $\left(-\infty, \frac{11}{18}\right)$
63. $\left\{x \mid x \ge -\frac{51}{31}\right\}$, or $\left[-\frac{51}{31}, \infty\right)$ **65.** $\{a \mid a \le 2\}$, or $(-\infty, 2]$
67. **(a)** 28.11; **(b)** $\{W \mid W <$ (approximately) 190.3 lb$\}$
69. $\{S \mid S \ge 84\}$ **71.** $\{B \mid B \ge \$11{,}500\}$
73. $\{S \mid S > \$7000\}$ **75.** $\{n \mid n > 25\}$ **77.** $\{p \mid p > 80\}$
79. $\{s \mid s > 8\}$ **81.** **(a)** 8.398 gal, 12.063 gal, 15.728 gal; **(b)** years after 1999 **83.** $3x^2 + 20x - 32$
84. $6r^2 - 23rs - 4s^2$ **85.** $6a^2 + 7a - 55$
86. $t^2 - 7st - 18s^2$ **87.** $\{x \mid x$ is a real number *and* $x \ne -8\}$, or $(-\infty, -8) \cup (-8, \infty)$
88. All real numbers **89.** All real numbers
90. $\left\{x \mid x \text{ is a real number } \textit{and } x \ne \frac{2}{3}\right\}$, or $\left(-\infty, \frac{2}{3}\right) \cup \left(\frac{2}{3}, \infty\right)$
91. ◆ **93.** **(a)** $\{p \mid p > 10\}$; **(b)** $\{p \mid p < 10\}$
95. True **97.** All real numbers **99.** All real numbers

Margin Exercises, Section 9.2, pp. 561–567

1. $\{0, 3\}$ **2.**

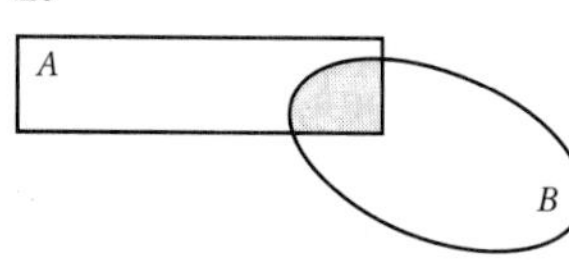

3. ; $(-1, 4)$
4. $\{x \mid -5 < x \le 10\}$, or $(-5, 10]$;
5. $\{x \mid x < -3\}$;
6. $\varnothing$ **7.** $\{x \mid 2 \le x \le 6\}$, or $[2, 6]$
8. $\{0, 1, 3, 4, 7, 9\}$ **9.**

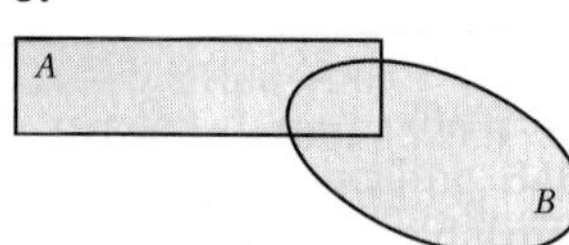

10. ; $(-\infty, -2] \cup (4, \infty)$
11. $\{x \mid x < 1$ *or* $x \ge 6\}$, or $(-\infty, 1) \cup [6, \infty)$;
12. $\left\{x \mid x \ge \frac{7}{2} \textit{ or } x < -2\right\}$, or $(-\infty, -2) \cup \left[\frac{7}{2}, \infty\right)$;
13. $\{x \mid x > -2$ *or* $x < -5\}$, or $(-\infty, -5) \cup (-2, \infty)$
14. All real numbers; **15.** $\{s \mid 8 < s < 19\}$

Exercise Set 9.2, p. 569

1. $\{9, 11\}$ **3.** $\{b\}$ **5.** $\{9, 10, 11, 13\}$
7. $\{a, b, c, d, f, g\}$ **9.** $\varnothing$ **11.** $\{3, 5, 7\}$
13. ; $(-4, 1]$
15. ; $(1, 6)$
17. $\{x \mid -4 \le x < 5\}$, or $[-4, 5)$;
19. $\{x \mid x \ge 2\}$, or $[2, \infty)$;
21. $\varnothing$ **23.** $\{x \mid -8 < x < 6\}$, or $(-8, 6)$
25. $\{x \mid -6 < x \le 2\}$, or $(-6, 2]$ **27.** $\{y \mid -1 < y \le 5\}$, or $(-1, 5]$ **29.** $\left\{x \mid -\frac{5}{3} \le x \le \frac{4}{3}\right\}$, or $\left[-\frac{5}{3}, \frac{4}{3}\right]$
31. $\{x \mid -1 < x \le 6\}$, or $(-1, 6]$
33. $\left\{x \mid -\frac{3}{2} \le x < \frac{9}{2}\right\}$, or $\left[-\frac{3}{2}, \frac{9}{2}\right)$ **35.** $\{x \mid 10 < x \le 14\}$, or $(10, 14]$ **37.** $\left\{x \mid -\frac{7}{2} < x < \frac{37}{2}\right\}$, or $\left(-\frac{7}{2}, \frac{37}{2}\right)$
39. ; $(-\infty, -2) \cup (1, \infty)$
41. ; $(-\infty, -3] \cup (1, \infty)$
43. $\{x \mid x < -5$ *or* $x > -1\}$, or $(-\infty, -5) \cup (-1, \infty)$;
45. $\left\{x \mid x \le \frac{5}{2} \textit{ or } x \ge 4\right\}$, or $\left(-\infty, \frac{5}{2}\right] \cup [4, \infty)$;
47. $\{x \mid x \ge -3\}$, or $[-3, \infty)$;
49. $\left\{x \mid x \le -\frac{5}{4} \textit{ or } x > -\frac{1}{2}\right\}$, or $\left(-\infty, -\frac{5}{4}\right] \cup \left(-\frac{1}{2}, \infty\right)$
51. All real numbers, or $(-\infty, \infty)$
53. $\{x \mid x > 2$ *or* $x < -4\}$, or $(-\infty, -4) \cup (2, \infty)$
55. $\left\{x \mid x < \frac{79}{4} \textit{ or } x > \frac{89}{4}\right\}$, or $\left(-\infty, \frac{79}{4}\right) \cup \left(\frac{89}{4}, \infty\right)$
57. $\left\{x \mid x \le -\frac{13}{2} \textit{ or } x \ge \frac{29}{2}\right\}$, or $\left(-\infty, -\frac{13}{2}\right] \cup \left[\frac{29}{2}, \infty\right)$
59. $\{d \mid 0 \le d \le 198\}$ **61.** $\{t \mid 14 \le t \le 19\}$, or from 2005 to 2010 **63.** Rounding to the nearest pound, $\{W \mid 152 \text{ lb} < W < 228 \text{ lb}\}$ **65.** $(-1, 2)$ **66.** $(-3, -5)$
67. $(4, -4)$ **68.** $y = -11x + 29$ **69.** $y = -4x + 7$
70. $y = -x + 2$ **71.** $6a^2 + 7ab - 5b^2$
72. $25y^2 + 35y + 6$ **73.** $21x^2 - 59x + 40$
74. $13x^2 + 37xy - 6y^2$ **75.** ◆
77. All real numbers; $\varnothing$ **79.** $\{x \mid -4 < x \le 1\}$, or $(-4, 1]$ **81.** $\left\{x \mid \frac{2}{5} \le x \le \frac{4}{5}\right\}$, or $\left[\frac{2}{5}, \frac{4}{5}\right]$

83. $\{x \mid -\frac{1}{8} < x < \frac{1}{2}\}$, or $\left(-\frac{1}{8}, \frac{1}{2}\right)$
85. $\{x \mid 10 < x \leq 18\}$, or $(10, 18]$

Margin Exercises, Section 9.3, pp. 573–578

1. $7|x|$ **2.** x^8 **3.** $5a^2|b|$ **4.** $\dfrac{7|a|}{b^2}$ **5.** $9|x|$ **6.** 29
7. 5 **8.** $|p|$
9. $\{6, -6\}$; **10.** $\varnothing$
11. $\{0\}$ **12.** $\{2, -2\}$ **13.** $\left\{\frac{17}{4}, -\frac{17}{4}\right\}$ **14.** $\{4, -4\}$
15. $\{3, 5\}$ **16.** $\left\{-\frac{13}{3}, 7\right\}$ **17.** $\varnothing$ **18.** $\left\{\frac{7}{4}, -\frac{1}{6}\right\}$
19. $\left\{-\frac{7}{2}\right\}$
20. $\{5, -5\}$; 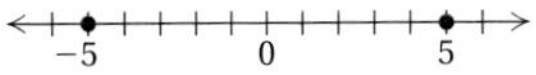
21. $\{x \mid -5 < x < 5\}$, or $(-5, 5)$;
22. $\{x \mid x \leq -5 \text{ or } x \geq 5\}$, or $(-\infty, -5] \cup [5, \infty)$;
23. $\{x \mid -2 < x < 5\}$, or $(-2, 5)$;
24. $\left\{x \mid 1 \leq x \leq \frac{11}{3}\right\}$, or $\left[1, \frac{11}{3}\right]$
25. $\left\{x \mid x \leq -\frac{7}{3} \text{ or } x \geq 1\right\}$, or $\left(-\infty, -\frac{7}{3}\right] \cup [1, \infty)$;

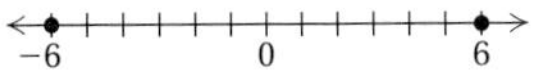

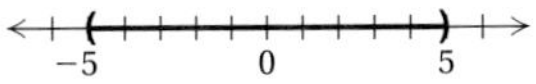

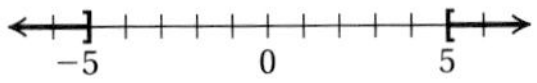

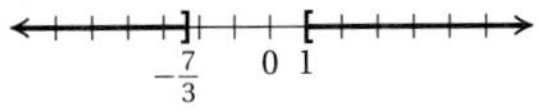

Exercise Set 9.3, p. 579

1. $9|x|$ **3.** $2x^2$ **5.** $2x^2$ **7.** $6|y|$ **9.** $\dfrac{2}{|x|}$ **11.** $\dfrac{x^2}{|y|}$
13. $4|x|$ **15.** 38 **17.** 19 **19.** 6.3 **21.** 5
23. $\{3, -3\}$ **25.** $\varnothing$ **27.** $\{0\}$ **29.** $\{15, -9\}$
31. $\left\{\frac{7}{2}, -\frac{1}{2}\right\}$ **33.** $\left\{\frac{23}{4}, -\frac{5}{4}\right\}$ **35.** $\{11, -11\}$
37. $\{291, -291\}$ **39.** $\{8, -8\}$ **41.** $\{7, -7\}$
43. $\{2, -2\}$ **45.** $\{8, -7\}$ **47.** $\{2, -12\}$ **49.** $\left\{\frac{7}{2}, -\frac{5}{2}\right\}$
51. $\varnothing$ **53.** $\left\{-\frac{13}{54}, -\frac{7}{54}\right\}$ **55.** $\left\{\frac{3}{4}, -\frac{11}{2}\right\}$ **57.** $\left\{\frac{3}{2}\right\}$
59. $\left\{5, -\frac{3}{5}\right\}$ **61.** All real numbers **63.** $\left\{-\frac{3}{2}\right\}$
65. $\left\{\frac{24}{23}, 0\right\}$ **67.** $\left\{32, \frac{8}{3}\right\}$ **69.** $\{x \mid -3 < x < 3\}$, or $(-3, 3)$
71. $\{x \mid x \leq -2 \text{ or } x \geq 2\}$, or $(-\infty, -2] \cup [2, \infty)$
73. $\{x \mid 0 < x < 2\}$, or $(0, 2)$ **75.** $\{x \mid -5 \leq x \leq -3\}$, or $[-5, -3]$ **77.** $\left\{x \mid -\frac{1}{2} \leq x \leq \frac{7}{2}\right\}$, or $\left[-\frac{1}{2}, \frac{7}{2}\right]$
79. $\left\{y \mid y < -\frac{3}{2} \text{ or } y > \frac{17}{2}\right\}$, or $\left(-\infty, -\frac{3}{2}\right) \cup \left(\frac{17}{2}, \infty\right)$
81. $\left\{x \mid x \leq -\frac{5}{4} \text{ or } x \geq \frac{23}{4}\right\}$, or $\left(-\infty, -\frac{5}{4}\right] \cup \left[\frac{23}{4}, \infty\right)$
83. $\{y \mid -9 < y < 15\}$, or $(-9, 15)$
85. $\left\{x \mid -\frac{7}{2} \leq x \leq \frac{1}{2}\right\}$, or $\left[-\frac{7}{2}, \frac{1}{2}\right]$
87. $\left\{y \mid y < -\frac{4}{3} \text{ or } y > 4\right\}$, or $\left(-\infty, -\frac{4}{3}\right) \cup (4, \infty)$
89. $\left\{x \mid x \leq -\frac{5}{4} \text{ or } x \geq \frac{23}{4}\right\}$, or $\left(-\infty, -\frac{5}{4}\right] \cup \left[\frac{23}{4}, \infty\right)$
91. $\left\{x \mid -\frac{9}{2} < x < 6\right\}$, or $\left(-\frac{9}{2}, 6\right)$
93. $\left\{x \mid x \leq -\frac{25}{6} \text{ or } x \geq \frac{23}{6}\right\}$, or $\left(-\infty, -\frac{25}{6}\right] \cup \left[\frac{23}{6}, \infty\right)$
95. $\{x \mid -5 < x < 19\}$, or $(-5, 19)$
97. $\left\{x \mid x \leq -\frac{2}{15} \text{ or } x \geq \frac{14}{15}\right\}$, or $\left(-\infty, -\frac{2}{15}\right] \cup \left[\frac{14}{15}, \infty\right)$
99. $\{m \mid -12 \leq m \leq 2\}$, or $[-12, 2]$
101. $\left\{x \mid \frac{1}{2} \leq x \leq \frac{5}{2}\right\}$, or $\left[\frac{1}{2}, \frac{5}{2}\right]$
103. $\{x \mid 0.49705 \leq x \leq 0.50295\}$, or $[0.49705, 0.50295]$
105. All real numbers **106.** $\left\{x \mid x \text{ is a real number and } x \neq -\frac{5}{2}\right\}$, or $\left(-\infty, -\frac{5}{2}\right) \cup \left(-\frac{5}{2}, \infty\right)$ **107.** $\{x \mid x \text{ is a real number and } x \neq 5\}$, or $(-\infty, 5) \cup (5, \infty)$ **108.** All real numbers **109.** $(6, -5)$ **110.** $(2, -4)$ **111.** ◈
113. $\left\{d \mid 5\frac{1}{2} \text{ ft} \leq d \leq 6\frac{1}{2} \text{ ft}\right\}$ **115.** $\{x \mid x \geq -5\}$, or $[-5, \infty)$
117. $\left\{1, -\frac{1}{4}\right\}$ **119.** $\varnothing$ **121.** All real numbers
123. All real numbers **125.** $|x| < 3$ **127.** $|x| \geq 6$
129. $|x + 3| > 5$

Margin Exercises, Section 9.4, pp. 583–590

1. No **2.** Yes
3.
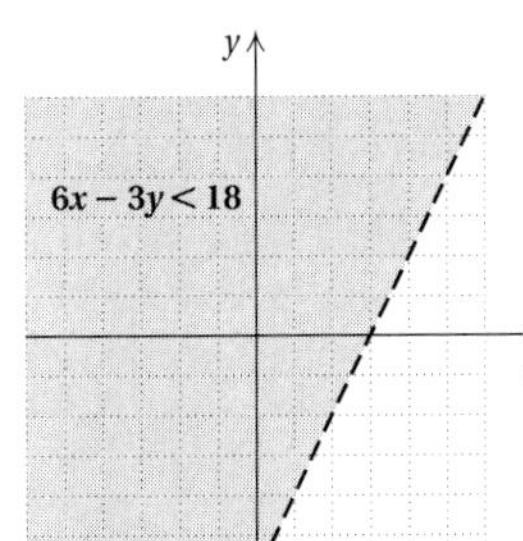

4.
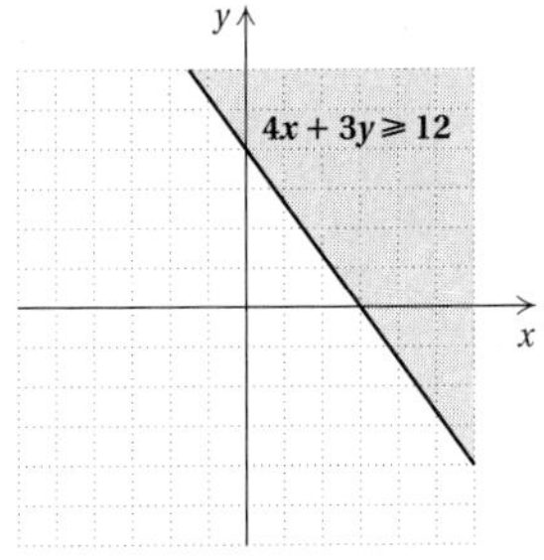

5.
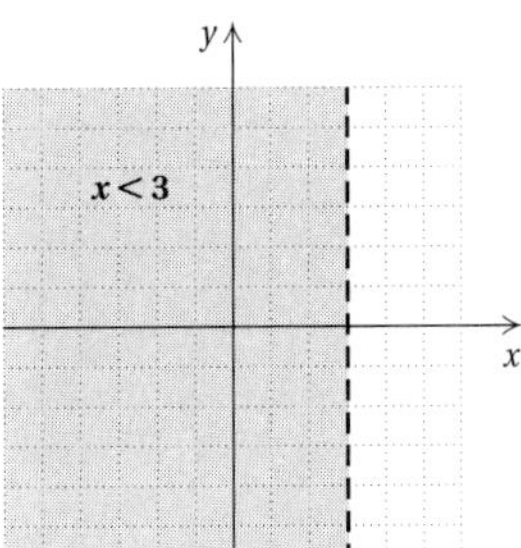

6.
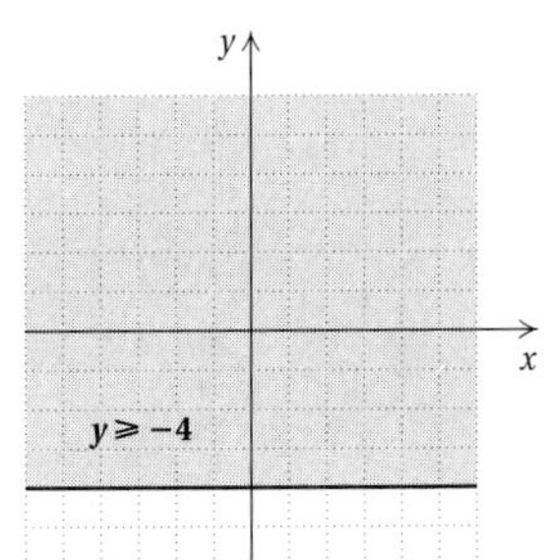

7.
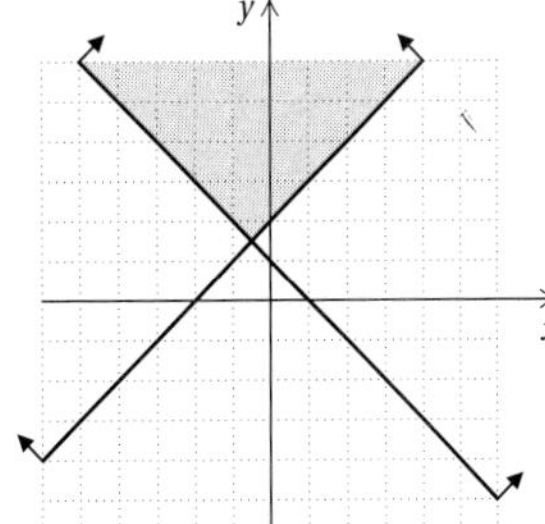

8.
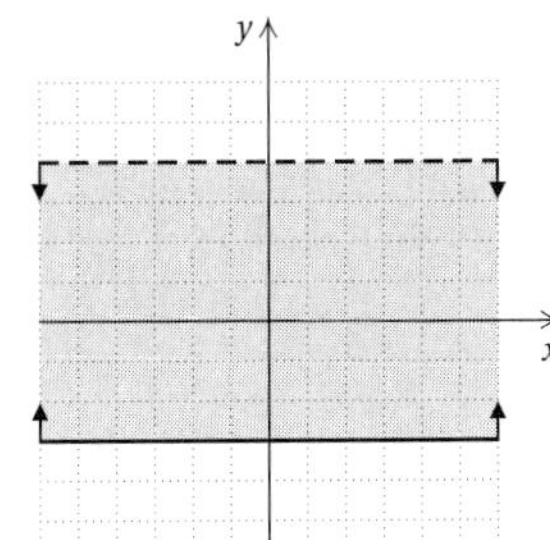

9.

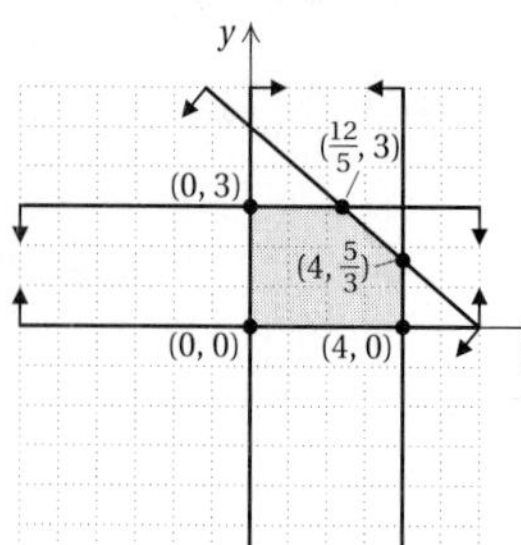

10.

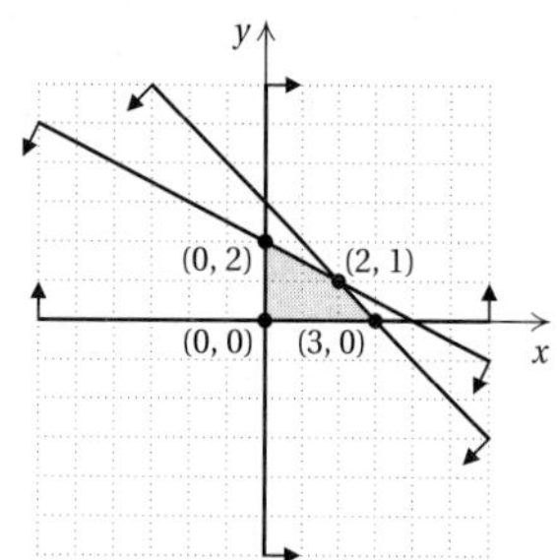

Exercise Set 9.4, p. 591

1. Yes **3.** Yes

5.

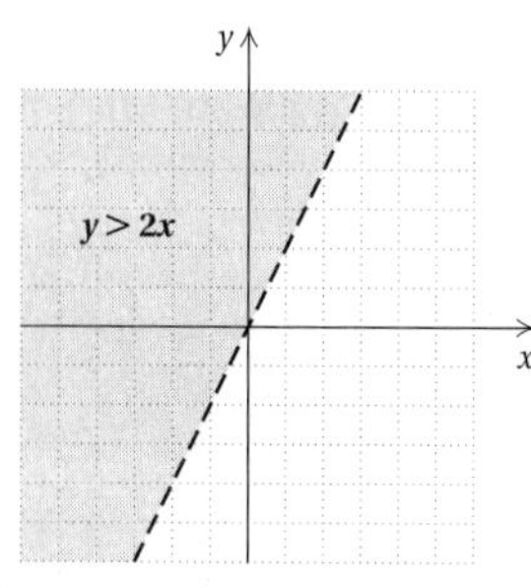

7.

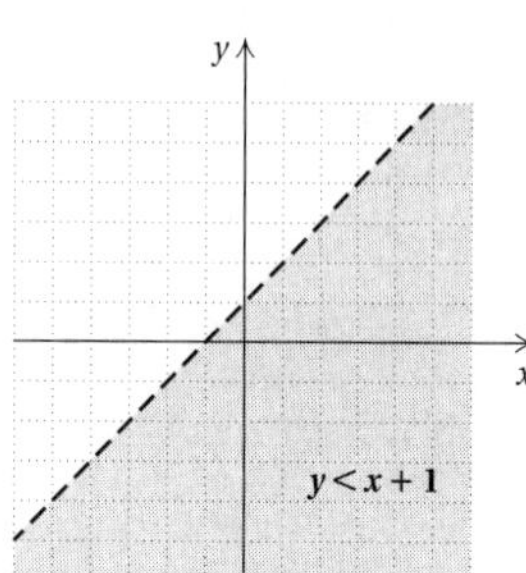

9.

$y > x - 2$

11.

$x + y < 4$

13.

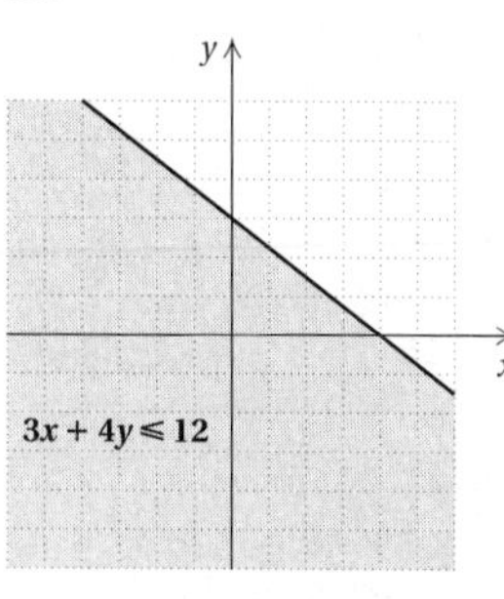

15.

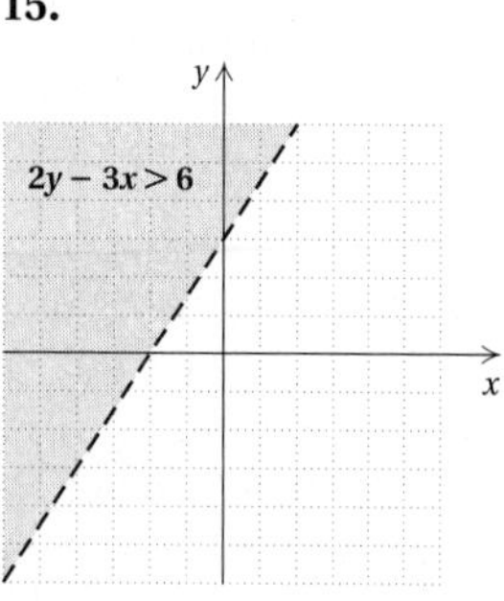

17.

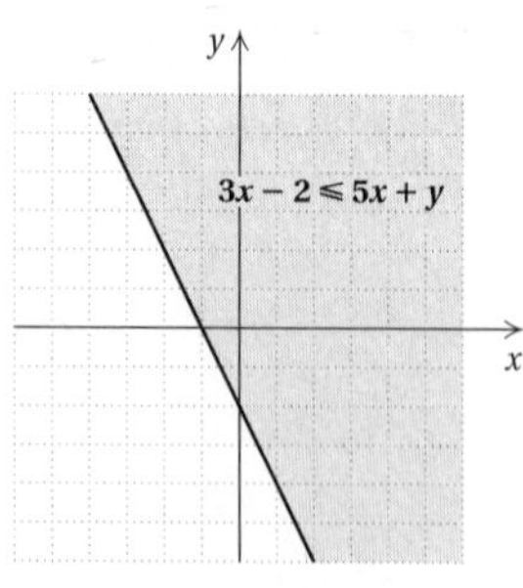

19.

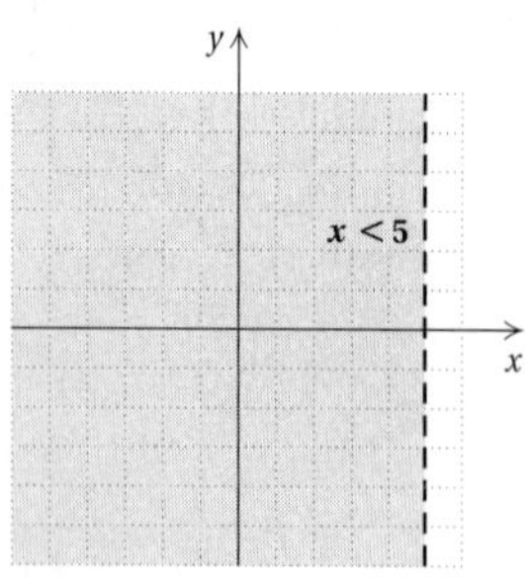

21.

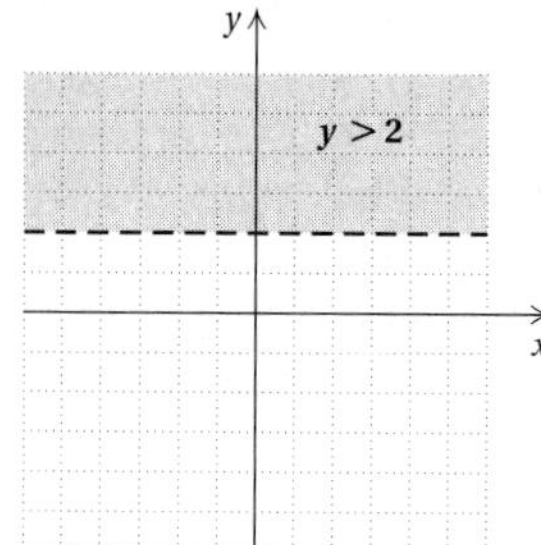

23.

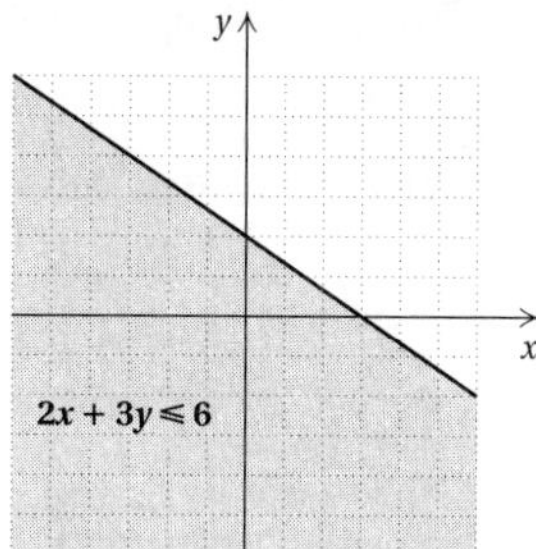

25.

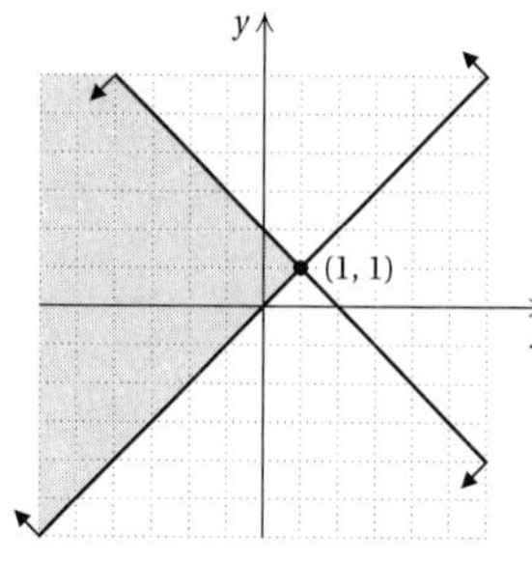

27.

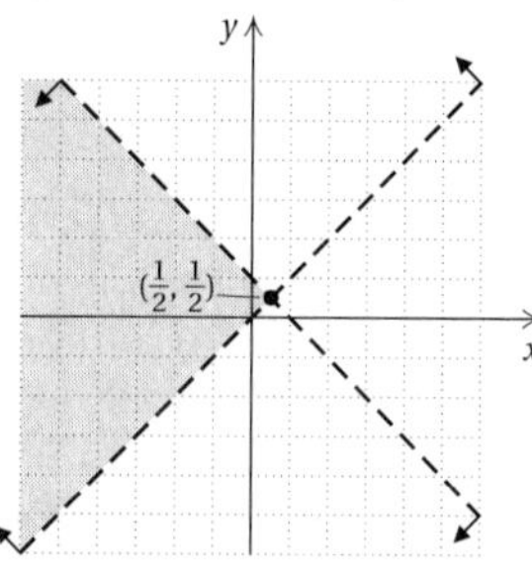

29.

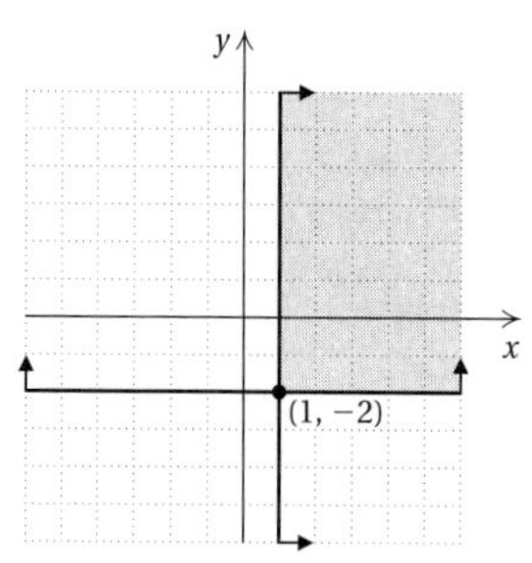

31.

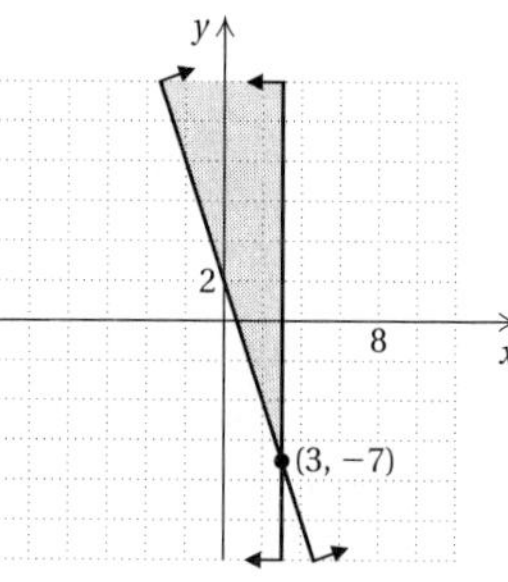

33.

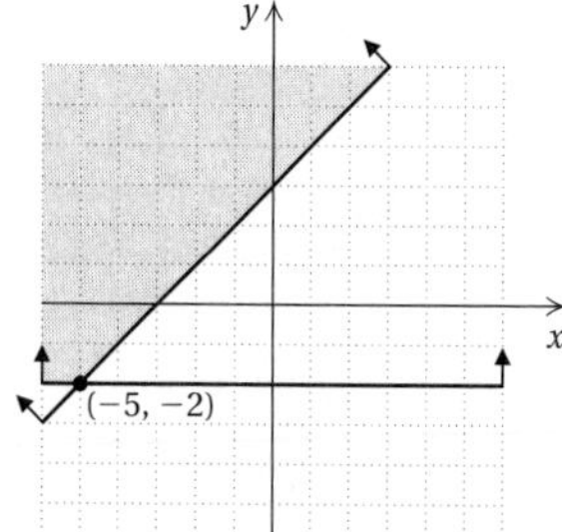

35.

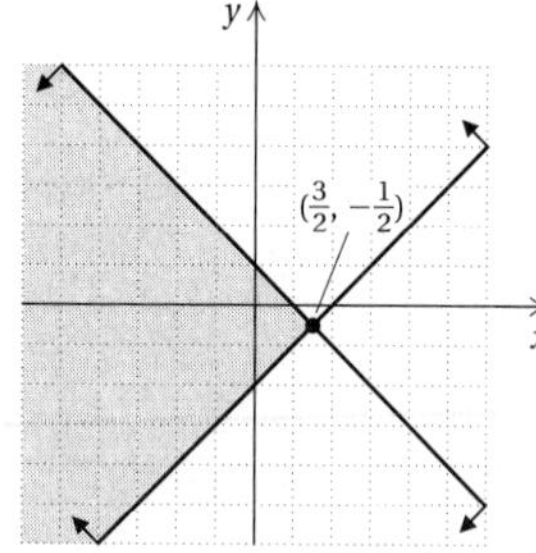

37.

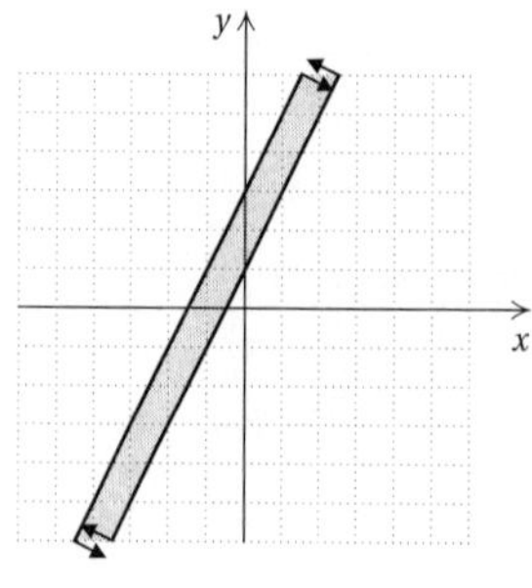

39.

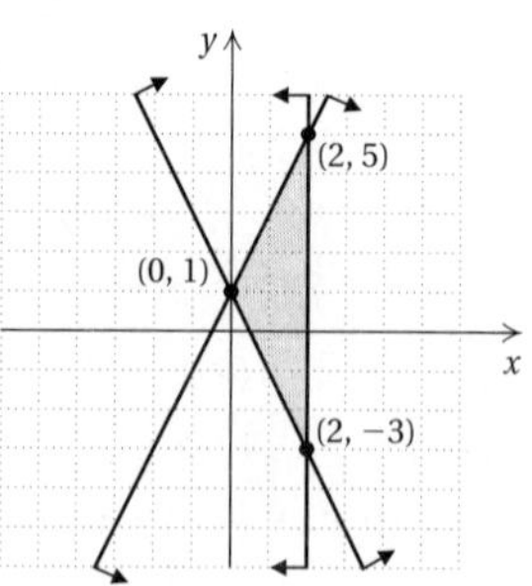

41.

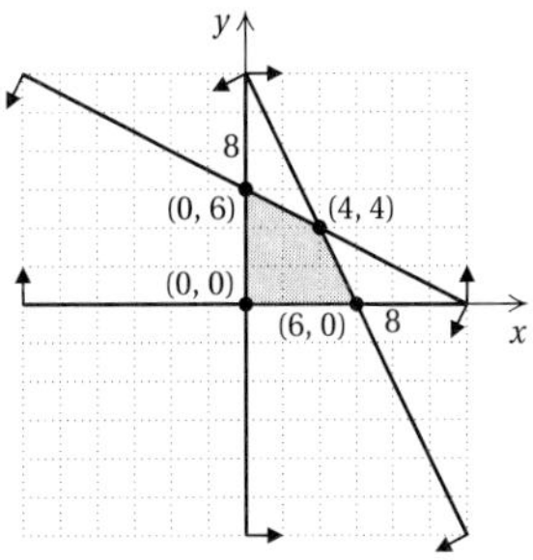

43.

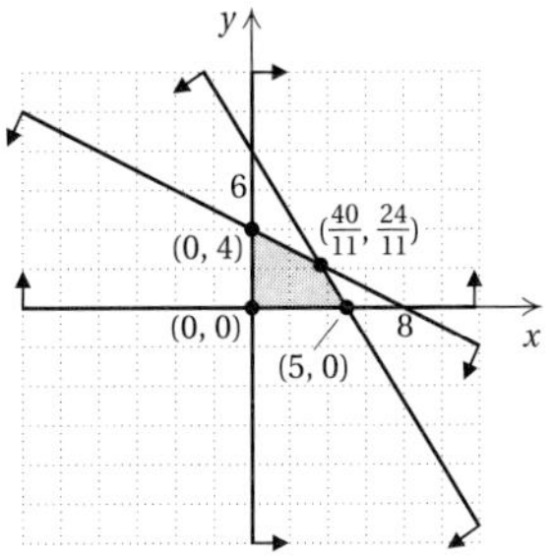

45. $y = \frac{1}{2}x - \frac{7}{2}$ **46.** $y = \frac{3}{8}x$ **47.** $y = -3x + 15$

48. $y = \frac{9}{2}x + 9$ **49.** $y = -\frac{3}{2}x - \frac{5}{2}$ **50.** $y = 6$

51. 2 **52.** 3 **53.** 1 **54.** 8 **55.** 4 **56.** $|2 - 2a|$, or $2|1 - a|$ **57.** 6 **58.** 0.2 **59.** ◈

61. $0 < w \le 62$,
$0 < h \le 62$,
$62 + 2w + 2h \le 108$, or $w + h \le 23$

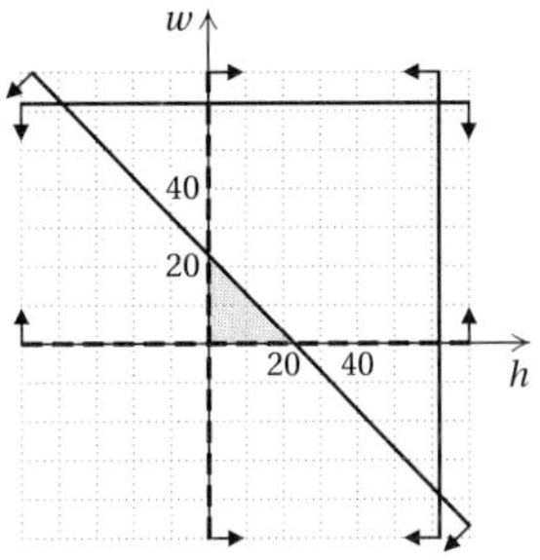

63.–67. Left to the student.

Summary and Review: Chapter 9, p. 595

1. $[-8, 9)$ **2.** $(-\infty, 40]$

3. 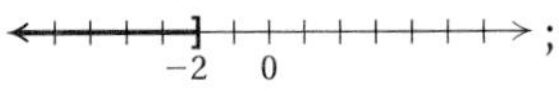 ; $(-\infty, -2]$

4. 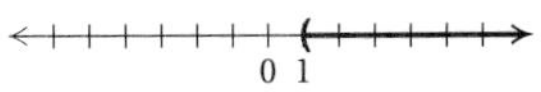 ; $(1, \infty)$

5. $\{a \mid a \le -21\}$, or $(-\infty, -21]$

6. $\{y \mid y \ge -7\}$, or $[-7, \infty)$ **7.** $\{y \mid y > -4\}$, or $(-4, \infty)$

8. $\{y \mid y > -30\}$, or $(-30, \infty)$

9. $\{x \mid x > -3\}$, or $(-3, \infty)$

10. $\left\{y \mid y \le -\frac{6}{5}\right\}$, or $\left(-\infty, -\frac{6}{5}\right]$

11. $\{x \mid x < -3\}$, or $(-\infty, -3)$ **12.** $\{y \mid y > -10\}$, or $(-10, \infty)$ **13.** $\left\{x \mid x \le -\frac{5}{2}\right\}$, or $\left(-\infty, -\frac{5}{2}\right]$

14. $\left\{t \mid t > 4\frac{1}{4} \text{ hr}\right\}$ **15.** \$10,000

16. 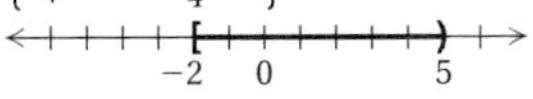 ; $[-2, 5)$

17. 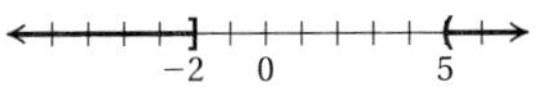 ; $(-\infty, -2] \cup (5, \infty)$

18. $\varnothing$ **19.** $\{x \mid -7 < x \le 2\}$, or $(-7, 2]$

20. $\left\{x \middle| -\frac{5}{4} < x < \frac{5}{2}\right\}$, or $\left(-\frac{5}{4}, \frac{5}{2}\right)$

21. $\{x \mid x < -3 \text{ or } x > 1\}$, or $(-\infty, -3) \cup (1, \infty)$

22. $\{x \mid x < -11 \text{ or } x \ge -6\}$, or $(-\infty, -11) \cup [-6, \infty)$

23. $\{x \mid x \le -6 \text{ or } x \ge 8\}$, or $(-\infty, -6] \cup [8, \infty)$

24. $\{t \mid 3.23 < t < 7.85\}$, or approximately between 1991 and 1996 **25.** $\dfrac{3}{|x|}$ **26.** $\dfrac{2|x|}{y^2}$ **27.** $\dfrac{4}{|y|}$ **28.** 62

29. 6, −6 **30.** $\{9, -5\}$ **31.** $\left\{-14, \frac{4}{3}\right\}$ **32.** $\varnothing$

33. $\left\{x \middle| -\frac{17}{2} < x < \frac{7}{2}\right\}$, or $\left(-\frac{17}{2}, \frac{7}{2}\right)$

34. $\{x \mid x \le -3.5 \text{ or } x \ge 3.5\}$, or $(-\infty, -3.5] \cup [3.5, \infty)$

35. $\left\{x \middle| x \le -\frac{11}{3} \text{ or } x \ge \frac{19}{3}\right\}$, or $\left(-\infty, -\frac{11}{3}\right] \cup \left[\frac{19}{3}, \infty\right)$

36. $\varnothing$ **37.** $\{1, 5, 9\}$ **38.** $\{1, 2, 3, 5, 6, 9\}$

39.

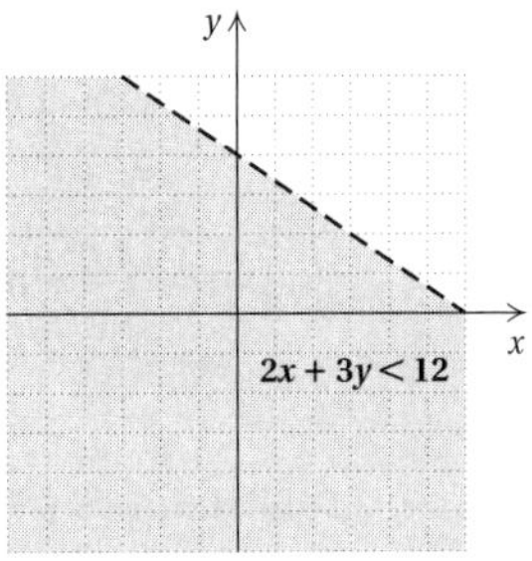

40.

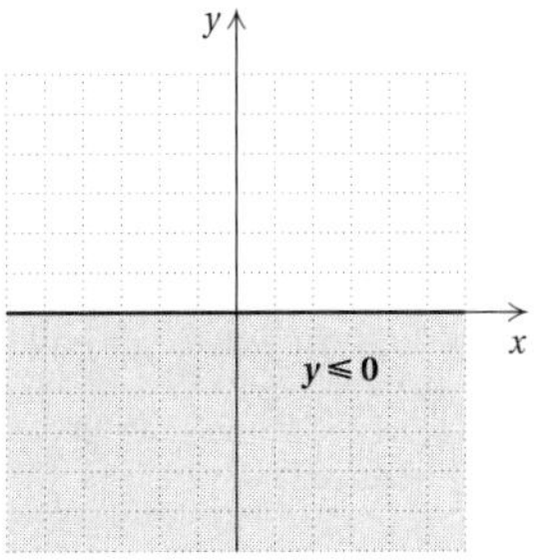

41.

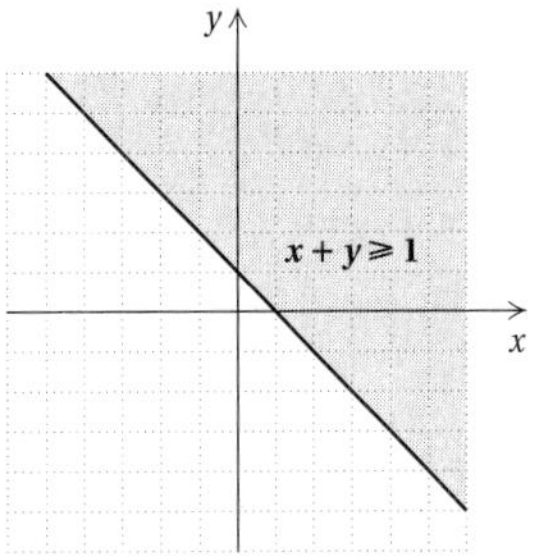

42.

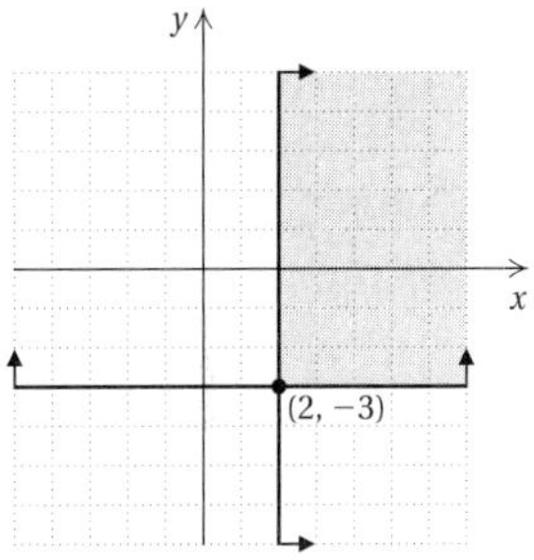

43.

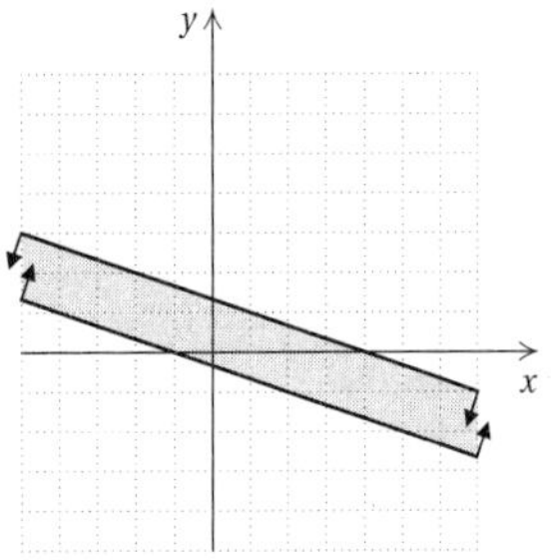

44.

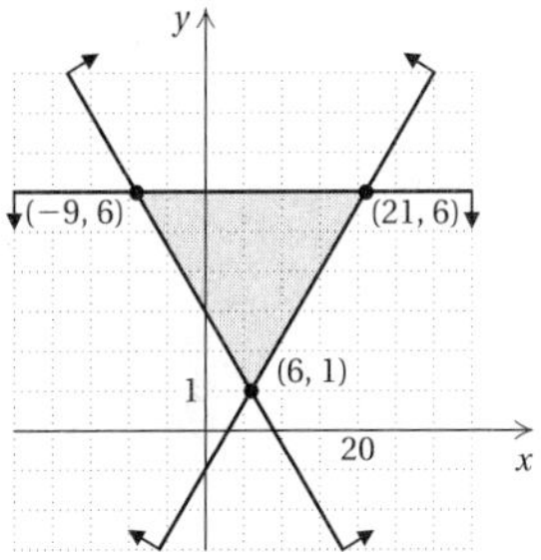

45. $24y^2 - 41y - 35$

46. $\{x \mid x \text{ is a real number } and\ x \ne -9\}$, or $(-\infty, -9) \cup (-9, \infty)$ **47.** $(2, -1)$ **48.** $y = -\frac{2}{5}x - \frac{9}{5}$

49. ◈ **(1)** $-9(x + 2) = -9x - 18$, not $-9x + 2$. **(2)** This would be correct if (1) were correct. **(3)** If (2) were correct, the right-hand side would be −5, not 8. **(4)** The inequality symbol should be reversed. The correct solution is

$$7 - 9x + 6x < -9(x + 2) + 10x$$
$$7 - 9x + 6x < -9x - 18 + 10x$$
$$7 - 3x < x - 18$$
$$-4x < -25$$
$$x > \frac{25}{4}.$$

50. "Solve" can mean to find all the replacements that make an equation or inequality true. It can also mean to express a formula as an equivalent equation with a given variable alone on one side.

51. $\left\{x \middle| -\frac{8}{3} \le x \le -2\right\}$, or $\left[-\frac{8}{3}, -2\right]$

Test: Chapter 9, p. 597

1. [9.1b] $(-3, 2]$ **2.** [9.1b] $(-4, \infty)$

3. [9.1c] 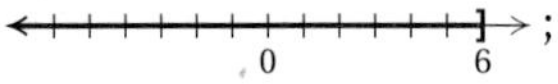

; $(-\infty, 6]$

4. [9.1c] 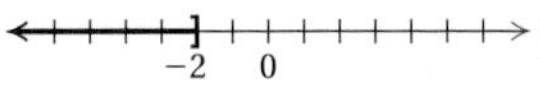

; $(-\infty, -2]$

5. [9.1c] $\{y|y > -50\}$, or $(-50, \infty)$

6. [9.1c] $\left\{a \middle| a \le \frac{11}{5}\right\}$, or $\left(-\infty, \frac{11}{5}\right]$ **7.** [9.1c] $\{y|y > 1\}$, or $(1, \infty)$ **8.** [9.1c] $\left\{x \middle| x > \frac{5}{2}\right\}$, or $\left(\frac{5}{2}, \infty\right)$

9. [9.1c] $\left\{x \middle| x \le \frac{7}{4}\right\}$, or $\left(-\infty, \frac{7}{4}\right]$ **10.** [9.1d] $\left\{h \middle| h > 2\frac{1}{10}\text{hr}\right\}$

11. [9.1d] $\{d|33 \text{ ft} \le d \le 231 \text{ ft}\}$

12. [9.2a] 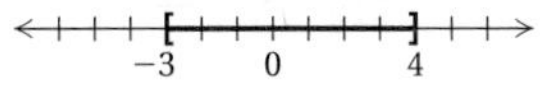

; $[-3, 4]$

13. [9.2b] 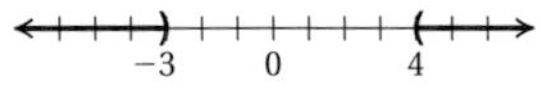

; $(-\infty, -3) \cup (4, \infty)$

14. [9.2a] $\{x|x \ge 4\}$, or $[4, \infty)$

15. [9.2a] $\{x|-1 < x < 6\}$, or $(-1, 6)$

16. [9.2a] $\left\{x \middle| -\frac{2}{5} < x \le \frac{9}{5}\right\}$, or $\left(-\frac{2}{5}, \frac{9}{5}\right]$

17. [9.2b] $\left\{x \middle| x < -4 \text{ or } x > -\frac{5}{2}\right\}$, or $(-\infty, -4) \cup \left(-\frac{5}{2}, \infty\right)$

18. [9.2b] All real numbers, or $(-\infty, \infty)$

19. [9.2b] $\{x|x < 3 \text{ or } x > 6\}$, or $(-\infty, 3) \cup (6, \infty)$

20. [9.3a] $\dfrac{7}{|x|}$ **21.** [9.3a] $2|x|$ **22.** [9.3b] 8.4

23. [9.3c] $\{9, -9\}$ **24.** [9.3e] $\{x|x < -3 \text{ or } x > 3\}$, or $(-\infty, -3) \cup (3, \infty)$ **25.** [9.3e] $\left\{x \middle| -\frac{7}{8} < x < \frac{11}{8}\right\}$, or $\left(-\frac{7}{8}, \frac{11}{8}\right)$ **26.** [9.3e] $\left\{x \middle| x \le -\frac{13}{5} \text{ or } x \ge \frac{7}{5}\right\}$, or $\left(-\infty, -\frac{13}{5}\right] \cup \left[\frac{7}{5}, \infty\right)$ **27.** [9.3d] $\{1\}$ **28.** [9.3c] $\varnothing$

29. [9.3e] $\{x|-99 \le x \le 111\}$, or $[-99, 111]$

30. [9.2a] $\{3, 5\}$ **31.** [9.2b] $\{1, 3, 5, 7, 9, 11, 13\}$

32. [9.4b]

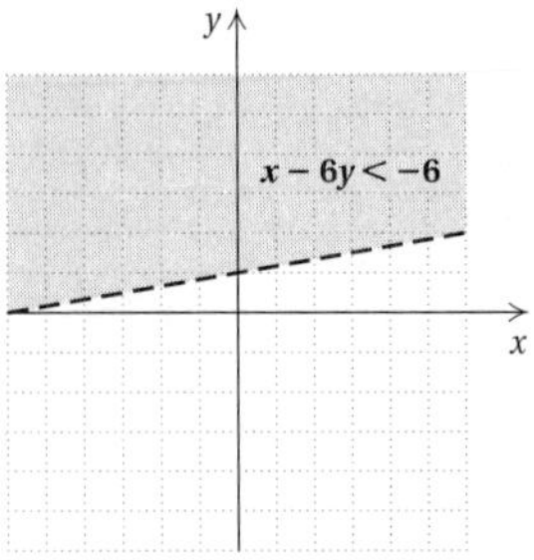

33. [9.4b]

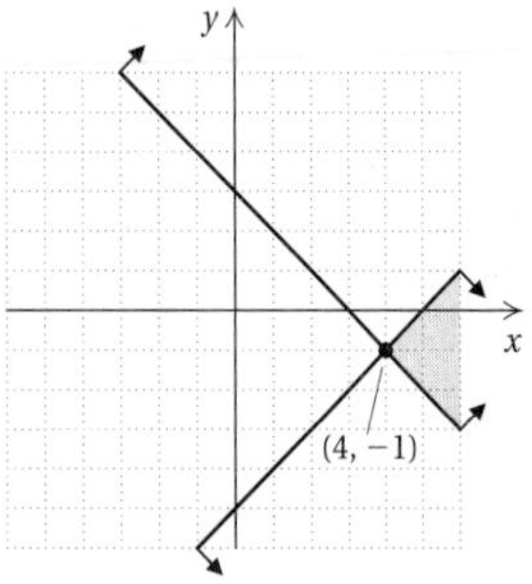

34. [9.4b]

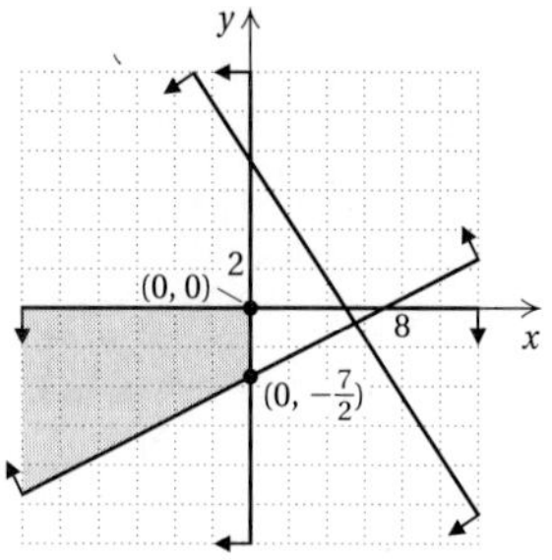

35. [4.6a] $72x^2 - 11x - 35$ **36.** [7.5c] $y = -2x + 4$

37. [8.2b] $(-8, -4)$ **38.** [7.2a] $\{x|x$ is a real number *and* $x \ne 20\}$, or $(-\infty, 20) \cup (20, \infty)$ **39.** [9.3e] $\varnothing$

40. [9.2a] $\left\{x \middle| \frac{1}{5} < x < \frac{4}{5}\right\}$, or $\left(\frac{1}{5}, \frac{4}{5}\right)$

Cumulative Review: Chapters 1–9, p. 599

1. [1.1a] $\frac{4}{3}$ **2.** [1.2e] 12 **3.** [1.6c] $-\frac{8}{9}$

4. [1.8b] $47b - 51$ **5.** [1.8d] 224 **6.** [4.2a, b] $\dfrac{x^6}{4y^8}$

7. [4.4a] $16p^2 - 8p$ **8.** [4.6c] $36m^2 - 12mn + n^2$

9. [4.6a] $15a^2 - 14ab - 8b^2$ **10.** [6.1d] $\dfrac{y-2}{3}$

11. [6.1f] $\dfrac{3x-5}{x+4}$ **12.** [6.4a] $\dfrac{6x+13}{20(x-3)}$

13. [6.4b] $\dfrac{4x+1}{(x+2)(x-2)}$ **14.** [6.8a] $\dfrac{y+2x}{3y-x}$

15. [6.8a] $\dfrac{y^3-2y}{y^3-1}$ **16.** [4.8b] $2x^2 - 11x + 23 + \dfrac{-49}{x+2}$

17. [2.3a] $\frac{15}{2}$ **18.** [2.6a] $C = \frac{5}{9}(F - 32)$

19. [9.2a] $\left\{x \middle| -3 < x < -\frac{3}{2}\right\}$, or $\left(-3, -\frac{3}{2}\right)$

20. [9.3e] $\{x|x \le -2.1 \text{ or } x \ge 2.1\}$, or $(-\infty, -2.1] \cup [2.1, \infty)$ **21.** [8.2b] Infinite number of solutions **22.** [8.4a] $(3, 2, -1)$ **23.** [5.8b] $\frac{1}{4}$

24. [5.8b] $-2, \frac{7}{2}$ **25.** [6.5a] No solution **26.** [6.5a] -1

27. [6.7a] $a = -\dfrac{bP}{P-3}$, or $\dfrac{bP}{3-P}$ **28.** [2.3a] 13

29. [9.1c] $\left\{x \middle| x \ge -\frac{7}{9}\right\}$, or $\left[-\frac{7}{9}, \infty\right)$

30. [9.3e] $\{x|-2 < x < 5\}$, or $(-2, 5)$ **31.** [9.3c] $\frac{4}{3}, \frac{8}{3}$

32. [8.2a] $(-2, 1)$ **33.** [8.4a] $\left(\frac{5}{8}, \frac{1}{16}, -\frac{3}{4}\right)$

34. [5.1b] $2x^2(2x + 9)$

35. [5.1c] $(2a - 1)(4a^2 - 3)$ **36.** [5.2a] $(x - 6)(x + 14)$

37. [5.3a], [5.4a] $(2x + 5)(3x - 2)$

38. [5.5c] $(4y + 9)(4y - 9)$ **39.** [5.5b] $(t - 8)^2$

40. [5.6a] $8(2x + 1)(4x^2 - 2x + 1)$

41. [5.6a] $(0.3b - 0.2c)(0.09b^2 + 0.06bc + 0.04c^2)$

42. [5.7a] $x^2(x^2 + 1)(x + 1)(x - 1)$

43. [5.3a], [5.4a] $(4x - 1)(5x + 3)$

44. [3.2b]

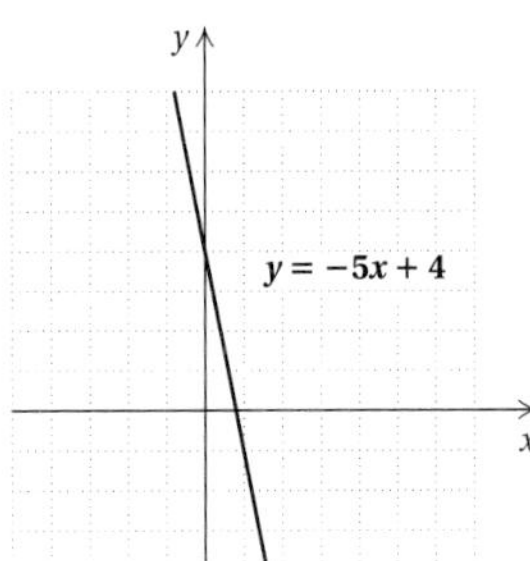

45. [3.3b]

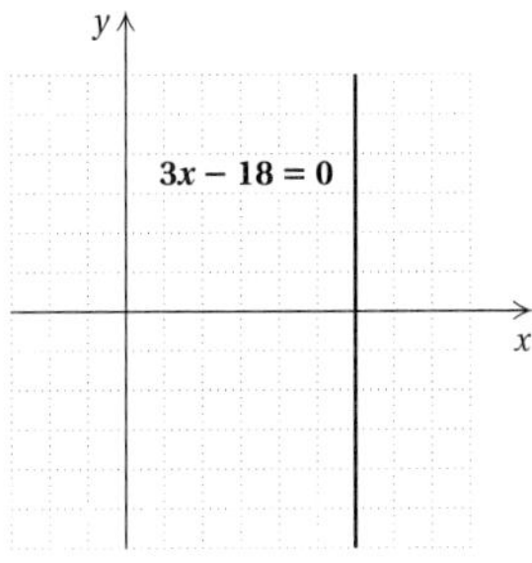

46. [9.4b]

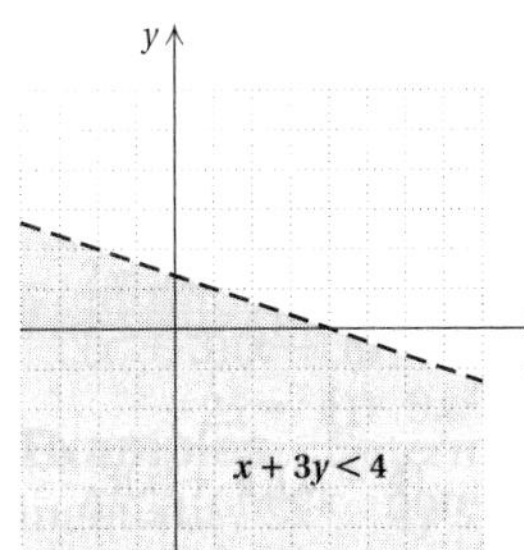

47. [9.4b]

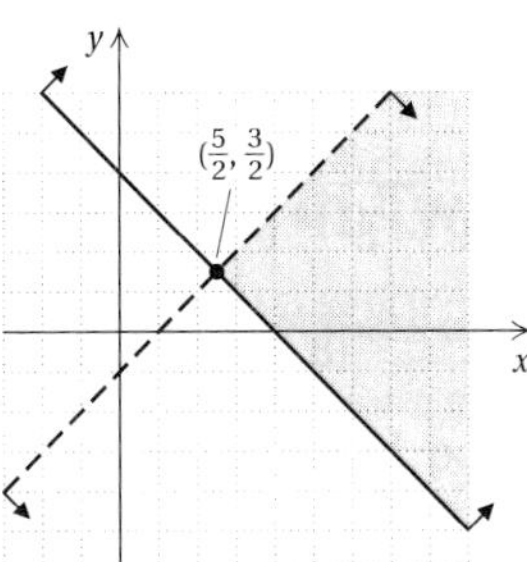

48. [7.5d] $y = -\frac{1}{2}x + \frac{17}{2}$ **49.** [7.5d] $y = \frac{4}{3}x - 6$
50. [7.5c] $x = -1$ **51.** [7.5b] $y = -3x + 7$
52. [7.3b] $m = \frac{3}{2}$; y-intercept $= (0, -4)$
53. [8.3a] Finland: \$7.99; Sweden: \$6.45
54. [8.2c] \$1170 for service; \$1716 for disposables
55. [2.4a] Yes **56.** [2.5a] $\$45{,}333.3\overline{3}$ **57.** [2.5a] 20.3%
58. [8.3a] Detroit: \$4.09; Chicago: \$1.66
59. [8.5a] Win: 38; lose: 30; tie: 13 **60.** [6.6a] $6\frac{2}{3}$ hr
61. [6.9b] 60 yr old **62.** [7.1e] **(a)** 4.8 lb; **(b)** 2030;
(c) $\{t \mid 34 \le t \le 39\}$, or from 2025 to 2030
63. [7.6b] **(a)** $S(x) = \frac{11}{15}x + \frac{143}{30}$;
(b) \$12.1 billion; \$15.0 billion
64. [7.2a] Domain $= \{-5, -4, -2, 0, 1, 2, 3, 4, 5\}$;
range $= \{-3, -2, 3, 4\}$ **65.** [7.2a] Domain $= [-5, 5]$;
range $= [-3, 4]$ **66.** [7.2a] Domain $= (-\infty, \infty)$;
range $= [-5, \infty)$ **67.** [7.2a] Domain $= [-5, 5]$;
range $= [0, 3]$ **68.** [8.5a] $a = 1$, $b = -5$, $c = 6$
69. [6.5a] All real numbers except 9 and -5
70. [5.8b] $0, \frac{1}{4}, -\frac{1}{4}$

Chapter 10

Pretest: Chapter 10, p. 604

1. [10.1b] $|t|$ **2.** [10.1c] $3x$ **3.** [10.1d] $|y|$
4. [10.2d] $3a\sqrt[3]{3ab^2}$ **5.** [10.4a] $4\sqrt{5}$
6. [10.3a] $12x^2\sqrt{x}$ **7.** [10.4b] $25 - 4\sqrt{6}$
8. [10.3b] $2\sqrt{a}$ **9.** [10.5b] $-6 + 3\sqrt{5}$
10. [10.6a] 13 **11.** [10.6a] $-\frac{7}{6}$ **12.** [10.6b] 0, 8
13. [10.7a] $\sqrt{39}$; 6.245 **14.** [10.7a] $\sqrt{32}$ ft; 5.657 ft
15. [10.2d] $\sqrt[6]{x^5}$ **16.** [10.8f] No
17. [10.8b] $-5 + 13i$ **18.** [10.8c] $24 - 7i$
19. [10.8e] $-\frac{14}{13} - \frac{5}{13}i$ **20.** [10.8d] $-i$

Margin Exercises, Section 10.1, pp. 605–610

1. 3, −3 **2.** 6, −6 **3.** 11, −11 **4.** 1 **5.** 6 **6.** $\frac{9}{10}$
7. 0.08 **8.** **(a)** 4; **(b)** −4; **(c)** does not exist as a real number **9.** **(a)** 7; **(b)** −7; **(c)** does not exist as a real number **10.** **(a)** 12; **(b)** −12; **(c)** does not exist as a real number **11.** 4.123 **12.** 6.325 **13.** 33.734
14. −29.455 **15.** 0.793 **16.** −5.569 **17.** $28 + x$
18. $\dfrac{y}{y + 3}$ **19.** 2; $\sqrt{22} \approx 4.690$; does not exist as a real number **20.** −2; $-\sqrt{7}$; does not exist as a real number **21.** Domain $= \{x \mid x \ge 5\} = [5, \infty)$
22. Domain $= \left\{x \mid x \ge -\frac{3}{2}\right\} = \left[-\frac{3}{2}, \infty\right)$
23.

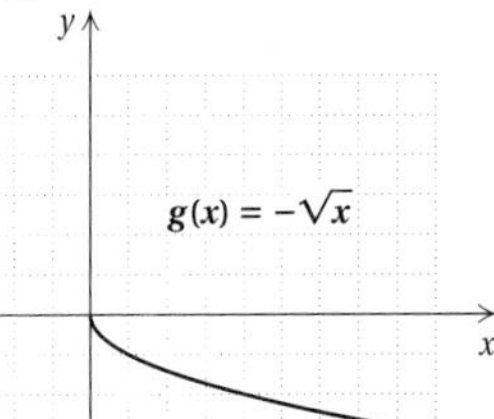

24.

25. $|y|$ **26.** 24 **27.** $5|y|$ **28.** $4|y|$ **29.** $|x + 7|$
30. $2|x - 2|$ **31.** $7|y + 5|$ **32.** $|x - 3|$ **33.** −4
34. $3y$ **35.** $2(x + 2)$ **36.** $-\frac{7}{4}$ **37.** −3; 0;
$\sqrt[3]{-5} \approx -1.710$; $\sqrt[3]{7} \approx 1.913$ **38.** 3 **39.** −3 **40.** x
41. y **42.** 0 **43.** $-2x$ **44.** $3x + 2$ **45.** 3
46. −3 **47.** Does not exist as a real number
48. 0 **49.** $2|x - 2|$ **50.** $|x|$ **51.** $|x + 3|$

Calculator Spotlight, p. 609

1. 2.802039331 **2.** −3.50339806 **3.** 3.601131332
4. −12.68265141

Calculator Spotlight, p. 610

1. Domain $= [0, \infty)$; range $= [0, \infty)$
2. Domain $= [-2, \infty)$; range $= [0, \infty)$
3. Domain $= (-\infty, \infty)$; range $= (-\infty, \infty)$
4. Domain $= (-\infty, \infty)$; range $= (-\infty, \infty)$
5. Domain $= [1, \infty)$; range $= [0, \infty)$
6. Domain $= (-\infty, \infty)$; range $= (-\infty, \infty)$
7. Domain $= [-3, \infty)$; range $= (-\infty, 5]$
8. Domain $= (-\infty, \infty)$; range $= (-\infty, \infty)$
9. Incorrect **10.** Correct

Exercise Set 10.1, p. 611

1. 4, −4 **3.** 12, −12 **5.** 20, −20 **7.** $-\frac{7}{6}$ **9.** 14
11. 0.06 **13.** 18.628 **15.** 1.962 **17.** $y^2 + 16$
19. $\dfrac{x}{y - 1}$ **21.** $\sqrt{20}$; 0; does not exist as a real number; does not exist as a real number **23.** $\sqrt{11}$; does not exist as a real number; $\sqrt{11}$; 12
25. Domain $= \{x \mid x \ge 2\} = [2, \infty)$
27. About 24.5 mph; about 54.8 mph

29.

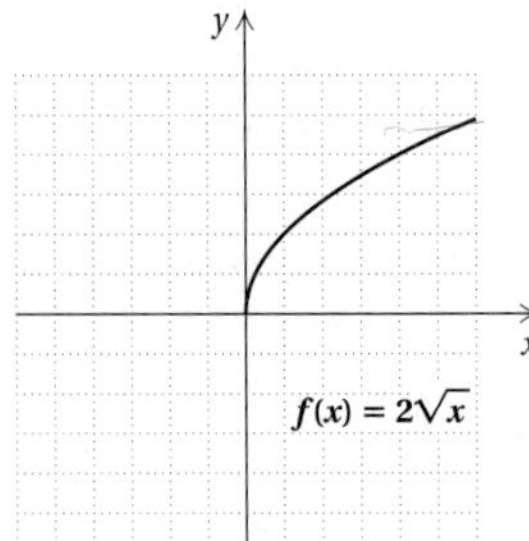

31.

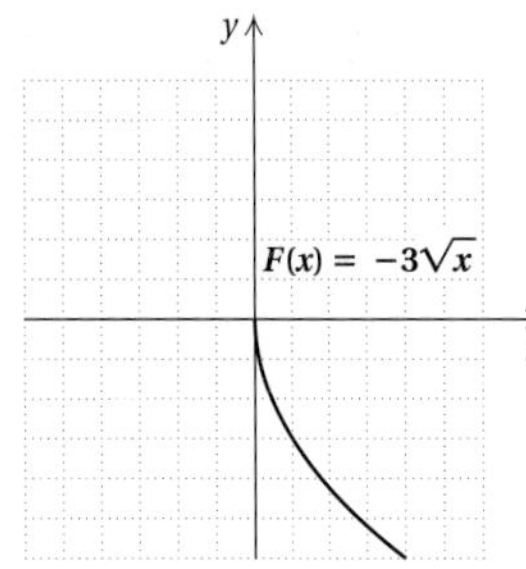

33.

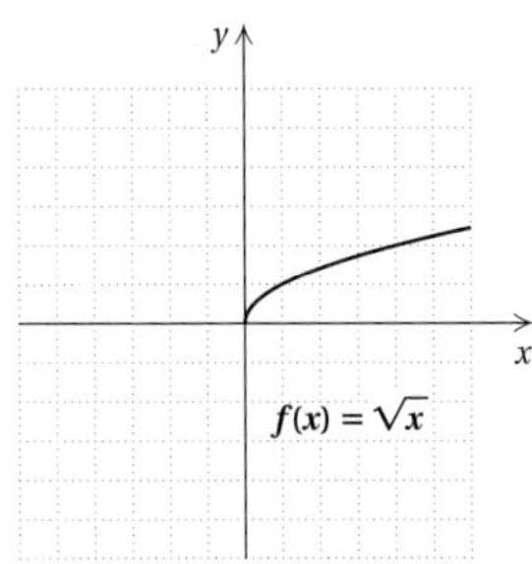

35.

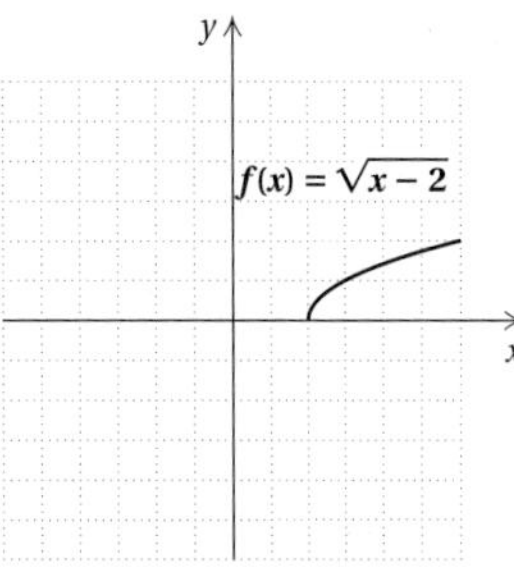

37.

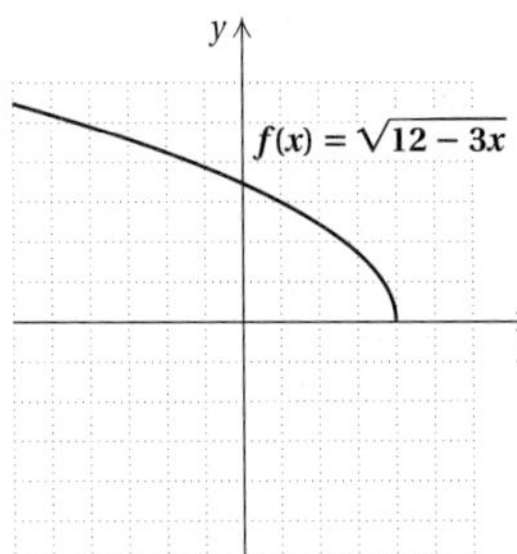

39.

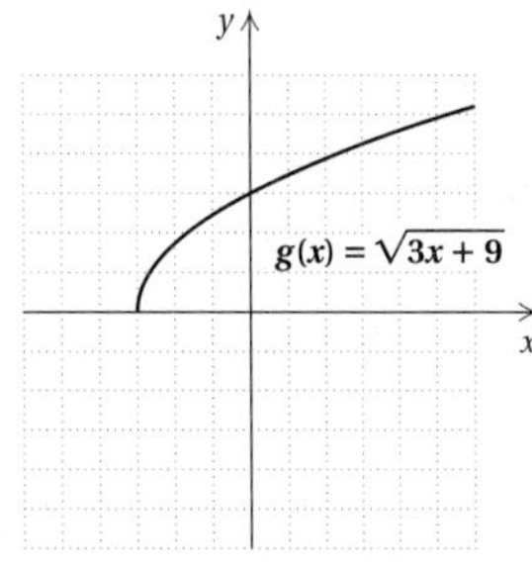

41. $4|x|$ **43.** $12|c|$ **45.** $|p + 3|$ **47.** $|x - 2|$ **49.** 3 **51.** $-4x$ **53.** -6 **55.** $0.7(x + 1)$ **57.** 2; 3; -2; -4 **59.** -1; approximately 2.7144; -4; -10 **61.** 5 **63.** -1 **65.** $-\frac{2}{3}$ **67.** $|x|$ **69.** $5|a|$ **71.** 6 **73.** $|a + b|$ **75.** y **77.** $x - 2$ **79.** -2, 1 **80.** -1, 0 **81.** $-\frac{7}{2}, \frac{7}{2}$ **82.** 4, 9 **83.** $-2, \frac{5}{3}$ **84.** $\frac{5}{2}$ **85.** $0, \frac{5}{2}$ **86.** 0, 1 **87.** $a^9b^6c^{15}$ **88.** $10a^{10}b^9$ **89.** ◈ **91.** $[-3, 2)$ **93.** 1.7; 2.2; 3.2 **95.** **(a)** Domain $= (-\infty, \infty)$, range $= (-\infty, \infty)$; **(b)** domain $= (-\infty, \infty)$, range $= (-\infty, \infty)$; **(c)** domain $= [-3, \infty)$, range $= (-\infty, 2]$; **(d)** domain $= [0, \infty)$, range $= [0, \infty)$; **(e)** domain $= [3, \infty)$, range $= [0, \infty)$

Margin Exercises, Section 10.2, pp. 615–618

1. $\sqrt[4]{y}$ **2.** $\sqrt{3a}$ **3.** 2 **4.** 5 **5.** $\sqrt[5]{a^3b^2c}$ **6.** $(19ab)^{1/3}$ **7.** $19(ab)^{1/3}$ **8.** $\left(\frac{x^2y}{16}\right)^{1/5}$ **9.** $7(2ab)^{1/4}$ **10.** $\sqrt{x^3}$ **11.** 4 **12.** 32 **13.** $(7abc)^{4/3}$ **14.** $6^{7/5}$ **15.** $\frac{1}{2}$ **16.** $\frac{1}{(3xy)^{7/8}}$ **17.** $\frac{1}{27}$ **18.** $\frac{7p^{3/4}}{q^{6/5}}$ **19.** $\left(\frac{7n}{11m}\right)^{2/3}$ **20.** $7^{14/15}$ **21.** $5^{1/3}$ **22.** $9^{2/5}$ **23.** $\frac{q^{1/8}}{p^{1/3}}$ **24.** $\sqrt{a}$ **25.** x **26.** $\sqrt{2}$ **27.** $\sqrt[4]{xy^2}$ **28.** $a^2\sqrt{b}$ **29.** ab^2 **30.** $\sqrt[4]{63}$ **31.** $\sqrt[6]{x^4y^3z^5}$ **32.** $\sqrt[4]{ab}$ **33.** $\sqrt[7]{5m}$ **34.** $\sqrt[6]{m}$ **35.** $a^{20}b^{12}c^4$ **36.** $\sqrt[10]{x}$

Calculator Spotlight, p. 616

1. 2.924 **2.** 5.278 **3.** 0.283 **4.** 11.053 **5.** 0.00006 **6.–11.** Left to the student.

Exercise Set 10.2, p. 619

1. $\sqrt[7]{y}$ **3.** 2 **5.** $\sqrt[5]{a^3b^3}$ **7.** 8 **9.** 343 **11.** $17^{1/2}$ **13.** $18^{1/3}$ **15.** $(xy^2z)^{1/5}$ **17.** $(3mn)^{3/2}$ **19.** $(8x^2y)^{5/7}$ **21.** $\frac{1}{3}$ **23.** $\frac{1}{1000}$ **25.** $\frac{1}{x^{1/4}}$ **27.** $\frac{1}{(2rs)^{3/4}}$ **29.** $\frac{2a^{3/4}c^{2/3}}{b^{1/2}}$ **31.** $\left(\frac{8yz}{7x}\right)^{3/5}$ **33.** $x^{2/3}$ **35.** $\frac{x^4}{2^{1/3}y^{2/7}}$ **37.** $\frac{7x}{z^{1/3}}$ **39.** $\frac{5ac^{1/2}}{3}$ **41.** $5^{7/8}$ **43.** $7^{1/4}$ **45.** $4.9^{1/2}$ **47.** $6^{3/28}$ **49.** $a^{23/12}$ **51.** $a^{8/3}b^{5/2}$ **53.** $\frac{1}{x^{2/7}}$ **55.** $\sqrt[3]{a}$ **57.** x^5 **59.** $\frac{1}{x^3}$ **61.** a^5b^5 **63.** $\sqrt[4]{2^5}$ **65.** $\sqrt[3]{2x}$ **67.** x^2y^3 **69.** $2c^2d^3$ **71.** $\sqrt[12]{7^4 \cdot 5^3}$ **73.** $\sqrt[20]{5^5 \cdot 7^4}$ **75.** $\sqrt[6]{4x^5}$ **77.** a^6b^{12} **79.** $\sqrt[18]{m}$ **81.** $\sqrt[12]{x^4y^3z^2}$ **83.** $\sqrt[30]{\frac{d^{35}}{c^{99}}}$ **85.** $a = \frac{Ab}{b - A}$ **86.** $s = \frac{Qt}{Q - t}$ **87.** $t = \frac{Qs}{s + Q}$ **88.** $b = \frac{ta}{t - a}$ **89.** ◈ **91.** Left to the student.

Margin Exercises, Section 10.3, pp. 621–624

1. $\sqrt{133}$ **2.** $\sqrt{21pq}$ **3.** $\sqrt[4]{2821}$ **4.** $\sqrt[3]{\frac{10}{pq}}$ **5.** $\sqrt[6]{500}$ **6.** $\sqrt[6]{25x^3y^2}$ **7.** $4\sqrt{2}$ **8.** $2\sqrt[3]{10}$ **9.** $10\sqrt{3}$ **10.** $6y$ **11.** $2a\sqrt{3b}$ **12.** $2bc\sqrt{3ab}$ **13.** $2\sqrt[3]{2}$ **14.** $3xy^2\sqrt[3]{3xy^2}$ **15.** $3\sqrt{2}$ **16.** $6y\sqrt{7}$ **17.** $3x\sqrt[3]{4y}$ **18.** $7\sqrt{3ab}$ **19.** 5 **20.** $56\sqrt{xy}$ **21.** $5a$ **22.** $\frac{20}{7}$ **23.** $\frac{5}{6}$ **24.** $\frac{x}{10}$ **25.** $\frac{3x\sqrt[3]{2x^2}}{5}$ **26.** $\sqrt[12]{xy^2}$

Exercise Set 10.3, p. 625

1. $2\sqrt{6}$ **3.** $3\sqrt{10}$ **5.** $5\sqrt[3]{2}$ **7.** $6x^2\sqrt{5}$ **9.** $3x^2\sqrt[3]{2x^2}$ **11.** $2t^2\sqrt[3]{10t^2}$ **13.** $2\sqrt[4]{5}$ **15.** $4a\sqrt{2b}$ **17.** $3x^2y^2\sqrt[4]{3y^2}$ **19.** $2xy^3\sqrt[5]{3x^2}$ **21.** $5\sqrt{2}$ **23.** $3\sqrt{10}$ **25.** 2 **27.** $30\sqrt{3}$ **29.** $3x^4\sqrt{2}$ **31.** $5bc^2\sqrt{2b}$ **33.** $a\sqrt[3]{10}$ **35.** $2y^3\sqrt[3]{2}$ **37.** $4\sqrt[4]{4}$ **39.** $4a^3b\sqrt{6ab}$ **41.** $\sqrt[6]{200}$ **43.** $\sqrt[4]{12}$ **45.** $a\sqrt[4]{a}$ **47.** $b\sqrt[10]{b^9}$ **49.** $xy\sqrt[6]{xy^5}$ **51.** $3\sqrt{2}$ **53.** $\sqrt{5}$ **55.** 3 **57.** $y\sqrt{7y}$ **59.** $2\sqrt[3]{a^2b}$ **61.** $4\sqrt{xy}$ **63.** $2x^2y^2$ **65.** $\frac{1}{\sqrt[6]{a}}$ **67.** $\sqrt[12]{a^5}$ **69.** $\sqrt[12]{x^2y^5}$ **71.** $\frac{5}{6}$ **73.** $\frac{4}{7}$ **75.** $\frac{5}{3}$ **77.** $\frac{7}{y}$

79. $\frac{5y\sqrt{y}}{x^2}$ **81.** $\frac{3a\sqrt[3]{a}}{2b}$ **83.** $\frac{3x}{2}$ **85.** $\frac{2x\sqrt[5]{x^3}}{y^2}$
87. $\frac{x^2\sqrt[6]{x}}{yz^2}$ **89.** $-10, 9$ **90.** Height: 4 in.; base: 6 in.
91. 8 **92.** $\frac{15}{2}$ **93.** No solution **94.** No solution
95. ◆ **97.** **(a)** 1.62 sec; **(b)** 1.99 sec; **(c)** 2.20 sec
99. $2yz\sqrt{2z}$

Margin Exercises, Section 10.4, pp. 629–630

1. $13\sqrt{2}$ **2.** $10\sqrt[4]{5x} - \sqrt{7}$ **3.** $19\sqrt{5}$
4. $(3y + 4)\sqrt[3]{y^2} + 2y^2$ **5.** $2\sqrt{x - 1}$ **6.** $5\sqrt{6} + 3\sqrt{14}$
7. $a\sqrt[3]{3} - \sqrt[3]{2a^2}$ **8.** $-4 - 9\sqrt{6}$
9. $3\sqrt{ab} - 4\sqrt{3a} + 6\sqrt{3b} - 24$ **10.** -3 **11.** $p - q$
12. $20 - 4y\sqrt{5} + y^2$ **13.** $58 + 12\sqrt{6}$

Exercise Set 10.4, p. 631

1. $11\sqrt{5}$ **3.** $\sqrt[3]{7}$ **5.** $13\sqrt[3]{y}$ **7.** $-8\sqrt{6}$ **9.** $6\sqrt[3]{3}$
11. $21\sqrt{3}$ **13.** $38\sqrt{5}$ **15.** $122\sqrt{2}$ **17.** $9\sqrt[3]{2}$
19. $29\sqrt{2}$ **21.** $(1 + 6a)\sqrt{5a}$ **23.** $(2 - x)\sqrt[3]{3x}$
25. $15\sqrt[3]{4}$ **27.** $4\sqrt{5} - 10$ **29.** $\sqrt{6} - \sqrt{21}$
31. $2\sqrt{15} - 6\sqrt{3}$ **33.** -6 **35.** $3a\sqrt[3]{2}$ **37.** 1
39. -12 **41.** 44 **43.** 1 **45.** 3 **47.** -19
49. $a - b$ **51.** $1 + \sqrt{5}$ **53.** $7 + 3\sqrt{3}$ **55.** -6
57. $a + \sqrt{3a} + \sqrt{2a} + \sqrt{6}$ **59.** $2\sqrt[3]{9} - 3\sqrt[3]{6} - 2\sqrt[3]{4}$
61. $7 + 4\sqrt{3}$ **63.** $3 - \sqrt[5]{24} - \sqrt[5]{81} + \sqrt[5]{72}$
65. $\frac{x(x^2 + 4)}{(x + 4)(x + 3)}$ **66.** $\frac{(a + 2)(a + 4)}{a}$ **67.** $a - 2$
68. $\frac{(y - 3)(y - 3)}{y + 3}$ **69.** $\frac{4(3x - 1)}{3(4x + 1)}$ **70.** $\frac{x}{x + 1}$
71. $\frac{pq}{q + p}$ **73.** ◆ **75.** Left to the student.
77. $2x - 2\sqrt{x^2 - 4}$ **79.** $\sqrt[3]{y} - y$ **81.** $12 + 6\sqrt{3}$

Margin Exercises, Section 10.5, pp. 635–637

1. $\frac{\sqrt{10}}{5}$ **2.** $\frac{\sqrt[3]{10}}{2}$ **3.** $\frac{2\sqrt{3ab}}{3b}$ **4.** $\frac{\sqrt[4]{56}}{2}$
5. $\frac{x\sqrt[3]{12x^2y^2}}{2y}$ **6.** $\frac{7\sqrt[3]{2x^2y}}{2y^2}$ **7.** $c^2 - b$ **8.** $a - b$
9. $\frac{\sqrt{3} + y}{\sqrt{3} + y}$ **10.** $\frac{\sqrt{2} - \sqrt{3}}{\sqrt{2} - \sqrt{3}}$ **11.** $-7 - 6\sqrt{2}$
12. $6 - 2\sqrt{2}$

Exercise Set 10.5, p. 639

1. $\frac{\sqrt{15}}{3}$ **3.** $\frac{\sqrt{22}}{2}$ **5.** $\frac{2\sqrt{15}}{35}$ **7.** $\frac{2\sqrt[3]{6}}{3}$ **9.** $\frac{\sqrt[3]{75ac^2}}{5c}$
11. $\frac{y\sqrt[3]{9yx^2}}{3x^2}$ **13.** $\frac{\sqrt[4]{s^3t^3}}{st}$ **15.** $\frac{\sqrt{15x}}{10}$ **17.** $\frac{\sqrt[3]{100xy}}{5x^2y}$
19. $\frac{\sqrt[4]{2xy}}{2x^2y}$ **21.** $\frac{54 + 9\sqrt{10}}{26}$ **23.** $-2\sqrt{35} - 2\sqrt{21}$
25. $\frac{\sqrt{15} + 20 - 6\sqrt{2} - 8\sqrt{30}}{-77}$ **27.** $\frac{6 - 5\sqrt{a} + a}{9 - a}$
29. $\frac{3\sqrt{6} + 4}{2}$ **31.** $\frac{x - 2\sqrt{xy} + y}{x - y}$ **33.** 30 **34.** $-\frac{19}{5}$

35. 1 **36.** $\frac{x - 2}{x + 3}$ **37.** ◆ **39.** Left to the student.
41. $-\frac{3\sqrt{a^2 - 3}}{a^2 - 3}$

Margin Exercises, Section 10.6, pp. 642–646

1. 100 **2.** No solution **3.** 1 **4.** 2, 5 **5.** 4 **6.** 17
7. 9 **8.** 27 **9.** 5 **10.** About 276 mi **11.** About 8.9 mi **12.** About 169 ft

Calculator Spotlight, p. 643

1. 9 **2.** 1 **3.** 2, 5 **4.** 15.6 **5.** 9.5 **6.** 2.0, 7.2

Exercise Set 10.6, p. 647

1. $\frac{19}{2}$ **3.** $\frac{49}{6}$ **5.** 57 **7.** $\frac{92}{5}$ **9.** -1 **11.** No solution
13. 3 **15.** 19 **17.** -6 **19.** $\frac{1}{64}$ **21.** 5 **23.** 9
25. 7 **27.** $\frac{80}{9}$ **29.** 6, 2 **31.** -1 **33.** No solution
35. 3 **37.** About 232 mi **39.** 16,200 ft
41. 211.25 ft; 281.25 ft **43.** 49 **45.** 5 **47.** $4\frac{4}{9}$ hr
48. Jeff: $1\frac{1}{3}$ hr; Grace: 4 hr **49.** 2808 mi **50.** 84 hr
51. 0, -2.8 **52.** 0, $\frac{5}{3}$ **53.** -8, 8 **54.** -3, $\frac{7}{2}$
55. ◆ **57.** Left to the student. **59.** 6912 **61.** 0
63. -3, -6 **65.** 2 **67.** 0, $\frac{125}{4}$ **69.** 2 **71.** $\frac{1}{2}$ **73.** 3

Margin Exercises, Section 10.7, pp. 651–652

1. $\sqrt{41}$; 6.403 **2.** $\sqrt{6}$; 2.449 **3.** $\sqrt{200}$; 14.142
4. $\sqrt{8200}$; 90.554 ft **5.** 7.4 ft

Exercise Set 10.7, p. 653

1. $\sqrt{34}$; 5.831 **3.** $\sqrt{450}$; 21.213 **5.** 5 **7.** $\sqrt{43}$; 6.557
9. $\sqrt{12}$; 3.464 **11.** $\sqrt{n - 1}$ **13.** $\sqrt{10{,}561}$; 102.767 ft
15. $\sqrt{116}$; 10.770 ft **17.** $s + s\sqrt{2}$ **19.** 7.1 ft
21. $\sqrt{181}$; 13.454 cm **23.** 12 in. **25.** $(3, 0)$, $(-3, 0)$
27. $\sqrt{340} + 8$; 26.439 ft **29.** 20.497 in.
31. Flash: $67\frac{2}{3}$ mph; Crawler: $53\frac{2}{3}$ mph **32.** $3\frac{3}{4}$ mph
33. -7, $\frac{3}{2}$ **34.** 3, 8 **35.** 1 **36.** -2, 2 **37.** 13
38. 7 **39.** ◆ **41.** 9 **43.** $\sqrt{75}$ cm

Margin Exercises, Section 10.8, pp. 657–661

1. $i\sqrt{5}$, or $\sqrt{5}\,i$ **2.** $5i$ **3.** $-i\sqrt{11}$, or $-\sqrt{11}\,i$
4. $-6i$ **5.** $3i\sqrt{6}$, or $3\sqrt{6}\,i$ **6.** $15 - 3i$ **7.** $-12 + 6i$
8. $3 - 5i$ **9.** $12 - 7i$ **10.** -10 **11.** $-\sqrt{34}$
12. 42 **13.** $-9 - 12i$ **14.** $-35 - 25i$
15. $-14 + 8i$ **16.** $11 + 10i$ **17.** $5 + 12i$ **18.** $-i$
19. 1 **20.** i **21.** -1 **22.** $8 - i$ **23.** 3
24. $-7 - 6i$ **25.** $-1 + i$ **26.** $6 - 3i$ **27.** $-9 + 5i$
28. $\pi + \frac{1}{4}i$ **29.** 53 **30.** 10 **31.** $2i$ **32.** $\frac{10}{17} - \frac{11}{17}i$
33. Yes **34.** Yes

Calculator Spotlight, p. 662

1. $-2 - 9i$ **2.** $20 + 17i$ **3.** $-47 - 161i$
4. $-0.5207 + 0.2517i$ **5.** $4i$ **6.** -20 **7.** -28.3725

8. $-0.64 - 0.02i$ **9.** 81 **10.** $-i$ **11.** -2
12. $-0.6358 - 0.4227i$ **13.** $-i$

Exercise Set 10.8, p. 663

1. $i\sqrt{35}$, or $\sqrt{35}\,i$ **3.** $4i$ **5.** $-2i\sqrt{3}$, or $-2\sqrt{3}\,i$
7. $i\sqrt{3}$, or $\sqrt{3}\,i$ **9.** $9i$ **11.** $7i\sqrt{2}$, or $7\sqrt{2}\,i$
13. $-7i$ **15.** $4 - 2\sqrt{15}\,i$ **17.** $(2 + 2\sqrt{3})i$
19. $12 - 4i$ **21.** $9 - 5i$ **23.** $7 + 4i$ **25.** $-4 - 4i$
27. $-1 + i$ **29.** $11 + 6i$ **31.** -18 **33.** $-\sqrt{14}$
35. 21 **37.** $-6 + 24i$ **39.** $1 + 5i$ **41.** $18 + 14i$
43. $38 + 9i$ **45.** $2 - 46i$ **47.** $5 - 12i$
49. $-24 + 10i$ **51.** $-5 - 12i$ **53.** $-i$ **55.** 1
57. -1 **59.** i **61.** -1 **63.** $-125i$ **65.** 8
67. $1 - 23i$ **69.** 0 **71.** 0 **73.** 1 **75.** $5 - 8i$
77. $2 - \frac{\sqrt{6}}{2}i$ **79.** $\frac{9}{10} + \frac{13}{10}i$ **81.** $-i$ **83.** $-\frac{3}{7} - \frac{8}{7}i$
85. $\frac{6}{5} - \frac{2}{5}i$ **87.** $-\frac{8}{41} + \frac{10}{41}i$ **89.** $-\frac{4}{3}i$ **91.** $-\frac{1}{2} - \frac{1}{4}i$
93. $\frac{3}{5} + \frac{4}{5}i$ **95.** Yes **97.** No **99.** 7 **100.** $\frac{70}{29}$
101. $-\frac{29}{3}, 5$ **102.** $\{x \mid -\frac{29}{3} < x < 5\}$, or $(-\frac{29}{3}, 5)$
103. $\{x \mid x \le -\frac{29}{3}$ *or* $x \ge 5\}$, or $(-\infty, -\frac{29}{3}] \cup [5, \infty)$
104. $-\frac{2}{5}, -12$ **105.** ◆ **107.** $-4 - 8i$; $-2 + 4i$; $8 - 6i$ **109.** $-3 - 4i$ **111.** $-88i$ **113.** 8
115. $\frac{3}{5} + \frac{9}{5}i$ **117.** 1

Summary and Review: Chapter 10, p. 667

1. 27.893 **2.** 6.378 **3.** $f(0)$, $f(-1)$, and $f(1)$ do not exist as real numbers; $f\left(\frac{41}{3}\right) = 5$
4. Domain $= \{x \mid x \ge \frac{16}{3}\}$, or $[\frac{16}{3}, \infty)$ **5.** $9|a|$ **6.** $7|z|$
7. $|c - 3|$ **8.** $|x - 3|$ **9.** -10 **10.** $-\frac{1}{3}$ **11.** 2; -2; 3
12. $|x|$ **13.** 3 **14.** $\sqrt[5]{a}$ **15.** 512 **16.** $31^{1/2}$
17. $(a^2b^3)^{1/5}$ **18.** $\frac{1}{7}$ **19.** $\frac{1}{4x^{2/3}y^{2/3}}$ **20.** $\frac{5b^{1/2}}{a^{3/4}c^{2/3}}$
21. $\frac{3a}{t^{1/4}}$ **22.** $\frac{1}{x^{2/5}}$ **23.** $7^{1/6}$ **24.** x^7 **25.** $3x^2$
26. $\sqrt[12]{x^4y^3}$ **27.** $\sqrt[12]{x^7}$ **28.** $7\sqrt{5}$ **29.** $-3\sqrt[3]{4}$
30. $5b^2\sqrt[3]{2a^2}$ **31.** $\frac{7}{6}$ **32.** $\frac{4}{3}x^2$ **33.** $\frac{2x^2}{3y^3}$ **34.** $\sqrt{15xy}$
35. $3a\sqrt[3]{a^2b^2}$ **36.** $\sqrt[15]{a^5b^9}$ **37.** $y\sqrt[3]{6}$ **38.** $\frac{5}{2}\sqrt{x}$
39. $\sqrt[12]{x^5}$ **40.** $7\sqrt[3]{x}$ **41.** $3\sqrt{3}$ **42.** $(2x + y^2)\sqrt[3]{x}$
43. $15\sqrt{2}$ **44.** $-43 - 2\sqrt{10}$ **45.** $8 - 2\sqrt{7}$
46. $9 - \sqrt[3]{4}$ **47.** $\frac{2\sqrt{6}}{3}$ **48.** $\frac{2\sqrt{a} - 2\sqrt{b}}{a - b}$ **49.** 13
50. 4 **51.** 9 cm **52.** $\sqrt{24}$; 4.899 ft **53.** 25
54. $\sqrt{46}$; 6.782 **55.** $(5 + 2\sqrt{2})i$ **56.** $-2 - 9i$
57. $1 + i$ **58.** 29 **59.** i **60.** $9 - 12i$ **61.** $\frac{2}{5} + \frac{3}{5}i$
62. 3 **63.** No **64.**

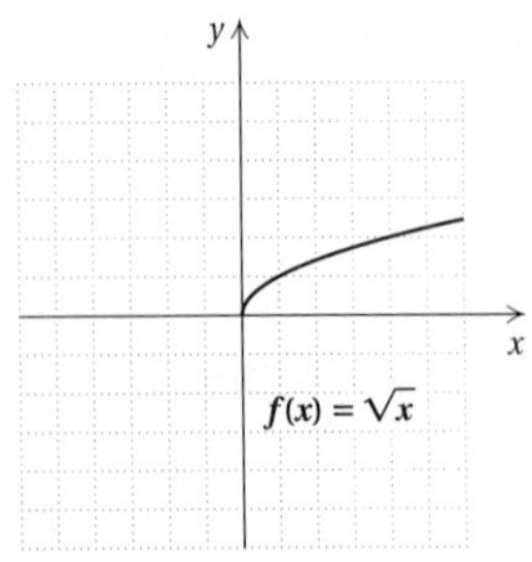

65. $\frac{23}{7}$ **66.** $-\frac{4}{3}, 3$ **67.** $\frac{x(x + 2)}{(x + y)(x - 3)}$
68. Motorcycle: 62 mph; SUV: 70 mph
69. ◆ The procedure for solving radical equations is to isolate one of the radical terms, use the principle of powers, repeat these steps if necessary until all radicals are eliminated, and then check the possible solutions. A check is necessary since the principle of powers does not always yield equivalent equations. **70.** -1 **71.** 3

Test: Chapter 10, p. 669

1. [10.1a] 12.166 **2.** [10.1a] 2; does not exist as a real number **3.** [10.1a] Domain $= \{x \mid x \le 2\}$, or $(-\infty, 2]$
4. [10.1b] $3|q|$ **5.** [10.1b] $|x + 5|$ **6.** [10.1c] $-\frac{1}{10}$
7. [10.1d] x **8.** [10.1d] 4 **9.** [10.2a] $\sqrt[3]{a^2}$
10. [10.2a] 8 **11.** [10.2a] $37^{1/2}$ **12.** [10.2a] $(5xy^2)^{5/2}$
13. [10.2b] $\frac{1}{10}$ **14.** [10.2b] $\frac{8a^{3/4}}{b^{3/2}c^{2/5}}$ **15.** [10.2c] $\frac{x^{8/5}}{y^{9/5}}$
16. [10.2c] $\frac{1}{2.9^{31/24}}$ **17.** [10.2d] $\sqrt[4]{x}$
18. [10.2d] $2x\sqrt{x}$ **19.** [10.2d] $\sqrt[15]{a^6b^5}$
20. [10.2d] $\sqrt[12]{8y^7}$ **21.** [10.3a] $2\sqrt{37}$ **22.** [10.3a] $2\sqrt[4]{5}$
23. [10.3a] $4a\sqrt[3]{4ab^2}$ **24.** [10.3b] $-\frac{2}{x^2}$ **25.** [10.3b] $\frac{5x}{6y^2}$
26. [10.3a] $\sqrt[3]{10xy^2}$ **27.** [10.3a] $xy\sqrt[4]{x}$
28. [10.3b] $\sqrt[5]{x^2y^2}$ **29.** [10.3b] $2\sqrt{a}$
30. [10.4a] $38\sqrt{2}$ **31.** [10.4b] -20
32. [10.4b] $9 + 6\sqrt{x} + x$ **33.** [10.5b] $\frac{13 + 8\sqrt{2}}{-41}$
34. [10.6a] 35 **35.** [10.6b] 7 **36.** [10.7a] 7 ft
37. [10.7a] $\sqrt{98}$; 9.899 **38.** [10.7a] 2 **39.** [10.8a] $11i$
40. [10.8b] $7 + 5i$ **41.** [10.8c] $37 + 9i$ **42.** [10.8d] $-i$
43. [10.8e] $-\frac{77}{50} + \frac{7}{25}i$ **44.** [10.8f] No
45. [10.1a]

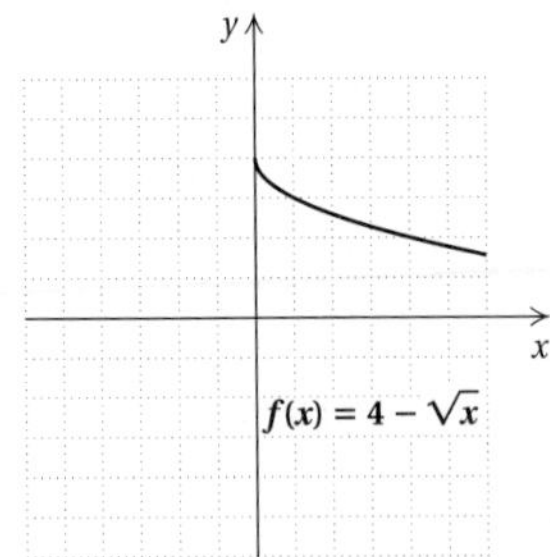

46. [6.5a] No solution **47.** [5.8b] $-\frac{1}{3}, \frac{5}{2}$
48. [6.1f] $\frac{x - 3}{x - 4}$ **49.** [6.6a] Fran: 36 hr; Juan: 45 hr
50. [10.8c, e] $-\frac{17}{4}i$ **51.** [10.6b] 3

Chapter 11

Pretest: Chapter 11, p. 672

1. [11.1b], [11.2a] 0, -3 **2.** [11.1b], [11.2a] $-2 + 2i$, $-2 - 2i$ **3.** [11.1a, b], [11.2a] 5

4. [11.4c] $\frac{\sqrt{3}}{3}, -\frac{\sqrt{3}}{3}, \sqrt{2}, -\sqrt{2}$ **5.** [11.2a] $-1, \frac{5}{6}$
6. [11.4c] 1 **7.** [11.2a] $\frac{-2 \pm \sqrt{6}}{2}$; 0.225, −2.225
8. [11.3b] $T = \frac{1}{RW^2}$ **9.** [11.4b] $3x^2 + 4x - 4 = 0$
10. [11.6a] **(a)** (3, −2); **(b)** $x = 3$;
(c)

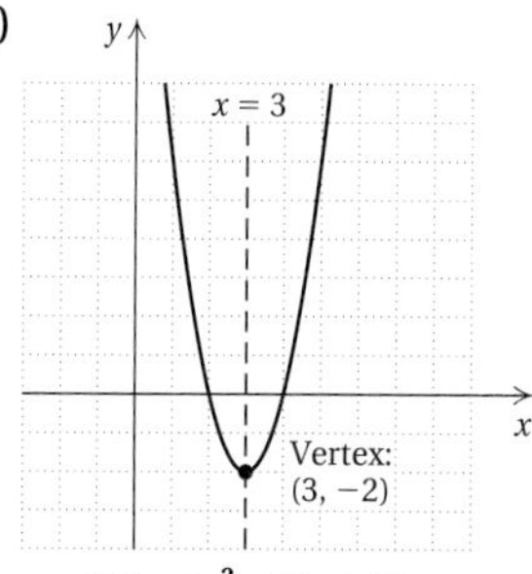

11. [11.6b] $(3 + \sqrt{5}, 0), (3 - \sqrt{5}, 0)$
12. [11.7b] $f(x) = 4x^2 - 5x + 1$ **13.** [11.3a] 12, 14, 16
14. [11.3a] 40 mph; 48 mph **15.** [11.7a] 5625 ft^2
16. [11.8a] $\{x|x < -5 \text{ or } 0 < x < 3\}$, or $(-\infty, -5) \cup (0, 3)$
17. [11.8b] $\{x|x < -3 \text{ or } x \geq 2\}$, or $(-\infty, -3) \cup [2, 0)$

Margin Exercises, Section 11.1, pp. 673–682

1. **(a)** (2, 0), (4, 0); **(b)** 2, 4; **(c)** The solutions of $x^2 - 6x + 8 = 0$, 2 and 4, are the first coordinates of the x-intercepts, (2, 0) and (4, 0), of the graph of $f(x) = x^2 - 6x + 8$. **2.** **(a)** $\frac{3}{5}$, 1; **(b)** $\left(\frac{3}{5}, 0\right)$, (1, 0)
3. 4 and −4, or ±4 **4.** $0, -\frac{7}{2}$ **5.** $\sqrt{3}$ and $-\sqrt{3}$, or $\pm\sqrt{3}$; 1.732 and −1.732, or ±1.732 **6.** $\frac{2\sqrt{6}}{3}$ and $-\frac{2\sqrt{6}}{3}$, or $\pm\frac{2\sqrt{6}}{3}$; 1.633 and −1.633, or ±1.633 **7.** $\frac{\sqrt{2}}{2}i$ and $-\frac{\sqrt{2}}{2}i$, or $\pm\frac{\sqrt{2}}{2}i$ **8.** **(a)** $1 \pm \sqrt{5}$; **(b)** $(1 + \sqrt{5}, 0)$, $(1 - \sqrt{5}, 0)$ **9.** $-8 \pm \sqrt{11}$ **10.** −2, −4 **11.** 10, −2
12. $-3 \pm \sqrt{10}$ **13.** $\frac{1 \pm \sqrt{22}}{3}$ **14.** \$3603.65
15. 20% **16.** About 36.0 in. **17.** 1 sec

Calculator Spotlight, p. 676

1. 1.633, −1.633 **2.** 0.650, −0.650

Calculator Spotlight, p. 677

The graphs of $y_1 = 4x^2 + 9$ and $y_2 = 0$ in Example 6 do not intersect. The graphs of $y_1 = 2x^2 + 1$ and $y_2 = 0$ in Margin Exercise 7 do not intersect. The calculator indicates an error since there are no real-number solutions.

Exercise Set 11.1, p. 683

1. **(a)** $\sqrt{5}, -\sqrt{5}$, or $\pm\sqrt{5}$; **(b)** $(\sqrt{5}, 0), (-\sqrt{5}, 0)$
3. **(a)** $\frac{5}{3}i, -\frac{5}{3}i$, or $\pm\frac{5}{3}i$; **(b)** no x-intercepts

5. $\pm\frac{\sqrt{6}}{2}$; ±1.225 **7.** 5, −9 **9.** 8, 0
11. $11 \pm \sqrt{7}$; 13.646, 8.354 **13.** $7 \pm 2i$ **15.** 18, 0
17. $\frac{3 \pm \sqrt{14}}{2}$; 3.371, −0.371 **19.** 5, −11 **21.** 9, 5
23. $-2 \pm \sqrt{6}$ **25.** $11 \pm 2\sqrt{33}$ **27.** $\frac{-1 \pm \sqrt{5}}{2}$
29. $\frac{5 \pm \sqrt{53}}{2}$ **31.** $\frac{-3 \pm \sqrt{57}}{4}$ **33.** $\frac{9 \pm \sqrt{105}}{4}$
35. 2, −8 **37.** $-11 \pm \sqrt{19}$ **39.** $5 \pm \sqrt{29}$
41. **(a)** $\frac{-7 \pm \sqrt{57}}{2}$; **(b)** $\left(\frac{-7 + \sqrt{57}}{2}, 0\right), \left(\frac{-7 - \sqrt{57}}{2}, 0\right)$
43. **(a)** $\frac{5}{4} \pm i\frac{\sqrt{39}}{4}$; **(b)** no x-intercepts **45.** $\frac{3 \pm \sqrt{17}}{4}$
47. $\frac{3 \pm \sqrt{145}}{4}$ **49.** $\frac{2 \pm \sqrt{7}}{3}$ **51.** $-\frac{1}{2} \pm i\frac{\sqrt{7}}{2}$
53. $2 \pm 3i$ **55.** 10% **57.** 12.5% **59.** 4%
61. 0.866 sec **63.** About 7.3 sec **65.** About 6.8 sec
67. **(a)** $f(x) = -1.175x + 12$;
(b)

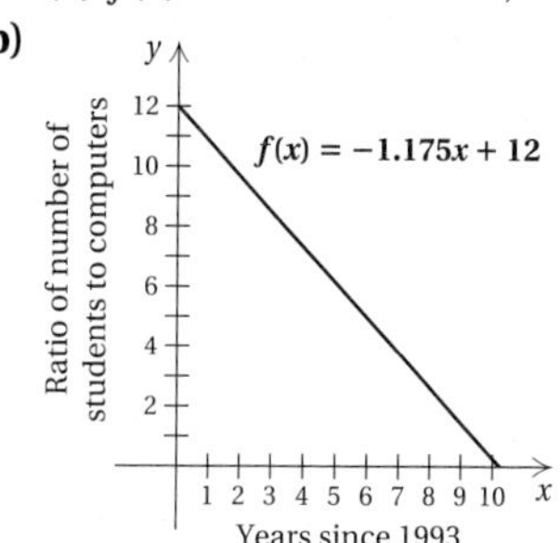

(c) 6.1; 3.8
68. **(a)** $f(x) = 0.675x + 3.6$;
(b)

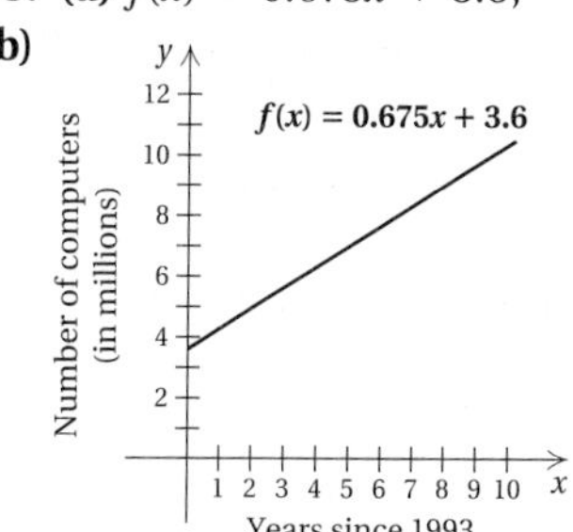

(c) 7.0 million; 8.3 million

69.

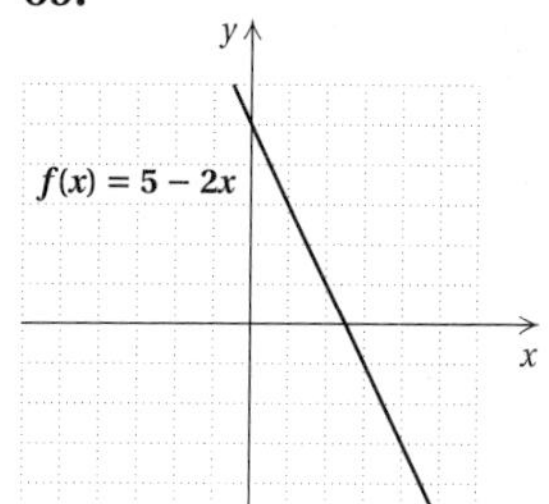

70.

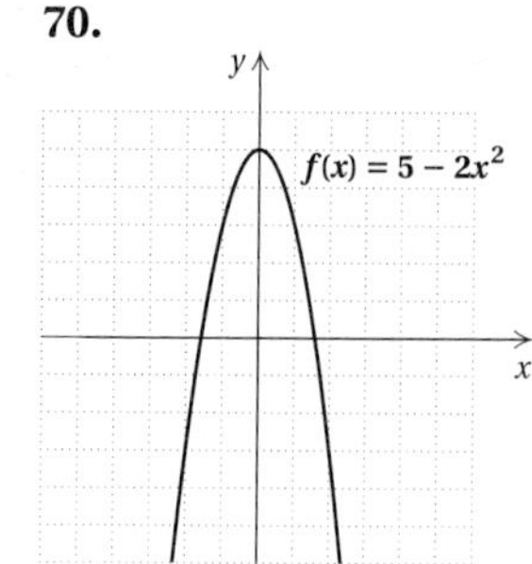

71.

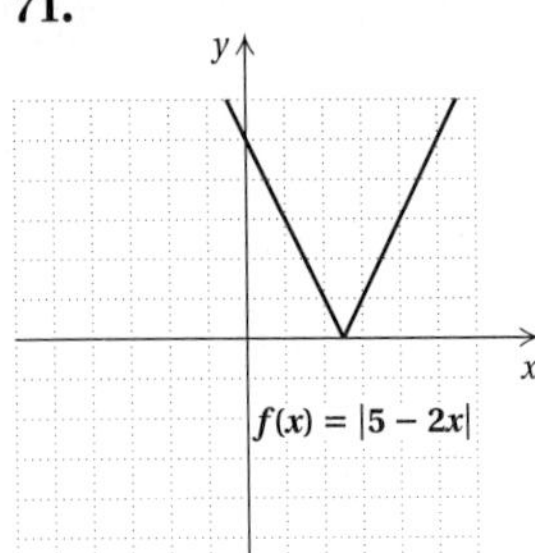

72.

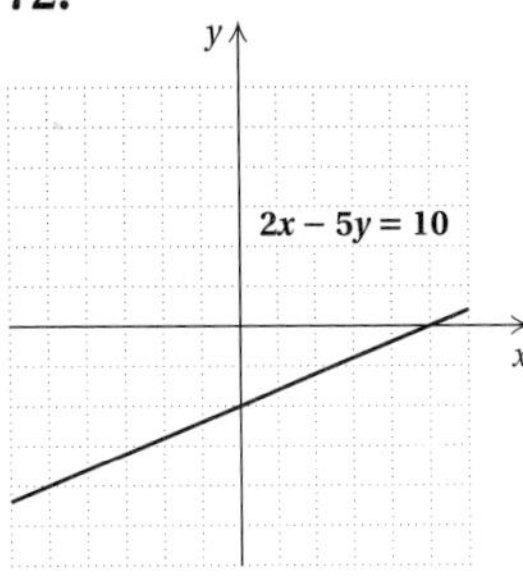

73. $2\sqrt{22}$ **74.** $\frac{\sqrt{10}}{5}$ **75.** ◈ **77.** Left to the student. **79.** 16, −16 **81.** $0, \frac{7}{2}, -8, -\frac{10}{3}$
83. Fishing boat: 15 km/h; barge: 8 km/h

Margin Exercises, Section 11.2, pp. 688–690

1. **(a)** $-\frac{1}{2}, 4$; **(b)** $-\frac{1}{2}, 4$ **2.** $\frac{-1 \pm \sqrt{22}}{3}$; 1.230, −1.897
3. $\frac{1 \pm i\sqrt{7}}{2}$, or $\frac{1}{2} \pm i\frac{\sqrt{7}}{2}$ **4.** $\frac{5 \pm \sqrt{73}}{6}$; 2.257, −0.591

Calculator Spotlight, p. 689

1. −1.897, 1.230 **2.** −0.309, 1.946

Exercise Set 11.2, p. 691

1. $-3 \pm \sqrt{5}$ **3.** $\frac{-4 \pm \sqrt{13}}{3}$ **5.** $\frac{1}{2} \pm i\frac{\sqrt{3}}{2}$ **7.** $2 \pm 3i$
9. $\frac{-3 \pm \sqrt{41}}{2}$ **11.** $-1 \pm 2i$ **13.** **(a)** 0, −1;
(b) (0, 0), (−1, 0) **15.** **(a)** $\frac{3 \pm \sqrt{229}}{22}$;
(b) $\left(\frac{3 + \sqrt{229}}{22}, 0\right), \left(\frac{3 - \sqrt{229}}{22}, 0\right)$ **17.** **(a)** $\frac{2}{5}$;
(b) $\left(\frac{2}{5}, 0\right)$ **19.** −1, −2 **21.** 5, 10 **23.** $\frac{17 \pm \sqrt{249}}{10}$
25. $2 \pm i$ **27.** $\frac{2}{3}, \frac{3}{2}$ **29.** $2 \pm \sqrt{10}$ **31.** $\frac{3}{4}, -2$
33. $\frac{1}{2} \pm \frac{3}{2}i$ **35.** $1, -\frac{1}{2} \pm i\frac{\sqrt{3}}{2}$ **37.** $-3 \pm \sqrt{5}$; −0.764, −5.236 **39.** $3 \pm \sqrt{5}$; 5.236, 0.764 **41.** $\frac{3 \pm \sqrt{65}}{4}$; 2.766, −1.266 **43.** $\frac{4 \pm \sqrt{31}}{5}$; 1.914, −0.314 **45.** 2 **46.** 3
47. 10 **48.** 8 **49.** No solution **50.** No solution
51. $\frac{15}{4}$ **52.** $\frac{17}{3}$ **53.** ◈ **55.** −0.797, 0.570
57. $\frac{1 \pm \sqrt{1 + 8\sqrt{5}}}{4}$ **59.** $\frac{-i \pm i\sqrt{1 + 4i}}{2}$
61. $\frac{-1 \pm 3\sqrt{5}}{6}$ **63.** $3 \pm \sqrt{13}$

Margin Exercises, Section 11.3, pp. 693–698

1. 2 ft **2.** A to B is 6 mi; A to C is 8 mi **3.** A to B is $\sqrt{31} - 1$, or 4.57 mi; A to C is $1 + \sqrt{31}$, or 6.57 mi

4.

	Distance	Speed	Time
Faster Ship	3000	$r + 10$	$t - 50$
Slower Ship	3000	r	t

20 knots, 30 knots
5. $w_2 = \frac{w_1}{A^2}$ **6.** $r = \sqrt{\frac{V}{\pi h}}$ **7.** $t = \frac{-g + \sqrt{g^2 + 64s}}{32}$
8. $b = \frac{pa}{\sqrt{1 + p^2}}$

Exercise Set 11.3, p. 699

1. Length: 9 ft; width: 2 ft **3.** Length: 18 yd; width: 9 yd **5.** 3 cm **7.** 6 ft, 8 ft
9. 28 and 29 **11.** Length: $2 + \sqrt{14} \approx 5.742$ ft; width: $\sqrt{14} - 2 \approx 1.742$ ft **13.** $\frac{17 - \sqrt{109}}{2} \approx 3.280$ in.
15. $7 + \sqrt{239} \approx 22.460$ ft; $\sqrt{239} - 7 \approx 8.460$ ft
17. First part: 60 mph; second part: 50 mph
19. 40 mph **21.** Cessna: 150 mph, Beechcraft: 200 mph; or Cessna: 200 mph, Beechcraft: 250 mph
23. Hillsboro: 10 mph; return trip: 4 mph
25. About 11 mph **27.** $s = \sqrt{\frac{A}{6}}$ **29.** $r = \sqrt{\frac{Gm_1m_2}{F}}$
31. $c = \sqrt{\frac{E}{m}}$ **33.** $b = \sqrt{c^2 - a^2}$
35. $k = \frac{3 + \sqrt{9 + 8N}}{2}$ **37.** $r = \frac{-\pi h + \sqrt{\pi^2h^2 + 2\pi A}}{2\pi}$
39. $g = \frac{4\pi^2L}{T^2}$ **41.** $H = \sqrt{\frac{700W}{I}}$, or $10\sqrt{\frac{7W}{I}}$
43. $v = \frac{c\sqrt{m^2 - m_0^2}}{m}$ **45.** $\frac{1}{x - 2}$
46. $\frac{x^3 + x^2 + 2x + 2}{(x - 1)(x^2 + x + 1)}$ **47.** $\frac{-x}{(x + 3)(x - 1)}$
48. $3x^2\sqrt{x}$ **49.** $2i\sqrt{5}$ **50.** $\frac{3x + 3}{3x + 1}$ **51.** $\frac{4b}{a(3b^2 - 4a)}$
53. ◈ **55.** $\pm\sqrt{2}$ **57.** $A(S) = \frac{\pi S}{6}$
59. $l = \frac{w + w\sqrt{5}}{2}$

Margin Exercises, Section 11.4, pp. 704–707

1. Two real **2.** One real **3.** Two nonreal
4. $x^2 - 5x - 14 = 0$ **5.** $3x^2 + 7x - 20 = 0$
6. $x^2 + 25 = 0$ **7.** $x^2 + \sqrt{2}x - 4 = 0$
8. ±3, ±1 **9.** 4 **10.** (−3, 0), (−1, 0), (2, 0), (4, 0)
11. $-\frac{1}{3}, \frac{1}{2}$

Exercise Set 11.4, p. 709

1. One real **3.** Two nonreal **5.** Two real **7.** One real **9.** Two nonreal **11.** Two real **13.** Two real

15. One real **17.** $x^2 - 16 = 0$ **19.** $x^2 + 9x + 14 = 0$
21. $x^2 - 16x + 64 = 0$ **23.** $25x^2 - 20x - 12 = 0$
25. $12x^2 - (4k + 3m)x + km = 0$
27. $x^2 - \sqrt{3}x - 6 = 0$ **29.** $\pm\sqrt{3}$ **31.** 1, 81
33. $-1, 1, 5, 7$ **35.** $-\frac{1}{4}, \frac{1}{9}$ **37.** 1 **39.** $-1, 1, 4, 6$
41. $\pm\sqrt{3 + \sqrt{13}}, \pm\sqrt{3 - \sqrt{13}}$ **43.** $-1, 2$
45. $\pm\frac{\sqrt{15}}{3}, \pm\frac{\sqrt{6}}{2}$ **47.** $-1, 125$ **49.** $-\frac{11}{6}, -\frac{1}{6}$
51. $-\frac{3}{2}$ **53.** $\frac{9 \pm \sqrt{89}}{2}, -1 \pm \sqrt{3}$ **55.** $\left(\frac{4}{25}, 0\right)$
57. $(4, 0), (-1, 0), \left(\frac{3 + \sqrt{33}}{2}, 0\right), \left(\frac{3 - \sqrt{33}}{2}, 0\right)$
59. Kenyan: 30 lb; Peruvian: 20 lb **60.** 4 L of A; 8 L of B **61.** $4x$ **62.** $3x^2$ **63.** $3a\sqrt[4]{2a}$ **64.** 4

65.

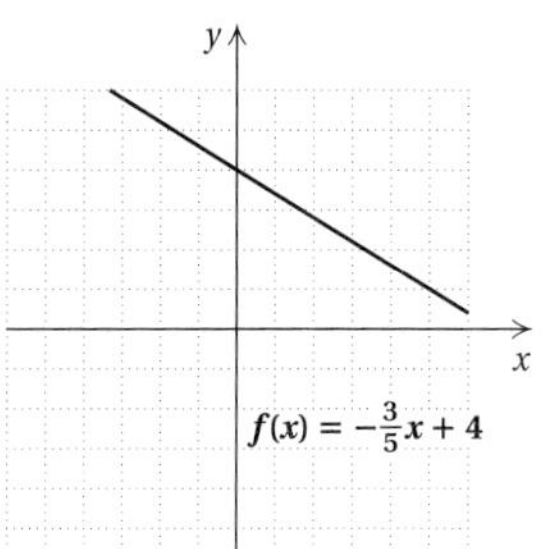

66.

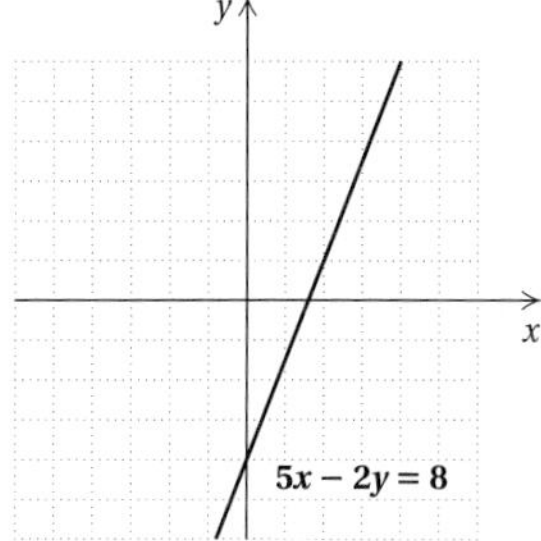

67.

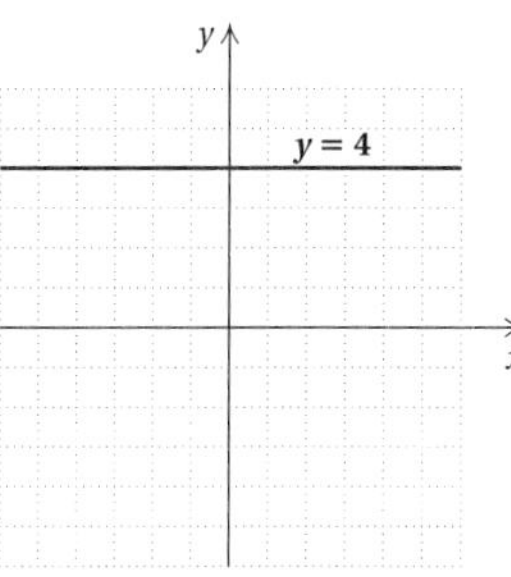

68.

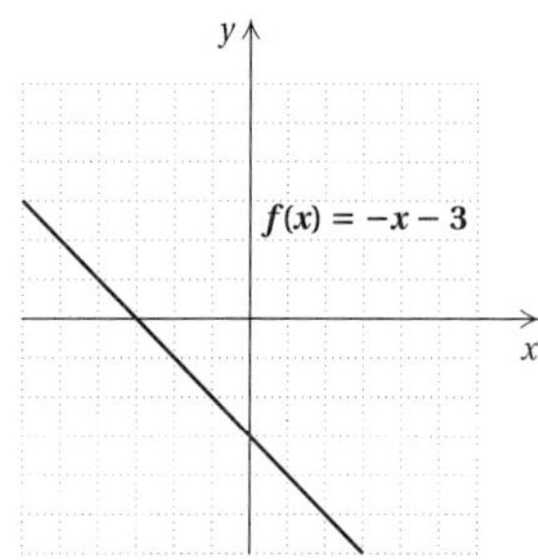

69. ◈ **71.** Left to the student. **73.** **(a)** $-\frac{3}{5}$; **(b)** $-\frac{1}{3}$
75. $x^2 - \sqrt{3}x + 8 = 0$ **77.** $a = 1, b = 2, c = -3$
79. $\frac{100}{99}$ **81.** 259 **83.** 1, 3

Margin Exercises, Section 11.5, pp. 713–718

1.

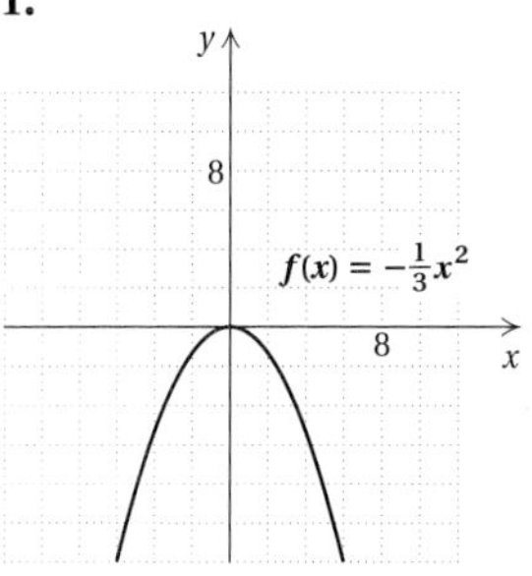

2.

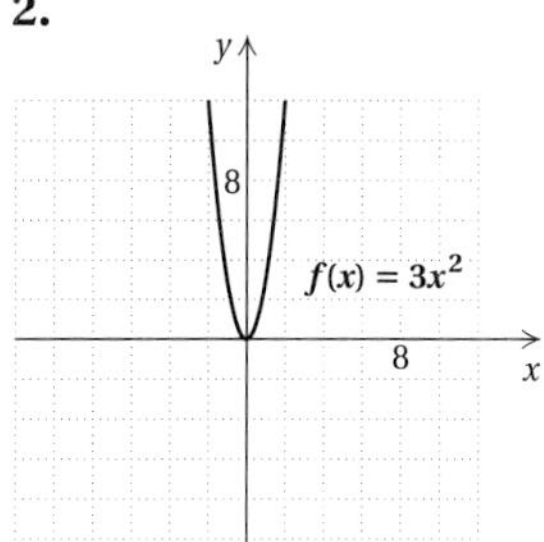

3.

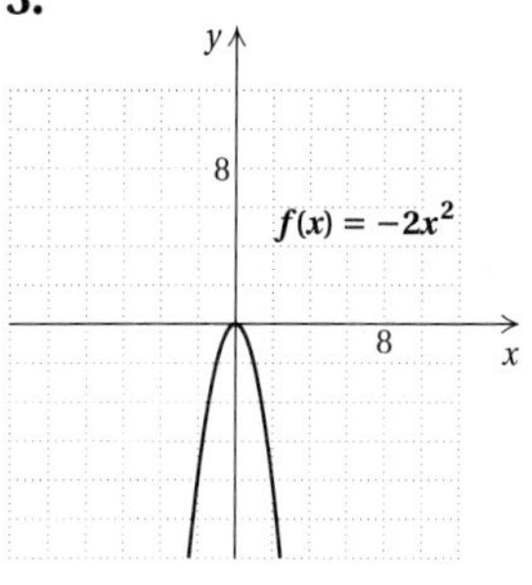

4.

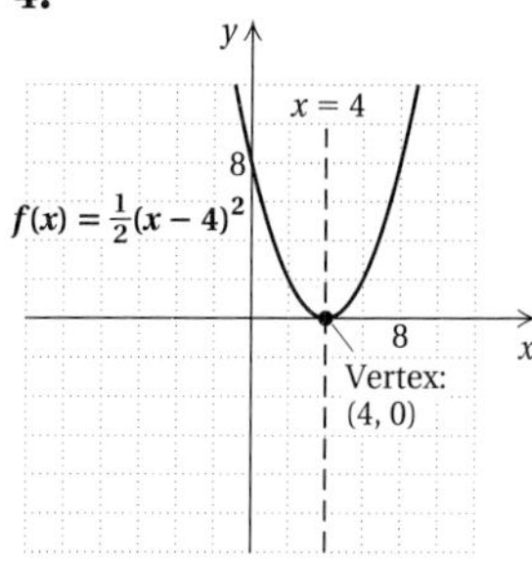

5.

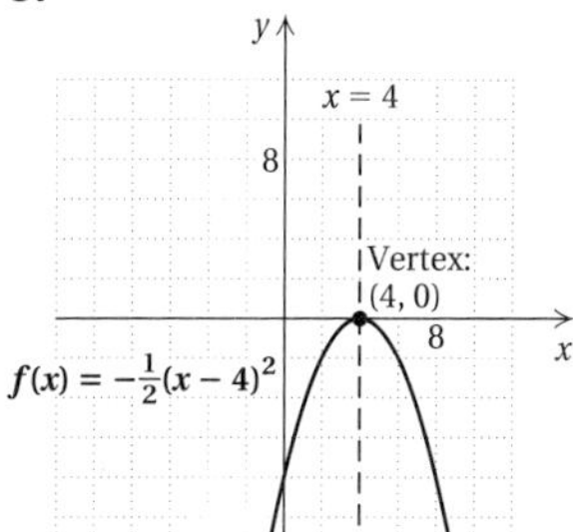

6.

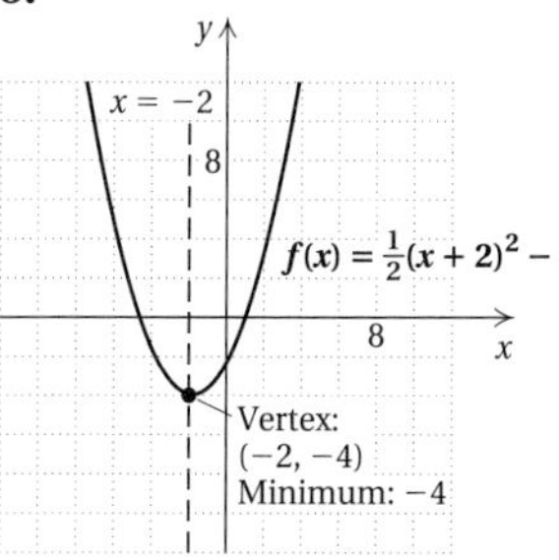

7.

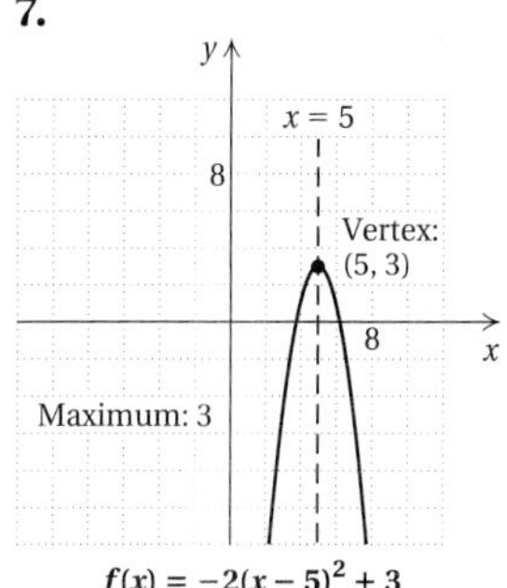

Exercise Set 11.5, p. 719

1.

x	$f(x)$
0	0
1	4
−1	4
2	16
−2	16

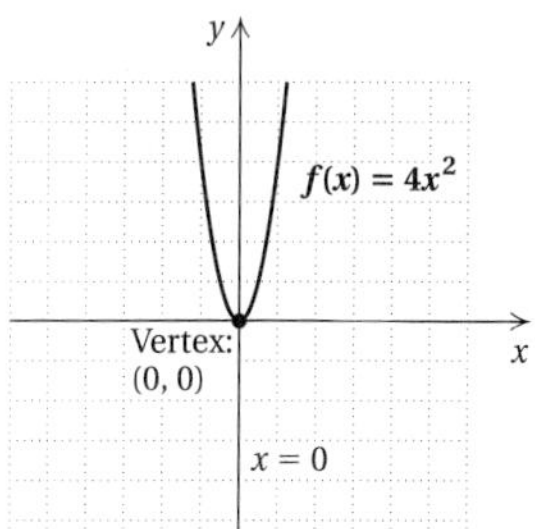

3.

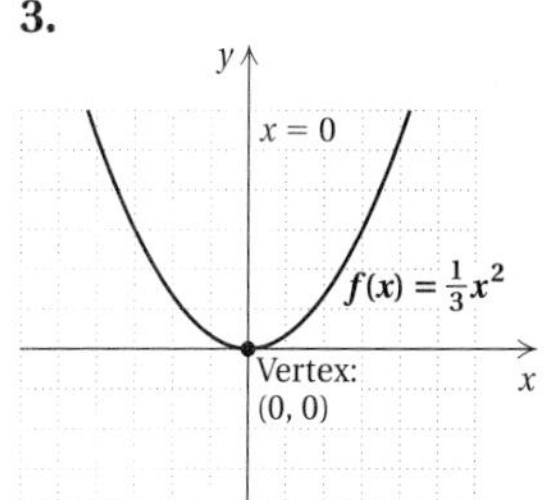

5.

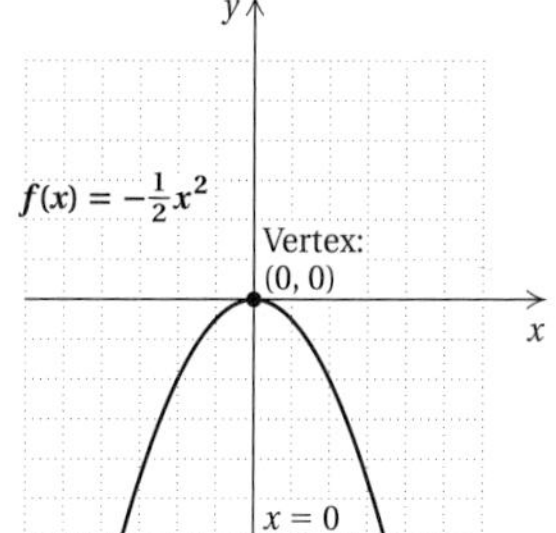

7.

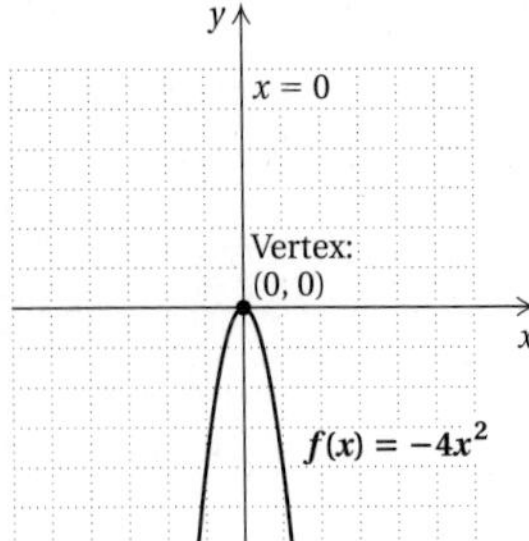

9.

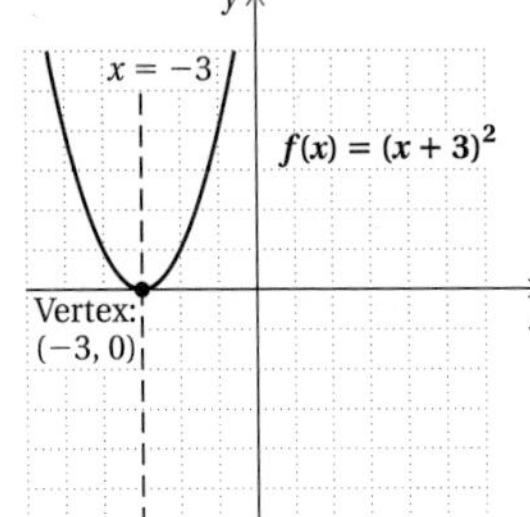

11.

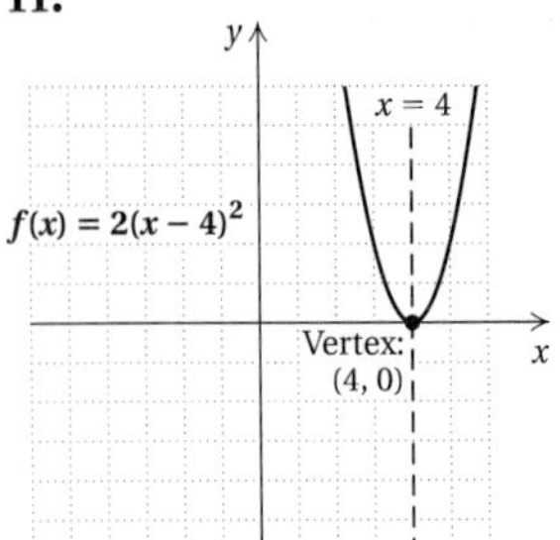

13.

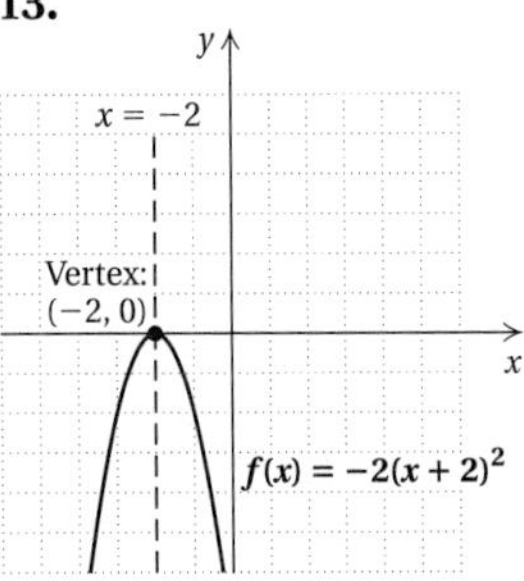

15.

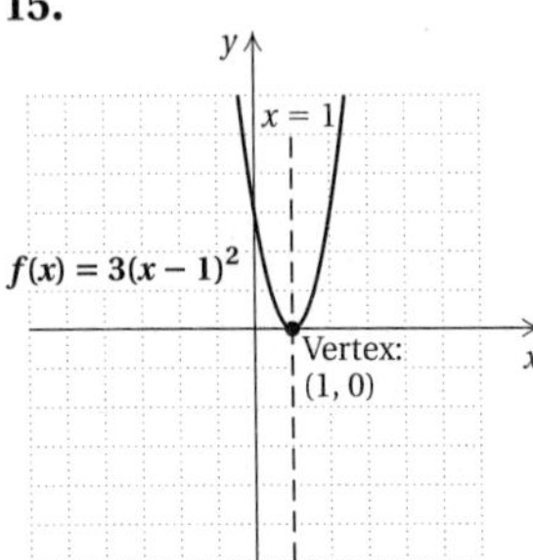

17.

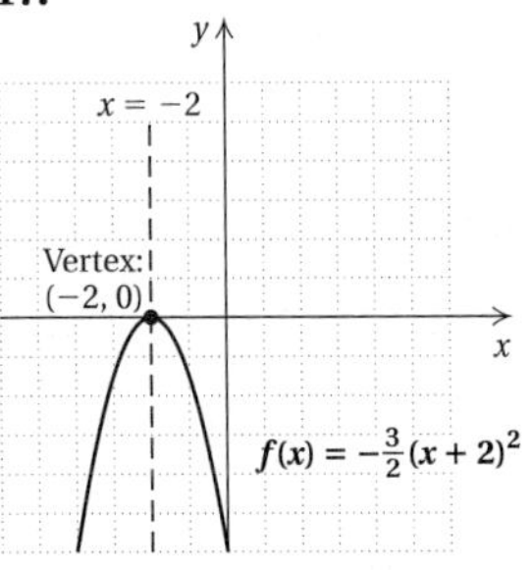

19.

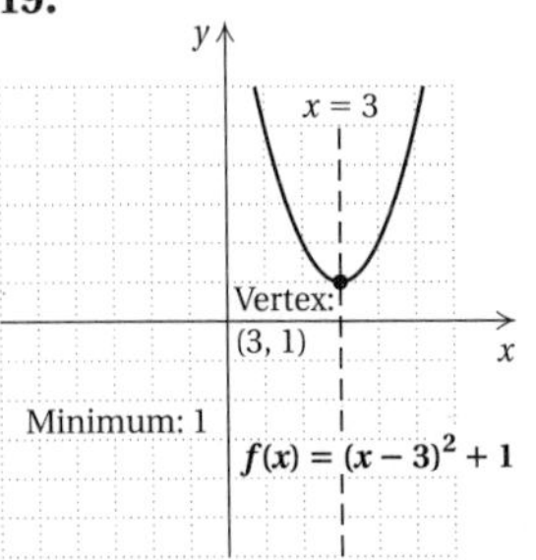

21.

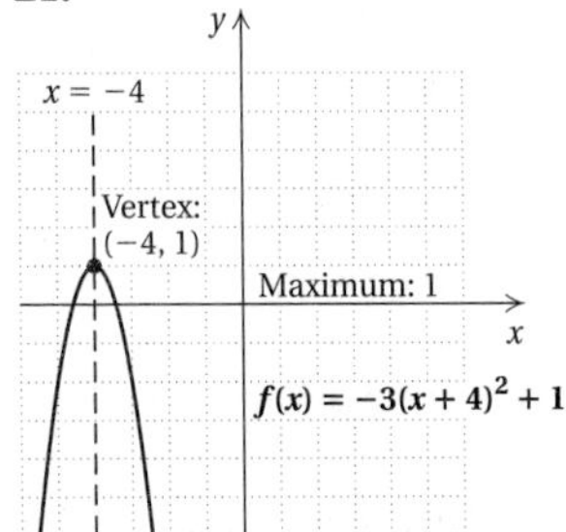

23.

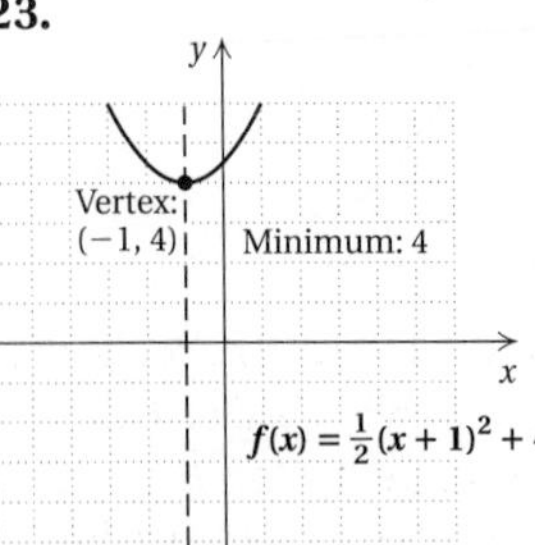

25.

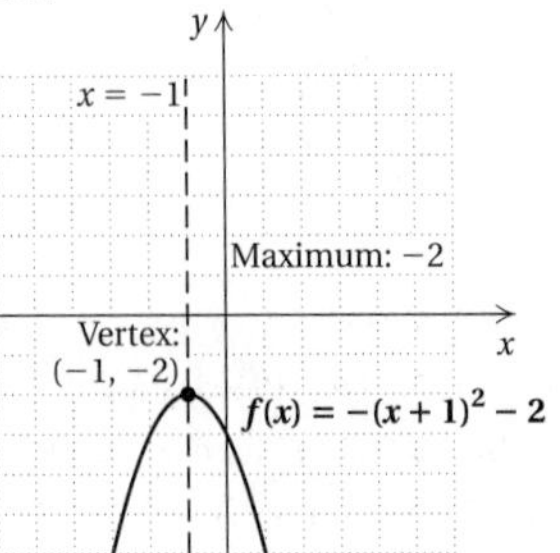

27. 9 **28.** $\frac{11}{3}$ **29.** No solution **30.** 15
31. ◈
33. Left to the student.

Margin Exercises, Section 11.6, pp. 721–724

1.

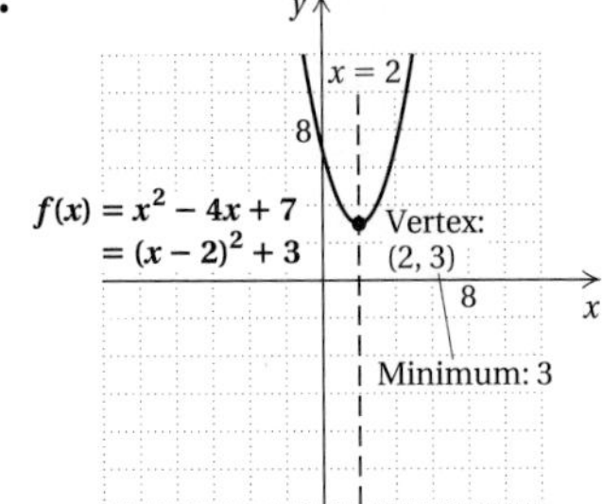

2.

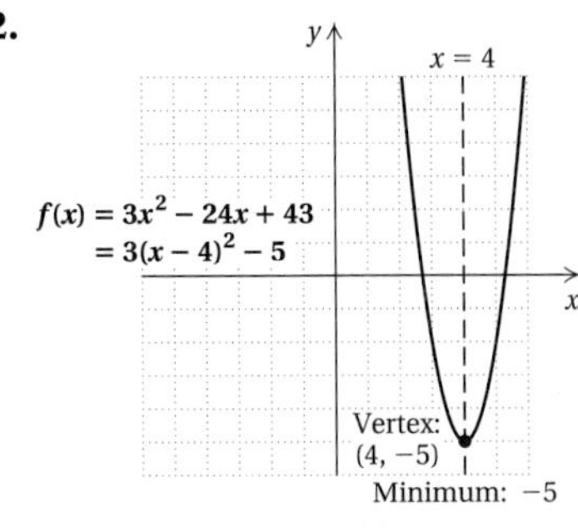

3.

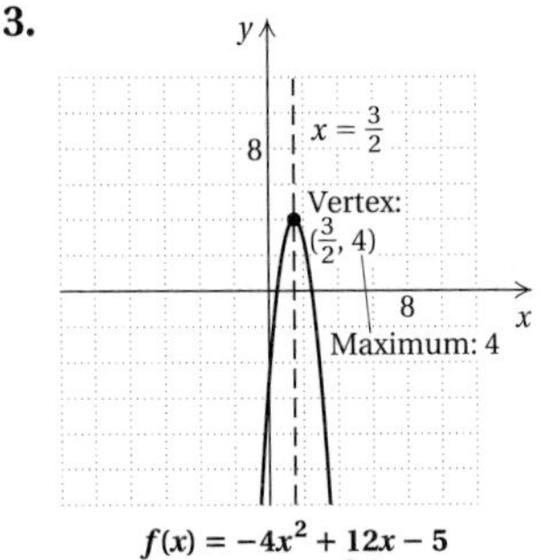

$f(x) = -4x^2 + 12x - 5$
$= -4(x - \frac{3}{2})^2 + 4$

4. $(3, -5)$ **5.** $(4, -5)$ **6.** $\left(\frac{3}{2}, 4\right)$
7. y-intercept: $(0, -3)$; x-intercepts: $(-3, 0)$, $(1, 0)$
8. y-intercept: $(0, 16)$; x-intercept: $(-4, 0)$
9. y-intercept: $(0, 1)$; x-intercepts: $(2 + \sqrt{3}, 0)$, $(2 - \sqrt{3}, 0)$, or $(3.732, 0)$, $(0.268, 0)$

Exercise Set 11.6, p. 725

1.

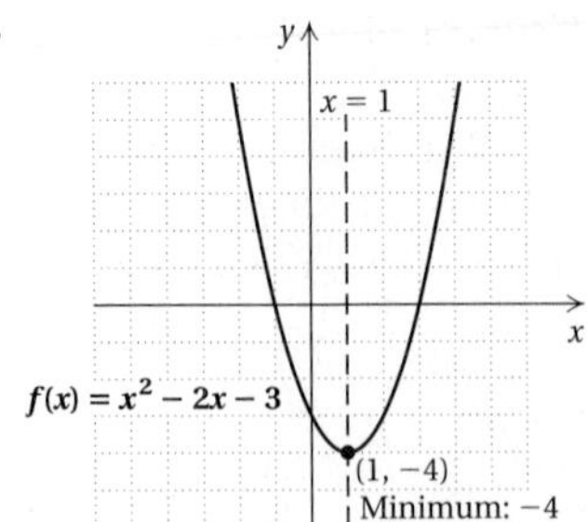

3.

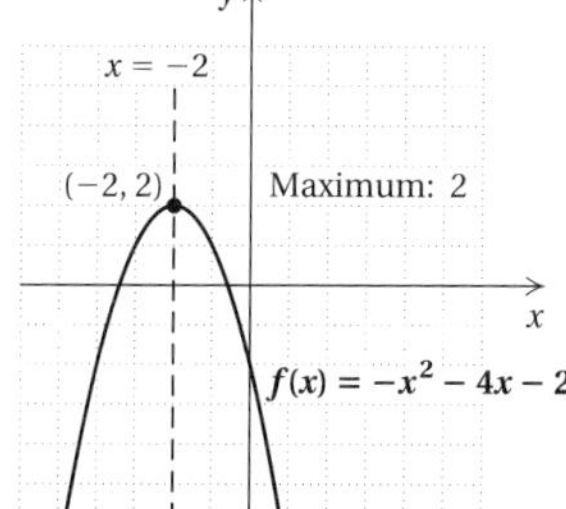

5.

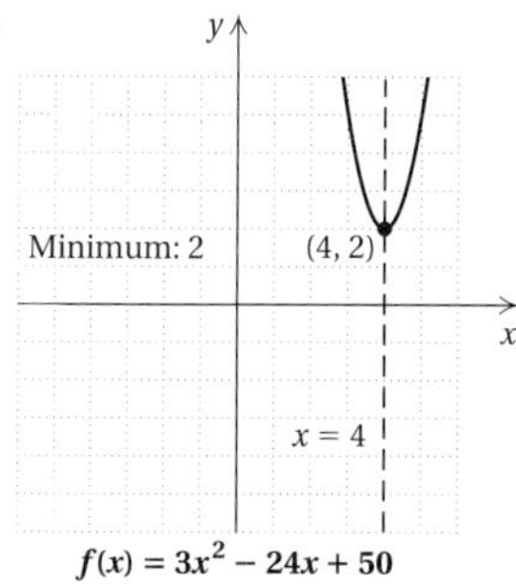

7.

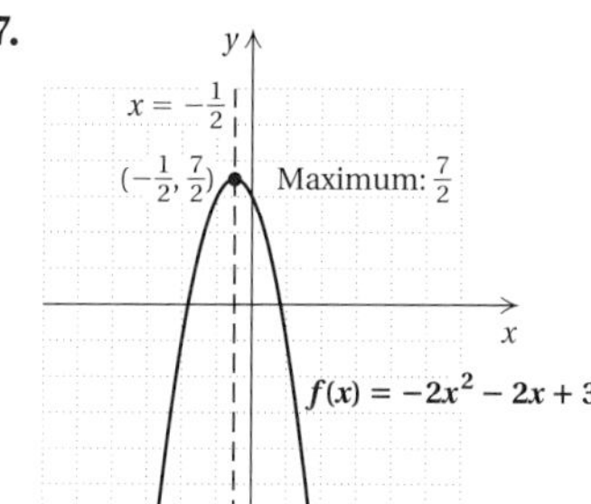

9.

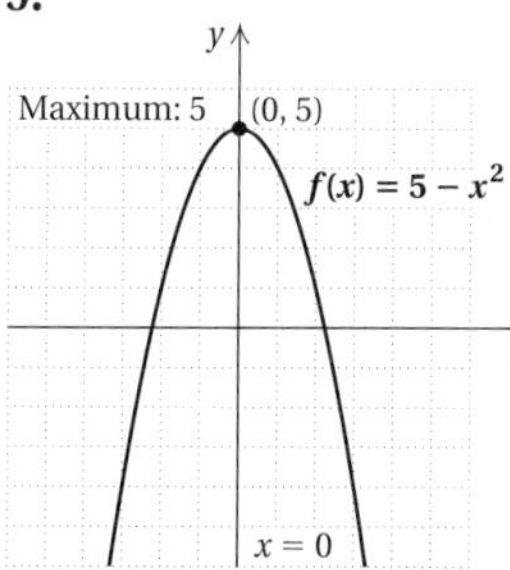

11.

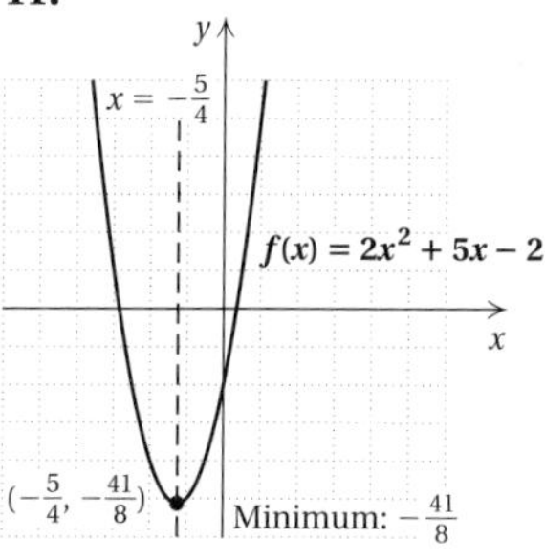

13. y-intercept: (0, 1); x-intercepts: $(3 + 2\sqrt{2}, 0)$, $(3 - 2\sqrt{2}, 0)$ **15.** y-intercept: (0, 20); x-intercepts: (5, 0), (−4, 0) **17.** y-intercept: (0, 9); x-intercept: $\left(-\frac{3}{2}, 0\right)$ **19.** y-intercept: (0, 8); x-intercepts: none **21.** **(a)** $D = 15w$; **(b)** 630 mg **22.** **(a)** $C = \frac{89}{6}t$; **(b)** 712 calories **23.** 250; $y = \frac{250}{x}$ **24.** 250; $y = \frac{250}{x}$ **25.** $\frac{125}{2}$; $y = \frac{125}{2}x$ **26.** $\frac{2}{125}$; $y = \frac{2}{125}x$ **27.** **29.** **(a)** Minimum: −6.954; **(b)** maximum: 7.014

31.

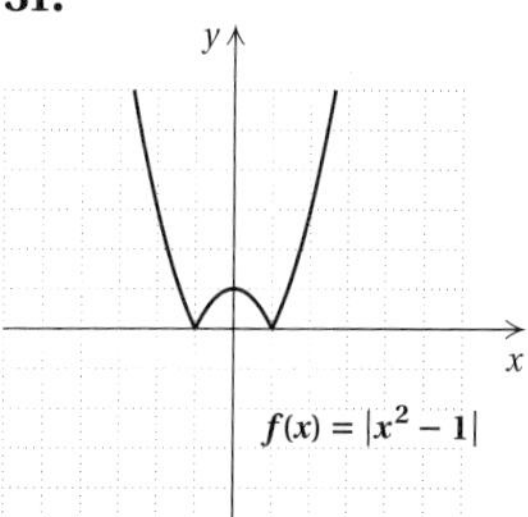

33.

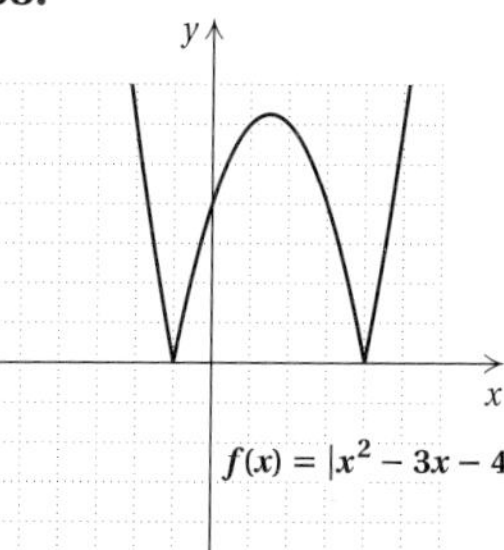

35. $f(x) = \frac{5}{16}x^2 - \frac{15}{8}x - \frac{35}{16}$ **37.** **(a)** Vertex: $\left(-1, -\frac{1}{2}\right)$; **(b)** line of symmetry: $x = -1$; **(c)** minimum: $-\frac{1}{2}$;
(d)

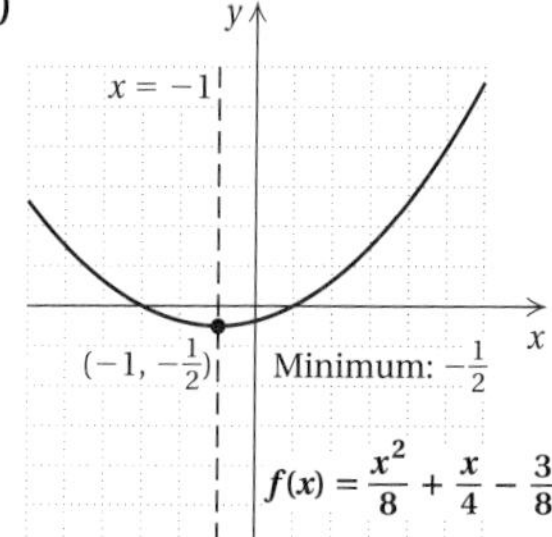

39. (0, 2), (5, 7)

Margin Exercises, Section 11.7, pp. 728–732

1. 25 yd by 25 yd **2.** $f(x) = ax^2 + bx + c,\ a > 0$ **3.** $f(x) = mx + b$ **4.** Polynomial, neither quadratic nor linear **5.** $f(x) = ax^2 + bx + c,\ a > 0$
6. **(a)**

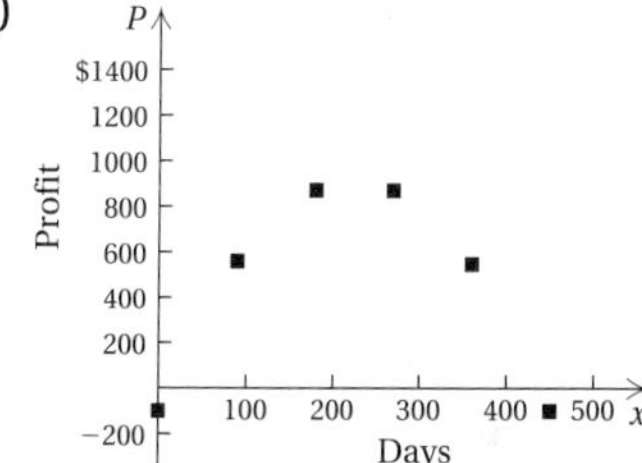

(b) yes; **(c)** $f(x) = -0.02x^2 + 9x - 100$; **(d)** $912.50

Calculator Spotlight, p. 733

1. **(a)**

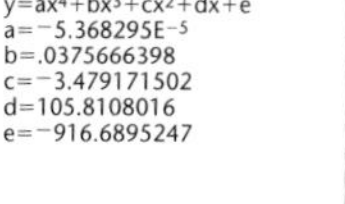

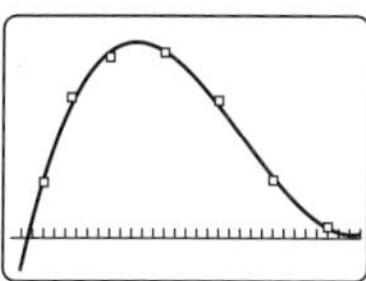

The quartic function is better than the quadratic, but it seems to fit equally as well as the cubic. **(b)** For the cubic function, domain values less than 15 give negative range values (negative number of live births). As domain values increase beyond 45, range values increase indefinitely (continual increase of live births). Neither of these facts is true for those ages.

2. **(a), (b)**

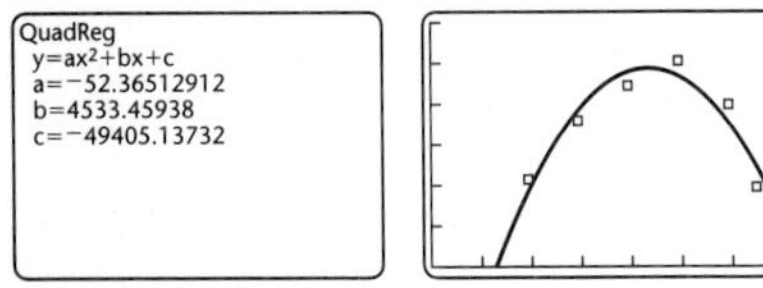

(c), (d)

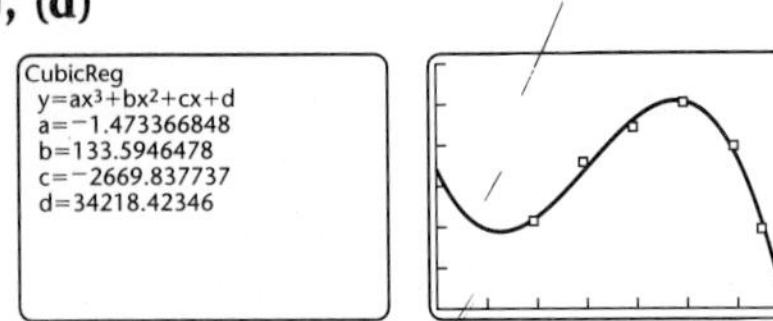

(e), (f)

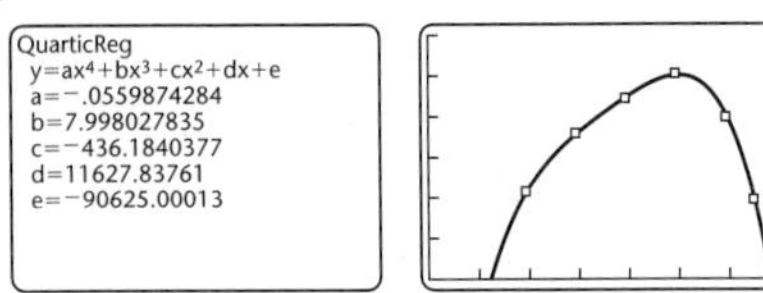

(g) quartic; **(h)** \$30,675; \$48,581
3. Left to the student.

Exercise Set 11.7, p. 735

1. 180 ft by 180 ft **3.** 121; 11 and 11 **5.** -4; 2 and -2 **7.** 36; -6 and -6 **9.** 200 ft^2; 10 ft by 20 ft **11.** 3.5 in. **13.** 3.5 hundred, or 350 **15.** $P(x) = -x^2 + 980x - 3000$; \$237,100 at $x = 490$ **17.** $f(x) = mx + b$ **19.** $f(x) = ax^2 + bx + c,\ a > 0$ **21.** Polynomial, neither quadratic nor linear **23.** Polynomial, neither quadratic nor linear **25.** $f(x) = 2x^2 + 3x - 1$ **27.** $f(x) = -\frac{1}{4}x^2 + 3x - 5$ **29.** **(a)** $A(s) = \frac{3}{16}s^2 - \frac{135}{4}s + 1750$; **(b)** about 531 per 200,000,000 km driven **31.** **(a)** $f(x) = 65.6x^2 - 143.1x + 131.5$; **(b)** \$1056 million, \$1634.5 million **33.** $5xy^2\sqrt[4]{x}$ **34.** $12a^2b^2$ **35.** 5 **36.** 4 **37.** No solution **38.** 4 **39.** ◈ **41.** $f(x) = -0.087x^4 + 1.837x^3 - 12.036x^2 + 23.694x + 41.232$

Margin Exercises, Section 11.8, pp. 739–744

1. $\{x \mid x < -3 \text{ or } x > 1\}$, or $(-\infty, -3) \cup (1, \infty)$
2. $\{x \mid -3 < x < 1\}$, or $(-3, 1)$ **3.** $\{x \mid -3 \le x \le 1\}$, or $[-3, 1]$ **4.** $\{x \mid x < -4 \text{ or } x > 1\}$, or $(-\infty, -4) \cup (1, \infty)$
5. $\{x \mid -4 \le x \le 1\}$, or $[-4, 1]$
6. $\{x \mid x < -1 \text{ or } 0 < x < 1\}$, or $(-\infty, -1) \cup (0, 1)$
7. $\left\{x \mid 2 < x \le \frac{7}{2}\right\}$, or $\left(2, \frac{7}{2}\right]$
8. $\{x \mid x < 5 \text{ or } x > 10\}$, or $(-\infty, 5) \cup (10, \infty)$

Calculator Spotlight, p. 742

1. $\{x \mid x < -5 \text{ or } x > 2\}$, or $(-\infty, -5) \cup (2, \infty)$
2. $\{x \mid x \le -5 \text{ or } x \ge 2\}$, or $(-\infty, -5] \cup [2, \infty)$
3. $\{x \mid -5 \le x \le 2\}$, or $[-5, 2]$ **4.** $\{x \mid -5 < x < 2\}$, or $(-5, 2)$ **5.** $\{x \mid -0.78 \le x \le 1.59\}$, or $[-0.78, 1.59]$
6. $\{x \mid x \le -0.21 \text{ or } x \ge 2.47\}$, or $(-\infty, -0.21] \cup [2.47, \infty)$ **7.** $\{x \mid x < -1.26 \text{ or } x > 2.33\}$, or $(-\infty, -1.26) \cup (2.33, \infty)$ **8.** $\{x \mid x > -1.37\}$, or $(-1.37, \infty)$

Exercise Set 11.8, p. 745

1. $\{x \mid x < -2 \text{ or } x > 6\}$, or $(-\infty, -2) \cup (6, \infty)$
3. $\{x \mid -2 \le x \le 2\}$, or $[-2, 2]$ **5.** $\{x \mid -1 \le x \le 4\}$, or $[-1, 4]$ **7.** $\{x \mid -1 < x < 2\}$, or $(-1, 2)$ **9.** All real numbers, or $(-\infty, \infty)$ **11.** $\{x \mid 2 < x < 4\}$, or $(2, 4)$
13. $\{x \mid x < -2 \text{ or } 0 < x < 2\}$, or $(-\infty, -2) \cup (0, 2)$
15. $\{x \mid -9 < x < -1 \text{ or } x > 4\}$, or $(-9, -1) \cup (4, \infty)$
17. $\{x \mid x < -3 \text{ or } -2 < x < 1\}$, or $(-\infty, -3) \cup (-2, 1)$
19. $\{x \mid x < 6\}$, or $(-\infty, 6)$ **21.** $\{x \mid x < -1 \text{ or } x > 3\}$, or $(-\infty, -1) \cup (3, \infty)$ **23.** $\left\{x \mid -\frac{2}{3} \le x < 3\right\}$, or $\left[-\frac{2}{3}, 3\right)$
25. $\left\{x \mid 2 < x < \frac{5}{2}\right\}$, or $\left(2, \frac{5}{2}\right)$
27. $\{x \mid x < -1 \text{ or } 2 < x < 5\}$, or $(-\infty, -1) \cup (2, 5)$
29. $\{x \mid -3 \le x < 0\}$, or $[-3, 0)$ **31.** $\{x \mid 1 < x < 2\}$, or $(1, 2)$ **33.** $\{x \mid x < -4 \text{ or } 1 < x < 3\}$, or $(-\infty, -4) \cup (1, 3)$ **35.** $\left\{x \mid 0 < x < \frac{1}{3}\right\}$, or $\left(0, \frac{1}{3}\right)$
37. $\{x \mid x < -3 \text{ or } -2 < x < 1 \text{ or } x > 4\}$, or $(-\infty, -3) \cup (-2, 1) \cup (4, \infty)$ **39.** $\frac{5}{3}$ **40.** $\dfrac{5}{2a}$
41. $\dfrac{4a}{b^2}\sqrt{a}$ **42.** $\dfrac{3c}{7d}\sqrt[3]{c^2}$ **43.** $\sqrt{2}$ **44.** $17\sqrt{5}$
45. $(10a + 7)\sqrt[3]{2a}$ **46.** $3\sqrt{10} - 4\sqrt{5}$ **47.** ◈
49. Left to the student.
51. $\{x \mid 1 - \sqrt{3} \le x \le 1 + \sqrt{3}\}$, or $[1 - \sqrt{3}, 1 + \sqrt{3}]$
53. All real numbers except 0, or $(-\infty, 0) \cup (0, \infty)$
55. $\left\{x \mid x < \frac{1}{4} \text{ or } x > \frac{5}{2}\right\}$, or $\left(-\infty, \frac{1}{4}\right) \cup \left(\frac{5}{2}, \infty\right)$
57. **(a)** $\{t \mid 0 < t < 2\}$, or $(0, 2)$; **(b)** $\{t \mid t > 10\}$, or $(10, \infty)$

Summary and Review: Chapter 11, p. 747

1. **(a)** $\pm\dfrac{\sqrt{14}}{2}$; **(b)** $\left(\dfrac{\sqrt{14}}{2}, 0\right), \left(-\dfrac{\sqrt{14}}{2}, 0\right)$ **2.** $0, -\dfrac{5}{14}$
3. 3, 9 **4.** $-\dfrac{3}{8} \pm i\dfrac{\sqrt{7}}{8}$ **5.** $\dfrac{7}{2} \pm i\dfrac{\sqrt{3}}{2}$ **6.** 3, 5
7. $-2 \pm \sqrt{3}$; $-0.268, -3.732$ **8.** 4, -2 **9.** $4 \pm 4\sqrt{2}$
10. $\dfrac{1 \pm \sqrt{481}}{15}$ **11.** $-2 \pm \sqrt{3}$ **12.** 0.90 sec
13. Length: 14 cm; width: 9 cm **14.** 1 in.
15. First part: 50 mph; second part: 40 mph
16. Two real **17.** Two nonreal
18. $25x^2 + 10x - 3 = 0$ **19.** $x^2 + 8x + 16 = 0$
20. $p = \dfrac{9\pi^2}{N^2}$ **21.** $T = \sqrt{\dfrac{3B}{2A}}$ **22.** 2, -2, 3, -3
23. 3, -5 **24.** $\pm\sqrt{7}, \pm\sqrt{2}$ **25.** 81, 16
26. **(a)** $(1, 3)$; **(b)** $x = 1$; **(c)** maximum: 3;
(d)

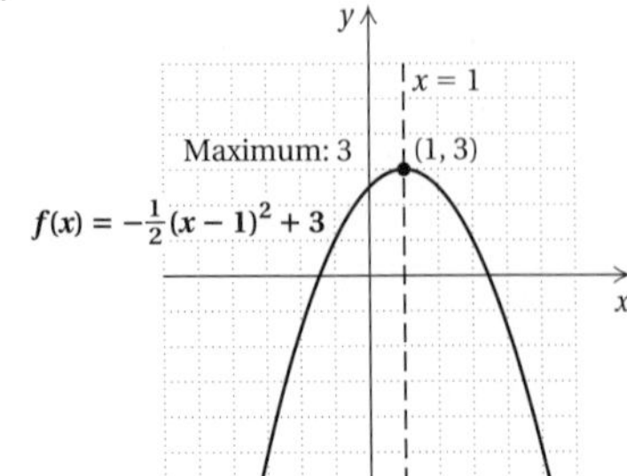

27. **(a)** $\left(\frac{1}{2}, \frac{23}{4}\right)$; **(b)** $x = \frac{1}{2}$; **(c)** minimum: $\frac{23}{4}$;
(d)

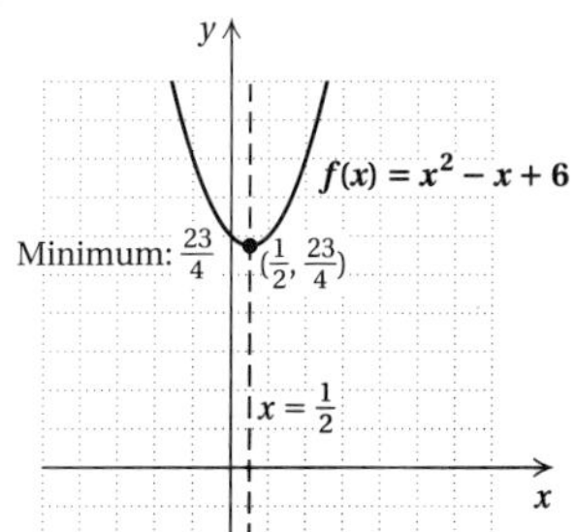

28. **(a)** $(-2, 4)$; **(b)** $x = -2$; **(c)** maximum: 4;
(d)

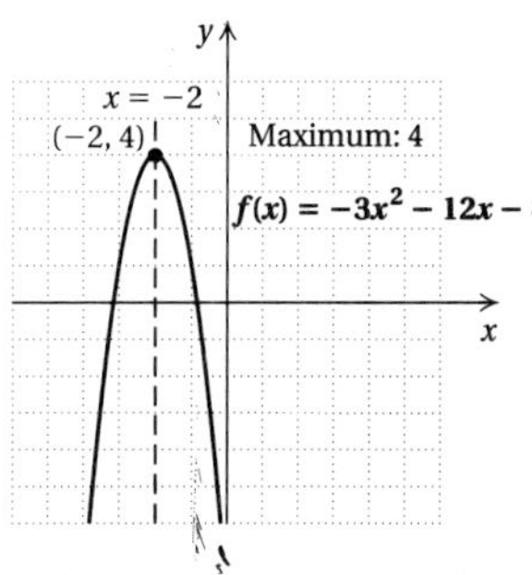

29. y-intercept: $(0, 14)$; x-intercepts: $(2, 0)$, $(7, 0)$
30. -121; 11 and -11 **31.** $f(x) = -x^2 + 6x - 2$
32. **(a)** $f(x) = -0.170x^2 + 1.846x + 60.124$; **(b)** 61.6%; 52.6% **33.** $\{x \mid -2 < x < 1 \text{ or } x > 2\}$, or $(-2, 1) \cup (2, \infty)$ **34.** $\{x \mid x < -4 \text{ or } -2 < x < 1\}$, or $(-\infty, -4) \cup (-2, 1)$ **35.** **(a)** $f(x) = 1.2x + 497.8$;
(b) $f(10) = 509.8$, $f(14) = 514.6$ **36.** $\dfrac{x + 1}{(x - 3)(x - 1)}$
37. $3t^5s\sqrt[3]{s}$ **38.** 2 **39.** ◆ The graph of $f(x) = ax^2 + bx + c$ is a parabola with vertex $\left(-\dfrac{b}{2a}, \dfrac{4ac - b^2}{4a}\right)$. The line of symmetry is $x = -\dfrac{b}{2a}$. If $a < 0$, then the parabola opens down and $\dfrac{4ac - b^2}{4a}$ is the maximum function value. If $a > 0$, the parabola opens up and $\dfrac{4ac - b^2}{4a}$ is the minimum function value. The x-intercepts are $\left(\dfrac{-b - \sqrt{b^2 - 4ac}}{2a}, 0\right)$ and $\left(\dfrac{-b + \sqrt{b^2 - 4ac}}{2a}, 0\right)$, for $b^2 - 4ac > 0$. When $b^2 - 4ac = 0$, there is just one x-intercept, $\left(-\dfrac{b}{2a}, 0\right)$. When $b^2 - 4ac < 0$, there are no x-intercepts.
40. ◆ The x-coordinate of the maximum or minimum point lies halfway between the x-coordinates of the x-intercepts. The function must be evaluated for this value of x in order to determine the maximum or minimum value. **41.** $f(x) = \frac{7}{15}x^2 - \frac{14}{15}x - 7$; minimum: $-\frac{112}{15}$ **42.** $h = 60$, $k = 60$ **43.** 18 and 324

Test: Chapter 11, p. 749

1. [11.1a] **(a)** $\pm\dfrac{2\sqrt{3}}{3}$; **(b)** $\left(\dfrac{2\sqrt{3}}{3}, 0\right), \left(-\dfrac{2\sqrt{3}}{3}, 0\right)$
2. [11.2a] $-\dfrac{1}{2} \pm i\dfrac{\sqrt{3}}{2}$ **3.** [11.4c] 49, 1 **4.** [11.2a] 9, 2
5. [11.4c] $\pm\sqrt{\dfrac{5 + \sqrt{5}}{2}}, \pm\sqrt{\dfrac{5 - \sqrt{5}}{2}}$
6. [11.2a] $-2 \pm \sqrt{6}$; 0.449, -4.449 **7.** [11.2a] 0, 2
8. [11.1b] $2 \pm \sqrt{3}$ **9.** [11.1c] About 6.7 sec
10. [11.3a] About 2.89 mph **11.** [11.3a] 7 cm by 7 cm
12. [11.1c] About 0.85 sec **13.** [11.4a] Two nonreal
14. [11.4b] $x^2 - 4\sqrt{3}x + 9 = 0$
15. [11.3b] $T = \sqrt{\dfrac{V}{48}}$, or $\dfrac{1}{4}\sqrt{\dfrac{V}{3}}$
16. [11.6a] **(a)** $(-1, 1)$; **(b)** $x = -1$; **(c)** maximum: 1; **(d)**

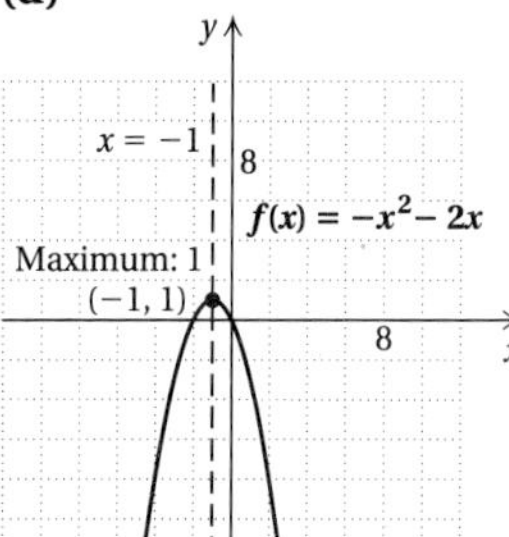

17. [11.6a] **(a)** $(3, 5)$; **(b)** $x = 3$; **(c)** minimum: 5; **(d)**

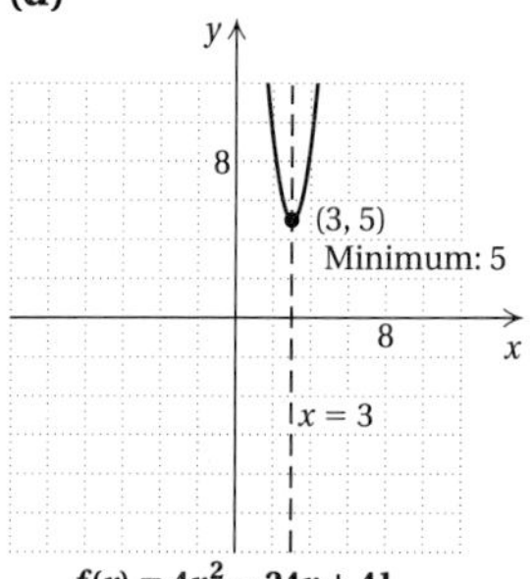

18. [11.6b] y-intercept: $(0, -1)$; x-intercepts: $(2 - \sqrt{3}, 0)$, $(2 + \sqrt{3}, 0)$ **19.** [11.7a] -16; 4 and -4
20. [11.7b] $f(x) = \frac{1}{5}x^2 - \frac{3}{5}x$
21. [11.7b] **(a)** $f(x) = -7x^2 + 105.35x - 266.95$; **(b)** 86.550; 44,900 **22.** [11.8a] $\{x \mid -1 < x < 7\}$, or $(-1, 7)$ **23.** [11.8b] $\{x \mid -3 < x < 5\}$, or $(-3, 5)$
24. [7.6b] **(a)** $f(x) = 1.6x + 498.4$; **(b)** 514.4, 520.8
25. [6.4a] $\dfrac{x - 8}{(x + 6)(x + 8)}$ **26.** [10.3a] $ab\sqrt[4]{2a^2}$
27. [10.6a] 6 **28.** [11.6a, b], [11.7b] $f(x) = -\frac{4}{7}x^2 + \frac{20}{7}x + 8$; maximum: $\frac{81}{7}$
29. [11.2a] $\frac{1}{2}$ **30.** [11.4c] $\pm 2, \pm\sqrt[4]{4}$

Chapter 12

Pretest: Chapter 12, p. 752

1. [12.1a]

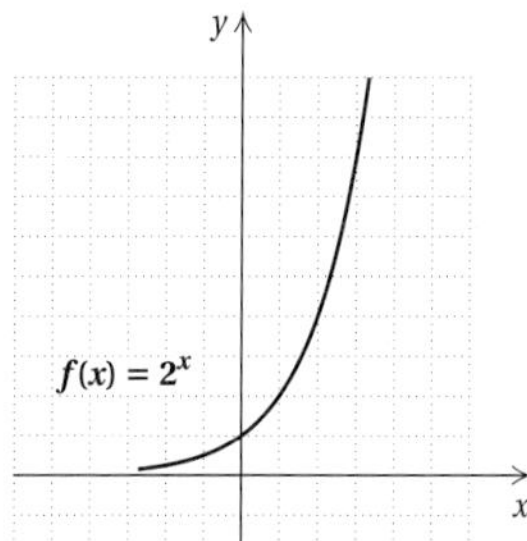

2. [12.3a]

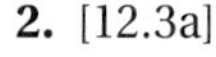

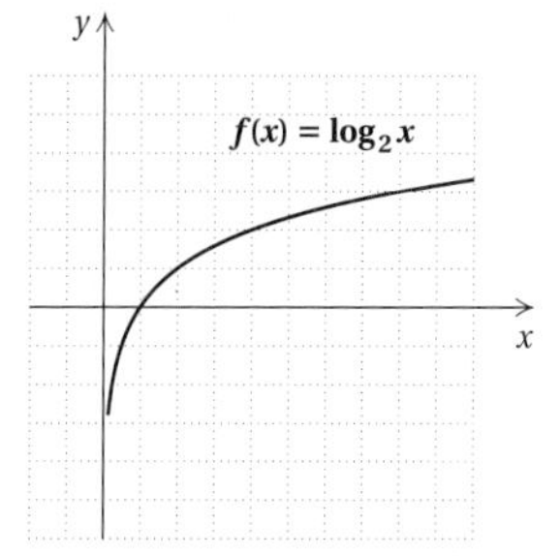

3. [12.3b] $\log_a 1000 = 3$ **4.** [12.3b] $2^t = 32$
5. [12.4d] $\log_a \frac{M^2N}{\sqrt[3]{Q}}$ **6.** [12.4d] $\frac{3}{4}\log_a x + \frac{1}{4}\log_a y - \frac{1}{2}\log_a z$ **7.** [12.6b] 2 **8.** [12.6b] 25
9. [12.6a] $\frac{\log 8.6}{\log 3} \approx 1.9586$ **10.** [12.6a] $\frac{7}{2}$
11. [12.6b] 8 **12.** [12.6b] 1 **13.** [12.3d] 2.8537
14. [12.5a] −7.0745 **15.** [12.5a] 0 **16.** [12.5a] 66.6863
17. [12.5a] 0.1188 **18.** [12.3d] −1.6308
19. [12.1c] $724.27 **20.** [12.7b] About 11,893 yr old

Margin Exercises, Section 12.1, pp. 754–759

1. (a)

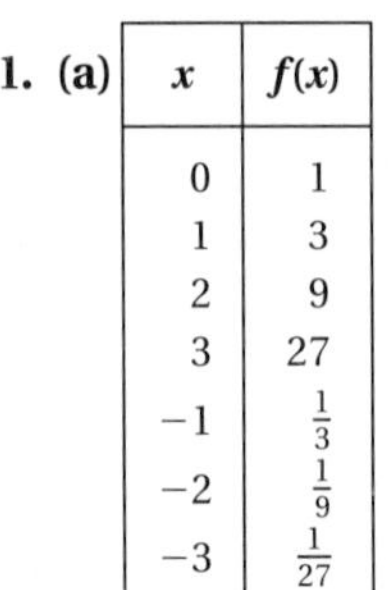

x	$f(x)$
0	1
1	3
2	9
3	27
−1	$\frac{1}{3}$
−2	$\frac{1}{9}$
−3	$\frac{1}{27}$

(b)

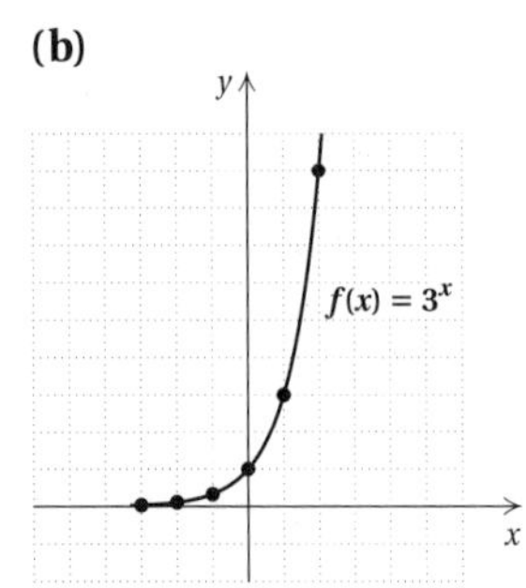

2. (a)

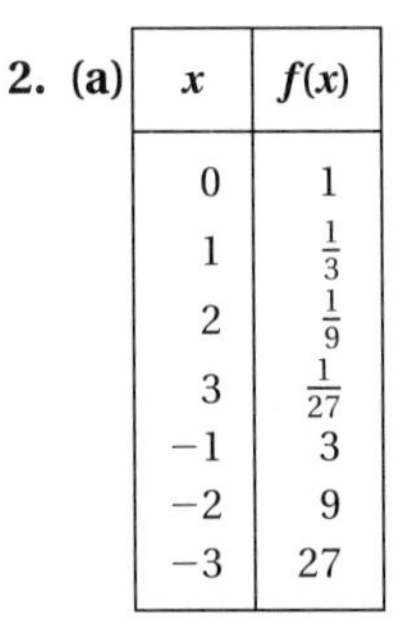

x	$f(x)$
0	1
1	$\frac{1}{3}$
2	$\frac{1}{9}$
3	$\frac{1}{27}$
−1	3
−2	9
−3	27

(b)

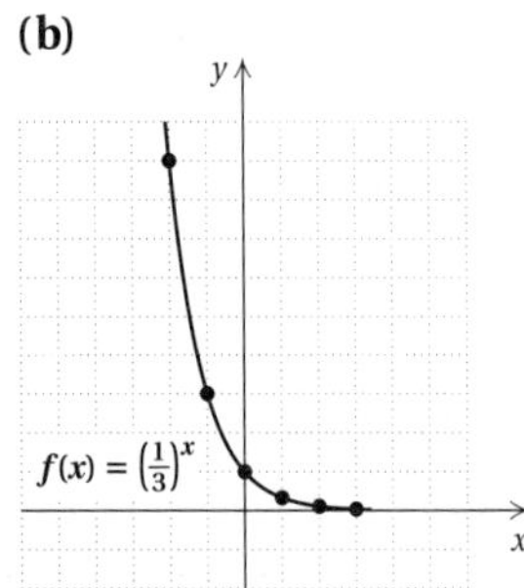

3.

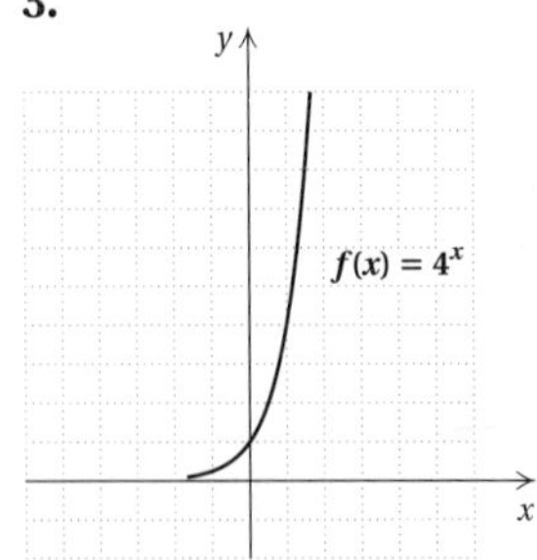

4.

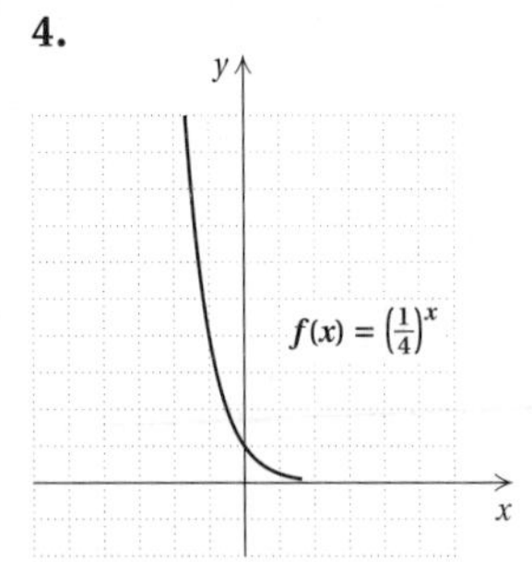

5.

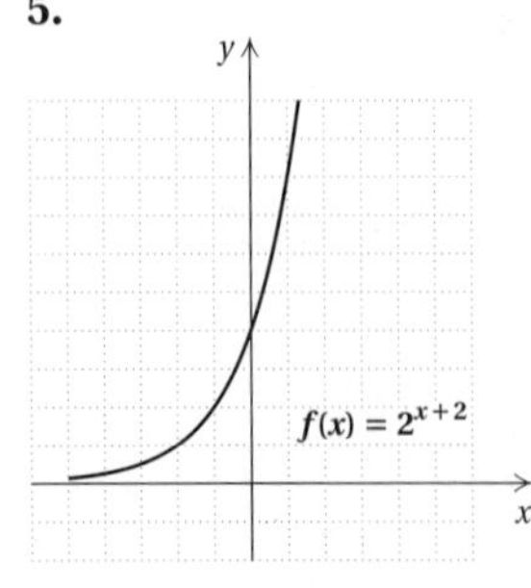

6.

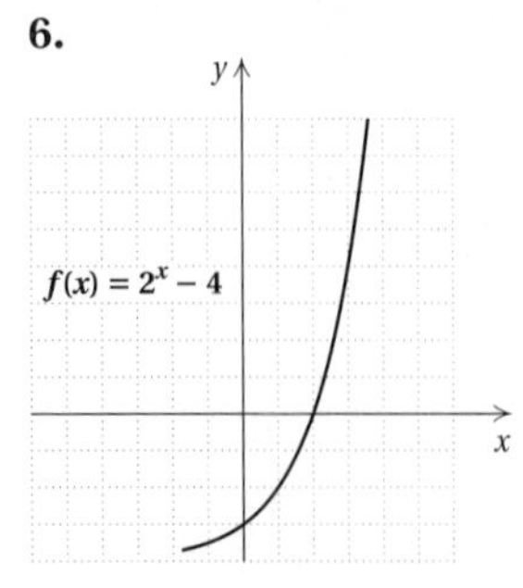

7.

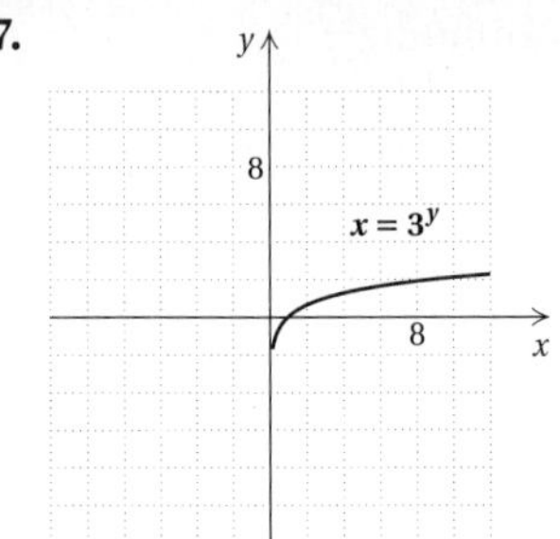

8. (a) $A(t) = \$40{,}000(1.05)^t$; **(b)** $40,000, $48,620.25, $59,098.22, $65,155.79;
(c)

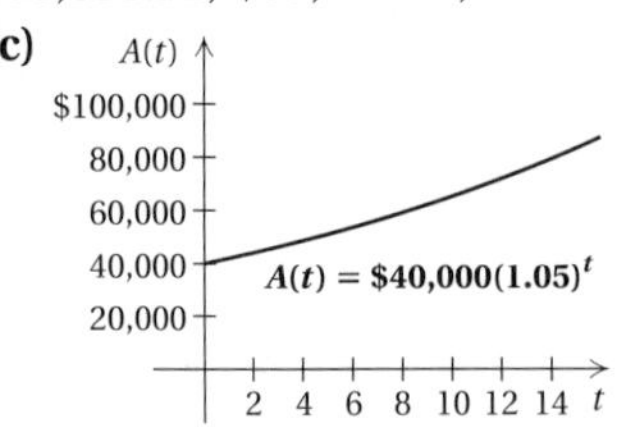

Calculator Spotlight, p. 753

1. 16.24 **2.** 451.81 **3.** 0.045

Calculator Spotlight, p. 760

1. $1125.51 **2.** $1127.16 **3.** $30,372.65
4. $1,480,569.70 **5. (a)** $10,540; **(b)** $10,547.29; **(c)** $10,551.03; **(d)** $10,554.80; **(e)** $10,554.84
6. (a) $10,300; **(b)** $10,302.25; **(c)** $10,303.39; **(d)** $10,304.53; **(e)** $10,304.54

Exercise Set 12.1, p. 761

1.

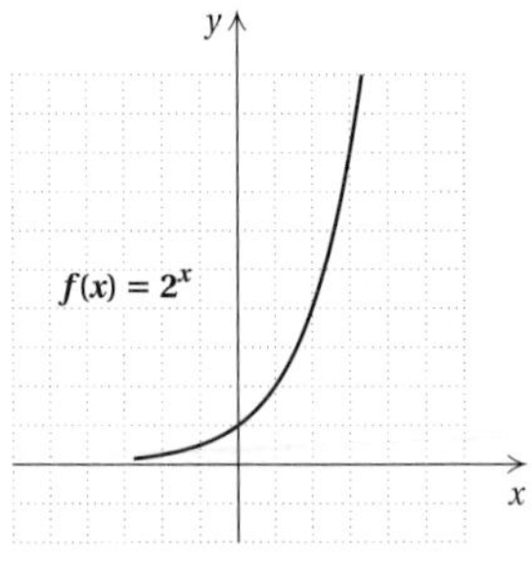

3.

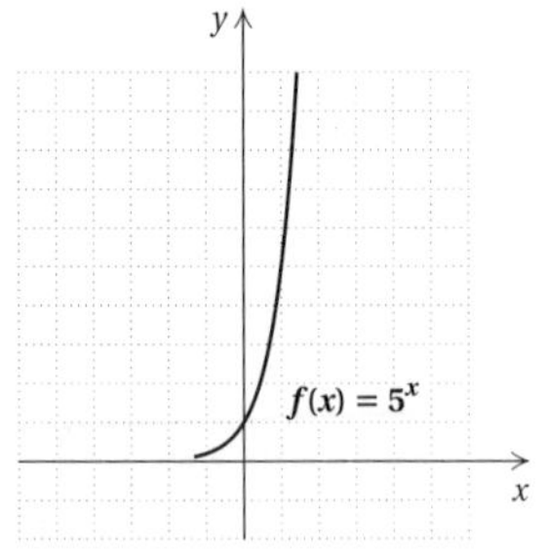

5.

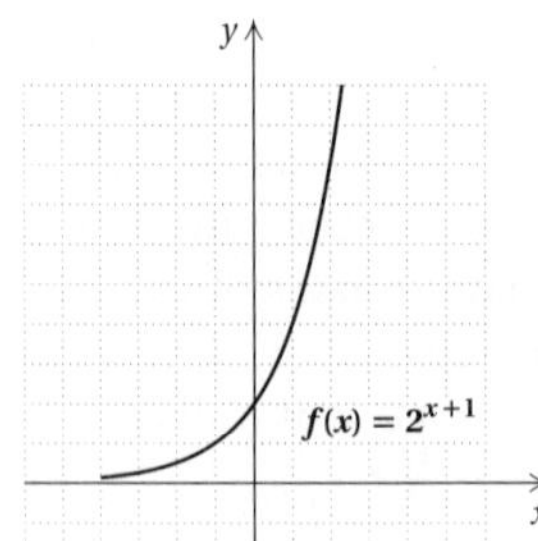

7.

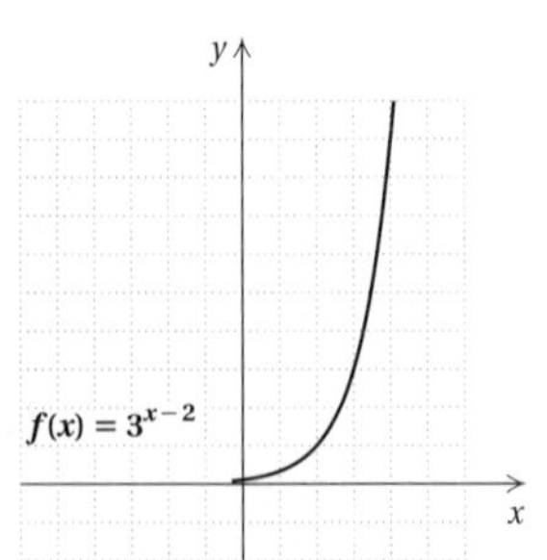

9.

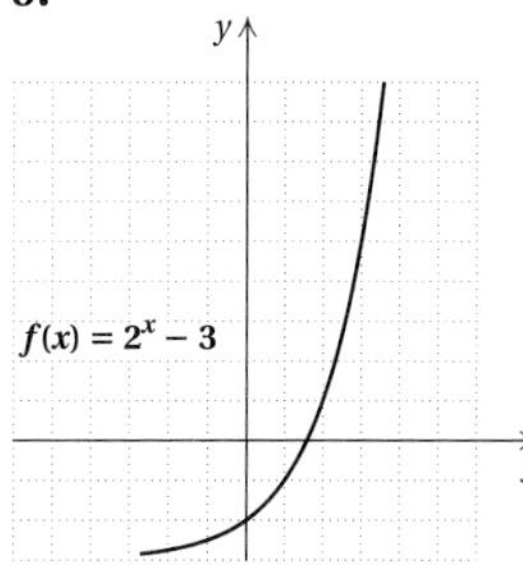

11.

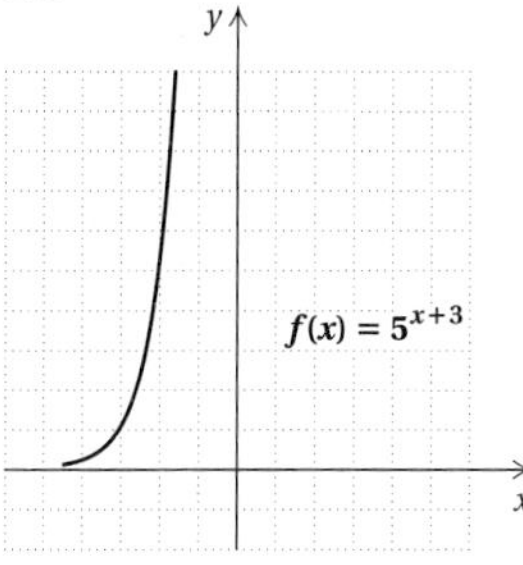

13.

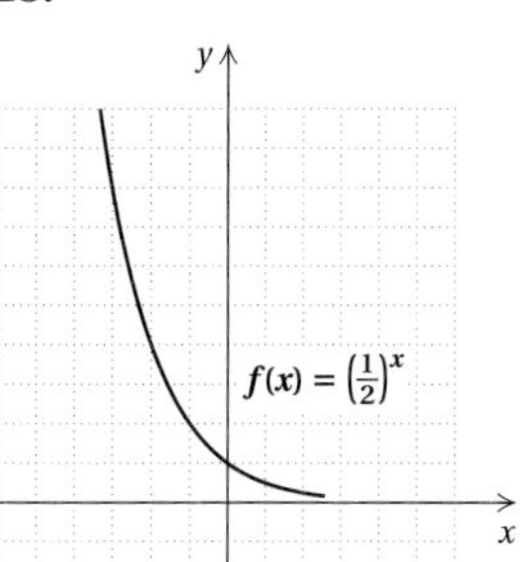

15.

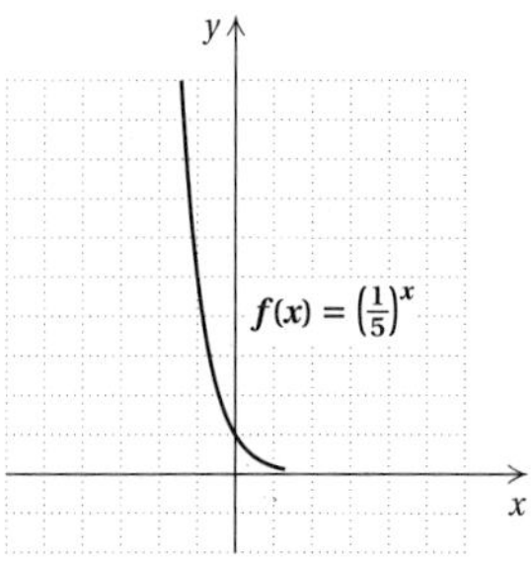

17.

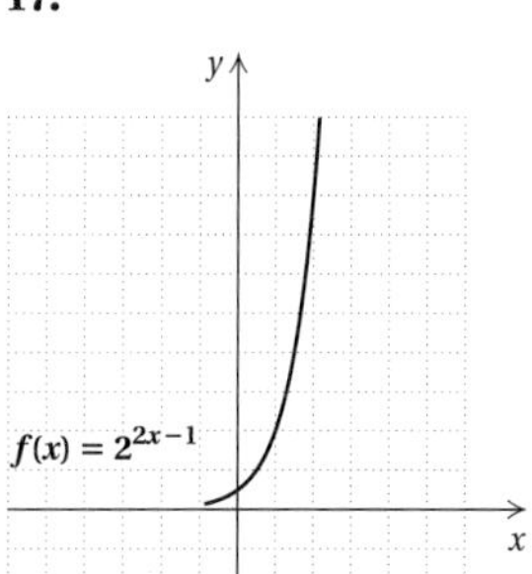

19.

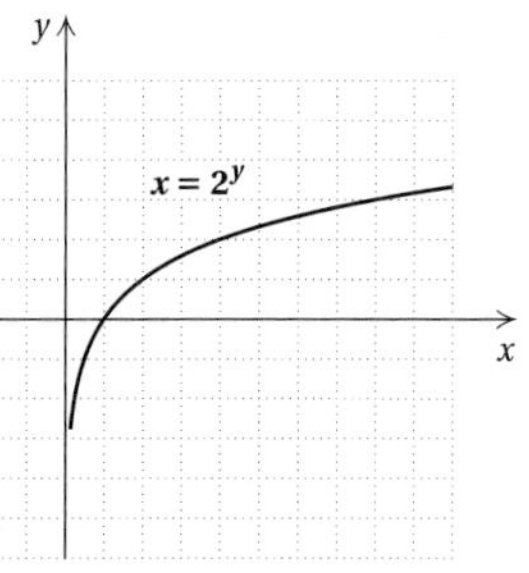

21.

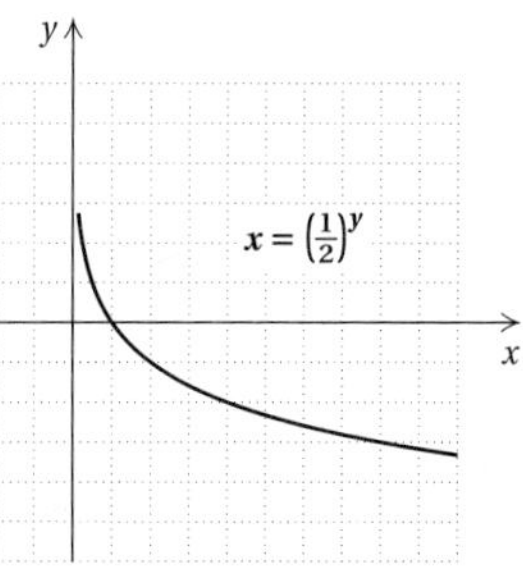

23.

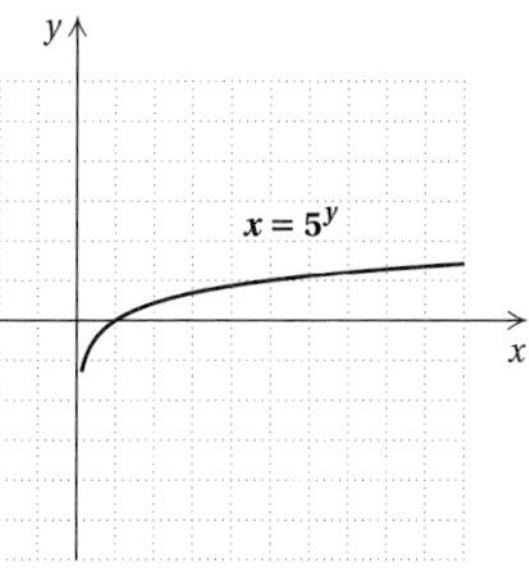

25.

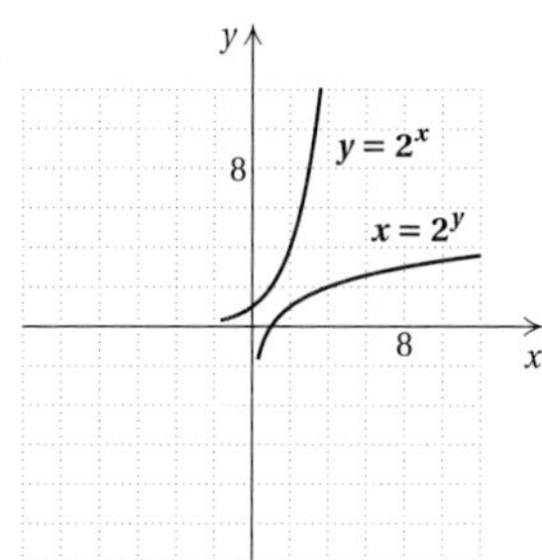

27. **(a)** 62 billion ft^3, 63.116 billion ft^3; **(b)** 65.409 billion ft^3, 78.183 billion ft^3;

(c)

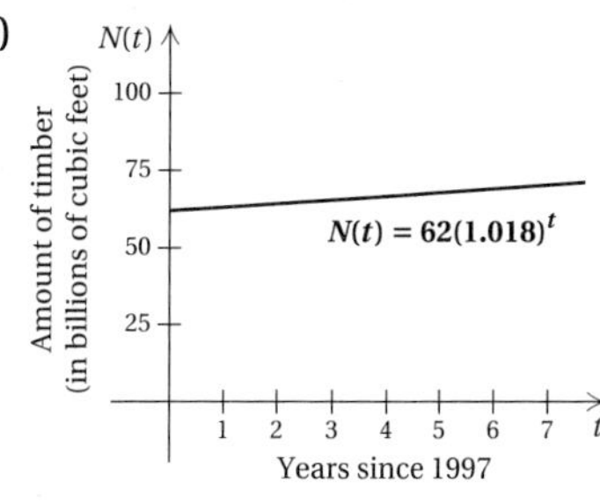

29. **(a)** 4243, 6000, 8485, 12,000, 24,000

(b)

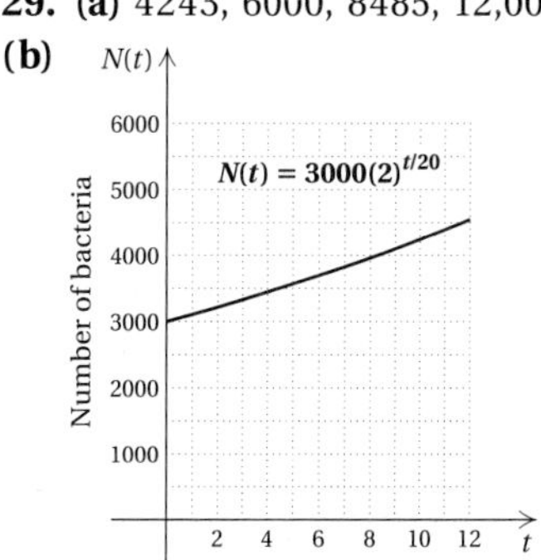

31. **(a)** 454,354,240 cm^2, 525,233,501,400 cm^2;

(b)

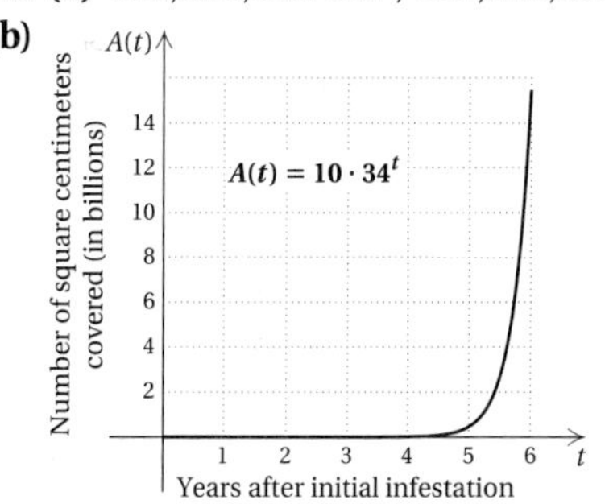

33. $\frac{1}{x^2}$ **34.** $\frac{1}{x^{12}}$ **35.** 1 **36.** 1 **37.** $\frac{2}{3}$ **38.** 2.7

39. $\frac{1}{x^7}$ **40.** $\frac{1}{x^{10}}$ **41.** x **42.** x **43.** ◈

45. 5^4, or 625

47.

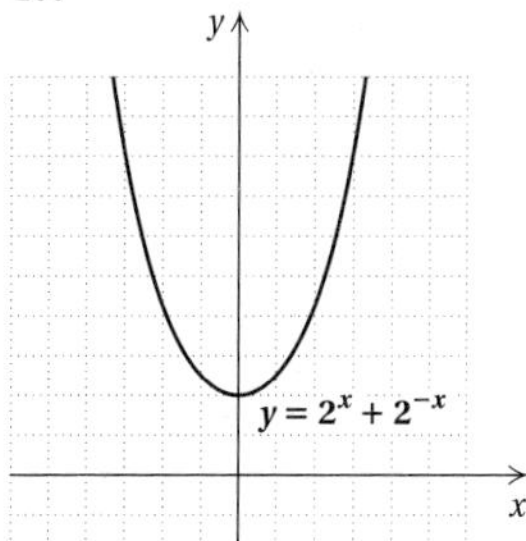

49.

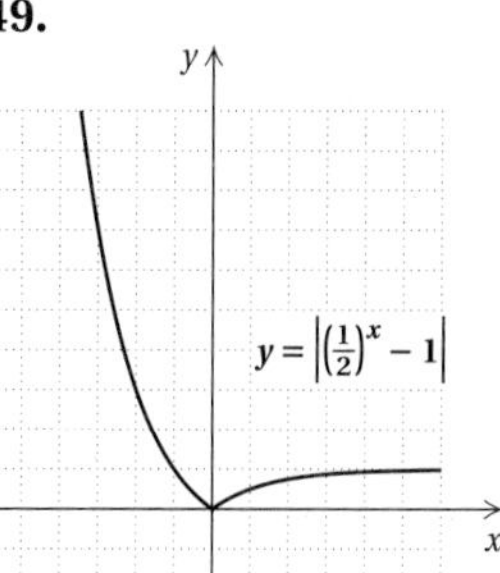

51.

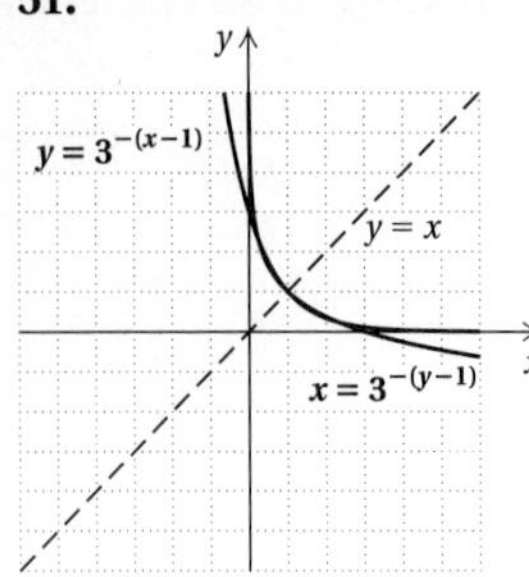

53. Left to the student.

Margin Exercises, Section 12.2, pp. 765–776

1. Inverse of $g = \{(5, 2), (4, -1), (1, -2)\}$

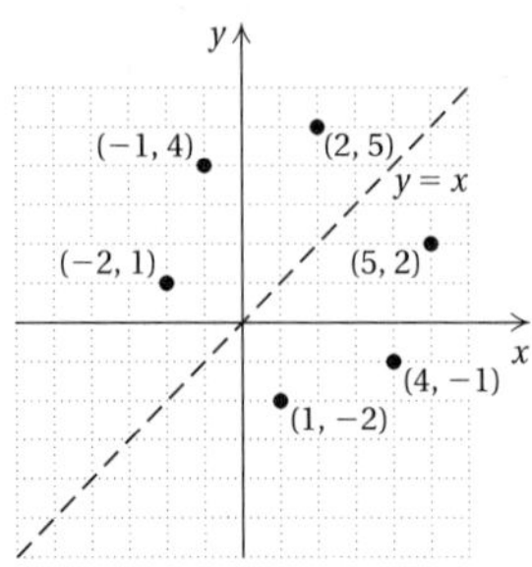

2. Inverse: $x = 6 - 2y$;

x	y
6	0
2	2
0	3
−4	5

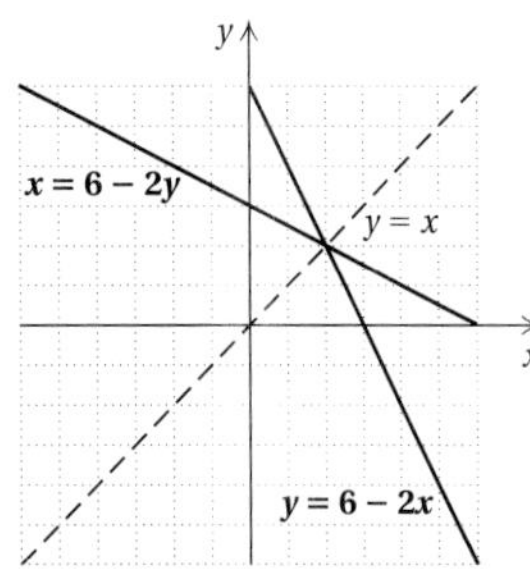

3. Inverse: $x = y^2 - 4y + 7$;

x	y
7	0
4	1
3	2
4	3
7	4

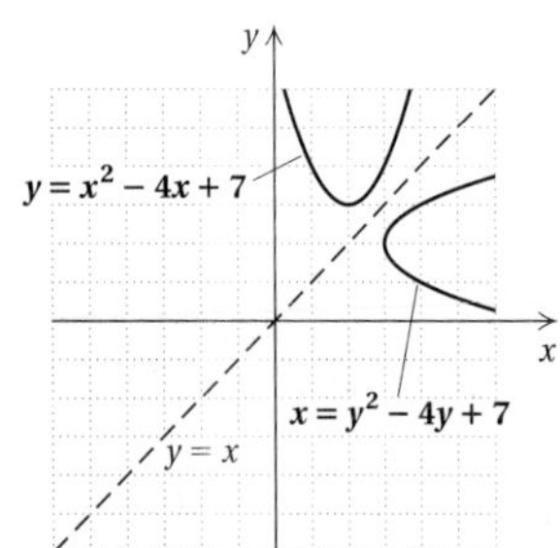

4. Yes **5.** No **6.** Yes **7.** No **8.** **(a)** Yes; **(b)** $f^{-1}(x) = 3 - x$ **9.** **(a)** Yes; **(b)** $g^{-1}(x) = \dfrac{x + 2}{3}$

10.

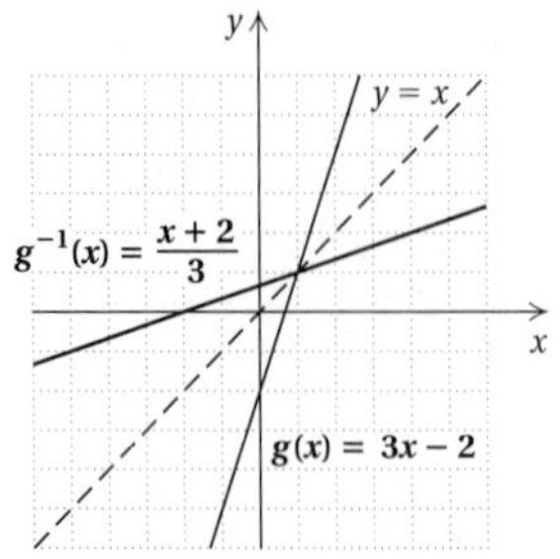

11. **(a)** Yes; **(b)** $f^{-1}(x) = \sqrt[3]{x - 1}$;
(c)

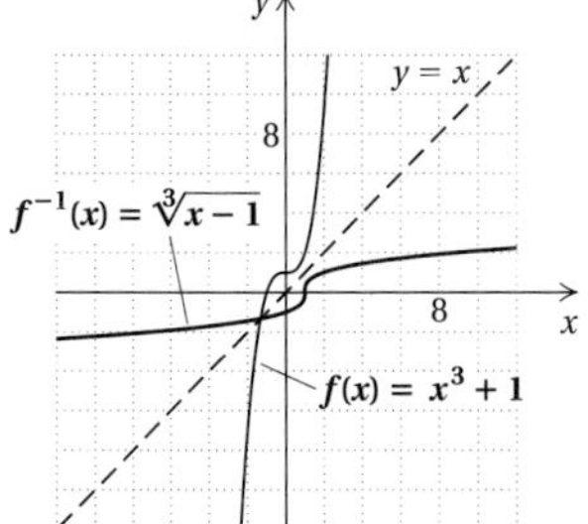

12. $x^2 + 4$; $x^2 + 10x + 24$ **13.** $4\sqrt[3]{x} + 5$; $\sqrt[3]{4x + 5}$

14. **(a)** $f(x) = \sqrt[3]{x}$, $g(x) = x^2 + 1$; **(b)** $f(x) = \dfrac{1}{x^4}$, $g(x) = x + 5$

15.
$$f^{-1} \circ f(x) = f^{-1}(f(x)) = f^{-1}\left(\tfrac{2}{3}x - 4\right)$$
$$= \frac{3\left(\frac{2}{3}x - 4\right) + 12}{2} = \frac{2x - 12 + 12}{2}$$
$$= \frac{2x}{2} = x;$$
$$f \circ f^{-1}(x) = f(f^{-1}(x)) = f\left(\frac{3x + 12}{2}\right)$$
$$= \frac{2}{3}\left(\frac{3x + 12}{2}\right) - 4 = \frac{6x + 24}{6} - 4$$
$$= x + 4 - 4 = x$$

Calculator Spotlight, p. 773

1. Left to the student. **2.** No **3.** Yes **4.** Yes **5.** Yes

Calculator Spotlight, p. 776

1. Left to the student. **2.** No **3.** No **4.** Yes **5.** No

Exercise Set 12.2, p. 777

1. Inverse:
$\{(2, 1), (-3, 6), (-5, -3)\}$

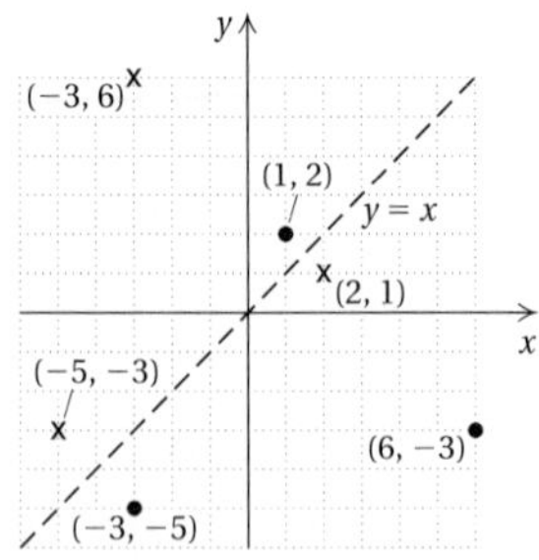

3. Inverse: $x = 2y + 6$

x	y
4	−1
6	0
8	1
10	2
12	3

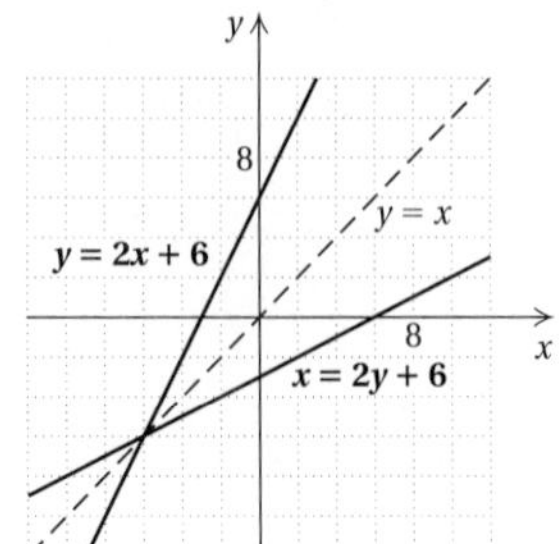

5. Yes **7.** No **9.** No **11.** Yes **13.** $f^{-1}(x) = \dfrac{x+2}{5}$

15. $f^{-1}(x) = \dfrac{-2}{x}$ **17.** $f^{-1}(x) = \frac{3}{4}(x - 7)$

19. $f^{-1}(x) = \dfrac{2}{x} - 5$ **21.** Not one-to-one

23. $f^{-1}(x) = \dfrac{1 - 3x}{5x - 2}$ **25.** $f^{-1}(x) = \sqrt[3]{x + 1}$

27. $f^{-1}(x) = x^3$

29.

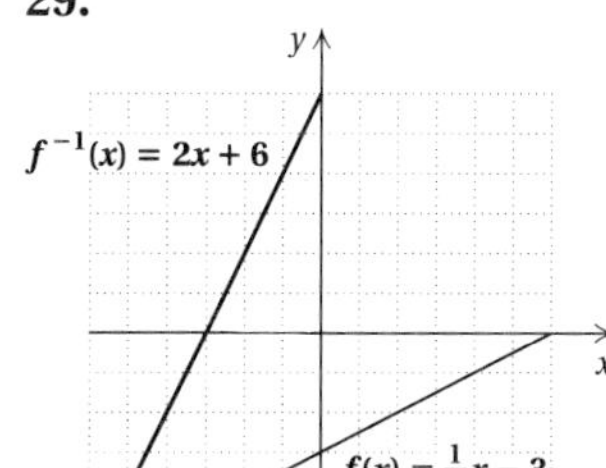

31.

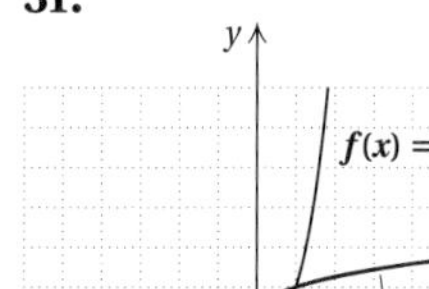

33. $-8x + 9$; $-8x + 18$ **35.** $12x^2 - 12x + 5$; $6x^2 + 3$

37. $\dfrac{16}{x^2} - 1$; $\dfrac{2}{4x^2 - 1}$ **39.** $x^4 - 10x^2 + 30$; $x^4 + 10x^2 + 20$ **41.** $f(x) = x^2$; $g(x) = 5 - 3x$

43. $f(x) = \sqrt{x}$; $g(x) = 5x + 2$ **45.** $f(x) = \dfrac{1}{x}$; $g(x) = x - 1$ **47.** $f(x) = \dfrac{1}{\sqrt{x}}$; $g(x) = 7x + 2$

49. $f(x) = x^4$; $g(x) = \sqrt{x} + 5$

51. $f^{-1} \circ f(x) = f^{-1}(f(x)) = f^{-1}\left(\frac{4}{5}x\right)$

$= \frac{5}{4}\left(\frac{4}{5}x\right) = x$;

$f \circ f^{-1}(x) = f(f^{-1}(x)) = f\left(\frac{5}{4}x\right)$

$= \frac{4}{5}\left(\frac{5}{4}x\right) = x$

53. $f^{-1} \circ f(x) = f^{-1}(f(x)) = f^{-1}\left(\dfrac{1 - x}{x}\right)$

$= \dfrac{1}{\dfrac{1 - x}{x} + 1} = \dfrac{1}{\dfrac{1}{x}} = x$;

$f \circ f^{-1}(x) = f(f^{-1}(x)) = f\left(\dfrac{1}{x + 1}\right)$

$= \dfrac{1 - \dfrac{1}{x + 1}}{\dfrac{1}{x + 1}} = \dfrac{\dfrac{x}{x + 1}}{\dfrac{1}{x + 1}} = x$

55. $f^{-1}(x) = \frac{1}{3}x$ **57.** $f^{-1}(x) = -x$

59. $f^{-1}(x) = x^3 + 5$ **61.** **(a)** 40, 42, 46, 50; **(b)** $f^{-1}(x) = x - 32$; **(c)** 8, 10, 14, 18 **63.** $\sqrt[3]{a}$

64. $\sqrt[3]{x^2}$ **65.** a^2b^3 **66.** $2t^2$ **67.** $\sqrt{3}$ **68.** $2\sqrt[4]{2}$

69. $\sqrt{2xy}$ **70.** $\sqrt[4]{p^2t}$ **71.** $2a^3b^8$ **72.** $10x^3y^6$

73. $3a^2b^2$ **74.** $3pq^3$ **75.** ◈ **77.** No **79.** Yes

81. **(1)** C; **(2)** A; **(3)** B; **(4)** D

Margin Exercises, Section 12.3, pp. 781–786

1. $\log_2 64$ is the power to which we raise 2 to get 64; 6

2.

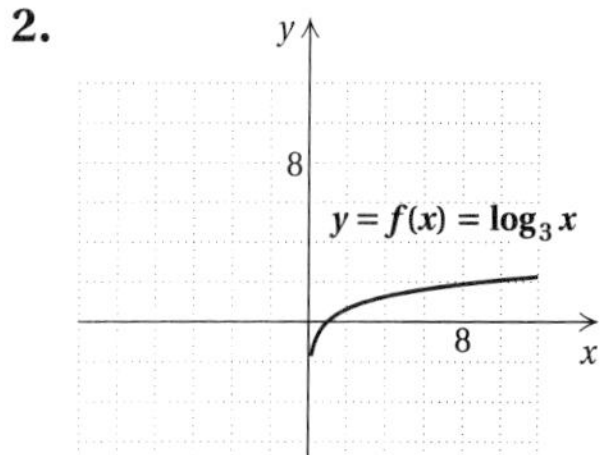

3. $0 = \log_6 1$ **4.** $-3 = \log_{10} 0.001$

5. $0.25 = \log_{16} 2$ **6.** $T = \log_m P$ **7.** $2^5 = 32$

8. $10^3 = 1000$ **9.** $a^7 = Q$ **10.** $t^x = M$ **11.** 10,000

12. 3 **13.** $\frac{1}{4}$ **14.** 4 **15.** -4 **16.** 0 **17.** 0

18. 1 **19.** 1 **20.** 0 **21.** 4.8934 **22.** -4.5100

23. Does not exist **24.** log 1000 = 3, log 10,000 = 4; between 3 and 4; 3.9945 **25.** 78,234.8042

Exercise Set 12.3, p. 787

1.

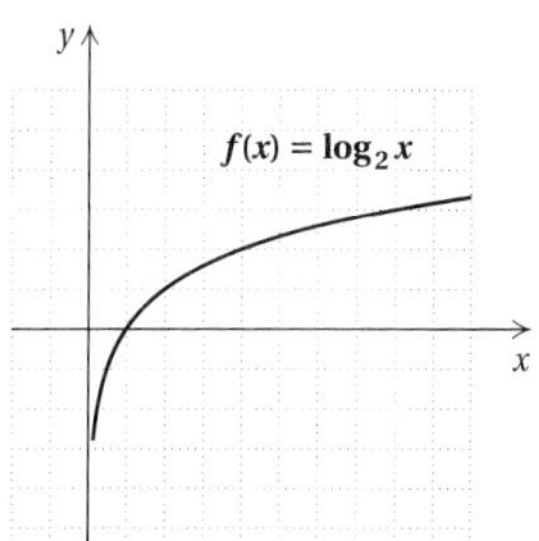

3.

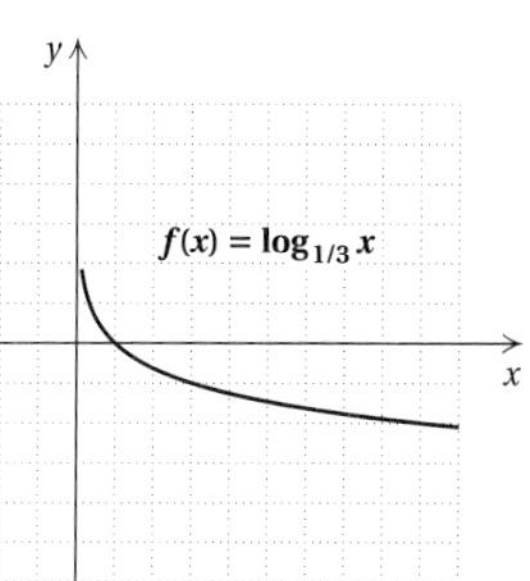

5.

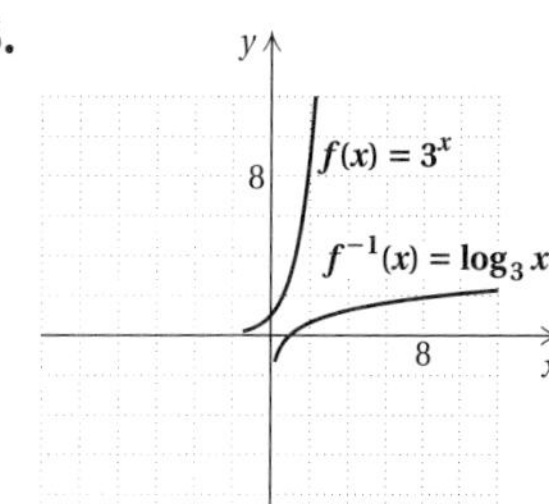

7. $3 = \log_{10} 1000$

9. $-3 = \log_5 \frac{1}{125}$

11. $\frac{1}{3} = \log_8 2$

13. $0.3010 = \log_{10} 2$

15. $2 = \log_e t$

17. $t = \log_Q x$

19. $2 = \log_e 7.3891$

21. $-2 = \log_e 0.1353$

23. $4^w = 10$ **25.** $6^2 = 36$ **27.** $10^{-2} = 0.01$

29. $10^{0.9031} = 8$ **31.** $e^{4.6052} = 100$ **33.** $t^k = Q$ **35.** 9

37. 4 **39.** 4 **41.** 3 **43.** 25 **45.** 1 **47.** $\frac{1}{2}$ **49.** 2

51. 2 **53.** -1 **55.** 0 **57.** 4 **59.** 2 **61.** 3

63. -2 **65.** 0 **67.** 1 **69.** $\dfrac{2}{3}$ **71.** 4.8970

73. -0.1739 **75.** Does not exist **77.** 0.9464

79.

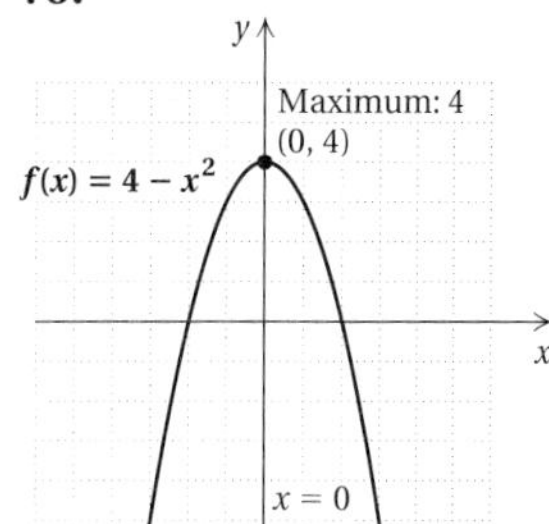

80.

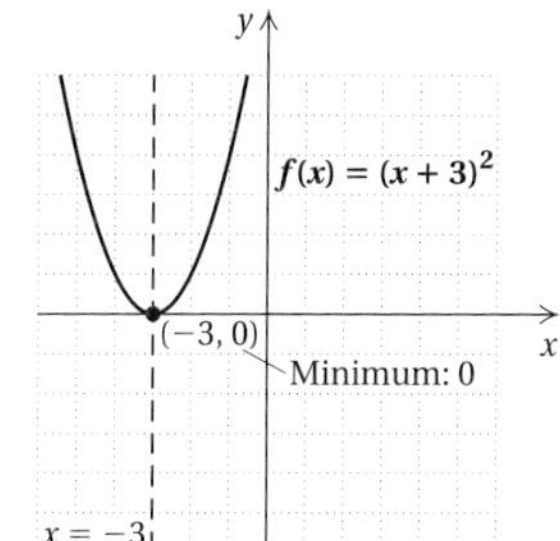

81.

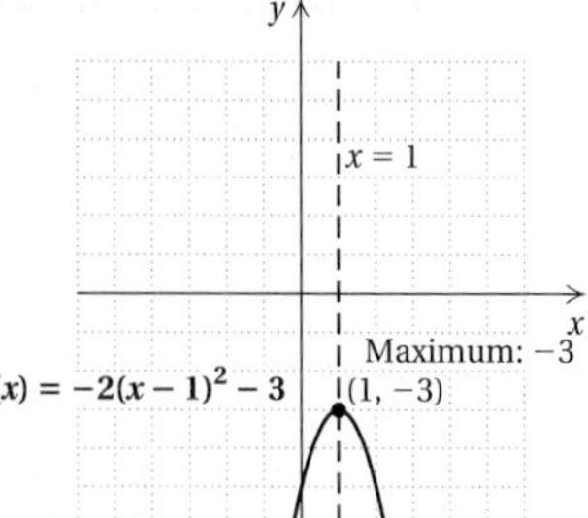

82.

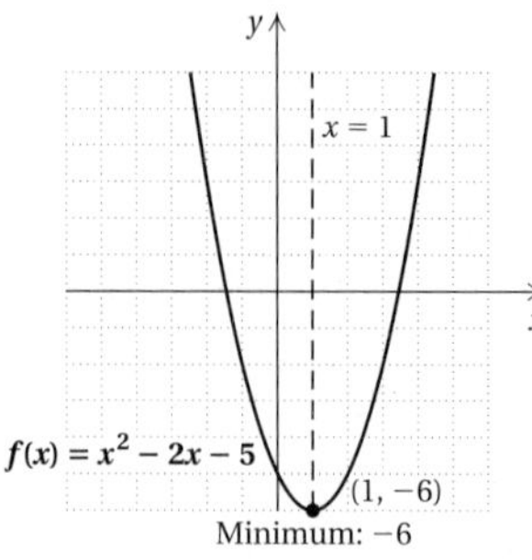

83. $c = \sqrt{\dfrac{E}{m}}$ **84.** $a = \dfrac{2 + \sqrt{4 + 3B}}{3}$ **85.** $b = \dfrac{A^2}{3a}$

86. $L = \dfrac{T^2 g}{4\pi^2}$ **87.** **(a)** $f(x) = -\frac{39}{280}x^2 + \frac{963}{140}x - \frac{356}{7}$;
(b) 30%, 13% **89.** ◈

91.

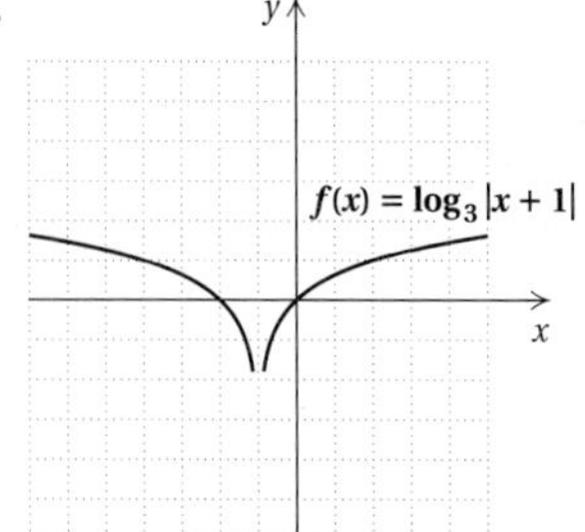

93. 25 **95.** $-\frac{7}{16}$ **97.** 3 **99.** 0 **101.** −2

Margin Exercises, Section 12.4, pp. 791–794

1. $\log_5 25 + \log_5 5$ **2.** $\log_b P + \log_b Q$ **3.** $\log_3 35$
4. $\log_a (JAM)$ **5.** $5 \log_7 4$ **6.** $\frac{1}{2} \log_a 5$
7. $\log_b P - \log_b Q$ **8.** $\log_2 5$
9. $\frac{3}{2} \log_a z - \frac{1}{2} \log_a x - \frac{1}{2} \log_a y$
10. $2 \log_a x - 3 \log_a y - \log_a z$
11. $3 \log_a x + 4 \log_a y - 5 \log_a z - 9 \log_a w$
12. $\log_a \dfrac{x^5 z^{1/4}}{y}$ **13.** $\log_a \dfrac{1}{b\sqrt{b}}$, or $\log_a b^{-3/2}$
14. 0.602 **15.** 1 **16.** −0.398 **17.** 0.398
18. −0.699 **19.** $\frac{3}{2}$ **20.** 1.699 **21.** 1.204 **22.** 6
23. 3.2 **24.** 12

Calculator Spotlight, p. 791

1. Correct **2.** Correct **3.** Correct **4.** Not correct

Exercise Set 12.4, p. 795

1. $\log_2 32 + \log_2 8$ **3.** $\log_4 64 + \log_4 16$
5. $\log_a Q + \log_a x$ **7.** $\log_b 252$ **9.** $\log_c Ky$
11. $4 \log_c y$ **13.** $6 \log_b t$ **15.** $-3 \log_b C$
17. $\log_a 67 - \log_a 5$ **19.** $\log_b 2 - \log_b 5$ **21.** $\log_c \frac{22}{3}$
23. $2 \log_a x + 3 \log_a y + \log_a z$
25. $\log_b x + 2 \log_b y - 3 \log_b z$
27. $\frac{4}{3} \log_c x - \log_c y - \frac{2}{3} \log_c z$
29. $2 \log_a m + 3 \log_a n - \frac{3}{4} - \frac{5}{4} \log_a b$ **31.** $\log_a \dfrac{x^{2/3}}{y^{1/2}}$,
or $\log_a \dfrac{\sqrt[3]{x^2}}{\sqrt{y}}$ **33.** $\log_a \dfrac{2x^4}{y^3}$ **35.** $\log_a \dfrac{\sqrt{a}}{x}$ **37.** 2.708
39. 0.51 **41.** −1.609 **43.** $\frac{3}{2}$ **45.** 2.609 **47.** t
49. 5 **51.** 7 **53.** −7 **55.** i **56.** −1 **57.** 5
58. $\frac{3}{5} + \frac{4}{5}i$ **59.** $23 - 18i$ **60.** $10i$ **61.** $-34 - 31i$
62. $3 - 4i$ **63.** ◈ **65.** Left to the student.
67. $\log_a (x^6 - x^4y^2 + x^2y^4 - y^6)$
69. $\frac{1}{2} \log_a (1 - s) + \frac{1}{2} \log_a (1 + s)$ **71.** False
73. True **75.** False

Margin Exercises, Section 12.5, pp. 798–800

1. 11.2675 **2.** −10.3848 **3.** Does not exist **4.** Does not exist **5.** 2.3026 **6.** 78,237.1596 **7.** 0.1353
8. 1.086 **9.** 5.5236

10.

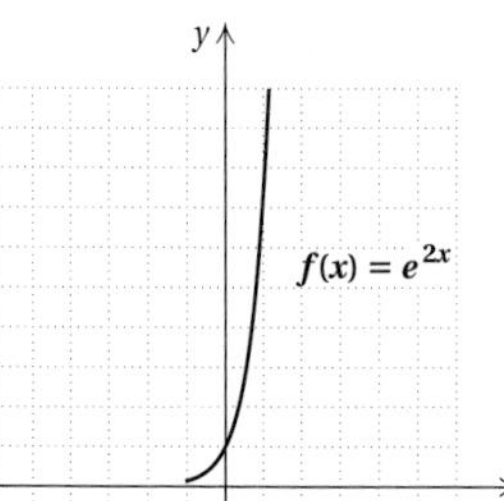

11.

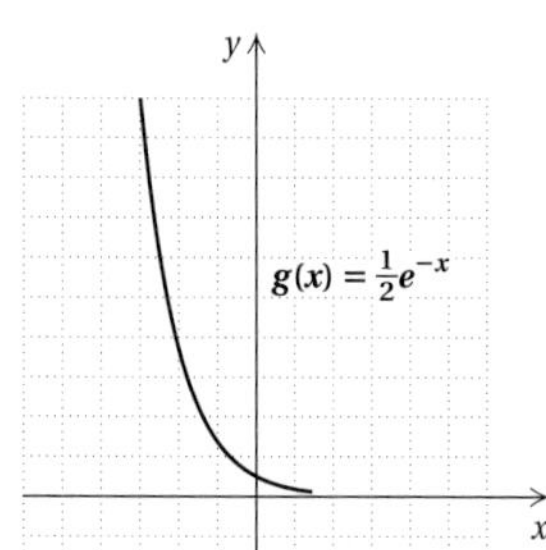

12.

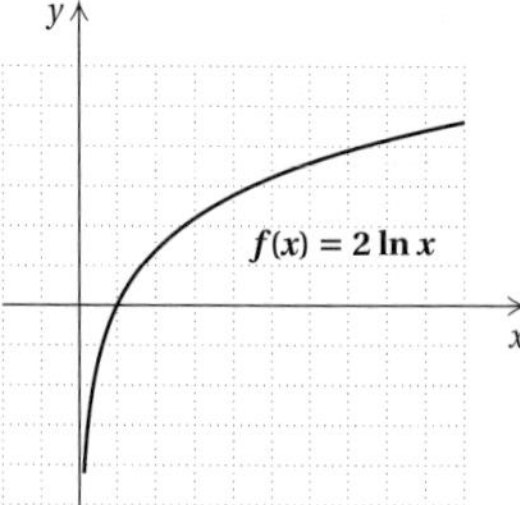

13.

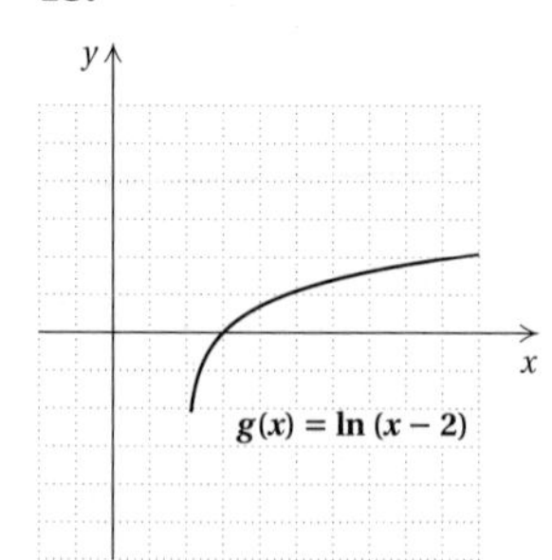

Exercise Set 12.5, p. 801

1. 0.6931 **3.** 4.1271 **5.** 8.3814 **7.** −5.0832
9. −1.6094 **11.** Does not exist **13.** −1.7455 **15.** 1
17. 15.0293 **19.** 0.0305 **21.** 109.9472 **23.** 5
25. 2.5702 **27.** 6.6439 **29.** 2.1452 **31.** −2.3219
33. −2.3219 **35.** 4.6284

37.

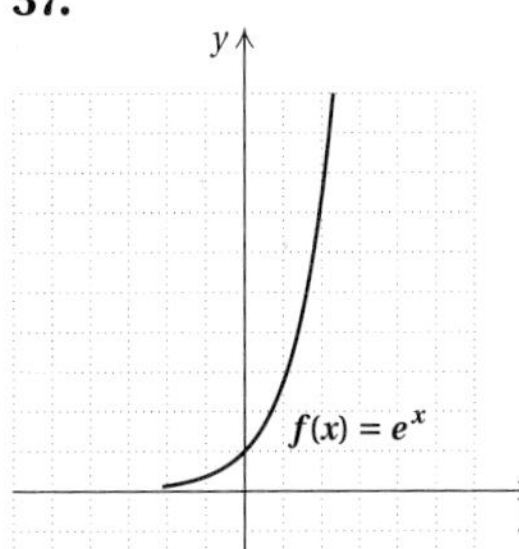

39.

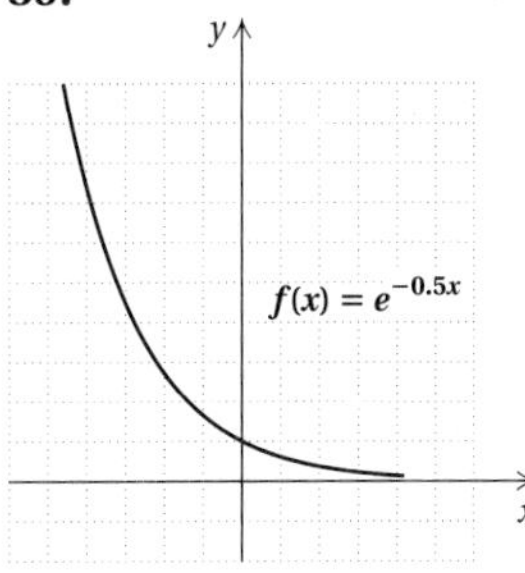

41.

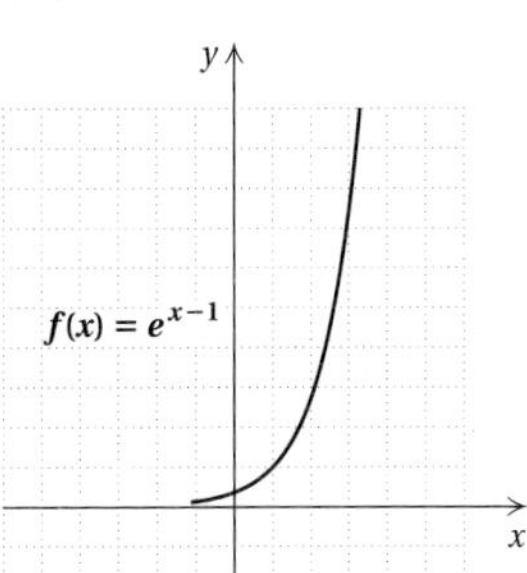

43.

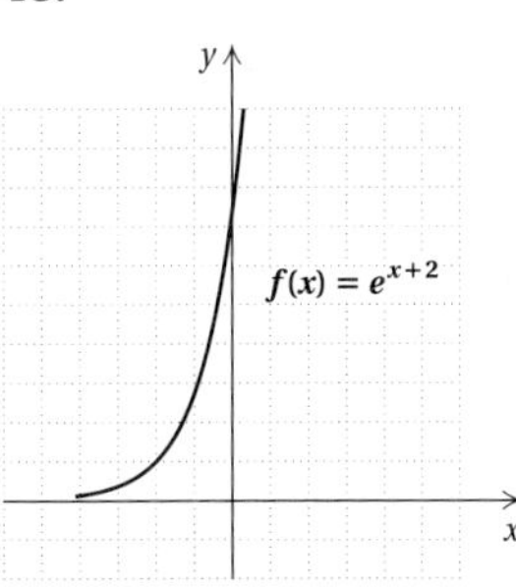

45.

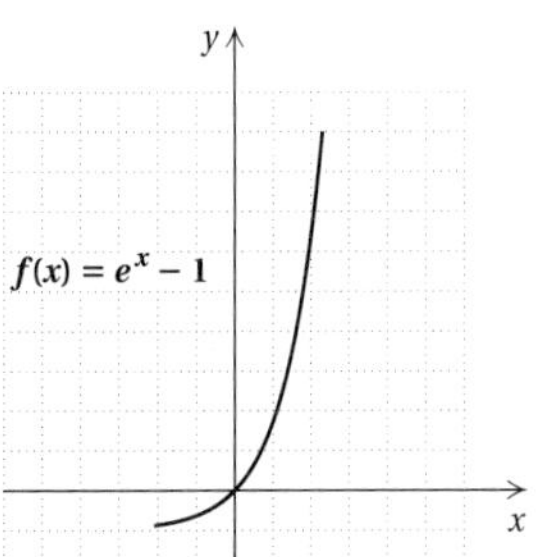

47.

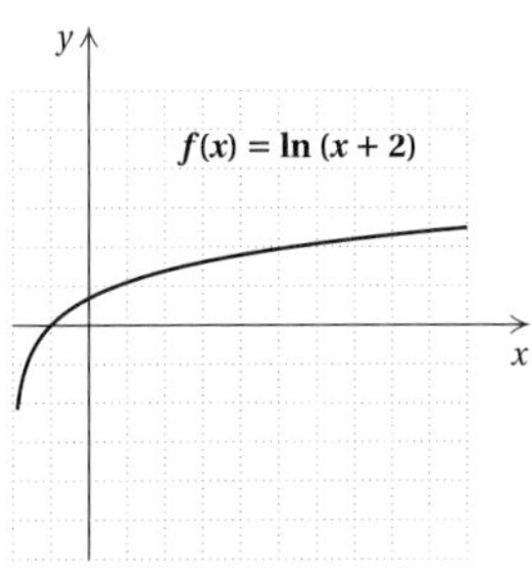

49.

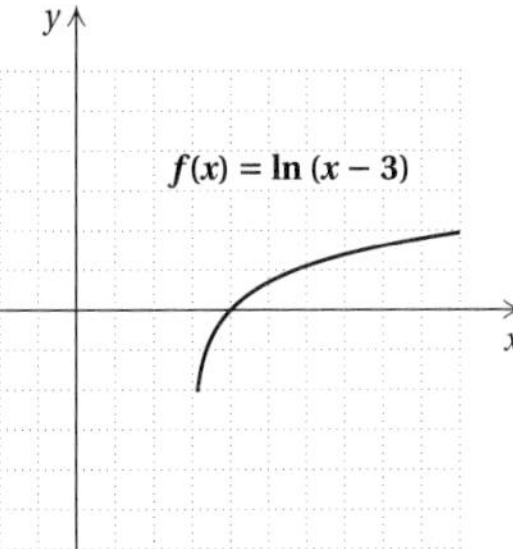

51.

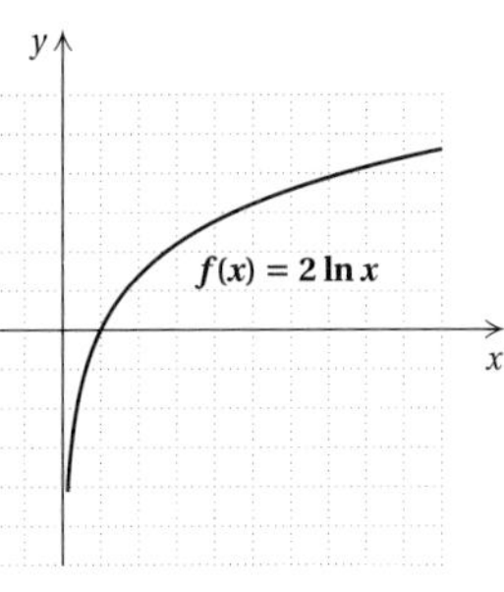

53.

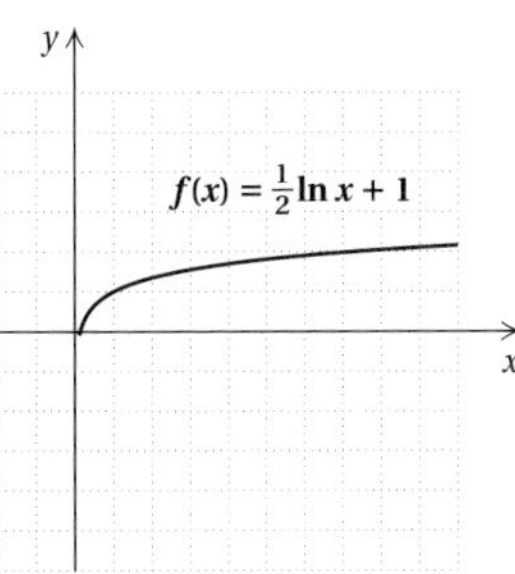

55.

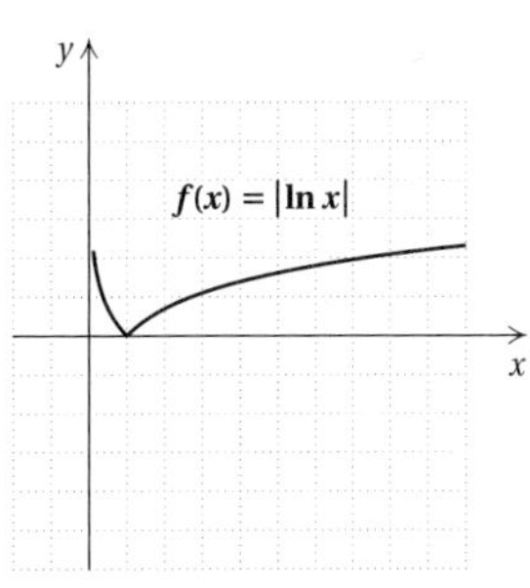

57. 16, 256 **58.** $\frac{1}{4}$, 9 **59.** 49, 121 **60.** $\pm 3, \pm 4$
61. ◆ **63.** Domain: $(-\infty, \infty)$; range: $[0, \infty)$
65. Domain: $(-\infty, \infty)$; range: $(-\infty, 100)$
67. Domain: $\left(\frac{5}{2}, \infty\right)$

Margin Exercises, Section 12.6, pp. 803–806

1. 1 **2.** 3 **3.** 1.5395 **4.** 14.6068 **5.** 25 **6.** $\frac{2}{5}$
7. 2 **8.** 5

Exercise Set 12.6, p. 807

1. 3 **3.** 4 **5.** $\frac{5}{2}$ **7.** $\frac{3}{5}$ **9.** 3.4594 **11.** 5.4263
13. $\frac{5}{2}$ **15.** $-3, -1$ **17.** $\frac{3}{2}$ **19.** 4.6052 **21.** 2.3026
23. 140.6705 **25.** 2.7095 **27.** 3.2220 **29.** 256
31. $\frac{1}{32}$ **33.** 10 **35.** $\frac{1}{100}$ **37.** $e^2 \approx 7.3891$
39. $\frac{1}{e} \approx 0.3679$ **41.** 121 **43.** 10 **45.** $\frac{1}{3}$ **47.** 3
49. $\frac{2}{5}$ **51.** 5 **53.** No solution **55.** $\pm 10, \pm 2$
56. $-64, 8$ **57.** $-2, -3, \frac{-5 \pm \sqrt{41}}{2}$ **58.** $-\frac{1}{10}, 1$
59. $\frac{y^{4/3}}{25x^2z^4}$ **60.** $-i$ **61.** ◆ **63.** 1 **65.** **(a)** 0.377;
(b) -1.962; **(c)** 0.904; **(d)** -1.532 **67.** 3, 4 **69.** -4
71. 2 **73.** $\pm\sqrt{34}$ **75.** $10^{100,000}$ **77.** 1, 100
79. 3, -7 **81.** 1, 1.465

Margin Exercises, Section 12.7, pp. 810–816

1. About 65 decibels **2.** $10^{-9.2}$ W/m^2 **3.** About 4.9
4. 10^{-7} moles per liter **5.** **(a)** 27.1 yr; **(b)** 10.2 yr
6. About 271 million; about 282 million
7. **(a)** $k \approx 6\%$, $P(t) = 5000e^{0.06t}$; **(b)** \$5309.18, \$5637.48, \$9110.59; **(c)** about 11.6 yr **8.** **(a)** $k \approx 0.415$, $P(t) = 91e^{0.415t}$; **(b)** \$30,359 million, or \$30.359 billion; **(c)** 2001 **9.** 2972 yr

Calculator Spotlight, p. 816

1. **(a)** $y = 77.694(1.589)^x$;
(b)

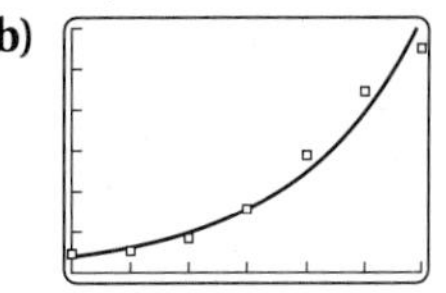

(c) \$50,750 million, or \$50.75 billion
2. **(a)** $y = 706{,}278.268(1.075)^x$;
(b)

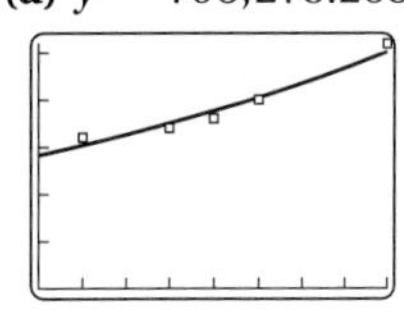

(c) \$1,449,669; \$2,975,514; **(d)** 2017

Exercise Set 12.7, p. 819

1. About 95 dB **3.** $10^{-1.5}$, or about 3.2×10^{-2} W/m^2
5. About 6.8 **7.** 1.58×10^{-8} moles per liter
9. 3.00 ft/sec **11.** 1.88 ft/sec **13.** **(a)** 4860;
(b) 3.6 days; **(c)** 0.63 day **15.** **(a)** \$231 billion; **(b)** 2010;
(c) 4.7 yr **17.** **(a)** $P(t) = 6e^{0.015t}$; **(b)** 7.2 billion; **(c)** 2032;
(d) 46.2 yr **19.** **(a)** $P(t) = P_0e^{0.06t}$; **(b)** \$5309.18,

\$5637.48, \$9110.59; **(c)** in 11.6 yr **21.** **(a)** 7%, $P(t) = 40{,}000e^{0.07t}$; **(b)** \$42,900.33, \$46,010.95, \$80,550.11; **(c)** 9.9 yr **23.** **(a)** $k \approx 0.034$, $P(t) = 20e^{0.034t}$; **(b)** 38¢; **(c)** 2028 **25.** About 2103 yr **27.** About 7.2 days **29.** 69.3% per year **31.** **(a)** $V(t) = 451{,}000e^{0.07t}$, where t is the number of years since 1991; **(b)** \$1,201,670; **(c)** 9.9 yr; **(d)** 2002 **33.** -1 **34.** i **35.** $-1 - i$ **36.** -2 **37.** $\frac{63}{65} - \frac{16}{65}i$ **38.** $-\frac{2}{41} + \frac{23}{41}i$ **39.** 41 **40.** $91 + 60i$ **41.** ◆ **43.** -0.937, 1.078, 58.770 **45.** -0.767, 2, 4

Summary and Review: Chapter 12, p. 823

1.

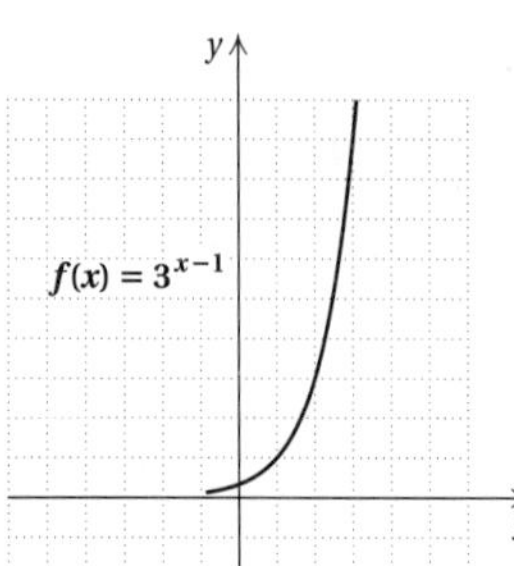

2.

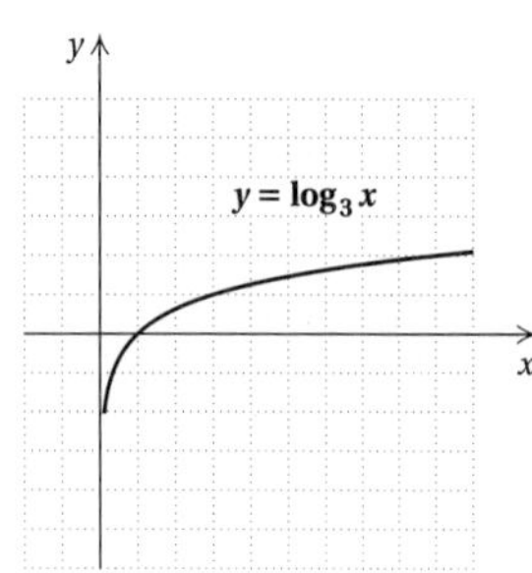

3.

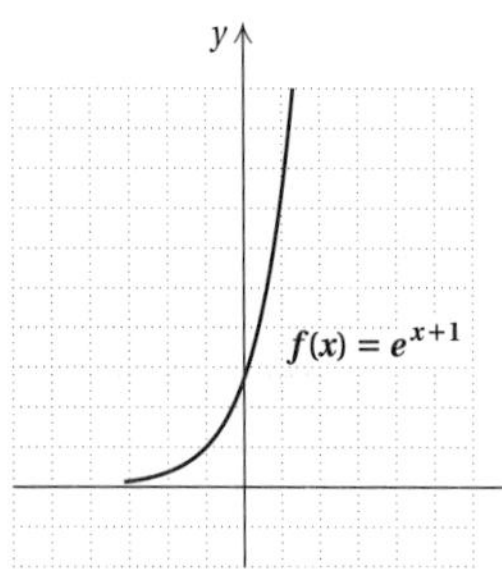

4.

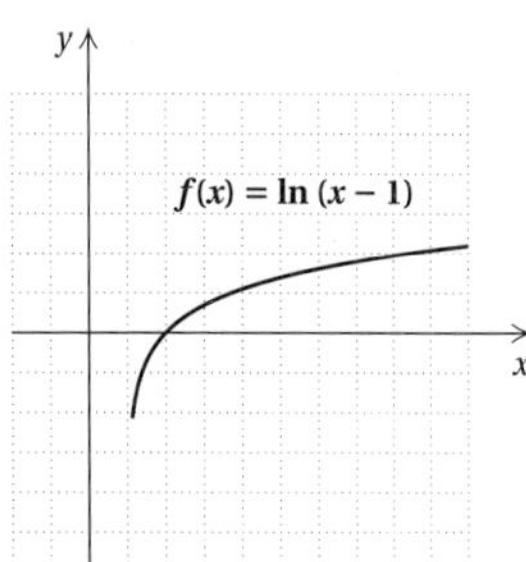

5. $\{(2, -4), (-7, 5), (-2, -1), (11, 10)\}$

6. Not one-to-one. **7.** $g^{-1}(x) = \dfrac{7x + 3}{2}$

8. $f^{-1}(x) = \dfrac{1}{2}\sqrt[3]{x}$ **9.** $f^{-1}(x) = \dfrac{3x - 4}{2x}$

10.

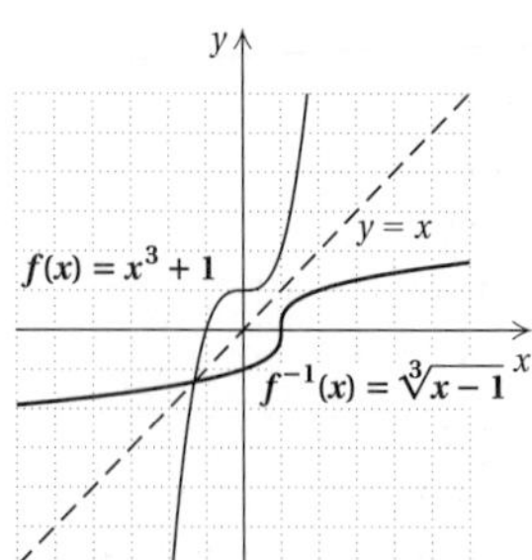

11. $f \circ g(x) = 9x^2 - 30x + 25$; $g \circ f(x) = 3x^2 - 5$
12. $f(x) = \sqrt{x}$, $g(x) = 4 - 7x$; answers may vary
13. $\log 10{,}000 = 4$ **14.** $\log_{25} 5 = \frac{1}{2}$ **15.** $4^x = 16$
16. $\left(\frac{1}{2}\right)^{-3} = 8$ **17.** 2 **18.** -1 **19.** 1 **20.** 0
21. -2.7425 **22.** Does not exist
23. $4 \log_a x + 2 \log_a y + 3 \log_a z$
24. $\frac{1}{2} \log z - \frac{3}{4} \log x - \frac{1}{4} \log y$ **25.** $\log_a 120$

26. $\log \dfrac{a^{1/2}}{bc^2}$ **27.** 17 **28.** -7 **29.** 8.7601
30. 3.2698 **31.** 2.54995 **32.** -3.6602 **33.** -2.6921
34. 0.3753 **35.** 18.3568 **36.** 0 **37.** Does not exist
38. 1 **39.** 0.4307 **40.** 1.7097 **41.** $\frac{1}{9}$ **42.** 2
43. $\frac{1}{10{,}000}$ **44.** $e^{-2} \approx 0.1353$ **45.** $\frac{7}{2}$ **46.** 1, -5
47. 1.5266 **48.** 35.0656 **49.** 2 **50.** 8 **51.** 20
52. $\sqrt{43}$ **53.** 90 dB **54.** **(a)** \$341.4 billion; **(b)** 1997; **(c)** 1.6 yr **55.** **(a)** $k \approx 0.347$, $N(t) = 6e^{0.347t}$, where t is the number of years since 1994; **(b)** 136; **(c)** 2004
56. $k \approx 0.043$, or 4.3% **57.** About 8.25 yr **58.** About 3463 yr **59.** **(a)** $f(x) = \frac{1}{64}x^2 + \frac{5}{16}x + \frac{5}{2}$; **(b)** \$19
60. ± 4, $\pm i\sqrt{5}$
61.

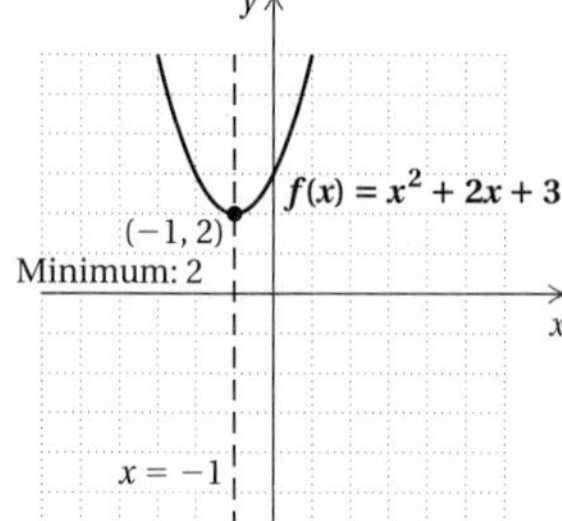

62. $11 - 10i$ **63.** 1 **64.** $-\frac{11}{10} - \frac{17}{10}i$ **65.** $5 - 2i$
66. ◆ You cannot take the logarithm of a negative number because logarithm bases are positive and there is no power to which a positive number can be raised to yield a negative number. **67.** ◆ $\log_a 1 = 0$ because $a^0 = 1$. **68.** e^{e^3} **69.** $\left(\frac{8}{3}, -\frac{2}{3}\right)$

Test: Chapter 12, p. 825

1. [12.1a]

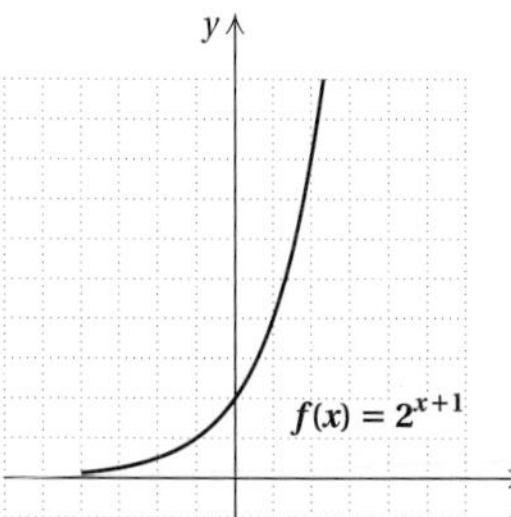

2. [12.3a]

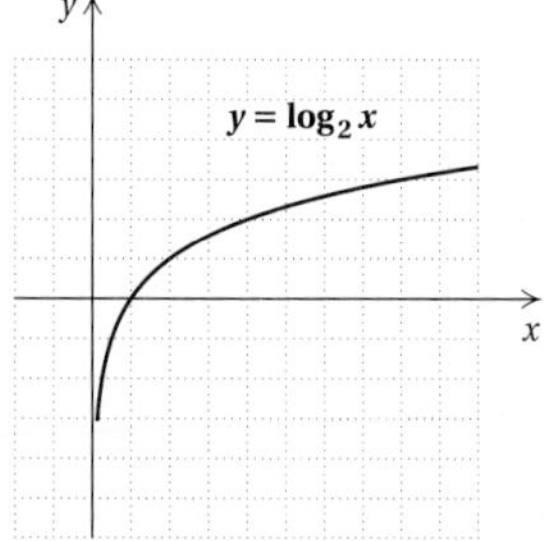

3. [12.5c]

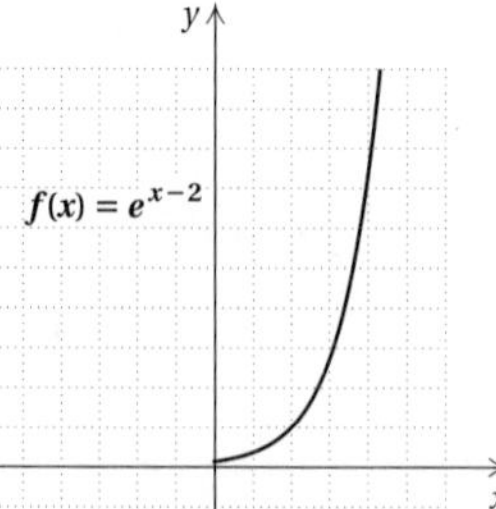

4. [12.5c]

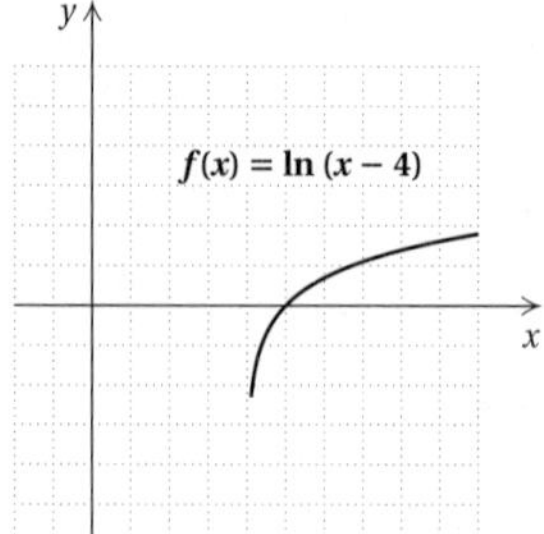

5. [12.2a] $\{(3, -4), (-8, 5), (-3, -1), (12, 10)\}$

6. [12.2b, c] $f^{-1}(x) = \dfrac{x + 3}{4}$

7. [12.2b, c] $f^{-1}(x) = \sqrt[3]{x} - 1$

8. [12.2b, c] Not one-to-one
9. [12.2e] $f \circ g(x) = 25x^2 - 15x + 2$, $g \circ f(x) = 5x^2 + 5x - 2$
10. [12.3b] $\log_{256} 16 = \frac{1}{2}$ **11.** [12.3b] $7^m = 49$
12. [12.3c] 3 **13.** [12.3c] 23 **14.** [12.3c] 0
15. [12.3e] -1.9101 **16.** [12.3e] Does not exist
17. [12.4d] $3 \log a + \frac{1}{2} \log b - 2 \log c$
18. [12.4d] $\log_a \dfrac{x^{1/3}z^2}{y^3}$ **19.** [12.4d] -0.544
20. [12.4d] 1.079 **21.** [12.5a] 6.6938
22. [12.5a] 107.7701 **23.** [12.5a] 0 **24.** [12.5b] 1.1881
25. [12.6b] 5 **26.** [12.6b] 2 **27.** [12.6b] 10,000
28. [12.6b] $e^{1/4} \approx 1.2840$ **29.** [12.6a] 0.0937
30. [12.6b] 9 **31.** [12.6b] 1 **32.** [12.7a] 4.2
33. [12.7b] **(a)** \$39,103; **(b)** 2010; **(c)** 11.9 yr
34. [12.7b] **(a)** $k \approx 0.006$, $N(t) = 30e^{0.006t}$, where t is the number of years since 1996 and N is in millions; **(b)** 31.287 million, 32.629 million; **(c)** 2081; **(d)** 115.5 yr
35. [12.7b] About 4.6% **36.** [12.7b] About 4684 yr
37. [11.7b] **(a)** $\frac{1}{33{,}600}x^2 + \frac{17}{336}x$; **(b)** about 11.3 min
38. [11.4c] 64, 1
39. [11.6a]

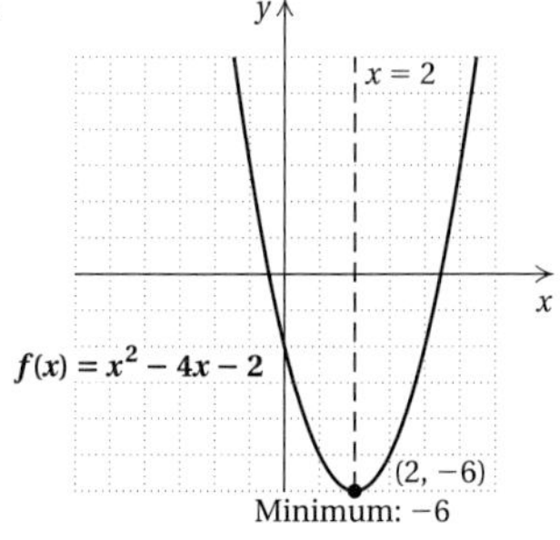

40. [10.8d] i **41.** [10.8e] $\frac{3}{2} - \frac{5}{2}i$ **42.** [12.6b] 44, -37
43. [12.4d] 2

Cumulative Review: Chapters 1–12, p. 827

1. [1.2e] $\frac{2}{15}$ **2.** [4.1e, f] $-9x^6y^2$ **3.** [1.8d] 62.5
4. [1.8c] $-x + 30$ **5.** [9.1c] $\{x \mid x \geq -1\}$, or $[-1, \infty)$
6. [9.3e] $\{x \mid x < -6.4$ *or* $x > 6.4\}$, or $(-\infty, -6.4) \cup (6.4, \infty)$ **7.** [9.2a] $\{x \mid -1 \leq x < 6\}$, or $[-1, 6)$ **8.** [8.2b] $\left(-\frac{1}{3}, 5\right)$ **9.** [8.4a] $(-1, 2, 3)$
10. [5.8b] $-1, \frac{3}{2}$ **11.** [6.5a] $-\frac{2}{3}, 3$ **12.** [10.6a] 4
13. [5.8b] $-7, -3$ **14.** [11.2a] $-\dfrac{1}{4} \pm i\dfrac{\sqrt{7}}{4}$
15. [11.4c] $-3, -2, 2, 3$ **16.** [6.5a] -16
17. [12.6b] 1 **18.** [12.6a] 1.748 **19.** [12.6b] 9
20. [11.8a] $\{x \mid x \leq -1$ *or* $x \geq 1\}$, or $(-\infty, -1] \cup [1, \infty)$
21. [11.8b] $\{x \mid x < -1$ *or* $x > 2\}$, or $(-\infty, -1) \cup (2, \infty)$
22. [2.6a] $N = \dfrac{4P - 3M}{6}$ **23.** [6.7a] $p = \dfrac{qf}{q - f}$
24. [4.5d] $2x^3 - x^2 - 8x - 3$
25. [4.4b, c] $-x^3 + 3x^2 - x - 6$ **26.** [6.1d] $\dfrac{2m - 1}{m + 1}$
27. [6.4b] $\dfrac{x + 2}{x + 1}$ **28.** [6.8a] $\dfrac{1}{x + 1}$
29. [4.8b, c] $x^3 + 2x^2 - 2x + 1 + \dfrac{3}{x + 1}$

30. [10.3b] $5x^2\sqrt{y}$ **31.** [10.4a] $11\sqrt{2}$ **32.** [10.2d] 8
33. [10.8c] $16 + i\sqrt{2}$ **34.** [10.8e] $\frac{3}{10} + \frac{11}{10}i$
35. [7.4a]

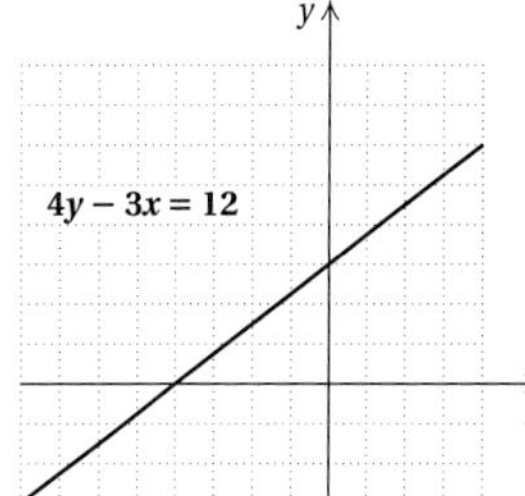

36. [9.4b]

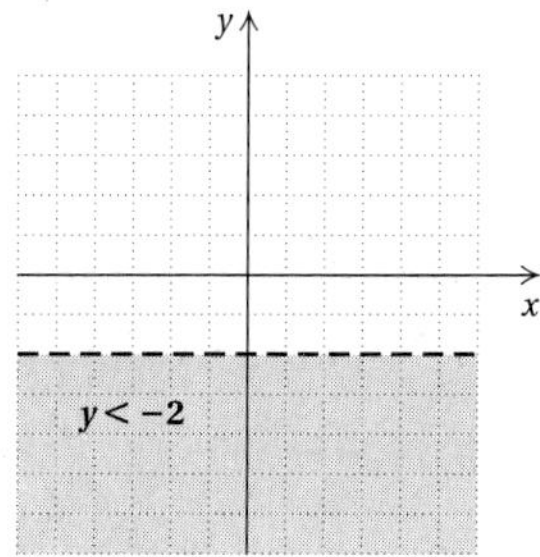

37. [9.4c]

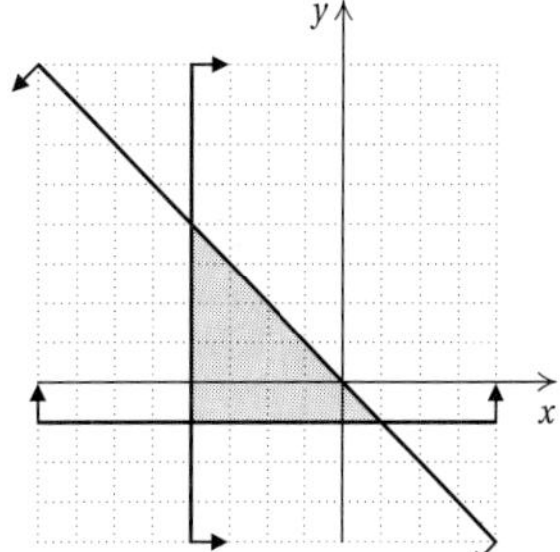

38. [11.6a]

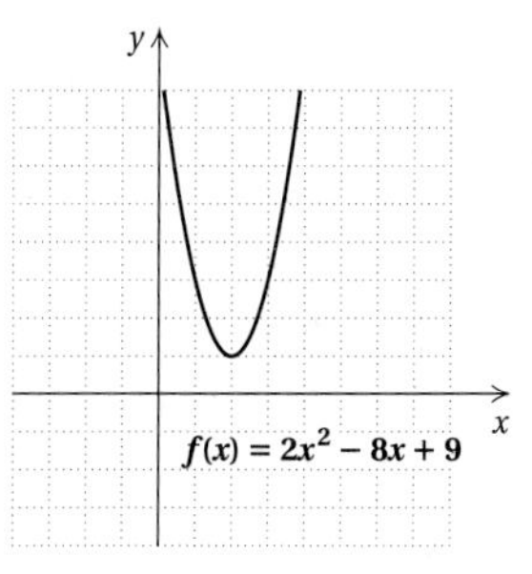

39. [12.5c]

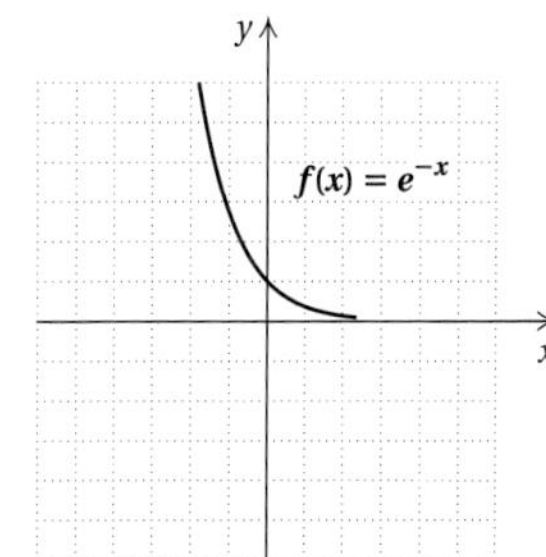

40. [12.3a]

y / x / $f(x) = \log_2 x$

41. [5.1c] $(2x^3 + 1)(x - 6)$
42. [5.2a] $3(a - 9b)(a + 5b)$
43. [5.2a] $(x - 8)(x - 9)$
44. [5.5c] $(9m^2 + n^2)(3m + n)(3m - n)$
45. [5.5b] $4(2x - 1)^2$
46. [5.6a] $3(3a - 2)(9a^2 + 6a + 4)$
47. [5.3a], [5.4a] $2(5x - 2)(x + 7)$
48. [5.3a], [5.4a] $3x(2x - 1)(x + 5)$
49. [7.5c] $y = 2x + 2$ **50.** [7.5d] $y = -\frac{1}{2}x + \frac{5}{2}$
51. [12.2c] $f^{-1}(x) = \frac{1}{2}(x + 3)$ **52.** [6.9e] $\frac{4}{5}$
53. [7.1b] -10 **54.** [10.5b] $\dfrac{15 + 8\sqrt{a} + a}{9 - a}$
55. [7.2a] All real numbers **56.** [7.2a] $\left\{x \mid x \text{ is a real number } and\ x \neq -\frac{1}{2}\right\}$, or $\left(-\infty, -\frac{1}{2}\right) \cup \left(-\frac{1}{2}, \infty\right)$
57. [11.3a] $-13, 12$ **58.** [6.6a] 1 hr **59.** [11.3a] True for any three consecutive even integers
60. [12.7b] **(a)** $P(t) = 102e^{0.92t}$; **(b)** \$4044 million, or \$4.044 billion; **(c)** 0.75 yr **61.** [6.6c] 612 mi
62. [12.7b] 2397 yr **63.** [6.9d] 3360 kg
64. [7.4b] $f(x) = -\frac{1}{3}x - \frac{7}{3}$

65. [11.7b] $f(x) = -\frac{17}{18}x^2 - \frac{59}{18}x + \frac{11}{9}$
66. [12.3b] $\log r = 6$ **67.** [12.3b] $3^x = Q$
68. [12.4d] $\log_b \left(\frac{x^7}{yz^8}\right)^{1/5}$
69. [12.4d] $-6 \log_b x - 30 \log_b y + 6 \log_b z$
70. [11.7a] 169 **71.** [12.2b] No **72.** [7.2a] **(a)** -5; **(b)** $(-\infty, \infty)$; **(c)** $-2, -1, 1, 2$; **(d)** $[-7, \infty)$
73. [5.8b] $0, \frac{b}{a}$ **74.** [1.8d], [4.1f] $\frac{328}{3}$
75. [10.6a] $\frac{2}{51 + 7\sqrt{61}}$, or $\frac{-51 + 7\sqrt{61}}{194}$
76. [5.6a] $\left(\frac{a}{2} + \frac{2b}{9}\right)\left(\frac{a^2}{4} - \frac{ab}{9} + \frac{4b^2}{81}\right)$
77. [6.6a] **(a)** $t = \frac{ab}{a+b}$; **(b)** $a = \frac{bt}{b-t}$; **(c)** $b = \frac{at}{a-t}$

Appendixes

Margin Exercises, Appendix A, pp. 831–832

1. $70 \frac{\text{mi}}{\text{hr}}$ **2.** $25 \frac{\text{m}}{\text{sec}}$ **3.** 16 yd **4.** 72 in^2
5. 66 man-hours **6.** $800 \frac{\text{kW-hr}}{\text{da}}$ **7.** 84 in.
8. $132 \frac{\text{ft}}{\text{sec}}$

Exercise Set A, p. 833

1. 68 ft **3.** 45 g **5.** $15 \frac{\text{mi}}{\text{hr}}$ **7.** $3.3 \frac{\text{m}}{\text{sec}}$ **9.** $4 \frac{\text{in.-lb}}{\text{sec}}$
11. 12 yd **13.** 16 ft^3 **15.** $\frac{\$970}{\text{day}}$ **17.** 51.2 oz
19. 3080 ft/min **21.** 96 in. **23.** Approximately 0.03 yr **25.** Approximately 31,710 yr **27.** $80 \frac{\text{oz}}{\text{in.}}$
29. 172,800 sec **31.** $\frac{3}{2}$ ft^2 **33.** $1.08 \frac{\text{ton}}{\text{yd}^3}$

Margin Exercises, Appendix B, pp. 834–837

1. -5 **2.** -7 **3.** -6 **4.** 93 **5.** $(3, -1)$
6. $(1, 3, -2)$

Exercise Set B, p. 838

1. 10 **3.** 0 **5.** -48 **7.** 0 **9.** -10 **11.** -3
13. 5 **15.** 0 **17.** $(2, 0)$ **19.** $\left(-\frac{25}{2}, -\frac{11}{2}\right)$ **21.** $\left(\frac{3}{2}, \frac{5}{2}\right)$
23. $(-4, 3)$ **25.** $(2, -1, 4)$ **27.** $(1, 2, 3)$
29. $\left(\frac{3}{2}, -4, 3\right)$ **31.** $(2, -2, 1)$

Margin Exercises, Appendix C, pp. 840–841

1. $(-8, 2)$ **2.** $(-1, 2, 3)$

Exercise Set C, p. 843

1. $\left(\frac{3}{2}, \frac{5}{2}\right)$ **3.** $(-4, 3)$ **5.** $\left(\frac{1}{2}, \frac{3}{2}\right)$ **7.** $(10, -10)$
9. $\left(\frac{3}{2}, -4, 3\right)$ **11.** $(2, -2, 1)$ **13.** $(0, 2, 1)$
15. $\left(4, \frac{1}{2}, -\frac{1}{2}\right)$ **17.** $(w, x, y, z) = (1, -3, -2, -1)$

Margin Exercises, Appendix D, p. 845

1. **(a)** $2x^2$; **(b)** 6; **(c)** $x^4 - 9$; **(d)** $\frac{x^2 + 3}{x^2 - 3}$; **(e)** $x^4 + 6x^2 + 9$
2. **(a)** -5; **(b)** 15; **(c)** -54; **(d)** 6

Exercise Set D, p. 846

1. 1 **3.** -41 **5.** 12 **7.** $\frac{13}{18}$ **9.** 5 **11.** 2
13. $x^2 - x + 1$ **15.** 21 **17.** 5
19. $-x^3 + 4x^2 + 3x - 12$ **21.** 42 **23.** $-\frac{3}{4}$ **25.** $\frac{1}{6}$
27. $x^2 + 3x - 4$; $x^2 - 3x + 4$; $3x^3 - 4x^2$; $\frac{x^2}{3x - 4}$
29. $\frac{1}{x-2} + 4x^3$; $\frac{1}{x-2} - 4x^3$; $\frac{4x^3}{x-2}$; $\frac{1}{4x^3(x-2)}$
31. $\frac{3}{x-2} + \frac{5}{4-x}$; $\frac{3}{x-2} - \frac{5}{4-x}$; $\frac{15}{(x-2)(4-x)}$; $\frac{3(4-x)}{5(x-2)}$

Index

M

N

R

S

T

U

V

W

X

Y

Z

Geometric Formulas

Plane Geometry

Rectangle
Area: $A = l \cdot w$
Perimeter: $P = 2 \cdot l + 2 \cdot w$

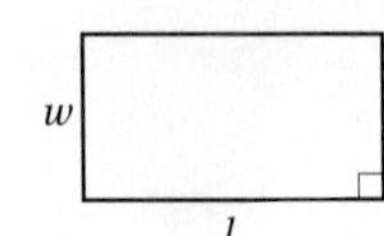

Square
Area: $A = s^2$
Perimeter: $P = 4 \cdot s$

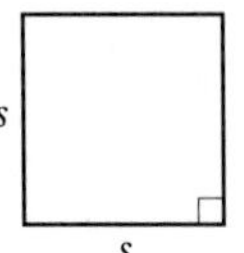

Triangle
Area: $A = \frac{1}{2} \cdot b \cdot h$

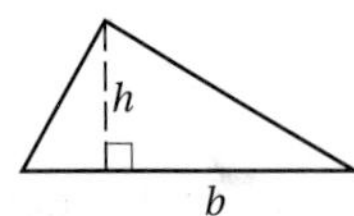

Sum of Angle Measures
$A + B + C = 180°$

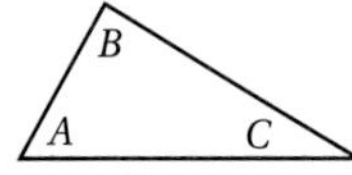

Right Triangle
Pythagorean Theorem:
$a^2 + b^2 = c^2$

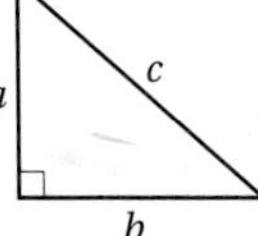

Parallelogram
Area: $A = b \cdot h$

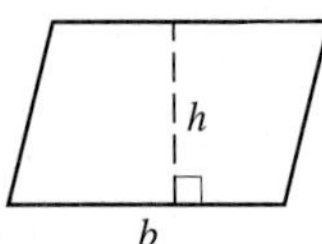

Trapezoid
Area: $A = \frac{1}{2} \cdot h \cdot (a + b)$

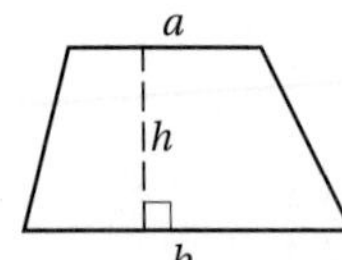

Circle
Area: $A = \pi \cdot r^2$
Circumference:
$C = \pi \cdot d = 2 \cdot \pi \cdot r$
($\frac{22}{7}$ and 3.14 are different approximations for π)

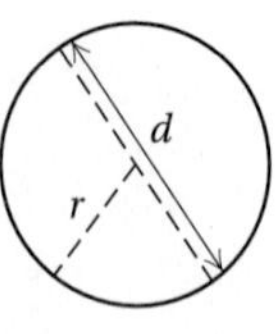

Solid Geometry

Rectangular Solid
Volume: $V = l \cdot w \cdot h$

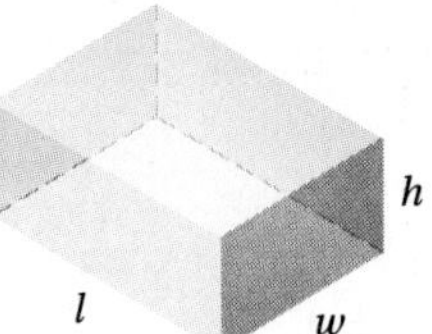

Cube
Volume: $V = s^3$

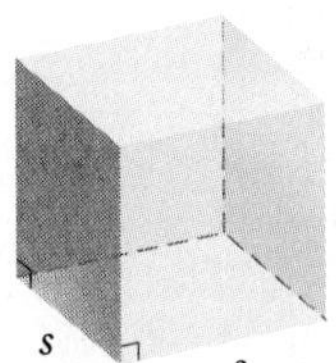

Right Circular Cylinder
Volume: $V = \pi \cdot r^2 \cdot h$
Surface Area:
$S = 2 \cdot \pi \cdot r \cdot h + 2 \cdot \pi \cdot r^2$

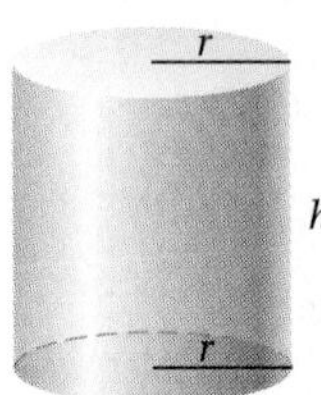

Right Circular Cone
Volume: $V = \frac{1}{3} \cdot \pi \cdot r^2 \cdot h$
Surface Area: $S = \pi \cdot r^2 + \pi \cdot r \cdot s$

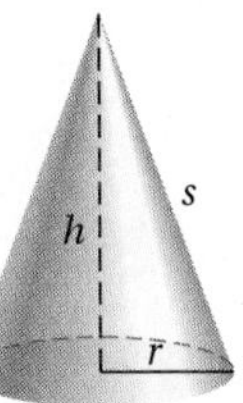

Sphere
Volume: $V = \frac{4}{3} \cdot \pi \cdot r^3$
Surface Area: $S = 4 \cdot \pi \cdot r^2$

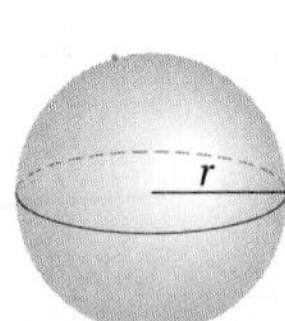